BIOLOGY
HOW LIFE WORKS

BIOLOGY
HOW LIFE WORKS

THIRD EDITION

James Morris
BRANDEIS UNIVERSITY

Daniel Hartl
HARVARD UNIVERSITY

Andrew Knoll
HARVARD UNIVERSITY

Robert Lue
HARVARD UNIVERSITY

Melissa Michael
UNIVERSITY OF ILLINOIS AT URBANA-CHAMPAIGN

Andrew Berry, Andrew Biewener,
Brian Farrell, N. Michele Holbrook
HARVARD UNIVERSITY

Jean Heitz, Mark Hens, Elena R. Lozovsky,
John Merrill, Randall Phillis, Debra Pires

w.h.freeman
Macmillan Learning
New York

VICE PRESIDENT, STEM Daryl Fox
PROGRAM DIRECTOR Andrew Dunaway
EXECUTIVE PROGRAM MANAGER Lisa Lockwood
DIRECTOR OF DEVELOPMENT Lisa Samols
SENIOR DEVELOPMENT EDITOR Susan Moran
SENIOR TEACHING AND LEARNING CONSULTANT Elaine Palucki
DEVELOPMENT EDITOR Erin Mulligan
ASSESSMENT DEVELOPMENT EDITORS Susan Teahan, Jennifer Hart, Anna Bristow
EDITORIAL ASSISTANTS Alexandra Hudson, Justin Jones, Casey Blanchard
EXECUTIVE MARKETING MANAGER Will Moore
MARKETING ASSISTANT Maddie Inskeep
PROJECT MANAGER Karen Misler
DIRECTOR OF CONTENT, LIFE SCIENCES Jennifer Driscoll Hollis
CONTENT DEVELOPMENT MANAGER, BIOLOGY Amber Jonker
LEAD CONTENT DEVELOPER, BIOLOGY Mary Tibbets
CREATIVE DIRECTOR Emiko Paul
MEDIA EDITORS Michael Jones, Amy Thorne, Jennifer Compton

DIRECTOR, CONTENT MANAGEMENT ENHANCEMENT Tracey Kuehn
SENIOR MANAGING EDITOR Lisa Kinne
SENIOR CONTENT PROJECT MANAGERS Liz Geller, Martha Emry
DIRECTOR OF DIGITAL PRODUCTION Keri deManigold
SENIOR MEDIA PROJECT MANAGER Chris Efstratiou
COPYEDITOR Jill Hobbs
MEDIA PERMISSIONS MANAGER Christine Buese
PHOTO RESEARCHER Lisa Passmore
DIRECTOR OF DESIGN, CONTENT MANAGEMENT Diana Blume
DESIGN SERVICES MANAGER Natasha A. S. Wolfe
TEXT DESIGNER Hespenheide Design
COVER DESIGNER John Callahan
ART MANAGER Matthew McAdams
ILLUSTRATOR Imagineering
SENIOR CONTENT WORKFLOW MANAGER Paul Rohloff
COMPOSITION Lumina Datamatics, Inc.
PRINTING AND BINDING LSC Communications
COVER PHOTO George Steinmetz/Getty Images

Library of Congress Control Number: 2018961482
ISBN-13: 978-1-319-01763-7
ISBN-10: 1-319-01763-0
© 2019, 2016, 2013 by W. H. Freeman and Company
All rights reserved.
Printed in the United States of America
1 2 3 4 5 6   22 21 20 19 18

W. H. Freeman and Company
One New York Plaza
Suite 4500
New York, NY 10004-1562
www.macmillanlearning.com

To all who are curious about life
and how it works

ABOUT THE AUTHORS

JAMES R. MORRIS is Professor of Biology and Chair of the Program in Health: Science, Society, and Policy at Brandeis University. He teaches a wide variety of courses for majors and non-majors, including introductory biology, evolution, genetics and genomics, epigenetics, comparative vertebrate anatomy, and a first-year seminar on Darwin's *On the Origin of Species*. He is the recipient of numerous teaching awards from Brandeis and Harvard. His research focuses on the rapidly growing field of epigenetics, making use of the fruit fly *Drosophila melanogaster* as a model organism. He currently pursues this research with undergraduates in order to give them the opportunity to do genuine, laboratory-based research early in their scientific careers. Dr. Morris received a PhD in genetics from Harvard University and an MD from Harvard Medical School. He was a Junior Fellow in the Society of Fellows at Harvard University, and a National Academies Education Fellow and Mentor in the Life Sciences. He also writes short essays on science, medicine, and teaching at his Science Whys blog (http://blogs.brandeis.edu/sciencewhys).

DANIEL L. HARTL is Higgins Professor of Biology in the Department of Organismic and Evolutionary Biology at Harvard University and Professor of Immunology and Infectious Diseases at the Harvard T. H. Chan School of Public Health. He has taught highly popular courses in genetics and evolution at both the introductory and advanced levels. His lab studies molecular evolutionary genetics and population genetics and genomics. Dr. Hartl is the recipient of the Samuel Weiner Outstanding Scholar Award as well as the Gold Medal of the Stazione Zoologica Anton Dohrn, Naples. He is a member of the National Academy of Sciences and the American Academy of Arts and Sciences. He has served as President of the Genetics Society of America and President of the Society for Molecular Biology and Evolution. Dr. Hartl's PhD is from the University of Wisconsin, and he did postdoctoral studies at the University of California, Berkeley. Before joining the Harvard faculty, he served on the faculties of the University of Minnesota, Purdue University, and Washington University Medical School. In addition to publishing more than 400 scientific articles, Dr. Hartl has authored or coauthored 30 books.

ANDREW H. KNOLL is Fisher Professor of Natural History in the Department of Organismic and Evolutionary Biology at Harvard University. He is also Professor of Earth and Planetary Sciences. Dr. Knoll teaches introductory courses in both departments. His research focuses on the early evolution of life, Precambrian environmental history, and the interconnections between the two. He has also worked extensively on the early evolution of animals, mass extinction, and plant evolution. He currently serves on the science team for NASA's mission to Mars. Dr. Knoll received the Phi Beta Kappa Book Award in Science for *Life on a Young Planet*. In 2018, he was awarded the International Prize for Biology. He is a member of the National Academy of Sciences and a foreign member of the Royal Society of London. He received his PhD from Harvard University and then taught at Oberlin College before returning to Harvard.

ROBERT A. LUE is Professor of Molecular and Cellular Biology Harvard University and the Richard L. Menschel Faculty Direct the Derek Bok Center for Teaching and Learning. Dr. Lue has a longstanding commitment to interdisciplinary teaching and resea and chaired the faculty committee that developed the first integra science foundation in the country to serve science majors as well pre-medical students. The founding director of Life Sciences Educ at Harvard, Dr. Lue led a complete redesign of the introductory curriculum, redefining how the university can more effectively fc new generations of scientists as well as science-literate citizens. I Lue has also developed award-winning multimedia, including the animation "The Inner Life of the Cell." He has coauthored unde graduate biology textbooks and chaired education conferences college biology for the National Academies and the National Scie Foundation and on diversity in science for the Howard Hughes Medical Institute and the National Institutes of Health. In 2012, D Lue's extensive work on using technology to enhance learning toc new direction when he became faculty director of university-wide online education initiative HarvardX; he now helps to shape Harv engagement in online learning to reinforce its commitment to tea excellence. Dr. Lue earned his PhD from Harvard University.

MELISSA MICHAEL is Director for Core Curriculum and Assista Director for Undergraduate Instruction for the School of Mole and Cellular Biology at the University of Illinois at Urbana-Champaign. A cell biologist, Dr. Michael primarily focuses on the continuing development of the School's undergraduate curricula. is engaged in several projects aimed at improving instruction ar assessment at the course and program levels. Her research focu primarily on how creative assessment strategies affect student lea outcomes and how outcomes in large-enrollment courses can be improved through the use of formative assessment in active classrc

ANDREW BERRY is Lecturer in the Department of Organismic a Evolutionary Biology and an undergraduate advisor in the Life Sciences at Harvard University. With research interests in evolutio biology and history of science, he teaches courses that either foc on one of the areas or combine the two. He has written two book *Infinite Tropics*, a collection of the writings of Alfred Russel Wall and, with James D. Watson, *DNA: The Secret of Life*, which is p history, part exploration of the controversies surrounding DNA-b technologies.

[N]DREW A. BIEWENER is Charles P. Lyman Professor of Biology in [the] Department of Organismic and Evolutionary Biology at Harvard [Uni]versity and Director of the Concord Field Station. He teaches [both] introductory and advanced courses in anatomy, physiology, and [bio]mechanics. His research focuses on the comparative biomechanics [and] neuromuscular control of mammalian and avian locomotion, with [rele]vance to biorobotics. He is currently Deputy Editor-in-Chief for the [Jou]rnal of Experimental Biology. He also served as President of the [Am]erican Society of Biomechanics.

[BRI]AN D. FARRELL is Director of the David Rockefeller Center for [Lat]in American Studies and Professor of Organismic and Evolutionary [Bio]logy and Curator in Entomology at the Museum of Comparative [Zoo]logy at Harvard University. He is an authority on coevolution [bet]ween insects and plants and a specialist on the biology of beetles. [He] is the author of many scientific papers and book chapters on the [evo]lution of ecological interactions between plants, beetles, and other [ins]ects in the tropics and temperate zone. Dr. Farrell also spearheads [initi]atives to repatriate digital information from scientific specimens of [ins]ects in museums to their tropical countries of origin. In 2011–2012, [he] was a Fulbright Scholar to the Universidad Autónoma de Santo [Do]mingo in the Dominican Republic. Dr. Farrell received a BA in [Zoo]logy and Botany from the University of Vermont and MS and PhD [fro]m the University of Maryland.

[N.] MICHELE HOLBROOK is Charles Bullard Professor of Forestry in [the] Department of Organismic and Evolutionary Biology at Harvard [Uni]versity. She teaches an introductory course on plant biology as well [as a]dvanced courses in plant physiology. Her research focuses on the [phy]sics and physiology of vascular transport in plants with the goal of [un]derstanding how constraints on the movement of water and solutes [bet]ween soil and leaves influences ecological and evolutionary [pro]cesses. Dr. Holbrook received her PhD from Stanford University.

ASSESSMENT AUTHORS

[JEA]N HEITZ is a Distinguished Faculty Associate at the University [of W]isconsin in Madison, WI. She has worked with the two-semester [intr]oductory sequence for biological sciences majors for over 40 years. [Her] primary roles include developing both interactive discussion/ [reci]tation activities designed to uncover and modify misconceptions in [biol]ogy and open-ended investigative labs designed to give students a [mor]e authentic experience with science. The lab experience includes [eng]aging all second-semester students in independent research, [eith]er mentored research or a library-based meta-analysis of an open [que]stion in the literature. She is also the advisor to the Peer Learning [Ass]ociation and is actively involved in TA training. She has taught a [gra]duate course in Teaching College Biology, has presented active-[lear]ning workshops at a number of national and international meetings, [and] has published a variety of lab modules, workbooks, and articles [rela]ted to biology education.

[MA]RK HENS is Associate Professor of Biology at the University of [Nor]th Carolina Greensboro, where he has taught introductory biology [sin]ce 1996. He is a National Academies Education Mentor in the [Life] Sciences and is the director of his department's Introductory [Bio]logy Program. In this role, he guided the development of a [com]prehensive set of assessable student learning outcomes for the [two]-semester introductory biology course required of all science majors at UNCG. In various leadership roles in general education, both on his campus and statewide, he was instrumental in crafting a common set of assessable student learning outcomes for all natural science courses for which students receive general education credit on the 16 campuses of the University of North Carolina system.

ELENA R. LOZOVSKY is Principal Staff Scientist in the Department of Organismic and Evolutionary Biology at Harvard University. She received her PhD in genetics from Moscow State University in Russia and before joining Harvard carried out research at the Institute of Molecular Biology in Moscow, Cornell University, and Washington University School of Medicine in St. Louis. Her research has focused on transposable elements and genome evolution in eukaryotes and on the evolution of drug resistance in malaria parasites. She has also had extensive experience in teaching genetics and evolution.

JOHN MERRILL is Director of the Biological Sciences Program in the College of Natural Science at Michigan State University. This program administers the core biology course sequence required for all science majors. He is a National Academies Education Mentor in the Life Sciences. In recent years he has focused his research on teaching and learning with emphasis on classroom interventions and enhanced assessment. A particularly active area is the NSF-funded development of computer tools for automatic scoring of students' open-ended responses to conceptual assessment questions, with the goal of making it feasible to use open-response questions in large-enrollment classes.

RANDALL PHILLIS is Associate Professor of Biology at the University of Massachusetts Amherst. He has taught in the majors introductory biology course at this institution for 19 years and is a National Academies Education Mentor in the Life Sciences. With help from the Pew Center for Academic Transformation (1999), he has been instrumental in transforming the introductory biology course to an active learning format that makes use of classroom communication systems. He also participates in an NSF-funded project to design model-based reasoning assessment tools for use in class and on exams. These tools are being designed to develop and evaluate student scientific reasoning skills, with a focus on topics in introductory biology.

DEBRA PIRES is an Academic Administrator at the University of California, Los Angeles. She teaches the introductory courses in the Life Sciences Core Curriculum. She is also the Instructional Consultant for the Center for Education Innovation & Learning in the Sciences (CEILS). Many of her efforts are focused on curricular redesign of introductory biology courses. Through her work with CEILS, she coordinates faculty development workshops across several departments to facilitate pedagogical changes associated with curricular developments. Her current research focuses on what impact the experience of active learning pedagogies in lower-division courses have on student performance and concept retention in upper-division courses.

BIOLOGY: HOW LIFE WORKS

has been a revolutionary force for both instructors and students in the majors biology course. It was the first truly comprehensive set of integrated tools for introductory biology, seamlessly incorporating powerful text, media, and assessment to create the best pedagogical experience for students.

THE VISUAL PROGRAM The already impressive visual program has been greatly improved and expanded. The powerful **Visual Synthesis tools** have been reimagined, allowing for more flexibility for both students and instructors. A new **Tour Mode** allows for learning objective–driven tours of the material. We've also added deep-linking from the eText to encourage the student to jump immediately from the reading experience into a more interactive visual representation of the content. Instructors can also create customized tours to use for engaging in-class presentation and active learning opportunities. And finally, new animations have been added to the library, including a new 3D animation to support the animal form and function chapters.

A FOCUS ON SCIENTIFIC SKILLS The third edition has an increased focus on helping students develop the skills they will need to be successful scientists. We've designed skills learning experiences not separate from the core content, but aligned to it. New **Skills Primers** are self-paced tutorials that guide students to learn, practice, and use skills like data visualization, experimental design, working with numbers, and more. New **How Do We Know Activities** are digital extensions of the application-based learning tool found in the text, and focus on further developing students' scientific inquiry skills.

THE HUB The best teaching resources in the world aren't of use if instructors can't find them. The **HUB** provides a one-stop destination for valuable teaching and learning resources, including all of our well-vetted in-class activities.

IMPROVED ORGANIZATION OF TOPICS We implemented several organizational changes based on extensive user feedback with the goal of creating an improved narrative for students and a more flexible teaching framework for instructors.

A new chapter on **Animal Form, Function, and Evolutionary History** leads off the animal anatomy and physiology chapters to provide a whole-body view of structure and function and to provide better context for the more specific systems in following chapters.

The ecology coverage has been enriched and reorganized for a more seamless flow. A new chapter on **Ecosystem Ecology** combines ecosystem concepts formerly housed in separate chapters to present a more cohesive view of the flow of matter and energy in ecosystems.

All of these changes and improvements represent the next step in the evolution of *Biology: How Life Works*. We think we have created the best learning resource for introductory biology students, and we think instructors will find joy in the improvements they can make in their classes with these materials.

DEAR STUDENTS AND INSTRUCTORS,

A new edition of *How Life Works* provides an opportunity for us to continue to innovate. It gives us a chance not just to update and organize, but also to rethink and reimagine the resources we develop to support teaching and learning in introductory biology. Many of our revisions are based on what we've heard from you—the growing community of students and instructors who use *How Life Works*.

From the start, we have written *How Life Works* to be a streamlined text that focuses on concepts and skills relevant to introductory biology. We take an integrated approach that aims to connect concepts and processes, so students can see their interrelationships. Instead of presenting vast amounts of content, we organize our presentation around a few key core concepts. And we recognize that the text, media, and assessment have to work together to provide a rich and meaningful learning experience for students. Our digital platform is designed to provide maximum flexibility and ease of use. With each edition, we keep these ideas front and center.

The third edition continues and expands on this approach. We are particularly excited about new resources that help to develop skills that students will encounter in introductory biology and beyond, including how to ask questions, develop hypotheses, interpret data, read graphs, use quantitative reasoning, and apply statistical and evolutionary thinking. In short, our new resources emphasize the skills that enable students to think like scientists.

Every chapter includes at least one **How Do We Know Figure**, which takes the time to describe how the scientific community came to know a piece of information, process, or concept presented in the text. These were so well received in the first and second editions that we decided to connect them to **How Do We Know Activities** that ask students to answer questions and explore the figure actively and in depth.

In addition, we have revised and expanded the **Primers** that we developed in earlier editions so that they focus on skills and include media and assessment. In this way, they are interactive and serve as short tutorials on fundamental skills.

We also took a fresh look at the chapters on animal anatomy and physiology. A new case focuses on the exciting field of biology-inspired design. This interdisciplinary field taps into biology, engineering, and material science to build tools and devices inspired by all types of organisms, including animals. In addition, a new chapter on animal form, function, and evolution provides a way to orient students to main concepts before delving into detail in the subsequent chapters. Finally, we have new media and assessment that work closely with this section.

In response to thoughtful feedback from instructors, we reorganized our coverage of ecological systems. We now bring together in a single chapter an integrated discussion of flow of matter and energy in ecosystems. This new arrangement allows us to move seamlessly from organisms to populations to species interactions to interactions with the physical environment to global ecology, ending with a discussion of the impact of human activities on the biosphere.

The pace of change in biology is matched by our ever-growing understanding of how students learn, drawing on the latest findings in neuroscience, psychology, and education. We think about ways to create a classroom that is active, inclusive, and evidence-based. We ask ourselves questions as we work to make our material more student-focused. Can we replace a short lecture with an activity in which students construct their own knowledge? What are some common misconceptions that students might encounter along the way? Do all students feel supported and encouraged?

Toward this end, in our new edition we have added a **Learning Objective Framework** while continuing to diversify our assessments. We focus on high-order questions and visual interpretation questions that ask students to make sense of a graph or figure. We continue to expand our rich collection of in-class activities to provide support for instructors who are leading student-centered classrooms. **Assessments** are found everywhere in *How Life Works*—associated with the text, figures, animations, simulations, maps, and in-class activities.

Finally, our media program continues to be fresh and creative, while staying aligned with our other resources. In this edition, we have redesigned our **Visual Synthesis Maps** to provide new functionality. Students can still freely explore spaces and processes in these maps, but we now offer "tours" based on learning objectives so that students can follow a path as a way to focus their exploration.

As one of our recent adopters said, hearteningly, **"Who is the book for?" if not for students**. That simple question guides everything we do. With every edition and update, we aim to improve the learning experience as students engage in discovering how life works.

Sincerely,
The *How Life Works* team

CONNECTED VISUAL TOOLS

The art in the text of **Biology: How Life Works** and the associated media in LaunchPad were developed in coordination with the text and assessments to present an integrated and engaging visual experience for students.

Two of the biggest challenges introductory biology students face are connecting concepts across chapters, and building a contextual picture, or visual framework, of a complex process. To help students think like biologists, we provide **Visual Synthesis Figures** at key points in the text with associated **Visual Synthesis Maps** in the media package. These figures bring together multiple images students have already seen into a visual summary, helping students see how individual concepts connect to tell a single story.

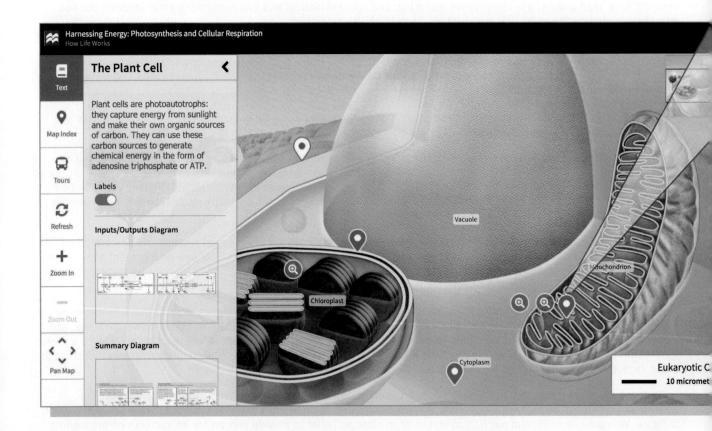

REIMAGINED VISUAL SYNTHESIS PROGRAM

The **Visual Synthesis Tool's** user experience has been reimagined to provide students a more effective visual learning environment, helping them synthesize the connections between biological concepts.

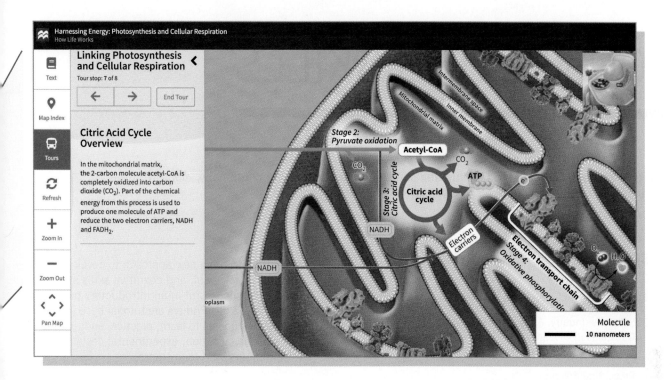

New Tour Mode Functionality: As in previous editions of *How Life Works*, students will still have the opportunity to freely explore these visual concept maps. However, we have now rebuilt the visual synthesis tool to provide students with learning objective–driven tours of the major concepts being visualized in the map. Through these synthesis-focused tours, students follow defined pathways through the core concepts of the maps, as well as measure their understanding of the tours through aligned and assignable assessment opportunities. Instructors will also be able to create their own tours for students to explore in assignments, or for use as in-class teaching resources.

Deep-link from the eText to the Visual Synthesis Map: Students, with one click of a button, will be taken from their digital reading experience into the exact spot in our **Visual Synthesis Maps** where a particular figure/visualization from the book appears. These deep links will provide students with additional information about the figure, an enhanced visual opportunity to learn the concept, and the unique experience of seeing how this figure connects into the larger conceptual framework of the map.

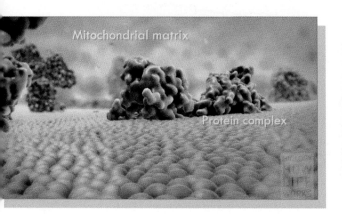
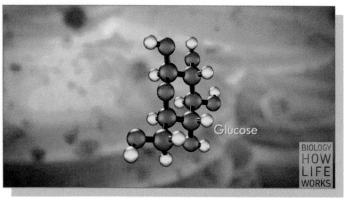

REIMAGINED 3D ANIMATIONS

The powerful **3D Animations** have been reimagined as more flexible tools for students. While still available as full-length videos, they are also available split into shorter clips of more manageable length with added annotations and an improved alignment of the assessments. New animations have been added to the library, including a new 3D animation to support the revised animal form and function topics.

CONNECTING SKILLS

How Life Works is dedicated to teaching the skills students need to think critically, along with the concepts and content they'll need for success in later biology courses.

SKILLS PRIMERS

1. Quantitative Reasoning: Working with Numbers
2. Scientific Inquiry: Asking and Answering Questions in Biology
3. Data and Data Visualization: Interpreting and Presenting Results
4. Phylogenetics: Reading and Building Evolutionary Trees
5. Probability and Statistics: Predicting and Analyzing Data
6. Models: Visualizing Biological Structures and Processes

Skills Primers are a new approach to developing the scientific, biological, and cognitive skills students need to be successful in an introductory biology course. The Skills Primers are tutorial-based resources that guide students to learn, practice, and use skills like data visualization, experimental design, working with numbers, and many more.

How Do We Know Activities ask students questions connected to How Do We Know Figures in the text. In answering these questions, students explore the figures in depth and develop the skills they need to understand scientific inquiry. These online companion exercises are assignable and gradable.

HOW DO WE KNOW?

FIG. 33.18

Do animals tend to get bigger over time?

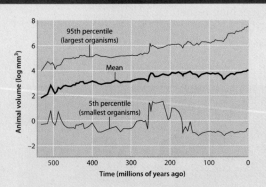

BACKGROUND Cope's rule, named after the American paleontologist Edward Cope, suggests that there is a trend through time toward increasing size among animals. However, it is unclear whether such a general trend actually exists. Large body size can be associated with increased fitness—for example, by making it easier to capture prey or compete with other organisms. Notable examples of very large animals that lived long ago are the dinosaurs. There are also exceptions to this trend in some groups of animals.

HYPOTHESIS Animals tend to increase in size over evolutionary time.

EXPERIMENT Jonathan Payne and his colleagues at Stanford University decided to test the hypothesis that animals tend to get bigger over time, focusing on marine animals over the last 541 million years. They measured the three major body axes of images of animal specimens to determine the mean volume (as a measure of body size) for more than 17,000 genera of animals, including arthropods, brachiopods, echinoderms, mollusks, and chordates.

RESULTS The researchers found that the mean volume for all genera increased over time (the dark black line in the graph). In addition, they found that the range in sizes increased dramatically (the difference between the two light black lines in the graph). Most of this expansion resulted from an increase in the maximum size of animals (the upper light black line in the graph).

CONCLUSION The data support the hypothesis that there is an evolutionary trend toward larger body size in many groups of animals.

FOLLOW-UP WORK The researchers next asked whether the increase in size is the result of natural selection or genetic drift (Chapter 20). They compiled observed trends in the mean, maximum, and minimum sizes of a number of animal groups and compared them to the predictions of models based on selection and drift. They found that their observations are consistent with models based on selection, and not consistent with models based on drift. These findings suggest that increased body size has selective advantages.

SOURCE Heim, N. A., et al. 2015. "Cope's Rule in the Evolution of Marine Mammals." *Science* 347:867–870.

CONNECTED LEARNING TOOLS

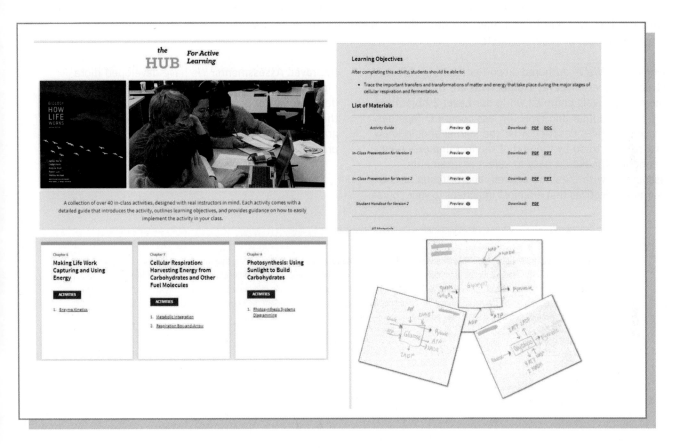

How Life Works has always been committed to providing powerful and effective resources for instructors and pedagogical support for students. The third edition expands this support with even better tools for self-study ahead of class and active work in the classroom.

The Hub is the teaching and learning destination of *How Life Works*. The Hub provides an easy way to find teaching and learning resources, including all of our in-class activities.

Chapter Reading Guides: Designed to improve and focus the reading experience, the reading guides provide students with easy access to chapter resources, including the **Learning Objective Framework** and answers to the **Self-Assessment Questions**.

In-Class Activities: Active learning exercises provide students with hands-on exploration and help correct common misconceptions. In-class activities contain all of the resources to engage students in an active classroom, including a preparation guide for instructors, in-class presentation slides with assessment questions, and student handouts. The third edition of *How Life Works* includes 10 new in-class activities for the second half of the text.

CONNECTED ASSESSMENT

Well-designed assessment is a tremendous tool for instructors preparing students for class, engaging students during class, and gauging student understanding. The *How Life Works* assessment author team applied decades of experience researching and implementing assessment practices to create a variety of questions and activities for pre-class, in-class, homework, and exam settings. All assessment items are carefully aligned with the text and media and have the flexibility to meet the needs of instructors with any experience level, classroom size, or teaching style.

Objectives: We've authored a two-tier learning objective structure wherein **Level 1 objectives** are broader and more cognitively complex while our **Level 2 objectives** are more granular and typically less complex. Together these objectives describe what the learner should know and be able to do by the end of the learning experience. LaunchPad makes it easy to select assessments associated with these objectives, and to focus on the learning objectives that are most important to your course.

Improved Assessments: We have developed more high-order questions, more questions that support the development of quantitative skills, and more questions that require students to engage with figures or consider graphically represented data. Improved data analytics provide insight into student performance.

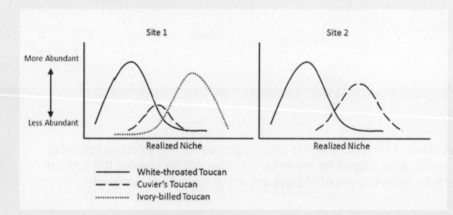

Toucans are frugivorous (fruit-eating) birds that inhabit tropical forests in Central and South America where they can potentially consume any fruit that is available (generalists). The realized niches of toucans are determined by competition in the consumption of different species of fruits (niche overlap). White-throated toucans are the largest species and very strong competitors who easily scare the mid-sized Cuvier's and Ivory-billed toucans off the fruiting trees, altering access to the resources among the three species. The graphs shown represent the occupation of niche space at two different sites, where two or three species coexist, in different abundances.

Which of the graphs shown best characterizes the fundamental niche of any of these species of toucans? Assume the X-axis is the niche space available.

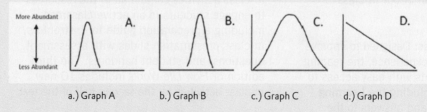

a.) Graph A   b.) Graph B   c.) Graph C   d.) Graph D

CONNECTING THROUGH LAUNCHPAD

Ordinarily, textbooks are developed by first writing chapters, then making decisions about art and images, and finally assembling a test bank and ancillary media. The authors of *How Life Works* developed the text, visual program, and assessment at the same time. These three threads are tied to the same set of core concepts, share a common language, and use the same visual palette, which ensures a seamless learning experience for students throughout the course.

The text, visuals, and assessments come together most effectively through **LaunchPad, Macmillan's integrated learning management system**. In LaunchPad, students and instructors can access all components of *How Life Works*.

LaunchPad resources for *How Life Works* are flexible and aligned. Instructors have the ability to select the visuals, assessments, and activities that best suit their classroom and students. All resources are aligned to one another as well as to the text to ensure effectiveness in helping students build skills and develop knowledge necessary for a foundation in biology.

LearningCurve's game-like quizzing motivates each student to engage with the content, and reporting tools help teachers get a handle on their class needs.

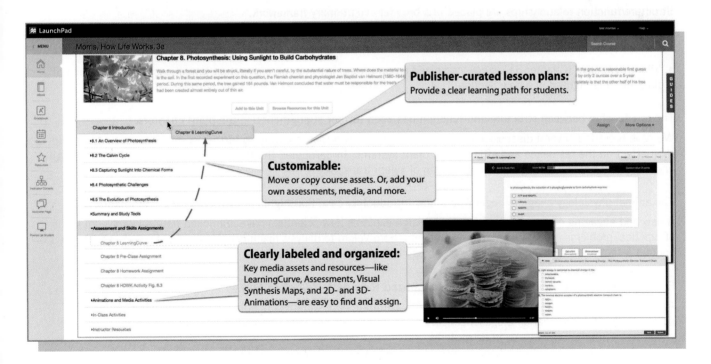

WHAT'S NEW IN THE THIRD EDITION?

MAJOR CHANGES AND UPDATES TO THE BOOK

In developing the third edition of *Biology: How Life Works*, we focused particularly on the form and function chapters and the ecology chapters.

New introductions set the scene for the plant and animal form and function chapters and highlight key themes in structure/function relationships.

- A new chapter, "Animal Form, Function, and Evolutionary History" (Chapter 33), leads off the animal physiology chapters. This chapter provides a whole-body view of structure and function that provides context for the specific systems discussed in the chapters that follow. It focuses on animal body plans and tissue types and introduces homeostasis as the major regulatory theme of the animal physiology chapters.

- The first section of "Plant Form, Function, and Evolutionary History" (Chapter 27) is a completely reconceived introduction to the plant form and function chapters. This section highlights major structure/function differences distinguishing bryophytes and vascular plants. It focuses on how the two groups maintain hydration, specifically on how the reliance on diffusion by bryophytes and bulk flow by vascular plants is reflected in overall structure and cell properties.

Structure/function relationships are placed in a broader evolutionary framework.

- The new "Animal Form, Function, and Evolutionary History" chapter (Chapter 33) concludes with an overview of the history of animal evolution, placing major anatomical and physiological innovations in an evolutionary context.

- "Plant Diversity" (Chapter 31) is now organized around four major structure/function transitions in the evolution of plant life, highlighted in a new Section 31.1.

The relationship between structure and function has been further strengthened in the plant chapters.

We have recrafted several discussions of vascular structure and root structure to further clarify these structures and their effect on the resilience and efficiency of plant systems. In particular, the third edition provides a more thorough and insightful understanding of the mechanism of xylem transport.

The animal physiology chapters begin with a new introductory case that highlights structure/function relationships.

A new and engaging case on Biology-Inspired Design explores how scientists have mimicked nature to solve all kinds of practical problems of real-life interest to students, from Velcro to dialysis machines. Most animal physiology chapters contain a section discussing an example of biology-inspired design.

Ecology coverage has been enriched and reorganized for a more seamless flow.

A new chapter on Ecosystem Ecology, Chapter 46, combines ecosystem concepts such as food webs and trophic pyramids with the material on biogeochemical cycles formerly in separate chapters to present a more cohesive view of the flow of matter and energy in ecosystems. This new arrangement allows us to move seamlessly from organisms to populations to species interactions to interactions with the physical environment to global ecology, ending with a discussion of the impact of human activities on the biosphere.

We continue to expand our treatment of ecological systems, one of our six grand themes. Chapters 44 and 45 ("Population Ecology" and "Species Interactions and Communities") have been enriched by the addition of new concepts and examples to deepen the discussions of life histories and tradeoffs, island biogeography, the niche, biodiversity, and succession, among other topics.

NEW CASE ON BIOLOGY-INSPIRED DESIGN

the internal organs against hard blows to the body and enables the body to turn without twisting these organs. Once molecular studies were conducted, however, researchers discovered that they did not support this traditional phylogenetic division of bilaterians into acoelomate, coelomate, and pseudocoelomate groups. Similarly, molecular sequence comparisons have not supported the once widespread view that segmented bodies indicate a close relationship between earthworms and lobsters.

CASE 7 BIOLOGY-INSPIRED DESIGN: USING NATURE TO SOLVE PROBLEMS

Can we mimic the form and function of animals to build robots?

Scientists who design robots often try to mimic the form and function of animals. Fishes are one of the most popular types of robotic animals. We can learn a great deal about innovations by studying fishes, whose remarkable adaptations have been shaped by natural selection over more than 500 million years.

The first robotic fish was created in the 1980s by a team of scientists and engineers at Massachusetts Institute of Technology. It was called RoboTuna because it mimicked the form and function of a tuna, allowing it to swim effectively through water. Since then, many more fish-like robots have been made, so that today there are more than 400 different types of robotic fishes. Some, like RoboTuna, have one joint; others have many joints; and still others are made of dynamic materials that can bend fluidly.

Why build a robotic fish? The first and most basic reason is to understand how fish move. Fish are spectacular swimmers. If we can build a robotic fish that swims like a living one, it suggests that we understand how a fish is able to move forward, turn, and navigate its surroundings. Robotic fishes can also be used in studies of fish behavior. A robotic fish that looks and even smells like a real fish, and that has sensors to detect and record its surroundings, provides a way to quietly "spy" on fish in a way that is not possible by direct observation or by the use of cameras. There are all kinds of practical applications of robotic fish as well, from mapping coastlines, coral reefs, and the sea floor, to monitoring underwater cables and pipelines, to exploring marine life.

Scientists and engineers have copied the form and function of many animals, including cockroaches, geckos, jellyfish, hummingbirds, and bats, to name just a few. One of the most recent and exciting robots is based on an octopus. Octopuses show incredible dexterity and strength, all without an internal skeleton.

The robot octopus, called an octobot and designed by researchers at Harvard University, has several remarkable characteristics (**Fig. 33.9**). It is entirely soft, with no rigid components

FIG. 33.9 Octobot. Octobot is a soft robot inspired by the octopus.
Source: Lori K. Sanders, Ryan Truby, Michael Wehner, Jennifer Lewis & Robert Wood, Harvard University.

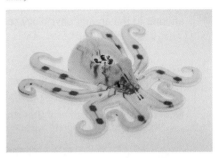

like metal or even batteries. It is powered by a simple chemical reaction: liquid hydrogen peroxide is converted to a gas in the presence of a catalyst (in this case, platinum), providing pressure that powers movement. Finally, the robot is produced with a 3D printer.

The octobot brings together engineers, material scientists, and chemists to create a robot that is inspired by biology. This truly interdisciplinary effort is the first step toward more flexible and innovative robots for research and diverse applications.

Self-Assessment Questions

1. Draw a simplified animal tree of life, indicating the relationships among sponges, cnidarians, protostomes, and deuterostomes.
2. Which features distinguish sponges, cnidarians, protostomes, and deuterostomes? Place these features on the phylogeny that you drew for Self-Assessment Question 1.
3. How is cephalization an adaption for forward locomotion?
4. Which animals are more closely related to sponges: cnidarians or bilaterians?

33.2 TISSUES AND ORGANS

As we have seen, animals show a diversity of form. Nevertheless, the cells of most animals are organized into just four types of tissues: epithelial, connective, muscle, and nervous. **Tissues** are collections of cells that carry out a specific

---

FIG. 33.2 Symmetry in animal form. (a) Radial versus (

a. Radial symmetry

The jellyfish displays radial symmetry: it has a body axis that runs from mouth to base, but many planes of symmetry around the axis.

Planes of symmetry

the eyespots of flatworms are located at one end of as are the brain and sense organs of earthworms, in vertebrates (**Fig. 33.3**). The concentration of nervo components at one end of the body, defined as the

FIG. 33.3 Cephalization. (a) Flatworms, (b) earthworms
Photo sources: (left to right) Science History Images/Alamy Stock Photo; bazilfo

(a)

(b)

(c)

(d)

NEW CHAPTER 33
ANIMAL FORM, FUNCTION, AND EVOLUTIONARY HISTORY

LIST OF NEW TOPICS & OTHER REVISIONS

The following is a detailed list of content changes. These range from small to quite substantial (an entire new chapter in the animal physiology section). Especially important changes are indicated with an asterisk ✽.

- Snottites story replaced with life on Mars story (Case 1)
- ✽ A rearranged discussion of DNA structure, DNA function, and process of transcription (Chapter 3)
- Chromatin structure preview removed and discussed in full in Chapter 13 (Chapter 3)
- A new **How Do We Know Figure** on Anfinsen's experiment on primary and tertiary structure (Chapter 4)
- An expanded discussion of osmotic pressure (Chapter 5)
- Simplified Figure 7.3, The Four Stages of Cellular Respiration, to show only glucose and glycolysis (Chapter 7)
- A new Figure 7.18 giving an overview of the catalysis of fuel molecules other than glucose (Chapter 7)
- A redesigned bottom panel of the Visual Synthesis Figure 8.20 for clearer and more accurate inputs and outputs (Chapter 8)
- A new section illustrating how individual responses create a whole body response (Chapter 9)
- An expanded discussion of cilia (Chapter 10)
- ✽ The section on nondisjunction now moved to Chapter 11 and condensed (Chapter 11)
- Supercoils moved to Section 13.4 from Chapter 3 (Chapter 13)
- ✽ Chapters on Mutation and DNA Repair and on Genetic Variation combined and condensed to more fully integrate mutations and their consequences (genotype and phenotype) (Chapter 14)
  - 14.1 Genotype and phenotype: mostly new section
  - 14.2 Nature of mutations
  - 14.3 Small scale mutations
  - 14.4 Chromosomal mutations
  - 14.5 DNA damage and repair
- In human genetics section, *BRCA* example replaced by *TCF7L2* diabetes example (Chapter 15)
- Extra explanation of walking through combinatorial control in plants (Chapter 19)
- Fruit fly example of Hardy-Weinberg equilibrium replaced by a continuation of Mendel's pea example (Chapter 20)
- Greenish warbler example of populations that differ in reproductive isolation replaced by briefer towhee example (Chapter 21)
- A new example and figure on the formation of a new species by hybridization: *Geospiza fortis* finches (Chapter 21)
- Figure 23.9 from the second edition removed and replaced by three new figures on different ways of using sequence data for phylogenetic reconstruction (Chapter 23)
- A new figure on the carbon cycle (Chapter 24)
- Updated bacterial and archaeal phylogenies (Chapter 24)
- Updated section on the origin of the eukaryotic cell (Chapter 25)
- Discussion of eukaryotic diversity revised in light of recent discoveries from the monumental Tara Ocean Expedition (Chapter 25)
- ✽ Overall focus changed to feeding a growing population (Case 6)
- ✽ A new Section 27.1 compares bryophyte and vascular plant structure and function, with a focus on hydration (Chapter 27)
- A new, clearer description of xylem conduits (Chapter 27)
- ✽ A completely revised and expanded explanation of the mechanism of xylem transport (Chapter 27)
- A revised description of the mechanism of filtering by the Casparian strip (Chapter 27)
- A new discussion of the unwanted effects of nitrogen fertilization (Chapter 27)
- A new paragraph on electrical signaling by plants (Chapter 30)
- ✽ A new Section 31.1 on the four major transitions in the evolution of plant life (Chapter 31)

- A new table on the "Diversity of Plant Life" (Chapter 31)

- Two new figures helping students integrate information from earlier chapters (Chapter 31)

- Revised phylogenies and text to reflect the uncertain relationship among the bryophytes (Chapter 31)

- ✪ A new section on the evolutionary relationship of angiosperms and gymnosperms and their relative diversity (Chapter 31)

- ✪ A new section on how innovations increased the efficiency of the angiosperm life cycles and xylem transport (Chapter 31)

- A revised explanation of diameter increase and vascular bundle distribution in monocot stems (Chapter 31)

- More on the release of digestive enzymes from hyphae tips (Chapter 32)

- More on spore dispersal mechanisms (Chapter 32)

- ✪ A new Case on biology-inspired design (Case 7)

- ✪ A new chapter "Animal Form, Function, and Evolutionary History" providing an introduction to the animal physiology chapters (Chapter 33)
  - animal body plans
  - tissues and organs
  - introduction to homeostasis
  - evolutionary history of animals

- A new Case section on robots inspired by animals (Chapter 33)

- ✪ The chapters on Animal Nervous Systems and Animal Sensory Systems now combined into a single chapter, with some material removed or condensed (Chapter 34)

- A new Case section on cochlear implants (Chapter 34)

- A new Case section on smart materials to heal damaged bones and improve artificial joints (Chapter 35)

- A new Case section on artificial heart valves (Chapter 37)

- New coverage of insect mandibles (Chapter 38)

- A new Case section on dialysis (Chapter 39)

- A new Case section on treating sepsis with an artificial spleen (Chapter 41)

- Two sections moved to the new chapter on Animal Form, Function, and Evolutionary History (Chapter 42)

- A new Table 43.1 summarizing Tinbergen's four questions (Chapter 43)

- A new discussion of altruistic behavior by non-haplodiploid and non-eusocial species (Chapter 43)

- Population dynamic equations standardized and further explained (Chapter 44)

- A new density-independent factor example: replaced song sparrow example with Scottish Red Deer example (Chapter 44)

- Further explanation of life histories and tradeoffs (Chapter 44)

- ✪ Rearrangements and significant revisions to island biogeography explanation (Chapter 44)

- Significant revisions to Case section on island populations and diversity (Chapter 44)

- Revision of niche concept figure and textual explanation—replaced generic pond with Connell experiment (Chapter 45)

- New example illustrating phylogenetic niche conservation: *Anolis* and *Ameiva* lizards (Chapter 45)

- New concepts of species richness and evenness in biodiversity section (Chapter 45)

- New concepts of primary and secondary succession in succession section (Chapter 45)

- ✪ A new chapter on Ecosystem Ecology, Chapter 46, combines ecosystem concepts such as food webs and trophic pyramids with the material on biogeochemical cycles formerly in separate chapters to present a more cohesive view of the flow of matter and energy in ecosystems.

- New discussion of lake turnover (Chapter 47)

TABLE OF CONTENTS

Chapters are arranged in a familiar way to allow easy use in a range of introductory biology courses. On closer look, there are significant differences that aim to help students understand key concepts in modern biology and think like a scientist.

Key differences are identified by ● and **unique chapters** by ✪.

PART 1 FROM CELLS TO ORGANISMS

CHAPTER 1 Life: Chemical, Cellular, and Evolutionary Foundations

CASE 1 Life's Origins: Information, Homeostasis, and Energy

CHAPTER 2 The Molecules of Life
CHAPTER 3 Nucleic Acids and Transcription
CHAPTER 4 Translation and Protein Structure
CHAPTER 5 Organizing Principles: Lipids, Membranes, and Cell Compartments
CHAPTER 6 Making Life Work: Capturing and Using Energy
CHAPTER 7 Cellular Respiration: Harvesting Energy from Carbohydrates and Other Fuel Molecules
CHAPTER 8 Photosynthesis: Using Sunlight to Build Carbohydrates

CASE 2 Cancer: Cell Signaling, Form, and Division

CHAPTER 9 Cell Signaling
CHAPTER 10 Cell and Tissue Architecture: Cytoskeleton, Cell Junctions, and Extracellular Matrix
CHAPTER 11 Cell Division: Variations, Regulation, and Cancer

CASE 3 Your Personal Genome: You, from A to T

CHAPTER 12 DNA Replication and Manipulation
CHAPTER 13 Genomes
CHAPTER 14 Mutation and Genetic Variation
CHAPTER 15 Mendelian Inheritance
CHAPTER 16 Inheritance of Sex Chromosomes, Linked Genes, and Organelles
✪ CHAPTER 17 The Genetic and Environmental Basis of Complex Traits
CHAPTER 18 Genetic and Epigenetic Regulation
CHAPTER 19 Genes and Development

● **SIX GRAND THEMES:** Chapter 1 introduces themes that act as core concepts for introduc biology. They help students to see biology no set of disparate facts, but instead as an integ whole. They connect and unite the many dime of biology, from molecules to the environme These themes continue to be explored throu *How Life Works*. They include scientific inqui chemistry and physics, the cell, evolution, ecological systems, and human impacts.

● **CASES:** Biology is best understood when pres using real and engaging examples as a frame for synthesizing information. Eight carefully positioned Cases help provide this framework example, the case about life's origins is intro before the first set of chapters to help stude understand the fundamental features of a cel therefore of life itself. It is then revisited in subsequent set of chapters. This case has be updated to include recent advances in our se for life on Mars.

● **CHEMISTRY:** After a brief chapter focusing c chemical principles, chemistry is taught in t context of biological processes in later chapt This approach helps students to see in actio key principle.

● **THE CELL:** The cell is the fundamental unit c This concept is introduced in Chapter 1. The Chapters 3–8, it is explored by highlighting th fundamental aspects of a cell—information (Chapters 3 and 4), homeostasis (Chapter 5), harnessing energy (Chapters 6–8). Placing th basic points upfront allows students to start v solid, integrated understanding of the cell.

● **CELL DIVISION:** Mitosis and meiosis are disc in one chapter to allow students to see these forms of cell division side-by-side and unders their similarities, differences, and evolutiona origins.

● **GENETICS:** The genetics section begins with timely case on personal genomics that is revis in the following chapters. The chapters start v genomes and move to inheritance to provide modern, molecular look at genetic variation an traits are transmitted. Current techniques suc the gene editing tool CRISPR are discussed.

● **VARIATION:** Mutation and genetic variation a combined into a single chapter to help studen the interrelationship of mutation, genetic vari and phenotypic variation.

CASE 4 Malaria: Coevolution of Humans and a Parasite

CHAPTER 20	Evolution: How Genotypes and Phenotypes Change Over Time
CHAPTER 21	Species and Speciation
CHAPTER 22	Evolutionary Patterns: Phylogeny and Fossils
★ CHAPTER 23	Human Origins and Evolution

PART 2 FROM ORGANISMS TO THE ENVIRONMENT

CASE 5 The Human Microbiome: Diversity Within

CHAPTER 24	Bacteria and Archaea
CHAPTER 25	Eukaryotic Cells: Origins And Diversity
★ CHAPTER 26	Being Multicellular

CASE 6 Agriculture: Feeding a Growing Population

CHAPTER 27	Plant Form, Function, and Evolutionary History
CHAPTER 28	Plant Reproduction: Finding Mates and Dispersing Offspring
CHAPTER 29	Plant Growth and Development
★ CHAPTER 30	Plant Defense
CHAPTER 31	Plant Diversity
CHAPTER 32	Fungi

CASE 7 Biology-Inspired Design: Using Nature to Solve Problems

CHAPTER 33	Animal Form, Function, and Evolutionary History
CHAPTER 34	Animal Nervous Systems
CHAPTER 35	Animal Movement: Muscles and Skeletons
CHAPTER 36	Animal Endocrine Systems
CHAPTER 37	Animal Cardiovascular and Respiratory Systems
CHAPTER 38	Animal Metabolism, Nutrition, and Digestion
CHAPTER 39	Animal Renal Systems: Water and Waste
CHAPTER 40	Animal Reproduction and Development
CHAPTER 41	Animal Immune Systems
CHAPTER 42	Animal Diversity

CASE 8 Conserving Biodiversity: Rainforest and Coral Reef Hotspots

CHAPTER 43	Behavior and Behavioral Ecology
CHAPTER 44	Population Ecology
CHAPTER 45	Species Interactions and Communities
CHAPTER 46	Ecosystem Ecology
CHAPTER 47	Biomes and Global Ecology
★ CHAPTER 48	The Anthropocene: Humans as a Planetary Force

- **EVOLUTION:** Evolution provides a thread throughout all of the chapters and resources in *How Life Works*. It is introduced in Chapter 1 as one of six core concepts of biology, allowing us to revisit evolutionary concepts and phylogenetic trees throughout. In this way, evolution provides a way for students to think broadly and in an integrated way about biology. Chapters 20–23 provide a focused discussion of key evolutionary processes, such as natural selection and speciation, and include a chapter devoted to human origins and evolution.

- **PROKARYOTIC DIVERSITY:** The diversity of Bacteria and Archaea is discussed in the context of the carbon, nitrogen, and sulfur cycles, providing a conceptual backbone for understanding the tremendous diversity and ecological importance of these organisms.

- **UPDATED CASE:** Case 6 has been substantially revised to highlight the challenge of feeding an ever-growing human population.

- **PLANT DEFENSE:** The chapter on plant defense provides a strong ecological and case-based perspective on the strategies plants use to survive their exploitation by pathogens and herbivores.

- **DIVERSITY:** Diversity follows chapters on anatomy, development, physiology, and life cycles, here and elsewhere, in order to provide a basis for understanding the groupings of organisms and to avoid presenting diversity as a list of names to memorize. When students understand the structure and function of organisms, they can understand the different groups in depth and organize them intuitively. To give instructors flexibility, brief descriptions of unfamiliar organisms and the major groups of organisms are included in the physiology chapters, and the diversity chapters include a brief review of organismal form and function.

- **NEW CASE:** Biology-inspired design looks to nature to build a wide variety of tools and devices, from Velcro to prosthetic limbs. This fascinating area combines biology, engineering, and computer science, allowing students to see how different fields come together to solve complex problems. It also serves to emphasize key concepts about the anatomy and physiology of animals.

- **NEW CHAPTER 33 — ANIMAL FORM, FUNCTION, AND EVOLUTIONARY HISTORY:** This chapter is new to the third edition and provides an overview of the structure, function, and evolution of animals.

- **NERVOUS SYSTEMS:** The biology of nerve cells, sensory systems, and the brain are combined in a single chapter to give students an integrated perspective on these topics and concepts.

- **EXPANDED ECOLOGY COVERAGE:** We continue to expand our treatment of ecological systems, one of our six grand themes. In this edition, we progress seamlessly from organisms, to populations and their interactions with other populations, to interactions among species and between species and the physical world, to global patterns of ecology, culminating in an exploration of how human activities are affecting the biosphere in the 21st century.

PRAISE FOR *HOW LIFE WORKS*

THE BOOK

Simply, the only textbook for this generation of students. We are no longer bearers of knowledge; the internet and Google have replaced us. So, we must teach students how science works and how we put concepts together into a coherent and meaningful way called biology. This text is designed to teach students within this new paradigm.

— Paul Moore, Bowling Green State University

Students tell me (unsolicited) that they like the textbook and that they can actually "read" it. Some say this is the first text they have actually liked reading.

— Tonya Kane, University of California, Los Angeles

This textbook does a good job of connecting concepts together and provides the student with a logical story throughout a chapter. I appreciate that and so do many of my students.

— Matthew Nusnbaum, Georgia State University

THE CASES

The cases... provide an insightful way to introduce the topics of a particular section. In many instances they show how the text information is applicable and relevant to real life situations.

— Nilo Marin, Broward College

ASSESSMENTS

The LearningCurve assignments have really benefited the students. The feedback I have gotten from students is extremely positive.

— Sally Harmych, University of Toledo

The assessment questions in general are very good and the entire system is pretty flexible and allows instructors to build a learning path and assessments that reflects the way they want their course to be taught.

— Mark Mort, University of Kansas

VISUAL SYNTHESIS MAPS

I love the VSMs! The conceptual depth of each is really quite impressive... I am continually impressed with how deeply students dive into each VSM.

— Jeremy Rose, Oregon State University

I really like the VS maps and the ability to drill down to the level of the atom/molecular group.

— Karen Smith, University of British Columbia

Acknowledgments

Biology: How Life Works is not just a book. Instead, it is an integrated set of resources to support student learning and instructor teaching in introductory biology. As a result, we work closely with a large community of authors, publishers, instructors, reviewers, and students. We would like to sincerely thank this dedicated group.

First and foremost, we thank the thousands of students we have collectively taught. Their curiosity, intelligence, and enthusiasm have been sources of motivation for all of us.

Our teachers and mentors have provided us with models of patience, creativity, and inquisitiveness that we strive to bring into our own teaching and research. They have encouraged us to be lifelong learners, teachers, and scholars.

We feel very lucky to be a partner with Macmillan Learning. From the start, our publisher has embraced our project, giving us the space to achieve something unique, while at the same time providing guidance, support, and input from a broad and diverse community of instructors and students.

We especially thank Andrew Dunaway, our program director. He has been a genuine and tireless partner in our efforts. He keeps a watchful eye on important trends in science, education, and technology, and we all deeply appreciate his enthusiasm, creativity, and vision.

Lead developmental editor Lisa Samols continues to have just the right touch—she's a good listener, serious with a touch of humor, and quiet but persistent. Senior developmental editor Susan Moran has an eye for detail and always offers thoughtful suggestions. Both Lisa and Susan have been with us from the very start, and we really value their dedication to the project and continued partnership. Developmental editor Erin Mulligan brings a fresh eye to this edition, and has the uncanny ability to read the text like a student.

Karen Misler kept us all on schedule in a clear and firm, but also understanding and compassionate way.

Will Moore, our marketing manager, refined the story of the third edition of *How Life Works* and works tirelessly with our sales teams to bring the third edition to instructors and students everywhere.

We thank Elizabeth Geller and Martha Emry for coordinating the move from manuscript to the page, and Jill Hobbs for helping to even out the prose. We also thank Diana Blume, our design director; Hespenheide Design, our text and cover designer; and Natasha Wolfe, Design Services Manager. Together, they managed the look and feel of the book, coming up with creative solutions for page layout.

The creation and preparation of the media resources was overseen by Jennifer Driscoll Hollis, Director of Content, Life Sciences, and her team: Amber Jonker, Content Development Manager, Biology; and Mary Tibbets, Lead Content Developer, Biology. They kept all of the pieces of our ever-expanding suite of resources moving smoothly through the creation and implementation process. Michael Jones, Media Editor, ably oversaw the updating of the animations and Visual Synthesis Maps, as well as the creation of our new 3D physiology animation and two new Visual Synthesis Maps. Emiko Paul, Creative Director, gave the Visual Synthesis Figures and Maps a sharp visual design.

Elaine Palucki, Senior Teaching and Learning Consultant, was the guiding light behind our Skills Primers and *How Do We Know* Activities. She and development editor Erin Mulligan steered the transformation of these activities into interactive learning tools.

We thank Keri deManigold and Chris Efstratiou for managing and coordinating the media and websites. They each took on this project with dedication, persistence, enthusiasm, and attention to detail that we deeply appreciate.

We would also like to thank Alexandra Hudson, Justin Jones, and Casey Blanchard for their consistent and tireless support.

Imagineering, under the patient and intelligent guidance of Mark Mykytuik, continues to provide creative, insightful art to complement, support, and reinforce the text. We also thank our illustration coordinator, Matt McAdams, for skillfully guiding our collaboration with Imagineering. Christine Buese, our photo editor, and Lisa Passmore, our photo researcher, provided us with a steady stream of stunning photos, and never gave up on those hard-to-find shots. Paul Rohloff, our workflow manager, ensured that the journey from manuscript to printing was seamless.

We would also like to acknowledge Daryl Fox, Vice President, STEM; Susan Winslow, Managing Director; and Ken Michaels, Chief Executive Officer, for their support of *How Life Works* and our unique approach.

We are extremely grateful for all of the hard work and expertise of the sales representatives, regional managers, and regional sales specialists. We have enjoyed meeting and working with this dedicated sales staff, who are the ones that ultimately put the book in the hands of instructors.

Countless reviewers made invaluable contributions to this book and deserve special thanks. From catching mistakes to suggesting new and innovative ways to organize the content, they provided substantial input to the book. They brought their collective years of teaching to the project, and their suggestions are tangible in every chapter.

Finally, we would like to thank our families. None of this would have been possible without their support, inspiration, and encouragement.

CONTRIBUTORS, THIRD EDITION

Thank you to all the instructors who worked in collaboration with the authors and assessment authors to help write *Biology: How Life Works* assessments, activities, and other media resources.

Cynthia Giffen, University of Michigan, Skills Primers
Christine Griffin, Ohio University, Skills Primers
Sven-Uwe Heinrich, Harvard University, Visual Synthesis Maps and Timelines, content creator/editor
Benedict J. Kolber, Duquesne University, *How Do We Know* Activities, content creator/editor
Jessica Liu, Harvard University, Visual Synthesis Maps and Timelines, content creator/editor
Fulton Rockwell, Harvard University, assessment contributor

REVIEWERS

Thank you to all the instructors who reviewed chapters, art, assessment questions, and other *Biology: How Life Works* materials.

First Edition

Thomas Abbott, University of Connecticut
Tamarah Adair, Baylor University
Sandra Adams, Montclair State University
Jonathon Akin, University of Connecticut
Eddie Alford, Arizona State University
Chris Allen, College of the Mainland
Sylvester Allred, Northern Arizona University
Shivanthi Anandan, Drexel University
Andrew Andres, University of Nevada, Las Vegas
Michael Angilletta, Arizona State University
Jonathan Armbruster, Auburn University
Jessica Armenta, Lone Star College System
Brian Ashburner, University of Toledo
Andrea Aspbury, Texas State University
Nevin Aspinwall, Saint Louis University
Felicitas Avendano, Grand View University
Yael Avissar, Rhode Island College
Ricardo Azpiroz, Richland College
Jessica Baack, Southwestern Illinois College
Charles Baer, University of Florida
Brian Bagatto, University of Akron
Alan L. Baker, University of New Hampshire
Ellen Baker, Santa Monica College
Mitchell Balish, Miami University
Teri Balser, University of Florida
Paul Bates, University of Minnesota, Duluth
Jerome Baudry, University of Tennessee, Knoxville
Michel Baudry, University of Southern California
Mike Beach, Southern Polytechnic State University
Andrew Beall, University of North Florida
Gregory Beaulieu, University of Victoria
John Bell, Brigham Young University
Michael Bell, Richland College
Rebecca Bellone, University of Tampa
Anne Bergey, Truman State University
Laura Bermingham, University of Vermont
Aimee Bernard, University of Colorado, Denver
Annalisa Berta, San Diego State University
Joydeep Bhattacharjee, University of Louisiana, Monroe
Arlene Billock, University of Louisiana, Lafayette
Daniel Blackburn, Trinity College
Mark Blackmore, Valdosta State University
Justin Blau, New York University
Andrew Blaustein, Oregon State University
Mary Bober, Santa Monica College
Robert Bohanan, University of Wisconsin, Madison
Jim Bonacum, University of Illinois at Springfield
Laurie Bonneau, Trinity College
David Bos, Purdue University
James Bottesch, Brevard Community College
Jere Boudell, Clayton State University
Nancy Boury, Iowa State University
Matthew Brewer, Georgia State University
Mirjana Brockett, Georgia Institute of Technology
Andrew Brower, Middle Tennessee State University
Heather Bruns, Ball State University
Jill Buettner, Richland College
Stephen Burnett, Clayton State University
Steve Bush, Coastal Carolina University
David Byres, Florida State College at Jacksonville
James Campanella, Montclair State University
Darlene Campbell, Cornell University
Jennifer Campbell, North Carolina State University
John Campbell, Northwest College
David Canning, Murray State University
Richard Cardullo, University of California, Riverside
Sara Carlson, University of Akron
Jeff Carmichael, University of North Dakota
Dale Casamatta, University of North Florida
Anne Casper, Eastern Michigan University
David Champlin, University of Southern Maine
Rebekah Chapman, Georgia State University
Samantha Chapman, Villanova University
Mark Chappell, University of California, Riverside
P. Bryant Chase, Florida State University
Young Cho, Eastern New Mexico University
Tim Christensen, East Carolina University
Steven Clark, University of Michigan
Ethan Clotfelter, Amherst College
Catharina Coenen, Allegheny College
Mary Colavito, Santa Monica College
Craig Coleman, Brigham Young University
Alex Collier, Armstrong Atlantic State University
Sharon Collinge, University of Colorado, Boulder
Jay Comeaux, McNeese State University
Reid Compton, University of Maryland
Ronald Cooper, University of California, Los Angeles

Victoria Corbin, University of Kansas
Asaph Cousins, Washington State University
Will Crampton, University of Central Florida
Kathryn Craven, Armstrong Atlantic State University
Scott Crousillac, Louisiana State University
Kelly Cude, College of the Canyons
Stanley Cunningham, Arizona State University
Karen Curto, University of Pittsburgh
Bruce Cushing, University of Akron
Rebekka Darner, University of Florida
James Dawson, Pittsburg State University
Jennifer Dechaine, Central Washington University
James Demastes, University of Northern Iowa
D. Michael Denbow, Virginia Polytechnic Institute and State University
Joseph Dent, McGill University
Terry Derting, Murray State University
Jean DeSaix, University of North Carolina at Chapel Hill
Elizabeth De Stasio, Lawrence University
Donald Deters, Bowling Green State University
Hudson DeYoe, University of Texas, Pan American
Leif Deyrup, University of the Cumberlands
Laura DiCaprio, Ohio University
Jesse Dillon, California State University, Long Beach
Frank Dirrigl, University of Texas, Pan American
Kevin Dixon, Florida State University
Hartmut Doebel, George Washington University
Jennifer Doll, Loyola University, Chicago
Logan Donaldson, York University
Blaise Dondji, Central Washington University
Christine Donmoyer, Allegheny College
James Dooley, Adelphi University
Jennifer Doudna, University of California, Berkeley
John DuBois, Middle Tennessee State University
Richard Duhrkopf, Baylor University
Kamal Dulai, University of California, Merced
Arthur Dunham, University of Pennsylvania
Mary Durant, Lone Star College System
Roland Dute, Auburn University
Andy Dyer, University of South Carolina, Aiken
William Edwards, Niagara University
John Elder, Valdosta State University
William Eldred, Boston University
David Eldridge, Baylor University
Inge Eley, Hudson Valley Community College
Lisa Elfring, University of Arizona
Richard Elinson, Duquesne University
Kurt Elliott, Northwest Vista College
Miles Engell, North Carolina State University
Susan Erster, Stony Brook University
Joseph Esdin, University of California, Los Angeles
Jean Everett, College of Charleston
Brent Ewers, University of Wyoming
Melanie Fierro, Florida State College at Jacksonville
Michael Fine, Virginia Commonwealth University

Jonathan Fingerut, St. Joseph's University
Ryan Fisher, Salem State University
David Fitch, New York University
Paul Fitzgerald, Northern Virginia Community College
Jason Flores, University of North Carolina at Charlotte
Matthias Foellmer, Adelphi University
Barbara Frase, Bradley University
Caitlin Gabor, Texas State University
Michael Gaines, University of Miami
Jane Gallagher, City College of New York, City University of New York
J. Yvette Gardner, Clayton State University
Kathryn Gardner, Boston University
Gillian Gass, Dalhousie University
Jason Gee, East Carolina University
Topher Gee, University of North Carolina at Charlotte
Vaughn Gehle, Southwest Minnesota State University
Tom Gehring, Central Michigan University
John Geiser, Western Michigan University
Alex Georgakilas, East Carolina University
Peter Germroth, Hillsborough Community College
Arundhati Ghosh, University of Pittsburgh
Phil Gibson, University of Oklahoma
Cindee Giffen, University of Wisconsin, Madison
Matthew Gilg, University of North Florida
Sharon Gillies, University of the Fraser Valley
Leonard Ginsberg, Western Michigan University
Florence Gleason, University of Minnesota
Russ Goddard, Valdosta State University
Miriam Golbert, College of the Canyons
Jessica Goldstein, Barnard College, Columbia University
Steven Gorsich, Central Michigan University
Sandra Grebe, Lone Star College System
Robert Greene, Niagara University
Ann Grens, Indiana University, South Bend
Theresa Grove, Valdosta State University
Stanley Guffey, University of Tennessee, Knoxville
Nancy Guild, University of Colorado, Boulder
Lonnie Guralnick, Roger Williams University
Laura Hake, Boston College
Kimberly Hammond, University of California, Riverside
Paul Hapeman, University of Florida
Luke Harmon, University of Idaho
Sally Harmych, University of Toledo
Jacob Harney, University of Hartford
Sherry Harrel, Eastern Kentucky University
Dale Harrington, Caldwell Community College and Technical Institute
J. Scott Harrison, Georgia Southern University
Diane Hartman, Baylor University
Mary Haskins, Rockhurst University
Bernard Hauser, University of Florida
David Haymer, University of Hawaii
David Hearn, Towson University
Marshal Hedin, San Diego State University

Paul Heideman, College of William and Mary
Gary Heisermann, Salem State University
Brian Helmuth, University of South Carolina
Christopher Herlihy, Middle Tennessee State University
Albert Herrera, University of Southern California
Brad Hersh, Allegheny College
David Hicks, University of Texas at Brownsville
Karen Hicks, Kenyon College
Alison Hill, Duke University
Kendra Hill, South Dakota State University
Jay Hodgson, Armstrong Atlantic State University
John Hoffman, Arcadia University
Jill Holliday, University of Florida
Sara Hoot, University of Wisconsin, Milwaukee
Margaret Horton, University of North Carolina at Greensboro
Lynne Houck, Oregon State University
Kelly Howe, University of New Mexico
William Huddleston, University of Calgary
Jodi Huggenvik, Southern Illinois University
Melissa Hughes, College of Charleston
Randy Hunt, Indiana University Southeast
Tony Huntley, Saddleback College
Brian Hyatt, Bethel College
Jeba Inbarasu, Metropolitan Community College
Colin Jackson, University of Mississippi
Eric Jellen, Brigham Young University
Dianne Jennings, Virginia Commonwealth University
Scott Johnson, Wake Technical Community College
Mark Johnston, Dalhousie University
Susan Jorstad, University of Arizona
Stephen Juris, Central Michigan University
Jonghoon Kang, Valdosta State University
Julie Kang, University of Northern Iowa
George Karleskint, St. Louis Community College at Meramec
David Karowe, Western Michigan University
Judy Kaufman, Monroe Community College
Nancy Kaufmann, University of Pittsburgh
John Kauwe, Brigham Young University
Elena Keeling, California Polytechnic State University
Jill Keeney, Juniata College
Tamara Kelly, York University
Chris Kennedy, Simon Fraser University
Bretton Kent, University of Maryland
Jake Kerby, University of South Dakota
Jeffrey Kiggins, Monroe Community College
Scott Kight, Montclair State University
Stephen Kilpatrick, University of Pittsburgh, Johnstown
Kelly Kissane, University of Nevada, Reno
David Kittlesen, University of Virginia
Jennifer Kneafsey, Tulsa Community College
Jennifer Knight, University of Colorado, Boulder
Ross Koning, Eastern Connecticut State University
David Kooyman, Brigham Young University
Olga Kopp, Utah Valley University
Anna Koshy, Houston Community College

Todd Kostman, University of Wisconsin, Oshkosh
Peter Kourtev, Central Michigan University
Carol Gibbons Kroeker, University of Calgary
William Kroll, Loyola University, Chicago
Dave Kubien, University of New Brunswick
Allen Kurta, Eastern Michigan University
Troy Ladine, East Texas Baptist University
Ellen Lamb, University of North Carolina at Greensboro
David Lampe, Duquesne University
Evan Lampert, Gainesville State College
James Langeland, Kalamazoo College
John Latto, University of California, Santa Barbara
Brenda Leady, University of Toledo
Jennifer Leavey, Georgia Institute of Technology
Hugh Lefcort, Gonzaga University
Brenda Leicht, University of Iowa
Craig Lending, College at Brockport, State University of New York
Nathan Lents, John Jay College of Criminal Justice, City University of New York
Michael Leonardo, Coe College
Army Lester, Kennesaw State University
Cynthia Littlejohn, University of Southern Mississippi
Zhiming Liu, Eastern New Mexico University
Jonathan Lochamy, Georgia Perimeter College
Suzanne Long, Monroe Community College
Julia Loreth, University of North Carolina at Greensboro
Jennifer Louten, Southern Polytechnic State University
Janet Loxterman, Idaho State University
Ford Lux, Metropolitan State College of Denver
Elaine Dodge Lynch, Memorial University of Newfoundland
José-Luis Machado, Swarthmore College
C. Smoot Major, University of South Alabama
Charles Mallery, University of Miami
Mark Maloney, Spelman College
Carroll Mann, Florida State College at Jacksonville
Carol Mapes, Kutztown University of Pennsylvania
Nilo Marin, Broward College
Diane Marshall, University of New Mexico
Heather Masonjones, University of Tampa
Scott Mateer, Armstrong Atlantic State University
Luciano Matzkin, University of Alabama in Huntsville
Robert Maxwell, Georgia State University
Meghan May, Towson University
Michael McGinnis, Spelman College
Kathleen McGuire, San Diego State University
Maureen McHale, Truman State University
Shannon McQuaig, St. Petersburg College
Lori McRae, University of Tampa
Susan McRae, East Carolina University
Mark Meade, Jacksonville State University
Brad Mehrtens, University of Illinois at Urbana–Champaign
Michael Meighan, University of California, Berkeley
Douglas Meikle, Miami University
Richard Merritt, Houston Community College
Jennifer Metzler, Ball State University

James Mickle, North Carolina State University
Allison Miller, Saint Louis University
Brian Miller, Middle Tennessee State University
Yuko Miyamoto, Elon University
Ivona Mladenovic, Simon Fraser University
Marcie Moehnke, Baylor University
Chad Montgomery, Truman State University
Jennifer Mook, Gainesville State College
Daniel Moon, University of North Florida
Jamie Moon, University of North Florida
Jeanelle Morgan, Gainesville State College
David Morgan, University of West Georgia
Julie Morris, Armstrong Atlantic State University
Becky Morrow, Duquesne University
Mark Mort, University of Kansas
Nancy Morvillo, Florida Southern College
Anthony Moss, Auburn University
Mario Mota, University of Central Florida
Alexander Motten, Duke University
Tim Mulkey, Indiana State University
John Mull, Weber State University
Michael Muller, University of Illinois at Chicago
Beth Mullin, University of Tennessee, Knoxville
Paul Narguizian, California State University, Los Angeles
Jennifer Nauen, University of Delaware
Paul Nealen, Indiana University of Pennsylvania
Diana Nemergut, University of Colorado, Boulder
Kathryn Nette, Cuyamaca College
Jacalyn Newman, University of Pittsburgh
James Nienow, Valdosta State University
Alexey Nikitin, Grand Valley State University
Tanya Noel, York University
Fran Norflus, Clayton State University
Celia Norman, Arapahoe Community College
Eric Norstrom, DePaul University
Jorge Obeso, Miami Dade College
Kavita Oommen, Georgia State University
David Oppenheimer, University of Florida
Joseph Orkwiszewski, Villanova University
Rebecca Orr, Collin College
Don Padgett, Bridgewater State College
Joanna Padolina, Virginia Commonwealth University
One Pagan, West Chester University
Kathleen Page, Bucknell University
Daniel Papaj, University of Arizona
Pamela Pape-Lindstrom, Everett Community College
Bruce Patterson, University of Arizona, Tucson
Shelley Penrod, Lone Star College System
Roger Persell, Hunter College, City University of New York
John Peters, College of Charleston
Chris Petrie, Brevard Community College
Patricia Phelps, Austin Community College
Steven Phelps, University of Texas at Austin
Kristin Picardo, St. John Fisher College
Aaron Pierce, Nicholls State University

Debra Pires, University of California, Los Angeles
Thomas Pitzer, Florida International University
Nicola Plowes, Arizona State University
Crima Pogge, City College of San Francisco
Darren Pollock, Eastern New Mexico University
Kenneth Pruitt, University of Texas at Brownsville
Sonja Pyott, University of North Carolina at Wilmington
Rajinder Ranu, Colorado State University
Philip Rea, University of Pennsylvania
Amy Reber, Georgia State University
Ahnya Redman, West Virginia University
Melissa Reedy, University of Illinois at Urbana–Champaign
Brian Ring, Valdosta State University
David Rintoul, Kansas State University
Michael Rischbieter, Presbyterian College
Laurel Roberts, University of Pittsburgh
George Robinson, University at Albany, State University of New York
Peggy Rolfsen, Cincinnati State Technical and Community College
Mike Rosenzweig, Virginia Polytechnic Institute and State University
Doug Rouse, University of Wisconsin, Madison
Yelena Rudayeva, Palm Beach State College
Ann Rushing, Baylor University
Shereen Sabet, La Sierra University
Rebecca Safran, University of Colorado
Peter Sakaris, Southern Polytechnic State University
Thomas Sasek, University of Louisiana, Monroe
Udo Savalli, Arizona State University
H. Jochen Schenk, California State University, Fullerton
Gregory Schmaltz, University of the Fraser Valley
Jean Schmidt, University of Pittsburgh
Andrew Schnabel, Indiana University, South Bend
Roxann Schroeder, Humboldt State University
David Schultz, University of Missouri, Columbia
Andrea Schwarzbach, University of Texas at Brownsville
Erik Scully, Towson University
Robert Seagull, Hofstra University
Pramila Sen, Houston Community College
Alice Sessions, Austin Community College
Vijay Setaluri, University of Wisconsin
Jyotsna Sharma, University of Texas at San Antonio
Elizabeth Sharpe-Aparicio, Blinn College
Patty Shields, University of Maryland
Cara Shillington, Eastern Michigan University
James Shinkle, Trinity University
Rebecca Shipe, University of California, Los Angeles
Marcia Shofner, University of Maryland
Laurie Shornick, Saint Louis University
Jill Sible, Virginia Polytechnic Institute and State University
Allison Silveus, Tarrant County College
Kristin Simokat, University of Idaho
Sue Simon-Westendorf, Ohio University
Sedonia Sipes, Southern Illinois University, Carbondale
John Skillman, California State University, San Bernardino
Marek Sliwinski, University of Northern Iowa

Felisa Smith, University of New Mexico
John Sollinger, Southern Oregon University
Scott Solomon, Rice University
Morvarid Soltani-Bejnood, University of Tennessee
Chrissy Spencer, Georgia Institute of Technology
Vladimir Spiegelman, University of Wisconsin, Madison
Kathryn Spilios, Boston University
Ashley Spring, Brevard Community College
Bruce Stallsmith, University of Alabama in Huntsville
Jennifer Stanford, Drexel University
Barbara Stegenga, University of North Carolina, Chapel Hill
Patricia Steinke, San Jacinto College, Central Campus
Asha Stephens, College of the Mainland
Robert Steven, University of Toledo
Eric Strauss, University of Wisconsin, La Crosse
Sukanya Subramanian, Collin College
Mark Sugalski, Southern Polytechnic State University
Brad Swanson, Central Michigan University
Ken Sweat, Arizona State University
David Tam, University of North Texas
Ignatius Tan, New York University
William Taylor, University of Toledo
Christine Terry, Lynchburg College
Sharon Thoma, University of Wisconsin, Madison
Pamela Thomas, University of Central Florida
Carol Thornber, University of Rhode Island
Patrick Thorpe, Grand Valley State University
Briana Timmerman, University of South Carolina
Chris Todd, University of Saskatchewan
Gail Tompkins, Wake Technical Community College
Martin Tracey, Florida International University
Randall Tracy, Worcester State University
James Traniello, Boston University
Bibit Traut, City College of San Francisco
Terry Trier, Grand Valley State University
Stephen Trumble, Baylor University
Jan Trybula, State University of New York at Potsdam
Alexa Tullis, University of Puget Sound
Marsha Turell, Houston Community College
Mary Tyler, University of Maine
Dirk Vanderklein, Montclair State University
Marcel van Tuinen, University of North Carolina at Wilmington
Jorge Vasquez-Kool, Wake Technical Community College
William Velhagen, New York University
Dennis Venema, Trinity Western University
Laura Vogel, North Carolina State University
Jyoti Wagle, Houston Community College
Jeff Walker, University of Southern Maine
Gary Walker, Appalachian State University
Andrea Ward, Adelphi University
Fred Wasserman, Boston University
Elizabeth Waters, San Diego State University
Douglas Watson, University of Alabama at Birmingham
Matthew Weand, Southern Polytechnic State University
Michael Weber, Carleton University

Cindy Wedig, University of Texas, Pan American
Brad Wetherbee, University of Rhode Island
Debbie Wheeler, University of the Fraser Valley
Clay White, Lone Star College System
Lisa Whitenack, Allegheny College
Maggie Whitson, Northern Kentucky University
Stacey Wild East, Tennessee State University
Herbert Wildey, Arizona State University and Phoenix College
David Wilkes, Indiana University, South Bend
Lisa Williams, Northern Virginia Community College
Elizabeth Willott, University of Arizona
Ken Wilson, University of Saskatchewan
Mark Wilson, Humboldt State University
Bob Winning, Eastern Michigan University
Candace Winstead, California Polytechnic State University
Robert Wise, University of Wisconsin, Oshkosh
D. Reid Wiseman, College of Charleston
MaryJo Witz, Monroe Community College
David Wolfe, American River College
Kevin Woo, University of Central Florida
Denise Woodward, Penn State
Shawn Wright, Central New Mexico Community College
Grace Wyngaard, James Madison University
Aimee Wyrick, Pacific Union College
Joanna Wysocka-Diller, Auburn University
Ken Yasukawa, Beloit College
John Yoder, University of Alabama
Kelly Young, California State University, Long Beach
James Yount, Brevard Community College
Min Zhong, Auburn University

Second Edition

Barbara J. Abraham, Hampton University
Jason Adams, College of DuPage
Sandra D. Adams, Montclair State University
Richard Adler, University of Michigan, Dearborn
Nancy Aguilar-Roca, University of California, Irvine
Shivanthi Anandan, Drexel University
Lynn Anderson-Carpenter, University of Michigan
Christine Andrews, University of Chicago
Peter Armbruster, Georgetown University
Jessica Armenta, Austin Community College
Brian Ashburner, University of Toledo
Ann J. Auman, Pacific Lutheran University
Nicanor Austriaco, Providence College
Felicitas Avendano, Grand View University
J. P. Avery, University of North Florida
Jim Bader, Case Western Reserve University
Ellen Baker, Santa Monica College
Andrew S. Baldwin, Mesa Community College
Stephen Baron, Bridgewater College
Paul W. Bates, University of Minnesota, Duluth
Janet Batzli, University of Wisconsin, Madison
David Baum, University of Wisconsin, Madison
Kevin S. Beach, University of Tampa

Philip Becraft, Iowa State University
Alexandra Bely, University of Maryland
Lauryn Benedict, University of Northern Colorado
Anne Bergey, Truman State University
Joydeep Bhattacharjee, University of Louisiana, Monroe
Todd Bishop, Dalhousie University
Catherine Black, Idaho State University
Andrew R. Blaustein, Oregon State University
James Bolton, Georgia Gwinnett College
Jim Bonacum, University of Illinois at Springfield
Laurie J. Bonneau, Trinity College
James Bottesch, Eastern Florida State College
Lisa Boucher, University of Texas at Austin
Nicole Bournias-Vardiabasis, California State University, San Bernardino
Nancy Boury, Iowa State University
Matthew Brewer, Georgia State University
Christopher G. Brown, Georgia Gwinnett College
Jill Buettner, Richland College
Sharon K. Bullock, University of North Carolina at Charlotte
Lisa Burgess, Broward College
Jorge Busciglio, University of California, Irvine
Stephen Bush, Coastal Carolina University
David Byres, Florida State College at Jacksonville
Gloria Caddell, University of Central Oklahoma
Guy A. Caldwell, University of Alabama
Kim A. Caldwell, University of Alabama
John S. Campbell, Northwest College
Jennifer Capers, Indian River State College
Joel Carlin, Gustavus Adolphus College
Sara G. Carlson, University of Akron
Dale Casamatta, University of North Florida
Merri Lynn Casem, California State University, Fullerton
Anne Casper, Eastern Michigan University
David Champlin, University of Southern Maine
Rebekah Chapman, Georgia State University
P. Bryant Chase, Florida State University
Thomas T. Chen, Santa Monica College
Young Cho, Eastern New Mexico University
Sunita Chowrira, University of British Columbia
Tim W. Christensen, East Carolina University
Steven Clark, University of Michigan
Beth Cliffel, Triton College
Liane Cochran-Stafira, Saint Xavier University
John G. Cogan, Ohio State University
Reid Compton, University of Maryland
Ronald H. Cooper, University of California, Los Angeles
Janice Countaway, University of Central Oklahoma
Joseph A. Covi, University of North Carolina at Wilmington
Will Crampton, University of Central Florida
Kathryn Craven, Armstrong State University
Lorelei Crerar, George Mason University
Kerry Cresawn, James Madison University
Richard J. Cristiano, Houston Community College Northwest
Cynthia K. Damer, Central Michigan University

David Dansereau, Saint Mary's University
Mark Davis, Macalester College
Tracy Deem, Bridgewater College
Kimberley Dej, McMaster University
Terrence Delaney, University of Vermont
Tracie Delgado, Northwest University
Mark S. Demarest, University of North Texas
D. Michael Denbow, Virginia Polytechnic Institute and State University
Jonathan Dennis, Florida State University
Elizabeth A. De Stasio, Lawrence University
Brandon S. Diamond, University of Miami
AnnMarie DiLorenzo, Montclair State University
Frank J. Dirrigl, Jr., University of Texas, Pan American
Christine Donmoyer, Allegheny College
Samuel Douglas, Angelina College
John D. DuBois, Middle Tennessee State University
Janet Duerr, Ohio University
Meghan Duffy, University of Michigan
Richard E. Duhrkopf, Baylor University
Jacquelyn Duke, Baylor University
Kamal Dulai, University of California, Merced
Rebecca K. Dunn, Boston College
Jacob Egge, Pacific Lutheran University
Kurt J. Elliott, Northwest Vista College
Miles Dean Engell, North Carolina State University
Susan Erster, Stony Brook University
Barbara I. Evans, Lake Superior State University
Lisa Felzien, Rockhurst University
Ralph Feuer, San Diego State University
Ginger R. Fisher, University of Northern Colorado
John Flaspohler, Concordia College
Sam Flaxman, University of Colorado, Boulder
Nancy Flood, Thompson Rivers University
Arthur Frampton, University of North Carolina at Wilmington
Caitlin Gabor, Texas State University
Tracy Galarowicz, Central Michigan University
Raul Galvan, South Texas College
Deborah Garrity, Colorado State University
Jason Mitchell Gee, East Carolina University
T. M. Gehring, Central Michigan University
John Geiser, Western Michigan University
Susan A. Gibson, South Dakota State University
Cynthia J. Giffen, University of Michigan
Matthew Gilg, University of North Florida
Sharon L. Gillies, University of the Fraser Valley
Leslie Goertzen, Auburn University
Marla Gomez, Nicholls State University
Steven Gorsich, Central Michigan University
Daniel Graetzer, Northwest University
James Grant, Concordia University
Linda E. Green, Georgia Institute of Technology
Sara Gremillion, Armstrong State University
Ann Grens, Indiana University, South Bend
John L. Griffis, Joliet Junior College

Nancy A. Guild, University of Colorado, Boulder
Lonnie Guralnick, Roger Williams University
Valerie K. Haftel, Morehouse College
Margaret Hanes, Eastern Michigan University
Sally E. Harmych, University of Toledo
Sherry Harrel, Eastern Kentucky University
J. Scott Harrison, Georgia Southern University
Pat Harrison, University of the Fraser Valley
Diane Hartman, Baylor University
Wayne Hatch, Utah State University Eastern
David Haymer, University of Hawaii at Manoa
Chris Haynes, Shelton State Community College
Christiane Healey, University of Massachusetts, Amherst
David Hearn, Towson University
Marshal Hedin, San Diego State University
Triscia Hendrickson, Morehouse College
Albert A. Herrera, University of Southern California
Bradley Hersh, Allegheny College
Anna Hiatt, East Tennessee State University
Laura Hill, University of Vermont
Jay Hodgson, Armstrong State University
James Horwitz, Palm Beach State College
Sarah Hosch, Oakland University
Kelly Howe, University of New Mexico
Kimberly Hruska, Langara College
William Huddleston, University of Calgary
Pat Uelmen Huey, Georgia Gwinnett College
Carol Hurney, James Madison University
Brian A. Hyatt, Bethel University
Bradley C. Hyman, University of California, Riverside
Anne Jacobs, Allegheny College
Robert C. Jadin, Northeastern Illinois University
Rick Jellen, Brigham Young University
Dianne Jennings, Virginia Commonwealth University
L. Scott Johnson, Towson University
Russell Johnson, Colby College
Susan Jorstad, University of Arizona
Matthew Julius, St. Cloud State University
John S. K. Kauwe, Brigham Young University
Lori J. Kayes, Oregon State University
Todd Kelson, Brigham Young University, Idaho
Christopher Kennedy, Simon Fraser University
Jacob Kerby, University of South Dakota
Stephen T. Kilpatrick, University of Pittsburgh, Johnstown
Mary Kimble, Northeastern Illinois University
Denice D. King, Cleveland State Community College
David Kittlesen, University of Virginia
Ann Kleinschmidt, Allegheny College
Kathryn Kleppinger-Sparace, Tri-County Technical College
Daniel Klionsky, University of Michigan
Ned Knight, Reed College
Benedict Kolber, Duquesne University
Ross Koning, Eastern Connecticut State University
John Koontz, University of Tennessee, Knoxville
Peter Kourtev, Central Michigan University

Elizabeth Kovar, University of Chicago
Nadine Kriska, University of Wisconsin, Whitewater
Carol A. Gibbons Kroeker, Ambrose University
Tim L. Kroft, Auburn University at Montgomery
William Kroll, Loyola University of Chicago
Dave Kubien, University of New Brunswick
Jason Kuehner, Emmanuel College
Josephine Kurdziel, University of Michigan
Troy A. Ladine, East Texas Baptist University
Diane M. Lahaise, Georgia Perimeter College
Janice Lai, Austin Community College, Cypress Creek
Kirk Land, University of the Pacific
James Langeland, Kalamazoo College
Neva Laurie-Berry, Pacific Lutheran University
Brenda Leady, University of Toledo
Adrienne Lee, University of California, Fullerton
Chris Levesque, John Abbott College
Bai-Lian Larry Li, University of California, Riverside
Cynthia Littlejohn, University of Southern Mississippi
Jason Locklin, Temple College
Xu Lu, University of Findlay
Patrice Ludwig, James Madison University
Ford Lux, Metropolitan State University of Denver
Morris F. Maduro, University of California, Riverside
C. Smoot Major, University of South Alabama
Barry Margulies, Towson University
Nilo Marin, Broward College
Michael Martin, John Carroll University
Heather D. Masonjones, University of Tampa
Scott C. Mateer, Armstrong State University
Robert Maxwell, Georgia State University
Joseph McCormick, Duquesne University
Lori L. McGrew, Belmont University
Peter B. McIntyre, University of Wisconsin, Madison
Iain McKinnell, Carleton University
Krystle McLaughlin, Lehigh University
Susan B. McRae, East Carolina University
Mark Meade, Jacksonville State University
Richard Merritt, Houston Community College Northwest
James E. Mickle, North Carolina State University
Chad Montgomery, Truman State University
Scott M. Moody, Ohio University
Daniel Moon, University of North Florida
Jamie Moon, University of North Florida
Jonathan Moore, Virginia Commonwealth University
Paul A. Moore, Bowling Green State University
Tsafrir Mor, Arizona State University
Jeanelle Morgan, University of North Georgia
Mark Mort, University of Kansas
Anthony Moss, Auburn University
Karen Neal, Reynolds Community College
Kimberlyn Nelson, Pennsylvania State University
Hao Nguyen, California State University, Sacramento
John Niedzwiecki, Belmont University
Alexey G. Nikitin, Grand Valley State University

Matthew Nusnbaum, Georgia State University
Robert Okazaki, Weber State University
Nathan Opolot Okia, Auburn University at Montgomery
Tiffany Oliver, Spelman College
Jennifer S. O'Neil, Houston Community College
Kavita Oommen, Georgia State University
Robin O'Quinn, Eastern Washington University
Sarah A. Orlofske, Northeastern Illinois University
Don Padgett, Bridgewater State University
Lisa Parks, North Carolina State University
Samantha Terris Parks, Georgia State University
Nilay Patel, California State University, Fullerton
Markus Pauly, University of California, Berkeley
Daniel M. Pavuk, Bowling Green State University
Marc Perkins, Orange Coast College
Beverly Perry, Houston Community College
John S. Peters, College of Charleston
Chris Petrie, Eastern Florida State College
John M. Pleasants, Iowa State University
Michael Plotkin, Mt. San Jacinto College
Mary Poffenroth, San Jose State University
Dan Porter, Amarillo College
Sonja Pyott, University of North Carolina at Wilmington
Mirwais Qaderi, Mount Saint Vincent University
Nick Reeves, Mt. San Jacinto College
Adam J. Reinhart, Wayland Baptist University
Stephanie Richards, Bates College
David A. Rintoul, Kansas State University
Trevor Rivers, University of Kansas
Laurel Roberts, University of Pittsburgh
Casey Roehrig, Harvard University
Jennifer Rose, University of North Georgia
Michael S. Rosenzweig, Virginia Polytechnic Institute and State University
Caleb M. Rounds, University of Massachusetts, Amherst
Yelena Rudayeva, Palm Beach State College
James E. Russell, Georgia Gwinnett College
Donald Sakaguchi, Iowa State University
Thomas Sasek, University of Louisiana at Monroe
Leslie J. Saucedo, University of Puget Sound
Udo M. Savalli, Arizona State University West Campus
Smita Savant, Houston Community College Southwest
Leena Sawant, Houston Community College Southwest
H. Jochen Schenk, California State University, Fullerton
Aaron E. Schirmer, Northeastern Illinois University
Mark Schlueter, Georgia Gwinnett College
Gregory Schmaltz, University of the Fraser Valley
Jennifer Schramm, Chemeketa Community College
Roxann Schroeder, Humboldt State University
Tim Schuh, St. Cloud State University
Kevin G. E. Scott, University of Manitoba
Erik P. Scully, Towson University
Sarah B. Selke, Three Rivers Community College
Pramila Sen, Houston Community College
Anupama Seshan, Emmanuel College

Alice Sessions, Austin Community College
Vijay Setaluri, University of Wisconsin
Timothy E. Shannon, Francis Marion University
Wallace Sharif, Morehouse College
Mark Sherrard, University of Northern Iowa
Cara Shillington, Eastern Michigan University
Amy Siegesmund, Pacific Lutheran University
Christine Simmons, Southern Illinois University, Edwardsville
S. D. Sipes, Southern Illinois University, Carbondale
John Skillman, California State University, San Bernardino
Daryl Smith, Langara College
Julie Smith, Pacific Lutheran University
Karen Smith, University of British Columbia
Leo Smith, University of Kansas
Ramona Smith-Burrell, Eastern Florida State College
Joel Snodgrass, Towson University
Alan J. Snow, University of Akron–Wayne College
Judith Solti, Houston Community College, Spring Branch
Ann Song, University of California, Fullerton
C. Kay Song, Georgia State University
Chrissy Spencer, Georgia Institute of Technology
Rachel Spicer, Connecticut College
Ashley Spring, Eastern Florida State College
Bruce Stallsmith, University of Alabama in Huntsville
Maria L. Stanko, New Jersey Institute of Technology
Nancy Staub, Gonzaga University
Barbara Stegenga, University of North Carolina
Robert Steven, University of Toledo
Lori Stevens, University of Vermont
Mark Sturtevant, Oakland University
Elizabeth B. Sudduth, Georgia Gwinnett College
Mark Sugalski, Kennesaw State University, Southern Polytechnic State University
Fengjie Sun, Georgia Gwinnett College
Bradley J. Swanson, Central Michigan University
Brook O. Swanson, Gonzaga University
Ken Gunter Sweat, Arizona State University West Campus
Annette Tavares, University of Ontario Institute of Technology
William R. Taylor, University of Toledo
Jessica Theodor, University of Calgary
Sharon Thoma, University of Wisconsin, Madison
Sue Thomson, Auburn University at Montgomery
Mark Tiemeier, Cincinnati State Technical and Community College
Candace Timpte, Georgia Gwinnett College
Nicholas Tippery, University of Wisconsin, Whitewater
Chris Todd, University of Saskatchewan
Kurt A. Toenjes, Montana State University, Billings
Jeffrey L. Travis, University at Albany, State University of New York
Stephen J. Trumble, Baylor University
Jan Trybula, State University of New York at Potsdam
Cathy Tugmon, Georgia Regents University
Alexa Tullis, University of Puget Sound
Marsha Turell, Houston Community College
Steven M. Uyeda, Pima Community College
Rani Vajravelu, University of Central Florida

Dirk Vanderklein, Montclair State University
Moira van Staaden, Bowling Green State University
William Velhagen, Caldwell University
Sara Via, University of Maryland
Christopher Vitek, University of Texas, Pan American
Neal J. Voelz, St. Cloud State University
Mindy Walker, Rockhurst University
Andrea Ward, Adelphi University
Jennifer Ward, University of North Carolina at Asheville
Pauline Ward, Houston Community College
Alan Wasmoen, Metropolitan Community College
Elizabeth R. Waters, San Diego State University
Matthew Weand, Southern Polytechnic State University
K. Derek Weber, Raritan Valley Community College
Andrea Weeks, George Mason University
Charles Welsh, Duquesne University
Naomi L. B. Wernick, University of Massachusetts, Lowell
Mary E. White, Southeastern Louisiana University
David Wilkes, Indiana University, South Bend
Frank Williams, Langara College
Kathy S. Williams, San Diego State University
Lisa D. Williams, Northern Virginia Community College
Christina Wills, Rockhurst University
Brian Wisenden, Minnesota State University, Moorhead
David Wolfe, American River College
Ramakrishna Wusirika, Michigan Technological University
G. Wyngaard, James Madison University
James R. Yount, Eastern Florida State College
Min Zhong, Auburn University

Third Edition

Danyelle Aganovic, University of North Georgia
Adam Aguiar, Stockton University
Nancy Aguilar-Roca, University of California, Irvine
Christine Andrews, University of Chicago
Jenny Archibald, University of Kansas
Jessica Armenta, Austin Community College
Hilal Arnouk, Saint Louis College of Pharmacy
Andrea Aspbury, Texas State University
Ann Auman, Pacific Lutheran University
Felicitas Avendano, Grand View University
Ellen Baker, Santa Monica College
Gerard Beaudoin, Trinity University
Lauryn Benedict, University of Northern Colorado
Joydeep Bhattacharjee, University of Louisiana, Monroe
Andrew Blaustein, Oregon State University
Joel A. Borden, University of South Alabama
Andrew Bouwma, Oregon State University
Christopher Brown, Georgia Gwinnett College
Jill Buettner, Richland College
Sharon Bullock, University of North Carolina, Charlotte
Denise Carroll, Georgia Southern University
Dale Casamatta, University of North Florida
Anne Casper, Eastern Michigan University
Michelle Cawthorn, Georgia Southern University

Rebekah Chapman, Georgia State University
Steven Clark, University of Michigan
Kimberley Clawson, Temple College
John Cogan, Ohio State University
Brett Couch, University of British Columbia
Jennifer Cymbola, Grand Valley State University
Cari Deen, University of Central Oklahoma
Tracie Delgado, Northwest University
Stephanie DeVito, University of Delaware
Sunethra Dharmasiri, Texas State University
Christine Donmoyer, Allegheny College
Janet Duerr, Ohio University
Laura Eidietis, University of Michigan
Bert Ely, University of South Carolina
Miles Engell, North Carolina State University
Jean Everett, College of Charleston
David Fitch, New York University
Matthias Foellmer, Adelphi University
Caitlin Gabor, Texas State University
Jason Gee, East Carolina University
Cynthia Giffen, University of Michigan
Sharon Gillies, University of the Fraser Valley
Marcia Graves, University of British Columbia
Linda Green, Georgetown University
Kathryn Gronlund, Lone Star College, Montgomery
Nancy Guild, University of Colorado, Boulder
Susan R. Halsell, James Madison University
Sally Harmych, University of Toledo
Phillip Harris, University of Alabama
Joseph Harsh, James Madison University
Mary Haskins, Rockhurst University
Christiane Healey, University of Massachusetts, Amherst
David Hearn, Towson University
Susan Herrick, University of Connecticut
Brad Hersh, Allegheny College
James Hickey, Miami University
Sarah E. Hosch, Oakland University
Brian Hyatt, Bethel University
Joanne K. Itami, University of Minnesota, Duluth
Jamie Jensen, Brigham Young University
Michele Johnson, Trinity University
Seth Jones, University of Kentucky
Tonya Kane, University of California, Los Angeles
Lori Kayes, Oregon State University
Benedict Kolber, Duquesne University
Ross Koning, Eastern Connecticut State University
Peter Kourtev, Central Michigan University
Nadine Kriska, University of Wisconsin, Whitewater
Tim Kroft, Auburn University at Montgomery
Troy A. Ladine, East Texas Baptist University
James Langeland, Kalamazoo College
Neva Laurie-Berry, Pacific Lutheran University
Christian Levesque, John Abbott College
B. Larry Li, University of California, Riverside
Jason Locklin, Temple College

Dale Lockwood, Colorado State University
David Lonzarich, University of Wisconsin, Eau Claire
Kari Loomis, University of Massachusetts, Amherst
Patrice Ludwig, James Madison University
Nilo Marin, Broward College
Timothy McCay, Colgate University
Lori McGrew, Belmont University
Michael Meighan, University of California, Berkeley
James Mickle, North Carolina State University
Chad Montgomery, Truman State University
Scott Moody, Ohio University
Paul Moore, Bowling Green State University
Mark Mort, University of Kansas
Kimberlyn Nelson, Pennsylvania State University
Judy Nesmith, University of Michigan, Dearborn
Alexey Nikitin, Grand Valley State University
Matthew Nusnbaum, Georgia State University
Mary Olaveson, University of Ontario Institute of Technology
Jennifer O'Neil, Houston Community College
Samantha Parks, Georgia State University
Thomas Peavy, California State University, Sacramento
John Peters, College of Charleston
Michael Plotkin, Mt. San Jacinto College
Mirwais Qaderi, Mount Saint Vincent University
Raul Ramirez, Oklahoma City Community College
Amy Reber, Georgia State University
Trevor Rivers, University of Kansas
Jeremy Rose, Oregon State University
Caleb Rounds, University of Massachusetts, Amherst
Jaime L. Sabel, University of Memphis
Donald Sakaguchi, Iowa State University
Leslie Saucedo, University of Puget Sound
Erik Scully, Towson University

Miriam Segura-Totten, University of North Georgia
Justin Shaffer, University of California, Irvine
Mark Sherrard, University of Northern Iowa
James Shinkle, Trinity University
Marcia Shofner, University of Maryland, College Park
Karen Smith, University of British Columbia
William Smith, University of Kansas
Kay Song, Georgia State University
Frederick Spiegel, University of Arkansas, Fayetteville
Barbara Stegenga, University of North Carolina
Robert Steven, University of Toledo
Tara Stoulig, Southeastern Louisiana University
Mark Sturtevant, Oakland University
Elizabeth Sudduth, Georgia Gwinnett College
Brad Swanson, Central Michigan University
Ken Sweat, Arizona State University, West Campus
Jonathan Sylvester, Georgia State University
Annette Tavares, University of Ontario Institute of Technology
Casey terHorst, California State University, Northridge
Sharon Thoma, University of Wisconsin, Madison
Gail Tompkins, Wake Technical Community College
Ron Vanderveer, Eastern Florida State College
Daryle Waechter-Brulla, University of Wisconsin, Whitewater
Stacey Weiss, University of Puget Sound
Naomi L. B. Wernick, University of Massachusetts, Lowell
Alan White, University of South Carolina
Mary White, Southeastern Louisiana University
Frank Williams, Langara College
Lisa Williams, Northern Virginia Community College
Paul Wilson, Trent University
Jennifer Zettler, Georgia Southern University, Armstrong Campus
Hongmei Zhang, Georgia State University

Brief Contents

PART I — FROM CELLS TO ORGANISMS

1 Life: Chemical, Cellular, and Evolutionary Foundations — 2

CASE 1 Life's Origins: Information, Homeostasis, and Energy — 25
2 The Molecules of Life — 28
3 Nucleic Acids and Transcription — 50
4 Translation and Protein Structure — 70
5 Organizing Principles: Lipids, Membranes, and Cell Compartments — 92
6 Making Life Work: Capturing and Using Energy — 118
7 Cellular Respiration: Harvesting Energy from Carbohydrates and Other Fuel Molecules — 134
8 Photosynthesis: Using Sunlight to Build Carbohydrates — 156

CASE 2 Cancer: Cell Signaling, Form, and Division — 180
9 Cell Signaling — 184
10 Cell and Tissue Architecture: Cytoskeleton, Cell Junctions, and Extracellular Matrix — 202
11 Cell Division: Variation, Regulation, and Cancer — 224

CASE 3 Your Personal Genome: You, from A to T — 252
12 DNA Replication and Manipulation — 256
13 Genomes — 280
14 Mutation and Genetic Variation — 300
15 Mendelian Inheritance — 322
16 Inheritance of Sex Chromosomes, Linked Genes, and Organelles — 342
17 The Genetic and Environmental Basis of Complex Traits — 360
18 Genetic and Epigenetic Regulation — 376
19 Genes and Development — 398

CASE 4 Malaria: Coevolution of Humans and a Parasite — 422
20 Evolution: How Genotypes and Phenotypes Change over Time — 426
21 Species and Speciation — 446
22 Evolutionary Patterns: Phylogeny and Fossils — 466
23 Human Origins and Evolution — 490

PART II — FROM ORGANISMS TO THE ENVIRONMENT

CASE 5 The Human Microbiome: Diversity Within — 514
24 Bacteria and Archaea — 518
25 Eukaryotic Cells: Origins and Diversity — 544
26 Being Multicellular — 568

CASE 6 Agriculture: Feeding a Growing Population — 586
27 Plant Form, Function, and Evolutionary History — 590
28 Plant Reproduction: Finding Mates and Dispersing Young — 616
29 Plant Growth and Development — 638
30 Plant Defense — 664
31 Plant Diversity — 684
32 Fungi — 712

CASE 7 Biology-Inspired Design: Using Nature to Solve Problems — 734
33 Animal Form, Function, and Evolutionary History — 738
34 Animal Nervous Systems — 756
35 Animal Movement: Muscles and Skeleton — 786
36 Animal Endocrine Systems — 808
37 Animal Cardiovascular and Respiratory Systems — 828
38 Animal Metabolism, Nutrition, and Digestion — 854
39 Animal Renal Systems: Water and Waste — 876
40 Animal Reproduction and Development — 896
41 Animal Immune Systems — 922
42 Animal Diversity — 944

CASE 8 Conserving Biodiversity: Rainforest and Coral Reef Hotspots — 974
43 Behavior and Behavioral Ecology — 978
44 Population Ecology — 1000
45 Species Interactions and Communities — 1018
46 Ecosystem Ecology — 1042
47 Climate and Biomes — 1066
48 The Anthropocene: Humans as a Planetary Force — 1092

Glossary — G-1
Index — I-1

Contents

About the Authors — vi
Preface — viii
Acknowledgments — xxiv

PART 1 FROM CELLS TO ORGANISMS

CHAPTER 1 LIFE CHEMICAL, CELLULAR, AND EVOLUTIONARY FOUNDATIONS — 2

Ralph Lee Hopkins/Getty Images

1.1 Scientific Inquiry — 4

Observation allows us to draw tentative explanations called hypotheses. — 4

A hypothesis makes predictions that can be tested by observation and experiments. — 4

HOW DO WE KNOW? What caused the extinction of the dinosaurs? — 6

A theory is a general explanation of natural phenomena supported by many experiments and observations. — 7

1.2 Chemical and Physical Principles — 8

The living and nonliving worlds follow the same chemical rules and obey the same physical laws. — 8

Scientific inquiry shows that living organisms come from other living organisms. — 10

HOW DO WE KNOW? Can living organisms arise from nonliving matter? — 10

1.3 The Cell — 11

HOW DO WE KNOW? Can microscopic life arise from nonliving matter? — 11

Nucleic acids store and transmit information needed for growth, function, and reproduction. — 13

Membranes define cells and spaces within cells. — 13

Metabolism converts energy from the environment into a form that can be used by cells. — 14

A virus is genetic material that requires a cell to carry out its functions. — 15

1.4 Evolution — 15

Variation in populations provides the raw material for evolution. — 15

Evolution predicts a nested pattern of relatedness among species, depicted as a tree. — 16

Evolution can be studied by means of experiments. — 18

HOW DO WE KNOW? Can evolution be demonstrated in the laboratory? — 18

1.5 Ecological Systems — 19

Basic features of anatomy, physiology, and behavior shape ecological systems. — 19

Ecological interactions play an important role in evolution. — 20

1.6 The Human Footprint — 21

CASE 1 Life's Origins: Information, Homeostasis, and Energy — 25

CHAPTER 2 THE MOLECULES OF LIFE — 28

LOOK Die Bildagentur der Fotografen GmbH / Alamy

2.1 Properties of Atoms — 29

Atoms consist of protons, neutrons, and electrons. — 29

Electrons occupy regions of space called orbitals. — 30

Elements have recurring, or periodic, chemical properties. — 31

2.2 Molecules and Chemical Bonds — 32

A covalent bond results when two atoms share electrons. — 32

A polar covalent bond is characterized by unequal sharing of electrons. — 33

An ionic bond forms between oppositely charged ions. — 33

A chemical reaction involves breaking and forming chemical bonds. — 34

2.3 Water — 35

Water is a polar molecule. — 35

A hydrogen bond is an interaction between a hydrogen atom and an electronegative atom. — 35

Hydrogen bonds give water many unusual properties. — 36

pH is a measure of the concentration of protons in solution. — 36

2.4	**Carbon**	37		What kinds of nucleic acids were present in the earliest cells?	60
	Carbon atoms form four covalent bonds.	37		In transcription, DNA is used as a template to make complementary RNA.	61
	Carbon-based molecules are structurally and functionally diverse.	38		Transcription starts at a promoter and ends at a terminator.	61
2.5	**Organic Molecules**	39		RNA polymerase adds successive nucleotides to the 3' end of the transcript.	62
	Functional groups add chemical character to carbon chains.	39		The RNA polymerase complex is a molecular machine that opens, transcribes, and closes duplex DNA.	64
	Proteins are composed of amino acids.	39	3.4	**RNA Processing**	65
	Nucleic acids encode genetic information in their nucleotide sequence.	40		Primary transcripts in prokaryotes are translated immediately.	65
	Complex carbohydrates are made up of simple sugars.	41		Primary transcripts in eukaryotes undergo several types of chemical modification.	65
	Lipids are hydrophobic molecules.	43		Some RNA transcripts are processed differently from protein-coding transcripts and have functions of their own.	67
2.6	**Life's Origins**	45			
	How did the molecules of life form?	45			
	The building blocks of life can be generated in the laboratory.	45			
	Experiments show how life's building blocks can form macromolecules.	45			
	HOW DO WE KNOW? Could the building blocks of organic molecules have been generated on the early Earth?	46	**CHAPTER 4**	**TRANSLATION AND PROTEIN STRUCTURE**	70

CHAPTER 3 NUCLEIC ACIDS AND TRANSCRIPTION — 50

Pasieka/Science Source

LAGUNA DESIGN/Getty Images

4.1	**Molecular Structure of Proteins**	71
	Amino acids differ in their side chains.	71
	Successive amino acids in proteins are connected by peptide bonds.	73
	The sequence of amino acids dictates protein folding, which determines function.	74
	Secondary structures result from hydrogen bonding in the polypeptide backbone.	74
	Tertiary structures result from interactions between amino acid side chains.	76
	Polypeptide subunits can come together to form quaternary structures.	77
	Chaperones help some proteins fold properly.	77
	HOW DO WE KNOW? What determines secondary and tertiary structure of proteins?	78
4.2	**Protein Synthesis**	80
	Translation uses many molecules found in all cells.	80
	The genetic code shows the correspondence between codons and amino acids.	82
	HOW DO WE KNOW? How was the genetic code deciphered?	83

3.1	**Chemical Composition and Structure of DNA**	52
	HOW DO WE KNOW? Can genetic information be transmitted between two strains of bacteria?	52
	HOW DO WE KNOW? Which molecule carries genetic information?	53
	A DNA strand consists of subunits called nucleotides.	53
	DNA is a linear polymer of nucleotides linked by phosphodiester bonds.	54
	Cellular DNA molecules take the form of a double helix.	55
3.2	**DNA Structure and Function**	58
	DNA molecules are copied in the process of replication, which relies on base pairing.	58
	RNA is an intermediary between DNA and protein.	58
3.3	**Transcription**	59
	RNA is a polymer of nucleotides in which the 5-carbon sugar is ribose.	60

	Translation consists of initiation, elongation, and termination.	84	
	How did the genetic code originate?	85	
4.3	**Protein Origins and Evolution**	86	
	Most proteins are composed of modular folding domains.	86	
	Amino acid sequences evolve through mutation and selection.	87	
	VISUAL SYNTHESIS Gene Expression	88	

CHAPTER 5 ORGANIZING PRINCIPLES — 92

Dr. Jeremy Burgess/Science Source.

5.1	**Structure of Cell Membranes**	94
	Cell membranes are composed of two layers of lipids.	94
	How did the first cell membranes form?	95
	Cell membranes are dynamic.	95
	Proteins associate with cell membranes in different ways.	97
5.2	**Movement In and Out of Cells**	98
	HOW DO WE KNOW? Do proteins move in the plane of the membrane?	98
	The plasma membrane maintains homeostasis.	99
	Passive transport involves diffusion.	100
	Primary active transport uses the energy of ATP.	101
	Secondary active transport is driven by an electrochemical gradient.	102
	Many cells maintain size and composition using active transport.	103
	The cell wall provides another means of maintaining cell shape.	103
5.3	**Internal Organization of Cells**	104
	Eukaryotes and prokaryotes differ in internal organization.	105
	Prokaryotic cells lack a nucleus and extensive internal compartmentalization.	105
	Eukaryotic cells have a nucleus and specialized internal structures.	105
5.4	**The Endomembrane System**	107
	The endomembrane system compartmentalizes the cell.	108
	The nucleus houses the genome and is the site of RNA synthesis.	108
	The endoplasmic reticulum is involved in protein and lipid synthesis.	109
	The Golgi apparatus modifies and sorts proteins and lipids.	110
	Lysosomes degrade macromolecules.	111
	Protein sorting directs proteins to their proper location in or out of the cell.	112
5.5	**Mitochondria and Chloroplasts**	114
	Mitochondria provide the eukaryotic cell with most of its usable energy.	114
	Chloroplasts capture energy from sunlight.	115

CHAPTER 6 MAKING LIFE WORK — 118

Purestock/Getty Images

6.1	**An Overview of Metabolism**	119
	Organisms can be classified according to their energy and carbon sources.	119
	Metabolism is the set of chemical reactions that sustain life.	121
6.2	**Kinetic and Potential Energy**	121
	Kinetic energy and potential energy are two forms of energy.	121
	Chemical energy is a form of potential energy.	122
	ATP is a readily accessible form of cellular energy.	122
6.3	**Laws of Thermodynamics**	123
	The first law of thermodynamics: energy is conserved.	123
	The second law of thermodynamics: energy transformations always result in an increase in disorder in the universe.	123
6.4	**Chemical Reactions**	124
	A chemical reaction occurs when molecules interact.	124
	The laws of thermodynamics determine whether a chemical reaction requires or releases energy available to do work.	125
	The hydrolysis of ATP is an exergonic reaction.	126
	Non-spontaneous reactions are often coupled to spontaneous reactions.	127

6.5	**Enzymes and the Rate of Chemical Reactions**	127
	Enzymes reduce the activation energy of a chemical reaction.	127
	Enzymes form a complex with reactants and products.	128
	Enzymes are highly specific.	129
	Enzyme activity can be influenced by inhibitors and activators.	129
	Allosteric enzymes regulate key metabolic pathways.	129
	HOW DO WE KNOW? Do enzymes form complexes with substrates?	130
	What naturally occurring elements might have spurred the first reactions that led to life?	131

CHAPTER 7 CELLULAR RESPIRATION 134

Steve Satushek/Getty Images

7.1	**An Overview of Cellular Respiration**	135
	Cellular respiration uses chemical energy stored in molecules such as carbohydrates and lipids to produce ATP.	135
	ATP is generated by substrate-level phosphorylation and oxidative phosphorylation.	136
	Redox reactions play a central role in cellular respiration.	136
	Cellular respiration occurs in four stages.	137
7.2	**Glycolysis**	139
	Glycolysis is the partial breakdown of glucose.	139
7.3	**Pyruvate Oxidation**	141
	The oxidation of pyruvate connects glycolysis to the citric acid cycle.	141
7.4	**The Citric Acid Cycle**	142
	The citric acid cycle produces ATP and reduced electron carriers.	142
	What were the earliest energy-harnessing reactions?	143
7.5	**The Electron Transport Chain and Oxidative Phosphorylation**	144
	The electron transport chain transfers electrons and pumps protons.	144
	The proton gradient is a source of potential energy.	144
	ATP synthase converts the energy of the proton gradient into the energy of ATP.	144
	HOW DO WE KNOW? Can a proton gradient drive the synthesis of ATP?	147
7.6	**Anaerobic Metabolism**	148
	Fermentation extracts energy from glucose in the absence of oxygen.	148
	How did early cells meet their energy requirements?	149
7.7	**Metabolic Integration**	150
	Excess glucose is stored as glycogen in animals and starch in plants.	150
	Sugars other than glucose contribute to glycolysis.	150
	Fatty acids and proteins are useful sources of energy.	151
	The intracellular level of ATP is a key regulator of cellular respiration.	152
	Exercise requires several types of fuel molecules and the coordination of metabolic pathways.	153

CHAPTER 8 PHOTOSYNTHESIS 156

ooyoo/Getty Images

8.1	**An Overview of Photosynthesis**	157
	Photosynthesis is widely distributed.	157
	Photosynthesis is a redox reaction.	158
	HOW DO WE KNOW? Does the oxygen released by photosynthesis come from H_2O or CO_2?	159
	The photosynthetic electron transport chain takes place in specialized membranes.	160
8.2	**The Calvin Cycle**	160
	The incorporation of CO_2 is catalyzed by the enzyme rubisco.	160
	NADPH is the reducing agent of the Calvin cycle.	161
	HOW DO WE KNOW? How is CO_2 incorporated into carbohydrates?	162
	The regeneration of RuBP requires ATP.	163
	The steps of the Calvin cycle were determined using radioactive CO_2.	163
	Carbohydrates are stored in the form of starch.	163
8.3	**Capturing Sunlight into Chemical Forms**	164
	Chlorophyll is the major entry point for light energy in photosynthesis.	164

xxxix

Antenna chlorophyll passes light energy to reaction centers.		165
The photosynthetic electron transport chain connects two photosystems.		166
HOW DO WE KNOW? Do chlorophyll molecules operate on their own or in groups?		166
The accumulation of protons in the thylakoid lumen drives the synthesis of ATP.		167
Cyclic electron transport increases the production of ATP.		169
8.4	**Photosynthetic Challenges**	169
	Excess light energy can cause damage.	169
	Photorespiration leads to a net loss of energy and carbon.	171
	Photosynthesis captures just a small percentage of incoming solar energy.	172
8.5	**The Evolution of Photosynthesis**	173
	How did early cells use sunlight to meet their energy requirements?	173
	The ability to use water as an electron donor in photosynthesis evolved in cyanobacteria.	173
	Eukaryotic organisms are believed to have gained photosynthesis by endosymbiosis.	174
	VISUAL SYNTHESIS Harnessing Energy: Photosynthesis and Cellular Respiration	176

CASE 2 Cancer: Cell Signaling, Form, and Division 180

CHAPTER 9 CELL SIGNALING 184

Quest/Science Source

9.1	**Principles of Cell Signaling**	185
	Cells communicate using chemical signals that bind to receptors.	186
	Signaling involves receptor activation, signal transduction, response, and termination.	187
	The response of a cell to a signaling molecule depends on the cell type.	188
9.2	**Distance Between Cells**	188
	Endocrine signaling acts over long distances.	188
	Signaling can occur over short distances.	188
	Signaling can occur by direct cell–cell contact.	189

	HOW DO WE KNOW? Where do growth factors come from?	190
9.3	**Signaling Receptors**	191
	Receptors for polar signaling molecules are located on the cell surface.	191
	Receptors for nonpolar signaling molecules are located in the interior of the cell.	192
	Cell-surface receptors act like molecular switches.	192
9.4	**G Protein-Coupled Receptors**	193
	The first step in cell signaling is receptor activation.	193
	Signals are often amplified in the cytosol.	194
	Signals lead to a cellular response.	194
	Signaling pathways are eventually terminated.	196
9.5	**Receptor Kinases**	197
	Receptor kinases phosphorylate each other, activate intracellular signaling pathways, lead to a response, and are terminated.	198
	How do cell signaling errors lead to cancer?	198
	Signaling pathways are integrated to produce a response in a cell.	199

CHAPTER 10 CELL AND TISSUE ARCHITECTURE 202

F. Fox/AGE Fotostock

10.1	**Tissues and Organs**	203
	Tissues and organs are communities of cells.	203
	The structure of skin relates to its function.	204
10.2	**The Cytoskeleton**	205
	Microtubules and microfilaments are polymers of protein subunits.	205
	Microtubules and microfilaments are dynamic structures.	205
	Motor proteins associate with microtubules and microfilaments to cause movement.	208
	Intermediate filaments are polymers of proteins that vary according to cell type.	210
	The cytoskeleton is an ancient feature of cells.	210
10.3	**Cell Junctions**	212
	Cell adhesion molecules allow cells to attach to other cells and to the extracellular matrix.	212
	Anchoring junctions connect adjacent cells and are reinforced by the cytoskeleton.	213

Tight junctions prevent the movement of substances through the space between cells. 215

Molecules pass between cells through communicating junctions. 215

10.4 The Extracellular Matrix 216

The extracellular matrix of plants is the cell wall. 216

The extracellular matrix is abundant in connective tissues of animals. 217

How do cancer cells spread throughout the body? 219

Extracellular matrix proteins influence cell shape and gene expression. 220

HOW DO WE KNOW? Can extracellular matrix proteins influence gene expression? 221

CHAPTER 11 CELL DIVISION 224

Dr. Torsten Wittmann/Science Source

11.1 Cell Division 225

Prokaryotic cells divide by binary fission. 225

Eukaryotic cells divide by mitotic cell division. 226

The cell cycle describes the life cycle of a eukaryotic cell. 227

11.2 Mitotic Cell Division 227

The DNA of eukaryotic cells is organized as chromosomes. 228

Prophase: Chromosomes condense and become visible. 228

Prometaphase: Chromosomes attach to the mitotic spindle. 229

Metaphase: Chromosomes align as a result of dynamic changes in the mitotic spindle. 230

Anaphase: Sister chromatids fully separate. 230

Telophase: Nuclear envelopes re-form around newly segregated chromosomes. 230

The parent cell divides into two daughter cells by cytokinesis. 230

11.3 Meiotic Cell Division 231

Pairing of homologous chromosomes is unique to meiosis. 231

Crossing over between DNA molecules results in exchange of genetic material. 232

The first meiotic division reduces the chromosome number. 233

The second meiotic division resembles mitosis. 234

Division of the cytoplasm often differs between the sexes. 235

Meiosis is the basis of sexual reproduction. 237

11.4 Nondisjunction 238

Nondisjunction in meiosis results in extra or missing chromosomes. 238

Some human disorders result from nondisjunction. 239

Extra or missing sex chromosomes have fewer effects than extra autosomes. 240

11.5 Cell Cycle Regulation 241

Protein phosphorylation controls passage through the cell cycle. 241

HOW DO WE KNOW? How is progression through the cell cycle controlled? 242

Different cyclin–CDK complexes regulate each stage of the cell cycle. 243

Cell cycle progression requires successful passage through multiple checkpoints. 243

11.6 Cancer 245

What genes are involved in cancer? 245

Oncogenes promote cancer. 245

Proto-oncogenes are genes that when mutated may cause cancer. 245

Tumor suppressors block specific steps in the development of cancer. 245

HOW DO WE KNOW? Can a virus cause cancer? 246

Most cancers require the accumulation of multiple mutations. 247

VISUAL SYNTHESIS Cellular Communities 248

CASE 3 Your Personal Genome: You, from A to T 252

CHAPTER 12 DNA REPLICATION AND MANIPULATION 256

Andrea Danti/Shutterstock

12.1 DNA Replication 257

During DNA replication, the parental strands separate and new partners are made. 257

New DNA strands grow by the addition of nucleotides to the 3′ end. 258

HOW DO WE KNOW? How is DNA replicated? 258

xli

In replicating DNA, one daughter strand is synthesized continuously and the other in a series of short pieces. 261

A small stretch of RNA is needed to begin synthesis of a new DNA strand. 261

Synthesis of the leading and lagging strands is coordinated. 261

DNA polymerase is self-correcting because of its proofreading function. 263

12.2 Replication of Chromosomes 264

Replication of DNA in chromosomes starts at many places almost simultaneously. 264

Telomerase restores tips of linear chromosomes shortened during DNA replication. 264

12.3 DNA Techniques 266

The polymerase chain reaction selectively amplifies regions of DNA. 267

Electrophoresis separates DNA fragments by size. 269

Restriction enzymes cleave DNA at particular short sequences. 269

DNA strands can be separated and brought back together again. 270

DNA sequencing makes use of the principles of DNA replication. 272

What new technologies are used to sequence your personal genome? 273

12.4 Genetic Engineering 274

Recombinant DNA combines DNA molecules from two or more sources. 274

Recombinant DNA is the basis of genetically modified organisms. 275

DNA editing can be used to alter gene sequences almost at will. 276

CHAPTER 13 GENOMES 280

CSP_iDesign/AGE Fotostock

13.1 Genome Sequencing 281

Complete genome sequences are assembled from smaller pieces. 282

Sequences that are repeated complicate sequence assembly. 282

HOW DO WE KNOW? How are whole genomes sequenced? 282

Why sequence your personal genome? 284

13.2 Genome Annotation 284

Genome annotation identifies various types of sequence. 285

Genome annotation includes searching for sequence motifs. 285

Comparison of genomic DNA with messenger RNA reveals the intron–exon structure of genes. 286

An annotated genome summarizes knowledge, guides research, and reveals evolutionary relationships among organisms. 287

The HIV genome illustrates the utility of genome annotation and comparison. 287

13.3 Genes, Genomes, and Organismal Complexity 288

Gene number is not a good predictor of biological complexity. 288

Viruses, bacteria, and archaeons have small, compact genomes. 289

Among eukaryotes, no relationship exists between genome size and organismal complexity. 289

About half of the human genome consists of transposable elements and other types of repetitive DNA. 290

13.4 Organization of Genomes 292

Bacterial cells package their DNA as a nucleoid composed of many loops. 292

Eukaryotic cells package their DNA as one molecule per chromosome. 293

The human genome consists of 22 pairs of chromosomes and two sex chromosomes. 294

Organelle DNA forms nucleoids that differ from those in bacteria. 295

13.5 Viruses and Viral Genomes 296

Viruses can be classified by their genomes. 296

The host range of a virus is determined by viral and host surface proteins. 297

Viruses have diverse sizes and shapes. 298

CHAPTER 14 MUTATION AND GENETIC VARIATION 300

Tetra Images/Getty Images

14.1 Genotype and Phenotype 301

Genotype is the genetic makeup of a cell or organism, and the phenotype is its observed characteristics. 302

Some genetic differences are harmful. 302

Some genetic differences are neutral. 302

A few genetic differences are beneficial. 303

The effect of a mutation may depend on the genotype and environment. 303

14.2 The Nature of Mutations 305

Mutation of individual nucleotides is rare, but mutation across the genome is common. 305

Only germ-line mutations are transmitted to progeny. 306

What can your personal genome tell you about your genetic risk factors for cancer? 307

Mutations are random with regard to an organism's needs. 308

HOW DO WE KNOW? Do mutations occur randomly, or are they directed by the environment? 308

14.3 Small-Scale Mutations 309

Point mutations are changes in a single nucleotide. 309

The effect of a point mutation depends in part on where in the genome it occurs. 310

Small insertions and deletions involve several nucleotides. 311

Some mutations are due to the insertion of a transposable element. 312

HOW DO WE KNOW? What causes sectoring in corn kernels? 313

14.4 Chromosomal Mutations 314

Duplications and deletions result in gain or loss of DNA. 314

Gene families arise from gene duplication and divergence. 315

Copy-number variation constitutes a significant proportion of genetic variation. 315

Tandem repeats are useful in DNA typing. 316

An inversion has a chromosomal region reversed in orientation. 316

A reciprocal translocation joins segments from nonhomologous chromosomes. 317

14.5 DNA Damage and Repair 317

DNA damage can affect both DNA backbone and bases. 317

Most DNA damage is corrected by specialized repair enzymes. 318

CHAPTER 15 MENDELIAN INHERITANCE 322

Edoma/Getty Images

15.1 Early Theories of Inheritance 323

Early theories of heredity predicted the transmission of acquired characteristics. 323

Belief in blending inheritance discouraged studies of hereditary transmission. 324

15.2 Foundations of Modern Transmission Genetics 325

Mendel's experimental organism was the garden pea. 325

In crosses, one of the traits was dominant in the offspring. 326

15.3 Segregation 328

Genes come in pairs that segregate in the formation of reproductive cells. 328

The principle of segregation was tested by predicting the outcome of crosses. 329

A testcross is a mating to an individual with the homozygous recessive genotype. 330

Segregation of alleles reflects the separation of chromosomes in meiosis. 330

Dominance is not universally observed. 331

The principles of transmission genetics are statistical and are stated in terms of probabilities. 331

Mendelian segregation preserves genetic variation. 332

15.4 Independent Assortment 333

Independent assortment is observed when genes segregate independently of one another. 333

Independent assortment reflects the random alignment of chromosomes in meiosis 334

Phenotypic ratios can be modified by interactions between genes. 334

HOW DO WE KNOW? How are single-gene traits inherited? 335

15.5 Human Genetics 337

Dominant traits appear in every generation. 337

Recessive traits skip generations. 337

Many genes have multiple alleles. 338

Incomplete penetrance and variable expression can obscure inheritance patterns. 339

How do genetic tests identify disease risk factors? 339

xliii

CHAPTER 16 INHERITANCE OF SEX CHROMOSOMES, LINKED GENES, AND ORGANELLES 342

FatCamera/Getty Images

16.1 The X and Y Chromosomes 343

In many animals, sex is genetically determined and associated with chromosomal differences. 343

Segregation of the sex chromosomes predicts a 1 : 1 ratio of females to males. 344

16.2 Inheritance of Genes in the X Chromosome 345

X-linked inheritance was discovered through studies of male fruit flies with white eyes. 345

Genes in the X chromosome exhibit a crisscross inheritance pattern. 346

X-linkage provided the first experimental evidence that genes are in chromosomes. 348

Genes in the X chromosome show characteristic patterns in human pedigrees. 349

16.3 Genetic Linkage 350

Nearby genes in the same chromosome show linkage. 350

The frequency of recombination is a measure of the genetic distance between linked genes. 352

Genetic mapping assigns a location to each gene along a chromosome. 353

Genetic risk factors for disease can be localized by genetic mapping. 353

HOW DO WE KNOW? Can recombination be used to construct a genetic map of a chromosome? 354

16.4 Inheritance of Genes in the Y Chromosome 355

Y-linked genes are transmitted from father to son. 355

How can the Y chromosome be used to trace ancestry? 356

16.5 Inheritance of Mitochondrial and Chloroplast DNA 357

Mitochondrial and chloroplast genomes often show uniparental inheritance. 357

Maternal inheritance is characteristic of mitochondrial diseases. 358

How can mitochondrial DNA be used to trace ancestry? 358

CHAPTER 17 THE GENETIC AND ENVIRONMENTAL BASIS OF COMPLEX TRAITS 360

Ryan McVay/Getty Images

17.1 Heredity and Environment 361

Complex traits are affected by the environment. 363

Complex traits are affected by multiple genes. 363

The relative importance of genes and environment can be determined by differences among individuals. 365

Genetic and environmental effects can interact in unpredictable ways. 365

17.2 Resemblance Among Relatives 366

For complex traits, offspring resemble parents but show regression toward the mean. 366

Heritability is the proportion of the total variation due to genetic differences among individuals. 367

17.3 Twin Studies 369

Twin studies help separate the effects of genes and environment in differences among individuals. 369

HOW DO WE KNOW? What is the relative importance of genes and of the environment for complex traits? 370

17.4 Complex Traits in Health and Disease 371

Most common diseases and birth defects are affected by many genes, each with a relatively small effect. 371

Human height is affected by hundreds of genes. 373

Can personalized medicine lead to effective treatments of common diseases? 373

CHAPTER 18 GENETIC AND EPIGENETIC REGULATION 376

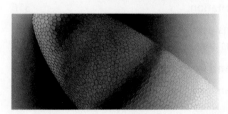

Lawrence Berkeley National Laboratory/Science Source

18.1 Chromatin to Messenger RNA in Eukaryotes 377

Gene expression can be influenced by chemical modification of DNA or histones. 378

xliv

Gene expression can be regulated at the level of an entire chromosome. 380

Transcription is a key control point in gene expression. 381

RNA processing is also important in gene regulation. 382

18.2 Messenger RNA to Phenotype in Eukaryotes 384

Small regulatory RNAs inhibit translation or promote mRNA degradation. 384

Translational regulation controls the rate, timing, and location of protein synthesis. 384

Protein structure and chemical modification modulate protein effects on phenotype. 385

How do lifestyle choices affect expression of your personal genome? 386

18.3 Transcriptional Regulation in Prokaryotes 386

Transcriptional regulation can be positive or negative. 386

Lactose utilization in *E. coli* is the pioneering example of transcriptional regulation. 388

HOW DO WE KNOW? How does lactose lead to the production of active β-galactosidase enzyme? 388

The repressor protein binds with the operator and prevents transcription, but not in the presence of lactose. 389

The function of the lactose operon was revealed by genetic studies. 389

The lactose operon is also positively regulated by CRP–cAMP. 390

Transcriptional regulation determines the outcome of infection by a bacterial virus. 391

VISUAL SYNTHESIS Virus: A Genome in Need of a Cell 394

CHAPTER 19 GENES AND DEVELOPMENT 398

Perennou Nuridsany/Science Source

19.1 Genetic Basis of Development 399

The fertilized egg is a totipotent cell. 399

Cellular differentiation increasingly restricts alternative fates. 400

HOW DO WE KNOW? How do stem cells lose their ability to differentiate into any cell type? 401

Can cells with your personal genome be reprogrammed for new therapies? 403

19.2 Hierarchical Control 403

Drosophila development proceeds through egg, larval, and adult stages. 403

The egg is a highly polarized cell. 404

Development proceeds by progressive regionalization and specification. 406

Homeotic genes determine where different body parts develop in the organism. 407

19.3 Master Regulators 409

Animals have evolved a wide variety of eyes. 409

Pax6 is a master regulator of eye development. 410

19.4 Combinatorial Control 412

Floral differentiation is a model for plant development. 412

The identity of the floral organs is determined by combinatorial control. 412

19.5 Cell Signaling in Development 414

A signaling molecule can cause multiple responses in the cell. 414

Developmental signals are amplified and expanded. 415

VISUAL SYNTHESIS Genetic Variation and Inheritance 418

CASE 4 Malaria: Coevolution of Humans and a Parasite 422

CHAPTER 20 EVOLUTION 426

Andrew Berry

20.1 Genetic Variation 427

Population genetics is the study of patterns of genetic variation. 427

Mutation and recombination are the two sources of genetic variation. 428

20.2 Measuring Genetic Variation 429

To understand patterns of genetic variation, we require information about allele frequencies. 429

Early population geneticists relied on observable traits and gel electrophoresis to measure variation. 429

xlv

DNA sequencing is the gold standard for measuring genetic variation. 430

HOW DO WE KNOW? How did gel electrophoresis allow us to detect genetic variation? 431

20.3 Evolution and the Hardy–Weinberg Equilibrium 432

Evolution is a change in allele or genotype frequency over time. 432

The Hardy–Weinberg equilibrium describes situations in which allele and genotype frequencies do not change. 432

The Hardy–Weinberg equilibrium relates allele frequencies and genotype frequencies. 433

The Hardy–Weinberg equilibrium is the starting point for population genetic analysis. 434

20.4 Natural Selection 434

Natural selection brings about adaptations. 434

The Modern Synthesis combines Mendelian genetics and Darwinian evolution. 436

Natural selection increases the frequency of advantageous mutations and decreases the frequency of deleterious mutations. 436

Which genetic differences have made some individuals more and some less susceptible to malaria? 437

Natural selection can be stabilizing, directional, or disruptive. 437

HOW DO WE KNOW? How far can artificial selection be taken? 438

Sexual selection increases an individual's reproductive success. 439

20.5 Non-Adaptive Mechanisms of Evolution 440

Genetic drift is a change in allele frequency due to chance. 440

Genetic drift has a large effect in small populations. 441

Migration reduces genetic variation between populations. 441

Mutation increases genetic variation. 442

Nonrandom mating alters genotype frequencies without affecting allele frequencies. 442

20.6 Molecular Evolution 442

The molecular clock relates the amount of sequence difference between species and the time since the species diverged. 443

The rate of the molecular clock varies. 443

CHAPTER 21 SPECIES AND SPECIATION 446

21.1 The Biological Species Concept 447

Species are reproductively isolated from other species. 447

The BSC is more useful in theory than in practice. 448

The BSC does not apply to asexual or extinct organisms. 449

Hybridization complicates the BSC. 449

Ecology and evolution can extend the BSC. 450

21.2 Reproductive Isolation 451

Pre-zygotic isolating factors occur before egg fertilization. 451

Post-zygotic isolating factors occur after egg fertilization. 452

21.3 Speciation 452

Speciation is a by-product of the genetic divergence of separated populations. 452

Allopatric speciation is speciation that results from the geographical separation of populations. 452

Dispersal and vicariance can isolate populations from each other. 453

HOW DO WE KNOW? Can vicariance cause speciation? 454

Co-speciation is speciation that occurs in response to speciation in another species. 457

How did malaria come to infect humans? 457

Sympatric populations—those in the same place—may undergo speciation. 458

Speciation can occur instantaneously. 459

Speciation can occur with or without natural selection. 460

VISUAL SYNTHESIS Speciation 462

CHAPTER 22 EVOLUTIONARY PATTERNS 466

Dudley Museums Service

xlvi

22.1	**Reading a Phylogenetic Tree**	467
	Phylogenetic trees provide hypotheses of evolutionary relationships.	468
	The search for sister groups lies at the heart of phylogenetics.	469
	A monophyletic group consists of a common ancestor and all its descendants.	470
	Taxonomic classifications are information storage and retrieval systems.	470
22.2	**Building a Phylogenetic Tree**	472
	Homology is similarity by common descent.	472
	Shared derived characters enable biologists to reconstruct evolutionary history.	473
	The simplest tree is often favored among multiple possible trees.	473
	Molecular data complement comparative morphology in reconstructing phylogenetic history.	475
	Phylogenetic trees can help solve practical problems.	477
	HOW DO WE KNOW? Did an HIV-positive dentist spread the AIDS virus to his patients?	478
22.3	**The Fossil Record**	479
	Fossils provide unique information.	479
	Fossils provide a selective record of past life.	479
	Geologic data indicate the age and environmental setting of fossils.	481
	Fossils can contain unique combinations of characters.	484
	HOW DO WE KNOW? Do fossils bridge the evolutionary gap between fish and tetrapod vertebrates?	486
	Rare mass extinctions have altered the course of evolution.	486
22.4	**Comparing Evolution's Two Great Patterns**	487
	Phylogeny and fossils complement each other.	487
	Agreement between phylogenies and the fossil record provides strong evidence of evolution.	488

CHAPTER 23 HUMAN ORIGINS AND EVOLUTION 490

23.1	**The Great Apes**	491
	Comparative anatomy shows that the human lineage branches off the great apes tree.	491
	Molecular analysis reveals that the human lineage split from the chimpanzee lineage about 5–7 million years ago.	493
	HOW DO WE KNOW? How closely related are humans and chimpanzees?	493
	The fossil record gives us direct information about our evolutionary history.	494
23.2	**African Origins**	497
	Studies of mitochondrial DNA reveal that modern humans evolved in Africa relatively recently.	497
	HOW DO WE KNOW? When and where did the most recent common ancestor of all living humans live?	498
	Studies of the Y chromosome provide independent evidence for a recent origin of modern humans.	499
	Neanderthals disappear from the fossil record as modern humans appear, but have contributed to the modern human gene pool.	500
23.3	**Human Traits**	500
	Bipedalism was a key innovation.	501
	Adult humans share many features with juvenile chimpanzees.	501
	Humans have large brains relative to body size.	502
	The human and chimpanzee genomes help us identify genes that make us human.	503
23.4	**Human Genetic Variation**	504
	Humans have very little genetic variation.	504
	The prehistory of humans influenced the distribution of genetic variation.	504
	The recent spread of modern humans means that there are few genetic differences between groups.	505
	Some human differences have likely arisen by natural selection.	506
	Which human genes are under selection for resistance to malaria?	506
23.5	**Culture, Language, and Consciousness**	507
	Culture changes rapidly.	507
	Is culture uniquely human?	508
	Is language uniquely human?	509
	Is consciousness uniquely human?	509

PART 2 FROM ORGANISMS TO THE ENVIRONMENT

CASE 5 The Human Microbiome: Diversity Within — 514

CHAPTER 24 BACTERIA AND ARCHAEA — 518

STEVE GSCHMEISSNER/SCIENCE PHOTO LIBRARY/Getty Images

24.1 Two Prokaryotic Domains — 519
- The bacterial cell is small but powerful. — 519
- Diffusion limits cell size in bacteria. — 520
- Horizontal gene transfer promotes genetic diversity in bacteria. — 521
- Archaea form a second prokaryotic domain. — 523

24.2 An Expanded Carbon Cycle — 524
- Many photosynthetic bacteria do not produce oxygen. — 525
- Many bacteria respire without oxygen. — 526
- Photoheterotrophs obtain energy from light but obtain carbon from preformed organic molecules. — 527
- Chemoautotrophy is a uniquely prokaryotic metabolism. — 527

24.3 Sulfur and Nitrogen Cycles — 528
- Bacteria and archaeons dominate Earth's sulfur cycle. — 528
- The nitrogen cycle is also driven by bacteria and archaeons. — 529

24.4 Bacterial Diversity — 530
- Bacterial phylogeny is a work in progress. — 530
- **HOW DO WE KNOW?** How many kinds of bacterium live in the oceans? — 531
- What, if anything, is a bacterial species? — 533
- Proteobacteria are the most diverse bacteria. — 534
- The gram-positive bacteria include organisms that cause and cure disease. — 534
- Photosynthesis is widely distributed on the bacterial tree. — 534

24.5 Archaeal Diversity — 535
- The archaeal tree has anaerobic, hyperthermophilic organisms near its base. — 536
- The Archaea include several groups of acid-loving microorganisms. — 537
- Only Archaea produce methane as a by-product of energy metabolism. — 537
- One group of the Euryarchaeota thrives in extremely salty environments. — 537
- Thaumarchaeota may be the most abundant cells in the deep ocean. — 537
- **HOW DO WE KNOW?** How abundant are archaeons in the oceans? — 538

24.6 The Evolutionary History of Prokaryotes — 539
- Life originated early in our planet's history. — 539
- Prokaryotes have coevolved with eukaryotes. — 540
- How do intestinal bacteria influence human health? — 541

CHAPTER 25 EUKARYOTIC CELLS — 544

Lebendkulturen.de/Shutterstock

25.1 A Review of the Eukaryotic Cell — 545
- Internal protein scaffolding and dynamic membranes organize the eukaryotic cell. — 545
- In eukaryotic cells, energy metabolism is localized in mitochondria and chloroplasts. — 547
- The organization of the eukaryotic genome also helps explain eukaryotic diversity. — 547
- Sex promotes genetic diversity in eukaryotes and gives rise to distinctive life cycles. — 547

25.2 Eukaryotic Origins — 549
- What role did symbiosis play in the origin of chloroplasts? — 549
- **HOW DO WE KNOW?** What is the evolutionary origin of chloroplasts? — 550
- What role did symbiosis play in the origin of mitochondria? — 551
- How did the eukaryotic cell originate? — 553
- In the oceans, many single-celled eukaryotes harbor symbiotic bacteria. — 553

25.3 Eukaryotic Diversity — 554
- Our own group, the opisthokonts, is the most diverse eukaryotic superkingdom. — 555
- Amoebozoans include slime molds that produce multicellular structures. — 556
- Archaeplastids, which include land plants, are photosynthetic organisms. — 558

	Stramenopiles, alveolates, and rhizarians dominate eukaryotic diversity in the oceans.	560	26.5	**The Evolution of Complex Multicellularity** 580
	HOW DO WE KNOW? How did photosynthesis spread through the Eukarya?	562		Fossil evidence of complex multicellular organisms is first observed in rocks deposited 575–555 million years ago. 580
	Photosynthesis spread through eukaryotes by repeated endosymbioses involving eukaryotic algae.	563		Oxygen is necessary for complex multicellular life. 582
25.4	**The Fossil Record of Protists**	564		Land plants evolved from green algae that could carry out photosynthesis on land. 583
	Fossils show that eukaryotes existed at least 1800 million years ago.	564		Regulatory genes played an important role in the evolution of complex multicellular organisms. 583
	Protists have continued to diversify during the age of animals.	565		**HOW DO WE KNOW?** What controls color pattern in butterfly wings? 584

CHAPTER 26 BEING MULTICELLULAR 568

CASE 6 Agriculture: Feeding a Growing Population 586

CHAPTER 27 PLANT FORM, FUNCTION, AND EVOLUTIONARY HISTORY 590

26.1	**The Phylogenetic Distribution of Multicellular Organisms**	569
	Simple multicellularity is widespread among eukaryotes.	569
	Complex multicellularity evolved several times.	571
26.2	**Diffusion and Bulk Flow**	572
	Diffusion is effective only over short distances.	572
	Animals achieve large size by circumventing limits imposed by diffusion.	573
	Complex multicellular organisms have structures specialized for bulk flow.	573
26.3	**How to Build a Multicellular Organism**	574
	Complex multicellularity requires adhesion between cells.	574
	HOW DO WE KNOW? How do bacteria influence the life cycles of choanoflagellates?	575
	How did animal cell adhesion originate?	575
	Complex multicellularity requires communication between cells.	576
	Complex multicellularity requires a genetic program for coordinated growth and cell differentiation.	577
26.4	**Plants Versus Animals**	578
	Cell walls shape patterns of growth and development in plants.	578
	Animal cells can move relative to one another.	579

27.1	**Photosynthesis on Land**	591
	Land plants share many cellular features with green algae.	591
	Bryophytes rely on surface water for hydration.	592
	Water pulled from the soil moves through vascular plants by bulk flow.	593
	Vascular plants produce four major organ types.	594
27.2	**Carbon Dioxide Gain and Water Loss**	595
	CO_2 uptake results in water loss.	595
	The cuticle restricts water loss from leaves but inhibits the uptake of CO_2.	596
	Stomata allow leaves to regulate water loss and carbon gain.	596
	CAM plants use nocturnal CO_2 storage to avoid water loss during the day.	597
	C_4 plants suppress photorespiration by concentrating CO_2 in bundle-sheath cells.	598
	HOW DO WE KNOW? Does C_4 photosynthesis suppress photorespiration?	600
27.3	**Water Transport**	601
	Xylem provides a low-resistance pathway for the movement of water.	601

	Water is pulled through xylem by an evaporative pump.	602	

Water is pulled through xylem by an evaporative pump. 602

HOW DO WE KNOW? Do plants generate negative pressures? 603

Xylem transport is at risk of conduit collapse and cavitation. 605

27.4 Transport of Carbohydrates 606

Phloem transports carbohydrates from sources to sinks. 606

Carbohydrates are pushed through phloem by an osmotic pump. 607

Phloem feeds both the plant and the rhizosphere. 607

27.5 Uptake of Water and Nutrients 608

Plants obtain nutrients from the soil. 608

Nutrient uptake by roots is highly selective. 608

Nutrient uptake requires energy. 610

Mycorrhizae enhance nutrient uptake. 611

Symbiotic nitrogen-fixing bacteria supply nitrogen to both plants and ecosystems. 612

How has nitrogen availability influenced agricultural productivity? 613

CHAPTER 28 PLANT REPRODUCTION 616

Rogelio Moreno/SCIENCE PHOTO LIBRARY/Science Source

28.1 Alternation of Generations 617

The algal sister groups of land plants have one multicellular generation in their life cycle. 618

Land plants have two multicellular generations in their life cycle. 618

Bryophytes illustrate how the alternation of generations allows the dispersal of spores in the air. 619

Dispersal enhances reproductive fitness in several ways. 620

Spore-dispersing vascular plants have free-living gametophytes and sporophytes. 620

28.2 Seed Plants 622

The seed plant life cycle is distinguished by four major steps. 622

Pine trees illustrate how the transport of pollen in air allows fertilization to occur in the absence of external sources of water. 623

Seeds enhance the establishment of the next sporophyte generation. 624

28.3 Flowering Plants 626

Flowers are reproductive shoots specialized for the transfer and receipt of pollen. 626

The diversity of floral morphology is related to modes of pollination. 629

HOW DO WE KNOW? How do long nectar spurs evolve? 630

Angiosperms have mechanisms to increase outcrossing. 630

Angiosperms delay provisioning their ovules until after fertilization. 631

Fruits enhance the dispersal of seeds. 633

How did scientists increase crop yields during the Green Revolution? 635

28.4 Asexual Reproduction 635

Asexually produced plants disperse with and without seeds. 635

CHAPTER 29 PLANT GROWTH AND DEVELOPMENT 638

Dietrich Rose/Getty Images

29.1 Shoot Growth and Development 639

Stems grow by adding new cells at their tips. 639

Stem elongation occurs just below the apical meristem. 641

Stems branch by producing new apical meristems. 641

The shoot apical meristem controls the production and arrangement of leaves. 642

Young leaves develop vascular connections to the stem. 643

Flower development terminates the growth of shoot meristems. 644

29.2 Plant Hormones 644

Hormones affect the growth and differentiation of plant cells. 645

Polar transport of auxin guides the placement of leaf primordia and the development of vascular connections with the stem. 646

What is the developmental basis for the shorter stems of high-yielding rice and wheat? 646

Cytokinins, in combination with other hormones, control the outgrowth of branches. 647

29.3	**Secondary Growth**	648
	Shoots produce two types of lateral meristem.	648
	The vascular cambium produces secondary xylem and phloem.	648
	The cork cambium produces an outer protective layer.	650
	Wood has both support and transport functions.	650
29.4	**Root Growth and Development**	651
	Roots grow by producing new cells at their tips.	651
	Root elongation and vascular development are coordinated.	652
	The formation of new root apical meristems allows roots to branch.	653
	The structures and functions of root systems are diverse.	653
29.5	**The Environmental Context of Growth and Development**	654
	Plants orient the growth of their stems and roots by light and gravity.	654
	HOW DO WE KNOW? How do plants grow toward light?	655
	Seeds can delay germination if they detect the presence of plants overhead.	656
	HOW DO WE KNOW? How do seeds detect the presence of plants growing overhead?	657
	Plants grow taller and branch less when growing in the shade of other plants.	658
	Roots elongate more and branch less when water is scarce.	658
	Exposure to wind results in shorter and stronger stems.	659
29.6	**Timing of Developmental Events**	660
	Flowering time is affected by day length.	660
	Plants use their internal circadian clock and photoreceptors to determine day length.	661
	Vernalization prevents plants from flowering until winter has passed.	662
	Plants use day length as a cue to prepare for winter.	662

CHAPTER 30 PLANT DEFENSE 664

30.1	**Protection Against Pathogens**	665
	Plant pathogens infect and exploit host plants by a variety of mechanisms.	665
	Plants are able to detect and respond to pathogens.	666
	Plants respond to infections by isolating infected regions.	668
	Mobile signals trigger defenses in uninfected tissues.	668
	HOW DO WE KNOW? Can plants develop immunity to specific pathogens?	669
	Plants defend against viral infections by producing siRNA.	670
	A pathogenic bacterium provides a way to modify plant genomes.	670
30.2	**Defense Against Herbivores**	671
	Plants use mechanical and chemical defenses to avoid being eaten.	672
	Diverse chemical compounds deter herbivores.	673
	Some plants provide food and shelter for ants, which actively defend them.	674
	Grasses can regrow quickly following grazing by mammals.	675
30.3	**Allocating Resources to Defense**	676
	Some defenses are always present, whereas others are turned on in response to a threat.	676
	Plants can sense and respond to herbivores.	676
	Plants produce volatile signals that attract insects that prey upon herbivores.	677
	Nutrient-rich environments select for plants that allocate more resources to growth than to defense.	677
	HOW DO WE KNOW? Can plants communicate?	678
	Exposure to multiple threats can lead to trade-offs.	679
30.4	**Defense and Plant Diversity**	679
	The evolution of new defenses may allow plants to diversify.	680
	Pathogens, herbivores, and seed predators can increase plant diversity.	680
	Can modifying plants genetically protect crops from herbivores and pathogens?	680

CHAPTER 31 PLANT DIVERSITY — 684

Alfredo Maiquez/Getty Images

31.1 Major Themes in the Evolution of Plant Diversity — 685

Four major transformations in life cycle and structure characterize the evolutionary history of plants. — 685

Plant diversity has changed over time. — 688

31.2 Bryophytes — 689

Bryophytes are small and tough. — 689

The small gametophytes and unbranched sporophytes of bryophytes are adaptations for reproducing on land. — 690

Bryophytes exhibit several cases of convergent evolution with the vascular plants. — 690

Sphagnum moss plays an important role in the global carbon cycle. — 691

31.3 Spore-Dispersing Vascular Plants — 692

Rhynie chert fossils provide a window into the early evolution of vascular plants. — 692

Lycophytes evolved leaves and roots independently from all other vascular plants. — 693

Ancient lycophytes included giant trees that dominated coal swamps about 320 million years ago. — 694

HOW DO WE KNOW? Did woody plants evolve more than once? — 695

Ferns and horsetails are morphologically and ecologically diverse. — 696

Fern diversity has been strongly affected by the evolution of angiosperms. — 697

31.4 Gymnosperms — 697

Seed plants have been the dominant plants on land for more than 200 million years. — 698

Cycads and ginkgos were once both diverse and widespread. — 698

Conifers are woody plants that thrive in dry and cold climates. — 700

Gnetophytes are gymnosperms that have independently evolved xylem vessels and double fertilization. — 700

31.5 Angiosperms — 701

Several innovations increased the efficiency of the angiosperm life cycle and xylem transport. — 701

Angiosperm diversity may result in part from coevolutionary interactions with animals and other organisms. — 702

Monocots are diverse in shape and size despite not forming a vascular cambium. — 703

HOW DO WE KNOW? When did grasslands expand over the land surface? — 704

Eudicots are the most diverse group of angiosperms. — 705

? What can be done to protect the genetic diversity of crop species? — 706

VS **VISUAL SYNTHESIS** Angiosperms: Structure and Function — 708

CHAPTER 32 FUNGI — 712

Chris Ted/Getty Images

32.1 Growth and Nutrition — 713

Hyphae permit fungi to explore their environment for food resources. — 713

Fungi transport materials within their hyphae. — 714

Not all fungi produce hyphae. — 715

Fungi are principal decomposers of plant tissues. — 715

Fungi are important plant and animal pathogens. — 716

Many fungi form symbiotic associations with plants and animals. — 717

Lichens are symbioses between a fungus and a green alga or a cyanobacterium. — 718

32.2 Reproduction — 719

Fungi proliferate and disperse using spores. — 719

Multicellular fruiting bodies facilitate the dispersal of sexually produced spores. — 720

Sexual reproduction in fungi often includes a stage in which haploid cells fuse, but nuclei do not. — 721

HOW DO WE KNOW? What determines the shape of fungal spores that are ejected into the air? — 721

Genetically distinct mating types promote outcrossing. — 723

Parasexual fungi generate genetic diversity by asexual means. — 723

32.3 Diversity — 724

Fungi are highly diverse. — 724

Chytrids are aquatic fungi that lack hyphae. — 724

	Zygomycetes produce hyphae undivided by septa.	725		The animal body plans we see today emerged during the Cambrian Period.	750

Zygomycetes produce hyphae undivided by septa. 725

Glomeromycetes form endomycorrhizae. 726

The Dikarya produce regular septa during mitosis. 726

Ascomycetes are the most diverse group of fungi. 726

Basidiomycetes include smuts, rusts, and mushrooms. 728

HOW DO WE KNOW? Can a fungus influence the behavior of an ant? 728

How do fungi threaten global wheat production? 731

CASE 7 Biology-Inspired Design: Using Nature to Solve Problems 734

CHAPTER 33 ANIMAL FORM, FUNCTION, AND EVOLUTIONARY HISTORY 738

4FR/Getty Images

33.1 Animal Body Plans 739

What is an animal? 739

Animals can be classified based on type of symmetry. 740

Many animals have a brain and specialized sensory organs at the front of the body. 740

Some animals also show segmentation. 742

Animals can be classified based on the number of their germ layers. 742

Molecular sequence comparisons have confirmed some relationships and raised new questions. 744

Can we mimic the form and function of animals to build robots? 745

33.2 Tissues and Organs 745

Most animals have four types of tissues. 746

Tissues are organized into organs that carry out specific functions. 747

33.3 Homeostasis 748

Homeostasis is the active maintenance of stable conditions inside of cells and organisms. 748

Homeostasis is often achieved by negative feedback. 748

33.4 Evolutionary History 749

Fossils and phylogeny show that animal forms were initially simple but rapidly evolved complexity. 749

The animal body plans we see today emerged during the Cambrian Period. 750

Five mass extinctions have changed the trajectory of animal evolution during the past 500 million years. 751

Animals began to colonize the land 420 million years ago. 752

Some animals show a trend toward increased body size over time. 752

HOW DO WE KNOW? Do animals tend to get bigger over time? 753

CHAPTER 34 ANIMAL NERVOUS SYSTEMS 756

Callista Images/AGE Fotostock

34.1 Nervous System Function and Evolution 757

Animal nervous systems have three types of nerve cells. 757

Nervous systems range from simple to complex. 758

34.2 Neuron Structure 759

Neurons share a common organization. 759

Neurons differ in size and shape. 760

Neurons are supported by other types of cells. 761

34.3 Signal Transmission 761

The resting membrane potential is negative and results in part from the movement of potassium and sodium ions. 761

Neurons are excitable cells that transmit information by action potentials. 762

Neurons propagate action potentials along their axons by sequentially opening and closing adjacent Na^+ and K^+ ion channels. 764

HOW DO WE KNOW? How does electrical activity change during an action potential? 766

Neurons communicate at synapses. 766

Signals between neurons can be excitatory or inhibitory. 768

34.4 Nervous System Organization 769

Nervous systems are organized into peripheral and central components. 769

Peripheral nervous systems have voluntary and involuntary components. 770

Simple reflex circuits provide rapid responses to stimuli. 772

liii

34.5	**Sensory Systems**	773
	Sensory receptor cells detect diverse stimuli.	773
	Sensory transduction converts a stimulus into an electrical impulse.	773
	Chemoreceptors respond to chemical stimuli.	773
	Mechanoreceptors detect physical forces.	775
	How do cochlear implants work?	778
	Electromagnetic receptors sense light.	778
34.6	**Brain Organization and Function**	780
	The brain processes and integrates information received from different sensory systems.	780
	The brain is divided into lobes with specialized functions.	781
	Information is topographically mapped into the vertebrate cerebral cortex.	782
	The brain allows for memory, learning, and cognition.	782

CHAPTER 35 ANIMAL MOVEMENT: MUSCLES AND SKELETONS — 786

SPL/Science Source

35.1	**How Muscles Work**	787
	Muscles use chemical energy to produce force and movement.	788
	Muscles can be striated or smooth.	788
	Skeletal and cardiac muscle fibers are organized into repeating contractile units called sarcomeres.	789
	Muscles contract by the sliding of myosin and actin protein filaments.	790
	Calcium regulates actin–myosin interaction through excitation–contraction coupling.	792
	Calmodulin regulates calcium activation and relaxation of smooth muscle.	793
35.2	**Muscle Contractile Properties**	794
	Antagonist pairs of muscles produce reciprocal motions at a joint.	794
	HOW DO WE KNOW? How does actin–myosin filament overlap affect force generation in muscles?	795
	Muscle length affects actin–myosin overlap and generation of force.	795
	Muscle force and shortening velocity are inversely related.	796
	Muscle force is summed by an increase in stimulation frequency and the recruitment of motor units.	796
	Skeletal muscles have slow-twitch and fast-twitch fibers.	798
	How do different types of muscle fibers affect the speed of animals?	799
35.3	**Animal Skeletons**	799
	Hydrostatic skeletons support animals by muscles that act on a fluid-filled cavity.	799
	Exoskeletons provide hard external support and protection.	801
	The rigid bones of vertebrate endoskeletons are jointed for motion and can be repaired if damaged.	802
35.4	**Vertebrate Skeletons**	803
	Vertebrate bones form directly or by forming a cartilage model first.	803
	The two main types of bone are compact bone and spongy bone.	803
	Bones grow in length and width, and can be repaired.	804
	Can we design smart materials to heal bones that are damaged and improve artificial joints?	805
	Joint shape determines range of motion and skeletal muscle organization.	805

CHAPTER 36 ANIMAL ENDOCRINE SYSTEMS — 808

FLPA/Chris Mattison/AGE Fotostock

36.1	**Endocrine Function**	809
	The endocrine system helps to regulate an organism's response to its environment.	809
	The endocrine system regulates growth and development.	810
	HOW DO WE KNOW? How are growth and development controlled in insects?	811
	The endocrine system underlies homeostasis.	812
36.2	**Hormones**	815
	Hormones act specifically on cells that bind the hormone.	815
	Two main classes of hormones are peptide and amines, and steroid hormones.	815

Hormonal signals are amplified to produce a strong effect. 816

Hormones are evolutionarily conserved molecules with diverse functions. 817

36.3 The Vertebrate Endocrine System 820

The pituitary gland integrates diverse bodily functions by secreting hormones in response to signals from the hypothalamus. 820

Many targets of pituitary hormones are endocrine tissues that also secrete hormones. 822

Other endocrine organs have diverse functions. 822

The fight-or-flight response represents a change in set point in many different organs. 823

36.4 Other Forms of Chemical Communication 824

Local chemical signals regulate neighboring target cells. 824

Pheromones are chemical compounds released into the environment that signal physiological and behavioral changes. 824

CHAPTER 37 ANIMAL CARDIOVASCULAR AND RESPIRATORY SYSTEMS 828

Power and Syred/Science Source

37.1 Delivery of Oxygen and Elimination of Carbon Dioxide 829

Diffusion governs gas exchange over short distances. 829

Bulk flow moves fluid over long distances. 830

37.2 Respiratory Gas Exchange 831

Many aquatic animals breathe through gills. 832

Insects breathe air through tracheae. 834

Most terrestrial vertebrates breathe by tidal ventilation of internal lungs. 834

Mammalian lungs are well adapted for gas exchange. 835

The structure of bird lungs allows unidirectional airflow for increased oxygen uptake. 837

Voluntary and involuntary mechanisms control breathing. 837

37.3 Oxygen Transport by Hemoglobin 838

Blood is composed of fluid and several types of cell. 838

Hemoglobin is an ancient molecule with diverse roles related to oxygen binding and transport. 838

Hemoglobin reversibly binds oxygen. 839

HOW DO WE KNOW? What is the molecular structure of hemoglobin and myoglobin? 839

Myoglobin stores oxygen, enhancing oxygen delivery to muscle mitochondria. 840

Many factors affect hemoglobin–oxygen binding. 840

37.4 Circulatory Systems 842

Circulatory systems have vessels of different sizes. 843

Arteries are muscular vessels that carry blood away from the heart under high pressure. 844

Veins are thin-walled vessels that return blood to the heart under low pressure. 845

Compounds and fluid move across capillary walls by diffusion, filtration, and osmosis. 845

Hormones and nerves provide homeostatic regulation of blood pressure. 845

37.5 Structure and Function of the Heart 846

Fishes have two-chambered hearts and a single circulatory system. 847

Amphibians and reptiles have three-chambered hearts and partially divided circulations. 847

Mammals and birds have four-chambered hearts and fully divided pulmonary and systemic circulations. 848

How can we engineer replacement heart valves? 849

Cardiac muscle cells are electrically connected to contract in synchrony. 850

Heart rate and cardiac output are regulated by the autonomic nervous system. 851

CHAPTER 38 ANIMAL METABOLISM, NUTRITION, AND DIGESTION 854

Thomas Kokta/Getty Images

38.1 Patterns of Animal Metabolism 855

Animals rely on anaerobic and aerobic metabolism. 855

Metabolic rate varies with activity level. 857

Metabolic rate is affected by body size. 858

Metabolic rate is linked to body temperature. 858

lv

	HOW DO WE KNOW? How is metabolic rate affected by running speed and body size?	859	**39.2 Excretion of Wastes**	881

HOW DO WE KNOW? How is metabolic rate affected by running speed and body size? 859

VISUAL SYNTHESIS Homeostasis and Thermoregulation 860

38.2 Nutrition and Diet 862
Energy balance is a form of homeostasis. 862

An animal's diet must supply nutrients that it cannot synthesize. 862

38.3 Adaptations for Feeding 864
Suspension filter feeding is common in many aquatic animals. 864

Large aquatic animals apprehend prey by suction feeding and active swimming. 864

Specialized structures allow for capture and mechanical breakdown of food. 865

38.4 Regional Specialization of the Gut 867
Most animal digestive tracts have three main parts: a foregut, midgut, and hindgut. 867

Digestion begins in the mouth. 868

Further digestion and storage of nutrients take place in the stomach. 868

Final digestion and nutrient absorption take place in the small intestine. 870

The large intestine absorbs water and stores waste. 872

The lining of the digestive tract is composed of distinct layers. 873

Plant-eating animals have specialized digestive tracts adapted to their diets. 873

CHAPTER 39 ANIMAL RENAL SYSTEMS: WATER AND WASTE 876

Steve Allen/Getty Images

39.1 Water and Electrolyte Balance 877
Osmosis governs the movement of water across cell membranes. 877

Osmoregulation is the control of osmotic pressure inside cells and organisms. 879

Osmoconformers match their internal solute concentration to that of the environment. 879

Osmoregulators have internal solute concentrations that differ from that of their environment. 879

39.2 Excretion of Wastes 881
The excretion of nitrogenous wastes is linked to an animal's habitat and evolutionary history. 882

Excretory organs work by filtration, reabsorption and secretion. 883

Animals have diverse excretory organs. 884

Vertebrates filter blood under pressure through paired kidneys. 885

39.3 The Mammalian Kidney 887
The mammalian kidney has an outer cortex and inner medulla. 887

Glomerular filtration isolates wastes carried by the blood along with water and small solutes. 888

The proximal convoluted tubule reabsorbs solutes by active transport. 888

The loop of Henle acts as a countercurrent multiplier to create a concentration gradient from the cortex to the medulla. 889

HOW DO WE KNOW? How does the mammalian kidney produce concentrated urine? 891

The distal convoluted tubule secretes additional wastes. 891

The final concentration of urine is determined in the collecting ducts and is under hormonal control. 892

The kidneys help regulate blood pressure and blood volume. 893

How does dialysis work? 894

CHAPTER 40 ANIMAL REPRODUCTION AND DEVELOPMENT 896

SPL/Science Source

40.1 The Evolutionary History of Reproduction 897
Asexual reproduction produces clones. 897

Sexual reproduction involves the formation and fusion of gametes. 899

Many species reproduce both sexually and asexually. 899

Exclusive asexuality is often an evolutionary dead end. 900

HOW DO WE KNOW? Do bdelloid rotifers only reproduce asexually? 902

40.2 Movement onto Land and Reproductive Adaptations 903

Fertilization can take place externally or internally. 903

r-strategists and *K*-strategists differ in number of offspring and parental care. 903

Animals either lay eggs or give birth to live young. 904

40.3 Human Reproductive Anatomy and Physiology 905

The male reproductive system is specialized for the production and delivery of sperm. 905

The female reproductive system produces eggs and supports the developing embryo. 907

Hormones regulate the human reproductive system. 908

40.4 Gamete Formation to Birth in Humans 911

Male and female gametogenesis have both shared and distinct features. 911

Fertilization occurs when a sperm fuses with an oocyte. 912

The first trimester includes cleavage, gastrulation, and organogenesis. 914

The second and third trimesters are characterized by fetal growth. 916

Childbirth is initiated by hormonal changes. 917

VISUAL SYNTHESIS Reproduction and Development 918

CHAPTER 41 ANIMAL IMMUNE SYSTEMS 922

SPL/Science Source

41.1 An Overview of the Immune System 923

Pathogens cause disease. 923

The immune system distinguishes self from nonself. 924

The immune system consists of innate and adaptive immunity. 925

41.2 Innate Immunity 925

The skin and mucous membranes provide the first line of defense against infection. 925

White blood cells provide a second line of defense against pathogens. 926

Phagocytes recognize foreign molecules and send signals to other cells. 927

Inflammation is a coordinated response to tissue injury. 928

The complement system participates in the innate and adaptive immune systems. 929

How can we use a protein that circulates in the blood to treat sepsis? 930

41.3 B Cells and Antibodies 930

B cells produce antibodies. 931

Mammals produce five classes of antibodies with different functions. 932

Clonal selection is the basis for antibody specificity. 932

Clonal selection explains immunological memory. 933

HOW DO WE KNOW? How is antibody diversity generated? 934

Genomic rearrangement generates antibody diversity. 935

41.4 T Cells and Cell-Mediated Immunity 936

T cells include helper and cytotoxic cells. 936

T cells have T cell receptors on their surface that recognize an antigen in association with MHC proteins. 936

The ability to distinguish between self and nonself is acquired during T cell maturation. 938

41.5 Three Pathogens 939

The flu virus evades the immune system by antigenic drift and shift. 939

Tuberculosis is caused by a slow-growing, intracellular bacterium. 940

The malaria parasite changes surface molecules by antigenic variation. 941

CHAPTER 42 ANIMAL DIVERSITY 944

Brian Farrell

42.1 Sponges, Cnidarians, Ctenophores, and Placozoans 945

Sponges share some features with choanoflagellates but also exhibit adaptations conferred by multicellularity. 946

Cnidarians are the architects of life's largest constructions: coral reefs. 948

Ctenophores and placozoans represent the extremes of body organization among phyla that branch from early nodes on the animal tree. 949

Branching relationships among early nodes on the animal tree remain uncertain. 951

| 42.2 | **Protostome Animals** | 952 |

Lophotrochozoans account for nearly half of all animal phyla, including the diverse and ecologically important annelids and mollusks. 952

Ecdysozoans are animals that episodically molt their external cuticle during growth. 956

| 42.3 | **Arthropods** | 957 |

The Arthropoda is the most diverse animal phylum. 957

HOW DO WE KNOW? How did the diverse feeding structures of arthropods arise? 958

Insects make up the majority of all known animal species and have adaptations that allow them to live in diverse habitats. 959

| 42.4 | **Deuterostome Animals** | 960 |

Hemichordates include acorn worms and pterobranchs, and echinoderms include sea stars and sea urchins. 961

Chordates include vertebrates, cephalochordates, and tunicates. 961

| 42.5 | **Vertebrates** | 963 |

Fish are the most diverse vertebrate animals. 964

The common ancestor of tetrapods had four limbs. 966

Amniotes evolved terrestrial eggs. 967

VISUAL SYNTHESIS Diversity Through Time 970

CASE 8 Conserving Biodiversity: Rainforest and Coral Reef Hotspots 974

CHAPTER 43 BEHAVIOR AND BEHAVIORAL ECOLOGY 978

Dan L. Perlman/EcoLibrary.org

| 43.1 | **Tinbergen's Questions** | 979 |

Tinbergen asked proximate and ultimate questions about behavior. 980

| 43.2 | **Dissecting Behavior** | 981 |

The fixed action pattern is a stereotyped behavior. 981

The nervous system processes stimuli and evokes behaviors. 981

Hormones can trigger certain behaviors. 982

Breeding experiments can help determine the degree to which a behavior is genetic. 984

Molecular techniques provide new ways of testing the role of genes in behavior. 984

HOW DO WE KNOW? Can the same gene influence behavior differently in different species? 986

| 43.3 | **Learning** | 987 |

Non-associative learning occurs without linking two events. 987

Associative learning occurs when two events are linked. 987

HOW DO WE KNOW? To what extent are insects capable of learning? 988

Learning is an adaptation. 988

| 43.4 | **Information Processing** | 989 |

Orientation involves a directed response to a stimulus. 989

Navigation is demonstrated by the remarkable ability of homing in birds. 990

Biological clocks provide important time cues for many behaviors. 990

HOW DO WE KNOW? Does a biological clock play a role in birds' ability to orient? 991

| 43.5 | **Communication** | 991 |

Communication is the transfer of information between a sender and a receiver. 992

Some forms of communication are complex and learned during a sensitive period. 993

Various forms of communication can convey specific information. 994

| 43.6 | **Social Behavior** | 994 |

Group selection is a weak explanation of altruistic behavior. 995

Reciprocal altruism is one way that altruism can evolve. 995

Kin selection is based on the idea that it is possible to contribute genetically to future generations by helping close relatives. 996

CHAPTER 44 POPULATION ECOLOGY 1000

Eladio M. Fernandez

| 44.1 | **Populations and Their Properties** | 1001 |

A population includes all the individuals of a species in a particular place. 1002

Three key features of a population are its size, range, and density.	1002
Ecologists estimate population size by sampling.	1003
HOW DO WE KNOW? How many butterflies are there in a given population?	1004

44.2 Population Growth and Decline — 1005

- Population size is affected by birth, death, immigration, and emigration. — 1005
- Population size increases rapidly when the per capita growth rate is constant over time. — 1005
- Carrying capacity is the maximum number of individuals a habitat can support. — 1007
- Logistic growth produces an S-shaped curve and describes the growth of many natural populations. — 1007
- Factors that influence population growth can be dependent on or independent of population density. — 1008

44.3 Age-Structured Population Growth — 1009

- Birth and death rates vary with age and environment. — 1009
- Survivorship curves record changes in survival probability over an organism's life-span. — 1010
- Patterns of survivorship vary among organisms. — 1011
- Reproductive patterns reflect the predictability of a species' environment. — 1012
- The life history of an organism shows trade-offs among physiological functions. — 1012

44.4 Metapopulation Dynamics — 1013

- A metapopulation is a group of populations linked by immigrants. — 1013
- How do populations colonize islands? — 1015

CHAPTER 45 SPECIES INTERACTIONS AND COMMUNITIES — 1018

Charles J. Smith

45.1 The Niche — 1019

- The niche is a species' place in nature. — 1019
- The realized niche of a species is more restricted than its fundamental niche. — 1020
- Niches are shaped by evolutionary history. — 1021

45.2 Antagonistic Interactions — 1021

- Limited resources foster competition. — 1022
- Species compete for resources other than food. — 1022
- Competitive exclusion prevents two species from occupying the same niche at the same time. — 1022
- Can competition drive species diversification? — 1023
- Predation, parasitism, and herbivory are interactions in which one species benefits at the expense of another. — 1024
- **HOW DO WE KNOW?** Can predators and prey coexist stably in certain environments? — 1025

45.3 Mutualistic Interactions — 1026

- Mutualisms are interactions between species that benefit both participants. — 1027
- Mutualisms may evolve increasing interdependence. — 1027
- Digestive symbioses recycle plant material. — 1027
- **HOW DO WE KNOW?** Have aphids and their symbiotic bacteria coevolved? — 1028
- Mutualisms may be obligate or facultative. — 1029
- In some interactions, one partner is unaffected by the interaction. — 1029
- Facilitation occurs when one species indirectly benefits another. — 1030
- The costs and benefits of species interactions can change over time. — 1030

45.4 Communities — 1031

- Species that live in the same place make up communities. — 1031
- How is biodiversity measured? — 1032
- Species influence each other in a complex web of interactions. — 1032
- Keystone species have disproportionate effects on communities. — 1033
- Disturbance can modify community composition. — 1034
- Succession describes the community response to new habitats or disturbance. — 1035
- Island biogeography explains species diversity on habitat islands. — 1036
- **VISUAL SYNTHESIS** Succession: Ecology in Microcosm — 1038

CHAPTER 46 ECOSYSTEM ECOLOGY — 1042

Andre Gallant/Getty Images

46.1 The Short-Term Carbon Cycle — 1043

- The Keeling curve records changing levels of carbon dioxide in the atmosphere over time. — 1044

Photosynthesis and respiration are key processes in short-term carbon cycling.	1044
The regular oscillation of CO_2 reflects the seasonality of photosynthesis in the Northern Hemisphere.	1045
Human activities play an important role in the modern carbon cycle.	1045
HOW DO WE KNOW? How much CO_2 was in the atmosphere 1000 years ago?	1046
Carbon isotopes show that much of the CO_2 added to air over the past half century comes from burning fossil fuels.	1046
HOW DO WE KNOW? What is the major source of the CO_2 that has accumulated in Earth's atmosphere over the past two centuries?	1047
46.2 The Long-Term Carbon Cycle	**1048**
Records of atmospheric composition over 400,000 years show periodic shifts in CO_2 levels.	1049
Reservoirs and fluxes are key in long-term carbon cycling.	1051
46.3 Food Webs and Trophic Pyramids	**1053**
Food webs trace carbon and other elements through communities and ecosystems.	1053
Energy as well as carbon is transferred through ecosystems.	1054
46.4 Other Biogeochemical Cycles	**1057**
The nitrogen cycle is closely linked to the carbon cycle.	1057
Phosphorus cycles through ecosystems, supporting primary production.	1058
46.5 The Ecological Framework of Biodiversity	**1059**
Biological diversity reflects the many ways that organisms participate in biogeochemical cycles.	1059
Biological diversity can influence primary production and therefore the biological carbon cycle.	1059
Biogeochemical cycles weave together biological evolution and environmental change through Earth history.	1059
HOW DO WE KNOW? Does species diversity promote primary productivity?	1060
VISUAL SYNTHESIS Flow of Matter and Energy Through Ecosystems	1062

CHAPTER 47 CLIMATE AND BIOMES	**1066**
Mike Hill/Getty Images	
47.1 Climate	**1067**
The principal control on Earth's surface temperature is the angle at which solar radiation strikes the surface.	1067
Heat is transported toward the poles by wind and ocean currents.	1068
Global circulation patterns determine patterns of rainfall, but topography also matters.	1070
47.2 Biomes	**1071**
Terrestrial biomes reflect the distribution of climate.	1072
Aquatic biomes reflect climate, the availability of nutrients and oxygen, and the depth to which sunlight penetrates through water.	1074
Marine biomes cover most of our planet's surface.	1083
47.3 Global Patterns	**1086**
Global patterns of primary production reflect variations in climate and nutrient availability.	1086
HOW DO WE KNOW? Does iron limit primary production in some parts of the oceans?	1087
Biodiversity is highest at the equator and lowest toward the poles.	1088
How do evolutionary and ecological history explain biodiversity?	1090
CHAPTER 48 THE ANTHROPOCENE	**1092**
Michael Utech/Getty Images	
48.1 The Anthropocene Epoch	**1093**
Humans are a major force on the planet.	1093
48.2 Human Influence on the Carbon Cycle	**1095**
As atmospheric carbon dioxide levels have increased, so has mean surface temperature.	1095

Changing environments affect species distribution and community composition. 1098

How has global environmental change affected coral reefs around the world? 1100

What can be done? 1101

HOW DO WE KNOW? How will rising twenty-first-century CO_2 levels affect coral reefs? 1102

48.3 Human Influence on the Nitrogen and Phosphorus Cycles 1104

Nitrogen fertilizer transported to lakes and the sea causes eutrophication. 1104

Phosphate fertilizer is also used in agriculture, but has finite sources. 1105

What can be done? 1105

48.4 Human Influence on Evolution 1107

Human activities have reduced the quality and size of many habitats, decreasing the number of species they can support. 1107

Overexploitation threatens species and disrupts ecological relationships within communities. 1108

Humans play an important role in the dispersal of species. 1108

Humans have altered the selective landscape for many pathogens. 1109

Are amphibians ecology's "canary in the coal mine"? 1110

48.5 Conservation Biology 1111

What are our conservation priorities? 1111

Conservation biologists have a diverse toolkit for confronting threats to biodiversity. 1112

Climate change provides new challenges for conservation biology in the twenty-first century. 1113

Sustainable development provides a strategy for conserving biodiversity while meeting the needs of the human population. 1113

48.6 Scientists and Citizens in the Twenty-First Century 1114

Glossary G-1

Index I-1

BIOLOGY
HOW LIFE WORKS

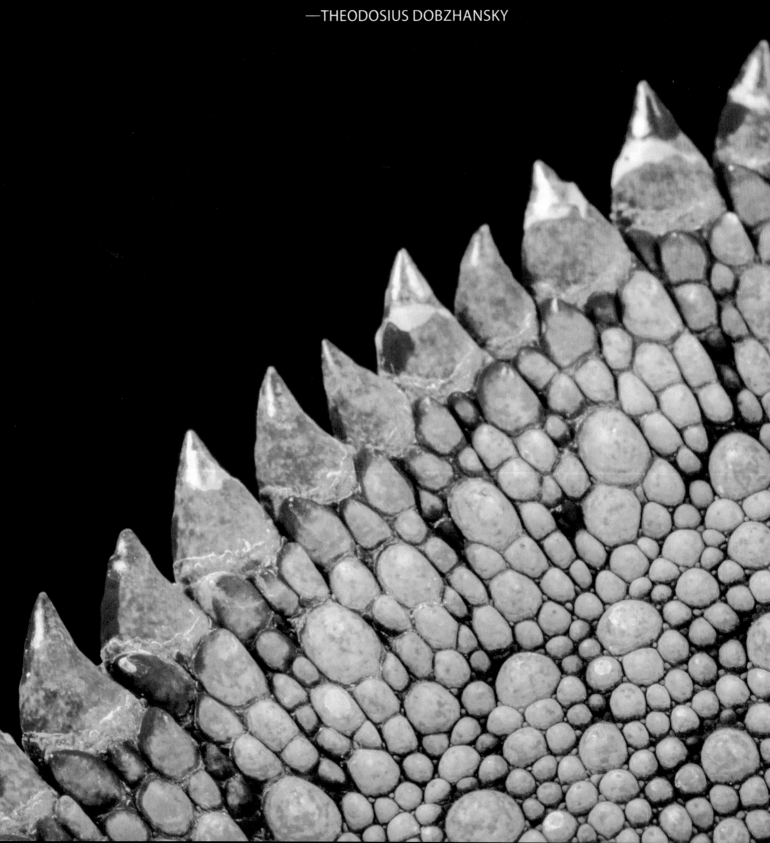

"Nothing in biology makes sense except in the light of evolution."
—THEODOSIUS DOBZHANSKY

PART 1

FROM CELLS TO ORGANISMS

CHAPTER 1 Life
Chemical, Cellular, and Evolutionary Foundations

CORE CONCEPTS

1.1 SCIENTIFIC INQUIRY: Scientific inquiry is a deliberate way of asking and answering questions about the natural world.

1.2 CHEMICAL AND PHYSICAL PROPERTIES: Life works according to fundamental principles of chemistry and physics.

1.3 THE CELL: The fundamental unit of life is the cell.

1.4 EVOLUTION: Evolution explains the features that organisms share and those that set them apart.

1.5 ECOLOGICAL SYSTEMS: Organisms interact with one another and with their physical environment, shaping ecological systems that sustain life.

1.6 THE HUMAN FOOTPRINT: In the twenty-first century, humans have become major agents in ecology and evolution.

Every day, remarkable things happen within and around you. Strolling through a local market, you come across a bin full of crisp apples, pick one up, and take a bite. Underlying this unremarkable occurrence is an extraordinary series of events. Your eyes sense the apple from a distance, and nerves carry that information to your brain, permitting identification. Biologists call this cognition, an area of biological study. Stimulated by the apple and recognizing it as ripe and tasty, your brain transmits impulses through nerves to your muscles. How we respond to external cues motivates behavior, another biological discipline. Grabbing the apple requires the coordinated activities of dozens of muscles that move your arm and hand to a precise spot. These movements are described by biomechanics, yet another area of biological research. And, as you bite down on the apple, glands in your mouth secrete saliva, starting to convert energy stored in the apple as sugar into energy that you will use to fuel your own activities. Physiology, like biomechanics, lies at the heart of biological function.

Studies of cognition, behavior, biomechanics, and physiology are all ways of approaching **biology,** the science of how life works. **Biologists,** scientists who study life, have come to understand a great deal about these and other processes at levels that run the gamut from molecular mechanisms within the cell, through the integrated actions of many cells within an organ or body, to the interactions among different organisms in nature. We don't know everything about how life works—in fact, it seems as if every discovery raises new questions. But biology provides us with an organized way of understanding ourselves and the world around us.

Why study biology? The example of eating an apple was deliberately chosen because it is an everyday occurrence that we ordinarily wouldn't think about twice. The scope of modern biology, however, is vast, raising questions that can fire our imaginations, affect our health, and influence our future. How, for example, will our understanding of the human genome change the way that we fight cancer? How do bacteria in our digestive system help determine health and well-being? Will expected changes in the temperature and pH of seawater doom coral reefs? Is there, or has there ever been, life on Mars? And, to echo the great storyteller Rudyard Kipling, why do leopards have spots, and tigers stripes?

We can describe six grand themes that connect and unite the many dimensions of life science, from molecules to the biosphere. These six themes are stated as Core Concepts for this chapter and are introduced in the following sections. Throughout the book, these themes will be visited again and again. We view them as the keys to understanding the many details in subsequent chapters and relating them to one another. Our hope is that by the time you finish this book, you will have an understanding of how life works, from the molecular machines inside cells and the metabolic pathways that cycle carbon through the biosphere to the process of evolution, which has shaped the living world that surrounds (and includes) us. You will, we hope, see the connections among these different ways of understanding life, and come away with a greater understanding of how scientists think about and ask questions about the natural world. How, in fact, do we know what we think we know about life? We also hope you will develop a basis for making informed decisions about your life, career, and the actions you take as a citizen.

1.1 SCIENTIFIC INQUIRY

How do we go about trying to understand the vastness and complexity of nature? For most scientists, studies of the natural world involve the complementary processes of observation and experimentation. **Observation** is the act of viewing the world around us. **Experimentation** is a disciplined and controlled way of asking and answering questions about the world in an unbiased manner.

Observation allows us to draw tentative explanations called hypotheses.

Observations allow us to ask focused questions about nature. Let's say you observe a hummingbird like the one pictured in **Fig. 1.1** hovering near a red flower, occasionally dipping its long beak into the bloom. What motivates this behavior? Is the bird feeding on some substance within the flower? Is it drawn to the flower by its vivid color? What benefit, if any, does the flower derive from this busy bird?

Observations such as these, and the questions they raise, allow us to propose tentative explanations, or **hypotheses**. We might, for example, hypothesize that the hummingbird is carrying pollen from one flower to the next, facilitating reproduction in the plant. Or we might hypothesize that nectar produced deep within the flower provides nutrition for the hummingbird—that the hummingbird's actions reflect its need to take in food. Both hypotheses provide a reasonable explanation of the behavior we observed, but they may or may not be correct. To find out, we have to test them.

Charles Darwin's classic book, *On the Origin of Species*, published in 1859, beautifully illustrates how we can piece together individual observations to construct a working hypothesis. In this book, Darwin discussed a wide range of observations, from pigeon breeding to fossils and from embryology to the unusual animals and plants found on islands. Darwin noted the success of animal breeders in selecting specific individuals for reproduction, thereby generating new breeds for agriculture or show. He appreciated that selective breeding is successful only if specific features of the animals can be passed from one generation to the next by inheritance. Reading economic treatises by the English clergyman Thomas Malthus, he understood that limiting environmental resources could select among the different individuals in populations in much the way that breeders select among cows or pigeons.

Gathering together all these seemingly disparate pieces of information, Darwin argued that life has evolved over time by means of natural selection. Since its formulation, Darwin's initial hypothesis has been tested by experiments, many thousands of them. Our knowledge of many biological phenomena, ranging from biodiversity to the way the human brain is wired, depends on direct observation followed by careful inferences that lead to models of how things work.

FIG. 1.1 A hummingbird visiting a flower. This simple observation leads to questions: why do hummingbirds pay so much attention to some flowers? Why do they hover near red flowers? *Source: Charles J. Smith.*

A hypothesis makes predictions that can be tested by observation and experiments.

Not just any idea qualifies as a hypothesis. Two features set hypotheses apart from other ways of attacking problems. First, a good hypothesis makes predictions about observations not yet made or experiments not yet run. Second, because hypotheses make predictions, we can test them. That is, we can devise an experiment to see whether the predictions made by the hypothesis actually occur, or we can go into the field to try to make further observations predicted by the hypothesis. A hypothesis, then, is a statement about nature that can be tested by experiments or by new observations. Hypotheses are testable because, even as they suggest an explanation for observations made previously, they make predictions about observations yet to be made.

Once we have a hypothesis, we can test it to see if its predictions are accurate (**Fig. 1.2**). Returning to the hummingbird and flower, we can test the hypothesis that the bird is transporting pollen from one flower to the next, enabling the plant to reproduce. Observation provides one type of test: if we catch and examine the bird just after it visits a flower, do we find pollen stuck to its beak or feathers? If so, our hypothesis survives the test.

Note, however, that we haven't proved the case. Pollen might be stuck on the bird for a different reason—perhaps it provides food for the hummingbird. However, if the birds *didn't* carry pollen from flower to flower, we would reject the hypothesis that they facilitate pollination. In other words, a single observation or experiment can lead us to reject a hypothesis, or it can support the hypothesis, but it cannot prove that a hypothesis is correct. To move forward, then, we might make a second set of observations. Does pollen that adheres to the hummingbird rub off when the bird visits a second flower of the same species? If so, we have stronger support for our hypothesis.

We might also use observations to test a more general hypothesis about birds and flowers. Does red color generally attract birds and so facilitate pollination in a wide range of flowers? To answer this question, we might catalog the pollination of many red flowers and ask whether they are pollinated mainly by birds. Or we might go the opposite direction and catalog the flowers visited by many different birds—are they more likely to be red than chance alone might predict?

Finally, we can test the hypothesis that the birds visit the flowers primarily to obtain food, spreading pollen as a side effect of their feeding behavior. We can measure the amount of nectar in the flower before and after the bird visits and calculate how much energy has been consumed by the bird during its visit. Continued observations over the course of the day will tell us whether the birds gain the nutrition they need by drinking nectar, and whether the birds have other sources of food.

In addition to observations, in many cases we can design experiments to test hypotheses (Fig. 1.2). One of the most powerful types of experiment is called a controlled experiment. In a controlled experiment, the researcher sets up several groups to be tested, keeping the conditions and setup as similar as possible from one group to the next. Then, the researcher deliberately introduces something different, known as a **variable,** into one group that he or she hypothesizes might have some sort of an effect. This is called the **test group.** In another group, the researcher does not introduce this variable. This is a **control group,** and the expectation is that no effect will occur in this group.

Controlled experiments are extremely powerful. By changing just one variable at a time, the researcher is able to determine whether that variable is important. If many variables were changed at once, it would be difficult, if not impossible, to draw conclusions from the experiment because the researcher would not be able to figure out which variable caused the outcome. The control group plays a key role as well. Having a group in which no change is expected ensures that the experiment works as it is supposed to and provides a baseline against which to compare the results of the test groups.

For example, we might test the hypothesis that hummingbirds facilitate pollination by doing a controlled experiment. In this case, we could set up groups of red flowers that are all similar to one another. For one group, we could surround the flowers with a fine mesh that allows small insects access to the plant but keeps hummingbirds away. For another group, we would not use a mesh. The variable, then, is the presence of a mesh; the test group is the flowers with the mesh; and the control group is the flowers without the mesh since the variable was not introduced in this group.

Will the flowers be pollinated? If only the group without the mesh is pollinated, this result lends support to our initial hypothesis. In this case, the hypothesis becomes less tentative and more certain. If both groups are pollinated, our hypothesis

FIG. 1.2 Scientific inquiry.

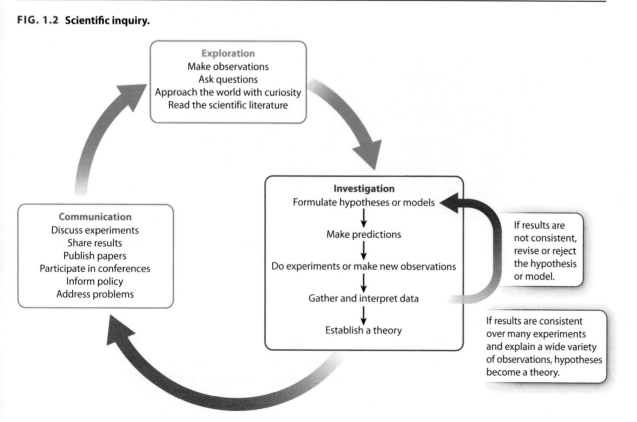

is not supported, in which case we may discard it for another explanation or change it to account for the new information.

Using observations to generate a hypothesis and then making predictions based on that hypothesis are two steps in what is sometimes called the scientific method. In reality, however, there is no single pathway to new knowledge about the world. To convey the full range of activities by which science works, we instead use the term **scientific inquiry** to describe the process outlined in Fig. 1.2. Scientific inquiry encompasses several careful and deliberate ways of asking and answering questions about the unknown. We ask questions, make observations, collect field or laboratory samples, and design and carry out experiments or analyses to make sense of things we initially do not understand. We then communicate what we find with other scientists and the public. Discussing and sharing ideas and results often lead to new questions, which in turn can be tested by observations and experiments. Scientific inquiry has been spectacularly successful in helping us to understand the world around us. We explore several aspects of scientific inquiry, including skills that scientists use to ask and answer questions, in your online courseware.

To emphasize the power of scientific inquiry, we turn to a famous riddle drawn from the fossil record (**Fig. 1.3**). Since the nineteenth century, paleontologists have known that before mammals expanded to their current ecological importance, other large animals dominated Earth. Dinosaurs evolved about 240 million years ago and disappeared abruptly 66 million years ago, along with many other species of plants, animals, and microscopic organisms. Because in many cases the skeletons and shells of these creatures were buried in sediment and became fossilized, we can read a record of their evolutionary history from layers of sedimentary rock deposited one atop another.

Working in Italy, the American geologist Walter Alvarez collected samples from the precise point in the rock layers that corresponds to the time of the extinction. Careful chemical analysis showed that rocks at this level are unusually enriched in the element iridium. Iridium is rare in most rocks on continents

HOW DO WE KNOW?

FIG. 1.3
What caused the extinction of the dinosaurs?

BACKGROUND Dinosaurs dominated the Earth for nearly 150 million years but became extinct about 66 million years ago.

OBSERVATION

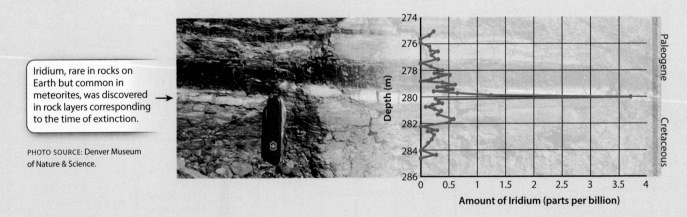

Iridium, rare in rocks on Earth but common in meteorites, was discovered in rock layers corresponding to the time of extinction.

PHOTO SOURCE: Denver Museum of Nature & Science.

HYPOTHESIS Geologist Walter Alvarez and his colleagues hypothesized that a large meteorite struck the Earth, disrupting communities on land and in the sea and causing the extinction of the dinosaurs and many other species.

PREDICTIONS A hypothesis makes predictions that can be tested by observation and experiments. To support the hypothesis, independent evidence of a meteor impact should be found in rock layers corresponding to the time of the extinction and be rare or absent in older and younger layers.

and the seafloor, but is relatively common in rocks that fall from space—that is, in meteorites. From these observations, Alvarez and his colleagues developed a remarkable hypothesis: 66 million years ago, a large meteor slammed into Earth, and in the resulting environmental havoc, dinosaurs and many other species became extinct. This hypothesis makes specific predictions, described in Fig. 1.3, which turn out to be supported by further observations.

A theory is a general explanation of natural phenomena supported by many experiments and observations.

As already noted, a hypothesis may initially be tentative. Indeed, it often provides only one of several possible ways of explaining existing data. With repeated observation and experimentation, however, a good hypothesis gathers strength, and we have more and more confidence in it. When a number of related hypotheses survive repeated testing and come to be accepted as good bases for explaining what we see in nature, scientists articulate a broader explanation that accounts for all the hypotheses and the results of their tests. We call this statement a **theory**, a general explanation of the world supported by a large body of experiments and observations (see Fig. 1.2).

Note that scientists use the word "theory" in a very particular way. In general conversation, "theory" is often synonymous with "hypothesis," "idea," or "hunch"—"I've got a theory about that." But in a scientific context, the word "theory" has a specific meaning. Scientists speak in terms of theories only if hypotheses have withstood testing to the point where they provide a general explanation for many observations and experimental results. Just as a good hypothesis makes testable predictions, a good theory both generates good hypotheses and predicts their outcomes. Thus, scientists talk about the theory of gravity—a set of hypotheses you test every day by walking down the street or dropping a fork. Similarly, the theory of evolution is not one explanation among many for the unity and diversity of life. Instead, it is a set of hypotheses that has been tested for more than a century and shown to provide an extraordinarily powerful explanation of biological observations that range from

FURTHER OBSERVATIONS

Quartz crystals that form only at high temperature and pressure—conditions met by giant meteors as they crash into Earth—occur abundantly in rock layers dated to the time of the extinction.

PHOTO SOURCE: Dr. David Kring/Science Source.

By 1990, geologists located the "smoking gun"—a crater (image on the right) of just the right age and size in the Yucatán Peninsula of Mexico (image on the left).

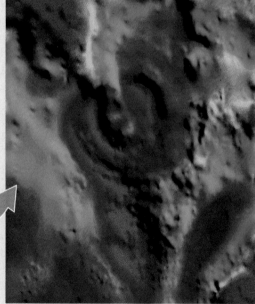

PHOTO SOURCE: Image courtesy of V. L. Sharpton/Lunar and Planetary Institute.

CONCLUSION A giant meteor struck the Earth 66 million years ago, causing the extinction of many species, including the dinosaurs.

FOLLOW-UP WORK Researchers have documented other mass extinctions, but the event that eliminated the dinosaurs appears to be the only one associated with a meteorite impact.

SOURCE Alvarez, W. 1998. *T. rex and the Crater of Doom*. New York: Vintage Press.

the amino acid sequences of proteins to the diversity of ants in a rain forest. In fact, as we discuss throughout this book, evolution is the single most important theory in all of biology. It provides the most general and powerful explanation of how life works.

> **Self-Assessment Questions**
>
> 1. How does a scientist turn an observation into a hypothesis and investigate that hypothesis?
> 2. What are the differences among a guess, hypothesis, and theory?
> 3. Mice that live in sand dunes commonly have light tan fur. Develop a hypothesis to explain this coloration.
> 4. What are the differences between a test group and a control group, and why is it important for an experiment to have both types of groups?
> 5. Design a controlled experiment to test the hypothesis that cigarette smoke causes lung cancer.

1.2 CHEMICAL AND PHYSICAL PRINCIPLES

We stated earlier that biology is the study of life. But what exactly *is* life? As simple as this question seems, it is frustratingly difficult to answer. We all recognize life when we see it, but coming up with a definition is harder than it first appears.

Living organisms are clearly different from nonliving things. But just how different is an organism from the rock shown in **Fig. 1.4**? On one level, the comparison is easy: the rock is much simpler than any living organism we can think of. It has far fewer components, and it is largely static, with no apparent response to environmental change on timescales that are readily tracked.

In contrast, even an organism as relatively simple as a bacterium contains many hundreds of different chemical compounds organized in a complex manner. The bacterium is also dynamic in that it changes continuously, especially in response to the environment. Unlike rocks and minerals, organisms metabolize—that is, they take in energy and materials from their environment and use those resources to grow and do work. Organisms reproduce, but minerals do not. And organisms do something else that rocks and minerals don't: they evolve. Indeed, the molecular biologist Gerald Joyce has defined life as a chemical system capable of undergoing Darwinian evolution.

From these simple comparisons, we can highlight key characteristics of living organisms: complexity, with precise spatial organization on several scales; the ability to change in response to the environment; the abilities to metabolize and to reproduce; and the capacity to evolve. Nevertheless, the living and nonliving worlds share an important attribute: both are subject to the laws of chemistry and physics.

The living and nonliving worlds follow the same chemical rules and obey the same physical laws.

The chemical elements found in rocks and other nonliving things are no different from those found in living organisms. In other words, all the elements that make up living things

FIG. 1.4 A climber scaling a rock. Living organisms like this climber contain chemicals that are found in rocks, but only living organisms reproduce in a manner that allows for evolution over time. *Source: Scott Hailstone/Getty Images.*

can be found in the nonliving environment—there is nothing special about our chemical components when taken individually. That said, the *relative* abundances of elements in organisms differ greatly from those in the nonliving world. In the universe as a whole, hydrogen and helium make up more than 99% of known matter, while Earth's crust contains mostly oxygen and silicon, with significant amounts of aluminum, iron, and calcium (**Fig. 1.5a**). In organisms, by contrast, oxygen, carbon, and hydrogen are by far the most abundant elements (**Fig. 1.5b**). As discussed more fully in Chapter 2, carbon provides the chemical backbone of life. The particular properties of carbon make possible a wide diversity of molecules that, in turn, support a wide range of functions within cells.

All living organisms are subject to the physical laws of the universe. Physics helps us to understand how animals move and why trees don't fall over; it explains how redwoods conduct water upward through their trunks and how oxygen gets into the cells that line your lungs. Indeed, two laws of thermodynamics, both of which describe how energy is transformed in any system, determine how living organisms are able to do work and maintain their spatial organization.

The **first law of thermodynamics** states that energy can neither be created nor destroyed; it can only be transformed from one form into another. In other words, the total energy in the universe is constant, but the form that energy takes can change. Living organisms are energy transformers. They acquire energy from the environment and transform it into a chemical form that cells can use. All organisms obtain energy from the sun or from chemical compounds. Some of this energy is used to do work—such as moving, reproducing, and building cellular components—and the rest is dissipated as heat. The energy that is used to do work plus the heat that is generated is the total amount of energy, which is the same as the input energy (**Fig. 1.6**). In other words, the total amount of energy remains constant before and after energy transformation.

The **second law of thermodynamics** states that the degree of disorder (or the number of possible positions and motions of molecules) in the universe tends to increase. Think about a box full of marbles distributed more or less randomly; if you want to line up all the red ones or blue ones in a row, you have to do work—that is, you have to add energy. In this case, the addition of energy increases the order of the system, or, put another way, decreases its disorder. Physicists quantify the amount of disorder (or the number of possible positions and motions of molecules) in a system as the **entropy** of the system.

Living organisms are highly organized. As with lining up marbles in a row, energy is needed to maintain this organization. Given the tendency toward greater disorder, the high level of organization of even a single cell would appear to violate the second law. But it does not. The key is that a cell is not an isolated system and therefore cannot be considered on its own; it exists in an environment. So we need to take into account the whole system, the cell plus the environment that surrounds it.

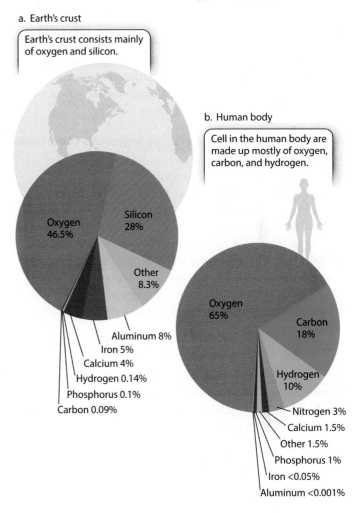

FIG. 1.5 Composition by mass of (a) Earth's crust and (b) cells in the human body. The Earth beneath our feet is made up of the same elements found in our feet, but in strikingly different proportions.

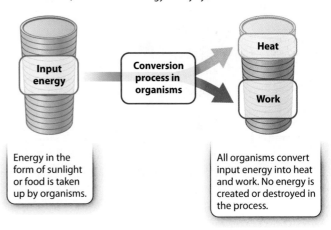

FIG. 1.6 Energy transformation and the first law of thermodynamics. The first law states that the total amount of energy in any system remains the same. Organisms transform energy from one form to another, but the total energy in any system is constant.

Only some of the energy harnessed by cells is used to do work; the rest is dissipated as heat (Fig. 1.6). That is, conversion of energy from one form to another is never 100% efficient. Heat is a form of energy, so the total amount of energy is conserved, as dictated by the first law. In addition, heat corresponds to the motion of small molecules—the greater the heat, the greater the motion, and the greater the motion, the greater the degree of disorder. Therefore, the release of heat as organisms harness energy means that the total entropy for the combination of the cell and its surroundings increases, in keeping with the second law (**Fig. 1.7**).

Scientific inquiry shows that living organisms come from other living organisms.

Life is made up of chemical components that also occur in the nonliving environment and obey the same laws of chemistry and physics. Can life spontaneously arise from these nonliving materials? We all know that living organisms come from other living organisms, but it is worth asking *how* we know this. Direct observation can be misleading here. For example, raw meat, if left out on a plate, will rot and become infested with maggots (fly larvae). It might seem as though the maggots appear spontaneously. In fact, the question of where maggots come from was a matter of vigorous debate for centuries. In the 1600s, the Italian physician and naturalist Francesco Redi conducted an experiment to test the hypothesis that maggots (and hence flies) in rotting meat come from other flies that laid their eggs in the meat. He found that living organisms come from other organisms and do not appear spontaneously (**Fig. 1.8**).

FIG. 1.7 Energy transformation and the second law of thermodynamics. The second law states that disorder in any system tends to increase. Entropy can decrease locally (inside a cell, for example) because the heat released increases disorder in the environment.

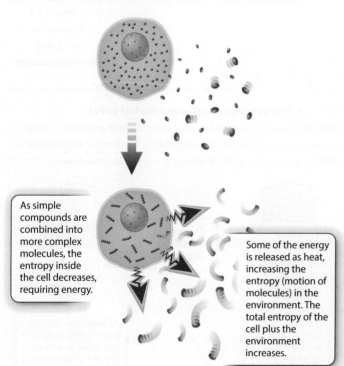

As simple compounds are combined into more complex molecules, the entropy inside the cell decreases, requiring energy.

Some of the energy is released as heat, increasing the entropy (motion of molecules) in the environment. The total entropy of the cell plus the environment increases.

HOW DO WE KNOW?

FIG. 1.8

Can living organisms arise from nonliving matter?

BACKGROUND Until the 1600s, many people believed that rotting meat spontaneously generates maggots (fly larvae).

HYPOTHESIS Francesco Redi hypothesized that maggots come from flies and are not spontaneously generated.

EXPERIMENT Redi placed meat in three jars. He left one jar open; he covered one with gauze; and he sealed one with a cap. The three jars were subject to the same conditions—the only variable was the opening. The open jar allowed for the passage of flies and air; the jar with the gauze allowed for the passage of air but not flies; and the sealed jar did not allow the passage of air or flies.

RESULTS Redi observed that maggots appeared only on the meat in the open jar. No maggots appeared in the other two jars, which did not allow access to the meat by flies.

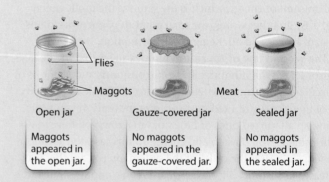

CONCLUSION The presence of maggots in the open jar and the absence of maggots in the gauze-covered and sealed jars supported the hypothesis that maggots come from flies.

FOLLOW-UP WORK Redi's experiment argued against the idea of spontaneous generation for insects. However, it was unclear whether his results could be extended to microbes. Applying the scientific process, Louis Pasteur used a similar approach about 200 years later to investigate this question (Fig. 1.9).

Some argued, however, that Redi's results might only apply to larger organisms; microscopic life might be another matter entirely. In the nineteenth century, the French chemist and biologist Louis Pasteur conducted another set of experiments using bacteria and came to the same conclusion as Redi, further supporting the hypothesis that life does not arise spontaneously (**Fig. 1.9**).

Redi's and Pasteur's experiments demonstrated that living organisms come from other living organisms. However, this conclusion raises the question of how life arose in the first place. If life comes from life, where did the first living organisms come from? Although today all organisms are produced by parental organisms, early in Earth's history this was not the case. Scientists hypothesize that life initially emerged from chemical compounds and reactions on early Earth about 4 billion years ago. We'll return to the great question of life's origin in Case 1 Life's Origins and in Chapters 2 through 8.

> **Self-Assessment Questions**
>
> 6. What are the first and second laws of thermodynamics, and how do they apply to living organisms?
> 7. What experimental evidence demonstrates that living organisms come from other living organisms?

1.3 THE CELL

The **cell** is the simplest entity that can exist as an independent unit of life. Every known living organism is either a single cell or an ensemble of a few to many cells (**Fig. 1.10**). Most bacteria (like those in Pasteur's experiment), yeasts, and the tiny algae that float in oceans and ponds spend their lives as single cells. In contrast, plants and animals contain billions to trillions of cells that function in a coordinated fashion.

HOW DO WE KNOW?

FIG. 1.9

Can microscopic life arise from nonliving matter?

BACKGROUND Educated people in the mid-nineteenth century knew that microbes (microorganisms like bacteria) grow well in nutrient-rich liquids such as broth. It was also known that boiling sterilizes broth, killing the microbes.

HYPOTHESIS Chemist and biologist Louis Pasteur hypothesized that microbes come from microbes and are not spontaneously generated. He argued that if microbes were generated spontaneously from nonliving matter, they would reappear in sterilized broth without the addition of microbes.

EXPERIMENT Pasteur filled two flasks with broth and boiled the broth to sterilize it and kill all the microbes. One flask had a straight neck and one had a swan neck. The straight-neck flask allowed dust particles with microbes to enter. The swan-neck flask did not. As in Redi's experiments (Fig 1.8), there was only one variable—in this case, the shape of the neck of the flask.

RESULTS Microbes grew in the broth inside the straight-neck flask but not in the swan-neck flask.

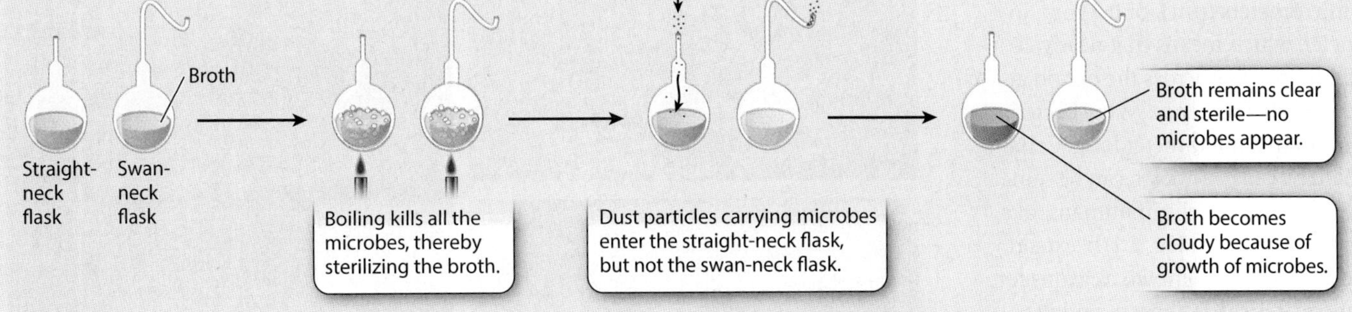

CONCLUSION The presence of microbes in the straight-neck flask and the absence of microbes in the swan-neck flask supported the hypothesis that microbes come from other microbes and are not spontaneously generated.

DISCUSSION Redi's and Pasteur's research illustrate classic attributes of well-designed experiments. Multiple treatments are set up, and nearly all conditions are the same in all of them. These common conditions are constant, so they cannot be the cause of different outcomes of the experiment. One feature—the variable—is changed from one treatment to the next. This is a place to look for explanations of different experimental outcomes.

FIG. 1.10 Unicellular and multicellular organisms. All living organisms are made up of cells: (a) bacteria; (b) brewer's yeast; (c) algae; (d) cheetahs; (e) humans. *Sources: a. Steve Gschmeissner/Science Source; b. Steve Gschmeissner/Science Source; c. Michael Abbey/Getty Images; d. Sven-Olof Lindblad/Science Source; e. Megapress/Alamy.*

Most cells are tiny, with dimensions that are well below the threshold of detection by the naked eye (**Fig. 1.11**). The cells that make up the layers of your skin (Fig. 1.11a) average about 100 micrometers (μm), or 0.1 mm, in diameter, which means that nearly 20 would fit in a row across the period at the end of this sentence. Many bacteria are less than a micrometer long. Certain specialized cells, however, can be quite large. Some nerve cells in humans, like the ones pictured in Fig. 1.11b, extend slender projections known as axons for distances as great as a meter, and the cannonball-size egg of an ostrich in Fig. 1.11c is a single giant cell.

The types of cell just mentioned—bacteria, yeasts, skin cells, nerve cells, and an egg—seem very different, but all are organized along broadly similar

FIG. 1.11 Cell diversity. Cells vary greatly in size and shape: (a) skin cells; (b) nerve cells; (c) ostrich egg. *Sources: a. Biophoto Associates/Science Source; b. Dr. Jonathan Clarke. Wellcome Images; c. Hemis/Alamy.*

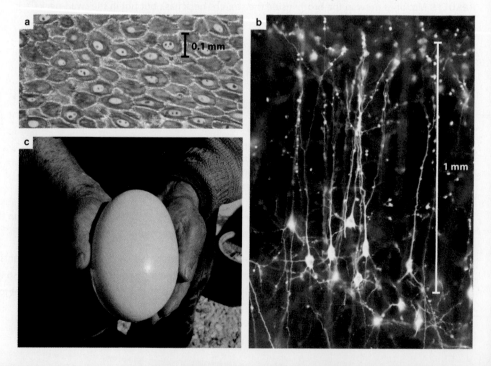

lines. In general, all cells contain a stable blueprint of information in molecular form; they have a discrete boundary that separates the interior of the cell from its external environment; and they have the ability to harness materials and energy from the environment.

Nucleic acids store and transmit information needed for growth, function, and reproduction.

The first essential feature of a cell is its ability to store and transmit information. To accomplish this, cells require a stable archive of information that encodes and helps determine their physical attributes. Just as the construction and maintenance of a house requires a blueprint that defines the walls, plumbing, and electrical wiring, organisms require an accessible and reliable archive of information that helps determine their structure and metabolic activities. Another hallmark of life is the ability to reproduce. To reproduce, cells must be able to copy their archive of information rapidly and accurately. In all organisms, the information archive is a remarkable molecule known as **deoxyribonucleic acid**, or **DNA** (Fig. 1.12).

DNA is a double-stranded helix, with each strand made up of varying sequences of four different kinds of molecules connected end to end. It is the arrangement of these molecular subunits that makes DNA special; in essence, they provide a four-letter alphabet that encodes cellular information. Notably, the information encoded in DNA directs the formation of **proteins**, the key structural and functional molecules that do the work of the cell. Virtually every aspect of the cell's existence—its internal architecture, its shape, its ability to move, and its various chemical reactions—depends on proteins.

FIG. 1.12 A molecule of DNA. DNA is a double helix made up of varying sequences of four different subunits.

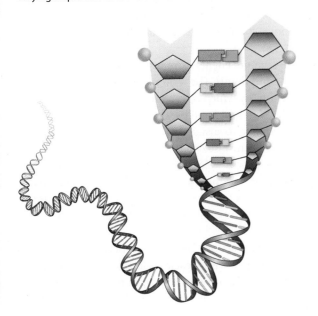

How does the information stored in DNA direct the synthesis of proteins? First, existing proteins create a copy of the DNA's information in the form of a closely related molecule called **ribonucleic acid**, or **RNA**. The synthesis of RNA from a DNA template is called **transcription**, a term that describes the copying of information from one form into another. Specialized molecular structures within the cell then "read" the RNA molecule to determine which building blocks to use to create a protein. This process, called **translation**, converts information stored in the language of nucleic acids to information in the language of proteins.

The pathway from DNA to RNA (specifically to a form of RNA called messenger RNA, or mRNA) to protein is known as the **central dogma** of molecular biology (**Fig. 1.13**). The central dogma describes the basic flow of information in a cell and, while there are exceptions, it constitutes a fundamental principle in biology. As proteins are ultimately encoded by DNA, we can define specific stretches or segments of DNA according to the proteins that they encode. This is the simplest definition of a **gene**: the DNA sequence that corresponds to a specific protein product.

FIG. 1.13 The central dogma of molecular biology. The central dogma defines the flow of information in all living organisms from DNA to RNA to protein.

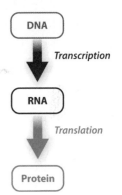

DNA has another remarkable feature. In addition to storing information, it is easily copied, or **replicated,** allowing genetic information to be passed from cell to cell or from an organism to its progeny. Each organism's DNA archive can be stably and reliably passed from generation to generation in large part because of its double-stranded helical structure. During replication, each strand of the double helix serves as a template for a new strand. Replication is necessarily precise and accurate because mistakes introduced into the cell's information archive may be lethal to the cell. That said, errors in DNA can and do occur during the process of replication, and environmental insults can damage DNA as well. Such changes to DNA sequence are known as **mutations**; they can spell death for the cell, or they can lead to the variations that underlie the diversity of life and the process of evolution.

Membranes define cells and spaces within cells.

The second essential feature of all cells is a **plasma membrane** that separates the living material within the cell from the nonliving environment around it (**Fig. 1.14**). This boundary between inside and outside does not mean that cells are closed

FIG. 1.14 The plasma membrane. The plasma membrane surrounds every cell and controls the exchange of material with the environment.

Sources: (top) Dr. Gopal Murti/SPL/Science Source; (bottom) Don W. Fawcett/Science Source.

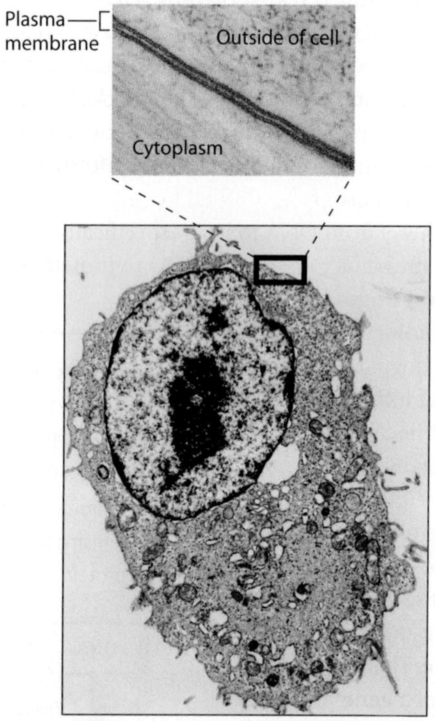

Transmission electron micrograph of a cell

systems independent of the environment. On the contrary, there is an active and dynamic interplay between cells and their surroundings that is mediated by the plasma membrane. All cells require sustained contributions from their surroundings, both simple ions and the building blocks required to manufacture macromolecules. They also release waste products into the environment. As discussed more fully in Chapter 5, the plasma membrane controls the movement of materials into and out of the cell.

In addition to the plasma membrane, many cells have internal membranes that divide the cell into discrete compartments, each specialized for a particular function. A notable example is the **nucleus**, which houses the cell's DNA. Like the plasma membrane, the nuclear membrane selectively controls movement of molecules into and out of it. As a result, the nucleus occupies a discrete space within the cell, separate from the space outside the nucleus, called the **cytoplasm**.

Not all cells have a nucleus. In fact, cells can be grouped into two broad classes depending on whether they have a nucleus. Cells without a nucleus are called **prokaryotes**, and cells with a nucleus are **eukaryotes**.

The first cells that emerged about 4 billion years ago were prokaryotic. Their descendants include the familiar bacteria, found today nearly everywhere that life can persist.

Some prokaryotes live in peaceful coexistence with humans, inhabiting our gut and aiding digestion (Case 5 The Human Microbiome). Others cause disease—salmonellosis, tuberculosis, and cholera are familiar examples of such bacterial diseases. The success of prokaryotes depends in part on their small size, their ability to reproduce rapidly, and their ability to obtain energy and nutrients from diverse sources. Most prokaryotes live as single-celled organisms, but some have simple multicellular forms.

Eukaryotes evolved much later, roughly 2 billion years ago, from prokaryotic ancestors. They include familiar groups such as animals, plants, and fungi, along with a wide diversity of single-celled microorganisms called protists. Eukaryotic organisms exist as single cells like yeasts or as multicellular organisms like humans. In multicellular organisms, cells may specialize to perform different functions. For example, in humans, muscle cells contract; red blood cells carry oxygen to tissues; and skin cells provide an external barrier.

The terms "prokaryotes" and "eukaryotes" are useful in drawing attention to a fundamental distinction between these two groups of cells. However, today, biologists recognize three domains of life—**Bacteria**, **Archaea**, and **Eukarya** (Chapters 24 and 25). Bacteria and Archaea are mostly single-celled microorganisms that lack a nucleus and are therefore prokaryotes, whereas Eukarya have a nucleus and are eukaryotic. Many Archaea flourish under seemingly hostile conditions, such as those found in the hot springs of Yellowstone National Park.

Metabolism converts energy from the environment into a form that can be used by cells.

A third key feature of cells is the ability to harness energy from the environment. Let's go back to our introductory example of eating an apple. The apple contains sugars, which store energy in their chemical bonds. By breaking down sugar, our cells harness this energy and convert it into a form that can be used to do the work of the cell. Energy from the food we eat allows us to grow, move, communicate, and do all the other things that we do.

Organisms acquire energy from just two sources—the sun and chemical compounds. The term **metabolism** describes chemical reactions by which cells convert energy from one form to another and build and break down molecules. These reactions are required to sustain life. Regardless of their source of energy, all organisms use chemical reactions to break down molecules, in the process releasing energy that is stored in a chemical form called **adenosine triphosphate**, or **ATP**. This molecule enables cells to carry out all sorts of work, including growth, division, and moving substances into and out of the cell.

Many metabolic reactions are highly conserved between organisms, meaning the same reactions are found in many different organisms. This observation suggests that the reactions evolved early in the history of life and have been maintained for billions of years because of their fundamental importance to cellular biochemistry.

A virus is genetic material that requires a cell to carry out its functions.

It's worth taking a moment to consider viruses. A virus is an agent that infects cells. It is usually smaller and simpler than cells. Why, then, aren't viruses the smallest unit of life? We just considered three essential features of cells—the capacity to store and transmit information, a membrane that selectively controls movement in and out, and the ability to harness energy from the environment. Viruses have a stable archive of genetic information, which can be RNA or DNA, surrounded by a protein coat and sometimes a lipid envelope. But viruses cannot harness energy from the environment. Therefore, on their own, viruses cannot read and use the information contained in their genetic material, nor can they regulate the passage of substances across their protein coats or lipid envelopes the way that cells do. To replicate, they require a cell.

A virus infects a cell by binding to the cell's surface, inserting its genetic material into the cell, and, in most cases, using the cellular machinery to produce more viruses. In essence, the virus "hijacks" the cell. Having used the cell's machinery to make more of themselves, viruses can then lyse, or break, the cell, enabling the new viruses to seek out and infect more cells. In some cases, the genetic material of the virus becomes integrated into the DNA of the host cell.

We discuss viruses many times throughout the book. Each species of Bacteria, Archaea, and Eukarya is susceptible to many types of virus that are specialized to infect its cells. Several hundred types of virus are known to infect humans, and the catalog is still incomplete. Useful tools in biological research, viruses have provided a model system for many problems in biology, including how genes are turned on and off and how cancer develops.

> ### Self-Assessment Questions
>
> 8. What does it mean to say that a cell is life's functional unit?
> 9. How does the central dogma help us to understand how mutations in DNA can result in disease?

1.4 EVOLUTION

The themes introduced in the last two sections stress life's unity: cells form the basic unit of all life; DNA, RNA, and proteins carry out the molecular functions of all cells; and metabolic reactions build and break down macromolecules. We need only look around us, however, to recognize that for all its unity, life displays a remarkable degree of diversity. We don't really know how many species share our planet, but reasonable estimates run to 10 million or more. Both the unity and the diversity of life are explained by the process of **evolution**, or change over time.

Variation in populations provides the raw material for evolution.

Described in detail, evolution by **natural selection** calls on complex mathematical formulations, but at heart its main principles are simple, indeed unavoidable. When there is variation within a population of organisms, and when that variation can be inherited (that is, when it can be passed from one generation to the next), the variants best suited for growth and reproduction in a given environment will contribute disproportionately to the next generation. As Darwin recognized, farmers have used this principle for thousands of years to select for crops with high yield or improved resistance to drought and disease. It is how people around the world have developed breeds of dog ranging from terriers to huskies (**Fig. 1.15**). And it is why antibiotic resistance is on the rise in many disease-causing microorganisms. Life has been shaped by evolution since its origin, and the capacity for Darwinian evolution may be life's most fundamental property.

FIG. 1.15 Artificial selection. Selection over many centuries has resulted in remarkable variations among dogs. Charles Darwin called this "selection under domestication" and noted that it resembles selection that occurs in nature. *Source: © Rob Brodman 2011.*

The apples in the bin from which you made your choice didn't all look alike. Had you picked your apple in an orchard, you would have seen that different apples on different trees, or even the same tree, looked different—some smaller, some greener, some misshapen, a few damaged by worms. Such variation is so commonplace that we scarcely pay attention to it. Variation is observed among individuals in virtually every species of organism. Variation that can be inherited provides the raw material on which evolution acts.

The causes of variation among individuals within a species are usually grouped into two broad categories. Variation among individuals is sometimes due to differences in the environment; this is called **environmental variation**. Among apples on the same tree, some may have good exposure to sunlight; some may be hidden in the shade; some were lucky enough to escape the female codling moth, whose egg develops into a caterpillar that eats its way into the fruit. These are all examples of environmental variation.

The other main cause of variation among individuals is differences in the genetic material that is transmitted from parents to offspring; this is known as **genetic variation**. Differences among individuals' DNA can lead to differences among the individuals' RNA and proteins, which affect the molecular functions of the cell and ultimately can lead to physical differences that we can observe. Genetic differences among apples produce varieties whose mature fruits differ in taste and color, such as the green Granny Smith, the yellow Golden Delicious, and the scarlet Red Delicious.

But even on a single tree, each apple contains seeds that are genetically distinct because the apple tree is a sexual organism. Bees carry pollen from the flowers of one tree and deposit it in some of the flowers of another, enabling the sperm inside pollen grains to fertilize egg cells within that single flower. All the seeds on an apple tree contain shared genes from one parent, the tree on which they developed. But they contain distinct sets of genes contributed by sperm transported in pollen from other trees. In all sexual organisms, fertilization produces unique combinations of genes, which explains in part why sisters and brothers with the same parents can be so different from one another.

Genetic variation ultimately stems from mutations. Mutations arise either from random errors during DNA replication or from environmental factors such as ultraviolet (UV) radiation, which can damage DNA. If these mutations are not corrected, they are passed on to the next generation. To put a human face on this, consider that lung cancer can result from an environmental insult, such as cigarette smoking, or from a genetic susceptibility inherited from the parents.

In nature, most mutations that harm growth and reproduction die out after a handful of generations. Those that are neither harmful nor beneficial can persist for hundreds or thousands of generations. And those that are beneficial to growth and reproduction can gradually become incorporated into the genetic makeup of every individual in the species. That is how evolution works: the genetic makeup of a population changes over time.

Evolution predicts a nested pattern of relatedness among species, depicted as a tree.

Evolutionary theory predicts that new species arise by the divergence of populations through time from a common ancestor. As a result, closely related species are likely to resemble each other more closely than they do more distantly related species. You know this to be true from common experience. All of us recognize the similarity between a chimpanzee's face and body and our own (**Fig. 1.16a**), and biologists have long known that we humans share more features with chimpanzees than we do with any other species.

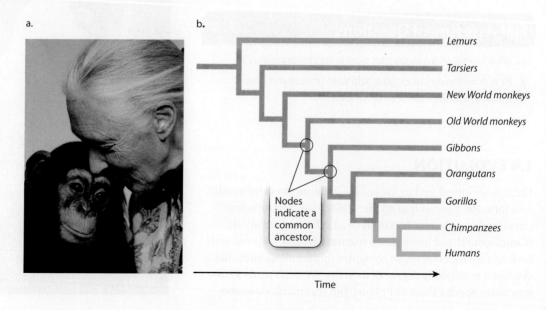

FIG. 1.16 Phylogenetic relationships among primates. (a) Humans share many features with chimpanzees. (b) Humans and chimpanzees, in turn, share more features with gorillas than they do with other species, and so on down through the evolutionary tree of primates. Treelike patterns of nested similarities are the predicted result of evolution. *Source: AP Photo/Bela Szandelszky.*

Humans and chimpanzees, in turn, share more features with gorillas than they do with any other species. And humans, chimpanzees, and gorillas share more features with orangutans than they do with any other species. And so on. We can continue to include a widening diversity of species, successively adding monkeys, lemurs, and other primates, to construct a set of evolutionary relationships that can be depicted as a tree (**Fig. 1.16b**). In this tree, the tips or branches on the right represent different groups of organisms, nodes (where lines split) represent the most recent common ancestor, and time runs from left to right.

Evolutionary theory predicts that primates should show a nested pattern of similarity, and this is indeed what morphological and molecular observations reveal. We can continue to add other mammals and then other vertebrate animals to our comparison, in the process generating a pattern of evolutionary relationships that forms a larger tree, with the primates confined to one limb. Using comparisons of DNA among species, we can then generate still larger trees, ones that include plants as well as animals and the full diversity of microscopic organisms. Biologists call the full set of evolutionary relationships among all organisms the tree of life.

This tree, illustrated in **Fig. 1.17,** has two major branches, one made up of Bacteria and the other consisting of Archaea and Eukarya. Although eukaryotes include large multicellular organisms such as humans, most branches on the tree of life consist of microorganisms. The last common ancestor of all living organisms, which forms a root to the tree, is thought to lie between the branch leading to Bacteria and the branch leading to Archaea and Eukarya. The tree shows that eukaryotes arose from within the Archaea, but the origin of eukaryotic cells may be more complicated than that. The Eukarya is now thought by many biologists to have originated from a partnership, or symbiosis, between an archaeon and a bacterium. (We'll explore this hypothesis more fully in Chapter 25.) One feature of the tree of life that everyone accepts is that the plants and animals so conspicuous in our daily existence make up only two branches on the eukaryotic limb of the tree.

The tree of life makes predictions for the order of appearance of different life-forms in the history of life, documented in rock in the fossil record. For example, Fig. 1.16—one small branch on the greater tree—predicts that humans should appear later in the fossil record than monkeys, and that primates more akin to lemurs and tarsiers should appear even earlier. The greater tree also predicts that all records of animal life should be preceded by a long interval of microbial evolution. As we will see in subsequent chapters, these predictions are confirmed by the geologic record.

Shared features, then, sometimes imply inheritance from a common ancestor. In combination, fossils and molecular studies such as comparisons of DNA sequences show that the close similarity between humans and chimpanzees reflects descent from a common ancestor that lived about 6 million years ago. Their differences reflect what Darwin called "descent with modification"—evolutionary changes that have accumulated over time since the two lineages split. For example, as discussed in Chapter 23, the flat face of humans, our small teeth, and our upright posture all evolved within our ancestors after they diverged from the ancestors of chimpanzees. At a broader scale, the fundamental features shared by all organisms reflect inheritance from a common ancestor that lived billions of years ago. Moreover, the differences that characterize the many branches on the tree of life have formed through the continuing action of evolution *since* the time of our earliest ancestors.

Four decades ago, the geneticist Theodosius Dobzhansky wrote, "Nothing in biology makes sense except in the light of evolution." For this reason, evolution permeates discussions throughout this book, whether we are explaining the molecular biology of cells, how organisms function and reproduce, how species

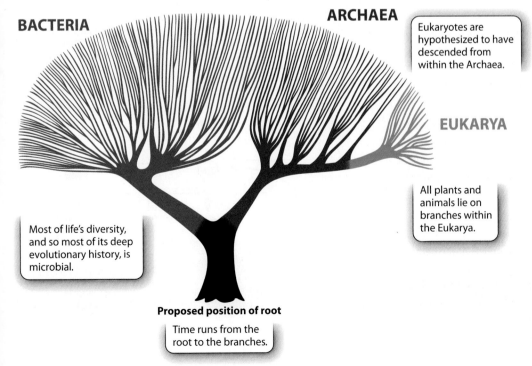

FIG. 1.17 The tree of life. *Information from Spang, A., and T. J. G. Ettema. 2016. "The Tree of Life Comes of Age."* Nature Microbiology *1:16056.*

interact in nature, or the remarkable biological diversity of our planet.

Evolution can be studied by means of experiments.

Both the nested patterns of similarity among living organisms and the succession of fossils in the geologic record fit the predictions of evolutionary theory. But can we capture evolutionary processes in action? One way to accomplish this goal is in the laboratory. Bacteria are ideal for these experiments because they reproduce rapidly and can form populations with millions of individuals. Large population size means that mutations are likely to form in nearly every generation, even though the probability that any individual cell will acquire a mutation is small. (In contrast to bacteria, think about trying evolutionary experiments on elephants!)

One such experiment is illustrated in **Fig. 1.18**. Beginning in 1988, microbiologist Richard Lenski grew populations of the common intestinal bacterium *Escherichia coli* in liquid medium containing small amounts of the sugar glucose as the only source of food. Lenski and his colleagues hypothesized that any bacterium with a mutation that increased its ability to use glucose would grow and reproduce at a faster rate than other bacteria in the population. In this experiment, then, they asked: did bacteria from later generations grow and reproduce more rapidly than those of earlier generations—that is, did evolution occur?

In fact, the bacteria did evolve an improved ability to use glucose, as demonstrated by the results shown in Fig. 1.18. This experiment both illustrates how experiments can be used to test hypotheses and shows evolution in action. Furthermore, follow-up studies of the bacterial DNA identified differences in genetic makeup that resulted in the improved ability of *E. coli* to use glucose. Many experiments of this general type have been carried out, applying scientific inquiry to demonstrate how bacteria adapt through mutation and natural selection to any number of environments.

Experiments in laboratory evolution help us understand how life works, and they have an immensely important practical side. They allow biologists to develop new and beneficial strains of microorganisms that, for example, remove toxins from lakes and rivers. In addition, they show how some of our worst pathogens develop resistance to drugs designed to eliminate them.

HOW DO WE KNOW?

FIG. 1.18
Can evolution be demonstrated in the laboratory?

BACKGROUND *Escherichia coli* is an intestinal bacterium commonly used in the laboratory. In 1988, the biologist Richard Lenski asked whether *E. coli* grown generation after generation on a medium containing limiting amounts of glucose as food would evolve in ways that improved their ability to metabolize this compound.

HYPOTHESIS Any bacterium with a random mutation that increases its ability to take up and utilize glucose will reproduce at a faster rate than other bacteria in the population. Over time, such a mutant will increase in frequency relative to other types of bacteria, thereby demonstrating evolution in a bacterial population.

EXPERIMENT Cells of *E. coli* can be frozen at −80 degrees Celsius, which keeps them in a sort of suspended animation in which no biological processes take place, but the cells survive. At the beginning of the experiment, Lenski and his students froze samples of the starting bacteria ("Ancestral"). As the experiment progressed, they took samples of the bacterial populations every 500 generations ("Later") and grew them together with a thawed sample of the starting bacteria on glucose-containing medium. They then compared the rate of growth of the ancestral bacteria with that of the bacteria samples taken at later time points.

RESULTS Through time, the cells from later time points grew increasingly more rapidly than the ancestral cells when the two populations were grown together on the glucose medium.

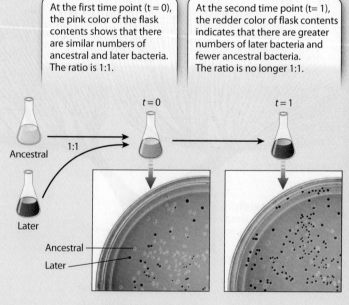

At the first time point (t = 0), the pink color of the flask contents shows that there are similar numbers of ancestral and later bacteria. The ratio is 1:1.

At the second time point (t = 1), the redder color of flask contents indicates that there are greater numbers of later bacteria and fewer ancestral bacteria. The ratio is no longer 1:1.

PHOTOS SOURCE: Neerja Hajela, Michigan State University.

Self-Assessment Questions

10. How does evolution account for both the unity and the diversity of life?

11. How might the heavy-handed use of antibiotics result in the increase of antibiotic-resistant cells in bacterial populations?

1.5 ECOLOGICAL SYSTEMS

When watching a movie, we don't need a narrator to tell us whether the action is set in a tropical rain forest, the African savanna, or the Arctic tundra. The plants in the scene make it obvious. Plants can be linked closely with environment—palm trees with the tropics, for example, or cacti with the desert—because the environmental distributions of different species reflect the sum of their biological features. Palms cannot tolerate freezing and so are confined to warmer environments; the cactus can store water in its tissues, enabling it to withstand prolonged drought. But the geographic ranges of palms and cacti reflect more than just their physical tolerances. They are also strongly influenced by interactions with other species, including other plants that compete for the limited resources available for growth, animals that feed on them or spread their pollen, and microorganisms that infest their tissues. **Ecology** is the study of how organisms interact with one another and with their physical environment in nature.

Basic features of anatomy, physiology, and behavior shape ecological systems.

Apple trees, we noted, reproduce sexually. How is this accomplished, given that the trees have no moving parts? The answer is that bees carry pollen from one flower to the next, enabling sperm carried within the pollen grain to fuse with an egg cell protected within the flower (**Fig. 1.19**). The process is much like the one discussed at the beginning of the chapter in which hummingbirds carry pollen from flower to flower. In both examples, the plants complete their life cycles by exploiting close interactions with animal species. Birds and bees visit flowers, attracted, as we are, by their color and odor. Neither visits flowers with the intent to pollinate; they come in expectation of a meal. As the birds and bees nestle into flowers to collect nutritious nectar, pollen rubs off on their bodies.

But if animals help apples to fertilize their eggs, how do apple trees disperse from one site to another? Again, the plants rely on animals—the apple's flesh attracts mammals that eat the fruit, seeds and all, and spread the seeds through defecation. Humans tend to be finicky—we carefully eat only the sweet outer flesh, discarding the seedy core. Nonetheless, even as humans have selected apples for their quality of fruit, we have dispersed apple trees far beyond their natural range.

The many ways that organisms interact in nature reflect their basic functional requirements. If a plant is to grow and

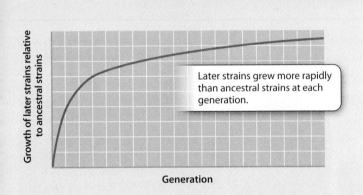

CONCLUSION Evolution occurred in the population: the *E. coli* bacteria evolved an improved ability to metabolize glucose.

FOLLOW-UP WORK As of 2017, the experiment has continued for more than 60,000 generations. Taking advantage of new genomic tools, Lenski and colleagues have been able to characterize increases in fitness in terms of specific mutations and follow the fate of individual genes through time.

SOURCES (experiment diagram) Elena, S. F., and R. E. Lenski. 2003. "Evolution Experiments with Microorganisms." *Nature Review Genetics* 4:457–469; (graph) Lenski, R. E. 2017. "Experimental Evolution and the Dynamics of Adaptation and Genome Evolution in Microbial Populations." *The ISME Journal* 11:2181–2194.

FIG. 1.19 Bee pollination. Many plants reproduce sexually by exploiting the behavior of animals. Here a honeybee pollinates an apple flower. *Source: Donald Specker/Animals Animals – Earth Scenes.*

reproduce, for example, it must have access to light, carbon dioxide, water, and basic nutrients. All plants have these requirements, so one plant's success in gathering nutrients may mean that fewer nutrients are available for its neighbor—the plants may compete for limited resources needed for growth (**Fig. 1.20a**). Animals in the same area require organic molecules for nutrition, and some of them may eat the plant's leaves or bark (**Fig. 1.20b**). Predation also benefits some organisms at the expense of others. In combination, the physical tolerances of organisms and the ways that organisms interact with one another determine the structure and diversity of communities.

In short, ecological relationships reflect the biomechanical, physiological, and behavioral traits of organisms in nature. Form and function, in turn, arise from molecular processes within cells, governed by the expression of genes.

Ecological interactions play an important role in evolution.

G. Evelyn Hutchinson, one of the founders of modern ecology, wrote a book called *The Ecological Theater and the Evolutionary Play* (1965). This wonderful title succinctly captures a key feature of biological relationships: it suggests that ecological communities provide the stage on which the play of evolution takes place.

As an example, let's look again at plants competing for resources. As a result of this competition, natural selection may favor plants that have more efficient uptake of nutrients or water. In the example of animals eating plants, natural selection may favor animals with greater jaw strength or more efficient extraction of nutrients in the digestive system. In turn, plants that avoid predation by synthesizing toxic compounds in their leaves may gain the upper hand. In each case, interactions between organisms lead to the evolution of particular traits.

Consider another example from mammal–plant interactions: selection for fleshy fruits improved the dispersal of apples because it increased the attractiveness of these seed-bearing structures to hungry mammals. Tiny yeasts make a meal on sugary fruits as well, in the process producing alcohol that deters potential competitors for the food. Humans learned to harness this physiological capability of yeasts in prehistoric times, and for this reason the total abundance and distribution of *Vitis vinifera*, the wine grape, has increased dramatically through time.

Self-Assessment Question

12. What are the features that determine the shape of ecological systems?

FIG. 1.20 Ecological relationships. (a) The trees in a rain forest all require water and nutrients. Neighboring plants compete for the limited supplies of these materials. (b) Plants provide food for many animals. Leaf-cutter ants cut slices of leaves from tropical plants and transport them to their nests, where the leaves grow fungi that the ants eat. *Sources: a. Louise Murray/Getty Images; b. Gail Shumway/Getty Images.*

1.6 THE HUMAN FOOTPRINT

The story of life has a cast of millions, with humans playing only one of the many roles in an epic 4 billion years in the making. *Homo sapiens* has existed for only the most recent 1/200 of 1% of life's history, yet there are compelling reasons to pay special attention to our own species. We want to understand how our own bodies work and how humans came to be: curiosity about ourselves is, after all, a deeply human trait.

We also want to understand how biology can help us conquer disease and improve human welfare. Epidemics have decimated human populations throughout history. The Black Death—bubonic plague caused by the bacterium *Yersinia pestis*—is estimated to have killed half the population of Europe in the fourteenth century. Casualties from the flu pandemic of 1918 exceeded those of World War I. Even King Tut, we now know, suffered from malaria in his Egyptian palace approximately 3300 years ago. Throughout this book, we discuss how basic biological principles are helping scientists to prevent and cure the great diseases that have persisted since antiquity, as well as modern ones such as AIDS and Ebola.

Furthermore, we need to understand the evolutionary and ecological consequences of a human population that now exceeds 7 billion. All species affect the world around them. However, in the twenty-first century human activities have taken on special importance because our numbers and technological abilities make our footprint on Earth's ecology so large. Human activities now emit more carbon dioxide than do volcanoes, through industrial processes we convert more atmospheric nitrogen to ammonia than nature does, and we commandeer, either directly or indirectly, as much as 25% of all photosynthetic production on land. To chart our environmental future, we need to understand our role in the Earth system as a whole.

And, as we have become major players in ecology, humans have become important agents of evolution. As our population has expanded, some species have expanded along with us (**Fig. 1.21**). We've seen how humans' pursuit of agriculture sharply increased the abundance and distribution of grapes,

FIG. 1.21 Humans as agents of evolution. Species that have benefited from human activity include (a) corn, (b) rats, and (c) cockroaches. *Sources: a. Fred Dimmick/iStockphoto; b. Arndt Sven-Erik/age footstock; c. Nigel Cattlin/Alamy.*

and the same is true for corn, cows, and apple trees. At the same time, we have inadvertently helped other species to expand—the crowded and not always clean environments of cities provide excellent habitats for cockroaches and rats.

Other species are in decline, their populations reduced by hunting and fishing, changes in land use, and other human activities (**Fig. 1.22**). When Europeans first arrived on the Indian Ocean island of Mauritius, large flightless birds called dodos were plentiful. Within a century, the dodo was extinct. Early Europeans in North America were greeted by vast populations of passenger pigeons, more than a million in a single flock. By the early twentieth century, the species had disappeared, a victim of hunting and habitat change through expanding agriculture. Other organisms both great (the Bali tiger) and small (the dusky seaside sparrow) have become extinct in recent decades. Still others are imperiled by human activities: victimized by habitat destruction and poaching, the last surviving male of the Northern White Rhinoceros died in 2018. Whether any rhinos will exist at the end of this century will depend almost entirely on decisions we make today.

FIG. 1.22 Extinct and endangered species. Humans have caused many organisms to become extinct, such as (a) the dodo, (b) passenger pigeon, (c) Bali tiger, and (d) dusky seaside sparrow. Burgeoning human populations have diminished the ranges of many others, including (e) the white rhinoceros.
Sources: a. Photo Researchers/Science Source; b. G. I. Bernard/Science Source; c. Look and Learn/Bridgeman Images; d. P. W. Sykes/U.S. Fish and Wildlife Service; e. James Warwick/Science Source.

Throughout this book, we return to the practical issues of life science. How can we use the principles of biology to improve human welfare, and how can we live our lives in ways that control our impact on the world around us? The answers to the questions critically depend on understanding biology in an *integrated* fashion. While it is tempting to consider molecules, cells, organisms, and ecosystems as separate entities, they are inseparable in nature. To tackle biological problems—whether building an artificial cell, stopping the spread of infectious diseases such as HIV or malaria, feeding a growing population, or preserving endangered habitats and species—we need an integrated perspective. In decisively important ways, our future welfare depends on improving our knowledge of how life works.

Self-Assessment Questions

13. What are three ways that humans have affected life on Earth?
14. What are the six themes that are discussed in this chapter? Summarize each one.

CORE CONCEPTS SUMMARY

1.1 SCIENTIFIC INQUIRY: Scientific inquiry is a deliberate way of asking and answering questions about the natural world.

Observations are used to generate a hypothesis, a tentative explanation that makes predictions that can be tested. page 4

On the basis of a hypothesis, scientists design experiments and make additional observations that test the hypothesis. page 4

A controlled experiment typically involves several groups in which all the conditions are the same and one group where a variable is deliberately introduced to determine whether that variable has an effect. page 5

Scientific inquiry involves making observations, asking questions, doing experiments or making further observations, and drawing and sharing conclusions. page 6

If a hypothesis is supported through continued observation and experiments over long periods of time, it is elevated to a theory, a sound and broad explanation of some aspect of the world. page 7

1.2 CHEMICAL AND PHYSICAL PROPERTIES: Life works according to fundamental principles of chemistry and physics.

The living and nonliving worlds follow the same chemical rules and obey the same physical laws. page 8

Experiments by Redi in the 1600s and Pasteur in the 1800s demonstrated that organisms come from other organisms and are not spontaneously generated. page 10

Life originated on Earth about 4 billion years ago, arising from nonliving matter. page 11

1.3 THE CELL: The fundamental unit of life is the cell.

The cell is the simplest biological entity that can exist independently. page 11

Information in a cell is stored in the form of the nucleic acid DNA. page 13

The central dogma describes the usual flow of information in a cell, from DNA to RNA to protein. page 13

The plasma membrane is the boundary that separates the cell from its environment. page 13

Cells with a nucleus are eukaryotes; cells without a nucleus are prokaryotes. page 14

Metabolism is the set of chemical reactions in cells that build and break down macromolecules and harness energy. page 14

A virus is an infectious agent composed of a genome and protein coat that uses a host cell to replicate. page 15

1.4 EVOLUTION: Evolution explains the features that organisms share and those that set them apart.

When variation is present within a population of organisms, and when that variation can be inherited, the variants best able to grow and reproduce in a particular environment will contribute disproportionately to the next generation, leading to a change in the population over time, or evolution. page 15

Variation can be genetic or environmental. The ultimate source of genetic variation is mutation. page 16

Organisms show a nested pattern of similarity, with humans being more similar to primates than other organisms, primates being more similar to mammals, mammals being more similar to vertebrates, and so on. page 16

Evolution can be demonstrated by laboratory experiments. page 18

1.5 ECOLOGICAL SYSTEMS: Organisms interact with one another and with their physical environment, shaping ecological systems that sustain life.

Ecology is the study of how organisms interact with one another and with their physical environment in nature. page 19

These interactions are driven in part by the anatomy, physiology, and behavior of organisms—that is, the basic features of organisms shaped by evolution. page 19

1.6 THE HUMAN FOOTPRINT: In the twenty-first century, humans have become major agents in ecology and evolution.

Humans have existed for only the most recent 1/200 of 1% of life's 4-billion-year history. page 21

In spite of humans' recent arrival on the planetary scene, our growing numbers are leaving a large ecological and evolutionary footprint. page 21

Solving biological problems requires an integrated understanding of life, with contributions from all the fields of biology, including molecular biology, cell biology, genetics, organismal biology, and ecology, as well as from chemistry, physics, and engineering. page 23

Log in to LaunchPad to check your answers to the Self-Assessment Questions and to access additional learning tools.

CASE 1

Life's Origins

Information, Homeostasis, and Energy

The walls of Endurance crater expose sandstones deposited on the surface of Mars more than 3 billion years ago. These sandstones provide information about Mars' surface environment around the same time that life was gaining a foothold on Earth. Most of the sands were deposited by wind, but thin rippled beds near the top of the crater indicate that some of the sands were transported by water. Further evidence for water on Mars comes from the chemistry of the beds. Most grains are made of salts, indicating that volcanic rocks interacted with water for many years, dissolving some minerals and precipitating others. Among the salts is a mineral called jarosite, known on Earth from acid mine waste and other environments where the pH is low. Iron oxides also tell us that a small amount of oxygen gas existed in the ancient atmosphere of Mars.

Could life have existed on the young Mars? On Earth, some microorganisms thrive in environments that are similar to the environment recorded by rocks at Endurance crater. As yet we have no evidence of ancient organisms on Mars, but maybe—just maybe—life could have subsisted in the arid, acidic, and oxidizing setting recorded by Endurance rocks. At the time the Endurance rocks formed, Mars was growing drier and colder, but older sedimentary rocks found elsewhere on the planet indicate that there were once lakes and streams that persisted for many thousands of years. Whether organisms could *thrive* in the environments that existed on early Mars is one question; whether life could *originate* under these conditions is another.

When and how life originated are some of the biggest questions in biology. They are as central to understanding our own planet as they are for Mars. Earth is nearly 4.6 billion years old. Chemical evidence from 3.5-billion-year-old rocks in Australia indicates that biologically based carbon and sulfur cycles existed at the time those rocks formed. Because living organisms drive these cycles, life must have originated earlier. Evidence from other continents appears to push the origin of life back to 3.8 billion years or even earlier. In the eons since then, the first primitive life-forms have evolved into the millions of different species that populate Earth today.

How did the first living cell arise? Scientists generally accept that life arose from nonliving materials—a process called abiogenesis—and thousands of laboratory experiments performed over the last sixty years provide glimpses of how this might have occurred. In our modern world, the features that separate life from nonlife are relatively easy to discern. However, Earth's first organisms were almost certainly much less complex than even the simplest bacteria alive today. And before those first living organisms appeared, molecular systems presumably existed that hovered somewhere between the living and the nonliving.

All cells require an archive of information, a membrane to separate the inside of the cell from its surroundings, and the ability to gather materials and harness energy from the environment. In modern organisms, the cell's information archive is deoxyribonucleic acid (DNA), the double-stranded molecule that contains the instructions needed for cells to

> *All cells require an archive of information, a membrane to separate the inside of the cell from its surroundings, and the ability to gather materials and harness energy from the environment.*

CASE 1

Endurance crater. Analysis of ancient sedimentary rocks exposed in the walls of this crater and others on Mars provides clues to what the environment was like on early Mars. Could life thrive under such conditions? Could it originate under such conditions? *Source: NASA/JPL/Cornell*

grow, differentiate, and reproduce. Without nucleic acids like DNA, life as we know it would not exist.

DNA is critical to life as we know it, and it is complex. Among the organisms alive today, the smallest known genome belongs to the bacterium *Carsonella rudii*. Even that genome contains nearly 160,000 DNA base pairs. How could such sophisticated molecular systems have arisen?

The likely answer to that question is "step by step." Laboratory experiments have shown how precursors to nucleic acids might have formed and interacted under the chemical conditions present on the young Earth. It's exceedingly unlikely that a molecule as complex as DNA was used by the very first living cells. As you'll see in the chapters that follow, scientists have gathered evidence suggesting that RNA, rather than DNA, stored information in early cells. Very likely, RNA performed other functions in early cells as well, catalyzing chemical reactions much as proteins do today.

While some kind of information archive was necessary for life to unfold, there is more to the story. Living things must have a barrier that separates them from their environment. All cells, whether found as single-celled bacteria or by the trillions in trees or humans, are individually encased in a cell membrane.

Once again, scientists can only guess at how the first cell membranes came about, but research shows that the molecules that make up modern membranes possess properties that may have led them to form spontaneously on the early Earth. At first, the membranes were probably quite simple—straightforward (but leaky) barriers that kept the contents of early cells separated from the world at large. Over time, as chance variations arose, those membranes that provided a better barrier were favored by natural selection. Moreover, proteins became embedded in membranes, providing gates or channels that regulated the transport of ions and small molecules into and out of the cell. In this way, early organisms gained the capacity to regulate cell interiors—what biologists call homeostasis.

A third essential characteristic of living things is the ability to harness energy from the environment. Here, too, a series of natural chemical processes might have led to cells that could achieve this feat. Simple reactions that produced carbon-bearing molecular by-products would have enabled more complex reactions down the road. Ultimately, that collection of reactions—combined with an archive of information and enclosed in some kind of primitive membrane—evolved into individual units that could grow, reproduce, and evolve.

Such a series of events may sound unlikely. However, some scientists argue that given the chemicals present on the early Earth, it was likely—perhaps even inevitable—that they would come together in such a way that life would emerge. Indeed, relatively simple, naturally occurring materials such as metal ions have been shown to play a role in key cellular reactions. Billions of years after the first cells arose, some of those metal ions—such as complexes of iron and sulfur—still play a critical role in cells.

One way to investigate how life originated is to mimic in the laboratory environmental conditions of the early Earth and analyze what compounds are produced. A different approach involves the study of extremophiles—organisms that live and even thrive in extreme environments, such as deep-sea hydrothermal vents. These studies can help us understand how life might have originated and evolved in early environments quite different from those we typically experience today.

Did life arise just once? Or could it have started up and died out several times before it finally got a firm foothold? If, given Earth's early chemistry, life on our planet was inevitable, or at least probable, could it have arisen elsewhere in the universe? Indeed, might life have originated on our neighbor Mars? The study of life's origins produces many more questions than answers—and not just for biologists. The mystery of life spans the fields of biology, chemistry, physics, and planetary science. Though the questions are vast, our understanding of life's origins is likely to come about the same way life itself arose: step by step.

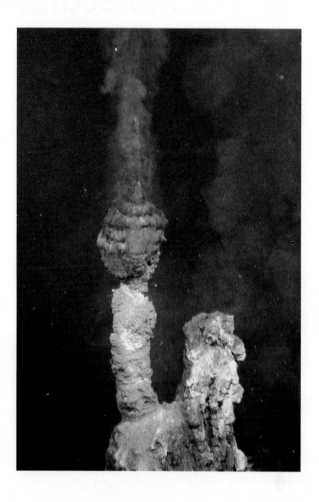

A hydrothermal vent. Some scientists think that this type of vent provided a favorable environment for chemical reactions that led to the origin of life on Earth. *Source: Image courtesy of New Zealand American Submarine Ring of Fire 2007 Exploration, NOAA Vents Program, the Institute of Geological & Nuclear Sciences and NOAA-OE.*

DELVING DEEPER INTO CASE 1

For each of the Cases, we provide a set of questions to pique your curiosity. These questions cannot be answered directly from the Case itself, but instead introduce issues that will be addressed in chapters to come. We revisit this Case in Chapters 2–8 on the pages indicated below.

1. How did the molecules of life form? See page 45.
2. What kinds of nucleic acids were present in the earliest cells? See page 60.
3. How did the genetic code originate? See page 85.
4. How did the first cell membranes form? See page 95.
5. Which naturally occurring elements might have spurred the reactions that led to the emergence of life? See page 131.
6. What were the earliest energy-harnessing reactions? See page 143.
7. How did early cells meet their energy requirements? See page 149.
8. How did early cells use sunlight to meet their energy requirements? See page 173.

CHAPTER 2: The Molecules of Life

CORE CONCEPTS

2.1 PROPERTIES OF ATOMS: The atom is the fundamental unit of matter.

2.2 MOLECULES AND CHEMICAL BONDS: Atoms can combine to form molecules linked by chemical bonds.

2.3 WATER: Water is essential for life.

2.4 CARBON: Carbon is the backbone of organic molecules.

2.5 ORGANIC MOLECULES: Organic molecules include proteins, nucleic acids, carbohydrates, and lipids, each of which is built from simpler units.

2.6 LIFE'S ORIGINS: Life likely originated on Earth by a set of chemical reactions that gave rise to the molecules of life.

When biologists speak of diversity, they commonly point to the 2 million or so species named and described to date, or to the 10–100 million living species thought to exist in total. Life's diversity can also be found at a very different level of observation: in the molecules within cells. Life depends critically on many essential functions, including storing and transmitting genetic information, establishing a boundary to separate cells from their surroundings, and harnessing energy from the environment. These functions ultimately depend on the chemical characteristics of the molecules that make up cells and organisms.

In spite of the diversity of molecules and functions, the chemistry of life is based on just a few types of molecule, which in turn are made up of just a few elements. Of the 100 or so chemical elements, only about a dozen are found in more than trace amounts in living organisms. These elements interact with one another in only a limited number of ways. So, the question arises: how is diversity generated from a limited suite of chemicals and interactions? The answer lies in some basic features of chemistry.

2.1 PROPERTIES OF ATOMS

Since antiquity, it has been accepted that the materials of nature are made up of a small number of fundamental substances combined in various ways. Today, we call these substances **elements.** From the seventeenth century through the end of the nineteenth century, elements were defined as pure substances that could not be broken down further by the methods of chemistry. In time, it was recognized that each element contains only one type of **atom,** the basic unit of matter. By 1850, about 60 elements were known, including such common ones as oxygen, copper, gold, and sodium. Today, 118 elements are known. Of these, 94 occur naturally and 24 have been created artificially in the laboratory. Elements are often indicated by a chemical symbol that consists of a one- or two-letter abbreviation of the name of the element. For example, carbon is represented by C, hydrogen by H, and helium by He.

Atoms consist of protons, neutrons, and electrons.

Elements are composed of atoms. The atom contains a dense central **nucleus** made up of positively charged particles called **protons** and electrically neutral particles called **neutrons.** A third type of particle, the negatively charged **electron,** moves around the nucleus at some distance from it. Carbon, for example, typically has six protons, six neutrons, and six electrons (**Fig. 2.1**). The number of protons, or the atomic number, specifies an atom as a particular element. An atom with one proton is hydrogen, for example, and an atom with six protons is carbon.

Together, the protons and neutrons determine the **atomic mass,** the mass of the atom. Each proton and neutron, by definition, has a mass of 1, whereas an electron has negligible mass. The number of neutrons in atoms of a particular element can differ, changing its mass. **Isotopes** are atoms of the same element that have different numbers of neutrons. For example, carbon has three isotopes: approximately 99% of carbon atoms have six neutrons and six protons, for an atomic mass of 12; approximately 1% have seven neutrons and six protons, for an atomic mass of 13; and only a very small fraction have eight neutrons and six protons, for

FIG. 2.1 A carbon atom. Most carbon atoms have six protons, six neutrons, and six electrons. The net charge of any atom is neutral because there are the same numbers of electrons and protons.

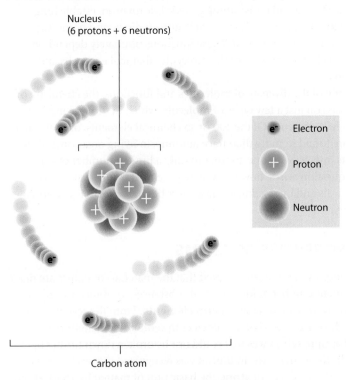

an atomic mass of 14. The atomic mass is sometimes indicated as a superscript to the left of the chemical symbol. For instance, ^{12}C is the isotope of carbon with six neutrons and six protons.

Typically, an atom has the same number of protons and electrons. Because a carbon atom has six protons and six electrons, the positive and negative charges cancel each other out and the carbon atom is electrically neutral. Some chemical processes cause an atom to either gain or lose electrons. An atom that has lost an electron is positively charged, and one that has gained an electron is negatively charged. Electrically charged atoms are called **ions**. The charge of an ion is specified as a superscript to the right of the chemical symbol. Thus, H^+ indicates a hydrogen ion that has lost an electron and is positively charged.

Electrons occupy regions of space called orbitals.

Electrons move around the nucleus, but not in the simplified way shown in Fig. 2.1. The exact path that an electron takes cannot be known, but it is possible to identify a region in space, called an **orbital**, where an electron is present most of the time. For example, **Fig. 2.2a** shows the orbital for hydrogen, which is simply a sphere occupied by a single electron. Most of the time, the electron is found within the space defined by the sphere, although its exact location at any instant is unpredictable.

The maximum number of electrons in any orbital is two. Most atoms have more than two electrons, so they have several

FIG. 2.2 Electron orbitals and energy levels (shells) for hydrogen and carbon. The orbital of an electron can be visualized as a cloud of points that is denser where the electron is more likely to be. The hydrogen atom contains a single orbital, in a single energy level (a and c). The carbon atom has five orbitals, one in the first energy level and four in the second energy level (b and c). In the second energy level, electrons in the spherical orbital have slightly less energy than those in the dumbbell-shaped orbitals.

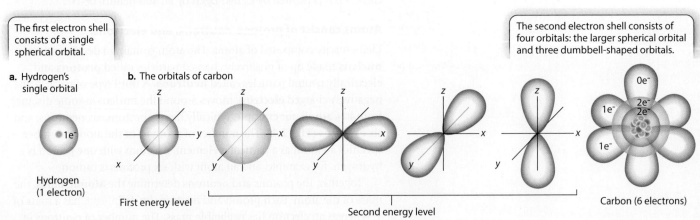

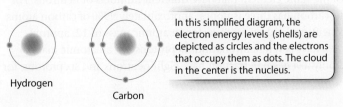

orbitals occupied by electrons that are positioned at different distances from the nucleus. These orbitals differ in size and shape. Electrons in orbitals close to the nucleus have less energy than do electrons in orbitals farther away. An analogy can help to explain this concept. When you bike up a hill, part of the energy you expend is stored in the increased elevation of you and the bike, and you get some of that stored energy back again when you coast down the hill. The elevation of the bike is analogous to the distance of an electron from the nucleus. An electron gives up energy when it moves to a closer orbital, which is why electrons fill up orbitals close to the nucleus before occupying those farther away. Several orbitals can exist at a given energy level, or **shell**. The first shell consists of the spherical orbital shown in Fig. 2.2a.

Fig. 2.2b shows the electron orbitals for carbon. Of carbon's six electrons, two occupy the small spherical orbital representing the lowest energy level. The remaining four are distributed among four possible orbitals at the next highest energy level: one of these four orbitals is a sphere (larger in diameter than the orbital at the lowest energy level) and three are dumbbell-shaped. In carbon, the outermost spherical orbital has two electrons, two of the dumbbell-shaped orbitals have one electron each, and one of the dumbbell-shaped orbitals is empty. Because a full orbital contains two electrons, it would take a total of four additional electrons to completely fill all of the orbitals at this energy level. Therefore, after the first shell, the maximum number of electrons per energy level is eight. **Fig. 2.2c** shows simplified diagrams of atoms in which the highest energy level, or shell, of hydrogen, represented by the outermost circle, contains one electron, and that of carbon contains four electrons.

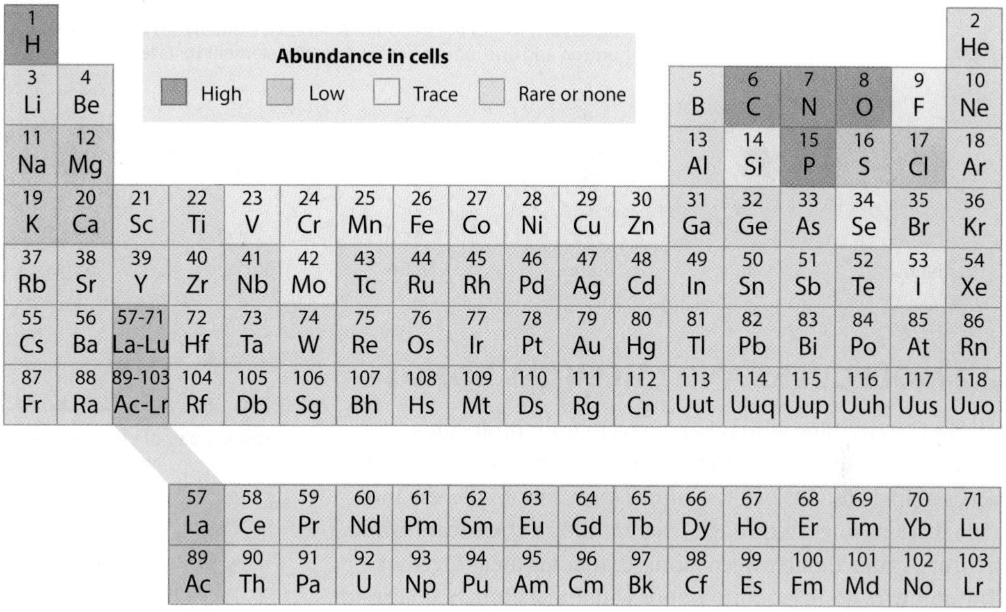

Elements have recurring, or periodic, chemical properties.

The chemical elements are often arranged in a tabular form known as the **periodic table of the elements,** shown in **Fig. 2.3**. The development of the periodic table is generally credited to the nineteenth-century Russian chemist Dmitri Mendeleev. The table provides a way to organize all the chemical elements in terms of their chemical properties.

In the periodic table, the elements are indicated by their chemical symbols and arranged in order of increasing atomic number. For example, the second row of the periodic table begins with lithium (Li) with 3 protons and ends with neon (Ne) with 10 protons.

For the first three horizontal rows in the periodic table, elements in the same row have the same number of shells, so they also have the same number and types of orbitals available to be filled by electrons. Across a row, therefore, electrons fill the shell until a full complement of electrons is reached on the right-hand side of the table. **Fig. 2.4** shows the filling of the shells for elements in the second row of the periodic table.

The elements in a vertical column are called a group or family. Members of a group all have the same number of electrons in their outermost shell. For example, carbon (C) and lead (Pb) both have four electrons in their outermost shell. The number of electrons in the outermost shell determines in large part how elements interact with other elements to form a diversity of molecules, as we will see in the next section.

FIG. 2.3 The periodic table of the elements. Elements are arranged by increasing number of protons, the atomic number. The elements in a column share similar chemical properties.

Self-Assessment Questions

1. In the early 1900s, Ernest Rutherford produced a beam of very small positive particles and directed it at a thin piece of gold foil just a few atoms thick. Most of the particles passed through the foil without changing their path; very rarely, a particle was deflected. What conclusions can you draw from this experiment about the structure of an atom?

2. What are atoms made up of? Describe each component.

3. What is the logic behind how the periodic table of the elements is organized?

FIG. 2.4 Energy levels (shells) of row 2 of the periodic table. The complete complement of electrons in the outer shell of this row of elements is eight.

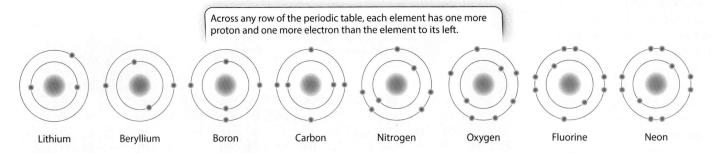

2.2 MOLECULES AND CHEMICAL BONDS

Atoms can combine with other atoms to form **molecules**, which are groups of two or more atoms attached together that act as a single unit. When two atoms form a molecule, the individual atoms interact through what is called a **chemical bond**, a form of attraction between atoms that holds them together. The ability of atoms to form bonds with other atoms explains in part why just a few types of element can come together in many different ways to make a variety of molecules that can carry out diverse functions in a cell. There are several ways in which atoms can interact with one another, and therefore many different types of chemical bond.

A covalent bond results when two atoms share electrons.

The ability of atoms to combine with other atoms is determined in large part by the electrons farthest from the nucleus—those in the outermost orbitals of an atom. These electrons are called **valence electrons**, and they are at the highest energy level of the atom. When atoms combine with other atoms to form a molecule, the atoms share valence electrons with each other. Specifically, when the outermost orbitals of two atoms come into proximity to each other, two atomic orbitals each containing one electron merge into a single orbital containing a full complement of two electrons. The merged orbital is called a **molecular orbital**, and each shared pair of electrons constitutes a **covalent bond** that holds the atoms together.

We can represent a specific molecule by its chemical formula, which is written as the letter abbreviation for each element followed by a subscript giving the number of that type of atom in the molecule. Among the simplest molecules is hydrogen gas, illustrated in **Fig. 2.5**, which consists of two covalently bound hydrogen atoms as indicated by the chemical formula H_2. Each hydrogen atom has a single electron in a spherical orbital. When the atoms join into a molecule, the two orbitals merge into a single molecular orbital containing two electrons that are shared by the hydrogen atoms. A covalent bond between atoms is denoted by a single line connecting the two chemical symbols for the atoms, as shown in the structural formula in Fig. 2.5.

Two adjacent atoms can sometimes share two pairs of electrons, forming a **double bond** denoted by a double line connecting the two chemical symbols for the atoms. In this case, four orbitals, each occupied by a single electron, merge to form two molecular orbitals.

Molecules tend to be most stable when the two atoms forming a bond share enough electrons to completely occupy the outermost energy level or shell. The outermost shell of a hydrogen atom can hold two electrons; by comparison, this shell can hold eight electrons in carbon, nitrogen, and oxygen atoms. This simple rule for forming stable molecules is known as the octet rule and applies to many, but not all, elements. For example, as shown in **Fig. 2.6**, one carbon atom (C, with four valence electrons) combines with four hydrogen atoms (H, with one valence electron each) to form CH_4 (methane); nitrogen

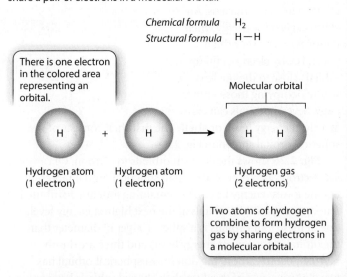

FIG. 2.5 A covalent bond. A covalent bond is formed when two atoms share a pair of electrons in a molecular orbital.

FIG. 2.6 Four molecules. Atoms tend to combine in such a way as to complete the complement of electrons in the outer shell.

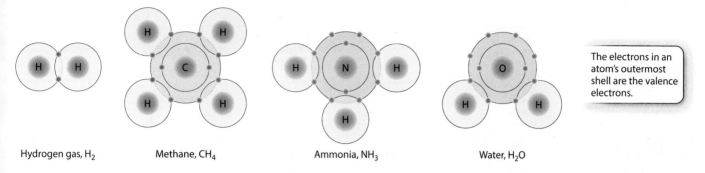

Hydrogen gas, H₂   Methane, CH₄   Ammonia, NH₃   Water, H₂O

The electrons in an atom's outermost shell are the valence electrons.

(N, with five valence electrons) combines with three H atoms to form NH_3 (ammonia); and oxygen (O, with six valence electrons) combines with two H atoms to form H_2O (water). Interestingly, the elements of the same column in the next row behave similarly. This is just one example of the recurring, or periodic, behavior of the elements.

A polar covalent bond is characterized by unequal sharing of electrons.

In hydrogen gas (H_2), the electrons are shared equally by the two hydrogen atoms. In many bonds, however, the electrons are not shared equally by the two atoms. A notable example is the bonds in a water molecule (H_2O): it consists of two hydrogen atoms, each of which covalently bound to a single oxygen atom (**Fig. 2.7**).

In a molecule of water, the electrons are more likely to be located near the oxygen atom. The unequal sharing of electrons results from a difference in the ability of the atoms to attract electrons, a property known as **electronegativity**. Electronegativity tends to increase across a row in the periodic table; as the number of positively charged protons across a row increases, negatively charged electrons are held more tightly to the nucleus. Therefore, oxygen is more electronegative than hydrogen and attracts electrons more readily than does hydrogen. In a molecule of water, oxygen has a slight negative charge, while the two hydrogen atoms have a slight positive charge (Fig. 2.7). When electrons are shared unequally between the two atoms, the resulting interaction is described as a **polar covalent bond**.

A covalent bond between atoms that have the same, or nearly the same, electronegativity is described as a **nonpolar covalent bond**, which means that the atoms share the bonding electron pair almost equally. Nonpolar covalent bonds include those in gaseous hydrogen (H_2) and oxygen (O_2), as well as carbon–carbon (C—C) and carbon–hydrogen (C—H) bonds. Molecules held together by nonpolar covalent bonds are important in cells because they do not mix well with water.

An ionic bond forms between oppositely charged ions.

In a molecule of water, the difference in electronegativity between the oxygen and hydrogen atoms leads to unequal sharing of electrons. In more extreme cases, when an atom of very high electronegativity is paired with an atom of very low electronegativity, the difference in electronegativity is so great that the electronegative atom "steals" the electron from its less electronegative partner. In this case, the atom with the extra electron has a negative charge and is a negative ion. The atom that has lost an electron has a positive charge and is a positive ion. The two ions are not covalently bound, but because opposite charges attract, they associate with each other in an **ionic bond**. An example of a compound formed by the attraction of a positive ion and a negative ion is sodium chloride (NaCl), commonly known as table salt (**Fig. 2.8a**).

When sodium chloride is placed in water, the salt dissolves to form sodium ions (written as "Na^+") that have lost an electron and so are positively charged, and chloride ions (Cl^-) that have gained an electron and so are negatively charged. In solution, the two ions are pulled apart and become surrounded

FIG. 2.7 A polar covalent bond. In a polar covalent bond, the two atoms do not share the electrons equally.

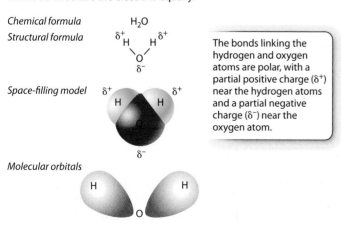

Chemical formula: H_2O

Structural formula

Space-filling model

Molecular orbitals

The bonds linking the hydrogen and oxygen atoms are polar, with a partial positive charge (δ^+) near the hydrogen atoms and a partial negative charge (δ^-) near the oxygen atom.

FIG. 2.8 An ionic bond. (a) Sodium chloride (salt) is formed by the attraction of two ions. (b) In solution, the ions are surrounded by water molecules.

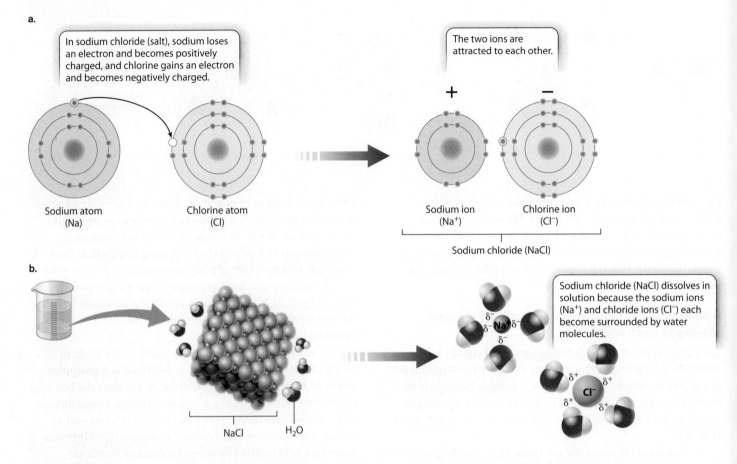

by water molecules. The negatively charged ends of water molecules are attracted to the positively charged sodium ion, and the positively charged ends of other water molecules are attracted to the negatively charged chloride ion (**Fig. 2.8b**). Only as the water evaporates do the concentrations of Na^+ and Cl^- increase to the point where the ions join and precipitate as salt crystals.

A chemical reaction involves breaking and forming chemical bonds.

The chemical bonds that link atoms in molecules can change in a **chemical reaction**, a process by which atoms or molecules, called **reactants**, are transformed into different molecules, called **products**. During a chemical reaction, atoms keep their identity but change which atoms they are bonded to.

For example, two molecules of hydrogen gas ($2H_2$) and one molecule of oxygen gas (O_2) can react to form two molecules of water ($2H_2O$), as shown in **Fig. 2.9**. In this reaction, the numbers of each type of atom are conserved, but their arrangement is different in the reactants and the products. Specifically, the H—H bond in hydrogen gas and the O=O bond in oxygen are broken. At the same time, each oxygen atom forms

FIG. 2.9 A chemical reaction. During a chemical reaction, atoms retain their identity, but their connections change as bonds are broken and new bonds are formed.

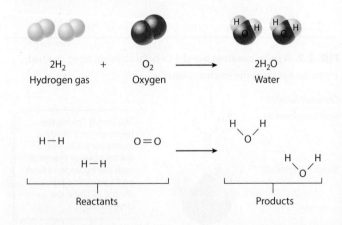

new covalent bonds with two hydrogen atoms, forming two molecules of water. In fact, this reaction is the origin of the word "hydrogen," which literally means "water former." The reaction releases a good deal of energy and is used in some rockets as a booster in satellite launches.

In biological systems, chemical reactions provide a way to build and break down molecules for use by the cell, as well as to harness energy, which can be held in chemical bonds (Chapter 6).

Self-Assessment Questions

4. From their positions in the periodic table (see Fig. 2.3), can you predict how many lithium atoms and hydrogen atoms can combine to form a molecule?
5. What are the differences between covalent bonds and polar covalent, hydrogen, and ionic bonds?

2.3 WATER

On Earth, all life depends on water. Indeed, life originated in water, and the availability of water strongly influences the environmental distributions of different species. Furthermore, water is the single most abundant molecule in cells, so water is the medium in which the molecules of life interact. So important to life is water that, in the late 1990s, the National Aeronautics and Space Administration (NASA) announced that the search for extraterrestrial life would be based on a simple strategy: follow the water. NASA's logic makes sense because, within our own solar system, Earth stands out both for its abundance of water and for the life that water supports. In addition, as we saw in Case 1 Life's Origins, life might even have been present on ancient Mars when liquid water was still plentiful. What makes water so special as the medium of life?

Water is a polar molecule.

As we saw earlier, water molecules have polar covalent bonds, characterized by an uneven distribution of electrons. A molecule like water that has regions of positive and negative charge is called a **polar** molecule. Polar molecules tend to interact with other polar molecules, whereas nonpolar molecules tend not to interact with polar molecules. Therefore molecules, or even different regions of the same molecule, can be organized into two general classes, depending on how they interact with water: **hydrophilic** ("water loving") and **hydrophobic** ("water fearing").

Hydrophilic compounds are polar; they dissolve readily in water. That is, water is a good **solvent**, capable of dissolving many substances. Think of what happens when you stir a teaspoon of sugar into water: the sugar seems to disappear as it dissolves. The sugar molecules disperse through the water and separate from one another, forming a solution in the watery, or **aqueous**, environment.

By contrast, hydrophobic compounds are **nonpolar**. Nonpolar compounds do not have regions of positive and negative charge, so they arrange themselves to minimize their contact with water. For example, oil molecules are hydrophobic. Thus, when oil and water are mixed, the oil molecules organize themselves into droplets that limit the oil–water interface. This **hydrophobic effect**, in which polar molecules like water exclude nonpolar ones, drives such biological processes as the folding of proteins (Chapter 4) and the formation of cell membranes (Chapter 5).

A hydrogen bond is an interaction between a hydrogen atom and an electronegative atom.

Because the oxygen and hydrogen atoms have slight charges, water molecules orient themselves to minimize the repulsion of like charges so that positive charges are near negative charges. For example, because of its slight positive charge, a hydrogen atom in a water molecule tends to orient itself toward the slightly negatively charged oxygen atom in another molecule.

A **hydrogen bond** is the name given to this interaction between a hydrogen atom with a slight positive charge and an electronegative atom of another molecule. In fact, any hydrogen atom covalently bound to an electronegative atom (such as oxygen or nitrogen) will have a slight positive charge and can form a hydrogen bond. In the case of water, the result of many such interactions is a molecular network stabilized by hydrogen bonds. Typically, a hydrogen bond is depicted by a dotted line, as in **Fig. 2.10**.

Hydrogen bonds are much weaker than covalent bonds, but it is hydrogen bonding that gives water many interesting properties, described next. In addition, the presence of many weak hydrogen bonds can help stabilize biological molecules, as in the case of nucleic acids and proteins.

FIG. 2.10 Hydrogen bonds in liquid water. Because of thermal motion, hydrogen bonds in water are continually breaking and forming between different pairs of molecules.

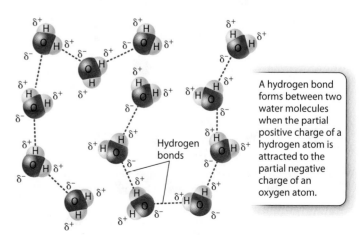

A hydrogen bond forms between two water molecules when the partial positive charge of a hydrogen atom is attracted to the partial negative charge of an oxygen atom.

Hydrogen bonds give water many unusual properties.

Hydrogen bonds influence the structure of both liquid water (**Fig. 2.11a**) and ice (**Fig. 2.11b**). When water freezes, most water molecules become hydrogen bonded to four other water molecules, forming an open crystalline structure we call ice. As the temperature increases and the ice melts, some of the hydrogen bonds are destabilized and break, allowing the water molecules to pack more closely and explaining why liquid water is denser than ice. As a result, ice floats on water, and ponds and lakes freeze from the top down and so do not freeze completely. This property allows fish and aquatic plants to survive winter in the cold water under the layer of ice.

Hydrogen bonds also give water molecules the property of **cohesion**, meaning that they tend to stick to one another. A consequence of cohesion is high surface tension, a measure of the difficulty of breaking the surface of a liquid. Cohesion between molecules contributes to water movement in plants. As water evaporates from leaves, water is pulled upward through the plant's vessels. Sometimes this water may rise as high as 100 meters above the ground in giant sequoia and coast redwood trees, which are among the tallest trees on Earth.

The hydrogen bonds of water also influence how water responds to heating. Molecules are in constant motion, and this motion increases as the temperature increases. When water is heated, some of the energy added by heating is used to break hydrogen bonds instead of causing more motion among the molecules. As a consequence, the temperature increases less in these situations than if there were no hydrogen bonding. The abundant hydrogen bonds make water more resistant to temperature changes than other substances, a property that is important for living organisms on a variety of scales. In the cell, water resists temperature variations that would otherwise result from numerous biochemical reactions. On a global scale, the oceans minimize temperature fluctuations, stabilizing the temperature on Earth in a range compatible with life.

In short, water is clearly the medium of life on Earth. But does it fill this critical role because water is uniquely suited for life, or is it because life on Earth has adapted through time to a watery environment? We don't know the answer to this question, but probably both explanations are partly true. Chemists have proposed that under conditions of high pressure and temperature, other small molecules, including ammonia (NH_3) and some simple carbon-containing molecules, might display similar characteristics friendly to life. However, under the conditions that exist on Earth, water is the only molecule uniquely suited to life. Water is a truly remarkable substance, and life on Earth would not be possible without it.

pH is a measure of the concentration of protons in solution.

In any solution of water, a small proportion of the water molecules exist as protons (H^+) and hydroxide ions (OH^-). The pH of a solution measures the proton concentration ($[H^+]$), which is important because the pH influences many chemical reactions and biological processes. It is calculated by the following formula:

$$pH = -\log[H^+]$$

The pH of a solution can range from 0 to 14. Since the pH scale is logarithmic, a difference of one pH unit corresponds to a tenfold difference in hydrogen ion concentration. A solution

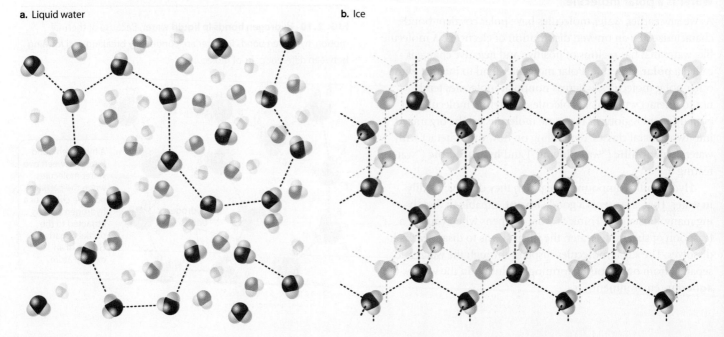

FIG. 2.11 Liquid water and ice. Hydrogen bonds create (a) a dense structure in water and (b) a highly ordered, less dense, crystalline structure in ice.

a. Liquid water

b. Ice

is neutral (pH = 7) when the concentrations of protons (H^+) and hydroxide ions (OH^-) are equal. When the concentration of protons is higher than that of hydroxide ions, the pH is lower than 7 and the solution is **acidic**. When the concentration of protons is lower than that of hydroxide ions, the pH is higher than 7 and the solution is **basic**. In turn, an acid can be described as a molecule that releases a proton (H^+), and a base is a molecule that accepts a proton in aqueous solution.

Pure water has a pH of 7—that is, it is neutral, with an equal concentration of protons and hydroxide ions. The pH of most cells is approximately 7 and is tightly regulated, as most chemical reactions can be carried out only in a narrow pH range. Certain cellular compartments, however, have a much lower pH. The pH of blood is slightly basic, with a pH around 7.4. This value is sometimes referred to in medicine as physiological pH. Freshwater lakes, ponds, and rivers tend to be slightly acidic because carbon dioxide from the air dissolves in the water and forms carbonic acid (Chapter 6).

Self-Assessment Questions

6. Why do containers of water, milk, soda, or other liquids sometimes burst when frozen?
7. What are three unusual properties of water, and why do these properties make water conducive to life?

2.4 CARBON

Hydrogen and helium are far and away the most abundant elements in the universe. In contrast, the solid Earth is dominated by silicon, oxygen, aluminum, iron, and calcium (Chapter 1). In other words, Earth is not a typical sample of the universe. Nor is the cell a typical sample of the solid Earth. **Fig. 2.12** shows the relative abundance by mass of chemical elements present in human cells after all the water has been removed. Note that just four elements—carbon (C), oxygen (O), hydrogen (H), and nitrogen (N)—account for approximately 90% of the total dry mass, and that the most abundant element is carbon.

The elemental composition of human cells is typical of all cells. Human life, and all life as we know it, is based on carbon. Carbon molecules play such an important role in living organisms that carbon-containing molecules have a special name—they are called **organic molecules**. Carbon has the ability to combine with many other elements to form a wide variety of molecules, each specialized for the functions it carries out in the cell.

Carbon atoms form four covalent bonds.

One of the special properties of carbon is that, in forming molecular orbitals, a carbon atom behaves as if it had four unpaired electrons. This behavior occurs because one of the electrons in the outermost spherical orbital moves into the

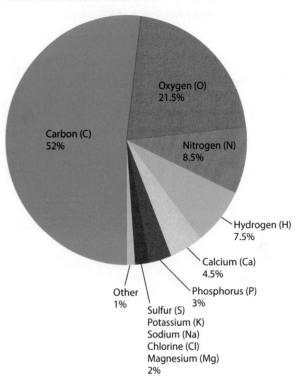

FIG. 2.12 Approximate proportions by dry mass of chemical elements found in human cells.

empty dumbbell-shaped orbital (see Fig. 2.2). In this process, the single large spherical orbital and three dumbbell-shaped orbitals change shape, becoming four equivalent hybrid orbitals, each with one electron.

Fig. 2.13 shows the molecular orbitals that result when one atom of carbon combines with four atoms of hydrogen to

FIG. 2.13 A carbon atom with four covalent bonds. In each molecule of methane, a carbon atom shares electrons with four hydrogen atoms in hybrid orbitals, forming a tetrahedron.

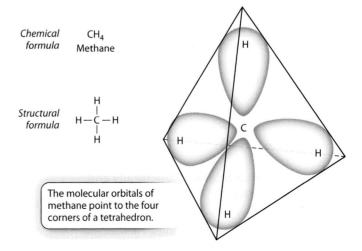

FIG. 2.14 Diverse carbon-containing molecules. (a) The molecule ethane contains one C—C bond. (b) A linear chain of carbon atoms and a ring structure that also contains oxygen contain multiple C—C bonds.

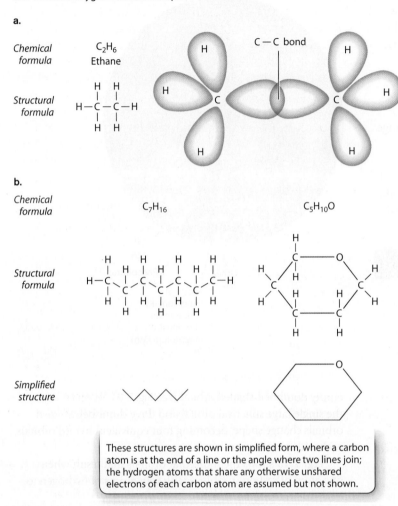

These structures are shown in simplified form, where a carbon atom is at the end of a line or the angle where two lines join; the hydrogen atoms that share any otherwise unshared electrons of each carbon atom are assumed but not shown.

form the gas methane (CH_4). Each of the four valence electrons of carbon shares a new molecular orbital with the electron of one of the hydrogen atoms. These bonds can rotate freely about their axis. Furthermore, because of the shape of the orbitals, the carbon atom lies at the center of a three-dimensional structure called a tetrahedron, and the four molecular orbitals point toward the four corners of this structure. The ability of carbon to form four covalent bonds, the spatial orientation of these bonds in the form of a tetrahedron, and the ability of each bond to rotate freely all contribute in important ways to the structural diversity of carbon-based molecules.

Carbon-based molecules are structurally and functionally diverse.

Carbon has other properties that contribute to its ability to form a diversity of molecules. For example, carbon atoms can link with other carbon atoms through covalent bonds, forming long chains. These chains can be branched, or two carbons at the ends of the chain or within the chain can link to form a ring.

Among the simplest chains is ethane, shown in **Fig. 2.14a**. Ethane forms when two carbon atoms become connected by a covalent bond. In this case, the orbitals of unpaired electrons in two carbon atoms form the covalent bond. Each carbon atom is also bound to three hydrogen atoms. Some more complex examples of carbon-containing molecules are shown in **Fig. 2.14b**, as both structural formulas and in simplified form.

Two adjacent carbon atoms can also share two pairs of electrons, forming a double bond, as shown in **Fig. 2.15**. Note that each carbon atom has exactly four covalent bonds, but in this case two are shared between adjacent carbon atoms. The double bond is shorter than a single bond and is not free to rotate, so all of the covalent bonds formed by the carbon atoms connected by a double bond are in the same geometrical plane. As with single bonds, double bonds can be found in chains of atoms or ring structures.

While the types of atoms making up a molecule help characterize the molecule, the spatial arrangement of atoms is also important. For example, 6 carbon atoms, 13 hydrogen atoms, 2 oxygen atoms, and 1 nitrogen atom can join covalently in many different arrangements to produce molecules with different structures. Two of these many arrangements are shown in **Fig. 2.16**. Note that some of the connections between atoms are identical in the two molecules (black) and some are different (green), even though the chemical formulas are the same ($C_6H_{13}O_2N_1$). Molecules that have the same chemical formula but different structures are known as **isomers**.

We have seen that carbon-containing molecules can adopt a wide range of arrangements, a versatility that helps us to understand how a limited number of elements can create an astonishing variety of molecules. We might ask whether carbon is uniquely versatile. Put another way, if we ever discover life on a distant planet, will it be based on carbon, like humans? Silicon, which is found just below carbon in the

FIG. 2.15 Molecules containing double bonds between carbon atoms.

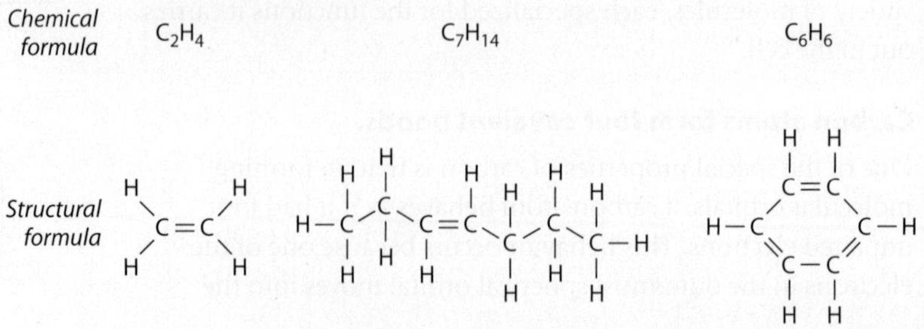

FIG. 2.16 The isomers isoleucine and leucine. Isoleucine and leucine have the same chemical formula, but their structures differ.

Chemical formula: $C_6H_{13}O_2N_1$ Isoleucine $C_6H_{13}O_2N_1$ Leucine

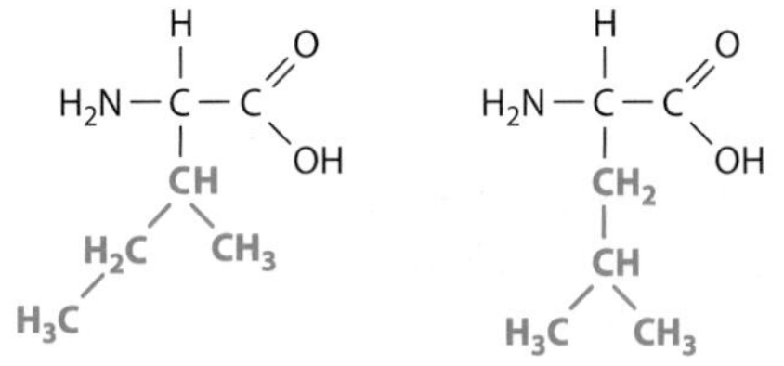

periodic table (see Fig. 2.3), is the one other element that is both reasonably abundant and characterized by four atomic orbitals with one electron each. Some scientists have speculated that silicon might provide an alternative to carbon as a chemical basis for life. However, silicon readily binds with oxygen. On Earth, nearly all of the silicon atoms found in molecules are covalently bound to oxygen. Studies of Mars and meteorites show that silicon is tightly bound to oxygen throughout our solar system, and that relationship is likely to be repeated everywhere we might explore. There are approximately 1000 different silicate minerals on Earth, but this diversity pales before the millions of known carbon-based molecules. If we ever discover life beyond Earth, very likely its chemistry will be based on carbon.

Self-Assessment Questions

8. What are the four most common elements in organic molecules and which common macromolecules always contain all four of these elements?
9. What features of carbon allow it to form diverse structures?

2.5 ORGANIC MOLECULES

Chemical processes in the cell depend on just a few classes of carbon-based molecules. **Proteins** provide structural support and act as catalysts that facilitate chemical reactions. **Nucleic acids** encode and transmit genetic information. **Carbohydrates** provide a source of energy and make up the cell wall in bacteria, plants, and algae. **Lipids** make up cell membranes, store energy, and act as signaling molecules.

These molecules are all large, consisting of hundreds or thousands of atoms, and many are **polymers,** complex molecules made up of repeated simpler units connected by covalent bonds. Proteins are polymers of **amino acids,** nucleic acids are polymers of **nucleotides,** and carbohydrates such as starch are polymers of simple **sugars.** Lipids are a bit different, as we will see, in that they are defined by a property rather than by their chemical structure. The lipid membranes that define cell boundaries consist of **fatty acids** bonded to other organic molecules.

Building macromolecules from simple, repeating units provides a means of generating an almost limitless chemical diversity. Indeed, in macromolecules, the building blocks of polymers play a role much like that of the letters in words. In written language, a change in the content or order of letters changes the meaning of the word (or renders it meaningless). For example, by reordering the letters of the word SILENT you can write LISTEN, a word with a different meaning. Similarly, rearranging the building blocks that make up macromolecules provides an important way to make a large number of diverse macromolecules whose functions differ from one to the next.

In the following sections, we focus on the building blocks of these four key molecules of life, reserving a discussion of the structure and function of the macromolecules for later chapters.

Functional groups add chemical character to carbon chains.

The simple repeating units of polymers are often based on a nonpolar core of carbon atoms. Attached to these carbon atoms are **functional groups,** groups of one or more atoms that have particular chemical properties on their own, regardless of what they are attached to. Among the functional groups frequently encountered in biological molecules are amine (=NH), amino (—NH_2), carboxyl (—COOH), hydroxyl (—OH), ketone (=O), phosphate (—O—PO_3H_2), sulfhydryl (—SH), and methyl (—CH_3). The nitrogen, oxygen, phosphorus, and sulfur atoms in these functional groups are more electronegative than the carbon atoms, and functional groups containing these atoms are polar. The methyl group (—CH_3), in contrast, is nonpolar.

Because many functional groups are polar, molecules that contain these groups—molecules that would otherwise be nonpolar—become polar. As a result, these molecules become soluble in the cell's aqueous environment. In other words, they disperse in solution throughout the cell. Moreover, because many functional groups are polar, they are reactive. Notice in the following sections that the reactions joining simpler molecules into polymers usually take place between functional groups.

Proteins are composed of amino acids.

Proteins do much of the cell's work. Some function as catalysts that accelerate the rates of chemical reactions (in which case they are called **enzymes**); others act as structural components necessary for cell shape and movement. The human body contains many thousands of distinct types of protein that perform a wide range of functions. Since proteins consist of amino acids linked covalently to form a chain, we need to examine the chemical features of amino acids to understand the diversity and versatility of proteins.

The general structure of an amino acid is shown in **Fig. 2.17a.** Each amino acid contains a central carbon atom, called the α **(alpha) carbon,** covalently linked to four groups: an **amino group** (—NH_2; blue), a **carboxyl group** (—COOH; brown), a hydrogen atom (H), and an **R group,** or **side chain** (green), that differs from one amino acid to the next. The identity of each amino acid is determined by the structure and composition of the

FIG. 2.17 Amino acids and peptide bonds. (a) An amino acid contains four groups attached to a central carbon atom. (b) In the environment of a cell, the amino group gains a proton and the carboxyl group loses a proton. (c) Peptide bonds link amino acids to form a protein.

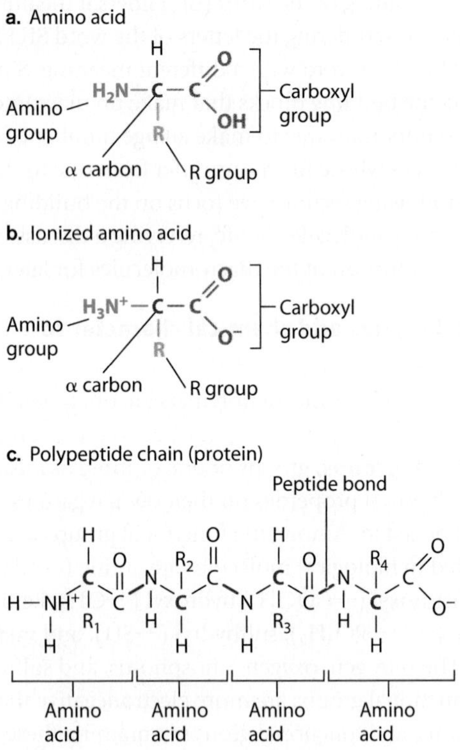

a. Amino acid

b. Ionized amino acid

c. Polypeptide chain (protein)

side chain. The side chain of the amino acid glycine is simply H, for example, and that of alanine is CH_3. In most amino acids, the α carbon is covalently linked to four different groups. Glycine is the exception, since its R group is a hydrogen atom.

At the pH commonly found in a cell (pH 7.4), the amino and carboxyl groups are ionized (charged) owing to interactions with the surrounding medium. The amino group gains a proton ($—NH_3^+$; blue) and the carboxyl group loses a proton ($—COO^-$; brown), as shown in **Fig. 2.17b**.

Amino acids are linked in a chain to form a protein (**Fig. 2.17c**). The carbon atom in the carboxyl group of one amino acid is joined to the nitrogen atom in the amino group of the next by a covalent linkage called a **peptide bond**. In Fig. 2.17c, the chain of amino acids includes four amino acids, and the peptide bonds are indicated in red. The formation of a peptide bond involves the loss of a water molecule. That is, to form a C—N bond, the carbon atom of the carboxyl group must release an oxygen atom and the nitrogen atom of the amino group must release two hydrogen atoms. These oxygen and hydrogen atoms can then combine to form a water molecule (H_2O). The loss of a water molecule also occurs in the linking of subunits to form polymers such as nucleic acids and complex carbohydrates.

Cellular proteins are composed of combinations of 20 different amino acids, each of which can be classified according to the chemical properties of its R group. The particular sequence, or order, in which amino acids are present in a protein determines how it folds into its three-dimensional structure. The three-dimensional structure, in turn, determines the protein's function. In Chapter 4, we examine how the sequence of amino acids in a particular protein is specified and discuss how proteins fold into their three-dimensional shapes.

Nucleic acids encode genetic information in their nucleotide sequence.

Nucleic acids are examples of informational molecules—that is, large molecules that carry information in the sequence of nucleotides that make them up. This molecular information is much like the information carried by the letters in an alphabet, but in the case of nucleic acids, the information is in chemical form.

The nucleic acid **deoxyribonucleic acid (DNA)** is the genetic material in all organisms. It is transmitted from parents to offspring, and it contains the information needed to specify the amino acid sequence of all the proteins synthesized in an organism. The nucleic acid **ribonucleic acid (RNA)** has multiple functions; it is a key player in protein synthesis and the regulation of gene expression.

DNA and RNA are long molecules consisting of nucleotides bonded covalently one to the next. Nucleotides, in turn, are composed of three components: a 5-carbon sugar, a nitrogen-containing compound called a **base**, and one or more phosphate groups (**Fig. 2.18**). The sugar in RNA is ribose,

FIG. 2.18 A ribonucleotide and a deoxyribonucleotide, the units of RNA and DNA.

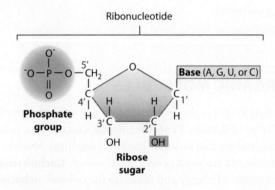

and the sugar in DNA is deoxyribose. The sugars differ in that ribose has a hydroxyl (OH) group on the second carbon (designated the 2 carbon), whereas deoxyribose has a hydrogen atom at this position (hence, *deoxy*ribose). (By convention, the carbons in the sugar are numbered with primes—1′, 2′, and so on—to distinguish them from carbons in the base—1, 2, and so on.)

The bases are built from nitrogen-containing rings and are of two types. The **pyrimidine** bases (**Fig. 2.19a**) have a single ring and include **cytosine (C), thymine (T),** and **uracil (U)**. The **purine** bases (**Fig. 2.19b**) have a double-ring structure and include **guanine (G)** and **adenine (A)**. DNA contains the bases A, T, G, and C, whereas RNA contains the bases A, U, G, and C. Just as the order of amino acids provides the information carried in proteins, so, too, does the sequence of nucleotides determine the information in DNA and RNA molecules.

In DNA and RNA, each adjacent pair of nucleotides is connected by a **phosphodiester bond,** which forms when a phosphate group in one nucleotide is covalently joined to the sugar unit in another nucleotide (**Fig. 2.20**). Like the formation of a peptide bond, the formation of a phosphodiester bond involves the loss of a water molecule.

DNA in cells usually consists of two strands of nucleotides twisted around each other in the form of a **double helix** (**Fig. 2.21a**). The sugar–phosphate backbones of the strands wrap like a ribbon around the outside of the double helix, and the bases point inward. The bases form specific purine–pyrimidine pairs that are **complementary**: where one strand carries an A, the other carries a T; and where one strand carries a G, the other carries a C. Base pairing results from hydrogen bonding between the bases (**Fig. 2.21b**).

FIG. 2.19 Pyrimidine bases and purine bases. (a) Pyrimidines have a single-ring structure, whereas (b) purines have a double-ring structure.

a. Pyrimidine bases

Cytosine (C)   Thymine (T)   Uracil (U)

b. Purine bases

Guanine (G)   Adenine (A)

In a nucleic acid, each base is attached to either a ribose or a deoxyribose by the bond indicated in red.

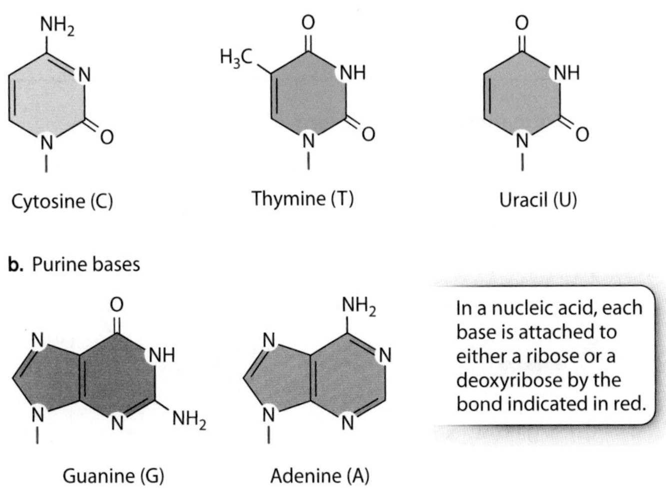

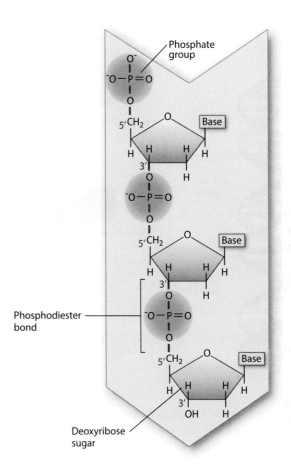

FIG. 2.20 The phosphodiester bond. Phosphodiester bonds link successive deoxyribonucleotides, forming the backbone of the DNA strand.

The genetic information in DNA is contained in the sequence, or order, in which successive nucleotides occur along the molecule. Successive nucleotides along a DNA strand can occur in *any* order, so a long molecule could contain any of an immense number of possible nucleotide sequences. This is one reason why DNA is an efficient carrier of genetic information. In Chapter 3, we consider the structure and function of DNA and RNA in greater detail.

Complex carbohydrates are made up of simple sugars.

Many of us, when we feel tired, reach for a candy bar for a quick energy boost. The energy in a candy bar comes from sugars, which are quickly broken down to release energy. Sugars belong to a class of molecules called **carbohydrates,** distinctive molecules composed of C, H, and O atoms, usually in the ratio 1:2:1. Carbohydrates are a major source of energy for metabolism.

The simplest carbohydrates are sugars (also called **saccharides**). Simple sugars are linear or, far more commonly, cyclic molecules containing five or six carbon atoms. All 6-carbon

FIG. 2.21 The structure of DNA. (a) DNA most commonly occurs in the form of a double helix, with the sugar and phosphate groups forming the backbone and the bases oriented inward. (b) The bases are complementary: A is always paired with T, and G is always paired with C. Base pairing results from hydrogen bonds.

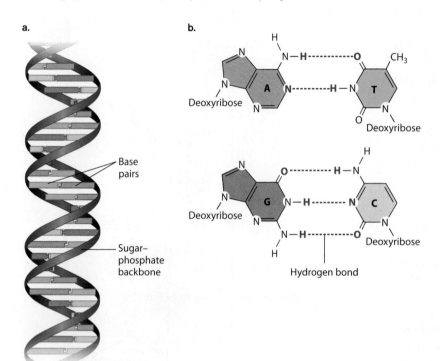

sugars have the same chemical formula ($C_6H_{12}O_6$) and differ only in configuration. Glucose (the product of photosynthesis), galactose (found in dairy products), and fructose (a commercial sweetener) are examples; they share the same formula ($C_6H_{12}O_6$) but differ in the arrangement of their atoms (**Fig. 2.22**).

A simple sugar is also called a **monosaccharide** (*mono* means "one"), and linking two simple sugars together by a covalent bond forms a disaccharide (*di* means "two"). Sucrose ($C_{12}H_{22}O_{11}$), or table sugar, is a disaccharide that combines one molecule each of glucose and fructose. Simple sugars combine in many ways to form polymers called **polysaccharides** (*poly* means "many") that provide long-term energy storage (starch and glycogen) or structural support (cellulose in plant cell walls). Long, branched chains of monosaccharides are called **complex carbohydrates**.

Let's take a closer look at monosaccharides, the simplest sugars. Monosaccharides are unbranched carbon chains with either an aldehyde (HC=O) or a ketone (C=O) group (Fig. 2.22). Monosaccharides with an aldehyde group are called aldoses and those with a ketone group are known as ketoses. In both types of monosaccharide, the other carbons each carry one hydroxyl (—OH) group and one hydrogen (H) atom. When the linear structure of a monosaccharide is written with the aldehyde or ketone group at the top, the carbons are numbered from top to bottom.

Almost all of the monosaccharides in cells are in ring form (**Fig. 2.23**), rather than linear structures. To form a ring, one end of the chain bonds to another part of the chain: the carbon in the aldehyde or ketone group forms a covalent bond with the oxygen of a hydroxyl group carried by another carbon in the same molecule.

For example, cyclic glucose is formed when the oxygen atom of the hydroxyl group on carbon 5 forms a covalent bond with carbon 1, which is part of an aldehyde group. The cyclic structure is approximately flat, and you can visualize it as being perpendicular to the plane of the paper, with the covalent bonds indicated by the thick lines in the foreground of Fig. 2.23. The groups attached to any carbon project either above or below the ring. When the ring is formed, the aldehyde oxygen becomes a hydroxyl group. The presence of the polar hydroxyl groups through the sugar ring makes these molecules highly soluble in water.

FIG. 2.22 Structural formulas for some 6-carbon sugars.

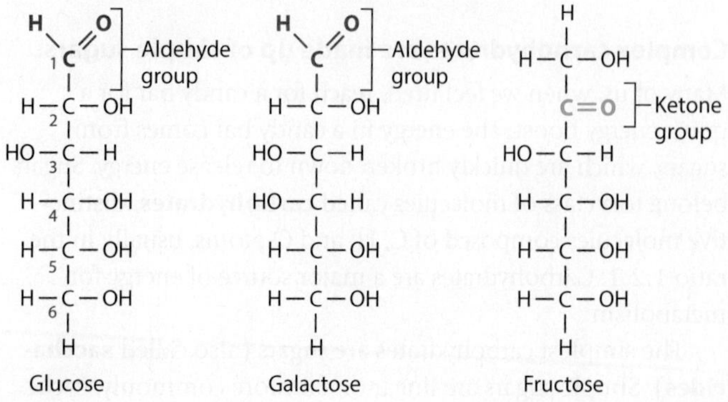

FIG. 2.23 Formation of the cyclic form of glucose.

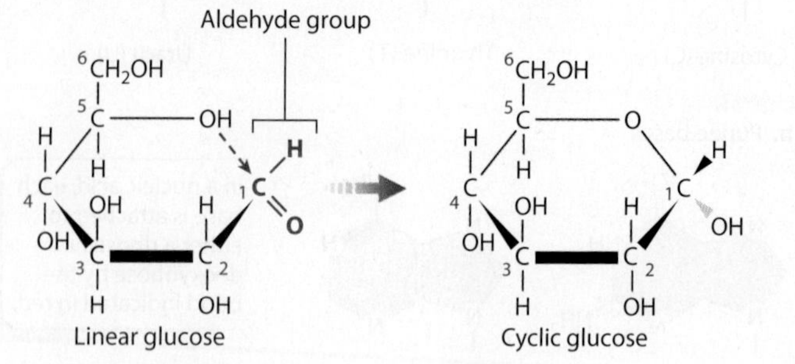

Monosaccharides, especially 6-carbon sugars, are the building blocks of complex carbohydrates. Monosaccharides are attached to each other by covalent bonds called **glycosidic bonds** (**Fig. 2.24**). As with peptide bonds, the formation of glycosidic bonds involves the loss of a water molecule. A glycosidic bond is formed between carbon 1 of one monosaccharide and a hydroxyl group carried by a carbon atom in a different monosaccharide molecule.

Carbohydrate diversity stems in part from the monosaccharides that make up carbohydrates, similar to the way that protein and nucleic acid diversity stems from the sequence of their subunits. Some complex carbohydrates are composed of a single type of monosaccharide, while others are a mix of different kinds of monosaccharide. Starch, for example, is a sugar storage molecule in plants composed completely of glucose molecules, whereas pectin, a component of the cell wall, contains up to five different types of monosaccharide.

Lipids are hydrophobic molecules.

Proteins, nucleic acids, and carbohydrates all are polymers made up of smaller, repeating units with a defined structure. Lipids, however, are different. Instead of being defined by a chemical structure, they share a particular property: lipids are all hydrophobic. Because they share a property rather than a structure, lipids are a chemically diverse group of molecules. They include the familiar fats that make up part of our diet, components of cell membranes, and signaling molecules. Let's briefly consider each in turn.

Triacylglycerol is an example of a lipid that is used for energy storage. It is the major component of animal fat and vegetable oil. A triacylglycerol molecule is made up of three fatty acids joined to glycerol (**Fig. 2.25**). A **fatty acid** is a long chain of carbon atoms attached to a carboxyl group (—COOH) at one end (Figs. 2.25a and 2.25b). **Glycerol** is a 3-carbon molecule

FIG. 2.24 Glycosidic bonds. Glycosidic bonds link carbohydrate molecules together. In this example, they link glucose monomers together to form the polysaccharide starch.

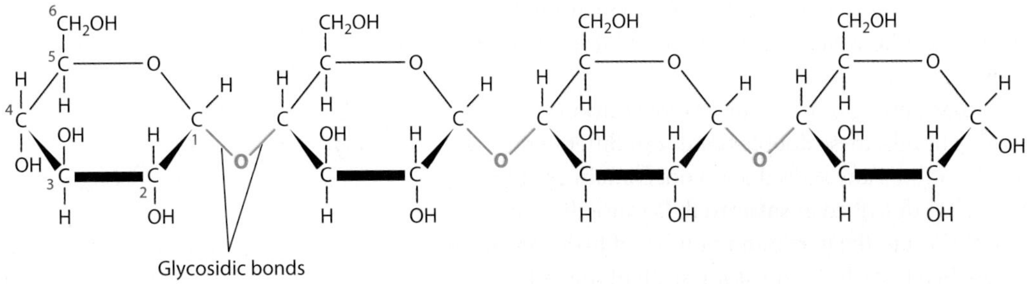

FIG. 2.25 Triacylglycerol and its components.

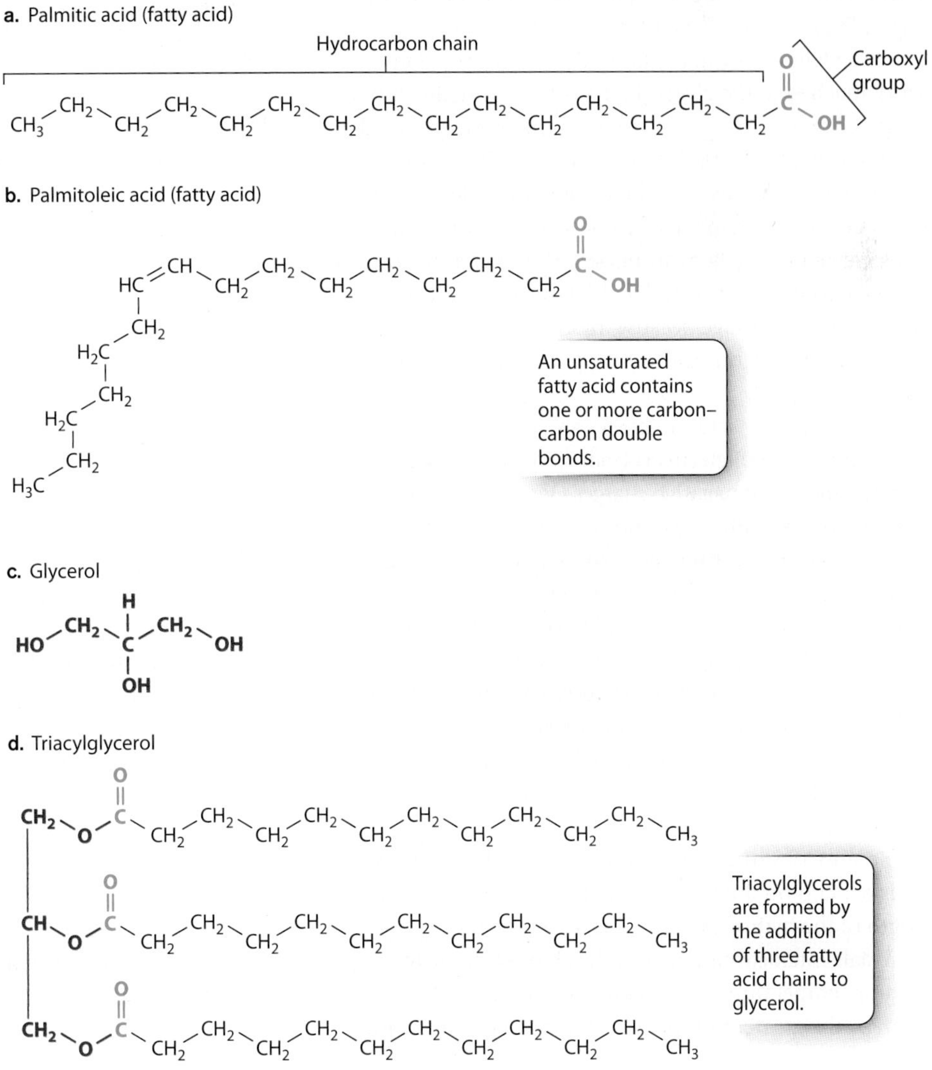

a. Palmitic acid (fatty acid)

b. Palmitoleic acid (fatty acid)

An unsaturated fatty acid contains one or more carbon–carbon double bonds.

c. Glycerol

d. Triacylglycerol

Triacylglycerols are formed by the addition of three fatty acid chains to glycerol.

with OH groups attached to each carbon (Fig. 2.25c). The carboxyl end of each fatty acid chain attaches to glycerol at one of the OH groups (2.25d), releasing a molecule of water.

Fatty acids differ in the length of their hydrocarbon chain—that is, they differ in the number of carbon atoms in the chain. (A hydrocarbon is a molecule composed entirely of carbon and hydrogen atoms.) Most fatty acids in cells contain an even number of carbon atoms because they are synthesized by the stepwise addition of 2-carbon units.

Some fatty acids have one or more carbon–carbon double bonds; these double bonds can differ in number and location. Fatty acids that do not contain double bonds are described as **saturated**. Because there are no double bonds, the maximum number of hydrogen atoms is attached to each carbon atom, so all of the carbon atoms are said to be "saturated" with hydrogen atoms (Fig. 2.25a). Fatty acids that contain carbon–carbon double bonds are **unsaturated** (Fig. 2.25b). The chains of saturated fatty acids are straight, while the chains of unsaturated fatty acids have a kink at each double bond.

Triacylglycerols can contain different types of fatty acids attached to the glycerol backbone. The hydrocarbon chains of fatty acids do not contain polar covalent bonds like those in a water molecule. Instead, their electrons are distributed uniformly over the whole molecule, so these molecules are uncharged. As a consequence, triacylglycerols are all extremely hydrophobic and, therefore, form oil droplets inside the cell. Triacylglycerols are an efficient form of energy storage because, by excluding water molecules, a large number can be packed into a small volume.

Although fatty acid molecules are uncharged, the constant motion of electrons leads to regions of slight positive and slight negative charges (**Fig. 2.26**). These charges, in turn, either attract or repel electrons in neighboring molecules, setting up areas of positive and negative charge in those molecules as well. The temporarily polarized molecules weakly bind to one another because of the attraction of opposite charges. These interactions are known as **van der Waals forces**. The van der Waals forces come into play only when atoms are sufficiently close to one another, and they are weaker than hydrogen bonds. Even so, many van der Waals forces acting together help to stabilize molecules.

Because of van der Waals forces, the melting points of fatty acids depend on their length and level of saturation. As the length of the hydrocarbon chains increases, the number of van der Waals interactions between the chains also increases. The melting temperature increases because more energy is needed to break the greater number of van der Waals interactions.

FIG. 2.26 Van der Waals forces. Transient asymmetry in the distribution of electrons along fatty acid chains leads to asymmetry in neighboring molecules, resulting in weak attractions.

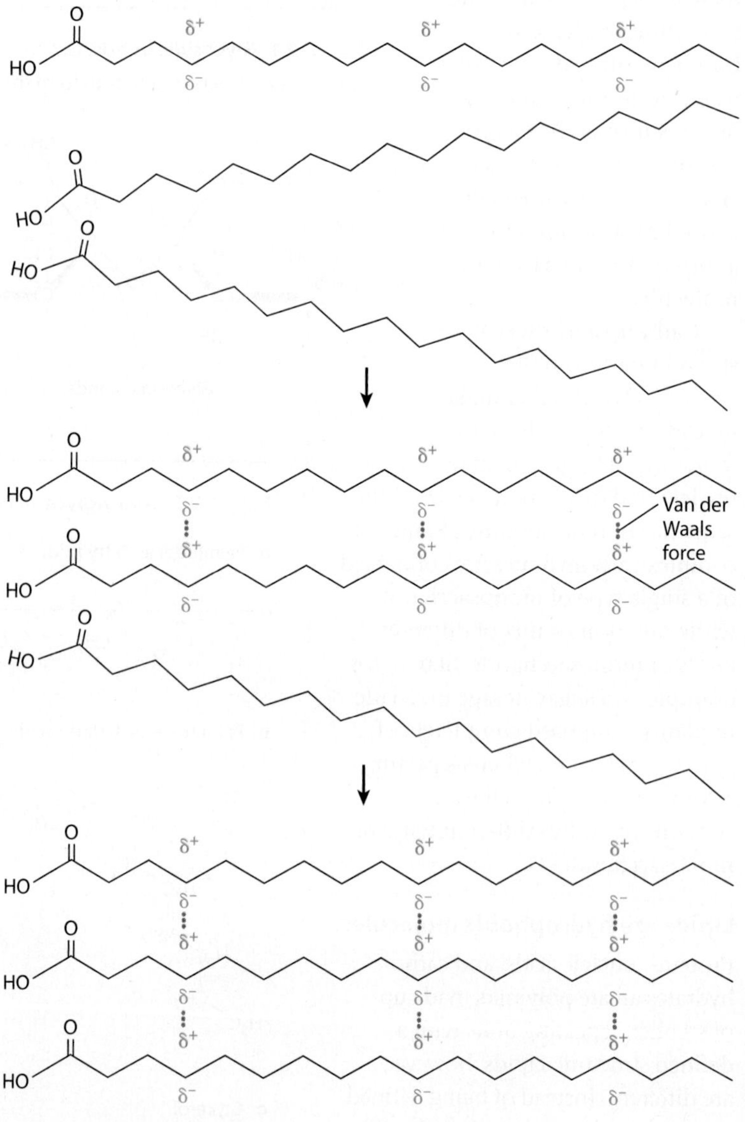

Kinks introduced by double bonds reduce the tightness of the molecular packing and, therefore, the number of intermolecular interactions. As a result, the melting temperature is lower. Thus, an unsaturated fatty acid has a lower melting point than a saturated fatty acid of the same length. Animal fats such as butter are composed of triacylglycerols with saturated fatty acids and are solid at room temperature, whereas plant fats and fish oils are composed of triacylglycerols with unsaturated fatty acids and are liquid at room temperature.

Steroids such as cholesterol are a second type of lipid (**Fig. 2.27**). Like other steroids, cholesterol has a core composed of 20 carbon atoms bonded to form four fused rings, and it

FIG. 2.27 The chemical structure of cholesterol.

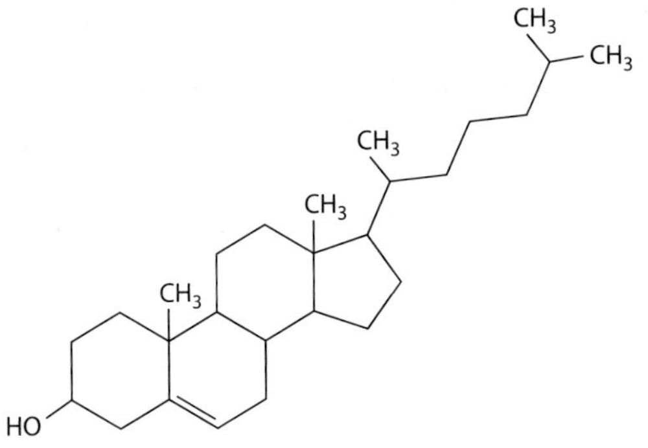

is hydrophobic. Cholesterol is a component of animal cell membranes (Chapter 5) and serves as a precursor for the synthesis of steroid hormones such as estrogen and testosterone (Chapter 36).

Phospholipids are a third type of lipid. They are a major component of the cell membrane and are described in Chapter 5.

> **Self-Assessment Questions**
>
> 10. Take a close look at Fig. 2.22. How is glucose different from galactose?
> 11. What are essential functions of proteins, nucleic acids, carbohydrates, and lipids?
> 12. How is diversity achieved in polymers? Use proteins as an example.
> 13. What are the basic structures of amino acids, nucleotides, monosaccharides, and fatty acids? Sketch your answer.

2.6 LIFE'S ORIGINS

 CASE 1 LIFE'S ORIGINS: INFORMATION, HOMEOSTASIS, AND ENERGY

How did the molecules of life form?

In Chapter 1, we considered the similarities and differences between living and nonliving things. Four billion years ago, however, the differences may not have been so pronounced. Life originated early in our planet's history, created by a set of chemical processes that, through time, produced organisms that could be distinguished from their nonliving surroundings. How can we think scientifically about one of biology's deepest and most difficult problems? One idea is to approach life's origins experimentally, asking whether the chemical reactions likely to have taken place on the early Earth can generate the molecules of life. It is important to note that even the simplest organisms living today are far more complicated than our earliest ancestors. No one suggests that cells as we know them emerged directly from primordial chemical reactions. Rather, the quest is to discover simple molecular systems that are able to replicate themselves and are subject to natural selection.

A key starting point is the observation, introduced earlier in this chapter, that the principal macromolecules found in organisms are themselves made of simpler molecules joined together. Thus, if we want to understand how proteins might have emerged on the early Earth, we should begin with the synthesis of amino acids, and if we are interested in nucleic acids, we should focus on nucleotides.

The building blocks of life can be generated in the laboratory.

Research into the origins of life was catapulted into the experimental age in 1953 with an elegant experiment carried out by Stanley Miller, then a graduate student in the laboratory of Nobel laureate Harold Urey. Miller started with water vapor, methane, ammonia, and hydrogen gas—all of which are thought to have been present in the early atmosphere. He put these gases into a sealed flask and then passed a spark through the mixture (**Fig. 2.28**). On the primitive Earth, lightning might have supplied the energy needed to drive chemical reactions, and the spark was meant to simulate its effects. Analysis of the contents of the flask showed that a number of amino acids were generated.

Miller and others conducted many variations on his original experiment, all with similar results. Today, many scientists doubt that the early atmosphere had the composition found in Miller's experimental apparatus, but amino acids and other biologically important molecules can form in a variety of simulated atmospheric compositions. If oxygen gas (O_2) is absent and hydrogen is more common in the mixture than carbon, the addition of energy generates diverse amino acids. The absence of oxygen gas is critical because these types of reactions cannot run to completion in modern air or seawater. Here, however, geology supports the experiments: chemical analyses of Earth's oldest sedimentary rocks indicate that, for the first 2 billion years of our planet's history, Earth's surface contained little or no oxygen.

Later experiments have shown that other chemical reactions can generate simple sugars, the bases found in nucleotides, and the lipids needed to form primitive membranes. Independent evidence that simple chemistry can form the building blocks of life comes from certain meteorites, which provide samples of the early solar system and contain diverse amino acids, lipids, and other organic molecules.

Experiments show how life's building blocks can form macromolecules.

From the preceding discussion, we have seen that life's simple building blocks can be generated under conditions likely to

HOW DO WE KNOW?

FIG. 2.28

Could the building blocks of organic molecules have been generated on the early Earth?

BACKGROUND In the 1950s, it was widely believed that Earth's early atmosphere was rich in water vapor, methane, ammonia, and hydrogen gas, with no free oxygen.

EXPERIMENT Stanley Miller built an apparatus, shown here, designed to simulate Earth's early atmosphere: a sealed flask filled with water vapor (H_2O), methane (CH_4), ammonia (NH_3), and hydrogen gas (H_2), all thought to have been present in the early atmosphere. To simulate lightning, Miller passed a spark through the mixture.

RESULTS As the experiment proceeded, reddish material accumulated in the flask. Analysis showed that this material included a number of amino acids, the building blocks that make up proteins, which are the key structural and functional molecules that do much of the work of the cell.

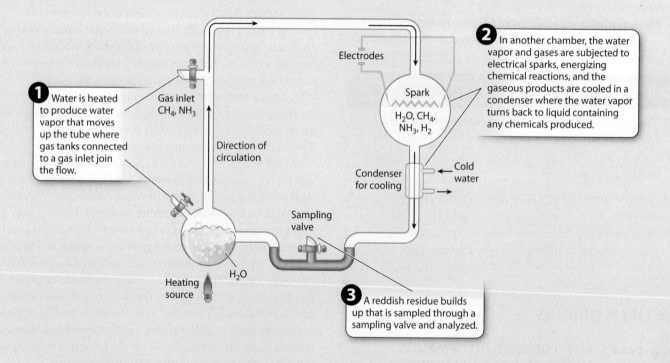

CONCLUSION Amino acids can be generated in conditions that mimic those of the early Earth.

FOLLOW-UP WORK Recent analysis of the original extracts, saved by Miller, shows that the experiment actually produced about 20 different amino acids, not all of them found in organisms.

SOURCES Miller, S. L. 1953. "Production of Amino Acids under Possible Primitive Earth Conditions." *Science* 117:528–529; Johnson, A. P., et al. 2008. "The Miller Volcanic Spark Discharge Experiment." *Science* 322:404.

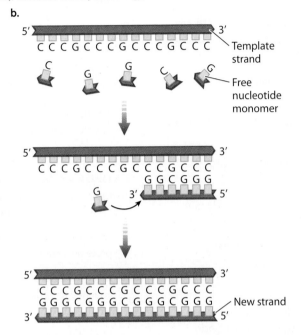

FIG. 2.29 Spontaneous polymerization of nucleotides. (a) Clays may have played an important role in the origin of life by providing surfaces for nucleotides to form nucleic acids. (b) Addition of a short RNA molecule to a flask containing modified nucleotides results in the formation of a complementary RNA strand. *Source: (a) Ray L. Frost, Professor of Physical Chemistry, Queensland University of Technology, Australia.*

have been present on the early Earth. But can these simple units be linked together to form polymers? Once again, careful experiments have shown how polymers could have formed in the conditions of the early Earth. Clay minerals that form from volcanic rocks can bind nucleotides on their surfaces (**Fig. 2.29a**). The clays provide a surface that places the nucleotides near one another, making it possible for them to join to form chains or simple strands of nucleic acid.

In a classic experiment, biochemist Leslie Orgel placed a short nucleic acid sequence into a reaction vessel and then added individual chemically modified nucleotides. The nucleotides spontaneously joined into a polymer, forming the sequence complementary to the nucleic acid already present (**Fig. 2.29b**).

Such experiments show that nucleic acids can be synthesized experimentally from nucleotides, but until recently the synthesis of nucleotides themselves presented a formidable problem for researchers seeking to uncover the origins of life. Many researchers tried to generate nucleotides from their sugar, base, and phosphate components, but no one succeeded until 2009. That year, John Sutherland and his colleagues showed that nucleotides can be synthesized under conditions thought to resemble those on the young Earth. These chemists showed how simple organic molecules likely to have formed in abundance on the early Earth react in the presence of phosphate molecules, yielding the long-sought nucleotides.

Such humble beginnings eventually gave rise to the abundant diversity of life we see around us, described memorably by Charles Darwin in the final paragraph of *On the Origin of Species*:

It is interesting to contemplate an entangled bank, clothed with many plants of many kinds, with birds singing on the bushes, with various insects flitting about, and with worms crawling through the damp earth, and to reflect that these elaborately constructed forms, so different from each other, and dependent on each other in so complex a manner, have all been produced by laws acting around us. . . . There is grandeur in this view of life . . . from so simple a beginning endless forms most beautiful and most wonderful have been, and are being, evolved.

Self-Assessment Question

14. What evidence supports the hypothesis that life originated on Earth by the generation and polymerization of small organic molecules by natural processes?

CORE CONCEPTS SUMMARY

2.1 PROPERTIES OF ATOMS: The atom is the fundamental unit of matter.

Atoms consist of positively charged protons and electrically neutral neutrons in the nucleus, as well as negatively charged electrons moving around the nucleus. page 29

The number of protons determines the identity of an atom. page 29

The number of protons and neutrons together determines the mass of an atom. page 29

The number of protons versus the number of electrons determines the charge of an atom. page 30

Negatively charged electrons travel around the nucleus in regions called orbitals. page 30

The periodic table of the elements reflects a regular and repeating pattern in the chemical behavior of elements. page 31

2.2 MOLECULES AND CHEMICAL BONDS: Atoms can combine to form molecules linked by chemical bonds.

Valence electrons occupy the outermost energy level (shell) of an atom and determine the ability of an atom to combine with other atoms to form molecules. page 32

A covalent bond results from the sharing of electrons between atoms to form a molecular orbital. page 32

A polar covalent bond results when two atoms do not share electrons equally as a result of a difference in the ability of the atoms to attract electrons, a property called electronegativity. page 33

An ionic bond results from the attraction of oppositely charged ions. page 33

2.3 WATER: Water is abundant and essential for life.

Water is a polar molecule because shared electrons are distributed asymmetrically between the oxygen and hydrogen atoms. page 35

Hydrophilic molecules dissolve readily in water, whereas hydrophobic molecules in water tend to associate with one another, minimizing their contact with water. page 35

A hydrogen bond results when a hydrogen atom covalently bonded to an electronegative atom interacts with an electronegative atom of another molecule. page 35

Water forms hydrogen bonds, which help explain its high cohesion, surface tension, and resistance to rapid temperature change. page 36

The pH of an aqueous solution is a measure of the acidity of the solution. page 36

2.4 CARBON: Carbon is the backbone of organic molecules.

A carbon atom can form up to four covalent bonds with other atoms. page 37

The geometry of these covalent bonds helps explain the structural and functional diversity of organic molecules. page 38

2.5 ORGANIC MOLECULES: Organic molecules include proteins, nucleic acids, carbohydrates, and lipids, each of which is built from simpler units.

Amino acids are linked by covalent bonds to form proteins. page 39

An amino acid consists of a carbon atom (the α carbon) attached to a carboxyl group, an amino group, a hydrogen atom, and a side chain. page 39

The side chain determines the properties of an amino acid. page 40

Nucleotides assemble to form nucleic acids, which store and transmit genetic information. page 40

Nucleotides are composed of a 5-carbon sugar, a nitrogen-containing base, and a phosphate group. page 40

Nucleotides in DNA incorporate the sugar deoxyribose, and nucleotides in RNA incorporate the sugar ribose. page 40

The bases are pyrimidines (cytosine, thymine, and uracil) and purines (guanine and adenine). page 40

Sugars are carbohydrates, molecules composed of C, H, and O atoms, usually in the ratio 1:2:1, and are a source of energy. page 41

Monosaccharides assemble to form disaccharides or longer polymers called complex carbohydrates. page 42

Lipids are hydrophobic. page 43

Triacylglycerols store energy and are made up of glycerol and fatty acids. page 43

Fatty acids consist of a linear hydrocarbon chain of variable length with a carboxyl group at one end. page 43

Fatty acids are either saturated (no carbon–carbon double bonds) or unsaturated (one or more carbon–carbon double bonds). page 44

The tight packing of fatty acids in lipids is the result of van der Waals forces, a type of weak, noncovalent bond. page 44

2.6 LIFE'S ORIGINS: Life likely originated on Earth by a set of chemical reactions that gave rise to the molecules of life.

In 1953, Stanley Miller and Harold Urey demonstrated that amino acids can be generated in the laboratory in conditions that mimic those found on the early Earth. page 45

Other experiments have shown that sugars, bases, and lipids can be generated in the laboratory. page 45

Once the building blocks were synthesized, they could join together in the presence of clay minerals to form polymers. page 47

Log in to LaunchPad to check your answers to the Self-Assessment Questions and to access additional learning tools.

CHAPTER 3 Nucleic Acids and Transcription

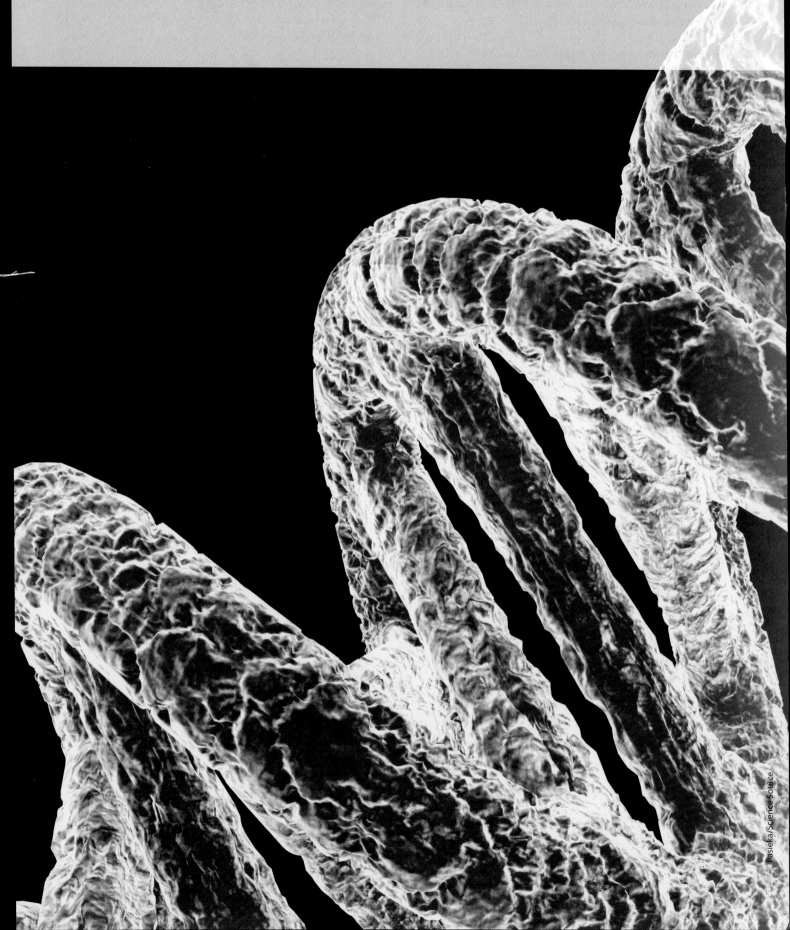

CORE CONCEPTS

3.1 CHEMICAL COMPOSITION AND STRUCTURE OF DNA: DNA is a polymer of nucleotides that forms a double helix.

3.2 DNA STRUCTURE AND FUNCTION: The structure of DNA allows it to store and transmit information.

3.3 TRANSCRIPTION: Genetic information is retrieved by transcription, the process by which RNA is synthesized from a DNA template.

3.4 RNA PROCESSING: The primary transcript is processed to become messenger RNA.

So much in biology depends on shape. Take your hand, for example. You can pick up a pin, text on a smartphone, or touch your pinky to your thumb. These activities are made possible by the coordinated movement of dozens of bones, muscles, nerves, and blood vessels that give your hand its shape. The functional abilities of your hand emerge from its structure. A close connection between structure and function exists in many molecules, too. Proteins are a good example. Composed of long, linear strings of 20 different kinds of amino acids in various combinations, each protein folds into a specific three-dimensional shape due to chemical interactions between the amino acids along the chain. The three-dimensional structure of the protein determines its functional properties and enables the protein to carry out its job in the cell.

Another notable example of the relationship between structure and function is the macromolecule **deoxyribonucleic acid (DNA)**, a linear polymer of four different subunits. DNA molecules from all cells and organisms have a very similar three-dimensional structure, reflecting their shared ancestry. This structure, called a **double helix,** is composed of two strands coiled around each other to form a molecule that resembles a spiral staircase. The banisters of the spiral staircase are formed by the linear backbone of the paired strands, and the steps are formed by the pairing of the subunits at the same level in each strand.

The spiral-staircase structure is common to all cellular DNA molecules, and its structure gave clues to its function. First, DNA *stores* **genetic information** that is encoded in the sequence of subunits along its length, much as textual information is encoded in the sequence of letters along this sentence. Some of the information in DNA encodes proteins that provide structure and do much of the work of the cell. Genetic information in DNA is organized in the form of **genes,** as textual information is organized in the form of words. Genes can exist in different forms in different individuals, even within a single species. Differences in genes can affect the shape of the hand, for example, yielding long or short fingers, or extra or missing fingers. As we will see, it is the linear order of individual subunits (bases) of DNA that accounts for differences in genes. Genes usually have no effect on the organism unless they are "turned on" and their product is made. The turning on of a gene is called **gene expression.** The molecular processes that control whether gene expression occurs at a given time, in a given cell, or at what level are collectively called **gene regulation.**

Second, DNA *transmits* genetic information to other molecules and from one generation to the next. The transmission of genetic information from parents to their offspring enables species of organisms to maintain their identity through time. The genetic information in DNA guides the development of the offspring, ensuring that parental apple trees give rise to apple seedlings and parental geese give rise to goslings. As we will see, determining the double-helical structure of DNA provided one of the first hints of how genetic information could be faithfully copied from cell to cell, and from one generation to the next.

In this chapter, we examine the structure of DNA in more detail and show how its structure is well suited to its function as the carrier and transmitter of genetic information.

3.1 CHEMICAL COMPOSITION AND STRUCTURE OF DNA

DNA is the molecule by which hereditary information is transmitted from generation to generation. Today the role of DNA is well known, but at one time hardly any biologist would have bet on it. Any poll of biologists before about 1950 would have shown overwhelming support for the idea that proteins are life's information molecules. Compared with the seemingly monotonous, featureless structure of DNA, the three-dimensional structures of proteins are highly diverse. Proteins carry out most of the essential activities in a cell, so it seemed logical to assume that they would play a key role in heredity, too. But while proteins do play a role in heredity, they do so by supporting replication, error correction, and readout of the information encoded in DNA.

The first hint that DNA might be the genetic material came in 1928, when Frederick Griffiths conducted studies on the transmission of genetic information and showed that macromolecules in extracts from bacteria could transmit genetic information from one bacterial call to another (**Fig. 3.1**). Almost 20 years later, experiments carried out by Oswald Avery, Colin MacLeod, and Maclyn McCarty showed that this information-carrying macromolecule is DNA, rather than RNA or protein (**Fig. 3.2**).

HOW DO WE KNOW?

FIG. 3.1
Can genetic information be transmitted between two strains of bacteria?

BACKGROUND In the 1920s, it was not clear which biological molecule carries genetic information. Fred Neufeld, a German microbiologist, identified several strains of the bacterium *Streptococcus pneumoniae,* which causes pneumonia in mice. One of these strains is virulent and causes death when injected into mice (Fig. 3.1a). A second strain is nonvirulent and does not cause illness when injected into mice (Fig. 3.1b).

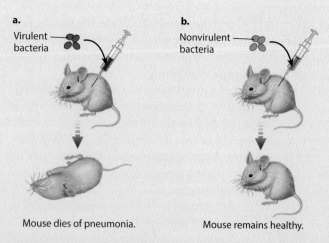

EXPERIMENT In 1928, Frederick Griffith, another microbiologist who was also interested in bacterial virulence, made a puzzling observation. He noted that live nonvirulent bacteria, when injected into mice, do not cause mice to get sick (Fig. 3.1b). He also noted that dead virulent bacteria, when injected into mice, do not cause mice to get sick (Fig. 3.1c). However when Griffith mixed live nonvirulent bacteria and dead virulent bacteria and injected the mice with this mixture, they became sick and died (Fig. 3.1d). Furthermore, when Griffith isolated bacteria from the dead mice, the isolated bacteria appeared to be virulent, even though the mice had been injected with live nonvirulent bacteria.

RESULTS

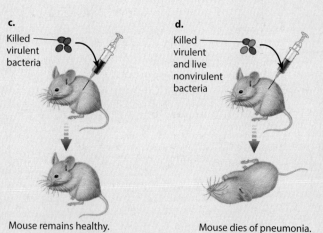

CONCLUSION Griffith concluded that the virulent bacteria, although dead, had somehow caused the nonvirulent bacteria to become virulent. He asserted that a molecule that is present in the debris of dead virulent bacteria carries the genetic information for virulence, but he did not identify the molecule. He based this conclusion on the observation that one strain of bacteria (nonvirulent) was transformed into another (virulent) by an unknown molecule from the virulent cells.

FOLLOW-UP WORK Griffith's experiments were followed up by many researchers, most notably Oswald Avery, Colin MacLeod, and Maclyn McCarty, who identified DNA as the molecule responsible for transforming the bacteria from a nonvirulent strain to a virulent strain (Fig. 3.2).

SOURCE Griffith, F. 1928. "The Significance of Pneumococcal Types." *Journal of Hygiene* 27:113–159.

CHAPTER 3 NUCLEIC ACIDS AND TRANSCRIPTION 53

HOW DO WE KNOW?

FIG. 3.2
Which molecule carries genetic information?

BACKGROUND Following Frederick Griffith, researchers Oswald Avery, Colin MacLeod, and Maclyn McCarty also studied virulence in pneumococcal bacteria. In the early 1940s, they recognized the significance of Griffith's experiments (see Fig. 3.1) and set out to identify the molecule that is responsible for transforming nonvirulent bacterial cells into virulent bacterial cells.

EXPERIMENT Avery, MacLeod, and McCarty killed virulent bacterial cells with heat, and then purified the remains of the dead virulent cells to make a solution. In their control experiment, they found that this solution of dead virulent cells did indeed transform nonvirulent bacteria into virulent ones. To identify what caused the transformation, they treated the solution with three different enzymes. Each enzyme (DNase, RNase, and protease) destroyed one of the three types of molecules (DNA, RNA, and protein) that they believed might be responsible for transformation. Their hypothesis was that transformation would not occur if the molecule responsible for transformation was destroyed by the enzymes.

RESULTS When the researchers treated the solution of dead virulent cells with enzymes that destroy RNA (RNase) or protein (protease), the solution was still able to transform nonvirulent bacteria into virulent cells. In contrast, when the solution was treated with the enzyme that destroys DNA (DNase), the extract was not able to transform nonvirulent bacteria into virulent cells. In other words, the solution was no longer lethal to the mice after it was treated with DNase.

CONCLUSION DNA is the molecule responsible for transforming nonvirulent bacteria into virulent bacteria. This experiment provided a key piece of evidence for the idea that DNA is the genetic material.

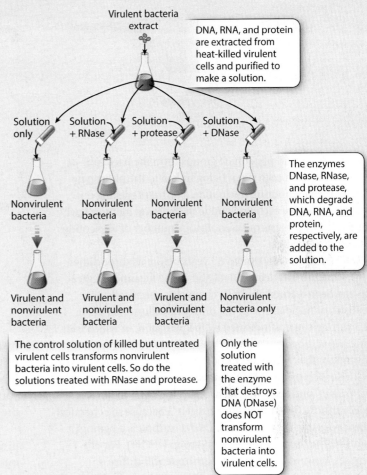

FOLLOW-UP WORK These experiments were followed up by Alfred Hershey and Martha Chase, who used a different system to confirm that DNA is the genetic material.

SOURCE Avery, O., C. MacLeod, and M. McCarty. 1944. "Studies on the Chemical Nature of the Substance Inducing Transformation of Pneumococcal Types." *Journal of Experimental Medicine* 79:137–158.

To serve as the genetic material, DNA would have to be able to store genetic information, make copies of itself, and direct the synthesis of other macromolecules in the cell. How could one molecule do all this? Part of the answer emerged when James D. Watson and Francis H. C. Crick of Cambridge University announced a description of the three-dimensional structure of DNA. This discovery marked a turning point in modern biology, as the structure of DNA revealed a great deal about its function.

A DNA strand consists of subunits called nucleotides.

In the time since the publication of Watson and Crick's paper, we have all become familiar with the iconic double helix of DNA. The elegant shape of the twisting strands relies on the structure of DNA's subunits, called **nucleotides**. As we saw in Chapter 2, nucleotides consist of three components: a 5-carbon **sugar**, a **base**, and one or more **phosphate groups (Fig. 3.3)**. Each component plays an important role in DNA structure. The

FIG. 3.3 Nucleotide structure.

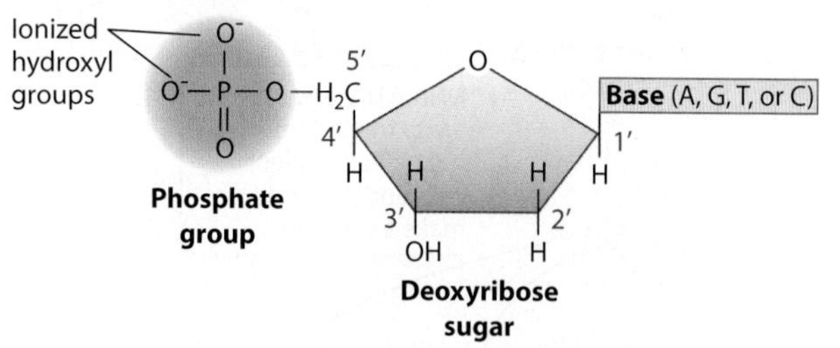

5-carbon sugars and phosphate groups form the backbone of the molecule, with each sugar being linked to the phosphate group of the neighboring nucleotide. The bases sticking out from the sugar give each nucleotide its chemical identity. Each strand of DNA consists of an enormous number of nucleotides linked one to the next.

DNA had been discovered 85 years before its three-dimensional structure was determined, and in the meantime a great deal had been learned about its chemistry, specifically, the chemistry of nucleotides. In the nucleotide illustrated in Fig. 3.3, the 5-carbon sugar is indicated by the pentagon, in which four of the five vertices represent the position of a carbon atom. By convention, the carbon atoms of the sugar ring are numbered clockwise with primes (1′, 2′, and so forth, read as "one prime," "two prime," and so forth). Technically, the sugar in DNA is 2′-deoxyribose (literally "minus oxygen") because the chemical group projecting downward from the 2′ carbon is a hydrogen atom (—H) rather than a hydroxyl group (—OH). For our purposes, however, the term **deoxyribose** will suffice.

Note in Fig. 3.3 that the phosphate group attached to the 5′ carbon has negative charges on two of its oxygen atoms. These charges are present because at cellular pH (around 7), the free

FIG. 3.4 Bases normally found in DNA.

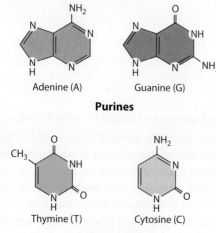

FIG. 3.5 Nucleosides and nucleotides.
A nucleoside is a sugar attached to a base, and nucleotides are nucleosides with one, two, or three phosphate groups attached. Thus, a nucleotide is a nucleoside phosphate.

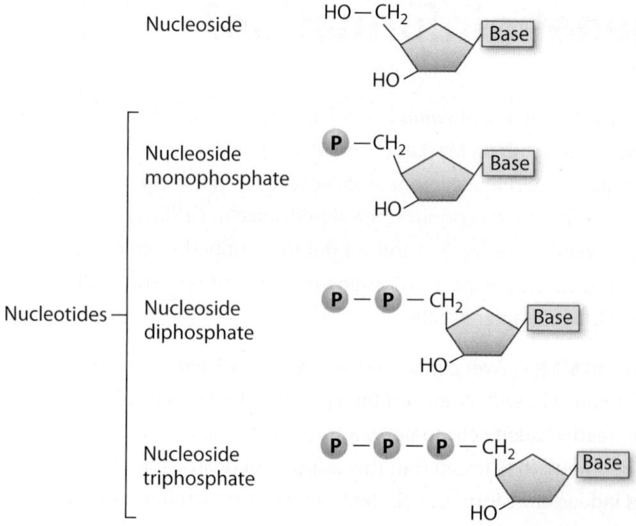

hydroxyl groups attached to the phosphorus atom are ionized by the loss of a proton and, therefore, are negatively charged. It is these negative charges that make DNA a mild acid, which you will recall from Chapter 2 is a molecule that tends to lose protons to the aqueous environment.

Each base is attached to the 1′ carbon of the sugar and projects above the sugar ring. A nucleotide normally contains one of four kinds of bases, denoted A, G, T, and C (**Fig. 3.4**). Two of the bases are double-ring structures known as **purines**; these are the bases **adenine (A)** and **guanine (G)**, shown across the top of the figure. The other two bases are single-ring structures known as **pyrimidines**; these are the bases **thymine (T)** and **cytosine (C)**, shown across the bottom.

The combination of sugar and base is known as a **nucleoside**, which is shown in simplified form in **Fig. 3.5**. A nucleoside with one or more phosphate groups is termed a nucleotide. More specifically, a nucleotide with one, two, or three phosphate groups is called a nucleoside monophosphate, diphosphate, or triphosphate, respectively (Fig. 3.5). The nucleoside triphosphates are particularly important because, as we will see later in this chapter, they are the molecules that are used to form DNA and RNA. In addition, nucleoside triphosphates have other functions in the cell, notably as carriers of chemical energy in the form of ATP and GTP.

DNA is a linear polymer of nucleotides linked by phosphodiester bonds.

Not only were the nucleotide building blocks of DNA known before the structure was discovered, but it was also known how they were linked into a long chain or polymer. The chemical linkages between nucleotides in DNA are shown in **Fig. 3.6.** The characteristic covalent bond that connects one nucleotide to the next is indicated by the vertical red lines that connect the 3′ carbon of one nucleotide to the 5′ carbon of the next nucleotide in line through the 5′-phosphate group. This C—O—P—O—C linkage is known as a **phosphodiester bond.** In DNA, it is a relatively stable bond that can withstand stresses such as heat and substantial changes in pH that would break weaker bonds. The succession of phosphodiester bonds traces the backbone of the DNA strand.

The phosphodiester linkages in a DNA strand give it **polarity,** which means that one end differs from the other. In Fig. 3.6, the nucleotide at the top has a free 5′ phosphate, and is known as the **5′ end** of the molecule. The nucleotide at the bottom has a free 3′ hydroxyl and is known as the **3′ end.** The DNA strand in Fig. 3.6 has the sequence of bases AGCT from top to bottom, but because of strand polarity we need to specify which end is which. For this strand of DNA, we could say that the base sequence is 5′-AGCT-3′ or, equivalently, 3′-TCGA-5′. When a base sequence is stated without specifying the 5′ end, by convention the end at the left is the 5′ end. Therefore, we can also say the sequence in Fig. 3.6 is AGCT, which means 5′-AGCT-3′.

Cellular DNA molecules take the form of a double helix.
With the knowledge of the chemical makeup of the nucleotides and their linkages in a DNA strand, Watson and Crick set out to build a molecular model of the structure of DNA. To do this, they combined three critical pieces of information. The first consisted of results from X-ray crystallography of DNA carried out by Rosalind Franklin and Maurice Wilkins, also at Cambridge University. Franklin's results were the most clear in indicating the likelihood that DNA had some sort of helical structure with a simple repeating structure all along its length, but they were incomplete and did not contain enough information for crystallographers to build models of the molecule.

The second piece of information consisted of published results of Erwin Chargaff, a biochemist at Columbia University. Chargaff had shown that DNA extracted from cells of a wide variety of organisms all have a characteristic feature: the number of molecules of the nucleotide base adenine (A) always equals the number of molecules of thymine (T), and the number of molecules of guanine (G) always equals the number of molecules of cytosine (C). This is true even though the relative amounts of A and G vary widely from one organism to the next.

The third piece of information came from Jerry Donohue and John Griffith, colleagues of Watson and Crick at Cambridge University. They determined that, if the bases were to pair in some way, the most likely way would be that A paired with T and that G paired with C.

An accurate model of DNA would have to account for the results of all previous chemical and physical experiments. Watson and Crick went to work, using sheet metal cutouts of the bases and wire ties for the sugar–phosphate backbone. After many false starts, they finally found a structure that worked: a double-helical structure with the backbones on the outside, the bases pointing inward, and A paired with T and G paired with C.

The pair realized immediately that they had made one of the most important discoveries in all of biology. That day, February 28, 1953, they lunched at the Eagle, a pub across the street from their laboratory, where Crick loudly pronounced, "We have discovered the secret of life." The Eagle is still there in Cambridge, England, and sports on its wall a commemorative plaque marking the table where the two ate. Watson and Crick published the structure of DNA in 1953. Knowing the structure of DNA opened the door to understanding how genetic information is stored, faithfully replicated, and able to direct the synthesis of other macromolecules. For this discovery, Watson,

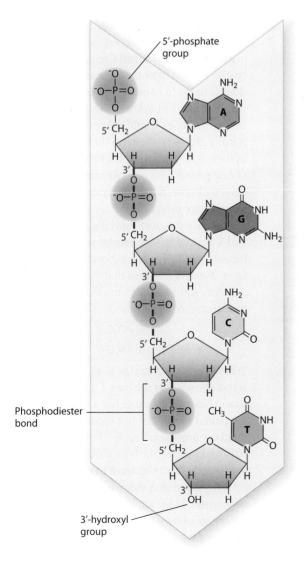

FIG. 3.6 Nucleotides linked by phosphodiester bonds to form a DNA strand.

Crick, and Wilkins were awarded the Nobel Prize in Physiology or Medicine.

The Watson–Crick structure is shown in **Fig. 3.7**. Fig. 3.7a is a space-filling model, in which each atom is represented as a color-coded sphere. The big surprise of the structure is that it consists of two DNA strands like those in Fig. 3.6, each wrapped around the other in the form of a helix coiling to the right, with the sugar–phosphate backbones winding around the outside of the molecule and the bases pointing inward. In the double helix, there are 10 base pairs per complete turn, and the diameter of the molecule is 2 nm. This measurement is hard to relate to everyday objects, but it might help to know that the cross section of a bundle of 100,000 DNA molecules would be about the size of the period at the end of this sentence. The outside contours of the twisted strands form an uneven pair of grooves, called the **major groove** and the **minor groove**. These grooves are important because proteins that interact with DNA often recognize a particular sequence of bases by making contact with the bases by the major or minor groove or both.

Importantly, the individual DNA strands in the double helix are **antiparallel**, which means that they run in opposite directions. That is, the 3′ end of one strand is opposite the 5′ end of the other. In Fig. 3.7a, the strand that starts at the bottom left and coils upward begins with the 3′ end and terminates at the top with the 5′ end. Its partner strand begins with its 5′ end at the bottom and terminates with the 3′ end at the top.

Fig. 3.7b shows a different depiction of double-stranded DNA, called a ribbon model, in which the sugar–phosphate backbones wind around the outside with the bases paired between the strands. The ribbon model of the structure closely resembles a spiral staircase, with the backbones forming the banisters and the base pairs the steps. If the amount of DNA in a human egg or sperm (3 billion base pairs) were scaled to the size of a real spiral staircase, it would reach from the Earth to the moon.

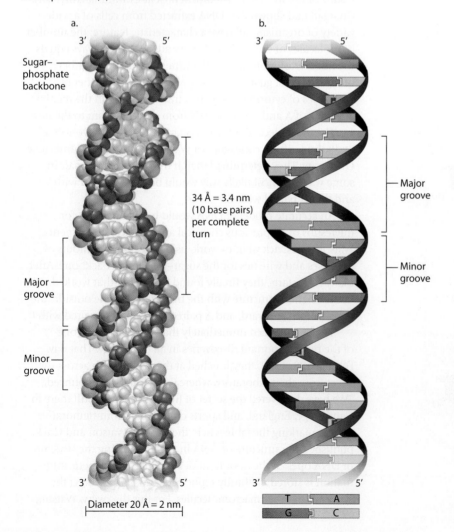

FIG. 3.7 Structure of DNA. The DNA double helix can be shown with (a) the atoms as solid spheres or (b) the backbones as ribbons.

As shown in Fig. 3.7b, an A in one strand pairs only with a T in the other strand, and G pairs only with C. Each base pair contains a purine and a pyrimidine. The pairing of A with T and of G with C nicely explains Chargaff's observations, now called Chargaff's rule. This precise pairing maintains the structure of the double helix. In contrast, pairing two purines would cause the backbones to bulge and pairing two pyrimidines would cause them to narrow, putting excessive strain on the covalent bonds in the sugar–phosphate backbone. The pairing of one purine with one pyrimidine preserves the distance between the backbones along the length of the entire molecule.

Because they form specific pairs, the bases A and T are said to be **complementary**, as are the bases G and C. The formation of only A—T and G—C base pairs means that the paired strands in a double-stranded DNA molecule have different base sequences. The strands are paired like this:

5′-ATGC-3′

3′-TACG-5′

Where one strand has the base A, the other strand across the way has the base T. Likewise, where one strand has a G, the other has a C. In other words, the paired strands are not identical but complementary. Because of the A—T and G—C base pairing, knowing the base sequence in one strand tells you the base sequence in its partner strand.

Why is it that A pairs only with T, and G only with C? **Fig. 3.8** illustrates the answer. The specificity of base pairing is brought about by hydrogen bonds that form between A and T (two hydrogen bonds) and between G and C (three

FIG. 3.8 Base pairing. Adenine (A) pairs with thymine (T), and guanine (G) pairs with cytosine (C). These base pairs differ in the number of hydrogen bonds.

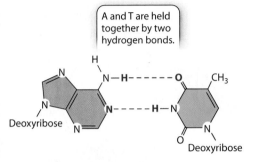

Self-Assessment Questions

1. What are three key structural features of DNA?
2. The letter R is conventionally used to represent any purine base (A or G) and Y to represent any pyrimidine base (T or C). In double-stranded DNA, what is the relation between the number of molecules of R and the number of molecules of Y?

FIG. 3.9 Interactions stabilizing the double helix. Hydrogen bonds between the bases in opposite strands and base stacking of bases within a strand contribute to the stability of the DNA double helix.

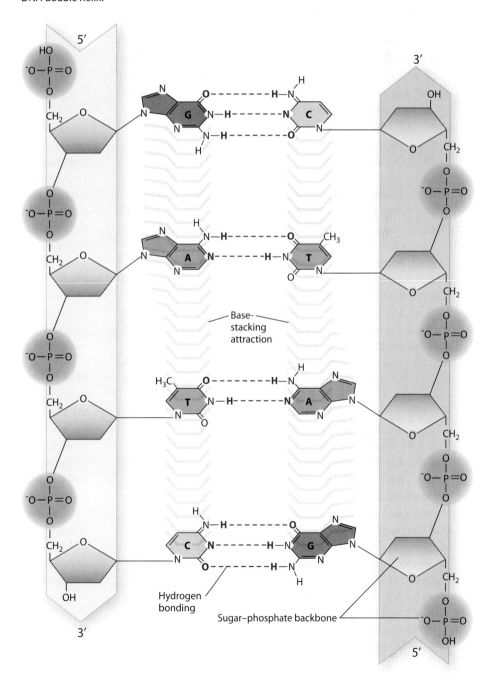

hydrogen bonds). A hydrogen bond in DNA is formed when an electronegative atom (O or N) in one base shares a hydrogen atom (H) with another electronegative atom in the base across the way. Hydrogen bonds are relatively weak bonds, typically 5% to 10% of the strength of covalent bonds, and can be disrupted by high pH or heat. However, added together, millions of these weak bonds along the molecule contribute to the stability of the DNA double helix.

An almost equally important factor contributing to the stability of the double helix is the interactions between bases in the same strand (**Fig. 3.9**). This stabilizing force, known as **base stacking**, occurs because the nonpolar, flat surfaces of the bases tend to group together away from water molecules, and hence stack on top of one another as tightly as possible.

3.2 DNA STRUCTURE AND FUNCTION

The double helix is a good example of how chemical structure and biological function come together. The structure of the molecule itself suggested how genetic information is stored in DNA—in the linear order or sequence of the base pairs. One of the most important features of DNA structure is that there is no restriction on the sequence of bases along a DNA strand. Any A, for example, can be followed by another A or by T, C, or G. The lack of sequence constraint suggested that the genetic information in DNA could be encoded in the sequence of bases along the DNA. With any of four possible bases at each nucleotide site, the information-carrying capacity of a DNA molecule is unimaginable. The number of possible base sequences of a DNA molecule only 133 nucleotides in length is equal to the estimated number of electrons, protons, and neutrons in the entire universe! This is how DNA can carry the genetic information for so many different types of organisms, and how variation in DNA sequence even within a single species can underlie genetic differences among individuals.

In addition, the structure of DNA gave important hints about other functions of DNA. For example, the base pairing of A with T and of G with C suggests how DNA is copied (replicated), as we discuss next.

DNA molecules are copied in the process of replication, which relies on base pairing.

DNA can serve as the genetic material because it is unique among cellular molecules in being able to specify exact copies of itself, a process known as **replication**. Faithful replication is critical because it enables DNA to pass genetic information from cell to cell and from parent to offspring.

Even to those seeing Watson and Crick's double-helix model for the first time, the complementary, double-stranded nature of the double helix suggests a mechanism by which DNA replication takes place. In their paper on the structure of DNA, Watson and Crick noted with dry wit, "It has not escaped our notice that the specific pairing we have postulated immediately suggests a possible copying mechanism for the genetic material."

A simplified diagram of DNA replication is shown in **Fig. 3.10**. The two strands of a parental double helix unwind and separate into single strands. As they do, each of the parental strands serves as a template, or pattern, for the synthesis of a complementary daughter strand. When the process is complete, there are two molecules, each containing one parental strand and one daughter strand, and each of which is identical in sequence to the original molecule, except possibly for rare errors that cause one base pair to be replaced with another. The sequence of bases along either strand determines that of the other because wherever one strand carries an A, the other must carry a T, and wherever one carries a G, the other must carry a C.

Reproducing the sequence of nucleotides as precisely as possible is important because, as we will see in Chapter 14,

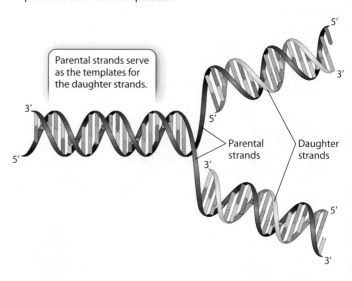

FIG. 3.10 DNA replication. The structure of the double helix gave an important clue to how it replicates.

mistakes that go unrepaired may be harmful to the cell or organism. An unrepaired error in DNA replication results in a **mutation**, which is a change in the genetic information in DNA. A mutation in DNA causes the genetic difference between virulent and nonvirulent *Streptococcus pneumoniae*, as discussed in Fig. 3.1. While most mutations in genes are harmful, rare favorable mutations are essential in the process of evolution because they allow populations of organisms to change through time and adapt to their environment.

The molecular details of DNA replication are discussed in Chapter 12 in the context of cell division and inheritance, two processes that fundamentally depend on DNA replication. Here, we focus on how genetic information stored in DNA is used to make RNA and proteins.

RNA is an intermediary between DNA and protein.

Biologists often say that genetic information in DNA dictates the activities in a cell or guides the development of an organism, but these effects of DNA are actually indirect. Most of the active molecules in cells and developmental processes are proteins, including the enzymes that convert energy into usable forms and the proteins that provide structural support for the cell. The genetic information in DNA is contained in the sequence of A's, T's, G's, and C's along its length. In a protein-coding gene, DNA acts indirectly by specifying the sequence of amino acid subunits of which each protein is composed, and this sequence in turn determines the three-dimensional structure of the protein, its chemical properties, and its biological activities (Chapter 4).

To specify the amino acid sequence of proteins, DNA acts through an intermediary molecule known as **ribonucleic acid (RNA)**, another type of nucleic acid. As we saw in Chapter 1, the usual flow of genetic information in a cell is from DNA to RNA to protein, which has come to be known as the **central dogma** of molecular biology (**Fig. 3.11**).

The first step in decoding DNA is **transcription**, in which the genetic information in a molecule of DNA is used as a template to generate a molecule of RNA. Base pairing between a strand of DNA and RNA means that the information stored in DNA is transferred to RNA. The term "transcription" is used because it emphasizes that the information is copied from DNA to RNA in the same language of nucleic acids. Transcription is the first step in gene expression, which is the production of a functional gene product. The second step in the readout of genetic information is **translation**, in which a molecule of RNA is used as a code for the sequence of amino acids in a protein. The term "translation" is used to indicate a change of languages, from nucleotides that make up nucleic acids to amino acids that make up proteins.

Through the years, some exceptions to the usual flow from DNA to RNA to protein have been discovered, including the transfer of genetic information from RNA to DNA (as in the human immunodeficiency virus [HIV], which causes acquired immune deficiency syndrome [AIDS]) and from RNA to RNA (as in replication of the genetic material of influenza virus). Nevertheless, in most cases the flow of information is from DNA to RNA to protein.

The processes of transcription and translation are regulated, meaning that they do not occur at all times in all cells, even though all cells in an individual contain the same DNA (Chapter 18). Genes are expressed, or "turned on," only at certain times and places, and not expressed, or "turned off," at other times and places. In multicellular organisms, for instance, cells are specialized for certain functions, and these different functions depend on which genes are on and which genes are off in specific cells. Muscle cells express genes that encode for proteins involved in muscle contraction, but these genes are not expressed in skin cells or liver cells, for example. Similarly, during development of a multicellular organism, genes may be required at certain times but not at others. In this case, the timing of expression is carefully controlled.

The flow of genetic information from DNA to RNA to protein applies to both prokaryotes and eukaryotes, but the differences in cell structure between the two groups mean the details of the processes differ. In prokaryotes, transcription and translation occur in the cytoplasm. In contrast, in eukaryotes, the two processes are spatially separated from each other, with transcription occurring in the nucleus and translation in the cytoplasm. The separation of transcription and translation in time and space in eukaryotic cells allows for additional levels of gene regulation that are not possible in prokaryotic cells. In spite of this and other differences in the details of transcription and translation between prokaryotes and eukaryotes, the processes are sufficiently similar that they must have evolved early in the history of life.

Self-Assessment Questions

3. How does the sequence of a molecule of DNA, made up of many monomers of only four nucleotides, encode the enormous amount of genetic information stored in the chromosomes of living organisms?

4. How are the parental strands in a DNA molecule used in DNA replication?

5. What is the usual flow of genetic information in a cell?

3.3 TRANSCRIPTION

Although the three-dimensional structure of DNA gave important clues about how DNA stores and transmits information, it left open many questions about how the genetic information in DNA is read out to control cellular processes. In 1953, when the double helix was discovered, almost nothing was known about these processes. Within a few years, however, an accumulating body of evidence began to show that DNA carries the genetic information for proteins, and that proteins are synthesized on particles called **ribosomes**. But in eukaryotes, DNA is located in the nucleus and ribosomes are located in the cytoplasm. Thus, there must be an intermediary molecule by which the genetic information is transferred from the DNA in the nucleus to ribosomes in the cytoplasm, and some researchers began to suspect that this intermediary was RNA.

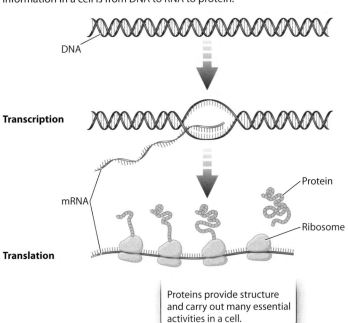

FIG. 3.11 The central dogma of molecular biology. The usual flow of information in a cell is from DNA to RNA to protein.

Proteins provide structure and carry out many essential activities in a cell.

The hypothesis of an RNA intermediary that carries genetic information from DNA to the ribosomes was supported by a clever experiment carried out in 1961 by Sydney Brenner, François Jacob, and Matthew Meselson. They used the virus T2, which infects cells of the bacterium *Escherichia coli* and hijacks the cellular machinery to produce its own viral proteins. The researchers found that while T2 DNA never associates with bacterial ribosomes, the bacteria infected with viruses produce a burst of RNA molecules shortly after infection and before they make viral proteins. This finding and others suggested that ribosomes do not synthesize proteins directly from DNA, but rather that RNA transmits the genetic information stored in DNA for use in protein synthesis. The transfer of genetic information from DNA to RNA constitutes the key process of transcription, which we examine more closely in this section.

RNA is a polymer of nucleotides in which the 5-carbon sugar is ribose.

RNA is a polymer of nucleotides linked by phosphodiester bonds, similar to DNA. Each RNA strand therefore has a polarity determined by which end of the chain carries the 3′ hydroxyl group (—OH) and which end carries the 5′ phosphate group. A number of important differences distinguish RNA from DNA, however (**Fig. 3.12**). First, the sugar in RNA is **ribose**, which carries a hydroxyl group on the 2′ carbon (highlighted in pink in Fig. 3.12a). Hydroxyls are reactive functional groups, so the additional hydroxyl group on ribose in part explains why RNA is a less stable molecule than DNA. Second, the base **uracil (U)** in RNA replaces thymine in DNA (Fig. 3.12b). The groups that participate in hydrogen bonding are identical, so that uracil pairs with adenine in RNA (U—A) just as thymine pairs with adenine in DNA (T—A). Third, while the 5′ end of a DNA strand is typically a monophosphate, the 5′ end of an RNA molecule is typically a triphosphate.

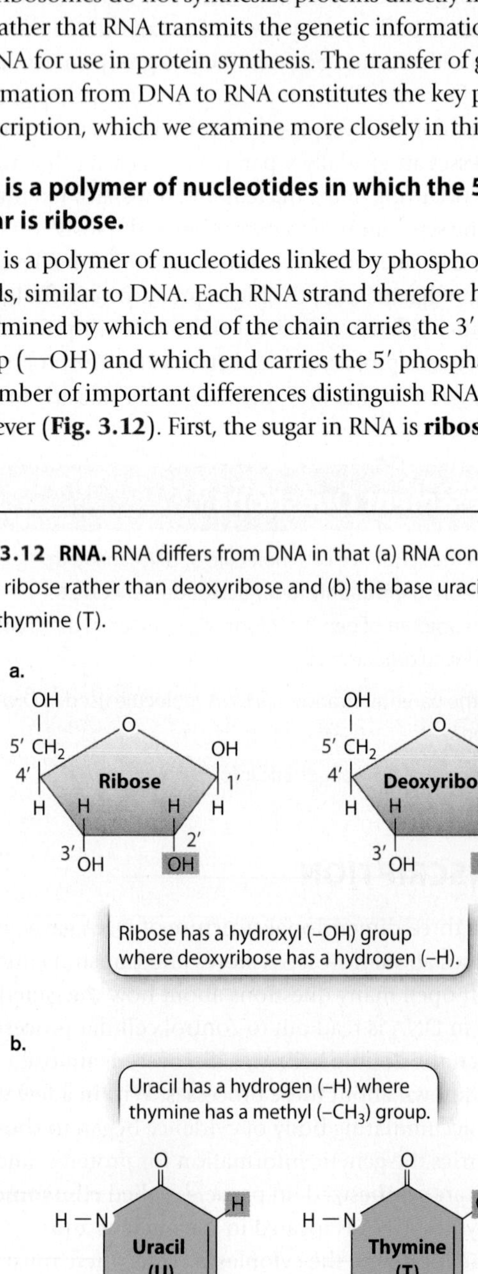

Two other features that distinguish RNA from DNA are physical rather than chemical. First, RNA molecules are usually much shorter than DNA molecules. A typical RNA molecule used in protein synthesis consists of a few thousand nucleotides, whereas a typical DNA molecule consists of millions or tens of millions of nucleotides. Second, most RNA molecules in the cell are single stranded, whereas DNA molecules, as we saw, are double stranded. Although RNA molecules are synthesized as single stranded, they can fold into complex, three-dimensional structures containing one or more double-stranded regions where the RNA pairs with itself. Folded RNA structures can have a three-dimensional complexity rivaling that of proteins, and some can even serve as catalysts in biochemical reactions.

 CASE 1 LIFE'S ORIGINS: INFORMATION, HOMEOSTASIS, AND ENERGY

What kinds of nucleic acids were present in the earliest cells?

RNA is a remarkable molecule. Like DNA, it can store information in its sequence of nucleotides. In addition, some RNA molecules can actually act as enzymes that facilitate chemical reactions. Because RNA has properties of both DNA (information storage) and proteins (enzymes), many scientists think that RNA, not DNA, was the original information-storage molecule in the earliest forms of life on Earth. This idea, sometimes called the **RNA world hypothesis**, is supported by other evidence as well. Notably, as we will see, RNA is involved in key cellular processes, including DNA replication, transcription, and translation. Many scientists believe that this involvement is a remnant of a time when RNA played a more central role in life's fundamental processes.

Ingenious experiments carried out by Jack W. Szostak and collaborators show how RNA could have evolved the ability to catalyze a simple reaction. These researchers synthesized a strand of RNA in the laboratory and then replicated it many times to produce a large population of identical RNA molecules. Next, they exposed the RNA to a chemical that induced random changes in the identity of some of the nucleotides in these molecules. These random changes were mutations that created a population of diverse RNA molecules, much in the way that mutation builds genetic variation in cells.

FIG. 3.12 RNA. RNA differs from DNA in that (a) RNA contains the sugar ribose rather than deoxyribose and (b) the base uracil (U) rather than thymine (T).

Next, the researchers placed all of these RNA molecules into a tube, and those RNA variants that successfully catalyzed a simple reaction—cleaving a strand of RNA, for example, or joining two strands together—were isolated, and the cycle was repeated. In each round of the experiment, the RNA molecules that functioned best were retained, replicated, subjected to treatments that induced additional mutations, and then tested for the ability to catalyze the same reaction. With each generation, the RNA catalyzed the reaction more efficiently, and after only a few dozen rounds of the procedure, very efficient RNA catalysts had evolved. Experiments such as this one suggest that RNA molecules can evolve over time and act as catalysts. Therefore, many scientists believe that RNA, with its dual functions of information storage and catalysis, was a key molecule in the very first forms of life.

If RNA played a key role in the origin of life, why do cells now use DNA for information storage and proteins to carry out other cellular processes? RNA is much less stable than DNA, and proteins are more versatile, so a plausible explanation is that life evolved from an RNA-based world to one in which DNA, RNA, and proteins are specialized for different functions.

In transcription, DNA is used as a template to make complementary RNA.

Conceptually, the process of transcription is straightforward. As a region of the DNA duplex unwinds, one strand is used as a template for the synthesis of an **RNA transcript** that is complementary in sequence to the template according to the base-pairing rules, except that the transcript contains U (uracil) where the template contains A (adenine) (**Fig. 3.13**). The transcript is produced by polymerization of ribonucleoside triphosphates. The enzyme that carries out the polymerization is known as **RNA polymerase**, which acts by adding successive nucleotides to the 3′ end of the growing transcript (Fig. 3.13). Only the template strand of DNA is transcribed. Its partner, called the **nontemplate strand**, is not transcribed.

Although transcription is conceptually simple, in practice it takes place in stages. The first stage is **initiation**, in which RNA polymerase and other proteins are attracted to double-stranded DNA, the DNA strands are separated, and transcription of the template strand actually begins. The second stage is **elongation**, in which successive nucleotides are added to the 3′ end of the growing RNA transcript as the RNA polymerase proceeds along the template strand. The third stage is **termination**, in which the RNA polymerase encounters a sequence in the template strand that causes transcription to stop and the RNA transcript to be released.

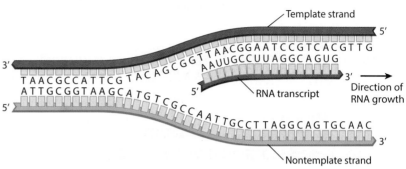

FIG. 3.13 Transcription. The DNA double helix unwinds for transcription. For each gene, usually only one strand, the template strand, is transcribed.

It is important to keep in mind the direction of growth of the RNA transcript and the direction that the DNA template is read. All nucleic acids are synthesized by addition of nucleotides to the 3′ end. That is, they grow in a 5′-to-3′ direction, also described simply as the 3′ direction. Just like the two strands of DNA in a double helix, the DNA template and the RNA strand transcribed from it are antiparallel, meaning the DNA template runs in the opposite direction from the RNA (Fig. 3.13). This means that, as the RNA transcript is synthesized in the 5′-to-3′ direction, the DNA template is read in the opposite direction (3′-to-5′).

Transcription starts at a promoter and ends at a terminator.

A long DNA molecule typically contains thousands of genes, most of them coding for proteins or RNA molecules with specialized functions, and hence thousands of different transcripts are produced. For example, the DNA molecule in the bacterium E. coli has about 4.6 million base pairs and produces RNA transcripts for about 4300 proteins. A typical map of a small part of a long DNA molecule is shown in **Fig. 3.14**. Each dark green segment indicates the position where transcription is initiated, and each dark purple segment indicates the position where it ends.

The green segments are **promoters**, regions of typically a few hundred base pairs where RNA polymerase and associated proteins bind to the DNA duplex. Although the term "promoter" refers to a region in double-stranded DNA because both strands are needed to recruit these proteins, transcription is initiated on only one strand. In many eukaryotic and archaeal promoters, the promoter sequence in the nontemplate strand includes 5′-TATAAA-3′ or something similar, which is known as a **TATA box** because of the sequence 5′-TATA-3′ (Fig. 3.14). Transcription takes place using the opposite strand as the template strand, usually starting about 25 nucleotides from the 3′ end of the complement to the TATA box, and elongation

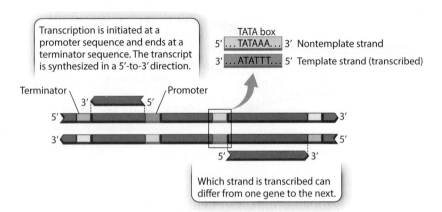

FIG. 3.14 **Transcription along a stretch of DNA.** A DNA molecule usually contains many genes that are transcribed individually and at different times, often from opposite strands.

occurs as the RNA polymerase moves along the template strand in the 3'-to-5' direction.

Transcription continues until the RNA polymerase encounters a sequence known as a **terminator** (dark purple in Fig. 3.14). Transcription stops at the terminator, and the transcript is released. A long DNA molecule contains the genetic information for hundreds or thousands of genes. For any one gene, usually only one DNA strand is transcribed; however, different genes in the same double-stranded DNA molecule can be transcribed from opposite strands. Which strand is transcribed depends on the orientation of the promoter. When the promoters are in opposite orientation (Fig. 3.14), transcription occurs in opposite directions because, as noted earlier, transcription can proceed only by successive addition of nucleotides to the 3' end of the transcript.

Transcription does not take place indiscriminately from promoters, but rather is a regulated process. For the so-called housekeeping genes, whose products are needed at all times in all cells, transcription takes place continually. Most genes, however, are transcribed only at certain times, under certain conditions, or in certain cell types. In *E. coli*, for example, the genes that encode proteins needed to utilize the sugar lactose (milk sugar) are transcribed only when lactose is present in the environment. For such genes, regulation of transcription often depends on whether the RNA polymerase and associated proteins are able to bind with the promoter (Chapter 18).

In bacteria, promoter recognition is mediated by a protein called **sigma factor,** which associates with RNA polymerase and facilitates its binding to specific promoters. Sigma binding is transient. Once transcription is initiated, the sigma factor dissociates and the RNA polymerase continues transcription on its own. One type of sigma factor is used for transcription of housekeeping genes and many others, but there are other sigma factors for genes whose expression is needed under special environmental conditions such as lack of nutrients or excess heat.

Promoter recognition in eukaryotes is considerably more complicated. Transcription requires the combined action of at least six proteins known as **general transcription factors** that assemble at the promoter of a gene. Assembly of the general transcription factors is necessary for transcription to occur, but not sufficient. Also needed is the presence of one or more types of **transcriptional activator protein**, each of which binds to a specific DNA sequence known as an **enhancer** (**Fig. 3.15**). Transcriptional activator proteins help control when and in which cells transcription of a gene will occur. They are able to bind with enhancer DNA sequences as well as with proteins that allow transcription to begin. As a consequence, the presence of the transcriptional activator proteins that bind with enhancers controlling the expression of the gene is required for transcription of any eukaryotic gene to begin (Fig. 3.15).

Once transcriptional activator proteins have bound to enhancer DNA sequences, they can attract, or recruit, a **mediator complex** of proteins, which in turn recruits the RNA polymerase complex to the promoter. Because enhancers can be located almost anywhere, sometimes far from the gene, the recruitment of the mediator complex and the RNA polymerase complex may require the DNA to loop around as shown in Fig. 3.15. Cells have several different types of RNA polymerase enzymes, but in both prokaryotes and eukaryotes all protein-coding genes are transcribed by just one of them. In eukaryotes, the RNA polymerase complex responsible for transcription of protein-coding genes is called **Pol II**. Once the mediator complex and Pol II complex are in place, transcription is initiated.

RNA polymerase adds successive nucleotides to the 3' end of the transcript.

Once transcriptional initiation takes place, successive ribonucleotides are added to grow the transcript in the process of elongation. Transcription takes place in a sort of bubble in which the strands of the DNA duplex are separated and the growing end of the RNA transcript is paired with the template strand, creating an RNA–DNA duplex (**Fig. 3.16**). In bacteria, the total length of the transcription bubble is about 14 base pairs, and the length of the RNA–DNA duplex in the bubble is about 8 base pairs.

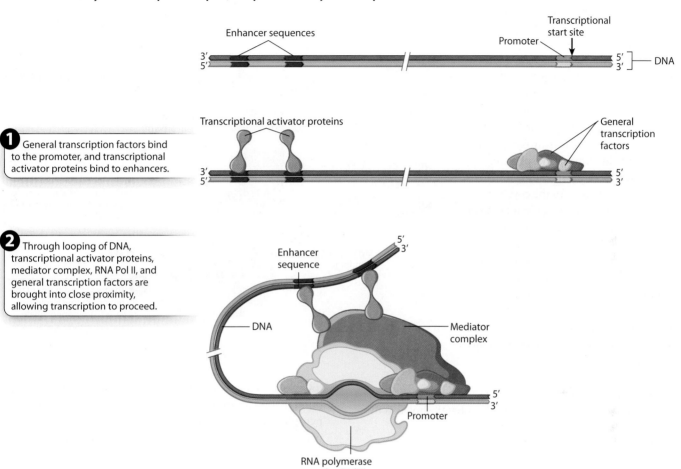

FIG. 3.15 The eukaryotic transcription complex, composed of many different proteins.

1. General transcription factors bind to the promoter, and transcriptional activator proteins bind to enhancers.

2. Through looping of DNA, transcriptional activator proteins, mediator complex, RNA Pol II, and general transcription factors are brought into close proximity, allowing transcription to proceed.

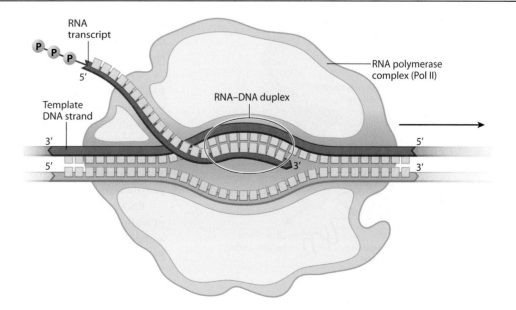

FIG. 3.16 Transcription bubble. Within the RNA polymerase, the two strands of DNA separate and the growing RNA strand forms a duplex with the DNA template.

Details of the polymerization reaction are shown in **Fig. 3.17**. The incoming ribonucleoside triphosphate, shown at the bottom right, is accepted by the RNA polymerase only if it undergoes proper base pairing with the base in the template DNA strand. In Fig. 3.17, there is a proper match because U pairs with A. At this point, the RNA polymerase orients the 3′ end of the growing strand so that the oxygen in the hydroxyl group can attack the innermost phosphate of the triphosphate of the incoming ribonucleoside, competing for the covalent bond. The bond connecting the innermost phosphate to the next is a high-energy phosphate bond, which when cleaved provides the energy to drive the reaction that creates the phosphodiester bond attaching the incoming nucleotide to the 3′ end of the growing chain. The term "high-energy" here refers to the amount of energy released when the phosphate bond is broken; this energy can be used to drive other chemical reactions.

The polymerization reaction releases a phosphate–phosphate group (pyrophosphate), shown at the

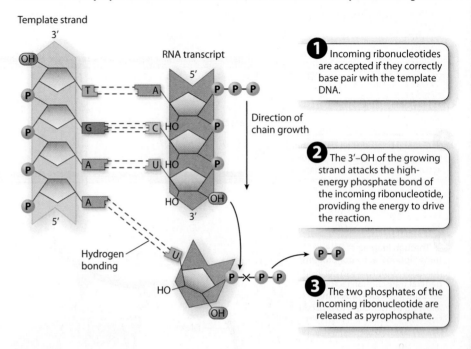

FIG. 3.17 **The polymerization reaction that allows the RNA transcript to be elongated.**

lower right in Fig. 3.17, which also has a high-energy phosphate bond that is cleaved by another enzyme. Cleavage of the pyrophosphate molecule makes the polymerization reaction irreversible, and the next ribonucleoside triphosphate that complements the template is brought into line.

The RNA polymerase complex is a molecular machine that opens, transcribes, and closes duplex DNA.

Transcription does not take place spontaneously. It requires template DNA, a supply of ribonucleoside triphosphates, and RNA polymerase, the large multiprotein complex in which transcription occurs. To illustrate RNA polymerase in action, we consider here bacterial RNA polymerase because of its relative simplicity.

The transcription bubble forms and transcription takes place inside the polymerase (**Fig. 3.18**). RNA polymerase contains structural features that separate the DNA strands, allow an RNA–DNA duplex to form, elongate the transcript nucleotide by nucleotide, release the finished transcript, and restore the original DNA double helix. Fig. 3.18 shows how structure and function come together in the bacterial RNA polymerase.

RNA polymerase is a remarkable molecular machine capable of adding thousands of nucleotides to a transcript before dissociating from the template. It is also very accurate, with fewer than 1 incorrect nucleotide incorporated per 10,000 nucleotides.

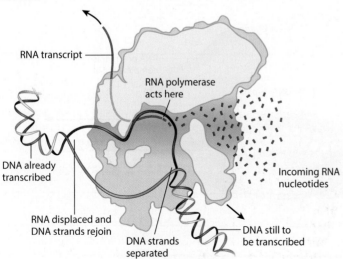

FIG. 3.18 **The RNA polymerase complex in prokaryotes.** This molecular machine has channels for DNA input and output, nucleotide input, and RNA output, and features that disrupt the DNA double helix, stabilize the RNA–DNA duplex, and allow the DNA double helix to re-form.

Self-Assessment Questions

6. What are four differences between the structure of DNA and RNA?
7. A segment of one strand of a double-stranded DNA molecule has the sequence 5'-ACTTTCAGCGAT-3'. What is the sequence of an RNA molecule synthesized from this DNA template?
8. What is the consequence for a growing RNA transcript if an abnormal nucleotide with a 3'-H is incorporated rather than a 3'-OH? How about a 2'-H rather than a 2'-OH?

3.4 RNA PROCESSING

The RNA transcript that comes off the template DNA strand is known as the **primary transcript,** and it contains the complement of every base that was transcribed from the DNA template. For protein-coding genes, this means that the primary transcript includes the information needed to direct the ribosome to produce the protein corresponding to the gene (Chapter 4). The RNA molecule that combines with the ribosome to direct protein synthesis is known as **messenger RNA (mRNA)** because it carries the genetic "message" (information) from the DNA to the ribosome. As we will see in this section, there is a major difference between prokaryotes and eukaryotes in how the primary transcript and mRNA differ from each other. We will also see that some genes do not code for proteins, but rather for RNA molecules that have functions of their own.

Primary transcripts in prokaryotes are translated immediately.

In prokaryotes, the relation between the primary transcript and mRNA is as simple as can be: the primary transcript *is* the mRNA. Even as the 3' end of the primary transcript is still being synthesized, ribosomes bind with special sequences near its 5' end and begin the process of protein synthesis (**Fig. 3.19**).

This intimate connection between transcription and translation can take place because prokaryotes have no nuclear envelope to spatially separate transcription from translation. Thus, the two processes are coupled, which means that they are connected in space and time.

Primary transcripts for protein-coding genes in prokaryotes have another feature not shared with those in eukaryotes: they often contain the genetic information for the synthesis of two or more different proteins. Usually these proteins code for successive steps in the biochemical reactions that produce small molecules needed for growth, or for successive steps

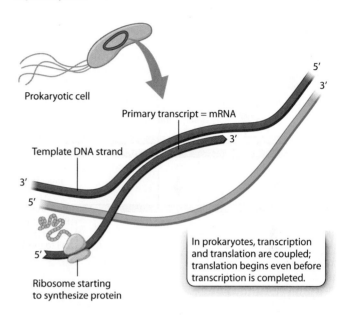

FIG. 3.19 Fate of the primary transcript for protein-coding genes in prokaryotes.

needed to break down a small molecule used for nutrients or energy. Molecules of mRNA that code for multiple proteins are known as **polycistronic mRNA** because the term "cistron" was once widely used to refer to a protein-coding sequence in a gene.

Primary transcripts in eukaryotes undergo several types of chemical modification.

In eukaryotes, the nuclear envelope is a barrier between the processes of transcription and translation. Transcription takes place in the nucleus, and translation in the cytoplasm. The separation allows for a complex chemical modification of the primary transcript, known as **RNA processing,** which converts the primary transcript into the finished mRNA, which can then be translated by the ribosome.

RNA processing consists of three principal types of chemical modifications, illustrated in **Fig. 3.20**. First, the 5' end of the primary transcript is modified by the addition of a special nucleotide attached in an unusual linkage. This addition, which is called the **5' cap,** consists of a modified nucleotide called 7-methylguanosine. An enzyme attaches the modified nucleotide to the 5' end of the primary transcript essentially backward: in a normal linkage between two nucleotides, the phosphodiester bond forms between the 3'-OH group of one nucleotide and the 5' carbon of the next nucleotide, but here the cap is linked to the RNA transcript by a triphosphate bridge between the 5' carbons of both ribose

FIG. 3.20 Fate of the primary transcript for protein-coding genes in eukaryotes.

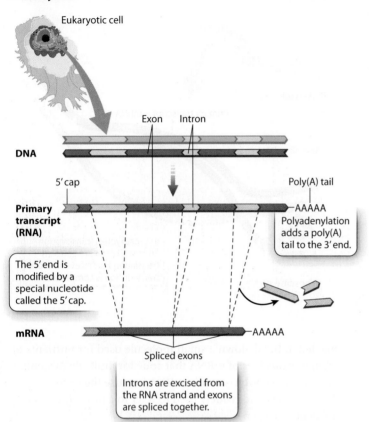

Not every stretch of the RNA transcript ends up being translated into protein. Transcripts in eukaryotes often contain regions of protein-coding sequence, called **exons**, interspersed with noncoding regions called **introns**. The third type of modification of the primary transcript is the removal of the noncoding introns (see Fig. 3.20). The process of intron removal is known as **RNA splicing**, which is catalyzed by a complex of RNA and protein known as the **spliceosome**.

The mechanism of splicing is outlined in **Fig. 3.22**. In the first step, the spliceosome brings a specific sequence within the intron into proximity with the 5' end of the intron, at a site known as the 5' splice site (Fig. 3.22a). The proximity enables a reaction that cuts the RNA at the 5' splice site, and the cleaved end of the intron connects back on itself, forming a loop and tail called a **lariat** (Fig. 3.22b). In the next step, the spliceosome brings the 5' splice site close to the splice site at the 3' end of the intron. The 5' splice site attacks the 3' splice site (Fig. 3.22c), cleaving the bond that holds the lariat on the transcript and attaching the ends of the exons to each other. The result is that the exons are connected and the lariat is released (Fig. 3.22d). The lariat making up the intron is quickly broken down into its constituent nucleotides.

The presence of introns has important implications for gene expression. Approximately 90% of all human

sugars (**Fig. 3.21**). The 5' cap is essential for translation because in eukaryotes, the ribosome recognizes an mRNA by its 5' cap. Without the cap, the ribosome would not attach to the mRNA and translation would not occur.

The second major modification of eukaryotic primary transcripts is **polyadenylation**, the addition of a string of about 250 consecutive A-bearing ribonucleotides to the 3' end, forming a **poly(A) tail** (see Fig. 3.20). Polyadenylation plays an important role in the export of mRNA into the cytoplasm. In addition, both the 5' cap and the poly(A) tail help to stabilize the RNA transcript. Single-stranded nucleic acids can be unstable and are even susceptible to enzymes that break them down. In eukaryotes, the 5' cap and poly(A) tail protect the two ends of the transcript and increase the stability of the RNA transcript until it is translated in the cytoplasm.

FIG. 3.21 Structure of the 5' cap in eukaryotic messenger RNA.

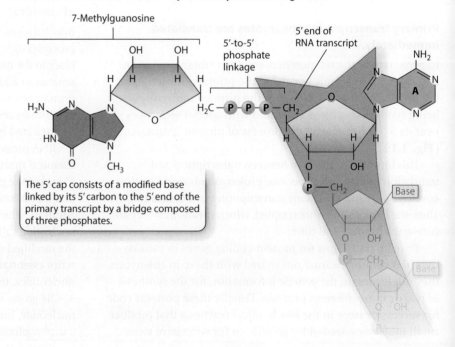

FIG. 3.22 RNA splicing.

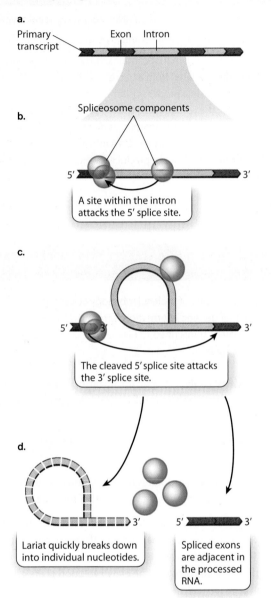

FIG. 3.23 Alternative splicing. A single primary transcript can be spliced in different ways.

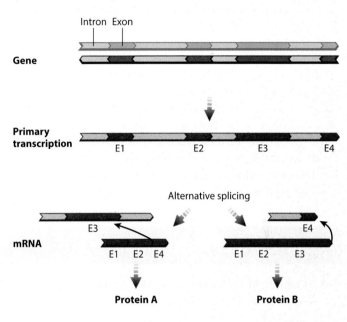

allows the same transcript to be processed in diverse ways to produce mRNA molecules with different combinations of exons coding for different proteins.

Some RNA transcripts are processed differently from protein-coding transcripts and have functions of their own.

Not all primary transcripts are processed into mRNA. Some RNA transcripts have functions of their own, and many of these transcripts are produced by RNA polymerases other than Pol II. These primary transcripts undergo different types of RNA processing, and their processed forms include some important noncoding types of RNA:

- **Ribosomal RNA (rRNA),** which makes up the bulk of ribosomes and is essential in translation. In eukaryotic cells, the genes and transcripts for rRNA are concentrated in the **nucleolus,** a distinct, dense, non–membrane-bound spherical structure observed within the nucleus. The structure and function of ribosomes are examined in Chapter 4.

- **Transfer RNA (tRNA),** which carries individual amino acids for use in translation. The way in which amino acids become attached to tRNA and are then transferred to a growing polypeptide chain is discussed in Chapter 4.

- **Small nuclear RNA (snRNA),** which is an essential component of the spliceosome required for RNA processing.

genes contain at least one intron. Although most genes contain 6 to 9 introns, the largest number is 147 introns, found in a muscle gene. Most introns are just a few thousand nucleotides in length, but roughly 10% are longer than 10,000 nucleotides. The presence of multiple introns in most genes allows for a process known as **alternative splicing,** in which primary transcripts from the same gene can be spliced in different ways to yield different mRNAs and, therefore, different protein products (**Fig. 3.23**). More than 80% of human genes are alternatively spliced. In most cases, the alternatively spliced forms differ in whether a particular exon is or is not removed from the primary transcript along with its flanking introns. Alternative splicing

- Small, regulatory RNA molecules that can inhibit translation or cause destruction of an RNA transcript. Two major types of small regulatory RNA are known as **microRNA (miRNA)** and **small interfering RNA (siRNA)**. The roles of these types of RNA in gene regulation are examined in Chapter 18.

By far, the most abundant transcripts in mammalian cells are those for ribosomal RNA and transfer RNA. In a typical mammalian cell, approximately 80% of the RNA consists of ribosomal RNA, and another approximately 10% consists of transfer RNA. Why are these types of RNA so abundant? The answer is that they are needed in large amounts to synthesize the proteins encoded in the messenger RNA, which is the subject of the next chapter.

Self-Assessment Questions

9. What are three mechanisms of RNA processing in eukaryotes?
10. What are five types of noncoding RNA and their functions?
11. When a region of DNA that contains the genetic information for a protein is isolated from a bacterial cell and inserted into a eukaryotic cell in a proper position between a promoter and a terminator, the resulting cell usually produces the correct protein. But when the experiment is done in the reverse direction (inserting eukaryotic DNA into a bacterial cell), the correct protein is often not produced. Why is this the case?

CORE CONCEPTS SUMMARY

3.1 CHEMICAL COMPOSITION AND STRUCTURE OF DNA: DNA is a polymer of nucleotides that forms a double helix.

Experiments on *Streptococcus pneumoniae* conducted by F. Griffith showed that a nonvirulent strain of bacteria could be transformed into a virulent one by some unknown substance. page 52

Experiments on pneumococcal bacteria conducted by Avery, MacLeod, and McCarty provided evidence that DNA is the genetic material. page 52

A nucleotide consists of a 5-carbon sugar, a phosphate group, and a base. page 53

The four bases of DNA are adenine (A), guanine (G), cytosine (C), and thymine (T). page 54

Successive nucleotides are linked by phosphodiester bonds to form a linear strand of DNA. page 54

DNA strands have polarity, with a 5′-phosphate group at one end and a 3′-hydroxyl group at the other end. page 55

In a DNA double helix, A pairs with T, and G pairs with C. page 55

3.2 DNA STRUCTURE AND FUNCTION: The structure of DNA allows it to store and transmit information.

Information is coded in the sequence of bases of DNA. page 58

DNA structure suggests a mechanism for replication, in which each parental strand serves as a template for a daughter strand. page 58

Ribonucleic acid (RNA) is synthesized from a DNA template in the process of transcription. page 58

The central dogma of molecular biology states that the usual flow of genetic information is from DNA to RNA to protein. DNA is transcribed to RNA, and RNA is translated to protein. page 58

3.3 TRANSCRIPTION: Genetic information is retrieved by transcription, the process by which RNA is synthesized from a DNA template.

RNA, like DNA, is a polymer of nucleotides linked by phosphodiester bonds. page 60

Unlike DNA, RNA incorporates the sugar ribose instead of deoxyribose and the base uracil instead of thymine. page 60

Some types of RNA can store genetic information, while other types can catalyze chemical reactions. These characteristics have led to the RNA world hypothesis, the idea that RNA played a critical role in the early evolution of life on Earth. page 60

RNA is synthesized from one of the two strands of DNA, called the template strand. page 61

RNA is synthesized by RNA polymerase in a 5′-to-3′ direction, starting at a promoter and ending at a terminator in the DNA template. page 61

Specific DNA sequences such as enhancers, promoters, and terminators participate in transcriptional regulation. page 62

RNA polymerase separates the two DNA strands, allows an RNA–DNA duplex to form, elongates the transcript, releases the transcript, and restores the DNA duplex. page 62

3.4 RNA PROCESSING: The primary transcript is processed to become messenger RNA.

In prokaryotes, the primary transcript is immediately translated into protein. page 65

In eukaryotes, transcription and translation are separated in time and space, with transcription occurring in the nucleus and then translation in the cytoplasm. page 65

In eukaryotes, there are three major types of modification to the primary transcript—the addition of a 5′ cap, polyadenylation, and splicing. page 65

Alternative splicing is a process in which primary transcripts from the same gene are spliced in different ways to yield different protein products. page 66

Some RNAs, called noncoding RNAs, do not code for proteins, but instead have functions of their own. page 67

Log in to **LaunchPad** to check your answers to the Self-Assessment Questions and to access additional learning tools.

CHAPTER 4: Translation and Protein Structure

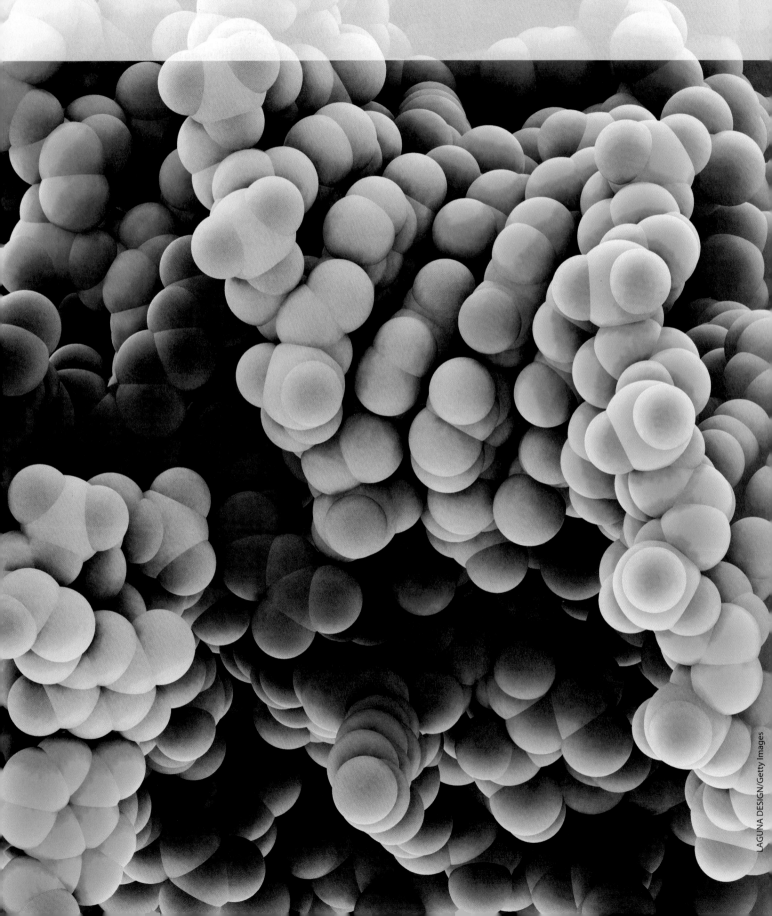

CORE CONCEPTS

4.1 MOLECULAR STRUCTURE OF PROTEINS: Proteins are linear polymers of amino acids that form three-dimensional structures with specific functions.

4.2 PROTEIN SYNTHESIS: Translation is the process in which the sequence of bases in messenger RNA specifies the order of successive amino acids in a newly synthesized protein.

4.3 PROTEIN ORIGINS AND EVOLUTION: Proteins evolve through mutation and selection and by combining functional units.

Hardly anything happens in the life of a cell that does not require proteins. They are the most versatile of macromolecules, each with its own built-in ability to carry out a cellular function. Some proteins aggregate to form relatively stiff filaments that help define the cell's shape and hold organelles in position. Others span the cell membrane and form channels or pores through which ions and small molecules can move. Many others are enzymes that catalyze the thousands of chemical reactions needed to maintain life. Still others are signaling proteins that enable cells to coordinate their internal activities or to communicate with other cells.

Want to see some proteins? Look at the white of an egg. Apart from the 90% or so that is water, most of what you see is protein. The predominant type of protein is ovalbumin. Easy to obtain in large quantities, ovalbumin was one of the first proteins studied through scientific inquiry (Chapter 1). In the 1830s, ovalbumin was shown to consist largely of carbon, hydrogen, nitrogen, and oxygen. Each molecule of ovalbumin was estimated to contain at least 400 carbon atoms. Leading chemists of the time scoffed at this number, believing that no organic molecule could possibly be so large. Little did they know: the number of carbon atoms in ovalbumin is actually closer to 2000 than to 400!

The reason that proteins can be such large organic molecules began to become clear only about a hundred years ago, when scientists hypothesized that proteins are polymers (large molecules made up of repeated subunits). Now we know that proteins are linear polymers of any combination of 20 amino acids, each of which differs from the other proteins in its chemical characteristics. In size, ovalbumin is actually an average protein, consisting of a chain of 385 amino acids.

In Chapter 3, we discussed how genetic information flows from the sequence of bases of DNA into the sequence of bases in a transcript of RNA. For protein-coding genes, the transcript is processed into messenger RNA (mRNA). In this chapter, we examine how amino acid polymers are assembled by ribosomes by means of a template of messenger RNA and transfer RNAs. We also discuss how the amino acid sequences of proteins help determine their three-dimensional structures and diverse chemical activities, as well as how proteins change through evolutionary time.

4.1 MOLECULAR STRUCTURE OF PROTEINS

If you think of a protein as analogous to a word in the English language, then the amino acids are like letters. The comparison is not altogether fanciful, as there are nearly as many amino acids in proteins as there are letters in the alphabet, and the order of both amino acids and letters is important. For example, the word PROTEIN has the same letters as POINTER, but the two words have completely different meanings. Similarly, the exact order of amino acids in a protein determines the protein's shape and function.

Amino acids differ in their side chains.

The general structure of an amino acid was discussed in Chapter 2 and is shown again in **Fig. 4.1**. It consists of a central carbon atom, called the **α (alpha) carbon**, connected by covalent bonds to four different chemical groups (Fig. 4.1a): an **amino group** ($-NH_2$, shown in dark blue), a **carboxyl group** ($-COOH$, shown in brown), a hydrogen atom

FIG. 4.1 Structure of an amino acid. The central carbon atom is attached to an amino group, a carboxyl group, an R group or side chain, and a hydrogen atom.

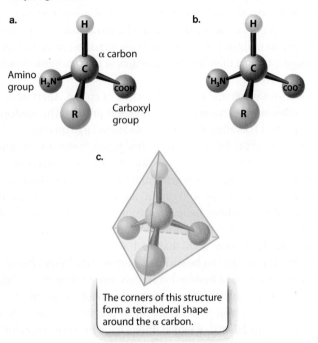

The corners of this structure form a tetrahedral shape around the α carbon.

(—H, shown in light blue), and a variable **side chain** or **R group** (shown in green). In the environment of a cell, where the pH is in the range 7.35–7.45 (called physiological pH), the amino group gains a proton to become —NH_3^+ and the carboxyl group loses a proton to become —COO^- (Fig. 4.1b). The four covalent bonds from the α carbon are at equal angles. As a result, an amino acid forms a tetrahedron, a pyramid with four triangular faces (Fig. 4.1c).

The R groups of the amino acids differ from one amino acid to the next. They are what make the "letters" of the amino acid "alphabet" distinct from one another. Just as letters differ in their shapes and sounds—vowels like E, I, and O and consonants like B, P, and T—amino acids differ in their chemical and physical properties.

The chemical structures of the 20 amino acids commonly found in proteins are shown in **Fig. 4.2**. The R groups (shown in green) are chemically diverse and are grouped according to their properties, with a particular emphasis on whether they are hydrophobic or hydrophilic, or have special characteristics that might affect a protein's structure. These properties strongly influence how a polypeptide folds, and hence affect the three-dimensional shape of the protein.

Hydrophobic amino acids do not readily interact with water or form hydrogen bonds. Most hydrophobic amino acids have nonpolar R groups composed of hydrocarbon chains or uncharged carbon rings. Because water molecules in the cell form hydrogen bonds with each other instead of with the hydrophobic R groups, the hydrophobic R groups tend to aggregate with each other. Their aggregation is also stabilized by weak van der Waals forces (Chapter 2), in which asymmetries in electron distribution create temporary charges in the interacting molecules, which are then attracted to each other. The tendency for hydrophilic water molecules to interact with each other and for hydrophobic molecules to interact with each other is the very same tendency that leads to the formation of oil droplets in water. It is also the reason why most hydrophobic amino acids tend to be buried in the interior of folded proteins, where they do not interact with water.

Amino acids with polar R groups have a permanent charge separation, in which one end of the R group is slightly more negatively charged than the other. As we saw in Chapter 2, polar molecules are hydrophilic, and they tend to form hydrogen bonds with each other or with water molecules.

The R groups of the basic and acidic amino acids are typically charged and, therefore, are strongly polar. At the pH of a cell, the R groups of the basic amino acids gain a proton and become positively charged, whereas those of the acidic amino acids lose a proton and become negatively charged. Because the R groups of these amino acids are charged, they are usually located on the outside surface of the folded molecule. The charged groups can also form ionic bonds with each other and with other charged molecules in the environment, with a negatively charged group or molecule bonding with a positively charged group or molecule. This ability to bind another molecule of opposite charge is an important way in which proteins can associate with each other or with other macromolecules such as DNA.

The properties of several amino acids are noteworthy because of their effect on protein structure. These amino acids include glycine, proline, and cysteine. Glycine is different from the other amino acids because its R group is hydrogen, exactly like the hydrogen on the other side of the α carbon, and therefore the glycine molecule is not asymmetric. All of the other amino acids have four different groups attached to the α carbon and are asymmetric. In addition, glycine is nonpolar and small enough to tuck into spaces where other R groups would not fit. The small size of glycine's R group also allows for freer rotation around the C—N bond since its R group does not get in the way of the R groups of neighboring amino acids. Thus, glycine increases the flexibility of the polypeptide backbone, which can be important in the folding of the protein.

Proline is also distinctive, but for a different reason. Note how its R group is linked back to the amino group. This linkage creates a kink or bend in the polypeptide chain and restricts rotation of the C—N bond, thereby imposing constraints on protein folding in its vicinity, an effect that is the very opposite of glycine's.

FIG. 4.2 Structures of the 20 amino acids commonly found in proteins.

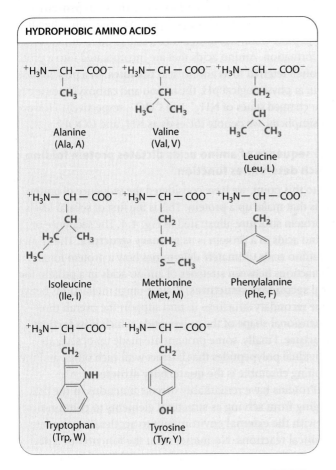

Cysteine makes a special contribution to protein folding through its —SH group. When two cysteine side chains in the same or different polypeptides come into proximity, they can react to form an S—S disulfide bond, which covalently joins the side chains. Such disulfide bonds are stronger than the ionic interactions of other pairs of amino acid, and form cross-bridges that can connect different parts of the same protein or even different proteins. This property contributes to the overall structure of single proteins or combinations of proteins.

Successive amino acids in proteins are connected by peptide bonds.

Amino acids are linked together to form proteins (**Fig. 4.3**). The bond formed between the two amino acids is a **peptide bond**,

FIG. 4.3 Formation of a peptide bond. A peptide bond forms between the carboxyl group of one amino acid and the amino group of another.

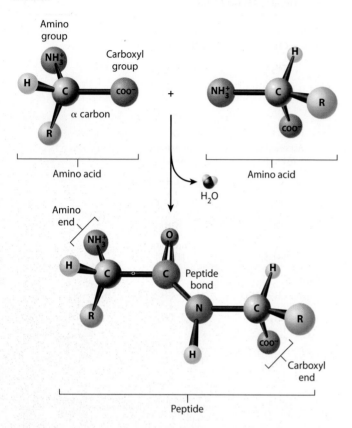

shown in red in Fig. 4.3. In forming the peptide bond, the carboxyl group of one amino acid reacts with the amino group of the next amino acid in line, and a molecule of water is released. In the resulting molecule, the R groups of each amino acid point in different directions.

The C=O group in the peptide bond is known as a carbonyl group, and the N—H group is an amide group. Note in Fig. 4.3 that these two groups appear on either side of the peptide bond. The electrons of the peptide bond are more attracted to the C=O group than to the NH group because of the greater electronegativity of the oxygen atom. The result is that the peptide bond has some of the characteristics of a double bond. The peptide bond is shorter than a single bond, for example, and it is not free to rotate like a single bond. The other bonds are free to rotate around their central axes.

Polymers ranging from as few as two to many hundreds of amino acids linked together share a chemical feature common to individual amino acids: namely, the ends are chemically distinct from each other. One end, shown at the left in Fig. 4.3, has a free amino group; this is the **amino end** of the molecule. The other end has a free carboxyl group; it is the **carboxyl end** of the molecule. More generally, a polymer of amino acids connected by peptide bonds is known as a **polypeptide**. Typical polypeptides produced in cells consist of a few hundred amino acids. In human cells, the shortest polypeptides are approximately 100 amino acids in length; the longest is the muscle protein titin, with 34,350 amino acids. The term **protein** is often used as a synonym for polypeptide, especially when the polypeptide chain has folded into a stable, three-dimensional conformation. Amino acids that are incorporated into a protein are often referred to as amino acid **residues**. In a polypeptide chain at physiological pH, the amino and carboxyl ends are in their charged states of NH_3^+ and COO^-, respectively. However, for simplicity, we denote the ends as NH_2 and COOH.

The sequence of amino acids dictates protein folding, which determines function.

Up to this point, we have considered the sequence of amino acids that make up a protein. This is the first of several levels of protein structure, illustrated in **Fig. 4.4**. The sequence of amino acids in a protein is its **primary structure**. The sequence of amino acids ultimately determines how a protein folds. Interactions between stretches of amino acids in a protein form local **secondary structures**. Longer-range interactions between these secondary structures in turn support the overall three-dimensional shape of the polypeptide, which is its **tertiary structure**. Finally, some proteins are made up of several individual polypeptides that interact with each other, and the resulting ensemble is the **quaternary structure**.

Proteins have remarkably diverse functions in the cell, ranging from serving as structural elements to communicating with the external environment to accelerating the rate of chemical reactions. No matter what the function of a protein is, the ability to carry out this function depends on the three-dimensional shape of the protein. When fully folded, some proteins contain pockets with positively or negatively charged side chains at just the right positions to trap small molecules; others have surfaces that can bind another protein or a sequence of nucleotides in DNA or RNA; some form rigid rods for structural support; and still others keep their hydrophobic side chains away from water molecules by inserting into the cell membrane.

The sequence of amino acids in a protein (its primary structure) is usually represented by a series of three-letter or one-letter abbreviations for the amino acids (abbreviations for the 20 common amino acids are given in Fig. 4.2). By convention, the amino acids in a protein are listed in order from left to right, starting at the amino end and proceeding to the carboxyl end. The amino and the carboxyl ends are different, so the order matters. Just as TIPS is not the same word as SPIT, the sequence Thr–Ile–Pro–Ser is not the same polypeptide as Ser–Pro–Ile–Thr.

Secondary structures result from hydrogen bonding in the polypeptide backbone.

Hydrogen bonds can form between the carbonyl group in one peptide bond and the amide group in another, thus allowing localized regions of the polypeptide chain to fold. This localized

FIG. 4.4 Levels of protein structure. The manner in which a polypeptide folds determines its shape and function.

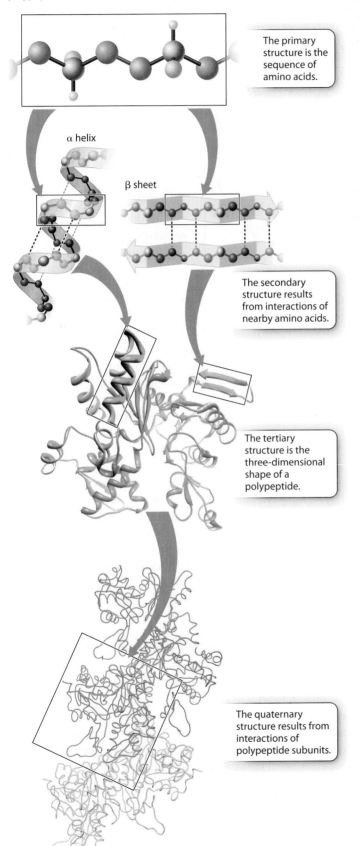

folding is a major contributor to the secondary structure of the protein.

In the early 1950s, American structural biologists Linus Pauling and Robert Corey used X-ray crystallography to study the structure of proteins. In this procedure, researchers first form a crystal in which the atoms of the protein are in an ordered pattern in three dimensions, and then rotate the crystal in a beam of X-rays. Some X-rays pass through the crystal, while the atoms of the proteins scatter other X-rays in different directions. A film or other detector records the pattern of scattered X-rays as a series of spots, and the locations and intensities of these spots can be used to infer the position and arrangement of the atoms in the molecule. Pauling and Corey studied crystals of highly purified proteins and discovered that two types of secondary structure are found in many different proteins: the *α* **(alpha) helix** and the *β* **(beta) sheet**. Both of these secondary structures are stabilized by hydrogen bonding along the polypeptide backbone.

In *α* helices, like the one shown in **Fig. 4.5**, the polypeptide backbone is twisted tightly in a right-handed coil with 3.6 amino acids per complete turn. The helix is stabilized by hydrogen bonds that form between each amino acid's carbonyl group (C=O) and the amide group (N—H) four residues ahead in the sequence, as indicated by the dashed lines in Fig. 4.5. Note that the R groups project outward from the *α* helix. The

FIG. 4.5 An *α* helix. Hydrogen bonds between carbonyl and amide groups in the backbone stabilize the helix.

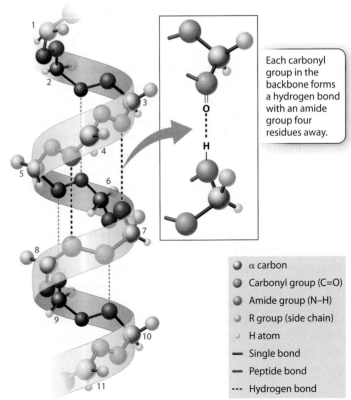

FIG. 4.6 A β sheet. Hydrogen bonds between neighboring strands stabilize the structure.

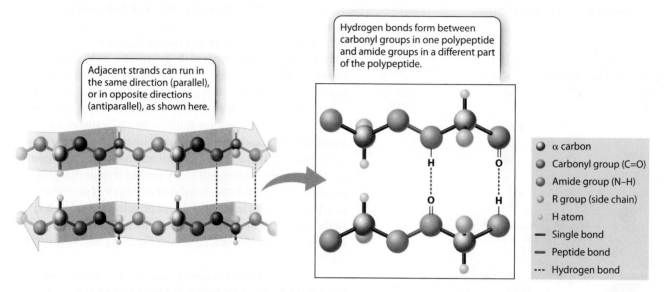

chemical properties of the projecting R groups largely determine where the α helix is positioned in the folded protein, and how it might interact with other molecules.

The other secondary structure that Pauling and Corey found is the β sheet, depicted in **Fig. 4.6**. In a β sheet, the polypeptide folds back and forth on itself, forming a pleated sheet that is stabilized by hydrogen bonds between carbonyl groups in one chain and amide groups in the other chain across the way (dashed lines). The R groups project alternately above and below the plane of the β sheet. β sheets typically consist of 4 to 10 polypeptide chains aligned side by side, with the amide groups in each chain hydrogen-bonded to the carbonyl groups on either side (except for those at the ends of each strand).

The β sheets are typically denoted by broad arrows, where the direction of the arrow runs from the amino end of the polypeptide segment to the carboxyl end. In Fig. 4.6, the arrows run in opposite directions, and the polypeptide chains are said to be antiparallel. β sheets can also be formed by hydrogen bonding between polypeptide chains that are parallel (pointing in the same direction). However, the antiparallel configuration is more stable because the carbonyl and amide groups are more favorably aligned for hydrogen bonding.

Tertiary structures result from interactions between amino acid side chains.

The tertiary structure of a protein is the three-dimensional conformation of a single polypeptide chain, usually made up of several secondary structure elements. Tertiary structure is defined largely by interactions between the amino acid R groups. By contrast, the formation of secondary structures relies on interactions in the polypeptide backbone and is relatively independent of the R groups. Tertiary structure is determined by the spatial distribution of hydrophilic and hydrophobic R groups along the molecule, as well as by different types of chemical bonds and interactions (ionic, hydrogen, and van der Waals) that form between various R groups. The amino acids whose R groups form bonds with each other may be far apart in the polypeptide chain, but can end up near each other in the folded protein. Hence, the tertiary structure usually includes loops or turns in the backbone that allow these R groups to sit near each other in space and for bonds to form.

The three-dimensional shapes of proteins can be illustrated in different ways, as shown in **Fig. 4.7**. A ball-and-stick model (Fig. 4.7a) draws attention to the atoms in the amino acid chain. A ribbon model (Fig. 4.7b) emphasizes secondary structures, with α helices depicted as twisted ribbons and β sheets as broad arrows. Finally, a space-filling model (Fig. 4.7c) shows the overall shape and contour of the folded protein.

Recall that the folding of a polypeptide chain is determined by the sequence of amino acids. The primary structure determines the secondary and tertiary structures. Furthermore, tertiary structure determines function because it is the three-dimensional shape of the molecule—the contours and distribution of charges on the outside of the molecule and the presence of pockets that might bind with smaller molecules on the inside—that enables the protein to serve as structural support, membrane channel, enzyme, or signaling molecule. **Fig. 4.8** shows the tertiary structure of a bacterial protein that contains a pocket in the center in which certain R groups can form hydrogen bonds with a specific small molecule and hold it in place.

The principle that structure determines function can be demonstrated by many observations. For example, most proteins can be unfolded, or **denatured**, by chemical treatment

FIG. 4.7 Three ways of showing the structure of the protein tubulin: (a) ball-and-stick model; (b) ribbon model; (c) space-filling model.

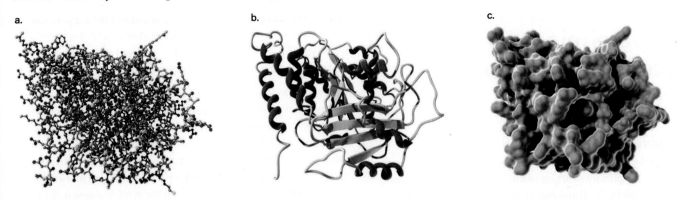

or high temperature that disrupts the hydrogen and ionic bonds holding the tertiary structure together. Under these conditions, the proteins lose their functional activity. Similarly, mutant proteins containing an amino acid that prevents proper folding are often inactive or do not function properly.

The first demonstration that the primary structure of a protein (its amino acid sequence) determines its folding into secondary and tertiary structures was based on studies of the enzyme ribonuclease A. In 1961, Christian Anfinsen and colleagues chemically denatured ribonuclease A and completely inactivated the enzyme. They then removed the denaturing agents and found that the enzyme completely recovered its activity (**Fig. 4.9**). These experiments demonstrated that the secondary and tertiary structures are formed by the primary sequence. For these experiments, Anfinsen was awarded a Nobel Prize in 1972.

Polypeptide subunits can come together to form quaternary structures.

Although many proteins are complete and fully functional as a single polypeptide chain with a tertiary structure, many other proteins are composed of two or more polypeptide chains or subunits with a tertiary structure that come together to form a higher-order quaternary structure. In the case of a multi-subunit protein, the activity of the complex depends on the quaternary structure formed by the combination of the various tertiary structures.

In a protein with quaternary structure, the polypeptide subunits may be either identical or different (**Fig. 4.10**). Fig. 4.10a shows an example of a protein produced by the human immunodeficiency virus (HIV) that consists of two identical polypeptide subunits. By contrast, many proteins, such as hemoglobin (shown in Fig. 410b), are composed of different subunits. In either case, the subunits can influence each other in subtle ways and influence their function. For example, the hemoglobin in red blood cells that carries oxygen has four subunits. When one of these subunits binds to oxygen, a slight change in its structure is transmitted to the other subunits, making it easier for them to take up oxygen. In this way, oxygen transport from the lungs to the tissues is improved.

Chaperones help some proteins fold properly.

The amino acid sequence (primary structure) of a protein determines how it forms its secondary, tertiary, and quaternary structures. For approximately 75% of proteins, the folding process takes place within milliseconds as the molecule is synthesized. Some proteins fold more slowly, however, and for these molecules folding is a dangerous business. The longer these polypeptides remain in a denatured (unfolded) state, the longer their hydrophobic groups are exposed to other

FIG. 4.8 Tertiary structure determines function. This bacterial protein has a cavity that can bind with a small molecule (shown as a ball-and-stick model in the center).

HOW DO WE KNOW?

FIG. 4.9
What determines secondary and tertiary structure of proteins?

BACKGROUND Natural selection influences the function that derives from the folded secondary and tertiary structure of a protein. However, only the primary structure (amino acid sequence) is encoded by a protein-coding gene. This seeming paradox was reconciled by the demonstration that the folded secondary and tertiary structure is determined by the primary structure.

EXPERIMENT Christian B. Anfinsen and his colleagues at the National Institutes of Health conducted experiments on the structure of ribonuclease A, a relatively small protein of 124 amino acids that acts as an enzyme. Anfinsen and colleagues measured the activity of ribonuclease A after treating it with various combinations of two chemicals, 2-mercaptoethanol (ME) and urea. In ribonuclease A, as in many other proteins, the sulfhydryl groups (—SH) of pairs of cysteine residues form a covalent disulfide bond (—S—S—) that links the cysteine side chains (Fig. 4.9a). These disulfide bonds are relatively weak covalent bonds and they break in the presence of the mild reducing agent ME. Urea forms stronger hydrogen bonds than do water molecules, and for this reason urea disrupts the hydrogen bonds that help stabilize a polypeptide chain as well as the hydrophobic core of the protein.

RESULTS When the researchers treated the purified ribonuclease A enzyme with a mixture of urea and ME, the enzyme completely lost activity (Fig. 4.9a). When the ME was removed and the enzyme was treated with urea only, the enzyme was misfolded and also inactive (Fig. 4.9b). However, when the urea was removed and only a trace of ME retained, the protein regained activity (Fig. 4.9c).

INTERPRETATION Treating the enzyme with urea and ME results in the loss of all disulfide bonds and complete unfolding of the protein (Fig. 4.9a). As a result, the enzyme loses its activity. When the ME is removed but urea retained, the protein folds in such a way that the disulfide bonds form incorrectly between the cysteine residues (Fig. 4.9b). The improper disulfide bonds result in loss of enzyme activity. When the urea is removed and only a trace of ME is retained, the trace ME allows the incorrect disulfide bonds to break and re-form with the proper partner, so that the protein folds into the correct structure, regaining its activity (Fig. 4.9c).

CONCLUSION These findings supported the hypothesis that the primary sequence dictates protein folding. Under normal conditions in the cell, the folding takes place spontaneously as the protein reaches a state of minimum free energy.

FOLLOW-UP WORK Some proteins fold relatively slowly and require the help of other proteins called chaperones. Chaperones sequester the folding proteins so that the unfolded intermediates do not aggregate to form more complex structures. Many diseases, including Alzheimer's disease, Parkinson's disease, and mad cow disease (bovine spongiform encephalopathy), result from improperly folded protein aggregates.

SOURCE Anfinsen, C. B., Haber, E., Sela, M., and White, F. H., Jr. 1961. The kinetics of formation of native ribonuclease during oxidation of the reduced polypeptide chain. *Proceedings of the National Academy of Sciences U.S.A.* 47:1309–1314.

FIG. 4.10 Quaternary structure.
Polypeptide units of proteins may be identical, as in (a) an enzyme from HIV, or different, as in (b) hemoglobin.

a.

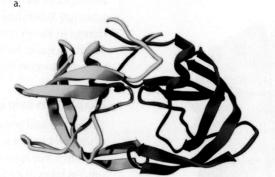

This enzyme consists of two identical polypeptide subunits, shown in light green and dark green.

b.

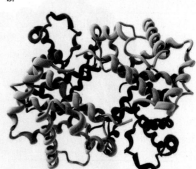

Hemoglobin is made up of four subunits: two copies of the polypeptide depicted in magenta and two copies of the polypeptide depicted in blue.

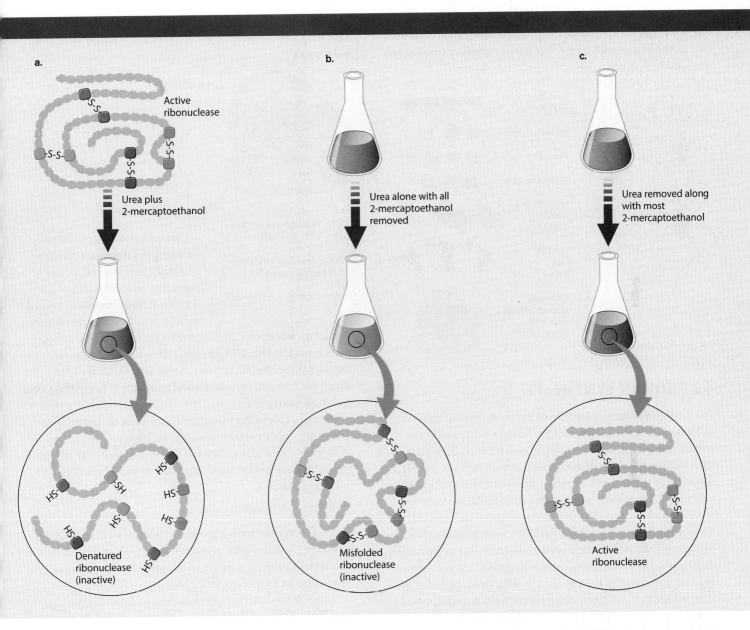

macromolecules in the crowded cytoplasm. The hydrophobic effect, along with van der Waals interactions, tends to bring the exposed hydrophobic groups together, and their inappropriate aggregation may prevent proper folding. Correctly folded proteins can sometimes unfold because of elevated temperature, for example, and in the denatured state they are subject to the same risks of aggregation.

Cells have evolved proteins called **chaperones** that help protect slow-folding or denatured proteins until they can attain their proper three-dimensional structure. Chaperones bind with hydrophobic groups and nonpolar R groups to shield them from inappropriate aggregation, and in repeated cycles of binding and release they give the polypeptide time to find its correct shape.

Self-Assessment Questions

1. Draw one of the 20 amino acids and label the amino group, the carboxyl group, the R group (side chain), and the α carbon.
2. What are the three major groups of amino acids as categorized by the properties of their R groups? How do the chemical properties of each group affect protein shape?
3. How do peptide bonds, hydrogen bonds, ionic bonds, disulfide bridges, and noncovalent interactions (van der Waals forces and the hydrophobic effect) define a protein's four levels of structure?
4. What ultimately determines the three-dimensional shape of a protein?
5. A mutation leads to a change in one amino acid in a protein. The result is that the protein no longer functions properly. How is this possible?

FIG. 4.11 The central dogma, showing how information flows from DNA to RNA to protein. Note the large number of cellular components required for translation.

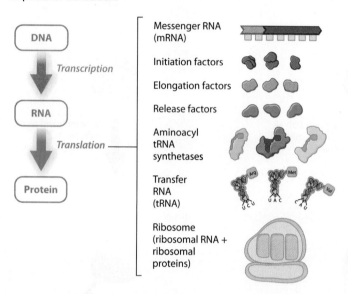

FIG. 4.12 Simplified structure of ribosomes in prokaryotes and eukaryotes.

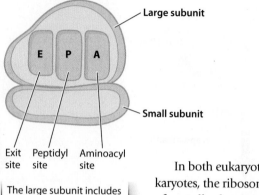

The large subunit includes three binding sites for tRNAs.

4.2 PROTEIN SYNTHESIS

The three-dimensional structure of a protein determines what it can do and how it works, and the immense diversity in the tertiary and quaternary structures among proteins explains their wide range of functions in cellular processes. Yet it is the sequence of amino acids along a polypeptide chain—its primary structure—that governs how the molecule folds into a stable three-dimensional configuration. This sequence is specified by the sequence of nucleotides in the DNA, in coded form. The decoding of the information takes place according to the central dogma of molecular biology, which defines information flow in a cell from DNA to RNA to protein (**Fig. 4.11**). In transcription, the sequence of bases along part of a DNA strand is used as a template in the synthesis of the complementary sequence of bases in a molecule of RNA, as described in Chapter 3. In **translation,** the sequence of bases in an RNA molecule known as **messenger RNA** (mRNA) is used to specify the order in which successive amino acids are added to a newly synthesized polypeptide chain.

Translation uses many molecules found in all cells.

Translation requires many components. More than 100 genes encode components needed for translation, some of which are shown in Fig. 4.11. What are these needed components? First, the cell needs **ribosomes,** which are complex structures of RNA and protein that bind with mRNA and are the site of translation. In prokaryotes, translation occurs as soon as the mRNA comes off the DNA template. In eukaryotes, the processes of transcription and translation are physically separated: transcription takes place in the nucleus, and translation takes place in the cytoplasm.

In both eukaryotes and prokaryotes, the ribosome consists of a small subunit and a large subunit, each composed of 1 to 3 types of ribosomal RNA and 20 to 50 types of ribosomal protein. Eukaryotic ribosomes are larger than prokaryotic ribosomes. As indicated in **Fig. 4.12**, the large subunit of the ribosome includes three binding sites for molecules of transfer RNA, which are called the **A (aminoacyl) site,** the **P (peptidyl) site,** and the **E (exit) site.**

A major role of the ribosome is to ensure that, when the mRNA is in place on the ribosome, the sequence in the mRNA coding for amino acids is read in successive, non-overlapping groups of three nucleotides, much as you would read the following sentence:

THEBIGBOYSAWTHEBADMANRUN

Each non-overlapping group of three adjacent nucleotides (like THE or BIG or BOY in our sentence analogy) constitutes a **codon.** In turn, each codon in the mRNA codes for a single amino acid in the polypeptide chain.

In the preceding example, it is clear that the sentence begins with THE. However, in a long linear mRNA molecule, the ribosome could begin at any nucleotide. As an analogy, if we knew that the letters THE were the start of the phrase, then we would know immediately how to read

ZWTHEBIGBOYSAWTHEBADMANRUN

However, without knowing where to start reading this string of letters, we could find three ways to break the sentence into three-letter words:

ZWT HEB IGB OYS AWT HEB ADM ANR UN

Z WTH EBI GBO YSA WTH EBA DMA NRU N

ZW THE BIG BOY SAW THE BAD MAN RUN

The different ways of parsing the string into three-letter words are known as **reading frames.** The protein-coding sequence in an mRNA consists of its sequence of bases, and just as in our sentence analogy, it can be translated into the correct protein only if it is translated in the proper reading frame.

While the ribosome establishes the correct reading frame for the codons, the actual translation of each codon in the mRNA into one amino acid in the polypeptide is carried out by means of transfer RNA (tRNA). Transfer RNAs are small RNA molecules consisting of 70 to 90 nucleotides (**Fig. 4.13**). Each has a characteristic self-pairing structure that can be drawn as a cloverleaf, as in Fig. 4.13a, in which the letters indicate the bases common to all tRNA molecules. The actual structure, though, is more like that in Fig. 4.13b. Three bases in the anticodon loop make up the **anticodon**; these are the three nucleotides that undergo base pairing with the corresponding codon.

Each tRNA has the nucleotide sequence CCA at its 3′ end (Fig. 4.13), and the

FIG. 4.14 Function of aminoacyl tRNA synthetase enzymes.
Aminoacyl tRNA synthetases attach specific amino acids to tRNAs. In this way, they are responsible for translating the codon sequence in an mRNA into an amino acid sequence in a protein.

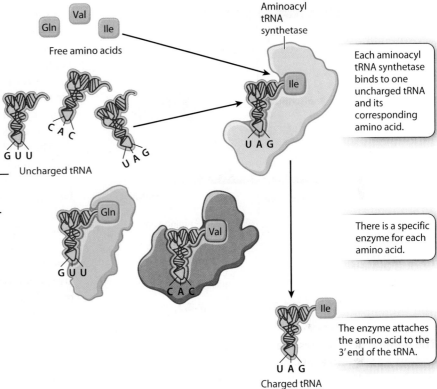

FIG. 4.13 Transfer RNA structure depicted in (a) a cloverleaf configuration and (b) a more realistic three-dimensional structure.

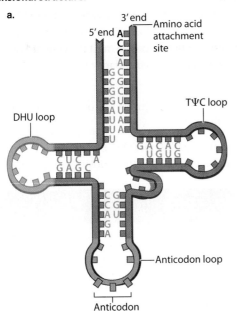

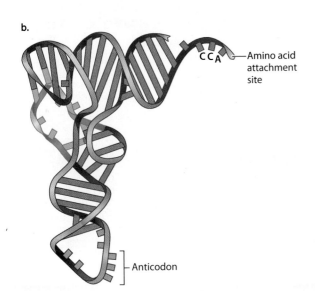

3′ hydroxyl of the A is the attachment site for the amino acid corresponding to the anticodon. Enzymes called **aminoacyl tRNA synthetases** connect specific amino acids to specific tRNA molecules (**Fig. 4.14**). Therefore, these enzymes are directly responsible for actually translating the codon sequence in a nucleic acid to a specific amino acid in a polypeptide chain. Most organisms have one aminoacyl tRNA synthetase for each amino acid. The enzyme binds to multiple sites on any tRNA that has an anticodon corresponding to the amino acid, and it catalyzes formation of the covalent bond between the amino acid and tRNA. A tRNA that has no amino acid attached is said to be uncharged, whereas one with its amino acid attached is said to be charged. Amino acid tRNA synthetases are very accurate and attach the wrong amino acid far less often than 1 time in 10,000.

Although the aminoacyl tRNA synthetase has the specificity necessary for attaching an amino acid to the correct tRNA, the specificity of DNA–RNA and codon–anticodon interactions result from base pairing. **Fig. 4.15** shows the relationships for one codon in double-stranded DNA, in the corresponding mRNA, and in the codon–anticodon pairing between the mRNA and the tRNA. Note that the first (5′) base in the codon in mRNA pairs with the last (3′) base in the anticodon because, as noted in Chapter 3, nucleic acid strands that undergo base pairing must be antiparallel.

FIG. 4.15 Role of base pairing. Base pairing determines the relationship between a three-base triplet in double-stranded DNA and a codon in mRNA, and between a codon in mRNA and an anticodon in tRNA.

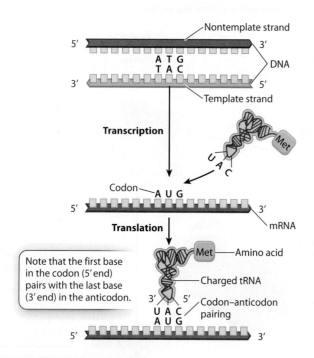

The genetic code shows the correspondence between codons and amino acids.

Fig. 4.15 shows how the codon AUG specifies the amino acid methionine (Met) by base pairing with the anticodon of a charged tRNA, denoted tRNA^{Met}. Most codons specify an amino acid according to a **genetic code**. This code is sometimes called the "standard" genetic code because, while it is used by almost all cells, some minor differences are found in a few organisms as well as in mitochondria.

The codon at which translation begins is called the initiation codon. It is coded by AUG, which specifies Met. The polypeptide is synthesized from the amino end to the carboxyl end, so Met forms the amino end of any polypeptide being synthesized; in many cases, however, the Met is cleaved off by an enzyme after synthesis is complete. The AUG codon also specifies the incorporation of Met at internal sites within the polypeptide chain.

As is apparent in Fig. 4.15, the AUG codon that initiated translation is preceded by a region in the mRNA that is not translated. The position of the initiator AUG codon in the mRNA establishes the reading frame that determines how the downstream codons (those following the AUG) are to be read.

Once the initial Met creates the amino end of a new polypeptide chain, the downstream codons are read one by one in non-overlapping groups of three bases. At each step, the ribosome binds to a tRNA with an anticodon that can base pair with the codon, and the amino acid on that tRNA is attached to the growing chain to become the new carboxyl end of the polypeptide chain. This process continues until one of three "stop" codons is encountered: UAA, UAG, or UGA. (The stop codons are also called termination codons or sometimes nonsense codons.) At this point, the polypeptide is finished and released into the cytosol.

The standard genetic code was deciphered in the 1960s by a combination of techniques, but among the most ingenious were the chemical methods for making synthetic RNAs of known sequence developed by American biochemist Har Gobind Khorana and his colleagues. This experiment is illustrated in **Fig. 4.16**.

The standard genetic code shown in **Table 4.1** has 20 amino acids specified by 61 codons. Thus, many amino acids are specified by more than one codon, and the genetic code is redundant, or degenerate. The redundancy has strong patterns, however:

- The redundancy results almost exclusively from the third codon position.

- When an amino acid is specified by two codons, they differ either in whether the third position is a U or a C

TABLE 4.1 The standard genetic code.

First position (5' end)	Second position				Third position (3' end)
	U	C	A	G	
U	UUU Phe ⎤ F UUC Phe ⎦ UUA Leu ⎤ L UUG Leu ⎦	UCU Ser ⎤ UCC Ser ⎥ S UCA Ser ⎥ UCG Ser ⎦	UAU Tyr ⎤ Y UAC Tyr ⎦ UAA Stop UAG Stop	UGU Cys ⎤ C UGC Cys ⎦ UGA Stop UGG Trp W	U C A G
C	CUU Leu ⎤ CUC Leu ⎥ L CUA Leu ⎥ CUG Leu ⎦	CCU Pro ⎤ CCC Pro ⎥ P CCA Pro ⎥ CCG Pro ⎦	CAU His ⎤ H CAC His ⎦ CAA Gln ⎤ Q CAG Gln ⎦	CGU Arg ⎤ CGC Arg ⎥ R CGA Arg ⎥ CGG Arg ⎦	U C A G
A	AUU Ile ⎤ AUC Ile ⎥ I AUA Ile ⎦ AUG Met M	ACU Thr ⎤ ACC Thr ⎥ T ACA Thr ⎥ ACG Thr ⎦	AAU Asn ⎤ N AAC Asn ⎦ AAA Lys ⎤ K AAG Lys ⎦	AGU Ser ⎤ S AGC Ser ⎦ AGA Arg ⎤ R AGG Arg ⎦	U C A G
G	GUU Val ⎤ GUC Val ⎥ V GUA Val ⎥ GUG Val ⎦	GCU Ala ⎤ GCC Ala ⎥ A GCA Ala ⎥ GCG Ala ⎦	GAU Asp ⎤ D GAC Asp ⎦ GAA Glu ⎤ E GAG Glu ⎦	GGU Gly ⎤ GGC Gly ⎥ G GGA Gly ⎥ GGG Gly ⎦	U C A G

Nonpolar · Polar · Basic · Acidic · Stop codon

HOW DO WE KNOW?

FIG. 4.16
How was the genetic code deciphered?

BACKGROUND The genetic code shows the correspondence between three-letter nucleotide codons in RNA and the amino acids in a protein. American biochemist Har Gobind Khorana performed key experiments that helped to crack the code. For this work he shared the Nobel Prize in Physiology or Medicine in 1968 with Robert W. Holley and Marshall E. Nirenberg.

METHOD Khorana and his group constructed synthetic RNAs with predetermined sequences. They then added these synthetic RNAs to a solution containing all of the other components needed for translation. The researchers used laboratory conditions that allowed ribosomes to initiate synthesis with any codon, even when the codon was not AUG.

EXPERIMENT 1 AND RESULTS When Khorana's group constructed an mRNA with just repeating uracils [called poly(U)], it was translated into a polyphenylalanine polypeptide (Phe–Phe–Phe…):

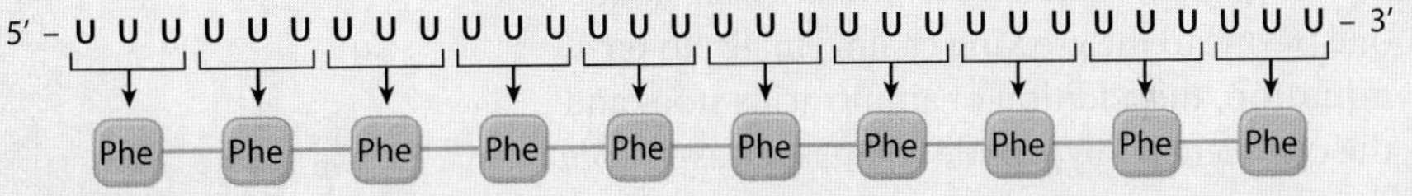

CONCLUSION Ribosomes translate the codon UUU into the amino acid Phe. The poly(U) mRNA can be translated in three possible reading frames, depending on which U is the 5' end of the start codon. In each of those reading frames, all the codons are UUU.

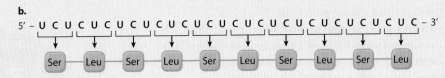

EXPERIMENT 2 AND RESULTS When Khorana's group synthesized an mRNA with alternating U and C ribonucleotides, the resulting polypeptide had alternating serine (Ser) and leucine (Leu):

CONCLUSION Here again are three possible reading frames, but each has alternating UCU and CUC codons. The researchers could not deduce from this result whether UCU corresponds to Ser and CUC to Leu, or the other way around; the correct assignment came from experiments using other synthetic mRNA molecules.

EXPERIMENT 3 AND RESULTS When Khorana's group synthesized an mRNA with repeating UCA, three different polypeptides were produced from multiple rounds of translation: polyserine (Ser), polyhistidine (His), and polyisoleucine (Ile).

CONCLUSION This mRNA could be read by the ribosome as repeated UCAs, repeated CAUs, or repeated AUCs, depending on which bases the ribosome read first. The results do not reveal which of the three reading frames corresponds to which amino acid, but this was sorted out by studies of other synthetic polymers.

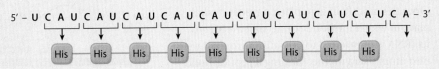

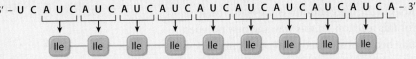

SOURCE Khorana, H. G. 1972. "Nucleic Acid Synthesis in the Study of the Genetic Code." In *Nobel Lectures, Physiology or Medicine 1963–1970*. Amsterdam: Elsevier.

(both pyrimidine bases), or in whether the third position is an A or a G (both purine bases).

- When an amino acid is specified by four codons, the identity of the third codon position does not matter; it could be U, C, A, or G.

The chemical basis of these patterns results from two features of translation. First, in many tRNA anticodons, the 5' base that pairs with the 3' (third) base in the codon is chemically modified into a form that can pair with two or more bases at the third position in the codon. Second, in the ribosome, there is less than perfect alignment between the third position of the codon and the base that pairs with it in the anticodon, so the requirements for base pairing are somewhat relaxed. This feature of the codon–anticodon interaction is referred to as wobble.

Translation consists of initiation, elongation, and termination.

Translation is usually divided into three separate processes: initiation, elongation, and termination. First, in **initiation,** the initiator AUG codon is recognized and Met is established as the first amino acid in the new polypeptide chain. Next, in **elongation,** successive amino acids are added one by one to the growing chain. Finally, in **termination,** the addition of amino acids stops and the completed polypeptide chain is released from the ribosome.

Initiation of translation (**Fig. 4.17**) requires a number of protein **initiation factors** that bind to the mRNA. In eukaryotes, one group of initiation factors binds to the 5′ cap that is added to the mRNA during processing. These factors recruit a small subunit of the ribosome, and other initiation factors bring up a transfer RNA charged with Met (Fig. 4.17a). The initiation complex then moves along the mRNA until it encounters the first AUG triplet. The position of this AUG establishes the translational reading frame.

When the first AUG codon is encountered, a large ribosomal subunit joins the complex, the initiation factors are released, and the next tRNA is ready to join the ribosome (Fig. 4.17b). Note in Fig. 4.17b that the tRNA^Met binds with the P (peptidyl) site in the ribosome and that the next tRNA in line comes in at the A (aminoacyl) site. Once the new tRNA is in place, a reaction takes place in which the bond connecting the Met to its tRNA is transferred to the amino group of the next amino acid in line, forming a peptide bond between the two amino acids (Fig. 4.17c). An RNA in the large subunit is the actual catalyst for this reaction. In this way, the new polypeptide becomes attached to the tRNA in the A site. The ribosome then shifts one codon to the right (Fig. 4.17d). With this move, the uncharged tRNA^Met shifts to the E site and is released into the cytoplasm, and the peptide-bearing tRNA shifts to the P site. The movement of the ribosome also empties the A site, making it available for the next charged tRNA in line to come into place.

At this point, steps (d) and (e) in Fig. 4.17 are repeated over and over again in the process

FIG. 4.17 Translational initiation and elongation. In eukaryotes, the codon used for initiation is the AUG codon nearest the 5′ end of the mRNA. Note that the growing polypeptide chain remains attached to a tRNA throughout translation.

a.
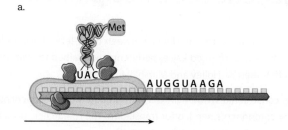

Initiation factors recruit the small ribosomal subunit and tRNA^Met and scan the mRNA for an AUG codon.

b.
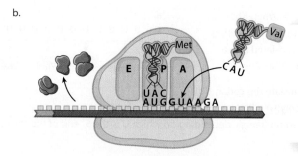

When the complex reaches an AUG, the large ribosomal subunit joins, the initiation factors are released, and a tRNA complementary to the next codon binds to the A site.

c.
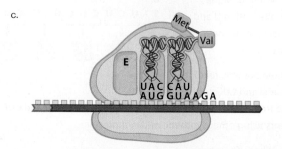

A reaction transfers the Met to the amino acid on the tRNA in the A site, forming a peptide bond.

d.
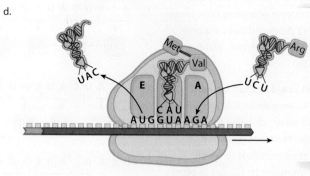

The ribosome moves down one codon, which puts the tRNA carrying the polypeptide into the P site and the now-uncharged tRNA into the E site, where it is ejected. A new tRNA complementary to the next codon binds to the A site.

e.
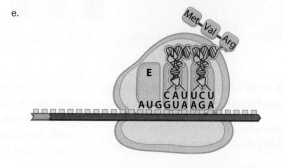

The polypeptide transfers to the amino acid on the tRNA in the A site. The polypeptide is elongated by repeating steps (d) and (e).

of elongation, so that the polypeptide chain grows in length through the addition of successive amino acids. Ribosome movement along the mRNA and formation of the peptide bonds require energy, which is obtained with the help of proteins called **elongation factors**. Elongation factors are bound to GTP molecules, and break their high-energy bonds to provide energy for the elongation of the polypeptide.

The elongation process shown in Fig. 4.17 continues until the ribosome encounters one of the stop codons (UAA, UAG, or UGA); these codons signal termination of polypeptide synthesis. Termination takes place because the stop codons do not have corresponding tRNA molecules. Rather, when the ribosome encounters a stop codon, a protein **release factor** binds to its A site (**Fig. 4.18**). The release factor causes the bond connecting the polypeptide to the tRNA to break, creating the carboxyl terminus of the polypeptide and completing the chain. Once the finished polypeptide is released, the small and large ribosomal subunits disassociate both from the mRNA and from each other.

Although elongation and termination are very similar in prokaryotes and in eukaryotes, translation initiation differs between the two (**Fig. 4.19**). In eukaryotes, the initiation complex forms at the 5′ cap and scans along the mRNA until the first AUG is encountered (Fig. 4.19a). In prokaryotes, the mRNA molecules have no 5′ cap. Instead, the initiation complex is formed at one or more internal sequences present in the mRNA known as a Shine–Dalgarno sequence (Fig. 4.19b). In *Escherichia coli*, the Shine–Dalgarno sequence is 5′-AGGAGGU-3′, and it is followed by an AUG codon eight nucleotides farther downstream that serves as an initiation codon for translation. The ability to initiate translation internally allows prokaryotic mRNAs to code for more than one protein. Such an mRNA is known as a **polycistronic mRNA**. In Fig. 4.19b, the polycistronic mRNA codes for three different polypeptide chains, each with its own AUG initiation codon preceded eight nucleotides upstream by its own Shine–Dalgarno sequence. Each Shine–Dalgarno sequence can serve as an initiation site for translation, so all three polypeptides can be translated.

A polycistronic mRNA results from transcription of a group of functionally related genes located in tandem along the DNA

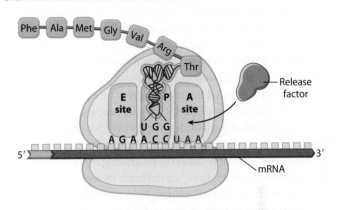

FIG. 4.18 Translational termination. When the ribosome encounters a termination codon (UAA, UAG, or UGA), a protein release factor combines with the A site. This binding triggers release of the polypeptide chain from the final tRNA and dissociation of the ribosomal subunits.

and transcribed as a single unit from one promoter. This type of gene organization is known as an **operon**. Prokaryotes have many of their genes organized into operons because the production of a polycistronic mRNA allows all the protein products to be expressed together whenever they are needed. Typically, the genes organized into operons are those whose products are needed either for successive steps in the synthesis of an essential small molecule, such as an amino acid, or for successive steps in the breakdown of a source of energy, such as a complex carbohydrate.

 CASE 1 LIFE'S ORIGINS: INFORMATION, HOMEOSTASIS, AND ENERGY

How did the genetic code originate?

During transcription and translation, proteins and nucleic acids work together to convert the information stored in DNA into proteins. If we think about how such a system might have originated, however, we immediately confront a chicken-and-egg problem: cells need nucleic acids to make proteins, but proteins are required to make nucleic acids. Which came first?

FIG. 4.19 Initiation in eukaryotes and in prokaryotes. (a) In eukaryotes, translation is initiated only at the 5′ cap. (b) In prokaryotes, initiation takes place at any Shine–Dalgarno sequence; this mechanism allows a single mRNA to include coding sequences for multiple polypeptides.

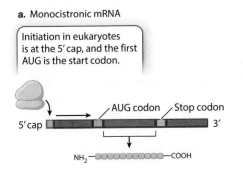
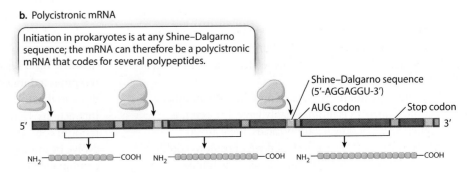

In Chapter 3, we discussed the special features that make RNA an attractive candidate for both information storage and catalysis in early life. Early in evolutionary history, then, proteins had to be added to the mix. No one fully understands how they were incorporated, but researchers are looking closely at tRNA, the molecule involved in the "translating" step of translation.

In modern cells, tRNA shuttles amino acids to the ribosome, but an innovative hypothesis suggests that in early life tRNA-like molecules might have served a different function. This proposal holds that the early precursors of the ribosome were RNA molecules that facilitated the replication of other RNAs, not proteins. In this version of an RNA world, precursors to tRNA would have shuttled nucleotides to growing RNA strands. Researchers hypothesize that tRNAs bound to amino acids may have acted as simple catalysts, facilitating more accurate RNA synthesis. Through time, amino acids brought into close proximity in the process of building RNA molecules might have polymerized to form polypeptide chains. From there, natural selection would favor the formation of polypeptides that enhanced replication of RNA molecules, bringing proteins into the chemistry of life.

Self-Assessment Questions

6. What are the relationships among the template strand of DNA, the codons in mRNA, anticodons in tRNA, and amino acids?

7. Which polypeptide sequences would you expect to result from a synthetic mRNA with the repeating sequence 5'-UUUGGGUUUGGGUUUGGG-3'?

8. What are the steps of translation? Name and describe each one.

9. Bacterial DNA containing an operon encoding three enzymes is introduced into chromosomal DNA in yeast (a eukaryote) in such a way that it is properly flanked by a promoter and a transcriptional terminator. The bacterial DNA is transcribed and the RNA correctly processed, but only the protein nearest the promoter is produced. Why?

4.3 PROTEIN ORIGINS AND EVOLUTION

The amino acid sequences of more than 1 million proteins are known, and the particular three-dimensional structure has been determined for each of more than 10,000 proteins. While few of the sequences and structures are identical, many are sufficiently similar that the proteins can be grouped into approximately 25,000 **protein families**. A protein family is a group of proteins that are structurally and functionally related as a result of their shared evolutionary history.

Why are there not more types of proteins? The number of possible sequences is unimaginably large. For example, for a polypeptide of only 62 amino acids, there are 20^{62} possible sequences (because each of the 62 positions could be occupied by any of the 20 amino acids). The number 20^{62} equals approximately 10^{80}; this number is also the estimated total number of electrons, protons, and neutrons in the entire universe! So why are there so few protein families? The most likely answer is that the chance that any random sequence of amino acids would fold into a stable configuration and carry out some useful function in the cell is very close to zero.

Most proteins are composed of modular folding domains.

If functional proteins are so unlikely, how could life have evolved? The answer is that the earliest proteins were probably much shorter than modern proteins and needed only a trace of function. Only as proteins evolved through billions of years did they become progressively longer and more specialized in their functions. Many protein families that exist today exhibit small regions of three-dimensional structure in which the protein folding is similar. These regions range in length from 25 to 100 or more amino acids. A region of a protein that folds in a similar way relatively independently of the rest of the protein is known as a **folding domain**.

Two examples of folding domains are illustrated in **Fig. 4.20**. Many folding domains are functional units in themselves. The folding domain called a TIM barrel (Fig. 4.20a) is named after the enzyme triose phosphate isomerase, in which it is a prominent feature. The TIM barrel consists of alternating α helices and parallel β sheets connected by loops. In many enzymes with a TIM barrel, the active site is formed by the loops at the carboxyl ends of the sheets. Fig. 4.20b is a β barrel formed from antiparallel β sheets. β barrel structures occur in proteins in some types of bacteria, usually in proteins that span the cell membrane, where the β barrel provides a channel that binds hydrophilic molecules.

Approximately 2500 folding domains are known, far fewer than the number of protein families. The reason for the discrepancy is that different protein families contain different combinations of folding domains. Modern protein families are composed of different combinations of a number of folding domains, each of which contributes some structural or functional feature of the protein. Different types of protein folds occur again and again in different contexts and combinations. The earliest proteins may have been little more than single folding domains that could aggregate to form more complex functional units. As life evolved, the proteins became longer by joining the DNA coding for the individual folding units together into a single molecule.

For example, human tissue plasminogen activator, a protein that is used in treating strokes and heart attacks because it dissolves blood clots, contains domains shared with cell-surface receptors, a domain shared with cellular growth factors, and a domain that folds into large loops facilitating protein–protein interactions. Hence, novel proteins do not always evolve from random combinations of amino acids; instead, they often evolve by joining already functional folding domains into novel combinations.

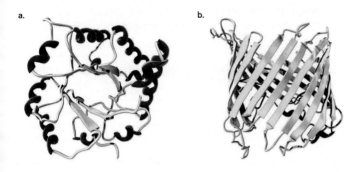

FIG. 4.20 Examples of folding domains: (a) TIM barrel; (b) β barrel.

Amino acid sequences evolve through mutation and selection.

Another important reason that complex proteins can evolve against seemingly long odds is that evolution proceeds stepwise through the processes of mutation and selection. A **mutation** is a change in the sequence of a gene. The process of mutation is discussed in Chapter 14, but for now all you need to know is that mutations affecting proteins occur at random in regard to their effects on protein function. In protein-coding genes, some mutations may affect the amino acid sequence; others might change the level of protein expression or the time in development or type of cell in which the protein is produced. Here, we will consider only those mutations that change the amino acid sequence.

By way of analogy, we can use a simple word game. The object of the game is to change an ordinary English word into another meaningful English word by changing exactly one letter. Consider the word GONE. To illustrate "mutations" of the word that are random with respect to function (that is, random with respect to whether the change will yield a meaningful new word), we wrote a computer program that would choose one letter in GONE at random and replace it with a different random letter. The first 24 "mutants" of GONE are:

UONE	GNNE	GONJ	GOZE
GONH	GOLE	GFNE	XONE
NONE	GKNE	GJNE	DONE
GCNE	GONB	GOIE	GGNE
GONI	GFNE	GPNE	GENE
BONE	GOWE	OONE	GYNE

Most of the mutant words are gibberish, corresponding to the biological reality that most random amino acid replacements impair protein function to some extent. Nevertheless, some mutant proteins function just as well as the original, and a precious few change function. In the word-game analogy, the mutants that can persist correspond to meaningful words, those words shown in red.

In a population of organisms, random mutations are retained or eliminated through the process of **selection** among individuals on the basis of their ability to survive and reproduce. This process was introduced in Chapter 1 and is considered in greater detail in Chapter 20, but the principle is straightforward. Most mutations that impair protein function will be eliminated because, if the function of the nonmutant protein contributes to survival and reproduction, the individuals carrying these mutations will leave fewer offspring than others. Mutations that do not impair function may remain in the population for long periods because their carriers survive and reproduce in normal numbers; a mutation of this type has no tendency to either increase or decrease in frequency over time. In contrast, individuals that carry the occasional mutation that improves protein function will survive and reproduce more successfully than others. Because of the enhanced reproduction, the mutant gene encoding the improved protein will gradually increase in frequency and spread throughout the entire population.

In the word game, any of the mutants in red may persist in the population. Suppose that one of them, GENE, is actually superior to GONE (considered more euphonious, perhaps). Then GENE will gradually displace GONE, and eventually GONE will be gone. Similar to the way in which one meaningful word may replace another, one amino acid sequence may be replaced with a different one over the course of evolution.

A real-world example that mirrors the word game is found in the evolution of resistance in the malaria parasite to the drug pyrimethamine. This drug inhibits an enzyme known as dihydrofolate reductase, which the parasite needs to survive and reproduce inside red blood cells. Resistance to pyrimethamine is known to have evolved through a stepwise sequence of four amino acid replacements. In the first replacement, serine (S) at the 108th amino acid in the polypeptide sequence (position 108) was replaced with asparagine (N); then cysteine (C) at position 59 was replaced with arginine (R); asparagine (N) at position 51 was then replaced with isoleucine (I); and finally, isoleucine (I) at position 164 was replaced with leucine (L). If we list the amino acids according to their single-letter abbreviation in the order of their occurrence in the protein, the evolution of resistance followed this pathway:

$$NCSI \rightarrow NCNI \rightarrow NRNI \rightarrow IRNI \rightarrow IRNL$$

where the mutant amino acids are shown in red. Each successive amino acid replacement increased the level of resistance so that a greater concentration of drug was needed to treat the disease. The quadruple mutant IRNL is resistant to such high levels of pyrimethamine that the drug is no longer useful.

Depicted according to stepwise amino acid replacements, the analogy between the evolution of pyrimethamine resistance and the word game is clear. It should also be clear from our earlier discussion that hundreds of other mutations causing amino acid replacements in the enzyme must have occurred in the parasite during the course of evolution, but only these amino acid changes occurring in this order persisted and increased in frequency because they conferred greater survival and reproduction of the parasite under treatment with the drug.

All of the steps in gene expression, including transcription and translation, are summarized in **Fig. 4.21**.

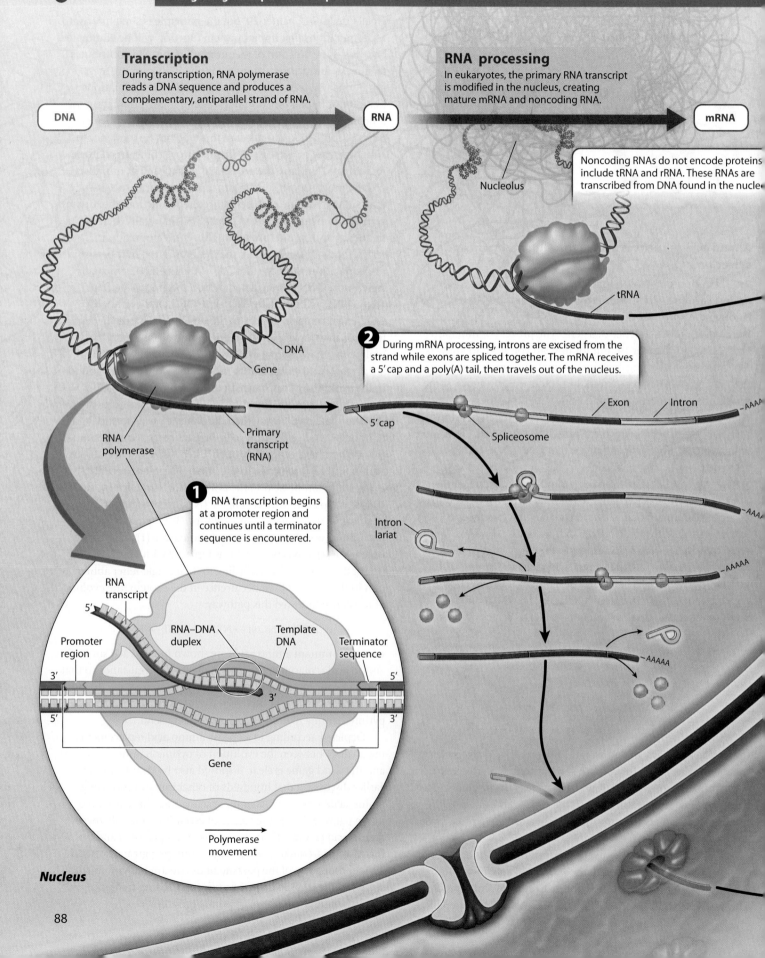

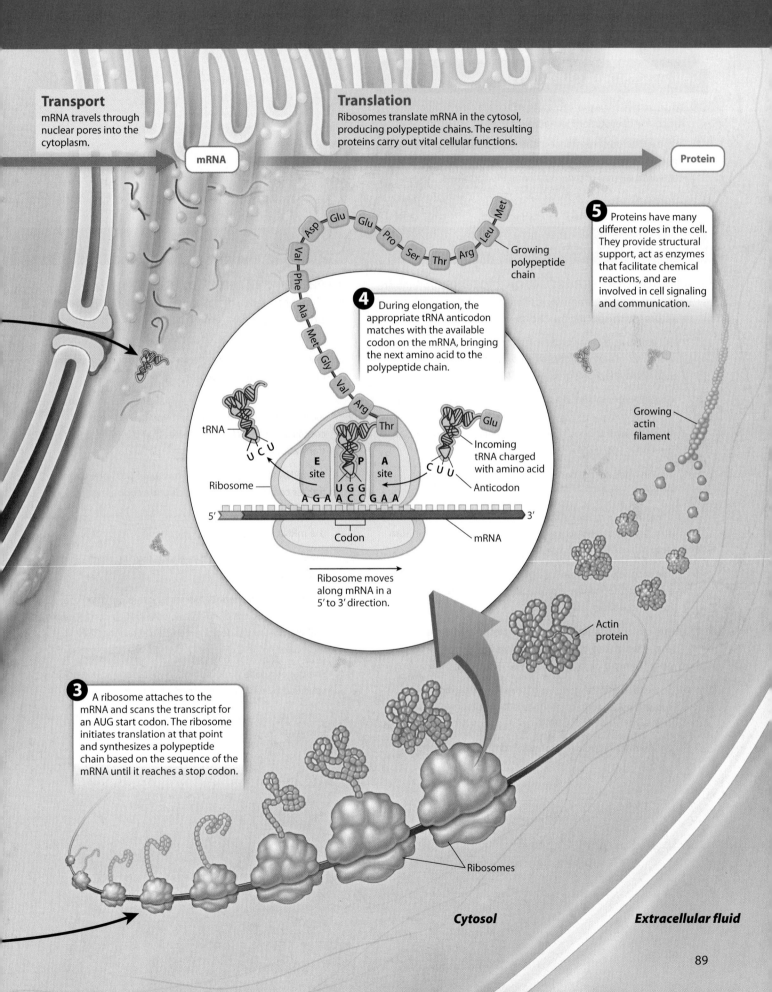

Self-Assessment Questions

10. What are two ways that proteins can acquire new functions in the course of evolution? Explain each one.

11. In the evolution of resistance in the malaria parasite to pyrimethamine, what do you think happened to the mutations that decreased survival or reproduction of the parasites?

CORE CONCEPTS SUMMARY

4.1 MOLECULAR STRUCTURE OF PROTEINS: Proteins are linear polymers of amino acids that form three-dimensional structures with specific functions.

An amino acid consists of an α carbon connected by covalent bonds to an amino group, a carboxyl group, a hydrogen atom, and a side chain or R group. page 71

The 20 common amino acids differ in their R groups. Amino acids are categorized by the chemical properties of their R groups—hydrophobic, basic, acidic, polar—and by special structures. page 72

Amino acids are connected by peptide bonds to form proteins. page 73

The primary structure of a protein is its amino acid sequence. This structure determines how a protein folds, which in turn determines how it functions. page 74

The secondary structure of a protein results from the interactions of nearby amino acids. Examples include the α helix and β sheet. page 74

The tertiary structure of a protein is its three-dimensional shape, which results from long-range interactions of amino acid R groups. page 76

Some proteins are made up of several polypeptide subunits; this group of subunits is the protein's quaternary structure. page 77

Chaperones help some proteins fold properly. page 77

4.2 PROTEIN SYNTHESIS: Translation is the process by which the sequence of bases in messenger RNA specifies the order of successive amino acids in a newly synthesized protein.

Translation requires many cellular components, including ribosomes, tRNAs, and proteins. page 80

Ribosomes are composed of a small and a large subunit, each consisting of RNA and protein; the large subunit contains three tRNA-binding sites that play different roles in translation. page 80

An mRNA transcript of a gene has three possible reading frames composed of three-nucleotide codons. page 80

tRNAs have an anticodon that base pairs with the codon in the mRNA and carries a specific amino acid. page 80

Aminoacyl tRNA synthetases attach specific amino acids to tRNAs. page 81

The genetic code defines the relationship between the three-letter codons of nucleic acids and their corresponding amino acids. It was deciphered using synthetic RNA molecules. page 82

The genetic code is redundant, in that many amino acids are specified by more than one codon. page 82

Translation consists of three steps: initiation, elongation, and termination. page 84

4.3 PROTEIN ORIGINS AND EVOLUTION: Proteins evolve through mutation and selection and by combining functional units.

- A protein family is a group of proteins that are structurally and functionally related. page 86

- There are far fewer protein families than the total number of possible proteins because the probability that a random sequence of amino acids will fold properly to carry out a specific function is very small. page 86

- A region of a protein that folds in a particular way and that carries out a specific function is called a folding domain. page 86

- Proteins evolve by combining different folding domains. page 87

- Proteins also evolve through changes in their amino acid sequence, which occurs by mutation and selection. page 87

Log in to **LaunchPad** to check your answers to the Self-Assessment Questions and to access additional learning tools.

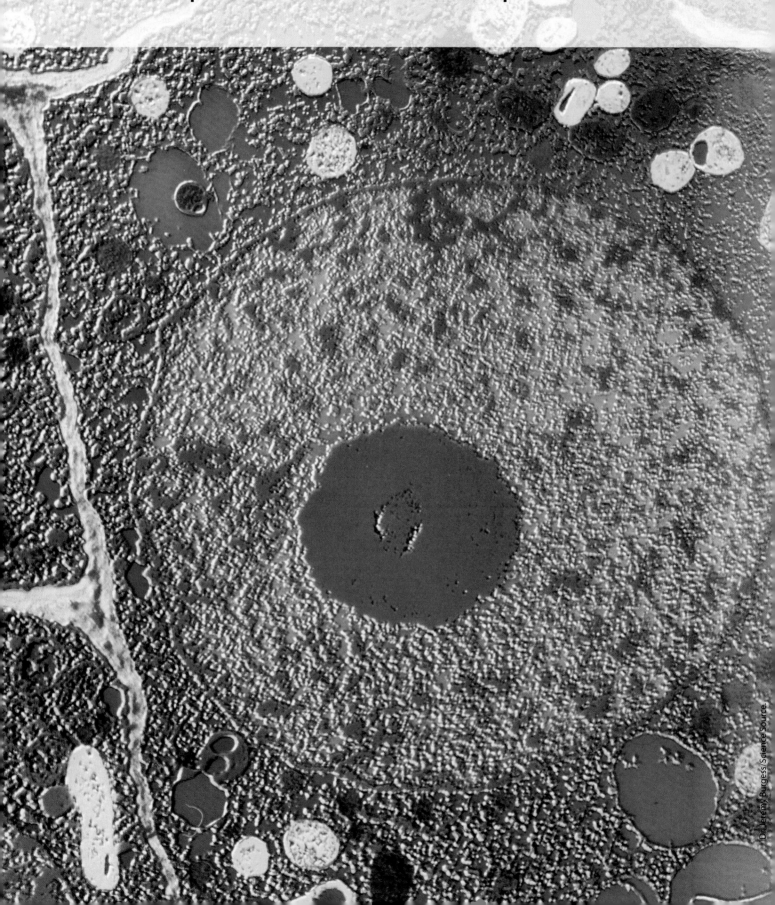

CHAPTER 5   Organizing Principles
Lipids, Membranes, and Cell Compartments

CORE CONCEPTS

5.1 STRUCTURE OF CELL MEMBRANES: Cell membranes are composed of lipids, proteins, and carbohydrates.

5.2 MOVEMENT IN AND OUT OF CELLS: The plasma membrane is a selective barrier that controls the movement of molecules between the inside and the outside of the cell.

5.3 INTERNAL ORGANIZATION OF CELLS: Cells can be classified as prokaryotes or eukaryotes, which differ in their degree of internal compartmentalization.

5.4 THE ENDOMEMBRANE SYSTEM: The endomembrane system is an interconnected system of membranes that defines spaces in the cell, synthesizes important molecules, and traffics and sorts these molecules in and out of the cell.

5.5 MITOCHONDRIA AND CHLOROPLASTS: Mitochondria and chloroplasts are organelles that harness energy, and likely evolved from free-living prokaryotes.

"With the discovery of the cell, biologists found their atom." So stated François Jacob, the French biologist who shared the Nobel Prize in Physiology or Medicine in 1965. Just as the atom is the smallest, most basic unit of matter, the cell is the smallest, most basic unit of living organisms. All organisms, from single-celled algae to complex multicellular organisms like humans, are made up of cells. Therefore, essential properties of life, including growth, reproduction, and metabolism, must be understood in terms of cell structure and function.

Cells were first seen sometime around 1665, when the English scientist Robert Hooke built a microscope that he used to observe thin sections of dried cork tissue derived from plants. In these sections, Hooke observed arrays of small cavities and named them "cells" (**Fig. 5.1**). Although Hooke was probably looking at the cell walls of empty (rather than living) cells, his observations nevertheless led to the concept that cells are the fundamental unit of life. Later in the seventeenth century, the Dutch microbiologist Anton van Leeuwenhoek greatly improved the magnifying power of microscope lenses, enabling him to see and describe unicellular organisms, including bacteria, protists, and algae. Today, modern microscopy provides unprecedented detail and a deeper understanding of the inner architecture of cells.

FIG. 5.1 The first observation of cells. Robert Hooke used a simple microscope to observe small chambers in a sample of cork tissue that he described as "cells."

Sources: (left) Science Museum/SSPL/The Image Works; (right) Ted Kinsman/Science Source.

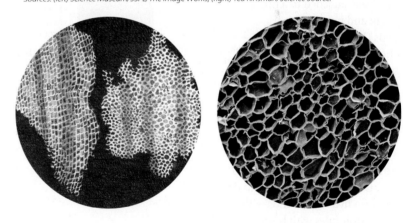

Cells differ in size and shape, but they share many features. This similarity in the microscopic organization of all living organisms led to the development in the middle of the nineteenth century of one of the pillars of modern biology: **cell theory**. Based on the work and ideas of Matthias Schlieden, Theodor Schwann, Rudolf Virchow, and others, cell theory states that all organisms are made up of cells, that the cell is the fundamental unit of life, and that cells come from preexisting cells. There is no life without cells, and the cell is the smallest unit of life.

This chapter focuses on cells and their internal organization. We pay particular attention to the key role that membranes play in separating a cell from the external environment and in defining structural and functional spaces within cells.

5.1 STRUCTURE OF CELL MEMBRANES

Cells are defined by membranes. Membranes physically separate cells from their external environment and define spaces within many cells that allow them to carry out their diverse functions.

Lipids are the main component of cell membranes. They have properties that allow them to form a barrier in an aqueous (watery) environment. Membranes are not, however, made up of only lipids. Proteins are often embedded in or associated with the membrane, where they perform important functions such as transporting molecules. Carbohydrates can also be found in cell membranes, usually attached to lipids (glycolipids) and proteins (glycoproteins).

Cell membranes are composed of two layers of lipids.

The major types of lipids found in cell membranes are phospholipids, which were introduced in Chapter 2. Most phospholipids are made up of a glycerol backbone attached to a phosphate group and two fatty acids (**Fig. 5.2**). The phosphate head group is hydrophilic ("water-loving") because it is polar, enabling it to form hydrogen bonds with water. By contrast, the two fatty acid tails are hydrophobic ("water-fearing") because they are nonpolar and do not form hydrogen bonds with water. Molecules with both hydrophilic and hydrophobic regions in a single molecule are termed **amphipathic**.

In an aqueous environment, amphipathic molecules such as phospholipids behave in an interesting way. Namely, they spontaneously arrange themselves into various structures in which the polar head groups on the outside interact with water and the nonpolar tail groups come together on the inside away from water. This arrangement results from the tendency of polar molecules like water to exclude nonpolar molecules or nonpolar groups of molecules.

The shape of the structure is determined by the bulkiness of the head group relative to the hydrophobic tails. For example, lipids with bulky heads and a single hydrophobic fatty acid tail are wedge-shaped and pack into spherical structures called **micelles** (**Fig. 5.3a**). By contrast, lipids with less bulky head groups and two hydrophobic tails are roughly rectangular and form a **bilayer** (**Fig. 5.3b**). A lipid bilayer is a structure formed of two layers of lipids in which the hydrophilic heads are the

FIG. 5.2 Phospholipid structure. Phospholipids, the major component of cell membranes, are made up of glycerol attached to a phosphate-containing head group and two fatty acid tails. They are amphipathic because they have both hydrophilic and hydrophobic domains.

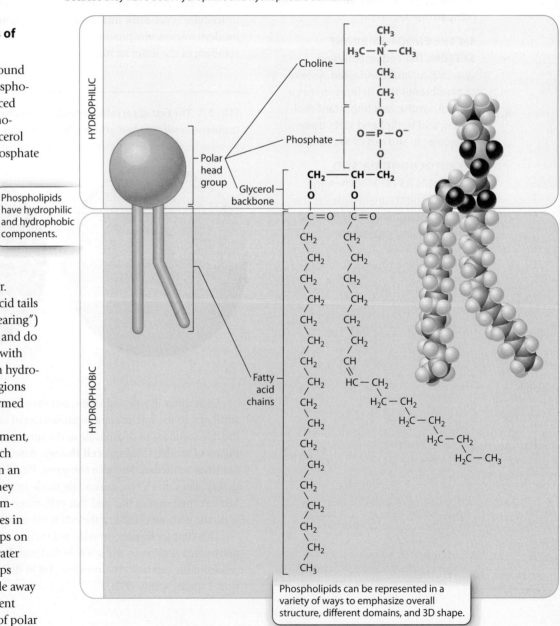

Phospholipids have hydrophilic and hydrophobic components.

Phospholipids can be represented in a variety of ways to emphasize overall structure, different domains, and 3D shape.

FIG. 5.3 Phospholipid structures. Phospholipids can form (a) micelles, (b) bilayers, or (c) liposomes when placed in water.

a. Micelle

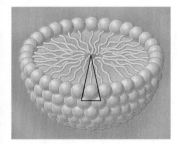

b. Bilayer

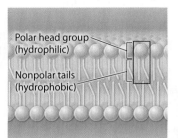

Polar head group (hydrophilic)

Nonpolar tails (hydrophobic)

c. Liposome

outside surfaces of the bilayer and the hydrophobic tails are sandwiched in between, isolated from contact with the aqueous environment.

The bilayers form closed structures with an inner space because free edges would expose the hydrophobic chains to the aqueous environment. This organization in part explains why bilayers are effective cell membranes. It also explains why membranes are self-healing. Small tears in a membrane are rapidly sealed by the spontaneous rearrangement of the lipids surrounding the damaged region because of the tendency of water to exclude nonpolar molecules.

 CASE 1 LIFE'S ORIGINS: INFORMATION, HOMEOSTASIS, AND ENERGY

How did the first cell membranes form?

What are the consequences of the ability of phospholipids to form a bilayer when placed in water? The bilayer structure forms spontaneously, dependent solely on the properties of the phospholipid and without the action of an enzyme, as long as the concentration of free phospholipids is high enough and the pH of the solution is similar to that of a cell. The pH is important because it ensures that the head groups are in their ionized (charged) form and suitably hydrophilic. Thus, if phospholipids are added to a test tube of water at neutral pH, they spontaneously form spherical bilayer structures called **liposomes** that surround a central space (**Fig. 5.3c**). As the liposomes form, they may capture macromolecules present in solution.

Such a process may have been at work in the early evolution of life on Earth. Experiments show that liposomes can form, break, and re-form in environments such as tidal flats that are repeatedly dried and flooded with water. The liposomes can even grow, incorporating more and more lipids from the environment, and capture nucleic acids and other molecules in their interiors. Depending on their chemical composition, early membranes might have been either leaky or almost impervious to the molecules of life. Over time, they evolved in such a way as to allow at least limited molecular traffic between the environment and the cell interior. At some point, new lipids no longer had to be incorporated from the environment. Instead, proteins guided lipid synthesis within the cell, although how this switch to protein-mediated synthesis happened remains uncertain.

All evidence suggests that membranes formed originally by straightforward physical processes, but that their composition and function evolved over time. François Jacob once said that evolution works more like a tinkerer than an engineer, modifying already existing materials rather than designing systems from scratch. It seems that the evolution of membranes is no exception to this pattern.

Cell membranes are dynamic.

Lipids freely associate with one another because of the extensive van der Waals forces between their fatty acid tails (Chapter 2). These weak interactions are easily broken and re-formed, so lipid molecules are able to move within the plane of the membrane, sometimes very rapidly: a single phospholipid can move across the entire length of a bacterial cell in less than 1 second. Lipids can also rapidly rotate around their vertical axis, and individual fatty acid chains are able to flex, or bend. As a result, membranes are dynamic: they are continually moving, forming, and re-forming during the lifetime of a cell.

Because membrane lipids are able to move in the plane of the membrane, the membrane is said to be **fluid**. The degree of membrane fluidity depends on which types of lipid make up the membrane. In a single layer of the lipid bilayer, the strength of the van der Waals interactions between the lipids' tails depends on the length of the fatty acid tails and the presence of double bonds between neighboring carbon atoms. The longer the fatty acid tails, the more surface is available to participate

FIG. 5.4 Saturated and unsaturated fatty acids in phospholipids. The composition of cell membranes affects the tightness of packing.

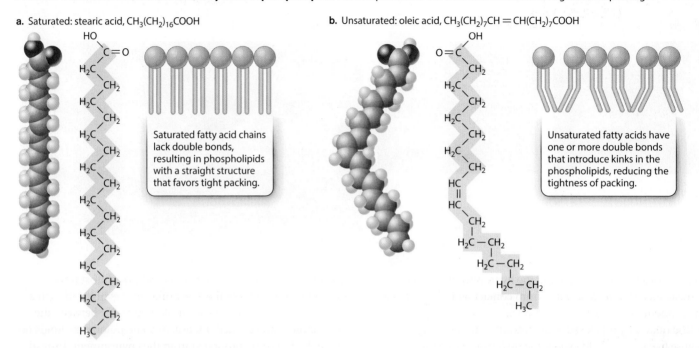

a. Saturated: stearic acid, $CH_3(CH_2)_{16}COOH$

Saturated fatty acid chains lack double bonds, resulting in phospholipids with a straight structure that favors tight packing.

b. Unsaturated: oleic acid, $CH_3(CH_2)_7CH=CH(CH_2)_7COOH$

Unsaturated fatty acids have one or more double bonds that introduce kinks in the phospholipids, reducing the tightness of packing.

in van der Waals interactions. The tighter packing that results tends to reduce lipid mobility. Saturated fatty acid tails, which have no double bonds, are straight and tightly packed—again reducing mobility (**Fig. 5.4a**). The double bonds in unsaturated fatty acids introduce kinks in the fatty acid tails, reducing the tightness of packing and enhancing lipid mobility in the membrane (**Fig. 5.4b**).

In addition to phospholipids, cell membranes often contain other types of lipids, which can also influence membrane fluidity. For example, **cholesterol** is a major component of animal cell membranes, representing approximately 30% by mass of the membrane lipids. Like phospholipids, cholesterol is amphipathic, with both hydrophilic and hydrophobic groups being present in the same molecule. In cholesterol, the hydrophilic region is simply a hydroxyl group (—OH) and the hydrophobic region consists of four interconnected carbon rings with an attached hydrocarbon chain (**Fig. 5.5**). This structure allows a cholesterol molecule to insert itself into the lipid bilayer so that its head group interacts with the hydrophilic head group of phospholipids, while the ring structure participates in van der Waals interactions with the fatty acid chains.

FIG. 5.5 Cholesterol in the lipid bilayer. Cholesterol molecules embedded in the lipid bilayer affect the fluidity of the membrane.

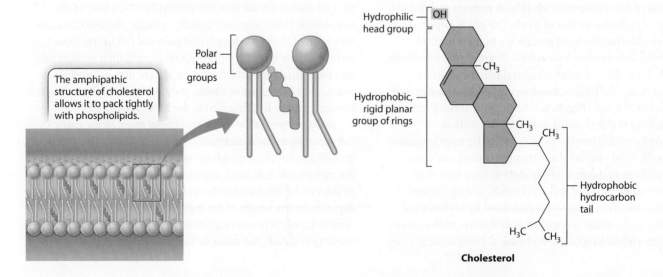

FIG. 5.6 Functions of membrane proteins.

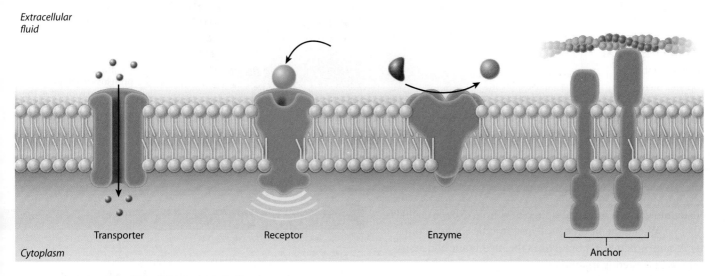

Cholesterol increases or decreases membrane fluidity depending on the temperature. At the temperatures typically found in a cell, cholesterol decreases membrane fluidity because the interaction of the rigid ring structure of cholesterol with the phospholipid fatty acid tails reduces the mobility of the phospholipids. However, at low temperatures, cholesterol increases membrane fluidity because it prevents phospholipids from packing tightly with other phospholipids. Thus, cholesterol helps maintain a consistent state of membrane fluidity by preventing dramatic transitions from a fluid to a solid state.

For many decades, it was thought that the various types of lipids found in the membrane were randomly distributed throughout the bilayer. More recent studies show that specific types of lipids, such as sphingolipids, sometimes assemble into defined patches called **lipid rafts.** Cholesterol and other membrane components such as proteins also appear to accumulate in some of these regions. Thus, membranes are not always a uniform fluid bilayer, but instead can contain regions with discrete components.

Although lipids are free to move in the plane of the membrane, the spontaneous transfer of a lipid between layers of the bilayer, known as lipid flip-flop, is very rare. This is not surprising because flip-flop requires the hydrophilic head group to pass through the hydrophobic interior of the membrane. As a result, little exchange of components occurs between the two layers of the membrane, which in turn allows the two layers to differ in composition. In fact, in many membranes, different types of lipids are present primarily in one layer or the other.

Proteins associate with cell membranes in different ways.

Most membranes contain proteins as well as lipids. For example, proteins represent as much as 50% by mass of the membrane of a red blood cell. Membrane proteins serve different functions (**Fig. 5.6**). Some act as **transporters,** moving ions or other molecules across the membrane. Other membrane proteins act as **receptors,** which allow the cell to receive signals from the environment. Still others are **enzymes** that catalyze chemical reactions or **anchors** that attach to other proteins and help to maintain cell structure and shape.

The various membrane proteins can be classified into two groups depending on how they associate with the membrane (**Fig. 5.7**). **Integral membrane proteins** are permanently associated with cell membranes and cannot be separated from the membrane experimentally without destroying the membrane itself. **Peripheral membrane proteins** are temporarily associated with the lipid bilayer or with integral membrane proteins through weak noncovalent interactions. They are easily separated from the membrane by simple experimental procedures that leave the structure of the membrane intact.

Most integral membrane proteins are **transmembrane proteins** that span the entire lipid bilayer, as shown in Fig. 5.7. These proteins are composed of three regions: two hydrophilic regions, one protruding from each face of the membrane, and a connecting hydrophobic region that spans the membrane. This structure allows for separate functions and capabilities of each end of the protein. For example, the hydrophilic region on the external side of a receptor can interact with signaling molecules, whereas the hydrophilic region of that receptor protein on the internal side of the membrane often interacts with other proteins in the cytoplasm of the cell to pass along the message.

Peripheral membrane proteins may be associated with either the internal or external side of the membrane (Fig. 5.7). These proteins interact either with the polar heads of lipids or with integral membrane proteins by weak noncovalent interactions such as hydrogen bonds. Peripheral membrane proteins are only transiently associated with the membrane and can play a

FIG. 5.7 Integral and peripheral membrane proteins. Integral membrane proteins are permanently associated with the membrane. Peripheral membrane proteins are temporarily associated with one or other of the two lipid bilayers or with an integral membrane protein.

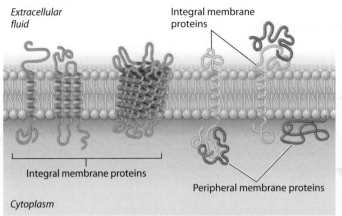

role in transmitting information received from external signals. Other peripheral membrane proteins limit the ability of transmembrane proteins to move within the membrane and assist proteins in clustering in lipid rafts.

Proteins, like lipids, are free to move in the membrane. How do we know this? The mobility of proteins in the cell membrane can be demonstrated using an elegant experimental technique called fluorescence recovery after photobleaching (FRAP), described in **Fig. 5.8**.

The idea that lipids and proteins coexist in the membrane, and that both are able to move in the plane of the membrane, led American biologists S. Jonathan Singer and Garth Nicolson to propose the **fluid mosaic model** in 1972. According to this model, the lipid bilayer is a fluid structure within which molecules move laterally, and is a mosaic (a mixture) of two types of molecules, lipids and proteins.

Self-Assessment Questions

1. How do lipids with hydrophilic and hydrophobic regions behave in an aqueous environment?
2. Like cell membranes, many fats and oils are made up in part of fatty acids. Most animal fats (like butter) are solid at room temperature, whereas plant fats (like canola oil) tend to be liquid. Can you predict which type of fat contains saturated fatty acids, and which type contains unsaturated fatty acids?
3. What are two ways in which proteins associate with membranes?
4. What would happen in the FRAP experiment if proteins did not move in the plan of a membrane?

5.2 MOVEMENT IN AND OUT OF CELLS

Phospholipids with embedded proteins make up the membrane surrounding all cells. This membrane, called the **plasma membrane**, is a fundamental, defining feature of all

HOW DO WE KNOW?

FIG. 5.8

Do proteins move in the plane of the membrane?

BACKGROUND Fluorescent recovery after photobleaching (FRAP) is a technique used to measure mobility of molecules in the plane of the membrane. A fluorescent dye is attached to proteins embedded in the cell membrane in a process called labeling. Labeling all the proteins in a membrane creates a fluorescent cell that can be visualized with a fluorescence microscope. A laser is then used to bleach a small area of the membrane, leaving a nonfluorescent spot on the surface of the cell.

HYPOTHESIS If membrane components such as proteins move in the plane of the membrane, the bleached spot should become fluorescent over time as unbleached fluorescent molecules move into the bleached area. If membrane components do not move, the bleached spot should remain intact.

EXPERIMENT AND RESULTS

Over time, fluorescence appears in the bleached area, telling us that fluorescent proteins that were not bleached moved into the bleached area.

CONCLUSION The gradual recovery of fluorescence in the bleached area indicates that proteins move in the plane of the membrane.

SOURCE Peters, R., et al. 1974. "A Microfluorimetric Study of Translational Diffusion in Erythrocyte Membranes." *Biochimica et Biophysica Acta* 367:282–294.

PHOTO SOURCE: FRAP of cytoplasmic EGFP in living HeLa cells, performed using an UltraVIEW® spinning disk confocal system (PerkinElmer Inc.). HeLa cells were transfected with pEGFP-C1 (Clontech Laboratories, Inc.) using GeneJuice® transfection reagent (Novagen®). A region of interest in the cytoplasm was photobleached using the UltraVIEW® photokinesis unit, and the recovery of fluorescence in this region was observed. (Fluorescence Recovery After Photobleaching (FRAP) using the UltraVIEW PhotoKinesis accessory, PerkinElmer Technical Note).

cells. It is the boundary that defines the space of the cell, separating its internal contents from the surrounding environment. But the plasma membrane is not simply a passive boundary or wall. Instead, it serves an active and important function. The environment outside the cell is constantly changing. In contrast, the internal environment of a cell operates within a narrow window of conditions, such as a particular pH range or salt concentration. It is the plasma membrane that actively maintains intracellular conditions compatible with life.

The cells of many organisms also have a **cell wall** external to the plasma membrane. The cell wall plays an important role in maintaining the shape of these cells. In this section, we consider the functions performed by these two key structures.

The plasma membrane maintains homeostasis.

The active maintenance of a constant environment within the cells, known as **homeostasis,** is a critical attribute of cells and of life itself. Chemical reactions and protein folding, for example, are carried out efficiently only within a narrow range of conditions. How does the plasma membrane maintain homeostasis? The answer is that it is **selectively permeable** (or semipermeable). This means that the plasma membrane lets some molecules in and out freely; it lets others in and out only under certain conditions; and it prevents still other molecules from passing through at all.

The membrane's ability to act as a selective barrier is the result of the combination of lipids and embedded proteins of

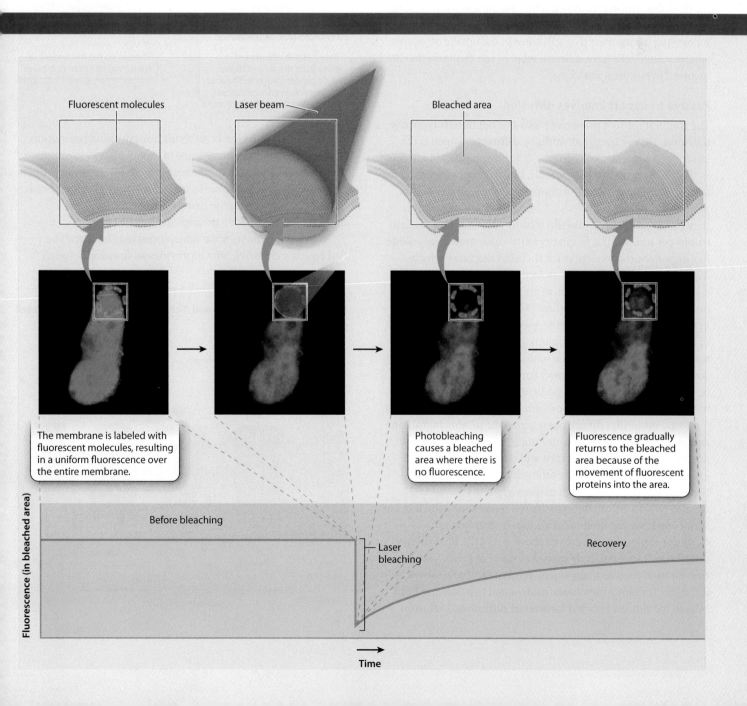

which it is composed. The hydrophobic interior of the lipid bilayer prevents ions and charged polar molecules from moving across it. Furthermore, many macromolecules such as proteins and polysaccharides are too large to cross the plasma membrane on their own. By contrast, gases such as oxygen and carbon dioxide, and nonpolar molecules such as lipids, can move across the lipid bilayer. Small uncharged polar molecules, such as water, are able to move through the lipid bilayer to a very limited extent, but this movement is not biologically significant. However, protein channels and transporters in the membrane can greatly facilitate the movement of molecules, including ions, water, and nutrients, that cannot cross the lipid bilayer on their own.

The identity and abundance of these membrane-associated proteins vary among cell types, reflecting the specific functions of different cells. For example, cells in your gut contain membrane transporters that specialize in the uptake of glucose, whereas nerve cells have different types of ion channels that are involved in electrical signaling.

Passive transport involves diffusion.

The simplest form of movement into and out of cells is passive transport. Passive transport works by **diffusion**, which is the random movement of molecules. Molecules are always moving in their environments. For example, molecules in water at room temperature move around at about 500 m/sec, which means that they can move only about 3 molecular diameters before they run into another molecule, leading to about 5 trillion collisions per second. The frequency with which molecules collide has important consequences for chemical reactions, which depend on the interaction of molecules (Chapter 6).

Diffusion leads to a net movement of the substance from one region to another when there is a concentration gradient in the distribution of a molecule, meaning that there are areas of higher and lower concentrations. In this case, diffusion results in net movement of the molecule from an area of higher concentration to an area of lower concentration. When there is no longer a concentration gradient, net movement stops but movement of molecules in both directions continues (**Fig. 5.9**).

Some molecules diffuse freely across the plasma membrane as a result of differences in concentrations between the inside and the outside of a cell. Oxygen and carbon dioxide, for example, move into and out of the cell in this way. Certain hydrophobic molecules, such as triacylglycerols (Chapter 2), are also able to diffuse through the cell membrane, which is not surprising since the lipid bilayer is also hydrophobic.

Some molecules that cannot move across the lipid bilayer directly can move passively toward a region of lower concentration through protein transporters. When a molecule moves by diffusion through a membrane protein and bypasses the lipid bilayer, the process is called **facilitated diffusion**. Diffusion

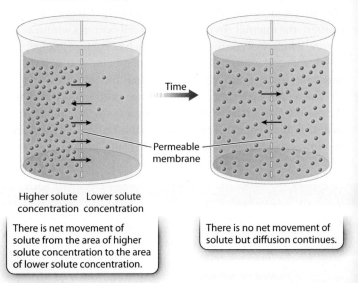

FIG. 5.9 Diffusion, the movement of molecules due to random motion. Net movement of molecules occurs when there are concentration differences.

and facilitated diffusion both result from the random motion of molecules, and net movement of the substance occurs when there are concentration differences (**Fig. 5.10**). In the case of facilitated diffusion, the molecule moves through a membrane transporter, whereas in the case of simple diffusion, the molecule moves directly through the lipid bilayer.

Two types of membrane transporters may be used. The first type is a **channel,** which provides an opening between

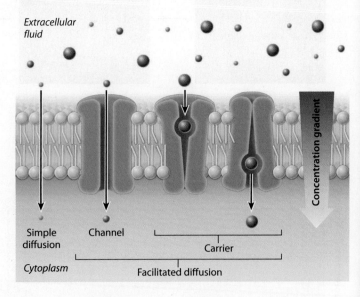

FIG. 5.10 Simple diffusion and facilitated diffusion through the cell membrane.

the inside and outside of the cell through which certain molecules can pass, depending on their shape and charge. Some membrane channels are gated, which means that they open in response to some sort of signal, which may be chemical or electrical (Chapter 9). The second type of membrane transporter is a **carrier**, which binds to and then transports specific molecules. Membrane carriers exist in two conformations: one that is open to one side of the cell, and another that is open to the other side of the cell. Binding of the transported molecule induces a conformational change in the membrane protein, allowing the molecule to be transported across the lipid bilayer, as shown on the right in Fig. 5.10.

Up to this point, we have focused our attention on the movement of molecules (the solutes) in water (the solvent). We now take a different perspective and focus instead on water movement. Water moves into and out of cells by passive transport. Although the central part of the phospholipid bilayer is hydrophobic, water molecules are small enough to move passively through the bilayer to a very limited extent by simple diffusion. However, this movement does not appear to be biologically important. Instead, water moves through the plasma membrane by protein channels called **aquaporins**. These channels allow water to move much more readily across the plasma membrane by facilitated diffusion than is possible by simple diffusion.

The net movement of a solvent such as water across a selectively permeable membrane such as the plasma membrane is known as **osmosis**. As in any form of diffusion, water moves from regions of higher *water* concentration to regions of lower *water* concentration (**Fig. 5.11**). Because water is a solvent within which nutrients such as glucose or ions such as sodium or potassium are dissolved, water concentration decreases as solute concentration increases. Therefore, it is sometimes easier to think about water moving from regions of lower *solute* concentration toward regions of higher *solute* concentration. Either way, the direction of water movement is the same.

During osmosis, the net movement of water toward the side of the membrane with higher solute concentration continues until a concentration gradient no longer exists or until the movement is opposed by another force. This force could be pressure due to gravity (in the case of Fig. 5.11) or the cell wall (in the case of plants, fungi, and bacteria, as described later in this chapter). Osmosis can therefore be prevented by applying a force to the compartment with the higher solute concentration. The pressure that would need to be applied to stop water from moving into a solution by osmosis is referred to as the **osmotic pressure**. Osmotic pressure is a way of describing the tendency of a solution to "draw water in" by osmosis (Chapter 39).

Primary active transport uses the energy of ATP.

Passive transport works to the cell's advantage only if the concentration gradient is in the right direction, from higher on the outside to lower on the inside for nutrients that the cell needs to take in, and from higher on the inside and lower on the outside for wastes that the cell needs to export. However, many of the molecules that cells require are not highly concentrated in the environment. Although some of these molecules can be synthesized by the cell, others must be taken up from the environment. In other words, cells have to move these substances from areas of lower concentration to areas of higher concentration. The "uphill" movement of substances against a concentration gradient, called **active transport**, requires energy. The transport of many kinds of molecules across membranes requires energy, either directly or indirectly. In fact, most of the energy used by a cell goes into keeping the inside of the cell different from the outside, a function carried out by proteins in the plasma membrane.

During active transport, cells move substances through transport proteins embedded in the cell membrane. Some of these proteins act as pumps, using energy directly to move a substance into or out of a cell. A good example is the sodium–potassium pump (**Fig. 5.12**). Within cells, sodium is kept at concentrations much lower than in the exterior environment; the opposite is true of potassium. Therefore, both sodium and potassium have to be moved against a concentration gradient. The sodium–potassium pump actively moves sodium out of

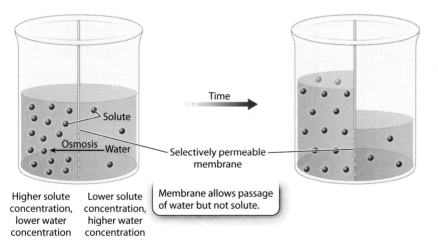

FIG. 5.11 Osmosis. Osmosis is the net movement of a solvent such as water across a selectively permeable membrane from an area of lower solute concentration to an area of higher solute concentration.

FIG. 5.12 Primary active transport. The sodium–potassium pump is a membrane protein that uses the energy stored in ATP to move sodium and potassium ions against their concentration gradients.

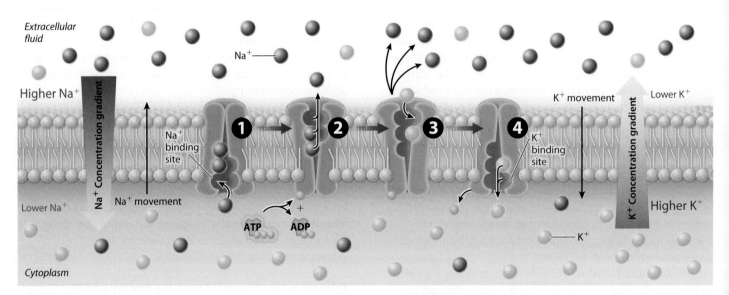

the cell and potassium into the cell. This movement of ions takes energy, which comes from the chemical energy stored in adenosine triphosphate (ATP). Active transport that uses energy directly in this manner is called **primary active transport**. Note that the sodium ions and potassium ions move in opposite directions; protein transporters that work in this way are referred to as antiporters. Other transporters move two molecules in the same direction, and are referred to as symporters or cotransporters.

Secondary active transport is driven by an electrochemical gradient.

Active transport can work in yet another way. Because small ions cannot cross the lipid bilayer, many cells use a transport protein to build up the concentration of a small ion on one side of the membrane. The resulting concentration

FIG. 5.13 Secondary active transport. Protons are pumped across a membrane by (a) primary active transport, resulting in (b) an electrochemical gradient, which drives (c) the movement of another molecule against its concentration gradient.

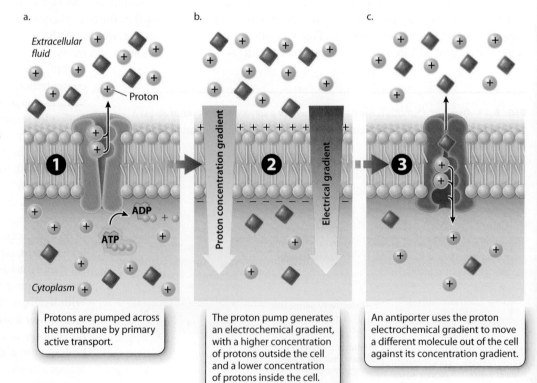

Protons are pumped across the membrane by primary active transport.

The proton pump generates an electrochemical gradient, with a higher concentration of protons outside the cell and a lower concentration of protons inside the cell.

An antiporter uses the proton electrochemical gradient to move a different molecule out of the cell against its concentration gradient.

gradient stores potential energy that can be harnessed to drive the movement of other substances across the membrane against their concentration gradient.

For example, some cells actively pump protons (H^+) across the cell membrane using ATP (**Fig. 5.13a**). As a result, in these cells the concentration of protons is higher on one side of the membrane and lower on the other side. In other words, the pump generates a concentration gradient, also called a chemical gradient because the entity forming the gradient is a chemical (**Fig. 5.13b**). We have already seen that concentration differences favor the movement of protons back to the other side of the membrane. By blocking the movement of protons to the other side, the lipid bilayer creates a store of potential energy, just as a dam or battery does.

In addition to the chemical gradient, another force favors the movement of protons back across the membrane: a difference in charge. Because protons carry a positive charge, the side of the membrane with more protons is more positive than the other side. This difference in charge is called an electrical gradient. Protons (and other ions) move from areas of like charge to areas of unlike charge, driven by an electrical gradient. A gradient that has both charge and chemical components is known as an **electrochemical gradient** (Fig. 5.13b).

If protons are then allowed to pass through the cell membrane by a transport protein, they will move down their electrochemical gradient toward the region of lower proton concentration. These transport proteins can use the movement of protons to drive the movement of other molecules against their concentration gradient (**Fig. 5.13c**). The movement of protons is always from regions of higher to lower concentration, whereas the movement of the coupled molecule is from regions of lower to higher concentration. Because the movement of the coupled molecule is driven by the movement of protons and not by ATP directly, this form of transport is called s**econdary active transport.** Secondary active transport uses the potential energy of an electrochemical gradient to drive the movement of molecules; by contrast, primary active transport uses the chemical energy of ATP directly.

The use of an electrochemical gradient as a temporary energy source is a common cellular strategy. For example, cells use a sodium electrochemical gradient generated by the sodium–potassium pump to transport glucose and amino acids into cells. In addition, cells use a proton electrochemical gradient to move other molecules and, as we discuss later in this chapter and in Chapter 7, to synthesize ATP.

Many cells maintain size and composition using active transport.

Many cells use active transport to maintain their size. Consider human red blood cells placed in a variety of solutions (**Fig. 5.14**). If a red blood cell is placed in a hypertonic solution

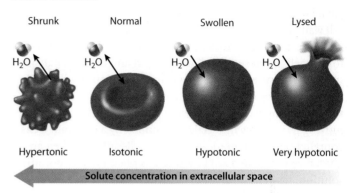

FIG. 5.14 Changes in red blood cell shape due to osmosis. Red blood cells shrink, swell, or burst because of net water movement driven by differences in solute concentration between the inside and the outside of the cell.

(one with a higher solute concentration than that inside the cell), water leaves the cell by osmosis and the cell shrinks. By contrast, if a red blood cell is placed in a hypotonic solution (one with a lower solute concentration than that inside the cell), water moves into the cell by osmosis and the cell lyses, or bursts. Animal cells solve the problem of water movement in part by keeping the intracellular fluid isotonic (that is, at the same solute concentration) with the extracellular fluid. Cells use the active transport of ions to maintain equal concentrations inside and out, and the sodium–potassium pump plays an important role in keeping the inside of the cell isotonic with the extracellular fluid.

Human red blood cells avoid shrinking or bursting by maintaining an intracellular environment isotonic with the extracellular environment, meaning the blood. But what about a single-celled organism, like *Paramecium*, swimming in a freshwater lake? In this case, the extracellular environment is hypotonic compared with the concentration in the cell's interior. As a result, *Paramecium* faces the risk of bursting from water moving in by osmosis. *Paramecium* and some other single-celled organisms contain **contractile vacuoles** that solve this problem. Contractile vacuoles are compartments that take up excess water from inside the cell and then, by contraction, expel it into the external environment. The mechanism by which water moves into the contractile vacuoles differs depending on the organism. The contractile vacuoles of some organisms take in water through aquaporins, while the contractile vacuoles of other organisms first take in protons through proton pumps, with water following by osmosis.

The cell wall provides another means of maintaining cell shape.

As we have seen, animals often maintain cell size by using active transport to keep the inside and outside of the cell isotonic.

How do organisms such as plants, fungi, and bacteria maintain cell size and shape? A key feature of these organisms is the cell wall surrounding the plasma membrane (**Fig. 5.15**). The cell wall plays a critical role in the maintenance of cell size and shape. When Hooke looked at cork through his microscope, what he saw was not living cells but rather the remains of cell walls.

The cell wall provides structural support and protection for the cell. Because the cell wall is rigid and resists expansion, it allows pressure to build up when water enters a cell. The force exerted by water pressing against an object is called **turgor pressure**. Turgor pressure builds as a result of water moving by osmosis into cells surrounded by a cell wall. Cells contain high concentrations of solutes. Recall that when an animal cell, such as a red blood cell, is placed in a hypotonic solution, it swells until it bursts. By contrast, when a plant cell is placed in a hypotonic solution, water enters the cell by osmosis until the turgor pressure created by the cell wall increases to a level to stop osmosis. Turgor pressure develops because the cell wall resists being stretched and pushes back on the interior of the cell, just as a balloon pushes back on the air inside.

The pressure exerted by water inside the cell on the cell wall provides structural support for many organisms that is similar in function to the support provided by animals' skeletons. In addition, plant and fungal cells have another structure, called the **vacuole** (different from the contractile vacuole), that absorbs water and contributes to turgor pressure (Fig. 5.15). Its functioning explains why plants wilt when dehydrated: the loss of water from the vacuoles reduces turgor pressure, so the cells can no longer maintain their shape within the cell wall. Plant vacuoles have many other functions and are often the most conspicuous feature of plant cells. They also explain in part why plant cells are typically larger than animal cells and can store water, as well as nutrients, ions, and wastes.

The cell wall is made up of many different components, including carbohydrates and proteins. The specific components differ depending on the organism. The plant cell wall is composed of polysaccharides, including cellulose, a polymer of the sugar glucose. Cellulose is the most abundant biological material in nature. Many types of algae have cell walls made up of cellulose, as in plants, but others have cell walls made of silicon or calcium carbonate. Fungi have cell walls made of chitin, another polymer of sugars. In bacteria, the cell wall is made up primarily of peptidoglycan, a polymer of amino acids and sugars.

Self-Assessment Questions

5. What are the roles of lipids and proteins in maintaining the selective permeability of membranes?

6. A container is divided into two compartments by a membrane that is fully permeable to water and small ions. Water is added to one side of the membrane (side A), and a 5% solution of sodium chloride (NaCl) is added to the other (side B). In which direction will water molecules move? In which direction will sodium and chloride ions move? When the concentration is equal on both sides, will diffusion stop?

7. What is the difference between passive and active transport?

8. In the absence of the sodium–potassium pump, the extracellular solution becomes hypotonic relative to the inside of the cell. Poisons such as the snake venom ouabain can interfere with the action of the sodium–potassium pump. What are the consequences for the cell?

9. What are three different ways in which cells maintain size and shape?

5.3 INTERNAL ORGANIZATION OF CELLS

In addition to the plasma membrane, many cells contain membrane-bound regions within which specific functions are carried out. Such a cell can be compared to a large factory with many different rooms and departments. Each department has a specific function and internal organization that contribute to the work of the factory. In this section, we give an overview of two broad classes of cells that can be distinguished by the presence or absence of these membrane-enclosed compartments.

FIG. 5.15 The plant cell wall and vacuoles. The plant cell wall is a rigid structure that maintains the shape of the cell. The pressure exerted by water absorbed by vacuoles also provides support. *Source: Biophoto Associates/Getty Images.*

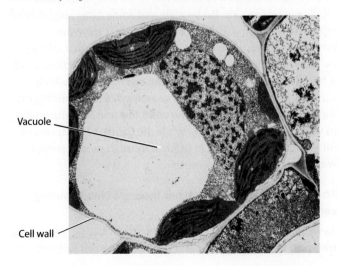

Vacuole

Cell wall

Eukaryotes and prokaryotes differ in internal organization.

All cells have a plasma membrane and contain genetic material. In some cells, this genetic material is housed in a membrane-bound space called the **nucleus**. Cells can be divided into two classes based on the absence or presence of a nucleus (Chapter 1). **Prokaryotes**, including bacteria and archaeons, lack a nucleus; **eukaryotes**, including animals, plants, fungi, and protists, have a nucleus (**Fig. 5.16**). The presence of a nucleus in eukaryotes allows for the processes of transcription and translation to be separated in time and space. This separation, in turn, allows for more complex ways to regulate gene expression than are possible in prokaryotes (Chapter 18).

There are other differences between prokaryotes and eukaryotes. For example, we saw in Chapter 3 that transcription and translation differ in some ways in prokaryotes and eukaryotes. In addition, there are differences in the types of lipids that make up their cell membranes. In mammals, cholesterol is present in cell membranes. Cholesterol belongs to a group of chemical compounds known as sterols, which are molecules containing a hydroxyl group attached to a four-ringed structure. In eukaryotes other than mammals, diverse sterols are synthesized and present in cell membranes. Most prokaryotes do not synthesize sterols, but some synthesize compounds called hopanoids. These five-ringed structures are thought to serve a function similar to that of cholesterol in mammalian cell membranes.

In spite of these and other differences between prokaryotes and eukaryotes, it is the absence or presence of a nucleus that defines the two groups. However, from an evolutionary perspective, archaeons and eukaryotes are more closely related to each other than either are to bacteria. For example, archaeons share with eukaryotes many genes involved in transcription and translation, and DNA is packaged with histones in both groups.

Prokaryotic cells lack a nucleus and extensive internal compartmentalization.

Prokaryotes do not have a nucleus—that is, there is no physical barrier separating the genetic material from the rest of the cell. Instead, the DNA is concentrated in a discrete region of the cell interior known as the **nucleoid**. Bacteria often contain additional small circular molecules of DNA known as **plasmids** that carry a few genes. Plasmids are commonly transferred between bacteria through the action of threadlike structures known as **pili** (singular, **pilus**), which extend from one cell to another. Genes for antibiotic resistance are commonly transferred in this way, which accounts for the quick spread of antibiotic resistance among bacterial populations.

Although the absence of a nucleus is a defining feature of prokaryotes, other features also stand out. For example, prokaryotes are small, typically just 1–2 micrometers (a micrometer is 1/1,000,000 of a meter) in diameter or smaller. By contrast, eukaryotic cells are commonly much larger, on the order of 10 times larger in diameter and 1000 times larger in volume. The small size of prokaryotic cells means that they have a relatively high ratio of surface area to volume, which makes sense for an organism that absorbs nutrients from the environment. In other words, a large amount of membrane surface area is available for absorption relative to the volume of the cell that it serves. In addition, most prokaryotes lack the extensive internal organization characteristic of eukaryotes.

Eukaryotic cells have a nucleus and specialized internal structures.

Eukaryotes are defined by the presence of a nucleus, which houses the vast majority of the cell's DNA. The nuclear membrane allows for more complex regulation of gene expression than is possible in prokaryotic cells (Chapters 3 and 18). In eukaryotes, DNA is transcribed to RNA inside the nucleus, but the RNA molecules carrying the genetic message travel from inside to outside the nucleus, where they instruct the synthesis of proteins.

Eukaryotes have a remarkable internal array of membranes. These membranes define compartments, called

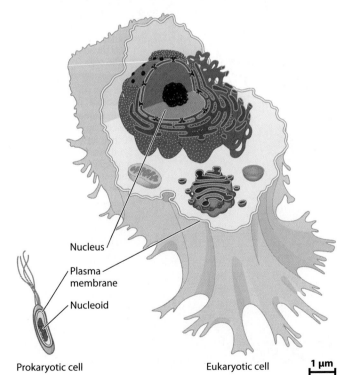

FIG. 5.16 Prokaryotic and eukaryotic cells. Prokaryotic cells lack a nucleus and extensive internal compartmentalization. Eukaryotic cells have a nucleus and extensive internal compartmentalization.

organelles, that divide the cell contents into smaller spaces specialized for different functions. **Fig. 5.17a** shows a macrophage, a type of animal cell, with various organelles and their functions. **Fig. 5.17b** shows a typical plant cell. In addition to the organelles described in Fig. 5.17a, plant cells have a cell wall outside the plasma membrane, vacuoles specialized for water uptake, and **chloroplasts** that convert energy of sunlight into chemical energy. Plant cells also have intercellular connections called plasmodesmata, channels that allow the passage of large molecules such as mRNA and proteins between neighboring cells.

The entire contents of a cell other than the nucleus make up the **cytoplasm**. The jelly-like internal environment of the cell that surrounds the organelles inside the plasma

FIG. 5.17 An animal cell and a plant cell. Animal and plant cells have many cell components in common.

a. Typical features of an animal cell

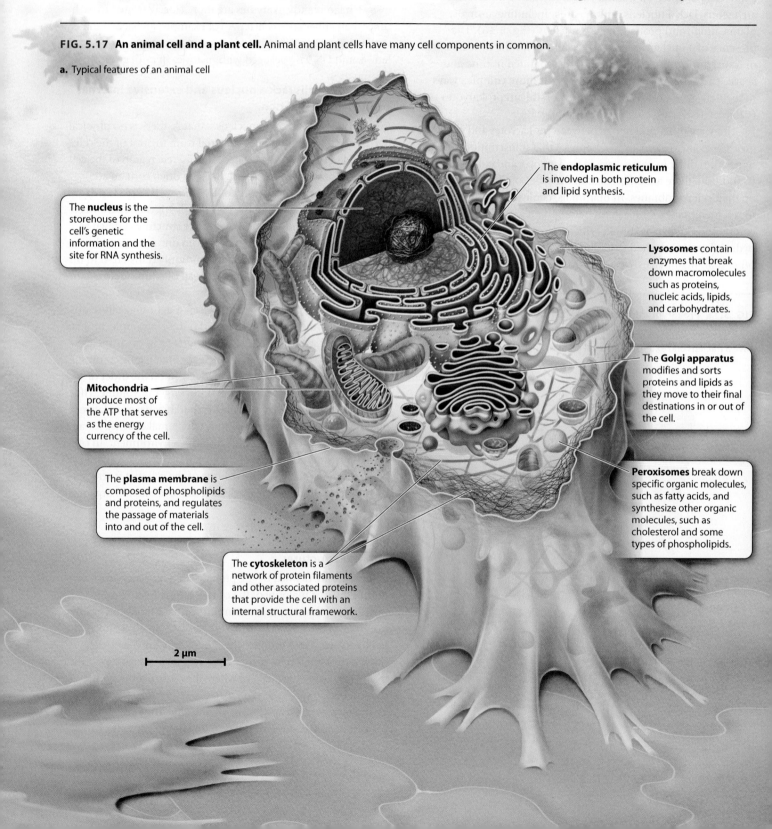

The **nucleus** is the storehouse for the cell's genetic information and the site for RNA synthesis.

The **endoplasmic reticulum** is involved in both protein and lipid synthesis.

Lysosomes contain enzymes that break down macromolecules such as proteins, nucleic acids, lipids, and carbohydrates.

Mitochondria produce most of the ATP that serves as the energy currency of the cell.

The **Golgi apparatus** modifies and sorts proteins and lipids as they move to their final destinations in or out of the cell.

The **plasma membrane** is composed of phospholipids and proteins, and regulates the passage of materials into and out of the cell.

Peroxisomes break down specific organic molecules, such as fatty acids, and synthesize other organic molecules, such as cholesterol and some types of phospholipids.

The **cytoskeleton** is a network of protein filaments and other associated proteins that provide the cell with an internal structural framework.

2 μm

membrane is referred to as the **cytosol**. In the next two sections, we consider these organelles in more detail, focusing on the role of membranes in forming distinct compartments within the cell.

Self-Assessment Question

10. Compare the organization, degree of compartmentalization, and size of prokaryotic and eukaryotic cells.

5.4 THE ENDOMEMBRANE SYSTEM

In eukaryotes, the total surface area of intracellular membranes is about tenfold greater than that of the plasma membrane. This high ratio of internal membrane area to plasma membrane area underscores the significant degree to which a eukaryotic cell is divided into internal compartments. Many of the organelles inside cells are not isolated entities, but instead communicate with one another. In fact, the membranes of these organelles are either physically connected by membrane "bridges" or transiently connected by **vesicles,** small membrane-enclosed sacs that transport substances within a cell or from the interior to the exterior of the cell. These

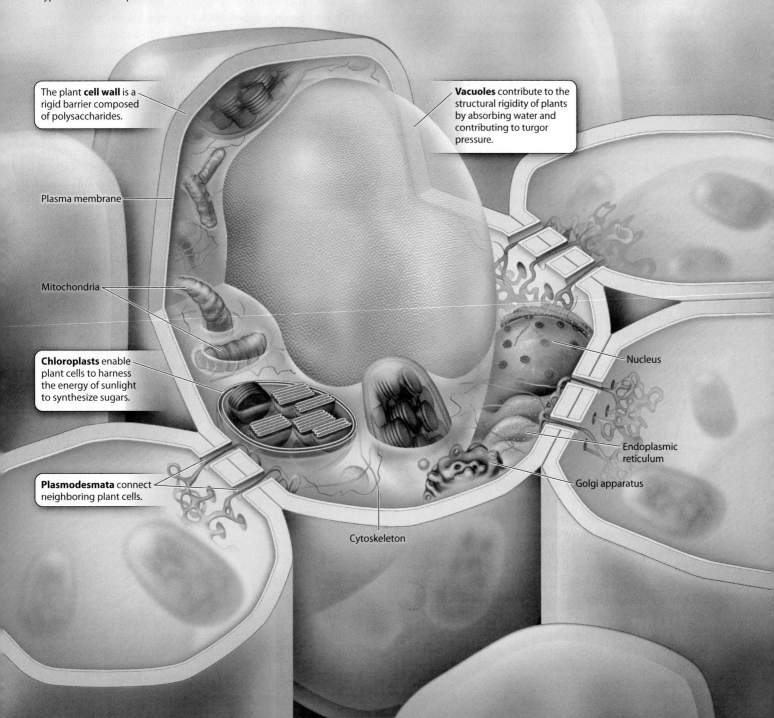

b. Typical features of a plant cell

The plant **cell wall** is a rigid barrier composed of polysaccharides.

Plasma membrane

Mitochondria

Chloroplasts enable plant cells to harness the energy of sunlight to synthesize sugars.

Plasmodesmata connect neighboring plant cells.

Vacuoles contribute to the structural rigidity of plants by absorbing water and contributing to turgor pressure.

Nucleus

Endoplasmic reticulum

Golgi apparatus

Cytoskeleton

vesicles form by budding off an organelle, taking with them a piece of the membrane and internal contents of the organelle from which they derive. They then fuse with another organelle or the plasma membrane, re-forming a continuous membrane and unloading their contents.

In total, these interconnected membranes make up the **endomembrane system**. The endomembrane system includes the nuclear envelope, endoplasmic reticulum, Golgi apparatus, lysosomes, the plasma membrane, and the vesicles that move between them (**Fig. 5.18**). In plants, the endomembrane system is actually continuous between cells through plasmodesmata (see Fig. 5.17; Chapters 10 and 26).

Extensive internal membranes are not common in prokaryotic cells. However, photosynthetic bacteria have internal membranes that are specialized for harnessing light energy (Chapters 8 and 25).

The endomembrane system compartmentalizes the cell.

Because many types of molecules are unable to cross cell membranes on their own, the endomembrane system divides the interior of a cell into two distinct "worlds," one inside the spaces defined by these membranes and one outside these spaces. A molecule within the interior of the endoplasmic reticulum (ER) can stay in the ER or end up in the interior of the Golgi apparatus or even outside the cell by the budding off and fusing of a vesicle between these organelles. Similarly, a molecule associated with the ER membrane can move to the Golgi membrane or the plasma membrane by vesicle transport. Molecules in the cytosol are in a different physical space, separated by membranes of the endomembrane system. This physical separation allows specific functions to take place within the spaces defined by the membranes and within the membrane itself.

In spite of forming a continuous and interconnected system, the various compartments have unique properties and maintain distinct identities determined in part by which lipids and proteins are present in their membranes.

Vesicles bud off from and fuse not only with organelles but also with the plasma membrane. When a vesicle fuses with the plasma membrane, the process is called **exocytosis**. It provides a way for a vesicle to empty its contents to the extracellular space or to deliver proteins embedded in the vesicle membrane to the plasma membrane (Fig. 5.18). This process also works in reverse: A vesicle can bud off from the plasma membrane, enclosing material from outside the cell and bringing it into the cell interior, in a process called **endocytosis**. Together, exocytosis and endocytosis provide a way to move material into and out of cells without passing through the cell membrane.

The nucleus houses the genome and is the site of RNA synthesis.

The innermost organelle of the endomembrane system is the nucleus, which stores DNA, the genetic material that encodes the information for all the activities and structures of the cell. The **nuclear envelope** defines the boundary of the nucleus (**Fig. 5.19**). It actually consists of two membranes, the inner and outer membranes. Each is a lipid bilayer with associated proteins.

The inner and outer membranes are continuous with each other at openings called **nuclear pores**. These large protein complexes allow molecules to move into and out of the nucleus, so they are essential for the nucleus to communicate with the rest of the cell. For example, some proteins that are synthesized in the cytosol, such as transcription factors, move through nuclear pores to enter the nucleus, where they control how and when genetic information is expressed.

In addition, the transfer of information encoded by DNA depends on the movement of mRNA (messenger RNA)

FIG. 5.18 The endomembrane system. The endomembrane system is a series of interconnected membrane-bound compartments in eukaryotic cells.

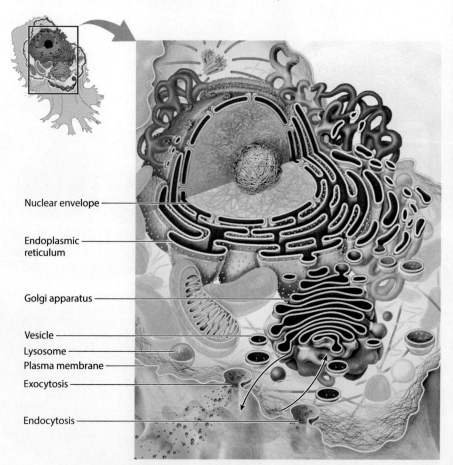

FIG. 5.19 A surface view of the nuclear envelope. The nucleus is surrounded by a double membrane and houses the cell's DNA.

Source: Don W. Fawcett/Science Source.

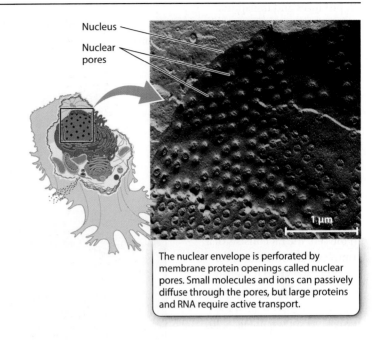

The nuclear envelope is perforated by membrane protein openings called nuclear pores. Small molecules and ions can passively diffuse through the pores, but large proteins and RNA require active transport.

molecules out of the nucleus through these pores. After exiting the nucleus, mRNA binds to free ribosomes in the cytosol or ribosomes associated with the endoplasmic reticulum (ER). **Ribosomes** are the sites of protein synthesis, in which amino acids are assembled into polypeptides guided by the information stored in mRNA (Chapter 4). In this way, the nuclear envelope and its associated protein pores regulate which molecules move into and out of the nucleus.

The endoplasmic reticulum is involved in protein and lipid synthesis.

The outer membrane of the nuclear envelope is physically continuous with the **endoplasmic reticulum**, an organelle bounded by a single membrane (**Fig. 5.20**). The ER is a conspicuous feature of many eukaryotic cells, accounting in some cases for as much as half of the total amount of membrane. The

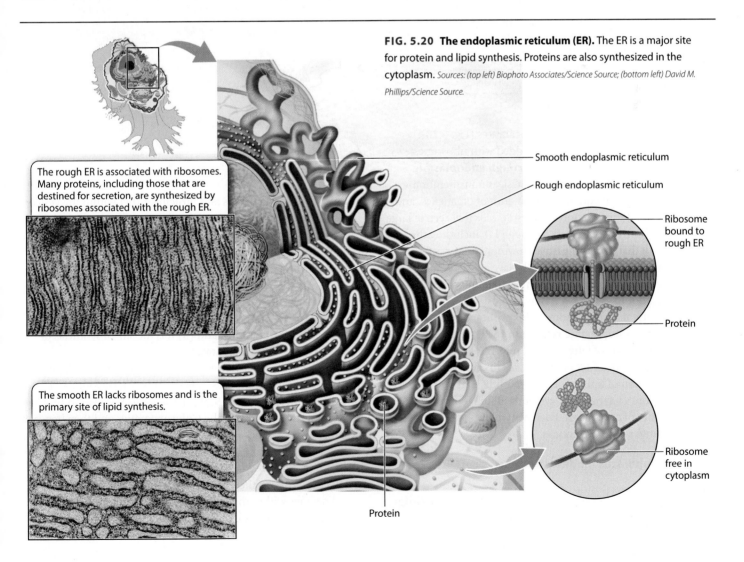

FIG. 5.20 The endoplasmic reticulum (ER). The ER is a major site for protein and lipid synthesis. Proteins are also synthesized in the cytoplasm. *Sources: (top left) Biophoto Associates/Science Source; (bottom left) David M. Phillips/Science Source.*

The rough ER is associated with ribosomes. Many proteins, including those that are destined for secretion, are synthesized by ribosomes associated with the rough ER.

The smooth ER lacks ribosomes and is the primary site of lipid synthesis.

ER produces and transports many of the proteins and lipids used inside and outside the cell, including all transmembrane proteins, as well as proteins destined for the Golgi apparatus, lysosomes, or export out of the cell. The ER is also the site of production of most of the lipids that make up the various internal and external cell membranes.

Unlike the nucleus, which is a single spherical structure in the cell, the ER consists of a complex network of interconnected tubules and flattened sacs. Its interior is continuous throughout and is called the **lumen**. As shown in Fig. 5.20, the ER has an almost mazelike appearance when sliced and viewed in cross section. Its membrane is extensively convoluted, allowing a large amount of membrane surface area to fit within the cell.

When viewed through an electron microscope, ER membranes have two different appearances (Fig. 5.20). Some look rough because they are studded with ribosomes; this portion of the ER is referred to as **rough endoplasmic reticulum (RER)**. The rough ER synthesizes transmembrane proteins, proteins that end up in the interior of organelles, and proteins destined for secretion. As a result, cells that secrete large quantities of protein have extensive rough ER, including the cells of the gut that secrete digestive enzymes and the cells of the pancreas that produce insulin. All cells have at least some rough ER for the production of transmembrane and organelle proteins.

A small amount of ER membrane in most cells appears smooth because it lacks ribosomes; this portion of the ER is called **smooth endoplasmic reticulum (SER)** (Fig. 5.20). Smooth ER is the site of fatty acid and phospholipid biosynthesis. Thus, this type of ER predominates in cells specialized for the production of lipids. For example, cells that synthesize steroid hormones have a well-developed SER that produces large quantities of cholesterol. Enzymes within the SER convert cholesterol into steroid hormones.

The Golgi apparatus modifies and sorts proteins and lipids.

Although it is not physically continuous with the ER, the **Golgi apparatus** is often the next stop for vesicles that bud off the ER.

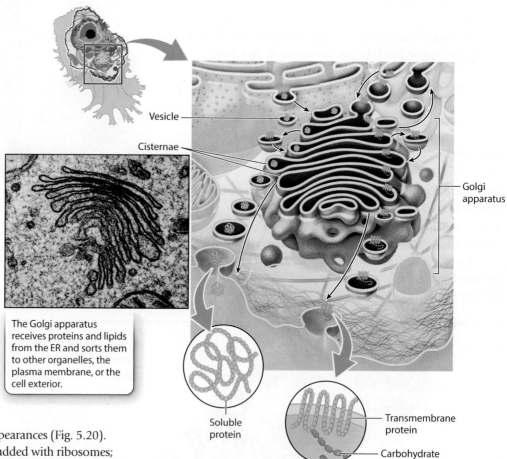

FIG. 5.21 The Golgi apparatus. The Golgi apparatus sorts proteins and lipids to other organelles, the plasma membrane, or the cell exterior. *Source: Biophoto Associates/Science Source.*

The Golgi apparatus receives proteins and lipids from the ER and sorts them to other organelles, the plasma membrane, or the cell exterior.

These vesicles carry lipids and proteins, either within the vesicle interior or embedded in their membranes. The movement of these vesicles from the ER to the Golgi apparatus and then to the rest of the cell is part of a biosynthetic pathway in which lipids and proteins are sequentially modified and delivered to their final destinations. The Golgi apparatus has three primary roles: (1) it further modifies proteins and lipids produced by the ER; (2) it acts as a sorting station as these proteins and lipids move to their final destinations; and (3) it is the site of synthesis of most of the cell's carbohydrates.

Under the microscope, the Golgi apparatus looks like stacks of flattened membrane sacs, called **cisternae,** surrounded by many small vesicles (**Fig. 5.21**). These vesicles transport proteins and lipids from the ER to the Golgi apparatus, and then between the various cisternae, and finally from the Golgi apparatus to the plasma membrane or other organelles. Vesicles are therefore the primary means by which proteins and lipids move through the Golgi apparatus to their final destinations.

Enzymes within the Golgi apparatus chemically modify proteins and lipids as they pass through it. These modifications take place in a sequence of steps, each performed in a different region of the Golgi apparatus, because each region contains a different set of enzymes that catalyzes specific reactions. As a result, there is a general movement of vesicles from the ER through the Golgi apparatus and then to their final destinations.

An example of a chemical modification that occurs predominantly in the Golgi apparatus is glycosylation, in which sugars are covalently linked to lipids or specific amino acids of proteins, forming glycoproteins. As these lipids and proteins move through the Golgi apparatus, they encounter different enzymes in each region that add or trim sugars. Glycoproteins are important components of the eukaryotic cell surface. The sugars attached to the protein can protect the protein from enzyme digestion by blocking access to the peptide chain. As a result, glycoproteins form a relatively flexible and protective coating over the plasma membrane. The distinctive shapes that sugars contribute to glycoproteins and glycolipids also allow them to be recognized specifically by other cells and molecules in the external environment. For example, human blood types (A, B, AB, and O) are defined by the particular sugars that are linked to proteins and lipids on the surface of red blood cells.

While traffic usually travels from the ER to the Golgi apparatus, a small amount of traffic moves in the reverse direction, from the Golgi apparatus to the ER. This reverse pathway is important to retrieve proteins in the ER or Golgi that were accidentally moved forward and to recycle membrane components.

Lysosomes degrade macromolecules.

The ability of the Golgi apparatus to sort and dispatch proteins to particular destinations is dramatically illustrated by lysosomes. **Lysosomes** are specialized vesicles derived from the Golgi apparatus that degrade damaged or unneeded macromolecules (**Fig. 5.22**). They contain a variety of enzymes that break down macromolecules such as proteins, nucleic acids, lipids, and complex carbohydrates. The Golgi apparatus packages macromolecules destined for degradation into vesicles. These vesicles then fuse with lysosomes, delivering their contents to the lysosome interior. The Golgi is also responsible for delivering the enzymes that break down macromolecules to the lysosomes.

The formation of lysosomes also illustrates the ability of the Golgi apparatus to sort key proteins. The enzymes inside the lysosomes are synthesized in the RER, sorted in the Golgi apparatus, and then packaged into lysosomes. In addition, the Golgi apparatus sorts and delivers specialized proteins that become embedded in lysosomal membranes. These include proton pumps that keep the internal environment at an acidic pH of about 5, the optimal pH for the activity of the enzymes inside. Other proteins in the lysosomal membranes transport the breakdown products of macromolecules, such as amino acids and simple sugars, across the membrane to the cytosol for use by the cell.

The function of lysosomes underscores the importance of having separate compartments within the cell bounded by selectively permeable membranes. Lysosomal enzymes cannot function in the normal cellular environment, which has a pH of about 7, and many of a cell's enzymes and proteins would unfold and degrade if the entire cell were at the pH of the inside of a lysosome. By restricting the activity of these enzymes to the lysosome, the cell protects proteins and organelles in the cytosol from degradation.

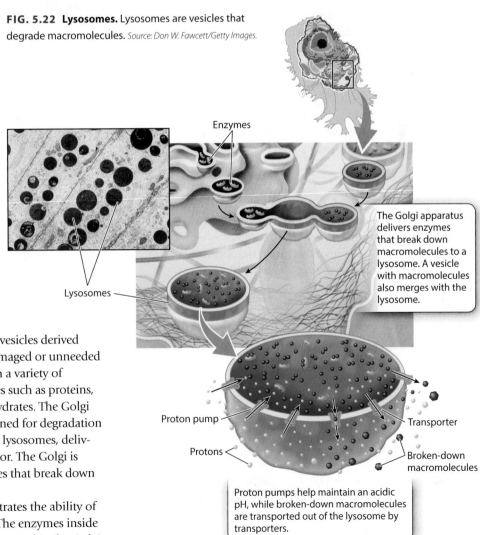

FIG. 5.22 Lysosomes. Lysosomes are vesicles that degrade macromolecules. *Source: Don W. Fawcett/Getty Images.*

The Golgi apparatus delivers enzymes that break down macromolecules to a lysosome. A vesicle with macromolecules also merges with the lysosome.

Proton pumps help maintain an acidic pH, while broken-down macromolecules are transported out of the lysosome by transporters.

Protein sorting directs proteins to their proper location in or out of the cell.

As we have seen, eukaryotic cells have many compartments, and different proteins function in different places, such as enzymes in lysosomes or transmembrane proteins embedded in the plasma membrane. **Protein sorting** is the process by which proteins end up where they need to be to perform their function. Protein sorting directs proteins to the cytosol, the lumen of organelles, or the membranes of the endomembrane system, or even out of the cell entirely.

Recall that proteins are produced in two places: free ribosomes in the cytosol and membrane-bound ribosomes on the rough ER. Proteins produced on free ribosomes are sorted after they are translated. These proteins often contain amino acid sequences, called **signal sequences,** that allow them to be recognized and sorted. As shown in **Fig. 5.23**, several types of signal sequences direct proteins synthesized on free ribosomes to different cellular compartments. Proteins with no signal sequence remain in the cytosol (Fig. 5.23a). Proteins destined for mitochondria or chloroplasts often have a signal sequence at their amino ends (Fig. 5.23b). Proteins targeted to the nucleus usually have signal sequences located internally (Fig. 5.23c). These nuclear signal sequences, called **nuclear localization signals,** enable proteins to move through pores in the nuclear envelope.

In contrast to proteins synthesized by free ribosomes, proteins produced on the rough ER end up in the lumen of the endomembrane system, secreted out of the cell, or as transmembrane proteins (**Fig. 5.24**). These proteins are sorted as they are translated.

Proteins destined for the ER lumen or secretion have an amino-terminal signal sequence that directs the ribosome

FIG. 5.24 Pathways for proteins destined (left) to be secreted or (right) to be transported to the plasma membrane.

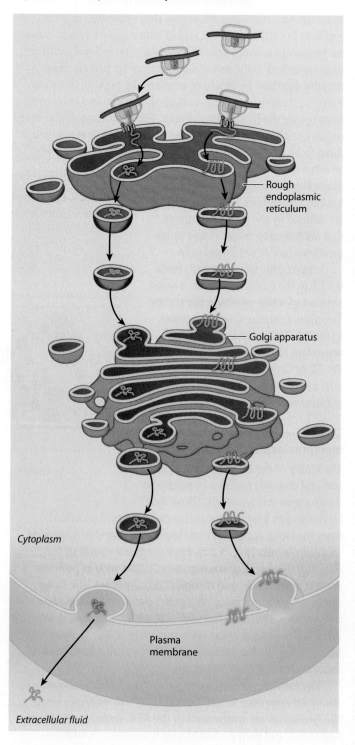

FIG. 5.23 Signal sequences on proteins synthesized by free ribosomes. (a) Most proteins synthesized by free ribosomes with no signal peptide remain in the cytosol. Other signal sequences direct proteins (b) to mitochondria and chloroplasts or (c) to the nucleus.

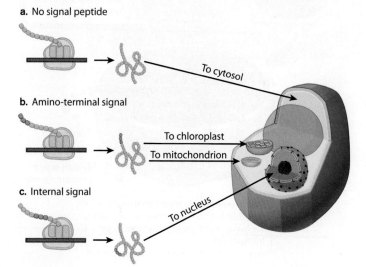

FIG. 5.25 Interaction of a signal sequence, signal-recognition particle (SRP), and SRP receptor. Binding of a signal sequence with the SRP halts translation, followed by docking of the ribosome on the ER membrane, release of the SRP, and continuation of translation.

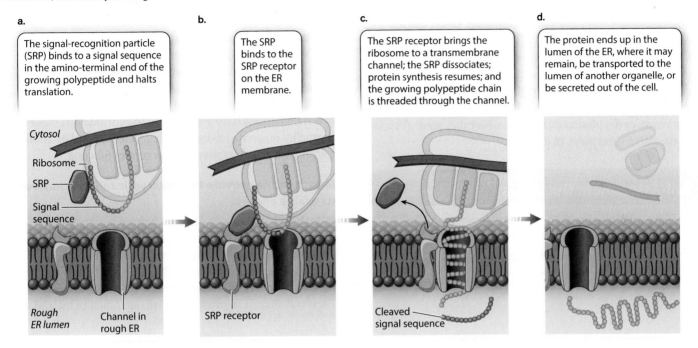

to the rough ER (**Fig. 5.25**). As a free ribosome translates the protein in the cytosol, this sequence is recognized by an RNA–protein complex known as a **signal-recognition particle (SRP)**. The SRP binds to both the signal sequence and the free ribosome, and brings about a pause in translation (Fig. 5.25a). The SRP then binds with a receptor on the RER so that the ribosome becomes associated with the RER (Fig. 5.25b). The SRP receptor brings the ribosome to a channel in the membrane of the RER. The SRP then dissociates and translation continues, allowing the growing polypeptide chain to be threaded through the channel (Fig. 5.25c). A specific protease cleaves the signal sequence as it emerges in the lumen of the ER (Fig. 5.25d). Some proteins are retained in the interior of the ER, whereas others are transported in vesicles to the interior of the Golgi apparatus. Some of these proteins are secreted by exocytosis.

Proteins destined for cell membranes contain a **signal-anchor sequence** in addition to the amino-terminal signal sequence (**Fig. 5.26**). After the growing polypeptide chain and its ribosome are brought to the ER, this complex is threaded

FIG. 5.26 Targeting of a transmembrane protein by a signal-anchor sequence. Proteins with a hydrophobic signal-anchor sequence end up embedded in the membrane.

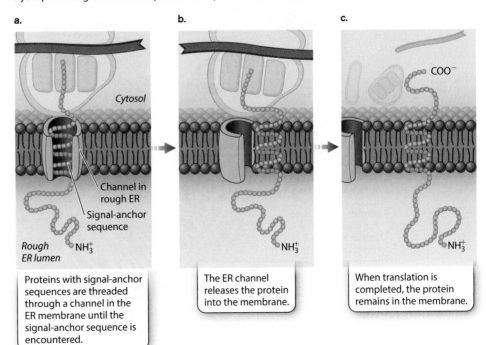

through the channel in the ER membrane until the signal-anchor sequence is encountered (Fig. 5.26a). The signal-anchor sequence is hydrophobic, so it is able to diffuse laterally in the lipid bilayer (Fig. 5.26b). At this point, the ribosome dissociates from the channel while translation continues. When translation is complete, the amino end of the polypeptide is in the ER lumen, the carboxyl end remains on the cytosolic side of the ER membrane, and the region between them resides in the membrane (Fig. 5.26c). Transmembrane proteins such as these may stay in the membrane of the ER or end up in other internal membranes or the plasma membrane, where they serve as transporters, pumps, receptors, or enzymes.

Self-Assessment Questions

11. What are the names and functions of the major organelles of the endomembrane system in eukaryotic cells?
12. How does a protein end up free in the cytosol, embedded in the plasma membrane, or secreted from the cell?

5.5 MITOCHONDRIA AND CHLOROPLASTS

The membranes of two organelles, mitochondria and chloroplasts, are not part of the endomembrane system. Both of these organelles are specialized to harness energy for the cell. Interestingly, they are both semi-autonomous organelles that grow and multiply independently of the other membrane-bound compartments, and they contain their own circular genomes. As we discuss in Chapters 8 and 25, the similarities between the DNA of these organelles and the DNA of certain bacteria have led scientists to conclude that these organelles originated as bacteria that were captured by another cell and, over time, evolved to their current function.

Mitochondria provide the eukaryotic cell with most of its usable energy.

Mitochondria are organelles that harness energy from chemical compounds such as sugars and convert it into ATP, which serves as the universal energy currency of the cell. ATP is able to drive the many chemical reactions in the cell.

Mitochondria are present in nearly all eukaryotic cells. These rod-shaped organelles have an outer membrane and a

FIG. 5.27 Mitochondria. Mitochondria synthesize most of the ATP required to meet the cell's energy needs. *Source: Keith R. Porter/Science Source.*

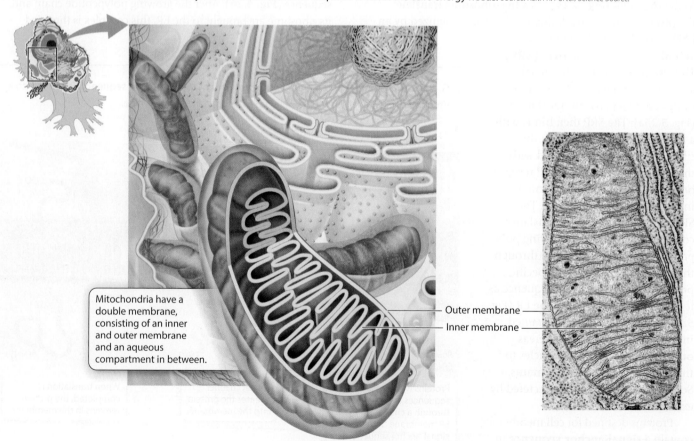

Mitochondria have a double membrane, consisting of an inner and outer membrane and an aqueous compartment in between.

Outer membrane
Inner membrane

highly convoluted inner membrane whose folds project into the interior (**Fig. 5.27**). A proton electrochemical gradient is generated across the inner mitochondrial membrane, and the energy stored in the gradient is used to synthesize ATP for use by the cell. In the process of breaking down sugar and synthesizing ATP, oxygen is consumed and carbon dioxide is released.

Does this process sound familiar? It also describes your own breathing, or respiration. Mitochondria are the site of cellular respiration, and the oxygen that you take in with each breath is used by mitochondria to produce ATP. Cellular respiration is discussed in greater detail in Chapter 7.

Chloroplasts capture energy from sunlight.

Both animal and plant cells have mitochondria to provide them with life-sustaining ATP. In addition, plant cells and green algae have organelles called **chloroplasts** that capture the energy of sunlight to synthesize simple sugars (**Fig. 5.28**). This process, called **photosynthesis**, results in the release of oxygen as a waste product. Like the nucleus and mitochondria, chloroplasts are surrounded by a double membrane. They also have a third, internal membrane, which defines a separate internal compartment called the **thylakoid**. The thylakoid membrane contains specialized light-collecting molecules called pigments, of which **chlorophyll** is the most important. The green color of chlorophyll explains why so many plants have green leaves.

Chlorophyll plays a key role in the chloroplast's ability to capture energy from sunlight. Using the light energy collected by this pigment, enzymes present in the chloroplast use carbon dioxide as a carbon source to produce carbohydrates. Photosynthesis is discussed in greater detail in Chapter 8.

Self-Assessment Questions

13. Are mitochondria present in animal cells, plant cells, or both?
14. What are three similarities between mitochondria and chloroplasts?

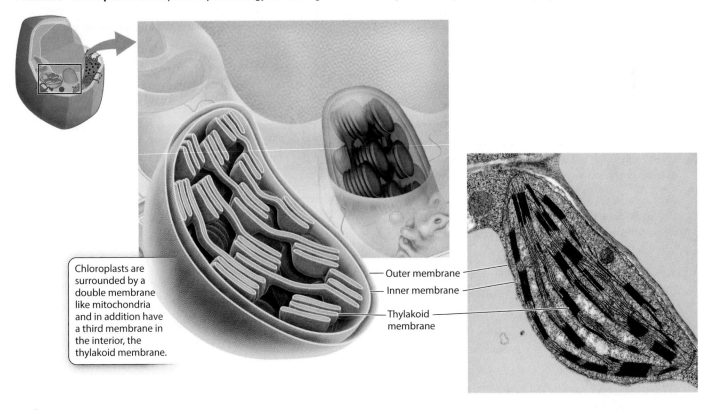

FIG. 5.28 Chloroplasts. Chloroplasts capture energy from sunlight and use it to synthesize sugars. *Source: Dr. Jeremy Burgess/Science Source.*

Chloroplasts are surrounded by a double membrane like mitochondria and in addition have a third membrane in the interior, the thylakoid membrane.

- Outer membrane
- Inner membrane
- Thylakoid membrane

CORE CONCEPTS SUMMARY

5.1 STRUCTURE OF CELL MEMBRANES: Cell membranes are composed of lipids, proteins, and carbohydrates.

Phospholipids have both hydrophilic and hydrophobic regions. As a result, they spontaneously form structures such as micelles and bilayers when placed in an aqueous environment. page 94

Membranes are fluid, meaning that membrane components are able to move laterally in the plane of the membrane. page 95

Membrane fluidity is influenced by the length of fatty acid chains, the presence of carbon–carbon double bonds in fatty acid chains, and the amount of cholesterol. page 95

Many membranes also contain proteins that span the membrane (transmembrane proteins) or are temporarily associated with one or other layer of the lipid bilayer (peripheral proteins). page 97

5.2 MOVEMENT IN AND OUT OF CELLS: The plasma membrane is a selective barrier that controls the movement of molecules between the inside and the outside of the cell.

Selective permeability results from the combination of lipids and proteins that makes up cell membranes. page 99

Passive transport is the movement of molecules by diffusion, the random movement of molecules. There is a net movement of molecules from regions of higher concentration to regions of lower concentration. page 100

Passive transport can occur by the diffusion of molecules directly through the plasma membrane (simple diffusion) or be aided by protein transporters (facilitated diffusion). page 100

Active transport moves molecules from regions of lower concentration to regions of higher concentration and requires energy. page 101

Primary active transport uses energy stored in ATP; secondary active transport uses the energy stored in an electrochemical gradient. page 101

Animal cells often maintain their size and shape through the actions of protein pumps that actively move ions in and out of the cell. page 103

Plants, fungi, and bacteria have a cell wall outside the plasma membrane that maintains cell size and shape. page 103

5.3 INTERNAL ORGANIZATION OF CELLS: Cells can be classified as prokaryotes or eukaryotes, which differ in their degree of internal compartmentalization.

Prokaryotic cells lack a nucleus and other internal membrane-enclosed compartments. page 105

Prokaryotic cells include bacteria and archaeons and are typically much smaller than eukaryotes. page 105

Eukaryotic cells have a nucleus and other internal compartments called organelles. page 105

Eukaryotes include animals, plants, fungi, and protists. page 105

5.4 THE ENDOMEMBRANE SYSTEM: The endomembrane system is an interconnected system of membranes that defines spaces in the cell, synthesizes important molecules, and traffics and sorts these molecules in and out of the cell.

The endomembrane system includes the nuclear envelope, endoplasmic reticulum, Golgi apparatus, lysosomes, vesicles, and plasma membrane. page 107

The nucleus, which is enclosed by a double membrane called the nuclear envelope, houses the genome. page 108

The endoplasmic reticulum is continuous with the outer nuclear envelope and manufactures proteins and lipids for use by the cell or for export out of the cell. page 109

The Golgi apparatus communicates with the endoplasmic reticulum by transport vesicles. It receives proteins and lipids from the endoplasmic reticulum and directs them to their final destinations. page 110

Lysosomes break down macromolecules such as proteins into simpler compounds that can be used by the cell. page 111

Protein sorting directs proteins to their final destinations inside or outside of the cell. page 112

Proteins synthesized on free ribosomes are sorted after translation, whereas proteins synthesized on ribosomes associated with the rough endoplasmic reticulum are sorted during translation. page 112

Proteins synthesized on free ribosomes are often sorted by means of a signal sequence and are destined for the cytosol, mitochondria, chloroplasts, or nucleus. page 112

Proteins synthesized on ribosomes on the rough endoplasmic reticulum have a signal sequence that is recognized by a signal-recognition particle. These proteins end up as transmembrane proteins, reside in the interior of various organelles, or are secreted. page 112

5.5 MITOCHONDRIA AND CHLOROPLASTS:
Mitochondria and chloroplasts are organelles that harness energy, and likely evolved from free-living prokaryotes.

Mitochondria harness energy from chemical compounds for use by both animal and plant cells. page 114

Chloroplasts harness the energy of sunlight to build sugars. page 115

Log in to LaunchPad to check your answers to the Self-Assessment Questions and to access additional learning tools.

CHAPTER 6
Making Life Work
Capturing and Using Energy

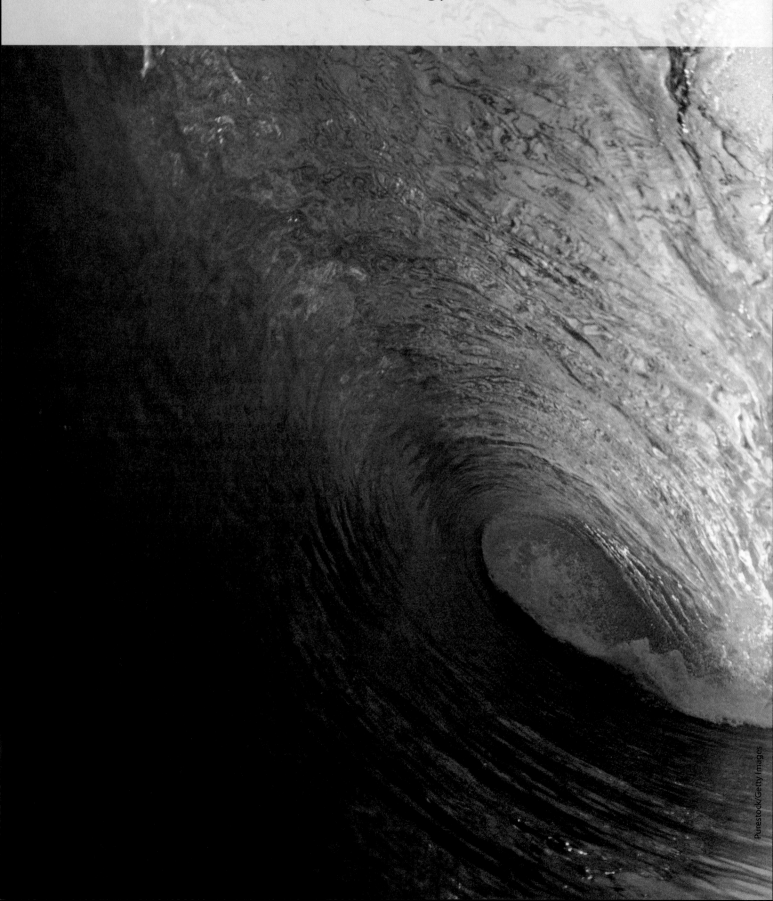

CORE CONCEPTS

6.1 AN OVERVIEW OF METABOLISM: Metabolism is the set of biochemical reactions that transforms molecules and transfers energy.

6.2 KINETIC AND POTENTIAL ENERGY: Kinetic energy is energy of motion, and potential energy is stored energy.

6.3 LAWS OF THERMODYNAMICS: The first and second laws of thermodynamics govern energy flow in biological systems.

6.4 CHEMICAL REACTIONS: Chemical reactions involve the breaking and forming of bonds.

6.5 ENZYMES AND THE RATE OF CHEMICAL REACTIONS: The rate of chemical reactions is increased by protein catalysts called enzymes.

We have seen that cells require a way to encode and transmit information and a membrane to separate inside from out. The third requirement of a cell is energy. Cells need energy to do work: to grow and divide, move, change shape, pump ions in and out, transport vesicles, and synthesize macromolecules such as DNA, RNA, proteins, and complex carbohydrates. All of these activities are considered work, and they all require energy.

We are all familiar with different forms of energy—the sun and wind are sources of energy, as are fossil fuels such as oil and natural gas. We have learned to harness the energy from these sources and convert it to other forms, such as electricity, to provide needed power to our homes and cities.

Cells face a similar challenge. They must harness energy from the environment and convert it to a form that allows them to do the work necessary to sustain life. Cells harness energy from the sun and from chemical compounds, including carbohydrates, lipids, and proteins. Although the source of energy may differ among cells, all cells convert energy to a form that can be easily used to drive cellular processes. All cells use energy in the form of a molecule called **adenosine triphosphate (ATP)**.

ATP is often called the universal "currency" of cellular energy to indicate that ATP provides energy in a form that all cells can readily use to perform work. Although "currency" provides a useful analogy for the role that ATP plays, there is an important distinction between actual currency, such as a dollar bill, and ATP. A dollar bill represents a certain value but does not have any value in itself. By contrast, ATP does not represent energy; it actually contains energy in its chemical bonds. Nevertheless, the analogy is useful because ATP, like a dollar bill, engages in a broad range of energy "transactions" in the cell.

In this chapter, we consider energy in the context of cells. What exactly is energy? What principles govern its flow in biological systems? And how do cells make use of it?

6.1 AN OVERVIEW OF METABOLISM

When considering a cell's use of energy, it is helpful to also consider the cell's sources of carbon. Carbon is the backbone of the organic molecules that make up cells and cells often use carbon-based compounds as a stable form of energy storage. How organisms obtain the energy and carbon they need is so fundamental that it is sometimes used to provide a metabolic classification of life (**Fig. 6.1**). Simply put, organisms have two ways of harvesting energy from their environment and two sources of carbon. This means that, together, there are four principal ways in which organisms acquire the energy and materials needed to grow, function, and reproduce.

Organisms can be classified according to their energy and carbon sources.

Organisms have two ways of harvesting energy from their environment: they can obtain energy either from the sun or from chemical compounds (Fig. 6.1). Organisms that capture energy from sunlight are called **phototrophs**. Plants are the most familiar example. Plants use the energy

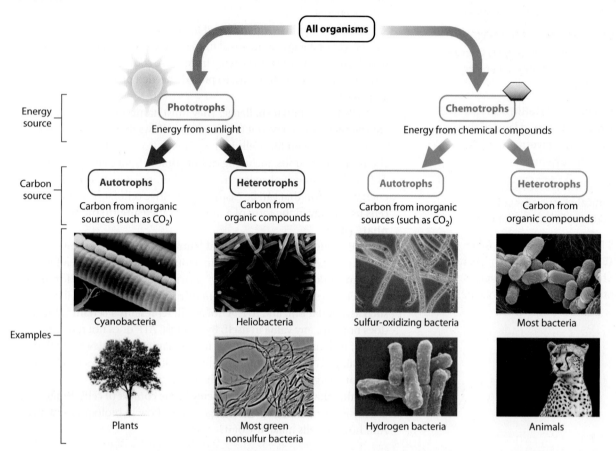

FIG. 6.1 A metabolic classification of organisms. Organisms can be classified according to their energy and carbon sources.

Photo sources: (clockwise from top-left) Dr. Tony Brain/Science Source; Dr. André Kempe/ScienceFoto.de; Copyright 1997 Microbial Diversity, Rolf Schauder; Martin Oeggerli/Science Source; DNY59/iStockphoto; David J. Patterson; EM image by Manfred Rohde, Helmholtz Centre for Infection Research, Braunschweig, Germany; Anna Omelchenko/Dreamstime.com.

of sunlight to convert carbon dioxide and water into sugar and oxygen (Chapter 8). Sugars, such as glucose, contain energy in their chemical bonds that can be used to synthesize ATP, which in turn can power the work of the cell.

Other organisms derive their energy directly from chemical compounds. These organisms are called **chemotrophs**, and animals are familiar examples. Animals ingest other organisms, obtaining glucose and other organic molecules that they break down in the presence of oxygen to produce carbon dioxide and water. In this process, the energy in the chemical bonds of the organic molecules is converted to energy carried in the bonds of ATP (Chapter 7).

In drawing this distinction between plants and animals, we have to be careful. Although sunlight provides the energy that plants use to synthesize glucose, plants still power most cellular processes by breaking down the sugar they make, just as animals do. And although animals harness energy from organic molecules, the energy in these organic molecules is generally derived from the sun.

Organisms can also be classified in terms of where they get their carbon (Fig. 6.1). Some organisms are able to convert carbon dioxide (an inorganic form of carbon) into glucose (an organic form of carbon). These organisms are **autotrophs**, or "self feeders." Plants again are an example, so plants are both phototrophs and autotrophs, or photoautotrophs.

Other organisms do not have the ability to convert carbon dioxide into organic forms of carbon. Instead, they obtain their carbon from organic molecules synthesized by other organisms. These organisms eat other organisms or molecules derived from other organisms. Such organisms are **heterotrophs**, or "other feeders." Animals obtain carbon in this way, and so animals are both chemotrophs and heterotrophs, or chemoheterotrophs. In fact, animals get their energy and carbon from the same molecule. That is, a molecule of glucose supplies both energy and carbon.

As we explore the diversity of life in Part 2, we will see that not all organisms fit into the two categories of photoautotrophs and chemoheterotrophs (Fig. 6.1). Some microorganisms gain energy from sunlight but obtain their carbon by ingesting organic molecules; such organisms are called photoheterotrophs. Other microorganisms extract energy from inorganic molecules but build their own organic molecules; these organisms are called chemoautotrophs. They are often found in extreme environments, such as deep-sea vents, where sunlight is absent and inorganic compounds such as hydrogen sulfide are plentiful.

Metabolism is the set of chemical reactions that sustain life.

The breaking down of organic molecules such as glucose and the harnessing and release of energy in the process are driven by chemical reactions in the cell. The term **metabolism** encompasses the entire set of chemical reactions that convert molecules into other molecules and transfer energy in living organisms. These chemical reactions are occurring all the time in your cells. Many of these reactions are linked, in that the products of one are the reactants of the next, forming long pathways and intersecting networks.

Metabolism is divided into two branches: **catabolism** is the set of chemical reactions that break down molecules into smaller units and, in the process, produce ATP, and **anabolism** is the set of chemical reactions that build molecules from smaller units and require an input of energy, usually in the form of ATP (**Fig. 6.2**). For example, carbohydrates can be broken down, or catabolized, into sugars, fats into fatty acids and glycerol, and proteins into amino acids. These initial products can be broken down further to release energy stored in their chemical bonds. By contrast, the synthesis of macromolecules such as carbohydrates and proteins is anabolic.

> ### Self-Assessment Questions
>
> 1. What are the ways that organisms obtain energy and carbon from the environment? What are the names used to describe these organisms?
> 2. What is the difference between catabolism and anabolism?

6.2 KINETIC AND POTENTIAL ENERGY

There are many different sources of energy, including sunlight, wind, electricity, and fossil fuels. The food we eat also contains energy. **Energy** is important in biological systems because it is needed to do work. A cell is doing work when it carries out the processes we discussed in earlier chapters, such as synthesizing DNA, RNA, and proteins, pumping substances across the plasma membrane, and moving vesicles between various compartments of a cell.

Here, we consider two forms of energy: kinetic energy and potential energy (**Fig. 6.3**). Kinetic energy is energy associated with movement, and potential energy is energy that is stored, for example in the chemical bonds of molecules such as glucose and ATP.

Kinetic energy and potential energy are two forms of energy.

Kinetic energy is the energy of motion, and it is perhaps the most familiar form of energy. A moving object, such as a ball bouncing down a set of stairs, possesses kinetic energy. Kinetic energy is associated with any kind of movement, such as a person running or a muscle contracting. Light, electricity, and thermal energy (perceived as heat) are also forms of kinetic energy, because light is associated with the movement of photons, electricity with the movement of electrons, and thermal energy with the movement of molecules.

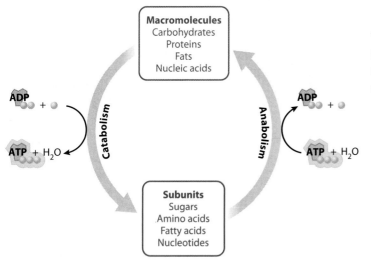

FIG. 6.2 Catabolism and anabolism. The energy harvested as ATP during the breakdown of molecules in catabolism can be used to synthesize molecules in anabolism.

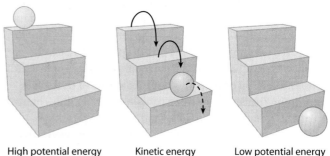

FIG. 6.3 Potential and kinetic energy. Some of the high potential energy of a ball at the top of a set of stairs is transformed to kinetic energy as the ball rolls down the stairs.

Energy is not always associated with motion. An immobile object can still possess a form of energy called **potential energy**, or stored energy. Potential energy depends on the structure of the object or its position relative to a field (gravitational, electrical, magnetic), and it is released by a change in the object's structure or position. For example, the potential energy of a ball is higher at the top of a flight of stairs than at the bottom (Fig. 6.3). If it were not blocked by the floor, it would move from the position of higher potential energy (the top of the stairs) to the position of lower potential energy (the bottom of the stairs) because of the force of gravity. Similarly, an electrochemical gradient of molecules across a cell membrane is a form of potential energy. Given a pathway through the membrane, the molecules move down their concentration and electrical gradients from higher to lower potential energy (Chapter 5).

Energy can be converted from one form to another. The ball at the top of the stairs has a certain amount of potential energy because of its position. As it rolls down the stairs, this potential energy is converted to kinetic energy associated with movement of the ball and the surrounding air. When the ball reaches the bottom of the stairs, the remaining energy is stored as potential energy. Conversely, it takes an input of energy to move the ball back to the top of the stairs, and this input of energy is stored as potential energy.

Chemical energy is a form of potential energy.

We obtain the energy we need from the food we eat, which contains chemical energy. **Chemical energy** is a form of potential energy held in the chemical bonds between pairs of atoms in a molecule. Recall from Chapter 2 that a covalent bond results from the sharing of electrons between two atoms. Covalent bonds form when the sharing of electrons between two atoms results in a more stable configuration than if they were not shared—that is, than if the orbitals of the two atoms did not overlap.

The more stable configuration will always be the one with lower potential energy. As a result, energy is required to break a covalent bond because going from a lower energy state to a higher one requires an input of energy. Conversely, energy is released when a covalent bond forms.

Some bonds are stronger than other ones. A strong bond is hard to break because the arrangement of orbitals in these molecules is much more stable than if the two atoms do not share electrons. Because strong bonds have a very stable arrangement of orbitals, they do not require a lot of energy to remain intact. As a result, strong bonds do not contain very much chemical energy, similar to the potential energy of a ball at the bottom of a flight of stairs (Fig. 6.3). Examples of molecules with strong covalent bonds that contain relatively little chemical energy are carbon dioxide (CO_2) and water (H_2O).

Conversely, some covalent bonds are relatively weak. These bonds are easily broken because the arrangement of orbitals in these molecules is only somewhat more stable than if the two atoms did not share electrons. As a result, weak covalent bonds require a lot of energy to stay intact and contain a lot of chemical energy, similar to the potential energy of a ball at the top of a flight of stairs (Fig. 6.3). Organic molecules such as carbohydrates, lipids, and proteins contain relatively weak covalent bonds, including many carbon–carbon (C—C) bonds and carbon–hydrogen (C—H) bonds. Therefore, organic molecules are rich sources of chemical energy and are sometimes called fuel molecules.

ATP is a readily accessible form of cellular energy.

The chemical energy carried in carbohydrates, lipids, and proteins is harnessed by cells to do work. But cells do not use this energy all at once. Instead, through the series of chemical reactions described in Chapter 7, they package this energy into a chemical form that is readily accessible to the cell. One form of chemical energy is adenosine triphosphate, or ATP, shown in **Fig. 6.4**. The chemical energy in the bonds of ATP is used in turn to drive many cellular processes, such as muscle contraction, cell movement, and membrane pumps. In this way, ATP serves as a go-between, acting as an intermediary between fuel molecules that store a large amount of potential energy in their bonds and the activities of the cell that require an input of energy.

The core of ATP is adenosine, which is made up of the base adenine and the five-carbon sugar ribose. To complete the molecule, the ribose is attached to triphosphate, or three phosphate groups (Fig. 6.4). ATP's chemical relatives are ADP (adenosine diphosphate) and AMP (adenosine monophosphate), with two phosphate groups and one phosphate group,

FIG. 6.4 The structure of ATP.

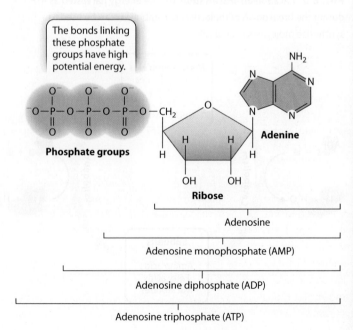

respectively. The use of ATP as an energy source in nearly all cells reflects its use early in the evolution of life on Earth.

The accessible chemical energy of ATP is held in the bonds connecting the phosphate groups. At physiological pH, these phosphate groups are negatively charged and have a tendency to repel each other. The chemical bonds connecting the phosphate groups therefore contain a lot of chemical energy to keep the phosphate groups connected. This energy is released when new, more stable bonds are formed that contain less chemical energy (section 6.4). The released energy in turn can be harnessed to power the work of the cell.

Self-Assessment Questions

3. What are the two forms of energy? Provide an example of each.
4. What is the relationship between strength of a covalent bond and the amount of chemical energy it contains?
5. Draw the structure of ATP, indicating the bonds that are broken during hydrolysis.

6.3 LAWS OF THERMODYNAMICS

Plants and other phototrophs convert energy from the sun into chemical potential energy held in the bonds of molecules. Chemical reactions can transfer this chemical energy between different molecules. And energy from these molecules can be used by the cell to do work. In all these instances, energy is subject to the laws of thermodynamics. In fact, all processes in a cell, from diffusion to osmosis to the pumping of ions to cell movement, are subject to these laws. There are four laws of thermodynamics, but the first and second laws are particularly relevant to biological processes.

The first law of thermodynamics: energy is conserved.

The **first law of thermodynamics** is the law of conservation of energy, which states that the universe contains a constant amount of energy. Therefore, energy is neither created nor destroyed. New energy is never formed, and energy is never lost. Instead, energy changes from one form to another. For example, kinetic energy can change to potential energy and vice versa, but the total amount of energy always remains the same (**Fig. 6.5**).

In the simple example of a ball rolling down a set of stairs (see Fig. 6.3), the potential energy stored at the top of the stairs is converted to kinetic energy associated with the movement of the ball and the surrounding air. The total amount of energy in this transformation is constant—that is, the total potential energy of the ball at the top of the stairs equals the amount of kinetic energy associated with the movement of the ball and air molecules plus the potential energy of the ball at the bottom of the stairs.

FIG. 6.5 The first law of thermodynamics. When energy is transformed from one form to another, the total amount of energy remains the same.

The second law of thermodynamics: energy transformations always result in an increase in disorder in the universe.

When energy changes forms, the total amount of energy remains constant. However, in going from one form of energy to another, the energy available to do work decreases (**Fig. 6.6**). How can the total energy remain the same, but the energy available to do work decrease? The answer is that some of the energy is no longer available to do work. Energy transformations are never 100% efficient because the amount of energy available to do work decreases every time energy changes forms. Thus, there is a universal price to pay in transforming energy from one form to another.

The energy that is not available to do work takes the form of an increase in disorder. This principle is summarized by the **second law of thermodynamics**, which states that the transformation of energy is associated with an increase in disorder of the universe. For example, when kinetic energy is changed into potential energy, the amount of disorder always increases. The degree of disorder is called **entropy**.

The degree of disorder is just one way to describe entropy. Another way to think about entropy is to consider the number

FIG. 6.6 The second law of thermodynamics. Because the amount of disorder increases when energy is transformed from one form to another, some of the total energy is not available to do work.

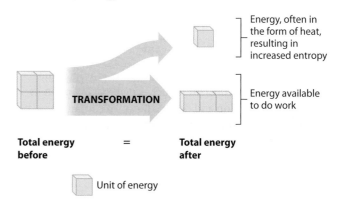

of possible positions and motions (collectively called microstates) a molecule can take on in a given system. As entropy increases, so does the number of positions and motions available to the molecule. Consider the expansion of a gas after the lid is removed from its container. Given more space, the molecules of the gas are less constrained and able to move about more freely. They have more positions available to them and move about at a larger range of speeds, so they have more entropy.

In chemical reactions, most of the entropy increase occurs through the transformation of various forms of energy into thermal energy, which we experience as heat. Thermal energy is a type of kinetic energy corresponding to the random motion of molecules and results in a given temperature. The higher the temperature, the more rapidly the molecules move and the higher the disorder.

Consider the contraction of a muscle, a form of kinetic energy associated with the shortening of muscle cells. The contraction of muscle is powered by chemical potential energy in fuel molecules. Some of this potential energy is transformed into kinetic energy (movement), and the rest is dissipated as thermal energy (which is why your muscles warm as you exercise). Consistent with the first law of thermodynamics, the amount of chemical potential energy expended is equal to the amount of kinetic energy plus the amount of thermal energy. Consistent with the second law, thermal energy flowing as heat is a necessary by-product of the transformation of chemical potential energy into kinetic energy.

In living organisms, catabolic reactions result in an increase of entropy as a single ordered molecule is broken down into several smaller ones with more freedom to move around. Anabolic reactions, by contrast, might seem to decrease entropy because they use individual building blocks to synthesize more ordered molecules such as proteins or nucleic acids. Do these anabolic reactions violate the second law of thermodynamics? No, because the second law of thermodynamics always applies to the universe as a whole, not to a chemical reaction in isolation. A local decrease in entropy is always accompanied by an even higher increase in the entropy of the surroundings. The production of heat in a chemical reaction increases entropy because heat is associated with the motion of molecules. The combination of the decrease in entropy associated with the building of a macromolecule and the increase in entropy associated with heat always results in a net increase in entropy.

The key point here is that a single cell or a multicellular organism requires a constant input of energy to maintain its high degree of function and organization. This input, as we have seen, comes either from the sun or from the energy stored in chemical compounds. This energy allows molecules to be built and other work to be carried out, but it also leads to greater entropy in the surroundings.

Self-Assessment Questions

6. What are the first and second laws of thermodynamics and how do they relate to chemical reactions?

7. Cold air has less entropy than hot air. The second law of thermodynamics states that entropy always increases. Do air conditioners violate this law?

6.4 CHEMICAL REACTIONS

Living organisms build and break down molecules, pump ions across membranes, move, and function largely through chemical reactions. As we have seen, organisms break down food molecules, such as glucose, storing energy in the bonds of ATP. That ATP then powers chemical reactions that sustain life. Chemical reactions are therefore central to life processes. In this section, we describe what a chemical reaction is, how the laws of thermodynamics apply to chemical reactions, and how ATP drives many chemical reactions in cells.

A chemical reaction occurs when molecules interact.

As we saw in Chapter 2, a chemical reaction is the process by which molecules, called reactants, are transformed into other molecules, called products. During a chemical reaction, atoms keep their identity, but the atoms that share bonds change. For example, carbon dioxide (CO_2) and water (H_2O) can react to produce carbonic acid (H_2CO_3), as shown in **Fig. 6.7**.

This reaction is common in nature. For example, it occurs when carbon dioxide in the air enters ocean waters, and it explains why the oceans are becoming more acidic as carbon dioxide levels in the atmosphere increase. This reaction also takes place in the bloodstream. When cells break down glucose,

FIG. 6.7 A chemical reaction. Atoms retain their identity during a chemical reaction as bonds are broken and new bonds are formed to yield new molecules.

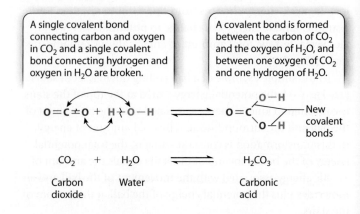

they produce carbon dioxide. In animals, this carbon dioxide diffuses out of cells and into the blood. Rather than remaining dissolved in solution, most of the carbon dioxide is converted into carbonic acid by the preceding reaction. In an aqueous (watery) environment like the ocean or the blood, carbonic acid exists as bicarbonate ions (HCO_3^-) and protons (H^+).

Many chemical reactions in cells are readily reversible: the products can react to form the reactants. For example, carbon dioxide and water react to form carbonic acid, and carbonic acid can dissociate to produce carbon dioxide and water. This reverse reaction also occurs in the blood, allowing carbon dioxide to be removed from the lungs or gills (Chapter 37). The reversibility of the reaction is indicated by a double arrow (Fig. 6.7).

The way the reaction is written defines forward and reverse reactions: a forward reaction proceeds from left to right and the reactants are located on the left side of the arrow; a reverse reaction proceeds from right to left and the reactants are located on the right side of the arrow.

The direction of a reaction can be influenced by the concentrations of reactants and products. For example, increasing the concentration of the reactants or decreasing the concentration of the products favors the forward reaction. This effect explains how many reactions in metabolic pathways proceed: the products of many reactions are quickly consumed by the next reaction, helping to drive the first reaction forward.

The laws of thermodynamics determine whether a chemical reaction requires or releases energy available to do work.

We have seen that cells use chemical reactions to perform much of their work. The amount of energy available to do work is called **Gibbs free energy (G)**. In a chemical reaction, we can compare the free energy of the reactants and products to determine whether the reaction releases energy that is available to do work. The difference between two values is denoted by the Greek letter delta (Δ). In this case, ΔG is the free energy of the products minus the free energy of the reactants (**Fig. 6.8**). If the products of a reaction have more free energy than the reactants, then ΔG is positive and a net input of energy is required to drive the reaction forward (Fig. 6.8a). By contrast, if the products of a reaction have less free energy than the reactants, ΔG is negative and energy is released and available to do work (Fig. 6.8b).

Reactions with a negative ΔG that release energy and proceed spontaneously are called **exergonic**, and reactions with a positive ΔG that require an input of energy and are not spontaneous are called **endergonic**. Note that the term "spontaneous" does not imply instantaneous or even rapid. "Spontaneous" in this context means that a reaction releases energy; "non-spontaneous" means that a reaction requires a sustained input of energy.

Recall that the total amount of energy is equal to the energy available to do work plus the energy that is not available to

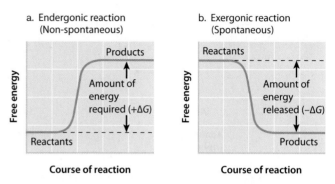

FIG. 6.8 Endergonic and exergonic reactions. An endergonic reaction requires an input of energy; an exergonic reaction releases energy.

do work because of the increase in entropy. We can write this relationship as an equation in which the total amount of energy is **enthalpy (H)**, the energy available to do work is Gibbs free energy (G), and the energy lost to entropy is **entropy (S)** multiplied by the **absolute temperature (T**, measured in degrees Kelvin), because temperature influences the movement of molecules (and hence the degree of disorder). Therefore,

Total amount of energy (H) =
    energy available to do work (G) + energy lost to entropy (TS)

We are interested in the energy available for a cell to do work, or G. Therefore, we express G in terms of the other two parameters, H and TS, as follows:

$$G = H - TS$$

In a chemical reaction, we can compare the total energy and entropy of the reactants with the total energy and entropy of the products to see if there is energy available to do work. As a result, we get

$$\Delta G = \Delta H - T\Delta S$$

This equation is a useful way to see if a chemical reaction takes place spontaneously and whether net energy is required or released. Let's consider a few examples. The value of ΔG depends on both the change in enthalpy and the change in disorder. Consider a reaction in which the products have less chemical energy (lower enthalpy) in their bonds than the reactants have, and the products are more disordered (higher entropy) than the reactants are. This reaction has a negative value of ΔH and a positive value of ΔS. Notice that, in the equation ΔG = ΔH − TΔS, there is a negative sign in front of ΔS, so both the negative ΔH and positive ΔS contribute to a negative ΔG. Therefore, this reaction proceeds spontaneously.

Now consider a reaction in which chemical energy increases (positive ΔH) and the degree of disorder decreases (negative ΔS). This reaction has a positive value of ΔG and is not spontaneous.

FIG. 6.9 ATP hydrolysis. ATP hydrolysis is an exergonic reaction that releases free energy.

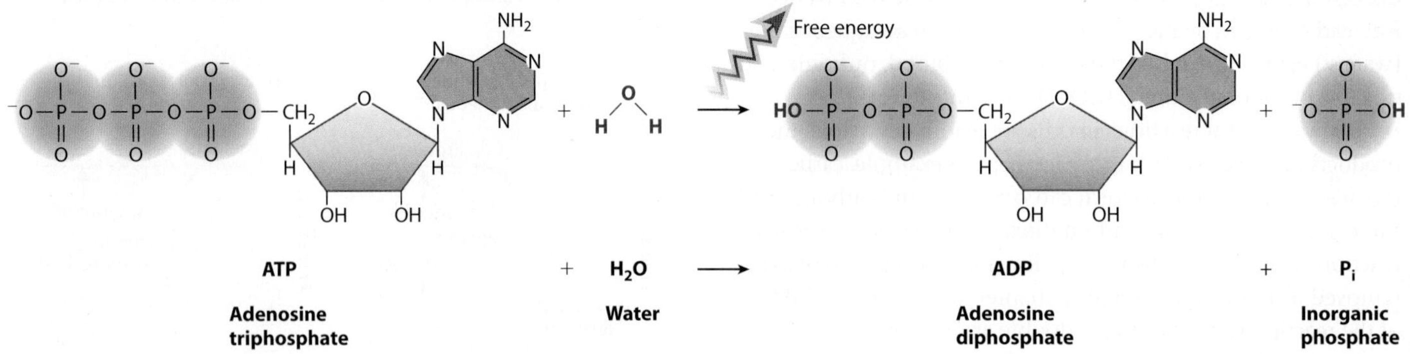

ATP	+	H₂O	→	ADP	+	P$_i$
Adenosine triphosphate		Water		Adenosine diphosphate		Inorganic phosphate

In other cases, the change in enthalpy and the change in entropy are both positive or both negative. In these cases, the absolute values of the energy and enthalpy changes determine whether ΔG is positive or negative and therefore whether a reaction is spontaneous or not.

The hydrolysis of ATP is an exergonic reaction.

Let's apply these concepts to a specific chemical reaction. Earlier, we introduced ATP, the molecule that drives many cellular processes using the chemical potential energy in its chemical bonds. ATP reacts with water to form ADP and inorganic phosphate, $P_i(HPO_4^{2-})$, as shown here and in **Fig. 6.9**:

$$ATP + H_2O \longrightarrow ADP + P_i$$

This is an example of a hydrolysis reaction, a chemical reaction in which a water molecule is split into a proton (H^+) and a hydroxyl group (OH^-). Hydrolysis reactions often break down polymers into their subunits, and in the process one product gains a proton and the other gains a hydroxyl group.

The reaction of ATP with water is an exergonic reaction because there is less free energy in the products than in the reactants. To explain the free energy difference, we can refer to the formula we derived in the last section:

$$\Delta G = \Delta H - T\Delta S$$

Recall that the phosphate groups of ATP are negatively charged at physiological pH and repel each other. ATP has three phosphate groups, and ADP has two. Therefore, ADP is more stable (contains less chemical energy in its bonds) than ATP and the value of ΔH is negative. In addition, a single molecule of ATP is broken down into two molecules, ADP and P_i. Therefore, entropy increases and the value of ΔS is positive. Because $\Delta G = \Delta H - T\Delta S$, ΔG is negative and the reaction is a spontaneous one that releases energy available to do work.

The free energy difference for ATP hydrolysis is approximately −7.3 kcal per mole (kcal/mol) of ATP. This value is influenced by several factors, including the concentration of reactants and products, the pH of the solution in which the reaction occurs, and the temperature and pressure. The value −7.3 kcal/mol is the value under standard laboratory conditions in which the concentrations of reactants and products are equal and pressure is held constant. In a cell, the free energy difference for ATP hydrolysis is likely higher, on the order of −12 kcal/mol.

The release of free energy during ATP hydrolysis comes from breaking weaker bonds (with more chemical energy) in the reactants and forming more stable bonds (with less chemical energy) in the products. The release of free energy then drives chemical reactions and other processes that require a net input of energy, as we discuss next.

FIG. 6.10 Energetic coupling. A spontaneous (exergonic) reaction drives a non-spontaneous (endergonic) reaction. (a) The hydrolysis of ATP drives the formation of glucose 6-phosphate from glucose. (b) The hydrolysis of phosphoenolpyruvate drives the synthesis of ATP.

The coupled reaction proceeds because ΔG is negative and P_i is shared between the two reactions.

a.

ATP	+ H₂O	→	ADP	+ P$_i$	$\Delta G_1 = -7.3$ kcal/mol	**Exergonic reaction**
Glucose	+ P$_i$	→	Glucose 6-phosphate	+ H₂O	$\Delta G_2 = +3.3$ kcal/mol	**Endergonic reaction**
Glucose	+ ATP	→	Glucose 6-phosphate	+ ADP	$\Delta G = -4$ kcal/mol	**Coupled reaction**

b.

Phosphoenolpyruvate	+ H₂O	→	Pyruvate	+ P$_i$	$\Delta G_1 = -14.8$ kcal/mol	**Exergonic reaction**
ADP	+ P$_i$	→	ATP	+ H₂O	$\Delta G_2 = +7.3$ kcal/mol	**Endergonic reaction**
Phosphoenolpyruvate	+ ADP	→	Pyruvate	+ ATP	$\Delta G = -7.5$ kcal/mol	**Coupled reaction**

Non-spontaneous reactions are often coupled to spontaneous reactions.

If the conversion of reactant A into product B is spontaneous, the reverse reaction converting reactant B into product A is not. The ΔG's for the forward and reverse reactions have the same absolute value but opposite signs. In this situation, you might expect that the direction of the reaction would always be from A to B. However, in living organisms, not all chemical reactions are spontaneous. Anabolic reactions are a good example; they require an input of energy to drive them in the right direction. This raises the question: what drives non-spontaneous reactions?

Energetic coupling is a process in which a spontaneous reaction (negative ΔG) drives a non-spontaneous reaction (positive ΔG). It requires that the net ΔG of the two reactions be negative. In addition, the two reactions must occur together. In some cases, this coupling can be achieved if the two reactions share an intermediate.

For example, a spontaneous reaction, ATP hydrolysis, can be used to drive a non-spontaneous reaction, the addition of a phosphate group to glucose to produce glucose 6-phosphate, as shown in **Fig. 6.10a**. In this case, the phosphate group transferred to glucose is released during ATP hydrolysis. The net ΔG for the two reactions is negative. So ATP hydrolysis provides the thermodynamic driving force for the non-spontaneous reaction, and the shared phosphate group couples the two reactions together.

Following ATP hydrolysis, the cell needs to replenish its ATP so that it can carry out additional chemical reactions. The synthesis of ATP from ADP and P_i is an endergonic reaction with a positive ΔG, requiring an input of energy. In some cases, exergonic reactions can drive the synthesis of ATP by energetic coupling (**Fig. 6.10b**). The sum of the ΔG's of the two reactions is negative and the reactions share a phosphate group, allowing the two reactions to proceed.

Like ATP, other phosphorylated molecules can be hydrolyzed, releasing free energy. Hydrolysis reactions can be ranked by their free energy differences (ΔG). **Fig. 6.11** shows that, compared with the hydrolysis of other phosphorylated molecules, ATP hydrolysis has an intermediate free energy difference. Those reactions that have a ΔG more negative than that of ATP hydrolysis transfer a phosphate group to ADP by energetic coupling, and those reactions that have a less negative ΔG receive a phosphate group from ATP by energetic coupling. Therefore, ADP is an energy acceptor and ATP an energy donor. The ATP–ADP system is at the core of energetic coupling between catabolic and anabolic reactions.

> ### Self-Assessment Questions
>
> 8. For a chemical reaction with a positive value of ΔG, what are the possible relative values for ΔH and TΔS? Is the reaction spontaneous or not? Answer the same questions for a chemical reaction with a negative value of ΔG.
> 9. How does increasing the temperature affect the change in free energy (ΔG) of a chemical reaction?
> 10. How can the hydrolysis of ATP drive non-spontaneous reactions in a cell?

6.5 ENZYMES AND THE RATE OF CHEMICAL REACTIONS

Up to this point, we have focused on the spontaneity and direction of chemical reactions. Now we address their rate. As mentioned earlier, a spontaneous reaction is not necessarily a fast one. For example, the breakdown of glucose into carbon dioxide and water is spontaneous with a negative ΔG, but the rate of the reaction is close to zero and the breakdown of glucose is imperceptible. However, glucose is readily broken down inside cells all the time. How is this possible? The answer is that chemical reactions in a cell are accelerated by chemical catalysts.

The rate of a chemical reaction is defined as the amount of product formed (or reactant consumed) per unit of time. Catalysts are substances that increase the rate of chemical reactions without themselves being consumed. In biological systems, the catalysts are usually proteins called **enzymes**, although, as we saw in Chapter 2, some RNA molecules have catalytic activity as well. Enzymes can increase the rate of chemical reactions dramatically. Moreover, because they are highly specific, acting only on certain reactants and catalyzing only some reactions, enzymes play a critical role in determining which chemical reactions take place from all the possible reactions that could occur in a cell. In this section, we discuss how enzymes increase the rate of chemical reactions and how this ability gives them a central role in metabolism.

Enzymes reduce the activation energy of a chemical reaction.

Earlier, we saw that exergonic reactions release free energy and endergonic reactions require free energy. Nevertheless, all chemical reactions

FIG. 6.11 Values of ΔG of common hydrolysis reactions in a cell. ATP hydrolysis has an intermediate value of ΔG compared to those of other common hydrolysis reactions.

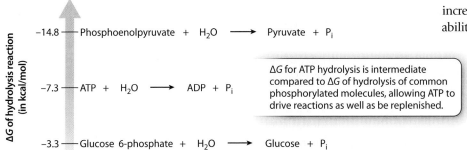

require an initial input of energy to proceed, even exergonic reactions that release energy. For an exergonic reaction, the energy released is more than the initial input of energy, so there is a net release of energy.

Why is an input of energy needed for all chemical reactions? As a chemical reaction proceeds, existing chemical bonds break and new bonds form. For an extremely brief period of time, a compound is formed in which the old bonds are breaking and the new ones are forming. This intermediate between reactants and products is called the **transition state**. It is highly unstable and therefore has a large amount of free energy.

In all chemical reactions, reactants adopt at least one transition state before their conversion into products. The graph in **Fig. 6.12** represents the free energy levels of the reactant, transition state, and product. This reaction is spontaneous because the free energy of the reactant is higher than the free energy of the product and ΔG is negative. However, the highest free energy value corresponds to the transition state.

To reach the transition state, the reactant must absorb energy from its surroundings (the uphill portion of the curve in Fig. 6.12). As a result, all chemical reactions, even spontaneous ones that release energy, require an input of energy that we can think of as an "energy barrier." The energy input necessary to reach the transition state is called the **activation energy (E_A)**. Once the transition state is reached, the reaction proceeds, products are formed, and energy is released into the surroundings (the downhill portion of the curve in Fig. 6.12). There is an inverse correlation between the rate of a reaction and the height of the energy barrier: the lower the energy barrier, the faster the reaction and the higher the barrier, the slower the reaction.

In chemical reactions that take place in the laboratory, heat is a common source of energy used to overcome the energy barrier. In living organisms, enzymes accelerate chemical reactions. Enzymes do not act by supplying heat. Instead, they reduce the activation energy by stabilizing the transition state and decreasing its free energy (the red curve in Fig. 6.12). As the activation energy decreases, the speed of the reaction increases.

Although an enzyme reduces the activation energy, the difference in free energy between reactants and products (ΔG) does not change. In other words, an enzyme changes the path of the reaction between reactants and products, but not the starting or end point (Fig. 6.12). Consider the breakdown of glucose into carbon dioxide and water. The ΔG of the reaction is the same whether it proceeds by combustion or by the action of multiple enzymes in a metabolic pathway in a cell.

Enzymes form a complex with reactants and products.

As catalysts, enzymes participate in a chemical reaction but are not consumed in the process. They emerge from a chemical reaction unchanged, ready to catalyze the same reaction again. Enzymes accomplish these functions by forming a complex with the reactants and products.

In a chemical reaction catalyzed by an enzyme, the reactant is often referred to as the **substrate**. In such a reaction, the substrate (S) is converted to a product (P):

$$S \rightleftharpoons P$$

In the presence of an enzyme (E), the substrate first forms a complex with the enzyme (enzyme–substrate, or ES). While still part of the complex, the substrate is converted to product (enzyme–product, or EP). Finally, the complex dissociates, releasing the enzyme and product. Therefore, a reaction catalyzed by an enzyme can be described as

$$S + E \rightleftharpoons ES \rightleftharpoons EP \rightleftharpoons E + P$$

The formation of this complex is critical for accelerating the rate of a chemical reaction. Recall from Chapter 4 that proteins adopt three-dimensional shapes and that the shape of a protein is linked to its function. Enzymes are folded into three-dimensional shapes that bring particular amino acids into close proximity to form an **active site**. The active site is the part of the enzyme that binds substrate and catalyzes its conversion to the product (**Fig. 6.13**). In the active site, the enzyme and substrate form transient covalent bonds, weak noncovalent interactions, or both. Together, these interactions stabilize the transition state and decrease the activation energy.

Enzymes also reduce the energy of activation by positioning two substrates to react: within the active site, their reactive chemical groups are aligned and their motion relative to each other restrained.

The size of the active site is extremely small compared with the size of the enzyme. If only a small fraction of the enzyme is necessary for the catalysis of a reaction, why are enzymes so large? Of the many amino acids that form the active site, only a few actively contribute to catalysis. Each of these amino acids has to occupy a very specific spatial position to align with the correct reactive group on the substrate. If the few essential

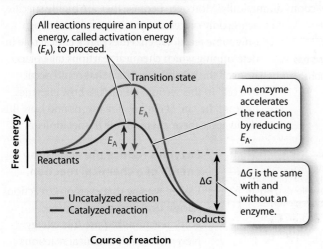

FIG. 6.12 An enzyme-catalyzed reaction. An enzyme accelerates a reaction by lowering the activation energy, E_A.

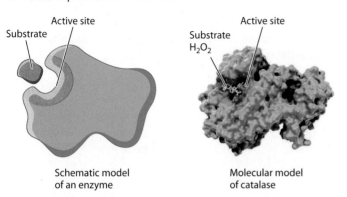

FIG. 6.13 The active site of the enzyme catalase, shown schematically and as a surface model. Substrate binds and is converted to product at the active site.

amino acids were part of a short peptide, the alignment of chemical groups between the peptide and the substrate would be difficult or even impossible because the length of the bonds and the bond angles in the peptide would constrain its three-dimensional structure. In many cases, the catalytic amino acids are spaced far apart in the primary structure of the enzyme, but brought close together by protein folding (**Fig. 6.14**). In other words, the large size of many enzymes is required at least in part to bring the catalytic amino acids into very specific positions in the active site of the folded enzyme.

The formation of a complex between enzymes and substrates can be demonstrated experimentally, as illustrated in **Fig. 6.15**.

Enzymes are highly specific.

Enzymes are remarkably specific both for the substrate and the reaction that is catalyzed. In general, enzymes recognize either a unique substrate or a class of substrates that share common chemical structures. In addition, enzymes catalyze only one reaction or a very limited number of reactions.

For example, the enzyme succinate dehydrogenase acts only on succinate. Even more remarkably, the enzyme β-(beta-) galactosidase cleaves the glycosidic bond that links galactose to glucose in the disaccharide lactose, but it does not cleave an almost identical molecule that differs from lactose only in the orientation of the glycosidic bond. In this case, the enzyme does not recognize the whole substrate but a particular structural motif within it, and very small differences in the structure of this motif affect the activity of the enzyme. The specificity of enzymes can be attributed to the structure of their active sites. The active site interacts only with substrates having a precise three-dimensional structure.

Enzyme activity can be influenced by inhibitors and activators.

The activity of enzymes can be influenced by **inhibitors** and **activators**. Inhibitors decrease the activity of enzymes, whereas activators increase the activity of enzymes. Enzyme inhibitors are quite common. They are synthesized naturally by many plants and animals as a defense against predators. Similarly, pesticides and herbicides often target enzymes to inactivate them. Many drugs used in medicine are enzyme inhibitors, including drugs used to treat infections as well as drugs used to treat cancer and other diseases. Given the importance of chemical reactions and the role of enzymes in metabolism, it is not surprising that enzyme inhibitors have such widespread applications.

There are two classes of inhibitors. Irreversible inhibitors usually form covalent bonds with enzymes and irreversibly inactivate them. Reversible inhibitors form weak bonds with enzymes and therefore easily dissociate from them.

Inhibitors can act in many different ways, two of which are shown in **Fig. 6.16**. In some cases, an inhibitor is similar in structure to the substrate and therefore is able to bind to the active site of the enzyme (Fig. 6.16a). Binding of the inhibitor prevents the binding of the substrate. In other words, the inhibitor competes with the substrate for the active site of the enzyme. These types of inhibitors can often be overcome by increasing the concentration of substrate. Other inhibitors bind to a site other than the active site of the enzyme but still inhibit the activity of the enzyme (Fig. 6.16b). In this case, binding of the inhibitor changes the shape and therefore the activity of the enzyme. This type of inhibitor usually has a structure very different from that of the substrate.

Allosteric enzymes regulate key metabolic pathways.

Enzymes that are regulated by molecules that bind at sites other than their active sites are called **allosteric enzymes.** The activity of allosteric enzymes can be influenced by both inhibitors and activators. They play a key role in the regulation of

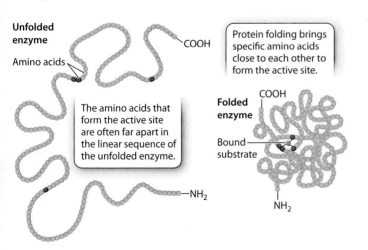

FIG. 6.14 Formation of the active site of an enzyme by protein folding.

HOW DO WE KNOW?

FIG. 6.15

Do enzymes form complexes with substrates?

BACKGROUND The idea that enzymes form complexes with substrates to catalyze a chemical reaction was first proposed in 1888 by the Swedish chemist Svante Arrhenius. In the 1930s, American chemist Kurt Stern performed one of the earliest experiments that supported Arrhenius' idea. Stern studied an enzyme called catalase, which is very abundant in animal and plant tissues. In the conclusion of his paper, he wrote, "It remains to be seen to which extent the findings of this study apply to enzyme action in general." To illustrate that Stern's findings do indeed apply to enzymes in general, another experiment that analyzes a different enzyme and technique are described here.

Lactose belongs to a family of molecules called β-galactosides. The enzyme β-galactosidase catalyzes the cleavage (splitting) of the glycosidic bond in β-galactoside. In β-thiogalactoside, the oxygen atom in the glycosidic bond is replaced by sulfur. As a result of this difference, the enzyme β-galactosidase binds β-thiogalactoside but cannot cleave or break the glycosidic bond in β-thiogalactoside as it can in β-galactoside.

β-galactoside   β-thiogalactoside

METHOD A container is separated into two compartments by a selectively permeable membrane. The membrane is permeable to β-galactoside and β-thiogalactoside, but it is not permeable to the enzyme β-galactosidase.

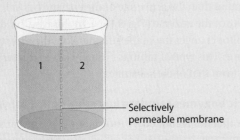

EXPERIMENT 1 AND RESULTS Radioactively labeled β-thiogalactoside (the substrate, S) is added to compartment 1 and the movement of S is followed by measuring the level of radioactivity in the two compartments. Over time, the level of radioactivity becomes the same in the two compartments.

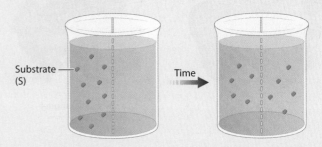

EXPERIMENT 2 AND RESULTS Radioactively labeled β-thiogalactoside (S) is added to compartment 1, the enzyme β-galactosidase (E) is added to compartment 2, and the movement of S is followed by measuring the level of radioactivity in the two compartments. Over time, the level of radioactivity becomes greater in compartment 2 than in compartment 1.

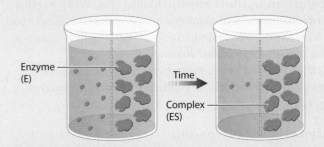

INTERPRETATION The substrate β-thiogalactoside (S) diffuses from compartment 1 to compartment 2, forms a complex (ES) with the enzyme β-galactosidase (E), and is not released because the enzyme (E) does not catalyze the conversion of substrate (S) to product (P). The level of radioactivity is greater in compartment 2 compared to the level in compartment 1 because E and S form a complex (ES) and the radioactive substrate is not released from the complex to diffuse back across the membrane.

SOURCES Adapted from Doherty, D. G., and F. Vaslow. 1952. "Thermodynamic Study of an Enzyme–Substrate Complex of Chymotrypsin." *J. Am. Chem. Soc.* 74:931–936. Stern, K. G. 1936. "On the Mechanism of Enzyme Action: A Study of the Decomposition of Monoethyl Hydrogen Peroxide by Catalase and of an Intermediate Enzyme–Substrate Compound." *J. Biol. Chem.* 114:473–494.

metabolic pathways. Consider the synthesis of isoleucine from threonine, a pathway found in some bacteria. This conversion requires five reactions, each catalyzed by a different enzyme (**Fig. 6.17**).

Once the bacterium has enough isoleucine for its needs, it would be a waste of energy to continue synthesizing the amino acid. To shut down the pathway once it is no longer needed, the cell relies on an enzyme inhibitor. The inhibitor is isoleucine, the final product of the five reactions. Isoleucine binds to the first enzyme in the pathway, threonine dehydratase, at a site distinct from the active site. Threonine dehydratase is thus an example of an allosteric enzyme. The binding of isoleucine changes the shape of the enzyme and in this way inhibits its function.

The isoleucine pathway provides an example of **negative feedback**, in which the final product inhibits the first step of the reaction. This is a common mechanism used widely in organisms to maintain homeostasis—that is, to actively maintain stable conditions or steady levels of a substance (Chapter 5).

The activity of threonine dehydratase, and therefore the rate of the reaction, depends on the concentration of substrate. At a

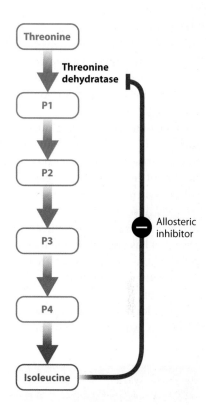

FIG. 6.17 Regulation of threonine dehydratase, an allosteric enzyme. The enzyme is inhibited by the final product, isoleucine, which binds to a site distinct from the active site.

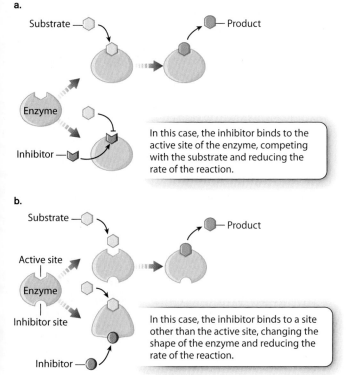

FIG. 6.16 Two mechanisms of inhibitor function. (a) Some inhibitors bind to the active site of the enzyme and (b) other inhibitors bind to a site that is different from the active site. Both types of inhibitor reduce the activity of an enzyme and therefore decrease the rate of the reaction.

low concentration of threonine, the rate of the reaction is very slow. As the concentration of threonine increases, the activity of the enzyme increases. At a particular threshold, a small increase in threonine concentration results in a large increase in reaction rate. Finally, when there is excess substrate, the reaction rate slows down.

In the next two chapters, we examine more closely two key metabolic processes, cellular respiration and photosynthesis. Both of these processes require many chemical reactions acting in a coordinated fashion. Allosteric enzymes catalyze key reactions in these and other metabolic pathways. These enzymes are usually found at or near the start of a metabolic pathway or at the crossroads between two metabolic pathways. Allosteric enzymes are one way that the cell coordinates the activity of multiple metabolic pathways.

CASE 1 LIFE'S ORIGINS: INFORMATION, HOMEOSTASIS, AND ENERGY

What naturally occurring elements might have spurred the first reactions that led to life?

Many enzymes contain metal ions in addition to amino acids. These ions are one type of **cofactor**, a substance that associates with an enzyme and plays a key role in its function. Metallic cofactors, especially iron, magnesium, manganese, cobalt, copper, zinc, and molybdenum, bind to diverse proteins, including enzymes used in DNA synthesis and nitrogen metabolism. With this in mind, scientists have asked whether metal ions might, by themselves, catalyze chemical reactions thought to have played a role in the origin of life. They do. For example, magnesium and zinc ions added to solutions can accelerate the linking of nucleotides to form RNA and DNA molecules.

Metallic cofactors also bind to enzymes used in the transport of electrons for cellular respiration and photosynthesis, as discussed in the next two chapters. Some particularly important enzymes used for electron transport contain iron and sulfur clustered together. Iron–sulfur minerals, especially pyrite (or fool's gold, FeS_2), form commonly in mid-ocean hydrothermal vent systems and other environments where oxygen is absent. It has been proposed that reactions now carried out in cells by iron–sulfur proteins are the evolutionary descendants of chemical reactions that took place spontaneously on the early Earth.

The idea that your cells preserve an evolutionary memory of ancient hydrothermal environments may seem like science fiction, but it finds support in laboratory experiments. For example, the reaction of H_2S and FeS to form pyrite has been shown to catalyze a number of plausibly pre-biotic chemical reactions, including the formation of pyruvate (a key intermediate in energy metabolism discussed in Chapter 7). Thus, the metals in enzymes help connect the chemistry of life to the chemistry of Earth.

Self-Assessment Questions

11. What are three characteristics of enzymes, and how does each permit chemical reactions to occur in cells?
12. Which of the following do enzymes change? ΔG, reaction rate, types of product, activation energy, the laws of thermodynamics.
13. How does protein folding allow for enzyme specificity?

CORE CONCEPTS SUMMARY

6.1 AN OVERVIEW OF METABOLISM: Metabolism is the set of biochemical reactions that transforms molecules and transfers energy.

Organisms can be grouped according to their source of energy: phototrophs obtain energy from sunlight and chemotrophs obtain energy from chemical compounds. page 119

Organisms can also be grouped according to the source of carbon they use to build organic molecules: heterotrophs obtain carbon from organic molecules, and autotrophs obtain carbon from inorganic sources, such as carbon dioxide. page 120

Catabolism is the set of chemical reactions that break down molecules and release energy, and anabolism is the set of chemical reactions that build molecules and require energy. page 121

6.2 KINETIC AND POTENTIAL ENERGY: Kinetic energy is energy of motion and potential energy is stored energy.

Kinetic energy is energy associated with movement. page 121

Potential energy is stored energy; it depends on the structure of an object or its position relative to its surroundings. page 122

Chemical energy is a form of potential energy held in the bonds of molecules. page 122

The bonds linking phosphate groups in ATP have high potential energy that can be harnessed for use by the cell. page 122

6.3 LAWS OF THERMODYNAMICS: The first and second laws of thermodynamics govern energy flow in biological systems.

The first law of thermodynamics states that energy cannot be created or destroyed. page 123

The second law of thermodynamics states that there is an increase in entropy in the universe over time. page 123

6.4 CHEMICAL REACTIONS: Chemical reactions involve the breaking and forming of bonds.

In a chemical reaction, atoms themselves do not change, but which atoms are linked to each other changes, forming new molecules. page 124

The direction of a chemical reaction is influenced by the concentration of reactants and products. page 125

Gibbs free energy (G) is the amount of energy available to do work. page 125

Three thermodynamic parameters define a chemical reaction: Gibbs free energy (G), enthalpy (H), and entropy (S). page 125

Exergonic reactions are spontaneous ($\Delta G < 0$) and release energy. page 125

Endergonic reactions are non-spontaneous ($\Delta G > 0$) and require energy. page 125

The change of free energy in a chemical reaction is described by $\Delta G = \Delta H - T\Delta S$. page 125

The hydrolysis of ATP is an exergonic reaction that drives many endergonic reactions in a cell. page 126

In living systems, non-spontaneous reactions are often coupled to spontaneous ones. page 127

6.5 ENZYMES AND THE RATE OF CHEMICAL REACTIONS: The rate of chemical reactions is increased by protein catalysts called enzymes.

Enzymes reduce the free energy level of the transition state between reactants and products, thereby reducing the energy input, or activation energy, required for a chemical reaction to proceed. page 128

During catalysis, the substrate and product form a complex with the enzyme. Transient covalent bonds, weak noncovalent interactions, or both stabilize the complex. page 128

The size of the active site of an enzyme is small compared to the size of the enzyme as a whole and the active site amino acids occupy a very specific spatial arrangement. page 128

An enzyme is highly specific for its substrate and for the types of reaction it catalyzes. page 129

Inhibitors reduce the activity of enzymes and can act irreversibly or reversibly. page 129

Activators increase the activity of enzymes. page 129

Allosteric enzymes bind activators and inhibitors at sites other than the active site, resulting in a change in their shape and activity. page 129

Allosteric enzymes are often found at or near the start of a metabolic pathway or at the crossroads of multiple pathways. page 130

Log in to **LaunchPad** to check your answers to the Self-Assessment Questions and to access additional learning tools.

CHAPTER 7: Cellular Respiration
Harvesting Energy from Carbohydrates and Other Fuel Molecules

CORE CONCEPTS

7.1 AN OVERVIEW OF CELLULAR RESPIRATION: Cellular respiration is a series of catabolic reactions that converts the energy in fuel molecules into energy in ATP.

7.2 GLYCOLYSIS: Glycolysis is the partial oxidation of glucose and results in the production of pyruvate, as well as ATP and reduced electron carriers.

7.3 PYRUVATE OXIDATION: Pyruvate is oxidized to acetyl-CoA, connecting glycolysis to the citric acid cycle.

7.4 THE CITRIC ACID CYCLE: The citric acid cycle results in the complete oxidation of fuel molecules and the generation of ATP and reduced electron carriers.

7.5 THE ELECTRON TRANSPORT CHAIN AND OXIDATIVE PHOSPHORYLATION: The electron transport chain transfers electrons from electron carriers to oxygen, using the energy released to pump protons and synthesize ATP by oxidative phosphorylation.

7.6 ANAEROBIC METABOLISM: Glucose can be broken down in the absence of oxygen by fermentation, producing a modest amount of ATP.

7.7 METABOLIC INTEGRATION: Metabolic pathways are integrated, allowing control of the energy level of cells.

All living organisms have the ability to harness energy from the environment. We saw in Chapter 6 that energy is needed for all kinds of functions, including cell movement and division, muscle contraction, growth and development, and the synthesis of macromolecules. Organic molecules such as carbohydrates, lipids, and proteins are good sources of chemical energy. Some organisms, such as humans and other heterotrophs, consume organic molecules in their diet. Others, such as plants and other autotrophs, synthesize these molecules from inorganic molecules such as carbon dioxide. Regardless of how they obtain organic molecules, nearly all organisms—animals, plants, fungi, and microbes—break them down in the process of **cellular respiration**, releasing energy that can be used to do the work of the cell. Cellular respiration is a series of chemical reactions that convert the chemical energy in fuel molecules into the chemical energy of **adenosine triphosphate (ATP)**. ATP can be readily used as a source of energy by all cells.

It is tempting to think that organic molecules are converted into energy in this process, but this is not the case. Recall from Chapter 6 that the first law of thermodynamics states that energy cannot be created or destroyed. Rather than creating energy, cellular respiration converts energy from one form to another. Specifically, it converts the chemical potential energy stored in organic molecules to the chemical potential energy in ATP.

It is also easy to forget that organisms other than animals, such as plants, use cellular respiration. If plants use sunlight as a source of energy, why would they need cellular respiration? As we will see in the next chapter, plants use the energy of sunlight to make carbohydrates. Plants then break down these carbohydrates in the process of cellular respiration to produce ATP.

In this chapter, we discuss the metabolic pathways of cellular respiration that supply the energy needs of a cell: the breakdown, storage, and mobilization of sugars such as glucose, the synthesis of ATP, and the coordination and regulation of these metabolic pathways.

7.1 AN OVERVIEW OF CELLULAR RESPIRATION

In Chapter 6, we saw that catabolism describes the set of chemical reactions that breaks down molecules into smaller units. In the process, these reactions release chemical energy that can be stored in molecules of ATP. Anabolism, by contrast, is the set of chemical reactions that build molecules from smaller units. Anabolic reactions require an input of energy, usually in the form of ATP.

Cellular respiration is one of the major sets of catabolic reactions in a cell. During cellular respiration, fuel molecules such as glucose, fatty acids, and proteins are catabolized into smaller units, releasing the energy stored in their chemical bonds to power the work of the cell.

Cellular respiration uses chemical energy stored in molecules such as carbohydrates and lipids to produce ATP.

Cellular respiration is a series of catabolic reactions that converts the energy stored in food molecules, such as glucose, into the energy stored in ATP and that produces carbon dioxide as a waste or by-product. Cellular respiration can occur in the presence of oxygen (termed aerobic respiration) or in its absence (termed anaerobic respiration). Most organisms that you are familiar with are capable of aerobic respiration; some bacteria respire anaerobically (Chapter 24). Here we focus on aerobic respiration. Oxygen is consumed in aerobic respiration, and carbon dioxide and water are produced:

$$C_6H_{12}O_6 + 6O_2 \rightarrow 6CO_2 + 6H_2O + \text{energy}$$
Glucose Oxygen Carbon dioxide Water

In Chapter 6, we saw that molecules such as carbohydrates and lipids have a large amount of potential energy in their chemical bonds. In contrast, molecules such as carbon dioxide and water have less potential energy in their bonds. Cellular respiration releases a large amount of energy because the sum of the potential energy in all of the chemical bonds of the reactants (glucose and oxygen) is higher than that of the products (carbon dioxide and water). The maximum amount of free energy—energy available to do work—released during cellular respiration is −686 kcal per mole of glucose. Recall from Chapter 6 that when $\Delta G < 0$, energy is released.

The overall reaction for cellular respiration helps us focus on the starting reactants, final products, and release of energy. However, it misses the many intermediate steps. Tossing a match into the gas tank of a car releases a tremendous amount of energy in the form of an explosion, but this energy is not used to do work. Instead, it is released as light and heat. Similarly, if all the energy stored in glucose were released at once, most of it would be released as heat and the cell would not be able to harness it to do work.

In cellular respiration, energy is released gradually in a series of chemical reactions (**Fig. 7.1**). As a result, some of this energy can be used to form ATP. On average, 32 molecules of ATP are produced from the aerobic respiration of a single molecule of glucose. The energy needed to form one mole of ATP from ADP and P_i is at least 7.3 kcal. Thus, cellular respiration harnesses at least $32 \times 7.3 = 233.6$ kcal of energy in ATP for every mole of glucose that is broken down in the presence of oxygen.

About 34% of the total energy released by aerobic respiration is harnessed in the form of ATP ($233.6/686 = 34\%$), with the remainder of the energy given off as heat. This degree of efficiency compares favorably with the approximately 25% efficiency of a gasoline engine.

ATP is generated by substrate-level phosphorylation and oxidative phosphorylation.

In cellular respiration, ATP is produced in two ways (Fig. 7.1). In the first, an organic molecule transfers a phosphate group directly to ADP, as we saw in Chapter 6. In this case, a single enzyme carries out two coupled reactions: the hydrolysis of an organic molecule to yield a phosphate group and the addition of that phosphate group to ADP. The hydrolysis reaction releases enough free energy to drive the synthesis of ATP. This way of generating ATP is called **substrate-level phosphorylation** because a phosphate group is transferred to ADP from an enzyme substrate, in this case an organic molecule.

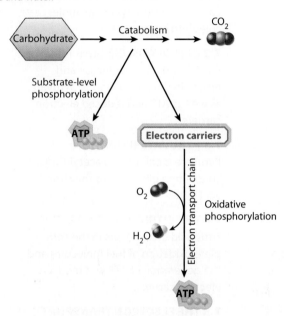

FIG. 7.1 An overview of cellular respiration. In most organisms, cellular respiration consumes oxygen and produces ATP by substrate-level phosphorylation and oxidative phosphorylation, as well as carbon dioxide and water.

Substrate-level phosphorylation produces only a small amount of the total ATP generated in cellular respiration, about 12% if the fuel molecule is glucose. The remaining 88% is produced by **oxidative phosphorylation** (Fig. 7.1). In this case, the chemical energy of organic molecules is transferred first to **electron carriers**. The role of these electron carriers is exactly what their name suggests—they carry electrons (and energy) from one set of reactions to another. In cellular respiration, electron carriers transport electrons released during the catabolism of organic molecules to the respiratory **electron transport chain**. Electron transport chains in turn transfer electrons along a series of membrane-associated proteins to a final electron acceptor and in the process, harness the energy released to produce ATP. In aerobic respiration, oxygen is the final electron acceptor, resulting in the formation of water. Electron transport chains are central to the energy economy of most cells. In respiration, electron transport chains harness energy from fuel molecules such as glucose, and in photosynthesis they harness energy from sunlight (Chapter 8). In section 7.5, we discuss how electron movement through the electron transport chain is coupled to ATP synthesis.

Redox reactions play a central role in cellular respiration.

Chemical reactions in which electrons are transferred from one atom or molecule to another are called **oxidation–reduction reactions** ("redox reactions" for short). Oxidation is the loss of electrons, and reduction is the gain of electrons. The loss and gain of electrons always occur together in a coupled oxidation–reduction reaction: electrons are transferred from one molecule to another so that one molecule loses electrons and one molecule gains those electrons. The molecule that loses electrons is oxidized and the molecule that gains electrons is reduced.

We can illustrate oxidation–reduction reactions by looking at the role played by electron carriers in cellular respiration. Two important electron carriers are nicotinamide adenine dinucleotide and flavin adenine dinucleotide. These electron carriers exist in two forms—an oxidized form (NAD^+ and FAD) and a reduced form (NADH and $FADH_2$). When fuel molecules such as glucose are catabolized, some of the steps are oxidation reactions. These oxidation reactions are coupled with the reduction of electron carrier molecules. These reduction reactions can be written as:

$$NAD^+ + 2e^- + H^+ \rightarrow NADH$$
$$FAD + 2e^- + 2H^+ \rightarrow FADH_2$$

Here, NAD^+ and FAD accept electrons and are converted to their reduced forms, NADH and $FADH_2$. Note that in redox reactions involving organic molecules such as NAD^+ or FAD, the gain (or loss) of electrons is often accompanied by the gain (or loss) of protons (H^+). Therefore, you can easily recognize reduced molecules by an increase in C—H bonds and the corresponding oxidized molecules by a decrease in C—H bonds.

In their reduced forms, NADH and $FADH_2$ can donate electrons. As they are oxidized to NAD^+ and FAD, they transfer electrons (and energy) to the electron transport chain:

$$NADH \rightarrow NAD^+ + 2e^- + H^+$$
$$FADH_2 \rightarrow FAD + 2e^- + 2H^+$$

NAD^+ and FAD can then accept electrons from the breakdown of fuel molecules. In this way, electron carriers act as shuttles, transferring electrons derived from the oxidation of fuel molecules such as glucose to the electron transport chain.

Cellular respiration can itself be understood as a redox reaction, even though it consists of many steps. In aerobic respiration, glucose is oxidized, releasing carbon dioxide, and at the same time oxygen is reduced, forming water:

Oxidation
$$C_6H_{12}O_6 + 6O_2 \rightarrow 6CO_2 + 6H_2O + energy$$
Reduction

Let's first consider the oxidation reaction (**Fig. 7.2a**). In glucose, there are many C—C and C—H covalent bonds, in which electrons are shared about equally between the two atoms. By contrast, in carbon dioxide, electrons are not shared equally. The oxygen atom is more electronegative than the carbon atom, so the electrons are more likely to be found near the oxygen atom. As a result, carbon has partially lost electrons to oxygen and is oxidized.

FIG. 7.2 Oxidation and reduction in the overall reaction for cellular respiration. Electrons are partially lost or gained in the formation of polar covalent bonds.

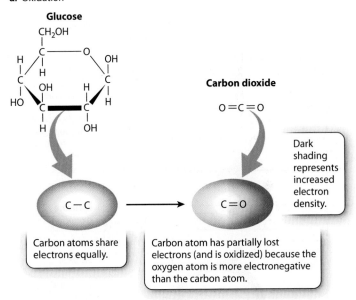

a. Oxidation

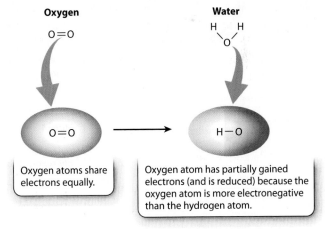

b. Reduction

Let's now consider the reduction reaction (**Fig. 7.2b**). In oxygen gas, electrons are shared equally between two oxygen atoms. In water, the electrons that are shared between hydrogen and oxygen are more likely to be found near oxygen because oxygen is more electronegative than hydrogen. As a result, oxygen has partially gained electrons and is reduced.

Cellular respiration occurs in four stages.

Up to this point, we have focused on the overall reaction of cellular respiration and the chemistry of redox reactions. Let's now look at the entire process, starting with glucose. Cellular respiration occurs in four stages (**Fig. 7.3**).

FIG. 7.3 The four stages of cellular respiration. Cellular respiration consists of glycolysis, pyruvate oxidation, the citric acid cycle, and oxidative phosphorylation.

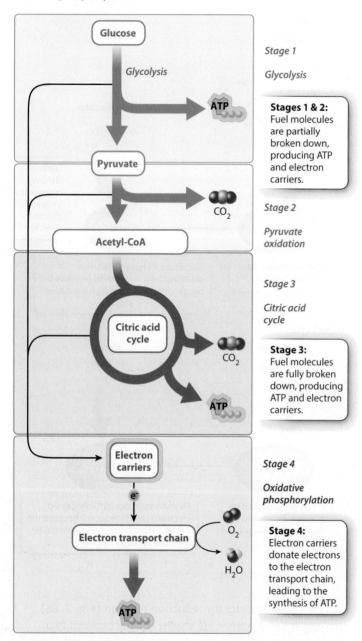

In stage 1, glucose is partially broken down to produce pyruvate, and energy is transferred to ATP and reduced electron carriers, a process known as **glycolysis**.

In stage 2, pyruvate is oxidized to another molecule called acetyl-coenzyme A (acetyl-CoA), producing reduced electron carriers and releasing carbon dioxide.

Acetyl-CoA enters stage 3, the **citric acid cycle,** also called the tricarboxylic (TCA) cycle or the Krebs cycle. In this series of chemical reactions, the acetyl group is completely oxidized to carbon dioxide and energy is transferred to ATP and reduced electron carriers. The amount of energy transferred to ATP and reduced electron carriers in this stage is nearly twice that transferred by stages 1 and 2 combined.

Stage 4 is **oxidative phosphorylation**. In this series of reactions, reduced electron carriers generated in stages 1–3 donate electrons to the electron transport chain and a large amount of ATP is produced.

Each stage of cellular respiration consists of a series of reactions, some of which are redox reactions. Therefore, glucose is not oxidized all at once to carbon dioxide. Instead, it is oxidized slowly and in a controlled manner, allowing for energy to be harnessed. Some of this energy is used to synthesize ATP directly and some of it is stored temporarily in reduced electron carriers and then used to generate ATP by oxidative phosphorylation.

Fig. 7.4 shows the change of free energy at each step in the catabolism of glucose. As you can see, the individual reactions allow the initial chemical energy present in a molecule of glucose to be "packaged" into molecules of ATP and reduced electron carriers. Note that the change in free energy is much greater for the steps that generate reduced electron carriers than for those that produce ATP directly.

In eukaryotes, glycolysis takes place in the cytoplasm, and pyruvate oxidation, the citric acid cycle, and oxidative phosphorylation all take place in mitochondria. The electron transport chain is made up of proteins and small molecules associated with the inner mitochondrial membrane (Chapter 5). In some bacteria, these reactions take place in the cytoplasm, and the electron transport chain is located in the plasma membrane.

Self-Assessment Questions

1. What are the four major stages of cellular respiration? Briefly describe each one.
2. What is an oxidation–reduction reaction? Why is the breakdown of glucose in the presence of oxygen to produce carbon dioxide and water an example of an oxidation–reduction reaction?
3. For each of the following pairs of molecules, indicate which member of the pair is reduced and which is oxidized, and which has more chemical energy and which has less chemical energy: NAD^+/NADH; FAD/$FADH_2$; CO_2/$C_6H_{12}O_6$.
4. What are two different ways in which ATP is generated in cellular respiration?

FIG. 7.4 Change in free energy in cellular respiration. Glucose is oxidized through a series of chemical reactions, releasing energy in the form of ATP and reduced electron carriers.

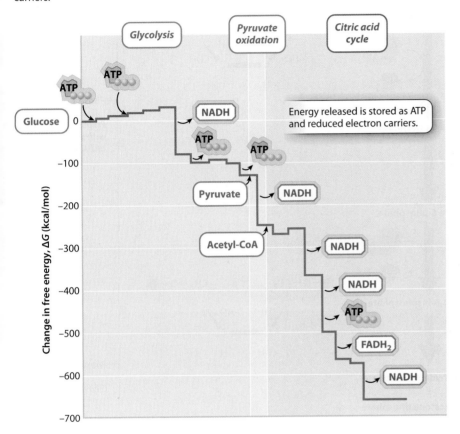

7.2 GLYCOLYSIS

Glucose is the most common fuel molecule in animals, plants, and microbes. It is the starting molecule for glycolysis, which results in the partial oxidation of glucose and the synthesis of a relatively small amount of both ATP and reduced electron carriers. Glycolysis literally means "splitting sugar," an apt name because a 6-carbon sugar (glucose) is split in two in this process, yielding two 3-carbon molecules. Glycolysis is anaerobic because oxygen is not consumed. It evolved very early in the evolution of life, when oxygen was not present in Earth's atmosphere. It occurs in nearly all living organisms and is therefore probably the most widespread metabolic pathway among organisms.

Glycolysis is the partial breakdown of glucose.

Glycolysis begins with a molecule of glucose and produces two 3-carbon molecules of pyruvate and a net total of two molecules of ATP and two molecules of the electron carrier NADH. ATP is produced directly by substrate-level phosphorylation.

Glycolysis is a series of 10 chemical reactions (**Fig. 7.5**). These reactions can be divided into three phases.

The first phase prepares glucose for the next two phases by the addition of two phosphate groups to glucose. This phase requires an input of energy. To supply that energy and provide the phosphate groups, two molecules of ATP are hydrolyzed per molecule of glucose. In other words, the first phase of glycolysis is an endergonic process. The phosphorylation of glucose has two important consequences. Whereas glucose enters and exits cells through specific membrane transporters, phosphorylated glucose is trapped inside the cell. In addition, the presence of two negatively charged phosphate groups in proximity destabilizes the molecule so that it can be broken apart in the second phase of glycolysis.

The second phase is the cleavage phase, in which the 6-carbon molecule is split into two 3-carbon molecules. For each molecule of glucose entering glycolysis, two 3-carbon molecules enter the third phase of glycolysis.

The third and final phase of glycolysis is sometimes called the payoff phase because ATP and the electron carrier NADH are produced. Later, NADH will contribute to the synthesis of ATP during oxidative phosphorylation. This phase ends with the production of two molecules of pyruvate.

In summary, glycolysis begins with a single molecule of glucose (six carbons) and produces two molecules of pyruvate (three carbons each). These reactions yield four molecules of ATP and two molecules of NADH. However, two ATP molecules are consumed during the initial phase of glycolysis, resulting in a net gain of two ATP molecules and two molecules of NADH (Fig. 7.5).

Self-Assessment Questions

5. What is the overall chemical equation for glycolysis?
6. At the end of glycolysis, but before the subsequent stages in cellular respiration, which molecules contain some of the chemical energy held in the original glucose molecule?

FIG. 7.5 Glycolysis. Glucose is partially oxidized to pyruvate, with the net production of two ATP and two NADH.

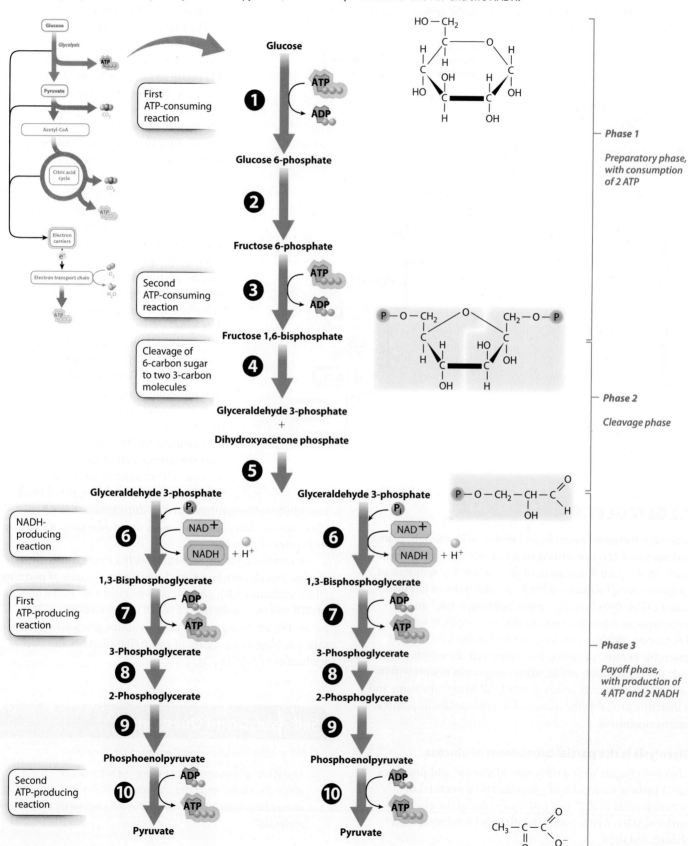

7.3 PYRUVATE OXIDATION

Glycolysis occurs in almost all living organisms, but it does not generate very much energy in the form of ATP. The end product, pyruvate, still contains a good deal of chemical potential energy in its bonds. In the presence of oxygen, pyruvate can be further oxidized to release more energy, first to acetyl-CoA and then even further in the series of reactions in the citric acid cycle. Pyruvate oxidation to acetyl-CoA is a key step that links glycolysis to the citric acid cycle.

The oxidation of pyruvate connects glycolysis to the citric acid cycle.

In eukaryotes, pyruvate oxidation is the first step that takes place inside the mitochondria. Mitochondria are rod-shaped organelles surrounded by a double membrane (**Fig. 7.6**; Chapter 5). The inner membrane has folds that project inward. These membranes define two spaces: the space between the inner and outer membranes is called the **intermembrane space**, and the space enclosed by the inner membrane is called the **mitochondrial matrix**.

Pyruvate is transported into the mitochondrial matrix, where it is converted into acetyl-CoA (**Fig. 7.7**). First, part of the pyruvate molecule is oxidized and splits off to form carbon dioxide, the most oxidized (and therefore the least energetic) form of carbon. The electrons lost in this process are donated to NAD^+, which is reduced to NADH. The remaining part of the pyruvate molecule, an acetyl group ($COCH_3$), still contains a large amount of potential energy. It is transferred to coenzyme A (CoA), a molecule that carries the acetyl group to the next set of reactions.

Overall, the synthesis of one molecule of acetyl-CoA from pyruvate results in the formation of one molecule of carbon dioxide and one molecule of NADH. Recall, however, that a single molecule of glucose forms two molecules of pyruvate during glycolysis. Therefore, two molecules of carbon dioxide, two molecules of NADH, and two molecules of acetyl-CoA are produced from a single glucose molecule in this stage of cellular respiration. Acetyl-CoA is the substrate of the first step in the citric acid cycle.

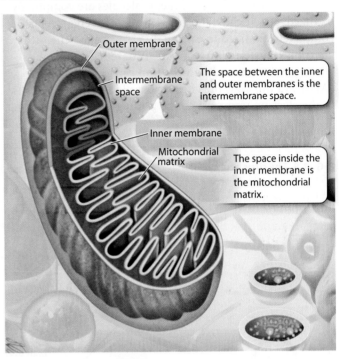

FIG. 7.6 Mitochondrial membranes and compartments.

- Outer membrane
- Intermembrane space: The space between the inner and outer membranes is the intermembrane space.
- Inner membrane
- Mitochondrial matrix: The space inside the inner membrane is the mitochondrial matrix.

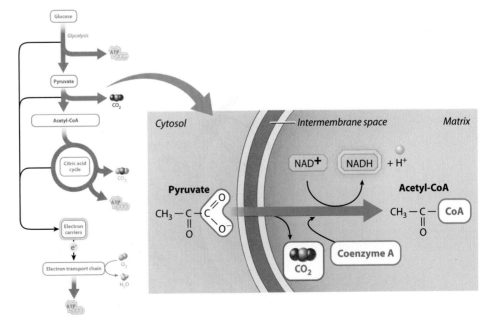

FIG. 7.7 Pyruvate oxidation. Pyruvate is oxidized in the mitochondrial matrix, forming acetyl-CoA, the first substrate in the citric acid cycle.

Self-Assessment Question

7. At the end of pyruvate oxidation, but before the subsequent stages of cellular respiration, which molecules contain the energy held in the original glucose molecule?

7.4 THE CITRIC ACID CYCLE

During the citric acid cycle, fuel molecules are completely oxidized. Specifically, the acetyl group of acetyl-CoA is completely oxidized to carbon dioxide and the chemical energy is transferred to ATP by substrate-level phosphorylation and to the reduced electron carriers NADH and $FADH_2$. In this way, the citric acid cycle supplies electrons to the electron transport chain, leading to the production of much more energy in the form of ATP than is obtained by glycolysis alone.

The citric acid cycle produces ATP and reduced electron carriers.

Like the synthesis of acetyl-CoA, the citric acid cycle takes place in the mitochondrial matrix. It is composed of eight reactions and is called a cycle because the starting molecule, oxaloacetate, is regenerated at the end (**Fig. 7.8**).

FIG. 7.8 The citric acid cycle. The acetyl group of acetyl-CoA is completely oxidized, with the net production of one ATP, three NADH, and one $FADH_2$.

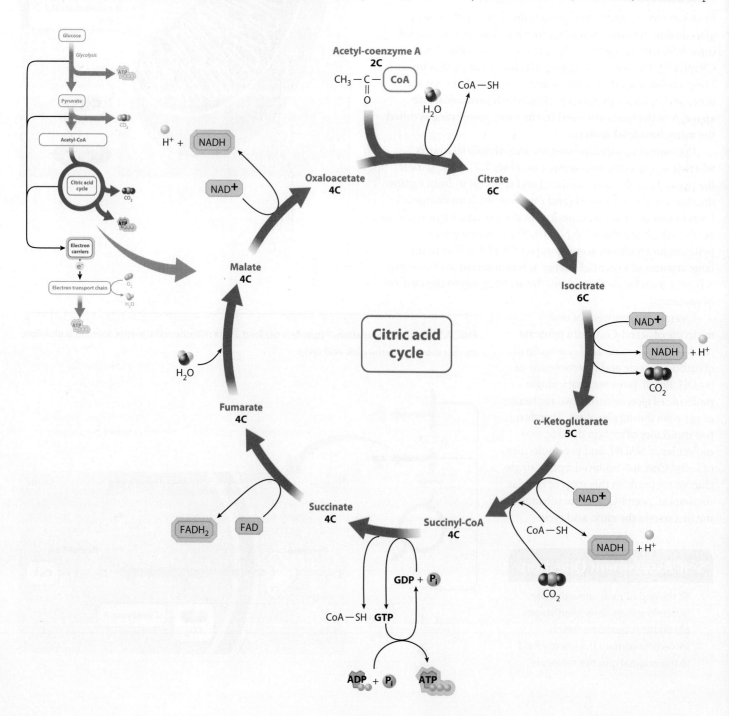

In the first reaction, the 2-carbon acetyl group of acetyl-CoA is transferred to a 4-carbon molecule of oxaloacetate to form the 6-carbon molecule citric acid or tricarboxylic acid (hence the alternative names "citric acid cycle" and "tricarboxylic acid cycle"). The molecule of citric acid is then oxidized in a series of reactions. The last reaction of the cycle regenerates a molecule of oxaloacetate, which can join to a new acetyl group and allow the cycle to continue.

The citric acid cycle results in the complete oxidation of the acetyl group of acetyl-CoA. Because the first reaction creates a molecule with six carbons and the last reaction regenerates a 4-carbon molecule, two carbons are eliminated during the cycle. These carbons are released as carbon dioxide. Along with the release of carbon dioxide from pyruvate during pyruvate oxidation, these reactions are the sources of carbon dioxide released during cellular respiration and therefore the sources of the carbon dioxide that we exhale when we breathe.

Four redox reactions, including the two that release carbon dioxide, produce the reduced electron carriers NADH and $FADH_2$. In this way, energy released in the oxidation reactions is transferred to a large quantity of reduced electron carriers: three molecules of NADH and one molecule of $FADH_2$ per turn of the cycle. These electron carriers donate electrons to the electron transport chain.

A single substrate-level phosphorylation reaction generates a molecule of GTP (Fig. 7.8). GTP can transfer its terminal phosphate to a molecule of ADP to form ATP.

Overall, two molecules of acetyl-CoA produced from a single molecule of glucose yield two molecules of ATP, six molecules of NADH, and two molecules of $FADH_2$ in the citric acid cycle.

CASE 1 LIFE'S ORIGINS: INFORMATION, HOMEOSTASIS, AND ENERGY

What were the earliest energy-harnessing reactions?
Some bacteria run the citric acid cycle in reverse, incorporating carbon dioxide into organic molecules instead of liberating it. Running the citric acid cycle in reverse requires energy, which is supplied by sunlight (Chapter 8) or chemical reactions (Chapter 24).

Why would an organism run the citric acid cycle backward? The answer is that running the cycle in reverse allows an organism to build, rather than break down, organic molecules. Whether the cycle is run in the reverse or forward direction, the intermediates generated step by step as the cycle turns provide the building blocks for synthesizing the cell's key organic molecules (**Fig. 7.9**). Pyruvate, for example, is the starting point for the synthesis of sugars and the amino acid alanine; acetate is the starting point for the synthesis of the cell's lipids; oxaloacetate is modified to form different amino acids and pyrimidine bases; and α-(alpha-)ketoglutarate is modified to form other amino acids.

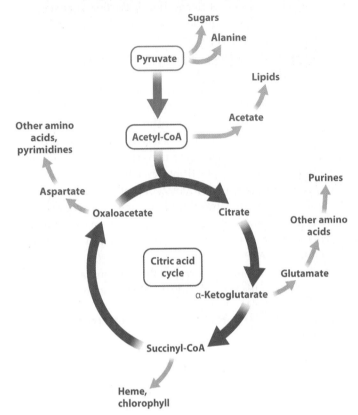

FIG. 7.9 Molecules produced by the citric acid cycle. Many organic molecules can be synthesized from citric acid cycle intermediates.

Therefore, for organisms that run the cycle in the forward direction, the citric acid cycle is used to generate both energy-storing molecules (ATP and reduced electron carriers) and intermediates in the synthesis of other molecules. For organisms that run the cycle in reverse, it is used to generate intermediates in the synthesis of other molecules and also to incorporate carbon into organic molecules.

The citric acid cycle is thus central to both the synthesis of organic molecules and to meeting the energy requirements of cells, suggesting that it evolved early in some of the first cells to feature metabolism. The early appearance of the citric acid cycle, in turn, implies that the great variety of biosynthetic and energy-yielding pathways found in modern cells evolved through the extension and modification of this deeply rooted cycle. As typical of evolution, new cellular capabilities arose by the modification of simpler, more general sets of reactions.

Self-Assessment Question

8. At the end of the citric acid cycle, but before the subsequent stages of cellular respiration, which molecules contain the energy held in the original glucose molecule?

7.5 THE ELECTRON TRANSPORT CHAIN AND OXIDATIVE PHOSPHORYLATION

The complete oxidation of glucose during the first three stages of cellular respiration results in the production of two kinds of reduced electron carriers: NADH and $FADH_2$. We are now going to see how the energy stored in these electron carriers is used to synthesize ATP.

The energy in these electron carriers is released in a series of redox reactions that occur as electrons pass through a chain of protein complexes in the inner mitochondrial membrane to the final electron acceptor, oxygen, which is reduced to water. The energy released by these redox reactions is not converted directly into the chemical energy of ATP, however. Instead, the passage of electrons is coupled to the transfer of protons (H^+) across the inner mitochondrial membrane, creating a concentration and charge gradient (Chapter 5). This electrochemical gradient provides a source of potential energy that is then used to drive the synthesis of ATP.

We next explore the properties of the electron transport chain, the role of the proton gradient, and the synthesis of ATP.

The electron transport chain transfers electrons and pumps protons.

The electron transport chain is made up of four large protein complexes (known as complexes I to IV) that are embedded in the inner mitochondrial membrane (**Fig. 7.10**). The inner mitochondrial membrane contains one of the highest concentrations of proteins found in eukaryotic membranes.

Electrons donated by NADH and $FADH_2$ are transported along this series of protein complexes. Electrons enter the electron transport chain at either complex I or II. Electrons donated by NADH enter through complex I, and electrons donated by $FADH_2$ enter through complex II. (Complex II is the same enzyme that catalyzes step 6 in the citric acid cycle.) These electrons are transported through either complex I or II to complex III and then through complex IV.

Within each protein complex of the electron transport chain, electrons are passed from electron donors to electron acceptors. Each donor and acceptor is a redox couple, consisting of an oxidized and a reduced form of a molecule. The electron transport chain contains many of these redox couples. When oxygen accepts electrons at the end of the electron transport chain, it is reduced to form water:

$$O_2 + 4e^- + 4H^+ \rightarrow 2H_2O$$

This reaction is catalyzed by complex IV.

Electrons also must be transported between the four complexes (Fig. 7.10). **Coenzyme Q (CoQ)**, also called ubiquinone, accepts electrons from both complexes I and II. In this reaction, two electrons and two protons are transferred to CoQ from the mitochondrial matrix, forming $CoQH_2$. Once formed, $CoQH_2$ diffuses in the inner membrane to complex III.

In complex III, electrons are transferred from $CoQH_2$ to **cytochrome c** and protons are released into the intermembrane space. When it accepts an electron, cytochrome c is reduced, diffuses in the intermembrane space, and passes the electron to complex IV.

The electron transfer steps within complexes are each associated with the release of energy. Some of this energy is used to reduce the next electron acceptor in the chain, but in complexes I, III, and IV, some of it is used to pump protons (H^+) across the inner mitochondrial membrane, from the mitochondrial matrix to the intermembrane space (Fig. 7.10). Thus, the transfer of electrons through complexes I, III, and IV is coupled with the pumping of protons. The result is an accumulation of protons in the intermembrane space.

The proton gradient is a source of potential energy.

Like all membranes, the inner mitochondrial membrane is selectively permeable: protons cannot passively diffuse across this membrane, and the movement of other molecules is controlled by transporters and channels (Chapter 5). We have just seen that the movement of electrons through membrane-embedded protein complexes is coupled to the pumping of protons from the mitochondrial matrix into the intermembrane space. The result is a proton gradient, a difference in proton concentration across the inner membrane.

The proton gradient has two components: a chemical gradient that results from the difference in concentration and an electrical gradient that results from the difference in charge between the two sides of the membrane. To reflect the dual contribution of the two gradients, the proton gradient is also called an electrochemical gradient.

The proton gradient is a source of potential energy, as discussed in Chapters 5 and 6. It stores energy in much the same way that a battery or a dam does. Protons in this gradient have a high concentration in the intermembrane space and a low concentration in the mitochondrial matrix. As a result, there is a tendency for protons to diffuse back to the mitochondrial matrix, driven by a difference in concentration and charge on the two sides of the membrane. This movement, however, is blocked by the membrane. The gradient stores potential energy because, if a pathway is opened through the membrane, the resulting movement of the protons through the membrane can be used to perform work.

In sum, the oxidation of the electron carriers NADH and $FADH_2$ leads to the generation of a proton electrochemical gradient. This gradient is a source of potential energy used to synthesize ATP.

ATP synthase converts the energy of the proton gradient into the energy of ATP.

In 1961, Peter Mitchell proposed a hypothesis to explain how the energy stored in the proton electrochemical gradient is used

FIG. 7.10 The electron transport chain. (a) The electron transport chain consists of four complexes (I to IV) in the inner mitochondrial membrane. (b) Electrons flow from electron carriers to oxygen, the final electron acceptor. (c) The proton gradient formed from the electron transport chain has potential energy that is used to synthesize ATP.

a. The electron transport chain in cellular respiration

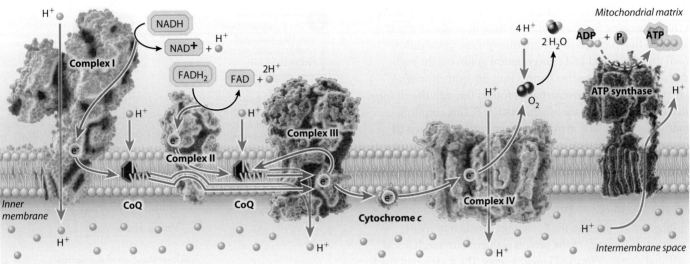

b. Electron transport

c. Proton transport and ATP synthesis

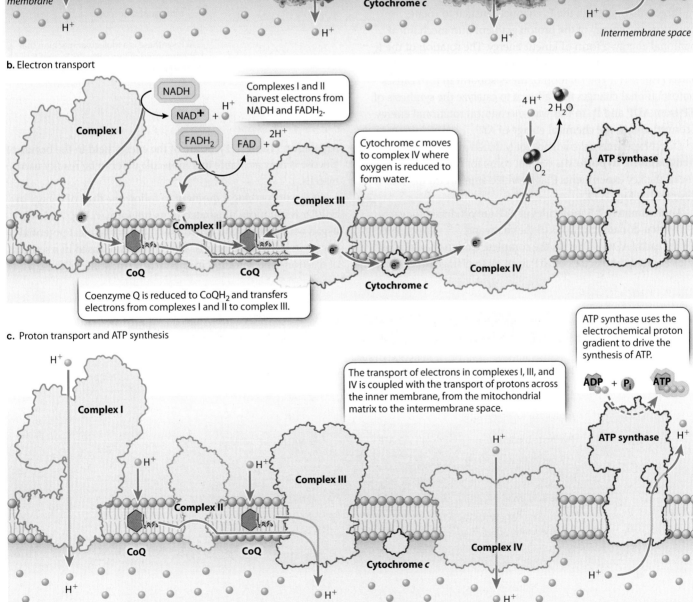

to synthesize ATP. In 1978, he was awarded the Nobel Prize in Chemistry for work that fundamentally changed the way we understand how a cell harnesses energy.

According to Mitchell's hypothesis, the gradient of protons provides a source of potential energy that is converted into chemical energy stored in ATP. First, for the potential energy of the proton gradient to be released, there must be an opening in the membrane for the protons to flow through. Mitchell suggested that protons in the intermembrane space diffuse down their electrical and concentration gradients through a transmembrane protein channel into the mitochondrial matrix. Second, the movement of protons through the channel must be coupled with the synthesis of ATP. This coupling is made possible by **ATP synthase**, an enzyme composed of two distinct subunits called F_o and F_1 (**Fig. 7.11**). F_o forms the channel in the inner mitochondrial membrane through which protons flow; F_1 is the catalytic unit that synthesizes ATP. Proton flow through the channel makes it possible for the enzyme to synthesize ATP.

Proton flow through the F_o channel causes it to rotate, converting the energy of the proton gradient into mechanical rotational energy, a form of kinetic energy. The rotation of the F_o subunit leads to rotation of the F_1 subunit in the mitochondrial matrix (Fig. 7.11). The rotation of the F_1 subunit in turn causes conformational changes that allow it to catalyze the synthesis of ATP from ADP and P_i. In this way, mechanical rotational energy is converted into the chemical energy of ATP.

Direct experimental evidence for Mitchell's idea, called the **chemiosmotic hypothesis**, did not come for over a decade. One of the key experiments that provided support for his idea is illustrated in **Fig. 7.12**.

Approximately 2.5 molecules of ATP are produced for each NADH that donates electrons to the chain and 1.5 molecules of ATP for each $FADH_2$. Overall, the complete oxidation of glucose yields about 32 molecules of ATP from all four stages of cellular respiration (**Table 7.1**). Some of the energy held in the bonds of glucose is now available in a molecule that can be readily used by cells.

It is worth taking a moment to follow the flow of energy in cellular respiration, illustrated in its full form in **Fig. 7.13**. We began with glucose and noted that it holds chemical potential energy in its covalent bonds. This energy is released in a series of reactions and captured in chemical form. Some of these

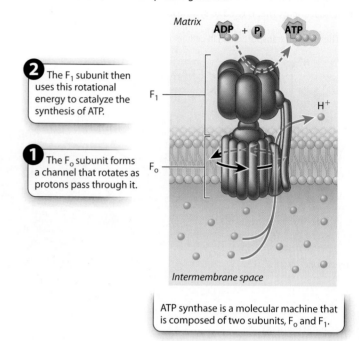

FIG. 7.11 ATP synthase. ATP synthase drives the synthesis of ATP by means of an electrochemical proton gradient.

① The F_o subunit forms a channel that rotates as protons pass through it.

② The F_1 subunit then uses this rotational energy to catalyze the synthesis of ATP.

ATP synthase is a molecular machine that is composed of two subunits, F_o and F_1.

TABLE 7.1	Approximate Total ATP Yield in Cellular Respiration		
PATHWAY	SUBSTRATE-LEVEL PHOSPHORYLATION	OXIDATIVE PHOSPHORYLATION	TOTAL ATP
Glycolysis (glucose → 2 pyruvate)	2 ATP	2 NADH = 5 ATP	7
Pyruvate oxidation (2 pyruvate → 2 acetyl-CoA)	0 ATP	2 NADH = 5 ATP	5
Citric acid cycle (2 turns, 1 for each acetyl-CoA)	2 ATP	6 NADH = 15 ATP 2 $FADH_2$ = 3 ATP	20
Total	4 ATP	28 ATP	32

HOW DO WE KNOW?

FIG. 7.12

Can a proton gradient drive the synthesis of ATP?

BACKGROUND Peter Mitchell's hypothesis that a proton gradient can drive the synthesis of ATP was met with skepticism because he proposed the idea before experimental evidence existed to support it. In the 1970s, biochemist Efraim Racker and his collaborator Walther Stoeckenius tested the hypothesis.

EXPERIMENT Racker and Stoeckenius built an artificial system consisting of a collection of vesicle-shaped membranes, each with ATP synthase and a bacterial proton pump that was activated by light.

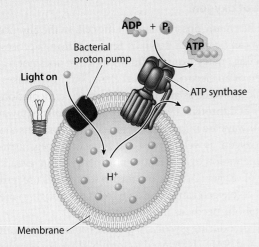

They measured the concentration of protons (pH) in the external medium and the amount of ATP produced in the presence and absence of light.

RESULTS In the presence of light, the pH of the external medium increased. An increase in pH indicates a decrease in concentration of protons as a result of movement of protons from the outside to the inside of the vesicles. This result demonstrates that the proton pump was activated and the vesicles were taking up protons. In the dark, the pH returned to the starting level.

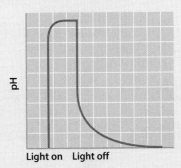

Furthermore, ATP was generated in the light, but not in the dark.

CONDITION	RELATIVE LEVEL OF ATP
Light	594
Dark	23

INTERPRETATION In the presence of light, the proton pump was activated and protons were pumped to one side of the membrane, leading to the formation of a proton gradient. The proton gradient, in turn, powered synthesis of ATP by ATP synthase.

CONCLUSION A membrane, proton gradient, and ATP synthase are sufficient to synthesize ATP. This result provided experimental evidence for Mitchell's hypothesis.

SOURCES Mitchell, P. 1961. "Coupling of Phosphorylation to Electron and Hydrogen Transfer by a Chemiosmotic Type of Mechanism." *Nature* 191:144–148; Racker, E., and W. Stoeckenius. 1974. "Reconstitution of Purple Membrane Vesicles Catalyzing Light-Driven Proton Uptake and Adenosine Triphosphate Formation." *J. Biol. Chem.* 249:662–663.

---

reactions generate ATP directly by substrate-level phosphorylation. Others are redox reactions that transfer energy to the electron carriers NADH and $FADH_2$. These electron carriers donate electrons to the electron transport chain.

The electron transport chain uses the energy stored in the electron carriers to pump protons across the inner membrane of the mitochondria. In other words, the energy of the reduced electron carriers is transformed into energy stored in a proton electrochemical gradient. ATP synthase then converts the energy of the proton gradient to rotational energy, which drives the synthesis of ATP. The cell now has a form of energy that it can use in many ways to perform work.

Self-Assessment Questions

9. Animals breathe in air that contains more oxygen than the air they breathe out. Where is oxygen consumed?
10. How does the movement of electrons along the electron transport chain lead to the generation of a proton gradient?
11. How is a proton gradient used to generate ATP?
12. Uncoupling agents are proteins spanning the inner mitochondrial membrane that allow protons to pass through the membrane and bypass the channel of ATP synthase. Describe the consequences for the proton gradient and ATP production.

FIG. 7.13 The flow of energy in cellular respiration. A single glucose molecule yields 32 ATP molecules.

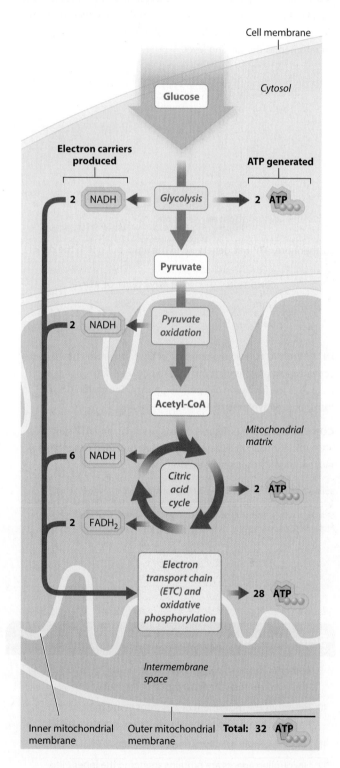

7.6 ANAEROBIC METABOLISM

Up to this point, we have followed a single metabolic path: the breakdown of glucose in the presence of oxygen to produce carbon dioxide and water. However, metabolic pathways more often resemble intersecting roads rather than a single, linear path. As an example, we saw earlier that intermediates of the citric acid cycle often feed into other metabolic pathways.

One of the major forks in the metabolic road occurs at pyruvate, the end product of glycolysis (section 10.2). When oxygen is present, it is converted to acetyl-CoA, which then enters the citric acid cycle. When oxygen is not present, pyruvate is metabolized along a number of different pathways. These pathways occur in many living organisms today and played an important role in the early evolution of life on Earth.

Fermentation extracts energy from glucose in the absence of oxygen.

Pyruvate has many possible fates in the cell. In the absence of oxygen, it can be broken down by **fermentation**, a process for extracting energy from fuel molecules that does not rely on oxygen or an electron transport chain, but instead uses an organic molecule as an electron acceptor. Fermentation is accomplished through a wide variety of metabolic pathways. These pathways are important for anaerobic organisms that live without oxygen, as well as yeast and some other organisms that favor fermentation over oxidative phosphorylation, even in the presence of oxygen. Aerobic organisms sometimes use fermentation when oxygen cannot be delivered fast enough to meet the cell's metabolic needs, as in exercising muscle.

Recall that during glycolysis, glucose is oxidized to form pyruvate and NAD^+ is reduced to form NADH. For glycolysis to continue, NADH must be oxidized to NAD^+; otherwise, glycolysis would grind to a halt. In the presence of oxygen, NAD^+ is regenerated when NADH donates its electrons to the electron transport chain. In the absence of oxygen during fermentation, NADH is oxidized to NAD^+ when pyruvate or a derivative of pyruvate is reduced.

Two of the major fermentation pathways are **lactic acid fermentation** and **ethanol fermentation** (**Fig. 7.14**). Lactic acid fermentation occurs in animals and bacteria. During lactic acid fermentation, electrons from NADH are transferred to pyruvate to produce lactic acid and NAD^+ (Fig. 7.14a). The overall chemical reaction is written as follows:

$$\text{Glucose} + 2\ \text{ADP} + 2\ P_i \rightarrow 2\ \text{lactic acid} + 2\ \text{ATP} + 2H_2O$$

Ethanol fermentation occurs in plants and fungi. During ethanol fermentation, pyruvate releases carbon dioxide to form acetaldehyde, and electrons from NADH are transferred to acetaldehyde to produce ethanol and NAD^+ (Fig. 7.14b). The overall chemical reaction is written as follows:

$$\text{Glucose} + 2\ \text{ADP} + 2\ P_i \rightarrow 2\ \text{ethanol} + 2CO_2 + 2\ \text{ATP} + 2H_2O$$

FIG. 7.14 Lactic acid and ethanol fermentation pathways.

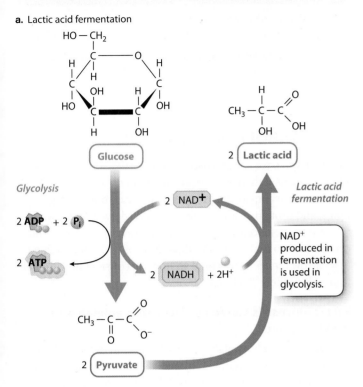

In both fermentation pathways, NADH is oxidized to NAD$^+$. However, NADH and NAD$^+$ do not appear in the overall chemical equations because there is no net production or loss of either molecule. NAD$^+$ molecules that are reduced during glycolysis are oxidized when lactic acid or ethanol is formed.

The breakdown of a molecule of glucose by fermentation yields only two molecules of ATP. The energetic gain is relatively small compared with the yield of aerobic respiration because the end products, lactic acid and ethanol, are not fully oxidized and still contain a large amount of chemical energy in their bonds. Organisms that produce ATP by fermentation therefore must consume a large quantity of fuel molecules to power the cell.

CASE 1 LIFE'S ORIGINS: INFORMATION, HOMEOSTASIS, AND ENERGY

How did early cells meet their energy requirements?

The four stages of cellular respiration lead to the full oxidation of glucose and the release of a large amount of energy stored in its chemical bonds. In the first stage, glycolysis, glucose is only partially oxidized, so just some of the energy held in its chemical bonds is released. Nearly all organisms are capable of partially breaking down glucose, suggesting that glycolysis evolved very early in the history of life.

The atmosphere contained no oxygen when life first evolved about 4 billion years ago. The earliest organisms probably used one of the fermentation pathways to generate ATP. Fermentation occurs in the cytoplasm, does not require atmospheric oxygen, and does not require proteins embedded in specialized membranes.

As we have seen, cellular respiration makes use of an electron transport chain, composed of proteins embedded in a membrane and capable of transferring electrons from one protein to the next and pumping protons. The resulting proton gradient powers the synthesis of ATP. Like fermentation, cellular respiration can occur in the absence of oxygen, but in that case molecules other than oxygen, such as sulfate and nitrate, are the final electron acceptor (Chapter 26). This form of respiration is known as anaerobic respiration and occurs in some present-day bacteria. The electron transport chain in these bacteria is located in the plasma membrane, not in an internal membrane.

How might such a system have evolved? An intriguing possibility is that early prokaryotes evolved pumps to drive protons out of the cell in response to an increasingly acidic environment (**Fig. 7.15**). Some pumps might have used the energy of ATP to pump protons, while others used the energy from electron transport to pump protons (Fig. 7.15a). At some point, proton pumps powered by electron transport might have generated a large enough electrochemical gradient that the protons could pass back through the ATP-driven pumps, running them in reverse to synthesize ATP (Fig. 7.15b).

FIG. 7.15 The possible evolution of the electron transport chain and oxidative phosphorylation.

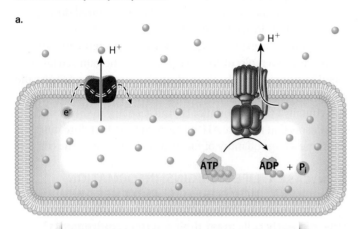

Early cells evolved mechanisms to pump protons out of the cell, powered by ATP and electron transport.

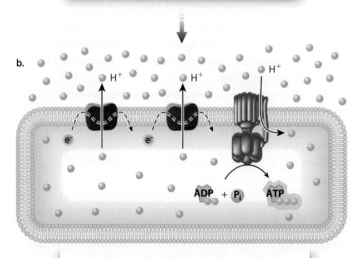

Eventually, electron-transport-powered pumps became efficient enough to run the ATP-driven pump in reverse.

Organisms capable of producing oxygen, the cyanobacteria, did not evolve until about 2.5 billion years ago, or maybe earlier. The evolution of this new form of life introduced oxygen into Earth's atmosphere. This dramatic change led to the evolution of new life-forms with new possibilities for extracting energy from fuel molecules such as glucose. Aerobic respiration, in which oxygen serves as the final electron acceptor in the electron transport chain, generates much more energy than does anaerobic respiration or fermentation.

The evolution of cellular respiration illustrates that evolution often works in a stepwise fashion, building on what is already present. In this case, aerobic respiration picked up where anaerobic respiration left off, making it possible to harness more energy from organic molecules to power the work of the cell.

Self-Assessment Questions

13. Bread making involves ethanol fermentation and typically uses yeast, sugar, flour, and water. Why are yeast and sugar used?

14. What are two different metabolic pathways that pyruvate can enter?

7.7 METABOLIC INTEGRATION

In this chapter, we have focused on the breakdown of glucose. What happens if there is more glucose than the cell needs? As well as glucose, organisms consume diverse carbohydrates, lipids, and proteins. How are these broken down? And how are the various metabolic pathways that break down fuel molecules coordinated so that the intracellular level of ATP is maintained in a narrow range? In this final section, we consider how the cell responds to these challenges.

Excess glucose is stored as glycogen in animals and starch in plants.

Glucose is a readily available form of energy in organisms, but it is not always broken down immediately. Excess glucose can be stored in cells and then mobilized when necessary. Glucose can be stored in two major forms: as **glycogen** in animals and as **starch** in plants (**Fig. 7.16**). Both these molecules are large branched polymers of glucose.

Carbohydrates in the diet are broken down into glucose and other simple sugars and circulate in the blood. The level of glucose in the blood is tightly regulated (Chapter 37). When the blood glucose level is high, as it is after a meal, glucose molecules that are not consumed by glycolysis are linked together to form glycogen in liver and muscle. Glycogen stored in muscle is used to provide ATP for muscle contraction. By contrast, the liver does not store glycogen primarily for its own use, but is a glycogen storehouse for the whole body. Glucose molecules located at the end of glycogen chains can be cleaved one by one and released in the form of glucose 1-phosphate. Glucose 1-phosphate is converted into glucose 6-phosphate, an intermediate in glycolysis (see Fig. 7.5). One glucose molecule cleaved off a glycogen chain produces three and not two molecules of ATP by glycolysis because the ATP-consuming step 1 of glycolysis is bypassed.

Sugars other than glucose contribute to glycolysis.

The carbohydrates in your diet are digested to produce a variety of sugars. Some of these are disaccharides (maltose, lactose, and sucrose) with two sugar units; others are monosaccharides (fructose, mannose, and galactose) with one sugar unit (**Fig. 7.17**). The disaccharides are hydrolyzed into monosaccharides, which are transported into cells.

FIG. 7.16 Storage forms of glucose. (a) Glycogen is a storage form of glucose in animal cells, and (b) starch is a storage form of glucose in plant cells.

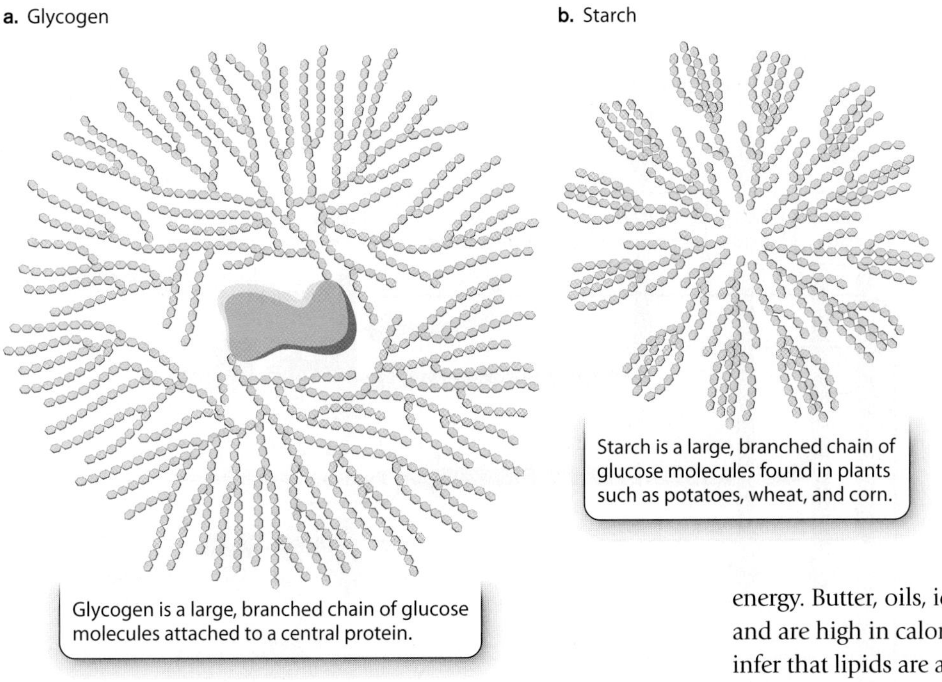

a. Glycogen

b. Starch

Starch is a large, branched chain of glucose molecules found in plants such as potatoes, wheat, and corn.

Glycogen is a large, branched chain of glucose molecules attached to a central protein.

Glucose molecules released during digestion directly enter glycolysis. What happens to other sugars? They, too, enter glycolysis, although not as glucose. Instead, they are converted into intermediates of glycolysis that come later in the pathway (**Fig. 7.18**). For example, fructose is produced by the hydrolysis of sucrose (table sugar) and receives a phosphate group to form either fructose 6-phosphate or fructose 1-phosphate. In the liver, fructose 1-phosphate is cleaved and converted into glyceraldehyde 3-phosphate, which enters glycolysis at reaction 6 (see Fig. 7.5).

Fatty acids and proteins are useful sources of energy.

We know from common experience that lipids are also a good source of energy. Butter, oils, ice cream, and the like all contain lipids and are high in calories, which are units of energy. We can also infer that lipids are a good source of energy from their chemical structure. Recall from Chapter 2 that a type of fat called triacylglycerol is composed of three fatty acid molecules bound to a

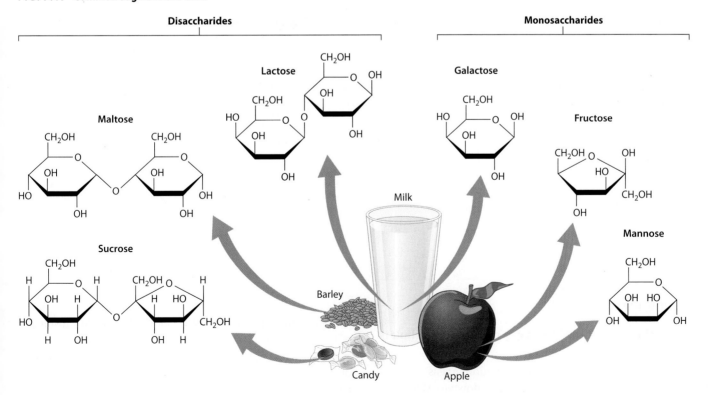

FIG. 7.17 Common sugars in the diet.

glycerol backbone (see Fig. 2.25). These fatty acid molecules are rich in carbon–carbon and carbon–hydrogen bonds, which, as we saw earlier, carry chemical potential energy.

Following a meal, the small intestine very quickly absorbs triacylglycerols, which are then transported by the bloodstream and either consumed or stored in fat (adipose) tissue. Triacylglycerols are broken down inside cells to glycerol and fatty acids (Fig. 7.18). Then, the fatty acids themselves are shortened by a series of reactions that sequentially remove two carbon units from their ends. This process is called *β-*(**beta-**)**oxidation**. It does not generate ATP, but produces NADH and $FADH_2$ molecules that provide electrons for the synthesis of ATP by oxidative phosphorylation. In addition, the end product of the reaction is acetyl-CoA, which feeds the citric acid cycle and leads to the production of additional reduced electron carriers.

The oxidation of fatty acids produces a large amount of ATP. For example, the complete oxidation of a molecule of palmitic acid, a fatty acid containing 16 carbons, yields about 106 molecules of ATP. By contrast, glycolysis yields just 2 molecules of ATP, and the complete oxidation of a glucose molecule produces about 32 molecules of ATP (Table 7.1). Fatty acids therefore are a useful and efficient source of energy, but they cannot be used by all tissues of the body. Notably, the brain and red blood cells depend primarily on glucose for energy.

Proteins, like fatty acids, are a source of chemical energy that can be broken down, if necessary, to power the cell. Proteins are typically first broken down to amino acids, which enter at various points in glycolysis, pyruvate oxidation, and the citric acid cycle (Fig. 7.18).

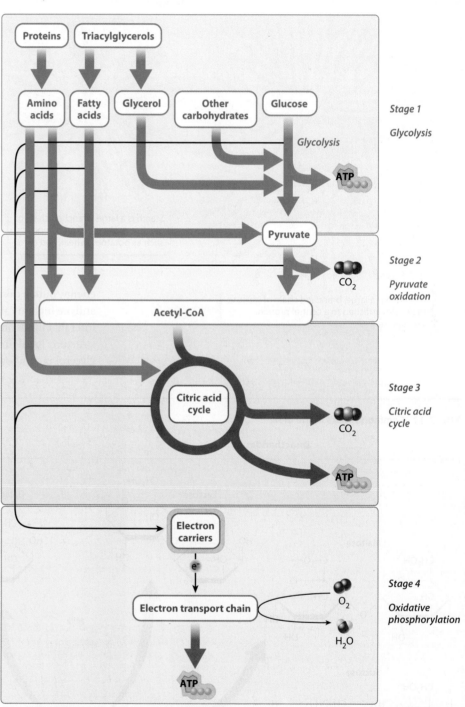

FIG. 7.18 Other fuel molecules. Sugars other than glucose, fats such as triacylglycerol, and proteins are all fuel molecules that are broken down and enter the cellular respiration pathway at different places.

The intracellular level of ATP is a key regulator of cellular respiration.

ATP is the key end product of cellular respiration. ATP is constantly being turned over in a cell, broken down to ADP and P_i to supply the cell's energy needs, and re-synthesized by fermentation and cellular respiration. The level of ATP inside a cell can therefore be an indicator of how much energy a cell has available. When ATP levels are high, the cell has a high amount of free energy and is poised to carry out cellular processes. In this case, pathways that generate ATP are slowed, or down-regulated. By contrast, when ATP levels are low, pathways that generate ATP

FIG. 7.19 Regulation of cellular respiration.
Cellular respiration is inhibited by its products, including ATP and NADH, and activated by its substrates, including ADP and NAD$^+$.

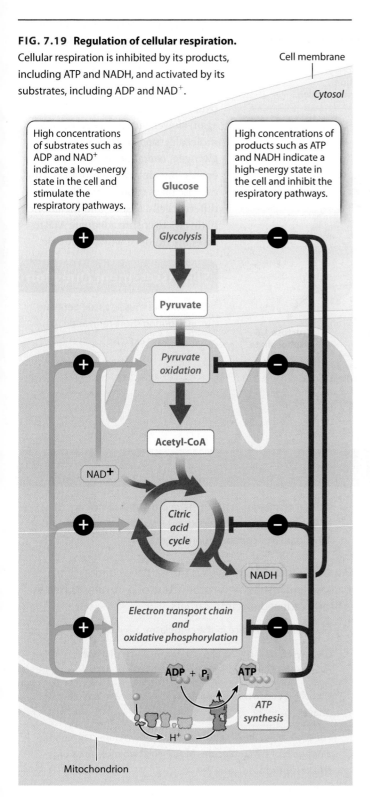

are activated, or up-regulated. Other intermediates of cellular respiration, such as NADH, have a similar effect in that high NAD$^+$ levels stimulate cellular respiration, whereas high NADH levels inhibit it (**Fig. 7.19**).

How is this kind of coordinated response of the cell possible? The cell uses several mechanisms, one of which is the regulation of enzymes that control key steps of the pathway. One of these key reactions is reaction 3 of glycolysis. In this reaction, fructose 6-phosphate is converted to fructose 1,6-bisphosphate, and a molecule of ATP is consumed. This is a key step in glycolysis because it is highly endergonic and irreversible. As a result, it is considered a "committed" step and is subject to tight control. This reaction is catalyzed by the enzyme phosphofructokinase-1 (PFK-1). You can think of this enzyme as a metabolic valve that regulates the rate of glycolysis.

PFK-1 is an allosteric enzyme with many activators and inhibitors (**Fig. 7.20**). Recall from Chapter 6 that an allosteric enzyme changes its shape and activity in response to the binding of molecules at a site other than the active site. ADP and AMP are allosteric activators of PFK-1. When ADP and AMP are abundant, one or the other binds to the enzyme and causes the enzyme's shape to change. The shape change activates the enzyme, increasing the rate of glycolysis and the synthesis of ATP. When ATP is in abundance, it binds to the same site on the enzyme as ADP and AMP, but in this case binding inhibits the enzyme's catalytic activity. As a result, glycolysis and the rate of ATP production slow down.

PFK-1 is also regulated by one of its downstream products, citrate, an intermediate in the citric acid cycle (see Fig. 7.8). Citrate acts as an allosteric inhibitor of the enzyme, slowing its activity. High levels of citrate indicate that it is not being consumed by the citric acid cycle and glucose breakdown can be slowed. The role of citrate in controlling glycolysis is an example of how the activity of glycolysis and the citric acid cycle is coordinated.

Exercise requires several types of fuel molecules and the coordination of metabolic pathways.

In the last two chapters, we considered what energy is and how it is harnessed by cells. Let's apply the concepts we discussed to a familiar example: exercise. Exercise such as running, walking, and swimming is a form of kinetic energy, powered by ATP in muscle cells. Where does this ATP come from?

Muscle cells, like all cells, do not contain a lot of ATP, and stored ATP is depleted by exercise in a matter of seconds. As a result, muscle cells rely on fuel molecules to generate ATP. For a short sprint or a burst of activity, muscle can convert stored glycogen to glucose, and then break down glucose anaerobically to pyruvate and lactic acid by lactic acid fermentation. This pathway is rapid, but it does not generate a lot of ATP.

For longer, more sustained exercise, other metabolic pathways come into play. Muscle cells contain many mitochondria, which produce ATP by aerobic respiration. The energy yield of aerobic respiration is much greater than that of fermentation, but the process is slower. This slower production of ATP by aerobic respiration in part explains why runners cannot maintain the pace of a sprint for longer runs.

For even longer exercise, liver glycogen supplements muscle glycogen: the liver releases glucose into the blood that is taken up by muscle cells and oxidized to produce ATP. In addition,

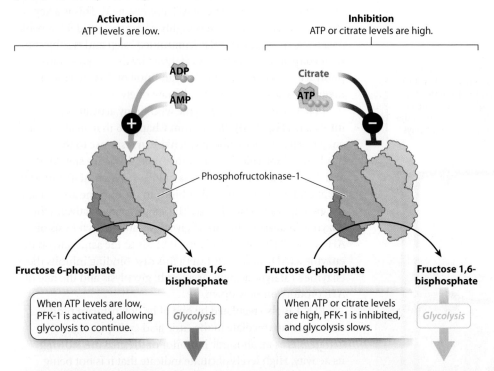

FIG. 7.20 Regulation of PFK-1. The regulation of the glycolytic enzyme phosphofructokinase-1 (PFK-1) is an example of integrated metabolic control.

fatty acids are released from adipose tissue and taken up by muscle cells, where they are broken down by β-oxidation. β-oxidation yields even more ATP than does the complete oxidation of glucose, but the process is again slower. Storage forms of energy molecules, such as fatty acids and glycogen, contain large reservoirs of energy, but are slow to mobilize. Thus, exercise takes coordination between different cells, tissues, and metabolic pathways to ensure adequate ATP to meet the needs of working muscle.

Self-Assessment Question

15. How does muscle tissue generate ATP during short-term and long-term exercise?

CORE CONCEPTS SUMMARY

7.1 AN OVERVIEW OF CELLULAR RESPIRATION: Cellular respiration is a series of catabolic reactions that converts the energy in fuel molecules into energy in ATP.

During cellular respiration, sugar molecules such as glucose are broken down in the presence of oxygen to produce carbon dioxide and water. page 135

Cellular respiration releases energy because the potential energy of the reactants is greater than that of the products. page 136

ATP is generated in two ways during cellular respiration: substrate-level phosphorylation and oxidative phosphorylation. page 136

Cellular respiration is an oxidation–reduction reaction. page 136

In oxidation–reduction reactions, electrons are transferred from one molecule to another. Oxidation is the loss of electrons, and reduction is the gain of electrons. page 136

Electron carriers transfer electrons to an electron transport chain, which harnesses the energy of these electrons to generate ATP. page 137

Cellular respiration is a four-stage process that includes (1) glycolysis; (2) pyruvate oxidation; (3) the citric acid cycle; and (4) oxidative phosphorylation. page 137

7.2 GLYCOLYSIS: Glycolysis is the partial oxidation of glucose and results in the production of pyruvate, as well as ATP and reduced electron carriers.

Glycolysis takes place in the cytoplasm. page 138

Glycolysis is a series of 10 reactions in which glucose is oxidized to pyruvate. page 139

Glycolysis consists of preparatory, cleavage, and payoff phases. page 139

For each molecule of glucose broken down during glycolysis, a net gain of two molecules of ATP and two molecules of NADH is produced. page 139

The synthesis of ATP in glycolysis results from the direct transfer of a phosphate group from a substrate to ADP, a process called substrate-level phosphorylation. page 139

7.3 PYRUVATE OXIDATION: Pyruvate is oxidized to acetyl-CoA, connecting glycolysis to the citric acid cycle.

The conversion of pyruvate to acetyl-CoA results in the production of one molecule of NADH and one molecule of carbon dioxide. page 141

Pyruvate oxidation occurs in the mitochondrial matrix. page 141

7.4 THE CITRIC ACID CYCLE: The citric acid cycle results in the complete oxidation of fuel molecules and the generation of ATP and reduced electron carriers.

The citric acid cycle takes place in the mitochondrial matrix. page 142

The acetyl group of acetyl-CoA is completely oxidized in the citric acid cycle. page 142

The citric acid cycle is a cycle because the acetyl group of acetyl-CoA combines with oxaloacetate, and then a series of reactions regenerates oxaloacetate. page 143

A complete turn of the citric acid cycle results in the production of one molecule of GTP (which is converted to ATP), three molecules of NADH, and one molecule of $FADH_2$. page 143

Citric acid cycle intermediates are starting points for the synthesis of many different organic molecules. page 143

7.5 THE ELECTRON TRANSPORT CHAIN AND OXIDATIVE PHOSPHORYLATION: The electron transport chain transfers electrons from electron carriers to oxygen, using the energy to pump protons and synthesize ATP by oxidative phosphorylation.

NADH and $FADH_2$ donate electrons to the electron transport chain. page 144

In the electron transport chain, electrons move from one redox couple to the next. page 144

The electron transport chain is made up of four complexes. Complexes I and II accept electrons from NADH and $FADH_2$, respectively. The electrons are transferred from these two complexes to coenzyme Q. page 144

Reduced coenzyme Q transfers electrons to complex III and cytochrome *c* transfers electrons to complex IV. Complex IV reduces oxygen to water. page 144

The transfer of electrons through the electron transport chain is coupled with the movement of protons across the inner mitochondrial membrane into the intermembrane space. page 144

The buildup of protons in the intermembrane space results in a proton electrochemical gradient, which stores potential energy. page 144

The movement of protons back into the mitochondrial matrix through the F_o subunit of ATP synthase is coupled with the formation of ATP, a reaction catalyzed by the F_1 subunit of ATP synthase. page 146

Complete oxidation of one glucose molecule by the four stages of aerobic cellular respiration nets approximately 32 molecules of ATP. page 146

7.6 ANAEROBIC METABOLISM: Glucose can be broken down in the absence of oxygen by fermentation, producing a modest amount of ATP.

Pyruvate, the end product of glycolysis, is processed differently in the presence and the absence of oxygen. page 148

In the absence of oxygen, pyruvate enters one of several fermentation pathways. page 148

In lactic acid fermentation, pyruvate is reduced to lactic acid. page 148

In ethanol fermentation, pyruvate is converted to acetaldehyde, which is reduced to ethanol. page 148

During fermentation, NADH is oxidized to NAD^+, allowing glycolysis to proceed. page 148

Glycolysis and fermentation are ancient biochemical pathways and were likely used in the common ancestor of all organisms living today. page 149

7.7 METABOLIC INTEGRATION: Metabolic pathways are integrated, allowing control of the energy level of cells.

Excess glucose molecules are linked together and stored in polymers called glycogen (in animals) and starch (in plants). page 150

Other monosaccharides derived from the digestion of dietary carbohydrates are converted into intermediates of glycolysis. page 151

Fatty acids contained in triacylglycerols are an important form of energy storage in cells. The breakdown of fatty acids is called β-oxidation. page 151

Phosphofructokinase-1 controls a key step in glycolysis. It has many allosteric activators, including ADP and AMP, and allosteric inhibitors, including ATP and citrate. page 153

The ATP in muscle cells used to power exercise is generated by lactic acid fermentation, aerobic respiration, and β-oxidation. page 153

Log in to **LaunchPad** to check your answers to the Self-Assessment Questions and to access additional learning tools.

CHAPTER 8

Photosynthesis
Using Sunlight to Build Carbohydrates

CORE CONCEPTS

8.1 AN OVERVIEW OF PHOTOSYNTHESIS: Photosynthesis is the major pathway by which energy and carbon are incorporated into carbohydrates.

8.2 THE CALVIN CYCLE: The Calvin cycle is a three-step process that uses carbon dioxide to synthesize carbohydrates.

8.3 CAPTURING SUNLIGHT INTO CHEMICAL FORMS: The light-harvesting reactions use sunlight to produce the ATP and NADPH required by the Calvin cycle.

8.4 PHOTOSYNTHETIC CHALLENGES: Challenges to the efficiency of photosynthesis include excess light energy and the oxygenase activity of rubisco.

8.5 THE EVOLUTION OF PHOTOSYNTHESIS: The evolution of photosynthesis had a profound impact on life on Earth.

Walk through a forest and you will be struck, literally if you aren't careful, by the substantial nature of trees. Where does the material to construct these massive organisms come from? Because trees grow upward from a firm base in the ground, a reasonable first guess is the soil. In the first recorded experiment on this question, the Flemish chemist and physiologist Jan Baptist van Helmont (1580–1644) found that the 200 pounds of dry soil into which he had planted a small willow tree decreased by only 2 ounces over a 5-year period. During this same period, the tree gained 164 pounds. Van Helmont concluded that water must be responsible for the tree's growth. He was, in fact, half right: a tree is roughly half liquid water. However, what he missed completely is that the other half of his tree had been created almost entirely out of thin air.

The process that allowed Van Helmont's tree to increase in mass using material pulled from the air is called **photosynthesis**. Photosynthesis is a biochemical process for building carbohydrates using energy from sunlight and carbon dioxide (CO_2) taken from the air. These carbohydrates are used both as starting points for the synthesis of other molecules and as a means of storing energy in chemical bonds. As we saw in Chapter 7, plants and animals alike need energy in the form of ATP to power the work of the cell. The energy from sunlight is converted into the energy in the chemical bonds of carbohydrates, and that energy, in turn, can be converted into ATP through cellular respiration.

8.1 AN OVERVIEW OF PHOTOSYNTHESIS

Photosynthesis is the major entry point for energy into biological systems. It is the source of all of the food we eat, both directly when we consume plant material and indirectly when we consume meat. It is also the source of all the oxygen that we breathe, as well as the fuels we use for heating and transportation. Fossil fuels are the legacy of ancient photosynthesis: oil has its origin in the bodies of marine algae and the organisms that grazed on them, whereas coal represents the geologic remains of terrestrial (land) plants. Thus, one motivation to understand photosynthesis is the sheer magnitude and importance of this process for life on Earth. Before exploring the details of how photosynthesis actually occurs, let's look at what types of organism carry out photosynthesis and where they live, as well as what structural components allow cells to capture energy in this remarkable way.

Photosynthesis is widely distributed.

The photosynthesis that is most evident to us is carried out by plants on land. Trees, grasses, and shrubs are all examples of photosynthetic organisms. However, photosynthesis is carried out by prokaryotic as well as eukaryotic organisms, on land as well as in the sea. Approximately 60% of global photosynthesis is carried out by terrestrial organisms, with the remaining 40% taking place in the ocean. The majority of photosynthetic organisms in marine environments are unicellular. About half of oceanic photosynthesis is carried out by single-celled marine eukaryotes, while the other half is carried out by photosynthetic bacteria.

Photosynthesis takes place almost everywhere sunlight is available to serve as a source of energy. In the ocean, photosynthesis occurs in the surface layer extending to about 100 m deep, called the **photic zone**, through which enough sunlight penetrates to enable photosynthesis. On land, photosynthesis occurs most readily in environments that are both moist and warm. Tropical rain forests have high photosynthetic productivity, as do grasslands and forests in the temperate zone. However, photosynthetic organisms have evolved adaptations that allow them to tolerate a wide range of environmental conditions, like those illustrated in **Fig. 8.1**. In very dry regions, a combination of photosynthetic bacteria and unicellular algae forms an easily disturbed layer on the surface of the soil known as desert crust. Photosynthetic bacteria are also found in the hot springs of Yellowstone National Park at temperatures up to 75°C. At the other extreme, unicellular algae can grow on the surfaces of glaciers, causing the snow to appear red. In section 8.4, we discuss why photosynthetic organisms growing in such extreme environments often appear red or yellow rather than green.

FIG. 8.1 Photosynthesis in extreme environments. (a) Desert crust in the Colorado Plateau formed by photosynthetic bacteria and algae. (b) A hot spring in Yellowstone National Park. The yellow color is due to photosynthetic bacteria. (c) The surface of a permanent snow pack. The red color is due to photosynthetic algae. *Sources: a. Jayne Belnap/USGS; b. f11photo/Shutterstock; c. Shattil & Rozinski/Naturepl.com.*

Photosynthesis is a redox reaction.

Carbohydrates are synthesized from CO_2 molecules during photosynthesis, yet they have more energy stored in their chemical bonds than is contained in the bonds of CO_2 molecules. Therefore, to build carbohydrates using CO_2 requires an input of energy. This energy comes from sunlight.

How does energy from sunlight become incorporated into chemical bonds? The answer to this question arises from the fact that the synthesis of carbohydrates from CO_2 and water is a reduction–oxidation, or redox, reaction. In Chapter 7, we saw that **reduction** reactions are reactions in which a molecule acquires electrons and gains energy, whereas **oxidation** reactions are reactions in which a molecule loses electrons and releases energy. During photosynthesis, CO_2 molecules are reduced to form higher-energy carbohydrate molecules (**Fig. 8.2**).

Where do the electrons used to reduce CO_2 come from? In photosynthesis carried out by plants and many algae, the ultimate electron donor is water. The oxidation of water results in the production of electrons, protons, and O_2. Thus, oxygen is formed in photosynthesis as a by-product of water's role as a source of electrons. We can demonstrate that water is the source

FIG. 8.2 Overview of photosynthesis. In photosynthesis, energy from sunlight is used to synthesize carbohydrates from CO_2.

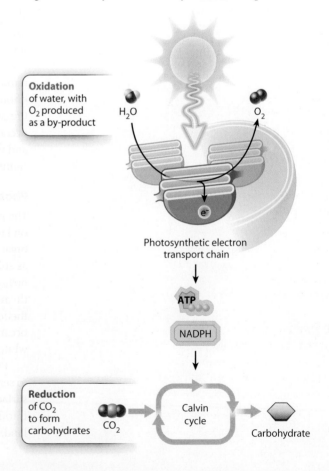

of the oxygen released during photosynthesis using isotopes, which are elements that have different numbers of neutrons and thus can be distinguished on the basis of their mass (**Fig. 8.3**).

Overall, then, the equation for photosynthesis leading to the synthesis of glucose ($C_6H_{12}O_6$) can be described as follows:

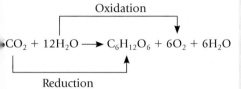

The oxidation of water is linked with the reduction of CO_2 through a series of redox reactions in which electrons are passed from one compound to another. This series of reactions constitutes the **photosynthetic electron transport chain**. The process begins with the absorption of sunlight by protein–pigment complexes. The absorbed sunlight provides the energy that drives electrons through the photosynthetic electron transport chain. In turn, the movement of electrons through this transport chain is used to produce ATP and the electron carrier **nicotinamide adenine dinucleotide phosphate (NADPH)**. Finally, ATP and NADPH are the energy sources needed to synthesize carbohydrates using CO_2 in a process called the **Calvin cycle** (see Fig. 8.2). In eukaryotes, all of these processes, both the capture of light energy and the synthesis of carbohydrates, take place in chloroplasts.

Although photosynthetic organisms are correctly described as autotrophs because they can form carbohydrates from CO_2, they also require a constant supply of ATP to meet each cell's energy requirements. ATP is produced within chloroplasts, but only carbohydrates (and not ATP) are exported from chloroplasts to the cytosol. This explains why cells that

HOW DO WE KNOW?

FIG. 8.3

Does the oxygen released by photosynthesis come from H_2O or CO_2?

BACKGROUND The reactants in photosynthesis are water and carbon dioxide. Both contain oxygen, so it is unclear which one is the source of the oxygen gas that is produced in the reaction. How did scientists determine whether the source of oxygen is water or carbon dioxide?

METHOD Isotopes were used to identify the source of oxygen in photosynthesis. Most of the oxygen in the atmosphere is ^{16}O, a stable isotope containing 8 protons and 8 neutrons. A small amount (0.2%) is ^{18}O, a stable isotope with 8 protons and 10 neutrons. The relative abundance of molecules containing ^{16}O versus ^{18}O can be measured using a mass spectrometer. Water and carbon dioxide containing a high percentage of ^{18}O were prepared, and separate solutions of the algae *Chlorella* were exposed to each one.

EXPERIMENT

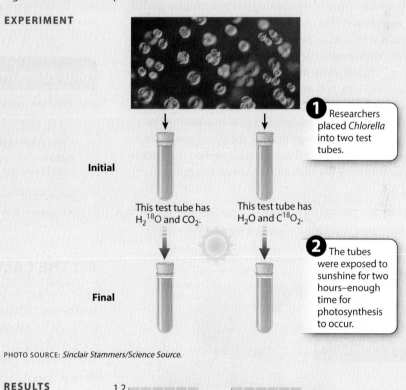

PHOTO SOURCE: *Sinclair Stammers/Science Source.*

CONCLUSION The percentage of $^{18}O_2$ produced by photosynthesis in the algae increased only when water contained a high percentage of ^{18}O, but not when carbon dioxide contained a high percentage of ^{18}O. This finding indicates that the oxygen produced in photosynthesis comes from water, not carbon dioxide.

FOLLOW-UP WORK Carbon also has several isotopes. The measurement of the relative abundance of these isotopes has been used to determine the source of increased carbon dioxide in the atmosphere today (Chapter 46).

SOURCE Adapted from Ruben, S., M. Randall, M. Kamen, and J. L. Hyde. 1941. "Heavy Oxygen (O^{18}) as a Tracer in the Study of Photosynthesis." *Journal of the American Chemical Society* 63:877–879.

have chloroplasts also contain mitochondria. In mitochondria, carbohydrates are broken down to generate ATP (Chapter 7). Cellular respiration is therefore one of several features that heterotrophic organisms like ourselves share with photosynthetic organisms.

The photosynthetic electron transport chain takes place in specialized membranes.

Electron transport chains play a key role in both photosynthesis and respiration (Chapter 7). In both cases, electrons move within and between large protein complexes embedded in specialized membranes. In photosynthetic bacteria, the photosynthetic electron transport chain is located in membranes within the cytoplasm or, in some cases, directly in the plasma membrane. In eukaryotic cells, photosynthesis takes place in chloroplasts. In the center of the chloroplast is the highly folded **thylakoid membrane** (**Fig. 8.4**). The photosynthetic electron transport chain is located in the thylakoid membrane.

The name "thylakoid" is derived from *thylakois*, the Greek word for "sac." Thylakoid membranes form structures that resemble flattened sacs, and these sacs are grouped into structures called **grana** (singular, granum) that look like stacks of interlinked pancakes. Grana are connected to one another by membrane bridges in such a way that the thylakoid membrane encloses a single interconnected compartment called the **lumen**. The region surrounding the thylakoid membrane is called the **stroma**. Carbohydrate synthesis takes place in the **stroma**, whereas sunlight is captured and transformed into chemical energy by the photosynthetic electron transport chain in the thylakoid membrane.

Next, we consider the underlying biochemistry of photosynthesis. In this chapter, we focus on photosynthesis as carried out by eukaryotic cells, but remarkable diversity exists in the way it is carried out in bacteria. We discuss this diversity in Chapter 24. Because carbohydrates are the major product of photosynthesis, we first examine the Calvin cycle, the biochemical pathway used in photosynthesis to synthesize carbohydrates from CO_2. Once we understand what energy forms are needed to drive this autotrophic pathway, we turn our attention to how energy is captured from sunlight. We then examine some of the challenges of coordinating these two metabolic stages and consider ways in which photosynthetic organisms cope with the inherent biochemical challenges of photosynthesis. Finally, we provide a short overview of the evolutionary history of photosynthesis.

Self-Assessment Questions

1. Write the overall reaction for photosynthesis. Which molecules are oxidized and which molecules are reduced?
2. If you want to produce carbohydrates that contain the heavy oxygen (^{18}O) isotope, should you water your plants with $H_2^{18}O$ or inject $C^{18}O_2$ into the air?
3. How are the overall reactions for photosynthesis and cellular respiration similar and how are they different?

8.2 THE CALVIN CYCLE

The Calvin cycle consists of 15 chemical reactions that synthesize carbohydrates from CO_2. These reactions can be grouped into three main steps: (1) **carboxylation**, in which CO_2 is added to a 5-carbon molecule; (2) **reduction**, in which energy and electrons are transferred to the compounds formed in step 1; and (3) **regeneration** of the 5-carbon molecule needed for carboxylation (**Fig. 8.5**). Carbon and oxygen make up more than 90% of the mass of a carbohydrate molecule, such that step 1 represents the major input of raw materials. The major input of energy occurs in step 2. Without this input of energy from ATP and NADPH, carbohydrates could not be produced from CO_2. Step 3 is essential because without the 5-carbon molecule needed for carboxylation, the Calvin cycle cannot proceed.

The incorporation of CO_2 is catalyzed by the enzyme rubisco.

In the first step of the Calvin cycle (Fig. 8.5), CO_2 is added to a 5-carbon sugar called **ribulose 1,5-bisphosphate (RuBP)**. This step is catalyzed by the enzyme **ribulose bisphosphate carboxylase oxygenase,** or **rubisco** for short. An enzyme that

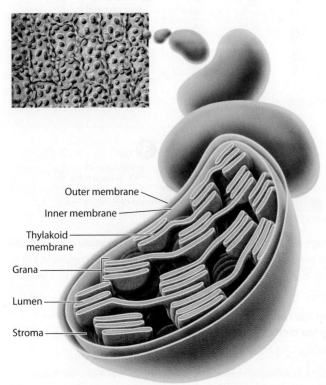

FIG. 8.4 Chloroplast structure. Chloroplasts contain highly folded thylakoid membranes. *Source: Biophoto Associates/Science Source.*

adds CO_2 to another molecule is called a carboxylase, explaining part of rubisco's long name.

Before rubisco can act as a carboxylase, RuBP and CO_2 must diffuse into its active site. Once the active site is occupied, the addition of CO_2 to RuBP proceeds spontaneously in the sense that no addition of energy is required. The product is a 6-carbon compound that immediately breaks into two molecules of **3-phosphoglycerate (3-PGA)**. These 3-carbon molecules are the first stable products of the Calvin cycle.

NADPH is the reducing agent of the Calvin cycle.

Rubisco is responsible for the addition of the carbon atoms needed for the formation of carbohydrates, but rubisco alone does not increase the amount of energy stored within the newly formed bonds. For this energy increase to take place, the carbon compounds formed by rubisco must be reduced. NADPH is the reducing agent used in the Calvin cycle. NADPH transfers the electrons that allow carbohydrates to be synthesized from CO_2 (Fig. 8.5).

In the Calvin cycle, the reduction of 3-PGA involves two reactions: (1) ATP donates a phosphate group to 3-PGA, and (2) NADPH transfers two electrons plus one proton (H^+) to the phosphorylated compound, which releases one phosphate group (P_i). Because two molecules of 3-PGA are formed each time rubisco catalyzes the incorporation of one molecule of CO_2, two ATP and two NADPH are required for each molecule of CO_2 incorporated by rubisco. NADPH provides most of the energy incorporated in the bonds of the carbohydrate molecules produced by the Calvin cycle. Nevertheless, ATP plays an essential role in preparing 3-PGA for the addition of energy and electrons from NADPH.

These energy transfer steps result in the formation of 3-carbon carbohydrate molecules known as **triose phosphates**. Triose phosphates are the true products of the Calvin cycle and the principal form of carbohydrate exported from the chloroplast during photosynthesis. However, if every triose phosphate molecule produced by the Calvin cycle were exported from the chloroplast, RuBP could not be regenerated and the Calvin cycle would grind to a halt. In fact, most of the triose phosphate molecules must be used to regenerate RuBP. For every six triose phosphate molecules produced, only one can be withdrawn from the Calvin cycle.

FIG. 8.5 The Calvin cycle. CO_2 is the input and triose phosphate is the output of the Calvin cycle.

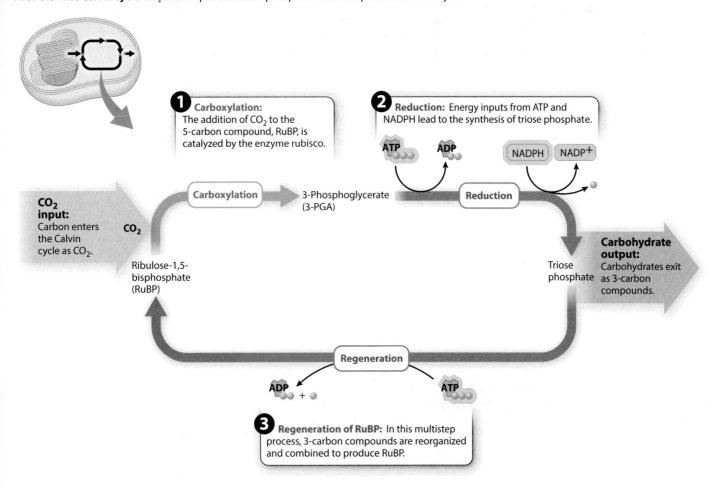

HOW DO WE KNOW?

FIG. 8.6
How is CO_2 incorporated into carbohydrates?

BACKGROUND In the 1940s, radioactive $^{14}CO_2$ became available in quantities that allowed scientists to use it for experiments. Melvin Calvin and Andrew Benson used $^{14}CO_2$ to follow the incorporation of CO_2 into carbohydrates.

EXPERIMENTS AND RESULTS

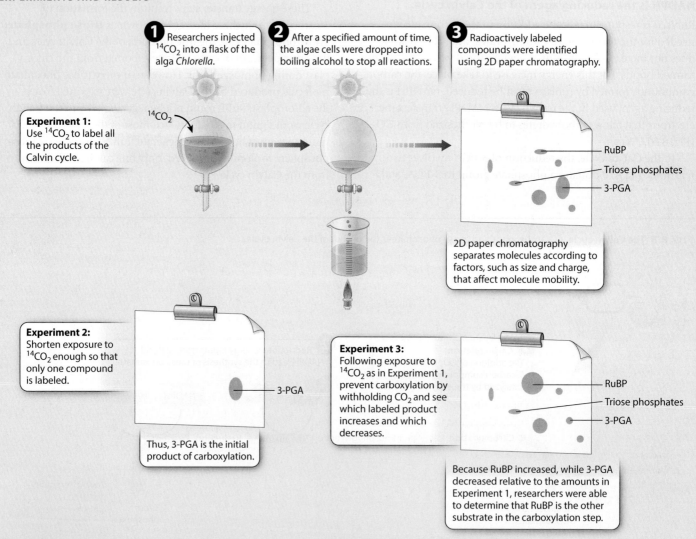

1 Researchers injected $^{14}CO_2$ into a flask of the alga *Chlorella*.

2 After a specified amount of time, the algae cells were dropped into boiling alcohol to stop all reactions.

3 Radioactively labeled compounds were identified using 2D paper chromatography.

Experiment 1: Use $^{14}CO_2$ to label all the products of the Calvin cycle.

2D paper chromatography separates molecules according to factors, such as size and charge, that affect molecule mobility.

Experiment 2: Shorten exposure to $^{14}CO_2$ enough so that only one compound is labeled.

Thus, 3-PGA is the initial product of carboxylation.

Experiment 3: Following exposure to $^{14}CO_2$ as in Experiment 1, prevent carboxylation by withholding CO_2 and see which labeled product increases and which decreases.

Because RuBP increased, while 3-PGA decreased relative to the amounts in Experiment 1, researchers were able to determine that RuBP is the other substrate in the carboxylation step.

CONCLUSION Preventing carboxylation resulted in an excess of RuBP and less 3-PGA, leading researchers to determine that the initial step in the Calvin cycle unites the 5-carbon RuBP with CO_2, resulting in the production of two molecules of 3-PGA.

FOLLOW-UP WORK In the 1950s, Marshall Hatch and colleagues showed that some plants, including corn and sugarcane, accumulate a 4-carbon compound as the first product in photosynthesis. In Chapter 27, we explore how C_4 photosynthesis allows plants to avoid the reaction of rubisco with oxygen.

SOURCE Calvin, M., and H. Benson. 1949. "The Path of Carbon in Photosynthesis IV: The Identity and Sequence of the Intermediates in Sucrose Synthesis." *Science* 109:140–142.

The regeneration of RuBP requires ATP.

Of the 15 chemical reactions that make up the Calvin cycle, 12 occur in the last step, the regeneration of RuBP (Fig. 8.5). Many reactions are needed to rearrange the carbon atoms from five 3-carbon triose phosphate molecules into three 5-carbon RuBP molecules. ATP is required for the regeneration of RuBP, raising the Calvin cycle's total energy requirements to two molecules of NADPH and three molecules of ATP for each molecule of CO_2 incorporated by rubisco.

The Calvin cycle does not use sunlight directly. For this reason, this pathway is sometimes referred to as the light-independent or even the "dark" reactions of photosynthesis. However, this pathway cannot operate without the energy input provided by a steady supply of NADPH and ATP. Both are supplied by the photosynthetic electron transport chain, in which light is captured and transformed into chemical energy. In addition, several Calvin cycle enzymes are regulated by cofactors that must be activated by the photosynthetic electron transport chain. Thus, in a photosynthetic cell, the Calvin cycle occurs only in the light.

The steps of the Calvin cycle were determined using radioactive CO_2.

In a series of experiments conducted between 1948 and 1954, the American chemist Melvin Calvin and colleagues identified the carbon compounds produced during photosynthesis (**Fig. 8.6**). They supplied radioactively labeled CO_2 ($^{14}CO_2$) to the unicellular green alga *Chlorella*, and then plunged the cells into boiling alcohol, thereby halting all enzymatic reactions. The carbon compounds produced during photosynthesis were thus radioactively labeled and could be identified by their radioactivity (Experiment 1 in Fig. 8.6).

Additional experiments were required, however, to determine the chemical reactions that connected these labeled compounds. For example, by using a very short exposure to $^{14}CO_2$, Calvin and colleagues determined that 3-PGA was the first stable product of the Calvin cycle (Experiment 2 in Fig. 8.6).

To determine how 3-PGA is formed, they first supplied $^{14}CO_2$ so that all of the molecules in the Calvin cycle became radioactively labeled. When they then cut off the supply of CO_2, the amount of RuBP increased relative to the amount observed in the first experiment. Based on this buildup of RuBP, they concluded that the first step in the Calvin cycle was the addition of CO_2 to RuBP (Experiment 3 in Fig. 8.6).

Carbohydrates are stored in the form of starch.

The Calvin cycle is capable of producing more carbohydrates than the cell needs or, in a multicellular organism, more than the cell is able to export. If carbohydrates accumulated in the

FIG. 8.7 A chloroplast containing starch granules, shown here in yellow. *Source: Biophoto Associates/Science Source.*

cell, they would cause water to enter the cell by osmosis, perhaps damaging the cell. Instead, excess carbohydrates are converted to starch, a storage form of carbohydrates discussed in Chapter 2. Because starch molecules are not soluble, they provide a means of carbohydrate storage that does not lead to osmosis. Starch formed during the day provides photosynthetic cells with a source of carbohydrates that they can use during the night (**Fig. 8.7**).

Self-Assessment Questions

4. What are the major inputs and outputs of the Calvin cycle?
5. What are the three major steps in the Calvin cycle and the role of the enzyme rubisco?
6. The Calvin cycle requires both ATP and NADPH. Which of these molecules provides the major input of energy needed to synthesize carbohydrates?

8.3 CAPTURING SUNLIGHT INTO CHEMICAL FORMS

In order for sunlight to power the Calvin cycle, the cell must be able to use light energy to produce both NADPH and ATP. In photosynthesis, light energy absorbed by pigment molecules drives the flow of electrons through the photosynthetic electron transport chain, and this flow of electrons leads to the formation of both NADPH and ATP.

Chlorophyll is the major entry point for light energy in photosynthesis.

To understand how energy from sunlight is captured and stored by photosynthesis, we need to know a little about light. The sun, like all stars, produces a broad spectrum of electromagnetic radiation, ranging from gamma rays to radio waves. Each point along the electromagnetic spectrum has a different energy level and a corresponding wavelength. **Visible light** is the portion of the electromagnetic spectrum apparent to our eyes, and it includes the range of wavelengths used in photosynthesis. The wavelengths of visible light range from 400 nm to 700 nm. Approximately 40% of the sun's energy that reaches Earth's surface is in this range.

Pigments are molecules that absorb some wavelengths of visible light. Pigments look colored because they reflect light enriched in the wavelengths that they do not absorb.

Chlorophyll is the major photosynthetic pigment. The graph in **Fig. 8.8** shows the extent to which wavelengths of visible light are absorbed by pigments in an intact leaf. Leaves appear green because they are poor at absorbing green wavelengths.

The chlorophyll molecule consists of a large, light-absorbing "head" that contains a magnesium atom at its center and a long hydrocarbon "tail" (**Fig. 8.9**). The large number of alternating single and double bonds in the head region explains why chlorophyll is so efficient at absorbing visible light. Chlorophyll molecules are bound to integral membrane proteins in the thylakoid membrane. These protein–pigment complexes, referred to as **photosystems**, are the functional and structural units that absorb light energy and use it to drive electron transport.

Photosystems contain pigments other than chlorophyll, called **accessory pigments**. The most notable are the

FIG. 8.9 Chemical structure of chlorophyll. Shown is chlorophyll *a*, found in all photosynthetic eukaryotes and cyanobacteria.

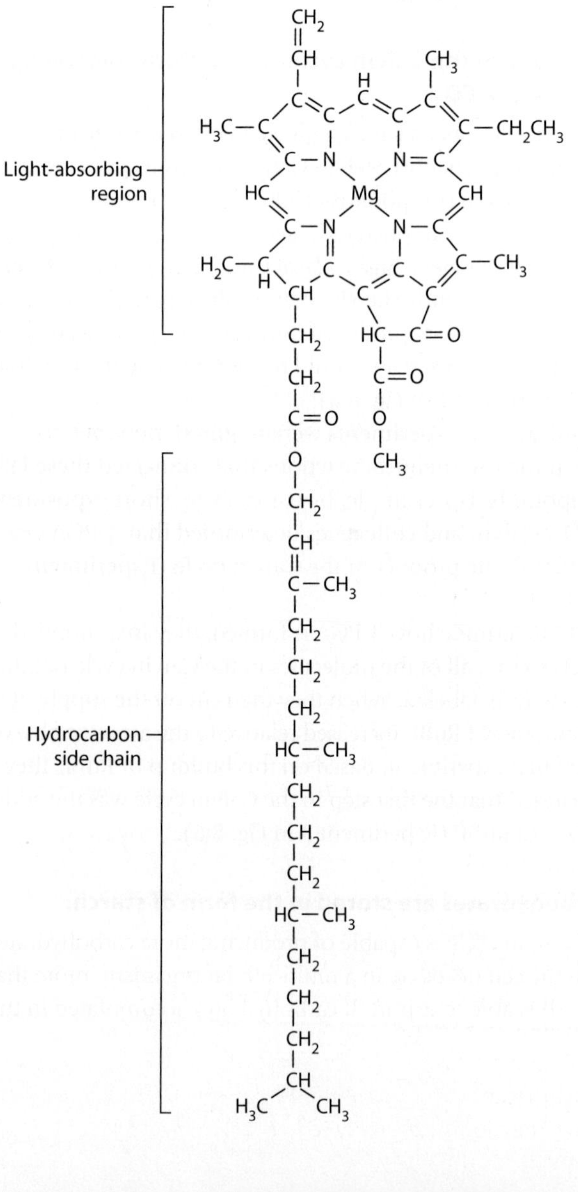

FIG. 8.8 Light absorbed by leaves. The graph shows the extent to which wavelengths of visible light are absorbed by pigments in an intact leaf.

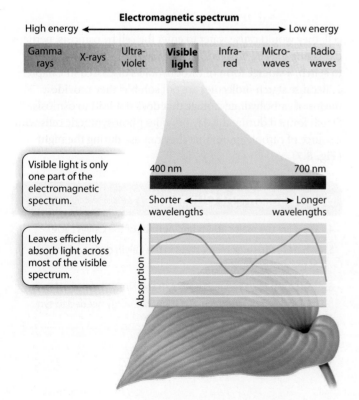

FIG. 8.10 Absorption of light energy by chlorophyll. Absorption of light energy by (a) an isolated chlorophyll molecule in the lab and (b) an antenna chlorophyll molecule.

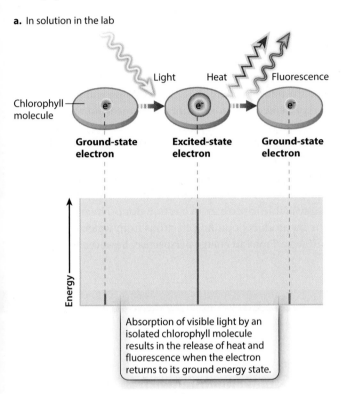

Absorption of visible light by an isolated chlorophyll molecule results in the release of heat and fluorescence when the electron returns to its ground energy state.

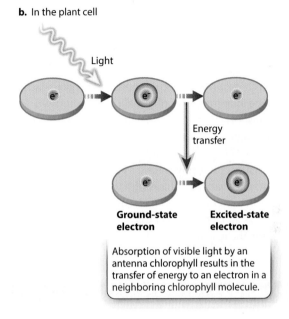

Absorption of visible light by an antenna chlorophyll results in the transfer of energy to an electron in a neighboring chlorophyll molecule.

orange-yellow carotenoids, which can absorb light from regions of the visible spectrum that are poorly absorbed by chlorophyll. Thus, the presence of these accessory pigments allows photosynthetic cells to absorb a broader range of visible light than would be possible with just chlorophyll alone. As we will see in section 8.4, carotenoids play an important role in protecting the photosynthetic electron transport chain from damage.

Antenna chlorophyll passes light energy to reaction centers.

When a chlorophyll molecule absorbs visible light, one of its electrons is elevated to a higher energy state (**Fig. 8.10**). For chlorophyll molecules that have been extracted from chloroplasts in the laboratory, this absorbed light energy is rapidly released, allowing the electron to return to its initial "ground" energy state (Fig. 8.10a). Most of the energy (more than 95%) is converted into heat; a small amount is reemitted as light (fluorescence).

By contrast, chlorophyll molecules within an intact chloroplast can transfer energy to an adjacent chlorophyll molecule instead of losing the energy as heat (Fig. 8.10b). Energy is still released as an excited electron returns to its ground state, but the release of this energy raises the energy level of an electron in an adjacent chlorophyll molecule. This mode of energy transfer is extremely efficient (that is, very little energy is lost as heat), allowing energy initially absorbed from sunlight to be transferred from one chlorophyll molecule to another and then on to another.

Most of the chlorophyll molecules in the thylakoid membrane function as an antenna: energy is transferred between chlorophyll molecules until it is finally transferred to a specially configured pair of chlorophyll molecules known as the **reaction center (Fig. 8.11)**.

FIG. 8.11 The reaction center. (a) Antenna chlorophylls deliver absorbed light energy to the reaction center, allowing an electron to be transferred to an electron acceptor. (b) After the reaction center has lost an electron, it must gain an electron from an electron donor before it can absorb additional light energy.

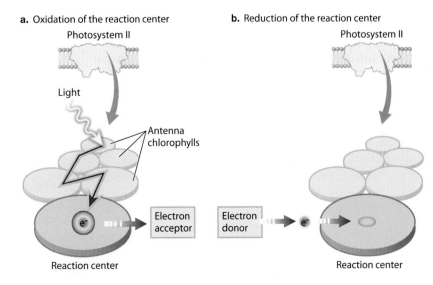

The reaction center is where light energy is converted into chemical energy: the excited electron itself, not just its energy, is transferred to an adjacent molecule. This division of labor among chlorophyll molecules was discovered in the 1940s in a series of experiments by the American biophysicists Robert Emerson and William Arnold, who showed that only a small fraction of chlorophyll molecules are directly involved in electron transport (**Fig. 8.12**). We now know that several hundred antenna chlorophyll molecules transfer energy to each reaction center. The antenna chlorophylls allow the photosynthetic electron transport chain to operate efficiently. Without the antennae to gather light energy, reaction centers would sit idle much of the time, even in bright sunlight.

The reaction center chlorophylls have a configuration distinct from that of the antenna chlorophylls. Therefore, when excited, the reaction center transfers an electron to an adjacent molecule that acts as an electron acceptor (Fig. 8.11a). When the transfer takes place, the reaction center becomes oxidized and the adjacent electron-acceptor molecule is reduced. The result is the conversion of light energy into a chemical form. This electron transfer initiates a light-driven chain of redox reactions that leads ultimately to the formation of NADPH.

Once the reaction center has lost an electron, it can no longer absorb light or contribute additional electrons. Thus, for the photosynthetic electron transport chain to continue, another electron must be delivered to take the place of the one that has entered the transport chain (Fig. 8.11b). As we will see below, these replacement electrons ultimately come from water.

The photosynthetic electron transport chain connects two photosystems.

In many ways, water is an ideal source of electrons for photosynthesis. Water is so abundant within cells that it is always available to serve as an electron donor. In addition, O_2, the by-product of pulling electrons from water, diffuses readily away. From an energy perspective, however, water is

HOW DO WE KNOW?

FIG. 8.12

Do chlorophyll molecules operate on their own or in groups?

BACKGROUND By about 1915, scientists knew that chlorophyll was the pigment responsible for absorbing light energy in photosynthesis. However, it was unclear how these pigments contributed to the reduction of CO_2. The American physiologists Robert Emerson and William Arnold set out to determine the nature of the "photochemical unit" by quantifying how many chlorophyll molecules are needed to produce one molecule of O_2.

EXPERIMENT Emerson and Arnold exposed flasks of the green alga *Chlorella* to flashes of light of very short duration (10^{-5} s) and high intensity (several times more intense than full sunlight) in order to "excite" each chlorophyll molecule just one time. They also made sure that the time between flashes was long enough to allow the reactions resulting from each flash to run to completion. They then determined the number of O_2 molecules produced per flash of high-intensity light and measured the concentration of chlorophyll present in their solution of cells.

RESULTS

Measurement 1 (rate of O_2 production):

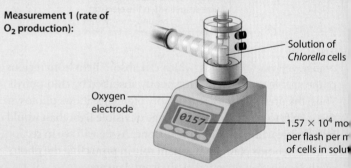

1.57×10^4 mol per flash per m of cells in solut

Measurement 2 (chlorophyll concentration):

3.9×10^7 mol chlorophyll per mm³ of solution

CONCLUSION Because the number of chlorophyll molecules in their flasks was much greater the number of O_2 molecules produced per flash, Emerson and Arnold concluded that many chlorophyll molecules are needed to produce one O_2 molecule and that each photochemical unit must contain many chlorophyll molecules.

FOLLOW-UP WORK Emerson and Arnold's work was followed by studies that demonstrated that the photosynthetic electron transport chain contains two photosystems arranged in series.

SOURCE Emerson, R., and W. Arnold. 1932. "The Photochemical Reaction in Photosynthesis." *Journal of General Physiology* 16:191–205.

a challenging electron donor: it takes a great deal of energy to pull electrons from water. The amount of energy that a single photosystem can capture from sunlight is not enough to simultaneously pull an electron from water and produce an electron donor capable of reducing $NADP^+$ to NADPH. The solution is to use two photosystems, each with distinct chemical properties, arranged in series (**Fig. 8.13**). The energy supplied by the first photosystem allows electrons to be pulled from water, and the energy supplied by the second photosystem allows electrons to be transferred to $NADP^+$ to form NADPH.

If you follow the flow of electrons from water through both photosystems and on to $NADP^+$, as shown in Fig. 8.13, you can see a large increase in energy as the electrons pass through each of the two photosystems. You can also see that an overall decrease in energy occurs as electrons move between the two photosystems. This decrease in energy explains why electrons move in one "direction" through the photosynthetic electron transport chain. To run these reactions in the opposite direction would require an input of energy. Because the overall energy trajectory has an up-down-up configuration resembling a "Z," the photosynthetic electron transport chain is sometimes referred to as the **Z scheme**.

The major protein complexes of the photosynthetic electron transport chain include the two photosystems as well as the **cytochrome–b_6f complex (Cyt)** in between (**Fig. 8.14**). Small, relatively mobile compounds convey electrons between these protein complexes: plastoquinone (Pq) diffuses through the membrane from photosystem II (PSII) to the cytochrome–b_6f complex, and plastocyanin (Pc) diffuses through the thylakoid lumen from the cytochrome–b_6f complex to photosystem I (PSI).

Water donates electrons to one end of the photosynthetic electron transport chain, whereas $NADP^+$ accepts electrons at the other end. The enzyme that pulls electrons from water, releasing both H^+ and O_2, is located on the lumen side of photosystem II. NADPH is formed when electrons are passed from photosystem I to a membrane-associated protein called ferredoxin (Fd) on the stromal side of the thylakoid membrane (Fig. 8.14b). The enzyme ferredoxin–$NADP^+$ reductase then catalyzes the formation of NADPH by transferring two electrons from two molecules of reduced ferredoxin to $NADP^+$ as well as a proton from the surrounding solution:

$$NADP^+ + 2e^- + H^+ \longrightarrow NADPH$$

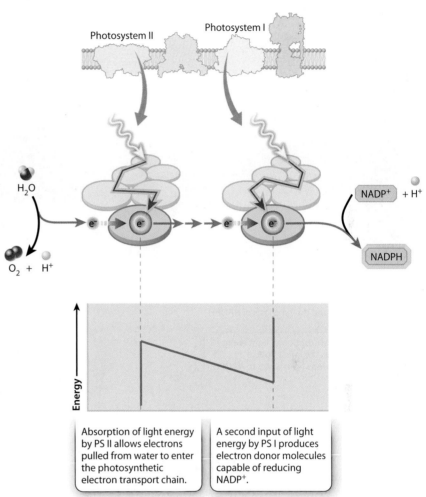

FIG. 8.13 The Z scheme. The use of water as an electron donor requires input of light energy at two places in the photosynthetic electron transport chain.

The accumulation of protons in the thylakoid lumen drives the synthesis of ATP.

So far, we have considered only how the photosynthetic electron transport chain leads to the formation of NADPH. However, we know that the Calvin cycle also requires ATP. In chloroplasts, as in mitochondria, ATP is synthesized by ATP synthase, a transmembrane protein powered by a proton gradient (Chapter 7). In chloroplasts, ATP synthase is oriented such that the synthesis of ATP is the result of the movement of protons from the thylakoid lumen to the stroma, as shown in Fig. 8.14c.

How do protons accumulate in the thylakoid lumen? Two features of the photosynthetic electron transport chain are responsible for the buildup of protons in the thylakoid lumen (Fig. 8.14c). First, the oxidation of water releases protons and O_2 into the lumen. Second, the cytochrome–b_6f complex and plastoquinone together act as a proton pump that functions much like proton pumping in the electron transport chain of cellular respiration (Chapter 7). In fact, the two proton pumping systems are evolutionarily related.

FIG. 8.14 The photosynthetic electron transport chain. (a) An overview of the production of NADPH and ATP. (b) The linear flow of electrons from H_2O to NADPH. (c) The use of a proton electrochemical gradient to synthesize ATP.

a. The production of NADPH and ATP by photosynthesis

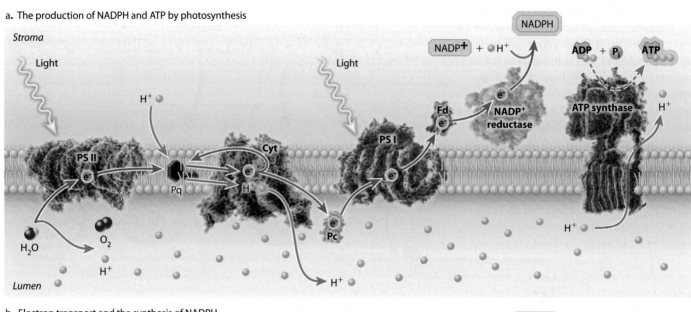

b. Electron transport and the synthesis of NADPH

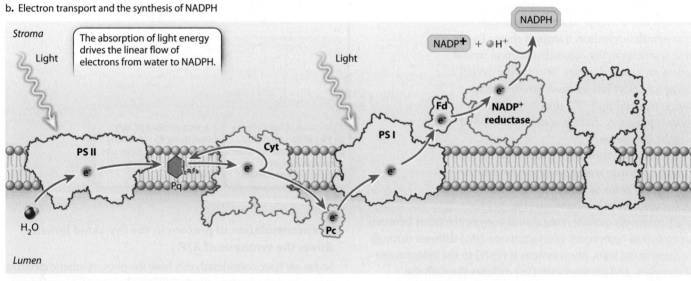

c. Proton transport and ATP synthesis

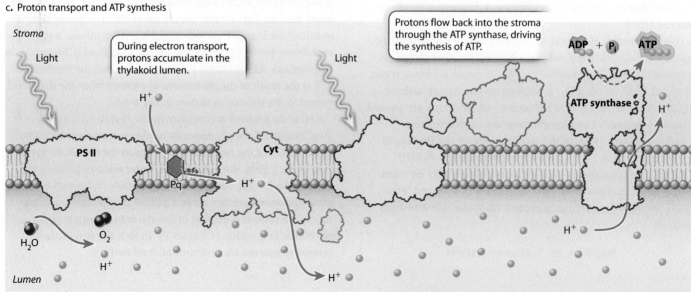

Together, these mechanisms are quite powerful. When the photosynthetic electron transport chain is operating at full capacity, the concentration of protons in the lumen can be more than 1000 times greater than their concentration in the stroma (equivalent to a difference of 3 pH units). This accumulation of protons on one side of the thylakoid membrane can then be used to power the synthesis of ATP by oxidative phosphorylation as described in Chapter 7.

Cyclic electron transport increases the production of ATP.

The Calvin cycle requires two molecules of NADPH and three molecules of ATP for each CO_2 incorporated into carbohydrates. However, the transport of four electrons through the photosynthetic electron transport chain, which is needed to reduce two $NADP^+$ molecules, does not transport enough protons into the lumen to produce the required three ATPs. An additional pathway for electrons is needed to increase the production of ATP.

In **cyclic electron transport**, electrons from photosystem I are redirected from ferredoxin back into the electron transport chain (**Fig. 8.15**). These electrons reenter the photosynthetic electron transport chain by plastoquinone. Because these electrons eventually return to photosystem I, this alternative pathway is cyclic in contrast to the linear movement of electrons from water to NADPH.

How does cyclic electron transport lead to the production of ATP? As the electrons from ferredoxin are picked up by plastoquinone, additional protons are transported from the stroma to the lumen. As a result, there are more protons in the lumen that can be used to drive the synthesis of ATP.

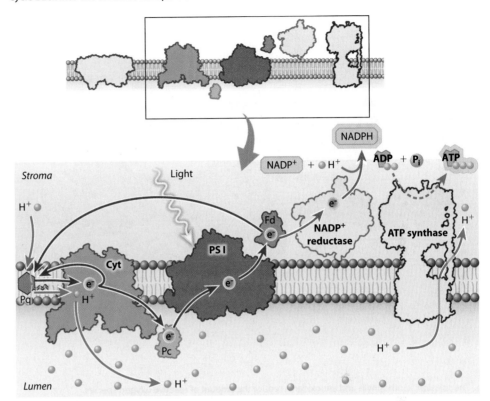

FIG. 8.15 Cyclic electron transport. To increase ATP production, some electrons from photosystem I cycle back into the electron transport chain.

8.4 PHOTOSYNTHETIC CHALLENGES

Two major challenges can hinder photosynthesis from functioning efficiently. The first is that if more light energy is absorbed than the Calvin cycle can use, excess energy can damage the cell. The second challenge stems from a property of rubisco: this enzyme can catalyze the addition of *either* carbon dioxide *or* oxygen to RuBP. The addition of oxygen instead of carbon dioxide can substantially reduce the amount of carbohydrate produced.

Excess light energy can cause damage.

Photosynthesis is an inherently dangerous enterprise. Unless the photosynthetic reactions are carefully controlled, molecules will form that can damage cells by oxidizing lipids, proteins, and nucleic acids indiscriminately (**Fig. 8.16**).

Under normal conditions, the photosynthetic electron transport chain proceeds in an orderly fashion from the absorption of light to the formation of NADPH. When $NADP^+$ is in short supply, however, the electron transport chain "backs up" and leaves electrons and energy with no "safe" place to go. This greatly increases the probability of creating reactive forms of oxygen known collectively as **reactive oxygen species** (Fig. 8.16a). These highly reactive molecules can be formed either by the transfer of absorbed light energy from antenna chlorophyll directly to O_2 or by the transfer of an electron,

> ### Self-Assessment Questions
>
> 7. How do antenna chlorophylls differ from reaction center chlorophylls?
> 8. Why are two photosystems needed when H_2O is used as an electron donor?
> 9. How is energy from sunlight used to produce ATP?
> 10. What are the products of linear electron transport and cyclic electron transport? What is the role of cyclic electron transport?

FIG. 8.16 Defenses against reactive oxygen species. (a) Reactive oxygen species are generated when light energy or electrons are transferred to oxygen. (b) Defenses against reactive oxygen species include antioxidants that neutralize these highly reactive molecules.

a. The problem: harmful reactive oxygen species are generated.

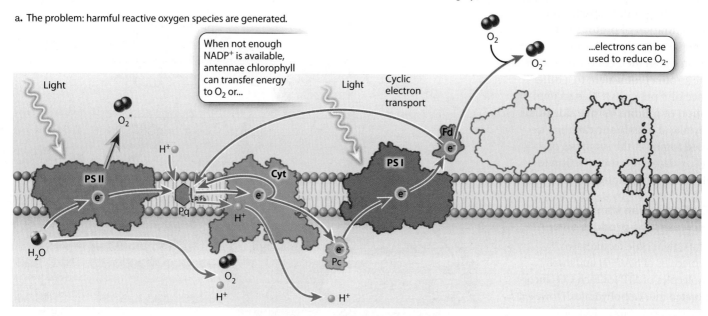

b. The solution: xanthophylls and antioxidants reduce the amount of reactive oxygen species.

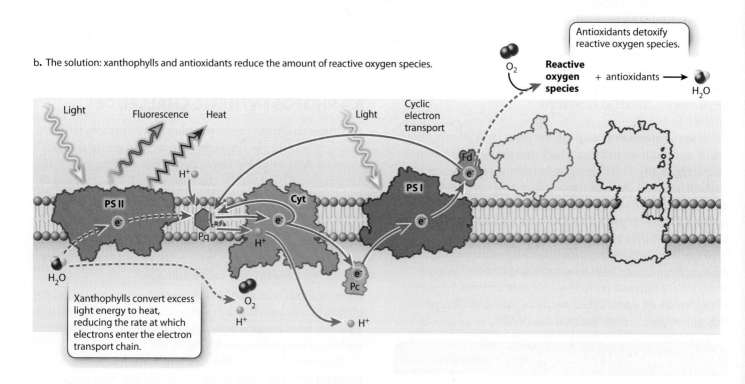

forming O_2^-. Both forms of O_2 can cause substantial damage to the cell.

$NADP^+$ is returned to the photosynthetic electron transport chain by the Calvin cycle's use of NADPH. Thus, any factor that causes the rate of NADPH use to fall behind the rate of light-driven electron transport can potentially lead to damage. Such an imbalance is likely to occur, for example, in the middle of the day when light intensity is highest.

Could the cell right the balance by supplying $NADP^+$ more quickly? Photosynthetic cells could speed up the resupply of $NADP^+$ by synthesizing more Calvin cycle enzymes. This strategy, however, would be energetically expensive. When light levels are low, such as in the morning and late afternoon, Calvin cycle enzymes would sit idle. An alternative strategy of reducing the amount of chlorophyll in the leaf runs into similar problems. Thus, excess light energy is an everyday event for photosynthetic cells, rather than something that occurs only in extreme environments.

The rate at which the Calvin cycle can make use of NADPH is also influenced by a number of factors that are independent of light intensity. For example, at cold temperatures, the enzymes of the Calvin cycle function more slowly, but temperature has little impact on the absorption of light energy. On a cold, sunny day, more light energy is absorbed than can be used by the Calvin cycle.

Photosynthetic organisms employ two major lines of defense to avoid the stresses that occur when the Calvin cycle cannot keep up with light harvesting (Fig. 8.16b). First among these are chemicals that detoxify reactive oxygen species. Ascorbate (vitamin C), β-(beta-)carotene, and other antioxidants are able to neutralize reactive oxygen species. These compounds exist in high concentration in chloroplasts. Some of these antioxidant molecules are brightly colored, like the red pigments found in algae that live on snow shown in Fig. 8.1c. The presence of antioxidant compounds is one of the many reasons why eating green, leafy vegetables is good for your health.

A second line of defense is to prevent reactive oxygen species from forming in the first place. **Xanthophylls** are yellow-orange pigments that slow the formation of reactive oxygen species by reducing excess light energy. These pigments accept absorbed light energy directly from chlorophyll, and then convert this energy to heat (Fig. 8.16b). Photosynthetic organisms that live in extreme environments often appear brown or yellow because they contain high levels of xanthophyll pigments, as seen in Figs. 8.1a and 8.1b. Plants that lack xanthophylls grow poorly when exposed to moderate light levels and die in full sunlight.

Converting absorbed light energy into heat is beneficial at high light levels, but at low light levels it would decrease the production of carbohydrates. Therefore, this capability is switched on only when the photosynthetic electron transport chain is working at high capacity.

Photorespiration leads to a net loss of energy and carbon.

A second challenge to photosynthetic efficiency is rubisco's ability to use both CO_2 and O_2 as substrates. If O_2 instead of CO_2 diffuses into the active site of rubisco, the reaction can still proceed, although O_2 is added to RuBP in place of CO_2. An enzyme that adds O_2 to another molecule is called an oxygenase. Recall that rubisco is shorthand for RuBP carboxylase oxygenase, reflecting rubisco's ability to catalyze two different reactions.

When rubisco adds O_2 instead of CO_2 to RuBP, the result is one molecule with three carbon atoms (3-PGA) and one molecule with only two carbon atoms (2-phosphoglycolate). A serious problem is that 2-phosphoglycolate cannot be used by the Calvin cycle to either produce triose phosphate or regenerate RuBP.

A metabolic pathway to recycle 2-phosphoglycolate is present in photosynthetic cells. Some of the carbon atoms in 2-phosphoglycolate are converted into 3-PGA, which can reenter the Calvin cycle. However, other carbon atoms in 2-phosphoglycolate are released as CO_2. Because the overall effect is the consumption of O_2 and release of CO_2 in the presence of light, this process is referred to as **photorespiration** (Fig. 8.17). However, whereas respiration produces ATP, photorespiration consumes ATP. In photorespiration, ATP drives the reactions that recycle 2-phosphoglycolate into 3-PGA. Thus, photorespiration represents a net energy drain in two ways: first, it results in the oxidation and loss, in the form of CO_2, of carbon atoms that had already been incorporated and reduced by the Calvin cycle; second, it consumes ATP.

The Calvin cycle originated long before the accumulation of oxygen in Earth's atmosphere, which provides an explanation why an enzyme with these properties might have initially evolved. Still, why would photorespiration persist in the face of what must be strong evolutionary pressure to reduce or eliminate the unwanted reaction with oxygen?

The difficulty is that in order for rubisco to favor the addition of CO_2 over O_2, the enzyme must be highly selective—and the price of high selectivity is speed. CO_2 and O_2 are similar in size and chemical structure, and for this reason rubisco can achieve selectivity only by binding more tightly with the transition state (Chapter 6) of the carboxylation reaction. As a result, the better rubisco is at discriminating between CO_2 and O_2, the slower its catalytic rate.

Nowhere is this trade-off more evident than in land plants, whose photosynthetic cells acquire CO_2 from an O_2-rich and CO_2-poor atmosphere. The rubiscos of plants are highly selective: if they are exposed to equal concentrations of CO_2 and O_2, the ratio of CO_2 addition to O_2 addition is approximately 80:1. As a result, rubisco is a very slow enzyme, with catalytic rates on the order of three reactions per second. To put this rate in perspective, it is common for metabolic enzymes to achieve a catalytic rate of tens of thousands of reactions per second.

This trade-off between selectivity and speed is a key constraint for photosynthetic organisms. For plants, rubisco's low catalytic rate means that photosynthetic cells must produce huge amounts of this enzyme. As much as 50% of

FIG. 8.17 Photorespiration. Carbon and energy are lost when rubisco acts as an oxygenase in photosynthesis.

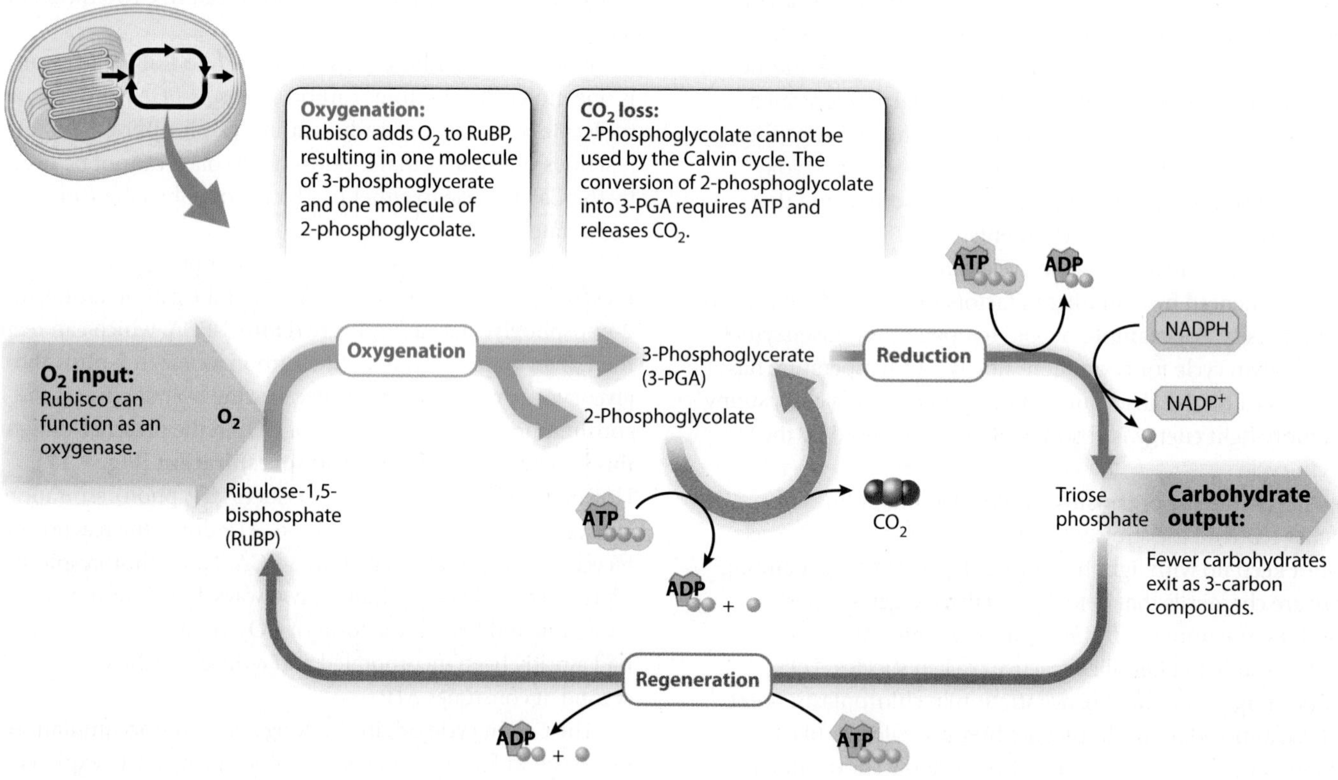

Oxygenation:
Rubisco adds O2 to RuBP, resulting in one molecule of 3-phosphoglycerate and one molecule of 2-phosphoglycolate.
CO2 loss:
2-Phosphoglycolate cannot be used by the Calvin cycle. The conversion of 2-phosphoglycolate into 3-PGA requires ATP and releases CO2.
O2 input:
Rubisco can function as an oxygenase.
O2
Oxygenation
3-Phosphoglycerate (3-PGA)
2-Phosphoglycolate
Reduction
ATP
ADP
NADPH
NADP+
Ribulose-1,5-bisphosphate (RuBP)
ATP
ADP +
CO2
Triose phosphate
Carbohydrate output:
Fewer carbohydrates exit as 3-carbon compounds.
Regeneration
ADP +
ATP

the total protein within a leaf is rubisco, and it is estimated to be the most abundant protein on Earth. At the same time, because O_2 is approximately 500 times more abundant in the atmosphere than CO_2, as much as one-quarter of the reduced carbon formed in photosynthesis can be lost through photorespiration.

Photosynthesis captures just a small percentage of incoming solar energy.

Typically, only 1% to 2% of the sun's energy that lands on a leaf ends up in carbohydrates. Does this mean that photosynthesis is incredibly wasteful? Or is this process, which is the product of billions of years of evolution, surprisingly efficient? This is not an idle question. Photosynthesis is relevant to solving several pressing global issues: the effects of rising CO_2 concentrations on Earth's climate; the search for renewable, carbon-neutral fuels to power our transportation sectors; and the agricultural demands of our skyrocketing human population.

Photosynthetic efficiency is typically calculated relative to the total energy output of the sun (**Fig. 8.18**). However, only visible light has the appropriate energy levels to raise the energy state of electrons in chlorophyll. Most of the sun's output (~60%) is not absorbed by chlorophyll and thus cannot be used in photosynthesis. In addition, leaves are not perfect at absorbing visible light—about 8% is either reflected or passes through the leaf. Finally, even under the best conditions, not all of the light energy absorbed by chlorophyll is transferred to the reaction center; some is given off as heat (also ~8%). This percentage increases when light levels are high, because excess light is actively converted into heat by xanthophyll pigments.

The photosynthetic electron transport chain therefore captures, at most, about 24% of the sun's energy arriving at the surface of a leaf (100% − 60% − 8% − 8% = 24%). Although this number may appear low, it is on a par with the efficiency of high-performance photovoltaic cells in solar panels, which convert sunlight into electricity. However, energy is also lost at a later step. The incorporation of CO_2 into carbohydrates results in considerable energy loss—equivalent to about 20% of the total incoming solar radiation. Some of this energy loss is due to photorespiration.

In total, the maximum energy conversion efficiency of photosynthesis is calculated to be around 4% (24% – 20%). Efficiencies achieved by real plants growing in nature, however, are typically much lower, on the order of 1% to 2%. In Chapter 27, we explore the many factors that can constrain the photosynthetic output of plants and see how some plants have

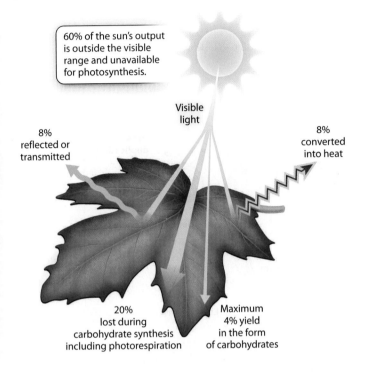

FIG. 8.18 Photosynthetic efficiency. Maximum photosynthetic efficiency is theoretically about 4% of incoming solar energy, but actual yields are closer to 1% to 2%.

evolved ways to minimize losses in productivity due to drought and photorespiration.

Self-Assessment Questions

11. What are two strategies that plants use to limit the formation and effects of reactive oxygen species?
12. How is photorespiration similar to cellular respiration and how does it differ?
13. Why does rubisco have such a low catalytic rate (that is, why is it so slow)?
14. What is the overall efficiency of photosynthesis, and where in the pathway is energy dissipated?

8.5 THE EVOLUTION OF PHOTOSYNTHESIS

The evolution of photosynthesis had a profound impact on the history of life on Earth. Not only did photosynthesis provide organisms with a new source of energy, but it also released oxygen into the atmosphere. As discussed in the previous chapter, evolution often works in a stepwise fashion, building on what is already present. Here, we consider hypotheses for how the photosynthetic pathways that are the dominant entry point for energy into the biosphere today may have evolved.

CASE 1 LIFE'S ORIGINS: INFORMATION, HOMEOSTASIS, AND ENERGY

How did early cells use sunlight to meet their energy requirements?

Sunlight is a valuable source of energy, but it can also cause damage. Ultraviolet wavelengths, in particular, can damage DNA and other macromolecules. Thus, the earliest interactions with sunlight may have been the evolution of UV-absorbing compounds that could shield cells from the sun's damaging rays. Over time, random mutations could have produced chemical variants of these UV-absorbing molecules. One or more of these variant compounds might have been capable of using sunlight to meet the energy needs of the cell, perhaps by transferring electrons to another molecule as a present-day reaction center does.

The earliest reaction centers may have used light energy to drive the movement of electrons from an electron donor outside the cell in the surrounding medium to an electron-acceptor molecule within the cell. In this way, energy from sunlight could have been used to synthesize carbohydrates. The first electron donor could have been a soluble inorganic ion like reduced iron, Fe^{2+}, which is thought to have been abundant in the early ocean. Alternatively, the first forms of light-driven electron transport may have been cyclic and thus did not require an electron donor. In either configuration, light-driven electron transport could have been coupled to the net movement of protons across the membrane, allowing for the synthesis of ATP.

Similarly, it is unlikely that these first photosynthetic organisms employed chlorophyll as a means of absorbing sunlight for the simple reason that the biosynthetic pathway for chlorophyll is complex, consisting of at least 17 enzymatic steps. Yet some of the intermediate compounds leading to chlorophyll are capable of absorbing light. Perhaps each of these now-intermediate compounds was, at one time, a functional end product used as a pigment by an early photosynthetic organism. The biosynthetic pathway may have gained steps as chemical variants produced by random mutations were selected because they were more efficient or more able to absorb new portions of the visible spectrum. Selection would have eventually resulted in the chlorophyll pigments that are used by photosynthetic organisms today.

The ability to use water as an electron donor in photosynthesis evolved in cyanobacteria.

The most ancient forms of photosynthesis have only a single photosystem in their photosynthetic electron transport chains.

As we have seen, a single photosystem cannot capture enough energy from sunlight to both pull electrons from water as well as raise their energy level enough that they can be used to reduce CO_2. Instead, photosynthetic organisms with a single photosystem must use more easily oxidized compounds, such as hydrogen sulfide (H_2S), as electron donors. These organisms can exist only in environments where the electron-donor molecules are abundant. Because these organisms do not use water as an electron donor, they do not produce O_2 during photosynthesis.

A major event in the history of life was the evolution of photosynthetic electron transport chains that use water as an electron donor. The first organisms to accomplish this feat were the cyanobacteria. These photosynthetic bacteria incorporated two different photosystems into a single photosynthetic electron transport chain: one to pull electrons from water molecules and one to raise the energy level of the electrons so that they can be used to reduce CO_2.

How did cyanobacteria end up with two photosystems? We cannot say for sure, but one clue is that each of the two photosystems present in cyanobacteria is similar in structure to photosystems found in groups of photosynthetic bacteria that contain only a single photosystem. Thus, it is highly unlikely that the photosystems in cyanobacteria evolved independently. One hypothesis is that the genetic material associated with one photosystem was transferred to a bacterium that already had the other photosystem, resulting in a single bacterium with the genetic material to produce both types of photosystems (shown on the left in **Fig. 8.19**). The mechanisms by which genetic material is transferred between bacteria are discussed more fully in Chapter 24. Another hypothesis is that the genetic material associated with one photosystem underwent duplication. Over time, one of the two photosystems diverged slightly in sequence and function through mutation and selection, giving rise to two distinct but related photosystems (shown on the right in Fig. 8.19). This mechanism, called duplication and divergence, is discussed more fully in Chapter 14.

The ability to use water as an electron donor in photosynthesis had two major impacts on life on Earth. First, photosynthesis could now occur anywhere there was both sunlight and sufficient water for cells to survive. Second, water releases oxygen as it donates electrons. Before the evolution of oxygenic photosynthesis, little or no free oxygen existed in Earth's atmosphere. All the oxygen in Earth's atmosphere results from photosynthesis by organisms containing two photosystems.

Eukaryotic organisms are believed to have gained photosynthesis by endosymbiosis.

Photosynthesis is hypothesized to have gained a foothold among eukaryotic organisms when a free-living cyanobacterium took up residence inside a eukaryotic cell (Fig. 8.19). Over time, the cyanobacterium lost its ability to survive outside its host cell and evolved into the chloroplast. The outer membrane of the chloroplast is thought to have originated from the plasma

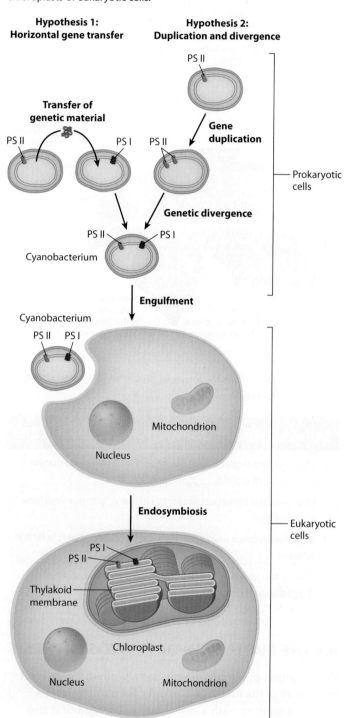

FIG. 8.19 The evolutionary history of photosynthesis. (a) Hypotheses for the origin of the two photosystems in cyanobacteria. (b) Later, by endosymbiosis, free-living cyanobacteria became the chloroplasts of eukaryotic cells.

membrane of the ancestral eukaryotic cell, which surrounded the ancestral cyanobacterium as it became incorporated into the cytoplasm of the eukaryotic cell. The inner chloroplast membrane is thought to correspond to the plasma membrane of the ancestral free-living cyanobacterium. The thylakoid membrane then corresponds to the internal photosynthetic membrane found in cyanobacteria. Finally, the stroma corresponds to the cytoplasm of the ancestral cyanobacterium.

The process in which one cell takes up residence inside of another cell is called endosymbiosis. Therefore, the idea that chloroplasts and mitochondria arose in this way is called the endosymbiotic hypothesis. It is discussed in Chapter 25.

Cellular respiration and photosynthesis are complementary metabolic processes. Cellular respiration breaks down carbohydrates in the presence of oxygen to supply the energy needs of the cell and produces carbon dioxide and water as by-products, whereas photosynthesis uses carbon dioxide and water in the presence of sunlight to build carbohydrates and releases oxygen as a by-product. We summarize the two processes in **Fig. 8.20**.

Self-Assessment Questions

15. What are two hypotheses to explain how cyanobacteria ended up with two photosystems?

16. What are three major steps in the evolutionary history of photosynthesis?

VISUAL SYNTHESIS
FIG. 8.20

Harnessing Energy: Photosynthesis and Cellular Respiration
Integrating concepts from Chapters 6–8

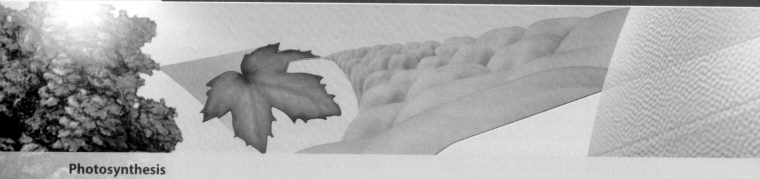

Photosynthesis
Energy captured from sunlight is stored in carbohydrates.

Photosynthesis requires light energy, CO_2, and H_2O, and produces carbohydrates and O_2. The **photosynthetic electron transport chain** uses energy from sunlight to drive the movement of electrons from water to $NADP^+$ and to produce ATP.

Carbon dioxide is reduced by NADPH and, using the energy provided by ATP, is incorporated into a carbohydrate molecule in the **Calvin cycle**.

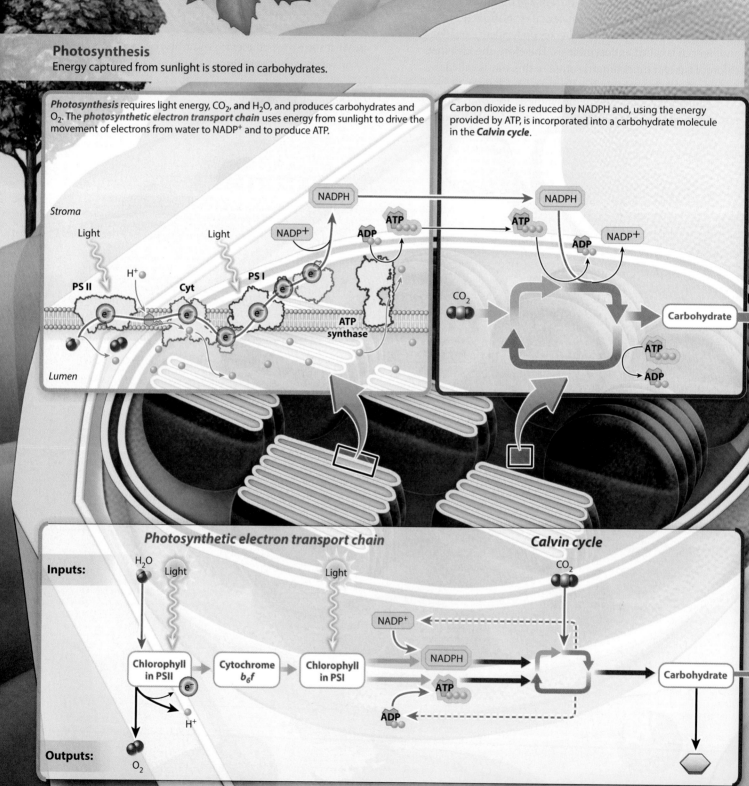

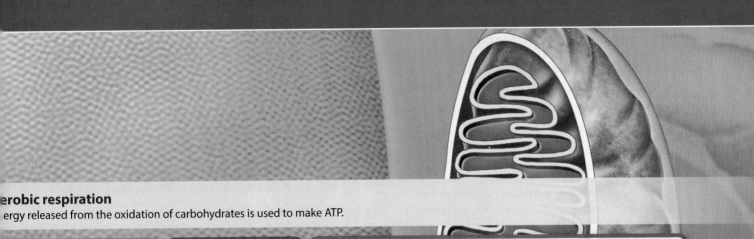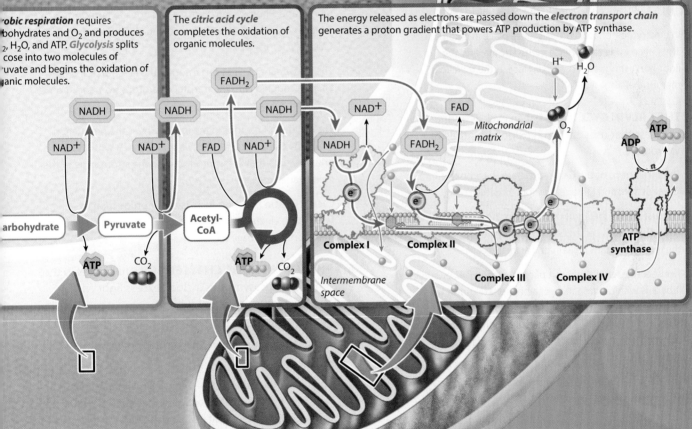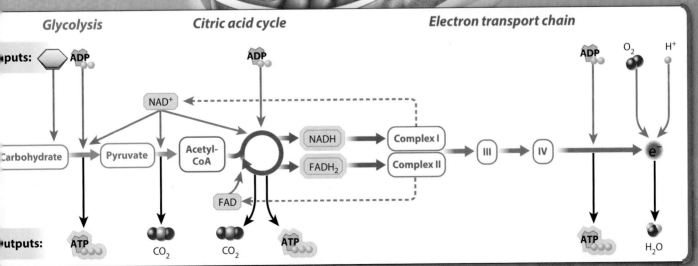

CORE CONCEPTS SUMMARY

8.1 AN OVERVIEW OF PHOTOSYNTHESIS: Photosynthesis is the major pathway by which energy and carbon are incorporated into carbohydrates.

In photosynthesis, water is oxidized, releasing oxygen, and carbon dioxide is reduced, forming carbohydrates. page 158

Photosynthesis consists of two sets of reactions: (1) the Calvin cycle, in which carbon dioxide is reduced to form carbohydrates, and (2) light-harvesting reactions, in which ATP and NADPH are generated to drive the Calvin cycle. page 159

In eukaryotes, photosynthesis takes place in chloroplasts: The Calvin cycle takes place in the stroma, and the light-harvesting reactions take place in the thylakoid membrane. page 160

8.2 THE CALVIN CYCLE: The Calvin cycle is a three-step process that uses carbon dioxide to synthesize carbohydrates.

The three steps of the Calvin cycle are (1) carboxylation; (2) reduction; and (3) regeneration. page 160

The first step is the addition of CO_2 to the 5-carbon sugar RuBP. This step is catalyzed by the enzyme rubisco, considered the most abundant protein on Earth. page 160

The second step is the input of energy captured by sunlight. It begins with the donation of a phosphate group by ATP, followed by the transfer of electrons from NADPH to produce 3-carbon triose phosphate molecules. Some of these triose phosphates are exported from the chloroplast to the cytosol. page 161

The third step is the regeneration of RuBP from five 3-carbon triose phosphates. page 163

Starch formation provides chloroplasts with a way of storing carbohydrates that will not cause water to enter the cell by osmosis. page 163

8.3 CAPTURING SUNLIGHT INTO CHEMICAL FORMS: The light-harvesting reactions use sunlight to produce the ATP and NADPH required by the Calvin cycle.

Chlorophyll absorbs visible light. page 164

Antenna chlorophyll molecules transfer absorbed light energy to the reaction center. page 165

Reaction centers are located within pigment–protein complexes known as photosystems. Special chlorophyll molecules in the reaction center transfer excited-state electrons to an electron-acceptor molecule, thus initiating the photosynthetic electron transport chain. page 165

The electron transport chain consists of a series of electron transfer or redox reactions that take place within both protein complexes and diffusible compounds. Water is the electron donor and $NADP^+$ is the final electron acceptor. page 166

The linear transport of electrons from water to NADPH requires the energy input of two photosystems. page 166

Photosystem II pulls electrons from water, resulting in the production of oxygen and protons on the lumen side of the membrane. Photosystem I passes electrons to ferredoxin, which then reduces $NADP^+$, producing NADPH for use in the Calvin cycle. page 167

The movement of electrons through the electron transport chain is coupled with the transfer of protons from the stroma to the lumen. page 169

The buildup of protons in the lumen drives the production of ATP by oxidative phosphorylation. The ATP synthase is oriented such that ATP is produced on the stroma side of the membrane. page 169

Cyclic electron transport increases ATP production by redirecting electrons from ferredoxin back into the electron transport chain. page 169

8.4 PHOTOSYNTHETIC CHALLENGES: Challenges to the efficiency of photosynthesis include excess light energy and the oxygenase activity of rubisco.

An imbalance between the light-harvesting reactions and the Calvin cycle can lead to the formation of reactive oxygen species. page 170

Protection from excess light energy includes antioxidant molecules that neutralize reactive oxygen species and xanthophyll pigments that dissipate excess light energy as heat. page 171

Rubisco can act catalytically on oxygen as well as on carbon dioxide. When it acts on oxygen, energy is lost and CO_2 is released. page 171

Rubisco has evolved to favor carbon dioxide over oxygen, but the cost of this selectivity is reduced speed. page 171

The synthesis of carbohydrates through the Calvin cycle results in significant energy losses, which are due in part to photorespiration. page 171

The maximum theoretical efficiency of photosynthesis is approximately 4% of total incoming solar energy. page 172

8.5 THE EVOLUTION OF PHOTOSYNTHESIS: The evolution of photosynthesis had a profound impact on life on Earth.

The ability to use water as an electron donor in photosynthesis evolved in cyanobacteria. page 174

Cyanobacteria evolved two photosystems either by the transfer of genetic material or by gene duplication and divergence. page 174

Photosynthesis in eukaryotes likely evolved by endosymbiosis. page 174

All of the oxygen in Earth's atmosphere results from photosynthesis by organisms containing two photosystems. page 174

Log in to LaunchPad to check your answers to the Self-Assessment Questions and to access additional learning tools.

CASE 2

Cancer

Cell Signaling, Form, and Division

Imagine a simple vaccine that could prevent about 500,000 cases of cancer worldwide every year. You'd think such a discovery would be hailed as a miracle. Not quite. A vaccine to prevent cervical cancer, which affects about half a million women and kills as many as 275,000 annually, has been available since 2006. In the United States, however, this vaccine (and others) has stirred controversy.

Cancer is uncontrolled cell division. Cell division is a normal process that occurs during development of a multicellular organism and subsequently as part of the maintenance and repair of adult tissues. It is carefully regulated so that it occurs only at the right time and place. Sometimes, however, this careful regulation can be disrupted. When the normal checks on cell division become derailed, cancer can result.

Most cancers are caused by inherited or acquired mutations, but some are caused by viruses. In fact, nearly all cases of cervical cancer are caused by a virus called human papillomavirus (HPV). There are hundreds of strains of HPV. Some of these strains cause minor problems such as common warts and plantar warts. Other strains are sexually transmitted. Some can cause genital warts but aren't associated with cancer. However, a handful of "high-risk" HPV strains are strongly tied to cancer of the cervix. More than 99% of cervical cancer cases are believed to arise from HPV infections.

HPV infects epithelial cells, a type of cell that lines the body cavities and covers the outer surface of the body. Once inside a cell, the virus hijacks the cellular machinery to produce new viruses. The high-risk strains of HPV that infect the cervix go one step further: those viruses that aren't fought off by the immune system can permanently integrate their own eight-gene DNA sequence into the host cell's DNA.

Once integrated into the DNA of human epithelial cells, the virus produces the proteins E6 and E7. These viral proteins, in turn, inhibit two cellular proteins, p53 and Rb, respectively. Both of these proteins are tumor suppressors. Tumor suppressors prevent cell division in healthy cells when DNA damage occurs or in other circumstances when cell division is harmful. In the presence of E6 and E7, p53 and Rb are unable to put the brakes on cell division, so cervical cells divide uncontrollably and cancer can result.

As they grow and multiply, the abnormal cells push through the basal lamina, a thin layer that separates the epithelial cells lining the cervix from the connective tissue beneath. Untreated, invasive cancer can spread to other organs. Once cancer travels beyond its primary location, the situation is grim. Most cancers cannot be cured after they have spread.

The U.S. Food and Drug Administration (FDA) has approved two vaccines for HPV, both of which protect against the high-risk strains responsible for cervical cancer. Medical groups such as the American Academy of Pediatrics and government organizations such as the Centers for Disease Control and Prevention (CDC) recommend vaccinating girls at age 11 or 12, before they become sexually active. In addition, recent CDC guidelines also recommend vaccinating boys because they can transmit the virus to women when they become sexually active. On the basis of these recommendations, many states are considering legislation that would require middle-school girls to receive the vaccine.

However, there are critics of these recommendations and mandates. Some say the government shouldn't be making medical decisions. Others argue that the use of the vaccine encourages sexual activity among young girls and boys. Still others have expressed concern over the vaccine's safety.

> *When the normal checks on cell division become derailed, cancer can result.*

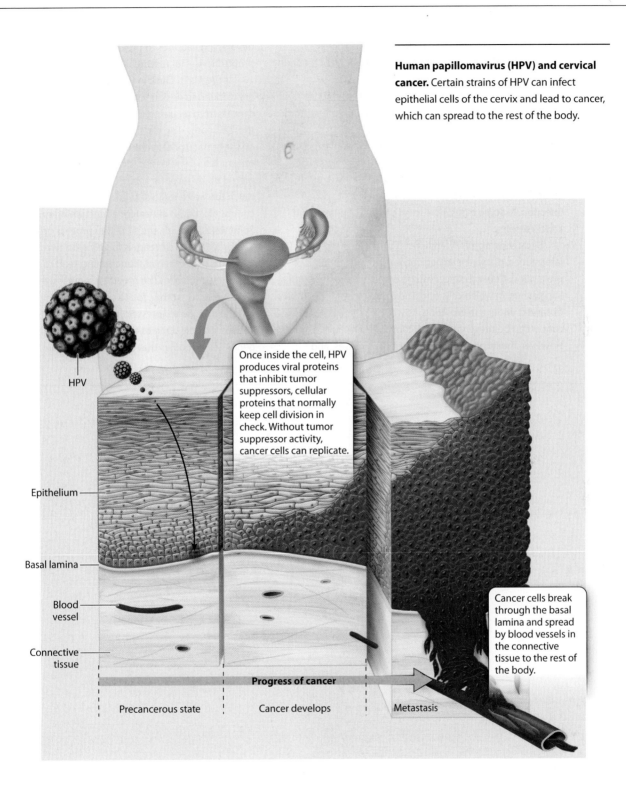

Human papillomavirus (HPV) and cervical cancer. Certain strains of HPV can infect epithelial cells of the cervix and lead to cancer, which can spread to the rest of the body.

Once inside the cell, HPV produces viral proteins that inhibit tumor suppressors, cellular proteins that normally keep cell division in check. Without tumor suppressor activity, cancer cells can replicate.

Cancer cells break through the basal lamina and spread by blood vessels in the connective tissue to the rest of the body.

Studies have found the HPV vaccine to be safe, a conclusion backed by a 2010 report from the Institute of Medicine, an independent nonprofit organization that advises the U.S government on issues of health. In addition, after 10 years of use, the HPV vaccine has reduced the number of cases of cervical cancer by half. Despite these findings, the HPV vaccine is slow to be accepted in the United States. The CDC found that by 2010, a full 5 years after the vaccine was introduced, only 32% of teenage girls had been fully vaccinated. In 2016, the percentage increased

CASE 2

to 50%. Vaccination rates are lower in teenage boys compared to those of teenage girls; in 2016, 38% of teenage boys had been fully vaccinated.

Proponents of HPV vaccination say those numbers should cause concern. HPV is common: a 2007 study found that nearly 27% of American women ages 14–59 were infected with the virus. Among 20- to 24-year-olds, the infection rate was nearly 45%. Certainly, not everyone who contracts HPV will develop cancer. Most people manage to clear the virus from their bodies within 2 years of infection. In some cases, however, the virus hangs on and cancer results.

Because cancer is often difficult to treat, especially in its later stages, many researchers focus on early detection or prevention. Cervical cancer can be detected early by routine Pap smears, in which cells from the cervix are collected and observed under a microscope. As we have seen, cervical cancer can be prevented with a conventional vaccine. However, most cancers are not caused by viruses, and developing vaccines to prevent those cancers is a trickier proposition—but researchers are pushing ahead.

At the Mayo Clinic in Rochester, Minnesota, researchers are working to design vaccines to prevent breast and ovarian cancers from recurring in women who have been treated for these diseases. The researchers have zeroed in on proteins on the surface of cancerous cells. Cells communicate with one another by releasing signaling molecules that are picked up by receptor molecules on another cell's surface, much the way a radio antenna picks up a signal. Cellular communication is critical for a functioning organism. Sometimes, though, the signaling process goes awry.

The Mayo Clinic team is focusing on two cell-surface receptors that, when malfunctioning, lead to cancer. One, Her2/neu, promotes the growth of aggressive breast cancer cells. The other, folate receptor α (alpha) protein, is frequently overexpressed in breast and ovarian tumors. The researchers hope to train patients' immune systems to generate antibodies that recognize these proteins, and then destroy the cancerous cells. The approach has successfully prevented tumors in mice. Currently, the researchers are testing the vaccines in humans.

Even if these new therapies are successful, cancer researchers have much more work to do. Cancer is not one disease but many, and most cancers are caused by a complex interplay of genetic and environmental factors. Many things can go wrong as cells communicate, grow, and divide. However, as researchers learn more about the cellular processes involved, they can step in to prevent or treat cancer. As the HPV vaccine shows, there is significant progress to be made.

 DELVING DEEPER INTO CASE 2

For each of the Cases, we provide a set of questions to pique your curiosity. These questions cannot be answered directly from the Case itself, but instead introduce issues that will be addressed in chapters to come. We revisit this Case in Chapters 9–11 on the pages indicated below.

1. How do cell signaling errors lead to cancer? See page 198.
2. How do cancer cells spread throughout the body? See page 219.
3. Which genes are involved in cancer? See page 245.

CHAPTER 9 Cell Signaling

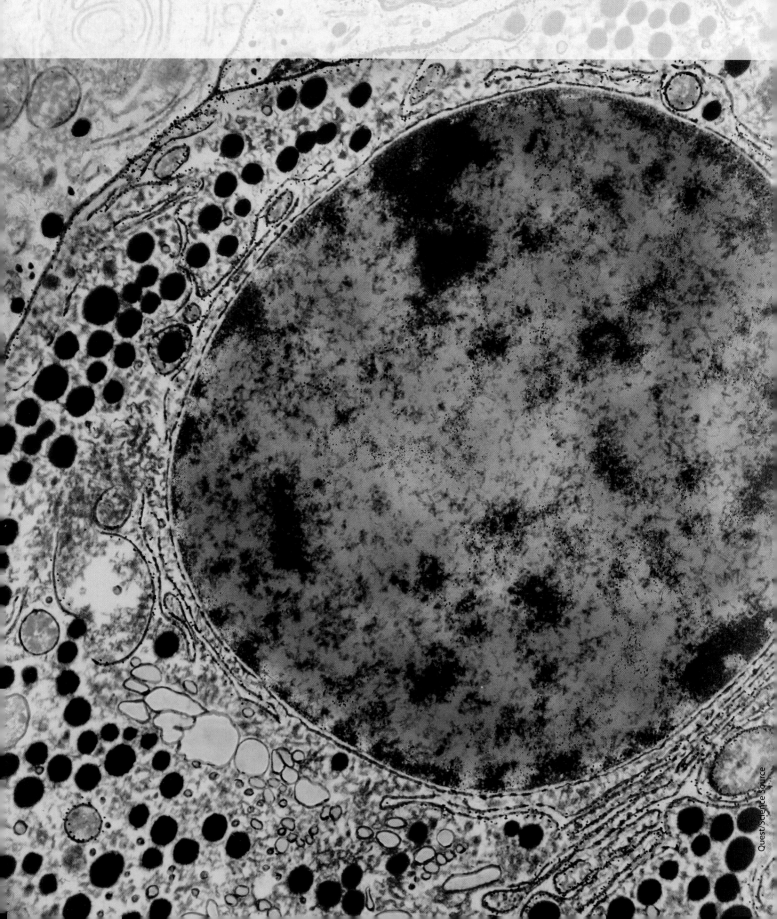

CORE CONCEPTS

9.1 PRINCIPLES OF CELL SIGNALING: Cells communicate primarily by sending and receiving chemical signals.

9.2 DISTANCE BETWEEN CELLS: Cells communicate over long and short distances.

9.3 SIGNALING RECEPTORS: Signaling molecules bind to and activate cell-surface and intracellular receptors.

9.4 G PROTEIN-COUPLED RECEPTORS: G protein-coupled receptors are a large, conserved family of receptors that often lead to short-term responses, such as activating enzymes or opening ion channels.

9.5 RECEPTOR KINASES: Receptor kinases are widespread and often lead to long-term responses, such as changes in gene expression.

Up to this point, we have considered how life works by looking mainly at what happens inside individual cells. We have seen how a cell uses the information coded in genes to synthesize the proteins necessary to carry out diverse functions. We have also seen how the plasma membrane actively keeps the environment inside the cell different from the environment outside it. Finally, we have explored how a cell harvests and uses energy from the environment.

In the next three chapters, we zoom out from individual cells and consider cells in context. Cells always exist in a particular environment, and their behavior is often strongly influenced by this environment. For example, the environment of unicellular organisms consists of their physical surroundings and other cells nearby. The ability of cells to sense and respond to this environment is critical for survival. A unicellular organism senses information that signals it when to feed, when to move, and when to divide.

In multicellular organisms, cells do not exist by themselves, but instead are part of cellular communities. These cells may physically adhere to one another, forming tissues and organs. Often multicellular organisms are highly complex, made up of cells of many types with specialized functions. These cells respond to signals, which may come from the outside environment or from other cells of the organism. Cell communication helps to coordinate the activities of the thousands, millions, and trillions of cells that make up the organism. These cells all come about through the process of cell division. Cells divide as part of growth and development, as well as tissue maintenance and repair. Cell division is an example of a cellular activity that is often regulated by signals—in this case, the signals tell a cell when to start dividing and when to stop dividing.

In Chapter 9, we look at how cells send, receive, and respond to signals. In Chapter 10, we focus on how cells maintain their shape and, in some cases, physically adhere to one another. As we saw in the case of molecules, form is inextricably linked to function. Finally, in Chapter 11, we consider how one cell becomes two through the process of cell division.

9.1 PRINCIPLES OF CELL SIGNALING

The activities of virtually all cells are influenced by their surroundings. Cells receive large amounts of information from numerous sources in their surroundings. One key source of information is the physical environment. Another is other cells—neighboring cells, nearby cells, and even very distant cells. Both unicellular and multicellular organisms sense information from their surroundings and respond to that information by changing their activity or even dividing.

In this section, we consider the general principles of cell communication, including how cells send and receive signals and how a cell responds after it receives a signal. These mechanisms first evolved in unicellular organisms as a way to sense and respond to their environment, and also as a way to influence the behavior and activity of other cells. These basic principles apply to all cells, both prokaryotic and eukaryotic, and to all organisms, both unicellular and multicellular.

Cells communicate using chemical signals that bind to receptors.

Cellular communication consists of four essential elements: a **signaling cell**, a **signaling molecule**, a **receptor protein,** and a **responding cell** (**Fig. 9.1**). The signaling cell is the source of a signaling molecule. Signaling molecules vary immensely and include peptides, lipids, and gases. In all cases, the signaling molecule carries information from one cell to the next. The signaling molecule binds to a receptor protein on or in the responding cell; in turn, the responding cell changes its activity or behavior.

Let's consider a simple, all-too-familiar example. You're startled or scared, and you experience a strange feeling in the pit of your stomach and your heart beats faster. Collectively, these physiological changes are known as the "fight or flight" response. For these changes to occur, a signal must go from signaling cells to responding cells in the heart, lungs, stomach, and many other organs (**Fig. 9.2**). This signal is the hormone adrenaline (also called epinephrine). In response to stress, adrenaline is released from the adrenal glands, located above the kidneys. Adrenaline circulates through the body and acts on many types of cells, including the cells of your heart, causing it to beat more strongly and quickly. As a result, the heart is able to deliver oxygen more effectively to the body.

This example illustrates the four basic elements involved in cellular communication. In this case, the signaling cells are specific cells within the adrenal glands. The signaling molecule is the hormone adrenaline released by these cells. Adrenaline is carried in the bloodstream and binds to receptor proteins located on the surface of responding cells, like those of the heart.

Let's look at a second example, this time focusing on communication among bacteria (**Fig. 9.3**). *Streptococcus pneumoniae* (also known as pneumococcus) is a bacterium that causes pneumonia, meningitis, and some kinds of arthritis. Many bacteria, including

FIG. 9.1 The four elements required for cellular communication: a signaling cell, a signaling molecule, a receptor protein, and a responding cell.

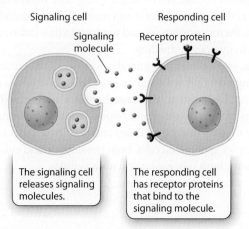

FIG. 9.2 Communication among cells in a multicellular organism. Adrenaline binds to receptors on many cells in many organs, leading to the "fight or flight" response.

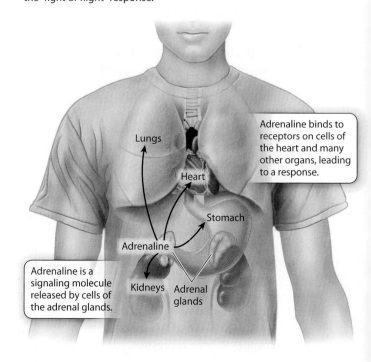

pneumococci, are able to take up DNA from the environment and incorporate it into their genome (Chapter 24). By this means, individual bacterial cells can acquire genes with advantageous properties, including antibiotic resistance.

In the 1960s, it was observed that the rate of DNA uptake by pneumococcal cells increased sharply once the bacterial population reached a certain density. Scientists concluded that the bacteria were able to coordinate DNA uptake across the population so that it occurred only at high population density. How is it possible that these bacteria "know" how many other bacteria are present? It turns out that they are able to communicate this information to one another through the release of a small peptide.

In the 1990s, scientists discovered a short peptide consisting of 17 amino acids that is continuously synthesized and released by pneumococcal cells. Not long after, a receptor for this peptide was discovered on the surface of these pneumococcal cells. The binding of this peptide to its receptor causes a bacterium to express the genes required for DNA uptake. When the bacteria are present at only a low density, the peptide is at too low a concentration to bind to the receptor (Fig. 9.3a). As the population density increases, so does the concentration of the peptide, until it reaches a level high enough that the peptide binds to enough receptors to cause the cells to turn on genes necessary for DNA uptake (Fig. 9.3b).

This is an example of quorum sensing, a process used by bacteria to determine whether they are at low or high population density and then turn on specific genes across the entire

FIG. 9.3 Communication among bacterial cells. The binding of receptors by the signaling molecule stimulates uptake of DNA.

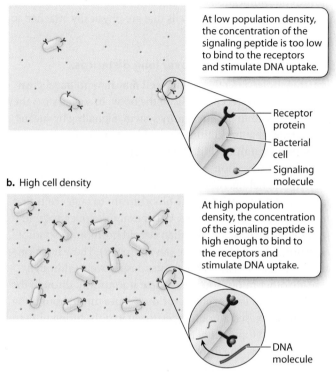

a. Low cell density

At low population density, the concentration of the signaling peptide is too low to bind to the receptors and stimulate DNA uptake.

b. High cell density

At high population density, the concentration of the signaling peptide is high enough to bind to the receptors and stimulate DNA uptake.

community. It is used to control and coordinate many different types of bacterial behaviors, such as bioluminescence, antibiotic production, and biofilm formation.

This example also illustrates the four essential elements involved in communication between cells. In this case, each bacterial cell is both a signaling cell and a responding cell because each cell makes the signaling molecule and its receptor. The general idea that cells are able to communicate by sending a signaling molecule that binds to a receptor on a responding cell is universal among prokaryotes and eukaryotes. These four elements act in very much the same way in myriad types of cellular communication.

Signaling involves receptor activation, signal transduction, response, and termination.

What happens when a signaling molecule binds to a receptor on a responding cell? The first step is **receptor activation**. On binding the signal, the receptor is turned on, or activated (**Fig. 9.4**). The signal usually activates the receptor by causing a conformational change in the receptor. As a result, some receptors bind to and activate other proteins located inside the cell. Other receptors are themselves enzymes, and binding of the signal changes the shape and activity of the enzyme. Still other receptors are channels that open or close in response to binding a signaling molecule.

Once activated, the receptor often triggers a series of downstream events in a process called **signal transduction**. During signal transduction, one molecule activates the next molecule, which activates the next, and so on. In this way, signal transduction can be thought of as a chain reaction or cascade of biochemical events set off by the binding and activation of the receptor. An important aspect of signal transduction is that the signal is often amplified at each step in the pathway. As a result, a low signal concentration can have a large effect on the responding cell.

Next, there is a cellular **response**, which can take different forms depending on the nature of the signal and the type of responding cell. For example, signaling pathways can activate enzymes involved in metabolic pathways, or turn on genes that cause the cell to divide, change shape, or signal other cells.

The last step is **termination**, in which the cellular response is stopped. The response can be terminated at any point along the signaling pathway. Termination protects the cell from overreacting to existing signals, thereby helping the cell have an appropriate level of response. It also allows the cell to respond to new signals.

In the case of pneumococcal cells, the peptide binds to and activates a receptor on the cell surface. When enough receptors are bound by the signaling molecule, the message is relayed by signal transduction pathways to the nucleoid. There, genes are turned on that express proteins involved in DNA uptake from the environment. Eventually, when the density of bacteria is

FIG. 9.4 Steps in cell signaling: receptor activation, signal transduction, response, and termination.

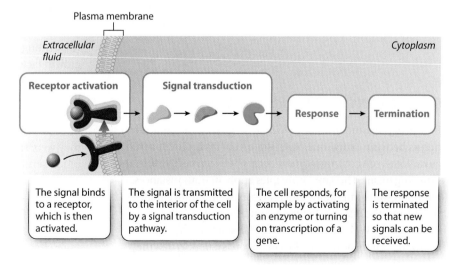

low, the initiating signal falls below a critical threshold and gene expression is turned back off.

This example is relatively simple. Remarkably, other more complex signaling pathways in a wide range of organisms involve the same four elements and steps. Indeed, in many cases they utilize similar signaling molecules, receptors, and signal transduction systems that have been evolutionarily conserved over long periods of time.

The response of a cell to a signaling molecule depends on the cell type.

Cells are typically exposed to many different types of signaling molecules. What determines whether a cell responds to a signaling molecule? A cell responds to a signal only if it has receptors that are able to bind to the signaling molecule. If specific receptors are present, binding of a signaling molecule leads to downstream effects; if they are not, the cell is not able to respond. In the case of adrenaline, only cells with receptors for adrenaline are able to respond. Adrenaline receptors are widespread in the body, including in the heart, lungs, stomach, and many other organs. The presence of these receptors allows cells of these organs to respond to adrenaline.

When a signaling molecule binds to a receptor, what determines the specific response of the cell? The cell's response depends on the set of proteins that is found in the cell, as different cell types have different sets of intracellular proteins and signaling pathways. As a result, the same signaling molecule can have different effects in different types of cells. Adrenaline, for example, can cause heart muscle cells to contract more quickly and forcefully, whereas it causes cells lining the airways of the lung to relax.

Self-Assessment Questions

1. What are the four essential elements of cell signaling?
2. What are the steps that occur when a signaling molecule binds to a receptor on a responding cell?
3. If a hormone is released into the bloodstream so that it comes into contact with many cells, what determines which cells in the body respond to the hormone?

9.2 DISTANCE BETWEEN CELLS

In prokaryotes and unicellular eukaryotes, cell communication occurs between individual organisms. In complex multicellular eukaryotes, cell communication occurs between cells within the same organism. The same principles apply in both instances, although some important differences exist. In multicellular organisms, the distance between communicating cells varies considerably. When the two cells are far apart, the signaling molecule is transported by the circulatory system. When they are close, the signaling molecule simply moves by diffusion. In addition, many cells in multicellular organisms are physically attached to one another, in which case the signaling molecule is not released from the signaling cell at all. In this section, we explore communication over long and short distances, as well as communication between cells that are physically attached to one another.

Endocrine signaling acts over long distances.

Signaling molecules released by a cell may have to travel great distances to reach receptor cells in the body. In such a case, they are often carried in the circulatory system. Signaling by means of molecules that travel through the bloodstream is called **endocrine signaling** (**Fig. 9.5a**; Chapter 36).

Adrenaline (section 9.1) provides a good example of endocrine signaling. Adrenaline is produced in the adrenal glands, and then carried by the bloodstream to target cells that are far from the signaling cells. Other examples of endocrine signaling involve the mammalian steroid hormones estradiol (an estrogen) and testosterone (an androgen). These hormones travel from the ovaries and the testes, respectively (although there are other minor sources of these hormones), through the bloodstream, to target cells in various tissues throughout the body. The increased amount of estrogen in girls during puberty causes the development of breast tissue and the beginning of menstrual cycles. The increased amount of testosterone in boys during puberty causes the growth of muscle cells, deepening of the voice, and growth of facial hair (Chapter 40).

Signaling can occur over short distances.

Signaling can also occur between two cells that are close to each other (**Fig. 9.5b**). In this case, movement of the signaling molecule through the circulatory system is not needed. Instead, the signaling molecule can simply move by diffusion between the two cells. This form of signaling is called **paracrine signaling**. Paracrine signaling molecules travel distances of approximately 20 cell diameters, or a few hundred micrometers.

In paracrine signaling, the signal usually takes the form of a small, water-soluble molecule such as a **growth factor**. A growth factor is a type of signaling molecule that causes the responding cell to grow, divide, or differentiate. One of the first growth factors discovered was found by scientists who were attempting to understand how to maintain cultures of cells in the laboratory. For decades, medical research has worked with cells maintained in culture. Initially, these cultured cells had limited use because they failed to divide outside the body unless they were supplied with unidentified factors from mammalian blood serum. In 1974, American scientists Nancy Kohler and Allan Lipton discovered that one of these factors is secreted by platelets, leading it to be named platelet-derived growth factor (PDGF) (**Fig. 9.6**). We now know of scores of molecules secreted by cells that function as growth factors; in most cases, their effects are confined to neighboring cells.

FIG. 9.5 Signaling distances. Cell communication can be classified according to the distance between the signaling and responding cells.

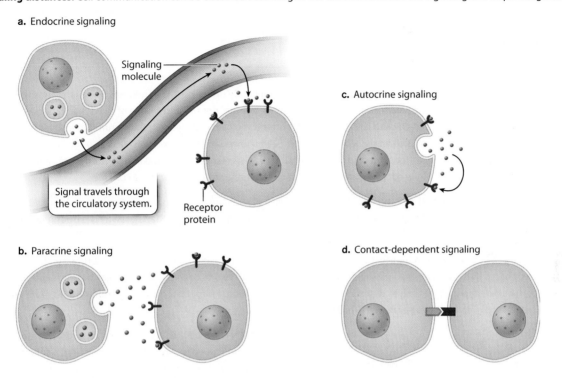

Growth factors secreted by cells in an embryo work over short distances to influence the kind of cells their neighbors will become. In this way, they help shape the structure of the adult's tissues, organs, and limbs. For example, in developing vertebrates, paracrine signaling by the growth factor Sonic Hedgehog (yes, it's named after a video game character) ensures that the motor neurons in your spinal cord are in the proper locations, that the bones of your vertebral column form correctly, and that your thumb and pinky fingers are on the correct sides of your hands.

A specialized form of short-range signaling is the communication between neurons (nerve cells), and between neurons and muscle cells (Chapters 34 and 35). Neurotransmitters are a type of signaling molecule released from a neuron. After release, they diffuse across a small space, called a synapse, between the signaling cell and the responding cell. If the adjacent cell is a neuron, it often responds by transmitting a nerve impulse and then releasing additional neurotransmitters. If the responding cell is a muscle cell, it may respond by contracting.

Sometimes signaling molecules may be released by a cell and then bind to receptors on the very same individual cell. Such cases, where signaling cell and responding cell are one and the same, are examples of **autocrine signaling (Fig. 9.5c)**. Autocrine signaling is especially important to multicellular organisms during the development of the embryo (Chapters 19 and 40). For example, once a cell differentiates into a specialized cell type, autocrine signaling is sometimes used to maintain this developmental decision. In addition, autocrine signaling can be used by cancer cells to promote cell division.

Signaling can occur by direct cell–cell contact.

Sometimes a cell communicates with another cell through direct contact, without diffusion or circulation of the signaling molecule. This form of signaling requires that the two communicating cells be in physical contact with each other. A transmembrane protein on the surface of one cell acts as the signaling molecule, and a transmembrane protein on the surface of an adjacent cell acts as the receptor (**Fig. 9.5d**). In this case, the signaling molecule is not released from the cell, but instead remains associated with the plasma membrane.

This form of signaling is important during embryonic development. As an example, consider the development of the central nervous system of vertebrate animals. In the brain and spinal cord, neurons transmit information in the form of electrical signals that travel from one part of the body to another. The neurons in the central nervous system are greatly outnumbered by supporting cells, called glial cells, which nourish and insulate the neurons. Both the neurons and the glial cells start out as similar cells in the embryo, but some of these undifferentiated cells become neurons and many more become glial cells.

During brain development, the amount of a transmembrane protein called Delta dramatically increases on the surface of

HOW DO WE KNOW?

FIG. 9.6
Where do growth factors come from?

BACKGROUND Cells grown outside the body in culture survive and grow well only under certain conditions. For a long time, researchers hypothesized that certain substances are required for cells to grow in culture, but the identity and source of these substances remained a mystery. A key insight came from the observation that chicken cells grow much better if they are cultured in blood serum rather than in blood plasma. Blood serum is the liquid component of blood that is collected after blood has been allowed to clot. Blood plasma is also the liquid component of blood, but it is collected from blood that has not clotted. In the 1970s, American biologists Nancy Kohler and Allan Lipton were interested in identifying the source of the factor in blood serum (that is not present in blood plasma) that allows cells to survive in culture.

HYPOTHESIS Kohler and Lipton knew that clotting depends on the release of substances from platelets. Because cells grow better in serum (which has been allowed to clot) compared to plasma (which has not clotted), they hypothesized that a growth-promoting factor was released by platelets during clotting.

EXPERIMENT 1 AND RESULTS The researchers first confirmed earlier observations using mouse cells called fibroblasts that are easily grown in culture. They grew two sets of fibroblasts in small plastic dishes. They added serum to one of the dishes and plasma to the other. They counted the number of cells in both culture dishes over time to determine the rate of cell division. Fibroblasts grown in serum divided more rapidly than those grown in plasma, as expected from earlier experiments (Fig. 9.6a).

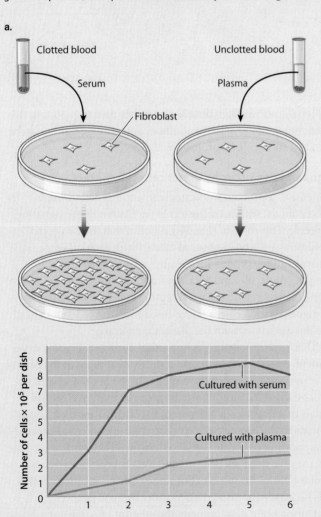

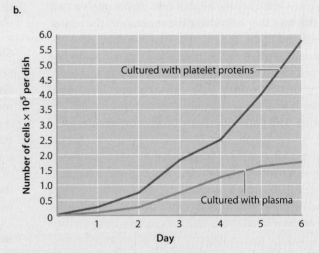

EXPERIMENT 2 AND RESULTS To see if the growth-promoting factor is released from platelets, the researchers prepared a solution of proteins made from purified platelets, added these proteins to cultured fibroblasts, and counted the number of cells over time. They found that fibroblasts grown in the presence of platelet proteins divided more rapidly than fibroblasts grown in plasma (Fig. 9.6b).

CONCLUSION Kohler and Lipton concluded that the growth-promoting factor is a protein that is released by platelets on clot formation, and that is normally present in serum but absent in plasma.

FOLLOW-UP WORK Over the next few years, Kohler, Lipton, and other researchers purified and characterized the growth factor we know today as platelet-derived growth factor, or PDGF.

SOURCE Kohler, N., and A. Lipton. 1974. "Platelets as a Source of Fibroblast Growth-Promoting Activity." *Experimental Cell Research* 87:297–301.

some of these undifferentiated cells. These cells will become neurons. Delta proteins on each new neuron bind to transmembrane proteins called Notch on the surface of adjacent, undifferentiated cells. In this case, the signaling cell is the cell with elevated levels of Delta protein. The Delta protein, in turn, is the signaling molecule, and Notch is its receptor. Cells with activated Notch receptors become glial cells and not neurons. Because one signaling cell sends this same message to all the cells it contacts, it is easy to understand how there can be so many more glial cells than neurons in the central nervous system.

As you can see from these examples, the same fundamental principles are at work when signaling guides a developing embryo, allows neurons to communicate with other neurons or muscles, triggers DNA uptake by pneumococcal cells, or allows your body to respond to stress. All of these forms of communication are based on signals that are sent from a signaling cell to a responding cell. These signaling molecules are the language of cellular communication.

Self-Assessment Questions

4. What is one way in which endocrine, paracrine, autocrine, and contact-dependent signaling are similar to one another?

5. What is one way in which endocrine, paracrine, autocrine, and contact-dependent signaling are different from one another?

9.3 SIGNALING RECEPTORS

Receptors are proteins that receive and interpret information carried by signaling molecules. Regardless of the distance between communicating cells, a responding cell receives a message when the signaling molecule binds to a receptor protein on or in that cell. For this reason, the signaling molecule is often referred to as a **ligand** (from the Latin *ligare*, which means "to bind"). The signaling molecule binds to a specific part of the receptor protein called the **ligand-binding site**. The bond formed is noncovalent and highly specific: the signaling molecule binds only to a receptor with a ligand-binding site that recognizes the molecule.

The binding of a signaling molecule to the ligand-binding site of a receptor typically causes a conformational change in the receptor. We say that the conformational change "activates" the receptor because it is through this change that the receptor passes the message from the signaling molecule to the interior of the cell. In many ways, this change in receptor shape is similar to the change that occurs when a substrate binds to the active site of an enzyme (Chapter 6). The conformational change in the receptor ultimately triggers chemical reactions or other changes in the cytosol, so it serves as a crucial step in the reception and interpretation of communications from other cells.

Receptors for polar signaling molecules are located on the cell surface.

The location of a particular receptor in a cell depends largely on whether the signaling molecule is polar or nonpolar (**Fig. 9.7**). Many signaling molecules, such as the growth factors we just discussed, are small, polar proteins that cannot pass through the hydrophobic core of the plasma membrane. The receptor proteins for these signals are located on the outside surface of the responding cell (Fig. 9.7a).

Receptor proteins for growth factors and other polar ligands are transmembrane proteins with an extracellular domain, a transmembrane domain, and a cytoplasmic domain. When a signaling molecule binds to the ligand-binding site in the extracellular domain, the entire molecule, including the cytoplasmic

FIG. 9.7 Cell-surface and intracellular receptors. (a) Cell-surface receptors interact with polar signaling molecules that cannot cross the plasma membrane. (b) Intracellular receptors interact with nonpolar signaling molecules that can cross the plasma membrane.

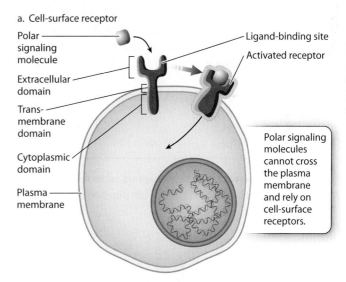

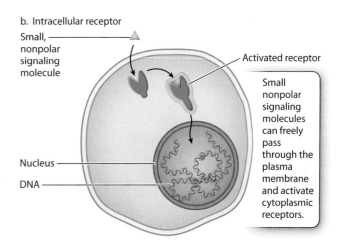

domain of the receptor, undergoes a conformational change, and as a result the molecule is activated. In this way, the receptor acts as a bridge between the inside and the outside of the responding cell that carries the message of the hydrophilic signal across the hydrophobic core of the plasma membrane.

Receptors for nonpolar signaling molecules are located in the interior of the cell.

The receptors for nonpolar signaling molecules, such as the steroid hormones involved in endocrine signaling, are not located on the surface of the cell. Since steroids are hydrophobic, they pass easily through the hydrophobic core of the phospholipid bilayer and into the target cell. Once inside, steroid hormones bind to receptor proteins located in the cytosol or in the nucleus to form receptor–steroid complexes (Fig. 9.7b). Steroid–receptor complexes formed in the cytosol enter the nucleus, where they act to control the expression of specific genes. Steroid receptors located in the nucleus are often already bound to DNA and need only to bind to their steroid counterpart to turn on gene expression.

Many examples of steroid hormones exist, including sex hormones, glucocorticoids (which raise blood glucose levels), and ecdysone (involved in insect molting). However, because much of the information received by cells is transmitted across the plasma membrane through transmembrane receptors, we focus our attention here on the sequence of events that takes place when receptors on the surface of cells bind their ligands.

Cell-surface receptors act like molecular switches.

As we saw earlier, a receptor is activated after a signaling molecule binds to its ligand-binding site. Many receptors act as binary molecular switches, existing in two alternative states, either "on" or "off" (**Fig. 9.8**). In this way, receptors behave similarly to a light switch (Fig. 9.8a). When bound to their signaling molecule, the molecular switch is turned on. When the signaling molecule is no longer bound, the switch is turned off.

Thousands of different receptor proteins are present on the surface of any given cell. Most of them can be placed into one of three groups on the basis of their structures and what occurs immediately after the receptor binds its ligand. One type of cell-surface receptor is called a **G protein-coupled receptor** (Fig. 9.8b; section 9.4). When a ligand binds to a G protein-coupled receptor, the receptor couples to, or associates with, a **G protein**, as its name suggests. G protein-coupled receptors are evolutionary conserved and all have a similar molecular structure.

A second group of cell-surface receptors are themselves enzymes, which become catalytically active when the receptor binds its ligand. Most of these are **receptor kinases** (Fig. 9.8c and section 9.5). A kinase is an enzyme that catalyzes the transfer of a phosphate group from ATP to a substrate. To catalyze this reaction, it binds both ATP and the substrate in a process called phosphorylation (**Fig. 9.9**). Phosphorylation is important because it alters the activity of the substrate: when a protein is phosphorylated by a kinase, it typically becomes active and is switched on. The addition of a phosphate group

---

FIG. 9.8 Three types of cell-surface receptors. (a) Receptors act like a light switch. (b) G protein-coupled receptors, (c) receptor kinases, and (d) ligand-gated ion channels can be either "off" (inactive) or "on" (active).

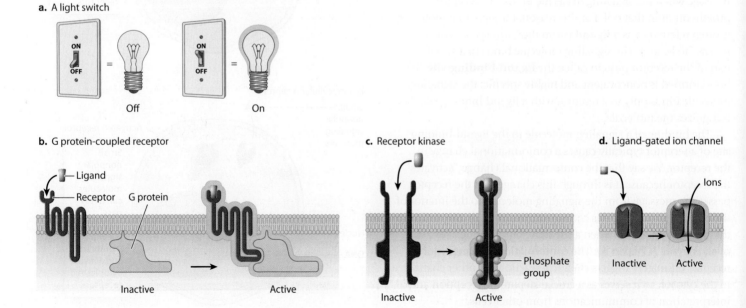

FIG. 9.9 Phosphorylation and dephosphorylation. A kinase transfers a phosphate group from ATP to a protein, typically activating the protein. A phosphatase removes a phosphate group from a protein, typically deactivating the protein.

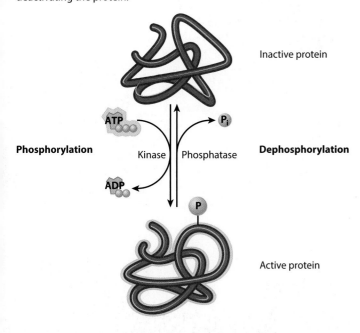

can activate a protein by altering its shape or providing a new site for other proteins to bind. **Phosphatases** remove a phosphate group, a process called dephosphorylation (Fig. 9.9). When a protein is dephosphorylated by a phosphatase, it typically becomes inactive and is switched off.

Receptors in the third group, **ion channels,** alter the flow of ions across the plasma membrane. Recall from Chapter 5 that channel proteins help ions and other molecules diffuse into and out of the cell by providing a hydrophilic pathway through the hydrophobic core of the plasma membrane. Most of the time, the channels are closed. However, when a signal arrives, the channel undergoes a conformational change that opens it and allows ions to flow in and out. These channels can be opened in different ways. Some open in response to changes in voltage across the membrane; these are called voltage-gated ion channels and are discussed in Chapter 34. Other ion channels open when bound by their ligand; these are called ligand-gated ion channels (Fig. 9.8d) and are discussed in Chapter 35. Signaling using ion-channel receptors is especially important for nerve and muscle cells because their functions depend on a rapid change in ion flow across the plasma membrane.

What happens after a signaling molecule binds to its receptor and flips a molecular switch? Following receptor activation, signaling pathways transmit the signal to targets in the interior of the cell, the cell responds, and eventually the signal is terminated. In the next two sections, we examine the signaling pathways activated by G protein-coupled receptors and receptor kinases.

Self-Assessment Questions

6. How are receptors that bind polar and nonpolar signaling molecules similar, and how are they different?
7. How can cells respond to external signaling molecules, even when those signaling molecules cannot enter the cell?
8. What are three possible initial steps following the binding of a signaling molecule to a receptor?

9.4 G PROTEIN-COUPLED RECEPTORS

G protein-coupled receptors, introduced in section 9.3, are a very large family of cell-surface molecules. They are found in almost every eukaryotic organism. In humans, for example, approximately 800 different G protein-coupled receptors have been identified. All of these receptors have three characteristics in common. First, they have a similar structure, consisting of a single polypeptide chain that has seven transmembrane spanning regions, with the ligand-binding site on the outside of the cell and the portion that binds to the G protein on the inside of the cell. Second, when activated, they associate with a G protein. In this way, they are able to transmit the signal from the outside to the inside of the cell. Third, the cellular responses to the activation of G protein-coupled receptors all tend to be rapid and short-lived. These characteristics result from their shared evolutionary history.

In spite of their similarity, different G protein-coupled receptors are able to respond to different signaling molecules. Hormones, neurotransmitters, and small molecules are among the diverse set of signaling molecules that bind to this group of receptors. In addition, the effects of these receptors are quite diverse. For example, signaling through these receptors is responsible for our senses of sight, smell, and taste (Chapter 34). In fact, it is thought that G protein-coupled receptors evolved from sensory receptors in unicellular eukaryotes.

The first step in cell signaling is receptor activation.

A G protein-coupled receptor can be inactive (off) or active (on) depending on whether it is bound to a ligand. In the absence of a ligand, it is inactive. When a ligand binds to a G protein-coupled receptor, the receptor is active. In its active state, it couples to (that is, it binds to) a G protein located on the cytoplasmic side of the plasma membrane.

The G protein also has two states, off or on. A G protein is able to bind to the guanine nucleotides GDP and GTP. When it is bound to GDP, it is off, or inactive; when it is bound to GTP, it is on, or active.

When a G protein-coupled receptor binds to a G protein, it activates the G protein by causing it to release GDP and bind GTP. As long as the G protein is bound to GTP, it is in the on position. The activated G protein goes on to activate additional proteins in the signaling pathway.

Some G proteins are composed of three subunits: α (alpha), β (beta), and γ (gamma) (**Fig. 9.10**). The α subunit is the part of the G protein that binds to either GDP or GTP. When GDP bound to the α subunit is replaced by GTP, the α subunit separates from the β and γ subunits. In most cases, the isolated GTP-bound α subunit becomes active and is able to bind to target proteins in the cell.

Signals are often amplified in the cytosol.

Let's follow the steps in cell signaling by examining a specific example. Adrenaline, discussed earlier, binds to a G protein-coupled receptor. When adrenaline binds to its receptor on cardiac muscle cells, GDP in the G protein is replaced by GTP and the G protein is activated.

The GTP-bound α subunit then binds to and activates an enzyme in the cell membrane called adenylyl cyclase

FIG. 9.10 Activation of a G protein by a G protein-coupled receptor. A G protein-coupled receptor is activated when it binds a signaling molecule, which leads to exchange of GDP for GTP on the G protein's α subunit, separation of the α subunit, and downstream effects.

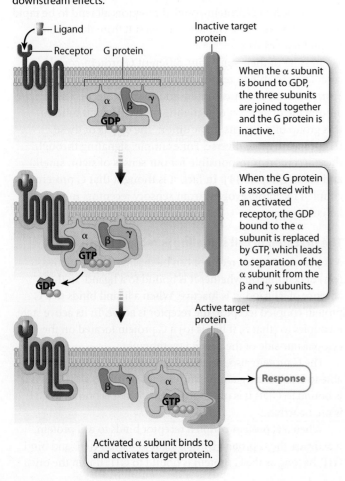

FIG. 9.11 Adrenaline signaling in heart muscle. Adrenaline binds to a G protein-coupled receptor, leading to production of the second messenger cAMP and activation of protein kinase A. The cell's response increases the heart rate.

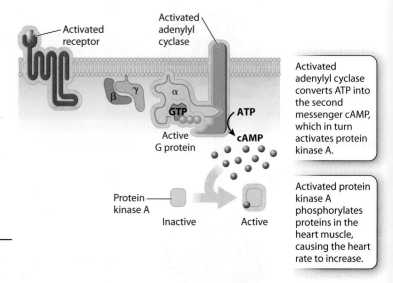

(**Fig. 9.11**). Adenylyl cyclase converts the nucleotide ATP into cyclic AMP (cAMP). Cyclic AMP is known as a **second messenger**. Second messengers are signaling molecules found inside cells that relay information to the next target in the signal transduction pathway. For example, the second messenger cAMP binds to and activates protein kinase A (PKA). The term "second messenger" was coined to differentiate these signaling molecules from those like adrenaline and growth factors, which are considered first messengers.

A little adrenaline goes a long way as a result of signal amplification (**Fig. 9.12**). First, a single receptor bound to adrenaline can activate several G protein molecules. Next, each molecule of adenylyl cyclase catalyzes the production of large amounts of the second messenger cAMP. Then, each PKA molecule activates multiple protein targets by phosphorylation. These sequential molecular changes in the cytosol amplify the signal so that a very small amount of signaling molecule has a large effect on a responding cell.

Signals lead to a cellular response.

As we have seen, the response of a cell to a signaling molecule depends on which type of cell it is and, therefore, which types of proteins are present within the cell. G protein-coupled receptors, for example, have different effects in different cell types. In the case of heart cells, activated PKA leads to the opening of calcium channels that are present in heart muscle cells. The resulting influx of calcium ions results in shorter intervals between muscle contractions, leading to a faster heart rate. As long as adrenaline is bound to its receptor, the heart rate remains rapid. The rapid heart rate, in turn, increases blood flow to the brain

FIG. 9.12 Amplification of G protein-coupled signaling. Signaling through G protein-coupled receptors is amplified at several places, so a small amount of signal can produce a large response in the cell.

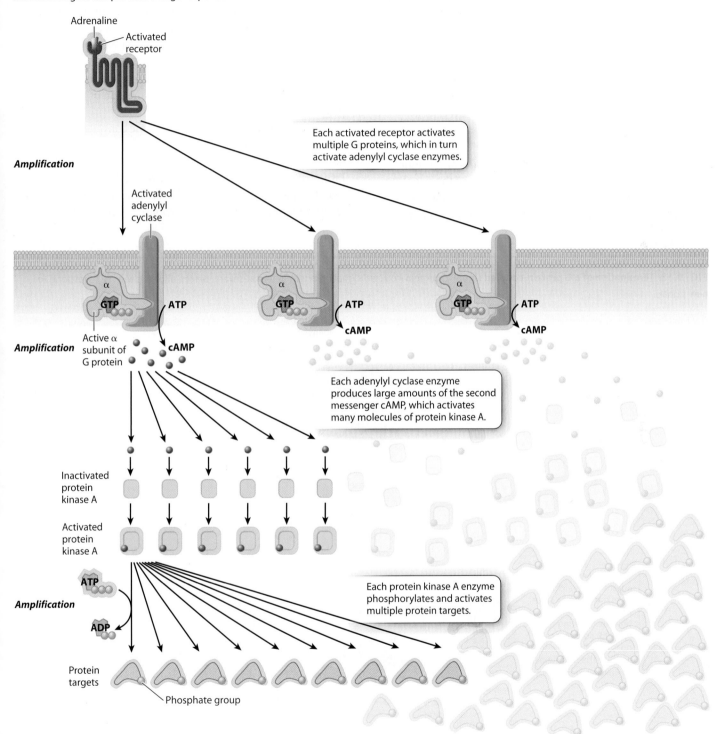

and skeletal muscles to deal with the stress that triggered the signal in the first place.

This example is typical of signaling through G protein-coupled receptors. These receptors tend to activate downstream enzymes or, in some cases, open ion channels. Because they often modify proteins that are already synthesized in the cell, their effects tend to be rapid, short-lived, and easily reversible, as we will see next.

Signaling pathways are eventually terminated.

After a good scare, we eventually calm down and our heartbeat returns to normal. This change means that the signaling pathway initiated by adrenaline has been terminated. How does this happen? First, most ligands, including adrenaline, do not bind to their receptors permanently. How long a signaling molecule remains bound to its receptor depends on how tightly the receptor holds on to it, a property called **binding affinity**. Once adrenaline leaves the receptor, the receptor reverts to its inactive conformation and no longer activates G proteins (**Fig. 9.13**).

Even when a receptor is turned off, a signal will continue to be transmitted unless the other components of the signaling pathway are also inactivated. A second place where the signal is terminated is at the G protein (Fig. 9.13). G proteins can catalyze the hydrolysis of GTP to GDP and inorganic phosphate. This means that an active, GTP-bound α subunit in the "on" position automatically turns itself "off" by converting GTP to GDP. In fact, the α subunit converts GTP to GDP almost as soon as a molecule of GTP binds to it. Thus, a G protein is able to activate adenylyl cyclase, and adenylyl cyclase is able to make cAMP only during the very short time it takes the α subunit to convert GTP to GDP. Without an active receptor to generate more active G protein α subunits, transmission of the signal quickly comes to a halt.

Farther down the pathway, an enzyme converts the second messenger cAMP to AMP, which no longer activates protein kinase A. Phosphatases remove the phosphate groups added by PKA, inactivating PKA's target proteins (Fig. 9.13). In fact, most signaling pathways are counteracted at one or more points as a means of terminating the cell's response to the signal.

FIG. 9.13 Termination of a G protein-coupled signal. G protein-coupled signaling is terminated at several places, allowing the cell to respond to new signals.

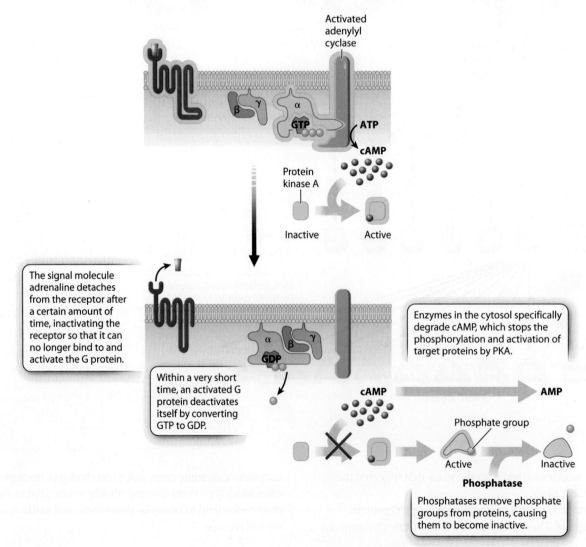

> **Self-Assessment Questions**
>
> 9. Is the term "G protein" just a shorter name for a G protein-coupled receptor?
> 10. What are three ways in which a signal is amplified in a G protein-coupled receptor signaling pathway?
> 11. What are three ways in which the response of a cell to a signal can be terminated?

9.5 RECEPTOR KINASES

Like the cells responding to G protein-coupled receptors, cells respond to signaling through receptor kinases in many ways. In response to receptor kinase signaling during embryonic development, limb buds form and eventually elongate into our arms and legs. When we cut a finger, platelets in the blood release platelet-derived growth factor, a receptor kinase ligand that triggers the cell division necessary to repair the wound.

Signaling through receptor kinases takes place in most eukaryotic organisms, and the structure and function of these receptors have been conserved over hundreds of millions of years. A well-studied receptor kinase called Kit provides an example. In vertebrates, signaling through the Kit receptor kinase leads to the production of pigment in skin, feathers, scales, and hair. As you can see from **Fig. 9.14**, mammals, reptiles, birds, and fish with a mutation in the *kit* gene have similar large white patches with no pigment. This observation indicates that the function of this gene has remained fairly constant since the appearance of the last common ancestor of these groups, more than 500 million years ago. (By convention,

FIG. 9.14 Mutations in the Kit receptor kinase. Similar patterns of incomplete pigmentation are present in mammals, reptiles, birds, and fish that have this mutated receptor. *Sources: (clockwise) Mark Boulton/Science Source; Mark Smith/Science Source; CID/ISM/Science Source; Werner Bollmann/Getty Images; Phillip Colla/Oceanlight.*

the name of a protein, like Kit, is capitalized and in roman type. The name of the gene that encodes the protein, like *kit*, is lowercase and italicized.)

The cellular responses that result from receptor kinase activation tend to be more long-lasting than the shorter-term changes that result from G protein-coupled receptors. That difference reflects the fact that receptor kinase activation often leads to changes in gene expression that allow cells to grow, divide, differentiate, or change shape.

Receptor kinases phosphorylate each other, activate intracellular signaling pathways, lead to a response, and are terminated.

Signaling through receptor kinases follows the same basic sequence of events that we saw in signaling by G protein-coupled receptors, including receptor activation, signal transduction, cellular response, and termination. Let's consider an example. Think about the last time you got a cut. The cut likely bled for a minute or two, and then the bleeding stopped. The signaling molecule platelet-derived growth factor helps start the healing process. When platelets in the blood encounter damaged tissue, they release a number of proteins, including PDGF. This signaling molecule binds to PDGF-specific receptor kinases on the surface of cells at the site of a wound.

Receptor kinases have an extracellular portion that binds the signaling molecule and an intracellular portion that is a kinase, an enzyme that transfers a phosphate group from ATP to another molecule. A single molecule of PDGF binds to the extracellular portion of two receptors at once, causing the receptors to partner, or dimerize, with each other. As a consequence, the kinase domains of the paired receptors are activated and phosphorylate each other at multiple sites on their tails (**Fig. 9.15**). The addition of these phosphate groups provides places on the receptor where other proteins in the cytoplasm can bind and become active.

One of the downstream targets of an activated receptor kinase is Ras. Ras is a G protein that consists of a single subunit, similar to the α subunit of the three-subunit G proteins. In the absence of a signal, Ras is bound to GDP and is inactive. However, when Ras is activated by a receptor kinase, it exchanges GDP for GTP. Activated GTP-bound Ras triggers the activation of a protein kinase that is the first in a series of kinases that are activated in turn, as each kinase phosphorylates the next in the series. Collectively, the series of kinases are called the mitogen-activated protein (MAP) kinase pathway (**Fig. 9.16**). The final activated kinase in the series enters the nucleus, where it phosphorylates target proteins. Some of these proteins include transcription factors that turn on genes needed for cell division so that your cut can heal.

The signal is amplified many times as it is passed from kinase to kinase. Each phosphorylated kinase in the series activates many molecules of the downstream kinase, and the downstream kinase in turn activates many molecules of another kinase still farther downstream. In this way, a very small amount of signaling molecule (PDGF, in our example) can cause a large-scale response in the cell.

Receptor kinase signaling is terminated by the same basic mechanisms that are involved in G protein-coupled receptor pathways. For example, protein phosphatases inactivate receptor kinases and the enzymes of the MAP kinase pathway. Furthermore, Ras quickly converts GTP to GDP and becomes inactive, just like the G protein α subunit. Without an active receptor kinase to generate more active Ras, activation of the MAP kinase pathway stops.

 CASE 2 CANCER: CELL SIGNALING, FORM, AND DIVISION

How do cell signaling errors lead to cancer?
Many cancers arise when something goes wrong with the way a cell responds to a signal for cell division or, in some

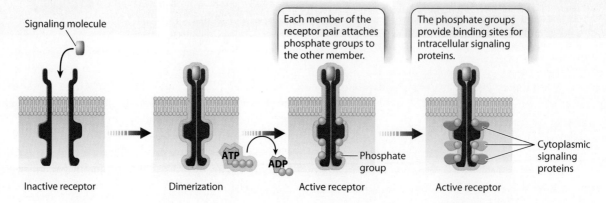

FIG. 9.15 Receptor kinase activation and signaling. Receptor kinases bind signaling molecules, dimerize, phosphorylate each other, and activate intracellular signal molecules.

FIG. 9.16 The MAP kinase pathway. Some receptor kinases signal through Ras, which in turn activates the MAP kinase pathway, leading to changes in gene expression.

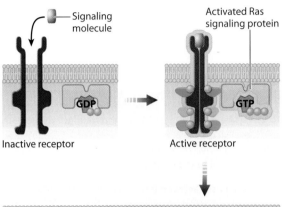

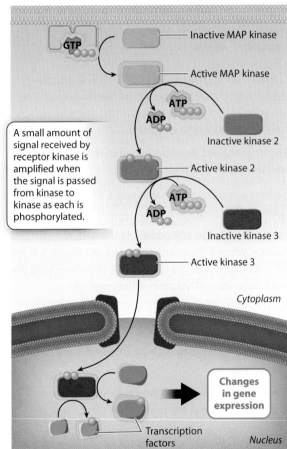

A small amount of signal received by receptor kinase is amplified when the signal is passed from kinase to kinase as each is phosphorylated.

cases, when a cell behaves as if it has received a signal for cell division when in fact it hasn't. Defects in cell signaling that lead to cancer can take place at just about every step in the cell signaling process.

In some cases, a tumor may form when a signaling molecule is overproduced or produced in an altered form. In other cases, the source of the problem is the receptor. For example, individuals with some forms of cancer have from 10 to 100 times the normal number of receptors for a signaling molecule called epidermal growth factor (EGF). The EGF receptor is a receptor kinase, like the PDGF receptor. Under normal conditions, binding of EGF to its receptor leads to the controlled division of cells. However, in cancer, the presence of excess receptors heightens the response of the signaling pathway, leading to abnormally high gene expression and excess cell division. In certain breast cancers, for example, an EGF receptor called HER2/neu is overexpressed.

Farther down the pathway, mutant forms of the Ras protein are often present in cancers. One especially harmful mutation prevents Ras from converting its bound GTP to GDP. The protein remains locked in the active GTP-bound state, causing the sustained activation of the MAP kinase pathway. More than 30% of all human cancers involve abnormal Ras activity as a result of one or more mutations in the *ras* gene.

Signaling pathways are integrated to produce a response in a cell.

In this chapter, we have focused on individual signaling pathways to illustrate the general principles of communication between cells. In each case, we saw that a cell releases a signaling molecule or in some cases retains the signaling molecule attached to its surface. The signaling molecule binds to a cell-surface or intracellular receptor. Binding of the signaling molecule induces a conformational change in the receptor, causing it to become active. The activated receptor then causes changes in the interior of the cell, frequently turning on a signal transduction cascade that is amplified and leads to a cellular response. Eventually, the response terminates, and the cell is ready to receive new signals.

Focusing on each pathway one at a time allowed us to understand how each pathway operates. But in the context of an entire organism or even a single cell, cell signaling can be quite complex, with multiple pathways acting at once and interacting with one another.

Multiple types of signaling molecules can bind to receptors on a single cell and activate several signaling pathways simultaneously. In this case, a cell's final response depends on how the pathways intersect with one another. The integration of different signals gives cells a wide range of possible responses to their environment. For example, receiving two different signals may enhance a particular response, such as cell growth.

Alternatively, one signal may inhibit the signaling pathway triggered by the other signal, weakening the response. For example, studies have shown that activated PKA can inhibit enzymes in the MAP kinase pathway. Recall that PKA is activated by the G protein-coupled receptor pathway. So, activation of the G protein-coupled pathway can inhibit the receptor kinase pathway.

Researchers have started to make use of this molecular cross talk in studying cancer. For example, many patients with breast cancer have elevated MAP kinase activity in their tumor cells.

When human breast cancer cells are genetically modified to express an activated G protein α subunit and then transplanted into mice, their growth is significantly reduced. In addition, in cell culture, elevated cAMP levels block cells from responding to growth factors that use the MAP kinase pathway. As we better understand how different signaling pathways interact in specific cell types, we may become able to alter the activity of particular pathways and ultimately the response of the cell.

Self-Assessment Question

12. How are receptor kinase pathways similar to G protein-coupled receptor pathways, and how are they different?

CORE CONCEPTS SUMMARY

9.1 PRINCIPLES OF CELL SIGNALING: Cells communicate primarily by sending and receiving chemical signals.

There are four essential players in communication between two cells: a signaling cell, a signaling molecule, a receptor protein, and a responding cell. page 186

The signaling molecule binds to its receptor on the responding cell, leading to receptor activation, signal transduction and amplification, a cellular response, and eventually termination of the response. page 187

9.2 DISTANCE BETWEEN CELLS: Cells communicate over long and short distances.

Endocrine signaling takes place over long distances and often relies on the circulatory system for transport of signaling molecules. page 188

Paracrine signaling takes place over short distances between neighboring cells and relies on diffusion. page 188

Autocrine signaling occurs when a cell signals itself. page 189

Some forms of cell communication depend on direct contact between two cells. page 189

9.3 SIGNALING RECEPTORS: Signaling molecules bind to and activate cell-surface and intracellular receptors.

A signaling molecule, or ligand, binds specifically to the ligand-binding site of the receptor. Binding causes the receptor to undergo a conformational change that activates the receptor. page 191

Receptors for polar signaling molecules, including growth factors, are located on the plasma membrane. page 191

Receptors for nonpolar signaling molecules, such as steroid hormones, are located in the cytosol or in the nucleus. page 192

There are three major types of cell-surface receptors: G protein-coupled receptors, receptor kinases, and ion channels. All act as molecular switches. page 192

G protein-coupled receptors associate with G proteins, which relay the signal to the interior of the cell. page 192

Receptor kinases activate target proteins. page 192

Ligand-gated ion channels open in response to a signal, allowing the movement of ions across the plasma membrane. page 193

9.4 G PROTEIN-COUPLED RECEPTORS: G protein-coupled receptors are a large, conserved family of receptors that often lead to short-term responses, such as activating enzymes or opening ion channels.

G protein-coupled receptors bound to signaling molecules associate with G proteins. page 193

G proteins are active when bound to GTP and inactive when bound to GDP. page 193

Some G proteins are composed of three subunits, denoted α, β, and γ. When a G protein encounters an activated receptor, the α subunit exchanges GDP for GTP, dissociates from the β and γ subunits, and becomes active. page 194

The GTP-bound α subunit activates adenylyl cyclase, which converts ATP into the second messenger cAMP. page 194

Signal transduction cascades are amplified in the cytosol. page 194

Cellular responses to G protein-coupled receptor signals tend to be short-lived and are quickly terminated. page 195

9.5 RECEPTOR KINASES: Receptor kinases are widespread and often lead to long-term responses, such as changes in gene expression.

Ligand binding to receptor kinases causes them to dimerize. The paired receptor kinases phosphorylate each other's cytoplasmic domains, leading to activation of intracellular signaling pathways. page 198

The phosphorylated receptors bind other proteins, which in turn activate other cytosolic signaling molecules, such as Ras. page 198

When Ras is activated, the GDP to which it is bound is released and is replaced by GTP. The active, GTP-bound Ras binds to and activates the first in a series of kinases. page 198

Cancer can be caused by errors in any step of the signaling pathways involved in cell division. page 198

Mutations in receptor kinases and Ras are frequently associated with human cancers. page 199

Several signaling pathways can take place simultaneously in a single cell, and the cellular response depends on how these pathways interact. page 199

Log in to LaunchPad to check your answers to the Self-Assessment Questions and to access additional learning tools.

CHAPTER 10 Cell and Tissue Architecture
Cytoskeleton, Cell Junctions, and Extracellular Matrix

CORE CONCEPTS

10.1 TISSUES AND ORGANS: Tissues and organs are communities of cells that perform specific functions.

10.2 THE CYTOSKELETON: The cytoskeleton is composed of microtubules, microfilaments, and intermediate filaments that help maintain cell shape.

10.3 CELL JUNCTIONS: Cell junctions connect cells to one another to form tissues.

10.4 THE EXTRACELLULAR MATRIX: The extracellular matrix provides structural support and informational cues.

A recurring theme in this book, and in all of biology, is that form and function are inseparably linked. This is true at every level of structural organization, from molecules to organelles to cells and to organisms themselves. We saw in Chapter 4 that the function of proteins depends on their shape. In Chapters 7 and 8, we saw that the function of organelles like mitochondria and chloroplasts depends in part on the large surface areas of internal membranes.

The functions of different cell types are also reflected in their shape and internal structural features (**Fig. 10.1**). Consider a red blood cell. It is shaped like a disk that is slightly indented in the middle, and it lacks a nucleus and other organelles (Fig. 10.1a). This unusual shape and internal organization allow it to deform readily, so it can pass through blood vessels with diameters smaller than that of the red blood cell itself. Spherical cells in the liver (Fig. 10.1b) that synthesize proteins and glycogen look and function very differently from the long, slender muscle cells (Fig. 10.1c) that contract to exert force. A neuron (Fig. 10.1d), has long and extensively branched extensions that communicate with other cells. It is structurally and functionally quite distinct from a cell lining the intestine (Fig. 10.1e) that absorbs nutrients.

In multicellular organisms, cells often exist in communities, forming tissues and organs. Again, form and function are intimately linked. Think of the branching structure and large surface area of the mammalian lung, which is adapted for gas exchange, or the muscular heart, which is adapted for pumping blood through large circulatory systems.

In this chapter, we look at the internal structures of cells that determine their shapes. We also examine the different ways cells connect to one another to build tissues and organs. Finally, we discuss the physical environment outside cells that is synthesized by cells and influences their behavior.

10.1 TISSUES AND ORGANS

Multicellular organisms consist of communities of cells with several notable features (discussed both in this section and in Chapter 26). First, cells adhere to one another. Most cells in multicellular organisms are physically attached to other cells or to substances in their surroundings. Second, cells communicate with one another. As we saw in Chapter 9, cells respond to signals from neighboring cells and the physical environment. In addition, as we will see, cells in tissues have pathways for the movement of molecules from one cell to another. Third, cells are specialized to carry out different functions as a result of their development and differentiation. The different shapes of cells often reflect these specialized functions.

Tissues and organs are communities of cells.

A biological **tissue** is a collection of cells that work together to perform a specific function. Animals and plants have tissues that allow them to carry out the various processes necessary to sustain them. In animals, for example, four types of tissues—epithelial, connective, nervous, and muscle—combine to make up all the organs of the body (Chapter 33). Two or more tissues often combine and function together as an **organ**, such as a heart or lung.

FIG. 10.1 Diverse cell types. Cells differ in shape and are well adapted for their various functions. *Sources: a. Cheryl Power/Science Source; b. Professors Pietro M. Motta & Tomonori Naguro/Science Source; c. Innerspace Imaging/Science Source; d. Biophoto Associates/Science Source; e. Don W. Fawcett/Science Source.*

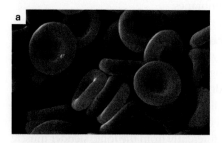

The biconcave shape of a red blood cell maximizes its surface area for gas exchange and allows it to deform as it passes through the circulatory system.

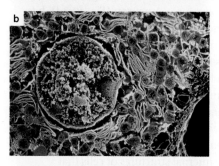

Hepatocytes (liver cells) contain large amounts of rough ER, shown here surrounding the central nucleus, needed for protein synthesis.

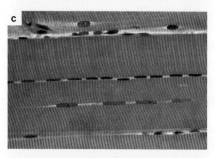

Long, multinucleated muscle cells have a striped appearance and are specialized for contraction.

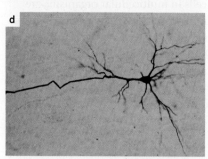

Long, slender extensions of the plasma membrane allow neurons to communicate with other cells.

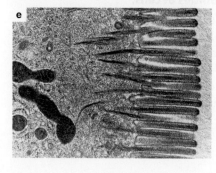

Thin, comb-like projections, called microvilli, on intestinal cells increase their absorptive surface area.

Tissues and organs have distinctive shapes that reflect how they work and what they do. In the same way, the different cell types that make up these organs have distinctive shapes that reflect what they do in the organ. In animals, the shape of cells is determined and maintained by structural protein networks in the cytoplasm called the **cytoskeleton** (section 10.2). The shape and structural integrity of tissues and organs depend on the ability of cells to connect to one another. In turn, the connection of cells to one another depends on structures called **cell junctions** (section 10.3). Outside many cells and tissues is a meshwork of proteins and polysaccharides called the **extracellular matrix** (section 10.4). The ability of cells to adhere to this meshwork is important to ensure a strong, properly shaped tissue or organ.

The structure of skin relates to its function.

To understand the role of the cytoskeleton, cell junctions, and the extracellular matrix, let's consider a community of cells very familiar to all of us—our own skin (**Fig. 10.2**). The structure of mammalian skin is tied to its function. Skin has two main layers. The outer layer, the **epidermis**, serves as a water-resistant, protective barrier. The inner layer, the **dermis**, supports the epidermis, both physically and by supplying it with nutrients. It also provides a cushion surrounding the body.

As you can see in Fig. 10.2, the epidermis is several cell layers thick. Cells arranged in one or more layers are called epithelial cells and together make up a type of animal tissue called epithelial tissue. Epithelial tissue covers the outside of the body and lines many internal structures, such as the digestive tract and vertebrate blood vessels. The epidermal layer of skin is primarily composed of epithelial cells called keratinocytes. The epidermis also contains melanocytes, which produce the pigment that gives skin its color.

Keratinocytes in the epidermis are specialized to protect the underlying tissues and organs. They are able to perform this function in part because of their elaborate system of cytoskeletal filaments. These filaments are often connected to the cell junctions that hold adjacent keratinocytes together. Cell junctions also connect the bottom layer of keratinocytes to a specialized form of extracellular matrix called the **basal lamina** (also called the basement membrane, although it is not really a membrane). The basal lamina underlies and supports all epithelial tissues.

The dermis is made up mostly of connective tissue, a type of tissue characterized by few cells and substantial amounts of extracellular matrix. The main type of cell in the dermis is the fibroblast, which synthesizes the extracellular matrix. The dermis is strong and flexible because its extracellular matrix is composed of tough protein fibers. The dermis also contains many blood vessels and nerve endings.

We will come back to the skin several times in this chapter to show more precisely how these multiple connections among

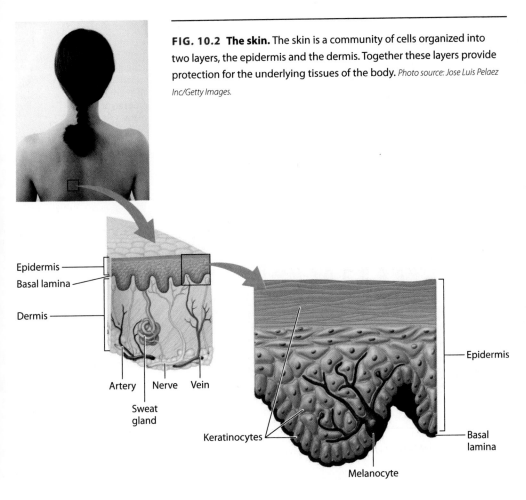

FIG. 10.2 The skin. The skin is a community of cells organized into two layers, the epidermis and the dermis. Together these layers provide protection for the underlying tissues of the body. *Photo source: Jose Luis Pelaez Inc/Getty Images.*

cytoskeleton, cell junctions, and the extracellular matrix make the skin a watertight and strong protective barrier.

Self-Assessment Questions

1. What are the four basic tissue types in animals?
2. How does the structure of the epidermis of the skin relate to its function?

10.2 THE CYTOSKELETON

Just as the bones of vertebrate skeletons provide internal support for the body, the protein fibers of the cytoskeleton provide internal support for cells (**Fig. 10.3**). All eukaryotic cells have at least two cytoskeletal elements, **microtubules** and **microfilaments**. Animal cells also have a third element, **intermediate filaments**. All three of these cytoskeletal elements are long chains, or polymers, made up of protein subunits. In addition to providing structural support, microtubules and microfilaments enable cells to change shape, move about, and transport substances.

Microtubules and microfilaments are polymers of protein subunits.

Microtubules are hollow tubelike structures with the largest diameter of the three cytoskeletal elements, about 25 nm (Fig. 10.3a). They are polymers of protein dimers. Each dimer is made up of two slightly different **tubulin** proteins, called α (alpha) and β (beta) tubulin. The tubulin dimers are assembled to form the microtubule.

Microtubules help maintain cell shape and internal structure. In animal cells, these structures radiate outward to the cell periphery from a microtubule organizing center called the **centrosome**. This spokelike arrangement of microtubules helps cells withstand compression. Many organelles are tethered to microtubules, so that microtubules guide the arrangement of organelles in the cell.

Microfilaments are polymers of **actin** monomers, arranged to form a helix. The thinnest of the three cytoskeletal fibers, they are approximately 7 nm in diameter. Microfilaments are present in various locations in the cytoplasm (Fig. 10.3b). They are relatively short and extensively branched in the cell cortex, the area of the cytoplasm just beneath the plasma membrane. In the cortex, microfilaments reinforce the plasma membrane and help to organize proteins associated with it.

These cortical microfilaments also play important roles in maintaining the shape of a cell, such as the biconcave shape of red blood cells discussed earlier (see Fig. 10.1a). In addition, long bundles of microfilaments form a band that extends around the circumference of epithelial cells. Microfilaments also maintain the shape of absorptive epithelial cells such as those in the small intestine. In these cells, bundles of microfilaments are found in microvilli, hairlike projections that extend from the surface of the cell (**Fig. 10.4**).

Microtubules and microfilaments are dynamic structures.

When we think of the parts of a skeleton, we are inclined to think of our bones. Unlike our bones, which undergo remodeling but do not change rapidly, microtubules and microfilaments are dynamic. They become longer by adding subunits to their ends, and become shorter by losing subunits.

FIG. 10.3 Three types of cytoskeletal elements. (a) Microtubule; (b) microfilament; and (c) intermediate filament.

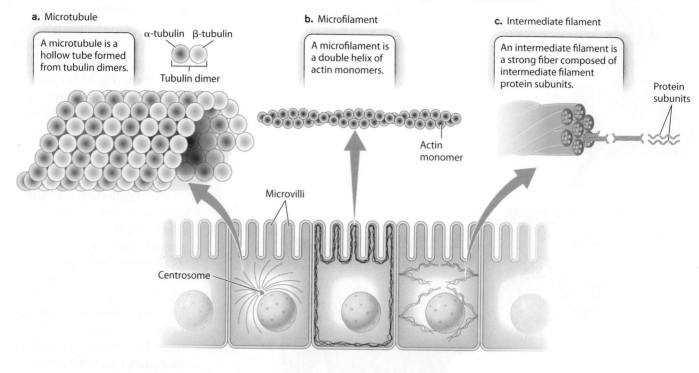

FIG. 10.4 Microfilaments in intestinal microvilli. Actin microfilaments help to maintain the structure of microvilli in the intestine. *Photo source: ©1982 Hirokawa, N. et al. The Journal of Cell Biology. 94:425-443. doi: 10.1083/jcb.94.2.425.*

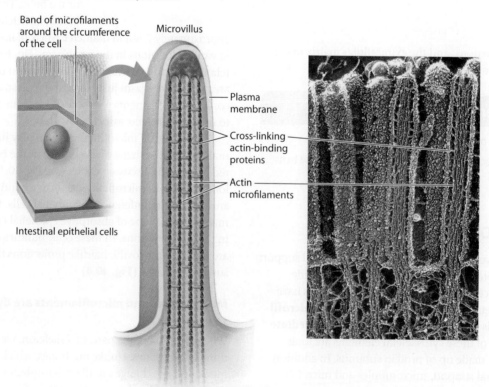

FIG. 10.5 Plus and minus ends of microtubules and microfilaments. The two ends assemble at different rates.

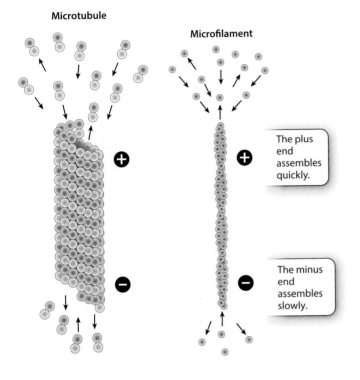

The rate at which protein subunits are added depends on the concentrations of tubulin and actin in that region of the cell. When high concentrations of subunits are present, microtubules and microfilaments can become longer at both ends, although the subunits are assembled more quickly on one end than on the other. The faster-assembling end is called the plus end, and the slower-assembling end is called the minus end (**Fig. 10.5**).

The ability of microfilaments and microtubules to lengthen and shorten is important for some of their functions. For example, some forms of cell movement, such as a single-celled amoeba foraging for food or a mammalian white blood cell chasing down foreign bacteria, depend on actin polymerization and depolymerization. The lengthening/shortening ability is also required when a single cell divides in two during cytokinesis (Chapter 11).

Microtubules make up the spindles that attach to chromosomes during cell division. In this case, the ability of spindle microtubules to "explore" the space of the cell and encounter chromosomes is driven by a unique property of microtubules: their plus ends undergo seemingly random cycles of rapid depolymerization followed by slower polymerization. These cycles of depolymerization and polymerization, which are called **dynamic instability** (**Fig. 10.6**), allow spindle microtubules to quickly find and attach to chromosomes during cell division.

FIG. 10.6 Dynamic instability. The plus ends of microtubules undergo cycles of rapid disassembly, followed by slow assembly.

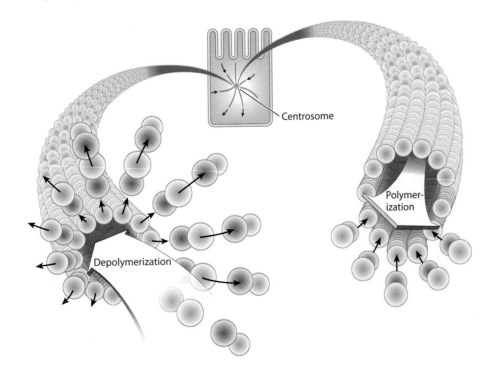

Motor proteins associate with microtubules and microfilaments to cause movement.

As noted earlier, microtubules and microfilaments have some capacity to lengthen and shorten by polymerization and depolymerization. When joined by small accessory proteins called **motor proteins**, microtubules and microfilaments are capable of causing amazing movements. A motor is a device that imparts motion.

For example, microtubules function as tracks for transport within the cell. Two motor proteins that associate with these microtubule tracks are **kinesin** and **dynein**. Kinesin transports cargo toward the plus end of microtubules, located at the periphery of the cell (**Fig. 10.7**). By contrast, dynein carries its load away from the plasma membrane toward the minus end, located at the centrosome in the interior of the cell. Movement along microtubules by kinesin and dynein is driven by conformational changes in the motor proteins and is powered by energy harvested from ATP.

Let's look at an especially striking example of this system at work. Specialized skin cells called melanophores are present in some vertebrates. Melanophores are similar to the melanocytes in our own skin that produce the pigment melanin. However, rather than hand off their melanin to other cells (as occurs in humans), melanophores keep their pigment granules and move them around inside the cell in response to hormones or neuronal signals. This redistribution of melanin within the cell allows animals such as fish or amphibians to change color. For example, at night the melanin granules in the skin of a zebrafish embryo are dispersed throughout the melanophores, making the skin darkly colored. As morning comes and the day brightens, the pigment granules aggregate at the center of the cell around the centrosome, causing the embryo's color to lighten (**Fig. 10.8**).

The melanin granules in the melanophores move back and forth along microtubules, transported by kinesin and dynein. Kinesin moves the granules out toward the plus end of the microtubule during dispersal, and dynein moves them back toward the minus end during aggregation. The color change provides daytime and nighttime camouflage: hungry predators lurking in the water below are less likely to spot the young, developing zebrafish embryo.

In addition to providing tracks for the transport of material within the cell, microtubules are found in **cilia** (singular cilium). Cilia are rod-like structures that extend from the surface of cells and are well conserved across eukaryotes. There are two types of cilia: those that don't move (called nonmotile cilia) and those that move (called motile cilia). Nonmotile cilia are very common and can be found on many eukaryotic cells. In these cells, they often serve a sensory function, taking in environmental signals and transducing them to the cell interior. For example, nonmotile cilia are found in mammalian olfactory neurons in the nose and photoreceptors in the eyes.

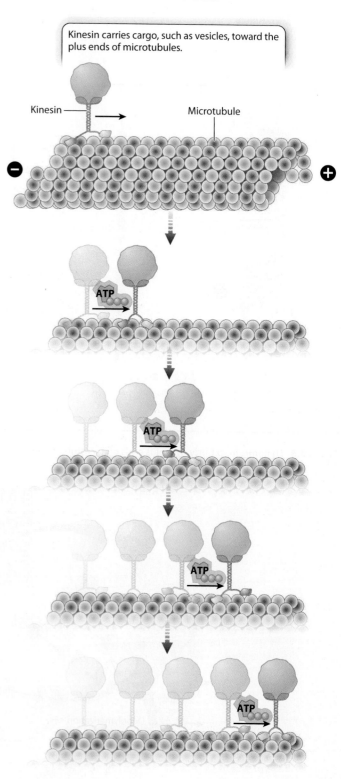

FIG. 10.7 Intracellular transport. The motor protein kinesin interacts with microtubules to move vesicles in the cell.

FIG. 10.8 Color change in zebrafish embryos driven by motor proteins kinesin and dynein. Melanin granules are redistributed along microtubules in the melanophores of the skin. *Photo sources: (zebrafish) Daisuke Kojima and Tomoya Shiraki; (redistribution of melanin) Courtesy Darren Logan, Wellcome Trust Sanger Institute.*

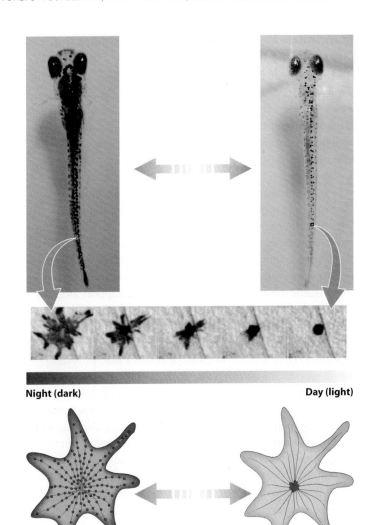

Night (dark) — Day (light)

In the dark, melanin granules are dispersed outward by kinesin, causing the embryo to be darkly colored.

In the light, melanin granules are aggregated toward the center by dynein, causing the embryo to be lightly colored.

Motile cilia propel the movement of cells or fluid surrounding the cell (**Fig. 10.9**). They can be found in single-celled eukaryotes, such as the green alga *Chlamydomonas* (Fig. 10.9a) and the protist *Paramecium* (Fig. 10.9b), where they propel the organisms through the water. They are also found in sperm cells (Fig. 10.9c). Long, elaborate cilia present in *Chlamydomonas* and sperm cells are sometimes called flagella, but they are similar in structure to the shorter cilia and are unrelated to the flagella found on bacterial cells, so most biologists now prefer the term "cilia."

Motile cilia are also present on epithelial cells, such as those that line the upper respiratory tract (Fig. 10.9d). These cilia move the fluid above the surface of the cell layer, carrying away foreign particles and pathogens. In motile cilia, microtubules associate with the motor protein dynein. Because the microtubules are fixed in place, dynein causes the microtubules to bend, causing a wave-like motion of the cilia. In turn, the motion of the cilia leads to movement of the cell or surrounding fluid.

Like microtubules, microfilaments associate with motor proteins to produce movement. Actin microfilaments associate with the motor protein **myosin** to transport various types of cellular cargo, such as vesicles, within cells. Microfilaments associated with myosin are also responsible for changes in the shape of many types of cell. One of the most dramatic examples of cell shape change is the shortening (contraction) of a muscle cell. Muscle contraction depends on the interaction of myosin with microfilaments, and is powered by ATP (Chapter 35).

FIG. 10.9 Motile cilia. Motile cilia move cells or allow cells to propel substances. (a) The green alga *Chlamydomonas* moves by the motion of two long cilia; (b) the protist *Paramecium* is covered with short cilia that beat in a coordinated fashion, allowing the organism to move through its environment; (c) sperm cells swim by movement of one long cilium, which is sometimes called a flagellum; (d) epithelial cells lining the upper respiratory tract have cilia that sweep fluid with debris and pathogens out of the airway. *Sources: (a) Andrew Syred/Science Source; (b) SPL/Science Source; (c) Juergen Berger/Science Source; (d) Eye of Science/Science Source.*

a. *Chlamydomonas*

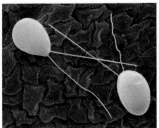

b. *Paramecium*

c. Sperm cells

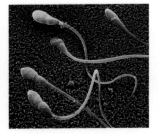

d. Airway epithelial cells

Intermediate filaments are polymers of proteins that vary according to cell type.

The intermediate filaments of animal cells have a diameter that is intermediate between those of microtubules and microfilaments, approximately 10 nm (see Fig. 10.3c). They are polymers of intermediate filament proteins that combine to form strong, cable-like structures in the cell. In this way, they provide cells with mechanical strength.

We have seen that different cell types use tubulin dimers to form microtubules and actin monomers to form microfilaments. By contrast, the proteins making up intermediate filaments differ from one cell type to another. For example, in epithelial cells, these protein subunits are keratins; in fibroblasts, they are vimentins; and in neurons, they are neurofilaments. Some intermediate filaments, called lamins, are even found inside the nucleus, where they provide support for the nuclear envelope (**Fig. 10.10**). More than 100 different kinds of intermediate filaments have been identified to date.

Once assembled, many intermediate filaments become attached to cell junctions at their cytoplasmic side, providing strong support for the cells (**Fig. 10.11**). In the case of epithelial cells, this anchoring results in structural continuity from one cell to another that greatly strengthens the entire epithelial tissue. This is especially important for tissues that are regularly subject to physical stress, such as the skin and the lining of the intestine.

Genetic defects that disrupt the intermediate filament network can have severe consequences. For example, some individuals with epidermolysis bullosa, a group of rare genetic diseases, have defective keratin genes. Intermediate filaments do not polymerize properly in these individuals, so they form weaker connections between the layers of cells that make up the epidermis. As a consequence, the outer layers can detach, resulting in extremely fragile skin that blisters in response to the slightest trauma (Fig. 10.11). The sensitivity to physical stress is so extreme that infants with epidermolysis bullosa often suffer significant damage to the skin during childbirth. Therefore, cesarean section is sometimes recommended in cases where the disease is diagnosed during pregnancy.

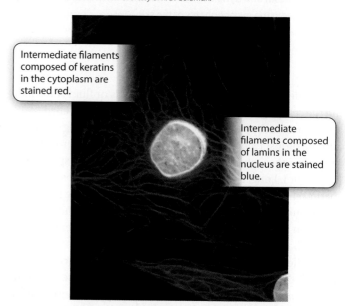

FIG. 10.10 Intermediate filaments in the cytoplasm and the nucleus of a cell. *Source: Courtesy of R. D. Goldman.*

Intermediate filaments composed of keratins in the cytoplasm are stained red.

Intermediate filaments composed of lamins in the nucleus are stained blue.

The cytoskeleton is an ancient feature of cells.

Actin and tubulin are found in all eukaryotic cells, and their structure and function have remained relatively unchanged throughout the course of evolution. The amino acid sequences of yeast tubulin and human tubulin are 75% identical. In fact, a mixture of yeast and human actin monomers forms hybrid microfilaments that are able to function normally in the cell.

TABLE 10.1	Major Functions of Cytoskeletal Elements	
CYTOSKELETAL ELEMENT	**SUBUNITS**	**MAJOR FUNCTIONS**
Microtubules	Tubulin dimers	Cell shape and support Cell movement (by cilia, flagella) Cell division (chromosome segregation) Vesicle transport Organelle arrangement
Microfilaments	Actin monomers	Cell shape and support Cell movement (by crawling) Cell division (cytokinesis) Vesicle transport Muscle contraction
Intermediate filaments	Diverse	Cell shape and support

FIG. 10.11 Intermediate filaments in the epidermis of the skin. Intermediate filaments bind to cell junctions called desmosomes, forming a strong, interconnected network. Epidermolysis bullosa is a group of blistering diseases of the skin that can be caused by a defect in intermediate filaments. *Source: Helen Osler/Northscot/REX/Shutterstock.*

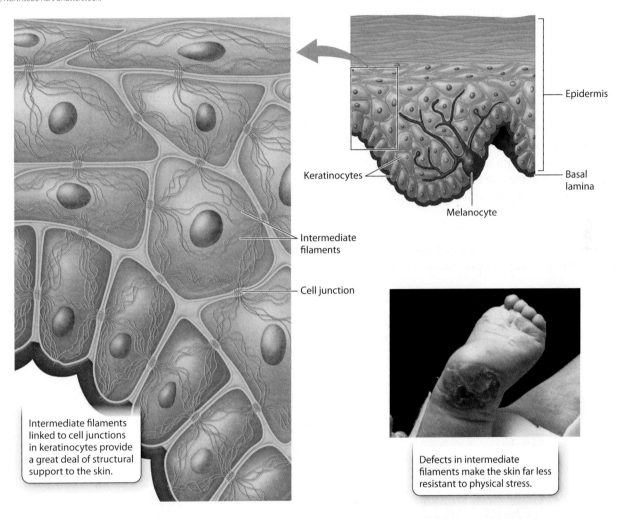

Not long ago, it was believed that cytoskeletal proteins were present only in eukaryotic cells. Recently, a number of studies have shown that many prokaryotes also have a system of proteins similar in structure to the cytoskeletal elements of eukaryotic cells. These bacterial proteins are involved in similar processes as those in eukaryotic cells, including the separation of daughter cells during cell division. Interestingly, at least one of these prokaryotic cytoskeleton-like proteins is expressed in the mitochondria and chloroplasts of some eukaryotic cells. The presence of this protein in these organelles lends support to the theory that mitochondria and chloroplasts were once independent prokaryotic cells that developed a symbiotic relationship with another cell. This idea, called the endosymbiotic theory, is discussed in Chapter 25.

The subunits and major functions of microtubules, microfilaments, and intermediate filaments are summarized in **Table 10.1**.

Self-Assessment Questions

3. What are three types of cytoskeletal elements? For each, indicate the subunits they are composed of, their relative size, and major functions.
4. How is the dynamic nature of microtubules and microfilaments important for their functions?
5. What are three types of motor proteins? For each, indicate which cytoskeletal element it interacts with and its function.
6. Would a defect in dynein or in kinesin cause a zebrafish embryo to remain darkly colored after daybreak?

10.3 CELL JUNCTIONS

The cell is the fundamental unit of living organisms (Chapter 1). Recent estimates place the number of cells in an adult human at approximately 30–40 trillion. Although it is difficult to determine this number exactly, it is clear that humans, as well as other multicellular organisms, are made up of a lot of cells! But what keeps us from slumping into a pile of cells? And what keeps cells organized into tissues, and tissues into organs?

Tissues are held together and function as a unit because of cell junctions. Cell junctions physically connect one cell to the next and anchor cells to the extracellular matrix. Some tissues have cell junctions that perform roles other than adhesion. For example, cell junctions in the outer layer of the skin and the lining of intestine provide a seal so that the epithelial sheet can act as a selective barrier. Other cell junctions allow adjacent cells to communicate so that they work together as a unit. In this section, we look more closely at the roles of the various types of cell junctions in tissues, as well as their interactions with the cytoskeleton.

Cell adhesion molecules allow cells to attach to other cells and to the extracellular matrix.

In 1907, American embryologist H. V. Wilson discovered that if he pressed a live sponge through fine cloth, he could break up the sponge into individual cells. If he then swirled the cells together, they would coalesce back into a group resembling a sponge. If he swirled the cells from sponges of two different species together, the cells sorted themselves out—that is, cells from one species of sponge associated only with cells from that same species (**Fig. 10.12a**). Fifty years later, German-born

FIG. 10.12 Cell type–specific cell adhesion. Experiments showed that cells from (a) sponges and from (b) amphibian embryos are in each case able to adhere to one another in a specific manner. *Photo sources: a. (top) Borut Furlan/WaterF/AGE Fotostock; (bottom) Franco Banfi/WaterF/AGE Fotostock.*

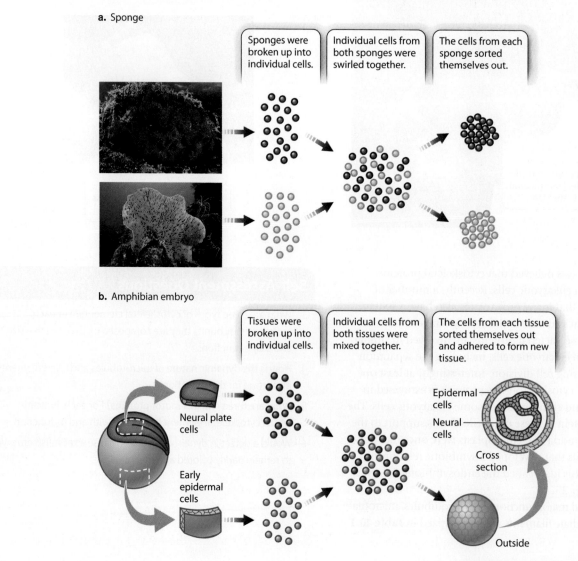

FIG. 10.13 Cell adhesion by cadherins and integrins. (a) Cadherins are transmembrane proteins that connect cells to other cells; (b) integrins are transmembrane proteins that connect cells to the extracellular matrix.

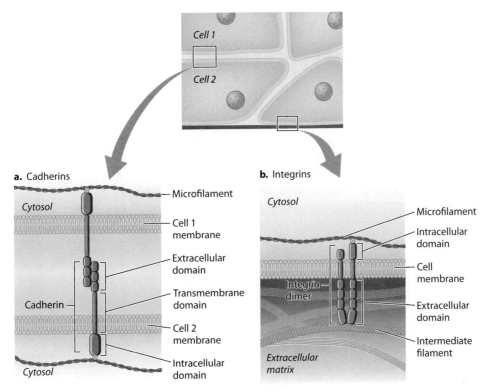

embryologist Johannes Holtfreter took neuronal cells and skin cells from an amphibian embryo and treated them the same way that Wilson had treated sponge cells. He observed that the embryonic cells sorted themselves according to tissue type (**Fig. 10.12b**).

Cells are able to sort themselves because proteins on the surfaces of cells of the same type recognize each other. Various proteins on the cell surface called **cell adhesion molecules** attach cells to one another or to the extracellular matrix. While a number of cell adhesion molecules are now known, the **cadherins** (calcium-dependent adherence proteins) are especially important. Many different kinds of cadherins exist, but a given cadherin may bind only to another cadherin of the same type. This property explains Holtfreter's observations that cells from amphibian embryos sorted themselves by tissue type. E-cadherin ("epidermal cadherin") is present on the surface of embryonic epidermal cells, and N-cadherin ("neural cadherin") is present on neuronal cells. The epidermal cells adhere to one another through E-cadherin, and the neuronal cells adhere to one another through N-cadherin.

Cadherins are transmembrane proteins (Chapter 5). The extracellular domain of a cadherin molecule binds to the extracellular domain of a cadherin of the same type on an adjacent cell. The cytoplasmic portion of the protein is linked to the cytoskeleton, including microfilaments and intermediate filaments (**Fig. 10.13a**). This arrangement essentially connects the cytoskeleton of one cell to the cytoskeleton of another, increasing the strength of tissues and organs.

As well as being stably connected to other cells, cells attach to proteins of the extracellular matrix through cell adhesion molecules called **integrins**. Like cadherins, integrins are transmembrane proteins, and their cytoplasmic domain is linked to microfilaments or intermediate filaments (**Fig. 10.13b**). Also like cadherins, integrins are of many different types, each binding to a specific extracellular matrix protein. Integrins are present on the surface of virtually every animal cell. In addition to their role in adhesion, integrins act as receptors that communicate information about the extracellular matrix to the interior of the cell (section 10.4).

Anchoring junctions connect adjacent cells and are reinforced by the cytoskeleton.

Cadherins and integrins are often organized into cell junctions, complex structures in the plasma membrane that allow cells to adhere to one another. These anchoring cell junctions are of two types: adherens junctions and desmosomes (**Fig. 10.14**).

In section 10.2, we saw that a long bundle of actin microfilaments forms a band that extends around the circumference of epithelial cells, such as those lining the intestine. This band

FIG. 10.14 Cell junctions. Cell junctions connect cells to other cells or to the basal lamina. They are reinforced by the cytoskeleton.

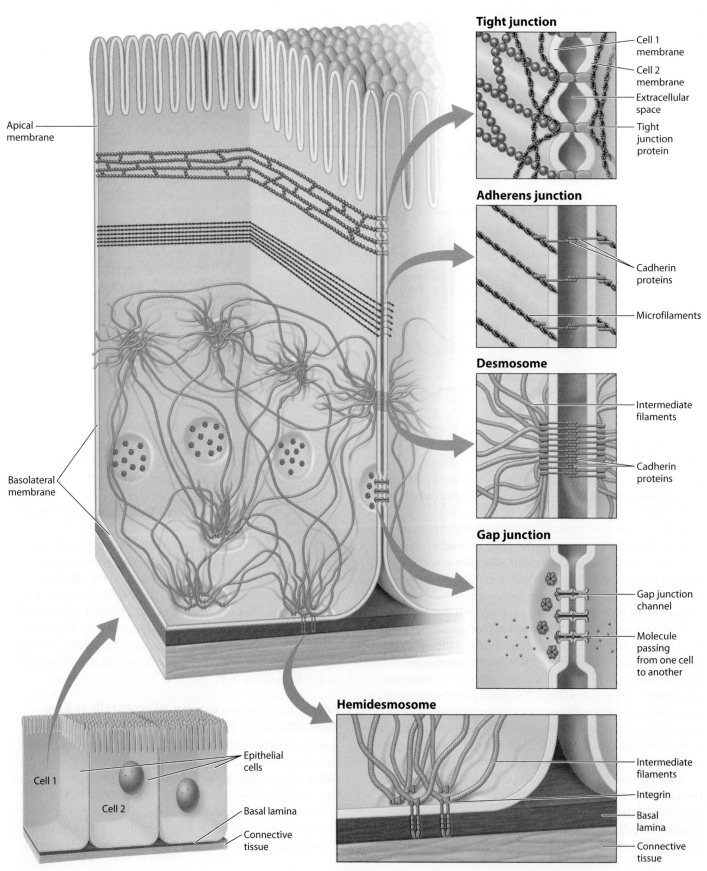

of actin is attached to the plasma membrane by cadherins in a beltlike structure called an **adherens junction** (Fig. 10.14). The cadherins in the adherens junction of one cell attach to the cadherins in the adherens junctions of adjacent cells. This arrangement establishes a physical connection among the actin cytoskeletons of all cells present in an epithelial layer.

Like adherens junctions, **desmosomes** are cell junctions that allow cells to adhere to one another. Whereas adherens junctions form a belt around the circumference of cells, desmosomes are buttonlike points of adhesion (Fig. 10.14). Cadherins are at work here, too, strengthening the connection between cells. Cadherins in the desmosome of one cell bind to cadherins in the desmosomes of adjacent cells. The cytoplasmic domain of these cadherins connects to intermediate filaments in the cytoskeleton. This second type of physical connection among neighboring cells greatly enhances the strength of epithelial cell layers.

Epithelial cells are attached not only to one another, but also to the underlying extracellular matrix (specifically, the basal lamina). In this case, the cells are firmly anchored to the extracellular matrix by a type of desmosome called a **hemidesmosome** (Fig. 10.14). Integrins are the most prominent cell adhesion molecules in hemidesmosomes. The extracellular domains bind extracellular matrix proteins, and the cytoplasmic domains connect to intermediate filaments. These intermediate filaments connect to desmosomes in other parts of the plasma membrane. The result is a firmly anchored and reinforced layer of cells.

Tight junctions prevent the movement of substances through the space between cells.

Epithelial cells form sheets or boundaries that line tissues and organs, including the digestive tract, respiratory tract, and outer layer of the skin. Like any effective boundary, a layer of epithelial cells must limit or control the passage of material across it. Adherens junctions and desmosomes provide strong adhesion between cells, but they do not prevent materials from passing freely through the spaces between the cells. This function is provided by a different type of cell junction. In vertebrates, these are called **tight junctions** (Fig. 10.14). Tight junctions establish a seal between cells that prevents molecules from moving through this channel. For example, the only way a substance can travel from one side of a sheet of epithelial cells to the other is by moving *through* the cells by means of one of the cellular transport mechanisms discussed in Chapter 5.

A tight junction is a band of interconnected strands of integral membrane proteins, particularly proteins called claudins and occludins. Like adherens junctions, tight junctions encircle the epithelial cell. The proteins forming the tight junction in one cell bind to the proteins forming the tight junctions in adjacent cells. Also like adherens junctions, tight junctions connect to actin microfilaments.

The tight junction divides the plasma membrane into two distinct regions (Fig. 10.14). Thus cells that have tight junctions have two sides. The portion of the plasma membrane in contact with the lumen, or the inside of any tubelike structure such as the gut, is called the apical membrane. The apical membrane defines the "top" side of the cell. The rest of the plasma membrane is the basolateral membrane, which defines the bottom ("baso") and sides ("lateral") of the cell.

A tight junction prevents lipids and proteins in the membrane on one side of the junction from diffusing to the other side. As a result, the apical and basolateral membranes of a cell are likely to have different integral membrane proteins, which causes them to be functionally different as well. In the small intestine, for example, glucose is transported from the lumen into intestinal epithelial cells by transport proteins on the apical side of the cells, and is transported out of the cells into the circulation by facilitated diffusion through a different type of glucose transporter restricted to the basolateral sides of the cells (Chapter 38).

Molecules pass between cells through communicating junctions.

Not all cell junctions are involved in the adhesion of cells to each other or in sealing a layer of cells. **Gap junctions** (Fig. 10.14) of animal cells and **plasmodesmata** of plant cells (see Fig. 5.17b) permit materials to pass directly from the cytoplasm of one cell to the cytoplasm of another, allowing cells to communicate with one another.

Gap junctions are a complex of integral membrane proteins called connexins arranged in a ring. The ring of connexin proteins connects to a similar ring of proteins in the membrane of an adjacent cell. Together, the two connexin rings form a channel that connects the cytoplasm of adjacent cells. Ions and signaling molecules pass through these junctions, allowing cells to act in unison. In the heart, for example, ions pass though gap junctions connecting cardiac muscle cells. This rapid electrical communication allows the muscle cells to beat in a coordinated fashion (Chapter 37).

Plasmodesmata (the singular form is "plasmodesma") are passages through the cell walls of adjacent plant cells. Like gap junctions, they allow cells to exchange ions and small molecules, but the similarity ends there. In plasmodesmata, the plasma membranes of the two connected cells are actually continuous. The size of the opening is considerably larger than that in gap junctions, large enough for cells to transfer RNA molecules and proteins, an ability that is especially important during embryonic development. Plant cells can send signals to one another through plasmodesmata despite being enclosed within rigid cell walls.

In summary, cell junctions interact to create stable communities of cells in the form of tissues and organs. These cell junctions are important for the functions of tissues, allowing cells to adhere to each other and the extracellular matrix, act as a barrier, and communicate rapidly. The types and functions of the cell junctions are summarized in **Table 10.2**.

TABLE 10.2 Types and Functions of Cell Junctions

CELL JUNCTION	MAJOR COMPONENT	CYTOSKELETAL ATTACHMENT	PRIMARY FUNCTION
Anchoring			
Adherens junction	Cadherins	Microfilaments	Cell–cell adhesion
Desmosome	Cadherins	Intermediate filaments	Cell–cell adhesion
Hemidesmosome	Integrins	Intermediate filaments	Cell–extracellular matrix adhesion
Barrier			
Tight junction	Claudins, occludins		Epithelial boundary
Communicating			
Gap junction	Connexins		Communication between animal cells
Plasmodesma	Cell membrane		Communication between plant cells

Self-Assessment Questions

7. What are three major types and functions of cell junctions?
8. Adherens junctions and desmosomes both attach cells to other cells and are made up of cadherins. How, then, are they different?
9. What do you think would happen if you interfered with the function of cadherins and integrins?

10.4 THE EXTRACELLULAR MATRIX

Up to this point, we have looked at how the cytoskeleton maintains the shape of cells. We have also seen how the stable association of animal cells with one another and with the extracellular matrix is made possible by cell junctions, and noted that these junctions are reinforced by the cytoskeleton. As important as the cytoskeleton and cell junctions are to the structure of cells and tissues, it is the extracellular matrix that provides the molecular framework that helps determine the structural architecture of plants and animals.

The extracellular matrix is an insoluble meshwork composed of proteins and polysaccharides. Its components are synthesized, secreted, and modified by many different cell types. The many different forms of extracellular matrix differ in the amount, type, and organization of the proteins and polysaccharides that make them up. In both plants and animals, the extracellular matrix not only contributes structural support, but also provides informational cues that determine the activity of the cells that are in contact with it.

The extracellular matrix of plants is the cell wall.

The paper we write on, the cotton fibers in the clothes we wear, and the wood in the chairs we sit on are all composed of the extracellular matrix of plants. In plants, the extracellular matrix forms the cell wall, and the main component of the plant cell wall is the polysaccharide cellulose (Chapters 2 and 5). Cellulose is, in fact, the most widespread organic macromolecule on Earth.

The plant cell wall is possibly one of the most complex examples of an extracellular matrix, and certainly one of the most diverse in terms of the functions it performs. Cell walls maintain the shape and turgor pressure of plant cells and act as a barrier that prevents foreign materials and pathogens from reaching the plasma membrane. In many plants, cell walls collectively serve as a support structure for the entire plant, much like the skeleton of an animal.

The plant cell wall is composed of as many as three layers: the outermost middle lamella, the primary cell wall, and the secondary cell wall, located closest to the plasma membrane (**Fig. 10.15**). The middle lamella is synthesized first, during the late stages of cell division. It is composed of a gluelike complex carbohydrate, and is the main mechanism by which plant cells adhere to one another. The primary cell wall is formed next. It consists mainly of cellulose, but also contains a number of other molecules, including pectin. The primary cell wall is laid down while the cells are still growing. It is assembled by enzymes on the surface of the cell and remains thin and flexible.

When cell growth has stopped, the secondary cell wall is constructed in many, but not all, plant cells. It is made largely of cellulose, but also contains a substance called lignin. Lignin hardens the cell wall and makes it water resistant. In woody plants, the cell wall can be as much as 25% lignin. The rigid secondary cell wall permits woody plants to grow to tremendous heights. Giant sequoia trees, for example, grow to more than 300 feet, supported entirely by the lignin-reinforced cellulose fibers of the interconnected cell walls.

As a plant cell grows, it must synthesize additional cell wall components to expand the area of the wall. Unlike the

FIG. 10.15 The three layers of the plant cell wall: middle lamella, primary cell wall, and secondary cell wall. The major component of the plant cell wall is cellulose, a polymer of glucose. *Photo source: Biophoto Associates/Science Source.*

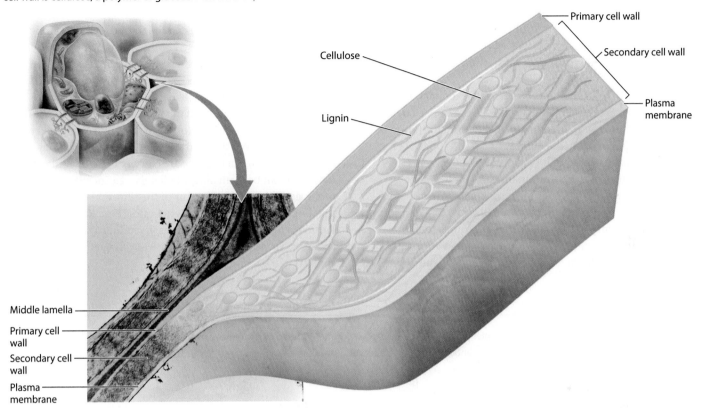

extracellular matrix components that are secreted by animal cells, the cellulose polymer is assembled outside the cell, on the extracellular surface of the plasma membrane. Both the glucose monomers that form the polymer and the enzymes that attach them are delivered to the cell surface by arrays of microtubules. This is yet another example of how the cytoskeleton plays an indispensable role in regulating the shape of a cell.

The extracellular matrix is abundant in connective tissues of animals.

The extracellular matrix of animals, like that of plants, is a mixture of proteins and polysaccharides secreted by cells. The animal extracellular matrix is composed of large fibrous proteins, including collagen, elastin, and laminin, which impart tremendous tensile strength. These fibrous proteins are embedded in a gel-like polysaccharide matrix. The matrix is negatively charged, attracting positively charged ions and water molecules that provide protection against compression and other physical stress.

The extracellular matrix can be found in abundance in animal connective tissue (**Fig. 10.16**). Connective tissue has two functions, both of which are necessary for multicellularity. First, it physically connects various parts of the body. For example, the tendons that connect your muscles to bones

FIG. 10.16 Animal connective tissue. Connective tissue is composed of protein fibers in a gel-like polysaccharide matrix. *Photo source: Biophoto Associates/Science Source.*

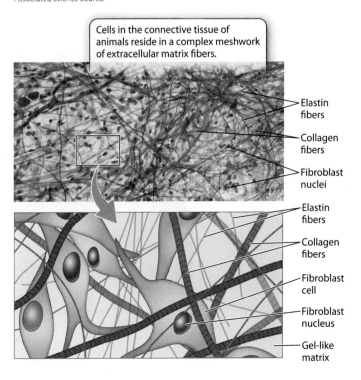

and the ligaments that connect your bones to other bones are connective tissue. Second, connective tissue supports various parts of the body. It underlies all epithelial tissues, as we have seen (section 10.1). For example, the dermis of the skin is a connective tissue that provides support and nutrients to the overlying epidermis. The main type of cell in the dermis is the fibroblast, which synthesizes most of the extracellular matrix proteins.

Connective tissue is an unusual tissue type in that it is dominated by the extracellular matrix and has a low cell density. Consequently, the extracellular matrix determines the properties of different types of connective tissue.

All animals express similar connective tissue proteins, highlighting those proteins' importance and evolutionarily conserved function. Collagen is the most abundant protein in the extracellular matrix of animals. More than 20 different forms of collagen exist, and in humans collagen accounts for almost 25% of all protein present in the body. More than 90% of this collagen is type I collagen. Present in the dermis of the skin, type I collagen provides strong, durable support for the overlying epidermis. Tendons and ligaments are able to withstand the physical stress placed on them because they are made up primarily of collagen.

Collagen's strength is related to its structure. Like a rope or a cable, this protein is composed of intertwined fibers that make it much stronger than if it were a single fiber of the same diameter. A collagen molecule consists of three polypeptides wound around one another in a triple helix. A bundle of collagen molecules forms a fibril, and the fibrils are assembled into fibers (**Fig. 10.17**). Once multiple collagen fibers are assembled into a ligament or tendon, the final structure is incredibly strong.

The basal lamina is a specialized layer of extracellular matrix that is present beneath all epithelial tissues, including the lining of the digestive tract, epidermis of the skin, and endothelial cells that line the blood vessels of vertebrates (**Fig. 10.18**). The basal lamina is made of several proteins, including a special type of collagen. The triple-helical structure of collagen provides flexible support to the epithelial sheet.

FIG. 10.17 Collagen. Type I collagen molecules are organized in a triple helix and grouped into bundles called fibrils, which in turn are grouped into bundles called fibers. This type of arrangement, seen in fibers and steel cables, imparts tremendous strength. *Photo sources: (top to bottom) Tom Grundy/Alamy; Egon Bömsch/imageBROKER/age fotostock; Eye of Science/Science Source.*

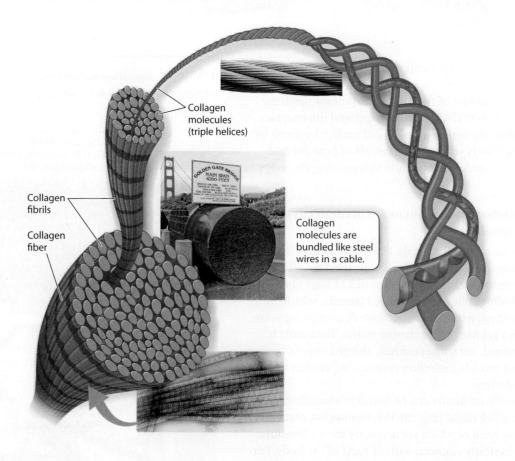

FIG. 10.18 The basal lamina. The basal lamina, found beneath epithelial tissue, is a specialized form of extracellular matrix. *Photo source: ISM/Jean-Claude RÉVY/Diomedia.*

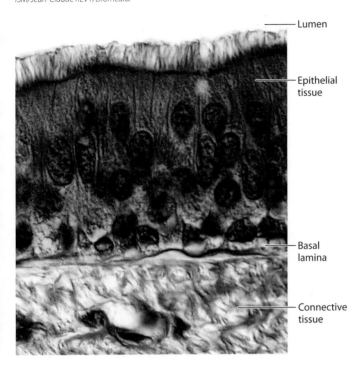

FIG. 10.19 Metastatic cancer cells. Some cancer cells spread from the original site of cancer formation to the bloodstream and then to distant organs of the body.

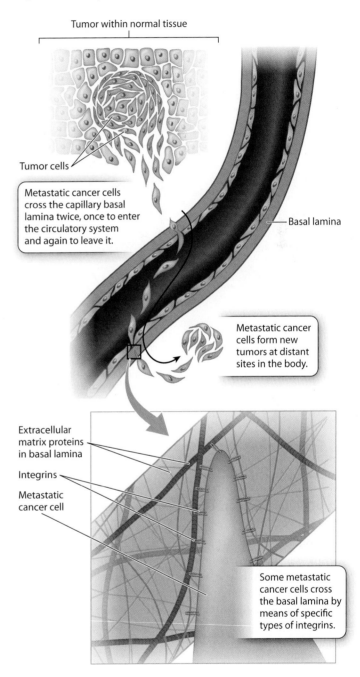

 CASE 2 CANCER: CELL SIGNALING, FORM, AND DIVISION

How do cancer cells spread throughout the body?

Nonmalignant, or benign, tumors are encapsulated masses of cells that divide continuously because regulation of cell division has gone awry (Chapter 11). As the tumor grows, it pushes outward against adjacent tissues. Benign tumors are rarely life threatening unless the tumor interferes with the function of a vital organ.

Malignant tumors are more dangerous. They contain some cells that can metastasize—that is, break away from the main tumor and travel to distant sites in the body. Metastatic tumor cells have an enhanced ability to adhere to extracellular matrix proteins, especially those in the basal lamina. This is significant because for a cell to metastasize, it must enter and leave the bloodstream through capillaries or other vessels. Since all blood vessels, including capillaries, have a basal lamina, a metastatic tumor cell needs to cross a basal lamina at least twice—once on the way into the bloodstream and again on the way out (**Fig. 10.19**). Since cells attach to basal lamina proteins by means of integrins, many studies have compared the integrins in metastatic and non-metastatic cells in the search for potential targets for treatment.

In some types of cancer, the number of specific integrins on the cell surface is an indicator of metastatic potential. Melanoma provides an example. A specific type of integrin is present in high amounts on metastatic melanoma cells, but is absent on non-metastatic cells from the same tumor. In laboratory tests, blocking these integrins eliminates the melanoma cell's ability

FIG. 10.20 Influence of the extracellular matrix on cell shape. (a) Fibroblasts adopt different shapes depending on whether they are grown on a two-dimensional or three-dimensional matrix. (b) Neurons adopt different shapes depending on whether they are cultured with the extracellular matrix protein laminin. *Sources: (a) Republished with permission of American Physiological Society, from Modulation of smooth muscle phenotype in vitro by homologous cell substrate, F. Tao, S. Chaudry, B. Tolloczko, J. G. Martin, and S. M. Kelly, Am J Physiol Cell Physiol 284:(6) C1531-C1541, 2003; permission conveyed through Copyright Clearance Center, Inc. (b) Courtesy Motoyoshi Nomizu.*

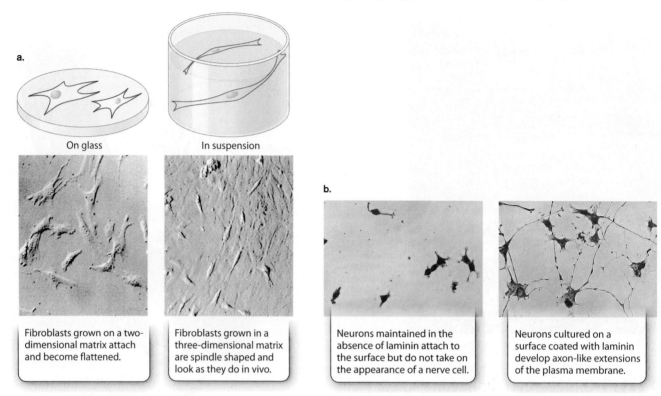

to cross an artificial basal lamina. Drugs targeting this integrin protein are currently in clinical trials.

Extracellular matrix proteins influence cell shape and gene expression.

Cells continue to interact with the extracellular matrix long after they have synthesized it or moved into it, and these interactions can have profound effects on cell shape and gene expression. Some of these cellular responses are the result of interactions between the extracellular matrix and integrins on the surface of cells. Integrins act as receptors that relay the signal to the cell interior as the first step in this signal transduction pathway (Chapter 9).

Biologists have studied how the extracellular matrix affects cell shape using cells grown in culture in the laboratory (**Fig. 10.20**). For example, fibroblasts cultured on a two-dimensional surface coated with extracellular matrix proteins attach to the matrix and flatten out as they maximize their adhesion to the matrix. By contrast, the same cells cultured in a three-dimensional gel of extracellular matrix look and behave like the spindle-shaped, highly migratory fibroblasts present in living connective tissue (Fig. 10.20a). Similarly, when nerve cells are grown in culture on a plastic surface, they attach to the surface of the dish but do not take on a neuron-like shape. However, when these cells are grown on the same surface coated with the extracellular matrix protein laminin, they develop long extensions that resemble the axons and dendrites of normal nerve cells (Fig. 10.20b).

In addition to influencing cell shape, the structure and composition of the extracellular matrix can influence gene expression, as described in **Fig. 10.21**. This experiment and the others described here demonstrate that there is a dynamic interplay between the extracellular matrix and the cells that synthesize it.

Self-Assessment Questions

10. Where can extracellular matrix be found in plants and in animals?
11. What are two effects that the extracellular matrix can have on cells?
12. Do you think cadherins or integrins are responsible for the changes in gene expression described in Fig. 10.21? Why?

HOW DO WE KNOW?

FIG. 10.21

Can extracellular matrix proteins influence gene expression?

BACKGROUND Adhesion of cells to the extracellular matrix is required for proper cell shape, as we discussed. In the 1980s, research by Iranian American cell biologist Mina Bissell and colleagues demonstrated that a cell's interaction with extracellular matrix proteins also influences gene expression. Bissell discovered that mammary cells synthesize and secrete high levels of the milk protein β-casein when grown in a three-dimensional collagen matrix but not in a two-dimensional collagen matrix. In the 1990s, American cell biologist Joan Caron followed up these studies using liver cells (hepatocytes).

HYPOTHESIS Caron hypothesized that a specific protein in the extracellular matrix is necessary for the expression of the protein albumin from hepatocytes grown in culture. Albumin is a major product of hepatocytes.

EXPERIMENT Caron cultured hepatocytes on a thin layer of type I collagen, which does not induce albumin synthesis. Next, she added a mixture of several different extracellular matrix proteins to the culture and looked for changes in albumin gene expression and protein secretion. She then tested individual extracellular matrix proteins from the mixture to see which one was responsible for the increase in albumin gene expression.

RESULTS Caron found that when she cultured cells on type I collagen with a combination of three extracellular matrix proteins—laminin, type IV collagen, and heparin sulfate proteoglycan (HSPG)—the cells synthesized high levels of albumin mRNA (top graph, blue line) and albumin protein (middle graph, blue line). However, if she cultured the cells on type I collagen alone, they did not synthesize high levels of albumin mRNA (top graph, red line) or albumin protein (middle graph, red line). When she tested the three extracellular matrix proteins individually, she found that laminin, but neither of the other proteins (type IV collagen and HSPG), caused an increase in synthesis of albumin mRNA (bottom bar graph).

CONCLUSION The experiments supported Caron's hypothesis. A specific extracellular matrix protein, laminin, influences the expression of albumin by hepatocytes.

FOLLOW-UP WORK Bissell continued her work with mammary cells. She found that laminin increases the expression of the β-casein gene just as it increases expression of the albumin gene in hepatocytes.

SOURCES Lee, E. Y., et al. 1985. "Interaction of Mouse Mammary Epithelial Cells with Collagen Substrata: Regulation of Casein Gene Expression and Secretion." *Proceedings of the National Academy of Sciences, USA* 82:1419–1423; Caron, J. M. 1990. "Induction of Albumin Gene Transcription in Hepatocytes by Extracellular Matrix Proteins." *Molecular and Cellular Biology* 10:1239–1243.

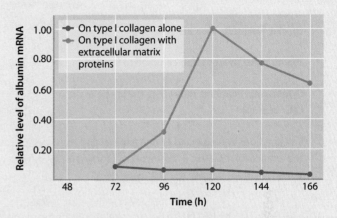

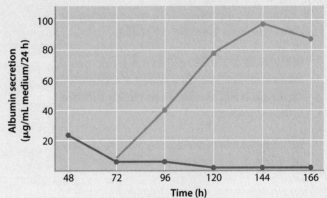

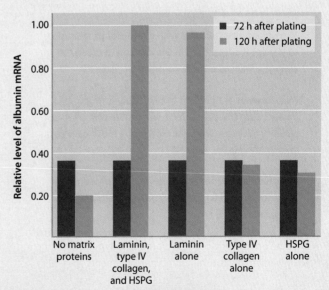

CORE CONCEPTS SUMMARY

10.1 TISSUES AND ORGANS: Tissues and organs are communities of cells that perform specific functions.

A tissue is a collection of cells that work together to perform a specific function. page 203

Two or more tissues often combine to form an organ. page 203

Cytoskeletal elements determine the shape of the cell. page 204

Cell junctions connect cells to one another and to the extracellular matrix, a meshwork of proteins and polysaccharides outside the cell. page 204

10.2 THE CYTOSKELETON: The cytoskeleton is composed of microtubules, microfilaments, and intermediate filaments that help maintain cell shape.

Microtubules are hollow polymers of tubulin dimers, and microfilaments are helical polymers of actin monomers. Both microtubules and microfilaments provide structural support to the cell. page 205

Microtubules and microfilaments are dynamic structures and can assemble and disassemble rapidly. page 205

Microtubules associate with the motor proteins dynein and kinesin to transport substances in the cell. page 206

Microfilaments associate with the motor protein myosin to transport substances in the cell and to cause cell shape changes, such as muscle contraction. page 207

Microtubules are present in cilia, rodlike structures that extend from the cell surface. Nonmotile cilia often serve a sensory function, whereas motile cilia move cells or fluid surrounding cells. page 208

Intermediate filaments are polymers of proteins that differ depending on cell type. They provide stable structural support for many types of cells. page 209

Some prokaryotic cells have protein polymers that function similarly to microtubules and microfilaments. page 210

10.3 CELL JUNCTIONS: Cell junctions connect cells to one another to form tissues.

Cell junctions anchor cells to each other and to the extracellular matrix, allow sheets of cells to act as a barrier, and permit communication between cells in tissues. page 211

Anchoring cell junctions include adherens junctions and desmosomes. page 213

Adherens junctions form a belt around the circumference of a cell. They are composed of cell adhesion molecules called cadherins and connect to microfilaments inside the cell. page 213

Desmosomes are buttonlike points of adhesion between cells. They are composed of cadherins and connect to intermediate filaments inside the cell. page 213

Hemidesmosomes connect cells to the extracellular matrix. They are composed of integrins and connect to intermediate filaments inside the cell. page 213

Tight junctions prevent the passage of substances through the space between cells and divide the plasma membrane into apical and basolateral regions. page 213

Gap junctions (in animals) and plasmodesmata (in plants) allow cells to communicate rapidly with one another. page 215

10.4 THE EXTRACELLULAR MATRIX: The extracellular matrix provides structural support and informational cues.

The extracellular matrix is an insoluble meshwork of proteins and polysaccharides secreted by the cells it surrounds. It provides structural support to cells, tissues, and organs. page 216

In plants, the extracellular matrix is found in the cell wall, and the main component of the plant cell wall is the polysaccharide cellulose. page 216

In animals, the extracellular matrix is found in abundance in connective tissue. page 217

The exceptionally strong protein fiber collagen is the primary component of connective tissues in animals. page 217

A specialized extracellular matrix called the basal lamina is present underneath all epithelial cell layers. page 218

The extracellular matrix can influence cell shape and gene expression. page 219

Log in to **LaunchPad** to check your answers to the Self-Assessment Questions and to access additional learning tools.

CHAPTER 11 Cell Division
Variation, Regulation, and Cancer

CORE CONCEPTS

11.1 CELL DIVISION: During cell division, a single parental cell produces two daughter cells.

11.2 MITOTIC CELL DIVISION: Mitotic cell division produces genetically identical daughter cells and is the basis for asexual reproduction and development.

11.3 MEIOTIC CELL DIVISION: Meiotic cell division produces genetically unique daughter cells and is the basis for sexual reproduction.

11.4 NONDISJUNCTION: Nondisjunction results in extra or missing chromosomes.

11.5 CELL CYCLE REGULATION: The cell cycle is regulated so that cells divide only at appropriate times and places.

11.6 CANCER: Cancer is uncontrolled cell division that can result from mutations in genes that regulate cell division.

Cells come from preexisting cells. This is one of the fundamental principles of biology and a key component of the cell theory, which was introduced in Chapter 5. **Cell division** is the process by which cells make more cells. Multicellular organisms begin life as a single cell, and then cell division produces the millions, billions, or in the case of humans, trillions of cells that make up the fully developed organism. Even after a multicellular organism has achieved its adult size, cell division continues. In plants, cell division is essential for continued growth. In many animals, cell division replaces worn-out blood cells, skin cells, and cells that line much of the digestive tract. If you fall and scrape your knee, the cells at the site of the wound begin dividing to replace the damaged cells and heal the scrape.

Cell division is also important in reproduction. In bacteria, for example, a new generation is produced when the parent cell divides and forms two daughter cells. The parent cell first makes two copies of its genetic material. Then the parent cell divides in two so that each of the two daughter cells receives one copy of the genetic material. Reproduction that occurs when offspring receive genetic material from a single parent is called **asexual reproduction**.

By contrast, reproduction that combines the genetic material from two parents is called **sexual reproduction**. Half the genetic material is supplied by the female parent and is contributed by the egg and the other half is supplied by the male parent and is contributed by the sperm. Eggs and sperm are specialized cells called **gametes**. A female gamete and male gamete fuse during fertilization to form a new organism (Chapter 40). Gametes are produced by a form of cell division that results in daughter cells with half the number of chromosomes as the parent cell. So when fertilization occurs, the combination of genetic material from the egg and the sperm results in a new organism with the same number of chromosomes as the parents.

What determines when cells divide and, importantly, when they should not? And what determines which cells divide? To answer these questions, we need to understand the process of cell division and how it is controlled. We will then be able to explore how cancer results from a loss of control of cell division.

11.1 CELL DIVISION

Cell division is the process by which a single cell produces two daughter cells. In order for a cell to divide successfully, it must be large enough to divide in two and contribute sufficient nuclear and cytoplasmic components to each daughter cell. Before the cell divides, therefore, key cellular components are duplicated. This duplication of material followed by cell division is achieved in a series of steps that constitutes the life cycle of a cell. When you think of a life cycle, you might think of various stages beginning with birth and ending with death. In the case of a cell, the life cycle begins and ends with cell division. In this section, we explore the different ways in which prokaryotic and eukaryotic cells divide.

Prokaryotic cells divide by binary fission.

Prokaryotic cells produce daughter cells by **binary fission**. In this form of cell division, a cell replicates its DNA, increases in size, and divides into two daughter cells. Each daughter cell receives one copy of the replicated parental DNA. Binary fission has been studied most extensively in bacteria. The process of binary fission is similar in archaeons as well as in chloroplasts and mitochondria, organelles within plant, fungal, and animal cells that evolved from free-living prokaryotic cells (Chapters 5 and 25).

Let's consider the process of binary fission in the intestinal bacterium *Escherichia coli* (**Fig. 11.1**). The circular genome of *E. coli* is attached by proteins to the inside of the plasma membrane (Fig. 11.1, step 1). DNA replication is initiated at a specific location on the circular DNA molecule, called the origin of replication, and proceeds in opposite directions around the circle. The result is two DNA molecules, each attached to the plasma membrane at a different site. The two attachment sites are initially close together (Fig. 11.1, step 2). The cell then elongates and, as it does so, the two DNA attachment sites move apart (Fig. 11.1, steps 3 and 4). When the cell is about twice its original size and the DNA molecules are well separated, a constriction forms at the midpoint of the cell (Fig. 11.1, step 5). Eventually, new membrane and cell wall are synthesized at the site of the constriction, dividing the single cell into two (Fig. 11.1, step 6). The result is two daughter cells, each having the same genetic material as the parent cell.

Like most cellular processes, binary fission requires the coordination of many components. Recent research has identified several genes whose products play a key role in bacterial cell division. One of these genes, called *FtsZ*, has been especially well studied. *FtsZ* encodes a protein that forms a ring at the site of constriction where the new cell wall forms between the two daughter cells. *FtsZ* is present in the genomes of diverse bacteria and archaeons, suggesting that it plays a fundamental role in binary fission. Interestingly, it appears to be evolutionarily related to the protein tubulin. Recall from Chapter 10 that tubulin makes up microtubules found in eukaryotic cells that are important in intracellular transport, cell movement, and cell division.

Eukaryotic cells divide by mitotic cell division.

The basic steps of binary fission that we just saw—replication of DNA, segregation of replicated DNA to daughter cells, and division of one cell into two—occur in all forms of cell division. However, cell division in eukaryotes is more complicated than cell division in prokaryotes. When eukaryotic cells divide, they first divide the nucleus by **mitosis** and then divide the cytoplasm into two daughter cells by **cytokinesis**.

The genome of a prokaryotic cell is a single, relatively small, circular DNA molecule. In comparison, the genome of a eukaryotic cell is typically much larger and is organized into one or more linear chromosomes. Each of these chromosomes must be replicated and separated into daughter cells. The DNA of prokaryotes is attached to the inside of the plasma membrane, allowing replicated DNA to be separated into daughter cells by cell growth. In contrast, the DNA of eukaryotes is located in the nucleus. As a result, eukaryotic cell

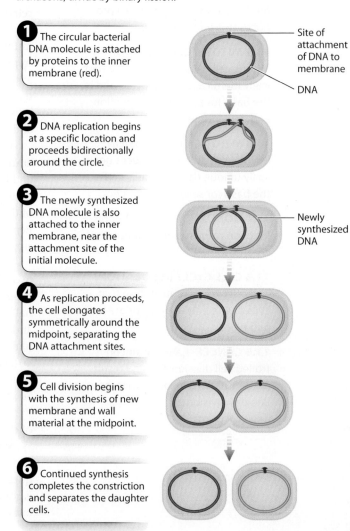

FIG. 11.1 Binary fission. Prokaryotic cells, such as bacteria and archaeons, divide by binary fission.

① The circular bacterial DNA molecule is attached by proteins to the inner membrane (red).

② DNA replication begins at a specific location and proceeds bidirectionally around the circle.

③ The newly synthesized DNA molecule is also attached to the inner membrane, near the attachment site of the initial molecule.

④ As replication proceeds, the cell elongates symmetrically around the midpoint, separating the DNA attachment sites.

⑤ Cell division begins with the synthesis of new membrane and wall material at the midpoint.

⑥ Continued synthesis completes the constriction and separates the daughter cells.

division requires first the breakdown and then the re-formation of the nuclear envelope, as well as mechanisms other than cell growth to separate replicated DNA. As we saw in Chapter 10 and discuss in more detail in section 11.2, chromosomes of dividing eukaryotic cells attach to the mitotic spindle, which separates them into daughter cells.

Interestingly, some unicellular eukaryotes exhibit forms of cell division that have characteristics of both binary fission and mitosis. For example, dinoflagellates, like all eukaryotes, have a nucleus and linear chromosomes. However, unlike in most eukaryotes, the nuclear envelope does not break down but stays intact during cell division. Furthermore, the replicated DNA is attached to the nuclear envelope. The nucleus then grows and divides in a manner reminiscent of binary fission. These and other observations of intermediate forms of cell division strongly suggest that mitosis evolved from binary fission.

The cell cycle describes the life cycle of a eukaryotic cell.

Cell division in eukaryotic cells proceeds through a number of steps that make up the **cell cycle** (**Fig. 11.2**). The cell cycle consists of two stages: **M phase** and **interphase.** During M phase, the parent cell divides into two daughter cells. M phase consists of two different events: (1) mitosis, the separation of the chromosomes into two nuclei, and (2) cytokinesis, the division of the cell itself into two separate cells. Typically, these two processes go hand in hand, with cytokinesis beginning even before mitosis is complete. In most mammalian cells, M phase lasts about an hour.

The second stage of the cell cycle, called interphase, is the time between two successive M phases (Fig. 11.2). For many years, it was thought that the relatively long period of interphase is uneventful. Today, we know that the cell makes many preparations for division during this stage. These preparations include DNA replication and cell growth. The DNA in the nucleus first replicates so that each daughter cell receives a copy of the genetic material. The cell then increases in size so that each daughter cell receives sufficient amounts of cytoplasmic and membrane components to allow it to survive on its own.

Interphase can be divided into three phases, as shown in Fig. 11.2. DNA replication occurs during S phase. This phase is called **S phase** because DNA replication involves the synthesis of DNA. In most cells, S phase does not immediately precede or follow mitosis but is separated from it by two gap phases: **G_1 phase** between the end of M phase and the start of S phase, and **G_2 phase** between the end of S phase and the start of M phase. Many essential processes occur during both "gap" phases, despite the name. For example, during G_1, regulatory proteins, such as kinases, are made. These proteins activate enzymes that synthesize DNA. In G_2, both the size and protein content of the cell increase. Thus, G_1 is a time of preparation for S-phase DNA synthesis, and G_2 is a time of preparation for M-phase mitosis and cytokinesis.

How long a cell takes to pass through the cell cycle depends on the type of cell and the organism's stage of development. Most actively dividing cells in your body take about 24 hours to complete the cell cycle. Cells in your skin and intestine that require frequent replenishing usually take about 12 hours. A unicellular eukaryote such as yeast can complete the cell cycle in just 90 minutes. The embryonic cells of some frog species complete the cell cycle even faster. Early cell divisions divide the cytoplasm of the large frog egg cell into many smaller cells and so no growth period is needed between cell divisions. Consequently, there are virtually no G_1 and G_2 phases, and as little as 30 minutes pass between cell divisions.

Not all the cells in your body actively divide. Instead, many cells pause in the cell cycle somewhere between M phase and S phase for periods ranging from days to more than a year. This period is the **G_0 phase** (Fig. 11.2). It is distinguished from G_1 by the absence of preparations for DNA synthesis. Liver cells remain in G_0 for as much as a year. Other cells such as nerve cells and those that form the lens of the eye enter G_0 permanently; these cells are nondividing. Thus, many brain cells lost to disease or damage cannot be replaced. Although cells in G_0 have exited the cell cycle, they are active in other ways—in particular, cells in G_0 still perform their specialized functions. For example, liver cells in G_0 still carry out metabolism and detoxification.

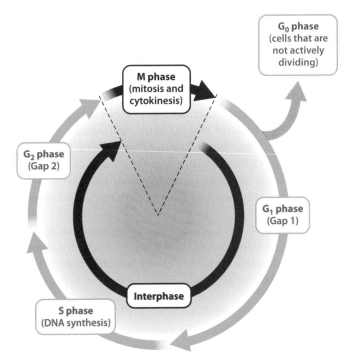

FIG. 11.2 The cell cycle. The eukaryotic cell cycle consists of M phase (mitosis and cytokinesis) and interphase.

Self-Assessment Questions

1. How is cell division similar and different in prokaryotic cells and eukaryotic cells?
2. What do you predict would be the consequence of a mutation in *FtsZ* that disrupts the function of the protein it encodes?

11.2 MITOTIC CELL DIVISION

Mitotic cell division (mitosis followed by cytokinesis) is the basis for asexual reproduction in single-celled eukaryotes, and the means by which an organism's cells, tissues, and organs develop and are maintained in multicellular eukaryotes. During mitosis and cytokinesis, the parental cell's DNA is replicated and passed on to two daughter cells. This process is continuous, but it is divided into discrete steps. These steps are marked by dramatic changes in the cytoskeleton and in the packaging and movement of the chromosomes.

The DNA of eukaryotic cells is organized as chromosomes.

One of the key challenges faced by a dividing eukaryotic cell is ensuring that the daughter cells receive an equal and complete set of chromosomes. The length of DNA contained in the nucleus of an average eukaryotic cell is on the order of 1 to 2 meters, well beyond the diameter of a cell. The DNA therefore is condensed to fit into the nucleus, and then the DNA is further condensed during cell division so that it does not become tangled as it segregates into daughter cells.

In eukaryotic cells, DNA is organized with histones and other proteins into chromatin, which is packaged to form the structures we know as chromosomes (Chapters 3 and 13). During interphase, chromosomes are long, thin, threadlike structures. One of the earliest events in mitosis is the condensing of chromosomes to short, dense forms that are identifiable under the microscope during M phase.

Every species is characterized by a specific number of chromosomes, and each chromosome contains a single molecule of DNA carrying a specific set of genes. When chromosomes condense and become visible during mitosis, they adopt characteristic shapes and sizes that allow each chromosome to be identified by its appearance in the microscope. The portrait formed by the number and shapes of chromosomes representative of a species is called its **karyotype**. Most of the cells in the human body, with the exception of the gametes, contain 46

FIG. 11.3 A human karyotype. This karyotype shows 22 pairs of chromosomes plus 2 sex chromosomes, or 46 chromosomes in total.
Source: ISM/Sovereign/Medical Images.

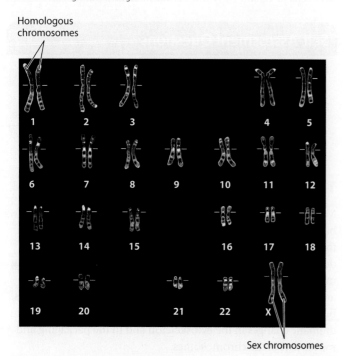

FIG. 11.4 Homologous chromosomes and sister chromatids. Sister chromatids result from the duplication of chromosomes. They are held together at the centromere.

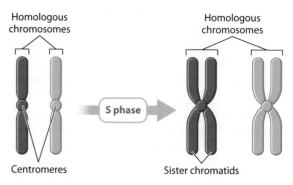

chromosomes (**Fig. 11.3**). In contrast, cells of horses have 64 chromosomes, and cells of corn have 20.

In a normal human karyotype, the 46 chromosomes can be arranged into 23 pairs, 22 pairs of chromosomes numbered 1 to 22 from the longest to the shortest chromosome and 1 pair of sex chromosomes (Fig. 11.3). Pairs of chromosomes of the same type carrying the same set of genes are said to be **homologous**. One chromosome in each pair was received from the mother and the other from the father. The sex chromosomes are the X and Y chromosomes. Individuals with two X chromosomes are female, and those with an X and a Y chromosome are male.

The number of complete sets of chromosomes in a cell is known as its ploidy. A cell with one complete set of chromosomes is **haploid**, and a cell with two complete sets of chromosomes is **diploid**. Some organisms, such as plants, can have four or sometimes more complete sets of chromosomes. Such cells are polyploid.

In order for cell division to proceed normally, every chromosome in the parent cell must be duplicated so that each daughter cell receives a full set of chromosomes. This duplication occurs during S phase. Even though the DNA in each chromosome duplicates, the two identical copies, called **sister chromatids**, do not separate. They stay side by side, physically held together at a constriction called the **centromere**. At the beginning of mitosis, the nucleus of a human cell still contains 46 chromosomes, but each chromosome is a pair of identical sister chromatids linked together at the centromere (**Fig. 11.4**). Thus, counting chromosomes is simply a matter of counting centromeres.

During mitosis, the sister chromatids separate from each other and go to opposite ends of the cell, so that each daughter cell receives the same number of chromosomes as is present in the parent cell, as we describe now.

Prophase: Chromosomes condense and become visible.

Mitosis takes place in five stages, each of which is easily identified by events that can be observed in the microscope

(**Fig. 11.5**). When you look in a microscope at a cell in interphase, you cannot distinguish specific chromosomes because they are long and thin. As the cell moves from G_2 phase to the start of mitosis, the chromosomes condense and become visible in the nucleus. The first stage of mitosis is known as **prophase** and is characterized by the appearance of visible chromosomes (Fig. 11.5, step 1).

Outside the nucleus, in the cytosol, the cell begins to assemble the **mitotic spindle**, a structure made up predominantly of microtubules that pull the chromosomes to opposite ends of the dividing cell. Recall from Chapter 10 that the **centrosome** is a compact structure that is the microtubule organizing center for animal cells. The centrosome is thus the structure from which the microtubules radiate. Plant cells also have a mitotic spindle formed of microtubules, but they lack centrosomes.

During S phase in animal cells, the centrosome duplicates and each one begins to migrate around the nucleus. The two centrosomes ultimately halt at opposite poles in the cell at the start of prophase. The final locations of the centrosomes define the opposite ends of the cell that will eventually be separated into two daughter cells. As the centrosomes make their way to the poles of the cell, tubulin dimers assemble around them, forming microtubules that radiate from each centrosome. These radiating filaments form the mitotic spindle and later serve as the guide wires for chromosome movement.

Prometaphase: Chromosomes attach to the mitotic spindle.

In the next stage of mitosis, known as **prometaphase**, the nuclear envelope breaks down and the microtubules of the mitotic spindle attach to chromosomes (Fig. 11.5, step 2). The microtubules radiating from the centrosomes grow and shrink as they explore the region of the cell where the nucleus once was. This process of growing and shrinking depends on the dynamic instability of microtubules, discussed in Chapter 10.

As the ends of the microtubules encounter chromosomes, they attach to the

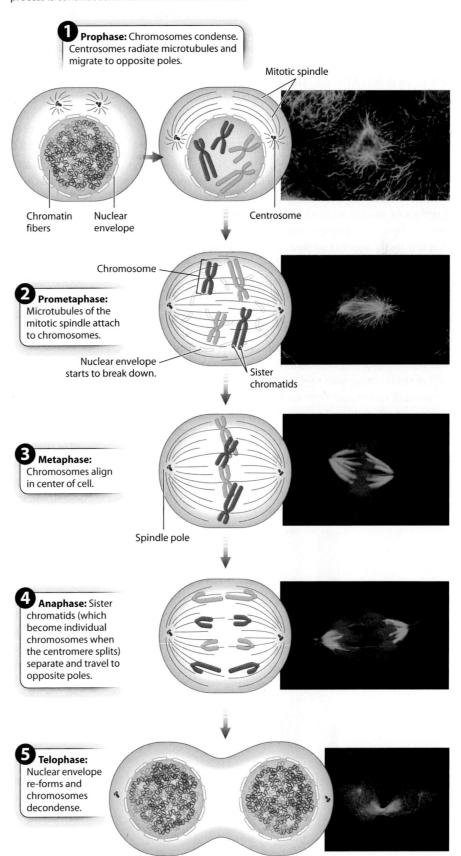

FIG. 11.5 Mitosis, or nuclear division. Mitosis can be divided into separate steps, but the process is continuous. *Source: Jennifer Waters/Science Source.*

chromosomes at their centromeres. Associated with the centromere of each chromosome are two protein complexes called **kinetochores**, one located on each side of the constriction (**Fig. 11.6**). Each kinetochore is associated with one of the two sister chromatids and forms the site of attachment for a single spindle microtubule. This arrangement ensures that each sister chromatid is attached to a spindle microtubule radiating from one of the poles of the cell. The symmetrical tethering of each chromosome to the two poles of the cell is essential for proper chromosome segregation.

Metaphase: Chromosomes align as a result of dynamic changes in the mitotic spindle.

Once each chromosome is attached to the mitotic spindles from both poles of the cell, the microtubules of the mitotic spindle lengthen or shorten to move the chromosomes to the middle of the cell. There the chromosomes are lined up in a single plane that is roughly equidistant from both poles of the cell. This stage of mitosis, when the chromosomes are aligned in the middle of the dividing cell, is called **metaphase** (see Fig. 11.5, step 3). It is one of the most visually distinctive stages under the microscope.

Anaphase: Sister chromatids fully separate.

In the next stage of mitosis, called **anaphase**, the sister chromatids separate (see Fig. 11.5, step 4). The centromere holding a pair of sister chromatids together splits, allowing the two sister chromatids to separate from each other. After separation, each chromatid is considered to be a full-fledged chromosome. The spindle microtubules attached to the kinetochores gradually shorten, pulling the newly separated chromosomes to the opposite poles of the cell.

After this event, the chromosomes are equally segregated between the two daughter cells. During S phase in a human cell, each of the 46 chromosomes is duplicated to yield 46 pairs of sister chromatids. Then, when the chromatids are separated at anaphase, an identical set of 46 chromosomes arrives at each spindle pole, the complete genetic material for one of the daughter cells.

Telophase: Nuclear envelopes re-form around newly segregated chromosomes.

Once a complete set of chromosomes arrives at a pole, the chromosomes have entered the area that will form the cytosol of a new daughter cell. This event marks the beginning of **telophase**. During this stage, the cell prepares for its division into two new cells (see Fig. 11.5, step 5). The microtubules of the mitotic spindle break down and disappear, while a nuclear envelope re-forms around each set of chromosomes, creating two new nuclei. As the nuclei become increasingly distinct in the cell, the chromosomes contained within them decondense, becoming less visible under the microscope. This stage marks the end of mitosis.

The parent cell divides into two daughter cells by cytokinesis.

Usually, as mitosis is nearing its end, cytokinesis begins and the parent cell divides into two daughter cells (**Fig. 11.7**). In animal cells, this stage begins when a ring of actin filaments, called the **contractile ring**, forms against the inner face of the cell membrane at the equator of the cell perpendicular to the axis of what was the spindle (Fig. 11.7a). As if pulled by a drawstring, the ring contracts, pinching the cytoplasm of the cell and dividing it in two. This process is similar to what occurs in binary fission, although in the case of binary fission, the process is driven by FtsZ protein, a homolog of tubulin, not by actin. The constriction of the contractile ring is driven by motor proteins that slide bundles of actin filaments in opposite directions. Successful division results in two daughter cells, each with its own nucleus. The daughter cells then enter G_1 phase and start the process again.

For the most part, mitosis is similar in animal and in plant cells, but cytokinesis is different (Fig. 11.7b). Since plant cells have a cell wall, the cell divides in two by constructing a new cell wall. During telophase, dividing plant cells form a structure called the **phragmoplast** in the middle of the cell. The phragmoplast consists of overlapping microtubules that guide vesicles containing cell wall components to the middle of the cell. During late anaphase and telophase, these vesicles fuse to form a new cell wall, called the cell plate, in the middle of the dividing cell. Once this developing cell wall is large enough, it fuses with the original cell wall at the perimeter of the cell. Cytokinesis is then complete and the plant cell has divided into two daughter cells.

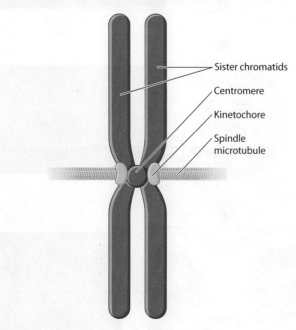

FIG. 11.6 Kinetochores. Kinetochores are sites of spindle attachment. There is one kinetochore on each side of the centromere.

FIG. 11.7 Cytokinesis, or cytoplasmic division. (a) In animal cells, a contractile ring made of actin constricts, pinching the cell in two. (b) In plant cells, a new cell wall called a cell plate grows between the two new cells. *Sources: a. Dr. Paul Andrews, University of Dundee/Science Source; b. Carolina Biological/ Medical Images.*

a. Animal cell cytokinesis

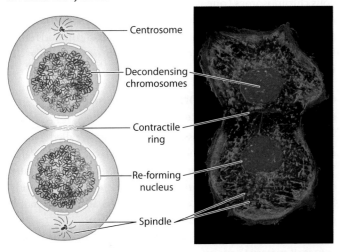

b. Plant cell cytokinesis

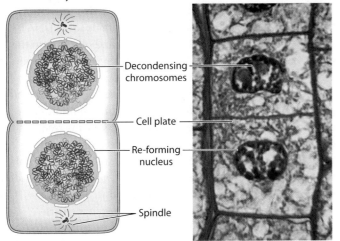

Self-Assessment Questions

3. Which DNA sequences are more alike: a pair of sister chromatids or a pair of homologous chromosomes?
4. What are three situations in which mitotic cell division occurs?
5. What are the five stages of mitosis? Draw the structure and position of the chromosomes at each stage.
6. What would be the consequence if a cell underwent mitosis but not cytokinesis?
7. How does cytokinesis differ between animal and plant cells?

11.3 MEIOTIC CELL DIVISION

As we discussed, mitotic cell division is important in the development of a multicellular organism and in the maintenance and repair of tissues and organs. Mitotic cell division is also the basis of asexual reproduction in unicellular eukaryotes. We now turn to the basis of sexual reproduction. In sexual reproduction, two gametes fuse during fertilization to form a new organism. This new organism has the same number of chromosomes as each of its parents because the two gametes each contain half the number of chromosomes as each diploid parent. Gametes are produced by **meiotic cell division,** a form of cell division that includes two rounds of nuclear division. By producing haploid gametes, meiotic cell division makes sexual reproduction possible.

There are several major differences between meiotic cell division and mitotic cell division. First, meiotic cell division results in four daughter cells instead of two. Second, each of the four daughter cells contains half the number of chromosomes as the parent cell. (The word "meiosis" is from the Greek for "diminish" or "lessen.") Third, each of the four daughter cells is genetically unique. In other words, they are genetically different from each other and from the parental cell.

In multicellular animals, the cells produced by meiosis are haploid eggs and sperm. In other organisms, such as fungi, the products are spores, and in some unicellular eukaryotes, the products are new organisms. In this section, we consider the steps by which meiosis occurs, its role in sexual reproduction, and how it likely evolved.

Pairing of homologous chromosomes is unique to meiosis.

Like mitotic cell division, meiotic cell division follows one round of DNA synthesis, but, unlike mitotic cell division, meiotic cell division consists of two successive cell divisions. The two cell divisions are called **meiosis I** and **meiosis II**, and they occur one after the other. Each cell division results in two cells, so that by the end of meiotic cell division a single parent cell has produced four daughter cells. During meiosis I, homologous chromosomes separate from each other, reducing the total number of chromosomes by half. During meiosis II, sister chromatids separate, as in mitosis.

Meiosis I begins with **prophase I**, illustrated in **Fig. 11.8**. The beginning of prophase I is when the condensing chromosomes first become visible. The chromosomes first appear as long, thin threads present throughout the nucleus. Because DNA replication has already taken place, each chromosome has become two sister chromatids held together at the centromere.

What happens next is an event of enormous importance, and it is unique to meiosis. The homologous chromosomes pair with each other, coming together to lie side by side, gene for gene, in a process known as **synapsis.** Even the *X* and *Y* chromosomes pair, but only at the tip, where their DNA sequences

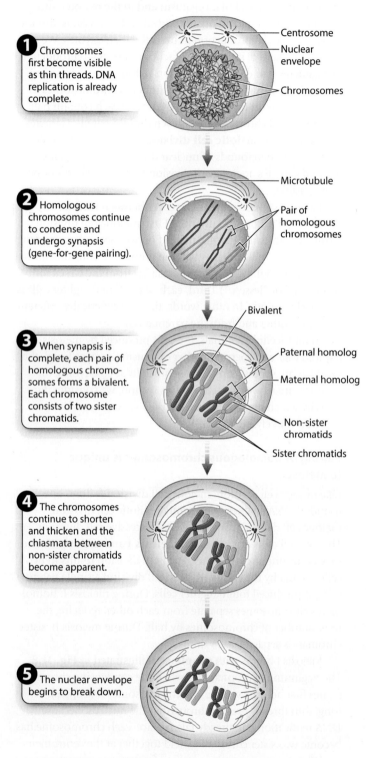

FIG. 11.8 Prophase I of meiosis. Chromosomes condense and homologous chromosomes pair.

① Chromosomes first become visible as thin threads. DNA replication is already complete.

② Homologous chromosomes continue to condense and undergo synapsis (gene-for-gene pairing).

③ When synapsis is complete, each pair of homologous chromosomes forms a bivalent. Each chromosome consists of two sister chromatids.

④ The chromosomes continue to shorten and thicken and the chiasmata between non-sister chromatids become apparent.

⑤ The nuclear envelope begins to break down.

Because each homologous chromosome is a pair of sister chromatids attached to a single centromere, a pair of synapsed chromosomes creates a four-stranded structure: two pairs of sister chromatids aligned along their length. The whole unit is called a **bivalent**, and the chromatids attached to different centromeres are called **non-sister chromatids** (Fig. 11.8). Non-sister chromatids result from the replication of homologous chromosomes (one is maternal and the other is paternal in origin), so they have the same set of genes in the same order, but are not genetically identical. By contrast, sister chromatids result from replication of a single chromosome, so are genetically identical.

Crossing over between DNA molecules results in exchange of genetic material.

Within the bivalents are crosslike structures, each called a **chiasma** (from the Greek meaning a "cross piece"; the plural is "chiasmata") (**Fig. 11.9**). Each chiasma is a visible manifestation of a **crossover**. A crossover results when non-sister chromatids break and then join in such a way that maternal and paternal genetic material is exchanged.

Through the process of crossing over, homologous chromosomes of maternal origin and paternal origin exchange DNA segments. The positions of these crossovers along the chromosome are essentially random, and therefore each chromosome that emerges from meiosis is unique: it contains some DNA segments from the maternal chromosome and others from the paternal chromosome. The process is very precise: usually, no nucleotides are gained or lost as homologous chromosomes exchange material. Occasionally, the exchange is imprecise and portions of the chromatids may be gained or lost, resulting in duplication or deletion of genetic material. Note the results of crossing over as shown in Fig. 11.9: the recombinant chromatids are those that carry partly paternal and partly maternal segments. By creating new combinations of genes in a chromosome, crossing over increases genetic diversity.

The number of chiasmata that form during meiosis depends on the species. In humans, the usual range is 50–60

FIG. 11.9 Chiasmata. Crossing over at chiasmata between non-sister chromatids results in recombinant chromatids.

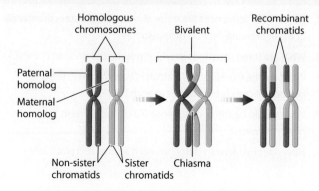

are nearly identical. Because one of each pair of homologs is maternal in origin and the other is paternal in origin, chromosome pairing provides an opportunity for the maternal and paternal chromosomes to exchange genetic information, as described in the next section.

chiasmata per meiosis. Most bivalents have at least one chiasma. Even the *X* and *Y* chromosomes are joined by a chiasma in the small region where they are paired. In addition to their role in exchanging genetic material, the chiasmata also hold the bivalents together while they become properly oriented in the center of the cell during metaphase, the stage we turn to next.

The first meiotic division reduces the chromosome number.

At the end of prophase I, the chromosomes are fully condensed and have formed chiasmata, the nuclear envelope has begun to disappear, and the meiotic spindle is forming. We are now ready to move through the remaining stages of meiosis I, which are illustrated in **Fig. 11.10**.

In **prometaphase I**, the nuclear envelope breaks down and the meiotic spindles attach to kinetochores on chromosomes (Fig. 11.10, step 2).

In **metaphase I**, the bivalents move until they end up on an imaginary plane cutting transversely across the spindle (Fig. 11.10, step 3). Each bivalent lines up so that its two centromeres lie on opposite sides of this plane, pointing toward opposite poles of the cell. Importantly, the orientation of these bivalents is random with respect to each other. For some, the maternal homolog is attached to the spindle radiating from one pole and the paternal homolog is attached to the spindle originating from the other pole. For others, the orientation is reversed. As a result, when the homologous chromosomes separate from each other in the next stage, a complete set of chromosomes moves toward each pole, and that chromosome set is a random mix of maternal and paternal homologs. The random alignment of maternal and paternal chromosomes in metaphase I further increases genetic diversity in the products of meiosis.

At the beginning of **anaphase I**, the two homologous chromosomes of each bivalent separate as they are pulled in opposite directions (Fig. 11.10, step 4). The key feature of anaphase I is that the centromeres do not split and the two chromatids that make up each chromosome remain together. Chromatids remain paired because spindle microtubules from one pole of the cell attach to both kinetochores of a given chromosome during prometaphase I. Anaphase I is thus very different from anaphase of mitosis, in which the centromeres split and each pair of chromatids separates.

Anaphase I ends as the chromosomes arrive at the poles of the cell. Only one of the two homologous chromosomes goes to each pole, so in human cells there are 23 chromosomes (the haploid number of chromosomes) at each pole at the end of meiosis I. Each of these chromosomes consists of two chromatids attached to a single centromere. Meiosis I is sometimes called the **reductional division** because it reduces the number of chromosomes in daughter cells by half.

In **telophase I**, the chromosomes may uncoil slightly and a nuclear envelope briefly reappears (Fig. 11.10, step 5). In many

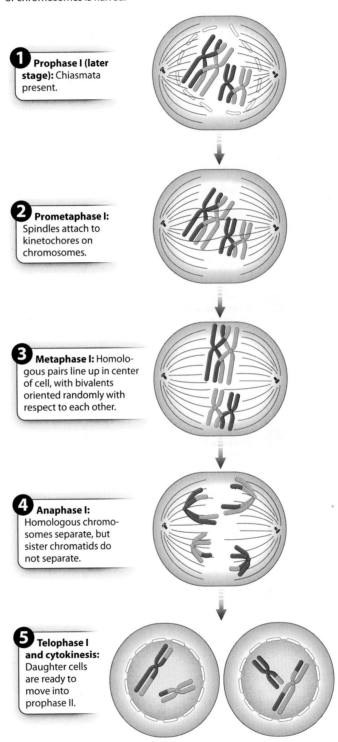

FIG. 11.10 Meiosis I. Meiosis I is the reductional division: the number of chromosomes is halved.

1. **Prophase I (later stage):** Chiasmata present.

2. **Prometaphase I:** Spindles attach to kinetochores on chromosomes.

3. **Metaphase I:** Homologous pairs line up in center of cell, with bivalents oriented randomly with respect to each other.

4. **Anaphase I:** Homologous chromosomes separate, but sister chromatids do not separate.

5. **Telophase I and cytokinesis:** Daughter cells are ready to move into prophase II.

species (including humans), the cytoplasm divides, producing two separate cells. The chromosomes do not completely decondense, however, and so telophase I blends into the first stage of the second meiotic division. Importantly, there is no DNA synthesis between the two meiotic divisions.

The second meiotic division resembles mitosis.

Now let's turn to the second meiotic division, meiosis II, shown in **Fig. 11.11**. In this division, sister chromatids separate. Starting with **prophase II**, the second meiotic division is in many respects like mitotic cell division, except that the nuclei in prophase II have the haploid number of chromosomes, not the diploid number. In prophase II, the chromosomes recondense to their maximum extent. Toward the end of prophase II, the nuclear envelope begins to disappear (in those species in which it has formed), and the spindle begins to be set up (Fig. 11.11, step 1).

FIG. 11.11 Meiosis II. Meiosis II is the equational division: the number of chromosomes stays the same.

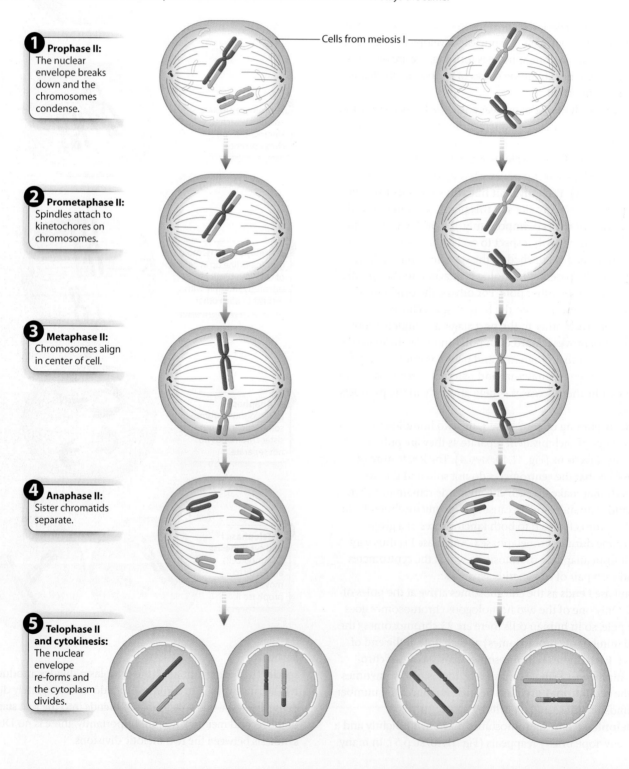

In **prometaphase II**, spindles attach to kinetochores (Fig. 11.11, step 2) and, in **metaphase II**, the chromosomes line up so that their centromeres lie on an imaginary plane cutting across the spindle (Fig. 11.11, step 3).

In **anaphase II**, the centromere of each chromosome splits. The separated chromatids, now each regarded as a full-fledged chromosome, are pulled toward opposite poles of the spindle (Fig. 11.11, step 4). In this way, anaphase II resembles anaphase of mitosis.

Finally, in **telophase II**, the chromosomes uncoil and become decondensed and a nuclear envelope re-forms around each set of chromosomes (Fig. 11.11, step 5). The nucleus of each cell now has the haploid number of chromosomes. Because cells in meiosis II have the same number of chromosomes at the beginning and at the end of the process, meiosis II is also called the **equational division**. Telophase II is followed by the division of the cytoplasm in many species.

A comparison of mitosis and meiosis gives us hints about how meiosis might have evolved (**Fig. 11.12** and **Table 11.1**). During meiosis I, maternal and paternal homologs separate from each other, whereas during meiosis II, sister chromatids separate from each other, similar to mitosis. The similarity of meiosis II and mitosis suggests that meiosis likely evolved from mitosis. Mitosis occurs in all eukaryotes and was certainly present in the common ancestor of all living eukaryotes. Meiosis is present in most, but not all, eukaryotes. Because the steps of meiosis are the same in all eukaryotes, meiosis is thought to have evolved in the common ancestor of all eukaryotes and has been subsequently lost in some groups.

Division of the cytoplasm often differs between the sexes.

In multicellular organisms, division of the cytoplasm in meiotic cell division differs between the sexes. In female mammals (**Fig. 11.13a**), the cytoplasm is divided very unequally in both meiotic divisions. Most of the cytoplasm is retained in one meiotic product, a very large cell called the oocyte. This large cell can develop into the functional egg cell. The other meiotic products receive only small amounts of cytoplasm and are smaller cells called **polar bodies**. In male mammals (**Fig. 11.13b**), the cytoplasm divides about equally in both meiotic divisions, and each of the resulting meiotic products goes on to form a functional sperm. During the development of sperm, most of the cytoplasm is eliminated, and what is left is essentially a nucleus in the sperm head equipped with a long whiplike flagellum to help propel it toward the egg.

TABLE 11.1 Comparison of Mitosis and Meiosis

	MITOSIS	MEIOSIS
Function	Asexual reproduction in unicellular eukaryotes Development in multicellular eukaryotes Tissue regeneration and repair in multicellular eukaryotes	Sexual reproduction Production of gametes and spores
Organisms	All eukaryotes	Most eukaryotes
Number of rounds of DNA synthesis	1	1
Number of cell divisions	1	2
Number of daughter cells	2	4
Chromosome complement of daughter cell compared with parent cell	Same	Half
Pairing of homologous chromosomes	No	Meiosis I—Yes Meiosis II—No
Crossing over	No	Meiosis I—Yes Meiosis II—No
Separation of homologous chromosomes	No	Meiosis I—Yes Meiosis II—No
Centromere splitting	Yes	Meiosis I—No Meiosis II—Yes
Separation of sister chromatids	Yes	Meiosis I—No Meiosis II—Yes

FIG. 11.12 Comparison of mitosis and meiosis.

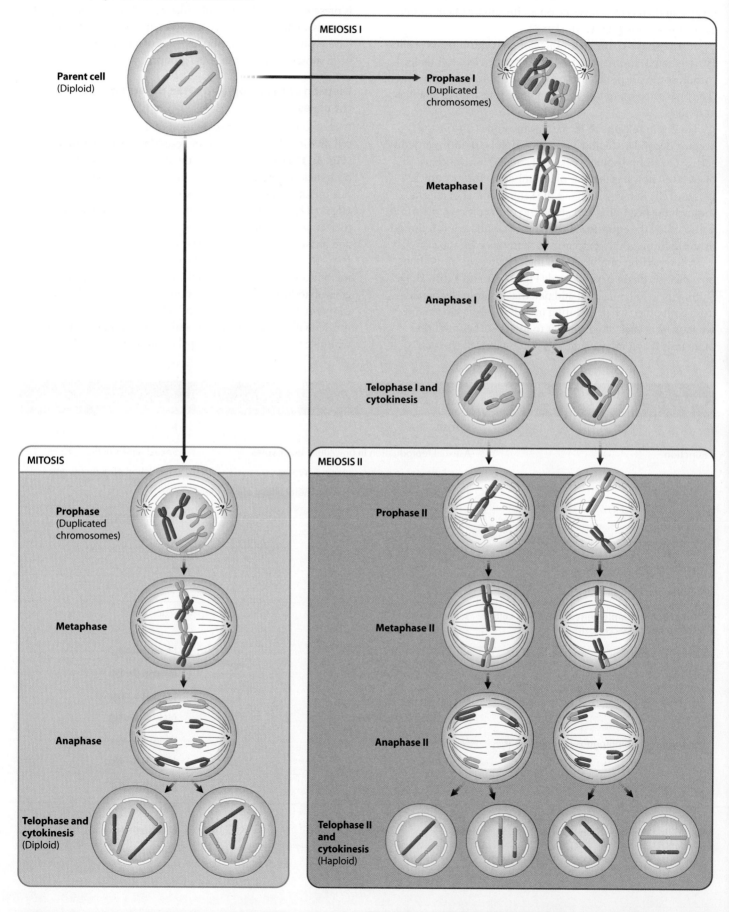

FIG. 11.13 Cytoplasmic division in females and in males. (a) In females, cytoplasmic division results in one oocyte and three polar bodies; (b) in males, it results in four sperm cells.

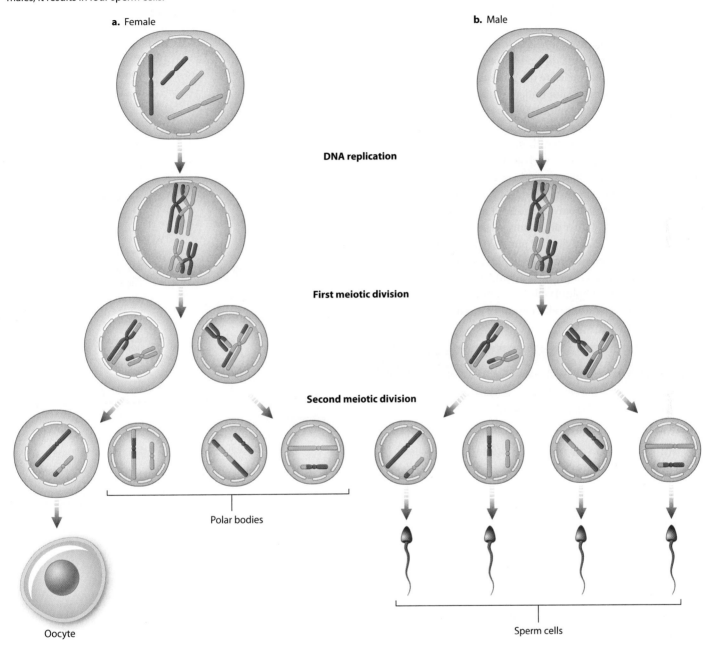

Meiosis is the basis of sexual reproduction.

Sexual reproduction involves two processes: meiotic cell division and fertilization. Meiotic cell division, as we just saw, produces cells with half the number of chromosomes present in the parent cell. In multicellular animals, the products of meiotic cell division are gametes: an egg cell is a gamete and a sperm cell is a gamete. Each gamete is haploid, containing a single set of chromosomes. In humans, meiosis takes place in the ovaries of the female and in the testes of the male. Each resulting gamete contains 23 chromosomes, including one each of the 22 numbered chromosomes plus either an *X* or a *Y* chromosome.

During fertilization, these gametes fuse to form a single cell called a **zygote**. The zygote is diploid, having two complete sets of chromosomes, one from each parent. Therefore, fertilization restores the original chromosome number.

As we discuss further in Chapters 15 and 40, sexual reproduction plays a key role in increasing genetic diversity. Genetic diversity results from both meiotic cell division and fertilization. The cells produced by meiotic cell division are each genetically different from one another as a result of crossing over and the random alignment and segregation of maternal and paternal chromosomes. In fertilization, different gametes are combined to produce a new, unique individual. Sexual reproduction therefore increases genetic diversity, allowing organisms to evolve and adapt more quickly to their environment than is possible with asexual reproduction.

In the life cycle of multicellular animals such as humans, then, the diploid organism produces single-celled haploid gametes that fuse to make a diploid zygote. In this case, the only haploid cells in the life cycle are the gametes, and the products of meiotic cell division do not undergo mitotic cell division but instead fuse to become a diploid zygote. However, there are a number of life cycles in other organisms that differ in the timing of meiotic cell division and fertilization, discussed more fully in Chapter 25. Some organisms, such as most fungi, are haploid. These haploid cells can fuse to produce a diploid zygote, but this cell immediately undergoes meiotic cell division to produce haploid cells, so that the only diploid cell in the life cycle is the zygote (Chapter 32). Other organisms, such as plants, have both multicellular haploid and diploid phases (Chapter 28). In this case, meiotic cell division produces haploid spores that divide by mitotic cell division to produce a multicellular haploid phase; subsequently, haploid cells fuse to form a diploid zygote that also divides by mitotic cell division to produce a multicellular diploid phase.

FIG. 11.14 Nondisjunction. (a) Chromosomes separate evenly into gametes in normal meiotic cell division. (b) Homologous chromosomes fail to separate in first-division nondisjunction. (c) Sister chromatids fail to separate in second-division nondisjunction.

a. Normal meiosis

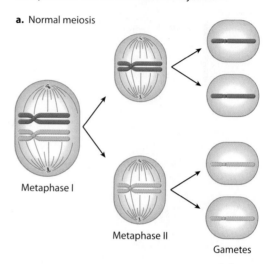

b. First-division nondisjunction

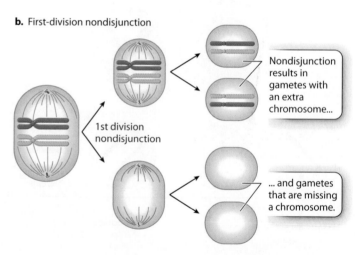

c. Second-division nondisjunction

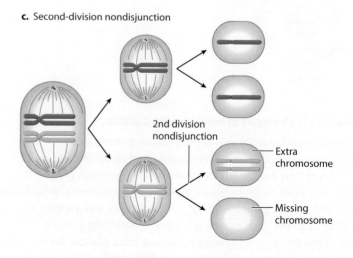

Self-Assessment Questions

8. In a human cell at the end of prophase I, how many chromatids, centromeres, and bivalents are present?
9. What are three ways in which meiosis I differs from mitosis?
10. What two processes during meiosis result in genetically unique daughter cells?

11.4 NONDISJUNCTION

In some cases, errors during mitosis and meiosis can lead to cells with extra or missing chromosomes. These occur quite commonly in humans. They are infrequently observed, however, because most of them are lethal. Nevertheless, a few major chromosomal differences are found in the general population. Some are common enough that their effects are familiar. Others are less common, and a few have no major effects at all. In this section, we focus on the types of difference that are observed, what causes them, and their effects.

Nondisjunction in meiosis results in extra or missing chromosomes.

Nondisjunction is the failure of a pair of chromosomes to separate during anaphase of cell division. The result is that one daughter cell receives an extra copy of one chromosome, and the other daughter cell receives no copy of that chromosome. Nondisjunction can take place in mitosis, and it leads to cell lineages with extra or missing chromosomes, which are often observed in cancer cells (section 11.6).

When nondisjunction takes place in meiosis, the result is reproductive cells (gametes) that have either an extra or a missing chromosome. **Fig. 11.14** illustrates nondisjunction in meiosis. Fig. 11.14a shows key stages in a normal meiosis, in which pairs of homologous chromosomes separate in the first division and sister chromatids separate in the second, resulting in gametes that each have one and only one copy of each chromosome.

Failure of chromosome separation can occur in either of the two meiotic divisions. Both types of nondisjunction result in gametes with an extra chromosome or a missing chromosome, but there is an important difference. In **first-division non-disjunction** (Fig. 11.14b), which is much more common, the homologous chromosomes fail to separate and all of the resulting gametes have an extra or missing chromosome. On the other hand, in **second-division nondisjunction** (Fig. 11.14c), sister chromatids fail to separate, giving rise to gametes with an extra chromosome, a missing chromosome, and normal numbers of chromosomes.

Some human disorders result from nondisjunction.

Approximately 1 out of 155 live-born children (0.6%) has a major chromosome abnormality of some kind. About half of these have a major abnormality in chromosome structure. The others have a chromosome that is present in an extra copy or a chromosome that is missing because of nondisjunction—a parent has produced an egg or a sperm with an extra or a missing chromosome.

Perhaps the most familiar of the conditions resulting from nondisjunction is **Down syndrome,** which results from the presence of an extra copy of chromosome 21. Down syndrome is also known as **trisomy 21** because affected individuals have three copies of chromosome 21. The extra chromosome can be seen in the chromosome layout called a karyotype (**Fig. 11.15**). The shorthand for the condition is 47,+21, meaning that the individual has 47 chromosomes with an extra copy of chromosome 21.

The presence of an extra chromosome 21 leads to a set of traits that, taken together, constitute a syndrome. These traits result from an increase in the number of genes and other elements present in chromosome 21, and some of these genetic elements have been identified. Children with Down syndrome are mentally disabled to varying degrees; most are in the mild to moderate range and with special education and support they can acquire basic communication, self-help, and social skills. People with Down syndrome are short because of delayed maturation of the skeletal system, their muscle tone is low, and they have a characteristic facial appearance. About 40% of children with Down syndrome have major heart defects, and life expectancy is currently about 55 years. Among adults with Down syndrome, the risk of dementia is about 15% to 50%, which is about 5 times greater than the risk in the general population.

Although Down syndrome affects approximately 1 in 750 live-born children overall, there is a pronounced effect of mother's age on the risk of Down syndrome. For mothers of ages 45–50, the risk of having a baby with Down syndrome is approximately 50 times greater than it is for mothers of ages 15–20. The increased risk is the reason that many physicians recommend tests

FIG. 11.15 Trisomy 21. This karyotype shows three copies of chromosome 21, which causes Down syndrome. *Source: Biophoto Associates/Science Source.*

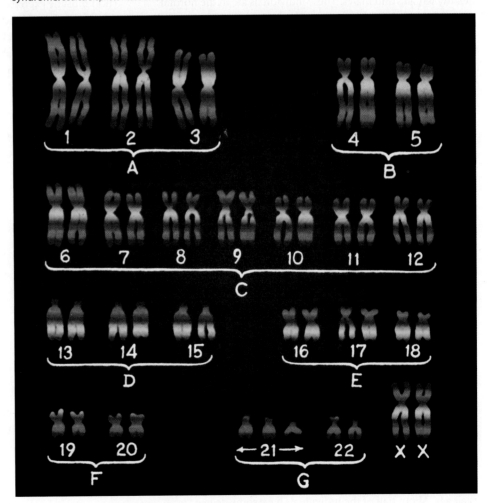

to determine abnormalities in the fetus for pregnant women over the age of 35.

Two other trisomies of autosomes (chromosomes other than the X and Y chromosomes) are sometimes found in live births. These are trisomy 13 and trisomy 18. Both are much rarer than Down syndrome and the effects are more severe. In both cases, the developmental abnormalities are so profound that newborns with these conditions usually do not survive the first year of life. Other trisomies of autosomes also occur, but they are not compatible with life.

Extra or missing sex chromosomes have fewer effects than extra autosomes.

Children born with an abnormal number of autosomes are rare, but extra or missing sex chromosomes (X or Y) are relatively common among live births. In most mammals, females have two X chromosomes (XX), and males have one X and one Y chromosome (XY). The presence of the Y chromosome, not the number of X chromosomes, leads to male development. The female karyotype 47, XXX (47 chromosomes, including three X chromosomes) and the male karyotype 47, XYY (47 chromosomes, including one X and two Y chromosomes) are found among healthy females and males. These individuals are in the normal range of physical development and mental capability, and usually their extra sex chromosome remains undiscovered until their chromosomes are examined for some other reason.

The reason that 47, XYY males show no detectable effects is the unusual nature of the Y chromosome. The Y chromosome contains the gene that stimulates the embryo to take the male developmental pathway and few other genes, so individuals do not have many extra genes, as is the case in Down syndrome. The absence of detectable effects in 47, XXX females has a completely different explanation that has to do with the manner in which the activity of genes in the X chromosome is regulated. In the cells of female mammals, all X chromosomes except one are inactivated and gene expression is largely repressed. (This process, called X-inactivation, is discussed in Chapter 19.) Because of X-inactivation, a 47, XXX female has one active X chromosome per cell, the same number as in a 46, XX female.

Two other sex chromosome abnormalities do have effects, especially on sexual development (**Fig. 11.16**). The more common is 47, XXY (47 chromosomes, including two X chromosomes and one Y chromosome). Individuals with this karyotype are male because of the presence of the Y chromosome, but they have a distinctive group of symptoms known as **Klinefelter syndrome** (Fig. 11.16a). Characteristics of the syndrome are very small testes but normal penis and scrotum. Growth and physical development, for the most part, are normal, although affected individuals tend to be tall for their age. About half of the individuals have some degree of mental impairment. When a person with Klinefelter syndrome reaches puberty, the testes fail to enlarge, the voice remains high pitched, pubic and facial hair remains sparse, and there is some enlargement of the breasts. Affected males do not produce sperm and hence are sterile; up to 10% of men who seek aid at infertility clinics turn out to be XXY.

Much rarer than all the other sex chromosome abnormalities is 45, X (45 chromosomes, with just one X chromosome), the karyotype associated with characteristics known as **Turner syndrome** (Fig. 11.16b). Affected females are short and often have a distinctive webbing of the skin between the neck and shoulders. There is no sexual maturation. The external sex organs remain immature, the breasts fail to develop, and pubic hair fails to grow. Internally, the ovaries are small or absent, and menstruation does not occur. Although the mental abilities of affected females are very nearly normal, they have specific defects in spatial abilities and arithmetical skills.

Earlier we saw that in XX females one X chromosome undergoes inactivation. Why, then, do individuals with just one X chromosome show any symptoms at all? The explanation is that some genes on the inactive X chromosome in XX females are not completely inactivated. Evidently, for normal development to take place, both copies of some genes that escape complete inactivation must be expressed.

Trisomies and the sex-chromosome abnormalities account for most of the simpler chromosomal abnormalities that are found among babies born alive. However, these chromosomal abnormalities represent only a minority of those that actually occur. In

FIG. 11.16 Symptoms associated with (a) Klinefelter syndrome and (b) Turner syndrome.

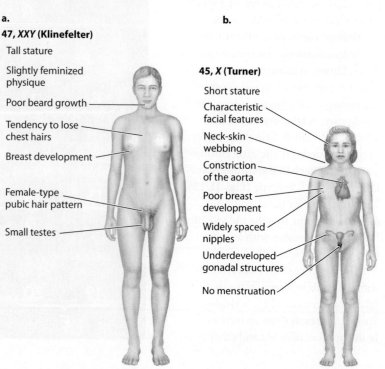

fertilized eggs with extra or missing chromosomes, the embryos usually fail to complete development. At some time during pregnancy—in some cases very early, in other cases relatively late—the chromosomally abnormal embryo or fetus undergoes spontaneous abortion.

> **Self-Assessment Questions**
>
> 11. How does nondisjunction in meiosis I or II result in extra or missing chromosomes? Make a diagram to answer the question.
>
> 12. A male baby is born with the sex-chromosome constitution *XYY*. Both parents have normal sex chromosomes (*XY* in the father, *XX* in the mother). In which meiotic division of which parent did the nondisjunction take place that produced the *XYY* baby?

11.5 CELL CYCLE REGULATION

Cell division occurs only at certain times and places. Mitotic cell division, for example, occurs as a multicellular organism grows, a wound heals, or cells are replaced in actively dividing tissues such as the skin or lining of the intestine. Similarly, meiotic cell division occurs only at certain times during development. Even for unicellular organisms, cell division takes place only when conditions are favorable—for example, when enough nutrients are present in the environment. In all these cases, a cell may have to receive a signal before it divides. In Chapter 9, we saw how cells respond to signals. Growth factors, for example, bind to cell-surface receptors and activate intracellular signaling pathways that lead to cell division.

Even when a cell receives a signal to divide, it does not divide until it is ready. Has all of the DNA been replicated? Has the cell grown large enough to support division into viable daughter cells? If these and other preparations have not been accomplished, the cell halts its progression through the cell cycle.

So, cells have regulatory mechanisms that initiate cell division, as well as mechanisms for spotting faulty or incomplete preparations and arresting cell division. When these mechanisms fail—for example, dividing in the absence of a signal or when the cell is not ready—the result may be uncontrolled cell division, a hallmark of cancer. In this section, we consider how cells control their passage through the cell cycle. Our focus is on control of mitotic cell division. However, many of the same factors also regulate meiotic cell division, a reminder of the close evolutionary connection between these two forms of cell division.

Protein phosphorylation controls passage through the cell cycle.

Early animal embryos, such as those of frogs and sea urchins, are useful model organisms for studying cell cycle control because they are large and undergo many rapid mitotic cell divisions following fertilization. During these rapid cell divisions, mitosis and S phase alternate with virtually no G_1 and G_2 phases in between. Studies of animal embryos revealed two interesting patterns. First, as the cells undergo this rapid series of divisions, several proteins appear and disappear in a cyclical fashion. Researchers hypothesized that these proteins might play a role in controlling the progression through the cell cycle. Second, several enzymes become active and inactive in cycles. These enzymes are kinases, proteins that phosphorylate other proteins (Chapter 9). The timing of kinase activity is delayed slightly relative to the appearance of the cyclical proteins.

These and many other observations led to the following view of cell cycle control. Proteins are synthesized that activate the kinases. These regulatory proteins are called **cyclins** because their levels rise and fall with each turn of the cell cycle. Once activated by cyclins, the kinases phosphorylate target proteins that promote cell division (**Fig. 11.17**). These kinases, called **cyclin-dependent kinases** (CDKs), are always present within the cell but are active only when bound to the appropriate

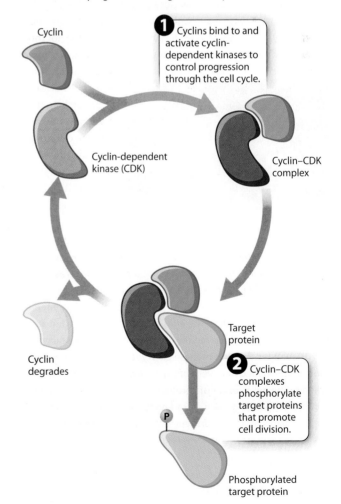

FIG. 11.17 Cyclins and cyclin-dependent kinases (CDKs). Cyclins and CDKs control progression through the cell cycle.

cyclin. It is the kinase activity of the cyclin–CDK complexes that triggers cell cycle events. In sum, the cyclical activity of the cyclin–CDK complexes depends on the cyclical levels of the cyclins. Cyclins in turn may be synthesized in response to signaling pathways that promote cell division.

These cell cycle proteins, including cyclins and CDKs, are widely conserved across eukaryotes, reflecting their fundamental role in controlling cell cycle progression, and have been extensively studied in yeast, sea urchins, mice, and humans (**Fig. 11.18**).

HOW DO WE KNOW?

FIG. 11.18
How is progression through the cell cycle controlled?

BACKGROUND Many components of the cell undergo cyclical changes during the cell cycle: chromosomes condense and decondense, the mitotic spindle forms and breaks down, and the nuclear envelope breaks down and re-forms. How these regular and cyclical changes are controlled is the subject of active research.

The first clues came from studies of yeast. Mutations in certain yeast genes blocked progression through the cell cycle, suggesting that these genes encode proteins that normally promote progression through the cell cycle. Additional clues came from studies of embryos of the sea urchin *Arbacia punctulata*. These embryos are large and divide rapidly by mitosis, so they are a good model system for the study of cell division. In the early 1980s, it was known that inhibition of protein synthesis blocks cell division in sea urchins. It was also known that an enzyme called MPF, or M-phase promotion factor, is necessary for the transition from G_2 to M phase.

SOURCE: *SeaPics.com*.

EXPERIMENT In the 1980s, English biochemist Tim Hunt and colleagues measured protein levels in sea urchin embryos as they divide rapidly by mitosis. The researchers added radioactive methionine (an amino acid) to eggs, which became incorporated into newly synthesized proteins. The eggs were then fertilized and allowed to develop. Samples of the rapidly dividing embryos were taken every 10 minutes. Extracts of these embryos were made. These extracts were loaded into wells of a gel. An electrical charge was then applied to the gel, causing the proteins to migrate toward one end of the gel and separating them according to their charge and size. This technique allowed the researchers to visualize the levels of different proteins.

RESULTS Most protein bands became darker as cell division proceeded, indicating more and more protein synthesis. However, the level of one protein band oscillated, increasing in intensity and then decreasing with each cell cycle, as shown in the graph.

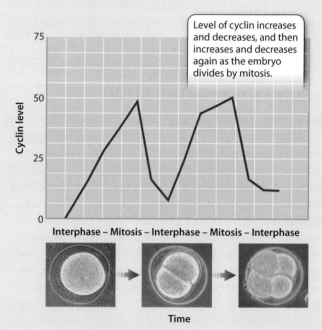

PHOTOS SOURCE: *Biology Pics/Science Source*.

CONCLUSION Hunt and colleagues called this oscillating protein "cyclin." Although they did not know its function, its fluctuating levels suggested it might play a role in the control of the cell cycle. More work was needed to figure out if cyclins actually cause progression through the cell cycle or if cyclin levels fluctuate in response to progression through the cell cycle.

FOLLOW-UP WORK Hunt and colleagues found a similar rise and fall in the levels of certain proteins in the sea urchin *Lytechinus pictus* and surf clam *Spisula solidissima* that they had found in *Arbacia punctulata*. MPF was found to consist of a cyclin protein and a cyclin-dependent kinase (CDK) that play a key role in the G_2–M transition. Together, these experiments suggested that cyclins control progression through the cell cycle by their interactions with cyclin-dependent kinases. Hunt shared the Nobel Prize in Physiology or Medicine in 2001 for his work on cyclins.

SOURCE T. Evans et al. 1983. "Cyclin: A Protein Specified by Maternal mRNA in Sea Urchin Eggs that is Destroyed at each Cleavage Division." *Cell* 33:389–396.

Different cyclin–CDK complexes regulate each stage of the cell cycle.

In mammals, there are several different cyclins and CDKs that act at specific steps of the cell cycle. Three steps in particular are regulated by cyclin–CDK complexes in all eukaryotes (**Fig. 11.19**).

During G_1, levels of cyclin D and cyclin E rise and activate CDKs, which allows the cell to enter S phase. These cyclin–CDK complexes activate transcription factors that lead to the expression of histone proteins needed for packaging the newly replicated DNA strands. They also result in expression and activation of DNA polymerase and the other enzymes that function during DNA replication.

During S phase, levels of cyclin A rise, activating CDKs that inhibit activity of the DNA synthesis enzymes once DNA replication is complete, which prevents the replication proteins from reassembling at the same place and re-replicating the same DNA sequence. Synthesizing the same sequence repeatedly would be dangerous because cells with too much DNA could die or become cancerous.

During G_2, cyclin B levels rise and activate CDKs that activate enzymes that initiate multiple events associated with mitosis. For example, cyclin B–CDK complex phosphorylates structural proteins in the nucleus, triggering the breakdown of the nuclear envelope in prometaphase. Cyclin B–CDK also phosphorylates proteins that regulate the assembly of tubulin into microtubules, promoting the formation of the mitotic spindle.

Cell cycle progression requires successful passage through multiple checkpoints.

Cyclins and CDKs not only allow a cell to progress through the cell cycle, but also give the cell opportunities to halt the cell cycle should something go wrong. For example, the preparations for the next stage of the cell cycle may be incomplete or there may be some kind of damage. In these cases, there are mechanisms that block the cyclin–CDK activity required for the next step, pausing the cell cycle until preparations are complete or the damage is repaired. Each of these mechanisms is called a **checkpoint**.

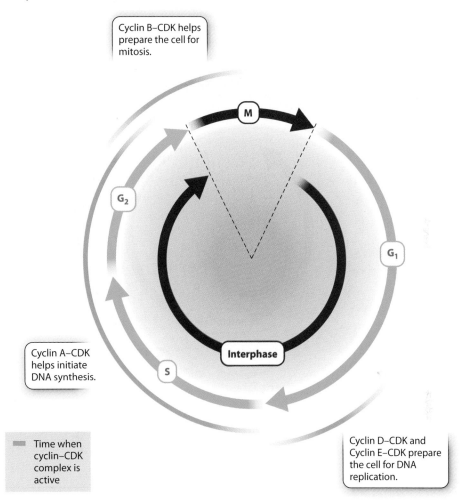

FIG. 11.19 Cyclin–CDK complexes. Cyclin–CDK complexes are active at different times in the cell cycle.

Cells have many cell cycle checkpoints. **Fig. 11.20** shows three major checkpoints. The presence of damaged DNA arrests the cell at the end of G_1 before DNA synthesis, the presence of unreplicated DNA arrests the cell at the end of G_2 before the cell enters mitosis, and abnormalities in chromosome attachment to the spindle arrest the cell in early mitosis. By way of illustration, we focus on the checkpoint that occurs at the end of G_1 in response to the presence of damaged DNA.

DNA can be damaged by environmental insults such as ultraviolet radiation or chemical agents (Chapter 14). Typically, damage takes the form of double-stranded breaks in the DNA. If the cell progresses through mitosis with DNA damage, daughter cells might inherit the damage, or the chromosomes might not segregate normally. The checkpoint late in G_1 delays progression through the cell cycle until DNA damage is repaired. This DNA damage checkpoint depends on several regulatory proteins: some of these recognize damaged DNA, whereas others arrest cell cycle progression before S phase.

FIG. 11.20 Cell cycle checkpoints. Cell cycle checkpoints monitor key steps in the cell cycle.

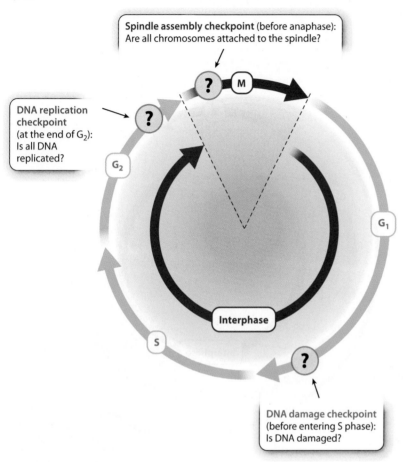

When DNA is damaged by radiation, a specific protein kinase is activated that phosphorylates a protein called p53. Phosphorylated p53 binds to DNA, where it turns on the expression of several genes. One of these genes codes for a protein that binds to and blocks the activity of the G_1/S cyclin–CDK complexes (**Fig. 11.21**). In so doing, p53 arrests the cell at the G_1/S transition, giving the cell time to repair the damaged DNA. The p53 protein therefore is a good example of a protein involved in halting the cell cycle when the cell is not ready to divide. Because of its role in protecting the genome from accumulating DNA damage, p53 is sometimes called the "guardian of the genome." Mutations in p53 are common in cancer, a topic we turn to next.

Self-Assessment Questions

13. What are the roles of cyclins and cyclin-dependent kinases in the cell cycle?

14. What are three examples of checkpoints that the cell monitors before proceeding through the cell cycle?

FIG. 11.21 DNA damage checkpoint controlled by p53.

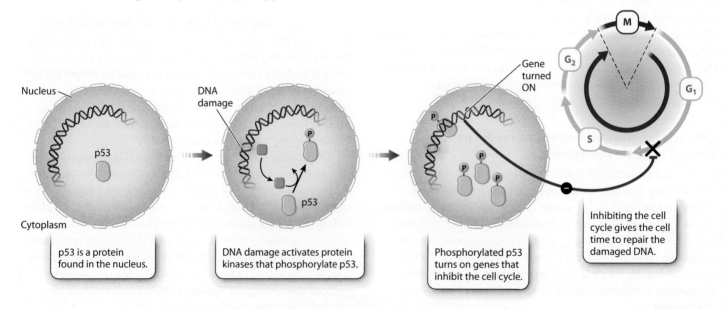

11.6 CANCER

CASE 2 CANCER: CELL SIGNALING, FORM, AND DIVISION

What genes are involved in cancer?

As we have seen, cells have evolved mechanisms that promote passage through the cell cycle, such as the cyclin–CDK complexes, as well as mechanisms that halt the cell cycle when the cell is not ready to divide, such as the DNA damage checkpoint that depends on the p53 protein. When cellular mechanisms that promote cell division are activated inappropriately or the normal checks on cell division are lost, cells may divide uncontrollably. The result can be cancer, a group of diseases characterized by uncontrolled cell division.

In this section, we explore various ways in which cell cycle control can fail and lead to cancer. We begin with a discussion of a cancer caused by a virus. Although viruses are not the cause of most cancers, the study of these cancers helped us to understand how cancers develop.

Oncogenes promote cancer.

Our understanding of cancer is based partly on early observations of cancers in animals. In the first decade of the twentieth century, Peyton Rous studied cancers called sarcomas in chickens (**Fig. 11.22**). His work and that of others led to the discovery of the first virus known to cause cancer in animals, named the Rous sarcoma virus.

All viruses have a protein coat surrounding a core of either RNA or DNA (Chapter 1). Viruses multiply by infecting cells and using the biochemical machinery of the host cell to synthesize viral components, which are assembled to make more copies of themselves. The investigation of cancer-causing viruses provided major insights into how cancers form. Because viruses typically carry only a handful of genes, it is relatively easy to identify which of those genes is a factor in causing cancer.

As discussed in Chapter 9, growth factors activate several types of proteins inside the cell that promote cell division. The gene from the Rous sarcoma virus that promotes uncontrolled cell division encodes an overactive protein kinase similar to the kinases in the cell that function as signaling proteins. This viral gene is named *v-src,* for *viral-src* (pronounced "sarc" and short for "sarcoma," the type of cancer it causes).

The *v-src* gene is one of several examples of an **oncogene,** or cancer-causing gene, found in viruses. A real surprise was the discovery that the *v-src* oncogene is found not just in the Rous sarcoma virus. It is an altered version of a gene normally found in the host animal cell, known as *c-src* (*cellular-src*). The *c-src* gene plays a role in the normal control of cell division during embryonic development.

Proto-oncogenes are genes that when mutated may cause cancer.

The discovery that the *v-src* oncogene has a normal counterpart in the host cell was an important step toward identifying the cellular genes that participate in cell growth and division. These normal cellular genes are called **proto-oncogenes.** They are involved in cell division. Nearly every protein that performs a key step in a signaling cascade that promotes cell division can be the product of a proto-oncogene. These include growth factors, cell-surface receptors, G proteins, and protein kinases. Each of these can be mutated to become oncogenes. However, proto-oncogenes do not themselves cause cancer. Only when they are mutated do they have the potential to cause cancer. Today, we know of scores of proto-oncogenes. Most were identified through the study of cancer-causing viruses in chickens, mice, and cats.

Oncogenes also play a major role in human cancers. Most human cancers are not caused by viruses. Instead, human proto-oncogenes can be mutated into cancer-causing oncogenes by environmental agents such as chemical pollutants. For example, organic chemicals called aromatic amines present in cigarette smoke can enter cells and damage DNA, resulting in mutations that can convert a proto-oncogene into an oncogene.

Let's consider an example of a proto-oncogene. In Chapter 9, we discussed platelet-derived growth factor, or PDGF. This protein binds to and dimerizes a receptor kinase in the membrane of the target cell. The dimerized receptor activates several signaling pathways that together promote cell division. In one type of leukemia (a cancer of blood cells), a mutation occurs in the gene that encodes the PDGF receptor. The mutant receptor is missing the extracellular portion needed to bind the growth factor and dimerizes on its own, independent of PDGF binding. The receptor is therefore always turned on. The overactive PDGF receptor activates too many target proteins over too long a time period, leading to uncontrolled proliferation of blood cells.

Tumor suppressors block specific steps in the development of cancer.

Up to this point, we have considered what happens when mechanisms that promote cell division are inappropriately activated. Now let's consider what happens when mechanisms that usually prevent cell cycle progression are removed.

Earlier, we discussed cell cycle checkpoints that halt the cell cycle until the cell is ready to divide. One of these checkpoints depends on the p53 protein, which normally arrests cell division in response to DNA damage (section 11.4). When the p53 protein is mutated or its function is inhibited, the cell can divide before the DNA damage is repaired. Such a cell continues to divide in the presence of damaged DNA, leading to the accumulation of mutations that promote cell division. The p53 protein is mutated in many types of human cancer, highlighting its critical role in regulating the cell cycle.

The p53 protein is one example of a **tumor suppressor.** Tumor suppressors are proteins whose normal activities inhibit cell division. Some tumor suppressors participate in cell cycle checkpoints, as is the case for p53. Other tumor suppressors

HOW DO WE KNOW?

FIG. 11.22

Can a virus cause cancer?

BACKGROUND In the early 1900s, little was known about the cause of cancer or the nature of viruses. Peyton Rous, an American pathologist, studied a form of cancer called a sarcoma in chickens. He moved a cancer tumor from a diseased chicken to a healthy chicken and found that the cancer could be transplanted. Rous then decided to conduct a further experiment to determine if a cell-free extract of the tumor injected into a healthy chicken could also induce cancer.

HYPOTHESIS Experiments in other organisms, such as mice, rats, and dogs, showed that an extract free of cells from a tumor does not cause cancer. Therefore, Rous hypothesized that a cell-free extract of the chicken sarcoma would not induce cancer in healthy chickens.

EXPERIMENT Rous took a sample of the sarcoma from a chicken, ground it up, suspended it in solution, and centrifuged it to remove the debris. In his first experiment, he injected this extract into a healthy chicken. In a second experiment, he passed the extract through a filter to remove all cells, including cancer cells and bacterial cells, and then injected this cell-free extract into a healthy chicken.

RESULTS In both cases, healthy chickens injected with the extract from the sarcoma developed cancer at the site of injection. Microscopic examination of the cancer showed it to be the same type of cancer as the cancer from the original chicken.

CONCLUSION AND FOLLOW-UP WORK Because the extract that did not contain any cells did cause cancer in chickens, Rous rejected his hypothesis. He concluded that an agent smaller than a cell, such as a virus or chemical, is capable of causing cancer. Later experiments confirmed that the cause of chicken sarcoma is a virus. This result was surprising, controversial, and dismissed at the time. A second cancer-causing virus was not found until the 1930s. Although most cancers are not caused by viruses, work with cancer-causing viruses helped to identify cellular genes that, when mutant, can lead to cancer. In 1966, Rous shared the Nobel Prize in Physiology or Medicine for his discovery.

SOURCES Rous, P. 1910. "A Transmissable Avian Neoplasm (Sarcoma of the Common Fowl)." *J. Exp. Med.* 12:696–705; Rous, P. 1911. "Transmission of a malignant new growth by means of a cell-free filtrate." *JAMA* 56:198.

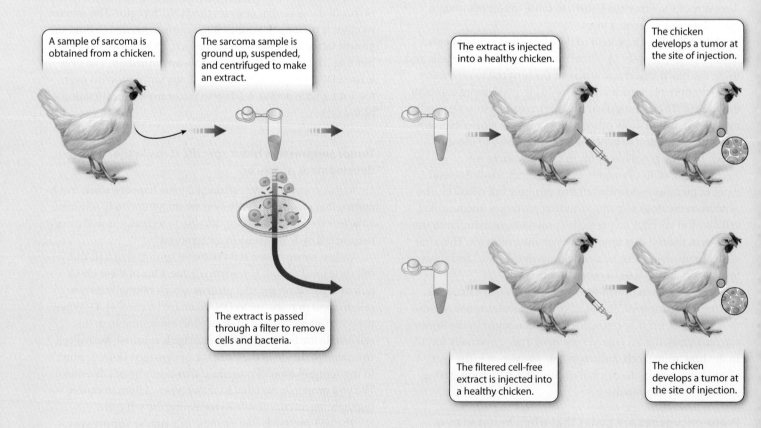

repress the expression of genes that promote cell division, and still others trigger cell death.

Tumor suppressors act in opposition to proto-oncogenes. Therefore, whether a cell divides or not depends on the activities of both proto-oncogenes and tumor suppressors:
proto-oncogenes must be turned on and tumor suppressors must be turned off for a cell to divide. Given the importance of controlling cell division, it is not surprising that cells have two counterbalancing systems that must be in agreement before cell division takes place.

Most cancers require the accumulation of multiple mutations.

Most human cancers require more than the overactivation of one oncogene or the inactivation of a single tumor suppressor. Given the multitude of different tumor suppressor proteins that are produced in the cell, it is likely that one will compensate for even the complete loss of another. Cells have evolved a redundant arrangement of these control mechanisms to ensure that cell division is properly regulated.

When several different cell cycle regulators fail, leading to both the overactivation of oncogenes and the loss of tumor suppressor activity, cancer will likely develop. The cancer may be either benign or malignant. A benign cancer is relatively slow growing and does not invade the surrounding tissue, whereas a malignant cancer grows rapidly and invades surrounding tissues. In many cases of malignant colon cancer, for example, tumor cells contain at least one overactive oncogene and several inactive tumor suppressor genes (**Fig. 11.23**). The gradual accumulation of these mutations over a period of years can be correlated with the stepwise progression of the cancer from a benign to a malignant cancer.

Taken together, we can now define some of the key characteristics that make a cell cancerous. Uncontrolled cell division is certainly important, but given the communities of cells and extracellular matrix we have discussed over the last few chapters, we should also consider additional characteristics. In 2000, American biologists Douglas Hanahan and Robert Weinberg highlighted key features of cancer cells. These include the ability to divide on their own in the absence of growth signals, resistance to signals that inhibit cell division or promote cell death, the ability to invade local and distant tissues (metastasis), and the production of signals to promote new growth of blood vessels for delivering nutrients to support cell division.

Considered in this light, a cancer cell is one that no longer plays by the "rules" of a normal cellular community. Cancer therefore serves to remind us of the normal controls and processes that allow cells to exist in a community. These processes, which work together, are summarized in **Fig. 11.24** on the following pages.

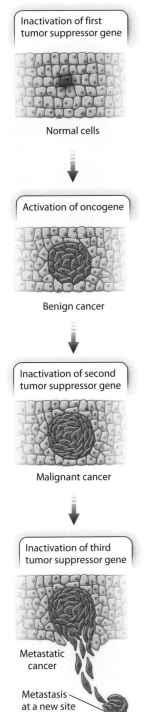

FIG. 11.23 Multiple-mutation model for the development of cancer. Cells that become cancerous typically carry mutations in several different genes.

> ### Self-Assessment Questions
>
> 15. How do oncogenes, proto-oncogenes, and tumor suppressor genes differ from one another?
>
> 16. What are two ways in which the function of p53 can be disrupted?

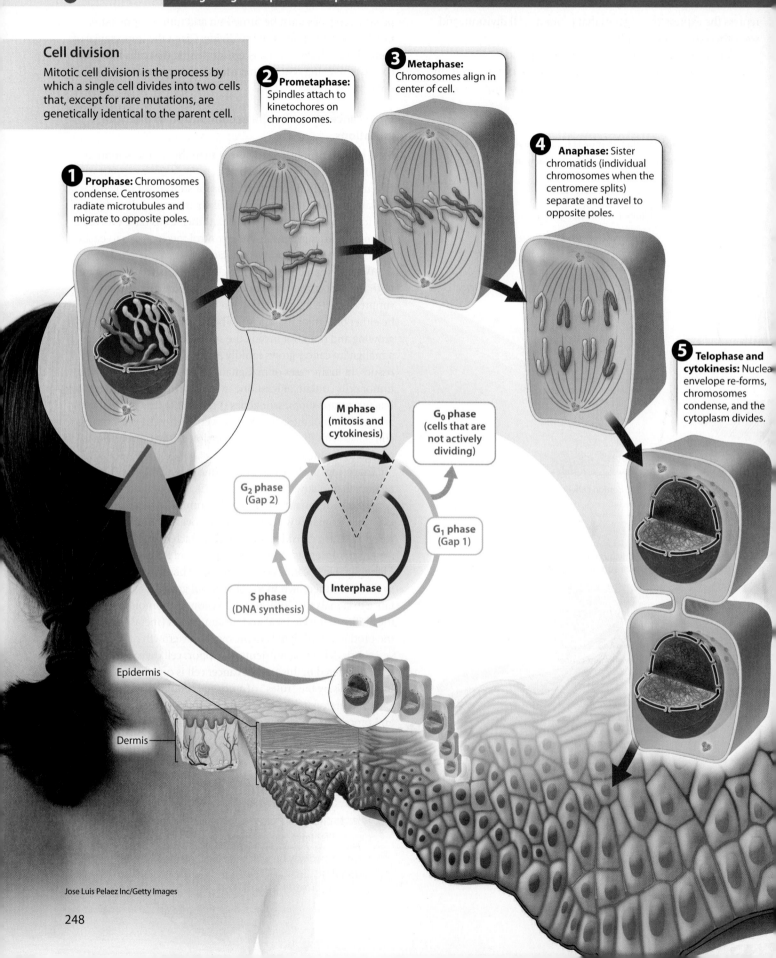

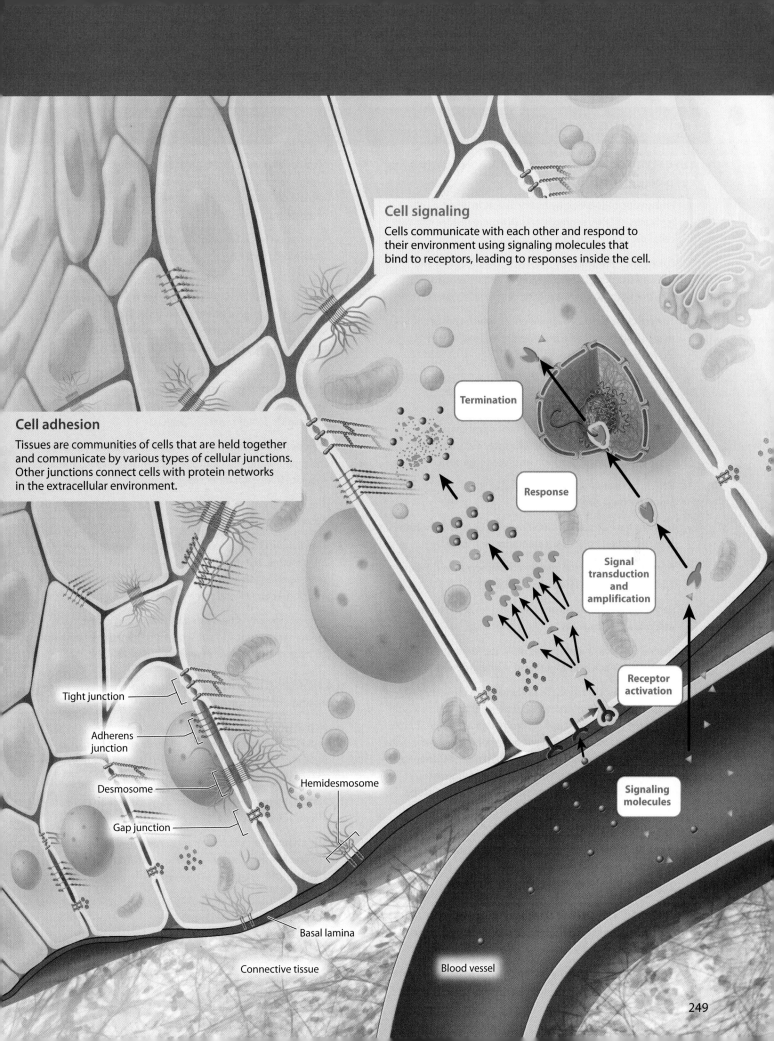

CORE CONCEPTS SUMMARY

11.1 CELL DIVISION: During cell division, a single parental cell produces two daughter cells.

Prokaryotic cells divide by binary fission, in which a cell replicates its DNA, segregates its DNA, and divides into two cells. page 225

Eukaryotic cells divide by mitosis (nuclear division) and cytokinesis (cytoplasmic division). Together, mitosis and cytokinesis are known as mitotic cell division. page 226

The cell cycle is divided into M phase, which consists of mitosis and cytokinesis, and interphase, which consists of G_1, S (synthesis), and G_2 phases. page 227

Cells that do not need to divide exit the cell cycle and are in G_0. page 227

11.2 MITOTIC CELL DIVISION: Mitotic cell division produces genetically identical daughter cells and is the basis for asexual reproduction and development.

DNA in a eukaryotic cell is packaged as linear chromosomes. page 228

Humans have 23 pairs of chromosomes, or 46 chromosomes total, including 1 pair of sex chromosomes. page 228

Homologous chromosomes are chromosomes of the same type with the same set of genes. page 228

Each parent contributes one complete set of 23 chromosomes at fertilization. page 228

During S phase, chromosomes replicate, forming a pair of sister chromatids held together at the centromere. page 228

Mitosis consists of five steps following DNA replication: (1) prophase—the chromosomes condense and become visible under the light microscope, (2) prometaphase— the spindle microtubules attach to the centromeres, (3) metaphase— the chromosomes line up in the middle of the cell, (4) anaphase— the centromeres split and the chromosomes move to opposite poles, and (5) telophase—the nuclear envelope re-forms and chromosomes decondense. page 228

Mitosis is followed by cytokinesis, in which one cell divides into two. In animals, a contractile ring of actin pinches the cell in two. In plants, a new cell wall, called the cell plate, is synthesized between the daughter cells. page 230

11.3 MEIOTIC CELL DIVISION: Meiotic cell division produces genetically unique daughter cells and is the basis for sexual reproduction.

Sexual reproduction involves meiosis and fertilization, both of which are important in increasing genetic diversity. page 231

Meiotic cell division is a form of cell division that reduces the number of chromosomes by half. It produces haploid gametes or spores that have one copy of each chromosome. page 231

Fertilization is the fusion of haploid gametes to produce a diploid cell. page 231

Meiosis consists of two successive cell divisions: the first is reductional (the chromosome number is halved), and the second is equational (the chromosome number stays the same). Each division consists of prophase, prometaphase, metaphase, anaphase, and telophase. page 231

In meiosis I, homologous chromosomes pair and exchange genetic material at chiasmata, or regions of crossing over. In contrast to mitosis, centromeres do not split and sister chromatids do not separate. page 232

Meiosis II is similar to mitosis in that chromosomes align in the center of the cell, centromeres split, and sister chromatids separate from each other. page 234

The similarity of meiosis II and mitosis suggests that meiosis evolved from mitosis. page 235

The division of the cytoplasm differs between the sexes: male meiotic cell division produces four functional sperm cells, whereas female meiotic cell division produces a single functional egg cell and three polar bodies. page 235

Genetic diversity is generated by (1) crossing over and (2) random alignment and subsequent segregation of maternal and paternal homologs in meiosis I. page 237

11.4 NONDISJUNCTION: Nondisjunction results in extra or missing chromosomes.

Nondisjunction is the failure of a pair of chromosomes to separate during anaphase of cell division. It can occur in mitosis or meiosis and results in daughter cells with extra or missing chromosomes. page 238

Nondisjunction in meiosis leads to gametes with extra or missing chromosomes. page 239

Most human fetuses with extra or missing chromosomes spontaneously abort, but some are viable and exhibit characteristic syndromes. page 239

Trisomy 21 (Down syndrome) is usually characterized by three copies of chromosome 21. page 239

Extra or missing sex chromosomes (X or Y) are common and result in syndromes such as Klinefelter syndrome (47, XXY) and Turner syndrome (45, X). page 240

11.5 CELL CYCLE REGULATION: The cell cycle is regulated so that cells divide only at appropriate times and places.

- Cells have regulatory mechanisms for promoting and preventing cell division. page 241

- Levels of proteins called cyclins increase and decrease during the cell cycle. page 241

- Cyclins form complexes with cyclin-dependent kinases (CDKs), activating the CDKs to phosphorylate target proteins that promote cell division. page 241

- Cyclin–CDK complexes are activated by signals that promote cell division. page 242

- Different cyclin–CDK complexes control progression through the cell cycle at key steps, including G_1/S phase, S phase, and M phase. page 243

- The cell has checkpoints that halt progression through the cell cycle if something is not right. page 243

- The p53 protein is an example of a checkpoint protein; it prevents cell division in the presence of DNA damage. page 244

11.6 CANCER: Cancer is uncontrolled cell division that can result from mutations in genes that regulate cell division.

- Cancer results when mechanisms that promote cell division are inappropriately activated or the normal checks on cell division are lost. page 245

- Cancers can be caused by certain viruses carrying oncogenes that promote uncontrolled cell division. page 245

- Viral oncogenes have cellular counterparts called proto-oncogenes that play normal roles in cell growth and division and that, when mutated, can cause cancer. page 245

- Oncogenes and proto-oncogenes often encode proteins involved in signaling pathways that promote cell division. page 245

- Tumor suppressors encode proteins, such as p53, that block cell division and inhibit cancers. page 245

- Cancers usually result from several mutations in proto-oncogenes and tumor suppressor genes that have accumulated over time within the same cell. page 247

Log in to **LaunchPad** to check your answers to the Self-Assessment Questions and to access additional learning tools.

CASE 3

Your Personal Genome

You, from A to T

Answers in the genetic sequence. Claudia Gilmore found that her genome contains a mutation that can lead to breast cancer. *Source: Photo by Dominic Gutierrez, courtesy Claudia Gilmore.*

> *Mutations that increase the risk of developing a particular disease are called genetic risk factors.*

As an American studies major at Georgetown University, Claudia Gilmore had plenty of experience taking exams. But at the age of 21, she faced an altogether different kind of test—and no amount of studying could have prepared her for the result. Gilmore had decided to be tested for a mutation in a gene known as *BRCA1*.

Specific mutations in the *BRCA1* and *BRCA2* genes are associated with an increased risk of breast and ovarian cancers.

In 2003, scientists published the first draft of the complete human genome. The human genome consists of a complete set of 23 chromosomes containing about 3 billion base pairs of DNA. Since then, genetic testing has become increasingly common and in some cases routine.

Gilmore's grandmother had battled breast cancer and later died from ovarian cancer. Before her death, she had tested positive for the *BRCA1* mutation. Her son, Gilmore's father, was tested and discovered he had inherited the mutation. After talking with a genetic counselor to help her understand the implications of the test, Gilmore gave a sample of her own blood and crossed her fingers.

"I had a fifty-fifty chance of inheriting the mutation. I knew there was a great possibility it would be a part of my future," she says. "But I was 21, I was healthy. A part of me also thought this could never really happen to me."

Unfortunately, it could. Two weeks after her blood was drawn, she learned that she, too, carried the mutated gene.

In reality, there isn't one single human genome. Everyone on Earth (with the exception of identical twins) has his or her own unique genetic sequence. Your personal genome, consisting of two sets of chromosomes, is the blueprint that codes for your hair color, the length of your nose, and your susceptibility to certain diseases. On average, any two human genomes are 99.9% identical, meaning that they differ at about 3 million base pairs.

Oftentimes, those differences have no impact on health. In some cases, however, a particular genetic signature is associated with disease. Sometimes a mutation makes a given illness inevitable. Certain mutations in a gene called *HTT*, for instance, always result in Huntington's disease, a degenerative brain condition that usually appears in middle age.

The link between mutations and disease isn't always so clear-cut, however. Most genetic diseases are complex, and it may take multiple mutations as well as other nongenetic factors for the disease to develop. Mutations that increase the risk of developing a particular disease are called genetic risk factors.

Certain mutations in the *BRCA1* and *BRCA2* genes are known genetic risk factors for breast and ovarian cancers, for instance. But not everyone with these mutations develops cancer. According to the National Cancer Institute, about 60% of women with a harmful *BRCA1* mutation will develop breast cancer in her lifetime, compared with about 12% of women in the general population. And 15% to 40% of women with a *BRCA1* mutation will be diagnosed with ovarian cancer, compared with just 1.4% of women without that genetic change.

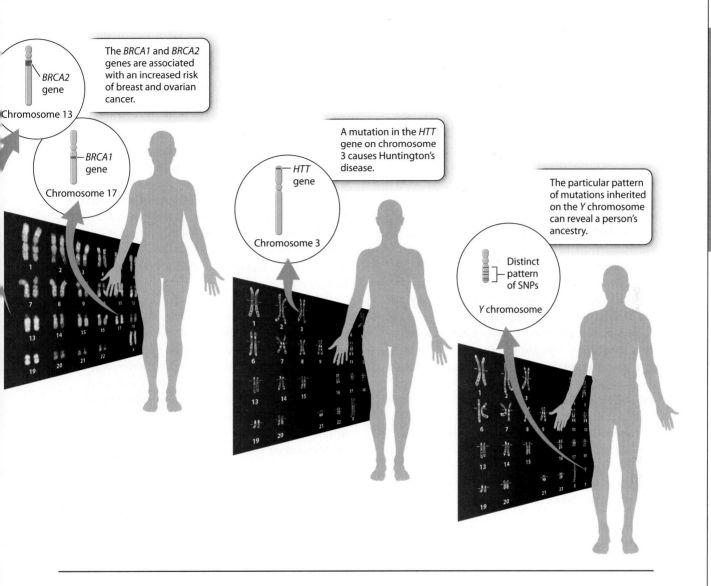

What does your genome say about you? Researchers have identified genes that, when mutated, contribute to disease, as well as differences that help reveal who you are and where you come from. Sources: (left to right) Willatt/Science Source; ISM/Sovereign/Medical Images, ISM/Sovereign/Medical Images.

The *BRCA1* mutations are just some of the thousands of harmful genetic changes that geneticists have identified so far. Other common mutations have been shown to elevate one's risk of developing heart disease, diabetes, various cancers, and numerous other common illnesses. Often, these mutations involve changes to just a single base pair of DNA. These common single-letter mutations are known as single nucleotide polymorphisms, or SNPs.

Several companies now offer genetic tests directly to the public. Unlike tests such as the *BRCA1* blood test that Gilmore was given, these direct-to-consumer (DTC) tests are offered to customers without any involvement of a medical professional, at a cost of a few hundred dollars.

Some of these tests aren't related to health. Your personal genome contains sequences that help to determine many unique features—from the shape of your fingernails to the shade of your skin—that don't affect your health but make you in part the person you are. Some DTC testing services aim to tell customers about their history by screening genes to identify variations that are more common in certain geographical regions or among members of certain ethnic groups. One such company offers genetic tests to African-Americans to determine in what part of the African continent their ancestors originated.

Other popular DTC tests inspect DNA samples for SNPs associated with certain diseases and physical

CASE 3

traits—everything from Parkinson's disease and age-related macular degeneration to earwax type and propensity for baldness.

Advocates of the tests say the technology puts the power of genetic information in the hands of consumers. Critics, on the other hand, argue that the information provided by DTC tests isn't always very meaningful. The presence of some SNPs might raise the risk of a rare disease by just 2% or 3%, for example. In many cases, the precise link between mutation and disease is still being sorted out. Also, without input from a genetic counselor or medical professional, consumers may not know how to interpret the information revealed by the tests. The American Medical Association has recommended that a doctor always be involved when any genetic testing is performed.

Moreover, knowing your genetic risk factors isn't the whole story. When it comes to your health and well-being, the environment also plays a significant role. Someone might override a genetic predisposition for skin cancer by using sunscreen faithfully every day. On the other hand, a person might have a relatively low genetic risk for type 2 diabetes, but still boost the odds of developing the disease by eating a poor diet and getting little physical exercise.

Claudia Gilmore is especially careful to exercise regularly and eat a healthy diet. Still, she can only control her environment to a degree. She knew that, given her genetic status, her risk of breast cancer remained high. She made the extraordinary decision to eliminate that risk by undergoing a mastectomy at the age of 23. It wasn't an easy decision, she says, but she feels privileged to have been able to take proactive steps to protect her health. "I'll be a 'previvor' instead of a survivor," she says.

Two years later the same medical choice was made by the American actress, film director, screenwriter, and author Angelina Jolie. She recounted her experience in *The New York Times,* writing, "Once I knew that this [*BRCA1* mutation] was my reality, I decided to be proactive and to minimize the risk as much I could. I made a decision to have a preventive double mastectomy. I started with the breasts, as my risk of breast cancer is higher than my risk of ovarian cancer, and the surgery is more complex. . . . I am writing about it now because I hope that other women can benefit from my experience."[1]

An unintended consequence of Jolie's public disclosure was that some women interpreted it as an endorsement. Immediately afterward, there was a brief surge in *BRCA1* and *BRCA2* testing. Most of the women tested had no family history of breast cancer and were unlikely to carry either the *BRCA1* or *BRCA2* mutation. Indeed, most of the tests were negative. The cost to the women was about $3000 per test, but a more important cost may have been a false sense of security. Even in the absence of the *BRCA1* or *BRCA2* mutations, women have about a 12% lifetime risk of breast cancer.[2]

We have entered an era of personal genomics. Already, doctors can design medical treatments based on a patient's personal genome. For example, people with a certain genetic profile are less likely than others to benefit from statins, medications prescribed to lower cholesterol. Doctors are also choosing which cancer drugs to prescribe based on the unique genetic signatures of patients and their tumors. Although there is much left to decipher, it's clear that each of our individual genomes contains a wealth of biological information. As Gilmore says, "I've always been taught that knowledge is power."

[1] Jolie, A. 2013. "My Medical Choice." *New York Times*, May 14.

[2] Johnson, C. Y. 2016. "The Unintended Consequence of Angelina Jolie's Viral Breast Cancer Essay." *Washington Post*, Wonkblog, December 15.

 DELVING DEEPER INTO CASE 3

For each of the Cases, we provide a set of questions to pique your curiosity. These questions cannot be answered directly from the Case itself, but instead introduce issues that will be addressed in chapters to come. We revisit this Case in Chapters 12–19 on the pages indicated below.

1. What new technologies are used to sequence your personal genome? See page 273.
2. Why sequence your personal genome? See page 284.
3. What can your personal genome tell you about your genetic risk factors for cancer? See page 307.
4. How do genetic tests identify disease risk factors? See page 339.
5. How can the *Y* chromosome be used to trace ancestry? See page 356.
6. How can mitochondrial DNA be used to trace ancestry? See page 358.
7. Can personalized medicine lead to effective treatments of common diseases? See page 373.
8. How do lifestyle choices affect expression of your personal genome? See page 386.
9. Can cells with your personal genome be reprogrammed for new therapies? See page 403.

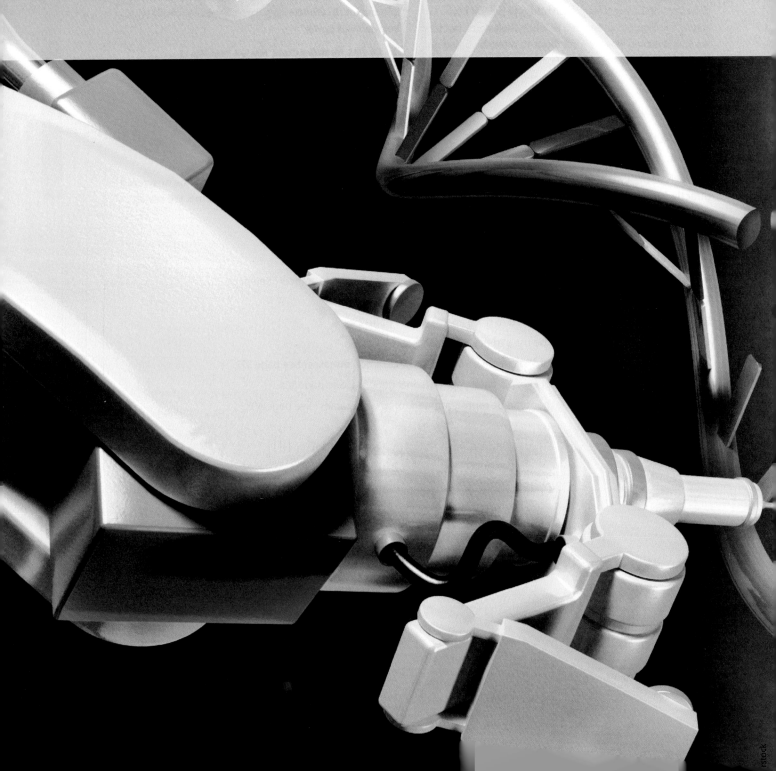

CHAPTER 12   DNA Replication and Manipulation

CORE CONCEPTS

12.1 DNA REPLICATION: In DNA replication, a single parental molecule of DNA produces two daughter molecules.

12.2 REPLICATION OF CHROMOSOMES: The replication of linear chromosomal DNA requires mechanisms that ensure efficient and complete replication.

12.3 DNA TECHNIQUES: Techniques for manipulating DNA follow from the basics of DNA structure and replication.

12.4 GENETIC ENGINEERING: Genetic engineering allows researchers to alter DNA sequences in living organisms.

One of the overarching themes of biology is that the functional unit of life is the cell, a theme that rests, in turn, on the fundamental concept that all cells come from preexisting cells (Chapter 1). In Chapter 11, we considered the mechanics of cell division and how it is regulated. Prokaryotic cells divide by binary fission, whereas eukaryotic cells divide by mitosis and cytokinesis. These processes ensure that cellular reproduction results in daughter cells that are like the parental cell. In other words, in cell division, like begets like. The molecular basis for the resemblance between parental cells and daughter cells is that a double-stranded DNA molecule in the parental cell duplicates and gives rise to two double-stranded daughter DNA molecules that are identical to each other, except for rare, uncorrected mutations that may have taken place.

The process of duplicating a DNA molecule is called **DNA replication**, and it occurs in virtually the same way in all organisms, reflecting its evolution very early in life's history. The process is conceptually simple—the parental strands separate and new complementary partner strands are made—but the molecular details are more complicated. Once scientists understood the molecular mechanisms of DNA replication, they could devise methods for manipulating and studying DNA. An understanding of DNA replication is therefore fundamental to understanding not only how cells and organisms produce offspring like themselves, but also some of the key experimental techniques in modern biology.

12.1 DNA REPLICATION

You may recall from Chapter 3 that double-stranded DNA consists of a pair of deoxyribonucleotide polymers wound around each other in antiparallel helical coils in such a way that a purine base (A or G) in one strand is paired with a pyrimidine base (T or C, respectively) in the other strand. To say that the strands are antiparallel means that they run in opposite directions: one strand runs in the 5'-to-3' direction and the other runs in the 3'-to-5' direction. These key elements of DNA structure are the only essential pieces of information needed to understand the mechanism of DNA replication.

During DNA replication, the parental strands separate and new partners are made.

When Watson and Crick published their paper describing the structure of DNA, they also coyly laid claim to another discovery: "It has not escaped our notice that the specific pairing we have postulated [A with T, and G with C] immediately suggests a copying mechanism for the genetic material." The copying mechanism they had in mind is exquisitely simple. The two strands of the parental duplex molecule separate (**Fig. 12.1**), and each individual parental strand serves as a model, or **template strand**, for the synthesis of a **daughter strand**. As each daughter strand is synthesized, the order of the bases in the template strand determines the order of the complementary bases added to the daughter strand. For example, the sequence 5'-ATGC-3' in the template strand specifies the sequence 3'-TACG-5' in the daughter strand because A pairs with T and G pairs with C. (The designations 3' and 5' convey the antiparallel orientation of the strands.)

A key prediction of the model shown in Fig. 12.1 is **semiconservative replication**. That is, after replication, each new DNA duplex consists of one

FIG. 12.1 DNA replication. During DNA replication, each parental strand serves as a template for the synthesis of a complementary daughter strand.

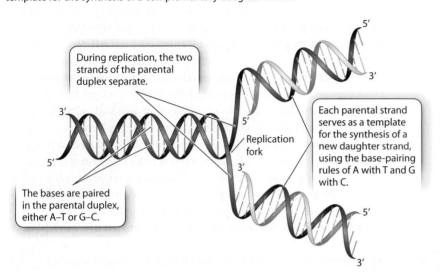

During replication, the two strands of the parental duplex separate.

Each parental strand serves as a template for the synthesis of a new daughter strand, using the base-pairing rules of A with T and G with C.

Replication fork

The bases are paired in the parental duplex, either A–T or G–C.

by Meselson and Stahl after two rounds of replication and exactly as predicted by the semiconservative replication model. The result also demonstrates that each eukaryotic chromosome contains a single DNA molecule that runs continuously all along its length.

New DNA strands grow by the addition of nucleotides to the 3′ end.

Although replication is semiconservative, this model alone does not tell us the details of replication. For example, the model implies that both daughter strands should grow in length by the addition of nucleotides near the site where the parental strands separate, a site called the **replication fork**. As more and more parental DNA is unwound and the replication fork moves forward (to the left in Fig. 12.1), both new strands would also grow in the

strand that was originally part of the parental duplex and one newly synthesized strand. An alternative model is conservative replication, which proposes that the original DNA duplex remains intact and the daughter DNA duplex is completely new. Which model is correct? If there were a way to distinguish newly synthesized daughter DNA strands ("new strands") from previously synthesized parental strands ("old strands"), the products of replication could be observed and the mode of replication determined.

American molecular biologists Matthew S. Meselson and Franklin W. Stahl carried out an experiment to determine how DNA replicates. This experiment, described in **Fig. 12.2**, has been called "the most beautiful experiment in biology" because it so elegantly demonstrates the process of generating hypotheses, making predictions, and performing an experiment to test the hypotheses (Chapter 1). Meselson and Stahl found that DNA replicates semiconservatively.

Important as it was in demonstrating semiconservative replication in bacteria, the Meselson–Stahl experiment left open the possibility that DNA replication in eukaryotes might be different. It was only a few years after the Meselson–Stahl experiment that methods for labeling DNA with fluorescent nucleotides were developed. These methods allowed researchers to visualize entire strands of eukaryotic DNA and follow each strand through replication.

Fig. 12.3 shows a human chromosome with unlabeled DNA that subsequently underwent two rounds of replication in medium containing a fluorescent nucleotide. The chromosome was photographed at metaphase of the second round of mitosis, after chromosome duplication but before the separation of the chromatids into the daughter cells. Notice that one chromatid contains hybrid DNA with one labeled strand and one unlabeled strand, which fluoresces faintly (light); the other chromatid contains two strands of labeled DNA, which fluoresces strongly (dark). This result is conceptually the same as what was seen

HOW DO WE KNOW?

FIG. 12.2

How is DNA replicated?

BACKGROUND Watson and Crick's discovery of the structure of DNA in 1953 suggested a mechanism by which DNA is replicated. Experimental evidence of how DNA is replicated came from research by American molecular biologists Matthew Meselson and Franklin Stahl in 1958.

HYPOTHESIS DNA replicates in a semiconservative manner, meaning that each new DNA molecule consists of one parental strand and one newly synthesized strand.

ALTERNATIVE HYPOTHESIS DNA replicates in a conservative manner, meaning that after replication, one DNA molecule consists of two parental strands and the other consists of two newly synthesized strands.

METHOD Meselson and Stahl distinguished parental strands ("old strands") from daughter strands ("new strands") using two isotopes of nitrogen atoms. Old strands were labeled with a heavy form of nitrogen with an extra neutron (^{15}N), and new strands were labeled with the normal, lighter form of nitrogen (^{14}N).

EXPERIMENT The researchers first grew bacterial cells on medium containing only the heavy ^{15}N form of nitrogen. As the cells grew, ^{15}N was incorporated into the DNA bases, resulting, after several generations, in DNA containing only ^{15}N. They then transferred the cells into medium containing only light ^{14}N nitrogen. After one

FIG. 12.3 Eukaryotic DNA replication. Further evidence that DNA replication is semiconservative came from observing the uptake of fluorescent nucleotides into chromosomal DNA. *Source: Daniel Hartl.*

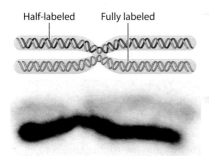

After two rounds of DNA replication in a labeled medium, one daughter molecule is half-labeled, and the other is fully labeled (compare with Figure 12.2).

direction of replication fork movement. However, it turns out that this scenario is impossible.

We have seen that the two DNA strands in a double helix run in an antiparallel fashion: one of the template strands (the strand that starts on the bottom in Fig. 12.1) has a left-to-right 5′-to-3′ orientation, whereas the other template strand (the strand that starts on the top in Fig. 12.1) has a left-to-right 3′-to-5′ orientation. Therefore, the new daughter strands also have opposite orientations so that near the replication fork, the daughter strand in the bottom duplex terminates in a 3′ hydroxyl, whereas that in the top duplex terminates in a 5′ phosphate. There's the rub: the strand that terminates in the 5′ phosphate cannot grow in the direction of the replication fork because new DNA strands can grow only by the addition of successive nucleotides to the 3′ end. That is, DNA always grows in the 5′-to-3′ direction.

DNA polymerization occurs only in the 5′-to-3′ direction because of the chemistry of nucleic acid synthesis, discussed in Chapter 3. The building blocks of DNA (Chapter 2) are nucleotides, each consisting of a deoxyribose sugar with three phosphate groups attached to the 5′ carbon (making them triphosphates), a nitrogenous base (A, T, G, or C) attached to the 1′ carbon, and a free hydroxyl (–OH) group attached to the

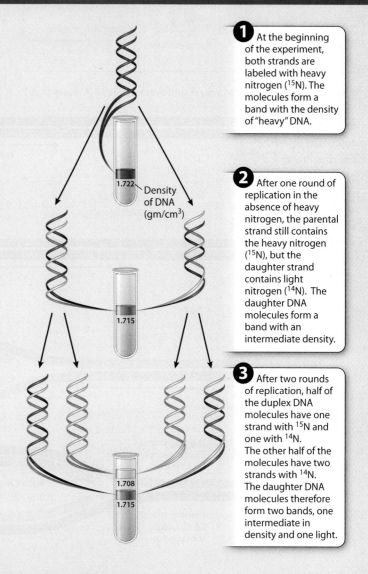

round of replication in this medium, cell replication was halted. The researchers could not observe the DNA directly, but they were able to measure the density of the DNA by spinning it in a high-speed centrifuge in tubes containing a solution of cesium chloride.

PREDICTION If DNA replicates conservatively, half of the DNA in the cells should be composed of two heavy parental strands containing ^{15}N, and half should be composed of two light daughter strands containing ^{14}N. If DNA replicates semiconservatively, after one round of replication, the daughter DNA molecules should each consist of one heavy strand and one light strand.

RESULTS When the fully ^{15}N-labeled parental DNA was spun, it concentrated in a single heavy band with a density of 1.722 gm/cm³ (Point 1 in the figure). This is the density of a duplex molecule containing two heavy (^{15}N-labeled) strands. After one round of replication, the DNA formed a band at a density of 1.715 gm/cm³, which is the density expected of a duplex molecule containing one heavy (^{15}N-labeled) strand and one light (^{14}N-labeled) strand (Point 2). DNA composed only of ^{14}N has a density of 1.708 gm/cm³. After two rounds of replication, half the molecules exhibited a density of 1.715 gm/cm³, indicating a duplex molecule with one heavy strand and one light strand, and the other half exhibited a density of 1.708 gm/cm³, indicating a duplex molecule containing two light strands (Point 3).

CONCLUSION DNA replicates semiconservatively, supporting the first hypothesis.

SOURCE Meselson, M., and F. W. Stahl. 1958. "The Replication of DNA in *Escherichia coli*." *PNAS* 44:671–82.

❶ At the beginning of the experiment, both strands are labeled with heavy nitrogen (^{15}N). The molecules form a band with the density of "heavy" DNA.

❷ After one round of replication in the absence of heavy nitrogen, the parental strand still contains the heavy nitrogen (^{15}N), but the daughter strand contains light nitrogen (^{14}N). The daughter DNA molecules form a band with an intermediate density.

❸ After two rounds of replication, half of the duplex DNA molecules have one strand with ^{15}N and one with ^{14}N. The other half of the molecules have two strands with ^{14}N. The daughter DNA molecules therefore form two bands, one intermediate in density and one light.

FIG. 12.4 DNA synthesis by nucleotide addition to the 3′ end of a growing DNA strand.

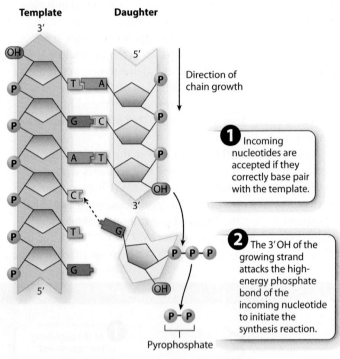

3′ carbon. DNA polymerization occurs when the 3′ hydroxyl at the growing end of the polynucleotide chain attacks the triphosphate group at the 5′ end of an incoming nucleotide triphosphate (**Fig. 12.4**). As the incoming nucleotide triphosphate is added to the growing DNA strand, one of the nucleotide's high-energy phosphate bonds is broken to release the outermost two phosphates (called pyrophosphate), and immediately the high-energy phosphate bond in the pyrophosphate is cleaved to drive the polymerization reaction forward and make it irreversible.

The polymerization reaction is catalyzed by **DNA polymerase,** an enzyme that is a critical component of a large protein complex that carries out DNA replication. DNA polymerases exist in all organisms and are highly conserved, meaning that they vary little from one species to another because they carry out an essential function. A cell typically contains several different DNA polymerase enzymes, each specialized for a particular situation. However, all DNA polymerases share the same basic function in that they synthesize a new DNA strand from an existing template. Most, but not all, also correct mistakes in replication, as we will see. DNA polymerases also have many practical applications in the laboratory, which we will discuss later in this chapter.

FIG. 12.5 Leading and lagging strand synthesis. In DNA replication, because the template strand is made of two antiparallel strands, one daughter strand (the leading strand) is synthesized continuously and the other daughter strand (the lagging strand) is synthesized in smaller pieces.

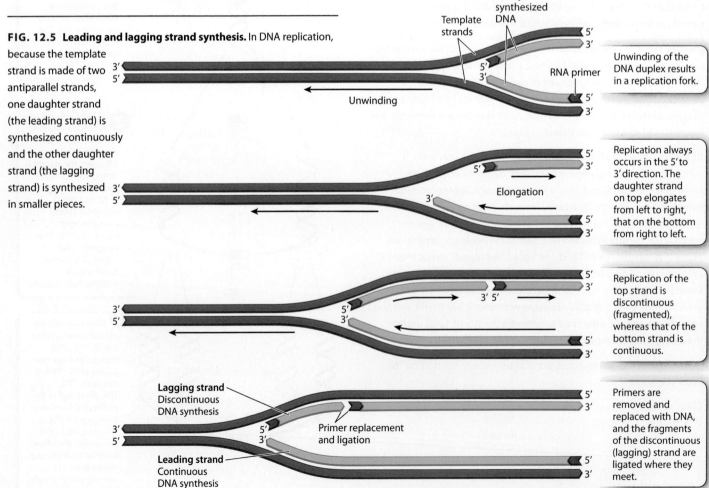

In replicating DNA, one daughter strand is synthesized continuously and the other in a series of short pieces.

Because a new DNA strand can be elongated only at the 3′ end, the two daughter strands are synthesized in quite different ways (**Fig. 12.5**). The daughter strand shown at the bottom of Fig. 12.5 has its 3′ end pointed toward the replication fork so that as the parental double helix unwinds, nucleotides can be added onto the 3′ end, and this daughter strand can be synthesized as one long, continuous polymer. This daughter strand is called the **leading strand**.

The situation is different for the daughter strand shown at the top in Fig. 12.5. Its 5′ end is pointed toward the replication fork, but the strand cannot grow in that direction. Instead, as the replication fork unwinds, it grows away from the fork and forms a stretch of single-stranded DNA of a few hundred to a few thousand nucleotides. Then, as the parental DNA duplex unwinds further, a new daughter strand is initiated with its 5′ end near the replication fork, and this strand is elongated at the 3′ end as usual. The result is that the daughter strand shown at the top in Fig. 12.5 is actually synthesized in relatively short, discontinuous pieces. As the parental double helix unwinds, a new piece is initiated at intervals, and each new piece is elongated at its 3′ end until it reaches the piece in front of it. This daughter strand is called the **lagging strand**. The short pieces in the lagging strand are sometimes called **Okazaki fragments** after their discoverer, Japanese molecular biologist Reiji Okazaki.

The presence of leading and lagging strands during DNA replication is a consequence of the antiparallel nature of the two strands in a DNA double helix and the fact that DNA polymerase can synthesize DNA in only one direction.

A small stretch of RNA is needed to begin synthesis of a new DNA strand.

Each new DNA strand must begin with a short stretch of RNA that serves as a **primer**, or starter, for DNA synthesis. The primer is needed because the DNA polymerase complex cannot begin a new strand on its own; it can only elongate the end of an existing piece of DNA or RNA. The primer is made by an RNA polymerase called **RNA primase**, which synthesizes a short piece of RNA complementary to the DNA template. Once the RNA primer has been synthesized, the DNA polymerase takes over and elongates the primer, adding successive DNA nucleotides to the 3′ end of the growing strand.

Because the DNA polymerase complex extends an RNA primer, all new DNA strands have a short stretch of RNA at their 5′ end. For the lagging strand, there are many such primers, one for each of the discontinuous fragments of newly synthesized DNA. As each of these fragments is elongated by DNA polymerase, it grows toward the primer of the fragment in front of it (**Fig. 12.6**). When the growing fragment comes into contact with the primer of the fragment synthesized earlier, a different DNA polymerase complex takes over, removing the earlier RNA

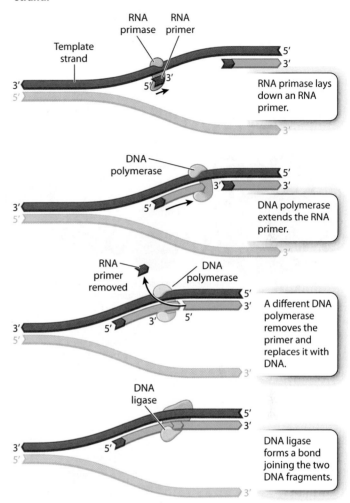

FIG. 12.6 Synthesis and removal of RNA primers in the lagging strand.

primer and extending the growing fragment with DNA nucleotides to fill the space left by its removal. When the replacement is completed, the adjacent fragments are joined, or ligated, by an enzyme called **DNA ligase**.

Synthesis of the leading and lagging strands is coordinated.

In addition to RNA primase and DNA ligase, many other enzymes play a role in DNA replication to ensure accurate and efficient synthesis of daughter strands. **Fig. 12.7a** shows enzymes that work at or near the replication fork. One of these, **topoisomerase II**, works upstream from the replication fork to relieve the stress on the double helix that results from its unwinding at the replication fork. In the meantime, **helicase** separates the strands of the parental double helix at the replication fork. Then, **single-strand binding protein** binds to these single-stranded regions to prevent the template strands from coming back together. These enzymes are evolutionarily conserved, and the process of replication is the same in all organisms. The two strands of the parental DNA duplex

FIG. 12.7 The replication fork. (a) Many proteins play different roles in replication, including separating and stabilizing the strands of the double helix. (b) In the cell, proteins involved in replication come together to form a complex that replicates both strands at the same pace.

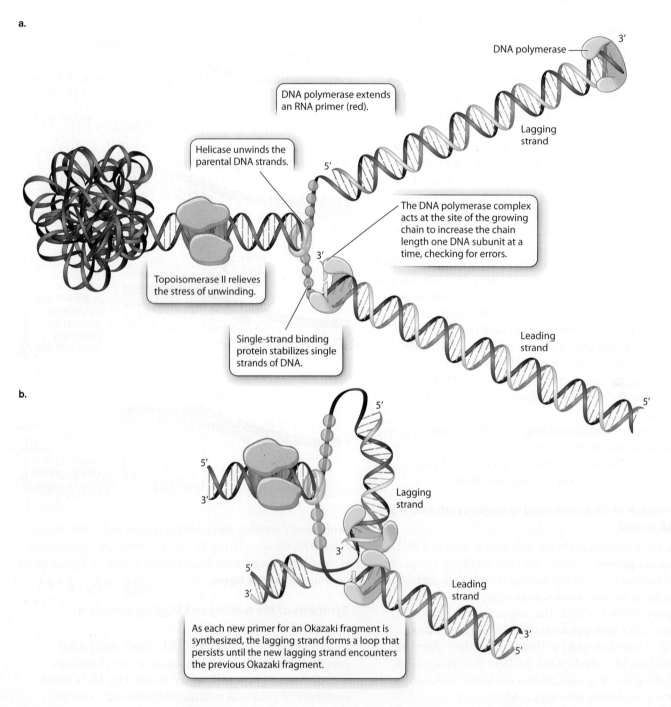

separate, and each serves as a template for the synthesis of a daughter strand according to the base-pairing rules of A–T and G–C, with each successive nucleotide being added to the 3′ end of the growing strand.

Replication does not occur exactly as it is shown in Fig. 12.7a. The DNA polymerase complexes for each strand stay in contact with each other so that synthesis of the leading and lagging strands is coordinated to occur at the same time and rate (**Fig. 12.7b**). In this arrangement, the lagging strand's polymerase releases and retrieves the lagging strand for the synthesis of each new RNA primer. The positioning of the polymerases is such that both the leading strand and the lagging strand pass through in the same direction, which requires that the lagging strand be looped around as shown in Fig. 12.7b. In this configuration, the 3' end of the lagging strand and the 3' end of the leading strand are elongated together, which ensures that neither strand outpaces the other in its rate of synthesis.

As we discuss in Chapter 14, when DNA damage occurs during replication, the rate of synthesis slows down so that the DNA can be repaired. If synthesis of one strand slows down to repair DNA damage, synthesis of the other strand slows down, too. The pairing of the replication complexes is disrupted when the RNA primer of the previous Okazaki fragment is encountered, and then takes place again when a new lagging-strand primer is formed. The lagging-strand loop is sometimes called a "trombone loop."

DNA polymerase is self-correcting because of its proofreading function.

Most DNA polymerases can correct their own errors in a process called **proofreading**, which is a separate enzymatic activity from strand elongation (synthesis) (**Fig. 12.8**).

When each new nucleotide comes into line in preparation for attachment to the growing DNA strand, the nucleotide is temporarily held in place by hydrogen bonds that form between the base in the new nucleotide and the base across the way in the template strand (see Fig. 12.4). The strand being synthesized and the template strand therefore have complementary bases—A paired with T, or G paired with C. However, on rare occasions, an incorrect nucleotide is attached to the new DNA strand. When this happens, DNA polymerase can correct the error because it detects mispairing between the template and the most recently added nucleotide. Mispairing between a base in the parental strand and a newly added base in the daughter strand activates a DNA-cleavage function of DNA polymerase that removes the incorrect nucleotide and inserts the correct one in its place.

Mutations resulting from errors in nucleotide incorporation still occur, but proofreading reduces their number. In the bacterium *E. coli,* for example, about 99% of the incorrect nucleotides that are incorporated during replication are removed and repaired by the proofreading function of DNA polymerase. Those that slip past proofreading and other repair systems lead to mutations, which are then faithfully copied and passed on to daughter cells. Some of these mutations may be harmful, but others are neutral and a rare few may be beneficial. These mutations are the ultimate source of genetic variation that we see among individuals of the same species and among species (Chapter 14).

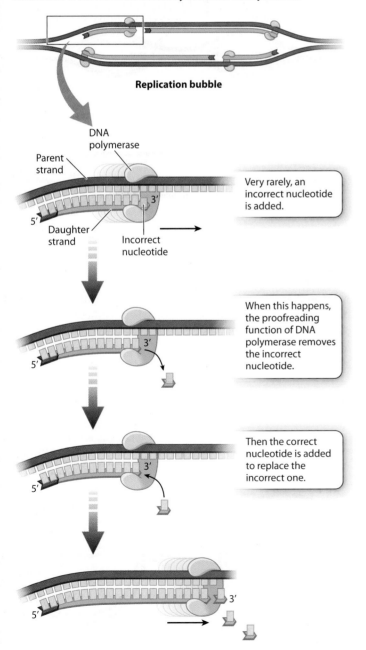

FIG. 12.8 Proofreading, the process by which an incorrect nucleotide is removed immediately after it is incorporated.

Self-Assessment Questions

1. Suppose Meselson and Stahl had done their experiment (Fig. 12.2) the other way around, starting with cells fully labeled with ^{14}N light DNA and then transferring them to medium containing only ^{15}N heavy DNA. What density of DNA molecule would you predict after one and two rounds of replication?

2. How does the structure of DNA itself suggest a mechanism of replication?

3. How does the chemical structure of deoxynucleotides determine the orientation of the DNA strands, and how does this affect the direction of DNA synthesis?

4. What are the differences and similarities in the way the two daughter strands of DNA are synthesized at a replication fork?

12.2 REPLICATION OF CHROMOSOMES

The basic steps involved in semiconservative DNA replication are universal, suggesting that they evolved in the common ancestor of all living organisms. In addition, they are the same whether they occur in a test tube or in a cell, or whether the segment of DNA being replicated is short or long. However, the replication of an entire linear chromosome poses particular challenges. Here, we consider two such challenges encountered by cells in replicating their chromosomes: how replication starts and how it ends.

Replication of DNA in chromosomes starts at many places almost simultaneously.

DNA replication is relatively slow. In eukaryotes, it occurs at a rate of about 50 nucleotides per second. At this rate, replication from end to end of the DNA molecule in the largest human chromosome would take almost two months. In fact, it takes only a few hours. This fast pace is possible because in a long DNA molecule, replication begins almost simultaneously at many places. Each point at which DNA synthesis is initiated is called an **origin of replication**. The opening of the double helix at each origin of replication forms a **replication bubble** with a replication fork on each side, each with a leading strand and a lagging strand with topoisomerase II, helicase, and single-strand binding protein playing their respective roles (**Fig. 12.9**). DNA synthesis takes place at each replication fork, and as the replication forks move in opposite directions, the replication bubble increases in size. When two replication bubbles meet, they fuse to form one larger replication bubble.

Note in Fig. 12.9 that within a single replication bubble, the same daughter strand is the leading strand at one replication fork and the lagging strand at the other replication fork. This situation results from the fact that the replication forks in each replication bubble move away from each other. When two replication bubbles fuse and the leading strand from one meets the lagging strand from the other, the ends of the strands that meet are joined by DNA ligase, just as happens when the discontinuous fragments within the lagging strand meet (see Fig. 12.6).

Some DNA molecules, including most of the DNA molecules in bacterial cells and the DNA in mitochondria and chloroplasts (Chapter 3) are small circles, not long linear molecules. Such circular DNA molecules typically have only one origin of replication (**Fig. 12.10**). Replication takes place at both replication forks, and the replication forks proceed in opposite directions around the circle until they meet and fuse on the opposite side, completing one round of replication.

Telomerase restores tips of linear chromosomes shortened during DNA replication.

A circular DNA molecule can be replicated completely because it has no ends, and the replication forks can move completely around the circle (Fig. 12.10). Linear DNA molecules have

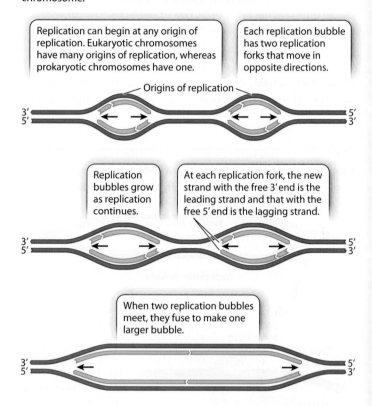

FIG. 12.9 Origins of replication. The replication of eukaryotic chromosomal DNA is initiated at many different places on the chromosome.

ends, however, and at each round of DNA replication, the ends become slightly shorter. The reason for the shortening is illustrated in **Fig. 12.11**.

Recall that each fragment of newly synthesized DNA starts with an RNA primer. On the leading strand, the only primer required is at the origin of replication when synthesis begins. The leading strand is elongated in the same direction as the moving replication fork and is able to replicate the template

FIG. 12.10 Replication of a circular bacterial chromosome.

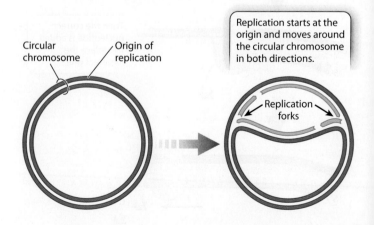

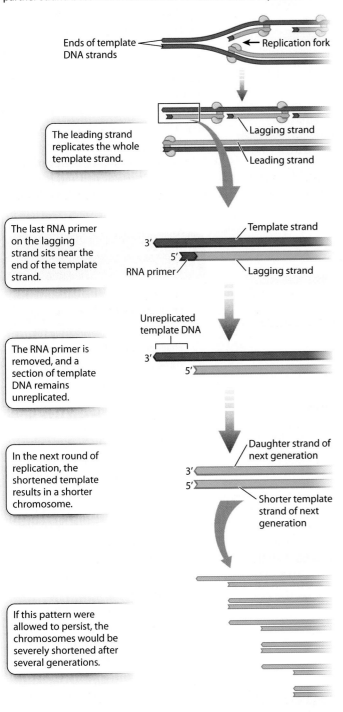

FIG. 12.11 **Shortening of linear DNA at the ends.** Because an RNA primer cannot be synthesized precisely at the 3' end of a DNA strand, its partner strand becomes a little shorter in each round of replication.

strand all the way to the end. On the lagging strand, however, which grows away from the replication fork, many primers are required, and the final RNA primer is synthesized about 100 nucleotides from the 3' end of the template. Because this primer initiates synthesis of the final Okazaki fragment of the lagging strand, no other Okazaki fragment is available to synthesize the missing 100 base pairs and remove the primer. Therefore, when DNA replication is complete, the new daughter DNA strand (light blue in Fig. 12.11) will be missing about 100 base pairs from the tip (plus a few more owing to the length of the primer). When this new daughter strand is replicated, the newly synthesized strand must terminate at the shortened end of the template strand, and so the new duplex molecule is shortened by about 100 base pairs from the original parental molecule. The strand shortening in each round of DNA replication is a problem because without some mechanism to restore the tips, the DNA in the chromosome would eventually be nibbled away to nothing.

Eukaryotic organisms have evolved mechanisms that solve the problem of shortened ends (**Fig. 12.12**). Each end of a eukaryotic chromosome is capped by a repeating sequence called the **telomere**. The repeating sequence that constitutes the telomere differs from one group of organisms to the next, but in the chromosomes of humans and other vertebrates, it consists of the sequence 3'-GGGATT-5' repeated over and over again in about 1500–3000 copies. For simplicity, just four copies of this telomere sequence are shown in Fig. 12.12. The telomere is slightly shortened in each round of DNA replication, as shown in Fig. 12.11, but the shortened end is quickly restored by an enzyme known as **telomerase**, which contains an RNA molecule complementary to the telomere sequence.

As shown in Fig. 12.12, the telomerase replaces the lost telomere repeats by using its RNA molecule as a guide to add successive DNA nucleotides to the 3' end of the template strand. Once the 3' end of the template strand has been elongated, an RNA primer and the complementary DNA strand are synthesized. In this way, the original telomere is completely restored. Because no genes are present in the telomere, the slight shortening and subsequent restoration that take place have no harmful consequences.

Telomerase activity differs from one cell type to the next. It is fully active in **germ cells**, which produce sperm or eggs, and also in **stem cells**, which are undifferentiated cells that can undergo an unlimited number of mitotic divisions and can differentiate into any of a large number of specialized cell types. Stem cells are found in embryos, where they differentiate into all the various cell types of the body. Stem cells are also found in some tissues of the body after embryonic development, where they replenish cells that have a high rate of turnover, such as blood and intestinal cells, and play a role in tissue repair.

In contrast to the high activity of telomerase in germ cells and stem cells, telomerase is almost inactive in most cells in the adult body. In these cells, the telomeres are actually shortened by about 100 base pairs in each mitotic division. Telomere shortening limits the number of mitotic divisions that the cells can undergo because human cells stop dividing when their chromosomes have telomeres with fewer than about 100 copies of the

FIG. 12.12 Telomerase. Telomerase prevents successive shortening of linear chromosomes.

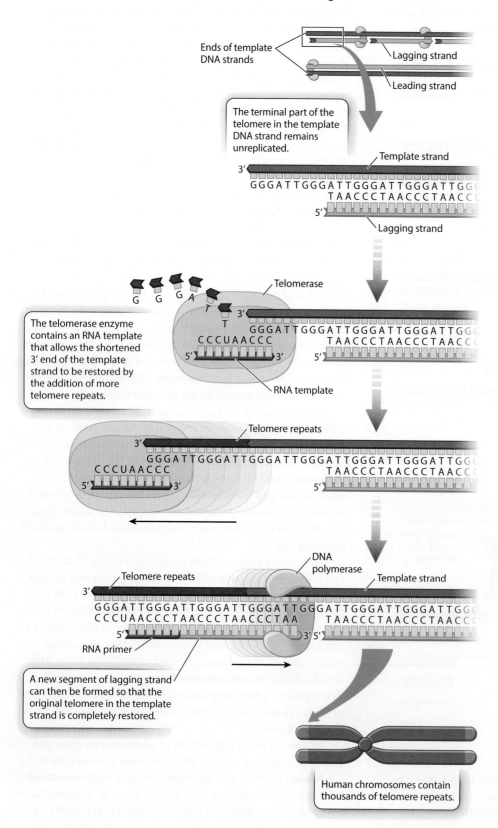

telomere repeat. Adult somatic cells can therefore undergo only about 50 mitotic divisions until the telomeres are so short that the cells stop dividing.

Many biologists believe that the limit on the number of cell divisions explains, in part, why our tissues become less youthful and wounds heal more slowly with age. The telomere hypothesis of aging is still controversial, but increasing evidence suggests that it is one of several factors that lead to aging. The flip side of the coin is observed in cancer cells, in which telomerase is reactivated and helps support the uncontrolled growth and division of abnormal cells.

Self-Assessment Questions

5. Why is replicating the ends of linear chromosomes problematic?

6. How does the cell overcome the challenge of replicating chromosome ends?

12.3 DNA TECHNIQUES

Watson and Crick's discovery of the structure of DNA and knowledge of the mechanism of replication allowed biologists not only to understand some of life's central processes, but also to create tools to study how life works. Biologists often need to isolate, identify, and determine the nucleotide sequence of particular DNA fragments. Such procedures can determine whether a genetic risk factor for diabetes has been inherited, blood at a crime scene matches that of a suspect, a variety of rice or wheat carries a genetic factor for insect resistance, or two species of organisms are closely related. Many of the experimental procedures for the isolation, identification, and sequencing of DNA are based on knowledge of DNA structure and physical properties of DNA. Others make use of the principles of DNA replication.

In this section, we discuss how particular fragments of double-stranded DNA can be produced, how DNA fragments of different sizes can be physically separated, and how the nucleotide sequence of a piece of DNA can be determined.

The polymerase chain reaction selectively amplifies regions of DNA.

DNA that exists in the nuclei of your cells is present in two copies per cell, except for DNA in the X and Y chromosomes in males, which are each present in only one copy per cell. In the laboratory, it is very difficult to manipulate or visualize a sample containing just one or two copies of a DNA molecule. Instead, researchers typically work with many identical copies of the DNA molecule they are interested in studying. A common method for making copies of a piece of DNA is the **polymerase chain reaction (PCR)**, which allows a targeted region of a DNA molecule to be replicated (or **amplified**) into as many copies as desired. PCR is both selective and highly sensitive, so it is used to amplify and detect small quantities of nucleic acids, such as HIV in blood-bank supplies, or to study DNA samples as minuscule as those left by a smoker's lips on a cigarette butt dropped at the scene of a crime. The starting sample can be as small as a single molecule of DNA.

The principles of PCR are illustrated in **Fig. 12.13**. Because the PCR reaction is essentially a DNA synthesis reaction, it requires the same basic components used by the cell to replicate its DNA. In this case, the procedure takes place in a small plastic tube containing a solution that includes four essential components:

1. Template DNA. At least one molecule of double-stranded DNA containing the region to be amplified serves as the template for amplification.

2. DNA polymerase. The enzyme DNA polymerase is used to replicate the DNA.

3. All four deoxynucleoside triphosphates. Deoxynucleoside triphosphates with the bases A, T, G, or C are needed as building blocks for the synthesis of new DNA strands.

4. Two primers. Two short sequences of single-stranded DNA are required for the DNA polymerase to start synthesis. Enough primer is added so that the number of primer DNA molecules is much greater than the number of template DNA molecules.

The primer sequences are **oligonucleotides** (*oligos* is the Greek word for "few") produced by chemical synthesis and are typically 20–30 nucleotides long. Their base sequences are chosen to be complementary to the ends of the region of DNA to be amplified. In other words, the primers flank the specific region of DNA to be amplified. The 3' end of each primer must be oriented toward the region to be amplified so that, when DNA polymerase extends the primer, it creates a new DNA strand complementary to the targeted region. Because the 3' ends of the primers both point toward the targeted region, one of the primers pairs with one strand of the template DNA and the other pairs with the other strand of the template DNA.

PCR creates new DNA fragments in a cycle of three steps, as shown in Fig. 12.13a. The first step, **denaturation**, involves heating the solution in a plastic tube to a temperature just short of boiling so that the individual DNA strands of the template separate (or "denature") as a result of the breaking of hydrogen bonds between the complementary bases. The second step, **annealing**, begins as the solution is cooled. Because of the great excess of primer molecules, the two primers bind (or "anneal") to their complementary sequence on the DNA (rather than two strands of the template duplex coming back together). In the final step, **extension**, the solution is heated to the optimal temperature for DNA polymerase and the polymerase elongates (or "extends") each primer with the deoxynucleoside triphosphates.

After sufficient time to allow new DNA synthesis, the solution is heated again, and the cycle of denaturation, annealing, and extension is repeated over and over, as indicated in Fig. 12.13b, usually for 25–35 cycles. In each cycle, the number of copies of the targeted fragment is doubled. The first round of PCR amplifies the targeted region into 2 copies, the next into 4, the next into 8, then 16, 32, 64, 128, 256, 512, 1024, and so forth. By the third round of amplification, the process begins to produce molecules that are only as long as the region of the template duplex flanked by the sequences complementary to the primers. After several more rounds of replication, the majority of molecules are of this type. The doubling in the number of amplified fragments in each cycle justifies the term "chain reaction."

Although PCR is elegant in its simplicity, the DNA polymerase enzymes from many species (including humans) irreversibly lose both structure and function at the high temperature required to separate the DNA strands. At each cycle, you would have to open the tube and add fresh DNA polymerase. This is possible, and, in fact, it was how PCR was performed when the technique was first developed, but the procedure is time-consuming and tedious. To solve this problem, we now use DNA polymerase enzymes that are heat-stable, such as those from the bacterial species *Thermus aquaticus*, which lives at the near–boiling point of water in natural hot springs, including those at Yellowstone National Park. This polymerase, called *Taq* polymerase, remains active at high temperatures. Once the reaction mixtures are set up, the entire procedure is carried out in a fully automated machine. The time of each cycle, temperatures, number of cycles, and other variables can all be programmed. The fact that DNA polymerase from a bacterium that lives in hot springs can be used to amplify DNA from any organism reminds us again of the conserved function and evolution of this enzyme early in the history of life.

FIG. 12.13 The polymerase chain reaction (PCR). PCR results in amplification of the DNA sequence flanked by the two primers.

a. Each cycle of amplification includes three steps.

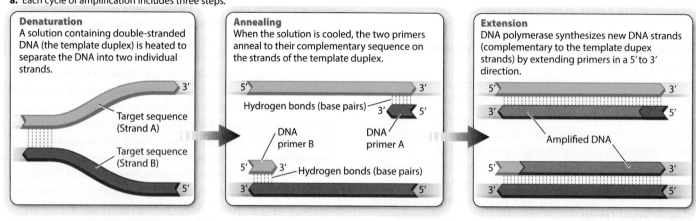

Denaturation
A solution containing double-stranded DNA (the template duplex) is heated to separate the DNA into two individual strands.

Annealing
When the solution is cooled, the two primers anneal to their complementary sequence on the strands of the template duplex.

Extension
DNA polymerase synthesizes new DNA strands (complementary to the template dupex strands) by extending primers in a 5' to 3' direction.

b. PCR repeats the cycle of amplification.

1 The template duplex is often longer than the amplified region.

2 Primers for PCR are typically 20–30 nucleotides in length. Their sequences are used for amplification. The primers are chosen to base pair with one or the other strand of the template duplex, with their 3' ends oriented toward each other. The primers are added to the reaction mixture in great excess to ensure pairing with any complementary sequence.

3 Each round of amplification doubles the number of molecules that have the same sequence as the template duplex.

4 After n cycles of amplification there are 2^n copies of the target sequence. When $n = 30$, there are $2^{30} < 10^9$ copies.

Electrophoresis separates DNA fragments by size.

PCR amplification does not always work as you might expect. Sometimes the primers have the wrong sequence and thus fail to anneal properly; sometimes they anneal to multiple sites and several different fragments are amplified. To determine whether or not PCR has yielded the expected product, a researcher must determine the size of the amplified DNA molecules. Usually, the researcher knows what the size of the correctly amplified fragment should be, making it possible to compare the expected size to the actual size.

One way to determine the size of a DNA fragment is by **gel electrophoresis (Fig. 12.14)**, a procedure in which DNA fragments are separated by size as they migrate through a porous substance in response to an electrical field. In this procedure, DNA samples are first inserted into slots or wells near the edge of a rectangular slab of porous material that resembles solidified agar (the "gel"). The porous nature of the gel is due to its composition of a tangle of polymers. This tangle of polymers makes it difficult for large molecules to pass through. The gel is then inserted into an apparatus and immersed in a solution that allows an electric current to be passed through it, from negative at the top of the gel where the DNA samples are initially placed, to positive at the bottom of the gel (Fig. 12.14a). This electrical field is what causes the DNA to migrate through the porous gel. Fragments of double-stranded DNA in this solution are negatively charged because of the ionized (negative) phosphate groups along the backbone in the solution. Therefore, the DNA molecules move toward the positive pole of the electric field.

The DNA molecules move according to their size. Short fragments pass through the pores of the gel more readily than large fragments, and so in a given interval of time, short fragments move a greater distance in the gel than large fragments. The rate of migration is dependent only on size, not on sequence, and so all fragments of a given size move together at the same rate in a discrete band, which can be made visible by dyes that bind to DNA and fluoresce under ultraviolet light (Fig. 12.14b). A solution of DNA fragments of known sizes is usually placed in one of the wells, resulting in a series of bands, called a ladder, which can be used for size comparison.

Consider a PCR experiment with 25 base-pair (bp) primers flanking a 250-bp length of DNA. PCR amplifies this region, generating many millions of copies of a 300-bp DNA fragment consisting of the 250-bp target sequence and the primer sequences on both ends. To determine if the PCR experiment worked as expected, a sample of the product is checked by gel electrophoresis. This sample is loaded into the well of the gel, a current is applied for a short period of time, and DNA fragments migrate to positions in the gel that correspond to their size, creating bands of DNA that are visualized with a dye. If a single band of 300 bp is seen on the gel, then the experiment worked as expected. Sometimes, however, the reaction might yield no bands, a single band of the incorrect size, or multiple bands. The researcher would then need to go back and investigate why the experiment did not work as expected.

Gel electrophoresis can separate DNA fragments produced by any means, not only PCR. Genomic DNA can be cut with certain enzymes and the resulting fragments separated by gel electrophoresis, as described in the next section. Similar procedures can also be used to separate protein molecules.

Restriction enzymes cleave DNA at particular short sequences.

In addition to amplifying segments of DNA, researchers often also make use of techniques that cut DNA at specific sites. Cutting DNA molecules allows pieces from the same or different organisms to be brought together in recombinant DNA technology, which is discussed in section 12.4. It also is a way to

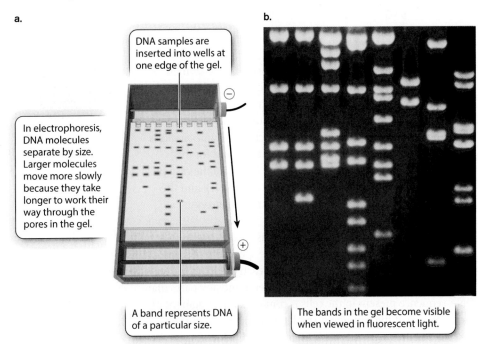

FIG. 12.14 Gel electrophoresis. (a) A typical electrophoresis apparatus consists of a plastic tray, gel, and solution. (b) DNA bands can be visualized after staining with a dye that fluoresces under ultraviolet light. *Photo source: Guy Tear/Wellcome Images.*

determine whether or not specific sequences are present in a segment of DNA, because techniques for cutting DNA depend on specific DNA sequences. Finally, cutting DNA allows whole genomes to be broken into smaller pieces for further analysis, such as DNA sequencing.

The method for cutting DNA makes use of enzymes that recognize specific, short nucleotide sequences in double-stranded DNA and cleave the DNA at these sites. These enzymes are known as **restriction enzymes.** There are about 1000 different kinds of restriction enzymes isolated from bacteria and other microorganisms, but they cut DNA from any organism whatsoever. The recognition sequences that the enzymes cleave, called **restriction sites**, are typically four or six base pairs long. Most restriction enzymes cleave double-stranded DNA at or near the restriction site. For example, the enzyme *Eco*RI recognizes the following restriction site:

$$\downarrow$$
$$5'\text{-GAATTC-}3'$$
$$3'\text{-CTTAAG-}5'$$
$$\uparrow$$

Wherever the enzyme finds this site in a DNA molecule, it cleaves each strand exactly at the position indicated by the vertical arrows. Note that the *Eco*RI restriction site is symmetrical: reading from the 5' end to the 3' end, the sequence of the top strand is exactly the same as the sequence of the bottom strand. This kind of symmetry is **palindromic** (it reads the same in both directions) and is typical of restriction sites. Note also that the site of cleavage is not in the center of the recognition sequence. The cleaved double-stranded molecules, therefore, each terminate in a short single-stranded overhang. In this case, the overhang is at the 5' end, as shown below:

$$\downarrow \qquad \qquad \downarrow$$
$$5'\text{-G}\qquad -3' \quad + \quad 5'\text{- AATTC-}3'$$
$$3'\text{-CTTAA -}5' \qquad 3'\text{-} \qquad \text{G-}5'$$
$$\uparrow \qquad \qquad \uparrow$$

Table 12.1 shows more examples of restriction enzymes. Some cleave their restriction site to produce a 5' overhang, others produce a 3' overhang, and still others cleave in the middle of their restriction site and leave blunt ends with no overhang. The standard symbols for restriction enzymes include both italic and roman letters. The italic letters stand for the species from which the enzyme is derived (*E. coli* in the case of *Eco*RI, *Bacillus amyloliquefaciens* in the case of *Bam*HI), and the Roman letters designate the particular restriction enzyme isolated from that species.

When a particular restriction enzyme is used to break up a whole genome into smaller fragments, the specificity of restriction enzymes ensures that the DNA from each cell in an individual organism yields the same set of fragments and any particular DNA sequence present in the cells is contained in a fragment of the same size. If the genome is small enough that the number and sizes of the fragments are limited, the individual fragments can be visualized directly in a gel (**Fig. 12.15**). These fragments can then be extracted from the gel for further analysis or manipulation. For instance, the extracted fragments can be sequenced or ligated to other fragments.

However, most genomes are too large to give such a simple picture. For example, the human genome is cleaved into more than a million different fragments by *Eco*RI, and these fragments appear as one big smear in a gel rather than as a series of discrete bands. In the next section, we discuss how individual bands containing a sequence of interest can be detected in such a gel.

DNA strands can be separated and brought back together again.

A researcher may wish to know whether a gene in one species is present in the DNA of a related species, but the nucleotide sequence of the gene is unknown. In such a case, techniques like PCR that require knowledge of the target DNA sequence cannot be used. Instead, the researcher can determine whether a DNA strand that contains the gene in one species can base pair with the complement of a similar gene in a DNA strand from the other species. The base pairing of complementary single-stranded nucleic acids is known as **renaturation** or hybridization, and the process is the opposite of denaturation. In denaturation, the DNA strands in a duplex molecule are separated, and in renaturation, complementary strands come together again. Denaturation and renaturation are also opposites in regard to the experimental conditions in which they occur. When a solution containing DNA molecules is gradually heated to a sufficiently high temperature, the strands denature. When the solution is allowed to cool, the complementary strands renature.

Denatured DNA strands from one source can renature with DNA strands from a different source if their sequences

TABLE 12.1	Examples of Restriction Enzymes with Six-Base Cleavage Sites		
BamHI	$\downarrow$ 5'-GGATCC-3' 3'-CCTAGG-5' $\uparrow$	HindIII	$\downarrow$ 5'-AAGCTT-3' 3'-TTCGAA-5' $\uparrow$
KpnI	$\downarrow$ 5'-GGTACC-3' 3'-CCATGG-5' $\uparrow$	SstI	$\downarrow$ 5'-GAGCTC-3' 3'-CTCGAG-5' $\uparrow$
HpaI	$\downarrow$ 5'-GTTAAC-3' 3'-CAATTG-5' $\uparrow$	SmaI	$\downarrow$ 5'-CCCGGG-3' 3'-GGGCCC-5' $\uparrow$

FIG. 12.15 *Eco*RI restriction fragments (a) on a circular DNA plasmid and (b) separated on a gel.

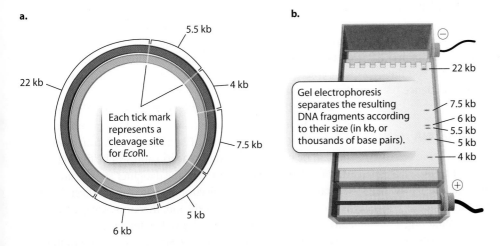

are precisely or mostly complementary. Two very closely related sequences will have more perfectly matched bases and thus more hydrogen bonds holding them together than two sequences that are less closely related. The amount of base pairing between two sequences affects the temperature at which they renature: very closely related sequences renature at a higher temperature than less closely related sequences. Evolutionary biologists have used this principle to estimate the proportion of perfectly matched bases present in the DNA of different species, which is used as a measure of how closely those species are related (Chapter 22). In general, the more closely two species are related, the more similar their DNA sequences. Knowledge of species relatedness is important in many applications, including conservation, identifying endangered species, and tracing the evolutionary history of organisms.

Renaturation makes it possible to use a small DNA fragment as a **probe**. This fragment is usually attached to a light-emitting or radioactive chemical that serves as a label. The probe can be used to determine whether or not a sample of double-stranded DNA molecules contains sequences that are complementary to it. Any DNA fragment can be used as a probe, and a probe can be obtained in any number of ways, such as by chemical isolation of a DNA fragment, amplification by PCR, or synthesizing a nucleic acid from free nucleotides.

Let's say you are interested in determining the number of copies of a particular DNA sequence in a genome or determining whether a given gene in human genomic DNA is intact. Recall that human and other genomic DNA cut with a restriction enzyme yield a large number of DNA fragments that look like a smear on a gel after electrophoresis. A labeled probe can be used to determine the size and number of a DNA sequence of interest in a gel. The method is known as a **Southern blot** (**Fig. 12.16**) after its inventor, the British molecular biologist Edwin M. Southern.

In a Southern blot, single-stranded DNA molecules that have migrated to bands on a gel are transferred to a filter paper, such that they are on the same spot on the paper as they were

FIG. 12.16 A Southern blot. *Photo source: Biochemical Journal: Reviews 1998 by Portland Press, Ltd. Reproduced with permission of Portland Press, Ltd. in the format Book via Copyright Clearance Center.*

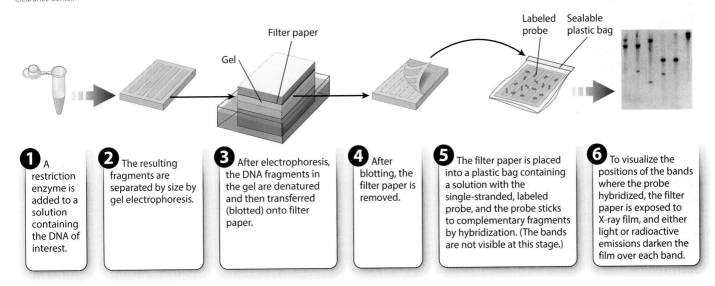

1 A restriction enzyme is added to a solution containing the DNA of interest.

2 The resulting fragments are separated by size by gel electrophoresis.

3 After electrophoresis, the DNA fragments in the gel are denatured and then transferred (blotted) onto filter paper.

4 After blotting, the filter paper is removed.

5 The filter paper is placed into a plastic bag containing a solution with the single-stranded, labeled probe, and the probe sticks to complementary fragments by hybridization. (The bands are not visible at this stage.)

6 To visualize the positions of the bands where the probe hybridized, the filter paper is exposed to X-ray film, and either light or radioactive emissions darken the film over each band.

on the gel. The filter paper is washed with a solution containing single-stranded probes that hybridize to whichever pieces of DNA on the filter paper are complementary. The paper is then exposed to X-ray film (in the case of radioactively labeled probes) so the bands can be visualized.

The presence of a band or bands on the film, therefore, indicates the number and size of DNA fragments that are complementary to the probe. This information, in turn, tells you the sizes and the number of copies of a particular DNA sequence present in the starting sample.

DNA sequencing makes use of the principles of DNA replication.

The ability to determine the nucleotide sequence of DNA molecules has given a tremendous boost to biological research. Techniques used to sequence DNA follow from our understanding of DNA replication. Consider a solution containing identical single-stranded molecules of DNA, each being used as the template for the synthesis of a complementary daughter strand that originates at a short primer sequence. The problem is to determine the nucleotide sequence of the template strand. A brilliant answer to this problem was developed by the English geneticist Frederick Sanger, an achievement rewarded with a share in the Nobel Prize in Chemistry in 1980. (It was his second Nobel; he had also been honored with the award in 1958 for his discovery of a method for determining the sequence of amino acids in a polypeptide chain.)

Recall that a free 3′ hydroxyl group is essential for each step in elongation because that is where the incoming nucleotide is attached (**Fig. 12.17a**). Making use of this fact, Sanger synthesized **dideoxynucleotides**, in which the 3′ hydroxyl group on the sugar ring is absent (**Fig. 12.17b**). Whenever a dideoxynucleotide is incorporated into a growing daughter strand, there is no hydroxyl group to attack the incoming nucleotide, and strand growth is stopped dead in its tracks (**Fig. 12.18a**). For this reason, a dideoxynucleotide is known as a **chain terminator**. By including a small amount of each of the chain terminators (A, G, C, and T) in a reaction tube along with larger quantities of all four normal nucleotides, a DNA primer, a DNA template, and DNA polymerase, Sanger was able to produce a series of interrupted daughter strands, each terminating at the site at which a dideoxynucleotide was incorporated.

Fig. 12.18b shows how the interrupted daughter strands help us to determine the DNA sequence by the procedure now called **Sanger sequencing**. In a tube containing dideoxy-A and all the other elements required for many rounds of DNA replication, a strand of DNA is synthesized complementary to the template until, when it reaches a T in the template strand, it incorporates an A. Only a small fraction of the A nucleotides in the sequencing reaction is in the dideoxy form, so only a fraction of the daughter strands incorporates a dideoxy-A at that point, resulting in termination. The rest of the strands incorporate a normal deoxy-A and continue synthesis, although most of these will be stopped at some point farther along the line when a T is reached again. Similarly, in a reaction containing dideoxy-C, DNA fragments will be produced whose sizes correspond to the positions of the Cs, and likewise for dideoxy-T and dideoxy-G.

Each of the four dideoxynucleotides is chemically labeled with a different fluorescent dye, as indicated by the different colors of A, C, T, and G in Fig. 12.18b, and so all four terminators can be present in a single reaction and still be distinguished. After DNA synthesis is complete, the daughter strands are separated by size with gel electrophoresis. The smallest daughter molecules migrate most quickly and, therefore, are the first to reach the bottom of the gel, followed by the others in order of increasing size. A fluorescence detector at the bottom of the gel "reads" the colors of the fragments as they exit the gel. What the scientist sees is a trace (or graph) of the fluorescence intensities, such as the one shown in **Fig. 12.18c**. The differently colored peaks, from left to right, represent the order of fluorescently tagged DNA fragments emerging from the gel. Thus, a trace showing peaks colored green-purple-red-green-purple-purple-blue-green-blue-red corresponds to a daughter strand having

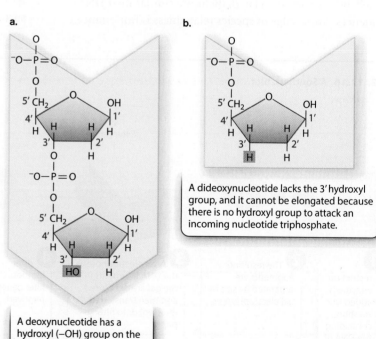

FIG. 12.17 (a) A deoxynucleotide and (b) a dideoxynucleotide. Incorporation of a dideoxynucleotide prevents strand elongation.

A deoxynucleotide has a hydroxyl (–OH) group on the 3′ carbon, allowing this end to be elongated.

A dideoxynucleotide lacks the 3′ hydroxyl group, and it cannot be elongated because there is no hydroxyl group to attack an incoming nucleotide triphosphate.

FIG. 12.18 Sanger sequencing. (a) The incorporation of an incoming dideoxynucleotide stops the elongation of a new strand. (b) Dideoxynucleotides terminate strands at different points in the template sequence. (c) Separation of interrupted daughter strands by size shows where each terminator was incorporated and hence the identity of the corresponding nucleotide in the template strand.

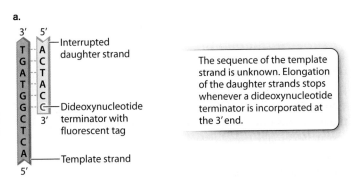

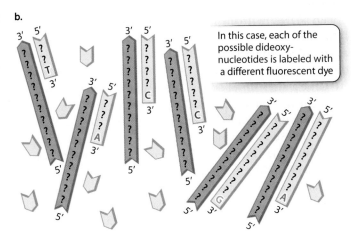

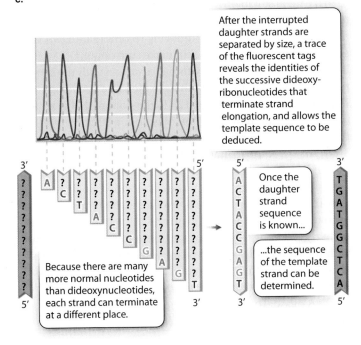

the sequence 5′-ACTACCGAGT-3′ (Fig. 12.18c). In the Sanger sequencing method, each sequencing reaction can determine the sequence of about 1000 nucleotides in the template DNA molecule.

CASE 3 YOUR PERSONAL GENOME: YOU, FROM A TO T

What new technologies are used to sequence your personal genome?

One of the high points of modern biology has been the determination of the complete nucleotide sequence of the DNA in a large number of species, including ours. The human genome and many others were sequenced by Sanger sequencing. This technique works well and is still the gold standard for accuracy, but it takes time and is expensive for large genomes like the human genome. Since it was first put into practice, Sanger sequencing has undergone many improvements, including the use of four-color fluorescent dyes to label the DNA fragments, capillary electrophoresis to separate the fragments, photocells to read the fluorescent signals automatically as the products run off the gel, and highly efficient enzymes. Together, these methods have increased the speed and decreased the cost of DNA sequencing considerably.

However, the ability to sequence everyone's genome, including yours, will require new technologies to further reduce the cost and increase the speed of sequencing. The first human genome sequence, completed in 2003 at a cost of approximately $2.7 billion, stimulated great interest in large-scale sequencing and the development of devices that increased scale and decreased cost. As in the development of computer hardware, emphasis was on making the sequencing devices smaller while increasing their capacity through automation.

Modern methods of sequencing include the use of fluorescent nucleotides or light-detection devices that reveal the identity of each base as it is added to the growing end of a DNA strand. Another approach passes individual DNA molecules through a pore in a charged membrane; as each nucleotide passes through the pore, it is identified by means of the tiny difference in charge that occurs. These miniaturized and automated devices carry out what is often called massively parallel sequencing, which can determine the sequence of hundreds of millions of base pairs in a few hours.

The goal of technology development was captured in the catchphrase "the $1000 genome," a largely symbolic target indicating a reduction in sequencing cost from about $1 million per megabase to less than $1 per megabase. For all practical purposes, the $1000-genome target has been achieved, but $1000 is still too costly to make genome sequencing a routine diagnostic procedure. The technologies are still evolving and the costs decreasing, and so it is a fair bet that in the coming years, your own personal genome can be sequenced quickly and cheaply.

Self-Assessment Questions

7. You have determined that the newly synthesized strand of DNA in your sequencing reaction has the sequence 5'-ACTACCGAGT-3'. What is the sequence of the template strand?

8. What does PCR do? Name and explain its three steps, and describe at least two uses for the PCR technique.

9. How do the properties of DNA determine how it moves through a gel, is cut by restriction enzymes, and hybridizes to other DNA strands?

10. How are DNA molecules sequenced?

12.4 GENETIC ENGINEERING

Along with methods to manipulate DNA fragments, the capability to isolate genes from one species and introduce them into another also emerged. This type of genetic engineering is called **recombinant DNA** technology because it literally recombines DNA molecules from two (or more) different sources into a single molecule. Recombinant DNA technology involves cutting DNA by restriction enzymes, isolating the resulting DNA fragments by gel electrophoresis, and ligating the fragments with enzymes used in DNA replication. This technology is possible because the DNA of all organisms is the same, differing only in sequence and not in chemical or physical structure. When DNA fragments from different sources are combined into a single molecule and incorporated into a cell, they are replicated and transcribed just like any other DNA molecule.

Recombinant DNA technology can combine DNA from any two sources, including different species. DNA from one species of bacteria can be combined with another, or a human gene can be combined with bacterial DNA, or the DNA from a plant and a fungus can be combined into a single molecule. These new sequences may be unlike any found in nature, raising questions about their possible effects on human health and the environment.

This section discusses one of the basic methods for producing recombinant DNA, some of the important applications of recombinant DNA technology such as genetically modified organisms, and a new method that can be used to change the DNA sequence of any organism into any other desired sequence.

Recombinant DNA combines DNA molecules from two or more sources.

The first application of recombinant DNA technology was the introduction of foreign DNA fragments into the cells of bacteria in the early 1970s. The method is simple and straightforward, and remains one of the mainstays of modern molecular research. It can be used to generate a large quantity of a protein for study or therapeutic use.

The method requires a fragment of double-stranded DNA that serves as the donor. The donor fragment may be a protein-coding gene, a regulatory part of a gene, or any DNA segment of interest. If you are interested in generating bacteria that could produce human insulin, you might use the coding region of the human insulin gene as your donor DNA molecule. Many genes in humans and other eukaryotes are very long—in some cases, more than a million base pairs. The messenger RNA from such long protein-coding genes is often much shorter than the gene itself because much of the primary transcript is composed of introns that are removed during RNA splicing (Chapter 3). To obtain a DNA fragment that contains only the protein-coding region of the gene, researchers make use of reverse transcriptase, which is an enzyme that produces a double-stranded DNA molecule from a single-stranded RNA template. For example, researchers can start with insulin mRNA, add reverse transcriptase, and obtain a DNA molecule that can be used as the donor. A DNA molecule produced with reverse transcriptase is known as **complementary DNA (cDNA)**. Reverse transcriptase is discussed further in Chapter 13.

Once a molecule of donor DNA has been isolated, another requirement for recombinant DNA is a **vector** sequence into which the donor fragment is inserted. The vector is the carrier of the donor fragment, and it must have the ability to be maintained in bacterial cells. A frequently used vector is a bacterial **plasmid**, a small circular molecule of DNA found naturally in certain bacteria that can replicate when the bacterial genomic DNA replicates and be transmitted to the daughter bacterial cells when the parental cell divides. Many naturally occurring plasmids have been modified by genetic engineering to make them suitable for use as vectors in recombinant DNA technology.

A common method for producing recombinant DNA is shown in **Fig. 12.19**. In order to make sure donor DNA can be fused with vector DNA, both pieces are cut with the same restriction enzyme so they both have the same overhangs. In the example shown in Fig. 12.19, the donor DNA is a fragment produced by digestion of genomic DNA with the restriction enzyme *Eco*RI, resulting in four-nucleotide 5' overhangs at the fragment ends. The vector is a circular plasmid that contains a single *Eco*RI cleavage site, so digestion of the vector with *Eco*RI opens the circle with a single cut that also has four-nucleotide 5' overhangs. Note that the overhangs on the donor fragment and the vector are complementary, allowing the ends of the donor fragment to renature with the ends of the opened vector when the two molecules are mixed. Once this renaturation has taken place, the ends of the donor fragment and the vector are covalently joined by DNA ligase. The joining of the donor DNA to the vector creates the recombinant DNA molecule.

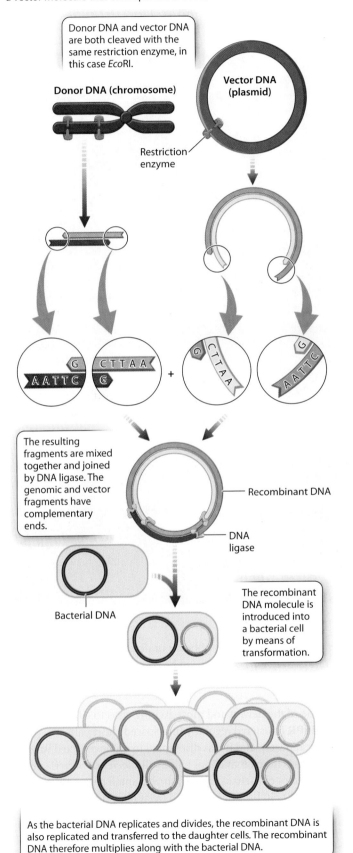

FIG. 12.19 Recombinant DNA. Donor DNA fragments are joined with a vector molecule that can replicate in bacterial cells.

The next step in the procedure is **transformation**, in which the recombinant DNA is mixed with bacteria that have been chemically coaxed into a physiological state in which they take up DNA from outside the cell. Having taken up the recombinant DNA, the bacterial cells are transferred into growth medium, where they multiply. Because the vector part of the recombinant DNA molecule contains all the DNA sequences needed for its replication and partition into the daughter cells, the recombinant DNA multiplies as the bacterial cell multiplies. If the recombinant DNA functions inside the bacterial cell, then new genetic characteristics may be expressed by the bacteria. For example, the recombinant DNA may allow the bacterial cells to produce a human protein, such as insulin or growth hormone.

Recombinant DNA is the basis of genetically modified organisms.

Applications of recombinant DNA have gone far beyond genetically engineered bacteria. Using methods that are conceptually similar to those described for bacteria but differing in many details, scientists have been able to produce varieties of genetically engineered viruses and bacteria, laboratory organisms, agricultural crops, and domesticated animals (**Fig. 12.20**). Examples include sheep that produce a human protein in their milk that is used to treat emphysema, chickens that produce eggs containing human antibodies to help fight harmful bacteria, and salmon with increased growth hormone for rapid growth. Plants such as corn, canola, cotton, and many others have been engineered to resist insect pests. Additional engineered products include rice with a high content of vitamin A, tomatoes with delayed fruit softening, potatoes with waxy starch, and sugarcane with increased sugar content. To model disease, researchers have used recombinant DNA to produce organisms such as laboratory mice that have been engineered to develop heart disease and diabetes. By studying these organisms, researchers can better understand human diseases and begin to find new treatments for them.

Genetically engineered organisms are known as **transgenic organisms** or **genetically modified organisms (GMOs)**. Transgenic laboratory organisms are indispensable in the study of gene function and regulation and for the identification of genetic risk factors for disease. In crop plants and domesticated animals, GMOs promise enhanced resistance to disease, faster growth and higher yields, more efficient utilization of fertilizer or nutrients, and improved taste and quality. However, there are concerns about unexpected effects on human health or the environment, the increasing power and influence of agribusiness conglomerates, and ethical objections to tampering with the genetic makeup of animals and plants. Nevertheless, more than 250 million acres of GMO crops are grown annually in more than 20 countries. The majority of this acreage is in the United States and South America. Resistance to the use of GMOs in Europe remains strong and vocal.

FIG. 12.20 Genetically modified organisms (GMOs). Organisms can be genetically modified so that (a) wheat resists weed-killing herbicides, (b) soybeans resist insects and have improved oil quality, (c) sheep produce more healthy fats, (d) chickens are unable to spread bird flu, (e) salmon have faster growth, and (f) pigs digest plant phosphorus more efficiently. *Sources: a. Adam Hart-Davis/Science Source; b. Steve Percival/Science Source; c. meirion matthias/Shutterstock; d. John Daniels/Ardea.com; e. AquaBounty Technologies; f. Bloomberg/Getty Images.*

DNA editing can be used to alter gene sequences almost at will.

Recombinant DNA technology combines existing DNA from two or more different sources. Its usefulness in research, medicine, and agriculture, however, is limited by the fact that it can make use only of existing DNA sequences. Therefore, scientists have also developed many different techniques to alter the nucleotide sequence of almost any gene in a deliberate, targeted fashion. In essence, these techniques allow researchers to "rewrite" the nucleotide sequence so that specific mutations can be introduced into genes to better understand their function, or mutant versions of genes can be corrected to restore normal function. Collectively, these techniques are known as **DNA editing.**

One of the newest and most exciting ways to edit DNA is **CRISPR** (clustered regularly interspaced short palindromic repeats), and it was discovered in an unexpected way. Researchers noted that about half of all species of bacteria and most species of Archaea contain similar small segments of DNA of about 20–50 base pairs derived from plasmids or viruses, but their function was, at first, a mystery. Later, it was discovered that they play a role in bacterial defense. When a bacterium is infected by a virus for the first time, it makes a copy of part of the viral genome and incorporates it into its genome. On subsequent infection by the same virus, the DNA copy of the viral genome is transcribed to RNA that combines with a protein that has a DNA-cleaving function. The RNA serves as a guide to identify target DNA in the virus by complementary base pairing, and the protein cleaves the target DNA, thus ending the viral threat. In this way, bacteria "remember" and defend themselves from past infections. The phrase "clustered regularly interspaced short palindromic repeats" describes the organization of the viral DNA segments in the bacterial genome.

In modern genetic engineering, the CRISPR mechanism is put to practical use to alter the nucleotide sequence of almost any gene in any kind of cell. One method is outlined in **Fig. 12.21**. Once researchers have identified a gene or DNA sequence they want to edit, they need to introduce, at different points in the process, three types of molecules into a cell containing that target DNA: a guide RNA that has been engineered to be complementary to the target DNA; a gene for a protein called Cas9 that cleaves DNA when it associates with the guide RNA; and a piece of DNA that acts as a template, with the desired new sequence for the target DNA. The target DNA will be identified by the guide RNA, cleaved by Cas9, and then replaced with the sequence of the template DNA.

The first step is to transform a cell with a plasmid that contains sequences that code for a CRISPR guide RNA as well as for Cas9. The guide RNA contains a region that can form a hairpin-shaped structure, as well as a region that researchers previously engineered to have bases complementary to the target DNA (Fig. 12.21a). When the guide RNA undergoes base pairing with the target DNA, Cas9 cleaves the target DNA (Fig. 12.21b). Exonucleases in the cell then expand the gap (Fig. 12.21c). The gap can be repaired using the newly introduced template DNA for editing the target DNA (Fig. 12.21d). This editing template DNA contains a sequence of interest to replace the degraded sequence of the target DNA, and is flanked by sequences complementary to the target. The strands of the gapped target DNA undergo base pairing with the complementary ends of the editing template, and DNA synthesis elongates the target DNA strands and closes the gap (Fig. 12.21d). The result is that the target DNA is restored, but its sequence has been altered according to the sequence present in the editing template (Fig. 12.21e).

FIG. 12.21 DNA editing. In this example, CRISPR guide RNA and its associated protein are used to cleave target DNA, and double-stranded template DNA is used in DNA repair to alter its sequence.

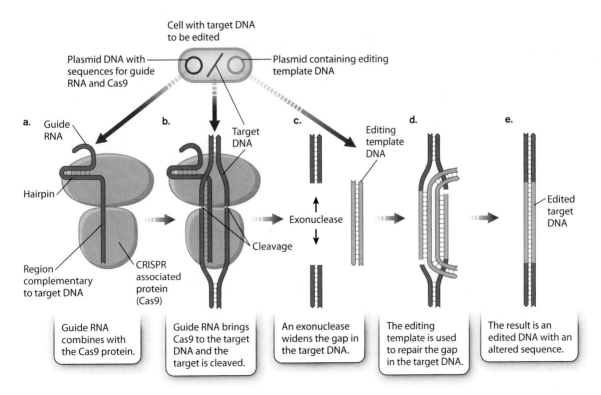

DNA editing by CRISPR is technically straightforward and highly efficient. The method has generated great interest because of its potential to correct genetic disorders of the blood, immune system, or other tissues and organs in which only a subset of cells with restored function can alleviate symptoms. CRISPR technology is so powerful that it can be used to insert new genes or change existing genes in virtually any cell type from any type of organism. In principle, the method could also be used to manipulate human germ cells, thereby affecting future generations. The uses and potential risks of human gene editing prompted a 2015 meeting of experts from around the world to discuss the science, ethics, and governance of the research, which resulted in a statement urging caution in the use of the technology to alter human germ cells.

Self-Assessment Questions

11. In making recombinant DNA molecules that combine restriction fragments from different organisms, researchers usually prefer restriction enzymes such as *Bam*HI or *Hin*dIII that generate fragments with "sticky ends" (ends with overhangs) rather than enzymes such as *Hpa*I or *Sma*I that generate fragments with "blunt ends" (ends without overhangs). Can you think of a reason for this preference?
12. How can recombinant DNA techniques be used to express a mammalian gene in bacteria?

CORE CONCEPTS SUMMARY

12.1 DNA REPLICATION: In DNA replication, a single parental molecule of DNA produces two daughter molecules.

DNA replication involves the separation of the two strands of the double helix at a replication fork and the use of these strands as templates to direct the synthesis of new strands. page 257

DNA replication is semiconservative, meaning that each daughter DNA molecule consists of a newly synthesized strand and a strand that was present in the parental DNA molecule. page 257

Nucleotides are added to the 3' end of the growing strand. Therefore, synthesis proceeds in a 5'-to-3' direction. page 258

Because synthesis occurs in a 5'-to-3' direction, at the replication fork, one new strand is synthesized continuously (the leading strand) and the other is synthesized in small pieces that are ligated together (the lagging strand). page 261

DNA polymerase requires RNA primers for DNA synthesis. page 261

DNA polymerase can correct its own mistakes by detecting a pairing mismatch between a template base and an incorrect new base. page 263

12.2 REPLICATION OF CHROMOSOMES: The replication of linear chromosomal DNA requires mechanisms that ensure efficient and complete replication.

Linear chromosomal DNA has many origins of replication, and replication proceeds from all of these almost simultaneously. page 264

Telomerase prevents chromosomes from shortening after each round of replication by adding a short stretch of DNA to the ends of chromosomes. page 264

12.3 DNA TECHNIQUES: Techniques for manipulating DNA follow from the basics of DNA structure and replication.

The polymerase chain reaction (PCR) is a technique for amplifying a segment of DNA. page 267

PCR requires a DNA template, DNA polymerase, the four nucleoside triphosphates, and two primers. It is a repeated cycle of denaturation, annealing, and extension. page 267

Gel electrophoresis allows DNA fragments to be separated according to size, with small fragments migrating farther than big fragments in a gel. page 269

Restriction enzymes cut DNA at specific recognition sequences called restriction sites, which cut DNA and leave a single-stranded overhang or a blunt end. page 269

In DNA denaturation, the two strands of a single DNA molecule separate from each other. In DNA renaturation or hybridization, two complementary strands come back together again. page 270

A Southern blot involves using a filter paper as a substrate for DNA fragments that have been cut up by restriction enzymes and hybridized to a probe. page 271

In Sanger sequencing of DNA, dideoxynucleotide chain terminators are used to stop the DNA synthesis reaction and produce a series of short DNA fragments from which the DNA sequence can be determined. page 272

New DNA sequencing technologies are being developed to increase the speed and decrease the cost of sequencing, perhaps making it possible to sequence everyone's personal genomes. page 273

12.4 GENETIC ENGINEERING: Genetic engineering allows researchers to alter DNA sequences in living organisms.

A recombinant DNA molecule can be made by cutting DNA from two organisms with the same restriction enzyme and then using DNA ligase to join them. page 274

Recombinant DNA is the basis for genetically modified organisms (GMOs), which offer both potential benefits and risks. page 275

Almost any DNA sequence in an organism can be altered by means of a form of DNA editing called CRISPR, which uses modified forms of molecules found in bacteria and archaeons that can cleave double-stranded DNA at a specific site. page 276

Log in to LaunchPad to check your answers to the Self-Assessment Questions and to access additional learning tools.

CHAPTER 13 Genomes

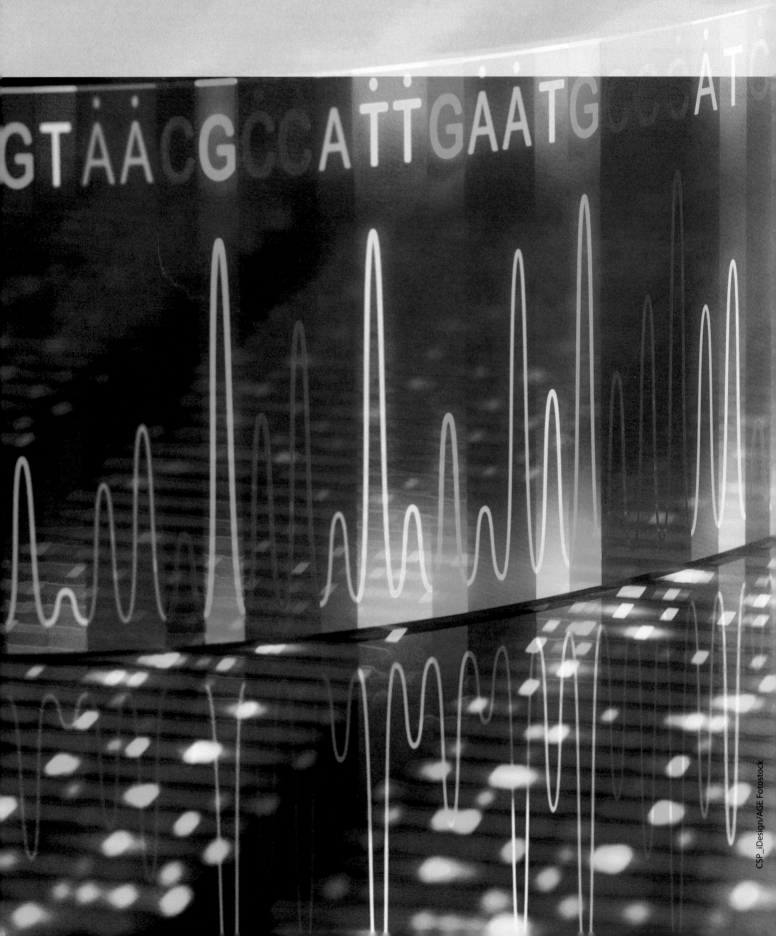

CORE CONCEPTS

13.1 GENOME SEQUENCING: A genome is the genetic material of a cell, organism, organelle, or virus, and its sequence is the order of bases along the DNA or (in some viruses) RNA.

13.2 GENOME ANNOTATION: Researchers annotate genome sequences to identify genes and other functional elements.

13.3 GENES, GENOMES, AND ORGANISMAL COMPLEXITY: The number of genes in a genome and the size of a genome do not correlate well with the complexity of an organism.

13.4 ORGANIZATION OF GENOMES: The orderly packaging of DNA allows it to carry out its functions and fit inside the cell.

13.5 VIRUSES AND VIRAL GENOMES: Viruses have diverse genomes, but all require a host cell to replicate.

In Chapter 12, we saw how small pieces of DNA are isolated, identified, and sequenced. This technology has advanced to the point where the complete genome sequences for thousands of species have been determined, including those of humans and our closest primate relatives, as well as dozens of other mammals. The term **genome** refers to the genetic material of an organism, cell, nucleus, organelle, or virus. Some genomes, like that of HIV, are small, whereas others, like the human genome, are large. Technically, the human genome refers to the DNA in the chromosomes present in a sperm or egg. However, the term is often used informally to mean all of the genetic material in an organism. So your personal genome consists of the DNA in two sets of chromosomes, one inherited from each parent, plus a much smaller mitochondrial genome inherited from your mother. The human genome sequence is so long that printing it in the size of the type used in this book would require 1.5 million pages. As we will see, however, the human genome is far from the largest among organisms.

The sequence of a genome is a long string of A's, T's, G's, and C's, which represent the order of bases present in successive nucleotides along the DNA molecules in the genome. However, a genome sequence, on its own, is not very useful to scientists. Additional research is required to understand what proteins and other molecules are encoded in the genome sequence, and to learn when these molecules are produced during an organism's lifetime and what they do.

In this chapter, we discuss how the sequence of a genome is determined and analyzed to reveal its key biological features, such as the protein-coding genes. We also examine what other kinds of DNA sequences are present in genomes and how these sequences are organized, with special emphasis on the human genome. Finally, we explore the diversity of genome types present in viruses.

13.1 GENOME SEQUENCING

What exactly is a genome? Originally, the term referred to the complete set of chromosomes present in a reproductive cell, like a sperm or an egg, which in the human genome is 23 chromosomes. The word "genome" is almost as old as the word "gene," and it was coined at a time when chromosomes were thought to consist of densely packed genes lined up one after another. A gene, as you may recall from Chapter 3, is a region of DNA associated with some function in the cell or organism. Almost all genes are initially transcribed into RNA, which may be chemically modified or processed. For some genes, the processed RNA can function on its own, but for most genes the processed RNA codes for protein (Chapter 4).

Chromosomes consist primarily of DNA and associated proteins, and the genetic information in the chromosomes resides in the DNA. One might therefore define the genome as the DNA molecules that are transmitted from parents to offspring. This definition has the advantage of including the DNA in organisms that lack true chromosomes, such as bacteria and archaeons, as well as eukaryotic organelles that contain their own DNA, such as mitochondria and chloroplasts. However, a definition restricted to DNA is too narrow because it excludes viruses like HIV, whose genetic material consists of RNA. Defining a genome as the genetic material transmitted from parent to offspring therefore embraces all known cellular forms of life, all known organelles, and all viruses.

In this first section, we focus on how genomes are sequenced, building on the DNA sequencing technology introduced in Chapter 12.

Complete genome sequences are assembled from smaller pieces.

In Chapter 12, we discussed a method of DNA sequencing known as Sanger sequencing. Sequencing methods are now automated to the point where specialized machines can determine the sequence of billions of DNA nucleotides in a single day. However, even with recent advances, the data are obtained in the form of short sequences, typically less than a few hundred nucleotides long. If you are interested in sequencing a short DNA fragment, these technologies work well. However, let's say you are interested in the sequence of human chromosome 1, which includes a DNA molecule approximately 250 million nucleotides long. How can you sequence a DNA molecule as long as that?

In one approach, the single long DNA molecule is first broken into small fragments, each of which is short enough to be sequenced by existing technologies. Even though the sequence data obtained from a small DNA fragment is only a minuscule fraction of the length of the DNA molecules in most genomes, each run of an automated sequencing machine yields hundreds of millions of these short sequences from random locations throughout the genome. To sequence a whole genome, researchers typically sequence such a large number of random DNA fragments that, on average, any particular small region of the genome is sequenced 10–50 times. This redundancy is necessary to minimize both the number of errors present in the final genome sequence and the number and size of gaps where the genome sequence is incomplete.

When the sequences of a sufficient number of short stretches of the genome have been obtained, the next step is **sequence assembly**: The short sequences are put together in the correct order to generate the long, continuous sequence of nucleotides in the DNA molecule present in each chromosome.

Assembly is accomplished using complex computer programs, but the principle is simple. The short sequences are assembled according to their overlaps, as illustrated in **Fig. 13.1**, which uses a sentence to represent the nucleotide sequence. This approach is called **shotgun sequencing** because the sequenced fragments do not originate from a particular gene or region but from sites scattered randomly across the chromosome.

Sequences that are repeated complicate sequence assembly.

Sequence assembly is not quite as straightforward as Fig. 13.1 suggests. Real sequences are composed of the nucleotides A, T, G, and C, and any given short sequence could come from either strand of the double-stranded DNA molecule. Therefore, the overlaps between fragments must be long enough both to ensure that the assembly is correct and to determine from which strand of DNA the short sequence originated.

HOW DO WE KNOW?

FIG. 13.1

How are whole genomes sequenced?

BACKGROUND DNA sequencing technologies can only determine the sequence of DNA fragments far smaller than the genome itself. How can the sequences of these small fragments be used to determine the sequence of an entire genome? In the early years of genome sequencing, many researchers thought that it would be necessary to know first where in the genome each fragment originated before sequencing it. A group at Celera Genomics reasoned that if so many fragments were sequenced that the ends of one would almost always overlap with those of others, then a computer program with sufficient power might be able to assemble the short sequences to reveal the sequence of the entire genome. By January 2000, Celera Genomics announced that it had assembled 90% of the human genome using its new computational methods.

HYPOTHESIS A genome sequence can be determined by sequencing small, randomly generated DNA fragments and assembling them into a complete sequence by matching regions of overlap between the fragments.

EXPERIMENT Hundreds of millions of short sequences from the genome of the fruit fly, *Drosophila melanogaster*, were sequenced. Fig. 13.1a shows examples of overlapping fragments, using a sentence from Watson and Crick's original paper on the chemical structure of DNA as an analogy.

RESULTS The computer program the group had written to assemble the fragments worked. The researchers were able to sequence the entire *Drosophila* genome by piecing together the fragments according to their overlaps. In the sentence analogy, the fragments (Fig. 13.1a) can be assembled into the complete sentence (Fig. 13.1b) by matching the overlaps between the fragments.

CONCLUSION The hypothesis was supported: Celera Genomics was able to determine the entire genomic sequence of an organism by sequencing small, random fragments and piecing them together at their overlapping ends.

Some features of genomes present additional challenges to sequence assembly, and the limitations of the computer programs for handling such features require hands-on assembly. Chief among these complicating features is the problem of repeated sequences known collectively as **repetitive DNA**.

There are many types of repeated sequences in eukaryotic genomes, and some are shown in **Fig. 13.2**. The repeated sequence may be several thousand nucleotides long and present in multiple identical or nearly identical copies. These long repeated sequences may be dispersed throughout the genome (Fig. 13.2a), or they may be in tandem, meaning that they are next to each other (Fig. 13.2b).

The difficulty with long repeated sequences is that they typically are much longer than the short fragments sequenced by automated sequencing. As a result, the repeat may not be detected at all. If the repeat is actually detected, there is no easy way of knowing the number of copies of the repeat, that is, whether the DNA molecule includes two, three, four, or any number of copies of the repeat. Sometimes, researchers can use the ends of repeats, where the fragments overlap with an adjacent, nonrepeating sequence, as a guide to the position and number of repeats.

To illustrate the assembly problems caused by repeated sequences, let's consider an analogy. In Shakespeare's play *Hamlet*, the word "Hamlet" occurs about 500 times throughout the text, similar to a dispersed repeat (Fig. 13.2a). If you chose short sequences of letters from the play at random and

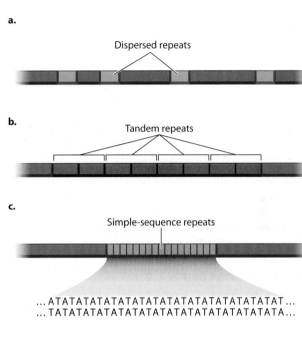

FIG. 13.2 Principal types of sequence repeats found in eukaryotic genomes. Repeats often pose problems in DNA sequencing. (Repeats are not drawn to scale.)

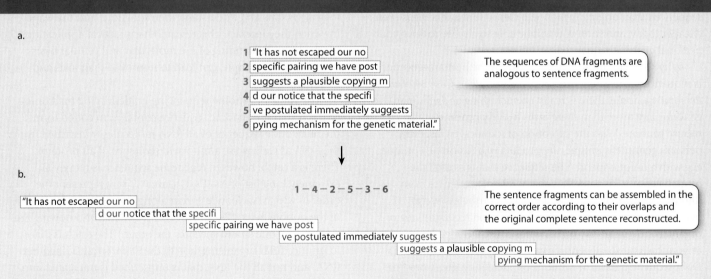

FOLLOW-UP WORK Today, computational methods are routinely used to assemble genome sequences. Computer assembly is also used to infer the genome sequences of hundreds of bacterial species simultaneously—for example, in bacterial communities sampled from seawater or from the human gut.

SOURCE Adams, M. D., et al. 2000. "The Genome Sequence of *Drosophila melanogaster*." *Science* 287:2185–2195.

found the sequence "Hamlet" or "amlet" or "Haml," you would have no way of knowing which of the 500 "Hamlets" these letters came from. Only if you found sequences that contained "Hamlet" overlapping with adjacent unique text could you identify their origin.

In another type of repeat, the repeating sequence is short. It could be as short as two nucleotides, such as AT, repeated over and over again in a stretch of DNA (Fig. 13.2c). Short repeating sequences of this kind are troublesome for sequencing machines because any single-stranded fragment consisting of alternating AT bases can fold back upon itself to form a double-stranded structure in which A is paired with T. Such structures are more stable than the unfolded single-stranded structures and are not easily sequenced. Similarly, the sequence 5'-AAAAAATTTTT-3' can also fold back itself to form a hairpin because the 5'-AAAAAA-3' is complementary to the 3'-TTTTTT-5'.

 CASE 3 YOUR PERSONAL GENOME: YOU, FROM A TO T

Why sequence your personal genome?

The goal of the Human Genome Project, which began in 1990, was to sequence the human genome as well as the genomes of certain organisms used as models in genetic research. The model organisms chosen are the mainstays of laboratory biology—a species each of bacteria, yeast, nematode worm, fruit fly, and mouse. By 2003, the genome sequences of these model organisms had been completed, as well as that of the human genome. By then, the cost of sequencing had become so low and the sequence output so high that many more genomes were sequenced than originally planned. Genome sequencing is even cheaper today, and soon you will be able to choose to have your personal genome sequenced.

Why sequence more genomes? And if the human genome is sequenced, why sequence yours? As we saw in Case 3 Your Personal Genome, there is really no such thing as *the* human genome, any more than there is *the* fruit fly genome or *the* mouse genome. With the exception of identical twins, every person's genome is unique, the product of a fusion of a unique egg with a unique sperm. The sequence that is called "the human genome" is actually a composite of sequences from different individuals. This sequence is nevertheless useful because most of us share the same genes and regulatory regions, organized the same way on chromosomes. Detailed knowledge of your own personal genome can be valuable. Our individual DNA sequences differ at millions of nucleotide sites from one person to the next. Some of these differences account in part for the physical differences we see among us; others have the potential to predict susceptibility to disease and response to medication. For Claudia Gilmore, knowledge of the sequence of her *BRCA1* gene had a significant impact on her life.

Determining these differences is a step toward **personalized medicine**, in which an individual's genome sequence, by revealing his or her disease susceptibilities and drug sensitivities, allows treatments to be tailored to that individual. There may come a time, perhaps within your lifetime, when personal genome sequencing becomes part of routine medical testing. Information about a patient's genome will bring benefits, but it will also raise ethical concerns and pose risks to confidentiality and insurability.

Self-Assessment Questions

1. DNA sequencing technology has been around since the late 1970s. Why did sequencing whole genomes present a challenge?

2. What is the shotgun method for determining the complete genome sequence of an organism?

3. Repeated sequences can be classified according to their organization in the genome as well as according to their function. What are two examples of each?

4. Let's say that a stretch of repeated AT is successfully sequenced. From what you know of the difficulties of sequencing long repeated sequences, what other problems might you encounter in assembling these fragments?

13.2 GENOME ANNOTATION

A goal of biology is to identify all the macromolecules in biological systems and understand their functions and the ways in which they interact. This research has practical applications: increased understanding of the molecular and cellular basis of disease, for example, can lead to improved diagnosis and treatment.

The value of genome sequencing in identifying macromolecules is that the genome sequence contains, in coded form, the nucleotide sequence of all RNA molecules transcribed from the DNA as well as the amino acid sequence of all proteins. There is a catch, however. A genome sequence is merely an extremely long list of A's, T's, G's, and C's that represent the order in which nucleotides occur along the DNA in one strand of the double helix. (Because of complementarity, knowing the sequence of one strand specifies the other.) The catch is that in multicellular organisms, not all the DNA is transcribed into RNA, and not all the RNA that is transcribed is translated into protein. Therefore, genome sequencing is just the first step in understanding the function of any particular DNA sequence. Following genome sequencing, the next step is to identify the locations and functions of the various types of sequence present in the genome.

Genome annotation identifies various types of sequence.

Genomes contain many different types of sequence, among them protein-coding genes (Chapters 3 and 4). Protein-coding genes are themselves composed of different regions, including transcribed and untranscribed regions. Transcribed regions contain exons with codons that specify the amino acid sequence of a polypeptide chain and introns that are removed during RNA processing. Untranscribed regions include promoters indicating where RNA transcription begins and other elements that specify when and where the RNA transcript is produced. Genomes also contain coding sequences for RNAs that are not translated into protein (noncoding RNAs), such as ribosomal RNA, transfer RNA, and other types of small RNA molecule. Although much of the DNA in the genomes of multicellular organisms is transcribed at least in some cell types, the functions of a large portion of these transcripts in metabolism, physiology, development, or behavior are unknown.

Genome annotation is the process by which researchers identify the various types of sequence present in genomes. Genome annotation is essentially an exercise in adding commentary to a genome sequence that identifies which types of sequence are present and where they are located. It can be thought of as a form of pattern recognition, where the patterns are regularities in a sequence that are characteristic of protein-coding genes or other types of sequence.

An example of genome annotation is shown in **Fig. 13.3**. Genes present in one copy per genome are indicated in orange. Most single-copy genes are protein-coding genes. In some computer programs that analyze genomes, the annotation of a single-copy gene also specifies any nearby regulatory elements that control transcription, the intron–exon boundaries in the gene, and any known or predicted alternative forms in which the introns and exons are spliced (Chapter 3). Each single-copy gene is given a unique name and its protein product is identified. Note in Fig. 13.3 that single-copy genes can differ in size from one gene to the next. The annotations in Fig. 13.3 also specify the locations of sequences that encode RNAs that are not translated into proteins, as well as various types of repeated sequence. Most genomes also contain some genes that are present in multiple copies that originate from gene duplication. However, these are usually identifiable in the same way as single-copy genes because the copies often differ enough in sequence that they can be distinguished.

Small genomes, such as that of HIV and other viruses, can be annotated by hand, but for large genomes like the human genome, computers are essential. In the human genome, some protein-coding genes extend for more than a million nucleotides. Roughly speaking, if the sequence of the approximately 3 billion nucleotides in a human egg or sperm was printed in normal-sized type like this, the length of the ribbon would stretch 4000 miles (6440 kilometers), which is about the distance from Fairbanks, Alaska, to Miami, Florida. By contrast, for the approximately 10,000 nucleotides in the HIV genome, the ribbon would extend a mere 70 feet (21 meters).

Genome annotation is an ongoing process because, as macromolecules and their functions and interactions become better understood, the annotations to the genome must be updated. A sequence that is annotated as nonfunctional today may be found to have a function tomorrow. For this reason, the annotation of certain genomes—including the human genome—will certainly continue to change.

Genome annotation includes searching for sequence motifs.

Because genome annotation is essentially pattern recognition, researchers begin annotating a genome by identifying patterns called **sequence motifs**, which are telltale sequences of nucleotides that indicate what types of function (or absence of function) may be encoded in a particular region of the genome (**Fig. 13.4**). Researchers can find sequence motifs in the DNA itself or in the RNA sequence inferred from the DNA sequence. Once identified, sequence motifs are typically confirmed by experimental methods. One sequence motif we have already encountered in Chapter 3 is a promoter, a DNA sequence where RNA polymerase and associated proteins bind to the DNA to initiate transcription.

Another example of a sequence motif is an **open reading frame (ORF)**, which researchers can identify in the DNA sequence or in the RNA sequence inferred from the DNA sequence (Fig. 13.4a). An ORF is a long string of nucleotides

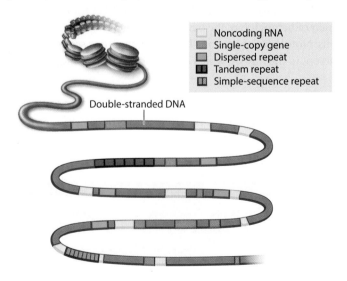

FIG. 13.3 Genome annotation. Given the DNA sequence of a genome, researchers can pinpoint locations of various types of sequence.

FIG. 13.4 Some common sequence motifs useful in genome annotation. (a) An open reading frame; (b) a noncoding RNA molecule; (c) transcription factor binding sites.

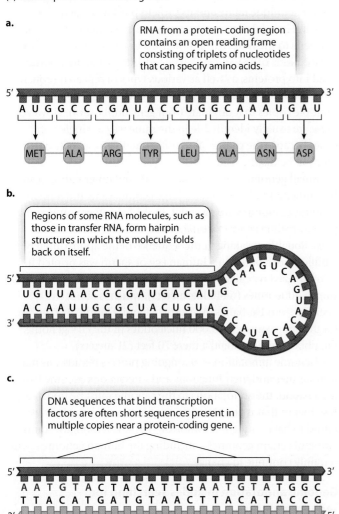

that, if transcribed and processed into messenger RNA, results in a set of codons for amino acids that does not contain a stop codon. The presence of an ORF motif alone is enough for a researcher to annotate the DNA segment as potentially protein coding. The qualifier "potentially" is necessary because ORFs identified in a DNA or RNA sequence may not actually produce a protein. For this reason, they are often called putative ORFs. A region containing a putative ORF may exist merely by chance; a putative ORF may not be transcribed; or a putative ORF might be transcribed but not translated if it produces a noncoding RNA or is found in what turns out to be in an intron. In the next section, we discuss how the analysis of messenger RNA sequences can determine whether a putative ORF is an actual ORF in DNA coding for protein.

Fig. 13.4b shows another sequence motif present in a hypothetical RNA transcript inferred from the DNA sequence. The nucleotide sequence at one end of the RNA is complementary to that at the other end, so the single-stranded molecule is able to fold back on itself and undergo base pairing to form a hairpin-shaped structure. Such hairpin structures are characteristic of certain types of RNA that function in gene regulation (Chapter 18). The DNA from which this RNA is transcribed also has complementary sequences on either end.

Fig. 13.4c shows one more type of sequence motif—a short sequence that is a binding site for DNA-binding proteins called transcription factors, which initiate transcription (Chapter 3). Transcription factor binding sites are often present in multiple copies (two copies, in red in Fig. 3.14c) and in either strand of the DNA. Sometimes they are located near the region of a gene where transcription is initiated because the transcription factor helps determine when the gene will be transcribed. However, they can also be located far upstream of the gene, downstream of the gene, or in introns, so their identification is difficult.

Comparison of genomic DNA with messenger RNA reveals the intron–exon structure of genes.

In annotating an entire genome, researchers typically make use of information outside the genome sequence itself. This information may include sequences of messenger RNA molecules that are isolated from various tissues or various stages of development of the organism. Recall from Chapter 3 that messenger RNA (mRNA) molecules have undergone processing and are therefore usually simpler than the DNA sequences from which they are transcribed. For example, introns are removed and exons are spliced together. The resulting mature mRNA therefore contains a long sequence of codons uninterrupted by a stop codon—in other words, an ORF. The ORF in an mRNA is the region that is actually translated into protein.

One aspect of genome annotation is the determination of which portions of the genome sequence correspond to sequences in mRNA transcripts. An example is shown in **Fig. 13.5**, which compares the DNA and mRNA for the beta (β) chain of hemoglobin, the oxygen-carrying protein in red blood cells. Note that the genomic DNA contains some sequences present in the mRNA, which correspond to exons, and some

FIG. 13.5 Identification of exons and introns by comparison of genomic DNA with mRNA sequence.

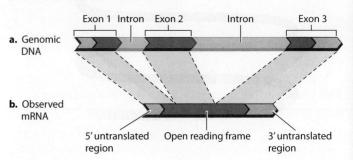

sequences that are not present in the mRNA, which correspond to introns. Comparison of mRNA with genomic DNA therefore reveals the intron–exon structure of protein-coding genes. In fact, introns were first discovered by comparing β-globin mRNA with genomic DNA.

An annotated genome summarizes knowledge, guides research, and reveals evolutionary relationships among organisms.

Genome annotation, which aims to identify all the functional and repeat sequences present in the genome, is an imperfect science. Even in a well-annotated genome, some protein-coding sequences or other important features may be overlooked, and occasionally the annotation of a sequence motif is incorrect.

Because researchers often have to rely on sequence motifs alone and not experimental data, their descriptions may be vague. For example, a common annotation in large genomes is "hypothetical protein." In some cases, such as the genome of the malaria parasite, this type of annotation accounts for about 50% of the possible protein-coding genes. There is no hint of what a hypothetical protein may do or even whether it is actually produced because it is determined solely by the presence of a putative ORF in the genomic sequence and not by the presence of actual mRNA or protein. Other annotations might be "DNA-binding protein," "possible hairpin RNA," or "tyrosine kinase"—with no additional detail about these motifs' functions or relationships to other sequences. In short, although some genome annotations summarize experimentally verified facts, many others are hypotheses and guides to future research.

Genome sequences contain information about ancestry and evolution, and so comparisons among genomes can reveal how different species are related. For example, the sequence of the human genome is significantly more similar to that of the chimpanzee than to that of the gorilla, indicating a more recent common ancestry of humans and chimpanzees (Chapter 23).

Analysis of the similarities and differences in protein-coding genes and other types of sequence in the genomes of different species is an area of study called **comparative genomics**. Such studies help us understand how genes and genomes evolve. They can also guide genome annotation because the sequences of important functional elements are often very similar among genomes of different organisms. Sequences that are similar in different organisms are said to be **conserved**. If a conserved sequence is found in several different organisms, researchers might look for that sequence in the genome they are annotating. A sequence motif that is conserved is likely to be important, even if its function is unknown, because it has changed very little over evolutionary time.

The HIV genome illustrates the utility of genome annotation and comparison.

HIV and related viruses provide an example of how genomes are annotated and how comparisons among genome sequences can reveal evolutionary relationships. A **virus** is a small infectious agent that contains a nucleic acid genome packaged inside a protein coat called a **capsid**. Viruses can bind to receptor proteins on cells of the host organism, enabling the viral genome to enter the cell. Infection of a host cell is essential to viral reproduction because viruses use cellular ATP and hijack cellular machinery to replicate, transcribe, and translate their genome in order to make more viruses. Later in this chapter, we will see that viruses can differ in the types of nucleic acid that make up their genome and how their messenger RNA is produced, but here we consider only the genome of HIV.

Whereas the genome of all cells consists of double-stranded DNA, the genome of HIV is single-stranded RNA. The sequence of the HIV genome identifies it as a retrovirus that replicates by a DNA intermediate that can be incorporated into the host genome. More narrowly, the sequence of the HIV genome groups it among the mammalian lentiviruses, so named because of the long lag between the initial time of infection and the appearance of symptoms (*lenti* means "slow").

Fig. 13.6 shows the evolutionary relationships among a sample of lentiviruses, which are grouped according to the similarity of their genome sequences. The evolutionary tree shows that closely related viruses have closely related hosts. For example, simian lentiviruses are more closely related to one another than they are to cat lentiviruses. This observation implies that the genomes of the viruses evolve along with the genomes of their hosts. A second feature shown by the evolutionary tree is that human HIV originated from at least two

FIG. 13.6 Evolutionary relationships among viruses related to HIV.
Data from A. Katzourakis, M. Tristem, O. G. Pybus, and R. J. Gifford, 2007, "Discovery and Analysis of the First Endogenous Lentivirus," Proceedings of the National Academy of Sciences USA 104(15): 6261–6265.

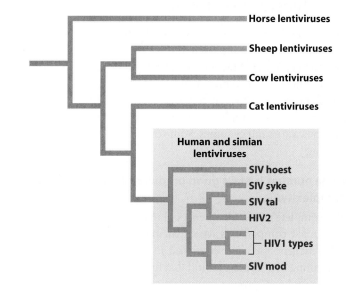

separate simian viruses that switched hosts from simians (most likely chimpanzees) to humans.

The annotated sequence of the HIV genome tells us a lot about the biology of this virus (**Fig. 13.7**). Many details are left out here, but the main point is to show the functional elements of HIV in the form of an annotated genome. The open reading frame denoted *gag* encodes protein components of the capsid, *pol* encodes proteins needed for reverse transcription of the viral RNA into DNA and incorporation into the host genome, and *env* encodes proteins that are embedded in the lipid envelope. The annotation in Fig. 13.7 also includes the genes *tat* and *rev*, which are encoding proteins that are essential for the HIV life cycle, as well as the genes *vif, vpr, vpu,* and *nef*, which encode proteins that enhance virulence in organisms. Identification of the genes necessary to complete the HIV cycle is the first step to finding drugs that can interfere with the cycle and prevent infection.

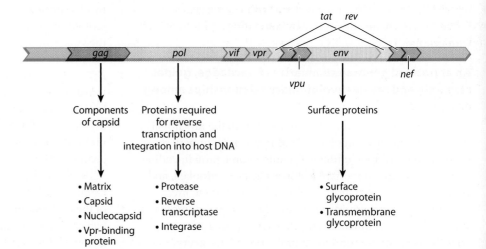

FIG. 13.7 The annotated genome of HIV.

Self-Assessment Questions

5. What is the purpose of genome annotation?
6. What does a comparison of the sequence of genomic DNA to messenger RNA indicate about the structure of a gene?
7. How can comparing the sequences of two genomes help to infer evolutionary relationships?

13.3 GENES, GENOMES, AND ORGANISMAL COMPLEXITY

Before whole-genome sequencing, molecular biology and evolution research tended to focus on single genes. Researchers studied how individual genes are turned on and off, and how a gene in one organism is related to a gene in another organism. Now, with the availability of genome sequences from multiple species, it is possible to make comparisons across full genomes. Some of the results have been striking.

Gene number is not a good predictor of biological complexity.

The complete genome sequences of many organisms allow us to make comparisons among them (**Table 13.1**). As different genomes were sequenced and annotated, it came as something of a surprise to find that humans have about the same number of protein-coding genes as many organisms with much smaller genomes. For example, the small flowering plant *Arabidopsis thaliana* has a genome only about 5% as large as that of humans, but the plant has at least as many protein-coding genes. The nematode worm *Caenorhabditis elegans* contains only 959 cells, of which 302 are nerve cells that form the worm's brain. Humans have 100 trillion cells, of which 100 billion are nerve cells. Despite having 100 million times as many cells as the worm, we have roughly the same number of protein-coding genes. Another example of the disconnect between level of complexity and genome size is the observation that the genome size of the mountain grasshopper *Podisma pedestris* is more than 100 times that of the fruit fly *Drosophila melanogaster*. Although there is no agreed-upon definition of organismal complexity, it would be hard to make a case that grasshoppers are any more or less complex than fruit flies.

TABLE 13.1 Gene Numbers of Several Organisms. Humans have more genes than fruit flies and nematode worms, but not as many as might be expected on the basis of complexity.

COMMON NAME	SPECIES NAME	APPROXIMATE NUMBER OF PROTEIN-CODING GENES
Mustard plant	*Arabidopsis thaliana*	27,000
Human	*Homo sapiens*	25,000
Nematode worm	*Caenorhabditis elegans*	22,000
Fruit fly	*Drosophila melanogaster*	17,000
Baker's yeast	*Saccharomyces cerevisiae*	6000
Gut bacterium	*Escherichia coli*	4000

Data from "Sequence Composition of the Human Genome" in International Genome Sequencing Consortium, 2001, "Initial Sequencing and Analysis of the Human Genome" (PDF). *Nature* 409(6822): 860–921, doi:10.1038/35057062, PMID 11237011.

On the other hand, some organisms are able to do more things with the genes they have than other organisms. For example, the expression of protein-coding genes can be regulated in many subtle ways, causing different gene products to be made in different amounts in different cells at different times (Chapters 18 and 19). Differential gene expression allows the same protein-coding genes to be deployed in different combinations to yield a variety of distinct cell types. In addition, proteins can interact with one another so that, even though there are relatively few types of protein, they are capable of combining in many different ways to perform different functions. We also have seen that a single gene may yield multiple proteins, either because of alternative splicing (different exons are spliced together to make different proteins) or posttranslational modification (proteins undergo biochemical changes after they have been translated).

Overall, the data in Table 13.1 pose many tantalizing evolutionary questions and suggest that major evolutionary changes can be accomplished not only by the acquisition of whole new genes, but also by modifying existing genes and their regulation in subtle ways.

Viruses, bacteria, and archaeons have small, compact genomes.

Not only can we compare numbers of genes, we can also compare sizes of genomes in different organisms. Before making such comparisons, we need to understand how genome size is measured. Genomes are measured in numbers of base pairs, and the yardsticks of genome size are a thousand base pairs (a kilobase, kb), a million base pairs (a megabase, Mb), and a billion base pairs (a gigabase, Gb).

Most viral genomes range in size from 3 kb to 300 kb, but a few are very large. Giant viruses, which infect certain species of amoebae, have genomes of approximately 1.5 Mb. These genomes contain approximately 1500 genes, including genes for nearly every component needed for translation except for ribosomes, as well as genes for sugar, lipid, and amino acid metabolism that are not found in other viruses.

The largest viral genome is three times as large and contains three times as many genes as the genome of the bacterium *Mycoplasma genitalium*. At 580 kb and encoding only 471 genes, the genome of *M. genitalium* is the smallest known among free-living bacteria—those capable of living entirely on their own. The complete sequence of small bacterial genomes has allowed researchers to define the smallest, or minimal, genome (and, therefore, the minimal set of proteins) necessary to sustain life. Current findings suggest that the minimum number of genes necessary to encode all the functions essential to life is approximately 500.

The genomes of bacteria and archaeons are information dense, meaning that most of the genome has a defined function. Roughly speaking, 90% or more of their genomes consist of protein-coding genes (although, in many cases, the protein has an unknown function). Bacterial genomes range in size from 0.5 to 10 Mb. The bigger genomes have more genes, allowing these bacteria to synthesize small molecules that other bacteria have to scrounge for, or to use chemical energy in the covalent bonds of substances that other bacteria cannot. Archaeons, whose genomes range in size from 0.5 to 5.7 Mb, have similar capabilities.

Among eukaryotes, no relationship exists between genome size and organismal complexity.

In eukaryotes, just as the number of genes does not correlate well with organismal complexity, the size of the genome is unrelated to the metabolic, developmental, and behavioral complexity of the organism (**Table 13.2**). The range of genome sizes is huge, even among similar organisms (**Fig. 13.8**). For comparison, Fig. 13.8 also shows the range of genome sizes among bacteria and archaeons. The largest eukaryotic genome exceeds the size of the smallest by a factor of more than 500,000—and both the smallest and the largest are found among protozoa. The range among flowering plants (angiosperms) is about three orders of magnitude, and the range among animals is about seven orders of magnitude. One species of lungfish has a genome size more than 45 times larger than the human genome (Table 13.2). Clearly, no relationship exists between the size of the genome and the complexity of the organism.

The disconnect between genome size and organismal complexity is called the **C-value paradox**. The C-value is the

TABLE 13.2 Genome Sizes of Several Organisms. Genome size varies tremendously among eukaryotic organisms, and there is no correlation between genome size and the complexity of an organism.

COMMON NAME	SPECIES NAME	APPROXIMATE GENOME SIZE (Mb)
Fruit fly	*Drosophila melanogaster*	180
Fugu fish	*Fugu rubripes*	400
Boa constrictor	*Boa constrictor*	2100
Human	*Homo sapiens*	3100
Locust	*Schistocerca gregaria*	9300
Onion	*Allium cepa*	18,000
Newt	*Amphiuma means*	84,000
Lungfish	*Protopterus aethiopicus*	140,000
Fern	*Ophioglossum petiolatum*	160,000
Amoeba	*Amoeba dubia*	670,000

Data from "Sequence Composition of the Human Genome" in International Human Genome Sequencing Consortium, 2001, "Initial Sequencing and Analysis of the Human Genome" (PDF). *Nature* 409(6822): 860–921, doi:10.1038/35057062, PMID 11237011.

FIG. 13.8 Ranges of genome size. Even within a single group, genome sizes of different species can vary by several factors of 10. *After Fig. 1, p. 186 in T. R. Gregory, 2004, "Macroevolution, Hierarchy Theory, and the C-Value Enigma," Paleobiology 30:179–202.*

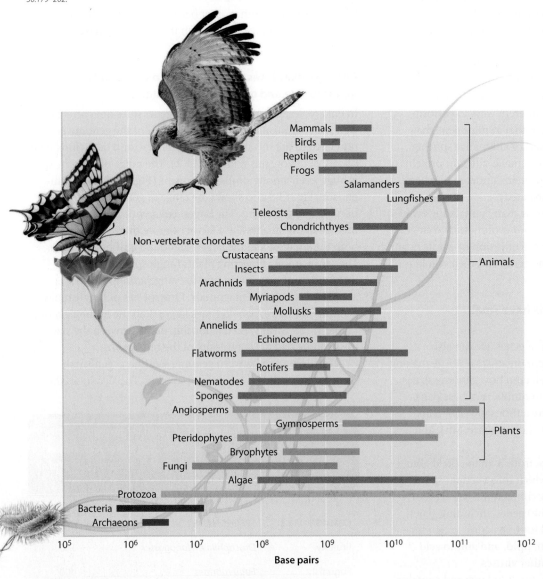

amount of DNA in a reproductive cell, and the paradox is the apparent contradiction between genome size and organismal complexity, leading to the difficulty of predicting one based on the other.

Why are some eukaryotic genomes so large? One reason is **polyploidy,** or having more than two sets of chromosomes in the genome. Polyploidy is especially prominent in many groups of plants. Humans have two sets of 23 chromosomes, giving us 46 chromosomes in total. The polyploid bread wheat *Triticum aestivum*, for example, has six sets of seven chromosomes.

Polyploidy has played an important role in plant evolution. Many agricultural crops are polyploid, including wheat, potatoes, olives, bananas, sugarcane, and coffee. Among flowering plants, it is estimated that 30% to 80% of existing species have polyploidy in their evolutionary histories, either because of the duplication of the complete set of chromosomes in a single species, or because of hybridization, or crossing, between related species followed by duplication of the chromosome sets in the hybrid (**Fig. 13.9**). Some ferns take polyploidy to an extreme: one species has 84 copies of a set of 15 chromosomes, or 1260 chromosomes altogether.

However, the principal reason for large genomes among some eukaryotes is that their genomes contain large amounts of DNA that do not code for proteins, such as introns and DNA sequences that are present in many copies. These repeated sequences are discussed next.

About half of the human genome consists of transposable elements and other types of repetitive DNA.

Complete genome sequencing has allowed the different types of noncoding DNA to be specified more precisely in a variety of organisms. It came as a great surprise to learn that in the human genome, only about 2.5% of the genome actually codes for proteins. The other 97.5% includes sequences we have encountered earlier, including introns, noncoding DNA, and other types of repetitive sequences.

In **Fig. 13.10**, we see the composition of the human genome, including the principal types of repeated sequences. Earlier, we saw how repeated sequences could be classified as dispersed or tandem according to their organization in the genome. In Fig. 13.10, the repeated sequences are classified according to their function. Among the repeated sequences is alpha (α) satellite DNA, which consists of tandem copies of a 171-bp sequence repeated near each centromere an average of 18,000 times. The α satellite DNA is essential for attachment of spindle fibers to the centromeres during cell division (Chapter 11).

Fig. 13.10 also shows the proportions of several types of repeated sequence collectively known as **transposable elements** (abbreviated *TE*; also called **transposons**), which are DNA sequences that can replicate and insert themselves into new positions in the genome. As a result, they have the potential to increase their copy number in the genome over time. Transposable elements are sometimes referred to as "selfish" DNA because it seems that their only function is to duplicate themselves and proliferate in the genome, making them the ultimate parasite.

Transposable elements make up about 45% of the DNA in the human genome. They can be grouped into two major classes based on the way they replicate. One class consists of **DNA transposons**, which replicate and transpose by DNA replication and repair. The other class consists of elements that transpose by means of an RNA intermediate. These are called **retrotransposons** because their RNA is used as a template to synthesize complementary strands of DNA, a process that reverses the usual flow of genetic information from DNA into RNA (*retro* means "backward"). More than 40% of the human genome consists of various types of retrotransposons, whereas only about 3% of human DNA consists of DNA transposons.

FIG. 13.9 Polyploidy in plants. A type of crocus known as Golden Yellow was formed by hybridization between two different species, and it contains a full set of chromosomes from each parent. The parental chromosomes in the hybrid are indicated in orange and blue in the diagram on the right. *Photo source: McPHOTO/age fotostock.*

Over the course of evolutionary time, the amount of repetitive DNA in a genome can change drastically, in large part because of the accumulation of transposable elements (Fig. 13.10). In some genomes, repetitive DNA is maintained over long periods of time even as new species evolve and diversify, whereas in other genomes the amount of repetitive DNA is held in check because of deletion and other processes. The result is that the genomes of different species can contain vastly differing amounts of repetitive DNA, and this accounts in large part for the C-value paradox.

FIG. 13.10 Sequence composition of the human genome.

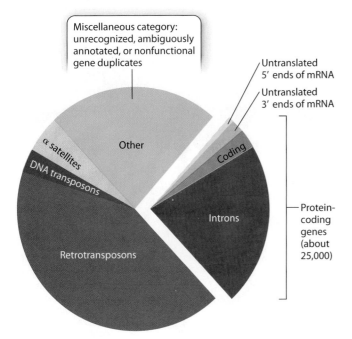

Self-Assessment Questions

8. Given our knowledge of genome sizes in different organisms, would you predict that *Homo sapiens* or the two-toed salamander (*Amphiuma means*) has the larger genome?

9. What are some reasons why, in multicellular eukaryotes, genome size is not necessarily related to the number of protein-coding genes or organismal complexity?

FIG. 13.11 Supercoils. A highly twisted rubber band forms coils of coils (supercoils), much as a circular DNA molecule does when it contains too many base pairs per helical turn.

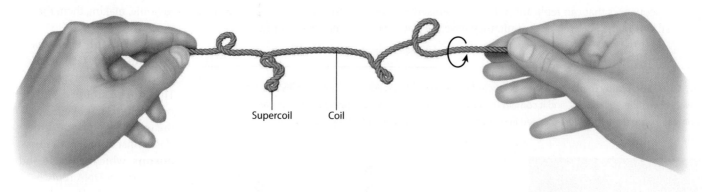

13.4 ORGANIZATION OF GENOMES

Regardless of whether an organism has a large or a small genome, the genomes of all organisms are large relative to the size of the cell. For example, if the circular genome of the intestinal bacterium *Escherichia coli* were fully extended, its length would be 200 times greater than the diameter of the cell itself. The fully extended length of DNA in human chromosome 1, our longest chromosome, would be 10,000 times greater than the diameter of the average human cell. Consequently, an enormous length of DNA must be packaged into a form that will fit inside the cell while still allowing the DNA to replicate and carry out its coding functions. The mechanism of packaging differs substantially in bacteria, archaeons, and eukaryotes. We focus here on bacteria and eukaryotes, primarily because less is known about how DNA is packaged in archaeons.

Bacterial cells package their DNA as a nucleoid composed of many loops.

Bacterial genomes are circular, and the DNA double helix is underwound, which means that it makes fewer turns in going around the circle than would allow every base in one strand to pair with its partner base in the other strand. Underwinding is caused by an enzyme, **topoisomerase II**, which breaks the double helix, rotates the ends to unwind the helix, and then seals the break. Underwinding creates strain on the DNA molecule, which is relieved by the formation of **supercoils**, in which the DNA molecule coils on itself. Supercoiling allows all the base pairs to form, even though the molecule is underwound. You can make your own supercoil by stretching and twirling the ends of a rubber band, and then relaxing the stress slightly to allow the twisted part to form coils around itself (**Fig. 13.11**). Supercoils that result from underwinding are called negative supercoils, and those that result from overwinding are positive supercoils. In most organisms, DNA is negatively supercoiled.

In bacteria, the supercoils of DNA form a structure with multiple loops called a **nucleoid** (**Fig. 13.12**). The supercoil loops are bound together by proteins. In *E. coli,* the nucleoid

FIG. 13.12 Bacterial nucleoid. The circular bacterial chromosome twists on itself to form supercoils, which are anchored by proteins. *Source: Dr. Klaus Boller/Science Source.*

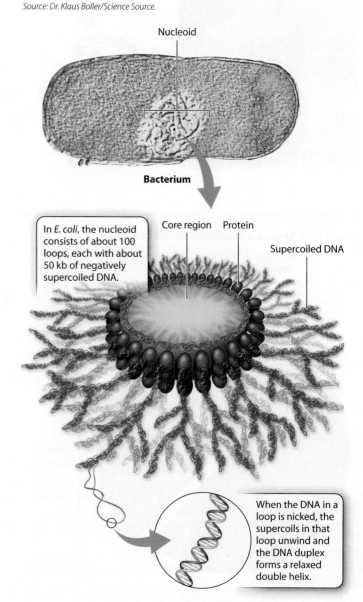

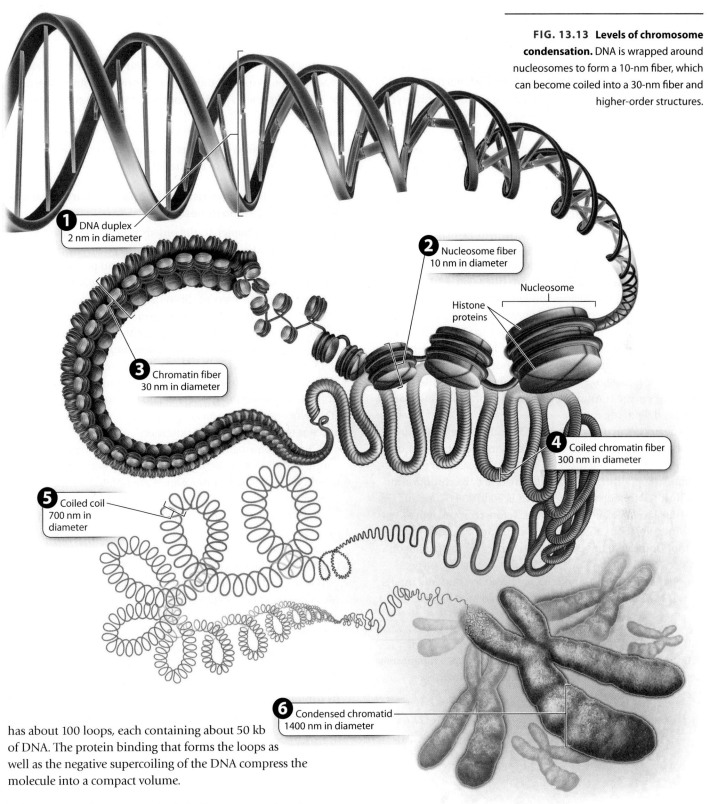

FIG. 13.13 Levels of chromosome condensation. DNA is wrapped around nucleosomes to form a 10-nm fiber, which can become coiled into a 30-nm fiber and higher-order structures.

has about 100 loops, each containing about 50 kb of DNA. The protein binding that forms the loops as well as the negative supercoiling of the DNA compress the molecule into a compact volume.

Eukaryotic cells package their DNA as one molecule per chromosome.

Eukaryotic cells also have topoisomerase II enzymes and their DNA is usually negatively supercoiled, but the DNA in chromosomes is packaged very differently from the way it is organized in prokaryotes. Eukaryotic DNA is linear and each DNA molecule forms a single chromosome. In a chromosome, DNA is packaged with proteins to form a DNA-protein complex called **chromatin**.

There are several levels of chromatin packaging (**Fig. 13.13**). First, eukaryotic DNA winds around histone proteins. Histone proteins are found in all eukaryotes, and they interact with any double-stranded DNA, no matter what its sequence is. Histone proteins are evolutionarily conserved, which means that they are very similar in sequence from one organism to the next, and histone proteins from one eukaryote can associate with DNA from another. The histone proteins form the core of a

nucleosome, which is made up of eight histone proteins: two each of histones H2A, H2B, H3, and H4. Each nucleosome also includes a stretch of DNA wrapped twice around the histone core. The histone proteins are rich in the positively charged amino acids lysine and arginine, which are attracted to the negatively charged phosphate groups in each DNA strand.

This first level of packaging of the DNA is sometimes referred to as "beads on a string," where the nucleosomes are the beads and the DNA is the string. It is also called a **10-nm fiber** in reference to its diameter, which is about five times the diameter of the DNA double helix (Fig. 13.13). In general, these are areas of the genome that are transcriptionally active.

The next level of packaging occurs when the chromatin is more tightly coiled, forming a **30-nm fiber** (Fig. 13.13). As the chromosomes in the nucleus condense in preparation for cell division, each chromosome becomes progressively shorter and thicker as the 30-nm fiber coils onto itself to form a 300-nm coil, a 700-nm coiled coil, and finally a 1400-nm condensed chromosome in a manner that is still not fully understood. The progressive packaging constitutes **chromosome condensation**, an active, energy-consuming process requiring the participation of several types of proteins.

Greater detail of the structure of a fully condensed chromosome is revealed when the histones are chemically removed (**Fig. 13.14**). Without histones, the DNA spreads out in loops around a supporting protein structure called the chromosome **scaffold**. Each loop of relaxed DNA is 30 to 90 kb long and anchored to the scaffold at its base. Before removal of the histones, the loops are compact and supercoiled. Each human chromosome contains 2000–8000 such loops, depending on its size.

Despite intriguing similarities between the nucleoid model in Fig. 13.12 and the chromosome scaffold model in Fig. 13.14, the structures evolved independently and make use of different types of protein to bind the DNA and form the folded structure of DNA and protein. Furthermore, the size of the eukaryotic chromosome is vastly greater than the size of the bacterial nucleoid. To appreciate the difference in scale, keep in mind that the volume of a fully condensed human *chromosome* is five times larger than the volume of a bacterial *cell*.

The human genome consists of 22 pairs of chromosomes and two sex chromosomes.

As discussed in Chapter 11, the orderly process of meiosis is possible because chromosomes come in pairs. The pairs usually match in size, general appearance, and position of the centromere, but there are exceptions, such as the *X* and *Y* sex chromosomes. The pairs of chromosomes that match in size and appearance are called **homologous chromosomes.** The members of each pair of homologous chromosomes have the same genes arranged in the same order along their length. If the DNA duplexes in each pair of homologs were denatured, each DNA strand could form a duplex with its complementary strand from the other homolog. There would be some differences in DNA sequence due to genetic variation, but not so many differences as to prevent DNA hybridization.

Chromosome painting illustrates the nearly identical nature of the DNA molecules in each pair of homologs. In this technique, individual chromosomes are isolated from cells in metaphase of mitosis. Metaphase of mitosis is the easiest stage in which to isolate chromosomes because of the availability of chemicals that prevent the spindle from forming. These chemicals block the cell cycle at metaphase, so cells progress to metaphase and then stop.

Once the chromosomes have been isolated, the DNA from each chromosome is fragmented, denatured, and labeled with a unique combination of fluorescent dyes. Under fluorescent light, the dyes give the DNA in each type of chromosome a different color. The fluorescently labeled DNA fragments are then mixed with and hybridized to intact metaphase chromosomes from another cell. Each labeled fragment hybridizes to its complementary sequence in the metaphase chromosomes, "painting" each metaphase chromosome with dye-labeled DNA fragments.

A chromosome paint of the chromosomes in a human male is shown in **Fig. 13.15**. Fig. 13.15a shows the chromosomes in the random orientation in which they were found in the metaphase cell, and Fig. 13.15b shows them arranged in a standard form called a **karyotype** (Chapter 11). To make a karyotype, the images of the homologous chromosomes are arranged in pairs from longest to shortest, with the sex chromosomes placed at the lower right. In this case, the sex chromosomes are *XY* (one *X* chromosome and one *Y* chromosome), indicating that the individual is male; in a female, the sex chromosomes are *XX* (a pair of *X* chromosomes). Including the sex chromosomes, humans have 23 pairs of chromosomes.

FIG. 13.14 (a) A chromosome with histones and (b) a chromosome depleted of histones, showing the underlying scaffold. *Source: Courtesy of Ulrich Laemmli.*

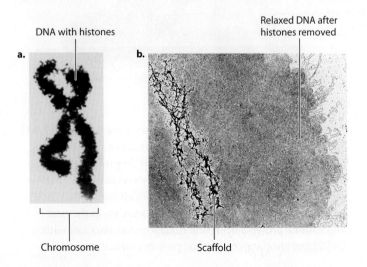

FIG. 13.15 A chromosome paint of chromosomes from a human male. (a) Condensed chromosomes are "painted" with fluorescent dyes. (b) Chromosomes are arranged in the standard form of a karyotype. *Source: NHGRI, www.genome.gov.*

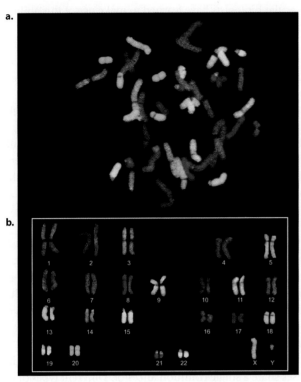

FIG. 13.16 Giemsa bands in a karyotype of a human male. Each dashed line marks the position of the centromere. *Source: Darryl Leja, NHGRI. www.genome.gov.*

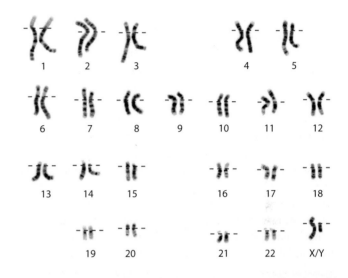

An important observation from the chromosome paint shown in Fig. 13.15 is that the two members of each pair of homologous chromosomes show the same pattern of fluorescent color. This means that a particular labeled DNA fragment hybridized only with the two homologs of one chromosome. Thus, the DNA in each pair of homologous chromosomes is different from that in any other pair of homologous chromosomes.

Higher resolution of human chromosomes can be obtained by the use of stains that bind preferentially to certain chromosomal regions and produce a visible pattern of bands, or crosswise striations, in the chromosomes. One such stain is the Giemsa stain; a Giemsa-stained karyotype is shown in Fig. 13.16. Note that each chromosome has a unique pattern of bands and that homologous chromosomes can readily be identified by their identical banding patterns. Procedures using the Giemsa stain yield about 300 bands that are used as landmarks for describing the location of genes along the chromosome. Such landmarks are also important in identifying specific chromosomal abnormalities associated with diseases, including some childhood leukemias.

Every species of eukaryote has its characteristic number of chromosomes. The number differs from one species to the next, with little relation between chromosome number and genome size. Despite variation in number, the rule that chromosomes come in pairs holds up pretty well, with the exception of the sex chromosomes. Polyploids, too, are an exception, but even in naturally occurring polyploids the number of copies of each homologous chromosome is usually an even number, so there are pairs of homologs after all.

The occurrence of chromosomes in pairs allows eukaryotes to reproduce sexually. When reproductive cells are formed during meiotic cell division, each cell receives one and only one copy of each of the pairs of homologous chromosomes (Chapter 11). When reproductive cells from two individuals fuse to form an offspring cell, the chromosome number characteristic of the species is reconstituted.

Organelle DNA forms nucleoids that differ from those in bacteria.

Most eukaryotic cells contain mitochondria, and many contain chloroplasts. Each type of organelle has its own DNA, meaning that eukaryotic cells have multiple genomes. Each eukaryotic cell has a **nuclear genome** consisting of the DNA in the chromosomes. Cells with mitochondria also have a **mitochondrial genome**, and those with chloroplasts also have a **chloroplast genome**.

Because the genome organization and mechanisms of protein synthesis in these organelles resemble those of bacteria, most biologists subscribe to the theory that the organelles originated as free-living bacterial cells that were engulfed by primitive eukaryotic cells billions of years ago (Chapter 25). In Chapters 7 and 8, we saw that the likely ancestor of mitochondria resembled a group of today's non-photosynthetic bacteria (a group that includes *E. coli*), and the likely ancestor of chloroplasts resembled today's photosynthetic cyanobacteria. In both cases, the DNA of the organelles became smaller during the course of evolution because most of the genes were transferred to the DNA in the nucleus. If the products of these transferred genes are needed in the organelles, they are synthesized in the cytoplasm and targeted for entry into the organelles by signal sequences (Chapter 5).

DNA in mitochondria and chloroplasts is usually circular and exists in multiple copies per organelle. Among animals, the size of a mitochondrial DNA molecule ranges from 14 to 18 kb. Mitochondrial DNA in plants is generally much larger, up to 100 kb. Plant chloroplast DNA is more uniform in size than mitochondrial DNA, with a range of 130 to 200 kb.

On the basis of size alone, some kind of packaging of organelle DNA is necessary. For example, the mitochondrial DNA in human cells is a circular molecule of 16 kb. Fully extended, it would have a circumference about as large as that of the mitochondrion itself. From their bacterial origin, you might expect organelle DNA to be packaged as a nucleoid rather than as a chromosome, and indeed it is. However, the structures of the nucleoid in mitochondria and chloroplasts differ from each other, and they also differ from those in free-living bacteria.

Self-Assessment Questions

10. What are the different mechanisms by which bacterial cells and eukaryotic cells package their DNA?
11. Draw a nucleosome, indicating the positions of DNA and proteins.
12. What are "homologous chromosomes"? What technique could you use to show their similarity?

13.5 VIRUSES AND VIRAL GENOMES

Now we discuss viral genomes, but not because they are particularly complex. Viral genomes are actually rather small and compact with little or no repetitive DNA, and their sequences are relatively easy to assemble, as we have seen in Fig. 13.7 for HIV. What sets viral genomes apart from those of cellular organisms are the types of nucleic acid that make up their genomes and the manner in which these genomes are replicated in the infected cells. In their nucleic-acid composition and manner of replication, viral genomes are far more diverse than the genomes of cellular organisms: all cellular organisms have genomes of double-stranded DNA that replicate by means of the processes described in Chapter 12. This is not so with viruses. Furthermore, whereas biologists classify cellular organisms according to their degree of evolutionary relatedness, they classify viruses based on their type of genome and mode of replication.

We mostly think of viruses as causing disease, and indeed they do. Familiar examples include influenza ("flu"), polio, and HIV. In some cases, they cause cancer (Case 2: Cancer). In fact, the term "virus" comes from the Latin for "poison." Yet viruses play other roles, too. Some viruses transfer genetic material from one cell to another. This process is called horizontal gene transfer to distinguish it from parent-to-offspring (vertical) gene transfer. Horizontal gene transfer has played a major role in the evolution of bacterial and archaeal genomes, as well as in the origin and spread of antibiotic resistance genes (Chapter 24). Molecular biologists have learned to make use of this ability of viruses to deliver genes into cells.

In this section, we focus on viral diversity with an emphasis on viral genomes. Thousands of viruses have been described in detail, and probably millions more have yet to be discovered. It has been estimated that life on Earth is host to 10^{31} virus particles—ten hundred thousand million more virus particles than grains of sand! Most of these viruses infect bacteria, archaeons, and unicellular eukaryotes, and they are especially abundant in the ocean (10^{11} viruses per liter). Viruses are small, consisting of little more than a genome in a package, and they can reproduce only by hijacking host-cell functions, but as a group they have had amazing evolutionary success.

Viruses can be classified by their genomes.

The genomes of viruses are diverse. Some genomes are composed of RNA and others of DNA. Some are single stranded, others are double stranded, and still others have both single- and double-stranded regions. Some are circular and others are made up of a single piece or multiple linear pieces of DNA (called linear and segmented genomes, respectively).

Unlike forms of cellular life, there is no evidence that all viruses share a single common ancestor. Different types of virus may have evolved independently more than once. Because classification of viruses based on evolutionary relatedness is not possible, other criteria are necessary. One of the most useful classifications is the type of nucleic acid the virus contains and how the messenger RNA, which produces viral proteins, is synthesized. The classification is called the Baltimore system, which was named after David Baltimore, who devised it.

According to the Baltimore system, there are seven major groups of viruses, designated I–VII, as shown in **Fig. 13.17**. These groups are largely based on aspects of their nucleic acid. Viral genomes can be double-stranded DNA, double-stranded RNA, partially double-stranded and partially single-stranded, or single-stranded RNA or DNA with a positive (+) or negative (−) sense. The sense of a nucleic acid molecule is positive if its sequence is the same as the sequence of the mRNA that is used for protein synthesis, and negative if it is the complementary sequence. An example of a (+)RNA virus (Group IV) is the Zika virus, which recently invaded South and Central America from Africa and is inexorably spreading northward into the United States. Zika is associated with severe neurological abnormalities in the fetuses of women who were infected when pregnant.

In a similar way that (+) and (−) RNA strands are defined in RNA viruses, a virus whose genome is (+)DNA strand has the same sequence as the mRNA, and one whose genome consists of a (−)DNA strand has the complementary sequence (except that U in RNA is replaced with T in DNA). Because mRNA is

FIG. 13.17 The Baltimore system of virus classification. This system classifies viruses by type of nucleic acid and the way mRNA is produced.

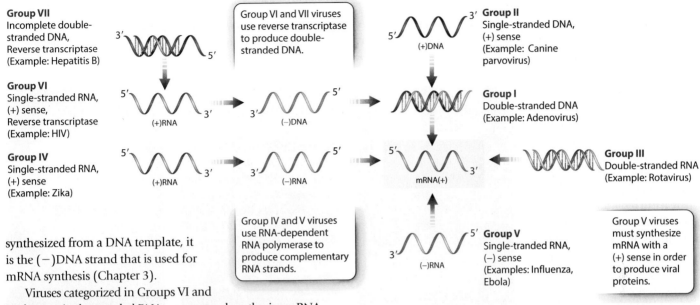

Group VII Incomplete double-stranded DNA, Reverse transcriptase (Example: Hepatitis B)

Group VI Single-stranded RNA, (+) sense, Reverse transcriptase (Example: HIV)

Group IV Single-stranded RNA, (+) sense (Example: Zika)

Group VI and VII viruses use reverse transcriptase to produce double-stranded DNA.

Group IV and V viruses use RNA-dependent RNA polymerase to produce complementary RNA strands.

Group II Single-stranded DNA, (+) sense (Example: Canine parvovirus)

Group I Double-stranded DNA (Example: Adenovirus)

Group III Double-stranded RNA (Example: Rotavirus)

Group V Single-tranded RNA, (−) sense (Examples: Influenza, Ebola)

Group V viruses must synthesize mRNA with a (+) sense in order to produce viral proteins.

synthesized from a DNA template, it is the (−)DNA strand that is used for mRNA synthesis (Chapter 3).

Viruses categorized in Groups VI and VII have a single-stranded RNA genome and synthesize mRNA by the enzyme reverse transcriptase. **Reverse transcriptase** is an RNA-dependent DNA polymerase that uses a single-stranded RNA as a template to synthesize a complementary DNA strand (Fig. 13.17). The reverse transcriptase then displaces the RNA template and replicates the DNA strand to produce a double-stranded DNA molecule that can be incorporated into the host genome. In synthesizing DNA from an RNA template, the enzyme reverses the usual flow of genetic information from DNA to RNA. This capability is so unusual and was so unexpected that many molecular biologists at first doubted whether such an enzyme could exist. Finally, the enzyme was purified and its properties verified, for which its discoverers, Howard Temin and David Baltimore, were awarded the Nobel Prize in Physiology or Medicine in 1975, which was shared with Renato Dulbecco.

As we saw earlier, genome size varies greatly among different viruses. RNA viral genomes tend to be smaller than DNA viral genomes. Most eukaryotic viruses that have RNA genomes and most plant viruses, including tobacco mosaic virus, are in Group IV. Among bacterial and archaeal viruses, most genomes consist of double-stranded DNA.

The host range of a virus is determined by viral and host surface proteins.

Although viruses show great diversity in their nucleic-acid composition and manner of replication, they all consist of some type of nucleic acid, a capsid, and sometimes a lipid **envelope** (Chapter 1). Viruses can reproduce only by infecting living cells and exploiting the cell's metabolism and protein synthesis to produce more viruses. Most viruses are tiny—some hardly larger than a ribosome—measuring 25–30 nm in diameter. Roughly speaking, the average size of a virus, relative to that of the host cell it infects, may be compared to the size of an average person relative to that of a commercial airliner.

All known cells and organisms are susceptible to viral infection, including bacteria, archaeons, and eukaryotes.

A cell in which viral reproduction occurs is called a **host cell.** Some viruses kill the host cell; others do not. Although viruses can infect all types of living organism, a given virus can infect only certain species or types of cell. At one extreme, a virus can infect just a single species. Smallpox, which infects only humans, is a good example. In cases like this, we say that the virus has a narrow host range. For other viruses, the host range can be broad. For example, rabies infects many different types of mammal, including squirrels, dogs, and humans. Similarly, tobacco mosaic virus, a plant virus, infects more than 100 different species of plants (**Fig. 13.18**). No matter how broad the host range, however, plant viruses cannot infect bacteria or animals, and bacterial viruses cannot infect plants or animals.

Host specificity results from the way that viruses gain entry into cells. Proteins on the surface of the capsid or envelope (if present) bind to proteins on the surface of host cells. These proteins interact in a specific manner, so the presence of the host

FIG. 13.18 A plant infected by tobacco mosaic virus. *Courtesy H. D. Shew; Reproduced, by permission, from Shew, H. D., and Lucas, G. B., eds. 1991. Compendium of Tobacco Diseases. American Phytopathological Society, St. Paul, MN.*

protein on the cell surface determines which cells a virus can infect. For example, a protein on the surface of the HIV envelope called gp120 (encoded by the *env* gene; see Fig. 13.7) binds to a protein called CD4 on the surface of certain immune cells (Chapter 41), so HIV infects these cells but not other cells.

Following attachment, the virus gains entry into the cell, where it can replicate using the host cell machinery to make new viruses. In some cases, the viral genome integrates into the host cell genome (Chapter 18).

The host range of a virus can change because of mutation and other mechanisms (Chapter 41). For example, avian, or bird, flu is a type of influenza virus that was once restricted primarily to birds but now can infect humans. The first reported case in humans was in 1996, and the disease has since spread widely. Similarly, HIV once infected only nonhuman primates, but its host range expanded in the twentieth century to include humans. Canine distemper virus, which infects dogs, expanded its host range in the early 1990s, leading to the infection and death of lions in Tanzania.

Viruses have diverse sizes and shapes.

Viruses show a wide variety of shapes and sizes, which are determined in some cases by the type of genome they carry. Three examples are shown in **Fig. 13.19**. T4 virus (Fig. 13.19a) infects cells of the bacterium *Escherichia coli*. Viruses that infect bacterial cells are called **bacteriophages**, which literally means "bacteria eaters." The T4 bacteriophage has a complex structure that includes a head composed of protein surrounding a molecule of double-stranded DNA, a tail, and tail fibers. In infecting a host cell, the T4 tail fibers attach to the surface, and the DNA and some proteins are injected into the cell through the tail.

Most viruses are not structurally so complex. Consider the tobacco mosaic virus, which infects plants (see Fig. 13.18). It has helical shape formed by the arrangement of protein subunits entwined with a molecule of single-stranded RNA (Fig. 13.19b). Tobacco mosaic virus was the first virus discovered. It was revealed in experiments showing that the infectious agent causing brown spots and discoloration of tobacco leaves was so small that it could pass through the pores of filters that could trap even the smallest bacterial cells.

Many viruses have an approximately spherical shape formed from polygons of protein subunits that come together at their edges to form a polyhedral capsid. Among the most common polyhedral shapes is an icosahedron, which has 20 identical

FIG. 13.19 Virus structures. Viruses come in a variety of shapes, including (a) the head-and-tail structure of a bacteriophage, (b) the helical shape of tobacco mosaic virus, and (c) the icosahedral shape of adenovirus.
Sources: a. Dept. of Microbiology, Biozentrum/Getty Images; b. Biology Pics/Science Source; c. BSIP/Science Source.

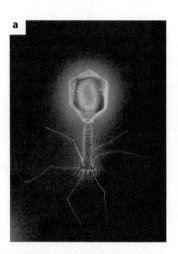
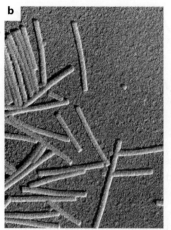
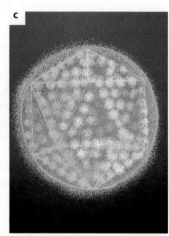

triangular faces. The example in Fig. 13.19c is adenovirus, a common cause of upper respiratory infections in humans. Many viruses that infect eukaryotic cells, such as adenovirus, are surrounded by a glycoprotein envelope composed of a lipid bilayer with embedded proteins and glycoproteins that recognize and attach to host cell receptors.

The genome of a virus contains the genetic information needed to specify all the structural components of the virus. Progeny virus particles are formed according to **molecular self-assembly**: when the viral components are present in the proper relative amounts and under the right conditions, the components interact spontaneously to assemble themselves into the mature virus particle.

Tobacco mosaic virus illustrates the process of self-assembly. In the earliest stages, the coat-protein monomers assemble into

FIG. 13.20 Tobacco mosaic virus. The protein and RNA components assemble themselves inside a cell.

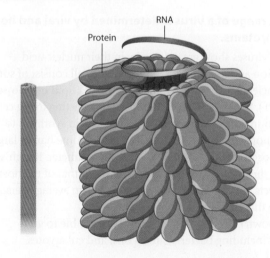

two circular layers and form a cylindrical disk. This disk binds with the RNA genome, and the combined structure forms the substrate for polymerization of all the other protein monomers into a helical filament that incorporates the rest of the RNA as the filament grows (**Fig. 13.20**). The mature virus particle consists of 2130 protein monomers and 1 single-stranded RNA molecule of 6400 ribonucleotides.

Self-Assessment Question

13. What are the steps necessary to synthesize mRNA from each of the following: double-stranded DNA, single-stranded (+)DNA, single-stranded (−)DNA, single-stranded (+)RNA, and single-stranded (−)RNA?

CORE CONCEPTS SUMMARY

13.1 GENOME SEQUENCING: A genome is the genetic material of a cell, organism, organelle, or virus, and its sequence is the order of bases along the DNA or (in some viruses) RNA.

The sequence of an organism's genome can be determined by breaking the genome into small fragments, sequencing these fragments, and then putting the sequences together at their overlaps. page 282

Sequences that are repeated in the genome can make sequence assembly difficult. page 282

13.2 GENOME ANNOTATION: Researchers annotate genome sequences to identify genes and other functional elements.

Genome annotation is the process by which the types and locations of the different kinds of sequences, such as protein-coding genes, are identified. page 284

Genome annotation sometimes involves scanning the DNA sequence for characteristic sequence motifs. page 285

Comparison of DNA sequences with messenger RNA sequences reveals the intron–exon structure of protein-coding genes. page 286

By comparing annotated genomes of different organisms, we can gain insight into their ancestry and evolution. page 287

The annotated HIV genome shows it is a retrovirus, and it contains the genes *gag, pol,* and *env.* page 287

13.3 GENES, GENOMES, AND ORGANISMAL COMPLEXITY: The number of genes in a genome and the size of a genome do not correlate well with the complexity of an organism.

The C-value paradox describes the observation that the size of a genome (measured by its C-value) does not correlate with an organism's complexity. page 288

In eukaryotes, the C-value paradox can be explained by differences in the amount of noncoding DNA, including repetitive sequences and transposons. Some genomes have a lot of noncoding DNA; others do not. page 289

13.4 ORGANIZATION OF GENOMES: The orderly packaging of DNA allows it to carry out its functions and fit inside the cell.

Bacteria package their circular DNA in a structure called a nucleoid. page 292

Eukaryotic cells package their DNA into linear chromosomes. page 293

DNA in eukaryotes is wound around groups of histone proteins called nucleosomes to form a 10-nm fiber, which in turn coils to form higher-order structures, such as the 30-nm fiber. page 293

Diploid organisms have two copies of each chromosome, called homologous chromosomes. page 294

Humans have 23 pairs of chromosomes, including the *X* and *Y* sex-chromosome pair. Females are *XX* and males are *XY.* page 294

The genomes of mitochondria and chloroplasts are organized into nucleoids that resemble, but are distinct from, those of bacteria. page 295

13.5 VIRUSES AND VIRAL GENOMES: Viruses have diverse genomes, but all require a host cell to replicate.

Viruses can be classified by the Baltimore system, which defines seven groups on the basis of type of nucleic acid and the way mRNA is synthesized. page 296

Viruses can infect all types of organism, but a given virus can infect only some types of cell. page 297

The host range of a virus is determined by proteins on viral and host cells. page 297

Viruses have diverse shapes, including head-and-tail, helical, and icosahedral. page 298

Viruses are capable of molecular self-assembly under the appropriate conditions. page 298

Log in to **LaunchPad** to check your answers to the Self-Assessment Questions and to access additional learning tools.

CHAPTER 14: Mutation and Genetic Variation

CORE CONCEPTS

14.1 GENOTYPE AND PHENOTYPE: The genetic makeup of a cell or organism is its genotype, and an observable characteristic is its phenotype.

14.2 THE NATURE OF MUTATIONS: Mutations are very rare for any given nucleotide and occur randomly without regard to the needs of an organism.

14.3 SMALL-SCALE MUTATIONS: Small-scale mutations include point mutations, insertions and deletions, and movement of transposable elements.

14.4 CHROMOSOMAL MUTATIONS: Chromosomal mutations involve large regions of one or more chromosomes.

14.5 DNA DAMAGE AND REPAIR: Mutations occur spontaneously and are induced by mutagens, but most DNA damage is repaired.

The sequencing of the human genome was a great step forward in our understanding of ourselves. However, as we saw in the previous chapter, it is an oversimplification to speak of *the* human genome. Almost all species have abundant amounts of **genetic variation,** which refers to the genetic differences that exist among individuals in a population at a particular time. Genetic variation accounts in part for the physical differences we see among individuals, such as differences in hair color, eye color, and height. And, on a much larger scale, genetic differences among species result in the diversity of organisms on this planet, from bacteria to blue whales.

Variation among different individuals' genomes arise from **mutations**. Any heritable change in the nucleotide sequence of DNA is a mutation. A change might be a substitution of one base for another, a small deletion or insertion of a few bases, or a larger alteration affecting a segment of a chromosome. By "heritable," we mean that the mutation persists through DNA replication and is transmitted during cell division (meiotic cell division, mitotic cell division, or binary fission). Most mistakes in DNA replication or chemical damage to DNA are immediately repaired by specialized enzymes in the cell. Some, however, escape these repair mechanisms and are transmitted from parental cell to daughter cell.

In this chapter, we examine some of the basic principles of mutation and genetic variation: different types of mutation, how and when they occur, and how errors in DNA replication and various types of DNA damage are repaired. In Chapters 15 and 16, we look at how this variation is inherited from one generation to the next.

14.1 GENOTYPE AND PHENOTYPE

A mutation occurs in an individual cell at a particular point in time. Through evolution (Chapters 1 and 20), the proportion of individuals in a population carrying this mutation may increase or decrease. Therefore, if we look at any present-day population of organisms, such as the human population, we will find that it harbors lots of genetic differences, all of which resulted from mutations that occurred sometime in the past.

We tend to think of mutations as something negative or harmful because, in everyday language, the term "mutant" suggests something abnormal. In reality, because mutations result in genetic variation among individuals, we are all "mutants" that differ from one another in our DNA. Although it is true that many mutations are harmful, other mutations have negligible effects on the organism, and a precious few are beneficial. Without beneficial mutations, evolution would not be possible. Mutations generate the occasional favorable variants that allow organisms to evolve and become adapted to their environment over time (Chapter 20).

In the human population, you can observe the effects of common genetic differences or mutations by looking at the people around you. They differ in height, weight, facial features, skin color, eye color, hair color, hair texture, and many other ways. These traits differ in part because of genetic variation, and in part because of the environment. Weight is affected by diet, for example, and skin color by exposure to sunlight.

Genotype is the genetic makeup of a cell or organism, and the phenotype is its observed characteristics.

The genetic makeup of a cell or organism constitutes its **genotype**. A population with a gene pool that has many variants in many different genes will consist of organisms with many different genotypes. For example, any two human genomes are likely to differ from each other at about 3 million nucleotide sites, or about one difference per thousand nucleotides across the genome. In other words, we differ from one person to the next but only in a very small fraction of nucleotides.

Mutations are the ultimate source of differences among genotypes. If we consider any present-day population of organisms, we will find that some genetic differences are especially common. Geneticists use the term **polymorphism** to refer to any genetic difference among individuals that is present in multiple individuals in a population. For example, if some people have an A—T base pair at a particular site in the genome, but others have a G—C base pair at the same site, this difference is a polymorphism. The polymorphism is the result of a mutation that occurred in the past, spread through the population, and is now shared among a subset of individuals in a population.

The different forms of any gene are called **alleles**, and they correspond to different DNA sequences (polymorphisms) in the genes. An individual who inherits the same allele of a gene from each parent is **homozygous**. By contrast, an individual who inherits a different allele of a gene from each parent is **heterozygous**.

An individual's **phenotype** is the individual's observable characteristics or traits, such as height, weight, eye color, and so forth. It may be visible, as in these characteristics, or it may be seen in the development, physiology, or behavior of a cell or organism. For example, color blindness and lactose intolerance are phenotypes. The phenotype results in part from the genotype. For example, a genotype with a mutation that affects expression of an enzyme that normally metabolizes lactose can lead to the phenotype of lactose intolerance. However, the environment also commonly plays an important role in the expression of the genotype, so it is most accurate to say that a phenotype results from an interaction between the genotype and the environment. These genotype–environment interactions are discussed in Chapter 18.

Some genetic differences are harmful.

Mutations affect not only observable characteristics of an organism, but also the probability that an organism is able to survive and reproduce in a particular environment. Viewed in this way, mutations can be harmful, neutral, or even beneficial. Harmful mutations are often eliminated in one or a few generations because they decrease the survival or reproduction of the individuals that carry them. Less harmful mutations may persist in a population.

For example, many polymorphisms increase susceptibility to particular diseases but do not cause the disease themselves. One such polymorphism increases the risk of emphysema, a condition marked by loss of lung tissue. Patients with emphysema suffer from shortness of breath even when at rest, a wheezy cough, and increased blood pressure in the arteries of the lungs. Without proper treatment, the disease progresses to respiratory failure or congestive heart failure, and eventually to death. Onset of the disease usually occurs in middle age, and the life-span of affected individuals is shortened by 10 to 30 years depending on the effect of treatment.

Emphysema results from reduced activity of the enzyme alpha-1 antitrypsin ($\alpha 1 AT$). The main function of $\alpha 1 AT$ is to inhibit another enzyme, known as elastase, which breaks down the connective-tissue protein elastin in lungs (**Fig. 14.1a**). The elasticity of the lung, which allows normal breathing, requires a balance between the production and the breakdown of elastin. Too-rapid breakdown by elastase is normally prevented by the inhibition of elastase by $\alpha 1 AT$.

Although approximately 80% of all cases of emphysema can be traced to cigarette smoking, in some affected individuals genetics plays a role. An individual's risk of emphysema is significantly increased by inheriting a mutation in the gene that encodes $\alpha 1 AT$. The mutation results in an enzyme with reduced activity (**Fig. 14.1b**). Among the many different alleles of this gene, a mutant allele denoted *PiZ* is particularly common in populations of European descent. Any mutation that increases the risk of disease in an individual is known as a **genetic risk factor** for that disease. A risk factor does not *cause* the disease, but rather makes the disease *more likely* to occur.

Some genetic differences are neutral.

Mutations that have negligible effects on survival or reproduction are considered **neutral.** Many mutations are neutral because they occur in noncoding DNA. Neutral mutations are therefore especially likely to occur in organisms with large genomes and abundant noncoding DNA (Chapter 13).

One example in human populations is the taster phenotype associated with perception of a bitter taste from certain chemicals, including phenylthiocarbamide (PTC). The ability to taste PTC is due largely, but not exclusively, to alleles of a single gene called *TAS2R38* that encodes a taste receptor in the tongue. Humans who are homozygous for an allele of this gene called *PAV* are almost all tasters, while most humans with at least one copy of another allele called *AVI* are almost all nontasters.

The ability to taste PTC is a polymorphism that differs in frequency among populations. Around the world, the frequency of nontasters varies from a low of 3% in West Africa to a high of 40% in India. Tasters often perceive a disagreeable bitterness in broccoli, cauliflower, kale, Brussels sprouts, and related vegetables.

FIG. 14.1 A harmful mutation. (a) The enzyme alpha-1 antitrypsin (α1AT) normally inhibits the activity of elastase, which breaks down elastin in the lung. (b) A mutation in α1AT called *PiZ* is harmful because it reduces α1AT activity and leads to increased risk of emphysema.

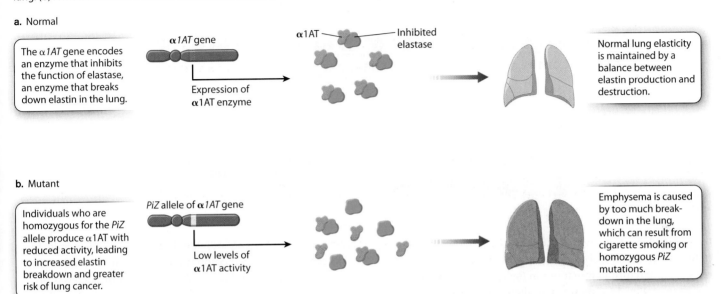

A few genetic differences are beneficial.

While many mutations are neutral and many others are harmful, a few mutations are beneficial. In human populations, beneficial mutations are often discovered through their effects in protecting against infectious disease. Here we consider a mutation that protects against acquired immune deficiency syndrome (AIDS), which is caused by the human immuno deficiency virus (HIV).

FIG. 14.2 HIV infection of T cells. HIV gains entry into a T cell by interacting with a CD4 protein and a CCR5 receptor on the surface of T cells.

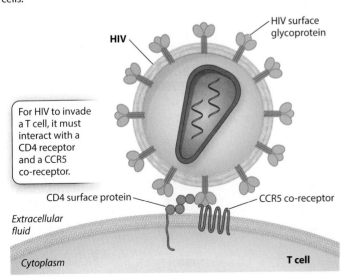

A glycoprotein on the surface of HIV interacts with a cell-surface receptor called CD4 to gain entry into immune cells called T cells (**Fig. 14.2**). Interaction with CD4 alone, however, does not enable the virus to infect the T cell. The HIV surface glycoprotein must also interact with another receptor on the T cell, which is denoted CCR5 (the red surface protein in Fig. 14.2), in the early stages of infection. The normal function of CCR5 is to bind certain small secreted proteins that promote tissue inflammation in response to infection. But because CCR5 is also an HIV receptor, cells lacking CCR5 are more difficult to invade.

A beneficial effect of a particular mutation in the *CCR5* gene was discovered in studies focusing on HIV-infected patients whose infection had not progressed to full-blown AIDS after 10 years or more. The protective allele is denoted as $\Delta 32$ (*delta 32*) because the mutation is a deletion that removes 32 base pairs in the coding sequence of the *CCR5* gene (**Fig. 14.3**). The mutant protein that results from this deletion is completely inactive.

The effect of the $\Delta 32$ allele is pronounced. In individuals with the homozygous $\Delta 32/\Delta 32$ genotype, HIV progression to AIDS is rarely observed. There is some protective effect from this mutation even in individuals with heterozygous $\Delta 32$ genotypes, in whom progression to AIDS is delayed by an average of about 2 years after infection by HIV.

The effect of a mutation may depend on the genotype and environment.

Whether a mutation is harmful, neutral, or beneficial often depends on factors other than the mutation itself. For example, some mutations can be harmful in some environments, but are neutral or even beneficial in others. Let's consider mutations in

FIG. 14.3 A beneficial mutation in the human population. Mutant *CCR5* has a 32-nucleotide deletion that results in defective CCR5 protein and, therefore, slows the progression of HIV to AIDS.

a. Nonmutant *CCR5* allele

b. Δ32 mutant *CCR5* allele

The deletion results in a mutant protein that is completely inactive.

the human DNA sequence coding for β-(beta-)globin, a subunit of the protein hemoglobin, which carries oxygen in red blood cells. We will return to mutations in β-globin several times in this chapter. **Fig. 14.4** shows three alleles of the β-globin gene: *A*, *S*, and *C*. These three alleles are relatively common in certain African populations.

The most frequent allele is the *A* allele, which has a GAG codon in the position indicated. This codon translates to glutamic acid (Glu) in the resulting polypeptide. The *S* and *C* alleles contain mutations that change the Glu found in the *A* allele to valine (Val) or lysine (Lys), respectively (red nucleotides in Fig. 14.4). Although in each case only one amino acid of the β-globin protein is affected, this change can have a dramatic effect on the function of the protein, since the amino acid sequence determines how a protein folds, and protein folding in turn determines the protein's function (Chapter 4).

For the hemoglobin *A*, *S*, and *C* alleles, there are three possible homozygous genotypes: *AA*, *SS*, and *CC*. The first letter or symbol in each pair indicates the allele inherited from one parent, and the second indicates the allele inherited from the other parent. By convention, alleles and genotypes are designated by italic letters.

There are also three possible heterozygous genotypes: *AS*, *AC*, and *SC*. While each individual can have only two alleles of a gene, many more alleles can exist in an entire population. In this example, a person can be homozygous, with two identical alleles, or heterozygous, with two different alleles, though there are three different alleles in the population.

What are the effects of these mutations? Are they beneficial, harmful, or neutral? The short answer is, it depends. Let's consider the *S*

allele. Homozygous *SS* individuals have sickle-cell anemia. In this condition, the hemoglobin molecules tend to bind together, or polymerize, when exposed to lower than normal levels of oxygen. The polymerization of hemoglobin causes the cell to change from its normal oval shape into the shape of a half-moon, or "sickle." In this form, the red blood cell is unable to carry the normal amount of oxygen. In addition, the sickled cells tend to block small capillaries, interrupting the blood supply to vital tissues and organs and resulting in severe pain. In the absence of proper medical care, patients with sickle-cell anemia usually die before adulthood.

Heterozygous *AS* individuals have only a mild anemia called the sickle-cell trait. Furthermore, in Africa, where malaria is

FIG. 14.4 Three alleles of the gene encoding β-globin, a subunit of hemoglobin. Both the *S* allele and the *C* allele are associated with increased resistance to malaria.

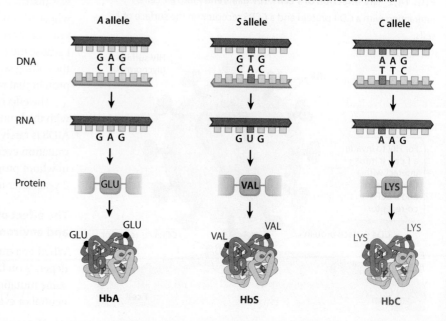

widespread, being a heterozygote is actually beneficial because it provides some protection against malaria.

This example illustrates two important principles about the connection between genotype and phenotype. First, the effect of a mutation often depends on whether the mutation is homozygous or heterozygous. In areas with malaria, the *S* allele is harmful as a homozygous genotype, but is beneficial as a heterozygous genotype. Second, the effect of a mutation may depend on the environment. The *S* allele, when heterozygous, is beneficial only in malarial-prone regions, where it offers protection from the disease that outweighs its other effects. In areas without malaria, it is harmful.

What about the *C* allele? Heterozygous *AC* individuals have partial protection against malaria, and homozygous *CC* individuals are not only more protected from malaria but also have a very mild anemia that usually needs no medical treatment. As with the *S* allele, the phenotype of the *C* allele depends on whether the individual is heterozygous or homozygous for the allele and whether malaria is present or absent.

> ### Self-Assessment Questions
>
> 1. What determines a genotype? What determines a phenotype?
> 2. With regard to mutations, what is meant by the terms "harmful," "beneficial," and "neutral"? Why it is sometimes an oversimplification to consider a mutation as harmful, beneficial, or neutral?
> 3. A mutation arises in a bacterium that confers antibiotic resistance. Is this mutation harmful, beneficial, or neutral?

14.2 THE NATURE OF MUTATIONS

Mutations result from mistakes in DNA replication or from unrepaired damage to DNA. The damage may result from reactive molecules produced in the normal course of metabolism, chemicals in the environment, or radiation of various types, including X-rays and ultraviolet light. Most genomes also contain DNA sequences that can "jump" from one position to another in the genome, and their insertion into or near genes is a source of mutation.

Most mutations are **spontaneous,** occurring by chance in the absence of any assignable cause. They occur randomly, unconnected to an organism's needs: it makes no difference whether a given mutation would benefit the organism. Whether a favorable mutation does or does not occur is purely a matter of chance. This key principle, that mutations are spontaneous and random, is the focus of the following section.

Mutation of individual nucleotides is rare, but mutation across the genome is common.

Later in this chapter, we'll distinguish among several different types of mutations. For now, let's consider the most common mutation: the substitution of one nucleotide for a different nucleotide. Nucleotide-substitution mutations are nevertheless relatively rare, and their rate of occurrence differs among organisms. **Fig. 14.5a** compares the rates of newly arising mutations in a given base pair in a single round of replication in viruses and several types of organisms. As seen in the graph, the nucleotide mutation rates for different organisms range

FIG. 14.5 Mutation rates. (a) Rate per nucleotide per round of replication. (b) Rate per genome per generation. *Data from J. W. Drake, B. Charlesworth, D. Charlesworth, J. F. Crow, 1998, "Rates of Spontaneous Mutation," Genetics 148:1667–1686.*

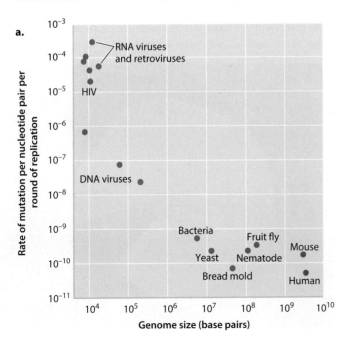
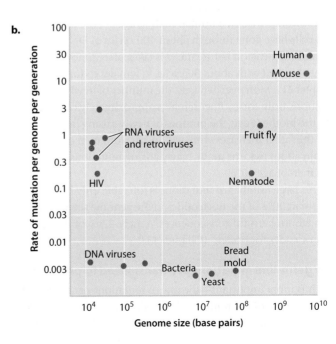

across almost eight orders of magnitude. Differences among organisms in the fidelity of replication and various other repair mechanisms account for much of this variation. For instance, polymerases in RNA viruses and retroviruses have no proofreading function, whereas in the cellular organisms in Fig. 14.5a, only one nucleotide in every 10 billion is mistakenly substituted for another and slips through the proofreading process.

However, averaging across an entire genome conceals many details. Certain nucleotides are especially prone to mutation and can exhibit rates of mutation that are greater than the average by a factor of 10 or more. Sites in the genome that are especially mutable are called **hotspots**. The rate of mutation also depends on whether the cells are **germ cells** (haploid gametes and the diploid cells that give rise to them) or **somatic cells** (the other cells of the body). In mammals, the rate of mutation per nucleotide per replication is greater in somatic cells than in germ cells because DNA repair mechanisms are more efficient in germ cells.

While the rate of mutation per nucleotide per replication in most multicellular organisms is low, the rate of mutation in a whole organism depends on the size of the genome and the number of cell divisions per generation of the organism (**Fig. 14.5b**). A comparison of the two graphs in Fig. 14.5 shows that, while humans have the lowest rate of mutation per nucleotide per replication (Fig. 14.5a), humans also have the highest *number* of mutations per genome per generation (Fig. 14.5b). This seeming paradox arises because humans have a large genome and undergo many cell divisions over their lifetime, providing more opportunities for mutations to occur.

In humans, the average number of newly arising nucleotide-substitution mutations per genome in one generation is about 30, or about 60 per diploid zygote. Approximately 80% of the newly arising mutations in a zygote come from the father. The main reason is that, in a human male at age 30, the diploid germ cells have gone through about 400 cycles of DNA replication and cell division before meiosis takes place to form sperm cells, compared with about 30 cycles in females to form oocytes (Chapter 11). Moreover, whereas the number of newly arising mutations from the mother remains approximately constant with the mother's age, the number of newly arising mutations from the father increases with age. Sperm from men of age 40 contain about twice as many newly arising mutations as sperm from men of age 20.

This large number of mutations is tolerable only because more than 90% of the nucleotides in the genome seem free to vary without harmful consequences. A mere 2.5% of the human genome codes for protein (Chapter 13).

Only germ-line mutations are transmitted to progeny.

Which is more important—the rate of mutations per nucleotide per replication or the number of mutations per genome per generation? The answer depends on whether we are interested in the cells of individual organisms or the transmission of mutations from parent to offspring. Mutations can take place in any type of cell. Those that occur in eggs and sperm and the cells that give rise to them are called **germ-line mutations**, and for mutations in these reproductive cells, researchers are most interested in the number of mutations per genome per generation. Germ-line mutations are important to the evolutionary process because, as they are passed from one generation to the next, they may eventually come to be present in many individuals descended from the original carrier.

In contrast, mutations in nonreproductive cells are called **somatic mutations**, and for these mutations, what matters is the rate of mutations per nucleotide per replication. Although somatic mutations are not transmitted to future generations, they are transmitted to daughter cells in mitotic cell divisions (Chapter 11). Hence, a somatic mutation affects not only the cell in which it occurs, but also all the cells that descend from it. The areas of different colors or patterns that appear in "sectored" flowers, valued as ornamental plants (**Fig. 14.6**), are usually due to somatic mutations in flower-color genes. A mutation in a flower-color gene occurs in one cell, and as the cell replicates during development of the flower, all its descendants in the cell lineage—the generations of cells that originate from a single ancestral cell—carry that mutation, producing a sector with altered coloration.

Most cancers result from mutations in somatic cells (Chapter 11). In some cases, the mutation increases the activity of a gene that promotes cell growth and division, whereas in other cases it decreases the activity of a gene that restricts cell growth and division. In either event, the mutant cell and its descendants

FIG. 14.6 Somatic mutation. Somatic mutations in the Japanese morning glory (*Ipomoea nil*) in cell lineages that differ in their ability to make purple pigment cause sectors of different pigmentation in the flower. *Source: Courtesy Atsushi Hoshino, National Institute for Basic Biology, Japan.*

escape from one of the normal control processes. Fortunately, a single somatic mutation is usually not sufficient to cause cancer—usually two or three or more mutations in different genes are required to derail control of normal cell division so extensively that cancer results.

In some cancers, several mutations must occur sequentially in a single cell lineage for the disease to become manifest. **Fig. 14.7** shows three key mutations that have been implicated in the origin of one type of invasive colon cancer: *p53*, *Ras*, and *APC*. Each mutation occurs randomly, but if by chance a mutation in the *Ras* gene occurs in a cell in which the *APC* gene is already mutated, that cell's progeny forms a polyp. Another chance mutation in the same cell lineage, which now carries mutations in both the *APC* and *Ras* genes, could lead to malignant cancer. Normally, the occurrence of multiple mutations in a single cell lineage is rare, but in people exposed to chemicals that cause mutations or who carry mutations in DNA repair processes, multiple mutations in a single cell lineage are more likely to occur, and so the risk of cancer is increased.

CASE 3 YOUR PERSONAL GENOME: YOU, FROM A TO T

What can your personal genome tell you about your genetic risk factors for cancer?

Cancer usually occurs because of a series of mutations that occur sequentially in a single lineage of somatic cells, as illustrated for colon cancer in Fig. 14.7. In most individuals with cancer, all the sequential mutations that cause the cancer are spontaneous mutations that take place in somatic cells. They are not transmitted through the germ line, so there is little or no increased risk of cancer in the offspring. In some families, however, a germ-line mutation in one of the genes implicated in cancer is transmitted from parents to their children. In any child who inherits the mutation, all cells in the body contain the defective gene, and hence the cells already have taken one of the mutational steps that lead to cancer.

For colon cancer, the major genetic risk factors are mutations in *APC*, *Ras*, and *p53*. For breast cancer, the major genetic risk factors are mutations in the *BRCA1* and *BRCA2* genes. For these genes, each is a risk factor because when it is mutated, fewer additional mutations are needed to bring about tumor growth.

The DNA sequence of each of our personal genomes can reveal the genetic risk factors that each of us carries, not only for cancer but for many other diseases as well. Not all genetic risk factors are known for all diseases, but many genetic risk factors are already known for a large number of common diseases, including high blood pressure, diabetes, inflammatory bowel disease, age-related macular degeneration, Alzheimer's disease, and many forms of cancer.

Our personal genomes can identify only genetic risk factors, however. In many cases, disease risk is substantially increased by environmental risk factors as well, especially lifestyle choices. While there are genetic risk factors for lung cancer, for example, the single biggest environmental risk factor is tobacco smoking. For skin cancer, the greatest environmental risk factor is exposure to the damaging ultraviolet rays in sunlight or in the sunlamps used in tanning beds. For heart disease, the key environmental risk factors are smoking, lack of physical activity, and obesity. For diabetes, it is an unhealthy diet.

For breast cancer, the environmental risk factors include certain forms of hormone therapy, lack of physical activity, and use of alcohol. While we may not be able to do much about the genetic risk factors for any of these conditions, knowing that we have them may make us more careful about the lifestyle choices that we make.

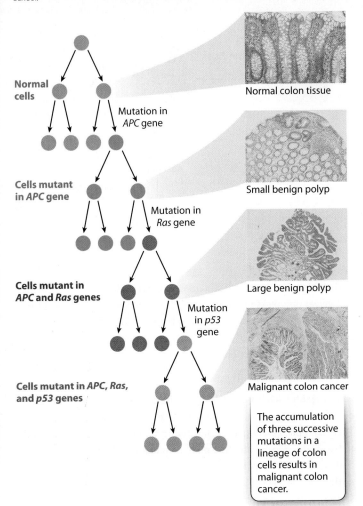

FIG. 14.7 Three somatic mutations implicated in the origin of invasive colon cancer. These mutations must occur in the same cell lineage for cancer to develop. *Source: Kathleen R. Cho, University of Michigan Medical School.*

The accumulation of three successive mutations in a lineage of colon cells results in malignant colon cancer.

308 SECTION 14.2 THE NATURE OF MUTATIONS

Mutations are random with regard to an organism's needs.

Consider the following experiment: an antibiotic is added to a liquid culture of bacterial cells that are growing and dividing. Most of the cells are killed, but a few survivors continue to grow and divide. These survivors are found to contain mutations that confer resistance to the antibiotic. This simple observation raises a profound question: were the antibiotic-resistant mutations already present before the antibiotic was added, implying that they arise spontaneously, or do the antibiotic-resistant mutations arise in response to the antibiotic?

These alternative hypotheses have deep implications for all of biology because they suggest two very different ways in which mutations might arise. The first suggests that mutations occur without regard to the needs of an organism. According to this hypothesis, the presence of the antibiotic in the experiment with bacterial cells does not direct or induce antibiotic resistance in the cells, but instead kills off all but the small number of preexisting antibiotic-resistant mutants, allowing them to flourish. The second hypothesis suggests that some sort of feedback occurs between the needs of an organism and the process of mutation, such that the environment directs specific mutations that are beneficial to the organism.

To distinguish between these two hypotheses, Joshua and Esther Lederberg in 1952 carried out a now-famous experiment, described in **Fig. 14.8**. Bacterial cells were grown and formed colonies on agar plates in the absence of antibiotic. Then, using a technique called replica plating, the Lederbergs transferred these colonies to new plates containing antibiotic. Only bacteria that were resistant to the antibiotic grew on the new plates. Because replica plating preserved the physical arrangement of the colonies, the Lederbergs were able to go back to the original plate and identify the colony that produced the antibiotic-resistant colony on the replica plate. From that original colony, they were then able to isolate a pure culture of antibiotic-resistant bacteria.

In the Lederbergs' experiments, isolating antibiotic-resistant bacteria from the original plate showed that the mutation for

HOW DO WE KNOW?

FIG. 14.8

Do mutations occur randomly, or are they directed by the environment?

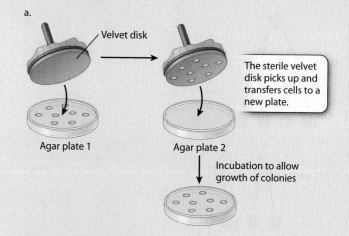

BACKGROUND Researchers have long observed that beneficial mutations tend to persist in environments where they are useful. For example, in the presence of antibiotics, bacterial populations become antibiotic resistant; in the presence of insecticides, insect populations become insecticide resistant.

HYPOTHESIS These observations lead to two hypotheses about how a mutation, such as one that confers antibiotic resistance to bacteria, might arise. The first suggests that mutations occur randomly in bacterial populations and over time become more common in the population in the presence of an antibiotic (which kills bacteria that do not have the mutation). In other words, they occur randomly with respect to the needs of an organism. The second hypothesis suggests that the environment, in this case the application of antibiotic, induces or directs mutations conferring antibiotic resistance.

METHOD To distinguish between these two hypotheses, Joshua and Esther Lederberg developed a technique called replica plating. In this technique, bacteria are grown on plates of a solid gel-like medium containing agar and nutrients, where they form colonies of cells (Fig. 14.8a). The cells in any one colony result from the division of a single original cell, so all individual bacteria in a single colony are genetically identical except for rare mutations that occur in the course of growth and division. Then a disk of sterilized velvet is pressed onto the plate. The velvet disk picks up cells from each colony and preserves in mirror image their geometrical positions on the plate. The velvet disk is pressed onto the surface of a fresh plate, transferring to the new plate some of the cells that originate from the colony on the first agar plate, in their original positions.

EXPERIMENT AND RESULTS First, the Lederbergs grew bacterial colonies on a master plate containing medium without antibiotic. Then, by replica plating, they transferred some cells from each colony to two plates (Fig. 14.8b). One plate contained medium without antibiotic. This first plate was the nonselective medium and served as the control; all cells were able to grow and form colonies on it. The second plate contained medium with antibiotic. This was the selective medium: only antibiotic-resistant cells multiplied and formed colonies

antibiotic resistance existed in the bacteria before they were ever exposed to antibiotic. This result supported the hypothesis that mutations occur spontaneously without regard to the needs of the organisms. The environment does not create specific mutations, but instead selects for them. The principle the Lederbergs demonstrated is true of all organisms so far examined.

Self-Assessment Questions

4. Mutations in which types of cell are most likely to contribute to evolutionary change in a population of organisms? Why?
5. What is the difference between the mutation rate for a given nucleotide and the mutation rate for a given cell?
6. If mutations occur at random with respect to an organism's needs, how does a species become more adapted to its environment over time?

14.3 SMALL-SCALE MUTATIONS

At the molecular level, a mutation is a change in the nucleotide sequence of a genome. Such changes can be small, affecting one or a few bases, or large, affecting entire chromosomes. In this section, we consider the origin and effects of small-scale changes to the DNA sequence. While mutation provides the raw material that allows evolution to take place, it can play this role only because of an important feature of living systems: once a mutation has taken place in a gene, the mutant genome is replicated as faithfully as the nonmutant genome.

Point mutations are changes in a single nucleotide.

Most DNA damage or errors in replication are immediately removed or corrected by specialized enzymes in the cell (section 14.5). We have already seen one type of DNA repair, the proofreading function of DNA polymerase during the process of replication, which acts to remove an incorrect nucleotide from the 3' end of the growing DNA strand (see Fig. 12.7). Damage

on this second plate. It served as the experimental treatment.

Because replica plating preserves the arrangement of the colonies, the location of an antibiotic-resistant colony that grew on the selective medium revealed the location of its parental colony on the control plate. In the final step of the experiment, the Lederbergs went back to the parental colony and plated cultures of just the antibiotic-resistant parent colony, the location of which had been revealed on the selective medium experimental treatment plate (Fig. 14.8c).

CONCLUSION The Lederbergs' replica-plating experiments showed that antibiotic-resistant mutants can arise in the absence of antibiotic, because antibiotic-resistant cells were isolated in those experiments even though the parental cells on nonselective medium never came into contact with the antibiotic. Only the daughter cells carried over to selective medium by replica plating were exposed to the antibiotic. In the final step of the experiment, the Lederbergs were able to isolate pure colonies of antibiotic-resistant cells.

FOLLOW-UP WORK These results have been extended to other types of mutation and other organisms, demonstrating that mutations are random and not directed by the environment.

SOURCE Lederberg, J., and E. M. Lederberg. 1952. "Replica Plating and Indirect Selection of Bacterial Mutants." *Journal of Bacteriology* 63:399–406.

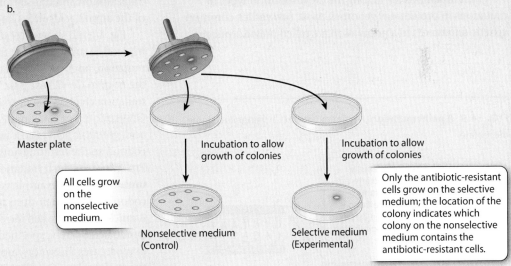

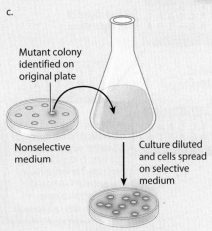

that is corrected is not regarded as a mutation because the DNA sequence is immediately restored to its original state and the mutation is not replicated in the daughter DNA.

If a mutation in DNA is to be inherited through mitotic or meiotic cell divisions, it must escape correction by the DNA repair systems. The example in **Fig. 14.9** shows how a mutation incorporated during replication can become a permanent change to the genome. Recall that in a double-stranded DNA molecule, the A's in one strand pair with T's in the other, and the G's in one strand pair with C's in the other. If, during replication, DNA polymerase mistakenly adds a G instead of a T (shown in red in Fig. 14.9) to a growing strand across from an A and its proofreading function does not correct the error, the result is a T—G mismatch in the double-stranded DNA. At the next replication, the G in the new template strand specifies a C in the daughter strand, with the result that the daughter DNA duplex has a perfectly matched C—G base pair. From this point forward, the DNA molecule containing the mutant C—G base pair will replicate as faithfully as the original molecule bearing the nonmutant T—A base pair. A mutation in which one base pair (in this example, T—A) is replaced by a different base pair (in this example, C—G) is called a nucleotide substitution or **point mutation.**

Point mutations are among the most common types of mutations in present-day genomes. As we saw earlier, common genetic differences in a population are called polymorphisms, so a difference between two sequences at a single nucleotide position in a DNA sequence is called a **single-nucleotide polymorphism** (abbreviated as **SNP** and pronounced "snip"). A SNP is the result of a point mutation that occurred sometime in the past and then increased in frequency so that many individuals in the population now carry it. For example, the differences among the A, S, and C alleles of the gene for β-globin are SNPs because they differ from one another at just one nucleotide site (see Fig. 14.4).

The effect of a point mutation depends in part on where in the genome it occurs.

In many multicellular eukaryotes, including humans, the vast majority of DNA in the genome does not code for protein or for RNAs with known function (Chapter 13). Most of the sequences in noncoding DNA have no known function, which may explain why many point mutations in noncoding DNA have no detectable effects on the organism.

In contrast, point mutations in coding sequences do have predictable consequences in an organism. **Fig. 14.10** shows an example in the human DNA sequence coding for β-globin, a subunit of hemoglobin, which we discussed in section 14.1. Fig. 14.10a shows a small part of the nonmutant DNA sequence, transcription and translation of which results in incorporation of the amino acids Pro–Glu–Glu in the β-globin polypeptide.

Fig. 14.10b shows an example of a neutral, or harmless, nucleotide substitution in this coding sequence. In this mutation, an A—T base pair (shown in red) is substituted for the normal G—C base pair. When transcribed into mRNA, the mutation changes the normal GAG codon into the mutant GAA codon. But GAG and GAA both code for the same amino acid, glutamic acid (Glu). In other words, they are synonymous codons, so the resulting amino acid sequences are the same: Pro–Glu–Glu. Such mutations are called **synonymous (silent) mutations.** This example is typical in that the synonymous codons differ at their third position (the 3' end of the codon). A quick look at the genetic code (see Table 4.1) shows that most amino acids can be specified by synonymous codons, and that in most cases the synonymous codons differ in the identity of the nucleotide at the third position.

A point mutation in a coding sequence can sometimes have harmful effects on the organism. Fig. 14.10c shows such an example, in which a T—A base pair (shown in red) is substituted for the normal A—T base pair. The result is a change in the mRNA from GAG, which specifies Glu (glutamic acid), to GUG, which specifies Val (valine). The resulting protein therefore contains the amino acids Pro–Val–Glu instead of Pro–Glu–Glu. This is the S allele that we discussed earlier (see Fig. 14.4). As we saw, this single amino acid change affects the ability of hemoglobin to function normally. In this case, the nucleotide substitution results in an **amino acid replacement.** Point mutations that cause amino acid replacements are called **nonsynonymous (missense) mutations.**

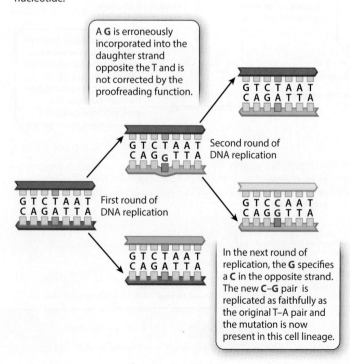

FIG. 14.9 A point mutation. A point mutation is a change in a single nucleotide.

FIG. 14.10 Synonymous and nonsynonymous mutations. Synonymous mutations do not change the amino acid sequence of the resulting protein. Nonsynonymous mutations change the amino acid sequence.

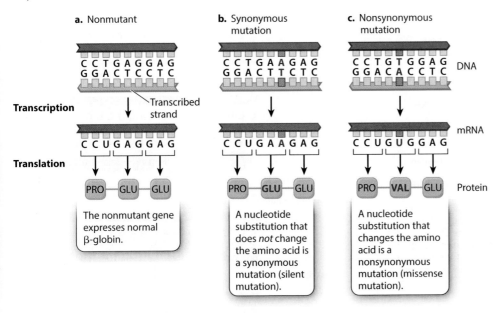

Fig. 14.11 shows a third way that a point mutation can affect a protein: a **nonsense mutation,** which creates a stop codon that terminates translation. A nonsense mutation nearly always has harmful effects. In Fig. 14.11, the mutation creates a UAG codon in the mRNA. Because UAG is a translational stop codon, the resulting polypeptide terminates after Pro. Polypeptides that are truncated are nearly always nonfunctional, are unstable, and are quickly destroyed. Eukaryotic cells have mechanisms to destroy most mRNA molecules that contain premature stop codons.

Small insertions and deletions involve several nucleotides.

Another relatively common type of mutation is the deletion or insertion of a small number of nucleotides. In noncoding DNA, such mutations have little or no effect. In protein-coding regions, their effects depend on their size. A small deletion or insertion that is an exact multiple of three nucleotides results in a polypeptide with fewer (in the case of a deletion) or more (in the case of an insertion) amino acids as there are codons deleted or inserted. For example, a deletion of three nucleotides eliminates one amino acid, and an insertion of six nucleotides adds two amino acids.

The effects of a deletion of three nucleotides can be seen in cystic fibrosis. This disease is characterized by the production of abnormal secretions in the lungs, liver, and pancreas and other glands. Patients with cystic fibrosis have an accumulation of thick, sticky mucus in their lungs, which often leads to respiratory complications, including recurrent bacterial infections. With proper medical care, including regular physical therapy to clear the lungs, antibiotics, pancreatic enzyme supplements, and good nutrition, the average life expectancy of a person with cystic fibrosis is currently 35 to 40 years.

The mutations responsible for cystic fibrosis occur in the gene encoding the cystic fibrosis transmembrane conductance regulator (CFTR) (**Fig. 14.12**). The CFTR protein is a chloride channel, which acts as a transporter to pump chloride ions out of the cell. Malfunction of CFTR causes ion imbalances that result in abnormal secretions from the many cell types in which the *CFTR* gene is expressed. Many different mutations can contribute to cystic fibrosis, including a specific mutation known as Δ508 (*delta 508*), which is a deletion of three nucleotides that eliminates a phenylalanine normally present at position 508 in the protein. The missing amino acid results in a CFTR protein that does not fold properly and is degraded before reaching the membrane. Researchers have created drugs that stabilize the Δ508 mutant protein and improve its function.

Small deletions or insertions that are not exact multiples of 3 can cause major changes in amino acid sequence of the corresponding proteins because they do not insert or delete entire codons. The effect of such a mutation can be appreciated by seeing how deletion of a single letter turns a perfectly

FIG. 14.11 A nonsense mutation. A nonsense mutation can change an amino acid to a stop codon, resulting in a shortened and unstable protein.

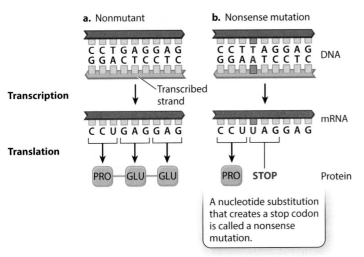

FIG. 14.12 Effect of a single amino acid deletion in CFTR, the cystic fibrosis transmembrane conductance regulator.

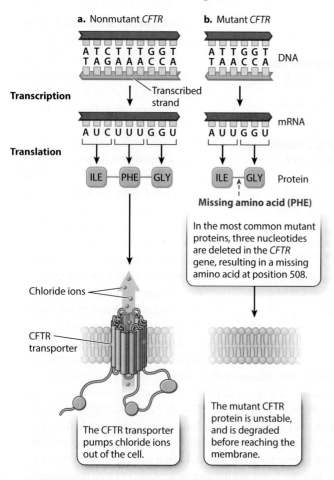

FIG. 14.13 A frameshift mutation. A frameshift mutation changes the translational reading frame.

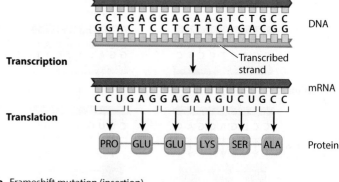

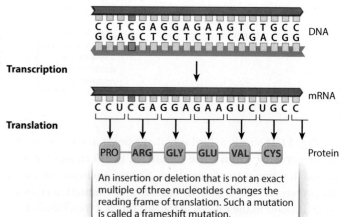

An insertion or deletion that is not an exact multiple of three nucleotides changes the reading frame of translation. Such a mutation is called a frameshift mutation.

sensible sentence of three-letter words into gibberish. Consider the sentence:

THE BIG BOY SAW THE CAT EAT THE BUG

If the red E is deleted, the new reading frame for three-letter words is as follows:

THB IGB OYS AWT HEC ATE ATT HEB UG

The result is unintelligible. Similarly, an insertion of a single nucleotide causes a one-nucleotide shift in the reading frame of the mRNA, and it changes all codons following the site of insertion. For this reason, such mutations are called **frameshift mutations**. Because frameshift mutations so profoundly alter the amino acid sequence, the mutant protein does not fold properly into its tertiary structure and becomes nonfunctional.

Fig. 14.13 shows the consequences of a frameshift mutation in the β-globin gene. The normal sequence in Fig. 14.13a corresponds to amino acids 5–10. The frameshift mutation in Fig. 14.13b is caused by the insertion of a C—G base pair. In turn, the mRNA transcript of the DNA also has a single-base insertion. When this mRNA is translated, the one-nucleotide shift in the reading frame

results in an amino acid sequence that bears no resemblance to the original protein. All amino acids downstream of the site of insertion are changed, resulting in loss of protein function.

Some mutations are due to the insertion of a transposable element.

An important source of new mutations in many organisms is the insertion of movable DNA sequences into or near a gene. Such movable DNA sequences are called **transposable elements** or **transposons**. As we saw in Chapter 13, the genomes of virtually all organisms contain several types of transposable element, each present in multiple copies per genome.

Transposable elements were discovered by American geneticist Barbara McClintock in the 1940s (**Fig. 14.14**). She studied corn (maize) because genetic changes that affect pigment formation can be observed directly in the kernels, which are normally purple. Since McClintock's work, biologists have discovered many different types of transposable elements, most ranging in length from a few hundred to a few thousand base pairs. When such a large piece of DNA inserts into a gene, it can interfere with transcription, cause errors in RNA processing, or disrupt the open

HOW DO WE KNOW?

FIG. 14.14

What causes sectoring in corn kernels?

BACKGROUND In the late 1940s, Barbara McClintock studied corn (*Zea mays*). Nonmutant corn has purple kernels, resulting from synthesis of purple anthocyanin pigment in every cell of the kernel (Fig. 14.14a). Mutant cells produce no purple pigment, resulting in yellow kernels. McClintock noticed that, in some mutant corn, speckles or streaks of purple pigment could be seen in many otherwise yellow kernels, and some kernels were entirely purple (Fig. 14.14b).

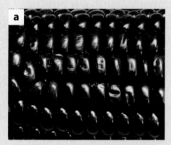
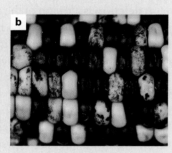

SOURCES: a. photo_journey/Shutterstock; b. Robert Martienssen, Cold Spring Harbor Laboratory.

HYPOTHESIS McClintock hypothesized that the yellow mutation resulted from insertion of a transposable element, which she called *Dissociation* (*Ds*), that had jumped into a site near or in the anthocyanin pigment gene and disrupted its function. She attributed the speckles of purple color in the yellow kernels to cell lineages in which the transposable element had jumped out again, restoring the nonmutant function of the anthocyanin gene.

EXPERIMENT AND RESULTS By a series of genetic crosses, McClintock showed that the jumping, or transposition, of *Ds* was due to a different genetic element, which she called *Activator* (*Ac*), located on another chromosome. She set up crosses in which she could track the *Ac*-bearing chromosome. She observed that in the presence of *Ac*, cells in mutant yellow kernels could revert to normal purple, resulting in purple speckles in an otherwise yellow kernel. From this observation, she inferred that the *Ds* element had jumped out of the anthocyanin gene in these cells, restoring its function. She also demonstrated that in cells in which the original purple color was restored, there were mutations elsewhere in the genome. From this observation, she inferred that after jumping out of the anthocyanin gene, the *Ds* element had become integrated elsewhere in the genome, where it disrupted the function of a different gene.

CONCLUSION McClintock's conclusion is illustrated in Fig. 14.14c: transposable elements can be excised from their original position in the genome and inserted into another position.

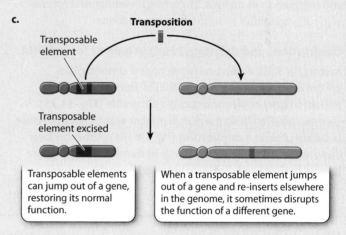

FOLLOW-UP WORK McClintock won the Nobel Prize in Physiology or Medicine in 1983. Later experiments showed that *Ds* is a transposable element that lacks a functional gene for transposase, the enzyme needed for the element to move, and *Ac* is a transposable element that encodes transposase. *Ac* produces active transposase that allows *Ds* to move. Much additional work has shown that many different types of transposable elements exist and that they are ubiquitous among organisms.

SOURCE McClintock, B. 1950. "The Origin and Behavior of Mutable Loci in Maize." *Proceedings of the National Academy of Sciences of the USA.* 36:344–355.

reading frame. The result in the case of maize is that the cell is unable to produce pigment, so the kernels will be yellow.

Self-Assessment Questions

7. What are the effects of nonsynonymous, synonymous, and nonsense mutations on a protein? Are these mutations usually harmful, beneficial, or neutral?

8. What are the effects of small insertions or deletions on a protein? Are these mutations usually harmful, beneficial, or neutral?

9. The coding sequences of genes that specify proteins with the same function in related species sometimes differ from each other by an insertion or deletion of contiguous nucleotides. The number of nucleotides that are inserted or deleted is almost always an exact multiple of 3. Why is this expected?

10. Given the discussion of the human genome in Chapter 13, would you predict that most mutations in humans would be harmful, beneficial, or neutral?

14.4 CHROMOSOMAL MUTATIONS

Whereas most mutations involve only one or a few nucleotides, some affect larger regions extending over hundreds of thousands or millions of nucleotides and have effects on chromosome structure that are often large enough to be visible through an ordinary optical microscope.

Chromosomal mutations can delete or duplicate regions of a chromosome containing several or many genes, and the resulting change in gene copy number also changes the amount of the products of these genes in the cell. Chromosomal mutations can also alter the linear order of genes along a chromosome or interchange the arms of nonhomologous chromosomes. While these types of chromosomal mutations do not change gene copy number, they do affect chromosome pairing and segregation in meiosis. These effects distinguish chromosome abnormalities from small-scale mutations.

Duplications and deletions result in gain or loss of DNA.

Among the most common chromosomal abnormalities are those in which a segment of the chromosome either is present in extra copies or is missing altogether (**Fig. 14.15**). A chromosome in which a region is present twice, instead of once, is said to contain a **duplication** (Fig. 14.15a). Although large duplications that include hundreds or thousands of genes are usually harmful and quickly eliminated from the population,

FIG. 14.15 Duplication and deletion. (a) In a duplication, a segment of chromosome is repeated. (b) In a deletion, a segment of chromosome is missing.

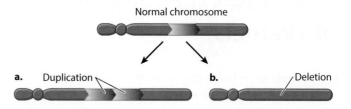

small duplications including only one or a few genes can be maintained over many generations. Usually, duplication of a region of the genome is less harmful than deletion of the same region.

A **deletion** is a region of the chromosome that is missing. An example is shown in Fig. 14.15b. A deletion can result from an error in replication or from the joining of breaks in a chromosome that occur on either side of the deleted region. Even though a deletion may eliminate a gene that is essential for survival, the deletion can persist in the population because chromosomes usually occur in homologous pairs (Chapter 13). If one member of a homologous pair has a deletion of an essential gene but the gene is present in the other member of the pair, that one copy of the gene is often sufficient for survival

FIG. 14.16 Origin of the β-globin gene family. The β-globin gene family evolved through several episodes of duplication and divergence. *Data from Y. A. Trusov and P. H. Dear, 1996, "A Molecular Clock Based on the Expansion of Gene Families," Nucleic Acids Research 24(6):995–999.*

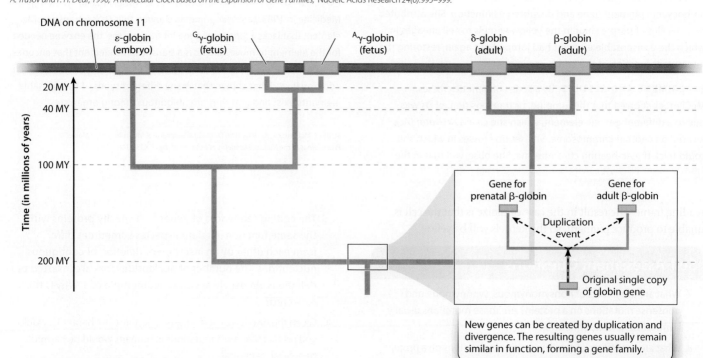

and reproduction. In these cases, the deletion can be transmitted from generation to generation, as long as the chromosome with the deletion is present along with a normal chromosome.

Deletions or duplications that include the **centromere,** the site associated with attachment of the spindle fibers that move the chromosome during cell division (Chapter 11), are rarely observed. An abnormal chromosome without a centromere, or one with two centromeres, is usually lost within a few cell divisions because it cannot be directed properly into the daughter cells during cell division.

Gene families arise from gene duplication and divergence.

Small duplications play an important role in the origin of new genes in the course of evolution. In most cases, when a gene is duplicated, one of the copies is free to change without harming the organism because the other copy continues to carry out the normal function of the gene. Occasionally, a mutation in the "extra" copy of the gene may result in a beneficial effect on survival or reproduction, so that gradually a new gene is formed from the duplicate. These new genes usually have a function similar to that of the original gene.

This process of creating new genes from duplicates of old ones is known as **duplication and divergence.** The term "divergence" refers to the slow accumulation of different mutations in duplicate copies of a gene that occurs on an evolutionary time scale. Multiple rounds of duplication and divergence can give rise to a group of genes with related functions known as a **gene family.** The largest gene family in the human genome includes approximately 400 genes and encodes proteins that detect odors. These proteins are structurally very similar, but differ in the region that binds small odor molecules. It is the diversity of the odorant binding sites that allows us to identify so many different smells.

Most gene families are not as large and diverse as that for odor detection. **Fig. 14.16** shows the evolutionary origin of the family of globin genes, which in humans are spread across about 50 kb of chromosome 11. The globin gene family consists of several genes that are expressed at various times during development (embryo, fetus, or adult). The sequences are all similar enough to imply that the genes arose through duplication and divergence.

Copy-number variation constitutes a significant proportion of genetic variation.

In the modern human population, a common form of genetic variation among individuals is differences in the number of copies of a region of the genome due to duplication and deletion. These differences are called **copy-number variations (CNVs).** The regions involved in CNVs are large and may include one or more genes. An example is shown in **Fig. 14.17.** The multiple copies of the CNV region are usually located adjacent to one another along the chromosome. Across the

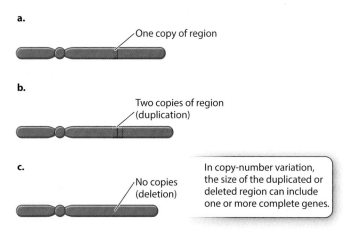

FIG. 14.17 Copy-number variation in a region of a chromosome.

In copy-number variation, the size of the duplicated or deleted region can include one or more complete genes.

genome as a whole, approximately 10% to 15% of the genome is subject to copy-number variation, and the average CNV length is 200 kb to 300 kb.

Some CNVs occur in noncoding regions, but others consist of genes that are present in multiple tandem copies along the chromosome. An example is the human gene *AMY1* for the salivary gland enzyme amylase, which aids in the digestion of starch. This gene is located on chromosome 1, and the *AMY1* copy number differs from one chromosome 1 to the next. **Fig. 14.18** shows the distribution of the *AMY1* copy number along chromosome 1 in two groups of people: societies with

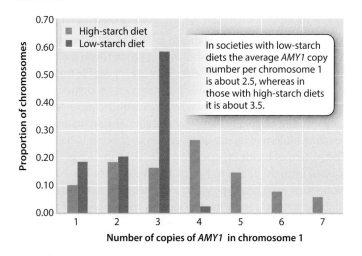

FIG. 14.18 Copy number variation in a gene that helps break down starch. Copy numbers of *AMY1* vary between societies with high-starch and low-starch diets. *Data from G. H. Perry, N. J. Dominy, K. G. Claw, A. S. Lee, H. Fiegler, R. Redon, J. Werner, F. A. Villanea, J. L. Mountain, R. Misra, N. P. Carter, C. Lee, and A. C. Stone, 2007, "Diet and the Evolution of Human Amylase Copy Number Variation,"* Nature Genetics *39:1256–1260.*

In societies with low-starch diets the average *AMY1* copy number per chromosome 1 is about 2.5, whereas in those with high-starch diets it is about 3.5.

a long history of a high-starch diet, and societies with a long history of a low-starch diet. There is a clear tendency for chromosomes from the latter group to have fewer copies of *AMY1* than those from the former group. A plausible hypothesis is that extra copies were selected in groups with a high-starch diet because of the advantage the extra copies conferred in digesting starch.

Tandem repeats are useful in DNA typing.

A special type of CNV consists of short sequences (2–50 base pairs) that are repeated in multiple tandem copies, called a **tandem repeat**. At each location, the number of copies of the short sequence may differ from chromosome to chromosome. The human genome contains tens of thousands of locations where one chromosome differs from another in the size of the tandem repeat (the number of copies of the short sequence). Just as an individual has two copies of each gene, possibly with two different alleles, so each individual has two copies of each of these tandem repeats. Each location with a tandem repeat typically has many alleles differing in copy number, so that, except for identical twins, an individual's genotype at 6–8 different locations is usually sufficient to identify the individual uniquely. Such differences among individuals are the basis of **DNA typing,** in which the analysis of a small quantity of DNA can be as reliable for identifying individuals as fingerprints. As a consequence, DNA typing is often called **DNA fingerprinting.**

Although the number of alleles of a particular tandem repeat at a given location in the genome can be very large in the population as a whole, any one diploid individual can have only 2 alleles inherited one each from the mother and the father. An example with 5 alleles is shown in **Fig. 14.19,** where the repeated sequence (AGT) is shown in green between the flanking unique sequences. With 5 alleles, there are 5 possible homozygous genotypes and 10 possible heterozygous genotypes. The genotypes can be distinguished by DNA sequencing or other methods, but originally they were visualized by separating the resulting molecules by size using gel electrophoresis (Chapter 12). The pattern of bands expected from the DNA in each genotype formed from the alleles in Fig. 14.19a is shown in Fig. 14.19b. Notice that each genotype yields a distinct pattern of bands, so these patterns can be used to identify the genotype of any individual for a particular region of DNA.

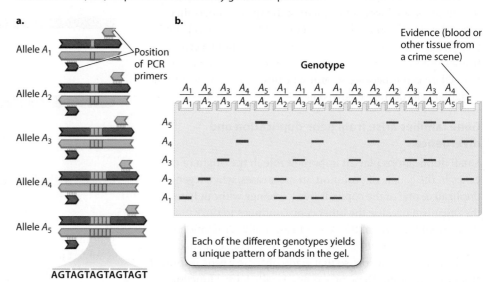

FIG. 14.19 Tandem repeats in DNA typing. (a) The number of short, repeated sequences at a given site in a human chromosome can vary. (b) These differences can be visualized using polymerase chain reaction (PCR) amplification followed by gel electrophoresis.

The lane labeled "E" at the far right in Fig. 14.19b shows the DNA bands for this same polymorphism from biological material collected at the scene of a crime. In this case, the bands in lane E imply that the source of the evidence is an individual of genotype A_2/A_4 because this is the only genotype with bands with vertical positions in the gel that exactly match those of the evidence sample. A single polymorphism of the type shown in Fig. 14.19 is not enough to say for sure that two pieces of DNA come from the same person, but because tandem repeat sites are scattered throughout the human genome, many multiple-allele tandem repeats can be examined. If six or eight of these polymorphisms match between two samples, it is very likely that the samples come from the same individual (or from identical twins). On the other hand, if any of the polymorphisms fails to match band for band, then it is almost certain that the samples come from different individuals (barring technical mishaps that lead to contaminated, degraded, or mislabeled DNA).

One of the advantages of DNA typing is that only a small amount of DNA is needed. Even minuscule amounts of DNA can be typed, so tiny spots of blood, semen, or saliva are sufficient samples. A cotton swab wiped across the inside of your cheek will contain enough DNA for typing; so will a discarded paper cup or a cigarette butt.

An inversion has a chromosomal region reversed in orientation.

Chromosomes in which the normal order of a block of genes is reversed contain an **inversion** (**Fig. 14.20**). An inversion

FIG. 14.20 Inversion. A chromosome with an inversion has a segment of chromosome present in reverse orientation.

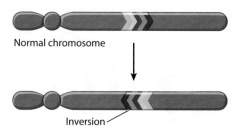

is typically produced when the region between two breaks in a chromosome is flipped in orientation before the breaks are repaired. Especially in large genomes, the breaks are likely to occur in noncoding DNA rather than within a gene. Whereas large inversions can cause problems in meiosis, small inversions are common in many populations and play an important role in chromosome evolution. A small inversion may have almost no effect on the organism because it still contains all of the genes present in the original chromosome, merely flipped in their order, and is too small to cause problems in meiosis. The accumulation of inversions over evolutionary time explains in part why the order of genes along a chromosome can differ even among closely related species.

A reciprocal translocation joins segments from nonhomologous chromosomes.

A **reciprocal translocation** (Fig. 14.21) occurs when two different (nonhomologous) chromosomes undergo an exchange of parts. In the formation of a reciprocal translocation, both chromosomes are broken and the terminal segments are exchanged before the breaks are repaired. In large genomes, the breaks are likely to occur in noncoding DNA, so the breaks themselves do not usually disrupt gene function.

Because reciprocal translocations change only the arrangement of genes and not their number, most reciprocal translocations do not affect the survival of organisms. Proper gene dosage requires the presence of both parts of the reciprocal translocation, however, as well as one copy of each of the normal homologous chromosomes. Problems can arise in meiosis because both chromosomes involved in the reciprocal translocation may not move together into the same daughter cells, resulting in gametes with only one part of the reciprocal translocation. This inequality does upset gene dosage, because these gametes have extra copies of genes in one of the chromosomes and are missing copies of genes in the other.

> ### Self-Assessment Questions
>
> 11. Which type of chromosomal abnormality—deletion, duplication, inversion, or translocation—would you expect to have the greatest effect on the organisms that carry it, and why?
>
> 12. How do gene families, such as the odorant receptor gene family, arise in evolution?

14.5 DNA DAMAGE AND REPAIR

Mutations result from unrepaired errors in DNA replication or from damage to DNA that is not repaired. DNA is a fragile molecule, prone to damage of many different kinds. Every day, in each cell in our bodies, the DNA is damaged in some way at tens of thousands of places along the molecule. Fortunately, cells have evolved mechanisms for repairing various types of damage and restoring the DNA to its original condition, which helps explain the low mutation rate in most organisms. The importance of DNA repair was recognized by the 2015 Nobel Prize in Chemistry awarded to Tomas Lindahl, Paul Modrich, and Aziz Sancar for their discoveries of key repair mechanisms. In this section, we examine some of the major types of DNA damage and some of the ways in which it is repaired.

DNA damage can affect both DNA backbone and bases.

Most mutations, whether small-scale or chromosomal, are spontaneous and occur naturally. However, mutations can also be induced by radiation or chemicals. **Mutagens** are agents that increase the probability of mutation. The presence of a mutagen can increase the probability of mutation by a factor of 100 or more.

Some of the most frequent types of DNA damage induced by mutagens are illustrated in **Fig. 14.22**, which shows a highly mutated DNA molecule. Some types of damage affect the structure of the DNA double helix. These include breaks in the sugar–phosphate backbone, one of the main mutagenic effects of X-rays. Breaks can occur in just one or both strands of a DNA molecule and can result in the types of chromosomal mutations described earlier. Ultraviolet light can cause cross-links between adjacent pyrimidine bases, especially thymine, resulting in the formation of thymine dimers that warp the backbone.

FIG. 14.21 Reciprocal translocation. A reciprocal translocation results from an exchange of parts between nonhomologous chromosomes.

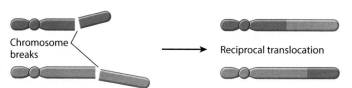

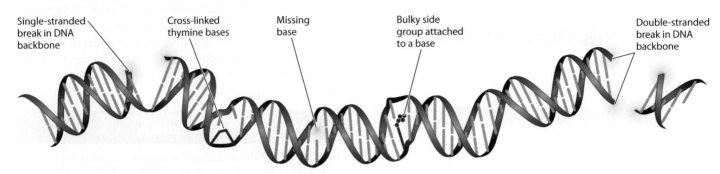

FIG. 14.22 **Major types of DNA damage.** Most types of DNA damage can be repaired by specialized enzymes.

Yet another type of structural damage is loss of a base from one of the deoxyribose sugars, resulting in a gap in one strand where no base is present. Spontaneous loss of a purine base is one of the most common types of DNA damage, occurring at the rate of approximately 13,000 purines lost per human cell per day. Most of these mutations result from the interaction between DNA and normal metabolic by-products. The mutation rate increases with age, and it can also be increased by exposure to oxidizing agents such as household bleach or hydrogen peroxide.

Other types of damage alter the chemical structure of the bases themselves, making them prone to mispairing. Chemicals that are highly reactive also tend to be mutagenic, often because they add bulky side groups to the bases that hinder proper base pairing. The main environmental source of such chemicals is tobacco smoke. Other chemicals can perturb the DNA replication complex and cause the insertion or deletion of one or occasionally several nucleotides.

Most DNA damage is corrected by specialized repair enzymes.

Cells contain many specialized DNA-repair enzymes that correct specific kinds of damage, and in this section we examine a few examples. Perhaps the simplest is the repair of breaks in the sugar–phosphate backbone, which are sealed by **DNA ligase,** an enzyme that can repair the break by using the energy in ATP to join the 3′ hydroxyl group of one end to the 5′ phosphate group of the other end. Most organisms have multiple different types of DNA ligase. One type of ligase seals single-stranded breaks in DNA, and a different type seals double-stranded breaks. Double-stranded breaks are less likely to be repaired than single-stranded breaks, in many cases resulting in chromosomal rearrangements. In addition to their role in DNA replication and repair, ligases are an important tool in research in molecular biology because they allow DNA molecules from different sources to be joined to produce recombinant DNA (Chapter 12).

As we saw earlier in this chapter, mispairing of bases during DNA replication leads to the incorporation of incorrect nucleotides that can become nucleotide substitutions in the next round of replication if they are not corrected. Approximately 99% of the mispaired bases are corrected immediately by the proofreading function of DNA polymerase, in which the mispaired nucleotide is removed immediately after incorporation and replaced by the correct nucleotide (Chapter 12).

Even after proofreading by DNA polymerase, the proportion of mismatched nucleotides in newly replicated DNA is about 10^{-6}, which is at least 1000 times greater than the actual frequency of errors in cellular organisms (see Fig. 14.5a). What accounts for the difference is a second-chance mechanism for catching mismatches that is known as **mismatch repair** (**Fig. 14.23**).

In the bacterium *E. coli*, mismatch repair begins when a protein known as MutS recognizes and binds to the site of a mismatch (Fig. 14.23a) and brings two other proteins (MutL and MutH) to the site. MutL determines which DNA backbone will be cleaved, and MutH cleaves the backbone in the vicinity of the mismatch (Fig. 14.23b). An exonuclease degrades the cleaved strand to a distance beyond the site of the mismatch (Fig. 14.23c). A DNA polymerase then synthesizes new DNA to fill the gap, correcting the mismatch, and a DNA ligase seals the remaining nick in the DNA backbone (Fig. 14.23d). Mismatch repair usually, but not always, cleaves the daughter strand, so the corrected strand matches the parental template strand.

Eukaryotes have proteins corresponding to those in Fig. 14.23. The importance of mismatch repair in these organisms is underlined by the finding that a genetic predisposition to certain types of colon cancer results from mutations in the human counterparts of the *MutS* and *MutL* genes.

The result of proofreading by DNA polymerase and mismatch repair is increased fidelity of DNA replication. Cells have evolved many other systems that repair different types of

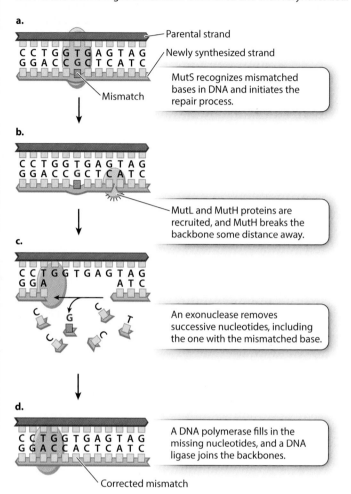

FIG. 14.23 Mismatch repair. In mismatch repair, the segment of a DNA strand containing the mismatch is removed and then resynthesized.

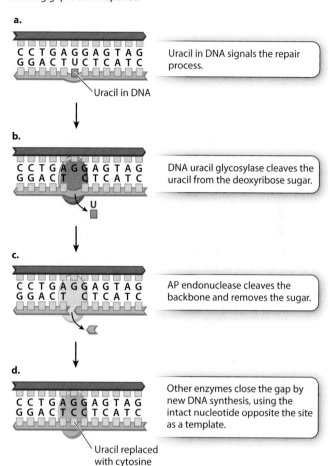

FIG. 14.24 Base excision repair. In base excision repair, an improper base in DNA and its deoxyribose sugar are both removed, and the resulting gap is then repaired.

DNA damage that can occur, even in nondividing cells such as those in the adult brain, heart, liver, and kidney. One example is **base excision repair,** which corrects abnormal or damaged bases. In the first step of base excision repair, an abnormal or damaged base is cleaved from the sugar in the DNA backbone. Then, the baseless sugar is removed from the backbone, leaving a gap of one nucleotide. Finally, a repair polymerase inserts the correct nucleotide into the gap **(Fig. 14.24)**.

Cells have also evolved mechanisms to repair short stretches of DNA containing mismatched or damaged bases. One of these is known as **nucleotide excision repair,** which has a similar mechanism of action to mismatch repair but uses different enzymes **(Fig. 14.25)**. Instead of degrading a DNA strand in a nucleotide by nucleotide fashion until the mismatch is removed, nucleotide excision repair removes an entire damaged section of a strand at once. Nucleotide excision repair is also used to remove nucleotides with bulky side groups, as well as thymine dimers resulting from ultraviolet light.

The importance of nucleotide excision repair is illustrated by the disease xeroderma pigmentosum (XP), in which nucleotide excision repair is defective. People with XP are exquisitely sensitive to the UV radiation in sunlight. Because of the defect in nucleotide excision repair, damage to DNA resulting from UV light is not corrected, leading to the accumulation of mutations in skin cells. The result is high rates of skin cancer. People with XP must minimize their exposure to sunlight and in extreme cases stay out of the sun altogether.

When DNA repair mechanisms function properly, they reduce the rate of mutation to a level that is compatible with life. But they are not perfect, and they do not catch all the mistakes. The result is genetic variation, the raw material of evolution.

FIG. 14.25 Nucleotide excision repair. In nucleotide excision repair, a damaged segment of a DNA strand is removed and resynthesized using the undamaged partner strand as a template.

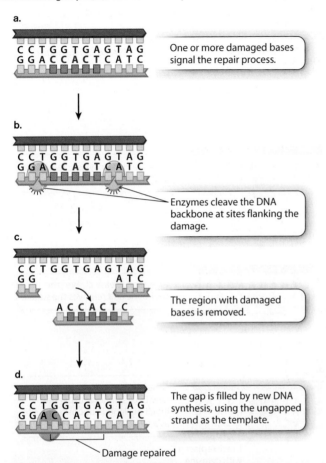

a. One or more damaged bases signal the repair process.

b. Enzymes cleave the DNA backbone at sites flanking the damage.

c. The region with damaged bases is removed.

d. The gap is filled by new DNA synthesis, using the ungapped strand as the template.

Damage repaired

Self-Assessment Questions

13. Earlier, we described the Lederbergs' experiment, which demonstrated that mutations are not directed by the environment. But mutagens, which are environmental factors, can lead to mutations. What's the difference?

14. What is a mutagen? Name two common mutagens and their effects on DNA.

15. What's the difference between nucleotide excision repair and mismatch repair?

CORE CONCEPTS SUMMARY

14.1 GENOTYPE AND PHENOTYPE: The genetic makeup of a cell or organism is its genotype, and an observable characteristic is its phenotype.

Polymorphisms result from mutations that persist through time and become common in the population. page 302

The genotype of an organism is its genetic makeup. page 302

The phenotype of an organism is any observable characteristic of that organism, such as its appearance, physiology, or behavior. The phenotype results from a complex interplay of the genotype and the environment. page 302

Different forms (sequences) of genes are called alleles. page 302

A person's genotype can be homozygous, with two alleles of the same type (one from each parent), or heterozygous, with a different type of allele from each parent. page 302

Mutations can be neutral (no effect on survival or reproduction), harmful (associated with decreased survival or reproduction), or beneficial (associated with increased survival or reproduction). page 302

14.2 THE NATURE OF MUTATIONS: Mutations are very rare for any given nucleotide and occur randomly without regard to the needs of an organism.

For an individual nucleotide, the frequency of mutation is very rare, though it differs among species. page 305

Mutations are common across an entire genome and across many cell divisions. page 305

Germ-line mutations occur in haploid gametes as well as in the diploid cells that give rise to them, whereas somatic mutations occur in nonreproductive cells. Only mutations in germ cells are transmitted to progeny and play a role in genetic variation and evolution. page 306

Mutations are random, in that they occur without regard to the needs of an organism. This principle can be demonstrated by replica plating of bacterial colonies in the absence and presence of an antibiotic. page 308

14.3 SMALL-SCALE MUTATIONS: Small-scale mutations include point mutations, insertions and deletions, and movement of transposable elements.

A point mutation, or nucleotide substitution, is a change of one base for another. page 310

A single-nucleotide polymorphism (SNP) is a difference in a single nucleotide. page 310

The effect of a point mutation depends on where it occurs. If it arises in noncoding DNA, it will likely have little or no effect on an organism. If it arises in a protein-coding gene, it can result in a change in the amino acid sequence (nonsynonymous mutation), no change in the amino acid sequence (synonymous mutation), or the introduction of a stop codon (nonsense mutation). page 310

Small insertions (or deletions) in DNA add (or remove) one base or a few contiguous bases. Their effect depends on where in the genome they occur and on their size. An insertion or deletion of a single nucleotide in a protein-coding gene results in a frameshift mutation, in which all the codons downstream of the insertion or deletion are changed. page 311

Transposable elements are DNA sequences that can jump from one place in a genome to another. They can affect the expression of a gene if they are inserted into or near a gene. page 312

14.4 CHROMOSOMAL MUTATIONS: Chromosomal mutations involve large regions of one or more chromosomes.

A duplication is a region of the chromosome that is present two times, whereas a deletion is the loss of part of a chromosome. page 314

Duplications and deletions both affect gene dosage, which can have important effects on a cell or organism. page 314

Gene duplication followed by evolutionary divergence results in gene families made up of genes with related, but not identical, functions. page 314

Copy-number variation (CNV) is a difference in the number of copies of a particular DNA sequence among chromosomes. page 314

DNA typing (DNA fingerprinting) analyzes genetic polymorphisms at multiple genes with multiple alleles, such as tandem repeat polymorphisms. Because there are so many possible genotypes in the population, this technique can uniquely identify an individual with high probability. page 316

An inversion is a segment of a chromosome in reverse orientation. page 316

A reciprocal translocation involves the exchange of a part of one chromosome with another. Reciprocal translocations do not affect gene dosage, but can lead to problems during meiosis. page 317

14.5 DNA DAMAGE AND REPAIR: Mutations can occur spontaneously or be induced by mutagens, but most DNA damage is repaired.

Some types of damage alter the structure of DNA, such as single-stranded and double-stranded breaks, cross-linked thymine dimers, and missing bases. page 317

Other types of DNA damage, such as changes in the side groups that form hydrogen bonds and the addition of side groups that interfere with base pairing, affect the bases themselves. page 318

DNA ligase seals breaks in DNA; thus, it is an important tool for molecular biologists. page 318

Mismatch repair provides a backup mechanism for mistakes not caught by the proofreading function of DNA polymerase. page 318

Base excision repair corrects individual nucleotides and involves several DNA repair enzymes working together. page 319

Nucleotide excision repair functions similarly to mismatch repair but excises longer stretches of damaged nucleotides. page 319

Log in to **LaunchPad** to check your answers to the Self-Assessment Questions and to access additional learning tools.

CHAPTER 15 Mendelian Inheritance

CORE CONCEPTS

15.1 EARLY THEORIES OF INHERITANCE: Early theories of heredity incorrectly assumed the inheritance of acquired characteristics and blending of parental traits in the offspring.

15.2 FOUNDATIONS OF MODERN TRANSMISSION GENETICS: The study of modern transmission genetics began with Gregor Mendel, who used the garden pea and studied traits with contrasting characteristics.

15.3 SEGREGATION: Mendel's first key discovery was the principle of segregation, which states that members of a gene pair separate equally into gametes.

15.4 INDEPENDENT ASSORTMENT: Mendel's second key discovery was the principle of independent assortment, which states that different gene pairs segregate independently of one another.

15.5 HUMAN GENETICS: The patterns of inheritance that Mendel observed in peas can also be seen in humans.

Except for identical twins, each of us has his or her own personal genome, a unique human genome differing from all that have existed before and from all that will come after. Genetic variation from one person to the next leads in part to our individuality, from differences in appearance to differences in the ways our bodies work. Examples of genetic variation in the human population range from harmless curiosities, such as the genetic difference in taste receptors that determines whether you taste broccoli as being unpleasantly bitter, to mutant forms of genes resulting in serious diseases such as sickle-cell anemia or emphysema.

This chapter focuses on how that genetic variation is inherited. **Transmission genetics** deals with the manner in which genetic differences among individuals are passed from generation to generation. We are all aware of genetics. We know that children resemble their parents and that there are sometimes uncanny similarities among even distant relatives. Some patterns, however, are more difficult to discern. Traits such as eye color, nose shape, or risk for a particular disease may be passed down faithfully generation after generation, but sometimes they are not, and sometimes they appear and disappear in seemingly random ways.

As a modern science, transmission genetics began with the pea-breeding experiments carried out by the monk Gregor Mendel in the 1860s. However, even before then, people understood enough about inheritance that they were able to select crops and livestock with particular characteristics.

15.1 EARLY THEORIES OF INHERITANCE

Thousands of years before Mendel, many societies carried out practical plant and animal breeding. The ancient practices were based on experience rather than on a full understanding of the rules of genetic transmission, but they were nevertheless highly successful. In Mesoamerica, for example, Native Americans chose corn (maize) plants for cultivation that had the biggest ears and softest kernels. Over many generations, their cultivated corn came to have increasingly less physical resemblance to its wild ancestral species. Similarly, people in the Eurasian steppes selected their horses for a docile temperament suitable for riding or for hitching to carts or sleds. Practical breeding of this kind provided the crops and livestock from which most of our modern domesticated animals and plants derive.

Early theories of heredity predicted the transmission of acquired characteristics.

The first written speculations about mechanisms of heredity were made by the ancient Greeks. Hippocrates (460–377 BCE), considered the founder of Western medicine, proposed that each part of the body in a sexually mature adult produces a substance that collects in the reproductive organs and that determines the inherited characteristics of the offspring. An implication of this theory is that any **trait,** or characteristic, of an individual can be transmitted from parent to offspring. Even traits that are acquired during the lifetime of an individual, such as muscle strength or bodily injury, were thought to be heritable because of the substance supposedly passed from each body part to the reproductive organs. The related term "phenotype" (Chapter 14) refers to the particular manifestation of a trait in an individual.

The theory that acquired characteristics can be inherited was invoked to explain such traits as the webbed feet of ducks, which were thought to result from many successive generations in which adult ducks stretched the skin between their toes while swimming and then passed this trait to their offspring. A few decades after Hippocrates, however, Aristotle (384–322 BCE) emphasized several observations that the theory of inheritance of acquired characteristics cannot account for:

- Traits such as hair color can be inherited, but it is difficult to see how hair—a nonliving tissue—could send substances to the reproductive organs.

- Traits that are not yet present in an individual can be transmitted to the offspring. For example, a father and his adult son can both be bald, even if the son was born before the father became bald.

- Parts of the body that are lost as a result of surgery or accident are not missing in the offspring.

From these and other observations, Aristotle concluded that the process of heredity transmits only the potential for producing traits present in the parents, and not the traits themselves. Nevertheless, Hippocrates' theory influenced biology until well into the 1800s. It was incorporated into an early theory of evolution proposed by the French biologist Jean-Baptiste Lamarck around 1800. Later, Charles Darwin developed an alternative theory of evolution: the theory of evolution by natural selection (Chapter 20).

While traits acquired during the lifetime of a parent are not transmitted to the offspring, parental misfortune or misbehavior can nevertheless result in impaired fetal development and in some cases permanent damage. For example, maternal malnutrition, drug addiction, alcoholism, infectious disease, and other conditions can severely affect the fetus, but these effects are due to disruption of fetal development and not to changes in the genome.

Belief in blending inheritance discouraged studies of hereditary transmission.

Darwin subscribed to the now-discredited model of **blending inheritance,** in which traits in the offspring resemble the average of those in the parents. For example, this model predicts that the offspring of plants with blue flowers and those with red flowers will have purple flowers. While traits of offspring are sometimes the average of those of the parents (think of certain cases of human height), the idea of blending inheritance—which implies the blending of the genetic material—as a general rule presents problems. For example, it cannot explain the reappearance of a trait several generations after it apparently "disappeared" in a family, such as red hair or blue eyes.

Another difficulty with the concept of blending inheritance is that variation is lost over time. Consider an example where black rabbits are rare and white rabbits are

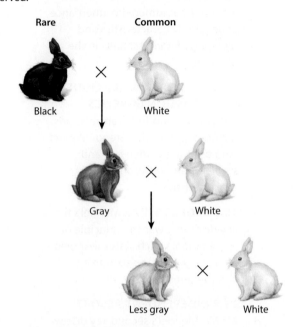

FIG. 15.1 Blending inheritance. This now-discredited model predicts loss of variation over time and blending of genetic material, which is not observed.

common (**Fig. 15.1**). Black rabbits, being rare, are most likely to mate with the much more common white rabbits. If blending inheritance occurs, the result will be gray rabbits. These will also mate with white rabbits, producing lighter-gray rabbits. Over time, the population will end up being all white, or very close to white, and there will be less variation. The only way that black rabbits will be present is if they are reintroduced into the population through mutation or migration.

Whatever the trait may be, blending inheritance predicts that inheritance will tend to be a homogenizing force, producing in each generation a blend of the original phenotypes. Of course, we know from common experience that variations in most populations are plentiful.

It is ironic that Darwin believed in blending inheritance because this mechanism of inheritance is incompatible with his theory of evolution by means of natural selection. The incompatibility was pointed out by some of Darwin's contemporaries, and Darwin himself recognized it as a serious problem. The problem with blending inheritance is that rare variants, such as the black rabbits in the previously mentioned example, will have no opportunity to increase in frequency because in each generation they must (most likely) mate with one of the superabundant white rabbits.

Although Darwin was convinced that he was right about natural selection, he was never able to reconcile his theory with the concept of blending inheritance. Unbeknownst to him, the solution had already been discovered in experiments carried out by Gregor Mendel (1822–1884), an Augustinian monk in a monastery in the city of Brno in what is now the Czech

Republic. Mendel's key discovery was this: traits are not transmitted in inheritance, but rather genes are transmitted.

> **Self-Assessment Question**
>
> 1. What are two observations that argue against the idea of blending inheritance?

15.2 FOUNDATIONS OF MODERN TRANSMISSION GENETICS

Mendel's scientific fame rests on the pea-breeding experiments that he carried out in the years 1856 to 1864 (**Fig. 15.2**), in which he demonstrated the basic principles of transmission genetics. As we will see in Chapter 17, most genetic variation in populations is due to multiple genes interacting with one another and with the environment, while traits due to single genes, such as those that Mendel studied, tend to be rare. Nevertheless, Mendel's experiments were so simple and presented with such clarity that they still serve as examples of the scientific method at its best (Chapter 1).

Mendel's experimental organism was the garden pea.

For his experimental material, Mendel used several strains of ordinary garden peas (*Pisum sativum*) that he obtained from local seed suppliers. His experimental approach was similar to that of a few botanists of the eighteenth and nineteenth centuries who studied the results of **hybridization,** or interbreeding between two different varieties or species of an organism. Where Mendel differed from his predecessors was in paying close attention to a small number of easily classified traits with contrasting characteristics. For example, where one variety, or strain, of peas had yellow seeds, another had green seeds; and where one had round seeds, another had wrinkled seeds (**Fig. 15.3**). Altogether Mendel studied seven physical features expressed in contrasting fashion among the strains: seed color, seed shape, pod color, pod shape, flower color, flower position, and plant height.

The expression of each trait in each of the original strains that Mendel obtained was **true breeding,** which means that the physical appearance of the offspring in each successive generation is identical to the previous one. For example, plants of the strain with yellow seeds produced only yellow seeds, and those of the strain with green seeds produced only green seeds. Likewise, plants of the strain with round seeds produced only round seeds, and those of the strain with wrinkled seeds produced only wrinkled seeds.

The objective of Mendel's experiments was simple. By means of crosses between the true-breeding strains and crosses among their progeny, Mendel hoped to determine whether statistical patterns could be seen in the occurrence of the contrasting characteristics, such as yellow seeds or green seeds. If such patterns could be found, he would seek to devise a hypothesis to explain them and then use his hypothesis to predict the outcome of further crosses.

In designing his experiments, Mendel departed from other plant hybridizers of the time in three important ways:

1. Mendel studied true-breeding strains, unlike many other plant hybridizers, who used complex and poorly defined material.

FIG. 15.2 Gregor Mendel and his experimental organism, the pea plant. *Sources: Juliette Wade/Getty Images; Authenticated News/Getty Images.*

FIG. 15.3 Contrasting traits. Mendel focused on seven contrasting traits.

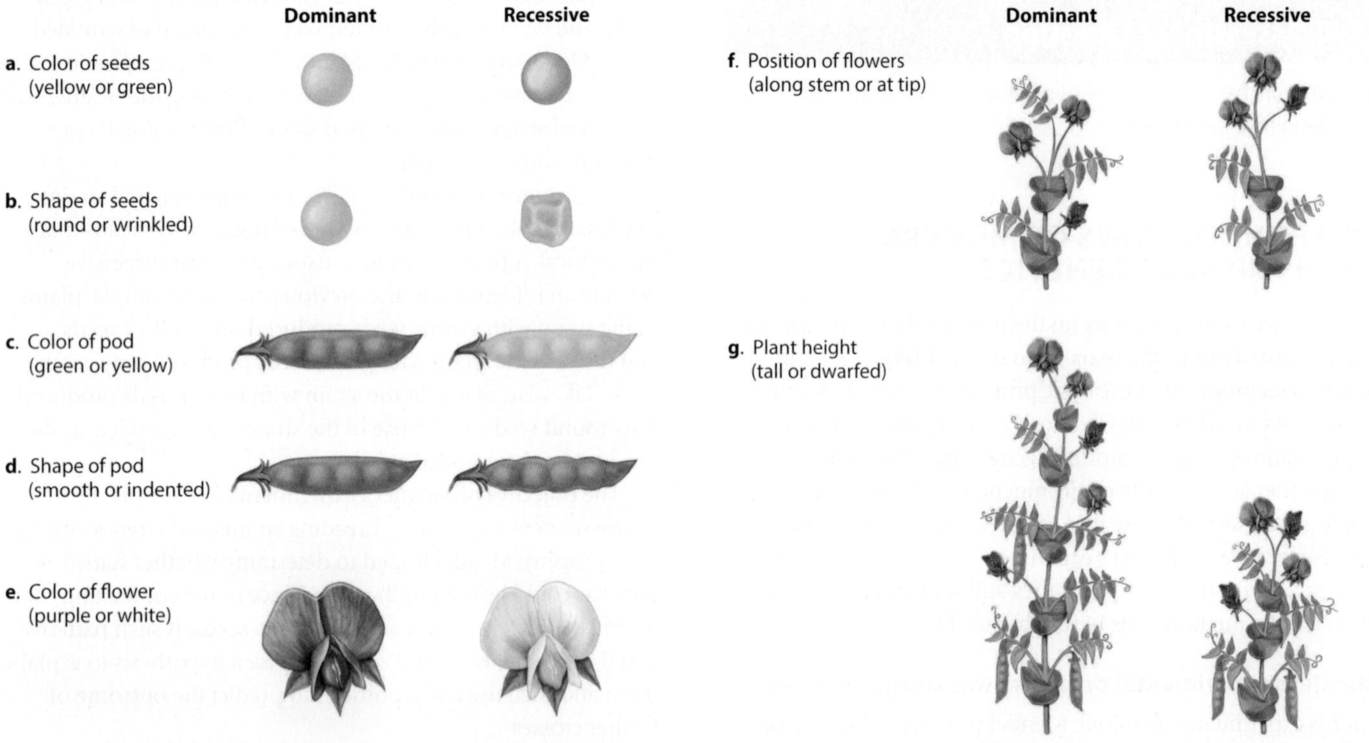

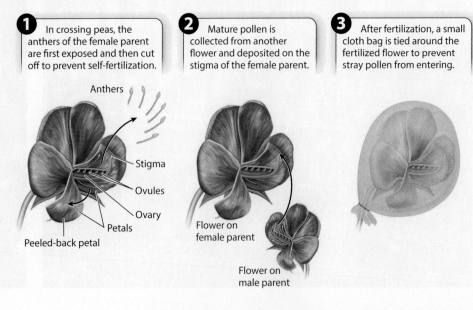

Dominant
Recessive
a. Color of seeds (yellow or green)
b. Shape of seeds (round or wrinkled)
c. Color of pod (green or yellow)
d. Shape of pod (smooth or indented)
e. Color of flower (purple or white)
Dominant
Recessive
f. Position of flowers (along stem or at tip)
g. Plant height (tall or dwarfed)

2. Mendel focused on one trait, or a small number of traits, at a time, with characteristics that were easily contrasted among the true-breeding strains. Other plant hybridizers crossed strains differing in many traits and tried to follow all the traits at once. This resulted in amazingly complex inheritance, and no underlying patterns could be discerned.

3. Mendel counted the progeny of his crosses, looking for statistical patterns in the offspring of his crosses. Others typically noted only whether offspring with a particular characteristic were present or absent, but did not keep track of and count all the progeny of a particular cross.

In crosses, one of the traits was dominant in the offspring.

Mendel began his studies with crosses between true-breeding strains that differed in a single contrasting characteristic. Crossing peas is not as easy as it may sound. Pea flowers include both female and male reproductive organs enclosed together within petals (**Fig. 15.4**). Because of this arrangement, pea plants usually fertilize themselves (self-fertilize). Performing crosses between different plants is a tedious and painstaking process. First, the flower of the designated female parent must be opened at an early stage and the immature anthers (male reproductive

FIG. 15.4 Crossing pea plants. Crossing plants in a way that isolates either the ovules or the pollen controls which genetic material each parent contributes to the offspring.

1 In crossing peas, the anthers of the female parent are first exposed and then cut off to prevent self-fertilization.
2 Mature pollen is collected from another flower and deposited on the stigma of the female parent.
3 After fertilization, a small cloth bag is tied around the fertilized flower to prevent stray pollen from entering.
Anthers
Stigma
Ovules
Ovary
Petals
Peeled-back petal
Flower on female parent
Flower on male parent

organs) clipped off and discarded, so the plant cannot self-fertilize. Then mature pollen from the designated male parent must be collected and deposited on the stigma (female reproductive organ) of the female parent, from where it travels to the reproductive cells in the ovule at the base of the flower. Finally, a cloth bag must be tied around the female flower to prevent stray pollen from entering. After a few days, a pea pod forms from the fertilized flower, with each ovule having developed into one pea in the pod. No wonder Mendel complained that his eyes hurt!

For each of the seven pairs of contrasting traits, true-breeding strains differing in the trait were crossed. **Fig. 15.5** illustrates a typical result, in this case for a cross between a plant producing yellow seeds and a plant producing green seeds. In these kinds of crosses, the parental generation is referred to as the **P_1 generation**, and the first offspring, or filial, generation is referred to as the **F_1 generation**. In the cross of P_1 yellow × P_1 green, Mendel observed that all the F_1 progeny had yellow seeds. This result was shown to be independent of the seed color of the pollen donor (the male parent) because both of the crosses shown below yielded progeny plants with yellow seeds:

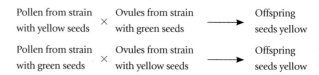

Crosses like these, in which the expressions of the trait in the female and male parents are interchanged, are known as **reciprocal crosses.** Mendel showed that reciprocal crosses yielded the same result for each of his seven pairs of contrasting traits.

Moreover, for each pair of contrasting traits, only one of the traits appeared in the F_1 generation. For simple crosses involving two parents that are true breeding for different traits, the trait that appears in the F_1 generation is said to be **dominant,** and the contrasting trait that does not appear is said to be **recessive.** For each pair of traits illustrated in Fig. 15.3, the dominant trait is shown on the left, and the recessive trait on the right. Thus, yellow seed is dominant to green seed, round seed is dominant to wrinkled seed, and so forth.

Mendel explained these findings by supposing that there is a hereditary factor for yellow seeds and a different hereditary factor for green seeds; likewise, there is a hereditary factor for round seeds and a different one for wrinkled seeds; and so on. We now know that the hereditary factors that result in contrasting traits are different forms of a gene that affect the trait, called **alleles.** The alleles of a gene are variations of the DNA sequence of a gene that occupies a particular region along a chromosome. The particular combination of alleles present in an individual constitutes its **genotype,** and the expression of the trait in the individual constitutes its **phenotype** (Chapter 14). As noted, Mendel found that, for each of his traits, one phenotype was dominant and the other phenotype was recessive.

In the cross between true-breeding plants that produce yellow seeds and true-breeding plants that produce green seeds, the genotype of each F_1 seed that grows from a female parent flower includes an allele for yellow from the yellow parent and an allele for green from the green parent. The phenotype of each F_1 seed is nevertheless yellow, because yellow is dominant to green.

The molecular basis of dominance in this case is the fact that, in diploid organisms, only one copy of a gene is needed to carry out its normal function. The yellow color in pea seeds results from an enzyme that breaks down green chlorophyll, allowing yellow pigments to show through. Green seeds result from a mutation in this gene that inactivates the enzyme, so that the green chlorophyll is retained. In an F_1 hybrid, such as that in Fig. 15.5, which receives a nonmutant gene from one parent and a mutant gene from the other, the seeds are yellow because one copy of the nonmutant gene produces enough of the enzyme to break down the chlorophyll to yield a yellow seed.

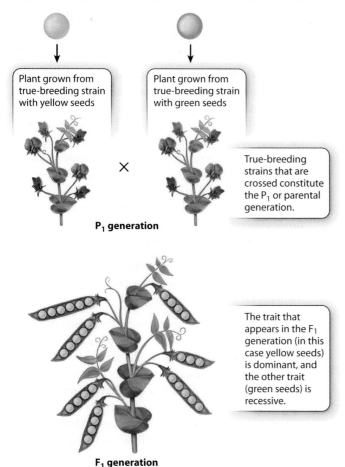

FIG. 15.5 The first-generation hybrid (F_1). A cross between two of Mendel's true-breeding plants (the parental, or P_1, generation) yielded first-generation hybrids displaying the dominant trait.

Self-Assessment Questions

2. In his famous paper, Mendel writes that he set out to "determine the number of different forms in which hybrid progeny appear" and to "ascertain their numerical interrelationships." How did his approach differ from earlier investigations in heredity?
3. What are the differences among gene, allele, genotype, and phenotype?

15.3 SEGREGATION

Mendel's most important discovery was that the F_1 progeny of a cross between plants with different traits did not breed true. In the **F_2 generation**, produced by allowing the F_1 flowers to undergo self-fertilization, the recessive trait reappeared (**Fig. 15.6**). Not only did the recessive trait reappear, it reappeared in a definite numerical proportion. Among a large number of F_2 progeny, Mendel found that the

TABLE 15.1 Observed F_2 Ratios in Mendel's Experiments

TRAIT	DOMINANT TRAIT	RECESSIVE TRAIT	RATIO
Seed color	6,022	2,001	3.01:1
Seed shape	5,474	1,850	2.96:1
Pod color	428	152	2.82:1
Pod shape	882	299	2.95:1
Flower color	705	224	3.15:1
Flower position	651	207	3.14:1
Plant height	787	277	2.84:1

dominant:recessive ratio was very close to 3:1. The results he observed among the F_2 progeny for each of the seven pairs of traits are given in **Table 15.1**. Across experiments for all seven traits, the ratio of dominant:recessive F_2 offspring was 14,949:5010. Although there is variation from one experiment to the next, the overall ratio of 14,949:5010 equals 2.98:1, which is a very close approximation to 3:1.

Genes come in pairs that segregate in the formation of reproductive cells.

The explanation for the 3:1 ratio in the F_2 and Mendel's observations in general can be summarized with reference to **Fig. 15.7**:

1. Except for cells involved in reproduction, each cell of a pea plant contains two alleles of each gene. In each true-breeding strain constituting the P_1 generation, the two alleles are identical. In Fig. 15.7, we designate the allele associated with yellow seeds as A and that associated with green seeds as a. The genotype of the true-breeding strain with yellow seeds can therefore be written as AA, and that of the true-breeding strain with green seeds as aa. The AA and aa genotypes are said to be **homozygous**, which means that both alleles inherited from the parents are the same.

2. Each reproductive cell, or **gamete**, contains only one allele of each gene. In this case, a gamete can contain the A allele or the a allele, but not both.

3. In the formation of gametes, the two members of a gene pair **segregate** (or separate) equally into gametes, so that half the gametes get one allele and half get the other allele. This separation of alleles into different gametes defines the **principle of segregation**. In the case of homozygous plants (such as AA or aa), all the gametes from an individual are the same. That is, the homozygous AA strain with yellow seeds produces gametes containing the A allele, and the homozygous

FIG. 15.6 Self-fertilization of the F_1 hybrid, resulting in seeds of the F_2 generation. The recessive trait appeared again in the F_2 generation.

Seeds from F_1 plants produced from a cross of true-breeding yellow-seed and green-seed plants are yellow because yellow is dominant and green is recessive in seed color.

F_1 generation

Peas are normally self-fertilizing and so, if they are left alone, the pollen produced in each flower fertilizes the ovules, producing a seed pod whose contents have genotypes that depend on the alleles in that individual flower's pollen and ovules.

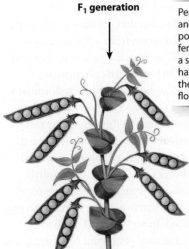

Because fertilization takes place at random, any individual pod can have a ratio of dominant:recessive that deviates from 3:1.

F_2 generation
(3 yellow seeds:1 green seed)

FIG. 15.7 The principle of segregation.

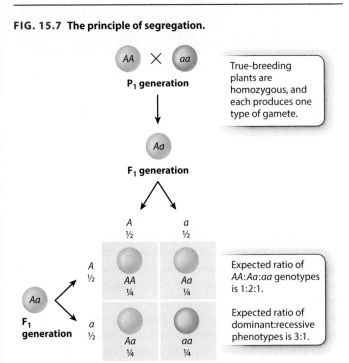

aa strain with green seeds produces gametes containing the *a* allele.

4. The fertilized egg cell, called the **zygote,** is formed from the random union of two gametes, one from each parent. For the cross $AA \times aa$, the zygote is an F_1 hybrid formed from the union of an *A*-bearing gamete with an *a*-bearing gamete. Each F_1 hybrid therefore has the genotype *Aa* (Fig. 15.7). The *Aa* genotype is **heterozygous,** which means that the two alleles for a given gene inherited from each parent are different. The seeds produced by the F_1 progeny are yellow because yellow is dominant to green. Note that each F_1 progeny contains an *a* allele because its genotype is *Aa*, but the phenotype of the seed is yellow and indistinguishable from that of an *AA* genotype.

5. When the F_1 progeny (genotype *Aa*) form gametes, by the principle of segregation the *A* and *a* alleles again separate equally so that half the gametes contain only the *A* allele and the other half contain only the *a* allele (Fig. 15.7).

6. In the formation of the F_2 generation, the gametes from the F_1 parents again combine at random. The consequences of the random union of gametes can be worked out by means of a checkerboard of the sort shown at the bottom of Fig. 15.7. In this kind of square, which is known as a **Punnett square** after its inventor, the British geneticist Reginald Punnett, the gametes from each parent, each with its respective frequency, are arranged along the top and sides of a grid. Each box in the grid represents the union of the gametes in the corresponding row and column, showing all the possible genotypes of offspring that can result from random fertilization.

7. The boxes of the Punnett square correspond to all of the possible offspring genotypes of the F_2 generation, with each possible genotype's frequency obtained by multiplying the gametic frequencies in the corresponding row and column. For the example illustrated in Fig. 15.7, the expected genotypes of the offspring are ¼ *AA*, ½ *Aa*, and ¼ *aa* (or 1 : 2 : 1). When there is dominance, however, as there is in this case, the *AA* and *Aa* genotypes have the same phenotype, so the ratio of dominant : recessive phenotypes is 3 : 1. The Punnett square in Fig. 15.7 illustrates the biological basis of the 3 : 1 ratio of phenotypes that Mendel observed in the F_2 generation. Nevertheless, the underlying ratio of *AA* : *Aa* : *aa* genotypes is 1 : 2 : 1.

The principle of segregation was tested by predicting the outcome of crosses.

The model of segregation of gene pairs (alleles) and their random combination in the formation of a zygote depicted in Fig. 15.7 was Mendel's hypothesis, an explanation he put forward to explain an observed result. Further experiments were necessary to support or disprove this hypothesis. The true test of a hypothesis is whether it can predict the results of experiments that have not yet been carried out (Chapter 1). If the predictions are correct, one's confidence in the hypothesis is strengthened. Mendel appreciated this relationship intuitively, even though the scientific method as understood today had not been formalized when he was doing his work.

The Punnett square in Fig. 15.7 makes two predictions. First, it predicts that the seeds in the F_2 generation showing the recessive green seed phenotype should be homozygous *aa*, in which case they should breed true. That is, when the seeds are grown into mature plants and self-fertilization is allowed to take place, the self-fertilized *aa* plants should produce only green seeds (*aa*). This prediction was confirmed by examining seeds actually produced by plants grown from the F_2 green seeds.

The second prediction from the Punnett square in Fig. 15.7 is more complex: it has to do with the seeds in the F_2 generation that show the dominant yellow phenotype. Although these seeds have the same phenotype, they have two different genotypes (*AA* and *Aa*). Among just the yellow seeds, ⅓ should have the genotype *AA* and ⅔ should have the genotype *Aa*, for a ratio of 1 *AA* : 2 *Aa*. (The proportions are ⅓ : ⅔ because we are considering *only* the seeds that are yellow.) Plants with the *AA* and *Aa* genotypes can be distinguished by the types of seed they produce when self-fertilized. The *AA* plants produce only

TABLE 15.2	Genetic Ratios from Self-Fertilization of Plants Showing the Dominant Phenotype		
TRAIT	HOMOZYGOUS DOMINANT	HETERO-ZYGOUS	RATIO
Yellow seeds	166	353	0.94:2
Round seeds	193	372	1.04:2
Green pods	40	60	1.33:2
Smooth pods	29	71	0.82:2
Purple flowers	36	64	1.13:2
Flowers along stem	33	67	0.99:2
Tall plants	28	72	0.78:2

TABLE 15.3	Phenotype of Progeny from Testcrosses of Heterozygotes		
TRAIT	DOMINANT TRAIT	RECESSIVE TRAIT	RATIO
Seed color	196	189	1.04:1
Seed shape	193	192	1.01:1
Flower color	85	81	1.05:1
Plant height	87	79	1.10:1

seeds with the dominant yellow phenotype (that is, they are true breeding), whereas the *Aa* plants yield dominant yellow and recessive green seeds in the ratio 3 : 1. Mendel performed such experiments, and the second prediction turned out to be correct. His data confirming the 1 : 2 ratio of *AA* : *Aa* among F_2 individuals with the dominant phenotype are shown in **Table 15.2**.

A testcross is a mating to an individual with the homozygous recessive genotype.

A more direct test of segregation is to cross the F_1 progeny with the true-breeding recessive strain instead of allowing them to self-fertilize. Any cross of an unknown genotype with a homozygous recessive genotype is known as a **testcross**.

The F_1 progeny show the dominant yellow seed phenotype. A yellow seed phenotype can result from either of two possible genotypes, *Aa* or *AA*. A testcross with plants of the homozygous recessive genotype (*aa*) can distinguish between these two possibilities (**Fig. 15.8**).

Let's first consider what happens in a testcross to an *Aa* individual (Fig. 15.8a). An *Aa* individual produces both *A*-bearing gametes and *a*-bearing gametes. As shown by the Punnett square, the predicted offspring from the testcross between an *Aa* parent and an *aa* parent are ½ *Aa* zygotes, which yield yellow seeds, and ½ *aa* zygotes, which yield green seeds. By contrast, an *AA* individual produces only *A*-bearing gametes, so all the zygotes resulting from a testcross with an *aa* individual have the *Aa* genotype, which yields yellow seeds (Fig. 15.8b). In other words, the testcross gives different results depending on whether the parent in question is heterozygous (*Aa*) or homozygous (*AA*).

Note that in a testcross the *phenotypes* of the progeny reveal the *alleles* present in the gametes from the tested parent. A testcross with an *Aa* individual yields ½ *Aa* (yellow seeds) and ½ *aa* (green seeds) because the *Aa* parent produces ½ *A*-bearing gametes and ½ *a*-bearing gametes. A testcross with an *AA* individual yields only *Aa* (yellow seeds) because the *AA* parent produces only *A*-bearing gametes. These results are a direct demonstration of the principle of segregation: the ratio of the phenotypes of progeny reflects the equal segregation of alleles into gametes. Some of Mendel's testcross data indicating 1 : 1 segregation in heterozygous genotypes are shown in **Table 15.3**.

Segregation of alleles reflects the separation of chromosomes in meiosis.

The principles of transmission genetics have a physical basis in the process of meiosis (Chapter 11). During meiosis I, maternal and paternal chromosomes (homologous chromosomes) align on the metaphase plate. Then, during anaphase I, the homologous chromosomes separate, and each chromosome goes to a different pole. Because gene pairs are carried on homologous chromosomes, the segregation of alleles observed by Mendel corresponds to the separation of chromosomes that takes place in anaphase I.

FIG. 15.8 A testcross. A cross with a homozygous recessive individual reveals the genotype of the other parent.

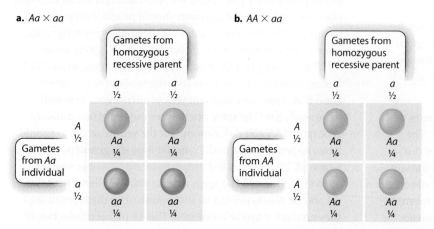

FIG. 15.9 Segregation of alleles of a single gene. Homologous chromosomes separate during meiosis, leading to segregation of alleles.

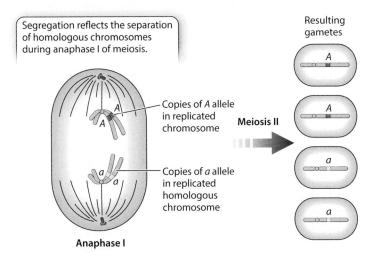

Fig. 15.9 illustrates the separation of a pair of homologous chromosomes in anaphase I. In the configuration shown, the copies of the *A* allele (dark blue) separate from the copies of the *a* allele (light blue) in anaphase I. The separation of chromosomes is the physical basis of the segregation of alleles.

Dominance is not universally observed.

Many traits do not show complete dominance such as Mendel observed with pea plants. Instead, most traits we see are determined by multiple genes or the interaction of genotype and the environment, so they do not display the expected 3 : 1 ratio of phenotypes. But even among traits that are determined by a single gene, the 3 : 1 phenotype ratio is not always observed. In some cases, the trait shows **incomplete dominance**, in which the phenotype of the heterozygous genotype is intermediate between those of the homozygous genotypes. In such cases, the result of segregation can be observed directly because each genotype has a distinct phenotype. An example is flower color in the snapdragon (*Antirrhinum majus*), in which the homozygous genotypes have red ($C^R C^R$) or white ($C^W C^W$) flowers, and the heterozygous genotype $C^R C^W$ has pink flowers (**Fig. 15.10**). In notating incomplete dominance, we use superscripts to indicate the alleles, rather than uppercase and lowercase letters, because neither allele is dominant to the other. A cross of homozygous $C^R C^R$ and $C^W C^W$ strains results in hybrid F_1 progeny that are pink ($C^R C^W$); when these are crossed, the resulting F_2 generation consists of ¼ red ($C^R C^R$), ½ pink ($C^R C^W$), and ¼ white ($C^W C^W$).

In this case, the genotype ratio and the phenotype ratio are both 1 : 2 : 1, because each genotype has a distinct phenotype. Such a direct demonstration of segregation makes one wonder whether Mendel's work might have been appreciated more readily if his traits had shown incomplete dominance!

FIG. 15.10 Incomplete dominance for flower color in snapdragons. In incomplete dominance, an intermediate phenotype is seen.

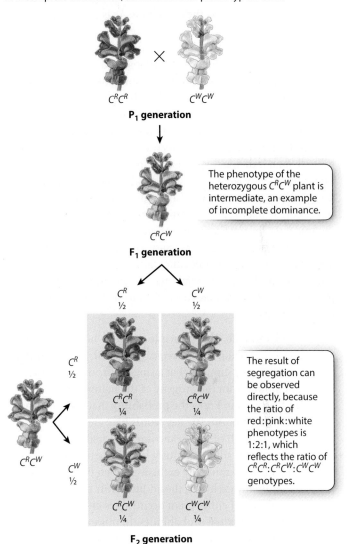

The principles of transmission genetics are statistical and are stated in terms of probabilities.

The element of chance in fertilization implies that the genotype of any particular progeny cannot be determined in advance. However, one can deduce the likelihood, or **probability**, that a specified genotype will occur. The probability of occurrence of a genotype must always lie between 0 and 1; a probability of 0 means that the genotype cannot occur, and a probability of 1 means that the occurrence of the genotype is certain. For example, in the cross $Aa \times AA$, no offspring can have the genotype aa, so in this mating the probability of aa is 0. Similarly, in the mating $AA \times aa$, all offspring must have the genotype Aa, so in this mating the probability of Aa is 1.

In many cases, the probability of a particular genotype is neither 0 nor 1, but some intermediate value. For one gene, the probabilities for a single individual can be deduced from the parental genotypes in the mating and the principle of segregation. For example, the probability of producing a homozygous recessive individual from the cross $Aa \times Aa$ is ¼ (see Fig. 15.7), and that from the cross $Aa \times aa$ is ½ (see Fig. 15.8a).

The genotype and phenotype probabilities for a single individual can also be inferred from observed data because the overall proportions of two (or more) genotypes among a large number of observations approximate the probabilities of each of the genotypes for a single observation. For example, in Mendel's F_2 data (see Table 15.1), the overall ratio of dominant:recessive is 2.98:1, or very nearly 3:1. This result implies that the probability that an individual F_2 plant has the homozygous recessive phenotype is very close to ¼, which is the value inferred from the principle of segregation. The reason large samples are important is that the average ratio is more likely to be close to the true ratio in larger samples compared to smaller ones.

Sometimes it becomes necessary to combine the probabilities of two or more possible outcomes of a cross, as in determining the probability of a genotype occurring based on the probabilities of the gametes occurring. In such cases, either of two rules is helpful:

1. **Addition rule.** This principle applies when the possible outcomes being considered cannot occur simultaneously. For example, suppose that a single offspring is chosen at random from the progeny of the mating $Aa \times Aa$, and we wish to know the probability that the offspring is either AA or Aa. The key words here are "either" and "or." Each of these outcomes is possible, but both cannot occur simultaneously in a single individual; the outcomes are mutually exclusive. When the possibilities are mutually exclusive, the addition rule states that the probability of either event occurring is given by the sum of their individual probabilities. In this example, the chosen offspring could have either genotype AA (with probability ¼, according to Fig. 15.7) or genotype Aa (with probability ½). Therefore, the probability that the chosen individual has either the AA or the Aa genotype is given by ¼ + ½ = ¾ (see Fig. 15.7). Alternatively, the ¾ probability could be interpreted to mean that, among a large number of offspring from the mating $Aa \times Aa$, the proportion exhibiting the dominant phenotype will be very close to ¾. This interpretation is verified by the data in Table 15.1.

2. **Multiplication rule.** This principle applies when outcomes can occur simultaneously, and the occurrence of one has no effect on the likelihood of the other. Events that do not influence one another are independent, and the multiplication rule states that the probability of two independent events occurring together is the product of their respective probabilities. This rule is widely used to determine the probabilities of successive offspring of a cross because each event of fertilization is independent of any other. For example, in the mating $Aa \times Aa$, one may wish to determine the probability that, among four peas in a pod, the one nearest the stem is green and the others yellow. Here, the word "and" is a simple indicator that the multiplication rule should be used. Because each seed results from an independent fusion of pollen and ovule, this probability is given by the product of the probability that the seed nearest the stem is aa and the probability that each of the other seeds is either AA or Aa; hence the probability is ¼ × ¾ × ¾ × ¾ = 27/256.

The addition and multiplication rules are very powerful when used in combination, as seen in **Fig. 15.11**. Consider the following question: in the mating $Aa \times Aa$, what is the probability that, among four seeds in a pod, exactly one is green? We have already seen that the multiplication rule gives the probability of the seed nearest the stem being green as 27/256, as shown for the top pod in Fig. 15.11. As the figure shows, there are only four possible ways in which exactly one seed can be green, each of which has a probability of 27/256, and these outcomes are mutually exclusive. Therefore, by the addition rule, the probability of there being exactly one green seed and three yellow seeds in a pod, occurring in any order, is given by 27/256 + 27/256 + 27/256 + 27/256 = 108/256, or approximately 42%.

Mendelian segregation preserves genetic variation.

As noted earlier, Darwin was befuddled because blending inheritance would make genetic variation disappear so rapidly that evolution by means of natural selection could not occur.

FIG. 15.11 Application of the multiplication and addition rules.

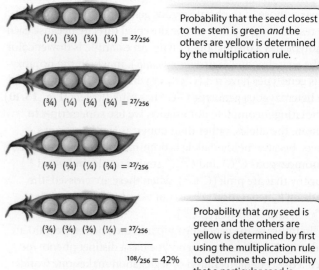

Probability that the seed closest to the stem is green *and* the others are yellow is determined by the multiplication rule.

Probability that *any* seed is green and the others are yellow is determined by first using the multiplication rule to determine the probability that a particular seed is green, then the addition rule to determine that any seed is green.

Although Darwin was completely unaware of Mendel's findings, segregation was the answer to his problem.

An important consequence of segregation is that it demonstrates that the alleles encoding a trait do not alter one another when they are present together in a heterozygous genotype (except in very rare instances). The recessive trait, which is masked in one generation, can appear in the next generation, looking exactly as it did in the true-breeding strains. Mendel fully appreciated the significance of this discovery. In one of his letters, he emphasized that "the two parental traits appear, separated and unchanged, and there is nothing to indicate that one of them has either inherited or taken over anything from the other." In other words, no hint of any sort of blending between the parental genetic material takes place, so there is no homogenization of the trait in the population. Because the individual genes maintain their identity down through the generations (except for rare mutations), genetic variation in a population also tends to be maintained through time. The maintenance of genetic variation is discussed further in Chapter 20.

Self-Assessment Questions

4. What are the genotypes and phenotypes for Mendel's true-breeding parent plants?
5. Is it possible for two individuals to have the same phenotype but different genotypes? The same genotype, but different phenotypes? How?
6. What is the purpose of a testcross?
7. What is the probability that *any* two seeds are green and two seeds are yellow in a pea pod with exactly four seeds?

15.4 INDEPENDENT ASSORTMENT

We have seen that segregation of different alleles of a single gene results in a 3:1 ratio of dominant:recessive phenotypes in the F_2 generation. What happens when the parental strains differ in two traits, such as when a strain having yellow and wrinkled seeds is crossed with a strain having green and round seeds? The results of these kinds of experiments constitute Mendel's second key discovery, the **principle of independent assortment**. This principle states that segregation of one set of alleles of a gene pair is independent of the segregation of another set of alleles of a different gene pair. That is, each pair of alleles assorts (segregates) into gametes without affecting or being affected by the assortment of any other pair of alleles.

Independent assortment is observed when genes segregate independently of one another.

In the cross between a strain of pea plants with seeds that are yellow and wrinkled and a strain with seeds that are green and round, the phenotype of the F_1 seeds is easily predicted. Because yellow is dominant to green, and round is dominant to wrinkled, the F_1 seeds are expected to be yellow and round, and in fact they are (**Fig. 15.12**). When these seeds are grown and the F_1 plants are allowed to undergo self-fertilization, the result

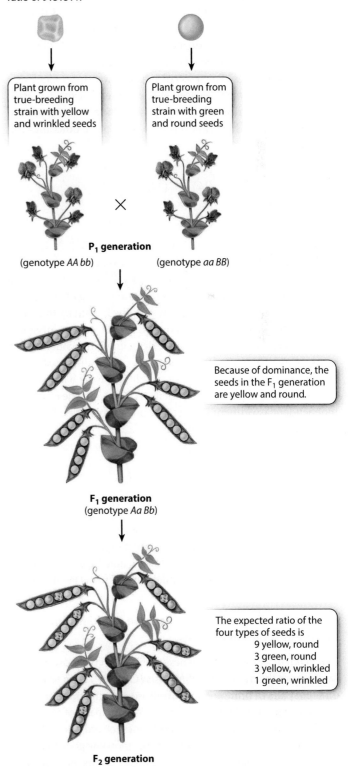

FIG. 15.12 Mendel's crosses with two traits. Plants heterozygous for two genes affecting different traits produce offspring with a phenotypic ratio of 9:3:3:1.

Plant grown from true-breeding strain with yellow and wrinkled seeds

Plant grown from true-breeding strain with green and round seeds

P_1 generation
(genotype *AA bb*) (genotype *aa BB*)

Because of dominance, the seeds in the F_1 generation are yellow and round.

F_1 generation
(genotype *Aa Bb*)

The expected ratio of the four types of seeds is
9 yellow, round
3 green, round
3 yellow, wrinkled
1 green, wrinkled

F_2 generation

is as shown in Fig. 15.12. Among 639 seeds from this cross, Mendel observed the following:

yellow round	367
green round	122
yellow wrinkled	113
green wrinkled	37

The ratio of these phenotypes is $9.9:3.3:3.1:1.0$, which Mendel realized is close to $9:3:3:1$. The latter ratio is that expected if the A and a alleles for seed color undergo segregation and form gametes independently of the B and b alleles for seed shape. How did Mendel come to expect a $9:3:3:1$ ratio of phenotypes? For seed color alone, we expect a ratio of ¾ yellow : ¼ green; for seed shape alone, we expect a ratio of ¾ round : ¼ wrinkled. If the traits are independent, then we can use the multiplication rule to predict the outcomes for both traits:

yellow round	(¾) × (¾)	=	9/16
green round	(¼) × (¾)	=	3/16
yellow wrinkled	(¾) × (¼)	=	3/16
green wrinkled	(¼) × (¼)	=	1/16

Note that $9/16:3/16:3/16:1/16$ is equivalent to $9:3:3:1$.

The underlying reason for the $9:3:3:1$ ratio of phenotypes in the F_2 generation is that the alleles for yellow versus green and those for round versus wrinkled are assorted into gametes independently of each other. In other words, the hereditary transmission of either gene has no effect on the hereditary transmission of the other gene. A Punnett square depicting independent assortment is shown in **Fig. 15.13**. The A and a alleles segregate equally into gametes as ½ A : ½ a, and likewise the B and b alleles segregate equally into gametes as ½ B : ½ b. The result of independent assortment is that the four possible gametic types are produced in equal proportions:

AB gametes	(½) × (½)	=	¼
Ab gametes	(½) × (½)	=	¼
aB gametes	(½) × (½)	=	¼
ab gametes	(½) × (½)	=	¼

As the Punnett square in Fig. 15.13 shows, random union of these gametic types produces the expected ratio of 9 yellow round, 3 green round, 3 yellow wrinkled, and 1 green wrinkled. **Fig. 15.14** summarizes how Mendel's experiments led him to formulate his two laws.

Independent assortment reflects the random alignment of chromosomes in meiosis.

Independent assortment of genes on different chromosomes results from the mechanics of meiosis (Chapter 11), in which different pairs of homologous chromosomes align randomly on the metaphase plate in meiosis I. For some pairs of chromosomes, the maternal chromosome goes toward one pole during anaphase I, and the paternal chromosome goes to the other pole; for other pairs, just the opposite occurs. Because the alignment is random, gene pairs on different chromosomes assort independently of one another.

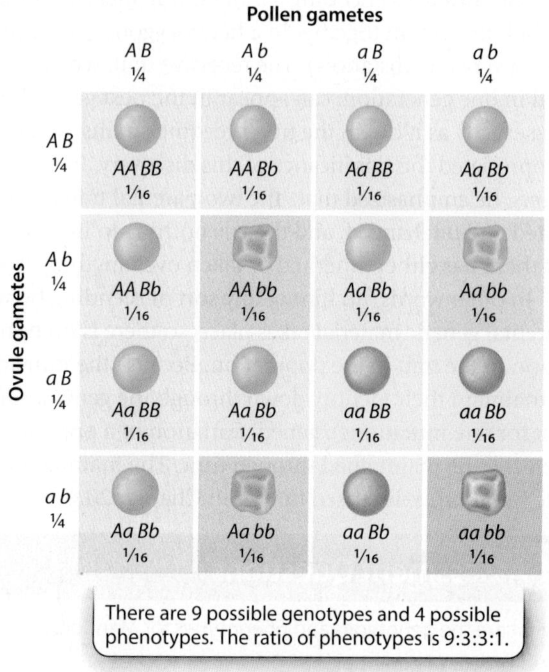

Fig. 15.15 illustrates two possible alignments that are equally likely. In one alignment, the B allele (dark red) goes to the same pole as the A allele (dark blue); in the other alignment, the b allele (light red) goes in the same direction as the A allele. The first type of alignment results in a $1:1$ ratio of $AB:ab$ gametes, and the second type of alignment results in a $1:1$ ratio of $Ab:aB$ gametes. Because the two orientations are equally likely, the overall ratio of $AB:ab:Ab:aB$ from a large number of cells undergoing meiosis is expected to be $1:1:1:1$. This is the principle of independent assortment for genes located in different chromosomes.

Not all genes undergo independent assortment. For example, genes near one another in the same chromosome do not assort independently of one another. Genes in the same chromosome that fail to show independent assortment are said to be linked and are discussed in Chapter 16.

Phenotypic ratios can be modified by interactions between genes.

The $9:3:3:1$ ratio of phenotypes results from independent assortment of two genes when one allele of each gene is dominant and when the two genes affect different traits. However, even with complete dominance, the ratio of phenotypes may be different if the two genes affect the same trait. This often happens when the genes code for proteins that act in the same biochemical pathway. In such cases, the gene products can interact to affect the phenotypic expression of the genotypes,

FIG. 15.13 Independent assortment of two genes. The Punnett square reveals the underlying phenotypic ratio of $9:3:3:1$.

	Pollen gametes			
Ovule gametes	AB ¼	Ab ¼	aB ¼	ab ¼
AB ¼	$AABB$ 1/16	$AABb$ 1/16	$AaBB$ 1/16	$AaBb$ 1/16
Ab ¼	$AABb$ 1/16	$AAbb$ 1/16	$AaBb$ 1/16	$Aabb$ 1/16
aB ¼	$AaBB$ 1/16	$AaBb$ 1/16	$aaBB$ 1/16	$aaBb$ 1/16
ab ¼	$AaBb$ 1/16	$Aabb$ 1/16	$aaBb$ 1/16	$aabb$ 1/16

There are 9 possible genotypes and 4 possible phenotypes. The ratio of phenotypes is 9:3:3:1.

HOW DO WE KNOW?

FIG. 15.14

How are single-gene traits inherited?

BACKGROUND Gregor Mendel's experiments, carried out in the years 1856–1863, are among the most important in all of biology. Mendel set out to improve upon previous research in heredity. He wrote that "among all the numerous experiments made, not one has been carried out to such an extent and in such a way as to make it possible to determine the number of different forms under which the offspring of the hybrids appear, or to arrange these forms with certainty according to their separate generations, or definitely to ascertain their numerical relations."

METHOD By studying simple traits across successive generations of crosses, Mendel observed how various traits are transmitted to find out whether there were regular and reproducible quantitative relations among the various phenotypes.

HYPOTHESIS By keeping track of the ancestral history of the parents in a mating, Mendel hypothesized that the phenotypes of the offspring could be predicted and their relative frequencies would obey certain "numerical relations."

RESULTS Crosses between plants that were hybrids of a single trait displayed two phenotypes in a ratio that averaged 3 : 1, with variation from experiment to experiment due purely to chance. Similarly, crosses between plants that were hybrids of two traits displayed four different phenotypes in a ratio averaging 9 : 3 : 3 : 1, again with variation among experiments due to chance.

CONCLUSION From observing these ratios among several different traits, Mendel came up with two conclusions about the inheritance of the genetic factors underlying traits, which are now called Mendel's laws:

1. The principle of segregation states that individuals inherit two copies (alleles) of each gene, one from the mother and one from the father. When the individual then forms reproductive cells, the two copies separate (segregate) equally in the eggs or sperm.

2. The principle of independent assortment states that the two copies of each gene segregate into gametes independently of the two copies of another gene.

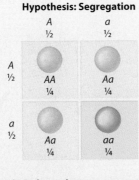

ANALYSIS Mendel found support for his two conclusions in the statistical analysis of the results of his meticulous crosses.

1. The principle of segregation: A prediction of the hypothesis of segregation is that, among seeds with the dominant phenotype, the ratio of homozygous to heterozygous genotypes should be 1 : 2. Mendel tested this prediction in several ways, one of which was simply to allow plants grown from F_2 seeds to self-fertilize. Any that were true breeding for the dominant phenotype he classified as homozygous, and any that segregated to yield both dominant and recessive phenotypes he classified as heterozygous. In the experimental test, the observed numbers fit the expected values within the margins that would be expected by chance.

Experimental test

	AA	Aa
Observed	166	353
Expected	173	346

2. The principle of independent assortment: Although the Punnett square for two pairs of alleles has 16 squares, there are only 9 genotypes, and the hypothesis of independent assortment predicts that these genotypes should appear in the ratios of 1 AA BB : 2 AA Bb : 1 AA bb : 2 Aa BB and 4 Aa Bb : 2 Aa bb : 1 aa BB : 2 aa Bb : 1 aa bb. As before, Mendel tested this hypothesis by self-fertilization of plants grown from the F_2 seeds, classifying any that bred true for either trait as homozygous and any that segregated to yield both dominant and recessive phenotypes as heterozygous. In the experimental test, once again the observed numbers fit the expected values within the margins that would be expected by chance.

Experimental test

	AA BB	AA Bb	AA bb
Observed	38	60	28
Expected	33	66	33

	Aa BB	Aa Bb	Aa bb
Observed	65	138	68
Expected	66	132	66

	aa BB	aa Bb	aa bb
Observed	35	67	30
Expected	33	66	33

FOLLOW-UP WORK Mendel's work was ignored during his lifetime, and its importance was not recognized until 1900, 16 years after his death. The rediscovery marks the beginning of the modern science of genetics.

SOURCE Mendel's paper in English is available at http://www.mendelweb.org/Mendel.html.

FIG. 15.15 Independent assortment of genes in different chromosomes. Chromosome pairs are sorted into daughter cells randomly during meiosis I, resulting in independent assortment of genes.

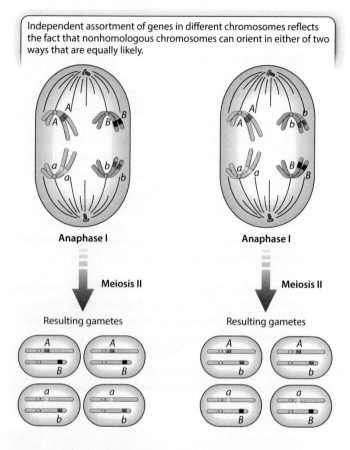

FIG. 15.16 Epistasis, the interaction of genes affecting the same trait. Epistasis can modify the 9 : 3 : 3 : 1 ratio of phenotypes, in this example to 13 : 3.

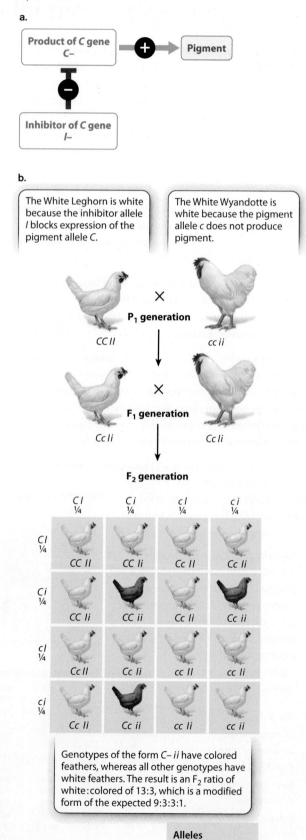

resulting in a modification of the expected ratio. Genes that modify the phenotypic expression of other genes are said to show **epistasis**.

There are many types of epistasis, leading to different modifications of the 9 : 3 : 3 : 1 ratio. Among the more common modified ratios are 12 : 3 : 1, as well as 9 : 3 : 4 and 13 : 3. **Fig. 15.16** shows one example, in which the F_2 generation of a cross between White Leghorn and White Wyandotte chickens displays the modified ratio 13 : 3. Both breeds are white, but for different genetic reasons. There are two genes involved in pigment production, each with two alleles. The C gene encodes a protein that affects coloration in feathers (Fig. 15.16a). The dominant allele C produces pigment, and the recessive allele c does not produce pigment. A different gene, I, codes for an inhibitor protein. The product of the dominant allele, I, inhibits the expression of C, whereas the recessive allele, i, does not produce the inhibitor and so does not inhibit the expression of C. Therefore, the White Leghorn (genotype $CC\,II$) is white because the product of the dominant allele I inhibits the pigment in the feathers, and the White Wyandotte (genotype $cc\,ii$) is white because the recessive allele c does not produce feather pigment to begin with. The F_1 generation has genotype $Cc\ Ii$ and is also white. With independent assortment, only the three C–ii offspring have colored feathers in the F_2 generation (the dash indicates that the second allele could be either C or c). The rest have white feathers, so the ratio of white : colored is 13 : 3 (Fig. 15.6b).

Self-Assessment Questions

8. What are Mendel's two laws?
9. How do the mechanics of meiosis and the movement of homologous chromosomes underlie Mendel's principles of segregation and independent assortment?
10. How can you predict the genotypes and phenotypes of offspring if you know the genotypes of the parents?
11. What are some reasons why a single trait might not show a 3:1 ratio of phenotypes in the F_2 generation of a cross between true-breeding strains, and why a pair of traits might not show a 9:3:3:1 ratio of phenotypes in the F_2 generation of a cross between true-breeding strains?

15.5 HUMAN GENETICS

Segregation of alleles takes place in human meiosis, just as it does in peas and most other sexual organisms. Nevertheless, the results are not so easily observed as in peas, for several reasons. First, humans do not choose their mating partners for the convenience of biologists, so experimental crosses are not possible. Second, the number of children in human families is relatively small, so the Mendelian ratios are often obscured by random fluctuations due to chance. Even so, in many cases segregation can be observed in human families.

In studying human families, the record of the ancestral relationships among individuals is summarized in a diagram of family history called a **pedigree**. Some typical symbols used in pedigrees are shown in **Fig. 15.17**. The same patterns of dominance and recessiveness that Mendel observed in his pea plants can be seen in some pedigrees. Over the next few sections, we'll see how to recognize telltale features of a pedigree that reveal the genotypic nature of a trait.

Dominant traits appear in every generation.

The pedigree shown in **Fig. 15.18** is for a rare dominant trait, brachydactyly, in which the middle long bone in the fingers fails to grow, such that the fingers remain very short. This pedigree, published in 1905, was the first demonstration of dominant Mendelian inheritance in humans. This particular form of brachydactyly results from a mutation in a gene whose normal product is a protein involved in cartilage formation, which is necessary for bone growth.

These are the features of the pedigree that immediately suggest dominant inheritance:

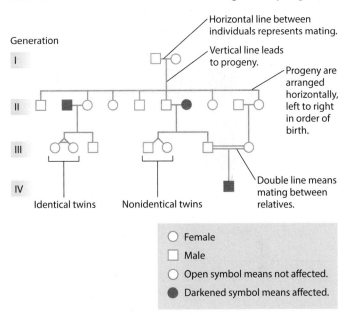

FIG. 15.17 Some conventions used in drawing human pedigrees.

1. Affected individuals are equally likely to be females or males.

2. Most matings that produce affected offspring have only one affected parent. The brachydactyly trait is rare, so a mating between two affected individuals is extremely unlikely.

3. Among matings in which one parent is affected, approximately half the offspring are affected.

If a dominant trait is rare, then affected individuals will almost always be heterozygous (Aa), not homozygous (AA). In a mating in which one parent is heterozygous for the dominant gene (Aa) and the other is homozygous recessive (aa), half the offspring are expected to be heterozygous (Aa) and the other half homozygous recessive (aa).

Recessive traits skip generations.

Recessive inheritance shows a pedigree pattern very different from that of dominant inheritance. The pedigree shown in **Fig. 15.19** pertains to albinism, in which the amount of melanin pigment in the skin, hair, and eyes is reduced. In most populations, the frequency of albinism is about 1 in 36,000, but it has a much higher frequency—about 1 in 200—among the Hopi and several other Native American tribes of the Southwest. (It is not unusual for genetic diseases to have elevated frequencies among isolated populations.) This type of albinism is due to a mutation in the gene *OCA2*, which encodes a membrane transporter protein thought to be important in transport of the amino acid tyrosine, which is used in the synthesis of the melanin pigment responsible for skin, hair, and eye color.

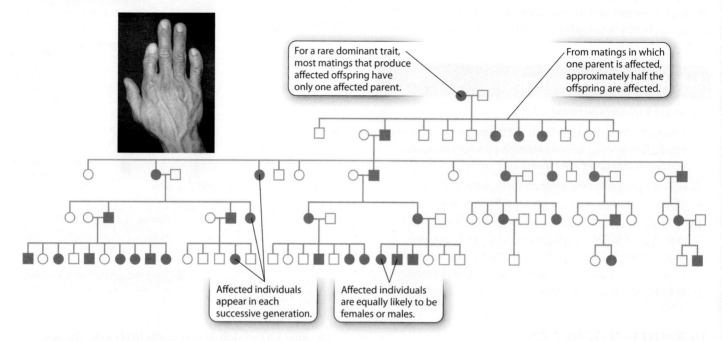

FIG. 15.18 Pedigree of a trait caused by a dominant allele. This pedigree shows the inheritance of shortened fingers associated with a form of brachydactyly (inset). *Stefan Mundlos, Institute for Medical Genetics, Charité, Berlin, Germany.*

Another type of mutation affecting expression of this same gene is associated with blue eyes. As shown in Fig. 15.19, double lines represent matings between relatives, and in both cases shown here the mating is between first cousins.

These are the principal pedigree characteristics of recessive traits:

1. The trait may skip one or more generations.
2. Affected individuals are equally likely to be females or males.
3. Affected individuals may have unaffected parents, as in the offspring of the second mating in the second generation in Fig. 15.19. For a recessive trait that is sufficiently rare, almost all affected individuals have unaffected parents.
4. Affected individuals often result from mating between relatives, typically first cousins.

Recessive inheritance has these characteristics because recessive alleles can be transmitted from generation to generation without manifesting the recessive phenotype. For an individual to be affected, that person must inherit the recessive allele from both parents. Mating between relatives often allows rare recessive alleles to become homozygous because an ancestor shared between the relatives may carry the gene (Aa). The recessive allele in the common ancestor can be transmitted to both parents, making them each a carrier of the allele as well. If both parents are unaffected carriers of the allele, they both have the genotype Aa, and ¼ of their offspring are expected to be homozygous aa and affected.

Many genes have multiple alleles.

The examples discussed so far in this chapter involve genes with only two alleles, such as A for yellow seeds and a for green seeds. Similarly, as noted in Chapter 14, most single-nucleotide polymorphisms (SNPs) have only two alleles, differing only in which particular base pair is present at a particular position in genomic DNA. In contrast, because a gene consists of a sequence of nucleotides, any nucleotide or set of nucleotides in the gene can undergo mutation. Because each of the mutant forms that exists in a population constitutes a different allele, a population of organisms may contain many different alleles of the same gene, which are called **multiple alleles.** Some genes have so many alleles that they can be used for individual identification by means of DNA typing, as discussed in Chapter 14.

In considering genetic diseases, multiple alleles are often grouped into categories such as "mutant" and "normal." But there are often many different "mutant" alleles and many different "normal" alleles in a population. For the "mutant" alleles, the DNA sequences are different from one another, but each produces a protein product whose function is impaired under the usual environmental conditions. For the "normal" alleles, the DNA sequences are also different, but they all are able to produce functional protein.

For example, more than 400 different recessive alleles of the gene that causes phenylketonuria (PKU) have been discovered

FIG. 15.19 Pedigree of a trait caused by a recessive allele. The photograph is of a group of Hopi males, three with albinism, which is caused by a recessive allele. *Source: BAE GN 02458C 06404400, National Anthropological Archives, Smithsonian Institution.*

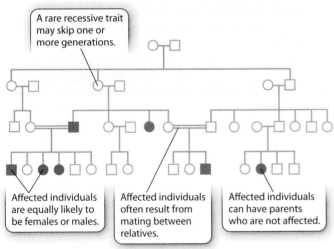

A rare recessive trait may skip one or more generations.

Affected individuals are equally likely to be females or males.

Affected individuals often result from mating between relatives.

Affected individuals can have parents who are not affected.

across the world. PKU is a moderate to severe form of mental retardation caused by mutations in the gene encoding the enzyme phenylalanine hydroxylase. Children affected with PKU are unable to break down the excess phenylalanine present in a normal diet, and the buildup impairs the development of neurons in the brain. About 1 in 10,000 newborns inherits two mutant alleles, which could be two copies of the same mutant allele or two different mutant alleles, and is affected. The "normal" form of the gene encoding phenylalanine hydroxylase also exists in the form of multiple alleles, each of which differs from the others, but nevertheless encodes a functional form of the enzyme.

Incomplete penetrance and variable expression can obscure inheritance patterns.

Many traits with single-gene inheritance demonstrate complications that can obscure the expected patterns in pedigrees. Chief among these are traits with **incomplete penetrance**, which means that individuals with a genotype corresponding to a trait do not actually show the phenotype, either because of environmental effects or because of interactions with other genes. Penetrance is the proportion of individuals with a particular genotype that show the expected phenotype. If the penetrance is less than 100%, then the trait shows reduced, or incomplete, penetrance.

For example, type 2 diabetes is associated with certain mutations in the gene *TCF7L2* that affect insulin secretion and glucose production, but the mutations are not completely penetrant. That is, some people who inherit a *TCF7L2* risk factor for type 2 diabetes do not actually develop the disease because environmental factors are also important in the causation of the disease. In other words, the genetic risk associated with the mutant forms of *TCF7L2* can often be offset by lifestyle choices such as proper diet and exercise. Similarly, familial cancers also often show incomplete penetrance.

Another common complication in human pedigrees is **variable expressivity,** which means that a particular phenotype is expressed with a different degree of severity in different individuals. Don't confuse variable expressivity with incomplete penetrance. With variable expressivity, the trait is always expressed, though the severity varies; with incomplete penetrance, the trait is sometimes expressed and sometimes not. Variation among individuals in the expression of a trait can result from the action of other genes, from effects of the environment, or both. An example is provided by deficiency of the enzyme alpha-1 antitrypsin (α1AT) discussed in Chapter 14, which is associated with loss of lung elasticity and emphysema. Among individuals with emphysema due to α1AT deficiency, the severity of the symptoms varies dramatically from one patient to the next. In this case, tobacco smoking is an environmental factor that increases the severity of the disease.

Incomplete penetrance and variable expressivity both provide examples where a given genotype does not always produce the same phenotype, since the expression of genes is often influenced by other genes, the environment, or a combination of the two.

 CASE 3 YOUR PERSONAL GENOME: YOU, FROM A TO T

How do genetic tests identify disease risk factors?

Your personal genome, as well as that of every human being, contains a unique combination of alleles of thousands of different genes. Most of these have no detectable effects on health or longevity, but as we saw in Chapter 14, many are risk factors for genetic diseases. Molecular studies have discovered particular alleles of genes associated with a large number of such conditions, and the presence of these alleles can be tested. More than a thousand genetic tests have already been deployed, and many more are actively being developed. A **genetic test** is a method of identifying the genotype of an individual. An example of genetic testing is the DNA fingerprinting used in criminal investigations (Chapter 14). In genetic tests for risk factors for disease, the tests may be carried out on entire populations or they may be restricted to high-risk individuals.

The benefits of genetic testing can be appreciated by an example. Screening of newborns for phenylketonuria identifies babies with high blood levels of phenylalanine. In the absence of treatment, 95% of such newborns will progress to moderate or severe mental retardation, whereas virtually all those placed on a special diet with a controlled amount of phenylalanine will have mental function within the normal range. For recessive conditions like phenylketonuria, tests can be carried out on people with affected relatives to identify the heterozygous genotypes. Testing can also identify genetic risk factors for disease, and carriers can take additional precautions. For example, individuals who are found to have an α1AT deficiency can prolong and improve the quality of their lives by not smoking tobacco, women with the genetic risk factors *BRCA1* and *BRCA2* for breast cancer can have frequent mammograms, and those with the *TCF7L2* risk factor for type 2 diabetes can decrease their risk by making lifestyle choices that include weight control and exercise.

While genetic testing has many potential benefits, it also has some perils. One major concern is maintaining the privacy of those persons who choose to be tested. With medical records increasingly going online, who will have access to your test results, and how will this information be used? Could your test results be used to deny you health or life insurance because you have a higher than average risk of some medical condition? Could an employer who got hold of your genetic test results decide to reassign you to another job, or even eliminate your position because of your genetic predispositions? Some safeguards have already been designed to protect you from such discrimination. The Genetic Information Nondiscrimination Act (GINA), which was signed into law in 2008, forbids the use of genetic information in decisions concerning employment and health insurance. The protection provided by GINA will, it is hoped, allow for the responsible and productive use of genetic information.

There is also increasing concern about the reliability and accuracy of genetic tests, especially direct-to-consumer (DTC) genetic tests. DTC tests can be purchased directly without the intervention of medical professionals. Because the consumer sends a biological sample and DTC tests are carried out by the provider, these tests are not regarded as medical devices and so are unregulated, which leads to some problems. Unfortunately, some DTC tests are based on flimsy and unconfirmed evidence connecting a gene with a disease. In addition, the link between genotype and risk may sometimes be exaggerated by the testing company for marketing purposes. Yet another problem is lack of information on quality control in the DTC laboratories. Finally, consumer misinterpretation may regard genotype as destiny, at one extreme descending into depression and despair, and at the other extreme using a low-risk genotype to justify an unhealthy lifestyle.

Self-Assessment Questions

12. How is it possible that there are multiple different alleles in a population, yet any individual can have only two alleles?
13. Construct a human pedigree for a dominant trait and a recessive trait. What are the patterns of inheritance in each?
14. What are some of the benefits and risks of genetic testing and personal genomics?

CORE CONCEPTS SUMMARY

15.1 EARLY THEORIES OF INHERITANCE: Early theories of heredity incorrectly assumed the inheritance of acquired characteristics and blending of parental traits in the offspring.

The inheritance of acquired characteristics suggests that traits that develop during the lifetime of an individual can be passed on to offspring. With rare exceptions, this mode of inheritance does not occur. page 323

Blending inheritance is the incorrect hypothesis that characteristics in the parents are averaged in the offspring. This model predicts the blending of genetic material, which does not occur. Different forms of a gene maintain their separate identities even when present together in the same individual. page 324

15.2 FOUNDATIONS OF MODERN TRANSMISSION GENETICS: The study of modern transmission genetics began with Gregor Mendel, who used the garden pea and studied traits with contrasting characteristics.

Mendel started his experiments with true-breeding plants, ones whose progeny are identical to their parents. He followed just one or two traits at a time, allowing him to discern simple patterns, and he counted all the progeny of his crosses. page 325

In crosses of one true-breeding plant with a particular trait and another true-breeding plant with a contrasting trait, just one of the two characteristics appeared in the offspring. The trait that appeared in this generation is dominant, and the trait that is not seen is recessive. page 326

Mendel explained this result by hypothesizing that there is a hereditary factor for each trait (now called a gene); that each pea plant carries two copies of the gene for each trait; and that one of two different forms of the gene (alleles) is dominant to the other one. page 327

15.3 SEGREGATION: Mendel's first key discovery was the principle of segregation, which states that members of a gene pair separate equally into gametes.

When Mendel allowed the progeny of the first cross to self-fertilize, he observed a 3 : 1 ratio of the dominant and recessive traits among the progeny. page 328

Mendel reasoned that the parent in this generation must have two different forms of the same gene (*A* and *a*) and that these alleles segregate from each other during gamete formation, with each gamete getting *A* or *a*, but not both. When the gametes combine at random during fertilization, they produce progeny in the genotypic ratio 1 *AA* : 2 *Aa* : 1 *aa*, which yields a phenotypic ratio of 3 : 1 because *A* is dominant to *a*. This idea became known as the principle of segregation. page 328

The principle of segregation reflects the separation of homologous chromosomes that occurs in anaphase I of meiosis. page 330

Some traits show incomplete dominance, in which the phenotype of the heterozygous genotype is intermediate between the phenotypes of the two homozygous genotypes. page 331

The expected frequencies of progeny of crosses can be predicted using the addition rule, which states that when two possibilities are mutually exclusive, the probability of either event occurring is the sum of their individual probabilities; and the multiplication rule, which states that when two possibilities occur independently, the probability of both events occurring is the product of the probabilities of each of the two events. page 332

15.4 INDEPENDENT ASSORTMENT: Mendel's second key discovery was the principle of independent assortment, which states that different gene pairs segregate independently of one another.

In a cross with two traits, each with two contrasting characteristics, the two traits behave independently of each other. This idea became known as the principle of independent assortment. For example, in a self-cross of a double heterozygote, the phenotypic ratio of the progeny is 9 : 3 : 3 : 1, reflecting the independent assortment of two 3 : 1 ratios. page 333

Independent assortment results from chromosome behavior during meiosis, specifically from the random orientation of different chromosomes on the meiotic spindle. page 334

In some cases, genes interact with each other, modifying the expected ratios in crosses. Epistasis is a gene interaction in which one gene affects the expression of another. page 334

15.5 HUMAN GENETICS: The patterns of inheritance that Mendel observed in peas can also be seen in humans.

In a human pedigree, females are represented by circles and males as squares; affected individuals are shown as filled symbols and unaffected individuals as open symbols. Horizontal lines denote matings. page 337

Dominant traits appear in every generation and affect males and females equally. page 337

Recessive traits skip one or more generations and affect males and females equally. Affected individuals often result from matings between close relatives. page 337

Although a given individual has only two alleles of each gene, there can be many alleles of a particular gene in the population as a whole. page 339

Interpreting pedigrees can be difficult because of incomplete penetrance and variable expressivity. Penetrance is the percentage of individuals with a particular genotype who show the expected phenotype, and expressivity is the degree to which a genotype is expressed in the phenotype. page 339

Genetic testing enables the genotype of an individual to be determined for one or more genes. Such tests carry both benefits and risks. page 339

Log in to **LaunchPad** to check your answers to the Self-Assessment Questions and to access additional learning tools.

CHAPTER 16: Inheritance of Sex Chromosomes, Linked Genes, and Organelles

CORE CONCEPTS

16.1 THE X AND Y CHROMOSOMES: Many organisms have a distinctive pair of chromosomes that differ between the sexes and show different patterns of inheritance from other chromosomes.

16.2 INHERITANCE OF GENES IN THE X CHROMOSOME: X-linked genes show a crisscross inheritance pattern.

16.3 GENETIC LINKAGE: Genes that are close together in the same chromosome tend to remain together in inheritance.

16.4 INHERITANCE OF GENES IN THE Y CHROMOSOME: Most Y-linked genes are passed from father to son.

16.5 INHERITANCE OF MITOCHONDRIAL AND CHLOROPLAST DNA: Mitochondria and chloroplast DNA follow their own inheritance pattern.

Mendel's principles of segregation and independent assortment are the foundation of transmission genetics (Chapter 15). For traits such as pea color and seed shape that are encoded by single genes and display simple dominance, these principles predict simple phenotypic ratios in the progeny from self-crosses of heterozygous genotypes.

However, we also saw in Chapter 15 that not all crosses are as simple as those for Mendel's pea plants. For example, in the case of alleles that show incomplete dominance, the $3:1$ phenotypic ratio in progeny of self-crosses of heterozygotes is $1:2:1$. The $9:3:3:1$ ratio of phenotypes observed for two genes that show independent assortment can be altered by epistasis, producing ratios such as $12:3:1$ or $13:3$. None of these is an exception to Mendel's laws because his laws reflect chromosome movement during meiosis (Chapter 11). Instead, they reflect how genes are expressed or how different genes interact to produce a phenotype.

This chapter highlights additional patterns of inheritance that Mendel did not observe because of his choice of experimental organism and the traits he studied. None of these patterns undermines or invalidates his insights, nor do they contradict later discoveries that genes in the nucleus are present in homologous chromosomes that pair and segregate in meiosis. Since Mendel's time, researchers have observed many inheritance patterns that seem to defy one or both of Mendel's laws. What such examples reveal is that the location of a gene is as important to our predictions about the inheritance of a trait as whether its alleles are dominant or recessive.

In this chapter, we discuss how genes carried in the sex chromosomes are transmitted differently in males and in females and how genes close to each other in the same chromosome do not undergo independent assortment and therefore violate Mendel's second law. Genes located in the genomes of mitochondria and chloroplast appear to defy Mendel's laws altogether because the organelles are inherited differently from the way chromosomes are inherited. Such unique patterns of inheritance expand the types of ratios and predictions we saw in Chapter 15 and draw our attention to the location and organization of genes—in specific chromosomes, relative to other genes, or outside the nucleus altogether.

16.1 THE X AND Y CHROMOSOMES

Mendel concluded from his experiments that reciprocal crosses yield the same types of progeny in the same proportions. This is in most cases true, but an important exception occurs with the X and Y sex chromosomes. For example, red–green color blindness in humans is due to mutant alleles of a gene in the X chromosome. Reciprocal crosses do not produce the same types and numbers of progeny. When a color-blind man mates with a woman who is not color blind, all of the sons and daughters have normal color vision. However, in the reciprocal cross, when a color-blind woman mates with a man who is not color blind, all of the daughters have normal color vision but all of the sons are color blind. For the X and Y sex chromosomes, reciprocal crosses are not equivalent.

In many animals, sex is genetically determined and associated with chromosomal differences.

Most chromosomes come in pairs that match in shape and size. The members of each pair are known as homologous chromosomes because they have the same genes along their length (Chapter 11). One member

FIG. 16.1 Human sex chromosomes. The human X and Y sex chromosomes differ in size and number of genes. Source: Science Photo Library/Science Source.

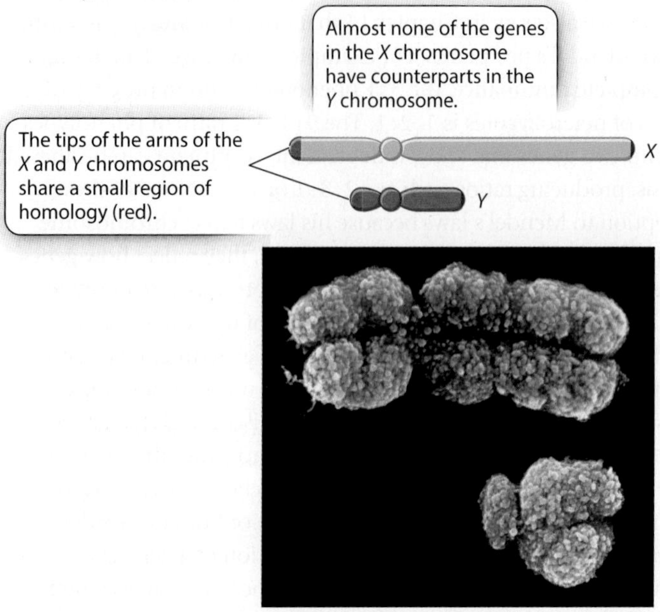

The tips of the arms of the X and Y chromosomes share a small region of homology (red).

Almost none of the genes in the X chromosome have counterparts in the Y chromosome.

of each pair of homologous chromosomes is inherited from the mother and the other from the father. In many animal species, however, the sex of an individual is determined by a distinctive pair of unmatched chromosomes known as the **sex chromosomes**, which are usually designated as the **X chromosome** and the **Y chromosome** (Chapter 13). Chromosomes other than the sex chromosomes are known as **autosomes**.

In humans, a normal female has two copies of the X chromosome (a sex-chromosome constitution denoted XX), and a normal male has one X chromosome and one Y chromosome (XY). As seen in **Fig. 16.1**, the sizes of the human X and Y chromosomes are very different from each other. The X chromosome DNA molecule is more than 150 Mb long, whereas the Y chromosome is only about 50 Mb long. Except for a small region near each tip (red in Fig. 16.1), the gene contents of the X and Y chromosomes differ from each other. The X chromosome, which includes more than 1000 genes, has a gene density similar to that of most autosomes. Most of these genes have no counterpart in the Y chromosome. In contrast to the X chromosome, the Y chromosome contains only about 50 protein-coding genes.

The regions of homology between the X and Y chromosomes consist of about 2.7 Mb of DNA near the tip of the short arm and about 0.3 Mb of DNA near the tip of the long arm. These regions of homology allow the chromosomes to pair during meiosis (Chapter 11). In most cells undergoing meiosis, a **crossover** (physical breakage, exchange of parts, and rejoining of the DNA molecules) occurs in the larger of these regions. The crossover allows the chromosomes to move as a unit to align properly at metaphase I so that, when their centromeres separate from each other at anaphase I, the X and Y chromosomes go to opposite poles.

The relative size and gene content of the X and Y chromosomes differ greatly among species. In some species of mosquitoes, the X and Y chromosomes are virtually identical in size and shape, and the regions of homology include almost the entire chromosome. This situation is unusual, however. In most species, the Y chromosome contains many fewer genes than the X chromosome. Some species, such as grasshoppers, have no Y chromosome: females in these species have two X chromosomes and males have only one X chromosome. The total number of chromosomes in grasshoppers, therefore, differs between females and males. In birds, moths, and butterflies, the sex chromosomes are reversed: Females have two different sex chromosomes and males have two of the same sex chromosome.

Segregation of the sex chromosomes predicts a 1 : 1 ratio of females to males.

It is ironic that Mendel did not interpret sex as an inherited trait. If he had, he might have realized that sex itself provides one of the most convincing demonstrations of segregation. In human males, segregation of the X chromosome from the Y chromosome during anaphase I of meiosis results in half the

FIG. 16.2 Inheritance of the X and Y chromosomes. Segregation of the sex chromosomes in meiosis and random fertilization result in a 1 : 1 ratio of female : male embryos.

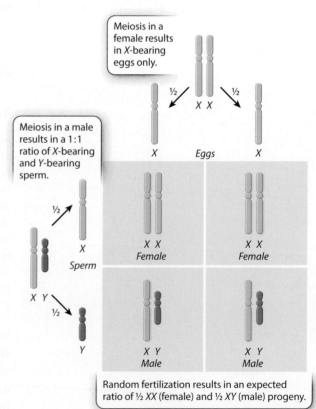

sperm bearing an *X* chromosome and the other half bearing a *Y* chromosome (**Fig. 16.2**). Meiosis in human females results in eggs that each contains one *X* chromosome. With random fertilization, as shown in Fig. 16.2, half of the fertilized eggs are expected to be chromosomally *XX* (and therefore female) and half are expected to be chromosomally *XY* (and therefore male).

The expected ratio of 1:1 of females:males refers to the sex ratio at the time of conception. This is the primary sex ratio, and it is not easily observed because the earliest stages of human fertilization and development are inaccessible to large-scale study. What can be observed is the sex ratio at birth, called the secondary sex ratio, which differs among populations but usually shows a slight excess of males. In the United States, for example, the secondary sex ratio is approximately 100 females:105 males. The explanation seems to be that, for reasons that are not entirely understood, female embryos are slightly less likely to survive from conception to birth. On the other hand, males are slightly less likely to survive from birth to reproductive maturity, and so, at the age of reproductive maturity, the sex ratio is very nearly 1:1. Male mortality continues to be greater than that of females throughout life, so that by age 85 and over, the sex ratio is about 2 females:1 male.

The sex of each birth appears to be random relative to previous births—that is, there do not seem to be tendencies for some families to have boys or for others to have girls. Many people are surprised by this fact, for they know of one or more large families consisting mostly of boys or mostly of girls. But the occasional family with children of predominantly one sex is expected simply by chance. When one occurs, its unusual sex distribution commands attention out of proportion to the actual numbers of such families. In short, families with unusual sex distributions are not more frequent than would be expected by chance.

Self-Assessment Questions

1. How can the human *X* and *Y* chromosomes pair during meiosis even though they are of different lengths and most of their genes are different?
2. What is the biological basis for the 1:1 ratio of males and females at conception in mammals?

16.2 INHERITANCE OF GENES IN THE *X* CHROMOSOME

Genes in the *X* chromosome are called **X-linked genes**. These genes have a unique pattern of inheritance first discovered by Thomas Hunt Morgan in 1910. Morgan's pioneering studies of genetics of the fruit fly *Drosophila melanogaster* helped bring Mendelian genetics into the modern era. The discovery of *X*-linked inheritance was not only important in itself, but it also provided the first experimental evidence that chromosomes contain genes.

X-linked inheritance was discovered through studies of male fruit flies with white eyes.

Morgan's discovery of *X*-linked genes began when he noticed a white-eyed male in a bottle of fruit flies in which all the others had normal, or wild-type, red eyes. (The most common phenotype in a population is often called the **wild type**.) This was the first mutant he discovered, and finding it was a lucky break.

Morgan's initial crosses are outlined in **Fig. 16.3**. In the first generation, he crossed the mutant white-eyed male with a wild-type red-eyed female. All of the progeny (F_1) fruit flies had wild-type red eyes, as you would expect from a cross with any recessive mutation. Morgan then carried out matings between brothers and sisters among the F_1 generation, and he found that the mutant phenotype of white eyes reappeared among the progeny. This result, too, was expected. However, there was a surprise: Morgan observed that the white-eye phenotype was associated with the sex of the fly. In the F_2 generation, all the white-eyed fruit flies were male, and the white-eyed males appeared along with red-eyed males in a ratio of 1:1. No

FIG. 16.3 Morgan's white-eyed fly. Morgan observed that the recessive white-eyed phenotype reappeared in the F_2 generation, but in males only.

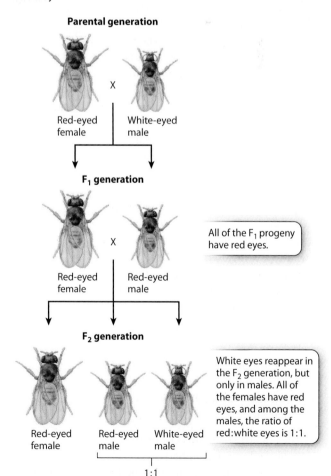

females with white eyes were observed; all the females had red eyes.

Genes in the *X* chromosome exhibit a crisscross inheritance pattern.

When Morgan did his crosses with the white-eyed male, the *X* chromosome had only recently been discovered by microscopic examination of the chromosomes in male and female grasshoppers. Morgan was the first to understand that the pattern of inheritance of the *X* chromosome would differ from that of the autosomes, and he proposed the hypothesis that the white-eyed phenotype was due to a mutation in a gene in the *X* chromosome. This hypothesis could explain the pattern of inheritance shown in Fig. 16.3.

The key features of *X*-linked inheritance are shown in **Fig. 16.4**. In *Drosophila*, as in humans, females are *XX* and males are *XY*. Fig. 16.4a shows an *XY* male in which the *X* chromosome contains a recessive mutation. Because the *Y* chromosome does not carry an allele of this gene, the recessive mutation will be reflected in the male's phenotype—in this case, white eyes.

The Punnett square in Fig. 16.4a explains why all the offspring had red eyes when Morgan crossed the white-eyed male with a red-eyed female in the parental generation. During meiosis in the male, the mutant *X* chromosome segregates from the *Y* chromosome, and each type of sperm, *X* or *Y*, combines with a normal *X*-bearing egg. The result is that the female progeny are heterozygous. They have only one copy of the mutant allele, and because the mutant allele is recessive, the heterozygous females do not express the mutant white-eye trait. The male progeny are also red-eyed because they receive their *X* chromosome from their wild-type red-eyed mother.

The Punnett square in Fig. 16.4a illustrates two important principles governing the inheritance of *X*-linked genes:

1. The phenotypes of the *XX* offspring indicate that a male transmits his *X* chromosome only to his daughters. In this case, the *X* chromosome transmitted by the male carries the white-eye mutation.

2. The phenotypes of the *XY* offspring indicate that a male inherits his *X* chromosome from his mother. In this

FIG. 16.4 X-linkage. (a) Cross between a homozygous nonmutant female and a male carrying an *X*-linked recessive allele (red). (b) Cross between a heterozygous female and a nonmutant male. *X*-linked recessive alleles are expressed in males because males have only one *X* chromosome.

a.

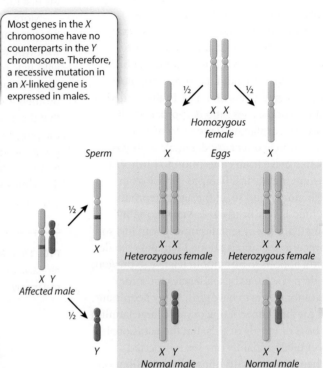

b.

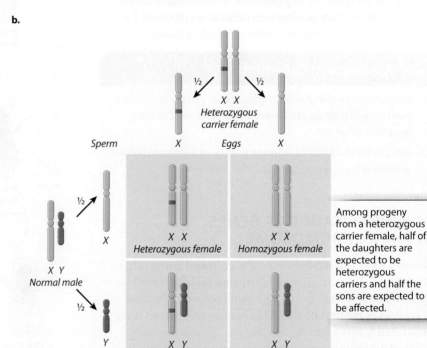

case, the X chromosome transmitted by the mother carries the nonmutant allele of the gene.

These principles underlie a pattern often referred to as **crisscross inheritance**: an X chromosome present in a male in one generation must be transmitted to a female in the next generation, and in the generation after that can be transmitted back to a male. Therefore, an X chromosome can crisscross, or alternate, between the sexes in successive generations.

The mating illustrated in Fig. 16.4b shows another important feature of X-linked inheritance, one that explains the results of Morgan's F_1 cross. In this case, the mother is a heterozygous female. During meiosis, the mutant allele and the nonmutant allele undergo segregation. Half of the resulting eggs contain an X chromosome with the mutant allele, and half contain an X chromosome with the nonmutant allele. These combine at random with either X-bearing sperm or Y-bearing sperm. Among the female progeny in the next generation, half are heterozygous for the mutant allele and the other half are homozygous for the normal allele. Thus, none of the females exhibits the white-eye trait because the mutation is recessive. Among the male progeny, half receive the mutant allele and have white eyes, whereas the other half receive the nonmutant allele and have red eyes. The expected progeny from the cross in Fig. 16.4b therefore consist of all red-eyed females, and there is a 1:1 ratio of red-eyed to white-eyed males, which is what Morgan observed in the F_2 generation of his crosses, as shown in Fig. 16.3.

Knowing the patterns revealed by the Punnett squares, we can now assign genotypes to Morgan's original crosses (**Fig. 16.5**). In Fig. 16.5, the white-eyed males that Morgan used in his parental generation crosses are given the genotype of w^-Y, and the red-eyed female has a genotype of w^+w^+. The symbol "w^-" stands for the recessive white-eye mutation in one X chromosome, and the symbol "w^+" stands for the dominant nonmutant allele (red eyes) in the other X chromosome. The symbol "Y" stands for the Y chromosome, and it is important to remember that the Y chromosome does not contain an allele of the white-eye gene.

In the cross between a wild-type red-eyed female and a white-eyed male, illustrated in Fig. 16.5a, the male offspring have red eyes because they receive their X chromosome from their mother. The female offspring from this cross receive one of their X chromosomes from their father, and hence they are heterozygous, w^+w^-. When the male and female F_1 progeny mate, their F_2 offspring consist of all red-eyed females (half of which are heterozygous) and a 1:1 ratio of red-eyed to white-eyed males, exactly as Morgan had observed.

The hypothesis of X-linkage not only explained the original data, but it also predicted the results of other crosses. One

FIG. 16.5 Genotypes of Morgan's white-eyed fruit fly crosses.

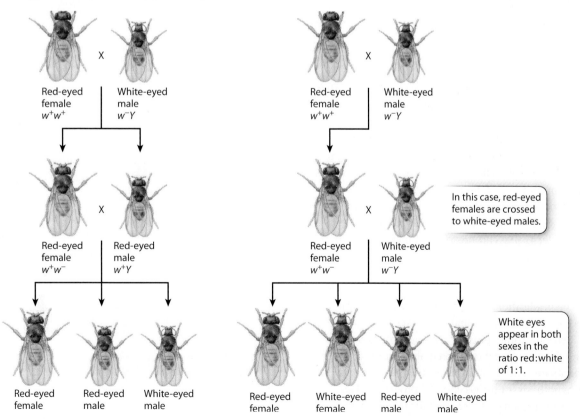

important test is outlined in Fig. 16.5b. Here, the parental cross is the same as that in Fig. 16.5a, but instead of mating the F_1 females to their brothers, they are mated to white-eyed males. The prediction is that there should be a 1 : 1 ratio of red-eyed females to white-eyed females as well as a 1 : 1 ratio of red-eyed males to white-eyed males. Again, these were the results observed. By the results of these crosses and others, Morgan demonstrated that the pattern of inheritance of the white-eye mutation parallels the pattern of inheritance of the X chromosome.

X-linkage provided the first experimental evidence that genes are in chromosomes.

Morgan's original experiments indicated that the white-eye mutation showed a pattern of inheritance like that expected of the X chromosome. However, it was one of Morgan's students who showed experimentally that the white-eye mutation was actually a physical part of the X chromosome. Today, it seems obvious that genes are in chromosomes because we know that genes consist of DNA and that DNA in the nucleus is found in chromosomes. But in 1916, when Calvin B. Bridges, who had joined Morgan's laboratory as a freshman, was working on his PhD research under Morgan's direction, neither the chemical nature of the gene nor the chemical composition of chromosomes was known.

In one set of experiments, Bridges crossed mutant white-eyed females with wild-type red-eyed males (**Fig. 16.6**). Usually, the progeny consisted of red-eyed females and white-eyed males (Fig. 16.6a). This is the result expected when the X chromosomes in the mother separate normally at anaphase I in meiosis because all the daughters receive a w^--bearing X chromosome from their mother as well as a w^+-bearing X chromosome from their father, whereas all the sons receive a w^--bearing X chromosome from their mother and a Y chromosome from their father.

But Bridges noted a few rare exceptions among the progeny. He saw that about 1 offspring in 2000 from the cross was "exceptional"—either a female with white eyes or a male with red eyes. The exceptional females were fertile, and the exceptional males were sterile. To explain these exceptional progeny, Bridges proposed the hypothesis diagrammed in Fig. 16.6b: the X chromosomes in a female occasionally fail to separate in anaphase I in meiosis, and both X chromosomes go to the same pole. Recall from Chapter 11 that chromosomes sometimes fail to separate normally in meiosis, a process known as **nondisjunction**. Nondisjunction of X chromosomes results in eggs containing either two X chromosomes or no X chromosome. Figure 16.6b shows the implications for eye color in the progeny if the hypothesis is correct. The exceptional white-eyed females would contain two X chromosomes plus a Y chromosome (genotype $w^-/w^-/Y$), and the exceptional red-eyed males would contain a single X chromosome and no Y chromosome (genotype w^+).

Bridges's hypothesis for the exceptional progeny in Fig. 16.6b was bold, as it assumed that *Drosophila* males could

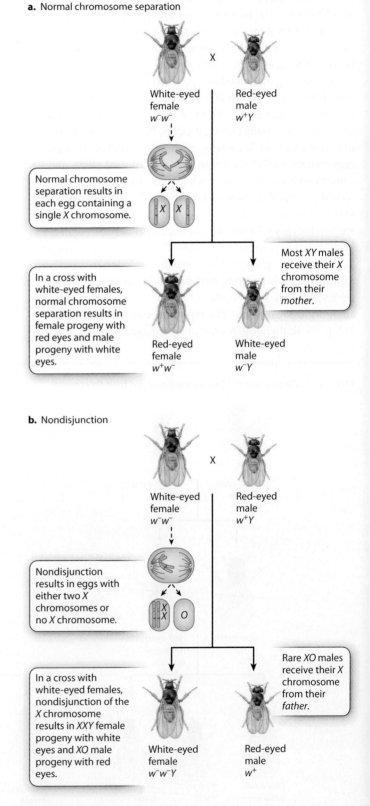

FIG. 16.6 Nondisjunction as evidence that genes are present in chromosomes. (a) Normal chromosome separation yields expected progeny. (b) Nondisjunction yields exceptional progeny.

develop in the absence of a *Y* chromosome (*XO* embryos yielding sterile males, where "*O*" indicates absence of a chromosome), and that females could develop in the presence of a *Y* chromosome (*XXY* embryos yielding fertile females). The hypothesis was accurate as well as bold. Microscopic examination of the chromosomes in the exceptional fruit flies confirmed that the exceptional white-eyed females had *XXY* sex chromosomes and that the exceptional sterile red-eyed males had an *X* but no *Y*. Because fruit flies with three *X* chromosomes (*XXX*) or no *X* chromosome (*OY*) were never observed, Bridges concluded that embryos with these chromosomal constitutions are unable to survive. Bridges also conducted crosses that showed that nondisjunction can take place in males as well as in females. From the phenotypes of these exceptional fruit flies and their chromosome constitutions, Bridges concluded that the white-eye gene (and by implication any other *X*-linked gene) is physically present in the *X* chromosome.

Bridges's demonstration that genes are present in chromosomes was also the first experimental evidence of nondisjunction. *Drosophila* differ from humans in that the *Y* chromosome is necessary for male fertility but not for male development. As we will see later in this chapter, a gene in the *Y* chromosome itself is the trigger for male development in humans and other mammals, and so for these organisms, the *Y* chromosome is needed both for male development and male fertility. Nondisjunction occasionally takes place in meiosis in humans as well as in fruit flies. When nondisjunction takes place in the human sex chromosomes, it results in chromosomal constitutions such as 47, *XXY* and 47, *XYY* males as well as 47, *XXX* and 45, *X* females (Chapter 11). Nondisjunction of autosomes can also occur, resulting in fetuses that have extra copies or missing copies of entire chromosomes.

Genes in the *X* chromosome show characteristic patterns in human pedigrees.

The features of *X*-linked inheritance can be seen in human pedigrees for traits caused by an *X*-linked recessive mutation. These are illustrated in **Fig. 16.7** for red–green color blindness, a condition that affects about 1 in 20 males. An individual with red–green color blindness will have difficulty seeing the number in the colored dots in Fig. 16.7.

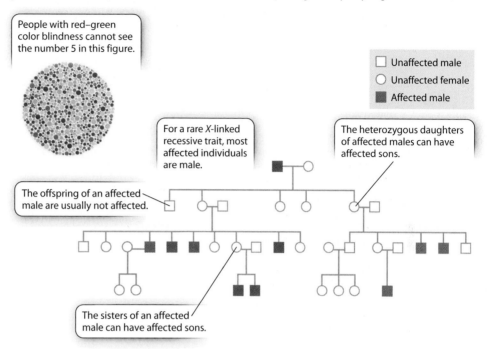

FIG. 16.7 Inheritance of an *X*-linked recessive mutation. This pedigree shows the inheritance of red–green color blindness, an *X*-linked recessive mutation. *Source: Dorling Kindersley/Getty Images.*

The key features of the inheritance of traits due to rare *X*-linked recessive alleles, which are noted in the pedigree, are listed here:

1. Affected individuals are usually males because males need only one copy of the mutant gene to be affected, whereas females need two copies to be affected.

2. Affected males have unaffected sons because males transmit their *X* chromosome only to their daughters.

3. A female whose father is affected can have affected sons because such a female must be a heterozygous carrier of the recessive mutant allele.

An additional feature worth mentioning is that the sisters of an affected male each have a 50% chance of being a heterozygous carrier because when a brother is affected the mother must be heterozygous for the recessive allele.

A pedigree for one of the most famous examples of human *X*-linked inheritance is shown in **Fig. 16.8**. The trait is a form of **hemophilia**, which results from a recessive mutation in a gene encoding a protein necessary for blood clotting. Affected individuals bleed excessively from even minor cuts and bruises, and internal bleeding can cause excruciating pain. Affecting about 1 in 7000 males, hemophilia is famous because of its presence in many members of European royalty descended from Queen Victoria of England (1819–1901), whose pedigree later revealed her to be a heterozygous carrier of the gene. By the marriages of her carrier granddaughters, the gene was introduced into the

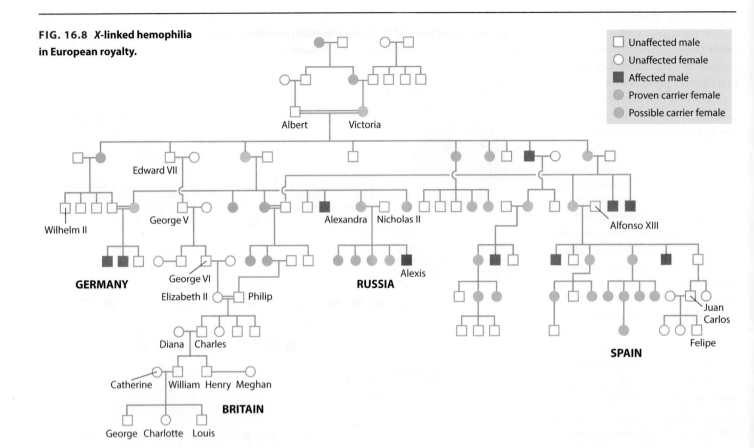

FIG. 16.8 *X*-linked hemophilia in European royalty.

royal houses of Germany, Russia, and Spain. The mutant allele is not present in the present royal family of Britain because this family descends from King Edward VII, one of Victoria's four sons, who was not himself affected and therefore passed only a normal X chromosome to his descendants.

The source of Queen Victoria's hemophilia mutation is not known. None of her ancestors is reported as having a bleeding disorder. Quite possibly the mutation was present for a few generations before Victoria was born but remained hidden because it was passed from heterozygous female to heterozygous female.

Self-Assessment Questions

3. For a recessive *X*-linked mutation, such as color blindness, what is the pattern of inheritance from an affected male through his daughters into her children?
4. Is it possible for an unaffected female to have female offspring with red–green color blindness?

16.3 GENETIC LINKAGE

Mendel was fortunate not only because peas do not possess sex chromosomes, but also because the genes that influenced the traits he studied, such as round/wrinkled and yellow/green seeds, are on separate chromosomes or far apart on the same chromosome. What happens when genes are close to each other in the same chromosome? We explore the answer to this question in this section.

Nearby genes in the same chromosome show linkage.

Genes that are sufficiently close together in the same chromosome are said to be **linked**. That is, they tend to be transmitted together in inheritance and do not assort independently of each other as Mendel observed. Note that "linked genes" refers to two genes that are close together in the same chromosome, which may be an autosome or sex chromosome. This is not to be confused with an "*X*-linked gene," which is one that is present anywhere in the *X* chromosome.

Linkage was discovered in *Drosophila* by Alfred H. Sturtevant, another of Morgan's students. Once again, genes in the *X* chromosome played a key role in the discovery because the phenotypes of the male offspring of a female reveal the female's genotype for *X*-linked genes. An example is shown in **Fig. 16.9**. Sturtevant worked with male fruit flies that have an *X* chromosome carrying two recessive mutations. One is in the *white* gene (*w*) discussed earlier, which when nonmutant (w^+) results in fruit flies with red eyes and when mutant (w^-) results in fruit flies with white eyes. The other recessive mutation is in a gene called *crossveinless* (*cv*), which when

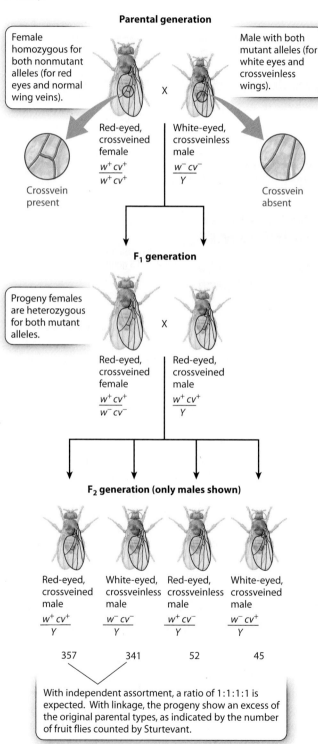

FIG. 16.9 Linkage of the *white* (*w*) and *crossveinless* (*cv*) genes in the *X* chromosome. Genes are written according to their order along the chromosome, with a horizontal line between homologous chromosomes, in this case the *X* and *Y*. Crossveins are shown in red for clarity.

nonmutant (cv^+) results in fruit flies with tiny crossveins in the wings and when mutant (cv^-) results in the absence of these crossveins.

Sturtevant crossed this doubly mutant male with a female carrying the nonmutant forms of the genes (w^+ and cv^+) in both *X* chromosomes. He saw that the offspring consist of phenotypically wild-type females that are heterozygous for both genes, and phenotypically wild-type males. When these are crossed with each other, the female F_2 progeny do not tell us anything because they are all wild type; each female receives the $w^+ cv^+$ *X* chromosome from her father and therefore has red eyes and normal crossveins. In the male F_2 progeny, however, the situation is different: each male progeny receives its *X* chromosome from the mother and its *Y* chromosome from the father, and so the phenotype of each male immediately reveals the genetic constitution of the *X* chromosome that the male inherited from the mother.

As shown in Fig. 16.9, the male F_2 progeny consist of four types:

Genotype of F_2 Progeny	Number of Fruit Flies
$w^+ cv^+/Y$ (red eyes, normal crossveins)	357
$w^- cv^-/Y$ (white eyes, missing crossveins)	341
$w^+ cv^-/Y$ (red eyes, missing crossveins)	52
$w^- cv^+/Y$ (white eyes, normal crossveins)	45

Although all four possible classes of maternal gametes are observed in the male progeny, they do not appear in the ratio 1 : 1 : 1 : 1 expected when gametes contain two independently assorting genes (Chapter 15). The lack of independent assortment means that the genes show linkage.

The male progeny fall into two groups. One group, represented by larger numbers of progeny, derives from maternal gametes containing either $w^+ cv^+$ or $w^- cv^-$. These are called **nonrecombinants** because the alleles are present in the same combination as that in the parent. The other group of male progeny consists of $w^+ cv^-$ and $w^- cv^+$ combinations of alleles. These are called **recombinants,** and they result from a crossover, the physical exchange of parts of homologous chromosomes, which takes place in prophase I of meiosis (Chapter 11).

Crossing over is a key process in meiosis (Chapter 11). Most chromosomes have one or more crossovers that form between the homologous chromosomes as they pair and undergo meiosis. Human females average about 2.75 crossovers per chromosome pair, and human males average about 2.50 crossovers per chromosome pair. As noted earlier for the *X* and *Y* chromosomes, crossovers between homologous chromosomes are important mechanically because they help hold the homologs together so they can align properly at metaphase I and segregate to opposite poles at anaphase I.

FIG. 16.10 Linkage and recombination. (a) Crossing over between genes results in recombination between the alleles of the genes. (b) When crossing over occurs outside the interval between genes, there is no recombination between the alleles of the genes.

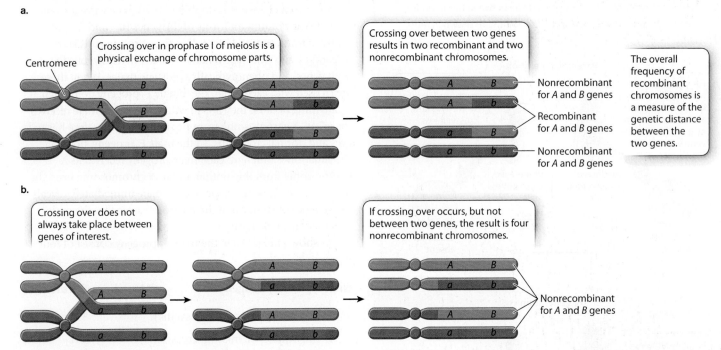

Fig. 16.10 shows how chromosomes with recombined alleles arise from crossing over between genes, using the hypothetical genes *A* and *B*. In a cell undergoing meiosis in which a crossover takes place between the genes, the allele combinations are broken up in the chromatids involved in the exchange, and the resulting gametes are *AB*, *Ab*, *aB*, and *ab* (Fig. 16.10a).

Note that crossing over does not result only in recombinant chromosomes. Fig. 16.10a shows that, even when a crossover occurs in the interval between the genes, two of the resulting chromosomes contain the nonrecombinant configuration of alleles. These nonrecombinant configurations occur because crossing over occurs at the four-strand stage of meiosis (when each homologous chromosome is a pair of sister chromatids), but only two of the four strands (one sister chromatid from each homologous chromosome) are included in any crossover.

Fig. 16.10b shows a second way in which nonrecombinant chromosomes originate. When two genes are close together in a chromosome, a crossover may occur not in the interval between the genes but at some other location along the chromosome. In this example, the crossover occurs in the region between gene *A* and the centromere, and therefore the combination of alleles *A* and *B* remains intact in one pair of chromatids and the combination of alleles *a* and *b* remains intact in the homologous chromatids. The resulting gametes are equally likely to have either *AB* or *ab*, both of which are nonrecombinant.

Nonrecombinant chromosomes can also be the result of two crossover events occurring between two genes, in which the effects of the first crossover (creating recombinants) is reversed by a second crossover (re-creating nonrecombinants).

The frequency of recombination is a measure of the genetic distance between linked genes.

When two genes are on separate chromosomes, a ratio of 1:1:1:1 is expected for the nonrecombinant (parental) and recombinant (nonparental) gametic types, as described by the principle of independent assortment (Chapter 15). For two genes present in the same chromosome, we can consider two extreme situations. If they are located very far apart from each other, one or more crossovers will almost certainly occur between them, and there will be a 1:1:1:1 ratio of nonrecombinant and recombinant gametes (as shown for the case of a single crossover in Fig. 16.10a). At the other extreme, if two genes are so close together that crossing over never takes place between them, we would expect only nonrecombinant chromosomes.

What happens in between these extremes? In these cases, in some cells undergoing meiosis, no crossover takes place between the genes, in which case all the resulting chromosomes are nonrecombinant (Fig. 16.10b); in other cells undergoing meiosis, a crossover occurs between the genes, in which case half the resulting chromosomes are nonrecombinant and half are recombinant (Fig. 16.10a).

The actual frequency of recombinants depends on the distance between the genes. The distance between the genes is important because whether or not a crossover occurs between the genes is a matter of chance, and the closer the genes are along the chromosome, the less likely it is that a crossover will take place in the interval between them. Because the formation of recombinant chromosomes requires at least one crossover between the genes, genes that are close together (more tightly linked) show less recombination than genes that are far apart. In fact, the proportion of recombinant chromosomes observed among the total, which is called the **frequency of recombination**, is a measure of genetic distance between the genes along the chromosome.

In the example with the genes w and cv, the total number of chromosomes observed among the progeny is $357 + 341 + 52 + 45 = 795$, and the number of recombinant chromosomes (w^+cv^-/Y and w^-cv^+/Y) is $52 + 45 = 97$. The frequency of recombination between w and cv is therefore $97/795 = 0.122$, or 12.2%, and this serves as a measure of the genetic distance between the genes. In studies of genetic linkage, the distance between genes is not measured directly by physical distance between them, but rather by the frequency of recombination.

The frequency of recombination between any two genes on the same chromosome ranges from 0% (when genes are so close together that crossing over never takes place between them) to 50% (when the genes are so far apart that a crossover between the genes almost always takes place). Genes that are linked have a recombination frequency somewhere between 0% and 50%. The maximum frequency of recombination is 50% provided that the non-sister chromatids involved in any crossover are chosen at random. A frequency of recombination of 50% yields the same ratio of gametic types as observed with independent assortment, which means that genes that are far enough apart in the same chromosome show independent assortment.

Recombination plays an important role in creating new combinations of alleles in each generation and in ensuring the genetic uniqueness of each individual. If there were no recombination (that is, if all the alleles in each chromosome were completely linked), any individual human would be able to produce only $2^{23} = 8.4$ million types of reproductive cells. Although this is a large number, the average number of sperm per ejaculate is much larger—approximately 350 million. Because recombination does occur, and because the crossovers resulting in recombination can occur at any of thousands of different positions in the genome, each of the 350 million sperm is virtually certain to carry a different combination of alleles.

Genetic mapping assigns a location to each gene along a chromosome.

With the exception of a few regions, such as the area near the centromere, the likelihood of a crossover occurring somewhere between two points on a chromosome is approximately proportional to the length of the interval between the points. Therefore, the frequency of recombination can be used as a measure of the physical distance between genes. These distances are used in the construction of a **genetic map**, which is a diagram showing the relative position of genes along a chromosome. The maps are drawn using a scale in which one unit of distance (called a **map unit**) is the distance between genes resulting in 1% recombination. Thus, in a *Drosophila* genetic map containing the genes w and cv that have a recombination frequency of 12.2%, the distance between the genes is 12.2 map units.

Genetic maps are built up step by step as new genes are discovered that are genetically linked, as shown in **Fig. 16.11**. Across distances that are less than about 15 map units, the map distances are approximately additive, which means that the distances between adjacent genes can be added to get the distance between the genes at the ends. For example, in Fig. 16.11, there are two genes between w and cv. The map distance between w and the next gene, ec, is 4.0, the distance between ec and the next gene, rb, is 2.0, and the distance between rb and cv is 6.2. The map distance between w and cv is therefore $4.0 + 2.0 + 6.2 = 12.2$ map units, and hence the expected frequency of recombination between these genes is 12.2%, which is the value observed.

However, for two genes that are farther apart than about 15 map units, the observed recombination frequency is somewhat smaller than the sum of the map distances between the genes. The reason is that, with greater distances, two or more crossovers between the genes may occur in the same chromosome, and thus an exchange produced by one crossover may be reversed by another crossover farther along the way.

Genetic risk factors for disease can be localized by genetic mapping.

The discovery of abundant genetic variation in DNA sequences in human populations, such as single-nucleotide polymorphisms (Chapter 14), made it possible to study genetic linkage in the human genome. At first, the focus was on finding mutations that cause disease, such as the mutation that causes cystic fibrosis. The method was to study large families extending over three or more generations in which the disease was present and then to identify the genotypes of each of the individuals for thousands of genetic markers (previously discovered DNA polymorphisms) throughout the genome. The goal was to find genetic markers that showed a statistical association with the disease gene, which would indicate genetic linkage and reveal the approximate location of the disease gene along the chromosome.

HOW DO WE KNOW?

FIG. 16.11

Can recombination be used to construct a genetic map of a chromosome?

BACKGROUND In 1910, American geneticist Thomas Hunt Morgan discovered X-linkage by studying the white-eye mutation in *Drosophila*. Soon other X-linked mutations were found. Alfred H. Sturtevant, Morgan's student, decided to test whether mutant genes in the same X chromosome were inherited together, that is, linked. He found that genes in the X chromosome were linked, but not completely, and that different pairs of genes showed great differences in their linkage. Some genes showed almost no recombination, whereas others underwent so much recombination that they showed independent assortment.

HYPOTHESIS Sturtevant hypothesized that recombination was due to crossing over between the genes, and that genes farther apart in the chromosome would show more recombination.

EXPERIMENT Taking this idea a step further, Sturtevant reasoned that if one knew the frequency of recombination between genes *a* and *b*, between *b* and *c*, and between *a* and *c*, then one should be able to deduce the order of these three genes along the chromosome. The logic is that the two genes showing the highest frequency of recombination are farthest apart, and the third gene is in the middle. He also predicted that, if the order of genes were known to be *a–b–c* and if the genes were close enough, the frequency of recombination between *a* and *c* should equal the sum of the frequency of recombination between *a* and *b* and the frequency of recombination between *b* and *c*.

RESULTS Sturtevant studied the frequencies of recombination between many pairs of genes along the X chromosome, including some of those in the illustration shown here.

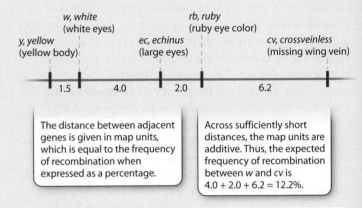

The distance between adjacent genes is given in map units, which is equal to the frequency of recombination when expressed as a percentage.

Across sufficiently short distances, the map units are additive. Thus, the expected frequency of recombination between *w* and *cv* is 4.0 + 2.0 + 6.2 = 12.2%.

CONCLUSION The results supported Sturtevant's hypothesis and showed that genes could be arranged in the form of a genetic map, depicting their linear order along the chromosome, with the distance between any pair of genes proportional to the frequency of recombination between them. Across sufficiently short regions, the frequencies of recombination are additive.

FOLLOW-UP WORK Genetic mapping remains a cornerstone of genetic analysis, allowing researchers to determine which chromosome contains a mutant gene and where along the chromosome the gene is located. The method helped to identify the genes responsible for many single-gene inherited disorders, including Huntington's disease, cystic fibrosis, and muscular dystrophy.

SOURCE Sturtevant, A. H. 1913. "The Linear Arrangement of Six Sex-Linked Factors in *Drosophila*, as Shown by Their Mode of Association." *Journal of Experimental Zoology* 14:43–59.

Fig. 16.12 illustrates the underlying concept. It assumes 100 chromosomes observed among different individuals in a pedigree, of which 50 carry a mutant allele of a gene and 50 carry a nonmutant allele. These chromosomes are tested for a marker of known location, in this case a single-nucleotide polymorphism (SNP) in which one of the nucleotide pairs in the DNA is a G—C base pair in some chromosomes and A—T in others. In Fig. 16.12a, there is clearly an association between the disease gene and the SNP. Almost all of the chromosomes that carry the mutant allele show the G—C nucleotide pair, whereas almost all of the chromosomes that carry the nonmutant allele show the A—T nucleotide pair. The two chromosomes in which the mutant and nonmutant genes are associated with the other SNPs can be attributed to recombination. Associations between disease genes and SNPs arise because of ancestry. When a new mutation takes place, it occurs in a single chromosome surrounded by a particular combination of SNP alleles. If the new mutation increases in frequency and establishes itself in the population, the particular combination of SNP alleles is carried along. The tighter the linkage between the mutation and the surrounding SNP alleles, the longer the association persists.

The association in Fig. 16.12a may be contrasted with the pattern in Fig. 16.12b, in which there is no association. In Fig. 16.12b, each of the alleles of the disease gene is equally likely to carry either form of the SNP. The failure to find an association means that the SNP is not closely linked to the disease gene, and may in fact be in a different chromosome. In actual studies, an association is almost never observed, but the lucky find of

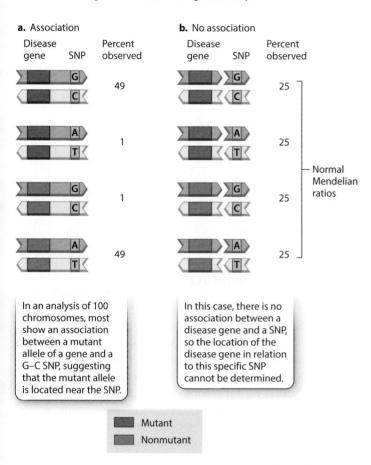

FIG. 16.12 Genetic mapping. SNPs associated with a mutant gene show where that gene is located in the genetic map.

In an analysis of 100 chromosomes, most show an association between a mutant allele of a gene and a G–C SNP, suggesting that the mutant allele is located near the SNP.

In this case, there is no association between a disease gene and a SNP, so the location of the disease gene in relation to this specific SNP cannot be determined.

an association helps identify the location of the disease gene in the genetic map. Using such association methods, hundreds of important disease genes have been located by genetic mapping. Once the location of the disease gene is known, the identity and normal function of the gene can be determined.

Self-Assessment Questions

5. Why don't linked genes exhibit independent assortment?
6. Why is the upper limit of recombination 50% rather than 100%?
7. For two genes that show independent assortment, what is the frequency of recombination?
8. How can recombination frequency be used to build a genetic map?

16.4 INHERITANCE OF GENES IN THE Y CHROMOSOME

Like the X chromosome, the Y chromosome exhibits a particular pattern of inheritance because of its association with the male sex. In humans and other mammals, all embryos initially develop immature internal sexual structures of both females and males. *SRY*, a gene in the Y chromosome, encodes a protein that is the trigger for male development. ("*SRY*" stands for "sex-determining region in the Y chromosome.") In the presence of *SRY*, male structures complete their development and female structures degenerate. In the absence of *SRY*, male embryonic structures degenerate and female structures complete their development. The *SRY* gene is therefore the male-determining gene in humans and other mammals. Because they are linked to *SRY*, most genes in the Y chromosome show a distinctive pattern of inheritance in pedigrees, different from the patterns of autosomal genes Mendel observed in his pea plants, in that they are transmitted only from father to son.

Y-linked genes are transmitted from father to son.

Genes that are present in the unique region of the Y chromosome (the part that cannot cross over with the X) are known as **Y-linked genes**. In addition to the *SRY* male-determining gene, the small number of Y-linked genes include several in which mutations are associated with impaired fertility and low sperm count.

The pedigree characteristics of Y-linked inheritance are striking (**Fig. 16.13**):

1. Only males are affected with the trait.
2. Females never inherit or transmit the trait, regardless of how many affected male relatives they have.
3. All sons of affected males are also affected.

Because the Y chromosome is always transmitted from father to son (and never transmitted to daughters), a trait determined by a Y-linked gene will occur in fathers, sons, grandsons, and so forth. Traits resulting from Y-linked genes cannot be present in females nor can they be transmitted by females. However, other than maleness itself and some types of impaired fertility, no physical traits are known that follow a

FIG. 16.13 Inheritance of Y-linked traits. The pedigree pattern is that of father to son to grandson to great-grandson, and so forth.

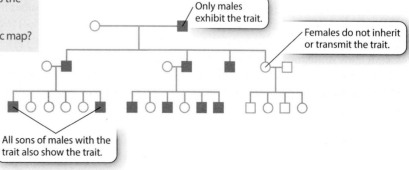

strict Y-linked pattern of inheritance. This observation emphasizes the extremely low density of functional genes in the Y chromosome.

 CASE 3 YOUR PERSONAL GENOME: YOU, FROM A TO T

How can the Y chromosome be used to trace ancestry?

Your personal genome not only can tell you about your genetic risk factors for disease, but it also contains important information about your genetic ancestry. For example, your male ancestors can be traced through your Y chromosome. The regions at the tips in which the X and Y chromosomes share homology is only about 6% of the entire length of the Y chromosome. This means that 94% of the Y chromosome consists of sequences that are completely linked with one another because that portion of the chromosome does not pair with another chromosome and does not undergo crossing over.

Because of this complete linkage, each hereditary lineage of Y chromosomes is separate from every other lineage. As mutations occur along the Y chromosome, they are completely linked to any past mutations that may be present and also completely linked to any future mutations that may take place. The mutations therefore accumulate, allowing the evolutionary history of a set of sequences to be reconstructed.

Fig. 16.14 shows an evolutionary tree based on the accumulation of mutations at a set of nucleotide sites along the Y chromosome. Each unique combination of nucleotides constitutes a Y-chromosome **haplotype**, or haploid genotype. In the figure, the most ancient Y chromosomes are at the top, and the accumulation of new mutations as the generations proceed results in the successive creation of new haplotypes. Each Y-chromosome lineage may leave some nonmutant descendants as well as some mutant descendants, and hence any or all of the sequences shown may coexist in a present-day population.

In human history, the mutations creating new Y-chromosome haplotypes were occurring at the same time as populations were migrating and founding new settlements across the globe, and so each geographically distinct population came to have a somewhat different set of Y-chromosome haplotypes. The differences among populations are offset to some extent by migration among populations, which mixes the geographical locations of various haplotypes. Nevertheless, the fact that the mutations accumulate through time and are completely linked allows the evolutionary history of the haplotypes to be reconstructed. It also enables Y chromosomes to be traced to their likely ethnic origin.

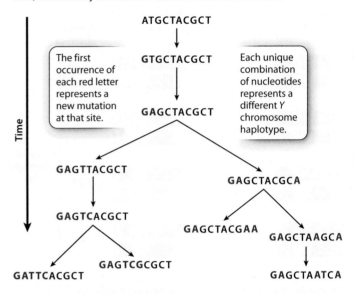

FIG. 16.14 Y-chromosome haplotypes. Because sequences in the Y chromosome are completely linked and mutations accumulate over time, the ancestry of a male's Y chromosome can be traced.

The worldwide distribution of real Y-chromosome lineages among human populations is shown in **Fig. 16.15**. The different colors in the pie charts represent different haplotypes. Neighboring populations tend to have more closely related Y chromosomes than more distant populations, as seen in the similarly colored sectors of neighboring pie charts. Four major clusters of Y-chromosome lineages can be recognized in Fig. 16.15. One is concentrated in Africa, another in Southeast Asia and Australia, a third in Europe and central and western Asia, and the fourth in North and South America. These clusters correspond roughly with the spread of human settlements around the globe inferred from archaeological evidence.

The implication of Fig. 16.15 is that the haplotype of your Y chromosome (if you have a Y chromosome) contains genetic information about its origin. And you can learn what this information is from genetic testing companies that sell direct-to-consumer (DTC) services. Their tests are not regarded as medical devices and so are unregulated, and quality control is sometimes uncertain. Nevertheless, you can send saliva or other biological samples to a DTC provider, which will (for a fee) test your Y chromosome and send you a report that details its possible origin. Of course, because of recombination and independent assortment of genes in other chromosomes, the ethnic origin of your Y chromosome may have little or nothing to do with the ethnic origins of genes in any of your other chromosomes.

FIG. 16.15 Geographical distribution of Y-chromosome haplotypes among native populations. The evolutionary trees of Y-chromosome haplotypes reflect the origin and movement of different Y chromosomes over time. *Data from M. A. Jobling and C. Tyler-Smith, 2003, "The Human Y Chromosome: An Evolutionary Marker Comes of Age," Nature Reviews Genetics 4:598–612.*

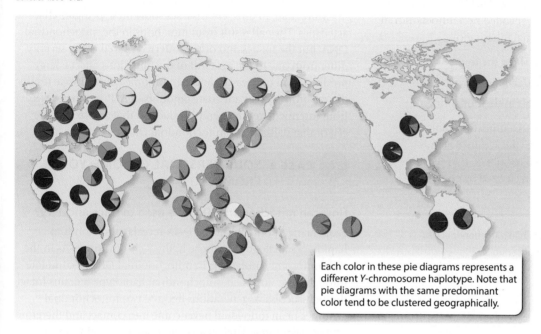

Each color in these pie diagrams represents a different Y-chromosome haplotype. Note that pie diagrams with the same predominant color tend to be clustered geographically.

Self-Assessment Questions

9. What is the pattern of inheritance expected from a Y-linked gene in a human pedigree?
10. How can Y-chromosome data can be used to trace ancestry?

16.5 INHERITANCE OF MITOCHONDRIAL AND CHLOROPLAST DNA

Sex chromosomes and linked genes are not the only genes that are inherited in ways that are unlike the patterns of inheritance that Mendel observed in peas. Genes in mitochondria and chloroplasts also show distinct inheritance patterns. Mitochondria and chloroplasts are organelles of eukaryotic cells originally acquired by the engulfing of prokaryotic cells (Chapter 5). Mitochondria generate ATP that cells use for their chemical energy. Chloroplasts are found only in plant cells and eukaryotic algae. These organelles carry out photosynthesis, producing sugars that are essential for growth (Chapter 8). Mitochondria and chloroplasts have their own genomes that contain genes for many of the enzymes that carry out the organelles' functions. Genes present in these genomes move with the organelle during cell division, independent of the segregation of chromosomes in the nucleus.

Mitochondrial and chloroplast genomes often show uniparental inheritance.

During sexual reproduction, organelles do not show the regular, highly choreographed movements that chromosomes undergo during cell division. Organelles are partitioned to the gametes along with other cytoplasmic components, and therefore their mode of inheritance in different species depends on how the gametes are formed, how much cytoplasm is included in the gametes, and the fate of the cytoplasm in each parental gamete after fertilization.

Considering the great diversity in the details of reproduction in different groups of organisms, it is not surprising that there is a diversity of types of inheritance of cytoplasmic organelles. The three most important types are:

- **Maternal inheritance**, in which the organelles in the offspring cells derive from those in the mother.
- **Paternal inheritance**, in which the organelles in the offspring cells derive from those in the father.
- **Biparental inheritance**, in which the organelles in the offspring cells derive from those in both parents.

In most organisms, either maternal inheritance or paternal inheritance predominates, but sometimes there is variation from one offspring to the next. For example, transmission of chloroplasts ranges from strictly paternal in the giant redwood *Sequoia*, to strictly maternal in the sunflower *Helianthus*, to either maternal or paternal (or less frequently biparental) in the fern *Scolopendrium*, to mostly maternal but sometimes paternal or biparental in the snapdragon *Antirrhinum*.

There is likewise great diversity among organisms in the inheritance of mitochondrial DNA. Most animals show maternal transmission of mitochondria, as would be expected from their large, cytoplasm-rich eggs and the small, cytoplasm-poor sperm. Among other organisms, there is again much variation, including maternal transmission of mitochondria in flowering plants and paternal transmission in the green alga *Chlamydomonas*.

Maternal inheritance is characteristic of mitochondrial diseases.

In humans and other mammals, mitochondria normally show strictly maternal inheritance—the mitochondria in the offspring cells derive from those in the mother. **Fig. 16.16** shows the characteristic pedigree patterns of a trait encoded by a mitochondrial genome transmitted through maternal inheritance:

1. Both males and females can show the trait.
2. All offspring from an affected female show the trait.
3. Males do not transmit the trait to their offspring.

The pedigree in Fig. 16.16 follows a mitochondrial disease known as MERRF syndrome (the acronym stands for "myoclonic epilepsy with ragged red fibers"). As its name suggests, the syndrome is characterized by epilepsy, a neurological disease characterized by seizures, as well as by the accumulation of abnormal mitochondria in muscle fibers. This extremely rare disease is associated with a single point mutation in a mitochondrial gene involved in protein synthesis that affects oxidative phosphorylation (Chapter 7).

More than 40 different diseases show these pedigree characteristics. They all result from mutations in the mitochondrial DNA, but the tissues and organs affected as well as the severity differ from one to the next. All of the mutations affect energy production in one way or another. In some cases, disease results directly from lack of adequate amounts of ATP, in other cases, from intermediates in energy production that are toxic to the cell or that damage the mitochondrial DNA.

CASE 3 YOUR PERSONAL GENOME: YOU, FROM A TO T

How can mitochondrial DNA be used to trace ancestry?

Sequencing of mitochondrial DNA reveals mitochondrial haplotypes analogous to those in the Y chromosome. As in the Y chromosome, mutations in mitochondrial DNA accumulate through time, and each mitochondrial haplotype remains intact through successive generations because two mitochondrial genomes in an individual never come into contact and therefore cannot recombine. This means that mitochondrial DNA can be used to trace ancestry and population history, much as described previously for the Y chromosome. The mitochondrial DNA is actually more informative than Y-chromosomal DNA: it shows substantially more genetic variation, so its ancestry can be tracked on a finer scale.

The use of mitochondrial DNA to trace human origins and migration is discussed in greater detail in Chapter 23. As with the Y chromosome, your personal mitochondrial genome includes information about its origin. If you want to learn about your mitochondrial DNA, you can send a tissue sample to any of a number of direct-to-consumer genotyping services that will, for a fee, analyze the DNA and provide you with a report.

FIG. 16.16 Maternal inheritance. Human mitochondrial DNA is transmitted from a mother to all of her offspring. *Source: Courtesy of Dr. Kurenai Tanji, Columbia University Medical Center, New York, NY.*

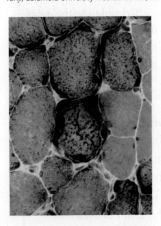

Inherited mitochondrial diseases are often associated with muscle weakness reflecting deficient production of ATP. The red patches in the microscopic image result from clumps of defective mitochondria in muscle fibers observed in one form of epilepsy that results from a mutation in mitochondrial DNA.

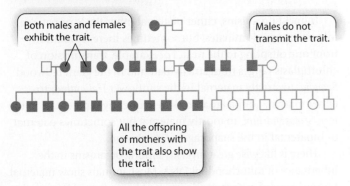

Both males and females exhibit the trait.

Males do not transmit the trait.

All the offspring of mothers with the trait also show the trait.

Self-Assessment Questions

11. A man affected with a mitochondrial disease mates with a woman who does not have a mitochondrial disease. What is the probability that their offspring will be affected?
12. How can mitochondrial DNA data be used to trace ancestry?

CORE CONCEPTS SUMMARY

16.1 THE X AND Y CHROMOSOMES: Many organisms have a distinctive pair of chromosomes that differ between the sexes and show different patterns of inheritance from other chromosomes.

In humans and other mammals, *XX* individuals are female and *XY* individuals are male. page 344

The human *X* and *Y* chromosomes are different lengths and contain different genes, except for small regions of homology that allow the two chromosomes to pair in meiosis. page 344

Segregation of the *X* and *Y* chromosomes during male meiosis results in half of the sperm receiving an *X* chromosome and half a *Y* chromosome so that random union of gametes predicts a 1:1 female:male sex ratio at the time of fertilization. page 344

16.2 INHERITANCE OF GENES IN THE X CHROMOSOME: *X*-linked genes show a crisscross inheritance pattern.

Thomas Hunt Morgan studied a mutation in the fruit fly *Drosophila melanogaster* that resulted in fruit flies with white eyes rather than wild-type red eyes. In this species, as in mammals, females are *XX* and males are *XY*. page 345

In a cross of a normal red-eyed female with a mutant white-eyed male, all of the male and female progeny had red eyes. When brothers and sisters of this cross were mated with each other, all of the females had red eyes, but males were red-eyed and white-eyed in a 1:1 ratio. page 345

This pattern of inheritance is observed because the gene Morgan studied is located in the *X* chromosome. The nonmutant w^+ allele is dominant to the mutant w^- allele, and the gene is present only in the *X* chromosome and not in the *Y* chromosome. page 346

X-linked genes show a crisscross inheritance pattern, in which the *X* chromosome with the mutant gene that is present in males in one generation is present in females in the next generation. page 347

Calvin Bridges observed rare fruit flies that did not follow the usual pattern for *X*-linked inheritance and inferred that these exceptional fruit flies resulted from nondisjunction, or failure of homologous chromosomes to segregate, in male or female meiosis. His observations provided evidence that genes are carried in chromosomes. page 348

In humans, *X*-linked inheritance shows a pattern in which affected individuals are almost always males, affected males have unaffected sons, and a female whose father is affected can have affected sons. page 349

16.3 GENETIC LINKAGE: Genes that are close together in the same chromosome tend to remain together in inheritance.

Genes that are close together in the same chromosome are linked and do not undergo independent assortment. page 350

Recombinant chromosomes result from crossing over between genes on the same chromosome and show a nonparental combination of alleles. page 351

Nonrecombinant chromosomes have the same configuration of alleles as one of the parental chromosomes. page 351

In genetic mapping, the observed proportion of recombinant chromosomes is the frequency of recombination and can be used as a measure of distance along a chromosome. A recombination frequency of 1% is 1 map unit. page 351

Gene linkage and mapping are used to identify the locations of disease genes in the human genome. page 353

16.4 INHERITANCE OF GENES IN THE Y CHROMOSOME: Most *Y*-linked genes are passed from father to son.

In humans and other mammals, the *Y* chromosome contains a gene called *SRY* that results in male development. page 355

In *Y*-linked inheritance, only males are affected and all sons of an affected male are affected. Females are never affected and do not transmit the trait. page 355

Most *Y*-linked genes show complete linkage, which allows their evolutionary history to be traced. page 356

16.5 INHERITANCE OF MITOCHONDRIAL AND CHLOROPLAST DNA: Mitochondria and chloroplast DNA follow their own inheritance pattern.

Mitochondria and chloroplasts have their own genomes, which reflect their evolutionary history as free-living prokaryotes. page 357

Mitochondria in humans and other mammals show maternal inheritance, in which individuals inherit their mitochondrial DNA from their mother. page 358

Because mitochondrial DNA does not undergo recombination and is maternally inherited, it can be used to trace human ancestry and migration. page 358

Log in to **LaunchPad** to check your answers to the Self-Assessment Questions and to access additional learning tools.

CHAPTER 17
The Genetic and Environmental Basis of Complex Traits

CORE CONCEPTS

17.1 HEREDITY AND ENVIRONMENT: Complex traits are those influenced both by the action of many genes and by environmental factors.

17.2 RESEMBLANCE AMONG RELATIVES: Genetic effects on complex traits are reflected in resemblance among relatives.

17.3 TWIN STUDIES: Twin studies help separate the effects of genotype and environment on variation in a trait.

17.4 COMPLEX TRAITS IN HEALTH AND DISEASE: Many common diseases and birth defects are affected by multiple genetic and environmental risk factors.

Biologists initially had a hard time accepting the principles of Mendelian inheritance because they seem so at odds with everyday observations. Common and easily observed traits such as height, weight, hair color, and skin color give no clear-cut phenotypic ratios such as 3:1 or 9:3:3:1 observed for simple Mendelian traits. The lack of these characteristic ratios raised serious doubt whether Mendel's principles are valid for common traits. Some biologists concluded that they apply only to seemingly trivial traits such as round and wrinkled seeds in peas.

At about the time that Mendel was studying inheritance in garden peas, the biologist Francis Galton, a friend and cousin of Charles Darwin, was studying common traits including human height. From studies of height and other common traits in parents and their offspring, Galton discovered general principles in the inheritance of such traits. For example, parents who are tall tend to have offspring who are taller than average but not as tall as themselves. Galton's principles for the inheritance of common traits did not refer to genes, segregation, independent assortment, or other features of Mendelian inheritance, but they did describe Galton's observations. Genes—what Mendel called "hereditary factors"—were unnecessary in Galton's theory of inheritance, and many biologists thought that Galton's observations and Mendel's laws were incompatible.

This, it turned out, was not the case. The traits that Mendel studied are now called **single-gene traits** because each one is determined by variation at a single gene and the traits for the most part are not influenced by the environment. By focusing on single-gene traits, Mendel was able to infer underlying mechanisms of inheritance based on physical factors we now call genes. By contrast, **complex traits,** such as human height, are influenced by multiple genes as well as by the environment. As a result, their inheritance patterns are more difficult to follow and simple phenotypic ratios are not observed.

In many ways, complex traits are more important than single-gene Mendelian traits. One reason is their prevalence—complex traits are found in all organisms and include most of the traits we can see around us. By contrast, there are relatively few examples of common single-gene traits. Complex traits are also important in human health and disease. In the most common disorders—among them heart disease, diabetes, autism, and schizophrenia—single-gene Mendelian inheritance is seldom found. It is therefore important to understand the inheritance of complex traits and common disorders, which is the subject of this chapter. We begin by describing some of the features of complex traits, and then show how these principles are not only compatible with, but are in fact predicted by, Mendelian inheritance. Finally, we describe how modern molecular genetics and genomics have allowed the identification of genes affecting complex traits.

17.1 HEREDITY AND ENVIRONMENT

Complex traits are important not only in humans, but in all organisms—including agricultural plants and animals. We will examine human height in some detail because it has been widely studied, but equally well-known complex traits are number of eggs laid by hens, milk production in dairy cows, and yield per acre of grain (**Fig. 17.1**).

FIG. 17.1 Examples of complex traits. (a) Human height, (b) egg number, (c) milk production, and (d) grain yield. *Sources: a. Randy Faris/Corbis/VCG/Getty Images; b. muratart/Shutterstock; c. smereka/Shutterstock; d. Radius Images/Getty Images.*

Many common human diseases, including high blood pressure, obesity, diabetes, and depression, are complex traits (**Fig. 17.2**). High blood pressure, for example, affects about one-third of the U.S. population, and obesity another third. Type 2 diabetes affects around 8% of the U.S. population, and an estimated 15% will suffer at least one episode of severe depression in the course of a lifetime. Taken together, about 200 million Americans—two-thirds of the entire population—suffer from one or more of these common disorders. None of these traits shows single-gene Mendelian inheritance.

In many complex traits, the phenotype of an individual is determined by measurement: human height is measured in inches, milk yield by the gallon, grain yield by the bushel, egg production by the number of eggs, blood pressure by millimeters of mercury, and blood sugar by millimoles per liter. Because the phenotype of complex traits such as these is measured along

FIG. 17.2 Examples of human diseases that are complex traits. (a) High blood pressure, (b) obesity, (c) diabetes, and (d) depression. *Sources: a. burwellphotography/iStockphoto; b. George Doyle/Getty Images; c. Junophoto/Getty Images; d. Imago/ZUMApress.com.*

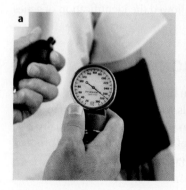
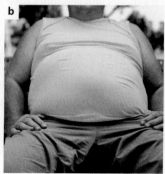
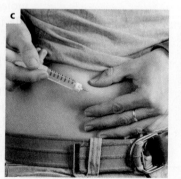

FIG. 17.3 Phenotypic variation due to variation in the environment. Genetically identical corn plants vary in height because of variation in exposure to sun and in soil composition at different locations in the same field. *Source: Bruce Leighty/Getty Images.*

a continuum with only small intervals between similar individuals, complex traits such as these are often called **quantitative traits**. By contrast, single-gene traits often appear in one of two or more different phenotypes, such as round versus wrinkled seeds, or green versus yellow seeds.

Complex traits are affected by the environment.

Expression of complex traits is notoriously susceptible to lifestyle choices and other environmental factors. Inadequate nutrition is linked to slow growth rate and short stature in adults. Salt intake is associated with an increased likelihood of high blood pressure and is therefore an **environmental risk factor** for this common disorder. An environmental risk factor is a characteristic in a person's surroundings that increases the likelihood of developing a particular disease. For example, a junk-food diet high in fat and carbohydrates is an environmental risk factor for obesity and diabetes.

Environmental effects are also important in agriculture. Farmers are well aware that adequate nutrition is essential to normal growth of chicks, lambs, piglets, and calves, and that continued high-quality feed is necessary for high egg production and milk yield when the animals reach adulthood. In crop plants such as grains, adequate soil moisture and nutrients are necessary for sustained high yields.

Environmental factors not only affect the average phenotype for complex traits, but they also affect the variation in phenotype from one individual to the next. For example, most fields of corn you see planted along the roadside come from seeds that are genetically identical to one another, yet there is phenotypic variation in complex traits like plant height (**Fig. 17.3**). Because the plants are genetically identical, the differences in phenotype result from differences in the environment. Some parts of the field may receive more sunlight than others, and some parts may have better water drainage. No matter how uniform an environment may seem, there are always minor differences from one area to the next, and these differences can result in variation in complex traits. In the example in Fig. 17.3, the plants in the foreground are shorter than the others because that corner of the field has poorer drainage, and the plants growing there have to compete for nutrients with the grass growing nearby.

Environmental effects on complex traits in animals can be observed in true-breeding, homozygous strains such as those used by Mendel in his experiments with pea plants. Such true-breeding strains are called **inbred lines**, and they are often used for research. Even though all animals in any inbred line are genetically identical and are caged in the same facility and fed the same food, there is variation from one animal to the next. In one study of cholesterol levels in an inbred line of mice, for example, average serum cholesterol was 120 mg/dl (milligrams per deciliter), but the range was 60–180 mg/dl. Because the mice are genetically identical, the variation in serum cholesterol resulted entirely from variation in the environment.

Complex traits are affected by multiple genes.

Most complex traits are affected by many genes, in contrast to Mendel's traits, which are primarily affected by only one. Therefore, in complex traits the familiar phenotypic ratios such as 3 : 1 that Mendel saw when he crossed two true-breeding strains with contrasting characters are not observed. For complex traits, the effects of individual genes are obscured by variation in phenotype that is due to multiple genes affecting the trait and also due to the environment. The number of genes affecting complex traits is usually so large that different genotypes can have very similar phenotypes, which also makes it difficult to see the effects of individual genes on a trait.

In a few traits, however, the effects of the environment are minor and the number of genes is small, and in these cases, the genetic basis of the trait can be analyzed. A classic example, studied by Herman Nilsson-Ehle about a century ago, concerns the color of seed casing in wheat (which give the seeds their color), which ranges from nearly white to dark red (**Fig. 17.4**). His experiment demonstrated that complex traits are subject to the same laws that Mendel worked out for single-gene traits, but that the inheritance patterns are more difficult to see because of the number of genes involved.

In studying seed color in true-breeding varieties and their first-generation and second-generation hybrids, Nilsson-Ehle realized that the relative frequencies of different shades of red color could be explained by the effects of three genes that undergo independent assortment, as illustrated by the Punnett square shown in Fig. 17.4. Each of the genes has two alleles, designated by combinations of upper-case and lower-case letters. In the development of seed color, each upper-case allele in a genotype intensifies the red coloration in an additive fashion. Altogether there are seven possible phenotypes, ranging from a phenotype of 0 (nearly colorless, genotype *aa bb cc*) to 6 (dark red, genotype *AA BB CC*).

Nilsson-Ehle first crossed true-breeding dark red plants (*AA BB CC*) with true-breeding colorless plants (*aa bb cc*) to obtain the hybrids with genotype *Aa Bb Cc*. He then crossed the hybrids together, which is the cross shown at the top of Fig. 17.4. Because the three genes show independent assortment (Chapter 15), the distribution of phenotypes expected in progeny from the cross *Aa Bb Cc* × *Aa Bb Cc* is as shown in the Punnett square. The bar graph below the Punnett square shows the seed-color phenotypes and their relative proportions, from which Nilsson-Ehle inferred three genes with independent assortment. The distribution of seed-color phenotypes is approximated by a bell-shaped curve known as a **normal distribution**. The phenotypes of many complex traits, including human height, conform to the normal distribution. Nilsson-Ehle's seed-color case is exceptional in that virtually all complex traits are affected by many more genes than three (in the case of human height, hundreds, as discussed below).

When differences in phenotype due to the environment can be ignored, the genetic variation affecting complex traits can be detected more easily. And when studying inbred lines, differences in phenotype due to genotype can be ignored because all individuals have the same genotype, and the effects of environment can be observed. In most cases, however, both genetic variation and environmental variation among individuals are present, and it is difficult to quantify how much variation in phenotype is due to genes and how much is due to environment.

Keep in mind that complex traits are not really more "complex" than any other biological trait. The term is used merely to imply that both genetic factors and environmental

FIG. 17.4 Multiple genes contributing to a complex trait. Three unlinked genes, each with two alleles, influence the intensity of red coloration of the seed casing in wheat.

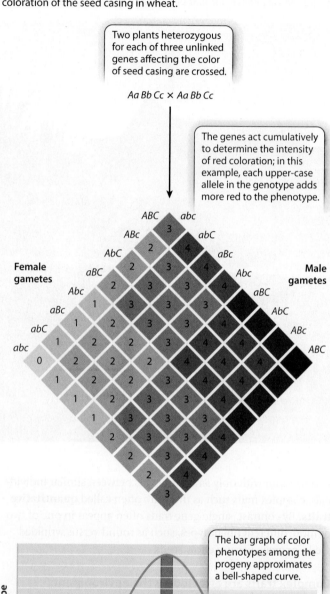

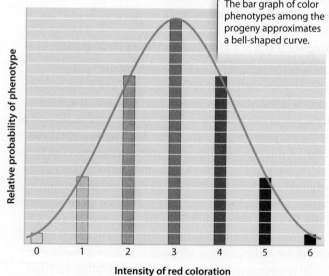

factors contribute to variation in phenotype among individuals. Just as there are environmental factors that affect complex traits such as height, so there are genetic factors that affect height. Similarly, just as an environmental risk factor increases the likelihood of a common disease, so does a genetic risk factor predispose an individual to the condition. For example, the human gene *ApoE* encodes a protein that helps transport fat and cholesterol. Certain alleles of *ApoE* are associated with high levels of cholesterol, and one particular allele is a genetic risk factor for Alzheimer's disease, the most common serious form of age-related loss of cognitive ability. Lifestyle environmental factors such as diet, physical exercise, and mental stimulation are also thought to play a role in the risk of Alzheimer's disease.

The relative importance of genes and environment can be determined by differences among individuals.

For any one individual, it is impossible to specify the relative roles of genes and environment in the expression of a complex trait. For example, in an individual 66 inches tall, it would be meaningless to attribute 33 inches of height to parentage (genes) and 33 inches of height to nutrition (environment). This kind of partitioning makes no sense because genes and environment act together so intimately in each individual that their effects are inseparable. To attempt to separate them for any individual would be like asking to what extent it is breathing or oxygen that keeps us alive. Both are important, and both must take place together if we are to live.

It is nevertheless possible to separate genes and environment in regard to their effects on the *differences,* or variation, among individuals within a particular population. For some traits, the variation seen among individuals is due largely, if not exclusively, to differences in the environment. For other traits, the variation is due mainly to genetic differences. In the case of human height, roughly 60% of the variation among individuals of the same sex is due to genetic differences, and the remaining 40% to differences in their environment, primarily differences in nutrition during their years of growth.

Genetic and environmental effects can interact in unpredictable ways.

One of the features of complex traits is that genetic and environmental effects may interact, often in unpredictable ways. Consider the example shown in **Fig. 17.5**. In this experiment, two genetically different strains of corn were each grown in a series of soils in which the amount of nitrogen had been enriched by the growth of legumes. The yield of strain 1 varies little across these different environments. The yield of strain 2, however, increases dramatically with soil nitrogen. Each of the lines is known as a **norm of reaction,** which for any genotype graphically depicts how the environment (shown on the *x*-axis) affects phenotype (shown on the *y*-axis) across a range of environments. Note that in one case in Fig. 17.5, the environment

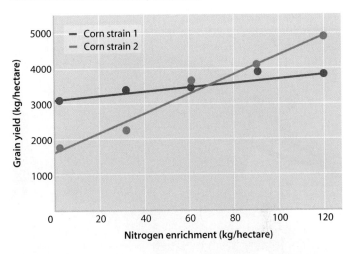

FIG. 17.5 Genotype-by-environment interaction for grain yield in corn. The effect of soil nitrogen on grain yield is minimal in strain 1, but dramatic in strain 2. *Data courtesy of J. W. Dudley.*

has little to no effect on the phenotype (the norm of reaction is nearly flat), but in the other it has a very noticeable effect. In cases such as strain 2, it is impossible to predict the phenotype of a given genotype without knowing what the environmental conditions are and how the phenotype changes in response to variation in the environment.

Such variation in the effects of the environment on different genotypes is known as **genotype-by-environment interaction.** This type of interaction is important because it implies that the effect of a genotype cannot be specified without knowing the environment, and the other way around. A further implication is that there may be no genotype that is the "best" across a broad range of environments, and likewise no environment that is "best" for all genotypes. For complex traits, phenotype depends on both genotype *and* environment. For the strains of corn in Fig. 17.5, for example, which strain is "better" depends on the soil nitrogen. With little nitrogen enrichment, strain 1 is better; at intermediate values, both strains yield about the same; and with high nitrogen enrichment, strain 2 is better. In this case, knowing the norms of reaction and recognizing the magnitude and direction of genotype-by-environment interaction, allows each farmer to use the strains likely to perform best under the available conditions of cultivation.

Genotype-by-environment interaction makes the interplay between genes and the environment difficult to predict. An example of genotype-by-environment interaction affecting obesity is shown in **Fig. 17.6**. The animals shown are adults of two inbred lines of mice. When fed a normal diet, both inbred lines grow to about the same size. When fed a high-fat diet, however, the inbred line A becomes obese, but inbred line B does not. Hence, it is not genotype alone that causes obesity in line A because with a normal diet, line A does not become

FIG. 17.6 Genotype-by-environment interaction for obesity in mice. When fed a normal diet, both inbred lines grow to about the same size. When fed a high-fat diet, mouse A becomes obese but mouse B does not. *Source: Phil Jones/Georgia Regents University.*

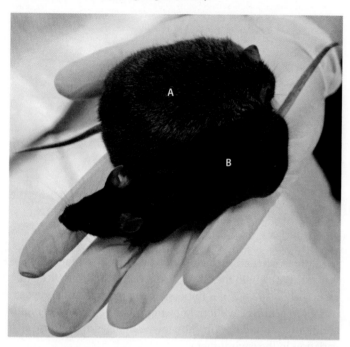

obese. Nor is it a high-fat diet alone that causes obesity, because the high-fat diet does not cause obesity in line B. Rather, the obesity in line A results from a genotype-by-environment interaction (which in this case is a genotype-by-diet interaction).

Self-Assessment Questions

1. Why are some complex traits also called quantitative traits?
2. What is an example of a complex trait?
3. What are some factors that influence variation in complex traits?
4. Why does it not make sense to try to separate the effects of genes ("nature") and the environment ("nurture") in a single individual, whereas it does make sense to separate genetic and environmental effects on differences among individuals?
5. Why can the effect of a genotype on a phenotype not always be determined without knowing what the environment is? Why can the effect of a particular environment on a phenotype not always be determined without knowing what the genotype is?

17.2 RESEMBLANCE AMONG RELATIVES

Mendel had many advantages over Galton in his studies of inheritance. Mendel's peas were true breeding, produced a new generation each year, and yielded large numbers of progeny. The environment had a negligible effect on the traits, and alleles of a single gene were responsible for variation of each trait. Mendel's crosses yielded simple ratios such as 3 : 1 or 1 : 1, which could be interpreted in terms of segregation of dominant and recessive alleles.

By contrast, Galton studied variation in such traits as height in humans. Humans are obviously not true breeding, have a long generation time, and produce few offspring. Most importantly, for complex traits such as height, both genetics and environment affect the phenotype. Two individuals with the same genotype may have a different phenotype because of differences in the environment, and two individuals in exactly the same environment may differ in phenotype because of differences in genotype. In one respect Galton had an advantage, though: whereas most simple Mendelian traits are relatively uncommon, Galton's traits are readily observed in everyday life. What did Galton discover?

For complex traits, offspring resemble parents but show regression toward the mean.

Galton's observations are as important in understanding complex traits as Mendel's are in understanding single-gene traits. He studied many complex traits, including human height, strength, and various other physical characteristics, such as the number of fingerprint ridges. The discovery he regarded as fundamental resulted from his data on adult height of parents and their progeny (**Fig. 17.7**). Galton noted that each category of parent (tall or short) produced a range of progeny forming a distribution with its own **mean** (the average value).

The bar graph in Fig. 17.7a shows the distribution of height among the progeny of the tallest parents in the population Galton studied, whose mean height is 72 inches. The mean height of the offspring is 71 inches, which is greater than the mean height of the whole population of the study (68.25 inches) but less than that of the parents. The bar graph in Fig. 17.7b is the distribution of height of progeny of the shortest parents, whose mean equals 66 inches. In this case, the mean height of the progeny is 67 inches, which is less than the mean height of the population but greater than that of the parents. Note, however, that *some* of the offspring of tall parents are taller than their parents, and likewise *some* of the offspring of short parents are shorter than their parents. It is only *on average* that the height of the offspring is less extreme than that of the parents.

Galton regarded this observation as his most important discovery and reported it in 1886. Today, we call it **regression toward the mean**. The offspring exhibit a mean phenotype that is closer to the population mean than the phenotype of the parents. In other words, when the mean height of the parents is *smaller* than the population mean, then the mean height of the offspring is *greater* than that of the parents (but smaller than the population mean). Likewise, when the mean height of the parents is *greater* than the population mean, then the mean

FIG. 17.7 Galton's data showing distribution of height of offspring of (a) the tallest parents and (b) the shortest parents. *Data from F. Galton, 1888, "Co-Relations and Their Measurement, Chiefly from Anthropometric Data," Proceedings of the Royal Society, London 45:135–145, reprinted in K. Pearson, 1920, "Notes on the History of Correlation," Biometrika 13:25–45.*

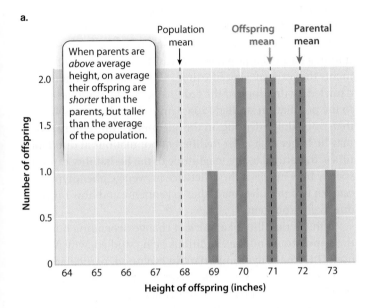

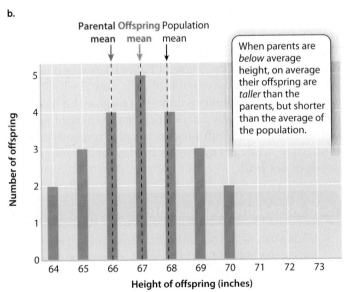

height of the offspring is *smaller* than that of the parents (but greater than the population mean).

Regression toward the mean is observed for two reasons. The first is that during meiotic cell division (Chapter 11), segregation and recombination break up combinations of genes that result in extreme phenotypes, such as very tall or very short, that are present in the parents. The second reason is that the phenotype of the parents results not only from genes but also from the environment. Environmental effects are not inherited, so any effect of the environment on the parents' phenotypes is not transmitted to the offspring. For example, if the parents are tall or short purely because of an environmental effect such as better or worse nutrition, then the mean height of the offspring will be equal to the population mean.

Heritability is the proportion of the total variation due to genetic differences among individuals.

The data points in red in **Fig. 17.8** show that regression toward the mean occurs across the full range of heights and not only at the extremes as in Fig. 17.7. For convenience in visualization, parents with similar heights have been grouped and averaged (*x*-axis), and for each group of parents the heights of their offspring have also been averaged (*y*-axis). The line in red is the *regression line* that best fits the data.

Let's contrast the regression line in Fig. 17.8 with two hypothetical extremes. One extreme is shown by the blue line with a slope of 1.0. This is the regression line expected when all the variation in height results from genetic differences among individuals and when height is determined by a large number of genes, each with small effect and with no dominance or gene interaction. (The genetic model is exactly like that shown in Fig. 17.4 but with many more genes.) Under these assumptions, the mean height of the offspring will equal the mean height of the parents, and any deviations from the blue line will be due entirely to chance.

The black line in Fig. 17.8 with a slope of 0.0 represents the opposite extreme when all variation in height results

FIG. 17.8 Regression toward the mean. Galton's data on the average adult height of parents and that of their offspring show that offspring mean height (red line) falls between the parental mean (blue line) and the population mean (black line).

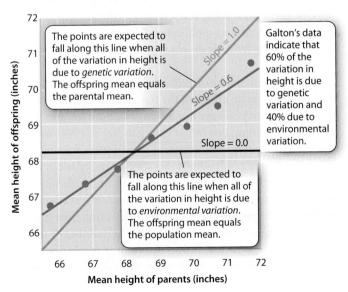

from environmental differences among individuals. As long as environmental effects are not transmitted from one generation to the next, the mean phenotype of the offspring will be equal to the mean of the population as a whole, irrespective of the phenotypes of the parents. Although environmental effects are usually not transmitted in plant and animal populations, human populations are exceptional in showing **cultural transmission** of some environmental effects. For example, the average wealth of the offspring of rich parents is greater than that of the offspring of poor parents, and this difference is obviously due to transmission of the parents' money, not their genes.

How much of the variation in height among individuals is due to genetic differences, and how much is due to environmental differences? The answer to this question is found in the slope of the regression line, which in Galton's data equals 0.60. Because a slope of 0.0 implies variation due only to the environment and a slope of 1.0 implies variation due only to genetics, a slope of 0.6 implies that 60 percent of the variation in height is due to genetic differences among individuals and 40 percent due to environmental differences among individuals.

The proportion of the total variation in the trait that is due to genetic differences among individuals is called the **heritability** of the trait. Thus, the slope of the regression line of progeny mean against parental mean is an estimate of the heritability of adult height (0.6 in Galton's data). For a complex trait, the heritability determines how closely the mean of the progeny resembles that of the parents. The similarity between first cousins and other types of relatives also depends on the heritability.

The term "heritability" is often misinterpreted. The problem is that, in nonscientific contexts, the word means "the capability of being inherited or being passed by inheritance." This definition suggests that heritability has something to do with the inheritance of a trait. But "heritability" as used for complex traits means no such thing. It refers only to the *variation* in a trait among individuals, and specifically to the proportion of the variation among individuals in a population due to differences in genotype.

Hence, a heritability of 100% does not imply that the environment has no effect on the trait. As emphasized earlier, the environment is always important—without oxygen, you wouldn't be reading this. What a heritability of 100% means is that *variation* in the environment does not contribute to differences among individuals in a specific population. For example, if genetically different strains of chrysanthemums are grown in a greenhouse and subject to identical environmental conditions, then differences in flowering time have to be due to genetic differences, and the heritability of the trait would be 100%. Similarly, a heritability of 0% does not imply that genotype has no effect on the trait. A heritability of 0% merely means that differences in genotype do not contribute to the *variation* in the trait among individuals in a specific population. If genetically identical strains of chrysanthemums are grown in different environments, then differences in flowering time must be due to the environment, and heritability would be 0%.

Heritability is not an intrinsic property of a trait. For chrysanthemum flowering time, the heritability in one case was 100% and in the second 0%. Heritability applies only to the trait in a particular population across the range of environments that exist at a specific time. Similarly, the heritability depicted by the slope of the red line in Fig. 17.8 applies only to the population studied (205 pairs of British parents and their 930 adult offspring, in the late nineteenth century) and may be larger or smaller in different populations at different times. In particular, the magnitude of the heritability cannot specify how much of the difference in average phenotype *between* two populations is due to genotype and how much due to environment.

If the heritability of a trait can change depending on the population and the conditions being studied, why is it useful? Heritability is important in evolution, particularly in studies of artificial selection, a type of selective breeding in which only certain chosen individuals are allowed to reproduce (Chapter 20). Practiced over many generations, artificial selection can result in considerable changes in morphology or behavior or almost any trait that is selected. The large differences among breeds of pigeons and other domesticated animals prompted Charles Darwin to point to artificial selection as an example of what natural selection could achieve. Heritability is important because this quantity determines how rapidly a population can be changed by artificial selection. A trait with a high heritability responds rapidly to selection, whereas a trait with a low heritability responds slowly or not at all.

Self-Assessment Questions

6. Graph a trait, such as human height, with height on the *x*-axis and number of individuals on the *y*-axis, and describe the shape of the resulting graph.

7. How might you go about determining the relative importance of genes and the environment for variation in risk for a complex trait such as type 2 diabetes?

8. What is "regression toward the mean" and why does it occur for complex traits?

9. Does regression toward the mean imply that the human population is getting shorter over time?

10. What is the heritability of a trait and why does the heritability of a given trait depend on the population being studied?

17.3 TWIN STUDIES

Studies of twins are one way to separate the effects of genotype and environment on phenotypic differences among individuals. Depending on whether they arise from one or from two egg cells, twins can be **monozygotic (MZ)** twins or **dizygotic (DZ)** twins. MZ twins arise from a single fertilized egg (the zygote), which, after several rounds of cell division, separates into two distinct but genetically identical embryos. Often strikingly similar in overall appearance (**Fig. 17.9a**), MZ twins are commonly known as identical twins. However, this is a misnomer. MZ twins differ from one another in many ways because of environmental effects. Nevertheless, MZ twins have stimulated the imagination since antiquity, inspiring stories by the Roman playwright Plautus, William Shakespeare (himself the father of twins), Alexander Dumas, Mark Twain, and many others.

In contrast, DZ twins (also called fraternal twins) result when two separate eggs, produced by a double ovulation, are fertilized by two different sperm. Whereas MZ twins are genetically identical, DZ twins are only as closely related as any other pair of siblings (**Fig. 17.9b**).

Twin studies help separate the effects of genes and environment in differences among individuals.

The utility of twins in the study of complex traits derives from the genetic identity of MZ twins. Whereas differences between DZ twins with regard to any trait may arise because of genetic or environmental factors (or genotype-by-environment interactions), differences between MZ twins must arise only from environmental factors because the twins are genetically identical. Consequently, if variation in a trait has an important genetic component, then MZ twins will be more similar to each other than DZ twins are similar to each other. But if the genetic contribution to variation is negligible, then MZ twins will not be more similar to each other than DZ twins are to each other. One complication of twin studies is that half of all DZ twins are of the opposite sex whereas MZ twins are always of the same sex. This issue is easily resolved by studying only same-sex DZ twins. Another caveat of twin studies is that the environment of MZ twins may be more similar than that of DZ twins; this can be remedied in some cases by studying MZ twins separated from each other shortly after birth and raised in different environments.

For complex traits, such as depression or diabetes, the extent to which twins are alike is measured according to the **concordance** of the trait, which is defined as the percentage of cases in which both members of a pair of twins show the trait when it is known that at least one member shows it. The relative importance of genetic and environmental factors in causing differences in phenotype can be estimated by comparing the concordances of MZ and DZ twins. A significantly higher concordance rate for a given trait for MZ twins compared with that for same-sex DZ twins suggests that the trait has a strong genetic component. Similar concordance rates for MZ and same-sex DZ twins suggest that the genetic component is less important.

In the study of concordance, twin pairs in which neither member is affected contribute no information, and these twin pairs must be removed from the analysis. The concordance is based solely on twin pairs in which one or both members express the trait. To take a concrete example, let us symbolize each member of a MZ twin pair as a closed square (■) if the

FIG. 17.9 Twins. (a) Monozygotic twins arise from a single fertilized egg and are genetically identical. (b) Dizygotic twins arise from two different fertilized eggs and are no more closely related than other pairs of siblings. *Sources: a. Craig Gaffield/AP Images; b. Wendy Connett/Getty Images.*

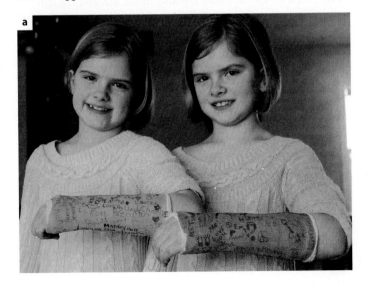

member shows the trait and as an open square (□) if the member does not show the trait. Then, in any sample of MZ twins, there are three possibilities, illustrated with data for type 2 diabetes:

- ■ ■ 20 MZ twin pairs, both members showing type 2 diabetes
- ■ □ 30 MZ twin pairs, only one member showing type 2 diabetes
- □ □ 500 MZ twin pairs, neither member showing type 2 diabetes

In this example, the 500 twin pairs who do not show the trait provide no information. The concordance is based only on the first two types of twin pairs, among which in 20 pairs both members show the trait and in 30 pairs only one shows the trait. In this case the concordance among MZ twins is 20/(20 + 30) = 40%. Among same-sex DZ twins, by contrast, the observed concordance for type 2 diabetes is only 10%. The difference in concordance between MZ and DZ twins (40% for MZ versus 10% for DZ) implies that differences in the risk of type 2 diabetes have an important genetic component. Furthermore, the observation that the concordance for MZ twins is much less than 100% implies that environmental factors unique to each twin in a pair (factors that we now know to include diet and exercise) are also important in the risk of type 2 diabetes. That the MZ concordance is not 100% implies that both genes and environment—"nature" *and* "nurture"—play important roles in the differences in risk.

Figure 17.10 uses the complex trait schizophrenia as an example to illustrate how the MZ and DZ concordance rates are affected by genetic factors, environmental factors common to members of each twin pair, and environmental factors that are unique to each twin in a pair.

Table 17.1 shows concordance rates for other complex traits. Note that for handedness, measles, or death from acute infection the concordance rates for MZ twins are very similar to those for same-sex DZ twins. Equality of the MZ and DZ concordance implies that there is no significant genetic component to variation in the trait, at least insofar as can be determined from twin studies. (Many students are surprised to learn that handedness does not have a detectable genetic component.) For infectious diseases, exposure to the disease agents in the environment is the determinant. With measles, for example, the data imply that when one twin catches measles, the other will very likely catch it, too, and it does not matter whether the twins are MZ or DZ.

TABLE 17.1 Twin Concordance Rates for Several Traits

DISORDER	CONCORDANCE IN MONOZYGOTIC TWINS (%)	CONCORDANCE IN DIZYGOTIC TWINS (%)
High blood pressure	25	7
Asthma	47	24
Rheumatoid arthritis	34	7
Epilepsy	37	10
Handedness (left or right)	79	77
Measles	95	87
Acute infection leading to death	8	9

Data from M. McGue and T. J. Bouchard, Jr., 1998, "Genetic and Environmental Influences on Human Behavioral Differences," Annual Review of Neuroscience 21:1–24.

HOW DO WE KNOW?

FIG. 17.10

What is the relative importance of genes and of the environment for complex traits?

BACKGROUND Twin studies remain important for assessing the relative importance of "nature" (genotype) and "nurture" (environment) in determining variation among individuals for complex traits. The idea of using twins to distinguish nature from nurture is usually attributed to English biologist Francis Galton because of an article he wrote about twins in 1875. In fact, the modern twin study does not trace to Galton but to Curtis Merriman in the United States and Hermann Siemens in Germany, who independently hit upon the idea in 1924. In Galton's time, it was not known that there are genetically two different kinds of twins.

EXPERIMENT The rationale of a twin study is to compare MZ twins with same-sex DZ twins. In principle, MZ twins differ only because of environment, whereas DZ twins differ because of genotype as well as environment. The extent to which DZ twins differ more from each other than MZ twins differ from each other measures the effect of fewer shared genes in the DZ twins. Some twin studies focus on twins reared apart in order to compensate for shared environmental influences that may be stronger for MZ twins than for DZ twins.

Self-Assessment Questions

11. What is twin concordance?
12. How can twin studies be used to investigate the importance of genetic and environmental factors in the expression of a trait?

RESULTS Part (a) in the illustration presents MZ and DZ data for schizophrenia. For simplicity, we have imagined a sample of 1000 MZ twins and 1000 DZ twins who show the same concordance rates as actual studies of tens of thousands of twin pairs. The concordance between twins is the fraction of twin pairs in which both twins show the trait among all those pairs in which at least one twin shows the trait. Part (b) shows the data as a bar graph. Roughly speaking, the difference in the concordance between MZ and DZ twins is a measure of the relative importance of genotype. The concordance for MZ twins is less than 1.0 because environmental factors unique to each twin in an MZ pair affect concordance.

CONCLUSION The example of schizophrenia shown here indicates large differences in the importance of genotype versus environment for this complex trait. This result is typical of the most common human diseases, all of which are complex traits.

FOLLOW-UP WORK Because of complications due to cultural inheritance, twin studies must always be interpreted in light of other studies comparing complex traits among individuals with various degrees of genetic relatedness. On the whole, twin studies of complex traits have yielded results that are consistent with other available evidence.

SOURCES Polderman, T. J. C. et al. 2015. "Meta-analysis of the Heritability of Human Traits Based on Fifty Years of Twin Studies." *Nature Genetics* 47:702–709; McGue, M., and T. J. Bouchard, Jr. 1998. "Genetic and Environmental Influences on Human Behavioral Differences." *Annual Review of Neuroscience* 21:1–24.

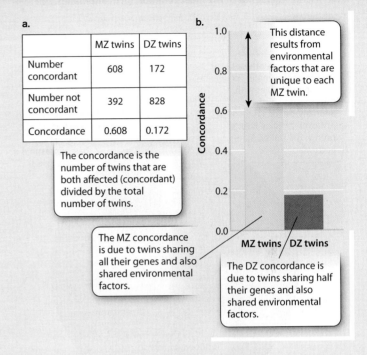

17.4 COMPLEX TRAITS IN HEALTH AND DISEASE

Much evidence beyond twin studies implies an important role for particular alleles as risk factors for diabetes, high blood pressure, asthma, rheumatoid arthritis, epilepsy, schizophrenia, clinical depression, autism, and many other conditions. These are all complex traits, affected by genotype, environment, and genotype-by-environment interactions.

Even the most common birth defects are affected by multiple genetic risk factors. The most common birth anomalies and the numbers of affected babies born in the United States each year—collectively more than 35,000—are indicated in **Fig. 17.11**. About 20% of these anomalies are due to an extra chromosome. Trisomy 21 (Down syndrome) is by far the most common of these chromosome-number abnormalities (Chapter 11). Another roughly 10% are due to defects in metabolism, many of which—among them cystic fibrosis, sickle-cell anemia, and phenylketonuria (the inability to break down the amino acid phenylalanine)—are rare simple Mendelian disorders. Birth anomalies due to extra or missing chromosomes and those due to single-gene Mendelian inheritance are not considered complex traits in terms of their inheritance. Together these account for about 30% of the of birth anomalies in Fig. 17.11. Most (70%) birth anomalies in Fig. 17.11 are inherited as complex traits that are affected by both genetic and environmental risk factors.

Most common diseases and birth defects are affected by many genes, each with a relatively small effect.

Because the most common birth anomalies as well as childhood and adult disorders are complex traits, biologists are keenly interested in identifying the genes that contribute to differences in risk, understanding what these genes do, and translating this knowledge into prevention or treatment. They have therefore begun to apply modern molecular methods—genome

372 SECTION 17.4 COMPLEX TRAITS IN HEALTH AND DISEASE

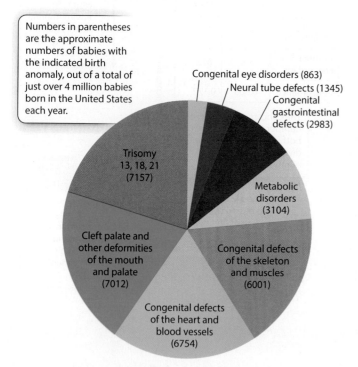

FIG. 17.11 Incidence of the most common birth anomalies. *Data from Centers for Disease Control and Prevention; National Newborn Screening and Genetics Resource Center; March of Dimes.*

Numbers in parentheses are the approximate numbers of babies with the indicated birth anomaly, out of a total of just over 4 million babies born in the United States each year.

- Congenital eye disorders (863)
- Neural tube defects (1345)
- Congenital gastrointestinal defects (2983)
- Metabolic disorders (3104)
- Congenital defects of the skeleton and muscles (6001)
- Congenital defects of the heart and blood vessels (6754)
- Cleft palate and other deformities of the mouth and palate (7012)
- Trisomy 13, 18, 21 (7157)

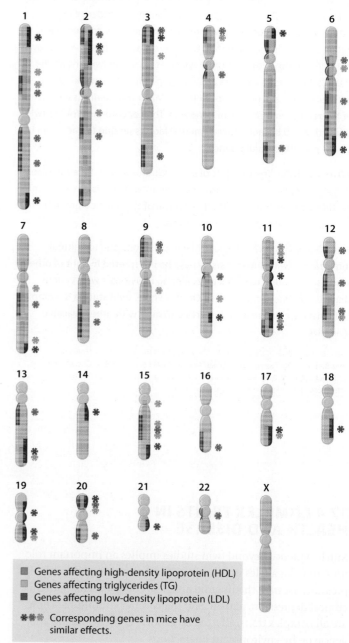

FIG. 17.12 Genes affecting the level of serum cholesterol in the human genome. *Data from X. Wang and B. Paigen, 2005, "Genetics of Variation in HDL Cholesterol in Humans and Mice," Circulation Research 96:27–42.*

- ■ Genes affecting high-density lipoprotein (HDL)
- ■ Genes affecting triglycerides (TG)
- ■ Genes affecting low-density lipoprotein (LDL)
- ✱✱✱ Corresponding genes in mice have similar effects.

sequencing (Chapter 13), genome annotation (Chapter 13), and genotyping (Chapter 14)—to identifying genes affecting complex traits.

The identification of genes affecting complex traits is not only a major goal of much current research in human genetics, but also in the genetics of domesticated animals and plants. Gene identification is also important in model organisms used in research, especially the laboratory mouse. Much is already known about these model organisms and they are easily manipulated, making it possible to discover the molecular mechanisms by which genes affecting complex traits exert their effects.

The study of complex traits has reached a stage at which patterns are beginning to emerge. Many of these patterns can be seen in the chromosome map in **Fig. 17.12**, which shows the location of genes in the human genome that affect cholesterol levels. The first observation is that many genes contribute to amounts of different types of cholesterol in humans—specifically, more than 50 genes contribute to the level of HDL (high-density lipoproteins, or "good cholesterol"), LDL (low-density lipoproteins, or "bad cholesterol"), and triglycerides. A second observation is that many of these genes affect two or even all three of the types of molecule. (A single gene that has multiple effects is said to show **pleiotropy**.) Third, many of the genes occur in clusters, being physically close together in the same chromosome. Clustering of genes shows that many genes arose through the process of duplication of a single gene followed by divergence over time, generating a family of genes near one another with related functions (Chapter 14).

Two other patterns are not visible in Fig. 17.12. One is that many of the genes show epistasis. Epistasis occurs when

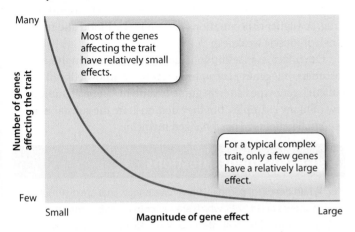

FIG. 17.13 Relationship between number of genes and their effect on complex traits.

multiple genes act in the same pathway to affect a trait (Chapter 15). Finally, the functions of many of the genes that contribute to a complex trait are often not known. Some of the genes in Fig. 17.12 affect serum cholesterol through known metabolic pathways, but many others work through unknown pathways. Many of the human genes shown have counterparts in the mouse genome that have similar effects on serum cholesterol, so these genes are open to direct experimental investigation.

A principle that is not evident in Fig. 17.12 is that the effects of each of the genes on cholesterol levels are very unequal. Some have relatively large effects, whereas others have small ones. The distribution of the magnitude of gene effects for complex traits resembles that shown in **Fig. 17.13**. The axes are labeled only in relative terms, because the actual numbers and effect sizes differ from one trait to the next. However, the main point is that for most genes that contribute to a complex trait, the magnitude of their individual effects is typically quite small. The magnitude of the effects of individual genes also often differs between the sexes, which helps explain why complex traits so often differ in prevalence or severity between males and females.

Human height is affected by hundreds of genes.

For most complex traits, the genes that have been identified to date account for only a relatively small fraction of the total variation in the trait. One extreme example is adult height. An enormous amount of data is available for height, not because height has been studied extensively for its own sake, but because in studies of common diseases the height of each individual is recorded, and hence data on height are available without additional effort or expense.

In one analysis, results from 79 separate studies were combined. These studies included more than 250,000 individuals of European ancestry who were genotyped for common nucleotide variants (SNPs) at about 3 million nucleotide sites.

The analysis identified 697 genes affecting height. Some of these genes are known to affect skeletal development, growth hormones, or other growth factors, but most have no obvious connection to the biology of growth. Some of the genes had previously been identified by studies of rare mutations that have pronounced effects on skeletal growth, either in human families or in laboratory mice. An unexpected finding was that a few genes affecting height are known to be associated with bone mineral density, obesity, and rheumatoid arthritis. These and many other of the genes identified might affect height indirectly.

The 697 genes for height identified among the 250,000 individuals account for about 60% of the genetic variation in height, a number that suggests that many more genes with still smaller effect also contribute to variation in height. And this analysis does not address at all the effects of the environment on human height, nor genotype-by-environment interactions, which likely play an important role as well. Complex traits therefore require sophisticated studies to tease apart all the genetic and environmental factors that play a role.

 CASE 3 YOUR PERSONAL GENOME: YOU, FROM A TO T

Can personalized medicine lead to effective treatments of common diseases?

The multiple genetic and environmental factors affecting complex traits imply that different people can have the same disease for different reasons. For instance, one person might develop breast cancer because of a mutation in the *BRCA1* or *BRCA2* genes, whereas another might develop breast cancer because of other genetic risk factors or even environmental ones. Similarly, one person might develop emphysema because of a mutation in the gene that encodes the enzyme alpha-1 antitrypsin (α1AT) (Chapter 14), and another might have emphysema as a result of cigarette smoking. Other examples where the same disease can be the result of different underlying genetic or environmental factors include elevated cholesterol levels, high blood pressure, and depression. Because the underlying genetic basis for the same disease may be different in different patients, some patients respond well to certain drugs and others do not.

The traditional strategy for treating diseases is to use the same medicine for the same disease. However, because we now know that the same disease may have different causes, another possibility has emerged: identifying each patient's genotype for each of the relevant genes and then matching the treatment to the genetic risk factors in each patient. The approach that matches the treatment to the patient, not the disease, is known as **personalized medicine** or **precision medicine**. (Prescription eyeglasses are a 400-year-old type of personalized medicine.)

Personalized medicine not only aims to identify ahead of time medicine that will work effectively, but also to avoid medicines that may lead to harmful side effects, even death. Advertisements and enclosures with prescription drugs enumerate long lists of side effects that you may get if you take a particular medicine. At present, it is difficult to know in advance who will get one or more of these side effects and who won't—that is, the side effects of taking medicine are themselves complex traits and the result of many underlying genetic and environmental factors. If we could identify ahead of time those patients who will respond negatively to a particular medicine and those who won't, we could minimize potentially harmful effects of medicines on certain individuals.

Someday, it may be possible to determine each patient's genome sequence quickly, reliably, and cheaply. At present, personalized medicine is restricted to studies of a few key genes known to have important effects on treatment outcomes. In the treatment of asthma, for example, some of the differences in response to albuterol inhalation have been traced to genetic variation in the gene *ADRB2*, encoding the β-(beta-)2-adrenergic receptor. Similarly, certain drugs used in the treatment of Alzheimer's disease are less effective in women with a particular apolipoprotein E (*APOE*) genotype than in other classes of patients. Also, more than half of the cases of muscle weakness occurring as an adverse effect of drug treatment used to control high cholesterol can be traced to genetic variation in the gene *SLCO1B1*, which encodes a liver transport protein. Certain variants of this protein have decreased ability to transport the drug, so higher concentrations remain in the blood and there is a risk of muscle weakness.

Other factors in addition to genes are involved in these disorders, but genes play an important role. Knowledge of a patient's genotype can help guide treatment. These are only a few of many examples, but they demonstrate the substantial potential benefits of personalized medicine.

Self-Assessment Questions

13. When genes for complex traits have effects that are distributed as shown in Fig. 17.13, which are easier to identify: those that are numerous with small effects, or those that are few with large effects?

14. What is personalized medicine and how does it relate to complex traits such as human diseases?

CORE CONCEPTS SUMMARY

17.1 HEREDITY AND ENVIRONMENT: Complex traits are those influenced both by the action of many genes and by environmental factors.

Complex traits that are measured on a continuous scale, such as human height, are called quantitative traits. page 363

It is usually difficult to assess the relative roles of genes and the environment ("nature" versus "nurture") in the production of a given trait in an individual, but it is reasonable to consider the relative roles of genetic and environmental variation in accounting for differences among individuals for a given trait. page 365

The relative roles of genes and environment in causing differences in phenotype among individuals differ among traits. For some traits (such as height), genetic differences are the more important source of variation, whereas for others (such as cancer), environmental differences can be the more important. page 365

Genetic and environmental factors can interact in unpredictable ways, resulting in genotype-by-environment interactions. page 365

17.2 RESEMBLANCE AMONG RELATIVES: Genetic effects on complex traits are reflected in resemblance between relatives.

In an analysis of heights of parents and offspring, Galton observed regression toward the mean, in which the offspring exhibit an average phenotype that is less different from the population mean than that of the parents. page 366

"Heritability" refers to the proportion of the total variation in a trait that can be attributed to genetic differences among individuals. page 368

The heritability of a given trait can differ among populations because of differences in genotype or environment. page 368

17.3 TWIN STUDIES: Twin studies help separate the effects of genotype and environment on variation in a trait.

Monozygotic (MZ), or identical, twins result from the fertilization of a single egg and are genetically identical. page 369

Dizygotic (DZ), or fraternal, twins result from the fertilization of two eggs and are genetically related to each other in the same way that other siblings are related to each other. page 369

"Concordance" refers to the percentage of cases in which both members of a pair of twins show the trait when it is known that at least one member shows it. page 369

Comparisons of concordance rates of MZ twins and concordance rates of DZ twins can help to determine to what extent variation in a particular trait has a genetic component. page 369

17.4 COMPLEX TRAITS IN HEALTH AND DISEASE: Many common diseases and birth defects are affected by multiple genetic and environmental risk factors.

Complex traits are often influenced by many genes with multiple, interacting, and unequal, effects. page 372

Hundreds of genes affect human height. page 373

Personalized medicine tailors treatment to an individual's genetic makeup. page 373

Log in to LaunchPad to check your answers to the Self-Assessment Questions and to access additional learning tools.

CHAPTER 18 Genetic and Epigenetic Regulation

CORE CONCEPTS

18.1 CHROMATIN TO MESSENGER RNA IN EUKARYOTES: Gene expression in eukaryotes can be regulated at the level of DNA packaging in chromosomes, transcription, and RNA processing.

18.2 MESSENGER RNA TO PHENOTYPE IN EUKARYOTES: After an mRNA is exported to the cytoplasm, gene expression in eukaryotes can be regulated at the level of mRNA stability, translation, and posttranslational modification of proteins.

18.3 TRANSCRIPTIONAL REGULATION IN PROKARYOTES: Gene expression in prokaryotes can be regulated at the level of transcription in response to environmental conditions.

In Chapter 17, we learned that complex traits are not determined by single genes with simple Mendelian inheritance. The traits we usually encounter in ourselves or in others, such as height, weight, diabetes, and high blood pressure, are influenced by multiple genes that interact with one another and with environmental factors, such as diet and exercise. The number of genes influencing complex traits can be large. For example, hundreds of genes contribute to adult height. The activities of these genes must be coordinated in time (at a particular point in development) and place (in a particular type of tissue). For example, growth to normal adult height requires that the production of growth hormone be coordinated in time and place with the production of growth-hormone receptor.

In Chapters 3 and 4, we outlined the basic steps of information flow in a cell, focusing on transcription and translation. In these processes, DNA is a template for the production of messenger RNA (mRNA), which itself is the template for the synthesis of a protein. For these processes to work in a living organism to produce the traits that we see, they must be coordinated so that genes are only **expressed**, or turned on, in the right place and time, and in the right amount. In a multicellular organism, for example, certain genes are expressed in some cells but not in others. Genes encoding muscle actin and myosin are turned on in muscle cells but not in liver or kidney cells. Even for single-celled organisms like bacteria, certain genes are expressed only in response to environmental signals, such as the availability of nutrients, as we discuss in section 18.3.

Gene regulation encompasses the ways in which cells control gene expression. It can be thought of as the *where? when?* and *how much?* of gene expression. Where (in which cells) are genes turned on? When (during development or in response to changes in the environment) are they turned on? How much gene product is made? This chapter provides an overview of the most common ways in which gene expression is regulated.

One of the important principles we discuss is that gene regulation can occur at almost any step in the path from DNA to mRNA to protein—at the level of the chromosome itself, by controlling transcription or translation, and, perhaps surprisingly, even after the protein is made. Each of these steps or levels of gene expression may be subject to regulation. In other words, each successive event that takes place in the expression of a gene is a potential control point for gene expression.

We begin by discussing gene regulation in eukaryotes, and then turn to gene regulation in prokaryotes and viruses. All life, from the simplest to the most complex, requires gene regulation. Gene regulation relies on similar processes in all organisms because all life shares common ancestry. Nevertheless, certain features of eukaryotes—the packaging of DNA into chromosomes, mRNA processing, and the separation in space of transcription and translation—provide additional levels of gene regulation in eukaryotes that are not possible in prokaryotes.

18.1 CHROMATIN TO MESSENGER RNA IN EUKARYOTES

Gene regulation in multicellular eukaryotes leads to cell specialization: different types of cell express different genes. The human body contains about 200 major cell types, and although for the most part they share the same genome, they look and function differently from one another because each type of cell expresses different sets of genes. For example, the insulin needed to regulate sugar levels in the blood is produced only by small patches of cells in the pancreas. Every cell in the body contains the genes that encode insulin, but only in these patches of pancreatic cells are they expressed. Insufficient insulin is one factor that causes type 2 diabetes, a complex trait in which levels of blood glucose are higher than normal. In this section, we explore gene regulation as it occurs in eukaryotic cells, focusing on regulation at the level of DNA, chromatin, and mRNA.

Gene expression can be influenced by chemical modification of DNA or histones.

Fig. 18.1 shows the major places where gene regulation in eukaryotes can take place. The first level of control is at the chromosome, even before transcription takes place. The manner in which DNA is packaged in the nucleus in eukaryotes provides an important opportunity for regulating gene expression. Regulation at this level of gene expression is determined, in part, by whether the proteins necessary for transcription can gain physical access to the genes they transcribe.

DNA in eukaryotes is packaged as **chromatin**, a complex of DNA, RNA, and proteins that gives chromosomes their structure (Chapter 13). When chromatin is in its coiled state, the DNA is not accessible to the proteins that carry out transcription. The chromatin must loosen to allow space for transcriptional enzymes and proteins to work. This is accomplished through **chromatin remodeling**, in which the nucleosomes are repositioned to expose different stretches of DNA to the nuclear environment.

One way in which chromatin is remodeled is by chemical modification of the histones around which DNA is wound (**Fig. 18.2**). Modification usually occurs on **histone tails**, which are strings of amino acids that protrude from the histone

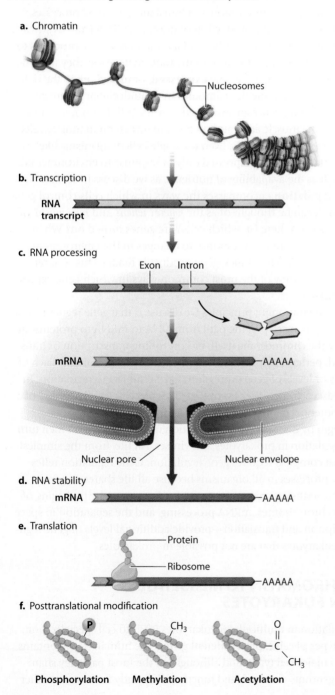

FIG. 18.1 Levels of gene regulation in eukaryotes.

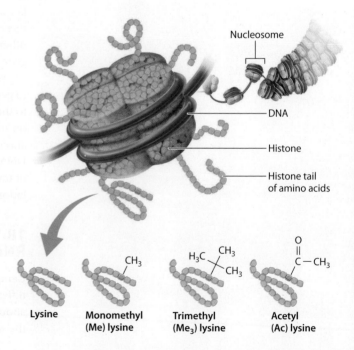

FIG. 18.2 Histone modifications. Typical modifications of the amino acid lysine observed in the histone tails of nucleosomes include addition of a methyl group (Me), addition of three methyl groups (Me$_3$), and acetylation (Ac).

proteins in the nucleosome. Individual amino acids in the tails can be modified by the addition (or later removal) of different chemical groups, including methyl groups (—CH$_3$) and acetyl groups (—COCH$_3$). Most often, methylation or acetylation occurs on the lysine residues of the histone tails. Some of these modifications tend to activate transcription and others to repress transcription. The pattern of modifications of the histone tails constitutes a **histone code** that affects chromatin structure and gene transcription. Modification of histones takes place at key times in development to ensure that the proper genes are turned on or off, as well as in response to environmental cues.

In many eukaryotic organisms, gene expression is also affected by chemical modification of certain bases in the DNA, the most common of which is the addition of a methyl group to the base cytosine (**Fig. 18.3**). DNA methylation recruits proteins that lead to changes in chromatin structure, histone modification, and nucleosome positioning that restrict access of transcription factors to promoters. Methylation of cytosines often occurs in **CpG islands**, which are clusters of adjacent CG nucleotides located in or near the promoter of a gene. Heavy cytosine methylation is associated with transcriptional repression of the gene near the CpG island. The methylation state of a CpG island can change over time or in response to environmental signals, providing a way to turn genes on or off. As a defense mechanism, cells sometimes heavily methylate CpG islands of transposable elements or viral DNA sequences that are integrated into the genome, thus preventing the expression of genes in the foreign DNA (Chapter 14).

Together, the modification of cytosine bases, changes to histones, and alterations in chromatin structure are often termed **epigenetic**, from the Greek *epi* ("over and above") and "genetic" ("inheritance"). That is, epigenetic mechanisms of gene regulation typically involve changes not to the DNA sequence itself but to the manner in which the DNA is packaged. Epigenetic modifications can, in some cases, affect gene expression. They can be inherited through mitotic cell divisions, just as genes are, but are often reversible and responsive to changes in the environment. Furthermore, epigenetic modifications present in sperm or eggs are sometimes transmitted from parent to offspring, meaning that these chemical modifications can be inherited.

In humans and other mammals, about 100 genes are repressed by chemical modifications such as DNA methylation in the germ line in a sex-specific manner. This sex-specific silencing of gene expression is known as **imprinting**. For some genes, the allele of the gene inherited from the mother is

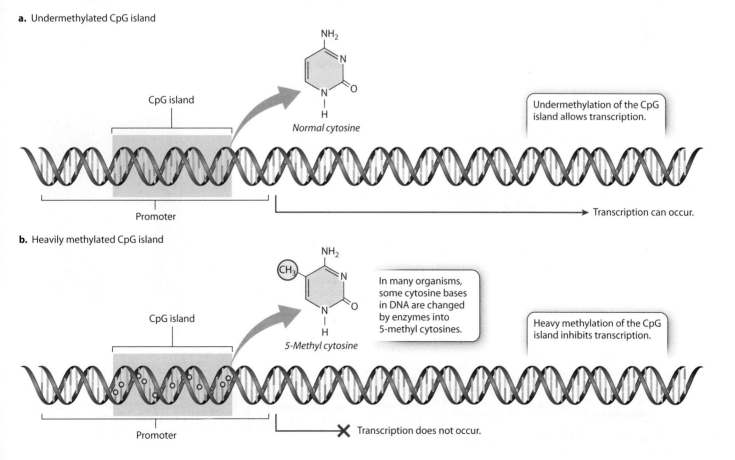

FIG. 18.3 Methylation states of CpG islands in or near the promoter of a gene.

imprinted and therefore silenced, so only the allele inherited from the father is expressed. For other imprinted genes, the allele of the gene inherited from the father is imprinted and therefore silenced, so only the allele inherited from the mother is expressed. The imprinting persists in somatic cells throughout the lifetime of the individual, but it is reset in its germ line according to the individual's sex.

Some of the genes that are imprinted affect the growth rate of the embryo. For example, in matings between lions and tigers that take place in captivity, the offspring of a male tiger with a female lion (a "tigon") is about the same size as its parents, whereas the offspring of a male lion and a female tiger (a "liger") is a giant cat—the largest of any big cat that exists. Much of the size difference between a liger and a tigon is thought to result from whether genes for rapid embryonic growth that are subject to imprinting are inherited from the mother or the father.

Gene expression can be regulated at the level of an entire chromosome.

A striking example of an epigenetic form of gene regulation is the manner in which mammals equalize the expression of X-linked genes in XX females and XY males. For most genes, a direct relation exists between the number of copies of the gene (the gene dosage) and the level of expression of the gene. An increase in gene dosage increases the level of expression because each copy of the gene is regulated independently of other copies. For example, as we saw in Chapter 14, the presence of an extra copy of most human chromosomes results in spontaneous abortion because of the increase in the expression of the genes in that chromosome.

XX females have twice as many X chromosomes as XY males. For genes located in the X chromosome, the dosage of genes is twice as great in females as it is in males. However, the level of expression of X-linked genes is about the same in both sexes. These observations imply that the regulation of genes in the X chromosome is different in females and in males. This differential regulation is called **dosage compensation**.

Different species have evolved different mechanisms of dosage compensation. In the fruit fly *Drosophila*, males double the transcription of the single X chromosome to achieve equal expression compared to the two X chromosomes in females. In the nematode worm *Caenorhabditis*, transcription of both X chromosomes in females is decreased to one-half the level of the single X chromosome in males. In mammals, including humans, dosage compensation occurs through the inactivation of one X chromosome in each cell in females. This process, known as **X-inactivation**, was first proposed by Mary F. Lyon in the early 1960s.

Soon after a fertilized egg with two X chromosomes implants in the mother's uterine wall (after several cell divisions have occurred), one X chromosome at random is inactivated (**Fig. 18.4**). In Fig. 18.4a, the inactive X chromosome is shown in gray. The inactive state persists through cell division, so in

FIG. 18.4 X-inactivation in female mammals. *X*-inactivation equalizes the expression of most genes in the *X* chromosome between *XX* females and *XY* males. *Source: b. Eyal Nahmias/Alamy.*

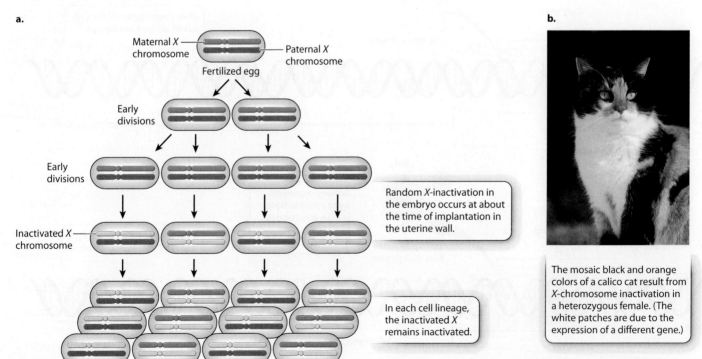

each cell lineage, the same X chromosome that was originally inactivated remains inactive. The result is that a normal female is a mosaic, or patchwork, of tissue. In some patches, the genes in the maternal X chromosome are expressed (and the paternal X chromosome is inactivated), whereas in other patches, the genes in the paternal X chromosome are expressed (and the maternal X chromosome is inactivated). The term "inactive X chromosome" is a slight exaggeration because a substantial number of genes are still transcribed, although usually at a low level.

As one argument for her X-inactivation hypothesis, Lyon called attention to calico cats, which are nearly always female (Fig. 18.4b). In calico cats, the orange and black fur colors are due to different alleles of a single gene in the X chromosome. In a heterozygous female, X-inactivation predicts discrete patches of orange and black, and this is exactly what is observed. (The white patches on a calico cat are due to an autosomal gene.)

How does X-inactivation work? Some of the details are still unknown, but the main features of the process are shown in **Fig. 18.5**. A key player is a small region in the X chromosome called the X-chromosome inactivation center (XIC), which contains a gene called *Xist* (X-inactivation specific transcript). The *Xist* gene is normally transcribed at a very low level, and the RNA is unstable. However, in an X chromosome that is about to become inactive, *Xist* transcription markedly increases. The transcript undergoes RNA splicing, but it does not encode a protein. Therefore, *Xist* RNA is an example of a noncoding RNA (Chapter 3). Instead of being translated, the processed *Xist* RNA coats the XIC region, and as it accumulates, the coating spreads outward from the XIC until the entire chromosome is coated with *Xist* RNA. The presence of *Xist* RNA along the chromosome recruits factors that promote DNA methylation, histone modification, and other changes associated with transcriptional repression.

The *Xist* gene is both necessary and sufficient for X-inactivation: if it is deleted, X-inactivation does not occur; if it is inserted into another chromosome, it inactivates that chromosome. Using recombinant DNA techniques such as those described in Chapter 12, researchers were able to insert *Xist* into one of the three copies of chromosome 21 present in cells taken from a patient with Down syndrome. Remarkably, *Xist* in its new location in chromosome 21 produced an RNA transcript that coated that copy of chromosome 21 and repressed most of the transcription of its genes! Whether all genes in chromosome 21 are repressed, and whether they are as fully repressed as those in the inactive X chromosome, are questions that are still being investigated, but this finding raises clear (but still speculative) ideas about how *Xist* might be used as a potential treatment for Down syndrome and various other chromosomal disorders.

Transcription is a key control point in gene expression.

Although access to DNA is necessary for transcription, it is not sufficient. The molecular machinery that actually carries out transcription is also required once the template DNA is made accessible through chromatin remodeling and histone modification. The mechanisms that regulate whether or not transcription

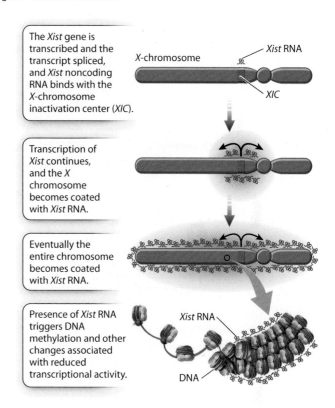

FIG. 18.5 The role of *Xist* in X-inactivation. The *Xist* transcript is a noncoding RNA that is expressed from the inactive X chromosome and coats the entire chromosome, leading to inactivation of most of the genes in the chromosome.

occurs are known collectively as **transcriptional regulation** (see Fig. 18.1b).

Transcriptional regulation in eukaryotic cells requires the coordinated action of many proteins that interact with one another and with DNA sequences near the gene. Let's first review the basic process of transcription in eukaryotes (Chapter 3). Initially, an important group of proteins called **general transcription factors** are recruited to the gene's promoter, which is the region of a gene where transcription is initiated. The transcription factors are brought there by one of the proteins that bind to a short sequence in the promoter called the TATA box, which is usually situated 25–30 nucleotides upstream of the nucleotide site where transcription begins. Once bound to the promoter, the transcription factors recruit the components of the **RNA polymerase complex**, which is the enzyme complex that synthesizes the RNA transcript complementary to the template strand of DNA.

Where in the many steps of transcription initiation does regulation occur? Transcription initiation can be regulated at the level of recruitment of the general transcription factors and components of the RNA polymerase complex. Recruitment of these elements is controlled by proteins called **regulatory transcription factors.** Some regulatory transcription factors

FIG. 18.6 Protein–DNA and protein–protein interactions in the eukaryotic transcription complex.

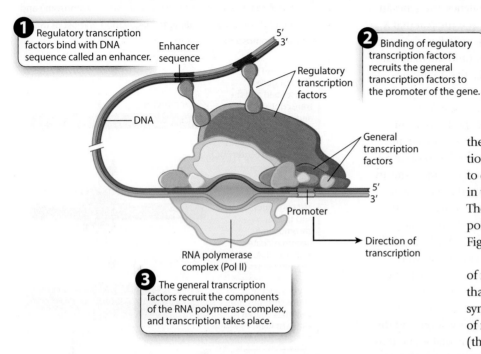

recruit chromatin remodeling proteins that allow physical access to a gene. Other regulatory transcription factors have two binding sites, one of which binds with a particular DNA sequence in or near a gene known as an **enhancer** and the other of which recruits one or more general transcription factors to the promoter region (**Fig. 18.6**). The general transcription factors then recruit the RNA polymerase complex, and transcription can begin (Chapter 3). Transcription does not occur if the regulatory transcription factors fail to recruit the general transcription factors to the gene.

Hundreds of different regulatory transcription factors control the transcription of thousands of genes. Some bind with enhancers and stimulate transcription; others bind with DNA sequences known as **silencers** and repress transcription. Enhancers and silencers are often in or near the genes they regulate, but, in some cases, they may be many thousands of nucleotides distant from the genes. A typical gene may be regulated by multiple enhancers and silencers of different types, each with one or more regulatory transcription factors that can bind with it. Transcription takes place only when the proper combination of regulatory transcription factors is present in the same cell, as shown in Fig. 18.6. Transcription of a gene with multiple silencers and enhancers depends on the presence of a particular combination of regulatory transcription factors, so this type of regulation is called **combinatorial control**.

RNA processing is also important in gene regulation.

A great deal happens in the nucleus after transcription takes place. The initial transcript, called the **primary transcript**, undergoes several modifications, which are collectively called **RNA processing** (Chapter 3) and include the addition of a nucleotide cap to the 5′ end and a string of tens to hundreds of adenosine nucleotides to the 3′ end to form the poly(A) tail. These modifications are necessary for the RNA molecule to be transported to the cytoplasm and recognized by the translational machinery. The poly(A) tail also helps to determine how long the RNA will persist in the cytoplasm before being degraded. Therefore, RNA processing is an important point where gene regulation can occur (see Fig. 18.1c).

In eukaryotes, the primary transcript of many protein-coding genes is far longer than the mRNA ultimately used in protein synthesis. The long primary transcript consists of regions that are retained in the mRNA (the **exons**) interspersed with regions that are excised and degraded (the **introns**). The introns are excised during **RNA splicing** (Chapter 3). The exons are joined together in their original linear order to form the processed mRNA.

RNA splicing provides an opportunity for regulating gene expression because the same primary transcript can be

FIG. 18.7 Alternative splicing of a mammalian insulin-receptor transcript. Alternative splicing generates different processed mRNAs and different proteins from the same primary transcript.

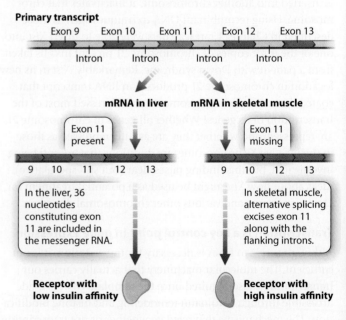

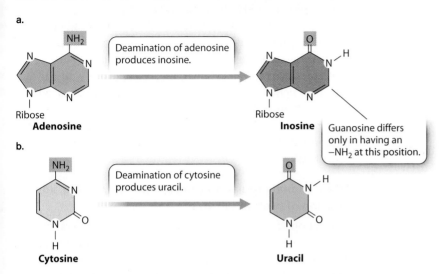

FIG. 18.8 RNA editing. RNA editing results in chemical modifications to the bases in mRNA, which can lead to changes in the amino acid sequence of the protein.

spliced in different ways to yield different proteins in a process called **alternative splicing**. This process takes place because what the spliceosome—the splicing machinery—recognizes as an exon in some primary transcripts, it also recognizes as part of an intron in other primary transcripts. The alternative-splice forms may be produced in the same cells or in different types of cell. Alternative splicing accounts, in part, for the observation that we produce many more proteins than our total number of genes. By some estimates, over 90% of human genes undergo alternative splicing.

Fig. 18.7 shows the primary transcript of a gene encoding an insulin receptor found in humans and other mammals. During RNA splicing in liver cells, exon 11 is included in the mRNA, and the insulin receptor produced from this mRNA has low affinity for insulin. In contrast, in cells of skeletal muscle, the 36 nucleotides of exon 11 are spliced out of the primary transcript along with the flanking introns. The resulting protein is 12 amino acids shorter, and this form of the insulin receptor has a high affinity for insulin. The different forms of the protein are important: the higher sensitivity of muscle cells to insulin enables them to absorb enough glucose to fulfill their energy needs.

Some RNA molecules can become a substrate for enzymes that modify particular bases in the RNA, thus changing its sequence and what it codes for. This process is known as **RNA editing** (**Fig. 18.8**). One type of editing enzyme removes the amino group (–NH$_2$) from adenosine and converts it to inosine, which in translation functions like guanosine (Fig. 18.8a). Another enzyme removes the amino group from cytosine and converts it to uracil (Fig. 18.8b). In the human genome, hundreds if not thousands of transcripts undergo RNA editing. In many cases, not all copies of the transcript are edited, and some copies may be edited more extensively than others.

The result is that transcripts from the same gene can produce multiple types of proteins, even in a single cell.

Transcripts from the same gene may undergo different editing in different cell types. An example of tissue-specific RNA editing is shown in **Fig. 18.9**. The mRNA fragments show part of the coding sequence for apolipoprotein B. The unedited mRNA in the liver (Fig. 18.9a) is translated into a protein that transports cholesterol in the blood. In contrast, RNA editing of the message occurs in the intestine (Fig. 18.9b). The cytosine nucleotide in codon 2153 is edited to uracil. The edited codon is UAA, which is a stop codon. Translation therefore terminates at this point, releasing a protein only about half as long as the liver form. This shorter form of the protein helps the cells of the intestine absorb lipids from the foods we eat.

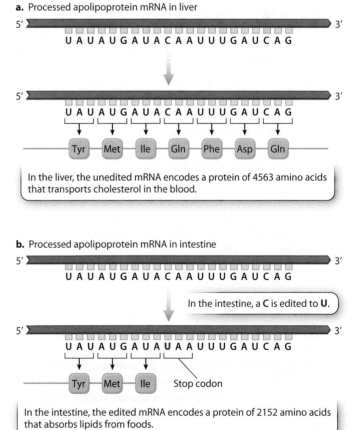

FIG. 18.9 Tissue-specific RNA editing of the human apolipoprotein B transcript. Different RNA editing in (a) the liver and (b) the intestine results in proteins with different functions.

> **Self-Assessment Questions**
>
> 1. What is the difference between gene expression and gene regulation?
> 2. What is meant by different "levels" of gene regulation?
> 3. How can DNA bases and chromatin can be modified to regulate gene expression? Does this modification result in increased or decreased gene expression?
> 4. How does *X*-inactivation in female mammals result in patchy coat colors in calico cats?
> 5. How can one protein-coding gene code for more than one polypeptide chain?

18.2 MESSENGER RNA TO PHENOTYPE IN EUKARYOTES

In eukaryotes, a processed mRNA must exit the nucleus before the translation step of gene expression can occur. The mRNA migrates to the cytoplasm through one of a few thousand nuclear pores, which are large protein complexes that span both layers of the nuclear envelope and regulate the flow of macromolecules in and out of the nucleus (Chapter 5). Once the mRNA is in the cytoplasm, there are multiple opportunities for gene regulation at the levels of mRNA stability, translation, and protein activity (see Figs. 18.1d–18.1f).

Small regulatory RNAs inhibit translation or promote mRNA degradation.

Regulatory RNA molecules known as **small regulatory RNAs** are among the most exciting recent discoveries in gene regulation. They are of exceptional interest to biologists and drug researchers because their small size allows easy synthesis in the laboratory, and researchers can design their sequences to target transcripts of interest.

Two important types of small regulatory RNAs are known as **siRNA (small interfering RNA)** and **miRNA (microRNA)**. Although they have somewhat different functions, they share similarities in how they are produced (**Fig. 18.10**). Both types of small regulatory RNA are transcribed from DNA and form hairpin, or stem-and-loop, structures in the nucleus, which are stabilized by base pairing in the stem. Enzymes in the cytoplasm specifically recognize these structures and cleave the stem from the hairpin, and then further cut the stem into small, double-stranded fragments that are 20–25 base pairs in length.

One of the two strands from each RNA fragment is incorporated into a protein complex known as **RISC (RNA-induced silencing complex)**. The small, single-stranded RNA targets the RISC to specific RNA molecules by base pairing with short regions in the target. Depending on the type of small regulatory RNA, the RNA sequence in the RISC, and the particular type of RISC, the small regulatory RNA may result in chromatin remodeling, degradation of RNA transcripts, or inhibition of mRNA translation (Fig. 18.10).

Regulation by small regulatory RNAs is widespread in eukaryotes, and it is thought to have evolved originally as a defense against viruses and transposable elements. These molecules also evolved to have important regulatory roles in their own right. Human chromosomes are thought to encode about 1000 microRNAs, each of which can target tens or hundreds of mRNA molecules. Half or more of human proteins may have their synthesis regulated, in part, by such small regulatory RNAs.

Translational regulation controls the rate, timing, and location of protein synthesis.

Translation of mRNA into protein provides another level of control of gene expression. **Fig. 18.11** shows the structure of a hypothetical mRNA molecule in a eukaryotic cell and highlights some of the features that help regulate its translation (Chapter 4). Not all mRNA molecules have all the features shown, but almost all mRNA molecules have a 5′ cap, a 5′ untranslated region (5′ UTR), an open reading frame (ORF) containing the codons that determine the amino acid sequence of the protein, a 3′ untranslated region (3′ UTR), and a poly(A) tail. The 5′ UTR and 3′ UTR may contain regions that bind with regulatory proteins. These RNA-binding proteins help control

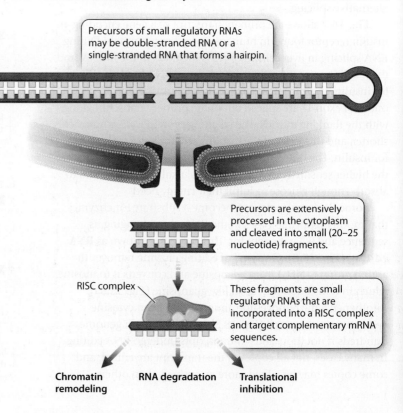

FIG. 18.10 Production of small regulatory RNAs and an overview of their functions.

FIG. 18.11 Some features of mRNA that affect gene expression. These include the 5′ cap, 5′ untranslated region (5′ UTR), open reading frame (ORF), 3′ untranslated region (3′ UTR), and poly(A) tail.

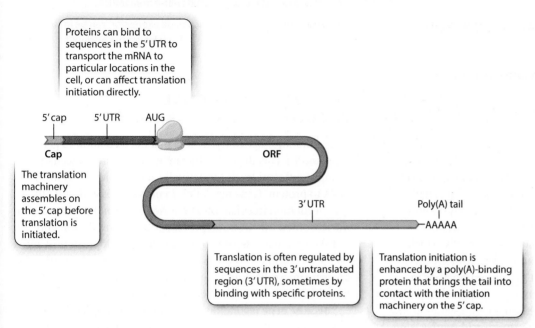

Protein structure and chemical modification modulate protein effects on phenotype.

Once translation is completed, the resulting protein can alter the phenotype of the cell or organism by affecting metabolism, signaling, gene expression, or cell structure. After translation, proteins are modified in multiple ways that regulate their structure and function. Collectively, these processes are called **posttranslational modification** (see Fig. 18.1e). Regulation at this level is essential because some proteins are downright dangerous. For example, proteases, such as the digestive enzyme trypsin,

mRNA translation and degradation. The UTRs may also contain binding sites for small regulatory RNAs. During development, as we will see in Chapter 19, some RNA-binding proteins interact with molecular motors that transport the mRNA to particular regions of the cell. In other cases, the proteins are only in particular locations in the cell and repress translation of the mRNAs that are transported there and to which they bind. By either transport or repression, these proteins cause the mRNA to be translated only in certain places in the cell.

The cap structure is one of the main recognition signals for translation initiation, which requires the coordinated action of about 25 proteins. These proteins are present in most cells in limited amounts. Therefore, at any given time, while some mRNAs from a gene are being translated, other mRNAs transcribed from the same gene may not have a translation initiation complex assembled.

The 3′ UTR and the poly(A) tail are also important in translation initiation. The efficiency of translation initiation is greatly increased by physical contact between a protein that binds the poly(A) tail and one that binds the 5′ cap. The physical contact creates a loop in the mRNA, bringing the 3′ end of the mRNA into proximity with the start site for translation. In fact, most mRNA sequences that regulate translation are present in the 3′ UTR.

Although translation initiation is the principal mode of translational regulation, not all mRNA molecules are equally accessible to translation. Among the key variables are the secondary (folded) structure of the 5′ UTR, the distance from the 5′ cap to the AUG initiation codon, and the sequences flanking the AUG initiation codon.

must be kept inactive until they are secreted out of the cell. If they were not kept inactive, their activity would kill the cell. These types of protein are often controlled by being translated in inactive forms that are made active by modification after secretion.

Folding and acquiring stability are key control points for some proteins (Chapter 4). Although many proteins fold properly as they come off the ribosome, others require help from proteins called chaperones that act as folding facilitators. Correct folding is important because improperly folded proteins may bind together to form aggregates that are destructive to cell function. Many diseases are associated with protein aggregates, including Alzheimer's disease, Huntington's disease, and the human counterpart of mad cow disease.

Posttranslational modification also helps regulate protein activity. Many proteins are modified by the addition of one or more sugar molecules to the side chains of some amino acids. This modification can alter the protein's folding and stability, or target the molecule to particular cellular compartments. Reversible addition of a phosphate group to the side chains of amino acids such as serine, threonine, or tyrosine is a key regulator of protein activity (Chapter 9). Introduction of the negatively charged phosphate group alters the conformation of the protein, in some cases switching it from an inactive state to an active state and in other cases the reverse. Because the function of a protein molecule results from its shape and charge (Chapter 4), a change in protein conformation affects protein function.

Marking proteins for enzymatic destruction by the addition of chemical groups after translation is also important in controlling their activity. For example, we have seen how the destruction of successive waves of cyclin proteins helps move the cell through its division cycle (Chapter 11).

 CASE 3 YOUR PERSONAL GENOME: YOU, FROM A TO T

How do lifestyle choices affect expression of your personal genome?

If you examine Fig. 18.1 as a whole and consider the DNA sequence shown as your personal genome, the situation looks pretty grim. You might be led to believe that genes dictate everything, and that biology is destiny. However, if you focus on the lower levels of regulation in Fig. 18.1, a different picture emerges. The picture is different because much of the regulation that occurs after transcription (regulation of mRNA stability, regulation of translation, posttranslational modifications) is determined by the physiological state of your cells, which, in turn, is strongly influenced by your lifestyle choices. For example, your cells can synthesize 12 of the amino acids in proteins, but if any of these is present in sufficient amounts in your diet, it is absorbed during digestion and not synthesized. The amino acid you ingest blocks the amino acid-synthesizing pathway through feedback effects.

The effect of an intervention in gene expression—genetic or environmental—at any given level can affect regulatory processes at both higher and lower levels. This cascade of regulatory effects in both directions can occur because the expression of any gene is regulated at multiple levels, and because much feedback and signaling occurs back and forth between nucleus and cytoplasm. It is because of these feedback and signaling mechanisms that the effects of lifestyle choices can be propagated up the regulatory hierarchy. For example, it has been shown that dietary intake of fats and cholesterol affects not only the activity of enzymes directly involved in the metabolism of fats and cholesterol, but also the levels of transcription of the genes encoding these enzymes by affecting the activity of their regulatory transcription factors. Similarly, lifestyles that combine balanced diets with exercise and stress relief have been shown to increase transcription of genes whose products prevent cellular dysfunction and decrease transcription of genes whose products promote disease.

The interplay of different levels of gene regulation with environmental factors underlies many complex traits of the type discussed in Chapter 17, which are affected by multiple genes and by multiple environmental factors as well as by genotype-by-environment interaction. For example, both genetic and environmental risk factors exist for breast and ovarian cancers, as we have seen. Simple Mendelian traits caused by mutations in single genes, such as cystic fibrosis and alpha-1 antitrypsin (α-1AT) deficiency (Chapter 16), are less responsive to lifestyle choices. Even in these cases, however, lifestyle matters. For example, people with α-1AT deficiency should not smoke tobacco and should avoid environments with low air quality.

Self-Assessment Questions

6. How do small regulatory RNAs differ from mRNA?
7. What are three ways in which gene expression can be influenced after mRNA is processed and leaves the nucleus?

18.3 TRANSCRIPTIONAL REGULATION IN PROKARYOTES

The central message of Fig. 18.1 is that the regulation of gene expression occurs through a hierarchy of regulatory mechanisms acting at different levels (and usually at multiple levels) from DNA to protein. Gene regulation in prokaryotes is simpler than gene regulation in eukaryotes because DNA is not packaged into nucleosomes, mRNA is not processed, and transcription and translation are not separated by a nuclear envelope. In prokaryotes, expression of a protein-coding gene entails transcription of the gene into mRNA and translation of the mRNA into protein. Each of these levels of gene expression is subject to regulation.

Because gene regulation in prokaryotes is simpler than gene regulation in eukaryotes, prokaryotes have served as model organisms for our understanding of how genes are turned on and off. In this section, we consider in more detail how gene expression is regulated at the level of transcription in bacteria and in viruses that infect bacteria. We focus on two well-studied systems: (1) the regulation of genes in the intestinal bacterium *Escherichia coli* (*E. coli*) that allows lactose-metabolizing proteins to be produced only when lactose is present in the environment and only when it is the best nutrient available, and (2) the regulation of genes in a virus that infects *E. coli* that controls whether the virus integrates its DNA into the bacterial host or lyses (breaks open) the cell. In both cases, specific genes are turned on and off in response to environmental conditions.

Transcriptional regulation can be positive or negative.

In both eukaryotes and prokaryotes, transcription can be positively or negatively regulated. In **positive regulation**, a regulatory molecule (usually a protein) must bind to the DNA at a site near the gene in order for transcription to take place. In **negative regulation**, a regulatory molecule (again, usually a protein) must bind to the DNA at a site near the gene in order for transcription to be prevented. Most promoters contain one or more short sequences that help recruit the proteins needed for transcription when transcription of the gene is activated. In eukaryotes, many promoters contain the sequence 5'-TATAAA-3' (or some variant of this sequence) about 25–35 base pairs upstream from the transcription start site, which helps recruit the proteins needed for transcription. Many prokaryotic promoters contain a pair of such regulatory sequences located about 10 and 35 base pairs upstream of the transcription start site. These promoter sequences are called the −10 and −35 sequence motifs.

Fig. 18.12 illustrates positive regulation. The main players are DNA, the RNA polymerase complex, and a regulatory protein called a transcriptional **activator**. The DNA contains two important binding sites: one for the activator protein, and one for the RNA polymerase complex. As shown in Fig. 18.12a, when the activator protein is present in a state that can interact with its binding site in the DNA, the RNA polymerase complex is recruited to the promoter of the gene and transcription takes place. When the activator is not present or is not able to bind with the DNA (Fig. 18.12b), the RNA polymerase is not recruited to the promoter and transcription does not occur. The binding site for the activator may be upstream of the promoter, as shown in the figure, be downstream of the promoter, or even overlap the promoter.

Sometimes, the activator protein combines with a small molecule in the cell and undergoes a change in shape that alters its binding affinity for DNA. The change in shape is an example of an **allosteric effect** (Chapter 6). In some cases, the activator on its own cannot bind with DNA, but combining with the small molecule changes its shape so that it can bind with particular sequences in the DNA. Therefore, the presence of the small molecule in the cell results in transcription of the gene. Genes subject to this type of positive control typically encode proteins that are needed only when the small molecule is present in the cell. For example, in *E. coli*, the genes for the breakdown of the sugar arabinose are normally repressed in the absence of the sugar because a regulatory protein binds the DNA in such a way as to prevent transcription. In the presence of arabinose,

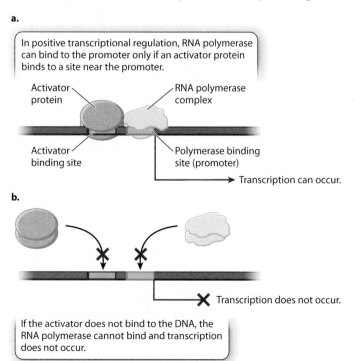

FIG. 18.12 Positive transcriptional regulation in prokaryotes. When an activator protein binds to DNA, it promotes transcription of a gene.

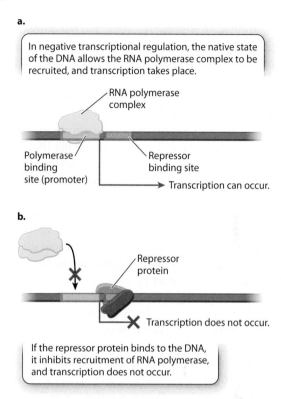

FIG. 18.13 Negative transcriptional regulation in prokaryotes. When a repressor protein binds to DNA, it prevents transcription of a gene.

however, the shape of the protein is altered so that it binds a different site in the DNA, allowing transcription of the genes. For other genes, activator proteins can bind DNA only when a small molecule is absent from the cell. These genes typically encode proteins needed for synthesis of the small molecule. In *E. coli*, the genes for synthesis of the amino acid cysteine are regulated in this fashion.

Fig. 18.13 illustrates negative regulation. In this case, the DNA in its native state can recruit the RNA polymerase complex, and transcription takes place at a constant rate unless something turns it off (Fig. 18.13a). Binding with a protein called a **repressor** turns off transcription (Fig. 18.13b). Again, the binding site for the repressor can be upstream of the promoter, downstream of the promoter, or overlapping with the promoter.

As with positive control, the ability of the repressor to bind DNA is often determined by an allosteric interaction with a small molecule. A small molecule that interacts with the repressor and prevents it from binding DNA and blocking transcription is called an **inducer**. In the next section, we will look closely at the regulation of genes for the breakdown of the sugar lactose in *E. coli*, and we will see that a derivative of lactose binds to and inhibits the repressor and is therefore an inducer of these genes.

In other cases, the small molecule changes the conformation of the repressor so that it can bind to the repressor binding site and inhibit transcription. Genes regulated in this way are often needed for synthesis of the small molecule. In *E. coli*, for example, the genes for the synthesis of the amino acid tryptophan are negatively regulated by tryptophan. When tryptophan is present in sufficient amounts, it binds with a regulatory protein to form the functional repressor, and transcription of the genes does not occur. When the level of tryptophan drops too low to form the repressor, transcription of the genes is initiated.

Lactose utilization in *E. coli* is the pioneering example of transcriptional regulation.

The principle that the product of one gene can regulate transcription of other genes was first discovered in the 1960s by François Jacob and Jacques Monod, who studied how the bacterium *E. coli* regulates production of the proteins needed for utilization of the sugar lactose. Lactose consists of one molecule each of the sugars glucose and galactose, which are covalently linked by a β-(beta-)galactoside bond. An enzyme called β-galactosidase cleaves lactose, releasing glucose and galactose. Both molecules can then be broken down and used as a source of carbon and energy (Chapter 7).

Jacob and Monod began their research with the observation that active β-galactosidase enzyme is observed in cells only in the presence of lactose or certain molecules that are chemically similar to lactose. Why is this so? **Fig. 18.14** describes and tests two hypotheses. One is that lactose stabilizes an unstable form of β-galactosidase that is produced by all cells all the time, thus activating the enzyme. The other is that lactose leads to the expression of the gene for β-galactosidase. The experiment shown in Fig. 18.14 supported the hypothesis that lactose turns on the β-galactosidase gene. The way in which lactose activates expression of the β-galactosidase gene was the subject of additional experiments. These Nobel Prize winning follow-up studies of Jacob and Monod gave the first demonstrated example of transcriptional gene regulation.

To understand how the genes for lactose utilization are regulated, you need to know the main players, and these are shown in **Fig. 18.15**. Let's start with the two coding sequences on the right side of the figure, in yellow and green:

- *lacZ* is the gene (coding sequence) for the enzyme β-galactosidase, which cleaves the lactose molecule into its glucose and galactose constituents. A functional, nonmutant form of the *lacZ* gene is denoted as *lacZ*$^+$.

- *lacY* is the gene (coding sequence) for the protein lactose permease, which transports lactose from the external medium into the cell. A functional, nonmutant form of the *lacY* gene is denoted as *lacY*$^+$.

These genes are called **structural genes** because they code for the sequence of amino acids making up the primary structure of each protein. Bacteria that contain mutations that eliminate function of the *lacZ* gene (denoted as *lacZ*$^-$ mutants) or those that eliminate function of the *lacY* gene (denoted as *lacY*$^-$ mutants), cannot utilize lactose as a source of energy.

HOW DO WE KNOW?

FIG. 18.14

How does lactose lead to the production of active β-galactosidase enzyme?

BACKGROUND Active β-galactosidase enzyme is observed only in *E. coli* cells that are growing in the presence of lactose.

HYPOTHESIS One hypothesis is that the enzyme is always being produced, but is produced in an unstable form that breaks down rapidly in the absence of lactose. An alternative hypothesis is that the enzyme is stable, but is produced only in the presence of lactose.

EXPERIMENT French biologists François Jacob and Jacques Monod exposed a culture of growing cells to lactose and later removed the lactose. The researchers measured the amount of β-galactosidase present in the culture during the exposure to lactose and its removal.

RESULTS Almost immediately upon addition of lactose, β-galactosidase began to accumulate, and its amount steadily increased. When lactose was removed, the enzyme did not disappear immediately (as would be the case if it were unstable, as predicted by the first hypothesis). Instead, the amount of enzyme remained the same as when lactose was present. This result is expected only if β-galactosidase is a stable enzyme that is synthesized when lactose is added and stops being synthesized when lactose is removed (as predicted by the second hypothesis).

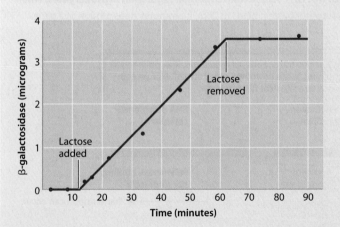

CONCLUSION Synthesis of β-galactosidase turns on when lactose is added and turns off when lactose is removed. These results support the second hypothesis, which asserts that the enzyme is stable but is produced only in the presence of lactose.

FOLLOW-UP WORK These pioneering experiments ultimately led to the discovery of the lactose operon and the mechanism of transcriptional regulation by a repressor protein.

SOURCE Monod, J. 1965. "From Enzymatic Adaption to Allosteric Transitions." Nobel Prize lecture. http://nobelprize.org/nobel_prizes/medicine/laureates/1965/monod-lecture.htm.

FIG. 18.15 Structural and regulatory elements of the lactose operon. The *lac* operon consists of the promoter, the operator, and all the structural genes that are transcribed into a single mRNA called a polycistronic mRNA.

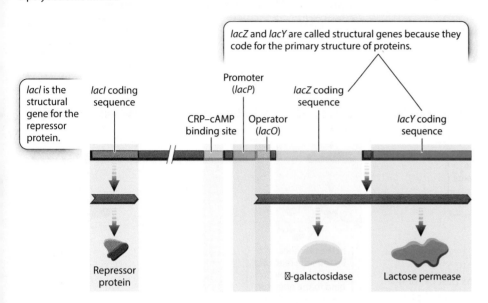

Without a functional product from *lacY*, lactose cannot enter the cell, and without a functional product from *lacZ*, lactose cannot be cleaved into its component sugars. A functional form of β-galactosidase ($lacZ^+$) and of permease ($lacY^+$) are both essential for the utilization of lactose and for cell growth.

Regulation of the *lacZ* and *lacY* structural genes is controlled by the product of another structural gene, called *lacI* (purple in Fig. 18.15), which encodes a repressor protein. Located between *lacI* and *lacZ* are a series of regulatory sequences for *lacZ* and *lacY* that include a promoter, *lacP* (light green), whose function is to recruit the RNA polymerase complex and initiate transcription, and an **operator**, *lacO* (orange), which is the binding site for the repressor. Another regulatory region is a binding site for a protein called CRP, which is discussed later.

The *lacZ* and *lacY* genes share a single set of regulatory elements and are transcribed together—a gene organization that is common in bacteria. Typically, a group of functionally related genes are located next to one another along the bacterial DNA and they share a promoter. Subsequently, when the coding sequences of the structural genes are transcribed from the promoter, they are transcribed together into a single molecule of mRNA. Such an mRNA is called a **polycistronic mRNA** ("cistron" is an old term for "coding sequence"). In Fig. 18.15, the polycistronic RNA includes the coding sequences for β-galactosidase and lactose permease. The region of DNA consisting of the promoter, the operator, and the coding sequence for the structural genes is called an **operon**.

Operons are found in bacteria and archaeons, whose cells can translate polycistronic mRNA molecules correctly because their ribosomes can initiate translation anywhere along an mRNA that contains a proper ribosome-binding site (Chapter 4). In a polycistronic mRNA, each of the coding sequences is preceded by a ribosome-binding site, so translation can be initiated at each coding sequence.

The repressor protein binds with the operator and prevents transcription, but not in the presence of lactose.

The lactose operon is negatively regulated by the repressor protein encoded by the *lacI* gene. That is, the structural genes of the lactose operon are always expressed unless the operon is turned off by a regulatory molecule, in this case the repressor. **Fig. 18.16a** shows what the operon looks like in the absence of lactose. The *lacI* gene, which is encoding the repressor protein, is expressed constantly at a low level. The repressor protein binds with the operator (*lacO*), the RNA polymerase complex is not recruited, and transcription does not take place.

The configuration of the lactose operon in the presence of lactose is shown in **Fig. 18.16b**. When lactose is present in the cell, the repressor protein is unable to bind to the operator, RNA polymerase is recruited, and transcription occurs. In other words, lactose acts as an inducer of the lactose operon because it prevents binding of the repressor protein and results in derepression. The inducer is not actually lactose itself, but rather an isomer of lactose called allolactose, which differs in the way the sugars are linked. Lactose in the cell is always accompanied by a small amount of allolactose, and so induction of the lactose operon occurs in the presence of lactose.

The binding of the inducer to the repressor results in an allosteric change in repressor structure that inhibits the protein's ability to bind to the operator. The absence of repressor from the operator allows the RNA polymerase complex to be recruited to the promoter, and the polycistronic mRNA is produced. The resulting lactose permease allows lactose to be transported into the cell on a large scale, and β-galactosidase cleaves the molecules to allow the constituents to be used as a source of energy and carbon. Therefore, the lactose operon is an example of negative regulation by a repressor, whose function is modulated by an inducer.

The function of the lactose operon was revealed by genetic studies.

Although the interactions shown in Fig. 18.16 have since been confirmed by direct biochemical studies, the inferences about how the repressor and operator work were originally

FIG. 18.16 Repression and induction of the lactose operon. (a) In the absence of lactose, transcription is repressed. (b) In the presence of lactose, transcription is induced.

a. In the absence of lactose, the lactose operon is repressed.

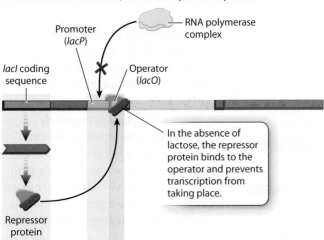

b. In the presence of lactose, the lactose operon is induced.

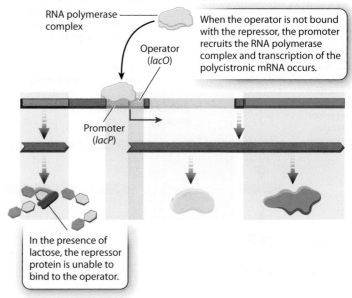

drawn from studies of mutations (**Fig. 18.17**). As part of their investigation, Jacob and Monod identified bacterial mutants that always expressed β-galactosidase and permease, even in the absence of lactose. The phenotype of a cell carrying such a mutation is said to be **constitutive** for production of the proteins. Constitutive expression means that expression occurs continuously. The most common constitutive phenotype resulted from a mutation in the *lacI* gene (*lacI*⁻ mutants) that produced a defective repressor protein (Fig. 18.17a).

Jacob and Monod also conducted experiments in which *E. coli* contained not one but two lactose operons (Fig. 18.17b). In bacterial cells containing one mutant and one nonmutant copy of the *lacI* repressor gene in the absence of lactose, expression of the structural genes was no longer constitutive but instead showed normal regulation. This finding is consistent with the idea that the nonmutant form of the gene produces a diffusible protein, which implies that the normal repressor is able to bind to and repress transcription from both operons, not just the one to which it is physically linked.

A much less common class of constitutive mutants identified the operator (Fig. 18.17c). Genetic studies of the mutations in these cells showed that they were not located in the *lacI* gene that encodes the repressor but, instead, a genetic element closer to the *lacZ* sequence, called the lactose operator (*lacO*). The constitutive mutations were designated *lacO*ᶜ ("c" for "constitutive"). When two different lactose operons, one normal and one with the *lacO*ᶜ mutation, were in the same cell in the absence of lactose, the operon carrying *lacO*ᶜ was transcribed constitutively, even in the presence of the normal repressor, because the repressor was unable to bind to the mutant operator site in *lacO*ᶜ (Fig. 18.17d). Jacob and Monod wrote that "to explain this effect, it seems necessary to invoke a new type of genetic entity, called an 'operator,' which would be: (a) adjacent to the group of [structural] genes and would control their [transcriptional] activity; and (b) would be sensitive to the repressor produced by a particular regulatory gene."

The lactose operon is also positively regulated by CRP–cAMP.

Fig. 18.16 shows how the ability of the repressor to bind with either the operator (in the absence of lactose) or with the inducer (in the presence of lactose) provides a simple and elegant way for the bacterial cell to transcribe the genes needed for lactose utilization only in the presence of lactose. The elucidation of these interactions was as far as the research tools used by Jacob and Monod could take them. Since the original experiments, the lactose operon has been studied in much greater detail and additional levels of regulation have been discovered.

One of these additional levels involves the CRP binding site shown in light blue in Fig. 18.15, which in **Fig. 18.18** is occupied by a protein called the CRP–cAMP complex. The CRP–cAMP complex is a positive regulator of the lactose operon, which you will recall is a protein that activates gene expression upon binding DNA. "CRP" stands for "cAMP receptor protein," and the CRP–cAMP complex is a molecule of CRP bound to a molecule of cAMP. The role of CRP–cAMP is to provide another level of control of transcription that is more sensitive to the nutritional needs of the cell than the level of control provided by the presence or absence of lactose. *E. coli* can utilize many kinds of molecules as sources of energy. When more than one type of energy source is available in the environment, certain sources are used before others. For example, glucose is preferred to lactose, and lactose is preferred to glycerol. The CRP–cAMP complex helps regulate which compounds are utilized.

FIG. 18.17 Lactose operon regulatory mutants. Jacob and Monod's results with repressor and operator mutants. (a) A cell with a mutant repressor protein produces the enzymes constitutively. (b) Addition of a nonmutant repressor restores normal regulation. (c) A cell with a mutant operator region produces the enzymes constitutively. (d) Addition of a nonmutant operator region fails to restore normal regulation, and the enzymes continue to be produced constitutively.

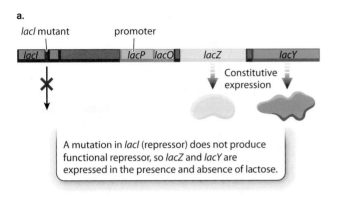

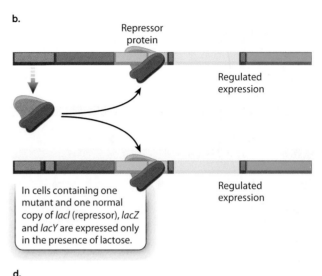

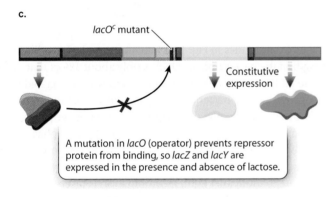

The concentration of the small molecule cAMP in the cell is a signal about the nutritional state of the cell. When there are low levels of glucose, cAMP levels are high. cAMP binds to CRP, changing the shape of CRP so that it can bind at a site near the operator and stimulate binding of RNA polymerase to transcribe *lacZ* and *lacY* when lactose is present in the cell (Fig. 18.18a). In this way, cAMP is an allosteric activator of CRP binding. However, if lactose is not present in the cell, the lactose repressor binds to the lactose operator and prevents transcription, even in the presence of the cAMP–CRP complex.

When glucose is abundant, cAMP levels are low, and the cAMP–CRP complex does not bind the lactose operon. As a result, even in the presence of lactose, the lactose operon is not transcribed to high levels (Fig. 18.18b). In this way, *E. coli* preferentially utilizes glucose when both glucose and lactose are present, and it utilizes lactose only when glucose is depleted.

Transcriptional regulation determines the outcome of infection by a bacterial virus.

Transcriptional regulation has also been well studied in viruses. Bacterial cells are susceptible to infection by a variety of viruses known as **bacteriophages** ("bacteriophage" literally means "bacteria-eater," and is often shortened to just "phage"). Among bacteriophages is a type that can undergo one of two fates when infecting a cell. The best-known example is bacteriophage λ (lambda), which infects cells of *E. coli*. The possible results of λ infection are illustrated in **Fig. 18.19**.

Upon infection, the virus injects its linear DNA genome into the bacterial cell, and almost immediately the ends of the molecule join to form a circle. In normal cells growing in nutrient medium, the usual outcome of infection is the **lytic pathway**, shown on the left side of Fig. 18.19. In the lytic pathway, the virus hijacks the cellular machinery to replicate the viral genome and produce viral proteins. After about an hour, the infected cell undergoes **lysis** and bursts open to release a hundred or more progeny phage that are capable of infecting other bacterial cells.

FIG. 18.18 The CRP–cAMP complex, a positive regulator of the lactose operon. (a) In the absence of large amounts of glucose, cAMP levels are high and the CRP–cAMP complex binds to a site near the promoter, where it activates transcription. (b) In the presence of large amounts of glucose, cAMP levels are low and the CRP–cAMP complex does not bind, so transcription is not induced to high levels, even in the presence of lactose.

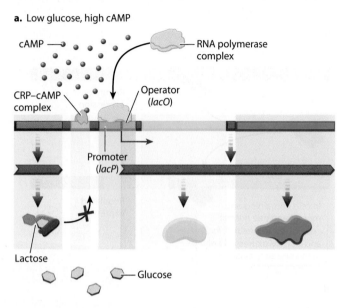

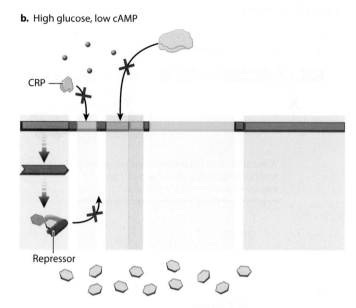

FIG. 18.19 Alternative outcomes of infection by bacteriophage λ. The bacteriophage can enter either the lytic or the lysogenic pathway.

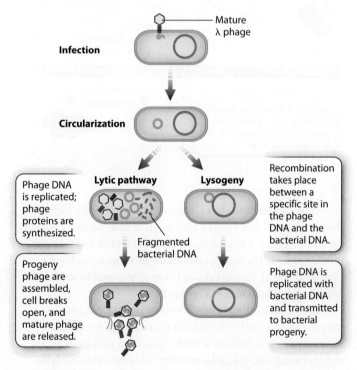

The alternative to the lytic pathway is **lysogeny**, shown on the right side of Fig. 18.19. In lysogeny, the bacteriophage DNA integrates into the bacterial DNA by undergoing recombination at a specific site in both molecules. Lysogeny often takes place in cells that are growing in poor conditions. The relative sizes of the DNA molecules in Fig. 18.19 are not to scale. In reality, the length of the bacteriophage DNA is only about 1% of that of the bacterial DNA. When the bacteriophage DNA is integrated by lysogeny, the only bacteriophage gene transcribed and translated is one that represses the transcription of other phage genes, preventing entry into the lytic pathway. The bacteriophage DNA is replicated along with the bacterial DNA and transmitted to the bacterial progeny when the cell divides. Under stress, such as exposure to ultraviolet light, recombination is reversed, freeing the phage DNA and initiating the lytic pathway.

At the molecular level, the choice between the lytic and lysogenic pathways is determined by the positive and negative regulatory effects of a small number of bacteriophage proteins produced soon after infection. Which pathway results depends on the relative production of a protein known as cro and that of another protein known as cI. If the production of cro predominates, the lytic pathway results; if cI predominates, the lysogenic pathway takes place.

Fig. 18.20a shows the small region of the bacteriophage DNA in which the key interactions take place. Almost immediately after infection and circularization of the bacteriophage DNA, transcription takes place from the promoters P_L and P_R. Transcription of genes controlled by the P_R promoter results in a transcript encoding the proteins cro and cII. The cro protein represses transcription of a gene controlled by another promoter P_M, whose transcript encodes the protein cI. In normal cells growing in nutrient medium, proteases present in the bacterial cell degrade cII and prevent its accumulation. With cro protein preventing *cI* expression and cII protein being unable to accumulate, transcription of bacteriophage genes in the lytic pathway takes place, including those genes

FIG. 18.20 Transcriptional regulation of *cI* and *cro* genes, which determines whether the viral pathway is lytic (a) or lysogenic (b).

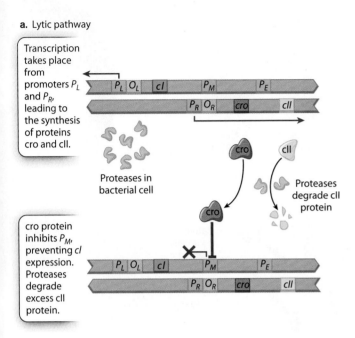
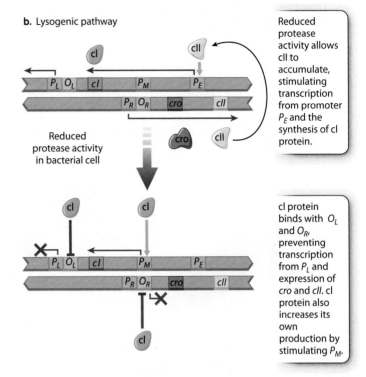

needed for bacteriophage DNA replication, those encoding proteins in the bacteriophage head and tail, and, finally, those needed for lysis.

Alternatively, in bacterial cells that are growing in poor conditions (**Fig. 18.20b**), reduced protease activity allows cII protein to accumulate. When cII protein reaches a high enough level, it stimulates transcription from the promoter P_E. Transcription from P_E proceeds through P_M and continues farther to include the coding sequence for cI protein. The cI protein has three functions:

- It binds with the operator O_R and prevents further expression of *cro* and *cII*.
- It stimulates transcription of its own coding sequence from the promoter P_M, establishing a positive feedback loop that keeps the level of cI protein high.
- It binds with the operator O_L and prevents further transcription from P_L.

The result is that cI production shuts down transcription of all bacteriophage genes except its own gene, and this is the regulatory state that produces lysogeny. (The protein needed for integration of the bacteriophage DNA into the bacterial DNA is produced by transcription from the P_L promoter before it is shut down by cI.) When cells that have undergone lysogeny are exposed to ultraviolet light or certain other stresses, the cI protein is degraded. In this case, cro and cII are produced again, and the lytic pathway follows.

Regulation of the lytic and lysogenic pathways works to the advantage of bacteriophage λ, but the process is not like something an engineer might design. That is because biological systems are not engineered—they evolve. Regulatory mechanisms are built up over time by the selection of successive mutations. Each evolutionary step refines the regulation in such a way as to be better adapted to the environment than it was previously. Each successive step occurs only because it increases survival and reproduction.

We summarize these two alternative outcomes of infection by a bacterial virus, and the earlier discussions of viral diversity, replication, host range, and effects on a cell, in **Fig. 18.21**.

Self-Assessment Questions

8. How can the binding of a molecule as small as a single amino acid change the ability of a large, multi-subunit protein molecule to bind a specific DNA sequence?
9. What would be the consequence of a mutation in the *lacI* repressor gene that produces repressor protein that is able to bind to the operator but is not able to bind allolactose?
10. What is the role of the CRP–cAMP complex in positive regulation of the lactose operon in *E. coli*?
11. How are the lytic and lysogenic pathways regulated?

VISUAL SYNTHESIS FIG. 18.21 — Virus: A Genome in Need of a Cell
Integrating concepts from Chapters 1, 4, 11, 13, 14, and 18

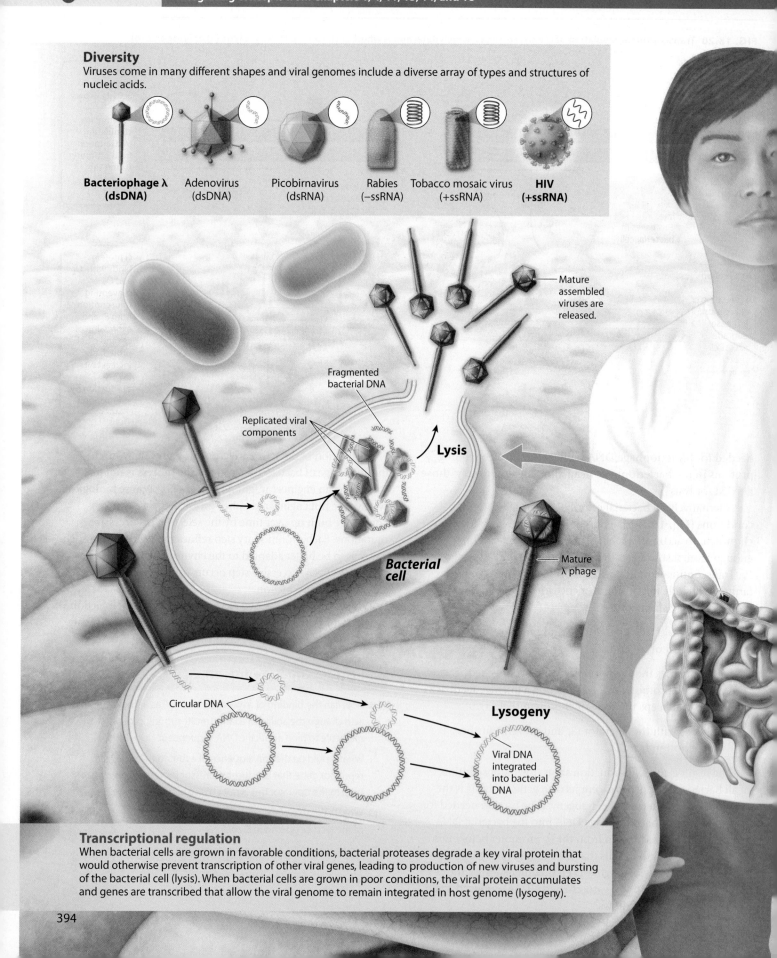

Diversity
Viruses come in many different shapes and viral genomes include a diverse array of types and structures of nucleic acids.

- Bacteriophage λ (dsDNA)
- Adenovirus (dsDNA)
- Picobirnavirus (dsRNA)
- Rabies (−ssRNA)
- Tobacco mosaic virus (+ssRNA)
- HIV (+ssRNA)

Mature assembled viruses are released.
Fragmented bacterial DNA
Replicated viral components
Lysis
Bacterial cell
Mature λ phage
Circular DNA
Lysogeny
Viral DNA integrated into bacterial DNA

Transcriptional regulation
When bacterial cells are grown in favorable conditions, bacterial proteases degrade a key viral protein that would otherwise prevent transcription of other viral genes, leading to production of new viruses and bursting of the bacterial cell (lysis). When bacterial cells are grown in poor conditions, the viral protein accumulates and genes are transcribed that allow the viral genome to remain integrated in host genome (lysogeny).

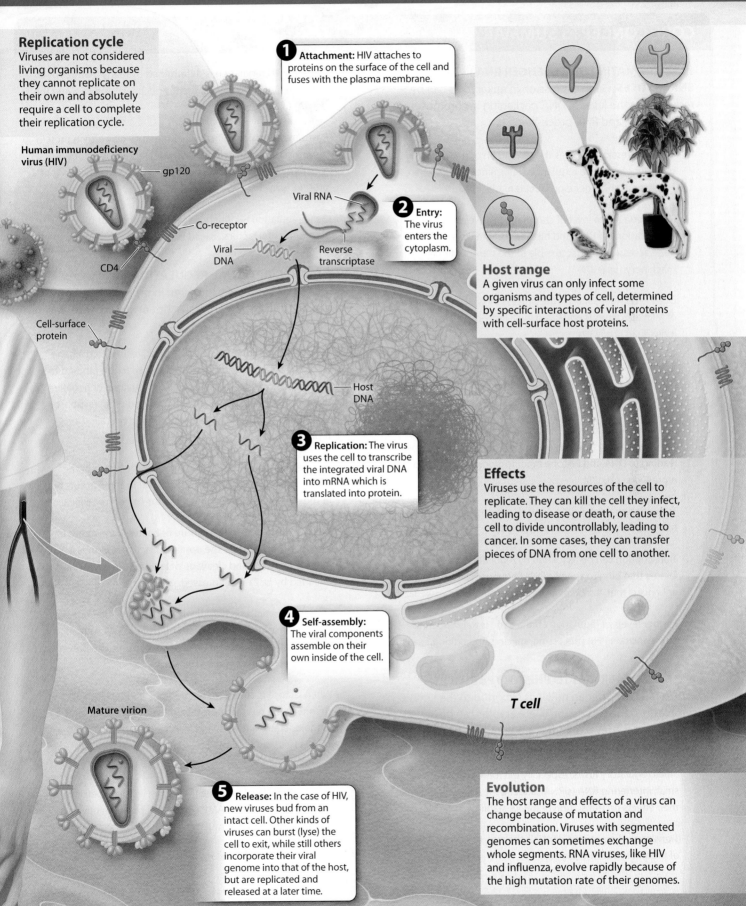

CORE CONCEPTS SUMMARY

18.1 CHROMATIN TO MESSENGER RNA IN EUKARYOTES: Gene expression in eukaryotes can be regulated at the level of DNA packaging in chromosomes, transcription, and RNA processing.

Gene expression involves the turning on or turning off of a gene, whereas gene regulation determines where, when, how much, and which gene product is made. page 377

Regulation at the level of chromatin involves chemical modifications of DNA and histones that make a gene accessible or inaccessible to the transcriptional machinery. page 378

Dosage compensation is the process by which the expression of X-linked genes is equalized in *XX* individuals and *XY* individuals. page 380

X-inactivation, the mechanism of dosage compensation for sex chromosomes in mammals, involves the inactivation of one of the two *X* chromosomes in females. page 380

X-inactivation occurs by the transcription of a noncoding RNA known as *Xist*, which coats the entire *X* chromosome, leading to DNA and histone modifications and transcriptional repression. page 381

Transcriptional regulation controls whether or not transcription of a gene occurs. page 381

Transcription can be regulated by regulatory transcription factors that bind to specific DNA sequences known as enhancers that can be near, in, or far from genes. page 382

Further levels of regulation after a gene is transcribed to mRNA include RNA processing, splicing, and editing. page 382

18.2 MESSENGER RNA TO PHENOTYPE IN EUKARYOTES: After an mRNA is exported to the cytoplasm, gene expression in eukaryotes can be regulated at the level of mRNA stability, translation, and posttranslational modification of proteins.

Small regulatory RNAs, especially microRNA (miRNA) and small interfering RNA (siRNA), affect gene expression through their effects on translation or mRNA stability. page 384

Translational regulation controls the rate, timing, and location of protein synthesis. page 384

Translational regulation is determined by many features of an mRNA molecule, including the 5′ and 3′ UTR, the cap, and the poly(A) tail. page 385

Posttranslational modification comes into play after a protein is synthesized, and includes chemical modification of side groups of amino acids, affecting the structure and activity of a protein. page 385

Gene regulation is influenced by both genetic and environmental factors. page 386

18.3 TRANSCRIPTIONAL REGULATION IN PROKARYOTES: Gene expression in prokaryotes can be regulated at the level of transcription in response to environmental conditions.

Transcriptional regulation can be positive, in which a gene is usually off and is turned on in response to the binding to DNA of a regulatory protein called an activator, or negative, in which a gene is usually on and is turned off in response to the binding to DNA of a regulatory protein called a repressor. page 386

Jacob and Monod demonstrated how production of the enzymes encoded in the lactose operon in *E. coli* occurs only in the presence of lactose in the culture medium. page 388

When lactose is added to culture of bacteria, the genes for enzymes that facilitate the uptake of lactose (permease, encoded by *lacY*) and cleavage of lactose (β-galactosidase, encoded by *lacZ*) are expressed. page 388

The lactose operon is negatively regulated by the repressor protein (encoded by *lacI*), which binds to a DNA sequence known as the operator when lactose is absent and prevents transcription of *lacY* and *lacZ*. page 389

When lactose is added to the medium, it induces an allosteric change in the repressor protein, preventing it from binding to the operator and allowing transcription of *lacY* and *lacZ*. In this way, lactose acts as an inducer of the lactose operon. page 389

An additional level of regulation of the lactose operon is provided by the CRP–cAMP complex, a positive activator of transcription that stimulates the operon when glucose is scarce. page 390

The lytic and lysogenic pathways of bacteriophage λ have also been well studied as a model of gene regulation. When bacteriophage λ infects *E. coli,* it can lyse the cell (the lytic pathway) or its DNA can become integrated into the bacterial genome (the lysogenic pathway). page 391

In infection of *E. coli* cells by bacteriophage λ, predominance of cro protein results in the lytic pathway, whereas predominance of the cI protein results in the lysogenic pathway. page 392

Log in to LaunchPad to check your answers to the Self-Assessment Questions and to access additional learning tools.

CHAPTER 19　Genes and Development

CORE CONCEPTS

19.1 GENETIC BASIS OF DEVELOPMENT: In the development of humans and other animals, stem cells become progressively more restricted in their possible pathways of cellular differentiation.

19.2 HIERARCHICAL CONTROL: The genetic control of development is a hierarchy in which genes are activated in groups that, in turn, regulate the next set of genes.

19.3 MASTER REGULATORS: Many master regulators that play key roles in development are evolutionarily conserved but can have dramatically different effects in different organisms.

19.4 COMBINATORIAL CONTROL: Combinatorial control is a developmental strategy in which cellular differentiation depends on the particular combination of transcription factors present in a cell.

19.5 CELL SIGNALING IN DEVELOPMENT: Ligand–receptor interactions activate signal transduction pathways that converge on transcription factors and genes that determine cell fate.

Altogether, the human body contains about 200 different types of cell, all of which derive from a single cell, the zygote. Some cells derived from the zygote become muscle cells, others nerve cells, and still others liver cells. All of these cells have almost exactly the same genome: they differ not in their content of genes, but instead in the groups of genes that are expressed or repressed. In other words, these cell types differ as a result of gene regulation, which was discussed in Chapter 18.

Gene regulation is especially important in any multicellular organism because it underlies **development**, the process in which a fertilized egg undergoes multiple rounds of cell division to become an embryo with specialized tissues and organs. During development, cells undergo changes in gene expression as genes are turned on and off at specific times and places. Gene regulation causes cells to become progressively more specialized, a process known as **differentiation**.

In this chapter, we focus on the general principles by which genes control development. We will see that, as cells differentiate along one pathway, they progressively lose their ability to differentiate along other pathways. Yet gene expression can sometimes be reprogrammed to reopen pathways of differentiation that had previously been shut off, a process that has important medical applications in therapeutic replacement of diseased or damaged tissue. From a broader perspective, the study of evolutionary changes in developmental processes constitutes the field of evolutionary developmental biology, which is often called **evo-devo**. We will see that, although some of the key molecular mechanisms of development are used over and over again in different organisms, they have evolved to yield such differences in shape and form as to conceal the underlying similarity in mechanism.

19.1 GENETIC BASIS OF DEVELOPMENT

Genetic programs and computer programs have some features in common. Computer code is written as a linear string of letters, analogous to the sequence of nucleotides in genomic DNA. Once a computer program is initiated, it automatically runs and performs its coded task. Small mistakes in the code, like mutations, can have big consequences and even cause the program to crash.

The analogy between genetic programs and computer programs has an important limitation. Computer programs are consciously written, whereas genetic programs evolve. The genetically encoded developmental programs of all living organisms emerged over billions of years through mutation and natural selection. These developmental programs changed through time, persisting only if they produced organisms that could successfully survive and reproduce in the existing environment. Here, we explore the genetic program of development—that is, the genetic instructions that lead a single fertilized egg to become a complex multicellular organism.

The fertilized egg is a totipotent cell.

In all sexually reproducing organisms, the fertilized egg is special because of its developmental potential. The fertilized egg is said to be **totipotent**, which means that it can give rise to a complete organism. In mammals, the egg also forms the membranes that surround and support the developing embryo (Chapter 40).

In humans, after fertilization, the fertilized egg, or **zygote**, undergoes successive mitotic cell divisions as it moves along a fallopian tube. One cell becomes two, two become four, four become eight, eight become sixteen, and so on, with all the cells contained within the egg's tough outer glycoprotein layer (**Fig. 19.1**). Within 4–5 days after fertilization, the zygote has turned into a ball of cells called the **morula** and has traveled from the site of fertilization in one of the fallopian tubes to the uterus.

These early cell divisions are different from the mitotic cell divisions that occur later in life because the cells do not grow between divisions; they merely replicate their chromosomes and divide again. The result is that the cytoplasm of the egg is partitioned into smaller and smaller cells, with the new cells all bunched together inside the membrane that surrounds the developing embryo.

Cell division continues in the morula until there are a few thousand cells. The cells then begin to move relative to one another, pushing against and expanding the membrane that encloses them and rearranging themselves to form a hollow sphere called a **blastocyst** (Fig. 19.1). Attached to the inside of the wall in one region of the blastocyst is a group of cells known as the **inner cell mass**, from which the body of the embryo develops. The wall of the blastocyst forms several membranes that envelop and support the developing embryo. Once the blastocyst forms, it implants in the uterine wall.

Once implanted in the uterine wall, the multiplying cells of the inner cell mass migrate and reorganize to form a **gastrula**, in which the cells of the blastula become organized into three **germ layers** (Fig. 19.1). The germ layers are sheets of cells called the ectoderm, mesoderm, and endoderm. These layers of cells differentiate further into specialized cells. Those formed from the **ectoderm** include the outer layer of the skin and nerve cells in the brain; cells formed from the **mesoderm** include cells that make up the inner layer of the skin, muscle cells, and red blood cells; and cells formed from the **endoderm** include cells of the lining of the digestive tract and lung, as well as liver cells and pancreas cells (Chapter 40).

Cellular differentiation increasingly restricts alternative fates.

At each successive stage in development in which the cells differentiate, they lose the potential to develop into any kind of cell. The fertilized egg is a totipotent cell because it can differentiate into both the inner cell mass and supporting membranes, and eventually into an entire organism. The cells of the inner cell mass, called embryonic stem cells, are **pluripotent** because they can give rise to any of the three germ layers, and therefore to any cell of the body. However, pluripotent cells cannot give rise to an entire organism on their own, as a totipotent cell can. Cells further along in differentiation are **multipotent**; these cells can form a limited number of types of specialized cell. Cells of the germ layers are multipotent because they can give rise only to the cell types specified for each germ layer in Figure 19.1. Totipotent, pluripotent, and multipotent cells are all **stem cells**, which are cells that are capable of differentiating into different cell types.

Why do differentiating cells increasingly lose their developmental potential? One hypothesis focuses on gene regulation. When cells become committed to a particular developmental pathway, genes that are no longer needed are turned off (that is, repressed) and are difficult to turn on again. Another hypothesis is genome reduction: as cells become differentiated, they delete DNA for genes they no longer need.

These hypotheses can be distinguished by an experiment in which differentiated cells are placed in an environment that reprograms them to mimic earlier states. If loss of developmental potential is due to gene regulation, then differentiated cells could be reprogrammed to become pluripotent or multipotent.

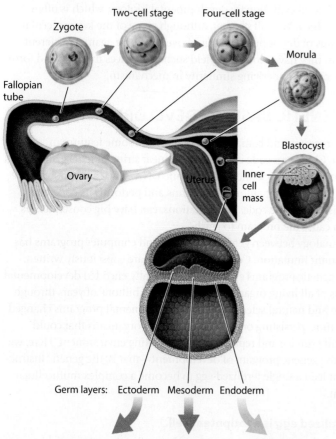

FIG. 19.1 Early development of a human embryo. The zygote is a totipotent cell because its daughter cells can develop into any cell type and eventually into a complete organism.

If loss of developmental potential is due to genome reduction, then differentiated cells could not be reprogrammed to become pluripotent or multipotent because the genes for alternate developmental routes are no longer present.

British developmental biologist John Gurdon carried out such experiments in the early 1960s (**Fig. 19.2**). Gurdon used a procedure called **nuclear transfer**, in which a hollow glass needle is used to insert the nucleus of a cell into the

HOW DO WE KNOW?

FIG. 19.2

How do stem cells lose their ability to differentiate into any cell type?

BACKGROUND During differentiation, cells become progressively more specialized and restricted in their fates. Early studies left the mechanisms of differentiation unclear.

HYPOTHESIS One hypothesis is that differentiation occurs when patterns of gene expression in the cell change as it develops and genes that are no longer needed are repressed. An alternative hypothesis is that differentiation occurs as a result of genome reduction, in which genes that are not needed are deleted.

EXPERIMENT John Gurdon carried out experiments in the amphibian *Xenopus laevis* to test these hypotheses. He transferred nuclei from differentiated cells (in this case, intestinal cells) into unfertilized eggs whose nuclei had been inactivated with ultraviolet light. Despite the nucleus being degraded, the egg cytoplasm still contains the molecules that would have kept the nucleus in an undifferentiated state. If differentiation is due to changes in gene expression, then the molecules in the egg cytoplasm should be able to reprogram the differentiated nucleus into an earlier state, and the reprogrammed nucleus should be able to differentiate again into all the cells of a tadpole. If differentiation is accompanied by loss of genes, then differentiation is irreversible and development cannot proceed.

RESULTS The experiment was carried out 726 times. In 716 cases, development did not occur; in 10 cases, development proceeded normally.

CONCLUSION Although the experiment succeeded in only 10 of 726 attempts, it demonstrated that the nucleus of a differentiated cell and the cytoplasm of the unfertilized egg were able to support complete development of a normal animal. This result allowed Gurdon to reject the hypothesis that differentiation occurs by the loss of genes. The first hypothesis—that cells become differentiated as a result of changes in gene expression—was supported. However, because of the small number of successes in reprogramming, additional experiments were needed to validate the conclusions.

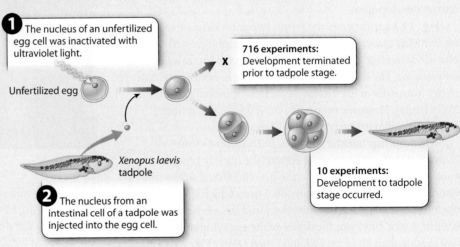

FOLLOW-UP WORK Gurdon's work was controversial. Some critics argued that the successful experiments resulted from a small number of undifferentiated cells present in intestinal epithelium. Others accepted the conclusion but expressed misgivings about possible applications to humans. Later experiments that succeeded in cloning mammals from fully differentiated cells confirmed Gurdon's original conclusion.

SOURCE Gurdon, J. B. 1962. "The Developmental Capacity of Nuclei Taken from Intestinal Epithelium Cells of Feeding Tadpoles." *Journal of Embryology & Experimental Morphology* 10:622–640.

FIG. 19.3 Results of nuclear transfer of differentiated cells. Data from: J. B. Gurdon and D. A. Melton, "Nuclear Reprogramming in Cells," 2008, Science 322:1811–1815.

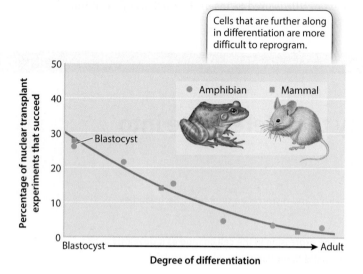

FIG. 19.4 Celebrity clones and their genetic mothers. (a) Dolly; (b) CopyCat. Sources: a. JOHN CHADWICK/AP Images; b. PAT SULLIVAN/AP Images.

cytoplasm of an egg whose own nucleus has been destroyed or removed. Previous nuclear transfer experiments had been carried out in the leopard frog, *Rana pipiens*. Whereas nuclei from pluripotent or multipotent cells could often be reprogrammed to develop into normal tadpoles, attempts with nuclei from fully differentiated cells failed. In other words, Gurdon's findings supported the first hypothesis: all of the same genes are present in intestinal cells as in early embryonic cells, but some of the genes are turned off, or repressed, during development.

Fig. 19.3 summarizes the results of many nuclear transfer experiments carried out in mammals and amphibians. The percentage of reprogramming experiments that fail increases as cells differentiate. The best chance of success is to use pluripotent nuclei from cells in the blastocyst (or its amphibian equivalent, the blastula). However, even some experiments using nuclei from fully differentiated cells have been successful.

When nuclear transfer succeeds, the result is a **clone**—an individual that carries an exact copy of the nuclear genome of another individual. In this case, the new individual shares the same genome as that of the individual from which the donor nucleus was obtained. (The mitochondrial DNA is not from the nuclear donor, but from the donor of the egg cytoplasm.) The first mammalian clone was a lamb called Dolly (**Fig. 19.4a**), born in 1996. She was produced from the transfer of the nucleus of a cell in the mammary gland of a sheep to an egg cell with no nucleus, and was the only successful birth among 277 nuclear transfers. Successful cloning in sheep soon led to cloning in cattle, pigs, and goats.

The first household pet to be cloned was a kitten named CopyCat (**Fig. 19.4b**), born in 2001 and derived from the nucleus of a differentiated ovarian cell. CopyCat was the only success among 87 attempts. As shown in Fig. 19.4b, the cat from which the donor nucleus was obtained was a calico, but CopyCat herself was not, even though the two cats were clones of each other. The reason for their different appearance has to do with X-inactivation, discussed in Chapter 18. Recall that the mottled orange and black calico pattern results from random inactivation of one of the two X chromosomes during development. The lack of a calico pattern in CopyCat implies that the X chromosomes in the transferred nucleus did not "reset" as they do in normal embryos. Instead, the inactive

X in the donor nucleus remained inactive in all the cells in the clone. Thus, although CopyCat and her mother share the same nuclear genome, the genes were not expressed in the same way because of irreversible epigenetic regulation in the donor nucleus.

CASE 3 YOUR PERSONAL GENOME: YOU, FROM A TO T

Can cells with your personal genome be reprogrammed for new therapies?

Stem cells play a prominent role in **regenerative medicine**, which aims to use the natural processes of cell growth and development to replace diseased or damaged tissues. Stem cells are already used in bone marrow transplantation and may someday be used to treat Parkinson's disease, Alzheimer's disease, heart failure, certain types of diabetes, severe burns and wounds, and spinal cord injury.

At first, it seemed as though the use of embryonic stem cells gave the greatest promise for regenerative medicine because of their pluripotency. This approach proved ethically controversial because obtaining embryonic stem cells requires the destruction of human blastocysts—that is, early-stage embryos. A major breakthrough occurred in 2006 when Japanese scientists demonstrated that adult cells can be reprogrammed by activation of just a handful of genes, most of them encoding transcription factors or chromatin proteins. The reprogrammed cells were pluripotent and were therefore called induced **pluripotent stem cells (iPS cells)**.

The success rate was only about one iPS cell per thousand, and the genetic engineering technique required the use of viruses that can sometimes cause cancer. Nevertheless, the result was regarded as spectacular. Other researchers soon found other genes that could be used to reprogram adult cells into pluripotent or multipotent stem cells, and still other investigators developed virus-free methods for delivering the genes. In recent years, researchers have even discovered small organic molecules that can reprogram adult cells.

This kind of reprogramming opens the door to personalized stem cell therapies. The goal is to create stem cells derived from the adult cells of the individual patient. Because these cells contain the patient's own genome, problems with tissue rejection are minimized or eliminated (Chapter 41). There remains much to learn before therapeutic use of induced stem cells becomes routine. Researchers will face challenges such as increasing the efficiency of reprogramming, verifying that reprogramming is complete, making sure that the reprogrammed cells are not prone to cancer, and demonstrating that the reprogrammed cells differentiate as they should. Nevertheless, researchers hope that someday soon, your own cells containing your personal genome could be reprogrammed to restore cells or organs damaged by disease or accident.

Self-Assessment Questions

1. The terms totipotent, pluripotent, and multipotent derive from Latin roots: *toti* ("all"), *pluri* ("more"), *multi* ("many"), and *potent* ("power"). Do these terms accurately convey the developmental potential of each type of cell?

2. From what you know about embryonic development, do you think that a cell from the inner cell mass or one from the ectoderm has more developmental potential? Why?

3. Explain how an individual's own cells might be used in stem cell therapy.

4. X-inactivation can result in two clones of a cell that differ in the genes they express. Can you think of other reasons why two genetically identical individuals might look different from each other?

19.2 HIERARCHICAL CONTROL

During development of a complex multicellular organism, many genes are activated and repressed at different times, thus restricting cell fates. One of the key principles of development is that genes expressed early in an organism's development control the activation of other groups of genes that act later in development. Gene regulation during development is therefore **hierarchical**: genes expressed at each stage in the process control the expression of genes that act later.

Drosophila development proceeds through egg, larval, and adult stages.

The fruit fly *Drosophila melanogaster* has played a prominent role in our understanding of the genetic control of early development, particularly the hierarchical control of development. Researchers have isolated and analyzed a large number of mutant genes that lead to a variety of defects at different stages in development. These studies have revealed many of the key genes and processes in development, which are the focus of the following sections.

The major events in *Drosophila* development are illustrated in **Fig. 19.5**. DNA replication and nuclear division begin soon after the egg and sperm nuclei fuse (Fig. 19.5a). Unlike in mammalian development, the early nuclear divisions in the *Drosophila* embryo occur without cell division; therefore, the embryo consists of a single cell with many nuclei in the center (Fig. 19.5b). When there are roughly 5000 nuclei, they migrate to the edge of the cell (Fig. 19.5c), where each nucleus becomes enclosed in its own cell membrane, and together they form the **cellular blastoderm** (Fig. 19.5d).

Then begins the process of **gastrulation**, in which the cells of the blastoderm migrate inward, creating layers of cells within the embryo. As in humans and most other animals (section 19.1

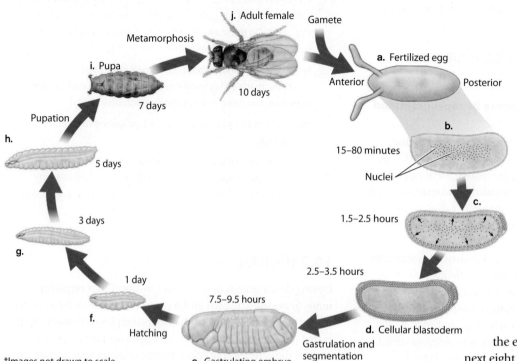

FIG. 19.5 Life cycle of the fruit fly *Drosophila melanogaster*. The life cycle begins with (a) a fertilized egg, followed by (b–e) a developing embryo, (f–h) larval stages, (i) pupa, and (j) adult.

and Chapter 40), gastrulation forms the three germ layers (ectoderm, mesoderm, and endoderm) that differentiate into different types of cell. A *Drosophila* embryo during gastrulation is shown in Fig. 19.5e. At this stage, the embryo already shows an organization into discrete parts or segments, the formation of which is known as **segmentation**. There are three cephalic segments, C1–C3 (the term "cephalic" refers to the head); three thoracic segments, T1–T3 (the thorax is the middle region of an insect); and eight abdominal segments (A1–A8). Each of these segments has a different fate in development.

About one day after fertilization, the embryo hatches from the egg as a larva (Fig. 19.5f). Over the next eight days, the larva grows and replaces its rigid outer shell, or cuticle, twice (Figs. 19.5g and 19.5h). After a week of further growth, the cuticle forms a casing—called the pupa (Fig. 19.5i)—in which the larva undergoes the dramatic developmental changes, known as metamorphosis, that give rise to the adult fruit fly.

The egg is a highly polarized cell.

The ways in which genes control development in *Drosophila* was inferred from systematic studies of mutants by Christiane Nusslein-Volhard and Eric F. Wieschaus—work for which they were awarded the 1995 Nobel Prize in Physiology or Medicine. One of their findings was that development starts even before a zygote is formed, in the maturation of the **oocyte**, which is the unfertilized egg cell produced by the mother. The oocyte, which matures under control of the mother's genes, is also important for normal embryonic development. This finding applies not only to insects like *Drosophila*, but also to many multicellular animals.

Among the striking mutants that Nusslein-Volhard and Wieschaus generated and investigated were ones that significantly affected early development (**Fig. 19.6**). These defects in very early development can be easily seen by the time the mutants reach the larval stage. In one class of mutants, called *bicoid*, larvae are missing segments at the anterior, or head, end. In another class of mutants, called *nanos*, larvae are missing segments at the posterior, or tail, end. The mutant larvae are grossly abnormal and do not survive. Nusslein-Volhard and Wieschaus were able to identify each segment that was missing

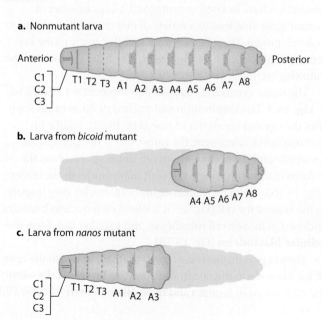

FIG. 19.6 Normal and mutant *Drosophila* larvae. (a) Nonmutant larva have anterior, middle, and posterior segments (b) *Bicoid* mutant larva lack anterior segments. (c) *Nanos* mutant larva lack posterior structures.

based on each segment's distinctive pattern of hairlike projections. In Fig. 19.6, the patterns are shown as dark shapes on each segment.

A distinguishing feature of *bicoid* and *nanos* mutants is that the abnormalities in the embryo depend on the genotype of the mother, not the genotype of the embryo. The reason the genotype of the mother can affect the phenotype of the developing embryo is that successful development requires the mother to produce a functioning oocyte before fertilization occurs. In *Drosophila* and many other organisms, the composition of the egg includes macromolecules (such as RNA and protein) synthesized by cells in the mother and transported into the egg. If the mother carries mutations in her genes that inhibit proper development of the oocyte, the offspring can be abnormal. Genes such as *bicoid* and *nanos* that are expressed by the mother but affect the phenotype of the offspring (in this case, the developing embryo and larva) are called **maternal-effect genes**.

A normal *Drosophila* oocyte is highly polarized, meaning that one end is distinctly different from the other. For example, gradients of macromolecules define the anterior–posterior (head-to-tail) axis of the embryo as well as the dorsal–ventral (back-to-belly) axis. The best known of these gradients are those of messenger RNAs (mRNAs) that are transcribed from the maternal-effect genes *bicoid* and *nanos*. **Fig. 19.7a** shows the gradient of *bicoid* mRNA across the oocyte. The bulk of *bicoid* mRNA comes from the mother, and is localized in the anterior of the egg by proteins that attach the mRNA to the cytoskeleton.

The mRNA corresponding to the maternal-effect gene *nanos* is also present in a gradient, but most of it is at the posterior end (**Fig. 19.7b**). Like *bicoid*, mRNA for *nanos* is synthesized by the mother's cells, and then imported into the oocyte during oogenesis. After fertilization, the zygote produces Bicoid and Nanos proteins from the localized mRNAs, and they have concentration gradients resembling those of the mRNAs in the oocyte (Fig. 19.7).

The anterior–posterior axis that is set up by the gradients of Bicoid and Nanos proteins is reinforced by gradients of two transcription factors called Caudal and Hunchback (**Fig. 19.8**). Like the mRNAs for Bicoid and Nanos, the mRNAs for Caudal and Hunchback are transcribed from the mother's genome and transported into the egg. As shown in Fig. 19.8a, the mRNAs for *caudal* and *hunchback* are spread uniformly in the cytoplasm of the fertilized egg. However, the mRNAs are not translated uniformly in the unfertilized egg. Bicoid protein represses translation of *caudal*, and Nanos protein represses translation of *hunchback* (Fig. 19.8b).

FIG. 19.7 Gradients of *bicoid* and *nanos* mRNA and protein in the developing embryo. (a) The mRNA and protein for *bicoid* are localized in the anterior end of the egg. (b) The mRNA and protein for *nanos* are localized in the posterior end of the egg.

a. Distribution of *bicoid* mRNA and protein in the egg

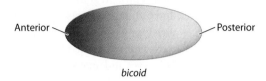

bicoid

b. Distribution of *nanos* mRNA and protein in the egg

nanos

FIG. 19.8 Caudal and Hunchback gradients in the developing embryo. (a) mRNA levels of *hunchback* and *caudal* are uniform across the embryo. (b) Hunchback and Caudal protein levels are localized to the anterior and posterior ends of the embryo, respectively, because Bicoid and Nanos control the translation of *hunchback* and *caudal* mRNA.

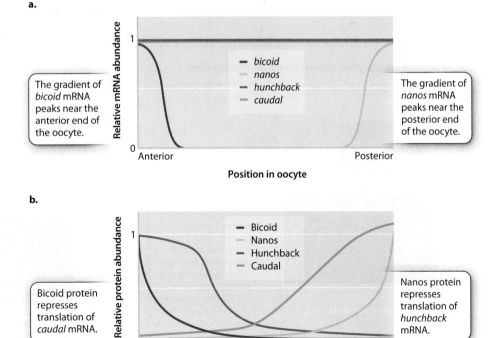

Caudal protein is therefore concentrated at the posterior end and Hunchback protein is concentrated at the anterior end. The expression of Caudal and Hunchback illustrates gene regulation at the level of translation, as discussed in Chapter 18. Bicoid protein is also a transcription factor that promotes transcription of the *hunchback* gene from zygotic nuclei, which reinforces the localization of Hunchback protein at the anterior end.

The Hunchback and Caudal gradients set the stage for the subsequent steps in development. The Hunchback transcription factor targets genes of the embryo needed for the development of anterior structures like eyes and antennae, and Caudal targets genes of the embryo needed for the development of posterior structures like genitalia. In this way, maternal genes expressed early in development influence the expression of genes of the embryo that are important in later development. Because the products of the *bicoid* and *nanos* mRNA are the ones initially responsible for organizing the anterior and posterior ends of the embryo, respectively, mothers that are mutant for *bicoid* have larvae that lack anterior structures, and mothers that are mutant for *nanos* have larvae that lack posterior structures.

Development proceeds by progressive regionalization and specification.

Nusslein-Volhard and Wieschaus also discovered mutants of the embryo's developmental genes. They discovered three classes of such mutants, and their analysis showed that genes in the embryo controlling its development are turned on in groups, and that each successive group acts to spatially restrict and further refine the pattern of differentiation generated by previous groups. This, too, is a general principle of development in many multicellular organisms.

The anterior–posterior gradient set up by the maternal-effect genes is first narrowed by genes called gap genes (**Fig. 19.9**), each of which is expressed in a broad region of the embryo. The name "gap gene" derives from the phenotype of mutant embryos, which are missing groups of adjoining segments, leaving a gap in the pattern of segments. Fig. 19.9 shows the expression pattern of the gap gene *Krüppel*, which is expressed in the middle region of the embryo. Mutants of *Krüppel* lack some thoracic and abdominal segments when they reach the larval stage. *Krüppel* is expressed in the pattern shown in Fig. 19.9 because it is under the control of the transcription factor Hunchback, which in this embryo is stained in green. High concentrations of Hunchback repress *Krüppel* transcription entirely, and low concentrations fail to induce *Krüppel* transcription. Because Hunchback is present in an anterior–posterior gradient with high concentration at the anterior end and low concentration at the posterior end (see Fig. 19.8), the pattern of Hunchback expression means that *Krüppel* is transcribed only in the middle region of the embryo where Hunchback is present but not too abundant.

The gap genes encode transcription factors that control genes in the next level of the regulatory hierarchy, which consists of pair-rule genes (**Fig. 19.10**). Pair-rule genes receive their

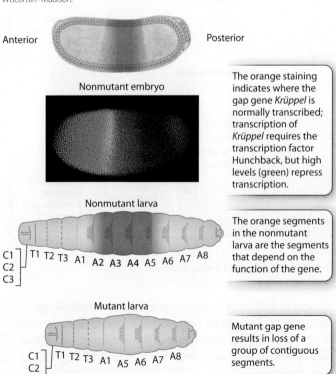

FIG. 19.9 Normal gap-gene expression pattern and mutant phenotype. *Source: James Langeland, Steve Paddock and Sean Carroll, HHMI, Univ. Wisconsin–Madison.*

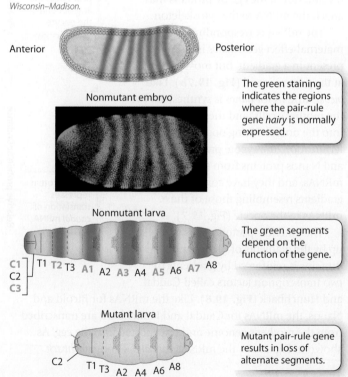

FIG. 19.10 Normal pair-rule gene expression pattern and mutant phenotype. *Source: James Langeland, Steve Paddock, and Sean Carroll, HHMI, Univ. Wisconsin–Madison.*

name because larvae with these mutations lack alternate body segments. The pair-rule gene illustrated in Fig. 19.10 is *hairy,* whose mutants lack the odd-numbered thoracic segments and the even-numbered abdominal segments. The pair-rule genes help to establish the uniqueness of each of seven broad stripes across the anterior–posterior axis.

The pair-rule genes encode a set of transcription factors that help to regulate the next level in the segmentation hierarchy, which consists of segment-polarity genes (**Fig. 19.11**). The segment-polarity genes refine the 7-striped pattern still further into a 14-striped pattern. Each of the 14 stripes has distinct anterior and posterior ends that are determined by the segment-polarity genes. Embryos with mutations in segment-polarity genes lose this anterior–posterior differentiation, with the result that the anterior and the posterior halves of each segment are mirror images. The example in Fig. 19.11 is *engrailed,* which eliminates the posterior pattern element in each stripe and replaces it with a mirror image of the anterior pattern element.

Homeotic genes determine where different body parts develop in the organism.

Together, the segment-polarity genes and other genes expressed earlier in the hierarchy control the pattern of expression of another set of genes called **homeotic (*Hox*) genes.** Originally discovered in *Drosophila,* homeotic genes encode some of the most important transcription factors in animal development. A homeotic gene is a gene that specifies the identity of a body part or segment during embryonic development. For example, homeotic genes instruct the three thoracic segments (T1, T2, and T3) each to develop a set of legs, and the second thoracic segment (T2) also to develop wings.

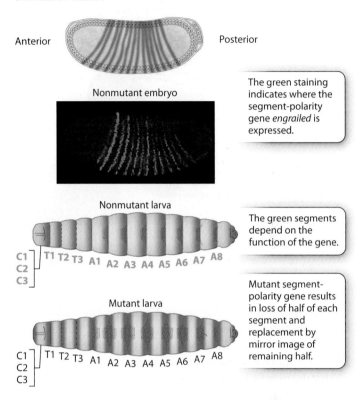

FIG. 19.11 Normal segment-polarity gene expression pattern and mutant phenotype. Source: James Langeland, Steve Paddock, and Sean Carroll, HHMI, Univ. Wisconsin–Madison.

Two classic examples of the consequences of mutations in homeotic genes in *Drosophila* are shown in **Fig. 19.12.** Fig. 19.12a shows what happens when the homeotic gene *Antennapedia,* which specifies the development of the leg, is

FIG. 19.12 Homeotic genes and segment identity. (a) Normal antennae are transformed into legs in an *Antennapedia* mutant. (b) Normal structures in the third thoracic segment are transformed into wings in a *Bithorax* mutant. Sources: a. (left and right) F. Rudolf Turner, Ph.D., Indiana University; b. (left) Thomas Deerinck, NCMIR/Science Source; (right) David Scharf/Science Source.

FIG. 19.13 Tissues controlled by homeotic genes. Homeotic genes specify the fate of clumps of tissue in the *Drosophila* larva.

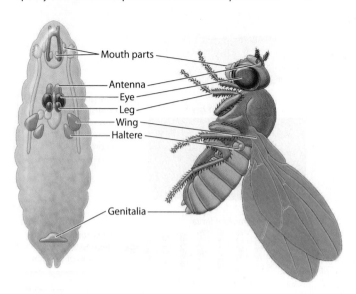

FIG. 19.14 Organization of the *Hox* gene clusters in *Drosophila* and the body parts that they affect. The order of genes along the chromosome corresponds to their positions along the anterior–posterior axis in the developing embryo.

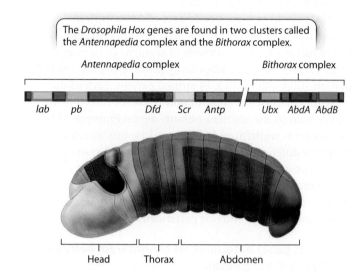

inappropriately expressed in anterior segments. The mutation causes legs to grow where antennae usually would. Similarly, a mutation in the homeotic gene *Bithorax* results in the transformation of thoracic segment 3 (T3) into thoracic segment 2 (T2), so that the fruit fly has two T2 segments in a row. As shown in the Fig. 19.12b, the result is a fruit fly with two complete sets of wings.

In *Drosophila*, the adult body parts such as legs, antennae, and wings are formed from specific tissues located throughout the larval body (**Fig. 19.13**). The development of these tissues and their metamorphosis into the adult body parts are regulated by the homeotic genes. First expressed at about the same time as the segment-polarity genes (see Fig. 19.11), the homeotic genes continue to be expressed even after the genes that regulate early development have shut down. Their continuing activity is due to the presence of chromatin remodeling proteins that keep the chromatin physically accessible to the transcription complex (Chapter 18).

Homeotic genes encode transcription factors. The domain of the homeotic transcription factors that binds DNA to promote or repress transcription is a sequence of 60 amino acids called a **homeodomain**, whose sequences are very similar from one homeotic protein to the next across different species.

The *Drosophila* genome contains eight *Hox* genes comprising two distinct clusters: the *Antennapedia* complex and the *Bithorax* complex (**Fig. 19.14**). The genes are arranged along the chromosome in the same order as their products function in anterior–posterior segments along the embryo. In addition, the timing of their expression corresponds to their order along the chromosome and location of expression, with genes that are expressed closer to the anterior end turned on earlier than genes that are expressed closer to the posterior end. The correlation among linear order along the chromosome, anterior–posterior position in the embryo, and timing of expression is observed in *Hox* clusters in almost all organisms studied.

Because the amino acid sequences of the homeodomains of *Hox* gene products are very similar from one organism to the next, *Hox* gene clusters have been identified in a wide variety of animals with bilateral symmetry (organisms in which both sides of the midline are mirror images), from insects to mammals. Comparison of the number and types of *Hox* genes in different species supports the hypothesis that the ancestral *Hox* gene cluster had an organization very similar to what we now see in most organisms with *Hox* gene clusters. In its evolutionary history, the vertebrate genome was duplicated in its entirety twice; thus, vertebrates have four copies of the *Hox* gene cluster (**Fig. 19.15**).

Unlike the *Hox* genes in *Drosophila*, mammalian *Hox* genes do not specify limbs but are important in the embryonic development of structures that become parts of the hindbrain, spinal cord, and vertebral column (Fig. 19.15). As in *Drosophila*, the genes in each cluster are expressed according to their linear order along the chromosome, which coincides with the head-to-tail order of regions that the genes affect in the embryo. Each gene helps to specify the identity of the region in which it is expressed. Many of the genes in the mammalian *Hox* clusters have redundant or overlapping functions so that learning exactly what each gene does continues to be a research challenge. The evolutionary and developmental study of *Hox* gene conservation and expression is a good example of recent evo-devo research.

FIG. 19.15 Organization and content of *Hox* gene clusters in the mouse and the regions of the embryo in which they are expressed.

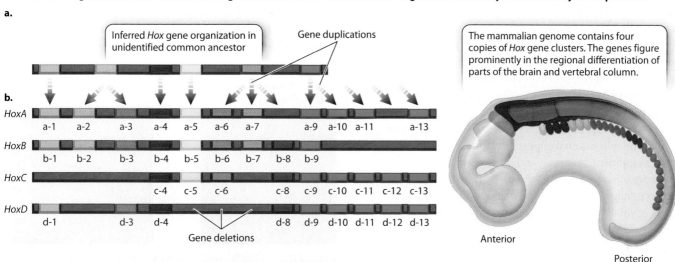

Self-Assessment Questions

5. Would development happen normally if the mother has normal *bicoid* function, but the embryo does not? Why or why not?

6. Draw a diagram to illustrate how a concentration gradient of a transcription factor along the anterior–posterior axis of a *Drosophila* embryo can create a region in the middle in which transcription of a target gene takes place without being transcribed in either the anterior or posterior region.

7. Would the pattern of segment-polarity gene expression be normal if one or more gap genes were not expressed properly? Why or why not?

8. Expression of a homeotic gene in the wrong tissue in *Drosophila* results in the development of an inappropriate body part from that tissue. Why does this happen?

19.3 MASTER REGULATORS

As we have seen, *Hox* genes and the proteins they encode are very similar in a wide range of organisms. As a result, they can be identified by their similarity in DNA or amino acid sequence. DNA or proteins that are similar in sequence among distantly related organisms are said to be **evolutionarily conserved**. Their similarity in sequence suggests that they were present in the most recent common ancestor and have changed very little over time because they carry out a vital function. Many transcription factors that are important in development are evolutionarily conserved. An impressive example is found in the development of animal eyes.

Animals have evolved a wide variety of eyes.

Animal eyes show an amazing diversity in their development and anatomy (**Fig. 19.16**). Among the simplest eyes are

FIG. 19.16 Diversity of animal eyes. *Sources: a. David M. Dennis/AGE Fotostock; b. Masa Ushioda/SeaPics.com; c. Reinhard Dirscherl/AGE Fotostock; d. Andrey Armyagov/iStockphoto; e. Julian Brooks/AGE Fotostock; f. Walter Geiersperger/AGE Fotostock.*

a. Planarian

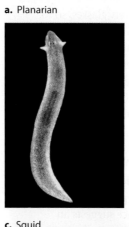

b. Jellyfish

c. Squid

d. Human

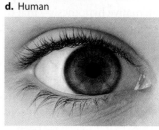

e. House fly

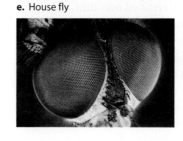

f. Trilobite

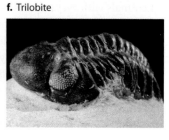

those of the planarian flatworm (Fig. 19.16a), which are only pit-shaped cells containing light-sensitive photoreceptors. The planarian eye has no lens to focus the light, but the animal can perceive differences in light intensity. The incorporation of a spherical lens, as in some jellyfish (Fig. 19.16b), improves the image.

Among the most complex eyes are the camera-type eyes of the squid (Fig. 19.16c) and human (Fig. 19.16d), which have a single lens to focus light onto a light-sensitive tissue, the retina (Chapter 35). Though the single-lens eyes of squids and vertebrates are similar in external appearance, they are vastly different in their development, anatomy, and physiology.

Some organisms, such as the house fly (Fig. 19.16e) and other insects, have compound eyes, that is, eyes consisting of hundreds of small lenses arranged on a convex surface and pointing in slightly different directions. This arrangement of lenses allows a wide viewing angle and detection of rapid movement.

In the history of life, eyes are very ancient. By the time of the extraordinary diversification of animals 542 million years ago (known as the Cambrian explosion), organisms such as trilobites (Fig. 19.16f) already had well-formed compound eyes. These differed greatly from the eyes in modern insects, however. For example, trilobite eyes had hard mineral lenses composed of calcium carbonate (the world's first safety glasses, so to speak). Altogether, about 96% of living animal species have true eyes that produce an image, as opposed to simply being able to detect differences in light intensity. Despite this common feature, the extensive diversity among the eyes of different organisms encouraged evolutionary biologists in the belief that the ability to perceive light may have evolved independently about 40 to 60 times.

Pax6 is a master regulator of eye development.

The wide diversity of structure and function of animal eyes suggests that eyes may have evolved independently in different organisms. However, more recent evidence suggests an alternative hypothesis—that they evolved once, very early in evolution, and subsequently diverged over time. One piece of evidence that supports the hypothesis of a single origin for light perception is the observation that the light-sensitive molecule in all light-detecting cells is the same, a derivative of vitamin A in association with a protein called opsin (Chapter 35). The presence of the same light-sensitive molecule in diverse eyes argues that it may have been present in the common ancestor of all animals with eyes and has been retained over time.

Another argument against multiple independent origins of eyes came from studies of eye development. Researchers identified *eyeless*, a gene in the fruit fly *Drosophila*. As its name implies, the phenotype of *eyeless* mutants is abnormal eye development (**Fig. 19.17a**). When the protein product of the *eyeless* gene was identified, it was found to be a transcription factor called Pax6. Mutant forms of a *Pax6* gene were already known to cause small

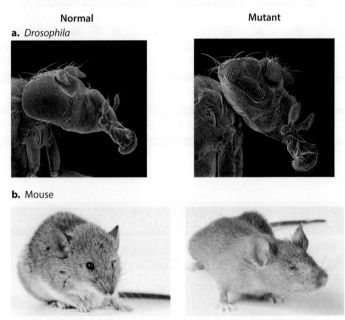

FIG. 19.17 Effect of *Pax6* mutations on eye development. (a) Normal and mutant eye phenotypes in *Drosophila*; (b) normal and mutant eye phenotypes in a mouse. *Sources: a. (left and right) David Scharf/Science Source; b. (left) INSADCO Photography / Alamy; (right) Jennifer L. Torrance, Photographer, The Jackson Laboratory.*

eyes in the mouse (**Fig. 19.17b**) and aniridia (absence of the iris) in humans.

The mutations in the *Pax6* gene that cause the development defects in the eye in fruit flies, mice, and humans are classified as **loss-of-function mutations**. A loss-of-function mutation is a mutation that inactivates the normal function of a gene. For example, the mutation that causes the green color in Mendel's peas is a loss-of-function mutant of an enzyme that normally breaks down the green pigment chlorophyll, so the peas remain green and obscure the yellow pigments that are also present (Chapter 15). In the case of eye development, loss-of-function mutations in *Pax6* make a defective version of the Pax6 transcription factor that is not able to carry out its function.

The similarity in phenotypes of *Pax6* mutants, along with the strong conservation of Pax6 in *Drosophila* and mouse, led Swiss developmental biologist Walter Gehring to hypothesize that Pax6 might be a master regulator of eye development. In other words, he hypothesized that Pax6 binds to regulatory regions of a set of genes that turns on a developmental program that induces eye development. In theory, this means that Pax6 can induce the development of an eye in any tissue in which it is expressed.

To test this hypothesis, Gehring and collaborators genetically engineered fruit flies that would produce the normal Pax6 transcription factor in the antenna, where it is not normally expressed (**Fig. 19.18a**). Any mutation in which a gene is expressed in the wrong place or at the wrong time is known as a **gain-of-function mutation**. (We saw an example of gain-of-function mutations when the *Hox* genes were expressed in the wrong segments of the fruit fly, producing legs where antennae normally develop and thus resulting in four-winged fruit flies.) For genes that control a developmental pathway, loss-of-function mutations and gain-of-function mutations often have opposite effects on phenotype.

Because loss-of-function mutations in *Pax6* result in an eyeless phenotype, a gain-of-function mutation should result in eyes developing in whatever tissue *Pax6* is expressed. In the gain-of-function mutation, the antenna developed into a miniature compound eye, and electrical recordings demonstrated that some of these antennal eyes were functional (**Fig. 19.18b**). The researchers also created other gain-of-function mutations that led to eyes on the legs, wings, and other tissues, which the *New York Times* publicized in an article headlined "With New Fly, Science Outdoes Hollywood."

Gehring and his group then went one step further. They took the *Pax6* gene from mice and expressed it in fruit flies to see whether the mouse *Pax6* gene is similar enough to the fruit fly version of the gene that it could induce eye development in the fruit fly. Specifically, they created transgenic fruit flies that expressed the *Pax6* gene from mice in the fruit fly antenna. The mouse gene induced a miniature eye in the fly (**Fig. 19.18c**). However, the *Pax6* gene from mice induced the development of a compound eye of *Drosophila*, not the single-lens eye of mouse.

The ability of mouse Pax6 to make an eye in fruit flies suggests that mouse and fruit fly *Pax6* are not only similar in DNA and amino acid sequence, but also similar in function, and indeed act as a master switch that can turn on a developmental program that leads to the formation of an eye. However, these observations led to another question: why did the mouse *Pax6* gene produce fruit fly eyes instead of mouse eyes?

The answer is that the fruit fly genome does not include the genes needed to make mouse eyes. Mouse Pax6 protein induces fly eyes because it affects genes in the fly genome that function later in the process of eye development. The later-acting genes are called **downstream genes.** Transcription factors like Pax6 interact with their target genes by binding with short DNA sequences that are adjacent to the gene, usually at the 5′ end, called **cis-regulatory elements**. These regulatory elements help determine whether the adjacent DNA is transcribed. When bound to cis-regulatory elements, some transcription factors act as repressors that prevent transcription of the target gene, and others serve as activators by recruiting the transcriptional machinery to the target gene (Chapter 18). A transcription factor can even repress some of its target genes and activate others.

In the fruit fly, Pax6 binds to cis-regulatory elements in many genes, turning some genes on and others off. The products of these downstream genes, in turn, affect the expression of further downstream genes. The total number of genes that are downstream of Pax6 and that are needed for eye development is estimated at about 2000. Most are not direct targets of Pax6, but they are activated indirectly through other transcription factors downstream of Pax6. When mouse Pax6 is expressed in fruit flies, it is similar enough in sequence to activate the genes involved in fruit fly eye development, so it makes sense that mouse Pax6 leads to the development of a fruit fly eye, not a mouse eye.

One scenario of how Pax6 became a master regulator for eye development in a wide range of organisms, yet produces a diversity of eyes in these organisms, is that Pax6 evolved early in the history of life as a transcription factor that was able to bind to and regulate genes involved in early eye development. Over time, different genes in different organisms acquired new Pax6-binding cis-regulatory elements by mutation, and if these were beneficial they persisted. Therefore, the downstream genes that are targets of Pax6 are different in different organisms, but they share two features: they are able to bind to and are regulated by Pax6, and they are involved in eye development. So, the early steps in eye development are conserved, but the later ones are not.

The *Pax6* gene and the *Hox* genes reveal an important principle in evo-devo, which is that master regulatory genes that control development are often evolutionarily conserved, whereas the downstream genes that they regulate may not be. Downstream genes can evolve new functions, or genes not originally controlled by the master regulator may evolve to come under its influence, or genes formerly controlled by the master regulator may evolve to be unresponsive. In this way, a conserved master regulatory mechanism may result in distinct developmental outcomes in different organisms. In the case of *Pax6*, for example, the master regulatory mechanism for eye development evolved early and is shared among a wide range of animals, even though the eyes that are produced are quite diverse.

FIG. 19.18 *Pax6*, **a master regulator controlling eye development.** (a) Normal fly antenna; (b) eye tissue induced by expression of fruit fly *Pax6* in the antenna; (c) eye tissue induced by expression of mouse *Pax6* in the antenna. *Sources: a. Cheryl Power/Science Source; b. Eye of Science/Science Source; c. Prof. Walter Gehring/Science Source.*

Normal antenna

Antennal eye induced by *Drosophila Pax6* gene

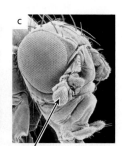

Antennal eye induced by mouse *Pax6* gene

Self-Assessment Questions

9. Why do master regulatory genes tend to be more strongly conserved in evolution than the downstream genes they regulate?

10. For genes that control pathways of development, loss-of-function mutations are usually recessive, whereas gain-of-function mutations are usually dominant. Why is this the case?

19.4 COMBINATORIAL CONTROL

Most cis-regulatory elements are composed of one or more binding sites for transcription factors, some of which are activators of transcription and others repressors of transcription. Therefore, the rate of transcription of any gene in any type of cell is determined by the combination of transcription factors that are present in the cell and by the relative balance of transcription factors that are activators versus those that are repressors. Regulation of gene transcription according to the mix of transcription factors in the cell is known as **combinatorial control**. Combinatorial control of transcription is another general principle often seen in all multicellular organisms at many stages of development. Here, we discuss flower development as an example of this general principle.

Floral differentiation is a model for plant development.

The plant *Arabidopsis thaliana*, a weed commonly called mouse-ear cress, is a model organism for developmental studies and presents a clear case of combinatorial control. As in all plants, *Arabidopsis* has regions of undifferentiated cells, called meristems, where growth of shoots, roots, and flowers can take place (Chapter 29). Meristem cells are similar to stem cells in animals because they consist of multipotent cells that can differentiate into different structures in the flower. In floral meristems of *Arabidopsis*, the flowers develop from a pattern of four concentric circles of cells, or whorls, each of which differentiates into a distinct type of floral structure.

From the outermost whorl (whorl 1) of cells to the innermost whorl (whorl 4), the floral organs are formed as shown in **Fig. 19.19**:

- Cells in whorl 1 form the green sepals, which are modified leaves forming a protective sheath around the petals in the flower bud that unfold like petals when the flower opens.
- Cells in whorl 2 form the petals.
- Cells in whorl 3 form the stamens, which are the male sexual structures in which pollen is produced. (The small granules on the stamens in Fig. 19.19 are pollen grains.)
- Cells in whorl 4 form the carpels, which are the female reproductive structures that receive pollen and contain the ovaries.

The identity of the floral organs is determined by combinatorial control.

The genetic control of flower development in *Arabidopsis* was discovered by the analysis of mutant plants, an investigation similar in principle to the way that genetic control of development in fruit flies was determined. The plant mutants fell into three classes showing characteristic floral abnormalities (**Fig. 19.20**). A nonmutant (wild-type) flower is shown in Fig. 19.20a. Mutations in the gene *apetala-2* inactivate the gene's function and affect whorls 1 and 2 (Fig. 19.20b); those in *apetala-3* or *pistillata* affect whorls 2 and 3 (Fig. 19.20c); and those in *agamous* affect whorls 3 and 4 (Fig. 19.20d). (Different standards are used for naming genes and proteins in different organisms. In *Arabidopsis*, wild-type alleles are written in capital letters with italics, and wild-type proteins in capital letters without italics. Mutant alleles and proteins follow the same rules but are written in lower-case letters.)

The observation that the mutant phenotypes affect development of pairs of adjacent whorls already hints at combinatorial control, which researchers incorporated into a model of floral development called the **ABC model**. The model invokes three activities arbitrarily called A, B, and C. These activities represent the function of one or more different protein products of different genes that are hypothesized to be present in cells of each whorl as shown in **Fig. 19.21**. In the course of floral differentiation, the cells of each whorl become different from those in the other whorls, and different combinations of genes are activated or repressed in each whorl. Activity A is present in whorls 1 and 2, activity B in whorls 2 and 3, and activity C in

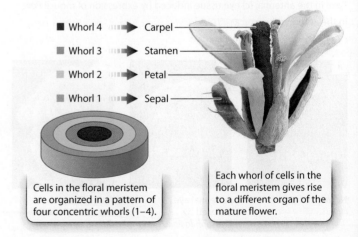

FIG. 19.19 Development of the *Arabidopsis* floral organs from concentric whorls of cells in the floral meristem. *Photo source: Juergen Berger/Max Planck Institute for Developmental Biology, Tuebingen, Germany.*

- Whorl 4 → Carpel
- Whorl 3 → Stamen
- Whorl 2 → Petal
- Whorl 1 → Sepal

Cells in the floral meristem are organized in a pattern of four concentric whorls (1–4).

Each whorl of cells in the floral meristem gives rise to a different organ of the mature flower.

FIG. 19.20 Phenotypes of the normal flower and floral mutants of *Arabidopsis*. Source: Courtesy John Bowman.

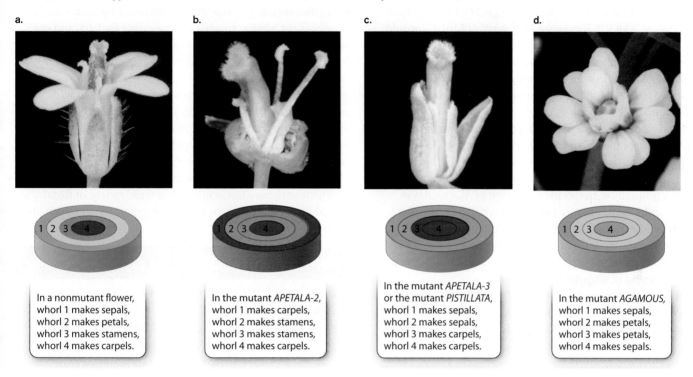

In a nonmutant flower, whorl 1 makes sepals, whorl 2 makes petals, whorl 3 makes stamens, whorl 4 makes carpels.

In the mutant *APETALA-2*, whorl 1 makes carpels, whorl 2 makes stamens, whorl 3 makes stamens, whorl 4 makes carpels.

In the mutant *APETALA-3* or the mutant *PISTILLATA*, whorl 1 makes sepals, whorl 2 makes sepals, whorl 3 makes carpels, whorl 4 makes carpels.

In the mutant *AGAMOUS*, whorl 1 makes sepals, whorl 2 makes petals, whorl 3 makes petals, whorl 4 makes sepals.

FIG. 19.21 The ABC model for flower development in *Arabidopsis*.

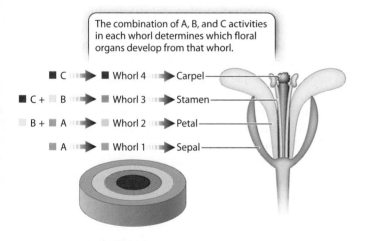

whorls 3 and 4. According to the ABC model, activity A alone results in the formation of sepals, A and B together result in petals, B and C together result in stamens, and C alone results in carpels. Like any good hypothesis (Chapter 1), this one makes specific predictions. Mutants lacking activity A will have defects in whorls 1 and 2; mutants lacking activity B will have defects in whorls 2 and 3; and mutants lacking activity C will have defects in whorls 3 and 4.

Matching these predictions with the mutants in Fig. 19.20 yields the following correspondence:

- The A activity is encoded by the gene *APETALA-2*.
- The B activity is encoded jointly by the genes *APETALA-3* and *PISTILLATA*.
- The C activity is encoded by the gene *AGAMOUS*.

Comparing the mutant phenotypes in Fig. 19.20 with the differentiation of the whorls in Fig. 19.21, it is clear that the normal product of *APETALA-2* (A activity) is needed for proper development of whorls 1 and 2; the normal products of *APETALA-3* and *PISTILLATA* (together constituting B activity) are required for proper development of whorls 2 and 3; and the normal product of *AGAMOUS* (C activity) is needed for proper development of whorls 3 and 4. These inferences suggest that *APETALA-2* is normally expressed in whorls 1 and 2, *APETALA-3* and *PISTILLATA* in whorls 2 and 3, and *AGAMOUS* in whorls 3 and 4.

To account for all of the mutant phenotypes in Fig. 19.20, you need only to postulate that when A is absent, C is expressed in whorls 1 and 2 in addition to its normal expression in whorls 3 and 4; and when C is absent, A expression expands into whorls 3 and 4 in addition to its normal expression in whorls 1 and 2. The domains of expression expand in this way because, in addition to its role in activating certain downstream genes, the product of *APETALA-2* represses the expression of *AGAMOUS* in whorls 1 and 2. As a result, *AGAMOUS* is normally not expressed in whorls 1 and 2. However, when *APETALA-2* is nonfunctional, *AGAMOUS* is expressed wherever *APETALA-2* would normally repress it. The result is activity C in whorls 1 and 2. Similarly, in addition to its role in activating other genes, the product of *AGAMOUS* represses the expression of *APETALA-2* in whorls 3 and 4, so *APETALA-2* is normally not expressed in whorls 3 and 4.

We can see this relationship between genes and activity, and the interaction of the activities, in the mutants in Fig. 19.20. In the *APETALA-2* mutant, for example, the *APETALA-2* gene is disrupted, meaning that there is no A activity in the plant at all. Activity A normally represses the expression of *AGAMOUS* in whorls 1 and 2, so its absence means that whorl 1 (normally just activity A) expresses only activity C, leading to the development of carpels in whorl 1; and whorl 2 (normally activities A and B, with C repressed) expresses activities B and C, leading to the development of stamens in whorl 2.

The identification of the products of these genes demonstrates combinatorial control and explains the mutant phenotypes. All of the genes encode transcription factors, which are called AP2, AP3, PI, and AG, corresponding to the four genes. The transcription factors bind to cis-regulatory elements of genes that encode proteins that are necessary for each whorl's development, similar to what we saw earlier for Pax6 and eye development. AP3 and PI are both necessary for the B activity because the proteins form a heterodimer, a protein made up of two different subunits.

Although other transcription factors associated with other activities that also contribute to flower development were discovered later, the original ABC model is still valid and serves as an elegant example of combinatorial control. As in the case of the *Pax6* gene in the development of animals' eyes, evolution of the cis-regulatory sequences in downstream genes has resulted in a wide variety of floral morphologies in different plant lineages, some of which are shown in **Fig. 19.22**.

Self-Assessment Questions

11. How does combinatorial control work in the context of floral development?

12. What is the phenotype of a flower in which *APETALA-2* (activity A) and *AGAMOUS* (activity C) are expressed normally, but *APETALA-3* and *PISTILLATA* (activity B) are expressed in all four whorls?

19.5 CELL SIGNALING IN DEVELOPMENT

As we have seen in the discussion of stem cells, differentiated cells can be reprogrammed by the action of only a few key genes or small organic molecules. In some cases, the processes that push differentiation in a forward direction are also quite simple. An important example is **signal transduction**, in which an extracellular molecule acts as a signal to activate a membrane protein that, in turn, activates molecules inside the cell that control differentiation (Chapter 9). The signaling molecule is called the **ligand** and the membrane protein that it activates is called the **receptor**. The following example shows how a simple ligand–receptor pair can have profound effects not only on the differentiation of the cell that carries the receptor, but also on its neighbors.

A signaling molecule can cause multiple responses in the cell.

Pioneering experiments on signal transduction in development were carried out in the soil nematode *Caenorhabditis elegans* (**Fig. 19.23**). This worm is a useful model organism for developmental studies because of its small size (the adult is about 1 mm long), short generation time (2½ days from egg to sexually mature adult), and process of development that is always the same in each individual. The adult animal consists of exactly 959 somatic (body) cells, and the cells' patterns of division,

FIG. 19.22 Floral diversity. (a) Colorado blue columbine (*Aquilegia coaerulea*); (b) slipper orchid (*Paphiopedilum holdenii*); (c) ginger flower (*Smithatris supraneanae*); (d) garden rose (hybrid). Sources: a. Ijh images/Shutterstock; b. Korkiat/Getty Images; c. Nonn Panitvong; d. Maria Mosolova/AGE Fotostock.

FIG. 19.23 The nematode worm *Caenorhabditis elegans*.

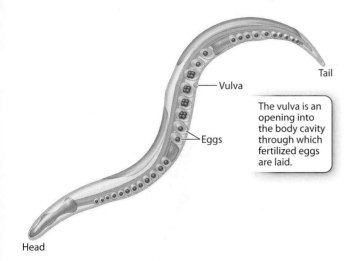

FIG. 19.24 Vulva development in *Caenorhabditis elegans*. The cells of the vulva differentiate in response to molecular signals from other cells.

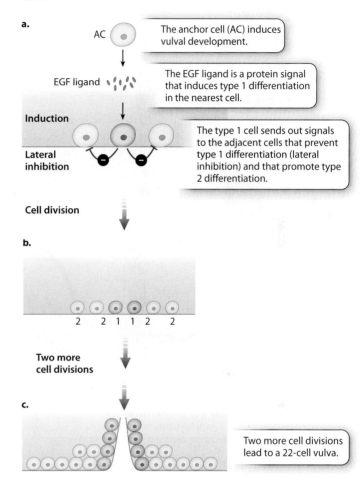

migration, differentiation, and morphogenesis are identical in each individual *C. elegans*. The adult is typically a hermaphrodite that produces both male and female reproductive cells. The eggs are fertilized internally by sperm, and after undergoing five or six cell divisions, the eggs are laid through an organ known as the vulva (Fig. 19.23).

Vulva development is relatively simple and therefore amenable to experimental studies. The entire structure arises from only three multipotent cells that undergo either of two types of differentiation and divide to produce a vulva consisting of exactly 22 cells. Cells resulting from type 1 differentiation form the opening of the vulva, and cells resulting from type 2 differentiation form a supporting structure around the vulval opening. In each of the two types of differentiation, different genes are activated or repressed, resulting in differentiated cells with different functions.

How each of the three original cells, called progenitor cells, differentiates is determined by its position in the developing worm (**Fig. 19.24**). The fates of the progenitor cells are determined by their proximity to another cell, called the anchor cell (the green cell labeled "AC" in Fig. 19.24a), which secretes a protein called epidermal growth factor (EGF) that binds to and activates a transmembrane EGF receptor (Chapter 9). The progenitor cell closest to the anchor cell receives the most amount of signal, and upon activation of its EGF receptor, the progenitor cell carries out three functions:

- It activates the genes for differentiation into a type 1 cell (Fig. 19.24a).
- It prevents the adjacent cells from differentiating as type 1 (a process called **lateral inhibition**).
- It induces the adjacent cells to differentiate as type 2 cells.

Developmental signals are amplified and expanded.

How can a single ligand–receptor pair cause so many changes in gene expression that it determines the pathway of differentiation not only of the cell itself but also of its neighbors? **Fig. 19.25** shows the mechanisms in simplified form. On the left in red is the progenitor cell nearest the anchor cell, which receives the strongest EGF signal. Activation of the EGF receptor by the ligand initiates a process of signal transduction in the cytoplasm. The signal received by the EGF receptor is transmitted from one protein to the next by proteins at each stage phosphorylating several others, which also amplifies the signal at each stage (Chapter 9). The result of signal transduction is that a set of transcription factors is activated.

In the nucleus, the transcription factors activate transcription of genes for type 1 differentiation. The transcription factors also activate transcription of genes that prevent type 1 differentiation in neighboring cells, including the

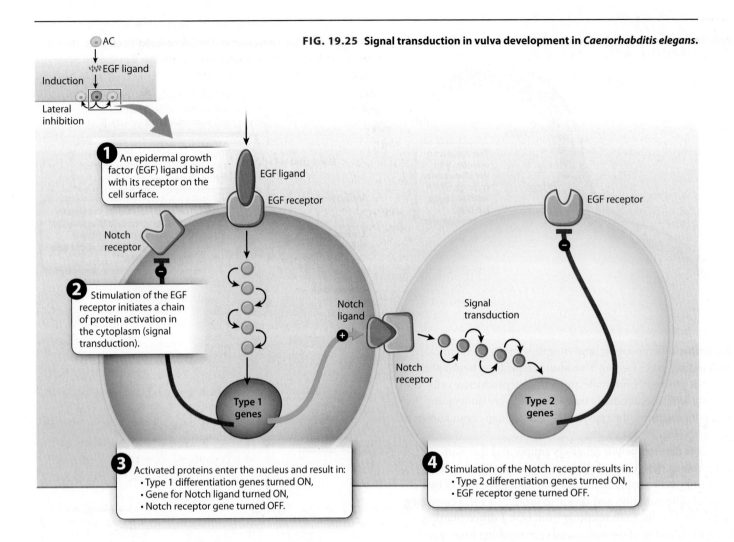

FIG. 19.25 Signal transduction in vulva development in *Caenorhabditis elegans*.

genes that produce another type of protein ligand, called Notch. The Notch ligand is embedded in the type 1 cell's membrane and activates Notch receptors in the neighboring cells. Activation of the Notch receptor in these cells activates a signal transduction cascade in these cells, which results in transcription of the genes for type 2 differentiation. The cascade started by the binding of Notch also activates transcription of other genes whose products inhibit the EGF receptor. Inhibiting the EGF receptor in type 2 cells prevents EGF from eliciting a type 1 response in the type 2 cell. In addition, at the same time that the type 1 cell produces the Notch ligand, it produces proteins that inhibit its own Notch receptors, and this prevents Notch from initiating a type 2 response in the type 1 cell.

Although vulva development in nematodes is a fairly simple example of the importance of ligand–receptor signaling in development, EGF and Notch ligands and their receptors are found in virtually all animals. They are among dozens of ligand–receptor pairs that have evolved as signaling mechanisms to regulate processes in cellular metabolism and development. Humans have several distinct but related gene families of EGF and Notch ligands and receptors. Human EGF is important in cell survival, proliferation, and differentiation. EGF functions in the differentiation and repair of multiple types of cells in the skin; it is present in all body fluids and helps regulate rapid metabolic responses to changing conditions. Human Notch ligands are involved in development of the nervous and immune systems as well as heart, pancreas, and bone. Abnormalities in EGF or Notch signaling are associated with many different types of cancer.

Therefore, just as we saw in our discussion of eye and flower development, the molecular players involved in development are often evolutionarily conserved across a wide range of organisms. This is true even of genes that we typically associate with disease, such as *BRCA1* and *BRCA2* and their link with cancer. These genes not only play a role in cell cycle control in the adult, but also in early development in many organisms. In fact, although heterozygous mutations in these genes predispose individuals to breast and ovarian cancers in humans, homozygous mutations are lethal in early embryonic stages. In **Fig. 19.26**, we focus on the *BRCA1* gene to summarize key concepts about DNA replication, mutation, genetic variation, inheritance, gene regulation, and development.

Self-Assessment Question

13. Diagram a pathway of signal transduction including a ligand, receptor, and ultimately a transcription factor that activates a gene that inhibits the receptor.

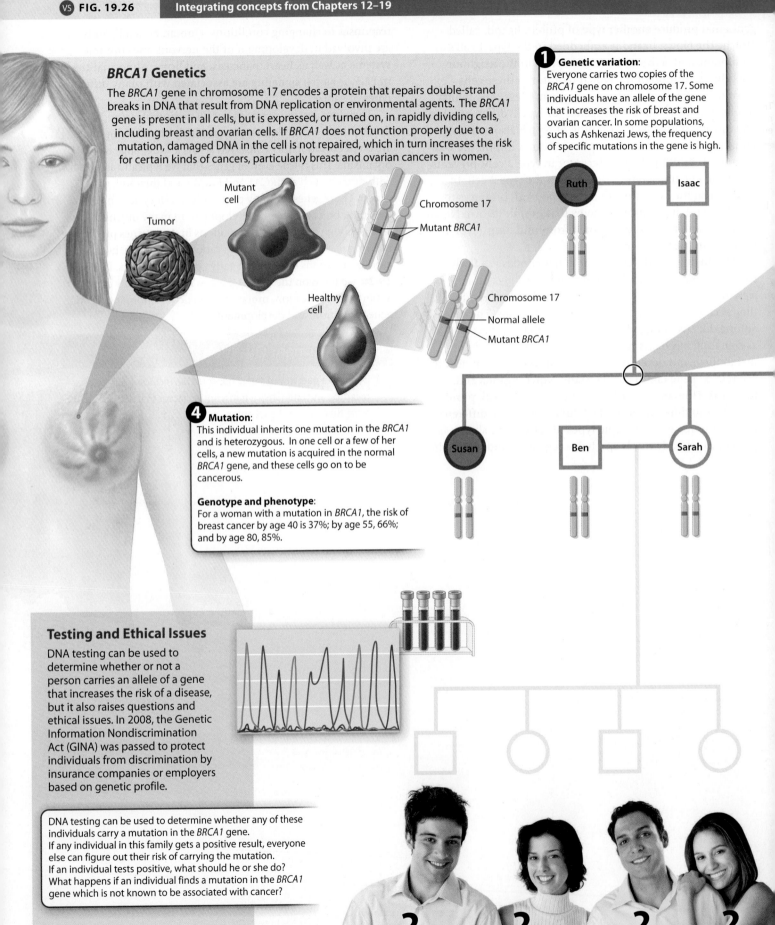

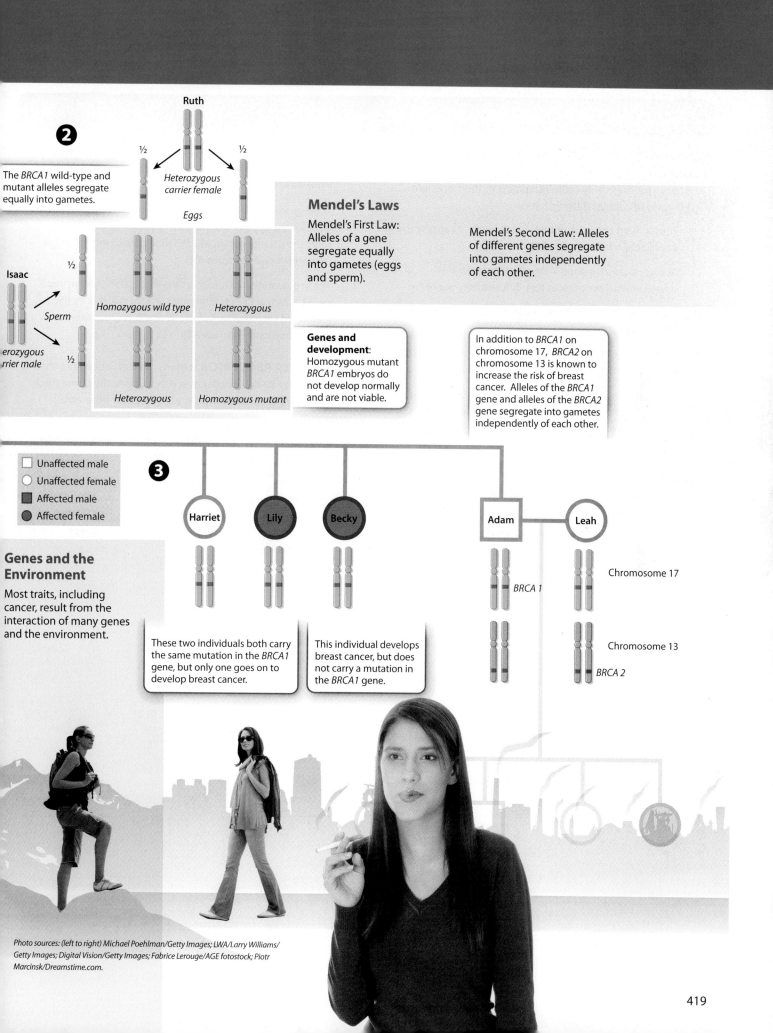

CORE CONCEPTS SUMMARY

19.1 GENETIC BASIS OF DEVELOPMENT: In the development of humans and other animals, stem cells become progressively more restricted in their possible pathways of cellular differentiation.

- The fertilized egg is totipotent, which means it can give rise to a complete organism. page 399

- At each successive stage in development, cells lose developmental potential as they differentiate. page 400

- Embryonic stem cells are pluripotent and can give rise to any of the three germ layers. page 400

- Multipotent cells can form only a limited number of specialized cell types. page 400

- Stem cells play a prominent role in regenerative medicine, in which stem cells—in some cases, reprogrammed cells from the patient's own body—are used to replace diseased or damaged tissues. page 400

19.2 HIERARCHICAL CONTROL: The genetic control of development is a hierarchy in which genes are activated in groups that, in turn, regulate the next set of genes.

- Hierarchical gene control can be seen in fruit fly (*Drosophila*) development. page 403

- The oocyte of a fruit fly is highly polarized, with gradients of maternal mRNA that set up anterior–posterior and dorsal–ventral axes. page 404

- These gradients, in turn, affect the expression of segmentation genes in the zygote, including the gap, pair-rule, and segment-polarity genes, which define specific regions in the developing embryo. page 406

- The segmentation genes direct the expression of homeotic genes, which are key transcription factors that specify the identity of each segment of the fly and that are conserved in animal development. page 407

19.3 MASTER REGULATORS: Many master regulators that play key roles in development are evolutionarily conserved but can have dramatically different effects in different organisms.

- Many proteins that are important in development are similar in sequence from one organism to the next. Such proteins are said to be evolutionarily conserved. page 409

- The downstream targets of homeotic genes are different in different animals, allowing homeotic genes to activate different developmental pathways in different organisms. page 411

Although animals exhibit an enormous diversity in eye morphology, the observation that the proteins involved in light perception are evolutionarily conserved suggests that the ability to perceive light may have evolved once, early in the evolution of animals. page 411

The Pax6 transcription factor is a master regulator of eye development. Loss-of-function mutations in *Pax6* result in abnormalities in eye development, whereas gain-of-function mutations result in eye development in tissues in which eyes do not normally form. page 410

19.4 COMBINATORIAL CONTROL: Combinatorial control is a developmental strategy in which cellular differentiation depends on the particular combination of transcription factors present in a cell.

By analyzing mutants that affect flower development in the plant *Arabidopsis*, researchers were able to determine the genes involved in normal flower development. page 412

The ABC model of flower development invokes three activities (A, B, and C) present in circular regions (whorls) of the developing flower, with the specific combination of factors determining the developmental pathway in each whorl. page 412

19.5 CELL SIGNALING IN DEVELOPMENT: Ligand–receptor interactions activate signal transduction pathways that converge on transcription factors and genes that determine cell fate.

Cell signaling involves a ligand, an extracellular molecule that acts as a signal to activate a membrane receptor protein, which, in turn, activates molecules inside the cell. page 414

Activation of a receptor sets off a pathway of signal transduction, in which a series of proteins in the cytoplasm become sequentially activated. page 414

Because signal transduction can amplify and expand a developmental signal, a single ligand–receptor pair can cause major changes in gene expression and ultimately determine the pathway of differentiation. page 414

An example of signal transduction in development is differentiation of the nematode vulva, which is determined by means of an EGF ligand and receptor. page 415

Log in to **LaunchPad** to check your answers to the Self-Assessment Questions and to access additional learning tools.

CASE 4

Malaria

Coevolution of Humans and a Parasite

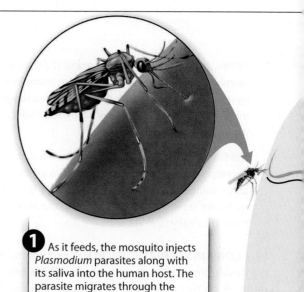

1 As it feeds, the mosquito injects *Plasmodium* parasites along with its saliva into the human host. The parasite migrates through the bloodstream to the host's liver.

Most people would say the world has more than enough mosquitoes, but in 2010 scientists at the University of Arizona conjured up a new variety. The high-tech bloodsuckers were genetically engineered to resist *Plasmodium*, the single-celled eukaryote that causes malaria.

> As humans have tried a succession of ways to defeat P. falciparum, the parasite has evolved, thwarting our efforts.

Normally, the *Plasmodium* parasite grows in the mosquito's gut and is spread to humans by the insect's bite. But by altering a single gene in the mosquito's genome, the researchers made the insects immune to the malaria parasite. As a consequence, these mosquitoes could not spread malaria to humans. The accomplishment is a noteworthy advance, the latest in a long line of efforts to stop *Plasmodium* in its tracks.

Malaria is one of the most devastating diseases on the planet. The World Health Organization estimates that 500 million people contract malaria annually, primarily in tropical regions. The disease is thought to claim about 1 million lives each year. Of those deaths, 85% to 90% occur in sub-Saharan Africa, mostly among children younger than 5 years of age.

Five species of *Plasmodium* can cause malaria in humans. One of them, *P. falciparum*, is particularly dangerous, accounting for the vast majority of malaria fatalities. Over thousands of years, this parasite and its human host have played a deadly game of tug-of-war. As humans have tried a succession of ways to defeat *P. falciparum*, the parasite has evolved, thwarting our efforts. In turn, the tiny organism has helped to shape human evolution.

Plasmodium is a wily and complicated parasite, requiring both humans and mosquitoes to complete its life cycle. The human part of the cycle begins with a single bite from an infected female mosquito. As the insect draws blood from a human, it releases *Plasmodium*-laden saliva into the human's bloodstream. Once inside their human host, the *Plasmodium* parasites invade the liver cells. There, the parasites undergo cell division for several days, so that their numbers increase. Eventually, the parasites infect red blood cells, where they continue to grow and multiply. The mature parasites burst from the red blood cells at regular intervals, triggering malaria's telltale cycle of fever and chills.

Some of the freed *Plasmodium* parasites go on to infect new red blood cells. Others divide to form gametocytes, the male and female reproductive cells, which travel through the victim's blood vessels. When another mosquito bites the infected individual, it takes up the gametocytes with its blood meal. Inside the insect, the parasite completes its life cycle. First the gametocytes fuse to form zygotes. Those zygotes then bore into the mosquito's stomach, where they form cells called oocysts that give rise to a new generation of parasites. When the infected mosquito sets out to feed, the cycle begins again.

The battle between malaria and humankind has raged through the ages. Scientists have recovered *Plasmodium* DNA from the bodies of 3500-year-old Egyptian mummies—evidence that ancient humans were infected with the malaria parasite. The close connection between humans, mosquitoes, and the malaria parasite almost certainly extends back much further.

In fact, people who hail from regions where malaria is endemic are more likely than others to have certain genetic signatures that offer some degree of protection from the parasite. That finding indicates that *Plasmodium* has been exerting evolutionary pressure on humankind for quite some time.

Meanwhile, we've done our best to fight back. For centuries, humans have fought the infection with quinine, a chemical found in the bark of the South American *Cinchona* tree. In the 1940s, scientists developed a more sophisticated drug based on the *Cinchona* compound. That drug, chloroquine, was effective, inexpensive, easy to administer, and caused few side effects. As a result, it was widely used. Unfortunately, in the late 1950s, *P. falciparum* began showing signs of resistance to chloroquine. Within 20 years, the resistance had spread to Africa, so that today the once-potent medication is no longer effective against most strains of *P. falciparum*.

Just as bacteria develop resistance to antibiotics, *Plasmodium* evolves resistance to the antiparasitic drugs designed to fight it. In poor, rural areas where malaria is prevalent, people often can't follow the recommended protocols for antimalarial treatment. Sick individuals may be able to afford only a few pills rather than the full recommended course. The strength and quality of those pills may be questionable, and the drugs are often taken without oversight from a medical professional.

This kind of inadequate use of the drugs fuels resistance. When pills are altered or the course of treatment is abbreviated, not all *Plasmodium* parasites are wiped out.

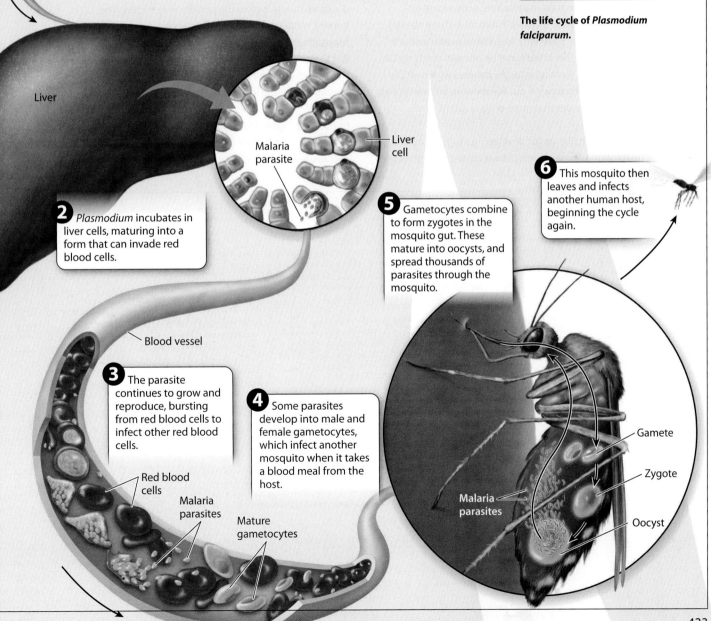

The life cycle of *Plasmodium falciparum*.

CASE 4

The parasites that survive in the presence of the drug are resistant to the drug. Because these resistant parasites have a survival advantage compared to parasites that succumb to the drug, the population evolves. As a result, the genes for drug resistance spread quickly through the population.

Since chloroquine resistance has emerged among *Plasmodium* species, pharmaceutical researchers have developed a variety of new medications to prevent or treat malaria. Most, however, are far too expensive for people in the poverty-stricken regions where malaria is rampant. And just as was the case with chloroquine, almost as quickly as new drugs are developed, *Plasmodium* evolves resistance to them.

One of the newest drugs used to treat malaria is artemisinin, a compound derived from the Chinese wormwood tree. Artemisinin is an effective and relatively inexpensive way to treat malaria infections. However, pockets of artemisinin-resistant *Plasmodium* have already been uncovered in Southeast Asia. Public health workers now recommend that artemisinin be given in combination with other drugs. Treating infected patients with multiple drugs is more likely to wipe out *Plasmodium* in their bodies, reducing the chances that more drug-resistant strains will emerge.

Given the challenges of developing drugs to prevent or treat malaria, some scientists have turned their attention to other approaches. One option is to produce a malaria vaccine. The parasite's complex life cycle makes that a complicated endeavor, though. Several vaccines are now in various stages of testing, and some show promise. Even so, researchers expect it will be years before a safe, effective vaccine for malaria could be available.

Other researchers are focusing their efforts on the mosquitoes that carry the parasite, rather than on *Plasmodium* itself. The genetically engineered insects created by the team in Arizona are a promising step in that direction.

The Arizona researchers set out to alter a cellular signaling gene that plays a role in the mosquito's life cycle. Mosquitoes normally live 2 to 3 weeks, and *Plasmodium* takes about 2 weeks to mature in the mosquito's gut. The researchers hoped to create mosquitoes that would die prematurely, before the parasite was mature. The genetic modification worked as planned: the engineered mosquitoes' life-spans were shortened by around 20%. But the genetic tweak also had a surprising side effect: the altered gene completely blocked the development of *Plasmodium* in the mosquitoes' guts. As a consequence, the engineered mosquitoes were incapable of spreading malaria to humans, regardless of how long they lived.

While the outcome of this work was a laboratory success, it will be much harder to translate the results to the real world. To create malaria-free mosquitoes in the wild, scientists would have to release the genetically modified mosquitoes and hope that their altered gene spreads through the wild mosquito population. But that gene would be passed on only if it gave the insects a distinct evolutionary advantage. The engineered mosquitoes may be malaria free, but so far they have no advantage over their wild counterparts.

It's clear that slashing malaria rates will not be an easy task. Despite decades of research, insecticide-treated bed nets are still the best method for preventing the disease in humans. For millennia, the malaria parasite has managed to withstand human efforts to squelch it, yet science continues to push the boundaries. Who will emerge the victor? Stay tuned.

DELVING DEEPER INTO CASE 4

For each of the Cases, we provide a set of questions to pique your curiosity. These questions cannot be answered directly from the Case itself, but instead introduce issues that will be addressed in chapters to come. We revisit this Case in Chapters 20–23 on the pages indicated below.

1. Which genetic differences make some individuals more and some less susceptible to malaria? See page 437.
2. How did malaria come to infect humans? See page 457.
3. Which human genes are under selection for resistance to malaria? See page 507.

CHAPTER 20

Evolution
How Genotypes and Phenotypes Change over Time

CORE CONCEPTS

20.1 GENETIC VARIATION: Genetic variation refers to differences in DNA sequences.

20.2 MEASURING GENETIC VARIATION: Patterns of genetic variation can be described by allele frequencies.

20.3 EVOLUTION AND THE HARDY–WEINBERG EQUILIBRIUM: Evolution is a change in the frequency of alleles or genotypes in a population over time.

20.4 NATURAL SELECTION: Natural selection leads to adaptations, which increase the fitness of an organism to its environment.

20.5 NON-ADAPTIVE MECHANISMS OF EVOLUTION: Genetic drift, migration, mutation, and non-random mating are also mechanisms of evolution.

20.6 MOLECULAR EVOLUTION: Molecular evolution is a change in DNA or amino acid sequences over time.

A walk down any street reveals how variable our species is: skin color and hair color, for example, vary from person to person. Until the publication in 1859 of Charles Darwin's *On the Origin of Species*, scientists tended to view all the variation we see in humans and other species as biologically unimportant. According to the prevailing view at the time, not only were species individually created in their modern forms by a divine Creator, but, because the Creator had a specific design in mind for each species, they were fixed and unchanging. In keeping with this perspective, departures or variations from this divinely ordained type were ignored.

Since Darwin, however, we have learned that a species consists of a range of variants. In our own species, we can observe that people may be tall, short, dark-haired, fair-haired, and so on. In fact, variation is an essential ingredient of Darwin's theory because the mechanism he proposed for evolution, natural selection, depends on the differential success—in terms of surviving and reproducing—of variants. Darwin changed how we view variation. Before Darwin, variation was irrelevant, something to be ignored; after Darwin, it was recognized as the key to the evolutionary process.

20.1 GENETIC VARIATION

Variation is a major feature of the natural world. We humans are particularly good at noticing phenotypic variation among individuals of our own species. As we discussed in Chapter 15, a phenotype is an observable trait, such as human height or butterfly wing color. Recall from Chapter 17 that two factors contribute to phenotype: an individual's genotype, which is the set of alleles possessed by the individual, and the environment in which the individual lives. By sequencing DNA regions in multiple individuals, we can take the environment out of the equation. Thus, this approach allows us to explore genetic variation directly by looking at differences at the DNA sequence level.

Population genetics is the study of patterns of genetic variation.

Remarkably, in spite of a high degree of phenotypic variation, humans actually rank low in terms of overall genetic variation compared with other species. Any two randomly selected humans differ from each other, on average, by one DNA base per thousand (the two genomes are 99.9% identical), whereas two fruit flies differ by ten bases per thousand (the two genomes are 99% identical). Even one of the most seemingly uniform species on the planet, the Adélie penguins seen in **Fig. 20.1**, is two to three times more genetically variable than humans are.

As we discuss in Chapter 21, a **species** is a group of individuals that are capable, through reproduction, of sharing alleles with one another. Individuals represent different combinations of alleles drawn from the species' **gene pool**, which consists of all the alleles present in all of the individuals in the species. The human gene pool includes alleles that cause differences in skin color, hair type, eye color, and so on. Each one of us has a different set of those alleles—alleles that cause brown hair and brown eyes, for example, or alleles that cause black hair and blue eyes—drawn from that gene pool.

Population genetics is the study of genetic variation in natural **populations**, which are interbreeding groups of organisms of the same species

FIG. 20.1 Genetic diversity in Adélie penguins. Adélie penguins are uniform in appearance but are actually more genetically diverse than humans.
Source: Ralph Lee Hopkins/Getty Images.

living in the same geographical area. Which factors determine the amount of variation in a population and in a species? Why are humans genetically less variable than fruit flies and Adélie penguins? Which factors affect the distribution of particular variations? Population genetics addresses detailed questions about patterns of variation. And small variations, given enough time, can accumulate to produce the major differences we see among species.

Mutation and recombination are the two sources of genetic variation.

Genetic variation has two sources. First, mutation generates new variation. Second, recombination followed by segregation of homologous chromosomes during meiotic cell division shuffles mutations to create new combinations. In both cases, new alleles are formed, as shown in **Fig. 20.2**.

As we saw in Chapter 14, mutations can be **somatic mutations**, which occur in the body's tissues in nonreproductive cells, or **germ-line mutations**, which occur in the reproductive cells. Germ-line mutations are passed on to the next generation, so, from an evolutionary standpoint, these are our primary focus. A somatic mutation affects only the cells descended from the one cell in which the mutation originally arose, so it affects only that one individual. In contrast, a germ-line mutation appears in every cell of an individual

FIG. 20.2 Mutation and recombination. The formation of new alleles occurs by mutation and recombination.

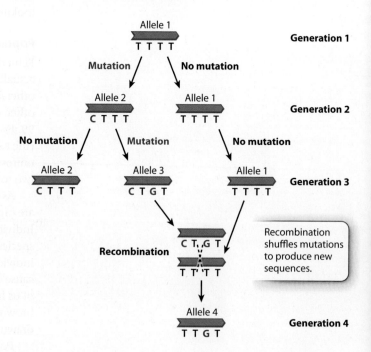

derived from the fertilization involving the mutation-bearing gamete, so it appears in its descendants.

We can also classify mutations by their effects on an organism (Chapter 14). Mutations occur randomly throughout the genome. Every human is born with about 60 new mutations, but, because most of the genome consists of noncoding DNA, most of these are **neutral** and have little or no effect. In contrast, a large proportion of those mutations that occur in protein-coding regions of the genome have a **deleterious** (harmful) effect on an organism. Rarely, a mutation occurs that has a beneficial effect. These kinds of mutations are **advantageous** if they improve their carriers' chances of survival or reproduction. Advantageous mutations, as we will see, can increase in frequency in a population until eventually they are carried by every member of a species. These mutations are the ones that result in a species that is adapted to its environment—better able to survive and reproduce in that environment.

Self-Assessment Questions

1. Why are germ-line mutations more important in evolution than somatic ones?
2. Why is recombination critical to generating genetic variation?

20.2 MEASURING GENETIC VARIATION

Mutations, whether deleterious, neutral, or advantageous, are sources of genetic variation. The goal of population genetics is to make inferences about the evolutionary process from patterns of genetic variation in nature. The raw information for this comes from the rates of occurrence of alleles in populations, which population geneticists refer to as **allele frequencies** in populations.

To understand patterns of genetic variation, we require information about allele frequencies.

The allele frequency of an allele x is the number of x's present in the population divided by the total number of alleles. In other words, allele frequency is the proportion, among all the alleles of a gene in a population, that consists of a specified allele. Consider, for example, pea color in Mendel's pea plants. In Chapter 15, we discussed how pea color (yellow or green) results from variation at a single gene. Two alleles of this gene are the dominant A (yellow) allele and the recessive a (green) allele. AA homozygotes and Aa heterozygotes produce yellow peas, whereas aa homozygotes produce green peas. Imagine that in a population every pea plant produces green peas, meaning that only one allele, a, is present. In this population, the allele frequency of a is 100%, and the allele frequency of A is 0%. When a population exhibits only one allele at a particular gene, we say that the population is **fixed** for that allele. Similarly, fixation is the change from a situation where there are at least two alleles to a situation where only one of the alleles remains.

Now consider another population of 100 pea plants with genotype frequencies of 50% aa, 25% Aa, and 25% AA. The **genotype frequency** is the proportion in a population of each genotype at a particular gene or set of genes. The genotype numbers for this population are 50 green pea homozygotes (aa), 25 yellow pea heterozygotes (Aa), and 25 yellow pea homozygotes (AA).

What is the allele frequency of a in this population? Each of the 50 aa homozygotes has two a alleles and each of the 25 heterozygotes has one a allele. There are no a alleles in AA homozygotes. The total number of a alleles is $(2 \times 50) + 25 = 125$. To determine the allele frequency of a, we divide the number of a alleles by the total number of alleles in the population, 200 (because each pea plant is diploid, meaning that it has two alleles): $125/200 = 62.5\%$. Because we are dealing with only two alleles in this example, the allele frequency of A is $100\% - 62.5\% = 37.5\%$.

The allele frequencies of A and a provide a measure of genetic variation at one gene in a given population. In this example, we were given the genotype frequencies, and from this information we determined the allele frequencies. But how are genotype and allele frequencies measured? In the following section, we consider three ways to measure genotype and allele frequencies in populations: observable traits, gel electrophoresis, and DNA sequencing.

Early population geneticists relied on observable traits and gel electrophoresis to measure variation.

It would be a simple matter to measure genetic variation in a population if we could always use observable traits, as we did with the pea plant example. In that case, we could simply count the individuals displaying variant forms of a trait and have a measure of the variation of that trait's gene. However, as we saw in Chapter 17, this approach works only rarely, for two important reasons. First, many traits are encoded by a large number of genes. In these cases, it is difficult, if not impossible, to make direct inferences from a phenotype to the underlying genotype. Even traits that seem to have a simple set of phenotypes often prove to have a complicated genetic basis. For instance, human skin color is determined by at least six different genes. Second, the phenotype is a product of both the genotype and the environment.

Until the 1960s, only one workable solution was available: to limit population genetics to the study of phenotypes that are encoded by a single gene. As these phenotypes are relatively rare, the number of genes that population geneticists could study was extremely small. Human blood groups, including the ABO system, provided an early example of a trait encoded by a single gene with multiple alleles. At this gene, there are three alleles in the population— A, B, and O—and therefore six possible genotypes, which result in four different phenotypes (**Table 20.1**).

TABLE 20.1	The ABO blood system.
PHENOTYPE	GENOTYPE
A	AA or AO
B	BB or BO
AB	AB
O	OO

Other instances in which phenotypic variation can be readily correlated with genotype include certain markings in invertebrates. For example, the coloring of the two-spot ladybug *Adalia bipunctata* is controlled by a single gene (**Fig. 20.3**). The genetic basis of most traits, however, is not so simple.

Single-gene variation became much easier to detect in the 1960s with the application of gel electrophoresis. In Chapter 12, we saw how gel electrophoresis separates segments of DNA according to their size. Before DNA technologies were developed, the same basic process was applied to proteins to separate them according to their electrical charge and their size. In gel electrophoresis, the proteins being studied migrate through a gel when an electrical charge is applied. The rate at which the proteins move from one end of the gel to the other is determined by their charge and their size.

Early studies of protein electrophoresis focused on enzymes that catalyze reactions that can be induced to produce a dye when the substrate for the enzyme is added. If researchers add the substrate, they can see the locations of the proteins in the gel. The bands in the gel provide a visual picture of genetic variation in the population, revealing which alleles are present and what their frequencies are. **Fig. 20.4** shows this sort of experiment.

DNA sequencing is the gold standard for measuring genetic variation.

Protein gel electrophoresis was a leap forward in our ability to detect genetic variation, but this technique had significant limitations. Researchers could study only enzymes because they needed to be able to stain specifically for enzyme activity. The technique also could detect only mutations that resulted in amino acid substitutions that changed a protein's mobility in the gel. Only with DNA sequencing did researchers finally have an unambiguous means of detecting all genetic variation in a stretch of DNA, whether in a coding region or not. The variations studied by modern population geneticists are differences in DNA sequence, such as a T rather than a G at a specified nucleotide position in a particular gene.

Calculating allele frequencies with DNA sequencing involves collecting a population sample and counting the number of occurrences of a given mutation. For example, researchers can look even more closely at the example of the *Drosophila Adh* gene from Fig. 20.4 to focus not on the amino acid difference between the *Fast* and *Slow* phenotypes, but on the A or G nucleotide difference underlying the two phenotypes. If they sequence the *Adh* gene from 50 individual flies, they then have 100 gene sequences from these diploid individuals. If their finding is that 70 sequences have an A and 30 have a G at the position in question, the allele frequency of A is $70/100 = 0.7$ and the allele frequency of G is 0.3. In general, in a sample of n diploid individuals, the allele frequency is the number of occurrences of that allele divided by twice the number of individuals.

FIG. 20.3 Single-gene variation. A genetic difference in color in the two-spot ladybug, *Adalia bipunctata*, results from variation in a single gene. *Sources: (left)* © *Biopix: G Drange; (right) Howard Marsh/Shutterstock.*

HOW DO WE KNOW?

FIG. 20.4

How did gel electrophoresis allow us to detect genetic variation?

BACKGROUND The introduction of protein gel electrophoresis in 1966 gave researchers the opportunity to identify differences in amino acid sequence in proteins both among individuals and, in the case of heterozygotes, within individuals. In gel electrophoresis, proteins with different amino acid sequences move at different rates through a gel in an electric field. Often, a single amino acid difference is enough to affect the mobility of a protein in a gel.

METHOD Researchers load material derived from crude tissue—in this case, the ground-up whole body of a fruit fly (*Drosophila melanogaster*)—on a gel and turn on the current. The rate at which any given protein migrates depends on its size and its charge, both of which may be affected by its amino acid sequence. To visualize the protein at the end of the gel run, researchers use a biochemical indicator that produces a stain when the protein of interest is active. In the experiment shown here, researchers used a biochemical agent that detects the activity of alcohol dehydrogenase (*Adh*) as the indicator. A series of bands on the gel shows where the *Adh* proteins in each lane migrated.

Drosophila. The body of the fly is ground up.

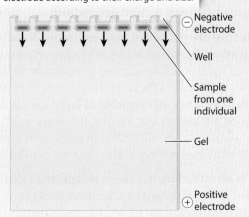

Each well of the gel is loaded with a sample from one individual, and an electric current is passed through the gel. The proteins in each sample migrate toward the positive electrode according to their charge and size.

- Negative electrode
- Well
- Sample from one individual
- Gel
- Positive electrode

RESULTS The *Adh* gene has two common alleles, which are distinguished by a single amino acid difference that changes the charge of the protein. One allele, *Fast* (*F*), runs faster than the other, *Slow* (*S*). In the diagram below, the experimental results show that four individuals are *S* homozygotes, two are *F* homozygotes, and two are *FS* heterozygotes. Note that the heterozygotes do not stain as strongly on the gel because each band has half the intensity of the single band in the homozygote. We can measure the allele frequencies by counting the alleles. Each homozygote has two of the same allele, and each heterozygote has one of each.

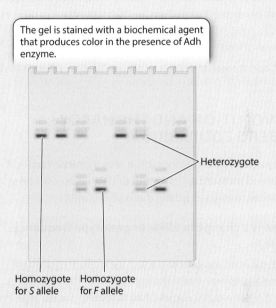

The gel is stained with a biochemical agent that produces color in the presence of Adh enzyme.

Heterozygote

Homozygote for *S* allele   Homozygote for *F* allele

Total number of alleles in the population of 8 individuals = $8 \times 2 = 16$

Number of *S* alleles in the population = $2 \times$ (number of *S* homozygotes) + (number of heterozygotes) = $8 + 2 = 10$

Frequency of $S = \frac{10}{16} = \frac{5}{8}$

Number of *F* alleles in the population = $2 \times$ (number of *F* homozygotes) + (number of heterozygotes) = $4 + 2 = 6$

Frequency of $F = \frac{6}{16} = \frac{3}{8}$

Note that the two allele frequencies add to 1.

CONCLUSION The experiment provides a profile of genetic variation at this gene for these eight individuals. Population genetics involves comparing data such as these with data collected from other populations to determine the forces shaping patterns of genetic variation.

FOLLOW-UP WORK This technique is seldom used now because it is easy to recover much more detailed genetic information about genetic variation from DNA sequencing.

ORIGINAL PAPER Lewontin, R. C., and J. L. Hubby. 1966. "A Molecular Approach to the Study of Genic Heterozygosity in Natural Populations. II. Amount of Variation and Degree of Heterozygosity in Natural Populations of *Drosophila pseudoobscura*." *Genetics* 54:595–609.

> **Self-Assessment Questions**
>
> 3. What is genetic variation, and how is it measured?
> 4. Using the example of pea color in Mendel's pea plants, devise equations to determine the allele frequencies of *A* and *a* from the genotype frequencies of *aa*, *Aa*, and *AA*.
> 5. Data on genetic variation in populations have become ever more precise over time, from phenotypes determined by a single gene to results obtained through gel electrophoresis that looks at variation among genes that encode for enzymes, to data generated directly from the DNA sequence. Has this increase in precision resulted in the discovery of more genetic variation or less?
> 6. What does it mean to say that an allele is "fixed" in a population?

20.3 EVOLUTION AND THE HARDY–WEINBERG EQUILIBRIUM

Allele frequencies give us information about genetic variation. Following and measuring changes in that variation over time is key to understanding the genetic basis of evolution.

Evolution is a change in allele or genotype frequency over time.

At the genetic level, evolution is a change in the frequency of an allele or a genotype from one generation to the next. For example, if there are 200 copies of an allele that causes blue eye color in a population in generation 1 and there are 300 copies of that allele in a population of the same size in generation 2, evolution has occurred. In principle, evolution may occur without allele frequencies changing. For instance, in our fruit fly example, even if the *A/G* allele frequencies stay the same from one generation to the next, the frequencies of the different genotypes (*AA*, *AG*, and *GG*) may change. This would be evolution *without* allele frequency change.

Evolution is therefore a change in the genetic makeup of a population over time. Note an important and often misunderstood aspect of this definition: *populations* evolve, not *individuals*. Note, too, that this definition does not specify a mechanism for this change. As we will see, many mechanisms can cause allele or genotype frequencies to change. Regardless of which mechanisms are involved, any change in allele frequencies, genotype frequencies, or both constitutes evolution.

The Hardy–Weinberg equilibrium describes situations in which allele and genotype frequencies do not change.

Allele and genotype frequencies change over time only if specific forces act on the population. This principle, which was demonstrated independently in 1908 by the English mathematician G. H. Hardy and the German physician Wilhelm Weinberg, has become known as the **Hardy–Weinberg equilibrium**.

In essence, the Hardy–Weinberg equilibrium describes the situation in which evolution does *not* occur. In the absence of evolutionary forces (such as natural selection), allele and genotype frequencies do not change.

To determine whether evolutionary forces are at work, we need to determine whether a population is in Hardy–Weinberg equilibrium. The Hardy–Weinberg equilibrium specifies the relationship between allele frequencies and genotype frequencies when several key conditions are met. In cases in which these conditions are met, we can conclude that evolutionary forces are not acting on the gene in the population we are studying. In many ways, then, the Hardy–Weinberg equilibrium is most interesting when we find instances in which allele or genotype frequencies *depart* from expectations. This finding implies that one or more of the conditions are *not* met and that evolutionary mechanisms are at work.

A population that is in Hardy–Weinberg equilibrium meets these conditions:

1. **There is no difference in the survival and reproductive success of individuals.** Let's examine what happens when this condition is *not* met. Given two alleles, *A* and *a*, consider what occurs when *a*, a recessive mutation, is lethal. In other words, all *aa* individuals die. In every generation, then, there is a selective elimination of *a* alleles, meaning that the frequency of *a* will gradually decline (and the frequency of *A* correspondingly increase) over the generations. As we discuss later in this chapter, we call this differential success of alleles **selection**.

2. **The population is sufficiently large to prevent sampling errors.** Small samples are often more misleading than large ones. On a campus-wide basis, a college's sex ratio may be close to 50:50, but it is not improbable that a small class of eight individuals might consist of six women and two men (a 75:25 ratio). Sample size, in the form of population size, also affects the Hardy–Weinberg equilibrium such that it technically holds only for infinitely large populations. A change in the frequency of an allele due to the random effects of small population size is called **genetic drift**.

3. **Populations are not added to or subtracted from by migration.** Again, let's see what happens when this condition is *not* met. Consider two populations, the first with only *A* alleles and the second with a mix of *A* and *a* alleles. If there is a sudden influx of individuals from the second population into the first, the frequency of *A* in the first population will decline in proportion to the number of immigrants as *a* alleles enter the population.

4. **There is no mutation.** If *A* alleles mutate into *a* alleles (or other alleles, if the gene has multiple alleles), and vice versa, then again we see changes in the

allele frequencies over the generations, meaning that this condition is *not* met and evolutionary change is occurring. In general, because mutation is so rare, it has a very small effect on changing allele frequencies on the timescales studied by population geneticists.

5. **Individuals mate at random.** For a population to be in Hardy–Weinberg equilibrium, mate choice must be made without regard to genotype. For example, an *AA* homozygote, when offered a choice of mate from among *AA*, *Aa*, or *aa* individuals, must choose at random. In contrast, **nonrandom mating** occurs when individuals do not mate randomly. For example, *AA* homozygotes might preferentially mate with other *AA* homozygotes. Nonrandom mating affects genotype frequencies from generation to generation, but does not affect allele frequencies.

The Hardy–Weinberg equilibrium relates allele frequencies and genotype frequencies.

Now that we have established the conditions required for a population to be in Hardy–Weinberg equilibrium, let us explore the idea in detail. Let's return to the Mendel's peas example we looked at earlier where there are two alleles, *A* and *a*, determining pea color. In a different population from the one we originally looked at, we find 70 *A* alleles and 30 *a* alleles. What are the genotype frequencies? How many *AA* homozygotes, *Aa* heterozygotes, and *aa* homozygotes are there in the population? The Hardy–Weinberg equilibrium predicts the expected genotype frequencies from allele frequencies as long as the Hardy–Weinberg conditions are met.

Recall that random mating is a condition that must be met if a population is in Hardy–Weinberg equilibrium. Random mating is the equivalent of putting all the population's gametes into a single pot and drawing out pairs of them at random to form zygotes. This is the same principle we saw in action in the discussion of independent assortment in Chapter 15. If there are 70 *A* alleles and 30 *a* alleles and mating is random, what is the probability of picking an *AA* homozygote? That is, what is the probability of picking an *A* allele followed by another *A* allele?

The probability of picking an *A* allele is the same as the *A* allele's frequency in the population. Therefore, the probability of picking the first *A* is $70/100 = 0.7$. What is the probability of picking the second *A*? Also 0.7. The probability of picking an *A* followed by another *A* is the product of the two probabilities: $0.7 \times 0.7 = 0.49$. There is a 49% probability of picking an *AA* homozygote. That means that the frequency of an *AA* genotype is 0.49. We can follow the same steps to determine the genotype frequency for the *aa* genotype. Its frequency is 0.3×0.3, or 0.09, so there is a 9% probability of picking an *aa* homozygote.

What about the genotype frequency of the heterozygote, *Aa*? If mating is random and a population is in Hardy–Weinberg equilibrium, what is the probability of an *Aa* heterozygote? We compute this value by determining the probability of drawing *a* followed by *A*, or *A* followed by *a*. In other words, there are two separate ways in which we can generate the heterozygote. Its frequency is therefore $(0.7 \times 0.3) + (0.3 \times 0.7) = 0.42$, so there is a 42% probability of picking an *Aa* heterozygote.

As seen in the Punnett square on the left in **Fig. 20.5**, we can generalize these calculations algebraically by substituting letters for the numbers to derive the relation defined by the Hardy–Weinberg equilibrium. If the allele frequency of one allele, *A*, is *p*, and the other allele, *a*, is *q*, then $p + q = 1$. This is because there are no other alleles at this gene and allele and genotype frequencies always sum to 1 (or 100%).

Genotypes	*AA*	*Aa*	*aa*
Frequencies	p^2	$2pq$	q^2

In the graph on the right-hand side of Fig 20.5, we have replaced the *p*'s and *q*'s with numbers. If no *a* alleles are present,

FIG. 20.5 Hardy–Weinberg relation. The Hardy–Weinberg relation predicts genotype frequencies from allele frequencies, and vice versa.

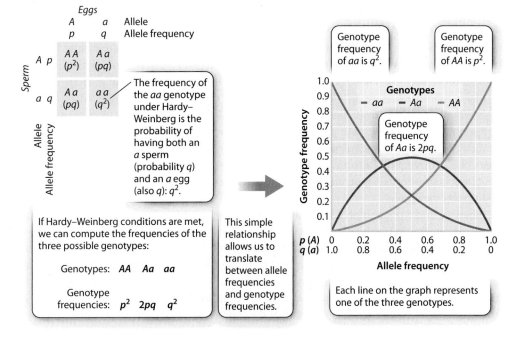

then $q = 0$ and $p = 1$ and the frequency of AA is 1. In other words, the allele frequency of a is 0, the allele frequency of A is 1, and the genotype frequency of AA is 1. In this case, all the genotypes in the population are AA. Accordingly, the blue line representing the frequency of AA in the population is at 1, and the red line representing the frequency of aa and the purple line representing the frequency of Aa are both at 0. When no A alleles are present, all individuals have genotype aa, and the red line is at 1 and the blue and purple lines are at 0.

Not only does the Hardy–Weinberg relation predict genotype frequency from allele frequencies, but it works in reverse, too: genotype frequencies predict allele frequencies. The graph in Fig. 20.5 shows how allele and genotype frequencies are related. We can use the graph to determine allele frequencies for given genotype frequencies, and genotype frequencies for given allele frequencies. For example, if the population we are examining has a 0.5 genotype frequency of heterozygotes, Aa, the purple line in the graph indicates that allele frequencies for p and q are both 0.5. Similarly, if we know that the allele frequency for $p = 0.5$ (and therefore $q = 0.5$), we can look at the lines to infer that the genotype frequency of heterozygotes is 0.5 and the genotype frequency of both homozygotes, AA and aa, is the same, 0.25.

We can do this mathematically as well. Knowing the genotype frequency of AA, for example, permits us to calculate allele frequencies: if, as in our pea example, p^2 is 0.49 (that is, 49% of the population has genotype AA), then p, the allele frequency of A, is 0.7. Because $p + q = 1$, then q, the allele frequency of a, is $1 - 0.7 = 0.3$.

Note that these relationships hold only if the Hardy–Weinberg conditions are met. If not, then allele frequencies can be determined only from genotype frequencies, as described earlier in section 20.2.

The Hardy–Weinberg equilibrium is the starting point for population genetic analysis.

Recall the definition of evolution: a change in allele or genotype frequency from one generation to the next. Given this definition, it might seem odd that we have spent so much time discussing factors necessary for allele frequencies to stay the same. Why study the Hardy–Weinberg equilibrium at all? The Hardy–Weinberg equilibrium provides a means of converting between allele and genotype frequencies, but, critically, when it is *not* upheld, this equilibrium also serves as an indicator that something interesting is happening in a population from an evolutionary viewpoint.

When we find a population whose allele or genotype frequencies are not in Hardy–Weinberg equilibrium, we can infer that evolution has occurred in that population. With further study of the population, we can consider, for the gene in question, whether the population is subject to selection, genetic drift, migration, mutation, or nonrandom mating. These are the primary mechanisms of evolution. (Recall that the conditions of the Hardy–Weinberg equilibrium are concerned with these five mechanisms.) Thus, the Hardy–Weinberg equilibrium gives us a baseline from which to explore the evolutionary processes affecting populations. In the next section, we start our exploration by considering one of the most important evolutionary mechanisms: natural selection.

Self-Assessment Questions

7. Can evolution occur without allele frequency changes? If not, why not? If so, how?

8. What can and can't we conclude about a population whose allele frequencies are not in Hardy–Weinberg equilibrium?

9. How would you calculate genotype frequencies of a population in Hardy–Weinberg equilibrium, given the allele frequencies for that trait?

20.4 NATURAL SELECTION

Natural selection results in allele frequencies changing from generation to generation according to the allele's impact on the survival and reproduction of individuals. New mutations that are deleterious and eliminated by natural selection have no long-term evolutionary impact; in contrast, mutations that are beneficial can result in adaptation to the environment over time.

Natural selection brings about adaptations.

The adaptations we see in the natural world—the exquisite fit of organisms to their environment—were typically taken by pre-Darwinian biologists as evidence of a divine Creator's existence. Each species, they argued, was so well adapted—the desert plant so physiologically adept at coping with minimal levels of rainfall and the fast-swimming fish so hydrodynamically streamlined—that it must have been designed by a Creator.

With the publication of *On the Origin of Species* in 1859, Darwin, pictured in **Fig. 20.6**, overturned the biological convention of his day on two fronts. First, he showed that species are not unchanging, but rather have evolved over time. Second, he suggested a mechanism, natural selection, that brings about adaptation. Natural selection is a brilliant solution to the central problem of biology: how organisms come to fit so well in their environments. From where does the woodpecker get its powerful chisel of a bill? And the hummingbird its long delicate bill for probing the nectar stores in flowers? Darwin showed how a simple mechanism, without foresight or intentionality, could result in the extraordinary range of adaptations that all of life is testimony to.

For 20 years after first conceiving the essence of his theory, Darwin collected supporting evidence for that theory. In 1858, he was finally spurred to begin writing *On the Origin of Species* by a letter from a little-known naturalist collecting specimens in what is today Indonesia. By a remarkable coincidence, Alfred

FIG. 20.6 Charles Darwin. This photograph was taken at about the time Darwin was writing *On the Origin of Species*. Source: AKG Images.

FIG. 20.7 Alfred Russel Wallace. This photograph was taken in Singapore during Wallace's expedition to Southeast Asia. Source: A. R. Wallace Memorial Fund & G. W. Beccaloni.

Russel Wallace, shown in **Fig. 20.7,** had also developed the theory of evolution by natural selection. Aware that Darwin was interested in the problem but having no idea that Darwin was working on the same theory, Wallace wrote to Darwin in 1858 to ask what Darwin thought of his idea.

Suddenly, Darwin was confronted with the prospect of losing his claim on the theory that he had been quietly nurturing for 20 years. But all was not lost. Darwin's colleagues arranged for the publication of a joint paper by Wallace and Darwin in 1858. This was done without consulting Wallace, who nonetheless never resented Darwin and afterward was careful to insist that the idea belonged to Darwin. It was Darwin's publication of *On the Origin of Species* in 1859 that brought both evolution and natural selection, its underlying mechanism of adaptation, to public attention. Wallace is only fleetingly mentioned in Darwin's great work, and Darwin, not Wallace, is now the name associated with the discovery.

Both Darwin and Wallace recognized their debt to the writings of a British clergyman, Thomas Robert Malthus. In his *An Essay on the Principle of Population,* first published in 1798, Malthus pointed out that natural populations have the potential to increase in size geometrically, meaning that populations can grow larger at an ever-increasing rate. For example, imagine that human couples each have just four children (two males and two females), so the population doubles every generation. Starting with a single couple, by the twentieth generation, the population will have grown to more than a million people—1,048,576, to be precise.

In reality, this geometric expansion of populations does not occur. In fact, population sizes are typically stable from generation to generation, largely because the resources upon which populations depend—food, water, places to live—are limited. In each generation, many individuals fail to survive or reproduce. Put simply, there are not enough food and other resources to go around, so individuals within a population must compete for survival.

Which individuals will win the competition? Darwin and Wallace suggested that those that are best adapted would most likely survive and leave more offspring. Genetic variation among individuals results in some individuals that are more likely than others to survive and reproduce, passing their genetic material to the next generation. As a result, the next generation has a higher proportion of these same advantageous alleles. Darwin used the term "natural selection" for the filtering process that acts against deleterious alleles and in favor of advantageous ones.

Competitive advantage is a function of how well an organism is adapted to its environment. A desert plant that is more efficient at minimizing water loss than another plant is better adapted to the desert environment. An organism that is better adapted to its environment is more fit. **Fitness**, in this context, is a measure of the extent to which the individual's genotype is represented in the next generation. We say that the first plant's fitness is higher than the second plant's fitness if the former leaves more surviving offspring either because that plant itself survives for longer, giving it greater opportunity to reproduce, or because it has some other reproductive advantage, such as the ability to produce more seeds.

Assuming that the trait maintains its advantage, natural selection acts over generations to increase the overall fitness of a population. A plant population newly arrived in a desert may be poorly adapted to its environment. Over time, however, alleles that minimize water loss increase under natural selection, resulting in a population that is better adapted to the desert.

Such changes in populations take time. Indeed, borrowing from the geologists of his day, Darwin recognized that time was a critical ingredient of his theory. Geologists had put forward a view of Earth's history that argued large geological changes—like the carving of the Grand Canyon—can be explained by simple day-to-day processes operating over vast timescales. Darwin applied this worldview to biology. He recognized that small changes, such as subtle shifts in the frequencies of alleles, could add up to major changes over long enough time periods. What might seem to us to be a trivial change over the short term can, over the long term, result in substantial differences among populations.

The Modern Synthesis combines Mendelian genetics and Darwinian evolution.

Darwinian evolution addresses the change over time of the genetic composition of populations. It is thus a genetic theory. Although Mendel published his genetic studies of pea plants in 1866, not long after *On the Origin of Species*, Darwin never saw Mendel's work, so a key component was missing from Darwin's theory.

The rediscovery of Mendel's work in 1900 unexpectedly provoked a major controversy among evolutionary biologists. Some argued that Mendel's discoveries did not apply to most genetic variation because the traits studied by Mendel were discrete, meaning that they had clear alternative states, such as either yellow or green color in peas. Most of the variation we see in natural populations, by contrast, is continuous, meaning that variation occurs across a spectrum (Chapter 17). Human height, for example, does not come in discrete classes. People are not either 5 feet tall or 6 feet tall and of no height in between. Instead, they may be any height within a certain range.

How could the factors that controlled Mendel's discrete traits account for the continuous variation seen in natural populations? This question was answered by the English theoretician Ronald Fisher, who realized that, instead of a single gene dictating a trait like human height, several genes might contribute to the trait. He argued that extending Mendel's theory to include multiple genes per trait could account for patterns of continuous variation that we see all around us.

Fisher's insight formed the basis of a synthesis between Darwin's theory of natural selection and Mendelian genetics that was forged during the middle part of the twentieth century. The product of this **Modern Synthesis** is our current theory of evolution.

Natural selection increases the frequency of advantageous mutations and decreases the frequency of deleterious mutations.

Natural selection increases the frequency of advantageous alleles, resulting in adaptation. In some cases, natural selection increases the frequency of an advantageous allele until the population is fixed for the allele (that is, 100% of the population has the allele). To start with, a new advantageous allele will exist as a single copy in a single individual (that is, as a heterozygote). Subsequently, under the influence of natural selection, the advantageous allele can eventually replace all the other alleles in the population. Natural selection that increases the frequency of an advantageous allele is called **positive selection**.

As we have noted, most mutations to genes are deleterious. In extreme cases, they are lethal to the individuals carrying them and, therefore, are eliminated from the population. Sometimes, however, natural selection is inefficient in getting rid of a deleterious allele. Consider a recessive lethal mutation, b (that is, one that is lethal only as a homozygote, bb, and has no effect as a heterozygote, Bb). When b first arises, all the other alleles in the population are B, which means that the first b allele that appears in the population must be paired with a B allele, resulting in a Bb heterozygote. Because natural selection does not act against heterozygotes in this case, the b allele may increase in frequency by chance alone (we discuss how this happens in the next section). Only when two b alleles come together to form a bb homozygote does natural selection act to rid the population of those two copies of the b allele. Natural selection that decreases the frequency of a deleterious allele is called **negative selection**.

Many human genetic diseases show this same pattern: the deleterious allele is rare and recessive. Because it is rare, homozygotes for it are formed only infrequently. Recall that the expected frequency of homozygotes in a population under the Hardy–Weinberg equilibrium is the square of the frequency of the allele in the population. Therefore, if the recessive lethal allele frequency is 0.01, we expect 0.01×0.01, or 1 in every 10,000 individuals, to be homozygous for it. Thus, the genetic disease occurs rarely, and the allele remains in the population because it is recessive and not expressed as a heterozygote.

CASE 4 MALARIA: COEVOLUTION OF HUMANS AND A PARASITE

Which genetic differences have made some individuals more and some less susceptible to malaria?

In addition to allowing alleles to be either eliminated or fixed, natural selection can maintain an allele at some intermediate frequency between 0 and 1. This form of natural selection, called **balancing selection**, acts to maintain two or more alleles in a population. For example, consider members of a species that face different conditions depending on where they live. One allele might be favored by natural selection in a dry area, but a different allele is favored in a wet area. In the species as a whole, these two alleles are both maintained by natural selection at intermediate frequencies.

Another example of balancing selection occurs when the heterozygote's fitness is higher than that of either of the homozygotes, resulting in selection that ensures both alleles remain in the population at intermediate frequencies. This form of balancing selection is called **heterozygote advantage**. Heterozygote advantage is illustrated by human populations in Africa affected by malaria. Because the malaria parasite spends part of its life cycle in human red blood cells, mutations in the hemoglobin molecule that affect the structure of the red blood cells can have a negative effect on the parasite. These hemoglobin mutations can reduce the severity of malarial attacks.

Two alleles of the gene for one of the subunits of hemoglobin are *A* and *S* (Chapter 14). The *A* allele codes for normal hemoglobin, resulting in fully functional, round red blood cells. The *S* allele encodes a polypeptide that differs from the *A* allele's polypeptide in just a single amino acid. This difference makes the red blood cell distort into a sickle shape and causes the disease called sickle-cell anemia.

In regions of the world with malaria, heterozygous individuals (*AS*) have an advantage over homozygous individuals (*AA* and *SS*). *AA* homozygotes lack sickle-cell anemia, but are vulnerable to malaria. *SS* homozygotes are protected against malaria, but are burdened with sickle-cell anemia. Sickle-shaped red blood cells can block capillaries, which leaves people with the *SS* genotype prone to debilitating, painful, and sometimes fatal episodes resulting from capillary blockage. *AS* heterozygotes do not have severe sickling disease and have some protection from malaria. As a result, natural selection maintains both the *A* and *S* alleles in the population at intermediate frequencies.

In areas where there is no malaria, this balance is shifted. Many African-Americans who are descended from Africans upon whom natural selection operated in favor of the heterozygote still carry the *S* allele, even though the allele is no longer useful to them in their current malaria-free environment. If natural selection were to run its course among African-Americans, the *S* allele would gradually be eliminated. This is a slow process, however, and many more people will continue to suffer from sickle-cell anemia before it is complete. Fortunately, medical interventions are available that can help those affected with the disease.

Natural selection can be stabilizing, directional, or disruptive.

Up to this point, we have followed the fate of individual mutations, which can increase, decrease, or be maintained at an intermediate frequency under the influence of natural selection. We can also look at the consequence of natural selection from a different perspective. Instead of following individual mutations, we can look at changes over time in a particular trait of an organism. For example, we might track the evolution of height in a population, despite not knowing the specifics of the genetic basis of height differences. When we look at natural selection from this perspective, we see three types of patterns: stabilizing, directional, and disruptive.

Stabilizing selection maintains the status quo and acts against extremes. A good example is provided by human birth weight, a trait affected by a number of factors, including many fetal genes (**Fig. 20.8**). If a baby is too small, its chances of survival after birth are low. However, if it is too big, there may be complications during delivery that endanger both mother and baby. The optimal birth weight is between these two extremes. In this case, natural selection acts against the extremes. The vast majority of natural selection is of this kind, meaning that deleterious mutations that cause a departure from the optimal phenotype are selected against.

Whereas stabilizing selection keeps a trait the same over time, **directional selection** leads to a change in a trait over time. A well-documented case of directional

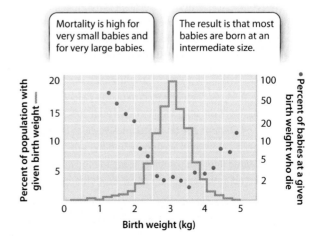

FIG. 20.8 Stabilizing selection. Stabilizing selection on human birth weight results in selection against babies that are either too small or too large. *Source: Data from L. L. Cavalli-Sforza and W. F. Bodmer, 1971, The Genetics of Human Populations, San Francisco, CA: W. H. Freeman, p. 613.*

FIG. 20.9 Directional selection. Directional selection in a population of Galápagos finches caused a shift toward larger bills in a single drought year, 1977. *Data from Freeman & Herron Evolutionary Analysis 2004.*

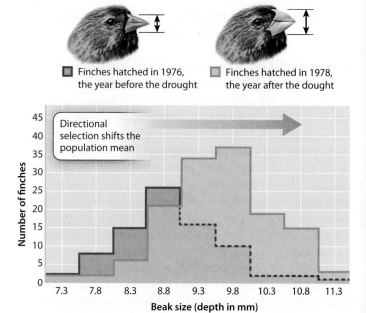

selection is found in Darwin's finches on the Galápagos Islands (**Fig. 20.9**). A severe drought in 1977 killed a significant amount of the vegetation that provided food for one island's population of seed-eating ground finches, *Geospiza fortis*, reducing that population's numbers from 750 to 90. Because plant species that produce big seeds fared better in drought conditions than other plant species did, the average size of seeds available to the remaining birds increased. Birds with bigger bills were better at handling the big seeds and, therefore, had higher survival rates than birds with smaller bills. Because bill size is genetically determined, the drought resulted in directional selection for increased bill size.

HOW DO WE KNOW?

FIG. 20.10

How far can artificial selection be taken?

BACKGROUND From the 1890s to the present day, an experiment at the University of Illinois has attempted to manipulate the properties of corn. This experiment has become one of the longest-running biological experiments in history.

HYPOTHESIS Researchers hypothesized that there is a limit to the extent to which a population can respond to continued directional selection.

EXPERIMENT Corn was artificially selected for either high oil content or low oil content. Every generation, researchers bred together just the plants that produced corn with the highest oil content, and did the same for the plants that produced corn with the lowest oil content. Each generation, kernels showed a range of oil levels, but only the 12 kernels with the highest or the lowest oil content were used for the next generation.

RESULTS In the line selected for high oil content, the percentage of oil more than quadrupled, from about 5% to more than 20%. In the line selected for low oil content, the oil content fell so close to zero that it could no longer be measured accurately, and the selection was terminated. Both selected lines are completely outside the range of any phenotype observed at the beginning of the experiment.

FOLLOW-UP WORK Genetic analysis of the selected lines indicates that the differences in oil content are due to the effects of at least 50 genes.

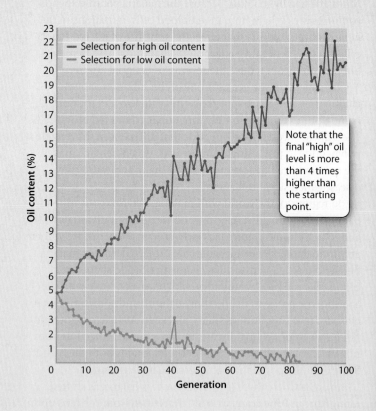

SOURCE Moose, S. P., J. W. Dudley, and T. R. Rocheford. 2004. "Maize Selection Passes the Century Mark: A Unique Resource for 21st Century Genomics." *Trends in Plant Science* 9:358–364.

FIG. 20.11 Disruptive selection. Disruptive selection has produced two genetically distinct populations of apple maggot fly, each one coordinated with the fruiting time of its host tree species. *Sources: (photo) Rob Oakleaf, Michigan State University, Tree Fruit Entomology Lab; Data from Filchak et al. 2000 Nature 407:739–742.*

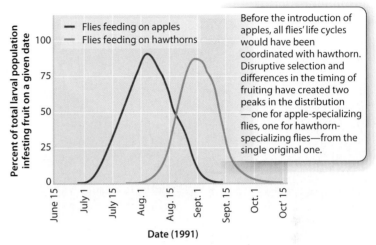

Before the introduction of apples, all flies' life cycles would have been coordinated with hawthorn. Disruptive selection and differences in the timing of fruiting have created two peaks in the distribution—one for apple-specializing flies, one for hawthorn-specializing flies—from the single original one.

Artificial selection, which has been practiced by humans since at least the dawn of agriculture, is a form of directional selection. Artificial selection is analogous to natural selection, but the competitive element is removed. Successful genotypes are selected by the breeder, not through competition. Because it can be carefully controlled by the breeder, artificial selection is astonishingly efficient at generating genetic change. When practiced over many generations, artificial selection can create a population in which the selected phenotype is quite different from that of the starting population. **Fig. 20.10** shows the result of artificial selection for oil content in kernels of corn.

A third mode of selection, known as **disruptive selection**, operates in favor of extremes and against intermediate forms. Apple maggot flies of North America, *Rhagoletis pomonella*, provide an example (**Fig. 20.11**). The larvae of these flies originally fed on the fruit of hawthorn trees. However, since the introduction of apples from Europe about 150 years ago, these flies have become pests of apples as well. Apple trees flower and produce fruit earlier every summer than hawthorns, so disruptive selection has resulted in the production of two genetically distinct groups of flies, one feeding on apple trees and the other on hawthorn trees. Disruptive selection acts against intermediates between the two groups, which would miss the peaks of both the apple and hawthorn seasons. We explore this mechanism, which can lead to the evolution of new species, in more detail in the next chapter.

Sexual selection increases an individual's reproductive success.

Initially, Darwin was puzzled by features of organisms that seemed to reduce an individual's chances of survival. In a letter dated a few months after the publication of *The Origin*, he wrote, "The sight of a feather in a peacock's tail, whenever I gaze at it, makes me sick!" The tail is metabolically expensive to produce; it is an advertisement to potential predators; and it is an encumbrance in any attempt to escape a predator. How could such a feature evolve under natural selection?

In his 1871 book, *The Descent of Man, and Selection in Relation to Sex*, Darwin introduced a solution to this problem. Natural selection is indeed acting to reduce the showiness and size of the peacock's tail, but another form of selection, **sexual selection**, is acting in the opposite direction. Sexual selection promotes traits that increase an individual's access to reproductive opportunities.

Darwin recognized that this could occur in two different ways (**Fig. 20.12**). In one form of sexual selection, members of one sex (usually the males) compete with one another for access to the other sex (usually the females). This form is called **intrasexual selection**, since it focuses on interactions between individuals of one sex. Because competition typically occurs among males, it is in males that we see physical traits such as large size and horns and other elaborate weaponry, as well as fighting ability. Larger, more powerful males tend to win more fights, hold larger territories, and have access to more females.

Darwin also recognized a second form of sexual selection. In this variant, males (typically) do not fight with one another, but instead compete for the attention of the female with bright colors or advertisement displays. In this case, females choose their mates. This form of selection is called **intersexual selection**, since it focuses on interactions between males and females. The peacock's tail is thought to be the product of intersexual selection: its evolution has been driven by a female preference for ever-showier tails. In the absence of sexual selection, natural selection would act to minimize the size of the peacock's tail. Presumably, the peacocks' tails we see are a compromise, a

FIG. 20.12 Sexual selection. (a) Intrasexual selection often involves competition between males, as in this battle between two male elk. (b) Intersexual selection often involves bright colors and displays by males to attract females, as shown by this male and female Japanese red-crowned crane. *Source: (a) Kelly Funk/Getty Images; (b) Steven Kaufman/Getty Images.*

trade-off between the conflicting demands of attracting a mate and survival.

Self-Assessment Questions

10. What is natural selection, and how is it different from other mechanisms of evolution?

11. Sexual selection tends to cause bigger size, more elaborate weaponry, or brighter colors in males. Is this an example of stabilizing, directional, or disruptive selection?

20.5 NON-ADAPTIVE MECHANISMS OF EVOLUTION

Selection is evolution's major driving force, preferentially passing on to each new generation those mutations that best fit organisms to their environments. However, natural selection is not the only evolutionary mechanism. Other mechanisms—genetic drift, migration, mutation, and nonrandom mating—can cause allele and genotype frequencies to change as well. Unlike natural selection, however, they do not lead to adaptations. In this section, we discuss each of these other evolutionary mechanisms.

Genetic drift is a change in allele frequency due to chance.

Genetic drift is the random change in allele frequencies from generation to generation. By "random," we mean that frequencies go up or down simply by chance.

Genetic drift dramatically affects small populations. Consider a rare allele, *A*, with a frequency of 1/1000. Imagine that habitat destruction reduces the population to just one pair of individuals, one of which, by chance, is carrying *A*. The frequency of *A* in this new population is 1/4 (0.25) because each individual has two alleles, giving a total of four alleles. So, we see that there was a dramatic change in allele frequency, from 1/1000 to 1/4. In addition, a marked loss of genetic variation typically occurs as much of the variation present in the original population is not present in the surviving pair. That is, the surviving pair carries only a few of the alleles that were present in their original population.

This is an example of a population **bottleneck**, which occurs when an originally large population is reduced to just a few individuals. Genetic drift often results from population bottlenecks. A population of Galápagos tortoises that has very low levels of genetic diversity probably went through just such a bottleneck about 100,000 years ago when a volcanic eruption eliminated most of the tortoises' habitat.

Genetic drift also occurs when a few individuals start a new population, in what is called a **founder event.** Such events occur, for example, when a small number of individuals arrive on an island and colonize it. Once again, relative to the parent population, allele frequencies are randomly changed and genetic variation is lost.

Earlier, we considered the fate of beneficial and harmful mutations under the influence of natural selection. What about neutral mutations? Natural selection, by definition, does not govern the fate of neutral mutations. Consider a neutral mutation, *m*, which occurs in a noncoding region of DNA and, therefore, has no effect on fitness. At first, it is present in just a single heterozygous individual. What happens if that individual fails to reproduce (for reasons unrelated to m)? In this case, m is lost from the population, but not by natural selection (which does not select against m). Rather, this change is due to random change, so it represents a case of genetic drift. Alternatively, the m-bearing individual might by chance leave many offspring (again for reasons unrelated to m), in which case the frequency of m increases. In principle, it is possible over a long period of time for m to become fixed in the population. At the end of the process, every member of the population is homozygous mm.

Like natural selection, genetic drift leads to allele frequency changes and, in turn, to evolution. Unlike natural selection, however, genetic drift does not lead to adaptations because the alleles whose frequencies are changing as a result of drift do not affect an individual's ability to survive or reproduce.

Genetic drift has a large effect in small populations.

The impact of genetic drift depends on population size (**Fig. 20.13**). If m arises in a very small population, its frequency will change rapidly, as shown in Fig. 20.13a and Fig. 20.13b. Imagine m arising in a population of just six individuals (or three pairs). Its initial frequency is 1 in 12, or about 8% (there are a total of 12 alleles because each individual is diploid). If, by chance, one pair fails to breed and the other two pairs (including the individual who is an Mm heterozygote) each produce three offspring, and all three of the Mm individual's offspring happen to inherit the m allele, then the frequency of m has increased to 3 in 12 (25%) in a single generation. Over subsequent generations, m is either fixed or eliminated. In effect, genetic drift is equivalent to a sampling error. In a small sample, extreme departures from the expected outcome are common.

In contrast, if the population is large, as in Fig. 20.13c and Fig. 20.13d, then changes in allele frequency from generation to generation will be much smaller, typically less than 1%. A large population is analogous to a large sample size, in which we tend not to see large departures from expectation. Toss a coin 1000 times, and you will end up with approximately 500 heads. Toss a coin 5 times, and you might well end up with zero heads. In other words, in a small sample of coin tosses, we are much more likely to see departures from our 50:50 expectation than in a large sample. The same is true of genetic drift: it is likely to be much more significant in small populations than in large ones.

Migration reduces genetic variation between populations.

Migration is the movement of individuals from one population to another. It results in **gene flow**, the movement of alleles from one population to another. It is relatively straightforward to see how movements of individuals and alleles can lead to changes in allele frequencies. As an example, consider two isolated

FIG. 20.13 Genetic drift. The fate of neutral mutations is governed by genetic drift, the effect of which is more extreme in small populations than in large populations.

a. Population size = 4

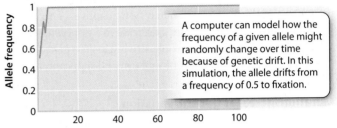

A computer can model how the frequency of a given allele might randomly change over time because of genetic drift. In this simulation, the allele drifts from a frequency of 0.5 to fixation.

b. Population size = 4

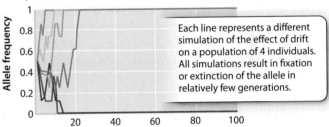

Each line represents a different simulation of the effect of drift on a population of 4 individuals. All simulations result in fixation or extinction of the allele in relatively few generations.

c. Population size = 40

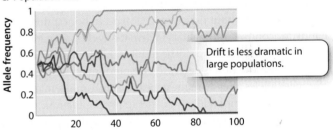

Drift is less dramatic in large populations.

d. Population size = 400

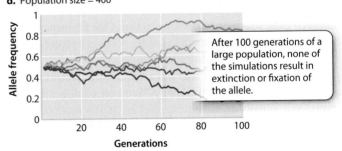

After 100 generations of a large population, none of the simulations result in extinction or fixation of the allele.

island populations of rabbits, one white and the other black. Now imagine that the isolation breaks down—a bridge is built between the islands—and migration occurs. Over time, black alleles enter the white population, and vice versa, and the allele frequencies of the two populations gradually become the same.

The consequence of migration is the homogenization of populations, making them more similar to each other and reducing genetic differences between them. Because populations are often adapted to their particular local conditions (for example, dark-skinned humans in regions of high sunlight versus fair-skinned humans in regions of low sunlight), migration may be worse than merely non-adaptive—it may be maladaptive, in that it causes a decrease in a population's average fitness. Fair-skinned people arriving in an equatorial region, for example, are at risk of sunburn and skin cancer.

Mutation increases genetic variation.

As we saw earlier in this chapter, mutation is a rare event. As a consequence, it is generally not important as an evolutionary mechanism that leads allele frequencies to change. Even so, as we have also seen, it is the source of new alleles and the raw material on which the other forces act. Without mutation, there would be no genetic variation and no evolution. Mutation is important in the long term but typically unimportant in the short term.

Nonrandom mating alters genotype frequencies without affecting allele frequencies.

As we saw earlier, another way that a population can evolve is through nonrandom mating. In random mating, individuals select mates without regard for genotype. In nonrandom mating, by contrast, individuals preferentially choose mates according to their genotypes. The result is that certain phenotypes increase and others decrease. Nonrandom mating just rearranges alleles already in the gene pool and, unlike migration or mutation, does not add new alleles to the population. As a result, genotype frequencies change in nonrandom mating, whereas allele frequencies do not.

Probably the most evolutionarily significant form of nonrandom mating is inbreeding, in which mating occurs between close relatives. Inbreeding increases the frequency of homozygotes and decreases the number of heterozygotes in a population without affecting allele frequencies.

Consider a rare allele b that has a frequency of 0.001 in a population. According to the Hardy–Weinberg equilibrium, the expected frequency of bb homozygotes is $0.001^2 = 0.000001$. What happens to the frequency of bb homozygotes when there is inbreeding? Suppose a father with one b allele has a son and a daughter who mate incestuously and have offspring. We know that the probability that the brother inherited the allele from his father is 0.5, and the same is true for his sister. The probability that the siblings' child is homozygous bb is the probability each sibling inherited a copy from their father (0.5×0.5) multiplied by the probability that their child inherited b from both of them (0.5×0.5), or $(0.5)^4 = 0.0625$. Clearly, 0.0625 is a considerably higher frequency than 0.000001.

If b is a deleterious recessive mutation, it may contribute to **inbreeding depression** in the child, a reduction in the child's fitness caused by homozygosity of deleterious recessive mutations. Inbreeding depression is a major problem in conservation biology, especially when endangered species are bred in captivity in programs starting with a just a small number of individuals.

> **Self-Assessment Questions**
>
> 12. Why, of all the evolutionary mechanisms, is selection the only one that can result in adaptation?
> 13. A female lizard floats on a log to an island where she lays her previously fertilized eggs and starts a new population on the island, where there are no other lizards. How do you think the genetic variation and allele frequencies of the island population will compare to those of the mainland population? Which mechanism of evolution is at work here?

20.6 MOLECULAR EVOLUTION

How do DNA sequence differences arise among species? Imagine starting with two pairs of identical twins, one pair male and the other female. We place one member of each pair together on two different islands (**Fig. 20.14**). If we assume the islands are completely isolated from each other, what, in genetic terms, happens over time? The original pairs are the founders of the populations on each island. The genetic starting point, in each case, is exactly the same, but, over time, differences accumulate between the two populations. Mutations occur in one population that have not arisen in the other population, and vice versa.

A mutation in either population has one of three fates: it goes to fixation (either through genetic drift or through positive selection); it is maintained at intermediate frequencies (by balancing selection); or it is eliminated (either through natural selection or genetic drift). Different mutations are fixed in each population. When we come back thousands of generations later and sequence the DNA of the original identical individuals' descendants, we will find that many differences have accumulated. The populations have diverged genetically. What we are seeing is evidence of **molecular evolution**.

Species are the biological equivalents of islands because they, too, are isolated. They are *genetically* isolated because, by definition, members of one species cannot exchange genetic material with members of another species (Chapter 21). The amount of time that two species have been isolated from each other is the time since their most recent common ancestor. For example, humans and chimpanzees, whose most recent common ancestor lived about 6–7 million years ago, have been isolated from each other for about 6–7 million years.

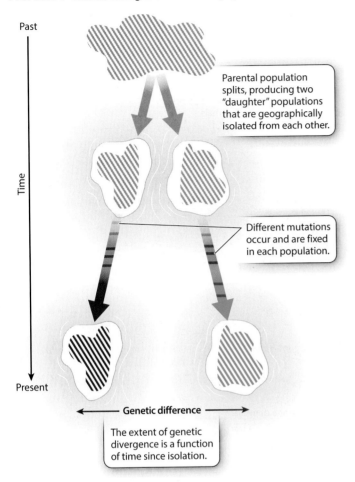

FIG. 20.14 Genetic divergence in isolated populations.

Parental population splits, producing two "daughter" populations that are geographically isolated from each other.

Different mutations occur and are fixed in each population.

Genetic difference

The extent of genetic divergence is a function of time since isolation.

Mutations arose and were fixed in the human lineage over that period; mutations, usually different ones, also arose and were fixed in the chimpanzee lineage over the same period. The result is the genetic difference between humans and chimpanzees.

The molecular clock relates the amount of sequence difference between species and the time since the species diverged.

The extent of genetic difference, or genetic divergence, between two species is a function of the time they have been genetically isolated from each other. The longer they have been apart, the greater the opportunity for mutation and fixation to occur in each population. This correlation between the time for which two species have been evolutionarily separated and the amount of genetic divergence between them is known as the **molecular clock**.

For a clock to function properly, it not only needs to keep time, but also needs to be set. We set the clock by using dates from the fossil record. For example, in a 1967 study, Vince Sarich and Allan Wilson determined from fossils that the lineages that gave rise to the two major groups of monkeys, the Old World and New World monkeys (Chapter 23), separated about 30 million years ago. Finding that the amount of genetic divergence between humans and chimpanzees was approximately one-fifth of that between Old World and New World monkeys, they concluded that humans and chimpanzees had been separated one-fifth as long, or about 6 million years. Although this is the generally accepted number today, Sarich and Wilson's result was revolutionary at the time, when it was thought that the two species had been separated for as long as 25 million years.

The rate of the molecular clock varies.

Molecular clocks can be useful for dating evolutionary events like the separation of humans and chimpanzees. However, because the rates of molecular clocks vary from gene to gene, clock data should be interpreted cautiously. These rate differences can be attributed largely to differences in intensity of negative selection (which results in the elimination of harmful mutations) among different genes.

The slowest molecular clock on record belongs to the histone genes, which encode the proteins around which DNA is wrapped to form chromatin (Chapters 3 and 13). These proteins are exceptionally similar in all organisms; only two amino acids (in a chain of about 100) distinguish plant and animal histones. Plants and animals last shared a common ancestor more than 1 billion years ago, which means, because each evolutionary lineage is separate, that there have been at least 2 billion years of evolution since they were in genetic contact. Despite this lengthy span, the histones have hardly changed at all. Almost any amino acid change fatally disrupts the histone protein, preventing it from carrying out its proper function. Negative selection has thus been extremely effective in eliminating just about every amino acid–changing histone mutation over 2 billion years of evolution. As a consequence, the histone molecular clock is breathtakingly slow.

Other proteins are less subject to such strong negative selection. Occasional mutations may become fixed in these proteins, either through drift (if they are neutral) or through selection (if they are beneficial). An extreme case of a fast molecular clock is that associated with a **pseudogene**, a gene that is no longer functional. Because *all* mutations in a pseudogene are by definition neutral—there is no function for a mutation to disrupt, so a mutation is neither deleterious nor beneficial—we expect to see a pseudogene's molecular clock tick at a very fast rate. In the histone genes, virtually all mutations are selected against, slowing the rate of evolution; in pseudogenes, no such negative selection is operating.

FIG. 20.15 The molecular clock. Different genes evolve at different rates because of differences in the intensity of negative selection. *Data from Fig. 20-3, p. 733, in A. J. F. Griffiths, S. R. Wessler, S. B. Carroll, and J. Doebley, 2012, Introduction to Genetic Analysis, 10th ed., New York, NY: W. H. Freeman.*

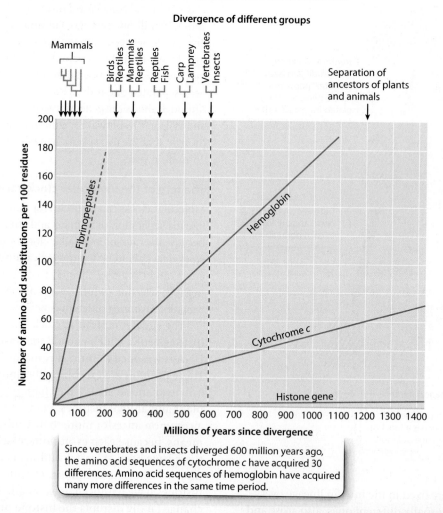

Since vertebrates and insects diverged 600 million years ago, the amino acid sequences of cytochrome *c* have acquired 30 differences. Amino acid sequences of hemoglobin have acquired many more differences in the same time period.

Fig. 20.15 shows the varying rates of the molecular clock for different genes.

In the next chapter, we examine how genetic divergence between populations can lead to the evolution of new species.

Self-Assessment Question

14. How is a molecular clock used to determine the time of divergence of two species?

CORE CONCEPTS SUMMARY

20.1 GENETIC VARIATION: Genetic variation refers to differences in DNA sequences.

Differences among members of a species (phenotypic variation) are the result of differences at the DNA level (genetic variation) as well as the influence of the environment. page 427

Mutation and recombination are the two sources of genetic variation, but all genetic variation ultimately comes from mutation. page 428

Mutations can be either somatic (in body tissues) or germ line (in gametes), but germ-line mutations are the only ones that can be passed on to the next generation. page 428

When a mutation occurs in a gene, it creates a new allele. page 428

Mutations can be deleterious, neutral, or advantageous. page 429

20.2 MEASURING GENETIC VARIATION: Patterns of genetic variation can be described by allele frequencies.

An allele frequency is the number of occurrences of a particular allele divided by the total number of all alleles of that gene in a population. page 429

In the past, population geneticists relied on observable traits determined by single genes and protein gel electrophoresis to measure genetic variation. page 429

DNA sequencing is now the standard technique for measuring genetic variation. page 430

20.3 EVOLUTION AND THE HARDY-WEINBERG EQUILIBRIUM: Evolution is a change in the frequency of alleles or genotypes in a population over time.

The Hardy–Weinberg equilibrium describes situations in which allele frequencies do not change. By determining whether a population is in Hardy–Weinberg equilibrium, we can determine whether evolution is occurring in a population. page 432

The Hardy–Weinberg equilibrium makes five assumptions: the population experiences no selection, no migration, no mutation, no sampling error due to small population size, and random mating. page 432

The Hardy–Weinberg equilibrium allows allele frequencies and genotype frequencies to be calculated from each other. page 433

20.4 NATURAL SELECTION: Natural selection leads to adaptations, which enhance the fit of an organism to its environment.

Independently conceived by Charles Darwin and Alfred Russel Wallace, natural selection is the differential reproductive success of genetic variants. page 434

Under natural selection, a harmful allele decreases in frequency and a beneficial one increases in frequency. Natural selection does not affect the frequency of neutral mutations. page 436

Natural selection can maintain alleles at intermediate frequencies by balancing selection. page 437

Changes in phenotype show that natural selection can be stabilizing, directional, or disruptive. page 437

In artificial selection, a form of directional selection, breeders govern the selection process. page 439

Sexual selection involves the evolution of traits that increase an individual's access to members of the opposite sex. page 439

In intrasexual selection, individuals of the same sex compete with one another, resulting in traits such as large size and horns. page 439

In intersexual selection, interactions between females and males result in traits such as elaborate plumage in male birds. page 439

20.5 NON-ADAPTIVE MECHANISMS OF EVOLUTION: Genetic drift, migration, mutation, and nonrandom mating are also mechanisms of evolution.

Genetic drift is a kind of sampling error that results in allele frequency change by chance. page 440

Genetic drift acts more strongly in small populations than in large ones. page 441

Migration involves the movement of alleles between populations (gene flow) and tends to have a homogenizing effect among populations. page 441

Mutation is the ultimate source of variation, but it also can change allele frequencies on its own. page 442

Nonrandom mating, such as inbreeding, results in an increase in homozygotes and a decrease in heterozygotes, but does not change allele frequencies. page 442

20.6 MOLECULAR EVOLUTION: Molecular evolution is a change in DNA or amino acid sequences over time.

The extent of sequence difference between two species is a function of the time the two species have been genetically isolated from each other. page 442

The correlation between the sequence differences among species and the time since the common ancestry of those species is known as the molecular clock. page 443

The rate of the molecular clock varies among genes because some genes are subject to more intense negative selection than others. page 443

Log in to LaunchPad to check your answers to the Self-Assessment Questions and to access additional learning tools.

CHAPTER 21: Species and Speciation

CORE CONCEPTS

21.1 THE BIOLOGICAL SPECIES CONCEPT: Reproductive isolation is the key to the biological species concept.

21.2 REPRODUCTIVE ISOLATION: Reproductive isolation is caused by barriers to reproduction before or after egg fertilization.

21.3 SPECIATION: Speciation underlies the diversity of life on Earth.

Imagine for a moment a world without **speciation,** the process that produces new and distinct forms of life. Life would have originated and natural selection would have done its job of winnowing advantageous mutations from disadvantageous ones, but the planet would be inhabited by a single kind of generally adapted organism. Instead of the staggering biological diversity we see around us—current estimates suggest that between 10 and 100 million species call Earth home—there would be just a single life-form. From a biological perspective, the planet would be a decidedly dull place. Evolution is as much about speciation, the engine that generates this breathtaking biodiversity, as it is about adaptation, the result of natural selection.

21.1 THE BIOLOGICAL SPECIES CONCEPT

The definition of **species** has been a long-standing problem in biology. In *On the Origin of Species,* Darwin wrote, "No one definition has as yet satisfied all naturalists; yet every naturalist knows vaguely what he means when he speaks of a species." The difficulty of defining species has come to be called the species problem.

Here is the problem in a nutshell: the species, as an evolutionary unit, must by definition be fluid and capable of changing, giving rise through evolution to new species. The whole point of the Darwinian revolution is that species are not fixed. How, then, can we define something that changes over time, and, by the process of speciation, even gives rise to two species from one?

Species are reproductively isolated from other species.

We can plainly see biodiversity, but are what we call "species" real biological entities? To test whether species are real, we can examine the natural world, measure some characteristic of the different living organisms we see, and then plot these measurements on a graph. **Fig. 21.1** shows such a plot, graphing antenna length and wing length of three different

FIG. 21.1 Species clusters on the basis of two characteristics. *Photo sources: Museum of Comparative Zoology, Harvard University.*

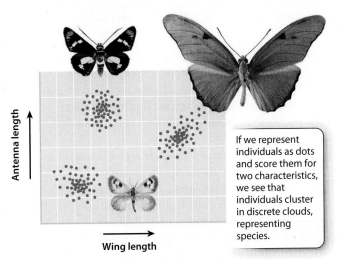

If we represent individuals as dots and score them for two characteristics, we see that individuals cluster in discrete clouds, representing species.

types of butterfly. Note that the dots, representing individual organisms, fall into non-overlapping clusters. Today, we can add a molecular dimension to this kind of analysis. When we compare genomes of multiple organisms, they, too, cluster on the basis of similarity. Each cluster is a species, and the fact that the clusters are distinct implies that species are biologically real.

The distances we see between the dots within a cluster reflect variation from one individual to the next within a species (Chapter 20). Humans are highly variable, but overall we are more similar to one another than to our most humanlike relative, the chimpanzee. On a plot such as the one in Fig. 21.1, we form a messy cluster, but that cluster does not overlap with the chimpanzee cluster.

Species, then, are real biological entities, not just a convenient way to group organisms. Whether two individuals are members of the same species is a reflection of their ability (or inability) to exchange genetic material by producing fertile offspring. Consider a new advantageous mutation that appears initially in a single individual. That individual and its offspring that inherit the mutation will have a competitive advantage over other members of the population, and the mutation will increase in frequency until it reaches 100%. Migration among populations causes the mutation to spread further until all individuals within the species have it. The mutation spreads within the species but, with some exceptions, cannot spread beyond it.

A species represents a closed gene pool, in which alleles are shared among members of that species but usually not with members of others. As a result, it is species that become extinct, and it is species that, through genetic divergence, give rise to new species.

The definition of "species" continues to be debated to this day. The most widely used and generally accepted definition of a species is known as the **biological species concept (BSC)**. The BSC was described by the great evolutionary biologist Ernst Mayr (1904–2005) as follows:

> Species are groups of actually or potentially interbreeding populations that are reproductively isolated from other such groups.

Let us look at this definition closely. At its heart is the idea of reproductive compatibility. Members of the same species are capable of producing offspring together, whereas members of different species are incapable of producing offspring together. In other words, members of different species are reproductively isolated from one another.

As Mayr and many others realized, however, reproductive compatibility entails more than just the ability to produce offspring. The offspring must be fertile and, therefore, capable of passing their genes on to their own offspring. For example, although a horse and a donkey—which are two different species—can mate to produce a mule, the mule is infertile. If a hybrid offspring is infertile, then it represents a genetic dead end.

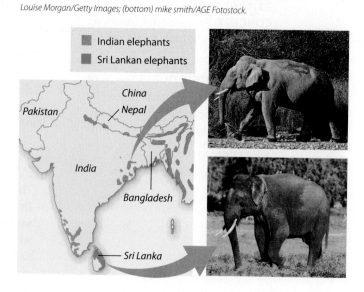

FIG. 21.2 *Elephas maximus* **in Sri Lanka and in India.** *Photo sources: (top) Louise Morgan/Getty Images; (bottom) mike smith/AGE Fotostock.*

Note, too, the "actually or potentially" part of Mayr's definition. An Asian elephant living on the island of Sri Lanka and one living in nearby India are considered members of the same species, *Elephas maximus*, even though Sri Lankan and Indian Asian elephants never have a chance to mate with each other in nature because they are geographically separated (**Fig. 21.2**).

The BSC is more useful in theory than in practice.

Species, then, are real biological entities and the BSC is the most accepted way to define them. Nevertheless, the BSC has shortcomings, the most important of which is that it can be difficult to apply. Imagine you are on a field expedition in a rain forest and you find two insects that look reasonably alike. Are they members of the same species? To use the BSC, you would need to test whether they are capable of producing fertile offspring. In practice, you probably will not have the time and resources to perform such a test. Even if you do attempt such a test, it is possible that members of the same species will not reproduce when you place them together in a laboratory setting or in your field camp. The conditions may be too unnatural for the insects to behave in a normal way. Finally, even if they do mate, you may not have time to determine whether their offspring are fertile.

Thus, we consider the BSC to be a valid framework for thinking about species, but one that is difficult to implement. As an alternative, on a day-to-day basis, biologists often use a rule of thumb called the **morphospecies concept**. Stated simply, the morphospecies concept holds that members of the same species usually look alike. For example, the shape, size, and coloration of the bald eagle make it reasonably straightforward to determine whether a bird observed in the wild is a member of the bald eagle species.

Today, the morphospecies concept has been extended to the molecular level: members of the same species usually have similar DNA sequences that are distinct from those of other species. A remarkable project called the Barcode of Life has established a database linking DNA sequences to species. Scientists can use it to identify species from DNA alone by sequencing the relevant segment of DNA and finding the matching sequence (and therefore the species) in the Barcode of Life database. Here, at both morphological and molecular levels, we return to those biological clusters of Fig. 21.1 in which members of a species fall close to one another within a single, discrete group.

Although the morphospecies concept is useful and generally applicable, it is not infallible. Members of a species may not always look alike, but instead may show different phenotypes called polymorphisms. For example, some species of birds display color differences. Sometimes, males of a species may look different from females—think of the showy peacock, which differs dramatically from the relatively drab peahen. Young can also be very different from old: caterpillars that mature into butterflies are a striking example.

As these cases suggest, members of the *same* species can look quite *different* from one another. But it is also true that members of *different* species can look quite *similar*. Some species of butterfly, for example, appear similar but have chromosomal differences that can be observed only with the aid of a microscope (**Fig. 21.3**). These chromosomal differences prevent the formation of interspecies hybrids, so those butterflies meet the BSC criterion—they are different species—but they do not meet the morphospecies criterion because they look so much alike.

With the application of ever more powerful DNA-based approaches to studies of natural populations, scientists frequently uncover what are called cryptic species. These species consist of organisms that had been traditionally considered as belonging to a single species because they look similar, but turn out to belong to two species because of differences at the DNA sequence level.

The BSC does not apply to asexual or extinct organisms.

In addition to the difficulties of putting the BSC into practice, a second problem is that it overlooks some organisms. For example, because it is based on the sexual exchange of genetic information, the BSC cannot apply to species that reproduce asexually, such as bacteria. Although some asexual species do occasionally exchange genetic information in a process known as conjugation (Chapter 24), true asexual organisms do not fit the BSC.

Furthermore, because it depends on reproduction, the BSC obviously cannot be applied to species that are extinct and known only through the fossil record, such as different "species" of trilobites and dinosaurs.

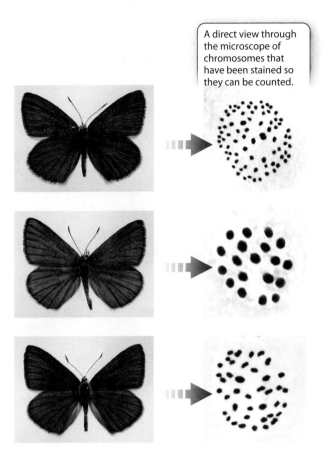

FIG. 21.3 Three species of *Agrodiaetus* butterflies. These similar-appearing butterflies are identifiable only by differences in their chromosome numbers. *Photos source: Vladimir Lukhtanov, Zoological Institute of Russian Academy of Sciences, St. Petersburg, Russia.*

A direct view through the microscope of chromosomes that have been stained so they can be counted.

Hybridization complicates the BSC.

Populations of one species may vary in the extent to which they are reproductively isolated from a closely related species. In one location, the two groups may be reproductively isolated from each other, while, in another location, they interbreed to produce **hybrid offspring**. Under the BSC, the first pair of populations represents two species, whereas the second pair represents just one. A classic example is shown in **Fig. 21.4**. Two closely related birds, called towhees, co-occur in various locations in Mexico. In some places they interbreed; in others they do not.

This kind of incomplete reproductive isolation seems to be especially common in plants. Despite apparently being good morphospecies that can be distinguished by appearance alone, many different species of willow (*Salix*), oak (*Quercus*), and dandelion (*Taraxacum*) are still capable of exchanging genes with other species in their genera through **hybridization**. According to the BSC, these different forms should be

FIG. 21.4 Hybridization in towhee species. Two species of closely related birds (*Pipilo maculatus* and *Pipilo ocai*) overlap in places in their distributions in central Mexico. In some locations, they hybridize; in others, they do not. *Source: Futuyma, Evolution, 2005 Sinauer.*

a.

Pipilo maculatus   *Pipilo ocai*   Hybrid   Hybrid

b.

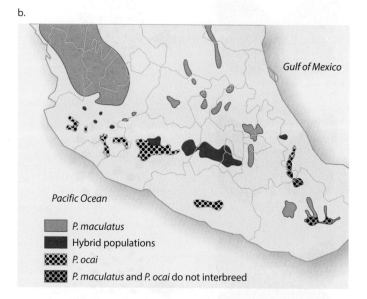

considered one large species because they are able to reproduce and produce fertile offspring. However, because the species maintain their distinct appearances, it appears that natural selection must work against the hybrid offspring.

With the application of modern comparative genomic approaches, we are discovering that the boundaries between many closely related species are also not as strictly drawn as traditionally supposed. For example, discussed in Chapter 23, our own species, *Homo sapiens*, interbred in the past with another species, *Homo neanderthalensis*.

Ecology and evolution can extend the BSC.

To remedy these problems with the application of the BSC, much work has been put into modifying and improving the BSC and many alternative definitions of species have been suggested. In general, these efforts highlight how difficult it is to make all species fit easily into one definition. The natural world truly defies neat categorization!

Even so, several useful ideas have come out of this literature. One of these is the notion that a species can sometimes be characterized by its ecological **niche**. A species' ecological niche (which we will discuss further in Chapter 45) is a complete description of the role the species plays in its environment—its habitat requirements, its nutritional and water needs, and the like. It turns out that it is impossible for two species to coexist in the same location if their niches are too similar because competition between them for resources inevitably leads to the extinction of one species. This observation has given rise to the **ecological species concept (ESC),** the idea that there is a one-to-one correspondence between a species and its niche. Applying this concept, we can determine whether asexual bacterial lineages are distinct species on the basis of differences or similarities in their ecological requirements. If two lineages have very different nutritional needs, for example, we can infer on ecological grounds that they are separate species.

Another species concept is the **phylogenetic species concept (PSC),** which emphasizes that members of a species all share a common ancestry and a common fate. It is, after all, species rather than individuals that become extinct. The PSC requires that all members of a species be descended from a single common ancestor, but does not specify on what scale this idea should be applied. All mammals derive from a single common ancestor that lived approximately 200 million years ago, but there are thousands of what we recognize as species of mammals that have evolved since that long-ago common ancestor. Under a strict application of the PSC, would we consider all mammals to be a single species? Similarly, in human families, siblings and cousins are all descended from a common ancestor, a grandmother, but that is surely not sufficient grounds to classify more-distant relatives as a distinct species. The PSC can be useful when thinking about asexual species, but, given the arbitrariness of the decisions involved in assessing whether the descendants of a single ancestor warrant the term "species," its utility is limited.

It is worth bearing these ecological and phylogenetic considerations in mind when thinking about species. Although the BSC remains our most useful definition of species, the ESC and PSC broaden and generalize the concept. For example, in the case of a normally asexual species that is difficult to study because conjugation is so infrequent, we can jointly apply the ESC and PSC. That is, we can use the ESC to loosely define the species in terms of its ecological characteristics (for example, its nutritional requirements), and we can refine that definition by using genetic analyses to determine whether the group is indeed a species by the PSC's standard (that is, all its members derive from a single common ancestor).

Despite the shortcomings of the BSC and the usefulness of alternative ideas, we stress that the BSC is the most constructive way to think about species. In particular, by focusing on reproductive isolation—the inability of different species to produce viable, fertile offspring—the BSC gives us a means of studying and understanding speciation, the process by which two populations, originally members of the same species, become distinct.

Self-Assessment Questions

1. Why has the scientific community been unable to come up with a single, comprehensive, and agreed-upon concept of a "species"?
2. How is the term "species" defined according to the biological species concept (BSC)?
3. Given a group of organisms, how would you test whether they all belong to one species or whether they belong to two separate species according to the morphospecies concept?
4. Which types of organisms do not fit easily into the biological species concept? Which species concept would work best for these organisms?
5. Why do morphospecies typically correspond to BSC-defined species?

21.2 REPRODUCTIVE ISOLATION

Factors that cause reproductive isolation are generally divided into two categories, depending on when they act. **Pre-zygotic** isolating factors act before the fertilization of an egg, and **post-zygotic** isolating factors come into play after fertilization. In other words, pre-zygotic factors prevent fertilization from taking place, whereas post-zygotic factors result in the failure of the fertilized egg to develop into a fertile individual.

Pre-zygotic isolating factors occur before egg fertilization.

Most species are reproductively isolated by pre-zygotic isolating factors, which can take many forms. Among animals, species are often **behaviorally isolated**, meaning that individuals mate only with other individuals based on specific courtship rituals, songs, or other behaviors. Chimpanzees may be humans' closest relative—and therefore the species we are most likely to confuse with our own—but a chimpanzee of the appropriate sex, however attractive to a chimpanzee of the opposite sex, fails to provoke even the faintest reproductive impulse in a human. In this case, the pre-zygotic reproductive isolation of humans and chimpanzees is behavioral.

Behavior does not play a role in plants, but pre-zygotic factors can still be important in their reproductive isolation. Pre-zygotic isolation in plants can take the form of incompatibility between the incoming pollen and the receiving flower, such that fertilization fails to take place. We see similar forms of isolation between members of marine species, such as abalone, which simply discharge their gametes into the water. In these cases, membrane-associated proteins on the surface of sperm interact specifically with membrane-associated proteins on the surface of eggs of the same species, but not with those of different species. These specific interactions ensure that a sperm from one abalone species, *Haliotis rufescens*, fertilizes only an egg of its own species and not an egg from *H. corrugata*, a closely related species. Incompatibilities between the gametes of two different species is called **gametic isolation**.

In some animals, especially insects, incompatibility arises earlier in the reproductive process. As an example, the genitalia of males of the fruit fly *Drosophila melanogaster* are configured in such a way that they fit only with the genitalia of females of the same species. Thus, attempts by males of *D. melanogaster* to copulate with females of another species of fruit fly, *D. virilis*, are prevented by **mechanical incompatibility**.

Both plants and animals may also be pre-zygotically isolated in time (**temporal isolation**). For example, closely related plant species may flower at different times of the year, so there is no chance that the pollen of one species will come into contact with the flowers of the other. Similarly, members of a nocturnal animal species are unlikely to encounter members of closely related species that are active only during the day.

Plants and animals can also be isolated in space (**geographic** or **ecological isolation**). This type of isolation can be subtle. For example, the two Japanese species of ladybug beetle shown in **Fig. 21.5** can be found living side by side in the same

FIG. 21.5 Ecological isolation. The ladybugs (a) *Henosepilachna yasutomii* and (b) *H. niponica* are reproductively isolated from each other because they feed and mate on different host plants. *Source: Courtesy Dr. Haruo Katakura.*

a.
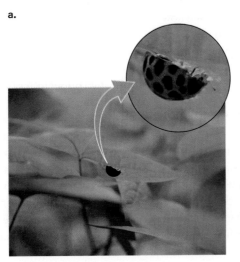

b.

field, but they feed on different plants. Because their life cycles are so intimately associated with their host plants (adults even mate on their host plants), these two species never breed with each other. This ecological separation is what leads to their pre-zygotic isolation.

Post-zygotic isolating factors occur after egg fertilization.

Post-zygotic isolating factors involve mechanisms that come into play after fertilization of the egg. Typically, they involve some kind of **genetic incompatibility**. One example, which we saw earlier in Fig. 21.3 and will explore later in the chapter, is the case of two organisms with different numbers of chromosomes.

In some instances, the effect can be extreme. For example, the zygote may fail to develop after fertilization because the two parental genomes are sufficiently different to prevent normal development. In others, the effect is less obvious. Some matings between different species produce perfectly viable adults, as in the case of the horse–donkey hybrid, the mule. As we have seen, though, all is not well with the mule from an evolutionary perspective. Because horses and donkeys have different chromosome numbers (haploid complements of 32 and 31, respectively), proper homologous chromosome pairing in meiosis in the mule cannot take place (Chapter 11), resulting in infertility of the mule. As a general rule, the more closely related—and therefore genetically similar—a pair of species, the less extreme the genetic incompatibility between their genomes.

Self-Assessment Questions

6. What is an example of a pre-zygotic isolating factor?
7. What is an example of a post-zygotic isolating factor?

21.3 SPECIATION

Recognizing that species are groups of individuals that are reproductively isolated from other such groups, we are now in a position to recast the key question addressed in this chapter. That is, instead of asking, "How do new species arise?" we can ask, "How does reproductive isolation arise between populations?"

Speciation is a by-product of the genetic divergence of separated populations.

The key to speciation is the fundamental evolutionary process of genetic divergence between genetically separated populations. As we saw in Chapter 20, if a single population is split into two populations that are isolated from each other, different mutations appear by chance in the two populations. Like all mutations, these are subject to genetic drift or natural selection (or both), resulting over time in the genetic divergence of the two populations. Two separate populations that are initially identical will, over long periods of time, gradually become distinct as different mutations are introduced, propagated, and fixed in each population. At some stage in the course of divergence, changes occur in one population that lead to its members becoming reproductively isolated from members of the other population (**Fig. 21.6**). It is this process that results in speciation.

Speciation—the development of reproductive isolation between populations—is, therefore, typically just a by-product of the genetic divergence of separated populations. As Fig. 21.6 shows, it is typically a gradual process. If we try to cross members of two populations that have genetically diverged but not yet diverged far enough for full reproductive isolation (that is, speciation) to have arisen, we may find that the populations are **partially reproductively isolated**. In other words, they are not truly separate species, but the genetic differences between them are extensive enough that the hybrid offspring they produce have reduced fertility or viability compared to offspring produced by crosses between individuals within each population.

Allopatric speciation is speciation that results from the geographical separation of populations.

As just noted, speciation is the process by which two groups of organisms become reproductively isolated from each other. Because this process requires genetic isolation between the diverging populations and because geography is the easiest way to ensure physical and therefore genetic isolation, many models of speciation focus on geography.

The process usually begins with the creation of **allopatric** (literally, "different place") populations, which are geographically separated from each other. Clearly, physical separation does not immediately cause reproductive isolation. If a single population is split in two by a geographic barrier, and, after a few generations, individuals from each population are allowed to interbreed again, they are still capable of producing fertile offspring and, therefore, are members of the same species. In addition to separation, the other requirement for speciation to occur is time. Time allows for different mutations to become fixed in the two separated populations so that eventually the two populations become reproductively isolated from each other.

Because genetic divergence is typically gradual, we often find allopatric populations that have yet to evolve even partial reproductive isolation but that have accumulated a few population-specific traits. This genetic distinctness is sometimes recognized by taxonomists, who call each geographic form a **subspecies** and add a further designation after its species name. For example, Sri Lankan Asian elephants, subspecies *Elephas maximus maximus*, are generally larger and darker than

FIG. 21.6 Speciation. Speciation occurs when two populations that are genetically diverging become reproductively isolated from each other.

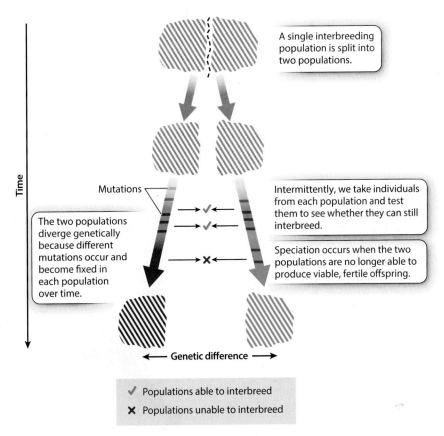

their Indian counterparts, subspecies *Elephas maximus indicus* (see Fig. 21.2).

Dispersal and vicariance can isolate populations from each other.

There are two ways in which populations may become allopatric. The first is by **dispersal**, in which some individuals colonize a distant place, such as an island, far from the main source population. The second is by **vicariance**, in which a geographic barrier arises within a single population, separating it into two or more isolated populations. For example, when sea levels rose at the end of the most recent ice age, new islands formed along coastlines as the low-lying land around them was flooded. The populations on those new islands suddenly found themselves isolated from other populations of their species. This kind of island formation is a vicariance event. Regardless of how the allopatric populations came about—whether through dispersal or vicariance—the outcome is the same: the two separated populations diverge genetically until speciation occurs.

Often, vicariance-derived speciation is the easier to study because we can date the time at which the populations were separated if we know when the vicariance occurred. One such event whose history is well known is the formation of the Isthmus of Panama between Central and South America, shown in **Fig. 21.7**. This event took place approximately 3.5 million years ago. In its wake, populations of marine organisms in the western Caribbean and eastern Pacific that had formerly been able to interbreed freely were separated from each other. The result was the formation of many distinct species, with each one's closest relative being on the other side of the isthmus.

Dispersal is important in a specific kind of allopatric speciation known as **peripatric speciation** (that is, in a peripheral place). In this model, a few individuals from a **mainland population** (the central population of a species) disperse to a new location remote from the original population and evolve separately. This may be an intentional act of dispersal, such as young mammals migrating away from where they were raised, or it could be an accident brought about by, for example, an unusual storm that blows migrating birds off their normal route. The result is a distant, isolated **island population**.

"Island" in this case may refer to a true island—such as Hawaii—or may simply refer to a patch of habitat on the

HOW DO WE KNOW?

FIG. 21.7
Can vicariance cause speciation?

BACKGROUND Three and a half million years ago, the Isthmus of Panama was not completely formed (top map). Several marine corridors remained open, allowing interbreeding between marine populations in the Caribbean and the eastern Pacific. Subsequently, the gaps in the isthmus were plugged, separating the Caribbean and eastern Pacific populations and preventing gene flow between them (bottom map).

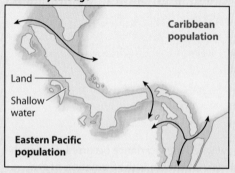

Interbreeding between Eastern Pacific and Caribbean populations of *Alpheus* was possible via the corridors that existed before the final formation of the Isthmus of Panama.

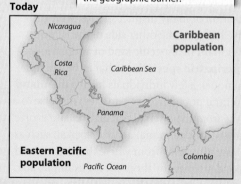

Interbreeding between Eastern Pacific and Caribbean populations is no longer possible because of the geographic barrier.

HYPOTHESIS In 1993, American marine biologist Nancy Knowlton and her colleagues hypothesized that patterns of speciation would reflect the impact of the vicariance resulting from the closing of the direct marine connections between the Pacific and the Caribbean. Specifically, they predicted that each ancestor species (from the time before the formation of the isthmus) split into two "daughter" species, one in the Caribbean and the other in the Pacific. The closest relative of each current Pacific species, then, is predicted to be a Caribbean species, and vice versa.

EXPERIMENT This study focused on 17 species of snapping shrimp in the genus *Alpheus*, a group that is distributed on both sides of the isthmus. The first step was to sequence the same segment of DNA from each species. The next step was to compare those sequences so as to reconstruct the phylogenetic relationships among the species.

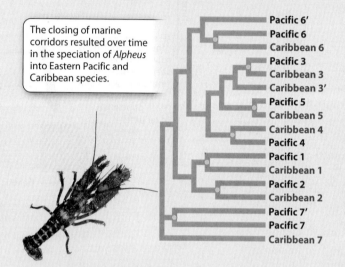

The closing of marine corridors resulted over time in the speciation of *Alpheus* into Eastern Pacific and Caribbean species.

Photo source: Dr. Arthur Anker, NUS, Singapore.

RESULTS The phylogeny reveals that species show a distinctly paired pattern of relatedness: the closest relative of each species is one from the other side of the isthmus.

CONCLUSION That we see these consistent Pacific/Caribbean sister species pairings strongly supports the hypothesis that the vicariance caused by the formation of the Isthmus of Panama has driven speciation in *Alpheus*. Each Pacific/Caribbean pairing is derived from a single ancestral species (indicated by a yellow dot in the phylogeny) whose continuous distribution between the Caribbean and eastern Pacific was disrupted by the formation of the isthmus. Here, we see striking evidence of the role of vicariance in multiple speciation events.

FOLLOW-UP WORK Speciation is about more than just genetic differences between isolated populations, as shown here. Knowlton and colleagues also tested the different species for reproductive isolation and found that there were high levels of isolation between Caribbean/Pacific pairs. Given that we know each species pair has been separated for at least 3 million years, we can use this information to calibrate the rate at which new species are produced.

SOURCE Knowlton, N., et al. 1993. "Divergence in Proteins, Mitochondrial DNA, and Reproductive Compatibility Across the Isthmus of Panama." *Science* 260 (5114):1629–1632.

mainland that is appropriate for the species but is geographically remote from the mainland population's habitat area. For a species adapted to life on mountaintops, a new island might be another previously uninhabited mountaintop. For a rain forest tree species, that new island might be a patch of lowland forest on the far side of a range of mountains that separates it from its mainland forest population.

The island population is classically small and often resides in an environment that is slightly different from that of the mainland population. The peripatric speciation model suggests that change accumulates faster in these peripheral isolated populations than in the large mainland populations for two reasons. First, genetic drift is more pronounced in smaller populations than in larger ones. Second, the environment may differ between the mainland and the island in a way that results in natural selection driving differences between the two populations. Together, these two mechanisms cause genetic divergence of the island population from the mainland one, ultimately leading to speciation.

It is possible to glimpse peripatric differentiation in action (**Fig. 21.8**). Studies of a kingfisher, *Tanysiptera galatea,* in New Guinea and nearby islands show the process under way. There are eight recognized subspecies of *T. galatea*: three on mainland New Guinea (where they exist in large populations separated by mountain ranges) and five on nearby islands (where, because the islands are small, the populations are correspondingly small). The mainland subspecies are still quite similar to one another, but the island subspecies are much more distinct, suggesting that genetic divergence is occurring faster in the small island populations. If we wait long enough, these subspecies will probably diverge into new species.

Because dating such dispersal events is tricky—newcomers on an island tend not to leave a record of when they arrived—we can sometimes use vicariance information to study the timing of peripatric speciation. For instance, consider the finches of the Galápagos Islands made famous by Charles Darwin. We know that the oldest of the Galápagos Islands were formed 4–5 million years ago by volcanic action. Some time early in the history of the Galápagos, individuals of a small South American bird species arrived there. Conditions on the Galápagos are very different from those on the South American continent, where the mainland population of this ancestral finch lived, so the isolated island population evolved to become distinct from its mainland ancestor and eventually became a new species.

The finches' subsequent dispersal among the other islands of the Galápagos promoted further peripatric speciation (**Fig. 21.9**). In the graph at the bottom of Fig. 21.9, we see that the number of finch species is correlated with the number of islands in the archipelago. This is clear indication of the

FIG. 21.8 Peripatric speciation in action among populations of New Guinea kingfishers. *Sources: Data from D. J. Futuyma, 2009, Evolution, 2nd ed., Sunderland, MA: Sinauer Associates, Fig. 18.7, p. 484; photo from C.H. Greenewalt/VIREO.*

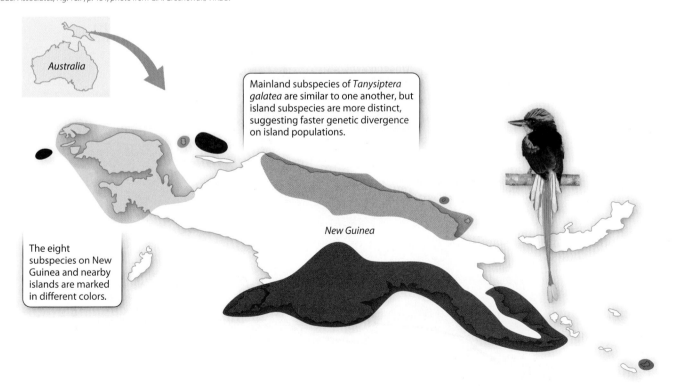

FIG. 21.9 Adaptive radiation in Darwin's finches. Today, there are 13 different species of Darwin's finches. *Source: Phylogenetic tree data from Fig. 2.1, p. 15, finch species plot data from Fig. 2.4, p. 23, in P. R. Grant and B. R. Grant, 2008, How and Why Species Multiply, Princeton, NJ: Princeton University Press.*

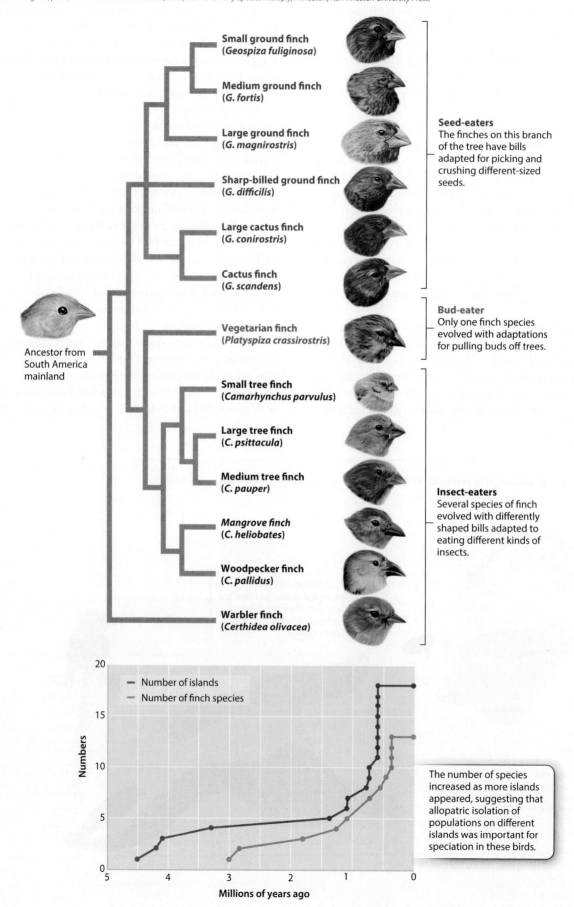

importance of geographic separation (allopatry) in speciation: the availability of islands provided opportunities for populations to become isolated from one another, allowing speciation. The result was the evolution of 13 different species of finches, collectively known today as Darwin's finches.

The Galápagos finches and their frenzy of speciation illustrate the important evolutionary idea of **adaptive radiation**, a bout of unusually rapid evolutionary diversification in which natural selection accelerates the rates of both speciation and adaptation. Adaptive radiation occurs when an environment offers many ecological opportunities for exploitation. Consider the finch ancestor immigrants arriving on the Galápagos. A wealth of ecological opportunities were open and available to them. Until the arrival of the colonizing immigrant birds, there were no birds on the islands to eat the plant seeds, or to eat the insects on the plants, and so on. Suppose that the finch ancestors fed specifically on medium-sized seeds on the South American mainland; that is, they were medium-seed specialists, with bills that are the right size for handling medium-sized seeds. On the mainland, they were constrained to that size of seed because any attempt to eat larger or smaller seeds brought them into conflict with other species—a large-seed specialist and a small-seed specialist—that already used these resources. In effect, stabilizing selection (Chapter 20) was operating on the mainland to eliminate the extremes of the bill-size spectrum in the medium-seed specialists.

When the medium-billed immigrants first arrived on the Galápagos, however, no such competition existed. Therefore, natural selection promoted the formation of new species of small- and large-seed specialists from the original medium-seed-eating ancestral stock. With the elimination of the stabilizing selection that had kept the medium-billed birds medium-billed on the mainland, selection operated in the opposite direction, favoring the large and small extremes of the bill-size spectrum because these individuals could take advantage of the abundance of unused resources, consisting of small and large seeds. It is this combination of emptiness—the availability of ecological opportunity—and the potential for allopatric speciation that results in adaptive radiation.

Co-speciation is speciation that occurs in response to speciation in another species.

As we have seen, physical separation is often a critical ingredient in speciation. Two populations that are not fully separated from each other—that is, gene flow exists between them—will typically not diverge from each other genetically, because genetic exchange homogenizes them. This is why most speciation is allopatric. Although allopatric speciation brings to mind populations separated from each other by stretches of ocean or deserts or mountain ranges, separation can be just as complete in the absence of such geographic barriers.

Consider an organism that parasitizes a single host species. Suppose that the host undergoes speciation, producing two daughter species. The original parasite population is also split into two populations, one for each host species. Thus, the two new parasite populations are physically separated from each other and diverge genetically, ultimately undergoing speciation. This divergence results in a pattern of coordinated host–parasite speciation called **co-speciation**, a process in which two groups of organisms speciate in response to each other and at the same time.

The phylogenetic trees of parasites and their hosts that undergo co-speciation are similar for each group. Each time a branching event—that is, speciation—occurs in one lineage, a corresponding branching event occurs in the other (**Fig. 21.10**).

 CASE 4 MALARIA: COEVOLUTION OF HUMANS AND A PARASITE

How did malaria come to infect humans?

Let's consider a human parasite, *Plasmodium falciparum,* the single-celled eukaryote that causes malaria. It had been suggested that *P. falciparum*'s closest relative is another *Plasmodium* species, *P. reichenowi,* found in chimpanzees, our closest living relative.

FIG. 21.10 Co-speciation. Parasites and their hosts often evolve together, and the result is similar phylogenies. *Source: Data from J. P. Huelsenbeck and B. Rannala, 1997, "Phylogenetic Methods Come of Age: Testing Hypotheses in an Evolutionary Context," Science 276:227, doi: 10.1126/science.276.5310.227; photo sources: (left) Photo by Alex Popinga, courtesy James Demastes; (right) Richard Ditch @ richditch.com.*

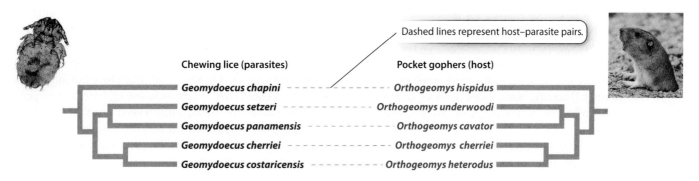

Were *P. falciparum* and *P. reichenowi* the products of co-speciation? When the ancestral population split millions of years ago to give rise to the human and chimpanzee lineages, that population's parasitic *Plasmodium* population could also have been split, ultimately yielding *P. falciparum* and *P. reichenowi*.

Recent studies, however, have disproved this hypothesis. We now know that *P. falciparum* was introduced to humans relatively recently from gorillas. Why doesn't the evolutionary history of *Plasmodium* follow the classical host–parasite co-speciation pattern? This history is complex and is still being unraveled. Moreover, the mosquito-borne phase of the parasite's life cycle facilitates its transfer to new hosts. Malaria parasites are not inextricably tied to their hosts, so their evolutionary history does not parallel that of their hosts as closely as the evolutionary history of the chewing lice follows that of their pocket gopher hosts.

Sympatric populations—those in the same place—may undergo speciation.

Can speciation occur without complete physical separation of populations? The answer to this question is "yes," although evolutionary biologists are still exploring how common this phenomenon is. Recall how separated populations inevitably diverge genetically over time (see Fig. 21.6). If a mutation arises in population A after it has separated from population B, that mutation is present only in population A. It may eventually become fixed in that population, either through natural selection, if it is advantageous, or through drift, if it is neutral. Once the mutation is fixed in population A, it represents a genetic difference between populations A and B. Repeated independent fixations of different mutations in the two populations result over time in the genetic divergence of separated populations.

Now imagine that populations A and B are not completely separated, such that there is some gene flow between them. The mutation that arose in population A can, in principle, appear in population B as members of the two populations interbreed. Gene flow effectively negates the genetic divergence of populations. If there is gene flow, a pair of populations may change over time, but they do so together. How, then, can speciation occur if gene flow exists? The term we use to describe populations that are in the same geographic location is **sympatric** (literally, "same place"). So we can rephrase the question in technical terms: how can speciation occur sympatrically?

For speciation to occur sympatrically, natural selection must act strongly to counteract the homogenizing effect of gene

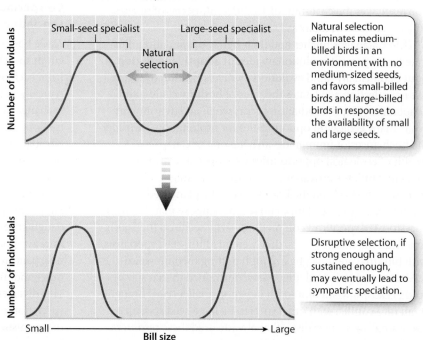

FIG. 21.11 Sympatric speciation by disruptive selection. Natural selection eliminates individuals in the middle of the spectrum.

flow. Consider two sympatric populations of finch-like birds, represented in the graph in **Fig. 21.11**. One population begins to specialize in feeding on small seeds and the other in feeding on large seeds. If the two populations freely interbreed, no genetic differences between the two occur, and speciation does not take place. Now suppose that the offspring produced by the pairing of a big-seed specialist with a small-seed specialist is an individual best adapted to eat medium seeds, and there are no medium-sized seeds available in the environment. Natural selection acts against the hybrids, which starve to death because there are no medium-sized seeds for them to eat and they are not well adapted to compete with the big- or small-seed specialists. Natural selection, in effect, eliminates the products of gene flow. So, although gene flow is occurring, it does not affect the divergence of the two populations because the hybrid individuals do not survive to reproduce. As discussed in Chapter 20, this form of natural selection, which operates against the middle of a spectrum of variation, is called disruptive selection.

It turns out to be difficult to find evidence of sympatric speciation in nature, though it may not be especially rare in plants, as we will see. We might find two very closely related species in the same location and argue that they must have arisen through sympatric speciation, but there is an alternative explanation. One species could have arisen elsewhere by, for example, peripatric speciation and subsequently moved into the environment of the other species. In other words, the speciation occurred in the past by allopatry, and the two species are only currently sympatric because migration after speciation occurred.

However, recent studies of plants on an isolated island have provided strong, if not definite, evidence of sympatric speciation. Lord Howe Island is a tiny island (about 10 km long and 2 km across at its widest) located approximately 600 km east of Australia. Two species of palm tree that are found only on this island are each other's closest relatives. Because the island is so small, there is little chance that the two species could be geographically separated from each other, meaning that the species are and were sympatric. Moreover, because of the distance between Lord Howe Island and Australia (or other islands), it is extremely likely that the palms evolved and speciated from a common ancestor on the island.

This and other evidence show that sympatric speciation can occur. We must recognize, however, that we do not yet know just how much of all speciation is sympatric and how much is allopatric. **Fig. 21.12** summarizes modes of speciation based on geography.

Speciation can occur instantaneously.

Although speciation is typically a lengthy process, it can occasionally occur in a single generation, making it sympatric by definition. Typically, cases of such **instantaneous speciation** are caused by hybridization between two species in which the offspring are reproductively isolated from both parents.

For example, hybridization in the past between two sunflower species, *Helianthus annuus* and *H. petiolaris* (the ancestor of the cultivated sunflower), has apparently given rise to three new sunflower species: *H. anomalus*, *H. paradoxus*, and *H. deserticola* (**Fig. 21.13**). *H. petiolaris* and *H. annuus* have probably formed innumerable hybrids in nature, virtually all of them inviable. However, a few hybrids—the ones with a workable genetic complement—survived to yield these three daughter species. Each of these new species acquired a different mix of parental chromosomes. It is this species-specific chromosome complement that makes all three distinct and reproductively isolated from the parent species and from one another.

In many cases of hybridization, chromosome numbers change in the offspring compared to the parents. Two diploid parent species with 5 pairs of chromosomes, for a total of 10 chromosomes each, may produce a hybrid with double the number of chromosomes (that is, the hybrid inherits a full paired set of chromosomes from each parental species)—a total of 20. In this case, the hybrid has 4 genomes rather than the diploid number of 2. We call such a double diploid a tetraploid. In general, animals cannot sustain this kind of expansion in chromosome complement, but plants are more likely to do so. As a result, the formation of new species through polyploidy—multiple chromosome sets (Chapter 13)—has been relatively common in plants.

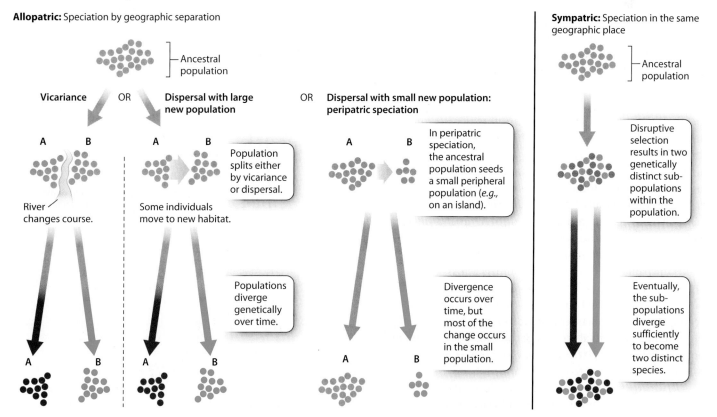

FIG. 21.12 Modes of speciation.

FIG. 21.13 Speciation by hybridization. In sunflowers, *Helianthus anomalus* (bottom) is the product of natural hybridization between *H. annuus* (top left) and *H. petiolaris* (top right). *Source: (top left) Gary A. Monroe, hosted by the USDA-NRCS PLANTS Database; (top right) Jason Rick; (bottom) Gerald J. Seiler, USDA-ARS.*

Helianthus annuus × *Helianthus petiolaris*

↓

Helianthus anomalus

Polyploids may be allopolyploids, meaning that they are produced from hybridization of two different species. For example, related species of *Chrysanthemum* appear, on the basis of their chromosome numbers, to be allopolyploids. Alternatively, polyploids may be autopolyploids, meaning that they are derived from an unusual reproductive event between members of a single species. In this case, through an error of meiosis in one or both parents in which homologous chromosomes fail to separate, a gamete may be produced that is not haploid. For example, *Anemone rivularis*, a plant in the buttercup family, has 16 chromosomes, and its close relative *A. quinquefolia* has 32 chromosomes. *Anemone quinquefolia* appears to be an autopolyploid derived from the joining of *A. rivularis* gametes with two full sets of chromosomes each.

So rampant is speciation by polyploidy in plants that it affects the pattern of chromosome numbers across all plants. In **Fig. 21.14**, the haploid chromosome numbers of thousands of plant species are plotted on a graph. Note that as the numbers get higher, even numbers tend to predominate, suggesting that the doubling of the total number of chromosomes (a form of polyploidy that always results in an even number of chromosomes) is an important factor in plant evolution.

Do we ever see instantaneous speciation *without* changes in chromosome complement? Analysis of Darwin's finches in the Galápagos Islands suggests that hybridization between closely related species may occasionally result in new species (**Fig. 21.15**). Detailed study of a population of *Geospiza fortis* on one small island, Daphne Major, allowed scientists to identify a hybridization event between a resident *G. fortis* female and an incoming male *G. conirostris* from another island, Española, 100 km away. Remarkably, the offspring of this pairing were able to breed to produce a population that is reproductively isolated from Daphne Major's resident *G. fortis*. Critically, the bill dimensions of the new species are different from those of other *Geospiza* species, so the new species can specialize on food resources not readily accessed by other species. Undoubtedly this was a rare and unusual event, but, given the vast periods of time over which the evolutionary process unspools, perhaps freak events such as this one play an important role in the generation of new species.

Speciation can occur with or without natural selection.

The association of both the origin of species and natural selection with Charles Darwin may make it seem that one

FIG. 21.14 Plant chromosome numbers. Plant chromosome numbers suggest that polyploidy has played an important role in plant evolution. *Data from Fig. 20-18, p. 752, in A. J. F. Griffiths, S. R. Wessler, S. B. Carroll, and J. Doebley, 2012, Introduction to Genetic Analysis, 10th ed., New York, NY: W. H. Freeman.*

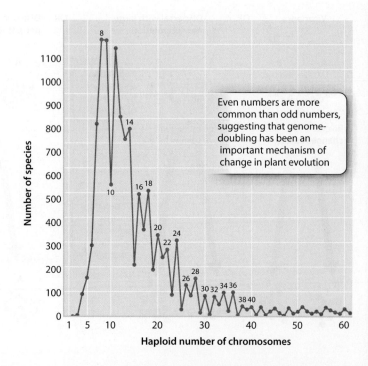

Even numbers are more common than odd numbers, suggesting that genome-doubling has been an important mechanism of change in plant evolution

FIG. 21.15 Speciation by hybridization. A new species of Darwin's finch derived from a mating between members of two closely related *Geospiza* species. *Modified by Sangeet Lamichhaney from Lamichhaney et al. 2018. Rapid hybrid speciation in Darwin's finches. Science 359:224–228. Photos by Peter & Rosemary Grant.*

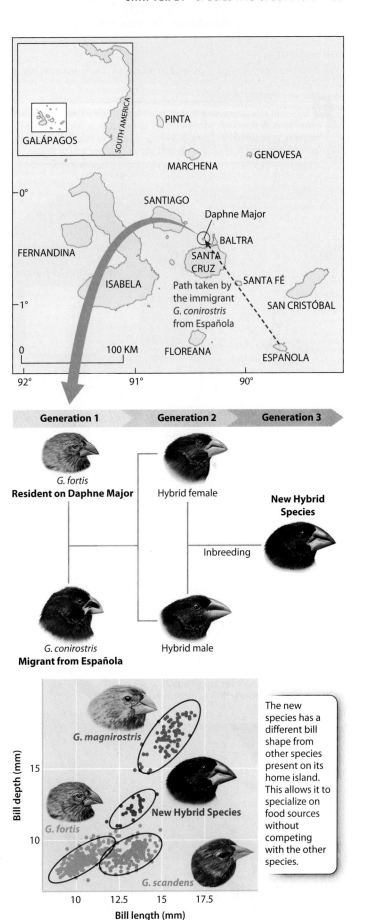

cannot occur without the other. In reality, speciation can occur in the presence or absence of natural selection, and natural selection does not always lead to speciation. The genetic divergence of two populations can be entirely due to genetic drift, for example, with no role for natural selection.

We have also seen two ways in which natural selection can be involved in speciation. First, sympatric speciation requires some form of disruptive selection, as when hybrid offspring are competitively inferior to nonhybrid offspring. Second, allopatric speciation (and adaptive radiation) may be facilitated by natural selection. For example, when a peripheral population enters a new environment, natural selection can act to promote its adaptation to the new conditions, accelerating the rate of genetic divergence between it and its parent population.

Fig. 21.16 summarizes the evolutionary mechanisms that lead to speciation. Speciation is caused by the accumulation of genetic differences between populations. Mutation is therefore a key component of the process, but so, too, is the fixation process. (Recall from Chapter 20 that fixation is the change from a situation where there are at least two variants of a particular allele present in a population to a situation where only one of the alleles remains.) A mutation can be fixed either by selection, if it is advantageous, or by genetic drift, if it is neutral. Multiple forces acting over long periods on each population produce the differences that accumulate between diverging populations.

Self-Assessment Questions

8. How are genetic divergence and reproductive isolation related to each other?

9. What is the difference between allopatric and sympatric speciation, and which is thought to be more common? Explain why.

10. What is the difference between allopatric speciation by dispersal and allopatric speciation by vicariance? Give an example of each.

11. Why are there so many examples of adaptive radiation on mid-ocean volcanic archipelagos such as the Galápagos?

12. There are hundreds of species of cichlid fish in Lake Victoria in Africa. Some scientists argue that they evolved sympatrically, but recent studies of the lake suggest that it periodically dried out, leaving a series of small ponds. How is this observation relevant to evaluating the hypothesis that these species arose by sympatric speciation?

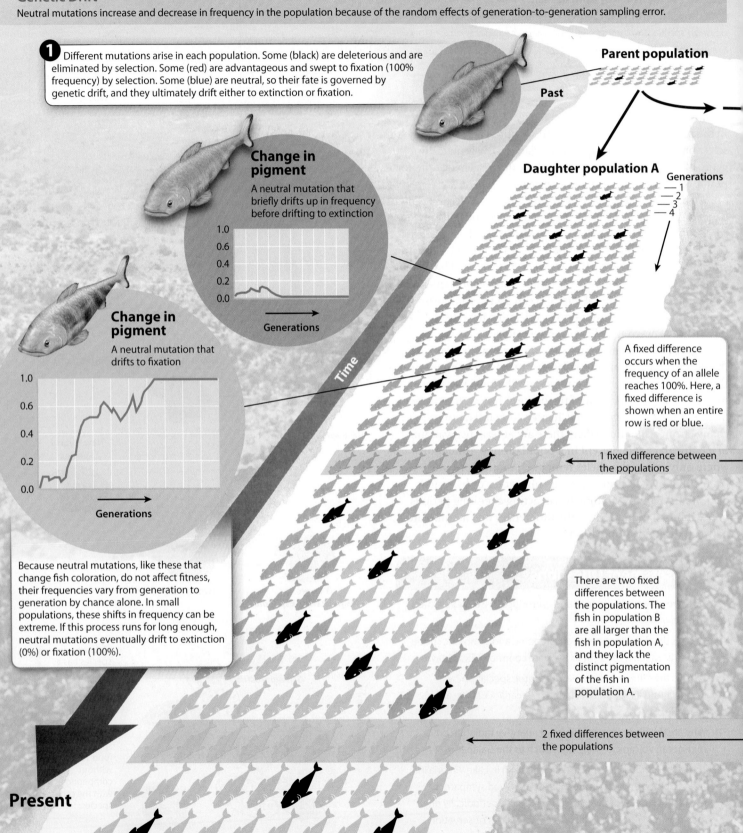

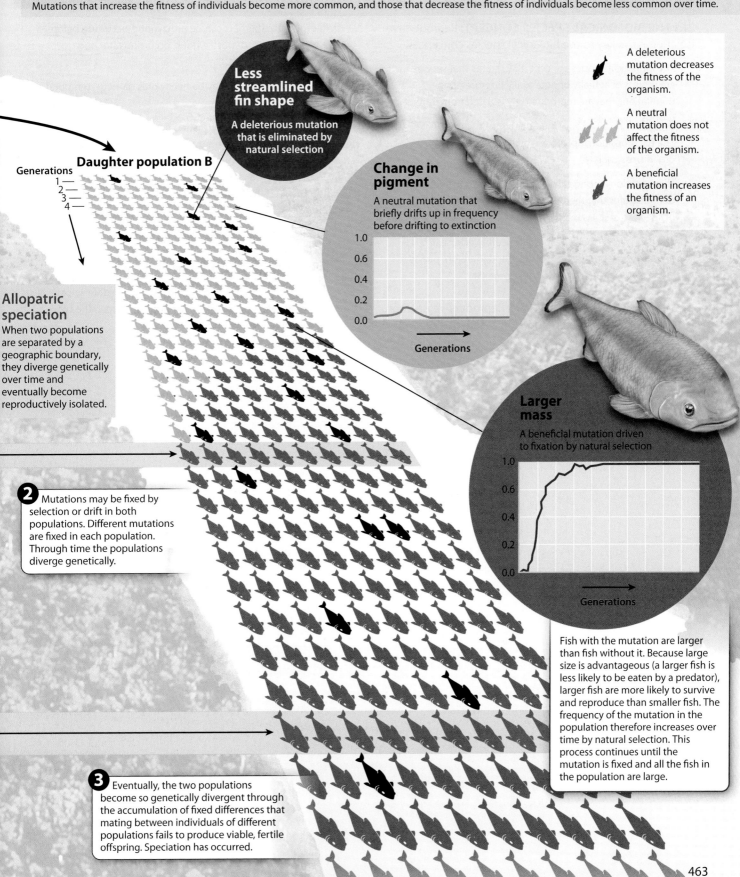

CORE CONCEPTS SUMMARY

21.1 THE BIOLOGICAL SPECIES CONCEPT: Reproductive isolation is the key to the biological species concept.

The biological species concept (BSC) states that species are groups of actually or potentially interbreeding populations that are reproductively isolated from other such groups. page 448

The BSC does not apply to asexual or extinct organisms. page 449

Hybridization among closely related species further demonstrates that the BSC is not a comprehensive definition of species. page 449

The BSC is nevertheless especially useful because it emphasizes reproductive isolation, and therefore provides studies of speciation with a focus. page 450

21.2 REPRODUCTIVE ISOLATION: Reproductive isolation is caused by barriers to reproduction before or after egg fertilization.

Reproductive barriers can be pre-zygotic, occurring before egg fertilization, or post-zygotic, occurring after egg fertilization. page 451

Pre-zygotic isolation may be behavioral, gametic, temporal, or ecological. page 451

In post-zygotic isolation, mating occurs but genetic incompatibilities prevent the development of viable, fertile offspring. page 452

21.3 SPECIATION: Speciation underlies the diversity of life on Earth.

Speciation is typically a by-product of genetic divergence that occurs as a result of the fixation of different mutations in two populations that are not regularly exchanging genes. page 452

If divergence continues long enough, chance differences will arise that result in reproductive barriers between the two populations. page 452

Most speciation is thought to be allopatric, involving two geographically separated populations. page 452

Geographic separation may be caused by dispersal, resulting in the establishment of a new and distant population, or by vicariance, in which the range of a species is split by a change in the environment. page 453

A special case of allopatric speciation by dispersal is peripatric speciation, in which the new population is small and outside the species' original range. page 453

Adaptive radiation, in which speciation occurs rapidly to generate a variety of ecologically diverse forms, is best documented on oceanic islands following the arrival of a single ancestral species. page 457

Co-speciation occurs when one species undergoes speciation in response to speciation in another species. In parasites and their hosts, co-speciation can result in phylogenies that have the same branching patterns. page 457

Speciation may be sympatric, meaning that there is no geographic separation between the diverging populations. page 458

Sympatric speciation can occur by hybridization or by disruptive selection against intermediate forms that is strong enough to offset the homogenizing effect of gene flow. page 460

Separated populations can diverge as a result of genetic drift, natural selection, or both. page 461

Log in to LaunchPad to check your answers to the Self-Assessment Questions and to access additional learning tools.

CHAPTER 22
Evolutionary Patterns
Phylogeny and Fossils

CORE CONCEPTS

22.1 READING A PHYLOGENETIC TREE: A phylogenetic tree is a hypothesis of the evolutionary relationships among organisms.

22.2 BUILDING A PHYLOGENETIC TREE: A phylogenetic tree is built on the basis of shared derived characters.

22.3 THE FOSSIL RECORD: The fossil record provides direct evidence of evolutionary history.

22.4 COMPARING EVOLUTION'S TWO GREAT PATTERNS: Phylogeny and fossils provide independent and corroborating evidence of evolution.

All around us, nature displays nested patterns of similarity among species. For example, as noted in Chapter 1, humans are more similar to chimpanzees than either humans or chimpanzees are to monkeys. Humans, chimpanzees, and monkeys, in turn, are more similar to one another than any one of them is to a mouse. And humans, chimpanzees, monkeys, and mice are more similar to one another than any of them is to a catfish. This pattern of nested similarity was recognized more than 200 years ago and used by the Swedish naturalist Carolus Linnaeus to classify biological diversity. A century later, Charles Darwin recognized this pattern as the expected outcome of a process of "descent with modification," or evolution.

Evolution produces two distinct but related patterns, both evident in nature. First is the nested pattern of similarities found among species on the present-day Earth. The second is the historical pattern of evolution recorded by fossils. Life, in its simplest form, originated more than 3.5 billion years ago. Today, an estimated 10 million species inhabit the planet. Short of inventing a time machine, how can we reconstruct those 3.5 billion years of evolutionary history so as to understand the extraordinary series of events that have ultimately resulted in the biological diversity we see around us today? These two great patterns provide the answer.

Darwin recognized that the species he observed must be the modified descendants of earlier species. Distinct populations of an ancestral species separate and diverge through time, again and again, giving rise to multiple descendant species through the processes we explored in Chapters 20 and 21. The result is the pattern of nested similarities observed in nature. This history of descent with branching is called a **phylogeny**, and is much like the genealogy that records our own family histories.

The evolutionary changes inferred from patterns of relatedness among present-day species make predictions about the historical pattern of evolution we should see in the fossil record. For example, groups with features that we infer to have evolved earlier than others should appear earlier in time as fossils. Paleontological research reveals that the history of life is, indeed, laid out in the chronological order predicted on the basis of comparative biology. How do we reconstruct the history of life from evolution's two great patterns, and how do they compare?

22.1 READING A PHYLOGENETIC TREE

Chapter 21 introduced the concept of speciation, the set of processes by which physically, physiologically, or ecologically isolated populations diverge from one another to the point where they can no longer produce fertile offspring. As illustrated in **Fig. 22.1,** speciation can be thought of as a process of branching. Now consider what happens as this process occurs over and over in a group through time. As species proliferate, their evolutionary relationships unfold in a treelike pattern, with individual species at the twig tips and their closest relatives connected to them at the nearest fork in the branch, called a **node**. Thus, a node represents the most recent common ancestor of two descendant species.

FIG. 22.1 The relationship between speciation and a phylogenetic tree. The phylogenetic tree, on the right, depicts the evolutionary relationships that result from the two successive speciation events diagrammed on the left.

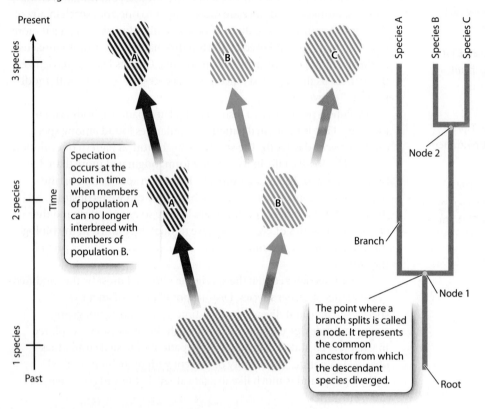

Phylogenetic trees provide hypotheses of evolutionary relationships.

Phylogenetics is one of two related disciplines within systematics, the study of evolutionary relationships among organisms. The other is taxonomy, the classification of organisms.

The aim of taxonomy is to recognize and name groups of individuals as species, and then to group closely related species into the more inclusive taxonomic group of the genus, and so on up through the taxonomic ranks—species, genus, family, order, class, phylum, kingdom, domain. Taxonomy, then, provides us with a hierarchical classification of species in groups that are more and more inclusive, giving us a convenient way to communicate information about the features each group possesses. For example, if we want to tell someone about a small animal we have seen that has fur, mammary glands, and extended finger bones that permit it to fly, we can provide this long description. Alternatively, we can just say we saw a bat, or a member of Order Chiroptera. All the rest is understood (or can be looked up).

Phylogenetics, by contrast, aims to discover the pattern of evolutionary relatedness among groups of species or other groups by comparing their anatomical or molecular features, and to depict these relationships as a **phylogenetic tree**. A phylogenetic tree is a hypothesis about the evolutionary history, or phylogeny, of the species. Phylogenetic trees are hypotheses because they represent the best model, or explanation, of the relatedness of organisms on the basis of the existing data. As with any model or hypothesis, the discovery of new evidence may suggest alternative relationships, leading to changes in the hypothesized pattern of branching on the tree. In recent years, biologists have worked to integrate phylogenetic branching patterns with taxonomic classification.

Many phylogenetic trees explore the relatedness of particular groups of individuals, populations, or species. We may, for example, want to understand how wheat is related to other, noncommercial grasses, or how disease-causing populations of *Escherichia coli* relate to more benign strains of the bacterium. At a much larger scale, universal similarities of molecular biology indicate that all living organisms are descended from a single common ancestor. This insight has inspired the goal of reconstructing phylogenetic relationships for all species as a means to understand how biological diversity has evolved since life originated. This universal tree is commonly referred to as the tree of life (Chapter 1). In Part 2 of this book, we will make use of the tree of life and many smaller-scale phylogenetic trees to understand our planet's biological diversity.

Fig. 22.2 shows a phylogenetic tree for vertebrate animals. The informal name at the end of each branch represents a group of organisms, many of them familiar. We sometimes find it useful to refer to groups of species this way (for example, "frogs," or "Class Anura"), rather than name all the individual species or list the characteristics they have in common. Remember, though, that such named groups represent a number of member species. If, for example, we could zoom in on the branch labeled "Frogs," we would see that it consists of many smaller branches, each representing a distinct species of frog, either living or extinct.

This tree provides information about evolutionary relationships among vertebrates. For example, it proposes that the closest living relatives of birds are crocodiles and alligators. The tree also proposes that the closest relatives of all tetrapod (four-legged) vertebrates are lungfish, which are fish with lobed limbs and the ability to breathe air. Phylogenetic trees are built from careful analyses of the morphological features (the organism's form) and molecular attributes of the species or other groups

FIG. 22.2 A phylogeny of vertebrate animals. The branching order constitutes a hypothesis of evolutionary relationships within the group.

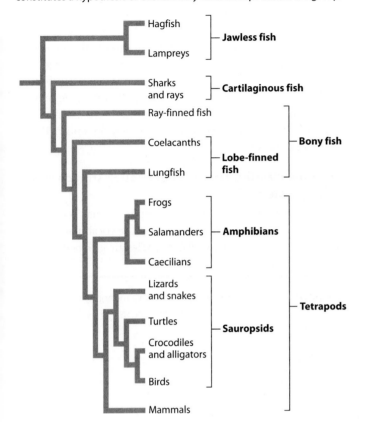

are the end products of the same interval of evolution since their divergence from a common ancestor more than 370 million years ago.

The search for sister groups lies at the heart of phylogenetics.

Two species, or groups of species, are considered to be closest relatives if they share a common ancestor not shared by any other species or group. Fig. 22.2, for example, illustrates that frogs are more closely related to salamanders than to any other group of organisms because frogs and salamanders share a common amphibian ancestor not shared by any other group. Similarly, lungfish are more closely related to tetrapods than to any other group. A lungfish may look more like a fish than it does an amphibian, but lungfish are more closely related to amphibians than they are to other fishes because lungfish share a common ancestor with amphibians (and other tetrapods) that was more recent than their common ancestor with other fishes (Fig. 22.2).

Groups that are more closely related to each other than either of them is to any other group, like lungfish and tetrapods, are called **sister groups**. Simply put, phylogenetic hypotheses amount to determining sister-group relationships because the simplest phylogenetic question we can ask is which two of any three species (or other groups) are more closely related to each other than either is to the third. In this light, we can see that a phylogenetic tree is simply a set of sister-group relationships; adding a species to the tree entails finding its sister group in the tree.

Closeness of relationship is determined by checking how recently two groups shared a common ancestor. Shared ancestry is indicated by a node, or branch point, on a phylogenetic tree. Branches, in turn, can be rotated around a node without changing the evolutionary relationships of the groups. **Fig. 22.3** shows four phylogenetic trees depicting evolutionary relationships among birds, crocodiles and alligators, and turtles. In all four trees, birds are a sister group to crocodiles and alligators because birds, crocodiles, and alligators share a common ancestor that is not shared by turtles. The more recent a common

under study. In essence, a tree is a hypothesis about the order of branching events in evolution, and that hypothesis can be tested by gathering more information about anatomical and molecular traits.

A phylogenetic tree does not in any way imply that more recently evolved groups are more advanced than groups that arose earlier. A modern lungfish, for example, is not more primitive or "less evolved" than an alligator, even though its group branches off the trunk of the vertebrate tree from an earlier node than the alligator group does. After all, both species

FIG. 22.3 Sister groups. The four trees illustrate the same set of sister-group relationships.

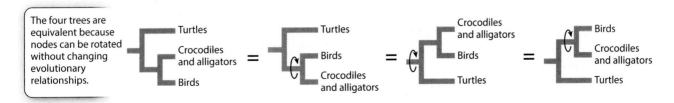

ancestor, the more closely related two groups are. Evolutionary relatedness, then, is determined by following nodes from the tips to the root of the tree, rather than by the order of the tips from the top to bottom of a page or across a page.

A monophyletic group consists of a common ancestor and all its descendants.

Up to this point, we have used the word "group" to mean all the species in some taxonomic entity, such as a genus or family. A more technical word for "group" is **taxon** (plural, **taxa**), with *taxonomy* providing a formal means of naming groups. Modern taxonomic classifications emphasize groups that are **monophyletic**, meaning that all members share a single common ancestor that is not shared with any other species or group of species. In Fig. 22.2, the tetrapods are monophyletic because all of the groups classified as tetrapods share a common ancestor not shared by any other taxa. Similarly, amphibians are monophyletic.

In contrast, consider the group of animals traditionally recognized as reptiles, which includes turtles, snakes, lizards, crocodiles, and alligators (Fig. 22.2). Birds are not considered reptiles, although they share a common ancestor with the included animals. The group "reptiles" is **paraphyletic**, meaning that it includes some, but not all, of the descendants of a common ancestor. Early zoologists did not recognize the existence of a common ancestor of birds and reptiles because birds are so distinctive. Nevertheless, many features of skeletal anatomy and DNA sequence strongly support the placement of birds as a sister group to the crocodiles and alligators.

There is a simple way to distinguish between monophyletic and paraphyletic groups, illustrated in **Fig. 22.4**. If to separate a group from the rest of the phylogenetic tree, you need to make only one cut, the group is monophyletic (blue in Fig. 22.4). If you need to make a second cut to trim away part of the separated branch, the group is paraphyletic (green in Fig. 22.4).

Groupings that do not include the last common ancestor of all members are called **polyphyletic**. Members of a polyphyletic group share traits that evolved independently by convergent evolution. For example, clustering bats and birds together as flying tetrapods results in a polyphyletic group (red in Fig. 22.4)

because wings evolved independently in the two groups and their common ancestor did not have wings.

Identifying monophyletic groups is a major goal of phylogenetics because monophyletic groups include all descendants of a common ancestor and only the descendants of that common ancestor. As a consequence, monophyletic groups alone show the evolutionary path a given group has taken since its origin. Omitting some members of a group, as in the case of reptiles and other paraphyletic groups, can provide a misleading sense of evolutionary history. By using monophyletic groups in taxonomic classification, we effectively convey our knowledge of their evolutionary history.

Taxonomic classifications are information storage and retrieval systems.

The nested pattern of similarities among species has been recognized by naturalists for centuries. In the vocabulary of formal classification, closely related species are grouped into

FIG. 22.4 Monophyletic, paraphyletic, and polyphyletic groups. Only monophyletic groups reflect evolutionary relationships because only they include all the descendants of a common ancestor.

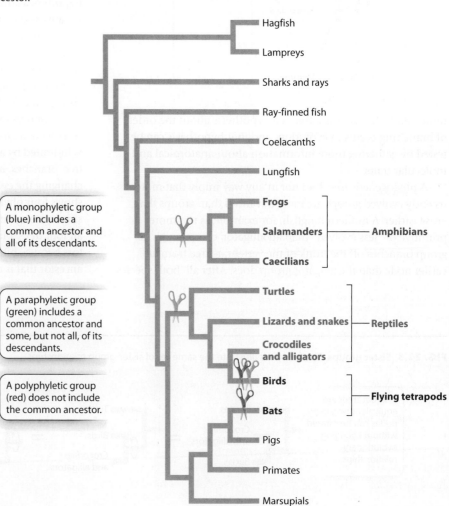

FIG. 22.5 Classification. Classification reflects our understanding of phylogenetic relationships, and the taxonomic hierarchy reflects the order of branching.

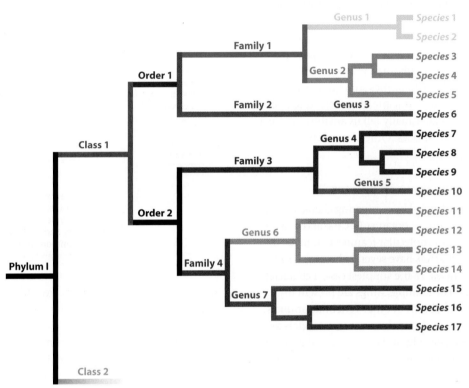

a **genus** (plural, **genera**). Closely related genera, in turn, belong to a larger, more inclusive branch of the tree, known as a **family**. Closely related families form an order, orders form a **class**, classes form a **phylum** (plural, **phyla**), and phyla form a **kingdom**, with each more inclusive taxonomic level occupying a successively larger limb on the tree (**Fig. 22.5**). Biologists today commonly refer to the three largest limbs of the entire tree of life as **domains** (Eukarya, or eukaryotes; Bacteria; and Archaea).

The ranks of classification form a nested hierarchy, but the boundaries of ranks above the species level are arbitrary. Put simply, there is nothing particular about a group that makes it, for example, a class rather than an order. A taxonomist examining the 17 species included in Fig. 22.5 might decide that species 3 and 4 are sufficiently distinct from species 5 to warrant placing them into a distinct genus. The same taxonomist might then decide that the grouping of the two new genera together should rank as a family. For this reason, it is not necessarily true that orders or classes of, for example, birds and ferns are equivalent in any meaningful way. In contrast, sister groups are equivalent in several ways. Notably, they diverged from a single ancestor at a single point in time. Therefore, if one branch is a sister group that contains 500 species and another branch has 6 species, the branches have experienced different rates of speciation, extinction, or both since they diverged. In Fig 22.5, Families 1 and 2 are sister groups, but Family 1 has five species while Family 2 has just one.

Self-Assessment Questions

1. In Fig. 22.3, why does the phylogeny indicate that birds and crocodiles are more closely related to each other than either is to turtles? On the phylogeny, label the common ancestor of crocodiles and birds, and the common ancestor of turtles, crocodiles, and birds.

2. Does either of the following phylogenetic trees indicate that humans are more closely related to lizards than to mice?

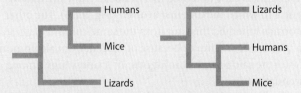

3. Look at Fig. 22.2. Are fish a monophyletic group?

4. In Fig 22.2, hagfish and lampreys (jawless fishes) are at the top of the tree and mammals at the bottom. Does this imply that jawless fishes are less evolved than mammals?

22.2 BUILDING A PHYLOGENETIC TREE

Up to this point, we have focused on how to interpret a phylogenetic tree, a diagram that depicts the evolutionary history of organisms. But how do we infer evolutionary history from a group of organisms? That is, how do we actually construct a phylogenetic tree? Biologists use characteristics of organisms to figure out their relationships. Similarities among organisms are particularly important, in that similarities sometimes suggest shared ancestry. However, a key principle of constructing trees is that only some similarities can actually tell us anything about a species' relationship to other species. Other shared characteristics can, in fact, be misleading.

Homology is similarity by common descent.

Phylogenetic trees are constructed by comparing character states shared among different groups of organisms. **Characters** are the anatomical, physiological, or molecular features that make up organisms. In general, characters have several observed conditions, called **character states**. In the simplest case, a character can be present or absent. For example, lungs are present in tetrapods and lungfish, but absent in other vertebrate animals. More commonly, there are multiple character states. Petals are a character of flowers, for example, and each observed arrangement—petals arranged in a helical pattern, petals arranged in a whorl, or petals fused into a tube—can be considered a state of the character of petal arrangement. All species contain some character states that are shared with other members of their group, some that are shared with members of other groups, and some that are unique.

Character states in different species can be similar for one of two reasons. First, the character state (for example, helically arranged petals) may have been present in the common ancestor of the two groups and retained over time (common ancestry). Alternatively, the character state may have evolved independently in the two groups as an adaptation to similar environments (convergent evolution, discussed earlier).

Consider two examples. Mammals and birds both produce amniotic eggs. Amniotic eggs occur only in groups descended from the common ancestor at the node connecting the mammal and sauropsid branches of the tree. Thus, we reason that birds and mammals each inherited this character from a common ancestor in which the trait first evolved (**Fig. 22.6**). The other sauropsids inherited this trait from the same common ancestor. Characters that are similar because of descent from a common ancestor are said to be **homologous**, or a **homology** among descendant species.

Not all similarities arise in this way, however. Think of wings, a character exhibited by both birds and bats. Much evidence supports the view that wings in these two groups do not reflect descent from a common, winged ancestor; instead, wings evolved independently in the two groups as adaptations to their environments. Similarities due to independent

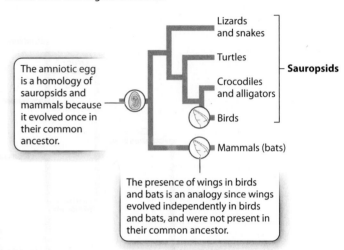

FIG. 22.6 Homology and analogy. A homology is a similarity that results from shared ancestry, whereas an analogy is a similarity that results from convergent evolution.

adaptation by different species are said to be **analogous**, or an **analogy** among different species. They are the result of convergent evolution.

Innumerable examples of convergent evolution less dramatic than wings are known. In some, we even understand the genetic basis of the convergence. For example, echolocation has evolved in both bats and dolphins, but not in most other mammals. Prestin is a protein in the hair cells of mammalian ears that is involved in hearing ultrasonic frequencies: bats and dolphins independently evolved similar changes in their Prestin genes, apparently convergent adaptations for echolocation. Similarly, unrelated fish that live in freezing water at the poles, Arctic and Antarctic, have evolved similar glycoproteins that act as molecular "antifreeze," preventing the formation of ice in their tissues.

In principle, two characters or character states are homologous if they are similar because of descent from a common ancestor that had the same character or character state; they are analogous if they arose independently because of similar selective pressures. In practice, to determine if characters observed in two organisms are homologous or analogous, we can weigh evidence that suggests where other traits place the two organisms on a phylogenetic tree, we can look at where on the organisms the trait occurs, and we can look at the anatomical or genetic details of how the trait is constructed. Wings in birds and bats are similar in morphological position (both are modified forelimbs), but differ in details of construction (the bat wing is supported by long fingers). Moreover, all other traits of birds and bats place them at the tips of different lineages, with many nonwinged species between them and their most recent common ancestor. Collectively, these pieces of information suggest that bats and birds did not retain wings from a common ancestor, but rather evolved wings independently.

Shared derived characters enable biologists to reconstruct evolutionary history.

Because homologies result from shared ancestry, only homologies, and not analogies, are useful in constructing phylogenetic trees. However, it turns out that only some homologies are useful. For example, character states that are unique to a given species or other monophyletic group can't tell us anything about its sister group. Because they evolved after the divergence of the group from its sister group, they can be used to characterize a group but not to relate it to other groups. Take wings again: we know that wings evolved in an ancestor of bats, but they do not occur in any other mammal. Because no other mammal has wings, this character cannot tell us which mammalian group is the sister group to bats. Similarly, homologies formed in the common ancestor of the entire group and therefore present in all its descendants do not help to identify sister-group relationships among the descendants of that common ancestor. All bats have wings, so wings by themselves do not help us to sort out evolutionary relationships among bat species.

What we need to build phylogenetic trees are homologies that are shared by some, but not all, of the members of the group under consideration. These shared derived characters are called **synapomorphies**. A derived character state is an evolutionary innovation, such as the change from five toes to a single toe—the hoof—in the ancestor of horses and donkeys. When this kind of novelty arises in the common ancestor of two taxa, it is shared by both; thus, the hoof is a synapomorphy defining horses and donkeys as sister groups.

Fig. 22.7 illustrates the major synapomorphies that have helped us construct the phylogeny of vertebrates. For example, the lung is a character present in lungfish and tetrapods, but absent in other vertebrates. The presence of lungs provides one piece of evidence that lungfish are the sister group of tetrapods. This approach to building a phylogenetic tree, or phylogenetic reconstruction, on the basis of synapomorphies is called **cladistics**.

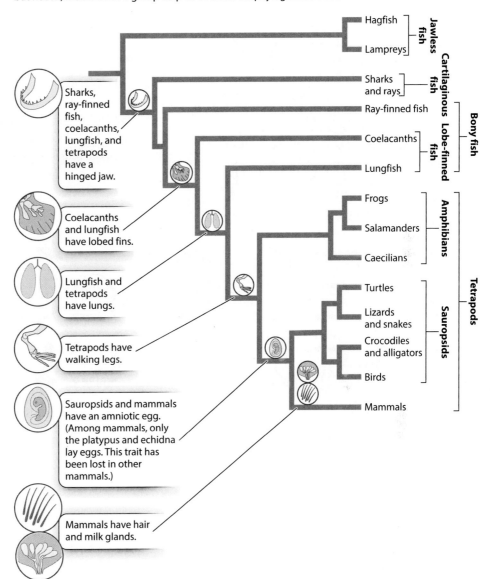

FIG. 22.7 Synapomorphies, or shared derived characters. Homologies that are present in some, but not all, members of a group help us to construct phylogenetic trees.

The simplest tree is often favored among multiple possible trees.

To see how synapomorphies help us chart out evolutionary relationships, let's consider the example in **Fig. 22.8**. We begin with four species of animals (labeled "A" through "D") for which we wish to build a phylogenetic tree. We will call this our ingroup, meaning it includes all of the groups for which we are interested in figuring out relationships. For comparison, we also include in our study a species that we believe is outside this ingroup—that is, it falls on a branch that splits off nearer the root of the tree—and so is called an outgroup (labeled "OG" in Fig. 22.8). Each species in the ingroup and the outgroup has a different combination of characters, such as leg number, presence or absence of wings, and whether development of young to adult is direct or goes through a pupal stage (Fig. 22.8a).

FIG. 22.8 Constructing a phylogenetic tree from shared derived traits. The strongest hypothesis of evolutionary relationships overall is the tree with the fewest number of changes because it minimizes the total number of independent origins of character states.

a. Character states

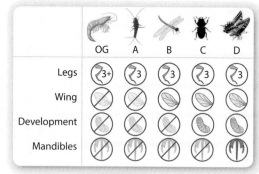

b. Possible phylogenetic trees

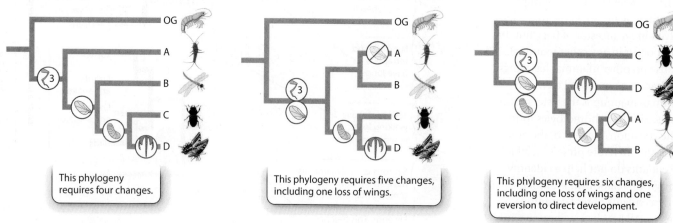

c. Phylogeny of Hexapoda (insects)

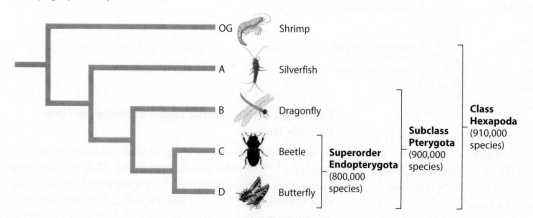

Since we are interested in the relationships among species A–D, we focus on potential synapomorphies, or character states shared by some, but not all, species within the group. For example, the table in Fig. 22.8a illustrating each group's character states shows that only C and D have pupae. This character suggests that C and D have a common ancestor that had pupae, and that the ancestor is not common to the other species. This would mean that C and D are more closely related to each other than either is to the other species. The alternatives to the explanation that C and D exclusively share a common ancestor are that C and D each evolved pupae independently or that pupae were present in the common ancestor of A–D but were lost in A and B.

How do we choose among the alternatives? Studies of the outgroup show that it does not form pupae, supporting the hypothesis that pupal development evolved within the ingroup. In practice, biologists examine multiple characters and choose the phylogenetic hypothesis that best fits all of the data.

How do we determine "best fit"? Fig. 22.8b illustrates three different hypotheses for the relationships among the species based on four characters and their various character states. Each reflects the sister-group relationship between C and D proposed earlier. The leftmost tree requires exactly four character-state changes during the evolution of these species: reduction of leg pairs to three in the group ABCD, wings in the group BCD, and pupae in group CD, plus a change of form of the mandible (jaw) in species D.

Now consider the middle tree in Fig. 22.8b. It groups A and B together with a common ancestor not shared by the other species. A does not have wings but B, C, and D do. If A and B share a common ancestor exclusively, it means that wings must have evolved in the common ancestor of A, B, C, and D and then were lost only in A, or that the common ancestor of A and B lost wings and then wings arose again in B—five changes in all.

The tree on the right groups A and B together, and requires two extra novel evolutionary changes for a total of six changes. No tree that we can construct from species A–D requires fewer than four evolutionary changes, so the left-hand version in Fig. 22.8b is the best available hypothesis of evolutionary relatedness. In fact, this is the phylogeny for a sample of species from the largest group of animals on Earth, the Hexapoda—insects and their closest relatives (Fig. 22.8c).

In general, trees with fewer character changes are preferred to ones that require more such changes because they provide the simplest explanation of the data. This approach is an example of **parsimony**, or choosing the simpler of two or more hypotheses to account for a given set of observations. When we use parsimony in phylogenetic reconstruction, we make the assumption that evolutionary change is typically rare. Over time, most features of organisms stay the same: we have the same number of ears as our ancestors, the same number of fingers, and so on. Thus, biologists commonly prefer the phylogenetic tree requiring the fewest evolutionary steps.

In systematics, parsimony suggests counting character changes on a phylogenetic tree to find the simplest tree for the data (the one with the fewest number of changes). Each change corresponds to a mutation (or mutations) in an ancestral species. Thus, the more changes or steps we propose, the more independent mutations we must also hypothesize.

Note also that it isn't necessary to make decisions in advance about which characters are homologies and which are analogies. We can construct all possible trees and then choose the one requiring the fewest evolutionary changes. This is a simple matter for the example in Fig. 22.8 because four species can be arranged into only 15 different trees. As the number of groups increases, however, the number of possible trees connecting them increases as well, and dramatically so. There are 105 possible trees for 5 groups, 945 trees for 6 groups, and nearly 2 million trees for 10 groups. For 50 groups, the possibilities balloon to 3×10^{76}! Clearly, computers are required to sort through all the possibilities.

As is true for all hypotheses, phylogenetic hypotheses can be supported strongly or weakly. Biologists use statistical methods to evaluate a given phylogenetic hypothesis. Available character data may not strongly favor any hypothesis. When support for a specific branching pattern is weak, biologists commonly depict the relationships as unresolved and show multiple groups diverging from one node, rather than just two. Such branching patterns are not meant to suggest that multiple species diverged simultaneously, but rather to indicate that we lack the data to choose unequivocally among several different hypotheses of relationship. In Part 2, we show unresolved branches in a number of groups. These shouldn't be read as admissions of defeat, but instead as problems awaiting resolution and opportunities for future research.

Molecular data complement comparative morphology in reconstructing phylogenetic history.

Trees can be built using anatomical features, but increasingly their construction relies on molecular data. The amino acids at particular positions in the primary structure of a protein can be used for this purpose, as can the nucleotides at specific positions along a strand of DNA.

From genealogy to phylogeny, tracing mutations in DNA or RNA sequences has revolutionized the reconstruction of historical genetic connections. Whether we are tracing the paternity of the children of Sally Hemings (mother of several of Thomas Jefferson's children), identifying the origin of a recent cholera epidemic in Haiti, or placing baleen whales near the hippopotamus family in a phylogenetic tree of mammals, molecular data are a rich source of phylogenetic insight.

There is nothing about molecular data that provides an intrinsically better record of history than does anatomical data; molecular data simply provide more details because there are more characters that can vary among the species. A sequence of DNA with hundreds or thousands of nucleotides can represent that many characters, as opposed to the tens of characters usually visible in morphological studies. For microbes and viruses, for example, very little morphology is available, so molecular information is critical for phylogenetic reconstruction. Once a gene or other stretch of DNA or RNA is identified that seems likely (based on previous studies of other species) to vary among the species to be studied, sequences are obtained and aligned to identify homologous nucleotide sites. Analyses of this kind commonly involve comparisons of sequences of about 1000 nucleotides from one or more genes. Increasingly, though, the availability of whole-genome sequences is changing the way we do molecular phylogenetics. With this approach, rather

FIG. 22.9 Phylogenetic reconstruction using molecular synapomorphies. We can use molecular synapomorphies to reconstruct phylogenies using sequence data.

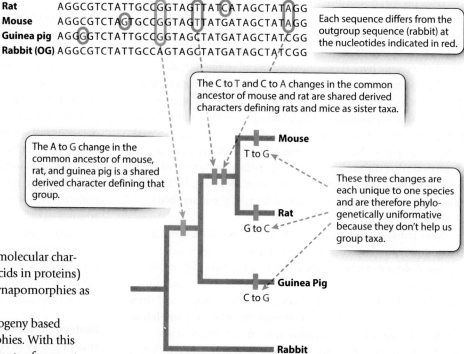

than comparing the sequences of a few genes, we compare the sequences of entire genomes.

The process of using molecular data is conceptually similar to the process described earlier for morphological data. Through comparison to an outgroup, we can identify shared derived molecular characters (whether DNA nucleotides or amino acids in proteins) and generate the phylogeny on the basis of synapomorphies as before (**Fig. 22.9**).

Alternatively, we can reconstruct the phylogeny based on overall similarity, rather than synapomorphies. With this approach, the assumption is that the descendants of a recent

FIG. 22.10 Phylogenetic reconstruction using distance methods. Sequence data can be used to reconstruct phylogeny on the basis of distance, where distance is measured in number of sequence differences between species.

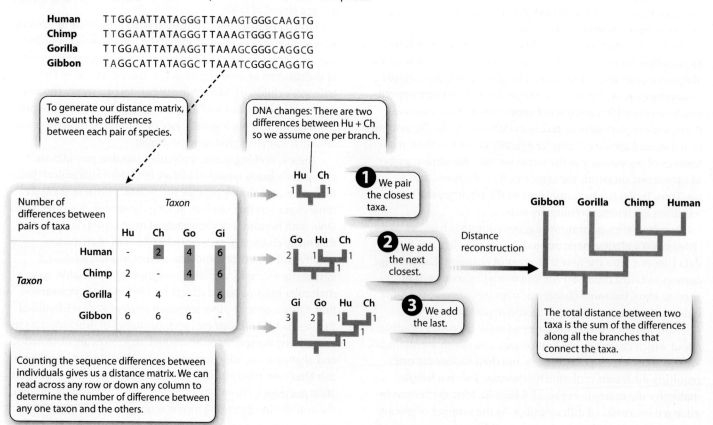

common ancestor will have had relatively little time to evolve differences, whereas the descendants of an ancient common ancestor have had a lot of time to do so. Thus, the extent of similarity (or distance) indicates how recently two groups shared a common ancestor. **Fig. 22.10** shows how distance methods work, using great apes as an example. We count the differences in DNA sequence for each pair of species. Those with the fewest differences are taken to be most closely related. In contrast, more distantly related species show a greater number of differences.

Underpinning this approach is the assumption that the rate of evolution is constant, or nearly so. If we find a total of six differences between two species, for example, this approach assumes that three arose in each lineage since the common ancestor (**Fig. 22.11a**). However, if the rate of evolution varied from one lineage to the next, few changes may have occurred within one lineage and many in the other. As shown in **Fig. 22.11b**, unequal rates of evolution in different lineages can result in inferring incorrect phylogenetic relationships. The rate-constancy assumption is less likely to be violated when we are using molecular data than when we are using morphological data. Recall from Chapter 20 that the molecular clock is based on the observation of constant accumulation of genetic divergence through time.

Molecular data are often combined with morphological data, and each can also serve as an independent assessment of the other. Not surprisingly, results from analyses of each kind of data are commonly compatible, at least for plants and animals rich in morphological characters.

The single largest library of taxonomic information is GenBank, the National Institutes of Health's genetic data storage facility. As of this writing, GenBank gives users access to more than 100 billion observations (mostly nucleotides) collected under more than 430,000 taxonomic names. A helpful Internet-based resource is the Encyclopedia of Life, which is gathering additional biological information about species, including ecology, geographic distributions, photographs, and sounds in pages for individual species that are easy to navigate. Another web resource, the Tree of Life, provides information on phylogenetic trees for many groups of organisms.

Phylogenetic trees can help solve practical problems.

The sequence of changes on a tree from its root to its tips documents evolutionary changes that have accumulated through time. Trees suggest which groups are older than others, and which traits came first and which followed later. Proper phylogenetic placement, therefore, reveals a great deal about evolutionary history. It can have practical consequences as well. For example, oomycetes, which are the microorganisms responsible for potato blight and other important diseases of food crops, were long thought to be fungi because they look like some fungal species. The discovery, using molecular characters, that oomycetes belong to a very different group of eukaryotic organisms has opened up new possibilities for understanding and controlling these plant pathogens. Similarly, in 2006, researchers used DNA sequences to identify the Malaysian parent population of a species of butterfly called lime swallowtails that had become an invasive species in the Dominican Republic, pinpointing the source populations from which natural enemies of this pest can be sought.

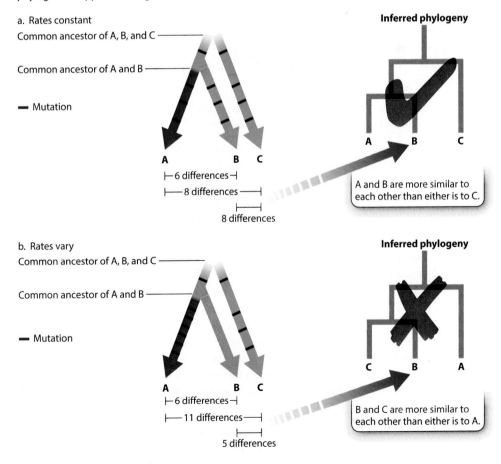

FIG. 22.11 Evolutionary rates. (a) When rates of evolution are constant, a distance-based phylogenetic method works. (b) When rates of evolution are not constant, a distance-based phylogenetic approach can give an incorrect tree.

Phylogenetics was employed to solves a famous case in which a dentist in Florida was accused of infecting his patients with human immunodeficiency virus (HIV) (**Fig. 22.12**). HIV nucleotide sequences evolve so rapidly that biologists can build phylogenetic trees that trace the spread of specific strains from one individual to the next. Phylogenetic study of HIV present in samples from several infected patients, the HIV-positive dentist, and other individuals provided evidence that the dentist had, indeed, infected his patients.

Similarly, phylogenetic studies of influenza virus strains show their origins and subsequent movements among geographic regions and individual patients. Today, there is a growing effort to use specific DNA sequences as a kind of fingerprint or barcode for tracking biological material. Such information could quickly identify samples of shipments of meat as being from endangered species, or track newly emerging pests. The Consortium for the Barcode of Life has already accumulated species-specific DNA barcodes for more than 100,000 species. Phylogenetic evidence provides a powerful tool for evolutionary analysis and is useful across timescales ranging from months to the entire history of life, from the rise of epidemics to the origins of metabolic diversity.

Self-Assessment Questions

5. Despite having many traits in common with fish, including a streamlined body and fins, dolphins are mammals, not fish. Are these similar traits homologous or analogous?

6. Which type of homology is useful in building phylogenetic trees? Why is this kind of homology, and not others, useful?

HOW DO WE KNOW?

FIG. 22.12
Did an HIV-positive dentist spread the AIDS virus to his patients?

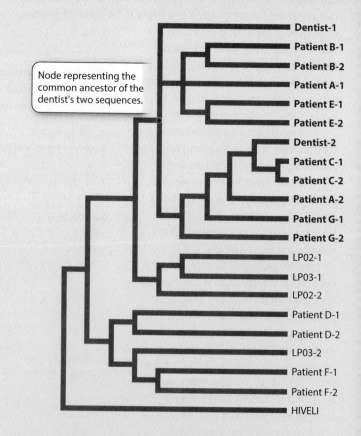

BACKGROUND In the late 1980s, several patients of a Florida dentist contracted acquired immunodeficiency syndrome (AIDS), the disease caused by HIV infection. Molecular analysis showed that the dentist was HIV-positive.

HYPOTHESIS It was hypothesized that the patients acquired HIV during dental procedures carried out by the infected dentist.

METHOD Researchers obtained two HIV samples each (denoted 1 and 2 in the figure) from several people, including the dentist (Dentist 1 and Dentist 2), several of his patients (Patients A through G), and other HIV-positive individuals chosen at random from the local population (LP). In addition, a strain of HIV from Africa (HIVELI) was included in the analysis as an outgroup.

RESULTS Biologists constructed a phylogeny based on the nucleotide sequence of a rapidly evolving gene in the genome of HIV. Because the gene evolves so quickly, its mutations preserve a record of evolutionary relatedness on a very fine scale. The HIV samples taken from some of the infected patients—patients A, B, C, E, and G—were more closely related to the dentist's HIV than they were to samples taken from other infected individuals. Some patients' sequences, however, were different enough from the dentist's that they were not on the same part of the tree, suggesting that these patients, D and F, had acquired their HIV infections from other sources.

CONCLUSION HIV phylogeny makes it highly likely that the dentist infected several of his patients. The details of how the patients were infected remain unknown, but the safety practices now rigidly observed in dentistry make it unlikely that such a tragedy could occur again.

FOLLOW-UP WORK Phylogenies based on molecular sequence characters are now routinely used to study the origin and spread of infectious diseases, such as swine flu and Ebola.

SOURCE Hillis, D. M., J. P. Huelsenbeck, and C. W. Cunningham. 1994. "Application and Accuracy of Molecular Phylogenies." *Science* 264:671–677.

22.3 THE FOSSIL RECORD

Phylogenies based on living organisms provide hypotheses about evolutionary history. Branching events toward the root of the tree occurred earlier than those near the tips, and characters change and accumulate along the path from the root to the tips. Fossils provide direct documentation of ancient life, and so, in combination, fossils and phylogenies provide strong complementary insights into evolutionary history.

Fossils provide unique information.

Fossils can and do provide evidence for phylogenetic hypotheses. For example, they show that groups that branch from early nodes in phylogenies appear early in the geologic record. But the fossil record does more than this. First, fossils enable us to calibrate phylogenies in terms of time. It is one thing to infer that mammals diverged from the common ancestor of birds, crocodiles, turtles, and lizards and snakes before crocodiles and birds diverged from their common ancestor (see Fig. 22.7); it is another matter to state that birds and crocodiles diverged from each other about 240 million years ago, whereas the group represented today by mammals branched from other vertebrates about 80 million years earlier. As we saw in Chapter 20, estimates of divergence time can be made using molecular sequence data, but all such estimates can be pinpointed to a specific time in history only by using fossils.

The evolutionary relationship between birds and crocodiles highlights a second kind of information provided by fossils. Not only do fossils record past life, but they also provide our only record of extinct species. The phylogeny in Fig. 22.7 contains a great deal of information, but it is silent about dinosaurs. Fossils demonstrate that dinosaurs once roamed Earth, and details of skeletal structure place birds among the dinosaurs in the vertebrate tree. Indeed, some remarkable fossils from China show that the dinosaurs most closely related to birds had feathers (**Fig. 22.13**).

A third, and also unique, contribution of fossils is that they place evolutionary events within the context of Earth's dynamic environmental history. Again, dinosaurs illustrate the point. As discussed in Chapter 1, geologic evidence from several continents suggests that a large meteorite triggered drastic changes in the global environment 66 million years ago, leading to the extinction of dinosaurs (other than birds). In fact, at five times in the past, large environmental disturbances sharply decreased Earth's biological diversity. These events, called mass extinctions, have played a major role in shaping the course of evolution.

Fossils provide a selective record of past life.

Fossils are the remains of once-living organisms, preserved through time in sedimentary rocks. If we wish to use fossils to complement phylogenies built from characters of modern organisms, we must understand how the processes of fossil formation govern what is and is not preserved.

For all its merits, the fossil record should not be thought of as a complete encyclopedia of everything that ever walked, crawled, or swam across our planet's surface. Fossilization requires burial, as when a clam dies on the seafloor and is quickly covered by sand, or when a leaf falls to the forest floor and ensuing floods cover it in mud. Through time, accumulating sediments from sand or mud or soil harden into sedimentary rocks, such as those exposed so dramatically in the walls of the Grand Canyon (**Fig. 22.14**). If they are not buried, the remains

FIG. 22.13 *Microraptor gui*, **a remarkable fossil discovered in approximately 125-million-year-old rocks from China.** The structure of its skeleton identifies *M. gui* as a dinosaur, yet it had feathers on its arms, tail, and legs. *Source: Getty Images/Getty Images.*

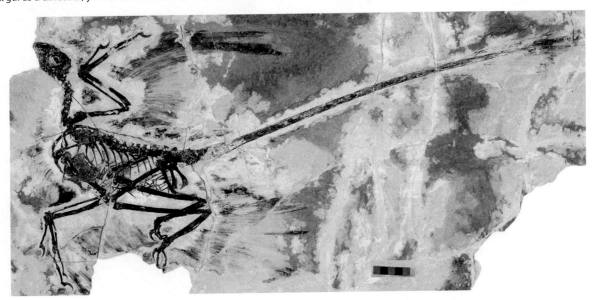

of organisms are eventually recycled by biological and physical processes, and no fossil forms. In general, the fossil record of marine life is more completely sampled than that for land-dwelling creatures because marine habitats are more likely than those on land to be places where sediments accumulate and become rock. Thus, trees and elks living high in the Rocky Mountains have a low probability of fossilization, whereas clams and corals on the shallow seafloor are commonly buried and become fossils.

Biological factors also contribute to the incompleteness of the fossil record. Most fossils preserve just the hard parts of organisms, those features that resist decay after death. For animals, this usually means mineralized skeletons. Clams and snails that secrete shells of calcium carbonate have excellent fossil records. More than 80% of the clam species found today along California's coast also occur as fossils in sediments deposited during the past million years. In contrast, nematodes, tiny worms that may be the most abundant animals on Earth, have no mineralized skeletons and are represented almost nowhere in the fossil record. The wood and pollen of plants, which are made in part of decay-resistant organic compounds, enter the fossil record far more commonly than do flowers. And, among unicellular organisms, the skeleton-forming diatoms, radiolarians, and foraminiferans have exceptionally good fossil records, whereas most amoebas are unrepresented.

Together, then, the properties of organisms (do they make skeletons or other features that resist decay after death?) and environment (did the organisms live in a place where burial was likely?) determine the probability that an ancient species will be represented in the fossil record.

Organisms that lack hard parts can leave a fossil record in two other distinctive ways. Many animals leave tracks and trails as they move about or burrow into sediments. These fossils of trails, rather than body parts, called **trace fossils**, range from dinosaur tracks to the feeding trails of snails and trilobites. In essence, they preserve a record of both anatomy and behavior (**Fig. 22.15**).

Organisms can also contribute **molecular fossils** to the rocks. Most biomolecules decay quickly after death. Proteins and DNA, for example, generally break down before they can be preserved, although, remarkably, the Neanderthal genome has been pieced together from DNA in 40,000-year-old bones (Chapter 23). Other molecules, especially lipids like cholesterol, are more resistant to decomposition. Sterols, bacterial

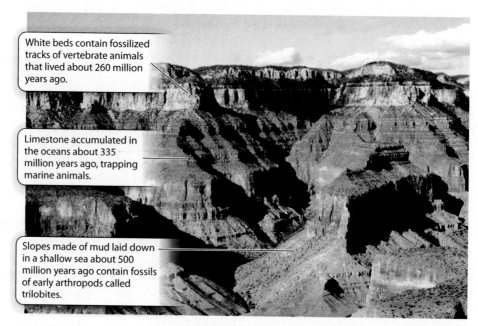

FIG. 22.14 The Grand Canyon. Erosion has exposed layers of sedimentary rock that record Earth history. *Source: National Park Service.*

FIG. 22.15 Trace fossils. These footprints in 150-million-year-old rocks record both the structure and the behavior of the dinosaurs that made them. *Source: José Antonio Hernaiz/AGE Fotostock.*

lipids, and some pigment molecules can accumulate in sedimentary rocks, documenting organisms that rarely form conventional fossils, especially bacteria and single-celled eukaryotes.

Rarely, unusual conditions preserve fossils of unexpected quality, including animals without shells, delicate flowers or mushrooms, fragile seaweeds or bacteria, and even the embryos of plants and animals. For example, 505 million years ago, during the Cambrian Period, a sedimentary rock formation called the Burgess Shale accumulated on a relatively deep seafloor covering what is now British Columbia. Waters just above the basin floor contained little or no oxygen. Thus, when mud swept into the basin, entombed animals were sealed off from scavengers, disruptive burrowing activity, and even bacterial decay. For this reason, Burgess rocks preserve a remarkable sampling of marine life during the initial diversification of animals (**Fig. 22.16**).

The Messel Shale formed more recently—about 50 million years ago—in a lake in what is now Germany. Release of toxic gases from deep within the Messel lake suffocated local animals, and their carcasses settled into muds on the lake floor that were oxygen-poor and sealed off from microbes that help decompose other organisms. In the Messel record, fish, birds, mammals, and reptiles are preserved as complete and articulated skeletons, and mammals retain impressions of fur and color patterning (**Fig. 22.17**). Plants were also beautifully preserved, as were insects, some with a striking iridescence still intact. Messel rocks provide a truly outstanding snapshot of life on land as the age of mammals began.

In general, the fossil record preserves some aspects of biological history well and others poorly. Fossils provide a good sense of how the forms, functions, and diversity of skeletonized animals have changed over the past 500 million years. The same is true for land plants and unicellular organisms that form mineralized skeletons. Collectively, these fossils shed light on major patterns of morphological evolution and diversity change through time; their geographic distributions record the movements of continents over millions of years; and the radiations and extinctions they document show how life responds to environmental change, both gradual and catastrophic.

FIG. 22.16 A Burgess Shale fossil. This fossil is *Opabinia regalis*, an extinct early relative of the arthropods, from the 505-million-year-old Burgess Shale. *Source: © Smithsonian Institution – National Museum of Natural History (USNM 57683). Photo: Jean-Bernard Caron.*

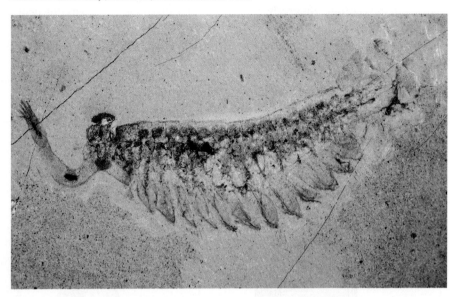

FIG. 22.17 A Messel Shale fossil. Even the furry tail of this extinct squirrel-like mammal is preserved in this remarkable 50-million-year-old fossil.

Geologic data indicate the age and environmental setting of fossils.

How do we know the age of a fossil? Beginning in the nineteenth century, geologists recognized that the types of fossils found in layers of rock change systematically from the bottom of a sedimentary rock formation to its top. As more of Earth's surface was mapped and studied, it became clear that certain fossils always occur in layers that lie beneath (and so are older than) layers that contain other species. From these patterns, geologists concluded that fossils mark time in Earth history. At first, geologists could not explain why fossils changed from one bed to the next, but after Darwin the reason became apparent: fossils record the evolution of life on Earth. Geologists eventually used these data to map out the **geologic timescale**, the series of time divisions that mark Earth's long history (**Fig. 22.18**).

The layers of fossils in sedimentary rocks can tell us that some rocks are older than others, but they cannot by themselves provide an absolute age. Calibration of the timescale became possible with the discovery of radioactive decay. In Chapter 2, we discussed isotopes, variants of an element that differ from one another in terms of the number of neutrons they contain. Many isotopes are unstable and spontaneously break down, ejecting neutrons, electrons, or other particles to form other, more stable isotopes. In the laboratory, scientists can measure how fast unstable isotopes decay. Then, by measuring the amounts of the unstable isotope and its stable daughter isotope inside a mineral, they can determine when the mineral formed.

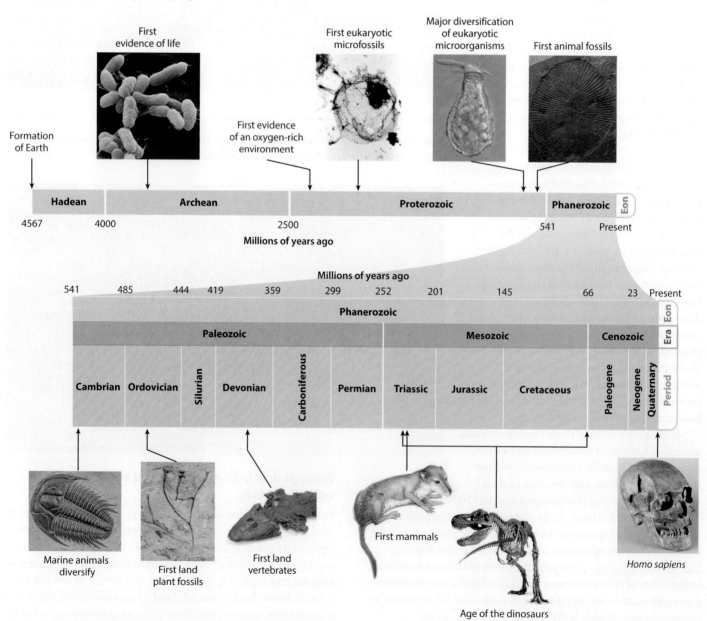

FIG. 22.18 The geologic timescale, showing major events in the history of life on Earth. Sources: (Top, left to right) Eye of Science/Science Source; Andrew Knoll, Harvard University; Antonio Guillén. Proyecto Agua Spain; Andrew Knoll, Harvard University. (Bottom, left to right) Scott Orr/iStockphoto.com; Hans Steur, The Netherlands; T. Daeschler/VIREO; Reconstruction illustration of Hadrocodium wui. Reconstruction artwork: Mark A. Klingler/Carnegie Museum of Natural History. From the cover of Science Vol. 292, no. 5521, 25 May 2001. Reprinted with permission from AAAS; dimair/Shutterstock; DEA/G. Cigolini/Getty Images.

Archaeologists commonly use the radioactive decay of the isotope carbon-14, or ^{14}C, to date wood and bone, a process called **radiometric dating**. ^{14}C is an isotope of carbon with eight neutrons and six protons instead of the more stable (and more common) configuration of six neutrons and six protons. As shown in **Fig. 22.19**, cosmic rays continually generate neutrons that collide with nitrogen in the atmosphere to form ^{14}C, much of which is incorporated into atmospheric carbon dioxide (CO_2). Through photosynthesis, carbon dioxide that contains ^{14}C is incorporated into wood, and animals incorporate small amounts of ^{14}C into their tissues when they eat plant material. The unstable ^{14}C in these tissues begins to break down immediately, losing an electron to form ^{14}N, a stable isotope of nitrogen. Once the organism dies, ^{14}C is no longer added, so the clock of radioactive carbon starts to tick. Laboratory measurements indicate that half of the ^{14}C in a given sample will decay to nitrogen in 5730 years, a period called its **half-life** (Fig. 22.19). All isotopes of elements have a unique half-life, which vary in length depending on their stability. Armed with this information, scientists can measure the amount of ^{14}C in an archaeological sample and, by comparing it to the amount of ^{14}C in a sample of known age (based on annual rings in trees, for example, or yearly growth of coral skeletons), determine the age of the sample.

Because its half-life is so short (by geologic standards), ^{14}C is useful only in dating materials younger than 50,000 to 60,000 years. Beyond that point, too little ^{14}C is left in materials to measure accurately. Older geologic materials are commonly dated using the radioactive decay of uranium (U) to lead (Pb): ^{238}U, which may be incorporated in trace amounts into the minerals of volcanic rocks, breaks down to ^{206}Pb with a half-life of 4.47 billion years; ^{235}U decays to ^{207}Pb with a half-life of 704 million years. Calibration of the geologic timescale is based mostly on the ages of volcanic ash found in sedimentary rocks that contain key fossils, as well as volcanic rocks that intrude into (and so are younger than) layers of rock containing fossils. In turn, the ages of those fossils provide calibration points for phylogenies.

The sedimentary rocks that contain fossils also preserve, encrypted in their physical features and chemical composition, information about the environment in which they formed. Sandstone beds, for example, may have rippled surfaces, like the ripples produced by currents that we see today in the sand of a seashore or lake margin. Pyrite (FeS_2), or fool's gold, forms when H_2S generated by anaerobic bacteria reacts with iron. Because these conditions generally occur where oxygen is absent, the presence of pyrite in ancient sedimentary rocks can indicate that the environment was depleted of oxygen at the time that the rock formed.

We might think our moment in geologic time is representative of Earth as it has always existed, but nothing could be further from the truth. In terms of the location and sizes of its continents, ocean chemistry, and atmospheric composition,

FIG. 22.19 ^{14}C **decay.** Scientists can determine the age of relatively young materials such as wood and bone from the amount of ^{14}C they contain.

the Earth we experience today is unlike any previous state of the planet. Today, for example, the continents are distributed widely over the planet's surface, but 280 million years ago they were clustered in a supercontinent called Pangaea (**Fig. 22.20**).

Oxygen gas permeates most surface environments of Earth today, but 3 billion years ago, there was no O_2 anywhere. And, just 20,000 years ago, 2 km of glacial ice stood where Boston lies today. Sedimentary rocks record the changing state of Earth's surface over billions of years and show that life and environment have changed together through time, each influencing the other.

Fossils can contain unique combinations of characters.

Phylogenies hypothesize impressive morphological and physiological shifts through time—amphibians from fish, for example, or land plants from green algae. Do fossils capture a record of these transitions as they took place?

Let's begin with an example introduced earlier in this chapter. Phylogenies based on living organisms generally place birds as the sister group to crocodiles and alligators, but birds and crocodiles are decidedly different from each other in structure. For example, birds have wings, feathers, toothless bills, and a number of other skeletal features distinct from those of crocodiles. In 1861, just two years after publication of *On the Origin of Species*, German quarry workers discovered a remarkable fossil that remains paleontology's most famous example of a form that is transitional between two groups. *Archaeopteryx lithographica*, now known from 11 specimens splayed for all time in fine-grained limestone, lived 150 million years ago. Its skeleton shares many characters with dromaeosaurs, a group of small, agile dinosaurs, but several features—its pelvis, its braincase, and, especially, its winglike forearms—are distinctly birdlike. Spectacularly, the fossils preserve evidence of feathers. *Archaeopteryx* clearly suggests a close relationship between birds and dinosaurs, and phylogenetic reconstructions that include information from fossils show that many of the characters found today in birds accumulated through time in their dinosaur ancestors. As noted earlier, even feathers first evolved in dinosaurs (**Fig. 22.21**).

Tiktaalik roseae and other skeletons in rocks deposited 375 to 362 million years ago record an earlier but equally fundamental transition: the colonization of land by vertebrates. Phylogenies built from characters of modern animals show that all land vertebrates, from amphibians to mammals, are descended from a common ancestral fish. As seen in **Fig. 22.22**, *Tiktaalik* had fins, gills, and scales like other fish of its day, but its skull was flattened, more like that of a crocodile than a fish, and it had a functional neck and ribs that could support its body—features today found only in tetrapods. Along with other fossils, *Tiktaalik* captures key moments in the evolutionary transition from water to land, confirming the predictions of phylogeny.

FIG. 22.20 Pangaea, 280 million years ago. Plate tectonics has shaped and reshaped Earth's geography through time. The white areas near the South Pole are glacial ice. *Source: Pangea from © Global Paleogeography and Tectonics © 2016 Colorado Plateau Geosystems Inc.*

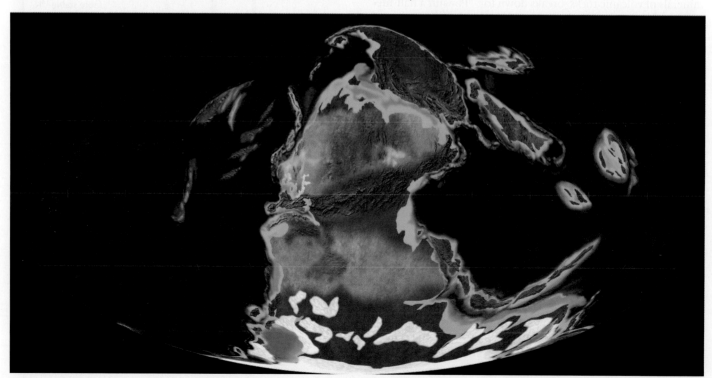

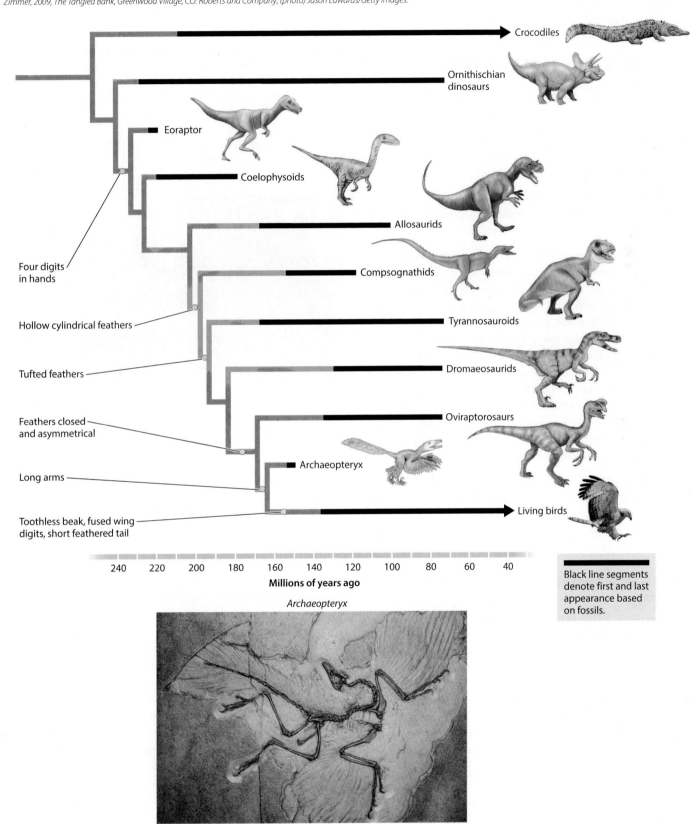

FIG. 22.21 Dinosaurs and birds. A number of dinosaur fossils link birds phylogenetically to their closest living relatives, the crocodiles. *Data from C. Zimmer, 2009, The Tangled Bank, Greenwood Village, CO: Roberts and Company; (photo) Jason Edwards/Getty Images.*

HOW DO WE KNOW?

FIG. 22.22

Do fossils bridge the evolutionary gap between fish and tetrapod vertebrates?

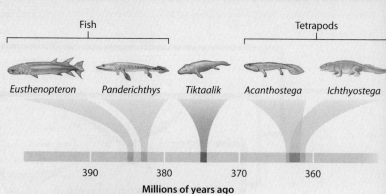

BACKGROUND Phylogenies based on both morphological and molecular characters of modern animals indicate that fish are the closest relatives of four-legged land vertebrates.

HYPOTHESIS Land vertebrates evolved from fish by modifications of the skeleton and internal organs that made it possible for them to live on land.

OBSERVATION Fossil skeletons 390 to 360 million years old show a mix of features seen in living fishes and amphibians. Older fossils have fins, fishlike heads, and gills, whereas younger fossils have weight-bearing legs, skulls with jaws able to grab prey, and ribs that help ventilate lungs. Paleontologists predicted that key intermediate fossils would be preserved in 380–370-million-year-old rocks.

In 2004, Edward Daeschler, Neil Shubin, and Farish Jenkins discovered the remarkable fossil *Tiktaalik* in rocks laid down by a meandering stream of just the right age. *Tiktaalik* had fins, scales, and gills like fishes, but an amphibian-like skull and a true neck (which fishes lack). Its limbs preserve wrist bones and fingers broadly similar to those in tetrapod vertebrates. Even so, as the reconstruction shows, its forelimbs were not completely adapted for movement on land.

CONCLUSION Fossils confirm the phylogenetic prediction that tetrapod vertebrates evolved from fish by the developmental modification of limbs, skulls, and other features.

Neil Shubin Lab

FOLLOW-UP WORK Research into the genetics of vertebrate development shows that the limbs of fish and amphibians are shaped by similar patterns of gene expression, providing further support for the phylogenetic connection between the two groups.

SOURCE Daeschler, E. B., N. H. Shubin, and F. A. Jenkins, Jr. 2006. "A Devonian Tetrapod-like Fish and the Evolution of the Tetrapod Body Plan." *Nature* 440:757–763.

Rare mass extinctions have altered the course of evolution.

Beginning in the 1970s, American paleontologist Jack Sepkoski scoured the paleontological literature, recording the first and last appearances in the geologic record for every genus of marine animals he could find. The results of this monumental effort are shown in **Fig. 22.23**. Sepkoski's diagram shows that animal diversity has increased over the past 540 million years: the biological diversity of animals in today's oceans may well be higher than it has ever been in the past. At the same time, it is clear that animal evolution is not simply a history of continual accumulation.

Repeatedly during the past 500 million years, animal diversity in the oceans has dropped both rapidly and substantially; waves of extinctions have also occurred on land. Known as **mass extinctions,** these events eliminated ecologically important taxa, thereby providing evolutionary opportunities for the survivors. In Chapter 21, we explored how species often radiate on islands where there is little or no competition or predation. For survivors, mass extinctions provided ecological opportunities on a grand scale.

The best-known mass extinction occurred 66 million years ago, at the end of the Cretaceous Period (rightmost drop in Fig. 22.23). On land, dinosaurs disappeared abruptly, following more than 150 million years of dominance in terrestrial ecosystems. In the oceans, cephalopod mollusks called ammonites, which had long been abundant predators, became extinct, and most skeleton-forming microorganisms in the oceans disappeared as well. As discussed in Chapter 1, a large body of geologic evidence supports the hypothesis that this biological catastrophe was caused by the impact of a giant meteorite. By eliminating dinosaurs and much more of the biological diversity that had built up over millions of years on land, the mass extinction at the end of the Cretaceous Period had the closely related

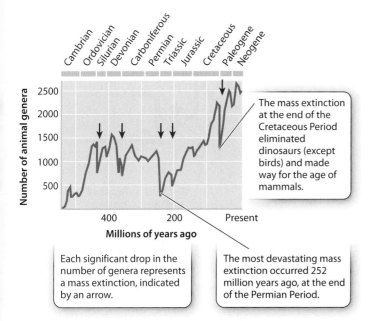

FIG. 22.23 Mass extinctions. Because they eliminate much of the life found on Earth at the time, mass extinctions allow new groups to proliferate and diversify. *Source: Sepkoski's Online Genus Database. http://strata.geology.wisc.edu/jack/.*

Each significant drop in the number of genera represents a mass extinction, indicated by an arrow.

The most devastating mass extinction occurred 252 million years ago, at the end of the Permian Period.

The mass extinction at the end of the Cretaceous Period eliminated dinosaurs (except birds) and made way for the age of mammals.

effect of generating new evolutionary possibilities for those terrestrial animals that survived the event, including mammals.

Before dinosaurs ever walked on Earth, the greatest of all mass extinctions occurred 252 million years ago, at the end of the Permian Period, when environmental catastrophe eliminated half of all families in the oceans and approximately 80% of all genera. Estimates of marine species losses run as high as 90%. Geologists have hypothesized that this mass extinction resulted from the catastrophic effects of massive volcanic eruptions. At the end of the Permian Period, most continents were gathered into the supercontinent Pangaea (see Fig. 22.20), and a huge ocean covered more than half of the Earth. Levels of oxygen in the deep waters of this ocean were low, as the result of sluggish circulation and warm seawater temperatures. Then, a massive outpouring of ash and lava—a million times larger than any volcanic eruption experienced by humans—erupted across what is now Siberia. Enormous emissions of carbon dioxide and methane from the volcanoes caused global warming (so even less oxygen reached deep oceans) and ocean acidification (making it difficult for animals and algae to secrete calcium carbonate skeletons).

The three-way insult of lack of oxygen, ocean acidification, and global warming doomed many species on land and in the seas. Seascapes dominated for 200 million years by corals and shelled invertebrates called brachiopods disappeared. As ecosystems recovered from this mass extinction, they came to be dominated by new groups descended from survivors of the extinction. Bivalve and gastropod mollusks diversified in the absence of their former competitors; new groups of arthropods radiated, including the ancestors of the crabs and shrimps we see today; and surviving sea anemones evolved a new capacity to make skeletons of calcium carbonate, resulting in the corals that build modern reefs. In short, mass extinction reset the course of evolution, much as it did and the end of the Cretaceous Period.

When you stroll through a zoo or snorkel above a coral reef, you are not simply seeing the products of natural selection played out over Earth history: you are encountering the descendants of Earth's biological survivors. Current biological diversity reflects the interplay through time of natural selection and rare massive perturbations to ecosystems on land and in the sea. The fossil record provides a silent witness to our planet's long and complex evolutionary history.

> **Self-Assessment Questions**
>
> 7. You have just found a novel vertebrate skeleton in 200-million-year-old rocks. How would you integrate this new fossil species into the phylogenetic tree depicted in Fig. 22.7?
> 8. Which of the following is most likely to have a good fossil record: earthworms, birds, corals, or tunicates (sea squirts)?
> 9. How can the fossil record be used to determine both the relative timescale and the absolute timescales of past events?
> 10. What is the significance of *Archaeopteryx* and *Tiktaalik*?
> 11. How have mass extinctions shaped the ecological landscape?

22.4 COMPARING EVOLUTION'S TWO GREAT PATTERNS

The diversity of life we see today is the result of evolutionary processes playing out over geologic time. Processes of evolution can be studied by experiment, both in the field and in the laboratory (Chapter 1), but experimentally exploring evolutionary history is another matter. There is no experiment we can do to determine why the dinosaurs became extinct—we cannot rerun the events of 66 million years ago, this time without the meteorite impact. Instead, the history of life must be reconstructed from evolution's two great patterns: the nested similarity observed in the forms and molecular sequences of living organisms, and the direct historical archive of the fossil record.

Phylogeny and fossils complement each other.

The great advantage of reconstructing evolutionary history from living organisms is that we can use a full range of features—skeletal morphology, cell structure, DNA sequence—to generate phylogenetic hypotheses. The disadvantage of using this kind of comparative biology is that we lack evidence of extinct species, any direct connection of phylogeny to time, and the environmental context of the species being studied. This is where the fossil record comes into play. Fossil evidence, in turn, has strengths and limitations that complement the evolutionary information in living organisms.

For example, we can use phylogenetic methods based on DNA sequences to infer that birds and crocodiles are closely related, but only fossils can show that the evolutionary link between birds and crocodiles runs through dinosaurs. Likewise, only the fossil record can show that mass extinction removed the dinosaurs, paving the way for the emergence of modern mammals. Paleontologists and biologists work together to understand evolutionary history. Biology provides a functional and phylogenetic framework for the interpretation of fossils, and fossils provide a record of life's history in the context of continually changing environments.

Agreement between phylogenies and the fossil record provides strong evidence of evolution.

Phylogenies based on morphological or molecular comparisons of living organisms make hypotheses about the timing of evolutionary changes through Earth history. We humans are a case in point. As we discuss in Chapter 23, comparisons of DNA sequences suggest that chimpanzees are our closest living relatives. This hypothesis is supported by observations that chimpanzees and humans share many features visible at a glance. However, among other differences, chimpanzees have smaller stature, smaller brains, long arms that facilitate knuckle-walking (the arms help support the body as the chimpanzee moves forward), a more prominent snout, and larger teeth.

How did these differences accumulate over the 7 million years or so since the human and chimpanzee lineages diverged? Fossils painstakingly unearthed over the past century show, for example, that the ability to walk upright evolved before humans evolved a larger-than-average brain. Lucy, a famous specimen of *Australopithecus afarensis* dating to approximately 3.2 million years ago, was fully bipedal but had a brain the size of a chimpanzee's. Fossils like Lucy that are a mix of ancestral (brain size) and derived (bipedalism) characters provide strong support for our phylogenetic tree that shows humans and chimpanzees as sister taxa.

The agreement of comparative biology and the fossil record can be seen at all scales of phylogeny. Humans form one tip of a larger branch that contains all members of the primate family. Primates, in turn, are nested within a larger branch occupied by mammals, and mammals are nested within a still larger branch containing all vertebrate animals, which include fish. This arrangement predicts that the earliest fossil fishes should be older than the earliest fossil mammals, the earliest fossil mammals older than the earliest primates, and the earliest primates older than the earliest humans. This is precisely what the fossil record shows (**Fig. 22.24**).

The agreement between fossils and phylogenies can be seen again and again when we examine different branches of the tree

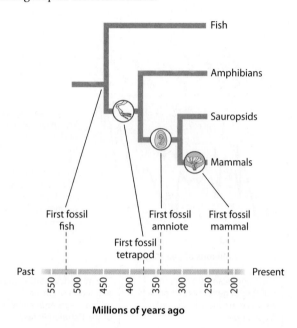

FIG. 22.24 Phylogeny and fossils of vertebrates. The branching order of the phylogeny corresponds to the order of appearance of each group in the fossil record.

of life or, for that matter, the tree as a whole. All phylogenies indicate that microorganisms diverged early in evolutionary history, whereas mammals, flowering plants, and other large complex organisms diverged more recently. The tree's shape implies that diversity has accumulated through time, beginning with simple organisms and later adding complex macroscopic forms.

The geologic record shows the same pattern. The vastness of the timescale may be daunting, but the message is simple: fossils show that life was, indeed, microbial for its first 3 billion years. Only at about 580 million years ago did the first animals arise, kicking off the long evolutionary journey that eventually resulted in our own species (see Fig. 22.18; Chapter 23).

In Part 2 of this book, we explore the evolutionary history of life in some detail. Here, it is sufficient to draw a key general conclusion: the fact that comparative biology and fossils, two complementary but independent approaches to reconstructing the evolutionary past, yield the same history is powerful evidence of evolution.

Self-Assessment Question

12. A popular creationist criticism of evolution is that the theory of evolution is impossible to falsify (that is, to prove wrong). Using information on phylogeny and the fossil record, imagine a discovery that would indeed falsify the theory.

CORE CONCEPTS SUMMARY

22.1 READING A PHYLOGENETIC TREE: A phylogenetic tree is a hypothesis of the evolutionary relationships among organisms.

The nested pattern of similarities seen among organisms is a result of descent with modification and can be represented as a phylogenetic tree. page 468

All organisms living today, regardless of their placement on a phylogenetic tree, have evolved for the same amount of time from the most recent common ancestor of all organisms, so none is more or less "evolved" than the others. page 469

Sister groups are more closely related to one another than they are to any other group. page 469

A node is a branching point on a phylogenetic tree, and it can be rotated without changing the interpretation of evolutionary relationships. page 469

Evolutionary relatedness is determined by following the sequence of nodes on a phylogenetic tree over time, rather than by assessing the order of the tips across a page. page 470

A monophyletic group includes all the descendants of a common ancestor. page 470

A paraphyletic group includes some, but not all, of the descendants of a common ancestor. page 470

A polyphyletic group includes organisms from distinct groups based on shared, convergent characters, and does not include a common ancestor. page 470

Organisms are classified into domain, kingdom, phylum, class, order, family, genus, and species. page 471

22.2 BUILDING A PHYLOGENETIC TREE: A phylogenetic tree is built on the basis of shared derived characters.

Characters, or traits, existing in different states are used to build phylogenetic trees. page 472

Homologies are similarities based on shared ancestry, while analogies are similarities based on independent convergently evolved adaptations. page 472

Homologies can be ancestral, unique to a particular group, or present in some, but not all, of the descendants of a common ancestor (shared derived characters). page 472

Only shared derived characters, or synapomorphies, are useful in constructing a phylogenetic tree. page 473

Molecular data, such as DNA sequences, provide a wealth of characters that complement other types of information in building phylogenetic trees. page 475

Phylogenetic trees can be used to understand evolutionary relationships of organisms and solve practical problems, such as how viruses evolve over time. page 478

22.3 THE FOSSIL RECORD: The fossil record provides direct evidence of evolutionary history.

Fossils are the remains of organisms preserved in sedimentary rocks. page 479

The fossil record is imperfect because fossilization requires burial in sediment, sediments accumulate episodically and discontinuously, and fossils typically preserve only the hard parts of organisms. page 479

Radioactive decay of unstable isotopes of elements provides a means of dating rocks. page 483

Transitional fossils, such as *Archaeopteryx* and *Tiktaalik*, help us understand important steps in evolution, including the bird–dinosaur transition and the fish–tetrapod transition, respectively. page 484

The history of life on Earth is characterized by five mass extinctions that changed the course of evolution by eliminating some groups and providing opportunities for other groups to diversify. page 486

Mass extinction at the end of the Cretaceous Period 66 million years ago included the extinction of the dinosaurs (other than birds). page 486

The extinction at the end of the Permian Period 252 million years ago is the largest documented mass extinction in the history of Earth. page 487

22.4 COMPARING EVOLUTION'S TWO GREAT PATTERNS: Phylogeny and fossils provide independent and corroborating evidence of evolution.

Phylogeny makes use of living organisms, while the fossil record supplies a record of species that no longer exist, absolute dates, and environmental context. page 487

Data from phylogeny and fossils are often in agreement, providing strong evidence for evolution. page 488

Log in to **LaunchPad** to check your answers to the Self-Assessment Questions and to access additional learning tools.

CHAPTER 23 Human Origins and Evolution

CORE CONCEPTS

23.1 THE GREAT APES: Anatomical, molecular, and fossil evidence shows that the human lineage branches off the great apes tree.

23.2 AFRICAN ORIGINS: According to phylogenetic analysis of mitochondrial and the Y chromosome DNA sequences, our species arose in Africa.

23.3 HUMAN TRAITS: The human lineage acquired a number of distinctive features in the 5–7 million years since the most recent common ancestor of humans and chimpanzees.

23.4 HUMAN GENETIC VARIATION: Human history has had an important impact on patterns of genetic variation in our species.

23.5 CULTURE, LANGUAGE, AND CONSCIOUSNESS: Culture, language, and consciousness are developed to a remarkable degree in humans.

Charles Darwin carefully avoided discussing the evolution of our own species in *On the Origin of Species.* He wrote only that he saw "open fields for far more important researches," and that "light will be thrown on the origin of man and his history." Darwin, an instinctively cautious man, realized that the ideas presented in *The Origin* were controversial enough without adding a discussion of humans into the mix. He presented his ideas on human evolution to the public only when he published *The Descent of Man* 12 years later, in 1871.

As it turned out, Darwin's delicate sidestepping of human origins had little effect. The initial print run of *The Origin* sold out on the day of publication, and the public was perfectly capable of reading between the lines. The Victorians found themselves wrestling with the book's revolutionary message: that humans are a species of ape.

Darwin's conclusions remain controversial to this day among some members of the general public, but they are not controversial among scientists. The evidence that humans are descended from a line of apes whose modern-day representatives include gorillas and chimpanzees is compelling. We know now that approximately 5–7 million years ago the family tree of the great apes split, with one branch ultimately giving rise to chimpanzees and the other to our species. Those 5–7 million years hold the key to our humanity. It was over this period—a brief span by evolutionary standards—that the attributes that make our species so remarkable arose. This chapter discusses what happened over those 5–7 million years and how humans came to be the way we are.

23.1 THE GREAT APES

We can approach the question of humans' place in the tree of life in three different ways: through comparative anatomy, through molecular analysis, and through the fossil record. In this section, we refer to data from all three sources as we apply the standard methods of phylogenetic reconstruction (Chapter 22) to figure out the evolutionary relationships between humans and other mammals.

Comparative anatomy shows that the human lineage branches off the great apes tree.

The approximately 400 species of **primates** include prosimians (lemurs, bushbabies), monkeys, and apes (**Fig. 23.1**). All primates share a number of general anatomical features that distinguish primates from other mammals: nails rather than claws, together with a versatile thumb, allow objects to be manipulated more dexterously, and eyes on the front of the face instead of the side allow stereoscopic (three-dimensional) vision.

Prosimians are thought to represent a separate primate group from the one that gave rise to humans. Lemurs, which today are found only on the island of Madagascar, are therefore only distantly related to humans. Monkeys underwent independent bouts of evolutionary change in the Americas and in Africa and Eurasia, so their family tree is split along geographic lines into New World and Old World monkeys. Although both New World and Old World groups have evolved similar habits, there are basic distinctions. Notably, the teeth of the two groups differ. Also, in New World monkeys the nostrils tend to be widely spaced, whereas in Old World species they are closer together.

FIG. 23.1 The primate family tree. This tree shows the evolutionary relationships of prosimians, monkeys, and apes. *Photo sources: (left to right) George Holton/Science Source; Penelope Dearman/Getty Images; Yellow Dog Productions/Getty Images; Patrick Shyu/Getty Images; Cavan Pawson/ANL/REX/Shutterstock.*

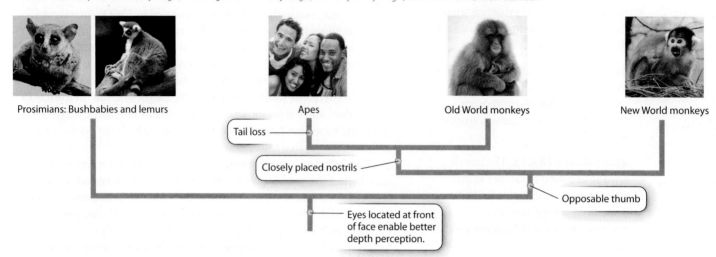

One line of Old World monkeys gave rise to the apes, which lack a tail and show more complex behaviors than other monkeys. The apes are split into two groups, the lesser apes and the great apes (**Fig. 23.2**). Lesser apes include the 14 species of gibbon, all of which are found in Southeast Asia. The great apes include orangutans (Southeast Asia), gorillas (Africa), chimpanzees (Africa), and humans. Taxonomists classify all the descendants of a specified common ancestor as belonging

FIG. 23.2 The family tree of the lesser and great apes. The apes consist of two major groups, the lesser apes and the great apes. The great apes group includes humans. *Sources: (left to right) Zoonar/K. Jorgensen/AGE Fotostock; S Sailer/A Sailer/AGE Fotostock; J & C Sohns/AGE Fotostock; Yellow Dog Productions/Getty Images; Robert HENNO/Alamy; FLPA/Jurgen & Christi/AGE Fotostock.*

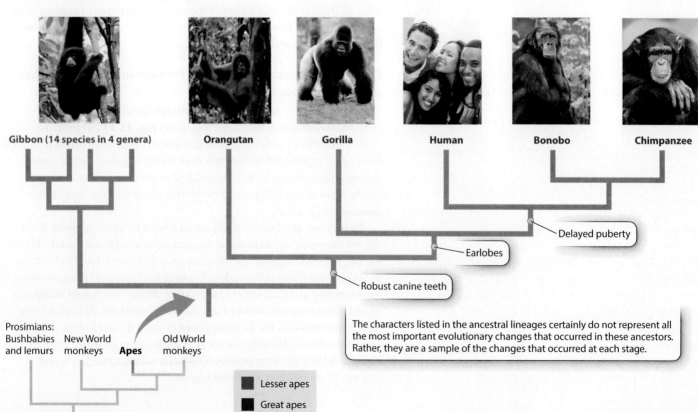

to a monophyletic group (Chapter 22). Thus, humans, blue whales, and hedgehogs are all mammals because all three are descended from the first mammal, the original common ancestor of all mammals. Because human ancestry can be traced to the common ancestor of orangutans, gorillas, and chimpanzees, humans, too, are members of the monophyletic great apes group.

Molecular analysis reveals that the human lineage split from the chimpanzee lineage about 5–7 million years ago.

Which great ape is most closely related to humans? That is, which is the sister group of humans? Traditional approaches of reconstructing evolutionary history by comparing anatomical features failed to determine which of two candidates, gorillas or chimpanzees, is the sister group to humans. It was only with the introduction of molecular methods of assessing evolutionary relationships—through the comparison of DNA and amino acid sequences from the different species—that the answer was determined. The closest relative to humans is the chimpanzee (Fig. 23.2), or, more accurately, the chimpanzees, plural, because there are two closely related chimpanzee species, the smaller of which is often called the bonobo.

Just how closely are humans and chimpanzees related? To answer this question, we need to know the timing of the evolutionary split that led to chimpanzees along one fork and to humans along the other. As discussed in Chapter 20, DNA sequence differences accumulate between isolated populations or species, and they do so at a more or less constant rate. As a result, the extent of sequence difference between two species is a good indication of the amount of time they have been separate, meaning the amount of time since their last common ancestor.

Mary-Claire King and Allan Wilson at the University of California at Berkeley carried out the first thorough comparison of DNA molecules between humans and chimpanzees before the advent of DNA sequencing methods (**Fig. 23.3**).

HOW DO WE KNOW?

FIG. 23.3
How closely related are humans and chimpanzees?

Perfect complementarity: 95°C denaturation

Some mismatch: 93°C denaturation

More mismatch: 91°C denaturation

BACKGROUND Phylogenetic analysis based on anatomical characteristics had established that chimpanzees are closely related to humans. In 1975, Mary-Claire King and Allan Wilson used molecular techniques to determine *how* closely related the two species are.

HYPOTHESIS Despite the marked anatomical and behavioral differences between chimpanzees and humans, the genetic distance between the two species is small, implying a relatively recent common ancestor.

METHOD When two complementary strands of DNA are heated, the hydrogen bonds between the two DNA strands break at approximately 95°C and the double helix denatures, or separates (Chapter 12). Complementary strands that have some mismatches separate at a temperature slightly lower than 95°C because fewer hydrogen bonds are holding the helix together. More mismatches between the two sequences results in an even lower denaturation temperature. By comparing hybrid DNA double helices with one strand contributed by each species—humans and chimpanzees, in this case—and determining their denaturation temperature, King and Wilson could infer the genetic distance (the extent of genetic divergence) between the two species.

RESULTS King and Wilson found that human–chimpanzee DNA molecules separated at a temperature approximately 1°C lower than the temperature at which human–human DNA molecules separate. They were able to calibrate this difference on the basis of studies of other species whose genetic distances were known from other methods. The DNA of humans differs from that of chimpanzees by approximately 1%.

CONCLUSION AND INTERPRETATION King and Wilson noted the discrepancy between the extent of genetic divergence (small) and the extent of anatomical and behavioral divergence (large) between humans and chimpanzees. They suggested that one way in which relatively little genetic change could produce extensive phenotypic change is through differences in gene regulation (Chapter 18). A small genetic change in a control region responsible for switching a gene on and off might have major consequences for the organism.

FOLLOW-UP WORK The genome sequences of the chimpanzee (completed in 2005) and the human (completed in 2003) allow us to compare the two sequences directly, and these data confirm King and Wilson's observations.

SOURCE King, M.-C., and A. C. Wilson. 1975. "Evolution at Two Levels in Humans and Chimpanzees." *Science* 188:107–116.

One of their methods of measuring molecular differences between species relied on DNA–DNA hybridization (Chapter 12). Two complementary strands of DNA in a double helix can be separated by heating the sample. If the two strands are not perfectly complementary, as occurs in the case of a base-pair mismatch (for example, a G paired with a T rather than with a C), less heat is required to separate the strands. King and Wilson used this observation to examine the differences between a human strand and the corresponding chimpanzee strand. They inferred the extent of DNA sequence divergence from the melting temperature and made a striking discovery: human DNA and chimpanzee DNA differ in sequence by just 1%.

King and Wilson's conclusions have proved robust as ever more powerful techniques have been applied to the comparison of human and chimpanzee DNA. Today, after the entire human and chimpanzee genomes have been sequenced and the differences between the two sequences have literally been counted, the figure of 1% still approximately holds. Genome sequencing studies give us superbly detailed information on the differences and similarities between the two genomes, revealing regions that are present in one species but not in the other, and which parts of the genomes have evolved more rapidly than others.

Because the amount of sequence difference is correlated with the length of time the two species have been isolated, we can estimate that the split between the human and chimpanzee lineages occurred approximately 5–7 million years ago. All of the extraordinary characteristics that set our species apart from the rest of the natural world—those attributes that are ours and ours alone—arose in just this short span of time.

The fossil record gives us direct information about our evolutionary history.

Molecular analysis is a powerful tool for comparing species and populations within species. It allows us to compare humans and chimpanzees and the differences among groups of humans or groups of chimpanzees. However, for a full picture of human evolution, we must turn to fossils (Chapter 22).

For the first several million years of human evolution, all the fossils from the human lineage are found in Africa. This fact is not surprising. Darwin himself noted that it was likely that the human lineage originated there, as humans' two closest relatives, chimpanzees and gorillas, live only in Africa. Fossil material varies in quality, and a great deal of ingenuity is often required to reconstruct the appearance and attributes of an individual from fragmentary fossil material. It is difficult to ascertain whether two fossil specimens with slight differences belong to the same or different species. And, of course, the ability to interbreed, the criterion that researchers typically use to define a species (Chapter 21), cannot be determined from fossils. As a result, experts often disagree over the details of the human fossil record, such as whether a specimen belongs to a particular

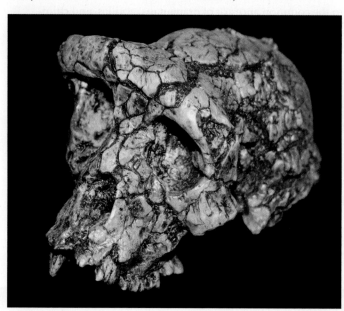

FIG. 23.4 The skull of *Sahelanthropus tchadensis*. This skull, which is about 7 million years old and was found in Chad, has both human and chimpanzee features. *Source: Sabena Jane Blackbird/Alamy.*

species. Despite these areas of uncertainty, we can draw several general conclusions.

The many different species, including modern humans, that have arisen on the human side of the split since the human/chimpanzee common ancestor are called **hominins**. The earliest known hominin is *Sahelanthropus tchadensis* (**Fig. 23.4**). Discovered in Chad in 2002, the skull of *S. tchadensis* combines both modern (human) and ancestral features. Dated to approximately 7 million years ago, this species had a chimpanzee-sized brain but hominin-type brow ridges. *S. tchadensis* probably lived shortly after the split between the hominin and chimpanzee lineages, although this conclusion is controversial because of difficulties in interpreting whether a fossil is part of a newly diverged lineage or just another specimen of the ancestral lineage.

Another important early hominin, dating from about 4.4 million years ago, is *Ardipithecus ramidus* from Ethiopia. This individual, known as **Ardi**, was capable of walking upright, using two legs on the ground but all four limbs in the trees.

An unusually complete early hominin fossil dating from around 3.2 million years ago is **Lucy**. She was found in 1974 at Hadar, Ethiopia, and was fully **bipedal**, habitually walking upright (**Fig. 23.5**). The fossil's name comes from the Beatles' song "Lucy in the Sky with Diamonds," which was playing in the paleontologists' field camp when she was unearthed. Lucy was a member of the species *Australopithecus afarensis* and was much smaller than modern humans, less than 4 feet tall, and

FIG. 23.5 Lucy, a specimen of *Australopithecus afarensis*. Lucy was fully bipedal, establishing that the ability to walk upright evolved at least 3.2 million years ago. *Source: The Natural History Museum, London/The Image Works.*

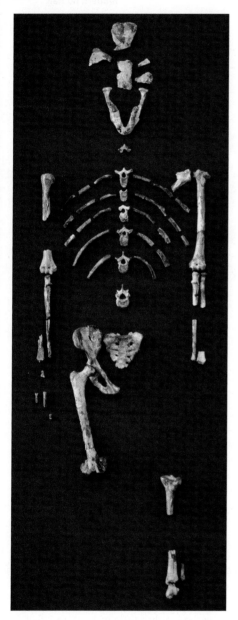

FIG. 23.6 Hominin lineages. Three main lineages, *Ardipithecus*, *Australopithecus*, and *Homo*, existed at various and sometimes overlapping times in history. *Sources: Data from R. G. Klein, 2009, The Human Career, Chicago: University of Chicago Press, p. 234; Homo floresiensis skull from R. D. Martin, A. M. MacLarnon, J. L. Phillips, L. Dussubieux, P. R. Williams, and W. B. Dobyns, 2006, Comment on "The Brain of LB1, Homo floresiensis," Science 312:999.*

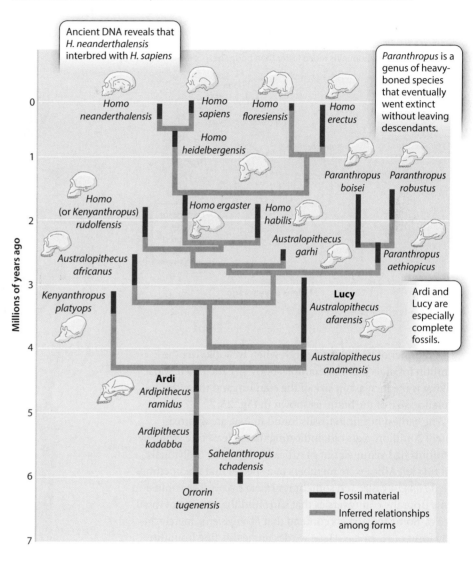

had a considerably smaller brain (even when the difference in body size is taken into consideration). In many ways, then, Lucy was similar to the common ancestor of humans and chimpanzees, except that she was bipedal.

Remarkably, many different hominin species were living in Africa at the same time. **Fig. 23.6** shows the main hominin species and the longevity of each in the fossil record, along with suggested evolutionary relationships among the species. Note that all hominins have a common ancestor, but not all of these groups lead to modern humans. The *Homo* lineage ultimately produced modern humans, *Homo sapiens*. Other branches of the hominin tree ultimately went extinct.

FIG. 23.7 Increase in brain size (as inferred from fossil cranium volume) over 3 million years. *Data from T. Deacon, "The Human Brain," pp. 115–123, in J. Jones, R. Martin, and D. Pilbeam (eds.), 1992, The Cambridge Encyclopedia of Human Evolution, Cambridge, UK: Cambridge University Press.*

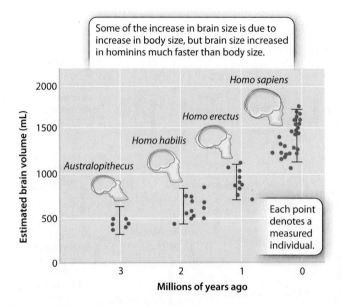

FIG. 23.8 Skeletons of *Homo neanderthalensis* and *Homo sapiens*. Neanderthals disappeared from Europe about 30,000 years ago, approximately 10,000 years after a group of modern humans called Cro-Magnon first appeared in Europe. *Source: Universal Images Group North America LLC/Alamy.*

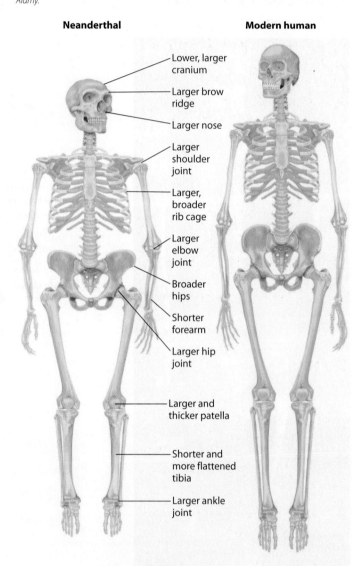

A number of trends can be seen when we look over the hominin fossil record. For example, body size increased. Most striking is the increase in size of the cranium and therefore, by inference, size of the brain, as shown in **Fig. 23.7**.

The earliest hominin fossils found in Asia are approximately 2 million years old, indicating that at least one group of hominins had ventured out of Africa by then. The individuals that first left Africa were members of a species that is sometimes called *Homo ergaster* and sometimes *Homo erectus*. The confusion stems from the controversies that surround the naming of fossil species. Some researchers contend that *H. ergaster* is merely an early form of *H. erectus*. Here, we designate this first hominin *Homo ergaster* and a later descendant species *H. erectus* (see Fig. 23.6). The naming details, however, are relatively unimportant. What matters is that some hominins first colonized areas beyond Africa about 2 million years ago.

Another hominin species was *Homo neanderthalensis*, whose fossils appear in Europe and the Middle East. **Neanderthals** represent a second hominin exodus from Africa dating from around 600,000 years ago (**Fig. 23.8**). Thicker boned than modern humans, and with flatter heads that contained brains about the same size as, or slightly larger than, ours, Neanderthals disappeared around 30,000 years ago. As we will see, genetic analysis suggests that this group likely interbred with our own group, *Homo sapiens*, so this disappearance was perhaps not as complete as the fossil record suggests.

Another hominin species became extinct only about 12,000 years ago: *H. floresiensis*, known popularly as the Hobbit. *H. floresiensis* is peculiar. Limited to the Indonesian island of Flores, adults of this species were only a little more than 3 feet tall. Some experts have suggested that *H. floresiensis* is not a genuinely distinct species, but rather an aberrant *H. sapiens*. Plenty of morphological evidence, however, suggests that *H. floresiensis* is a distinct species derived from an archaic *Homo* species, probably *H. erectus* (**Fig. 23.9**). Mammals often evolve small body size on islands because of the limited availability of food.

FIG. 23.9 Phylogenetic tree of recent hominins. When modern humans departed Africa about 60,000 years ago, they encountered and interbred with Neanderthals and Denisovans, groups derived from an earlier African exodus. *Photo sources (left to right): Fuse/Getty Images; INTERFOTO/Alamy; Sylvain Entressangle & Elisabeth Daynes/Science Source; Sebastien Plailly/Science Source; Sylvain Entressangle & Elisabeth Daynes/Science Source; Aila Images/Shutterstock.*

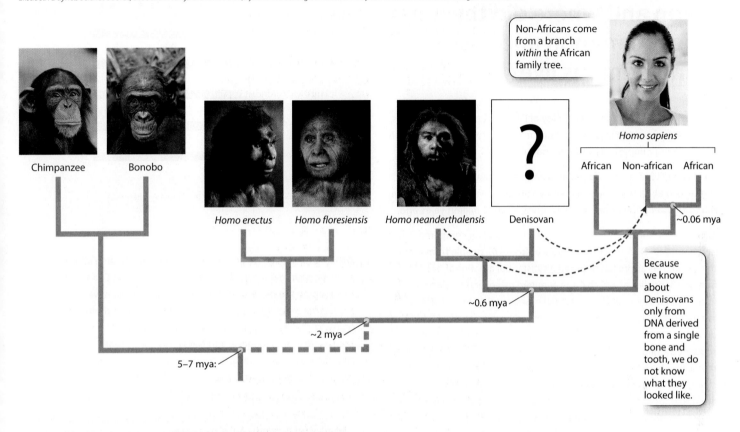

Self-Assessment Questions

1. Did humans evolve from chimpanzees? Explain.
2. Describe the evidence suggesting that chimpanzees are the closest living relatives of humans.
3. What can we learn about human evolution from the hominin fossil record that we *cannot* learn from analyses of living species?

23.2 AFRICAN ORIGINS

From the wealth of fossil evidence, it's clear that modern humans first evolved in Africa. But for a while the timing of this event was unclear. Can all modern humans, *Homo sapiens*, date their ancestry to early hominins in Africa that migrated out and spread around the world about 2 million years ago? Or did the groups that left Africa ultimately go extinct, so that modern humans evolved from a group of hominins in Africa that migrated out much more recently? Molecular studies helped to answer this question.

Studies of mitochondrial DNA reveal that modern humans evolved in Africa relatively recently.

For a long time, it was argued that modern humans derive from the *Homo ergaster* (or *H. erectus*) populations that spread around the world starting around the time of the early migration out of Africa 2 million years ago. This idea is called the **multiregional hypothesis** of human origins because it implies that different *Homo ergaster* populations throughout Africa and Eurasia evolved in parallel, with some limited gene flow among them, each producing modern *H. sapiens* populations. In short, the multiregional hypothesis asserts that racial differences among humans would have evolved over 2 million years in different geographic locations.

This idea was overturned in 1987 by a study that suggested instead that modern humans arose much more recently from *Homo ergaster* descendants that remained in Africa (sometimes called *Homo heidelbergensis*; see Fig 23.6). According to this hypothesis, all modern humans are descended from an African common ancestor dating from approximately 200,000 years ago. This newer idea is called the **out-of-Africa hypothesis** of human origins.

HOW DO WE KNOW?

FIG. 23.10

When and where did the most recent common ancestor of all living humans live?

BACKGROUND With the availability of DNA sequence analysis tools, it became possible to investigate human prehistory by comparing the DNA of living people from different populations.

HYPOTHESIS The multiregional hypothesis of human origins suggests that our most recent common ancestor was living at the time *Homo ergaster* populations first left Africa, about 2 million years ago.

METHOD In the late 1980s, American Rebecca Cann compared mitochondrial DNA (mtDNA) sequences from 147 people from around the world. This approach required a substantial amount of mtDNA from each individual, which Cann acquired by collecting placentas from women after childbirth. Instead of sequencing the mtDNA, Cann inferred differences among sequences by digesting the mtDNA with different restriction enzymes, each of which cuts DNA at a specific sequence (Chapter 12). If the specific sequence is present, the enzyme cuts. If any base in the recognition site of the restriction enzyme has changed in an individual and the specific sequence is *not* present, the enzyme does not cut. The resulting fragments were then separated by gel electrophoresis. By using 12 different enzymes, Cann was able to assay a reasonable proportion of all the mtDNA sequence variation present in each sample.

Fig. 23.10a is a sample dataset for four people and a chimpanzee. There are seven varying restriction sites.

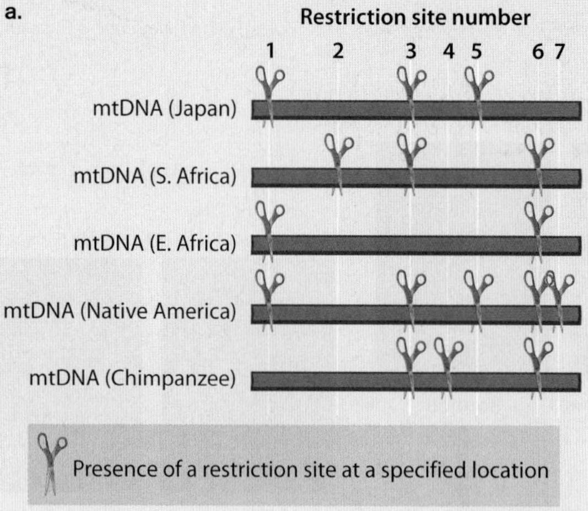

ANALYSIS Cann converted the data into a family tree (Fig. 23.10b) by using shared derived characters, as described in Chapter 22. By mapping the pattern of changes in this way, Cann was able to reconstruct the phylogenetic relationships of the mtDNA sequences. For example, the derived "cut" state at restriction site 1 shared by the East African, Japanese, and Native American sequences implies that the three groups have a common ancestor in which a mutation created the new restriction site.

RESULTS AND CONCLUSION A simplified version of Cann's phylogenetic tree based on mtDNA is shown in Fig. 23.10c. Modern humans arose relatively recently in Africa, and their most common ancestor lived about 200,000 years ago. The multiregional hypothesis is rejected. Also, Cann's data revealed that all modern humans who lived or live outside of Africa derive from within the African family

To test these two hypotheses about human origins, Rebecca Cann, a student in Allan Wilson's laboratory, analyzed DNA sequences to reconstruct the human family tree. Specifically, she studied sequences of a segment of mitochondrial DNA from people living around the world (**Fig. 23.10**).

Mitochondrial DNA (mtDNA) is a small circle of DNA, approximately 17,000 base pairs long, that is found in every mitochondrion (Chapter 16). Cann chose to study mtDNA for several reasons. Although a typical cell contains a single nucleus with just two copies of nuclear DNA, each cell has many mitochondria, each carrying multiple copies of mtDNA; thus, mtDNA is much more abundant than nuclear DNA and, therefore, easier to extract. More important, however, is mtDNA's mode of inheritance. All of a person's mtDNA is inherited from their mother as part of the egg she produces because sperm do not contribute mitochondria to the zygote. As a consequence, there is no opportunity for genetic recombination between different mtDNA molecules and the only way in which sequence variation can arise is through mutation. In nuclear DNA, by contrast, differences between two sequences can be introduced by both mutation and recombination (Chapter 20). Recombination obscures genealogical relationships because it mixes segments of DNA with different evolutionary histories.

Mitochondrial DNA sequences offered Cann a clear advantage. With sequence information from the mtDNA of 147 people from around the world, Cann reconstructed the human family tree (Fig. 23.10). The tree contained two major surprises. First, the two deepest branches—the ones that come off the tree earliest in time—are African. All non-Africans are branches off

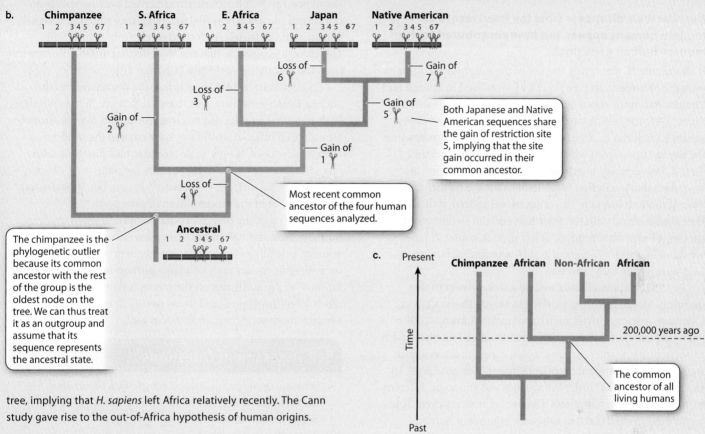

tree, implying that *H. sapiens* left Africa relatively recently. The Cann study gave rise to the out-of-Africa hypothesis of human origins.

FOLLOW-UP WORK This study set the stage for an explosion in genetic studies of human prehistory that used the same approach of comparing sequences so as to identify migration patterns. Today, with molecular tools vastly more powerful than those available to Cann, studies of the distribution of genetic variation are showing us a detailed pattern, even on relatively local scales, of demographic events.

SOURCE Cann, R. L., M. Stoneking, and A. C. Wilson. 1987. "Mitochondrial DNA and Human Evolution." *Nature* 325:31–36.

the African tree. The implication is that *Homo sapiens* evolved in Africa; only afterward did populations migrate out of Africa and become established elsewhere. This finding contradicted the expectations of the multiregional theory, which suggested that *Homo sapiens* evolved independently in different locations throughout the Old World.

Second, the tree is remarkably shallow. That is, even the most distantly related modern humans have a relatively recent common ancestor. By calibrating the rate at which mutations occur in mtDNA, Cann was able to estimate the time back to the common ancestor of all modern humans as about 200,000 years. Subsequent analyses have somewhat refined this estimate, but the key message has not changed: the data contradict the multiregional hypothesis, which predicted a number closer to 2 million years.

Studies of the Y chromosome provide independent evidence for a recent origin of modern humans.

Given that 200,000 years ago is 10 times more recent than 2 million years ago, this updated estimate represents a revolution in our understanding of human prehistory. Perhaps it is risky to make such major claims on the basis of a single study; maybe there's something peculiar about mtDNA, and it doesn't offer an accurate picture of the human past.

To find independent evidence for a recent origin of modern humans in Africa, another dataset was required. One such dataset comes from studies of the Y chromosome, another segment of human DNA that does not undergo recombination (Chapter 16). When an approach similar to that employed by Cann was used to reconstruct the human family tree using Y chromosome DNA sequences, the result was completely in agreement with the mtDNA result: the human family is young, and it arose in Africa.

Neanderthals disappear from the fossil record as modern humans appear, but have contributed to the modern human gene pool.

If we exclude *H. floresiensis,* the last of the nonmodern humans were the Neanderthals (see Fig. 23.9), who lived in Europe and Western Asia until about 30,000 years ago. What happened to them? Were they eliminated by the first population of *Homo sapiens* to arrive in Europe (a group named **Cro-Magnon** after the site in France from which specimens were first described)? Or did Neanderthals interbreed with the Cro-Magnons, and, if so, does that mean that Neanderthals should be included among modern humans' direct ancestors? Indeed, if modern humans interbred with Neanderthals and our definition of species is based on reproductive isolation (Chapter 21), perhaps we should consider *H. sapiens* and *H. neanderthalensis* as belonging to the same species.

In 1997, it appeared that we had a clear answer to this question. Matthias Krings, working in Svante Pääbo's lab in Germany, extracted intact stretches of mtDNA from 30,000-year-old Neanderthal bone. If Neanderthals had interbred with modern humans' ancestors, we would expect to see evidence in the form of Neanderthal mtDNA in our own gene pool. In fact, the Neanderthal mtDNA sequence was strikingly different from that of modern humans. That we see nothing even close to the Neanderthal mtDNA sequence in modern humans strongly suggests that Neanderthals did not interbreed with our ancestors.

This conclusion was subsequently reversed following remarkable technological advances that permitted Pääbo and others to sequence the entire Neanderthal genome (rather than just Neanderthal mtDNA). Careful population genetic analysis reveals that the ancestors of modern humans did interbreed with Neanderthals, and that 1% to 4% of the genome of every non-African is derived from that species.

A new scenario emerged to explain this discovery: as populations of our ancestors headed from Africa into the Middle East, they encountered Neanderthals, and it was then, before modern humans had spread out across the world, that interbreeding took place. Remarkably, ancient DNA analysis has further revealed that early non-African *Homo sapiens* did not interbreed only with Neanderthals. Another group of Eurasian hominins, relatives of Neanderthals known as Denisovans after the Siberian cave in which their remains were found, also contributed genetically to some modern human populations, especially indigenous people from Australasia (see Fig 23.9).

How can we reconcile the two apparently contradictory results about the contributions of Neanderthals to the modern human gene pool? The mtDNA study of Neanderthals indicates there was no genetic input from Neanderthals in modern humans, but the whole-genome study suggests that there was.

One possible solution lies in the difference in patterns of transmission between mtDNA and genomic material (Chapter 16). Recall that mtDNA is maternally inherited because sperm do not contribute mitochondrial material. Your mtDNA, whether you are male or female, is derived solely from your mother. Therefore, it is possible that the Neanderthal mtDNA lineage has been lost through genetic drift (Chapter 20).

An alternative explanation is that the discrepancy in the ancient DNA stems from a sex-based difference in interbreeding. If Neanderthal females did not interbreed with our ancestors, Neanderthal mtDNA would not have entered the modern human gene pool. If only male Neanderthals interbred with our ancestors, we would expect exactly the pattern we observe: no Neanderthal mtDNA in modern humans, but Neanderthal genomic DNA in the modern human gene pool.

Regardless of the details of what happened when modern humans, Neanderthals, and Denisovans encountered one another, one thing is clear: genomic sequencing demands that we reimagine the ancestry of a large portion of the human population. Every non-African on the planet is part Neanderthal, albeit a very small part, and many non-Africans can count Denisovans among their ancestors as well.

> **Self-Assessment Questions**
>
> 4. What are the multiregional and out-of-Africa hypotheses, and what are their respective predictions about the common ancestor of all humans that are living today?
> 5. How do studies of mitochondrial DNA and the *Y* chromosome support the out-of-Africa hypothesis of human origins?
> 6. How are genetic data consistent with the possibility that only male Neanderthals interbred with our ancestors?

23.3 HUMAN TRAITS

Many extraordinary changes in anatomy and behavior have occurred in the 5–7 million years since our human lineage split from the lineage that gave rise to the chimpanzees. Fossils tell us a great deal about those changes, especially high-quality material such as Lucy and Ardi. In general, though, hominin fossils are hard to come by and there is a lot of speculation. Speculation is especially rampant when we try to explain the reasons behind the evolution of a particular trait. Why, for example, did language evolve? It is easy enough to think of a scenario in which natural selection favors some ability to communicate—maybe language arose to facilitate group hunting. Plenty of plausible ideas on the subject have been proposed, but, in most cases, no evidence is available to support them. Thus, it is impossible to declare a winner among competing hypotheses. Nevertheless, we can be confident that the events that produced language occurred in Africa. Moreover, through paleontological studies of past environments, we can conclude that humans evolved in an environment similar in many ways to today's East African savanna.

Bipedalism was a key innovation.

The shift from walking on four legs to walking on two legs was probably one of the first changes in our lineage. Many primates are partially bipedal—chimpanzees, for example, may be observed knuckle-walking—but humans are the only living species that is wholly bipedal. Lucy, dating from about 3.2 million years ago, was already bipedal, and, 4.4 million years ago, Ardi was partially bipedal. We can further refine our estimate of when full bipedalism arose from the evidence of a trace fossil. A set of 3.5-million-year-old fossil footprints discovered in Laetoli, Tanzania, reveal a truly upright posture.

Becoming bipedal is not simply a matter of standing up on hind legs. The change required substantial shifts in a number of basic anatomical characteristics, described in **Fig. 23.11**.

Why did hominins become bipedal? Maybe it gave them access to berries and nuts located high on a bush. Maybe it allowed them to scan the vicinity for predators. Maybe it made long-distance travel easier. Whatever the reason for its emergence, bipedalism freed our ancestors' hands. No longer did they need their hands for locomotion, so for the first time there arose the possibility that specialized hand function could evolve. Most primates have some kind of opposable thumb, but the human version is much more dexterous. The human thumb has three muscles that are not present in the thumb of chimpanzees, which allow much finer motor control of the thumb. Tool use, which is present but crude in chimpanzees, can be much more subtle and sophisticated with a human hand.

Exactly when sophisticated tool use arose is controversial, but it is indisputable that bipedalism contributed to its development. Similarly, the ability to manipulate food with the hands, and to carry material using the hands rather than the mouth, likely permitted the evolution of the human jaw—indeed, the entire facial structure—in such a way that language became a possibility.

Adult humans share many features with juvenile chimpanzees.

King and Wilson's 1975 discovery that the DNA of humans and of chimpanzees are 99% identical has recently been confirmed by DNA sequence comparison of the human and chimpanzee genomes. It turns out that the two genomes are extraordinarily similar: all our genes are also present in the chimpanzee, and gene sequences in humans and in chimpanzees are extremely similar, suggesting that the functions of the proteins our genes code for are the same. If humans and chimpanzees are so similar, how can we account for the extensive differences between the two species?

In their original paper, King and Wilson suggested that much of the most significant evolution along the hominin lineage came about through changes in the regulation of genes (Chapter

FIG. 23.11 The shift from four to two legs. Anatomical changes were required for the shift to bipedalism, especially in the structure of the skull, spine, legs, and feet.

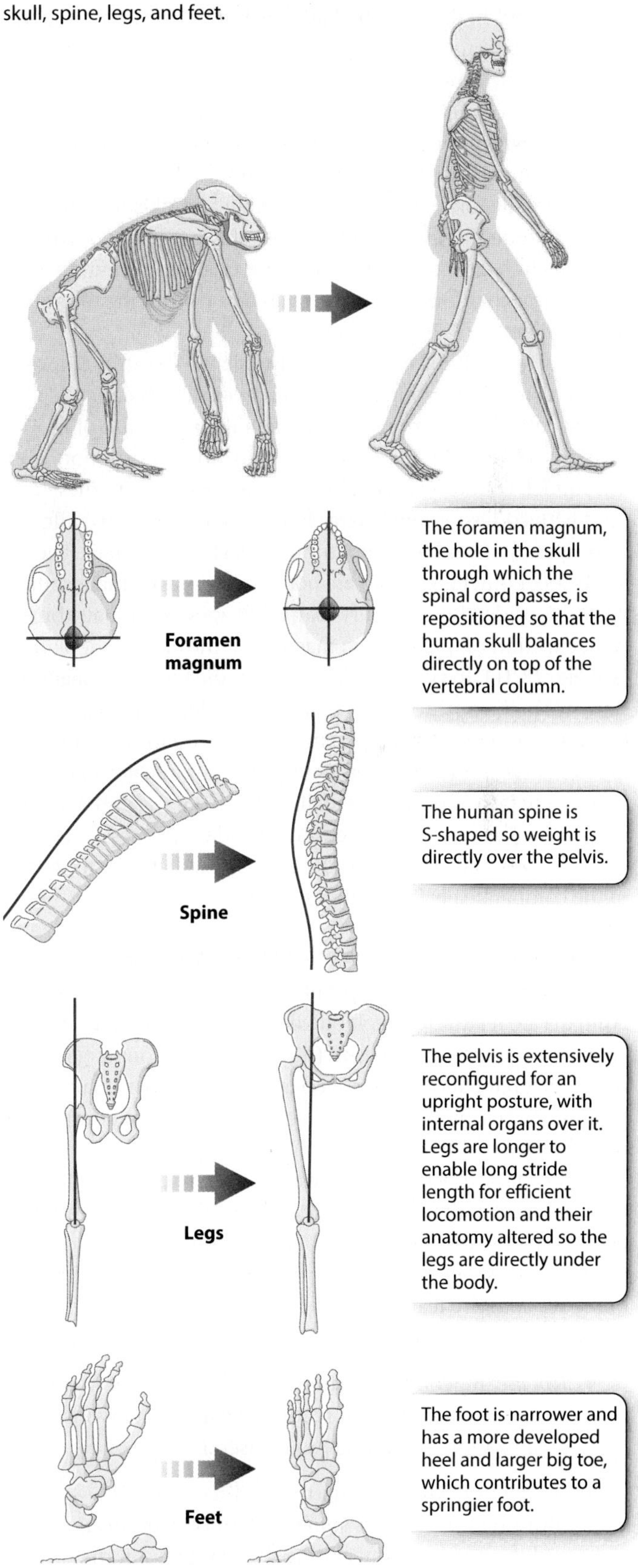

Foramen magnum

The foramen magnum, the hole in the skull through which the spinal cord passes, is repositioned so that the human skull balances directly on top of the vertebral column.

Spine

The human spine is S-shaped so weight is directly over the pelvis.

Legs

The pelvis is extensively reconfigured for an upright posture, with internal organs over it. Legs are longer to enable long stride length for efficient locomotion and their anatomy altered so the legs are directly under the body.

Feet

The foot is narrower and has a more developed heel and larger big toe, which contributes to a springier foot.

18). A small change—one that causes a gene to be transcribed at a different stage of development, for example—can have a major effect. In essence, King and Wilson introduced a model in which small changes in the software have a major impact even though the basic hardware is the same.

One of the gene-regulatory pathways that changed may be responsible for human **neoteny,** the long-term evolutionary process in which the timing of development is altered so that a sexually mature organism retains the physical characteristics of the juvenile form. In keeping with King and Wilson's idea, this shift in development could conceivably be achieved with relatively little genetic change. All that it might take would be a few changes to the regulatory switches that control the timing of development.

As early as 1836, French naturalist Étienne Geoffroy Saint-Hilaire noted that a young orangutan on exhibit in Paris looked considerably more like a human than the adult of its own species (**Fig. 23.12**). Several human attributes support this model, including heads that are large relative to bodies (and with correspondingly large brains), a feature of juvenile great apes. A second human attribute is lack of body hair. The juveniles of other great apes are not as hairless as humans, but they're considerably less hairy than adult apes. A third attribute is the position of the foramen magnum at the base of the skull (see Fig. 23.11). In primate development, the foramen magnum starts off in the position it occupies in the adult human and then, in nonhuman great apes, migrates toward the back of the skull. Adult humans have retained the juvenile great ape foramen magnum position. Finally, it has been suggested that human mentality, with its questioning and playfulness, is equivalent in many ways to that of a juvenile ape rather than to that of the comparatively inflexible adult ape.

Humans have large brains relative to body size.

In mammals, brain size is typically correlated with body size. Humans are relatively large-bodied mammals, but our brains are large even for our body size (**Fig. 23.13**). It is our large brains that have allowed our species' success, extraordinary technological achievements, and at times destructive dominion over the planet.

Which factors acted as selective pressure for the evolution of the large human brain? Again, speculation about this question runs rampant. However, because a large brain is metabolically expensive to produce and to maintain, we can conclude that the brain has adaptive features that natural selection acted to favor. What are these selective factors? Here are some possibilities:

- *Tool use.* Bipedalism permitted the evolution of manual dexterity, which in turn requires a complex nervous organization if hands are to be useful.

- *Social living.* Groups require coordination, and coordination requires some form of communication and the means of integrating and acting upon the

FIG. 23.12 A juvenile and an adult orangutan. Humans look more like juvenile orangutans than like adult orangutans, suggesting that humans may be neotenous great apes. *Sources: (top to bottom) McPHOTO/W. Layer/ AGE Fotostock; Russell Watkins/Shutterstock.*

information conveyed. One scenario, for example, sees group hunting as critical in the evolution of the brain: natural selection favored those individuals who cooperated best as they pursued large prey.

- *Language.* Did the evolution of language drive the evolution of large brains? Or did language arise as a result of having large brains? Again, we will probably

FIG. 23.13 Brain size plotted against body size for different species. Humans have large brains for their body size. *Data from Fig. 2.4, p. 44, in H. J. Jerison, 1973, Evolution of the Brain and Intelligence, New York: Academic Press.*

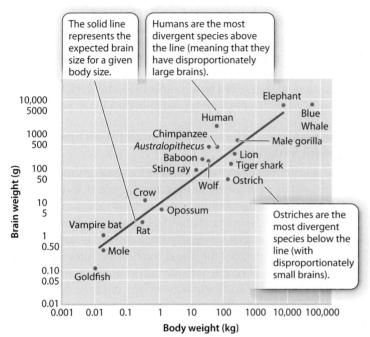

never know, but it is tempting to speculate that the brain and our extraordinary powers of communication evolved in concert.

Probably, as with bipedalism, no single factor drove the development of humans' large brain, but rather a combination of factors worked together to result in the evolution of large brain size. We tend to focus on brain size because it is a convenient stand-in for mental power and because we can measure it in the fossil record by making the reasonable assumption that the volume of a fossil's cranium reflects the size of its brain. Of course, our minds are not solely the products of larger brains. Instead, the reorganization of existing structures and pathways is the more important product of the evolution of the brain. Not only was the brain expanding over those 5–7 million years of human evolution, but it was also being rewired.

We can be sure that the human brain evolved through natural selection because it is such a metabolically costly organ. Although the brain contributes only 2% of a human's weight, it accounts for 20% of the body's metabolic budget. The evolutionary advantages of a large brain must be substantial enough to offset this cost. What we have today, as a result, is a learning machine capable of generating many more skills and abilities than just those that directly enhance evolutionary fitness. For example, let's assume that the brain evolved to make group hunting more efficient. Although the larger brain does indeed enhance our ability to hunt together, numerous by-products of this brain then take its abilities beyond just hunting. Making art, for instance, has nothing to do with group hunting, yet the brain allows us to do it. A brain that evolved for an essential but mundane task like group hunting lies at the heart of much that is wonderful about humanity.

The human and chimpanzee genomes help us identify genes that make us human.

The key genetic differences between humans and chimpanzees must lie in the approximately 1% of protein-coding DNA sequence that differs between the two genomes. What do we see when we compare human and chimpanzee genomes?

One gene that has attracted a lot of interest is *FOXP2*, a member of a large family of evolutionarily conserved genes that encode transcription factors that play important roles in development. People with mutations in *FOXP2* often have difficulty with speech and language. Interestingly, laboratory animals whose *FOXP2* gene has been knocked out also have communication impairments. Songbirds are less capable of learning new songs, and the high-frequency "songs" of mice are disrupted. In addition to its effects on the brain, *FOXP2* plays a role in the development of many tissues.

Studies of the *FOXP2* gene's amino acid sequence reveal a pattern of extreme conservation. These amino acid sequences in mice and chimpanzees, whose most recent common ancestor lived about 75 million years ago, differ by a single amino acid (**Fig. 23.14**). However, two amino acids are present in humans

FIG. 23.14 A family tree of mammals showing amino acid changes in *FOXP2*. The gene is highly conserved, but two changes are seen in the Neanderthals and modern humans.

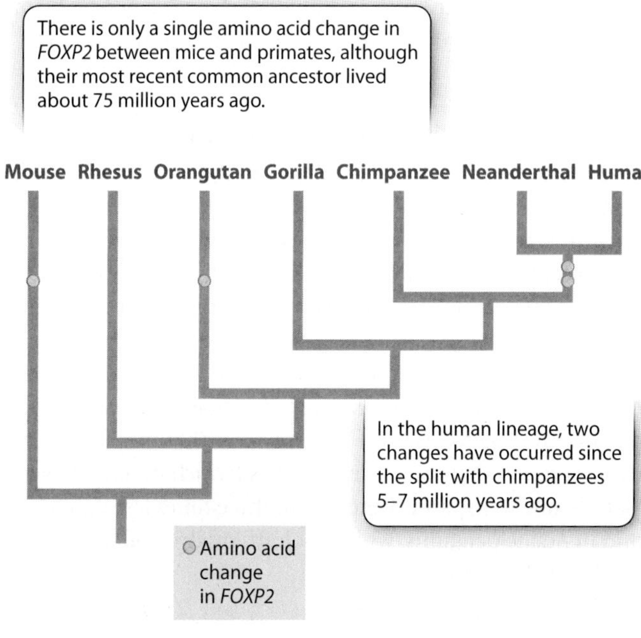

but absent in chimpanzees. Studies of Neanderthal DNA have shown that Neanderthals possessed the modern human version of *FOXP2*. The presence of the same version of the gene in both species suggests that the differences arose in the hominin line before the split between our species and Neanderthals, which occurred at least 600,000 years ago.

Studies of *FOXP2* give us a glimpse into the genetic architecture of the biological traits that are likely to be important in the determination of "humanness." Those two differences in amino acid sequence are intriguing, but they do not make an organism a human. Substitute the human *FOXP2* gene into a mouse and you will not get a talking mouse. The critical genetic differences are many, subtle, and interacting.

Self-Assessment Questions

7. What are four anatomical differences between chimpanzees and humans that facilitated walking upright?
8. Given the high genetic similarity of humans and chimpanzees, how might gene-regulatory pathways help us account for the differences we see between the two species?
9. What are three possible selective factors underlying the evolution of large brains in our ancestors?
10. The *FOXP2* gene is sometimes called the "language gene." Why is this name inaccurate?

23.4 HUMAN GENETIC VARIATION

So far, we have treated humans as all alike, and, in many ways, we are. But there are also many differences from one person to the next. Those differences ultimately have two sources: genetic variation and differences in environment (Chapter 17). For example, a person may be born with dark skin, or a person born with pale skin may acquire darker skin—a tan—in response to exposure to sun.

Humans have very little genetic variation.

The differences we see from one person to the next are deceptive. Despite appearances, ours is not the most genetically variable species on Earth. While it is certainly true that everyone alive today (except for identical twins) is genetically unique, our species is actually rather low in overall amounts of genetic variation. Modern estimates based on comparisons of many human DNA sequences indicate that, on average, about 1 in every 1000 base pairs differs among individuals (that is, our level of DNA variation is 0.1%). That's about 10 times less genetic variation than occurs in fruit flies (which nevertheless all look the same to us) and about two to three times less than is found in Adélie penguins, which look strikingly similar to one another (see Fig. 20.1).

Why, then, are humans all so phenotypically different if there is so little genetic variation in our species? Given the large size of our genome, a level of variation of 0.1% translates into a great many genetic differences. Our genome consists of approximately 3 billion base pairs, so 0.1% variation means that 3 million base pairs differ between any two people chosen at random. Many of those differences are in noncoding DNA, but some fall in regions of DNA that encode proteins or regulatory regions and, therefore, influence the phenotype. When those mutations are reshuffled by recombination, the result is the vast array of genetic combinations present in the human population.

The prehistory of humans influenced the distribution of genetic variation.

The reasons for our species' relative lack of genetic variation compared to other species lie in prehistory, and factors affecting the geographical distribution of that variation also lie in the past. Studies of the human family tree inspired by Rebecca Cann's original mtDNA analysis paint a detailed picture of how our ancestors colonized the planet and how that process affected the distribution of genetic variation across populations today. Detailed analyses of different populations, often using mtDNA or Y chromosomes, allow us to reconstruct the history of human population movements.

As we have seen, *Homo sapiens* arose in Africa. Genetic analyses indicate that perhaps 60,000 years ago, populations started to venture out through the Horn of Africa into the Middle East (**Fig. 23.15**). The first phase of colonization took our ancestors through Asia and into Australia about 50,000 years ago. Archaeological evidence indicates that it was not until about 15,000 years ago that the first modern humans crossed from Siberia to North America to populate the New World.

Genetic analyses indicate that other colonizations occurred even later. Despite its closeness to the African mainland, the first humans arrived in Madagascar only about 2000 years ago, and the colonists came from Southeast Asia, not Africa. Madagascar populations to this day bear the genetic imprint of this surprising Asian input. The Pacific Islands were among the last habitable places on Earth to be colonized during the Polynesians' extraordinary seaborne odyssey from Samoa, which began about 2000 years ago. Hawaii was colonized about 1500 years ago, and New Zealand only 1000 years ago.

By evolutionary standards, the beginning of the spread of modern humans out of Africa about 60,000 years ago is very recent. Consequently, differences have had relatively little time to accumulate among regional populations. Indeed, most of the variation in human populations today arose in ancestral populations before any humans left Africa. When we compare levels of variation in a contemporary African population with that in a non-African population, such as Europeans or Asians, we find there is more variation in the contemporary African population.

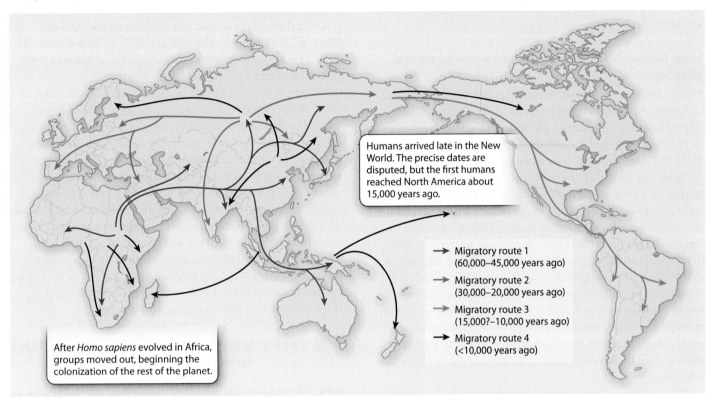

FIG. 23.15 Human migratory routes. Tracking the spread of mitochondrial DNA mutations around the globe allows us to reconstruct the colonization history of our species. *Data from Fig. 6.18, p 149, in D. J. Futuyma, 2009, Evolution, 2nd ed., Sunderland, MA: Sinauer Associates.*

The individuals who left Africa 60,000 years ago to found populations in the Middle East and beyond were a relatively small sample of the total amount of genetic variation then present in the human population. Contemporary descendants of these populations who left Africa early in human history therefore began with less genetic variation, and that initial lower variation is reflected in their genetic profiles today.

The recent spread of modern humans means that there are few genetic differences between groups.

Because the out-of-Africa migration was so recent, the genetic differences we see among geographical groups—sometimes called races—are actually quite minor. This fact is highly counterintuitive. People may note many superficial differences between a typical African and a typical European, such as skin color, facial form, and hair type, and assume that these superficial differences must reflect extensive genetic differences. This assumption made sense when the standard theory about the origin of modern humans was the multiregional one. If European and African populations had been geographically isolated from each other for as long as 2 million years, then it would make sense to expect significant genetic differences among populations.

Isolated populations diverge genetically over time as different mutations occur and are fixed in each population. The longer two populations have been isolated from each other, the more genetic differences between them we would expect to see (Chapter 20). Isolation lasting 2 million years implies that the differences are extensive, whereas isolation for just 60,000 years suggests differences are relatively few. Patterns of genetic variation among different human populations support the hypothesis that human populations dispersed as recently as 60,000 years ago. What we see when we look at genetic markers—variable A's, T's, G's, and C's in human DNA—is that there is actually very little genetic differentiation among human populations scattered across the globe.

In short, there's a disconnect: different groups may look different, but, from a genetic perspective, they're not very different at all. Any two humans differ from each other by, on average, only 3 million base pairs. Statistical analyses show that approximately 85% of that genetic variation occurs within a population (for example, the Yoruba in West Africa); 8% occurs between populations within so-called races (for example, between the Yoruba and the Kikuyu, another African group); and the remaining 7% occurs between so-called races. The characteristics used to assess an individual's ethnicity, such as skin color, eye type, and hair form, are encoded by genetic variants that lie in that 7%. If Earth were threatened with destruction and only one population—the Yoruba, for example—survived, 85% of the total amount of human genetic variation that exists today would still be present.

Some human differences have likely arisen by natural selection.

Patterns of genetic variation in human populations are shaped by the set of evolutionary forces discussed in Chapter 20. Neutral variants, for example, are subject to genetic drift, and, given the stepwise global colonization history of our species, it is likely that founder events have played a role as well.

Selection, too, has been important. For example, the genetic variants affecting traits we can easily see are an especially prominent feature of the 7% of human genetic variation that occurs between so-called races. If we look at other genetic variants, which don't affect traits that we can see, there is little or no racial pattern. A contemporary dark-skinned African is as likely to have a particular base-pair mutation in a randomly chosen gene as a contemporary fair-skinned European. So why are visible traits so markedly different among groups while other traits are not? Given the short amount of time (by evolutionary standards) since all *Homo sapiens* were in Africa, it is likely that the differences we see between groups are the product of selection.

People with dark skin tend to originate from areas in lower latitudes with high levels of solar radiation, whereas people with light skin tend to originate from areas in higher latitudes (near the equator) with low levels of solar radiation (toward the poles). It is likely that natural selection is responsible for the differences in pigmentation between these populations. Assuming that the ancestors of non-African populations were relatively dark-skinned, which selective factors might account for the loss of pigmentation among populations that now live at higher latitudes?

One likely factor is an essential vitamin, vitamin D, that is particularly important in childhood because it is needed for the production of bone. A deficiency of vitamin D can result in the skeletal malformation known as rickets. The body can synthesize vitamin D, but the process requires ultraviolet (UV) radiation. Heavily pigmented skin limits the entry of UV radiation into cells, thereby limiting the production of vitamin D. This does not present a difficulty in parts of the world where there is plenty of sunlight, but it can be problematic in regions of low sunlight. Presumably, natural selection favored lighter skin in the ancestors of populations from higher latitudes because lighter skin promotes the production of vitamin D.

A second likely factor is folate, another vitamin. Folate is necessary for normal embryonic development, and a deficiency of folate can lead to certain birth defects. UV radiation can break down folate, so natural selection favors darker skin in areas where there is a lot of sunlight, such as near the equator.

Here, too, there is the danger of UV damage both to skin cells and to the DNA molecules they contain, with cancers such as melanoma a possible outcome. Again, skin pigmentation plays a key protective role. Thus, UV radiation is a double-edged sword: it breaks down one essential vitamin (folate) and damages DNA, but is essential for the production of another vitamin, D. Skin color across the globe represents a balance between these opposing selective pressures.

Attempts have been made to identify the adaptive value of other visible differences among groups, such as facial features. It is possible that natural selection played a role in the evolution of these differences, but an alternative explanation, originally suggested by Charles Darwin, is more compelling: sexual selection (Chapter 20). As we have seen, there is an apparent mismatch between the extent of differences among groups in visible characters, such as facial features, and the overall level of genetic differences between human groups. Sexual selection can account for this mismatch because it operates solely on characteristics that can readily be seen—think of the peacock's tail. As we learn more about the genetic underpinnings of the traits in question, we will be able to investigate directly the factors responsible for the differences we see among groups.

 CASE 4 MALARIA: COEVOLUTION OF HUMANS AND A PARASITE

Which human genes are under selection for resistance to malaria?

We see evidence of regional genetic variation in response to local challenges, especially those posed by disease. Malaria, for example, is largely limited to warm climates because it is transmitted by a species of mosquito that can survive only in these regions. Historically, this disease has been devastating in Africa and the Mediterranean. As we saw in Chapter 20, the sickle allele of hemoglobin, *S*, has evolved to be present at high frequencies in these regions because in heterozygotes it confers some protection against the disease. But in homozygotes, the *S* allele is highly detrimental because it causes sickle-cell anemia. Homozygotes for the allele encoding normal hemoglobin are also at a disadvantage because they are entirely unprotected from the parasite.

The *S* allele is beneficial only in the presence of malaria. If there is no malaria in an area, the *S* allele is disadvantageous, so natural selection presumably acted rapidly to eliminate it in the ancestors of Europeans when they arrived in malaria-free regions. The continued high frequency of the *S* allele in contemporary Africans, in some Mediterranean populations, and in populations descended recently from Africans (such as African-Americans) is, however, a reflection of the response of natural selection to a regional disease.

The hemoglobin genes are not the only genes that are under selection for resistance to malaria. *Glucose-6-phosphate dehydrogenase* (*G6PD*), a gene involved in glucose metabolism, is one of several other genes implicated. People who are heterozygotes for a mutation in the *G6PD* gene—and therefore have a G6PD enzyme deficiency—can develop severe anemia when they eat certain foods, most notably fava beans (hence, the condition is called favism). People who are heterozygotes for a mutation in the *G6PD* gene also have increased resistance to malaria, apparently because they are better at clearing infected red blood cells from their bloodstream. In areas where malaria is common, the advantage of malaria resistance offsets the disadvantage of favism.

Detailed evolutionary analysis of mutations in *G6PD* shows that favism has arisen multiple times, each time selectively favored because of its role in the body's response to the malaria parasite. As expected, favism, like sickle-cell anemia, is mainly a feature of populations in malarial areas or of populations whose evolutionary roots lie in these areas.

Self-Assessment Questions

11. Why does the timing of the modern human out-of-Africa event affect the pattern of genetic variation we see among modern human groups?
12. How did phenotypic differences among different human populations arise?

23.5 CULTURE, LANGUAGE, AND CONSCIOUSNESS

The most remarkable outcome of the evolutionary process described in this chapter is the human brain. This brain allows us to do extraordinary things, such as play and appreciate music, write and read books, design and build skyscrapers, and invent and implant an artificial heart. But does this wonderful brain make us qualitatively different from other organisms? Does it in some way take us out of nature? Traditionally, the answer to these questions would have been a resounding "yes." More recently, research into the capabilities of other species has questioned this conclusion. The human brain is certainly remarkable, but, in essence, what we can do is merely an extension of what other animals can do.

Culture changes rapidly.

Culture is generally defined as a body of learned behavior that is socially transmitted among individuals and passed down from one generation to the next. Culture has permitted humans in part to transcend our biological limits. To take a simple example, clothing and ingeniously constructed shelters have enabled us to live in extraordinarily inhospitable parts of the planet, such as the Arctic (**Fig. 23.16**). The capacity to innovate coupled with the ability to transmit culture is the key to the success of humans.

FIG. 23.16 The power of culture. Inventions (such as clothing) have allowed our species to expand its geographic range. *Source: B&C Alexander/ArcticPhoto.*

Culture, of course, changes over time. In many ways, this kind of cultural change is responsible for our species' extraordinary achievements. Genetic evolution is slow because it involves mutation followed by changes in allele frequencies that take place over many generations. Cultural change, in contrast, can occur much more rapidly. Twenty years ago, no one had heard of smartphones; today, millions of people own them. Or consider the speed at which a change in clothing style—a shift from flared to straight-leg jeans, for example—spreads through a population. Today's human population as a whole is genetically almost identical to the population when your grandparents were young, but the cultural changes that have occurred in the 50 years or so between your grandparents' youth and your own are astounding.

Despite this clear contrast between biological evolution and cultural change, we should not necessarily think of the two processes as independent of each other. Sometimes cultural change drives biological evolution.

A good example of the interaction between cultural change and biological evolution is the evolution of lactose tolerance in populations for which domesticated animals became an important source of dairy product. Most humans around the world are lactose intolerant. Lactose, a sugar, is a major component of mammalian milk, including human breast milk. Humans have an enzyme, lactase, that breaks down lactose in the gut, but typically this enzyme is produced only in the first years of life, when we are breastfeeding. Once a child is weaned, lactase production is turned off. Lactase, however, is clearly a useful enzyme to have if there is a major dairy component to your diet.

Archaeological and genetic analyses indicate that cattle were domesticated probably three separate times in the past 10,000 years, in three separate places: in the Middle East, in East Africa, and in the Indus Valley. In at least two of these cases, subsequent human biological evolution has favored lactose tolerance, in the form of continued lactase production throughout life. Analysis of the gene region involved in switching lactase production on and off has revealed mutations in European lactose-tolerant people that are different from those in African lactose-tolerant people, implying independent, convergent evolution of this trait in the two populations. Furthermore, these changes appear to have evolved very recently, implying that they arose as a response to the domestication of cattle. This is an example of the interaction between cultural change and biological evolution. Biological evolution of continued lactase production has resulted from the change in a cultural practice, namely cattle domestication.

Is culture uniquely human?

Even nonprimates are known to be capable of learning from another member of their own species—culture is not uniquely human. Bird songs, for instance, may vary regionally because juveniles learn the song from local adults. In a famous case, illustrated in **Fig. 23.17a**, small birds called blue tits (*Parus caeruleus*) in Britain learned from each other how to peck through the aluminum caps of milk bottles left on doorsteps by milkmen to reach the rich cream at the top of the milk. Presumably, one individual discovered accidentally how to peck through a milk bottle cap and others then imitated the action.

Nor is teaching, one way in which culture is transmitted, limited to humans. In teaching, one individual tailors the information available to another so as to facilitate learning. Teaching and imitation together make learning highly efficient. For example, adult meerkats teach young meerkats how to handle prey by giving them the opportunity to interact with live prey (**Fig. 23.17b**).

The most sophisticated example of nonhuman culture involves chimpanzees. Detailed studies of several geographically isolated populations have revealed 39 culturally transmitted behaviors, such as ways of using tools to catch insects, which are

---

FIG. 23.17 Nonhuman culture. (a) English blue tits steal cream from a milk bottle on the doorstep. (b) Adult meerkats teach their young how to handle their prey. (c) Chimpanzees in different populations have devised different ways of using tools to hunt insects. *Sources: a. Colin F. Sargent; b. Dr. Alex Thornton, University of Exeter; c. FLPA/Peter Davey/AGE Fotostock.*

specific to particular populations (**Fig. 23.17c**). West African chimpanzees in both Guinea and the Ivory Coast use tools to help them harvest the same species of army ant for food (and avoid the painful defensive bites of the ant), but each population has its own distinct way of doing this. Chimpanzees, then, are like humans in demonstrating regional variations in culture.

Despite the cultural achievements of other species, it is clear that culture in *Homo sapiens* is very far removed from anything seen in the natural world. Humans are amazingly adept at imitation, learning, and acquiring culture. Teaching, learning, and cultural transmission all benefit from another extraordinary human attribute, language.

Is language uniquely human?

Language is a complex form of communication involving the transfer of information encoded in specific ways, such as speech, facial expressions, posture, and the like. Language is not unique to humans. The waggle dance of a worker bee on its return to the hive communicates information on the direction, distance, and nature of a food resource (Chapter 43). Vervet monkeys use warning vocalizations to specify the identity of a potential predator: they have one call for "leopard," for example, and another for "snake." But nonhuman animal languages are limited. Attempts to teach even our closest relatives, chimpanzees, to communicate using sign language have met with only limited success (**Fig. 23.18**).

Grammar provides a set of rules that allow us to combine words into a virtually infinite array of meanings. Noam Chomsky, father of modern linguistics, has pointed out that all human languages are basically similar from a grammatical viewpoint. Humans have what he has called a "universal grammar" that would lead a visiting linguist from another planet to conclude that all Earth's languages are dialects of the same basic language. This universal grammar is "hard-wired" into the human brain in such a way that every human infant spontaneously strives to acquire language. The specific attributes of the language depend on the baby's environment—a baby in France will learn French, and one in Japan will learn Japanese—but the basic process is similar in every case. Conversely, chimpanzees, and the entire natural world, lack the drive toward the acquisition of a grammatical language.

Is consciousness uniquely human?

Descartes famously wrote, "Cogito, ergo sum"—"I think, therefore I am." Can we legitimately rewrite his statement to declare, "Animals think, therefore they are"? With the growth of the animal rights movement, particularly in reference to the treatment of animals in factory farms, this question is of more than academic interest. We can cite many examples of animal thinking from a range of species, including, not surprisingly, chimpanzees and gorillas.

Perhaps even more remarkable are the examples that come from animal species that are not closely related to us. In experiments carried out by Alex Kacelnik in Oxford, England, a pair of New Caledonian crows was presented with two pieces of wire, one straight and the other hooked, and offered a food reward that could be obtained only by using the hooked wire.

FIG. 23.18 Dr. Susan Savage Rumbaugh with Panbanisha, a female bonobo that learned to communicate using sign language. Chimpanzees and bonobos are able to learn and use sign language to express words and simple sentences. *Source: Laurentiu Garofeanu/Barcroft USA/BarcoftMedia/Getty Images.*

One member of the pair, the male, disregarded the experiment and flew off with the hooked wire. The female, having discovered that she could not get the food reward with the straight wire, went to some considerable trouble to bend a hook into the straight wire. She succeeded in getting the food. It is difficult to deny that the crow thought about the problem and was able to solve it, perhaps in the same way that a human would. Definitions of consciousness are contested, but, as with language and culture, it seems clear that other species are capable of some form of conscious thought.

The evolutionary biologist Theodosius Dobzhansky once said, "All species are unique, but humans are uniquest." Our "uniquest" status is not derived from having attributes absent in other species, but rather from the extent to which those attributes are developed in us. Human language, culture, and consciousness are extraordinary products of our extraordinary brains. Nevertheless, as Darwin taught us, we are fully a part of the natural world. Today, humans dominate the planet more comprehensively than any single species ever has before, meaning that Darwin's lesson is crucially important: the future fate of the planet—home to us *and* millions of other species—is in our hands.

Self-Assessment Questions

13. What is an argument for the idea that culture, language, and consciousness are uniquely human?
14. What is an argument against the idea that culture, language, and consciousness are uniquely human?

CORE CONCEPTS SUMMARY

23.1 THE GREAT APES: Anatomical, molecular, and fossil evidence shows that the human lineage branches off the great apes tree.

Anatomical features indicate that primates are a monophyletic group that includes prosimians, monkeys, and apes. The apes, in turn, include the lesser and great apes. page 491

The great apes include orangutans, gorillas, chimpanzees, and humans. page 492

Analysis of sequence differences between humans and our closest relatives, chimpanzees, indicate that the human lineage split from chimpanzees 5–7 million years ago. page 493

Lucy, an unusually complete specimen of *Australopithecus afarensis*, demonstrates that our human ancestors were bipedal by about 3.2 million years ago. page 494

Hominin fossils occur only in Africa until about 2 million years ago, when *Homo ergaster* migrated out of Africa to colonize the Old World. page 496

23.2 HUMAN ORIGINS: Phylogenetic analysis of mitochondrial DNA and the *Y* chromosome shows that the human species arose in Africa.

Studies of mitochondrial DNA (mtDNA) suggest that the common ancestor of all modern humans lived approximately 200,000 years ago, implying that modern humans (*Homo sapiens*) arose in Africa (the out-of-Africa hypothesis). page 497

The mtDNA out-of-Africa pattern is supported by *Y* chromosome analysis, which also shows a recent African origin of modern humans. page 499

Analysis of Neanderthal DNA from 30,000-year-old material indicates that, as the ancestors of non-African humans emigrated from Africa, they interbred with the Neanderthals. page 500

Our species originated in Africa and subsequently colonized the rest of the planet, starting about 60,000 years ago. page 500

23.3 HUMAN TRAITS: During the 5–7 million years since the most recent common ancestor of humans and chimpanzees, our lineage acquired a number of distinctive features.

The development of bipedalism involved a wholesale restructuring of anatomy. page 501

Neoteny is the process in which the timing of development is altered over the course of evolution so that a sexually mature organism retains the physical characteristics of the juvenile form; humans are neotenous, exhibiting several traits as adults that chimpanzees exhibit as juveniles. page 502

Many possible selective factors may explain the evolution of humans' large brain, including tool use, social living, and language. page 502

FOXP2, a gene involved in brain development, may be important in language, as mutations in *FOXP2* are implicated in speech pathologies. page 503

23.4 HUMAN GENETIC VARIATION: Human history has had an important impact on patterns of genetic variation in our species.

Because our ancestors left Africa very recently in evolutionary terms, there has been little time for genetic differences to accumulate among geographically separated populations. page 504

Humans have very little genetic variation, with only about 1 in every 1000 base pairs varying among individuals. page 504

Most of the variation among humans occurs within populations. As much as 85% of the total amount of genetic variation in humans can be found within a single population. Only about 7% of human genetic variation segregates between groups that are so-called races. page 505

Some differences among groups, such as skin color and resistance to malaria, have probably arisen by natural selection. page 506

Other differences among groups have probably arisen by sexual selection. page 506

23.5 CULTURE, LANGUAGE, AND CONSCIOUSNESS: Culture, language, and consciousness are developed to a remarkable degree in humans.

Cultural evolution and biological evolution may interact, as in the case of the evolution of lactose tolerance in regions where cattle were domesticated. page 508

Other animals possess simple versions of culture, language, and even consciousness, but the advances of our species in all three are truly exceptional. page 508

Log in to LaunchPad to check your answers to the Self-Assessment Questions and to access additional learning tools.

"… from so simple a beginning endless forms most beautiful and most wonderful have been, and are being, evolved."

—CHARLES DARWIN

FROM ORGANISMS TO THE ENVIRONMENT

PART 2

CASE 5

The Human Microbiome

Diversity Within

Clostridium difficile doesn't grab many headlines, but perhaps it should. This bacterium is thought to be the most common cause of hospital-acquired infections in the United States. It sickens about half a million U.S. hospital patients each year, causing fevers, chills, diarrhea, and severe intestinal pain. It also adds more than $1 billion annually to the nation's health-care costs. Altogether, the infection kills nearly 30,000 people every year, according to the Centers for Disease Control and Prevention.

> *No one has yet identified all of the many microbes living on and within our bodies, but recent estimates suggest that each of us contains at least as many bacterial cells as human cells.*

Now a promising (if stomach-turning) new treatment is gaining traction: fecal transplants. The vast majority of *C. difficile* cases occur after patients have taken antibiotics for other infections. These drugs kill the beneficial bacteria normally found in a healthy gut, allowing *C. difficile* to proliferate. Recently, pioneering researchers have begun to replace the missing microbes by transplanting fecal material from healthy donors into patients sick with *C. difficile* infections. Some doctors have reported that the procedure alleviates symptoms in more than 80% of patients.

While transplanting fecal bacteria may not yet be mainstream medicine, there is no question that our gut, as well as our skin, mouth, nose, and urogenital tract, is home to a diversity of bacterial species. No one has yet identified all of the many microbes living on and within our bodies, but recent estimates suggest that each of us contains at least as many bacterial cells as human cells. For the most part, the relationship between "us and them" is a type of symbiosis, in which both humans and bacteria benefit from the close association. Indeed, the human *microbiome*, as our assemblage of bacterial houseguests is known, affects human health in many important ways.

Some of the most mysterious, and most intriguing, members of that microbiome live in the gut. On the one hand, bacteria found in our intestines help us break down food and synthesize vitamins. On the other hand, abnormal communities of microbes in the gut have been linked to Crohn's disease, colon cancer, and inflammatory bowel disease. The list of microbial residents in the human body grows longer each year.

Your own personal gut flora is a unique collection of species—even identical twins can have surprisingly different microbiomes. This variability has made it difficult for scientists to decipher exactly what role individual species play in health and in disease. The sheer diversity of bacteria in the gut merely compounds the problem. One international team of scientists recently reported that each human gut contains at least 160 different bacterial species.

Despite these challenges, researchers have made strides toward understanding the unique bacterial communities within our intestinal tract. In 2011, scientists compared all the DNA sequences from gut microbes of people living in Europe, America, and Japan. They found that each individual fell into one of three enterotypes, or "gut types." Most people had a bacterial mix dominated by organisms from the genus *Ruminococcus*. Others, however, had gut types dominated by the genus *Bacteroides* or the genus

Prevotella. The three different gut compositions appear to produce different vitamins and may even make a person more, or less, susceptible to certain diseases. Continuing research shows that the boundaries between enterotypes may be fuzzier that originally thought, but understanding the variations in gut biota among human populations may in time lead to novel and focused treatments for diseases of the intestinal tract.

Surprisingly, the researchers reported that enterotype was unrelated to gender, body mass index, or nationality. So what determines your gut type? A follow-up study suggested that enterotypes are associated with long-term dietary patterns. People who eat a diet high in animal protein and saturated fats are more likely to contain a gut community dominated by *Bacteroides*. In contrast, people who stick to a high-carbohydrate diet rich in plant-based nutrition are more likely to host an abundance of *Prevotella*.

The researchers don't know yet what a person's enterotype means for his or her health. But scientists are eager to explore the link between microbiome and physical well-being. Meanwhile, various researchers are exploring the microbiomes of other organisms—specifically, cows.

As it happens, there are good reasons to peer into a cow's gut. The cow's rumen (one of its four digestive

The human microbiome. Different groups of bacteria reside in or on each part of the body. When *Helicobacter pylori* is present in the stomach (*H. pylori* +), it can overwhelm other bacterial populations (*H. pylori* –), causing stomach ulcers and cancer. *Source: Mimi Haddon/Getty Images.*

CASE 5

chambers) contains microbes that break down the cellulose in plant material. Without them, cattle wouldn't be able to extract adequate nutrition from the grass on which they graze. Those microbes are the particular focus of scientists at the Department of Energy (DOE), who are interested in converting material containing cellulose, such as switchgrass, into biofuels.

As researchers look for more sustainable replacements for fossil fuels, many have pinned their hopes on biofuels. Unfortunately, the enzymes currently used commercially to break down cellulose aren't efficient or cost-effective enough to produce biofuels on an industrial scale. That's where the cows come in. When DOE scientists recently analyzed the microbial genes present in the cow rumen, they discovered many genes for proteins that metabolize carbohydrates, at least 50 of which were previously unknown. The hope is that some of these newly discovered proteins will help break down cellulose in ways that will produce biofuels more efficiently and cheaply.

Mammals such as cows and humans are hardly the only organisms that have symbiotic relationships with bacteria. All living animals have their own microbiomes. Even tiny

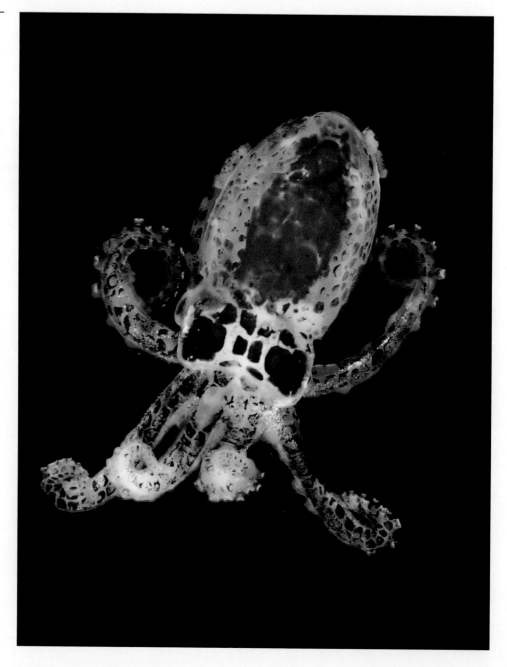

The bobtail squid and its luminescent bacteria. This squid contains specialized organs that harbor light-emitting bacteria. *Source: Gary Bell/OceanwideImages.com.*

single-celled eukaryotes can harbor still-tinier bacterial guests.

Symbiotic bacteria provide a variety of benefits to their hosts. Consider the bobtail squid, *Euprymna scolopes*, which possesses a specialized organ to house bacteria. The bacteria residing there, *Vibrio fischeri*, are luminescent—that is, capable of generating light. The squid use their *Vibrio*-filled light organs like spotlights, projecting light downward to match light from the surface as a way to camouflage themselves from would-be predators below.

In many cases, animals and their bacterial symbionts are so intimately connected that it's hard to separate one from the other. Many human gut bacteria are poorly studied because scientists haven't been able to grow them outside the body. In contrast, some other species have taken their intimate bacterial partnership a step further.

Aphids are small insects that feed on plant sap. They rely on the bacterium *Buchnera aphidicola* to synthesize certain amino acids for them that aren't available from their plant-based diet. The partnership between aphids and *Buchnera* dates back as far as 250 million to 150 million years ago. Over time, the organisms have become completely dependent on each other for survival. The *Buchnera* bacteria actually live inside the aphid's cells. In some ways, the bacteria resemble organelles rather than independent organisms. Intracellular symbionts may not be so unusual. In fact, mitochondria and chloroplasts—key organelles of eukaryotic cells—are thought to have originated from symbiotic bacteria. Over time, the progenitor bacteria became so intertwined with their host cells that they developed into an integral part of the cells themselves. Studying simple sap-sucking aphids, it turns out, may offer insights into the evolutionary history of all eukaryotic cells.

Clearly, there are many good reasons to learn more about our bacterial comrades. From evolution to energy to human health, our microbiomes hold great promise—and great mystery. We've evolved hand in hand with these bacteria for millions of years. Let's hope it takes less time to uncover their secrets.

 DELVING DEEPER INTO CASE 5

For each of the Cases, we provide a set of questions to pique your curiosity. The questions cannot be answered directly from the Case itself, but instead introduce issues that will be addressed in chapters to come. We revisit this Case in Chapters 24–25 on the pages indicated below.

1. How do intestinal bacteria influence human health? See page 541.
2. What role did symbiosis play in the origin of chloroplasts? See page 548.
3. What role did symbiosis play in the origin of mitochondria? See page 551.
4. How did the eukaryotic cell originate? See page 552.

CHAPTER 24 Bacteria and Archaea

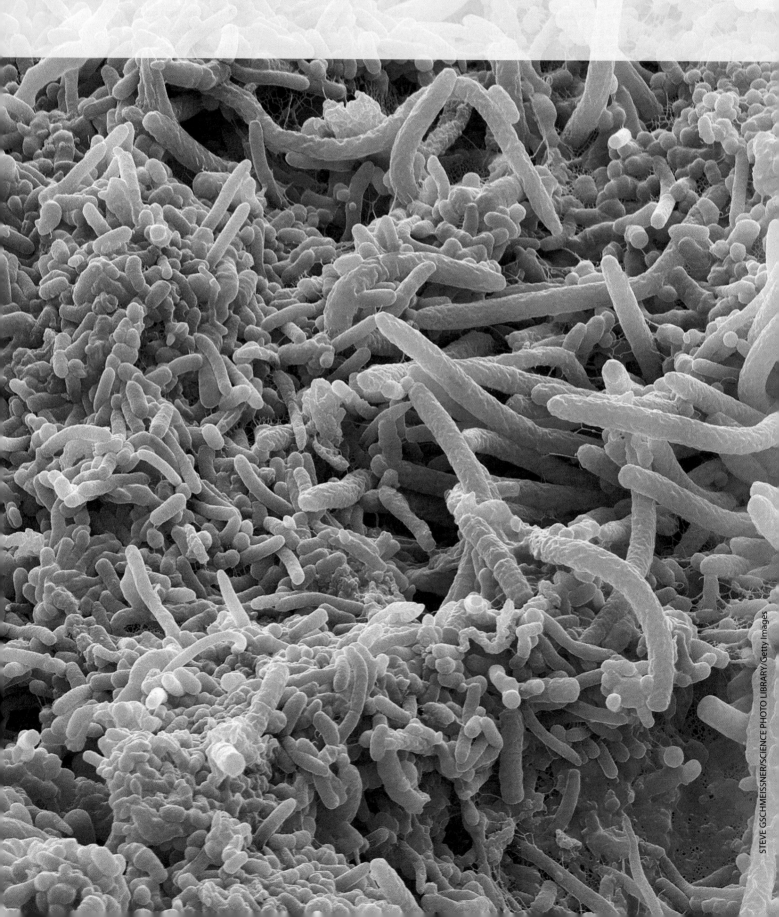

CORE CONCEPTS

24.1 TWO PROKARYOTIC DOMAINS: The two prokaryotic domains are Bacteria and Archaea.

24.2 AN EXPANDED CARBON CYCLE: Bacteria and Archaea cycle carbon in diverse ways.

24.3 SULFUR AND NITROGEN CYCLES: Bacteria and Archaea are critical to the biological cycling of sulfur and nitrogen.

24.4 BACTERIAL DIVERSITY: Genomic surveys of non-culturable bacteria reveal the extent of bacterial diversity.

24.5 ARCHAEAL DIVERSITY: The diversity of Archaea has only recently been recognized.

24.6 THE EVOLUTIONARY HISTORY OF PROKARYOTES: The earliest forms of life on Earth were Bacteria and Archaea.

In Chapters 7 and 8, we discussed the complementary metabolic processes of aerobic respiration and photosynthesis. Photosynthesis transforms CO_2 into organic molecules, and cellular respiration returns CO_2 to the environment. These two processes allow plants and animals to cycle carbon, which in turn is linked to their production and consumption of oxygen. But can carbon cycle through the deep waters of the Black Sea, within the black muds found beneath swamps and marshlands, or in other habitats where oxygen is limited or absent? Can biological communities even survive without oxygen? The answer, emphatically, is yes. Indeed, life existed on Earth for more than a billion years before oxygen-rich habitats first appeared on our planet.

The seemingly alien oxygen-poor environments on today's Earth are populated by organisms that neither produce nor consume oxygen, yet are able to cycle carbon. The organisms responsible for this expanded cycle share one fundamental feature: they have prokaryotic cell organization. Prokaryotes, which include bacteria and archaeons, lack a nucleus (Chapter 5). They inhabit the full range of environments present on Earth, including those rich in oxygen. Notably, however, their ability to live in habitats where plants and animals cannot suggests that some of these organisms have biological features quite unlike those of our familiar world.

The **carbon cycle** describes the path of carbon as it moves between organisms, the atmosphere, rocks, soil, and the ocean. We'll look at the carbon cycle again later in this chapter, but first we discuss **Bacteria** and **Archaea**, the prokaryotic domains of microscopic organisms so critical to the cycling of carbon and other elements essential to life. We examine their diversity and consider how they have evolved through time.

24.1 TWO PROKARYOTIC DOMAINS

Chapter 5 outlined the two distinct ways that cells are organized internally. Eukaryotic cells, which include the cells that make up our bodies, have a membrane-bounded nucleus and organelles that form separate compartments for distinct cell functions. Prokaryotic cells have a simpler organization. No membrane surrounds the cell's DNA, and there is little in the way of cell compartments. Prokaryotic cell organization is an ancestral character state for life as a whole—that is, this feature was present in the last common ancestor of all organisms alive today. Nevertheless, the group defined traditionally as the prokaryotes is paraphyletic, in that it excludes some descendants of the last common ancestor of all living organisms, namely eukaryotes (Chapter 22).

Bacteria and Archaea are the two domains characterized by prokaryotic cell structure. These organisms are present almost everywhere on Earth. What they lack in complexity of cell structure, these tiny cells more than make up for in their dazzling metabolic diversity. As will become clear over the course of this chapter, Bacteria and Archaea underpin the efficient operation of ecosystems on our planet.

The bacterial cell is small but powerful.

Because of their small size and deceptively simple cell organization, bacteria were long dismissed as primitive organisms, distinguished mostly by the eukaryotic features they lack: they have no membrane-bounded nuclei, no energy-producing organelles, no sex. This point of view turns out to be more than a little misleading. Bacteria are the diverse and remarkably successful products of nearly 4 billion years of evolution. Today, bacterial cells outnumber eukaryotic cells by several orders of magnitude. Even in your own body, bacteria equal or outnumber human cells.

Fig. 24.1 illustrates the bacterial cell, which was briefly introduced in Chapter 5. The cell's DNA is present in a single circular chromosome, in contrast to the multiple linear chromosomes characteristic of eukaryotic cells. Many bacteria carry additional DNA in the form of **plasmids,** small circles of DNA that replicate independently of the cell's circular chromosome. In general, plasmid DNA is not essential for the cell's survival, but it may contain genes that have adaptive value under specific environmental conditions. Because no nuclear membrane separates DNA from the surrounding cytoplasm, transcribed mRNA is immediately translated into proteins by ribosomes.

Bacteria lack the membrane-bounded organelles found in eukaryotic cells. Instead, cell processes such as metabolism are carried out by proteins that float freely in the cytoplasm or are embedded in the plasma membrane. A few bacteria, notably the photosynthetic bacteria, contain internal membranes similar to those found in chloroplasts and mitochondria. The light-triggered reactions of photosynthetic bacteria take place in association with membranes distributed within the cytoplasm.

Structural support for bacteria is provided by a cell wall made of **peptidoglycan,** a complex polymer of sugars and amino acids. Some bacteria have thick walls made up of multiple peptidoglycan layers, while others have thin walls surrounded by an outer layer of lipids. For many years, it was believed that bacteria lacked the cytoskeletal framework that organizes cytoplasm in eukaryotic cells. More recently, careful studies have shown that bacteria do possess an internal scaffolding of proteins that plays an important role in determining the shape, polarity, and other spatial properties of bacterial cells (Chapter 10).

Diffusion limits cell size in bacteria.

Most bacterial cells are tiny: the smallest are only 200–300 nanometers (nm) in diameter, and relatively few are more than 1–2 micrometers (μm) long. Why are bacteria so small? The answer has to do with diffusion, the random motion of molecules introduced in Chapter 5. If you could watch the movement of any particular molecule in air or water, you would see that its motion is random, sometimes going in one direction and sometimes in another. On average, however, more molecules

FIG. 24.1 A bacterial cell. The cells of Bacteria (and Archaea) do not have a membrane-bounded nucleus or other organelles.

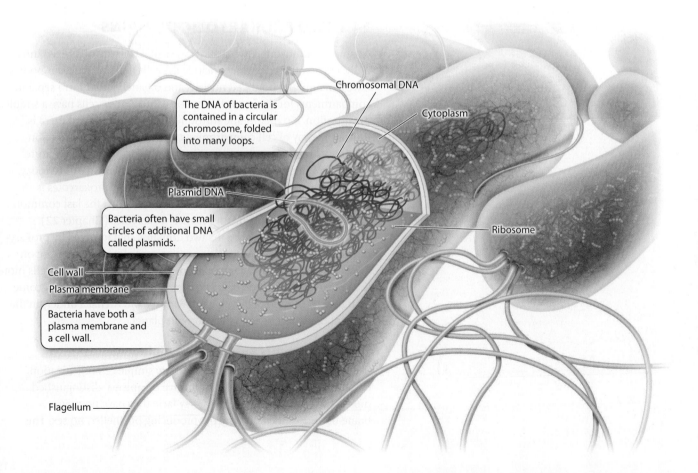

move from a region with a higher concentration of the molecule to a region with a lower concentration of the molecule than move in the opposite direction. Net movement stops only when the two regions achieve equal concentrations of the molecule, although diffusion continues. Photosynthetic bacteria gain the carbon dioxide they need by the diffusion of CO_2 from the environment into the cell; similarly, respiring bacteria take in small organic molecules and oxygen by diffusion.

Diffusion explains why bacterial cells tend to be small. A small cell has more surface area in proportion to its volume, and the interior parts of a small cell are closer to the surrounding environment than those of a larger cell. As a consequence, slowly diffusing molecules do not have to travel far to reach every part of a small cell's interior. The surface area of a spherical cell—the area available for taking up molecules from the environment—increases as the square of the radius. However, the cell's volume—the amount of cytoplasm that is supported by diffusion—increases as the cube of the radius. Therefore, as cell size increases, it becomes harder to supply the cell with the materials needed for growth. For this reason, most bacterial cells are tiny spheres, rods, spirals, or filaments—small enough for molecules to diffuse into the cell's interior (**Fig. 24.2**).

A few exceptional bacteria exceed 100 μm in length. The largest known bacterium, *Thiomargarita namibiensis*, lives in oxygen-poor sediments off the coast of southwestern Africa. Its total volume is approximately 100 million times larger than that of *Escherichia coli* (**Fig. 24.3**). But, in one sense, *T. namibiensis* cheats: 98% of its volume is taken up by a large vacuole, so the metabolically active cytoplasm is restricted to a thin film around the cell's periphery. Thus, the distance through which nutrients move by diffusion is only a few micrometers, as in many other bacteria.

Some bacteria are multicellular, forming simple filaments or sheets of cells. More unusual are myxobacteria, which aggregate to form multicellular reproductive structures that are composed of several distinct cell types (see Fig. 24.2e).

Horizontal gene transfer promotes genetic diversity in bacteria.

Bacterial genomes are generally smaller than the genomes of eukaryotes, in part because bacteria lack the large stretches of

FIG. 24.2 Cell shape and size in Bacteria and Archaea. *Sources: a. Eye of Science/Science Source; b. Scimat Scimat/Getty Images; c. Courtesy Mike Dyall-Smith; d. David Scharf/Science Source; e. © ANIMA RES – 3D Medical Animation.*

a. *Streptococcus*, strings of spheroidal or coccoidal bacteria

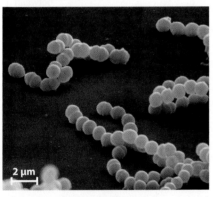

b. *E. coli*, bacterial rods

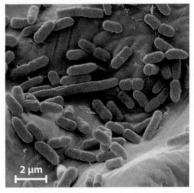

c. *Haloquadratum walsbyi*, a square archaeon that lives in salt ponds

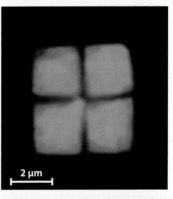

d. *Streptomyces*, helical bacteria that produce antibiotics

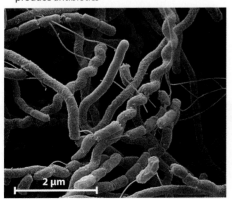

e. A myxobacterium, a bacterium in which cells aggregate to form fruiting bodies

noncoding DNA characteristic of eukaryotic chromosomes (Chapter 13). The streamlining of the bacterial genome confers certain benefits. For example, bacteria can reproduce rapidly when the nutrients required for their growth are available. Bacteria replicate their DNA from only one or a small number of initiation sites, so a smaller genome can be duplicated more quickly.

Bacteria do not undergo meiotic cell division and cell fusion. That is, they lack the sexual processes characteristic of eukaryotic organisms (Chapters 11 and 40). Despite this, bacterial populations display remarkable genetic diversity. For example, distinct populations of *Pseudomonas aeruginosa*, a common disease-causing organism, may differ in genome size by nearly a factor of 2 (3.7 million base pairs compared to 7.1 million base pairs), yet these strains are quite similar in function.

How does this variation arise? In eukaryotic organisms, genes generally pass from parent to offspring. Bacteria likewise inherit most of their genes from parental cells, but they can also obtain new genes from distant relatives by **horizontal gene transfer**. This process is a major source of genetic diversity in bacteria.

How does DNA move from one bacterial cell into another independently of cell division?

FIG. 24.3 ***Thiomargarita namibiensis*, the largest known bacterial cell.** A vacuole takes up most of the volume of the cell, so the active cytoplasm is only a few micrometers thick. The small spheroidal granules consist of sulfur, formed by the metabolism of hydrogen sulfide in their environment. *Source: Courtesy Heide Schulz-Vogt, Leibniz Institute for Baltic Sea Research, Department of Biological Oceanography.*

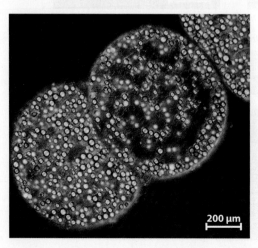

FIG. 24.4 Horizontal gene transfer in bacteria. DNA shown in red originates from the donor cell. DNA shown in blue is that of the recipient cell. *Photo source: Dr. Linda M. Stannard, University of Cape Town/Science Source.*

a. DNA transfer by conjugation

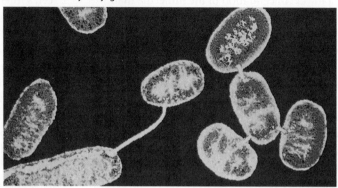

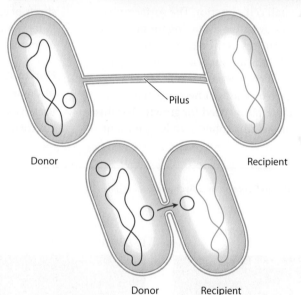

In conjugation, DNA (usually a plasmid) from a donor cell is transferred to an adjacent recipient cell. First, a pilus tethers the donor to the recipient and brings the cells together.

Once the cells are closely aligned, the DNA passes through a small opening formed between the cells.

b. DNA transfer by transformation

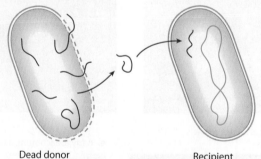

In transformation, DNA released into the environment by dead cells is taken up by a recipient cell.

c. DNA transfer by transduction

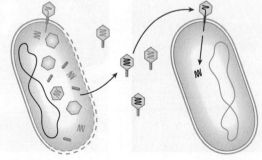

In transduction, DNA is transferred from a donor to a recipient cell by a virus.

Some bacteria synthesize thin strands of membrane-bound cytoplasm called pili (Chapter 5) that connect them to other cells (**Fig. 24.4a**). After joining to a second cell, the pilus contracts, drawing the two cells close together. A pore-like opening develops where the two cells are in close contact, providing a migration route for the direct cell-to-cell transfer of DNA. This process, called **conjugation,** commonly transfers plasmids from one cell to another, spreading novel genes throughout a population. Genes that confer resistance to antibiotics are a well-studied example of horizontal gene transfer by conjugation.

Genes can also be transferred from one cell to another without any direct bridge between cells. DNA released to the environment by cell breakdown can be taken up by other cells, in a process called **transformation** (**Fig. 24.4b**). Transformation was discovered when experiments showed that harmless strains of the bacteria causing pneumonia could be transformed into virulent strains by exposure to media containing dead cells of disease-causing strains (Chapter 3). Scientists reasoned that the transformation occurred because the living bacteria were taking up some substance from the dead cells. The "transforming substance" was later shown to be DNA. Today, biologists commonly use transformation in the laboratory to introduce genes into cells.

Viruses provide a third mechanism of horizontal gene transfer. Recall from Chapter 18 that viruses that invade bacterial cells sometimes integrate their DNA into the host's bacterial DNA. This viral DNA persists within the cells as they grow and divide. Before the virus leaves the cell to infect others, the viral DNA removes itself from the bacterial genome and is packaged in a protein capsule. The complete virus particle can then be released into the environment. This excision is not always precise, and sometimes genetic material from the bacterial host is incorporated into the virus. Viruses released from their host cell go on to infect others, bringing host-derived genes with them. Horizontal gene transfer by means of viruses is called **transduction** (**Fig. 24.4c**). It is common in nature and is also widely used in the laboratory to introduce novel genes into bacteria for medical research.

Horizontal gene transfer allows bacterial cells to gain beneficial genes from organisms distributed throughout the bacterial domain and beyond. Indeed, bacteria everywhere are constantly reshaping their genomes. Bacteria, therefore, are not merely the poor cousins of eukaryotes, unable to generate genetic diversity by sexual recombination. Instead, bacteria have their own highly efficient mechanisms for adding and subtracting genes that permit them to evolve and adapt rapidly to local conditions. Perhaps the most widely discussed and worrisome manifestation of horizontal gene transfer is the rapid spread of antibiotic resistance among pathogenic bacteria (Chapter 48).

Archaea form a second prokaryotic domain.

For many years, all prokaryotic cells were classified as bacteria. However, in 1977, George Fox and Carl Woese published a revolutionary hypothesis. Based on comparisons of RNA molecules from the small subunits of ribosomes (and later, comparisons of the genes for these RNAs), Fox and Woese argued that prokaryotic organisms actually fall into two distinct groups, as different from each other as either is from eukaryotes. If this is true, the tree of life must have three great branches: Bacteria, Eukarya, and the limb recognized by Fox and Woese, called Archaea (**Fig. 24.5**). Research since the late 1970s has confirmed this hypothesis. Like the cells of bacteria, archaeal cells are prokaryotic—they have no membrane-bounded nucleus and their genes are arrayed along a single circular chromosome. Moreover, cell size in Archaea is also limited by diffusion, and genetic diversity is promoted by horizontal gene transfer, much as in Bacteria.

In detail, however, Archaea are quite distinctive (**Table 24.1**). Their membranes are made from lipids that differ from the fatty acids found in bacterial and eukaryotic membranes. Archaea also show a diversity of molecules in their cell walls, but none has the peptidoglycan characteristic of Bacteria or the cellulose or chitin found in most eukaryotic cell walls.

Archaea differ from Bacteria in another intriguing way. Transcription in archaeons employs RNA polymerase and ribosomes more similar to those of eukaryotes than to the components of bacteria (Chapter 3). Moreover, many of the antibiotics that target protein synthesis in bacteria are ineffective against archaeons, suggesting fundamental differences in translation as well.

Many of the microorganisms first identified as Archaea have unusual physiological properties or inhabit extreme environments. For example, some archaeons live in acidic mine water at pH 1 (or less!). Others live in water salty enough to precipitate NaCl (**Fig. 24.6**), or in deep-sea hydrothermal vents where temperatures can exceed 100°C. We now know that many archaeons live under less extreme conditions in soils, lakes, and the sea. Indeed, they may be the most abundant organisms throughout much of the ocean.

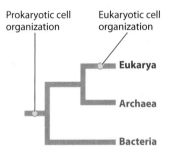

FIG. 24.5 A three-branched tree of life, based on sequence comparisons of the genes for the small subunit of ribosomal RNA. Archaea and Bacteria form distinct branches, although they inherited prokaryotic cell organization from a common ancestor. Recent data suggest that eukaryotes arose within the Archaea, making this domain paraphyletic.

TABLE 24.1 Principal Differences among Archaea, Bacteria, and Eukarya

CHARACTERISTIC	ARCHAEA	BACTERIA	EUKARYA
Cell contains a nucleus and other membrane-bound organelles	No	No	Yes
DNA occurs in a circular form*	Yes	Yes	No
Ribosome size	70S	70S	80S
Membrane lipids are ester-linked[†]	No	Yes	Yes
Photosynthesis with chlorophyll	No	Yes	Yes
Capable of growth at temperatures greater than 80°C	Yes	Yes	No
Histone proteins present in cell	Yes	No	Yes
Operons present in DNA	Yes	Yes	No
Introns present in most genes	No	No	Yes
Capable of methanogenesis	Yes	No	No
Sensitive to the antibiotics chloramphenicol, kanamycin, and streptomycin	No	Yes	No
Capable of nitrogen fixation	Yes	Yes	No
Capable of chemoautotrophy	Yes	Yes	No

* Eukaryote DNA is linear.
[†] Archaea membrane lipids are ether-linked.

FIG. 24.6 Archaeons in San Francisco Bay. These ponds are flooded and then evaporated to precipitate table salt. Salt-loving archaeons, conspicuous by their reddish purple pigment, thrive here. *Source: David Wall/Alamy.*

Self-Assessment Questions

1. How are bacterial and archaeal cells similar, and how are they different?
2. How do prokaryotic cells obtain nutrients, and how does this process put constraints on their size?
3. How does surface area and volume of a cell change with its size?
4. In eukaryotes, sexual reproduction is the main process that generates new gene combinations. How do bacteria generate new gene combinations in the absence of sexual reproduction?

24.2 AN EXPANDED CARBON CYCLE

In our discussion of metabolism in Chapters 7 and 8, we emphasized oxidation–reduction (redox) chemistry. In redox reactions, a pair of molecules reacts, with one molecule becoming more oxidized and the other becoming more reduced. If we follow the transfer of electrons in a redox reaction, we see that the molecule that is reduced gains electrons, whereas the molecule that is oxidized loses electrons. Thus, reduction requires a source of electrons, or an electron donor, while oxidation requires a sink for electrons, or an electron acceptor.

In photosynthesis carried out by plants, algae, and cyanobacteria (Chapter 8), carbon dioxide (CO_2) is reduced to form carbohydrates, and water is oxidized to oxygen gas (O_2). Water is the electron donor needed to reduce CO_2, with this process generating O_2 as a by-product. In respiration (Chapter 7), organic molecules are oxidized to CO_2, and O_2 is reduced to water. In this case, O_2 serves as the electron acceptor needed to oxidize organic molecules.

In all known photosynthetic eukaryotes, the photosynthetic reaction is **oxygenic,** or oxygen producing. There is a good reason for this behavior. Water occurs nearly everywhere and, in the oxygen-rich environments where most eukaryotic organisms thrive, no other molecule is available to donate electrons. Nonetheless, where oxygen is limited or absent, other electron donors such as hydrogen sulfide (H_2S) are available for photosynthesis and are used by photosynthetic microorganisms. Similarly, essentially all respiration in eukaryotic cells is **aerobic,** or oxygen using. Again, this makes good chemical sense because oxidation of carbohydrates using oxygen generates far more energy than can be obtained using other oxidants. Once again, where oxygen is absent, other electron acceptors can be used for respiration. The organisms that can use these alternative electron donors and electron acceptors are Bacteria and Archaea.

In oxygenic photosynthesis and aerobic respiration, carbon is cycled from the atmosphere to plants and animals, and back to the atmosphere, forming one circuit of the carbon cycle. Bacteria and Archaea can cycle carbon and harvest energy in unique ways, making them indispensable parts of nature (**Fig. 24.7**). Consequently, prokaryotic organisms are considered the real foundation of the carbon cycle, as they are capable of building and dismantling organic molecules throughout the full breadth of habitats on Earth.

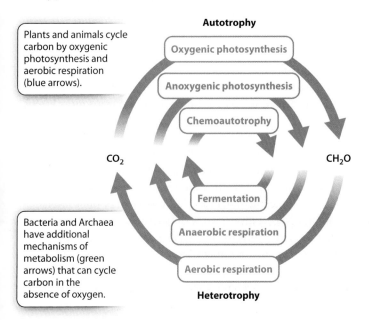

FIG. 24.7 The biological carbon cycle. Microbes have complementary metabolisms that enable them to cycle carbon in ways other than those used by plants and animals.

Many photosynthetic bacteria do not produce oxygen.

In saline bays along the west coast of Australia and many other low-lying tropical coastlines, carpets of a deep blue-green color cover tidal flats along restricted lagoons. High salinity and episodic exposure limit the colonization of these tidal flats by algae and animals (**Fig. 24.8**). The carpets shown in Fig. 24.8a comprise densely packed communities of (mostly) bacteria and archaeons called microbial mats. These communities have a predictable structure based on the depths to which light and oxygen penetrate into the mat (Figs. 24.8b and 24.8c).

In the upper layer, where both light and oxygen are present, the mat is dominated by cyanobacteria. These bacteria engage in photosynthesis just as plants and algae do, using water as an electron donor and releasing oxygen gas (the blue-green layer in Fig. 24.8b). It is the photosynthetic pigments in the cyanobacteria that impart the blue-green color to microbial mats. The cyanobacteria share this layer with heterotrophic bacteria, which get their carbon from preformed organic molecules. Both cyanobacteria and the heterotrophic bacteria respire aerobically, using O_2 to oxidize organic molecules to CO_2. Thus, at the mat surface, prokaryotic microorganisms carry out a biological carbon cycle much like plants and animals do.

What happens beneath the mat surface is a different story. Note in Fig. 24.8b that a distinct purple layer lies beneath the blue-green mat surface. Light penetrates into this layer, but oxygen does not, because aerobic respiration rapidly consumes O_2 within the mat (Fig. 24.8c). What kinds of organisms live where light is present and oxygen absent?

Like the cyanobacteria in the surface zone, many bacteria within the purple layer are photosynthetic, but they do not use water as an electron donor and so do not generate oxygen gas. Called **anoxygenic** because they do not produce oxygen, these photosynthetic bacteria use electron donors such as hydrogen sulfide (H_2S), hydrogen gas (H_2), ferrous iron (Fe^{2+}), and even the arsenic compound arsenite (AsO_3^{3-}). Where O_2 is present, these compounds oxidize quickly and so are not available for photosynthesis. Therefore, microorganisms that conduct anoxygenic photosynthesis are restricted to sunlit habitats where O_2 is absent.

The purple color in the mat layer comes from **bacteriochlorophyll,** a light-harvesting pigment that is closely related to the chlorophyll found in plants, algae, and cyanobacteria (**Fig. 24.9a**). Because of its chemical structure, bacteriochlorophyll absorbs light most strongly at wavelengths different from those absorbed by the cyanobacteria above them in the mat (**Fig. 24.9b**).

Anoxygenic photosynthesis differs from oxygenic photosynthesis in another important way: it employs only a single photosystem. Recall from Chapter 8 that a photosystem is a protein–pigment complex that absorbs light, which then powers photosynthesis. A single photosystem cannot capture enough energy to oxidize water and then use those electrons to reduce CO_2. Therefore, oxygenic photosynthesis relies on

FIG. 24.8 Microbial mats. Mats, like the example from (a) Western Australia, cover tidal flats and lagoons where salinity or other environmental factors inhibit animals and seaweeds. As shown by the example from (b) Cape Cod, microbial mats have sharp vertical gradients of light, oxygen, and other chemical compounds, so they support nearly all the energy metabolisms found in (c) the expanded carbon cycle. *Sources: a. Andrew Knoll, Harvard University; b. Republished with permission of Springer Science+Business Media, from The Prokaryotes, Vol. 2: Ecophysiology and Biochemistry 2006, The Phototrophic Way of Life, Jörg Overmann and Ferran Garcia-Pichel, Figure 5. Springer eBook. Photo courtesy Jörg Overmann.*

a.

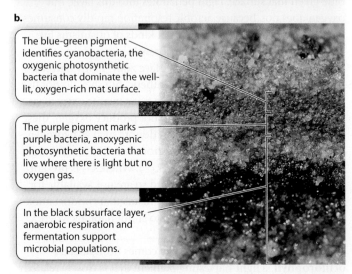

b.

The blue-green pigment identifies cyanobacteria, the oxygenic photosynthetic bacteria that dominate the well-lit, oxygen-rich mat surface.

The purple pigment marks purple bacteria, anoxygenic photosynthetic bacteria that live where there is light but no oxygen gas.

In the black subsurface layer, anaerobic respiration and fermentation support microbial populations.

c.

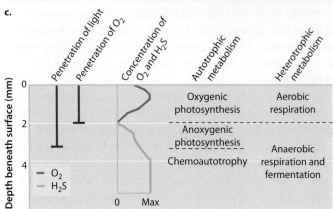

FIG. 24.9 Bacteriochlorophyll *a* and chlorophyll *a*. Note that bacteria absorb wavelengths of light that are outside the visible spectrum.

a. Structures of light-absorbing pigments

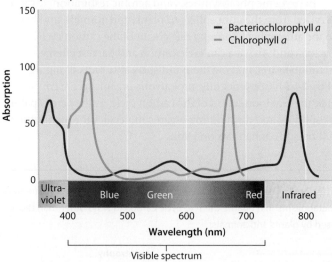

The differences in structure (shown in red) between the bacteriochlorophyll used by anoxygenic photosynthetic bacteria and the chlorophyll used by cyanobacteria and chloroplasts in phytosynthetic eukaryotes account for the differences in wavelengths absorbed by the two molecules.

b. Absorption spectra

two photosystems that operate in tandem. With only a single photosystem, bacteria that carry out anoxygenic photosynthesis use electron donors that donate their electrons more easily than water.

Many bacteria respire without oxygen.

Light, but not oxygen, penetrates into the purple layer in Fig. 24.8b, but neither light nor oxygen reaches the black layer deep within the microbial mat. Even so, both of these layers are inhabited by a diverse array of heterotrophic Bacteria and Archaea that metabolize the organic molecules originally synthesized by photosynthetic bacteria higher in the mat community.

Because no oxygen is present in these layers, the organisms found there must use other molecules as electron acceptors in cellular respiration, including oxidized forms of nitrogen (NO_3^-), sulfur (SO_4^{2-}), manganese (Mn^{4+}), iron (Fe^{3+}), and even arsenic (AsO_4^{3-}). Such microorganisms thrive in many environments where oxygen is lacking.

Still other types of heterotrophs live in deeper layers of microbial mats. These heterotrophs may rely on **fermentation** rather than cellular respiration as a way of extracting energy from organic molecules. Whereas cellular respiration entails the full oxidation of carbon compounds to CO_2, fermentation involves the partial oxidation of carbon compounds to molecules that are less oxidized than CO_2 (Chapter 7). On the one hand, fermentation has an advantage over cellular respiration in that it does not require an external electron acceptor, such as O_2. On the other hand, fermentation yields only a modest amount of energy.

Fermentation plays an important role in oxygen-poor environments that are rich in organic matter, such as landfills and the digestive tracts of animals. In these settings, the breakdown of organic molecules typically requires more than one organism, each of which is able to metabolize a different intermediate. These cooperating groups of fermenters can break down substances that could not be metabolized by any one organism alone. In essence, groups of fermenters form an ecological solution to the metabolically difficult problem of gaining energy from complex organic molecules. Fermentation is widespread in Bacteria and Archaea, but is of only minor importance in most organisms belonging to Eukarya. The exception is yeasts, which thrive in the sugar-rich environment of rotting fruit, as well as in breweries throughout the world.

In summary, where O_2 is present, many different kinds of organisms participate in the carbon cycle. Photosynthetic plants, algae, and cyanobacteria all transform CO_2 into organic molecules, and animals, fungi, single-celled eukaryotes, bacteria, and archaeons return CO_2 to the environment through aerobic respiration. However, only prokaryotic heterotrophs can complete the recycling of organic carbon in oxygen-poor environments. Thus, in carrying out anaerobic respiration or a diversity of fermentation reactions, prokaryotes play a major role in the carbon cycle.

Photoheterotrophs obtain energy from light but obtain carbon from preformed organic molecules.

In Chapter 6, we saw that organisms have two sources of energy: the sun (phototrophs) and chemical compounds (chemotrophs). Organisms also have two sources of the carbon needed for growth and other functions: inorganic molecules like CO_2 (autotrophs) and organic molecules like glucose (heterotrophs). Thus, organisms can acquire the energy and carbon they need in four different ways (see Fig. 6.1).

Up to this point, we have considered only two of these ways. Plants, algae, and cyanobacteria are photoautotrophs, gaining energy from sunlight and carbon from CO_2, whereas animals, fungi, and many prokaryotes are chemoheterotrophs, gaining both energy and carbon from organic molecules taken up from the environment. Among the Bacteria and Archaea, additional mechanisms of metabolism are possible. Some use the energy from sunlight to make adenosine triphosphate (ATP), just as plants do, but rather than reducing CO_2 to make their own organic molecules, they rely on organic molecules obtained from the environment as their source of carbon. These organisms are **photoheterotrophs**.

Photoheterotrophy can be advantageous in environments that are rich in dissolved organic compounds. In particular, organisms can use all of their absorbed light energy to make ATP while directing all absorbed organic molecules toward growth and reproduction. Photoheterotrophic bacteria have a single photosystem, so they can live where oxygen is limited or absent. The purple layer of a microbial mat very likely contains heliobacteria or green nonsulfur bacteria, as both are photoheterotrophs that take advantage of the large amounts of organic matter generated by photoautotrophs in the mat.

Chemoautotrophy is a uniquely prokaryotic metabolism.

On the deep seafloor, animals are relatively uncommon, their abundance limited by the slow descent of organic matter from the ocean's surface. Perhaps surprisingly, then, animal populations can be remarkably dense around hydrothermal vents (**Fig. 24.10**), places where hot water from the Earth's interior is expelled into the overlying sea. Why are animals so abundant here?

FIG. 24.10 Deep-sea hydrothermal vents. Chemoautotrophs fuel the carbon cycle in these environments. *Source: Science Source/Getty Images.*

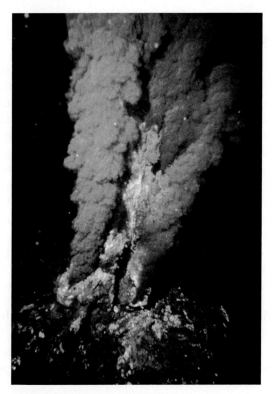

Many of the animals found around deep-sea hydrothermal vents subsist on a thriving community of microorganisms. These microorganisms gain carbon by reducing CO_2 to form carbohydrates. However, they obtain the energy to fuel this process not from sunlight, but rather from chemical reactions. Hence, these microorganisms are **chemoautotrophs**. They oxidize molecules and ions such as H_2, H_2S, and Fe^{2+} to generate the ATP and reducing power required to incorporate CO_2 into organic molecules. In chemoautotrophic metabolism, the electron acceptor is most commonly O_2 or nitrate (NO_3^-). Thus, chemoautotrophy requires access to both oxidized (O_2 and nitrate) and reduced (H_2, H_2S, and Fe^{2+}) substances.

Oxidized and reduced molecules react readily with each other, so they are rarely present in the same environment. For this reason, chemoautotrophs tend to live along the interface between an environment where oxygen is present and one in which oxygen is absent. Notably, chemoautotrophic prokaryotes thrive along hydrothermal vents, where reduced gases from Earth's interior meet oxygen-rich seawater.

> ### Self-Assessment Questions
>
> 5. How can photosynthesis occur without the production of oxygen, and how can respiration occur without requiring oxygen?
> 6. How do photoautotrophs such as cyanobacteria differ from photoheterotrophs such as heliobacteria?
> 7. How did the biological carbon cycle work on the early Earth, where oxygen gas was essentially absent from the atmosphere and oceans?

24.3 SULFUR AND NITROGEN CYCLES

Let's think further about prokaryotic metabolism. We saw that some photosynthetic and chemoautotrophic bacteria oxidize reduced sulfur compounds like H_2S to sulfate (SO_4^{2-}), whereas some heterotrophic bacteria reduce sulfate to H_2S. These reverse chemical transformations may remind you of the carbon cycle outlined in Chapters 7 and 8, in which plants generate oxygen and animals consume it. As Bacteria and Archaea have already shown us, mechanisms of metabolism at work where oxygen is absent can also complete the carbon cycle, taking in and releasing CO_2. They do so by coupling the cycling of carbon to the cycling of other elements—in this case, sulfur. Once again, we see complementary metabolic pathways. The expanded carbon cycle of Bacteria and Archaea promotes the cycling of sulfur, nitrogen, and other biologically important elements.

Bacteria and archaeons dominate Earth's sulfur cycle.

The human body is mostly carbon, oxygen, and hydrogen, but we also contain approximately 0.2% sulfur by weight. Sulfur is a component of the amino acids cysteine and methionine and, therefore, is present in many proteins (Chapter 2). Sulfur is also present in the iron–sulfur clusters that are key components of enzymes fundamental to electron transfer (Chapter 6) and in some vitamins. Where do we get the sulfur we need? From the food we eat. In particular, protein in steak or fish provides an especially rich source of sulfur. Where do cows and fish get the sulfur they need? Like humans, from the food they eat. Food chains don't go on forever, so somewhere in the system, some organisms must take up inorganic sulfur from the environment. **Primary producers**, the organisms that reduce CO_2 to form carbohydrates, fulfill this role.

Plants are the dominant primary producers on land. In the process called **assimilation**, plants take up sulfate (SO_4^{2-}) ions from the soil and reduce them within their cells to hydrogen sulfide (H_2S) that can be incorporated into cysteine and other biomolecules (**Fig. 24.11**). Algae and photosynthetic bacteria do much the same thing in lakes, rivers, and oceans. Why don't primary producers take up H_2S directly? There are two answers to this question. First, H_2S is rapidly oxidized in the presence of oxygen, so it does not occur in environments where oxygenic photosynthesis is common. Second, H_2S is toxic to most eukaryotic organisms, so plants and algae do not thrive where it is abundant. (The H_2S produced within eukaryotic cells has a short lifetime and is restricted to intracellular sites distant from those that are vulnerable to its toxic effects.)

We've now seen half the sulfur cycle, the conversion of sulfate to H_2S within cells. But how did sulfate molecules get into the soil in the first place? After cells die, fungi and bacteria decompose the cells, returning their carbon, sulfur, and other compounds to the environment. Reduced sulfur compounds released from decomposing cells are oxidized to sulfate by

FIG. 24.11 The sulfur cycle. Plants take up sulfur in the form of sulfate ions for incorporation into proteins and other compounds, and fungi release sulfur during decomposition. However, the major role in cycling sulfur through the biosphere is played by microbes that oxidize or reduce sulfur compounds to gain energy or carbon.

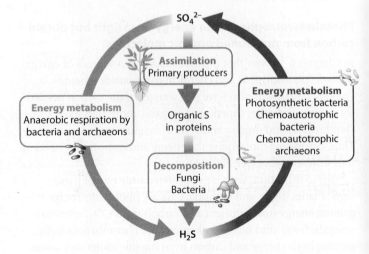

bacteria and archaeons, completing the cycle. These microbes are either chemoautotrophs that obtain energy by oxidizing H_2S or photosynthesizers that use H_2S as the electron donor. Some of this sulfur is assimilated by primary producers, but where oxygen is not available, much of it is consumed by heterotrophic bacteria. These bacteria use sulfate rather than oxygen as the final electron acceptor in respiration and reduce it to H_2S. In fact, most of the sulfur cycled biologically is used to drive energy metabolism in oxygen-poor environments, rather than to build proteins.

Note that eukaryotes use neither H_2S for photo- (or chemo-) synthesis nor sulfate in respiration. Thus, the biological sulfur cycle is completed by bacteria and archaeons alone.

The nitrogen cycle is also driven by bacteria and archaeons.

The model provided by the sulfur cycle can be extended to other biologically important elements. Nitrogen is the fourth most abundant element in the human body, contributing 3% by weight as a component of proteins, nucleic acids, and other compounds. As is true of sulfur, primary producers incorporate inorganic forms of nitrogen into biomolecules, and we get the nitrogen we need from our food. As we'll see, however, there is a twist.

Let's examine the biological nitrogen cycle (**Fig. 24.12**). Carbon is a minor component of the atmosphere and sulfur only a trace constituent; the air we breathe consists mostly (78%) of nitrogen gas (N_2). Indeed, most of the nitrogen at Earth's surface resides in the atmosphere. Although nitrogen is plentiful, the nitrogen available to primary producers in many environments is limited. The reason is that neither plants nor algae can make use of nitrogen gas. Indeed, only certain bacteria and archaeons can reduce N_2 to ammonia (NH_3), a form of nitrogen that can be incorporated into biomolecules.

The process of converting N_2 into a biologically useful form such as ammonia is called **nitrogen fixation,** and it is one of the most important reactions ever evolved. Farmers plant soybeans to regenerate nitrogen nutrients in soils, but it is not the soybean that fixes nitrogen. Instead, nodules on soybean roots harbor nitrogen-fixing bacteria (**Fig. 24.13**). Plants like soybean have entered into an intimate partnership with nitrogen-fixing bacteria to obtain the biologically useful forms of nitrogen needed for growth (Chapter 27).

The nitrogen cycle, then, begins with nitrogen fixation, and it continues as further chemical transformations are carried out by many additional metabolisms, most of them confined to bacteria and archaeons. The nitrogen taken up by plants and animals is released in the form of ammonia by the decomposition of dead cells. Ammonia can be taken up by primary producers, but more commonly it is oxidized to nitrate (NO_3^-) by chemoautotrophic bacteria, a process called **nitrification** (see Fig. 24.12). Because this process works efficiently in both soil and water, nitrate is the most common form of nitrogen available to life in both environments. Most primary producers can assimilate nitrate as well as ammonia.

Anaerobic respiration completes the nitrogen cycle by returning N_2 to the atmosphere. Some bacteria can use nitrate as an electron acceptor in respiration, a process known as **denitrification** (see Fig. 24.12). Accepting an electron sets off the transformation of nitrate into N_2. Because denitrification is so

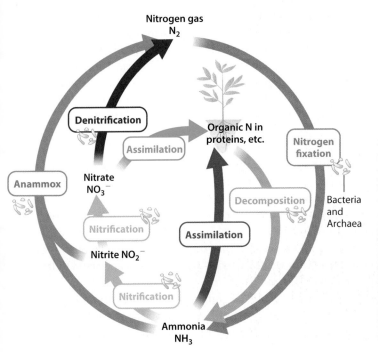

FIG. 24.12 The nitrogen cycle. Unique bacterial and archaeal metabolisms include nitrogen fixation, nitrification and denitrification, and anammox.

FIG. 24.13 Roots of a soybean plant. The nodules contain bacteria capable of reducing nitrogen gas to ammonia. *Source: Scimat/Getty Images.*

efficient at returning nitrogen to the atmosphere as nitrogen gas, nitrogen fixation must continually make nitrogen available for organisms. Without nitrogen fixation, the nitrogen cycle would slowly wind down, and the productivity of Earth's biota—that is, of all its organisms—would wind down with it. This is why nitrogen fixation is such an important process.

Until the 1990s, the nitrogen cycle was described—completely, it was thought—in terms of the metabolic reactions just discussed. Then, a new and unexpected form of nitrogen metabolism was discovered in the oceans. A number of prokaryotic microorganisms function as chemoautotrophs by reacting ammonium ion (NH_4^+) with nitrite (NO_2^-) to gain energy. The reaction is shown here:

$$NH_4^+ + NO_2^- \rightarrow N_2 + 2H_2O$$

Note that this is a redox reaction, like other reactions of energy metabolism. In this case, the ammonium ion is oxidized to nitrogen gas, and nitrite is reduced to nitrogen gas. Water is produced as a by-product. This reaction, called **anammox** (<u>an</u>aerobic <u>am</u>monia <u>ox</u>idation), provides only a small amount of energy, but it supports many bacteria and archaeons in oxygen-depleted parts of the ocean. In doing so, it provides a second major route by which fixed nitrogen returns to the atmosphere as N_2 (see Fig. 24.12).

The biological nitrogen cycle, then, begins and ends with prokaryotic metabolisms. Bacteria and archaeons fix N_2 into a biologically available form, and nitrifying bacteria oxidize ammonia to nitrate. Other prokaryotes gain energy from denitrification and anammox, returning N_2 to the air. Unlike the carbon and sulfur cycles, the nitrogen cycle does not rely at any point on elements stored as minerals in rocks and sediments. Most of Earth's nitrogen was released to the atmosphere as gas early in our planet's history, and today movement into sediments and out of volcanoes contributes relatively little to the cycle. Humans, however, contribute in important ways, most notably through the industrial synthesis of nitrate fertilizer (Chapter 48).

Self-Assessment Questions

8. What are the roles of bacteria and archaeons in the sulfur and nitrogen cycles?
9. How would the nitrogen cycle operate in the absence of bacteria and archaeons?

24.4 BACTERIAL DIVERSITY

Traditionally, bacterial taxonomy was a laborious process. Microbiologists grew bacteria in beakers or on petri dishes, carefully separating and enriching individual populations until only one type of bacterium—that is, a pure culture—was present. Species were identified on the basis of details of cell structure and division observed under the microscope, and metabolic capabilities were established through experiments and reactions to chemical stains or antibiotics. By the 1980s, a few thousand species had been described.

With the sequencing technologies now available, bacterial species are increasingly being characterized by their DNA sequences, especially the genes for the small subunit of ribosomal RNA (rRNA). A novel application of sequencing technology paved the way for modern studies of bacterial diversity and ecology. Microbiologists reasoned that the same procedures used to extract, amplify, and sequence genes from pure cultures in the lab could be used to study bacteria in nature. That is, they could extract and read gene sequences directly from soil or seawater, thereby revealing the diversity of the microbial communities in these environments.

The results were stunning. Most bacteria in nature—perhaps 99% or more—are species that have not been cultured in the lab. Indeed, species that have been studied in pure cultures are commonly only minor components of natural communities. To give just one illustration, a bacterium called SAR11 was originally identified on the basis of gene sequences amplified from seawater. SAR11 caught the attention of microbiologists because it makes up approximately one-third of all sequences identified in samples from the surface ocean. How could we be ignorant of a bacterium that must be among the most abundant organisms on Earth? The answer is simple: no one had cultured it. Recently, SAR11 has been cultured. Formally described as *Pelagibacter ubique* (*ubique* is Latin for "everywhere"), this bacterium is now known to be one of several closely related heterotrophs that play an important role in recycling organic molecules dissolved in the sea.

More recently, biologists have figured out how to sequence whole genomes from environmental samples. Using this sequence information, microbiologists have begun to match uncultured populations with specific metabolisms. We no longer need be content just to know from molecular sequences what is living in a given environment. Instead, we can now begin to understand what the different microorganisms are doing there. For this reason, another revolution in our understanding of the microbial world has begun (**Fig. 24.14**).

Bacterial phylogeny is a work in progress.

In Chapter 22, we introduced molecular phylogenies, which are constructed by comparing gene or protein sequences. Through such comparisons, we can draw conclusions about evolutionary relatedness among populations in nature. Gene sequencing made it possible to test traditional bacterial groupings based on form and physiology. Some traditionally defined groups—for example, the cyanobacteria—passed the test and stand today as well-defined bacterial groups. Many others did not.

Molecular sequence comparisons bring enormous opportunities to the study of prokaryotic diversity and evolution, but they introduce new problems as well. One set of problems

HOW DO WE KNOW?

FIG. 24.14

How many kinds of bacterium live in the oceans?

BACKGROUND The oceans are full of bacteria. By some estimates, the sea harbors more than 10^{29} bacterial cells. How diverse are these bacterial communities?

METHOD 1 Different types of bacterium can be recognized by the unique nucleotide sequences of their genes. In fact, individual types of bacterium can be identified on the basis of the nucleotide sequence in a relatively small region of the gene for the small subunit of ribosomal RNA (rRNA). This molecular "barcode" can be sequenced quickly and accurately for samples of DNA drawn from the environment.

Using this strategy, American biologist Mitchell Sogin and colleagues associated with the International Census of Marine Life obtained samples of seawater and seafloor sediment from deep-ocean environments. In the laboratory, the genes that code for rRNA were amplified by means of the polymerase chain reaction (PCR, a technique that allows a targeted region of a DNA molecule to be replicated into many copies), separated by gel electrophoresis, and sequenced. These rRNA sequences provide a library of tags that document bacterial diversity.

METHOD 2 In another set of experiments, scientists collected samples of seawater from the Sargasso Sea. From these samples, they amplified whole genomes—more than 1 billion nucleotides' worth. They used small-subunit rRNA and other genes to characterize species richness, and a large assortment of additional sequence data to understand genetic and physiological diversity.

RESULTS Both surveys found that the bacterial diversity of marine environments is much higher than had been thought. The barcode survey (method 1) found as many as 20,000 distinct types of bacteria in a single liter of seawater (bottom graph). From these results, it is estimated that as many as 5–10 million different microbes may live in the world's oceans. Some of these bacteria are abundant and widespread, but most are rare. The genomic survey (method 2) of the Sargasso Sea found more than 1800 distinct bacteria in sea-surface samples (top graph). Remarkably, this survey identified more than 1.2 million previously unknown genes within these genomes.

CONCLUSION The bacterial diversity of the oceans is huge and still largely unexplored. Both barcode and genomic surveys are continuing, and promise a more nearly complete accounting of marine diversity. Together, these surveys will help us understand how marine bacteria function and how they sort into communities.

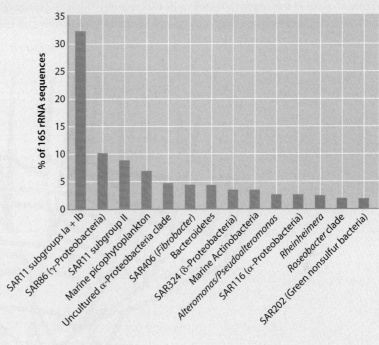

Relative abundance of different bacterial groups in samples from the Sargasso Sea. Each bar shows the percentage of all small-subunit rRNA sequences represented by a unique small-subunit rRNA sequence. Note the abundance of SAR11, which has only recently been characterized.

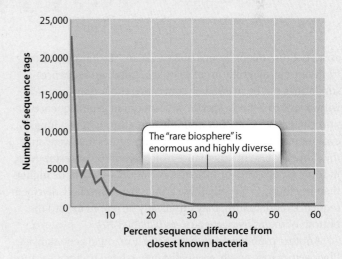

Computer-aided comparison of sequences from the world's oceans shows that approximately 75% of all sequences collected are similar to other sequences in the library of sequence tags. However, 25% are distinctive, forming a rare but enormously diverse pool of bacterial types.

SOURCES Venter, J. C., et al. 2004. "Environmental Genome Shotgun Sequencing of the Sargasso Sea." *Science* 304:66–74; Sogin, M. L., et al. 2006. "Microbial Diversity in the Deep Sea and the Underexplored 'Rare Biosphere'" *Proceedings of the National Academy of Sciences USA* 103:12115–12120.

FIG. 24.15 Bacterial and archaeal phylogeny. Some scientists argue that the evolution of prokaryotes should be viewed not as a tree, but rather as a series of branches that diverge and then come back together. This arrangement of branches represents genes that diverge from one another but then come to reside together in new organisms through horizontal gene transfer. *Data from W. F. Doolittle, 2000, "Uprooting the Tree of Life,"* Scientific American *282:90–95.*

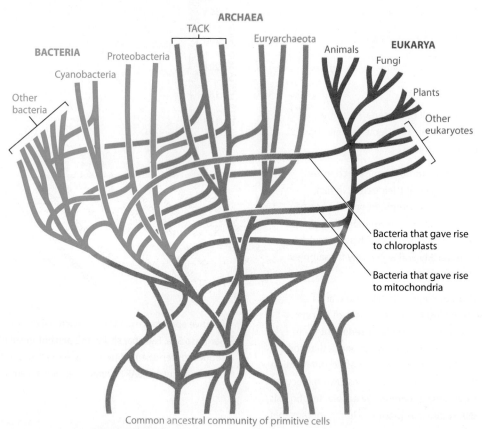

arises because major groups of Bacteria and Archaea separated from one another billions of years ago. Inherited sequence similarities can be masked by continuing molecular evolution within groups. That is, specific nucleotides in gene sequences may change not once, but multiple times throughout a long evolutionary history, erasing molecular features that were once the same because of inheritance from a common ancestor. Also, in some groups rates of sequence evolution appear to be higher than in others. As a result, these groups have especially divergent sequences and, in turn, may mistakenly be placed toward the bottom of their phylogenetic trees (Chapter 22).

Another problem arises because of horizontal gene transfer. Phylogenies may falsely group distantly related bacteria because they contain genes passed on by conjugation, transformation, or transduction. Statistical analyses of complete bacterial genomes suggest that at least 85% of all genes in bacterial genomes have been transferred horizontally at least once. Such apparently rampant gene exchange might well doom attempts to build a bacterial tree of life. As an alternative, some biologists argue that bacterial evolution is more properly depicted as a complex bush with many connecting branches (**Fig. 24.15**).

Other microbiologists are more optimistic. Small-subunit rRNA genes, for example, show little evidence of horizontal transfer, suggesting that they might faithfully record the evolutionary history of the organisms in which they occur. Nonetheless, we need to be cautious when using bacterial trees to study the evolution of specific metabolic or physiological characteristics. Nitrogen fixation, photosystems, drug tolerance—such features depend on genes known to jump from branch to branch on the tree.

To date, molecular signatures have been extremely useful in identifying major groups of bacteria, but their contributions are less clear in establishing branching order among these groups. Nearly 100 major groups of bacteria (each one very roughly equivalent to a eukaryotic phylum) have been recognized. Nearly half of these groups cannot now be cultured, but rather have been identified solely through the cloning of genes from environmental samples. To support our discussion of bacterial diversity, we present a phylogeny based on comparisons of whole genomes in **Fig. 24.16**. As you examine this figure, remember that microbiologists continue to debate branching order on the bacterial tree. Remarkably, the large branch labeled

FIG. 24.16 Evolutionary relationships among bacteria, inferred from data from 1000 complete genomes. Photosynthetic groups are shown in green (the Acidobacteria have one known photoheterotrophic member). Other groups discussed in the text are shown in red. The branch labeled "Candidate phyla" has only recently been discovered. *Data from A. Hug et al., 2016, Nature Microbiology 1:16048; Mukherjee et al., 2017, Nature Biotechnology 35:676–683.*

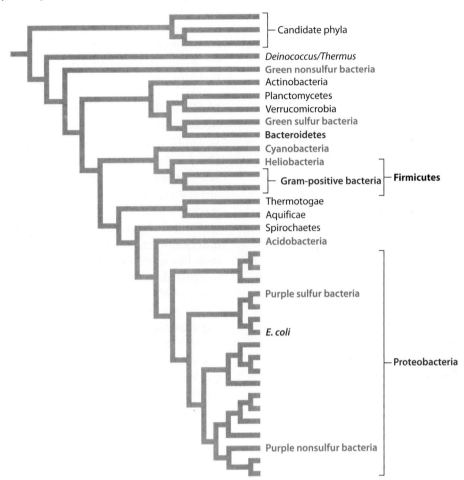

Candidate Phyla consists of bacteria that were unknown before 2015. Few of these microbes have been cultured, but their genomes suggest that most cannot synthesize nucleotides or amino acids; thus, they may live in symbiotic association with other organisms. We still have a lot to learn about bacterial diversity on land and in the sea.

What, if anything, is a bacterial species?

What do we mean when we talk about bacterial species? As we saw in Chapter 21, Ernst Mayr defined species as "groups of interbreeding natural populations that are reproductively isolated from other such groups." This widely accepted definition is called the biological species concept, but a moment's reflection will show that it can't be applied to the most abundant organisms on Earth—Bacteria and Archaea. Organisms in these domains cannot interbreed because there is no sexual reproduction. How, then, do we think about species in the prokaryotic domains?

Some microbiologists argue that we should abandon the concept of species when talking about bacteria, especially since distantly related bacteria can exchange genes. Nonetheless, nature is full of microbial populations with well-defined features of form and function.

In the age of molecular sequence comparisons, bacterial species have sometimes been defined as populations whose small-subunit rRNA sequences are more than 97% identical. It would be useful, however, to conceive of bacterial species in terms of a functional process, much as Ernst Mayr did for eukaryotic organisms. In plants, animals, and single-celled eukaryotes, interbreeding ensures that populations share the same pool of genes. That is, interbreeding provides a counterforce to the mutations and selection pressures that promote divergence.

A similar process exists in bacteria. In early research on the population genetics of bacteria, biologists observed that genetic diversity in laboratory cultures gradually increases through time and then rapidly decreases with the emergence of a successful variant that outcompetes the rest. This kind of episodic loss of diversity is called **periodic selection**. Some biologists argue that the periodic selection process provides a means of recognizing bacterial species: populations selected during the same episode of periodic selection belong to a single species. This is a useful way to define bacterial species, but other definitions are possible, and as yet there is no consensus among microbiologists.

Proteobacteria are the most diverse bacteria.

Given the immense and still unfolding diversity of bacteria, we cannot hope to discuss all major groups here. Instead, we introduce a few of the most diverse, ecologically interesting, or medically important groups of bacteria.

Proteobacteria are far and away the most diverse of all bacterial groups (Fig. 24.16). Defined largely by similarities in their rRNA sequences, Proteobacteria include many of the organisms that populate our expanded carbon cycle and other biogeochemical cycles. For example, they include anoxygenic photosynthetic bacteria and chemoautotrophs that oxidize NH_3, H_2S, and Fe^{2+}, as well as bacteria able to respire using SO_4^{2-}, NO_3^-, or Fe^{3+}.

Many Proteobacteria have evolved intimate ecological relationships with eukaryotic organisms. Some of these are mutually beneficial, such as the association between soybeans and nitrogen-fixing bacteria in root nodules. Other Proteobacteria are our worst pathogens: the *Rickettsia* bacteria that cause typhus and spotted fever, the *Vibrio* species that trigger cholera, the *Salmonella* bacteria responsible for typhoid fever and food poisoning, and the *Pseudomonas aeruginosa* strains that infect burn victims and others whose resistance to infection is reduced. *Escherichia coli*, famous as a laboratory model organism (see Fig. 24.4b), is usually a harmless presence in our intestinal tract, but pathogenic strains can cause serious diarrhea and even death.

The gram-positive bacteria include organisms that cause and cure disease.

In 1884, the Danish scientist Hans Christian Gram developed a diagnostic dye that, after washing, is retained by bacteria with thick peptidoglycan walls, but not by those with thin walls. Bacteria that retain the dye are now called **gram-positive bacteria**. Molecular sequence comparisons confirm that most **gram-positive bacteria** form a well-defined branch of the bacterial tree (Fig. 24.16). One species in this group, *Bacillus subtilis*, is the focus of many studies exploring basic molecular function in bacteria. All of us harbor *Streptococcus* species in our mouths, and many of us have suffered from strep throat, produced by streptococcal pathogens. More serious infections, including some forms of meningitis, are also associated with streptococci. *Staphylococcus* bacteria are commonly found on human skin, but only occasionally do we suffer from maladies such as staph infection, toxic shock syndrome, or food poisoning induced by these bacteria.

In contrast to these pathogens, gram-positive bacteria called streptomycetes have a remarkable property that has proved invaluable to medicine. More than half of the species in this group secrete compounds that kill other bacteria and fungi. Streptomycetes have provided us with tetracycline, streptomycin, erythromycin, and many other antibiotics that combat infectious disease.

Photosynthesis is widely distributed on the bacterial tree.

Molecular studies confirm that all bacteria capable of oxygenic photosynthesis form a single branch of the bacterial tree, known as the **cyanobacteria**. Cyanobacteria can be found in environments that range from deserts to the open ocean. Their diverse species include unicellular rods and spheroidal cells, as well as multicellular balls and filaments (**Fig. 24.17**). Some filamentous cyanobacteria can even form several different cell types, including specialized cells for nitrogen fixation and resting cells that provide protection when the local environment does not favor growth.

In contrast to the monophyly of cyanobacteria, molecular sequence comparisons show that *anoxygenic* photosynthesis is distributed widely on the bacterial tree (see Fig. 24.16). Think of the purple layer beneath the surface of many microbial mats (see Fig. 24.7). The vivid color comes from the light-sensitive pigments of the aptly named purple bacteria, found within the Proteobacteria. Purple bacteria are capable of photoautotrophic growth using bacteriochlorophyll and a single photosystem. Most show evidence of metabolic versatility: when light or appropriate electron donors are absent, they can grow as heterotrophs.

A green layer lies beneath the surface of other microbial mats, its color caused by light reflected by the pigments of another group of photosynthetic bacteria, called green sulfur bacteria because they commonly gain electrons from hydrogen sulfide and deposit elemental sulfur on their walls. Unlike most groups of photosynthetic bacteria, the green sulfur bacteria are intolerant of oxygen gas. Oddly, a species of green sulfur bacteria has been found in hydrothermal rift vents 2 km beneath the surface of the ocean. Light doesn't penetrate to these depths, so if these organisms harvest light, it must come from incandescent lava as it erupts on the bottom of the ocean.

Light-harvesting bacteria occur on several other branches of the bacterial tree—the photoheterotrophic heliobacteria, for example, and the green nonsulfur bacteria often found in freshwater hot springs. Despite a century of research, our knowledge of photosynthesis within the bacterial tree remains incomplete, as illustrated by the discovery in 2007 of unique light-harvesting bacteria in Yellowstone National Park. Genomic surveys of hot springs turned up evidence of a previously unknown microorganism containing bacteriochlorophyll, and

FIG. 24.17 Diversity of form among cyanobacteria. *Sources: a. Dr. Ralf Wagner; b. blickwinkel/Alamy; c. Hecker/Sauer/AGE Fotostock; d. Jason Oyadomari.*

a. *Aphanothece*, small single cells

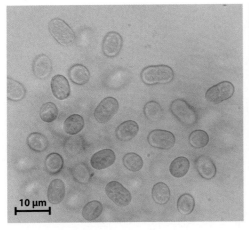

b. *Chroococcus*, larger cells bound into a colony by a common extracellular envelope made of polysaccharides

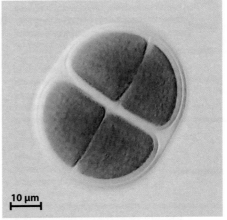

c. *Oscillatoria*, filaments with no cell differentiation

d. *Anabaena*, strings of undifferentiated cells at either end of the filament, flanking two elongated resting cells, and in the middle, a heterocyst specialized for nitrogen fixation

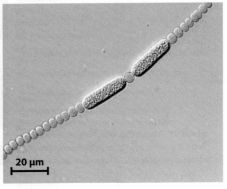

gene sequencing established that this organism belongs to the phylum Acidobacteria. There is no evidence for carbon fixation, however, and recent success in culturing this Yellowstone bacterium confirms that it is a photoheterotroph.

Self-Assessment Question

10. Name and describe three major groups of Bacteria.

24.5 ARCHAEAL DIVERSITY

We have already noted that Archaea and Bacteria are prokaryotic, but differ in their membrane and wall composition, molecular mechanisms for transcription and translation, and many other biological features (see Table 24.1). As a result, the two groups tend to thrive in very different environments. Many archaeons tolerate environmental extremes such as heat and acidity. In fact, for many environmental conditions, archaeons define the known limits of life. Do these disparate environments have anything in common that might help us to understand the biology of Archaea? One proposal is that archaeons thrive in environments where the energy available to fuel cell activities is limited—where there is no light, or a limited abundance of fuels and oxidants for chemoautotrophy or respiration, or both. That is, archaeons commonly live where energy resources are too low to support bacteria and eukaryotes.

Bacteria and Archaea diverged from their last common ancestor more than 3 billion years ago. Just as with the Bacteria, the diversity of Archaea defies easy categorization. Archaea include many distinct types of organisms, and new groups are being discovered every year. In turn, as new archaeons are discovered, the shape of the archaeal tree changes. **Fig. 24.18** shows a recent phylogeny, in which the three main groups are labeled Euryarchaeota, TACK, and DPANN. The names of the latter two are based on the initials of major subgroups that make

FIG. 24.18 Phylogeny of Archaea. Different branches of the domain have evolved distinct temperature preferences, salt tolerance, acid tolerance, and production of methane. *Data from Williams, T. A. et al., 2017, "Integrative Modeling of Gene and Genome Evolution Roots the Archaeal Tree of Life," Proceedings of the National Academy of Sciences USA 114:E4602–E4611.*

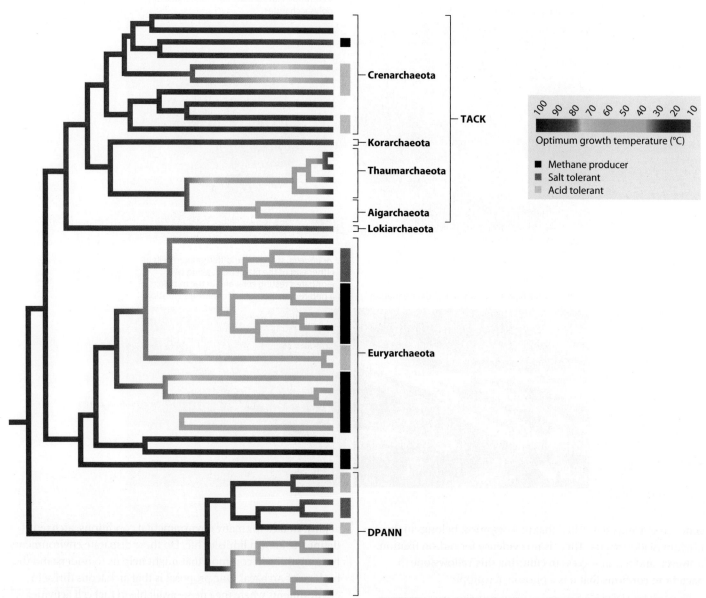

up these branches. Remarkably, the diverse DPANN group has come to light only over the past few years. Like the Bacteria, the Archaea include forms able to respire aerobically or anaerobically, as well as chemoautotrophs and fermenters. There are, however, metabolic differences between the two domains. No Archaea carry out photosynthesis using chlorophyll and related pigments, but some Archaea exhibit forms of energy metabolism unknown in bacteria or eukaryotes.

The archaeal tree has anaerobic, hyperthermophilic organisms near its base.

Archaeons that branch from lower nodes of the archaeal tree exhibit a remarkable ability to grow and reproduce at temperatures of 80°C or more. Called **hyperthermophiles**, these organisms do not simply tolerate high temperatures; they require them for survival. At present, the world record for high-temperature growth is held by *Methanopyrus kandleri*, an archaeon that grows at 122°C among hydrothermal vents on the seafloor beneath the Gulf of California. Most of these heat-loving archaeons are anaerobic. They live as anaerobic respirers or as chemoautotrophs that obtain energy from the oxidation of hydrogen gas.

Why do anaerobic hyperthermophiles sit on the lowest branches of the archaeal tree? Many biologists believe that Archaea first evolved in hot environments with no oxygen. As we discussed in Chapter 22, geologists agree that oxygen gas was not present on the early Earth, consistent with the inference we can make from the archaeal tree that early Archaea were anaerobic. A hot environment for early Archaea can be understood in two ways. Either the entire ocean was hot—for which there is little evidence—or Archaea first evolved in hot springs where they could live as chemoautotrophs deriving energy from local supplies of hydrogen and metals. Because some versions of the bacterial tree also show hyperthermophilic species branching from the earliest nodes, it has been suggested that the last common ancestor of *all* living organisms lived in a hot, oxygen-free environment.

The Archaea include several groups of acid-loving microorganisms.

In many parts of the globe where humans mine coal or metal ores, abandoned mines fill up with very acidic water. Called acid mine drainage, this water has been acidified through the oxidation of pyrite (FeS_2) and other sulfide minerals to produce sulfuric acid (H_2SO_4). The pH of these waters is commonly in the range of 1 to 2, and can actually fall to near 0. To us, acid mine drainage is an environmental catastrophe, but a number of archaeons thrive in it (**Fig. 24.19**). Their modified protein and membrane structures enable them to tolerate these harsh conditions, making it possible to use sulfur compounds in the environment for both chemoautotrophic and heterotrophic growth. What is the limit to growth at low pH? *Picrophilus torridus*, a euryarchaeote archaeon, grows optimally at 60°C and in environments with a pH of 0.7—the pH of battery acid. It is capable of growth in environments down to pH 0, which is a record, as far as we know. In fact, *P. torridus* requires an acidic environment. It will not grow in rainwater because the pH is too high.

Only Archaea produce methane as a by-product of energy metabolism.

Euryarchaeotes include Earth's only biological source of natural gas, or methane (CH_4). Known as **methanogens,** they generate methane by fermenting acetate (CH_3COO^-) or by chemoautotrophic pathways in which hydrogen gas (H_2) reduces CO_2 to CH_4. Methanogenic Archaea play an important role in the carbon cycle, recycling organic molecules in environments where the supplies of electron acceptors for respiration are limited.

Most methanogens live in soil or sediments. They are particularly prominent in peats and lake bottoms where sulfate levels are low and bacterial sulfate reduction is limited. Some, however, live in the rumens (a specialized chamber in the gut) of cows (Case 5), helping to maximize the energy yield from ingested grass. Methane doesn't last long in the atmosphere; it is quickly oxidized to carbon dioxide. Nonetheless, the steady supply of methane generated by archaeons makes a significant contribution to the greenhouse properties of the atmosphere. For this reason, biologists are interested in how methanogenic Archaea will respond to twenty-first-century climate change (Chapter 48).

FIG. 24.19 Headwaters of the Rio Tinto, southwestern Spain. Archaeons thrive in these highly acidic waters, pH 1–2. *Source: Andrew Knoll, Harvard University.*

The red color comes from iron sulfate ions that form in highly acidic water.

One group of the Euryarchaeota thrives in extremely salty environments.

Methanogens are not the only unique euryarchaeotes. A related group illustrates the limits of tolerance to another environmental condition—salinity, or saltiness. The Haloarchaea are halophilic (salt-loving) photoheterotrophs that use the protein bacteriorhodopsin to absorb energy from sunlight (see Fig. 24.6). They live in waters that are salty enough to precipitate table salt (NaCl), maintaining osmotic balance by accumulating ions (especially K^+) and organic solutes in their cytoplasm. Just as extreme acidophiles require acid conditions, these extreme halophiles require high salt conditions and cannot live in dilute water.

Thaumarchaeota may be the most abundant cells in the deep ocean.

Environmental sequencing has yielded another surprise discovery: the immense volumes of water beneath the surface waters of the oceans contain vast numbers of archaeons. In the deep ocean, these organisms may be nearly as abundant as the total number of bacteria in the oceans. Recently segregated as the Thaumarchaeota (the T in TACK), these archaeons are chemoautotrophs that derive energy from the oxidation of ammonia (**Fig. 24.20**). It is becoming clear that Thaumarchaeota play an important, if only recently recognized, role in the marine nitrogen cycle.

HOW DO WE KNOW?

FIG. 24.20

How abundant are archaeons in the oceans?

BACKGROUND Early exploration of microbial diversity in the oceans revealed that archaeons are tremendously abundant. Just how abundant are they, and which metabolisms do they employ to obtain energy and carbon?

METHOD Marine biologists sampled seawater throughout the depth of the ocean at a test site in the northern Pacific Ocean. To the samples, they added molecular tags bound to fluorescent molecules that are visible under the microscope. The tags were RNA sequences that bind to small-subunit ribosomal RNA genes known to be useful in identifying different types of bacteria and archaeons. In this way, separate and distinct fluorescent markers were attached to thaumarchaeote archaeons (a group within the TACK Archaea), euryarchaeote archaeons, and bacteria.

RESULTS By counting the cells marked by different fluorescent tags in seawater samples, the biologists showed that most cells near the surface are bacteria, but thaumarchaeotes make up approximately 40% of all cells in deeper waters (Fig. 24.20a).

CONCLUSION Marine Thaumarchaeota, unknown in 1990, are now known to be among the most abundant organisms in the oceans.

FOLLOW-UP WORK Archaea are among the most abundant of all cells in the world's oceans. How do these cells live? Biologists have taken microbial samples from water known to be sites of nitrification (that is, the conversion of ammonia to nitrite or nitrate). These samples contained an abundance of thaumarchaeote cells, supporting the hypothesis that thaumarchaeotes are nitrifiers. In fact, the higher the abundance of thaumarchaeote cells (Fig. 24.20b, dark blue line), the higher the amount of nitrite produced (Fig. 24.20b, light blue line). Thaumarchaeotes grown in culture were shown to grow by consuming ammonia (Fig. 24.20b, red line). In other words, these organisms oxidize ammonia to provide the ATP and reducing power needed to incorporate CO_2 into organic molecules. In doing so, they play a major role in the marine nitrogen cycle, thriving where sources of carbon and energy for other types of metabolism are scarce.

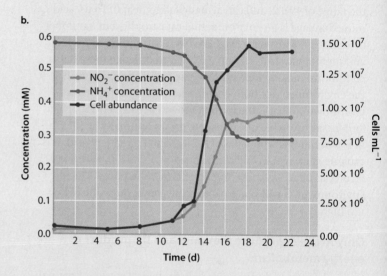

SOURCES Karner, M. B., E. F. DeLong, and D. M. Karl. 2001. "Archaeal Dominance in the Mesopelagic Zone of the Pacific Ocean." *Nature* 409:507–510 (The figure has been modified to reflect more recent nomenclature.); Könneke, K., et al. 2005. "Isolation of an Autotrophic Ammonia-Oxidizing Marine Archaeon." *Nature* 437:543–546.

Like the discovery of SAR11, the discovery of marine thaumarchaeotes makes it clear that before the still-emerging age of environmental genetics and genomics, we simply didn't have a very good idea of which microorganisms lived where and did what in the oceans. Fortunately, that situation is changing rapidly.

Self-Assessment Questions

11. What are three environmental extremes where Archaea thrive?
12. If early branches on the bacterial and archaeal trees are dominated by hyperthermophilic microorganisms, does this mean that the early oceans were very hot?

24.6 THE EVOLUTIONARY HISTORY OF PROKARYOTES

On the tree of life, the plants and animals of our familiar existence populate only the most recent branches. For most of Earth's history, life was entirely microbial. Perhaps surprisingly given their small size, Bacteria and Archaea have left a fossil record that allows paleontologists to reconstruct aspects of life's early evolution. Sedimentary rocks preserve microscopic fossils of ancient bacteria, including the photosynthetic cyanobacteria (**Fig. 24.21**). Limestones and related rocks also preserve **stromatolites,** layered structures that record sediment accumulation by microbial communities (**Fig. 24.22**). Some especially well-preserved ancient rocks also contain molecular fossils—biomolecules, usually lipids, that resist decay and so preserve a chemical signature of ancient life.

FIG. 24.21 Bacterial microfossils. (a) Bacteria that probably metabolized iron in 1.9-billion-year-old rock from the Gunflint Formation in northwest Ontario, Canada. (b) Cyanobacteria preserved in 800-million-year-old silicified marine carbonates from the Draken Formation in Spitsbergen, Norway. (c) A short cyanobacterial filament from silicified tidal flat carbonates of the 1.5-billion-year-old Bilyakh Group in northern Siberia. *Source: a–c. Andrew Knoll, Harvard University.*

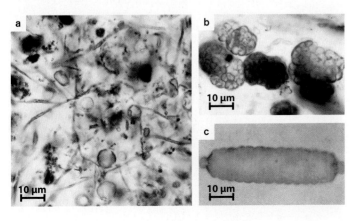

FIG. 24.22 Stromatolites. Modern-day stromatolites at Shark Bay, Western Australia (top), show microbial communities building the domed structures exhibited by stromatolites in ancient rocks (bottom). Stromatolites are among the earliest records of life on Earth. *Sources: (top) Andrew Knoll, Harvard University; (bottom) François Gohier/Science Source.*

Life originated early in our planet's history.

The Earth is approximately 4.6 billion years old, but few sedimentary rocks have survived destruction by geologic processes and remain to document our planet's earliest history. Sedimentary rocks deposited nearly 3.5 billion years ago in Western Australia and South Africa provide some of our earliest records of Earth's history. Remarkably, they contain a scarce but discernible signature of life. Tiny organic structures preserved in these rocks have been interpreted as microfossils, although this conclusion remains controversial.

More pervasive (and persuasive) evidence of life in early oceans is provided by stromatolites. Stromatolites in ancient rocks are strikingly similar to modern structures formed by microbial communities and so provide evidence of life that goes back about 3.5 billion years (Fig. 24.22). Also, because purely physical processes cannot explain the isotopic composition of carbon in very old limestones and organic matter, we can conclude that a biological carbon cycle existed 3.5 billion years ago (and, from the evidence of 3.8-billion-year-old rocks in Greenland, possibly earlier).

The chemistry of Earth's oldest sedimentary rocks also shows that the early atmosphere and ocean contained little or no free oxygen. What would the carbon cycle look like on an oxygen-free planet? We know the answer from our discussion of prokaryotic metabolism. Photosynthesis can incorporate CO_2 into organic molecules in oxygen-poor waters, using H_2S, H_2, or Fe^{2+} as electron donors, and fermenters and anaerobic respirers can recycle carbon, releasing CO_2 back into the environment, using electron acceptors such as SO_4^{2-} and Fe^{3+}. Geologic evidence suggests that iron played a particularly important role in cycling carbon for the first billion years of evolutionary history.

The evolution of cyanobacteria, with their capacity for oxygenic photosynthesis, made possible the accumulation of oxygen in the atmosphere and oceans. Oxygen began to accumulate in the atmosphere approximately 2.4 billion years ago, making possible aerobic respiration and, eventually, our present-day carbon cycle. Both microfossils and molecular fossils indicate that cyanobacteria and other photosynthetic bacteria continued to dominate primary production until about 650 million years ago. Thus, our modern carbon cycle, powered overwhelmingly by oxygenic photosynthesis and with major participation by eukaryotic organisms, may have existed for only the last 15% of Earth's history.

Prokaryotes have coevolved with eukaryotes.

Even after the rise of eukaryotic cells and, later, plants and animals, Bacteria and Archaea have remained essential to the functioning of biogeochemical cycles and hence to the maintenance of habitable environments on Earth. Prokaryotic organisms have also radiated into the many novel environments made possible by eukaryotes. We noted earlier that bacteria fix nitrogen in nodules formed on soybean roots. Soybeans and nitrogen-fixing bacteria have evolved together, in a process called **coevolution**.

Coevolutionary relationships between prokaryotic microorganisms and eukaryotes are common in nature. As noted earlier, cows metabolize grass with the help of methanogenic archaeons in the rumen, a specialized digestive organ. These microbes enable the cow to gain nutrition from plant materials that could otherwise not be digested; in turn, the microorganisms benefit from a steady supply of food. Similarly, clams that live near methane-rich vents in the Gulf of Mexico have evolved specialized organs that house methane-oxidizing bacteria. These chemoautotrophic bacteria excrete organic compounds that provide the clams with a reliable source of nutrition.

Bacteria can affect animal behavior as well as nutrition. For example, some squid species attract mates by emitting light from specialized organs full of bioluminescent bacteria (Case 5). The bacteria provide the squid with a mechanism for communication in the dark deep-sea environment, and the squid feeds the bacteria.

Interestingly, many different kinds of bacteria live in the waters that surround the squid, but only the bioluminescent species accumulates in the squid's light organ. Molecular signals enable the squid and bacteria to recognize and respond to each other.

Bacteria can also influence animal reproduction. Earth supports more than 1 million species of insects, and perhaps two-thirds of these harbor the proteobacterial parasite *Wolbachia* within their tissues. A heterotroph that obtains nutrition by metabolizing organic compounds supplied by its host, *Wolbachia* infects its host's reproductive tissues and passes from one generation to the next in the host's eggs (**Fig. 24.23**). Remarkably, many *Wolbachia* strains alter the host's reproductive cells so that the insect host can successfully reproduce only with partners infected by the same *Wolbachia* strain. Because *Wolbachia* is transferred by eggs and not sperm, these organisms have little use for males. Consequently, some cause male host embryos to die, increasing the proportion of females. Others cause genetically male embryos to develop as females, or enable eggs to develop into female embryos without fertilization.

Wolbachia has another effect on its insect hosts that is highly relevant to human health: it commonly inhibits infection by viruses and the eukaryotic parasites that cause malaria. Recent research shows that *Wolbachia* populations in some fly species protect their hosts against infection by RNA viruses. Scientists introduced these bacteria into mosquitos that transmit dengue fever, a tropical disease that infects 50–100 million people each year. The *Wolbachia*-infected mosquitos were strongly resistant to the dengue virus, which causes dengue fever. Moreover, when introduced into the wild, these virus-resistant forms rapidly expand to dominate regional mosquito populations. *Wolbachia* parasites present a possible path to controlling dengue fever, malaria, and other insect-borne diseases.

FIG. 24.23 A large population of *Wolbachia* bacteria (fluorescing green) in the embryo of a fruit fly. *Wolbachia* can influence the development and reproductive biology of their animal hosts. *Source: Image by Zoe Veneti and Kostas Bourtzis.*

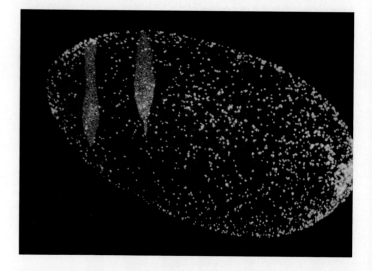

All around us, plants, animals, and microorganisms live in intimate association—sometimes to our detriment, but often in mutually beneficial relationships that influence how we grow, develop, eat, and behave. Humans are no exception.

 CASE 5 THE HUMAN MICROBIOME: DIVERSITY WITHIN

How do intestinal bacteria influence human health?

It has been estimated that the bacteria (and, to a much lesser extent, archaeons) in and on your body match or outnumber your own cells. Estimates of the number of microbial species that inhabit our bodies vary, but approximately 750 types have been identified in the mouth alone, and the list remains incomplete (**Fig. 24.24**). A comparable diversity resides in the colon, and still more bacteria live on the skin and elsewhere. Some are transients, entering and leaving the body within a few bacterial generations. Others are specifically adapted for life within humans, forming complex communities of interacting species.

At present, much research is focused on the bacteria within our intestinal tracts. We commonly think of gut bacteria as harmful, but this perception arises from only a small—albeit devastating—subset of our microbial guests. Illnesses known to be caused by bacteria within our bodies include cholera, dysentery, and tuberculosis. Once the leading causes of death in New York and London, these diseases remain major killers in Africa. Other diseases, not previously thought to be of microbial origin, are now known to be caused by bacteria—ulcers, for example, and stomach cancer, both mediated by the acid-tolerant bacterium *Helicobacter pylori*. More often than not, however, intestinal bacteria have a beneficial effect. They help to break down food in our digestive system and secrete vitamins and other biomolecules into the colon for absorption into our tissues. Molecular signals from bacteria guide the proper development of cells that line the interior of our intestines.

A fertilized human egg has no bacteria attached to it, so our microscopic passengers arrive by colonization—by infection, if you will. Research on children from Europe and rural Africa clearly shows the importance of diet in establishing our gut microbiota (**Fig. 24.25**). African children, who are typically

FIG. 24.25 Diet and the intestinal microbiota. Studies of gut bacteria in children from (a) rural Africa and (b) Western Europe show markedly different communities. *Data from C. De Filippo, D. Cavalieri, M. Di Paola, M. Ramazzotti, J. B. Poullet, S. Massart, S. Collini, G. Pieraccini, and P. Lionetti, 2010, "Impact of Diet in Shaping Gut Microbiota Revealed by a Comparative Study in Children from Europe and Rural Africa," Proceedings of the National Academy of Sciences USA 107(33):14693, doi:10.1073/pnas.1005963107.*

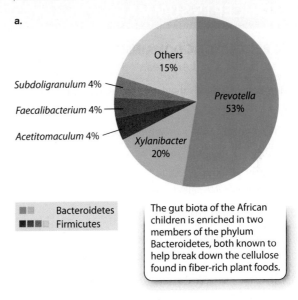

The gut biota of the African children is enriched in two members of the phylum Bacteroidetes, both known to help break down the cellulose found in fiber-rich plant foods.

FIG. 24.24 The human microbiome. This chart provides a rough guide to the distribution of the 3000 or more species of bacteria found on and within a healthy human adult, showing the locations of bacteria that had been sequenced as of 2010. *Data from J. Peterson et al., 2009, "The NIH Human Microbiome Project," Genome Research 19:2317–2323; originally published online October 9, 2009.*

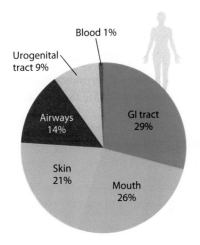

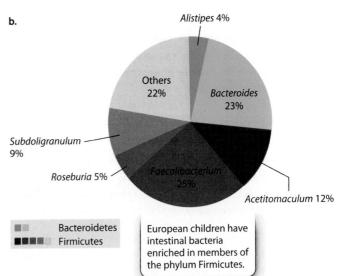

European children have intestinal bacteria enriched in members of the phylum Firmicutes.

raised on a diet low in fat and animal protein but rich in plant matter, have gut microbiotas enriched in species of the phylum Bacteroidetes that are known to help digest cellulose. In contrast, European children, when raised on a typical Western diet rich in sugar and animal fat, lack these bacteria, harboring instead diverse members of the phylum Firmicutes. We do not understand the full ramifications of these differences, but active research suggests that they may help to explain the relative prominence in Western societies of allergies and other disorders involving the immune system.

We also influence our gut biota by ingesting antibiotics. Although prescribed for the control of pathogens, most antibiotics kill a wide range of bacteria and, therefore, can change the balance of our intestinal microbiota. Clinical studies show that the incidence of inflammatory bowel disease, a disabling inflammation of the colon, increases in humans who have been treated with antibiotics.

To understand the relationships between the human microbiome and human health, we need to know what constitutes a healthy gut biota. This means studying the metabolism, ecology, and population genetics of the bacteria in our bodies. We need to know which bacteria have evolved to take advantage of the environments provided by the human digestive system, and which are "just passing through." Future visits to the doctor may include routine genetic fingerprinting of our bacterial biota, as well as treatments designed to keep our microbiome—and, so, ourselves—healthy.

Self-Assessment Question

13. What is the age of the Earth, and when is life thought to have first originated?

CORE CONCEPTS SUMMARY

24.1 TWO PROKARYOTIC DOMAINS: The two prokaryotic domains are Bacteria and Archaea.

- Prokaryotic cells are cells that lack a nucleus; they include Bacteria and Archaea. page 519

- Bacteria are small and lack membrane-bounded organelles. page 519

- The bacterial genome is circular. Some bacteria also carry smaller circles of DNA called plasmids. page 520

- Diffusion limits size in bacterial cells. page 520

- Bacteria can obtain DNA by horizontal gene transfer from organisms that may be distantly related. page 522

- Like Bacteria, Archaea lack a nucleus, but they collectively form a second prokaryotic domain that is distinct from Bacteria. page 523

- Some archaeons are extremophiles, living in environments characterized by low pH, high salinity, or high temperatures. Others live in less extreme environments such as the upper ocean or soil. page 523

24.2 AN EXPANDED CARBON CYCLE: Bacteria and Archaea cycle carbon in diverse ways.

- Bacteria are capable of oxygenic photosynthesis, using water as a source of electrons and producing oxygen as a by-product, and of anoxygenic photosynthesis, using electron donors other than water, such as H_2S, H_2, and Fe^{2+}. page 525

- Some bacteria and archaeons are capable of anaerobic respiration, in which NO_3^-, SO_4^{2-}, Mn^{4+}, and Fe^{3+} serve as the electron acceptor instead of oxygen gas. page 526

- Many bacteria and archaeons obtain energy from fermentation, which involves the partial oxidation of organic molecules and the production of ATP in limited quantities. page 527

- Some photosynthetic bacteria are photoheterotrophs, obtaining energy from sunlight but using preformed organic compounds rather than CO_2 as a source of carbon. page 527

- Chemoautotrophy, in which chemical energy is used to convert CO_2 to organic molecules, is unique to Bacteria and Archaea. page 527

24.3 SULFUR AND NITROGEN CYCLES: Bacteria and Archaea are critical to the biological cycling of sulfur and nitrogen.

- Plants and algae can take up sulfur and incorporate it into proteins, but bacteria and archaeons dominate the sulfur cycle by means of oxidation and reduction reactions that are linked to the carbon cycle. page 528

- Some bacteria and archaeons can reduce nitrogen gas to ammonia in a process called nitrogen fixation. page 529

- The nitrogen cycle also involves oxidation and reduction reactions by Bacteria and Archaea that are linked to the carbon cycle. page 530

24.4 BACTERIAL DIVERSITY: Genomic surveys of non-culturable bacteria reveal the extent of bacterial diversity.

- Traditionally, bacterial groups were recognized by their morphology, physiology, and ability to take up specific stains in culture. page 530

Direct sequencing of genes, and now genomes, from organisms in soil and seawater samples has revealed new groups of bacteria. page 530

Proteobacteria are the most diverse group of bacteria; they are involved in many of the biogeochemical processes that are linked to the carbon cycle. page 534

Gram-positive bacteria include important disease-causing strains as well as species that are principal sources of antibiotics. page 534

Photosynthetic bacteria are not limited to a single branch of the bacterial tree. page 534

24.5 ARCHAEAL DIVERSITY: The diversity of Archaea has only recently been recognized.

Archaeons tend to thrive where energy available for growth is limited. page 535

Archaea are commonly divided into three major groups, called DPANN, TACK, and Euryarchaeota. page 535

Archaeons that branch from the earliest nodes are hyperthermophiles, meaning that they grow only at high temperatures. page 536

A number of archaeons grow in highly acidic waters, such as those associated with acid mine drainage. page 537

Some archaeons (but no bacteria or eukaryotes) generate methane as a by-product of their energy metabolism, thereby making important contributions to climate and the carbon cycle. page 537

Haloarchaea are archaeons that can live only in extremely salty environments. page 537

Thaumarchaeotes are among the most abundant cells in the oceans. page 537

24.6 THE EVOLUTIONARY HISTORY OF PROKARYOTES: The earliest forms of life on Earth were Bacteria and Archaea

Evidence for the early history of life comes from microfossils, fossilized structures called stromatolites, and the isotopic composition of rocks and organic matter. page 539

Fossils indicate that life on Earth originated more than 3.5 billion years ago. page 539

The early atmosphere and ocean contained little or no free oxygen. page 540

Oxygen began to accumulate in the atmosphere and oceans approximately 2.4 billion years ago as a result of the success of cyanobacteria utilizing oxygenic photosynthesis. page 540

Prokaryotic metabolisms were essential in the early history of Earth and remain vital today, as many forms of life depend on biogeochemical cycles and metabolisms unique to Bacteria and Archaea. page 540

Most animals, including humans, live in intimate association with bacteria, which in turn affect health. page 541

Log in to LaunchPad to check your answers to the Self-Assessment Questions and to access additional learning tools.

CHAPTER 25 Eukaryotic Cells
Origins and Diversity

CORE CONCEPTS

25.1 A REVIEW OF THE EUKARYOTIC CELL: Eukaryotic cells are defined by the presence of a nucleus, but features such as a dynamic cytoskeleton and membrane system explain their evolutionary success.

25.2 EUKARYOTIC ORIGINS: The endosymbiotic hypothesis proposes that the chloroplasts and mitochondria of eukaryotic cells were originally free-living bacteria.

25.3 EUKARYOTIC DIVERSITY: Eukaryotes were formerly divided into four kingdoms, but are now divided into at least seven superkingdoms.

25.4 THE FOSSIL RECORD OF PROTISTS: The fossil record provides perspectives on the timing and environmental context of eukaryotic evolution.

Eukaryotic cells have a nucleus, which serves as their defining characteristic. As we will see, however, it is quite another set of features that distinguishes eukaryotes in terms of their function and their capacity for evolutionary innovation. Eukaryotic diversity—the animals, plants, fungi, and protists we are familiar with—does not reflect exceptional versatility in energy metabolism. In fact, compared with Bacteria and Archaea, the Eukarya obtain carbon and energy in only a limited number of ways. For example, oxygenic photosynthesis and aerobic respiration are widespread among eukaryotes, but anoxygenic photosynthesis is unknown. Furthermore, anaerobic respiration has been reported in only a handful of species, while chemoautotrophy has not been documented in eukaryotes at all. Many eukaryotes can ferment organic substrates, but with the exception of yeasts, which are particularly active fermenters, most use fermentation only as a supplementary metabolism.

Why then, have eukaryotes been so successful? The source of eukaryotic success lies in their remarkable capacity to form cells of diverse shapes and sizes and to produce multicellular structures in which multiple cell types act in concert. Together, these innovations opened up a range of possibilities for structure and function not found among Bacteria and Archaea. Eukaryotic diversity includes cells that can engulf other cells, as well as organisms as large as redwoods and as complex as humans.

25.1 A REVIEW OF THE EUKARYOTIC CELL

Chapter 5 introduced the fundamental features of eukaryotic cells. Here, we revisit those attributes (**Fig. 25.1**) and consider their consequences for function and diversity. How do eukaryotic cells differ from bacteria and archaeons, and how do these differences explain the roles that eukaryotes play in modern ecosystems?

Internal protein scaffolding and dynamic membranes organize the eukaryotic cell.

All cells require a mechanism to maintain spatial order in the cytoplasm. As we saw in Chapter 24, bacteria and archaeons rely primarily on walls that support the cell from the outside, along with a relatively rigid framework of proteins within the cytoplasm. Some eukaryotic cells have walls, but many others, including the cells in your body, do not. Eukaryotes rely mainly on an internal scaffolding of proteins to organize the cell, with most of those supports being microtubules composed of the protein tubulin and microfilaments of actin. This cytoskeleton, which is found in all eukaryotic cells, differs in one key property from the protein framework of bacteria: it can be remodeled quickly, enabling cells to change shape (Chapter 10).

Dynamic cytoskeletons require dynamic membranes that enable the cell to continue functioning even as it changes shape; eukaryotes maintain within their cells a remarkably dynamic network of membranes called the endomembrane system (Chapter 5). This network includes the nuclear envelope, an assembly of membranes that runs through the cytoplasm called the endoplasmic reticulum (ER) and Golgi apparatus, and a plasma membrane that surrounds the cytoplasm. All membranes of the endomembrane system are interconnected, either directly or by the movement of vesicles. Many are also capable of changing shape rapidly.

FIG. 25.1 A eukaryotic cell. Eukaryotic cells have a nucleus and extensive internal compartmentalization.

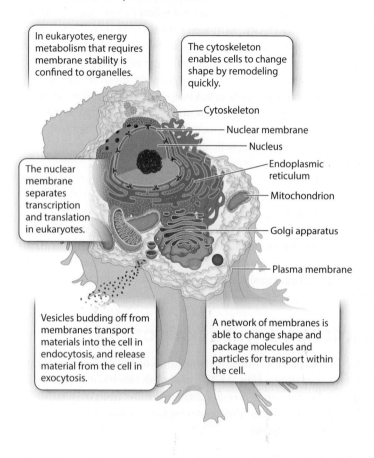

In eukaryotes, energy metabolism that requires membrane stability is confined to organelles.

The cytoskeleton enables cells to change shape by remodeling quickly.

- Cytoskeleton
- Nuclear membrane
- Nucleus
- Endoplasmic reticulum
- Mitochondrion
- Golgi apparatus
- Plasma membrane

The nuclear membrane separates transcription and translation in eukaryotes.

Vesicles budding off from membranes transport materials into the cell in endocytosis, and release material from the cell in exocytosis.

A network of membranes is able to change shape and package molecules and particles for transport within the cell.

In fact, the different membranes are interchangeable in the sense that material originally added to the endoplasmic reticulum may in time be transferred to the cell membrane or nuclear membrane. Biologists like to say that the membranes of eukaryotic cells are in dynamic continuity. Membranes that are stable, as required for energy metabolism, occur only in the mitochondria and chloroplasts.

In combination, the dynamic cytoskeleton and the membrane system provide eukaryotes with new possibilities for movement. For example, amoebas extend finger-like projections that pull the rest of the cell forward. The cytoskeleton and membrane system also enable eukaryotic cells to engulf molecules or particles, including other cells, in a process called endocytosis; prokaryotic cells cannot perform this activity. Phagocytosis is a specific form of endocytosis in which eukaryotic cells surround food particles and package them in vesicles that bud off from the cell membrane (**Fig. 25.2**). When packaged in this way, the particles can be transported into the cell interior for digestion. The same process also works in reverse: in exocytosis, molecules or cytoplasmic waste formed within the cell are packaged in vesicles and moved to the cell surface for removal (Chapter 5).

Intracellular vesicles and the molecules they carry are transported through the cytoplasm by means of molecular motors associated with the cytoskeleton (Chapter 10). In this way, both nutrients and signaling molecules move through the cell at speeds much greater than diffusion allows. A major consequence of this accelerated pace of transport is that eukaryotic cells can be much larger than most bacteria.

The cytoskeleton and membrane system are flexible in another way: a change in the expression of a few genes can

FIG. 25.2 Phagocytosis. The flexibility of the eukaryotic cytoskeleton and membrane system enables eukaryotic cells to engulf food particles, including other cells. This sequence shows an amoeba engulfing a yeast cell. Note how the cell membrane (green) folds inward locally to form a vesicle around the red food particle. *Source: Valentina Mercanti et al. Selective membrane exclusion in phagocytic and macropinocytic cups, Journal of Cell Science 119, 4079–4087. doi:10 1242/jcs.03190*

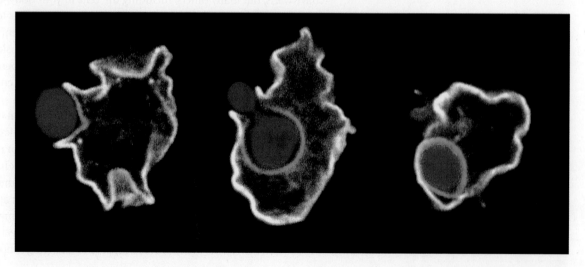

change their shape and organization. This flexibility makes possible yet another hallmark of eukaryotic evolution—complex multicellularity (Chapter 26). In complex multicellular organisms, patterns of gene expression can differ from cell to cell. A change in gene expression can modify the cytoskeleton and membranes of individual cells, enabling cells to function in different ways.

In eukaryotic cells, energy metabolism is localized in mitochondria and chloroplasts.

Relative to prokaryotic organisms, eukaryotes are fairly limited in the ways they obtain carbon and energy. Moreover, the metabolic processes that power eukaryotic cells take place only in specific organelles. Specifically, aerobic respiration occurs in the mitochondrion (Chapter 7), and photosynthesis in the chloroplast (Chapter 8). Only limited anaerobic processing of food molecules takes place within the cytoplasm.

As noted earlier, many eukaryotic cells engulf food particles and package them inside a vesicle, which is then transported into the cytoplasm. Within the cytoplasm, enzymes break down the particles into molecules that can be processed by the mitochondria. Many single-celled eukaryotes feed on bacteria or other eukaryotic cells. Animals, in turn, ingest larger foodstuffs, including other animals and plants. As a consequence, eukaryotes can exploit sources of food that are not readily available to bacterial heterotrophs, which feed on individual molecules. This ability opened up the great new ecological possibility of predation, which vastly increased the complexity of interactions among organisms.

The structural flexibility of eukaryotic cells also allows photosynthetic eukaryotes to interact with their environment in ways that photosynthetic bacteria cannot. Unicellular algae (which are eukaryotes) can move effectively through surface waters vertically as well as horizontally and, therefore, can seek and exploit local patches of nutrients. Diatoms, discussed shortly, go one step further: they store nutrients in large internal vacuoles for later use when nutrient levels in the environment become low. Plants have evolved multicellular bodies with many different cell types. For example, specialized cell types working together can provide leaves high in the canopy of trees with water and nutrients from the soil in which the plants are rooted. Thus supplied, leaves can capture sunlight many meters above the ground, giving plants a tremendous advantage on land.

The organization of the eukaryotic genome also helps explain eukaryotic diversity.

Bacteria and Archaea absorb available nutrients quickly, and both rapid deployment of metabolic enzymes and rapid reproduction are key to exploiting patchily distributed nutrients. DNA replication needs to be fast for reproduction to also be fast. In bacteria, the speed of DNA replication is limited because the majority of the cell's DNA is organized in a single circular chromosome, and replication begins from only one site. As a result, selection favors those strains of Bacteria and Archaea that retain only the genetic material vital to the organism.

Eukaryotes have multiple linear chromosomes and can begin replication from many sites on each one. On this basis, eukaryotes are able to replicate multiple strands of DNA simultaneously and rapidly (Chapter 12). This ability relieves the evolutionary pressure for streamlining, allowing eukaryotic genomes to build up large amounts of DNA that do not code for proteins. Most of this additional DNA was originally considered to have no function; indeed, it was traditionally called "junk DNA." That view has been modified in recent years, as the complete genome has become better understood (Chapter 13). Although eukaryotic genomes do appear to contain truly junky DNA, at least some of the DNA that does not code for proteins functions in gene regulation (Chapter 18). This regulatory DNA gives eukaryotes the fine control of gene expression required for both multicellular development and complex life cycles, two major features of eukaryotic diversity.

At this point, you can see that the evolutionary success of eukaryotes rests on a combination of features. The innovations of their dynamic cytoskeletal and membrane systems gave eukaryotes the structure required to support larger cells with complex shapes and the ability to ingest other cells. Thus, early unicellular eukaryotes did not gain a foothold in microbial ecosystems by outcompeting bacteria and archaeons, but rather succeeded by evolving novel functions such as the capacity to remodel cell shape. Along with these new functions, eukaryotes evolved complex patterns of gene regulation, which in turn enabled unicellular eukaryotes to evolve complex life cycles and multicellular eukaryotes to generate multiple, interacting cell types during growth and development. These abilities opened up still more possibilities for novel functions, which we explore in this and later chapters.

Sex promotes genetic diversity in eukaryotes and gives rise to distinctive life cycles.

In Chapter 24, we saw how Bacteria and Archaea generate genetic diversity by horizontal gene transfer. Horizontal gene transfer has been documented in eukaryotic species, though it is relatively uncommon. How, then, do eukaryotes generate and maintain genetic diversity within populations? The answer is quite simple: sex.

Sexual reproduction involves meiosis and the formation of gametes, and the subsequent fusion of gametes during fertilization (Chapters 11 and 40). Sex promotes genetic variation in two basic ways. First, meiotic cell division results in gametes that are genetically unique. Each gamete has a combination of alleles different from the other gametes and from the parental cell as a result of recombination and independent assortment. Second, in fertilization, new combinations of genes are brought together by the fusion of gametes. Interestingly, a few eukaryotic groups have lost the capacity for sexual reproduction. The best studied of these eukaryotes are tiny animals called bdelloid rotifers. The genetic diversity of these organisms is actually quite high, as it is maintained by high rates of horizontal gene transfer.

Meiotic cell division results in **haploid** cells, which have one set of chromosomes. Sexual fusion brings two haploid ($1n$) cells together to produce a **diploid** ($2n$) cell that has two sets of chromosomes. The life cycle of sexually reproducing eukaryotes, then, necessarily alternates between haploid and diploid states.

Many single-celled eukaryotes normally exist in the haploid stage and reproduce asexually by mitotic cell division (**Fig. 25.3a**). The green alga *Chlamydomonas* is one such organism. Nevertheless, under the right conditions, such as starvation or other environmental stress, two cells may fuse, forming a diploid cell, or **zygote**. The zygote formed by these single-celled eukaryotes commonly functions as a resting cell. It covers itself with a protective wall and then lies dormant until environmental conditions improve. In time, further signals from the environment induce meiotic cell division, resulting in four genetically distinct haploid cells that emerge from their protective coating to complete the life cycle.

As shown in **Fig. 25.3b**, some single-celled eukaryotes normally exist as diploid cells. An example is provided by the diatoms, single-celled eukaryotes commonly found in lakes, soils, and seawater. Most diatom cells are diploid and reproduce asexually by mitotic cell division to make more diploid cells. Because their external mineralized skeletons constrain their growth, diatoms become smaller with each asexual division. Once a critical size is reached, meiotic cell division is triggered, producing haploid gametes that fuse to regenerate the diploid state as a round, thick-walled cell. This cell eventually germinates to form an actively growing cell with a mineralized skeleton. In diatoms, then, short-lived gametes constitute the only haploid phase of the life cycle.

FIG. 25.3 Eukaryotic life cycles. (a and b) The life cycles of single-celled eukaryotes differ in the proportion of time spent as haploid ($1n$) versus diploid ($2n$) cells. (c) The life cycles of animals have many mitotic divisions between formation of the zygote and meiosis. (d) Vascular plants have two multicellular phases.

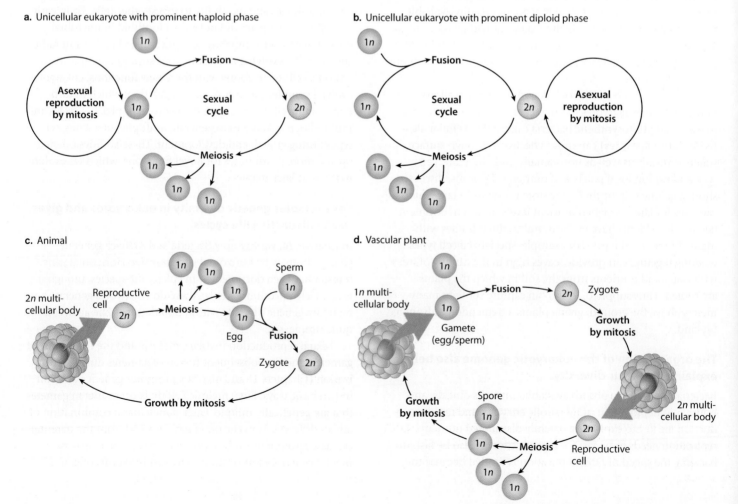

The two life cycles just introduced, of *Chlamydomonas* and diatoms, are similar in many ways. In both, haploid cells fuse to form diploid cells, and diploid cells undergo meiotic cell division to generate haploid cells. Both life cycles also commonly include cells capable of persisting in a protected form when the environment becomes stressful. Why some single-celled eukaryotes usually occur as haploid cells and others usually occur as diploid cells remains unknown—a good question for continuing research.

Sexual reproduction has never been observed in some eukaryotes, but most species appear to be capable of sex, even if they reproduce asexually most of the time. Variations on the eukaryotic life cycle can be complex, especially in parasitic microorganisms that have multiple animal hosts. All, however, have the same fundamental components as the two life cycles described here.

In chapters that follow, we discuss multicellular organisms in detail. Here we note only that animals, plants, and other complex multicellular organisms have life cycles with the same features as those just discussed. The big difference is that in animals, the zygote divides many times to form a multicellular diploid body before a small subset of cells within the body undergoes meiotic cell division to form haploid gametes (eggs and sperm). During fertilization, the egg and sperm combine sexually to form a zygote (**Fig. 25.3c**). In animals, as in diatoms, the only haploid phase of the life cycle is the gamete. As we will see in Chapter 28, plants have two multicellular phases in their life cycle, one haploid and one diploid (**Fig. 25.3d**). Like single-celled eukaryotes, many plants and animals can reproduce asexually. Humans and other mammals are unusual in that we cannot reproduce asexually.

Self-Assessment Questions

1. Which features distinguish a eukaryotic cell from a prokaryotic cell?
2. Which forms of energy metabolism are found in eukaryotes?
3. How did the evolutionary expansion of eukaryotic organisms change the way carbon is cycled through biological communities?

25.2 EUKARYOTIC ORIGINS

In the preceding section, we reviewed the basic features of eukaryotic cells. The distinctive cellular organization of Eukarya differs markedly from the cells of Bacteria and Archaea. How did the complex cells of eukaryotes come to exist? In Case 5: The Human Microbiome, we saw that symbioses are common, even in humans. Is the relationship between humans and bacteria unusual, or are symbioses so fundamental that they played key roles in the evolution of the eukaryotic cell?

CASE 5 THE HUMAN MICROBIOME: DIVERSITY WITHIN

What role did symbiosis play in the origin of chloroplasts?

The chloroplasts found in plant cells closely resemble certain photosynthetic bacteria, specifically cyanobacteria. The molecular workings of photosynthesis are nearly identical in the two, and internal membranes organize the photosynthetic machinery in similar ways (Chapter 8). The Russian botanist Konstantin Sergeevich Merezhkovsky recognized this similarity more than a century ago. He also realized that corals and some other organisms harbor algae within their tissues as symbionts that aid the growth of their host. A **symbiont** is an organism that lives in closely evolved association with another species. This association, called **symbiosis**, is discussed in Chapter 45.

Putting these two observations together, Merezhkovsky came up with a radical hypothesis. Chloroplasts, he argued, originated as symbiotic cyanobacteria that through time became permanently incorporated into their hosts. Such a symbiosis, in which one partner lives within the other, is called an **endosymbiosis**. Merezhkovsky's hypothesis of chloroplast origin by endosymbiosis was difficult to test with the tools available in the early twentieth century, and his idea was dismissed, more neglected than disproved, by most biologists.

In 1967, American biologist Lynn Margulis resurrected the endosymbiotic hypothesis. She supported her arguments with new types of data made possible by the then-emerging techniques of cell and molecular biology (**Fig. 25.4**). Transmission electron microscopy revealed that the structural similarities between chloroplasts and cyanobacteria extend to the submicrometer level, such as in the organization of the photosynthetic membranes. It also became clear that the chloroplasts in red and green algae (and in land plants) are separated from the cytoplasm that surrounds them by two membranes. This arrangement would be expected if a cyanobacterial cell had been engulfed by a eukaryotic cell. The inner membrane corresponds to the cell membrane of the cyanobacterium, and the outer one is part of the engulfing cell's membrane system, as shown in Fig. 25.2. In addition, the biochemistry of photosynthesis is essentially the same in cyanobacteria and chloroplasts. Both use two linked photosystems, including a common mechanism for extracting electrons from water, and both rely on the same reactions to reduce carbon dioxide (CO_2) to form organic matter (Chapter 8).

Such observations kindled renewed interest in the endosymbiotic hypothesis, but the decisive tests were made possible by another, and unexpected, discovery. It turns out that chloroplasts have their own DNA, which is organized into a single circular chromosome, like that of bacteria. Because of this, investigators were able to use the tools of molecular sequence comparison (Chapter 22) to study chloroplast genes. The sequences of nucleotides in chloroplast genes closely match

HOW DO WE KNOW?

Fig. 25.4

What is the evolutionary origin of chloroplasts?

BACKGROUND Transmission electron microscopy shows that cyanobacteria (top), red algal chloroplasts (middle), and plant chloroplasts (bottom) have very similar internal membranes. These structures, recognizable in the images as parallel laminae within cells or chloroplasts, organize the light reactions of photosynthesis. Other structural and biochemical features also indicate a strong similarity between cyanobacteria and the chloroplasts found in photosynthetic eukaryotes.

HYPOTHESIS The similarities between cyanobacteria and chloroplasts reflect descent from a common ancestor: chloroplasts evolved from cyanobacteria living as endosymbionts within a eukaryotic cell.

EXPERIMENT Like cyanobacteria, chloroplasts have DNA organized in a single circular chromosome. Biologists sequenced genes for small-subunit rRNA in chloroplasts and compared the sequences to those for small-subunit rRNA in the nuclei of various eukaryotes and in bacteria, including cyanobacteria. Phylogenies based on molecular sequence comparisons place chloroplasts unambiguously among the cyanobacteria.

CONCLUSION Molecular and electron microscope data support the hypothesis that chloroplasts originated as endosymbiotic cyanobacteria.

SOURCE Giovannoni, S. J., et al. 1988. "Evolutionary Relationships among Cyanobacteria and Green Chloroplasts." *Journal of Bacteriology* 170:3584–3592. Photos (top to bottom): Dr. Kari Lounatmaa/Science Source; Biology Pics/Getty Images; Dr. Jeremy Burgess/Science Source.

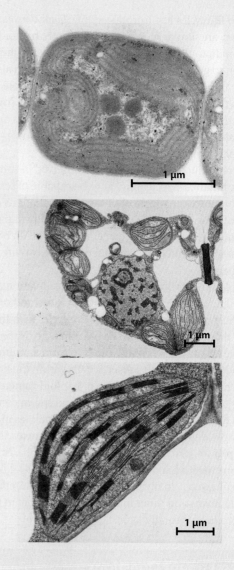

those of cyanobacterial genes but are strikingly different from the sequences of nucleotides in genes within the nuclei of photosynthetic eukaryotes. This finding provides strong support for the hypothesis of Merezhkovsky and Margulis. Chloroplasts are indeed the descendants of symbiotic cyanobacteria that lived within eukaryotic cells.

Eukaryotes were able to acquire the ability to perform photosynthesis because they could engulf and retain cyanobacterial cells. But engulfing another microorganism was just the beginning. The cyanobacteria most closely related to chloroplasts have 2000 to 3000 genes. In contrast, among photosynthetic eukaryotes, chloroplast gene numbers vary from just 60 to 200. Where did the rest of the cyanobacterial genome go? Some genes may simply have been lost if similar nuclear genes could supply the chloroplast's requirements. Many others were transported to the nucleus when chloroplasts broke or by hitching a ride with viruses. As an example, the nuclear genome of the flowering plant *Arabidopsis* contains several thousand genes of cyanobacterial origin, some of which code for proteins destined for use within the chloroplast.

Although foreign to mammals and other vertebrate animals, symbiosis between a heterotrophic host and a photosynthetic partner is common throughout the eukaryotic domain. For example, reef corals harbor photosynthetic cells that live symbiotically within their tissues. Some corals have even lost the capacity to capture food from surrounding waters. Giant clams of the genus *Tridacna*, found in tropical Pacific waters, also obtain some or even most of their nutrition from symbiotic algae that live within their tissues.

Until recently, most biologists agreed that chloroplasts originated from endosymbiotic cyanobacteria only once, in a common ancestor of the green algae and red algae. Now, remarkably, a second case of cyanobacterial endosymbiosis has come to light. *Paulinella chromatophora* is a photosynthetic amoeba (**Fig. 25.5**). Gene sequence comparisons show that its chloroplast originated in a branch of the cyanobacteria different

FIG. 25.5 A second example of chloroplast endosymbiosis. The photosynthetic amoeba *Paulinella chromatophora* acquired photosynthesis by endosymbiosis independently of, and more recently than, the endosymbiotic event that established chloroplasts in the algal ancestors of green plants. *Source: Courtesy of Hwan Su Yoon.*

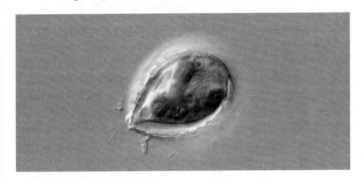

from the one that gave rise to chloroplasts in other photosynthetic eukaryotes. Approximately one-third to one-half of the ancestral genome remains in *P. chromatophora* chloroplasts, suggesting that in this organism chloroplast evolution is still in progress.

 CASE 5 THE HUMAN MICROBIOME: DIVERSITY WITHIN

What role did symbiosis play in the origin of mitochondria?

Like chloroplasts, mitochondria closely resemble free-living bacteria in organization and biochemistry. Also like chloroplasts, mitochondria contain DNA that confirms their close phylogenetic relationship to a form of bacteria—in this case, proteobacteria (Chapter 24). Like chloroplasts, then, mitochondria originated as endosymbiotic bacteria.

Mitochondria are also like chloroplasts in having a small genome. Indeed, the mitochondrial genome is dramatically reduced compared to the ancestral proteobacterial genome; in most eukaryotes, it contains only a handful of functioning genes. Human mitochondria, for example, code for just 13 proteins and 24 RNAs. Once again, many genes from the original bacterial endosymbiont migrated to the nucleus, where they still reside.

Most eukaryotic cells contain mitochondria, but a few single-celled eukaryotic organisms found in oxygen-free environments do not. Biologists had originally hypothesized that these eukaryotes evolved before the endosymbiotic event that established mitochondria in cells having a nucleus, but that proposal was abandoned when it was recognized that genes migrated from the endosymbiont to the host's nucleus. In fact, every mitochondria-free eukaryote examined to date has relict mitochondrial genes in its nuclear genome. This finding provides support for a second hypothesis, that eukaryotic cells without mitochondria had them once but have lost them.

Many eukaryotes that lack mitochondria contain small organelles called hydrogenosomes that generate ATP by anaerobic processes (**Fig. 25.6**). These organelles have little or no DNA, but genes of mitochondrial origin are found in the cell's nuclei. These genes code for proteins that function in the hydrogenosome. Thus, hydrogenosomes appear to be highly altered mitochondria adapted to life in oxygen-poor environments.

FIG. 25.6 Transmission electron microscope images comparing mitochondria within the aerobic protist *Euplotes* (left) and *Nyctotherus*, a close relative that lives in oxygen-free environments (right). The organelle in *Nyctotherus* is halfway between a normal mitochondrion and a hydrogenosome, supporting the view that these two organelles are descended from a common ancestor. *Source: Reprinted by permission from Macmillan Publishers Ltd: B. Boxma, et al., 2005, "An Anaerobic Mitochondrion That Produces Hydrogen," Nature 434:74–79. Copyright 2005.*

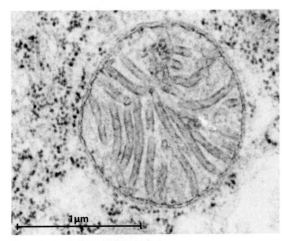

Normal mitochondrion in *Euplotes*

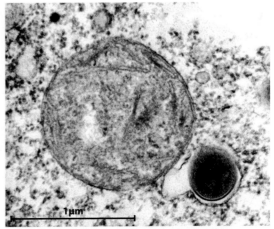

Altered mitochondrion in *Nyctotherus* produces ATP and generates hydrogen. The small dark body in lower right is an endosymbiotic archaeon.

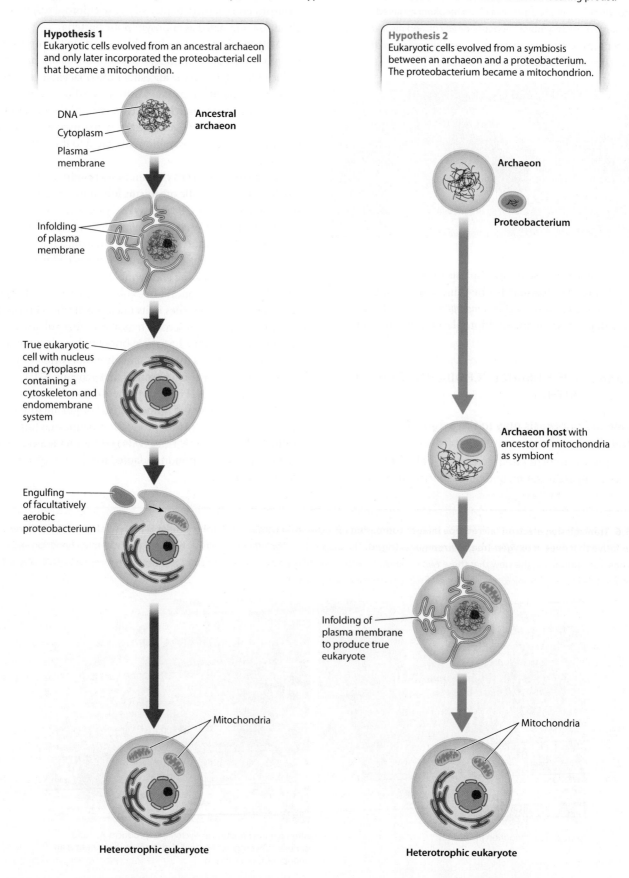

FIG. 25.7 Two hypotheses for the origin of the eukaryotic cell. Both hypotheses lead to the evolution of a mitochondrion-bearing protist.

Other anaerobic eukaryotes contain still smaller organelles called mitosomes that also appear to be remnant mitochondria. Interestingly, mitosomes have lost all capacity for metabolism, but they do retain genes needed for assembling the clusters of iron and sulfur found in many proteins. Why these genes are retained in a modified mitochondrion remains unknown.

CASE 5 THE HUMAN MICROBIOME: DIVERSITY WITHIN

How did the eukaryotic cell originate?

If mitochondria originated as proteobacteria and chloroplasts as cyanobacteria, where does the rest of the eukaryotic cell come from? Analysis of the nuclear genome alone provides no clear answer to this question, because genes of bacterial, archaeal, and purely eukaryotic origin are all present. As discussed, many nuclear genes originated with the mitochondria and chloroplasts acquired from specific bacteria. However, genes from other groups of bacteria also reside in the eukaryotic nucleus, recording multiple episodes of horizontal gene transfer through evolutionary history. In contrast, some genes are present only in eukaryotes and apparently evolved after the domain originated. Still others, including genes involved in DNA transcription and translation, are clearly related to the genes of Archaea.

For this reason, many biologists propose that Archaea-like microorganisms played a key role in the origin of the eukaryotic cell. Based on sequence comparisons of genes coding for the small subunit of ribosomal RNA, American biologists Carl Woese and George Fox hypothesized that the cells ancestral to the cytoplasm and nucleus of eukaryotes were once a distinct microbial lineage more closely related to Archaea than to Bacteria. Recent genome comparisons, however, increasingly place this ancestor within the Archaea, specifically alongside a recently discovered group called the Lokiarchaeota (Chapter 24). Studies of Lokiarchaeota living in hydrothermal vents beneath the Arctic Ocean reveal that the genome of these cells encodes actin-like proteins and other molecules fundamental to cell organization, vesicle trafficking, and phagocytosis in eukaryotes. Based on this information, it seems likely that some portion of the eukaryotic cell is archaeal.

Two starkly different hypotheses have been proposed to explain the origin of the eukaryotic cell. Some biologists believe that the host for mitochondrion-producing endosymbiosis was itself a true eukaryotic cell. This cell had a nucleus, cytoskeleton, and endomembrane system, but only limited ability to derive energy from organic molecules (hypothesis 1 in **Fig. 25.7**). According to this view, the Archaea-like component of the eukaryotic cell had already undergone massive changes in cell biology before it engulfed an endosymbiotic proteobacterium. As noted earlier, however, no one has ever found eukaryotes that do not have and never had mitochondria.

An alternative hypothesis is that no eukaryotic cell existed before mitochondria emerged. Some biologists propose that the eukaryotic cell as a whole began as a symbiotic association between a proteobacterium and an archaeon (hypothesis 2 in Fig. 25.7). The proteobacterium became the mitochondrion and provided many genes to the nuclear genome. The archaeon provided other genes, including those used to transcribe DNA and translate it into proteins.

Biologists continue to debate these alternatives. Both hypotheses explain the hybrid nature of the eukaryotic genome, but neither fully explains the origins and evolution of the nucleus, linear chromosomes, the eukaryotic cytoskeleton, or a cytoplasm subdivided by ever-changing membranes. There is no consensus on the origins of eukaryotes: it remains one of biology's deepest unanswered questions, awaiting novel observations by a new generation of biologists. Once a dynamic cytoskeleton became coupled to a flexible membrane system, however, the evolutionary possibilities of eukaryotic form were established.

In the oceans, many single-celled eukaryotes harbor symbiotic bacteria.

The evolution of the eukaryotic cell is marked by intimate associations between formerly free-living organisms. We might view this kind of arrangement as unusual, but symbioses between eukaryotic cells and bacteria are all around us. For example, an unexpected symbiosis was recently discovered in the Santa Barbara Basin, a local depression in the seafloor off the coast of southern California. Sediments accumulating in this basin contain little oxygen but large amounts of hydrogen sulfide (H_2S) generated by anaerobic bacteria within the sediments. We might predict that eukaryotic cells would be uncommon in these sediments. Not only is oxygen scarce, but in addition sulfide actively inhibits respiration in mitochondria. It came as a surprise, then, when samples of sediment from the Santa Barbara Basin were found to contain large populations of single-celled eukaryotes.

Some of these cells have hydrogenosomes, mitochondria altered by evolution to generate energy where oxygen is absent. Others appear to thrive by supporting populations of symbiotic bacteria on or within their cells. In one case, rod-shaped bacteria cover the surface of their eukaryotic host (**Fig. 25.8**). Research has shown that the bacteria metabolize sulfide in the local environment, thereby protecting their host from this toxic substance. Scientists hypothesize that the bacteria benefit from this association as well because they get a free ride as their host moves through the sediments, enabling these chemoautotrophs to maximize growth by remaining near the boundary between waters that contain oxygen and those rich in sulfide.

The diversity of these eukaryotic–bacterial symbioses is remarkable, yet poorly studied. The evidence gathered to date shows that single-celled eukaryotes have evolved numerous symbiotic relationships with chemoautotrophic bacteria. The bacteria feed and protect the eukaryotes, enabling them to colonize habitats where most eukaryotes cannot live. Clearly, then, the types of symbioses that led to mitochondria and

FIG. 25.8 A modern bacteria–eukaryotic cell symbiosis. A scanning electron microscope image shows bacterial cells on the surface of a single-celled eukaryote found in oxygen-depleted sediments in the Santa Barbara Basin, off the coast of California. The bacteria are hypothesized to metabolize the H_2S in this environment, thereby protecting the eukaryotes that they enclose. *Source: Reprinted by permission from Macmillan Publishers Ltd: V. P. Edgcomb, S. A. Breglia, N. Yubuki, D. Beaudoin, D. J. Patterson, B. S. Leander, and J. M. Bernhard, 2011, "Identity of Epibiotic Bacteria on Symbiontid Euglenozoans in O_2-Depleted Marine Sediments: Evidence for Symbiont and Host Co-evolution," ISME Journal 5:231–243. Copyright 2010. Courtesy of Naoji Yubuki.*

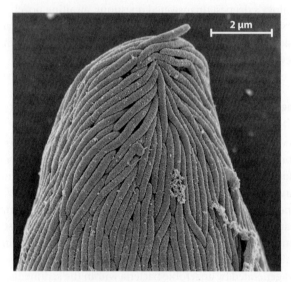

chloroplasts on the early Earth were not rare events, but basic associations between cells that continue to evolve today.

Self-Assessment Questions

4. What is the origin of the chloroplast and the mitochondrion?
5. What are two hypotheses for the origin of the eukaryotic cell?

25.3 EUKARYOTIC DIVERSITY

Historically, the domain Eukarya was divided into four kingdoms: plants, animals, fungi, and protists. Plants, animals, and fungi received special consideration from biologists for the obvious reason that they include large and relatively easily studied species. All remaining eukaryotes were grouped together as **protists**. These were defined as organisms having a nucleus but lacking other features specific to plants, animals, or fungi. The term "protist" is frowned on by some biologists because it does not refer to a monophyletic grouping of species—that is, an ancestral form and all its descendants (Chapter 22). Nonetheless, as an informal term that draws attention to a group of organisms sharing particular characteristics, "protist" usefully describes the diverse world of microscopic eukaryotes and seaweeds.

Two other terms scorned by some biologists but embraced by ecologists are "algae" and "protozoa." **Algae** are photosynthetic protists. They may be microscopic single-celled organisms or the highly visible, multicelled organisms we call seaweeds. **Protozoa** are heterotrophic protists, almost exclusively single-celled organisms. Although "algae" and "protozoa" may have little phylogenetic meaning, these simple and useful terms convey a great deal of information about the structure and function of these organisms.

Protist cells exhibit remarkable diversity. Some have cell walls, while others do not. Some are covered by skeletons of silica (SiO_2) or calcium carbonate ($CaCO_3$), while others are naked or live within **tests**, or "houses," constructed exclusively of organic molecules. Some are photosynthetic, others

FIG. 25.9 The eukaryotic tree of life showing major groups. Branches with photosynthetic species (shown in green) are distributed widely. Additional branches, including animals, foraminiferans, radiolarians, and ciliates, contain species that harbor photosynthetic symbionts. Dashed lines indicate a high degree of uncertainty.

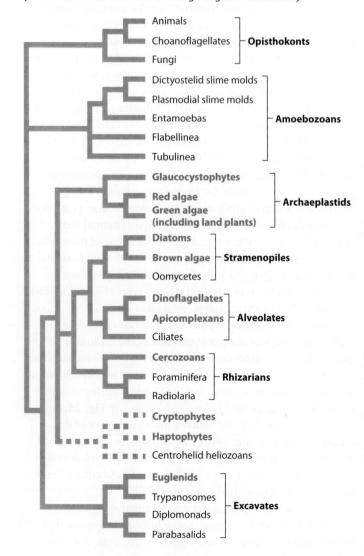

are heterotrophic, and still others are capable of both photosynthesis and heterotrophy. Some move by beating flagella; others, such as *Amoeba*, can extend fingers of cytoplasm, called pseudopodia, to move and capture food. Most are aerobic, but there are anaerobic protists as well. Surprisingly, most of these distinctive features have evolved multiple times and so do not define phylogenetically coherent groups. As a result, our understanding of evolutionary pattern in Eukarya had to wait for the molecular age.

Fig. 25.9 shows a phylogenetic tree of eukaryotes as biologists currently understand it. As in the case of Bacteria, it is easier to identify major groups of Eukarya than to establish their evolutionary relationships to one another. Molecular sequence comparisons consistently recognize seven major groups, called **superkingdoms**. These groups are noted in bold black type in Fig. 25.9. Animals fall within one of these superkingdoms, the Opisthokonta, whereas plants fall within another, the Archaeplastida. Most, but not all, of these major eukaryotic branches and their respective species are strongly supported by data from genes and genomes. A few groups, notably the cryptophytes, haptophytes, and centrohelid heliozoans, seem to wander from position to position from one phylogenetic analysis to the next, sometimes moving together, sometimes separately. Whether the well-resolved superkingdoms constitute the only major limbs on the eukaryotic tree is less certain.

Many microscopic protists remain unstudied, and we may yet be surprised by what they show.

In the following sections, we introduce the major features of eukaryotic diversity. The species numbers we present indicate only the taxonomic diversity that is known to us; most protistan diversity remains undiscovered. Many biologists estimate that only about 1 in 10 protist species has been described so far. As this is also the case for animals and fungi (but perhaps not for land plants, because they are all visible to the naked eye), the relative numbers of species within different superkingdoms may reflect something like the actual distribution of biodiversity among eukaryotes.

Our own group, the opisthokonts, is the most diverse eukaryotic superkingdom.

Over the past 250 years, biologists have described approximately 1.8 million species. Of these, 75% or so fall within the **Opisthokonta** (**Fig. 25.10a**), a group that encompasses animals, fungi, and some protists. The name for this superkingdom is derived from Greek words meaning "posterior pole." The name calls attention to the fact that cell movement within this group is propelled by a single flagellum attached to the posterior end of the cell. Not all cells in this superkingdom move around. In humans, for example, the flagellum is limited to sperm. Opisthokonts are

FIG. 25.10 Opisthokonts. This superkingdom includes animals and fungi, as well as several protistan groups. The closest protistan relatives of animals are the choanoflagellates. *Photo source: Nicole King, Howard Hughes Investigator and Professor of Genetics, Genomics and Development, University of California, Berkeley; Steve Paddock and Sean Carroll, HHMI, Univ. of Wisconsin.*

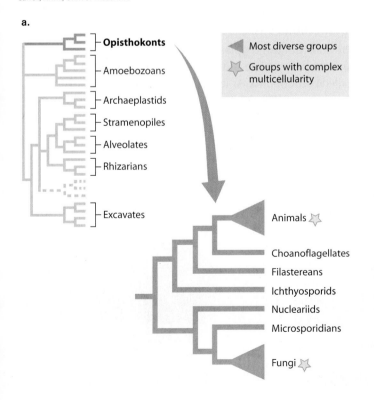

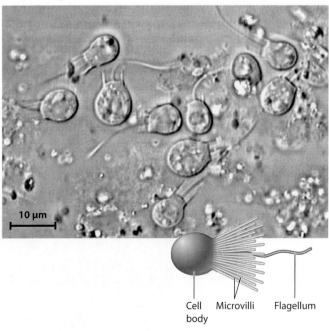

b. Choanoflagellates

heterotrophic, although some species harbor photosynthetic symbionts.

Animals are the most diverse and conspicuous opisthokonts. More than 1.3 million animal species, mostly insects and their relatives, have been described. Fungi are also diverse, with more than 75,000 described species, including the visually arresting mushrooms. Here, however, we focus on opisthokont protists, as these somewhat poorly known microorganisms hold clues to the origins of complex multicellularity in this superkingdom (Chapter 26).

Finding morphological characters that unite all opisthokonts is a challenge. One reason for this difficulty is that animals and fungi have diverged so strikingly from what must have been the ancestral condition of the group. However, as molecular sequence comparisons began to reshape our understanding of eukaryotic phylogeny, it soon became clear that fungi and animals are closely related.

Even more closely related to the animals are the **choanoflagellates**. This group of mostly unicellular protists is characterized by a ring of microvilli, fingerlike projections that form a collar around the cell's single flagellum (**Fig. 25.10b**). About 150 choanoflagellate species have been described from marine and freshwater environments, where they prey on bacteria. As early as 1841, the close similarity between choanoflagellates and the collared feeding cells of sponges suggested that these minute organisms might be our closest protistan relatives. This view gained widespread popularity with the discovery in 1880 of *Proterospongia haeckeli*, a choanoflagellate that lives in colonies of cells joined by adhesive proteins.

Molecular sequence comparisons now confirm the close relationship between animals and choanoflagellates. For example, the complete genome of the choanoflagellate *Monosiga brevicollis* shows that a number of signaling molecules known to play a role in animal development are present in our choanoflagellate relatives, although their function in these single-celled organisms remains largely unknown. More generally, many genes once thought to be unique to animals have now been identified in choanoflagellates and other unicellular opisthokonts, underscoring the deep evolutionary roots of animal biology.

Microsporidia form another group of single-celled opisthokonts. Microsporidia are parasites that live inside animal cells. Only their spores survive in the external environment, where they await the opportunity to infect a host and complete their life cycle. Microsporidia infect all animal phyla, so the approximately 1000 known species probably represent only the tip of the iceberg. More than a dozen species have been isolated from human intestinal tissues. Microsporidia infections can be particularly devastating in patients with AIDS, who have compromised immune systems as a result of their disease.

Beyond their importance in human health, microsporidians have attracted the attention of biologists because of features they *lack:* microsporidian cells have no aerobically respiring mitochondria, no Golgi apparatus, and no flagella. Furthermore, they have a highly reduced metabolism and among the smallest genomes of any known eukaryote. The cellular simplicity of microsporidians does not mean that they are early-evolved organisms, however. In fact, molecular sequencing studies show that microsporidians are the descendants of more complex organisms. Their simplicity is actually an adaptation for life as an intracellular parasite. It is now widely accepted that microsporidians are closely related to the fungi.

Other protistan opisthokonts have been identified in recent years. These organisms are mostly bacteria-eating unicells or parasites of aquatic animals. Nevertheless, as with the choanoflagellates described earlier, their biology is beginning to illuminate the evolutionary path to complex multicellularity in animals and fungi.

Amoebozoans include slime molds that produce multicellular structures.

Previously in this chapter, we learned one thing about amoebas: they move and feed by extending cytoplasmic fingers called pseudopodia. As its name implies, the superkingdom **Amoebozoa** (**Fig. 25.11a**) is a group of eukaryotes with

---

FIG. 25.11 Amoebozoans. This group includes protists with an amoeboid stage in their life cycle. *Photo sources: b. M. I. Walker/Science Source; c. AFIP/Science Source.*

a.

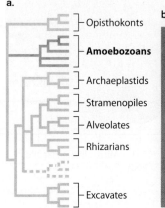

b. *Amoeba proteus*, with clearly visible pseudopodia

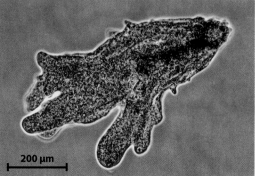

c. *Entamoeba histolytica*, the infectious agent in amoebic dysentery

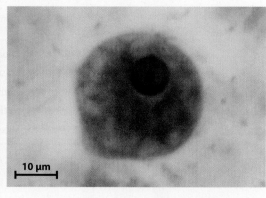

FIG. 25.12 Plasmodial slime molds. (a) Plasmodia, such as this pretzel slime mold, are coenocytic structures containing many nuclei. (b) Plasmodia generate sporangia, stalked structures that produce spores for dispersal. *Sources: a. Matt Meadows/Science Source; b. Ray Simons/Getty Images.*

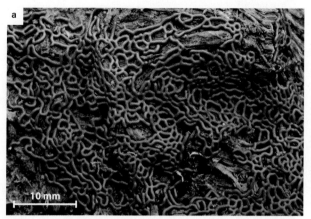

amoeba-like cells (illustrated by *Amoeba proteus* in **Fig. 25.11b**). More than 1000 amoebozoan species have been described. In some of them, large numbers of cells aggregate to form multicellular structures as part of their life cycle.

Although some amoeba-like cells are not members of the Amoebozoa superkingdom, all members of this group have cells similar to that of *Amoeba*, at least at some stage of their life cycle. Amoebozoans play an important role in soils as predators on other microorganisms. Some amoebozoans cause disease in humans. For example, *Entamoeba histolytica* (**Fig. 25.11c**), an anaerobic protist that causes amoebic dysentery, is responsible for 50,000 to 100,000 deaths every year.

Beyond considerations of human health, the amoebozoans of greatest biological interest are the slime molds. In plasmodial slime molds (**Fig. 25.12**), haploid cells fuse to form zygotes that subsequently undergo repeated rounds of mitosis but not cell division to form colorful, often lacy structures visible to the naked eye (Fig. 25.12a). These structures, called plasmodia, are **coenocytic**, which means they contain many nuclei within one giant cell. The plasmodia can crawl along surfaces and so are able to seek out and feed on the bacteria and small fungi commonly found on bark or plant litter on the forest floor. Triggered by poorly understood environmental signals, plasmodia eventually differentiate to form stalked structures called sporangia that can be 1 to 2 mm high (Fig. 25.12b). Within the mature sporangium, cell walls form around the many nuclei, producing discrete cells. These cells undergo meiosis, generating haploid spores that disperse into the environment. Fusion of these spores to form a diploid amoeba-like cell begins the life cycle anew.

Cellular slime molds (**Fig. 25.13**) are a second type of slime mold. These slime molds spend most of their life cycle as solitary amoeboid cells feeding on bacteria in the soil. If the food supply runs out, however, the cells undergo a dramatic transformation. The starving cells produce the chemical signal cyclic AMP (Chapter 9), which induces as many as 100,000 cells to aggregate into a large multicellular slug-like form (Fig. 25.13a). This "slug" can migrate by a coordinated movement of its cells governed by actin and myosin, the same proteins that control muscle movement in animals. Migrating "slugs" forage for food, leaving behind a trail of mucilage—which explains the "slime mold" name. Like plasmodial slime

FIG. 25.13 Cellular slime molds.
(a) Starvation causes amoeboid feeding cells of cellular slime molds to aggregate into a multicellular "slug" that can migrate along surfaces, leaving behind a trail of slime. (b) "Slugs" differentiate to form stalked sporangia that produce spores. *Source: DIOMEDIA/Carolina Biological.*

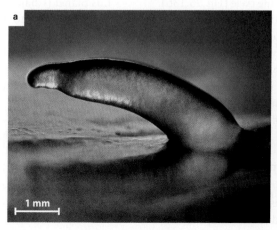

FIG. 25.14 Archaeplastids. (a) Archaeplastids are a photosynthetic group descended from the protist that acquired photosynthesis from an endosymbiotic cyanobacterium. They include the (b) glaucocystophytes, (c) red algae, and (d) green algae and their descendants, the land plants. *Photo sources: b. Jerome Pickett-Heaps/Science Source; c. Dr. D. P. Wilson/Science Source; d. D. J. Patterson.*

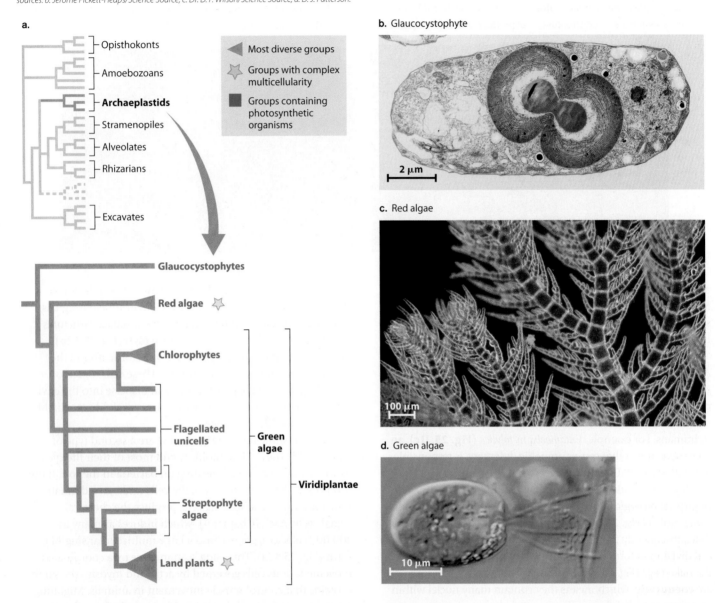

molds, the "slugs" of cellular slime molds eventually differentiate to form stalk-borne sporangia that produce the spores responsible for dispersal and for renewal of the life cycle (Fig. 25.13b).

Slime molds are not true intermediates in the evolution of multicellularity. In other words, they are not closely related to animals, fungi, or other groups in which complex multicellularity evolved. Nonetheless, biologists have studied the cellular slime mold *Dictyostelium* as a model organism because it has much to teach us about fundamental eukaryotic processes of cell signaling and differentiation, cellular motility, and mitosis.

Archaeplastids, which include land plants, are photosynthetic organisms.

After the opisthokonts, the most conspicuous and diverse eukaryotes fall within the superkingdom **Archaeplastida** (Fig. 25.14). This is where we find the land plants, whose 400,000 described species dominate eukaryotic biomass on this planet (Chapter 31). The archaeplastids comprise three major groups, and with the exception of a few parasitic and saprophytic (living on dead matter) species, all are photosynthetic (Fig. 25.14a). This ability indicates that the last common ancestor of living archaeplastids was itself photosynthetic. In fact, archaeplastids are the direct descendants of the protist that first evolved chloroplasts from endosymbiotic cyanobacteria. Some molecular analyses also place cryptophyte and haptophyte algae within this branch, but their location in the phylogenetic tree remains a topic of continuing research.

The least conspicuous archaeplastids are the glaucocystophytes (Fig. 25.14b), a small group of single-celled algae found in freshwater ponds and lakes. We might skip over

FIG. 25.15 Diversity of green algae. (a) *Halosphaera minor*, a small flagellated unicell found in the surface ocean; (b) *Spirogyra*, a filament-forming freshwater green alga with distinctive helical chloroplasts; (c) a desmid, among the most diverse and widespread of all unicellular green algae in fresh water; (d) *Volvox*, a simple multicellular organism; and (e) *Acetabularia*, a macroscopic (but single-celled!) green alga found in coastal marine waters in tropical climates. *Sources: a. Charles J. O'Kelly, Ph.D.; b. DIOMEDIA/Eric Grave; c. Gerd Guenther/Science Source; d. Frank Fox/Science Source; e. Wolfgang Poelzer /Wa/AGE Fotostock.*

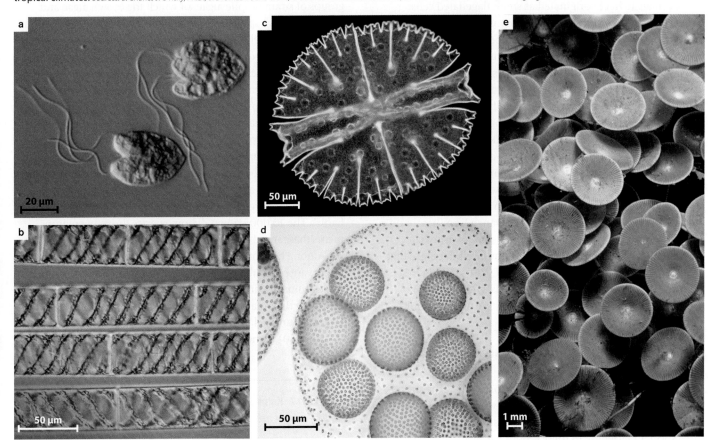

them entirely, except for one illuminating observation: glaucocystophyte chloroplasts retain more features of the ancestral cyanobacterial endosymbiont than the chloroplasts of any other algae. Their chloroplasts have walls of peptidoglycan, the same molecule found in cyanobacterial walls, and their photosynthetic pigments include biliproteins, also found in cyanobacteria.

More abundant and diverse is the second major archaeplastid group, the red algae (Fig. 25.14c). Approximately 5000 species of red algae are known, mostly from marine environments. Most are multicellular, and some have complex morphology that includes differentiated tissues such as sporangia and distinct surface and interior cell layers. Like other archaeplastids, most red algae have walls made of cellulose. The principal photosynthetic pigments in red algae are a form of chlorophyll called chlorophyll *a* and the biliproteins also found in glaucocystophytes and most cyanobacteria.

Red algae can be conspicuous as seaweeds on the shallow seafloor. One subgroup of red algae, known as the coralline algae, secretes skeletons of calcium carbonate ($CaCO_3$) that deter predators and help the algae withstand the energy of waves in shallow marine environments. If you examine the margin of coral reefs, where waves crash into the reef, you will often see that it is coralline algae and not the corals themselves that resist breaking waves, thereby enabling the reefs to expand upward into wave-swept environments.

Red algae are important as food in certain cultures. For example, they are consumed as laver in the British Isles (particularly in Wales, in cakes similar to oatcakes) and as nori in Japan. Without knowing it, you often ingest carageenan, a polysaccharide synthesized by red algae. Carageenan molecules stabilize mixtures of biomolecules and promote gel formation, so they are used in products ranging from ice cream to toothpaste. Agar, another red algal polysaccharide, is found universally in microbiology laboratories because the gel it forms at room temperature provides a nearly ideal substrate for bacterial growth.

The land plants and their algal relatives form Viridiplantae, the third branch of Archaeplastida. In this chapter, we are concerned with the protistan members of this group, the green algae (Fig. 25.14d). Green algae include a diverse assortment of species (about 10,000 have been described) that live in marine and, especially, freshwater environments. Green algae range from tiny single-celled flagellates to meter-scale seaweeds (**Fig. 25.15**). They all are united by two features: the presence of chlorophyll *a* and chlorophyll *b* in chloroplasts that have two membranes, and a unique attachment for flagella.

Green algae likely originated as small flagellated cells. Indeed, today many still live as photosynthetic cells in marine or fresh waters, like their ancestors, at nodes near the base of the green algal tree (Fig. 25.15a). The larger diversity of green algae, however, is found on two branches that diverged from these flagellated ancestors. One branch, the chlorophytes, radiated mostly in the sea and includes the common seaweeds seen along the world's seashores. Chlorophyte green algae have also played a major role in laboratory studies of photosynthesis (*Chlamydomonas*) and multicellularity (*Volvox*, Fig. 25.15d).

For humans, the more important branch of the green algal tree is the one that diversified in freshwater and, eventually, on land. Called streptophytes, the species on this branch show a progression of form from unicells that branch from basal nodes through simple cell clusters and filaments on intermediate branches (Figs. 25.15b and 25.15c), to multicellular algae that form some of the closest relatives of land plants (Chapter 31). As was true of choanoflagellates and animals, green algae display features of cell biology and genomics that tie them unequivocally to land plants.

Stramenopiles, alveolates, and rhizarians dominate eukaryotic diversity in the oceans.

The phylogenetically related superkingdoms called the stramenopiles, alveolates, and rhizarians may not be household names, but perhaps they should be. The surface ocean hosts three species belonging to these groups for every one marine animal species, and these protists also play important roles in the ecology of lakes, rivers, and soils.

The superkingdom **Stramenopila** (see Fig. 25.9) includes unicellular organisms and giant kelps, algae and protozoa, free-living cells and parasites. All have an unusual flagellum that bears two rows of stiff hairs (**Fig. 25.16**). Most also have a second, smooth flagellum.

FIG. 25.16 Stramenopiles. This electron micrograph of a unicellular stramenopile shows a flagellum with two rows of stiff hairs. Like this cell, most stramenopiles also have a second, smooth flagellum. *Source: D. J. Patterson.*

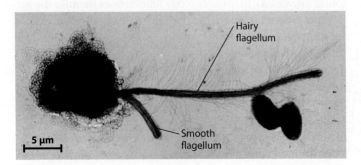

Despite the presence of heterotrophic species, most stramenopiles are photosynthetic. There are approximately a dozen groups of stramenopile algae, of which the brown algae and diatoms deserve special mention. Brown seaweeds are common along rocky shorelines across the world, and in the Sargasso Sea, huge masses of ropy brown algae (aptly named *Sargassum*) float at the surface. Easily the most impressive brown algae are the kelps, giant seaweeds that form forests above the seafloor (**Fig. 25.17**). Kelps not only develop functionally distinct tissues, but also have evolved distinct organs specialized for attachment, photosynthesis, flotation, and reproduction.

The most diverse stramenopiles are the diatoms (**Fig. 25.18**). Recognized by their distinctive external skeletons of silica, diatoms account for half of all photosynthesis in the sea (and, therefore, nearly one-fourth of all photosynthesis on Earth). The 10,000 known diatom species thrive in environments that range from wet soil to the open ocean.

Among heterotrophic stramenopiles, the oomycetes, or water molds, are particularly interesting. Originally classified as fungi, these multicellular pathogens cause a number of plant diseases, including the potato blight that devastated Ireland's major food crop and led to famine in that country during the 1840s (Chapter 30). Other species infect animals, including humans.

FIG. 25.17 Kelps. These complex multicellular brown algae can form "forests" tens of meters high in the ocean. *Source: Mark Conlin/Alamy.*

FIG. 25.18 Diatoms. Diatoms are the most diverse stramenopiles, and among the most diverse of all protists. Their shapes, outlined by their tiny skeletons made of silica (SiO_2), can be (a) like a pill-box, (b) elongated, or (c) twisted around the long axis of the cell. Some diatom species form colonies, like those shown in (d). *Sources: a. and b. Steve Gschmeissner/Science Source; c. Andrew Syred/Science Source; d. Biophoto Associates/Science Source.*

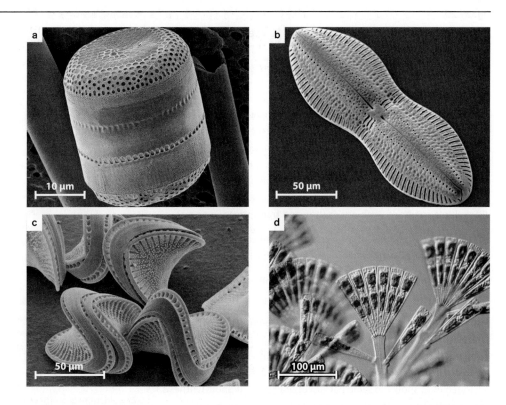

Like stramenopiles, the superkingdom **Alveolata** is a diverse group that includes both photosynthetic and heterotrophic species. Also like stramenopiles and many other eukaryotic groups, alveolates are united by a feature observed only with careful electron microscopy: small vesicles called cortical alveoli are packed beneath the cell surface and, in some species at least, store calcium ions for use by the cell (**Fig. 25.19a**).

Most photosynthetic alveolates belong to a large subgroup called dinoflagellates. Dinoflagellates are distinguished by two flagella that produce a rotary motion of the cell as it moves through its environment; many also have a distinctive covering of organic plates (**Fig. 25.19b**). Approximately 4000 dinoflagellate species have been described, but recent genomic surveys of the oceans make it clear that they are much more diverse, by nearly an order of magnitude. In fact, recent studies indicate that dinoflagellates and their poorly known parasitic relatives are the most diverse eukaryotes in the world's sunlit oceans. About half of the known dinoflagellate species are photosynthetic, but most of these are also capable of heterotrophic growth. Notably, dinoflagellates affect humans when they form red tides, large and toxic blooms in coastal waters. Red tides occur when nutrients are supplied to coastal waters in large amounts, sometimes for natural

FIG. 25.19 Alveolates. (a) The arrows point to alveoli, flattened vesicles just beneath the surface of the cell that give alveolates their name. (b) Dinoflagellates have cell walls made of interlocking plates of cellulose. (c) A ciliate has many short flagella, or cilia, that line the cell. *Sources: a. Gert Hansen, SCCAP, University of Copenhagen; b. Electron Microscopy Unit, Australian Antarctic Division © Commonwealth of Australia; c. Andrew Syred/Science Source.*

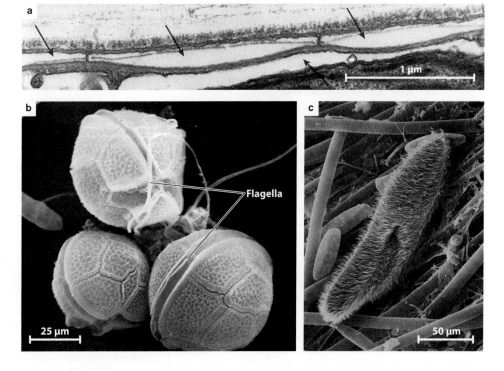

HOW DO WE KNOW?

Fig. 25.20

How did photosynthesis spread through the Eukarya?

BACKGROUND Many different branches on the eukaryotic tree include photosynthetic species, commonly interspersed with non-photosynthetic lineages (named groups in figure 25.20b). Did eukaryotes gain photosynthesis once, by the incorporation of a cyanobacterium, and then lose this capacity multiple times? Or is the history of photosynthetic endosymbioses more complicated? That is, did multiple events of symbiont capture and transformation occur?

HYPOTHESIS 1 Photosynthesis was established early in eukaryotic evolution by endosymbiosis and then lost in some lineages.

HYPOTHESIS 2 Eukaryotes acquired photosynthesis multiple times by repeated episodes of endosymbiosis.

PREDICTIONS The two hypotheses make different predictions. If photosynthesis were acquired in a single common ancestor of all living photosynthetic eukaryotes and then lost in some lineages, we would expect chloroplast gene phylogenies and nuclear gene phylogenies to show the same pattern of branching. If photosynthesis were acquired multiple independent times, we would expect chloroplast phylogenies and nuclear phylogenies to show different patterns.

EXPERIMENT Because chloroplasts have DNA, scientists developed a phylogenetic tree based on molecular sequence comparison of chloroplast genes (Fig. 25.20a). This chloroplast phylogeny was then compared with a second phylogeny, for all eukaryotes, based on molecular sequence comparison of nuclear genes (Fig. 25.20b).

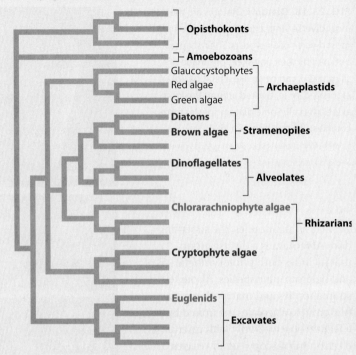

b. Phylogenetic relationships based on nuclear genes

RESULTS The molecular sequence comparisons show that evolutionary relationships among chloroplasts do not mirror those based on nuclear genes. The chlorarachniophyte algae and photosynthetic euglenids (shown in green) are classified into a monophyletic group with the green algae when chloroplast DNA is analyzed (see figure 25.20a). However, these groups lie far from green algae in the phylogeny based on nuclear genes (figure 25.20b). Similarly, brown algae, diatoms, most photosynthetic dinoflagellates, and cryptophyte algae (shown in red) form a monophyletic group with the red algae in the phylogeny based on chloroplast DNA (figure 25.20a), but analysis of nuclear genes places them in different groups (figure 25.20b).

CONCLUSION The chloroplast and nuclear phylogenies show different patterns of branching, supporting the hypothesis that photosynthesis spread through the Eukarya by means of multiple endosymbioses involving eukaryotic endosymbionts.

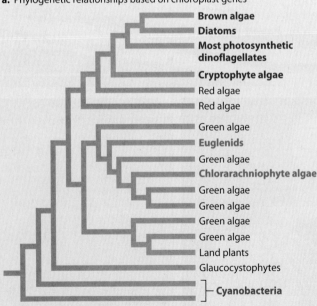

a. Phylogenetic relationships based on chloroplast genes

For branches in black, chloroplast relationships match the phylogeny indicated by nuclear genes—these groups diverged from the ancestor in which endosymbiosis first gave rise to chloroplasts.

Nuclear genes place brown algae and diatoms among the stramenopiles, dinoflagellates within the alveolates, and cryptophytes on their own poorly resolved branch. Chloroplast genes, however, relate them to red algae. These groups gained photosynthesis by incorporating red algal cells as endosymbionts.

Nuclear genes place chlorarachniophytes and euglenids among the rhizarians and excavates, respectively, but chloroplast genes relate them to green algae. Both groups gained photosynthesis by incorporating green algal cells as endosymbionts.

Chloroplasts form a monophyletic grouping nested within the cyanobacteria—a principal line of evidence supporting the endosymbiotic origin of chloroplasts.

SOURCE Hackett, J. D., et al. 2007. "Plastid Endosymbiosis: Sources and Timing of the Major Events." In *Evolution of Primary Producers in the Sea*, edited by P. G. Falkowski and A. H. Knoll, 109–132. Boston, MA: Elsevier Academic Press.

reasons but commonly because of human activities such as sewage seeps or runoff from highly fertilized croplands.

The alveolates also include a second diverse (8000 described species) group of protists called ciliates (**Fig. 25.19c**). Ciliates are also distinct at the cellular level, with two nuclei in each cell and numerous short flagella called cilia. These heterotrophs feed on bacteria or other protists and so play key roles as both predators and prey in soil and water.

The third principal group of alveolates, the Apicomplexans, perhaps attracts the most scientific attention. The reason is simple: the apicomplexan *Plasmodium falciparum* and related species cause malaria, a lethal disease that sickened millions and killed more than 400,000 people in 2015 (Case 4: Malaria). Although *Plasmodium* cells are heterotrophic, they contain vestigial chloroplasts that suggest evolutionary loss of photosynthesis. Recently, this hypothesis has been confirmed by the discovery in Sydney Harbour, Australia, of photosynthetic cells closely related to parasitic *Plasmodium*.

Photosynthesis spread through eukaryotes by repeated endosymbioses involving eukaryotic algae.

A curious feature of the eukaryotic tree is that photosynthetic eukaryotes are distributed widely but discontinuously among its branches (see Fig. 25.9). How can we explain this pattern?

As explored in **Fig. 25.20**, two hypotheses have been suggested: either photosynthesis was established early in eukaryotic evolution and then subsequently lost in some lineages, or eukaryotes acquired photosynthesis multiple times by repeated episodes of endosymbiosis. The second hypothesis, it turns out, is correct. The twist is that most of the symbionts involved in the spread of photosynthesis were photosynthetic eukaryotes, not cyanobacteria.

Once more, DNA sequence comparisons provide the decisive data. We can construct phylogenetic trees for photosynthetic eukaryotes using chloroplast genes and then repeat the exercise using nuclear genes. A comparison of the two trees, as shown in Figure 25.20, reveals that two groups, chlorarachniophytes and euglenids, gained photosynthesis by establishing green algal cells as endosymbionts. (The algal cells contained chloroplasts derived from an earlier symbiotic assimilation of cyanobacteria.) Members of other groups (including brown algae, diatoms, most photosynthetic dinoflagellates, and cryptophyte algae) gained photosynthesis as well, but most likely by ingesting red algal cells.

Several features of cell biology support the view that eukaryotes acquired photosynthesis multiple times by repeated episodes of endosymbiosis. The chloroplasts of euglenids and chloarachniophyte algae contain chlorophyll *a* and *b*, like green algae. In contrast, cryptophytes contain chlorophyll *a* and biliproteins, both of which are found in red algae, along with chlorophyll *c*, a form of chlorophyll not known from either red or green algae. (Diatoms, brown algae, and their relatives contain chlorophyll *a* and *c*, but not biliproteins.)

We noted earlier that chloroplasts in green and red algae are bounded by two membranes, but most of the algae shown in red and green in Fig. 25.20 have chloroplasts bounded by four membranes. The presence of four membranes is expected under the hypothesis of eukaryotic endosymbionts: the inner two membranes are those that bounded the chloroplast in the green or red algal endosymbiont; the third membrane is derived from the outer cell membrane of the symbiont; and the fourth, outer membrane is part of the cell membrane of the host (**Fig. 25.21**).

FIG. 25.21 Successive endosymbiotic events that established photosynthesis in chlorarachniophyte algae.

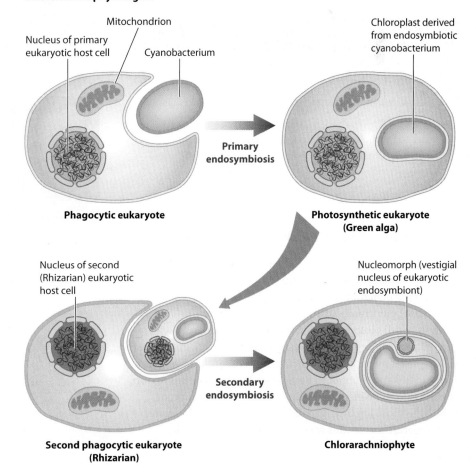

Chlorarachniophyte algae retain another feature that supports the eukaryotic origin of their chloroplasts: a small organelle called a nucleomorph (Fig. 25.21). The nucleomorph in these algae is the remnant nucleus of a green algal symbiont. It retains only a few genes, but those show close affinity with genes in green algae. Nucleomorph genes in cryptophyte algae similarly provide evidence for a red algal symbiont.

How many endosymbiotic events are required to account for the diversity of photosynthetic eukaryotes? The primary endosymbiosis gave rise to green and red algae, two further events established chloroplasts derived from green algae in euglenids and chlorarachniophytes, and at least one to three events established chloroplasts derived from red algae in cryptophytes, haptophytes, and stramenopiles.

The dinoflagellates add a remarkable coda to this accounting. Like stramenopiles, most dinoflagellates have chloroplasts derived from red algae. Three dinoflagellate groups, however, have chloroplasts that are clearly derived from green, cryptophyte, and haptophyte algae. Include *Paulinella* in the mix, and at least 8 to 10 endosymbiotic events are required to explain eukaryotic photosynthesis. Eukaryotes never evolved oxygenic photosynthesis—only the cyanobacteria accomplished that feat. Instead, eukaryotes appropriated the biochemistry of cyanobacteria and spread it throughout the domain by the horizontal transfer of entire cells.

Self-Assessment Questions

6. What are the names of the seven superkingdoms of eukaryotes? Name an organism in each one.
7. Was the common ancestor of plants and animals unicellular or multicellular?
8. How do the molecular sequences of genes in chloroplasts show how photosynthesis spread throughout the eukaryotic domain?

25.4 THE FOSSIL RECORD OF PROTISTS

Not all algae and protozoa fossilize easily, but those that do provide a striking record of evolution that both predates and parallels the better-known fossil record of animals. Fossilized protists do not contain DNA, nor do they preserve the features of cell biology that define eukaryotes, such as nuclear membranes and organelles. Thus, for a fossil to be identified as eukaryotic, it must preserve other features of morphology found today in Eukarya, but not in Bacteria or Archaea. For this reason, fossils that can be identified with certainty as eukaryotic are limited to those left by eukaryotes that synthesized distinctive and preservable cell walls at some stage of their life cycle.

Fossils show that eukaryotes existed at least 1800 million years ago.

Sedimentary rocks deposited in coastal marine environments 1800–1400 million years ago contain microscopic fossils (called **microfossils** for short) up to about 300 μm in diameter. Many of these have complicated wall structures that identify them as eukaryotic. Interlocking plates, long and branching arms, and complex internal layering have been imaged by both scanning and transmission electron microscopes (**Fig. 25.22a**). Comparison with living cells suggests that such fossils could be formed only by organisms with a cytoskeleton and endomembrane system, the hallmarks of eukaryotic biology.

FIG. 25.22 Eukaryotic fossils in Precambrian sedimentary rocks. (a) Simple protist in 1500-million-year-old rocks from Australia. (b) Simple multicellular red algae in 1050-million-year-old rocks from Canada. (c) Amoebozoan test in 730-million-year-old rocks from the Grand Canyon, Arizona. *Sources: a. Andrew Knoll, Harvard University; b. Nicholas J. Butterfield, University of Cambridge; c. Susannah Porter.*

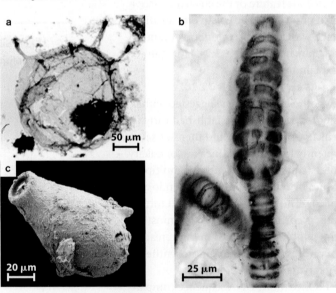

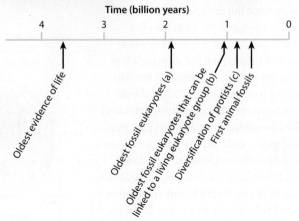

At present, it appears that the eukaryotic domain first appeared no later than 1800 million years ago and possibly earlier, although the record is difficult to trace in older rocks. None of these earliest eukaryotic microfossils can be placed with confidence into one of the living superkingdoms. Conceivably, at least some of the oldest protistan microfossils could predate the last common ancestor of living eukaryotes.

In sedimentary rocks dated at about 1050 million years old, we find the oldest fossils that can be linked clearly to a living group of eukaryotes (**Fig. 25.22b**). These fossils preserve features found today only in the red algae. Thus, by a billion years ago, Archaeplastida had already diverged from other eukaryotic superkingdoms. More generally, photosynthesis had become established in eukaryotes and simple multicellularity had already evolved within the domain.

Much more eukaryotic diversity is recorded in rocks 800–700 million years old, including green algae and the remains of test-forming amoebozoans (**Fig. 25.22c**). Molecular fossils further indicate that algae were expanding to become major photosynthesizers in the oceans. How can we explain this diversification? One hypothesis is that 800–700 million years ago, protists first evolved (or expanded) the ability to eat other protists rather than preying only on Bacteria and Archaea. The appearance of this ability initiated an evolutionary arms race that favored divergent means of preying on other organisms and avoiding predation. During this time interval, for reasons that are currently debated, nutrient levels of surface seawater also began to increase, providing conditions for the expansion of eukaryotic algae in the oceans. Both fossils and molecular clocks support these ecological hypotheses for eukaryotic diversification, perhaps assisted by a small increase in the oxygen content of seawater.

Protists have continued to diversify during the age of animals.

Modern protistan diversity took shape alongside evolving animals during the Phanerozoic Eon, 541 million years ago to the present. Two examples of these more recently emerging protist groups are foraminiferans and radiolarians, marine protozoans with shells of calcium carbonate and silica, respectively. These protist groups appear for the first time in rocks deposited near the beginning of the Phanerozoic Eon and have continued to diversify since that time (**Fig. 25.23**). Some seaweeds also evolved skeletons of calcium carbonate at this time, and diversified along with animals in Phanerozoic oceans. These mineralized skeletons presumably protect the organisms that form them from being eaten.

Marine protists underwent a new round of radiation during the Mesozoic Era, 252–66 million years ago. The most significant changes occurred among photosynthesizers. During the preceding Paleozoic Era, green algae and cyanobacteria were the primary photosynthesizers in the oceans. As the Mesozoic Era began, however, new groups of photosynthetic algae expanded, first dinoflagellates and coccolithophorid algae, followed by diatoms. Today, these three groups dominate photosynthesis in many parts of the sea.

Within the oceans, diatom and coccolithophorid evolution had another important consequence. The silica skeletons of diatoms are now the principal means by which SiO_2 is transferred from seawater to sediments. Similarly, the $CaCO_3$ scales made by coccolithophorids and the skeletons of planktonic foraminiferans are a primary source of carbonate sediments on the seafloor (**Fig. 25.24**). In this way, the ongoing evolution of protists in the age of animals has continued to transform marine ecosystems.

FIG. 25.23 (a) Foraminifera and (b) Radiolaria, showing skeletons of calcium carbonate and silica, respectively. *Sources: a. Astrid & Hanns-Frieder Michler/Science Source; b. M. I. Walker/Science Source.*

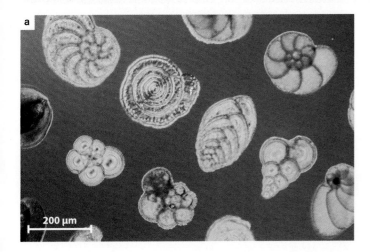
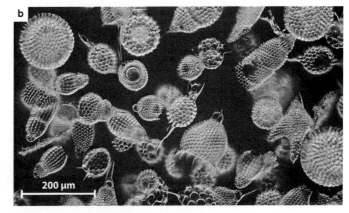

FIG. 25.24 Coccolithophorids. (a) Coccolithophorid algae, like *Emiliania huxleyi*, form tiny scales of calcium carbonate. (b) A satellite image captures a coccolithophorid bloom off the southwestern coast of England visible as light blue streamers of calcium carbonate in the otherwise dark blue ocean.
Sources: a. NHM Images, © The Trustees of the Natural History Museum, London; b. National Oceanography Centre, UK.

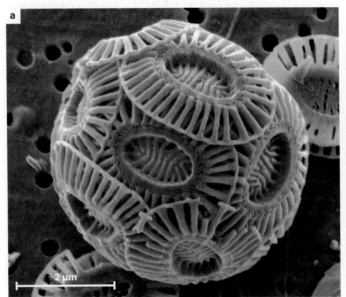

Today, microscopic eukaryotes dominate photosynthetic production in many parts of the oceans and play a major role in biogeochemical cycles. And for thousands of protists, plants and animals are simply habitats, ripe for colonization. For proof of this relationship, we need look no further than the many eukaryotic vectors of human disease. The key conclusion is that a remarkable, and still largely undocumented, diversity of eukaryotic organisms has evolved through time to form an integral part of ecosystems across our planet, shaping the world in which we live in ways both positive and negative.

Self-Assessment Questions

9. When did the first eukaryotic cell evolve?
10. What kind of evidence is used to date the timing of the first eukaryotic cell?

CORE CONCEPTS SUMMARY

25.1 A REVIEW OF THE EUKARYOTIC CELL: Eukaryotic cells are defined by the presence of a nucleus, but features such as a dynamic cytoskeleton and membrane system explain their evolutionary success.

Eukaryotic cells have a network of proteins inside the cell that allow them to change shape, move, and transfer substances in and out of the cell. page 545

Eukaryotic cells have dynamic membranes that provide new possibilities for movement and feeding. page 545

Eukaryotic cells compartmentalize their machinery for energy metabolism into mitochondria and chloroplasts, freeing the cell to interact with the environment in novel ways not available to prokaryotes. page 547

Eukaryotes have larger genomes than those of prokaryotes, as well as novel mechanisms of gene regulation. page 547

Sexual reproduction promotes genetic diversity in eukaryotic populations. page 547

25.2 EUKARYOTIC ORIGINS: The endosymbiotic hypothesis proposes that the chloroplasts and mitochondria of eukaryotic cells were originally free-living bacteria.

The endosymbiotic hypothesis is based on physical, biochemical, and genetic similarities between chloroplasts and cyanobacteria, and between mitochondria and proteobacteria. page 549

Chloroplasts and mitochondria have their own genomes, but their genomes are small relative to those of the free-living bacteria to which they are closely related, in part because of gene migration to the nuclear genome. page 550

Symbiosis between a heterotrophic host and a photosynthetic partner is common throughout the eukaryotic domain; reef-forming corals are an example. page 550

Most eukaryotic cells have mitochondria, but a few do not. Evidence suggests that cells lacking mitochondria once had them but subsequently lost them. page 551

The eukaryotic nuclear genome contains genes unique to Eukarya, but also genes related to Bacteria and Archaea. This suggests that the ancestor of the modern eukaryotic cell was either a primitive eukaryote descended from archaeal ancestors or an archaeon that engulfed a bacterium. page 552

25.3 EUKARYOTIC DIVERSITY: Eukaryotes were formerly divided into four kingdoms, but are now divided into at least seven superkingdoms.

The terms "protist," "algae," and "protozoa" are useful, but do not describe monophyletic groups. page 554

Protists are eukaryotes that do not have features of animals, plants, and fungi; algae are photosynthetic protists; and protozoa are heterotrophic protists. page 554

The seven superkingdoms of eukaryotes are Opisthokonta, Amoebozoa, Archaeplastida, Stramenopila, Alveolata, Rhizaria, and Excavata. page 555

Opisthokonts are the most diverse eukaryotic superkingdom. They include animals and fungi, as well as choanoflagellates (our closest protistan relatives). page 555

Amoebozoans produce multicellular structures by the aggregation of amoeba-like cells, and include organisms that cause human disease and those important in biological research. page 556

The archaeplastids, which include land plants, are photosynthetic, and are divided into three major groups. page 558

The related superkingdoms called the stramenopiles, alveolates, and rhizarians dominate eukaryotic diversity in the oceans. page 559

Photosynthesis spread through Eukarya by means of multiple independent events involving a protozoan host and a eukaryotic endosymbiont. page 562

25.4 THE FOSSIL RECORD OF PROTISTS: The fossil record provides perspectives on the timing and environmental context of eukaryotic evolution.

Fossils in sedimentary rocks as old as 1800 million years have unmistakable signs of eukaryotic cells, including complicated wall structures. page 564

The earliest fossil eukaryotes that can be placed into one of the present-day superkingdoms are red algae from rocks dated as 1050 million years old. page 565

Eukaryotic fossils diversified greatly about 800 million years ago, reflecting a new capacity for predation on other protists, increased nutrient levels in surface seawater, and, perhaps, a modest increase in oxygen. page 565

Protists continue to diversify even today. Green algae and cyanobacteria dominated primary production in earlier seas, but, since about 200 million years ago, dinoflagellates, coccolithophorids, and diatoms have become the primary photosynthetic organisms in the oceans. page 565

Protists have evolved to take advantage of the environments provided by animals and plants; these protists include many that cause disease in humans. page 566

Log in to **LaunchPad** to check your answers to the Self-Assessment Questions and to access additional learning tools.

CHAPTER 26 Being Multicellular

CORE CONCEPTS

26.1 THE PHYLOGENETIC DISTRIBUTION OF MULTICELLULAR ORGANISMS: Complex multicellularity arose several times in evolution.

26.2 DIFFUSION AND BULK FLOW: In complex multicellular organisms, bulk flow circumvents the limitations of diffusion.

26.3 HOW TO BUILD A MULTICELLULAR ORGANISM: Complex multicellularity depends on cell adhesion, communication, and a genetic program for development.

26.4 PLANTS VERSUS ANIMALS: Plants and animals evolved complex multicellularity independently of each other, and solved similar problems with different sets of genes.

26.5 THE EVOLUTION OF COMPLEX MULTICELLULARITY: The evolution of large and complex multicellular organisms, which required abundant oxygen, is recorded by fossils.

Chapters 24 and 25 introduced the extraordinarily diverse, but largely unseen world of microorganisms. Among the Bacteria, Archaea, and protists, each cell usually functions as an individual, growing and reproducing, moving from one place to another, taking in nutrients, and both sensing and transmitting molecular signals. In contrast, complex multicellular organisms can contain more than a trillion cells that work in close coordination. In your own body, for example, different cells are specialized for specific functions, so that while your body as a whole can perform the broad range of tasks accomplished by microorganisms, individual cells for the most part cannot. Cells lining the intestine absorb food molecules, but the bloodstream must then transport nutrients to other parts of the body. Cells lining the inner surface of lungs take up oxygen from inhaled air, but red blood cells must distribute it to other tissues and organs. And while cells at the body's surface sense signals from the environment, the signals affecting interior cells come mostly from surrounding cells.

The biological gulf between microorganisms and complex multicellular organisms is enormous, but complex multicellularity has evolved a half dozen times. In the chapters that follow, we explore the structure and diversity of plants, fungi, and animals. Here, we address the broader question of how complex multicellular organisms evolved in the first place.

26.1 THE PHYLOGENETIC DISTRIBUTION OF MULTICELLULAR ORGANISMS

Most prokaryotic organisms are composed of a single cell, although some form simple filaments or live in colonies. A few types of bacteria, notably some cyanobacteria, form simple multicellular structures with several distinct cell types. Nevertheless, no bacteria develop macroscopic bodies with differentiated tissues having specialized functions. Only eukaryotes have evolved those structures. Simple multicellular organisms, composed of multiple similar cells, occur widely within the eukaryotic tree of life, and a few of these evolved into complex multicellular organisms characterized by differentiated tissues and organs. How are these different types of organization distributed on the eukaryotic tree of life?

Simple multicellularity is widespread among eukaryotes.

A recent survey of eukaryotic organisms recognized 119 major groups within the superkingdoms discussed in Chapter 25. Of these, 83 contain only single-celled organisms. These groups consist of predominantly cells that engulf other microorganisms or ingest small organic particles, photosynthetic cells that live suspended in the water column, or parasitic cells that live within other organisms. Each of the 36 remaining branches exhibits some cases of simple multicellularity, mostly in the form of filaments, hollow balls, or sheets of little-differentiated cells (**Fig. 26.1**).

Simple multicellular eukaryotes share several properties. In many of these organisms, cell adhesion molecules cause adjacent cells to stick together, but there are few specialized cell types and relatively little communication or transfer of resources between cells. Most or all of the cells retain a full range of functions, including reproduction, such that the organism usually pays only a small penalty for individual cell death. Importantly, in simple multicellular organisms, nearly every

FIG. 26.1 **Simple multicellularity.** Cells in simple multicellular organisms show little differentiation and remain in close contact with the external environment. Many groups of eukaryotic organisms display simple multicellularity: (a) *Uroglena*, a stramenopile alga found commonly in lakes; (b) *Epistylis*, a stalked ciliate protozoan that lives attached to the surfaces of fish and crabs; and (c) *Prasiola*, a bladelike green alga only a single cell thick, found on tree trunks and rock surfaces. *Sources: a. Jason Oyadomari; b. D. J. Patterson; c. Fabio Rindi/British Phycological Society.*

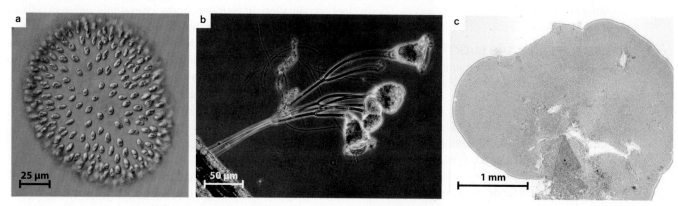

cell is in direct contact with the external environment, at least during those phases of the life cycle when the cells must acquire nutrients.

Simple multicellularity is most common among algae. In addition, stalk-like colonies of particle feeders have evolved in at least three groups of heterotrophic protists, and simple filamentous fungi absorb organic molecules as sources of carbon and energy. Four other eukaryotic groups have achieved simple multicellularity by a different route, aggregating during just one stage of the life cycle. Slime molds are the best-known example (Chapter 25).

Six groups (two algal, three protozoan, and one fungal) include species characterized by coenocytic organization. In coenocytic organisms, the nucleus divides multiple times, but the cell does not, so the nuclei are not partitioned into individual cells. The result is a large cell—sometimes even visible to the naked eye—with many nuclei. The green algae *Codium* and *Caulerpa*, found along the shorelines of temperate and tropical seas, respectively, are common examples of coenocytic organisms (**Fig. 26.2**). Coenocytic rhizarians on the deep seafloor can be 10 to 20 cm long. There is no evidence that any coenocytic organisms evolved from truly multicellular ancestors, nor has any given rise to complex multicellular descendants.

Which selection pressures favored the evolution of simple multicellular organisms from single-celled ancestors? One selective advantage is that multicellularity helps organisms avoid getting eaten. In an illuminating experiment, single-celled green algae were grown in the presence of a protistan predator. Within 10 to 20 generations, most of the algae were living in eight-cell colonies that were essentially invulnerable to predator attack.

FIG. 26.2 **Coenocytic organization.** Nuclei divide repeatedly but are not partitioned into individual cells. Two different groups of green algae have evolved large size as coenocytic organisms: (a) *Codium*, found abundantly along temperate coastlines, and (b) *Caulerpa*, common in shallow tropical seas and aquaria. *Sources: a. Marevision/AGE Fotostock; (b) Wolfgang Poelzer/Wa/AGE Fotostock.*

Another advantage is that multicellular organisms may be able to maintain their position on a surface or in the water column better than their single-celled relatives. Seaweeds, for example, live anchored to the seafloor in places where light and nutrients support growth. Feeding provides a third potential advantage: in colonial heterotrophs such as the stalked ciliate *Epistylis* (see Fig. 26.1b), the coordinated beating of flagella directs currents of food-laden water toward the cells.

Multicellularity also has some costs, particularly for complex multicellular organisms with differentiated reproductive tissues. In these organisms, most cells do not reproduce, instead supporting the few cells that do. This separation of function requires cooperation among cells, but also creates opportunities for cells to "cheat," by using nutrients for their own proliferation, rather than directing them toward the growth and reproduction of the organism as a whole. Cancer is the most conspicuous and lethal example of non-cooperation. Indeed, it might be considered the defining disease of complex multicellular organisms.

As we will see, complex multicellular organisms evolved from simple multicellular ancestors, but most taxonomic groups containing simple multicellular organisms never gave rise to complex descendants.

Complex multicellularity evolved several times.

Complex multicellular organisms are conspicuous parts of our daily existence. Plants, animals, and (if you look a bit more closely) mushrooms and complex seaweeds are what we notice when we view a landscape or the coastal ocean (**Fig. 26.3**). Complex multicellular organisms differ from one another in many ways, but they share three general features. First, they have highly developed molecular mechanisms for adhesion between cells. Second, they display specialized structures that allow cells to communicate with one another. Third, they display complex patterns of cellular and tissue differentiation, guided by networks of regulatory genes. Without these features, complex multicellularity would be impossible.

For example, both plants and animals have differentiated cells and tissues with specialized functions. Only some tissues photosynthesize or absorb organic molecules; other tissues transport food and oxygen through the body; and still others generate the molecular signals that govern development. Only a small subset of all cells contributes to reproduction. Because of this functional differentiation, cell or tissue loss can be lethal for the entire organism.

FIG. 26.3 Complex multicellular organisms. Complex multicellularity evolved independently in (a) red algae, (b) brown algae, (c) land plants, (d) animals, and (e) fungi (at least twice). *Sources: a. Dr. Keith Wheeler/Science Source; b. Dave Fleetham/Design Pics/Getty Images; c. Filmfoto/Dreamstime.com; d. Jack Milchanowski/AGE Fotostock; e. AntiMartina/Getty Images.*

One other feature of complex multicellular organisms is key to understanding their biology: they have a three-dimensional organization, so only some cells are in direct contact with the environment. Cells that are buried within tissues, relatively far from the exterior of the organism, do not have direct access to nutrients or oxygen. Therefore, interior cells cannot grow as fast as surface cells unless there is a way to transfer resources from one part of the body to another. Similarly, interior cells do not receive signals directly from the environment, even though all cells must be able to respond to environmental signals if the organism is to grow, reproduce, and survive. Complex multicellular organisms, therefore, require mechanisms for transferring environmental signals received by cells at the body's surface to interior cells, where genes will be activated or repressed in response. Development in complex multicellular organisms can, in fact, be defined as increasing or decreasing gene expression in response to molecular signals from surrounding cells (Chapter 19).

Complex multicellularity evolved at least six separate times in different eukaryotic groups (**Fig. 26.4**). Although complex multicellularity certainly characterizes animals, it also evolved at least twice in the fungi, once in the green algal group that gave rise to land plants, once in the red algae, and once in the brown algae, producing the giant kelps that form forests in the sea.

Self-Assessment Question

1. How do simple multicellular organisms differ from complex multicellular organisms?

26.2 DIFFUSION AND BULK FLOW

We already mentioned a key challenge of complex multicellularity: transporting food, oxygen, and molecular signals rapidly across large distances within the body. To illustrate the point, how does oxygen get from the air in your lungs into your bloodstream? How does carbon dioxide get from the atmosphere into leaves? And how does ammonia get from seawater into the cells of seaweeds? The answers to all three questions are the same: by diffusion. Even so, oxygen absorbed by your lungs doesn't reach your toes by diffusion alone. Instead, it is transported actively, and in bulk, by blood pumped through your circulatory system. Bulk flow of oxygen, nutrients, and signaling molecules, at rates and across distances far larger than diffusion alone can achieve, lies at the heart of how complex multicellular organisms function.

Diffusion is effective only over short distances.

Diffusion is the random motion of molecules, with net movement from areas of higher concentration to areas of lower concentration (Chapter 5). Diffusion is effective only over small

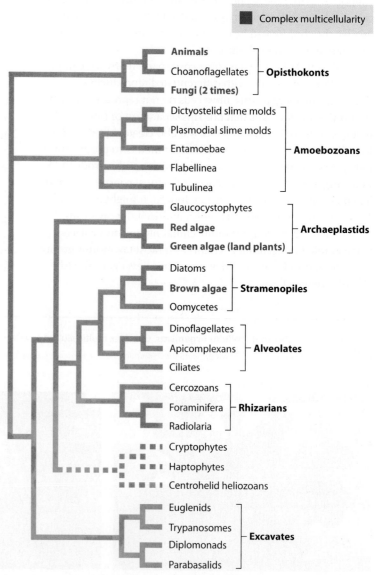

FIG. 26.4 Complex multicellularity within the eukaryotic tree of life. Complex multicellular organisms (found on branches marked by blue) evolved at least six times within three superkingdoms of eukaryotes.

distances. The distance over which molecules can diffuse quickly depends in part on how great the concentration difference is. Because diffusion supplies key molecules for metabolism, it exerts a strong constraint on the size, shape, and function of cells. It strictly limits the size and shape of bacterial cells, as discussed in Chapter 24, and also constrains how eukaryotic organisms function.

Oxygen provides a good example. Most eukaryotes require oxygen for respiration. If a cell or tissue must rely on diffusion for its oxygen supply, its thickness is limited by the difference in oxygen concentration between the cell or tissue and its environment. The concentration difference itself depends on the amount of oxygen in the environment and the rate at which oxygen is used for respiration inside the organism. In shallow water that is in direct contact with the atmosphere, animals relying on diffusion can reach thicknesses of about 1 mm to 1 cm, depending on their metabolic rate. Of course, a quick swim along the seacoast will reveal many animals that are much larger. And, obviously, you are larger. How do you and other large animals get enough oxygen to all of your cells?

Animals achieve large size by circumventing limits imposed by diffusion.

Sponges can reach overall dimensions of a meter or more, but they actually consist of only a few types of cell that line a dense network of pores and canals (**Fig. 26.5a**). As a result of this organization, their cells remain in close contact with circulating seawater. In essence, the large size of a sponge is achieved without placing metabolically active cells at any great distance from their environment. Similarly, in jellyfish, active metabolism is confined to thin tissues that line the inner and outer surfaces of the body. Essentially, a large flat surface is folded up to produce a three-dimensional structure (**Fig. 26.5b**). The jellyfish's bell-shaped body is often thicker than the metabolically active tissue, but its massive interior is filled by materials that are not metabolically active. This material constitutes the mesoglea, the jellyfish's "jelly." The mesoglea provides structural support but does not require much oxygen.

But what about us? How does the human body circumvent the constraints of diffusion? Our lungs gather the oxygen we need for respiration, but the lung is a prime example of diffusion in action, not a means of avoiding it. Because lung tissues have a very high surface area to volume ratio, oxygen can diffuse efficiently from the air you breathe into lung tissue (Chapter 37). A great deal of oxygen can be taken into the body in this way, but how does it get from the lungs to our brains or toes? The distances are far too large for diffusion to be effective. The answer is that oxygen binds to molecules of hemoglobin in red blood cells and then is carried through the bloodstream to distant sites of respiration. We circumvent diffusion by actively pumping oxygen-rich blood through our bodies.

Complex multicellular organisms have structures specialized for bulk flow.

We have seen how humans circumvent the limitations of oxygen diffusion. Many other animals similarly rely on the active circulation of fluids to transport oxygen and other essential molecules, including food and molecular signals, across distances far larger than those that could be traversed by diffusion alone. Indeed, without a mechanism like bulk flow, animals could not have achieved the range of size, shape, and function familiar to us.

Bulk flow is any means by which molecules move through organisms at rates beyond those possible by diffusion across a concentration gradient (**Fig. 26.6**). In humans and other vertebrate animals, the active pumping of blood through blood vessels supplies oxygen to tissues that may be more than a meter distant from the lungs (Fig. 26.6a). Most invertebrate animals lack well-defined blood vessels but pump fluids into the body cavity, where they circulate freely.

Bulk flow also distributes nutrients throughout the body. Intestinal cells absorb organic molecules required for respiration, but it is active transport through the bloodstream that distributes these molecules over large distances. Endocrine signaling molecules such as hormones (Chapters 9 and 36) also

---

FIG. 26.5 Circumventing limits imposed by diffusion. (a) Sponges can attain a large size because the many pores and canals in their bodies ensure that all cells are in close proximity to the environment. (b) Jellyfish also have thin layers of metabolically active tissue, but their familiar bell can be relatively thick because it is packed with metabolically inert molecules (the mesoglea, or "jelly"). *Sources: a. Andrew J. Martinez/SeaPics.com; b. D. R. Schrichte/SeaPics.com.*

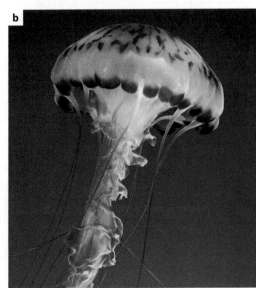

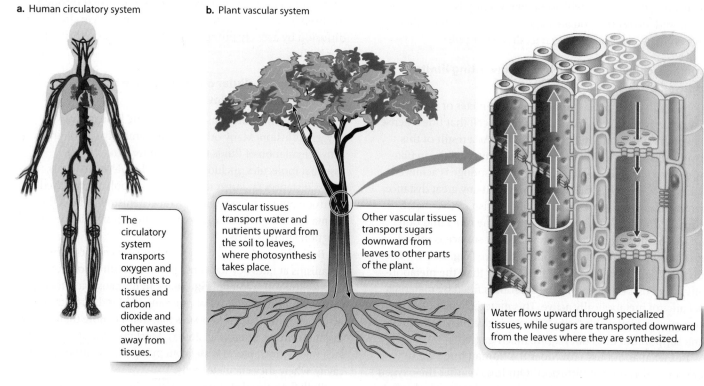

FIG. 26.6 Bulk flow. (a) The circulatory system in animals and (b) the vascular system in plants allow these organisms to get around the size limits of diffusion.

a. Human circulatory system

b. Plant vascular system

The circulatory system transports oxygen and nutrients to tissues and carbon dioxide and other wastes away from tissues.

Vascular tissues transport water and nutrients upward from the soil to leaves, where photosynthesis takes place.

Other vascular tissues transport sugars downward from leaves to other parts of the plant.

Water flows upward through specialized tissues, while sugars are transported downward from the leaves where they are synthesized.

move rapidly through the body by means of the blood and other fluids.

Complex organisms other than animals also rely on bulk flow. A redwood tree must transport water upward from its roots to leaves that may be 100 m above the soil. Plants that rely on diffusion to transport water are only a few millimeters tall. How, then, do most plants move water? Bulk flow operates in plants through a system of specialized tissues powered by the evaporation of water from leaf surfaces (Fig. 26.6b; Chapter 27). Vascular plants also have specialized tissues for the transport of nutrients and signaling molecules upward and downward through roots, stems, and leaves.

Still other complex multicellular organisms also have specialized tissues to transport essential materials over long distances. Fungi transport nutrients through networks of filaments that may be meters long, relying on osmosis to pump materials from sites of absorption to sites of metabolism (Chapter 32). The giant kelps have an internal network of tubular cells that transports molecules through a body that can be tens of meters long. We will discuss the biology of plants, fungi, and animals in subsequent chapters. For now, it is important to understand the functional similarities among these diverse multicellular organisms. In general, when some cells within an organism are buried within tissues, far from the external environment, bulk flow is required to supply those cells with the molecules needed for metabolism.

Self-Assessment Questions

2. How does diffusion limit the size of organisms?
3. How do multicellular organisms get around the size limits imposed by diffusion?

26.3 HOW TO BUILD A MULTICELLULAR ORGANISM

We have emphasized three general requirements for complex multicellular life: cells must stick together, they must communicate with one another, and they must participate in a network of genetic interactions that regulates cell division and differentiation. Once these elements are in place, the stage is set for the evolution of the specialized tissues and organs observed in plants, animals, and other complex multicellular organisms.

Complex multicellularity requires adhesion between cells.

If a fertilized egg is to develop into a complex multicellular organism, it must divide many times, and the cells produced from those divisions must stick together. In addition, they must retain a specific spatial relationship with one another for the developing organism to function properly. Chapter 10

HOW DO WE KNOW?

Fig. 26.7

How do bacteria influence the life cycles of choanoflagellates?

BACKGROUND Choanoflagellates are the closest protistan relatives of animals. Biologists are studying cell adhesion, cell differentiation, and multicellularity in these organisms to increase our understanding of early events in the evolution of animal multicellularity. American biologist Nicole King and her colleagues cultured the choanoflagellate species *Salpingoeca rosetta* in the laboratory. High-speed videos revealed that these choanoflagellates differentiate into several distinct cell types in the course of their life cycle. Actively swimming cells can be fast swimmers or slow swimmers that differ in shape and ornamentation (Fig. 26.7a). When swimming cells make contact with a solid surface, they can differentiate to form a stalk that tethers them to the substrate (Fig. 26.7b). Simple multicellular colonies in the shape of rosettes (Fig. 26.7c) were seen when the choanoflagellates were cultured in the presence of many bacteria, including *Algoriphagus machipongonesis*.

HYPOTHESIS The bacterium *Algoriphagus machipongonesis* induces the choanoflagellate to form rosette-like colonies.

EXPERIMENT The scientists grew the choanoflagellates in a medium containing only *A. machipongonesis*, as well as in a medium without this prey bacterium.

RESULTS The multicellular state formed only when *A. machipongonesis* was present (Fig. 26.7c).

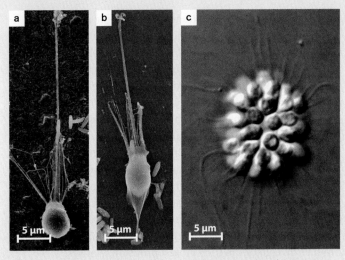

Photo source: Reprinted from M. Dayel et al., 2011, "Cell Differentiation and Morphogenesis in the Colony-Forming Choanoflagellate Salpingoeca rosetta," Developmental Biology, 357:73–82. Copyright 2011, with permission from Elsevier.

CONCLUSION Choanoflagellates differentiate into several cell types upon receiving cues from the environment. In particular, molecular signals from prey bacteria can induce a simple form of multicellularity. The direct ancestors of animals may have attained simple multicellularity by a similar route.

FOLLOW-UP WORK Researchers continue to investigate the molecular mechanisms of cell differentiation, cell adhesion, and signal reception in these distant relatives of ours.

SOURCE M. Dayel et al. 2011. "Cell Differentiation and Morphogenesis in the Colony-Forming Choanoflagellate *Salpingoeca rosetta*." Developmental Biology 357:73–82.

---

introduced the molecular mechanisms that hold multicellular organisms together. We begin by reviewing those mechanisms briefly before placing them in evolutionary context.

In animal development, proliferating cells commonly become organized into sheets of cells called epithelia that line the inside and cover the outside of the body. Within epithelia, adjacent cells adhere to each other by means of transmembrane proteins, especially cadherins, which form molecular attachments between cells. Epithelial cells also secrete an extracellular matrix called the basal lamina, made of proteins and glycoproteins, and these cells attach themselves to this matrix by means of other transmembrane proteins called integrins. In this way, cadherins, integrins, and additional transmembrane proteins provide the molecular mechanisms for adhesion in animals.

Cadherins and integrins do not exist in complex multicellular organisms other than animals, but all such organisms require some means of keeping their cells stuck together. How do they do it? Plants also synthesize cell adhesion molecules that bind cells into tissues, but, in this case, the molecules are polysaccharides called pectins (pectins are what make fruit jelly jell). Cell adhesion molecules are less well known in other complex multicellular organisms, but the general picture is clear. Without adhesion there can be no complex multicellularity, and different groups of eukaryotes have evolved distinct types of molecular "glue" for this purpose.

How did animal cell adhesion originate?

The genome of the choanoflagellate *Monosiga brevicollis* provides fascinating insight into the evolution of cell adhesion molecules. Choanoflagellates, the closest protistan relatives of animals, are unicellular microorganisms. Therefore, it came as a surprise that the genes of *M. brevicollis* code for many of the same protein families that promote cell adhesion in animals. In fact, the choanoflagellate genome contains genes for both cadherin and integrin proteins. Clearly, these proteins are not supporting epithelia in *M. brevicollis*, so what are they doing?

One possible answer comes from the observation that cells stick not only to one another but also to rock or sediment surfaces. Proteins that originally evolved to promote adhesion of individual cells to sand grains or a rock surface may have been modified during evolution to allow cell–cell adhesion. It is also possible that adhesion molecules were originally used in the capture of bacterial cells, to make prey adhere to the predator.

An intriguing clue to the role of these proteins in choanoflagellates comes from careful laboratory studies (**Fig. 26.7**). Although choanoflagellates are unicellular, molecular signals induce some choanoflagellates to form a simple multicellular structure. Unexpectedly, the source of these signals is a bacterium, the preferred prey of the choanoflagellates. The function of the multicellular structures in these choanoflagellates has not yet been determined, but these observations support the hypothesis that they aid predation.

To date, cadherins have been found only in choanoflagellates and animals, but proteins of the integrin complex extend even deeper into eukaryotic phylogeny: they are found in single-celled protists that branch from basal nodes of the opisthokont superkingdom (Chapter 25). Such a phylogenetic distribution provides strong support for the general hypothesis that cell adhesion in animals resulted from the redeployment of protein families that evolved to perform other functions before animals diverged from their closest protistan relatives.

Complex multicellularity requires communication between cells.

In complex multicellular organisms, it is not sufficient for cells to adhere to one another; they must also be able to communicate. Communication is important during development, guiding the patterns of gene expression that differentiate cells, tissues, and organs. The flow of information among cells is also necessary to coordinate the activity of cells within tissues and the activity of tissues within organs.

As discussed in Chapter 9, cells communicate by molecular signals. A signaling molecule (generally a protein) synthesized by one cell binds with a receptor protein on the surface of a second cell, essentially flipping a molecular switch that activates or represses enzymes or gene expression in the receptor cell's nucleus. Both plants and animals have receptors of the receptor kinase type, and these initiate similar signaling pathways within the cell. As in the case of adhesion molecules, these signaling pathways, or components of the pathways, have also been identified in their close protistan relatives. Evidently, many of the signaling pathways used for communication between cells in complex multicellular organisms first evolved in single-celled eukaryotes. Again we can ask, what function did molecular signals and receptors have in the protistan ancestors of complex organisms?

All cells have transmembrane receptors that respond to signals from the environment. In some cases, the signal is a molecule released by a food organism (such as the bacteria that induce simple multicellularity in choanoflagellates). In other cases, the cells sense nutrients, temperature, or oxygen level. Single-celled eukaryotes also communicate with other cells within the same species. For example, they need to communicate to ensure that two cells can find each other to fuse in sexual reproduction. Signaling between two cells within an animal's body can be seen as a variation on this more general theme of a cell responding to other cells and the physical environment.

While all eukaryotic cells have molecular mechanisms for communication between cells, complex multicellular organisms have distinct pathways for the movement of molecules from one cell to another. For example, animals more complex than sponges have **gap junctions**, protein channels that allow ions and signaling molecules to move from one cell into another (**Fig. 26.8**). Gap junctions not only help cells to communicate with their neighbors, but also allow *targeted* communication between a cell and specific cells adjacent to it (Chapter 10).

Plants, in contrast, have an intrinsic barrier to intercellular communication: the cell wall. Careful observation of plant cell

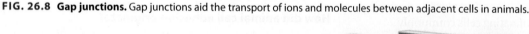

FIG. 26.8 **Gap junctions.** Gap junctions aid the transport of ions and molecules between adjacent cells in animals.

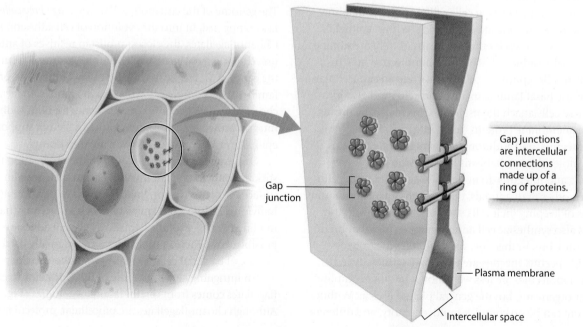

walls with the electron microscope reveals that these walls have tiny holes through which thin strands of cytoplasm extend from one cell to the next (**Fig. 26.9**). Called **plasmodesmata**, these intercellular channels are lined by extensions of the cell membrane (Chapter 10). Tubules of endoplasmic reticulum run through these channels, connecting the endomembrane systems of the two cells. Like gap junctions, plasmodesmata permit signaling molecules to pass between cells in such a way that they can be targeted to only one or a few adjacent cells.

Complex red and brown algae also have plasmodesmata, and complex fungi have pores between cells that enable cytoplasm that carries signals to flow between cells. As similar channels do not occur in most other eukaryotic organisms, they appear to represent an important step in the evolution of complex multicellularity.

Complex multicellularity requires a genetic program for coordinated growth and cell differentiation.

All of the cells in your body derive from a single fertilized egg, and most of those cells contain the same genes. Yet your body contains some 200 distinct cell types precisely arranged in a variety of tissues and organs. How can two cells with the same genes become different cell types? The answer lies in development, the system of gene regulation that guides growth from zygote to adult (Chapter 19).

Development is the result of molecular communication between cells. Cells have different fates depending on which genes are switched off or on, and genes are switched off and on by the molecular signals that cells receive. A signal commonly alters the production of proteins—inducing the reorganization of the cytoskeleton, for example. As a result, a stem cell may become an epithelial cell, or a muscle cell, or a neuron. This observation leads to another question: what causes the same gene to be turned on in one cell and turned off in another? The ultimate answer is that two cells in the same developing organism can be exposed to very different environments.

When we think about development as a process of programmed cell division and differentiation, the link to unicellular ancestors becomes clearer. Many biological innovations accompanied the evolution of complex multicellularity, but, as noted in Chapter 25, the differentiation of distinct cell types is not one of them. Many unicellular organisms have life cycles in which different cell types alternate in time, depending on environmental conditions. For example, if we experimentally starve dinoflagellate cells, two cells undergo sexual fusion to form resting cells protected by thick walls. That is, a nutrient shortage

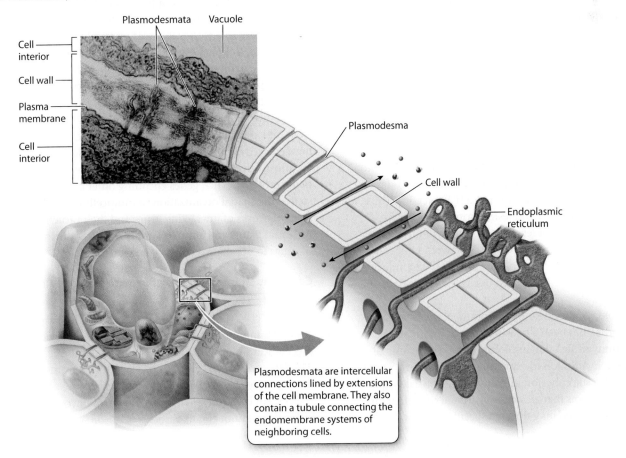

FIG. 26.9 Plasmodesmata. Plasmodesmata aid the movement of ions and molecules between cells in land plants and in complex red and brown algae. *Photo source: Biophoto Associates/Science Source.*

induces a change in gene expression that leads to the formation of a distinct type of cell. When food becomes available again, the cells undergo meiotic cell division to form new feeding cells. Many other single-celled eukaryotes form resting cells in response to environmental cues, especially deprivation of nutrients or oxygen.

The key innovation associated with complex multicellularity was the differentiation of cells in *space* instead of in *time*. In a three-dimensional multicellular organism, only surface cells are in direct contact with the outside environment. Interior cells are exposed to a different physical and chemical environment because nutrients, oxygen, and light become less abundant with increasing depth within tissues. In effect, a gradient of environmental signals exists within multicellular organisms. We might, therefore, hypothesize that in the earliest of these organisms, a nutrient or oxygen gradient triggered oxygen- or nutrient-starved interior cells to differentiate, much as these conditions trigger the formation of resting cells in their single-celled relatives. As cellular responses to signaling gradients came under genetic control, the seeds of complex development were sown.

Green algae provide a fascinating example that links cell differentiation in unicellular and multicellular organisms. The simple multicellular organism *Volvox* (see Fig. 25.15) has two types of cells: vegetative cells that photosynthesize and control movement of the organism, and reproductive cells. Cell differentiation in *Volvox* is regulated by a gene that is also involved in the formation of distinct cell types in the life cycle of *Volvox*'s single-celled relative *Chlamydomonas*. The similar function of this gene in the two organisms supports the hypothesis that the spatial differentiation of cells in multicellular organisms began with the redeployment of genes that regulate cell differentiation in single-celled relatives.

Bulk flow, which transports nutrients, oxygen, and water within complex multicellular organisms, also carries developmental signals. As with the molecular transport described earlier, bulk flow can carry signals for far greater distances through the body than is possible by diffusion alone. For example, in animals the endocrine system releases hormones directly into the bloodstream. These hormones can then act on cells far distant from those in which they formed (Chapter 36). Thus, the sex hormones estrogen and testosterone are synthesized in reproductive organs but regulate development throughout the body, contributing to the differences between males and females. In this way, signals carried by bulk flow can induce the formation of distinct cell types and tissues along the path of signal transport.

The genome of the choanoflagellate *Monosiga brevicollis*, discussed earlier, has been a treasure trove of information on the antiquity of signaling molecules deployed in animal development. In addition to expressing proteins that govern cell adhesion and tissue cohesion in animals, *M. brevicollis* expresses a number of proteins that are active in animal cell differentiation. For example, signaling through specific receptor kinases was long thought to be restricted to animals, but it also occurs in *M. brevicollis*. In other cases, individual components of signaling proteins are present in the *M. brevicollis* genome, but not the complex multidomain proteins formed by animals. This is true, for example, of several protein complexes that are important in development. Similarly, molecules that play an important role in plant development have been identified in the genomes of morphologically simple green algae. The key point is that genome sequences interpreted in light of eukaryotic phylogeny are now enabling biologists to piece together the patterns of gene evolution that accompanied the evolution of morphologic complexity and diversity in plants and animals.

Self-Assessment Questions

4. Which arose first: animals with many cells or cell adhesion molecules?

5. All cells in your body derive from a single fertilized egg, yet your body contains many different types of cells. How can we explain this phenomenon?

26.4 PLANTS VERSUS ANIMALS

Plants and animals are the best-known examples of complex multicellular organisms. The study of phylogenetic relationships makes it clear that complex multicellularity evolved independently in the two groups. In other words, they do not share a common ancestor that was multicellular. We can see the results of these separate evolutionary events by examining cell adhesion, signaling, and development in plants and animals.

Both plants and animals have evolved sophisticated systems for cell adhesion and molecular signaling. The underlying mechanisms must differ, however, because plant cells have cell walls and animal cells do not. Likewise, plants and animals have evolved similar genetic logic to govern development, but use mostly distinct sets of genes. In both plants and animals, many proteins switch genes encoding other proteins on or off, so that the spatial organization of multicellular organisms arises from networks of interacting genes and their protein products. Ancestral plants and animals simply recruited distinct families of genes to populate regulatory networks.

Cell walls shape patterns of growth and development in plants.

The plant cell wall (Chapter 5), made of cellulose, provides structural support to cells. In fact, the cell wall provides the mechanical support that allows plants to stand erect (Chapter 27). The presence of cell walls has largely determined the evolutionary fate of plants. For example, because all plant cells except eggs and sperm are completely surrounded by cell walls, they cannot engulf particles or absorb organic molecules.

Instead, most plants gain carbon and energy only through photosynthesis. Another consequence is that plant cells have no pseudopodia and no flagella (in conifers and flowering plants, even sperm have lost their flagella). This being the case, plant cells cannot move. Therefore, the plant must obtain nutrients, evade predators, and cope with environmental stress without any moving parts.

The inability of plant cells to move has major consequences for development. At the level of the cell, the entire program of growth and development involves cell division, cell expansion (commonly by developing large vacuoles in cell interiors), and cell differentiation. As a consequence, plant growth is confined to **meristems** (**Fig. 26.10**), populations of actively dividing cells at the tips of stems and roots (Chapter 29). The cells in meristem regions are more or less permanently undifferentiated cells that repeatedly undergo mitosis. A few millimeters from the region of active cell division in a stem or root meristem, cells stop dividing and begin to expand. Within another few millimeters, expansion is curtailed as well. There, signaling molecules induce cells to differentiate, forming the mature cells that will function in photosynthesis, storage, bulk flow, or mechanical support.

Because their cell walls render plants immobile, plants have evolved mechanisms to transport water and nutrients from the soil to leaves without the use of any moving parts or even the expenditure of ATP. Moreover, because plants are anchored in place, they are unable to move when growth conditions are unfavorable. Instead, plants respond to environmental signals by adjusting their meristem activity and, hence, growth pattern. Heat, drought, floods, and fire can all leave their mark in altered patterns of growth—just compare amply and poorly watered tomato plants. In addition, plants can't flee predators, so they have evolved mechanical structures (hairs and spines) and poisons to keep from being eaten. Indeed, it has been suggested that, in terms of responding to the environment, growth plays a role in plants similar to that played by behavior in animals.

The details of plant growth, development, reproduction, and function are presented in Chapters 27–31. Here, the key point to bear in mind is that the properties of a pine tree or a rice plant depend fundamentally on the basic features of cell adhesion, intercellular communication, and a regulatory network to guide development, all carried out under the constraints imposed by cell walls.

Animal cells can move relative to one another.

Having outlined the growth and development of plants, we can appreciate anew the remarkable ballet that characterizes development in animal embryos (**Fig. 26.11**; Chapter 40). Fertilized eggs undergo several rounds of mitosis to form a hollow ball called a **blastula**, formed of a single layer of undifferentiated cells. Then something happens that has no parallel in plant development: unconstrained by cell walls, animal cells move relative to one another. Blastula cells migrate, becoming reorganized into a hollow ball that folds inward at one location to form a second layer under the first. This multilayered structure is called the **gastrula**.

Gastrula formation brings new populations of cells into direct contact with one another, inducing patterns of molecular signaling and gene regulation that begin the long process of growth and tissue specification. As cells proliferate, molecular signals are expressed in some cells and diffuse into others, generating what amounts to a three-dimensional molecular map of the organism that guides cell differentiation. Gradients in signaling molecules define top, bottom, front, back, left, and right. Plants do much the same thing, but because animal cells can move during development, animal embryos are not restricted to growth only from localized regions like meristems. Instead, cell division and tissue differentiation occur throughout the developing animal body.

Because they are not constrained by cell walls, animals can form organs with moving parts—muscles that power active transport of food and fluids within the organism and allow movement. Thus, animals have possibilities for function that are far different from those of plants. If drought or predators strike, animals can respond by changing their behavior; for example, they can move to a new location (Chapter 35).

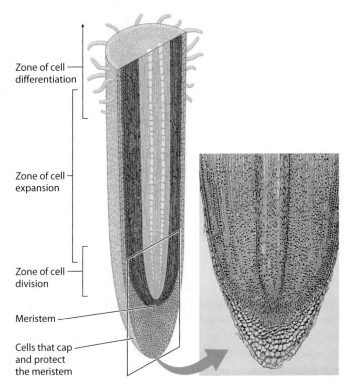

FIG. 26.10 A meristem. Because of their cell walls, plant cells can't move. For this reason, roots grow by cell division of the undifferentiated cells in the meristem region at the root tip. *Photo source: Walker Mi./Getty Images.*

Zone of cell differentiation

Zone of cell expansion

Zone of cell division

Meristem

Cells that cap and protect the meristem

FIG. 26.11 Gastrula formation. The movement of cells during embryogenesis in animals transforms a blastula into a gastrula.

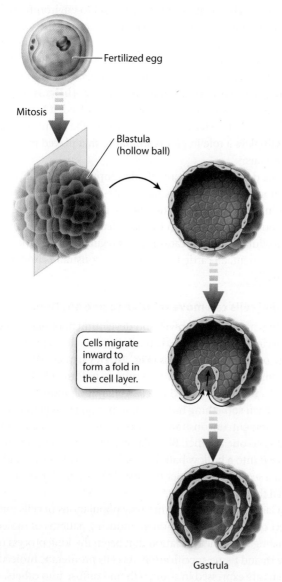

Plants and animals, then, display contrasting patterns of development and function that reflect both their independent origins from different groups of protists and the constraints imposed by cell walls in plants.

Self-Assessment Questions

6. How do plants and animals differ in the ways their cells adhere, communicate, and differentiate during development?

7. What is a key difference in structure between the cells of multicellular plants and the cells of multicellular animals?

26.5 THE EVOLUTION OF COMPLEX MULTICELLULARITY

The differences between an amoeba and a lobster might seem vast, but research over the past decade increasingly shows how this apparent gap was bridged through a series of evolutionary innovations. Throughout this chapter, we have emphasized that complex multicellular organisms have three requirements: mechanisms for cell adhesion, modes of communication between cells, and a genetic program to guide growth and development. These three requirements had to be acquired in a specific order. If the products of cell division don't stick together, there can be no complex multicellularity. Even so, adhesion is not sufficient: it must be followed by mechanisms for communication between cells. Moreover, cells must be able to send molecular messages to specific targets or regions, assisted in animals by gap junctions and in plants by plasmodesmata.

With these first two requirements in place, natural selection would favor the increase and diversification of genes that regulate growth and development, making possible more complex morphology and anatomy. Thus, the stage was set for the final key to complex multicellularity: the differentiation of tissues and organs that govern the bulk flow of fluids, nutrients, signaling molecules, and oxygen through increasingly large and complex bodies. This differentiation freed organisms from the tight constraints imposed by diffusion. When all these features are placed onto phylogenies, they show both the predicted order of acquisition and the tremendous evolutionary consequences of complex multicellularity (**Fig. 26.12**).

Fossil evidence of complex multicellular organisms is first observed in rocks deposited 575–555 million years ago.

In Chapter 22, we noted that phylogenies based on living organisms make predictions about the fossil record. Characters and groups associated with lower branches in phylogenies should appear as earlier fossils than those associated with later branches. Although fossils don't preserve molecular features, the record supports the phylogenetic pattern shown in Fig. 26.12.

Single-celled protists are found in sedimentary rocks as old as 1800 million years, and a number of simple multicellular forms have been discovered that are nearly as old (see Fig. 25.23). Nonetheless, the oldest fossils that record complex multicellularity are found in rocks that were deposited only 575–555 million years ago. **Fig. 26.13a** shows one of the oldest known animal fossils, a frondlike form preserved in 570-million-year-old rocks from Newfoundland, near the eastern tip of Canada. This fossil and many others found in rocks of this age are enigmatic. There is complex morphology to be sure, but it is difficult to understand these forms by comparing them to the body plans of living animals. The Newfoundland fossils show no evidence of head or tail, no limbs, and no opening that might have functioned as a mouth. They appear to be very simple organisms that obtained both carbon and oxygen by diffusion. These fossils are long and wide, but they are not thick. In fact, it is likely that active metabolism occurred only along the surfaces of these structures.

Such animals lie near the base of the animal tree (Chapter 42), and probably would have developed under the guidance of regulatory genes similar to those present in modern sponges or, perhaps, jellyfish. By 560–555 million years ago, more complex animals, with a distinct head and tail, top and bottom, left and right, enter the record (**Fig. 26.13b**). At the same time, we begin to see tracks and trails made by animals that moved across and through surface sediments with the aid of muscles (**Fig. 26.13c**). These tracks are an indirect record, complementary to fossils, of increasing animal complexity. These, then, are the first known occurrences of the types of animal that dominate ecosystems today, both in the sea and on land. These animals must have possessed a diverse array

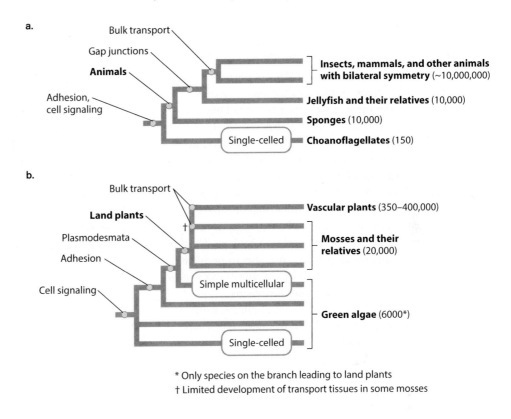

FIG. 26.12 The evolution of complex multicellularity. Phylogenetic relationships of complex (a) animals and (b) plants, with their close evolutionary relatives, show similar patterns of character accumulation and its consequences for biological diversity. Estimated species numbers are shown.

* Only species on the branch leading to land plants
† Limited development of transport tissues in some mosses

FIG. 26.13 The earliest fossils of animals. (a) Among the oldest known macroscopic animal fossils is a frond-like form made up of tubular structures, preserved in 570-million-year-old rocks from Newfoundland. (b) The oldest known animal fossil showing bilateral symmetry is a mollusk-like form preserved in 560–555-million-year-old rocks from northern Russia. (c) The oldest tracks and trails of mobile animals are found in rocks 560 million years old and younger. *Sources: a. Guy Narbonne; b. Mikhail Fedonkin; c. Andrew Knoll, Harvard University.*

of genes to guide development, something akin to those seen today in insects or snails.

Clearly, by 575–555 million years ago, the animal tree was beginning to branch. This development, however, raises a new question: why did complex multicellular animals appear so much later than simple multicellular organisms?

Oxygen is necessary for complex multicellular life.

Earlier in the chapter, we discussed the key biological requirements for complex multicellularity. There is a critical environmental requirement as well: the presence of oxygen. Other potential electron acceptors for respiration exist, such as sulfate and ferric iron. However, they not only occur in much lower abundances than oxygen, but also are not gases and so do not accumulate in air. In addition, only the oxidation of organic molecules by O_2 provides sufficient energy to support large and active predators like wolves and lions. And only oxygen in concentrations approaching those of the present day can diffuse into the interior cells of large, active organisms. On our planet—and, probably, on all planets—no other molecule is both sufficiently abundant and has the oxidizing power needed to support the biology of large complex organisms.

On the present-day Earth, lake or ocean bottoms with oxygen levels below approximately 8% of surface levels support only a limited diversity of animals, and those animals tend to be tiny. Large animals with energetic modes of life—carnivores, for example—live only in oxygen-rich environments. Clearly, then, clear restrictions on animal form and diversity exist at lower oxygen levels today. These restrictions have important implications for the distribution of large, active, and diverse animals through time. We noted in Chapter 24 that oxygen first began to accumulate in the atmosphere and surface oceans about 2400 million years ago, but the chemistry of sedimentary rocks deposited after that time tells us that oxygen levels remained low for a very long time. Our modern world of abundant oxygen came to exist only 580–560 million years ago, about the time when fossils first record large multicellular organisms (**Fig. 26.14**).

Scientists continue to debate the causes of this increase in oxygen levels, but the correspondence in time between oxygen enrichment and the first appearance of large, complex animals (and algae) suggests that more oxygen permitted animals to attain greater size and adopt more energetic modes of life. Their greater size created opportunities for tissue differentiation, leading to transport of nutrients and signaling molecules by bulk flow. The evolution of bulk flow in turn permitted still larger size, setting up a positive feedback that eventually resulted in the complex multicellular organisms we see today.

With morphologic complexity came new functions, including predation on other animals. Protozoan predators capture other microorganisms, but do so one cell at a time. In contrast, animals that obtain food by filtering seawater gather cells by the thousands, and larger animals can eat smaller ones. Predation by animals opened up endless possibilities for evolutionary specialization and, therefore, diversification. Among photosynthetic organisms, multicellular red and green algae were able to establish populations in wave-swept coastlines and other environments where simpler organisms

FIG. 26.14 Oxygen and the evolution of complex multicellularity.

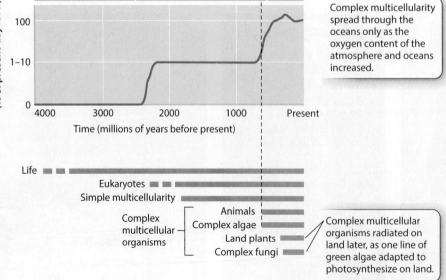

FIG. 26.15 Evolution of complex fungi. (a) This trunklike fossil, discovered in France, is actually a giant fungus that towered over early land plants 375 million years ago. (b) This image shows the interior of the fungus, which consists of tubes that transported nutrients through the large body. *Sources: Carol Hotton; b. Martha E. Cook, School of Biological Sciences, Illinois State University.*

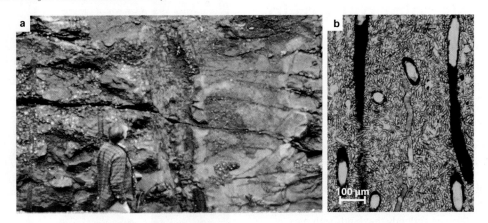

lacked the mechanical strength to hold their position along the shoreline.

The key point is that complex multicellular organisms didn't succeed by doing the same things as simpler eukaryotes. Instead, complex multicellularity spread through the oceans because it opened up new and unprecedented evolutionary possibilities.

Land plants evolved from green algae that could carry out photosynthesis on land.

Complex multicellular land plants originated about 465 million years ago, well after marine animals and seaweeds. Nonetheless, in some ways, the pattern of land plant evolution parallels that of animals. Ancestral green algae evolved molecular means for cell–cell adhesion and communication between cells, and the algae that form the sister group to land plants share with them the innovation of plasmodesmata. Then, as they developed larger and more complex three-dimensional bodies, the plant ancestors evolved a succession of genes and their protein products to guide cell division, expansion, and differentiation. Land plants, however, faced a different challenge from that of animals. They now had to carry out photosynthesis, which had evolved earlier in aquatic organisms, within their own tissues bathed by air. Their diversification across the land surface also required that these plants absorb nutrients and water from the soil and transport these substances throughout the plant, rather than simply taking them in by diffusion from the surrounding water.

Fossils indicate that by 400 million years ago, plants with specialized tissues for bulk flow of water and nutrients had begun to spread across the continents. As plant biomass built up on land, it created opportunities for additional types of complex multicellular organisms. In particular, the fungi, which decompose organic matter, radiated by taking advantage of this resource (**Fig. 26.15**), evolving complex multicellularity in two distinct lineages (Chapter 32).

Regulatory genes played an important role in the evolution of complex multicellular organisms.

Complex multicellular organisms account for a large proportion of all eukaryotic species. Thus, one consequence of complex multicellularity was an increase in biological diversity. Let's compare the diversity of complex multicellular groups with their simpler phylogenetic sisters. Choanoflagellates, for example, number about 150 species; simple animals such as sponges and jellyfish about 20,000; and complex animals with circulatory systems perhaps as many as 10 million. Plants and their relatives show a comparable pattern. There are about 6000 species of green algae on the branch that includes land plants, but about 400,000 species of anatomically complex land plants capable of bulk flow. Similarly, complex fungi and red algae include more species by an order of magnitude than their simpler relatives.

How did this immense diversity arise? At one level, the answer is functional and ecological. Complex multicellular organisms can perform a range of functions that simpler organisms cannot, so they have evolved many specific types of interaction with other organisms and the physical environment. Carnivory, for example, requires both sophisticated sensory systems and jaws or limbs to locate and capture prey. Once established, though, carnivorous animals radiated throughout the oceans and, eventually, on land, providing an ecological driver of diversification. This explanation, however, prompts another question: what is the genetic basis for the bewildering range of sizes and shapes displayed by complex multicellular organisms?

The answer has to do with the network of genes that guide development. Since the 1950s, biologists have learned a

HOW DO WE KNOW?

Fig. 26.16

What controls color pattern in butterfly wings?

BACKGROUND Butterfly wings commonly show a striking pattern of color, including circular features known as eyespots. Eyespots are adaptive, for example, in deterring predation by birds.

HYPOTHESIS The expression of regulatory genes during wing development governs eyespot formation on wing surfaces.

EXPERIMENT Paul Brakefield and his colleagues mapped the expression of a regulatory gene called *Distalless* in developing butterfly wings. They identified an antibody that binds with the protein expressed by *Distalless* genes and fused this antibody to green fluorescent protein (GFP). GFP glows vivid green when illuminated under blue light, enabling the biologists to visualize the spatial pattern of *Distalless* expression in developing wings.

RESULTS In both wild-type and mutant butterflies, the expression pattern of *Distalless* (on the left in each panel of the figure) in developing wings closely resembles the pattern of eyespots on the wing (on the right in each panel).

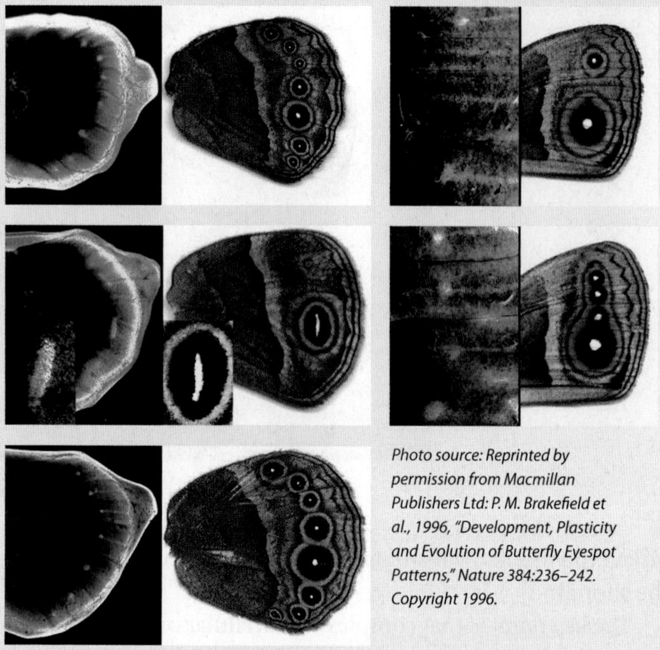

CONCLUSION Regulatory genes play an important role in butterfly wing coloration, and mutations in these genes can account for differences in wing color patterns among species. Continuing research using new tools for genomic analysis has enabled biologists to understand how interacting genes and their products give rise to morphologic pattern in butterfly wings and many other organ systems.

Photo source: Reprinted by permission from Macmillan Publishers Ltd: P. M. Brakefield et al., 1996, "Development, Plasticity and Evolution of Butterfly Eyespot Patterns," *Nature* 384:236–242. Copyright 1996.

SOURCE P. M. Brakefield et al. 1996. "Development, Plasticity and Evolution of Butterfly Eyespot Patterns." *Nature* 384:236–242.

---

remarkable amount about this genetic network and its host of interacting genes (Chapter 19). Many of the genes involved in development are what we might consider "middle managers": a molecular signal induces the expression of such a gene, and its protein product in turn prompts the expression or repression of another gene. It is the complex interplay of these genetic switches—turning specific genes on in one cell and off in another, depending on where those cells occur in the developing body—that results in crabs with large pincers and butterflies with patterned wings (**Fig. 26.16**).

It makes sense that if regulatory genes guide development of the body in each complex multicellular species, then mutations in regulatory genes may account for many of the differences we observe among different species. A whole new field of biology called evolutionary-developmental biology, or **evo-devo** for short, examines how developmental genes underpin evolutionary change. In evo-devo research, scientists compare the genetic programs for growth and development in species found on different branches of phylogenetic trees. One goal is to discover the molecular mechanisms that guide development in individual species. Another is to understand the genetic differences associated with differences in form and function among species.

Evo-devo is an exciting and rapidly expanding field of research because it is helping to illuminate long-suspected relationships between the development of individuals and patterns of evolutionary relatedness among species. Continuing research promises to shed new light on the similarities and differences among plants, animals, and other complex multicellular organisms.

Self-Assessment Questions

8. Which environmental change(s) are recorded by the sedimentary rocks that contain the oldest fossils of large active animals?

9. How does evo-devo research help us to understand how animals have diversified to form so many distinct species?

CORE CONCEPTS SUMMARY

26.1 THE PHYLOGENETIC DISTRIBUTION OF MULTICELLULAR ORGANISMS: Complex multicellularity arose several times in evolution.

Bacteria are unicellular or form simple multicellular structures. page 569

Eukaryotes can be unicellular or multicellular and exhibit both simple and complex multicellularity. page 569

Simple multicellularity requires the adhesion of cells although there is little cell differentiation; complex multicellularity requires cell adhesion, cell signaling, and differentiation and specialization among cells. page 569

Complex multicellular organisms have evolved independently at least six times: in animals, in vascular plants, in red algae, in brown algae (the kelps), and at least twice in the fungi. page 572

26.2 DIFFUSION AND BULK FLOW: In complex multicellular organisms, bulk flow circumvents the limitations of diffusion.

Diffusion is the random motion of molecules, with net movement occurring from regions of higher concentration to regions of lower concentration. It generally acts over small distances, placing limits on the size of multicellular organisms. page 572

Bulk flow is an active process that allows multicellular organisms to nourish or send signals to cells located far from the external environment, thereby circumventing the constraints imposed by diffusion. page 573

26.3 HOW TO BUILD A MULTICELLULAR ORGANISM: Complex multicellularity depends on cell adhesion, communication, and a genetic program for development.

Animal and plant cells are organized into tissues characterized by specific molecular attachments between cells. page 574

Choanoflagellates express some of the same proteins that permit cell adhesion in animals, even though choanoflagellates are unicellular. Experiments suggest that choanoflagellates may use these proteins to capture bacteria, not for cell adhesion. page 575

Gap junctions in animals and plasmodesmata in plants allow cells to communicate with each other in a targeted fashion. page 576

The cells of complex multicellular organisms are genetically programmed to differentiate into multiple cell types in space. page 578

A number of gene families known to play developmental roles in multicellular organisms are also present in their unicellular relatives, where at least some of them play a role in life cycle differentiation. page 578

26.4 PLANTS VERSUS ANIMALS: Plants and animals evolved complex multicellularity independently of each other and solved similar problems with different sets of genes.

The cell wall characteristic of plant cells provides structural and mechanical support, but does not allow plant cells to move. page 578

The cell wall of plants has led to distinct solutions to the problems of cell adhesion, cell communication, and development. For example, plants grow by the activity of meristems, populations of actively dividing cells at the tips of stems and roots. page 579

Animal cells do not have cell walls, allowing cell movement that is not possible in plants. For example, during animal development, cells of the embryo migrate inward to form a layered structure called a gastrula. page 579

26.5 THE EVOLUTION OF COMPLEX MULTICELLULARITY: The evolution of large and complex multicellular organisms, which required abundant oxygen, is recorded by fossils.

Oxygen may be required for the evolution of complex multicellularity because of its chemical properties and its abundance. page 582

The evolution of complex multicellularity observed in the fossil record correlates with increases in atmospheric oxygen that occurred 580–560 million years ago. page 582

Complex multicellular organisms evolved later on land, as ancestral plants evolved the capacity to photosynthesize surrounded by air rather than water. page 583

Evolutionary-developmental biology, or evo-devo, is a field of research that looks at both individual development and evolutionary patterns in an attempt to understand the developmental changes that allowed organisms to diversify and adapt to changing environments. page 584

Log in to LaunchPad to check your answers to the Self-Assessment Questions and to access additional learning tools.

CASE 6

Agriculture
Feeding a Growing Population

In 2008, people around the world, from South Asia to Africa to the Caribbean, took to the streets to protest rising food prices. Their fear and frustration were understandable: in the years between 2005 and 2008, food prices skyrocketed. The price of corn almost tripled, while the price of wheat and rice nearly doubled. The poor, who spend a large part of their income on food, were the hardest hit. The Food and Agricultural Organization of the United Nations estimates that higher prices pushed an additional 40 million people into hunger in 2008 alone.

> With the demand for food increasing, what will be the impact on our planet?

The sharp increase in food prices raised many questions. Had the human population outstripped its food supply? In 1798, Thomas Malthus's *An Essay on the Principle of Population* made the point that while populations have the potential to increase geometrically, their food supply is limited. Malthus's essay influenced Charles Darwin, who saw it as describing the great struggle of nature and, therefore, the survival of the fittest.

Yet there is no evidence that the 2008 spike in world food prices was due to overpopulation. Instead, a number of factors contributed to the rise in food prices, two of which especially stand out.

The first factor driving up food prices was drought in a number of key grain-producing regions. The lesson here is that even in developed countries, weather matters. In particular, drought and high temperature can slow or halt crop growth. Irrigation can help, but it's expensive, especially given the immense amounts of water that plants need. For example, a cow raised for meat drinks approximately 15,000 liters of water each year—a trivial amount compared to the roughly 3 million liters of water needed to grow the grain the cow will consume over the same time period.

The second factor creating the increased food prices was a steep rise in oil prices, which increased the cost of producing food. Historically, the sun's rays and the sweat of the farmer's brow were the major sources of energy for food production. Today, however, fossil fuels power tractors, are used to produce fertilizers and pesticides, and run irrigation pumps. Thus, when oil prices spike, the price of food also rises.

When oil prices began to fall during the second half of 2008, food prices declined. This provided relief for consumers, but can Malthus's concerns be ignored? Sometime in 2011, Earth's human population surpassed 7 billion. The United Nations estimates that by 2100 the human population will exceed 10 billion. Factor in the impacts of climate change, the spread of pests and pathogens, and the degradation of agricultural soils, and the challenge of producing enough food for all can seem daunting. With the demand for food increasing, what will be the impact on our planet?

Before humans began to cultivate plants, they hunted and gathered their food. These early hunter-gatherers favored plants that were easy to harvest, allowing them to gather large amounts of food in a short amount of time.

As humans began to settle in one place, they learned to cultivate food plants. The early agriculturalists needed to remove nonfood species (weeding), as well as save seeds to sow in the next season. Instead of pulling the seeds off in the field, they developed tools to cut stems and carry home entire seed heads. Harvesting became more efficient.

This new way of harvesting had a profound impact on plants. Only seeds that remained firmly attached to the stem were carried back to the village, so only these seeds would be replanted the next year. In this way, humans inadvertently became agents of artificial selection.

Consider wheat. In the wild, natural selection favors wheat whose seeds are protected by tough barbs and separate easily from the plant, allowing dispersal. Farmers, however, preferred wheat whose seeds remained attached to the plant and so were easier to harvest. Over time, crop plants lost the ability to disperse their own seeds. In this way, plant species became domesticated: they were no longer able to survive and reproduce on their own.

Humans have had a close association with wheat and other grasses for a long time. Today, more than 60% of the world's food energy comes from grasses, with wheat, rice, and corn (maize) being the three most important.

The earliest crops were domesticated a little more than 10,000 years ago. Humans made the transition from hunter-gatherers to agriculturists in at least six sites around the world. From these locations, agriculture spread along with technological innovations. For example, animal-drawn plows could break up heavy soils that could not be easily worked with hand tools. Irrigated crops could be grown during periods of low rainfall.

In late nineteenth century and early twentieth centuries, scientists began to recognize the role of microbes in plant diseases and the importance of nutrients such as nitrogen and phosphorus for plant growth. With the invention of the internal combustion engine, agriculture became mechanized. Yet increases in yield (measured in weight of crop output per area of land) remained modest.

By the 1940s, scientists had developed industrially produced nitrogen fertilizers that significantly boosted plant growth. They had also learned enough about genetics to begin applying that knowledge to plant breeding. These advances led to an amazingly productive period known as the Green Revolution. During the second half of the twentieth century, crop yields ballooned.

In the case of rice and wheat, both of which produce their seed heads at the tops of their stalks, a key factor was breeding for shorter and stronger stems. Plants

Domestication of wheat. Wild wheat (left) and modern domesticated wheat (right). The seeds of wild wheat are protected by tough barbs (the long spikes), and they fall easily from the plant, enhancing dispersal. The seeds of domesticated wheat remain attached to the plant, making the grain easier to harvest. *Sources: (left) Bob Gibbons/FLPA/Science Source; (right) Nigel Cattlin/Science Source.*

CASE 6

with short, strong stems could allocate more resources to seed production without causing the stems to bend over. In addition, rice varieties were developed that mature more quickly, so farmers could harvest several crops in a single year.

Today, we are entering a new phase of agriculture with the use of genetic engineering. Most of the corn and soy products now consumed in the United States come from plants that have been genetically engineered to resist certain pests and herbicides. Some of the genes come from different species altogether. For example, corn and cotton plants are often engineered to contain a gene from the bacterium *Bacillus thuringiensis* (*Bt*) that confers pest resistance.

Without question, humans have made great strides in increasing yields to feed our ever-growing population. At the same time, though, crop pests and pathogens have continued to evolve ways to evade our savviest plant breeders.

The nature of modern agriculture compounds this problem. Most modern farms consist largely of monocultures, single crop species grown over a large area. In a given field of corn or wheat, the individual plants are often genetically quite similar to one another. Growing only a single crop at a time makes it much easier to mechanize planting and harvesting. Unfortunately, it also means that a pathogen that can overcome a plant's natural defenses has the potential to wipe out an entire field.

In 2010, a new strain of yellow wheat rust fungus emerged in the Middle East. In its first year, the disease wiped out as much as half of Syria's wheat crop. It has since spread to other countries in the region. Meanwhile, crop scientists have warned that Ug99, a strain of an even more damaging fungal pathogen that first surfaced in eastern Africa, could devastate global wheat crops.

Today, yields continue to increase, although not as fast as they once did. If yield increases cannot be maintained, we may have to depend on the age-old method of increasing food production: we will have to add more agricultural land. While it might seem as if the planet has plenty of additional land that can be used to produce more food, some of this land is too dry, too cold, or too steep to be useful in agriculture. Other regions not under cultivation now are home to countless other species that would be vulnerable to the loss of their habitats.

One of humanity's greatest challenges is to produce enough food for all while saving room for nature. As plants are the cornerstone of agricultural systems, understanding what encourages plants' growth and reproduction and what constrains them is central to feeding us all in an increasingly crowded world.

DELVING DEEPER INTO CASE 6

For each of the Cases, we provide a set of questions to pique your curiosity. The questions cannot be answered directly from the Case itself, but instead introduce issues that will be addressed in chapters to come. We revisit this Case in Chapters 27–32 on the pages indicated below.

1. How has nitrogen availability influenced agricultural productivity? See page 613.
2. How did scientists increase crop yields during the Green Revolution? See page 635.
3. What is the developmental basis for the shorter stems of high-yielding rice and wheat? See page 646.
4. Can modifying plants genetically protect crops from herbivores and pathogens? See page 680.
5. What can be done to protect the genetic diversity of crop species? See page 706.
6. How do fungi threaten global wheat production? See page 731.

CHAPTER 27: Plant Form, Function, and Evolutionary History

CORE CONCEPTS

27.1 PHOTOSYNTHESIS ON LAND: A major challenge in the evolution of plants was being able to acquire carbon dioxide for photosynthesis without drying out.

27.2 CARBON DIOXIDE GAIN AND WATER LOSS: Leaves have a waxy cuticle and stomata, which allow plants to acquire carbon dioxide while limiting water loss.

27.3 WATER TRANSPORT: Xylem transports water and dissolved nutrients from the soil, allowing leaves to remain hydrated.

27.4 TRANSPORT OF CARBOHYDRATES: Phloem transports carbohydrates throughout the plant for use in growth and respiration.

27.5 UPTAKE OF WATER AND NUTRIENTS: Roots expend energy to obtain nutrients from the soil.

Single-celled organisms are the principal photosynthesizers of the seas, but plants carry out nearly all of the photosynthesis that occurs on land. This means that plants are the main entry point for both energy and carbon into terrestrial ecosystems. Directly or indirectly, plants are the source of most of what humans, and all other animals on land, eat. Only after land plants evolved did animals move out of the ocean and onto the land.

The closest living relatives to land plants are green algae that primarily grow in or close to streams and ponds. This relationship suggests that land plants originated in moist areas close to sources of fresh water. From this starting point, plants have spread out to colonize nearly all environments on land; some plants have even evolved to live in aquatic habitats, both fresh water and marine. But perhaps the most important event was the appearance of plants able to carry out photosynthesis far above the ground. The challenges of living on land and growing up into the air transformed plants from small, relatively simple forms into organisms with complex vascular tissues and organs specialized for nutrient uptake and photosynthesis.

In turn, the appearance of land plants had a profound impact on the environment. The terrestrial habitats of the first land plants were very different from today. In particular, there would have been no soil and very little shade. The presence of plants transformed the land: plants contributed to the formation of soil and, by growing tall, they created forests and other habitats in which plants and other land-dwelling organisms live. In this chapter, we focus on the structural and physiological innovations that allow plants to carry out photosynthesis so successfully on land. In later chapters, we explore how plants reproduce (Chapter 28), grow (Chapter 29), and defend themselves against pathogens and herbivores (Chapter 30).

27.1 PHOTOSYNTHESIS ON LAND

Organisms that live on land typically look nothing like those that live in water. The reason is that air and water differ in many fundamental properties. For example, water is denser than air, so organisms on land invest more in structures that can support the weight of their bodies. They are also exposed to higher amounts of ultraviolet radiation and wider swings in external temperatures. Nevertheless, the major risk of exposure to air is the potential to dry out.

Excessive water loss, or **desiccation,** is a constant hazard for photosynthetic organisms on land. Recall from Chapter 8 that photosynthesis uses energy from sunlight to synthesize organic (carbon-containing) molecules. Plants expose large surface areas to the air to obtain the sunlight and carbon dioxide (CO_2) needed for photosynthesis. Evaporation of water from these exposed surfaces can lead to desiccation. Consequently, many aspects of plant form and function can be understood in light of how they help sustain the hydration of photosynthetic cells.

Land plants share many cellular features with green algae.

Let's begin by revisiting where plants sit on the tree of life. In Chapter 25, we saw that "land plants" occur within the eukaryotic supergroup known as the Archaeplastids. Within this group, land plants are found within the Viridiplantae, or "green plants" (*viridi* = green, *plantae* = plants). The other members of the Viridiplantae are green algae. Land plants are monophyletic, meaning that they include all of the descendants of a single common ancestor (**Fig. 27.1**). Green algae are paraphyletic because the descendants of their last common ancestor also include the land plants. In this and succeeding chapters, we use "plants" and "land plants" interchangeably to denote the group that has diversified to become the dominant photosynthetic organisms on land. The scientific name for land plants (Embryophyta) highlights a feature shared by all plants: the formation of an embryo. We discuss the significance of this feature in Chapter 28.

Land plants are the only members of the Viridiplantae that have evolved complex multicellularity. Recall from Chapter 26 that organisms with complex multicellularity produce three-dimensional bodies with differentiated tissues and organs. In contrast, green algae are either unicellular or exhibit simple multicellularity, in which nearly every cell is in direct contact with the external environment.

Nevertheless, plant cells share many features with the cells of green algae. The cells of both plants and green algae contain chloroplasts with chlorophyll *a* and *b* and include a large membrane-bound vacuole that stores ions, metabolites, and water. Of particular importance, a cellulose-containing cell wall surrounds almost every cell (gametes are the exception). One consequence of this cell wall is that turgor pressure can develop within the cell. As water moves into a cell by osmosis, the cytoplasm pushes outward on the cell wall. At the same time, the cell wall resists being stretched and pushes back on the cytoplasm. The result is an increase in the pressure inside the cell (Chapter 5). Turgor pressure contributes to the mechanical support of plant tissues. In fact, many plant tissues visibly deform or "wilt" when excessive water loss results in a loss of turgor pressure.

In green algae, most cells have thin walls and contain chloroplasts for photosynthesis. Most likely the cells of the first plants did as well. As plants evolved, they produced a greater diversity of cell types. Many of these new cell types had thicker, and in some cases stiffer, walls that provided the mechanical support needed for plants to grow tall and to transport water and nutrients internally via bulk flow. As we will see, the evolution of specialized cell types played a key role in how plants have become so successful on land.

Bryophytes rely on surface water for hydration.

How have plants evolved the ability to survive and photosynthesize on land, even though the land surface is often dry? We can identify two main strategies. Plants called bryophytes rely on surface moisture for hydration, and they have evolved the ability to withstand intermittent drying. In contrast, vascular plants are able to pull water from the soil and to regulate their water loss. As a result, vascular plants maintain a more stable level of hydration compared to bryophytes.

The bryophytes include mosses, liverworts, and hornworts (Fig. 27.1). These three groups of plants share many features, including how they acquire water. The evolutionary relationships among the groups are unclear, however: we do not know which one is most closely related to vascular plants, or even if one or more might form a monophyletic group. Even so, the last common ancestor of mosses, liverworts and hornworts likely lived early in the evolutionary history of land plants.

Bryophytes are small plants, most less than 5 cm tall. Many look like small green tufts, and they are often found growing on rocks or logs (**Fig. 27.2**). A distinctive feature of

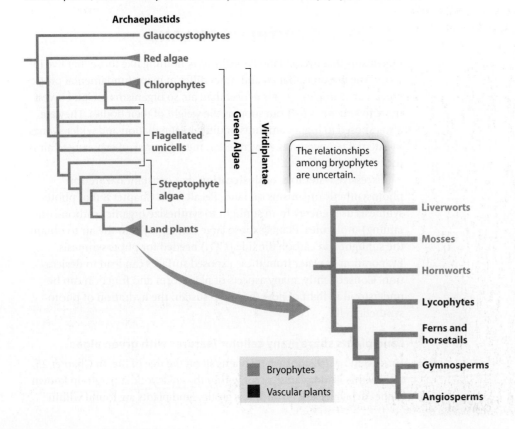

FIG. 27.1 The phylogenetic tree of land plants. Land plants can be divided into two types: the vascular plants, which actively control their hydration, and the bryophytes, which do not.

FIG. 27.2 Bryophytes. The understory of this northern forest is dominated by mosses and liverworts. *Source: Fulton Rockwell.*

bryophytes is that the majority of their aboveground surfaces are permeable to water. When it rains, these surfaces take up water and dissolved nutrients. Bryophytes do not produce roots. Instead, hair-like outgrowths from their lower surfaces anchor bryophytes in place. Like their green algal relatives, bryophytes lack specialized surfaces or organs for resource uptake.

In addition to being small, bryophytes have a high ratio of surface area to volume. This explains why most bryophytes rely entirely on diffusion to distribute water internally. Recall that diffusion is an effective transport mechanism only over short distances (Chapter 26). Because no cell is very far from an external surface in bryophytes, water is able to move into and through their cells by diffusion. A few moss do grow to more than 10 cm tall; these species produce elongated cells that allow fluids to be transported from one part of the plant to another by bulk flow.

Paradoxically, it is their dependence on the uptake of surface moisture that puts bryophytes at risk of drying out. Put simply, any surface that takes up water when wet will lose water to the surrounding air when dry. Because bryophytes do not have a way to regulate water loss, they desiccate quickly when their surfaces are dry. Thus, the water content of bryophyte cells fluctuates with that of their environment.

All cells require a high degree of hydration to function. As bryophyte cells dry out, these plants lose their ability to carry out photosynthesis. At the same time, many bryophytes exhibit **desiccation tolerance,** a set of biochemical traits that allows their cells to survive extreme dehydration without damage to membranes or macromolecules. As a result, they can resume photosynthesis when their surfaces once again come into contact with water.

Water pulled from the soil moves through vascular plants by bulk flow.

Vascular plants evolved a completely different approach to achieving the hydration needed to carry out photosynthesis on "dry" land. Their strategy has been highly successful: vascular plants are, by far, the dominant plant type found today. Ferns, sunflowers, and pine trees are all examples of vascular plants.

Vascular plants differ from bryophytes in several ways. The first one is obvious: vascular plants do not rely on diffusion to distribute materials internally, but instead transport water and other molecules by means of bulk flow through specialized, tube-like cells. This mechanism means that they can become much larger in size than bryophytes.

A second major difference is that water uptake primarily occurs in specialized organs that grow belowground—the roots. Through their roots, vascular plants can access the large amounts of water held by the soil. Because the plant's water comes from the soil, the aboveground surfaces of vascular plants can be relatively impermeable to water, limiting its loss.

The net result of these features is that vascular plants can carry out photosynthesis even when their aboveground surfaces are dry. Moreover, they can sustain the hydration of photosynthetic organs elevated as much as 100 m into the air. In many ways, the capacity to control the uptake, transport, and loss of water made possible the extraordinary evolutionary success of vascular plants.

More than 95% of all land plant species found today are vascular plants, and they provide almost all the photosynthetic output of terrestrial environments. Thus, the remainder of this chapter focuses on the structure and function of vascular plants.

We return to bryophytes when we examine plant reproduction (Chapter 28) and diversity (Chapter 31).

Vascular plants produce four major organ types.

Vascular plants are monophyletic, with four main subgroups (Fig. 27.1; **Fig. 27.3**). **Lycophytes** and **ferns and horsetails** are mostly small plants (Fig. 27.3a and b). The **gymnosperms** include pine trees and other conifers (Fig. 27.3c). The **angiosperms,** or flowering plants, include oak trees, grasses, and sunflowers (Fig. 27.3d). Many of the features that distinguish these groups are related to how they reproduce. For example, only gymnosperms and angiosperms produce seeds, while flowers and fruits occur only in angiosperms.

Vascular plants exhibit tremendous diversity of form, yet they are all built from the same four types of organs: leaves, reproductive organs, stems, and roots (**Fig. 27.4**). Leaves are specialized for photosynthesis. Reproductive organs promote genetic diversity and dispersal. Stems elevate and position both leaves and reproductive organs. These aboveground organs—leaves, reproductive organs, and stems—collectively form the **shoot.** Roots grow below ground, anchoring the plant and taking up nutrients and water. Variation in form arises from differences in length, size, and branching patterns of stems and roots, as well as in the structure of individual organs.

Although plant organs such as leaves come in all shapes and sizes, they are constructed from only three main types of tissues: epidermis, vascular, and ground tissues (Fig. 27.4). An outer layer of cells, the **epidermis,** forms the interface between the plant and the external environment. Epidermal cells provide protection to the plant. Some epidermal cells have specialized structures that enhance uptake of water and nutrients from the soil or uptake of CO_2 from the air. Fluids are transported in **vascular tissues,** from roots to leaves and also from leaves to roots. Vascular plants produce two types of vascular tissues: the **xylem,** whose primary function is to transport water, and the **phloem,** whose primary function is to transport carbohydrates. Both contain many specialized cell types that enable the long-distance transport of water, sugars, and signaling

FIG. 27.3 The four major groups of vascular plants. Example species include (a) a lycophyte (*Lycopodium annotinum*), (b) a fern (*Athyrium filix-femina*), (c) a gymnosperm (*Pseudotsuga menziesii*), and (d) an angiosperm (*Punica granatum*). *Sources: a. F. Teigler/AGE Fotostock; b. R A Kearton/Moment RF/Getty Images; c. petekarici/Getty Images; d. Design Pics Inc/Alamy.*

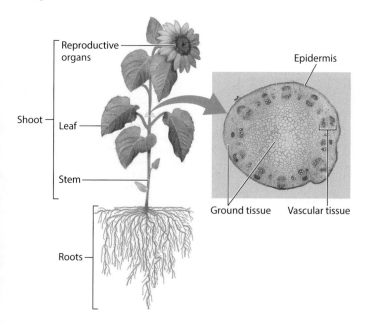

FIG. 27.4 Major organs and tissues of a vascular plant. The organs include leaves, stems, reproductive structures, and roots. The tissues include epidermis, vascular tissue, and ground tissue. *Photo source: Garry DeLong/Science Source.*

molecules. **Ground tissue** is the term for all of the remaining cells. In many cases, these will be **parenchyma,** thin-walled cells that are capable of further division and differentiation. Parenchyma cells perform most of a plant's metabolic functions, including photosynthesis and storage of nutrients, carbohydrates, and water.

We are now ready to answer the question of how vascular plants photosynthesize on land by examining how leaves, stems, and roots work together.

Self-Assessment Question

1. Why are vascular plants better able to sustain photosynthesis than are bryophytes?

27.2 CARBON DIOXIDE GAIN AND WATER LOSS

The **leaf** is the principal site of photosynthesis in most vascular plants. In this section, we explore how the structure of leaves allows vascular plants to photosynthesize in air without drying out.

CO_2 uptake results in water loss.

Plants use large amounts of water. In fact, water is the resource that most often limits a plant's ability to grow and function. This fact becomes readily apparent when you forget to water your houseplants or garden. And when you drive across a continent, you see that the plants growing in wet places look very different from those growing in dry ones. Why do plants require so much water?

The answer is *not* water's role as an electron donor in photosynthesis (Chapter 8): that process accounts for less than 1% of the water required by vascular plants. Instead, most of a plant's need for water is a consequence of the plant taking up the CO_2 needed for photosynthesis from the air. CO_2 is a minor constituent of air, present at about 400 parts per million. This low concentration limits the rate at which CO_2 can diffuse into the leaf. Therefore, how fast a leaf can take up CO_2 depends in large part on the extent to which its photosynthetic cells are exposed to the surrounding air.

A cross section shows the leaf's three major tissues: sheets of epidermal cells line the leaf's upper and lower surfaces; loosely packed photosynthetic cells make up the **mesophyll** (literally, "middle leaf"); and the system of vascular conduits called **veins** connects the leaf to the rest of the plant (**Fig. 27.5**). If we look inside a leaf, we see that each mesophyll cell is surrounded by air spaces. The mesophyll cells obtain CO_2 from these air spaces. If the air spaces within the leaf were completely sealed off, photosynthesis would quickly run out of CO_2 and come to a halt. But they're not sealed off; instead, the leaf's air spaces are connected to the air surrounding the leaf by pores in the epidermis. As photosynthesis uses up CO_2, the concentration of CO_2 molecules within the leaf's air spaces decreases relative to the concentration of CO_2 in the outside air. Because molecules diffuse from regions of higher concentration to regions of lower concentration, CO_2 diffuses into the leaf, replenishing the supply of CO_2 for photosynthesis.

The ability of leaves to draw in CO_2 comes at a price: as CO_2 diffuses into the leaf, water vapor diffuses out (Fig. 27.5). Furthermore, water vapor diffuses out of a leaf at a much faster rate than CO_2 diffuses inward. Recall that molecules diffuse from regions of higher concentration to regions of lower concentration and that the rate of diffusion is proportional to the difference in concentration. On a sunny summer day, the difference in water vapor concentration between the air spaces within a leaf and the air outside can be more than 100 times greater than the difference in concentration of CO_2. Add the fact that water is lighter than CO_2, and so diffuses 1.6 times faster *for the same concentration gradient,* and it becomes clear why, on a sunny summer day, several hundred water molecules can be lost for every molecule of CO_2 acquired for photosynthesis.

The loss of water vapor from leaves is referred to as **transpiration.** The rates at which plants transpire are often quite high. A sunflower leaf, for example, can transpire an amount equal to its total water content in as little as 20 minutes. To put this rate of water loss in perspective, you would have to drink about 2 liters per minute to survive a similar rate of loss. Vascular plants can sustain such high rates of water loss because they can access

FIG. 27.5 Leaf structure. The uptake of CO_2 by diffusion is accompanied by a much larger outward diffusion of water vapor.

Cuticle
Upper epidermis
Mesophyll cells
Lower epidermis
Cuticle
Guard cell
Stoma
Vein
CO_2
H_2O

~200 ppm
~40,000 ppm
CO_2 gradient
Water vapor gradient
~400 ppm
~20,000 ppm

Mesophyll cells near the upper surface of the leaf have a columnar arrangement that maximizes light interception.

Mesophyll cells close to the stomata have a honeycomb-like arrangement to allow CO_2 to spread easily throughout the leaf.

The concentration difference driving the diffusion of CO_2 into the leaf is typically 100 times smaller than the concentration difference driving the diffusion of water vapor out of the leaf.

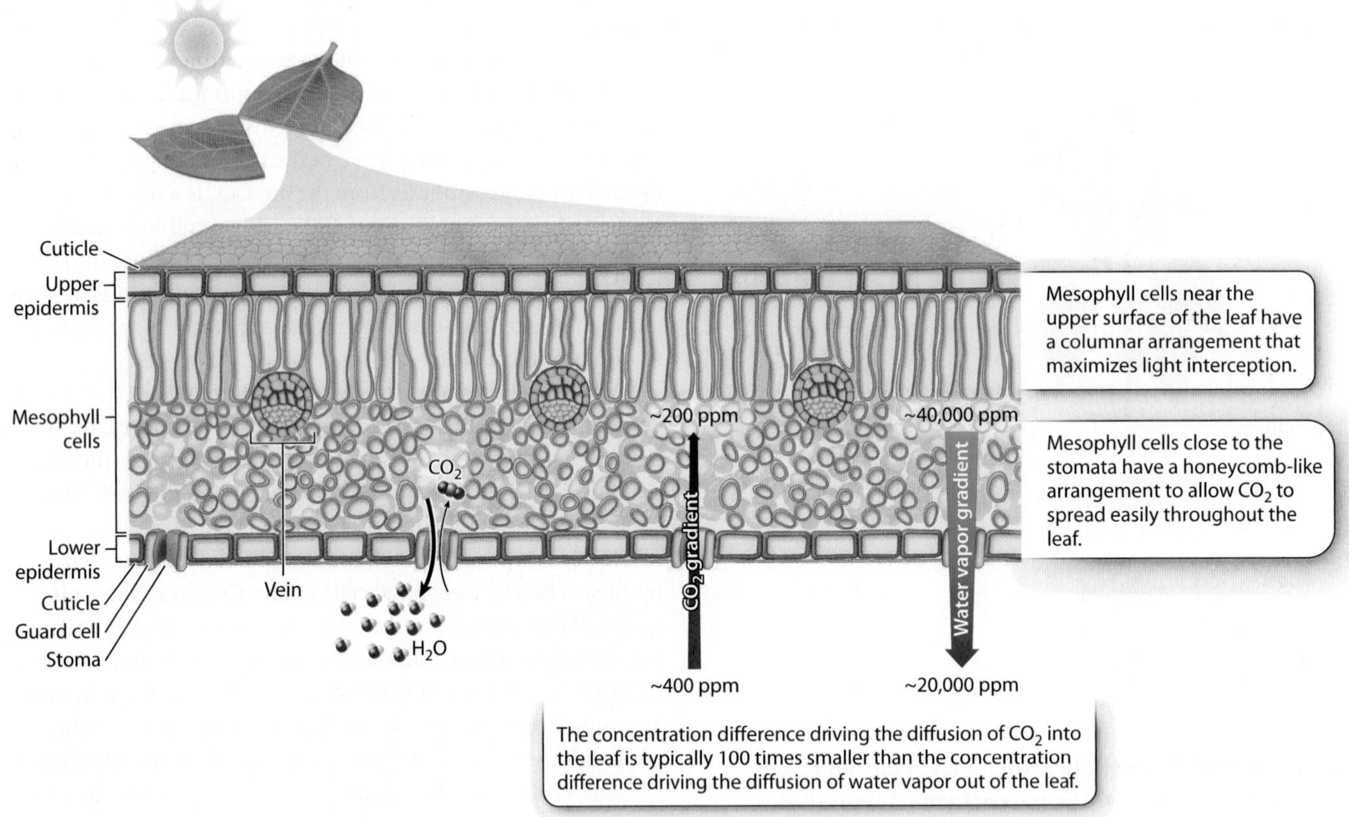

the only consistently available source of water on land: the soil. Water moves from the soil, through the bodies of vascular plants, and then, as water vapor, into the atmosphere. Therefore, the challenge of keeping photosynthetic cells hydrated is met partly by the continual supply of water from the soil. As we discuss next, it is also met by limiting the rate at which water already in leaf tissues is lost to the atmosphere.

The cuticle restricts water loss from leaves but inhibits the uptake of CO_2.

Epidermal cells secrete a waxy **cuticle** on their outer surface that limits water loss. Without a cuticle, the humidity within the internal air spaces would drop and the photosynthetic mesophyll cells would quickly dry out. However, the cuticle prevents CO_2 from diffusing into the leaf even as it restricts water vapor from diffusing out of it.

As noted earlier, small pores in the epidermis allow CO_2 to diffuse into the leaf (**Fig. 27.6a**). These pores are called **stomata** (singular, **stoma**). Stomata can be numerous, with many plants having hundreds per square millimeter on their epidermis. Yet because each stoma is small, less than 1% to 2% of the leaf surface is actually covered by these pores.

The epidermis with its stomatal pores represents a compromise between the challenges of providing food (CO_2 uptake) and preventing thirst (water loss). However, because both the dryness of the air and the wetness of the soil vary over time, leaves must be able to alter how much water and CO_2 pass through the leaf epidermis so that the plant can maintain a balance between the rates of water loss to the atmosphere and water delivery from the soil. Leaves can do this by opening and closing their stomata.

Stomata allow leaves to regulate water loss and carbon gain.

Stomata are more than just holes in the leaf epidermis: they are hydromechanical valves that can open and close. Each stoma consists of two **guard cells** surrounding a central pore. The guard cells can shrink or swell, changing the size of the pore between them (**Fig. 27.6b**). Let's consider how this valve system works.

Recall that cellulose provides strength to plant cell walls. In guard cells, the long cellulose molecules are oriented radially—that is, wrapped around the cell. Confined by this arrangement of cellulose, swelling guard cells tend to expand in length rather than in girth. Because the guard cells are firmly connected at their ends, an increase in length causes them to bow apart, opening the stomatal pore. Conversely, a decrease in guard cell volume causes the pore to close.

FIG. 27.6 Stomata. (a) A scanning electron microscope (SEM) image of stomata on a leaf surface; (b) guard cells of a stoma in open and closed positions. *Photo source: Dr. Jeremy Burgess/Science Source.*

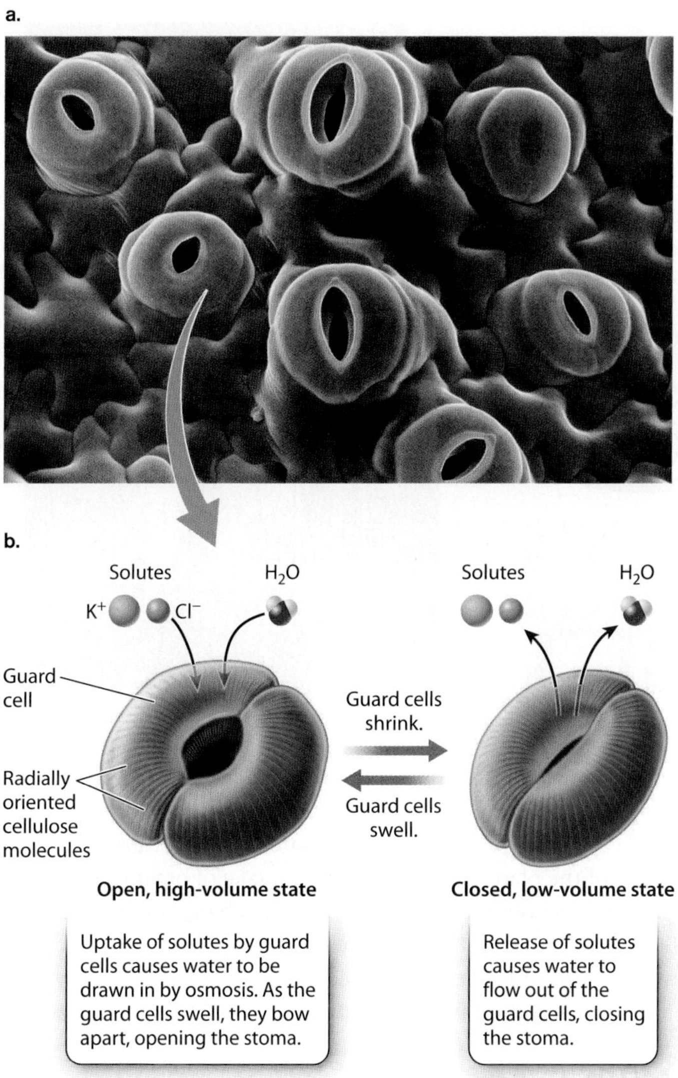

a.

b.

Open, high-volume state

Uptake of solutes by guard cells causes water to be drawn in by osmosis. As the guard cells swell, they bow apart, opening the stoma.

Closed, low-volume state

Release of solutes causes water to flow out of the guard cells, closing the stoma.

Guard cells control their volume by altering the concentration of solutes, such as potassium ions (K^+) and chloride ions (Cl^-), in their cytoplasm. ATP drives the uptake of solutes across the plasma membrane against their concentration gradient. An increase in solute concentration causes water to move into the cell by osmosis. Recall from Chapter 5 that osmosis is the net movement of water molecules across a selectively permeable membrane, such as the cell's plasma membrane, from a region of lower solute concentration to a region of higher solute concentration. As water enters the guard cells, their turgor pressure increases. Higher turgor pressure causes the walls to stretch, allowing the cells to swell.

Stomata close by the same process, only in reverse. That is, solutes leave the guard cells; as the solute concentration decreases, water exits the cells, causing the turgor pressure to decrease and the cells to shrink.

To function effectively, stomata must open and close in response to factors that affect CO_2 uptake and water loss. Stomata are thus key sites for the processing of physiological information. In most plants, light stimulates stomata to open, while high levels of CO_2 inside the leaf (a signal that CO_2 is being supplied faster than photosynthesis can take it up) cause stomata to close. Guard cell volume also changes in response to signaling molecules. For example, abscisic acid, a hormone produced during drought, causes stomata to close. Thus, stomata open when the conditions for photosynthesis are favorable, and close when a water shortage endangers the hydration of the leaf.

Our own experience leads us to think of evaporation (for example, of sweat) as a means of preventing overheating. Is there any evidence that plants control the temperature of their leaves by regulating transpiration? The answer, perhaps surprisingly, is no. Transpiration does cool leaves, but plants transpire whether their leaves are too hot, too cold, or at their optimal temperature. When leaves are above their optimal temperature for photosynthesis, cooling through transpiration is beneficial. But when temperatures are below the optimum, transpiration lowers leaf temperatures to become even less favorable. In addition, evaporative cooling would be most beneficial when temperatures spike above normal, yet often during heat waves, little water is available in the soil. For this reason, plants in arid regions tend to rely on mechanisms other than evaporative cooling, such as reflective hairs and waxes and small leaf size, to prevent their leaves from overheating.

CAM plants use nocturnal CO_2 storage to avoid water loss during the day.

In most plants, water loss and photosynthesis peak at the same time, when the sun is high overhead. What if a plant could open its stomata to capture CO_2 at night, when cool air limits rates of evaporation?

In fact, a number of plants have evolved precisely this mechanism to help balance CO_2 gain and water loss (**Fig. 27.7**). The mechanism is called **crassulacean acid metabolism (CAM)**; it is named after the Crassulaceae family of plants, which uses this pathway. CAM provides a system for storing CO_2 overnight by converting it into a form that will not diffuse away. The storage form of CO_2 is produced by the enzyme PEP carboxylase, which combines a dissolved form of CO_2 (bicarbonate ion, HCO_3^-) with a 3-carbon compound called phosphoenolpyruvate (PEP). The resulting product is a 4-carbon organic acid that is stored in the cell's vacuole.

When the sun comes up the next morning, the stomata close, conserving water (Fig. 27.7a). At the same time, the 4-carbon organic acids are transferred from the vacuole to the chloroplasts and their CO_2 is released. Because the stomata are now closed, this newly released CO_2 does not diffuse out of the

FIG. 27.7 CAM photosynthesis. (a) CAM plants store CO_2 at night in the vacuole; leaves can then photosynthesize during the day without opening their stomata. (b) A saguaro cactus, a CAM plant. (c) An epiphytic orchid (*Dendrobium* species), also a CAM plant. *Photo sources: b. Charles Harker/Getty Images; c. John Mason/ARDEA.*

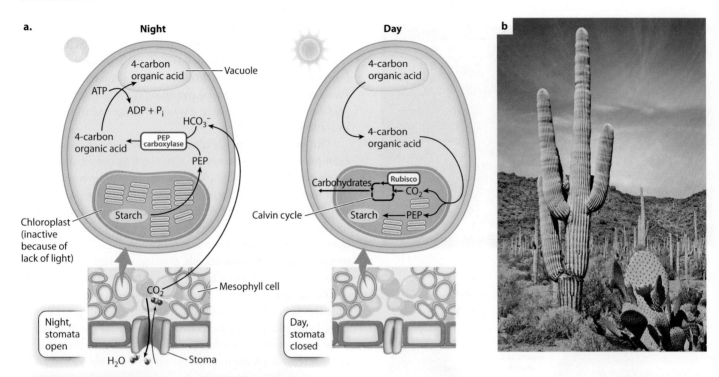

leaf. Instead, it becomes incorporated into carbohydrates by the Calvin cycle. As we saw in Chapter 8, the Calvin cycle can operate only in the light because it requires a continual supply of energy in the form of ATP and NADPH produced by the light reactions of photosynthesis. The 3-carbon PEP molecules remaining after CO_2 release are converted into starch and stored in the chloroplast until the sun goes down and they are again needed to capture and store CO_2 in the vacuole.

By opening their stomata only at night, CAM plants greatly increase the amount of CO_2 gain per unit of water loss. Indeed, the $CO_2:H_2O$ exchange ratio for CAM plants is in the range of 1:50, nearly 10 times higher than that of plants that open their stomata during the day. However, CAM has a drawback. CAM photosynthesis produces carbohydrates slowly because it uses some of the plant's ATP to drive the uptake of organic acids into the vacuole and because only so much organic acid can accumulate overnight in the vacuole. For these reasons, CAM is most commonly encountered in habitats such as deserts, where water conservation is crucial, and among **epiphytes,** plants that grow on the branches of other plants, without contact with the soil (Fig. 27.7b and 27.7c).

Because CAM uses enzymes (for example, PEP carboxylase) and compartments (for example, vacuoles) that exist in almost all plant cells, it is not surprising that it has evolved multiple times. CAM occurs in all four groups of vascular plants shown in Fig. 27.1. It is most widespread among the angiosperms, found in 5% to 10% of these species. Vanilla orchids and Spanish moss, two well-known epiphytes, are CAM plants, and so are many cacti.

C_4 plants suppress photorespiration by concentrating CO_2 in bundle-sheath cells.

Photorespiration adds another wrinkle to the challenge of acquiring CO_2 from the air. Recall from Chapter 8 that either CO_2 or O_2 can be a substrate for rubisco, the key enzyme in the Calvin cycle. When CO_2 is the substrate, the

Calvin cycle produces carbohydrates through photosynthesis. When O_2 is the substrate, there is a net loss of energy and a release of CO_2, through a process called photorespiration. (Photorespiration is similar to aerobic respiration only in the sense that it uses O_2 and releases CO_2. However, plants do not gain energy in photorespiration, but rather lose it.)

Photorespiration presents a significant challenge for land plants for two reasons. First, air contains approximately 21% O_2 but only 0.04% CO_2. Although rubisco reacts more readily with CO_2 than with O_2, the sheer abundance of O_2 means that without some means of concentrating CO_2, rubisco will use O_2 as a substrate some of the time. Second, air provides much less of a thermal buffer than does water. As a result, organisms on land experience higher and more variable temperatures than do organisms that live in water. Temperature has a major effect on photorespiration because rubisco becomes less able to distinguish between CO_2 and O_2 as temperatures increase. At moderate leaf temperatures, O_2 is the substrate instead of CO_2 as often as 1 time out of 4. At higher temperatures, O_2 is even more likely to be the substrate for rubisco.

Some plants have evolved a way to reduce the energy and carbon losses associated with photorespiration. These **C_4 plants** suppress photorespiration by increasing the concentration of CO_2 in the immediate vicinity of rubisco. C_4 plants take their name from the fact that they, like CAM plants, use PEP carboxylase to produce 4-carbon organic acids that subsequently supply the Calvin cycle with CO_2. The Calvin cycle produces 3-carbon compounds (Chapter 8), and plants that do not use 4-carbon organic acids to supply the Calvin cycle with CO_2 are called **C_3 plants.**

Both CAM and C_4 plants produce 4-carbon organic acids as the entry point for photosynthesis. However, in CAM plants, CO_2 capture and the Calvin cycle take place *at different times*; in C_4 plants, they take place *in different cells*.

C_4 plants initially capture CO_2 in mesophyll cells, whereas the Calvin cycle takes place in the bundle sheath, a cylinder of cells that surrounds each vein. The 4-carbon compounds produced in mesophyll cells diffuse through plasmodesmata into bundle-sheath cells (**Fig. 27.8**). Once inside these cells, the 4-carbon compounds release CO_2, which is then incorporated into carbohydrates through the Calvin cycle (Chapter 8). The C_4 cycle is completed as the 3-carbon molecules generated during CO_2 release diffuse back to the chloroplasts in the mesophyll cells. There, ATP is used to re-form the original 3-carbon compound (PEP) that combines with CO_2.

The significance of the C_4 cycle is that it operates much faster than the Calvin cycle. As a result, the concentration of CO_2 within bundle-sheath cells increases dramatically, reaching levels as much as 5 times higher than the levels in the air surrounding the leaf. The high concentration of CO_2 in bundle-sheath cells makes it unlikely that rubisco will use O_2 as a substrate. Thus, the C_4 cycle functions like a "fuel-injection" system that increases the efficiency of the Calvin cycle in bundle sheath cells.

FIG. 27.8 C_4 photosynthesis. C_4 plants suppress photorespiration by concentrating CO_2 in bundle-sheath cells.
Photo source: Medical Images RM/KEN WAGNER.

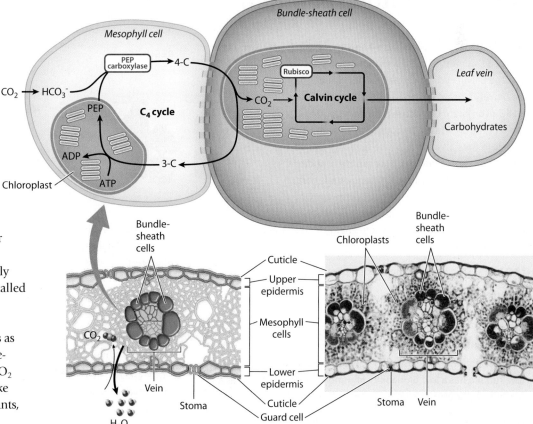

C_4 plants have high rates of photosynthesis because they do not suffer the losses in energy and reduced carbon associated with photorespiration (**Fig. 27.9**). At the same time, C_4 plants lose less water because they can restrict diffusion through their stomata to a greater extent than a C_3 plant can, while still maintaining high concentrations of CO_2 in bundle-sheath cells. However, C_4 photosynthesis consumes more energy than conventional (C_3) photosynthesis, because ATP must be used to regenerate PEP in the C_4 cycle. Thus, C_4 photosynthesis confers an advantage in hot, sunny environments where rates of photorespiration and transpiration would otherwise be high. This type of photosynthesis has evolved as many as 20 times, but is most common among tropical grasses and plants of open habitats with warm temperatures. C_4 plants include a number of important crops, such as maize (corn), sugarcane, and sorghum, as well as some of the most noxious agricultural weeds.

> **Self-Assessment Questions**
>
> 2. Why do plants transpire?
> 3. Draw and label the features of a leaf that are necessary to maintain a well-hydrated interior while still allowing CO_2 uptake from the atmosphere.
> 4. Diagram how the movement of solutes results in the opening and closing of stomata.
> 5. Diagram how the enzyme PEP carboxylase allows CAM plants to reduce transpiration rates and C_4 plants to suppress photorespiration.

HOW DO WE KNOW?

FIG. 27.9

Does C_4 photosynthesis suppress photorespiration?

BACKGROUND Before biologists knew about the C_4 pathway, studies had suggested there was something unusual about the Calvin cycle in certain species. Studies using radioactively labeled CO_2 showed that some species initially incorporate CO_2 into 4-carbon compounds instead of the 3-carbon compounds that are the first products in the Calvin cycle. These C_4 plants also have high rates of photosynthesis. How can these findings be explained? Are these 4-carbon compounds part of a new, more efficient photosynthetic pathway? Or do C_4 plants have high rates of photosynthesis because they are able to avoid the carbon and energy losses associated with photorespiration?

HYPOTHESIS C_4 plants do not exhibit photorespiration.

EXPERIMENT Olle Björkman and colleagues at the Carnegie Institution of Washington compared C_3 and C_4 species of the genus *Atriplex*. They measured rates of CO_2 uptake by plant leaves, first in normal air (21% O_2) and then in an experimental gas mixture in which the concentration of O_2 was only 1%. When the concentration of O_2 is low, rubisco has a low probability of using O_2 (instead of CO_2) as a substrate, so photorespiration does not occur.

Measurements of CO_2 uptake by leaves represent the sum of photosynthesis, which takes in CO_2, and photorespiration, which releases CO_2. Mitochondrial respiration also releases CO_2, but in an illuminated leaf the rates are low relative to photosynthesis and invariant across this range of O_2 concentrations. Consequently, the rate of CO_2 uptake measured in 1% O_2 tells us what the rate of photosynthesis is in normal air (21% O_2) under the same conditions (e.g., same light intensity, temperature).

RESULTS

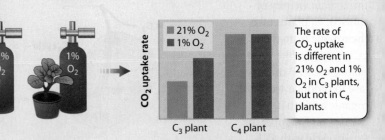

The rate of CO_2 uptake is different in 21% O_2 and 1% O_2 in C_3 plants, but not in C_4 plants.

CONCLUSION The rate of CO_2 uptake in C_4 plants is the same at the two different O_2 concentrations, indicating that significant photorespiration is not occurring in these plants. In contrast, CO_2 uptake of the C_3 plants increases in the low O_2 environment, indicating that photorespiration decreases net carbon gain in these plants in 21% O_2.

FOLLOW-UP WORK The higher photosynthetic efficiency of C_4 photosynthesis has prompted efforts, so far unsuccessful, to incorporate this pathway into C_3 crops such as rice.

SOURCE Björkman, O., and J. Berry. 1973. "High-Efficiency Photosynthesis." *Scientific American* 228:80–93.

27.3 WATER TRANSPORT

On a summer day, a tree can transport many hundreds of liters of water from the soil to its leaves. Impressively, this feat is accomplished without any moving parts. Even more remarkably, trees and other plants transport water without any direct expenditure of ATP. The structure of vascular plants allows them to use the evaporation of water from leaves to pull water from the soil.

Fig. 27.10 shows the cross section of a sunflower stem. Like a leaf, it has a surface layer of epidermal cells. This layer encloses thin-walled, undifferentiated parenchyma cells in the interior. Notice that the stem also contains differentiated tissues that lie in a ring near the outside of the stem. These are the vascular tissues, which form a continuous pathway that extends from near the tips of the roots, through the stem, and into the network of veins within leaves. The outer tissue, called phloem, transports carbohydrates from leaves to the rest of the plant body. The inner tissue, called xylem, transports water from the roots to the leaves.

In addition, both the xylem and the phloem transport dissolved nutrients and signaling molecules, such as hormones, throughout the plant. Importantly, the xylem is the pathway used to transport nutrients absorbed from the soil to the aboveground portions of the plant. Because the total concentration of solutes in the xylem is very low (typically less than 0.01%), here we will focus on the xylem as a means for transporting water from soil to leaves. We discuss the uptake of nutrients in section 27.5.

Xylem provides a low-resistance pathway for the movement of water.

Water travels with relative ease through xylem: it would take 10 orders of magnitude greater force to drive water through living parenchyma cells than it takes to transport the same amount of water through the xylem. This ease of flow is explained by the structure of the water-transporting cells within the xylem. These cells have cell walls but lack any cytoplasm or cell membranes at maturity, so water can flow unimpeded through their hollow centers (**Fig. 27.11**).

These open tubes do not extend the length of a plant. Instead, the xylem is made up of many hollow conduits, each of which tapers to a closed end. Thus, as water travels from the soil to the leaves, it must repeatedly flow from one conduit to another. The frequent transfer from one conduit to the next increases the resistance compared to an open tube. However, this arrangement has the advantage that if a conduit "fails" (a topic we discuss later), it eliminates only a small part of the xylem's capacity to transport water.

Xylem conduits have lignified secondary cell walls (Chapter 10). Lignin is a chemical compound that strengthens the cell wall, but also impedes the movement of water across the wall. As a result, water enters and exits xylem conduits primarily through **pits,** circular or ovoid regions lacking the secondary cell wall layer. The only barrier to water flow across pits is the water-permeable primary cell wall. As we will see, pits play an important role because they allow the passage of water, but not air, from one conduit to another.

A xylem conduit can be formed from a single cell or from multiple cells stacked to form a hollow tube. Unicellular conduits are called **tracheids** (Fig. 27.11a), whereas multicellular conduits are called **vessels** (Fig. 27.11b). Because tracheids are the product of a single cell, they are typically only 5–50 μm in diameter and less than 1 cm long. Vessels, which are made up of many cells called **vessel elements,** can be much wider and longer. Vessel diameters range from 10 to 500 μm, and their lengths can be up to several meters.

Fossils show that the first vascular plants produced only tracheids. Today, tracheids remain the principal conduit type in lycophytes, ferns and horsetails, and most gymnosperms. Xylem vessels have evolved several times. Some ferns, such as the widespread bracken fern, produce multicellular xylem conduits, as does one group of gymnosperms. Nearly all angiosperms produce vessels.

Water enters a tracheid through pits, travels upward through the conduit interior, and then flows outward through other pits into an adjacent, partially overlapping, tracheid. Water also enters and exits a vessel through pits. Once the water is inside the vessel, little or nothing blocks the flow of water from one

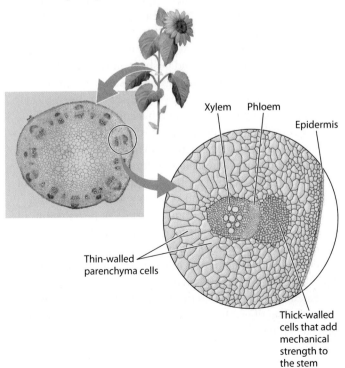

FIG. 27.10 Xylem and phloem. A cross section of the stem of a sunflower shows the two types of vascular tissues: xylem and phloem.
Photo source: Garry DeLong/Science Source.

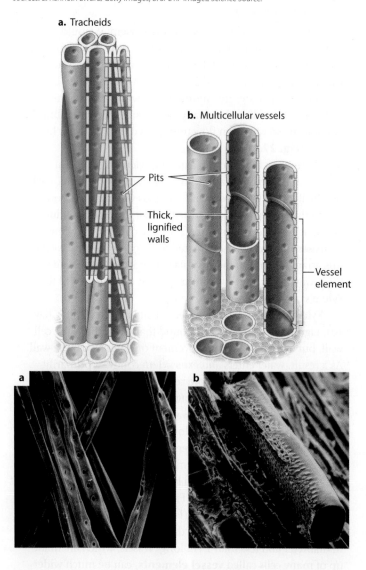

FIG. 27.11 Xylem structure. Xylem conduits are formed by hollow cells with stiff walls and no cell content or membrane. (a) Single-celled tracheids and (b) multicellular vessels are interconnected by pits. *Photo sources: a. Kenneth Eward/Getty Images; b. SPL RF Images/Science Source.*

vessel element to the next. During the development of a xylem vessel, the end walls of the vessel elements are digested away, which means that water is able to flow along the entire length of the vessel without having to cross any pits. At the end of a vessel, however, the water must flow through pits as it enters an adjacent vessel and thereby continues its journey from the soil to the leaves.

The rate at which water moves through xylem depends on both the number of conduits and their size. Water flows faster through longer conduits because it does not need to flow so often from conduit to conduit across pits, which exert a significant resistance to flow. Water flow through xylem conduits is also greater when the conduits are wider. For any open pipe, the flow rate is proportional to the radius of the conduit raised to the fourth power, so doubling the radius increases the flow 16-fold. Because vessels can be both longer and wider than tracheids, plants with vessels can achieve greater rates of water transport.

The xylem of many angiosperms has a much greater water transport capacity than the xylem of plants that have only tracheids. As a result, angiosperms can support the high rates of transpiration needed to obtain large amounts of CO_2 for photosynthesis. The evolution of angiosperms led to substantial increases in terrestrial productivity as well as an intensification of the water cycle on land, as higher rates of transpiration resulted in more water being returned to the atmosphere, where it became able to fall again as rain.

Water is pulled through xylem by an evaporative pump.

A leafy shoot placed in a sunny spot with its cut end in water will continue to transpire. This result demonstrates that the driving force for water transport is not generated in the roots, but instead comes from the leaves.

In the vascular systems of plants and animals, bulk flow is driven by differences in pressure. Pressure has units of force per area, and the direction of flow is always from higher to lower pressure. In your body, a mechanical pump—your heart—*raises* the pressure of the blood, causing it to flow from the heart through your arteries and veins. In plants, transpiration *lowers* the pressure of the water in the xylem near the sites of evaporation in the leaf, causing water to flow from the soil to the leaves. Plants face a challenge owing to this process: to transport water from the soil to the leaves at rates equal to the water loss from transpiration, the pressures in the leaf xylem must be very low, typically well below zero (**Fig. 27.12**).

We commonly think of pressure as the result of molecules bouncing into one another and pushing outward on the walls of a container. Negative pressures result when the forces between molecules act in the opposite direction. The molecules pull, rather than push, on one another and the forces on the container walls are oriented inward. For this reason, negative pressure in the xylem is sometimes described as "tension."

Water stands out among liquids for its ability to sustain large negative pressures. The strong hydrogen bonds that form between water molecules and between water molecules and many surfaces allow tensile (pulling) forces to be transmitted from one water molecule to another. In contrast, negative pressures cannot develop in gases because the molecules are so widely separated that they do not form intermolecular bonds and, therefore, cannot pull on one another.

Essentially, the water in the xylem functions as the "rope" in a game of tug-of-war between the soil and the leaves. Soils hold onto water due to surface tension at air–water interfaces. As soils dry out, they hold on even harder. If the leaf is to win the game,

HOW DO WE KNOW?

FIG. 27.12

Do plants generate negative pressures?

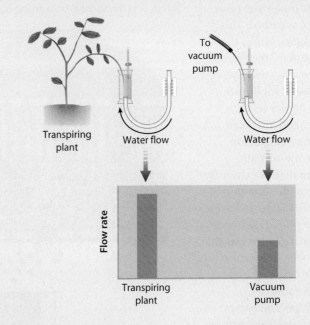

BACKGROUND The idea that water is pulled through the plant by negative pressures in leaves was first suggested in 1895. However, without a way to measure these pressures, there was no way to know how large they were.

HYPOTHESIS In 1912, German physiologist Otto Renner hypothesized that leaves are able to exert stronger suctions than could be generated with a vacuum pump.

EXPERIMENT Renner measured the rate at which water flowed from a reservoir into the cut tip of a branch of a transpiring plant. He next cut off about 10 cm of the branch and attached a vacuum pump in place of the transpiring plant. He compared the flow rate through the branch when attached to the transpiring plant with the flow rate through the branch generated by the vacuum pump.

RESULTS Renner found that the flow rates generated by transpiring plants were two to nine times greater than the flow rates generated by the vacuum pump.

CONCLUSION A transpiring plant pulls water through a branch faster than a vacuum pump. Therefore, the pulling force generated by a transpiring plant must be greater than that of a vacuum pump. Because vacuum pumps create suction by reducing the pressure in the air, a vacuum pump can reduce the pressure only to zero. For the transpiring leaves to pull even harder, the pressures in the xylem must be less than zero. Thus, transpiration generates negative pressures in the xylem.

The maximum height that one can lift water using a vacuum pump is 10 m (33 feet). The forces that pull water through plants have no comparable limit because the strong hydrogen bonds between water molecules mean that the xylem can sustain very large negative pressures. This is why plants can pull water from dry soils and why they can grow to more than 100 m, the height of a 30-story building.

FOLLOW-UP WORK In 1965, Per Fredrick Scholander developed a pressurization technique that uses reverse osmosis to measure the magnitude of the negative pressures exerted by leaves of intact plants. Using this technique, he demonstrated that the leaves at the top of a Douglas fir tree 100 m tall exert pressures that are even more negative than those exerted by leaves closer to the ground.

SOURCES Renner, O. 1925. "Zum Nachweis negativer Drucke im Gefässwasser bewurzeler Holzgewachse." *Flora* 118/119:402–408; Scholander, P. F., et al. 1965. "Sap Pressure in Vascular Plants." *Science* 148:339–346.

---

it has to pull harder than the soil, *and* it has to pull hard enough to overcome the force of gravity and the frictional resistance of moving water through xylem conduits. But how does transpiration generate tension in xylem?

As stomata open to allow CO_2 needed for photosynthesis to enter, water evaporates from the surface of the cells lining the leaf's air spaces (**Fig. 27.13**). The partial dehydration of these cell walls causes water to flow from the cytoplasm of nearby cells to the sites of evaporation, similar to how water flows into a freshly wrung-out sponge. The forces that pull water toward the sites of evaporation include surface tension at the air–water interface and interactions between water molecules and the components of the cell wall. The loss of water from the cytoplasm both increases the solute concentration and decreases the cell's turgor pressure. The net result is that water flows from adjacent, more hydrated cells toward the sites of evaporation, until eventually water is drawn from xylem conduits in leaf veins.

The loss of a water molecule from the xylem causes the pressure to decrease. Because the walls of xylem conduits are very stiff, the water molecules remaining in the xylem must stretch a tiny bit if the conduit is to remain full of water. Just as when you pull on a rubber band, the water molecules resist being stretched. This resistance causes the pressure in the xylem to decrease because the water molecules pull harder on each other and also exert an inward force on the conduit walls. Ultimately, the decrease in pressure causes another water molecule to be pulled up to take the place of the one that has been drawn into the leaf toward the sites of evaporation.

Water can enter the xylem from any part of the plant that is more hydrated than the leaf. Typically, roots are the wettest part of the plant because they are in contact with the soil. Thus, during transpiration, most of the water moving into the leaf comes from the roots, which in turn obtain water from the soil. If the soil becomes very dry, however, the leaf may be unable to generate sufficient tension without experiencing damaging levels of dehydration. This is why stomata close as soil water is depleted.

Many different processes affect the movement of water through plants. In soils, the surface tension of air–water interfaces dominates, whereas in living cells osmosis plays a central role. In the xylem, pressure gradients are the driving force for water movement, and everywhere gravity has an effect. **Water potential** is a parameter that combines all of the various factors that influence the movement of water, such as pressure, osmosis, and gravity. In a thermodynamic framework, water potential is the free energy of water and, therefore, its capacity to do work (Chapter 6). Water moves from areas with higher water potential to areas with lower water potential, just as happens when water flows downhill. Thus, if we know the water potential at any two points, we can say which way the water will flow. For example, in a transpiring plant, the water potential of the leaf is lower than the water potential of the soil, so water moves upward.

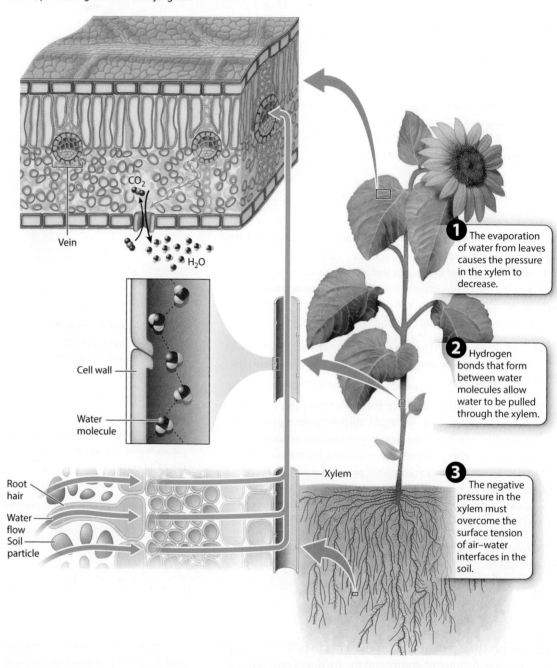

FIG. 27.13 Xylem transport from roots to leaves. Water is pulled through the plant by forces generated in the leaves, preventing them from drying out.

① The evaporation of water from leaves causes the pressure in the xylem to decrease.

② Hydrogen bonds that form between water molecules allow water to be pulled through the xylem.

③ The negative pressure in the xylem must overcome the surface tension of air–water interfaces in the soil.

Xylem transport is at risk of conduit collapse and cavitation.

Two distinct risks arise when the tensions in the xylem are large (**Fig. 27.14**). The first is the danger of collapse. If you suck too hard on a drinking straw, it collapses inward, blocking flow. Much the same thing can happen in xylem (Fig. 27.14a). To prevent collapse, xylem conduits have thick walls that contain lignin. Although lignin is more costly to produce than cellulose, lignin makes conduit walls rigid, reducing the risk of collapse.

A second way that xylem conduits can fail is through **cavitation**, which occurs when liquid water under tension is replaced by water vapor. Cavitation blocks flow because gases such as water vapor cannot transmit tensions. As a consequence, water can be pulled through the xylem only if a continuous column of water extends from the roots to the leaves. Cavitation in plants is thought to result from the expansion of tiny gas bubbles due to the tension in the water pulling outward on the bubble. Once a bubble starts to grow, it continues to expand until it fills the entire conduit. Initially, the gas is water vapor, but soon diffusion results in the conduit becoming filled with air.

One way to maintain liquid water in the xylem is to prevent bubbles from entering in the first place. Root tissues prevent the passage of any water bubbles that enter the xylem from the soil. However, bubbles can become introduced into the xylem in two other ways.

First, if xylem tensions are excessive, tiny air bubbles that can "seed" cavitation may be pulled in through a pit (Fig. 27.14b). Cavitation is more likely to develop in this way during drought, when the tensions in the xylem are large.

Second, bubbles can form in the xylem when gas comes out of solution during freezing (Fig. 27.14c). Gases are much less soluble in ice than in water, so as a conduit freezes, dissolved gas molecules aggregate into tiny bubbles. These bubbles can then expand when the conduit thaws and the water again experiences tensions. The wide vessels found in angiosperms are especially vulnerable to cavitation at freezing temperatures, which partly explains why the boreal forests of the subarctic contain few angiosperms.

The xylem is organized in ways that minimize the impact of cavitation. For example, xylem consists of many conduits in parallel, so the loss of any one conduit to cavitation does not result in a major loss of transport capacity. Similarly, as water flows from the soil to the leaves, it passes from one conduit to another, each time by flowing across pits. The likelihood that cavitation will spread is thereby reduced because only large pressure differences can pull air through pits.

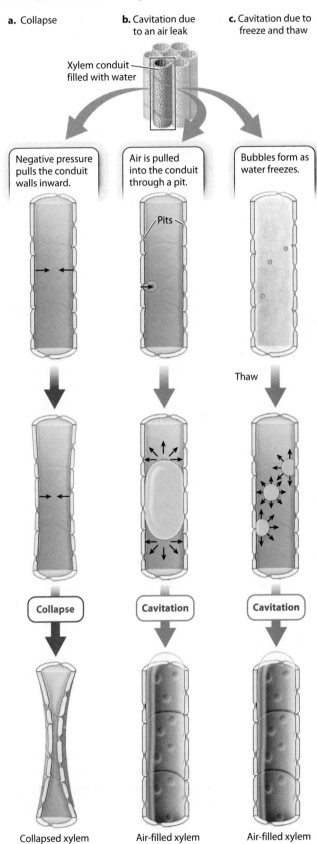

FIG. 27.14 Collapse and cavitation. The negative pressures exerted by transpiring leaves can cause xylem conduits to collapse inward or to cavitate, in both cases blocking flow.

Self-Assessment Question

6. How is the statement that water is transported through the xylem without requiring any input in energy by the plant both correct and incorrect?

27.4 TRANSPORT OF CARBOHYDRATES

Central to the success of vascular plants is their ability to pull water from the soil. But to do that, vascular plants need roots, and these roots need to be able to grow belowground. Without sunlight to power photosynthesis, growing roots depend on carbohydrates imported from aboveground. Vascular plants have evolved a second transport tissue, the phloem, capable of moving carbohydrates efficiently across the entire length of the plant. The phloem supports not only the growth of roots, but also the development of non-photosynthetic organs such as reproductive structures and stems.

Phloem transports carbohydrates from sources to sinks.

Whereas the direction of xylem transport is typically from the soil to the leaves, phloem moves carbohydrates from wherever they are synthesized or stored to wherever they are needed to support growth and respiration. In plants, regions that produce or store carbohydrates are referred to as **sources.** For example, leaves are sources because they produce carbohydrates by photosynthesis, and mature potato tubers (the part we eat) are sources because they have stored carbohydrates that can be supplied to the rest of the plant body when needed. **Sinks** are any portion of the plant that needs carbohydrates to fuel growth and respiration; examples are roots, young leaves, and developing fruits. Storage organs can also act as sinks—for example, developing potato tubers. Thus, the direction of phloem transport can be either up or down, depending on where the source and the sink are relative to each other.

In contrast to the xylem, phloem transport takes place through living cells containing cytoplasm. During development, these cells lose much of their intracellular structure, including the nucleus and the vacuole. At maturity, phloem transport cells retain an intact plasma membrane that encloses a modified cytoplasm containing only smooth endoplasmic reticulum and a small number of organelles, including mitochondria. Cellular functions such as protein synthesis are carried out by neighboring cells (called **companion cells** in angiosperms). Numerous plasmodesmata connect the phloem transport cells to these neighbors.

The cytoplasm of phloem transport cells is rich in sugars, 10% to 30% by weight. For comparison, a can of regular Coke has a sugar content of approximately 10%. These sugars and other solutes flow from one cell to another through pores connecting the cells. **Phloem sap** refers to the soluble components of the cytoplasm that can pass through these pores. In angiosperms, phloem sap moves though **sieve tubes** composed of cells that are connected by **sieve plates,** which are modified end walls with large pores (1–1.5 μm in diameter) (**Fig. 27.15**). The plasma membrane of adjacent cells is continuous through each of these pores, so each multicellular sieve tube can be considered a single cytoplasm-filled compartment.

Plants that are not angiosperms have narrow phloem conduits that lack well-defined sieve plates. Their capacity for

FIG. 27.15 Phloem structure. Phloem conduits are formed of living cells. (a) Sieve tubes and associated companion cells and (b) scanning electron microscope image showing sieve plates of squash (*Cucurbita maxima*). *Photo source: Courtesy D. L. Mullendore and M. Knoblauch, Washington State University.*

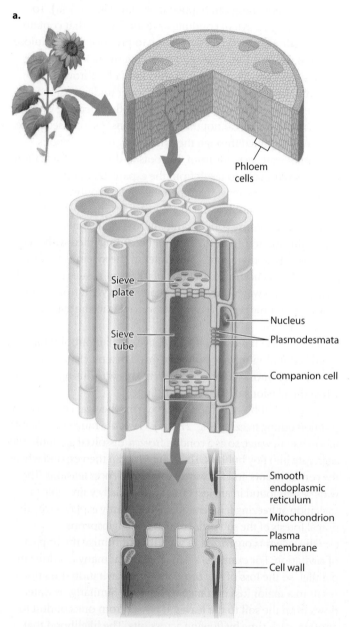

phloem sap transport appears to be much lower than that of angiosperms, paralleling the differences in the xylem between tracheids and vessels. We have seen that the more conductive xylem of angiosperms allows higher rates of photosynthesis. Their more conductive phloem allows angiosperms to transport the additional sugars produced at these higher rates to the rest of the plant.

Phloem transports carbohydrates as sucrose or larger sugars. Phloem also transports amino acids, inorganic forms of nitrogen, and ions including K^+, which are all present in much lower concentrations than the carbohydrates. Finally, phloem transports informational molecules such as hormones, protein signals, and even RNA. Thus, phloem forms a multicellular highway for moving raw materials and signaling molecules through the entire length of the plant.

Carbohydrates are pushed through phloem by an osmotic pump.

How does phloem transport sugars from source to sink? The overall mechanism results from the buildup of sugars in sources relative to a lower concentration of sugars in sinks. In sources, a high concentration of sugars in sieve tubes causes water to be drawn into the phloem by osmosis. Because the cell walls of the sieve tube resist being stretched outward, the turgor pressure (Chapter 5) at the source end increases. At sinks, sugars are transported out of the phloem into surrounding cells. This withdrawal of sugars causes water to leave the sieve tube, again by osmosis, reducing turgor pressure at the sink end. It is the difference in turgor pressure that drives the movement of phloem sap from source to sink (**Fig. 27.16**).

In some plants, sugars are actively transported into sieve tubes using energy from ATP. In others, sugars diffuse passively from the sites of photosynthesis into the sieve tubes as the sugar concentration builds up in photosynthesizing cells. These mechanisms are quite effective. Indeed, the pressures that develop in sieve tubes are approximately 100 times greater than the pressures that occur in our own circulatory system.

In sinks, sugars are thought to exit sieve tubes by diffusion, although this remains an area of active research. Sink cells use up sugars, either in respiration or to synthesize larger molecules such as cellulose or starch, so their sugar concentration is lower than that in the phloem. This difference in concentration provides a driving gradient for diffusion into sink cells. The water that exits with the sugars can remain within the sink or it can be carried back to the leaves in the xylem (Fig. 27.16). The volume of water that moves through the phloem, however, is tiny compared to the amount that must be transported through the xylem to replace water lost by transpiration. Therefore, the number and size of xylem conduits greatly exceeds the number and size of sieve tubes.

Like the xylem, the phloem is subject to risks that arise from the way the flow is generated. The contents of damaged sieve tubes can leak out, pushed by high turgor pressures in the phloem. Damage is an ever-present danger because the sugar-rich phloem is an attractive target for insects. Cell damage activates sealing mechanisms that block flow through sieve plates, thereby preventing phloem sap from leaking out. In some respects, these mechanisms are comparable to the formation of blood clots in humans, except that phloem can seal itself much more rapidly, typically in less than a second.

Phloem feeds both the plant and the rhizosphere.

With the exception of a few cell types that lack cytoplasm, all the cells in a plant's body contain mitochondria that carry out respiration to provide a constant supply of ATP. Typically, about 50% of the carbohydrates produced by photosynthesis in one day are converted back to CO_2 by respiration within 24 hours.

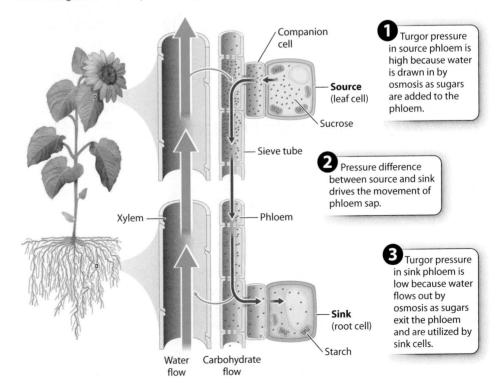

FIG. 27.16 Phloem transport from source to sink. Sources produce carbohydrates or supply them from storage, and sinks require carbohydrates for growth and respiration.

❶ Turgor pressure in source phloem is high because water is drawn in by osmosis as sugars are added to the phloem.

❷ Pressure difference between source and sink drives the movement of phloem sap.

❸ Turgor pressure in sink phloem is low because water flows out by osmosis as sugars exit the phloem and are utilized by sink cells.

Carbohydrates that are not immediately consumed in respiration can be used as raw materials for growth, or they can be stored for later use. Carbohydrates may be stored as starch within roots and stems or within tubers (specialized storage organs such as potatoes). Stored carbohydrates can support new growth in the spring or following a period of drought, and they can replace leaves consumed by insects or grazing mammals.

What determines how carbohydrates become distributed within the plant? Phloem transport to reproductive organs appears to have priority over transport to stems and developing leaves, and transport to stems and developing leaves has priority over transport to roots. In Chapter 29, we discuss the role that hormones play in controlling the growth and development of plants. These hormones may influence the ability of different sinks to compete successfully for resources transported in the phloem.

Phloem also supplies carbohydrates to fungi and bacteria that form symbiotic associations with roots (section 27.5). A fraction of the carbohydrates transported to the roots even spills out into the **rhizosphere,** the soil layer that surrounds actively growing roots. Nourished by these carbohydrates, a dense population of soil microbes flourishes near roots. These soil bacteria decompose soil organic matter rich in nutrients such as nitrogen and phosphorus. By releasing carbohydrates into the soil, roots are thought to acquire more nutrients from the soil.

> ### Self-Assessment Questions
>
> 7. How is phloem able to transport carbohydrates from the shoot to the roots, as well as from the roots to the shoot (but not at the same time)?
> 8. How are the xylem and the phloem similar? How are they different?
> 9. Why is phloem necessary to produce roots?

27.5 UPTAKE OF WATER AND NUTRIENTS

As anyone who has dug a hole in a forest or a field can tell you, plants make a lot of roots (**Fig. 27.17**). Laid end to end, the roots of a single corn plant would extend for more than 600 km. Why do plants make such an extensive investment belowground? A major reason is that, with the important exceptions of CO_2 and sunlight, everything that a vascular plant needs to build and sustain its body enters through its roots.

So far, we have focused on the importance of roots for obtaining water from the soil. To carry out photosynthesis, plants must also acquire the nutrients needed to produce the enzymes of the Calvin cycle and the components of the photosynthetic electron chain. Thus, as vascular plants evolved roots able to obtain water from the soil, they evolved mechanisms for these roots to acquire nutrients as well.

FIG. 27.17 Roots and root hairs. Extensive branching of roots (a) and the presence of root hairs (b) create a large surface area in contact with the soil. *Sources: a. Topic Photo Agency/AGE Fotostock; b. NHPA/Superstock.*

Plants obtain nutrients from the soil.

On average, plants are 85% water and only 15% dry material. Approximately 96% of the dry material, by weight, is made up of three elements: carbon, oxygen, and hydrogen. The remaining 4% comes from nutrients that plants obtain from the soil. Nutrients play essential roles in plant metabolism and structure (**Table 27.1**). A deficiency in any one of them results in a plant that grows poorly and has a low rate of photosynthesis.

The essential nutrients can be divided up into four groups. Nitrogen, phosphorus, and sulfur are components of many organic molecules, including amino acids, nucleic acids, and phospholipids. Another group of nutrients remain in an ionic (charged) form and play important roles as enzyme cofactors and second messengers (for example, calcium) and in maintaining the osmotic balance of cells (for example, potassium). The magnesium ions present in chlorophyll are critical for photosynthesis. Metal elements such as iron and copper are primarily electron donors and acceptors in reactions that transfer electrons from one molecule to another. The final group includes elements that contribute to the structural integrity of the cell wall, such as calcium, silica, and boron.

Nutrient uptake by roots is highly selective.

Nutrients from the soil enter roots dissolved in water and are transported from the roots to the rest of the plant in the xylem. Xylem, however, is never in direct contact with the soil. The living cells that surround the xylem allow roots to select which nutrients move from the soil into the xylem. Thus, the nutrient

TABLE 27.1 Nutrients Essential for Plant Metabolism and Structure

ELEMENT	CONCENTRATION IN DRY MATTER (%)	FUNCTIONS
Nutrients covalently bonded with carbon compounds		
Nitrogen	1.5	Component of amino acids, nucleic acids, nucleotides, and coenzymes
Phosphorus	0.2	Component of nucleic acids, nucleotides, coenzymes, and phospholipids; key role in reactions involving ATP
Sulfur	0.1	Component of two amino acids, coenzyme A, and other essential organic compounds
Nutrients that remain in ionic form		
Potassium	1.0	Cofactor for many enzymes, major cation in cell osmotic balance
Calcium	0.5	Cofactor for enzymes involved in hydrolysis of ATP and phospholipids, second messenger in metabolic regulation, important structural role in cell walls
Magnesium	0.2	Enzyme cofactor, component of chlorophyll
Chlorine, zinc, sodium	each ≤0.01	Cofactor for enzymes involved in photosynthesis and other reactions
Nutrients involved in redox reactions		
Iron, manganese, copper, nickel, molybdenum	each ≤0.01	Component of enzymes that catalyze electron transfer (for example, in the electron transport chains used in cellular respiration and photosynthesis)
Nutrients present in cell walls		
Silica, boron	0.1, <0.01	Contribute to the structural integrity of cell walls and defense against herbivores

Data from Table 5.2 in Lincoln Taiz et al., *Plant Physiology and Development*, 6th ed., 2015, Sinauer Associates, Inc.; Table 8.1 in Emmanuel Epstein and Arnold J. Bloom, *Mineral Nutrition of Plants: Principles and Perspectives*, 2nd ed., 2004, Sinauer Associates, Inc.

composition of the water in the xylem differs markedly from that of the water that can be easily extracted from the soil. Some elements, such as nitrogen and phosphorus, are more concentrated within the xylem, whereas others, such as sodium and calcium, are more abundant in the soil solution. Given that water travels in a continuous pathway from the soil to the leaves, how are roots able to concentrate some nutrients and exclude others?

The uptake of water and nutrients takes place primarily in young, actively growing roots. Like leaves and stems, newly formed roots have an outer cell layer of epidermis (**Fig. 27.18**). Epidermal cells near the root tip produce slender outgrowths known as **root hairs.** Root hairs greatly increase the surface area of the root available for absorbing nutrients. Inside the epidermis is the **cortex,** composed of parenchyma cells. Depending on the species, the cortex can vary in thickness from a single cell to many cells. Xylem and phloem are located at the center of the root and are surrounded by the **endodermis,** a layer of cells that acts as a gatekeeper controlling which nutrients move into the xylem.

In Chapter 5, we learned how cells control the movement of molecules across their plasma membranes. Ions and other charged molecules diffuse passively across cell membranes at a relatively slow rate. Alternatively, they can be transported across cell membranes by proteins embedded in the lipid bilayer. By coupling the hydrolysis of ATP to the activity of such transport proteins, cells can both accumulate desired nutrients and remove unwanted ones. These protein transporters form the basis for how the root functions as a living filter.

Epidermal and root cortex cells use protein transporters to increase the uptake of nutrients that the plant needs and to actively excrete undesired solutes. Once in the cytoplasm, nutrients can move from cell to cell via plasmodesmata. In addition, water and nutrients can move through the relatively porous walls that surround each cell. To prevent these "unfiltered" nutrients from reaching the xylem, each cell of the endodermis is encircled by a **Casparian strip,** a thin band of hydrophobic material that prevents water and any dissolved materials from moving through the cell wall. Water can still flow through

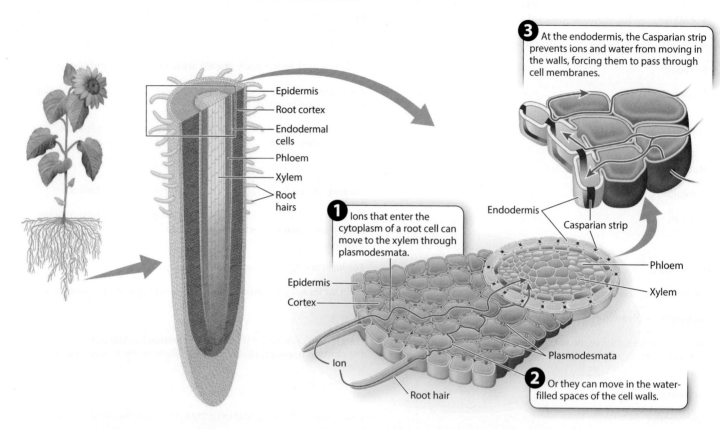

FIG. 27.18 Water movement through endodermal cells. As water carrying nutrients flows from the soil to the xylem of roots, it must pass through the plasma membranes of endodermal cells, which act as a selective barrier.

the endodermis by crossing the plasma membranes of the endodermal cells, but whether a specific solute will be taken up or excluded will depend on which protein transporters are present. As a result, roots can concentrate nutrients that are needed in large amounts (for example, nitrogen) in the xylem and exclude others that can damage plant cells (for example, sodium).

High concentrations of salts in the soil make it difficult for plants to obtain water. The reason is that to prevent salt from building up to toxic levels in leaves, plants must prevent salt from entering the xylem. As a result, salt concentrations build up in and around roots. Plants must then pull harder to overcome the osmotic effects of these salts if they are to obtain water from the soil. Thus, salt buildup in the soil resulting from poor irrigation practices may cause crop yields to plummet.

Nutrient uptake requires energy.

Roots expend a great deal of energy acquiring nutrients from the soil. We have already seen that energy is required to move nutrients across the cell membranes of the endodermis on their journey to the xylem. Energy is also required to make certain nutrients in the soil accessible to the root. The mitochondria of root cells must, therefore, respire at high rates to provide the energy needed for nutrient uptake. Their high respiration rates explain why plants are harmed by flooding: waterlogged soils slow the diffusion of oxygen, limiting respiration.

To understand how roots acquire nutrients and why this process requires energy, let's first look at how nutrients move through the soil. Nutrients move by diffusion through the films of water that surround individual soil particles. The active uptake of nutrients by root cells decreases nutrient concentrations at the root surface. As a result, a concentration gradient forms between the bulk soil and the root surface that drives the diffusion of nutrients to the root. However, because diffusion is efficient only over short distances, each root is able to acquire nutrients only from a small volume of soil. This limitation explains why plants produce so many roots and have so many root hairs: both structures increase the volume of soil from which a root can obtain nutrients. It also explains why roots elongate more or less continuously, exploiting new regions of soil to obtain nutrients required for growth.

Nutrients vary dramatically in the ease and speed with which they move through the soil. For example, nitrate (NO_3^-) and sulfate (SO_4^{2-}) are extremely mobile because they do not bind to soil particles. When transpiration rates are high, the flow of water through the soil can help to convey such mobile nutrients toward the root. In contrast, zinc and inorganic phosphate stay firmly attached to soil particles. One way that roots gain access to highly immobile nutrients is by releasing protons or compounds that make the soil environment close to the root more acidic. This requires energy in the form of ATP, but the decrease in pH weakens the attachment of immobile nutrients to soil particles. As a result, these formerly immobile nutrients are able to move by diffusion to the root.

Soil microorganisms—both fungi and bacteria—are more adept than plants at obtaining nutrients. Fungi (Chapter 32) are particularly good at mobilizing phosphorus and at decomposing organic matter, which releases nitrogen. Meanwhile, some bacteria can convert gaseous nitrogen (N_2) into a chemical form that can be incorporated into proteins and nucleic acids (Chapter 24). What plants have at their disposal is a renewable source of carbohydrates. Using these carbohydrates, plants support the growth of symbiotic fungi and bacteria, in effect exchanging carbon for phosphorus or nitrogen.

Mycorrhizae enhance nutrient uptake.

Plants can increase their access to soil nutrients by forming partnerships with fungi. **Mycorrhizae** are symbioses between roots and fungi that enhance nutrient uptake: the plant provides carbohydrates produced by photosynthesis, and the fungus provides nutrients it has obtained from the soil.

Mycorrhizae are of two main types (**Fig. 27.19**). **Ectomycorrhizae** produce a thick sheath of fungal cells that surrounds the root tip. Some of the fungus extends into the interior of the root, producing a dense network of filaments that surrounds individual root cells. Fungi and root cells exchange nutrients across this interface. In contrast, **endomycorrhizae**

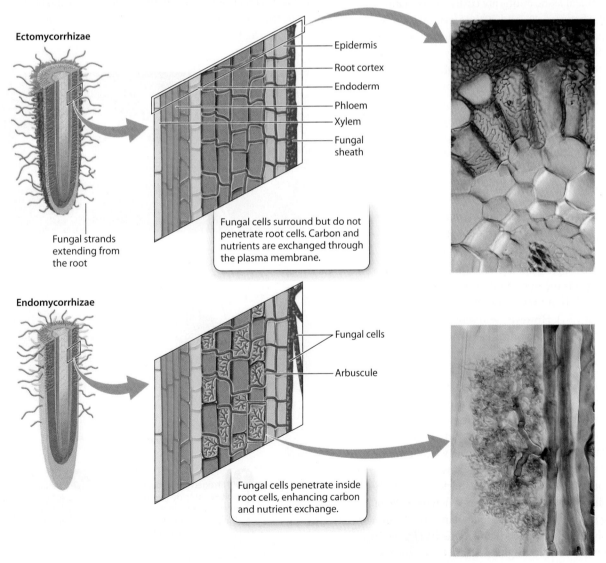

FIG. 27.19 Mycorrhizae. These fungus–root associations allow plants to gain nutrients from the soil and fungi to obtain carbohydrates. *Photo sources: Mark Brundrett.*

do not form structures that are visible on the outside of the root. Instead, they form highly branched structures, called arbuscules, that grow inside root cells and provide a large surface area for nutrient exchange. Both types of mycorrhizae produce an extensive network of hyphae that extend outward into the soil.

How do fungi enhance the ability of roots to obtain nutrients from the soil? Fungal networks can extend more than 10 cm from the root surface, greatly increasing the absorptive surface area in contact with the soil. And because fungal filaments are so thin, they can penetrate tiny spaces between soil particles and access nutrients that would otherwise be unavailable to roots. Another way fungi enhance nutrient uptake is by secreting enzymes that make soil nutrients more available. Endomycorrhizae are important in enhancing phosphorus uptake, while ectomycorrhizal fungi provide host plants with access to nitrogen that would be otherwise unavailable. Both kinds of mycorrhizae may also offer other benefits to plants. For example, mycorrhizal roots may be less susceptible to infection by pathogens.

Mycorrhizal associations are costly to the plant, consuming between 4% and 20% of its total carbohydrate production. The plant has some control over whether the roots become colonized by mycorrhizal fungi: when nutrients are readily available, as in a well-fertilized garden or field, the percentage of mycorrhizal roots decreases. In natural ecosystems, soil nutrients are usually not so abundant, and mycorrhizal roots are the norm rather than an exception. Fossils from more than 400 million years ago show that interactions between plants and fungi are ancient. Perhaps fungi helped plants gain a foothold on land.

Symbiotic nitrogen-fixing bacteria supply nitrogen to both plants and ecosystems.

Nitrogen is the nutrient that plants must acquire in greatest abundance from the soil, and in many ecosystems nitrogen is the nutrient that most limits growth. Nitrogen itself is not rare; 78% of Earth's atmosphere is nitrogen gas (N_2). However, as discussed in Chapter 24, plants and other eukaryotic organisms cannot use N_2 directly. Only certain bacteria and archaeons are able to convert N_2 into a biologically useful compound in the process of **nitrogen fixation**. Nitrogen-fixing bacteria transform N_2 into forms such as ammonia (NH_3) that can be used to build proteins, nucleic acids, and other compounds. The bacteria release a portion of these usable forms of nitrogen into the soil, where they are available to plants and other organisms.

Although many nitrogen-fixing bacteria live freely in the soil, some live in plant roots. When the level of available soil nitrogen is low, some plants form symbiotic relationships with nitrogen-fixing bacteria (**Fig. 27.20**). In response to chemical signals released from the root, bacteria multiply within the rhizosphere. At the same time, chemical signals from the bacteria trigger root cells to divide, leading to the formation of a pea-sized protrusion called a **root nodule**. Nitrogen-fixing bacteria enter the root either through root hairs or through breaks in the epidermis. They then move into the center of the nodule and begin to fix nitrogen. Even though the bacteria appear to have taken up residence within the cytoplasm of nodule cells, each bacterial cell is actually contained within a membrane-bound vesicle produced by the host plant cell.

The plant supplies its nitrogen-fixing partners with carbohydrates transported by the phloem and carries away the products of nitrogen fixation in the xylem. As in mycorrhizal symbioses, the carbon cost to the plant is high: as much as 25% of the plant's total carbohydrate goes to nitrogen-fixing bacteria. Relatively few plant species have evolved symbiotic associations with nitrogen-fixing bacteria, but those that have—especially

FIG. 27.20 Symbiosis between roots and nitrogen-fixing bacteria. (a) Symbiotic root nodules on alfalfa; (b) the formation of a root nodule. *Photo source: Inga Spence/Science Source.*

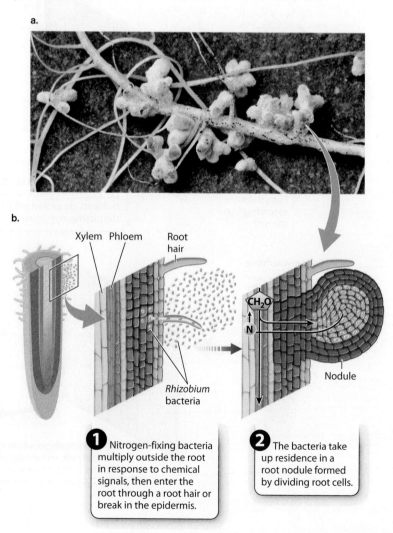

members of the legume (bean) family—represent an important way in which biologically available nitrogen enters into ecosystems. In addition, nitrogen fixation in legume roots helps to sustain fertility in many agricultural systems.

CASE 6 AGRICULTURE: FEEDING A GROWING POPULATION

How has nitrogen availability influenced agricultural productivity?

Harvesting crops removes nutrients, including nitrogen, from an ecosystem. These nutrients must be returned if the land is to remain fertile. In the early days of agriculture, farmers left fields unplanted for long periods between crops. During these periods, nitrogen-fixing bacteria replenished soil nitrogen, either in symbiotic association with roots or as free-living bacteria in the soil.

As populations grew, fields could not be left unplanted for long periods. Organic fertilizers such as crop residues, manure, and even human feces recycled some of the lost nitrogen. Unless they are brought in from off the farm, however, they cannot replace all the nitrogen removed during harvest.

Crop rotation is one way to sustain soil fertility. Every few years, a field is planted with legumes harboring nitrogen-fixing bacteria. The legume plants are not harvested, but instead are treated as "green manure" that is allowed to decay and replenish soil nitrogen. During the eighteenth and nineteenth centuries, legume rotation contributed to the higher crop yields needed to support increasing populations. Therefore, along with the mechanization of agriculture, nitrogen-fixing symbioses contributed to the social conditions that fueled the Industrial Revolution.

By the second half of the nineteenth century, more abundant and concentrated supplies of nitrogen fertilizers were sought. Mining of guano and sodium nitrate mineral deposits helped meet the demand for nitrogen, but with time these natural sources of nitrogen were depleted. Some way was needed to obtain nitrogen fertilizer from the vast storehouse of nitrogen in the atmosphere.

German chemist Fritz Haber developed a laboratory method for fixing nitrogen in the years 1908–1911, and Karl Bosch subsequently adapted it for industrial production. Their method allows humans to achieve what, until then, had been accomplished only by prokaryotes. Like nitrogen fixation by bacteria, the Haber–Bosch process requires large inputs of energy; unlike nitrogen fixation by bacteria, it occurs only at high temperatures and pressures. The prospect of an essentially unlimited supply of nitrogen in a form that plants can readily use has proved irresistible. Today, industrial fertilizers produced by the Haber–Bosch process account for more than 99% of all fertilizers used. Industrial fixation of nitrogen is approximately equal to the entire natural fixation of nitrogen, which is largely due to bacteria, although a small amount of nitrogen is fixed by lightning.

As we discuss in Chapter 48, human use of fertilizer is altering the nitrogen cycle on a massive scale, posing a threat to biodiversity and the stability of natural ecosystems. The problems arise when nitrogen added as fertilizers leaks out of agricultural fields and enters into either groundwater or, carried through the air, uncultivated areas. Nitrogen addition to terrestrial ecosystems can cause a variety of unwanted effects. For example, it may alter species composition by favoring the growth of nitrogen-demanding and often weedy species, or it may alter soil pH, leading to the loss of cations such as calcium and magnesium needed as nutrients. The compounds used in fertilizer may also release nitrogen-containing greenhouse gases. Nitrogen runoff into aquatic systems can stimulate algal blooms and reduce the quality of drinking water.

Nevertheless, nitrogen fertilizer is a cornerstone of modern agriculture. The high-yielding varieties of corn, wheat, and rice introduced in the mid-twentieth century result in high levels of grain production only when those crops are abundantly supplied with nitrogen. The fact that plants such as legumes (including beans, soybeans, alfalfa, and clover) form symbiotic relationships with nitrogen-fixing bacteria suggests a possible way around this problem. It might be possible to engineer nitrogen fixation into crops such as wheat and rice, either by inducing them to form symbiotic relationships with nitrogen-fixing bacteria or by transferring the genes that would enable them to fix nitrogen on their own. But would that eliminate the need for nitrogen fertilizers? The answer is yes, but with a caveat. As we have seen, nitrogen fixation requires a large amount of energy, diverting resources that plants could otherwise use to support growth and reproduction.

Self-Assessment Question

10. What are four adaptations of roots that enhance the uptake of nutrients from the soil?

CORE CONCEPTS SUMMARY

27.1 PHOTOSYNTHESIS ON LAND: A major challenge in the evolution of plants was being able to acquire carbon dioxide for photosynthesis without drying out.

Land plants evolved within the group of green algae. Their cells share many features with green algal cells, including cell walls containing cellulose, large membrane-bound vacuoles, and chloroplasts containing chlorophyll *a* and *b*. page 591

Bryophytes balance CO_2 gain and water loss passively, photosynthesizing in wet conditions and tolerating desiccation in dry ones. page 592

Vascular plants maintain the hydration of their photosynthetic cells with water pulled from the soil. page 593

Vascular plants produce four organ types: roots, stems, leaves, and reproductive organs. All four organ types are built from the same three tissue layers: epidermis, ground, and vascular. page 594

27.2 CARBON DIOXIDE GAIN AND WATER LOSS: Leaves have a waxy cuticle and stomata, which allow plants to acquire carbon dioxide while limiting water loss.

The low concentration of CO_2 in the atmosphere forces plants to expose their photosynthetic cells directly to the air. The outward diffusion of water vapor leads to a massive loss of water. page 595

The waxy cuticle on the outside of the epidermis slows rates of water loss from leaves but also slows the diffusion of CO_2 into leaves. page 596

Stomata are pores in the epidermis that allow CO_2 to diffuse into the leaf, but also allow water vapor to diffuse out of the leaf. page 596

Stomata can open or close to regulate the amount of water leaving the leaf. page 596

CAM plants capture CO_2 at night when evaporative rates are low. During the day, they close their stomata and use this stored CO_2 to supply the Calvin cycle, increasing the exchange ratio of CO_2 and water. page 597

C_4 plants suppress photorespiration by concentrating CO_2 in bundle-sheath cells. The buildup of CO_2 in bundle-sheath cells results from the rapid production of C_4 compounds in mesophyll cells. The C_4 compounds diffuse into the bundle-sheath cells and release CO_2 that can be used in the Calvin cycle. page 599

27.3 WATER TRANSPORT: Xylem transports water and dissolved nutrients from the soil, allowing leaves to remain hydrated.

Water flows through xylem conduits from the soil to the leaves. page 601

Xylem conduits are formed from cells that have thick, lignified cell walls and which lose their cell contents as they mature. Water flows into xylem conduits across small thin-walled regions called pits. page 601

Xylem conduits are of two types: tracheids and vessels. Tracheids are unicellular xylem conduits; vessels are formed from many cells. page 601

As water evaporates from leaves, water is drawn toward the sites of evaporation from the xylem. As water exits the xylem, the pressure falls, generating the driving force for water movement in the xylem. Because of the strong hydrogen bonds between water molecules, water can be pulled from the soil and transported through the xylem. page 602

Pulling water through the xylem creates the risk of mechanical failure, either by the collapse inward of conduit walls or by cavitation, in which a gas bubble expands to fill the entire conduit. page 605

27.4 TRANSPORT OF CARBOHYDRATES: Phloem transports carbohydrates throughout the plant for use in growth and respiration.

Carbohydrates are transported in the phloem through living cells. These cells lack a nucleus or vacuole at maturity. page 606

In angiosperms, phloem transport cells are connected end to end by sieve plates, forming multicellular sieve tubes. Phloem sap flows from one sieve tube cell to another through large pores in the sieve plates. page 606

High solute concentration brings water into sieve tubes by osmosis, increasing the turgor pressure. At sites of use, the removal of solutes leads to an outflow of water and a drop in turgor pressure. page 607

The difference in turgor pressure drives the movement of phloem sap from source to sink. page 607

Carbohydrates released from roots provide carbon and energy sources for the rhizosphere microbial community, stimulating decomposition of organic matter and increasing nutrient availability. page 608

27.5 UPTAKE OF WATER AND NUTRIENTS: Roots expend energy to obtain nutrients from the soil.

The endodermis allows roots to control which dissolved solutes from the soil enter the xylem. The uptake of solutes by roots is therefore highly selective. page 609

Plants exchange carbohydrates with symbiotic fungi in return for assistance obtaining nutrients from the soil. page 611

Plants supply carbohydrates to symbiotic nitrogen-fixing bacteria in exchange for usable forms of nitrogen. page 612

Agricultural productivity is closely linked to nitrogen supply. page 613

Log in to LaunchPad to check your answers to the Self-Assessment Questions and to access additional learning tools.

CHAPTER 28
Plant Reproduction
Finding Mates and Dispersing Young

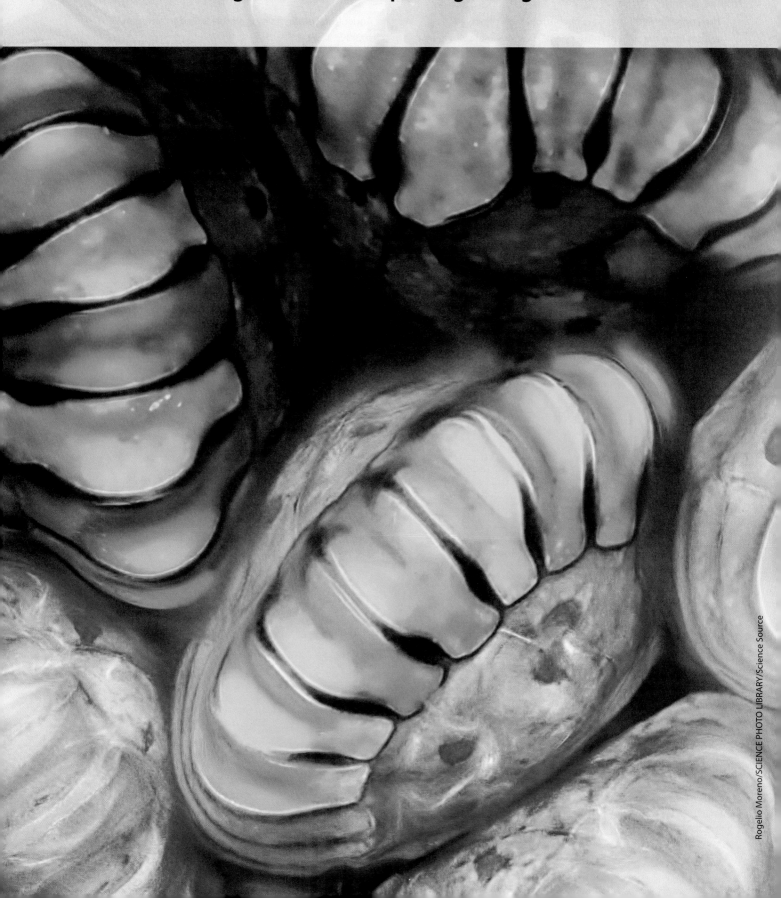

CORE CONCEPTS

28.1 ALTERNATION OF GENERATIONS: Alternation of generations evolved by the addition of a diploid sporophyte generation that allows plants to disperse spores through the air.

28.2 SEED PLANTS: Pollen allows the male gametophyte to be transported through the air, while resources stored in seeds support the development of the embryo.

28.3 FLOWERING PLANTS: Angiosperms produce flowers and fruits that increase the efficiency of pollination and seed dispersal, and provide resources for seeds after fertilization.

28.4 ASEXUAL REPRODUCTION: Many plants can reproduce asexually.

As plants evolved on land, the challenges of carrying out photosynthesis in air were matched by the difficulties of completing their life cycle. The algal ancestors of land plants relied on water currents to carry sperm to egg and to disperse their offspring. On land, however, plants are surrounded by air. Air is less buoyant than water, provides a poor buffer against changes in temperature and ultraviolet radiation, and increases the risk of drying out. Thus, as plants diversified on land, structures evolved that would allow gametes and offspring to survive being transported through air. For these structures to be useful, plant life cycles had to undergo radical change. The modification of land plant life cycles is a major theme in plant evolution.

The gametes and offspring of many land plants are carried passively by the movement of the air (or in a few cases, water—think coconuts). However, some plants evolved the ability to use animals as transport agents. In particular, the flowers and fruits of angiosperms (flowering plants) attract animals through their colors, scents, and food resources. Animals attracted by flowers transport male gametes from one flower to another, while animals that eat fruits are key agents of seed dispersal. Coevolution with animals is a second major theme in plant evolution.

28.1 ALTERNATION OF GENERATIONS

Land plants are rooted in one place, yet must be able to disperse their offspring to minimize competition for space and resources with the parent plant. Early land plants evolved the capacity to disperse their offspring through the air, but they still required water for fertilization. How could these plants exist in both water and air? They couldn't—at least not at the same time. Instead, early land plants evolved a life cycle with two distinct multicellular generations: one generation released sperm into a moist environment, and the following generation dispersed offspring through the air (**Fig. 28.1**). A pattern of alternating generations, one haploid and one diploid, still exists in all plants today.

FIG. 28.1 The phylogeny of land plants. This phylogenetic tree shows the evolutionary trajectory of the environmental setting for fertilization and dispersal.

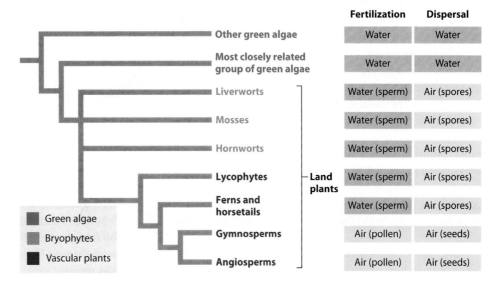

The algal sister groups of land plants have one multicellular generation in their life cycle.

Molecular sequence comparisons have identified three groups of green algae as being closely related to land plants. Although these three groups are diverse in form, their life cycles share many features. Here we use *Chara*, a member of one of these groups, to gain insights into the evolution of land plant life cycles (**Fig. 28.2**).

In animals, the multicellular body consists of diploid ($2n$) cells. *Chara* has a multicellular body as well, shown in Fig. 28.2, but it consists entirely of haploid ($1n$) cells. In animals, haploid gametes are formed by meiosis. In *Chara*, haploid eggs and sperm are formed instead by mitosis. The egg is retained within the female reproductive organ, while the sperm are released into the water (**Fig. 28.3**). Fertilization occurs when one egg and one sperm fuse to form a diploid zygote. In *Chara*, the zygotes are released and dispersed by water currents until they settle. The zygote then undergoes meiosis to produce four haploid cells. One of these haploid cells develops into a new multicellular (haploid) individual representing the beginning of the next generation.

In *Chara*, both fertilization and dispersal take place in water. Because they live along the margins of lakes and streams, however, *Chara* are occasionally left high and dry by drought. Their photosynthetic bodies shrivel quickly, but these algae survive because their zygotes are surrounded by a tough outer wall. When water once again covers the resting zygotes, they undergo meiosis and produce new haploid algae. This is the condition from which the more complicated life cycles of land plants evolved.

Land plants have two multicellular generations in their life cycle.

In Chapter 27, we saw that bryophytes—liverworts, mosses, and hornworts—share many features related to their dependence on surface moisture for hydration. They also share many reproductive traits. Here, we consider the life cycle of *Polytrichum commune*, a moss commonly found in forests and the edges of fields. *Polytrichum* illustrates how the evolution of two multicellular generations, one specialized for fertilization and the other for dispersal, contributed to the ability of plants to colonize the land.

Like *Chara*, *Polytrichum* has a photosynthetic body—the familiar green tufts observed on the forest floor—that is made of

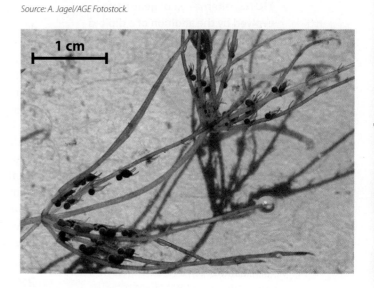

FIG. 28.2 *Chara*, **a green alga that is closely related to land plants.** Chara is an upright alga found along the margins of lakes and estuaries.
Source: A. Jagel/AGE Fotostock.

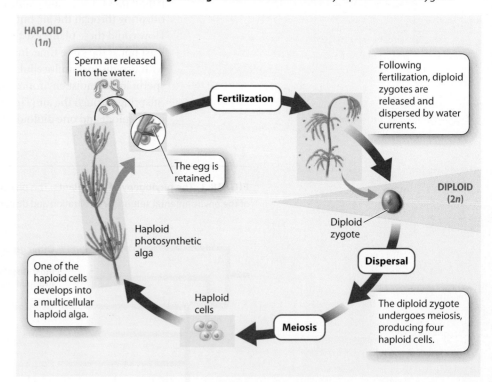

FIG. 28.3 The life cycle of the green alga *Chara corallina*. The only diploid cell is the zygote.

haploid cells and that forms gametes by mitosis. And like *Chara*, *Polytrichum* and other bryophytes release swimming sperm into the environment. For fertilization to occur, the sperm must swim through films of water that coat moist surfaces to reach eggs retained within reproductive organs.

FIG. 28.4 The life cycle of the moss *Polytrichum commune*. The multicellular sporophyte generation is a major innovation of land plants. *Photo source: Malcolm Storey, 2010, www.bioimages.org.uk.*

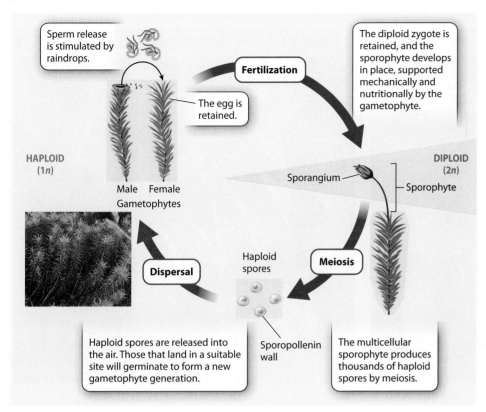

Bryophytes produce their gametes near the ground, where they are most likely to encounter a continuous film of water. They tend to be small, as sperm are able to swim only a few centimeters. Many bryophytes release their sperm only when agitated by raindrops. Raindrops signal the presence of surface water, and their impact can splash sperm farther than they could swim on their own.

It is after the fusion of egg and sperm gives rise to a diploid zygote that the life cycles of plants and their green algae relatives diverge. In plants, the zygote does not undergo meiosis, and it does not disperse. Instead, the zygote is retained within the female reproductive organ. There, it divides repeatedly by mitosis to produce a new multicellular plant—this one made of diploid cells (**Fig. 28.4**). Because the fertilized egg begins to develop while surrounded and supported by the maternal (that is, egg-producing) plant, it can be described as an embryo. Recall from Chapter 27 that the scientific name for land plants is Embryophyta. This name highlights the fact that in all land plants the fertilized egg is retained and develops in place, forming an embryo.

As the embryo develops into a mature diploid plant, some cells undergo meiosis. The haploid products of meiosis develop into **spores**, cells that disperse and give rise to new haploid individuals. Because the diploid multicellular plant produces spores, it is called the **sporophyte**, and this generation is called the sporophyte generation. In contrast, because the haploid multicellular plant produces gametes, it is called the **gametophyte**, and that generation is called the gametophyte generation. The resulting life cycle is called **alternation of generations** because a haploid gametophyte generation and a diploid sporophyte generation follow one after the other. All land plants exhibit alternation of generations.

Bryophytes illustrate how the alternation of generations allows the dispersal of spores in the air.

How might the new diploid generation have made it possible for plants to colonize land? To answer this question, let's look at the sporophyte of *Polytrichum*. Late in the growing season, the green tufts of this moss sprout an extension, consisting of a small brown capsule at the end of a cylindrical stalk a few centimeters high (**Fig. 28.5a**). This is the sporophyte, which originated from a fertilized egg. Because the fertilized egg was retained within the female reproductive organ, the sporophyte grows directly out of the gametophyte's body. Thus, the *Polytrichum* specimen pictured in Fig. 28.5a shows both a gametophyte (green) and several sporophytes (brown). The gametophyte is photosynthetically self-sufficient, but the sporophytes must obtain water and nutrients needed for their growth from the gametophyte.

The significance of the multicellular sporophyte is that it enhances the ability of plants to disperse on land. The capsule at the top of the *Polytrichum* sporophyte is a **sporangium**, a structure in which many thousands of diploid cells undergo meiosis, producing huge numbers of haploid spores. In contrast, fertilization in *Chara* is followed by meiosis in only a single cell (the zygote).

Spores are small and can be carried for thousands of kilometers by the wind. However, their tiny size puts spores at risk of being washed from the air by raindrops. The sporangia of many bryophytes release their spores only when the air is dry. *Polytrichum* releases spores through small openings in the sporangium the way a salt shaker releases salt. When the air is wet, the spores form clumps that are too big to fit through these openings. More commonly, a ring of teethlike projections curls over the top of the sporangia. These bend backward when air humidity is low, flinging the spores into the air (**Fig. 28.5b**).

FIG. 28.5 Moss sporangia. (a) When spores of *Polytrichum commune* are mature, a cap falls off, allowing the salt shaker–like sporangium to release the spores. (b) The sporangia of *Homalothecium sericeum* have a ring of teethlike projections that curl back as they dry, flinging spores into the air.
Sources: a. Keith Burdett/AGE Fotostock; b. Power and Syred/Science Source.

The spores are protected as they travel through the air. Earlier, we noted that the zygotes of *Chara* secrete a tough wall. This wall protects the zygote enclosed within it from environmental stresses such as ultraviolet radiation and desiccation. The wall contains **sporopollenin,** a complex mixture of polymers that is remarkably resistant to degradation. In mosses and other land plants, it is the spore, not the zygote, that secretes sporopollenin. This suggests that early land plants retained a capacity present in their algal ancestors, the ability to synthesize sporopollenin, but the timing of gene expression changed in these plants.

Following dispersal, most spores germinate within a few days. Some, however, can survive within the spore wall for months or even years. When conditions for growth are favorable, the spore wall ruptures, and the haploid cell inside can then develop into a multicellular and free-living gametophyte.

Dispersal enhances reproductive fitness in several ways.

With the evolution of the multicellular sporophyte generation, land plants were able to make many spores and disperse them across large distances on land. Dispersal is one of the most important stages in a plant's life cycle. To understand why, imagine what would happen if a plant's spores (or seeds) simply dropped straight down onto the ground below the parent plant. All the plant's offspring would then attempt to grow in a very small space that might contain enough nutrients and light for only a few to establish themselves. In contrast, offspring dispersed by wind or other means are more likely to settle in a spot where they will not be forced to compete with closely related individuals for resources.

In addition, dispersal allows offspring to avoid pathogens and parasites. Viruses and bacteria travel easily from individual to individual in a densely packed population. An individual that settles away from the parent plant is less likely to encounter a pathogen. Finally, dispersal allows offspring to colonize new habitats. The evolution of new structures to enhance dispersal on land is a major theme in plant evolution.

Spore-dispersing vascular plants have free-living gametophytes and sporophytes.

In Chapter 27, we saw that the evolution of vascular tissues allowed plants to grow tall. Xylem and phloem are present only in the sporophyte generation because spore dispersal is enhanced by height. In contrast, gametophytes must remain small and close to the ground to increase their chances of fertilization. A major change in the plant life cycle is that in vascular plants the sporophyte is the dominant generation, both because the sporophyte is physically larger than the gametophyte and because its photosynthetic production is much higher.

FIG. 28.6 The life cycle of the bracken fern *Pteridium aquilinum*. A free-living gametophyte generation, which remains small, alternates with a free-living vascularized sporophyte generation, which grows tall. *Photo sources: (left) Biophoto Associates/Science Source; (right) VOISIN/PHANIE/Getty Images.*

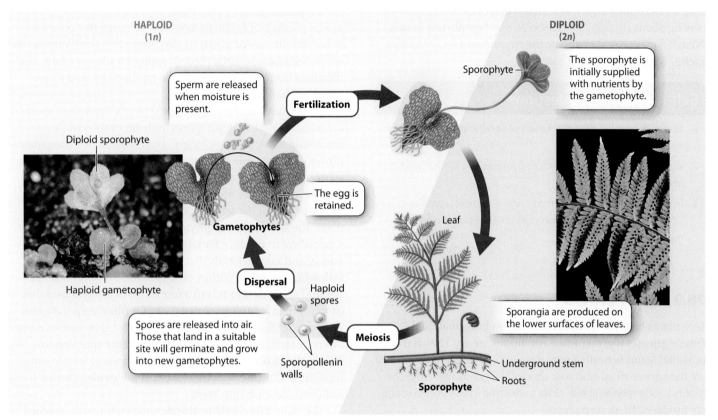

Two groups of vascular plants, the lycophytes and the ferns and horsetails, depend on swimming sperm for fertilization and disperse by spores that are released into the air (see Fig. 28.1). In these ways, their life cycle is similar to that of the bryophytes. Unlike bryophytes, however, both the gametophyte and the sporophyte generation are free-living, meaning that each is able to supply its own nourishment.

Let's examine the life cycle of the bracken fern *Pteridium aquilinum*. We begin with the largest component of the life cycle, the leafy diploid sporophyte generation (**Fig. 28.6**). If you examine a bracken leaf closely, you may find tiny brown packets along its lower margin. These are sporangia, and each contains diploid cells that undergo meiosis to generate haploid spores. Each spore becomes covered by a thick wall containing sporopollenin, making it well suited for dispersal through the air. The photo on the first page of this chapter shows fern sporangia (photographed through a microscope in ultraviolet light) that have not yet opened.

As the sporangia dry out, they break open, releasing the spores. Most spores travel for only short distances, but a few may end up far from their parent. Bracken ferns can be found on every continent except Antarctica, as well as on isolated oceanic islands. This distribution is testimony to the ability of spores to travel over long distances.

Spores that land in a favorable location can germinate and grow to form the haploid gametophyte generation. You have probably never seen a fern gametophyte—or never noticed one. Like most fern gametophytes, the bracken gametophyte is less than 2 cm long and only one to a few cells thick (Fig. 28.6). Because of its small size, one might assume that the haploid generation is a weak link in the fern life cycle. In reality, fern gametophytes often live longer and withstand stressful environmental conditions better than the much larger sporophytes. As discussed in Chapter 27, plants that lack roots must be able to tolerate drying out; this group includes both bryophytes and free-living fern gametophytes.

Fern gametophytes typically produce either male or female gametes. The female gametes (eggs) are retained within reproductive organs that remain attached to the gametophyte. The male gametes are released and swim to the egg of a nearby gametophyte through a film of water. The union of a male gamete and a female gamete forms a diploid zygote, which develops in place and is supported by the gametophyte as it begins to grow. Eventually, the developing sporophyte forms leaves and roots that allow it to become a physiologically independent diploid plant.

Ferns release swimming sperm, so they are able to reproduce only when conditions are wet. That is true of the other spore-dispersing plants (lycophytes, horsetails, and bryophytes) as well. What if a plant could eliminate the requirement for swimming sperm? Seed plants did just that.

> **Self-Assessment Questions**
>
> 1. How is the evolution of alternation of generations an adaptation for reproduction on land?
> 2. In what ways are spores similar to gametes, and in what ways do spores and gametes differ?
> 3. In what ways is fern reproduction similar to moss reproduction? In what ways is fern reproduction different from moss reproduction?

28.2 SEED PLANTS

Seed plants have the remarkable ability to bring male and female gametes together when conditions are dry. How is this possible? Stand beneath a pine tree on a spring day, and you will see the answer all around you: tiny yellow grains that cover the ground and everything else close to the tree. These pollen grains are produced by all seed plants.

In fact, in seed plants we encounter two new structures: pollen and seeds. Pollen is what liberates seed plants from the need to release swimming sperm into the environment, and seeds are multicellular structures adapted for dispersing offspring. The evolution of pollen and seeds required dramatic changes in the seed plant life cycle.

Seed plants form a monophyletic group. Their scientific name, Spermatophyta, comes from the Greek word for seed (*spermato* = seed, *phyte* = plant). There are two main types of seed plants: gymnosperms, which include conifers such as pine trees, and angiosperms, which are the flowering plants (see Fig. 28.1). In this section, we examine features of the life cycle that are common to all seed plants and use pine, a gymnosperm, to illustrate how seed plants reproduce without releasing swimming sperm into the environment. In section 28.3, we examine features of angiosperm reproduction, such as flowers and fruits, that distinguish their life cycle from that of other seed plants.

The seed plant life cycle is distinguished by four major steps.

Pollen and seeds are produced only by seed plants, with these structures taking on some of the roles previously played by sperm and spores. To understand them, we consider four signature steps of the life cycle of seed plants.

The first step is the formation of two types of spore, each produced in separate sporangia. One type develops into a male gametophyte and the other into a female gametophyte. Male gametophytes produce only male gametes, and female gametophytes produce only female gametes. Although this condition is not unique to seed plants (it also occurs in a small number of spore-dispersing vascular plants), without it the rest of the modifications that result in the life cycle exhibited by seed plants could not have evolved.

The second step is unique to seed plants. Seed plant spores are not dispersed. Instead, spores undergo mitosis while remaining inside their sporangia. As a consequence, the gametophytes develop while they are attached to and supported by the sporophyte.

The significance of this step is most apparent in the development of the female gametophyte. By developing in place, the female gametophyte can draw resources from the sporophyte to produce female gametes (eggs) and to store energy and raw materials that the embryo can make use of following dispersal. Within each sporangium, only a single haploid spore develops into a female gametophyte, and this gametophyte continues growing until it fills the sporangium. The haploid female gametophyte is protected by one or more layers of sporophyte tissue that surround the sporangium. This entire structure, consisting of sporangium, female gametophyte, and protective outer layers, is the **ovule**. Ovules, if fertilized, develop into seeds.

The male gametophyte also develops while still attached to the sporophyte. Many cells within each sporangium undergo meiosis, forming large numbers of haploid spores. Within each of these spores, a male gametophyte develops, producing a handful of cells by mitosis that all fit within the wall of the original spore. **Pollen** consists of the multicellular male gametophyte and a surrounding outer wall containing sporopollenin.

The third step begins when pollen is shed. **Pollination** is the transport of pollen, either in the air or by an animal, from the sporangium where it was produced to a location near an ovule. Because the male gametophyte produces male gametes, pollination eliminates the need to release swimming sperm into the environment. Nevertheless, pollination alone is not sufficient to ensure fertilization. The male gametophyte must germinate and grow through the tissues of the ovule to deliver the male gametes to the egg. For this reason, you should not think of pollen as flying sperm. The growth of the male gametophyte plays an essential role in bringing the male and female gametes together.

The fourth and final step is the maturation of a fertilized ovule into a **seed**. A seed consists of an embryo, stored resources, and an outer, protective coat. Seeds increase the probability that offspring will survive because their stored resources can be used to fuel the early growth of the next generation.

Pine trees illustrate how the transport of pollen in air allows fertilization to occur in the absence of external sources of water.

To illustrate how seed plants have freed themselves from a dependence on surface moisture for fertilization, let's look at the life cycle of a pine tree (**Fig. 28.7**).

In pines, as in all vascular plants, what we think of as "the plant" is the diploid sporophyte. However, a significant part of the pine's life cycle—from the production of haploid spores to the growth of the gametophyte from a spore, to the production of gametes, and ultimately to fertilization and the development of an embryo within a seed—takes place within reproductive structures called cones. Cones are compact shoots with modified leaves that produce sporangia on their surface. Pines develop two types of cones. **Ovulate cones** typically are located on upper branches and produce spores that develop into female gametophytes. **Pollen cones** are commonly found in clusters near the tips of branches lower in the tree; they produce spores that develop into male gametophytes.

In ovulate cones, each sporangium is surrounded by a jacket of protective tissue. Within each sporangium, a single cell undergoes meiosis, producing four haploid spores. Three of these spores degrade, leaving a single functional spore inside the sporangium. This spore is not dispersed, but instead undergoes

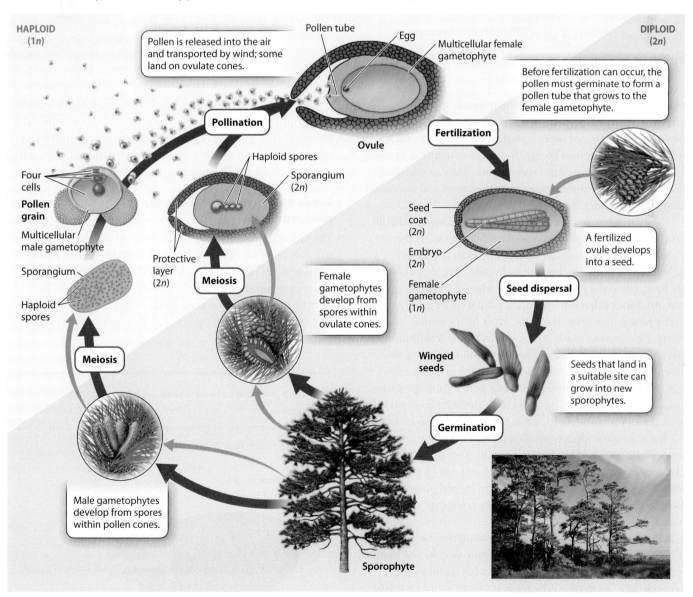

FIG. 28.7 The life cycle of the loblolly pine tree, *Pinus taeda*. *Photo source: Charles O. Cecil/Alamy.*

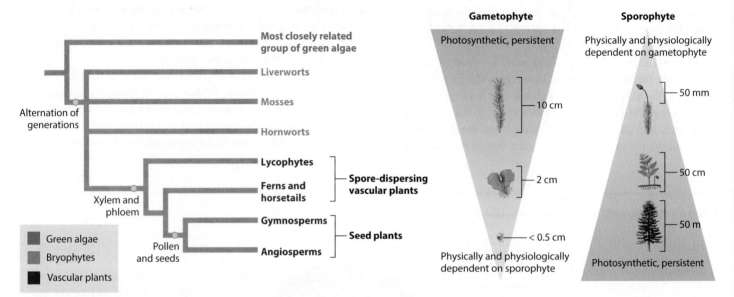

FIG. 28.8 Life-cycle evolution in land plants. A major trend in the evolution of land plants is a decrease in the size and independence of the gametophyte generation and a corresponding increase in the prominence of the sporophyte generation.

repeated mitoses to form a multicellular female gametophyte consisting of a few thousand haploid cells. The female gametophyte produces one or more reproductive organs in which a haploid cell develops into an egg. The entire structure of protective outer layers surrounding a sporangium, with a multicellular female gametophyte inside, constitutes the ovule, and it remains attached to the sporophyte.

In pollen cones, many cells within each sporangium divide by meiosis to form haploid spores. Each haploid spore then divides mitotically to form a multicellular male gametophyte inside the spore wall. In pine, the male gametophyte consists of only four cells at the time the pollen is released from the parent plant. After the pollen is released, the pollen cones are also shed.

The pine pollen is carried to the ovule by the wind. Pollination occurs as much as 15 months before fertilization. In fact, it is the arrival of the pollen that stimulates the development of the female gametophyte. When the pollen germinates, the male gametophyte produces a pollen tube that grows outward through an opening in the sporopollenin coat and into the tissues of the ovule. Then, when the egg is ready to be fertilized, the male gametophyte's pollen tube grows to the female gametophyte, attracted by chemical signals. Two sperm, which lack flagella, travel down the pollen tube, and one of them fuses with the egg to form a zygote. Fertilization in seed plants thus takes place without the male gametes ever being exposed to the external environment.

Following fertilization, the ovules develop into seeds, while the ovulate cone enlarges and becomes quite woody. The mature ovulate cone (also called a seed cone) is what is commonly referred to as a "pinecone." The seeds it contains, once released and dispersed, can give rise to a new, free-living sporophyte.

In seed plants, the relationship between sporophyte and gametophyte has been reversed completely from that found in bryophytes. The gametophyte, so prominent in mosses, is reduced to a small number of cells that depend entirely on their parent sporophyte for nutrition (**Fig. 28.8**).

Seeds enhance the establishment of the next sporophyte generation.

The union of gametes triggers the ovule to develop into a seed that can carry the new embryo away from the parent plant. Because seeds develop from fertilized ovules, they contain tissues produced by three generations (**Fig. 28.9**). On the outside is the protective **seed coat,** which is formed from tissues that surround the sporangium and, therefore, is a product of the

FIG. 28.9 Structure of a pine seed. A pine seed contains tissues from three generations: the seed coat formed from diploid sporophyte layers, the haploid female gametophyte, and the diploid embryo, which will grow into the next sporophyte generation. *Source: Jim French.*

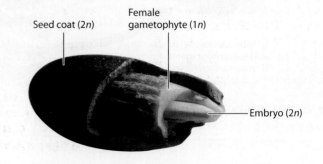

FIG. 28.10 Seed diversity. Seeds vary tremendously in size, from (a) the dustlike orchid seeds to (b) the large seeds of the coco-de-mer.
Sources: a. Charles Marden Fitch/SuperStock; b. Kevin O'Hara/AGE Fotostock.

diploid sporophyte. At the center is the embryo, which develops from the zygote and represents the next sporophyte generation. In gymnosperms such as pine trees, the embryo is surrounded by the haploid female gametophyte, which provides the raw materials that support the growth of the embryo. In angiosperms, the origin of this storage tissue differs, but it serves the same role of nourishing the embryo.

Compared with a unicellular spore, a seed is able to store more resources to support the growth and establishment of a new sporophyte generation. Many seeds contain high levels of protein and energy-rich oils (Chapter 2). As a result, seeds are an important food resource for animals and are the cornerstone of agricultural production.

Because seeds are larger than spores, they are not as easily dispersed by wind. As a result, seed plants have evolved structures to enhance seed dispersal. For example, many pine seeds have flattened extensions, referred to as "wings," that increase the distances over which they can be carried by a breeze. Other seed plants produce brightly colored and/or fleshy structures to attract and reward animals that may inadvertently carry seeds away from the parent plant. Many plants dispersed by animals develop seed coats that allow their seeds to pass unharmed through an animal's digestive system.

As seeds mature, most lose water. The seeds' metabolic activity drops to extremely low levels and the embryo ceases to grow. In this state, seeds can survive for long periods, in many cases one to several years, and sometimes much longer. Some seeds exhibit **dormancy**, meaning that they can delay germination even when conditions for growth, notably temperature and moisture, are favorable. Dormancy prevents seeds from germinating at the wrong time, such as on an uncharacteristically warm day at a time of year when the seedlings would be unable to survive. It also prevents seeds from germinating all at once, thereby spreading the risk of making the transition from embryo to seedling over a larger time span and range of conditions.

Seeds span more than 11 orders of magnitude in terms of their size (**Fig. 28.10**). The 100-nanogram (ng) seeds of orchids are so small (100 ng = 0.0001 mg) that they cannot germinate successfully without obtaining nutrients from a symbiotic fungus. At the other end of the spectrum are the 30-kilogram (kg) seeds of the coco-de-mer, a palm tree found on islands in the Indian Ocean. Large seeds, being better provisioned, are well adapted for the deep shade of a forest understory. However, large seeds typically disperse over shorter distances and often have little or no dormancy. In contrast, for the same investment in resources, a plant can make many small seeds, which can persist in the soil until the right combination of conditions triggers germination. As a result, most weeds produce small seeds.

Self-Assessment Questions

4. How do the male and female gametophyte generations differ in seed plants?

5. What are two advantages of seeds over spores in terms of the probability that the next sporophyte generation will become successfully established?

28.3 FLOWERING PLANTS

Suppose you want to send a postcard to a friend, but your only means of doing so is to throw the card into the air and hope the wind carries it to your friend's front door. How many cards would you have to send aloft to achieve a high probability of success? The answer, of course, is *many*, and this is precisely the reason that pines produce huge numbers of pollen grains each year. Now suppose instead that you can give your card to the letter carrier whose route goes by your friend's house: the odds of successful delivery are now much greater. This analogy gives some idea of a key evolutionary innovation that appeared in the angiosperms. The "letter carriers" in angiosperm fertilization are animals, and the evolution of flowers is what allowed angiosperms to use animals to transport pollen.

Animal pollination is far more efficient than pollination by wind; that is, a higher proportion of pollen reaches an egg when it is dispersed by this means. Thus, a plant that relies on animal pollination can afford to produce less pollen. Flowers may have initially evolved to make animal pollination possible in dense tropical forests. However, as angiosperms diversified into many habitats, some species reverted to wind pollination. Today, angiosperms make use of a wide range of pollination strategies. Here, we examine how angiosperms reproduce, beginning with flowers, which when fertilized develop into fruits.

Flowers are reproductive shoots specialized for the transfer and receipt of pollen.

Angiosperms are seed plants whose life cycle is similar to that of pine trees (see Fig. 28.7), with a few important differences. One of these differences is that in angiosperms pollen and ovules are produced in a single structure, the flower (**Fig. 28.11**), rather than in separate cones. The significance of this arrangement is that, because pollen and ovules are so close together, an animal pollinator can deliver pollen to one plant and take up pollen to carry to another plant in a single visit. Approximately 12% of flowering plant species produce pollen and ovules in separate flowers, but all of these species descend from ancestors whose flowers produced both. Such "unisexual" flowers are often found in the approximately 20% of angiosperm species that have reverted to wind pollination.

Flowers are spectacularly diverse in size, color, scent, and form, but they all have the same basic organization: concentric whorls of floral organs (**Fig. 28.12**; Chapter 19). The outer whorls consist of **sepals** and **petals**, while the inner whorls are made up of pollen-producing **stamens** and ovule-producing **carpels**. Let's look at the floral organs in more detail, starting with the carpels at the center of the flower.

Carpels are modified leaves that have become folded over and sealed along the edges to form a hollow chamber. Ovules

FIG. 28.11 Flower diversity. Shown here are (a) a lady's-slipper orchid (*Cypripedium reginae*); (b) a magnolia (*Magnolia grandiflora*); (c) French lavender (*Lavandula stoechas*); and (d) a tropical tree (*Brownea grandiceps*). *Sources: a. Barrett & MacKay/AGE Fotostock; b. Florida Images/Alamy; c..Jonathan Buckley/AGE Fotostock; d. Tim Laman/Getty Images.*

FIG. 28.12 Flower organization. The four whorls of organs in a flower are carpels, stamens, petals, and sepals. *Photo source: Cora Niele/Getty Images.*

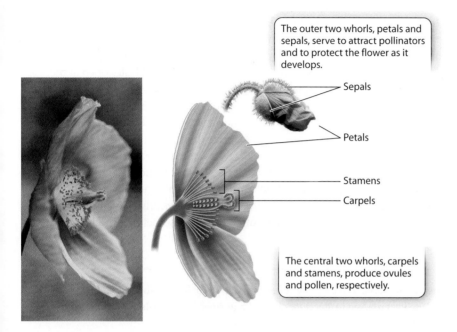

are produced on the inner surface of this cavity (**Fig. 28.13a**). Ovule development begins with the growth of sporangia, each of which is surrounded, typically, by two protective layers. Within each sporangium a single cell undergoes meiosis, forming four haploid spores. One of these spores develops into a multicellular female gametophyte. Recall that ovules, when fertilized, develop into seeds. The fact that the ovules develop within the carpel is what gives rise to the name "angiosperm," which is derived from Greek words meaning "vessel" and "seed." In gymnosperms, the ovules are not enclosed, and their name, again from the Greek, literally means "naked seeds."

Each flower produces one or more carpels (typically 3 to 5, but sometimes more than 20). Because the carpels are often fused, however, there may be only a single structure at the center of the flower. At the base is the **ovary,** in which one to many ovules develop. The ovary protects the ovules from being eaten

FIG. 28.13 Spore formation and gametophyte development in (a) carpels and (b) stamens.

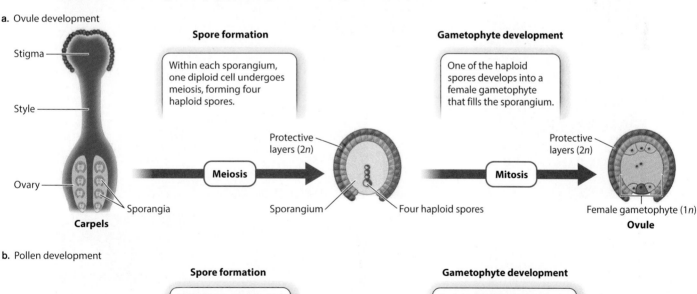

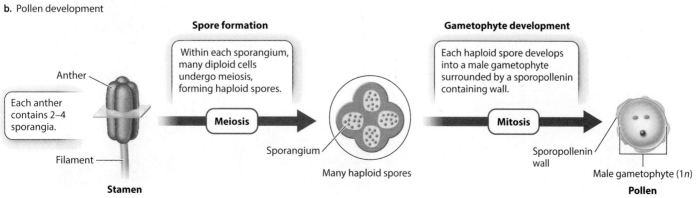

or damaged by animals, but also makes it impossible for pollen to land directly on the surface of the ovule. Instead, to reach the ovules, pollen must land on the **stigma**, a sticky or feathery surface at the top of the carpel(s), and germinate to produce a pollen tube that grows down through the stalk-like **style**. The female gametophytes produce chemicals that guide the pollen tubes toward unfertilized ovules. Because animal pollinators can deposit many pollen grains at once, the fastest-growing pollen tubes are the most likely to deliver sperm to an egg. In some plants, the style can be more than 10 cm long—corn silks, for instance, are styles—creating the opportunity for competition, and thus natural selection, between genetically distinct male gametophytes.

Immediately surrounding the carpels are the pollen-producing stamens. A stamen can have a leaflike structure bearing sporangia on its surface, but more commonly it consists of a filament that supports a structure known as the **anther**, which contains several sporangia (**Fig. 28.13b**). Within each sporangium, many cells undergo meiosis to form haploid spores. The male gametophyte then develops within the spore wall, forming pollen. In most flowers, the anther splits open, exposing the pollen grains. Once exposed, the pollen can come into contact with the body of a visiting pollinator or, in the case of a wind-pollinated species, be carried off by the wind. In other plants, such as the tomato, small holes open at the top of anthers. To extract the pollen, bees land on the flower and vibrate at just the right frequency to shake loose the pollen inside. Orchids and milkweeds do not release individual pollen grains. Instead, they disperse their pollen all together in a package with a sticky tag that attaches to a visiting pollinator.

The outer whorls of the flower produce neither pollen nor ovules, but instead contribute to reproductive success in other ways. Most flowers have two outer whorls. The outermost is made up of sepals, often green but sometimes brightly colored.

FIG. 28.14 Flowers providing food or other rewards for their pollinators. (a) A butterfly collecting pollen from flowers; (b) a hummingbird visiting a flower for nectar; (c) a bat visiting *Agave palmeri* for nectar; and (d) flies attracted to *Rhizanthes lowii* with its smell of rotting flesh. *Sources: a. Sue Kennedy/Flowerphotos/ardea.com; b. Glenn Bartley/AGE Fotostock; c. Rolf Nussbaumer/imageBROKER/AGE Fotostock; d. David M. Dennis/AGE Fotostock.*

Sepals protect the flower during its development. In contrast, petals are frequently brightly colored and distinctively shaped. Their role is to attract and orient animal pollinators. In addition to serving as visual cues, petals in many flowers produce volatile oils. These are the source of the distinctive odors—some pleasant, some decidedly not—that many flowers use to advertise their presence to pollinators.

The diversity of floral morphology is related to modes of pollination.

The evolutionary histories of the angiosperms and their animal pollinators are closely intertwined. A rapid increase in angiosperm diversity occurred between 100 and 65 million years ago, about the same time as bees (Hymenoptera) and butterflies (Lepidoptera) diversified. The evolution of flowers allowed animals to specialize on new food resources. At the same time, animal pollinators greatly assisted the movement of pollen between plants within the same population.

Flowers communicate their presence through both scent and color. For pollination to be reliable, however, it must be in the interest of the animal pollinator to move repeatedly between flowers of the same species. Flowers earn fidelity from their pollinators by providing rewards, frequently in the form of food (**Fig. 28.14**). Many insects consume some of the pollen itself, and many flowers also produce nectar, a sugar-rich solution that is attractive to bats and birds as well as to insects.

Some species provide rewards other than food. For example, many orchids secrete chemicals that male bees need to make their own sexual pheromones. Other flowers provide an enclosed and sometimes heated chamber in which insects can aggregate. Still other flowers do not provide rewards at all; instead, they "trick" their pollinators into visiting. For example, flowers that look and smell like rotting flesh attract flies, and orchids that look like a female bee even emit chemicals that mimic sex pheromones to attract male bees (**Fig. 28.15**). As male bees attempt to copulate with the flower, they deliver and receive pollen.

Producing rewards and attractants such as nectar and petals requires resources. Thus, while animals can provide more efficient pollination than can the wind, the resources expended may be wasted if nonpollinating visitors "steal" pollen or nectar. For this reason, many flowers have mechanisms to protect their investments. For example, tubular cardinal flowers provide rewards only to hummingbirds able to probe their depths, and snapdragons provide rewards only to insects that are strong enough to push apart their petals.

Many plants have evolved flowers that match rewards and attractants to the metabolic needs and sensory capabilities of their pollinators. For example, flowers pollinated by larger pollinators such as bats and birds produce copious amounts of nectar. Bat-pollinated flowers tend to be large, white or cream colored, and strongly scented; these traits are all good matches to the size and nocturnal habits of their pollinators. In addition,

FIG. 28.15 *Ophrys speculum*, **a Mediterranean orchid, which mimics the shape and color of female bees.** These flowers induce males to attempt mating. Through repeated attempts on different flowers, the males transfer pollen. *Source: José Antonio Jiménez/AGE Fotostock.*

these flowers are positioned among branches and leaves in such a way that they can be found by echolocation. In contrast, bird-pollinated flowers are often red, a color most insects cannot see, and have no scent. Because interactions between plants and pollinators affect gene flow, they can affect rates of speciation. A recent radiation in columbine occurred as the plants shifted from bumblebees to hummingbirds to hawkmoths as their primary pollinator (**Fig. 28.16**).

Angiosperms are thought to have evolved in tropical forests. Pollen cannot move freely through the air in the high density of foliage that is present year round in these forests, favoring the evolution of animal pollination. However, as angiosperms diversified and spread around the globe, some evolved wind pollination once again. The reversal to wind pollination is associated with habitats where pollinators are not reliably abundant, such as dry regions.

Wind pollination is most effective in species that are locally abundant. For example, many grasses rely on wind for pollination. Wind pollination is also common in temperate-zone trees, which produce flowers in the early spring, before leaves expand and before pollinators are active. Wind-pollinated species produce large amounts of pollen, and their petals tend to be tiny. Hay fever is an allergic reaction to pollen, almost all of which comes from wind-pollinated species.

HOW DO WE KNOW?

FIG. 28.16

How do long nectar spurs evolve?

BACKGROUND Some flowers produce nectar in tubular outgrowths of their petals, called nectar spurs. Pollinators pick up and deliver pollen as they access nectar in the spur with their tongue. In some species, nectar spurs can be up to 40 cm long. Scientists have long wondered which processes led to the evolution of such long nectar spurs.

HYPOTHESIS 1 Charles Darwin proposed that long nectar spurs result from a coevolutionary "race" between plants and pollinators that favors gradual increases in both spur and tongue lengths. Flowers with longer nectar spurs gain a reproductive advantage because pollinators must come into close contact with the flower if they are to reach the nectar. Such close contact increases the likelihood that pollen will be transferred from the body of the pollinator to the stigma and from the anthers to the pollinator. At the same time, pollinators with longer tongues are able to obtain a larger food reward. Thus, plants and pollinators coevolve toward increasingly longer nectar spurs and tongues.

HYPOTHESIS 2 An alternative hypothesis is that longer nectar spurs evolve as plants adapt to a series of unrelated pollinators, each with longer tongues. According to this "pollinator shift" model, nectar spur length evolves to match the tongue length of the new pollinator. Note that this model also predicts increases in nectar spur length because pollinators with short tongues will not visit plants that have nectar they cannot reach.

PREDICTIONS The two hypotheses make different predictions. Hypothesis 1 predicts that increases in nectar spur length occur in species with the same pollinator type. Hypothesis 2 predicts that increases in nectar spur length are associated with a change in pollinator type.

METHOD To determine which of these two hypotheses is correct, scientists studied columbines (members of the genus *Aquilegia*). Their visually striking flowers are pollinated by bumblebees (short tongues), hummingbirds (medium-length tongues), and hawkmoths (long tongues).

Justen Whittall of the University of California at Davis and Scott Hodges of the University of California at Santa Barbara mapped changes in pollinator type—specifically whether the flowers were adapted for pollination by bumblebees, hummingbirds, or hawkmoths—onto a phylogenetic tree of columbines. To do so, they assigned each of 25 species examined to one of three groups based on length of nectar spur, flower color, and most common pollinator. The group is indicated by the color circle at the tip of each branch in the phylogenetic tree in the figure. The researchers then conducted a statistical analysis, whose results suggested the most likely pollinator type at each node representing an ancestor (also indicated by color circles in the figure).

From this information, the researchers inferred the history of pollinator shifts (indicated by an *) during the radiation of this group. Next, they analyzed whether increases in nectar spur length occurred in species with the same pollinator type, as predicted by Darwin's coevolutionary race hypothesis, or instead were associated with a change in pollinator type, as predicted by the pollinator-shift hypothesis.

Angiosperms have mechanisms to increase outcrossing.

When pollen is deposited onto a stigma, the pollen must germinate and form a pollen tube that grows down through the style before fertilization can occur (**Fig. 28.17**). Typically, only pollen from the same species will germinate and grow successfully, thereby preventing gametes from two different species from coming into contact.

Plants do not have the elaborate courtship rituals that many animal species use to select an appropriate mate. Nevertheless, angiosperms have a wide range of mating systems. At one extreme are **self-compatible** species, in which gametes from male (pollen) and female (ovules) gametophytes produced by flowers on the same plant can form viable offspring. Self-compatible plants can reproduce even when they are physically isolated from other individuals of the same species or when pollinators are rare. Many weedy species and the majority of crops are self-compatible.

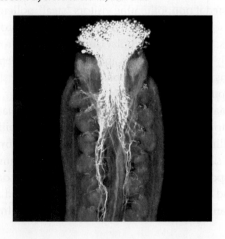

FIG. 28.17 Germination and growth of pollen tubes. The pollen tubes of *Arabidopsis thaliana* can be seen here; they have been stained with a fluorescent dye. *Source: Courtesy Keun Chae.*

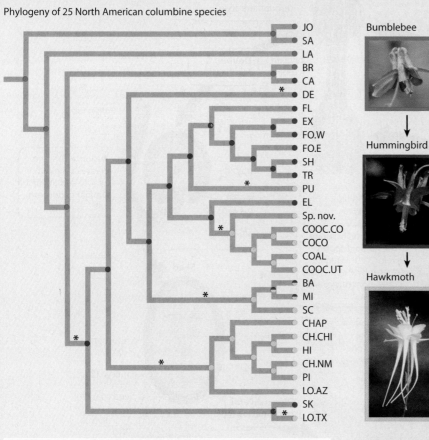

- Short nectar spurs, blue petals, visited by bumble bees
- Medium-length nectar spurs, red petals, visited by hummingbirds
- Long nectar spurs, white or yellow petals, visited by hawkmoths

* indicates a likely evolutionary shift in pollinator

RESULTS Within the recent radiation of columbines, there are multiple instances of shifts from a shorter-tongued pollinator to a longer-tongued pollinator. Most of the increases in nectar spur length are associated with changes in pollinator type, consistent with hypothesis 2, the pollinator-shift hypothesis.

CONCLUSION Spur length in columbines evolves by rapid increases to match the tongue length of a new pollinator type, indicating a key role for pollinator shifts in adaptive radiation.

FOLLOW-UP WORK Developmental studies show that the diversity of nectar spur length evolves as a result of changes in cell shape.

SOURCES Whittall, J. B., and S. A. Hodges. 2007. "Pollinator Shifts Drive Increasingly Long Nectar Spurs in Columbine Flowers." *Nature* 447:706–709; Puzey, J. R., et al. 2011. "Evolution of Spur Length Diversity in Aquilegia Petals Is Achieved Solely Through Cell Shape Anisotropy." *Proceedings of the Royal Society, London, Series B*, 279: 1640–1645.

Photo sources: (top) Paul Oomen/Getty Images; (middle) NNehring/Getty Images; (bottom) Daniela Duncan/Getty Images.

At the other extreme are species in which pollen must be transferred between different plants. Approximately half of all of angiosperm species are **self-incompatible,** meaning that pollination by the same or a closely related individual does not lead to fertilization. Self-incompatible plants must be able to recognize that a gamete comes from a closely related individual. Specifically, these plants recognize proteins produced from self-incompatibility genes, or *S*-genes. If the pollen's proteins match those of the carpel, the pollen either fails to germinate or germinates but the pollen tube grows slowly and eventually stops elongating. *S*-genes may have originated as a defense against fungal invaders, then later evolved as a means of promoting outcrossing. Because dozens of *S*-gene alleles can be present in a population, only pollen transfers between closely related individuals are blocked. Apples are examples of self-incompatible plants: apple orchards always contain a mixture of different varieties to ensure fertilization occurs.

Angiosperms delay provisioning their ovules until after fertilization.

As we have seen, a major trend in the evolution of plants is the decreasing size and independence of the gametophyte generation and the increasing prominence of the sporophyte generation. In angiosperms, this trend is taken even further (**Fig. 28.18**). The male gametophyte has only three cells: one cell controls the growth of the pollen tube, while the other two are male gametes or sperm. In most angiosperms, the female gametophyte contains only eight nuclei, arranged as six haploid cells (one of which gives rise to the egg) plus a central cell containing two nuclei. Thus, the female gametophyte of angiosperms is too small to support the growth of the embryo, as it does in gymnosperms. How, then, do flowering plants provide nutrition for the embryo to grow?

This is where the central cell of the female gametophyte, which contains two haploid nuclei, comes into play. During

FIG. 28.18 The life cycle of an angiosperm.

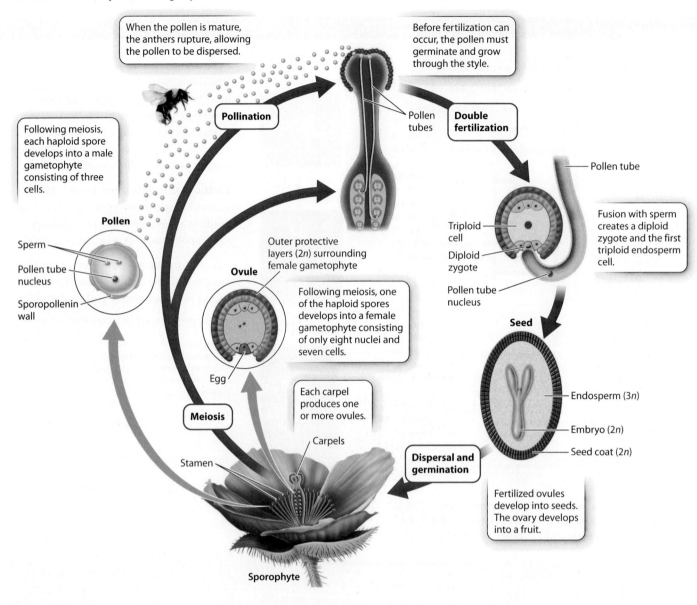

FIG. 28.19 Double fertilization. In double fertilization, one sperm fuses with the egg to form a zygote, while a second sperm fuses with the two haploid nuclei in the central cell of the female gametophyte, leading to the formation of triploid endosperm.

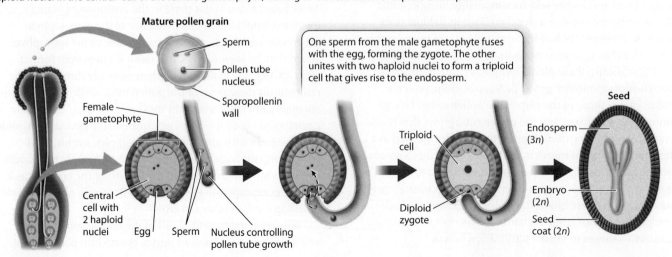

pollination, the two sperm travel down the pollen tube and enter the ovule (**Fig. 28.19**). One of the sperm fuses with the egg to form a zygote, just as occurs in all plants. The other sperm unites with the two haploid nuclei in the central cell to form a triploid (3*n*) cell. This triploid cell undergoes many mitotic divisions, forming a new tissue called **endosperm**. In angiosperm seeds, this endosperm supplies nutrition to the embryo. The process in which two sperm from a single pollen tube fuse with the egg and the two haploid nuclei in the central cell is called **double fertilization**.

Double fertilization and endosperm formation offer a distinct advantage in terms of plant reproduction. In pines and other gymnosperms, the female gametophyte provides food for the embryo. This provisioning occurs whether or not the egg is successfully fertilized, so it is potentially a significant waste of resources. In contrast, in angiosperms, the endosperm develops only when fertilization has occurred. In many flowering plants, including corn, wheat, and rice, the mature seed consists largely of endosperm (**Fig. 28.20**); the embryo itself is quite small. As the seeds germinate, the embryo draws resources from the endosperm to help it become established. It is no overstatement to suggest that the human population is sustained mostly by endosperm produced in the seeds of cereal crops.

In other angiosperms, the growing embryo consumes all the endosperm before the seed leaves the parent plant. For example, a peanut has virtually no endosperm left at the time it is dispersed. Almost all the resources that were previously held within the endosperm have been transferred into two large embryonic leaves, called cotyledons, that form the two halves of the part of the peanut we eat. Thus, the embryo itself contains all of the stored resources that will nourish the growth of the seedling at germination.

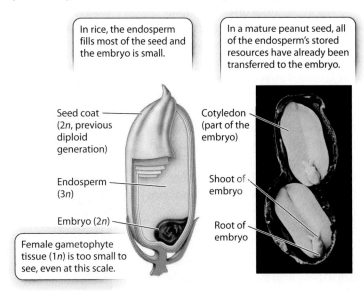

FIG. 28.20 Endosperm. Endosperm develops from the triploid cell produced by double fertilization. *Photo source: Greg Lawler/Alamy.*

Fruits enhance the dispersal of seeds.

The transformation of a flower into a **fruit** is as remarkable as the metamorphosis of a caterpillar into a butterfly (**Fig. 28.21**). As the embryo develops and the endosperm proliferates around it, the ovary wall develops into a fruit, stimulated by signaling molecules produced by the endosperm. For example, the fleshy fruit of a tomato is a mature ovary that encloses the seeds. In some plants, other parts of the flower may also become incorporated into the fruit. For instance, the fleshy part of an apple is formed from the outer part of the ovary combined with the base of the petals and sepals. Other fruits, such as pineapple, develop from multiple flowers. Fruits serve two functions: they

FIG. 28.21 Transition from flower to fruit. Fruits, in this case a plum, develop from ovaries and provide a means of dispersing seed. *Sources: (left to right) Gert Tabak The Netherlands/Getty Images; Trevor Sims/AGE Fotostock; McPHOTO/AGE Fotostock.*

enhance dispersal once the seeds are mature, and they protect immature seeds from being preyed on by animals.

The tremendous diversity of form seen in fruits reflects the many ways in which angiosperm seeds are dispersed (**Fig. 28.22**). The winged fruit of an ash tree is carried away on the wind, and a coconut can float on ocean currents for many months, protected from exposure to salt water by its thick outer husk. The burdock fruit has small hooks that attach to fur (or clothing): this small hitchhiker was the inspiration for Velcro. Other fruits are not dispersed, but instead open to expose their seeds. For example, as milkweed fruits mature and dry out, they split apart to release hundreds of seeds, each with a tuft of silky hairs that helps them float away on even a light breeze. Still others, such as *Impatiens* (jewelweed), produce fruits that forcibly eject their seeds.

Animals are important agents of seed transport, in many cases being attracted by the nutritious flesh of the fruit. In fleshy fruits, the seeds are protected by a hard seed coat and pass unharmed through the animals' digestive tract. In some species, this passage actually speeds up germination. In many other cases, the seed is consumed, and successful dispersal occurs only when the animals gather more seeds than they eat. Squirrels are a good example, storing many acorns that never get eaten. Humans are also an important dispersal agent: we can transport seeds long distances, sometimes far outside their usual range. In some cases, these kinds of accidentally introduced plant species can disrupt native plant communities.

It is essential that the fruits dispersed by animals not be consumed before the seeds are able to withstand a trip through their disperser's digestive tract. Thus, immature fleshy fruits are physically tough and their tissues highly astringent. As they ripen, the fruits are rapidly transformed in texture, palatability, and color. Ripening converts starches to sugars and loosens the connections between cell walls so that the fruit becomes softer.

In a number of species, including apples, bananas, and tomatoes, a gaseous hormone called ethylene triggers fruit ripening. It is the production of ethylene that explains why a banana ripens faster when placed with a ripe apple. Tomatoes and bananas are usually picked in an immature state and later ripened by exposure to ethylene. As a consequence, the fruits can be transported while they are still hard and less prone to damage. Because ethylene synthesis requires oxygen, the unripe fruits are typically stored in a low-oxygen environment. Without this precaution, the ethylene produced by the ripening of even a single fruit could set off a chain reaction in which all the stored fruit would ripen at once.

FIG. 28.22 Fruits and seed dispersal. (a) The seeds of the milkweed *Asclepias syriaca* are dispersed by wind; (b) a robin is eating berries of mountain ash (*Sorbus americana*); (c) cocklebur (*Xanthium strumarium*) fruits are hitching a ride on the fur of a cow; (d) a squirrel carries a fruit it will bury for later use. *Sources: a. Don Johnston/AGE Fotostock; b. David Norton/Alamy; c. Don Johnston/AGE Fotostock; d. blickwinkel/Alamy.*

CASE 6 AGRICULTURE: FEEDING A GROWING POPULATION

How did scientists increase crop yields during the Green Revolution?

Human civilization and plant reproduction are closely intertwined. Not only do seeds and fruits make up much of our diet, but the direct manipulation of plant reproduction also plays a critical role in agriculture. Plant breeding began as the artificial selection for plants with seeds that were easy to harvest. Today it has become a highly quantitative field in which controlled crosses between plants are used to combine a number of favorable traits into a single variety. The spectacular increases in productivity that are often referred to as the Green Revolution resulted, in part, from controlling the movement of pollen between plants.

Norman Borlaug, an American agronomist, has been called "the father of the Green Revolution" and "the man who saved a billion lives." In the mid-twentieth century, Borlaug led a project to make Mexico self-sufficient in bread wheat. Borlaug's aim was to breed wheat that could resist infection by a devastating class of fungal pathogens known as rusts. To do this, he and his group hand-pollinated plants that exhibited resistance to rusts with pollen from plants that grew well under field conditions in Mexico or had seeds that produced good flour.

Under natural conditions, wheat usually self-pollinates. To introduce new sources of pollen, Borlaug and his coworkers removed the anthers of each flower with tweezers before the flowers opened. Then they had to cover each flower to prevent unwanted pollen from coming into contact with the stigma. Finally, when the stigma was receptive, they dusted it with pollen collected from a plant with the desired characteristics. Under the hot sun, this is backbreaking work, but over a series of years the pollinations carried out by Borlaug and his coworkers achieved their goal of breeding disease-resistant wheat varieties.

Yet another problem remained to be solved. To obtain high yields requires large amounts of fertilizer (Chapter 27). Unfortunately, under these conditions, the new wheat varieties produced so many seeds that stalks became top heavy and fell over in the wind. Borlaug knew that an answer to this problem was to cross his plants with wheat varieties that had shorter, sturdier stems. Finding such a plant might be difficult, however, given the selective advantage of height in natural populations. Fortunately, Japanese wheat breeders had identified and preserved a mutant dwarf plant that had arisen spontaneously, and Borlaug obtained seeds of this dwarf variety.

Much hard work and many thousands of hand pollinations lay ahead. Even so, by 1963, more than 95% of Mexican wheat under cultivation was one of Borlaug's high-yielding semi-dwarf wheat varieties, and wheat yields were six times greater than they had been in 1944, the year he began work in Mexico. In 1965, the seeds were exported in large numbers, first to India and Pakistan and soon to the rest of the world. Used in combination with greater investments in fertilizer and irrigation, these varieties prevented the famines that many had predicted would result from growing human populations. For his work in alleviating world hunger, Borlaug was awarded the Nobel Peace Prize in 1970.

Self-Assessment Questions

6. Diagram the relationship between the sporophyte and gametophyte generations in bryophytes, ferns, gymnosperms, and angiosperms. Show the relative sizes and physical interactions (if any) of the two generations.
7. What are the names and functions of the structures in each whorl of a flower?
8. Contrast the investments that angiosperms and gymnosperms make and the structures that they produce to enhance pollination.
9. Diagram the structure of a mature angiosperm and a mature gymnosperm seed, indicating the ploidy (1n, 2n, 3n) of each tissue and its role in seed development and germination.
10. From which flower structure(s) is a fruit derived?

28.4 ASEXUAL REPRODUCTION

Many plants can also reproduce asexually, producing new individuals without the formation and fusion of gametes. While asexual reproduction avoids the challenges associated with fertilization, asexually formed plants must still disperse if they are to avoid competition for resources with the parent plant. Sexual reproduction relies on spores and seeds for dispersal. But how do asexually produced individuals move across the landscape?

Asexually produced plants disperse with and without seeds.

Some species of flowering plants are able to use seeds to disperse asexually formed individuals. In these species, seeds can develop even in the absence of fertilization, a process called **apomixis**. A variety of underlying mechanisms can result in "seeds without sex"; for example, an embryo can form directly from a diploid sporophyte cell. Dandelions are an example of plant species that produce seeds in the absence of sex. Apomixis would be of tremendous utility in agriculture if it could be turned on or off at will. The reason is that, once a desirable variety was developed, additional seeds could be produced without introducing new genetic variation. This ability to breed true would be useful in self-incompatible species such as apple and blueberry.

Most plants that reproduce asexually do so without producing seeds; instead, a new plant grows out of part of the

FIG. 28.23 Vegetative growth of aspen trees. Each distinct group of trees, distinguished by leaf color, is a group of genetically identical individuals produced by vegetative reproduction. *Source: Scott Smith/Getty Images.*

parent plant. In these cases, offspring are dispersed in one of several ways. In mosses and liverworts, for example, tiny tissue discs form by mitosis at the base of shallow cups. Raindrops landing in this cup can dislodge the tiny discs, splashing some away from the parent plant. If a disc lands in a suitable site, it can grow to produce a new individual.

Most plants that reproduce asexually do so by growing to a new location. You need look no further than the front lawn to find evidence of such **vegetative reproduction**. Grass forms horizontal stems underground from which new upright grass shoots are produced at a distance from the site where the parent plant originally germinated. Strawberries, bamboo, and spider plants (a common houseplant) also spread vegetatively by forming horizontal stems. In many cases, the connections that were needed initially to produce a new plant become severed. When this happens, genetic evidence is needed to determine whether a plant is the result of sexual or asexual reproduction.

Vegetative reproduction also occurs in woody plants. For example, when a redwood tree falls over, new upright stems can form along the now-horizontal trunk. Perhaps the most impressive example of the ability to spread vegetatively is the quaking aspen (*Populus tremuloides*), a common tree of the Rocky Mountains that produces brilliant yellow foliage in autumn (**Fig. 28.23**). Aspens produce upright stems from a spreading root system. Although each shoot lives for less than 200 years, a single individual can persist for tens of thousands of years by continuing to produce new stems. The current record holder is an aspen clone in Utah consisting of nearly 50,000 stems; it is estimated to be 80,000 years old.

Plants can reproduce vegetatively because of the way they build their bodies. As we will see in the next chapter, a plant continues to grow and develop throughout its life, and this is the plant characteristic that allows plants to produce new upright "individuals" from roots or horizontal stems.

Self-Assessment Question

11. Describe two ways that plants can reproduce asexually. How does each of these methods help to disperse offspring?

CORE CONCEPTS SUMMARY

28.1 ALTERNATION OF GENERATIONS: Alternation of generations evolved by the addition of a diploid sporophyte generation that allows plants to disperse spores through the air.

- The life cycle of the algal sister groups of land plants has one multicellular haploid generation. page 618

- The life cycle of all land plants is characterized by an alternation of haploid and diploid generations. page 619

- Spores are produced by meiosis in the multicellular diploid generation, which is referred to as the sporophyte. page 619

- Gametes are produced by mitosis in the multicellular haploid generation, which is referred to as the gametophyte. page 619

- Bryophytes have a free-living haploid gametophyte that makes gametes and a dependent diploid sporophyte that makes spores that are released into the air. page 619

- In spore-dispersing vascular plants, both the gametophyte generation and the sporophyte generation can survive on their own. page 621

- Bryophytes and spore-dispersing vascular plants rely for fertilization on swimming sperm released into the environment. page 621

28.2 SEED PLANTS: Pollen allows the male gametophyte to be transported through the air, while resources stored in seeds support the development of the embryo.

- Pollen is the male gametophyte that develops within the spore wall, a structure that enables it to be transported through the air. page 622

- The female gametophyte of seed plants is retained and nourished by the sporophyte. The female gametophyte, together with the surrounding layers of protective tissues, is called an ovule. page 622

- Pollination is the transport of pollen to the ovule, where fertilization takes place. page 622

- Ovules, when fertilized, develop into seeds. Seeds contain the diploid embryo, stored resources that support the growth of the embryo, and a protective seed coat. In gymnosperm seeds, it is the female gametophyte that provides nourishment to the embryo. page 624

- A trend in the evolution of plants is the progressive reduction in size and independence of the gametophyte and the increasing size and persistence of the sporophyte. page 624

28.3 FLOWERING PLANTS: Angiosperms produce flowers and fruits that increase the efficiency of pollination and seed dispersal, and provide resources for seeds after fertilization.

- The pollen and ovules of gymnosperms are produced in separate structures, whereas the pollen and ovules of angiosperms often form in one structure, the flower. page 626

- Flowers consist of four whorls of organs: ovule-bearing carpels, pollen-producing stamens, petals, and sepals. page 626

- Many flowers attract animals because they provide a reward, such as food, shelter, or chemicals. Animals, in turn, transfer pollen. page 629

- Many flowering plants have genetic and structural mechanisms to prevent self-fertilization. page 630

- Double fertilization leading to the formation of a diploid zygote and triploid endosperm is unique to angiosperms. The endosperm nourishes the embryo that develops from the zygote. page 633

- The fertilization of an ovule triggers the ovary to develop into a fruit, a structure that enhances seed dispersal. page 633

28.4 ASEXUAL REPRODUCTION: Many plants can reproduce asexually.

- Asexual reproduction produces new individuals without the need for uniting gametes. Dispersal of offspring, however, remains important. page 635

- Apomixis is the formation of seeds from diploid cells, in the absence of fertilization. Seeds formed by apomixis are therefore genetically identical to the parent plant. page 635

- Vegetative reproduction can occur when a plant stem or root spreads horizontally and develops new upright shoots. page 636

Log in to **LaunchPad** to check your answers to the Self-Assessment Questions and to access additional learning tools.

CHAPTER 29: Plant Growth and Development

CORE CONCEPTS

29.1 SHOOT GROWTH AND DEVELOPMENT: In plants, upward growth by stems occurs at shoot apical meristems, populations of totipotent cells that produce new cells for the lifetime of the plant.

29.2 PLANT HORMONES: Plant hormones are chemical signals that influence the growth and differentiation of plant cells.

29.3 SECONDARY GROWTH: Lateral meristems allow plants to grow in diameter, increasing their mechanical stability and the transport capacity of their vascular system.

29.4 ROOT GROWTH AND DEVELOPMENT: The root apical meristem produces new cells that allow roots to grow downward into the soil, enabling plants to obtain water and nutrients.

29.5 THE ENVIRONMENTAL CONTEXT OF GROWTH AND DEVELOPMENT: Plants respond to light, gravity, and wind through changes in internode elongation and the development of leaves, roots, and branches.

29.6 TIMING OF DEVELOPMENTAL EVENTS: Plants have sensory systems that control the timing of growth and development.

Plants build their bodies in ways that are fundamentally different from what occurs in animals. Consider briefly how our own bodies develop from fertilized eggs (Chapter 19). Animals grow by the repeated division of cells throughout the body. The first divisions of the fertilized egg form stem cells that are totipotent: each cell has the potential to give to give rise to all of the cell types found in a plant's body. As development proceeds, cells lose this totipotency and proceed down a specific developmental pathway. Once the body is mature, growth essentially stops.

In contrast to animals, cell division in plants persists in populations of totipotent cells called **meristems**, and growth commonly lasts throughout the life of the plant. A bristlecone pine in western North America may be several thousand years old, but it continues to sprout new branches, new leaves, and new cones every year.

Plants respond to the world around them not by moving about, but rather by modifying their size and shape. Plants acquire resources by growing to them, and they compete with neighboring plants by growing beyond them. Because development in plants can be modified in response to local environmental conditions, growth fulfills many of the same roles in plants that behavior does in animals (Chapter 43). And although plants can't move out of the way of danger, plants can rebuild their bodies if they become damaged by fire or frost, or if they are struck by a falling tree. Even reproduction requires the continuing formation of new organs.

This chapter considers how vascular plants build their bodies. We focus on the sporophyte generation because of its greater size and persistence. However, many of the principles of cell growth and differentiation discussed here also pertain to how gametophytes build their bodies. We begin by asking how the shoot system, consisting of leaves, stems, and reproductive structures, is formed. We then turn to the development of root systems, exploring how plants grow belowground. Finally, we consider how plants sense the environment and how they respond to what they sense by modifying their growth and development.

29.1 SHOOT GROWTH AND DEVELOPMENT

In tropical forests, vines weave their way upward through the branches of trees. Many of these vines have long, thin stems with widely spaced leaves (**Fig. 29.1a**). In contrast, a barrel cactus, living in the desert, has thick stems and leaves modified to form sharp spines (**Fig. 29.1b**). Although they look nothing alike, these two plants are constructed in the same way. Plant shoots are modular, meaning that they are formed of repeating units. Each unit consists of a **node**, the point where one or more leaves are attached, and an **internode**, the segment between two nodes (**Fig. 29.1c**). In vines, the internodes are long and the leaves large; by contrast, in cacti, the internodes are short and the leaves thin and sharp. The modular structure of plants helps us understand how the capacity for continued growth gives rise to the tremendous variation in plant form that we see around us.

Stems grow by adding new cells at their tips.

At the tip of each stem lies a tiny dome of cells called the shoot apical meristem (**Fig. 29.2**). The **shoot apical meristem** is a group of totipotent cells that, like embryonic stem cells in animals, gives rise to new tissues.

FIG. 29.1 Shoot organization. (a) A tropical vine and (b) barrel cacti look very different, but their shoot systems are built from (c) the same repeating units of nodes and internodes. *Sources: a. Le Do/Shutterstock; b. Liane M/Alamy.*

FIG. 29.2 The shoot apical meristem. The shoot apical meristem consists of undifferentiated cells that divide rapidly, giving rise to all the cells in stems and leaves. (a) The seedling is rowan (*Sorbus acuparia*). (b) The photo shows a longitudinal section of the shoot apical meristem of coleus (*Solenostemon scutellariodes*). (c) This scanning electron micrograph shows the shoot apical meristem of tomato (*Solanum lycopersicum*). *Sources: a. Ian Gowland/Science Source; b. M. I. Walker/Science Source; c. Siobhan Braybrook.*

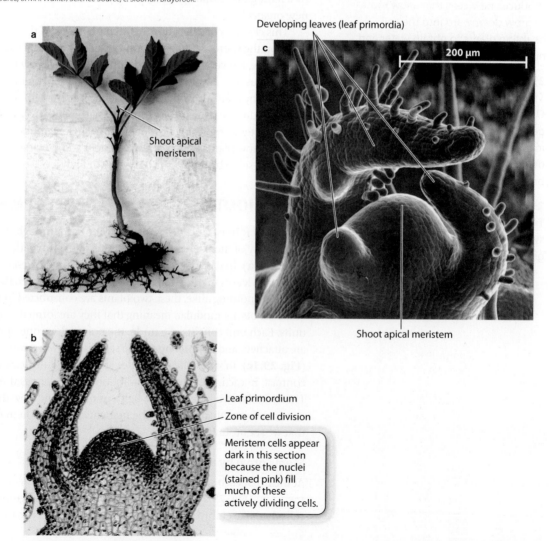

As we will see, plants have meristems in several parts of their bodies, but stem growth occurs exclusively at the shoot tip. This is where cell division occurs, generating all the new cells that elongate stems. The shoot apical meristem also initiates leaves and produces new meristems, which allow plants to branch.

The shoot meristem maintains a constant size even as its cells actively divide. As new cells are added near the shoot tip, cells that are farther from the shoot tip cease to divide. Cells near the shoot tip maintain their ability to divide through the expression of **meristem identity genes**. The expression of these genes is controlled by a network of chemical signals produced by cells at the very tip of the stem. Because these signals can be transmitted over a fairly limited distance, only the cells close to the shoot tip express meristem identity genes.

Stem elongation occurs just below the apical meristem.

In animals, cell division and cell enlargement typically go hand in hand. In plants, however, most of the increase in cell size occurs after cell division is complete. Cells that are too far from the shoot tip to express meristem identity genes cease to divide, yet they continue to grow. Most of the elongation of stems takes place in the internodes located immediately beneath the shoot apical meristem (**Fig. 29.3**). The increase in cell size is impressive. If all of the cells in the stem of a 100-m redwood tree remained the size they were when they left the shoot apical meristem, the tree would be less than 5 m tall.

Each elongating cell grows many times more in length than in width. The reason is the orientation of the cellulose molecules, which are primarily wrapped around the cell perpendicular to the axis of the stem rather than parallel with it. Because cellulose is very strong, having a tensile strength similar to that of steel, cells expand more easily in length than in girth. A large central vacuole forms during cell elongation. In fact, most of the increase in cell volume is due to the uptake of water and solutes that fill the vacuole. This explains in part why plant growth is markedly reduced during periods of drought.

As we move even farther from the shoot tip, cells reach their final size and complete their differentiation into mature cell types. This organization into successive zones of cell division, cell elongation, and cell maturation allows stems to grow without any predetermined limit to their length.

Once cells have matured, they no longer expand. Thus, during the upward growth of the shoot, it is the production and elongation of cells at the shoot tip that lift the meristem ever higher into the air. Imagine that, as a 10-year-old, you carved your initials into a tree. Twenty years later, your initials will be exactly the same distance from the ground as they were on the day you inscribed them. But the letters will be noticeably wider—a phenomenon we discuss later in this chapter.

Stems branch by producing new apical meristems.

Vascular plants evolved the ability to branch even before they evolved roots or leaves. Branching was important to these first plants because it allowed them to produce more sporangia. In present-day plants, branching enables them to support greater numbers of both reproductive structures and leaves.

In most spore-dispersing vascular plants, branches form when the shoot apical meristem divides in two, giving rise

---

FIG. 29.3 Stem elongation. Elongation of the internodes results from cells increasing many times in length, but only a small amount in width.

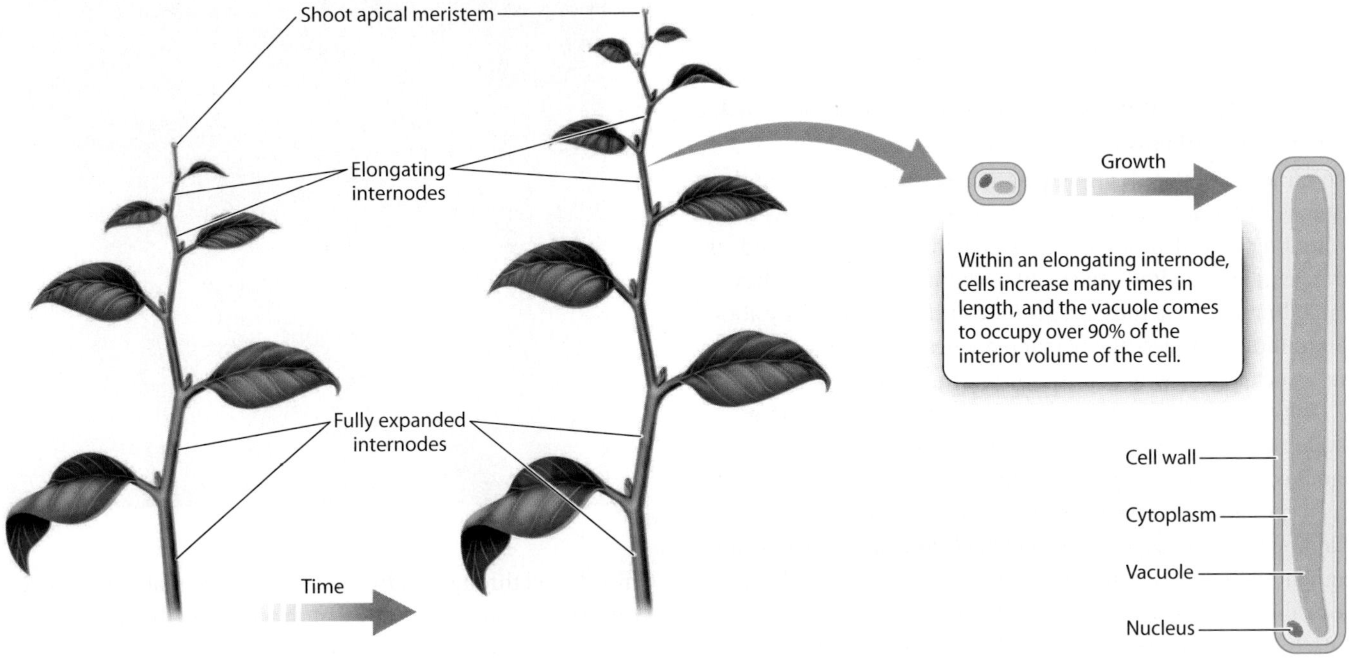

Within an elongating internode, cells increase many times in length, and the vacuole comes to occupy over 90% of the interior volume of the cell.

FIG. 29.4 Axillary buds.
In seed plants, branches grow out from axillary buds that are produced at nodes. *Photo source: Jacqueline S. Wong/ViewFinder Exis, LLC.*

FIG. 29.5 The evolution of leaves.
Fossils show that leaves evolved from branches that (1) lost the ability for continued growth, (2) developed in a plane, and (3) expanded the photosynthetic surface to fill the area between veins. The photo shows *Archaeopteris*, one of the earliest plants with flattened leaves. *Photo source: Topham/The Image Works.*

Early vascular plants lacked leaves. Their shoots consisted of photosynthetic stems.

In some of these early vascular plants, side branches lost their apical meristems and became flattened into a single plane.

The development of new meristems allowed tissues to form between these side branches, leading to the development of leaves.

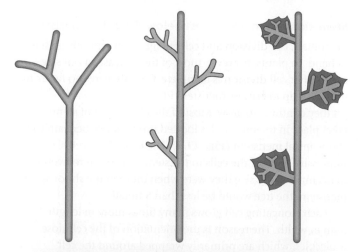

to two stems, each with its own shoot apical meristem. In seed plants, branches grow out from **axillary buds**, which are meristems that form at the base of each leaf (**Fig. 29.4**). Axillary buds have the same structure and developmental potential as the apical meristem and express the same meristem identity genes.

When a branch develops, the axillary bud becomes the shoot apical meristem of the new branch. Most plants have many shoot apical meristems: one at the tip of each branch. Axillary buds form at the same time as leaves do, but they remain dormant until triggered to grow, and they persist even after leaves are shed. Axillary buds provide seed plants with many points along their stem where new branches can form.

The shoot apical meristem controls the production and arrangement of leaves.

The earliest vascular plants were simple branching stems, and photosynthesis took place along the length of the stem. As evolution proceeded, some branch systems became flattened, an arrangement that allows the plant to capture more light. These planar branches lost the capacity for continued growth. By about 380 million years ago, the planar branches became modified into structures recognizable as leaves (**Fig. 29.5**).

The evolution of leaves required several changes in plant development. Apical meristem identity genes were down-regulated and leaf identity genes up-regulated, resulting in an organ specialized for photosynthesis. New meristems evolved, enabling leaves to expand into flattened structures for capturing sunlight. In fern leaves, these regions of cell division are located along the leaf margin. In pine needles, they occur at the base of each needle. In most flowering plants, meristem cells are distributed throughout the developing leaf, making possible a diversity of leaf shapes. In contrast to cells in the shoot apical meristems, leaf meristem cells divide only for a relatively short

FIG. 29.6 Diversity of leaf arrangements. *Photo sources: (left to right) Jonathan Buckley/GAP Gardens; Anneke Doorenbosch/Alamy; Christian Hütter/imageBROKER/ AGE Fotostock.*

Alternate   Opposite   Whorled

FIG. 29.7 Non-photosynthetic functions of leaves. (a) Protection (bud scales); (b) climbing (tendrils); (c) trapping insects (spines); and (d) attracting pollinators (color). *Sources: a. Dr. Keith Wheeler/Science Source; b. Marga Werner/AGE Fotostock; c. Ernie Janes/Alamy; d. Adisa/Dreamstime.com.*

period of time. This explains why, at a certain point, leaves stop growing.

Leaves begin as small bumps, called **primordia** (singular, **primordium**). These bumps form on the sides of the shoot apical meristem in a highly regular pattern (see Fig. 29.2). Each species has a characteristic number of leaves attached at each node along the stem. Some species have only a single leaf at each node, whereas others have two or more leaves at each site (**Fig. 29.6**). The regular placement of leaves reduces the shading of one leaf by another, thereby enhancing plants' ability to obtain sunlight.

When we think of leaves, it is the green photosynthetic ones that first come to mind. In reality, many plants produce leaves that are specialized for functions other than photosynthesis, including climbing, trapping insects, and attracting pollinators (**Fig. 29.7**). Most plants that overwinter produce small, hard **bud scales** that protect shoot apical meristems from water loss and damage due to cold. Bud scales may not look like leaves, but they form from leaf primordia and are arranged in the same way around the stem as the green leaves produced in spring.

Young leaves develop vascular connections to the stem.

During the initial growth of leaf primordia, cells rely on diffusion to obtain the resources needed to divide and expand. As the young leaves increase in size, their demand for water and carbohydrates quickly exceeds the amounts that diffusion can supply. If they are to grow to even a fraction of their full size, leaves must establish vascular connections with the xylem and phloem in the stem. As we saw in Chapter 27, the ability of leaves to carry out photosynthesis requires both a supply of water through the xylem and the ability to export carbohydrates to the rest of the plant through the phloem.

When a developing leaf is still very small, sets of cells within the leaf begin to elongate. These elongating cells form discrete strands of **procambial cells** that extend from near the tip of the leaf to the mature vascular tissues within the stem. Procambial cells ultimately give rise to both xylem and phloem within the leaf veins and in the vascular bundles in the stalk (petiole) attaching the leaf to the stem. The common origin of xylem and phloem from procambial cells explains why these two tissues are always found in close association throughout the length of the plant.

Spore-dispersing vascular plants typically have one or several vascular bundles in the center of their stems. Most commonly in seed plants, the vascular bundles form a ring near the outside of the stem. The xylem conduits of each bundle are located toward the center of the stem, while the phloem is found nearer the outside of the stem (**Fig. 29.8**). Typically, each leaf develops vascular connections with several of the stem's vascular bundles. The ground tissue between the epidermis and the vascular bundles is the **cortex**, and the ground tissue inside the ring of vascular bundles is the **pith**. The placement of the vascular bundles, which contain the stiff xylem conduits, near the outside of the stem increases the stem's ability to support its leaves without bending over.

Flower development terminates the growth of shoot meristems.

In vascular plants, sporangia are produced on the surfaces of leaves. In some plants, including many ferns, sporangia are located on leaves that also carry out photosynthesis. In conifers and some other plants, sporangia form on the surfaces of highly modified leaves and branches that form compact cones. In both cases, the arrangement of the sporangia-bearing leaves is the same as that of the leaves that do not bear sporangia.

Flowers develop from floral meristems formed by the conversion of shoot meristems. As a consequence, flowers can be produced at the tip of a shoot or at the base of leaves as the products of axillary buds. Floral meristems differ from shoot meristems in several ways. First, all of the cells within a floral meristem differentiate, so floral meristems lose the capacity for continued growth. Second, although floral organs (sepals, petals, stamens, and carpels) develop from primordia and are thought to be modified leaves, their arrangement differs from the placement of leaves along the stem. Floral organs occur in whorls, with each whorl consisting of only a single organ type (see Fig. 28.12).

Flowering is triggered by florigen, a protein produced in leaves and transported through the phloem to apical meristems and axillary buds. Florigen triggers the transition to floral meristems by initiating the down-regulation of meristem identity genes and the up-regulation of genes that govern floral identity. Once the meristem is launched along the trajectory for flower development, the identity of each whorl is controlled by the expression of homeotic genes (Chapter 19).

FIG. 29.8 Arrangement of vascular bundles. In most seed plants, the vascular bundles are arranged in a ring near the outside of the stem. *Photo source: Steve Gschmeissner/Science Source.*

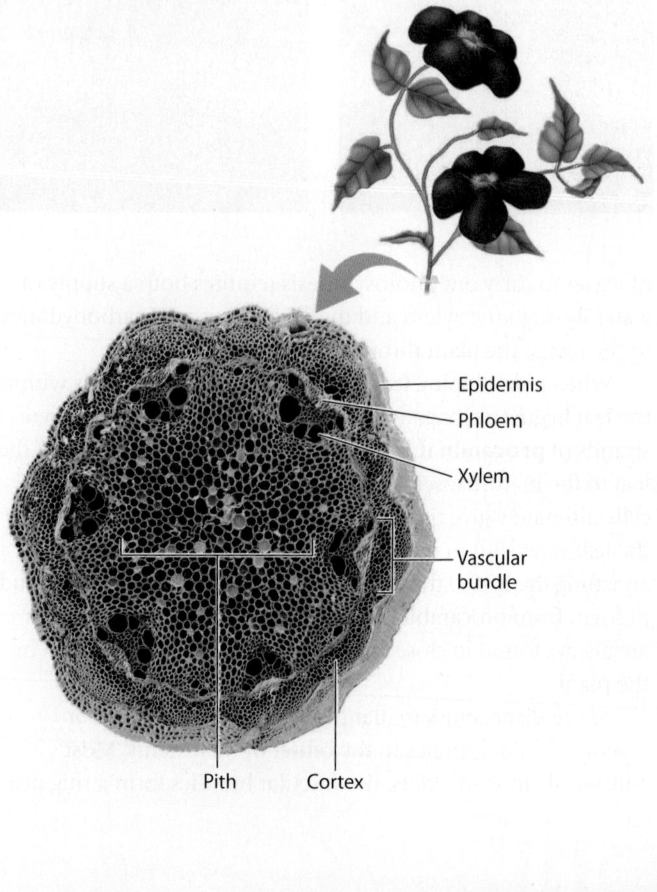

Self-Assessment Questions

1. What are three functions of the shoot apical meristem?
2. Diagram the zones of cell division, elongation, and maturation for plant stems. Why does this organization allow stems to grow without a predetermined limit to their length?

29.2 PLANT HORMONES

A key question in the development of multicellular organisms is what controls the identity, or "fate," of individual cells. In plants, each cell is bound to its neighbor by the cell wall and, therefore, is fixed in place. As a consequence, cell identity is determined by where a cell is located within the plant. For cells at the surface, the absence of neighboring cells on one side can be the signal to become an epidermal cell. For cells in the interior, chemical signals determine which sort of cell is formed. These signals provide information about both the identity of neighboring cells and the cell's position in the plant. In contrast, animal cells are not fixed in place and cell migration plays an important role in development (Chapter 19). For this reason, cell lineage, rather than position, often controls the identity of animal cells.

Some of the chemical signals that guide plant development affect only a small area. For example, as mentioned in section 29.1, the signals that cause cells to express meristem identity genes are restricted to the region close to the shoot tip. Other signaling molecules, or their precursors, enter the vascular system and are transported along the entire length of the plant. Because plants grow and develop in many places at once, signaling molecules that can move from one part of the plant to another play an important role in coordinating growth and development.

Hormones affect the growth and differentiation of plant cells.

Hormones are chemical signals that influence growth and development. The term "growth" refers to irreversible increases in size, whereas "development" encompasses the formation of new tissues and organs, as well as the differentiation of specific cell types. Hormones play two roles in plant growth and development. First, they help establish and maintain basic patterns of the plant body plan. For example, hormones determine how leaves are arranged, where branches form, and how much internodes elongate. Second, they coordinate growth in different parts of the plant in response to both internal and external environmental factors. For example, hormones control which axillary buds grow into branches and guide the growth of a stem toward sunlight. Hormones affect plant growth and development both by influencing physiological processes, such as the activity of membrane transport proteins, and by altering patterns of gene expression.

Traditionally, plant biologists have recognized five major plant hormones: auxin, gibberellic acid, cytokinins, ethylene, and abscisic acid (**Table 29.1**). How are such a small number of hormones able to shape the continuing growth and

TABLE 29.1 The Five Major Hormones

HORMONE	SYNTHESIS AND TRANSPORT	EFFECTS
Auxin (Indole-3-acetic acid)	Synthesized primarily in shoot apical meristems and young leaves. Transported by polar transport and through phloem.	Induces cell wall extensibility. Inhibits outgrowth of axillary meristems. Stimulates formation of root meristems. Stimulates differentiation of procambial cells. Involved in phototropism and gravitropism. Stimulates ethylene synthesis.
Gibberellic acid	Synthesized in the growing regions of both roots and shoots.	Stimulates stem elongation and cell division. Mobilizes seed resources for the developing embryo. Stimulates cytokinin synthesis.
Cytokinins	Synthesized in both root and shoot meristems. Transported in both xylem and phloem.	Stimulates cell division. Delays leaf senescence. Acts synergistically with auxin.
Ethylene	Synthesized in elongating regions of roots and stems, as well as in developing fruits. Gaseous; precursors can be transported in xylem.	Reduces cell elongation. Triggers fruit ripening. Stimulates formation of air spaces in roots.
Abscisic acid	Synthesized in root cap, developing seeds, and leaves. Transported from roots to leaves in xylem.	Maintains seed dormancy. Stimulates root elongation. Triggers closing of stomata. Antagonistic with ethylene.

development of plants? Each of these hormones triggers a wide range of developmental effects. For example, ethylene can lead to fruit ripening (Chapter 28), but it also slows the elongation of roots and stems. How a cell responds to a particular hormone depends on both its developmental stage and its position within the plant. In addition, the effect of any one hormone is often influenced by the presence of one or more other hormones.

Several additional chemicals with hormone-like properties have recently been discovered. Some of these have broad effects on plant development (for example, strigolactone). Others appear to be more specific (for example, florigen, whose only role is to stimulate flower development).

To illustrate how hormones influence plant growth and development, we consider the role of three hormones—auxin, gibberellic acid, and cytokinins—in establishing the basic patterns of shoot development. Later in the chapter, we consider the role of hormones in root development (section 29.4) and plant growth in response to light and gravity (section 29.5).

Polar transport of auxin guides the placement of leaf primordia and the development of vascular connections with the stem.

The plant hormone **auxin** was first identified when it was observed to cause shoots to elongate. As we will see in section 29.5, auxin's ability to trigger cell elongation underlies how shoots bend and grow toward light. Auxin also plays an important role in establishing spatial patterns of development. In shoots, auxin influences the spacing and placement of leaf primordia and the formation of the vascular bundles connecting young leaves to the mature vascular system of the stem.

Auxin's role in patterning results from the fact that its movement across a group of cells can be controlled. The net movement of this hormone in a single direction creates regions of high and low concentrations of auxin. These gradients can guide the growth and development of individual cells.

Let's consider first how auxin movement influences the placement of new leaf primordia. Auxin is synthesized in rapidly dividing cells within the shoot meristem and is transported in the outermost cell layer toward the shoot tip. High levels of auxin in the outermost layer of cells trigger the growth of new leaf primordia. Once a primordium begins to form, it becomes a sink for auxin. The movement of auxin into leaf primordia depletes this hormone from surrounding regions, which prevents new leaf primordia from forming nearby. As the shoot continues to grow, already formed leaf primordia become spread farther and farther apart. This creates spaces where auxin moving upward in the outermost cell layer can accumulate to the levels needed to trigger new leaf primordia.

Let's now consider how auxin transport guides the formation of vascular bundles that connect the leaf to the stem. As auxin is transported into leaf primordia, it promotes cell expansion, so that the young leaves grow. Auxin does not remain in the developing leaf, however. Instead, auxin that has accumulated at the tip of the young leaf is transported back toward the stem, but this time through cells in the interior of the leaf. As auxin moves through these cells, it causes the cells both to elongate and to produce more of the membrane transport proteins needed for auxin to exit the cell. The cells that initially, by chance, have the highest auxin levels become highly efficient at transporting this hormone toward the stem. This, in turn, creates distinct "channels" through which auxin drains from the leaf. These continuous strands of elongate procambial cells eventually develop into veins and vascular bundles containing mature xylem and phloem.

These two examples illustrate how the movement of auxin in a single direction establishes spatial patterns that can guide further differentiation and development. Nevertheless, they leave open the question of how undifferentiated cells control the movement of auxin in the first place.

The coordinated movement of auxin across many cells in a single direction is referred to as **polar transport** (Fig. 29.9). By means of polar transport, auxin can move independently of gravity at rates faster than the hormone could move by diffusion alone. Polar transport, however, requires an input of energy.

Polar transport depends on the difference in pH between the cell wall and the cytoplasm. In the cell wall, auxin carries no net charge, so it can enter surrounding cells by diffusing across the plasma membrane. Because the pH of the cytoplasm is higher than the pH of the cell wall, auxin loses a proton once it enters the cytoplasm. As a result, auxin has a net negative charge when inside the cell and cannot move easily across the plasma membrane. For auxin to leave the cell, an auxin-specific plasma membrane transport protein, known as PIN, is required. By restricting the placement of PIN proteins to only one face of each cell, plants can control the direction of auxin movement across an otherwise undifferentiated group of cells. For example, if PIN proteins are inserted only on those cell sides farthest away from the shoot tip, auxin will move toward the roots.

 CASE 6 AGRICULTURE: FEEDING A GROWING POPULATION

What is the developmental basis for the shorter stems of high-yielding rice and wheat?

In natural ecosystems, competition for sunlight creates a strong selective advantage for growing tall. Compare the slender vine in Fig. 29.1, which grows in a dense forest, with the compact barrel cactus that grows where there is little danger of being shaded by a neighboring plant. Whereas the ancestors of today's crops competed for sunlight by growing taller, a key property of the high-yielding rice and wheat varieties developed during the Green Revolution is their shorter stems. These shorter-stemmed plants can support the much-larger seed mass induced by high rates of fertilizer application (Chapter 27). Moreover, by investing less in stems, these plants can allocate more resources to the production of seeds. But what happens inside these plants to make the stems shorter?

FIG. 29.9 Polar transport of auxin. Auxin moves from the tip of each developing leaf to the base of the plant by polar transport.

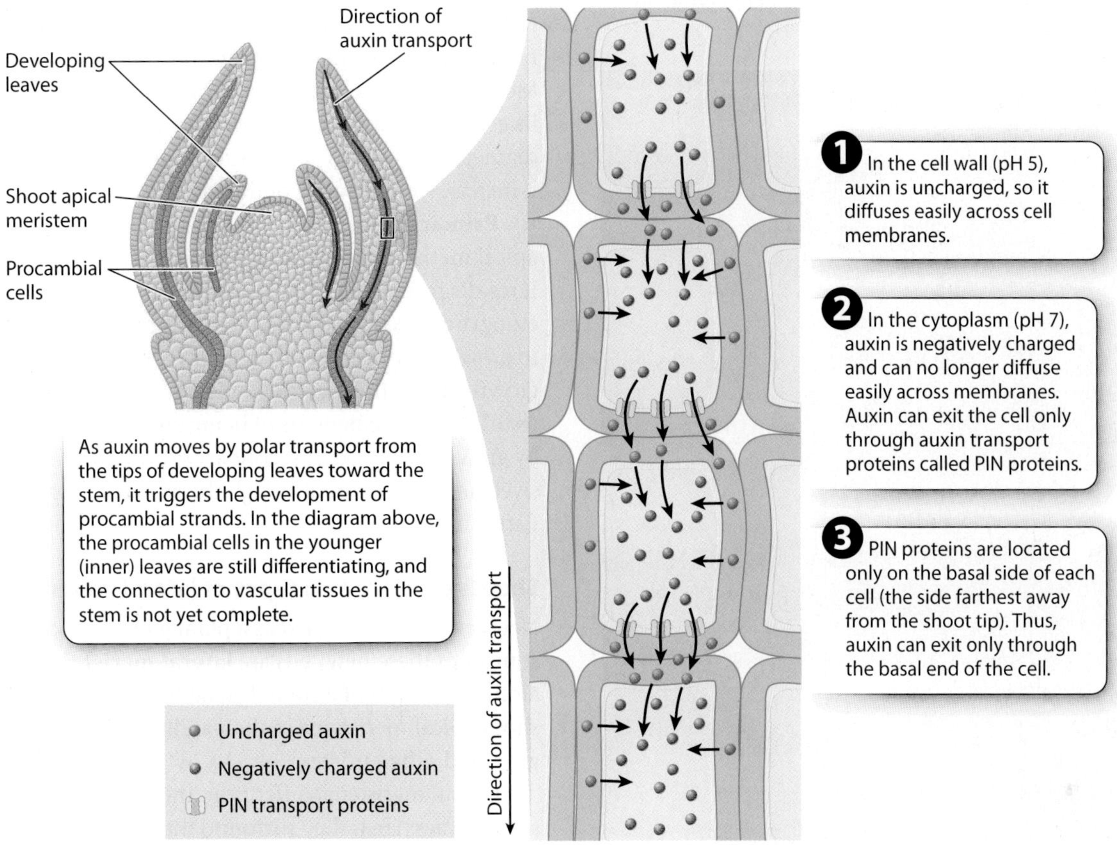
Direction of auxin transport
Developing leaves
Shoot apical meristem
Procambial cells
As auxin moves by polar transport from the tips of developing leaves toward the stem, it triggers the development of procambial strands. In the diagram above, the procambial cells in the younger (inner) leaves are still differentiating, and the connection to vascular tissues in the stem is not yet complete.
Uncharged auxin
Negatively charged auxin
PIN transport proteins
Direction of auxin transport
1 In the cell wall (pH 5), auxin is uncharged, so it diffuses easily across cell membranes.
2 In the cytoplasm (pH 7), auxin is negatively charged and can no longer diffuse easily across membranes. Auxin can exit the cell only through auxin transport proteins called PIN proteins.
3 PIN proteins are located only on the basal side of each cell (the side farthest away from the shoot tip). Thus, auxin can exit only through the basal end of the cell.

Gibberellic acid was first identified as a chemical produced by fungal pathogens that caused the stems of young rice plants to grow to many times their normal length. We now know that this hormone is produced naturally in plants, particularly in growing regions such as developing leaves and elongating stems. Gibberellic acid increases internode elongation by reducing the force needed to cause cell walls to expand. This raises the possibility that the high-yielding crop varieties with shorter stems might produce less gibberellic acid or be less sensitive to it.

At the time that the semidwarf varieties were developed, the developmental basis for their reduced stem growth was not known. Since then, the genes responsible for reduced stem growth have been identified in both wheat and rice. The semidwarf wheats are less sensitive to gibberellic acid, while the shorter rice varieties are deficient in the enzymes needed to synthesize this hormone. In both cases, the result is shorter but sturdier plants.

Cytokinins, in combination with other hormones, control the outgrowth of branches.

Plants with multiple branches can support more leaves, but if branches are too close together, leaves will overlap and shade one another. Seed plants have the potential to produce a new branch at each node. To avoid overlap, only a small number of axillary buds actually develop into branches. Which ones become branches is determined by the interaction of several hormones, including both cytokinins and auxin.

Cytokinins were discovered when they were seen to stimulate cell division in tissues grown in the laboratory. Not surprisingly, these hormones are produced in plant meristems, including the meristems within axillary buds. Cytokinins have many effects. Notably, when applied directly to axillary buds by researchers, they can stimulate these buds to develop into branches. This finding suggests that cytokinins could stimulate branching in nature as well, but leaves open the question of what determines whether an axillary bud remains dormant or begins to grow.

Removing the shoot tip provides a clue. With the shoot tip gone, branches begin growing from axillary buds along that branch or stem (**Fig. 29.10**). Evidently, the shoot apical meristem somehow suppresses the growth of axillary buds, a phenomenon called **apical dominance**. Apical dominance prevents branches from being formed too close to the shoot tip, yet new branches can still be formed if the shoot meristem is damaged.

Apical dominance occurs when chemical signals moving down from the shoot tip suppress the growth of axillary buds by inhibiting the synthesis of cytokinins. When the apical meristem

FIG. 29.10 Apical dominance. In the intact plant (left), the axillary buds are suppressed. Removal of the apical meristem (right) results in the outgrowth of two axillary buds. *Source: Nigel Cattlin/Alamy.*

is removed, cytokinin synthesis increases in the stem nodes, triggering the growth of axillary buds (Table 29.1).

The chemical that suppresses branching is auxin. Researchers have shown that if the shoot tip is cut off and a paste containing auxin is applied to the cut end of the stem, the axillary buds remain dormant. If instead the shoot tip is treated with a substance that inhibits auxin transport, branching is induced.

Recent studies indicate that this top-down control is complemented by the action of another hormone, **strigolactone**. This hormone is made in roots and transported upward in the xylem. Strigolactone inhibits the outgrowth of axillary buds: plants that are unable to synthesize this chemical produce many more branches than do wild-type plants. One hypothesis is that strigolactone is transported to the shoot when either water or nutrients limit growth. Thus, the plant would stop growing new branches whenever the root system is unable to supply the water and nutrients they require.

Self-Assessment Question

3. Name one role of the plant hormone auxin. How is auxin is transported within a plant?

29.3 SECONDARY GROWTH

Shoot apical meristems enable plants to grow in length and to branch. Even so, a plant unable to grow in diameter would not become very tall before buckling under its own weight. Moreover, a plant unable to add vascular tissue could not supply its ever-increasing number of leaves with water. Growth in diameter thus serves two functions: it strengthens the stem, and it increases the capacity of the vascular system.

Primary growth is the increase in length made possible by apical meristems. **Secondary growth** is an increase in diameter. It results from a new type of meristem, one that forms only after elongation is complete. The fossil record shows that the ability to grow in diameter evolved millions of years after primary growth, and it did so independently in several groups of plants, testimony to the benefits of being able to compete for sunlight by growing tall. Today, secondary growth occurs almost exclusively in seed plants, although this was not always the case in Earth's history.

Shoots produce two types of lateral meristem.

New cells must be added as a plant grows in diameter. The sources of these new cells are **lateral meristems**, which form along the length of a stem. Lateral meristems are similar to shoot apical meristems in their ability to produce new cells that grow and differentiate.

Lateral meristems differ from their apical counterparts in several ways. First, they surround the stem, rather than occur at its tip. Second, because lateral meristems form only after elongation is complete, the new cells they produce grow in diameter but not in length. Finally, as a stem becomes thicker, the number of meristem cells needed to encircle the stem also increases. Thus, lateral meristems become larger over time.

Plants produce two distinct lateral meristems that together result in secondary growth. One of these, the **vascular cambium** (plural, **cambia**), accounts for most of the increase in diameter. The vascular cambium is the source of new xylem and phloem. The xylem produced by secondary growth is familiar to you: it's wood. Plants with secondary growth are often described as woody, whereas plants that lack secondary growth are described as herbaceous. Most of the mass of a large tree consists of wood, generated from the vascular cambium.

As a plant's diameter increases, the epidermis formed during primary growth eventually ruptures. A second lateral meristem, the **cork cambium**, renews and maintains a protective outer layer. This outer layer protects the stem against herbivores, mechanical damage, desiccation, and, in some species, fire.

The vascular cambium produces secondary xylem and phloem.

The vascular cambium forms at the interface between the xylem and the phloem. In stems, the vascular cambium derives from procambial cells within each vascular bundle and parenchymal

FIG. 29.11 Secondary growth. The vascular cambium is the source of new xylem and phloem, while the cork cambium maintains an outer layer of bark.

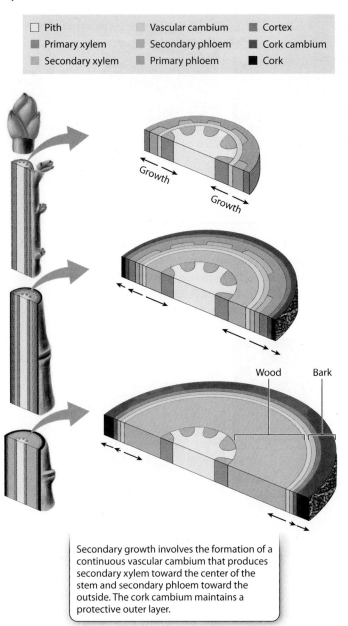

Secondary growth involves the formation of a continuous vascular cambium that produces secondary xylem toward the center of the stem and secondary phloem toward the outside. The cork cambium maintains a protective outer layer.

cells in between the bundles. In forming vascular cambium, both types of cell take on a new identity, becoming meristem cells.

The vascular cambium produces new cells on both its sides. Those to the inside of the vascular cambium become **secondary xylem**, while those on the outside become **secondary phloem** (**Fig. 29.11**). The vascular cambium forms a continuous layer that surrounds the xylem and runs nearly the entire length of the plant. The undifferentiated and actively dividing cells of the vascular cambium are weak and easily pulled apart. Thus, when you peel the bark off a stem, it separates from the wood at the vascular cambium. All of the material produced to the outside of the vascular cambium, including the secondary phloem, makes up the bark.

As the vascular cambium forms new secondary xylem cells along its inner face, the newly formed cells push the vascular cambium outward, much in the same way that the elongation of newly formed cells pushes the shoot apical meristem upward. The vascular cambium accommodates this outward movement by adding new cells that increase its circumference.

In contrast, secondary phloem is unable to produce any additional cells. As the vascular cambium is pushed outward, the phloem becomes stretched and, eventually, damaged. Consequently, the vascular cambium must continually produce new secondary phloem.

Over time, xylem conduits lose their ability to transport water. In long-lived trees, the xylem that is still active is located in a **sapwood** layer adjacent to the vascular cambium. The **heartwood**, in the center of the stem, does not conduct water. Heartwood is often darker in color than sapwood because it contains large amounts of chemicals such as resins. For this reason, heartwood is typically more resistant to decay than sapwood.

In many plants growing in seasonal climates, secondary xylem cells produced at the end of the growing season are smaller than ones produced earlier. If you cut down a tree and inspect the top surface of the stump, the layer of smaller cells is visible as the dark edge of a **growth ring**. Counting growth rings makes it possible to determine a tree's age.

Growth rings provide a window into the past. By measuring the width of each ring, it is possible to determine how much wood was laid down in a given year (**Fig. 29.12**). A tree produces more wood when moisture and temperature levels are

FIG. 29.12 Annual growth. In seasonal climates, tree rings provide a record of the annual accumulation of wood. The tree rings shown are from Douglas fir (*Pseudotsuga menziesii*). *Source: Henri D. Grissino-Mayer, The University of Tennessee, Knoxville.*

Wide growth rings indicate higher growth rates and thus suggest that the tree experienced more favorable conditions in these years.

Narrow growth rings indicate slower growth rates, perhaps because of low rainfall.

FIG. 29.13 Outer bark. (a) Outside and (b and c) inside, showing how successive cork cambia are produced. The growth of many, discontinuous cork cambia layers results in the patchy bark of pine and walnut trees. *Sources: a. Alan Majchrowicz/AGE Fotostock; b. Dr. Keith Wheeler/Science Source.*

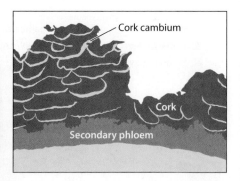

a. Outside of tree, ponderosa pine (*Pinus ponderosa*) **b.** Cross section of bark, walnut (*Juglans regia*) **c.**

favorable for photosynthesis. Thus, annual growth rings provide a record of past climatic conditions, as well as events such as fire.

The cork cambium produces an outer protective layer.

The cork cambium maintains a protective layer around a stem that is actively increasing in diameter (see Fig. 29.11). It forms initially from cortex cells that de-differentiate—that is, regress to their earlier state—to become meristem cells. With time, this layer of actively dividing cells becomes increasingly distant from its source of carbohydrates, the phloem. As the cork cambium becomes cut off from its carbohydrate supply, a new cork cambium forms within the secondary phloem. Thus, each cork cambium layer is relatively short-lived. Indeed, in many cases, the new cork cambia form in patches rather than as continuous layers, causing the bark of the tree to have a patchy and fissured appearance (**Fig. 29.13**).

The cells produced by the cork cambium are called cork. As cork cells mature, they become coated with **suberin**, a waxy compound that protects the cells against mechanical damage and the entry of pathogens, while also forming a barrier to water loss. However, this layer of waxy, nonliving cells impedes the inward diffusion of oxygen needed for cellular respiration. The outer bark cells are less tightly packed in regions called **lenticels**. Oxygen diffuses through lenticels into the stem (**Fig. 29.14**).

Wood has both support and transport functions.

We have seen that as a tree grows, its secondary xylem fulfills the two roles of secondary growth: to provide trees with the strength and stability to stand upright and support large crowns, and to transport water and nutrients from the roots to the leaves. In most gymnosperms, a single cell type, the tracheid, fulfills both of these functions (**Fig. 29.15a**). As discussed in Chapter 27, tracheids are single-celled xylem conduits that have thick lignified cell walls. They are nonliving and empty at maturity. The wood of a pine tree, for example, is relatively uniform in

FIG. 29.14 Lenticels in (a) water birch (*Betula occidentalis*) and (b) basswood (*Tilia americana*). Oxygen diffuses through lenticels in the outer bark to supply the respiratory needs of the cells within the stem.

Sources: a. Robert and Jean Pollock/Science Source; b. Garry DeLong/Science Source.

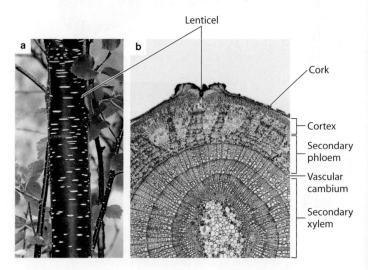

structure, consisting almost entirely of tracheids. This type of wood does not vary dramatically from species to species.

In contrast, the wood of angiosperms is diverse in structure. Xylem cells mature into several distinct types that separate the functions of water conduction and mechanical support (**Fig. 29.15b**). **Fibers** are narrow cells with extremely thick walls and almost no lumen; their primary function is support the plant as it grows taller. In contrast, **vessel elements** can be extremely wide cells; their primary function is water transport (Chapter 27). The largest vessel elements are nearly 0.5 mm in diameter and visible to the naked eye. Vessel elements can be much wider than tracheids because the fibers provide support. Because vessel elements develop open connections that link them end to end with other vessel elements, creating multicellular xylem vessels,

FIG. 29.15 Wood structure. These scanning electron micrographs show (a) a gymnosperm, Douglas fir (*Pseudotsuga menziesii*), and (b) an angiosperm, English elm (*Ulmus procera*). *Sources: a. Scanning electron micrograph courtesy of the N. C. Brown Center for Ultrastructure Studies, State University of New York College of Environmental Science and Forestry, Syracuse, NY; b. Andrew Syred/Science Source.*

a. Gymnosperms

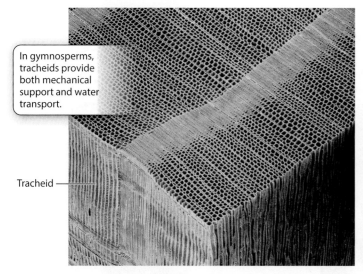

In gymnosperms, tracheids provide both mechanical support and water transport.

Tracheid

b. Angiosperms

In angiosperms, fibers provide mechanical support, allowing vessels to be specialized for water transport.

Vessels
Fiber

angiosperms can produce wood that transports water more efficiently than does the wood of gymnosperms.

The ability to form xylem vessels has a number of important consequences. For example, angiosperms can have higher rates of water flow through their stems than is possible with wood that contains only tracheids. As a result, angiosperms sustain higher rates of CO_2 uptake and, therefore, higher growth rates.

In addition, because vessels can transport water much more efficiently than do tracheids, angiosperms can devote a smaller fraction of their wood to cells for water conduction, and a larger fraction to other cell types. The wood of some angiosperms contains many fibers. In fact, in extreme cases, the fiber-heavy wood is so dense that it sinks. These dense woods are extremely strong, allowing trees to create spreading crowns with outstretched branches. Such "hardwoods" are useful as flooring materials and furniture. The wood of other angiosperms contains a high percentage of living cells instead of fibers, so it has a very low density. Balsa trees, whose wood is used in building model airplanes, are a good example.

Self-Assessment Questions

4. Why do plants have two types of lateral meristem?
5. Why does a plant that has a vascular cambium also have a cork cambium?
6. Why does the vascular cambium form a continuous sheath that runs from near the tips of the branches to near the tips of the roots, whereas the cork cambium is discontinuous in both space and time?

29.4 ROOT GROWTH AND DEVELOPMENT

Roots enable vascular plants to obtain water and nutrients from the soil and provide an anchor for plants capable of growing tall. As in the case of leaves and wood, the fossil record shows that the first vascular plants lacked roots. These plants survived only in extremely wet environments, where they absorbed water through horizontal, and possibly submerged, stems. Once roots evolved, vascular plants expanded into drier habitats and became able to support taller stems.

The fossil record indicates that roots evolved at least twice. Nevertheless, the roots of all vascular plants share many features. These commonalities suggest that the physical challenges of growing through dense soil and the demands of obtaining water and nutrients from that soil place strong constraints on root structure.

Roots grow by producing new cells at their tips.

In Chapter 27, we saw that plants produce large numbers of thin roots to obtain sufficient nutrients from the soil. Also, to obtain water during periods without rain, plants produce roots that penetrate deep into the soil. Growing roots into the soil presents new challenges. Although the dangers of drying out are much reduced belowground, the density of deeper soil layers can approach that of concrete. How are the thin and permeable roots needed for water and nutrient uptake able to penetrate and grow through the soil?

In one respect, roots grow in the same way as shoots: new cells are produced at their tip. The **root apical meristem** is the source of new cells. After cell division, these cells elongate and become differentiated. They complete their differentiation only after elongation has ceased (**Fig. 29.16**). Thus, as we move back from the tip of the elongating root, we encounter successive regions of cell division, elongation, and maturation, just as we saw in stems.

Roots also differ from stems in several important ways. One difference is that roots are much thinner than stems. Plants with thin roots can more efficiently produce the large surface area needed for uptake of water and nutrients from the soil. Thin roots can also more easily bend around soil particles and make use of channels burrowed through the soil by invertebrate animals.

A second difference is that the root apical meristem is covered by a **root cap** that protects the meristem. As the root elongates through the soil, root cap cells are rubbed off or damaged, while the meristem remains intact. The root cap is continually supplied with new cells produced by the root apical meristem.

A third difference between roots and stems, and one that can be seen without a microscope, is that the root apical meristem does not produce any lateral organs, whereas the shoot apical meristem makes leaves. Root hairs, the single-celled outgrowths of root epidermal cells that greatly increase the roots' surface area (Chapter 27), form only after elongation has ceased.

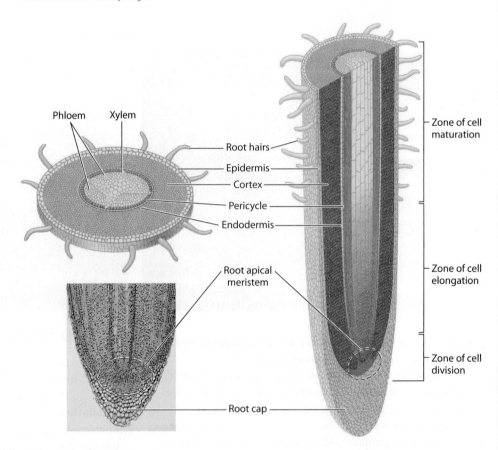

FIG. 29.16 The structure of a typical root, including the root apical meristem and root cap.
Photo source: Walker Mi./Getty Images.

Root elongation and vascular development are coordinated.

A cross section taken through the zone of cell maturation shows a series of concentric tissue layers: the epidermis is on the outside, followed by the root cortex, and then the endodermis (Fig. 29.16). Recall that the endodermis plays a key role in the selective uptake of nutrients (Chapter 27). The Casparian strip, a waxy band surrounding each endodermal cell, forces all materials that will enter the xylem to cross a plasma membrane. At the center of the root is a single vascular bundle in which the xylem is arranged as a fluted cylinder, with phloem nestled in each groove. In cross section, the xylem is arranged in a star shape; the number of points of the star varies among different plant species.

As in stems, the hormone auxin plays a central role in specifying which cells differentiate into vascular tissues. Auxin produced in shoot meristems is transported downward through the mature stem and into the roots by the phloem. At the end of the fully formed phloem, auxin moves by polar transport from cell to cell until it reaches the root apical meristem and the root cap. As auxin moves through the elongation zone, it triggers the formation of procambial cells that become phloem and xylem.

In this way, the production of auxin by developing leaves and the polar transport of auxin from cell to cell ensures that the vascular system is continuous between roots and leaves. Phloem extends closer to the root tip than do fully lignified xylem conduits. The root apical meristem does not need direct access to the xylem because it can obtain water and nutrients directly from the soil, but it depends entirely on the phloem for carbohydrates needed for growth.

So far, we have described only the primary growth of roots. Roots can undergo secondary growth as well: they increase in diameter and add new xylem and phloem. As in stems, the vascular cambium forms at the interface between the xylem and the phloem. Indeed, the vascular cambium must be continuous from the shoot to the roots to ensure that the secondary xylem and phloem provide an unbroken connection between the roots and the leaves.

FIG. 29.17 Branching in roots. New root meristems develop from the pericycle and must grow through the root cortex before reaching the soil. The light micrographs show willow (*Salix*) roots. *Source: Omikron/Science Source.*

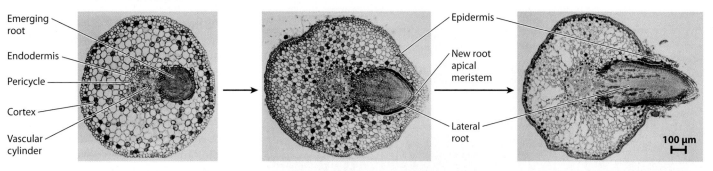

In older roots with extensive secondary growth, a cork cambium also forms. As in stems, the cork cambium renews the outer layers of cells that protect the living tissues from damage and predation.

The formation of new root apical meristems allows roots to branch.

Compared to stems, roots have extensive branching. Even a plant that has just a single stem aboveground will have many tens of thousands of elongating root tips. A major reason that roots branch so much is that most of the water and nutrients are taken up in a zone near the root tip where root hairs are most abundant. Branching is necessary to create enough root surface area to supply the water and nutrient needs of the shoot.

For a root to branch, it must form a new apical meristem. These new root meristems develop from the **pericycle**, a single layer of cells just inside the endodermis (**Fig. 29.17**). The pericycle is immediately adjacent to the vascular bundle, so the new root is connected to the vascular system from the beginning. Because the new root meristem develops internally, the new root must grow through the endodermis, cortex, and epidermis before it reaches the soil.

One consequence of initiating new root meristems from the pericycle is that new roots can form anywhere along the root. As a result, roots can proliferate locally where nutrients are abundant. For example, the presence of nitrate, the most common form of nitrogen in soil, triggers the formation of new roots.

Most new roots are active for only a short time, after which they die and are shed. Although this turnover of fine roots allows the root system to respond to changes in the availability of water and nutrients, it represents a major expenditure of carbon and energy.

Up to this point, we have described how new roots form from existing roots. New roots can also be produced by stems. Many plants that lack secondary growth produce horizontal stems that grow on or beneath the soil surface. In these plants, new roots form at each successive node, so the uptake capacity of the root system increases as the stem elongates.

When the stem of a plant is severed, new roots often form at the cut end. The formation of new root meristems is stimulated by auxin, which is produced by the young leaves and accumulates by polar transport at the cut end. Because their cut stems can form new roots, plants are able to survive damage and in some cases even proliferate afterward. Many commercially important plants are propagated from cut stems.

The structures and functions of root systems are diverse.

How a plant's roots are distributed in the soil has a tremendous impact on the ability of the root system to supply the shoot with water. Some plants produce very deep roots that provide access to groundwater that is independent of the variability of rainfall. For example, roots have been observed in mine shafts and caves more than 50 m below the soil surface. Producing and maintaining these non-photosynthetic structures is costly for the plant.

At the other extreme are plants that produce shallow and spreading root systems. For example, barrel cacti produce roots that penetrate less than 10 cm into the soil but extend out several meters from the base of the plant. These roots cast a wide but shallow net to capture as much water as possible from the intermittent rain showers that fall in desert environments.

Some plants produce distinctive roots whose principal role is *not* the absorption of water and nutrients from the soil (**Fig. 29.18**). For example, climbing plants such as poison ivy produce roots along their stems that adhere to the sides of trees (Fig. 29.18a). The conspicuous prop roots produced by some tropical trees provide mechanical stability that allows these plants to grow tall despite their slender stems (Fig. 29.18b).

Many plants produce swollen roots that store resources such as starch (Fig. 29.18c). Plants with belowground storage can produce new leaves and shoots quickly following either damage or a period of drought or cold that has killed the shoots. Cassava and sweet potato are examples of economically important species with storage roots.

FIG. 29.18 Root diversity. (a) Climbing roots of poison ivy; (b) prop roots of a *Pandanus* tree; (c) storage roots of yams; and (d) breathing roots of black mangrove (shown at low tide). *Sources: a. Doug Wechsler/AGE Fotostock; b. A Jagel/AGE Fotostock; c. PhilipYb Studio/Shutterstock; d. Patrick Lynch/Alamy.*

Perhaps the most unusual rooting structures are the "breathing" roots produced by trees that grow in water. For example, black mangroves produce pencil-sized roots that extend vertically upward out of the sandy soil (Fig. 29.18d). Breathing roots do not actually breathe; instead, they contain internal air spaces that provide an easier pathway for oxygen to diffuse into the roots than through the waterlogged soil. For this reason, breathing roots must extend above the surface of the water.

Self-Assessment Questions

7. How is root development similar to and different from stem development?
8. What are three structural differences between roots and shoots that allow roots to grow through the soil?
9. How would you tell whether an isolated piece of a plant came from the shoot or from the root?

29.5 THE ENVIRONMENTAL CONTEXT OF GROWTH AND DEVELOPMENT

Using their capacity for continued growth, plants often respond to changes in their environment by growing more in one direction than another. To respond in a manner that enhances fitness, however, plants must first gain information about the world around them. Plants rely on three types of sensory receptors to provide that information. Photoreceptors sense the availability of light needed to drive photosynthesis; mechanical receptors sense physical influences such as gravity and wind; and chemical receptors detect the presence of specific chemicals, as well as chemical gradients.

We saw in Chapter 9 how a signal can be conveyed to a cell by a change in the shape of a receptor molecule on the cell surface or within the cell. Sensory receptors in plants work in a similar manner. For example, the absorption of light by a photoreceptor changes the chemical properties of the photoreceptor, similar to what happens with the photoreceptors in your eye. When a plant's sensory receptor is activated, it produces a signal that triggers changes in the cell's metabolism or alters patterns of gene expression. Hormones play a key role in translating information gained by the plant's sensory receptors into an appropriate response.

Plants orient the growth of their stems and roots by light and gravity.

If you lay a plant on its side, its stems will turn upward and its roots will bend downward. Similarly, if you place a houseplant near a window, the shoot will grow toward the light. **Tropism** is the bending or turning of an organism in response to an external signal such as light or gravity. Plant stems are positively **phototropic**, meaning they bend toward the light. Plant stems are also negatively **gravitropic**, meaning they grow upward against the force of gravity. In contrast, plant roots grow down and away from the light. Thus, roots are positively gravitropic and negatively phototropic.

Observing that grass seedlings bend toward light, Charles Darwin and his son Francis conducted experiments to determine which part of the plant detects light. When they covered just the tip of the young plants, the plants no longer grew toward the light; in contrast, when the Darwins buried the plants in fine

HOW DO WE KNOW?

Fig. 29.19
How do plants grow toward light?

BACKGROUND In 1926, Frits Went, a Dutch graduate student, followed up on the experiments of Charles and Francis Darwin. Went sought to understand how the perception of light by the tip of the plant led to a growth response farther down the stem. He reasoned that if the signal from the shoot tip was a chemical, he could isolate the chemical signal by cutting the tip off of a growing plant and putting the tip on a block of gelatin. The gelatin would act like a trap, capturing any chemical that moved out of the shoot tip.

HYPOTHESIS A chemical signal links the perception of light by the shoot tip with the growth of the stem toward light.

EXPERIMENT Went cut off the tips of young oat plants (*Avena sativa*) and placed them on gelatin blocks for periods of 1 to 4 hours. Removing the tip of the plant caused the plants to stop elongating. When the gelatin block was placed on top of the cut plant (Experiment 1), growth resumed. This experiment demonstrated that a growth-inducing chemical was exported from the tip of the young oat plants to the gelatin block.

Interestingly, Went found that if the block of gelatin was not centered on the tip but instead was a bit off center, the plant bent away from the side of the gelatin block (Experiment 2). The side of the plant exposed to the gelatin block grew faster, causing bending, because it received more of the chemical signal.

CONCLUSION A signal is produced in the tip of the plant and travels downward through the stem, toward the roots. This signal stimulates growth. Light causes the signal to move to the opposite side, stimulating growth on that side, which in turn causes the plant to bend toward the light. This signal is now known to be a plant hormone, called auxin from the Latin word *augere*, "to increase."

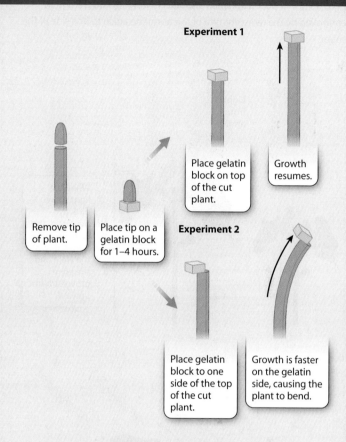

FOLLOW-UP WORK Went followed up his experiments by determining how the level of auxin relates to the degree of curvature. Other experiments showed that auxin moves in a directed fashion by polar transport. Later, auxin was shown to be indole-3-acetic acid (IAA).

SOURCES Darwin, C., and F. Darwin. 1880. *The Power of Movement in Plants*. London, UK: John Murray; Went, F. W. 1926. "On Growth-Accelerating Substances in the Coleoptile of *Avena sativa*." *Proceedings of the Royal Academy of Amsterdam* 30:10–19.

---

black sand so that only their tips were exposed, they did. This experiment established the shoot tip as the site where the light is perceived, but left open the question of which kind of signal caused the plants to bend (**Fig. 29.19**).

A plant's photoreceptors detect light and set off a response. The photoreceptors that trigger phototropism absorb blue wavelengths of light. If a plant is exposed to light from only one side, only the photoreceptors on the illuminated side are activated. Their absorption of blue light is thought to trigger a change in the placement of PIN proteins so that auxin is transported to the shaded side of the stem. As a result, concentrations of auxin are higher on the shaded side and lower on the illuminated side. Because auxin stimulates cell expansion in stems, the shoot grows faster on the shaded side than it does on the illuminated side, causing the plant to bend toward the light (**Fig. 29.20**).

Phototropism in roots is similar, in that illuminating a root from the side causes auxin to be redistributed toward the shaded side. In this case, auxin is transported to the shaded side within the root cap, and then moves upward in the outermost layer of cells into the zone of cell elongation. In contrast to what happens in shoots, higher auxin concentrations *decrease* the rate of cell elongation in roots, causing roots to bend *away* from light. In roots, high concentrations of auxin reduce cell

FIG. 29.20 Phototropism in shoots. Bending toward the light is controlled by the redistribution of the hormone auxin to one side of the shoot.

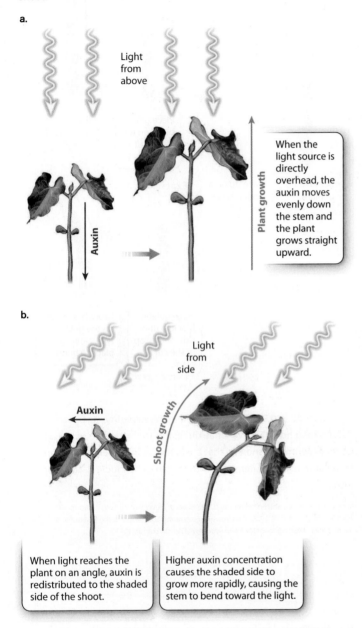

FIG. 29.21 Root orientation with respect to gravity. Starch-filled statoliths settle to the bottom of cells, triggering a redistribution of auxin to the lower side of the root. Auxin slows elongation on that side, causing the root to bend downward. *Photo source: Courtesy Dr. Christophe Jourdan.*

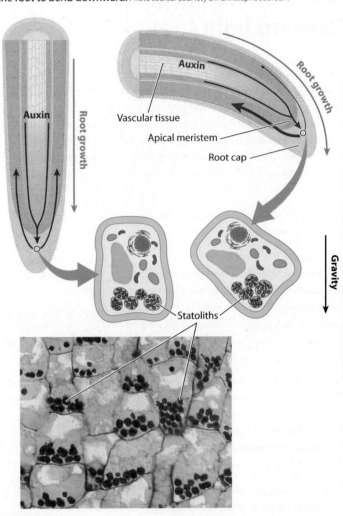

elongation and trigger the production of ethylene, a hormone that also reduces cell elongation.

A similar mechanism orients plants with respect to gravity. Moving a plant to a horizontal position results in the transport of auxin to the lower side. The accumulation of auxin on the lower surface causes stems to bend up and roots to turn down. How is the plant able to detect which way is up and which way is down? Recall that plants store carbohydrates as starch. Starch is a large molecule that has a density greater than water. As a consequence, starch-filled organelles sink to the bottom of the cell. The weight of these organelles pressing on the cytoskeleton or membranes is thought to be the signal that causes auxin to move toward the lower side of the plant. In roots, gravity is a critical source of information regarding orientation. Specialized gravity-sensing cells in the root cap contain large starch-filled organelles known as **statoliths** (**Fig. 29.21**).

Seeds can delay germination if they detect the presence of plants overhead.

For a small seed that has few stored reserves, germinating in the shade of another plant could be fatal. But how can a seed detect the presence of plants overhead?

As sunlight passes through leaves, the red wavelengths are absorbed by chlorophyll but the far-red wavelengths are not. For

HOW DO WE KNOW?

Fig. 29.22

How do seeds detect the presence of plants growing overhead?

BACKGROUND To study the effects of light on seed germination, American researchers Harry Borthwick and Sterling Hendricks exposed lettuce seeds to different wavelengths of light and then counted what fraction of the seeds germinated. Red light had the greatest ability to stimulate germination but, surprisingly, far-red light inhibited germination. That is, fewer of the seeds exposed to far-red light germinated than even control seeds kept in darkness. In the rush to conduct more experiments, petri dishes with the light-treated seeds piled up by the sink until someone noticed that the seeds that had been experimentally treated with far-red light were now germinating.

HYPOTHESIS This observation suggested that the inhibitory effect of far-red light can be overcome by a subsequent exposure to red light.

EXPERIMENT The scientists exposed lettuce seeds to red and far-red light in an alternating pattern, ending with either red light or far-red light. They then placed the seeds in the dark for two days and afterward counted the number of seeds that had germinated.

RESULTS When the lettuce seeds were exposed to red light last, approximately 75% of the seeds germinated. By contrast, when the lettuce seeds were exposed to far-red light last, the percentage of seeds that germinated was dramatically reduced.

Lettuce Seeds Germinating After Exposure to Red and Far-Red Light in Sequence

SEQUENCE OF LIGHT EXPOSURE	SEED GERMINATION (%)
Red	70
Red, far-red	6
Red, far-red, red	74
Red, far-red, red, far-red	6
Red, far-red, red, far-red, red	76
Red, far-red, red, far-red, red, far-red	7
Red, far-red, red, far-red, red, far-red, red	81
Red, far-red, red, far-red, red, far-red, red, far-red	7
Time →	

CONCLUSION Seed germination in lettuce is triggered by exposure to red light and is inhibited by exposure to far-red light in a reversible fashion. This finding indicates that plants are able to track changes in the relative amounts of red and far-red light, obtaining information on the presence or absence of plants overhead (see Fig. 29.23).

FOLLOW-UP WORK Additional studies showed that this result was due to a single photoreceptor, now known as phytochrome, that is converted into an active form by red light and reversibly converted into an inactive form by far-red light.

SOURCE Borthwick, H. A., et al. 1952. "A Reversible Photoreaction Controlling Seed Germination." *Proceedings of the National Academy of Sciences* 38:662–666 and Borthwick, H. A., et al. 1954 "Action of Light on Lettuce-Seed Germination." *Botanical Gazette* 115: 205–225.

---

this reason, the ratio of red to far-red light is much higher under an open sky than in the shade of another plant. When scientists first began to study the effects of light on seed germination, they found that red light is particularly good at stimulating germination, while far-red light inhibits germination. A simple "flip-flop" experiment by H. A. Borthwick and colleagues showed that the inhibitory effect of far-red light could be overcome by a subsequent exposure to red light (**Fig. 29.22**). Ultimately, it was shown that only a single photoreceptor was needed to detect both far-red and red light.

Phytochrome is a photoreceptor that switches back and forth between two stable forms depending on its exposure to light (**Fig. 29.23a**). Red light causes phytochrome to change into a form that absorbs primarily far-red light (P_{fr}); far-red light causes phytochrome to change back into the form that absorbs red light (P_r). Because red light stimulates seed germination, we know that P_{fr} is the active form of phytochrome. That is, red light causes phytochrome to change into its P_{fr} form. P_{fr} then moves from the cytoplasm to the nucleus, where it alters patterns of gene expression, leading to seed germination.

Phytochrome allows dormant seeds to detect the presence of plants overhead (**Fig. 29.23b**). Because the proportion of phytochrome that is in the active P_{fr} form depends on the ratio of red to far-red light, more P_{fr} will be produced in open environments than in the forest understory. Furthermore, because the ratio of red to far-red light is independent of light intensity, it provides a reliable signal that leaves are overhead, even when seeds are covered by a thin layer of soil. Independent of whether light is present, P_{fr} slowly converts back into the inactive P_r form, which allows the phytochrome system to reset itself overnight.

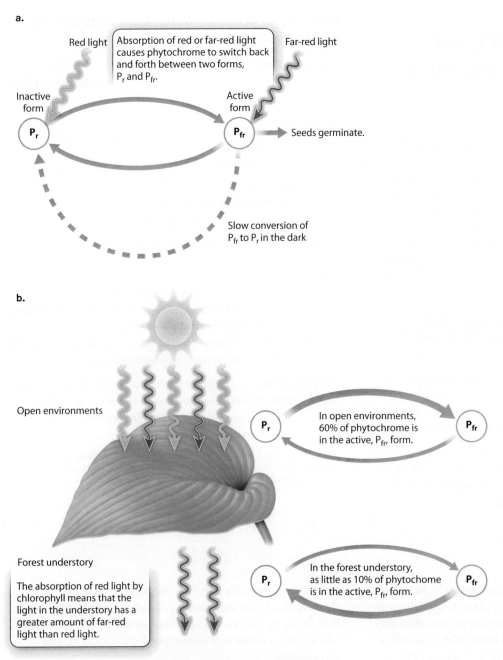

FIG. 29.23 Detection of shading. Phytochrome detects the depletion of red light under the canopy.

receives signals from several types of photoreceptors, including one type that absorbs blue wavelengths and another type, phytochrome, that absorbs both red and far-red wavelengths.

Even after a plant emerges above the soil, its ability to sense and respond to light can enhance its chances for survival. Some species are adapted for shade, producing thin leaves that efficiently use low levels of light; others grow best in direct sunlight. When such sun-adapted species are grown in the shade of another plant, they direct their resources primarily into growing tall. When shaded, these plant species produce stems that have long, thin internodes and fewer branches, and they invest less in root growth.

Interestingly, the effect is much less if the shade is produced by anything other than another plant. Plants can sense when shade is produced by neighboring plants because chlorophyll absorbs red, but not far-red, wavelengths of light. Thus, as light passes through leaves, it becomes depleted in red wavelengths. As a result, phytochrome is converted from the P_{fr} form to the P_r form. Furthermore, light reflected off the leaves of a plant's neighbors is also depleted in red wavelengths, so phytochrome allows a plant to detect not only when it is underneath another plant, but also when it is surrounded by other plants that could grow to shade it in the future. Thus, phytochrome provides an early-warning system that competitors are nearby.

Plants grow taller and branch less when growing in the shade of other plants.

A newly germinated seedling must reach the soil surface before using up its stored resources; otherwise, it will die. Seeds that germinate in the dark produce completely white seedlings with thin, highly elongated internodes and small leaves (**Fig. 29.24**). By putting all their resources into elongation, seedlings grown in the dark maximize their chances of reaching the light. If and when they do, internodal elongation slows markedly and green leaves expand. These dramatic changes are triggered by photoreceptors that signal the presence of light. The seedling

Roots elongate more and branch less when water is scarce.

The sensitivity of plant shoots to their environment is mirrored underground. When a plant experiences drought, it produces more roots, and these roots penetrate farther into the soil, in part because they produce fewer branches. Deeper roots increase a plant's chances of reaching moister soil.

Cells in the root sense and respond to the amount of water in the soil. Because the root cap is not well connected to the plant's vascular system, the water status of its cells provides a

FIG. 29.24 Seedlings grown in the light and the dark (left) or in full sun and shade (right). *Photo source: Nigel Cattlin/Science Source.*

Plants adapted to full sun develop long, thin internodes, and fewer leaves, when grown in shade.

Dark | Light

Full sun | Canopy shade

good indication of the moisture content of the soil. As soils dry out, root cap cells produce a hormone called **abscisic acid**, abbreviated as ABA (Table 29.1). ABA both stimulates root elongation and triggers stomata to close, reducing the demand for water.

Abscisic acid stimulates root elongation by suppressing ethylene synthesis. Ethylene slows root elongation by influencing the orientation of cellulose in the cell wall. Cells treated with ethylene expand in diameter, rather than in length. Thus, by inhibiting the production of ethylene, the production of ABA in drought-exposed roots leads to roots that penetrate deeper into the soil.

Exposure to wind results in shorter and stronger stems.

In the 1980s, scientists studied the model plant *Arabidopsis thaliana* to learn which genes are expressed in response to different hormones. The experimental treatment involved spraying plants with a variety of solutions containing different hormones; control plants were sprayed with water. The hormone treatments were successful, in that several genes were strongly up-regulated. Surprisingly, though, the same genes were turned on in the plants sprayed with water. Further experiments showed that the plants were responding to neither the hormones nor the water, but to their stems bending back and forth when the plants were sprayed. This finding led to the discovery of touch-sensitive genes in plants, which are activated by mechanical perturbation.

This discovery confirmed what foresters and horticulturists had long known. Plants exposed to wind produce stems that are shorter and wider than plants grown in more protected sites. In fact, commercial greenhouses often install fans, in part so that their plants will produce stems robust enough that the plants can thrive when moved outdoors. Flexing a stem back and forth triggers an increase in the synthesis of ethylene. As already noted, cells treated with ethylene expand more in diameter and less in length, resulting in shorter and thicker stems.

Self-Assessment Questions

10. What are the similarities and differences in the ways stems and roots respond to gravity and light?

11. Diagram how the amount of red light versus far-red light allows plants to determine whether they are in the open or in the forest understory.

12. What is an example of how a plant's ability to sense its environment improves the plant's chances for survival and reproduction?

29.6 TIMING OF DEVELOPMENTAL EVENTS

As plants colonized the land, they encountered environments with distinct seasons. In temperate regions, temperatures vary from summer to winter; in most tropical habitats, wet seasons alternate with dry ones. Thus, plants evolved mechanisms to control the timing of developmental transitions such as the production of flowers and other reproductive organs, the germination of spores or seeds, or the formation of overwintering buds. To time these events appropriately, plants need reliable information that tells them which season it is. While temperature tends to be unreliable—think of unusually warm or cold summers—day length is a dependable source of information about the seasons that works everywhere except close to the equator.

Flowering time is affected by day length.

In the 1910s, Wightman Garner and Henry Allard, scientists at the U.S. Department of Agriculture, set out to understand the unusual flowering schedule of Maryland Mammoth, a variety of tobacco. This plant does not produce flowers during the summer like other tobacco plants. Instead, these plants continue to grow throughout the summer, reaching heights several times that of a typical tobacco plant and producing many more leaves. Only in autumn does the Maryland Mammoth tobacco plant finally produce flowers, too late in the year for seeds to mature.

One possible explanation for the variety's delay in flowering time is that Maryland Mammoth must achieve a greater size before it produces flowers. However, when Garner and Allard grew Maryland Mammoth plants in a heated greenhouse during the winter, the plants flowered at the same time and size as other tobacco plants. Therefore, size does not determine when Maryland Mammoth flowers. Because the heated greenhouse was set to typical summer temperatures, temperature could also be eliminated as a factor. Garner and Allard were left with only one environmental variable that was different between the summer and the winter: the length of the day.

In a series of experiments, Garner and Allen artificially shortened the day length during summer and extended it during winter. Their results showed that in Maryland Mammoth, as well as in many other plants, flowering is controlled by day length. They coined the term **photoperiodism** to describe the effect of the "photo period," or day length, on flowering (**Fig. 29.25**). **Short-day plants**, like Maryland Mammoth, flower only when the day length is less than a critical value. When the light period exceeds this threshold, a short-day plant continues to produce new leaves, but no flowers are formed. In contrast, **long-day plants**, which include radish, lettuce, and some varieties of wheat, flower only when the light period exceeds a certain length. Flowering of **day-neutral plants** is independent of any change in day length.

Sensitivity to day length ensures that plants flower only when they are large enough to support the development of a large number of seeds *and* there is enough time for those seeds to mature before the onset of winter. A short-day plant that germinates in the spring will not flower until late summer, when the decreasing day length falls below the critical value for that species (**Fig. 29.26a**). This flowering strategy allows short-day plants to grow as large as possible and switch from producing leaves to producing flowers in time to complete seed development before winter. In contrast, many long-day plants produce flowers late spring or early summer. These plants germinate in spring and are triggered to flower when the day length exceeds a critical value (**Fig. 29.26b**).

FIG. 29.25 Control of flowering by day length. Short-day plants flower only when hours of daylight are less than a certain number. Long-day plants flower only when hours of daylight are greater than a certain number. *Photo sources: (top left) Darlyne A. Murawski/Getty Images; (bottom left) Liz Van Steenburgh/Shutterstock; (top right) DR JEREMY BURGESS/SCIENCE PHOTO LIBRARY; (bottom right) Dr Jeremy Burgess/Science Source.*

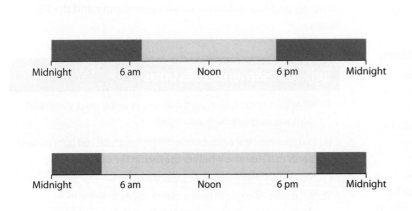

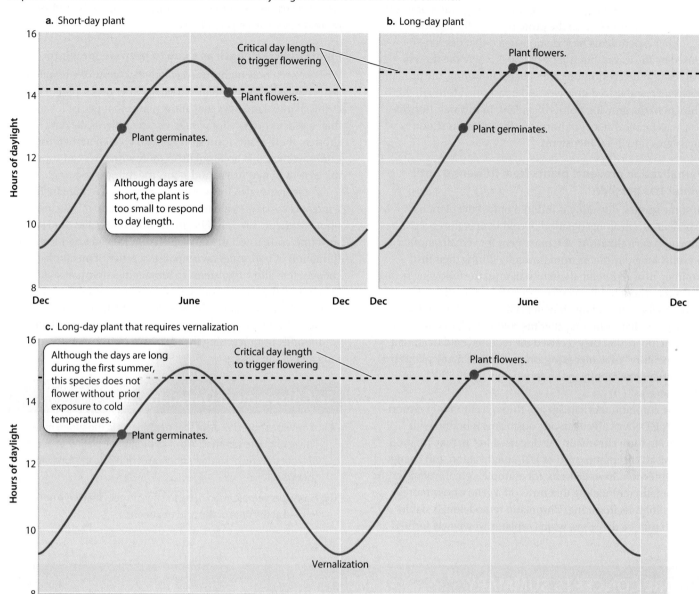

FIG. 29.26 Flowering strategies for short-day and long-day plants. (a) A short-day plant that germinates in the spring flowers in the late summer, when day length falls below a critical value. (b) A long-day plant flowers in the summer after a critical day length is reached. (c) A long-day plant that requires vernalization does not flower until the critical day length is reached in the second summer.

Plants use their internal circadian clock and photoreceptors to determine day length.

To determine day length, one first needs to know what time it is, specifically how long it has been since the sun came up, and then whether it is light or dark outside. Plants, like all organisms, have an internal circadian clock. Circadian clocks are biochemical mechanisms that oscillate with a 24-hour period and are coordinated with the day–night cycle. Phytochrome is the photoreceptor responsible for synchronizing the plant's internal oscillator with the first light of dawn. The circadian clock tells the plant how long it has been since the sun came up, while phytochrome tells the plant whether it is light or dark during a specific phase of the circadian cycle.

The actual mechanism used by plants to determine day length relies on the fact that the plant's circadian clock affects the transcription of many genes. For example, in *Arabidopsis thaliana,* a long-day plant, a gene whose protein product is critical for triggering flowering is transcribed at increasingly higher rates throughout the day, peaking approximately 16 hours after dawn. After that, the rate of transcription declines. The protein product of this gene is targeted for rapid degradation unless light is present. Thus, whether the protein builds up to a level sufficient to trigger flowering depends on day length. During long days, the protein reaches levels needed to trigger flowering because peak production occurs during daylight, when the protein is not targeted for destruction. During short days, the protein product does not reach levels needed to trigger flowering because the gene is transcribed at high rates only after the sun has gone down, so the protein is rapidly degraded.

Other experiments have demonstrated that day length is sensed by the leaves, but not by the buds where the flowers will form. When conditions are right for flowering, the protein florigen is synthesized in leaves and transported through the phloem to the growing points of a plant. In this way, florigen acts as a chemical signal or hormone that converts shoot meristems into floral meristems.

Vernalization prevents plants from flowering until winter has passed.

In some species, flowering is induced only if the plant has experienced a prolonged period of cold temperatures, a process known as **vernalization**. A requirement for vernalization prevents long-day plants from flowering during their first summer, instead forcing them to wait until the following spring to flower, when the chances for successful pollination and seed development are much better (**Fig. 29.26c**). Even parts of the plant that form long after the cold stimulus is past seem to "remember" that they received the required cold treatment. This "memory" of winter is important because many plants that overwinter grow rapidly in the spring. But what is the basis of this "memory"?

One common mechanism acts through chromatin remodeling. The DNA of all eukaryotic organisms is bound with proteins to form chromatin. As discussed in Chapter 18, modifications to chromatin, such as DNA methylation, can change gene expression. In *Arabidopsis*, for example, vernalization results in chromatin remodeling that turns off a gene whose protein product inhibits flowering. Chromatin remodeling is stable through mitotic divisions, which explains why newly formed parts of a plant "remember" winter. However, the slate is wiped clean during meiosis, and the requirement for vernalization is reinstated with each generation.

Plants use day length as a cue to prepare for winter.

We have seen how plants use day length to control when flowers are produced. Day length also triggers other developmental events, particularly ones that allow plants that persist for more than a year to prepare for winter. For example, as the days shorten, these plants form storage organs, often in their roots. Carbohydrates that accumulate in these structures can support the growth of new leaves and stems the following spring.

These plants also form overwintering buds to prepare for winter. As the days begin to shorten, plants stop producing photosynthetic leaves and begin forming bud scales. Bud scales surround and protect the meristem from ice and water loss. The formation of bud scales accompanies a series of metabolic changes that allow meristems to remain in a dormant state throughout the winter. For example, plugs are produced that block the plasmodesmata between the meristem and the rest of the plant. These plugs, which prevent any growth-stimulating compounds from reaching the meristem over the winter, are removed in the spring.

Self-Assessment Questions

13. Why will a short-day plant that germinates in the spring not flower until late summer? Why will a long-day plant that germinates at the end of summer not flower until late the following spring?

14. How does vernalization have an effect in cells that were not formed at the time of the cold treatment?

CORE CONCEPTS SUMMARY

29.1 SHOOT GROWTH AND DEVELOPMENT: In plants, upward growth by stems occurs at shoot apical meristems, populations of totipotent cells that produce new cells for the lifetime of the plant.

Stems are built from repeating modules consisting of nodes, where one or more leaves are attached, and internodes, which are the regions of stem between the nodes. page 639

The shoot apical meristem, located at the tip of each stem, is the source of new cells from which the stem and the leaves are formed. page 639

Stems contain zones of cell division, elongation, and maturation. Most of the increase in size occurs in the zone of cell elongation. page 641

In seed plants, branching occurs at axillary buds at the base of each leaf. page 642

Leaves differ from stems in two major ways: all cells differentiate fully, and leaves in most species have no capacity for continued growth. page 642

Leaves develop from leaf primordia produced by the shoot apical meristem. page 643

In most seed plants, the stem's vascular bundles are organized as a ring. page 643

Shoot apical meristems and axillary buds can be transformed into floral meristems, at which point they lose their capacity for continued growth. page 644

29.2 PLANT HORMONES: Plant hormones are chemical signals that influence the growth and differentiation of plant cells.

Traditionally, five plant hormones have been identified: auxin, gibberellic acid, cytokinins, ethylene, and abscisic acid. page 645

The transport of auxin guides the placement of leaf primordia and the development of vascular channels (xylem and phloem) that connect leaves and stems. page 646

The polar transport of auxin depends on a difference in pH between the cytoplasm and the cell walls, as well as the effect of this difference on the ionization of auxin. page 646

The polar transport of auxin through PIN proteins ensures the movement of auxin in a specific direction to establish the patterning of leaf primordia and the continuity of vascular channels. page 646

Gibberellic acid causes shoot internodes to elongate. page 647

The control of axillary bud growth by hormone interactions illustrates how growth and development are often regulated by interactions between two or more hormones. Page 647

29.3 SECONDARY GROWTH: Lateral meristems allow plants to grow in diameter, increasing their mechanical stability and the transport capacity of their vascular system.

The formation of two lateral meristems, the vascular cambium and the cork cambium, allows plants to grow in diameter. page 648

The vascular cambium produces secondary xylem toward the center and phloem toward the outside. page 649

The cork cambium produces a protective outer layer. page 650

Gymnosperm tracheids provide both mechanical support and water transport. Angiosperm xylem produces fibers that influence the strength of wood and vessel elements that function in water transport. page 650

29.4 ROOT GROWTH AND DEVELOPMENT: The root apical meristem produces new cells that allow roots to grow downward into the soil, enabling plants to obtain water and nutrients.

The root apical meristem is the source of new cells that allow roots to elongate. page 651

The root cap protects the root apical meristem from damage as the root grows through the soil. page 652

The root's vascular tissues are located in the center of the root, surrounded by the pericycle, the endodermis, the cortex, and the epidermis. page 652

Lateral roots develop from new root apical meristems that form in the pericycle. page 653

Roots alter their development in response to conditions in the soil. page 653

Some roots are involved in adhesion, storage, or gas diffusion. page 653

29.5 THE ENVIRONMENTAL CONTEXT OF GROWTH AND DEVELOPMENT: Plants respond to light, gravity, and wind through changes in internode elongation and the development of leaves, roots, and branches.

Plant stems grow toward the light and away from gravity as a result of redistribution of auxin. page 655

Roots grow away from light and toward gravity as a result of redistribution of auxin. page 655

Starch-filled statoliths allow roots to sense and respond to gravity. page 656

Seedlings germinating in the dark produce elongated internodes and suppress leaf expansion to increase their ability to reach the soil surface before their stored resources are depleted. page 657

Phytochrome is a photoreceptor that allows plants to determine the amount of red light versus far-red light. Because light filtered through leaves has lower levels of red wavelengths, phytochrome enables plants to detect the presence of plants overhead. page 658

Some plants respond to the presence of neighboring plants by elongating more rapidly and suppressing branching, thereby maximizing their height growth. page 657

Stems that experience mechanical stress, such as bending in the wind, become short and wide. page 659

29.6 TIMING OF DEVELOPMENTAL EVENTS: Plants have sensory systems that control the timing of developmental events.

Many plants are photoperiodic, meaning that their flowering is controlled by day length. page 660

Plants measure day length through an interaction between photoreceptors and their internal circadian clock. page 661

Vernalization is a prolonged period of exposure to cold temperatures necessary in some temperate-zone species before flowering can be induced. page 662

Decreases in the photoperiod trigger plants to prepare their meristems for winter by producing bud scales and reducing internode elongation. page 662

Log in to **LaunchPad** to check your answers to the Self-Assessment Questions and to access additional learning tools.

CHAPTER 30 Plant Defense

CORE CONCEPTS

30.1 PROTECTION AGAINST PATHOGENS: Plants have mechanisms to protect themselves from infection by pathogens.

30.2 DEFENSE AGAINST HERBIVORES: Plants use chemical, mechanical, and ecological defenses to protect themselves from being eaten by herbivores.

30.3 ALLOCATING RESOURCES TO DEFENSE: The production of defenses is costly, resulting in trade-offs between protection and growth.

30.4 DEFENSE AND PLANT DIVERSITY: Interactions among plants, pathogens, and herbivores contribute to the origin and maintenance of plant diversity.

Plants have complicated interactions with animals. As we saw in Chapter 28, many plants rely on animals for pollination or seed dispersal. In return, plant seeds provide a rich source of food for animals, and leaves furnish a vast, but lower-quality, food resource. The plant body also provides habitats for an array of viruses, bacteria, and protists, not to mention other plants. How do plants defend themselves against the animals that would eat them? How do they defend against pathogens? The answers to these questions help us understand how plant populations are distributed in nature, and they help us solve practical problems as well. In the United States alone, crop damage by herbivores and pathogens costs farmers more than $2 billion every year.

Plants can't run away and hide as animals can. Nonetheless, they can sense danger and are able to respond by deploying a diverse and powerful defensive arsenal. Their defenses include physical and chemical deterrents as well as aerial surveillance, armed guards, and sticky traps.

Plants produce a wide array of chemicals that alter the metabolism and influence the behavior of animals that seek to consume them. These molecules also interact with our own biochemistry, which explains why compounds derived from plants have been used since antiquity as sources of medicine. Our early ancestors turned to the leaves and bark of willow trees in much the same way that we rely on aspirin—and for good reason. Willows produce a compound that deters herbivores but in humans relieves aches and pain; this compound provides the chemical basis of aspirin. Quinine, derived from the tropical cinchona tree, is used in the treatment of malaria. Belladonna, a relative of the tomato, produces toxins that the ancient Romans used to eliminate rivals but in very low doses relieve gastrointestinal disorders. In fact, nearly 30% of all medicines used today are based on chemicals that plants produce to defend themselves.

30.1 PROTECTION AGAINST PATHOGENS

In September 1845, potato plants across Ireland began to die. Their leaves turned black and entire plants began to rot. When the potatoes were dug up, those that were not already rotten soon shriveled and decayed into a foul-smelling and inedible mush. The biological agent that had destroyed the potato crop looked like a fungus, but in fact was an oomycete called *Phytophthora infestans*, a type of protist (Chapter 25).

Phytophthora infestans is one example of the diverse pathogens that infect plants. Plant pathogens include viruses, bacteria, fungi, nematode worms, and even other plants. What all of these organisms have in common is the ability to grow on and in the tissues of plants, extracting resources to support their own growth. The outcome for the plant is often disease and death (**Fig. 30.1**). Most plants appear disease free only because they are able to defend themselves. Plants have an immune system that allows them to recognize and respond to pathogens that would otherwise exploit them. In the Irish potato famine, which lasted from 1845 to 1852, these defenses were breached. *P. infestans* devastated the potato crop, and many members of the Irish population starved to death.

Plant pathogens infect and exploit host plants by a variety of mechanisms.

The epidermis, with its thick walls and waxy cuticle, is a plant's first line of defense against pathogens. Viruses and bacteria cannot penetrate the cuticle, but instead commonly enter plants through wounds or on

FIG. 30.1 A potato plant infected with *Phytophthora infestans*, an oomycete that causes blight. *Source: Nigel Cattlin/Science Source.*

FIG. 30.2 Entry points for pathogens. A scanning electron micrograph shows *Phytophthora infestans* gaining access to the leaf interior by growing through stomata. *Source: Clouds Hill Imaging Ltd./Science Source.*

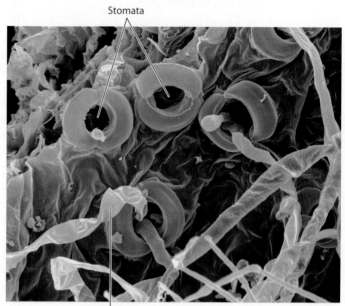

the piercing mouthparts of insects and nematodes. Stomata form a natural entry point for bacteria, oomycetes, and fungi (**Fig. 30.2**). Other pathogens, notably parasitic plants and some species of fungi, secrete enzymes that weaken epidermal cell walls. These invaders then force their way through the weakened epidermis. Crop plants damaged by farm machinery also contain entry points for pathogens.

Once inside the plant, fungi and bacteria can spread by secreting enzymes that degrade plant cell walls. In contrast, viruses move from one plant cell to another through plasmodesmata. The plant vascular system provides a highway that allows pathogens to move over long distances. Viruses and some bacteria are transported within the phloem, whereas fungi and other bacteria move through the xylem. During the 1950s, commercial banana production in Mesoamerica was nearly wiped out by *Fusarium oxysporum*, a fungal pathogen whose spores are carried in the xylem. More recently, the Florida citrus industry has been threatened by citrus greening disease, which is thought to result from infection by *Liberibacter*, a bacterium that moves through the phloem.

Biotrophic pathogens obtain resources from living cells, whereas **necrotrophic pathogens** kill cells before exploiting them. Most biotrophic pathogens do not enter into the host plant's cells. Instead, they produce chemicals that stimulate the host cells to secrete compounds that the pathogen can use as food. Viruses, however, are biotrophic pathogens that do enter the cell, where they hijack the living cell's machinery so that the virus can reproduce. Necrotrophic pathogens produce toxins that kill cells; the pathogens feed on compounds that leak from the dead and dying cells. Bacteria and fungal pathogens can be either biotrophic or necrotrophic. For example, wheat leaf rust (*Puccinia* species) is a biotrophic fungus that significantly reduces wheat yields (Chapter 32). Rice blast (*Magnaporthe grisea*) is a necrotrophic fungus that each year reduces rice production by an amount that would feed 60 million people.

Parasitic plants obtain resources by infecting other plants (**Fig. 30.3**). These plants produce structures that penetrate into the stems or roots of other plants, eventually tapping into the host plant's vascular system. Some, like some mistletoe (Fig. 30.3a), are xylem parasites: they obtain water and nutrients from the xylem of the host plant without expending the costs of producing roots themselves. Other plant pathogens target both the xylem and the phloem. In the most extreme cases, the parasite receives all of its carbohydrates from its host and carries out no photosynthesis of its own. *Rafflesia* (Fig. 30.3b), for example, grows entirely within the tissues of its host. It emerges only when it produces its giant flowers, the largest in the world. The ability to parasitize other plants has evolved independently multiple times. Currently, more than 4000 species of parasitic plants have been identified.

Plants are able to detect and respond to pathogens.

Most plants are at risk of infection by only a small subset of the many known plant pathogens. **Host plants** of a particular pathogen are the species that can be infected by that pathogen.

FIG. 30.3 Parasitic plants. (a) Mistletoe carries out photosynthesis, but relies on its host plant for water and nutrients. (b) *Rafflesia* produces the world's largest flower. It does not photosynthesize, but instead obtains everything it needs from its host. (c) *Cuscuta*, or dodder, parasitizes the stems of many plants. It also obtains everything it needs from its host. *Sources: a. Mark Boulton/NHPA/Photoshot; b. southeast asia/Alamy; c. Clive Varlack/clivevbugs/AGPix.*

Not all infections significantly damage the host plant. **Virulent** pathogens can overcome the host plant's defenses and lead to disease. In contrast, **avirulent** pathogens damage only a small part of the plant because the host plant is able to contain the infection. Whether an interaction is virulent or avirulent depends on the genotypes of both the pathogen and the host plant.

In the human body, the innate immune system provides a first line of defense against pathogens (Chapter 41). Plants, too, have an immune system that detects pathogens and mounts an appropriate response. A key feature of the immune system is that plants, like animals, can recognize a cell or virus as foreign. How do they accomplish this?

Plant cells have protein receptors (Chapter 9) that bind to molecules produced by pathogens and recognize them as foreign to the plant body. Binding changes the conformation of the receptor, triggering a cascade of reactions that enhance the plant's ability to resist infection.

The plant immune system has two components (**Fig. 30.4**). Both rely on receptors, but these receptors are in different

FIG. 30.4 The plant immune system. (a) In basal resistance, plasma membrane receptors recognize molecules produced by broad classes of pathogens. (b) Specific resistance depends on *R* genes that allow plant cells to identify and deactivate AVR proteins produced by specific pathogens.

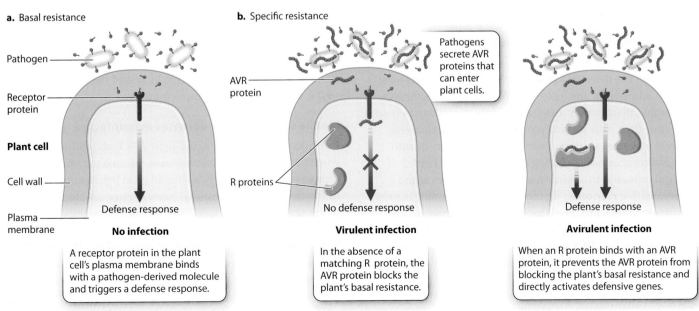

locations. The first component, known as basal resistance, consists of receptors located on the plasma membrane (Fig 30.4a). These receptors recognize highly conserved molecules generated by broad classes of pathogens. Examples of such molecules include flagellins, which are present in the flagella of bacteria, and chitin, which is a component of the fungal cell wall. When one of these molecules binds to the plant's receptor, an array of defense mechanisms is triggered that help protect the plant from infection.

Some pathogens have evolved a means of overcoming basal resistance. These pathogens produce proteins called AVR proteins that enter into plant cells and enable infection ("AVR" stands for "avirulence"). Some AVR proteins block defense responses triggered by the basal resistance of plants; others cause the cell to secrete molecules that the pathogen needs to support its own growth.

The second component of the plant immune system, known as specific resistance, targets these AVR proteins. In this case, the receptors are located inside the cell rather than on the plasma membrane (Fig. 31.4b). Plants harbor hundreds of these receptors, called **R proteins**. Each is expressed by a different gene, with the various genes being collectively called **R genes** ("R" stands for "resistance"). Each R protein recognizes a specific AVR protein. When an R protein binds with a pathogen-derived AVR protein, it both prevents the AVR protein from blocking the plant's defenses and activates additional defenses.

The large number of R genes is the result of an evolutionary arms race. Pathogens improve their success rate when mutation results in an AVR protein that does not bind with any of the R proteins produced by the host plant. For example, *P. infestans*, the pathogen responsible for the Irish potato famine, produces many more AVR proteins than related *Phytophthora* species. This large number of AVR proteins increases the probability that at least one of the proteins will escape detection by the infected plant. Conversely, when pathogens are successful, natural selection favors plant populations that produce new R proteins capable of binding to the AVR proteins of the invader.

Plants respond to infections by isolating infected regions.

Once a plant has detected a pathogen, how does it protect itself? One way is by reinforcing its natural barriers against pathogens. For example, plants under attack commonly strengthen their cell walls, close their stomata, and plug their xylem. Alternatively, the plant may counterattack. Plants produce a variety of antimicrobial compounds, including ones that attack bacterial or fungal cell walls. In a third type of defense, known as the **hypersensitive response,** uninfected cells surrounding the site of infection rapidly produce large numbers of reactive oxygen species. These species trigger cell wall reinforcement and cause the cells to die (**Fig. 30.5**). The resulting barrier of dead tissue prevents the spread of biotrophic pathogens.

FIG. 30.5 Hypersensitive response. A tobacco plant infected with tobacco mosaic virus produces a region of dead tissue that surrounds the site of infection and prevents the virus from spreading to other parts of the plant. *Source: Norm Thomas/Science Source.*

Site of initial infection

Plants can isolate infected regions to a far greater degree than is possible in most animals. For example, plants infected by xylem-transported pathogens often plug their xylem conduits. If the infection is caught in time, the pathogen is prevented from spreading throughout the plant. But if containment fails and the pathogen spreads, the plant may contribute to its own death by blocking its entire vascular system. Virulent pathogens that move through the xylem cause a variety of disorders collectively called vascular wilt diseases (**Fig. 30.6**). The widespread death of native elm trees in both Europe and North America is the result of a vascular wilt disease caused by a fungus introduced from Asia. This pathogen was first identified in the Netherlands, so it is commonly called Dutch elm disease.

Mobile signals trigger defenses in uninfected tissues.

In 1961, American plant pathologist A. F. Ross reported that when individual tobacco leaves were infected with the tobacco mosaic virus (TMV), they developed dead patches characteristic of a hypersensitive response. However, when uninfected leaves of the same plant were subsequently exposed to the virus, they suffered little or no damage (**Fig. 30.7**). Ross called this ability to resist future infections **systemic acquired resistance (SAR).** Initially reported for viral infections, SAR is now known to

FIG. 30.6 Vascular wilt diseases. Pathogens that are transmitted through the xylem can kill their host if plant defenses block water transport to the leaves. (a) Panama disease, which affects bananas, is caused by a fungal pathogen. (b) Pierce's disease, which affects grapes, is caused by a pathogenic bacterium. *Sources: a. Guy Blomme/Bioversity International; b. Bruce Fleming/Cephas Picture Library.*

HOW DO WE KNOW?

FIG. 30.7

Can plants develop immunity to specific pathogens?

BACKGROUND Tobacco is susceptible to an infectious disease that causes its leaves to turn a mottled yellow; hence the name "tobacco mosaic disease." In the 1890s, experiments with filters showed that the infectious agent was smaller than a bacterium, a finding that led to the discovery of viruses. Some plants succumb to the disease, but others do not. Are the plants that survive the initial infection immune from further attacks?

HYPOTHESIS Plants that survive infection with tobacco mosaic virus (TMV) have acquired immunity and, therefore, resist further attack.

EXPERIMENT American plant pathologist A. F. Ross infected one leaf on a tobacco plant with TMV. One week later, he exposed another leaf on the same plant to the virus.

RESULTS Plants that had been previously exposed to TMV showed only minor signs of infection when exposed to the virus a second time.

CONCLUSION Tobacco leaves that have never been exposed to TMV can acquire resistance to the pathogen if another leaf on the same plant has been previously exposed to the virus. This result indicates that a signal has been transmitted from the infected leaf to the undamaged parts of the plant, and that the transmitted signal

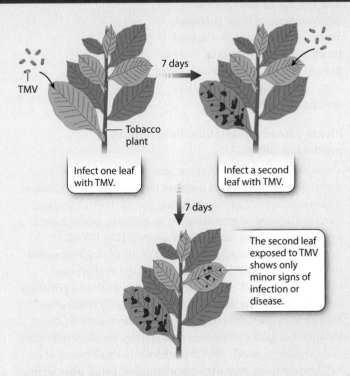

subsequently triggers an immune response that protects the plant from further infection.

FOLLOW-UP WORK Plants are now known to acquire immunity to a wide range of pathogens in addition to viruses.

SOURCE Ross, A. F. 1961. "Systemic Acquired Resistance Induced by Localized Virus Infections in Plants." *Virology* 14:340–358.

occur in response to a wide range of pathogens, including bacteria and fungi.

How do uninfected leaves acquire resistance to pathogens? One hypothesis is that a chemical signal is transported from the infected region through the phloem. This chemical signal then triggers the expression of genes encoding many of the same proteins that defend against the pathogen in infected cells. The identity of the mobile signal has proved elusive, but experiments show that salicylic acid (the chemical basis of aspirin) is required for SAR.

A plant steps up its production of salicylic acid when it is infected by a biotrophic pathogen that triggers a hypersensitive response. Plants produce different signaling molecules in response to necrotrophic pathogens, notably jasmonic acid and the plant hormone ethylene. Both of these chemicals activate defenses effective against broad classes of pathogens. Only in the case of viral infections, discussed next, is the response targeted to a specific pathogen.

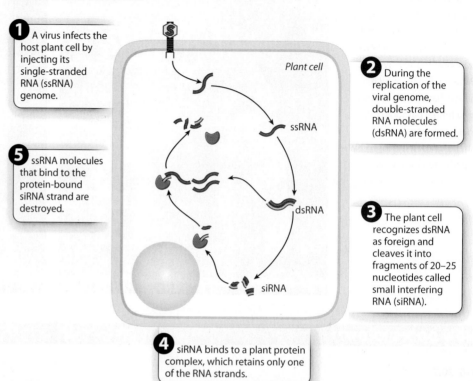

FIG. 30.8 Plant defense by small interfering RNAs (siRNAs). siRNAs target complementary single-stranded RNA for destruction.

1. A virus infects the host plant cell by injecting its single-stranded RNA (ssRNA) genome.
2. During the replication of the viral genome, double-stranded RNA molecules (dsRNA) are formed.
3. The plant cell recognizes dsRNA as foreign and cleaves it into fragments of 20–25 nucleotides called small interfering RNA (siRNA).
4. siRNA binds to a plant protein complex, which retains only one of the RNA strands.
5. ssRNA molecules that bind to the protein-bound siRNA strand are destroyed.

Plants defend against viral infections by producing siRNA.

Viruses can be potent infectious agents. To defend against them, plants have evolved multiple responses to viral infection, including a hypersensitive response, that are much like those mounted against other pathogens. In addition, plants have a form of defense that targets the virus itself (**Fig. 30.8**).

Most plant viruses have genomes made of single-stranded RNA (ssRNA). During the replication of the viral genome inside the plant cell, double-stranded RNA molecules (dsRNA) are formed. Because plant cells do not normally make double-stranded RNA, the replicating viral genomes are readily identified as foreign. Enzymes produced by the plant cell cleave these double-stranded RNA molecules into small pieces of 20 to 25 nucleotides, forming fragments called **small interfering RNA (siRNA)** (Chapter 3). The siRNA fragments then bind with protein complexes in the cell. The RNA–protein complexes subsequently play a role in targeting and destroying ssRNA molecules that have a complementary sequence—that is, the viral genome (Chapter 18). The damaged virus is unable to replicate and, therefore, cannot spread.

When a cell is attacked by a virus, the siRNA molecules that are produced in response to the initial infection can move through plasmodesmata, enter the phloem, and then spread throughout the plant. This movement allows the plant to acquire systemic immunity against specific viruses. Thus, the response to viral infection consists of both general defenses triggered by salicylic acid and the transport of highly specific RNA molecules that target and destroy viral genomes throughout the plant.

A pathogenic bacterium provides a way to modify plant genomes.

Some pathogenic bacteria alter their host plant's biology by inserting their own genes into the host's genome. One such pathogen, *Rhizobium radiobacter* (formerly known as *Agrobacterium tumifaciens*), is useful to humans, even though it can cause widespread damage to crops such as apples and grapes. The reason? This bacterium's naturally evolved ability to incorporate its DNA into the genome of its host plant provides a surprisingly easy way to genetically modify plants.

Rhizobium radiobacter is a soil bacterium with virulent strains that infect the roots and stems of many plants (**Fig. 30.9**). Virulent bacteria enter a plant through wounds and move through the plant's cell walls, propelled by flagella. A tumor,

FIG. 30.9 The formation of a crown gall tumor. *Rhizobium radiobacter,* a soil bacterium, can insert its genes into the plant genome, altering the growth and metabolism of the plant in ways that enhance the growth of *R. radiobacter*.

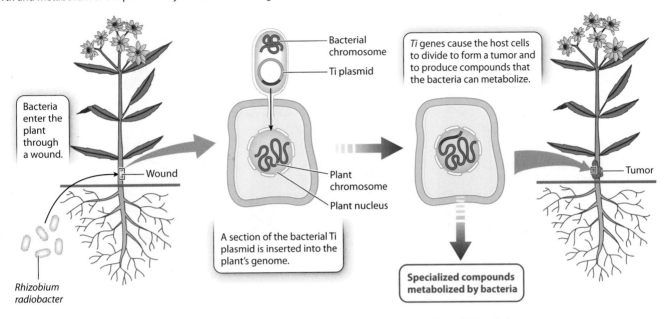

or gall, forms at the point of infection, often where the stem emerges from the soil. For this reason *R. radiobacter* infection is called crown gall disease.

Virulent strains of *R. radiobacter* contain a plasmid, called the **Ti plasmid**, which is a small (approximately 200 kb), circular DNA molecule. This plasmid contains genes that will become integrated into the host cell's genome, as well as all the genes needed to make this transfer. The integrated genes alter the growth and metabolism of infected cells. Some of the transferred genes—including ones that result in the synthesis of the two plant hormones auxin and cytokinin (Chapter 29)—cause infected cells to proliferate and form a gall. Other transferred genes induce cells in the gall to produce specialized compounds that the bacteria use as sources of carbon and nitrogen for growth. Note that because plant cells do not move (in contrast to animal cells), there is no possibility that the tumor-forming cells will spread to other parts of the plant.

Biologists make use of *R. radiobacter*'s ability to insert genes into the host genome to genetically modify crops and other plants. The first step is to replace the genes in the Ti plasmid that are inserted into the plant genome. Genes for gall formation are removed, and genes of interest are inserted. These genes may have products that increase the plant's nutritional value or confer resistance against disease. In addition, researchers need a way to identify which cells have taken up the genes, so a marker gene, such as one conferring antibiotic resistance, is typically included. Cells that survive an application of the antibiotic are known to have successfully incorporated the Ti plasmid into their genome.

Self-Assessment Questions

1. How do pathogens enter and move within the plant body?
2. What is the difference between biotrophic and necrotrophic plant pathogens?
3. What are the two components of the plant immune system? Describe them.
4. How does the plant hypersensitive response differ from systemic acquired resistance, and how are they alike?
5. What is a feature of *Rhizobium radiobacter* that makes it a useful tool in biotechnology?

30.2 DEFENSE AGAINST HERBIVORES

The first land plants were followed closely in time by the ancestors of spiders and scorpions. Fossils show that these earliest land animals soon began to feed on plant fluids and tissues. As plants diversified, so did herbivores, and the pressure on plants to defend themselves against organisms from caterpillars to cows only increased.

Given that they are unable to run away, plants would seem to be the ideal food. Plants, however, have evolved a diversity of mechanisms that deter would-be consumers. Because these mechanisms deplete resources that could otherwise be directed toward growth and reproduction, plants have also evolved means of deploying their defenses in a cost-effective manner. In

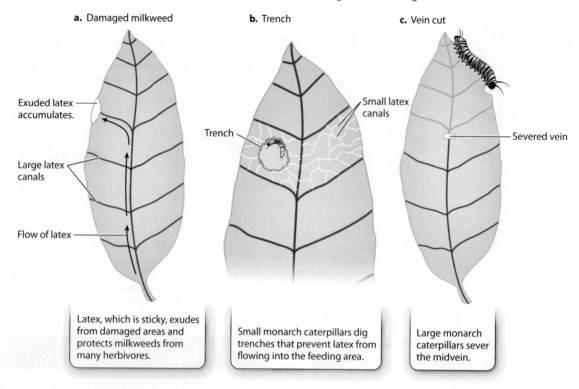

FIG. 30.10 How monarch caterpillars feed on milkweed. Young monarch caterpillars disarm milkweed defenses by digging trenches, while older, larger caterpillars sever major veins. Both methods prevent the sticky latex from flowing into the feeding area.

this section, we explore the mechanical, chemical, developmental, and even ecological means by which plants protect themselves in a world teeming with hungry herbivores.

Plants use mechanical and chemical defenses to avoid being eaten.

Milkweeds (*Asclepias* species) are common in open fields and roadsides across North America. They illustrate well how plants protect themselves from herbivorous animals. In fact, most generalist herbivores, referring to those herbivores that eat a diversity of plants, do not use milkweeds as a food source. Only specialists, such as caterpillars of the monarch butterfly (*Danaus plexippus*), can feed on these well-defended plants.

In part, milkweed's defenses are mechanical. Dense hairs cover the leaves of many milkweed species. An insect crawling on a leaf must make its way through as many as 3000 hairs per square centimeter, a journey likely more daunting to an insect than wading into a blackberry thicket is to a human. Monarch caterpillars may spend as much as an hour mowing off these hairs before they begin to feed. This long handling time represents a significant cost. To understand why monarch caterpillars make this investment, let's look at milkweed's other defenses.

A network of extracellular canals filled with a white, sticky liquid called **latex** runs through the milkweed leaf, both within the major veins and extending into the regions between veins (**Fig. 30.10**). Latex canals are under pressure in the intact leaf. When the leaf is damaged, the latex flows out, sticking to everything it comes into contact with and rapidly congealing on exposure to air. In particular, the tiny mouthparts of a feeding insect may become glued together by the latex, with potentially fatal consequences. Small monarch caterpillars cut trenches into the leaf that prevent latex from flowing onto the region where they are feeding. Larger caterpillars sever the midvein, relieving the pressure within the latex canals downstream of the cut. The caterpillars can then feed without risk on the part of the leaf that lies beyond the severed midvein.

Milkweed latex is not only sticky but toxic. In particular, it contains high concentrations of cardenolides, steroid compounds that cause heart arrest in animals. (In small doses, these compounds are used to treat irregular heart rhythms in humans.) Monarch caterpillars do not metabolize the cardenolides that they consume. Instead, they sequester them within specific regions of their bodies as a chemical defense against their own predators. As a consequence of this buildup of cardenolides, monarch caterpillars and even the adult butterflies are highly toxic to most birds and other animals that would otherwise eat them. The protection conferred by these chemicals helps explain why monarch caterpillars have evolved to feed on leaves that are time-consuming to eat.

FIG. 30.11 Mechanical defenses. (a) Spines on the peach palm (*Bactris gasipaes*) make it hard for animals to climb or penetrate its stem. Hairs on the leaves of (b) nettle (*Urtica dioica*) and (c) *Arabidopsis thaliana* deter much smaller herbivores. *Sources: a. Fletcher & Baylis/Science Source; b. DR KEITH WHEELER/ Science Source; c. Dr. John Runions/Science Source.*

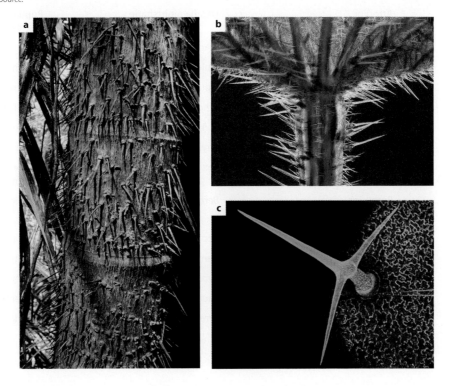

Most plant species have mechanical or chemical defenses against herbivores, or even both. Mechanical defenses vary. Hairs are common on leaves, and, as anyone who has ever rubbed against a stinging nettle knows, these hairs are sometimes armed with chemical irritants. Hard silica (SiO_2) plates form within the epidermal cells of grasses and some other plants. Silica wears down insect mouthparts, so the insects feed less efficiently and grow more slowly. On a larger scale, some trees have prickles, spines, or thorns (**Fig. 30.11**). Even something as simple as how tough a leaf is can have a large impact on the probability of the leaf's being eaten. In fact, leaves are most likely to be eaten while they are still growing and their tissues not fully hardened. Nonetheless, the most spectacular diversity in plant defenses lies in the realm of chemistry.

Diverse chemical compounds deter herbivores.

The cardenolides produced by milkweeds are just one example from the vast chemical arsenal available to plants for protection. Some plants produce **alkaloids,** nitrogen-bearing compounds that damage the nervous system of animals. Commonly bitter tasting, alkaloids include such well-known compounds as nicotine, caffeine, morphine, theobromine (found in chocolate), quinine (a treatment for malaria), strychnine, and atropine. Alkaloids are a costly defense because they are rich in nitrogen, an essential and often limiting element that plants need to build proteins for photosynthesis. However, even very small concentrations of alkaloids can deter herbivores. One of the first chemotherapy drugs (**Table 30.1**), vincristine, is an alkaloid extracted from the Madagascar periwinkle. It inhibits microtubule polymerization, thereby preventing cell division (Chapter 11). Vincristine does not harm the plant's own cells because it is stored within an extracellular system of latex-containing canals.

A second group of defensive compounds, the **terpenes**, do not contain nitrogen. As a result, plants can produce them for defense without having to make use of nitrogen that could otherwise be devoted to protein synthesis. Small terpenes are volatile and vaporize easily, so they make up many of the essential oils associated with plants. The distinctive smells of lemon peel, mint, sage, menthol, pine resins, and geranium leaves are all due to terpenes. These odors, while pleasant to us, are feeding deterrents to mammals such as squirrels and moose. In addition, terpenes obstruct the growth and metabolism of both fungi and insects. For this reason, pyrethrin, a terpene derivative extracted from chrysanthemums, is marketed commercially as an insecticide.

Phenols form the third main class of defensive compounds. Examples are the **tannins** found widely in plant tissues. Tannins bind with proteins, reducing their digestibility. Because plants store tannins in cell vacuoles, these compounds come into

TABLE 30.1 A Sampling of Plant Compounds Used in Medicine

CLASS	EXAMPLE	PLANT SOURCE	MEDICAL USE
Alkaloids (Morphine)	Atropine	Deadly nightshade	Pupil dilation; treatment of low heart rate
	Vincristine	Madagascar periwinkle	Cancer treatment
	Codeine	Opium poppy	Cough suppressant
	Morphine	Opium poppy	Pain treatment
	Quinine	Cinchona tree	Malaria treatment
	Ephedrine	*Ephedra* (Mormon tea)	Asthma treatment
	Turocurare (curare)	*Strychnos toxifera*	Muscle relaxant
	Digitoxin	Foxglove	Congestive heart failure treatment
	Berberine	*Berberis* (barberry)	Antibiotic, anti-inflammatory agent
Terpenoids (Menthol)	Menthol	Peppermint	Topical analgesic; mouthwash
	Camphor	Camphor laurel	Anti-itch cream
	Eucalyptol	*Eucalyptus* species	Mouthwash; cough suppressant
	Taxol	Pacific yew	Cancer treatment
	Artemisinin	*Artemisia annua*	Malaria treatment
Phenols (Salicylic acid)	Salicylic acid	Willow tree	Pain treatment

contact with the protein-rich cytoplasm only when cells are damaged. Herbivores attempting to feed on tannin-producing plants obtain a poor reward for their efforts. For this reason, natural selection favors individuals that avoid tannin-rich plants. Many unripe fruits are high in tannins: the unpleasant experience of biting into an unripe banana illustrates how tannins deter consumers. For thousands of years, humans have taken advantage of the protein-binding properties of tannins, using extracts from tree bark to process animal skins by "tanning" to produce leather.

Some chemical defenses found in plants are protein based. Plants and animals use the same 20 amino acids to construct proteins, but some plants produce additional amino acids as well. Plants do not incorporate these additional amino acids into their proteins, but herbivores that ingest them do. The resulting proteins can no longer fulfill their function. As a consequence, the insect herbivores that consume nonprotein amino acids grow slowly and often die early.

Another protein-based defense is the production of antidigestive proteins called **protease inhibitors**. These proteins bind to the active site of enzymes that break down proteins in the herbivore's digestive system. As a result, the enzymes can no longer break down proteins into their individual amino acids. Insects that feed on plants that produce protease inhibitors have reduced growth rates.

Some plants provide food and shelter for ants, which actively defend them.

While many plants have mechanical defenses against herbivores and nearly all have at least one form of chemical defense, a smaller number of species have evolved ecological defenses against herbivory. These plants "employ" animals as bodyguards in exchange for shelter or nourishment, much the way that many plants provide food or other rewards to animals that transfer pollen and disperse seeds (Chapter 28).

Nectar is typically associated with flowers, but many plants produce nectar in glands located on their leaves (**Fig. 30.12**). These extrafloral nectaries attract ants. As ants move actively throughout the plant in search of nectar, they may encounter the eggs or larvae of other insects, and they consume these as well. The value of this relationship to the plant is easily demonstrated: when ants are prevented experimentally from patrolling certain branches, those branches suffer higher rates of herbivore damage.

A small number of plants, several hundred in total, have evolved a much closer relationship with ants. These so-called ant-plants provide both food and shelter for an entire colony of ants, which then defend their host. The bullhorn acacia (*Acacia cornigera*), which grows in dry areas of Mexico and Central America, produces hollow spines at the base of each leaf, as well as protein- and lipid-rich food bodies on the tips of its expanding leaves. Symbiotic ants (*Pseudomyrmex ferruginea*) live in the spines and feed on the food bodies (**Fig. 30.13**). To protect this source of food and shelter, the ants actively defend the plant. When the ants detect an invader, they produce an alarm pheromone that causes the entire colony to swarm over the plant, attacking any insects or vertebrates (including scientists) that they encounter. The ants will attack the growing tip of a vine that might climb up the host. They even destroy plants growing in the soil beneath the host *Acacia*.

This system works well until it comes time to reproduce. Without some mechanism to control the hypervigilant ants, they would attack any visiting pollinator as though it were a truly unwanted guest. But the bullhorn acacia has a found a solution to this dilemma: to ensure that pollinators may visit its flowers unharmed, those flowers release a chemical that repels the resident ants.

The interactions between ant-plants and their ants are highly coevolved. The ant species are found only in association with their host plants. While ant-plants can be grown without ants, herbivores would rapidly consume an undefended plant in nature because ant-plants lack the chemical defenses present in related species that do not host ants.

Grasses can regrow quickly following grazing by mammals.

By far, the most important vertebrate herbivores are the bison, zebra, antelope, and their relatives that populate the great plains of East Africa and North America. Studies of herbivore damage in an East African savanna indicate that vertebrates annually consume approximately the same amount of leaf biomass as do insects.

Grasses are well adapted to cope with grazing mammals (**Fig. 30.14**). The vertically oriented leaves grow from stems with limited internodal elongation. The shoot apical meristem remains close to the ground, where it is less likely to be damaged. Grasses have persistent zones of cell division and

FIG. 30.12 Ants attracted by an extrafloral nectary. These ants receive a food reward and also consume any insect eggs or larvae they encounter on the plant. *Source: Alex Wild/alexanderwild.com.*

FIG. 30.13 The symbiosis between ants and the bullhorn acacia. (a) Ants live in the plant's thorns, and (b) the plant provides food (the yellow food bodies). The ants vigorously defend their home against animals and plants alike. *Source: Dan L. Perlman/EcoLibrary.org.*

FIG. 30.14 Adaptation for grazing. The ability of grass leaves to elongate from the base allows grasses to regrow following grazing.

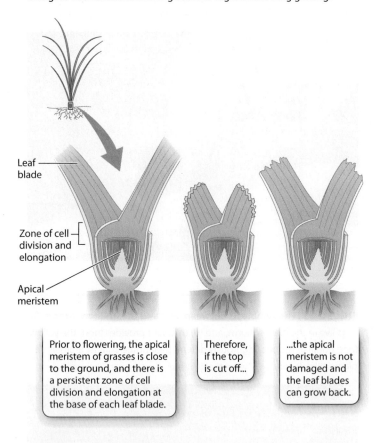

elongation at the base of each leaf. Because grass leaves elongate from their base, grasses can quickly replace leaf tissues removed by a grazing mammal, fire, or a lawn mower. Only when grasses produce their flowering stalks does the apical meristem emerge above the leaves.

Grasses are thought to have evolved in forest environments. In fact, grass-dominated ecosystems have become widespread only during the past 25 million years. The fossil record shows that as grasslands expanded, grazing mammals evolved along with them. Horses, for example, originated as small browsers that fed on the leaves of small trees and shrubs. As grasslands expanded, larger horse species with teeth adapted to eating grass evolved. Grasses have small plates of silica in their cell walls that increase their strength and may provide a mechanical defense against herbivores. Horses and other grazers that feed on these grasses wear down their tooth enamel grinding against the silica, so teeth with a thick enamel layer are favored by natural selection. It is not clear whether grasses evolved in response to grazing or to disturbances such as fire, but without question, their distinctive pattern of meristem activity minimizes the damage caused by grazing mammals.

> **Self-Assessment Question**
>
> 6. What are three ways that plants protect themselves from being eaten by herbivores?

30.3 ALLOCATING RESOURCES TO DEFENSE

By now, it should be clear that plants invest substantial resources in defense, resources that might otherwise be used for growth and reproduction. Given the considerable cost of plant defenses, their benefits must be correspondingly large. If a commonly encountered pathogen or herbivore is lethal, the resources invested in defense provide a clear reproductive advantage. But what if the pathogen or herbivore usually damages only half of a plant's leaves? Or 10%? Or encounters the plant only infrequently?

Some defenses are always present, whereas others are turned on in response to a threat.

In nature, a balance between cost and benefit is achieved in several different ways. When a threat is common, plants produce constitutive defenses—defenses that are always present. When a threat is uncommon, however, there is an advantage to mounting a defense only when the threat is encountered. A defense that is activated only when the plant senses the threat is called an inducible defense. An inducible defense can consist of either an increase in the level of an already present (constitutive) defense or the production of a new type of defense. The ability to sense and respond to the presence of herbivores allows plants to use resources efficiently.

When a plant is attacked by herbivores, it begins to express a new set of genes. Many of these genes increase the production of chemical defenses, while others stimulate the reinforcement of cell walls. In addition, herbivore damage triggers the synthesis of jasmonic acid, a signal that is transmitted through the phloem. Exposure to jasmonic acid induces the transcription of defensive genes even in parts of the plant untouched by herbivores.

Scientists are working to determine how information about herbivore damage is transmitted to distant parts of the plant. Chemical signals transported in the phloem may be the primary means of communicating with organs such as roots that are actively importing sugars. However, mature leaves export rather than import sugars, yet they also respond when a distant leaf is damaged. New research suggests that electrical signals transmitted within the plant vascular system may play an important role in alerting the plant to the presence of herbivores.

Plants can sense and respond to herbivores.

For more than 15 years, biologists have studied how coyote tobacco (*Nicotiana attenuata*), a relative of tobacco native to dry open habitats of the American West (**Fig. 30.15**), defends itself against herbivores. These studies provide a fascinating window into how plants allocate their resources toward defense under natural conditions.

Nicotiana attenuata is attacked by a wide variety of herbivores, both generalists and specialists. One of its major defenses against being eaten is the production of the alkaloid nicotine. Nicotine is an effective defense against many generalist herbivores, but it is expensive to produce because it diverts nitrogen away from the synthesis of proteins. Thus, *N. attenuata* plants that have not been exposed to herbivores produce only low levels of nicotine as a constitutive defense. When *N. attenuata* is attacked by a generalist herbivore, however, nicotine concentrations in leaves increase dramatically—typically fourfold, but in some cases as much as tenfold. Because nicotine is synthesized in roots and transported to leaves in the xylem, this inducible defense must be activated by signaling molecules transported in the phloem from the sites of insect damage to the roots.

Nicotiana attenuata is also attacked by a specialist herbivore, the tobacco hornworm caterpillar, *Manduca sexta*. These caterpillars, which can reach 7 cm in length, have evolved the ability to sequester and secrete nicotine. By this means, these caterpillars are able to consume leaves with high levels of nicotine. Since nicotine is not an effective defense against this voracious herbivore, how can *N. attenuata* protect itself?

It turns out that *N. attenuata* can recognize specific chemicals in the saliva of *M. sexta*. When attacked by *M. sexta*, *N. attenuata* does not waste time and resources synthesizing nicotine. Instead, it protects itself in an entirely different way: it attracts other insects that will attack the herbivore, as we discuss next.

Plants produce volatile signals that attract insects that prey upon herbivores.

How does *N. attenuata* attract its insect allies? A hint comes from the smell of a newly mown lawn. This distinctive smell originates from chemicals released from the cut blades of grass. Some of the chemicals released when plants are damaged are produced specifically in response to that damage. When *N. attenuata* is attacked by *M. sexta* caterpillars, it produces volatile signals that attract insects that prey on the caterpillar's eggs and larvae. The adage "The enemy of my enemy is my friend" seems to hold true when it comes to protecting plants from herbivore damage.

Studies first conducted in the 1980s showed that undamaged plants increase their synthesis of chemical defenses when neighboring plants are attacked by herbivores (**Fig. 30.16**). Although these plants were described in the popular press as "talking trees," the plants under attack are not altruistically warning their neighbors. Instead, subsequent studies suggested that the undamaged plants detect signals released by their neighbors that are directed at other insects, then ramp up their own defenses.

Nutrient-rich environments select for plants that allocate more resources to growth than to defense.

Some plant species allocate far more resources to defense than others. Because defenses are costly to produce, you might think

FIG. 30.15 *Nicotiana attenuata*, **the coyote tobacco.** This species has been developed as a model organism for studying how plants protect themselves against herbivores. (a) *Nicotiana attenuata* in a Utah desert; (b) *Manduca sexta*, a specialist herbivore, feeding on *N. attenuata*. *Sources: a. Danny Kessler, Max Planck Institute for Chemical Ecology, Jena, Germany; b. Celia Diezel, Max Planck Institute for Chemical Ecology, Jena, Germany.*

HOW DO WE KNOW?

FIG. 30.16

Can plants communicate?

BACKGROUND When damaged by herbivores, plants produce chemicals that make their tissues less palatable. Field experiments in which caterpillars were added to target plants suggested that undamaged neighboring plants increased their production of defensive chemicals, whereas plants farther away did not.

HYPOTHESIS Volatile chemicals released from plants that have been attacked by herbivores elicit a defensive response in undamaged plants.

EXPERIMENT One set of plants was placed downwind of other plants in which two leaves were torn, simulating herbivore damage, and a set of control plants was placed downwind of an empty chamber. The concentrations of defensive compounds in all sets of plants were measured and compared.

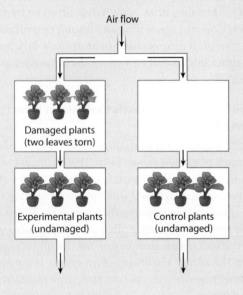

RESULTS When measured 52 hours after the initial damage, both the damaged plants and the undamaged plants downwind of them had elevated levels of defensive compounds in their leaves compared to control plants.

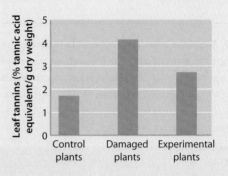

CONCLUSION Volatile chemicals released from damaged plant tissues can trigger the production of defensive chemicals in undamaged plants.

FOLLOW-UP WORK The volatile chemicals released from damaged plants have been identified and shown to attract predatory animals that feed on the herbivores, parasitize their eggs, or both. Thus, when plants respond to damaged neighbors, they are "eavesdropping" on signals that likely evolved to attract predatory insects, rather than to warn neighboring plants.

SOURCE Baldwin, I. T., and J. C. Schulze. 1983. "Rapid Changes in Tree Leaf Chemistry Induced by Damage: Evidence for Communication Between Plants." *Science* 221:277–279.

---

that plants growing in nutrient-rich habitats should be the best defended. But consider the counterargument: plants growing in nutrient-poor soils might invest heavily in defense because they cannot afford to replace tissues lost to herbivores. Such plants would grow relatively slowly because their resources are being used to build defenses rather than to produce new leaves and roots. By contrast, plants from nutrient-rich habitats would favor growth over defense because they can more readily replace lost or damaged tissues.

This situation is an example of a **trade-off**: something is gained and, at the same time, something is lost. You can't have your cake and eat it, too; you have to make a choice. With plants, there can be a trade-off between growth and defense: allocating resources for one means that those resources cannot be allocated for the other.

A recent study in the Amazon rain forest focused on a possible trade-off between growth and defense. Researchers compared plant species found growing in nutrient-rich clay soils with closely related species growing in nutrient-poor sandy soils. They found that species from the nutrient-rich clay soils invested less heavily in defenses than did species from the sandy sites. Because the clay-soil species divert fewer resources to defense, they grow faster in either soil type compared to sandy-soil species—as long as they are grown under a net that excludes herbivores (**Fig. 30.17**). When grown out in the open, each species grows best in the soil type in which it is naturally found.

FIG. 30.17 A trade-off between growth and defense. Plants from nutrient-rich clay habitats can overcome herbivory by growing fast, whereas plants from low-nutrient sandy habitats grow slowly and invest more in defense. The photos show that, without defenses, clay-soil plants are prone to being eaten by herbivores. *Source: Paul Fine*

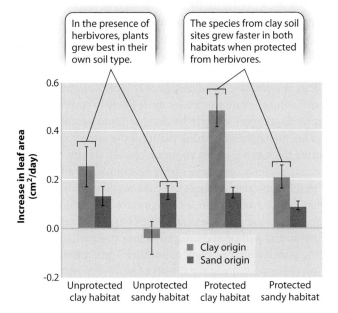

The clay-soil species grow poorly in sandy soils when not protected by a net because, without ample soil nutrients available, they are unable to replace tissues consumed by herbivores. Conversely, when grown in the nutrient-rich soils, species from the sandy habitats are poor competitors because they allocate resources that could have been used for growth to defensive compounds and structures. What we see is a classic example of a trade-off.

Exposure to multiple threats can lead to trade-offs.

In the real world, plants are routinely confronted with more than one threat. A plant may be attacked simultaneously by both a pathogen and an herbivore, or it might be in danger of being shaded by a fast-growing neighbor at the same time that it must defend its leaves from being eaten. Given that the resources a plant can draw on are limited, how do plants respond to multiple threats?

Often, the response to multiple threats suggests a trade-off. For example, a plant exposed to a pathogen may be less able to respond to a later herbivore attack. If tobacco plants are exposed to tobacco mosaic virus and systemic acquired resistance is activated, they become more susceptible to the hornworm caterpillar, *Manduca sexta*. Studies indicate that plants do not defend themselves as vigorously against subsequent herbivore attacks because of cross talk between the signaling pathways for responding to biotrophic pathogens and those for responding to herbivores.

There can also be trade-offs between the threat of competition with neighboring plants and the threat of attack by herbivores. For example, when the phytochrome receptor in plants detects the presence of neighboring plants (Chapter 29), plants allocate more resources to growing tall, but they also produce fewer defensive chemicals in response to herbivore damage. Interactions among the signaling molecules associated with each of these processes underlie this trade-off.

Self-Assessment Questions

7. What would happen to a plant in which jasmonic acid synthesis is blocked?
8. Why are there often trade-offs between plant growth and plant defense?
9. Does the trade-off between growth and defense favor a single species that dominates in all soil types or different species specialized for each habitat?

30.4 DEFENSE AND PLANT DIVERSITY

Many plants spend a significant fraction of their resource and energy budgets on defenses. Despite this commitment, insects consume approximately 20% of all new plant growth each year, while pathogens destroy a substantial additional amount. The magnitude of this loss means that pathogens and herbivores must be a potent force in both ecology and evolution, limiting the success of some species while allowing some of their competitors to prosper. In nature, plant fitness may be strongly influenced by the capacity to deter herbivores and resist pathogens.

Interactions between plants and their consumers are thought to have influenced patterns of plant diversification. Just like pollinating insects, plant-eating insects diversified along with angiosperms (the flowering plants). The coincidence of timing is consistent with an important role for plant–insect coevolution in generating diversity. Further evidence that defense has influenced plant diversification comes from plant genes and secondary compounds. For example, the *R* genes that are instrumental in pathogen recognition form one of the largest gene families in plants, while chemical defenses against herbivores include about 6000 alkaloids and more than 10,000 terpenoid compounds. Such observations suggest that plants are locked in an evolutionary arms race with herbivores and pathogens.

The evolution of new defenses may allow plants to diversify.

American scientists Paul Ehrlich and Peter Raven observed that closely related species of plants were fed upon by closely related species of butterflies. In 1964, they cited this observation as evidence that plants and their herbivores are caught up in an evolutionary arms race: plants evolve new forms of defense, and herbivores evolve mechanisms to overcome these defenses. In particular, Ehrlich and Raven hypothesized that the evolution of novel defenses has been an important force in the diversification of both plants and herbivorous insects.

A novel form of defense may allow a plant population to expand into new areas. If the population in a new area remains separated from the original population, it may evolve into a new species. Similarly, a novel means of overcoming plant defenses may allow an insect or pathogen population access to new plant resources. If these insects or pathogens became separated from their original population, then they might also evolve into new species. Thus, Ehrlich and Raven's "escape and radiate" hypothesis predicts a burst of diversification following the evolution of novel defenses.

This hypothesis makes intuitive sense, but it could be tested only with the advent of phylogenetic trees based on DNA sequence comparisons (Chapter 22). Biologists knew that latex or resin canals evolved independently in more than 40 different groups of plants. Reasoning that these features represent a novel defense, they asked whether the groups that evolved latex or resin canals were more diverse than closely related groups that lacked these forms of defense. In 14 of the 16 groups examined, the groups with the protective canals were significantly more diverse than their closest relatives that lacked the canals. This pattern supports the hypothesis that the evolution of latex and resin canals provided plants with the freedom to expand into new habitats.

Escalation of defenses is not the only possible evolutionary outcome. Phylogenetic research shows that, as milkweeds diversified, many of the newly evolved species produced *fewer* chemical defenses such as latex canals and cardenolides. Instead, these species grow in resource-rich habitats and replace tissues by growing quickly following damage. The presence of specialized herbivores such as monarch caterpillars that can disarm milkweeds' defenses may help explain why these more recent plant species reduce investments in defense in favor of more rapid growth.

Pathogens, herbivores, and seed predators can increase plant diversity.

The lowland rain forests of eastern Ecuador have the highest diversity of tree species recorded anywhere: more than 1000 species of trees were identified in a 500 m × 500 m plot. Because trees compete for sunlight, water, and nutrients, the species best able to acquire resources would be expected, over time, to outcompete all others. What allows so many tree species to coexist is one of the great questions of modern biology. In the 1970s, two ecologists, Daniel Janzen and Joseph Connell, independently proposed that interactions with host-specific pathogens, herbivores, or seed predators provide at least a partial answer.

The Janzen–Connell hypothesis, as it is now known, is based on the observation that seeds or seedlings often suffer high rates of mortality when the new plants grow directly beneath an adult tree of the same species (**Fig. 30.18**). Seed predators often gather in higher densities near trees with mature fruit, so seeds that disperse far from their maternal plant may escape notice by these predators. By a similar argument, pathogens are more common in the soil underneath an adult tree, so seedlings that grow farther away have a better chance of escaping these invaders. In both cases, high-density populations of seeds and seedlings are more at risk.

Rare species have an advantage over more common species: the offspring of a rare species are more likely to survive because their seeds are more likely to be dispersed into a location where there are no neighboring adult plants of the same species. By favoring the survival of less common species, density-dependent mortality prevents one species from outcompeting all others.

In the tropical rain forests of Ecuador and elsewhere, only a small number of tree species are common; the majority of tree species are rare. These rare species are able to escape herbivore and pathogen damage, but they need reliable ways of getting pollen from one plant to another. Thus, animal pollination (Chapter 28) and the threats of herbivory and infection by pathogens are thought to play complementary roles in promoting angiosperm diversity.

CASE 6 AGRICULTURE: FEEDING A GROWING POPULATION

Can modifying plants genetically protect crops from herbivores and pathogens?

The density-dependence of herbivore and pathogen damage is a particular challenge for agriculture as practiced in most countries. When acre after acre is covered by a single crop species, successful invaders are not slowed by low target density. On the contrary, pest populations build up to high levels, and the tendency to plant genetically uniform varieties, which play to the pests' strengths, increases the losses in yield. Not

FIG. 30.18 Density-dependent mortality promotes tree diversity. Seedlings of rare species have a greater probability of surviving because they are more likely to escape infection by species-specific pathogens.

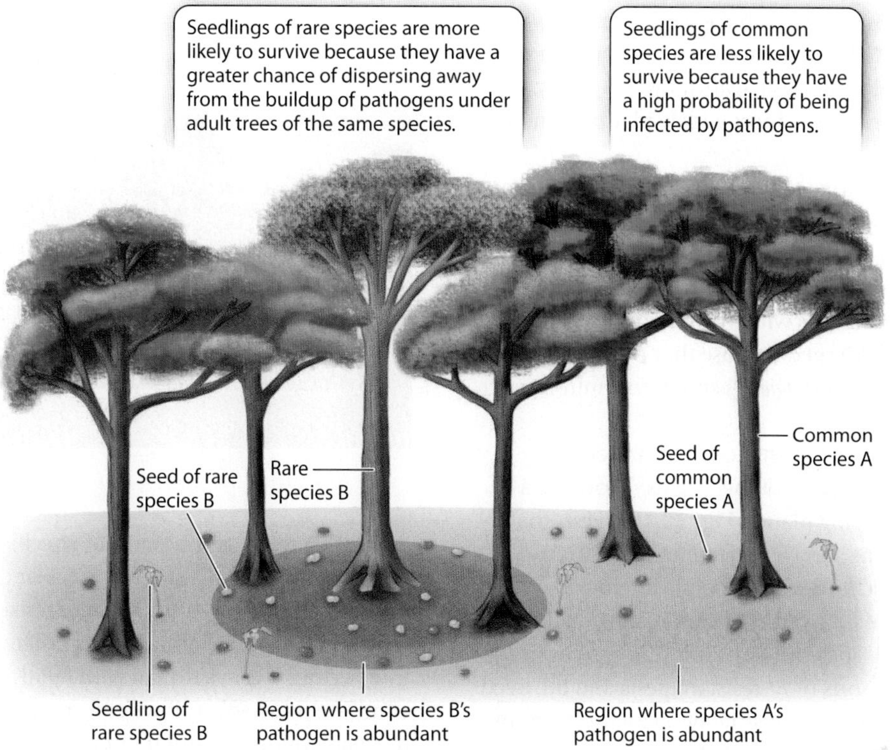

surprisingly, herbivores and pathogens exact a large toll on agricultural crops, and protecting crops from damage is a high priority for farmers.

The production of pesticides and herbicides increased dramatically in the second half of the twentieth century. Together with irrigation and fertilizers, these chemicals are important contributors to the increased yields of the Green Revolution. Unfortunately, large-scale application of pesticides has risks. In addition to the potential for toxic effects, widespread chemical treatments provide a strong selective force for the evolution of resistance. As pests become resistant to pesticides, one response is to apply more or stronger chemicals. A population of resistant pests can spiral out of control as chemical applications indiscriminately kill off their natural predators.

During the 1980s, farmers applied ever more pesticides to rice paddies in South Asia, yet crop losses continued to increase. This dangerous course was reversed by the introduction of a program that had originally been developed in the 1950s. Farmers in this program carefully determine which pests are actually present, whether they are at damaging levels, and whether there are natural predators of the pests. Only after understanding the ecology of the farm do farmers remove pests, starting with physical removal by hand or machine, followed by biological control using natural predators, and, only as a last resort, application of industrial pesticides. Farmers using this approach, called integrated pest management, can minimize routine pesticide use, thereby preventing pest species from evolving resistance.

Another approach is to improve crop defenses through breeding and genetic modification. Crop breeders must emulate, albeit on a faster timescale, the role played by evolution in selecting for resistance to newly evolving pests. For example, Norman Borlaug's original task was to develop wheat varieties for Mexico that were resistant to fungal pathogens. To do this, he obtained seeds of wheat varieties from around the world that showed evidence of pathogen resistance. He then crossed these disease-resistant varieties with varieties adapted to the local conditions. Although it took thousands of hand-pollinations, Borlaug successfully introduced resistance genes into wheat varieties that grew well in Mexico.

Crop breeding is a powerful way to alter the genetic makeup of cultivated species. However, because it relies on sexual reproduction, crop breeding is limited to the genetic diversity that exists within varieties that can interbreed. In contrast, *Rhizobium radiobacter* allows scientists to introduce specific genes found in distantly related organisms into cultivated species. One of the first commercial uses of this technology was to introduce chemical defenses from a bacterium into crop plants.

Bacillus thuringiensis (*Bt*) is a soil-dwelling bacterium that produces proteins toxic to insects. It has been used to control insect outbreaks in agriculture since the 1920s. At first, farmers applied spores containing the toxins or the toxins themselves directly to plants. In 1985, the genes that encode the toxins from *B. thuringiensis* were inserted into tobacco, where their expression conferred substantial protection against insects. Today, *Bt*-modified crops are some of the most widely planted genetically modified plants.

The introduction of *Bt* crops markedly reduced the application of pesticides, to the benefit of both farmworkers and the environment. Nevertheless, planting *Bt*-expressing crops widely has its own risk: constant exposure to *Bt* toxins increases the probability that pest populations will evolve resistance. Organic farmers are particularly concerned because they continue to use traditional application of *B. thuringiensis* spores to control insect outbreaks.

To prevent the evolution of resistance, U.S. farmers are required by law to plant a fraction of their fields, known as a refuge, with non–*Bt*-expressing plants (**Fig. 30.19**). Pests that are not resistant to *Bt* toxins survive by feeding on the non–*Bt*-expressing plants. These non-resistant pests interbreed with individuals feeding on the *Bt*-expressing plants, slowing the evolution of *Bt*-resistance. The efficacy of this practice has been called into question by reports of *Bt*-resistant pests and the need to apply increasing amounts of industrial pesticides to *Bt* crops.

If there is one practical lesson to be learned from our experience with *Bt*, it is that pathogens and herbivores will continue to evolve new ways of circumventing the defenses of crop plants. Thus, agriculture will require the ongoing development of new forms of crop protection. While widespread use of chemical pesticides is likely to continue, the application of genetics to protecting crops will increase, through both crop breeding and biotechnology. In either case, success will depend on the availability of existing genes honed by natural selection.

In some cases, crop protection will rely on genes from distantly related organisms, as illustrated by *Bt*. More commonly, genes are taken from other plants, often ones that

FIG. 30.19 ***Bt*-crops co-planted with non-*Bt* varieties to prevent the development of resistance.** *Source: John L. Obermeyer, IPM Specialist Department of Entomology, Purdue University.*

are closely related. For example, specific *R* genes typically are present only within a single species or closely related species. Thus, to develop potatoes that are resistant to the protist *Phytophthora infestans* that causes potato blight, scientists are transferring *R* genes from a wild relative of potato. We can continue these efforts if we commit to safeguarding the genetic diversity found in natural populations of crop species and their close relatives. Only by tapping into these natural genetic resources can breeders produce crops that stay one step ahead of the constantly evolving threats of pathogens and herbivores.

Self-Assessment Questions

10. Draw a phylogenetic tree that illustrates an "escape and radiate" pattern of diversification for plants that evolve novel defenses.

11. What is one benefit and one disadvantage of herbicide and pesticide use in agriculture?

CORE CONCEPTS SUMMARY

31.1 PROTECTION AGAINST PATHOGENS: Plants have mechanisms to protect themselves from infection by pathogens.

Pathogens enter plants through damaged tissue or stomata and spread by growing or moving through the plant's vascular system. Viruses and some bacteria move in the phloem, whereas fungi and other bacteria move in the xylem. page 665

Biotrophic pathogens obtain resources from living cells; necrotrophic pathogens kill cells before exploiting them. page 666

Plants have an immune system that allows them to detect and respond to pathogens. One component of the plant immune system acts generally and recognizes highly conserved molecules produced by broad classes of pathogens. page 667

A second component of the plant immune system acts on specific pathogens. It consists of "resistance" or *R* genes that encode R proteins, which recognize pathogen-derived AVR proteins. page 668

Plants may respond to pathogens by actively killing the cells surrounding the infection (the hypersensitive response). page 668

Infected plants may send a signal to uninfected tissues so that they can mount a defense (systemic acquired resistance). page 668

When attacked by a virus, plants produce siRNA that targets the viral genome for destruction. Because siRNA can move between cells and in the phloem, plants can acquire systemic immunity against specific viruses. page 670

The bacterium *Rhizobium radiobacter* infects plants by inserting some of its genes into the plant's genome, resulting in the formation of a tumor and providing a way to genetically engineer plants. page 670

30.2 DEFENSE AGAINST HERBIVORES: Plants use chemical, mechanical, and ecological defenses to protect themselves from being eaten by herbivores.

Plant defenses against herbivory include latex, chemicals, and mechanical defenses such as dense hairs. page 672

Plants produce a wide range of chemicals to deter herbivores, including alkaloids, terpenes, and tannins. page 673

Ant-plants, such as the bullhorn acacia, provide food and shelter for ants. The ants, in turn, defend their host plant. page 675

Grasses are well adapted to disturbances such as fire and grazing by mammals because their vertically oriented leaves can regrow from the base, while their apical meristems remain near the ground where they are less likely to be damaged. page 675

30.3 ALLOCATING RESOURCES TO DEFENSE: The production of defenses is costly, resulting in trade-offs between protection and growth.

Plants produce both constitutive defenses, which are produced whether or not a threat is present, as well as inducible defenses, which are triggered when a plant detects a specific attack. page 676

Plants produce volatile signals that attract insects that prey on herbivores. page 677

A trade-off is sometimes observed between plant growth and defense, in which allocating resources to defense results in slower growth rates. page 678

In nutrient-rich environments, species that allocate more resources to growth and less to defense can outcompete plants that invest more resources in defense and less in growth; the opposite is true in nutrient-poor soils. page 678

A trade-off can also occur when plants are confronted by multiple threats. For example, a plant attacked by a pathogen will allocate more resources to defend against this threat, but will also become more susceptible to a subsequent attack by a herbivore. page 679

30.4 DEFENSE AND PLANT DIVERSITY: Interactions among plants, pathogens, and herbivores contribute to the origin and maintenance of plant diversity.

The "escape and radiate" pattern of plant evolution suggests that plants undergo a burst of diversification following the evolution of a new form of defense against a pathogen or an herbivore. page 680

The Janzen–Connell hypothesis proposes that interactions with pathogens and herbivores increase plant diversity by favoring the survival of less common species. page 680

Herbivores and pathogens are a major concern for agriculture. page 680

Methods of crop protection include the use of chemical pesticides, integrated pest management, application of spores or toxins from *Bacillus thuringiensis* to plants, and insertion of genes that encode for toxins from *B. thuringiensis* to make *Bt*-modified plants. page 681

Log in to **LaunchPad** to check your answers to the Self-Assessment Questions and to access additional learning tools.

CHAPTER 31: Plant Diversity

CORE CONCEPTS

31.1 MAJOR THEMES IN THE EVOLUTION OF PLANT DIVERSITY: The evolutionary history of plants has been shaped by the challenges of fertilization, dispersal, and photosynthesis on land.

31.2 BRYOPHYTES: Bryophytes have a dominant gametophyte generation; they are small in stature and rely on surface moisture for hydration.

31.3 SPORE-DISPERSING VASCULAR PLANTS: Spore-dispersing vascular plants have small free-living gametophytes and much larger free-living sporophytes with xylem and phloem; they are most commonly found growing as epiphytes or in forest understories.

31.4 GYMNOSPERMS: Gymnosperms rely on pollen for fertilization and seeds for dispersal; they have woody stems and today are most common in seasonally cool or dry regions.

31.5 ANGIOSPERMS: Angiosperms are distinguished by flowers, fruits, double fertilization, and xylem vessels; they are the most diverse group of plants found today.

Five hundred million years ago, the land surface of Earth supported little more than a living crust made up of a mixture of photosynthetic and non-photosynthetic microbes. Then, through the process of natural selection, a group of freshwater green algae began to evolve new ways of surviving, reproducing, and carrying out photosynthesis on land. Their descendants are the organisms we call plants. Fast-forward 465 million years to the present time, and plants are everywhere. Much of what makes up plants, including roots, leaves, tree trunks, and seeds, evolved quickly. Within the first quarter of the time since plants diverged from their algal ancestors, all of these features were present. But one thing was missing: flowers. Flowering plants, also known as angiosperms, do not appear on the scene until the last quarter of this history, yet today they make up a remarkable 90% of all plant species. Many of the plant groups that were present before angiosperms appeared are still present, but where they live and grow has been shaped by the evolution of flowering plants.

31.1 MAJOR THEMES IN THE EVOLUTION OF PLANT DIVERSITY

The closest living relatives of plants are green algae that primarily grow in streams and ponds (**Fig. 31.1**). Some features found in plants are also present in the close algal relatives, including cellular structures such as plasmodesmata and the enzymes used to reduce CO_2 loss during photorespiration. But for the most part, plants are quite distinct from their aquatic ancestors. The reason is that, as plants evolved on land, they had to survive in a new medium, air. Algae can rely on water for hydration, support, and protection from ultraviolet radiation and temperature fluctuations, and they can use water currents to transport sperm and disperse offspring. Surviving on land required the evolution of new ways of growing and reproducing. We have seen the key adaptations that allow plants to photosynthesize and reproduce on land in Chapters 27–30. In this chapter, we bring these adaptations together to explore how plant diversity changed as new ways of growing and reproducing evolved.

Four major transformations in life cycle and structure characterize the evolutionary history of plants.

The phylogeny in Fig. 31.1 shows four major events that transformed how plants grow and reproduce. The first was alternation of generations (Chapter 28; **Fig. 31.2**). Every plant species alternates between two multicellular forms: a haploid gametophyte generation that makes gametes and a diploid sporophyte generation that makes spores (Fig. 31.2b). The green algae most closely related to plants have a life cycle in which there is only one multicellular form, which is made up entirely of haploid cells (Fig. 31.2a). It is thought that the addition of a second, diploid generation, extending upward above the ground-hugging gametophytes, gave newly evolving plants an advantage in dispersing their offspring through the air. Spores released from sporangia on the tips of these taller structures were more likely to be carried away by air currents. Meanwhile, the gametophyte generation grew low to the ground, where it could release swimming sperm into surface moisture layers. Thus, the evolution of the alternation of generations in plants allowed the gametophyte generation to have features that enhance fertilization and the sporophyte generation to have features that enhance dispersal.

FIG. 31.1 Phylogenetic tree of land plants. During the evolutionary history of land plants, there were four key innovations: alternation of generations, xylem and phloem, seeds and pollen, and flowers.

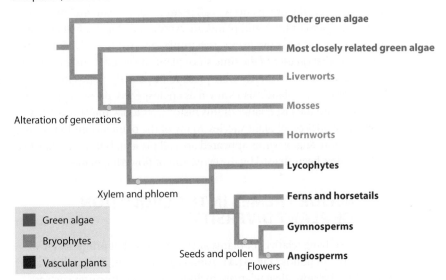

The second major event was the evolution of xylem and phloem in the sporophyte generation (see Fig. 31.1). These internal transport systems distinguish the vascular plants from the mosses, liverworts, and hornworts. Xylem transports water and dissolved nutrients, and phloem transports carbohydrates produced by photosynthesis (Chapter 27). Xylem and phloem transport water and carbohydrates efficiently over considerably longer distances than is possible with diffusion, and xylem allows vascular plants to pull water from the soil. Thus, these structures provided a competitive advantage in both height and sustained hydration over plants that lacked xylem and phloem.

The third major event was the evolution of seeds and pollen (see Fig. 31.1). Fertilization could now take place without releasing swimming sperm into the external environment. Natural selection for fertilization without surface moisture led to the gametophyte generation being retained within the tissues of the much larger and more productive sporophytes. Male gametophytes packaged into small structures called pollen could deliver sperm to the female reproductive organs of another plant by traveling through the air.

The third major event also resulted in a new dispersal unit: multicellular seeds packed with stored resources and surrounded by protective layers. Seeds provide new ways of increasing the probability of the next sporophyte generation becoming successfully established as a free-living plant (Chapter 28). Plants that produce seeds and pollen are called "seed plants." Their reliance on pollen for mating and seeds for dispersing offspring distinguishes them from all other plants, which release swimming sperm into the environment and disperse unicellular spores.

If we step back and look at these first three events, two major themes emerge (**Fig. 31.3**). The first theme is the transition from a gametophyte-dominated life cycle to a life

FIG. 31.2 Alternation of generations. The plants' life cycle (a) differs from that of their green algal relatives (b) by the addition of a multicellular sporophyte generation.

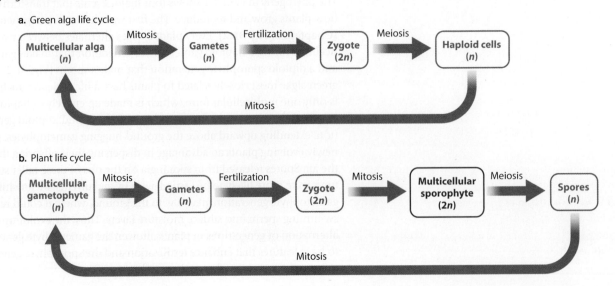

FIG. 31.3 Life cycle features of plants. *Photo sources: (top right) Keith Burdett/AGE Fotostock; (bottom right) VOISIN/PHANIE/Getty Images.*

	Gametophyte generation	**Sporophyte generation**
Bryophytes	Dependent on surface moisture for hydration and fertilization 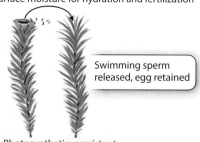 Swimming sperm released, egg retained; Photosynthetic, persistent, free-living, small stature	Obtains water and nutrients from the gametophyte  Aerial dispersal of haploid spores; Unbranched, often upright; remains supported by gametophyte
Spore-dispersing vascular plants	Dependent on surface moisture for hydration and fertilization 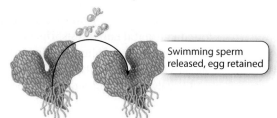 Swimming sperm released, egg retained; Tiny, free-living, often photosynthetic	Obtains water and nutrients from the soil  Aerial dispersal of haploid spores; Overgrows gametophyte to become free-living and photosynthetic; leaves and roots connected by xylem and phloem
Seed plants	Hydration and fertilization independent of surface moisture 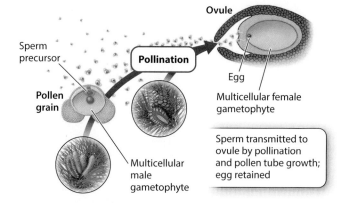 Sperm precursor, Pollen grain, Pollination, Ovule, Egg, Multicellular female gametophyte, Multicellular male gametophyte; Sperm transmitted to ovule by pollination and pollen tube growth; egg retained; Microscopic; develops while surrounded by sporophyte tissues. Gametophytes that produce sperm are shed as pollen; gametophytes that produce eggs are retained.	Obtains water and nutrients from the soil  Dispersal of multicellular seeds by wind, water, or animals; Grows from seed to become free-living and photosynthetic; leaves and roots connected by xylem and phloem

cycle dominated in size and photosynthetic output by the sporophyte. It is hypothesized that the diploid sporophyte generation, when it first evolved, was physically and physiologically dependent on the gametophyte. That is still true today in mosses, liverworts, and hornworts. In seed plants, the reverse is true: the gametophyte generation is physically and physiologically dependent on the sporophyte. The second theme is the increasing independence from the need for surface moisture for hydration and fertilization. With the evolution of lignified xylem, vascular plants gained the ability to pull water from the soil, meaning that they could engage in photosynthesis even when their surfaces were dry. Similarly, with the evolution of pollen, fertilization could occur without releasing swimming sperm into the environment.

The fourth major event was the evolution of flowering plants, the angiosperms (see Fig. 31.1). Flowers are the signature feature of this group, but other distinguishing characteristics include carpels, double fertilization leading to endosperm, xylem vessels, and fruits (Chapter 28). In section 31.5, we discuss how these and other features increase the efficiency with which angiosperms build their bodies and complete their life cycles, contributing to the evolutionary success and diversification of flowering plants.

Plant diversity has changed over time.

If we tabulate the numbers of species within each major branch on plant phylogenetic tree, one fact stands out: of the nearly 400,000 species of plants present today, approximately 90% are angiosperms (**Table 31.1**). Thus, the fourth major event in plant evolution gave rise to the most diverse group of plants.

Angiosperms are relative newcomers, first appearing in the fossil record approximately 140 million years ago (**Fig. 31.4**). By comparison, the oldest known evidence of land plants is found in rocks approximately 465 million years old. Thus, for more than 300 million years, terrestrial vegetation was made up of plants other than angiosperms. During this period, lycophytes, ferns and horsetails, and gymnosperms, as well as many groups of now-extinct plants, dominated the land, forming forests and landscapes unlike those that surround us today.

Once the flowering plants gained an ecological foothold, however, the number of angiosperm species increased at an unprecedented rate. As the number of angiosperm species rose, the number of species in other groups fell. At the same time, the total number of plant species increased dramatically. Thus, angiosperms' hold on plant diversity is the result of both squeezing out other groups and increasing the number of species that can coexist.

One way that angiosperm evolution may have increased the overall number of plant species on Earth was by transforming the environments into which they diversified. Climate models suggest that angiosperms may have been necessary for the formation of tropical rain forests as we know them today. Without the higher rates of transpiration exhibited by angiosperms, many tropical regions would have higher temperatures and lower rainfall amounts. The warmer, drier conditions would have been less conducive to the luxuriant plant growth found in tropical rain forests. The new tropical forests, with their dense shade and humid understories, provided new habitats into which both angiosperms and non-angiosperms could evolve. Thus, although early angiosperms outcompeted

TABLE 31.1	Diversity of Land Plants	
PHYLUM	**COMMON NAME**	**APPROXIMATE NUMBER OF SPECIES**
Bryophytes		
Marchantiophyta	Liverworts	8000
Bryophyta	Mosses	15,000
Anthocerophyta	Hornworts	100
Spore-Dispersing Vascular Plants		
Lycopodiophyta	Lycophytes	1200
Monilophyta	Ferns and horsetails	10,500
Gymnosperms		
Cycadophyta	Cycads	300
Ginkgophyta	Ginkgo	1
Coniferophyta	Conifers	630
Gnetophyta	Gnetophytes	90
Angiosperms		
Magnoliophyta	Flowering plants	>380,000

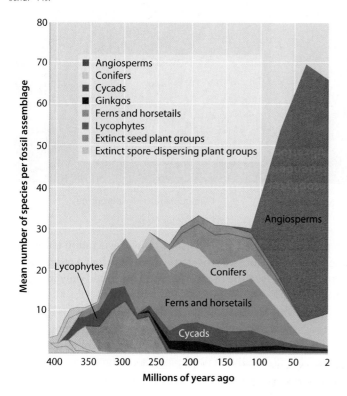

FIG. 31.4 The diversity of vascular plant groups through time. First appearing about 140 million years ago, angiosperms have come to dominate the fossil record. *Source: Adapted from A. H. Knoll and K. J. Niklas, 1987, "Adaptation, Plant Evolution, and the Fossil Record," Review of Palaeobotany and Palynology, 50:127–149.*

many species for space and light, their evolution also stimulated radiations of new species in at least some of the groups that preceded them.

Self-Assessment Questions

1. How did the addition of a sporophyte generation contribute to the success of plants on land?
2. Why did xylem and phloem evolve only in the sporophyte generation?
3. How was plant diversity affected by the evolution of angiosperms?

31.2 BRYOPHYTES

Liverworts, mosses, and hornworts share many features and are collectively referred to as bryophytes. Among their shared characteristics are a dependence on surface moisture for hydration and a life cycle dominated by a persistent, photosynthetic gametophyte. They are also united by the absence of key features of vascular plants, notably xylem and phloem.

The phylogenetic tree shown in Fig. 31.1 indicates that the relationships among liverworts, mosses, and hornworts remain uncertain. As a consequence, we do not know which of the three is most closely related to the vascular plants (that is, shares a more recent common ancestor) or even if two or more of them might form a monophyletic group. We do, however, believe that their last common ancestor lived long ago and that liverworts, mosses, and hornworts have evolved independently for hundreds of millions of years.

Bryophytes are small and tough.

Mosses are the most widely distributed of the bryophyte groups and the most diverse, encompassing approximately 15,000 species. You may have seen moss growing on a shady log or on top of rocks by a stream. In fact, mosses grow in all terrestrial environments, from deserts to tropical rain forests. Liverworts (about 8000 species) and hornworts (about 100 species) are less widespread and also less diverse. In considering the diversity of these three groups, let's start by reviewing some of their features that allow them to carry out photosynthesis on land.

A distinctive feature of bryophytes is the dominance of the gametophyte generation. Only the haploid gametophyte is free-living (capable of growing on its own). In contrast, the sporophyte remains attached to and dependent on the gametophyte. A second distinctive feature of bryophytes is their small stature. In most cases, the gametophyte is less than 5 cm tall (**Fig. 31.5**).

Bryophyte gametophytes come in two forms. Some species produce only a flattened photosynthetic structure called a **thallus**. Others have slender stalks with tiny leaf-like appendages. These leaf-like structures are quite different from the leaves of vascular plants, in that they are only one to several cells thick and lack internal air spaces. Liverworts can be either thalloid or leafy, whereas all mosses are of the leafy type and all hornworts are of the thalloid type.

Bryophytes produce neither roots nor xylem, so they are unable to pull water from the soil. Instead, these plants absorb moisture from droplets of rainfall or dew that land on their surfaces or are pulled by surface tension into the spaces between their closely packed stalks and leaf-like structures. Bryophytes can produce cuticle, a waxy layer that restricts water loss, but many of their surface cells, especially those involved in photosynthesis, lack this protection. As a result, these surfaces are capable of absorbing water and nutrients. Although hair-like structures, only a single cell wide, grow from the lower surfaces of bryophytes, they anchor the gametophyte to the substrate rather than transport resources from the soil.

Bryophytes can absorb enough water to remain metabolically active when the environment is wet, but they lose water quickly when their surfaces are dry. Without the ability to regulate water loss, their water content fluctuates in tandem with the environmental moisture. Some bryophytes grow only in habitats that are continually moist; others live in

FIG. 31.5 Bryophyte diversity. Bryophytes include (a) mosses (*Polytrichum commune* is shown); (b) liverworts (*Pellia epiphylla*, a thalloid liverwort, is shown); and (c) hornworts (*Anthoceros* species). *Sources: a. NHPA/Photoshot; b. ARCO/Reinhard H/AGE Fotostock; c. Daniel Vega/AGE Fotostock.*

Sporophyte   Gametophyte

Gametophyte   Sporophyte

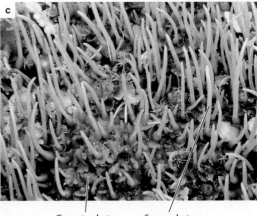
Gametophyte   Sporophyte

habitats that frequently dry out. The latter species typically exhibit desiccation tolerance (Chapter 27), which allows them to survive a nearly complete loss of water. Thus, while some bryophytes are delicate, others are extremely tolerant of stressful conditions. As a result, bryophytes can be found from the equator to the highest latitudes at which vegetation exists, from lowlands to the highest altitudes of vegetation, and from swamps to deserts.

Before the evolution of vascular plants, bryophytes may have covered much of the land surface. Bryophytes remain common today, although in most habitats they are dwarfed by the presence of the much larger vascular plants. Because bryophytes are small, they are poor competitors for light and do poorly when closely overtopped by a dense canopy of vascular plants or covered by a thick layer of fallen leaves. However, bryophytes can grow in places where plants with roots cannot—for example, on the surfaces of rocks. Many bryophytes found in rain forests live on the branches and trunks of trees rather than on the ground. Plants that grow on other plants are called **epiphytes** (from the Greek words *epi*, "on," and *phyton*, "plant"). Bryophytes are well suited to growing on other plants because they do not depend on the soil as a source of water.

The small gametophytes and unbranched sporophytes of bryophytes are adaptations for reproducing on land.

Bryophyte surfaces close to a solid substrate dry more slowly than those elevated in the air. Indeed, one reason that bryophytes are small is to sustain the hydration of their photosynthetic cells. A second reason for their diminutive stature is related to how they reproduce. Bryophytes release swimming sperm into the environment, and these sperm swim through films of surface water on their way to fertilize an egg or are transported by the splash of a raindrop. Sperm can travel in this manner only over a relatively limited distance and only if water is present. Consequently, the gametophyte generation tends to be small. A few mosses grow to be more than half a meter tall, but these giants of the bryophyte world are found only in the understory of very wet forests.

Following fertilization, the diploid zygote develops by mitosis into a multicellular sporophyte, while remaining physically attached to and nutritionally dependent upon the gametophyte. Recall that the diploid, spore-producing generation represents a new component of the life cycle as plants evolved on land (see Fig. 31.2). This sporophyte generation addresses the challenges plants face in dispersing offspring through the air. In bryophytes, the sporophyte produces many haploid spores by meiosis, with these spores then being dispersed to promote spread of offspring.

In most bryophytes, the unbranched sporophyte extends several centimeters above the gametophyte (Fig. 31.5). In mosses and liverworts, a capsule-like sporangium forms at the tip; spores are produced in this structure. After the spores mature, the sporophyte dries out and the sporangia open. Drying prevents spores from sticking together, thereby increasing the chances that a gust of wind will carry off a spore. In hornworts, the spores form at the top of the sporophyte, albeit within the stalk itself. The spores are released as the sporophyte stalk dries from the top down and splits open.

In mosses and liverworts, the sporophyte is short-lived, falling off after the spores are dispersed. In hornworts, the sporophyte can live nearly as long as the gametophyte because it can produce new cells at its base, similar to the way grass blades elongate (Chapter 29).

Bryophytes exhibit several cases of convergent evolution with the vascular plants.

Bryophytes are interesting partly because they are so different from the more familiar vascular plants. However, given that

FIG. 31.6 Moss specializations. (a) The yellow moose-dung moss, *Splachnum luteum*, attracts insects, which then transport its spores. (b) The internal transport tissues of *Dawsonia superba* allow it to grow more than 40 cm tall. *Sources: a. Biopix: A Neumann; b. blickwinkel/Alamy.*

they have long evolved in parallel with vascular plants, it is not surprising that bryophytes have developed similar solutions to environmental challenges, a process referred to as convergent evolution (Chapter 22). For example, some mosses depend on insects to transport their spores (**Fig. 31.6a**). The brightly colored sporangia of these mosses emit volatile chemicals that mimic compounds released by rotting flesh or herbivore dung. Attracted by these chemicals, insects land on the sporangia, and their legs become covered with spores. When the insects subsequently land on a real pile of dung, some of the spores fall off into this ideal site for supporting the growth of a new gametophyte.

Another example of convergent evolution is the presence of cells specialized for the internal transport of water and carbohydrates in some of the largest mosses and leafy liverworts (**Fig. 31.6b**). The structure of these cells indicates that they evolved independently from the xylem and phloem of vascular plants. In particular, the water-conducting cells in bryophytes do not have lignified cell walls. Nevertheless, water and carbohydrates can move efficiently from one part of the plant to the other through these elongated cells.

Stomata are present on the sporophytes of hornworts and some mosses, particularly in places where a waxy cuticle also develops. As discussed in Chapter 27, stomata are pores in the epidermis of vascular plants that open and close, providing an active means of controlling water loss. Bryophyte stomata, however, may play a different role. Bryophyte stomata open only once during development of the sporophyte; once open, they are unable to close. What purpose might these stomata serve? A sporophyte with open stomata will dry out quickly, so its spores are more easily released into the air. Further study is needed to determine whether these stomata are the result of convergent evolution.

Sphagnum moss plays an important role in the global carbon cycle.

In most ecosystems, bryophytes make only a small contribution to the total biomass. The major exception is in **peat bogs**. In these wetlands, dead organic matter accumulates rather than decomposes. A major component of peat bogs is sphagnum moss, any of the several hundred species of the genus *Sphagnum* (**Fig. 31.7**). These mosses play a key role in creating the wet and acidic conditions that slow rates of decomposition. They contain specialized cells that hold onto water, much like a sponge, and they secrete protons that acidify the surrounding water. Unlike many plants with roots, sphagnum moss thrives under wet, acidic conditions. In addition, these mosses produce phenols, organic compounds that slow decomposition under waterlogged conditions.

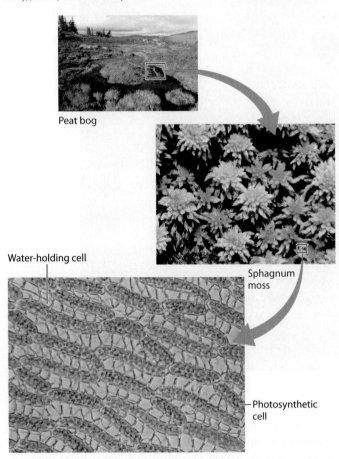

FIG. 31.7 Sphagnum moss. Peat bogs have low rates of organic decomposition due to the wet and acidic conditions created by sphagnum moss. *Sources: (top) Duncan Shaw/Science Source; (middle) All Canada Photos/Alamy; (bottom) blickwinkel/Alamy.*

Peat bogs occupy only 2% to 3% of the total land surface, but they store large amounts of organic carbon—on the order of 65 times the amount released each year from the combustion of fossil fuels. Thus, any increase in the decomposition rate of these naturally formed peats has the potential to increase significantly the CO_2 content of Earth's atmosphere.

What might cause an increase in the decomposition rate? Peat bogs are vulnerable to changes in the water table. As water levels fall, the peat becomes exposed to the air, allowing it to be broken down by microbes. Most peat bogs are located in northern latitudes, where the effects of climate change are predicted to be greatest. If rising summer temperatures result in higher rates of transpiration from surrounding forests, the result could be a lowering of the water table and a relatively rapid release of CO_2 from this moss-dominated ecosystem. Thus, destruction of peat bogs has the potential to accelerate rates of climate change over the next century.

> **Self-Assessment Questions**
>
> 4. Which factor is thought to limit the stature and productivity of the gametophyte generation?
> 5. Which features of the sporophyte generation enhance the likelihood of successful spore dispersal?
> 6. Are there any habitats where a bryophyte might be more productive than a vascular plant?
> 7. If the stomata of bryophytes and of vascular plants are not the result of convergent evolution, what is an alternative explanation for their appearance in these groups?

31.3 SPORE-DISPERSING VASCULAR PLANTS

Vascular plants first appear in the fossil record approximately 425 million years ago. The evolution of xylem and phloem gave vascular plants advantages in size and the ability to keep their photosynthetic cells hydrated using water pulled from the soil. As a result, they were able to outcompete bryophytes for light and other resources and to become the dominant plants on land.

Vascular plants can be divided into two groups according to how they complete their life cycle. Lycophytes, as well as ferns and horsetails, disperse by spores and rely on swimming sperm for fertilization, just like bryophytes. Gymnosperms and angiosperms are seed plants: they disperse seeds and rely on the aerial transport of pollen for fertilization. In this section, we focus on the spore-dispersing vascular plants, which include lycophytes and the group composed of ferns and horsetails. In both groups, the sporophyte dominates in physical size. The gametophyte is small, only a few centimeters across, and thalloid in structure.

Unique to the spore-dispersing vascular plants is that both the sporophyte and the gametophyte are free-living, capable of surviving on their own (see Fig. 31.3). In many cases, the gametophytes of spore-dispersing vascular plants are capable of photosynthesis and tolerant of desiccation. Because of their size and fossil record, our discussion emphasizes the sporophytes of these plants.

Rhynie chert fossils provide a window into the early evolution of vascular plants.

Fossils recovered from a single site in Scotland document key stages in the evolution of vascular plants. Four hundred million years ago, near the present-day village of Rhynie, a wetland community thrived in a volcanic landscape. Mineral-laden springs similar to those found today in Yellowstone National Park encased the newly deposited remains of local populations in chert, a mineral made of silica (SiO_2). These fossils preserve plants, fungi, protists, and small animals in remarkable detail.

FIG. 31.8 Rhynia, a fossil vascular plant from the 400-million-year-old Rhynie chert. All vascular plants are descended from these kinds of simple plants. *Photo sources: (top) Courtesy Hans Kerp © Palaeobotanical Research Group, University Münster; (middle & bottom) Hans Steur, The Netherlands.*

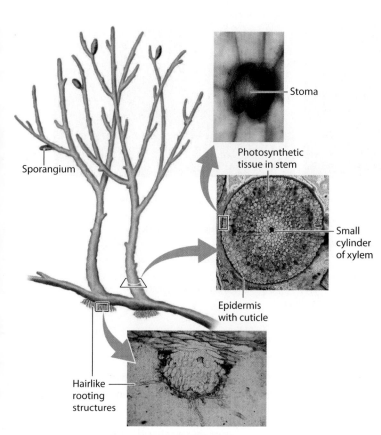

Lycophytes evolved leaves and roots independently from all other vascular plants.

One of the fossil populations in Rhynie cherts is notably larger and more complex than the others. These fossil sporophytes show leaflike structures arranged on the stem in a spiral pattern, and a thick-lobed cylinder of xylem runs through the stem (**Fig. 31.9**). These are among the earliest fossils of lycophytes, a group of vascular plants that today can be found in environments ranging from tropical rain forests to Arctic tundra. Because lycophytes diverged before any of the other groups of vascular plants, they are the sister group to all other vascular plants (see Fig. 31.1).

Fossils indicate that the leaves and roots of lycophytes evolved independently from those of ferns, horsetails, and seed plants. A distinctive feature of lycophyte leaves is that their small leaves have only a single vein. About 1200 species of lycophytes can be found today. Some grow low to the ground and produce branching stems covered by small leaves

FIG. 31.9 An early lycophyte preserved in the 400-million-year-old Rhynie chert. The fossil has helically arranged leaflike structures, adventitious roots (roots that arise from the shoot and not from the root system), and a central strand of xylem, features much like those of living lycophytes. *Sources: (top) photo by William Chaloner, Crown Copyright; (bottom) Photo by John Hall © Botanical Society of America.*

The fossils of Rhynie chert provide our best view of early vascular plants. These plants were small (maximum of 15 cm tall) and consisted primarily of photosynthetic stems that branched repeatedly, forming sporangia at the tips of short side branches (**Fig. 31.8**). They had no leaves, and their only root structures were small hairlike extensions on the lower parts of stems that ran along the ground. The stems had a cuticle with stomata, and a tiny cylinder of vascular tissue ran through the center. Many of the gametophytes also had erect cylindrical stems, not at all like the reduced gametophytes of living vascular plants.

The fossils of Rhynie chert provide another important glimpse into early plant evolution: some fossils have a branched sporophyte that includes cell walls reinforced with lignin, but there is no evidence of vascular tissue. Evidently, plants evolved the upright stature we associate with living vascular plants before they evolved the differentiated tissues of the vascular system (xylem and phloem). Once vascular tissues evolved, they played a key role in shaping the subsequent evolution of the plant sporophyte.

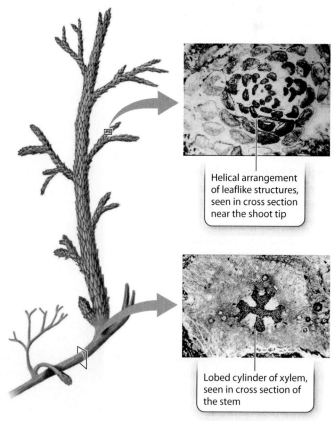

FIG. 31.10 Lycophyte diversity. Living lycophytes include about 1200 species, in three major groups: (a) *Lycopodium annotinum*, showing yellow sporangia-bearing leaves at the shoot tips of this 5- to 30-cm-tall plant; (b) *Selaginella wildenowii*, in a close-up showing the flattened arrangement of its 3-mm-long leaves; and (c) *Isoetes lacustris*, an aquatic lycophyte that grows from 8 to 20 cm tall. *Photo sources: a. Philippe Clement/Nature Picture Library; b. blickwinkel/Alamy; c. Biopix: JC Schou.*

(Fig. 31.10a) or leaves in rows held in a single plane (Fig. 31.10b), while others grow as epiphytes in tropical forests. The most distinctive living lycophytes are the quillworts (genus *Isoetes*), small plants that live along the margins of lakes and slow-moving streams (Fig. 31.10c). Seeing a quillwort today, you would never guess that its relatives once included enormous trees.

Ancient lycophytes included giant trees that dominated coal swamps about 320 million years ago.

Fossils show that ancient lycophytes evolved additional features convergently with seed plants, including a vascular cambium and cork cambium that enabled them to form trees up to 40 m tall (Fig. 31.11). Swamps that formed widely about 320 million years ago were dominated by tree-sized lycophytes. Unlike the vascular cambium of seed plants, the vascular cambium of these lycophytes

FIG. 31.11 Giant lycophytes. (a) Artist's depiction of giant lycophytes. (b) A fossil grove of lycophyte trees showing rooting structures and thick stems; the fossil trunks were preserved by burial during a flood, which molded their lower parts. *Sources: a. John Gurche; b. Biophoto Associates/Science Source.*

produced relatively little secondary xylem and no secondary phloem. Instead, the giant lycophytes relied on their thick bark for mechanical support. Tree-sized lycophytes were markedly different from the trees familiar to us today (**Fig. 31.12**).

These forest giants were not outcompeted by early seed plants, but persisted alongside them for millions of years in swampy environments. Environmental change, however, spelled their doom. When changing climate dried out the swamps, the

HOW DO WE KNOW?

FIG. 31.12
Did woody plants evolve more than once?

OBSERVATIONS Today, the vascular cambium is present almost exclusively among seed plants, with only limited development of secondary xylem in some adder's tongue ferns and quillwort lycophytes (*Isoetes* species). The fossil record, however, shows that other plants, now extinct, also had a vascular cambium.

HYPOTHESIS The vascular cambium evolved independently in several different groups of vascular plants.

EXPERIMENT AND RESULTS The vascular cambium is recorded in fossils by xylem cells in rows oriented radially in the stem or root. Thus, the giant tree lycophytes of Carboniferous coal swamps (the top photo) had a vascular cambium, as did extinct tree-sized relatives of the horsetails (the bottom photo) and other extinct horsetail relatives. Phylogenetic trees generated from morphological features preserved in fossils unambiguously show that woody lycophytes, woody horsetail relatives, and the group of seed plants and progymnosperms did not share a common ancestor that had a vascular cambium. Moreover, anatomical research shows that the vascular cambium of extinct woody lycophytes and the giant horsetails *Archaecalamites* and *Calamites* generated secondary xylem but not secondary phloem, unlike the vascular cambium of seed plants. Living horsetails do not have a vascular cambium, but fossils show that they are descended from ancestors that did make secondary xylem and have lost this trait through evolution.

CONCLUSION Fossils and phylogeny support the hypothesis that the vascular cambium and, hence, wood evolved more than once, reflecting a strong and persistent selection for tall sporophytes among vascular plants.

SOURCE Taylor, T. N., E. L. Taylor, and M. Krings. 2009. *Paleobotany*. Amsterdam: Elsevier.

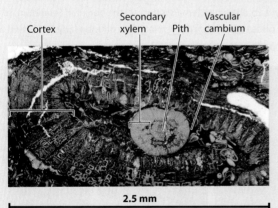

A small branch of *Lepidodendron*, an ancient giant lycophyte

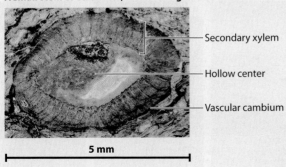

A small stem of *Calamites*, an ancient giant horsetail

Photos source: Andrew Knoll, Harvard University.

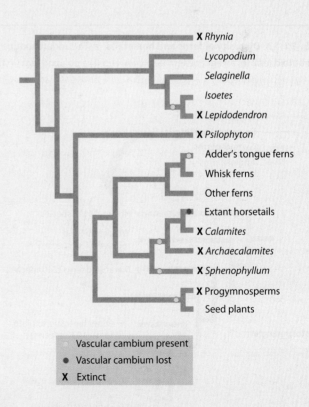

- ○ Vascular cambium present
- ● Vascular cambium lost
- X Extinct

large lycophytes disappeared, along with their habitat. Little *Isoetes* was left as the only living reminder of this once-dominant group of plants.

These ancient forests left one other legacy: the coal deposits we mine today. As trees died and fell over into the swamp, their bodies decomposed slowly. Over time, as material accumulated and became buried, the combined action of high temperature and pressure converted the dead organic matter first into peat and then into the carbon-rich material we call coal. Thus, a major source of energy we use today derives originally from photosynthesis carried out by lycophyte trees.

Ferns and horsetails are morphologically and ecologically diverse.

Ferns and horsetails form a monophyletic group that is the sister group to the seed plants (**Fig. 31.13**). Most ferns have distinctive leaves that uncoil during development from tightly wound "fiddleheads" (**Fig. 31.14a**). Horsetails, as well as whisk ferns and their relatives, lack this feature. In fact, horsetails and whisk ferns were once considered distinct groups because of their unique body organizations. Whisk ferns have photosynthetic stems without leaves and roots (see Fig. 31.13b), so they look similar to fossils of early vascular plants. Nevertheless, molecular sequence comparisons support the hypothesis that whisk ferns are the simplified descendants of plants that produced both leaves and roots.

The horsetails, represented by 15 living species of the genus *Equisetum*, produce tiny leaves arranged in whorls, giving the stem a jointed appearance (see Fig. 31.13c). Horsetail stems are hollow, and their cells accumulate high levels of silica, earning them the name "scouring rush." Today, horsetails are small plants, most less than a meter tall. In earlier times, however, tree-sized versions grew side by side with the ancient lycophyte trees (see Fig. 31.12).

In contrast, most ferns produce large leaves—in some cases up to 5 m long, although typically the photosynthetic surfaces are divided into smaller units called **pinnae**. Yet, because most ferns are not capable of secondary growth, their stems remain slender and frequently grow underground. Only the leaves emerge into the air. Many ferns exhibit little stem elongation and produce their leaves in tight clumps. Other species, such as bracken fern, produce underground stems with widely spaced leaves. Because bracken is toxic to cattle and other livestock, its tendency to invade pastures is a serious problem for farmers.

Despite the absence of secondary growth, some ferns can grow quite tall (**Fig. 31.14b**). Tree ferns can reach heights of more than 10 m by producing thick roots near the base of each leaf that descend parallel to the stem to the ground. These roots increase the mechanical stability of the plant and allow it to transport more water. Other ferns grow tall by producing leaves that twine around the stems of other plants, using those stems

FIG. 31.13 Diversity of ferns and horsetails. (a) The adder's tongue fern *Ophioglossum*, (b) the whisk fern *Psilotum*, (c) the horsetail *Equisetum*, (d) a marattioid fern, (e) an aquatic fern (*Salvinia*), and (f) a polypod fern (in the middle of the photo is *Athyrium filix-femina*, the common lady fern). *Photo sources: a. GAP Photos/Jonathan Buckley; b. Biophoto Associates/Science Source; c. REDA &CO srl / Alamy; d. Courtesy Gary Higgins; e. FLPA/Chris Mattison/AGE Fotostock; f. Ron Watts/Getty Images.*

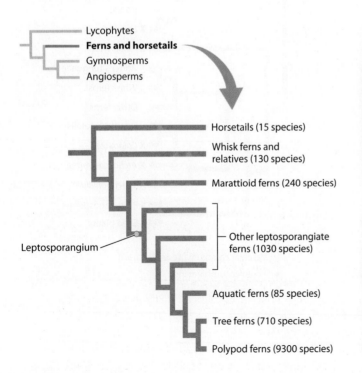

FIG. 31.14 Fern leaves and stems. (a) A fern leaf uncoiling from a "fiddlehead." (b) Silver tree ferns (*Cyathea meullaris*) grow 10 m tall or more. *Source: a. Andre Nantel/Shutterstock; b. AA World Travel Library/Alamy.*

for support. At the other extreme are tiny aquatic ferns such as *Salvinia* (see Fig. 31.13e) that float on the surface of the water.

Fern diversity has been strongly affected by the evolution of angiosperms.

The fossil record shows that ferns and horsetails originated more than 360 million years ago, and between 300 and 100 million years ago they were the most diverse group of plants (see Fig. 31.4). Then, as angiosperms evolved, their diversity markedly decreased. It is not clear why angiosperm evolution should have so affected ferns and horsetails, which had coexisted with a variety of seed plant groups for more than 200 million years.

The most diverse group of ferns today, the polypod ferns (about 9300 species; see Fig. 31.13f), evolved after the rise of the angiosperms. Thus, many of the fern species present today are likely to have evolved in habitats newly created by the development of angiosperm forests. Approximately 40% of living ferns are epiphytes, many of them growing in tropical rain forests dominated by angiosperm trees.

Polypod ferns may owe some of their evolutionary success to an enhanced capacity for spore dispersal. Horsetails, whisk ferns, and a few other fern groups make large sporangia similar to those produced by early vascular plants. These sporangia break open to release spores, but lack a mechanism for ejecting the spores into the air. In contrast, most ferns have a distinctive sporangium (called a **leptosporangium**) whose wall is only a single cell thick. In the leptosporangia of polypod ferns, a line of thick-walled cells runs along the sporangium surface (see the photo on the first page of Chapter 28). When the spores mature, the sporangium dries out and these cells contract like a catapult, forcibly flinging the spores into the air.

Self-Assessment Questions

8. Where are lycophytes found today?
9. What are three ways that ferns, which lack secondary growth, are able to elevate their leaves and thus access more sunlight?
10. How has fern diversity been affected by the evolution of the angiosperms?

31.4 GYMNOSPERMS

Gymnosperms are seed plants, as are angiosperms: both types of plants reproduce using pollen and seeds. Gymnosperms total only a bit more than 1000 species, whereas there are more than 380,000 species of angiosperms (Table 31.1). Despite their much lower diversity, gymnosperms are highly successful inhabitants of modern-day Earth. In cool or dry parts of the world, gymnosperms such as pine trees are often more common than angiosperms in the forest canopy.

The fossil record shows that there were once many more groups of gymnosperms than are found today. Taking these extinct groups into account, biologists hypothesize that gymnosperms are paraphyletic (**Fig. 31.15**). In this section, we examine the evolutionary history of living gymnosperms and explore how their diversity and distribution were affected by the

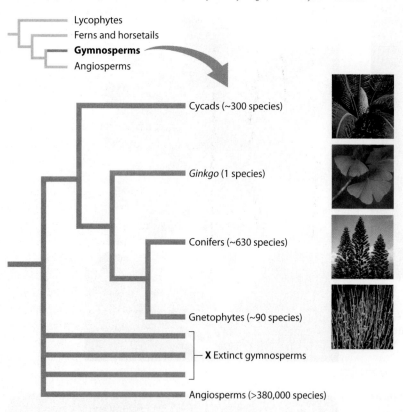

FIG. 31.15 Seed plant diversity. The phylogenetic tree shows a hypothesis of evolutionary relationships among gymnosperms. *Photo sources: (top to bottom) Robert Davis/AGE Fotostock; Michael P. Gadomski/Science Source; Pete Ryan/Getty Images; Patti Murray/AGE Fotostock.*

rise of the angiosperms. First, though, we review some of the features of seed plants that may have contributed to their evolutionary and ecological success.

Seed plants have been the dominant plants on land for more than 200 million years.

Seed plants have come to dominate land environments since their first appearance about 370 million years ago. While the coal-forming swamps of 300 million years ago were populated by spore-producing giants, the understory of these strange forests included plants that produced pollen and seeds. By 200 million years ago, seed plants had become the largest and most widespread plants on Earth.

Which factors have contributed to the success of seed plants? Notably, they do not require external water for fertilization. Instead, a tiny, sperm-producing male gametophyte is transported through the air in pollen; the female gametophyte develops while remaining hidden inside the sporophyte. Thus, the sporophyte generation dominates the seed plant's life cycle. Gametophytes still grow from spores produced inside a sporangium, but the spores are not dispersed. In gymnosperms, the sporangia develop in cones, whereas in angiosperms, they develop in flowers.

Pollination, if successful, leads to fertilization and the formation of a seed. Seeds contain an embryonic sporophyte, surrounded by food resources and a protective outer layer. The resources stored with each seed support the growth of the embryo following dispersal. In this way, seeds increase the likelihood that the embryo will grow into a photosynthetically self-sufficient new sporophyte.

Before these innovations, the ancestors of seed plants evolved the ability to produce additional or "secondary" xylem and phloem through the formation of a vascular cambium. This layer of actively dividing and differentiating cells surrounds stems and allows them to increase in diameter (Chapter 29). A second layer of dividing cells, the cork cambium, maintains an intact layer of protective outer bark. Vascular and cork cambia provide the support and water transport capacity needed for plants to grow tall and to support increasing numbers of leaves. Woody stems and roots are the ancestral condition in seed plants and all gymnosperms exhibit secondary growth.

Cycads and ginkgos were once both diverse and widespread.

Cycads typically have unbranched stems topped by a whorl of stiff leaves (**Fig. 31.16**). Although most cycads have thick stems, their vascular cambium produces little additional xylem, and most of their bulk is made up of unlignified (nonwoody) cells in the center of the stem (pith) and surrounding the vascular tissues (cortex). Cycads have low photosynthetic rates and, as a result, grow very slowly.

In nutrient-poor environments, cycads are able to grow by forming symbiotic associations with nitrogen-fixing cyanobacteria. They can survive wildfires because their vascular cambium is located far beneath the surface of their thick stems and their apical meristem is protected from heat by a dense covering of bud scales. Thus, cycads have traits that favor persistence, rather than high rates of photosynthesis.

Cycads are an ancient group, with fossils at least 280 million years old. During the time of the dinosaurs (about 230 to 66 million years ago), they were among the most common plants. As the angiosperms rose to ecological prominence, however, cycads' diversity and abundance declined markedly. The approximately 300 species of cycads found today occur most commonly in tropical and subtropical regions. Most species have a fragmented distribution, suggesting that they were much more widespread in the past.

Cycads may owe their persistence in part to their reliance on insects for pollination, as well as to their ability to live in nutrient-poor and fire-prone environments. Insects transfer pollen more efficiently than wind, so insect-pollinated species

FIG. 31.16 Cycads. (a) Most living cycads have thick, unbranched stems that support large leaves and cones. Shown here is *Encephalartos friderici-guilielmi*. (b) Many cycads, such as *Encephalartos transvenosus*, have large cones. (c) Mature seeds are large (the seeds of *Cycas circinalis* are shown here) and contain a relatively large female gametophyte that provides nutrition for the embryo and germinating sporophyte. *Sources: a. A. Laule/AGE Fotostock; b. & c. Andrew Knoll, Harvard University.*

with small populations are able to reproduce successfully. Many cycads are pollinated by beetles attracted by chemical signals released from the cones. The large seed cones provide shelter for the insects, and the pollen cones provide them with food. Following fertilization, the brightly colored, fleshy seed coats attract a variety of birds and mammals that serve as dispersal agents.

The Ginkgophyta, once a diverse lineage, today has only a single living representative: *Ginkgo biloba*. This species forms tall, branched trees, with fleshy seeds and fan-shaped leaves that turn brilliant yellow in autumn (**Fig. 31.17**). Once common in temperate forests, ginkgos declined in abundance, diversity, and geographic distribution over the past 100 million years, and it is unclear whether any truly natural populations exist in the wild. However, ginkgos have long been cultivated in China, especially near temples.

FIG. 31.17 *Ginkgo biloba*. (a) The fan-shaped leaves of *Ginkgo* turn yellow in autumn. (b) The fleshy seeds enclose a large female gametophyte, as in cycads. *Sources: a. JTB/Photoshot; b. Kathy Merrifield/Science Source.*

FIG. 31.18 Conifers. Giant sequoia (*Sequoiadendron giganteum*) can reach 85 m in height and have stems that are 8 m in diameter. *Source: Charles J. Smith.*

Today, gingkos are found all over the world. Their ability to tolerate the pollution and other stresses of urban environments have made them a tree of choice for street plantings.

Conifers are woody plants that thrive in dry and cold climates.

The approximately 630 species of conifers are, for the most part, trees. The tallest (more than 100 m) and oldest (more than 5000 years) trees on Earth are both conifers (**Fig. 31.18**). In addition, most conifers are evergreen, meaning that they retain their often needle-like leaves throughout the year and, in some species, for decades. Well-known conifers include pines, junipers, and redwoods.

Most conifers have strong apical dominance, so they develop a single straight stem from which extend much smaller horizontal branches. Conifers have traditionally served as masts for sailing ships, and today they are used for telephone poles. Conifer xylem consists almost entirely of tracheids, single-cell conduits that both support the tree and transport water. The abundance of similar-sized xylem conduits is why conifer wood is strong for its weight and well suited as a building material. Conifers supply much of the world's timber, as well as the raw material for producing paper.

Conifer reproduction tends to be slow. After pollination, it can take more than a year before the seeds are ready to be dispersed. Conifers are wind pollinated, and most species also rely on wind for seed dispersal. However, some conifers, such as junipers and yews, produce seeds surrounded by fleshy and often brightly colored tissues that attract birds and other animals.

Conifers dominate the vast boreal forests of Canada, Alaska, Siberia, and northern Europe. Conifers also tend to become more abundant as elevation increases, and they are common in dry areas, such as the western parts of North America and much of Australia. Only a small number of conifer species are found in tropical latitudes, most commonly at higher elevations. Before the rise of the angiosperms, conifers were more evenly distributed across the globe, including in the lowland tropics. One of the great questions in the evolutionary history of plants is why conifers were largely displaced from tropical latitudes by the angiosperms but continued to thrive in cold and dry environments.

An important difference between angiosperms and conifers is that angiosperms transport water in the xylem vessels, whereas conifers transport water in tracheids. In wet and warm environments, the xylem vessels of angiosperms can be both wide and long, increasing the efficiency of water transport through their stems. One hypothesis is that the evolution of xylem vessels gave angiosperms a competitive advantage in moist tropical environments. With more water flowing through their stems, angiosperms could have higher rates of photosynthesis and grow larger leaves. Once conifer populations began to dwindle, they could have disappeared completely, because wind pollination requires dense populations to be effective. In colder or drier regions, however the dangers of cavitation due to drought or freeze followed by thaw (Chapter 27) constrain the size of angiosperm vessels. This may explain the greater coexistence of conifers and angiosperms in cold or dry climates.

Gnetophytes are gymnosperms that have independently evolved xylem vessels and double fertilization.

Gnetophytes are a small group of gymnosperms (about 90 species) made up of three genera that look completely different from one another (**Fig. 31.19**). *Welwitschia mirabilis* is found only in the deserts of southwestern Africa. It produces just two straplike leaves that elongate continuously from their base. *Gnetum* grow in tropical rain forests, where they form woody vines and small trees with large broad leaves. *Ephedra,* which is native to arid regions, produces shrubby plants with photosynthetic stems and tiny leaves.

The fossil record shows that gnetophytes originated perhaps as early as 200 million years ago. They reached their greatest diversity at approximately the same time as the angiosperms began to diversify. Gnetophytes have long attracted the attention of plant biologists because they exhibit a number of traits that are typical of angiosperms, notably the formation of xylem vessels and double fertilization. Despite these shared features, DNA analysis indicates that gnetophytes are not closely related to the angiosperms but are, instead, more closely related to conifers. Thus, both xylem vessels and double fertilization evolved independently in this group.

FIG. 31.19 Gnetophytes. The three genera in this group include some of the most distinctive seed plants: (a) *Welwitschia mirabilis*; (b) *Gnetum* species; and (c) *Ephedra* species. *Sources: a. M. Philip Kahl/Science Source; b. Ahmad Fuad Morad [Flickr page] 072112; c. Bob Gibbons/Science Source.*

Terrestrial ecosystems became more productive as rates of photosynthesis increased in plants with the more efficient angiosperm xylem. As a result, more food became available for insects, birds, and mammals. All of these animal groups diversified along with the flowering plants.

Angiosperms are distinguished from other seed plants by a number of innovations that make pollination and provision of offspring more efficient. These innovations include the emblematic flower, but also closed carpels, double fertilization leading to the formation of endosperm, and fruits. Angiosperm vascular tissues are also distinct: water transport to photosynthetic leaves is more efficient in plants with multicellular xylem vessels. We briefly review some of these innovations in the next section and then consider how these new features may have contributed to the spectacular diversity of the flowering plants.

Several innovations increased the efficiency of the angiosperm life cycle and xylem transport.

Recall from Chapter 28 that flowers are a plant's way of enlisting animals to help them reproduce. The majority of angiosperms rely on animals to carry pollen from one plant to another, instead of depending on the wind for dispersal. Brightly colored petals and volatile scents advertise the presence of flowers, while rewards such as nectar encourage pollinators to return repeatedly to individuals of the same species. The movement of animals from one flower to another provides an efficient mode of pollen transfer. As a result, animal-pollinated plants are able to reduce their production of pollen compared to a wind-pollinated plant.

A second way in which angiosperms are more efficient at completing their life cycle than other seed plants is the process of double fertilization, in which one fertilization leads to the embryo and the other to the endosperm. In angiosperms, endosperm is the tissue that supports the growth of the embryo. Double fertilization allows angiosperms to delay forming nutritive tissue for the embryo until after fertilization occurs (Chapter 28). In contrast, in gymnosperms, resources for the embryo are committed before fertilization. Double fertilization prevents resources from being wasted when ovules are provisioned but never fertilized.

Fruits are another feature that contributes to the efficiency of angiosperm reproduction. Fruits result from the maturation

Self-Assessment Questions

11. How has the evolution of angiosperms affected the distribution of cycads and of conifers?

12. How does the xylem produced by conifers differ from that of angiosperms? How could this difference have influenced the present-day distribution of conifers?

13. Double fertilization evolved independently in gnetophytes and in flowering plants. What are two other features usually associated with angiosperms that evolved independently in gymnosperms?

31.5 ANGIOSPERMS

The evolution of angiosperms transformed life on land. The total number of plant species increased significantly as angiosperms diversified. Some non-angiosperm plant groups thrived as they expanded into new niches created by angiosperms, whereas others declined as many of their species went extinct.

of the ovary containing the seeds. By packaging seeds within a larger structure, angiosperms increase the ways in which seeds can be dispersed. In particular, because fruits provide rewards for animal dispersal agents, they make it more likely that seeds will be dispersed widely. This may be especially true for small seeds, which otherwise would be unlikely to attract an animal disperser.

In addition to reproducing efficiently, angiosperms are able to grow efficiently into a wide variety of forms, ranging from short-lived herbaceous plants to trees with large spreading crowns. One reason may be the structure of the angiosperm vascular system: instead of having a single cell type for mechanical support and water transport like gymnosperms, angiosperms produce thick-walled, elongate fibers that support the stem and xylem vessels that transport water. One outcome of this arrangement is that because xylem vessels are not a stem's sole source of support, they can be wide and long. Angiosperms with these wide xylem vessels are able to transport water through their stems at higher rates than plants with tracheids. Another outcome of producing efficient xylem is that angiosperms can grow quickly and with great flexibility of form. The diversity of shapes and sizes made possible by xylem vessels reduces competition for light and space, potentially allowing more species to coexist. Finally, because they can grow quickly, angiosperms may be able to make use of habitats and resources that are only fleetingly available, such as ephemeral pools or open areas that appear following a fire.

Angiosperm diversity may result in part from coevolutionary interactions with animals and other organisms.

Evidence in the fossil record shows that for the first 30 to 40 million years of their evolution, angiosperms were not particularly diverse. Nor were they the dominant plants of their ecosystems. In fact, fossil pollen indicates that, between about 140 and 100 million years ago, gnetophytes diversified just as much as angiosperms. Later in the Cretaceous Period, beginning about 100 million years ago, angiosperm diversity began to increase at a much higher rate. It is at this time that we first encounter fossils of the two groups that would come to dominate both angiosperm diversity and the ecology of many terrestrial environments: the monocots and the eudicots (**Fig. 31.20**).

Before looking at the diversity of these two groups, let's consider how characteristics of the angiosperms may have promoted their diversification. While many factors likely contributed to their success, here we focus on the ways in which angiosperms interact with other types of organisms.

The interaction of flowers and animal pollinators is thought to contribute to angiosperm diversity. Plants whose flowers attract different pollinators tend to more readily diverge to form new species than do wind-pollinated species. In addition, the carpel (the closed vessel in which seeds develop) is able to block

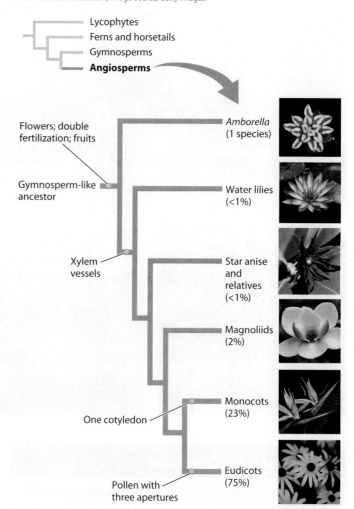

FIG. 31.20 Angiosperm phylogeny. Most angiosperm species belong to the monocots and eudicots. *Photo sources: (top to bottom) Kenneth Setzer; Tim Gainey/Alamy; Liz Cole/AGE Fotostock; GAP Photos/Jo Whitworth; Tropicals JR Mau/PhotoResourceHawaii.com; Image Source/Getty Images.*

the growth of pollen from different species and from closely related individuals, including pollen from the same plant. These genetic recognition systems may contribute to the reproductive isolation required for speciation (Chapter 21).

Diversity results from both high rates of species formation *and* low rates of species loss. The fossil record suggests that the persistence of species may play a particularly important role in angiosperm diversity. Wind-pollinated plants can reproduce only when their populations have a relatively high density. Since no wind-carried pollen is likely to fall on isolated individuals, species with low population density are more likely to go extinct. In contrast, animal pollinators actively searching for rare species are much more likely to find them, so animal-pollinated angiosperm populations can persist at low population densities.

Another reason rare species may be more persistent is that they are less likely to encounter soil pathogens and seed predators. These threats to survival tend to gather near adults of the same species. Because the seeds of rare species are more likely to land away from an individual of the same species, these seeds suffer less from density-dependent mortality than do those of more common species. Of course, this strategy works only if rare isolated individuals can remain in contact with a large enough population of potential mates. Once again, flowers and animal pollinators play a key role in species persistence.

If animal pollination has so many advantages, why did approximately 20% of angiosperm species subsequently evolve a dependence on wind for pollination? Wind-pollinated species grow primarily in seasonal environments. There, fewer animal pollinators are available than in tropical forests, and soil pathogens and seed predators are less abundant.

That angiosperm diversity is the result of multiple factors is reinforced by the observation that some of these "angiosperm" features are also present in other groups of plants. For example, cycads rely on animal pollination, and gnetophytes produce xylem vessels, yet neither group is particularly diverse. The initially slow but increasingly rapid buildup of angiosperm diversity may be explained by two factors: (1) the compounding effect of coevolutionary interactions with pollinators and dispersers and (2) adaptive radiations triggered by chemical arms races with pathogens and herbivores (Chapter 30). This is what is meant by the adage "Diversity begets diversity," a phenomenon further explored in Chapter 45.

Monocots are diverse in shape and size despite not forming a vascular cambium.

Monocots (more formally, monocotyledons) make up nearly one-fourth of all angiosperms. Monocots come in all shapes and sizes and are found in virtually every terrestrial habitat on Earth (**Fig. 31.21**). Some monocots grow in arid regions; an example is agave, which is used to make tequila. Others, such as orchids, grow primarily as epiphytes in tropical forests. A few monocots form vines—for example, the rattan palms, whose stems are used to produce rattan furniture. Monocots are the most important group of plants in our diets. Grasses such as corn, rice, wheat, and sugarcane are staples in the diets of most people around the world. Other monocots that make their way to our dinner table include banana, ginger, asparagus, pineapple, and vanilla.

Monocots take their name from the fact that they have one embryonic seed leaf, or cotyledon, whereas all other angiosperms have two. However, monocots are distinct in form in many other ways. For example, monocot flowers typically produce organs in multiples of three (for example, having three, six, or nine stamens), whereas eudicot flowers typically produce organs in multiples of four or five. Another difference is that in monocot leaves, the major veins are typically parallel and the base of the leaf surrounds the stem, forming a continuous sheath. However, the major difference is that monocots do not

FIG. 31.21 Monocot diversity and phylogeny. This diverse group of angiosperms includes (a) *Agave shawii*, a desert succulent; (b) *Leucojum vernum*, a spring wildflower; (c) bamboo, a forest grass; and (d) *Costus* species, which grow in the rain forest understory. (e) Monocot phylogeny indicates that orchids (approximately 22,000 species) and grasses (approximately 11,500) make up much of monocot diversity. *Photo sources: a. Biosphoto/François Gohier; b. Kerstin Hinze/naturepl.com/NaturePL; c. Axle71/Dreamstime.com; d. Tropicals JR Mau/PhotoResourceHawaii.com.*

HOW DO WE KNOW?

FIG. 31.22

When did grasslands expand over the land surface?

BACKGROUND Today, prairies, steppes, and other grasslands spread over broad areas in the interiors of continents. Grasses, however, do not fossilize readily. How can we understand how grasslands developed through time?

HYPOTHESIS Grasslands expanded as climate changed over the past 50 million years.

OBSERVATIONS AND EXPERIMENTS Many plants, including grasses, form small structures of silica (SiO_2) called phytoliths in their cells. Phytoliths provide a window into the past because they persist in the soil (i.e., they do not decay) and different groups of plants can be identified based on the structure of their phytoliths. Thus, phytolith assemblages in the soil can be used to distinguish between grasses growing in open grasslands (where grasses are present but there are few or no trees) and grasses growing in a forest understory. In addition, mammals that feed on grasses evolved high-crowned teeth, which are also preserved well, giving us an indirect record of grassland history. Finally, C_4 grasses, which are adapted to hot, sunny environments with limited rainfall, have a distinctive carbon isotopic composition imparted by the initial fixation of CO_2 by PEP

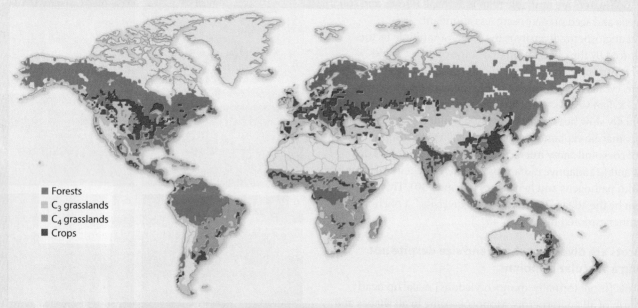

Data from Edwards, E. J., C. O. Osborne, C. A. E. Stromberg, S. A. Smith, and the C_4 Grasses Consortium. 2010. "The Origins of C_4 Grasslands: Integrating Evolutionary and Ecosystem Science." Science 328:587–591.

form a vascular cambium. In the stems of most monocots, the vascular bundles are scattered throughout the cross section, instead of being arranged in a ring like those of most other seed plants. The scattered arrangement of vascular bundles is thought to preclude the formation of a continuous layer of vascular cambium. As a result, monocots do not exhibit secondary growth.

Because monocots lack a vascular cambium, the vascular capacity of their stems and roots cannot increase even as the plants increase in size. Some monocots overcome this limitation by producing horizontal stems that continuously initiate new roots each time a new leaf is added. Others limit the production of new leaves that would need to be supported by an expanded vascular system: they form upright stems that are unbranched and, therefore, have a relatively fixed total leaf area. A few monocots, such as palm trees, have evolved alternative ways of producing larger-diameter stems and, in some cases, a small amount of additional xylem and phloem. Most monocots, though, lack the ability either to grow in diameter or to increase their vascular transport capacity.

Which factors might have led to such a radical revision in how monocots build their bodies? One hypothesis is that monocots evolved from ancestors that produced creeping, horizontal stems as they grew along the shores of lakes and other wetlands. Many monocots today grow in such habitats. In addition, many features of the monocot body plan are well adapted to environments with loose substrates, flowing water, and fluctuating water levels. For example, their leaf base provides a firm attachment that prevents leaves from being pulled off by flowing water. Furthermore, many monocots produce strap-shaped leaves that elongate from a persistent zone of cell division and expansion located at the base of the leaf blade. By continually elongating from the base, monocot leaves can extend above fluctuating water levels.

Although monocots may have originated in wet habitats, today they grow in a wide range of environments, including carboxylase (Chapter 27). Scientists can track C_4 grasslands through time by examining measurements of ^{13}C and ^{12}C in mammal teeth, soil carbonate minerals, and, more recently, tiny amounts of organic matter incorporated into phytoliths.

RESULTS Studies of mammal tooth structure (top graph), phytoliths (middle graph), and carbon isotopic composition of horse teeth enamel from the North American midcontinent (bottom graph) show that grasslands expanded 20 to 15 million years ago, and C_4 grasslands expanded later, about 8 to 6 million years ago.

CONCLUSION In North America, grasslands expanded as atmospheric CO_2 levels declined and climates became drier. Other continents show evidence of a similar linkage of grassland expansion to climate change.

FOLLOW-UP WORK At present, atmospheric CO_2 levels are increasing rapidly, which may affect the competitive abilities of C_3 and C_4 grasses. Scientists are working to understand how global change will influence grasslands and other vegetation.

SOURCE Strömberg, C. A. E. 2011. "Evolution of Grasses and Grassland Ecosystems." *Annual Review of Earth and Planetary Sciences* 39:517–544.

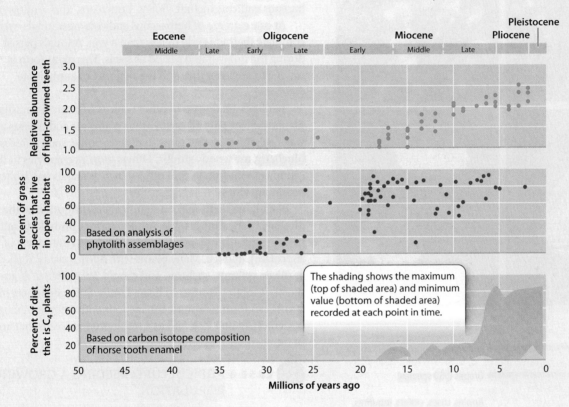

ones that are seasonally dry. One of the most diverse groups within the monocots is the grasses. Many grasses produce stems that grow horizontally and branch, allowing them to cover large areas. Grasses have linear leaves that elongate from the base, allowing them to survive grazing as well as fire and drought. Many grasses are able to tolerate dry environments by producing roots that extend deep into the soil. In addition, C_4 photosynthesis has evolved within the grasses multiple times. As described in Chapter 27, C_4 plants can avoid photorespiration, thereby photosynthesizing with greater efficiency. Grasses are among the most successful groups of plants, becoming widespread within the past 20 million years as climates changed (**Fig. 31.22**). Today, nearly 30% of terrestrial environments are grasslands.

Eudicots are the most diverse group of angiosperms.

Eudicots first appear in the fossil record about 125 million years ago. By 90 to 80 million years ago, most of the major groups we see today were present. Today, an estimated 160,000 species of eudicots are found on Earth, accounting for nearly three-fourths of all angiosperm species (**Fig. 31.23**). Eudicots are well represented in the fossil record, in part because their pollen is easily distinguished. Each eudicot pollen grain has three openings from which the pollen tube can grow. Pollen from all other seed plants has only a single opening.

Eudicots take their name from the fact that they produce two cotyledons, whereas monocots produce one. Because all other angiosperms except monocots also have two cotyledons, this largest of all angiosperm groups is referred to as the *eu-* or "true" dicots.

Many eudicots produce highly conductive xylem. Their high rates of water transport, and therefore their high rates of photosynthesis, may explain why eudicot trees are so abundant in many forests. In addition, many eudicot trees lack strong apical dominance and produce crowns with many spreading branches. The diversity of crown architecture exhibited by eudicot trees

FIG. 31.23 Eudicot diversity and phylogeny. Eudicots include (a) red oak trees (*Quercus rubra*); (b) tropical passionflower vines (*Passiflora caerulea*); (c) *Banksia* shrubs, native to Australia; and (d) beavertail cactus (*Opuntia basilaris*). (e) The tree shows a phylogeny of eudicots, with well-known species indicated. Groups with a small number of species are not shown. *Photo sources: a. Michael P. Gadomski/Getty Images; b. Sergio Hayashi/Dreamstime.com; c. Joanne Harris/Dreamstime.com; d. Danita Delimont/Getty Images.*

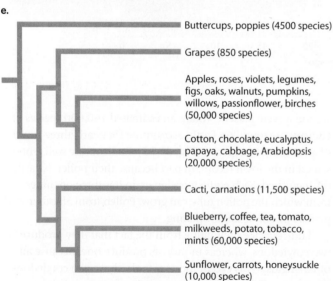

- Buttercups, poppies (4500 species)
- Grapes (850 species)
- Apples, roses, violets, legumes, figs, oaks, walnuts, pumpkins, willows, passionflower, birches (50,000 species)
- Cotton, chocolate, eucalyptus, papaya, cabbage, Arabidopsis (20,000 species)
- Cacti, carnations (11,500 species)
- Blueberry, coffee, tea, tomato, milkweeds, potato, tobacco, mints (60,000 species)
- Sunflower, carrots, honeysuckle (10,000 species)

reduces competition for light and space, potentially allowing more species to coexist. Today, tropical rain forest trees are an important component of the diversity of eudicots. Familiar eudicot trees in temperate regions include oaks, willows, and eucalyptus.

At the other end of the size spectrum are the herbaceous eudicots. These small plants make up a substantial fraction of eudicot diversity. Herbaceous plants do not form woody stems. Instead, the aboveground shoot dies back each year rather than withstand a period of drought or cold. The herbaceous growth form appears to have evolved many times within eudicots as tropical groups represented by woody plants expanded into temperate regions. Important to the success of herbaceous plants is the ability to produce efficient xylem: they can rebuild their stems year after year with minimal investment. Examples of herbaceous eudicots include violets, buttercups, and sunflowers.

At one extreme of herbaceous eudicots are annuals, which complete their life cycle in less than a year. Annuals persist during the unfavorable period as seeds. This plant form is unique to angiosperms, and the majority of annuals are eudicots.

Eudicots are diverse in other ways as well. Most parasitic plants and virtually all carnivorous plants are eudicots, as are water-storing cacti that grow in deserts. Some, such as roses and blueberry, are woody shrubs. Others grow as epiphytes in the canopy of rain forests. Still others, such as grapes and honeysuckle, are vines.

Finally, eudicots make a significant contribution to the diversity of the dinner table. Apples, carrots, pumpkins, and potatoes are all eudicots, as are coffee, cacao (the source of chocolate), and tea. Many plants with oil-rich seeds, such as olives, walnuts, soybean, and canola, are eudicots, as are members of the legume, or bean, family. Some members of this family can form symbiotic interactions with nitrogen-fixing bacteria (Chapter 27), and as a result produce seeds that are rich in nitrogen, such as lentils and soybean.

 CASE 6 AGRICULTURE: FEEDING A GROWING POPULATION

What can be done to protect the genetic diversity of crop species?

Of the more than 400,000 species of plants on Earth today, we eat surprisingly few. Of those we do eat, nearly all are angiosperms. Most of the plants we eat come from species that are grown in cultivation, and many of these cultivated species have, as a consequence of artificial selection, lost the ability to survive and reproduce on their own. Of the approximately 200 such domesticated species, 12 account for more than 80% of human caloric intake. Just three—wheat, rice, and corn—make up more than two-thirds of that intake.

Crop breeders select for varieties that grow well under cultivation and are easy to harvest, but a consequence of selective breeding is decreased genetic diversity of crop species. Because pathogens and pests continue to evolve, any loss in genetic diversity of crop species creates substantial risks. For example, before 1970, the fungal pathogen *Bipolaris maydis* had destroyed less than 1% of the U.S. corn crop annually. In 1970, however, it

FIG. 31.24 "Centers of origin" where different crop species were domesticated.

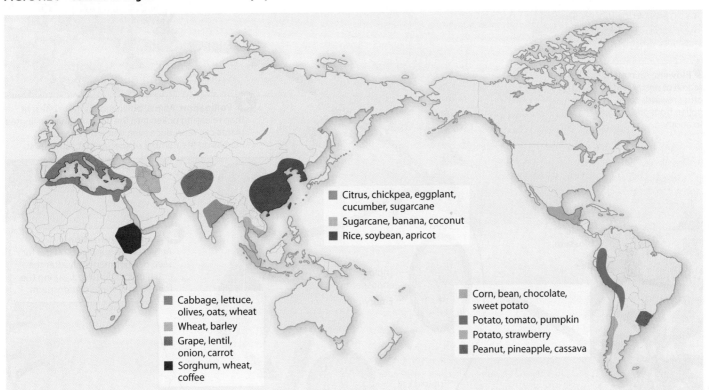

destroyed more than 15% of the corn crop in a single year. The severity of the epidemic was due to the genetic uniformity of the corn varieties planted at that time. In natural plant populations, genetic variation for pathogen resistance makes it highly unlikely that a newly evolved pathogen strain will be able to infect every plant.

Nicolai Vavilov, a Russian botanist and geneticist, was one of the first to recognize the importance of safeguarding the genetic diversity of both crop species and their wild relatives. In the early twentieth century, he mounted a series of expeditions to collect seeds from around the globe. Vavilov's observations of cultivated plants and their relatives led him to hypothesize that plants had been domesticated in specific locations. From these locations, plant domestication spread through human migration and commerce (**Fig. 31.24**). Vavilov called these regions "centers of origin." He believed that they coincided with the centers of diversity for both crop species and their wild relatives. In World War II, during the siege of Leningrad, in which more than 700,000 people perished, scientists at the Vavilov Institute sought to protect what was then the world's largest collection of seeds. At least one of the self-appointed caretakers died of starvation, despite being surrounded by vast quantities of edible seeds. The dedication of these scientists illustrates the priceless nature of the genetic diversity on which our food supply rests.

Today, seed banks help preserve the genetic diversity of crop species and their wild relatives. However, seed banks can store only a fraction of the genetic diversity present in nature. To help make up the difference, some have suggested establishing protected areas that coincide with Vavilov's centers of origin. As we discuss in Chapter 48, new threats lie just over the horizon. Meeting these threats will require that every genetic resource be brought to bear.

The reproduction, growth, and physiology of angiosperms are summarized in **Fig. 31.25** on pages 708 and 709.

Self-Assessment Questions

14. How does the movement of pollen from an animal-pollinated angiosperm compare with that from a wind-pollinated conifer?
15. What are two features (and their possible advantages) that have contributed to the diversity and success of angiosperms?
16. What are some of the distinctive features of monocots?
17. Explain how features found in angiosperms may have allowed the evolution of herbaceous plants, including annual plants.
18. Which three species make up more than two-thirds of human food consumption?

VISUAL SYNTHESIS
FIG. 31.25
Angiosperms: Structure and Function
Integrating concepts from Chapters 27–31

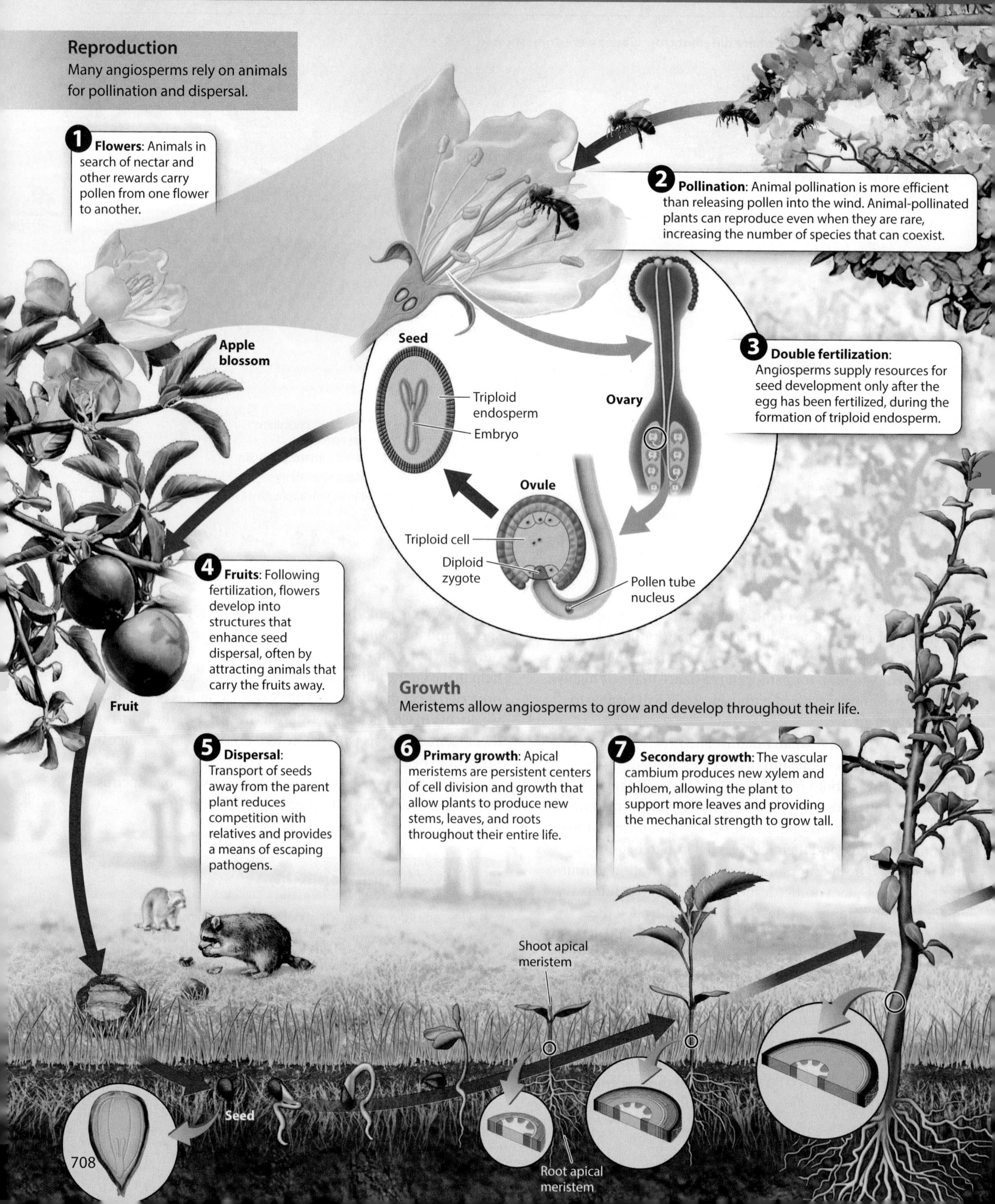

Reproduction
Many angiosperms rely on animals for pollination and dispersal.

1 Flowers: Animals in search of nectar and other rewards carry pollen from one flower to another.

2 Pollination: Animal pollination is more efficient than releasing pollen into the wind. Animal-pollinated plants can reproduce even when they are rare, increasing the number of species that can coexist.

3 Double fertilization: Angiosperms supply resources for seed development only after the egg has been fertilized, during the formation of triploid endosperm.

Apple blossom

Seed
— Triploid endosperm
— Embryo

Ovary

Ovule
— Triploid cell
— Diploid zygote
— Pollen tube nucleus

4 Fruits: Following fertilization, flowers develop into structures that enhance seed dispersal, often by attracting animals that carry the fruits away.

Fruit

Growth
Meristems allow angiosperms to grow and develop throughout their life.

5 Dispersal: Transport of seeds away from the parent plant reduces competition with relatives and provides a means of escaping pathogens.

6 Primary growth: Apical meristems are persistent centers of cell division and growth that allow plants to produce new stems, leaves, and roots throughout their entire life.

7 Secondary growth: The vascular cambium produces new xylem and phloem, allowing the plant to support more leaves and providing the mechanical strength to grow tall.

Seed

Shoot apical meristem

Root apical meristem

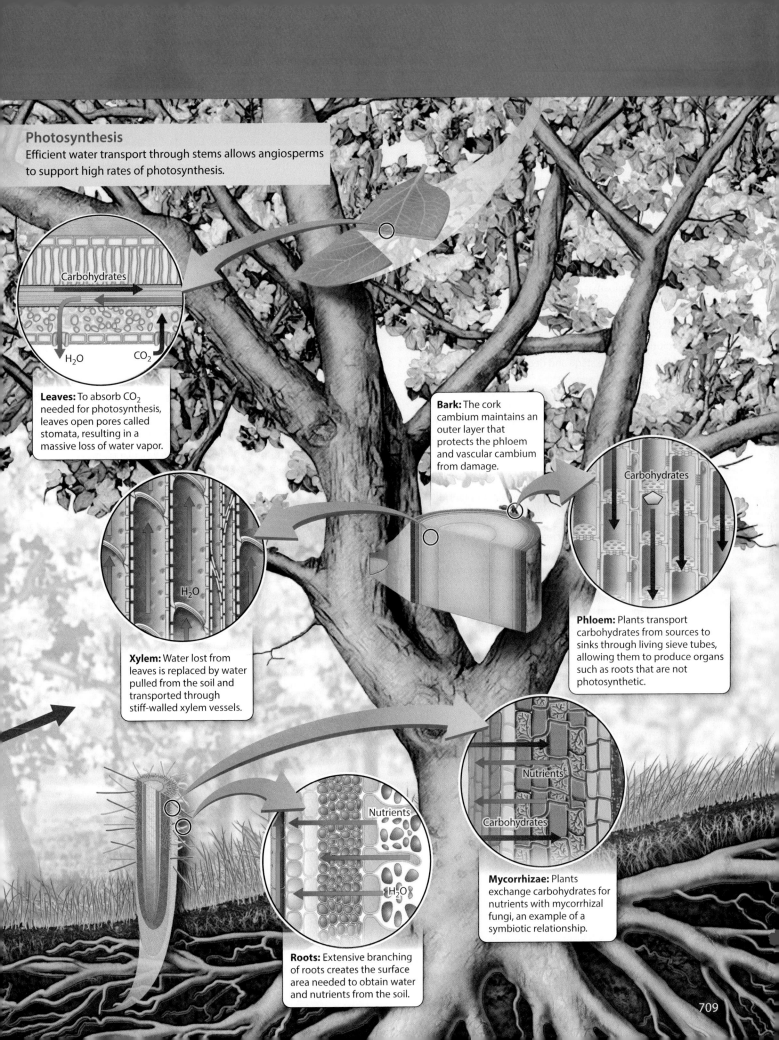

CORE CONCEPTS SUMMARY

31.1 MAJOR THEMES IN THE EVOLUTION OF PLANT DIVERSITY: The evolutionary history of plants has been shaped by the challenges of fertilization, dispersal, and photosynthesis on land.

Four major transformations in land plants have occurred, each of which defines a monophyletic group: alternation of generations is a characteristic of all land plants; formation of xylem and phloem is a trait common to all vascular plants; seeds and pollen are features of seed plants; and flowers (among other traits) define the flowering plants. page 685

Evolutionary trends in land plants are increasing dominance of the sporophyte stage and decreasing dependence on surface moisture. page 687

Nearly 400,000 species of plants are believed to exist today. Of these, 90% are angiosperms; the remaining 10% are distributed among the six other major groups of plants. page 688

Angiosperms first appear in the fossil record about 140 million years ago, more than 300 million years after plants first evolved on land. page 688

The rapid and dramatic increase in angiosperms starting 140 million years ago has resulted in an overall increase in plant diversity, but a decline in other plant groups such as gymnosperms and ferns. page 688

Moist tropical rain forests dominated by angiosperms provided new types of habitat into which other plants could evolve. page 688

31.2 BRYOPHYTES: Bryophytes have a dominant gametophyte generation; they are small in stature and rely on surface moisture for hydration.

The bryophytes are three groups of plants (mosses, liverworts, and hornworts) that share many features. The relationships among these three groups remain uncertain, but all diverged before the evolution of vascular plants. page 689

Bryophytes are small plants with a dominant gametophyte stage in the form of either a flattened thallus or a leafy stalk. page 689

Bryophytes do not form roots, but rather absorb water through their surfaces. Because they cannot restrict water loss, the hydration of bryophytes fluctuates with that of the environment. page 689

The sporophyte generation of bryophytes remains physically and physiologically dependent on the gametophyte. The sporophyte elevates the sites of spore production and, by drying out, enhances the likelihood of spores being dispersed by wind. page 690

Convergent evolution between bryophytes and vascular plants has occurred several times, resulting in insect dispersal of spores in some mosses and the evolution of internal transport cells in some mosses and liverworts. page 691

Sphagnum moss, the dominant plant of peat bogs, soaks up water and acidifies the environment, slowing decomposition. As a consequence, large amounts of organic carbon build up year after year in these bogs. page 691

31.3 SPORE-DISPERSING VASCULAR PLANTS: Spore-dispersing vascular plants have small free-living gametophytes and much larger free-living sporophytes with xylem and phloem; they are most commonly found growing as epiphytes or in forest understories.

Only two groups of spore-dispersing vascular plants have living relatives today: lycophytes, and ferns and horsetails. page 692

Three hundred million years ago, lycophytes included large trees that dominated swamp forests. Today, lycophytes are small plants that either grow in the forest understory as epiphytes or occur in shallow ponds. page 694

Ferns and horsetails are morphologically diverse. Ferns produce large leaves that uncoil as they grow; horsetails have tiny leaves; and whisk ferns have no leaves at all. page 696

Although ferns have a long history, most present-day species are the result of a radiation that occurred after the rise of the angiosperms. page 697

31.4 GYMNOSPERMS: Gymnosperms rely on pollen for fertilization and seeds for dispersal; they have woody stems and today are most common in seasonally cool or dry regions.

Seed plants, which include both gymnosperms and angiosperms, have dominated terrestrial environments for more than 200 million years. Gymnosperms were once much more diverse and widespread; today, there are fewer than 1000 species of these plants. page 697

Gymnosperms (including the many extinct groups) are paraphyletic. There are four groups of living gymnosperms: cycads, ginkgos, conifers, and gnetophytes. page 697

Cycads produce large leaves on thick, unbranched stems. Although they once were widely distributed, they now occur in small, fragmented populations, primarily in the tropics and subtropics. Many cycads are insect pollinated, and all form symbiotic associations with nitrogen-fixing bacteria. page 698

Ginkgo is the single living species of a group that was distributed globally before the evolution of the angiosperms. *Ginkgo* is wind pollinated and produces tall, branched trees. page 699

Conifers include the tallest and longest-lived trees on Earth. Wind-pollinated and largely evergreen, they are found primarily in cool to cold environments. page 700

Before the angiosperms appeared, conifers were widespread. Their rarity in the tropics and persistence in temperate regions may be due to their dependence on wind pollination and their xylem, which is formed entirely of tracheids. page 700

The gnetophytes are a small group, containing only three genera and few species, that has independently evolved xylem vessels and double fertilization. page 700

31.5 ANGIOSPERMS: Angiosperms are distinguished by flowers, fruits, double fertilization, and xylem vessels; they are the most diverse group of plants found today.

Angiosperms reproduce quickly and with more efficient use of resources as a result of pollination and double fertilization. page 701

Angiosperm diversity may result as much from low rates of extinction as from high rates of speciation. page 702

Flowering plants with animal pollinators can reproduce even if they are far apart, allowing rare species to persist and reproduce. page 702

The angiosperm phylogeny indicates a major split between two diverse groups, monocots and eudicots. page 702

Monocots have a single cotyledon, or embryonic seed leaf, and do not form a vascular cambium. page 703

Eudicots have two cotyledons. Many eudicots produce highly conductive xylem, and exhibit a wide range of growth forms. page 705

Herbaceous plants evolved multiple times within eudicots and make an important contribution to eudicot diversity. page 706

Modern agriculture is based on just a few plant species with low genetic diversity, making crops vulnerable to pests and pathogens. page 706

Log in to **LaunchPad** to check your answers to the Self-Assessment Questions and to access additional learning tools.

CHAPTER 32 Fungi

CORE CONCEPTS

32.1 GROWTH AND NUTRITION: Fungi are heterotrophic eukaryotes that feed by absorption.

32.2 REPRODUCTION: Fungi reproduce both sexually and asexually, and disperse by spores.

32.3 DIVERSITY: Other than animals, fungi are the most diverse group of eukaryotic organisms.

In tropical rain forests, vascular plants make up most of the biomass. Animals eat the plants, but they feed selectively. That is, animals consume leaves, fruits, and seeds, but largely avoid a much more abundant tissue—wood. In fact, most leaves also escape grazing and eventually fall, dead, onto the forest floor. Animals, too, may be eaten by predators, but some die of other causes and contribute their remains to the soil. Given this constant rain of biological materials, we might expect wood, leaves, and animal carcasses to accumulate in great piles, but a walk through the rain forest reveals that the forest floor is remarkably clean. What happened to all of the dead plant and animal tissues?

The reason we are not overwhelmed with dead plants and animals is that their bodies decompose. On land, the organisms principally responsible for the decomposition of plant and animal tissues are **fungi**, one of the most abundant and diverse groups of eukaryotic organisms. Many of us have seen mushrooms and toadstools in a meadow or woodland, but most fungal biomass lies within the soil, out of sight. Fungi are able to locate and break down the complex molecules and bulky tissues in plant and animal bodies. For this reason, fungi play an important role in the carbon cycle. Moreover, because they form intimate relationships with living plants and animals, fungi may either enhance or hinder the growth and reproduction of many other organisms. For example, fungi associated with plant roots dramatically increase plant growth by enhancing the uptake of mineral nutrients from the soil (Chapter 27). Other fungi are major agricultural pests (Chapter 30). Some helpful fungi enable us to turn wheat into bread, barley into beer, and milk into cheese. Other fungi cause athlete's foot, yeast infections, and, especially in patients with compromised immune systems, overwhelming infections that can lead to death.

32.1 GROWTH AND NUTRITION

Fungi are heterotrophs: they depend on preformed organic molecules for both carbon and energy. Unlike animals, fungi do not have organs that enable them to ingest food and break it down in a digestive cavity (Chapter 38). Instead, fungi absorb organic molecules directly through their cell walls. This mode of feeding presents two major problems.

First, although simple molecules like amino acids and sugars pass readily through the cell wall, more complicated molecules do not. To deal with this problem, fungi secrete a multitude of enzymes that break down complex organic molecules like starch or cellulose into simpler compounds that can be absorbed. Fungi, then, digest their food first and take it into the body afterward.

A second challenge is to find food in the environment. Other heterotrophic organisms such as bacteria, protists, and animals commonly move through their own habitats, actively searching for food. Fungi have no means of locomotion, so these organisms use the process of growth itself to find nourishment.

Hyphae permit fungi to explore their environment for food resources.

Most fungi consist of highly branched, multicellular filaments called **hyphae** (**Fig. 32.1**). The hyphae are slender, typically 10 to 50 times thinner than a human hair. The numerous long, thin hyphae provide

FIG. 32.1 Fungal hyphae. These thin filaments have an enormous surface area that increases absorption of nutrients. Shown here are hyphae of *Trichophyton*, a fungus that infects the skin of humans, as seen under a scanning electron microscope. *Source: Susumu Nishinaga/Science Source.*

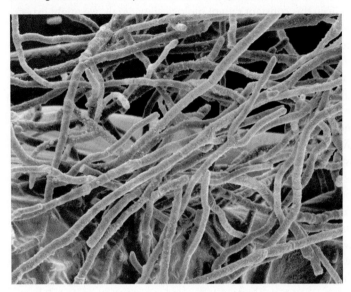

fungi with a large surface area for absorbing nutrients. The network of branching hyphae that forms the body of a fungus is called a **mycelium** (plural, **mycelia**).

Hyphae maintain their slender form by growing only at their tips. Elongating hyphae penetrate ever farther into their environment, encountering new food resources as they grow. Enzymes needed to break down complex food sources are secreted at the growing tip where new cell wall is being formed. In this way, hyphal growth is important for both finding and digesting food resources. Where resources are low, growth is slow or may stop entirely.

When fungi encounter a rich food resource, their hyphae grow rapidly and branch repeatedly. Mycelia can grow to be quite large—the largest known individual of the fungus *Armillaria ostoyae* covers more than 2000 acres in the Blue Mountains of Oregon and weighs many hundreds of tons. We may be impressed by blue whales and redwoods, but in fact some fungi are the largest living organisms on Earth.

A strong but flexible cell wall is key to hyphal growth. In fungi, cell walls are made of **chitin**, the same compound found in the exoskeletons of insects. Chitin is a modified polysaccharide that contains nitrogen. Because of their chemical makeup, fungal cell walls are thinner than plant cell walls. The fungal cell walls prevent hyphae from swelling and rupturing as water flows into the cytoplasm by osmosis, when cells are exposed to dilute solutions such as those in freshwater environments and in the soil. The inflow of water by osmosis provides the force that enables fungi to explore their surroundings: the inflow of water generates positive turgor pressure (Chapters 5 and 27), pushing the hyphal tip ever deeper into the local environment.

Fungi transport materials within their hyphae.

Fungi can transport food and signaling molecules across long distances in mycelia. As a result, they are able to grow between resource patches and to produce reproductive structures such as mushrooms that rise up above the ground. Molecules taken up actively from the environment drive water into the cell by osmosis, thereby increasing turgor pressure. At the same time, growth and respiration consume these molecules, resulting in a decrease in turgor pressure. These increases and decreases in turgor pressure along the hyphae drive bulk flow in a manner similar to phloem transport in plants (Chapter 27). Materials obtained in a nutrient-rich location are carried by bulk flow within fungi and used to fuel hyphal elongation across nutrient-poor locations. Transport by bulk flow also allows fungi to build relatively large reproductive structures aboveground. All the raw materials used to build a mushroom, for example, must be transported from hyphae in contact with a source of nutrients, such as the soil or a rotting log.

A continuous stream of cytoplasm, not divided by barriers, is essential for the movement of materials by bulk flow within mycelia. In some groups, the hyphae have many nuclei but no cell walls to separate them. In other groups, nuclear divisions are accompanied by the formation of **septa** (singular, **septum**),

FIG. 32.2 Septa separating individual cells in fungi. Septa, partitions between cells, have pores that allow transport of solutes along the hypha.

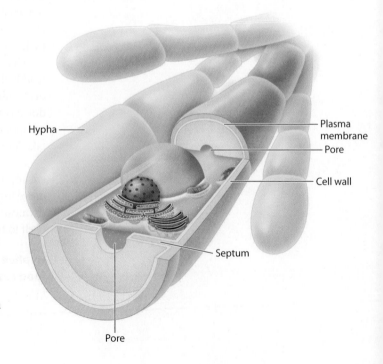

walls that partially divide the cytoplasm into separate cells (**Fig. 32.2**). Each septum contains one or more pores that allow water and solutes to move freely between cells. Septa play an important role when hyphae are damaged. Injury activates sealing mechanisms that plug pores in the septa, preventing the loss of pressurized cytoplasm.

Not all fungi produce hyphae.

Yeasts are single-celled fungi found in moist, nutrient-rich environments. In this kind of setting, the search for food does not require hyphae. All yeasts are descended from ancestors that developed hyphae, but hyphal development has been lost independently several times. Thus, the lack of hyphae is an example of convergent evolution, in which the same trait develops independently in different lineages in response to similar selective pressures (Chapter 22).

Most yeasts divide by budding. In this process, a small outgrowth increases in size and eventually breaks off to form a new cell (**Fig. 32.3**). The growth of the bud is similar to growth by elongation at hyphal tips. In fact, some yeasts can form hypha-like structures under certain conditions. For example, in *Candida albicans*, a yeast that infects humans, hyphae are produced during the transition of the fungus from a harmless variant to a pathogenic form.

Yeasts are commonly found on the surfaces of plants, and to a lesser extent on the surfaces and in the guts of animals. Humans have long used the yeast *Saccharomyces cerevisiae* to ferment plant carbohydrates as part of the process of producing leavened bread and alcoholic beverages. *Saccharomyces cerevisiae* is also an important model organism used in laboratory studies of genetics.

Fungi are principal decomposers of plant tissues.

Most fungi use dead organic matter as their source of energy and raw materials. On land, the organic matter in dead tissues on and within soils far exceeds the amount in the living biomass. In essence, the ground beneath our feet provides tremendous resources for heterotrophic growth. Bacteria and some protists can use this resource, but for the most part it is the fungi that convert dead organic matter back to carbon dioxide and water. Their activity helps to keep the biological carbon cycle in balance (Chapter 46) and returns nutrients to the soil, where they are available to support new plant growth.

A brief consideration shows why fungi have the advantage on a forest floor or in soil. Like fungi, bacteria secrete enzymes that break down organic molecules in their immediate environment, but in the soil bacteria have only a limited capacity to move from place to place. Their flagella can propel them a few micrometers through a thin film of water, but when local resources become depleted, bacteria must wait for new food to appear. In contrast, the growing hyphae of fungi can search actively through the soil for new food resources. Furthermore, because their hyphae can penetrate into the interior of a food source, fungi are particularly well suited for breaking down the bodies of multicellular organisms, such as tree stems.

Although fungi can obtain nutrition from dead animal, protozoan, or even bacterial cells, the most abundant biomolecule on and within soils is cellulose, the principal component of plant cell walls (Chapter 29). Cellulose is a polymer of the sugar glucose, so it is a particularly rich source of carbon and energy. But cellulose is also difficult to degrade because individual cellulose polymers bind tightly to one another. Fungi secrete enzymes that break the cellulose into individual glucose molecules, which are then absorbed and metabolized. Woody tissues, however, pose an additional challenge. In wood, cellulose is closely associated with lignin, which makes the cellulose even harder to get at. Lignin is difficult for enzymes to break apart, in part because it lacks a regular chemical structure. Wood-rotting fungi produce enzymes that degrade lignin, allowing them to access the cellulose polymers within the wood. A few bacteria are known to degrade lignin, but in nature, fungi account for most of the decomposition of wood (**Fig. 32.4**).

To break down lignified plant cell walls, fungi require oxygen. In water-logged environments, where oxygen concentrations are typically low, decay proceeds slowly, if at all. For this reason, woody biomass, once submerged in water, tends to accumulate, leading to the buildup of partially decayed plant material as peat. Over geologically long time intervals, peat becomes coal. Thus, this energy staple of the industrial world owes its existence to the inhibition of fungal decay.

FIG. 32.3 The yeast *Saccharomyces cerevisiae*. Yeasts are unicellular fungi that often divide by budding smaller cells off larger ones. *Source: Andrew Syred/Science Source.*

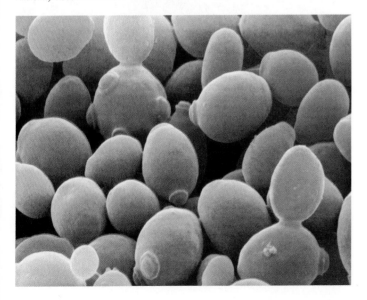

FIG. 32.4 Decomposition of wood by fungi. (a) A dead tree being decomposed by brown rot fungi. (b) These scanning electron microscope images show (left) normal cedar wood and (right) rotted cedar wood from the coffin in the tomb of King Midas in Gordion, Turkey. *Sources: a. Veronique Leplat/Science Source; b. Timothy R. Filley, Robert A. Blanchette, Elizabeth Simpson, and Marilyn L. Fogel. Nitrogen cycling by wood decomposing soft-rot fungi in the "King Midas tomb," Gordion, Turkey PNAS 2001 98 (23) 13346-13350. Copyright 2001 National Academy of Sciences, U.S.A. Photos courtesy of Robert A. Blanchette, University of Minnesota.*

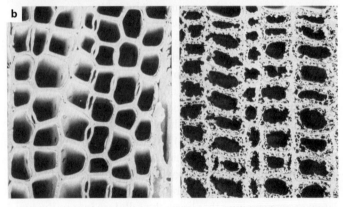

FIG. 32.5 Fungal infection of living tissue. (a) This stained section of heart tissue (pink) is infected with the fungus *Candida albicans* (the purple threadlike structures). (b) A scanning electron micrograph shows hyphae of *Ceratocystis ulmi*, a vascular wilt fungus, growing in xylem vessels of an elm tree. *Sources: a. Centers for Disease Control and Prevention/ScienceSource; b. Biophoto Associates/Science Source.*

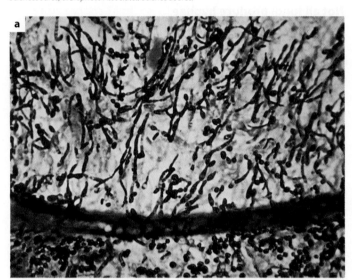

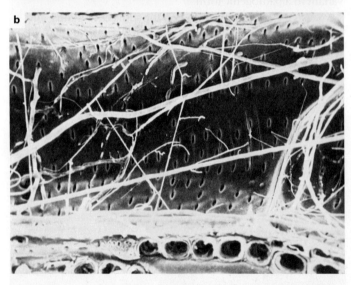

Fungi are important plant and animal pathogens.

Fungi not only feed on dead plants and animals, but also are well adapted to infect living tissues (**Fig. 32.5**). Plants, for example, are vulnerable to a diverse array of fungal pathogens—rusts, smuts, and molds—that cause huge losses in agricultural production, as we discuss in section 32.3.

Invasion is the key to success for plant pathogens (Chapter 30). Successful pathogens must be able to get past a plant's physical or chemical defenses. This explains why fungal pathogens exhibit host specificity, meaning that they are able to infect only one or a small number of host species. In many cases, fungi infect plants through wounds, which provide a route through a plant's outer defenses. Some fungi enter through stomata. Others penetrate epidermal cells directly, degrading the wall with enzymes and then using turgor pressure to push their hyphae into the plant's interior. The extent of infection usually depends on the ability of hyphae to expand into new tissues.

Vascular wilt fungi (Chapter 30) send their hyphae through xylem vessels, exploiting this route to spread throughout the entire plant. Old photographs of towns in the eastern United States show streets shaded by American elm trees, but today nearly all those trees are dead, victims of a vascular wilt fungus introduced into the United States early in the twentieth century.

Aboveground, plant infections are usually transmitted by fungal spores, carried either by the wind or on the bodies of insects. For example, many fungi that attack fruits are transmitted by insects also seeking those fruits; yeasts that feed on nectar within flowers may hitch rides on pollinators. Some fungi secrete sugar-rich droplets that attract insects. The insects then transport their fungal spores to other sites.

Belowground, infection is typically transmitted by hyphae that penetrate the roots of plants. Hyphae can feed on organic matter in the soil until they encounter the root of a suitable host plant. The persistence of fungal hyphae in soils means that populations of pathogenic fungi can build up in the soil near a host plant. As a consequence, seedlings germinating close to an adult tree of the same species are at risk of infection, whereas seedlings of distantly related species may be spared. By favoring the establishment of seedlings of less common species, fungal pathogens are thought to play an important role in maintaining high levels of species diversity in tropical rain forests (Chapter 30).

Insects and other invertebrates are also susceptible to fungal pathogens. Certain fungi exploit living animals in a unique way: by trapping them. Some fungal predators form sticky traps with their hyphae, whereas others actually lasso their prey (**Fig. 32.6**). The hyphae of these fungi form rings that can inflate within seconds, trapping anything within the ring. Nematodes, tiny worms common in soils, may inadvertently pass through the rings, setting off hyphal inflation. The nematodes are then trapped, providing food for the fungus.

Fungal infection is relatively rare in vertebrates. Severe infections are more frequent among fish and amphibians than in mammals, perhaps because fungi grow poorly at mammalian body temperatures. An apparent exception is the fungal infection that has caused dramatic declines in North American bat populations. These fungi tend to infect the bats during their winter hibernation, when the bats have lowered body temperatures that conserve energy. In humans, most fungal infections are annoying rather than life threatening; athlete's foot and yeast infections are prime examples. However, in individuals with immune systems compromised by human immunodeficiency virus (HIV) or illness, fungal infection can be fatal.

Many fungi form symbiotic associations with plants and animals.

While some interactions between species benefit the fungus at the expense of its host, others benefit both partners. In Chapter 27, we saw that mycorrhizal fungi supply plant roots with nutrients such as phosphorus and nitrogen from the soil and, in return, receive carbohydrates from their host (**Fig. 32.7**). Two main types of mycorrhizae exist. The hyphae of **ectomycorrhizal** fungi surround, but do not penetrate, root cells. In contrast, the hyphae of **endomycorrhizal** fungi penetrate into root cells, where they produce highly branched structures that provide a large surface area for nutrient exchange. As we discuss in section 32.3, the endomycorrhizal fungi constitute a monophyletic group dating back more than 400 million years. In contrast, ectomycorrhizae are present in multiple groups and appear to have evolved more than once.

Other fungi, called **endophytes**, live within leaves. Although endophytes are much less studied than mycorrhizae, they may be just as widespread. Endophytic hyphae grow within cell walls and in the spaces between cells. Long thought to be harmless, endophytes are now recognized as beneficial to the host plant. They may improve plant growth during stressful

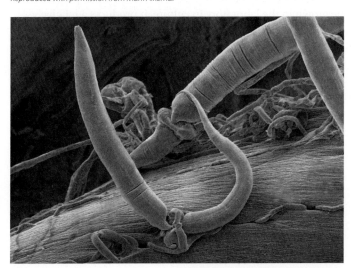

FIG. 32.6 A fungal lasso. Hyphae of *Drechslerella doedycoides* form inflated rings that capture nematode worms moving through the soil. *Source: Reprinted from Current Opinion in Neurobiology, 22, Jennifer K Pirri, Mark J Alkema, The neuroethology of C.elegans escape Fig. 3, Copyright 2011, with permission from Elsevier. Reproduced with permission from Mark Alkema.*

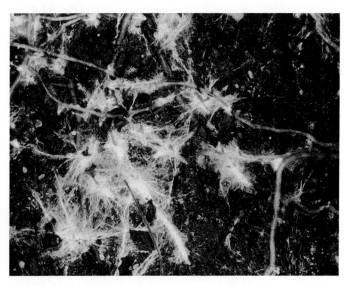

FIG. 32.7 Mutualistic relationship between fungi and plants. Mycorrhizal fungi (white) surround roots (brown) and produce extensively branched hyphae that extend into the soil. *Source: Dr. Jeremy Burgess/Science Source*

FIG. 32.8 Lichens. A lichen is a mutualistic association between a fungus and a photosynthetic microorganism. *Sources: (left to right) Wallace Garrison/Getty Images; Stephen Sharnoff; Stephen Sharnoff.*

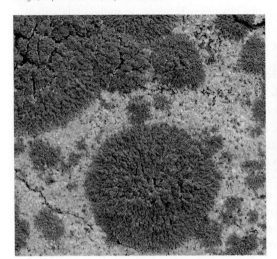

conditions, such as drought, or produce chemicals that deter invasion by pathogens and consumption by herbivorous insects.

Mutually beneficial associations between fungi and animals are much less common, but a few examples are known. None is more striking than the relationship in which insects grow fungi for food. The fungi benefit because the insects provide shelter, food, and protection from predators and pathogens. Insect–fungal agriculture has evolved at least three times: in leaf-cutter ants that live in tropical forests, in a group of African termites, and in some wood-boring beetles. These beetles maintain fungal gardens within tunnels that they excavate in damaged or dead trees.

Lichens are symbioses between a fungus and a green alga or a cyanobacterium.

Lichens are familiar sights in many environments, often forming colorful growths on rocks and tree trunks (**Fig. 32.8**). Lichens look, function, and even reproduce as if they are single organisms, but they are actually stable associations between a fungus and a photosynthetic microorganism. The microorganism is usually a green alga but sometimes a cyanobacterium. Nearly 15% of all known fungal species grow as lichens.

The dual nature of lichens was first proposed by a Swiss botanist, Simon Schwendener, in 1867, but his hypothesis found little favor at the time. The possibility of a composite "organism" challenged the idea that all life could be divided into discrete categories such as animals and plants. Biologists also questioned how an association of distinct species could function as an integrated whole. Given what we now understand about the widespread nature of mutualisms, the opposition to Schwendener's proposal may seem surprising. Yet lichens' dual nature was widely accepted only in 1939, after Eugen Thomas showed that lichens could be separated into their individual parts and then reassembled.

The bulk of a lichen consists of fungal hyphae. The photosynthetic algae or cyanobacteria form a thin layer just under the surface (**Fig. 32.9**). The hyphae anchor the lichen to a rock or tree. They also aid in the uptake and retention of water and nutrients, and produce chemicals that protect the organism against excess light and herbivorous animals. In turn, the photosynthetic partners provide a source of reduced carbon, such as carbohydrates. Cyanobacteria living in lichens can fix nitrogen (Chapter 24) using N_2 from the atmosphere. The two partners exchange nutrients through fungal hyphae that tightly encircle or even penetrate the walls of the photosynthetic cells. As a result of the association between the two partners, lichens can thrive where neither partner could exist on its own.

It remains an open question whether either partner can exist independently in nature. In cases where the fungal partner can be grown in the laboratory in culture, it produces a relatively undifferentiated hyphal mass. Thus, it appears that chemical signals from the photosynthetic partner influence the form and shape of the fungus.

Surprisingly, these associations are not species specific. There are approximately 13,500 known lichens, but only about 100 participating photosynthetic species. Thus, different lichens may include the same algae or cyanobacteria. Conversely, a given fungal partner may associate with more than one photosynthetic species without affecting the lichen's outward form. Because the lichen's shape and structure depend on the fungal species rather than the photosynthetic partner, the lichen and fungus are assigned the same scientific name.

Lichens spread asexually by fragmentation or through the formation of dispersal units consisting of a single

FIG. 32.9 Lichen anatomy. The fungus provides structure, and photosynthetic algae form a thin layer under the surface. The scanning electron micrograph shows a section through *Xanthoria flammea*. *Photo source: Eye of Science/Science Source.*

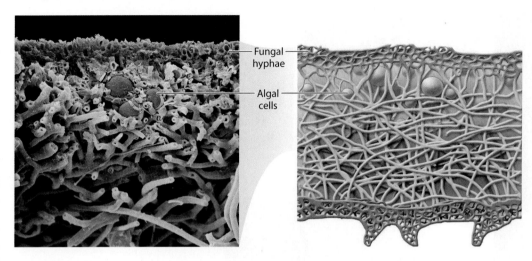

Lichens consist mostly of fungal hyphae that take up water and nutrients from the soil.

Photosynthetic algal cells form a thin layer just under the surface of the lichen.

photosynthetic cell surrounded by hyphae. Sexual reproduction, at least by the fungi, is common, whereas the photosynthetic cells reproduce asexually by mitotic cell division. Whether sexual or asexual reproduction is more important in allowing lichens to establish in new habitats is not known.

Lichens are remarkable for their ability to grow on the surfaces of rocks and tree trunks. They are among the first colonizers of lava flows and the barren land left after glacial retreat. In these habitats, lichens obtain nutrients from rainfall or by secreting organic acids that help release some nutrients from rocky substrates. Not surprisingly, lichens in these harsh environments grow slowly. Other lichens grow on soil. For example, reindeer "moss" grows in the Arctic tundra, where it is eaten by caribou. Reindeer moss is thought to compete successfully for space by secreting chemicals that hinder the growth of plants.

Given their small size and exposure to the environment, lichens must be able to tolerate drying out from time to time. All lichens have a high tolerance for desiccation, and can tolerate wide fluctuations in temperature and light. In contrast, lichens are quite sensitive to air pollution, particularly sulfur dioxide (SO_2). For this reason, lichen growth is sometimes used as an indicator of industrial pollution.

Self-Assessment Questions

1. What are two ways in which fungi differ from other heterotrophic organisms in how they obtain and digest their food?
2. What are two roles of hyphae?
3. How do fungi contribute to the carbon cycle?
4. Which types of organisms make up a lichen? How does each partner benefit from the association?

32.2 REPRODUCTION

Fungi face two challenges in completing their life cycles. First, to maintain genetic diversity within populations, they must find other individuals to mate with. Second, they must be able to disperse from one place to another. The majority of fungi reproduce both sexually and asexually. As we discuss in this section, one feature of the sexual life cycle of fungi distinguishes it from all other eukaryotes, and some asexual fungi have a unique way to generate genetic diversity. Fungal adaptations for dispersal broadly resemble those of plants. Fungi rely on wind, water, or animals to carry spores through the environment. Recall from Chapter 28 that spores are specialized cells well adapted for dispersal and long-term survival.

We most often come into contact with fungi through their reproductive structures. Fungal spores commonly cause respiratory illness, and mushrooms attract our attention because they are delicious or poisonous. Many other aspects of fungal reproduction are less easily observed. In this section, we emphasize the most general principles and patterns.

Fungi proliferate and disperse using spores.

The tissues, living or dead, that fungi feed on often have a patchy distribution. For this reason, fungi must be able to travel from one food source to another. The extensive networks of hyphae within soils or host organisms can spread locally but cannot disperse over great distances, so fungi produce spores that can be carried by the wind (**Fig. 32.10a**), in water, or attached to (or within) animals. In fungi that live in aquatic environments, spores have flagella that allow them to swim. The great majority of fungi, however, live on land, and their spores lack flagella. To overcome these obstacles to their proliferation, fungal spores are

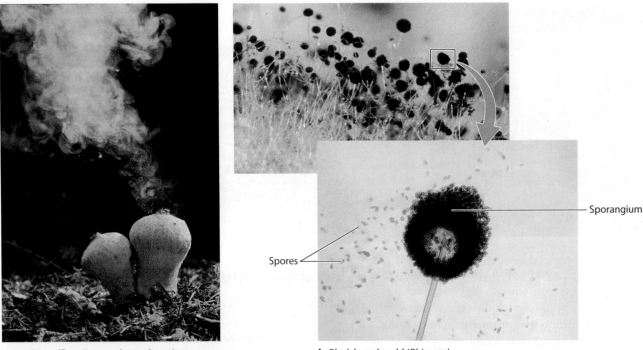

FIG. 32.10 Fungal spores dispersed by wind. Sources: a. Biosphoto / Yves Lanceau; b. (top) Gregory G. Dimijian/Science Source; b. (bottom) Garry DeLong/Getty Images.

a. Devil's snuffbox (*Lycoperdon perlatum*)

b. Black bread mold (*Rhizopus*)

encased in a thick wall that protects them as they are dispersed over habitats unsuitable for growth.

The probability that any given spore will come to rest in a favorable habitat is low, so fungi produce large numbers of spores. Fungal spores remain viable, able to grow if provided with an appropriate environment, for periods ranging from only a few hours in some species to many years in others. Thus, spores allow fungi to use resources that are patchy in time as well as in space. In fact, a shortage of resources is one of the cues that triggers spore formation.

Spores can form through either sexual or asexual reproduction. As part of sexual reproduction, spores form by meiotic cell division. In contrast, asexual spores form by mitotic cell division and, therefore, are genetically identical to their parent. In many species, asexual spores are produced within sporangia that form at the ends of erect hyphae, allowing the spores to be released into the air at the appropriate time. Take a close look at a moldy piece of bread: the surface is covered with hyphae carrying sporangia containing asexual spores (**Fig. 32.10b**).

Fungal species vary in the extent to which they rely on asexual versus sexual spore production. In some groups, including the most common mushrooms, sexual reproduction is the dominant means of spore production. In a small number of fungi, including the endomycorrhizal species, spores produced by meiotic cell division have never been observed. Although sexual reproduction appears to be absent in some fungal species, asexual fungi have other mechanisms for producing genetic diversity, as discussed later in this section.

Multicellular fruiting bodies facilitate the dispersal of sexually produced spores.

Fungi employ an astonishing array of mechanisms to enhance spore dispersal. You are probably most familiar with the multicellular **fruiting bodies** produced by some fungi. Fruiting bodies are structures that release sexually produced spores into the air for dispersal. (Fungal fruiting bodies should not be confused with the fruits of flowering plants, which are unrelated and very different structures.) Mushrooms, stinkhorns, puffballs, bracket fungi, and truffles are all fungal fruiting bodies.

Fruiting bodies are highly ordered and compact structures compared with the mycelia from which they grow, yet they are constructed entirely of hyphae (**Fig. 32.11**). In fact, in many cases their mechanisms of spore dispersal demand a high degree of structural precision. How tip-growing hyphae can produce such complex and regular structures remains largely a mystery.

The fruiting bodies of many fungi rise above the ground or grow outward from the trunks of dead trees, so the sexually produced spores are released high above the ground. Nevertheless, elevation by itself is not enough to ensure dispersal. To give their spores a boost on their travels, many fungi forcibly eject their spores. The tiny spores achieve velocities of more than 1 m/s, enough to penetrate the layer of unstirred

FIG. 32.11 Fruiting bodies. These complex multicellular structures are built from hyphae. Fruiting bodies of *Hygrophorus miniatus*, known commonly as the vermillion waxcap. *Photo source: Matt Meadows/Getty Images.*

air that surrounds the fruiting body (**Fig. 32.12**). Other fungi rely on external agents such as raindrops or animals to move their spores around. In section 32.3, we will encounter examples of the diverse dispersal mechanisms found in the fungi.

Sexual reproduction in fungi often includes a stage in which haploid cells fuse, but nuclei do not.

Like other sexually reproducing eukaryotes, fungi have life cycles that include haploid ($1n$) and diploid ($2n$) stages.

HOW DO WE KNOW?

FIG. 32.12
What determines the shape of fungal spores that are ejected into the air?

BACKGROUND Many fungi eject spores forcibly into the air. For spores to be picked up by the wind, however, they must escape a layer of still air, called the boundary layer, that lies close to solid surfaces. Escape from the boundary layer is easier if spores have a size and shape that minimize drag, the resistance of air to the movement of an object.

HYPOTHESIS In fungi that eject their spores into the atmosphere, natural selection favors spore shapes that minimize drag.

EXPERIMENT Working from computer models for minimizing drag on airplane wings, Marcus Roper and colleagues calculated which spore shapes would minimize drag. These models took into account spore size as well as the physical characteristics of the air through which spores travel. The scientists then measured spore shape for more than 100 species of spore-ejecting fungi.

RESULTS Nearly three-fourths of the examined spores had shapes that came within 1% of the shape calculated to minimize drag (shown

Fungal spores with outlines (blue) of computer-modeled shapes that minimize drag. *Source: Reprinted from Mycological Research, 102/11, R. Dulymamode, P.F. Cannon, A. Peerally, Fungi from Mauritius Anthostomella Species of Pandanus, 1998, with permission from Elsevier.*

in the figure as blue outlines). Related species that do not eject spores forcibly were less than half as likely to have drag-minimizing shapes.

CONCLUSION Natural selection has acted on fungi to make spore dispersal by wind efficient.

SOURCE Roper, M., et al. 2008. "Explosively Launched Spores of Ascomycete Fungi Have Drag-Minimizing Shapes." *Proceedings of the National Academy of Sciences, USA* 105:20583–20588.

The nuclei in fungal hyphae are haploid, so the fungal life cycle is similar to the life cycles of eukaryotic organisms that exist normally in the haploid stage (see Fig. 25.3a). Haploid-dominant organisms can multiply by either sexual or asexual reproduction. Asexual reproduction takes place through the production of haploid spores by mitosis, whereas sexual reproduction takes place as two haploid cells (often male and female gametes) fuse to form a diploid zygote. The zygote undergoes meiosis as its first division, giving rise to sexually produced haploid spores. Sexual reproduction in most fungi differs from all other haploid-dominant organisms in one important respect: in these fungi, the fusion of haploid cells is not immediately followed by the fusion of their nuclei.

In most fungi, the sexual phase of the life cycle takes place through the fusion of hyphal tips rather than the fusion of specialized reproductive cells, or gametes. During mating, two hyphae grow together and release enzymes that digest their cell walls at the point of contact. The contents of the two hyphal cells then merge, forming a single cell with two haploid nuclei.

In most sexually reproducing organisms, when two gametes merge, their nuclei fuse almost instantly to form a diploid zygote. In fungi, however, the cytoplasmic union of two cells (**plasmogamy**) is not always followed immediately by the fusion of their nuclei (**karyogamy**). Instead, the haploid nuclei retain their independent identities for a while. The period between cell fusion and nuclei fusion is known as the **heterokaryotic** ("different nuclei") stage (**Fig. 32.13**). During that stage, a cell contains nuclei from two parental hyphae, but the nuclei remain distinct and even continue to divide. The heterokaryotic stage ends when pairs of nuclei fuse to form diploid zygotes.

In some groups, the heterokaryotic stage consists of a single cell with many haploid nuclei. In other groups, new septa form as the genetically distinct nuclei replicate so that each cell contains two haploid nuclei, one from each parent. Heterokaryotic cells with two genetically distinct haploid nuclei are referred to as **dikaryotic**, or **n + n**, cells. In dikaryotic fungi (fungi with dikaryotic cells), there may be either a small number

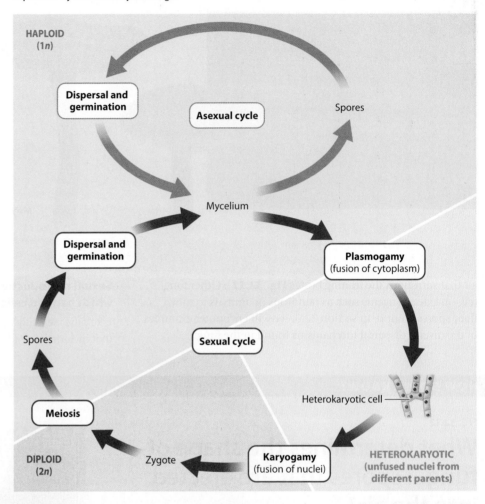

FIG. 32.13 Generalized fungal life cycle. In many fungi, plasmogamy and karyogamy are separated by a heterokaryotic stage.

of dikaryotic cells or extensively developed hyphae made up of dikaryotic cells. Some edible mushrooms found on market shelves consist entirely of dikaryotic hyphae.

Dikaryotic fungi account for more than 98% of all known fungal species. Their dominance suggests that the separation of plasmogamy and karyogamy in time and space may confer an evolutionary advantage. We can see one such advantage by thinking about the environments where fungi live. Mating takes place principally within the soil or inside the trunks of rotting trees, as those are the sites where hyphae are most likely to come into contact. Dispersal, however, is most effective when spores can be released into the air. The separation of plasmogamy and karyogamy allows mating and spore production to occur where each is most effective.

Nevertheless, in some dikaryotic fungi, the separation of plasmogamy and karyogamy is neither distant in space nor long in time. These fungi produce highly branched dikaryotic hyphae with many cells in which karyogamy will eventually take place.

In these fungi, the dikaryotic stage may serve primarily as a means of increasing the number of sexually produced spores that are formed.

At present, we really don't know why fungi proliferate $n + n$ cells rather than forming a multicellular diploid phase. There may be no single answer to the question of why the dikaryotic fungi have this unique cell type.

Genetically distinct mating types promote outcrossing.

Under most conditions found in nature, genetically diverse populations persist better than those lacking diversity. Indeed, sexual reproduction is widely viewed as a means of promoting genetic diversity for long-term ecological success in variable environments (Chapters 11 and 40). But with the exception of fungi that live in aquatic habitats, fungi do not produce male and female gametes. What prevents an individual from mating with itself?

Fungi have different **mating types** that are genetically determined and prevent self-fertilization. The mating type of an individual is determined by a mating-type gene. Fertilization can take place only between individuals that have different alleles at the mating-type gene. In some species, there are only two mating-type alleles. In this case, mating patterns are identical to those for species with male and female sexes. If the two mating-type alleles are spread evenly throughout the population, an individual should be able to mate with 50% of the general population.

Other fungi have more than two mating-type alleles. These fungi have a greater likelihood of encountering a compatible mating type. That likelihood is even higher for fungi in which mating type is determined by two different mating-type genes, each of which has multiple alleles. Fungi in the group that includes the common mushrooms can have as many as 20,000 different mating-type alleles. In this case, the odds of finding a compatible mate are close to 100%.

Genetic compatibility is a prerequisite for mating. How do fungi go about finding a suitable mate? The answer is that they secrete chemical signals that attract fungi with mating types different from their own. The chemical structure of these signals differs among groups.

Parasexual fungi generate genetic diversity by asexual means.

Approximately 20% of fungi appear to lack sexual reproduction altogether. Among these are such well-known groups as *Penicillium* (the source of the antibiotic penicillin), *Aspergillus* (the major industrial source of vitamin C), and all the

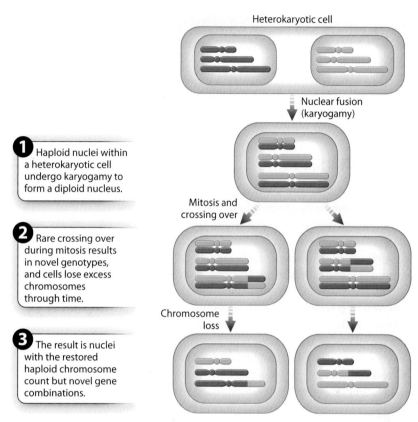

FIG. 32.14 Parasexuality. Crossing over during mitosis is thought to provide genetic diversity in asexual fungi.

1. Haploid nuclei within a heterokaryotic cell undergo karyogamy to form a diploid nucleus.
2. Rare crossing over during mitosis results in novel genotypes, and cells lose excess chromosomes through time.
3. The result is nuclei with the restored haploid chromosome count but novel gene combinations.

endomycorrhizal fungi. To proliferate and disperse, these fungi form spores by mitosis.

How are asexual species able to persist without a mechanism for generating genetic diversity? The short answer is, they don't. These fungal species are hypothesized to have another mechanism of generating genetic diversity: the crossing over of DNA during mitosis (**Fig. 32.14**). Such species are described as **parasexual**.

In parasexual fungi, the haploid nuclei within a heterokaryotic cell undergo karyogamy to form a diploid nucleus—so far, a familiar pattern. But then, during mitosis, a rare crossing over may occur between the two sets of chromosomes. Like sexual reproduction, crossing over produces novel gene combinations. The cells lose excess chromosomes over time, such that the haploid state is restored. The key difference between parasexuality and sexual reproduction is that parasexual species do not undergo meiosis.

Entirely asexual species of fungi posed a problem for early taxonomists. These species lack many of the morphological characters such as fruiting bodies or sporangium-bearing stalks that had been used to define different groups. With the advent of DNA sequence data, it has become clear that many asexual species are closely related to sexual forms, indicating that their evolutionary history as asexual species is not long. A major exception are the Glomeromycota, for which fossils as old as 450 million years have been reported.

> **Self-Assessment Questions**
>
> 5. How do fungi disperse?
> 6. What is the difference between a diploid cell and a dikaryotic cell?

32.3 DIVERSITY

When Aristotle classified the living world into animals and plants, he grouped fungi with plants because of their lack of motility. As discussed in Chapter 25, molecular sequence comparisons now make it clear that fungi are actually related more closely to animals. In fact, it would not be amiss to refer to mushrooms and molds as our "cousins."

Fungi share a number of features with animals: their motile cells, when present, have a single flagellum attached to their posterior ends; fungi and many animals synthesize chitin; and fungi store energy as glycogen, as do animals. Nevertheless, the last common ancestor of fungi and animals was a single-celled microorganism that lived in aquatic environments approximately 1 billion years ago. Multicellularity evolved independently in fungi and animals (Chapter 26), so the fungi have many features that are unique in the biological world.

Fungi are highly diverse.

By any measure, fungi are diverse, although just how diverse is not known. Approximately 75,000 species have been formally described, but estimates of true fungal diversity run as high as 5 million species. Among eukaryotes, the only organisms more diverse than fungi are the animals. The availability of DNA sequence data has greatly advanced our understanding of phylogenetic relationships within the fungi. The phylogeny depicted in **Fig. 32.15** clearly shows how the characters present in familiar mushrooms accumulated through the course of evolution: first cell walls of chitin, then hyphae, then regularly placed septa, and finally the complex multicellular reproductive bodies we call mushrooms.

The evolutionary history of fungi can be viewed as the transition from flagellated ancestral forms that lived in aquatic environments to soil-dwelling forms in which hyphae divided by septa develop complex structures visible to the naked eye. Moreover, in the ancestral forms, karyogamy is hypothesized to have followed quickly on the heels of plasmogamy without the formation of a heterokaryotic stage. Thus, another evolutionary trend is the progressive separation in time between plasmogamy and karyogamy in fungal life cycles. Many of the distinguishing features of fungi evolved as adaptations to the challenges faced

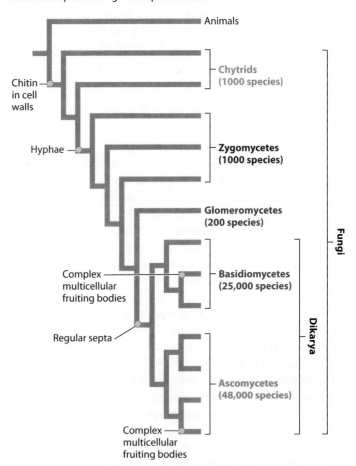

FIG. 32.15 A phylogeny of the fungi. Approximate numbers of described species are given in parentheses.

by land-dwelling nonmotile heterotrophs that absorb nutrients directly from their substrate.

We also see that the numbers of species are not spread evenly across the phylogeny. The two dikaryotic groups include more than 98% of known species, while the remaining 2% is divided among all other groups.

Chytrids are aquatic fungi that lack hyphae.

Fungi are opisthokonts, members of the eukaryotic superkingdom that includes animals (Chapter 25). Among fungi, the Chytridiomycota (commonly referred to as **chytrids**) were the first groups to diverge. Molecular clock analysis estimates that chytrids began to diversify 800 to 700 million years ago. Today, there are about 1000 species of chytrids found in aquatic or moist environments (**Fig. 32.16**).

Many chytrids are single cells with walls of chitin. They may also form short multinucleate structures, but they lack the well-defined hyphae characteristic of other fungi. Chytrids, therefore, do not form a true mycelium, although in some species elongated cellular outgrowths called rhizoids penetrate

FIG. 32.16 Chytrids. Chytrids are aquatic fungi that attach to decomposing organic matter by rhizoids. *Source: John Hardy, University of Wisconsin—Stevens PointDepartment of Biology.*

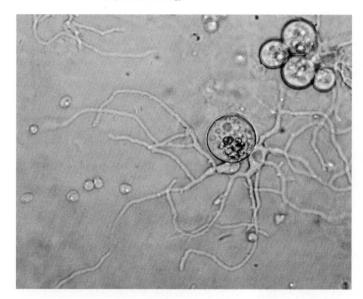

into organic substrates. Rhizoids anchor the organism in place and absorb food molecules.

Chytrids generally disperse by flagellated spores. Sexual reproduction appears to be rare, but there is substantial life-cycle diversity within this group. Chytrids lack a heterokaryotic stage but form flagellated gametes that swim through their aqueous environment.

Most chytrids are decomposers, and a few of them are pathogenic. Notably, infection by the chytrid *Batrachochytrium dendrobatidis* may be a cause of a widespread decline in amphibian populations (Chapter 48).

Zygomycetes produce hyphae undivided by septa.

Moving through the phylogenetic tree, we next encounter several groups traditionally united as the Zygomycota, or **zygomycetes** (see Fig. 32.15). These groups share a number of traits.

Collectively, zygomycetes account for less than 1% of the known fungal species. Some are decomposers, specializing on dead leaves, animal feces, and food. There are probably zygomycetes in your kitchen. Others live on and in plants, animals, and even other fungi. Zygomycetes have many typical fungal traits. For example, they produce mycelia and aerial spores, traits that may have evolved as adaptations for finding food and dispersing on land. Their spores lack flagella, possibly also an adaptation for life on land. Unlike more complex fungi, zygomycetes do not form regular septa along their hyphae, nor do they produce multicellular fruiting bodies.

Earlier, we introduced the black bread mold *Rhizopus*, which is a zygomycete (see Fig. 32.10b). *Rhizopus* and its relatives are specialists that grow on substrates containing abundant, easy-to-digest carbon compounds, such as bread, ripe fruits, and the dung of herbivorous animals. These fungi consume their substrates rapidly. Once their meal is finished, they release large numbers of aerial spores to locate another food source.

Sexual reproduction occurs when two compatible hyphal tips fuse to form a thick-walled structure containing many nuclei of each mating type. Karyogamy and meiosis follow, then the haploid cell germinates to form an elevated stalk. Each stalk develops a sporangium that contains spores produced asexually by mitotic cell division. If you look closely at a bread mold, you can see the white filamentous hyphae and small black sporangia on slender stalks. Each sporangium can produce as many as 100,000 spores that are dispersed by the wind, explaining how these organisms seem to get everywhere.

Some zygomycete fungi have evolved truly spectacular means of dispersal. *Pilobilus* is a zygomycete that consumes the dung of herbivorous animals. Its life cycle is similar to that of *Rhizopus* except that, instead of releasing individual spores, it forcibly ejects the entire sporangium (**Fig. 32.17**). Turgor pressure generated in the supporting stalk propels the sporangia as far as 2 m. Light-sensitive pigments in the stalk's hyphae control the orientation of this water cannon, sending the sporangia hurling toward light. Release in that direction ensures that the spores have the best chance of escaping the dung pile and landing on vegetation that is attractive to grazing herbivores. Feeding herbivores then disperse the spores further, while supplying them with fresh dung for food. The remarkable dispersal of *Pilobilus* sporangia benefits other species as well:

FIG. 32.17 Spore dispersal in a zygomycete. In *Pilobilus*, light-sensing pigments orient the turgor-propelled ejection of the sporangia. *Source: Carolina Biological/Medical Images.*

parasitic nematode worms hitch a ride on the sporangia to find a new host.

Glomeromycetes form endomycorrhizae.

The Glomeromycota, or **glomeromycetes**, are a monophyletic group of apparently low diversity (there are only some 200 described species) but tremendous ecological importance. All known glomeromycetes occur in association with plant roots. The most famous species are those that form endomycorrhizae. Molecular clock estimates suggest that glomeromycetes diverged from other fungi 600 to 500 million years ago, before the evolution of vascular plants. Establishment of glomerophyte–root symbiosis changed the course of evolution for both participants. Today, most vascular plant species harbor endomycorrhizae, which dramatically increase the plants' nutrient uptake from the soil.

Glomeromycetes are difficult to study because they cannot be grown independently of their plant partners, so many unanswered questions remain. For example, no evidence of sexual reproduction has ever been found in this group. Yet genetic studies of glomeromycete populations show some genetic diversity, which is thought to result from parasexual processes. Glomeromycetes produce extremely large (0.1 to 0.5 mm in diameter) multinucleate spores asexually. These large spores can persist in soils until they come into contact with an uninfected root.

The Dikarya produce regular septa during mitosis.

The final major node, or branching point, in the fungal phylogeny marks the origin of the **Dikarya**, the group that includes approximately 98% of all described fungal species. A key feature of the dikaryotic fungi is that every mitotic division is accompanied by the formation of a new septum. Once septa separated hyphae into many cells, these fungi could control the number of nuclei within each cell and, in turn, proliferate dikaryotic cells. The Dikarya include all the edible mushrooms; the yeast species used in the production of beer, bread, and cheese; the major wood-rotting fungi; and pathogens affecting both crops and humans.

The Dikarya contain two monophyletic subgroups (see Fig. 32.15), each named for the shape of the cell in which the key reproductive processes of nuclear fusion (karyogamy) and meiosis take place. The Ascomycota, or **ascomycetes**, are sometimes called sac fungi because nuclear fusion and meiosis take place in an elongated saclike cell called an ascus

FIG. 32.18 Two groups of dikaryotic fungi. (a) The morel (*Morchella esculenta*) is an ascomycete, whereas (b) the shiitake mushroom (*Lentinula edodes*) is a basidiomycete. *Sources: a. Matt Meadows/Getty Images; b. TOPIC PHOTO AGENCY IN/AGE Fotostock.*

(*askos* is Greek for "leather bag"). The Basidiomycota, or **basidiomycetes**, are popularly known as club fungi because the nuclear fusion and meiosis take place in a club-shaped cell called a basidium. ("Basidium" is derived from the Greek *basis*, which means "base" or "pedestal"; the reference is to the position of sexually produced spores at the tips of this club-shaped body.)

Although ascomycetes and basidiomycetes have the same basic life cycle, they diverged from each other as much as 600 to 500 million years ago and evolved complex multicellular structures independently of each other. The morel found in the supermarket produce section is an ascomycete, whereas the shiitake mushroom next to it is a basidiomycete (**Fig. 32.18**). In addition to asci and basidia, the two groups differ in several other features. For example, the septal pores that connect adjacent cells within hyphae are distinct in the two groups. So are the ways in which cells can plug the pores to minimize damage to a mycelium. When ascomycetes reproduce, mating cells form and their nuclei fuse in relatively close succession. Thus, the heterokaryotic stage is brief. In contrast, when basidiomycetes reproduce, the nuclei of two fused cells may remain separate for a long time. The dikaryotic hyphae, containing both types of nuclei, can continue to absorb nutrients from the substrate for a while before the nuclei fuse.

Ascomycetes are the most diverse group of fungi.

Ascomycetes make up 64% of all known fungal species. They include important wood-rotting fungi, many ectomychorrizal species, and significant pathogens of both animals and plants. Ascomycetes form the fungal partner in most lichens; approximately 40% of all ascomycete species occur in lichens.

Ascomycetes loom large in human history. They contribute the baker's and brewer's yeasts used to make bread and beer, respectively; the antibiotic penicillin; and model systems employed in laboratory investigations of eukaryotic cell biology and genetics. Ascomycetes are used to produce soy sauce, sake, rice vinegar, and miso, and to transform milk into Brie, Camembert, and Roquefort cheeses. They cause athlete's foot and other skin infections, as well as many more serious fungal diseases.

Ascomycetes include species that form multicellular fruiting bodies. A remarkable feature of these fruiting bodies is that they contain both dikaryotic and haploid cells. The proliferation of dikaryotic cells leads to the production of many asci, while the growth of haploid cells contributes to the bulk of the fruiting body.

Two ascomycete lineages consist largely of unicellular yeasts, but their close relatives all produce hyphae. Thus, the unicellular nature of yeasts is thought to be a derived feature evolved through loss of hyphae.

Fig. 32.19 illustrates the life cycle of a common ascomycete, the brown cup fungus. In ascomycetes, meiosis is followed by a single round of mitosis, producing eight haploid spores in each ascus. After they mature, the spores are ejected from the top of the asci, expelled by turgor pressure. In many ascomycetes,

FIG. 32.19 The sexual cycle of a common ascomycete, the brown cup fungus (*Peziza* species). *Photo source: Ed Reschke/Getty Images.*

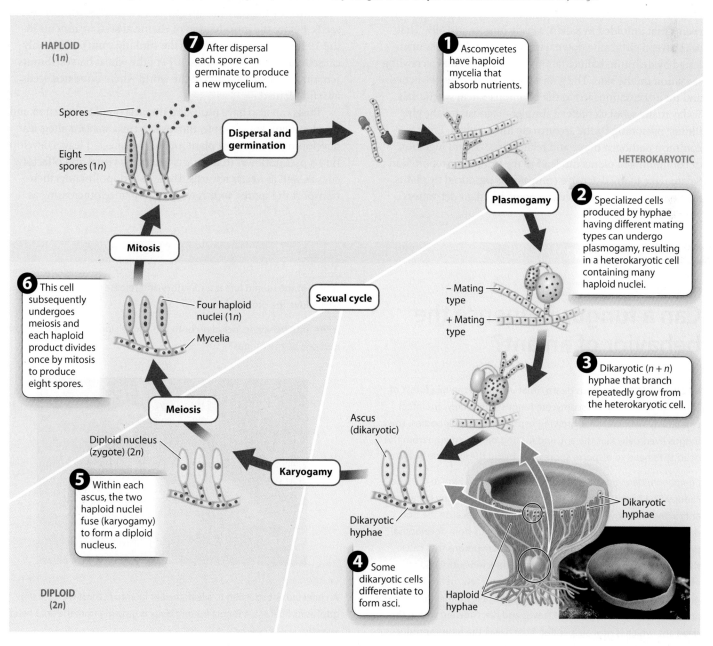

fruiting bodies elevate the asci on one or more cup-shaped surfaces. Spores released from these surfaces are easily caught and carried away by wind.

In the fruiting bodies of some ascomycetes, the asci are completely enclosed by a layer of tissue. The spores in these asci must be dispersed by other organisms rather than by the wind. The truffles prized in cooking provide an example. The edible truffle is the fruiting body of an ascomycete that grows as an ectomycorrhizal fungus on tree roots. Not only do truffles encase their spores in protective tissues, but they also develop underground. How, then, can their spores be dispersed? Developing truffles release androstenol, a hormone also produced by boars before mating. The hormone attracts female pigs, and the pigs unearth and consume the fruiting body. The spores pass through the pig's digestive tract without damage and are released into the environment in feces.

Another ascomycete is thought to have played a role in the events that unfolded in Salem, Massachusetts, in 1692. That year, several girls came down with an unknown illness manifested by delirium, hallucinations, convulsions, and a crawling sensation on the skin. The girls were thought to be bewitched, and their accusations led to the infamous Salem witch trials. Today many scholars prefer a simpler diagnosis for the girls' illness: poisoning by the ascomycete fungus ergot. Ergot is a common pathogen of rye and related grasses, and ingestion of contaminated plants can lead to the symptoms reported in Salem. Like many defensive compounds produced by plants, alkaloid molecules produced by ergot and their derivatives are used in medicine in low doses, such as in the treatment of migraines.

Ascomycetes are even known for producing so-called zombie ants, a topic explored in **Fig. 32.20**.

Basidiomycetes include smuts, rusts, and mushrooms.

The basidiomycetes make up 34% of all described fungal species. There are three major groups (**Fig. 32.21**). Two of these groups, the smuts and rusts, consist primarily of plant pathogens. The third includes many of the most commonly observed mushrooms.

Smut fungi, which infect the reproductive tissues of grasses and related plants, take their name from the black sooty spores that they produce. *Ustilago maydis,* the corn smut, turns developing corn kernels into soft gray masses that are a culinary delicacy in Mexico. Because smuts infect seeds, their spread is magnified by the harvest, storage, and eventual sowing of crop seeds. Before the introduction of chemical seed treatments in the 1930s, infection by *Tilletia* (the stinking smut) commonly caused the loss of as much as 50% of the wheat harvest. Smuts remain a problem in parts of the world where untreated seeds are still planted.

Rusts can also have profound impacts on plant function and development. For example, the rust *Puccinia monoica* alters leaf development in its host plant *Arabis*. The affected leaves develop into a pseudoflower that attracts insects by visual and olfactory cues as well as nectar rewards. These pseudopollinators then transport the spores. In this way they enhance outcrossing as

---

HOW DO WE KNOW?

FIG. 32.20

Can a fungus influence the behavior of an ant?

BACKGROUND A curious death ritual unfolds in the rain forest of Thailand. Spores of the ascomycete fungus *Ophiocordyceps* infect *Camponotus leonardi* ants, growing hyphae inside their bodies. The fungus eventually kills the ant, and fruiting bodies emerge from the dead ant's head to disperse spores that begin the life cycle anew.

Camponotus leonardi ants nest and forage for food high in the forest canopy. The fungus is unable to reproduce in the canopy because ants actively remove fungi from their nests and because the hot dry conditions of the canopy prevent the fungus from developing properly. Infected ants, however, undergo convulsions that cause them to fall to the forest floor. There, the infected ants wander erratically and are unable to climb more than a few meters above the ground before convulsions make them fall again. In their final act, the ants bite into the undersides of leaves and die. Growing within the dead ant, which is now suspended from a leaf, the fungus produces its spores in the humid forest understory and releases them into the air well above the ground.

Is the ants' behavior, including both the convulsions and leaf biting, induced by the fungus?

An infected ant attached to a leaf; the ant bit into the leaf with a "death grip" as its final act. A fungal fruiting body is growing from the ant's head.
Source: Alex Wild Photography.

FIG. 32.21 Basidiomycetes. (a) Corn smut (*Ustilago maydis*); (b) pseudoflowers that form when plants are infected by the rust *Puccinia monoica*; (c) a toadstool fungus, the highly toxic *Amanita*. *Sources: a. Inga Spence/Science Source; b. Bitty Roy, University of Oregon; c. David Clapp/Getty Images.*

well as dispersal of the fungi, but provide no service to the plant. In the next section we examine Ug99, a highly virulent wheat rust first reported in Uganda in 1999.

The third basidiomycete group is characterized by the formation of multicellular fruiting bodies. This group includes the iconic toadstools with their central stalk and umbrella-like cap, as well as the diverse shapes seen in stinkhorns, puffballs, and bracket fungi (**Fig. 32.22**).

Fig. 32.23 shows the life cycle of a typical basidiomycete mushroom. It is similar to the life cycle of ascomycetes (see Fig. 32.19), with a few differences. The fruiting body is made up entirely of dikaryotic hyphae, instead of a combination of dikaryotic and haploid hyphae as in ascomycetes. In addition, nuclear fusion (karyogamy) takes place in a specialized cell called a basidium rather than in an ascus. Finally, the haploid products of meiosis (the spores) do not undergo mitosis, so four spores are produced from each basidium rather than eight.

Most species in this third group form ectomycorrhizal associations or decompose wood and other substrates. The fruiting bodies that we see elevated above a rotting log or the

HYPOTHESIS A parasitic fungus manipulates ant behavior to complete its own reproductive cycle.

OBSERVATIONS Biologist David Hughes observed infected and uninfected ants for many hours, recording their behavior.

RESULTS Infected ants have repeated convulsions (indicated by vertical bars in the graph), but uninfected ants show no such behavior (not shown). Transitions from erratic wandering to a "death grip" moment in which the ant bites into a leaf (indicated by red triangles in the graph and shown in the photograph) occur at about the same time of day. In addition, dissections showed that the ants' death grip results from jaw-muscle wasting caused by the fungus.

CONCLUSION The fungi change the behavior of infected ants, inducing a stereotypical behavior that facilitates completion of the fungus's life cycle. The molecular basis of this manipulation is not yet known.

FOLLOW-UP WORK The zombie-like behavior induced by the fungus is an example of what has been called the extended phenotype: the idea that a phenotype should include the effects that an organism has on its environment—in this case, its effects on host behavior.

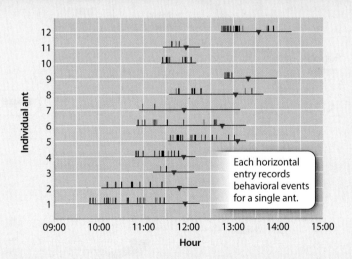

Each horizontal entry records behavioral events for a single ant.

SOURCE Hughes, D. P., et al. 2011. "Behavioral Mechanisms and Morphological Symptoms of Zombie Ants Dying from Fungal Infections." *BMC Ecology* 11:13–22.

FIG. 32.22 Diverse fruiting bodies of basidiomycetes. (a) Stinkhorn (*Dictyophora indusiata*); (b) puffball (*Calvatia gigantea*); (c) bracket fungi (*Ganoderma australe*). *Sources: a. Bill Gozansky/AGE Fotostock; b. Wally Eberhart/Getty Images; c. Bob Gibbons/Science Source.*

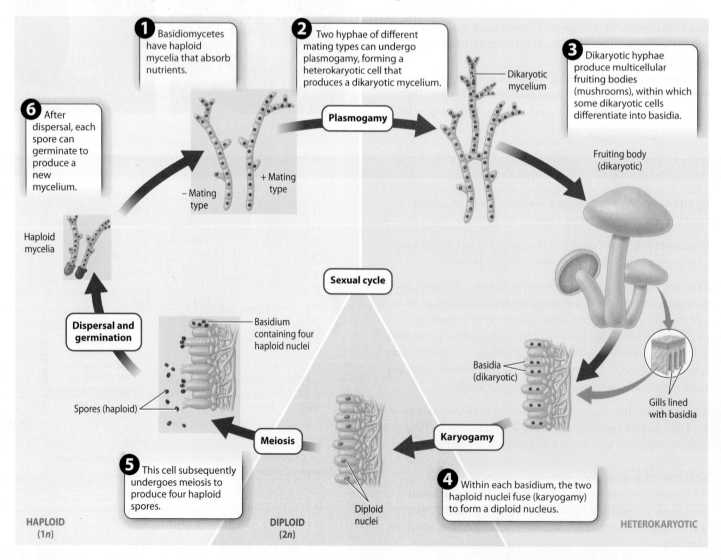

FIG. 32.23 The sexual cycle of a basidiomycete that forms multicellular fruiting bodies.

soil are connected to extensive mycelia that provide resources for fruiting-body development.

Basidiomycetes that form conspicuous fruiting bodies have diverse mechanisms for dispersing their spores. For example, puffballs function like bellows: the impact of raindrops forces the loose, dry spores out of a small hole at the tip of the ball (see Fig. 32.10a). Bird's nest fungi function like splash cups: raindrops displace groups of spores (the "eggs" in the nest) onto nearby vegetation. Herbivores then consume the spores along with the vegetation and deposit the spores in a new dung pile. Stinkhorns produce spores in a stinky, sticky mass that attracts insects; the spores stick to the insects' legs and bodies, and are carried off in this way to new locales.

Perhaps the most remarkable mechanism for spore dispersal occurs in the many basidiomycetes that depend on surface tension to catapult their spores through the air. The placement of their four spores on the ends of short stalks is the key to this mechanism. Both the spores and the supporting basidial cell actively secrete solutes that cause water to condense on their surfaces. At first, the droplets are independent, but as they grow they eventually come into contact with one another. At this point, the high surface tension of water causes them to merge into a single, smooth droplet. This change in shape shifts the center of mass of the water with enough force that it catapults the spore into the air.

Basidiomycetes that rely on surface tension produce their basidia on tightly packed vertical surfaces, such as the "gills" of common mushrooms (**Fig. 32.24**). The spores are flung from the basidia into the air between the vertical surfaces, where they sink due to gravity until they are outside of the fruiting body and can be carried off by the wind. For this mechanism to work, both the orientation of the basidia-lined surfaces and the catapult mechanism must be exquisitely coordinated. If the surfaces are not perfectly vertical, then the spores will not be able to fall from the mushroom. Similarly, if the surfaces are too close together or the force that ejects the spores is too great, the spores will crash into the opposing surface. This arrangement allows huge numbers of basidia to be produced within a single fruiting body, while providing a way for the spores to reach the surrounding air.

 CASE 6 AGRICULTURE: FEEDING A GROWING POPULATION

How do fungi threaten global wheat production?

Throughout much of human history, crops have been vulnerable to infection by *Puccinia graminis,* the black stem rust of wheat. *Puccinia graminis* first infects wheat leaves and stems through their stomata, and then extends within the plant's living tissues to fuel its own growth. Rust-colored pustules erupt along the stem. These pustules release a vast number of spores, leaving the host plant to wither and die.

Crop losses from *P. graminis* can be devastating. In the past, recognizing how badly food supplies could be disrupted using

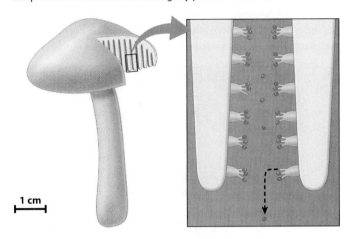

FIG. 32.24 Spore dispersal in gill-forming mushrooms. Spores are catapulted from basidia that line tightly packed vertical surfaces.

this fungus, both the United States and the Soviet Union tried to develop black stem rust as a biological weapon. The U.S. version was cluster bombs (since destroyed) containing turkey feathers smeared with rust spores. The aim was to disrupt crop yields in enemy countries.

Many rusts have a life history in which they alternate between different host species. For example, to complete its life cycle, *P. graminis* must infect first wheat and then barberry bushes (*Berberis* species). During a particularly severe outbreak of wheat rust during World War I, crop losses equaled one-third of total U.S. wheat consumption. In response, a major program of barberry eradication was started to bring stem rusts under control. This program reduced the threat, but did not totally eliminate it.

The Green Revolution seemed to give growers a decisive advantage over stem rust. During World War II, the Rockefeller Foundation sent Norman Borlaug, whose original training was in plant pathology, to Mexico to help combat crop losses due to stem rust. Borlaug was eventually able to breed resistant wheat varieties. Much of the resistance shown by his varieties could be attributed to variation in a small number of genes. Stem rusts were apparently vanquished by Borlaug's discoveries—that is, until the appearance of Ug99, a wheat stem rust first identified in Uganda in 1999 (hence the name). At that time, Ug99 was capable of defeating all known resistance genes. Although scientists have recently identified new genes that confer resistance to Ug99, it will take time to incorporate these into cultivated varieties. Given that wheat accounts for 20% of daily global calorie consumption by humans and that 90% of cultivated wheat has little or no resistance to Ug99, the appearance of this resistant fungus is cause for major concern.

At present, Ug99 shows every sign of spreading (**Fig. 32.25**). This fact is not surprising given that a single hectare of infected wheat produces upward of 1 billion spores. Ug99 currently devastates wheat production in Kenya and has been found in Yemen (2006), Iran (2007), and South Africa (2009). As you

FIG. 32.25 The spread of Ug99. Ug99 is a highly resistant stem rust of wheat and a major threat to global food production. *Sources: After Food and Agriculture Organization of the United Nations. 2010. "The Spread of Wheat Stem Rust UG99 Lineage." JPEG image, http://www.fao.org/agriculture/crops/rust/stem/rust-report/stem-ug99racettksk/en/; (photos) Yue Jin, USDA-ARS, Cereal Disease Laboratory.*

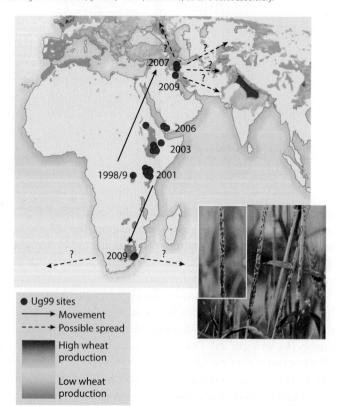

read this, Ug99 is poised to reach the major wheat-growing regions of Turkey and South Asia. Even more rapid spread is possible if spores accidentally lodge on cargo or airline passengers.

Just before his death in 2009, Norman Borlaug urged the world to take this threat seriously. Today, plant breeders are actively searching ancestral wheat varieties for genetic sources of resistance to Ug99, while other scientists try to understand how plants defend themselves against rusts and what makes Ug99 so virulent. The resurgence of *P. graminis* as a serious threat to world food production underscores the importance of safeguarding the genetic diversity of agricultural species so that the inevitable evolutionary battles with pathogenic species may be waged successfully.

Self-Assessment Questions

7. What are two key innovations in the evolutionary history of fungi that allowed them to move from water to land?

8. Based on the phylogenetic tree shown in Fig. 32.15, did complex multicellular fruiting bodies evolve once or twice (independently) in fungi?

9. Draw the life cycle of an ascomycete, and indicate the heterokaryotic stage.

10. Draw the life cycle of a basidiomycete, and indicate the heterokaryotic stage.

11. How did the evolution of vascular plants provide opportunities for fungal diversification?

CORE CONCEPTS SUMMARY

32.1 GROWTH AND NUTRITION: Fungi are heterotrophic eukaryotes that feed by absorption.

Most fungi produce slender, multicellular filaments called hyphae that form a network called a mycelium. Fungi rely on the growth of hyphae to find food and absorb nutrients. page 713

Fungi can transport food and signaling molecules within their hyphae using bulk flow, allowing them to grow between nutrient patches and to produce reproductive structures aboveground. page 714

Some fungi, such as yeasts, do not produce hyphae. page 715

Fungi are critical elements of the carbon cycle, feeding on dead organic matter and converting it back into carbon dioxide and water. page 715

Many fungi feed on living animals and plants, causing diseases. page 716

Fungi have repeatedly evolved mutually beneficial relationships with plants and animals. page 717

Lichens are stable associations between a fungus and a photosynthetic microorganism that look, function, and even reproduce as single organisms. page 718

32.2 REPRODUCTION: Fungi reproduce both sexually and asexually, and disperse by spores.

Fungi produce haploid spores that are dispersed by wind, water, or animals. Spores can be produced asexually (by mitosis) or sexually (by cell fusion and meiosis). page 719

During sexual reproduction, many fungi produce spores within multicellular fruiting bodies that enhance spore dispersal. page 720

The fungal life cycle is similar to the haploid-dominant life cycles found in many eukaryotic organisms, except that plasmogamy (cell fusion) and karyogamy (nuclear fusion) are often separated. page 722

During the period between cell fusion and nuclear fusion, called the heterokaryotic stage, genetically distinct haploid nuclei are found in the same cell. page 722

Genetically determined mating types prevent self-fertilization in many fungi. page 723

Parasexual fungi lack sexual reproduction. They are hypothesized to generate genetic diversity by fusion of haploid nuclei, followed by crossing over during mitosis and chromosome loss. page 723

32.3 DIVERSITY: Other than animals, fungi are the most diverse group of eukaryotic organisms.

Approximately 75,000 fungal species have been formally described, but the total diversity may be as high as 5 million species. page 724

Chytrids are fungi that reside in aquatic or moist habitats and lack well-defined hyphae. Only about 1000 species have been identified. page 724

Zygomycetes produce hyphae without septa and account for less than 1% of known fungal species. page 725

The glomeromycetes are a group of fungi of low diversity but tremendous ecological importance because of their association with plant roots. page 726

Most fungi belong to the Dikarya (ascomycetes and basidiomycetes), which form septa along their hyphae. During the heterokaryotic stage, these fungi produce cells that contain two haploid nuclei, one from each parent. page 726

Ascomycetes include wood-rotting fungi, many ectomycorrhizal species, plant and animal pathogens, the fungal partner in most lichens, and baker's and brewer's yeasts. page 726

Basidiomycetes include the familiar toadstools and puffballs, as well as plant pathogens page 728

Log in to **LaunchPad** to check your answers to the Self-Assessment Questions and to access additional learning tools.

CASE 7

Biology-Inspired Design

Using Nature to Solve Problems

What does Velcro have in common with the Eiffel Tower? At first, it might seem very little. Velcro is a common fastener, found in everything from sneakers to coats to tents. The Eiffel Tower is a landmark in Paris and symbol of France. It turns out, however, that both were inspired by nature.

> Biology-inspired design holds great promise for tackling all kinds of problems.

Velcro was invented in the 1940s by George de Mestral, a Swiss engineer. He realized that the burs of plants, which seem to stick everywhere, might provide a model for an effective way to hold things together. Taking a closer look, he noticed that their "stickiness" is due to the presence of small hooks that grab on to the loops of fabric, like pants, or to the fur of passing animals.

Why not use this same structure to build something that holds firmly, but releases easily? Taking his cue from nature, de Mestral constructed a fastener made up of two parts: one with small hooks and the other with small loops. He called this new material Velcro. Ever since its invention, Velcro has had many practical applications; it is used to fasten clothing, hang pictures, and close packages, to name just a few.

The engineers who built the Eiffel Tower in Paris also looked to nature to solve the problem of how to construct what was the tallest building in the world at the time. Built in 1889 for the World's Fair, the tower stands 324 m tall (approximately 1063 feet or 80 stories). The engineers who constructed the tower examined other structures notable for

Plant burs and Velcro. (a) A bur is a seed or fruit with small hooks that grab on to passing animals. (b) Velcro has two strips: one with loops, and the other with hooks similar to the burs of plants. *Sources: a. visual7/Getty Images; b. curtoicurto/Getty Images.*

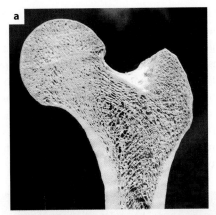
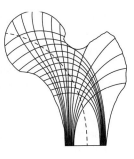

their strength, namely bones, and asked how these relatively light structures can support tremendous weight. Bones, like the human femur (thighbone), have an internal meshwork at the ends that is aligned in the direction of the major stresses, and strong support along the shafts.

The Eiffel Tower, too, is made up of crisscrossing beams that help to distribute the weight and includes strong lines of support running from the base to the tip. As a result, it is lightweight and uses very little material for its size. Many cranes, bridges, and skyscrapers use similar architectural principles, borrowed from nature.

The use of nature to tackle problems is known as biology-inspired design. It draws from many fields, including biology, engineering, physics, and materials science, providing an interdisciplinary way to solve all kinds of practical problems.

Take the problem of suction. Remoras (fishes in the family Echeneidae) are famous for their ability to hitch a ride on sharks, whales, rays, small boats, and occasionally scuba divers. A remora has a modified fin along its back that acts like a suction cup. It uses this structure to stick to other fishes and get a free ride, catch scraps of food, and aerate its gills as the host fish swims along.

Dr. Jason Nadler and his colleagues at the Georgia Tech Research Institute are studying the anatomy and physiology of the remora's "suction cup" in an effort to understand how it functions so effectively. The cup creates a tight seal, but it can also be released quickly and easily.

Bones and the Eiffel Tower. (a) Bones consist of a spongy meshwork of material with support along lines of stress. (b) The Eiffel Tower uses principles similar to those found in bones so that the structure is both strong and lightweight. *Sources: a. B Christopher/Alamy; b. Denitsa Ivanova/EyeEm/Getty Images.*

A remora. Remoras use a modified dorsal fin to stick to surfaces, such as this shark. *Source: Papilio/Alamy.*

CASE 7

When they understand how it functions, the researchers will try to mimic it so as to design, for example, underwater adhesives or medical bandages that hold firmly but can be removed easily.

Biology-inspired design sometimes uses biology to solve problems, as we have seen. Another aim of the field is to mimic nature to provide replacement parts, particularly in the field of medicine. Whether the replacement part is a mechanical valve to replace a damaged or diseased heart valve, a dialysis machine that functions like a kidney, a limb prosthesis for an arm or leg lost to trauma or disease, or a cochlear implant to provide a sense of hearing for individuals with hearing loss, many parts of the human body can now be built and replaced.

There is an old saying: If you can model something, you understand it. Another goal of biology-inspired design, then, is to test our understanding of a structure or process. Take the gravity-defying climbing ability of geckos (lizards in the group Gekkota). They are famous for running up smooth walls and even scurrying across ceilings. How do they do it?

Geckos have toe pads made up of many small hairs. It is thought that these hairs adhere to surfaces by van der Waals interactions (Chapter 2). Although these interactions are weak individually, the large number of hairs (an estimated 6.5 million in total!) results in a strong force overall.

We can test this idea by building it. In 2014, researchers at Stanford University's Biomimetic Dexterous Manipulation Laboratory engineered synthetic toe pad hairs made of silicone that allow humans to scale a glass wall. By examining gecko toe pads' adhesive properties, the researchers were able to copy it to provide suction that would support humans.

Sometimes, researchers don't just mimic part of an organism, like the toe pads of geckos. Instead, they build the entire animal. To date, researchers have developed bipedal human-like robots, as well as robotic dogs, hummingbirds, fishes, jellyfish, flies, grasshoppers, and even cockroaches. These devices are made in part to understand how the animals move, but also to take advantage of specific features of these animals. Cockroaches can move quickly and fit into tight spaces, so a robotic cockroach could be used to search for survivors after a building collapse, for example.

Recently, Dr. Joanna Aizenberg and her colleagues at Harvard University went beyond mimicking just one organism. Specifically, they combined parts of a desert beetle, cactus, and pitcher plant to make a material that collects water from the air. This material provides a way to collect water in deserts, where water is scarce. It also might prove useful in industries where it's important to keep the air dry and avoid condensation.

Biology-inspired design holds great promise for tackling all kinds of problems. In the end, what we build and what we find in nature may look quite similar to each other, but they arrived there along two different paths. We typically build things with a clear goal in mind, like the tallest tower or adhesive gloves. What we find in nature, by contrast, evolved over time without a goal. Evolution works by "tinkering," as the French biologist François Jacob famously said, using what it has and sometimes even modifying existing structures for new functions. After 4 billion years of life on Earth, it has stumbled across some beautiful and clever solutions that are worth mimicking.

Geckos and "gecko gloves." (a) Geckos can climb up walls. (b) Adhesive gloves modeled after gecko toe pads allow this man to scale a glass wall. Sources: a. nico99/Shutterstock; b. Eric Eason/Biomimetics and Dexterous Manipulation Lab, Stanford University.

DELVING DEEPER INTO CASE 7

For each of the Cases, we provide a set of questions to pique your curiosity. The questions cannot be answered directly from the Case itself, but instead introduce issues that will be addressed in chapters to come. We revisit this Case in Chapters 33–42 on the pages indicated below.

1. How can we mimic the form and function of animals to build robots? See page 607.
2. How do cochlear implants work? See page 774.
3. How can we design smart materials to heal damaged bones and improve artificial joints? See page 805.
4. How can we engineer replacement heart valves? See page 849.
5. How does dialysis work? See page 894.
6. How can we use a protein that circulates in the blood to treat sepsis? See page 930.

CHAPTER 33
Animal Form, Function, and Evolutionary History

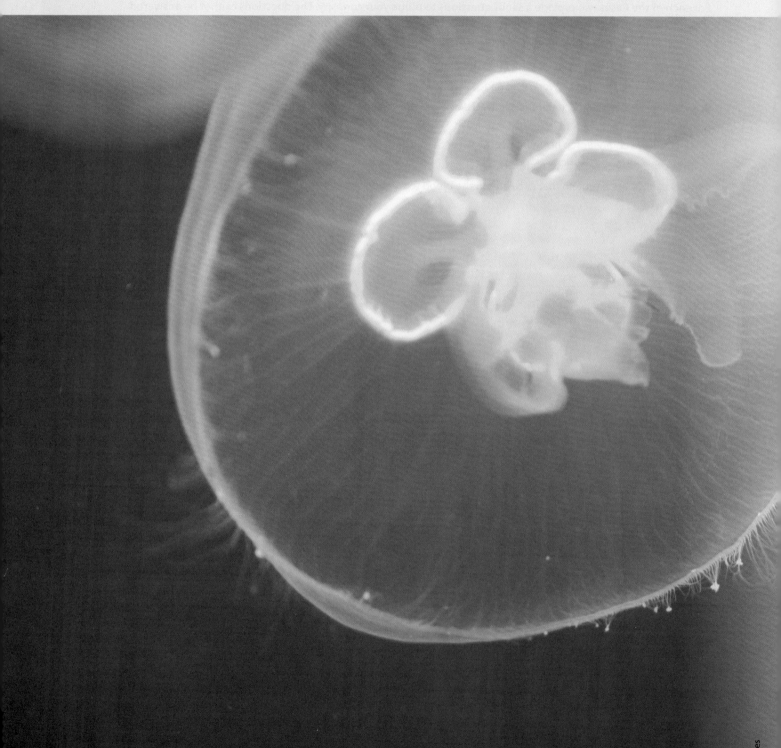

CORE CONCEPTS

33.1 ANIMAL BODY PLANS: Animals have a limited number of distinct body plans.

33.2 TISSUES AND ORGANS: Most animals have four different types of tissue, which combine to form organs.

33.3 HOMEOSTASIS: Homeostasis actively maintains a stable internal environment.

33.4 EVOLUTIONARY HISTORY: Animals evolved more than 600 million years ago in the oceans, and by 500 million years ago the major body plans were in place.

Animals move, feed, and behave in many different ways. Indeed, animals have evolved a remarkable variety of structures to carry out these and other functions. As an example, consider the ways in which animals move (Chapter 35). Sponges have no muscle cells at all; rooted to the ground, they don't move from place to place. Jellyfish have muscle fibers that contract to squeeze a fluid-filled cavity, powering movement by jet propulsion. Mammals have limbs for locomotion, which are powered by the coordinated actions of muscles attached to an internal skeleton. Insects move in much the same way, but their skeleton is external. Other organ systems show comparable variation among animals. The wide range of anatomical and physiological variations has permitted animals to diversify to a remarkable degree.

Biologists have described approximately 1.8 million species of eukaryotic organisms from the world's forests, deserts, grasslands, and oceans. Of these, some 1.3 million species are animals. There is reason to believe that animal diversification began in the oceans, but today most animal species are found on land. Indeed, the majority of all animal species are insects. In this chapter, we look at animals as a group, focusing on their basic form, function, and evolutionary history.

33.1 ANIMAL BODY PLANS

As discussed in Chapter 22, it is relatively easy to understand how the body plans of humans, chimpanzees, and gorillas show them to be more closely related to one another than any of them is to other animal species. Furthermore, it isn't hard to see that humans, chimpanzees, and gorillas are more closely related to monkeys than they are to lemurs, and that humans, chimpanzees, gorillas, and monkeys are more closely related to lemurs than they are to horses. Anatomy and morphology reveal key evolutionary relationships among vertebrate animals, but how do we come to understand the place of vertebrates in a broader tree that includes all animals? More generally, how can we construct an animal phylogeny that includes organisms with body plans as different as those of sponges, jellyfish, earthworms, mussels, sea stars, and humans?

What is an animal?

Let's begin by asking what an animal is. Which features differentiate animals from all other organisms? And, more specifically, which features differentiate animals from their nearest relatives?

As introduced in Chapter 25, the organisms most closely related to animals are choanoflagellates, a group of single-celled eukaryotes. This relationship was first proposed in the nineteenth century by scientists who noticed similarities in cell shape between choanoflagellates and the feeding cells of sponges; it was confirmed in the twenty-first century through comparisons of molecular sequences. Choanoflagellates are unicellular, whereas animals are multicellular. But multicellularity is not a unique feature of animals: plants and fungi are also multicellular. Thus, we need to find another feature to distinguish animals from other organisms.

Animals are heterotrophs, gaining energy and carbon from preformed organic molecules (Chapter 6). This character differentiates animals from plants, but not from fungi, which are also heterotrophs. Animal cells lack cell walls, allowing them to move during development and in the adult animal (Chapter 26). This character differentiates animals from both plants and fungi. Animals have a pattern of early embryological development that begins with a hollow ball of cells called a blastula and includes the movement of embryonic cells, usually to form a gastrula (Chapters 19, 26, and 40), a process unique to animals.

Animals, therefore, can be described as multicellular heterotrophic eukaryotes with a distinctive mode of early development. Other features also distinguish animals from other organisms, but as we learn more about choanoflagellates, some of these lines have blurred. For example, collagen, long considered to be unique to animals, is now known to occur in choanoflagellates.

Animals can be classified based on type of symmetry.

Early in the history of biology, taxonomists such as Carl Linnaeus recognized that for all their diversity, animals have a limited number of distinctive body plans. Animals with the same type of body plan can be placed into a group called a phylum. The problem for biologists has long been how to understand the evolutionary relationships among animal phyla. Indeed, this is still an active field of research.

In Chapter 21, we saw that the fundamental mechanism by which biological diversity increases is speciation, the divergence of two populations from a common ancestor. Played out repeatedly through time, speciation gives rise to a treelike pattern of evolutionary relatedness: more closely related groups diverge from branch points close to the tips of the tree, whereas more distantly related groups diverge from branch points nearer its base. Distinctive features of organisms, called characters, evolve throughout evolutionary history. We can estimate when a character arose from its shared presence in the descendants of the population in which the character first evolved. Often the characters analyzed are anatomical traits.

Fig. 33.1 shows a phylogenetic tree of animals. Most animals can be divided into three groups based on the symmetry of their bodies. **Porifera** (sponges) are irregular in form, with no clearly developed plane of symmetry. **Cnidaria** (the group that includes jellyfish, corals, and sea anemones) display radial symmetry. Their bodies have an axis that runs from mouth to base, with many planes of symmetry cutting through this axis (**Fig. 33.2a**). Because of this structure, jellyfish can move up and down in the water column by flexing the muscles around their bell-like bodies, and many sea anemones and corals can wave their ring of food-gathering tentacles in all directions at once.

Bilateria (most other animals, including humans, insects, and snails) show bilateral symmetry. Their bodies have a distinct front and back, top and bottom, and right and left, with a single plane of symmetry running between right and left at the midline (**Fig. 33.2b**). Animals with bilateral symmetry are able to move in a horizontal direction to capture prey, find shelter, and escape from enemies. These animals develop specialized sensory organs

FIG. 33.1 A phylogenetic tree of animals. The tree shows increasing complexity from the closest protistan relatives of animals, the choanoflagellates, to bilaterian animals such as insects and mammals.

Photo sources: (top to bottom) Mark Dayel (http://www.dayel.com/choanoflagellates) and Nicole King (http://kinglab.berkeley.edu); James D. Watt/SeaPics.com; UMI NO KAZE/a.collectionRF/Getty Images; British Antarctic Survey/Science Source.

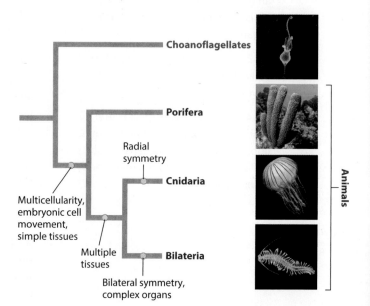

at the front end for guidance (Chapter 34), and specialized appendages along both sides for locomotion, grasping, and defense.

The phylogenetic tree in Fig. 33.1 also indicates that animals are closely related to choanoflagellates, but differ from them in having persistent multicellularity and in forming a blastula during early development. Poriferans have only a few different cell types, and these cells are organized into only simple tissues, not the well-defined tissues and complex organs found in other animals (Section 33.2). Cnidarians have well-defined tissues, but lack complex organs. Bilaterians have both well-defined tissues and complex organs. These organs are specialized for movement, digestion, gas exchange, and other functions.

Note that the phylogenetic tree shown in Fig. 33.1 does not tell us that sponges are older than other animals. Instead, it says that sponges and other animals diverged from a common ancestor, but it doesn't tell us what that common ancestor looked like. Similarly, cnidarians and animals with bilateral symmetry diverged from their last common ancestor at a single point in time and so are equally old, but structurally distinct.

Many animals have a brain and specialized sensory organs at the front of the body.

A notable feature of many bilaterian animals is that nervous system tissue, including specialized sense organs such as eyes, becomes concentrated at one end of the body. For example,

FIG. 33.2 Symmetry in animal form. (a) Radial versus (b) bilateral symmetry distinguishes cnidarians and bilaterian animals.

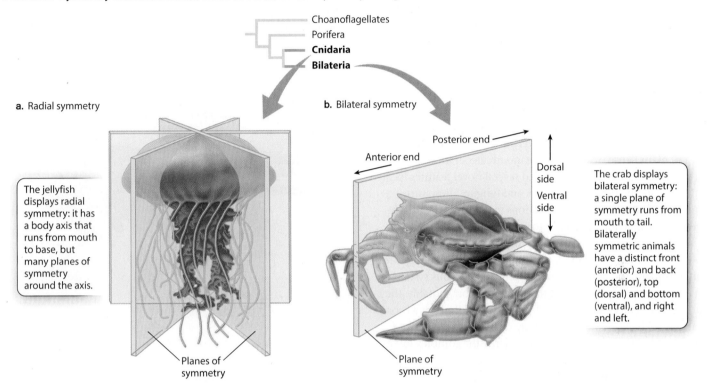

the eyespots of flatworms are located at one end of its body, as are the brain and sense organs of earthworms, insects, and vertebrates (**Fig. 33.3**). The concentration of nervous system components at one end of the body, defined as the "front," is referred to as **cephalization.** Cephalization is a key feature of the body plan of most animals, including vertebrates.

Cephalization is considered an adaptation for locomotion because it allows animals to take in sensory information

FIG. 33.3 Cephalization. (a) Flatworms, (b) earthworms, (c) insects, and (d) vertebrates all have a distinct head region with specialized sense organs.
Photo sources: (left to right) Science History Images/Alamy Stock Photo; bazilfoto/iStock/Getty Images; marcouliana/Getty Images; taviphoto/Getty Images.

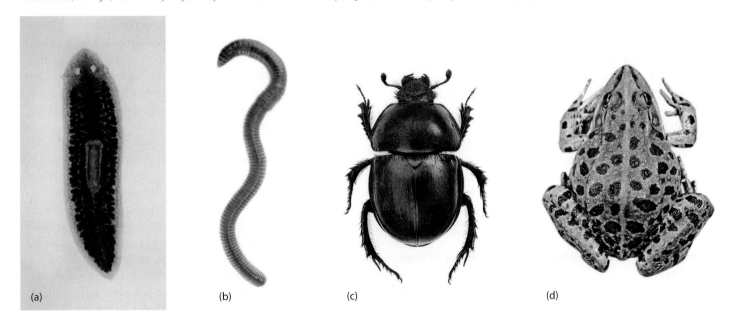

from the environment ahead of them as they move forward. In addition, the proximity of the sensory organs to the brain makes it possible to process this information quickly, thereby enabling a suitable behavioral response. As the quality and amount of sensory information taken in by animals increased, their brain size and complexity also increased. Cephalization is also considered to be an adaptation for predation, since animals with cephalization can better detect and capture prey, and for predator avoidance, providing a means of escaping capture.

Although cephalization is a feature of many animals, it has been particularly well studied in vertebrates. In vertebrates, the brain, many sense organs, and the mouth are all located in the head. Vertebrates have also evolved several novel features located in the head, including a jaw, teeth, and tongue. These structures are all thought to be adaptations for predation, or more generally the acquisition and processing of food.

As a result of evolutionary selection for enhanced sensory perception and the ability to respond to important cues in the environment, the brain, sensory organs, and nervous system of many animals are complex in their organization. Notably, they are linked to more sophisticated abilities that allow for a broad range of behaviors. Such abilities are critical to the success of both predators and prey, and they underlie the complex interactions that occur among members of a species when they mate, reproduce, disperse, and care for their young.

Some animals also show segmentation.

In addition to cephalization, another broad pattern that is evident among the body plans of bilaterian animals is segmentation. **Segmentation** is the organization of the body into units, or segments, that are repeated from front to back (along the anterior–posterior axis), but modified depending on where they are located in the body.

Insects, for example, have a head, thorax, and abdomen (**Fig. 33.4**). Each of these regions can be further subdivided into segments. For example, each of three pairs of legs grows from one of the three segments that make up the thorax. In Chapter 19, we considered some of the genetic mechanisms that generate this segmented pattern in the fruit fly *Drosophila*. Humans and other vertebrates also develop in a segmented fashion, as is evident in the repeated pattern of vertebrae and nerves that make up the backbone and spinal cord.

Three animal phyla are known for showing a segmented body plan: arthropods (including insects, spiders, and crustaceans), annelids (segmented worms), and chordates (including vertebrates). Segmentation allows for the differentiation of segments into distinct body parts and may have evolved as an adaptation for higher motility.

Animals can be classified based on the number of their germ layers.

New insights about animals became possible with the advent of microscopes that enabled the direct study of early animal

FIG. 33.4 Segmentation. Insects, such as the fruit fly *Drosophila melanogaster*, have a head, thorax, and abdomen, each of which is made up of segments. *Photo source: Herman Eisenbeiss/Getty Images.*

embryos. One important observation was that some animals that look very different as adults share patterns of early embryological development. For example, adult sea stars and catfish look very different from each other, but their embryos show a number of key similarities, including the way that the early cell divisions occur and the number of embryonic tissue layers they develop.

In cnidarians (radially symmetrical animals), the embryo has two germ layers, the endoderm and the ectoderm, from which the adult tissues develop. Because of this characteristic, these animals are called **diploblastic** (**Fig. 33.5**). In bilaterians (bilaterally symmetrical animals), the embryo has three germ layers, with the mesoderm between the endoderm and ectoderm. These animals are called **triploblastic** (Fig. 33.5). The evolution of the mesoderm in triploblasts allowed for the development of new types of tissues and organs, such as muscles and circulatory systems. These new types of tissues and organs, in turn, facilitated the evolution of new modes of locomotion, feeding, and behavior.

Comparative embryology also enabled biologists to divide bilaterian animals into the two groups shown in **Fig. 33.6**: the **protostomes** (from the Greek for "first mouth") and the **deuterostomes** (from the Greek for "second mouth"). In protostomes, the earliest-forming opening to the internal cavity of the developing embryo, called the blastopore, becomes the mouth. Protostomes include such groups as nematodes (roundworms), arthropods (including insects, spiders, and crustaceans), mollusks (gastropods like snails and slugs; bivalves like clams and mussels; and cephalopods like squids and octopus), annelids (segmented worms), and flatworms (**Fig. 33.7**).

FIG. 33.5 Variation in embryonic development. Early development of germ tissues in the embryo separates the diploblastic cnidarians from the triploblastic bilaterians.

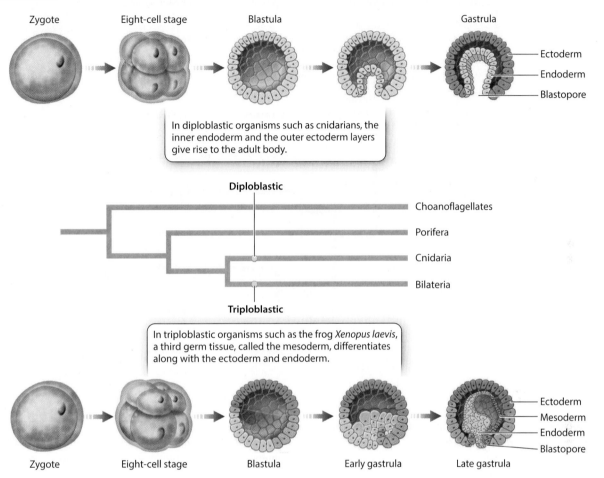

FIG. 33.6 A phylogenetic tree of bilaterian animals. Embryological features and molecular sequences of genes show that bilaterian animals can be organized into two major groups, deuterostomes and protostomes. *Photo sources: (top) Glenn Nagel/iStockphoto; (bottom) Lal/Getty Images.*

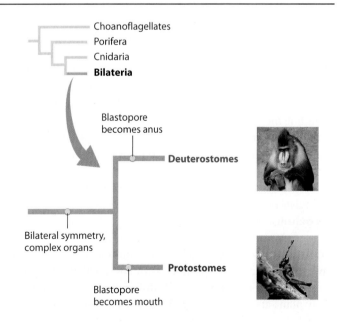

In deuterostomes, the blastopore becomes the anus, and the mouth forms second. Deuterostomes include chordates, such as vertebrates, and echinoderms, such as sea stars, sea urchins, and sand dollars (Fig. 33.7).

While the names of the protostomes and deuterostomes focus our attention on which structure the blastopore becomes, these groups exhibit a host of other differences. For example, they differ in the nature of the early divisions of the embryo, the origin of the mesoderm, and the position of the nerve cord running the length of the body. Only in the age of molecular sequence comparisons have the relationships among phyla within each of these two groups become clear.

FIG. 33.7 Phylogeny of animals, showing major phyla of bilaterians.

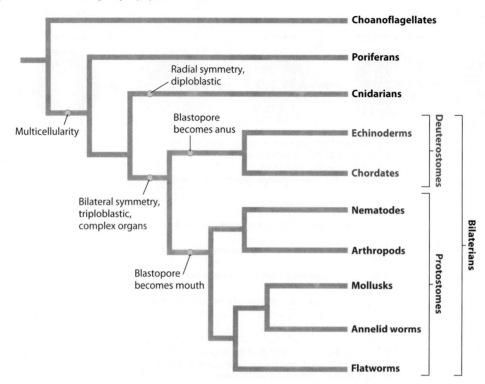

Molecular sequence comparisons have confirmed some relationships and raised new questions.

Early biologists laid the groundwork for understanding animal phylogeny by relying on their observations of comparative anatomy and embryonic development. Yet as was true for phylogenetic relationships in Bacteria, Archaea, and Eukarya as a whole, a modern understanding of evolutionary relationships among animals had to wait for the revolution in molecular sequencing. Over the past two decades, comparisons among DNA, RNA, and amino acid sequences have greatly improved our understanding of animal phylogeny. Molecular comparisons support many of the conclusions reached by the early biologists, including the early divergence of sponges and the separation of radially and bilaterally symmetrical animals. Moreover, molecular sequence comparisons confirm that choanoflagellates are the closest protistan relatives of animals.

Other hypotheses have been rejected. For example, some biologists pointed to the presence or absence of a cavity, or **coelom,** surrounding the gut, as a character that could be used to organize bilaterians into three groups: those without a body cavity (acoelomates) and those with a body cavity (coelomates and pseudocoelomates, which differ in the embryonic origin of the cells lining the cavity). These body plans are shown in **Fig. 33.8**. The coelom cushions

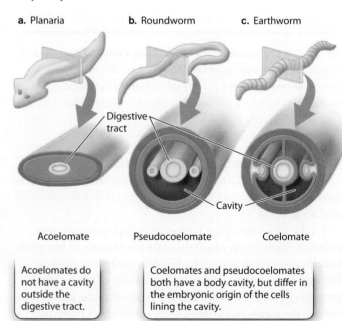

FIG. 33.8 Major types of anatomical organization of bilaterian animals. Bilaterians vary in the presence, absence, or structure of a body cavity.

the internal organs against hard blows to the body and enables the body to turn without twisting these organs. Once molecular studies were conducted, however, researchers discovered that they did not support this traditional phylogenetic division of bilaterians into acoelomate, coelomate, and pseudocoelomate groups. Similarly, molecular sequence comparisons have not supported the once widespread view that segmented bodies indicate a close relationship between earthworms and lobsters.

CASE 7 BIOLOGY-INSPIRED DESIGN: USING NATURE TO SOLVE PROBLEMS

Can we mimic the form and function of animals to build robots?

Scientists who design robots often try to mimic the form and function of animals. Fishes are one of the most popular types of robotic animals. We can learn a great deal about innovations by studying fishes, whose remarkable adaptations have been shaped by natural selection over more than 500 million years.

The first robotic fish was created in the 1980s by a team of scientists and engineers at Massachusetts Institute of Technology. It was called RoboTuna because it mimicked the form and function of a tuna, allowing it to swim effectively through water. Since then, many more fish-like robots have been made, so that today there are more than 400 different types of robotic fishes. Some, like RoboTuna, have one joint; others have many joints; and still others are made of dynamic materials that can bend fluidly.

Why build a robotic fish? The first and most basic reason is to understand how fish move. Fish are spectacular swimmers. If we can build a robotic fish that swims like a living one, it suggests that we understand how a fish is able to move forward, turn, and navigate its surroundings. Robotic fishes can also be used in studies of fish behavior. A robotic fish that looks and even smells like a real fish, and that has sensors to detect and record its surroundings, provides a way to quietly "spy" on fish in a way that is not possible by direct observation or by the use of cameras. There are all kinds of practical applications of robotic fish as well, from mapping coastlines, coral reefs, and the sea floor, to monitoring underwater cables and pipelines, to exploring marine life.

Scientists and engineers have copied the form and function of many animals, including cockroaches, geckos, jellyfish, hummingbirds, and bats, to name just a few. One of the most recent and exciting robots is based on an octopus. Octopuses show incredible dexterity and strength, all without an internal skeleton.

The robot octopus, called an octobot and designed by researchers at Harvard University, has several remarkable characteristics (**Fig. 33.9**). It is entirely soft, with no rigid components

FIG. 33.9 Octobot. Octobot is a soft robot inspired by the octopus. *Source: Lori K. Sanders, Ryan Truby, Michael Wehner, Jennifer Lewis & Robert Wood, Harvard University.*

like metal or even batteries. It is powered by a simple chemical reaction: liquid hydrogen peroxide is converted to a gas in the presence of a catalyst (in this case, platinum), providing pressure that powers movement. Finally, the robot is produced with a 3D printer.

The octobot brings together engineers, material scientists, and chemists to create a robot that is inspired by biology. This truly interdisciplinary effort is the first step toward more flexible and innovative robots for research and diverse applications.

Self-Assessment Questions

1. Draw a simplified animal tree of life, indicating the relationships among sponges, cnidarians, protostomes, and deuterostomes.
2. Which features distinguish sponges, cnidarians, protostomes, and deuterostomes? Place these features on the phylogeny that you drew for Self-Assessment Question 1.
3. How is cephalization an adaption for forward locomotion?
4. Which animals are more closely related to sponges: cnidarians or bilaterians?

33.2 TISSUES AND ORGANS

As we have seen, animals show a diversity of form. Nevertheless, the cells of most animals are organized into just four types of tissues: epithelial, connective, muscle, and nervous. **Tissues** are collections of cells that carry out a specific function (Chapter 10). They are combined in various ways to make **organs,** such as the heart, lung, or kidney. Most organs consist of several different types of tissues. For example, the skin is made up of two layers: the outer layer is the epidermis, which is epithelial tissue, and the inner layer is the dermis, which is connective tissue (**Fig. 33.10**). In this section, we discuss the four tissue types that combine to make organs in animals.

Most animals have four types of tissues.

Except in sponges, animal cells are organized into different kinds of well-defined tissue. **Epithelial tissue** provides a lining for all of the spaces inside and outside the body. The outer layer of the skin is epithelial tissue, as is the inner lining of the gut, bladder, and blood vessels. The cells that make up epithelial tissue are closely packed together and connected by cellular junctions (Chapter 10), forming a continuous sheet of cells. This sheet provides a boundary and controls the movement of substances into and out of the body.

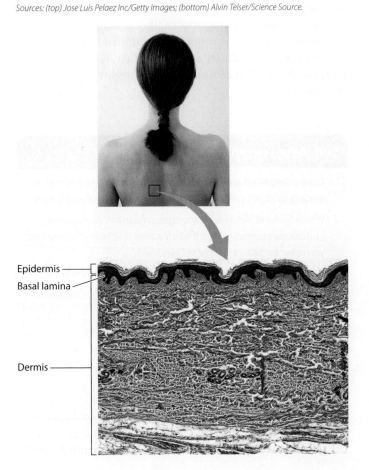

FIG. 33.10 Skin. The skin has two layers, the epidermis and dermis, that together provide protection for the underlying tissues of the body.

Sources: (top) Jose Luis Pelaez Inc/Getty Images; (bottom) Alvin Telser/Science Source.

Epithelial tissue is classified based on certain characteristics. The first characteristic is layering: a single layer of cells is considered simple, and more than one layer is labeled as stratified. The second characteristic is the shape of the cells: flat cells are squamous, round or square cells are cuboidal, and tall cells are columnar. For example, a single layer of flat cells is called simple squamous epithelium; this type of epithelial tissue lines the inside of blood vessels and the interior surface of the lung. Some epithelial tissues have special features, such as a layer of the protein keratin in the case of the skin, or cilia in the case of the upper airways.

Epithelial tissue not only forms a boundary, but also may absorb substances from and secrete substances into the space it surrounds. In the gut, for example, the epithelial lining absorbs nutrients. Notably, epithelial tissue has no blood vessels, so it gets its blood supply from the underlying tissue, which is connective tissue.

Connective tissue underlies epithelial tissues and is found elsewhere as well. In contrast to epithelial tissue, which is composed of closely packed cells, connective tissue has an extensive extracellular matrix and few cells. The extracellular matrix is an insoluble meshwork composed of proteins and polysaccharides. Its components are synthesized, secreted, and modified by the cells that reside within the connective tissue. Many different forms of extracellular matrix exist, which differ in the amount, type, and organization of the proteins and polysaccharides that compose them. The extracellular matrix not only contributes structural support, but also provides informational cues that determine the activity of the cells that are in contact with it (Chapter 10).

Connective tissue is characterized by the properties of the specific extracellular matrix. For example, two layers of connective tissue lie beneath the epidermis of the skin: first a specialized type of connective tissue called the basal lamina, and then the dermis, a type of connective tissue made up of cells that secrete the components of the extracellular matrix (Fig. 33.10). The dermis is strong and flexible because its extracellular matrix is composed of tough protein fibers. This layer contains blood vessels that nourish both the dermis and the overlying epidermis. The dermis also provides a cushion for the body.

The dermis of the skin is one of several types of connective tissue found in the body. Other connective tissues, such as bones, cartilage, tendons, and ligaments, help support the body and provide a system of levers for movement (Chapter 35). Finally, specialized connective tissues also include adipose tissue (fat) and blood.

Muscle tissue is made up of cells (called fibers) that are able to shorten or contract (Chapter 35). These specialized cells contain actin thin filaments and myosin thick filaments. Myosin is a motor protein that uses the energy of ATP to change conformation. This conformational change moves the thin filament relative to the thick filament. As a result, individual muscle

FIG. 33.11 Skeletal muscle. Viewed under the microscope, skeletal muscle cells have a striated, or striped, appearance. *Photo source: Photo Researchers/Getty Images.*

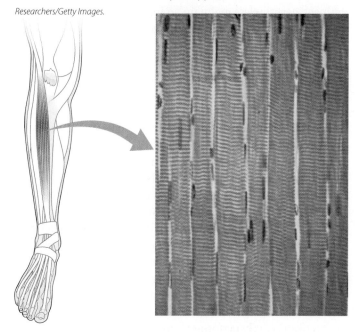

the other (**Fig. 33.12**). They communicate with each other at specialized junctions called synapses, where chemical signals are released by one nerve cell and received by another. The ability of nerve cells to communicate quickly and specifically and to form networks underlies rapid decision making and complex behaviors.

Tissues are organized into organs that carry out specific functions.

In bilaterians, multiple tissues can combine to make an organ. For example, the intestine is an organ that includes all four types of tissues: epithelial, connective, muscle, and nervous. In turn, organs may combine to form an **organ system.** For example, the intestine is one organ of the digestive system, which also includes the stomach, liver, pancreas, and other organs (Chapter 38).

Sponges have only simple epithelia that line the surface of the body. In contrast, jellyfish, corals, and sea anemones (cnidarians) have some tissues, but not true organs (see Fig. 33.1). These organisms have a set of nerves called a nerve net, but their nerves are not organized into a brain or central nervous system. Moreover, cnidarians have muscle cells, but not well-developed

cells shorten and the entire muscle tissue contracts. True muscle tissue is present in bilaterians but not cnidarians.

Vertebrate animals have three types of muscle tissues. Skeletal muscle attaches to bone and makes voluntary movements possible (**Fig. 33.11**). Cardiac muscle is found in the heart and contracts to make the heart beat. Smooth muscle can be found in such places as the gut, where it causes waves of contraction that push food along the digestive tract, and blood vessels, which constrict and relax to control blood flow.

Nervous tissue is the fourth type of animal tissue (Chapter 34). It can be found, for example, in the nerve nets of cnidarians and in the brain, spinal cord, and peripheral nerves of vertebrates. Nervous tissue takes in sensory information from the environment, processes information, and sends signals to target organs to elicit a response. For example, signals sent to muscles direct movement, and signals sent to organs help the body maintain a relatively constant internal state, called homeostasis (Section 33.3).

Nervous tissue is made up primarily of neurons (nerve cells) that send electrical impulses from one end of the cell to

FIG. 33.12 Nerve cells. Nerve cells, or neurons, send electrical impulses along their length. *Source: Eye of Science/Science Source.*

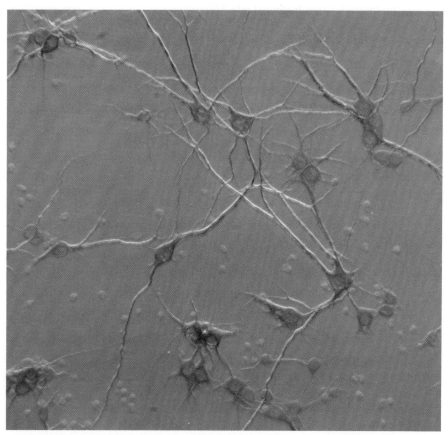

muscle tissues. They exchange gases by diffusion rather than by using lungs or gills, and they digest food in a central cavity with a single opening for both eating and excretion.

Mollusks, arthropods, mammals, and other bilaterians have true organs made up of one or more types of tissue (see Fig. 33.1). These include such organs as the heart for pumping blood (Chapter 37), gills or lungs for gas exchange (Chapter 37), stomach and intestine for nutrient absorption as substances pass through a digestive tract with two openings (Chapter 38), and kidneys or (in invertebrate bilaterians) nephridia for water balance and waste excretion (Chapter 39). Many of these organs play important roles in maintaining an internal environment compatible with life, a topic we turn to next.

Self-Assessment Question

5. What are the properties and functions of the four types of animal tissues?

33.3 HOMEOSTASIS

Animals commonly experience a range of environments during their lifetimes, yet many of them are able to maintain various physiological states such as body temperature and blood pH within a relatively narrow range. The active maintenance of stable internal conditions is called **homeostasis**. Homeostasis is a critical feature of cells and of life itself. In this section, we examine how animals are able to maintain homeostasis.

Homeostasis is the active maintenance of stable conditions inside of cells and organisms.

In Chapter 5, we looked at how cells maintain homeostasis. The environment outside of cells may change, but the environment inside of cells remains relatively constant. This consistency is critical for many functions that support life. Chemical reactions and protein folding, for example, are carried out efficiently only within a narrow range of conditions, such as pH range or salt concentration.

For cells, the selectively permeable plasma membrane actively maintains intracellular conditions compatible with life. The lipid bilayer with its embedded protein channels and pumps lets some molecules through freely but restricts the passage of others, allowing the cell to regulate what goes in and what goes out (Chapter 5). For example, the plasma membrane keeps ion concentrations within narrow ranges. The firing of action potentials by neurons is an example of a cell function that requires particular ion concentrations on either side of the membrane. Recent evidence suggests that nerve cell firing rates are maintained at a steady level in the brain of rats, regardless of sensory stimulus or deprivation, or whether the animals are awake or asleep. Temperature and pH are other parameters that do not vary much, because enzymes often work effectively only in narrow temperature and pH ranges. For example, cells found in the highly acidic waters associated with mine drainage nevertheless maintain an internal pH that is roughly neutral. To do so, they actively pump protons across the cell membrane.

Homeostasis is also maintained for the body as a whole. Many physiological parameters are maintained in a narrow range of conditions throughout the body, including temperature, heart rate, blood pressure, blood sugar, blood pH, and other ion concentrations. Similarly, the water content of the body as a whole is kept stable through the careful regulation of ions and other solutes, as discussed in Chapter 39.

The concept of homeostasis was first described as regulation of the body's "interior milieu" in the late 1800s by the French physiologist Claude Bernard, who is often credited with bringing scientific methods to the field of medicine. The term "homeostasis" was coined by the American physiologist Walter Cannon, whose book *The Wisdom of the Body* (first published in 1932) popularized the concept.

Maintaining steady and stable conditions takes work in the face of changing environmental conditions. That is, a cell or organism actively performs various processes that maintain homeostasis. For example, long periods of drought challenge an animal's ability to remain hydrated and maintain a stable water and ion balance. Animals facing drought must respond rapidly by changing the permeability of their skin and respiratory organs so that they can retain as much water as possible.

Homeostasis is often achieved by negative feedback.

How does the body maintain homeostasis? Homeostatic regulation often depends on **negative feedback** (Fig. 33.13). In negative feedback, a stimulus acts on a sensor that communicates with an effector, which produces a response that opposes the initial stimulus. For example, negative feedback is used to maintain a constant temperature in a house. Cool temperature (the stimulus) is detected by a thermostat (the sensor). The thermostat sends a signal to the furnace or other heating system (the effector), prompting it to generate heat (the response). The response (heat) opposes the initial stimulus (cool temperature) until the temperature setting of the thermostat is reached. No additional heat is produced until the temperature drops below the temperature setting. In this way, a stable temperature is maintained (Fig. 33.13a).

In a similar way, humans and other mammals maintain a steady body temperature even as the temperature outside fluctuates. Nerve cells in the hypothalamus (located in the base of the brain) act as the body's thermostat or sensor (Fig. 33.13b). When a decrease in the temperature in the environment causes a drop in body temperature (the stimulus), the hypothalamus (the sensor) activates the nervous system to induce shivering and the production of metabolic heat (the effectors), as discussed in Chapter 36. At the same time, the hypothalamus

FIG. 33.13 Temperature regulation by negative feedback in (a) a house and (b) a mammal. In negative feedback, a response (such as heat) opposes the stimulus (cold), leading to a stable state (a steady temperature).

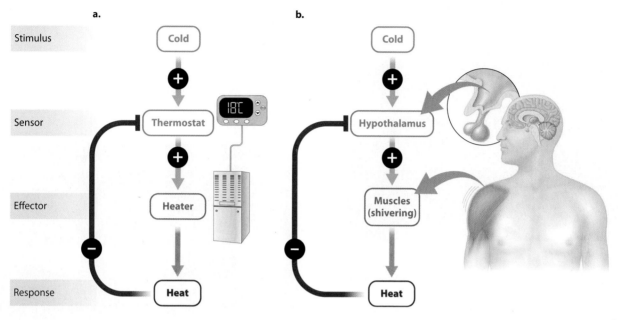

activates nerves that cause peripheral blood vessels to constrict (additional effectors). The reduction in blood flow near the body's surface reduces heat loss to the surrounding air. Conversely, an increase in temperature signals sweat glands to secrete moisture and peripheral blood vessels to dilate to aid heat loss from the skin.

Homeostatic regulation, therefore, relies on negative feedback to maintain a **set point,** or steady-state value. In the example, the set point is an animal's preferred body temperature. The ability to maintain a constant body temperature is known as thermoregulation, and preferred body temperature is just one of many physiological set points that the body actively maintains, as we discuss in subsequent chapters.

Self-Assessment Question

6. What is homeostasis? Provide one example of a condition that is maintained by homeostasis.

33.4 EVOLUTIONARY HISTORY

Because many animals form hard, mineralized skeletons, sedimentary rocks deposited throughout their evolutionary history over the past 541 million years contain a rich fossil record of shells and bones. Animals also leave sedimentary calling cards in the form of their tracks, trails, and burrows, which are also widespread in sedimentary rocks. Nevertheless, not all animal phyla are well represented in the fossil record. For example, fossil earthworms are rare and fossil flatworms are essentially unknown because they lack mineralized body parts (Chapter 22). In contrast, the fossilized shells and bones of bivalve mollusks (such as clams and mussels), brachiopods (marine shelled animals), echinoderms (such as sea stars and sea urchins), and mammals preserve an excellent record of evolutionary history within these groups.

Fossils also provide a record of now extinct species that often resemble, yet are distinct from, modern groups. They can show combinations of traits not seen in living animals—think, for example, of dinosaurs. In addition, they underscore the evolutionary importance of extinctions: a small number of events removed a majority of existing species, paving the way for renewed diversification among survivors. If we step back and look at the big picture, what do fossils tell us about the evolutionary history of the animal kingdom?

Fossils and phylogeny show that animal forms were initially simple but rapidly evolved complexity.

In Chapter 22, we discussed how to interpret and build phylogenetic trees based on morphological, developmental, and molecular traits of organisms. Phylogeny suggests that animals are relative latecomers in evolutionary history, and the fossil record supports this hypothesis. Life originated more than 3.5 billion years ago, but microorganisms were the only members of ecosystems for most of our planet's history. As noted in Chapter 26, macroscopic fossils of organisms thought to be animals first appear in rocks deposited only 575 million years ago. Called Ediacaran fossils after the Ediacara Hills of South Australia where they were discovered, these fossils have

FIG. 33.14 *Dickinsonia,* **an Ediacaran fossil.** *Dickinsonia* and similar fossils are structurally simple, showing that the earliest animals had not yet evolved the complex morphologies and organ systems seen today in bilaterians. *Source: O. Louis Mazzatenta/Getty Images.*

simple shapes that are not easily classified among living animal groups (**Fig. 33.14**).

Based on phylogenetic trees, we should look for sponges, cnidarians (jellyfish, corals, and sea anemones), and other diploblastic animals among the oldest animal fossils. Sponges, at least, are rare among Ediacaran fossils, probably because early sponges did not make mineralized parts. Instead, the majority of the organisms preserved as Ediacaran fossils show simple, fluid-filled tubes, without identifiable mouths or other organs. Many may have formed colonies, gaining complexity through colonial growth and differentiation, as some living cnidarians do. These early animals had epithelia and are thought to have obtained food by taking in dissolved organic matter or phagocytosing small particles. They would have exchanged gases by diffusion. Most Ediacaran fossils probably branched from an early node or nodes on the animal tree.

Why did the first observable animal radiation occur so late in the history of life? Scientists continue to debate this question, but part of the answer appears to lie in Earth's environmental history. Geochemical data suggest that only during the Ediacaran Period did the atmosphere and oceans come to contain sufficient amounts of oxygen to support the metabolism of large, active animals.

The animal body plans we see today emerged during the Cambrian Period.

Ediacaran fossils differ markedly from the shapes of living animals. In contrast, in the next interval of geologic history, the Cambrian Period (541–485 million years ago), we begin to see the fossilized remains of animals with familiar body plans (**Fig. 33.15**). Cambrian fossils commonly include skeletons made of the minerals silica, calcium carbonate, and calcium phosphate, and they record the presence of arthropods

FIG. 33.15 Fossils of the Cambrian Period. Cambrian rocks preserve early representatives of present-day phyla, including (a) early arthropods called trilobites, (b) annelid worms, (c) early deuterostomes, and (d) primitive vertebrates. *Sources: a. Sinclair Stammers/Science Source; b. Smithsonian Institution—National Museum of Natural History. Image: Jean-Bernard Caron; c. S. Bengtson, 2000, "Teasing Fossils out of Shales with Cameras and Computers," Palaeontologia Electronica 3, no. 1: art. 4:14 pp., 7.7 MBhttp://palaeo-electronica.org/2000_1/fossils/issue1_00.htm. Copyright: Palaeontological Association, 15 April 2000; d. Reprinted from Palaeoworld, 20, Jun-Yuan Chen, The Origins And Key Innovations Of Vertebrates And Arthropods. Fig. 5d. Copyright 2011 with permission from Elsevier.*

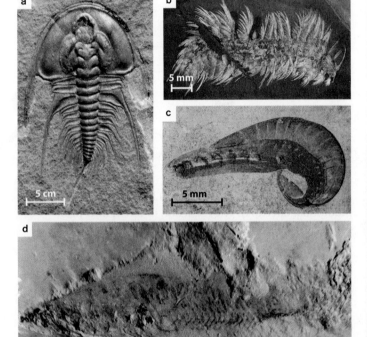

(the group that today includes insects, spiders, crabs, and their close relatives), echinoderms, mollusks, brachiopods, and other bilaterian animals in the oceans. The rocks also preserve complex tracks and burrows made by early organisms. And in a few places, notably at Chengjiang in China and the Burgess Shale in Canada, unusual environmental conditions have preserved a treasure trove of animals that did not form mineralized skeletons (Chapter 22).

These exceptional windows into early animal evolution show that, during the first 40 million years of the Cambrian Period, the body plans characteristic of most bilaterian phyla took shape. This period of rapid diversification in the fossil record is sometimes called the **Cambrian explosion.** Sponges and cnidarians radiated as well, producing through time the diverse habitats of reefs and imparting an ecological structure to life in the sea broadly similar to what we see today.

Scientists sometimes argue about whether the name "Cambrian explosion" is apt. The fossil record makes it clear that bilaterian body plans did not suddenly appear fully formed, so the event was not truly "explosive." Rather, fossils demonstrate a large accumulation of new characters in a relatively short period of time during which the key attributes of modern animal phyla emerged. For example, living arthropods have segmented bodies with a protective cuticle, jointed legs, other appendages specialized for feeding or sensing the environment, and compound eyes with many lenses. Cambrian fossils include the remains of organisms with some, but not all, of the major features present today in arthropods. In short, the first 40 million years of the Cambrian Period ushered in a world utterly distinct from anything known in the preceding 3 billion years.

Five mass extinctions have changed the trajectory of animal evolution during the past 500 million years.

The wealth of paleontological data we have show a dynamic history of animal radiations and extinctions through time (Chapter 22). **Fig. 33.16** is a graph of fossil occurrences through time, compiled from the paleontological literature by American paleontologist Jack Sepkoski. A notable feature of this graph is that great reductions in the number of animal genera appear to have occurred five times (indicated by the arrows), suggesting events that we call **mass extinctions.** A mass extinction is a large, widespread, and relatively rapid loss of life on Earth.

In Fig. 33.16, we also see that despite the burst of body plan evolution recorded by Cambrian fossils, there were still relatively few genera at the end of the Cambrian Period. The following Ordovician Period (485–444 million years ago) was a time of renewed animal diversification. In particular, heavily skeletonized animals evolved in the world's oceans. The number of genera recorded by fossils increased fivefold during this period, suggesting that species diversity might have increased by an order of magnitude (most genera contain multiple species). The Ordovician radiation established a marine ecosystem that persisted for more than 200 million years.

One interesting note: if you had walked along an Ordovician beach, the shells washing about your feet would have been far different from the ones you see today. The dominant shells were those of brachiopods, not clams **(Fig. 33.17)**. The broken corals in the Ordovician surf were the skeletons of now-extinct cnidarians only distantly related to modern reef-forming corals. And arthropod

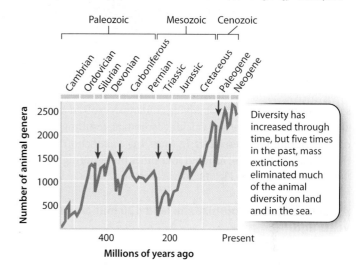

FIG. 33.16 Genus-level diversity of marine animals through time, compiled from the fossil record. Arrows indicate the five mass extinctions. *Source: Sepkoski's Online Genus Database. http://strata.geology.wisc.edu/jack/.*

Diversity has increased through time, but five times in the past, mass extinctions eliminated much of the animal diversity on land and in the sea.

exoskeletons, molted during growth, were those of now-extinct trilobites, not lobsters or crabs. Why was the Ordovician world so distinct from our own?

Another look at Fig. 33.16 provides the answer. At the end of the Permian Period, 252 million years ago, a mass extinction eliminated most genera in the oceans. Early coral-like cnidarians

FIG. 33.17 Brachiopod fossils. Brachiopods were common in marine communities during the Ordovician Period. *Source: Richard Paselk, HSU Natural Museum.*

become extinct, as did the trilobites. Brachiopods survived as a group, but most species disappeared. As noted in Chapter 22, the likely trigger for this devastation was a massive eruption of volcanos that unleashed global warming, ocean acidification, and oxygen loss from subsurface oceans.

As ecosystems recovered from this mass extinction, they came to be dominated by new groups descended from survivors of the extinction. Notably, bivalves and gastropods (two groups of mollusks) diversified. New groups of arthropods also radiated, including the ancestors of the crabs and shrimps we see today. In addition, the surviving cnidarians evolved a new ability to make skeletons of calcium carbonate, resulting in the corals that build modern reefs. In short, mass extinction reset the course of evolution, as it did four other times during the past 500 million years (see Fig. 33.16).

Animals began to colonize the land 420 million years ago.

The movement from water to land presented numerous challenges for animals, just as it did for plants (Chapter 27). In particular, animals had to develop ways to move about without the buoyancy provided by water (Chapter 35), extract oxygen from air rather than water (Chapter 37), and reproduce on land (Chapter 40).

Animals began to colonize land only after plants had already established themselves there. Land plants evolved during the Ordovician Period, about 460 million years ago. The first animals to colonize the land were arthropods, which moved onto land during the Silurian Period, about 420 million years ago. These arthropods included insects and chelicerates (the group that today includes spiders, mites, and scorpions).

The radiation of the major groups of insects began about 360 million years ago with a marked diversification of dragonflies and the ancestors of cockroaches and grasshoppers. Some of the dragonfly-like insects that darted among the plants of this period had bodies the size of a lobster and 75-cm wing spans! You might well wonder what prevents insects from attaining this size today. Scientists debate this question, but part of the answer probably implicates flying vertebrates—first now-extinct groups such as pterosaurs, and later, birds and bats—which competed with and preyed upon large insects.

Flies, beetles, bees, wasps, butterflies, and moths radiated later. Their rise in diversity parallels that of the flowering plants and reflects the coevolution of these pollinators and flowering plants.

Fossils of tetrapods (four-legged land vertebrates, such as amphibians) first appear in sedimentary rocks deposited near the end of the Devonian Period, about the same time as the early insect radiations of 360 million years ago. As discussed in Chapter 22 (see Fig. 22.22), the fossil record documents in some detail the shifts in skull, trunk, and limb morphology that allowed vertebrate animals to colonize land. These include the evolution of muscled, articulated legs from fins, together with a set of strong, articulated digits where the limbs meet the ground; lungs and a rib cage to support the muscles that control breathing; and an erect, elevated head with eyes oriented for forward vision.

Amniotes (the large group that includes reptiles, birds, and mammals) evolved later, about 310 million years ago. A key innovation that separates amniotes from amphibians is the evolution of a water-tight, or amniotic, egg (Chapter 40). This feature allowed land vertebrates to reproduce on land without returning to the water. The appearance in the fossil record of amphibians and then amniotes follows the predictions of phylogenies based on comparative biology.

During this time, new species of land animals continued to diversify, including small bipedal species that gave rise to some of the most remarkable creatures ever to walk on land: the dinosaurs. From their beginnings approximately 240 million years ago, dinosaurs radiated to produce many hundreds of species, dominating terrestrial ecosystems until the end of the Cretaceous Period, 66 million years ago. At that time, another mass extinction, caused by a catastrophic asteroid impact, eliminated nearly all dinosaur species (Chapter 1). We say "nearly all" because the fossil record documents the evolutionary divergence of birds from a specific subgroup of dinosaurs approximately 150 million years ago (Chapter 22).

Mammals have been the dominant vertebrates on land since the extinction of dinosaurs. This group actually originated much earlier, at least 210 million years ago. During the age of dinosaurs, most mammals were small nocturnal or tree-dwelling animals that stayed out of the way of large dinosaurs, although fossils from China show clearly that the largest mammals of this interval ate small dinosaurs and their eggs. Mammals are dominant components of terrestrial ecosystems today in part because they survived the end-Cretaceous mass extinction.

Some animals show a trend toward increased body size over time.

In our brief overview of animal evolution, we have seen giant dragonflies and large dinosaurs. A newly described dinosaur, *Patagotitan mayorum*, is thought to be the largest animal ever to have walked on land. The blue whale (*Balaenoptera musculus*) is the largest animal to have ever existed on Earth. These are just a few examples, but it turns out that many groups of animals show increases in body sizes over time (**Fig. 33.18**). These groups include the brachiopods, chordates (including vertebrates), and echinoderms (such as sea stars, sea urchins, and sand dollars).

Large body size presents distinct challenges. Within those large bodies, animals need to transport oxygen, nutrients, and signaling molecules over long distances. Diffusion, the random motion of molecules, is effective only over short distances. In Chapter 27, we discussed how bulk flow can transport

HOW DO WE KNOW?

FIG. 33.18
Do animals tend to get bigger over time?

BACKGROUND Cope's rule, named after the American paleontologist Edward Cope, suggests that there is a trend through time toward increasing size among animals. However, it is unclear whether such a general trend actually exists. Large body size can be associated with increased fitness—for example, by making it easier to capture prey or compete with other organisms. Notable examples of very large animals that lived long ago are the dinosaurs. There are also exceptions to this trend in some groups of animals.

HYPOTHESIS Animals tend to increase in size over evolutionary time.

EXPERIMENT Jonathan Payne and his colleagues at Stanford University decided to test the hypothesis that animals tend to get bigger over time, focusing on marine animals over the last 541 million years. They measured the three major body axes of images of animal specimens to determine the mean volume (as a measure of body size) for more than 17,000 genera of animals, including arthropods, brachiopods, echinoderms, mollusks, and chordates.

RESULTS The researchers found that the mean volume for all genera increased over time (the dark line in the graph). In addition, they found that the range in sizes increased dramatically (the difference between the two light lines in the graph). Most of this expansion resulted from an increase in the maximum size of animals (the upper light line in the graph).

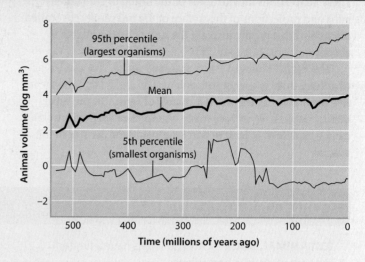

CONCLUSION The data support the hypothesis that there is an evolutionary trend toward larger body size in many groups of animals.

FOLLOW-UP WORK The researchers next asked whether the increase in size is the result of natural selection or genetic drift (Chapter 20). They compiled observed trends in the mean, maximum, and minimum sizes of a number of animal groups and compared them to the predictions of models based on selection and drift. They found that their observations are consistent with models based on selection, and not consistent with models based on drift. These findings suggest that increased body size has selective advantages.

SOURCE Heim, N. A., et al. 2015. "Cope's Rule in the Evolution of Marine Mammals." *Science* 347:867–870.

---

molecules over the long distances present in larger organisms. In some animals, for example, blood is circulated throughout the body by the pumping action of the heart, and oxygen is transported into the lungs by contraction of the muscular diaphragm (Chapter 37).

More generally, the form of animals is in part determined by their size. Therefore, as animals got bigger, their form changed as well. To understand why this is so, consider what happens if the size of an ant increases 10-fold without a change in its form. The strength of the legs would increase as the *square* of the scale factor, or 10^2 (100 times), because strength is proportional to the cross-sectional area of the legs. However, the weight of the ant would increase as the *cube* of the scale factor, or 10^3 (1000 times), because weight is proportional to volume. Therefore, as an ant gets bigger, its weight would quickly outstrip the strength of its legs, and it would not be able to support itself. So, an ant looks like an ant in part because of its size.

The issue with a large ant is that it increases in size but keeps its overall shape. This kind of change is called **isometry** ("same measure"). By contrast, increases in size are possible if they are accompanied by changes in shape. This kind of change is called **allometry** ("different measure"). For example, in the 1600s, Galileo Galilei noticed that the bones of larger animals are not simply scaled-up versions of the bones of smaller animals. Instead, in many cases, the bones of larger animals are disproportionately wider than those of smaller animals so as to support the increased weight.

As an animal (or any three-dimensional object) gets bigger, both surface area and volume increase, but surface area increases more slowly than does volume (Chapter 26). Therefore, shape changes often accompany size differences to balance the two. For example, nutrients are absorbed through the lining of the gut, which is one large surface area, but they support the entire body, which represents a volume. Larger organisms, therefore, have adaptations of the lining of the gut to increase the surface area, enabling them to "keep up" with the increased volume. These adaptations include folds of the lining of gut (called villi) and projections of the surface of the intestinal cells (called microvilli) that greatly increase the surface area available to absorb nutrients (Chapter 38).

The British biologist J. B. S. Haldane wrote, "For every type of animal there is a most convenient size, and a large change in size inevitably carries with it a change of form." In the subsequent chapters, we will see many examples of adaptations that can be explained at least in part by considering the size of the animal.

Self-Assessment Questions

7. What is the evolutionary significance of the Cambrian explosion?
8. What are two examples of how mass extinctions changed the ecological structure of life on Earth?
9. Which has more surface area relative to volume: a grape or a grapefruit?

CORE CONCEPTS SUMMARY

33.1 ANIMAL BODY PLANS: Animals have a limited number of distinct body plans.

Animals are multicellular heterotrophic eukaryotes that form a blastula and, commonly, a gastrula during development. page 740

Early biologists grouped animals on the basis of shared features of adult bodies, such as symmetry. page 740

Poriferans (sponges) have no well-defined plane of symmetry; cnidarians (jellyfish, corals, and sea anemones) are radially symmetric; and bilaterians (insects and vertebrates, for example) are bilaterally symmetric. page 740

Cephalization is the concentration of the nervous system and special sensory organs at the front of the body. page 741

Segmentation is the organization of the body into repeated units along the anterior–posterior (head–tail) axis. page 742

Cnidarians have two germ layers (ectoderm and endoderm), whereas bilaterians have three germ layers (ectoderm, mesoderm, and endoderm). page 742

33.2 TISSUES AND ORGANS: Most animals have four different types of tissues, which combine to form organs.

Most animals organize their cells into tissues. page 746

The four types of tissues found in animals are epithelial, connective, muscle, and nervous. page 746

Epithelial tissue lines spaces, such as cavities or the outside of the body. page 746

Connective tissue is characterized by an extensive extracellular matrix and few cells. It provides nutrients and a cushion for the overlying epithelial tissue, among other functions. page 746

Muscle tissue is able to shorten, or contract. page 746

Nervous tissue is specialized for transmitting electrical impulses, thereby allowing animals to take in information from the environment, process it, and elicit a response. page 747

Most animals with tissues, with the exception of cnidarians, have organs, which are collections of tissues that perform a specific function. page 747

Organs often work with other organs form an organ system. page 747

33.3 HOMEOSTASIS: Homeostasis actively maintains a stable internal environment.

Homeostasis is an active process that is often maintained for individual cells as well as for an organism as a whole. page 748

Many parameters, such as blood pH, ion concentrations, and body temperature, are actively maintained in a narrow range by homeostasis. page 748

Homeostasis is often achieved by negative feedback, in which the response inhibits the stimulus to maintain a set point. page 748

33.4 EVOLUTIONARY HISTORY: Animals evolved more than 600 million years ago in the oceans, and by 500 million years ago the major body plans were in place.

Ediacaran fossils from 575 million years ago provide evidence of early animals. page 749

Most of the animal body plans we see today first evolved during the Cambrian explosion, a time of rapid diversification that began 541 million years ago. page 750

Animal diversity has been shaped by five mass extinctions, which were followed by adaptive radiations, over the past 500 million years. page 751

Arthropods were the first animals to colonize the land, sometime around 420 million years ago. page 752

Four-legged vertebrates (tetrapods) first appear in the fossil record approximately 360 million years ago. page 752

Mammals originated at least 210 million years ago, but became dominant only after the extinction of the non-avian dinosaurs 66 million years ago. page 752

Some animal groups show a tendency toward larger sizes over time, resulting in changes in form. page 753

Log in to LaunchPad to check your answers to the Self-Assessment Questions and to access additional learning tools.

CHAPTER 34 Animal Nervous Systems

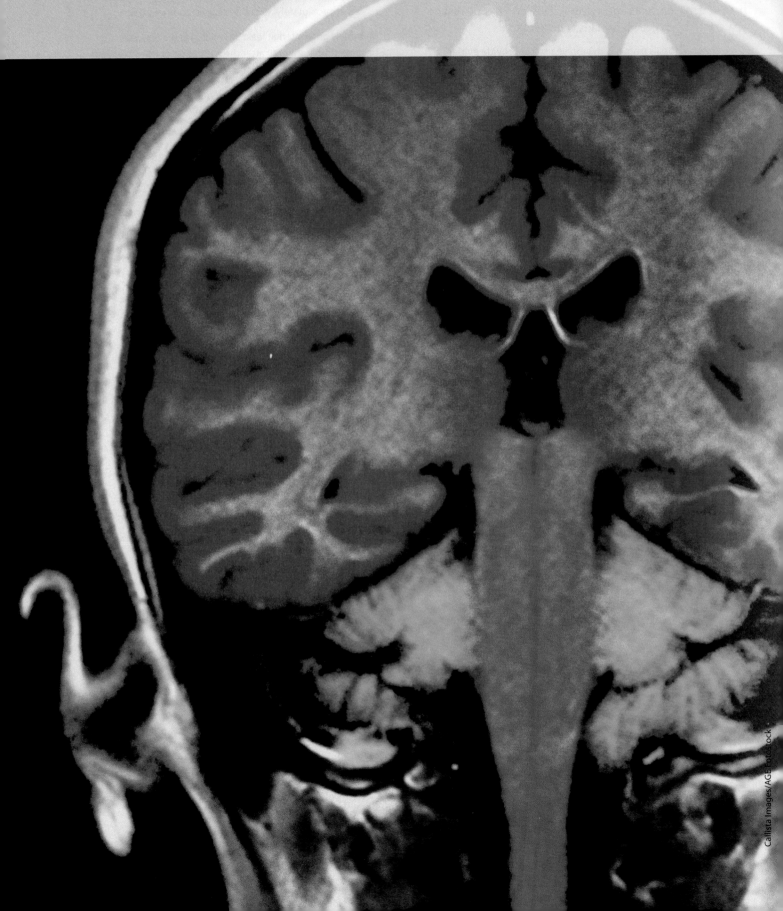

CORE CONCEPTS

34.1 NERVOUS SYSTEM FUNCTION AND EVOLUTION: Nervous systems allow animals to sense and respond to the environment, coordinate movement, and regulate internal functions of the body.

34.2 NEURON STRUCTURE: The functional unit of the nervous system is the neuron, which has dendrites that receive information and axons that transmit information.

34.3 SIGNAL TRANSMISSION: The electrical properties of neurons allow them to communicate rapidly with one another.

34.4 NERVOUS SYSTEM ORGANIZATION: Nervous systems are organized into peripheral and central components, and include voluntary and involuntary functions.

34.5 SENSORY SYSTEMS: Animal sensory receptors detect chemical, physical, and electromagnetic stimuli by changes in membrane potential.

34.6 BRAIN ORGANIZATION AND FUNCTION: The brain integrates information from diverse sources and is organized into distinct regions with specialized functions.

Gazelles grazing on an African plain are constantly alert to the threat of predators. Members of the herd scan the landscape for visual cues that might reveal the stealthy approach of a cheetah. Olfactory (smell) and auditory (sound) cues from the environment can also warn of the cheetah's approach. Herd members communicate the presence of the cheetah to one another by the directed attention of a dominant male or the pawing of a hoof. At what point is the threat real? When do the animals decide to stop grazing and turn to run? Which direction should the escape take to keep the herd together?

The cheetah scans the horizon, its keen vision spotting a grazing gazelle. The cheetah's ability to recognize wind direction enables it to move downwind of the gazelle to minimize auditory and olfactory cues that might otherwise be carried by the air toward suspecting prey. At what point does the cheetah decide to sprint in pursuit of its quick-running prey?

The perceptions and reactions of these animals depend on the activity of their nervous systems. The nervous system takes in information from the environment through sensory organs, processes that information by networks of nerve cells, and guides motor responses. In this chapter, we explore the organization of the nervous system, including the brain and the nerve cells that compose it.

34.1 NERVOUS SYSTEM FUNCTION AND EVOLUTION

Most multicellular animals possess a **nervous system**, a network of many interconnected nerve cells. Nervous systems first evolved in animals like jellyfish and flatworms. These early nervous systems could sense basic features of the environment, such as light, temperature, and chemical odors. Guided by these cues, nervous systems could trigger movements useful for obtaining food, finding a mate, or choosing a suitable habitat.

As animals evolved more complex bodies with increasingly specialized organ systems, their nervous systems also became more elaborate. These nervous systems could regulate internal body functions in response to sensory cues received from the environment. They also made possible more sophisticated behaviors that relied on improved decision making, learning, and memory.

Animal nervous systems have three types of nerve cells.

Nerve cells, or **neurons**, are the functional units of nervous systems. All animal nervous systems are made up of three types of neurons with different functions: **sensory neurons**, **interneurons**, and **motor neurons**. Sensory neurons receive and transmit information about an animal's environment or its internal physiological state. These neurons respond to physical features of the environment such as temperature, light, and touch, or to chemical signals such as molecules conveying odor and taste. Interneurons process information received by sensory neurons and transmit it to different body regions to produce a suitable response. Often, interneurons communicate with motor neurons. A motor neuron may, for example, stimulate a muscle to contract to produce movement. Other motor neurons may adjust an animal's internal physiology, constricting blood vessels to affect blood flow or causing wavelike contractions of the gut to aid digestion. In this way, nervous system function plays a fundamental role in maintaining homeostasis, the ability of animals, organs, and cells to actively regulate and maintain a stable internal state (Chapters 5 and 33).

Most nerve cells have fiberlike extensions, some of which receive information and others of which transit information. The result is a network of interconnected nerve cells that form a circuit. These circuits allow information to be received, processed, and delivered. In most animals, interneurons form circuits that may consist of a diverse array of nerve cells having different shapes, sizes, and chemical properties.

The ability to process information first evolved with the formation of **ganglia** (singular, **ganglion**). Ganglia are groups of nerve cell bodies that process sensory information received from a nearby region, then send signals to motor neurons that control some physiological function of the animal. They can be thought of as relay stations in nerve cell circuits. For example, ganglia assist in processing information from the eyes and controlling the digestive state of an animal's gut (Chapter 38).

Over time, a centralized concentration of neurons called a **brain** evolved. As a central nervous system organ, an animal brain is able to process complex sensory stimuli received from the environment or anywhere in the body. Neurons in the brain form complex circuits, allowing the brain to regulate a broad array of behaviors, as well as to learn and retain memories of past experiences.

Nervous systems range from simple to complex.

Single-celled organisms have cell-surface receptors that enable them to sense and respond to their environment. In contrast, most multicellular animals have a nervous system. In general, the organization and complexity of an animal's nervous system reflect its lifestyle. Some animals, such as sea anemones and clams, are sessile (immobile) and have relatively simple sense organs and nervous systems. By contrast, active animals such as arthropods (which include insects, spiders, and crustaceans), squid, and vertebrates have more sophisticated sense organs linked to a brain and more complex nervous systems. The nervous systems of these animals help to maintain homeostasis and coordinate complex behaviors that require decision making, forming memories, and learning.

Sponges are the only animals that lack a nervous system (**Fig. 34.1**). These marine animals lack any form of body symmetry. Many small pores along the side and a larger opening at the top draw nutrient-laden water into a central cavity of the sponge. The cells lining the cavity first capture food particles and dissolved organic matter by endocytosis, then metabolize these nutrients in the cell interior. The functions of different body regions are coordinated by groups of cells specialized for functions such as contracting or taking in water. These cells respond to local chemical and physical cues. For example, a

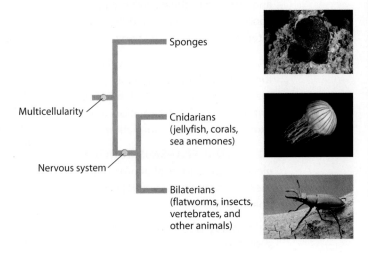

FIG. 34.1 A simplified phylogenetic tree of multicellular animals. All multicellular organisms except sponges have a nervous system. *Photo sources: (top) Andrew J. Martinez/Science Source; (middle) UMI NO KAZE/a.collectionRF/Getty Images; (bottom) Christof Wermter/AGE Fotostock.*

pore-like cell that allows water to enter the sponge closes when touched.

Among multicellular organisms, the simplest nervous systems are found in cnidarians, which are radially symmetric animals such as jellyfish and sea anemones (Fig. 34.1). At one end of a cnidarian's body is a mouth, surrounded by tentacles, that opens into a central body cavity, where food is digested. The nervous system of these animals is made up of neurons that are arranged like a net, but the system lacks ganglia and a central brain (**Fig. 34.2a**). The nerve net of a sea anemone is organized around an internal gut cavity. Sensory neurons sensitive to touch signal when one of the anemone's tentacles has captured food, and then motor neurons react to coordinate the tentacle's movement toward the mouth for feeding. Sensory neurons responding to a noxious stimulus signal the sea anemone to retract or detach from its substrate.

More complex nerve connections develop in the head end of flatworms, annelid worms, insects, and cephalopod mollusks, such as squid and octopuses. These animals are all bilaterians, meaning they have symmetrical right and left sides and distinct front and back ends (see Fig. 34.1). Flatworms were one of the earliest groups of bilaterians to evolve that were well adapted for forward locomotion. These elongated, flattened organisms glide along a streambed, trapping small animals under their bodies and sucking them in through a muscular opening in their underside. Sensory neurons are most numerous in the animal's front end, so the flatworm can perceive the environment ahead of it as it moves forward. The flatworm's nervous system can compare inputs from different sensory neurons and then send signals to the appropriate muscles to move toward a target. Paired ganglia (one on each side of the animal) in the head end of flatworms handle this processing task (**Fig. 34.2b**).

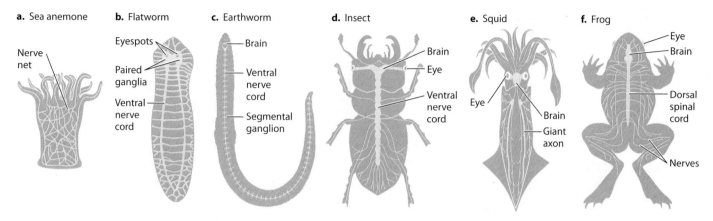

FIG. 34.2 **Animal nervous systems.** (a) Sea anemones have a nerve net; (b) flatworms have paired ganglia; (c) earthworms, (d) insects, (e) squid, and (f) frogs have a brain.

In animals with an organized nervous system, information is transmitted to different regions along the length of the animal's body by distinct **nerve cords** and **nerves**. These bundles contain long, fiberlike extensions from multiple nerve cells. Nerves connect the ganglia of a flatworm to the muscles along its body used for locomotion.

In earthworms, paired ganglia, each made up of many interneurons, are located in each of the body's segments (**Fig. 34.2c**). These segmental ganglia help regulate key processes in local regions and organs of the animal's body. For example, paired ganglia near the head regulate the activity of motor neurons that control movements of the earthworm's mouth for feeding.

Centralized collections of neurons forming a brain are present in earthworms and other annelid worms, insects, and mollusks (**Figs. 34.2d** and **34.2e**). Sense organs also developed in the head region, ranging from the eyespots in flatworms that respond to light to the more sophisticated image-forming eyes in octopuses and squid. The evolution of a brain in these invertebrate animals enabled them to learn and perform complex behaviors.

The evolution of a brain (**Fig. 34.2f**) also enabled vertebrates to evolve complex behaviors, learning, and memory. At the same time, animal brains became increasingly complex in terms of the number of nerve cells they contain (fruit fly, about 0.25 million; cockroach, about 1 million; mouse, about 75 million; humans, about 100 billion), the number of connections between neurons, and the variety of shapes and sizes of nerve cells.

Self-Assessment Questions

1. What are the three basic types of neurons?
2. What are the functions of these three types of neurons?

34.2 NEURON STRUCTURE

Although animal nervous systems differ in organization and complexity, their nerve cells are fundamentally alike: they all have the same basic features and they use the same mechanisms to communicate with neighboring cells. These mechanisms have been retained over the course of evolution to code and reliably transmit information. The ability to receive sensory information from the environment, and to transmit and process this information within a nervous system, is shared across a wide diversity of animal life and has greatly contributed to the success of animals.

Neurons share a common organization.

Neurons all share some basic features. For example, they have a cell body from which emerge two kinds of fiberlike extensions, **dendrites** and **axons** (**Fig. 34.3**). These extensions are the input and output ends of the nerve cell, respectively. Both types of cellular extensions can be highly branched, enabling neurons to communicate over large distances with many other cells.

A neuron's dendrites receive signals from other nerve cells or, in the case of sensory nerves, from specialized sensory endings. These signals travel along the dendrites to the neuron's cell body. At the junction of the cell body and its axon (an area known as the **axon hillock**), the signals from multiple dendrites are combined. If the sum of the signals is high enough, the neuron fires an **action potential**, or nerve impulse, that travels down the axon. An action potential is a brief electrical signal transmitted from the cell body along one or more axon branches.

Axons generally transmit signals away from the nerve's cell body. The end of each axon forms a swelling called the axon terminal. An axon terminal communicates with its neighboring cell through a junction called a **synapse**. A space, the **synaptic cleft**, separates the end of the axon of the presynaptic

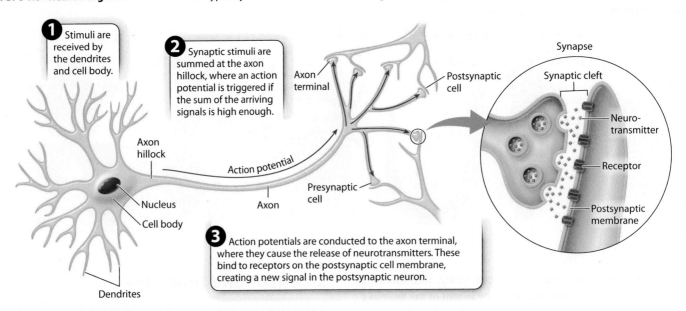

FIG. 34.3 Neuron organization. Nerve cells typically have dendrites that receive signals as well as axons that transmit signals.

cell and the neighboring postsynaptic cell. This synaptic cleft is commonly only 10 to 20 nm wide.

Molecules called **neurotransmitters** convey the signal from the presynaptic cell across the synaptic cleft to the postsynaptic cell. The arrival of a nerve signal at the axon terminal triggers the release of neurotransmitter molecules from vesicles located in the terminal. The vesicles fuse with the axon's membrane, releasing neurotransmitter molecules into the synaptic cleft. These molecules then diffuse across the synapse and bind to receptors on the plasma membrane of the target cell. The binding of neurotransmitters to these receptors causes a change in the electrical charge across the membrane of the receiving postsynaptic cell, which in turn propagates the signal in the second cell. In this way, most neurons communicate by forming hundreds to thousands of synapses with other cells. The massive number of interconnections enables the formation of complex nerve cell circuits and the transmission of information from one circuit to others.

Some neurons do not synapse with other neurons, but instead with other types of cell that produce some physiological response in the animal. Examples are muscle cells, which contract in response to nerve signals, and secretory cells in glands, which release hormones in response to nerve signals.

Neurons differ in size and shape.

In spite of their common organization, neurons show a remarkable diversity in terms of their size and shape. Some neurons, such as many interneurons, are very short, whereas others, like the motor neurons that extend from your spinal cord to muscles that control your toes, are several feet long. The number of extensions coming off the cell body can also vary (**Fig. 34.4**).

Some sensory neurons involved in smell, sight, and taste have just two extensions coming off the cell body (Fig. 34.4a), whereas other neurons have many more (Fig. 34.4b and 34.4c).

The degree of branching of a neuron's extensions varies from one neuron to the next. A neuron with more dendritic branches receives inputs from more sources than one with fewer branches. Some neurons receive information only from nearby regions through a few dendrites that branch close to the cell body. Other neurons receive information from many branching dendrites. Axons typically have fewer branches than dendrites.

Signals from all the dendrites converge on the neuron cell body. These signals determine the frequency and timing of signals carried by the neuron's axon. In turn, the frequency

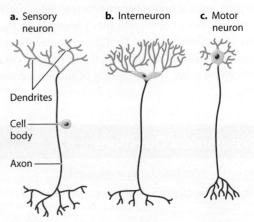

FIG. 34.4 Variation in neuron shape. Neurons display a diversity of shapes, reflecting their different functions.

of the signals carried by the neuron's axon determine their strength. The process of bringing together information gathered from different sources is referred to as information integration. The degree of branching and the number of synapses that a neuron makes with neighboring nerve cells, therefore, reflect how information is processed and integrated by the cell.

Neurons are supported by other types of cells.

Neurons in many body regions, and in particular within the brain, are supported by other types of cells that do not themselves transmit electrical signals. These include **glial cells**. Indeed, the human brain has more glial cells than neurons.

Glial cells have many functions. They surround neurons and provide them with nutrition and physical support. Some glial cells, called astrocytes, support endothelial cells that make up blood vessels in the brain. These endothelial cells are linked by tight junctions (Chapter 10) to form a selective barrier, called the blood–brain barrier, that limits the types of compounds that can diffuse from the blood into the brain. This barrier prevents pathogens and toxic compounds in the blood from entering the brain. Because nerve cells have limited capacity to regenerate after damage, protection of the brain and spinal cord is critically important. Nevertheless, lipid-soluble compounds such as alcohol and certain anesthetics readily diffuse across the blood–brain barrier, so they can affect an animal's mental state. They can also damage and even destroy neurons.

During development, glial cells help orient neurons as they develop their connections. They also provide electrical insulation to vertebrate neurons that allows nerve signals to be transmitted rapidly, as we discuss next.

Self-Assessment Questions

3. Diagram and label the basic features of a neuron, indicating where information is received and where it is sent.
4. How are signals transmitted from one end of a nerve cell to the other, and from one neuron to another?

34.3 SIGNAL TRANSMISSION

As described in the previous section, neurons send signals to other neurons by releasing neurotransmitters that diffuse across the synaptic cleft and bind to receptors. In Chapter 9, we introduced the binding of a chemical signal to receptors as a general mechanism of cell communication. In neurons, once the receptor has bound the neurotransmitter and become activated, neurons send electrical signals from one end of the cell to the other. These electrical signals travels at high speeds, up to 200 m/sec (450 mph). In this section, we discuss the electrical properties of neurons and consider how signals are propagated along neurons. We also describe how signals are transmitted between neurons.

The resting membrane potential is negative and results in part from the movement of potassium and sodium ions.

The inside surface of a neuron carries an electrical charge that is due to the presence of charged ions; the same is true of the outside surface. However, the electrical charge on the inside and the electrical charge on the outside are not the same because unequal numbers of charged ions are located inside and outside the cell. When no signal is present, there are more negative ions inside the cell, so the inside is negatively charged relative to the outside. Because the charge differs between the inside and the outside of the cell, the membrane is said to be **polarized**.

The charge difference between the inside and outside of a neuron due to differences in the numbers of charged ions is known as the cell's **membrane potential**, and is measured in volts. Other cells also have membrane potentials, but only nerve cells and muscle cells respond to changes in their membrane potential. Therefore, these cells are considered electrically excitable. The key to producing an electrical signal in a nerve cell is the movement of positively and negatively charged ions across the cell membrane. This movement creates a change in electrical charge across the membrane that constitutes the electrical signal.

Let's first consider the membrane potential when the neuron is at rest and no signal is being received or sent. Under these conditions, the cell's membrane potential is negative on its inside relative to its outside (**Fig. 34.5**). The negative membrane potential results from the buildup of negatively charged ions on the inside surface of the cell's plasma membrane and of positively charged ions on its outer surface. Notice in Fig. 34.5 that there are many more positively charged ions (colored purple and pink) outside the cell. The negative membrane potential at rest is the cell's **resting membrane potential**. This resting membrane potential ranges from −40 to −85 millivolts (mV) depending on the type of nerve cell. The charge of the cell's interior can be measured with respect to its outside charge by placing small glass electrodes on the inside and outside of the cell.

At rest, nerve (and muscle) cells have a greater concentration of sodium (Na^+) ions outside the cell than inside, and a greater concentration of potassium (K^+) ions inside the cell than outside. This distribution of ions results in part from the action of the sodium–potassium pump (Chapter 5). The pump uses the energy of ATP to move three Na^+ ions outside the cell for every two K^+ ions moved inside the cell (Fig. 34.5). The action of the pump makes the inside of the cell less positive, and therefore more negative, than the outside of the cell.

The exact value of the resting membrane potential depends on the movement of K^+ ions back out of the cell by passive diffusion through potassium ion channels (Fig. 34.5). As discussed in Chapter 5, ion channels are protein pores embedded in the cell membrane. The most important ions that move across the cell membrane are sodium (Na^+), potassium (K^+), chloride (Cl^-), and calcium (Ca^{2+}). When a nerve cell is at rest, the K^+ ion channels are open and K^+ ions move out of the

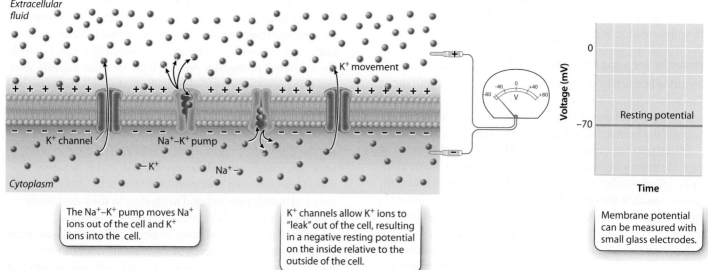

FIG. 34.5 **Neuron resting membrane potential.** The resting potential of a neuron is negative and results primarily from the movement of K+ ions out of the cell.

The Na+–K+ pump moves Na+ ions out of the cell and K+ ions into the cell.

K+ channels allow K+ ions to "leak" out of the cell, resulting in a negative resting potential on the inside relative to the outside of the cell.

Membrane potential can be measured with small glass electrodes.

cell. As a result, positive ions build up on the outside of the cell. Because negatively charged ions (largely proteins) remain on the inside of the cell, the inside of the nerve cell is more negative than its outside.

The relative proportion of ions does not by itself determine the resting membrane potential. In fact, the number of charged ions that build up at the cell's surface is just a tiny fraction of the total number of charged ions and proteins located inside and outside the cell. Instead, it is the *movement* of K+ ions relative to other ions, particularly Na+ ions, that largely determines the resting membrane potential.

Neurons are excitable cells that transmit information by action potentials.

When a nerve cell is excited, its membrane potential becomes less negative—that is, the inside becomes less negative, or more positive, than the outside of the cell. This increase in membrane potential is referred to as a **depolarization** of the membrane. Depolarization does not occur over the entire cell at once. Rather, it starts at the end of the dendrite, in response to neurotransmitter binding to membrane receptors, and travels to the cell body, losing strength along the way. If the depolarization is strong enough at the axon hillock of the cell body, the cell fires an action potential that carries the signal from the cell body to the ends of the axon.

The nerve fires an action potential at the axon hillock if the membrane at the hillock is depolarized to a voltage of approximately 15 mV above the resting membrane potential (about –55 mV). An action potential is a rapid, short-lasting rise and fall in membrane potential (**Fig. 34.6**). The critical depolarization voltage of –55 mV required for an action potential is called the cell's **threshold potential**. Nerve cells in different animals have slightly different resting and threshold potentials, but all operate similarly with respect to firing an action potential. When the threshold potential is exceeded (step 1 in Fig. 34.6), the nerve fires an action potential in an "all-or-nothing" fashion. As a consequence, once the threshold potential has been exceeded, the magnitude of the action potential is always the same independent of the strength of the stimulating input. At this point, a rapid spike in membrane potential to approximately +40 mV occurs over a very brief time period of 1 to 2 msec.

How is the action potential generated? As Fig. 34.6 shows, voltage-gated channels play a key role in how neurons fire action potentials. Recall that **voltage-gated channels** open and close in response to changes in membrane potential (Chapter 9).

As the cell membrane crosses its threshold potential, a dramatic change occurs in the state of the voltage-gated Na+ and voltage-gated K+ channels in the axon membrane, causing a large number of voltage-gated Na+ channels to open suddenly, and allowing Na+ ions to rush inside the cell (step 2). This influx of Na+ causes a flow of positive charge along the inside of the cell, toward more negatively charged nearby regions. The resulting rise in voltage causes additional, nearby voltage-gated Na+ channels to open. This is an example of **positive feedback**, in which a signal (depolarization) causes a response (open voltage-gated Na+ channels) that leads to an enhancement of the signal (more depolarization), which then leads to an even larger response (more open voltage-gated Na+ channels).

FIG. 34.6 An action potential. An action potential results from the opening and closing of voltage-gated ion channels and results in a positive spike in membrane potential.

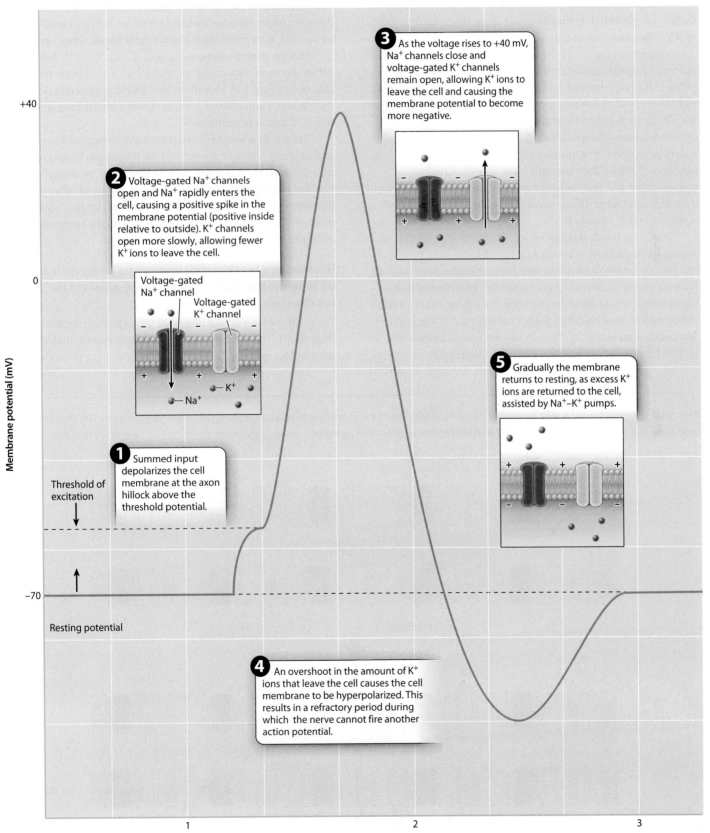

During the rising phase of the action potential, voltage-gated K^+ channels also begin to open. These channels are slow to activate relative to the rapidly opening voltage-gated Na^+ channels. As a result, few K^+ ions leave the cell, and the cell membrane potential continues to rise due to the large number of Na^+ ions entering the cell.

As shown in Fig. 34.6, the rising phase of the action potential (rapid depolarization) is followed by the falling phase (rapid repolarization). Two important factors cause this sudden reversal of the membrane potential. First, voltage-gated Na^+ channels begin to close once the membrane potential becomes positive (step 3). Second, voltage-gated K^+ channels continue to open in response to the change in membrane potential. Because they respond more slowly than the voltage-gated Na^+ channels, the membrane potential peaks and then falls as K^+ ions diffuse out of the axon through the open voltage-gated K^+ channels.

The charge inside the axon does not return immediately to resting levels. Instead, it briefly falls below the resting potential in a process known as hyperpolarization or undershoot (step 4). It then returns to the resting potential after another few milliseconds as K^+ channels close to restore the resting concentration of Na^+ and K^+ ions on either side of the cell membrane (step 5). The continuous action of the sodium–potassium pumps also helps to reestablish the resting membrane potential.

The period during which the membrane potential falls below and then returns to the resting potential is called the **refractory period** (Fig. 34.6). During the refractory period, a neuron cannot fire a second action potential. Although the duration of the refractory period varies for different kinds of nerve cells, it serves to limit a nerve cell's fastest firing frequency.

It might seem that many Na^+ and K^+ ions need to diffuse across the axon membrane to produce an action potential. In fact, only about 1 in 10 million Na^+ and K^+ ions need to cross the membrane. Thus, enough ions are always available to generate repeated action potentials.

Because all action potentials are alike in being all-or-nothing and uniform, the action potential itself contains little information. Instead, most neurons code information by changing the rate and timing of action potentials. Generally, a higher firing frequency codes for a more intense stimulus, such as a louder noise, or for a stronger signal transmitted by the nerve cell to other cells it contacts.

Neurons propagate action potentials along their axons by sequentially opening and closing adjacent Na^+ and K^+ ion channels.

Once initiated, an action potential propagates along the axon (**Fig. 34.7**). The local membrane depolarization initiated at the axon hillock triggers the opening of neighboring voltage-gated

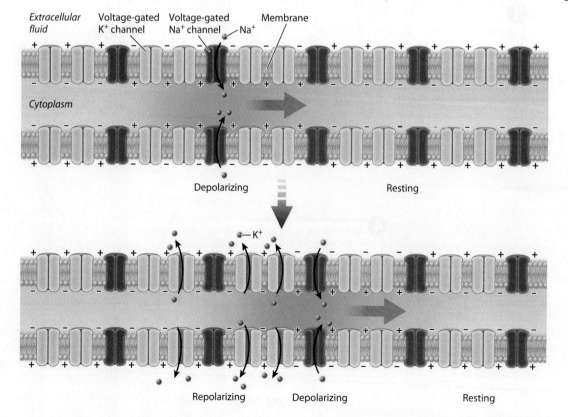

FIG. 34.7 Propagation of action potentials. Local depolarization of the membrane triggers the opening of nearby voltage-gated Na^+ channels, producing an action potential that spreads along the membrane. The action potential moves in only one direction (shown here as left to right).

FIG. 34.8 Saltatory propagation. Myelination insulates axons, enabling faster propagation of action potentials. *Photo source: Don W. Fawcett/Science Source.*

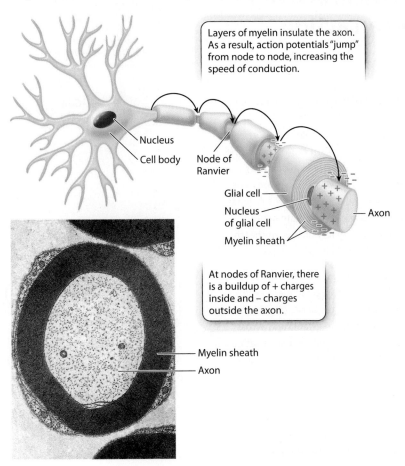

Na⁺ channels farther along the axon. The inward sodium current depolarizes the membrane above its threshold, triggering the opening of nearby voltage-gated Na⁺ channels still farther along the axon. By this means, the depolarization—the location of the action potential spike—spreads down the axon. Notice in Fig. 34.7 that the area of depolarization represented by the three Na⁺ ions (purple spheres) is farther along the axon in the bottom panel. Neighboring voltage-gated K⁺ channels subsequently open and close to reestablish a resting membrane potential after an action potential has fired. In this way, action potentials become **self-propagating**. They propagate in only one direction, usually from the cell body at the axon hillock to the axon terminal. The refractory period following an action potential prevents the membrane from reaching its threshold and firing an action potential in the reverse direction.

The conduction speed of action potentials is limited by the neuron's membrane properties and the diameter of its axon. Yet fast transmission speeds are essential for animals to respond quickly to stimuli. For example, when you inadvertently touch something very hot, your fingers sense and relay this information rapidly to ensure that you withdraw your hand quickly and avoid worse injury.

The neurons of vertebrate animals have an adaptation that lets them respond rapidly to stimuli in their environment. Glial cells form lipid-rich layers or sheaths called **myelin** that wrap around the axons of these neurons (**Fig. 34.8**). Myelin gives many nerves a glistening white appearance. The myelin sheath insulates the axon's membrane, spreading the charge from a local action potential over a much greater distance along the axon's length than would be possible in the absence of myelin.

At regular intervals, the axon membrane is exposed at sites called **nodes of Ranvier** that lie between adjacent segments wrapped with myelin (Fig. 34.8). Voltage-gated Na⁺ and K⁺ channels are concentrated at these nodes. As a result, rather than being conducted in a continuous fashion, as is the case for unmyelinated axons, the action potentials in myelinated axons "jump" from node to node. This **saltatory propagation** (from the Latin *saltus*, "jump") of action potentials by myelinated axons greatly increases the speed of signal transmission.

The British neurophysiologists Alan Hodgkin and Andrew Huxley first worked out the electrical properties of the axon plasma membrane by studying the giant axons in squid that stimulate contractions of these animals' body muscles for

HOW DO WE KNOW?

FIG. 34.9

How does electrical activity change during an action potential?

BACKGROUND To record the voltage of the inside of a nerve cell relative to the outside, a small electrical recording device called a microelectrode is inserted into a neuron. This technique is more easily performed with large cells, such as the squid giant axon. As its name suggests, this axon is quite large, measuring 0.5 mm in diameter (Fig. 34.9a), about 500 times larger than most vertebrate axons. The large diameter of the giant axon of the squid allows electrical signals to be propagated to the muscles very quickly. Vertebrate neurons (approximately 1 μm) are much smaller; they rely on myelin sheaths rather than large size for rapid conduction of electrical signals.

EXPERIMENT In 1939, British neurophysiologists Alan Hodgkin and Andrew Huxley inserted a microelectrode into a squid giant axon and placed a reference electrode outside the nerve cell (Fig. 34.9b) to test the hypothesis that action potentials are established across the membrane of a neuron's axon. They then used a separate set of electrodes (not shown) to depolarize the cell to threshold, triggering an action potential.

RESULTS Fig. 34.9c is a voltage trace as a function of time from Hodgkin and Huxley's experiment showing the resting membrane potential and the course of an action potential recorded by the electrode inside a giant squid axon. Note that the resting potential is negative on the inside of the axon relative to the outside, and that the action potential is the rapid spike in potential seen in the voltage trace. During the action potential, the inside of the cell quickly becomes positive, then negative again. The large size of the spike was a surprise.

FOLLOW-UP WORK This work was performed before anyone had identified the ion channels responsible for the changes in current across the membrane. Subsequent work focused on identifying these channels and their roles in generating the action potential.

SOURCES Hodgkin, A. L., and A. F. Huxley. 1939. "Action Potentials Recorded from Inside a Nerve Fibre." *Nature* 144:710–711; R. D. Keynes. 1989. "The Role of Giant Axons in Studies of the Nerve Impulse." *Bioessays* 10:90–94.

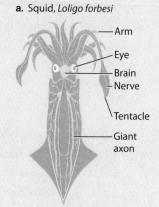

a. Squid, *Loligo forbesi*

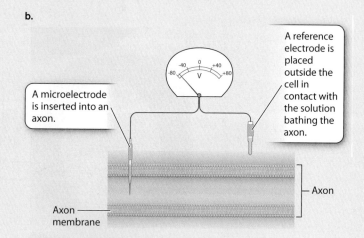

b.

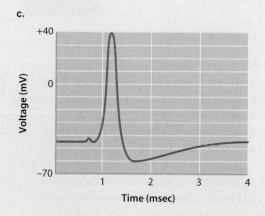

c.

swimming (**Fig. 34.9**). For this work, they shared the 1963 Nobel Prize in Physiology or Medicine, along with John Eccles.

Neurons communicate at synapses.

As described earlier, nerve cells communicate at specialized junctions called synapses. Synapses can be electrical or chemical. Electrical synapses provide direct electrical communication through gap junctions that form between neighboring cells (Chapters 10 and 26). Such electrical synapses are found in the giant axons of squids, as well as in the large motor nerves that fishes use for escape. Those found in the mammalian brain likely help to speed up information processing, but their role has not been well studied.

Chemical synapses are by far the more common type of synapse found in animal nervous systems. The signals conveyed at chemical synapses are neurotransmitters contained within small vesicles in the axon terminal. When an action potential reaches the end of an axon (**Fig. 34.10**, step 1), the resulting

FIG. 34.10 A chemical synapse. Neurons communicate with other neurons or with muscle cells by releasing neurotransmitters at a synapse.

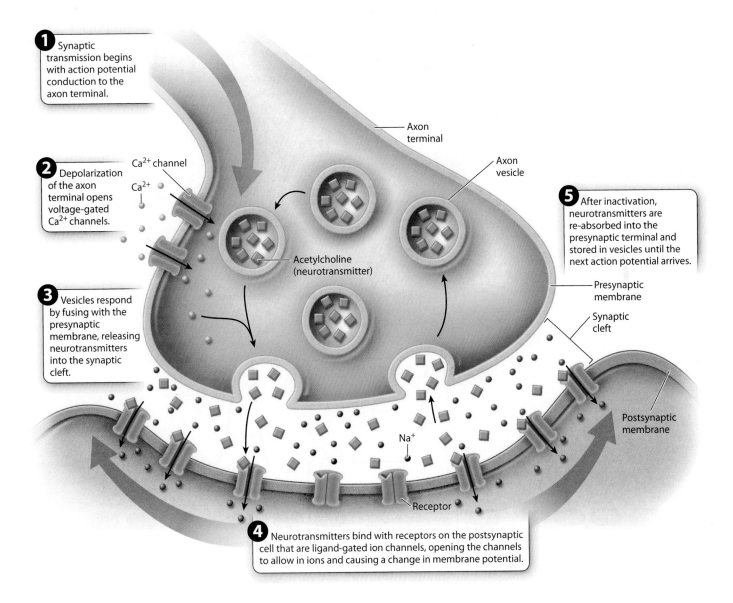

depolarization induces voltage-gated Ca^{2+} ion channels to open (step 2). These channels are found only in the axon terminal membrane. Because of their higher concentration outside the cell, Ca^{2+} ions diffuse through these channels into the axon terminal. In response to the rise in Ca^{2+} concentration, the vesicles fuse with the presynaptic membrane and release neurotransmitter molecules into the synaptic cleft by exocytosis (step 3). The neurotransmitters diffuse rapidly across the cleft and bind to postsynaptic membrane receptors of the neighboring cell. The binding of neurotransmitters opens or closes ion channels, causing a change in the postsynaptic cell membrane potential that allows the signal to propagate along the next neuron (step 4).

These postsynaptic membrane receptors are called **ligand-gated ion channels**, because when the neurotransmitter binds as a ligand, it causes the ion channel to open. Specific ions can then enter or leave the postsynaptic cell, changing its membrane potential. Ligand-gated ion channels underlie the signal transmission that occurs at synapses, regulating the activity of nerve and muscle cells.

The neurotransmitter's effect on the postsynaptic membrane ends shortly after binding. Neurotransmitters become unbound from the receptor, causing the ligand-gated ion channels to close, and neurotransmitters in the synaptic cleft are either rapidly broken down by deactivating enzymes or taken up again by the presynaptic cell and reassembled in newly formed vesicles (step 5). Because the synapse is rapidly cleared of neurotransmitters, the postsynaptic cell is ready to respond to a new action potential signal from the presynaptic cell shortly after responding to the first one.

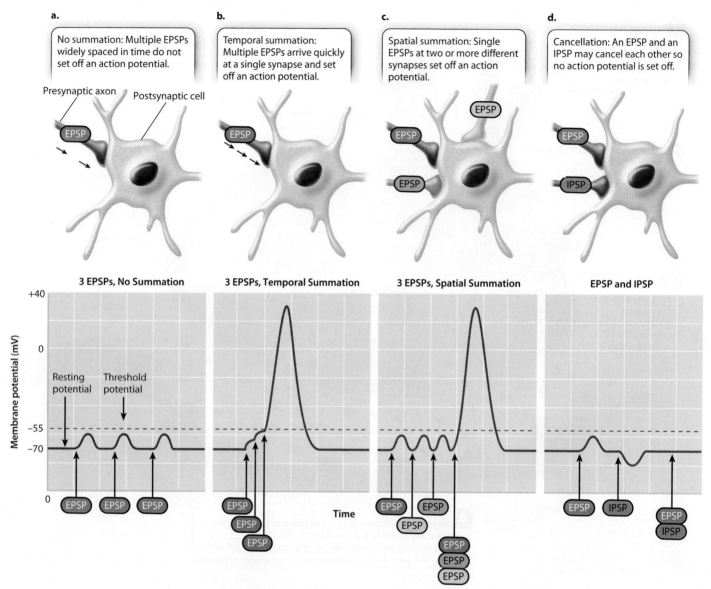

FIG. 34.11 Summation of excitatory and inhibitory postsynaptic potentials. (a) Widely spaced EPSPs do not sum. Other EPSPs and IPSPs can be summed (b) in time or (c) in space, or even (d) cancel each other out.

Signals between neurons can be excitatory or inhibitory.

Neurotransmitters bound to postsynaptic membrane receptors can either stimulate or inhibit the firing of action potentials in the postsynaptic neuron. The binding of some neurotransmitters depolarizes the postsynaptic membrane by opening ligand-gated Na^+ channels. As a result, Na^+ ions diffuse into the cell, making the membrane potential more positive. Such a positive change in membrane potential is termed an **excitatory postsynaptic potential (EPSP)**. Membrane depolarization makes a neuron more likely to respond and transmit an action potential.

In contrast, the binding of other neurotransmitters can make the postsynaptic membrane potential more negative, or hyperpolarized, through the opening of ligand-gated Cl^- or K^+ channels. In this process, Cl^- ions diffuse into the cell or K^+ ions diffuse out of the cell. The negative change in membrane potential is called an **inhibitory postsynaptic potential (IPSP)**. Membrane hyperpolarization inhibits the neuron from sending an action potential.

The postsynaptic nerve cell combines the excitatory and inhibitory synaptic inputs (the summed EPSPs and IPSPs) that it receives through its dendrites and cell body (**Fig. 34.11**). The neuron then fires an action potential only if the sum of synaptic stimulation results in a membrane potential that exceeds its threshold at the axon hillock (Fig. 34.11a).

EPSPs and IPSPs can be summed over time and space. When summed over time, the frequency of synaptic stimuli determines whether the postsynaptic cell fires an action potential (Fig. 34.11b). This is **temporal summation**. When summed over space, the number of synaptic stimuli received from different regions of the postsynaptic cell's dendrites determines whether the cell fires an action potential (Fig. 34.11c). This is **spatial summation**. Sometimes, excitatory and inhibitory signals may cancel each other out (Fig. 34.11d). Temporal and spatial summation of EPSPs and IPSPs are the fundamental forms of information processing carried out by the nervous system. They provide mechanisms for determining whether an organism responds to a particular sensory stimulus or ignores it.

Many different neurotransmitters exist. Although most nerve cells release only one type of neurotransmitter, some release two or more neurotransmitters. The amino acids glutamate (excitatory), glycine (inhibitory), and GABA (inhibitory) are key neurotransmitters that operate at synapses in the brain. The related amino acid derivatives dopamine, norepinephrine, and serotonin are also neurotransmitters acting in the brain. Simple peptides serve as neurotransmitters for sensory neurons. Acetylcholine is one of the key neurotransmitters produced in a variety of nerve cells: this excitatory neurotransmitter is released by motor neurons to cause muscle contraction (Chapter 35).

Excitatory synapses tend to transmit relevant information between nerve cells, and inhibitory synapses often serve to filter out unimportant information. For example, when a gazelle senses a predator, its nervous system transmits key sensory information about the presence of a predator, but filters out irrelevant information such as flies buzzing around its ears. Inhibitory synapses are also important for controlling the timing of muscle activity required for coordinated movement.

Self-Assessment Questions

5. Graph an action potential, showing the change in electrical potential on the *y*-axis and time on the *x*-axis. Indicate on the graph the phases when voltage-gated Na^+ and K^+ ion channels are opened and when they are closed.

6. What is meant by saying that action potentials are "all-or-nothing"?

7. Multiple sclerosis is a disease in which the immune system attacks and destroys the myelin sheath surrounding nerve cells. What effect would you expect the loss of myelin to have on the speed of nerve conduction?

8. How does neurotransmitter binding to receptors on a postsynaptic cell cause inhibition or excitation?

9. Acetylcholinesterase breaks down and inactivates acetylcholine. Some nerve gases used as chemical weapons block acetylcholinesterase. What effect would such nerve gases have on muscle contraction?

34.4 NERVOUS SYSTEM ORGANIZATION

The brain is the body's command center: it receives sensory information from the eyes, ears, and skin; from the antennae and legs of an insect; from the nose of a vertebrate, and from internal organs. Based on that information, it then sends signals to the rest of the body. Up to this point, we have examined how individual neurons send signals to other nerve and muscle cells. How are these neurons organized in the body to sense stimuli, process them, and issue an appropriate response? How does the brain coordinate the movement of limbs necessary to walk or run, and adjust heart rate and breathing in response to exercise? In this section, we look at how the nervous system is organized to allow the brain and the rest of the body to communicate with each other.

Nervous systems are organized into peripheral and central components.

As animals evolved the ability to sense and coordinate responses to increasingly complex stimuli, their nervous systems became organized into peripheral and central components (**Fig. 34.12**). Animals' eyes, receptors for the sense of touch, and other sensory organs are located on the surface of the body, where they can receive signals from the environment. In contrast, the brain and a main nerve cord are located in the interior. Among multicellular animals with nervous systems, only the diffuse nerve nets of cnidarians lack central and peripheral components.

Nerves form the lines of communication between these nervous system structures. Nerves are composed mainly of axons and dendrites from many different nerve cells. For example, the optic nerve contains axons that travel from nerve cells in the eye to the brain.

Nerves connecting to sensory organs and muscles make up the **peripheral nervous system (PNS)**. These nerves communicate with the **central nervous system (CNS)**, which is made up of the brain, a main nerve cord, or—in the case of animals such as flatworms that lack a brain—centralized ganglia. Sensory organs at or near the body surface transmit information by afferent neurons, which send information *toward* the CNS. After processing the information, the CNS sends signals through efferent neurons, which send information *away* from the CNS. The signals delivered by efferent neurons in peripheral nerves coordinate the activity of muscles and glands in different regions of the animal's body.

The central nervous system of vertebrates includes the brain and **spinal cord** (Fig. 34.12). The spinal cord is a central tract of neurons that passes through the vertebrae to transmit information between the brain and the periphery of the body. The vertebrate spinal cord is divided into segments, each of which controls body movement in a particular region along the animal's length. Each spinal cord segment contains axons from peripheral sensory neurons, a set of interneurons, and a set of motor neuron cell bodies.

FIG. 34.12 Human central (yellow) and peripheral (blue) nervous systems.

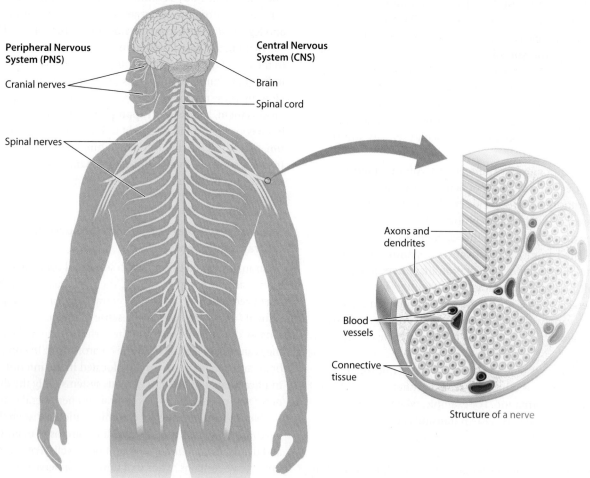

In humans and other vertebrate animals, the peripheral nervous system is organized into left and right sets of **cranial nerves** located within the head, and **spinal nerves** running from the spinal cord to the periphery. Most of the cranial nerves and all of the spinal nerves contain axons of both sensory and motor neurons. Cranial nerves link specialized sensory organs (eyes, ears, taste buds on the tongue) to the brain. Cranial nerves also control eye movement, facial expression, speech, and feeding. Spinal nerves exit from the spinal cord through openings located between adjacent vertebrae and then thread their way through the trunk and limbs of an animal's body. These nerves receive sensory information from receptors in nearby body regions along the length of the body and send motor signals from the spinal cord back to those regions (Fig. 34.12).

Peripheral nervous systems have voluntary and involuntary components.

As bodies with distinct internal organ systems evolved, two separate components of the peripheral nervous system emerged. When a gazelle senses a predator, some nerve circuits send a signal to run, an action that is under conscious control. Other nerve circuits signal the heart to beat faster and blood vessels supplying muscles to dilate, actions that occur unconsciously. Conscious reactions are under the control of the **voluntary component** of the peripheral nervous system, and unconscious ones are under the control of the **involuntary component**. Voluntary components mainly handle sensing and responding to external stimuli, whereas involuntary components typically regulate internal bodily functions. Both nervous system components are found in invertebrate and vertebrate animals.

In insects and crustaceans, for example, nerve circuits from the involuntary component regulate glands and muscles located in the animal's digestive system. Nerve circuits connecting sensory structures such as the antennae and eyes to the animal's brain, and connecting the brain to muscles, are part of the voluntary system.

In vertebrates, the peripheral nervous system is divided into **somatic** (voluntary) and **autonomic** (involuntary or visceral) components. The somatic nervous system is made up of sensory

FIG. 34.13 The human autonomic nervous system. The autonomic system, which controls involuntary functions, is divided into sympathetic and parasympathetic systems.

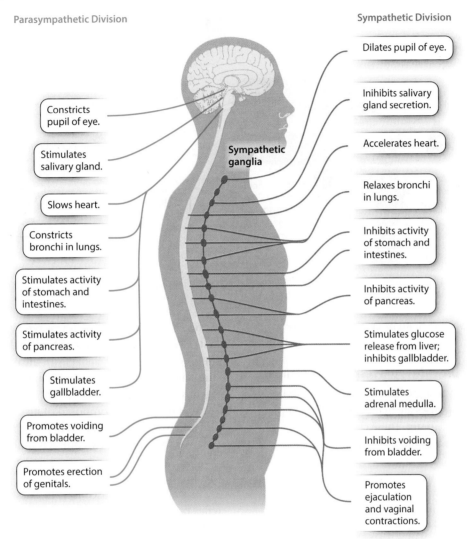

neurons that respond to external stimuli and motor neurons that synapse with voluntary muscles. This system is considered voluntary because it is under conscious control. The autonomic nervous system (**Fig. 34.13**) controls internal functions of the body such as heart rate, blood flow, digestion, excretion, and temperature. Thus, it is critical to the maintenance of homeostasis. The autonomic nervous system includes both sensory and motor nerves, which usually act without our conscious awareness.

The autonomic nervous system is organized into **sympathetic** and **parasympathetic** divisions. Both divisions continuously monitor and regulate internal functions of the body. Typically, sympathetic and parasympathetic nerves have opposite effects. For example, sympathetic neurons stimulate the heart to beat more rapidly, whereas parasympathetic neurons cause the heart to beat more slowly.

The sympathetic nervous system generally causes an increase in alertness and activity. This pathway is activated when animals are exposed to threatening conditions, resulting in the "fight or flight" response. As part of this response, the sympathetic nervous system increases heart rate and breathing rate, stimulates the liver to release glucose, and inhibits digestion. In contrast, the parasympathetic nervous system slows the heart and stimulates digestion as well as metabolic processes that store energy; it enables the body to "rest and digest."

In addition to having different functions, sympathetic and parasympathetic nerves are distributed in different regions of the central nervous system, as shown in Fig. 34.13. Sympathetic nerves leave the CNS from the middle (thoracic and lumbar) region of the spinal cord, forming ganglia along much of the length of the spinal cord that process information and coordinate activity within a local region of the body. Parasympathetic

FIG. 34.14 The knee-extension reflex. The knee-extension reflex involves a sensory neuron, a single synapse, and a motor neuron.

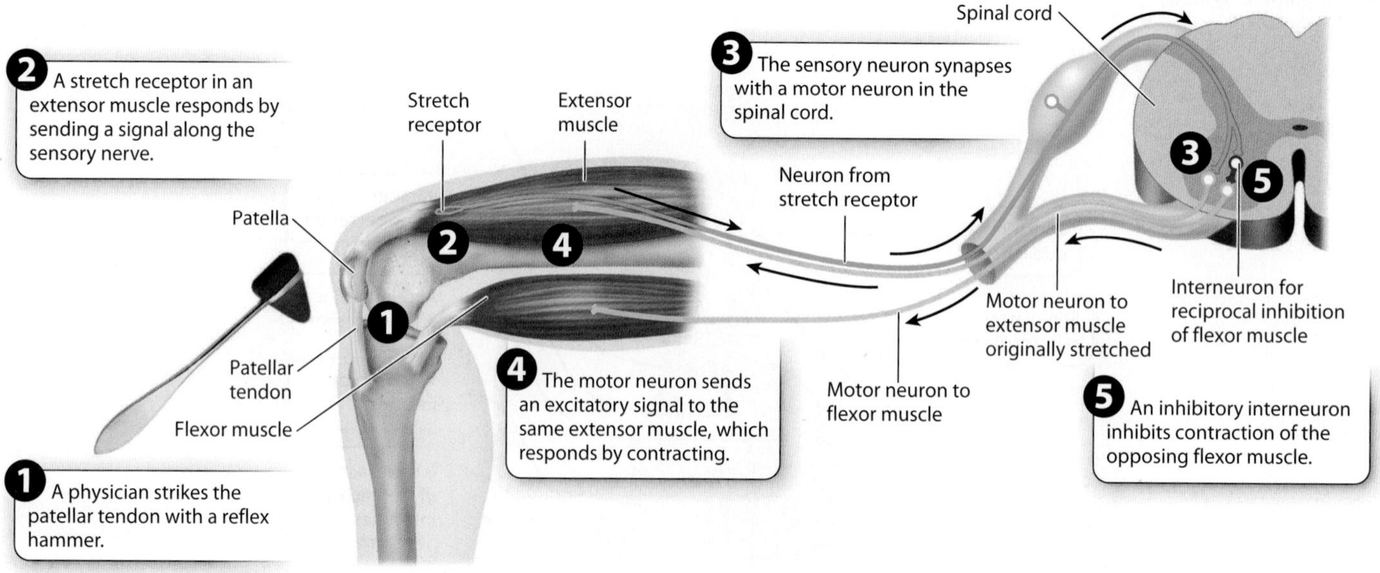

❶ A physician strikes the patellar tendon with a reflex hammer.

❷ A stretch receptor in an extensor muscle responds by sending a signal along the sensory nerve.

❸ The sensory neuron synapses with a motor neuron in the spinal cord.

❹ The motor neuron sends an excitatory signal to the same extensor muscle, which responds by contracting.

❺ An inhibitory interneuron inhibits contraction of the opposing flexor muscle.

nerves leave the CNS from the brain by cranial nerves and from lower (sacral) levels of the spinal cord.

Simple reflex circuits provide rapid responses to stimuli.

An animal that has perceived a predator has an advantage if it can move quickly. Fast responses are made possible by simple reflex circuits that bypass the brain. These circuits connect sensory neurons (detecting a stimulus) directly with motor neurons (providing a quick response to the stimulus). Reflex circuits are common in both the voluntary and involuntary components of the nervous system.

Simple spinal reflex circuits connect sensory and motor neurons directly in the spinal cord of vertebrates. The **knee-extension reflex** in humans is an example of a simple nerve circuit that includes only a single synapse between two neurons—a sensory neuron and a motor neuron (**Fig. 34.14**). Physicians commonly use this reflex to evaluate the performance of the peripheral nervous and muscular systems. The strike of the physician's hammer activates the extensor muscles on the front of the thigh, including the quadriceps muscle, causing the bent leg to straighten.

This pathway starts with specialized "stretch" receptors located on the dendrites of a sensory neuron in the extensor muscles of the leg. These dendrites extend to cell bodies in ganglia alongside the spinal cord. Axons extend from these cell bodies into the spinal cord. The stretch receptors sense the stretch of the muscle that occurs during movement or in response to a physician's strike of a reflex hammer. In response to a stretch, a signal is sent from the stretch receptor, through the dendrite and cell body, to the axon. In the spinal cord, the axon of the sensory neuron forms synapses with motor neurons that travel from the spinal cord back to the muscle where the stretch originated. The signal from the muscle stretch receptor stimulates the motor neurons: the muscle contracts, and the leg extends at the knee.

This reflex arc is composed of just two neurons and one synapse. Transmission is delayed by communication at synapses, but the muscle contracts quickly because just one synapse is required to relay the sensory information back to the muscle.

As well as being a useful medical test, the knee-extension reflex has a normal physiological role. In running or landing from a jump, the knee joint first flexes (bends), stretching the quadriceps muscles. This action sends the same sensory signal to the spinal cord as when a physician's hammer taps the muscle's tendon. As a result, after flexing, the knee reflexively extends (straightens). Knee flexion followed by extension allows rapid adjustments in muscle force and knee position that help stabilize the body during running and jumping.

Because muscles only contract, opposing movements such as flexion and extension require flexor and extensor muscles on either side of the joint (Chapter 35). These are often activated out of phase (at different times), so that when one set of muscles is activated (contracting), the other is inhibited (relaxed). The alternating extension and flexion of a leg during walking and running provide an example. This pattern of joint and limb movement is achieved by **reciprocal inhibition**: when stretch receptors of the knee extensor muscles are activated, they also inhibit opposing flexor muscles (Fig. 34.14). Reciprocal inhibition is possible because axons of the stretch receptor neurons synapse not only with motor neurons of extensor muscles, but also with interneurons that inhibit motor neuron stimulation of flexor muscles. This inhibitory pathway contains two synapses: one synapse between the sensory neuron and interneuron, and a second synapse between the interneuron and the motor neuron to the flexor muscle.

Self-Assessment Questions

10. Which functions of an animal are controlled by the voluntary component of the nervous system and which by the involuntary component of the nervous system?
11. Diagram a simple circuit that includes a sensory neuron that synapses with a motor neuron to produce a reflex. Indicate where in the nervous system this synapse is found.
12. Why is an interneuron needed to provide reciprocal inhibition of the flexor muscle when an extensor muscle is activated?

34.5 SENSORY SYSTEMS

To see, your eyes detect light, which is a form of electromagnetic radiation. To hear, your ears detect sound waves. To smell, your nose detects odor molecules present in the air. In all these cases, your sense organs detect a physical or chemical stimulus in the environment. In this section, we explore how specialized sensory cells detect these signals and code them as information that can be transmitted and processed by the nervous system.

Sensory receptor cells detect diverse stimuli.

Animals can sense many physical properties of their environment, including light, chemicals, temperature, pressure, and sound. These perceptions are useful to help animals find mates and food, and avoid predators and noxious environments. Early in evolutionary history, organisms evolved specialized protein receptors in the cell membrane that were able to detect these critical features in their environment. For example, bacteria evolved membrane receptors that sensed osmotic pressure that might otherwise rupture their cell membrane. Bacteria also evolved membrane receptors that detect chemicals, including nutrients, in their environment. Single-celled eukaryotes and sponges have these chemical-sensing membrane receptors as well.

Likewise, the senses of smell, taste, and sight in multicellular organisms with a nervous system rely on membrane receptors. These receptors are embedded in specialized **sensory receptor cells**. These cells either communicate with neurons, as for taste and sight, or are themselves neurons, as for smell. (The term "sensory receptor" refers to the entire sensory cell or sensory neuron, not just the membrane protein.)

In most multicellular animals, sensory receptors are organized into specialized **sensory organs** that convert physical or chemical stimuli into signals that are communicated to the brain. Cnidarians (including jellyfish, corals, and anemones) and roundworms (including the laboratory organism *Caenorhabditis elegans*) evolved sensory receptors that sense physical contact. Cnidarians and flatworms were among the first animals to evolve simple light-sensing organs.

Sensory transduction converts a stimulus into an electrical impulse.

The conversion of physical or chemical stimuli into nerve impulses is called **sensory transduction**. For example, receptors located in the ear convert the energy of sound waves into nerve impulses that allow an animal to distinguish loud versus soft sounds and high-pitched versus low-pitched sounds. Although the sense organs of different animals share many similar properties, some differences between them have also evolved. Consequently, different animals perceive the world differently. Many insects, for instance, are sensitive to ultraviolet light; nocturnal snakes can see at night by sensing infrared radiation; and dogs can distinguish odor compounds at concentrations as much as 100 million times lower than the concentration than humans can detect.

How is a physical phenomenon in the world outside an organism transformed into a nerve impulse? The initial transformation takes place inside the sensory receptor (**Fig. 34.15**). In many cases, the sound wave, touch, or other stimulus causes ion channels in the cell's plasma membrane to open. In most cases, the influx of ions depolarizes the membrane potential by altering the distribution of charged ions on either side of the membrane. That depolarization in membrane potential is the first step in firing an action potential. Sensory receptors either fire action potentials themselves or synapse with neurons that fire action potentials, which are then transmitted to the central nervous system.

Let's look at examples of how three different types of stimuli from the environment—chemicals, physical changes, and light—can be transformed into a nerve signal.

Chemoreceptors respond to chemical stimuli.

The most ancient type of sensory detection is chemoreception. All organisms respond to chemical cues in their environment. Even bacteria and archaea have protein receptors in their cell membranes that respond to molecules in the environment. Chemoreception underlies the senses of smell (**olfaction**) and taste (**gustation**).

In animals, **chemoreceptors** are sensory receptor cells that respond to molecules that bind to protein receptors on the cell membrane (Fig. 34.15a). Many animals detect food in their environment by sensing key molecules such as oxygen

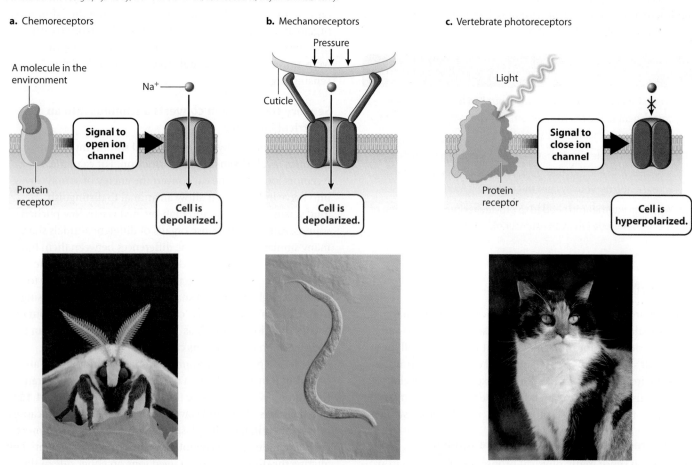

FIG. 34.15 Sensory transduction, which converts an external stimulus into a change in membrane potential. (a) Chemoreceptors are located in the antennae of the luna moth; (b) mechanoreceptors below the cuticle of a roundworm; and (c) photoreceptors in the eyes of a vertebrate. *Photo sources: a. Rolf Nussbaumer Photography/Alamy; b. Sinclair Stammers/Science Source; c. Eyal Nahmias/Alamy.*

(O_2), carbon dioxide (CO_2), glucose, and amino acids. Female mosquitoes track CO_2 levels to locate prey for blood meals, and coral polyps respond to simple amino acids in the water, extending their bodies and tentacles toward areas of greater concentration to feed. Other arthropods, such as flies and crabs, have chemosensory hairs on their legs and feet. These animals taste potential food sources by walking on them. Salmon detect chemical traces of the home waters of the river where they hatched, and where they will return to mate and spawn.

The sense of smell in mammals depends on specialized sensory neurons that line the nasal cavity (**Fig. 34.16**). These neurons have long, hairlike extensions containing membrane receptors that project into the nasal mucus. Odor molecules are captured by the mucus during inhalation and bind to membrane receptors. Each sensory receptor has one type of membrane receptor, which is able to detect one type of molecule, but there are typically many distinct sensory receptors, each detecting a different molecule. Mice have approximately 1500 genes that express odorant receptor proteins (about 7% of their genome). The olfactory system can detect specific odor molecules in quite low concentrations.

When bound to odor molecules, the sensory receptors produce excitatory postsynaptic potentials (EPSPs). If enough odor molecules bind to the receptor cell, the EPSPs are summed and trigger an action potential. That action potential travels down the axon to an interneuron in the olfactory bulb that receives information only from sensory receptors sensitive to a particular odor molecule. In Fig. 34.16, sensory neurons that bind the identical odor molecules shown in the nasal passage all synapse on the middle interneuron. The interneurons in the olfactory bulb, in turn, transmit the combined information from a number of chemoreceptors to the brain. These neurons communicate directly with the brain through the olfactory nerve, one of the cranial nerves.

FIG. 34.16 The senses of smell and taste. Olfactory receptors respond to chemical odors in the nasal passages. Taste cells are chemosensory receptors located in taste buds.

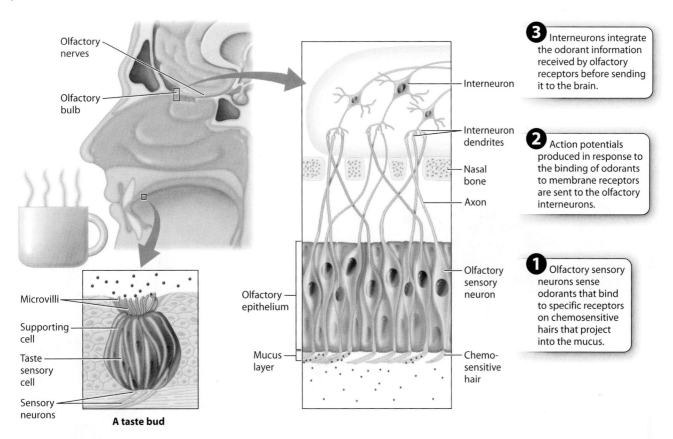

Clusters of chemosensory receptor cells are also located in specialized **taste buds**, the sensory organs for taste (Fig. 34.16). Some fishes have taste buds in their skin that sense amino acids in water, helping them to locate food. The taste buds of terrestrial vertebrates are found in the mouth; most of the taste buds in mammals are on the tongue. The human tongue has approximately 10,000 taste buds, which are contained within raised structures called papillae, giving the tongue its rough surface. Fingerlike projections called microvilli extend through a pore in each taste bud. These microvilli contact food items and provide a large surface area that is rich in membrane receptors.

In most cases, the binding of a molecule to a taste membrane receptor causes the receptor to change conformation, which in turn triggers the opening of Na^+ channels through G protein signal transduction pathways (Chapter 9). The influx of Na^+ ions depolarizes the receptor cell. Although no action potential fires, the depolarization travels far enough down the receptor's short axon to trigger the release of neurotransmitters. Taste bud sensory cells synapse with neurons that send axons to the brain. These neurons fire action potentials when molecules from a bite of food depolarize a sufficient number of sensory cells.

Mechanoreceptors detect physical forces.

Mechanoreceptors respond to physical deformations of their membrane produced by touch, stretch, pressure, motion, and sound. Mechanoreceptors are found in all multicellular animals. Even bacteria have pressure-sensitive protein receptors in their cell membrane.

In both bacteria and multicellular organisms, the protein receptor is a sodium ion channel. Thus, deformation of the receptor membrane opens sodium channels directly, causing the endings of the cell's dendrites to depolarize (see Fig. 34.15b). Pressure-sensitive protein receptors in bacteria sense internal cell pressure: when water moving into the cell by osmosis increases the risk of bursting (Chapter 5), the stretching of the membrane opens channels that let water leave the cell. Other mechanoreceptors in roundworms and anemones are linked to externally projecting cilia that sense forces at the animal's body surface.

One well-studied example of a mechanoreceptor is a touch receptor in the roundworm *Caenorhabditis elegans*. Roundworms, also called nematodes, are one of the most diverse and widespread groups of animals. Deformation of the surface of the worm exerts pressure on proteins connected to an ion channel. The mechanical force changes the shape of the ion

FIG. 34.17 Sensing gravity and motion. (a) The statocyst of invertebrates detects gravity and orientation by means of a statolith that deflects hair cells. (b) The vestibular system of vertebrates senses gravity by two otolith organs and senses angular rotations of the head by movements of fluid that deflect hair cells within three semicircular canals.

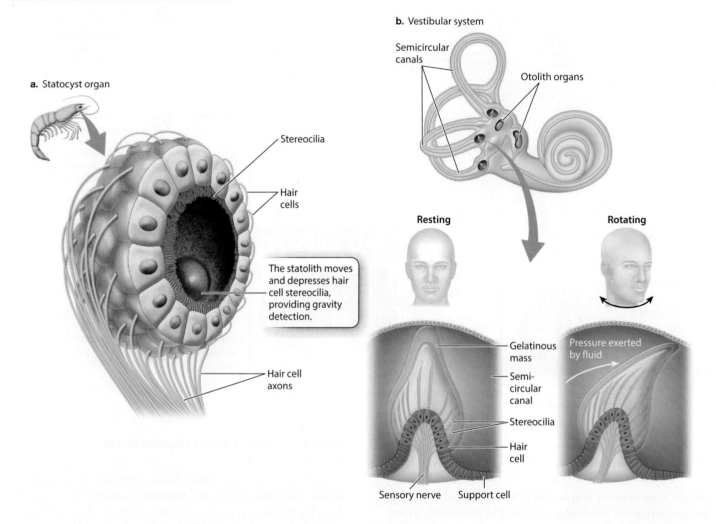

channel, causing it to open and produce a change in membrane potential.

One mechanoreceptor in humans and other mammals is the sensory receptor found in the skin that senses touch and pressure. The cell bodies of these neurons are located in ganglia near the spinal cord, with extensions going in two directions: to the skin and to the spinal cord. In the skin, these neurons have branched tips containing ion channels sensitive to deformations of the membrane. These branched tips act as the initial sensors of touch and pressure. If the stimulus is strong enough, depolarization at the tips leads to the firing of an action potential that travels to the spinal cord.

A very different group of mechanoreceptors called **hair cells** provide the ability to orient the animal's body with respect to gravity, detect motion, and hear. The hair cells of fishes and amphibians detect movement of the surrounding water, and those of many invertebrates sense gravity and other forces acting on the animal. Hair cells are also found in the ears of terrestrial vertebrates, where they sense body orientation, motion, and sound.

In all these cases, hair cells sense mechanical vibrations. Such vibrations move projections from the surface of the hair cell called **stereocilia**. In turn, the motion of stereocilia causes a depolarization of the cell's membrane by opening or closing ion channels. Hair cells themselves do not fire action potentials. Instead, when they become depolarized, the cells release neurotransmitters that alter the firing rate of adjacent neurons.

Unlike chemoreceptors, hair cells are generally similar in structure; there is not a specific hair cell for a particular body orientation or pitch of sound. Thus, for hair cells to provide information about mechanical stimuli, they must interact with structures of the sensory organ in which they reside. These structures influence how individual hair cells respond to body movement and sound waves.

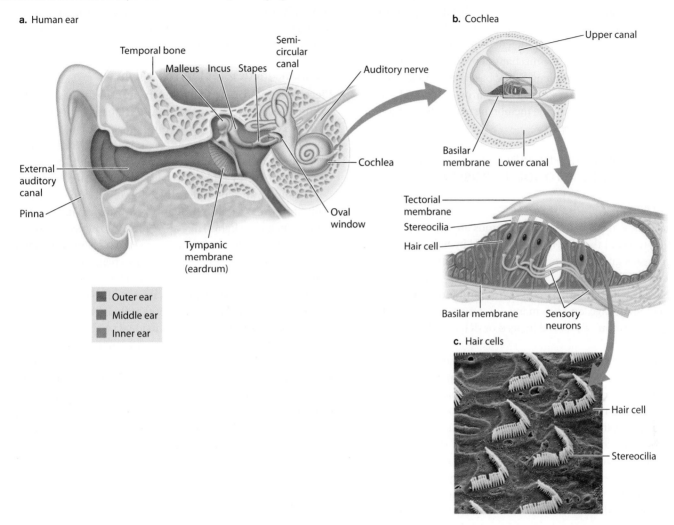

FIG. 34.18 The human ear. (a) The tympanic membrane and bones of the middle ear transmit sound waves to the inner ear. (b) Hair cells with stereocilia in the cochlea respond to the vibrations, sending signals to the brain. *Photo source: SPL/Science Source.*

For example, gravity-sensing organs called **statocysts** are found in most invertebrates (**Fig. 34.17a**). The statocyst is a chamber lined by hair cells with stereocilia that project into the chamber. Small granules of sand or other material form a dense particle called a **statolith** that is free to move within the statocyst organ. By pressing down on hair cells at the "bottom" of the chamber, the statolith activates those cells.

The mammalian inner ear contains organs that sense motions of the head and its orientation with respect to gravity. These organs make up the **vestibular system** (**Fig. 34.17b**), which consists of two otoliths and three **semicircular canals**. The otoliths are similar to the statocysts of invertebrates, providing a sense of gravity and body orientation. Hair cells in the semicircular canals sense angular motions of the head in three perpendicular planes, providing a sense of balance. When the head rotates, gelatinous fluid in the semicircular canals is accelerated, deflecting the stereocilia of the hair cells to activate sensory neurons. The brain interprets differences in the motion of the hair cells among the three semicircular canals to resolve angular motions of the head.

Hair cells in the ear also allow mammals to detect sound. What our ears interpret as sound are actually pressure waves traveling through the air. **Fig. 34.18** shows the structure of the human ear. The pinna (external ear), auditory canal, and tympanic membrane (eardrum) make up the outer ear (Fig. 34.18a). Sound waves enter the outer ear and lead to vibrations of the tympanic membrane. The middle ear contains three small bones: the **malleus**, **incus**, and **stapes**. These bones amplify the vibrations of the tympanic membrane and transmit them to a thin membrane (oval window) of the cochlea in the inner ear.

The **cochlea** (which means "snail" in Latin) is a coiled chamber in the skull (Fig. 34.18b). It contains two fluid-filled canals separated by a basilar membrane that supports hair cells (Fig. 34.18c). Stereocilia of the hair cells project into a rigid tectorial membrane. Vibrations of the oval window lead to fluid waves in the cochlea. These waves induce motions of the basilar membrane, bending the stereocilia and stimulating the hair cells to release excitatory neurotransmitters. The signal is sent by the auditory nerve to the brain where it is interpreted as sound.

The ability to discriminate different sound frequencies results from differences in the mechanical properties of the basilar membrane along its length. The basilar membrane is wide, thin, and flexible at one end, and narrow, thick, and stiff at the other end. As a result, the membrane responds to different frequencies in different regions.

CASE 7 BIOLOGY-INSPIRED DESIGN: USING NATURE TO SOLVE PROBLEMS

How do cochlear implants work?

Our ears allow us to detect sound. Sounds are pressure waves traveling through the air that are converted by hair cells into an electrical signal. Many people, however, are unable to detect sounds. Hearing loss, or deafness, affects millions of people worldwide. Deafness ranges from mild to profound, can occur at any age, and has many different causes. One of the most common causes is damaged or defective hair cells in the cochlea.

To make hearing possible for some deaf persons, scientists have developed cochlear implants. Cochlear implants are small neuro-prosthetic devices that bypass damaged hair cells in the cochlea. They work by electrically stimulating the auditory nerve, which transmits these signals to the brain. Sounds are received by an external microphone worn behind the ear (like a hearing aid) and then decoded by a speech processor that analyzes the sounds. The processor transmits the decoded signals to a receiver that is implanted under the skin. The receiver then sends these signals to electrodes that stimulate the auditory nerve. In this way, the microphone, speech processor, and electrodes mimic the function of the outer, middle, and inner ear that converts sound waves into electrical signals.

The cochlear implant allows deaf individuals to detect sounds, but the quality of the sounds is different from that of normal sounds. As a result, these individuals have to learn to make sense of the sounds they hear. Furthermore, persons who have never heard sounds need to learn how to interpret the sounds they are hearing for the first time. Thus, after undergoing surgery for a cochlear implant, a person must work with audiologists, counselors, and speech-language pathologists over several years. In addition, scientists continue to develop technology to improve the implants.

The cochlear implant is not without controversy. Many people in the Deaf community view deafness as a culture, with a unique identity, language (sign language), and history. From this perspective, the cochlear implant is a threat. Most deaf children have hearing parents, so decisions about inserting a cochlear implant in a child are often made by people with little or no knowledge of the Deaf community. At the same time, cochlear implants are more effective in younger patients, as the ability to hear and speak depends on early exposure to sounds. Some advocate for sign language to be taught alongside spoken language for children who receive an implant so they have the benefit of multiple forms of communication.

Electromagnetic receptors sense light.

Electromagnetic receptors are sensory cells that respond to electrical, magnetic, and light stimuli. Of these, light-detecting **photoreceptors** are the most common and diverse types (see Fig. 34.15c). Most animals sense light in their environment. They rely on the same light-sensitive protein, called **opsin**, to convert light energy into electrical signals in the receptor cell.

Opsins are ancient molecules that evolved early in animal evolution. These receptor proteins are sensitive to many different stimuli, including light. They are G protein-coupled receptors that activate G proteins, leading to a cellular response (Chapter 9). In this case, the cellular response is a change in membrane potential. The observation that different types of animal eye use a common light-sensitive protein suggests a common evolutionary origin for animal eyes about 500 million years ago.

Photoreceptors respond to individual photons of light energy. In vertebrates, for example, opsin molecules are arranged in cylindrical groups in the plasma membrane of photoreceptor cells. Each opsin protein contains a light-absorbing pigment molecule called **retinal**. Retinal is a derivative of vitamin A that changes conformation when it absorbs a photon (**Fig. 34.19**).

In most invertebrates, light causes Na^+ channels to open, depolarizing the cell. By contrast, vertebrate photoreceptor cells have leaky Na^+ channels that let some Na^+ ions into the cell even at rest. As a result, their resting membrane potential (in the dark) is less negative (about –35 mV) than that of nerve cells. When a photoreceptor is in the dark and its resting potential is –35 mV, it continuously releases the neurotransmitter glutamate. However, when retinal absorbs a photon of light, it undergoes a conformational change from a *cis* to a *trans* configuration, causing Na^+ channels to close. Without Na^+ ions entering the cell, the cell membrane becomes hyperpolarized and the cell's release of glutamate is reduced. The reduction in neurotransmitter triggers EPSPs in the dendrites of some connecting neurons, and IPSPs in others. Differences in the firing rates of these neurons provide information about the intensity and location of light.

Although opsins are the universal photoreceptor protein among animals, the way the visual world is perceived—what it looks like to the organism—depends in part on the structure of the eye in which the receptors are embedded. Here, we look at the three types of eyes shown in **Fig. 34.20**: **eyecups**, **compound eyes**, and **single-lens eyes**.

In flatworms, two eyecups on the back surface of the head detect the direction and intensity of light sources (Fig. 34.20a). Each eyecup contains photoreceptors that point up and to

FIG. 34.19 Opsin bound to the light-absorbing pigment retinal. Retinal changes conformation from *cis* to *trans* when it absorbs a photon of light. *Rendered by Dale Muzzey.*

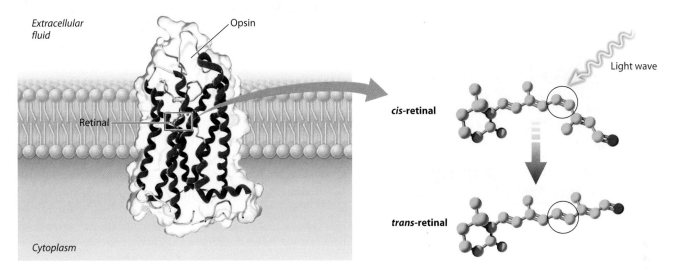

the left or right. In response to light, flatworms turn to move directly away from the light source, seeking a dark region that hides them from potential predators. To find a dark region, the flatworm's nervous system compares the light intensity received by photoreceptors of the two eyecups; the organism then moves in the direction of lower light intensity. Many invertebrates use simple light-sensitive photoreceptors in this way.

A key innovation for detecting images was the evolution of a lens. Two major types of image-forming eyes based on lenses evolved in other invertebrates: compound eyes and single-lens eyes. Insects and crustaceans have compound eyes that consist of individual light-focusing elements, each having a lens, called **ommatidia** (singular, **ommatidium**) (Fig. 34.20b). The number of ommatidia in the eye determines the resolution, or sharpness, of the image. The compound eyes of ants have only a few ommatidia, and those of fruit flies have about 800. In contrast, the eyes of dragonflies have 10,000 or more. Dragonflies are predators, and their high number of ommatidia is an adaptation for detecting motion and visually tracking their prey.

FIG. 34.20 Three types of eyes. *Photo sources: a. David M. Dennis/AGE Fotostock; b. Eye of Science/Science Source; c. PeopleImages.com/Getty Images.*

a. Eyecup

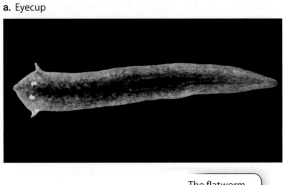

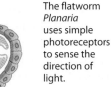

The flatworm *Planaria* uses simple photoreceptors to sense the direction of light.

b. Compound eye

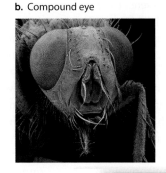

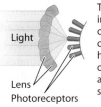

The compound eye of insects, such as the common housefly, are composed of hundreds of ommatidia, each with a lens, that individually sense light.

c. Single-lens eye

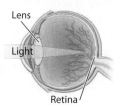

The single-lens eye of a human focuses light on a retina and allows for a high degree of acuity.

FIG. 34.21 Vertebrate photoreceptors: rods and cones. *Photo source: Omikron/Science Source.*

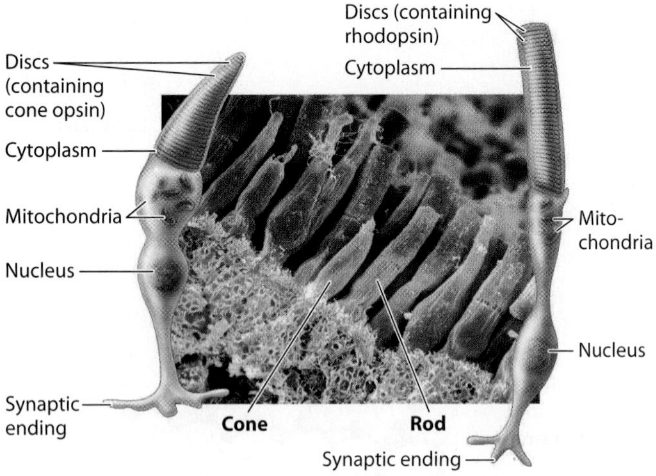

The single-lens eyes of vertebrates and cephalopod mollusks like squid and octopuses produce sharply defined images (Fig. 34.20c). In spite of their similar outward appearance, single-lens eyes evolved independently in vertebrates and in cephalopod mollusks. An advantage of single-lens eyes compared to compound eyes is that the single lens can focus light rays on a particular region of photoreceptors, improving image quality and light sensitivity.

To form an image, light is focused on a thin, light-sensitive tissue called the **retina** located in the back of the eye. The retina contains the photoreceptor cells that sense light stimuli and other nerve cells that initially process those stimuli. Vertebrates have two types of photoreceptor cells: **rod cells** and **cone cells** (**Fig. 34.21**). Rod cells are sensitive to blue-green light. Because all rod cells have the same opsin (called rhodopsin), the brain interprets the light detected by rod cells as shades of gray, not as blue-green. Rod cells are more numerous and sensitive than cone cells, so they allow animals to see in low light. Cone cells contain opsins sensitive to different wavelengths of light and are responsible for color vision.

Cone cells likely evolved from a rod cell precursor. Most non-mammalian vertebrates have four types of cone cells, each with a different opsin. In contrast, most mammals have two types of cone cells, and two opsins. It is likely that two opsins were lost early in mammalian evolution, when mammals were small, nocturnal, and burrowing. Old World primates, apes, and humans, as well as some New World primates, regained trichromatic (three-color) vision by means of gene duplication, likely in response to selection for better ability to locate fruit, a key part of their diet. Each human cone cell therefore has one of three opsins, which absorb light at blue, green, or red wavelengths. Stimulation of cone cells in varying combinations of these three opsins allows humans and other primates to see a full range of color.

Self-Assessment Questions

13. What are examples of chemosensory, mechanosensory, and electromagnetic sensory cells?
14. Diagram a generalized sensory receptor cell and show how it changes its firing rate in response to a stimulus.
15. State a hypothesis that explains why all animals have chemoreceptors.
16. Why are you unstable when you try to walk after you spin in place or take a ride on a merry-go-round?
17. Compare and contrast the roles of rod cells and cone cells in the retina.

34.6 BRAIN ORGANIZATION AND FUNCTION

Up to this point, we have looked at how sensory information, such as smell, taste, hearing, and vision, is detected by sensory receptors. In each case, the sensory information is sent to the brain, where it is processed. Much of what we know about how the brain functions has been learned over many years through studies of patients who suffered brain injuries and through studies in model organisms. The remainder of this chapter explores the functional organization of the vertebrate brain.

The brain processes and integrates information received from different sensory systems.

The vertebrate brain is organized into a **hindbrain**, **midbrain**, and **forebrain**, with a **cerebral cortex** formed from a portion of the forebrain (**Fig. 34.22a**). The hindbrain and midbrain control basic body functions and behaviors, whereas the forebrain, particularly the cerebral cortex, governs more advanced cognitive functions. The cerebral cortex of mammals—and particularly of primates and humans—is greatly expanded relative to that of other vertebrates.

In adult vertebrates, the hindbrain develops into the **cerebellum** and a portion of the **brainstem** (Fig. 34.22a). The remainder of the brainstem develops from the midbrain. The cerebellum coordinates the body's actions for complex motor tasks, such as catching a ball or writing. It integrates both motor and sensory information. The brainstem, consisting of the **medulla**, **pons**, and midbrain, initiates and regulates motor functions such as walking and controlling posture, and coordinates breathing and swallowing. The brainstem activates the forebrain by relaying information from lower spinal levels.

FIG. 34.22 Human brain development. (a) The brain develops from the forebrain, midbrain, and hindbrain. (b) Part of the forebrain develops into the limbic system, which controls drives and emotions.

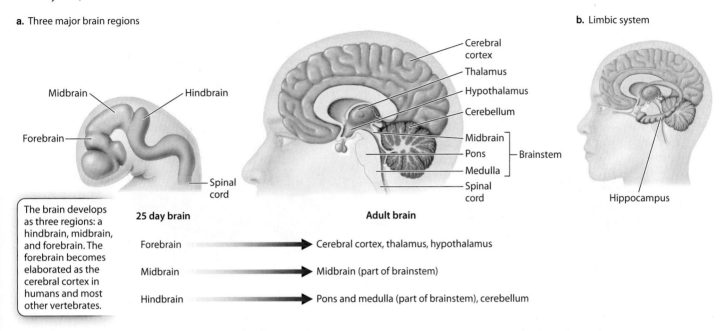

The brain is divided into lobes with specialized functions.

The cerebral hemispheres are the largest structures of the mammalian brain (**Fig. 34.23**). They contain a highly folded outer layer of **gray matter** that forms the cerebral cortex, which is made up of densely packed neuron cell bodies and their dendrites (Fig. 34.23a). The folds greatly increase the surface area of the cortex and are the result of selection for an increased number of cortical neurons within the limited volume of the skull. Deep inside the cerebral cortex is the **white matter**, which contains the axons of cortical neurons. The fatty myelin produced by glial cells surrounding the axon makes this region of the brain white. Neurons in different regions of the gray matter communicate with each other and with deeper regions of the forebrain, midbrain, and hindbrain.

Major regions of the cerebral hemispheres are defined by clearly visible anatomical lobes separated in most cases by deep crevices called **sulci** (singular, **sulcus**) (Fig. 34.23b). The **frontal lobe** is located in the front region of the cerebral cortex (behind your forehead). It is important in decision making and planning. The **parietal lobe** is located behind the frontal lobe and is separated from the frontal lobe by the central sulcus. The parietal lobe controls body awareness and the ability to perform complex tasks, such as animal courtship or human dressing. The **temporal lobe** lies below the parietal lobe. It is involved in processing sound, as well as performing other functions

The forebrain consists of an inner brain region that forms the **thalamus** and the underlying **hypothalamus**, and a more anterior region that develops into the **cerebrum**, the outer left and right hemispheres of the cerebral cortex (Fig. 34.22a). The thalamus is a central relay station for sensory information sent to higher brain centers of the cerebrum. The hypothalamus interacts closely with the autonomic and endocrine systems to regulate the general physiological state of the body. In humans and most other primates, the cerebral cortex is the largest part of the brain, overseeing sensory perception, memory, and learning.

Other inner components of the forebrain constitute the **limbic system**, which controls physiological drives, instincts, emotions, and motivation and, through interactions with midbrain regions, the sense of reward (**Fig. 34.22b**). Stimulation of the limbic system can induce strong sensations of pleasure, pain, or rage. A posterior region of the limbic system, the **hippocampus**, is involved in long-term memory formation, discussed in the next section.

Sensory information reaches the cerebral cortex from the cranial nerves and nerves passing through the spinal cord. This information passes first through the brainstem, and then through the thalamus, the central relay station for sensory information. From the thalamus, information for each of the senses goes to a different region of the brain specialized for further processing that kind of information.

FIG. 34.23 Human brain organization. (a) The folding of the brain and (b) the lobes of the cerebrum.

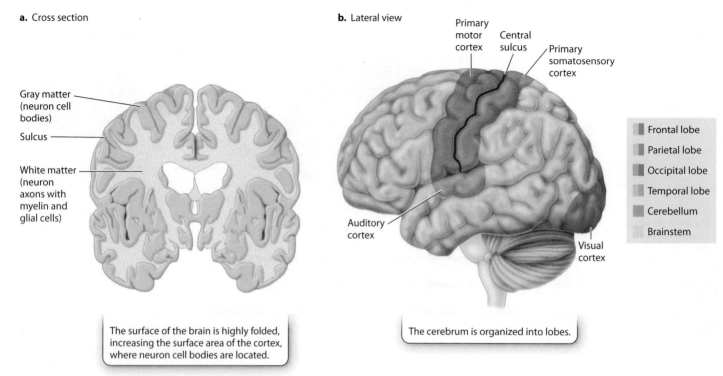

discussed later in this section. Located at the back of the brain is the **occipital lobe**, which processes visual information.

The central sulcus separates the **primary motor cortex** of the frontal lobe from the **primary somatosensory cortex** of the parietal lobe (Fig. 34.23b). "Command" neurons in the primary motor cortex produce complex coordinated behaviors by controlling skeletal muscle movements. The primary somatosensory cortex integrates information about pressure and touch from specific body regions and relays this information to the motor cortex. Both cortices make connections to the opposite side of the body by sending axons that cross over in the brainstem and spinal cord. As a result, the right cortex controls the left side of the body and the left cortex controls the right side of the body.

Information is topographically mapped into the vertebrate cerebral cortex.

The primary somatosensory cortex and motor cortex are organized as topographic maps that represent different body regions, as shown in **Fig. 34.24**. For example, the neurons that sense pressure and touch sensation of the foot and lower limb are located near the mid-axis of the somatosensory cortex. Those that sense motions of the face, jaws, and lips map to the side and farther down in the primary motor cortex. A similar topographic map exists in the motor cortex. Notice how large the hand and face are on the map: regions such as those representing the fingers, hands, and face, which must distinguish fine sensations or control fine motor movements, are represented by larger areas of the cortex.

The auditory cortex in the temporal lobe is similarly organized as a map. This region processes sound information transmitted from the cochlea of the ear. Neurons in the auditory cortex are organized by pitch: neurons sensitive to low frequencies are located at one end, while neurons sensitive to high frequencies are found at the other end.

The occipital lobe topographically maps visual information received from the optic nerves. The topographic mapping of information enables the brain to relate patterns and motion to precise regions of the visual field. As a result, highly visual animals, such as primates and birds, are able to detect complex patterns and movements in their environment.

The brain allows for memory, learning, and cognition.

Memory is the basis for learning. An animal that can remember its encounter with a predator or noxious plant will avoid that danger in the future. A memory is formed by changes in the synaptic connections between neurons in a neural circuit.

In humans and other primates, the hippocampus plays a special role in memory formation. We can remember much of the information we receive for several minutes. However, unless

FIG. 34.24 Somatosensory topographic map.

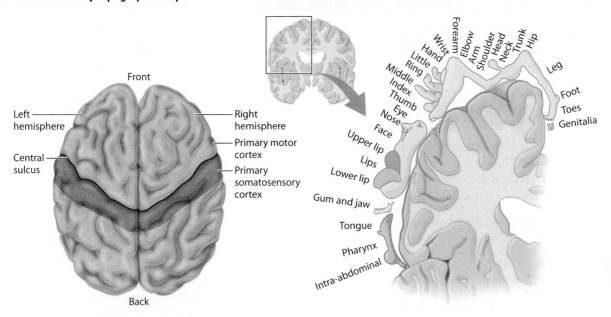

it is reinforced by repetition or by paying particular attention, this information is typically forgotten. The hippocampus transforms reinforced short-term memories into long-term memories.

The hippocampus forms long-term memories by repetitively relaying information to regions of the cerebral cortex. Although the details of these processes remain uncertain, memory and learning depend on establishing particular neural circuits in the hippocampus and cerebral cortex. These circuits can be activated to recall a memory by a particular sound, sight, or other relevant stimulus. Establishing a neural circuit requires establishing a particular set of connections between neurons. Thus, the formation of a memory circuit requires changes in synapses. Some synapses are weakened or removed altogether, whereas others are formed or strengthened. Synapses in the pathways that are repeatedly activated by the hippocampus are likely to be formed or strengthened.

Although the brain regions involved in memory and learning vary among different animals, the basic features of how memories are constructed through changes in synapses are likely broadly shared across animals capable of forming memories. For example, octopuses rival certain vertebrates in their ability to learn and remember motor tasks. Even the roundworm *C. elegans* is capable of memory and learning.

Human brains are able to process and integrate complex sources of information, interpret and remember past events, solve problems, reason, and form ideas. Collectively, these abilities are called **cognition**. Cognitive brain function ultimately gives rise to a state of consciousness—an awareness of oneself in relation to others.

The study of the brain is challenging because its circuits are so complex. Nevertheless, considerable progress is being made through the use of non-invasive neuroimaging techniques. Using these techniques, scientists observe the interactions of different brain regions within human and other animal subjects while the subjects perform a particular mental or motor task. The results of such neuroimaging studies can then be linked to studies of the molecular and cellular properties of nerve cells. Taken together, these studies advance our understanding of how nerve cells form and maintain neural circuits among different brain regions, and of how these circuits are linked to particular behaviors and cognitive processes.

Self-Assessment Questions

18. Draw the brain, label the lobes, and describe their primary functions.

19. What is topographic mapping of sensory input in the cortex? Use the primary somatosensory cortex as an example.

CORE CONCEPTS SUMMARY

34.1 NERVOUS SYSTEM FUNCTION AND EVOLUTION: Nervous systems allow animals to sense and respond to the environment, coordinate movement, and regulate internal functions of the body.

Nerve cells, or neurons, receive and send signals and are the functional unit of the nervous system. page 753

Animal nervous systems include three types of neurons: sensory neurons that respond to signals, interneurons that integrate and process sensory information, and motor neurons that produce a response from muscle. page 753

Ganglia are localized collections of nerve cell bodies that integrate and process information. page 754

Simply organized animals, such as cnidarians, have a nerve net to coordinate sensory and motor function. page 754

Forward locomotion favored the evolution of specialized sense organs in the head, along with concentrated groupings of nerve cells to form ganglia and a brain. page 754

34.2 NEURON STRUCTURE: The functional unit of the nervous system is the neuron, which has dendrites that receive information and axons that transmit information.

Neurons share a common organization: they have dendrites that receive inputs, a cell body that receives and sums the inputs, and axons that transmit signals to other nerve cells. page 755

Glial cells provide nutritional and physical support for neurons. page 757

34.3 SIGNAL TRANSMISSION: The electrical properties of neurons allow them to communicate rapidly with one another.

Neurons have electrically excitable membranes that code information by changes in membrane potential and transmit information in the form of electrical signals called action potentials. page 758

Ion channels open and close in response to changes in the membrane potential, underlying the production of action potentials in nerve cells. page 758

Action potentials fire in an all-or-nothing fashion; a brief refractory period follows their firing. page 758

In vertebrates, glial cells produce the myelin sheath that insulates axons, increasing the speed of nerve impulses. page 761

Action potentials are conducted in a saltatory fashion in myelinated axons, firing at nodes of Ranvier, where the axon membrane is exposed and not insulated by myelin. page 761

Most neurons communicate by chemical synapses formed between an axon terminal and a neighboring nerve or muscle cell. page 763

Communication across the synapse occurs when an arriving action potential triggers the release of neurotransmitters from vesicles within the axon terminal. page 763

Neurotransmitters released from presynaptic vesicles bind to receptors in the postsynaptic membrane, causing either an excitatory stimulus or an inhibitory stimulus. page 764

Excitatory stimuli depolarize the membrane, producing an excitatory postsynaptic potential (EPSP), whereas inhibitory stimuli hyperpolarize the membrane, producing an inhibitory postsynaptic potential (IPSP). page 764

EPSPs and IPSPs can be summed over time (temporal summation) and space (spatial summation). page 765

34.4 NERVOUS SYSTEM ORGANIZATION: Nervous systems are organized into peripheral and central components, and include voluntary and involuntary functions.

Nervous systems are divided into the peripheral nervous system (PNS) and central nervous system (CNS). page 765

The peripheral nervous system is distributed throughout the animal's body and is composed of sensory and motor nerve cells. page 765

The central nervous system includes the brain and one or more main trunks of nerve cells, such as the spinal cord. page 765

In many invertebrates and vertebrates, the peripheral nervous system is divided into voluntary and involuntary components. page 766

In vertebrates, the voluntary component is referred to as the somatic nervous system, and the involuntary component is referred to as the autonomic nervous system. page 766

The autonomic system regulates body functions through opposing actions of its sympathetic and parasympathetic divisions. page 767

Simple reflex circuits can involve as few as two neurons: a sensory neuron from the periphery, which synapses with a motor neuron in the spinal cord that sends a signal to a muscle. page 768

34.5 SENSORY SYSTEMS: Animal sensory receptors detect chemical, physical, and electromagnetic stimuli by changes in membrane potential.

Sensory receptors convert chemical or physical stimuli into nerve impulses in the process of sensory transduction. page 769

Chemoreceptors detect chemical stimuli and allow for the senses of taste and smell. page 769

Mechanoreceptors respond to physical deformations produced by touch, stretch, pressure, motion, and sound, and are important for the senses of balance, detecting the direction of gravity, and hearing. page 771

Hair cells are specialized mechanoreceptors that detect motion and vibration. page 772

Electromagnetic receptors are sensory cells that respond to electrical, magnetic, and light stimuli and provide the sense of vision. page 774

Opsin is a G protein-coupled receptor and the universal photoreceptor protein in animal eyes. page 774

Opsin converts light energy into chemical signals, altering the firing rate of neurons. page 774

Three types of eyes are the eyecups of flatworms, the compound eyes of many arthropods, and the single-lens eyes of vertebrates and cephalopod mollusks. page 774

Photoreceptor cells of the retina include rod cells, which detect light intensity, and cone cells, which detect different wavelengths of light and allow color vision. page 776

34.6 BRAIN ORGANIZATION AND FUNCTION: The brain integrates information from diverse sources and is organized into distinct regions with specialized functions.

The vertebrate brain is organized into a hindbrain, midbrain, and forebrain. The forebrain is elaborated into the cerebral cortex in birds and mammals. page 776

The cerebrum is divided into frontal, parietal, temporal, and occipital lobes, which are specialized for different functions. page 777

Somatosensory, motor, auditory, and visual information is topographically mapped to specific areas in the cerebral cortex. page 778

The brain stores memories of past experiences and enables learning through the creation of long-lasting neural circuits. page 778

Log in to **LaunchPad** to check your answers to the Self-Assessment Questions and to access additional learning tools.

CHAPTER 35 Animal Movement: Muscles and Skeletons

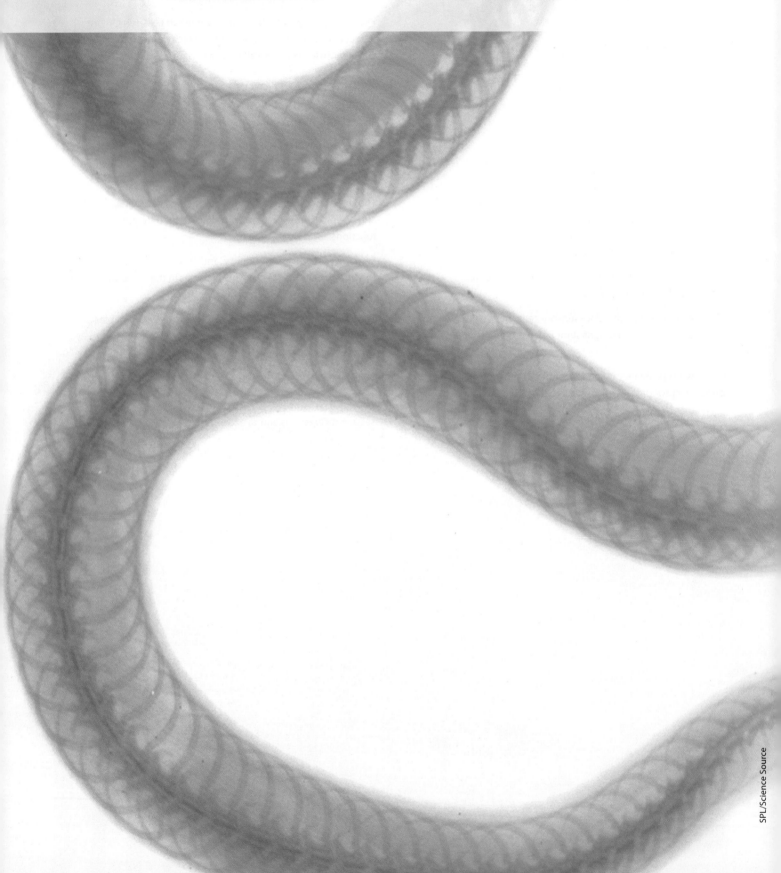

CORE CONCEPTS

35.1 HOW MUSCLES WORK: Muscles are biological motors composed of actin and myosin that generate force and produce movement for support, locomotion, and control of internal physiological functions.

35.2 MUSCLE CONTRACTILE PROPERTIES: The force generated by a muscle depends on muscle size, degree of actin–myosin overlap, shortening velocity, and stimulation rate.

35.3 ANIMAL SKELETONS: Hydrostatic skeletons, exoskeletons, and endoskeletons provide animals with mechanical support and protection.

35.4 VERTEBRATE SKELETONS: Vertebrate endoskeletons allow for growth and repair and transmit muscle forces across joints.

The ability to move is a defining feature of animal life. Movement allows animals to explore new environments and avoid inhospitable ones, escape predators, mate, feed, and play. An animal's motor and nervous systems work together to enable it to sense and respond to its environment. How are movements produced, and how are they controlled? What determines an athlete's performance? How do muscles and the skeleton work together to provide movement and support? How do muscles convert the chemical energy of ATP into force and movement during a contraction? This chapter explores the organization and function of the muscles and skeletons that power the movements of larger multicellular animals and provide mechanical support of their bodies (**Fig. 35.1**).

35.1 HOW MUSCLES WORK

We rely on our muscles for all kinds of movement. Muscles function as biological motors within the body because they generate force and produce movement. Some of these movements are obvious—running, climbing stairs, raising your hand, and playing the piano. We are less aware of others—moving food through the digestive tract, breathing in and out, pumping blood through the body. A muscle's ability to produce movement depends on electrically excitable muscle cells containing proteins that can be activated by the nervous system. As we will discuss, the geometry and organization of these proteins largely determine how muscles contract (shorten) to produce force and movement.

The contractile machinery of muscles has an ancient history. The proteins underlying muscle contraction have been found in eukaryotes that lived more than 1 billion years ago, and the first muscle fibers common to all animals are found in cnidarians (the group that includes jellyfish, corals, and sea anemones), which evolved at least 600 million years ago. Thus, basic features of muscle organization and function are conserved across the vast diversity of eukaryotes. Nevertheless, recent

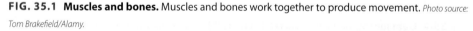

FIG. 35.1 Muscles and bones. Muscles and bones work together to produce movement. *Photo source: Tom Brakefield/Alamy.*

genetic analysis shows that certain features of muscle evolved independently in cnidarians, possibly ctenophores (comb jellies), and all other animals.

Muscles use chemical energy to produce force and movement.

Muscles are composed of elongated cells called muscle **fibers**. Muscle fibers use ATP generated through cellular respiration (Chapter 7) to generate force and change length during a contraction. This conversion of chemical energy into muscle work underlies much of the energy cost of animal movement discussed in Chapter 38.

A **force** is a push or pull by one object interacting with another object. Skeletal muscles, for example, generate pulling forces at specific sites of the skeleton. As a muscle shortens, it pulls against the part of the skeleton to which it is attached. The work performed by a muscle is equal to force times length change. For example, work is increased when either the force produced or the distance the skeleton is moved increases. Similarly, a muscle that produces a pulling force twice as great as another muscle but moves the skeletal segment only half the distance does the same work as that other muscle.

Because muscles exert only pulling forces, pairs of muscles are arranged to produce movements in two opposing directions at specific joints of the skeleton. For example, muscles pull on the bones of your leg and foot to swing them forward when you kick a ball. Opposing sets of muscles pull on the leg in the opposite direction to swing your leg back and flex your knee before and after the kick.

Muscles can shorten very quickly, and they can produce large forces for their weight. These forces can be many times an animal's body weight, allowing animals to move quickly. Some small muscles contract extremely quickly. For example, the muscles of many flying insects contract 200 to 500 times per second. Even in vertebrate animals, muscles can contract as many as 50 to 100 times per second. Other muscles, such as the closing muscle of a clam, contract slowly, but can continue contracting for prolonged periods without tiring. To see how muscles produce forces and contract to change length, we now consider how muscle fibers are organized at the molecular level.

Muscles can be striated or smooth.

All muscle fibers contain the same contractile proteins that enable them to shorten and produce force, namely actin and myosin (Chapter 10). These proteins are organized into long, narrow bundles called **filaments** that interact with one another to cause muscle fibers to shorten. Although all muscle fibers have filaments of actin and myosin, the filaments are arranged differently in different types of muscle.

Muscles are organized into two broad groups on the basis of their function and appearance under a light microscope (**Fig. 35.2**). As their name suggests, **striated muscles** appear striped under a light microscope because the actin and myosin filaments are arranged in a regularly repeating pattern. Striated muscles include **skeletal muscles** (Fig. 35.2a), which connect to the body skeleton to move the animal's limbs and torso, and **cardiac muscle** in the heart (Fig. 35.2b), which contracts to pump blood. Skeletal muscle cells are elongated and have many nuclei in each cell. By comparison, cardiac muscle cells are typically less elongated than skeletal muscle cells, often branched, and contain one or more nuclei per cell.

In contrast to skeletal and cardiac muscle, **smooth muscles** appear unstriated under the light microscope because the organization of actin and myosin filaments is irregular (Fig. 35.2c). In many vertebrates and invertebrates, smooth muscles are

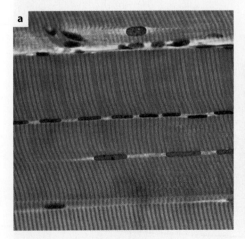
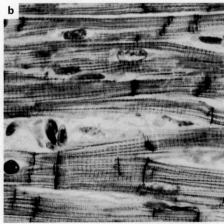
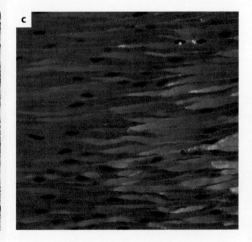

FIG. 35.2 Light micrographs of (a) skeletal muscle, (b) cardiac muscle, and (c) smooth muscle. Skeletal and cardiac muscles have a striated, or striped, appearance; smooth muscle does not. *Sources: a. Innerspace Imaging/Science Source; b. Manfred Kage/Science Source; c. SPL/Science Source.*

FIG. 35.3 Skeletal muscle organization. Muscles are made up of bundles of muscle fibers (cells), each of which contains myofibrils.

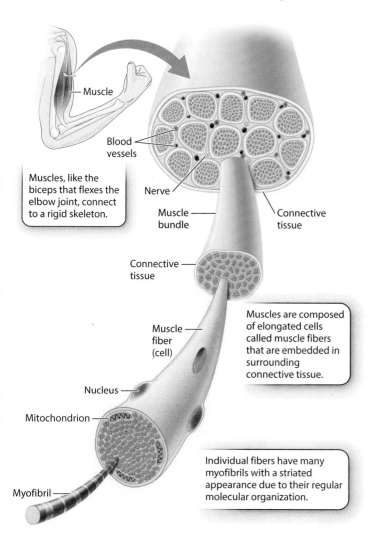

FIG. 35.4 Thin and thick filaments. A thin filament is made up of two actin subunits arranged in a double helix. A thick filament is made up of numerous myosin molecules arranged in parallel.

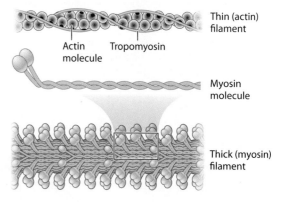

hundreds of long, rodlike structures called **myofibrils**. Myofibrils contain parallel arrays of actin and myosin filaments that cause a muscle to contract.

Let's consider the organization of actin and myosin filaments in skeletal muscles in more detail (**Fig. 35.4**). Actin molecules polymerize to create two long polypeptide chains, which are arranged as a double helix to form a **thin filament**. The protein **tropomyosin** runs in the grooves formed by the actin helices. Each myosin molecule consists of two long polypeptide chains coiled together, with each chain ending with a globular head. Consequently, the myosin molecule resembles a double-headed golf club. The myosin molecules are arranged in parallel to form a **thick filament**, with the numerous myosin heads extending out from their flexible necks along the myosin filament.

The thin filaments are attached to protein backbones called **Z discs** that are regularly spaced along the length of the myofibril (**Fig. 35.5a**). The region from one Z disc to the next is called the **sarcomere**. Sarcomeres are the functional units of muscles. In other words, the shortening (contraction) of a muscle is ultimately the result of the shortening of thousands of sarcomeres along a myofibril.

Sarcomeres are arranged in series along the length of the myofibril. Sets of actin thin filaments extend from both sides of the Z discs toward the midline of the sarcomere (Fig. 35.5a). In the middle of the sarcomere, but not directly contacting the Z discs, are myosin thick filaments. The thin filaments overlap with the myosin thick filaments, forming two regions of overlap within a sarcomere. Toward either end of the sarcomere, as well as in the middle, are regions where the actin and myosin filaments do not overlap. The regions that overlap appear dark under an electron microscope, while the regions of non-overlap appear lighter in color. A third large protein, called titin, links the myosin filaments to the Z discs at the ends of the

found in the walls of arteries to regulate blood flow, in the respiratory system to control air flow, and in the digestive and excretory systems to help transport food and waste products. Smooth muscles contract slowly compared with cardiac and skeletal muscles. Consequently, bivalve mollusks rely on smooth muscle fibers to keep their shells closed for long periods without having to expend much energy.

Skeletal and cardiac muscle fibers are organized into repeating contractile units called sarcomeres.

The visible bands in skeletal and cardiac muscles are an important clue to the mechanism that produces muscle contractions. Whole muscles are made up of parallel bundles of individual muscle fibers (**Fig. 35.3**). Recall that a muscle fiber is a muscle cell. Each muscle fiber, in turn, contains

FIG. 35.5 Muscle sarcomere. (a) Longitudinal and (b) cross-sectional views of a sarcomere, including two transmission electron micrographs. The sarcomere is the region between Z discs; it is the contractile unit of a muscle. *Photo sources: (a) Don W. Fawcett/Science Source; (b) Biophoto Associates/Science Source.*

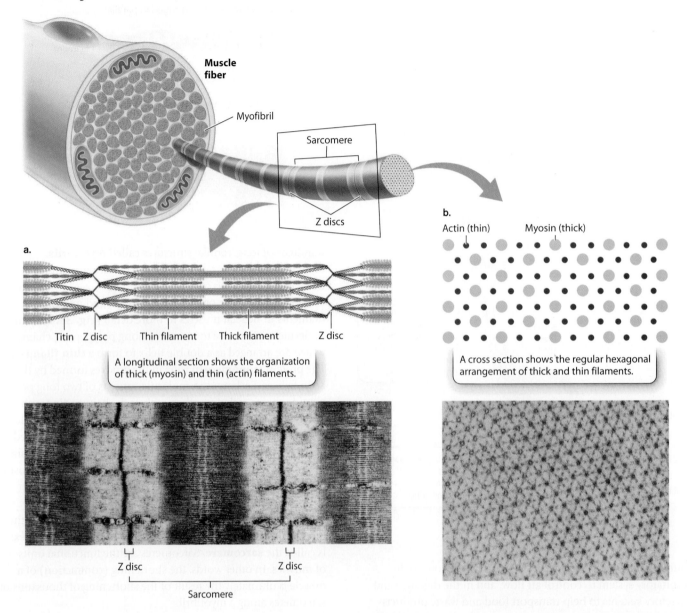

sarcomere. Titin is believed to help with assembly and protect the sarcomeres from being overstretched. Because it acts as a spring, titin also contributes to muscle elasticity.

As mentioned earlier, the regular pattern of actin and myosin filaments within sarcomeres along the length of the fiber gives skeletal muscles their striated appearance (see Fig. 35.2a). In cross section, thick and thin filaments are arranged in a hexagonal lattice. This arrangement allows each thick filament to interact with six adjacent thin filaments (**Fig. 35.5b**). The precise geometry of thin and thick filaments is critical to their interaction, as we discuss next.

Muscles contract by the sliding of myosin and actin protein filaments.

Using high-resolution light microscopy, two independent teams (physiologists Hugh Huxley and Jean Hanson in the United States and Andrew Huxley and Rolf Niedergerke in England) were able to quantify the changing banding patterns of sarcomeres when rabbit or frog myofibrils were stimulated to contract to different lengths. Their results led them to hypothesize that muscles change length and produce force by the sliding of actin filaments relative to myosin filaments. Their theory is known as the **sliding filament model** of muscle contraction.

FIG. 35.6 Sliding filament model. Muscle contraction, or shortening, results from the sliding of actin thin filaments relative to myosin thick filaments. Filament sliding changes the sarcomere banding pattern between Z discs. *Photo source: Reprinted by permission from Macmillan Publishers Ltd.: NATURE 173, H. Huxley and J. Hanson, Changes in the Cross-Striations of Muscle during Contraction and Stretch and Their Structural Interpretation, Copyright 1954.*

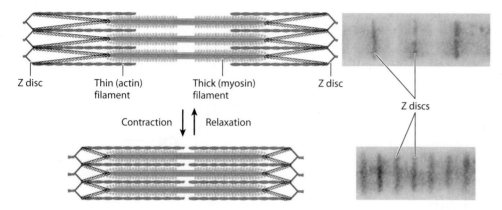

The banding patterns of sarcomeres revealed the extent of overlap between actin and myosin filaments. When myofibrils contracted to short lengths, the sarcomeres were observed to have increased actin–myosin overlap. When myofibrils were stretched to longer lengths, actin–myosin overlap decreased (**Fig. 35.6**). However, the lengths of the myosin and actin filaments never varied. Consequently, nearly all of the length change during a muscle contraction results from the sliding of actin filaments with respect to myosin filaments within individual sarcomeres. The length change of the whole muscle fiber, therefore, is a sum of the fractions by which each sarcomere shortens along the fiber's length.

A muscle's ability to change length and generate force is largely determined by the properties of its sarcomeres. Whereas sarcomere length is quite uniform in vertebrate animals (approximately 2.3 μm at rest), it is highly variable in invertebrate animals (ranging from as short as 1.3 μm to as long as 43 μm). Longer sarcomeres allow a greater degree of shortening. Thus, length changes are more limited in vertebrate muscles compared to those in some invertebrate muscles with long sarcomeres. For example, when the sarcomeres in an octopus tentacle contract, they can produce large motions of the tentacle. However, because vertebrates have shorter sarcomeres than an octopus, vertebrate muscles can shorten to produce movements more quickly.

Interactions between the myosin and actin filaments cause the muscle fiber to shorten and produce force. Along the myosin filament, one of the two heads of a myosin molecule at a given point in time binds to actin at a specific site to form a **cross-bridge** between the myosin and actin filaments. The myosin filaments pull the actin filaments toward each other by means of these cross-bridges.

The key to making the filaments slide relative to each other is the ability of the myosin head to undergo a conformational change and, therefore, pivot back and forth. The myosin head attaches to the actin filament and pivots forward, sliding the actin filament toward the middle of the sarcomere. The myosin head then detaches from the actin filament, pivots back, and reattaches, and the cycle repeats. The movement of the myosin head is powered by ATP. Together, these events constitute the **cross-bridge cycle**.

The cross-bridge cycle is just that—a cycle (**Fig. 35.7**). Let's look at it in more detail:

1. The myosin head binds ATP. Binding of ATP allows the myosin head to detach from actin and readies it for attachment to actin again.

2. The myosin head hydrolyzes ATP to ADP and inorganic phosphate (P_i). Hydrolysis of ATP results in a conformational change in which the myosin head is cocked back. ADP and P_i remain bound to the myosin head. Because ADP and P_i are bound rather than released, the myosin head is in a high-energy state.

3. The myosin head binds to actin, forming a cross-bridge.

4. When the myosin head binds to actin, the myosin head releases ADP and P_i. The result is another conformational change in the myosin head, called the **power stroke**. During the power stroke, the myosin head pivots forward and generates a force, causing the myosin and actin filaments to slide relative to each other over a distance of approximately 7 nm. The power stroke pulls the actin filaments toward the sarcomere midline.

The myosin head then binds a new ATP molecule (step 1 again), allowing it to detach from actin, and the cycle repeats, as the myosin binds to a new site farther along the actin filament.

Individual muscle contractions are the result of many successive cycles of cross-bridge formation and detachment among many myosin and actin filaments. During these cycles, muscle cells convert the chemical energy released by ATP into force and the kinetic energy of movement. Muscle fibers that contract especially quickly, such as those that power insect wing flapping or produce the sounds of cicadas and rattlesnake tails, express myosin molecules that have high rates of ATP hydrolysis. The muscle fibers have faster rates of cross-bridge cycling, force

FIG. 35.7 Cross-bridge cycle. The myosin head binds to actin and uses the energy of ATP to pull on the actin filament, causing muscle shortening.

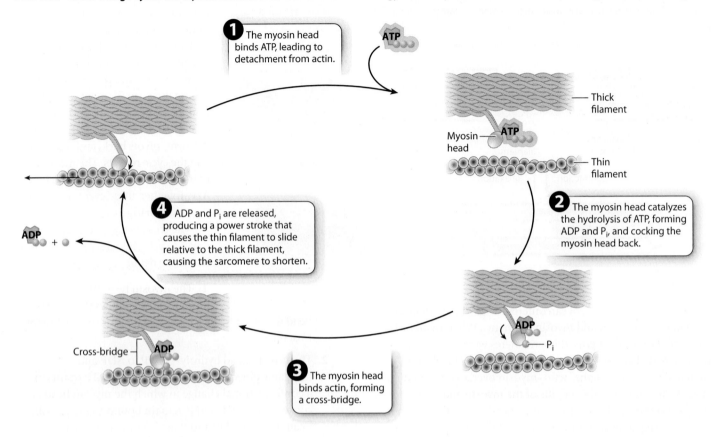

development, and shortening. Thus, myosin functions as both a structural protein and an enzyme.

Each thick filament can interact with as many as six actin filaments. Thus, at any one time numerous cross-bridges anchor the lattice of myosin filaments, while other myosin heads are detached to find new binding sites. When summed over millions of cross-bridges, these molecular events generate the force that shortens the whole muscle.

Calcium regulates actin–myosin interaction through excitation–contraction coupling.

We have discussed *what* makes muscles contract. We now consider *when* they contract. Skeletal and smooth muscle fibers are both activated by the nervous system. Whereas vertebrate skeletal muscles are activated by the somatic nervous system, smooth muscles are activated by the autonomic nervous system, as are cardiac muscles (Chapter 34).

Like nerve cells, muscle fibers are electrically excitable. Skeletal muscle fibers are activated by impulses transmitted by motor nerves (**Fig. 35.8**). Motor neuron axons have branches that allow each axon to synapse with multiple muscle fibers. When action potentials traveling down a motor neuron arrive at the neuromuscular junction, the neurotransmitter acetylcholine is released into the synaptic cleft. The neurotransmitter binds with postsynaptic receptors on the muscle cell at a region called the **motor endplate**, triggering the opening of Na^+ channels. The resulting influx of Na^+ ions initiates a wave of membrane depolarization that passes from the neuromuscular junction toward both ends of the muscle fiber (Chapter 34).

How does membrane depolarization lead to the cross-bridge cycle? Actin and myosin filaments can form cross-bridges only when the myosin-binding sites on actin are exposed. At rest, these binding sites are blocked by the protein tropomyosin. The wave of depolarization in the muscle cell initiates a chain of events that moves tropomyosin away from these binding sites, allowing cross-bridges between actin and myosin to form and the muscle to contract (Fig. 35.8).

Let's look at the chain of events that concludes with the exposure of the myosin-binding sites. You saw in Chapter 5 that eukaryotic cells contain several types of membrane-bound internal organelles. The myofibrils of muscle cells are surrounded by a highly branched membrane-bound organelle called the **sarcoplasmic reticulum (SR)**, a modified form of the endoplasmic reticulum. When the muscle is at rest, the SR contains a large internal concentration of calcium (Ca^{2+}) ions that are transported into the cell by calcium pumps in its membranes.

FIG. 35.8 Excitation–contraction coupling. Depolarization (excitation) leads to shortening (contraction) of the muscle. *Photo source: Don W. Fawcett/Science Source.*

a. Scanning electron micrograph of a motor nerve endplate

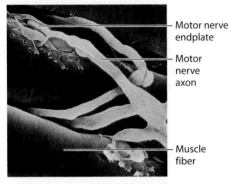

b. Muscle excitation–contraction coupling

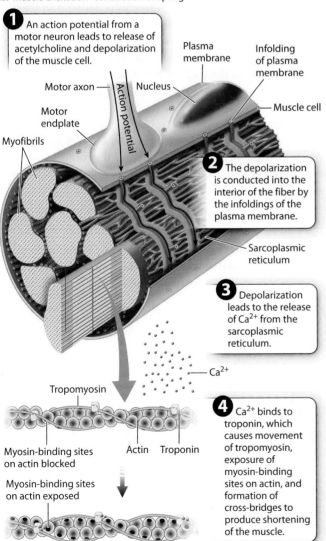

Depolarization initiated in the plasma membrane is conducted to the SR through infoldings of the plasma membrane (Fig. 35.8). A muscular contraction is initiated when depolarization reaches the SR and causes it to release Ca^{2+}. The Ca^{2+} diffuses into the myofibrils and binds to a protein called **troponin**, causing the troponin molecule to change shape. This conformational change of troponin, in turn, causes tropomyosin to move, exposing myosin-binding sites along the actin filament. Myosin cross-bridges then form with actin, producing a contraction.

Note that at this stage in the cross-bridge cycle, the myosin head has already hydrolyzed ATP, is bound to ADP and P_i, and is in the cocked-back "ready" position. Binding to actin then allows the power stroke to occur. In this way, contraction is initiated immediately following depolarization and release of Ca^{2+}.

The process by which membrane depolarization leads to Ca^{2+} release from the SR and the formation of myosin-actin cross-bridges is called **excitation–contraction coupling** because excitation of the muscle cell is coupled to contraction of the muscle. Together, these events are the molecular "switch" that causes a muscle to contract. The muscle relaxes when neural stimulation ends. Acetylcholine is broken down or reabsorbed, and Ca^{2+} is actively transported back into the sarcoplasmic reticulum. With Ca^{2+} gone, tropomyosin molecules can once again block myosin-binding sites along the actin filaments.

Calmodulin regulates calcium activation and relaxation of smooth muscle.

Smooth muscle controls many internal organs. How is the contraction of smooth muscle regulated? Smooth muscle can be activated by the autonomic nervous system, but it also responds to other factors. These include hormones, stretch of the muscle, and local factors such as pH, oxygen, carbon dioxide, and nitric oxide. For example, smooth muscle in the walls of arteries contracts when stimulated by the release of epinephrine to reduce blood flow to a body region (Chapter 37), and smooth muscle within the mammalian uterus contracts when oxytocin is released to stimulate labor (Chapter 40).

As in skeletal muscle, contraction of smooth muscle is regulated by Ca^{2+}. However, there are two sources of Ca^{2+} for smooth muscle contraction: intracellular (release of Ca^{2+} from the SR, as in skeletal muscle) and extracellular (opening of calcium channels in the plasma membrane). In addition, whereas release of Ca^{2+} in skeletal muscle is coupled to depolarization, Ca^{2+} channels in smooth muscle also open in response to intracellular signaling pathways triggered by hormones and other factors, as well as to stretch of the muscle cell.

Smooth muscle lacks the troponin–tropomyosin mechanism for regulating contraction. Instead, smooth muscle cells are activated when the protein **calmodulin** binds with Ca^{2+} released from the SR or entering through the cell's membrane.

The calmodulin–Ca^{2+} complex activates a myosin kinase that phosphorylates the smooth muscle myosin heads, causing them to bind actin and begin the cross-bridge cycle. Once the muscle has finished contracting, a second enzyme dephosphorylates the myosin heads, disrupting their ability to bind to actin, and the muscle relaxes.

Smooth muscle contracts slowly compared with skeletal muscle, using less ATP per unit time. The SR of smooth muscle cells is less extensive and has many fewer calcium pumps than the SR of skeletal muscle cells. As a result, calcium is returned more gradually from the myofibrils back to the SR, or pumped back out through the cell's membrane, slowing the relaxation of smooth muscle. Through these kinds of slow contractions of smooth muscle, clams are able to hold their shells closed for long periods, arteries can contract and restrict blood flow to particular body regions, and the gut can move the digestive contents through the gastrointestinal tract.

Self-Assessment Questions

1. Diagram a sarcomere, showing the basic organization of thin (actin) and thick (myosin) filaments. Indicate on your diagram the regions where myosin cross-bridges form with actin.
2. What is the sequence of events involved in the cross-bridge cycle? How do these events explain how muscles generate force and shorten when they are stimulated to contract?
3. When an animal dies, its limbs and body become stiff because its muscles go into *rigor mortis* (a term that means "stiffness of death"). Why would the loss of ATP following death cause this to happen?
4. What is the sequence of events that occurs when an action potential arrives at a muscle fiber's motor endplate, causing the muscle to be depolarized?
5. Curare is a paralyzing compound that blocks the action of the neurotransmitter acetylcholine at the muscle fiber's motor endplate. What effect does curare have on the release of calcium ions from the sarcoplasmic reticulum of the muscle cell?

35.2 MUSCLE CONTRACTILE PROPERTIES

Actin and myosin are evolutionarily conserved across all animal life, so all muscles have similar force-generating properties. The huge muscles of a weight lifter as well as the large closing muscle of a lobster claw are evidence that larger muscles can exert more force. Why is that? The force exerted by the whole muscle is the sum of the forces produced by the individual myofibrils of activated muscle fibers. As a consequence, larger muscles with more fibers produce greater forces than smaller muscles with fewer fibers. However, the force that a muscle produces also depends on its contraction length and the speed of contraction.

Antagonist pairs of muscles produce reciprocal motions at a joint.

Muscles can generate force only by pulling, not by pushing, on the skeleton. As described earlier, to produce reciprocal joint and limb movements, they are arranged as pairs of **antagonist muscles** that pull in opposite directions. **Flexion** is the joint motion that moves bone segments closer together, and **extension** is the joint motion that moves them farther apart. For example, muscles at the elbow joint are organized as flexors (the biceps muscle) that contract to bring the arm toward the shoulder, and extensors (the triceps muscle) that contract to extend the arm away from the shoulder (**Fig. 35.9**).

Antagonist muscles also control flexion and extension of the leg joints of jumping grasshoppers and locusts, as well as the muscles that elevate and depress the wings of flying insects. Additional pairs of antagonist muscles are present for movements in other directions. Muscles in the body wall of cnidarians (jellyfish and sea anemones) and segmented annelid worms are organized in longitudinal (lengthwise) and circumferential (circular) layers, as are the smooth muscles in the walls of intestines. These two muscle layers function as muscle antagonists that control movement and shape.

In contrast, when muscles combine to produce similar motions, they are termed muscle **agonists**. Muscles arranged as agonists increase the strength and improve the control of joint motion. For example, the three heads of the triceps are agonists of one another.

FIG. 35.9 Antagonist muscles. Skeletal muscles can produce force only by pulling on a skeletal attachment, so they are arranged as antagonists to produce opposing motions at a joint.

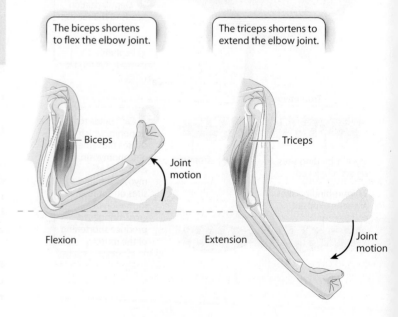

HOW DO WE KNOW?

FIG. 35.10

How does actin–myosin filament overlap affect force generation in muscles?

BACKGROUND From the 1930s through the 1960s, British and American muscle physiologists studied the properties of striated frog muscle fibers. In one set of studies, American physiologists Albert Gordon and Fred Julian, together with British physiologist Andrew Huxley, studied how a contracted muscle's length relates to the muscle's force-generating ability.

HYPOTHESIS Gordon, Huxley, and Julian hypothesized that changes in overlap between myosin and actin affect the number of cross-bridges, thereby determining the amount of force the muscle can produce.

EXPERIMENT The three scientists isolated a single muscle fiber and used optical microscopy to measure sarcomere length. They kept this length constant while stimulating the fiber to produce force. They then obtained measurements of force at various other sarcomere lengths. The length at which the muscle fiber generated the maximum force is indicated on the graphs as 100%. Other lengths are expressed relative to this length.

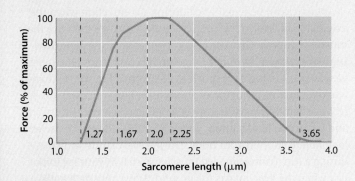

RESULTS The graph above shows the data obtained by Gordon, Huxley, and Julian for a single sarcomere of a single muscle fiber,

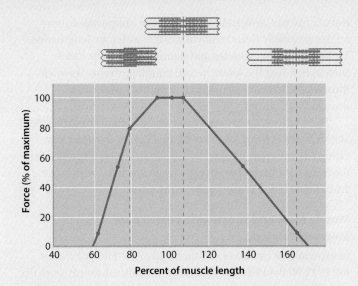

and the graph at the upper right reinterprets their data for a whole muscle. Note that these graphs have different x-axes. The lower left graph shows force (% maximum) plotted versus absolute length of the sarcomeres. The upper right graph shows force (% maximum) plotted against the percentage of muscle length, with 100% being the length at which the muscle generates maximum force. The researchers' results indicate that the muscle fiber produces its maximal force at an intermediate sarcomere length (approximately 2.3 μm). It is at this length, they hypothesized, that the greatest number of cross-bridges form, generating maximal force. As the sarcomere length is decreased or stretched, and the muscle fiber is then stimulated to contract, force declines. These results are consistent with a decrease in the number of cross-bridges that form as a result of decreased actin–myosin filament overlap.

FOLLOW-UP WORK Other studies confirmed that changes in myosin cross-bridge formation are linked to changes in actin and myosin filament overlap at different sarcomere lengths, supporting the sliding filament model of muscle contraction.

SOURCE Gordon, A. M., A. F. Huxley, and F. J. Julian. 1966. "The Variation in Isometric Tension with Sarcomere Length in Vertebrate Muscle Fibres." *Journal of Physiology* 184:170–192.

Muscle length affects actin–myosin overlap and generation of force.

Why is it difficult to jump while standing on your toes? A muscle's ability to generate force depends in part on how much it is stretched before contraction begins. The sliding filament model for muscle contraction, like any good model, makes specific predictions. In this case, it predicts that the amount of overlap between actin and myosin filaments within the sarcomere will determine the number of cross-bridges that can form. The more overlap, the more force that can be produced. Conversely, when a muscle fiber is pulled to long lengths and then contracts, it generates less force because the overlap between myosin and actin filaments is reduced. When a muscle fiber contracts at short lengths, it also produces less force because myosin filaments begin to run into the Z disc at each end of the sarcomere, disrupting the geometry of the myosin filament lattice and hindering cross-bridge formation. At intermediate lengths, the muscle fiber generates its maximum force because actin–myosin filament overlap is greatest. These predictions have been tested by experiments (**Fig. 35.10**).

You can't jump well if you are on your toes and your ankles are fully extended because this position shortens your calf muscle fibers. Conversely, you also can't jump well if your ankles are overly flexed and your calf muscles are lengthened. Through training, athletes and dancers learn techniques, such as bending the knees before jumping, so that their leg muscles contract at intermediate lengths to maximize muscle-force output.

The relationship between force and length of striated muscle also explains Starling's law for the function of the heart as a pump (discussed in Chapter 37). This law explains the fact that the heart contracts more strongly when it is filled with larger amounts of blood returning from the veins.

Muscle force and shortening velocity are inversely related.

In experiments carried out in the 1930s, A. V. Hill (who won the 1922 Nobel Prize in Physiology or Medicine for his work on muscle energy use) observed that a muscle shortens fastest when producing low forces (**Fig. 35.11**). To produce larger forces, the muscle must shorten at progressively slower velocities. For example, your arm moves faster when you throw a lightweight ball than when you throw a heavy one. The reason is that the faster a muscle fiber shortens, the fewer cross-bridges that can form within it, reducing the force that each fiber and the muscle as a whole can produce.

The term "muscle contraction" suggests that muscles shorten when stimulated to generate force. In reality, muscles may also exert force and remain a uniform length or even be lengthened. Consequently, the term "contraction" needs to be broadened beyond its familiar reference to muscle shortening to include those instances when a muscle exerts force while remaining the same length or being actively stretched. In the last section, we looked at what happens when you contract a muscle that has been stretched or shortened before the contraction begins. Now we look at what happens when you lengthen a muscle during the contraction or keep it the same length rather than shortening it.

When you carry a heavy bucket of water, your arm muscle exerts force, yet it stays the same length and does not shorten. In this case, the muscle's shortening velocity is zero and it generates an **isometric** force (*iso* means "same" and *metric* refers to "measure") that is greater than the force generated when it shortens. Isometric exercises take advantage of this type of muscle contraction: holding a weight in a fixed position or exerting force against an immovable object helps maintain muscle strength and may be beneficial in rehabilitation after injury by minimizing joint motion.

Muscles can be lengthened during a contraction when the external load against which they contract exceeds the force the muscle produces. This action is known as an active **lengthening contraction**. During a lengthening contraction, the muscle generates more force (as much as 75% more; Fig. 35.11) than when it remains isometric or shortens. When you lower a box from a shelf, your biceps muscle lengthens as it exerts force, and when you land a jump, your calf and quadriceps muscles lengthen to support your body. During every stride you take during a run, contracting leg muscles stretch as your leg joints flex when the leg contacts the ground. Active muscle lengthening reduces the muscle's energy consumption because fewer fibers are recruited to generate a given level of force.

Stretching a muscle too rapidly or repetitively however can cause injury. Rapid and repetitive stretching often occurs during prolonged downhill hiking, causing muscle damage and soreness, particularly in the quadriceps muscles in the front of the thigh. Suitable training and warm-up exercises can help muscles to stretch without damage.

Muscle force is summed by an increase in stimulation frequency and the recruitment of motor units.

How do animals vary the amount of force exerted by a muscle? As we have seen, skeletal muscles are stimulated by motor neurons, which conduct action potentials to the neuromuscular junction. The force exerted by a muscle depends on the frequency of stimulation by the motor neuron, as well as the

FIG. 35.11 Force–velocity relationship of muscle. Skeletal muscles exert greater force when they shorten more slowly; they produce even greater force when they are stretched relative to being isometric.

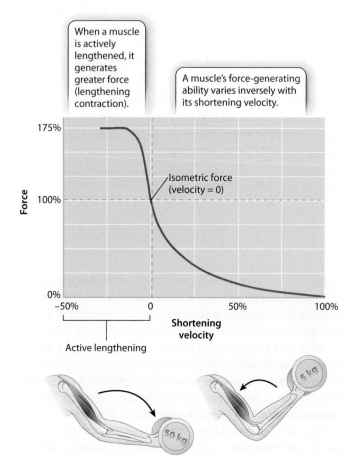

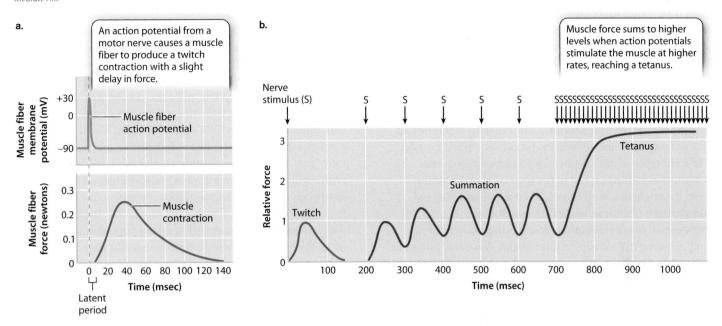

FIG. 35.12 (a) Twitch and (b) fused tetanus muscle contractions. *Source: Adapted from E. P. Widmaier, H. Raff, and K. T. Strang, 2011, Vander's Human Physiology, New York: McGraw-Hill.*

number of motor neurons that activate the muscle. As action potentials become more frequent, the amount of calcium that is released to activate cross-bridge formation in the muscle fibers also increases. Thus, the more frequent the action potentials, the more cross-bridges form.

A single action potential results in a **twitch** contraction of a certain force (**Fig. 35.12a**). If sufficient time is allowed for the muscle to pump Ca^{2+} back into the sarcoplasmic reticulum, the contraction ends, force falls to zero, and a second stimulus can elicit a twitch contraction of the same force. However, if a second action potential arrives before the muscle has relaxed, a greater force is produced because Ca^{2+} remains within the myofibrils to activate more cross-bridges. As the frequency of stimulation increases, muscle force increases as well (**Fig. 35.12b**). This property of muscle contraction is called force summation.

At a sufficiently high stimulation frequency, muscle force reaches a plateau and remains steady as stimulation continues. This muscle contraction of sustained force is called a **tetanus**. Measurements in animal muscles indicate that tetanic contractions are common, and that they generate steady and large forces. Tetanus is a normal physiological response of a muscle, but it can also be induced by a toxin from the bacterium *Clostridium tetani*, so named because it causes severe muscle spasms, particularly of the jaw, chest, back, and abdomen.

A motor neuron and the population of muscle fibers that it innervates are collectively termed a **motor unit** (**Fig. 35.13**). The number of muscle fibers in a motor unit can range from relatively few to several hundred. Invertebrate muscles typically have few motor units. Force is adjusted mainly by graded activation of the muscle's fibers rather than by adjusting the stimulation frequency of multiple motor units. In contrast, a vertebrate muscle's force output depends on the number of motor units (and therefore the number of muscle fibers) that are activated, as well as their stimulation frequency. The number of muscle fibers innervated by a given motor neuron, and

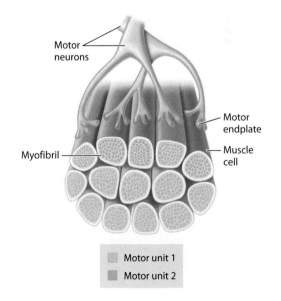

FIG. 35.13 Motor units. A motor unit consists of the motor neuron and the population of cells (fibers) it innervates.

hence the size of the motor unit, affects how finely force can be controlled. For example, finger muscles have small motor units allowing fine adjustments in muscle force, whereas leg muscles have large motor units providing larger adjustments in force. Within each motor unit, muscle fibers generally share similar contractile and metabolic properties. Differences in these properties can give muscles very different capabilities, as discussed in the next section.

Skeletal muscles have slow-twitch and fast-twitch fibers.

The difference between a sprinter and a marathoner is not just due to training; it also reflects the properties of their muscles. Different types of muscle fibers exist, allowing for greater variation in endurance and in the speed of muscle contraction. In vertebrates, such muscle fibers are classified as red **slow-twitch** fibers and white **fast-twitch** fibers (**Fig. 35.14**). Slow-twitch fibers are found in muscles that contract slowly and consume ATP slowly. An animal uses slow-twitch fibers to control its posture or move slowly and economically. Fast-twitch fibers generate force quickly, producing rapid movements, but they consume ATP more quickly. An animal depends on fast-twitch fibers to sprint or lunge at its prey. A key difference between the two types of fibers is that slow-twitch fibers obtain their energy through oxidative phosphorylation (aerobic respiration), whereas fast-twitch fibers obtain their energy mainly through glycolysis (Chapter 7).

Although slow-twitch fibers develop force more slowly, they have greater resistance to fatigue in response to repetitive stimulation; that is, their loss of force over time is less. Their greater endurance is explained by the ability of their mitochondria to supply ATP to muscle fibers by aerobic respiration. Slow-twitch fibers have many mitochondria and are well supplied by capillaries. They contain an abundance of **myoglobin**, an oxygen-binding protein related to hemoglobin that helps deliver oxygen to the mitochondria (Chapter 37). The iron in myoglobin gives muscles with these fibers their red appearance. Slow-twitch muscle fibers express a chemically "slow" form of myosin with a relatively low rate of ATP hydrolysis, limiting their speed of cross-bridge cycling and force development.

White fast-twitch muscle fibers have fewer mitochondria and capillaries, contain little myoglobin, and rely heavily on glycolysis to produce ATP. Fast-twitch fibers express a chemically "fast" form of myosin with a high rate of ATP hydrolysis, favoring rapid force development and movement. However, fast-twitch fibers also fatigue quickly. Because white fibers are larger than red fibers, they generate more force. In general, most skeletal muscles are composed of mixed populations of red and white fiber types, providing a broad range of speed, force, and endurance.

Athletes who have more oxidative slow-twitch fibers excel at endurance and long-distance competitions. In contrast, athletes who have larger numbers of fast-twitch fibers perform better in sprint races and weight-lifting competitions. Because the distribution and properties of muscle fiber types are strongly inherited, it is largely true that champions are born and not made. Nevertheless, aerobic training can significantly increase the energy output through oxidative phosphorylation of an individual's muscle fibers, just as weight lifting and speed training can enhance an individual's strength and sprinting ability. A popular belief is that weight lifting places strain on the muscle

FIG. 35.14 Slow-twitch (darkly stained) and fast-twitch (lightly stained) skeletal muscle fibers. These two types of fibers provide differences in endurance and the speed of contraction. (a) Mammalian limb muscle; (b) muscles of two different fish species. *Photo source: Biophoto Associates/Science Source.*

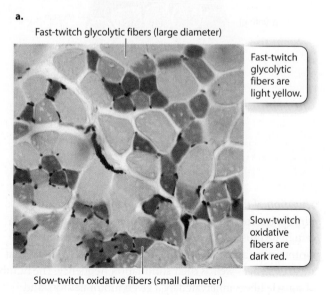
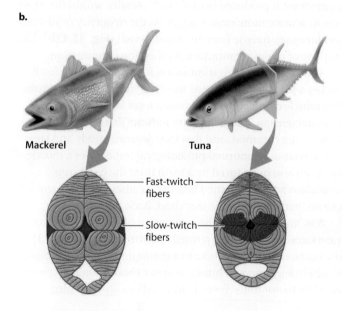

that builds new fibers. This is a myth: it actually increases the size of existing muscle fibers through the synthesis of additional myosin and actin filaments. This synthesis increases the cross-sectional area of the fiber and, in turn, its force-generating capacity.

How do different types of muscle fibers affect the speed of animals?

We have just seen that slow-twitch muscle fibers provide endurance for sustained activity, whereas larger fast-twitch fibers produce greater speed and strength but fatigue quickly. The relative distribution of these fiber types in the muscles of different animals is influenced mainly by genetic heritage. It affects the locomotive behavior and capacities of both predator and prey animals. Whereas cheetahs have exceptional speed, their muscles tire quickly. As their prey, antelope have muscles that enable endurance for long-distance running and contract rapidly for maneuvering.

Highly aerobic animals such as dogs and antelope have large fractions of slow-twitch fibers *and* specialized fast-twitch muscle fibers that have a high oxidative capacity and use glycolysis for rapid ATP synthesis. These muscle fibers allow the animals to move for extended periods of time at relatively fast speeds. In contrast, cats have mainly fast-twitch fibers, enabling them to sprint and pounce quickly. Nevertheless, cats and other animals use particular muscles with high concentrations of slow-twitch fibers (these are the dark-staining fibers shown in Fig. 35.14a) for slower postural movements and stealth. Animals such as sloths and lorises that move slowly have mostly oxidative slow-twitch fibers and few glycolytic fast-twitch fibers.

More generally, animals with high body temperatures and high metabolic rates (birds and mammals) have larger numbers of oxidative fibers compared with animals that have lower body temperatures and lower metabolic rates (reptiles). As a result, lizards use glycolytic fast-twitch fibers to sprint for brief periods but must then recover from the acid buildup produced by their muscles' anaerobic activity by resting for longer periods (Chapter 38). In contrast, birds and mammals can sustain activity over longer time periods because they have oxidative muscle fibers supported by an aerobic ATP supply.

Most fish have a narrow band of slow-twitch red muscle fibers that runs beneath their skin, which they use for slow, steady swimming. Tuna and some sharks have evolved deeper regions of red muscle that they can keep warm (Fig. 35.14b), enabling their muscles to contract longer and faster to enhance their swimming. When escaping from a predator or attacking prey, fish recruit their larger fast-twitch white trunk musculature for rapid swimming.

The evolutionary arms race between predators and prey has resulted in selection for particular muscle fiber types and other musculoskeletal specializations for rapid movement and maneuvering. Features of the skeleton that contribute to these specializations are discussed next.

Self-Assessment Questions

6. Why can you lift a larger load when your elbow is slightly flexed and your biceps muscle is at an intermediate length than you can when your elbow is fully extended?

7. Draw a graph of the isometric force–length relationship of striated muscle, indicating where maximal overlap between actin and myosin filaments occurs.

8. Draw a graph of the force–shortening velocity relationship of striated muscle.

9. Compare the force produced by a muscle over time when it is stimulated by a single twitch stimulus with the force produced by multiple stimuli at low frequency, and then with the force produced when the stimulation frequency is increased.

10. What are two mechanisms by which the force of contraction can be increased in vertebrate muscles?

35.3 ANIMAL SKELETONS

Most animal skeletons provide a rigid set of elements that meet at joints (articulate). These rigid elements transmit forces generated by muscles for movement and body support. They also enable many animals to manipulate their environment by digging burrows, building nests, or constructing webs to catch food. Animals have evolved three types of skeletal systems: **hydrostatic skeletons**, **exoskeletons**, and **endoskeletons**. All three types of skeletons have rigid elements that resist the pull of antagonist sets of muscles.

Hydrostatic skeletons support animals by muscles that act on a fluid-filled cavity.

Hydrostatic skeletons evolved early in multicellular animals, with the first cnidarians (jellyfish and sea anemones) approximately 600 million years ago. Today they are found in nearly all multicellular animals as well as in many vascular plants. In animals that depend on a hydrostatic skeleton, fluid contained within a body cavity serves as the supportive component of the skeleton. Muscles exert pressure against the fluid to produce movement.

Two sets of opposing muscles surround the fluid. By contracting and relaxing these opposing muscle sets, an animal can control the width and length of the body cavity and, in many cases, of the whole animal (**Fig. 35.15a**). Circular muscles reduce the diameter of the body cavity, and longitudinal muscles reduce its length. A sea anemone, for example, can bend its body in various directions to feed or to resist water currents by closing its mouth to keep its fluid volume constant and contracting its circular and longitudinal muscles. It extends itself into the water column by contracting its circular muscles and relaxing its longitudinal muscles. The sea anemone can also adjust its shape by altering the amount of fluid in its body cavity. For example,

FIG. 35.15 Diverse examples of hydrostatic skeletal support. (a) In sea anemone; (b) in earthworm; (c) intervertebral discs; (d) articular cartilage.
Photo sources: a. Carole Valkenier/Getty Images; b. Nigel Cattlin/Alamy.

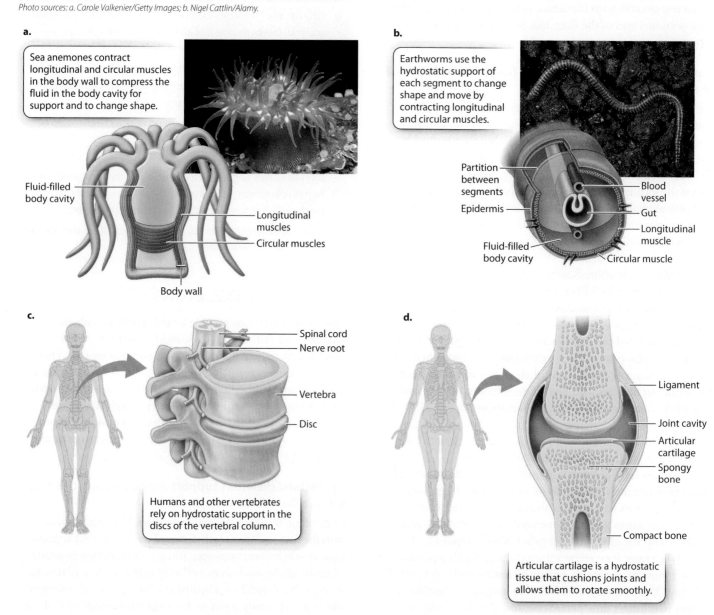

when threatened, a sea anemone rapidly retracts by contracting its longitudinal muscles and allowing water to escape from its body cavity.

Similarly, an earthworm burrows through the soil by causing movement in a series of hydrostatic segments along its body (**Fig. 35.15b**). By contracting longitudinal muscles in a few segments, the earthworm shortens those segments and also widens them to anchor those segments against the soil. It then extends intervening segments between anchor points by contracting circular muscles, causing the segments to lengthen and move the animal's body through the soil.

Because of their simple but effective organization, hydrostatic skeletons are well adapted for many uses. In animals such as squid and jellyfish, the hydrostatic skeleton has become specialized for jet-propelled locomotion. Large circular muscles contract to eject fluid from the body cavity, propelling the animal in the opposite direction. The muscles of other mollusks also exert pressure against a hydrostatic skeleton. By this means, for example, a clam uses its muscular "foot" to burrow into the sediment, and an octopus adeptly manipulates its flexible tentacles. (Clams also have shells that form an exoskeleton to protect the body.) These same principles allow an elephant to control its trunk, and the human tongue to manipulate food and assist in speech.

Even vertebrate animals with rigid endoskeletons have hydrostatic elements that provide flexibility and cushion loads transmitted by the skeleton. These elements include **intervertebral discs** and **cartilage.** Intervertebral discs are sandwiched between the bony vertebrae of the backbone (**Fig. 35.15c**). Each disc has a wall reinforced with connective tissue that surrounds a jelly-like fluid. Intervertebral discs enable the backbone to twist and bend. Disc walls damaged by repeated stress can rupture, causing significant lower back pain. Articular cartilage, another fluid-filled tissue, forms the joint surfaces between adjacent bones (**Fig. 35.15d**). The fluid within the cartilage resists the pressure between the ends of articulating bones, cushioning forces transmitted across the joint.

Exoskeletons provide hard external support and protection.

The first skeletons, consisting of needle-like elements called spicules hardened by calcium carbonate or silica, arose with sponges approximately 650 million years ago. Exoskeletons, rigid skeletons that lie outside the animal's soft tissues, subsequently evolved in other invertebrate groups adapted for life in aquatic and terrestrial environments (**Fig. 35.16**). Because exoskeletons are on the exterior of the animal, the muscles attach from the inside. Exoskeletons provide hard external support and protection. However, a main disadvantage is that exoskeletons limit growth of the organism.

Different invertebrates have different kinds of exoskeleton, but all types have relatively few cells for their volume of tissue. The shells of bivalve mollusks, such as clams and mussels, form a hard, mineralized calcium carbonate exoskeleton (Fig. 35.15a). So does the shell of the cephalopod mollusk *Nautilus,* a creature that has existed largely unchanged for nearly 500 million years (Fig. 35.15b). Mineralization is the process in which a mineral, in this case calcium carbonate, is laid down on an organic matrix, in this case a small amount of protein. The mixture of calcium carbonate and protein is an example of a composite material, one that combines substances with different properties. Its composite nature means the resulting shell is less brittle and more difficult to break than if it were made of just calcium carbonate. The shell expands as the organism grows and epidermal cells deposit new layers of mineralized protein in regular geometric patterns. Indeed, you can see these growth rings if you examine a mollusk shell. Although some marine bivalves achieve large size over many years, growth is limited by the rate at which new skeletal material can be deposited on the outside of the shell.

Arthropods have more complex exoskeletons. More than half of all known animal species, including insects and crustaceans, are arthropods. Arthropods are defined in part by their jointed legs and in part by a type of rigid exoskeleton called a **cuticle** that covers their entire body (Fig. 35.16c). Arthropod exoskeletons, which first evolved in aquatic crustaceans, protect animals from desiccation and physical insults. Consequently, exoskeletons were key to the success and diversification of insects in terrestrial environments approximately 450 million years ago.

The cuticle of insects is composed mainly of **chitin**, nitrogen-containing polysaccharide fibers that are reinforced by proteins. When initially formed, the cuticle of arthropods is soft, so the animal can grow before the cuticle hardens. Mature cuticle consists of two layers. A thin outer, waxy layer minimizes water loss. Water loss is especially dangerous for terrestrial

FIG. 35.16 Examples of exoskeletons. The shells of (a) a bivalve mollusk and (b) *Nautilus*, made of calcium carbonate, surround and protect internal soft body parts. The exoskeleton of arthropods, such as (c) a grasshopper, is made of stiff cuticle that surrounds internal muscles and tendons and other internal body parts. *Sources: a. Rita van den Broek/AGE Fotostock; b. Reinhard Dirscherl/Getty Images; c. Jim Simmen/Getty Images.*

arthropods, given their small size. The much thicker inner layer, whether flexible or stiff, is tough and hard to break. The cuticle remains flexible at joints, allowing motion between body segments. Because the rigid exoskeleton restricts their growth, arthropods shed their cuticle at intervals, a process called **molting**. After molting, arthropods expand and grow for a period before forming a new rigid exoskeleton.

Marine crustaceans, such as crabs and lobsters, incorporate calcium carbonate into their cuticle. Calcium carbonate makes the exoskeleton hard and stiff, much like the hardened cuticle of terrestrial insects, helping to protect these crustaceans from predators.

While offering several protective benefits, exoskeletons pose risks as well. Animals are vulnerable when the newly formed exoskeleton has not yet hardened. Exoskeletons are also hard to repair. If a skeleton is damaged while growing, the animal must produce an entirely new one. Because the exoskeleton is a thin-walled structure, it is prone to breaking if its surface area is very large. Consequently, the imperviousness of the monstrous insects depicted in science-fiction films is improbable—they would easily break from a sharp blow to their exoskeleton.

The bones of vertebrate skeletons consist of a variety of tubular, rodlike, and platelike elements that form a scaffold to which the muscles attach. Because bones are rigid and hard, there are joints between adjacent bones to enable movement. Muscles attach to the skeleton by connective tissue, often specialized as **tendons** made of collagen. Tendons transmit muscle forces, allowing the forces to be redirected and transmitted over a wide range of joint motion. Tendons, like a spring, also store and recover elastic energy.

The vertebrate skeleton can be divided into **axial** and **appendicular** regions (**Fig. 35.17**). The axial skeleton consists of the skull and jaws of the head, the vertebrae of the spinal column, and the ribs. Bones of the skull protect the brain and sense organs of the head. The appendicular skeleton consists of the bones of the limbs, including the shoulder and pelvis. The axial and appendicular regions reflect the evolutionary ancestry of vertebrates. The axial skeleton formed first, to protect the head and provide support and movement of the animal's body by bending its body axis. As a result, the axial skeleton is the main skeletal component of fishes. The appendicular skeleton includes the pectoral and pelvic fins of fishes, which evolved into limbs when vertebrate animals first moved onto land. In

The rigid bones of vertebrate endoskeletons are jointed for motion and can be repaired if damaged.

Endoskeletons first evolved approximately 350–500 million years ago, when the first vertebrates evolved endoskeletons made of cartilage, a gel-like tissue reinforced by protein and sugar polymer fibers. Lampreys are descendants of these early vertebrates. The cartilage endoskeleton became stiffened in other vertebrates through the incorporation of calcium phosphate mineral, or was replaced by bone. The calcium phosphate skeletons that evolved in marine vertebrates stand in contrast to the calcium carbonate skeletons that evolved in marine invertebrates.

An endoskeleton, made of either bone or calcified cartilage, lies internal to most of a vertebrate's soft tissues. In contrast to exoskeletons, vertebrate endoskeletons can grow extensively and, when broken, can be repaired. Endoskeletons also provide protection for certain key internal organs, such as the brain, lungs, and heart.

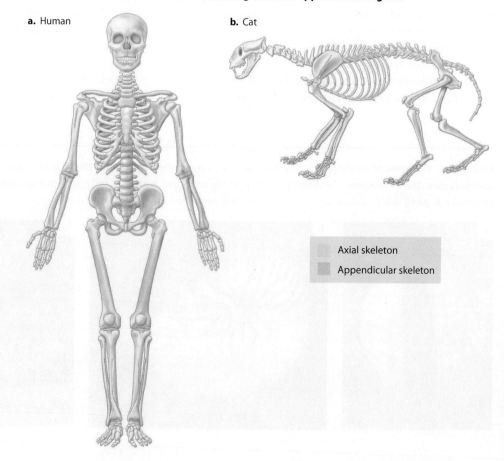

FIG. 35.17 The vertebrate skeleton, showing axial and appendicular regions.

a. Human

b. Cat

Axial skeleton
Appendicular skeleton

a special group of lobe-finned fishes, the fin bones became adapted for an amphibious lifestyle, evolving into the limb bones that terrestrial vertebrates use for support against gravity and for movement on land.

Skeletal tissues have relatively few cells compared to their volume because they consist mostly of an extensive **extracellular matrix** that is secreted by specialized cells. Thus, most of a skeletal tissue is connective tissue external to the cells (Chapter 10). Five main extracellular tissues are produced in the formation of the vertebrate skeleton: bone, tooth enamel, dentine, tendon, and cartilage.

Bone tissue is formed by cells called **osteoblasts**. These cells synthesize and secrete calcium phosphate as **hydroxyapatite** mineral crystals in close association with the protein **collagen**. Mineralized bone is removed by **osteoclasts** when necessary to reshape a bone or repair damaged regions. Generally, bone consists of two-thirds hydroxyapatite mineral and one-third type I collagen protein. The combination of mineral and protein makes bone a composite material, like an exoskeleton. Bone is rigid (because of the properties of the mineral), but also hard to break because it absorbs much energy before fracture (because of the properties of the collagen). During the normal process of aging, bone gradually becomes more mineralized and brittle. The genetic bone disorder osteogenesis imperfecta leads to bones that are brittle and fragile because the osteoblasts produce insufficient collagen in the extracellular matrix.

Articular cartilage is located at the joint surfaces of a bone. It forms a cushion between the bones meeting at a joint. As a gel-like matrix, cartilage resists fluid being squeezed out when forces press on the cartilage during movement. Articular cartilage consists of 70% water; 15% large, branched molecules called proteoglycans that bond with water molecules; and 15% type II collagen. The collagen fibers reinforce the cartilage, allowing the cartilage to cushion and distribute loads as the joint moves. Cartilage may degenerate for a variety of reasons, including rheumatoid arthritis and repetitive physical injury, but in general it lasts over an individual's lifetime.

Self-Assessment Questions

11. What are the two basic structural elements common to all animal musculoskeletal systems?
12. How are exoskeletons and endoskeletons alike? How are they different?
13. Glass is strong and rigid, but can be easily broken. In contrast, a bone is strong and rigid, but hard to break. How can this be?
14. What are the two primary components of bone tissue? How does each contribute to a bone's strength, stiffness, and resistance to fracture?

35.4 VERTEBRATE SKELETONS

Although both vertebrate skeletons and invertebrate exoskeletons are rigid and have flexible joints, certain features distinguish them. Vertebrate skeletons are adapted to grow quickly in the fetus and grow continuously, without molting, until adulthood. Even after bones stop growing, they can be repaired if damaged. Here we explore the biology of vertebrate skeletons, focusing on how bones develop and grow and how they are repaired and remodeled.

Vertebrate bones form directly or by forming a cartilage model first.

Vertebrate bones form and develop by two distinct processes: either they are produced directly as bone or they are generated by way of a cartilage model. Bones of the skull and the ribs form when precursor cells differentiate into osteoblasts that immediately begin producing bone as a skeletal sheet or membrane. These bones are often referred to as membranous bones. This method of forming bone is very slow, however. Indeed, if most bones in the body formed this way, the fetus would take a very long time to develop.

Most other bones of the body—the vertebrae, shoulder, pelvis, and limb bones—are first formed as cartilage (**Fig. 35.18**). The precursor cells initially become cartilage-producing cells, known as chondroblasts, which in the embryo produce cartilage models of most bones of the developing body (Fig. 35.18a). For example, at approximately 9 weeks of development, a human fetal hand consists mainly of cartilage models of the hand bones (Fig. 35.18b). Embryonic cartilage is the same as the articular cartilage that remains at the ends of bones to form the joint surfaces. Because cartilage is a pliable, fluid-filled matrix, it grows by expansion within the structure as well as at the surface, enabling the fetal skeleton to grow rapidly. As the fetus grows and its skeleton matures, the cartilage model of the bone is invaded by blood vessels. The presence of blood vessels triggers the transformation of cartilage into hard, mineralized bone tissue.

The two main types of bone are compact bone and spongy bone.

The cartilage template of a bone is replaced by one of the two main types of bone: compact or spongy (**Fig. 35.19**). Both types contribute to the overall form of many bones in the body.

Compact bone forms the walls of the bone's shaft. It consists mainly of dense, mineralized bone tissue containing bone cells derived from osteoblasts and the network of blood vessels supplying these cells with oxygen and nutrients.

A second type of bone, called **spongy bone**, consists of small plates and rods known as **trabeculae** with spaces between them. Spongy bone is found internally at the ends of limb bones and within the vertebrae. It reduces the weight of bone in these regions, but is also stiff so that forces are transmitted effectively

FIG. 35.18 Bone formation from a cartilage model. (a) Most bones other than the skull and ribs are formed as cartilage first, followed by invasion by blood vessels and the transformation into bone tissue by osteoblasts. (b) Cartilage model of a human fetal hand, showing initial sites of bone tissue formation at about 9 weeks of embryonic development. *Photo source: SPL/Science Source.*

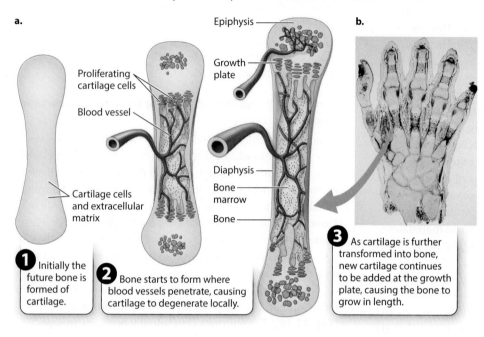

FIG. 35.19 Compact and spongy bone. *Source: Dave King/Getty Images.*

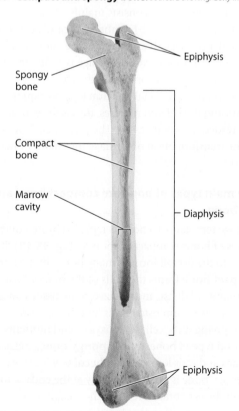

from the joint to the bone shaft. **Bone marrow** is a fatty tissue found between trabeculae and also within the bone's central cavity. It contains many important cells, including blood-forming cells, other stem cells (Chapter 19), immune system cells (Chapter 41), and fat cells.

Bones grow in length and width, and can be repaired.

Bones can grow in length, at least until an animal reaches maturity. Growth in length occurs at the **growth plate** (see Fig. 35.18), a region of cartilage between the middle region of the bone, called the **diaphysis**, and the end, called the **epiphysis**. The growth plate remains in place when the rest of the cartilage is transformed into bone. It adds new cartilage toward the bone's diaphysis, enabling bone length to continue to increase after birth (see Fig. 35.18). At maturity, each growth plate fuses as the remaining cartilage is replaced with bone, preventing further growth in length. The fusion of growth plates is typical of most mammals and birds. In amphibians and reptiles, the growth plates often do not fuse and growth may continue, albeit at a slower pace, over much of their lifetime.

Limb bones grow in diameter when osteoblasts deposit new bone on the bone's external surface. At the same time, bone is removed from the inner surface by osteoclasts, expanding the marrow cavity. Bone removal is slower than bone growth, thickening the walls during growth. Osteoclasts secrete digestive enzymes and acid to dissolve the calcium mineral and collagen. The dissolved compounds are reabsorbed and recycled for use in bone formation in other regions of the skeleton. Through this process, the vertebrate skeleton serves as an important store of calcium and phosphate ions. For example, in female birds, calcium removed from the skeleton is regularly used to form the eggshell.

Generally, physically active younger adults have thicker bones than less active younger adults because physical loading stimulates osteoblasts to produce more bone during growth. However, in older adults, bone tissue is gradually lost as osteoclasts remove more bone than is produced by osteoblasts. Bone loss is particularly severe in women after menopause, in part because of hormone shifts as well as reduced physical activity. This bone loss can lead to osteoporosis, a condition that significantly increases the risk of bone fracture.

A great advantage of endoskeletons compared with exoskeletons is that a damaged endoskeleton can be repaired. Osteoblasts and osteoclasts form and remove mineralized tissue as needed in damaged regions of the bone.

CASE 7 BIOLOGY-INSPIRED DESIGN: USING NATURE TO SOLVE PROBLEMS

Can we design smart materials to heal bones that are damaged and improve artificial joints?

Despite their elegant organization, which helped inspire the design of the Eiffel Tower, bones are often damaged due to fracture from accidents, extreme physical activity, and disease. Damaged bone is a major health problem for people of all ages, but particularly for older individuals.

New bone tissue engineering techniques that make use of three-dimensional printing hold promise to solve this problem. The strategy is to create artificial polymer–ceramic scaffolds that incorporate stem cells, growth factors, and bioactive molecules. With the help of recent advances in three-dimensional printing, such as solid free-form printing, researchers create complex geometries of multicomponent materials at microscopic scales that mimic the structure and chemical composition of bone tissue. After being implanted in the body, this scaffold is infiltrated with stem cells and growth factors to stimulate bone formation. Once the artificial scaffold is remodeled into natural bone tissue, the scaffold degrades.

As we have seen, the structure of bone is complex: it consists of a compact shell that is supported by plates and struts consisting of a more porous inner spongy bone. The challenge for bone tissue engineers and material scientists is to create three-dimensional scaffolds that mimic this complex structure. The scaffolds need to provide similar mechanical support when initially grafted to a damaged region of a bone, induce the formation of blood vessels for nutrient supply, and stimulate the cellular activity of osteoblasts. These processes of bone healing are critical to the remodeling of the artificial scaffold into natural bone tissue. An advantage of bioengineering artificial bone tissue, rather than relying on a bone graft from another donor, is that the risk of infection during surgical implantation is greatly reduced.

In addition to repairing damaged regions of bone, bone tissue engineering techniques are improving the long-term viability of joint prostheses, such as artificial hips and knees, which are commonly used in both humans and their pets. Often a joint prosthesis fails because its connection to the surrounding bone tissue loosens. By coating the surface of the prosthesis base with bioactive components, such as growth factors and stem cells, and by utilizing a three-dimensional porous surface, osteoblasts can be stimulated to form new bone tissue that grows into the prosthesis, reinforcing its interface to bone and greatly increasing the useful lifetime of an artificial joint.

Joint shape determines range of motion and skeletal muscle organization.

The shapes of the bone surfaces that meet at a joint determine the range of motion at that joint. Joints range from simple **hinge joints** to **ball-and-socket joints** (Fig. 35.20). The human elbow joint and the ankle joint of a dog are examples of hinge joints. The shoulder and hip joints are examples of ball-and-socket joints, which allow the widest range of motion. The joints at the base of each finger are intermediate: they allow you to flex and extend your fingers as well as spread them laterally or move them together when making a fist or grasping objects.

Joints with a larger range of motion are generally less stable than joints with a smaller range of motion. The shoulder is the most mobile joint in the human body, yet one of the joints that is most often dislocated or injured. In contrast, the ankle joint of dogs, horses, and other animals is a stable hinge joint that is unlikely to be dislocated, but its range of motion is limited to flexion and extension. Because muscles are arranged as paired sets of antagonists to move a joint in opposing directions, hinge

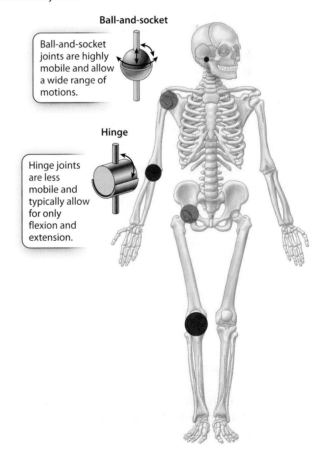

FIG. 35.20 Hinge and ball-and-socket joints. The elbow and knee are examples of hinge joints. The shoulder and hip are examples of ball-and-socket joints.

joints may be controlled by as few as two antagonist muscles (generally referred to as a flexor and an extensor). In contrast, ball-and-socket joints have at least three sets of muscle antagonists to control motion in many different planes. As a result, a more complex organization of muscles is needed to control the movements of the arm at the shoulder joint or the leg at the hip joint.

> **Self-Assessment Questions**
>
> 15. Diagram a limb bone, such as the tibia, showing the regions of articular cartilage, spongy trabecular bone, compact bone tissue, and marrow cavity.
>
> 16. The shoulder joint provides a good example of a trade-off, where a benefit comes with a cost. What is the trade-off?

CORE CONCEPTS SUMMARY

35.1 HOW MUSCLES WORK: Muscles are biological motors composed of actin and myosin that generate force and produce movement for support, locomotion, and control of internal physiological functions.

There are two main types of muscles: striated (skeletal and cardiac) and smooth. page 788

Skeletal and cardiac muscle fibers appear striated when viewed under the microscope because of the regular spacing of sarcomeres along their length. page 788

Smooth muscle fibers, which regulate air flow for breathing, blood flow through arteries, and the passage of food through the gut, lack regular sarcomere organization and appear smooth when viewed under the microscope. page 788

Skeletal muscle fibers are long, thin cells composed of parallel sets of myofibrils built up from smaller parallel arrays of actin and myosin filaments. page 789

The sarcomere is the basic contractile unit of a skeletal muscle. Sarcomeres are arranged in series along the length of a myofibril. page 789

Muscles change length and produce force by forming actin–myosin cross-bridges, which cause myosin and actin filaments to slide relative to each other. page 791

Muscles are stimulated by motor neurons at the fiber's motor endplate, leading to depolarization of the muscle cell that triggers the release of Ca^{2+} ions from the sarcoplasmic reticulum. Ca^{2+} binds troponin, which moves tropomyosin off the myosin-binding sites on actin, allowing myosin heads to form cross-bridges with actin. page 792

Excitation–contraction coupling is the process by which depolarization of the muscle cell leads to its shortening. page 793

35.2 MUSCLE CONTRACTILE PROPERTIES: The force generated by a muscle depends on muscle size, degree of actin–myosin overlap, shortening velocity, and stimulation rate.

Because muscles can transmit force only by pulling on the skeleton, they are arranged as antagonist pairs to produce reciprocal motions of a joint or limb. page 794

Muscles exert their greatest force when actin–myosin filaments have maximal overlap, allowing the largest number of cross-bridges to form. page 794

Muscles exert more force when they contract at slow velocities, compared with when they contract at high velocities. page 796

In vertebrates, a motor unit consists of a single motor neuron and the muscle fibers (cells) it innervates. page 797

Muscle force is increased by increasing the motor neuron firing rate and, in vertebrates, by increasing the number of motor units that are activated. page 797

Vertebrate muscles contain two types of fibers: red slow-twitch fibers that contract slowly over longer time periods and white fast-twitch fibers that contract rapidly but fatigue quickly. page 798

35.3 ANIMAL SKELETONS: Hydrostatic skeletons, exoskeletons, and endoskeletons provide animals with mechanical support and protection.

The rigid element of a hydrostatic skeleton is an incompressible fluid within a body cavity, used to support the body and change shape. page 799

Invertebrate exoskeletons form an external rigid support system, which protects the animal and limits water loss, but also limits growth and repair. page 801

Vertebrates have a bony endoskeleton that provides rigid support and protection of body organs, and that can grow and be repaired. page 802

The vertebrate endoskeleton is organized into axial (central) and appendicular (limb) components. page 802

Vertebrate endoskeletons consist of bone, which is a composite tissue of calcium phosphate mineral and type I collagen, and cartilage, which is a fluid-based gel that provides cushioning at joint surfaces. page 803

35.4 VERTEBRATE SKELETONS: Vertebrate endoskeletons allow for growth and repair and transmit muscle forces across joints.

Vertebrate bones develop by one of two processes: either directly or by way of a cartilage model. page 803

Cartilage forms much of the embryonic skeleton and remains as growth plates within the bone to provide rapid growth after birth, as well as forming the bone's joint (articular) surfaces. page 803

Bone has two basic structures: compact bone forms the solid walls of a bone's shaft, and spongy bone forms a mesh that supports the cartilage at the bone's ends. page 803

Bone formation by osteoblasts and removal by osteoclasts is a continuing process of growth, shape change, and repair. page 804

The shape of a bone's joint surfaces largely determines the range of motion and stability of a joint. page 805

Log in to LaunchPad to check your answers to the Self-Assessment Questions and to access additional learning tools.

CHAPTER 36 Animal Endocrine Systems

CORE CONCEPTS

36.1 ENDOCRINE FUNCTION: Animal endocrine systems release hormones into the bloodstream that respond to environmental cues, regulate growth and development, and maintain homeostasis.

36.2 HORMONES: Hormones bind to receptors on or inside target cells, and their signals are amplified to exert strong effects on target cells.

36.3 THE VERTEBRATE ENDOCRINE SYSTEM: In vertebrates, the hypothalamus and pituitary gland control and integrate diverse physiological functions and behaviors.

36.4 OTHER FORMS OF CHEMICAL COMMUNICATION: Chemical communication can occur locally between neighboring cells or between individuals, coordinating social interactions.

We have seen how the nervous system allows an animal to sense and respond to its environment, coordinating the animal's movement and behavior (Chapters 34 and 35). The nervous system also works closely with the **endocrine system** to regulate an animal's internal physiological functions. The endocrine system is an interacting set of secretory cells, glands, and organs, located in different regions of the body, which communicates using chemical signals. It responds to signals sent from the nervous system as well as to blood-borne signals from other organs to regulate internal bodily functions.

The signals communicated by the nervous and endocrine systems differ greatly in their modes of transmission and in the times over which they act. The nervous system sends signals rapidly by action potentials running along nerve axons, and adjacent nerve cells communicate by means of neurotransmitters. In contrast, cells and glands of the endocrine system secrete chemical signals called **hormones**, which influence the actions of other cells in the body. Whereas some hormones act locally, many others are released into the bloodstream. Circulating throughout the body, these hormones can influence distant target cells. As a result, endocrine communication is generally slower and more prolonged than the rapid and brief signals transmitted by nerve cells.

Hormones bind to receptors on or in target cells. Once bound, each hormone exerts a specific effect on those cells. Hormonal signals are amplified in a series of steps, so that even a small amount of hormone can have dramatic effects in the body (Chapter 9).

In this chapter, we explore the properties and actions of hormones that help to coordinate and maintain a broad set of physiological functions of the organism. In particular, we examine the mechanisms by which hormones trigger cellular responses in their target organs. Because most hormones are evolutionarily conserved, their chemical structure is similar across a diverse array of animals. However, their functions often evolve rapidly, enabling hormones to serve new and broader roles.

36.1 ENDOCRINE FUNCTION

The word "hormone" often popularly connotes teenagers and the changes that occur to their bodies as they grow and mature. Hormones do play a key role in animal growth and development, but that is just one of their many diverse roles. In particular, hormones regulate an organism's physiological response to the environment and help maintain stable physiological conditions within cells or within the animal as a whole. We start by highlighting these functions, before addressing the molecular mechanisms of how hormones work in section 36.2.

The endocrine system helps to regulate an organism's response to its environment.

Organisms face constant changes in their environment. These changes often present physiological challenges or stresses that the organism must respond to by altering the functional state of its body. Changes in light or temperature, the threat of a predator, or the presence of a potential mate stimulate responses of an animal's nervous system and endocrine system. Environmental cues activate the nervous system, which relays signals to the endocrine system, triggering a response. The endocrine system, with its slower and more prolonged signaling, reinforces physiological changes in the animal's body that better suit environmental conditions.

When a gazelle sees or smells a predator, its endocrine system helps to ready its body for rapid escape. When a female cardinal sees a bright red, singing male early in spring, endocrine signals released from within the female cardinal's brain initiate changes in its reproductive organs and increase its behavioral responsiveness, encouraging it to select the male as its mate. And when particular smells stimulate a sense of hunger and digestive function in animals such as dogs and humans, those animals are attracted to food. In each of these cases, the endocrine system helps the animal respond appropriately to environmental cues.

The endocrine system regulates growth and development.

Growth and development require broad changes in many different organ systems. The release of circulating hormones from the endocrine system accomplishes these changes, regulating how animals develop and grow. For example, in Chapter 40, we discuss the importance of the sex hormones estrogen and testosterone in determining female or male sexual characteristics both in the embryo and during puberty.

As another example, consider how the well-studied endocrine system of insects regulates their growth and development. Insects produce a rigid exoskeleton (Chapter 35), which must be periodically shed and replaced by a larger new one to enable the insect to grow (**Fig. 36.1**). Shedding of the exoskeleton is referred to as **molting**. Each molting produces a new larval stage. In some insects, such as moths, the animal's body also changes dramatically, undergoing **metamorphosis** at a key stage in development (Fig. 36.1a). In other insects, such as grasshoppers and crickets, the animal grows larger after each molt, but with little change in body form (Fig. 36.1b).

Molting and metamorphosis are regulated by hormones released from tissues in the insect's head. British physiologist Vincent Wigglesworth studied the effects of hormones on the blood-sucking insect *Rhodnius*. Normally, *Rhodnius* goes through five juvenile stages before developing into its final adult body form. Like grasshoppers and crickets, it does not undergo metamorphosis. Each of the developmental stages is triggered by a blood meal. Wigglesworth showed that he could prevent molting if the head of the insect were removed shortly after a blood meal (**Fig. 36.2**). *Rhodnius* can survive for long periods despite having its head removed. If the insect takes a blood meal and the head is removed a week later, the animal's body undergoes a molt and grows into its adult form. Wigglesworth hypothesized that a substance diffusing from the animal's head controls its molt. If the head is removed shortly after a blood meal, this substance does not have time to reach the insect's body to trigger a molt; if the removal is delayed, it does.

The German endocrinologist Alfred Kuhn obtained a similar result in experiments performed on moth caterpillars. Tying the caterpillar's head off from the rest of the body prevented the body from molting. Wigglesworth and Kuhn considered the diffusing substance to be a brain hormone. Subsequent work

FIG. 36.1 Growth and development in (a) moths and caterpillars and in (b) grasshoppers.

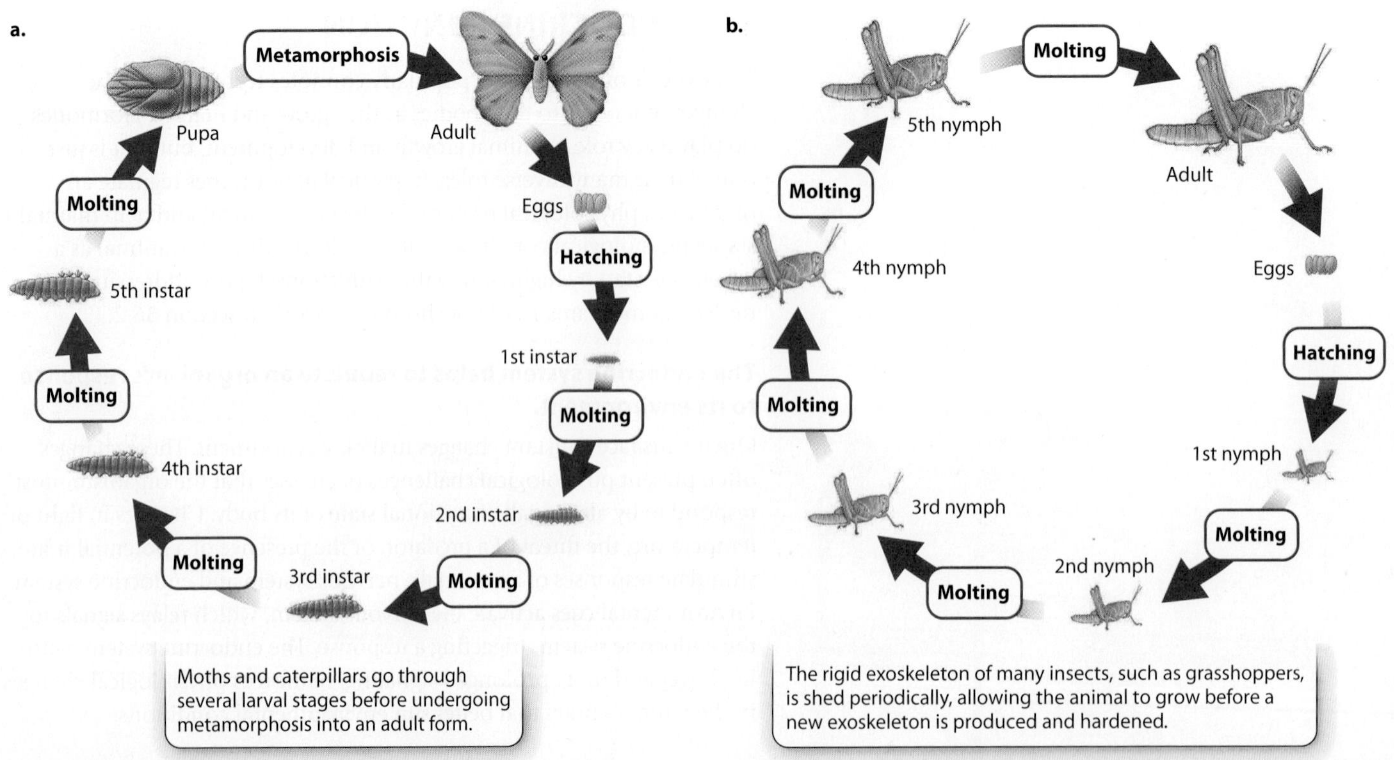

Moths and caterpillars go through several larval stages before undergoing metamorphosis into an adult form.

The rigid exoskeleton of many insects, such as grasshoppers, is shed periodically, allowing the animal to grow before a new exoskeleton is produced and hardened.

HOW DO WE KNOW?

FIG. 36.2

How are growth and development controlled in insects?

BACKGROUND During the 1930s, British physiologist Vincent Wigglesworth studied how a blood meal taken by the bug *Rhodnius* triggered molting and growth. This work pioneered the discovery of hormonal substances that stimulate the bug's growth. *Rhodnius* goes through five successive larval stages (called nymphs) before becoming a winged adult, as illustrated in Fig. 36.2a. Each developmental step is triggered by a blood meal.

HYPOTHESIS Wigglesworth hypothesized that a substance (specifically, a hormone) that diffuses from the head triggers molt in *Rhodnius*.

EXPERIMENT 1 AND RESULTS Wigglesworth decapitated juvenile bugs at different intervals of time after a blood meal and observed whether molting occurred. (Decapitation does not kill the insect.) He showed that if a bug is decapitated less than an hour after a blood meal, it fails to molt. If a bug is decapitated a week after the blood meal, it molts (Experiment 1; Fig. 32.2b).

EXPERIMENT 2 AND RESULTS Wigglesworth also found that if he used a fluid tube to join the body of a bug decapitated immediately after a blood meal with the body of one that wasn't decapitated until a week after feeding, the bodies of both bugs molted (Experiment 2; Fig. 32.2b). This experiment demonstrated that the diffusing hormone triggers molting in a bug that lacks the hormone because of an immediate decapitation.

CONCLUSION A substance diffuses from the head in *Rhodnius* and triggers the molting process.

FOLLOW-UP WORK In 1936, the German endocrinologist Alfred Kuhn confirmed Wigglesworth's result in experiments performed on moth caterpillars. More recently, Japanese molecular endocrinologists Atsushi Nagasawa and colleagues (1984) and Hiromichi Kawakami and colleagues (1990) isolated and sequenced the structure of the hormone from silkworm moths. It is now known as prothoracicotropic hormone (PTTH), because it acts on the prothoracic gland of the insect to release the molting hormone, ecdysone.

SOURCES Wigglesworth, V. B. 1934. "The Physiology of Ecdysis in *Rhodnius prolixus* (Hemiptera). II. Factors Controlling Moulting and 'Metamorphosis.'" *Quarterly Journal of Microscopical Sciences* 77:191–223; Nagasawa, H., et al. 1984. "Amino-Terminal Amino Acid Sequence of the Silkworm Prothoracicotropic Hormone: Homology with Insulin." *Science* 226:1344–1345; Kawakami, A., et al. 1990. "Molecular Cloning of the *Bombyx mori* Prothoracicotropic Hormone." *Science* 247:1333–1335.

a.
Development of *Rhodnius*

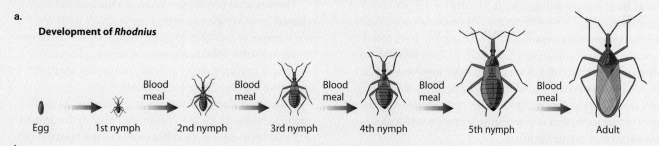

b.

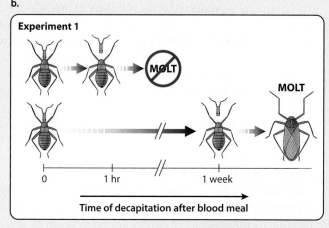

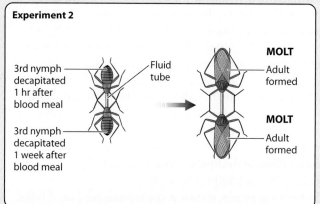

TABLE 36.1 Major Invertebrate Hormones

SECRETING TISSUE OR GLAND	HORMONE	TARGET GLAND OR ORGAN	ACTION
Brain	Brain hormone (PTTH) (peptide)	Prothoracic gland	Stimulates release of ecdysone to trigger molt and metamorphosis
Corpora allata	Juvenile hormone (peptide)	All body tissues	Inhibits metamorphosis to adult stages; stimulates retention of juvenile characteristics
Prothoracic gland	Ecdysone (steroid)	All body tissues	Stimulates molt and, in the absence or low levels of juvenile hormone, stimulates metamorphosis
Nervous system	Ecdysone (steroid)	All body tissues	Stimulates growth and regeneration in sea anemones, flatworms, nematodes, annelids, snails, sea stars
Brain	Melanocyte-stimulating hormone (peptide)	Chromatophores	Stimulates pigmentation changes in cephalopods (octopuses, squid, cuttlefish)
Brain and eye stalks	Chromatotropins (peptides)	Chromatophores	Stimulates pigmentation changes in crustaceans
Reproductive gland	Androgen (peptide)	Reproductive tract	Regulates development of testes and male secondary sexual characteristics of crustaceans

showed that cells in a specialized region of the insect brain secrete the peptide now known as prothoracicotropic hormone (PTTH) (**Table 36.1**). In *Rhodnius,* PTTH is released after a blood meal and acts to trigger the animal's molt by stimulating the release of a second hormone from the prothoracic gland.

Two decades later, the German endocrinologist Peter Karlson isolated and purified this second hormone, termed ecdysone. PTTH triggers molting by stimulating the release of ecdysone, a steroid hormone that coordinates the growth and reorganization of body tissues during a molt. The action of ecdysone is a good example of how a hormone can precisely coordinate broad changes in body organization and function, in contrast to the specific regulation that nerves provide. Karlson was able to purify a small amount of ecdysone from large numbers of silk moth *Bombyx* caterpillars. He ultimately isolated a mere 25-mg sample from 500 kg of moth larvae! This tiny amount highlights the general principle that relatively few hormone molecules can have a large effect on an organism.

In his studies of *Rhodnius,* Wigglesworth also noted that, whatever the larval stage, the decapitated bug always molted into an adult after its blood meal. By removing just the region behind the brain, Wigglesworth was able to show that paired endocrine glands, called corpora allata, in this region of the head release a hormone that normally prevents the earlier larval stages from molting into the adult form. This hormone, called juvenile hormone, is released in decreasing amounts during each successive larval molt. After the fifth and final stage, the level of juvenile hormone is so low that it no longer blocks maturation; thus, at the final molt the insect undergoes its final growth into the adult form (**Fig. 36.3**).

The brains of insects, similar to the nervous systems of other invertebrate and vertebrate animals, contain **neurosecretory cells**. These cells are neurons that release hormones, which act on endocrine glands or other targets within the insect, instead of secreting neurotransmitters that bind to another neuron or to muscle. In the case of insects, peptide hormones released from neurosecretory brain cells act on the corpora allata and the prothoracic gland (Fig. 36.3). The neurosecretory cells stimulate the corpora allata to produce juvenile hormone so that molting larvae retain their juvenile characteristics. At appropriate times, other neurosecretory brain cells secrete the peptide hormone PTTH, which stimulates the release of ecdysone from the prothoracic gland, triggering the transition from larval to pupal and adult forms.

These studies reveal how small amounts of hormones released from key glands within the animal's body regulate major stages of growth and changes in body form during metamorphosis. When an animal's body requires coordinated changes in multiple organ systems, hormonal regulation by the endocrine system plays a critical role.

Hormones also regulate growth in humans and other vertebrate animals. Growth hormone controls the growth of the skeleton and many other tissues in the human body. This hormone is produced by the **pituitary gland**, which is located beneath the brain. Tumors of the pituitary that cause an overproduction of growth hormone lead to gigantism, whereas tumors that result in too little growth hormone cause pituitary dwarfism. The role of the pituitary gland in regulating body function is described later in the chapter.

The endocrine system underlies homeostasis.

Endocrine control of internal body functions is also central to **homeostasis**, the maintenance of a steady physiological state within a cell or an organism (Chapters 5 and 33). Without some way to maintain a stable internal environment, changing environmental conditions would lead to dangerous shifts in an animal's physiological function. For example, an animal's body weight depends on the regulation of its energy intake relative to energy expenditure. Disruption of hormones that regulate appetite and food intake can lead to obesity on the one

FIG. 36.3 Hormonal control of growth and development in *Rhodnius*.

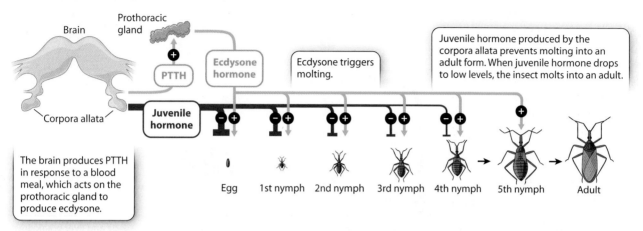

hand or to weakness and lethargy on the other hand. Similarly, hormones that regulate the concentrations of key ions in the body, such as Na⁺ and K⁺, are fundamental to healthy nerve and muscle function (Chapters 34 and 35) and fluid balance within the body (Chapter 39).

How does the body, and in particular the endocrine system, maintain homeostasis? Maintaining homeostasis depends on feedback from the target organ to the endocrine gland that secretes the hormone. Because hormones are transmitted through the bloodstream, this feedback can occur over varying distances within the body, coordinating the functions of several organs at any one time. In response to this feedback, the endocrine gland modifies its own subsequent production of hormone, either increasing or decreasing it. Two general types of feedback are provided: **negative feedback** and **positive feedback**.

FIG. 36.4 Negative feedback. Negative feedback results in steady conditions, or homeostasis.

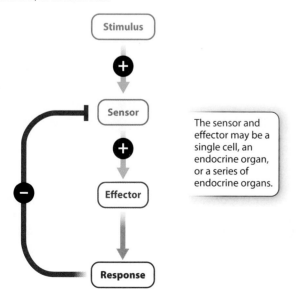

Homeostasis typically depends on negative feedback, in which a stimulus produces a response that in turn counteracts the initial stimulus. In Chapter 33, we saw how both the temperature in a house and the core body temperature of an animal are maintained at a constant level by this type of control.

Many physiological parameters—such as blood glucose and calcium levels—are maintained at relatively steady levels by negative feedback. For example, a change in the glucose or calcium level causes a response that brings the level back to the starting point, called the set point. In these cases, a change in level (the stimulus) is detected by a sensor in the endocrine organ (**Fig. 36.4**). The endocrine organ (the effector) releases a hormone, and the hormone causes a response that opposes the initial stimulus, so that the glucose or calcium level moves back toward the set point. As levels return to the set point, secretion of the hormone decreases, and there is a limit to further change. Constant feedback between the response and sensor maintains a set point. That is, the maintenance of homeostasis is an active process.

Let's examine the example of blood glucose levels in more detail. The amount of glucose in the blood of animals is maintained at a steady level (**Fig. 36.5**). If glucose levels are too low, cells of the body do not have a ready source of energy. Conversely, if glucose levels are too high for too long, they can damage organs. Maintaining steady blood glucose levels is challenging because glucose levels rise immediately after a meal when glucose is absorbed into the bloodstream from the intestine, then fall as glucose is taken up by cells to meet their energy needs.

Immediately after a meal, when blood glucose rises, β (beta) cells of the pancreas secrete the hormone insulin, which circulates in the blood (Fig. 36.5a). The presence of insulin triggers muscle and liver cells to take up glucose from the blood and either use it or convert it to a storage form called glycogen (Chapter 7). In this way, insulin guards against high levels of glucose in the blood.

Several hours after a meal, as blood glucose levels fall, a different population of cells in the pancreas, called α (alpha) cells, secrete the hormone glucagon. Its effects are roughly opposite to those of insulin (Fig. 36.5b): glucagon stimulates the breakdown of glycogen into glucose and its release from muscle and liver cells. The result is that blood glucose levels rise.

In both cases, the stimulus (either high or low blood glucose levels) is sensed by cells of the pancreas (β cells or α cells) and triggers a response (secretion of insulin or of glucagon) that brings blood glucose levels back to the set point. Note that in each case the response feeds back to the secreting cells (negative feedback) to reduce further hormone secretion.

When insulin fails to control blood glucose levels, a disease called diabetes mellitus results. When this disease is untreated, individuals with diabetes excrete excess glucose in their urine because their blood glucose levels are too high. Diabetes causes cardiovascular and neurological damage, including loss of sensation in the extremities, particularly the feet.

In some instances, it is necessary to accelerate the response of target cells for a period of time. That is the role of positive

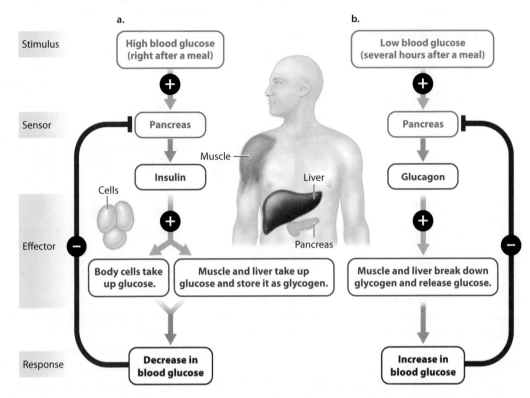

FIG. 36.5 Control of glucose levels in the blood by negative feedback.

feedback (**Fig. 36.6**). In positive feedback, a stimulus causes a response in the same direction as the initial stimulus, which leads to a further response, and so on. In the endocrine system, positive feedback to a stimulus leads to secretion of a hormone that causes a response, and the response causes the release of more hormone. The result is an escalation of the response. A positive feedback loop reinforces itself until it is interrupted or broken by some sort of external signal outside the feedback loop but from within the body.

Positive feedback occurs in mammals during birth (Chapter 40). In response to uterine contractions (the stimulus), the pituitary gland releases the hormone oxytocin. The release of oxytocin causes the uterine muscles (the effector) to contract more forcefully. The uterine contractions, in turn, stimulate (positive feedback) the pituitary gland to secrete more oxytocin, causing the uterine muscles to contract still more forcefully and frequently.

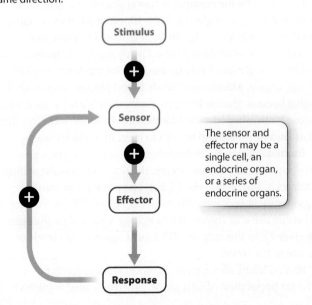

FIG. 36.6 Positive feedback. In positive feedback, a stimulus causes a response, and that response causes a further response in the same direction.

The sensor and effector may be a single cell, an endocrine organ, or a series of endocrine organs.

Self-Assessment Questions

1. What are three general functions of the endocrine system? Provide an example of each.

2. Diabetes mellitus is a disease characterized by high blood glucose levels. Given what you know about how insulin usually prevents high blood glucose levels, can you think of two different ways that diabetes can result?

3. Diagram an endocrine pathway, such as oxytocin regulation of labor during birth, that relies on positive feedback.

36.2 HORMONES

Because hormones are released into the bloodstream, they have the potential to affect many organ systems. How do hormones circulating in the blood target specific tissues? How can one hormone exert its effect on one cell type and another hormone exert its effect on a different cell type? The answer lies in the presence of receptors on cells within target organs—receptors that bind specific hormones.

Hormones act specifically on cells that bind the hormone.

Hormones circulating in the bloodstream bind to receptors located either on the surface of or inside target cells. Therefore, it is the presence or absence of a receptor for a given hormone that determines which cells respond and which do not. For example, when the hormone oxytocin is released into the bloodstream of mammals, it affects only cells that express a receptor on the cell surface capable of binding oxytocin as it flows by in the bloodstream, namely uterine muscle cells and secretory cells in breast tissue. When oxytocin binds to these cell-surface receptors, uterine muscle cells are stimulated to contract, and secretory cells in breast tissue release milk during breastfeeding. The cells of other organs do *not* express the receptor, so the hormone can exert its effect only on these specific tissues and no others.

The binding of a hormone to its receptor triggers changes in the target cell, resulting in a cellular response (Chapter 9). The specific action of a hormone depends on the kind of response that it triggers in the target cell. For example, the binding of a hormone can alter ion flow across the cell membrane, change the biochemical activity of a target cell, or initiate changes in gene expression.

Two main classes of hormones are peptide and amines, and steroid hormones.

Some hormone receptors are located on the cell surface; others are located inside the cell. Where a receptor is located depends on characteristics of the hormone molecules themselves. Hormones can be grouped into two general classes: hydrophilic molecules, which cannot diffuse across the plasma membrane, and hydrophobic molecules, which can. The hydrophilic hormones are **peptide hormones** and **amine hormones**, and the hydrophobic hormones are **steroid hormones**. **Fig. 36.7** shows examples of both types.

Peptide hormones and amine hormones are both derived from amino acids (Figs. 36.7a and 36.7b). Peptide hormones are short chains of amino acids, whereas amine hormones are

---

FIG. 36.7 (a) Peptide, (b) amine, and (c) steroid hormones.

a. Peptide hormones

Oxytocin

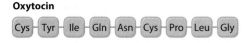

Antidiuretic hormone (ADH)

Oxytocin and ADH are peptide hormones that are 9 amino acids long. Both are released by the posterior pituitary gland and share a similar structure (only 2 amino acids differ).

b. Amine hormones

L-Dopa

Dopamine

Norepinephrine

Epinephrine

Amine hormones are derived from aromatic amino acids. All four shown here are derived from phenylalanine.

c. Steroid hormones

Progesterone

Testosterone

Cortisol

Cholesterol

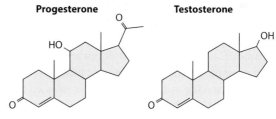

The steroid hormones all share a similar structure and are derived from cholesterol.

derived from a single aromatic amino acid, such as tyrosine or phenylalanine (Chapter 4). Steroid hormones are derived from cholesterol (Fig. 36.7c). Whereas peptide hormones (which are sometimes large enough to be considered proteins) can evolve through changes in their amino acid sequence, steroid hormones cannot. Evolutionary changes in steroid hormone function depend instead on changes in their synthetic enzymes or changes in the receptors they bind and the cellular responses they trigger.

Peptide and amine hormones are more abundant than steroid hormones and are more diverse in their actions. Because most peptide and amine hormones are hydrophilic and cannot diffuse across the plasma membrane (exceptions are thyroid hormones), nearly all of these hormones bind to receptors on the surface of the cell (**Fig. 36.8a**). Peptide and amine hormones alter the activity of the target by initiating signaling cascades within the target cell, as discussed in Chapter 9. The bound receptor activates an enzyme or other molecule within the cell, and it in turn activates another, and so on. Typically, the activated enzymes are protein kinases, which phosphorylate other proteins. These signaling cascades can alter metabolism by activating or inhibiting enzymes, or they can lead to changes in gene expression. For example, a peptide or amine hormone could trigger a cell to grow, divide, change shape, or release another hormone. Peptide and amine hormones act on timescales of minutes to hours.

There are many examples of peptide hormones. In vertebrates, for instance, growth hormone stimulates protein synthesis and the growth of many body tissues, particularly those in the musculoskeletal system (**Table 36.2**). As we saw, insulin and glucagon are peptide hormones released by the pancreas that regulate glucose metabolism by the liver, muscles, and other tissues. Gastrin and cholecystokinin are peptide hormones that regulate mammalian digestive function (Chapter 38). In addition to serving as a neurotransmitter released by the sympathetic nervous system (Chapter 34), epinephrine and norepinephrine (also known as adrenaline and noradrenaline, respectively) are amine hormones that support an animal's fight-or-flight response.

Because steroid hormones are hydrophobic, they diffuse freely across the cell membrane. Steroid hormones bind with receptors in the cytoplasm or nucleus, forming a steroid hormone–receptor complex (**Fig. 36.8b**). Hormone–receptor complexes that form in the cytoplasm are transported into the nucleus of the cell. These complexes most commonly act as transcription factors: they stimulate or repress gene expression, thereby altering the proteins produced by the target cell. Consequently, steroid hormones typically have profound and long-lasting effects on the cells and tissues they target, on timescales of days to months.

The sex hormones estrogen, progesterone, and testosterone (Table 36.2) are steroid hormones that regulate the

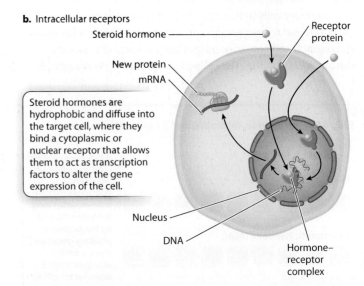

FIG. 36.8 Mechanism of action of (a) peptide and (b) steroid hormones.

development of the vertebrate reproductive organs. Cortisol is a steroid hormone that regulates a vertebrate animal's response to stress and inhibits inflammation. Steroid hormones similar to cortisol are used medically to reduce the symptoms of certain inflammatory disorders. However, they also carry risks; namely, they suppress the immune system, increasing the chance of infection and diminishing wound healing.

Hormonal signals are amplified to produce a strong effect.

Hormones are typically released in small amounts. As mentioned earlier, they can exert strong effects on the overall physiology of an organism by amplification (Chapter 9). Amplification occurs during two series of signaling steps, both before the hormone binds to a cell receptor and inside the cell afterward. The first set of signaling steps is the passing of a signal from one endocrine gland to the next in a hormonal pathway. The second set is signal transduction in the target cell. Signals are amplified at each step of the pathway, resulting in a large effect on the target cell or organ.

Some hormonal signaling pathways involve the release of a hormone by one gland that acts on another gland as its target. The target gland, in turn, releases another hormone that acts on a second gland or tissue in the pathway, and so on. Hormone signaling pathways between endocrine glands and tissues are often referred to as endocrine axes. In vertebrates, hormonal signals are amplified along a pathway called the hypothalamic–pituitary axis that goes from the hypothalamus to the anterior pituitary gland, and then from the anterior pituitary gland to target glands or tissues in the body. The hypothalamus initially releases trace amounts of peptide hormones called **releasing hormones** (Fig. 36.9). These releasing hormones bind to receptors on cells in the anterior pituitary gland, leading that organ to release a much larger amount of the associated hormones. For example, the hypothalamus secretes a corticotropin-releasing hormone, which stimulates the anterior pituitary gland to release a larger amount of adrenocorticotropic hormone (ACTH).

Hormones released by the anterior pituitary gland subsequently bind cell receptors in the target organ. For example, ACTH acts on cells of the adrenal cortex, stimulating their secretion of the hormone cortisol. Cortisol acts on many different cells and tissues in the body, causing an acute stress response. Among its effects, it causes the liver to convert glycogen and amino acids to glucose. This action yields a weight of glucose that is 56,000 times heavier than the initial weight of releasing hormones secreted by the hypothalamus.

We can see an example of amplification in invertebrates in the regulation of insect molting and metamorphosis (see Fig. 36.3). This pathway has two signaling steps in which an endocrine gland releases a hormone. First, the brain releases the hormone PTTH. In turn, PTTH signals the prothoracic gland, leading to the release of a much larger amount of ecdysone, the hormone that regulates growth and metamorphosis of the insect's body. Ecdysone then binds to cell receptors in the target organ. Signaling cascades within these cells further amplify the hormone signal. Thus, amplification applies both to signal transduction cascades within a cell and to the transmission of endocrine hormones between glands and tissues.

Hormones are evolutionarily conserved molecules with diverse functions.

Most hormones have an ancient evolutionary history and are evolutionarily conserved. For example, some vertebrate

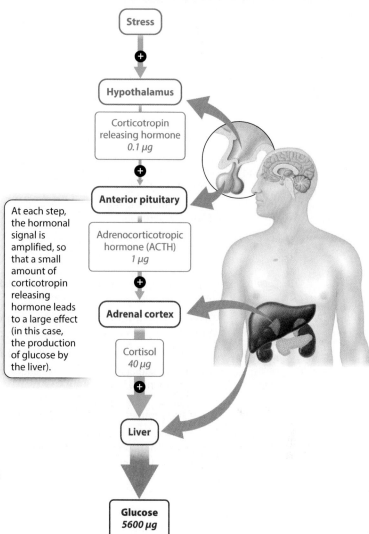

FIG. 36.9 Amplification of a hormonal signal.

At each step, the hormonal signal is amplified, so that a small amount of corticotropin releasing hormone leads to a large effect (in this case, the production of glucose by the liver).

hormones can also be found in many invertebrates. Given that the first vertebrate animals diverged from invertebrates more than 500 million years ago, some animal hormones must be even older. Their roles in many invertebrate animals have yet to be discovered.

Typically, the same hormone serves different functions in vertebrates and in invertebrates. The same hormone may even serve different functions within distinct groups of vertebrates. For example, thyroid-stimulating hormone (TSH) is released by the anterior pituitary gland and targets the thyroid gland. It regulates metabolism in mammals but triggers metamorphosis in amphibians and feather molt in birds. It has even been found in snails and other invertebrates that lack a thyroid gland. Its function in snails appears to be stimulation of sperm or egg production.

TABLE 36.2 Major Vertebrate Hormones

SECRETING TISSUE OR GLAND	HORMONE	TARGET GLAND OR ORGAN	ACTION
Hypothalamus	Releasing hormones (peptides)	Anterior pituitary gland	Stimulate secretion of anterior pituitary hormones
Anterior pituitary gland	Thyroid-stimulating hormone (TSH) (glycoprotein)	Thyroid gland	Stimulates synthesis and secretion of thyroid hormones by the thyroid gland
	Follicle-stimulating hormone (FSH) (glycoprotein)	Gonads	Stimulates maturation of eggs in females; stimulates sperm production in males
	Luteinizing hormone (LH) (glycoprotein)	Gonads	Stimulates production and secretion of sex hormones in ovaries (estrogen and progesterone) and testes (testosterone)
	Adrenocorticotropic hormone (ACTH) (peptide)	Adrenal glands	Stimulates production and release of cortisol
	Growth hormone (GH) (protein)	Bones, muscles, liver	Stimulates protein synthesis and body growth
	Prolactin (protein)	Mammary glands	Stimulates milk production
	Melanocyte-stimulating hormone (peptide)	Melanocytes	Regulates skin (and scale) pigmentation
Posterior pituitary gland	Oxytocin (peptide)	Uterus, breast, brain	Stimulates uterine contraction and release of milk; influences social behavior
	Antidiuretic hormone (ADH; vasopressin) (peptide)	Kidneys, brain	Stimulates uptake of water from the kidneys; involved in pair bonding
Thyroid gland	Thyroid hormones (peptides)	Many tissues	Stimulate and maintain metabolism for development and growth
	Calcitonin (peptide)	Bone	Stimulates bone formation by osteoblasts
Ovaries	Estrogen (steroid)	Uterus, breast, other tissues	Stimulates development of female secondary sexual characteristics and regulates reproductive behavior
	Progesterone (steroid)	Uterus	Maintains female secondary sexual characteristics and sustains pregnancy

Recent genomic analysis has shown that the receptors for many hormones evolved long before the hormones with which they now interact. Even though the structure of a hormone or its receptor is often largely unchanged across diverse groups of organisms, hormones and their receptors can readily be selected to take on new roles as organisms evolve new behaviors and exploit new environments. An example is the shift in function of TSH from its role of stimulating sperm or egg production in snails to regulating metabolism in mammals and feather molt in birds.

Another intriguing finding is that many peptides originally identified as hormones in various tissues also function as

TABLE 36.2 Major Vertebrate Hormones (continued)

SECRETING TISSUE OR GLAND	HORMONE	TARGET GLAND OR ORGAN	ACTION
Testes	Testosterone (steroid)	Various tissues	Stimulates development of male secondary sexual characteristics; regulates male reproductive behavior and stimulates sperm production
Adrenal cortex	Cortisol (steroid)	Liver, muscles, immune system	Regulates response to stress by increasing blood glucose levels and reduces inflammation
Adrenal medulla	Epinephrine (norepinephrine) (peptide)	Heart, blood vessels, liver	Stimulates heart rate, blood flow to muscles, and elevation of blood glucose level as part of fight-or-flight response
Parathyroid glands	Parathyroid hormone (PTH) (protein)	Bone	Stimulates bone resorption by osteoclasts to increase blood Ca^{2+} levels
Pancreas	Insulin (protein)	Liver, muscles, fat, other tissues	Stimulates uptake of blood glucose and storage as glycogen
	Glucagon (protein)	Liver	Stimulates breakdown of glycogen and glucose release into blood
	Somatostatin (peptide)	Digestive tract	Inhibits insulin and glucagon release; decreases digestive activity (secretion, absorption, and motility)
Stomach	Gastrin (peptide)	Stomach	Stimulates protein digestion by secretion of digestive enzymes and acid; stimulates contraction of smooth muscle in the stomach
Small intestine	Cholecystokinin (peptide)	Pancreas, liver, gallbladder	Stimulates secretion of digestive enzymes and products from liver and gallbladder
	Secretin (peptide)	Pancreas	Stimulates bicarbonate secretion from pancreas
Pineal gland	Melatonin (peptide)	Brain, various organs	Regulates circadian rhythms

neurotransmitters in the nervous system. For example, oxytocin, which stimulates uterine contraction and the release of milk during breastfeeding, also serves as a neurotransmitter in the brain. It is believed to influence social behavior, as well as to stimulate sexual arousal in mammals. Similarly, antidiuretic hormone, a peptide hormone that regulates water uptake in the kidneys (Chapter 39), also functions as a neurotransmitter in the brain, influencing mammalian mating and pair-bonding behavior. Hence, the same compound, expressed in two different cell types, may have entirely different functions in each cell type. The roles of hormones as chemical messengers are varied within an organism and easily changed over the course of evolution.

Self-Assessment Questions

4. How can hormones act specifically on only some cells or organs although they are transported by the bloodstream throughout the body?
5. What are the two general classes of hormone? Provide an example of each.
6. Where are receptors for peptides and steroid hormones located, and how does the difference in their location affect the resulting signaling pathways in the target cell?
7. How are hormones able to exert substantial and broad effects on their target cells even though they are released in small amounts by the initial cell, tissue, or gland?

36.3 THE VERTEBRATE ENDOCRINE SYSTEM

The vertebrate endocrine system responds to sensory cues received by its nervous system. Whereas certain sensory signals may be communicated directly to endocrine glands or tissues in the body, many are processed in the brain. After processing, these signals are transmitted to the endocrine system by the hypothalamus, a structure located in the forebrain. The hypothalamus is the main route by which nervous system signals are transmitted to the vertebrate endocrine system. The hypothalamus relays these signals to the pituitary gland, which lies just below it. The pituitary gland acts as a control center for many other endocrine glands in the body.

The pituitary gland integrates diverse bodily functions by secreting hormones in response to signals from the hypothalamus.

The pituitary gland is a central regulating gland of the vertebrate endocrine system. It releases several hormones that coordinate the action of many other endocrine glands and tissues. These glands, which are illustrated in **Fig. 36.10**, control the growth and maturation of the body, regulate reproductive development and metabolism, and control water balance. These effects reflect the broad range of roles that the endocrine and nervous systems perform in regulating body maturation and maintaining homeostasis.

The pituitary gland communicates with many cells, tissues, and organs of the body. Some of these, such as the thyroid and adrenal glands, secrete hormones in response to signals from the pituitary gland and, therefore, have exclusively endocrine functions. Others, such as the lungs, kidneys, and digestive tract, harbor endocrine cells that secrete hormones and also have other physiological roles in the body. Consequently, the vertebrate endocrine system is not localized in one part of the body, but rather is present throughout.

The pituitary gland is divided into anterior and posterior regions (**Fig. 36.11**). This division is not arbitrary: the two

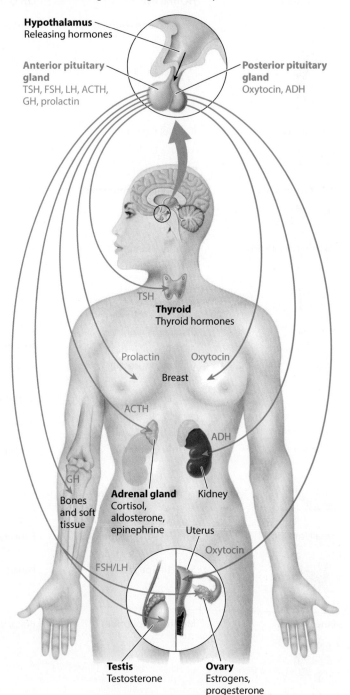

FIG. 36.10 The hypothalamus–pituitary axis. The hypothalamus sends signals to the pituitary gland, which in turn sends signals to diverse cells and organs throughout the body.

regions have distinct functions, organizations, and embryonic origins. The **anterior pituitary gland** forms from epithelial cells that develop and push up from the roof of the mouth, whereas the **posterior pituitary gland** develops from neural tissue at the base of the brain. Both sets of cells form pouches that develop into glands, which come to lie adjacent to each other. As a consequence of this development pathway, the

FIG. 36.11 The anterior and posterior pituitary glands. The hypothalamus communicates with the anterior pituitary gland by secreting releasing hormones into the blood. The hypothalamus communicates with the posterior pituitary gland by extending axons of neurosecretory cells into it.

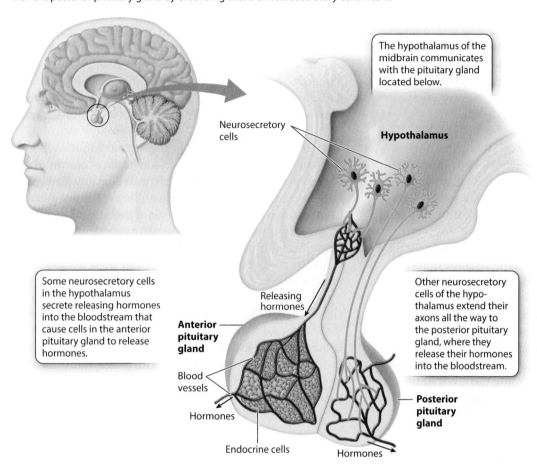

anterior and posterior pituitary glands should be thought of as two distinct glands, not one.

Because of their developmental differences, the anterior and posterior pituitary glands receive input from the hypothalamus in different ways. Like the brain of arthropods, the vertebrate hypothalamus contains neurosecretory cells, which are part of the brain and therefore neurons. Like the invertebrate neurosecretory cells, these cells release hormones into the bloodstream. Some of the neurosecretory cells communicate with the anterior pituitary gland. In this case, they secrete releasing hormones into small blood vessels that travel to and supply the anterior pituitary gland (Fig. 36.11). In response, cells of the anterior pituitary gland release hormones into the bloodstream. These hormones circulate throughout the body and bind to receptors on target cells, tissues, and organs.

In contrast, the hypothalamus and the posterior pituitary gland communicate without the use of releasing hormones. Instead, cell bodies of neurosecretory cells located in the hypothalamus extend axons into the posterior pituitary gland. These axons release hormones directly into the bloodstream for transport to distant sites (Fig. 36.11). Consequently, the posterior pituitary gland is part of the nervous system itself.

In response to signals from the hypothalamus, the anterior and posterior pituitary glands secrete distinct sets of hormones (Table 36.2). The hormones released by the anterior pituitary gland act on an endocrine gland to cause release of other hormones. Hormones that control the release of other hormones are called **tropic hormones**. For example, in response to thyroid-stimulating hormone, the thyroid gland releases thyroid hormones that regulate the metabolic state of the body. In response to follicle-stimulating hormone (FSH) and luteinizing hormone (LH), the ovaries release estrogen and progesterone and the testes release testosterone. In response to adrenocorticotropic hormone, the adrenal glands release cortisol.

The anterior pituitary gland also secretes two other hormones: growth hormone (GH) and prolactin. Growth hormone acts generally on the muscles, bones, and other body tissues to stimulate their growth. Prolactin stimulates milk production in the breasts of female mammals in response to an infant's suckling at the mother's nipple.

The hormones released by the posterior pituitary gland include **oxytocin** and **antidiuretic hormone** (ADH; also called **vasopressin**). These two evolutionarily related peptide hormones have similar structures. As discussed earlier, oxytocin plays several roles related to female reproduction. It causes uterine contraction during labor and stimulates the release of milk during breastfeeding. Antidiuretic hormone acts on the kidneys, regulating the concentration of urine that an animal excretes, which is critical to maintaining water and solute balance in the body (Chapter 39).

In addition to their roles in reproduction and kidney function, oxytocin and antidiuretic hormone are released by cells in the brain and may play roles in social behaviors. These hormones are able to act directly on the brain by binding to receptors on neurons. Recent evidence indicates that oxytocin may have important roles in regulating maternal behavior toward infants and in creating trust as a prelude to mating in certain groups of vertebrate animals. These functions reflect the long evolutionary history of oxytocin, which also plays a role in the social behavior of insects.

Recent evidence suggests that antidiuretic hormone plays a parallel role in regulating male social behavior and parental behavior in mammals. This is not surprising because oxytocin and antidiuretic hormone have very similar chemical structures (see Fig. 36.7). The integration of reproductive function with behavior for the care of young is likely favored strongly by natural selection.

Many targets of pituitary hormones are endocrine tissues that also secrete hormones.

As we have seen, some of the hormones released by the anterior pituitary gland act on endocrine organs, which then release hormones of their own. These tropic hormones are TSH, FSH and LH, and ACTH (see Fig. 36.10; Table 36.2). Here, we look at their target organs in more detail.

TSH acts on the **thyroid gland**, which is located in the front of the neck (see Fig. 36.10; Table 36.2). In response to TSH, the thyroid gland releases two peptide hormones, triiodothyronine (T_3) and thyroxine (T_4), that diffuse into cells and bind to intracellular receptors. These two hormones act to increase the resting metabolic rate of cells throughout the body. As levels of T_3 and T_4 rise, these hormones act as a brake on the release of TSH by the anterior pituitary gland. This regulation of TSH by negative feedback is critical to keeping TSH levels stable and maintaining the body as a whole in a stable metabolic state. Individuals who overproduce thyroid hormones (hyperthyroidism) experience increased appetite and weight loss caused by an overly active metabolic state. In contrast, individuals with thyroid hormone deficiency (hypothyroidism) experience fatigue and sluggishness caused by a metabolic state that is too low. Both conditions are diagnosed by blood tests that monitor circulating levels of thyroid hormones.

Because the thyroid hormones T_3 and T_4 require iodine for their production, individuals who do not acquire enough iodine from their diets produce too little thyroid hormone. In such a case, the anterior pituitary gland increases its production of TSH because it is no longer receiving negative feedback. Over time, in response to TSH the thyroid gland expands to form an enlargement of the throat called a goiter. The introduction of iodized salt has largely eliminated goiter formation and hypothyroidism in many countries, but in some underdeveloped areas these conditions remain significant public health problems.

The gonadotropic hormones are FSH and LH. They target the female and male gonads: the **ovaries** and **testes**, **respectively** (see Fig. 36.10; Table 36.2). In response to FSH and LH, both the ovaries and the testes secrete specific sex hormones that regulate their own development as well as the development of secondary sexual characteristics. Female sex hormones include the steroids estrogen and progesterone. The principal male sex hormone is the androgen testosterone. These sex hormones are common to a vast majority of vertebrates, regulating sexual differentiation, gonadal maturation, and reproductive behavior. The role of these sex hormones in regulating reproductive function is discussed in greater detail in Chapter 40.

Testosterone is a naturally occurring anabolic steroid that promotes protein synthesis to build body tissues and anabolic metabolism to store energy within cells. Thus, one role of testosterone is to stimulate male body tissue growth, particularly in muscles. A variety of synthetic anabolic steroids that mimic the action of testosterone have been developed to treat muscle wasting due to loss of appetite and diseases such as cancer and AIDS. Some male and female athletes use anabolic steroids illegally to promote muscle strengthening, undermining fair competition in their sports. The risks to health of continued use of anabolic steroids are considerable, including liver damage, heart disease, and high blood pressure. Tissues become damaged because the elevated circulating levels of testosterone produced by anabolic steroids inhibit the normal synthesis of tropic hormones by the anterior pituitary gland.

ACTH released by the anterior pituitary gland is also a tropic hormone. As noted in section 36.1, ACTH acts on the cortex (the outer portion) of the paired **adrenal glands** (see Fig. 36.10; Table 36.2). The adrenal glands are located adjacent to the kidneys (*ad-*, meaning "near"; *renal*, referring to the kidneys). In humans, each small adrenal gland lies just above each kidney. During times of stress, such as starvation, fear, or intense physical exertion, ACTH stimulates adrenal cortex cells to secrete cortisol. The release of cortisol affects a broad range of bodily functions. For example, it raises blood glucose levels, suppresses immune function (Chapter 41), and increases blood pressure (Chapter 37). Whereas acute (short-term) episodes of elevated cortisol are healthy physiological responses that enable animals to respond to stressful situations, chronic (long-term) stress often leads to a broad range of unhealthy conditions that include impaired cognitive and immune function, sleep disruption, and fatigue.

Other endocrine organs have diverse functions.

Although many endocrine organs receive signals from the hypothalamic–pituitary axis, some respond instead to internal physiological states of the body or to external environmental cues. One of these organs is the **parathyroid gland**, which is located on the thyroid gland. The parathyroid gland secretes parathyroid hormone (PTH). Working together with calcitonin secreted by the thyroid gland, PTH controls the level of calcium in the blood (Table 36.2). The two hormones act by regulating the actions of bone cells (osteoblasts and osteoclasts; Chapter 35) that control bone formation and bone removal.

When circulating levels of calcium fall too low, the parathyroid gland releases PTH. The release of PTH stimulates osteoclasts to reabsorb bone mineral, halting bone formation and triggering the release of calcium into the bloodstream. When calcium levels are too high, the production of PTH is inhibited and calcitonin is released. The release of calcitonin shifts bone metabolism toward net bone formation, building bone that stores calcium in the skeleton. As is true for blood glucose levels, calcium levels are maintained in a narrow range by negative feedback. The response to the hormone (increasing or decreasing the level of calcium in the blood) is in the opposite direction to the stimulus (low or high level of calcium in the blood), so that a stable set point is maintained.

Several organs of the digestive system, including the pancreas, stomach, and duodenum, produce hormones that regulate appetite and digestion. Three of these hormones are discussed in Chapter 38.

The **pineal gland** is located in the brain. In response to darkness, the pineal gland secretes melatonin, a hormone that helps control an animal's state of wakefulness (Table 36.2). Melatonin secretion is inhibited by light. Environmental light cues sensed by the retina are conveyed to the gland by the autonomic nervous system. In some vertebrates, such as lampreys, some fishes, and reptiles, the pineal gland responds to light sensed by a "third eye" composed of photosensitive cells that are located at the skull's surface between the animal's actual eyes. In both cases, melatonin levels rise at night. In diurnal animals (those that are active during the day), the rise in melatonin levels at night causes the animals to sleep. In nocturnal animals (those that are active during the night), the rise in melatonin levels at night causes the animals to become active.

The release of melatonin in response to daily (and seasonal) light cycles helps to maintain the body's circadian rhythms. Circadian rhythms are cycles of approximately 24 hours in which an animal's biochemical, physiological, and behavioral state shifts in response to changes in daily and seasonal environmental conditions. To overcome jet lag, travelers are advised to seek sunlight and often to take melatonin before sleeping to help reset their 24-hour circadian rhythm.

In several animals, including ground squirrels, bats, bears, and rattlesnakes, the pineal gland regulates hibernation, controlling the animal's metabolic state over longer time periods. In many animals, release of melatonin also regulates seasonal breeding cycles.

The fight-or-flight response represents a change in set point in many different organs.

The endocrine system works closely with the nervous system to enable animals to respond to external cues in the environment. In Chapter 34, we discussed how prey animals, such as gazelles, use visual and olfactory cues processed by the nervous system to detect the distant movement of a predator, such as a cheetah. We also saw how the sympathetic nervous system, in the fight-or-flight response, can trigger broad physiological changes, such as increased heart and breathing rates and changes in metabolism that ready the predator to attack or the prey to respond.

These actions are coordinated by the endocrine system. The sympathetic nervous system sends axons to the **adrenal medulla** (see Fig. 36.10), the inner part of the adrenal gland. In response to stimulation by the sympathetic nervous system, cells of the adrenal medulla secrete two hormones, epinephrine and norepinephrine (also known as adrenaline and noradrenaline, respectively). These hormones released by the adrenal gland have targets throughout the body, and they trigger the physiological changes of the fight-or-flight response. Interestingly, epinephrine and norepinephrine also act as neurotransmitters in the brain. In fact, cells of the adrenal medulla are modified nerve cells that have lost their axons and dendrites.

The same set of responses occurs when, walking along a deserted sidewalk, we note a shadowy movement or hear a footstep behind us. The perceived threat sets in motion a coordinated physiological reaction, mediated by the combined action of the nervous and endocrine systems. These changes make us alert and ready for action to avoid or resist a perceived threat. They also inhibit digestion and eliminate a sense of appetite.

Earlier, we considered the importance of homeostasis in maintaining a variety of parameters, such as temperature and blood glucose levels, at a steady level. In this case, there is an adjustment of the set points of different organs, which is critical in enabling an animal to respond to external cues and adjust its physiological response appropriately. When the real or perceived threat is gone, the body is able to return to its prior state by reestablishing earlier set points.

Set points are also changed when an animal acclimatizes to a new environment, such as when it moves from fresh water to a marine environment or to a new altitude. An adaptive change in body function to a new environment is referred to as **acclimatization**. Examples of acclimatization are the increase in the number of red blood cells and their capacity to carry more oxygen that occurs at higher altitude (Chapter 37) and the seasonal changes in the fur of mammals that achieve the insulation required to maintain a stable body temperature (Chapter 38). These physiological changes result from establishing new set points in a process coordinated by the endocrine system.

Self-Assessment Questions

8. What is the difference in how the hypothalamus controls the anterior pituitary gland and how it controls the posterior pituitary gland?

9. What is a tropic hormone? Give two examples, identifying which endocrine gland releases the tropic hormone and which endocrine gland or organ is its target.

10. Diagram an endocrine pathway, such as parathyroid regulation of blood calcium levels, that maintains homeostasis of a bodily function by means of negative feedback.

36.4 OTHER FORMS OF CHEMICAL COMMUNICATION

As we have seen, hormones enter the circulation and act on distant cells, namely those having receptors that bind the hormones. However, other chemical signals act over short distances, and still others are released into the environment to signal other individuals of the same species. In this section, we explore additional modes of chemical communication and the compounds that produce them.

Local chemical signals regulate neighboring target cells.

Unlike hormones, some chemical compounds act locally only, on neighboring cells (**Fig. 36.12**). These cells must have receptors that bind to the compound before it degrades or diffuses into the bloodstream.

Chemical compounds that act locally are said to have paracrine function (Fig. 36.12b). If a compound acts on the secreting cell itself, it is said to have autocrine function; that is, the chemical signal may stimulate or inhibit its own secretion. Growth factors (also referred to as cytokines and distinct from the growth hormone released by the anterior pituitary gland) and histamine are examples of paracrine chemical compounds released in small amounts that act locally on neighboring cells.

Growth factors (Chapter 9) enhance the differentiation and growth of particular kinds of tissues. For example, bone morphogenetic proteins (BMPs) promote the formation of bone in growing skeletons, and fibroblast growth factors (FGFs) stimulate the formation of connective tissue. Similarly, nerve growth factors (NGFs) promote the survival and growth of nerve cells.

When tissues are damaged, specialized cells nearby release histamine as a paracrine signaling molecule. This chemical triggers the dilation of blood vessels, allowing blood proteins and white blood cells to move into the region to fight infection and induce repair (Chapter 41). The release of histamine causes the swelling, redness, and warmth that we commonly associate with inflammation.

Another form of short-range chemical signaling is the release of neurotransmitters into the synapse between directly communicating nerve cells or between nerve cells and muscles at neuromuscular junctions (Fig. 36.12b). Synaptic signaling is extremely rapid and brief compared to the other modes of chemical communication.

Pheromones are chemical compounds released into the environment that signal physiological and behavioral changes.

Many animals release small chemical compounds into the environment to send a signal to other members of their species. These water-borne or airborne compounds, known as **pheromones**, influence the physiology and behavior of other members of the species. For example, bees and wasps release pheromones to signal the location of a nest or food

FIG. 36.12 Modes of chemical signaling. Chemical signaling can occur (a) over long distances (endocrine) or (b) locally (paracrine and synaptic).

a. Long-distance signaling

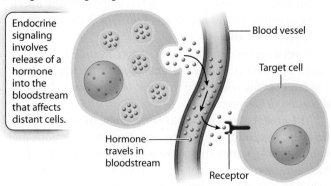

b. Local signaling

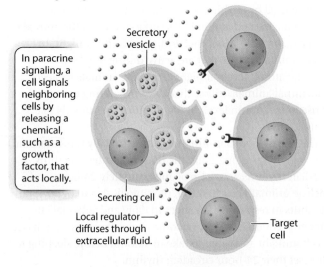

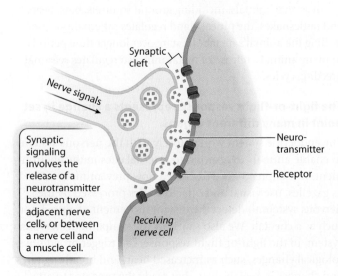

FIG. 36.13 Pheromone action. (a) Mating ladybugs are attracted by pheromones. (b) Wolves secrete pheromones in their urine that mark their territory. (c) Honeybees swarm to defend their nest after a bee that stings a threatening animal releases an alarm pheromone. *Sources: a. Krzysztof Bisztyga/AGE Fotostock; b. Jorg & Petra Werner/Animals Animals/Earth Scenes; c. Darios/Shutterstock.*

and to regulate the development and activities of other colony members. Even plants, such as orchids, release pheromones. Some plant pheromones attract pollinating bees or wasps by mimicking the chemical signals released by females of these species. There are many different pheromones, which serve very different functions, as shown in **Fig. 36.13**.

One of the most common types of pheromone signaling is the release of sex pheromones (Chapter 43). These pheromones play a key role in the mating behavior of many invertebrate and vertebrate species: they attract mates and trigger ritualistic social and mating behaviors. For example, the sex pheromone bombykol, released from a gland in the abdomen of the female silk moth, signals her location to males looking for a mate. In most mammals, amphibians, and reptiles, pheromones are detected by a vomeronasal organ with chemosensory neurons in the nasal region of the skull. Humans and nonhuman primates lack a vomeronasal organ, and most studies indicate that they do not produce pheromones.

Other pheromones signal territorial boundaries. Dogs and other canids (this family includes wolves and coyotes), as well as cats, secrete pheromones in their urine that mark their territories. Social seabirds, such as cormorants and boobies, mark their nesting sites with pheromones.

In many species, an animal under attack releases an alarm pheromone, warning its neighbors. Ants and other social insects deploy alarm pheromones to warn of an attack by an advancing army from another colony. The release of alarm pheromones from a stinging bee causes other bees to swarm and join the attack to defend their nest. Similarly, social mammals such as deer warn of an approaching predator by means of alarm pheromones.

Social insects, such as ants, deploy trail pheromones that provide chemical cues that other worker ants can follow to a food source (**Fig. 36.14**). An ant returning to the colony from

FIG. 36.14 Trail pheromones in ants. Here, a chemical trail has been laid down by worker ants signaling a food source for the colony. *Source: LatitudeStock–Patrick Ford/Getty Images.*

a food source releases trail pheromones from its abdomen onto the soil. Other worker ants then detect this trail with their antennae and track the concentration of trail pheromone, guiding them along the trail. If you have watched ants marching in line across a sidewalk, along a trail, or into and out of your house, you have seen worker ants using pheromone cues that serve as signposts for the way to food.

Aquatic species also depend on pheromone signaling. Fishes, salamanders, and tadpoles release alarm pheromones into the water to warn of a threatening predator. Prey species that share the same habitat may even sense and respond to the release of another species' alarm pheromone.

Fishes also release pheromones to coordinate mating and to regulate social interactions. In dense populations of cichlid fish living in Lake Tanganyika in central Africa, the ratio of brightly colored breeding males (**Fig. 36.15**) to females is highly regulated. Only approximately 10% of males are brightly colored and defend their territory to attract and mate with females. This restriction on the number of dominant breeding males is regulated by behavioral cues and chemical pheromones. When a dominant male dies, subordinate males fight for the vacated territory. When a new male assumes dominance, it becomes brightly colored to attract females. The color change is triggered by hormones released from the male fish's pituitary gland, which act on pigment cells called melanocytes that are located in the epidermis of its scales. The dominant breeding male reinforces its dominance by releasing pheromones in its urine that signal females and subordinate males in its breeding area. Thus, the fish makes use of both endocrine and pheromone signals to change its own behavior and the behavior of other members of its species.

FIG. 36.15 A brightly colored male cichlid fish. The behavior and markings of this fish are under hormonal control and triggered by social context. *Source: blickwinkel/Alamy.*

Self-Assessment Questions

11. How do hormones, paracrine signaling molecules, neurotransmitters, and pheromones differ?

12. The brown marmolated stink bug (*Halyomorpha halys*) emits at least two kinds of chemical compounds: one emits a noxious smell (explaining its name) that is used to ward off predators, and the other attracts other brown marmolated stink bugs. Which of these two chemicals is a pheromone and why?

CORE CONCEPTS SUMMARY

36.1 ENDOCRINE FUNCTION: Animal endocrine systems release hormones into the bloodstream that respond to environmental cues, regulate growth and development, and maintain homeostasis.

The nervous system responds rapidly to sensory information, providing immediate physiological responses, whereas the endocrine system acts more slowly and over prolonged periods to effect broad changes in the physiological and behavioral state of an animal. page 809

The endocrine system helps an animal respond to its environment. page 809

The endocrine system is involved in growth and development. page 810

The nervous and endocrine systems regulate physiological functions to maintain homeostasis. page 812

Homeostasis is often achieved by negative feedback, in which the response opposes the stimulus. page 813

A stimulus can be amplified by positive feedback, in which the response amplifies the initial stimulus. page 814

36.2 HORMONES: Hormones bind to receptors on or inside target cells, and their signals are amplified to exert strong effects on target cells.

Hormones are chemical compounds that are secreted and transported by the circulatory system, often to distant targets. page 815

Hormones bind to specific receptors located on the surface or inside the target cells that receive their signals. page 815

Two main classes of hormone are peptides and amines, and steroid hormones. page 815

Peptide and amine hormones are hydrophilic and bind to cell-membrane receptors, stimulating second-messenger pathways to alter the biochemical function of target cells. page 815

Steroids are hydrophobic and bind to intracellular or nuclear receptors, which function as transcription factors to regulate gene expression and protein synthesis. page 816

Hormones communicate by signaling cascades to amplify the strength of their downstream effects on target cells. page 816

The hypothalamic–pituitary axis forms the central endocrine pathway by which vertebrates regulate and integrate diverse bodily functions. page 816

Hormones are evolutionarily conserved molecules that are common to diverse groups of animals, but that have evolved novel functions. page 817

36.3 THE VERTEBRATE ENDOCRINE SYSTEM: In vertebrates, the hypothalamus and the pituitary gland control and integrate diverse functions and behaviors.

The anterior pituitary gland secretes hormones in response to releasing hormones secreted by the hypothalamus. page 821

Hormones released by the anterior pituitary gland include thyroid-stimulating hormone (TSH), the gonadotropic hormones follicle-stimulating hormone (FSH) and luteinizing hormone (LH), adrenocorticotropic hormone (ACTH), growth hormone (GH), and prolactin. page 821

The posterior pituitary gland contains the axons of neurosecretory cells that release the peptide hormones oxytocin and antidiuretic hormone into the bloodstream. page 821

Tropic hormones are hormones that act on endocrine glands to produce additional hormones. TSH acts on the thyroid gland to release thyroid hormones; FSH and LH act on the female ovaries to produce estrogen and progesterone and the male testes to produce testosterone; and ACTH acts on the adrenal gland to release cortisol. page 822

Some endocrine glands do not respond to signals from the pituitary gland. These include the parathyroid gland, which releases parathyroid hormone that helps to control calcium levels; the pancreas, which secretes insulin and glucagon to regulate blood glucose levels; and the pineal gland, which secretes melatonin in response to light and dark cycles. page 822

36.4 OTHER FORMS OF CHEMICAL COMMUNICATION: Chemical communication can occur locally between neighboring cells or between individuals, coordinating social interactions.

Paracrine signals act locally to affect neighboring cells, compared with the more distant signals transmitted by blood-borne hormones. page 824

Pheromones are chemical signals released into the air or water, signaling behavioral cues between individuals of a species. page 824

Sex pheromones attract mates and coordinate ritualized mating behavior; alarm pheromones signal predatory threats; trail pheromones signal the location of food sources; and territorial pheromones signal a male's dominance and breeding area. page 825

Log in to LaunchPad to check your answers to the Self-Assessment Questions and to access additional learning tools.

CHAPTER 37 Animal Cardiovascular and Respiratory Systems

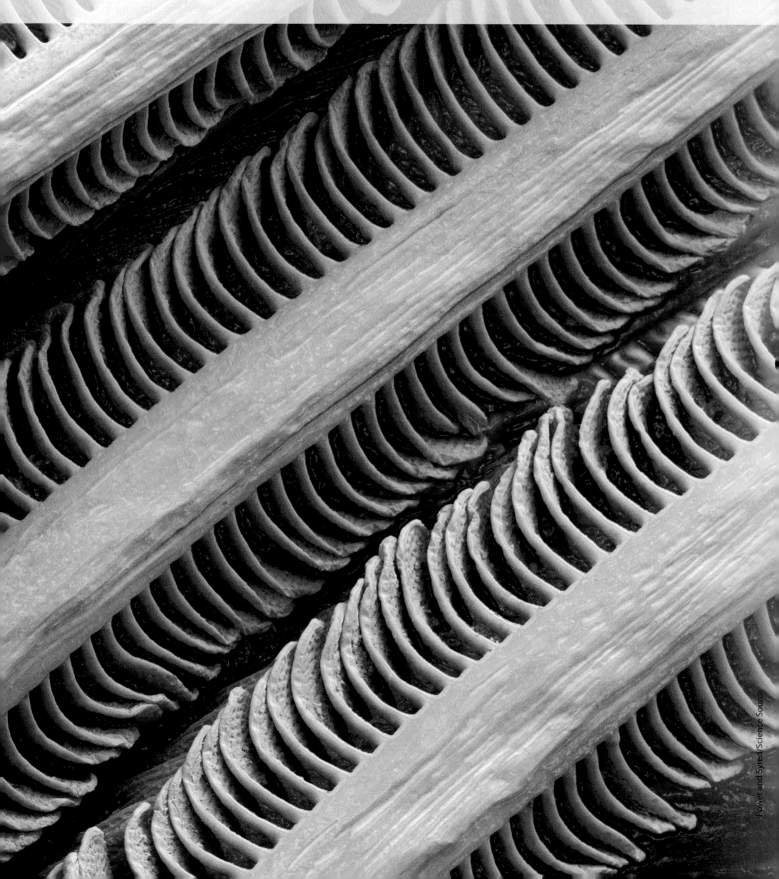

CORE CONCEPTS

37.1 DELIVERY OF OXYGEN AND ELIMINATION OF CARBON DIOXIDE: Respiration and circulation depend on diffusion over short distances and bulk flow over long distances.

37.2 RESPIRATORY GAS EXCHANGE: Respiration provides oxygen and eliminates carbon dioxide in support of cellular metabolism.

37.3 OXYGEN TRANSPORT BY HEMOGLOBIN: Red blood cells produce hemoglobin, greatly increasing the amount of oxygen transported by the blood.

37.4 CIRCULATORY SYSTEMS: Circulatory systems have vessels of different sizes that facilitate bulk flow and diffusion.

37.5 STRUCTURE AND FUNCTION OF THE HEART: The evolution of animal hearts reflects selection for a high metabolic rate, achieved by increasing the delivery of oxygen to metabolically active cells.

As animals evolved to be larger and more complex, they faced the challenge of obtaining an adequate supply of oxygen (O_2) and nutrients and, at the same time, eliminating carbon dioxide (CO_2) and other waste products. In this chapter, we explore how animal respiratory and circulatory systems evolved to meet the metabolic needs of an animal's cells. In particular, we focus on the uptake of O_2 and the release of CO_2, key gases in cellular respiration (Chapter 7).

Gases, nutrients, and wastes are continually being exchanged between an organism and its environment. Because of their small sizes or thin bodies, single-celled organisms and simple multicellular animals exchange compounds with the environment by diffusion. Diffusion works effectively only over very short distances, so larger and more complex animals cannot rely on diffusion as the sole means of exchanging compounds (Chapter 26). Instead, these animals evolved respiratory structures at the body's surface for taking up O_2 and releasing CO_2, and internal circulatory systems for long-distance transport to different body regions.

Because similar physical laws govern gas and liquid movement, the respiratory and circulatory systems of different animals have many features in common. However, fishes and other aquatic animals obtain O_2 from water, whereas land animals breathe air. Water and air have very different physical properties. Thus, aquatic and terrestrial environments pose distinct challenges for obtaining O_2. Because life first began in the water, the evolution of life on land required substantial changes in how animals exchange gases, take up nutrients, and eliminate wastes.

37.1 DELIVERY OF OXYGEN AND ELIMINATION OF CARBON DIOXIDE

Eukaryotic cells use O_2 to burn organic fuels—carbohydrates, lipids, and proteins—for adenosine triphosphate (ATP) production (Chapter 7). The increase in atmospheric O_2 with the evolution of photosynthesis was key to the evolution of O_2-based cellular respiration, which in turn contributed to the diversification of multicellular animal life during the Cambrian Period, approximately 500 million years ago. In the process of burning fuel by aerobic respiration, CO_2 is produced. Animals have evolved mechanisms to acquire O_2 from their environment and eliminate excess CO_2 from their bodies. These mechanisms make up the process known as animal respiration—or "breathing."

Diffusion governs gas exchange over short distances.

The transport of O_2 and CO_2 between an organism and its environment is referred to as **gas exchange**, and it is fundamental to all eukaryotic animals as well as to photosynthetic plants. Single-celled organisms and simple animals and plants exchange gases by diffusion. **Diffusion** is the random movement of individual molecules (Chapter 5). If there is a difference in the concentration of a substance between two regions, diffusion results in the net movement of molecules from regions of higher concentration to regions of lower concentration until the molecules are evenly distributed.

Given a large enough surface, diffusion is an extremely effective way to exchange gases and other substances over short distances between two compartments, such as between the lungs and the blood vessels supplying the lungs. The rate of diffusion is directly proportional to the surface area over which exchange occurs and to the concentration difference. It is inversely proportional to the distance over which the molecules move (or the thickness of the barrier). Diffusion is generally ineffective over distances exceeding about 100 μm (0.1 mm). Consequently, respiratory and circulatory structures have large surface areas and thin interfaces.

Because effective transport by diffusion is limited to short distances, some of the earliest and simplest invertebrate animals were, and still are, composed of thin sheets of cells with a large surface area. For example, sponges and sea anemones have a thin body with a central cavity through which water circulates, and flatworms have a flattened body surface (**Fig. 37.1**).

Diffusion results in the net movement of a substance when there is a concentration difference between two regions. When considering the diffusion of a gas, however, we usually refer to its partial pressure rather than its concentration. The **partial pressure (p)** of a gas is defined as its fractional concentration relative to other gases present, multiplied by the overall (that is, the atmospheric) pressure (**Fig. 37.2**). Pressure is measured in units of millimeters of mercury (mmHg). Oxygen makes up approximately 21% of the gases in air (nitrogen contributes about 78%, carbon dioxide about 0.03%, and the remainder

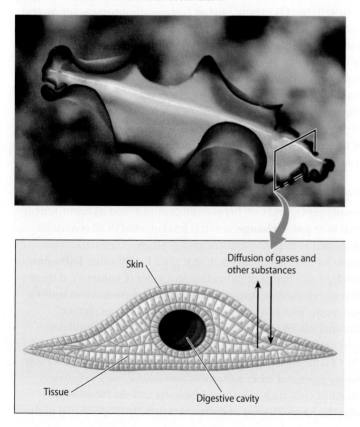

FIG. 37.1 The thin body of a flatworm. Gases can be exchanged by diffusion in flatworms because their bodies are only a few cell layers thick. *Photo source: Amar and Isabelle Guillen/SeaPics.com.*

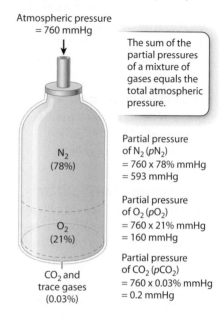

FIG. 37.2 Partial pressure. The partial pressure of a gas is its fractional concentration multiplied by the total atmospheric pressure.

consists of trace gases). At sea level, atmospheric pressure is approximately 760 mmHg, which means that O_2 in the atmosphere exerts a partial pressure (pO_2) of about $0.21 \times 760 = 160$ mmHg. For O_2 to diffuse from the air into cells, the partial pressure of O_2 inside the cells must be lower than the partial pressure of O_2 in the atmosphere.

Bulk flow moves fluid over long distances.

Larger, more complex animals transport O_2 and CO_2 over longer distances to cells within their bodies. These animals rely on **bulk flow** in addition to diffusion for this transport. Bulk flow is the physical movement of fluids, either a liquid or a gas, as a result of pressure differences. Specialized pumps, like the heart, generate the pressure that is required to move fluids.

Bulk flow occurs in two steps to meet the gas exchange needs of cells in larger animals. The first step is **ventilation**. Ventilation is the movement of the animal's respiratory medium—water or air—past a specialized respiratory surface. The second step is **circulation**. Circulation is the movement of a specialized body fluid that carries O_2 and CO_2. The circulatory fluid is called **hemolymph** in invertebrates and **blood** in vertebrates. This fluid delivers O_2 to cells within different regions of the body and carries CO_2 back to the respiratory exchange surface.

Both ventilation and circulation require a pump to produce a pressure (P) that drives flow (Q) against the resistance to flow (R). The rate of flow is governed by the following simple equation:

$$Q = P/R$$

FIG. 37.3 The four steps of gas transport and exchange.

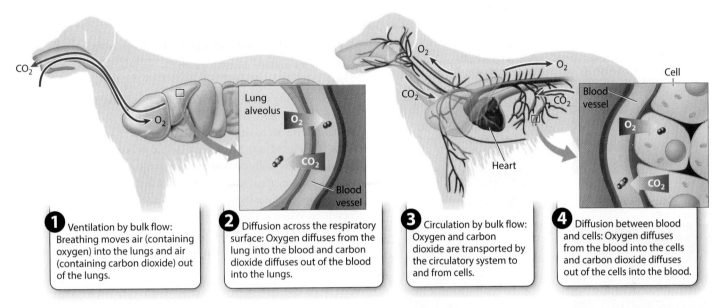

Resistance to flow measures the difficulty of pumping the fluid through a network of chambers and vessels located within the respiratory and circulatory systems. If the resistance to flow doubles, the flow rate is halved. The longer the network of vessels and the narrower the vessels within that network, the greater the vessels' resistance to the fluid (whether air, water, or blood) moving through them. As an illustration of that relationship, consider how much harder it is to blow fluid out of a long thin tube than a short wide one.

In summary, we can view gas exchange in complex multicellular organisms as four steps that are linked in series (**Fig. 37.3**). To deliver O_2 to the mitochondria in an animal's cells, (1) fresh air or water moves over the respiratory exchange surface in the process of ventilation (by bulk flow). Ventilation maximizes the concentration of O_2 in the air or water on the outside of the respiratory surface. (2) The buildup of O_2 then favors the diffusion of O_2 into the blood or hemolymph across its respiratory surface. (3) O_2 is transported by the circulation (by bulk flow) to the tissues. Internal circulation serves to maximize the concentration of O_2 outside cells. (4) Oxygen diffuses from the blood across the cell membrane and into the mitochondria, where fuel molecules are oxidized for ATP production.

The same four steps occur in reverse to remove CO_2 from the body: CO_2 (4) moves from cells to the blood or hemolymph by diffusion, (3) is carried in the circulation to the respiratory surface by bulk flow, (2) moves across the respiratory surface by diffusion, and (1) moves out of the animal by bulk flow. Because diffusion and bulk flow are coupled, it is important that each step in the transport chain (whether for a gas, nutrient, or waste) has a similar capacity—that is, the same amount of O_2 enters the blood through diffusion as is transported by the circulation.

Self-Assessment Questions

1. Which two structural features favor diffusion?
2. Individuals who experience an asthma attack have difficulty breathing. From the discussion of bulk flow, what do you think happens to these individuals' airways that makes it difficult for them to breathe?
3. Diagram the four basic steps of O_2 transport from an animal's respiratory medium (air or water) to its cells.

37.2 RESPIRATORY GAS EXCHANGE

All animals obtain O_2 from the surrounding air or water. Aquatic animals such as lobsters, aquatic salamanders, and fishes that take in O_2 from water breathe through **gills**, highly folded delicate structures that allow gas exchange with the surrounding water (**Fig. 37.4a**). Some animals, such as aquatic frogs and terrestrial adult salamanders, breathe through their skin. In contrast, most terrestrial vertebrates, such as reptiles, certain terrestrial salamanders, birds, and mammals, have internal **lungs** that perform gas exchange (**Fig. 37.4b**). Internal respiratory exchange surfaces are adaptations that protect the lung surfaces from drying out in a terrestrial environment. Instead of lungs, terrestrial insects evolved a system of air tubes called **tracheae** that branch from openings along their abdominal surface into smaller airways. These smaller airways, termed tracheoles, supply air directly to the cells within their bodies.

FIG. 37.4 Animal gas exchange organs. Gas exchange organs are shaped and folded to provide large surfaces for gas exchange. *Photo sources: (top) Reinhard Dirscherl/AGE Fotostock; (bottom) John Cancalosi/Alamy.*

a. Aquatic animals

Tube worm: External gills

Aquatic Salamander: External gills

b. Terrestrial animals

Grasshopper: Tracheae

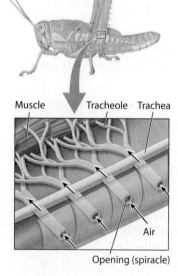

Dog: Internal lungs

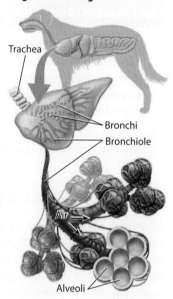

For both aquatic and terrestrial animals, ventilation supplies O_2 to gas exchange surfaces, and then O_2 crosses the surface through diffusion. How does the rate of diffusion through the gas exchange surface in lungs or gills keep up with the rate of O_2 supplied by ventilation? The respiratory surfaces of many animals are highly folded, creating a very large surface area. These surfaces are also exceedingly thin, just one or two cell layers thick, providing a diffusion distance of as little as 1 to 2 μm. In fact, this short distance is all that separates the air you breathe into your lungs from the bloodstream into which O_2 diffuses.

Many aquatic animals breathe through gills.

Most large aquatic animals (with the notable exception of marine mammals like dolphins and whales) breathe O_2-containing water through gills. Some sessile invertebrates have external gills; these organisms include tube worms and sea urchins, as well as nudibranch mollusks and aquatic amphibians (see Fig. 37.4a). However, external gills are easily damaged or may be eaten by predators. Most mollusks, crustaceans, and fishes have evolved internal gills that, although still formed as outfoldings of the body surface, are contained within protective cavities. Whereas the external gills of invertebrates often rely on the natural motions of the water to move water past the gills, invertebrates with internal gills often have cilia that direct water over the gill's surface.

The gills of fishes are located in a chamber behind the mouth cavity. Fishes need sufficient O_2 to meet the energy demands of swimming. To meet this need, they actively pump water through their mouth and over the gills (**Fig. 37.5**). The effort required to ventilate the gills could be high because of the density and viscosity of water, but fishes greatly reduce the energy cost of moving water by maintaining a continuous, unidirectional flow of water past their gills. In bony fishes, a protective flap overlying the gills, called the operculum, expands laterally to draw water over the gills while the mouth is refilling before its next pumping cycle.

The gills of fishes typically consist of a series of gill arches located on either side of the animal behind the mouth cavity and, in bony fishes, beneath the operculum. Each gill arch includes two stacked rows of flat leaf-shaped structures called gill filaments. Numerous **lamellae** (singular, **lamella**), thin sheetlike structures, are evenly but tightly spaced along the length of each gill filament and extend upward from the filament's surface. A series of blood vessels brings O_2-poor blood from the heart to the lamellar surfaces. The lamellae are composed of flattened epithelial cells and are extremely thin, so the water passing outside and blood passing inside the lamellae are separated by only about 1 to 2 μm. The lamellae give the gills an enormous surface area for their size. As a consequence, O_2 in the water passing over the gill can readily diffuse into the small blood vessels of the lamellae.

Blood flows through the lamellae in a capillary network. The lamellae are oriented so that blood flowing through the capillaries moves in a direction opposite to the flow of water past the gills (Fig. 37.5). This type of organization, in which fluids with different properties move in opposite directions, is an efficient way to exchange properties between the two fluids. The property exchanged can be a chemical property, such as O_2 concentration, or a physical property, such as heat. The movement of two fluids

FIG. 37.5 Respiration by fish gills. Fish gills extract O_2 from water efficiently through a unidirectional flow of water and countercurrent flow.

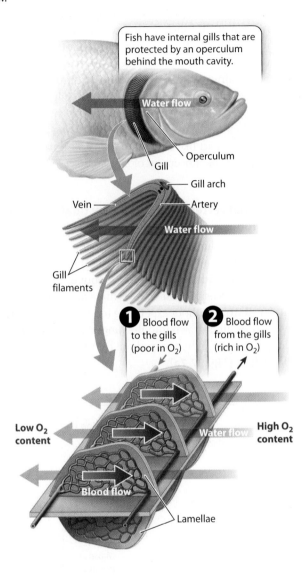

FIG. 37.6 Concurrent and countercurrent exchange. Countercurrent exchange is used widely in nature, as in fish gills, in cases where two fluids exchange properties (such as O_2, CO_2, or heat) efficiently.

a. Concurrent flow

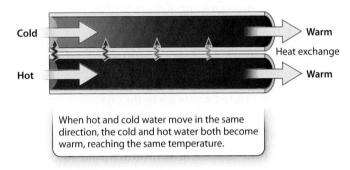

When hot and cold water move in the same direction, the cold and hot water both become warm, reaching the same temperature.

b. Countercurrent flow

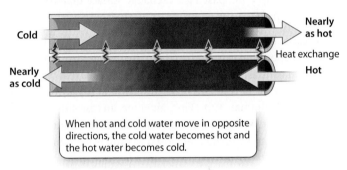

When hot and cold water move in opposite directions, the cold water becomes hot and the hot water becomes cold.

c. Countercurrent flow in fish gills

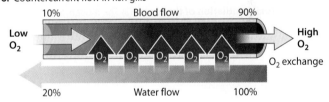

with different properties in opposite directions is called countercurrent flow.

Consider for a moment two tubes right next to each other, one carrying hot water and the other carrying cold water (**Fig. 37.6**). Heat is transferred between the two tubes: the hot water gets colder and the cold water gets hotter. The efficiency of heat transfer between the two tubes depends on the direction of flow. Suppose the water in the tubes flows in the same direction (concurrent flow; Fig. 37.6a). Given enough distance, the water will be at an intermediate (warm) temperature at the end of the tubes, reaching the same average temperature in both tubes.

Now consider what happens when the water in the two tubes flows in opposite directions (countercurrent flow). In this case, the cold water encounters increasingly hot temperatures and, given enough distance, becomes nearly as hot as the hot water entering the other tube (Fig 37.6b). Similarly, the hot water encounters increasingly cold temperatures, and becomes nearly as cold as the cold water entering the other tube. With countercurrent flow, the fluids in the two tubes essentially exchange properties (heat, in this example).

This mechanism, known as **countercurrent exchange**, is used widely in nature. As a result of countercurrent exchange, fish gills can extract nearly all the O_2 in the water that passes over them (Fig. 37.6c). The O_2 in the oxygen-rich water moves down its concentration gradient into the oxygen-poor blood of the gills along the entire length of the lamellae. At the same time, CO_2 readily diffuses out of the blood vessels and into the water that leaves the gill chamber. A set of blood vessels carries the O_2-rich blood away from the gills to supply the body.

Insects breathe air through tracheae.

With the evolution of life on land, animals that breathe air were able to increase their uptake of O_2 and achieve higher rates of metabolism. Oxygen uptake increased for the following reasons:

1. The O_2 content of air is typically 50 times greater than that of a similar volume of water. Fresh air contains approximately 210 mL O_2 per liter of air (21%), whereas the concentration of O_2 in well-mixed freshwater or salt water rarely exceeds 4 mL O_2 per liter of water (0.4%).

2. Oxygen diffuses nearly 8000 times faster in air than in water. This fact explains why the O_2 content can be close to zero a few centimeters beneath the surface of a stagnant pond.

3. Water is 800 times denser and 50 times more viscous than air. As a consequence, it requires more energy to pump water past a gas exchange surface than to pump air.

Because of their small size, insects can use a direct pathway of air transport that gets air directly to their tissues (see Fig. 37.4b). In contrast to terrestrial vertebrates, insects rely on a two-step process of ventilation and diffusion to supply their cells with O_2 and eliminate CO_2. First, air enters an insect through openings, called **spiracles**, along either side of its abdomen. The spiracles can be opened or closed to limit water loss and regulate O_2 delivery, much like the stomata of leaves in plants (Chapter 27). Inside the insect body, air is ventilated through a branching series of tiny air tubes, known as tracheae and tracheoles, directly to the cells. Second, diffusion occurs at the cell: O_2 supplied by the fine airways diffuses into the cells. CO_2 diffuses out and is eliminated through the insect's tracheae.

The mitochondria of the flight muscles and other metabolically active tissues of insects are located within a few micrometers of tracheole airways. Because O_2 and CO_2 diffuse rapidly in air, the tracheal system of insects delivers O_2 to cellular mitochondria at high rates. Active insects also have an air sac system connected to their tracheae. Powered by contracting abdominal muscles, the air sacs act like bellows to pump air through the tracheae, speeding the movement of air to tissues so that gas exchange is faster.

Most terrestrial vertebrates breathe by tidal ventilation of internal lungs.

Unlike fish gills, the lungs of most land vertebrates inflate and deflate to move fresh air with high levels of O_2 into the lungs and expire stale air with high levels of CO_2 out of the lungs. The low density and viscosity of air enables these animals to breathe by **tidal ventilation** without expending too much energy. In tidal respiration, air is drawn into the lungs during **inhalation** as the lungs inflate and then moved out during **exhalation** as the lungs deflate (**Fig. 37.7a**).

FIG. 37.7 Tidal ventilation of the lungs. (a) The movement of the diaphragm causes lung volume to increase and decrease during breathing, drawing in air and then expelling it. (b) The change in lung volume in humans is typically approximately 0.5 L when breathing at rest.

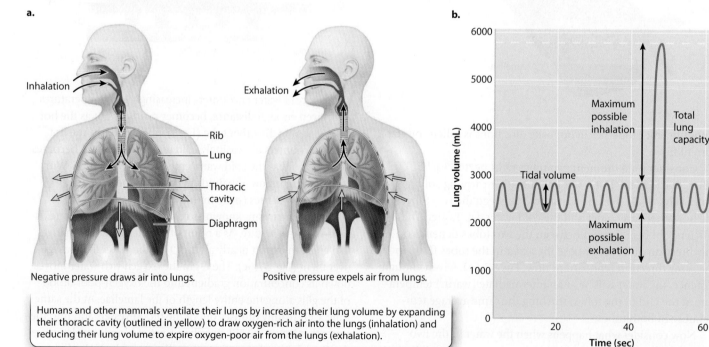

Humans and other mammals ventilate their lungs by increasing their lung volume by expanding their thoracic cavity (outlined in yellow) to draw oxygen-rich air into the lungs (inhalation) and reducing their lung volume to expire oxygen-poor air from the lungs (exhalation).

Mammals and reptiles expand their thoracic cavity to draw air inside their lungs on inhalation. The thoracic cavity is located in the front of the animal's trunk and contains the lungs. Expansion of the thoracic cavity causes the lungs to expand as well, which lowers the air pressure inside the lungs relative to the air pressure outside the lungs. The resulting negative pressure draws air into the lungs. By contrast, amphibians inflate their lungs by positive pressure produced by their mouth cavity. In most animals, exhalation is passively driven by elastic recoil of tissues that were previously stretched during inhalation. Elastic recoil compresses the lungs, causing the air pressure inside the lungs to become higher than the air pressure outside the lungs. The resulting positive pressure forces air out of the lungs.

In mammals, inhalation during normal, relaxed breathing is driven by contraction of the **diaphragm**, a domed sheet of muscle located at the base of the lungs that separates the thoracic and abdominal cavities (Fig. 37.7a). When the diaphragm contracts, it expands the thoracic cavity, drawing air into the lungs. Exhalation occurs passively by elastic recoil of the lungs and the chest wall surrounding the thoracic cavity. During exercise, other muscles come into play to assist with inhalation and exhalation. For example, **intercostal muscles**, which are attached to adjacent pairs of ribs, assist the diaphragm by elevating the ribs on inhalation and depressing them during exhalation. The action of the intercostal muscles helps to produce larger changes in the volume of the thoracic cavity, increasing the negative pressure that draws air into the lungs during inhalation, and assisting elastic recoil of the lungs and chest wall to pump air out of the lungs during exhalation.

At rest, humans inhale and exhale approximately 0.5 L of air every cycle (**Fig. 37.7b**). This amount represents the **tidal volume** of the lungs. With a breathing frequency of 12 breaths per minute, the ventilation rate (breathing frequency × tidal volume) is 6 L per minute. When more O_2 is needed during exercise, both breathing frequency and tidal volume increase to elevate ventilation rate.

The fresh air inhaled during tidal breathing mixes with O_2-depleted stale air that remains in the airways after exhalation. As a result, the pO_2 in the lungs (approximately 100 mmHg in humans) is lower than the pO_2 of freshly inhaled air (approximately 160 mmHg at sea level). Consequently, the fraction of O_2 that can be extracted is lower than the fraction that can be extracted by the countercurrent flow of fish gills, which can reach 90% or more. Typically, mammalian lungs extract less than 25% of the O_2 in the air, and reptile and amphibian lungs extract even less. Nevertheless, the disadvantages of tidal respiration are offset by the ease of ventilating the lung at high rates and by the high O_2 content of air.

Mammalian lungs are well adapted for gas exchange.

Despite the limitations of tidal respiration, mammalian lungs are well adapted for breathing air. They have an enormous surface area and a short diffusion distance for gas exchange. Consequently, the lungs of mammals supply O_2 quickly enough to support high metabolic rates.

Except in the case of birds (discussed in the next subsection), the lungs of air-breathing vertebrates are blind-ended sacs located within the thoracic cavity. Air is taken in through the mouth and nasal passages and then passes through the **larynx**, an organ in the throat made of muscle and cartilage that helps to isolate breathing from swallowing and within which the **vocal cords** are located. The vocal cords enable speech, song, and sound production. Air then enters the **trachea**, the central airway leading to the lungs (**Fig. 37.8**). The trachea divides into two airways, called the **primary bronchi**, one of which supplies each lung. The trachea and bronchi are supported by rings of cartilage that prevent them from collapsing during respiration, ensuring that airflow meets with minimal resistance. You can feel some of these cartilage rings at the front of your throat below your larynx (the part of the larynx you can feel is your Adam's apple).

The primary bronchi divide first into smaller secondary bronchi, and again into finer **bronchioles**. This branching continues until the terminal bronchioles have a diameter of less than 1 mm. The very fine respiratory bronchioles end with clusters of tiny thin-walled sacs, the **alveoli** (singular, **alveolus**), where gas exchange by diffusion takes place. As a result of this branching pattern, each lung consists of several million alveoli, providing a large surface area. Together, the two human lungs include 300 to 500 million alveoli with a combined surface area as large as a tennis court—about 100 m^2!

Small blood vessels, the **pulmonary capillaries**, supply the alveolar wall. Each alveolus is lined with thin epithelial cells in intimate contact with the endothelium of these small blood vessels (Fig. 37.8). As a result, the diffusion distance from the alveolus into the capillary is extremely short (approximately 2 μm), similar to that of gill lamellae and the terminal airways of bird lungs.

Keeping the surfaces of the airways and alveoli moist is critical. Moisture helps move O_2 molecules from the air into solution, thereby helping O_2 diffuse across the alveolar wall. A film of mucus secreted by specialized epithelial cells keeps the airways moist.

This film has surface tension, a cohesive force that holds the molecules of a liquid together, causing the surface to act like an elastic membrane. Surface tension is the reason why it requires greater pressure to inflate a small balloon than a large one. Specialized alveolar epithelial cells produce **surfactant**, a compound that acts to reduce the surface tension of the fluid film. Surfactant allows the lungs to be inflated easily at low volumes when the alveoli are partially collapsed and small. It also ensures that alveoli of different sizes inflate with similar ease, enabling more uniform ventilation of the lung.

FIG. 37.8 Human lung anatomy. (a) Millions of alveoli provide a vast surface for gas exchange. (b) A scanning electron micrograph of the lung alveolar surface. (c) A transmission electron micrograph showing a cross section of the alveolar wall and lung capillary. *Photo source: Reprinted from E. Weibel, B. Sapoval, and M. Filoche, 2005, "Design of Peripheral Airways for Efficient Gas Exchange," Respiratory Physiology & Neurobiology, 148:3–21. Copyright 2005, with permission from Elsevier.*

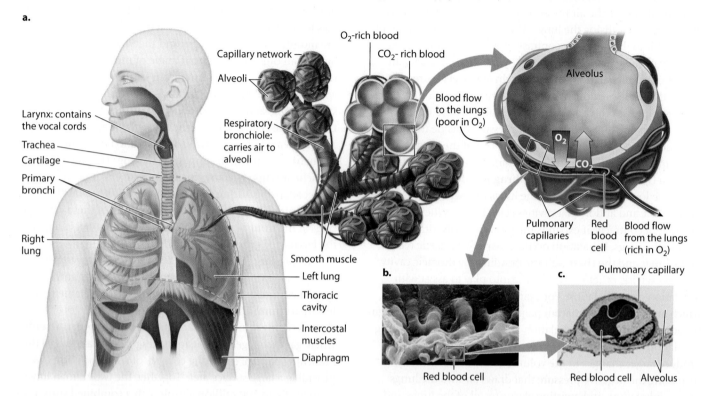

The mucus lining the lung airways not only keeps them moist, but also traps and removes foreign particles and microorganisms that an animal may breathe in with air. Beating cilia on the surface of other epithelial cells move the mucus and foreign debris out of the lungs and into the throat, where it is swallowed and digested. Smoking even one cigarette can stop the cilia's beating for several hours or even destroy them. This is one reason why smokers cough—they must clear the mucus that builds up.

FIG. 37.9 Respiration by bird lungs. Anterior and posterior air sacs move air unidirectionally through a rigid lung, increasing the uptake of O_2 at low pO_2.

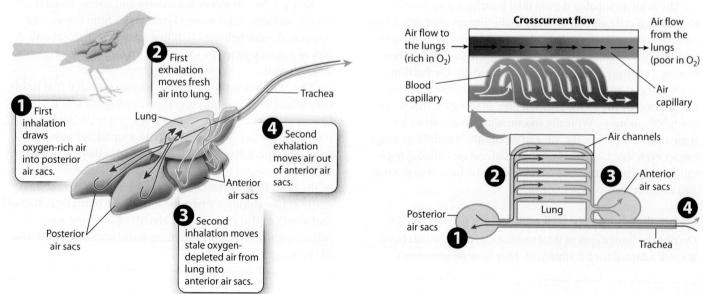

The structure of bird lungs allows unidirectional airflow for increased oxygen uptake.

Birds extract O_2 efficiently to supply the considerable energy needed for flight. An efficient O_2 supply is particularly important for migratory birds that fly at high altitude, where the partial pressure of O_2 is relatively low. The lungs of birds are unique in that they are rigid and do not inflate and deflate. As a consequence, birds do not use tidal breathing. Instead, they benefit from unidirectional airflow that enhances gas exchange by maintaining larger concentration gradients for diffusion.

A bird's respiratory system consists of two sets of air sacs: one located posterior to the lungs (behind them) and one located anterior to the lungs (in front of them). Air is pumped through these air sacs continuously in a bellows-like action (**Fig. 37.9**). As the posterior air sacs expand, air is drawn from the bird's mouth into those sacs. Surrounding muscles compress the sacs, pumping air through the bird's rigid lungs. The lungs consist of a series of larger air channels that branch into fine air capillaries having a large surface area. The air capillaries direct the air so that it moves **crosscurrent** (that is, at 90°) to the path of blood flow through the walls of the air capillaries. This crosscurrent flow is not as efficient as the countercurrent flow of fish gills, but it enables birds to extract more O_2 from the air than mammals can.

After losing O_2 and gaining CO_2, the air leaves the bird's lungs and enters the set of anterior air sacs. Fresh air is again inhaled into the posterior air sacs, and the stale air in the anterior air sacs is exhaled out of the bird's trachea and mouth. As a result, birds achieve a continuous supply of fresh air into the lung during both inhalation *and* exhalation. In contrast to other air-breathing vertebrates, therefore, birds move air through their respiratory system in two cycles of ventilation.

Voluntary and involuntary mechanisms control breathing.

Because an animal's need for O_2 varies with activity level, animals adjust their respiratory rate to meet their changing demand for O_2. Respiration is a unique physiological process in that it is controlled by both the voluntary and involuntary components of the nervous system (Chapter 34). In sleep and most other circumstances, breathing is controlled unconsciously.

The regulation of blood O_2 levels is a key example of homeostasis, like the maintenance of core body temperature (Chapter 33) and blood glucose levels (Chapter 36). Recall that homeostasis often depends on sensors that monitor the levels of the chemical being regulated. In the case of breathing, these sensors are chemoreceptors located within the brainstem and in sensory structures called the **carotid** and **aortic bodies** that are located in the neck and near the heart, respectively (**Fig. 37.10**). Together, they sense O_2, CO_2, and proton (H^+) concentrations of the blood. The most important factor in the control of breathing is the amount of CO_2 in the blood. When the concentration of CO_2 in the blood is too high, these chemoreceptors stimulate motor neurons that activate the respiratory muscles to contract

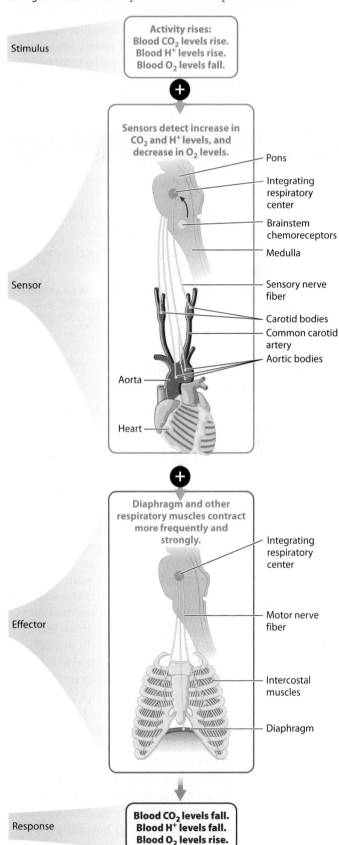

FIG. 37.10 Homeostatic control of breathing. Breathing becomes stronger or faster when CO_2 levels increase or O_2 levels decrease.

more strongly and frequently. Stronger, faster breathing rids the blood of excess CO_2 and increases the supply of O_2 to the body.

Breathing can also be controlled voluntarily. It is a simple matter to choose to hold your breath. Holding the breath makes it possible for humans and marine mammals to dive under water; it is also critical to the production of speech, song, and sound for communication. Sound is produced by voluntarily adjusting the magnitude and rate of airflow over the vocal cords of mammals, the syrinx of songbirds, or the glottal folds of calling amphibians, such as some toads and frogs.

Self-Assessment Questions

4. What are two features of the gills of aquatic animals that favor diffusion of O_2 and CO_2?
5. What are two differences in the physical properties of water and of air that affect gas exchange in aquatic animals?
6. How does the tracheal respiratory system of insects enable high metabolic rates?
7. What generates the pressure changes needed to draw air into the lungs and to expel air out of the lungs?

37.3 OXYGEN TRANSPORT BY HEMOGLOBIN

Once O_2 is extracted from the environment and diffuses into the animal, it is transported to respiring cells. Oxygen is transported in the circulation by specialized body fluids: the blood of vertebrates or the hemolymph of invertebrates.

Blood is composed of fluid and several types of cell.

Although this chapter focuses on the role of blood and hemolymph in O_2 and CO_2 transport, blood serves many other functions. These include heat transport for temperature regulation and nutrient transport to metabolically active tissues (Chapter 38), hormone transport (Chapter 36), waste transport to excretory organs (Chapter 39), and transport of immune cells involved in fighting pathogens (Chapter 41) as well as wound healing. Blood can be separated into a cellular fraction and a fluid fraction, called blood plasma. Red blood cells that carry **hemoglobin** constitute most of the cellular fraction of blood. In both vertebrates and invertebrates, hemoglobin evolved as a specialized iron-containing, or heme, molecule for O_2 transport.

The fraction of total blood volume that is red blood cells in vertebrates is defined as the **hematocrit**. Whereas a hematocrit of 45% is considered normal for humans (**Fig. 37.11**), fishes have hematocrits that range from 20% to 35%, crocodiles have hematocrits of approximately 20%, birds have hematocrits that vary from 35% to 58%, and diving mammals have hematocrits as high as 65%. Blood with a lower hematocrit flows with less resistance (that is, has a lower viscosity), but carries less oxygen. Humans with low hematocrits are considered to be iron

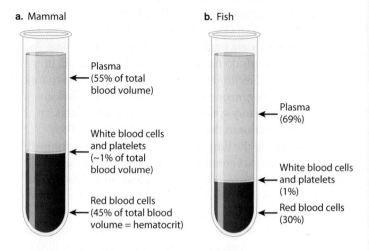

FIG. 37.11 Blood composition in (a) mammals and (b) fishes. Blood consists of a cellular fraction, which includes red blood cells, white blood cells, and platelets, and a fluid fraction called plasma.

deficient because their blood contains fewer red blood cells and, therefore, too little hemoglobin.

A much smaller fraction (approximately 1%) of cells are white blood cells, which help to defend the body against pathogens (Chapter 41). The remaining cells are platelets, which respond to damaged blood vessels by helping to form a clot to prevent blood loss.

Red blood cells are the densest component of blood. Therefore, when blood is spun in a centrifuge, the red blood cells form a layer in the bottom of a centrifuge tube, and a thin layer of white blood cells and platelets forms above them. The remaining components of the blood, the blood plasma and solutes, rise to the top of the centrifuge tube (Fig. 37.11).

Blood plasma can hold only as much O_2 or CO_2 as can be dissolved in solution. How much O_2 goes into solution at a given partial pressure is a measure of its **solubility**. Because O_2 is nearly 30 times less soluble than CO_2, only about 0.2 mL of O_2 can be carried in 100 mL of blood. By binding O_2 and removing it from solution, hemoglobin increases the amount of O_2 in the blood by a hundredfold. Because of its greater solubility, CO_2 is transported in solution within the blood. Rather than remaining dissolved in solution as a gas, most of the CO_2 (about 95%) is converted to carbonic acid, which dissociates to form bicarbonate ions (HCO_3^-) and protons (Chapter 6).

Whereas some invertebrate hemoglobins exist in solution in the hemolymph and others exist within cells carried by the hemolymph, all vertebrate hemoglobins are produced by and reside in red blood cells. Hemoglobin gives the cells and the blood their red appearance.

Hemoglobin is an ancient molecule with diverse roles related to oxygen binding and transport.

Hemoglobin is an ancient molecule that has been found in organisms from all kingdoms of life. In bacteria, archaeons, and fungi, similar forms of hemoglobin evolved as a terminal electron acceptor in cellular respiration. These hemoglobins may have also served a role in scavenging O_2 to avoid damage due to oxidative stress.

Plant hemoglobins bind and transport O_2 within the cell. In soybeans, hemoglobin binds O_2 in root nodule cells to keep concentrations low enough to avoid inhibiting nitrogenase, the enzyme used by symbiotic rhizobial bacteria in the root nodules for nitrogen fixation (Chapter 27).

Invertebrate hemoglobins found in roundworms, annelid worms, and arthropods bind O_2 in the circulatory fluid, as discussed earlier. They share considerable gene and amino acid sequence similarity with vertebrate hemoglobins, indicating that a common ancestor of invertebrate and vertebrate animals evolved a shared form of hemoglobin more than 670 million years ago.

Hemoglobin reversibly binds oxygen.

Hemoglobin exists in large concentrations within red blood cells. This globular protein consists of four polypeptide units (**Fig. 37.12**). Each of these four units surrounds a heme group that contains an iron atom, which reversibly binds one O_2 molecule. After O_2 diffuses into the blood, it diffuses into the red blood cells and binds to the heme groups in hemoglobin. Hemoglobin's binding of O_2 removes O_2 from solution, keeping the pO_2 of the red blood cell below that of the blood plasma. As a result, O_2 continues to diffuse into the cell. The removal of O_2 from the plasma, in turn, keeps the pO_2 of the plasma below that of the lung alveolus, so O_2 continues to diffuse from the lungs into the blood.

When more O_2 is present in blood plasma, we expect more O_2 to become bound to hemoglobin, and that is what happens. But an interesting wrinkle enables hemoglobin to readily bind O_2 leaving the lung. If we plot blood pO_2 against the percentage of O_2 bound to hemoglobin (the relation defined as "hemoglobin saturation"), we get the curve shown in **Fig. 37.13a**. As blood pO_2 increases, hemoglobin saturation rises slowly at first, then more steeply, and then more slowly again until it levels out. At 100% saturation, all hemoglobin molecules bind four O_2 molecules. The curve, which is called the hemoglobin's **oxygen dissociation curve**, has a sigmoidal shape—that is, a shape like an "S".

HOW DO WE KNOW?

FIG. 37.12

What is the molecular structure of hemoglobin and myoglobin?

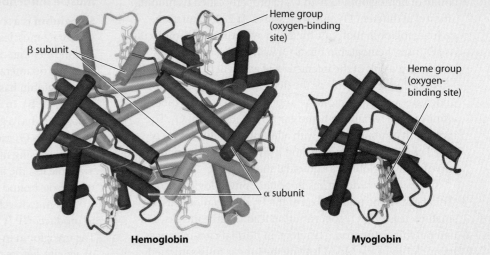

BACKGROUND In the 1950s, the Austrian scientist Max Perutz worked with British structural chemist John Kendrew in the Cavendish Laboratory of the University of Cambridge to determine the molecular structure of globular proteins. Perutz and Kendrew were interested in understanding how the structures of hemoglobin and myoglobin, the two O_2 transport proteins, enabled binding and transport of O_2.

EXPERIMENT Perutz and Kendrew developed and applied the new technique of X-ray crystallography to determine the three-dimensional molecular structures of hemoglobin and myoglobin.

RESULTS Perutz and Kendrew showed that an adult hemoglobin molecule consists of four polypeptide subunits: two α (alpha) and two β (beta) subunits. Each subunit contains a heme group that contains an iron atom. The iron atom is the site of O_2 binding. By contrast, a myoglobin molecule consists of only a single subunit with one heme group. Myoglobin is also more highly folded and compact than hemoglobin. These differences in molecular structure underlie the different O_2-binding and dissociation properties of the two O_2 transport proteins.

FOLLOW-UP WORK Work by Perutz in the 1960s showed how the molecular structure of hemoglobin changes when it binds O_2. Perutz also supervised Francis Crick as a doctoral student (Crick later worked with James Watson to determine the structure of DNA). Perutz and Kendrew received the 1962 Nobel Prize in Chemistry for their work.

SOURCES Kendrew, J. C., and M. F. Perutz. 1948. "A Comparative X-Ray Study of Foetal and Adult Sheep Haemoglobins." *Proceedings of the Royal Society, London, Series A* 194:375–398; Perutz, M. F. 1964. "The Hemoglobin Molecule." *Scientific American* 211:64–76; Perutz, M. F. 1978. "Hemoglobin Structure and Respiratory Transport." *Scientific American* 239:92–125.

FIG. 37.13 **(a) Hemoglobin and (b) myoglobin dissociation curves.**

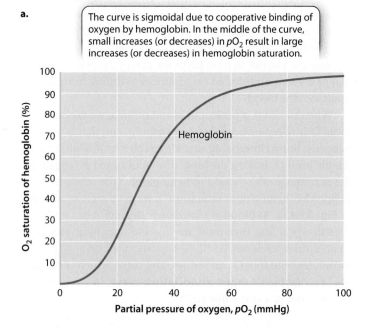
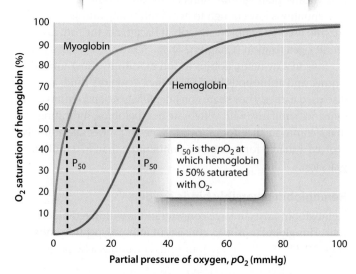

The shape of this curve can be explained by changes in the ability of hemoglobin to bind O_2 (a property called its binding affinity) at different O_2 partial pressures. At 25% saturation, each hemoglobin molecule binds, on average, one O_2 molecule. As pO_2 rises, hemoglobin binds O_2 with increasing binding affinity as a result of the interaction of adjacent heme groups. After one heme group binds the first O_2 molecule, hemoglobin undergoes a conformational change that increases the binding affinity of the remaining heme groups for additional O_2. The increase in binding affinity with additional binding of O_2 is called **cooperative binding**, and it gives the O_2 dissociation curve for hemoglobin its sigmoidal shape.

Cooperative binding has important physiological consequences. In the middle part of the O_2 dissociation curve, small increases in O_2 concentration lead to large increases in hemoglobin saturation. Therefore, in normal circumstances, hemoglobin in the blood leaving the lung is fully saturated with O_2; that is, each hemoglobin molecule is bound to four O_2 molecules.

The shape of the O_2 dissociation curve also helps to explain how O_2 is delivered to respiring cells. When the hemoglobin in red blood cells reaches tissues needing O_2 to supply their mitochondria, the O_2 is released from the hemoglobin. As active cells consume O_2, they reduce the local pO_2 of the cell and surrounding tissues to 40 mmHg or less. At these lower pO_2 values, the hemoglobin's O_2 dissociation curve has a steep slope (Fig. 37.13a). The steepness of the slope indicates that for a relatively small decrease in pO_2, large amounts of O_2 can be released from hemoglobin to diffuse into the cell.

Myoglobin stores oxygen, enhancing oxygen delivery to muscle mitochondria.

Myoglobin is a specialized O_2 carrier within the cells of vertebrate muscles. In contrast to hemoglobin, myoglobin is a monomer that contains only a single heme group. Because there are no interacting subunits, the O_2 dissociation curve for myoglobin has a different shape from that of hemoglobin (**Fig. 37.13b**). In fact, over the physiological range of pO_2, myoglobin has a greater affinity for O_2 than hemoglobin does, and it binds O_2 more tightly. As a result, hemoglobin releases O_2 to exercising muscles. As the exercising muscles consume the available O_2, the intracellular pO_2 drops, and the myoglobin releases its bound O_2 to the mitochondria.

Red muscle cells that depend mainly on aerobic respiration to produce ATP (Chapter 35) store large amounts of myoglobin. The myoglobin in these cells can release O_2 quickly at the onset of activity, before the respiratory and circulatory systems have had time to increase the supply of O_2. Diving marine mammals, such as whales and seals, have large amounts of myoglobin within their muscles. The myoglobin loads up with O_2 when the animals breathe at the surface before beginning a dive. The O_2 bound to the myoglobin is then used to supply ATP during the dive, when the animal cannot breathe. These marine mammals can stay under water for 30 minutes or longer and dive to considerable depths (**Fig. 37.14**).

Many factors affect hemoglobin–oxygen binding.

Obtaining enough O_2 can be a challenge. How does a mammalian fetus in its mother's uterus take up O_2 from its

FIG. 37.14 Myoglobin as an O_2 store for long-lasting dives. The graph plots the dives of a beaked whale (*Ziphius cavirostris*) fitted with a tracking device. *Source: Based on Fig. 3A in P. L. Tyack, M. Johnson, N. Aguilar de Soto, A. Sturlese, and P. T. Madsen, 2006, "Extreme Diving Behaviour of Beaked Whale Species Known to Strand in Conjunction with Use of Military Sonars," Journal of Experimental Biology 209:4238–4253. Photo source: Bill Curtsinger/Getty Images.*

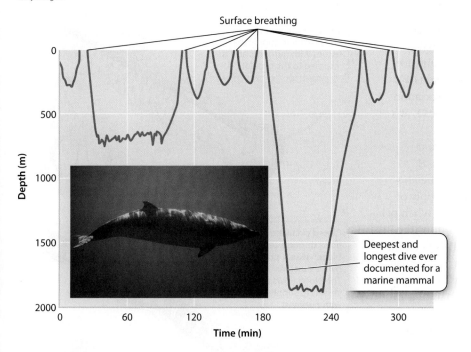

mother's blood? How do animals that live at high altitude obtain enough O_2 under conditions of low pO_2? In each case, the species adapts by evolving forms of hemoglobin with higher binding affinity.

In mammals, the mother's respiratory and cardiovascular systems must provide the fetus with O_2 and remove CO_2, as well as supply nutrients and remove wastes. Maternal and fetal blood do not mix, but rather exchange gases across the placenta. Yet the O_2 concentration gradient at this exchange point does not strongly favor the movement of O_2 from the mother's blood to the fetal blood. Fetal mammals solve the problem of extracting O_2 from their mother's hemoglobin by expressing a form of hemoglobin that has a higher affinity for O_2 than does their mother's hemoglobin. The O_2 dissociation curve for fetal hemoglobin is shifted to the left of the curve for maternal hemoglobin (**Fig. 37.15a**), allowing the fetal hemoglobin to extract O_2 from the mother's circulation. At birth, when the newborn begins to breathe, its red blood cells rapidly shift to

FIG. 37.15 Changes in hemoglobin's O_2 dissociation curve.

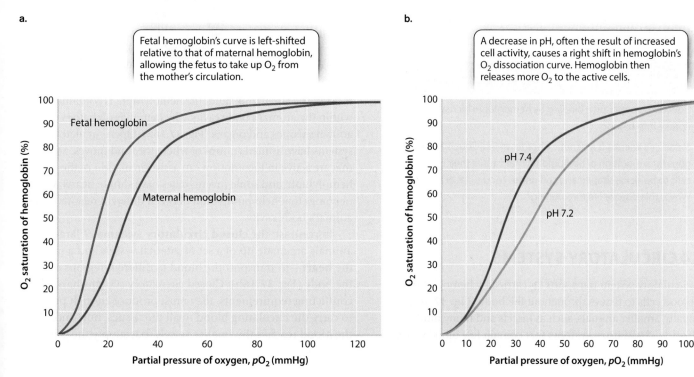

expressing the adult form of hemoglobin, maintaining this form throughout life.

Many animals live at high altitude. For example, birds have been observed flying over the highest peaks (taller than 9000 m), even though air at this altitude has less than one-third the O_2 content of air at sea level. Similarly, llamas inhabit the Andes at altitudes up to 5000 m even though the air has about one-half the O_2 content of air at sea level. Consequently, the lung pO_2 of a llama is only about 50 mmHg. How do these animals obtain enough O_2? Their hemoglobin has a higher affinity for O_2 and binds O_2 more readily than does the hemoglobin of animals living at sea level. Consequently, their O_2 dissociation curve is shifted to the left, favoring the binding of more O_2 at a given pO_2.

Shifts in blood and tissue pH cue the body that more oxygen is needed to fuel exercise. The pH falls (that is, H^+ concentration increases) when CO_2 is released from metabolizing cells during exercise, or when an inadequate supply of O_2 leads to the production of lactic acid (Chapter 7). When pH falls, the affinity of hemoglobin for O_2 *decreases*, shifting the O_2 dissociation curve to the right (**Fig. 37.15b**). This phenomenon is called the Bohr effect after Christian Bohr, the Danish physiologist who first described it in 1904. Because hemoglobin's affinity for O_2 is reduced, more O_2 is released to cells for aerobic ATP synthesis.

Carbon dioxide also reacts with the amine (NH_2) groups of hemoglobin, reducing hemoglobin's affinity for O_2. Thus, when released from respiring tissues, CO_2 promotes increased O_2 delivery both through its direct effect on hemoglobin and through its contribution to a decrease in blood pH by the Bohr effect. Whereas the production of CO_2 promotes O_2 release at the tissues, its elimination at the lung increases hemoglobin's affinity for O_2, thereby enhancing O_2 uptake.

Self-Assessment Questions

8. How does cooperative binding by hemoglobin increase O_2 uptake into blood?
9. Sickle-cell anemia is a genetic disorder of individuals homozygous for a mutation of hemoglobin that causes their red blood cells to be sickle shaped and stiff under conditions of low pO_2. Why is this disease life threatening?

37.4 CIRCULATORY SYSTEMS

The circulatory system transports O_2 carried by hemoglobin in red blood cells to tissues throughout the body (**Fig. 37.16**). Generally, smaller animals, such as insects and many mollusks, have **open circulatory systems** that contain few blood vessels:

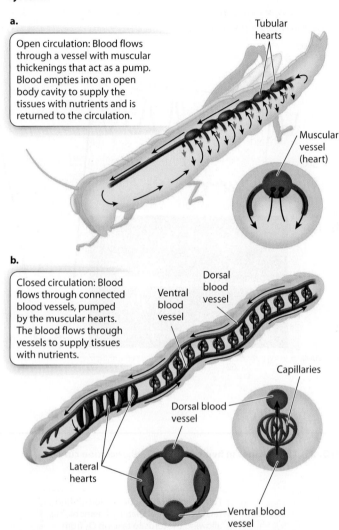

FIG. 37.16 (a) Open and (b) closed invertebrate circulatory systems.

a. Open circulation: Blood flows through a vessel with muscular thickenings that act as a pump. Blood empties into an open body cavity to supply the tissues with nutrients and is returned to the circulation.

b. Closed circulation: Blood flows through connected blood vessels, pumped by the muscular hearts. The blood flows through vessels to supply tissues with nutrients.

most of the circulating fluid, the hemolymph, is contained within the animal's body cavity. The hemolymph bathes the animal's tissues and organs (Fig. 37.16a). Open circulatory systems have limited control of where the fluid moves. Muscles that are active in locomotion can assist circulation of the hemolymph, and some invertebrates have simple hearts with openings that help pump fluid between different regions of the animal's body cavity.

In contrast, the **closed circulatory systems** of larger animals are made up of a set of internal vessels and a pump—the **heart**—to transport the blood to different regions of the body (Fig. 37.16b). Closed circulatory systems have two conflicting requirements: they must produce enough pressure to carry the circulating blood to all the tissues, but, once the blood reaches the smaller vessels that supply the cells within the tissues, the blood pressure and flow rate must not be too high. High pressure would push fluid through the walls of the smaller vessels, and it would cause the blood to flow so quickly that there would not be enough time to exchange gases.

About 500 million years ago, segmented worms first evolved closed circulatory systems that allow them to expand individual body segments to burrow underground. About 480 million years ago, the first cephalopods (squid and octopus) evolved closed circulatory systems that enabled them to be successful predators in Paleozoic oceans. A closed circulatory system delivers O_2 at high rates to exercising tissues, enabling their mitochondria to provide the energy needed to chase prey. In contrast, open circulatory systems generally operate under low pressure, and their transport capacity is limited. As a result, animals with open circulatory systems are less active. Insects can be very active despite having an open circulatory system because they obtain O_2 by their tracheal system independent of the circulatory system.

Open circulatory systems have limited ability to control the delivery of respiratory gases and metabolites to specific tissues and regions. By contrast, closed circulatory systems can control blood flow to specific regions of the body. For example, wading birds reduce blood flow to their legs when they are in cold water to reduce heat loss, and vertebrates can increase blood flow to their muscles to deliver O_2 and nutrients during exercise. Blood flow is controlled by varying the resistance to flow.

Circulatory systems have vessels of different sizes.

For a fluid like blood to flow through a set of pipes, pressure (P) is required to overcome the resistance to flow (R). To pump blood through a set of closed interconnected vessels, a muscular heart is needed to produce sufficient pressure to overcome the flow resistance of the vessels. The pressure developed by a heart when it contracts establishes a rate of blood flow that is governed by P/R. That is, the rate of blood flow increases with an increase in pressure and decreases with an increase in resistance. The resistance to flow is determined in part by the fluid's viscosity and the vessel's length. More viscous fluid and longer vessels of a given diameter have greater resistance. However, the main factor governing resistance to flow is the vessel's radius (r). Resistance is proportional to $1/r^4$, which means that if a vessel's radius is reduced by half, its resistance to flow increases 16 times. The high resistance of narrow vessels presents a challenge. To keep diffusion distances short, gas exchange between cells and blood requires narrow vessels. But how does the circulatory system overcome the dramatic increase in resistance that occurs with narrower diameters?

The organization of animal circulatory systems provides an answer (**Fig. 37.17**). Animal circulatory systems are organized to have blood flow over longer distances in a relatively few large-diameter vessels with low resistance to flow (Fig. 37.17a). **Arteries** are the large, high-pressure vessels that move blood flow away from the heart to the tissues. **Veins** are the large, low-pressure vessels that return blood to the heart. Arteries branch into blood

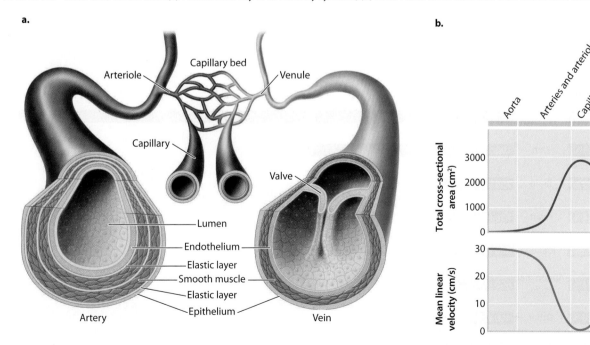

FIG. 37.17 Flow and vessel size. (a) Vessel hierarchy of circulatory systems; (b) total vessel cross-sectional area versus flow velocity.

vessels of progressively smaller diameter called **arterioles**. These arterioles ultimately connect to finely branched networks of very small blood vessels called **capillaries**, which are only one cell in diameter or less. Indeed, the diameter of a capillary is so small that red blood cells pass through one at a time, and must be flexible to squeeze through the capillaries.

Within the capillaries, gases are exchanged by diffusion with the surrounding tissues. The number of smaller-diameter vessels at each branching point in the circulatory system greatly exceeds the number of larger-diameter vessels. The reverse organization is found on the return side of circulation: numerous capillaries drain into vessels of progressively larger diameter called **venules**, and the venules drain into a few larger veins that return blood to the heart.

This organization has two advantages. First, it maintains the same volume of blood flow at all levels within the circulatory system: the increased resistance to flow in the capillaries is offset by the large increase in the number of capillaries. Second, it enables the blood to flow more slowly in smaller vessels (Fig. 37.17b), providing time for gases and metabolites to diffuse into and out of the neighboring cells.

Arteries are muscular vessels that carry blood away from the heart under high pressure.

When animals become active after a period of rest and feeding, blood flow must be increased to their muscles and reduced to their digestive organs (**Fig. 37.18**). Animals meet this challenge by changing the radius of vessels. Recall that resistance to flow is strongly affected by the vessel radius. Arterioles supplying regions of the body that need less blood become narrower by contracting the circular smooth muscle fibers in their walls. The narrowing of the blood vessels dramatically increases the resistance to flow, thereby reducing the rate of blood flow. For example, when you put your hands or feet into cold water, the arterioles constrict, reducing the loss of heat from blood flowing to your hands and feet.

Arterioles supplying regions of the body that need more blood become wider by relaxing the smooth muscles lining their walls. The larger vessels offer less resistance, so more blood can flow through them. As the rate of cellular respiration increases, cells release more CO_2 as well as other metabolites, such as lactate and hydrogen ions. The increase of these molecules in the bloodstream acts as a signal to the smooth muscle in the arteriole walls to relax. In this way, blood flow increases to the capillaries so as to match O_2 and nutrient delivery to the metabolic needs of the cells.

In humans, a strong pulse of flow pushes against the arterial walls about every second with each heartbeat that pumps blood through the arteries. This pulse of flow causes a momentary expansion of the arteries. Artery walls can withstand these repeated pressure pulses because they contain multiple elastic layers composed of two proteins: **collagen** and **elastin**. Collagen fibers are strong and resist the expansion of the arterial wall during pressure pulses. Collagen and elastin fibers together

FIG. 37.18 Changes in blood flow from resting to exercise as measured in humans. Numbers in parentheses indicate percent of total blood flow.
Source: Data from C. B. Chapman and J. H. Mitchell, 1965, "The Physiology of Exercise," Scientific American 212:88–96.

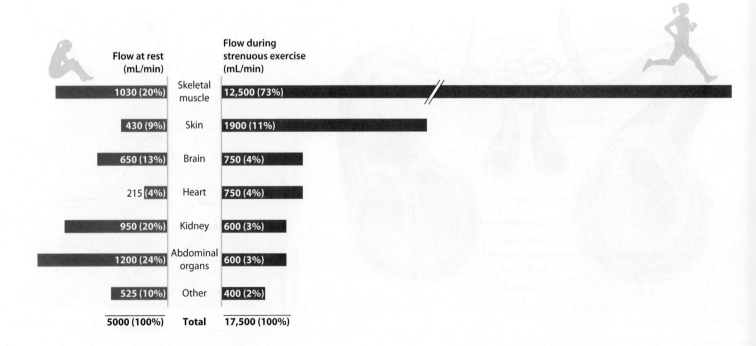

provide for an elastic rebound of the arterial wall once the pulse has passed, returning energy to help smooth out blood flow.

Both collagen and elastin resist overexpansion of the arterial wall. However, if an artery wall deteriorates, the collagen and elastin can become so thin that the artery bulges outward. This outward bulge, called an aneurysm, can lead to a life-threatening rupture.

Veins are thin-walled vessels that return blood to the heart under low pressure.

Blood collected from local capillary networks returns to the heart through progressively larger veins. These ultimately drain into the two largest veins, the **venae cavae** (singular, **vena cava**). The venae cavae drain blood from the head and body into the heart. Relatively little pressure is available to push the blood forward in the veins because pressure has been lost to the resistance of the arterioles and capillaries. Consequently, veins have thin walls combined with little smooth muscle or elastic connective tissue. Because of the low pressure, blood tends to accumulate within the veins: as much as 80% of your total blood volume resides in the venous side of your circulation at any one time.

Veins located in the limbs and in the body below the heart have one-way valves that help prevent blood from pooling owing to gravity (see Fig. 37.17a). However, the most important mechanism used to return blood to the heart is the voluntary muscle contractions that occur during walking and exercise, which exert pressure on the veins. Above the level of the heart, gravity assists the return of blood to the heart through the veins.

Compounds and fluid move across capillary walls by diffusion, filtration, and osmosis.

We have seen that O_2 and CO_2 move between capillaries and tissues by diffusion. In addition, blood pressure forces water, certain ions, and other small molecules to move from capillaries into the surrounding interstitial fluid (the extracellular fluid surrounding vessels). Proteins and blood cells remain in the capillaries. In this way, the blood is filtered as the plasma passes through the capillary wall.

Why doesn't the blood plasma lose all its water and ions over time? As water is filtered out of the capillaries, some ions, cells, proteins, and other large compounds remain inside; their concentration increases with the removal of the water (**Fig. 37.19**). As a result, water has a tendency to flow back into the capillaries by osmosis (Chapter 5). At the same time, the resistance to flow imposed by the capillaries decreases blood pressure. When the pressure is no longer high enough to counter osmosis, water moves back into the capillaries.

In vertebrates, excess interstitial fluid is also returned to the bloodstream through the **lymphatic system**, a network of vessels distributed throughout the body. The fluid that enters the lymphatic system is called **lymph**. Small lymphatic vessels

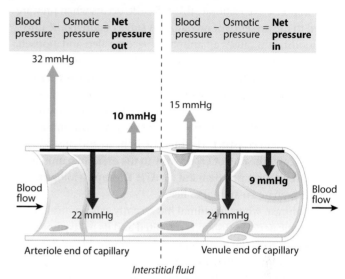

FIG. 37.19 Pressures pushing fluid out of blood vessels (blood pressure) and into them (osmotic pressure).

merge with progressively larger thin-walled vessels, draining the lymph into lymphatic ducts, which empty into the venous system and then the heart. Lymphatic vessels have one-way valves, similar to those in the veins, which assist the return of lymph to the circulatory system. Local muscle contractions also help pump the lymph toward the ducts that return lymph to the circulatory system of vertebrates. In addition to its role in returning plasma to the bloodstream, the lymphatic system has important functions in the immune system (Chapter 41).

Hormones and nerves provide homeostatic regulation of blood pressure.

An animal that is dehydrated or has lost blood following an injury may experience a decrease in blood pressure. In such cases, the posterior pituitary gland releases antidiuretic hormone (ADH) into the circulation (Chapter 36). ADH causes arteries to constrict, increasing their resistance to blood flow. The higher resistance increases blood pressure throughout the body.

Blood pressure is continually being adjusted to remain within normal physiological limits, a phenomenon that is an example of homeostasis. If blood pressure drops, animals reduce the supply of blood to the limbs by constricting arterioles that supply the limbs. This response helps maintain blood pressure to the heart, brain, and kidneys. In this case, sympathetic neurons synapsing on the smooth muscles of arterioles in the limbs stimulate these muscles to contract. In contrast, when blood pressure is too high, sympathetic neurons that synapse on the smooth muscles are inhibited. The smooth muscles relax, reducing resistance in the arterioles and increasing blood flow. These two responses are called **vasoconstriction** and **vasodilation**, respectively.

In some circumstances, it may be advantageous to an animal to have blood pressure rise higher than normal levels. Animals faced with a stressful situation exhibit the fight-or-flight response; for example, this response occurs when prey are alerted to the presence of predators. A surge of adrenaline into the bloodstream causes an animal to become alert and aroused. At the same time, the sympathetic nervous system increases the heart and respiratory rates and inhibits digestive processes. The increase in heart rate increases blood pressure. Arterioles supplying limb muscles dilate to enhance blood flow to exercising muscles, while arterioles supplying organs within the digestive system constrict. The increase in heart and breathing rates increases the rate of O_2 delivery to the exercising muscles to meet their increased demand for ATP. These physiological changes prepare an animal to flee.

In contrast, when an animal has recently eaten a meal, its sympathetic nervous system is inhibited and its parasympathetic nervous system is stimulated. In response, blood flow to the digestive system increases and smooth muscle in the gut becomes more active (Chapter 38). In this way, regulation of blood pressure is linked to the physiological state of the animal through feedback from the nervous and endocrine systems.

> **Self-Assessment Questions**
>
> 10. Which change in vessel shape most affects resistance to blood flow?
> 11. How does the branching of larger arteries into many smaller vessels affect the rate of and resistance to blood flow in the smaller vessels?

37.5 STRUCTURE AND FUNCTION OF THE HEART

The pressure required to drive blood through blood vessels is generated by the heart. Animal hearts are pumps made of muscle that rhythmically contracts, producing this pressure. Early hearts were as simple as a muscular thickening of a small region of a blood vessel. Such hearts are still found in insects (see Fig. 37.16a). More complex hearts have chambers that expand and fill with deoxygenated blood returning from the animal's tissues. After filling, the heart muscle contracts, pumping deoxygenated blood to the lungs or gills and the

FIG. 37.20 Fish heart and circulatory system. (a) Fishes have a two-chambered heart. (b) Blood flows in a single circuit from the heart, through gills, to the tissues for gas exchange, and back to the heart.

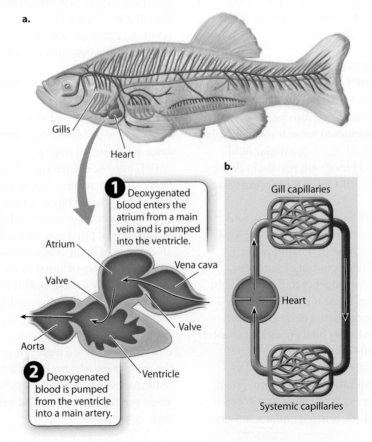

FIG. 37.21 Amphibian heart and circulatory system. (a) The frog heart has three chambers: two atria and one ventricle. (b) Mixed blood leaving the common ventricle goes either to the lung and back to the heart after being re-oxygenated, or to the tissues and back to the heart.

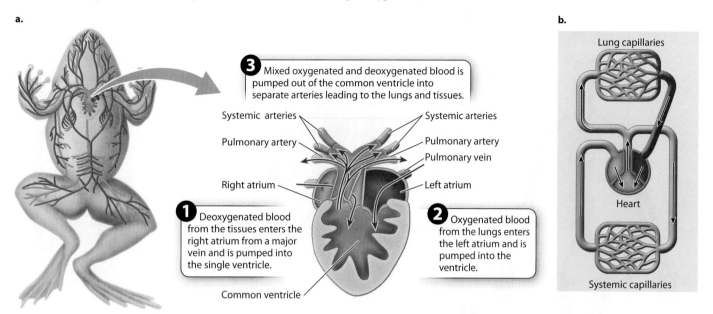

newly oxygenated blood onward to the animal's tissues. The closed circulatory system of cephalopod mollusks has three hearts: two hearts that pump deoxygenated blood through the gills and another heart that pumps oxygenated blood to the body. In contrast, vertebrates have a single heart with at least two chambers: one for receiving blood and the other for pumping blood to the body during each heartbeat cycle. Hearts also have one-way valves to ensure that blood does not flow backward when the heart contracts.

A theme running through the evolution of vertebrate circulatory systems is the progressive separation of circulation to the gas exchange organ from circulation to the rest of the body. In fishes, deoxygenated blood is first pumped to the gills to gain O_2; then the freshly oxygenated blood flows to the rest of the body and returns to the heart. This single circulation path limits the rate of blood flow to metabolically active body tissues. By contrast, birds and mammals evolved a separate **pulmonary circulation** to the lungs and **systemic circulation** to the rest of the body. This double circulation was made possible by the evolution of a four-chambered heart. This organization has two advantages: it increases the supply of oxygenated blood to active tissues, and it increases the uptake of O_2 at the gas exchange surface.

Fishes have two-chambered hearts and a single circulatory system.

Fish hearts have two chambers (**Fig. 37.20**), an **atrium** (plural, **atria**) and a **ventricle**. Deoxygenated blood returning from the fish's tissues enters the atrium, which fills and then contracts to move the blood into a thicker-walled ventricle. The muscular ventricle pumps the blood through a main artery to the gills for uptake of O_2 and elimination of CO_2. Oxygenated blood collected from the gills travels to the tissues through a large artery called the **aorta**.

The small gill capillaries impose a large resistance to flow. As a result, much of the blood pressure is lost in moving blood through the gills. This loss of pressure limits the flow of oxygenated blood to body tissues.

Amphibians and reptiles have three-chambered hearts and partially divided circulations.

As vertebrates moved onto land, the transition from breathing water to breathing air had important consequences for the organization of their circulatory systems. Land vertebrates evolved hearts that separated the circulation of deoxygenated blood pumped to their gas exchange organs from the circulation of oxygenated blood delivered to their body tissues. Gas exchange became more efficient and O_2 delivery increased. Metabolic rates rose, and animals became capable of greater activity.

Reflecting their aquatic and terrestrial lifestyle, amphibians evolved a variety of ways to breathe. Some retain gills, while others use lungs or their skin, or some combination of these, to breathe. Amphibian hearts evolved an additional atrium, becoming divided into two separate atria and a single ventricle (**Fig. 37.21**). This arrangement allows oxygenated blood from the gills, lungs, or skin to return to the heart separately from deoxygenated blood returning from metabolically active tissues. Such a partial separation of the pulmonary and

FIG. 37.22 The mammalian heart and circulatory system. (a) Mammals have a fully divided four-chambered heart with two atria and two ventricles. (b) Mammals have separate pulmonary and systemic circulations.

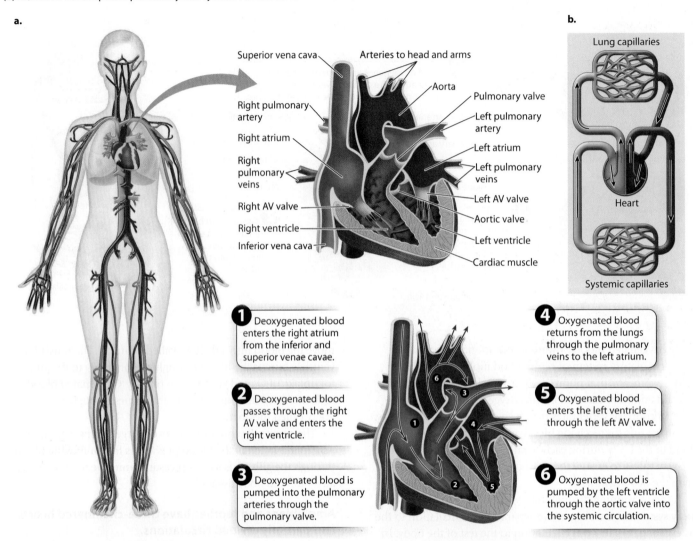

systemic circulations means that freshly oxygenated blood can be pumped under higher pressure to the body than is the case in fishes. However, because amphibians have a single ventricle, oxygenated blood returning to the heart from the gills, skin, or lungs mixes with deoxygenated blood returning from the animal's body before being pumped from the ventricle. Anatomical features of the ventricle limit this mixing, but mixing still hinders the supply of oxygenated blood to an amphibian's body.

Most reptiles also have a three-chambered heart, but their single ventricle has internal ridges to improve the separation of deoxygenated and oxygenated blood. As a result, reptiles are able to deliver more O_2 to their tissues than amphibians can. At the same time, the incompletely divided ventricle provides reptiles with versatile circulatory responses during exercise and when diving underwater. For example, when they dive, the blood of turtles and sea snakes bypasses their lungs, which they cannot use underwater. Notably, crocodiles and alligators have evolved four-chambered hearts, similar to those of mammals and birds.

Mammals and birds have four-chambered hearts and fully divided pulmonary and systemic circulations.

Mammals and birds evolved four-chambered hearts with separate atria and separate ventricles (**Fig. 37.22**). This arrangement completely separates blood flow to the lungs from blood flow to the tissues. As a result, these animals can pump blood to their lungs under lower pressure, allowing increased uptake of O_2, and at the same time supply blood gases and nutrients to their tissues at high pressure. Whereas the systemic blood pressure of fishes is only a few mmHg, in a bird or mammal it is 100 mmHg or higher. Moreover, blood flowing to the tissues is fully oxygenated at the outset, so it delivers more O_2 to body tissues at a higher rate. This change in heart structure and circulatory pattern was linked to an increase in the metabolic rates and the evolution of endothermy in birds and mammals (Chapter 38).

Following the pattern established in amphibians and reptiles, deoxygenated blood enters the right atrium from the venae cavae and, when the atrium contracts, moves through an **atrioventricular (AV) valve** into the right ventricle (Fig. 37.22). When the right ventricle contracts, blood is pumped through the **pulmonary valve** into the pulmonary trunk, which divides into the left and right **pulmonary arteries**, and then to the lungs for oxygenation. The atrioventricular and pulmonary valves ensure that blood flows in one direction into and then out of the right ventricle. The walls of the right ventricle are thinner than those of the left ventricle, and its weaker contractions eject blood at a lower pressure. As a result, the blood of the pulmonary circulation moves at a slower rate, allowing greater time for gases to diffuse into and out of the lungs.

The oxygenated blood returns from the lungs through the **pulmonary veins** and enters the left atrium of the heart. When the left atrium contracts, blood is pumped through a second atrioventricular valve into the left ventricle. The thick muscular walls of the left ventricle eject the blood under high pressure to the body. Oxygenated blood leaving the left ventricle passes through the **aortic valve** and then flows to the head and the rest of the body through the large artery called the aorta. The one-way aortic valve ensures that blood leaving the left ventricle does not flow back into the left ventricle as it relaxes.

The contraction of the two atria followed by the contraction of the two ventricles makes up the **cardiac cycle** (Fig. 37.23). The cardiac cycle is divided into two main phases: **diastole**, which is the relaxation of the ventricles, and **systole**, which is the contraction of the ventricles. Thus, diastole is the phase in which the atria contract and the ventricles fill with blood, and systole is the phase of the cardiac cycle in which blood is pumped from the ventricles of the heart into the pulmonary and systemic circulations.

 CASE 7 BIOLOGY-INSPIRED DESIGN: USING NATURE TO SOLVE PROBLEMS

How can we engineer replacement heart valves?

The human heart beats nearly 3 billion times in an average lifetime. This repeated pumping can lead to wear on the heart valves that regulate blood flow through the heart. As a result, damaged valves may need to be replaced or repaired. The aortic valve, which controls blood flow out of the heart, and the mitral (atrioventricular) valve on the left side of the heart are the valves most often in need of surgical replacement or repair. Approximately 2.5% of the U.S. population will eventually need heart valve replacement surgery (more than 80,000 people per year).

Researchers have developed two types of artificial heart valves, both modeled after their normal counterparts: mechanical valves and biological valves made from a pig's or cow's valve. Both types of valves open and close in response to pressure differences, just like regular valves. Each is associated with different risks and benefits. Mechanical heart valves last longer, often for the remainder of an individual's lifetime. However, because mechanical valves more frequently create blood clots, they carry the risk of stroke or heart attack. Once such a valve is installed, the person must remain on blood thinners the rest of his or her life, which affects the patient's lifestyle. Biological valves from pig or cow valve tissues do not carry this risk, but they deteriorate more quickly, lasting approximately 10 to 20 years. An individual with a biological valve is at risk of needing a second (or third) valve replacement surgery, with the challenges of surgery increasing later in life.

Because of these trade-offs in risk and lifestyle impact, mechanical valves are recommended for individuals younger than 50 years of age and biological valves are recommended for those older than 70 years of age. Nevertheless, in recent years biological valve replacements have become increasingly popular for younger individuals, in part due to their minimal impact on lifestyle. The trade-offs in longevity and health risk continue

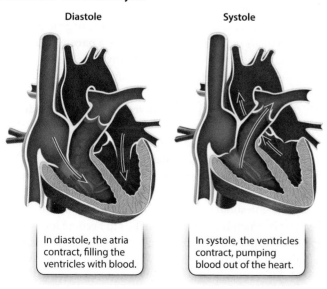

FIG. 37.23 The cardiac cycle.

In diastole, the atria contract, filling the ventricles with blood.

In systole, the ventricles contract, pumping blood out of the heart.

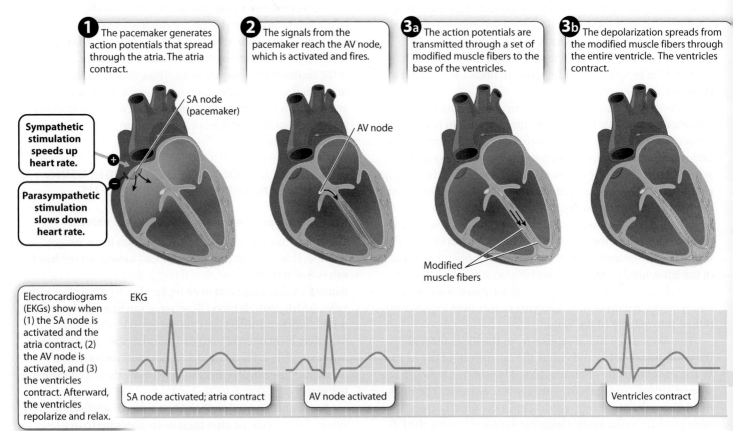

FIG. 37.24 Control of heart rate and rhythm. The pacemaker cells of the sinoatrial and atrioventricular nodes keep the heart beating with a basic rhythm. This rhythm can be assessed by recording the electrocardiogram (EKG) of the heart.

to stimulate further research with the aim of increasing the lifetime effectiveness of biological replacement valves and better assessing the long-term outcomes of mechanical heart valves relative to biological valves.

Cardiac muscle cells are electrically connected to contract in synchrony.

For the heart to function effectively as a pump, the **cardiac muscle** cells that make up the walls of the atria and ventricles must contract in a coordinated fashion. Atrial muscle cells contract in synchrony during diastole to fill the ventricles, and ventricular muscle cells contract in synchrony during systole to eject the blood from the heart. For cardiac muscle cells to pump in unison, they must be depolarized in unison by an action potential.

Cardiac muscle cells have contractile properties similar to those of skeletal muscle cells (Chapter 35), but are distinct in two ways. First, specialized cardiac muscle cells can generate action potentials on their own, independently of the nervous system. Second, cardiac muscle cells are in electrical continuity with one another through gap junctions (Chapter 10), meaning that they can pass their action potentials to adjacent cells. These two features of cardiac muscle cells ensure that all muscle cells in a region surrounding a heart chamber are activated and contract in unison.

The specialized cardiac muscle cells capable of generating action potentials independently function as a **pacemaker** that causes the heart to beat with a basic rhythm. These cells stimulate neighboring cells to contract in synchrony. Pacemaker cells are found in two specialized regions of the heart: the **sinoatrial (SA)** and **atrioventricular (AV) nodes** (**Fig. 37.24**).

Whereas most nerve cells maintain their resting membrane potential until they receive a signal from another cell, the critical feature of pacemaker cells is that their resting membrane potential gradually becomes less negative on its own until it reaches its threshold voltage and the cell fires an action potential. The membrane potential tends to become less negative over time due to slow leakage of sodium ions into the cell and decreased flow of potassium ions out of the cell. When the cell reaches its threshold, voltage-gated calcium ion channels open, causing more rapid depolarization. Like nerve cells, the pacemaker cells repolarize after firing an action potential to reach their resting potential, keeping the heart relaxed between contractions. Also, like nerve cells, pacemaker cells repolarize by opening voltage-gated potassium channels to allow potassium ions to leave the cell.

The heartbeat is initiated at the sinoatrial node, which is located at the junction of the vena cava and the right atrium (Fig. 37.24). Because adjacent cardiac muscle cells are in electrical contact, action potentials initiated at the sinoatrial node spread rapidly from one cell to the next. Cell depolarization spreads electrically throughout the right and left atria, causing them to contract in unison. Because there is no electrical contact between the atria and ventricles, the ventricles do not contract. Instead, the depolarization in the atria reaches a second set of pacemaker cells, located in the AV node.

Activation of the AV node transmits the action potential to the ventricles. A modified set of cardiac muscle fibers transmits the action potential from the AV node to the base of the ventricles. From there, the depolarization spreads throughout the ventricle walls, causing the ventricles to contract in unison. Transmission through the conducting fibers causes a delay that ensures the ventricles do not contract until they are fully filled with blood from the atria.

Electrodes placed over the surface of the chest and other body regions can record the electrical currents produced by the depolarization of the heart while it beats. This type of recording is called an electrocardiogram (abbreviated as "EKG" because *kardia* is Greek for "heart," although "ECG" is also used; Figure 37.24). Cardiologists use EKGs to diagnose heart problems or to determine which region of the heart has been damaged following a heart attack.

Heart rate and cardiac output are regulated by the autonomic nervous system.

The volume of blood pumped by the heart over a given interval of time is its **cardiac output (CO)**. This quantity is a key measure of heart function. Cardiac output is determined by calculating the product of the **heart rate (HR)** and the volume of blood pumped during each beat, the **stroke volume (SV)**: CO = HR × SV. As this equation indicates, animals can increase their cardiac output by increasing heart rate or stroke volume or both. Whereas humans and other mammals rely most heavily on increases in heart rate to increase cardiac output, fishes rely more on increases in stroke volume. Cardiac output rises or falls in response to the metabolic demand for O_2, which in turn depends on an animal's state of activity. When an animal sleeps, cardiac output is minimal; when it runs, cardiac output increases to meet the demand for O_2 of its active muscles.

The nervous system controls the heart rate. Stimulation by sympathetic nerves causes the pacemaker cells of the sinoatrial node to depolarize more rapidly and the heart rate to speed up. Stimulation by parasympathetic nerves causes these cells to depolarize more slowly and the heart rate to slow down. Adrenaline released by the adrenal gland into the circulation also increases the heart rate for more prolonged periods of time. This hormone is a key component of the fight-or-flight response of many vertebrate animals.

Stroke volume is adjusted automatically as activity level rises and falls. During exercise, when an animal's muscles regularly contract, the volume of blood returned to the heart is increased. The influx of additional blood stretches the walls of the atria and ventricles to a greater extent, causing them to contract more forcefully. As a result, more blood is ejected during each heartbeat. The adjustment of stroke volume to changes in activity ensures that heart emptying is matched to heart filling (that is, to venous return). The relationship between the volume of blood filling the heart and stroke volume is described by **Starling's Law**: a greater volume of blood returned to the heart increases stretching of the muscle, which leads to more forceful ejection of blood and a greater stroke volume.

As we have noted, circulation is linked to several other functions in addition to the transport of respiratory gases, nutrients, and wastes. The circulatory system plays a central role in the temperature regulation of many animals (Chapters 33 and 38). Changes in vasodilation or vasoconstriction are an important factor controlling heat loss or gain between an animal's body and its environment. The circulatory system also provides the route by which hormones (Chapter 36) and immune cells (Chapter 41) are distributed throughout the body, enabling the endocrine system to maintain homeostatic control of diverse bodily functions and the immune system to defend the body against pathogens. In doing so, the circulatory system serves as a general pathway by which hormonal and immune communication occurs to regulate, integrate, and protect the functional state of the whole organism.

Self-Assessment Questions

12. Diagram the path of blood flow through the heart, lungs, and body of a fish and of a mammal or bird.

13. Is the pressure in the atrium or ventricle higher when the atrioventricular valve closes?

14. What are two mechanisms by which the cardiac output of an animal's heart can be adjusted? Which is more important for increasing cardiac output in mammals and which is more important in fishes?

CORE CONCEPTS SUMMARY

37.1 DELIVERY OF OXYGEN AND ELIMINATION OF CARBON DIOXIDE: Respiration and circulation depend on diffusion over short distances and bulk flow over long distances.

Diffusion is effective only over short distances and requires large exchange surface areas with thin barriers. page 829

Ventilation and circulation provide bulk flow over long distances. page 830

Oxygen is delivered to tissues in four steps: (1) bulk flow of water or air past the respiratory surface (gills and lungs); (2) diffusion of O_2 across the respiratory surface into the circulatory system; (3) bulk flow through the circulatory system; and (4) diffusion of O_2 into tissues and cells. page 831

37.2 RESPIRATORY GAS EXCHANGE: Respiration provides oxygen and eliminates carbon dioxide in support of cellular metabolism.

Many aquatic animals exchange respiratory gases with water through gills. page 832

Countercurrent flow of water relative to blood in the gills enhances O_2 extraction from water. page 832

Terrestrial animals breathe air by means of internal tracheae or lungs. page 833

Terrestrial vertebrates inflate and deflate their lungs during bidirectional tidal ventilation driven by changes in pressure. page 834

Unidirectional flow of air maximizes O_2 uptake by bird lungs. page 837

Both the voluntary and involuntary nervous systems control breathing. page 837

37.3 OXYGEN TRANSPORT BY HEMOGLOBIN: Red blood cells produce hemoglobin, greatly increasing the amount of oxygen transported by the blood.

Hemoglobin greatly increases the capacity of the blood to transport O_2. page 838

Red blood cells contain hemoglobin with iron-containing heme groups that reversibly bind and release O_2. page 839

A sigmoidal O_2-dissociation curve describes the change in the binding affinity of hemoglobin for O_2 with changes in partial pressure. The shape of the curve results from cooperative binding of O_2 by hemoglobin. page 839

Myoglobin binds and stores O_2 in muscle cells, increasing the delivery of O_2 to muscle mitochondria for activity in general and for diving in marine mammals. page 840

Fetal hemoglobin expressed by the mammalian fetus has a higher binding affinity for O_2 to allow for O_2 uptake from the mother's blood. page 841

The affinity of hemoglobin for O_2 is reduced by a decrease in pH and an increase in CO_2 levels. page 842

37.4 CIRCULATORY SYSTEMS: Circulatory systems have vessels of different sizes that facilitate bulk flow and diffusion.

The fluid in an open circulatory system moves through only a few vessels and is mostly contained within the animal's body cavity. page 842

The fluid in a closed circulatory system travels through a set of internal vessels, moved by a pump, the heart. page 842

Resistance to blood flow depends most strongly on vessel radius. page 843

Arteries are thick-walled vessels that contain collagen and elastin fibers, which provide for support against pressure and elastic rebound. page 844

Capillaries are small-diameter, thin-walled vessels that facilitate diffusion of O_2 and CO_2. page 844

Arterioles control blood flow within the body by changing their resistance through contraction and relaxation of the smooth muscle in their walls. page 844

Veins are thin-walled vessels that return blood to the heart under low pressure. page 845

Liquids, gases, and other compounds move across capillary walls by diffusion, filtration, and osmosis. In vertebrates, some fluid is returned to the bloodstream by the lymphatic system. page 845

37.5 STRUCTURE AND FUNCTION OF THE HEART: The evolution of animal hearts reflects selection for a high metabolic rate, achieved by increasing the delivery of oxygen to metabolically active cells.

Fishes have two-chambered hearts and a single circulatory path. page 847

Amphibians and reptiles have three-chambered hearts and a partially divided circulatory path. page 847

Birds and mammals have four-chambered hears and a double circulatory path, consisting of a pulmonary circulation, from the heart to the lungs, and a systemic circulation, from the heart to the rest of the body. This arrangement enhances O_2 uptake and delivery to metabolizing tissues. page 848

One-way valves control blood flow through the heart. page 849

The cardiac cycle has discrete ventricular filling (diastole) and ventricular emptying (systole) phases. page 849

Cardiac muscle cells are electrically connected, allowing them to contract in synchrony as a unified heartbeat. page 850

Contraction of the heart is controlled by signals from pacemaker cells of the sinoatrial and atrioventricular nodes; these signals are transmitted by electrical conducting fibers to the ventricles. page 850

The vertebrate autonomic nervous system regulates the heart's cardiac output through changes in heart rate and stroke volume. page 851

Log in to LaunchPad to check your answers to the Self-Assessment Questions and to access additional learning tools.

CHAPTER 38 Animal Metabolism, Nutrition, and Digestion

CORE CONCEPTS

38.1 PATTERNS OF ANIMAL METABOLISM: Metabolic rate depends on activity level, body size, and body temperature.

38.2 NUTRITION AND DIET: An animal's diet supplies the energy the animal needs for homeostasis and essential nutrients it cannot synthesize on its own.

38.3 ADAPTATIONS FOR FEEDING: Animals have evolved different ways to capture food.

38.4 REGIONAL SPECIALIZATION OF THE GUT: The digestive tract has regions specialized for digestion, absorption, storage, and elimination.

It is literally true that "you are what you eat." Not only do animals build their bodies and obtain energy from the food they eat, but also their very appearance often reflects what they eat. Consider a cat. We know it's a carnivore—an organism that mainly eats animals—just from looking at it. It has sharp teeth and powerful jaws specialized for tearing meat; strong legs, sharp claws, and keen eyesight for hunting prey; and a short gut because breaking down animal protein is easier than breaking down plant material. Humans are omnivores, adapted for eating both other animals and plants, and this is reflected in our anatomy. Our teeth include sharp canines for tearing meat and flat molars for grinding plant material. Cows are herbivores, animals that eat plants. Cows have evolved a variety of adaptations, including a four-chambered stomach, that allow them to digest cellulose, which is tough but rich in energy.

Animals are heterotrophs, obtaining food from other organisms (Chapter 6). The energy animals obtain from food is essential for building bodies, moving, surviving, and reproducing. Much of an animal's activity therefore centers on obtaining food. In turn, the biology of an animal is in large part shaped by the type of food the animal consumes. For example, an animal's metabolism, including the chemical reactions by which it breaks down its food, has evolved in close association with the type of food available to the animal. Likewise, an animal's survival is closely linked to the energy and nutrients contained in that food.

The food that an animal eats also affects its place in an ecosystem. Primary producers, such as plants and some algae, are food for a great diversity of animals. Animals that feed on plants tend to exist in large numbers. Large animals that feed on other, smaller animals exist in smaller numbers and commonly represent fewer species.

This chapter examines how animals obtain and break down the food they eat. We begin at the cellular level, building on the biochemistry of metabolism discussed in Chapter 7. We then examine the dietary needs of animals, followed by the structure and function of animal digestive systems: how an animal eats, digests, and absorbs food to supply its cells with the energy required for its functions. Discussion of the relationship between an animal's diet and its place in an ecosystem is reserved until Chapter 46.

38.1 PATTERNS OF ANIMAL METABOLISM

The biochemical pathways that make up animal metabolism are remarkably conserved among diverse groups of organisms. These pathways evolved early in the history of life and have been retained throughout the evolution and diversification of life. In Chapter 7, we examined the pathway of glucose breakdown by cellular respiration, which provides ATP to fuel the work of a cell. Here, we put this and related pathways in an organismal context. We explore how carbohydrates, proteins, and fats are metabolized for building body structures and supplying energy needs, and how metabolic rate varies with an animal's level of activity, internal body temperature, and size.

Animals rely on anaerobic and aerobic metabolism.

Nearly all animals depend on three main classes of molecules as sources of energy and building blocks for growth and development: carbohydrates, fats, and proteins (Chapter 2). These molecules are broken down

FIG. 38.1 Cellular metabolism, which includes anaerobic and aerobic pathways. Animals obtain energy to fuel their activities and development from the breakdown of carbohydrates, fats, and proteins.

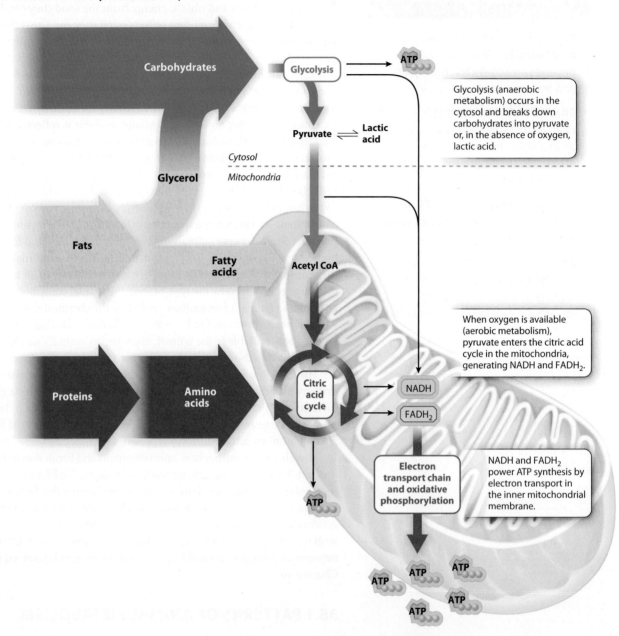

after ingestion, and in the process the energy in their chemical bonds is transferred to biologically useful forms, typically **adenosine triphosphate (ATP)**. While other nucleic acid triphosphates also serve as usable forms of energy, ATP is by far the one most commonly used by animals. The breakdown of carbohydrates, fats, and proteins to produce ATP takes place through a series of chemical reactions. Reactions that break down food sources to fuel the energy needs of a cell are termed **catabolic**. In contrast, reactions that result in net energy storage within cells and the organism are called **anabolic**.

Let's first review the breakdown of the carbohydrate glucose. Glucose can be partially broken down in the absence of oxygen by glycolysis (**Fig. 38.1**). Glycolysis occurs in the cytosol of the cell, and the breakdown of each glucose molecule results in the production of 2 molecules of pyruvate, 2 molecules of ATP, and 2 molecules of NADH. In animals, if oxygen is not present, or if it is present only in small amounts, pyruvate is converted by fermentation into lactic acid. The production of ATP by **anaerobic** ("absence of oxygen") **metabolism** provides rapid but short-term energy to the cell and organism.

In contrast, **aerobic** ("of oxygen") **metabolism**, which is carried out within the mitochondria of animals, provides a steady supply of ATP for longer-term, sustainable activity. When enough oxygen diffuses into the mitochondria, pyruvate can be further broken down by the **citric acid cycle** rather than converted to lactic acid (Fig. 38.1). In a series of oxidation–reduction reactions, energy-rich NADH and FADH$_2$ molecules are produced. These molecules then enter the electron transport chain in the inner mitochondrial membrane. The **electron transport chain** couples the transfer of electrons to the pumping of protons across the inner mitochondrial membrane, resulting in a proton electrochemical gradient that drives the synthesis of ATP by oxidative phosphorylation. Because oxygen is reduced to water in the process, oxygen is consumed and water is produced as a by-product. Aerobic metabolism is also called cellular respiration or aerobic respiration.

Overall, aerobic metabolism results in the formation of a total of 32 molecules of ATP when 1 molecule of glucose is broken down, much more than the 2 molecules of ATP from 1 molecule of glucose produced anaerobically. We can define metabolic efficiency as the amount of energy captured in a usable form (at least 234 kcal of useful ATP energy) per amount of energy in the starting molecule (686 kcal in 1 mol of glucose). The metabolic efficiency of aerobic respiration, then, is at least 34% (or 234/686 × 100). The remaining 66% of this energy is converted to heat and either lost to the environment or used to warm the animal. Glucose that is not metabolized is stored as glycogen, primarily in the liver and muscles.

Lipids are another important energy source for most animals. Lipids consumed in the diet, such as triacylglycerol, are broken down to glycerol and free fatty acids, which enter glycolysis or the citric acid cycle to yield ATP by oxidative phosphorylation (Chapter 7). Alternatively, they can be stored as fat (adipose tissue). Storing energy as fat is a particularly efficient mechanism because fat yields more than twice the energy supply per unit weight (9.4 kcal/g) as glycogen (4.2 kcal/g). When needed, stored fats can similarly be broken down into glycerol and free fatty acids, which yield ATP by the same pathways as lipids from the diet.

Proteins consumed in the diet can also be a useful energy source for animals. Carnivores such as cats get most of their energy from the protein and fat in the meat they eat. Proteins include enzymes and structural elements of cells and tissues, which are needed for building and maintaining the body. Only following prolonged food deprivation, when fat and carbohydrate reserves are depleted, do animals break down protein reserves (mainly from their muscles) to form ATP. Animals rely mainly on fats for long-term energy supply. Well-nourished humans have enough fat reserves to support their metabolic needs for 5 to 6 days before they start to use protein reserves. Because nerve cells depend primarily on glucose to produce ATP, carbohydrates are spared for the brain and nervous system when animals face starvation.

Metabolic rate varies with activity level.

An animal's overall rate of energy use is termed its **metabolic rate**. Metabolic rate can be measured by the animal's rate of oxygen consumption, or how much oxygen it consumes in a certain amount of time. The rate of oxygen consumption reflects the rate of aerobic production of ATP. (Recall the overall formula for cellular respiration: $C_6H_{12}O_6 + 6O_2 \rightarrow 6H_2O + 6CO_2 + ATP$.)

An animal's metabolic rate is affected by many factors, including its activity level. When an animal shifts from rest to activity, its metabolic rate and oxygen consumption rise to meet the increased demand for ATP. The onset of activity requires immediate energy, which in animals is provided by specialized energy stores in their tissues. In vertebrate muscle cells, for example, phosphocreatine is a ready source of energy. Phosphocreatine is hydrolyzed to provide high-energy phosphates that are used to synthesize ATP directly from ADP at the onset of activity. When the phosphocreatine is used up, the cell begins the relatively rapid production of ATP by glycolysis. Although glycolysis produces relatively few ATP molecules per molecule of glucose, the reactions are extremely fast, providing a rapid short-term supply of energy for animals. Animals rely on glycolysis for short bursts of intensive activity.

With longer activity, the ATP needs of the animal are met by aerobic respiration in mitochondria. The rate of oxygen consumption initially increases as aerobic respiration ramps up, then levels off (**Fig. 38.2**). At this point, the animal's need for energy is met entirely by aerobic respiration. The steady production of ATP can continue as long as adequate oxygen reaches the mitochondria, first by transport to the tissue through the cardiovascular system (Chapter 37) and then by diffusion into mitochondria.

Sprinters rely heavily on anaerobic ATP production for intense short-term bouts of activity. By comparison, longer races are run at progressively slower speeds for increased endurance, as a distance runner relies more heavily on aerobic metabolism to produce ATP. The world record for the 100-m sprint averages a little more than 10 m/s, but for the 1500-m race the average is about 6.6 m/s.

When activity ends, the animal's rate of oxygen consumption declines but does not immediately return to resting levels, as shown in Fig. 38.2. The elevated consumption of oxygen following activity is known as the animal's **recovery metabolism**. An animal continues to expend metabolic energy during this period in an effort to reestablish the resting metabolic state

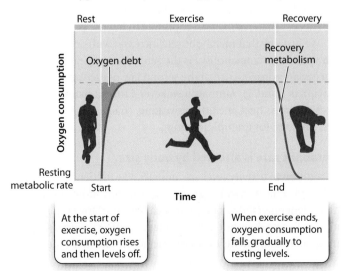

FIG. 38.2 Oxygen consumption during physical activity.

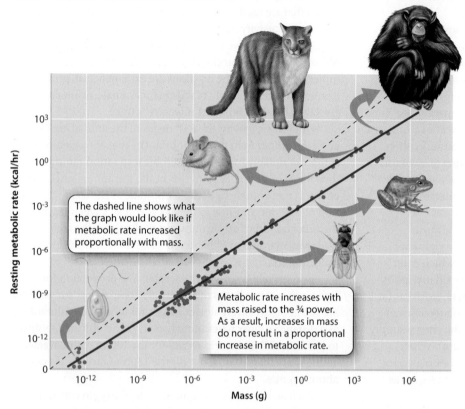

FIG. 38.3 Resting metabolic rate and body size in diverse organisms. The relationship plotted here shows that larger animals have lower metabolic rates per gram of tissue than smaller animals.
Source: Adapted from A. M. Hemingsen, 1960, "Energy Metabolism as Related to Body Size and Respiratory Surfaces, and Its Evolution." Re. Steno Memorial Hospital Nordisk Insulinlaboratorium 9:1–110.

range of organisms show that metabolic rate increases with mass raised to the ¾ power (**Fig. 38.3**). This means that the average rate at which each gram of body tissue consumes energy is *less* in larger organisms compared to smaller ones. Put another way, the larger the organism, the *lower* the metabolic rate *per gram* of body tissue. This scaling pattern of cellular energy metabolism is remarkable because it holds across a diverse size range of unicellular and multicellular organisms, as can be seen in Fig. 38.3.

In addition to comparing resting metabolic rates in different-sized animals, we can examine how metabolic rates vary with activity level in different-sized animals. For example, we can look at how much energy is used when animals move at different speeds. The American comparative physiologist C. Richard Taylor has extensively studied how metabolic rate varies with running speed in different-sized terrestrial animals (**Fig. 38.4**). He found that metabolic rate increases linearly with speed, and that larger animals expend less energy per unit mass than smaller ones.

Metabolic rate is linked to body temperature.

Metabolic rate is also affected by an animal's internal body temperature. Chemical reaction rates increase as temperature rises, so internal body temperature determines how fast fuel molecules can be mobilized and broken down to supply energy for a cell. Animals can be categorized as either endotherms or ectotherms based on the sources of most of their heat.

Animals that produce most of their own heat as by-products of metabolic reactions, including the breakdown of food, are **endotherms**. These animals usually, but not always, maintain a constant body temperature that is higher than that of their environment. As we saw in Chapter 33, the maintenance of a constant core body temperature is a form of homeostasis and requires balancing heat production and heat loss. Birds and mammals, including humans, are endotherms.

By contrast, animals that obtain most of their heat from the environment are **ectotherms**. Ectotherms often regulate their body temperatures by changing their behavior—moving into or out of the sun, for example. Think of a turtle or some other reptile sunning itself on a rock: it is raising its body temperature. The temperatures of these animals usually fluctuate with the outside environment. Most fish, amphibians, reptiles, and invertebrates are ectotherms.

of its cells. As part of recovery metabolism, cells re-synthesize depleted ATP stores and metabolize the end products of fermentation, particularly lactic acid. The difference between an animal's immediate energy need at the onset of activity and the energy supplied by aerobic metabolism is referred to as the animal's "oxygen debt." This debt is "paid back" following exercise by the animal's recovery metabolism. Recovery metabolism explains the elevated breathing and heart rate that you (and other animals) experience when resting after moderate to intense exercise.

We just saw that metabolic rate increases with activity level. A by-product of metabolism is the generation of heat, which is a necessary consequence of the second law of thermodynamics (Chapters 1 and 6). Animals have several mechanisms for dissipating excess heat, including sweating, changes in blood flow, and (in dogs, for example) panting.

Metabolic rate is affected by body size.

In addition to activity level, an animal's size influences its metabolic rate. At rest, larger animals consume more energy and have higher metabolic rates than smaller ones. However, resting metabolic rate does not increase linearly with an animal's mass. Instead, measurements of resting metabolic rate in a wide

HOW DO WE KNOW?

FIG. 38.4

How is metabolic rate affected by running speed and body size?

BACKGROUND In the 1970s and 1980s, the American physiologist C. Richard Taylor and his colleagues studied the relationship between metabolic rate and running speed in mammals of different sizes. They were interested in understanding the energetic costs of running in different animals.

EXPERIMENT Taylor and colleagues measured oxygen consumption during running for a wide range of organisms. They trained each animal to run on a treadmill at different speeds and measured oxygen consumption using either a face mask or enclosure. Oxygen consumption over time was used as a measure of metabolic rate.

RESULTS The researchers found that there is a linear increase in metabolic rate with speed in different-sized animals. They also found that larger animals expend less energy per unit body mass to move a given distance compared to smaller ones.

FOLLOW-UP WORK Similar studies were performed with kangaroos, which move by hopping rather than by running. Interestingly, it was found that it is "cheaper" (that is, it requires less energy) to hop than to run. This finding perhaps explains why kangaroos and other hopping animals survived, while many animals that run on four legs became extinct, in Australia 30,000 to 20,000 years ago with the arrival of early humans. Studies have also compared the energetic costs of running on two legs and running on four legs and have found that the cost is the same.

SOURCES Dawson, T. J., and C. R. Taylor. 1973. "Energetic Cost of Locomotion in Kangaroos." *Nature* 246:313–314; Taylor, C. R., N. C. Heglund, and G. M. O. Maloiy. 1982. "Energetics and Mechanics of Terrestrial Locomotion." *Journal of Experimental Biology* 97:1–21.

Photo source: Cary Wolinsky/Getty Images.

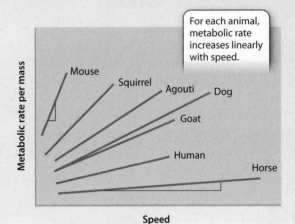

For each animal, metabolic rate increases linearly with speed.

The slopes for smaller animals, like mice, are greater than those for larger animals, like horses, indicating that the cost of transport is greater in smaller animals than larger ones.

Endotherms are sometimes referred to as "warm-blooded" and ectotherms as "cold-blooded," but these terms are misleading. Depending on the environmental conditions, cold-blooded animals can sometimes have core body temperatures that are higher than those of warm-blooded animals. Also keep in mind that these two ways of obtaining heat are not completely distinct, but instead represent extremes of a continuum.

An animal's major source of heat has profound implications for its metabolic rate. Endotherms have a higher metabolic rate than ectotherms. As a result, they are able to be active over a broader range of external temperatures than ectotherms. Both the activity level and metabolic rate of ectotherms increase with increasing body temperature. However, at similar body temperatures, ectotherms have metabolic rates that are approximately 25% those of endotherms of similar body mass. Although ectotherms can achieve activity levels similar to those of endotherms when their body temperatures are similar, ectotherms cannot sustain prolonged activity. Therefore, ectotherms such as lizards or insects demonstrate brief bouts of activity, followed by longer periods of inactivity compared to endotherms.

Endotherms benefit from having an active lifestyle. However, their high metabolic rate requires long periods of foraging to acquire enough food to keep warm and be active. By contrast, although ectotherms cannot sustain activity for long periods, they can survive for much longer periods without food.

As we have discussed, the control of core body temperature, or thermoregulation, is a form of homeostasis and requires the coordinated activities of the nervous, muscular, endocrine, circulatory, and digestive systems. We summarize how these systems work together in endotherms and ectotherms in **Fig. 38.5**.

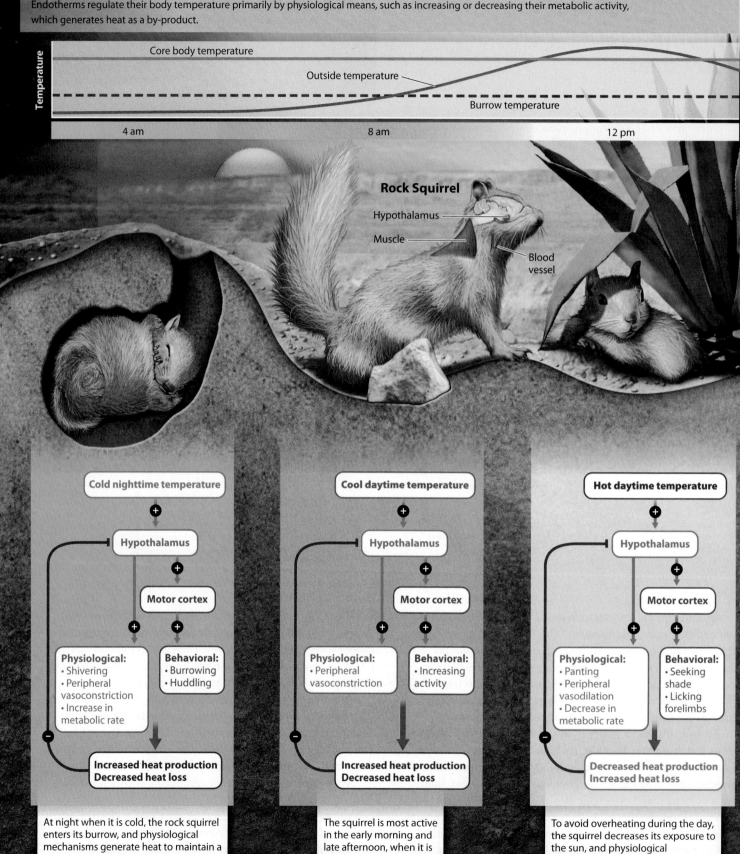

VISUAL SYNTHESIS — FIG. 38.5: Homeostasis and Thermoregulation
Integrating concepts from Chapters 33–38

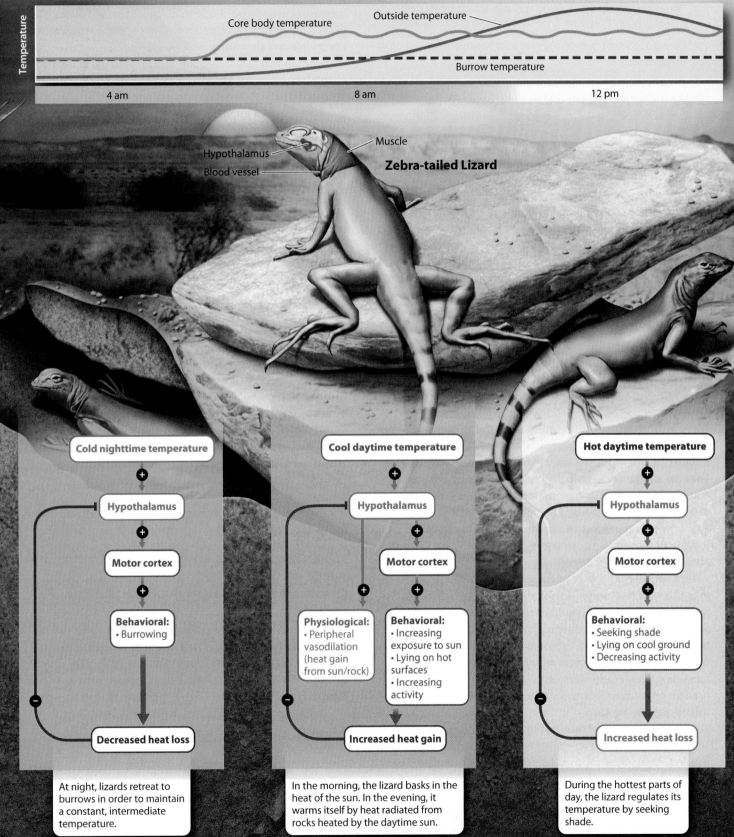

> **Self-Assessment Questions**
>
> 1. How can you measure the metabolic rate of an animal?
> 2. Why is the average world record speed for a 10-m race faster than that for a marathon?
> 3. Does a dog that is twice as heavy as a cat have twice the resting metabolic rate?
> 4. What happens in an endotherm and an ectotherm when outside temperature gets cold, and what happens when it gets hot?

38.2 NUTRITION AND DIET

Running, jumping, moving, growing, reproducing, and all other activities performed by animals require energy. Autotrophs like plants and certain microbes are able to capture energy from the sun or from inorganic compounds (Chapters 8 and 24), but animals must acquire food as organic compounds obtained from other living organisms, such as plants or other animals. Food does not consist just of carbohydrates, fats, and proteins, however. To enable various cell processes, animals must also obtain in their diet certain minerals and chemical compounds that they cannot synthesize on their own. In this section, we explore what animals need to obtain from their food, and how much they require, in order to sustain their cells and tissues.

Energy balance is a form of homeostasis.

Like core body temperature (Fig. 38.5), blood glucose levels (Chapter 36), and blood pressure (Chapter 37), the **energy balance** of an organism is often maintained at a constant level. An animal in energy balance takes in over time the same amount of calories of energy from food that it uses over time to meet its metabolic needs. Energy balance can be thought of as a form of homeostasis. Factors determining energy balance include sources of energy, or **energy intake**, and ways in which energy is expended, or **energy use**.

For animals, the source of energy is the diet. In turn, that energy is used to do work—maintain tissues, grow, move about, and the like. The energy required to support basic life processes accounts for the majority (approximately 70%) of energy use. Interestingly, digestion and absorption of food themselves require energy, and this use is included in the 70% figure. The remaining energy (approximately 30% of energy use) can be used for physical activity, with higher levels of activity using more energy, as we saw.

When energy intake does not equal energy use, there is an energy imbalance. If an animal eats more food than it requires, energy stores such as fat deposits grow over time. The result is that the body shifts its metabolism mostly to anabolic processes that build energy stores. Many animals achieve a net positive energy balance during the late summer and fall when food is plentiful, before it becomes scarce in winter. Other animals maintain a constant energy intake throughout the year by migrating to areas with more abundant food and warmer temperatures, thereby avoiding colder temperatures that require increased energy expenditure to remain active and warm. Still other animals hibernate, or become less active, to conserve their energy use over the winter.

Animals that cannot acquire enough food are in negative energy balance and become undernourished. During prolonged periods of inadequate food supply or starvation, an animal consumes its own internal fuel reserves. Starvation forces animals to deplete their glycogen and fat reserves first. Then, if no food is found, they resort to protein stores, primarily in muscle tissue. The consumption of muscle protein ultimately leads to muscle wasting. Undernourishment and starvation are particularly serious human health problems in many developing countries, especially those ravaged by war and political instability.

Humans, like most other animals, store excess food calories as fat. We do so because, over much of our evolutionary history, food was less abundant and more unpredictable in its availability than it is today. With the development of agriculture and domestication of livestock, food supplies rapidly increased, and many human populations were able to grow and consume increasing amounts of food. With the rise of mechanized agricultural food production and the availability of highly processed foods developed by the modern food industry, excessive intake of food calories has led to an increasing and now critical public health problem: obesity.

Obesity is now an epidemic in many industrialized nations. In the United States, approximately 36% of the adult population is considered obese. Obesity is a major public health concern because it increases the risk of diabetes, heart disease, and stroke, and contributes to a shorter life-span. For most animals, however, acquiring food and storing its products efficiently in the body allow them to have a fuel reserve to meet their energy requirements. This rationing of stored energy remains an essential part of their metabolic and digestive physiology.

An animal's diet must supply nutrients that it cannot synthesize.

Through their metabolic pathways, animals are able to obtain energy from the environment as well as to synthesize many of the compounds needed to sustain life. However, many nutrients necessary for life cannot be synthesized by an animal's metabolism and, therefore, must be acquired in the food that they eat.

Recall that amino acids are the basic units of proteins (Chapter 4). Twenty different amino acids are typically found in proteins (see Fig. 4.2). Although all of these amino acids are necessary for life, an **essential amino acid** is defined as one that cannot be synthesized by an organism's cellular biochemical pathways and instead must be ingested. Most animals can synthesize about half of their amino acids, but which of these each can synthesize varies from species to species.

Humans are unable to synthesize 8 of the 20 amino acids (**Table 38.1**). We have to obtain these eight essential amino acids in our diets. Infants require additional amino acids. The most reliable source of all eight essential amino acids is meat. Most plant proteins lack at least one of the eight amino acids essential to humans, and diets heavy in a particular food, such as corn (which lacks lysine), can lead to protein deficiency. Nevertheless, vegetarians can achieve a healthy intake of all essential amino acids by combining plant foods. For example, beans supply the lysine that corn lacks, and corn supplies the methionine that beans lack.

Dietary minerals are chemical elements other than carbon, hydrogen, oxygen, and nitrogen that are required in cells and must be obtained in the food that an animal eats (shown in red in **Fig. 38.6**). They include sodium (Na), chlorine (Cl), magnesium (Mg), phosphorus (P), potassium (K), calcium (Ca), iron (Fe), and zinc (Zn). Magnesium and zinc are sometimes associated with enzymes as cofactors (Chapter 6); calcium is required for neuron function (Chapter 34), muscle function (Chapter 35), and building skeletons (Chapter 35); and iron in hemoglobin binds oxygen and transports it in the blood (Chapter 37). Sodium and potassium are essential not only for nerve function, but also for supporting the sodium–potassium pumps characteristic of all cells. Humans typically obtain the minerals sodium and chloride through common table salt and other foods. Many animals seek exposed rock that they lick to obtain minerals and salts.

In addition to amino acids and minerals, animals must ingest essential **vitamins**, organic molecules that are required in very small amounts in the diet. Vitamins have diverse roles: some bind to and increase the activity of particular enzymes, whereas others act as antioxidants or chemical signals. Different animal species have different vitamin requirements, so a molecule that functions as a vitamin for one may not be a vitamin for another. Thirteen essential vitamins have been identified for humans (**Table 38.2**). Knowing which vitamins are required for an animal's health is critical to proper medical and veterinary care. Animals have evolved diets that ensure that their need for vitamins, as well as for essential amino acids and minerals, is met through the combination of foods that they eat.

In humans as well as in other animals, vitamin deficiency can have serious consequences. Whereas most mammals can synthesize ascorbic acid (vitamin C), primates (including

TABLE 38.1 Essential Amino Acids for Humans, Dogs, and Cats

ESSENTIAL AMINO ACID	HUMAN	DOG	CAT	COMMON SOURCES
Isoleucine	✓	✓	✓	Meat, dairy, lentils, wheat
Leucine	✓	✓	✓	Meat, cottage cheese, lentils, peanuts
Lysine	✓	✓	✓	Meat, beans, lentils, spinach
Methionine	✓	✓	✓	Meat, dairy, whole grains, corn
Phenylalanine	✓	✓	✓	Meat, dairy, peanuts, seeds
Threonine	✓	✓	✓	Meat, dairy, nuts, seeds
Tryptophan	✓	✓	✓	Meat, dairy, oats
Valine	✓	✓	✓	Meat, dairy, grains
Arginine		✓	✓	Meat, dairy, soy, peanuts
Histidine		✓	✓	Meat, dairy, soy, eggs
Taurine			✓	Meat, dairy, eggs

FIG. 38.6 Commonly recognized minerals required by humans. Dietary minerals, shown in red, are chemical elements other than carbon, hydrogen, oxygen, and nitrogen that are required in the diet.

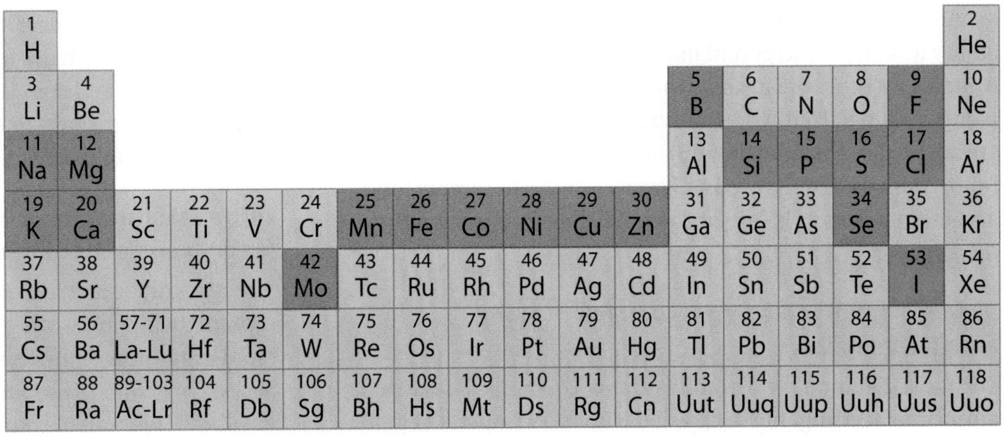

humans) cannot. Vitamin C is necessary for building connective tissue. Humans who do not ingest enough of this vitamin develop scurvy, a disease characterized by bleeding gums, loss of teeth, and slow wound healing. Until it was discovered that green vegetables and fruit supply vitamin C, scurvy was a common disease that could lead to death for sailors who were deprived of fresh produce for long periods of time while at sea.

Deficiencies of vitamins B_1, B_6, and B_{12} can cause nervous system disorders and various forms of anemia (low levels of red blood cells or hemoglobin in the blood). Vitamin D is essential for the absorption of calcium in the diet and, therefore, for skeletal growth and health. With adequate exposure to ultraviolet solar radiation, skin cells synthesize enough vitamin D to sustain a growing body. However, people inhabiting northern regions of the world do not get a lot of sunlight. As a result, they produce low levels of vitamin D and require more of it in their diet. Lack of adequate vitamin D can lead to rickets, a disease often seen in malnourished children in which bones do not mineralize properly. The role of vitamin E remains uncertain, but its absence is sometimes linked to anemia.

TABLE 38.2 Essential Vitamins for Humans

VITAMIN DESIGNATION	NAME	COMMON SOURCES
Vitamin A	Retinol	Green and orange vegetables, liver, milk
Vitamin B_1	Thiamine	Pork, grains
Vitamin B_2	Riboflavin	Dairy products, eggs, leafy green vegetables
Vitamin B_3	Niacin	Meat, dairy products, eggs, vegetables
Vitamin B_5	Pantothenic acid	Meat, grains
Vitamin B_6	Pyridoxine	Meat, grains
Vitamin B_7	Biotin	Meat, eggs
Vitamin B_9	Folic acid	Leafy green vegetables, eggs
Vitamin B_{12}	Cobalamin	Liver, eggs
Vitamin C	Ascorbic acid	Fruit, vegetables
Vitamin D	Calciferol	Fish, eggs
Vitamin E	Tocopherol	Fruit, vegetable oils
Vitamin K	Phylloquinone	Leafy green vegetables

Self-Assessment Questions

5. Muscle, fat, and glycogen are all reservoirs of energy. In which order are they used during a prolonged fast?
6. In the context of nutrition, what does "essential" mean?

38.3 ADAPTATIONS FOR FEEDING

Animals have evolved a variety of adaptations to enhance their ability to acquire energy and nutrients from the environment. Animals eat other organisms, whether they are other animals, plants, or fungi. Herbivores are adapted to eat plants; carnivores are adapted to eat animals; and omnivores are adapted to eat both. Animals have evolved a diversity of ways to capture food and mechanisms to break food down before its digestion in the gut.

Suspension filter feeding is common in many aquatic animals.

The most common form of food capture by animals is **suspension filter feeding,** in which water with food suspended in it is passed through a sievelike structure (**Fig. 38.7**). Suspension filter feeding is possible only in aquatic environments, and it is common in a great diversity of animals found in oceans and freshwater bodies, such as lakes and streams. This form of feeding evolved multiple times independently on many different branches of the tree of life, so these occurrences of filter feeding are examples of convergent evolution (Chapter 22).

Many worms and bivalve mollusks, such as scallops, clams, and oysters, pump water over their gills to trap food particles suspended in the water (Fig. 38.7a). Cells of the gills produce mucus that acts as an adhesive to trap food particles, and the cells' beating cilia sweep food trapped in the mucus to the mouth and gut.

Other aquatic organisms move water containing suspended food particles through filters in their oral cavity to separate the food from the water. The food is then conveyed into the gut. Large baleen whales have long comblike blades of baleen (which is made of keratin, the same material that makes up hair and nails) in their mouth to capture small shrimp called krill that are plentiful in ocean waters (Fig. 38.7b).

Large aquatic animals apprehend prey by suction feeding and active swimming.

Aquatic animals that do not rely on suspension filter feeding capture their prey in other ways. Many fish feed by suction. A rapid expansion of the fish's mouth cavity draws water and the desired prey into the mouth (**Fig. 38.8**). After the fish closes its mouth, the water is pumped out of the mouth past the gills.

FIG. 38.7 Suspension filter feeding by (a) a scallop and (b) a baleen whale. *Sources: a. Andrew J. Martinez/SeaPics.com; b. Jean-Marc Bour/Biosphoto.*

FIG. 38.8 Suction feeding. A rapid expansion of the fish's mouth draws in water and prey. *Source: Marevision/Getty Images.*

large size and formidable speed enable them to catch smaller or similar-sized prey. As discussed next, the evolution of jaws and teeth considerably enhanced the ability of these and other vertebrate animals to capture new kinds of food, allowing them to lead a more active and energetically demanding lifestyle.

Specialized structures allow for capture and mechanical breakdown of food.

Jaws and teeth were an important evolutionary innovation for active predators. Some of the first vertebrates, the jawed fishes, became dominant in their aquatic environment through the ability to swim and bite forcefully to obtain their food.

FIG. 38.9 Feeding by active swimming. Large marine predators, like this great white shark, use their size and speed to capture prey. *Source: C & M Fallows/SeaPics.com.*

Trapped inside the mouth, the prey moves into the pharynx (part of the throat) and is broken up by a second set of jaws in the pharynx before being swallowed. Suction feeding is the most common form of prey capture among fishes, as it allows them to be "sit-and-wait" predators; they can hide in a coral reef or under a rock before rapidly striking to capture prey moving in front of them. A wide variety of fishes use suction feeding, as do aquatic salamanders, and this feeding method has contributed to their evolutionary diversification and success.

Many insects that bite to obtain a blood meal draw the blood of their prey into their digestive system by suction. Young mammals feed by suckling milk from their mother's breast, using their tongue and generating suction as fishes do.

As top predators in marine and freshwater environments, larger fishes like sharks and marine mammals like whales and dolphins actively swim to capture their prey (**Fig. 38.9**). Their

FIG. 38.10 A phylogeny of vertebrates, showing the appearance of jaws.
Jaws, which evolved more than 400 million years ago from a group of jawless ancestors, contributed to the success of many groups of animals.

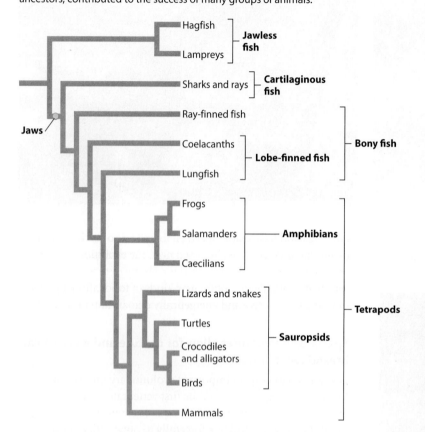

Fig. 38.10 shows a phylogenetic tree of vertebrates indicating the appearance of jaws. Jawed fishes evolved from jawless ancestors, and jaws can be found in present-day fishes, amphibians, reptiles, and mammals. As you can see from the large number of jawed groups, jaws are key to the evolutionary success of vertebrates. Jaws evolved from cartilage that supported the gills, providing a spectacular example of an organ adapted for one function (gill support) changing over time to become adapted for an entirely different function (predation).

Among vertebrates, mammals evolved a specialized jaw joint, the **temporomandibular joint**, as well as a great diversity of specialized forms of teeth. These teeth have cutting and crushing surfaces, enabling mammals to break down a variety of foods mechanically before swallowing those foods. **Fig. 38.11** illustrates the arrangement of specialized teeth in mammals with different diets. Teeth in the front of the mouth, the **incisors**, are specialized for biting. Other teeth, such as the **canines** of dogs, cats, and other carnivores, are specialized for piercing the body of their prey. Saber-toothed cats, which lived as recently as 10,000 years ago, had remarkably long canines (up to 16 inches [40 cm]!) specialized for killing large prey. **Premolars** and **molars** are teeth in the back of the mouth that are well adapted for crushing and shredding tougher foods, such as meat and fibrous plant material.

Herbivorous animals, such as cattle, sheep, and horses, have specialized premolars and molars with prominent surface ridges. These teeth can shred tough plant material before it is swallowed and digested. Mammalian herbivores use their front incisors and canines to bite grasses and leaves; they then move the food to the back of the mouth, where it is ground and crushed between their ridged premolars and molars.

Arthropods, such as insects and crustaceans, lack jaws and teeth, yet are also able to capture, manipulate, and break down food. Some of them use paired mouthparts called **mandibles**, one on each side of the head (**Fig. 38.12**). In contrast to the teeth of vertebrates, mandibles are located outside and in front of the mouth. They are used to capture and physically break down food before it enters the mouth and then the gut. Arthropod mandibles have evolved a great

FIG. 38.11 Jaws and teeth of carnivorous, herbivorous, and omnivorous mammals.
Mammals evolved a diverse arrangement of teeth for diverse diets.

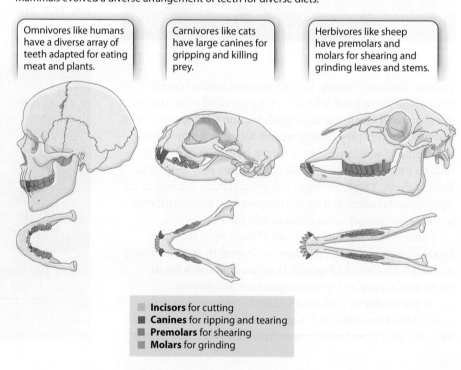

FIG. 38.12 Mandibles. (a) Ants and (b) beetles have jaw-like mandibles that are used to capture, hold, and tear apart food. *Sources: a. Mark Bowler/Nature Picture Library/Getty Images; b. Artverau/Getty Images.*

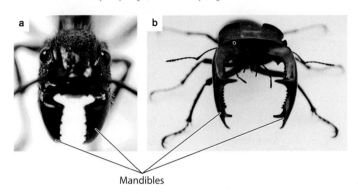

diversity of structures and are well adapted to different sources of food (Chapter 42).

Self-Assessment Questions

7. What are four ways in which animals capture prey? Provide an example of an animal that uses each one.

8. The jaws of vertebrates and the mandibles of insects have similar functions. Are they an example of analogous or homologous structures?

38.4 REGIONAL SPECIALIZATION OF THE GUT

After animals obtain food, they need to break it down. The process of breaking down food into smaller components is called **digestion**. Digestion can occur either mechanically or chemically. The food is isolated in a specialized organelle or compartment so that it can be broken down without damaging other organelles or structures of the body.

Organisms on different branches of the tree of life have evolved very different means of isolating and digesting food. Single-celled protists, which obtain food particles by phagocytosis, rely on intracellular digestion: their food is broken down within cells. In these organisms, lysosomes containing hydrolytic enzymes fuse with vacuoles containing food particles, thereby mixing enzymes and food (Chapter 5). Following the food particles' chemical breakdown, the products of intracellular digestion are available for use by the cell. Most animals, however, employ the process of extracellular digestion, in which food is isolated and broken down outside a cell in a body compartment. Following digestion, the breakdown products are taken up into the bloodstream in a process called **absorption**.

Most animal digestive tracts have three main parts: a foregut, midgut, and hindgut.

In extracellular digestion, food is broken down in a specialized cavity or body compartment. Some animals, such as sea anemones, carry out digestion in an internal cavity. Food flows directly into this cavity through the mouth. Cells lining the cavity secrete digestive enzymes that break down the food. These very same cells then absorb the smaller chemical compounds produced by digestion.

Other animals have evolved more elaborate digestive systems that allow the transport of food by a long tube that runs from one end of an animal to the other. This tube constitutes an animal's **gut** or **digestive tract**. Food is moved in a single direction through the gut, which contains regions specialized for different functions. These functions include digestion (the mechanical and chemical breakdown of food), absorption of released nutrients, storage, and elimination of waste products.

These various functions take place in the three main parts of the digestive tract: the **foregut**, **midgut**, and **hindgut** (**Fig. 38.13**). Most animals' digestive systems have the same basic plan, though some animals have evolved additional specialized structures, as discussed later in this section. The foregut includes the **mouth**, **esophagus**, and **stomach** or

FIG. 38.13 Organization of animal digestive tracts: foregut, midgut, and hindgut.

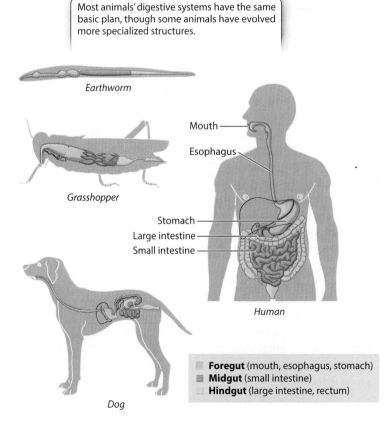

crop, which serves as an initial storage and digestive chamber. Next is the midgut, which includes the **small intestine,** where the remainder of digestion and most nutrient absorption takes place. Here, specialized organs secrete enzymes and other chemicals that aid in the breakdown of particular macromolecules, such as carbohydrates and fats. Finally, the digested material reaches the hindgut, which includes the **large intestine** and **rectum.** In the hindgut, water and inorganic molecules are absorbed, leaving the waste products, or feces. The feces are stored in the rectum until being eliminated from the body.

Food and its breakdown products move through the gastrointestinal tract by **peristalsis,** waves of smooth muscle contraction and relaxation. These waves keep substances moving from one end of the tract to the other, and help prevent backward movement.

TABLE 38.3 Major Sites of Chemical Digestion of Macromolecules

MACROMOLECULE	SITE OF DIGESTION	ENZYME
Protein	Stomach	Pepsin
	Small intestine	Trypsin
Carbohydrate	Mouth	Salivary amylase
	Small intestine	Pancreatic amylase, lactase
Lipid	Mouth	Lingual (tongue) lipase
	Stomach	Gastric (stomach) lipase
	Small intestine	Pancreatic lipase

Digestion begins in the mouth.

Most animals jump-start digestion by breaking down food mechanically through biting and chewing, as we saw in section 38.3. Breaking food into smaller pieces increases the surface area that is exposed to taste buds in the tongue and available for the action of enzymes.

After being ingested and mechanically broken down, the food is ready for chemical digestion, which is carried out by enzymes that break the bonds between the subunits of large molecules. These enzymes require the proper chemical environment to function properly. Some enzymes work best at neutral or near-neutral pH, like that found in the mouth or small intestine. Other enzymes work best in acidic environments, like that found in the stomach.

In some animals, such as mammals, chemical digestion begins in the mouth. Food entering the mouth is mixed with salivary secretions that contain **amylase,** an enzyme that breaks down carbohydrates, and secretions from the tongue that contain **lipase,** an enzyme that breaks down lipids (**Table 38.3**).

The muscular tongue of mammals and other land vertebrates manipulates food and transports it within the mouth cavity. The tongue moves the food into position within the mouth for effective cutting and chewing into smaller pieces by the teeth. When food has been chewed and is ready to be swallowed, the tongue moves it to the rear of the mouth cavity.

Swallowing is a complex set of motor reflexes carried out by several muscles and structures in the rear of the mouth and in the **pharynx,** the region of the throat that connects the mouth and nasal cavity (**Fig. 38.14**). Once initiated, swallowing reflexes are involuntary, under the control of the autonomic nervous system. During swallowing, food is pushed through the pharynx, over the **epiglottis** (a flap of tissue that protects food from entering the trachea and lungs), and into the esophagus (a tube that connects the pharynx to the stomach). In the esophagus, peristalsis moves the food toward the stomach.

Birds, alligators, crocodiles, and earthworms break down food into smaller pieces farther along their digestive tracts in the **gizzard,** a compartment with thick muscular walls. Birds and earthworms often ingest small rocks or sediment that help grind the food into smaller pieces within the muscular gizzard. Breaking food into smaller pieces aids the chemical digestion that follows.

Further digestion and storage of nutrients take place in the stomach.

Digestion, begun in the mouth, continues in the stomach. The stomach is one of the main sites of protein and lipid breakdown. The acidic environment of the stomach facilitates digestion. In addition, several digestive enzymes have evolved to act at the stomach's very low pH, in the range of 1–3. Most enzymes, by contrast, are denatured (that is, unfolded and thus made inactive) in acidic environments.

The stomach is maintained at a low pH through the secretion of hydrochloric acid (HCl) by specific cells lining the stomach. A low pH could damage the stomach when no food is present, so the amount of HCl is carefully regulated. The sight, smell, and taste of food send signals to the brain that stimulate a sense of appetite. Signals from the brain to the stomach then stimulate the secretion of HCl and digestive enzymes that break down proteins and lipids.

When food arrives in the stomach, cells lining the stomach secrete a peptide hormone called **gastrin.** Gastrin stimulates the cells lining the stomach to increase their production of HCl still further. If the pH of the stomach becomes too low, gastrin secretion is inhibited, an example of a negative feedback loop (Chapter 33).

The stomach has several mechanisms to protect itself from the acidic environment and the action of digestive enzymes. Glands in the lining of the stomach secrete mucus to protect the stomach wall. In addition, cells in the stomach secrete some digestive enzymes in an inactive form; otherwise, the cells themselves would be digested. After their secretion, the inactive enzymes are activated by a change in their chemical structure. The primary digestive enzyme produced in the stomach is

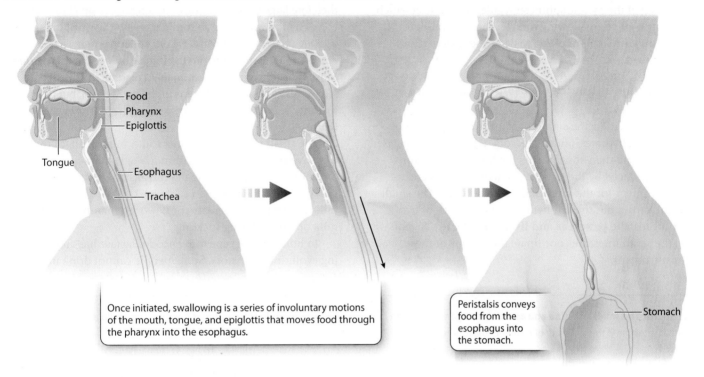

FIG. 38.14 **Swallowing.** Swallowing moves food from the mouth to the stomach.

Once initiated, swallowing is a series of involuntary motions of the mouth, tongue, and epiglottis that moves food through the pharynx into the esophagus.

Peristalsis conveys food from the esophagus into the stomach.

pepsin, an enzyme that breaks down proteins into amino acids (Table 38.3). Cells lining the stomach release pepsin in an inactive form called pepsinogen that is activated by the low pH of the stomach. The stomach also secretes lipases that break down lipids (Table 38.3). Like pepsin, gastric lipase works best in an acidic environment.

Animals often eat large amounts of food in a short time, but it takes much longer to digest that food. Therefore, the stomach serves as a storage compartment as well as a digestion compartment. In humans, it typically takes about 4 hours for the stomach to empty, so digestion and absorption of nutrients occur between meals. In other animals, stomach emptying can take much longer. Most carnivorous animals consume large and infrequent meals that are followed by long periods of digestion and nutrient absorption—much longer than the intervals in humans and other animals that eat more regularly. Snakes such as pythons that engulf large prey whole spend several days digesting and absorbing the nutrients from their meal (**Fig. 38.15**). To digest their single large meal, their gut undergoes extensive remodeling that produces new energy-demanding secretory and absorptive cell surfaces. As a result, their metabolic rate increases sharply during digestion. These extra cells are not retained between meals because of the high energy cost of maintaining them.

The stomach walls contract to mix the contents of the stomach, aiding their digestion. Waves of muscular contraction move the food toward the base of the stomach. There, the

FIG. 38.15 **Python and prey.** Because the prey is eaten whole, the python's metabolic rate is high during digestion, which takes place over several days. *Source: Biosphoto/Ken Griffiths/Photoshot.*

pyloric sphincter, a ring of muscle, opens and allows small amounts of digested food to enter the small intestine. Opening and closing of the pyloric sphincter regulates the rate at which the stomach empties, allowing time for the food products released into the small intestine to be further digested and then absorbed.

Final digestion and nutrient absorption take place in the small intestine.

After the food enters the small intestine, final digestion of protein, carbohydrates, and fats occurs. The nutrients released by digestion are then absorbed by the body. In addition, the acid produced by the stomach is neutralized to create an environment for the intestinal enzymes to act. The small intestine carries out these functions with the help of two other organs, the **pancreas** and **liver** (**Fig. 38.16**). Signals from the small intestine coordinate the actions of these accessory organs.

The small intestine is named for its small diameter, but it is actually a long (about 20 feet or 6 m in an adult human) and therefore large organ. Because of its length, the small intestine is coiled to fit within the abdominal cavity. The small intestine is made up of three sections (Fig. 38.16, upper right). Food enters from the stomach into the initial section, the **duodenum**. Most of the remaining digestion takes place in the duodenum. Nutrient absorption also begins in the duodenum, but the majority occurs in the next two sections, the **jejunum** and **ileum**.

The small intestine produces some, but not all, of the digestive enzymes that act within it. For example, its cells produce lactase, an enzyme that breaks down the sugar lactose into glucose and galactose (Table 38.3). Lactose is found in milk, so lactase is necessary for newborn mammals to digest milk obtained during suckling.

In humans, the gene that encodes lactase has an interesting evolutionary history. Some people cannot drink milk as

FIG. 38.16 The small intestine and associated organs. Digestion in the small intestine is aided by secretions from the liver, gallbladder, and pancreas. The small intestine is the main site of nutrient absorption.

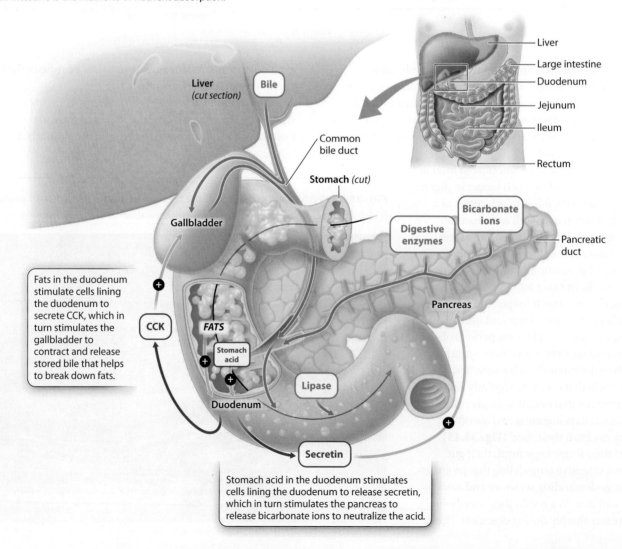

adults, a condition called lactose intolerance. In these individuals, the gene that encodes lactase is down-regulated and not expressed after weaning. Other people can continue to drink milk as adults. In these individuals, the gene for lactase continues to be expressed after weaning. Researchers have found that lactase persistence in these individuals is the result of mutations that occurred and then were selected for when cattle were domesticated and milk became part of the diet approximately 10,000 to 5,000 years ago.

Some of the digestive enzymes that act in the small intestine are produced by the pancreas, rather than by the small intestine. The pancreas is an organ that lies below the stomach, near the duodenum (Fig. 38.16). It secretes enzymes and other molecules into ducts that connect to the duodenum, as well as the hormones insulin and glucagon directly into the blood (Chapter 36). The pancreas produces a variety of digestive enzymes, including lipase, which breaks down fats; amylase, which breaks down carbohydrates; and **trypsin**, which breaks down proteins (Table 38.3). Like the stomach, which produces pepsin in an inactive form, the pancreas produces trypsin in an inactive form called trypsinogen to avoid digesting itself. Trypsinogen is converted to trypsin by an enzyme produced by intestinal cells.

The pancreas also secretes bicarbonate ions, which neutralize the acid produced by the stomach. The pancreas is stimulated to secrete bicarbonate ions by the hormone **secretin** (the first hormone to be identified). Cells lining the duodenum secrete the hormone in response to the acidic pH of the stomach contents entering the small intestine.

The liver participates in digestion by producing **bile**, which is composed of bile salts, bile acids, and bicarbonate ions. Bile aids in fat digestion by breaking large clusters of fats into smaller lipid droplets, a process called emulsification. Bile produced by the liver is stored in the **gallbladder** (Fig. 38.16). When fats enter the duodenum, cells lining the duodenum release a peptide hormone called **cholecystokinin (CCK)**, which causes the gallbladder to contract, leading to the release of bile into the duodenum. By breaking up the fats into smaller droplets, bile allows lipases to digest the lipids more effectively.

After large macromolecules have been broken down by digestive enzymes, the smaller molecules produced through this process are ready to be absorbed. The small intestine has highly folded inner surfaces, called **villi**, and the cells constituting the villi themselves have highly folded surfaces, called **microvilli** (**Fig. 38.17**). Together, these villi and microvilli greatly increase the surface area available for the absorption of nutrients. Cells that line the small intestine are connected by tight junctions (Fig. 38.17; Chapter 10) that prevent movement of substances between cells, instead directing the products of digestion to be absorbed across the microvilli surfaces. Membrane transporters can then control their movement through the cell and into the bloodstream.

Hydrophilic nutrient molecules, such as glucose and amino acids, need the help of a carrier protein to cross the plasma membrane into cells lining the intestine. These molecules are often co-transported into cells with sodium ions from gut contents. A sodium ion binds to a transmembrane protein on the surface of cells lining the intestine; this protein also binds the nutrient molecule. Absorption of the nutrient molecule into the cell is driven by the movement of the sodium ion down its concentration gradient (**Fig. 38.18**). The

FIG. 38.17 Intestinal villi and microvilli. These structures increase the surface area for absorption of nutrients. *Photo source: Biophoto Associates/Science Source.*

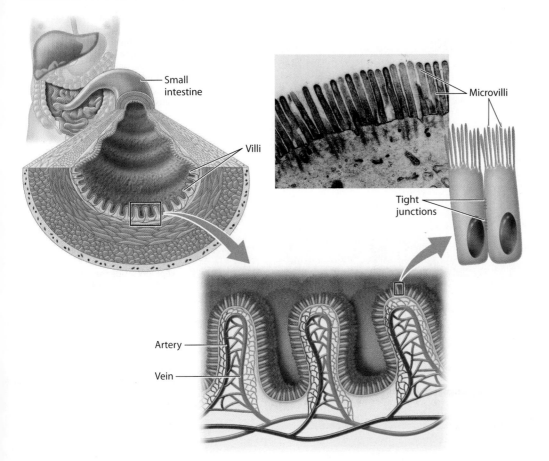

concentration of sodium ions inside the cell is kept low by the action of the sodium–potassium pump (Chapter 5). Molecules such as glucose then move passively down their concentration gradient out of the intestinal cell toward the blood through channels on the basal side of the cell.

The products of fat digestion—fatty acids and glycerol—are lipid soluble, so they readily diffuse across the plasma membrane into a cell without the assistance of a carrier protein. Eventually they are transported to the bloodstream. There, lipids bind molecules that make them water soluble, allowing for their transport in the aqueous (water-based) blood.

The large intestine absorbs water and stores waste.

Peristalsis moves the contents of the small intestine into the large intestine, or **colon**, of the hindgut at a pace that allows time for nutrient absorption in the small intestine. By the time the gut contents reach the large intestine, the nutrients have been absorbed into the body, but indigestible wastes, water, and inorganic ions remain. Most of the water and inorganic ions are absorbed in the colon (see Fig. 38.13), resulting in semisolid feces. Feces are stored in the final segment of the colon, the rectum, until they are periodically eliminated through the anus in the case of most mammals. In other vertebrates, including amphibians, reptiles, birds, and a few mammals, waste is eliminated through the cloaca, which is a shared opening for the digestive, urinary, and reproductive tracts.

Excess water absorption within the colon can cause constipation, whereas too little water absorption can cause diarrhea. Constipation and diarrhea can also be caused by toxins released by microorganisms in the food an animal eats. For example, the bacterium *Vibrio cholera* releases toxins that are the cause of cholera.

As we saw in Case 5 The Human Microbiome, digestion is not a solo endeavor. Large populations of bacteria reside in the small and large intestines and help extract nutrients that the animal's body cannot extract itself. The principal gut resident is *Escherichia coli*, the bacterium commonly used for research in laboratories. As they nourish themselves on the host's gut contents, the various bacteria provide nutrients and certain vitamins, such as biotin and vitamin K, that the animal cannot

FIG. 38.18 Glucose absorption by cells of the small intestine. Absorption of glucose into the cell is driven by the movement of the sodium ions down a concentration gradient.

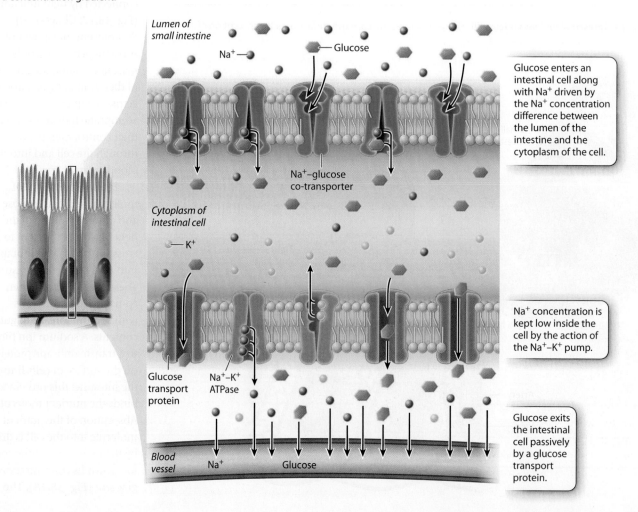

FIG. 38.19 Cross section of the mammalian intestine.

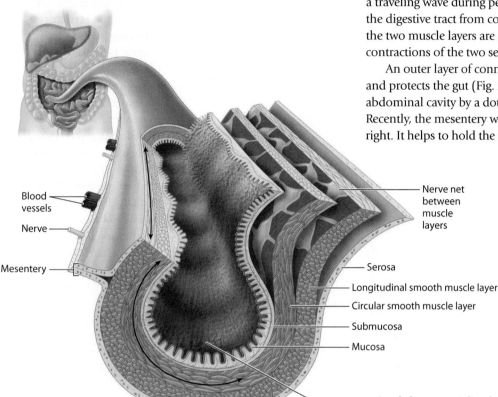

produce itself. Vitamin deficiency can sometimes result from prolonged antibiotic medication that kills large numbers of gut bacteria. Another by-product of bacterial action in the gut is gases such as methane and hydrogen sulfide. The mutual benefits to the host animal and the bacteria ensure the success of their symbiotic relationship.

The lining of the digestive tract is composed of distinct layers.

As we have seen, the digestive tract is not a passive tube. Indeed, it secretes enzymes and other chemical compounds for digestion, absorbs nutrients, and actively moves food through the body. How does a series of tubes accomplish all of this? As shown in **Fig. 38.19**, the digestive tract is made up of several layers of tissue, each with a specialized function. The central space, or **lumen**, through which the gut contents travel is surrounded by an inner tissue layer, the **mucosa**, which both secretes enzymes and absorbs nutrients. The cells of the mucosa secrete mucus to protect the gut wall from digestive enzymes and, in the stomach, from hydrochloric acid. Surrounding the mucosa is the **submucosa**, a layer containing blood vessels, lymph vessels, and nerves.

Outside these layers are two smooth muscle layers (Fig. 38.19). The inner **circular muscle** layer contracts to reduce the size of the lumen. The outer **longitudinal muscle** layer contracts to shorten small sections of the gut. Thus, these two muscle layers contract alternately to mix gut contents, producing a traveling wave during peristalsis that moves the contents along the digestive tract from compartment to compartment. Between the two muscle layers are autonomic nerves that control the contractions of the two sets of smooth muscle.

An outer layer of connective tissue called the **serosa** covers and protects the gut (Fig. 38.19). The gut is supported in the abdominal cavity by a double membrane called the **mesentery**. Recently, the mesentery was recognized as an organ in its own right. It helps to hold the intestines in place and provides a passageway for blood vessels and nerves to supply the gut.

Plant-eating animals have specialized digestive tracts adapted to their diets.

As noted earlier, herbivorous (that is, plant-eating) animals evolved specialized digestive tracts that enhance their ability to digest cellulose and other plant compounds. Animals, including herbivores, lack **cellulase**, the enzyme that breaks down cellulose. Instead, plant-eating animals have specialized compartments in their digestive tract that contain large bacterial populations, which do produce cellulase. As a result, the bacteria are able to digest the cellulose of plants. This association of herbivores and their gut microbes is another example of a symbiotic relationship.

Rather than the single stomach of other mammals, ruminants (cattle, sheep, and goats) have a four-chambered stomach that is highly specialized to enhance the ability of their gut bacteria to digest plants (**Fig. 38.20a**). The first two chambers—the **rumen** and the **reticulum**—harbor large populations of anaerobic bacteria that break down cellulose. To assist breakdown, the rumen contents are frequently regurgitated (in a process called rumination) for further chewing in the mouth. This repeated chewing decreases the size of the plant fibers, increasing the surface area available for bacterial digestion in the rumen. The bacteria break down cellulose by fermentation, and the host uses some of the products as nutrients. The site of fermentation in the foregut is the source of the name "foregut fermenters" for these mammals.

Bacterial fermentation produces carbon dioxide and methane gas. The amount of methane produced by domesticated ruminants is second only to the amount produced industrially. Since methane is a greenhouse gas that traps heat from the sun, methane production by cattle, sheep, and other animals contributes to atmospheric increases in methane and global climate change (Chapters 46 and 48).

After leaving the reticulum, the mixture of food and bacteria passes into the third chamber of the ruminant's

FIG. 38.20 Digestive systems of (a) a foregut fermenter and (b) a hindgut fermenter.

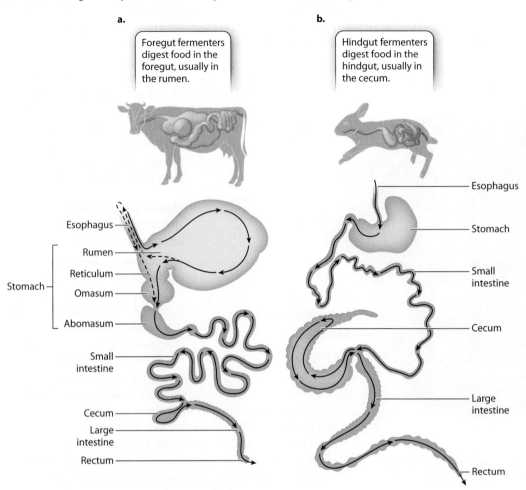

stomach, the **omasum,** where water is absorbed. It then enters the **abomasum,** the last of the four chambers, where protein digestion begins as it does in other mammals. The acid and protease secretions kill the bacteria, and the nutrients are absorbed in the small intestine. The bacteria, too, are an important protein source for the ruminant, yielding as much as 100 g of protein each day for a cow. This role is important because plants are a poor source of protein but contain considerable nitrogen that the bacteria use to synthesize their own amino acids. The bacteria lost to digestion are replaced by the continual reproduction of bacteria in the rumen.

In contrast to the foregut fermentation of cows and sheep, other mammalian herbivores, such as horses, rabbits, and koalas, digest the plant material they eat by hindgut fermentation (**Fig. 38.20b**). Hindgut fermentation occurs in the colon and in the **cecum,** a chamber that branches off the large intestine. Because the fermentation products released from the cecum have already passed through the small intestine, the main site of nutrient absorption, hindgut fermentation is less efficient at extracting nutrients than foregut fermentation.

In humans and other animals such as lemurs and squirrels, the cecum is present but small and includes the **appendix,** a narrow, tubelike structure that extends from the cecum. The appendix is an example of a **vestigial structure,** a structure that has lost its original function over evolutionary time and is now much reduced in size. Although the appendix has no clear function in nonherbivorous animals, evidence suggests it may play a role in the immune system (Chapter 41). The appendix can become infected when partially digested material enters or blocks it. Unless the infected appendix is removed, it can burst, releasing gut contents into the abdominal cavity; such a rupture can be life threatening.

Charles Darwin put forth a hypothesis explaining how the appendix might have evolved to take on its current form in humans. He suggested that in the ancestors of modern humans, the cecum harbored bacteria that aided in the digestion of plant material such as cellulose. However, as their diet became less reliant on plant material, the cecum became smaller and lost its function over time, becoming what we see today as the appendix. Other vestigial structures, such as the remnants of hind limbs in whales, small eyes in cave-dwelling animals, and wings in flightless birds, provide an illuminating window into the past and strong evidence for evolution.

Self-Assessment Questions

9. The rocks and sediments in the gizzards of birds break apart food mechanically. Which structures in vertebrates have a similar function?

10. What are the main sites of digestion of protein, carbohydrates, and fats, and which enzymes are involved in breaking down each of these?

11. What are the similarities and differences between foregut and hindgut fermentation?

12. Some animals, such as rabbits, whose digestive systems include hindgut fermentation, re-ingest the products of digestion. What function do you think this serves?

CORE CONCEPTS SUMMARY

38.1 PATTERNS OF ANIMAL METABOLISM: Metabolic rate depends on activity level, body size, and body temperature.

Animals acquire carbohydrates, fats, and protein from their diet for energy to fuel cellular processes and activities. page 855

Carbohydrates are broken down rapidly in the cytosol by glycolysis to produce a small amount of ATP in the absence of oxygen. page 856

Carbohydrates and fats produce large amounts of ATP by aerobic respiration through the citric acid cycle and oxidative phosphorylation. page 856

An animal's metabolic rate is its overall rate of energy use; this rate can be measured by oxygen consumption over time. page 857

Metabolic rate increases with increased physical activity. page 857

Larger animals consume more energy and have higher metabolic rates than smaller ones, but the increase in metabolic rate is less than proportional to mass. page 858

Endothermic animals maintain elevated body temperature by metabolic heat production and have high metabolic rates. page 858

Ectothermic animals depend on external heat sources to warm their bodies. They have lower metabolic rates compared to endothermic animals. page 858

38.2 NUTRITION AND DIET: An animal's diet supplies the energy the animal needs for homeostasis and essential nutrients it cannot synthesize on its own.

An animal is in energy balance when the amount of energy that it takes in is equal to the amount it uses to perform its functions and sustain life. page 862

Animals must ingest essential amino acids, minerals, and vitamins, those that cannot be synthesized in their cells. page 862

38.3 ADAPTATIONS FOR FEEDING: Animals have evolved different ways to capture food.

Aquatic animals take in food by suspension filter feeding, suction feeding, and active swimming. page 864

Vertebrates have specialized jaws and teeth for biting and mechanically processing food. page 865

Carnivorous animals have enlarged canines and slicing molars for catching prey and eating meat. page 866

Herbivorous animals have molars and premolars with surface ridges that enable them to grind plant matter. page 866

Many insects have paired mandibles for capturing, holding, and tearing prey. page 866

38.4 REGIONAL SPECIALIZATION OF THE GUT: The digestive tract has regions specialized for digestion, absorption, storage, and elimination.

The animal digestive tract is a long tubelike structure with distinct regions having different functions. page 867

Peristalsis mixes and moves food through the gut by means of smooth muscle contraction and relaxation. page 868

Most animals break down their food into small pieces in the mouth, making it easier to digest. page 868

Initial chemical digestion of carbohydrates and fat begins in the mouth. page 868

The stomach stores food and initiates protein and fat digestion through secretions of hydrochloric acid and digestive enzymes, including pepsin. page 868

Proteins, carbohydrates, and lipids are further broken down in the duodenum of the small intestine, assisted by the secretion of digestive enzymes by the pancreas and of fat-emulsifying bile by the liver. page 870

Nutrient absorption takes place in the small intestine. page 870

Nutrient absorption in the small intestine is enhanced by an enlarged surface area provided by villi and microvilli. page 871

The large intestine absorbs water and inorganic molecules before waste products are eliminated as semisolid feces from the rectum through the anus or cloaca. page 872

Herbivorous animals have specialized digestive chambers (the rumen and cecum) that enable bacterial fermentation to break down plant cellulose. page 873

Log in to LaunchPad to check your answers to the Self-Assessment Questions and to access additional learning tools.

CHAPTER 39 Animal Renal Systems: Water and Waste

CORE CONCEPTS

39.1 WATER AND ELECTROLYTE BALANCE: All animals regulate the water and electrolyte levels within their cells.

39.2 EXCRETION OF WASTES: Excretory organs eliminate nitrogenous wastes and regulate water and electrolyte levels.

39.3 THE MAMMALIAN KIDNEY: The mammalian kidney can produce urine that is more concentrated than blood as an adaptation for living on land.

In the 1960s, researchers at the University of Florida developed a new sports drink for players on the football team, the Gators. In addition to water and sugar, the drink contained electrolytes, ions such as sodium (Na^+) and potassium (K^+) that give the solution electrical properties. The researchers who came up with the formula for the drink reasoned that it was important to replace both water and electrolytes lost in sweat. The drink they created, called Gatorade, is now well known.

Water and electrolytes are two key substances that allow cells and organisms to maintain their size and shape, as well as to carry out essential functions. Life originated and diversified in a watery environment. Consequently, the chemical functions of cells and organisms crucially depend on the properties of water (Chapter 2) and the relative amounts of water and electrolytes inside a cell. In general, the internal environment of a cell, including the concentration of electrolytes, is maintained within a narrow window. The active maintenance of a stable internal environment in the face of a changing external environment is known as **homeostasis** (Chapters 5 and 33).

Animals must also maintain homeostasis in terms of the molecules produced by metabolic activity. Metabolism generates wastes that, in many cases, can be toxic to the cell. These wastes must somehow be eliminated. In most organisms, the elimination of wastes is closely tied to the maintenance of water and electrolyte balance.

In this chapter, we explore the physiological and cellular mechanisms that underlie water and electrolyte balance, as well as waste elimination. We examine how animals meet the demands for water and electrolyte balance in different environments. We also consider how water and electrolyte balance is linked to waste elimination and explore the diversity of excretory organs that animals have evolved to regulate these processes. The organization and function of the mammalian excretory organ, the **kidney**, is discussed in the last section of the chapter.

39.1 WATER AND ELECTROLYTE BALANCE

A healthy person can live for several weeks without food, but cannot survive for more than a few days without water. A loss of more than 10% of body water threatens survival. Homeostasis of water and electrolyte concentration is critical to normal physiological function, not just in humans, but in all organisms.

Osmosis governs the movement of water across cell membranes.

Water, like all molecules, moves by diffusion, or the random motion of molecules (Chapter 5). When there is a difference in water concentration between two regions, a net movement of water molecules occurs from the region of higher water concentration to the region of lower water concentration. Water often contains dissolved molecules, such as electrolytes, amino acids, and sugars, which are collectively called **solutes**. Because we typically describe solutions in terms of the concentration of dissolved solutes, it is sometimes easier to think about water moving from a region of lower solute concentration (higher water concentration) to a region of higher solute concentration (lower water concentration). Whether considered in terms of water concentration or solute concentration, the direction of net water movement is the same.

Sometimes a barrier separates two solutions. When that barrier allows water or solutes to diffuse through it freely, it is said to be permeable. When it blocks the diffusion of water or solutes entirely, it is described as impermeable. When the barrier allows the movement of some molecules but not others, it is **selectively permeable**.

In organisms, the plasma membrane of a cell serves as a barrier between solutions, as we saw in Chapter 5. The plasma membrane is a lipid bilayer with embedded proteins. It is selectively permeable: water and smaller solutes can diffuse freely through the membrane, but larger solutes, such as proteins, cannot. Water is able to move to a limited extent through the lipid bilayer by diffusion. However, it moves much more readily through channels called **aquaporins** by facilitated diffusion.

The process by which water moves from regions of higher water concentration to lower water concentration across a selectively permeable membrane is called **osmosis** (**Fig. 39.1**). Consider a situation in which there is a difference in solute concentration on the two sides of a selectively permeable membrane that is freely permeable to water but impermeable to the solute (Fig. 39.1a). Water moves by osmosis from regions of higher *water* concentration (lower *solute* concentration) to regions of lower *water* concentration (higher *solute* concentration) (Fig. 39.1b).

The tendency of water to move from one solution into another by osmosis is called **osmotic pressure**. It is the pressure needed to *prevent* water from moving by osmosis across a selectively permeable membrane. The higher the solute concentration, the higher the osmotic pressure of that solution (that is, there is a greater tendency for water to move *into* that solution).

Water movement by osmosis can be prevented by hydrostatic pressure that results from gravity or from the stiffness of the container walls. As water diffuses across a selectively permeable membrane into a region with a higher solute concentration (higher osmotic pressure), the hydrostatic pressure of that solution increases. When osmotic pressure equals hydrostatic pressure, the net movement of water stops and equilibrium is reached (Fig. 39.1c). At equilibrium, water molecules still move across the membrane in both directions by diffusion, but there is no net movement of water molecules.

Whether water moves by osmosis into or out of a cell depends on the solute concentration inside of the cell relative to the solute concentration outside of the cell. Similarly, whether water moves by osmosis into or out of an organism depends on the solute concentration inside an animal relative to its environment, as well as the permeability to water of the sheets of cells that make up the skin and other surfaces exposed to the environment.

Note that water movement is driven by differences in *total* solute concentration between the two sides of a cell membrane. That is, if a particular electrolyte such as potassium is present in a high concentration inside the cell and in a low concentration outside the cell, but another electrolyte, such as sodium, has exactly the opposite distribution, there will be no net water movement because the total concentrations of the solutes on the two sides of the membrane are the same.

Also note that water moves across selectively permeable membranes in response to concentration differences of *any* solute. That is, the solute can be an electrolyte, such as sodium, or a dissolved molecule, such as glucose, an amino acid, or even waste, as long as the membrane is impermeable to it.

FIG. 39.1 Osmosis and osmotic pressure.

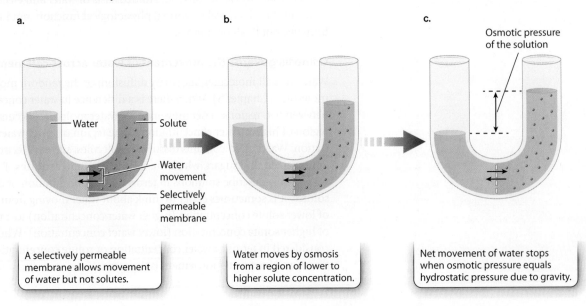

Osmoregulation is the control of osmotic pressure inside cells and organisms.

Animals on Earth can survive in a wide range of environments by adopting different means of osmoregulation. **Osmoregulation** is the regulation of osmotic pressure. It can be thought of as the regulation of water content, which keeps internal fluids from becoming too concentrated (high osmotic pressure) or too dilute (low osmotic pressure).

Osmoregulation, like energy balance (Chapter 38), is a form of homeostasis. High osmotic pressure inside a cell can lead to damage, sometimes even causing the cell to burst. Low osmotic pressure can lead to dehydration. Although cells and tissues can tolerate some degree of dehydration, often as high as 50% or more, excessive dehydration impairs a cell's metabolic function. Many chemical reactions depend on the presence of water, such as hydrolysis reactions that break down proteins and nucleic acids into individual subunits.

How do cells control their internal osmotic pressure? We have seen that water follows solutes across selectively permeable cell membranes. To regulate water and solute levels, and hence the osmotic pressure inside cells, the cell controls the solute concentration inside the cell relative to the solute concentration outside the cell. At the level of the cell, then, osmoregulation is achieved by the movement of solutes, particularly electrolytes.

At the level of the organism, osmoregulation is achieved by balancing input and output of water and electrolytes. Let's first consider how animals gain and lose water. Both terrestrial and freshwater animals gain water by drinking water that is less concentrated in solutes (hypotonic) than their body fluids. Animals also gain water through the food they eat. Finally, animals produce water as a product of cellular respiration (Chapter 7). This is the primary source of water for some animals that have become adapted to deserts and oceans, where drinkable water is not readily available. Animals lose water in their urine and feces. Terrestrial animals lose water through evaporation from their lungs, and, in the case of humans, by sweating. Freshwater fishes gain water through their gills, whereas most marine fishes lose water through their gills.

Now let's consider how animals gain and lose electrolytes. Animals gain electrolytes in the food they eat. Marine animals gain electrolytes by drinking salt water, which has a higher solute concentration than their body fluids. Marine animals also gain electrolytes as hypertonic water moves across their gills. In contrast, freshwater animals lose electrolytes through diffusion across their gills into the hypotonic watery environment. Humans lose electrolytes as well as water when they sweat. Drinking beverages high in electrolytes is helpful before engaging in demanding physical activities—including sports—because sweating during these activities can result in substantial water and electrolyte loss. Electrolytes are also lost by specialized glands (discussed later in this chapter) and in urine and feces.

Given these inputs and outputs, there are two ways in which animals maintain homeostasis of water and electrolytes: by matching their internal osmotic pressure to that of their external environment or by using energy to maintain an internal osmotic pressure different from that of their external environment.

Osmoconformers match their internal solute concentration to that of the environment.

Some animals keep their internal fluids at the same osmotic pressure as the surrounding environment. These animals are called **osmoconformers**. Because the solute concentrations inside and outside their bodies are similar, osmoconformers do not spend a lot of energy regulating osmotic pressure. However, they must adapt to the solute concentration of their external environment. Osmoconformers tend to live in environments that have stable solute concentrations, such as seawater, because maintaining stable internal solute concentrations is easier in stable environments.

Although osmoconformers generally match the overall concentration of solutes within their tissues with the overall concentration of solutes in their external environment, they often expend energy to regulate the concentrations of particular ions, such as sodium, potassium, and chloride, and particular molecules, such as amino acids and glucose. For example, whereas the intracellular space of nearly all animals has a relatively high concentration of potassium ions and a low concentration of sodium ions, the extracellular space has concentrations that are just the opposite: it is low in potassium and high in sodium. This means that these cells are actively pumping sodium out and potassium in (Chapter 5).

Most marine invertebrates, such as sea stars, mussels, lobsters, and scallops, are osmoconformers. These organisms maintain high concentrations of sodium and chloride in their cells to achieve an overall solute concentration close to that of seawater. Consequently, they have few specializations for osmoregulation beyond the need to regulate specific internal ion concentrations relative to their environment. Some marine vertebrates, such as hagfish and lampreys, are also osmoconformers that maintain high internal concentrations of particular electrolytes.

Other osmoconformers, including sharks, rays, and coelacanths, match seawater's solute concentration by maintaining a high internal concentration of a compound called urea. Urea is a waste product of protein metabolism that many animals excrete, thereby ensuring it does not build up to high concentrations (section 39.2). By retaining this solute, these marine vertebrates are able to achieve osmotic equilibrium with the surrounding seawater.

Osmoregulators have internal solute concentrations that differ from that of their environment.

The second way in which animals achieve water and electrolyte homeostasis is by maintaining an internal solute concentration

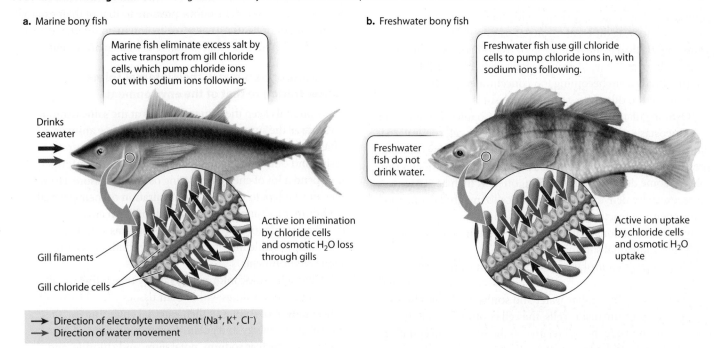

FIG. 39.2 Osmoregulators. Osmoregulators actively maintain an osmotic pressure different from that of their environment.

that is different from that of the environment. The animals that rely on this mechanism are called **osmoregulators**. Osmoregulators expend considerable energy pumping ions across cell membranes to regulate the movement of water into or out of their bodies. Some freshwater and marine fishes that are osmoregulators are estimated to expend 50% of their resting metabolic energy on osmoregulation. However, the ability to regulate osmotic pressure allows osmoregulators to live in diverse environments—in salt water, in fresh water, and on land.

Teleosts (also called bony fishes; they are the largest group of marine vertebrates) are osmoregulators. So are all freshwater and terrestrial animals. Marine bony fishes maintain concentrations of electrolytes much lower than those in the surrounding seawater, whereas freshwater fishes and amphibians maintain concentrations that are higher than those in the surrounding fresh water.

Let's consider how osmoregulators achieve osmotic balance in salt water, in fresh water, and on land. In salt water, marine organisms live in an environment that is hypertonic relative to their internal fluids, such as blood and the cytoplasm of the cell. These organisms maintain internal electrolyte concentrations lower than that of the surrounding salt water. As a consequence, they face the problem of water loss (across their gills, as well as in their urine and feces) and electrolyte gain (across their gills and in seawater that they drink). To manage this balance, they need to take in as much water as possible and eliminate excess electrolytes. They obtain water primarily through the food they eat and cellular respiration. They avoid water loss by producing concentrated urine (discussed in section 39.2).

Bony fishes have evolved specialized gills that allow them both to excrete excess electrolytes and to breathe. The thin gill filaments, with their extensive surface area, are useful sites for electrolyte transport out of the body in addition to being well adapted for gas exchange (**Fig. 39.2**). Specialized chloride cells in the gills pump chloride ions into the surrounding seawater. By pumping these negatively charged ions out of the body, the chloride cells create an electrical gradient that is balanced by positively charged sodium ions moving out of the body as well (Fig. 39.2a).

In fresh water, the environment is hypotonic relative to an organism's internal fluids. In this case, the problem is just the opposite of that faced by marine organisms: water gain and salt loss. To deal with this challenge, freshwater organisms tend to minimize their water intake and maximize water elimination. For example, in contrast to marine fishes, they do not drink water. In addition, freshwater fishes produce very dilute urine (discussed in section 39.2).

To maintain electrolyte concentrations higher than that of the surrounding water, freshwater fishes have gill chloride cells with reversed polarity relative to marine bony fishes. Their chloride cells pump chloride ions (with sodium ions following) into the body to counter their ongoing loss from the gills into the surrounding fresh water (Fig. 39.2b).

Nearly all amphibians also live in freshwater environments, so they face challenges similar to those experienced by freshwater fishes. To maintain homeostasis, amphibians take up electrolytes across their skin and produce dilute urine.

On land, the challenge is water loss. Terrestrial animals lose water in their urine and feces, through evaporation from their lungs, and, in humans, by sweating. To counter these losses, they drink water that is hypotonic relative to their body fluids. They also acquire water by eating food and by producing water during cellular respiration. Finally, terrestrial animals produce concentrated urine to minimize their water loss.

Many marine birds have evolved specialized nasal salt glands to rid themselves of the high salt content obtained through their diet (**Fig. 39.3**). These salt glands actively pump ions from the blood into the cells that make up the gland; the birds then excrete salt from their bodies through this gland and the nasal cavity. Their salt glands allow the birds to gain net water intake by drinking seawater. In contrast, humans and many other animals not adapted to marine life are incapable of deriving useful net water gain by drinking seawater because they cannot eliminate salt in high enough concentrations. The high concentration of magnesium sulfate in seawater results in diarrhea if too much seawater is consumed, exacerbating the net loss of water in the feces.

An extreme example of the ability of an animal to regulate water and electrolyte levels is provided by the life cycle of salmon. Mature salmon swim in the salt water of the open ocean but return upriver to spawn in fresh water. During the course of their return, they swim from high-salt ocean water through water of intermediate saltiness into freshwater rivers and streams that have little salt. After spawning, the young feed and grow in fresh water for up to 3 years, eventually returning to the open ocean, where they live until they become reproductively mature and return to their home river system to spawn again.

These changes in aquatic habitat demand that salmon make dramatic adjustments in the way they handle water and solutes. When in the ocean, salmon must avoid loss of water and uptake of electrolytes. The challenge is reversed when mature salmon migrate upriver to spawn and as young salmon continue to grow in freshwater streams and rivers: they must avoid gaining too much water and losing critical electrolytes across their gills. In salmon, radical changes in osmoregulation are required over the course of an animal's lifetime to enable it to reproduce in freshwater streams and return to the sea for further growth and maturation. Next, we explore how gills and kidneys meet these challenges.

Self-Assessment Questions

1. Given a selectively permeable membrane that is permeable to water but not to a particular solute, and given that there are different solute concentrations on the two sides of the membrane, draw a diagram showing the direction of water movement. Label the side with the higher osmotic pressure.

2. What happens when a cell is placed in a hypotonic solution (one with a solute concentration lower than that of the cell)? What happens when a cell is placed in a hypertonic solution (one with a solute concentration higher than that of a cell)?

3. A solution of water and dissolved ions is separated by a selectively permeable membrane that allows the passage of water but not small ions. The concentration of sodium is higher on one side of the membrane than on the other side. However, there is no net movement of water across the membrane. How is this possible?

4. How do animals gain and lose water and electrolytes?

5. Name two animals that are osmoconformers and two animals that are osmoregulators. What is the difference between the two types of animals?

39.2 EXCRETION OF WASTES

In most organisms, the regulation of water and solute levels is closely tied to waste excretion. Excretion is the elimination of waste products and toxic compounds from the body. Some of these wastes are generated as by-products of metabolism, particularly protein and nucleic acid breakdown. Others were originally ingested, like the undigested remains of food.

Most animals have evolved specialized excretory organs, such as the vertebrate kidney, to isolate, store, and eliminate

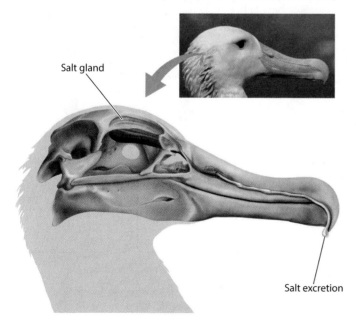

FIG. 39.3 Salt excretion from glands. Many marine birds have glands on the top of the skull that actively excrete salt. *Photo source: David Wall/Alamy.*

toxic compounds and metabolic waste products. Because these compounds are eliminated as solutes that have been dissolved and transported within the blood or other body fluids, their elimination from the body is intimately tied to water and electrolyte balance. The excretory organs of most animals therefore have two functions: they maintain water and electrolyte balance and they eliminate waste products.

Both the gastrointestinal tract (Chapter 38) and excretory organs like the vertebrate kidneys eliminate wastes from the body. However, the gastrointestinal tract is continuous with the outside world, so its role in the elimination of wastes is quite different from that of excretory organs that separate toxic compounds from essential nutrients, electrolytes, and cells circulating in the blood.

The excretion of nitrogenous wastes is linked to an animal's habitat and evolutionary history.

When proteins and nucleic acids are broken down, one of the by-products is ammonia (NH_3). Ammonia contains nitrogen and is toxic to organisms, so it is a form of **nitrogenous waste** (Fig. 39.4). By far, most ammonia is produced from the breakdown of proteins. Ammonia can disturb the pH balance of cells and damage neurons. At high levels, it is lethal to organisms. It is either eliminated from the body or converted into a less toxic form.

Some organisms, such as fishes, are able to excrete ammonia directly into the surrounding water, primarily through their gills. As a result, ammonia does not accumulate to toxic levels in their bodies. This process is not energetically expensive, but it requires high volumes of water in the surrounding environment to dilute the highly toxic ammonia.

When animals moved from water to land, they faced the challenge of how to eliminate waste. No longer could they rely on simple diffusion of ammonia into the surrounding water. Instead, they evolved biochemical pathways to convert ammonia to less toxic forms. One of these forms is urea (Fig. 39.4). Because urea is less toxic than ammonia, it can be stored in a fairly concentrated form before being excreted. Mammals (including humans), many amphibians, sharks, and some bony fishes excrete nitrogenous waste in the form of urea. In mammals, urea is produced in the liver and carried by the blood to the kidneys, from which it is eliminated. Urea is less toxic than ammonia but requires energy to produce and water to eliminate.

Ammonia can also be converted into uric acid (Fig. 39.4). Uric acid is the least toxic of the three forms of nitrogenous waste, so animals can store it at a higher concentration than urea. Birds and many reptiles, arthropods (including insects), and land snails excrete nitrogenous waste in the form of uric acid. These groups evolved the ability to convert ammonia to uric acid independently of one another as an adaptation to living on land, providing a good example of convergent evolution (Chapter 22). Uric acid precipitates from solution and

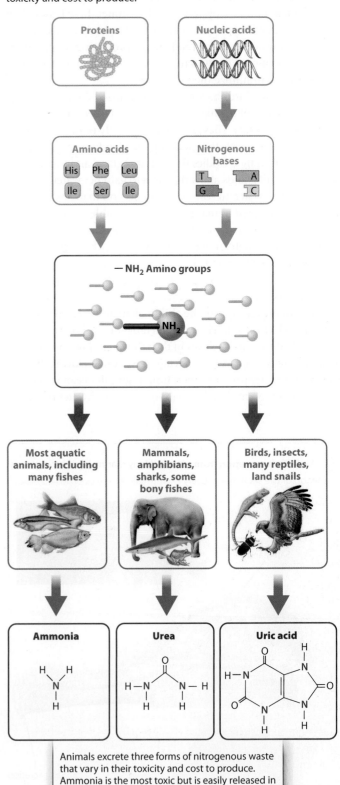

FIG. 39.4 Nitrogenous waste. Animals excrete nitrogenous waste in three major forms—ammonia, urea, and uric acid—that vary in their toxicity and cost to produce.

Animals excrete three forms of nitrogenous waste that vary in their toxicity and cost to produce. Ammonia is the most toxic but is easily released in water. Uric acid is the least toxic, but is energetically costly to produce.

forms a semisolid paste (familiar as bird droppings). Because this uric acid is not dissolved in water, it does not exert osmotic pressure and is eliminated with minimal water loss. By limiting water loss, many reptiles and insects are able to live in extremely hot and dry environments. However, uric acid is energetically expensive to produce.

Many amphibians, including frogs, excrete ammonia early in their development, when they live in water, then excrete urea when they move onto land after metamorphosis. Alligators and turtles also adjust what they excrete: ammonia when in the water and uric acid when on land.

Ammonia, urea, and uric acid are the three major forms of nitrogenous wastes, but they are not the only forms. Animals excrete nitrogen as trimethylamine oxide, creatine, creatinine, and even amino acids. The form of nitrogenous waste that an animal excretes is ultimately linked to its environment and evolutionary history.

Excretory organs work by filtration, reabsorption and secretion.

Organisms eliminate waste and other compounds from the blood through two mechanisms: filtration and secretion. **Filtration** is the process by which the blood is passed into an extracellular space, with some substances being allowed to pass through the blood vessel wall into the space but others being retained in the blood. **Secretion** is the active transport of molecules from the blood into the extracellular space. Filtration is energetically inexpensive and can be used to remove novel compounds, but it also removes compounds that the body needs, such as water, electrolytes, and essential nutrients. Secretion is specific but energetically expensive.

Initially, filtration is used to isolate most wastes, but **reabsorption** is required to retain filtered compounds that the body needs. Reabsorption is the process by which essential molecules are transported back into the blood. Reabsorption and secretion together help to balance water and electrolyte levels, producing dilute or concentrated urine as required by the animal.

The first step of excretion is to isolate the waste into an extracellular space. In the case of unicellular organisms such as *Paramecium* and simple multicellular organisms such as sponges and cnidarians (jellyfish and sea anemones), wastes are isolated in a cellular compartment called a **contractile vacuole** (**Fig. 39.5**). Fusion of the contractile vacuole with the cell membrane eliminates the waste contents by exocytosis from the cell (Chapter 5).

Animals with pressurized circulatory systems (annelid worms, mollusks, and vertebrates) isolate wastes from the blood by filtration (**Fig. 39.6**). Circulatory pressure pushes fluid containing the wastes through specialized filters into an extracellular space. The filtered fluid, called the filtrate, contains waste products along with water, electrolytes, and other solutes. The filtrate drains into **excretory tubules** that connect to the

FIG. 39.5 Contractile vacuole of a paramecium when (a) full and (b) after emptying. *Source: Michael Abbey/Science Source.*

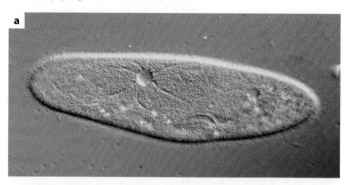

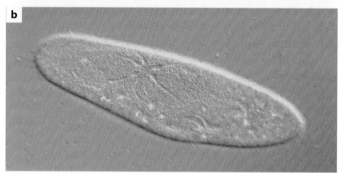

FIG. 39.6 Filtration, reabsorption, and secretion. These three processes eliminate wastes and provide osmoregulation in animals with pressurized circulatory systems.

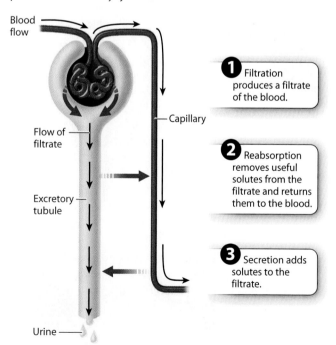

1. Filtration produces a filtrate of the blood.
2. Reabsorption removes useful solutes from the filtrate and returns them to the blood.
3. Secretion adds solutes to the filtrate.

outside. The excretory filter, like a cell's plasma membrane, is selectively permeable. It allows smaller ions and solutes to pass into the tubule along with water, but retains larger solutes such as proteins and cells.

Filtration is followed by reabsorption of key ions and solutes from the filtrate into the blood (Fig. 39.6). Cells lining the excretory tubule take up substances that are needed by the animal, such as electrolytes, amino acids, vitamins, and simple sugars, and return them to the bloodstream. Reabsorption can be an active process, requiring active transporters for some molecules, or it can be a passive process involving diffusion for other molecules. Water is reabsorbed into the cells by osmosis, driven by the osmotic pressure established by the active transport of solutes into the cell.

The final step in waste removal is the secretion of additional toxic compounds and excess ions (Fig. 39.6). Secretion is an active process, carried out by protein transporters in the plasma membranes of cells lining the tubules. Using the energy of ATP, these transporters pump molecules and ions from the blood into the excretory tubules. Secretion eliminates substances that were not filtered from the blood, and it helps to fine-tune electrolyte levels in the blood.

FIG. 39.7 Excretory organ of a flatworm.

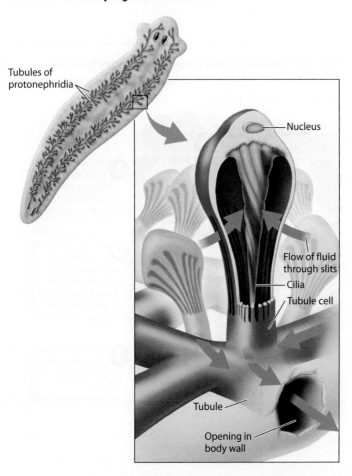

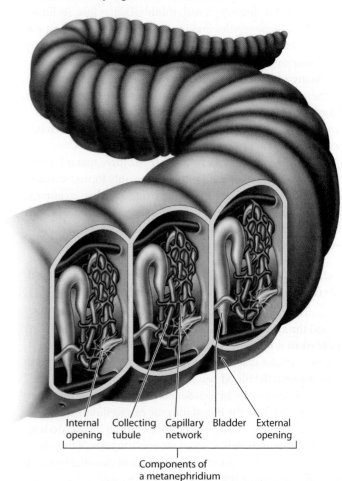

FIG. 39.8 Excretory organ of an earthworm.

In summary, filtration, reabsorption, and secretion are the three steps that almost all animal excretory systems perform to eliminate toxic wastes, retain essential nutrients, and regulate water and electrolyte levels.

Animals have diverse excretory organs.

As we have seen, the processes of filtration, reabsorption, and secretion serve two important and complementary functions: they isolate wastes that are then removed from the body, and they allow the organism to adjust the amounts of water and electrolytes required to meet its osmoregulatory needs. These two functions are carried out by diverse excretory organs adapted to the needs of specific organisms.

Consider waste excretion in flatworms, small flattened animals that glide along the streambed floor. Flatworms have no pressurized circulatory system, so waste is excreted without being filtered: the fluid from the body cavity is isolated in excretory organs called **protonephridia** (Fig. 39.7). Cells with beating cilia move the fluid down the lumen of a tubule to an excretory pore. Freshwater flatworms, such as *Planaria*, have an extensive network of tubules that eliminate excess fluid along with waste products. As the fluid passes down the tubule, cells lining the tubule modify the fluid's contents through reabsorption and secretion. As is typical of freshwater animals, the urine leaving a freshwater flatworm is less concentrated (that is, more dilute) than its internal body fluids to balance the water the animal ingests as it feeds.

Segmented annelid worms such as earthworms do have a pressurized circulatory system. In this case, the blood is filtered through small capillaries into a pair of excretory organs called **metanephridia** within each body segment (**Fig. 39.8**). The filtrate enters a metanephridium at a funnel-shaped opening surrounded by cilia. The cilia direct the filtrate into a series of tubes in the next body segment and then out of the animal through an excretory pore. Blood vessels and cells surrounding the tubules reabsorb and secrete additional compounds as urine is formed. Like freshwater flatworms, earthworms reabsorb electrolytes and eliminate dilute urine containing dissolved nitrogenous wastes.

In insects and other terrestrial arthropods, body fluid called hemolymph passes from the main body cavity into a series of tubes called **Malpighian tubules** that empty into the hindgut (**Fig. 39.9**). The cells lining the tubules actively secrete uric acid and excess electrolytes into the lumen of the tubules while retaining key solutes. Muscle fibers in the walls of the tubules help push the fluid to the hindgut. In the hindgut, the fluid becomes increasingly acidic as uric acid builds up, causing uric acid to precipitate from solution when it enters the rectum. Because uric acid is removed from the fluid, it does not influence the movement of water. Water is therefore free to diffuse back into the tissues as electrolytes are reabsorbed from the rectum. The uric acid and remaining dried digestive wastes are then eliminated. Malpighian tubules are one of the most successful adaptations for excreting nitrogenous wastes with minimal water loss. They enable insects to live in diverse terrestrial environments, including some extremely arid areas.

Vertebrates filter blood under pressure through paired kidneys.

The excretory organs of vertebrates are the kidneys. Like all excretory organs, they both eliminate wastes and play an important role in maintaining water and electrolyte balance. Moreover, vertebrate kidneys reflect the evolutionary transition of vertebrates from water to land, where there is a need to minimize water loss. Mammals and birds, for example, can excrete urine that is more concentrated in solute than their blood and tissues.

The kidneys lie posterior to the abdominal cavity. The kidneys of humans and other mammals are bean-shaped, whereas the kidneys of birds and other vertebrate animals are elongated and segmented (**Fig. 39.10**). Vertebrate kidneys receive blood from the heart through paired renal arteries ("renal" means "relating to the kidneys") that branch from the abdominal aorta.

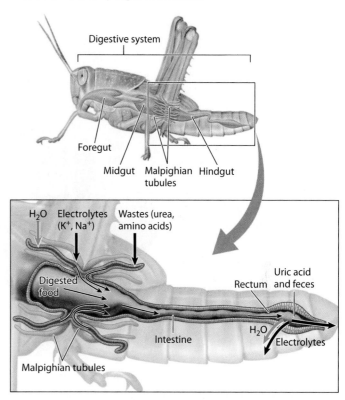

FIG. 39.9 Excretory organ of an insect.

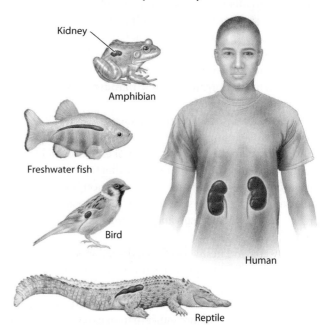

FIG. 39.10 Location and shape of kidneys in different vertebrates.

Vertebrates filter their blood through specialized capillaries that have openings in their walls. These porous capillaries form a tufted loop called a **glomerulus** (**Fig. 39.11**). There are thousands to millions of glomeruli in each kidney. Driven by circulatory pressure, nitrogenous wastes, electrolytes, other

FIG. 39.11 Vertebrate kidney and nephron.

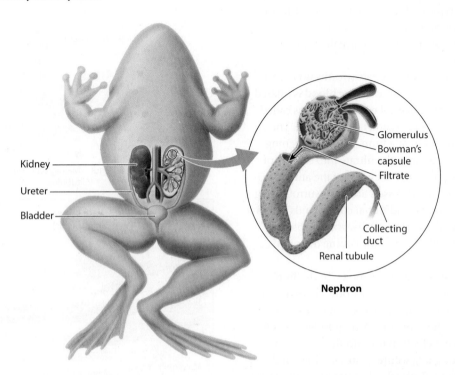

small solutes, and water move through small holes in the capillary walls into an extracellular space surrounded by a capsule (called Bowman's capsule after the surgeon William Bowman, who identified it in 1842). The filtrate then moves through a series of **renal tubules**, which further process the filtrate by reabsorption and secretion. Finally, the filtrate enters the **collecting ducts** as urine. The collecting ducts converge on a larger tube called the **ureter**, which brings urine from the kidneys to a hollow organ called the **bladder** for storage and elimination from the body.

The glomerulus, capsule, renal tubules, and collecting ducts make up a **nephron**, the functional unit of the kidney. Nephrons perform the three basic steps of excretion and osmoregulation: filtration of blood passing through the glomerulus, reabsorption of key electrolytes and solutes from the renal tubule and return of these substances to the bloodstream, and secretion of additional wastes and electrolytes by cells lining the renal tubules.

In saltwater and freshwater fishes, the kidneys run along the length of the body, and the nephrons are arranged segmentally. Because fishes excrete nitrogenous waste in the form of ammonia through their gills, the kidney is primarily an organ of water and electrolyte balance rather than one of waste elimination. Recall that saltwater and freshwater fishes face different challenges in achieving water balance. Saltwater fishes must eliminate excess electrolytes that enter through their gills and in the seawater that they drink, whereas freshwater fishes must eliminate excess water that enters through their gills. In saltwater fishes, the walls of the renal tubules are readily permeable to water: water diffuses out of the tubule by osmosis, and the urine that remains is relatively concentrated (high concentration of electrolytes and low concentration of water). By contrast, the walls of the renal tubules in freshwater fishes are relatively impermeable to water, but readily allow reabsorption of electrolytes, sugars, and amino acids. Because water is retained in the tubules, freshwater fishes excrete very dilute urine, thereby eliminating the excess water that continually enters their bodies through their gills.

The kidneys of amphibians are similar to those of freshwater fishes. Amphibians take in excess water across their gills and skin as they breathe. Amphibian kidneys therefore produce extremely dilute urine, enabling them to maintain an electrolyte concentration in their body that exceeds that of their freshwater environment.

Reptiles are found in a variety of habitats, from oceans to deserts, and are well adapted to their environments. Marine iguanas, for example, ingest salt water with their food, so they rely on specialized salt glands to eliminate excess electrolytes from their body. Terrestrial reptiles living in desert regions with limited water availability face the challenge of minimizing water loss. Because reptiles excrete nitrogenous waste in the form of uric acid, they are able to reabsorb most of the water from the urine.

Mammals also live on land, but they excrete most of their nitrogenous waste in the form of urea dissolved in water. Their kidneys are well adapted for excreting waste and balancing their water and electrolyte levels, as we discuss in the next section.

> **Self-Assessment Questions**
>
> 6. What are three forms of nitrogenous waste? How is each an adaptation for the environment in which an animal lives?
> 7. Why do mammals convert ammonia to urea rather than simply excreting it, as fishes do?
> 8. What are the three steps in which animals excrete wastes?

39.3 THE MAMMALIAN KIDNEY

Mammals excrete urea dissolved in solution, but they must be able to do so without losing too much water, an adaptation to living on land. To minimize the loss of water, they remove water from the waste-containing fluid by osmosis as it moves through the kidney. In this way, the urine is concentrated without the kidney ever actively transporting water itself.

The cells of the kidneys are surrounded by a solution called interstitial fluid. Mammals concentrate their urine by generating concentrated (hypertonic) interstitial fluid deep in the kidney, through which the collecting ducts pass. Then, as the filtrate passes through the collecting ducts, water moves out of collecting ducts and into the interstitial fluid by osmosis, leaving concentrated urine in the collecting ducts.

This process, which is quite efficient, allows animals to thrive in diverse environments. For example, the giant kangaroo rat shown in **Fig. 39.12** is a desert mammal; it reabsorbs so much water that it does not need to drink. It can obtain enough water from seeds and cellular respiration. The giant kangaroo rat is thought to produce the most concentrated urine of all mammals. Marine mammals, such as whales, dolphins, and seals, also have an exceptional ability to concentrate their urine. They can remain at sea for long periods of time, but cannot

FIG. 39.12 A giant kangaroo rat (*Dipodomys ingens*). Giant kangaroo rats do not drink water. Instead, they obtain water through cellular respiration and from the seeds they eat, which contain small amounts of water. *Source: Tom McHugh/Science Source.*

drink seawater for net water uptake. They, too, rely on water produced by cellular respiration in combination with water in the food they eat to remain hydrated.

In this section, we examine the mammalian kidney in more detail. The kidneys of birds share many features with those of mammals. These features arose independently in birds and mammals. Thus they are the result of convergent evolution, presumably occurring as adaptations for efficient processing of wastes in endothermic vertebrates with high metabolic rates (Chapter 38).

The mammalian kidney has an outer cortex and inner medulla.

The paired kidneys of mammals are organized into two layers: an outer layer called the **cortex** and a deeper layer, surrounded by the cortex, called the **medulla** (**Fig. 39.13**). Large numbers

FIG. 39.13 The organization of the mammalian kidney and nephron.

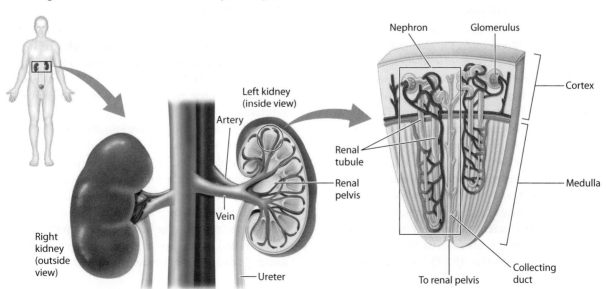

FIG. 39.14 The mammalian glomerulus and Bowman's capsule. Three layers create the filter: endothelial cells, basal lamina, and podocytes.

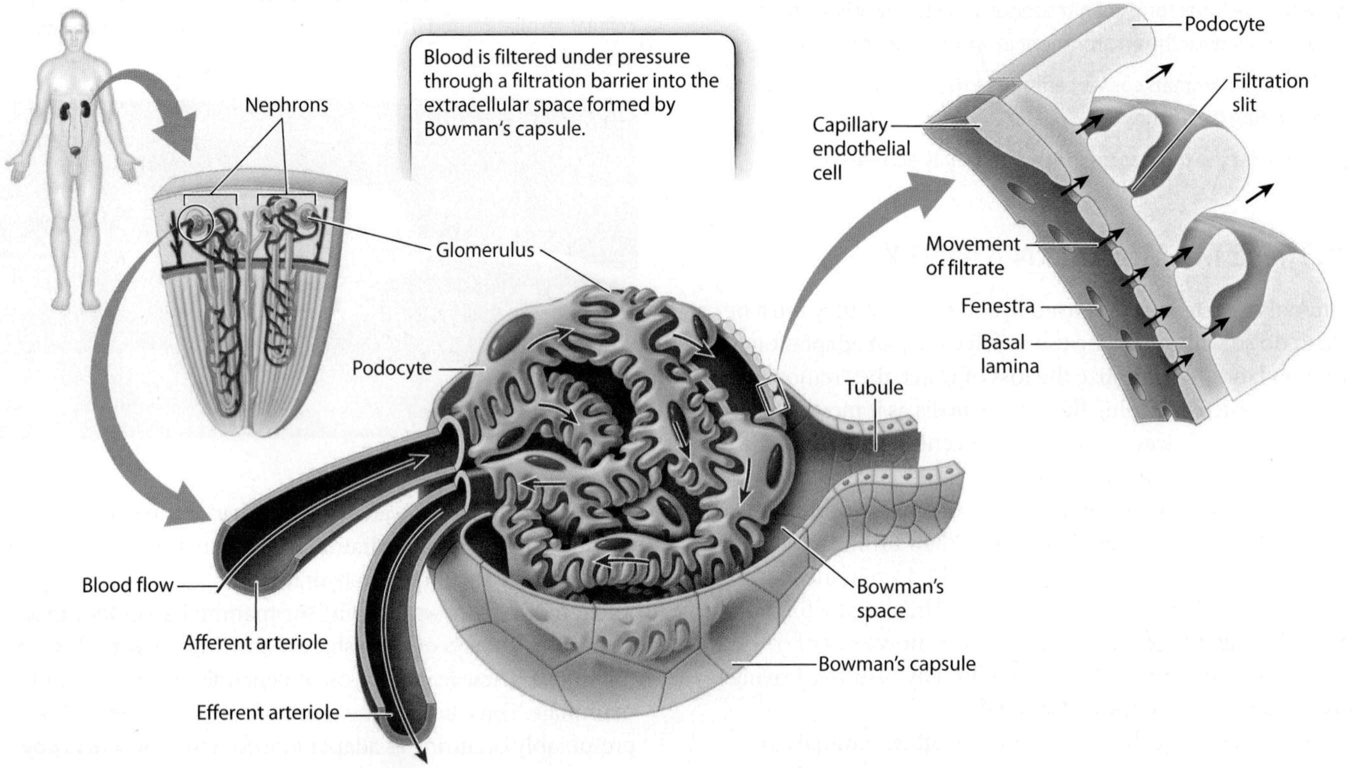

of nephrons are organized into wedge-shaped pyramids. Each nephron's glomerulus is located within the renal cortex or outer layer of the medulla and feeds into a renal tubule that starts in the renal cortex, then dips into the medulla at the base of the pyramid, and finally loops back up to the cortex. The renal tubules drain into the collecting ducts. Those ducts also descend into the pyramid's base, emptying into the renal pelvis and draining through the ureter into the bladder.

Glomerular filtration isolates wastes carried by the blood along with water and small solutes.

Blood is filtered as it passes through the tuft of capillaries forming the glomerulus. Blood enters the glomerulus by an afferent ("toward") arteriole and leaves by an efferent ("away") arteriole (**Fig. 39.14**). The tuft of capillaries is encased in a membranous sac called **Bowman's capsule**, and the space enclosed by the capsule is called Bowman's space. The capillary wall and Bowman's capsule together act as the filter: water, wastes, and solutes from the blood pass through the capillary wall and capsule into Bowman's space.

To filter the blood, the capillaries of the glomerulus have a different structure from that of capillaries elsewhere in the body. The cells that line the capillaries, known as endothelial cells, have pores called fenestrae (a term that means "windows") that allow fluid to pass through. The filtration barrier of the glomerulus is made up of three layers (Fig. 39.14): the endothelial cells, a basal lamina, and a layer of cells with footlike projections called **podocytes**, which loosely interlock to create slits in the cell layer. Together, these three layers create a filter that allows small molecules to pass through but blocks the passage of large proteins and cells, so that they remain in the blood. As a result, wastes such as urea, as well as water and solutes such as electrolytes, glucose, and amino acids, all end up in the filtrate. After passing into Bowman's space, the filtrate enters the renal tubule.

The proximal convoluted tubule reabsorbs solutes by active transport.

The renal tubule of the mammalian kidney is divided into three sections: the **proximal convoluted tubule**, the **loop of Henle**, and the **distal convoluted tubule** (**Fig. 39.15**). Although the renal tubule is one long tube, these three sections are specialized for different functions because they are permeable to different substances.

The filtrate initially moves into the first portion of the renal tubule, the proximal convoluted tubule (Fig. 39.15). Here, electrolytes and other nutrients that the animal requires are reabsorbed into the blood. To maximize absorption, the proximal convoluted tubule has thick walls composed of epithelial cells with fingerlike projections called **microvilli** on their surfaces, similar to those found in the small intestine (Chapter 38). The microvilli form a brush border along the inside (lumen) of the tubule, greatly increasing the surface area

FIG. 39.15 The organization of the mammalian renal tubules and collecting duct.

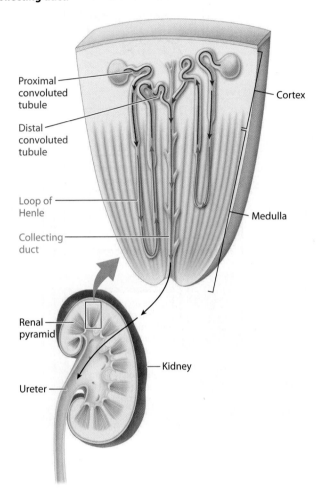

available for solute and water reabsorption. The epithelial cells also possess numerous mitochondria that produce the ATP needed to actively reabsorb solutes from the filtrate into the bloodstream.

The proximal tubule reabsorbs all the glucose and amino acids filtered by the glomerulus, as well as most sodium and chloride ions. Because the proximal tubule is permeable to water, water diffuses by osmosis in the same direction as the electrolytes and solutes. By the time the filtrate leaves the proximal convoluted tubule, approximately 75% of the water, along with most of the electrolytes and solutes, has been reabsorbed into the bloodstream. What is left is primarily water, urea, and electrolytes.

The loop of Henle acts as a countercurrent multiplier to create a concentration gradient from the cortex to the medulla.

The proximal convoluted tubule connects to the loop of Henle, the second portion of the renal tubule (Fig. 39.15). Although many nephrons contain short loops of Henle that are confined to the renal cortex, others descend all the way into the medulla before looping back up to the cortex. These longer loops of Henle play an important role: they create a concentration gradient in the interstitial fluid from the cortex to the medulla, so that the interstitial fluid of the cortex is less concentrated in solutes and the interstitial fluid of the medulla is more concentrated in solutes. This concentration gradient allows water passing through the downstream collecting duct to be reabsorbed by osmosis, producing concentrated urine.

How is the gradient generated? The key to understanding how the loop of Henle creates a concentration gradient is to recognize that the two limbs of the loop of Henle run in parallel but opposite directions, and that they differ in their permeability to water and ability to actively transport electrolytes out of the tubule. Although the filtrate enters the descending limb before looping around to the ascending limb, it is easier to consider the ascending limb first.

The thick portion of the ascending limb of the loop of Henle is impermeable to water but not electrolytes (**Fig. 39.16**). As the filtrate moves up the thick ascending limb of the loop of Henle, electrolytes are actively transported out of it into the surrounding interstitial fluid. As a result, the interstitial fluid becomes more concentrated, and the filtrate passing through the ascending limb becomes less concentrated.

In contrast to the ascending limb, the descending limb is permeable to water. Water moves passively out of the descending limb, driven by osmosis due to the electrolytes pumped into the interstitial fluid by the ascending limb. As water moves out of the descending limb, the filtrate becomes increasingly more concentrated, matching the concentration in the surrounding interstitial space. We have already seen that, as electrolytes are pumped out of the thick ascending limb, the filtrate becomes less concentrated. Thus, the filtrate enters and leaves the loop of Henle at approximately the same concentration. Although the filtrate concentration is unchanged, the trip around the loop creates a concentration gradient in the interstitial fluid surrounding the loop of Henle. Notably, fluid in the inner medulla is much more concentrated than fluid in the outer cortex.

The gradient results from the combination of active transport of electrolytes in the ascending limb, osmosis of water in the descending limb, and the movement of the filtrate through the loop of Henle. The filtrate is most concentrated at the base of the loop, and the concentration of solutes inside and outside the loop is roughly the same, leaving the medulla more concentrated than the cortex.

In Chapter 37, we saw how the movement of water and blood in opposite directions in fish gills provides an efficient way to extract oxygen from the water and move it into the blood. This mechanism, called countercurrent exchange, is an efficient mechanism by which two fluids can exchange properties such as amount of oxygen. In the loop of Henle, active transport of electrolytes creates a concentration gradient, which is then

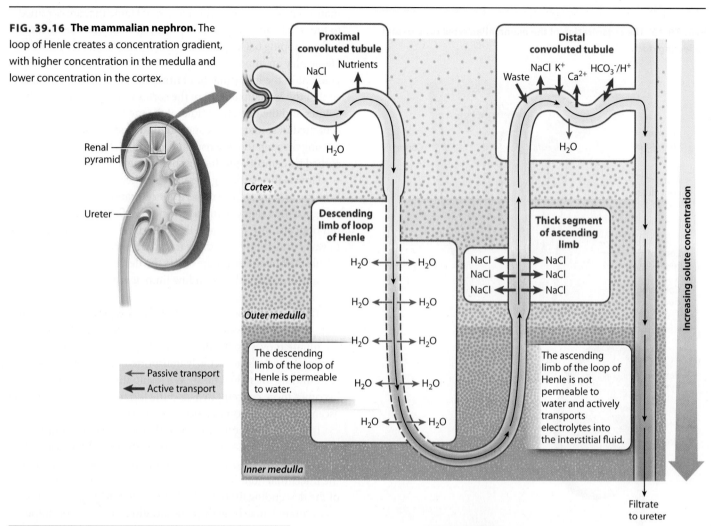

FIG. 39.16 The mammalian nephron. The loop of Henle creates a concentration gradient, with higher concentration in the medulla and lower concentration in the cortex.

FIG. 39.17 Vasa recta. The blood in the vasa recta moves in a countercurrent fashion, just like the filtrate in the loop of Henle, to maintain the concentration gradient from the cortex to the medulla.

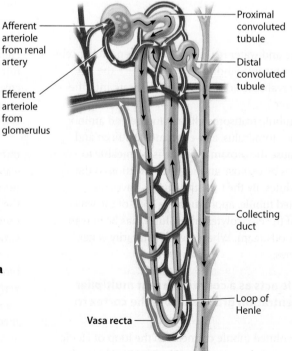

multiplied because the descending and ascending limbs move in parallel but opposite directions. This system is known as a **countercurrent multiplier**. While countercurrent exchangers maintain a concentration gradient, countercurrent multipliers generate them.

Typically, a high concentration of solutes in the interstitial fluid would be quickly lost as water flows out of blood vessels by osmosis. To avoid dissipating the concentration gradient, the blood vessels in the kidneys are organized in a unique pattern. The blood vessels, collectively called the **vasa recta**, have descending and ascending vessels arranged in a countercurrent organization, similar to the loops of Henle (**Fig. 39.17**). The blood within these vessels flows down into the medulla in one arm and back up through the cortex in the other. This arrangement maintains the concentration

HOW DO WE KNOW?

FIG. 39.18

How does the mammalian kidney produce concentrated urine?

BACKGROUND The mammalian kidney can produce urine that is more concentrated than the blood. How it concentrates urine was one of the most perplexing problems in renal physiology.

HYPOTHESIS One hypothesis is that the kidney actively transports water out of the collecting ducts as the filtrate leaves the kidney. Another hypothesis, first suggested by the Swiss chemist Werner Kuhn in the early 1950s, is that the loop of Henle creates a concentration gradient in the interstitial fluid surrounding the collecting ducts from the cortex to the medulla by a countercurrent multiplier mechanism. According to this model, the solute concentration of the interstitial fluid is low in the cortex and high in the medulla, so water leaves the collecting ducts passively as the filtrate moves from the cortex to the medulla.

PREDICTION Kuhn's hypothesis predicts that the solute concentrations of fluid taken at the same level from the medulla, loop of Henle, and collecting ducts should be similar to one another, and higher than that of the blood. If, instead, water is simply pumped out of the collecting ducts, then the fluid in the collecting ducts should be more concentrated than the fluid in the loop of Henle.

EXPERIMENT American physiologists Carl W. Gottschalk and Margaret Mylle tested Kuhn's hypothesis. They used innovative micropuncture techniques in hamsters to measure the solute concentration of fluid taken from the loop of Henle and the collecting ducts at the same level. They then compared the concentrations to that of blood plasma (the liquid portion of blood).

SOLUTE CONCENTRATION (MILLIOSMOLES PER KILOGRAM OF WATER)			
HAMSTER NUMBER	LOOP OF HENLE	COLLECTING DUCT	PLASMA
1	1391	1402	308
2	725	720	336
3	1270	1206	325
4	453	453	

RESULTS Their results, shown in the table, are consistent with the predictions of Kuhn's hypothesis. That is, the solute concentrations of the fluid in the loop of Henle and the fluid in the collecting ducts are similar to each other and higher than that of the blood plasma.

CONCLUSION These results are consistent with a model in which concentrated urine is produced by first generating a concentration gradient between the cortex and the medulla, and then allowing water to move out of the collecting ducts into the interstitial fluid by osmosis.

FOLLOW-UP WORK Similar measurements in other areas of the kidney and in other mammals and birds further supported the model.

SOURCES Gottschalk, C. W., and M. Mylle. 1958. "Evidence That the Mammalian Nephron Functions as a Countercurrent Multiplier System." *Science* 128:594; Gottschalk, C. W., and M. Mylle. 1959. "Micropuncture Study of the Mammalian Urinary Concentrating Mechanism: Evidence for the Countercurrent Hypothesis." *American Journal of Physiology* 196(4):927–936.

---

gradient. Blood flowing from the cortex to the base of the kidney encounters increasingly concentrated interstitial fluid, allowing water to diffuse passively out of the blood. The ascending vessels, which pass nearby, reabsorb the water. In effect, water passes straight from one arm of the vasa recta to the other, so the interstitial fluid and the bloodstream have no net gain or loss of water.

The generation of a concentration gradient by a countercurrent multiplier in the loop of Henle is a key step in producing urine that is more concentrated than the blood. It was first proposed by the Swiss chemist Werner Kuhn in the 1950s. Although his hypothesis was initially met with skepticism, experimental work confirmed its predictions (**Fig. 39.18**).

The distal convoluted tubule secretes additional wastes.

The distal convoluted tubule is located in the cortex. The filtrate entering the tube is dilute, having lost nearly all of its electrolytes in the ascending limb of the loop of Henle, as we have seen. Urea is the principal solute in the filtrate at this point.

The distal convoluted tubule is one of the main sites of secretion. Additional wastes not filtered by the glomerulus are actively secreted into the distal convoluted tubule from the bloodstream. In addition, the distal convoluted tubule participates in the regulation of key electrolytes, such as potassium, sodium, and calcium ions. Potassium ions are secreted from the blood into the tubule, whereas the few sodium and calcium ions left in the tubule are reabsorbed into the blood in response to hormones. Finally, the distal convoluted tubule helps to regulate blood pH: bicarbonate (HCO_3^-) and protons (H^+) are secreted into the tubule or reabsorbed out of it, as needed to maintain pH homeostasis.

As a result, the filtrate leaving the distal convoluted tubule and entering the collecting ducts is dilute and contains mainly urea and other wastes.

FIG. 39.19 Hormonal control of urine concentration by ADH. (a) In the presence of ADH, the collecting duct is more permeable to water than (b) in the absence of ADH.

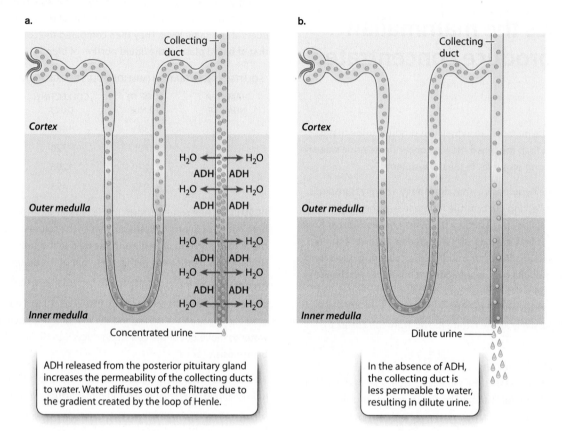

ADH released from the posterior pituitary gland increases the permeability of the collecting ducts to water. Water diffuses out of the filtrate due to the gradient created by the loop of Henle.

In the absence of ADH, the collecting duct is less permeable to water, resulting in dilute urine.

The final concentration of urine is determined in the collecting ducts and is under hormonal control.

The final segment of the nephron is the collecting duct. This is where the concentration gradient from cortex to medulla established by the loop of Henle becomes important: it allows the collecting ducts to retain or lose water as needed depending on the water levels of the organism. By this means, the kidney maintains water homeostasis.

The dilute filtrate enters the collecting ducts. These ducts pass from the cortex to the medulla and drain into the renal pelvis at the base of the kidney. In response to the concentration gradient from the cortex to the medulla, water moves out of the collecting ducts into the interstitial fluid by osmosis as long as the collecting ducts are permeable to water.

The permeability of the collecting ducts to water can be adjusted to make the urine either more dilute or more concentrated. The water permeability of the collecting ducts is controlled by **antidiuretic hormone (ADH)**, also called **vasopressin**. This peptide hormone is secreted by the posterior pituitary gland and is released when the body is dehydrated (Chapter 36). It acts to make the collecting ducts more permeable to water, so less water in lost in the urine.

The secretion of ADH is controlled by the hypothalamus. Receptors in the hypothalamus monitor the concentration of solutes in the blood. When a person is dehydrated, solute concentration is high, and the hypothalamus sends a signal to secrete ADH. In the presence of ADH, aquaporins are inserted into cell membranes (**Fig. 39.19a**). Water in the collecting duct diffuses into the interstitial fluid through the aquaporins, and the urine becomes more concentrated. When a person is dehydrated, the urine leaving the collecting ducts can become as concentrated as the interstitial fluid at the base of the kidney.

When a person is well hydrated, no ADH is released. The collecting ducts are then impermeable to water, and no water is reabsorbed (**Fig. 39.19b**). The result is dilute urine. Some diuretic substances (*diuretic* is from a Greek word meaning "to flow through"), such as alcohol, promote the dilution of urine by interfering with ADH production, thereby making the collecting ducts impermeable to water. When dilute urine is produced, there is net water loss. Consequently, diuretic drinks often cause dehydration.

The regulation of water and solute reabsorption by the kidney is remarkably precise. The kidneys of a human filter 180 L of blood each day (125 mL/min). The total blood volume of a human is about 5 L (of which 3 L is plasma), which means that the blood is filtered about 60 times per day. Humans produce, on average, 0.5 L of urine per day. Thus, under normal conditions, the kidneys reabsorb 99.9% of the water entering the renal tubules. In addition, the kidneys reabsorb all the glucose and amino acids that are filtered out of the blood, together with 99.9% of the electrolytes in the blood.

The exceptional ability of the kidneys to filter and reabsorb electrolytes, solutes, and water also explains why kidney damage can have such serious consequences. People with kidney disease must have their blood filtered by dialysis treatment as frequently as three times per week, with each treatment lasting several hours. Otherwise, toxic wastes will build up to dangerous levels in their bodies. Kidney disease remains a substantial health problem around the world, costing more than $34 billion in the United States alone.

The kidneys help regulate blood pressure and blood volume.

In another example of homeostasis, the body maintains blood pressure at relatively constant levels. Because the kidneys receive a large fraction of the blood leaving the heart, they are well suited for monitoring changes in blood pressure and helping to maintain blood pressure homeostasis. Low blood pressure can lead to dizziness and fainting, and high blood pressure contributes to heart disease, stroke, and kidney failure. The kidneys help to regulate blood pressure by secreting hormones that cause blood vessels to constrict and that result in the reabsorption of more electrolytes and water into the blood, increasing blood volume.

Specialized cells of the efferent arteriole leaving the glomerulus of each nephron are well positioned to sense and respond to changes in the blood pressure. These cells form the **juxtaglomerular apparatus (Fig. 39.20)**. In response to a drop in blood pressure, they secrete the hormone renin. Renin converts angiotensinogen, which is produced in the liver and

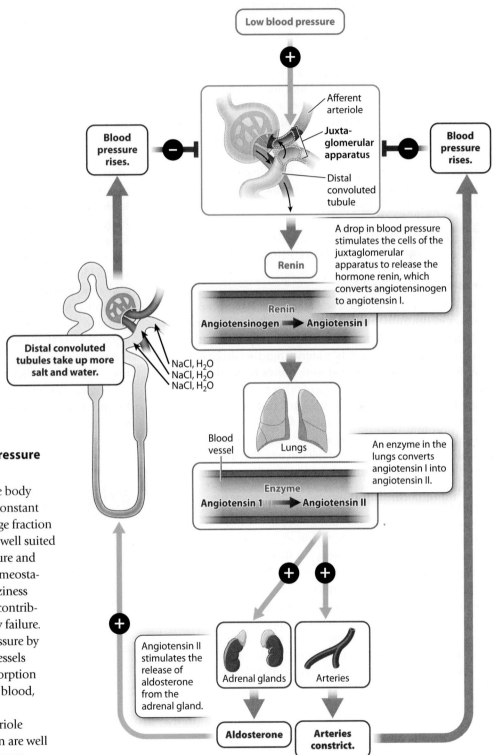

FIG. 39.20 Control of blood volume and blood pressure by the kidney. Cells of the efferent arteriole of the glomerulus sense blood pressure and start a hormone cascade when blood pressure gets too low.

circulates in the blood, into angiotensin I. Angiotensin I, in turn, is converted by an enzyme present in the lungs and elsewhere to its active form, **angiotensin II**. Angiotensin II acts on the smooth muscles of arterioles throughout the body, causing them to constrict; this action increases blood pressure and directs more blood back to the heart.

Angiotensin II also stimulates the release of the hormone **aldosterone** from the adrenal glands (Fig. 39.20). Aldosterone stimulates the distal convoluted tubules and collecting ducts to increase the reabsorption of electrolytes and water back into the blood. As blood volume increases, blood pressure increases, too.

Clearly, the kidneys play many roles in the body. In addition to eliminating wastes and other toxic compounds, they maintain homeostasis in several ways, including through the maintenance of water and electrolyte balance and the regulation of blood volume and blood pressure.

 CASE 7 BIOLOGY-INSPIRED DESIGN: USING NATURE TO SOLVE PROBLEMS

How does dialysis work?

As we have seen, the kidney serves many critical functions. It removes wastes from the bloodstream as well as maintains water and solute balance. These functions are so important that we cannot live without them. When kidney function drops to less than 10% to 15% of the normal level, wastes accumulate in the blood and disrupt the functioning of other organs, including the brain. In addition, water and solute homeostasis is not achieved, leading to high or low blood pressure, for example. The inability of the kidney to filter wastes and maintain water and solute balance is called renal failure, and it can be either a short-term (acute) or long-term (chronic) condition.

Chronic renal failure is currently treated in one of two ways: transplantation or dialysis. Transplantation is the surgical implantation of a healthy kidney in place of one that is damaged by disease. Donor kidneys can come from living or recently deceased donors, but their supply is limited. Transplantation also requires a proper immunological match between the donor and the recipient, so that the recipient's body will not reject the transplanted kidney (Chapter 41).

Many people with chronic renal failure depend on dialysis, or more specifically hemodialysis—a term that reflects the procedure's function of filtering (*dialysis*) of the blood (*hemo*). A dialysis machine is essentially a bioengineered apparatus that mimics the function of the kidney. Blood from a patient is passed through a filter, not unlike the natural filter of the glomerulus. The filter retains blood cells and large molecules that are returned to the patient. The filtrate is adjusted so that proper levels of water and solutes are also returned to the patient, while wastes such as urea are removed, similar to the function of the proximal convoluted tubules, distal convoluted tubules, and collecting ducts.

Dialysis is effective at replacing many of the functions of the kidney. Unfortunately, it requires treatment three times per week, and each treatment lasts several hours. In an effort to increase the convenience of therapy for patients with chronic renal failure, scientists and engineers are researching other kinds of treatments. For example, scientists at Vanderbilt University are working on what they call a bio-artificial kidney. This kidney is implanted in the body, so the individual does not require frequent and lengthy dialysis treatments. It also avoids the need for a supply of donor kidneys.

The researchers looked to nature in the design of the bio-artificial kidney. The filter does not rely on small circular holes as dialysis machines do, but instead on slit pores. These pores are similar to those of the glomerulus and other effective filters found in the natural world, such as the baleen of whales that separate food (krill) from water. Because it relies on slit pores, the bio-artificial kidney can use the pressure of the heart to push fluid through the filter in the first stage of filtering the blood.

Engineering a way to achieve proper water and solute balance is difficult, so the researchers incorporated living renal tubule cells into the device instead. Challenges remain, but an implantable artificial kidney would provide a much-needed alternative to current treatments of chronic renal failure.

Self-Assessment Questions

9. Draw a mammalian nephron, label and describe the primary function of each part, and show the direction of water and electrolytes in each part.

10. The filtrate is dilute at the start and the end of the loop of Henle. What, then, is the function of the loop of Henle?

11. What is the role of ADH in the regulation of urine concentration?

12. How do the kidneys help to regulate blood volume and blood pressure?

CORE CONCEPTS SUMMARY

39.1 WATER AND ELECTROLYTE BALANCE: All animals regulate the water and electrolyte levels within their cells.

Cell membranes act as selectively permeable membranes, allowing the passage of water but restricting or controlling the movement of many solutes. page 878

Water moves across selectively permeable membranes by osmosis from a region of lower solute concentration to a region of higher solute concentration. page 878

Osmotic pressure is a measure of the tendency for water to move by osmosis across a selectively permeable membrane into a solution with higher solute concentration. page 878

Osmoregulation is the regulation of osmotic pressure inside cells or organisms. page 879

Osmoconformers maintain an internal solute concentration similar to that of the environment, whereas osmoregulators have an internal solute concentration different from that of the environment. page 879

Osmoregulators that live in high-salt environments excrete excess electrolytes and minimize water loss. page 880

Osmoregulators that live in low-salt or freshwater environments excrete excess water and minimize electrolyte loss. page 880

Osmoregulators that live on land minimize water loss. page 881

39.2 EXCRETION OF WASTES: Excretory organs eliminate nitrogenous wastes and regulate water and electrolyte levels.

Osmoregulation and excretion are closely coordinated processes controlled by the excretory organs of animals. page 881

Ammonia, urea, and uric acid are three major forms of nitrogenous waste that animals excrete, depending on their evolutionary history and habitat. page 882

Ammonia is toxic but readily diffuses into water; urea is less toxic and can be stored before it is eliminated; uric acid is a semisolid and can be eliminated with minimal water loss as an adaptation for living on land. page 882

All animal excretory systems first isolate or filter fluid into an extracellular space. Key electrolytes, solutes, and water are then reabsorbed from the filtrate. Additional wastes are subsequently secreted into the remaining filtrate before it is eliminated from the body. page 883

Examples of excretory organs are the protonephridia of flatworms, the metanephridia of segmented annelid worms, the Malpighian tubules of insects, and the kidneys of vertebrates. page 884

The urine produced by the kidneys and other excretory organs is stored in a bladder until it is eliminated from the body. page 886

The functional unit of the vertebrate kidney is the nephron, which consists of a glomerulus, capsule, renal tubules, and collecting duct. page 886

39.3 THE MAMMALIAN KIDNEY: The mammalian kidney can produce urine that is more concentrated than blood as an adaptation for living on land.

The kidney has an outer cortex and an inner medulla. page 887

The renal tubules have specialized regions, including the proximal convoluted tubule, the loop of Henle, and the distal convoluted tubule. page 888

The loop of Henle of some nephrons extends into the medulla before looping back to the cortex, creating a concentration gradient from the cortex to the medulla by a countercurrent multiplier mechanism. page 889

The final concentration of urine is under the control of the hormone ADH, which regulates the water permeability of the walls of the collecting ducts as they pass from the cortex to the medulla through the concentration gradient generated by the loops of Henle. page 892

The kidneys help to regulate blood pressure by secreting the hormone renin, leading to the production of angiotensin II, which constricts blood vessels, and aldosterone, which increases reabsorption of electrolytes and water by the kidneys. page 893

Log in to **LaunchPad** to check your answers to the Self-Assessment Questions and to access additional learning tools.

CHAPTER 40 Animal Reproduction and Development

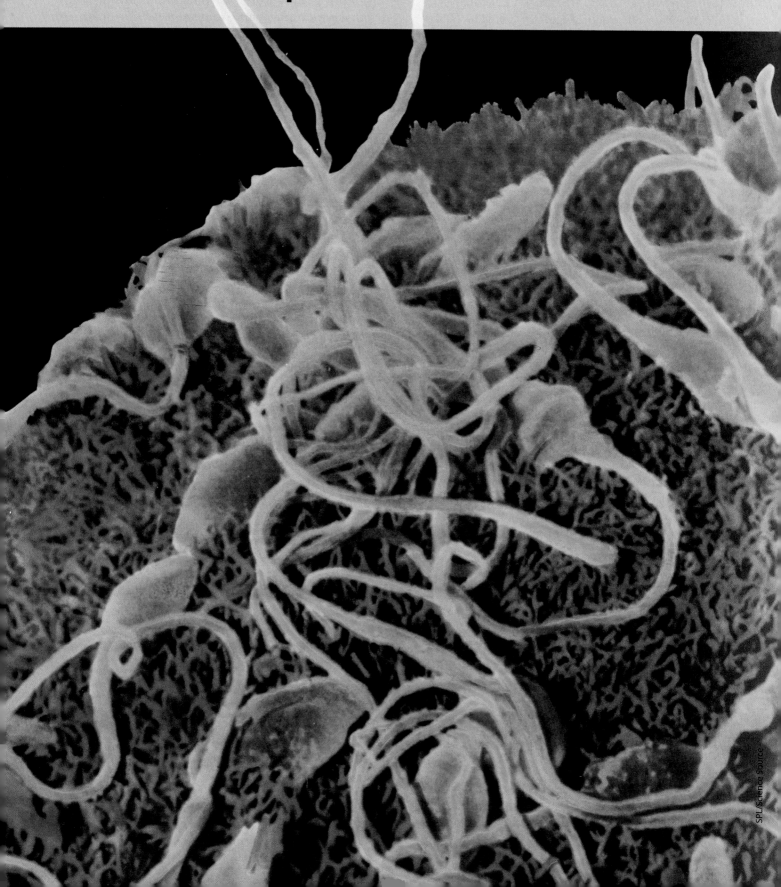

CORE CONCEPTS

40.1 THE EVOLUTIONARY HISTORY OF REPRODUCTION: Reproduction is a basic feature of living organisms and occurs both asexually and sexually.

40.2 MOVEMENT ONTO LAND AND REPRODUCTIVE ADAPTATIONS: The movement of vertebrates from water to land involved changes in reproduction, including internal fertilization and the amniotic egg.

40.3 HUMAN REPRODUCTIVE ANATOMY AND PHYSIOLOGY: The male reproductive system is adapted for the production and delivery of sperm, and the female reproductive system is adapted for the production of eggs and, in some cases, support of the developing fetus.

40.4 GAMETE FORMATION TO BIRTH IN HUMANS: Human reproduction involves the formation of gametes, fertilization, and growth and development.

Children are sometimes gently introduced to the subject of human reproduction by reference to "the birds and the bees." Since birds and bees reproduce, it is assumed that they can substitute for humans in discussions about reproduction. But how relevant is bird and bee reproduction to human reproduction? It might come as a surprise that most male birds don't have penises, and that male bees don't even have fathers. Perhaps, then, the lesson of "the birds and the bees" is not so much in what it teaches us about human reproduction, but, as we explore in this chapter, in showing us that reproductive strategies among organisms are spectacularly diverse.

Reproduction is a striking and conspicuous feature of the natural world. Flowers burst into bloom each spring to attract pollinators; male fireflies light up brilliantly on warm summer nights as they signal to females in the grass below; birds sing and display ornate plumage to attract mates. Most living organisms have the ability to reproduce—that is, to make new organisms like themselves. This remarkable ability is not unique to living organisms: after all, computer viruses and self-replicating molecules are capable of reproduction but are not themselves living. Nevertheless, reproduction is a key attribute of life. All living things came about by reproduction, and most are capable in turn of reproduction. Of course, this statement does not address the question of how life originated in the first place, a topic considered in Case 1 Life's Origins.

This chapter focuses on reproduction and development. We begin by considering reproduction broadly among diverse organisms. We then turn our attention to animal, and specifically vertebrate, reproduction and the surprise that even among vertebrates there is quite an array of reproductive strategies. In the last two sections, we focus on human reproduction, starting with reproductive anatomy and physiology, and then following reproduction from gamete formation to fertilization, pregnancy, and birth.

40.1 THE EVOLUTIONARY HISTORY OF REPRODUCTION

While the ability to reproduce is shared across all domains of life, the particular way in which organisms reproduce varies among species, from the familiar to what might strike us as truly bizarre. To take just one of many examples: soon after it is born, a male anglerfish finds a female anglerfish, bites her skin, and fuses with her. His body atrophies, leaving the male reproductive organs physically attached to the female and allowing the female's eggs to be fertilized when she is ready. To begin to make sense of this and other reproductive strategies, we start by considering how different forms of reproduction evolved, which organisms use them, and what their adaptive significance is.

Asexual reproduction produces clones.

Organisms have two potential ways to reproduce. One is the production of genetically identical cells or individuals called **clones**. This form of reproduction, which is called **asexual reproduction**, occurs in a wide variety of organisms (**Fig. 40.1**).

For example, bacteria (Fig. 40.1a) and archaeons reproduce by binary fission, a form of cell division in which the genome replicates and then the cell divides in two (Chapter 11). The result is two cells that are either

FIG. 40.1 Asexually reproducing organisms. (a) Bacteria divide by binary fission. (b) The yeast *Saccharomyces cerevisiae* reproduces by budding. (c) Corals are propagated by fragmentation. (d) Komodo dragons can reproduce by parthenogenesis. *Sources: a. CNRI/Science Source; b. SPL/Science Source; c. WaterFrame/Alamy; d. Mike Lane/Alamy.*

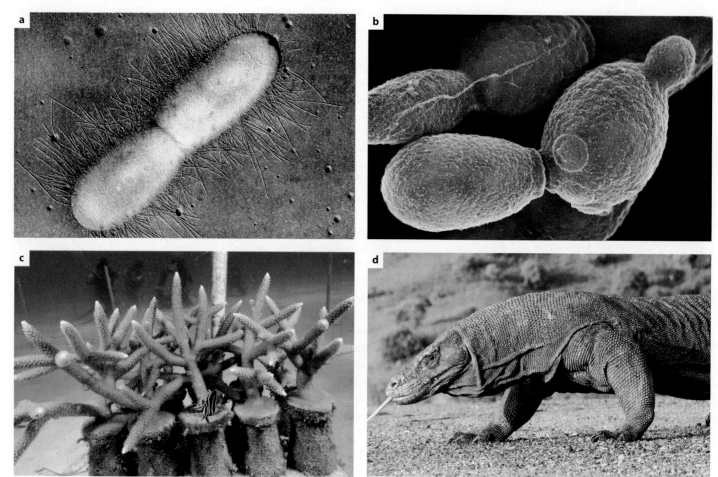

identical to the parent cell or nearly identical, differing only because of chance mutations.

Prokaryotes can nevertheless increase genetic variation by horizontal gene transfer, as described in Chapter 24. For example, they can transfer DNA from one individual to another, usually of the same species, and much more rarely to an individual of another species, during conjugation. They can also obtain DNA directly from the environment and by viral infection, incorporating genes that might have evolved in even distantly related species. These processes allow prokaryotes to acquire novel genetic sequences that can increase or decrease fitness.

Eukaryotic cells, in contrast, divide by mitotic cell division (Chapter 11), a process that evolved early and was present in the common ancestor of all living eukaryotes. Mitotic cell division is a mechanism of growth and development in all eukaryotes, and is the basis of asexual reproduction in many eukaryotes.

One form of asexual reproduction observed in fungi, plants, and some animals is **budding**. In this process, a bud, or protrusion, forms on an organism and eventually breaks off to form a new organism that is smaller than its parent. The budding yeast, *Saccharomyces cerevisiae*, which is used to make bread and beer, is so named because of its ability to produce new individuals by budding (Fig. 41.1b). Budding can also occur in animals: for example, the water-dwelling hydra can grow and subsequently break off part of itself to form a new, free-living individual. Whether it occurs in unicellular or multicellular organisms, budding involves an unequal division between a mother and daughter cell or organism. Budding is the result of mitosis, so, like binary fission, it results in an offspring that is genetically identical to the parent.

Another form of asexual reproduction is **fragmentation**. In this type of reproduction, a single individual splits into pieces, and each piece develops into a new individual by mitotic cell division. Certain molds, algae, worms, sea stars, and corals can reproduce by fragmentation. Fragmentation occurs naturally, but humans can use this mode of reproduction to propagate corals in the restoration of coral reefs (Fig. 40.1c). Once again, the new individual is a clone of the parent.

A final form of asexual reproduction is **parthenogenesis**, which literally means "virgin birth." In parthenogenesis, females produce eggs that are not fertilized by males, but rather divide by mitosis and develop into new individuals. Parthenogenesis explains why male bees don't have fathers. It occurs in invertebrates, including some insects and crustaceans, as well as in vertebrates, including some fishes and reptiles (Fig. 40.1d). The details of parthenogenesis differ depending on the organism. In most sexually reproducing animals, egg cells are usually the product of meiosis and are haploid, having one set of chromosomes. In some parthenogenic organisms, this haploid egg cell divides repeatedly by mitosis to become a haploid adult; in others, the DNA content doubles during development so the adult is diploid, having two sets of chromosomes; and in still others, the egg cell is not the product of meiosis (it is diploid) and divides by mitosis to become a diploid adult. In all cases, parthenogenesis involves a single maternal parent.

Sexual reproduction involves the formation and fusion of gametes.

The second way that organisms reproduce is by combining complete sets of genetic information from two individuals to make a new genetically unique individual. This form of reproduction is called **sexual reproduction**.

At the core of sexual reproduction are two basic biological processes: **meiotic cell division** and **fertilization** (**Fig. 40.2**). Meiotic cell division halves the number of chromosomes; during fertilization, products of meiotic cell division fuse to restore the original chromosomal content. Meiotic cell division evolved in the ancestors of modern eukaryotes, so sexual reproduction occurs only in eukaryotes.

As we saw in Chapter 11, meiotic cell division consists of one round of chromosome replication followed by two rounds of cell division, resulting in four cells each with half the number of chromosomes of the parent cell. So, in a diploid species with $2n$ chromosomes, the result of meiosis is four cells with n chromosomes. The resulting haploid cells are called **gametes** or **spores**. Human individuals, for example, have 23 pairs of chromosomes ($2n = 46$ chromosomes) and human gametes have half this number ($n = 23$ chromosomes).

Many species produce two types of gametes that differ in shape and size. The smaller, male gametes are called **spermatozoa** or **sperm**, and the larger, female gametes are called **ova** or **eggs**. For other organisms, including green algae, diatoms, slime molds, and most fungi, the two types of gametes are the same size. In spite of their similar size and appearance, they come in two (or more) distinct mating types and cannot reproduce sexually with individuals of the same type.

Fertilization takes place by the fusion of two gametes, one from each parent (Fig. 40.2). The result is a single cell called a

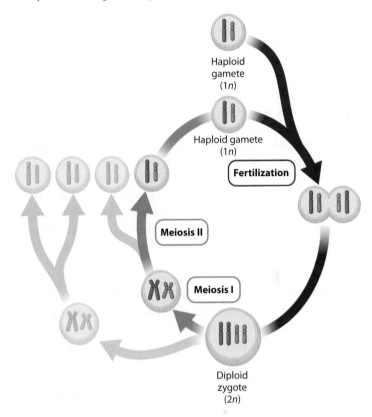

FIG. 40.2 Sexual reproduction. This form of reproduction involves the production of gametes by meiosis and their fusion by fertilization.

zygote. If each gamete has n chromosomes, the resulting zygote has $2n$ chromosomes. For diploid multicellular organisms, fertilization is the direct basis for reproduction. The zygote divides by mitosis and develops into an **embryo**, an early stage of multicellular development.

Meiotic cell division and fertilization have important genetic consequences. During meiotic cell division, pairs of homologous chromosomes undergo recombination, which shuffles the combination of alleles between chromosomes. In addition, meiotic cell division randomly sorts homologous chromosomes into different gametes. As a result, the products of meiotic cell division, the haploid gametes, are each genetically unique and represent a mixture of the parental genetic makeup. Then, during fertilization, two of these unique haploid gametes fuse, creating a new genetic combination and a unique diploid individual.

Many species reproduce both sexually and asexually.

Most organisms that reproduce asexually are also capable of reproducing sexually. For example, hydras reproduce asexually by budding, but are also able to reproduce sexually. Even budding yeast are capable of sexual reproduction. Likewise, most aphids, fishes, and reptiles capable of parthenogenesis can reproduce sexually.

What determines when an organism reproduces sexually and when it reproduces asexually? This question has been examined in the freshwater crustacean *Daphnia*. *Daphnia* reproduces asexually by parthenogenesis in the spring, when food is plentiful, perhaps because asexual reproduction is a rapid form of reproduction that allows the organism to take advantage of abundant resources (**Fig. 40.3**). Later in the season, conditions are less favorable because of increased crowding, decreased food, and cooler temperatures. At this time, *Daphnia* reproduces sexually, mating to produce a zygote that starts to form an embryo. At an early stage, however, development is arrested and metabolism slows dramatically. The embryo forms a thick-walled structure, called a cyst. In that form, it is able to resist freezing and drying, allowing it to survive the winter.

This link between environmental conditions and mode of reproduction has been observed in many other species, from algae to slime molds. The life cycle of *Daphnia* hints that asexual reproduction and sexual reproduction have both costs and benefits, a subject we address next.

Exclusive asexuality is often an evolutionary dead end.

Asexual reproduction has some advantages compared to sexual reproduction. It does not require taking time and energy to find and attract a mate. Certain birds, for example, go to elaborate lengths to attract mates. Male birds of paradise invest significant energy in growing colorful plumage, and male bowerbirds take time to build elaborate homes for potential mates (**Fig. 40.4**). For other organisms, finding a mate can be difficult because of very low population density. This is the case for anglerfish, and may help to explain their unusual reproductive strategy, described earlier.

In addition, asexual reproduction is rapid, allowing organisms to increase their numbers quickly. Consider rates

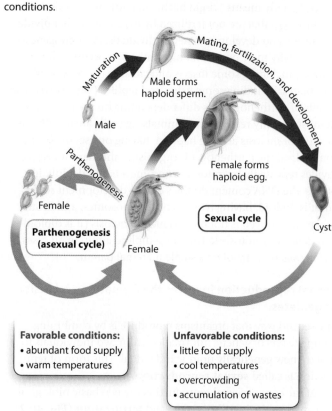

FIG. 40.3 Sexual and asexual reproduction in *Daphnia*. *Daphnia* can reproduce sexually or asexually, depending on environmental conditions.

of population growth for organisms that reproduce asexually and those that reproduce sexually. In asexual organisms, all offspring can produce more offspring, so population growth

FIG. 40.4 Time and energy costs of sexual reproduction. (a) A male bird of paradise has colorful plumage to attract a mate. (b) A bowerbird builds an elaborate hut to entice a female. *Source: Tim Laman Photography.*

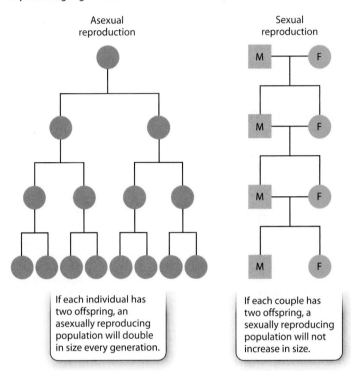

FIG. 40.5 The twofold cost of sex. Population size increases more rapidly in asexually reproducing organisms compared to sexually reproducing organisms.

If each individual has two offspring, an asexually reproducing population will double in size every generation.

If each couple has two offspring, a sexually reproducing population will not increase in size.

can be exponential if conditions are favorable. By contrast, in sexually reproducing organisms, it takes two individuals to produce new offspring. The consequence of this difference is that an asexual group can have a much greater rate of population growth than a sexually reproducing one. The British evolutionary biologist John Maynard Smith called this effect the **twofold cost of sex** (**Fig. 40.5**), and it is one of the key disadvantages of sexual reproduction.

In spite of the advantages of asexual reproduction, a survey of the natural world reveals a striking pattern: most organisms reproduce sexually at least part of the time. Even prokaryotes, as we have seen, have mechanisms for genetic exchange between individuals. The observation that the vast majority of organisms reproduce sexually at some point suggests that sexual reproduction offers advantages over asexual reproduction. Asexual reproduction is quick, but it produces offspring that are genetically identical to one another and to the parent. The only source of genetic variation with this means of reproduction is chance mutations. By contrast, one of the defining features of sexual reproduction is the production of offspring that are genetically different from one another and from their parents.

Many hypotheses have been developed to explain the advantages of producing genetically distinct offspring in spite of the costs of sexual reproduction. One proposes that sexually reproducing species adapt faster than asexually reproducing ones because rare beneficial mutations that arise in different organisms can be brought together, increasing the overall fitness of the population. Another set of hypotheses proposes that sexual reproduction allows a population to purge itself of harmful mutations more quickly than could a population of asexual individuals.

A third hypothesis proposes that sexual reproduction is a mechanism of parasite defense. In host–parasite interactions, the host evolves defenses against the parasite, and the parasite evolves new ways to infect the host. Thus, each partner evolves rapidly simply to maintain its ecological position. Called the Red Queen hypothesis, this idea takes its name from Lewis Carroll's *Through the Looking-Glass,* in which the Red Queen tells Alice that she needs to keep running to stay in the same place. Such rapid evolution requires lots of genetic variation on which natural selection can act, and sexual reproduction increases genetic variation among individuals faster than asexual reproduction does.

These different hypotheses for the prevalence of sexual reproduction are not mutually exclusive. It may be that several mechanisms work together, or that different ones are important for different organisms.

It is instructive to consider groups of eukaryotes that have lost the ability to reproduce sexually, and those that have lost the ability to reproduce asexually. While some organisms, such as the New Mexico whiptail lizard (*Cnemidophorus neomexicanus*), reproduce only asexually, these species are not very old in evolutionary terms. In other words, there are very few, if any, *ancient* asexual groups. It appears that asexual reproduction on its own rarely allows a species to persist for very long in evolutionary terms.

A possible exception to this rule may be bdelloid rotifers, a group of microscopic freshwater invertebrates that are thought to have reproduced asexually for at least 35 to 40 million years (**Fig. 40.6**). If bdelloid rotifers really have persisted for tens of millions of years without sexual reproduction, we would predict that they have evolved some process other than sexual reproduction for increasing their genetic variation. In fact, evidence suggests that these organisms are able to take up DNA from their environment, similar to what bacteria do through transformation (Chapter 24). The bdelloid rotifers seem to be fantastic "collectors" of DNA from other organisms, including bacteria, plants, and fungi. However, comparison of DNA sequences among bdelloids has recently demonstrated that different individuals can share genes that are identical or almost identical in sequence, which is not possible without at least occasional sexual reproduction. More research is needed to fully understand reproduction in this group.

In contrast to organisms that lose the ability to reproduce sexually, mammals are unusual in the opposite way: they have lost the ability to reproduce asexually. Parthenogenesis, as we have seen, occurs in some vertebrates, but does not occur in mammals. Although parthenogenesis may seem odd from our point of view, scientists have wondered just the opposite: why can't female mammals reproduce without a father? Asked another way, why do mammalian embryos need both

HOW DO WE KNOW?

FIG. 40.6

Do bdelloid rotifers only reproduce asexually?

BACKGROUND Many groups of animals reproduce asexually, but most of these organisms reproduce sexually at least some of the time. The bdelloid rotifer, one of several species of microscopic aquatic animals called rotifers, may be an exception.

OBSERVATION All bdelloid rotifers found to date are females. No males or hermaphrodites (organisms with both male and female reproductive organs) have ever been observed. Meiosis in bdelloid rotifers has never been documented.

HYPOTHESIS Bdelloid rotifers are asexual and reproduce solely by parthenogenesis, in which unfertilized eggs develop into adult females.

PREDICTION American biologists Mark Welch and Matthew Meselson reasoned that in the absence of meiosis, the two copies (alleles) of a gene do not recombine, and mutations that accumulate are not reshuffled between alleles. As a result, two alleles of a gene in an asexually reproducing organism should show a much higher degree of sequence difference compared with two alleles of a gene in a sexually reproducing organism.

RESULTS The researchers sequenced the two copies of the *hsp82* gene in a sexually reproducing rotifer species (Fig. 40.6a) and the two copies of the *hsp82* gene in an asexually reproducing rotifer species (Fig. 40.6b). As you can see, the two copies are very similar in Fig. 40.6a but very different in Fig. 40.6b.

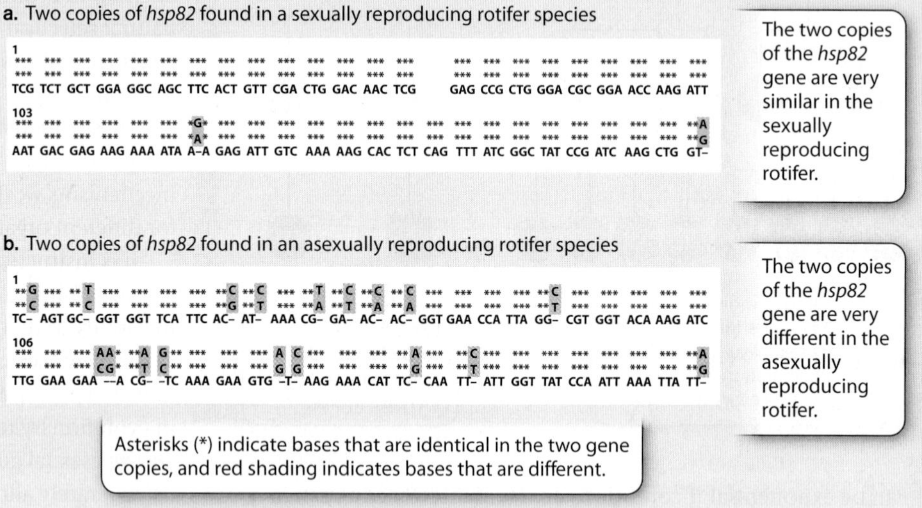

CONCLUSION The actual results are consistent with the predicted results and suggest that bdelloid rotifers are exclusively asexual.

FOLLOW-UP WORK In the absence of sexual reproduction, bdelloid rotifers might have evolved other mechanisms to increase genetic variation. In subsequent work, Meselson and his colleagues found genes in the bdelloid rotifer genome that originated in other organisms, such as bacteria, fungi, and plants. This observation suggests that these organisms might increase genetic variation by horizontal gene transfer, and it helps to explain how bdelloid rotifers might have persisted for extended evolutionary times without reproducing sexually.

a. Two copies of *hsp82* found in a sexually reproducing rotifer species

The two copies of the hsp82 gene are very similar in the sexually reproducing rotifer.

b. Two copies of *hsp82* found in an asexually reproducing rotifer species

The two copies of the hsp82 gene are very different in the asexually reproducing rotifer.

Asterisks (*) indicate bases that are identical in the two gene copies, and red shading indicates bases that are different.

SOURCES Welch, M. D., and M. S. Meselson. 2000. "Evidence for the Evolution of Bdelloid Rotifers Without Sexual Reproduction or Genetic Exchange." *Science* 288:1211–1215. Gladyshev, E. A., M. S. Meselson, and I. R. Arkhipova. 2008. "Massive Horizontal Gene Transfer in Bdelloid Rotifers." *Science* 320:1210–1213.

---

a maternal genome and a paternal genome? Wouldn't two maternal genomes from a single parent do the trick?

Evidently not. Scientists have known for some time that mammalian maternal and paternal genomes are not quite the same. Certain genes, called imprinted genes, are expressed differently depending on their parent of origin (Chapter 18). Imprinting ensures that mammals develop from embryos with one maternal and one paternal genome. When researchers manipulated key imprinted genes in a maternal genome, they were able to breed a parthenogenetic mouse, which had two copies of a maternal genome from a single parent. We can hypothesize that early mammals gained fitness advantages in the evolution of imprinted genes, but lost the ability to reproduce parthenogenetically.

Self-Assessment Questions

1. Sexual reproduction results in offspring that are genetically different from one another and from their parents. What are four mechanisms that produce this genetic variation?
2. How are asexual and sexual reproduction similar? How are they different?
3. What are the costs and benefits of asexual and of sexual reproduction?
4. How can you explain the observation that there are very few, if any, ancient asexual organisms?

40.2 MOVEMENT ONTO LAND AND REPRODUCTIVE ADAPTATIONS

We have considered the two modes of reproduction observed across all living organisms. In this section, we narrow our focus to vertebrates: fishes, amphibians, reptiles, birds, and mammals. Interestingly, even among vertebrates, there are many different reproductive strategies. Understanding these strategies not only gives us a picture of how these organisms reproduce, but also sheds light on how they live and behave.

How vertebrates reproduce is also intimately connected to the transition of vertebrates from water to land 400 to 375 million years ago. Moving from an aquatic environment to a terrestrial one involved many adaptations, among them changes in the way that reproduction took place, so it is helpful to consider reproductive strategies in an evolutionary context.

Fertilization can take place externally or internally.

To survive, eggs and sperm require a wet environment. Aquatic organisms such as fishes and amphibians can release eggs and sperm directly into the water by **external fertilization**. External fertilization, as you might imagine, is a chance affair: somehow, sperm must meet egg in a large environment. Given this reality, animals that release gametes externally have developed strategies to increase the probability of fertilization. Some species release large numbers of gametes. Many female fishes, including salmon and perch, lay eggs on or near a substrate, such as a rock, and males release their sperm onto the eggs to increase the probability of fertilization. Externally fertilizing animals may also come close together physically to improve chances that the sperm will fertilize the eggs. Male frogs grasp females, releasing sperm as the female lays her eggs (**Fig. 40.7**).

A challenge for land-dwelling organisms is keeping gametes from drying out. **Internal fertilization**, in which fertilization takes place inside the body of the female, is an adaptation for living on land. Reptiles, birds, and mammals all use internal fertilization. However, this mode of fertilization is not exclusive to land dwellers. Marine mammals such as whales and dolphins use internal fertilization, as did the land-dwelling ancestors from which they evolved. Even some fishes, such as guppies, reproduce by internal fertilization.

r-strategists and K-strategists differ in number of offspring and parental care.

While the terms "external" and "internal" fertilization focus our attention on where fertilization takes place, the two modes of reproduction are associated with a host of other behavioral and reproductive differences. One key difference is the number of offspring produced.

External fertilization is generally associated with the release of large numbers of gametes and the production of large numbers of offspring. An individual offspring has a low probability of survival in part because the offspring typically

FIG. 40.7 External fertilization. Tree frogs reproduce by external fertilization: the male frog holds on to the female and fertilizes her eggs as she lays them. *Source: National Geographic Creative/Alamy.*

receive little, if any, parental care. External fertilization, in other words, is a game of numbers. By comparison, when both fertilization and embryonic development take place inside the female, there is much more control over fertilization and subsequent development of the embryo. As a result, animals that use internal fertilization typically produce far fewer offspring and invest considerable time and energy into raising those offspring.

These two strategies represent two ends of a continuum of reproductive strategies that was first described by the American ecologists Robert H. MacArthur and E. O. Wilson in 1967. Organisms that produce large numbers of offspring without a lot of parental investment are called **r-strategists**, and those that produce few offspring but put in a lot of parental investment are called **K-strategists** (**Fig. 40.8**; Chapter 44). Fishes are r-strategists; humans are K-strategists.

According to this model, the environment places selective pressures on organisms to drive them in one or the other direction. In general, r-strategists evolve in unstable, changing, and unpredictable environments. In such an environment, there is an advantage to reproducing quickly and producing many offspring when conditions are favorable. By contrast, K-strategists often evolve in stable, unchanging, and predictable environments, where there tends to be more crowding and larger populations. In the face of intense competition for limited

FIG. 40.8 r-strategists and K-strategists. r-strategists, such as fishes, produce many offspring with low probability of survival, whereas K-strategists, such as primates, produce few offspring and put a lot of care into raising their young.

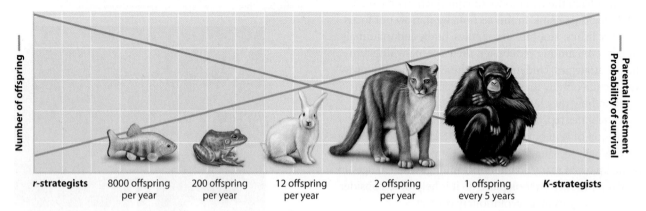

resources, traits such as increased parental care and few offspring are favored.

These two strategies represent the extremes of a spectrum, and there are notable exceptions. An octopus, for example, can lay up to 150,000 eggs, typical of r-strategists, but the female guards and takes care of them for a month or more, typical of K-strategists. Nevertheless, the two strategies highlight different reproductive patterns across animals and suggest important links between number of offspring, parental care, and the environment.

Animals either lay eggs or give birth to live young.

When we speak of parental care, it is natural to think of parents tending the young. However, parental care actually begins much earlier in development, in the provisioning of nutrients for the developing embryo. The eggs of animals that are externally fertilized have a substance called **yolk** that provides all the nutrients that the developing embryo needs until it hatches. The embryo obtains water and oxygen directly from the aquatic environment and metabolic wastes easily diffuse out.

The anatomy of the egg was transformed as vertebrates moved onto land. No longer could the developing embryo depend simply on diffusion of key substances into and out of a watery environment. A key innovation is the **amnion**, a membrane surrounding a fluid-filled cavity that allows the embryo to develop in a watery environment and prevents it from drying out. Reptiles, birds, and mammals are known as amniotes because of the presence of this key membrane in their developing embryos (**Fig. 40.9**). A second membrane, called the **allantois**, encloses a space where metabolic wastes collect. Finally, a third membrane, called the **chorion**, surrounds the entire embryo along with its yolk and allantoic sac. The embryo is further protected in many species by a hard shell just outside the chorion. Together, the chorion and shell limit water loss, and the chorion, shell, and allantois allow gas exchange with the environment. The yolk sac, amnion, allantois, and chorion are all **extraembryonic membranes**, sheets of cells that extend out from the developing embryo (**Fig. 40.10**).

Animals that are internally fertilized can either lay eggs or give birth to live young. Animals that lay eggs are oviparous (**oviparity** means "egg birth"). Oviparity is used by most insects, fishes, amphibians, and reptiles, as well as by all birds. Even some mammals—the monotremes, which include the spiny anteater and platypus—lay eggs. For oviparous animals, very little, if any, embryonic development occurs inside the mother, and all the nutrients for the developing embryo come from the yolk.

FIG. 40.9 The evolution of the amnion. The amnion is a membrane enclosing a fluid-filled cavity surrounding the developing embryo. It evolved during the transition of animals from water to land.

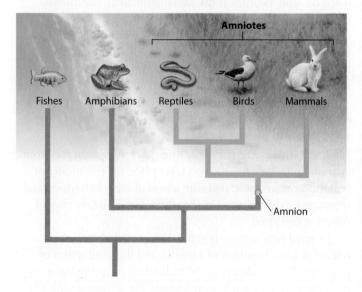

FIG. 40.10 The amniotic egg. The amniotic egg includes four extraembryonic membranes: the yolk sac, amnion, allantois, and chorion.
Photo source: Thanunkorn Klypaksi/iStockphoto.

Other animals give birth to live young. These animals differ in where the embryo gets its nourishment. In some animals, embryos develop inside eggs that are retained inside the mother until they hatch. These embryos receive nutritional support from the yolk, not from the mother. Such animals are ovoviviparous (**ovoviviparity** means "egg live birth"). Some sharks and snakes are ovoviviparous.

In other animals, the embryo develops inside the mother and receives all its nourishment from her. These animals are viviparous (**viviparity** means "live birth"). Most mammals and some sharks and snakes are viviparous. In fact, viviparity has evolved many times independently. In placental mammals, such as humans, the chorion and allantois fuse to form the **placenta**, an organ that allows nutrients to be obtained directly from the mother.

Self-Assessment Questions

5. What are three adaptations that allow reproduction to take place on land?
6. Do all mammals give birth to live young? Are mammals the only vertebrates that give birth to live young?

40.3 HUMAN REPRODUCTIVE ANATOMY AND PHYSIOLOGY

In this and the next section, we continue to narrow our focus so as to consider reproduction in more detail. In this section, we discuss human reproduction and development. Many features of human reproduction are seen in other mammals: humans are amniotes, reproduce sexually, use internal fertilization, and are viviparous. Other aspects are unique. Our focus on human reproduction allows us to provide a detailed treatment of the topic in one species and gives us a chance to explore ways we intervene in the process to promote, prevent, and follow pregnancy.

The male reproductive system is specialized for the production and delivery of sperm.

Let's start by considering the human male reproductive system. Male gametes, called spermatozoa or sperm, are produced in the two male **gonads**, or **testes** (singular, testis; **Fig. 40.11**). Over the course of his life, a man is capable of making hundreds of billions of sperm. In fact, a single ejaculate typically contains several hundred million sperm. If it takes just one sperm to fertilize an egg, why do males make so many? A likely explanation involves sperm competition. In species such as humans, where females can mate with more than one male, a male's reproductive success is at least partially correlated with the number of sperm: the more sperm, the better the chance of fertilizing the egg. A second explanation points out that the female reproductive tract is hostile to sperm; many sperm are needed to ensure that at least one makes it to the egg. Of the several hundred million sperm released by a male during a single ejaculation, only several hundred sperm make it to the egg.

Sperm are haploid cells with a unique shape and properties adapted for their function of delivering genetic material to the egg (**Fig. 40.12**). They have a small head containing minimal cytoplasm and a very densely packed nucleus. The head is surrounded by a specialized organelle, the **acrosome**, which contains enzymes that enable the sperm to traverse the outer coating of the egg. Sperm have a long tail, or **flagellum**, which moves the cell by a whipping motion powered by the sperm's mitochondria.

The testes are located outside the abdominal cavity in a sac called the **scrotum** (see Fig. 40.11). This location is critical for the production of sperm, which require cooler temperatures than are found in the abdominal cavity of humans. However, the testes of some mammals, such as elephants, whales, and bats, are located inside the abdominal cavity, indicating that there are species-specific differences in sperm temperature requirements.

Sperm are produced within the testes in a series of tubes called the **seminiferous** ("seed-bearing") **tubules** (**Fig. 40.13**). This is where diploid cells undergo meiosis to produce haploid sperm. When sperm are first produced, they are not fully mature: they are unable to move on their own and incapable of fertilizing an egg. Sperm mature as they move through the male reproductive system.

From the seminiferous tubules, sperm travel to the **epididymis**, where they become motile (Fig. 40.13). This is also where sperm are stored prior to ejaculation. From there, sperm enter

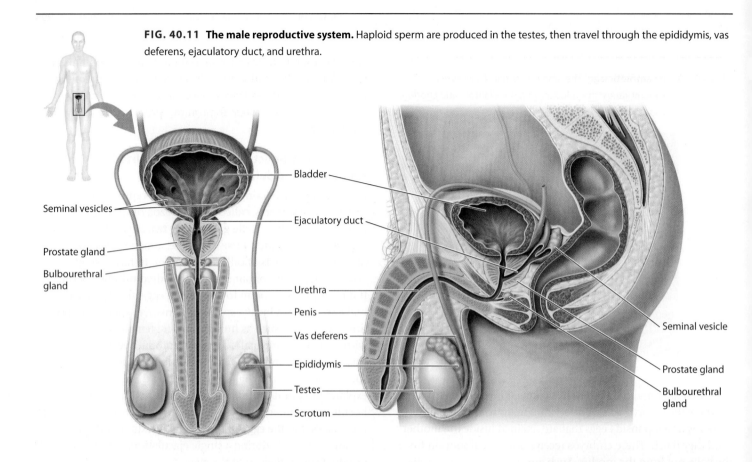

FIG. 40.11 The male reproductive system. Haploid sperm are produced in the testes, then travel through the epididymis, vas deferens, ejaculatory duct, and urethra.

FIG. 40.12 Sperm. Sperm are specialized cells with minimal cytoplasm, a densely packed nucleus, an acrosome with digestive enzymes, and a flagellum. *Photo source: Roland Birke/Getty Images.*

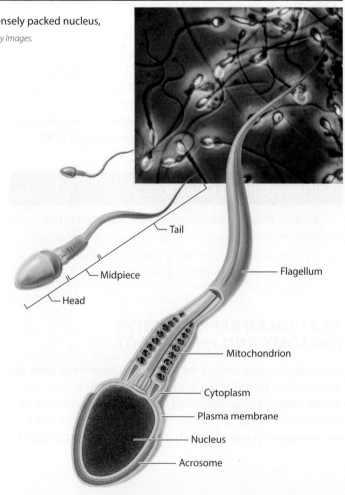

the **vas deferens**, a long, muscular tube that follows a circuitous path, starting off in the scrotum of each testis, passing into the abdominal cavity, running along the bladder, and then connecting with an **ejaculatory duct** (see Fig. 40.11). The two ejaculatory ducts merge at the **urethra**, just below the bladder. The urethra is used by both the urinary and reproductive systems in males.

Along this path, there are several exocrine glands that produce components of **semen**, the fluid that nourishes and sustains sperm as they travel first in the male reproductive tract and then in the female reproductive tract. Two **seminal vesicles**, located at the junction of the vas deferens and the prostate gland, secrete a protein- and sugar-rich fluid that makes up most of the semen and provides energy for sperm motility (see Fig. 40.11). The **prostate gland**, located just beneath the bladder, produces a thin, slightly alkaline fluid that helps maintain sperm motility and counteracts the acidity of the female reproductive tract. Finally, the **bulbourethral glands**, found below the prostate gland, produce a clear fluid that lubricates the urethra for passage of sperm.

The urethra is enclosed in the **penis**, the male copulatory organ (see Fig. 40.11). When a male is sexually aroused, the penis becomes erect as a result of changes in blood flow: the

FIG. 40.13 Seminiferous tubules, epididymis, and vas deferens.

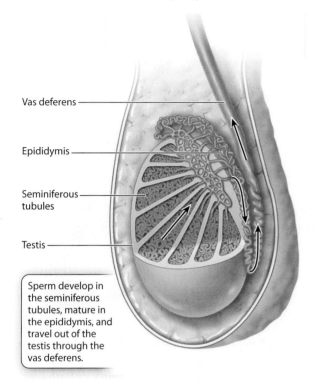

Sperm develop in the seminiferous tubules, mature in the epididymis, and travel out of the testis through the vas deferens.

arteries become dilated and the veins constricted. During ejaculation, semen is expelled from the penis.

The male reproductive system is well adapted for the production, storage, and delivery of sperm. Therefore, male contraceptive methods focus on blocking sperm. For example, a condom is a sheath of rubber or latex that covers the penis and prevents sperm from entering the female during sexual intercourse.

The human penis is a relatively simple organ, consisting of a shaft and a head called the **glans penis**. In other animals, the penis can be quite elaborate, with spikes, hooks, and grooves whose functions are to stimulate the female, grip the female, and in some cases even remove sperm from a previous copulation. In marsupial mammals such as kangaroos and koalas, males have forked penises. In snakes and lizards, males have two penises, called hemipenes. In insects, the external genitalia are often distinct enough to be used in species identification. By contrast, most species of birds do not have a penis.

The female reproductive system produces eggs and supports the developing embryo.

The female reproductive system, like the male reproductive system, produces gametes, but is also specialized to support the developing embryo. Developing female gametes, called **oocytes**, which mature into ova upon fertilization, are produced in the two female gonads, called **ovaries** (**Fig. 40.14**). Oocytes are released monthly in response to hormones in the process of menstruation, discussed in the next section.

FIG. 40.14 The female reproductive system. Oocytes are produced in the ovaries and then travel through the oviducts to the uterus, where they implant if they are fertilized. If not fertilized, the oocytes are shed during menstruation.

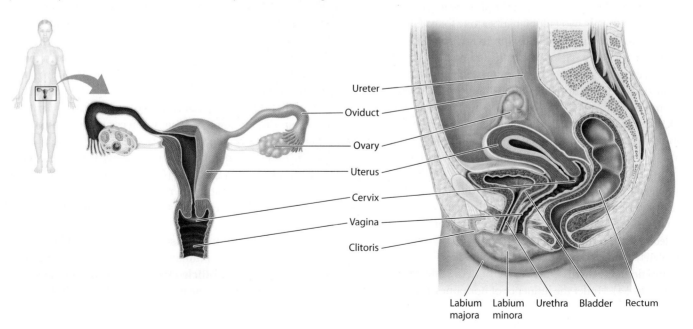

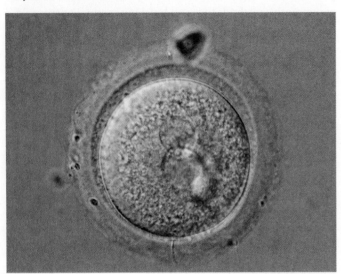

FIG. 40.15 **A female gamete.** Oocytes are produced in the ovaries, mature into ova (eggs), and are by volume the largest cells in the human body. *Source: Claude Cartier Science Source.*

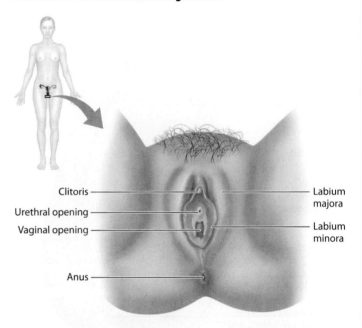

FIG. 40.16 The female external genitalia.

The oocyte is a remarkable and unique cell (**Fig. 40.15**). Whereas sperm are small and motile, the oocyte is the largest cell by volume in the human body and is visible to the naked eye; it is about the size of the period at the end of this sentence. Its large size reflects the presence of large amounts of cytoplasm containing molecules that direct the early cell divisions following fertilization. Whereas a male produces millions of sperm, a female usually produces just one oocyte each month, as she will nourish and support the developing embryo if the oocyte is fertilized.

Following its release from the ovary, the oocyte travels through the **oviduct**, also called a **fallopian tube** (see Fig. 40.14). There are two fallopian tubes, one for each ovary. The fallopian tubes have featherlike projections on their ends that help channel the egg into the tube, and not into the abdominal cavity.

The oocyte travels from the fallopian tube to the **uterus**, or womb (see Fig. 40.14). The uterus is a hollow organ with thick, muscular walls that is adapted to support the developing embryo if fertilization occurs and to deliver the baby during birth. In humans, the uterus is usually a single pear-shaped structure, but it can sometimes have two horns, which can interfere with pregnancy. Other mammals differ in the shape of the uterus, ranging from pear-shaped (humans and other primates), to two horns (pigs, horses), to more completely divided but still connected at the base (cats, dogs), to several completely separate structures (rodents, rabbits).

The end of the uterus is the **cervix**, or neck of the uterus (see Fig. 40.14). Under the influence of hormones, the cervix produces different kinds of mucus capable of either blocking or guiding sperm through the cervix.

The uterus is continuous with the **vagina**, or birth canal (see Fig. 40.14). The vagina is a tubular channel connecting the uterus to the exterior of the body. It is highly elastic, allowing it to stretch during sexual intercourse and birth.

The external genitalia of the female are collectively called the **vulva** (**Fig. 40.16**). The vulva includes two folds of skin, the **labia majora** and **labia minora**. The labia minora meet at the **clitoris**, the female homolog of the glans penis.

Hormones regulate the human reproductive system.

The male testes and female ovaries are not only the site of gamete production, but also part of the endocrine system (Chapter 36). That is, they respond to hormones and in turn secrete hormones. These hormones play key roles during embryonic development and puberty and regulate reproduction in the adult. Hormonal control of reproduction is critically important, as, for example, it allows an egg to be released only when the uterus is prepared to accept it.

The master regulator of the endocrine system is the hypothalamus, located in the brain (Chapter 36). The hypothalamus secretes hormones that act on the pituitary gland, which in turn releases hormones that act on target organs (**Fig. 40.17**). In the case of the reproductive system, the hypothalamus releases **gonadotropin-releasing hormone (GnRH)**, which stimulates the anterior pituitary gland to secrete **luteinizing hormone (LH)** and **follicle-stimulating hormone (FSH)**. These hormones act on the male and female gonads. In males,

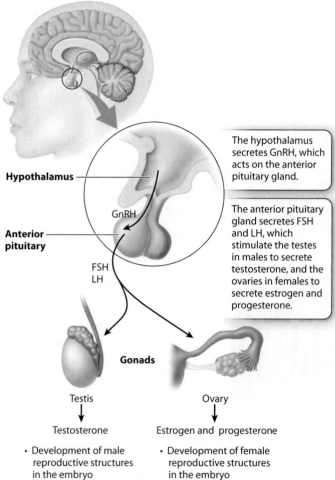

FIG. 40.17 Hormonal control of the reproductive system.

the hormones are secreted almost continually; in females, they are secreted cyclically. In response to the hormones, the testes secrete the hormone **testosterone** and the ovaries secrete the hormones **estrogen** and **progesterone**.

In males, LH acts on **Leydig cells**, cells in the testes that secrete testosterone. Testosterone is a steroid hormone derived from cholesterol that plays key roles in male growth, development, and reproduction. During the development of the embryo, testosterone directs development of the male reproductive organs. At puberty, hormones initiate the onset of sexual maturity. Levels of testosterone increase at this time, leading to the development of male **secondary sexual characteristics**, traits that characterize and differentiate the two sexes but that do not relate directly to reproduction. For males, these traits include a deep voice; facial, body, and pubic hair; and increased muscle mass. Testosterone and FSH also act on another group of cells in the testes, called **Sertoli cells**, to stimulate sperm production.

In females, an increase in the level of estrogen at puberty leads to the development of female secondary sexual characteristics, such as enlargement of the breasts, growth of body and pubic hair, and changes in the distribution of muscle and fat. Also at puberty, females begin a monthly cycle, the **menstrual cycle**, in which an oocyte matures and is released from the ovary under the influence of hormones (**Fig. 40.18**). The menstrual cycle consists of cyclical changes in the ovaries and uterus, coordinated by hormones, that time the release of an oocyte from the ovary with the growth of the uterine lining so that the uterus can support the developing embryo if fertilization occurs.

The menstrual cycle has two phases: the **follicular phase** and the **luteal phase**. In the follicular phase, an oocyte develops and is eventually released from the ovary. The oocyte is surrounded by a supporting shell of cells called a **follicle**. At the start of each menstrual cycle, FSH stimulates specific follicle cells, called granulosa cells, to secrete a form of estrogen called estradiol. In response to estradiol, oocytes in several follicles begin to mature each month, but usually only one becomes fully mature. Estradiol also acts on the lining of the uterus, causing it to thicken. Finally, estradiol acts on the hypothalamus and anterior pituitary gland, where it reduces the secretion of GnRH, FSH, and LH by negative feedback (Fig. 40.18).

At the end of the follicular phase, this negative feedback of estradiol changes to positive feedback (Fig. 40.18). The mechanism for this change is not well understood, but may be due in part to high levels of estradiol. Positive feedback of estradiol on the hypothalamus and anterior pituitary causes a rapid increase in the level of LH produced by the anterior pituitary gland. This LH surge causes ovulation, the release of the oocyte from the follicle in the ovary.

Ovulation marks the end of the follicular phase and the beginning of the luteal phase. The follicle, now devoid of the oocyte, is converted to a structure known as the **corpus luteum** (literally "yellow body," so named because of its appearance). The corpus luteum is a temporary endocrine structure that secretes the hormones progesterone and estradiol. Progesterone maintains the thickened and vascularized uterine lining, and estradiol inhibits the secretion of GnRH, FSH, and LH (Fig. 40.18).

The oocyte, meanwhile, is swept into the fallopian tube and travels to the uterus. If it is fertilized, the developing embryo implants into the uterine lining. In this case, the corpus luteum continues to secrete progesterone; the corpus luteum

FIG. 40.18 The menstrual cycle. Note the cyclical and coordinated changes in the anterior pituitary hormones, ovary, ovarian hormones, and uterine lining. These changes are regulated by hormonal feedback loops.

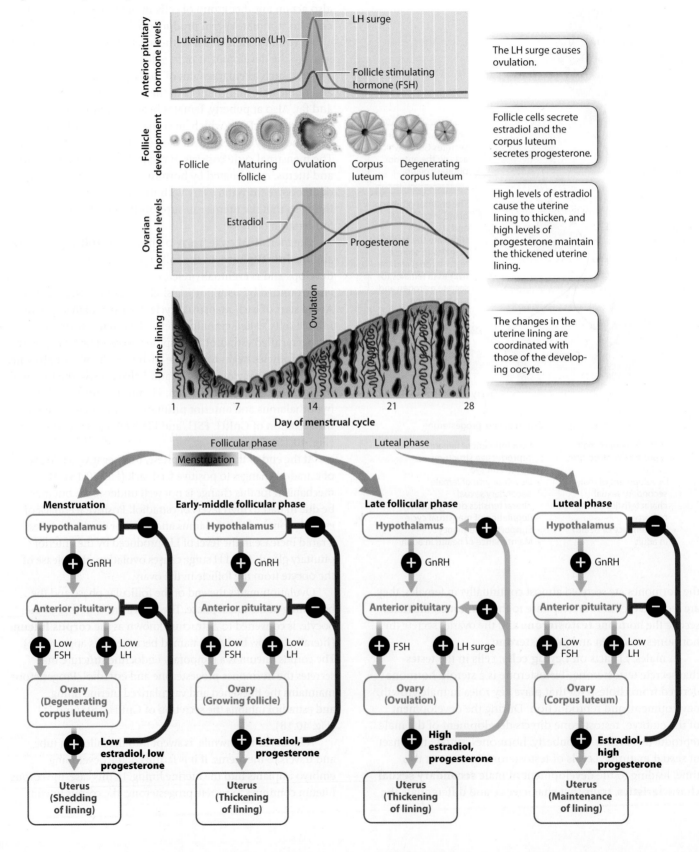

is maintained first by LH and then by the hormone **human chorionic gonadotropin (hCG)**, which is released by the developing embryo. Eventually, the placenta takes over estrogen and progesterone production to maintain the uterine lining and stimulate growth of the uterus. High levels of estrogen and progesterone during pregnancy also block ovulation because they suppress the release of GnRH, FSH, and LH.

If the oocyte is not fertilized, the corpus luteum degenerates, estrogen and progesterone levels drop, and the uterine lining is shed. The monthly shedding of the uterine lining is known as **menstruation**. Menstrual cycles usually start occurring around age 12 in the United States and continue until approximately age 45 to 55. The cessation of menstrual cycles, called **menopause**, results from decreasing production of estradiol and progesterone by the ovaries.

As we have seen, the brain, anterior pituitary gland, ovaries, uterus, and placenta pass hormonal messages back and forth to ensure that behavior, gamete release, the thickening of the uterine lining, and embryonic development are coordinated (Fig. 40.18). If hormone levels are low or absent, as occurs normally before puberty or as the result of malnutrition or some diseases, ovulation does not occur. Similarly, too much hormone at the wrong time can interfere with fertility, a factor that serves as the basis for oral contraceptive pills ("the pill"). Oral contraceptive pills contain various combinations of synthetic estrogens, progesterones, or both. Estrogens and/or progesterones in the pill suppress GnRH, FSH, and LH, thereby blocking oocyte development and ovulation.

Humans and chimpanzees both have menstrual cycles. Other placental mammals have an **estrus cycle**, which is characterized by phases in which females are sexually receptive. In these mammals, the uterine lining is reabsorbed instead of shed if fertilization does not occur.

Self-Assessment Questions

7. What is the pathway of sperm from its site of production to the urethra?
8. What effect would cutting the vas deferens on both sides and tying off the ends have on male fertility?
9. Describe the pathway of an oocyte (and the egg it develops into) from its site of production to the uterus.
10. What are the relationships among changes in levels of anterior pituitary and ovarian hormones, oocyte development, and changes in the uterine lining during a menstrual cycle?
11. A monthly course of an oral contraceptive pill usually consists of three weeks of hormone pills followed by one week of placebo (a sugar pill with no hormone). Why do you think placebos are sometimes used?

40.4 GAMETE FORMATION TO BIRTH IN HUMANS

With an understanding of male and female reproductive anatomy and endocrinology, we are ready to follow human reproduction and development from the formation of sperm and eggs, to fertilization, to subsequent development of the embryo and fetus, and finally to birth. Along the way, we consider how these processes vary in different organisms. Our understanding of these processes is in large part informed by studies in model organisms, including the sea urchin (*Strongylocentrotus purpuratus*), roundworm (*Caenorhabditis elegans*), fruit fly (*Drosophila melanogaster*), African clawed frog (*Xenopus laevis*), and house mouse (*Mus musculus*). These organisms have particular reproductive and developmental features that make them suitable for laboratory study.

Male and female gametogenesis have both shared and distinct features.

The formation of gametes is called **gametogenesis**. Gametogenesis follows the same steps in males and in females, but there are striking differences in its timing (**Fig. 40.19**).

Spermatogenesis is the formation of sperm. It occurs in the seminiferous tubules of the testes beginning at puberty and continuing throughout life. The process starts with spermatogonia, diploid stem cells that divide by mitosis to ensure a continuous population of precursor cells. During puberty and throughout adulthood, these cells differentiate into diploid **primary spermatocytes**, which undergo the first meiotic division to produce two **secondary spermatocytes** and the second meiotic division to produce four haploid spermatids joined by cytoplasmic bridges. The spermatids subsequently separate from each other to become distinct haploid cells. The maturation process, in which spermatids become spermatozoa or sperm, begins in the testes and continues in the epididymis, where they acquire motility. Spermatogenesis takes about 2 to 3 months.

Oogenesis is the formation of ova or eggs. The diploid precursor cells are called oogonia. These cells form during embryonic development and divide by mitosis. Some oogonia differentiate into diploid **primary oocytes** during fetal development. These enter the first meiotic division but their differentiation is arrested immediately in prophase I, the first stage of meiosis. All these steps take place before the female is even born. At birth, a female has approximately 1 to 2 million primary oocytes, all arrested in prophase I of meiosis. They do not resume the first meiotic division until a menstrual cycle begins. Therefore, remarkably, all of oocytes remain arrested for at least 12 years (from birth to the age of menarche) and some for as long as 50 years (to the age of menopause).

In the first phase of the menstrual cycle, approximately 10 to 20 follicles, each containing an oocyte, begin to mature

FIG. 40.19 Spermatogenesis in the male and oogenesis in the female.

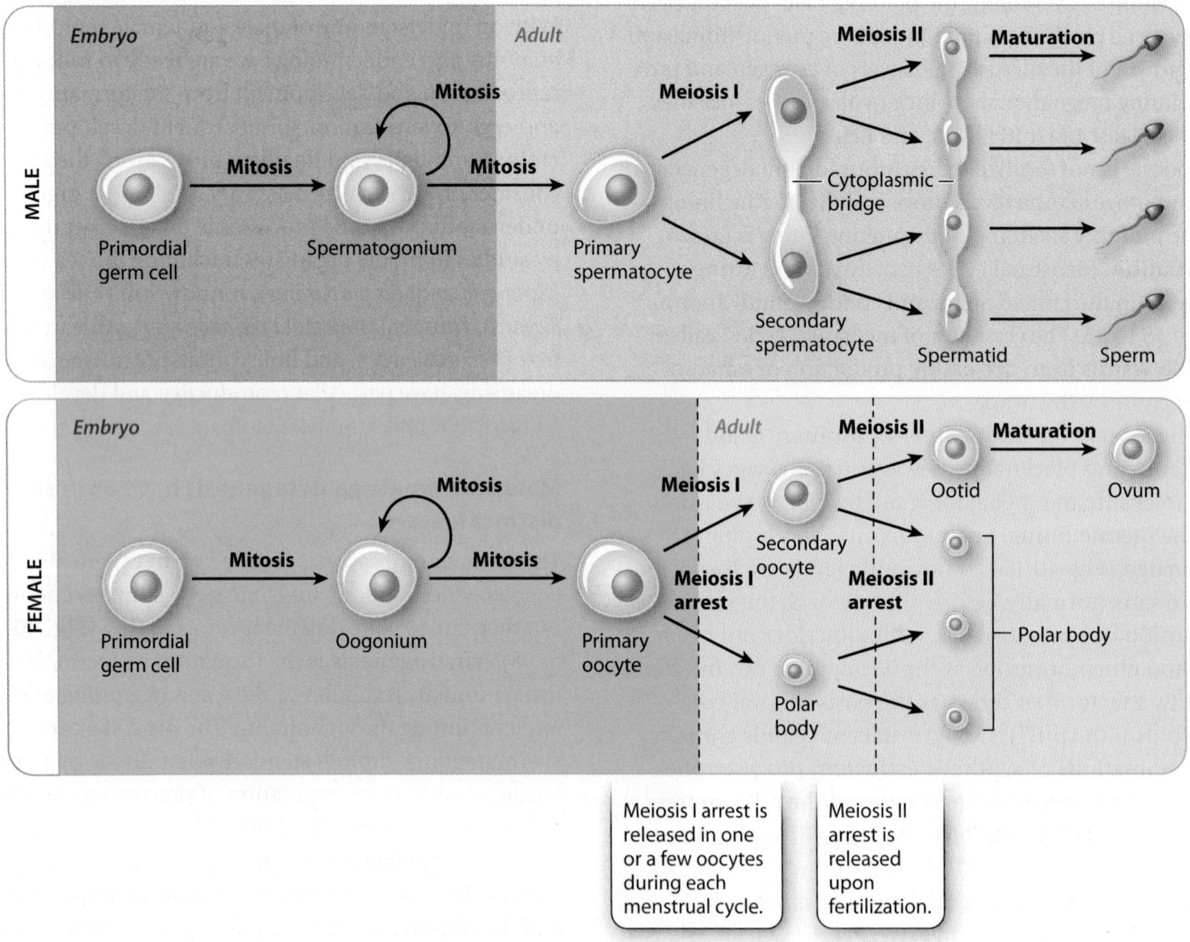

in response to rising FSH levels. One of these follicles becomes dominant, allowing the primary oocyte to complete the first meiotic division. This division is asymmetric, leading to the formation of a large **secondary oocyte** and a much smaller **polar body**. Most of the cytoplasm is partitioned into the secondary oocyte. The secondary oocyte is released from the ovary and immediately enters the second meiotic division, but again arrests, this time in metaphase II. The secondary oocyte remains arrested in metaphase II until fertilization takes place. Upon fertilization, the second meiotic division continues, producing an ootid and another polar body, again allowing most of the cytoplasm to end up in the developing egg. The ootid quickly develops into a mature ovum, or egg.

Fertilization occurs when a sperm fuses with an oocyte.

During fertilization, two gametes fuse, restoring the diploid chromosome content (**Fig. 40.20**). In humans, male and female gametes are usually brought together by sexual intercourse. Sexual arousal leads to a host of physiological changes, including erection of the penis, erection of the clitoris, and lubrication of the vagina. The erect penis is inserted into the vagina until orgasm occurs. Orgasm is characterized by feelings of pleasure and contractions of the pelvic muscles. In males, it is usually accompanied by ejaculation, in which semen, containing sperm, is delivered into the vagina.

Sperm travel through the cervix and into the uterus and fallopian tubes. Fertilization usually occurs in one of the fallopian tubes. Many details about how sperm travel through the female reproductive tract are poorly understood. Sperm are motile, but from studies of the time it takes sperm to reach the fallopian tubes, it appears that the female reproductive tract plays an active role in moving sperm to the oocyte, such as through muscular contractions.

In the female reproductive tract, sperm undergo **capacitation**, a key developmental step that is unique to

FIG. 40.20 Fertilization. Many sperm cluster around an oocyte, but usually just one fuses with it to form a zygote. *Photo source: David M. Phillips/Science Source.*

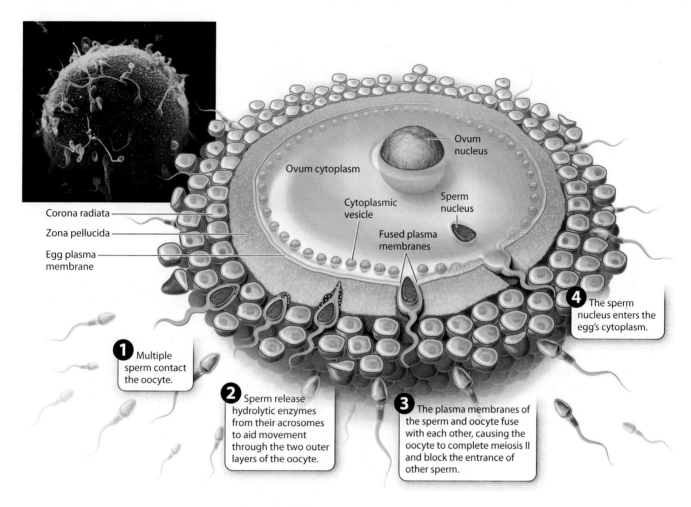

mammals. Capacitation is a series of physiological changes that allow the sperm to fertilize the egg. These changes include alterations in the fluidity of the plasma membrane, loss of some surface membrane proteins, and changes in the charge across the membrane, called the membrane potential (Chapter 34). As a result, sperm motility increases.

In many species, the oocyte releases soluble molecules, creating a concentration gradient that guides the sperm to the oocyte. On reaching the oocyte, the sperm passes through two layers—an outer layer of follicle cells called the corona radiata and an inner matrix of glycoproteins called the zona pellucida—before fusing with the plasma membrane of the oocyte (Fig. 40.20). In sea urchins, by contrast, a single vitelline membrane surrounds the egg. To facilitate its passage through these various outer layers, the head of the sperm releases enzymes from the acrosome. In addition, surface proteins on the gametes ensure that gametes interact only with other gametes from the same species. For organisms with external fertilization, such as sea urchins, these proteins are especially important.

After passing through the layers surrounding the oocyte, the plasma membranes of the sperm and oocyte fuse (Fig. 40.20). The oocyte then completes meiosis II and undergoes changes that prevent **polyspermy**, or fertilization by more than one sperm. The mechanisms for preventing polyspermy have been extensively studied in sea urchins and are classified according to how quickly they act. The first line of defense, or fast block to polyspermy, is caused by changes in the membrane potential of the egg: the membrane potential is usually negative but quickly becomes positive on fertilization. The fast block is quick but temporary and does not occur in mammals.

A second, more stable defense is slow block to polyspermy. In this process, vesicles in the egg fuse with the plasma membrane and release their contents, leading to modifications

FIG. 40.21 Cleavage divisions. *Photos source: Anatomical Travelogue/Science Source.*

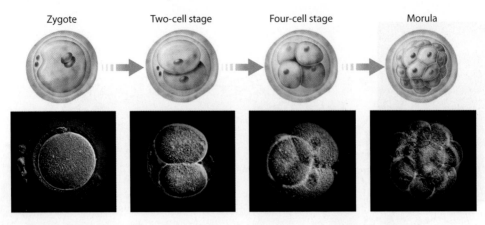

of the structure of the zona pellucida in mammals and the vitelline membrane in sea urchins that make this zone impassable to sperm.

Although sexual intercourse is one way in which gametes are brought together, there are now other methods to achieve fertilization and pregnancy in humans. The various techniques, which are collectively called assisted reproductive technologies, are typically used to treat infertility. For example, **in vitro fertilization (IVF)** is a process in which eggs and sperm are brought together in a petri dish, where fertilization and cell divisions occur. The developing embryos are then inserted into the uterus. The first IVF or "test tube" baby was born in 1978, and since then several million babies have been conceived in this way. Assisted reproductive technologies such as IVF have opened the door to pre-implantation genetic diagnosis, in which oocytes or embryos are screened for genetic diseases before implantation. This technique raises ethical concerns, as it can be used to determine and select the gender of the embryo and other traits not associated with disease.

The first trimester includes cleavage, gastrulation, and organogenesis.

How does a single, diploid cell (the zygote) become a complex, multicellular organism? Development depends on a precisely orchestrated set of cell divisions, movements, and specializations coordinated by both maternal and zygotic genetic instructions (Chapter 19). In this section, we focus on three key processes in early embryonic development: cleavage, gastrulation, and organogenesis.

After fertilization, the single-celled zygote divides by mitosis in a process called **cleavage** (**Fig. 40.21**). In cleavage, the single large egg divides into many smaller cells. In most organisms, these mitotic divisions are rapid, occur at the same time, and do not increase the overall size of the embryo. Furthermore, they occur in the absence of expression of the zygote's genes.

Instead, molecules in the egg's cytoplasm direct these early cell divisions. These molecules include mRNA and proteins encoded by maternal genes (Chapter 18) and stored in the egg.

Cleavage in mammals is unique in several respects. Notably, it is slow compared with cleavage in other organisms. The early divisions do not occur at the same time, so it is not unusual to find an odd number of cells in the developing embryo. Finally, transcription of zygotic genes begins very early, as early as the first mitotic division, or two-cell stage.

The pattern of cleavage is different for different organisms. For example, the presence of yolk inhibits cell division (**Fig. 40.22**). In animals with lots of yolk, such as insects, fishes, and birds, cell division does not extend through the entire egg (Fig. 40.22a). In animals with very little or no yolk, such as mammals and sea urchins, cell division extends completely through the egg (Fig. 40.22b). A developing embryo that lacks sufficient yolk depends on another source of nutrition; that source is the placenta in humans and other placental mammals.

Cleavage results in a solid ball of cells called a **morula** (see Fig. 40.21). Further cell divisions result in a fluid-filled ball of cells called a **blastula** (or a blastocyst in humans and other mammals). It is around this time, approximately 5 days after fertilization, that the developing embryo implants into the uterine lining, which is thickened and highly vascularized under the influence of progesterone from the corpus luteum.

Implantation marks the beginning of **pregnancy**, the carrying of one or more embryos in the uterus. In humans, pregnancy lasts approximately 38 weeks from fertilization. Because the time of fertilization is often not known, human pregnancy is typically measured not from implantation but as 40 weeks from the first day of the last menstrual period. Pregnancy is divided into three periods called **trimesters**, each lasting about 3 months. The divisions between the

FIG. 40.22 Cleavage patterns. (a) Partial cleavage, shown in this zebrafish embryo. (b) Full cleavage, shown in this sea urchin embryo.
Photo sources: a. © 2002 Steve Baskauf; b. DIOMEDIA/Carolina Biological.

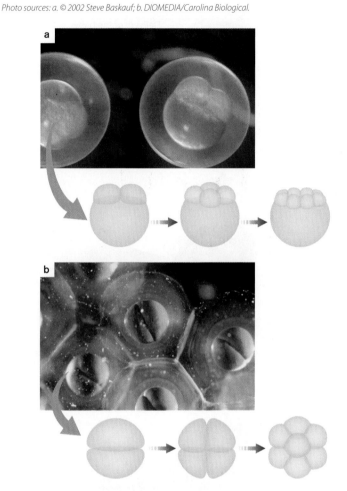

FIG. 40.23 Early human embryonic development.

a. Blastocyst

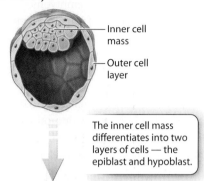

The inner cell mass differentiates into two layers of cells — the epiblast and hypoblast.

b. Bilaminar embryo

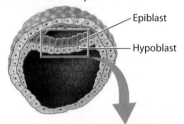

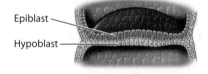

c. Gastrulation

The epiblast cells migrate inward, displace the hypoblast cells, and differentiate into all three germ layers — ectoderm, mesoderm, and endoderm.

d. Trilaminar embryo

Germ layers: Ectoderm   Mesoderm   Endoderm

- Cells in the outer layer of skin
- Pigment cells
- Nerve cells in the brain

- Cells in the inner layer of skin
- Muscle cells
- Bone cells
- Red blood cells

- Cells lining the inside of the digestive tract
- Cells lining the inside of the lung
- Liver cells
- Pancreas cells

three trimesters are arbitrary, but they provide a convenient way to follow the major changes that occur during pregnancy.

The first trimester is characterized by spectacular changes in the developing embryo under the influence of genes and developmental pathways that are largely conserved across vertebrates (Chapter 19). During this time, the embryo develops all its major organs and begins to be recognizably human. Toward the end of the first trimester, a human embryo is known as a **fetus**.

In mammals, such as humans, the blastocyst has an inner cell mass and an outer layer of cells (**Fig. 40.23a**). The inner cell mass becomes the embryo itself, and the outer cell layer forms part of the placenta. The placenta is composed of both maternal and embryonic cells and is the site of gas exchange, nutrient uptake, and waste excretion between maternal and fetal blood. The placenta also secretes the hormones estrogen and progesterone, which maintain the uterine lining during pregnancy. Meanwhile, the inner cell mass differentiates into a two-layered (bilaminar) structure consisting of the epiblast and the hypoblast (**Fig. 40.23b**).

Gastrulation is the second key process in embryonic development. It is a highly coordinated set of cell movements that leads to a fundamental reorganization of the embryo. In humans, cells of the epiblast divide and migrate inward (**Fig. 40.23c**). These migrating epiblast cells form three layers or sheets of cells called **germ layers**: the outer **ectoderm**, intermediate **mesoderm**, and inner **endoderm** (**Fig. 40.23d**). All adult tissues and organs are made from one or more of these three germ layers. At this stage, the embryo is known as a trilaminar (three-layered) embryo to reflect the presence of the three germ layers.

As with cleavage, the details of gastrulation vary in different organisms. One notable difference is the fate of the opening through which cells migrate during gastrulation. In protostomes ("mouth first"), such as worms, mollusks, and arthropods, it becomes the mouth. In deuterostomes ("mouth second"), such as vertebrates and sea urchins, it becomes the anus, with the mouth forming later (Chapter 42).

Organogenesis, the third key developmental process, is the transformation of the three germ layers into all the organ systems of the body. The ectoderm becomes the outer layer of the skin and nervous system; the mesoderm makes up the circulatory system, muscle, and bone; and the endoderm becomes the lining of the digestive tract and lungs (Fig. 40.23d). Because all of the basic structures and organ systems develop during the first trimester of pregnancy, it is during this time that the developing embryo is particularly vulnerable to toxins, drugs, and infections.

It is also during the first trimester that the embryo begins to develop male or female characteristics. In humans, sex is determined by the presence or absence of the Y chromosome. The Y chromosome contains a gene, *SRY*, that encodes a protein called testis-determining factor (TDF). TDF acts on the developing gonad to cause it to differentiate into a testis. In the absence of TDF in females, the immature gonad becomes an ovary. Once sex is determined, sexual differentiation follows under the influence of hormones produced by the gonads. In humans, then, sex determination is genetic, as is the case in other mammals and in birds. For other organisms, such as some reptiles and fishes, sex determination depends instead on environmental cues such as temperature and crowding.

The second and third trimesters are characterized by fetal growth.

During the second trimester, from around 13 to 27 weeks, the fetus continues to develop and grow. This is the time when the mother first detects fetal movements. Because of improvements in medical care, babies born in the last few weeks of the second trimester are often able to survive.

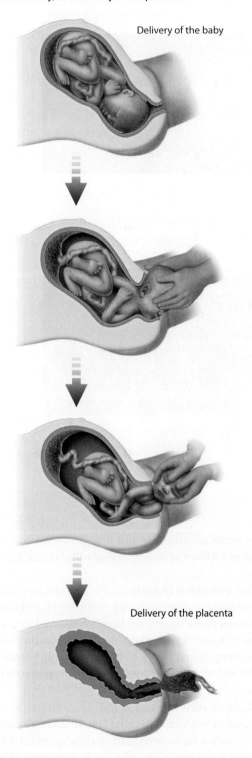

FIG. 40.24 Childbirth. Childbirth involves changes to the cervix, delivery of the baby, and delivery of the placenta.

The third trimester, from 28 to 40 weeks, is characterized by rapid growth, with the fetus doubling in size during the last 2 months. Weight gain is critical, as birth weight is a major factor in infant mortality. The fetus often moves less frequently during this trimester as it becomes increasingly compressed by limited space.

Childbirth is initiated by hormonal changes.

Childbirth marks the end of pregnancy. Although we tend to think of childbirth as the delivery of the baby, it actually consists of three events: changes in the cervix to allow passage of the baby, descent and delivery of the baby, and delivery of the placenta. These three events result from strong muscular contractions of the uterus stimulated by the hormone oxytocin, which is released by the posterior pituitary gland. The action of oxytocin is a good example of positive feedback: exposure to oxytocin causes uterine contractions, stimulating more oxytocin release, which in turn leads to further contractions, and so on.

During pregnancy, the cervix is a tube about 4 cm long closed with a mucus plug. To allow passage of the infant, the cervix shortens and widens. These changes are not enough to accommodate passage of the baby during delivery. Humans have a narrow pelvis relative to that of other primates as an adaptation for bipedalism. To ease its passage through the pelvis, the baby goes through a series of movements (**Fig. 40.24**). Babies are usually born head first with the face pointing down (toward the mother's back) when passing through the birth canal. Other positions make vaginal birth more difficult, and sometimes impossible.

The third and final stage of childbirth is the delivery of the placenta, which usually occurs approximately 15 to 30 minutes after the baby is born.

Although birth normally occurs vaginally, in some cases, such as when the baby is too large or positioned feet first, it is necessary to remove the baby by making an incision in the abdomen and uterus in a surgical procedure known as a caesarean section. Interventions such as caesarean sections, the use of antibiotics, and prenatal and postnatal care have greatly reduced the maternal and fetal mortality associated with childbirth.

Whichever way the baby is delivered, when the umbilical cord is cut and the baby takes its first breath, it begins its life independent of the mother—anatomically independent, that is. Otherwise, human newborns are completely dependent on others. Humans are unusual among animals in having an extended period of immaturity while their brains and bodies grow and develop. This long period of immaturity means that parents usually put considerable effort into raising and caring for their offspring, and it allows for extended periods of learning, the development of strong social bonds, and the transmission of culture to the next generation.

Human development from fertilization to the end of pregnancy is summarized in **Fig. 40.25**.

Self-Assessment Questions

12. When are primary oocytes and spermatocytes produced, and how many gametes result from one diploid cell in females and males?

13. How are male and female gametogenesis similar? How are they different?

14. What are three key steps in the development of a single-celled zygote into a multicellular individual?

15. A multicellular organism starts off as a single cell, which divides by mitosis to produce many cells. Therefore, for the most part, all the cells in a multicellular organism are genetically identical—they are clones—yet in time different groups of cells look different and function quite differently from one another. How can this be?

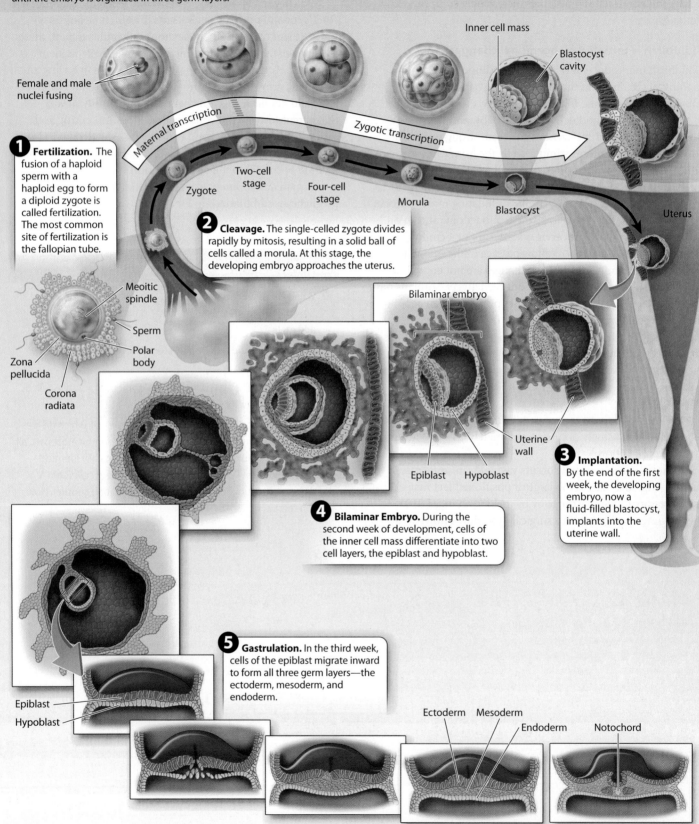

VISUAL SYNTHESIS FIG. 40.25
Reproduction and Development
Integrating concepts from Chapters 9, 10, 11, 18, 19, 26, and 40

Early development
The fertilized egg is a totipotent stem cell because it can give rise to a complete organism. During development, cells become increasingly restricted in their developmental potential as a result of changes in gene expression. During gastrulation, cells further differentiate and move relative to one another until the embryo is organized in three germ layers.

1 Fertilization. The fusion of a haploid sperm with a haploid egg to form a diploid zygote is called fertilization. The most common site of fertilization is the fallopian tube.

2 Cleavage. The single-celled zygote divides rapidly by mitosis, resulting in a solid ball of cells called a morula. At this stage, the developing embryo approaches the uterus.

3 Implantation. By the end of the first week, the developing embryo, now a fluid-filled blastocyst, implants into the uterine wall.

4 Bilaminar Embryo. During the second week of development, cells of the inner cell mass differentiate into two cell layers, the epiblast and hypoblast.

5 Gastrulation. In the third week, cells of the epiblast migrate inward to form all three germ layers—the ectoderm, mesoderm, and endoderm.

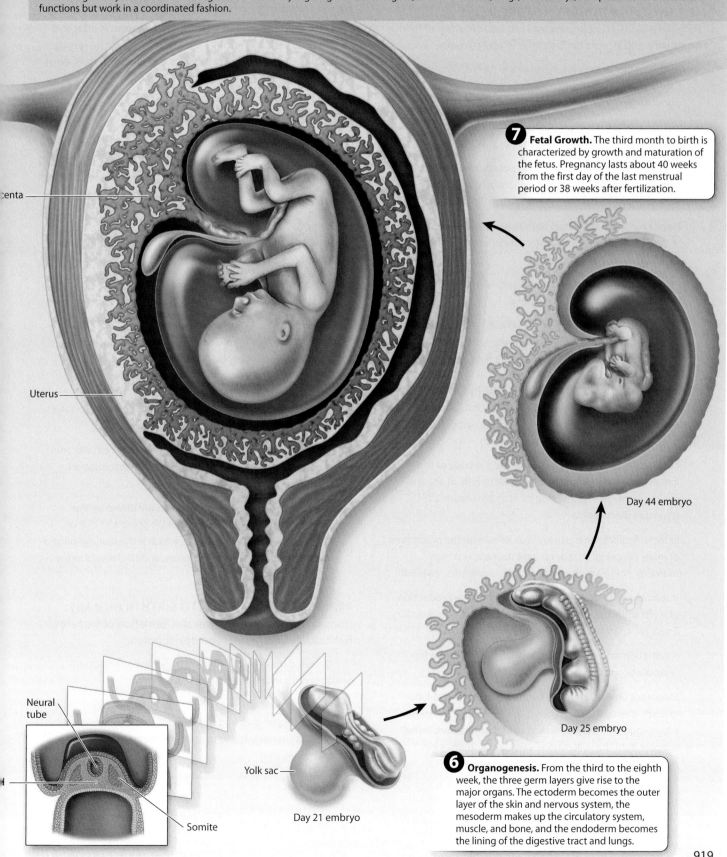

CORE CONCEPTS SUMMARY

40.1 THE EVOLUTIONARY HISTORY OF REPRODUCTION: Reproduction is a basic feature of living organisms and occurs both asexually and sexually.

Asexual reproduction involves a single parent. Offspring are produced by binary fission or mitotic cell division and are clones of the parent. page 897

Examples of asexual reproduction include budding, fragmentation, and parthenogenesis. page 898

Sexual reproduction involves two parents: two haploid gametes, produced by meiotic cell division, fuse to make a diploid zygote. page 899

Many organisms reproduce both sexually and asexually. For some organisms, environmental conditions favor one mode or the other. page 899

Asexually reproducing populations grow rapidly but are genetically uniform clones. page 900

Sexual reproduction produces genetically unique offspring, a feature that is thought to explain why most eukaryotes reproduce sexually at least some of the time. page 901

40.2 MOVEMENT ONTO LAND AND REPRODUCTIVE ADAPTATIONS: The movement of vertebrates from water to land involved changes in reproduction, including internal fertilization and the amniotic egg.

Fertilization can occur outside or inside the body of the female. External fertilization is possible only in an aquatic environment. Internal fertilization is an adaptation to terrestrial life. page 903

External fertilization is usually associated with the production of many offspring, little or no parental care, and high mortality. These are characteristics of *r*-strategists. page 903

Internal fertilization is often associated with the production of fewer offspring, increased parental care, and low mortality. These are characteristics of *K*-strategists. page 903

A watertight sac, called the amnion, is an adaptation for reproduction on land. page 904

Oviparous animals lay eggs. Ovoviviparous and viviparous animals give birth to live young, but ovoviviparous animals retain eggs until they hatch and commonly get nutritional support from the yolk, whereas viviparous animals get nutritional support from the mother. page 904

40.3 HUMAN REPRODUCTIVE ANATOMY AND PHYSIOLOGY: The male reproductive system is adapted for the production and delivery of sperm, and the female reproductive system is adapted for the production of eggs and, in some cases, support of the developing fetus.

Male gametes, called sperm, develop in the seminiferous tubules of the testes. page 905

Sperm travel from the seminiferous tubules to the epididymis to the vas deferens to the ejaculatory duct and finally to the urethra before being released during ejaculation. page 905

Semen is a mixture of sperm and fluid from the prostate gland, seminal vesicles, and bulbourethral glands. page 906

Developing female gametes, called oocytes, are produced in the ovaries. Mature oocytes are called eggs, or ova. page 907

An oocyte is released from the ovary, travels through the fallopian tube, and implants in the uterus if it is fertilized. page 908

The reproductive system responds to and produces hormones. The hypothalamus releases GnRH, which stimulates the anterior pituitary gland to release FSH and LH. These hormones, in turn, stimulate the ovaries to release estrogen and progesterone or the testes to secrete testosterone. page 908

In males, LH stimulates Leydig cells to secrete testosterone, and FSH and testosterone act on Sertoli cells to support sperm production. page 909

In females, monthly menstrual cycles are driven by the interplay of hypothalamic, anterior pituitary, and ovarian hormones, leading to ovulation, which is the maturation and release of an oocyte from the ovary, and to changes to the uterine lining. page 909

40.4 GAMETE FORMATION TO BIRTH IN HUMANS: Human reproduction involves the formation of gametes, fertilization, and growth and development.

Gametogenesis is called spermatogenesis in males and oogenesis in females. page 911

In males, spermatogonia differentiate into primary spermatocytes that undergo meiosis to make immature haploid spermatids and mature spermatozoa or sperm. page 911

Spermatogenesis begins at puberty and continues throughout life. page 911

In females, oogonia undergo differentiation into primary oocytes during fetal development. The primary oocytes begin meiosis but arrest in prophase I. During a menstrual cycle, these oocytes are released from arrest, finish meiosis I, but then arrest in metaphase II. They remain arrested in metaphase II until they are fertilized. page 911

Fertilization is the fusion of the oocyte and sperm plasma membranes. It usually occurs in the fallopian tube and results in the formation of a diploid zygote. page 912

Cleavage is the division of the zygote by mitotic cell division into smaller cells. At the blastocyst stage, the developing embryo implants in the uterus. page 914

Human pregnancy lasts approximately 38 weeks from fertilization until birth and is divided into three trimesters. page 914

Gastrulation results in the formation of the three germ layers: ectoderm, mesoderm, and endoderm. page 916

Organogenesis is the formation of organs. page 916

Childbirth, regulated by the posterior pituitary hormone oxytocin, involves changes in the cervix, delivery of the baby, and delivery of the placenta. page 917

Log in to LaunchPad to check your answers to the Self-Assessment Questions and to access additional learning tools.

CHAPTER 41 Animal Immune Systems

CORE CONCEPTS

41.1 AN OVERVIEW OF THE IMMUNE SYSTEM: The immune system protects organisms from pathogens.

41.2 INNATE IMMUNITY: The innate immune system acts generally against a diversity of pathogens.

41.3 B CELLS AND ANTIBODIES: The adaptive immune system includes B cells that produce antibodies against specific pathogens that reside outside cells in tissues.

41.4 T CELLS AND CELL-MEDIATED IMMUNITY: The adaptive immune system also includes helper and cytotoxic T cells that attack infected, diseased, and foreign cells.

41.5 THREE PATHOGENS: A virus, bacterium, and eukaryote illustrate how some pathogens have evolved mechanisms that enable them to evade the immune system.

All organisms have mechanisms that protect them against harmful pathogens. Pathogens evolve strategies to evade these mechanisms, and organisms in turn evolve new responses. In animals, the result of this evolutionary back and forth is an elaborate and sophisticated immune system. The immune system is remarkable in its ability, in most cases, to protect organisms from harmful pathogens. Consider how well your body is able to defend itself, without taking antibiotics, from the common cold and from bacterial infections that enter through cuts and scrapes. In fact, because our immune system is so effective, we don't even notice most of the pathogens we encounter every day. A key feature of the vertebrate immune system is that it has evolved both defenses that act against a broad range of pathogens and defenses that target specific pathogens.

We begin with an overview of the immune system and then discuss general defenses that protect against a diverse array of pathogens. We then turn to a powerful part of the immune system that is tailored to attack specific pathogens and that has the ability to learn and remember, attributes we usually associate with the nervous system. Finally, we consider three infections that illustrate how pathogens have evolved ways to evade these defenses.

41.1 AN OVERVIEW OF THE IMMUNE SYSTEM

The immune system differs in important ways from the other systems we have considered up to this point. Most organ systems are made up of discrete tissues or organs. Consider the cardiovascular system, which includes the heart and associated blood vessels. Or think of the gastrointestinal system, which is made up of a series of organs, such as the esophagus, stomach, and intestine, each with a specialized function for digesting food and absorbing nutrients.

By contrast, the immune system consists of organs, cells, proteins, and even biochemical reactions that are spread throughout the body. Its presence everywhere in the body makes sense, in that the immune system must be able to interact with pathogens wherever they might be found. Some organs are dedicated to the immune system, such as the thymus in many vertebrates, but otherwise the immune system is a diffuse system of cells, molecules, and processes. In fact, the phrase "immune system" was not even used until the late 1960s. It was introduced to draw attention to the fact that the immune system is truly a *system* made up of many different components. These components work by different mechanisms, but all share a common function: to protect an organism from pathogens.

Pathogens cause disease.

Pathogens are organisms and other agents that cause disease. For a bacterium, pathogens include viruses that infect bacteria, as we have seen (Chapter 18). These viruses are called bacteriophages (literally "bacteria eaters"). They can cause the bacterial cell to lyse, or burst, when viral progeny are produced inside the bacterium. In response, bacteria protect themselves with broadly effective mechanisms such as the cell wall or mechanisms such as CRISPR directed at specific pathogens. CRISPR allows bacteria to "remember" past viral infections and mount a response when subsequent infections by the same virus occur (Chapter 12).

For a fungus, pathogens include bacteria and viruses. In turn, fungi have evolved molecules that kill or inhibit the growth of bacteria. We have isolated many of these substances and learned how to use them as antibiotics. Penicillin is a famous example. It was discovered in 1928 by Alexander Fleming, who noticed that a fungus in the genus *Penicillium* interfered with the growth of bacteria in a petri dish.

For an animal, pathogens include viruses, bacteria, fungi, single-celled eukaryotes commonly called protists, and worms (**Fig. 41.1**). These organisms cause disease in diverse ways. In some cases, they infect a particular type of cell. For example, human immunodeficiency virus (HIV) destroys a specific type of cell involved in the immune system itself, making the body susceptible to other pathogens that it would usually be able to easily fight off. In other cases, pathogens proliferate in an organ and interfere with its function. Pneumonia, for example, can arise when bacteria or fungi proliferate in lung tissue and limit gas exchange. Sometimes, pathogens produce toxins that cause disease. Cholera, for instance, results from infection with a toxin-producing bacterium called *Vibrio cholera*. The toxin interferes with membrane transporters in the small intestine and causes diarrhea, resulting in dehydration and sometimes death.

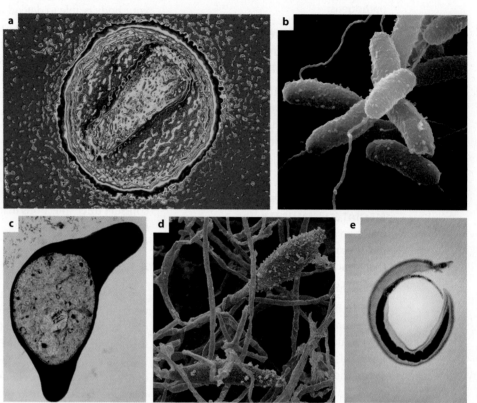

FIG. 41.1 Human pathogens. (a) Human immunodeficiency virus (HIV); (b) cholera bacteria; (c) a malaria parasite (green) inside a swollen red blood cell; (d) ringworm (a fungus); (e) schistosome (a worm). *Sources: a. James Cavallini/Science Source; b. Juergen Berger/Science Source; c. Dr. Tony Brain/Science Source; d. David Scharf/Science Source; e. Scott Camazine/Getty.*

The immune system distinguishes self from nonself.

Critical to the functioning of the immune system is the ability to recognize pathogens as foreign. That is, the immune system must be able to distinguish the molecules and cells of its own organism, described as host or **self**, from the molecules and cells of other organisms, described as foreign or **nonself**, and then respond only to those that are nonself.

The immune system is able to distinguish self from nonself largely by recognizing proteins that normally reside on the surface of cells. For example, in organ transplants, the immune system sometimes recognizes proteins on the surface of transplanted organs as nonself. The transplanted organs are then attacked by the immune system, leading to rejection of the transplant.

Sometimes, the immune system mistakes self and nonself. If the immune system reacts inappropriately to self, it can end up attacking its own cells or tissues. This type of reaction occurs in **autoimmune diseases**, such as rheumatoid arthritis, type 1 diabetes, and multiple sclerosis. Alternatively, the immune system may sometimes recognize as nonself harmless substances such as pollen or dust, leading to allergic reactions. In some cases, the immune system may not react when it should, so that it does not adequately protect an organism. Such a weak immune system is most likely to occur at the extremes of age or when infections or diseases compromise its function, a condition called **immunodeficiency**.

The activity of the immune system is carefully regulated. When functioning properly, immune cells and proteins remain in an inactive state until they come into contact with a foreign molecule or cell. In this way, the immune system reacts only when it is needed to attack a pathogen.

The distinction between self and nonself is a useful way to think about how the immune system functions, but it has limitations. For example, as we will see, the immune system is able to recognize and attack self cells that are diseased or infected, but it also tolerates bacteria and other nonself microorganisms that make up our microbiome (Case 5 The Human Microbiome). French immunologist Polly Matzinger proposed that another way to think about the immune system is that it recognizes and attacks agents that cause injury or damage, not necessarily those that are foreign. Her idea suggests that we have more work to do to fully understand how the immune system functions.

The immune system consists of innate and adaptive immunity.

The immune system consists of two parts that work in different ways and interact with each other to protect against infection. Each part is made up of different cell types and pathways. The first type of immunity, called **innate immunity** (or natural immunity), provides protection against all kinds of infection in a nonspecific manner. Innate immunity is present in fungi, plants, and animals and is an evolutionarily early form of immunity. It is considered *innate* because it does not depend on prior exposure to a pathogen.

By contrast, the second form of immunity, called **adaptive immunity** (or acquired immunity), is specific to a given pathogen. Adaptive immunity "remembers" past infections, meaning that subsequent encounters with the same pathogen generate a stronger response. In other words, the immune system *adapts* over time, and immunity is *acquired* after an initial exposure. Adaptive immunity evolved later than innate immunity and is unique to vertebrates.

Table 41.1 summarizes the major differences between these two forms of immunity.

Self-Assessment Questions

1. How are the innate and adaptive immune systems similar?
2. How are the innate and adaptive immune systems different?

41.2 INNATE IMMUNITY

The innate immune system provides general, nonspecific defenses against a broad array of pathogens. In animals, it includes tissues, such as the skin, that act as a barrier to infection, cells that act as sentries against pathogens, and proteins and other molecules circulating in the blood that signal the presence of infection. All these defenses

TABLE 41.1 Comparison of the Innate and Adaptive Immune Systems

	INNATE IMMUNE SYSTEM	ADAPTIVE IMMUNE SYSTEM
Targets	Diverse pathogens	Diverse pathogens
Ability to distinguish self from nonself?	Yes	Yes
Specificity?	No	Yes
Memory?	No	Yes
Organisms	All organisms	Vertebrates

are nonspecific, act immediately, and do not depend on past exposure to elicit a response.

The skin and mucous membranes provide the first line of defense against infection.

Organisms have a barrier that surrounds them. Examples include the cell wall of bacteria, the bark of trees, the cuticle of leaves, the exoskeleton of insects, the scales of fish, the shells of eggs, and the skin of mammals (**Fig. 41.2**). These structures act as physical barriers against infection.

FIG. 41.2 Physical barriers against infection. (a) Tree bark; (b) leaf cuticle; (c) insect exoskeleton; (d) egg shell; (e) skin. *Sources: a. Christian Beier/AGE Fotostock; b. Michael P. Gadomski/Science Source; c. Charles J. Smith; d. Inga Spence/Alamy; e. Janaka Dharmasena/Shutterstock.*

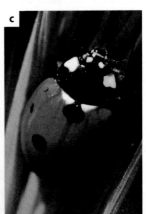

To understand the importance of this barrier, consider what happens when it is breached. Burn victims who lose the outer layer of their skin are particularly vulnerable to infection. The bite of a dog or the sting of an insect can insert viruses, bacteria, or parasites directly into the bloodstream, bypassing the protective layer of skin. Even a small puncture wound can provide entry for the bacterium *Clostridium tetani*, the cause of tetanus.

Any surface that comes in contact with the external environment is a potential entry site for pathogens. Like the skin, the mucous membranes of the respiratory, gastrointestinal, and genitourinary tracts act as physical barriers. However, they have additional protective mechanisms. For example, parts of the respiratory tract have **cilia**, hairlike projections that sweep mucus containing trapped particles and microorganisms out of the body. Saliva and tears contain enzymes that help to protect against bacteria. The stomach, with its low pH and enzymes that break down food, is inhospitable to many microorganisms.

In addition to acting as a physical and chemical barrier, the skin and mucous membranes provide a home to many nonpathogenic microorganisms. These microorganisms in some cases provide protection against harmful pathogens by competing with them for food and space. They may also maintain health in other ways, such as by aiding digestion. In this way, the human body engages in a mutualistic relationship with these microorganisms (Case 5 The Human Microbiome).

White blood cells provide a second line of defense against pathogens.

If a pathogen makes it past the skin and mucous membranes of an animal, a second line of defense may rise to combat it: **white blood cells** (leukocytes). Many different types of white blood cells are present in the body, but all arise through differentiation from stem cells in the bone marrow (**Fig. 41.3**). Just as the skin and mucous membranes are an external defense against pathogens, white blood cells are an internal defense. Some white blood cells mount an early response to a pathogen; these are nonspecific innate immune cells. If they are unable to clear the pathogen, they alert the more specialized white blood cells of the adaptive immune system. Here, we focus on the role of white blood cells in innate immunity.

Phagocytes (literally, "eating cells") are white blood cells that engulf and destroy foreign cells or particles. The engulfing of a cell or particle by another cell is called **phagocytosis**. The Russian microbiologist Ilya Mechnikov first described this remarkable process. For this discovery, he shared the Nobel Prize in Physiology or Medicine with the German scientist Paul Ehrlich in 1908.

The phagocytic process begins when a phagocyte encounters a cell or particle that it recognizes as foreign and binds to it. The phagocyte then extends its plasma membrane completely around the cell or particle until it is contained in a compartment or vesicle inside the phagocyte (**Fig. 41.4**). In some cases,

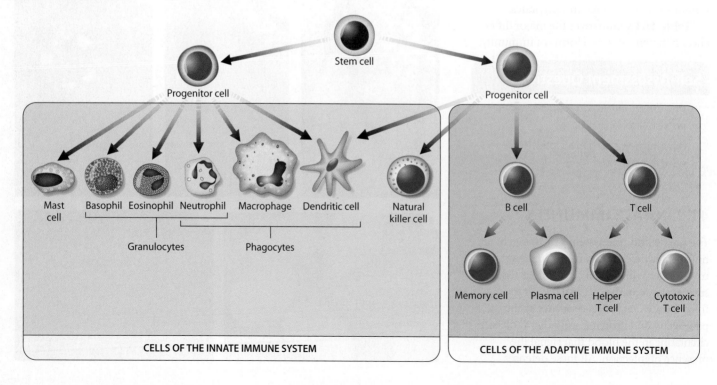

FIG. 41.3 White blood cells. Many different types of white blood cells function in the innate and adaptive immune systems. They develop from stem cells in the bone marrow.

FIG. 41.4 Phagocytosis. A phagocyte recognizes, engulfs, and digests foreign particles and cells. The scanning electron micrograph shows a macrophage engulfing tuberculosis bacteria. *Photo source: SPL/Science Source.*

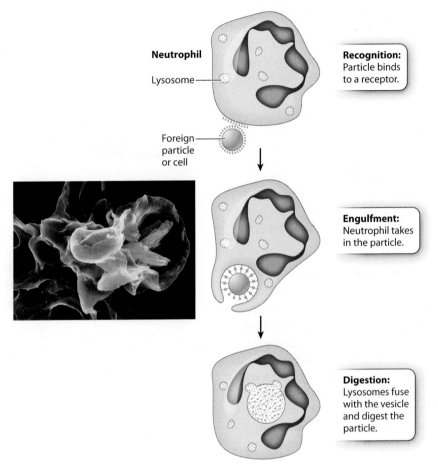

the vesicle merges with a lysosome, and enzymes within the lysosome digest the foreign cell or particle. Phagocytosis can also trigger a respiratory burst, a process that generates reactive oxygen species, such as superoxide radicals and hydrogen peroxide, and reactive nitrogen species, such as nitrogen oxide and nitrogen dioxide. These molecules have antimicrobial properties that directly damage pathogens.

Three major types of phagocytic cells are present in the body: **macrophages**, **dendritic cells**, and **neutrophils** (see Fig. 41.3). In each case, the name gives clues about the cell's shape and size. Macrophages are large cells that patrol the body (*macro* means "large"). Dendritic cells have long cellular projections that resemble the dendrites of neurons (Chapter 34). Both macrophages and dendritic cells are typically part of the natural defenses found in the skin and mucous membranes. Neutrophils are members of a group of cells called **granulocytes** because of the presence of granules in their cytoplasm. The granules of neutrophils take up both acidic and basic dyes, so their staining pattern is considered neutral. Neutrophils are abundant in the blood and are often the first cells to respond to infection.

The innate immune system includes several other white blood cell types in addition to phagocytic cells (see Fig. 41.3). Two types of granulocytes, **eosinophils** and **basophils**, defend against parasitic infections, but also contribute to allergies. **Mast cells** release **histamine**, an important contributor to allergic reactions and inflammation. Finally, **natural killer cells** do not recognize foreign cells, but instead identify and kill host cells that are infected by a virus or have become cancerous or otherwise abnormal.

Phagocytes recognize foreign molecules and send signals to other cells.

Phagocytes attack foreign cells that they encounter but leave host cells alone. How is a phagocyte able to distinguish a foreign cell from a host cell? All cells have on their surface an array of proteins and other molecules, as we saw in Chapters 5 and 9. Phagocytes detect surface molecules that are common to pathogens, but not to the host.

Toll-like receptors (TLRs) are proteins on the surface of phagocytes that recognize and bind to molecules on pathogens. First discovered in the fruit fly *Drosophila*, TLRs are found on cells in vertebrates and invertebrates, and similar molecules are found in plants and bacteria, indicating that they are an early, evolutionarily conserved part of the innate immune system. These proteins provide one of the earliest signals that an infection is present. TLRs recognize evolutionarily conserved surface molecules on a wide range of microorganisms. For example, one type of TLR recognizes a glycolipid on the cell wall of certain bacteria; another type recognizes a protein that is part of the bacterial flagellum.

Binding of TLRs to surface molecules on the pathogen is a signal to the phagocyte to engulf and destroy its target. Following binding of a foreign molecule to the TLR, phagocytes send a signal to the rest of the immune system that the body is under attack by releasing chemical messengers called **cytokines**. These chemical compounds recruit other immune cells to the site of injury or infection. Cytokines, like hormones, provide long-distance communication between cells (Chapters 9 and 36).

Inflammation is a coordinated response to tissue injury.

Although some immune cells are already present at the site of tissue injury or infection, many more must be recruited to fight infection successfully. That recruitment happens as part of a process called inflammation. The redness and swelling you see following an injury, a break in the skin, or an infection are signs of inflammation. More precisely, **inflammation** is a physiological response of the body to injury that removes the inciting agent if present and begins the healing process—for example, by initiating cell division to heal a wound. It works through the coordinated responses of the local tissue, the vascular system, and the innate immune system. Inflammation is typically beneficial, but if left unchecked or allowed to continue for a long time, it can be debilitating.

Inflammation is characterized by four signs, described by their Latin names: *rubor* (redness), *calor* (heat), *dolor* (pain), and *tumor* (swelling). The process begins following tissue injury, such as a cut or infection, as shown in **Fig. 41.5**. Certain cells in the tissue, such as dendritic cells and mast cells, recognize a pathogen as foreign and release cytokines and other chemical messengers that recruit more white blood cells to the site or make it easier for them to reach the site quickly. For example, histamine, which is released by mast cells and basophils, acts directly on blood vessels to cause vasodilation. Vasodilation increases the blood flow to the site of infection or injury, which leads to the characteristic redness and heat of inflammation. In addition, histamine increases the permeability of the blood vessel wall. Fluid leaks out of the blood vessel, carrying white blood cells into the damaged tissue. The increased fluid in the tissue surrounding the blood vessels is visible as swelling. Some chemical messengers act directly on nerve fibers, causing pain.

During inflammation, phagocytes in the blood often move from a blood vessel to the site of infection by **extravasation**

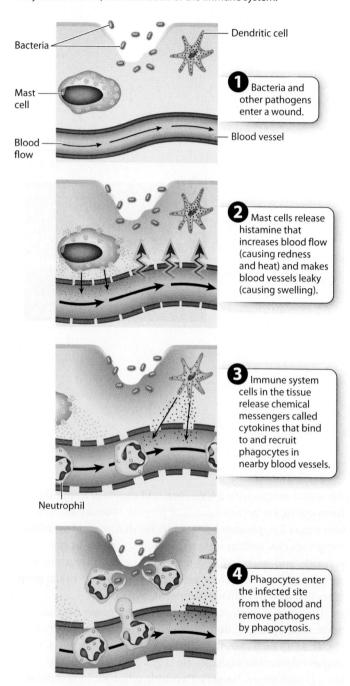

FIG. 41.5 Inflammation. Inflammation involves increased blood flow, leaky blood vessels, and activation of the immune system.

1. Bacteria and other pathogens enter a wound.
2. Mast cells release histamine that increases blood flow (causing redness and heat) and makes blood vessels leaky (causing swelling).
3. Immune system cells in the tissue release chemical messengers called cytokines that bind to and recruit phagocytes in nearby blood vessels.
4. Phagocytes enter the infected site from the blood and remove pathogens by phagocytosis.

(**Fig. 41.6**). In this process, the phagocyte travels along the vessel wall in a rolling motion, grabbing weakly onto the wall as glycoproteins on the phagocyte surface bind transiently to proteins on the endothelial cells. Stronger interactions put a brake on this rolling motion and allow the phagocyte to adhere more firmly to the vessel wall. As it nears the site of infection, the phagocyte encounters and binds to cytokines. The phagocyte

FIG. 41.6 Extravasation. In extravasation, a phagocyte moves from a blood vessel to the surrounding tissue.

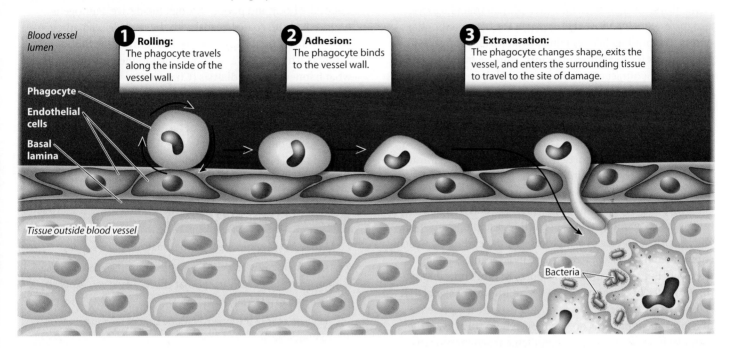

then changes shape and moves in between cells lining the blood vessel and into the surrounding tissue.

The complement system participates in the innate and adaptive immune systems.

We have seen that many white blood cells participate in innate immunity, providing a general defense against pathogens. Other white blood cells participate in adaptive immunity, and white cells from both the innate and adaptive immune systems can be activated by a set of proteins that circulate in the blood. Collectively, these proteins make up the **complement system**. First described by Paul Ehrlich in the late 1800s, this system is so named because it supplements or *complements* other parts of the immune system; it is considered to be part of the innate immune system.

The complement system consists of more than 25 proteins, many of which are enzymes, that circulate in the blood in an inactive form. The system is activated when one of these proteins binds to molecules found only on invading microorganisms or to antibodies (discussed later in this chapter). Activation sets off a biochemical cascade in which the product of one reaction is an enzyme that catalyzes the next. This kind of sequential activation leads to amplification of the response, as we saw earlier in cell signaling (Chapter 9) and in the endocrine system (Chapter 36), where a small stimulus can have a dramatic response.

Activation of the complement system has three effects (**Fig. 41.7**). First, and most dramatically, complement proteins circulating in the blood assemble to form a pore, called a

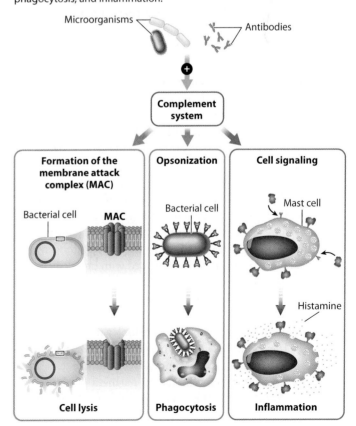

FIG. 41.7 The complement system. The proteins of the complement system contribute to the defense against pathogens by cell lysis, phagocytosis, and inflammation.

membrane attack complex (MAC), in bacterial cells. The MAC causes the bacterial cell to break open, or lyse. Second, the complement system coats bacteria with a protein that phagocytes recognize, allowing those cells to target the bacteria for phagocytosis. The binding of a protein to a pathogen to target it for phagocytosis is called **opsonization**. Third, activated complement proteins attract other components of the immune system. For example, some of these proteins stimulate mast cells, which release histamine and initiate the inflammatory response.

You can appreciate the importance of the complement system by considering what occurs in individuals who lack key components. Complement deficiencies are a form of immunodeficiency, a condition in which the immune system does not function properly. For example, one type of complement deficiency makes individuals susceptible to bacterial infections because fewer pathogens undergo cell lysis and opsonization. The intensity and type of symptoms vary depending on which component of the complement system is lacking, ranging from very mild, to recurrent infections, to other immunological disorders. Because antibodies are an important signal that activates the complement system, and because the presence of complement proteins on a pathogen helps to activate the adaptive immune system, the complement system provides an important bridge between the innate, nonspecific part of the immune system and the adaptive, specific part, a topic we consider in the next section.

 CASE 7 BIOLOGY-INSPIRED DESIGN: USING NATURE TO SOLVE PROBLEMS

How can we use a protein that circulates in the blood to treat sepsis?

One complement protein is known as mannose-binding lectin (MBL). MBL circulates in the blood, where it recognizes carbohydrates on the surface of diverse pathogens, including viruses, bacteria, fungi, and protists. MBL binds to carbohydrates on the surface of these pathogens and can activate the complement system.

Researchers at the Wyss Institute for Biologically Inspired Engineering at Harvard University made use of MBL to build a device to treat sepsis. Sepsis is a blood infection that can lead to organ failure and death; it is a common cause of patients' deaths in the hospital. Because the identity of the pathogen is often not known, sepsis is typically treated with broad-spectrum antibiotics, which act against a wide variety of common bacteria. However, the antibiotics do not always work and many patients do not recover.

The researchers at the Wyss Institute took a different approach. They built a device that makes use of MBL, which recognizes and binds to many common bacteria, like those that cause sepsis. They began by coating magnetic beads with a modified form of MBL. The beads were inserted into a device that mimics the spleen, which helps to remove old cells and pathogens from the blood. Blood from a patient was circulated out of the body and passed through the device. Pathogens bound the MBL-coated beads, with the complex of pathogen–MBL–bead subsequently being removed from the blood by magnets. The blood was then returned to the patient, similar to what happens during dialysis (Chapter 39).

In experiments with rats, the biospleen, as the device is called, was able to remove 90% of bacteria in the blood. Furthermore, the survival rate of rats with sepsis increased significantly when rats were treated in this way compared to untreated control rats. The biospleen, if it proves effective in humans, has the potential to revolutionize the treatment of sepsis.

Self-Assessment Questions

3. What are the components of the innate immune system?
4. Certain diseases, as well as cigarette smoke, can damage cilia lining the respiratory tract. What effects would you expect from damage to respiratory cilia?
5. How do cells of the innate immune system recognize pathogens?
6. How do the four signs of inflammation relate to the changes that occur in the underlying tissues?

41.3 B CELLS AND ANTIBODIES

The innate immune system can react to diverse pathogens and has the capacity to distinguish self from nonself. The adaptive immune system shares these features, but also has two additional ones. The first is specificity: the adaptive immune system produces an array of molecules, each of which has the potential to target a specific pathogen it has never before encountered. As a result, it is often more effective at eliminating a pathogen than the innate immune system is. The second is memory: the adaptive immune system remembers past infections and mounts a stronger response on re-exposure.

Consider for a moment the number of pathogens that we might encounter—on the order of millions—and remember that they, too, are evolving all the time. How can the adaptive immune system tailor its response to each of these pathogens? And how does it remember past infections? We explore answers to these questions in the next two sections.

While many cells and tissues play a role in adaptive immunity, two types of white blood cells are particularly important: **B lymphocytes** (also called **B cells**) and **T lymphocytes** (also called **T cells**) (see Fig. 41.3). B and T cells are named for their sites of maturation. In birds, B cells mature in the bursa of Fabricius, an organ located off the digestive tract (**Fig. 41.8**). In mammals, B cells mature in the bone marrow.

FIG. 41.8 Organs that participate in the immune system.

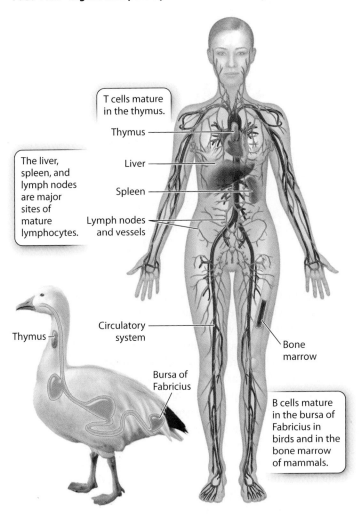

In most vertebrates, T cells mature in the thymus, an organ of the immune system located just behind the sternum, or breastbone. B and T cells circulate in blood vessels and lymphatic vessels, which are thin-walled vessels that absorb excess fluid from tissues (Chapter 37). B and T cells also travel through the spleen, liver, and lymph nodes, where they sample the body for pathogens. We begin with a discussion of the role of B cells in adaptive immunity.

B cells produce antibodies.

One of the hallmarks of adaptive immunity is the ability to target specific novel pathogens. Of course, many different viruses, bacteria, single-celled eukaryotes, fungi, and worms can lead to infection. How does the adaptive immune system recognize each one? The specificity of the adaptive immune system is in part the result of antibodies produced by B cells.

An **antibody** is a large protein found on the surface of B cells or free in the blood or tissues. An antibody binds to molecules on the surface of foreign cells, called **antigens**, which, by definition, are molecules that lead to the production of antibodies. (The word "antigen" is a contraction of "antibody" and "generator.") Any molecule that naturally occurs on or in a microorganism can be an antigen. The great diversity of antigens on different microorganisms is matched by a great diversity of antibodies produced by an individual. It is estimated that humans produce approximately 10 billion different antibodies, each with the ability to recognize one or a few antigens on microorganisms.

Antibodies have a distinctive structure that reflects their specificity for a given antigen. The simplest antibody molecule, shown in **Fig. 41.9**, is a Y-shaped protein made up of four polypeptide chains: two identical **light (L) chains** and two identical **heavy (H) chains**, so named because of their relative molecular weights. Covalent disulfide bonds hold the four chains together.

The light and heavy chains are both subdivided into **variable (V)** and **constant (C)** regions (Fig. 41.9). The constant region, as the name implies, is the same in all antibodies of a given class. In contrast, the variable region is unique to each antigen that the antibody targets. Within the variable regions of the L and H chains are regions that are even more variable, called hypervariable regions. These hypervariable regions interact in a specific manner with just a portion of the antigen.

Binding of antibodies to antigens is the first step in recognition and removal of pathogens. Binding alone is sometimes enough to disable a pathogen. More often, the pathogen is killed by a different component of the immune system, such as by a phagocyte or the complement system. In this case, the function of an antibody is to recognize the pathogen, then

FIG. 41.9 Structure of one type of antibody. An antibody has variable regions that distinguish it from the billions of other antibodies and allow it to bind a specific antigen.

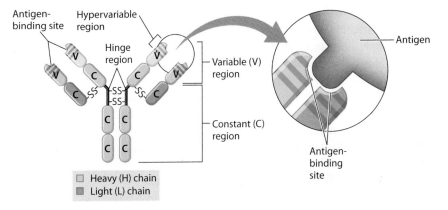

recruit other cells of the immune system or activate the complement system.

Mammals produce five classes of antibodies with different functions.

In mammals, there are five classes of antibodies, called immunoglobulins G, M, A, D, and E (abbreviated as **IgG**, **IgM**, **IgA**, **IgD**, and **IgE**, respectively), each with a different function (**Fig. 41.10**). IgG is the Y-shaped antibody depicted in Figs. 41.9 and 41.10. IgG circulates in the blood and is particularly effective against bacteria and viruses.

IgM is a pentamer in mammals and a tetramer in fishes. (A pentamer is a molecule made of five monomers; a tetramer is a molecule made of four monomers.) The individual units—the monomers—are linked by a joining chain. IgM also exists as a monomer on the surface of B cells. IgM is important in the early response to infection.

IgA is usually a dimer consisting of two antibody molecules linked by a joining chain. It is the major antibody on mucosal surfaces, such as those found in the respiratory, gastrointestinal, and genitourinary tracts. IgA is also present in secretions, including tears, saliva, and breast milk.

IgD and IgE are monomers. IgD is typically found on the surface of B cells. This immunoglobulin helps initiate inflammation. IgE plays a central role in inflammation, as well as in allergies, asthma, and other **immediate hypersensitivity reactions**, in which the body mounts a heightened or inappropriate immune response to common antigens. Immediate hypersensitivity reactions result from the binding of IgE to receptors on the surface of mast cells and basophils. Binding leads to the release of histamine and cytokines, resulting in inflammation that can be life threatening.

Clonal selection is the basis for antibody specificity.

A central question in immunology is how an individual can generate a great diversity of antibodies, each of which is specific for a given antigen. An organism might come across seemingly limitless types of antigens. How can B cells generate antibodies that interact only with specific antigens among this tremendous diversity?

Two alternative hypotheses have been proposed to explain antibody specificity. In 1900, Paul Ehrlich suggested that the body contains a large pool of B cells, each with a different antibody on its surface. The antigen binds to one or a few of these antibodies, stimulating the B cell to divide. In this model, specificity exists before any exposure to the antigen. Ehrlich's model drew on concepts from chemistry such as the binding of an enzyme to a substrate in the manner of a lock and key (Chapter 6).

Later, in the early 1900s, a different hypothesis was suggested. This hypothesis proposed that the antigen instructs the antibody to fold in a particular way so that the two interact

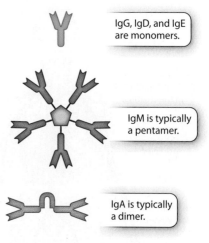

FIG. 41.10 Five classes of antibodies.

in a specific manner. In this model, the antigen plays a more active role in the process, and antibody specificity is determined only after interaction with the antibody.

Which of these two hypotheses is correct? Does the antigen *select* a B cell with a preexisting surface antibody with which it interacts specifically, as in the first hypothesis, or does it *instruct* the B cell to make a specific antibody, as in the second?

Experimental data on the genetic basis for antibody diversity (described later in this section) provided strong evidence for the first hypothesis. Every individual has a very large pool of B cells, each with a different antibody on its cell surface (**Fig. 41.11**). Collectively, these antibodies recognize a great diversity of antigens, but each individual antibody can recognize only one antigen or a few antigens. An antigen interacts with the antibody on the B cell, or a small set of B cells, that has the best fit for the antigen. Binding leads the B cell to divide and differentiate into two types of cells. Daughters of the B cell called **plasma cells** secrete antibodies that bind to the antigen and elicit an immune response against the pathogen. Long-lived daughters of the B cell called **memory cells** contain membrane-bound antibody having the same antigen specificity as the parent cell. Since the resulting population of B cells arose from one B cell, they are all clones of one another.

The process by which antigen binding generates a clone of B cells is called **clonal selection**, and it is one of the central principles of immunology. Almost counterintuitively, antibody specificity is achieved before any exposure to the antigen. In this way, clonal selection resembles natural selection (Chapter 20). In natural selection, genetic variation occurs in a population; the environment then selects among this variation, allowing some individuals to survive and reproduce more than others. Similarly, in clonal selection, there is variation in a population of B cells, and the antigen selects among this variation, allowing some cells to proliferate more than others.

FIG. 41.11 Clonal selection. Those B cells that recognize a foreign antigen are selected to divide and differentiate into plasma cells and memory cells.

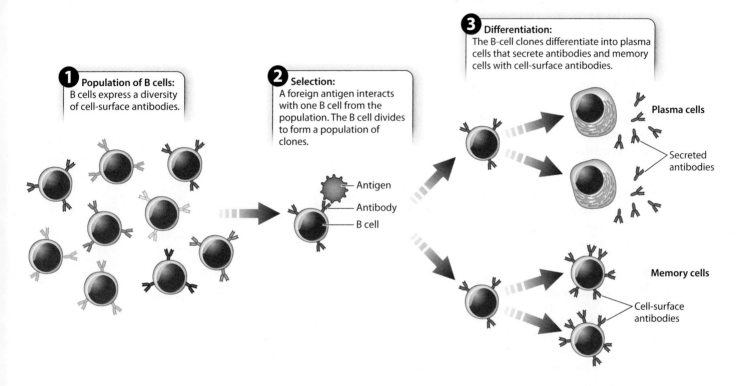

Clonal selection explains immunological memory.

Clonal selection is the basis for antibody specificity. It also explains the second feature of the adaptive immune system: the ability to remember past infections—that is, to react more vigorously on re-exposure. The first encounter with an antigen leads to a **primary response (Fig. 41.12)**. In this response, there is a short lag before antibody is produced. The lag is the time required for B cells to divide and form plasma cells and memory cells. The plasma cells secrete antibodies, typically IgM. The level of IgM in the blood increases, peaks, and then declines.

On re-exposure to the same antigen, even months or years after the first exposure, there is a **secondary response**, which is quicker, stronger, and longer-lasting than the primary response (Fig. 41.12). During this response, memory B cells formed during the primary response differentiate into plasma cells that release IgG antibodies. These memory B cells respond more quickly to antigen exposure than naïve B cells (B cells that have not been previously exposed to antigen). In addition, the pool of memory B cells is larger than the pool of naïve B cells for a given antigen, explaining why the secondary response produces more antibodies for a longer period of time.

The presence of long-lived memory cells provides long-lasting immunity after an infection such as chickenpox. We have learned to make use of immunological memory in **vaccination**. A vaccine deliberately exposes an individual to an antigen from a pathogen to induce a primary response but not induce the disease, thereby providing future protection from infection by the same pathogen.

Vaccination was discovered by a combination of observation and experiment. Until the early 1700s, it was common

FIG. 41.12 Primary and secondary responses. The immune system responds more strongly to the second exposure to an antigen than to the first exposure because of the presence of memory B cells.

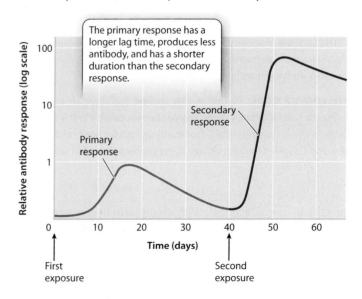

HOW DO WE KNOW?

FIG. 41.13

How is antibody diversity generated?

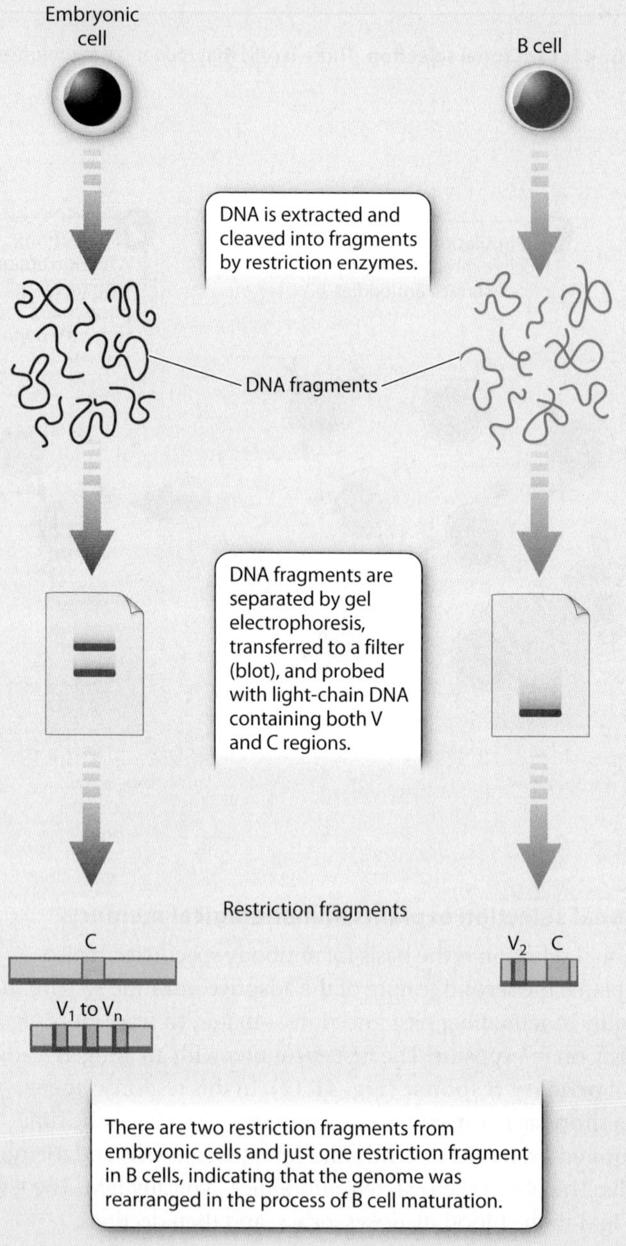

BACKGROUND Humans produce an estimated 10 billion different antibodies. Antibodies are proteins, encoded by genes. However, the human genome contains only approximately 25,000 protein-coding genes. How can humans (and other vertebrates) produce such a diversity of antibodies from this limited set of genes? This was a central, unanswered question in immunology until the 1960s and 1970s.

HYPOTHESIS In 1965, American biologists William Dreyer and J. Claude Bennett proposed a model in which there are many slightly different copies of two gene segments, which they called the V (variable) and C (constant) segments. According to this model, each unique antibody gene is assembled by recombining a single copy of each of these segments. At the time, there was no direct experimental evidence to support this hypothesis.

EXPERIMENT In 1976, Japanese immunologists Nobumichi Hozumi and Susumu Tonegawa tested the Dreyer and Bennett hypothesis. Using mice as a model system, they isolated DNA from embryonic cells (before recombination was thought to occur) and from adult B cell tumor lines that produce one type of antibody (after recombination was thought to occur). The procedure Hozumi and Tonegawa used was slightly different from what is described here, but conceptually the same. Extracted DNA is cut by restriction enzymes, the fragments are separated on a gel and transferred to a filter paper (blot), and light chain DNA (containing both V and C regions) is used as a radioactive probe to determine the size and positions of the V and C regions. (This technique, known as a Southern blot, is discussed in Chapter 12.)

RESULTS Hozumi and Tonegawa found that the V and C regions were in two different restriction fragments in embryonic cells and in one restriction fragment in adult B cells. Thus, the V and C regions are far apart in embryonic cells and close together in adult B cells.

CONCLUSION Antibody genes are assembled by recombination of individual gene segments during B cell differentiation, providing a mechanism for generating antibody diversity.

FOLLOW-UP WORK Further studies showed that there are multiple copies of several different gene segments, now called the V (variable), D (diversity), J (joining), and C (constant) regions for heavy chains, and V, J, and C regions for light chains. These are recombined in unique patterns in different B cells, so each B cell produces one antibody and a population of B cells produces a diversity of antibodies.

SOURCES Dreyer, W. J., and J. C. Bennett. 1965. "The Molecular Basis of Antibody Formation: A Paradox." *Proceedings of the National Academy of Sciences* 54:864–869; Hozumi, N., and S. Tonegawa. 1976. "Evidence for Somatic Rearrangement of Immunoglobulin Genes Coding for Variable and Constant Regions." *Proceedings of the National Academy of Sciences* 73:3628–3632.

---

practice to inoculate people with smallpox to induce a mild disease and prevent a more severe, even lethal, infection. It was also well known that milkmaids, exposed to the relatively benign cowpox virus from milking cows, were immune to the related but more deadly smallpox virus. In 1796, the English scientist Edward Jenner specifically tested the consequences of exposure to cowpox by inoculating a young boy with cowpox and demonstrating that he became immune to smallpox. In fact, the word "vaccine" is derived from the Latin *vacca*, which means "cow."

Today, vaccines take many forms: they can be a protein or a part of a protein from the pathogen, a live but weakened form of the pathogen, or a killed pathogen. They are among the most effective public health measures ever developed.

Genomic rearrangement generates antibody diversity.
We pointed out earlier that antibody specificity is determined before exposure to the antigen. Clonal selection therefore requires a mechanism for producing a great diversity of antibodies. How is this diversity achieved? Diversity would result if each antibody were encoded by a separate gene. However, this cannot be the case, as the number of antibodies produced by an organism far exceeds the total number of genes in the genome. For example, the human genome consists of approximately 25,000 protein-coding genes, but humans can produce an estimated 10 billion different antibodies.

In 1965, American biologists William Dreyer and J. Claude Bennett suggested a novel hypothesis to explain antibody diversity. As they put it, this hypothesis was "radically different from anything found in modern molecular genetics." Dreyer and Bennett proposed that a single antibody is made by separate gene segments that are brought together by recombination. According to this model, many copies of each gene segment are present, each slightly different from the others. Only one copy of each gene segment ends up in the final, recombined antibody gene (and hence in the antibody protein). Diversity is therefore achieved by combining different copies of each segment in different B cells as they mature.

More than 10 years after Dreyer and Bennett proposed their model, direct experimental evidence showed that their hypothesis is correct (**Fig. 41.13**). As a B cell differentiates, different gene segments are joined in a process called **genomic rearrangement** that produces a specific antibody.

Genes for H chains are composed of multiple different V (variable), D (diversity), J (joining), and C (constant) gene segments (**Fig. 41.14**). Genes for mice H chains, for example, can be composed of any of 100 different V segments, 30 D segments, 6 J segments, and 8 C segments. During B cell differentiation, the DNA for all of these gene segments undergoes recombination so that just one of each type of segment is present in the DNA, and the intervening DNA is deleted. The assembled VDJ segment encodes the variable portion, and the C segment encodes the constant region. L chain genes also contain multiple copies of gene segments, but they do not have a D gene segment, so the VJ segment encodes the variable portion and the C segment encodes the constant region. The result of this process is that each B cell encodes a single H chain and a single L chain, and each B cell expresses a unique antibody.

It is worth pausing here to consider the significance of this mechanism. We generally think of DNA as being a stable

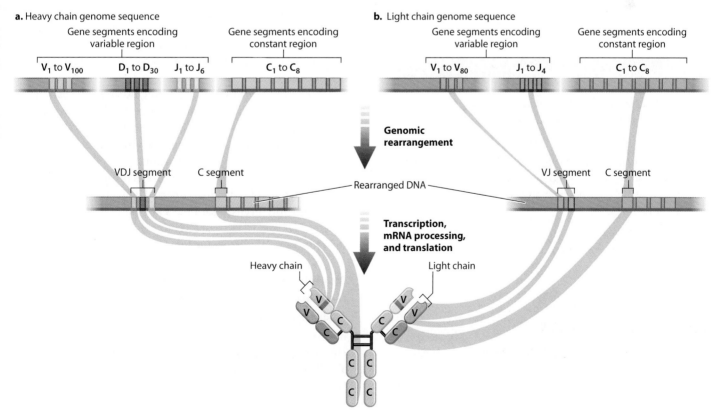

FIG. 41.14 Genomic rearrangement in mice. In genomic rearrangement, different gene segments recombine to generate a functional gene encoding the parts of an antibody.

blueprint present in identical copies in all of the somatic cells in our bodies. B cells are an exception. As a result of genomic rearrangement, the DNA in each mature B cell is different from the DNA in every other mature B cell, and is also different from the DNA in other cells in the body.

Genomic rearrangement is the primary mechanism that creates antibody diversity, but other mechanisms contribute as well. For example, the association of different L and H chains makes different kinds of functional antibodies. In addition, recombination is sometimes imprecise, creating different sequences at the junctions between the different gene segments. Other mechanisms add or replace nucleotides within VJ and VDJ segments. Finally, alternative splicing of mRNA (Chapter 3) allows a single B cell to express both membrane-bound and secreted antibodies.

As a group, B cells produce an enormous diversity of antibodies, whereas a given B cell makes one and only one antibody. But B cells are diploid, containing maternal and paternal copies of the genes that encode for antibodies. Maternal and paternal alleles of the antibody genes are similar, but not identical. Therefore, a B cell that is able to use both alleles would make two different antibodies, yet B cells always make one. Remarkably, once one allele undergoes genomic rearrangement, the other allele is prevented from doing so. As a result, B cells function as if they have a single allele for each antibody gene: they express the maternal gene or the paternal gene, but not both.

The mechanisms we have considered pertain to humans and mice. Whereas B cells mature in the bone marrow in humans and mice, they mature in gut-associated lymphoid tissue (GALT) in many other vertebrates. Furthermore, genomic rearrangement is less important in many vertebrates. Instead, other mechanisms appear to play a more important role.

Self-Assessment Questions

7. Draw a picture of an antibody molecule, labeling each of the components and the places where antigens bind.
8. How is immunological memory achieved?
9. Look again at Fig. 41.12. Imagine that a different novel antigen is added on day 40 rather than a second injection of the same antigen. Would there be a primary or a secondary response to this second antigen?
10. If a given B cell produces only one type of antibody, how do organisms produce a great diversity of antibodies?

41.4 T CELLS AND CELL-MEDIATED IMMUNITY

Antibodies are remarkable for their diversity and specificity. They provide protection against many kinds of bacteria, especially those enclosed in a polysaccharide capsule, such as the bacteria that cause pneumonia and meningitis. However, B cells and their antibodies have limitations. For example, B cells on their own can make antibodies against only some antigens; for other antigens, they require the assistance of other cells. Furthermore, they are sometimes ineffective against pathogens that take up residence inside a host cell, as do the bacteria that cause tuberculosis.

Another part of the adaptive immune system handles these and other kinds of pathogens. This aspect of the immune response depends on the second type of lymphocyte: the T cell. T cells do not secrete antibodies. Instead, they participate in **cell-mediated immunity**, so named because cells, rather than antibodies, recognize and act against pathogens. T cells target infected or diseased cells. Their importance is most dramatically demonstrated by individuals taking immunosuppressive drugs or infected by HIV, both of which suppress the activity of T cells. The result can be overwhelming infections and even tumors.

T cells include helper and cytotoxic cells.

There are two major subpopulations of T cells with different names, functions, and cell-surface markers: **helper T cells** and **cytotoxic T cells** (see Fig. 41.3). Helper T cells do just that: they *help* other cells of the immune system by secreting cytokines. One of their key roles is to activate B cells to secrete antibodies. Although B cells can work on their own, particularly against encapsulated bacteria, in most cases they require the participation of helper T cells. Helper T cells also activate macrophages, cytotoxic T cells, and other cells of the immune system.

Cytotoxic T cells *kill* other cells—that's what "cytotoxic" means. Like B cells, they are activated by cytokines released from helper T cells. Cytotoxic T cells are particularly effective against altered host cells, such as those infected with a virus or that have become cancerous.

These two types of T cells are distinguished by the presence of different glycoproteins on their surface: CD4 on helper T cells and CD8 on cytotoxic T cells. The ratio of these two classes of T cell can be used to assess immune function. In healthy individuals, the CD4 : CD8 ratio is usually about 2 : 1. A lower ratio, indicating the presence of fewer helper T cells relative to cytotoxic T cells, is typical of individuals infected with HIV. HIV infects and kills helper T cells, disarming a key player in the immune system. CD4 and CD8 assist receptors on T cells in carrying out their function. We discuss the function of T cells next.

T cells have T cell receptors on their surface that recognize an antigen in association with MHC proteins.

T cells, like B cells, originate in the bone marrow. However, unlike B cells, T cells mature in the thymus. A mature T cell is characterized by the presence on the plasma membrane of a **T cell receptor (TCR)**, a protein receptor that recognizes and binds to the antigen.

In many ways, TCRs are similar to antibodies on the surface of B cells. For example, TCRs recognize antigens with a specific structure. In addition, the many TCRs differ from one another, but each T cell has just one type of TCR on its surface. Binding of TCR to an antigen triggers the T cell to divide into clones, resulting in a pool of T cells that are each specific for a given antigen. Finally, as with the antibody diversity, the diversity of TCRs among T cells results from genomic rearrangement of V, D, J, and C gene segments.

However, TCRs are different from antibodies in important ways. First, they are composed of two, rather than four, polypeptide chains (**Fig. 41.15**). Second, they are not secreted like antibodies, but are always bound to the membrane on the T cell's surface. Third, the TCR does not recognize an antigen by itself. Instead, it recognizes an antigen in association with specific proteins that appear on the surface of most mammalian cells. These proteins are encoded by the **major histocompatibility complex (MHC)**. MHC proteins were first discovered in transplantation biology because their presence leads to acceptance or rejection of transplanted tissues.

The MHC is a cluster of genes in mammals that encode proteins on the surface of cells. This complex is composed of many genes, all having a high rate of polymorphism, meaning that there is a lot of variation in the gene sequence (and consequently in the protein sequence) among different individuals. In humans and mice, the genes are divided into three classes: **class I** genes are expressed on the surface of all nucleated cells; **class II** genes are expressed on the surface of macrophages, dendritic cells, and B cells; and **class III** genes encode several proteins of the complement system and proteins involved in inflammation.

How do helper and cytotoxic T cells recognize and respond to antigens? Let's consider the activation of helper T cells first (**Fig. 41.16**). When an antigen enters the body, it may be recognized by an antibody directly or be taken up by **antigen-presenting cells**. These cells, which include macrophages, dendritic cells, and B cells, take up the antigen and return portions of it to the cell surface bound to MHC class II proteins. Through their T cell receptors, helper T cells recognize processed antigen along with MHC class II proteins. When TCR binds to antigen and MHC class II proteins,

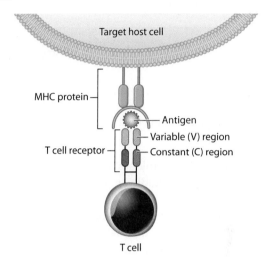

FIG. 41.15 The T cell receptor (TCR). The TCR is a protein on the surface of T cells that binds antigens found in association with MHC proteins.

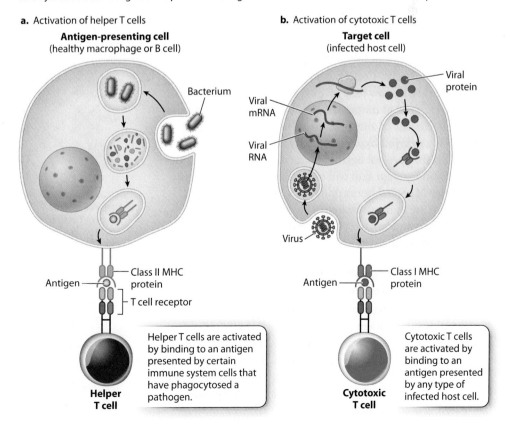

FIG. 41.16 Three-way interactions among the T cell receptor, MHC protein, and processed antigen. Helper T cells recognize the processed antigen found in association with class II MHC protein, and cytotoxic cells recognize the processed antigen found in association with class I MHC protein.

TABLE 41.2	Comparison of Helper and Cytotoxic T Cells	
	HELPER T CELL	**CYTOTOXIC T CELL**
Functions	Activation of macrophages Activation of B cells Activation of cytotoxic T cells Secretion of cytokines	Cell killing
Surface molecule	CD4	CD8
MHC protein recognized by the T cell	Class II	Class I

the helper T cells release cytokines that activate other parts of the immune system, including macrophages, B cells, and cytotoxic T cells.

Cytotoxic T cells also recognize antigens displayed by host cells, but only those antigens that are associated with MHC class I proteins (Fig. 41.16). Because class I proteins are present on almost all host cells, cytotoxic T cells recognize and kill any host cell that becomes abnormal in some way. For example, a virus-infected cell often expresses viral antigens with MHC class I proteins on its surface. Cytotoxic T cells recognize the antigen and MHC class I proteins and kill the virus-infected cell. Tumor cells express novel antigens along with MHC class I proteins, and in some cases may be eliminated by cytotoxic T cells. **Table 41.2** summarizes the major features of helper and cytotoxic T cells.

Once the helper or cytotoxic T cells become bound to the MHC–antigen complex, they divide and form clones. Some cells of each type are memory cells that provide long-lasting immunity following an initial infection, as in the case of B cells. Also like B cells, T cells can sometimes be activated too strongly. Earlier, we saw how IgE bound to mast cells and basophils can lead to immediate hypersensitivity reactions, like those characteristic of allergies and asthma. The counterpart in T cells is a **delayed hypersensitivity reaction**, which, as its name suggests, does not begin right away. For example, if you touch poison ivy, your skin will turn red and start itching only after a delay of several hours or days. Delayed hypersensitivity reactions are initiated by helper T cells. These cells release cytokines that attract macrophages to the site of exposure, which is typically the skin.

The ability to distinguish between self and nonself is acquired during T cell maturation.

Of the many T cell receptors that may potentially be generated by genomic rearrangement, only some are useful. Useful T cell receptors must be able to react with the host's own MHC proteins, but not with other molecules that are normally present in or on a host organism's own cells. In other words, T cells must respond only to self MHC proteins in association with nonself antigens.

A sorting process ensures that only properly functioning T cells mature and others are eliminated (**Fig. 41.17**). As T cells

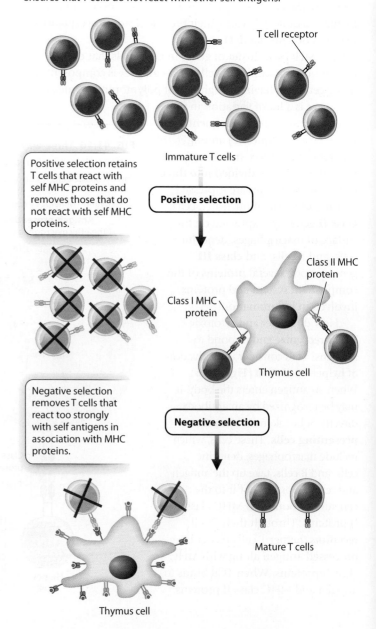

FIG. 41.17 Positive and negative selection. Positive selection ensures that T cells react with self MHC proteins, and negative selection ensures that T cells do not react with other self antigens.

mature in the thymus, they interact with cells of that organ's epithelium. The T cells that recognize self MHC proteins on epithelium cells are **positively selected** and continue to mature. The T cells that react too strongly to self antigens in association with MHC are **negatively selected** and eliminated through cell death. In spite of their names, both processes involve the elimination of some T cells.

The results of this sorting process are twofold. First, T cells become MHC restricted, meaning that they interact only with antigens that are associated with MHC proteins (helper T cells with antigen plus MHC class II proteins, and cytotoxic T cells with antigen plus MHC class I proteins). Second, T cells exhibit **tolerance**: they do not respond to self antigens, even though the immune system functions normally otherwise. Those antigens present as T cells mature are identified as self and do not elicit a response; those not present are nonself and, if encountered, do elicit a response.

B cells also exhibit tolerance to self antigens because they go through a similar process of negative selection in the bone marrow. During that process, B cells that react strongly to self antigens are eliminated through cell death. Although most self-reactive T and B cells are eliminated, some T and B cells that react to self antigens escape this developmental check and end up in the circulation. Additional mechanisms are in place to eliminate these self-reactive cells, either through the activity of T cells or through cell death.

The ability to distinguish self from nonself is critical. Failure to make this distinction leads to autoimmune disease, in which tolerance is lost and the immune system becomes active against antigens of the host. Autoimmune diseases can be debilitating, as T cells or antibodies attack cells and organs of the host. For example, in rheumatoid arthritis, self-reactive T cells attack the joints, causing chronic inflammation and tissue damage; in type 1 diabetes, the attack is directed against insulin-producing cells of the pancreas; and in multiple sclerosis, the target is the myelin sheath surrounding nerves of the central nervous system. It is not fully understood how these self-reactive T cells evade checks on their development.

Self-Assessment Questions

11. How does T cell activation differ from B cell activation?
12. How do positive and negative selection lead to a population of T cells that react to MHC molecules in association with nonself antigens, but not to self antigens?
13. What happens when tolerance to self antigens is lost?

41.5 THREE PATHOGENS

Although we discussed the different components of the immune system one at a time, they actually work together to provide a coordinated response to pathogens. We have seen how B cells often require T cell help, how the complement system can be activated by antibodies produced by B cells, and how macrophages present antigens in association with MHC proteins to T cells. In turn, pathogens have evolved means of evading the immune system. In this section, we consider three pathogens: the flu virus, the bacterium that causes tuberculosis, and the malaria parasite. These pathogens illustrate how pathogens evolve and adapt in response to the immune system, and how the immune system in turn evolves and adapts in response to pathogens.

The flu virus evades the immune system by antigenic drift and shift.

Everyone knows the symptoms: fever, chills, sore throat, cough, weakness, and fatigue, perhaps a headache and muscle pain. Many viruses cause these common flu-like symptoms. However, the common cold, stomach flu, and 24-hour flu are caused by unrelated viruses, not by the flu virus.

The flu (influenza) is caused by an RNA virus that infects mammals and birds. It is notable for causing seasonal outbreaks, or epidemics. Four times in the last century—in the Spanish flu of 1918, the Asian flu of 1957–1958, the Hong Kong flu of 1968–1969, and the swine flu of 2009—the virus spread more widely than usual and infected more people, and the disease reached pandemic levels. The flu is more than just a nuisance: during pandemics, it causes hundreds of thousands of deaths worldwide.

Part of the success of the virus stems from its ability to spread easily. It can spread in the air by small droplets produced when an infected person coughs or sneezes. It can also spread by direct contact or by touching a contaminated surface, such as a doorknob. Hand washing, therefore, is one of the most effective means of protecting against the flu.

Another characteristic of the flu is its ability to evade the immune system. The flu virus can change the antigens on its outer coat, so memory cells, produced from earlier infections, no longer recognize them. There are three major types of flu virus (A, B, and C), and many different strains. For example, the swine flu pandemic of 2009 was caused by a type A flu virus of the strain H1N1. These strains differ in several respects, including the structure of a cell-surface glycoprotein called hemagglutinin (HA). HA binds to epithelial cells and controls viral entry into these cells. It can bind only to cells that display a complementary cell-surface protein. As a result, hemagglutinin determines in part which organisms the flu virus infects—humans, pigs, or birds—and which part of the body it infects—the nose, throat, or lungs.

Cells infected by viruses secrete cytokines that bring macrophages, T cells, and B cells to the site of infection, as we saw earlier. These cytokines, which are produced in abundance, lead to many of the symptoms commonly associated with the flu. For example, the cytokine interleukin-1 causes fever.

Once in the cell, the virus replicates its genome and makes more virus particles (Chapter 18). However, viral replication is prone to error, so there is a high rate of mutation. As a result, the amino acid sequences of antigens present on the viral surface, including HA, change over time. This process, called **antigenic drift** (Fig. 41.18a), allows a population of viruses to evolve over time and evade memory B and T cells that remember past infections.

While antigenic drift leads to a gradual change of the virus over time, the flu virus is also capable of sudden changes by a process called **antigenic shift** (Fig. 41.18b). The viral genome consists of eight linear RNA strands. If a single cell is infected with two or more different flu strains at one time, the RNA strands can reassort to generate a new strain. H1N1, for example, has genetic elements from human, pig, and bird flu viruses. Antigenic drift and shift make it difficult to predict from year to year which strains will be most prevalent and, therefore, which vaccine will be most effective.

Tuberculosis is caused by a slow-growing, intracellular bacterium.

Our second example is tuberculosis (TB). Like the flu, it is spread through small droplets dispersed in the air when a person with active disease coughs or sneezes. Also like the flu, the TB pathogen replicates inside host cells. But unlike the flu, TB is caused by a bacterium, not a virus. TB results from infection by *Mycobacterium tuberculosis*, a bacterium with several unusual properties. It is very small and has a compact genome, lacks a plasma membrane outside the cell wall, and replicates exceedingly slowly. *M. tuberculosis* takes approximately 16–20 hours to divide, compared to 20 minutes for the common intestinal bacterium *Escherichia coli*.

Tuberculosis is very common, affecting an estimated one third of the world's population. Symptoms are absent in most cases because the bacteria are in a dormant state. Approximately 10% of cases, however, are active. Active cases are characterized by chronic cough, fever, night sweats, and the weight loss that gives TB its other name: consumption.

Tuberculosis primarily affects the lungs. The bacteria infect macrophages in the alveoli of the lungs and replicate. In response, infected macrophages release cytokines, recruiting T and B cells to the site of infection. These lymphocytes surround the infected macrophages, forming a structure called a **granuloma** (Fig. 41.19). The granuloma helps to prevent the spread of the infection and aids in killing infected cells.

Many TB cases remain asymptomatic. However, the bacteria can sometimes overcome host defenses, especially if the immune system is compromised or suppressed, as in the case of co-infection with HIV. Treatment for active TB consists of a long course of multiple antibiotics. In many areas of the world, however, TB has evolved antibiotic resistance, a growing and alarming problem.

Diagnosis of TB poses challenges, especially in the case of asymptomatic infections. The tuberculin skin test is performed by injecting protein from the bacteria under the skin and

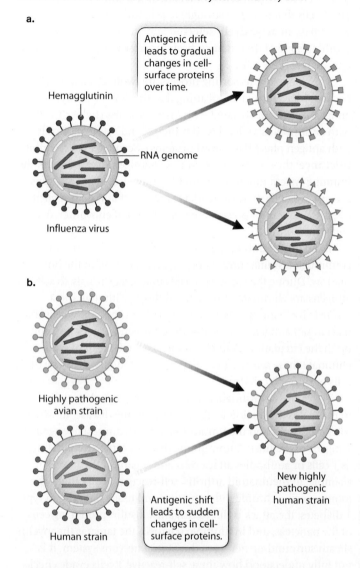

FIG. 41.18 (a) Antigenic drift and (b) antigenic shift. Virus populations evolve over time and evade the immune system's memory cells.

observing whether a reaction occurs, evidenced by the formation of a hard, raised area. A positive skin test is an example of a delayed hypersensitivity reaction. The reaction is caused by the recruitment of memory T cells to the site of inoculation, indicating that there had been an earlier infection with *M. tuberculosis*. Unfortunately, the tuberculin skin test is not always accurate, so newer tests are being developed.

Developing an effective vaccine to TB is challenging. The only vaccine currently available is BCG (Bacillus Calmette-Guérin, named for Albert Calmette, a French bacteriologist, and Camille Guérin, a French veterinarian, who developed it early in the twentieth century). BCG is a live but weakened strain of a related bacterium that infects cows. This vaccine was developed following the success of using cowpox to immunize against smallpox. However, BCG is only sometimes effective, and in 1930 a public health disaster in Germany resulted in the deaths

FIG. 41.19 A granuloma in a tubercular lung. *Photo source: Biophoto Associates/Science Source.*

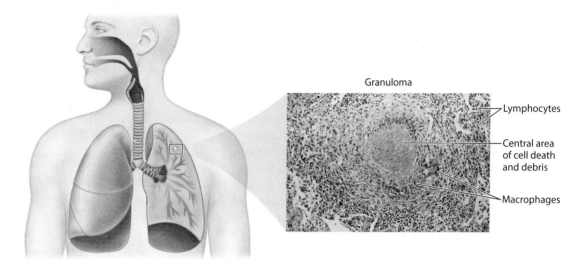

of 73 newborns from contaminated vaccine. Consequently, BCG is not used in the United States.

The malaria parasite changes surface molecules by antigenic variation.

"Bad air." That's the literal translation of "malaria," a contraction of the Italian *mala* ("bad") and *aria* ("air"). But malaria is not caused by the air, nor is it caused by a virus or bacterium. Instead, malaria is caused by a single-celled eukaryote of the genus *Plasmodium* (Case 4 Malaria). Several species infect humans, but *P. falciparum* is the most common and most virulent.

Malaria is both devastating and common. Approximately 500 million cases and 2 million deaths from this disease occur every year, mostly in sub-Saharan Africa, and mostly involving children. Malaria causes cyclical fevers and chills and can lead to coma and death.

The malaria parasite is transmitted to humans by the bite of an infected mosquito (**Fig. 41.20a**). Using mosquitoes as a vector, the malaria parasite deftly bypasses the natural protective barrier provided by the skin. The malaria parasite spreads to the liver and then to red blood cells, where it completes its life cycle. The progeny can be taken up again by mosquitoes.

Malaria is an ancient parasite that has coevolved with humans. We have already seen how altered hemoglobin provides some protection against infection (Chapter 20). But the malaria parasite, too, has evolved: it is able to evade the human immune system.

The malaria parasite resides in liver cells and red blood cells. Infected red blood cells can be removed by the spleen,

FIG. 41.20 Malaria. (a) The malaria parasite, a single-celled eukaryote, is transmitted by the bite of a mosquito. (b) Following infection, red blood cells adhere to blood vessel walls by interactions between a malaria-encoded cell-surface protein and receptors on the blood vessel wall. *Photo source: a. CDC/James Gathany/Science Source.*

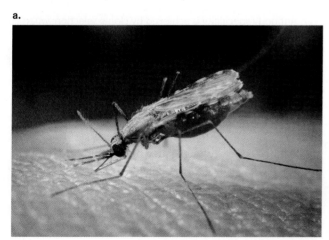
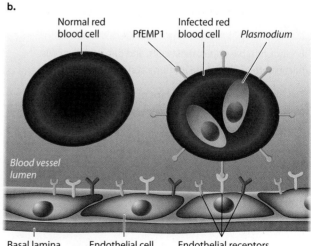

but the malaria parasite has evolved a way to avoid this filtering action. It expresses an adhesive protein, PfEMP1 (*Plasmodium falciparum* erythrocyte membrane protein 1), that inserts itself on the surface of red blood cells. This protein interacts with proteins on the surface of cells lining blood vessels, helping infected cells stick to blood vessel walls and keeping them in the circulatory system and out of the spleen (**Fig. 41.20b**). PfEMP1 could conceivably be a target for antibodies or TCRs. However, it is encoded by any one of approximately 60 genes, each slightly different from the others. The parasite expresses just one of these genes at a time and can change which one is expressed, in a process called **antigenic variation**. The great and ever-changing diversity of this protein, both in a single parasite and in the population as a whole, makes it a moving target for the immune system.

Despite continuing research, no vaccine against malaria has been developed. Researchers hope that the sequence of the malaria genome, completed in 2002, will provide new targets for vaccine research. In the meantime, efforts to control malaria focus on measures to control mosquitoes, such as insecticide-treated bed nets and spraying insecticides on areas where mosquitoes breed, and on treatment of patients with antimalarial medications. The evolution of resistance, however, is an all too familiar problem.

Self-Assessment Question

14. What are three mechanisms that allow pathogens to evade the immune system?

CORE CONCEPTS SUMMARY

41.1 AN OVERVIEW OF THE IMMUNE SYSTEM: The immune system protects organisms from pathogens.

Pathogens are agents or organisms that cause disease. page 923

The immune system is able to distinguish host molecules and cells, called self, from foreign molecules and cells, called nonself. page 924

The immune system consists of two parts: the innate immune system acts nonspecifically and the adaptive immune system acts specifically against pathogens. page 925

41.2 INNATE IMMUNITY: The innate immune system acts generally against a diversity of pathogens.

The skin and mucous membranes are physical barriers consisting of chemicals, cells, and microorganisms that provide protection against pathogens. page 925

Phagocytes are white blood cells that engulf other cells and debris. page 926

Eosinophils, basophils, mast cells, and natural killer cells participate in the innate immune system. page 927

Toll-like receptors on the surface of phagocytes recognize evolutionarily conserved molecules on the surface of foreign molecules. page 927

Inflammation is the body's response to infection or injury. It involves an increase in blood flow and blood vessel permeability, production of cytokines that signal phagocytes, and recruitment of phagocytes. page 928

The complement system consists of circulating proteins that activate one another. Activation results in the formation of a membrane attack complex that lyses bacteria, production of proteins that enhance phagocytosis, and generation of cytokines. page 929

41.3 B CELLS AND ANTIBODIES: The adaptive immune system includes B cells that produce antibodies against specific pathogens that reside outside cells in tissues.

B cells produce antibodies that bind to antigens. page 930

Antibodies have two light chains and two heavy chains joined by covalent disulfide bonds. page 931

Light and heavy chains have constant and variable regions. page 931

Mammals produce five types of antibodies: IgG, IgM, IgA, IgD, and IgE. page 932

Antibody specificity is achieved by clonal selection: when an antigen binds to antibodies on the surface of a B cell, the B cell differentiates into plasma and memory cells. page 932

The primary immune response is characterized by a long lag period and a short response. The secondary immune response is quicker and more robust, and provides the basis for immunity following infection or vaccination. page 933

Antibody diversity is generated by genomic rearrangement, in which individual gene segments are joined to make a functional gene. page 935

41.4 T CELLS AND CELL-MEDIATED IMMUNITY: The adaptive immune system includes helper and cytotoxic T cells that attack infected, diseased, and foreign cells.

- Two types of T cells are helper T cells and cytotoxic T cells. Helper T cells activate other cells of the immune system, including macrophages, B cells, and cytotoxic T cells. page 936

- Cytotoxic T cells kill altered host cells. page 936

- T cells are activated by the binding of a T cell receptor to an antigen in association with an MHC protein. page 937

- As T cells mature in the thymus, they undergo positive and negative selection, so only T cells that recognize MHC, but not self antigens, survive. page 938

- The ability to distinguish self from nonself antigens is critical for proper immune function; autoimmune diseases result when the immune system attacks cells of the host. page 939

41.5 THREE PATHOGENS: A virus, bacterium, and eukaryote illustrate how some pathogens have evolved mechanisms that enable them to evade the immune system.

- The flu is caused by an RNA virus that can change its surface proteins by antigenic drift and antigenic shift. page 939

- The bacterium that causes tuberculosis grows slowly and resides inside cells, particularly macrophages. page 940

- Malaria is caused by a single-celled eukaryote with a complex life cycle having stages in mosquitoes and in humans. The malaria parasite has evolved several mechanisms, including antigenic variation, that help it reproduce in humans without being eliminated by the immune system. page 941

Log in to **LaunchPad** to check your answers to the Self-Assessment Questions and to access additional learning tools.

CHAPTER 42 Animal Diversity

CORE CONCEPTS

42.1 SPONGES, CNIDARIANS, CTENOPHORES, AND PLACOZOANS: The simplest animals evolved multicellularity from single-celled ancestors.

42.2 PROTOSTOME ANIMALS: Protostomes are bilaterian animals and include lophotrochozoans and ecdysozoans.

42.3 ARTHROPODS: Arthropods, bilaterian animals named for their jointed legs, are extraordinarily diverse.

42.4 DEUTEROSTOME ANIMALS: Deuterostomes are bilaterian animals and include humans and other chordates, as well as acorn worms and sea stars.

42.5 VERTEBRATES: Vertebrates are deuterostomes with a bony cranium and typically a vertebral column, enabling them to become ecologically important in the water and on land.

The human body is complex, our diverse functions made possible by sophisticated, interacting organ systems: from muscle pairs attached to skeletons that allow us to run, jump, and dance, to lungs that exchange gases with the atmosphere and a brain and nervous system that enable us to sense our environment and respond to it. None of these organ systems is well developed in sponges, the simplest animals, and few can be found in sea anemones or jellyfish. How can we understand the evolutionary pathways that have given rise to animals as different as humans, jellyfish, and sponges?

In Chapter 33, we began our exploration of animal biology with a brief discussion of phylogeny, the evolutionary relationships among organisms. In this chapter, we return to questions of biological diversity, building a phylogeny of all animals based on features of anatomy, embryology, and molecular sequences. How do the combinations of characters exhibited by different animal groups reflect evolutionary history, and how do they relate to the distribution of species diversity among phyla?

Chapter 22 introduced the logic of phylogenetic reconstruction. Synapomorphies—shared derived characters, whether they be morphological or molecular—enable us to identify sister groups that diverged from a common ancestor at some point in the past. As we add morphological or molecular characters, we identify successively more inclusive sets of sister groups, eventually building a tree-like phylogeny that shows the evolutionary relationships among animals with diverse body plans and functional capabilities.

While phylogenies are built from the tips down, we want to interpret them as evolutionary histories, read from the root upward. Which features were present in the earliest animals, and how did successive evolutionary innovations give rise to the patterns of taxonomic diversity and ecological distribution we observe today? Biologists recognize approximately 36 distinct phyla (the singular is "phylum"), each with a fundamentally distinct body plan that marks a major branch of the animal tree. These body plans reflect adaptations that enabled different groups to persist and diversify through time, and they also impose constraints on the evolutionary pathways followed by different groups. For example, the external skeleton of arthropods has enabled them to diversify in aquatic and terrestrial environments, but prevents them from growing as large as the largest vertebrates. The animal tree illustrates how more than 600 million years of evolution have resulted in patterns of diversity, ecology, and geography that we see when we look at animals across the globe.

42.1 SPONGES, CNIDARIANS, CTENOPHORES, AND PLACOZOANS

Fig. 42.1 revisits the basic phylogeny introduced in Chapter 33. It shows that animals as a whole are sister to the choanoflagellates, a group of single-celled eukaryotes (introduced in Chapter 25). Choanoflagellates are mostly marine organisms that feed by ingesting bacteria, sometimes forming simple multicellular groups that appear to facilitate feeding by beating their flagella in a way that draws water carrying food particles past the cells. The close evolutionary relationship between choanoflagellates and animals suggests that the earliest animals were simple multicellular organisms that lived in the sea; they had little in the way of differentiated cell types, but were able to coordinate cell activity to facilitate feeding. This prediction is supported by **sponges** (Porifera; **Table 42.1**), in many ways the simplest of living animals.

Sponges share some features with choanoflagellates but also exhibit adaptations conferred by multicellularity.

The common ancestor of animals and choanoflagellates carried out all functions as individual cells, from metabolism to

TABLE 42.1 Animal Phyla

BRANCH OF THE ANIMAL TREE	PHYLUM	COMMON NAME	APPROXIMATE NUMBER OF DESCRIBED SPECIES
Basal nodes	Porifera	Sponges	9000
	Cnidaria	Corals, jellyfish	9000
	Ctenophora	Comb-jellies	100–150
	Placozoa	*Trichoplax adhaerens*	1
Basal bilaterians	Acoelomorpha	Acoels	350
Basal protostomes	Chaetognatha	Arrow worms	100
Protostomia/Lophotrochozoa	Acanthocephala	Thorny-headed worms	1000
	Annelida	Segmented worms	15,000
	Brachiopoda	Lamp shells	350
	Bryozoa	Moss animals	5000
	Cycliophora	*Symbion*	At least 3
	Echiura	Spoon worms	150
	Entoprocta	Goblet worms	150
	Gastrotricha	Hairy backs	700
	Gnathostomulida	Jaw worms	100
	Micrognathozoa	*Limnognathia maerski*	1
	Mollusca	Mollusks	80,000
	Nemertea	Ribbon worms	1200
	Phoronida	Horseshoe worms	20
	Platyhelminthes	Flatworms	25,000
	Rhombozoa	—	75
	Rotifera	Rotifers	2000
	Sipuncula	Peanut worms	350
Protostomia/Ecdysozoa	Arthropoda	Arthropods	800,000–1,000,000
	Kinorhyncha	Mud dragons	150
	Loricifera	Brush heads	120
	Nematoda	Roundworms	20,000
	Nematomorpha	Horsehair worms	320
	Onychophora	Velvet worms	75
	Priapulida	Priapulid worms	17
	Tardigrada	Water bears	More than 600
Deuterostomia	Chordata	Chordates	70,000
	Echinodermata	Echinoderms	7000
	Hemichordata	Acorn worms, pterobranchs	90
	Xenoturbellida	—	2

SOURCE: Jørgen Olesen.

FIG. 42.1 A simple phylogenetic tree of animals. The tree shows increasing complexity from the closest protistan relatives of animals, the choanoflagellates, to bilaterian animals such as insects and mammals. *Photo sources: (top to bottom) Mark Dayel (http://www.dayel.com/choanoflagellates) and Nicole King (http://kinglab.berkeley.edu); James D. Watt/SeaPics.com; Umi No Kaze/ a.collectionRF/Getty Images; British Antarctic Survey/Science Source.*

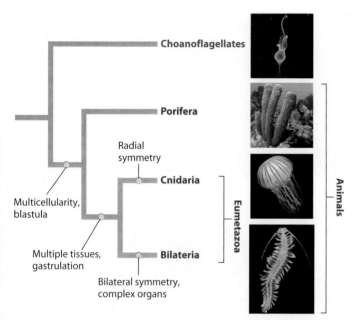

FIG. 42.2 Anatomy of a sponge. *Photo source: imageBROKER/Alamy.*

reproduction. Animals, however, are united by synapomorphies not shared by choanoflagellates, including persistent multicellularity and development from an embryo that includes a blastula (Chapter 33). The cells making up sponges retain much of the individuality of choanoflagllate cells while also having some benefits of multicellularity. Coordination among cells means that distinct kinds of cells can be specialized for different functions: some cells may form a protective skin, for example, while others secrete enzymes for digestion. Most dramatically, multicellularity permits sponges to develop bodies that extend above the seafloor, giving them access to food suspended in water currents.

The sponge body plan resembles a flower vase, with many small pores along its sides and a larger opening at the top (**Fig. 42.2**). The cells on the outer surface of the sponge form a tough layer that acts as the sponge's skin. The interior surface is lined by cells called **choanocytes**, which have flagella and function in nutrition and gas exchange. Choanocyte cells have a collar of small cilia around their flagellum, much like the cells of choanoflagellates. Between the interior and exterior cell layers lies a gelatinous mass called the **mesohyl** (Fig. 42.2). Mesohyl is mostly noncellular, but it contains some amoeba-like cells that function in skeleton formation, reproduction, and the dispersal of nutrients.

Sponges obtain nutrition through intracellular digestion (Chapter 38). The choanocytes that surround the interior chambers and passageways of the sponge beat their flagella, creating a current that draws water from outside the body, through the pores in its walls and upward through the central cavity of the sponge, where the water exits through the large opening at the top. The circulating water contains food particles and dissolved organic matter, which cells lining the cavity capture by endocytosis. Individual cells metabolize the food in their interiors, much in the same way that protozoans feed.

From this observation, we might conclude that a sponge is simply a group of uncoordinated cells, each working for itself, but that isn't the case. The choanocytes beat their flagella in a coordinated pattern, helping to draw water into the body interior through the pore system and outward again through the vase opening. Moreover, the shape of the sponge body itself directs water movement across feeding cell surfaces, facilitating food uptake. Many sponges obtain at least part of their nutrition from symbiotic microorganisms living within their bodies. In fact, sponge bodies are commonly full of bacterial cells. At least some of these bacteria are probably symbionts, but only a few experiments have demonstrated their function.

Sponge cells require oxygen for respiration and must get rid of the carbon dioxide that respiration generates. Gas exchange occurs across the cell membranes by diffusion, aided by the movement of water through the sponge cavity.

Sponges don't have highly developed reproductive organs. Instead, cells recruited from the choanocyte layer migrate into the mesohyl, where they undergo meiotic cell division and differentiate into sperm or eggs. Sperm released into the water from one sponge fuse with eggs in the mesohyl of other sponges.

Many sponges build skeletons of simple structures called spicules. Some sponges have spicules of glasslike silica (SiO_2). Other sponges make their spicules—and sometimes massive skeletons—of calcium carbonate ($CaCO_3$). Still other sponges have a skeleton made up of proteins (these are the sponges used in the bath). In sum, sponges are intermediate in the sophistication of their bodies: their cells function much as single-celled protozoans do, but they coordinate their activities and so are more efficient in extracting food and oxygen from seawater.

The features shared by sponges reflect both their divergence from choanoflagellate-like ancestors and the evolutionary innovations that have shaped their unique anatomy and functions. Despite their anatomic simplicity, sponges are major contributors to seafloor communities. Approximately 9000 species of sponges have been described; most of them are ocean dwelling, but a few live in fresh water. Over the past decade, biologists have debated whether sponges form a monophyletic group, which includes all the descendants of their last common ancestor, or whether they are paraphyletic, sharing their last common ancestor with other animal groups. At present, most data indicate that sponges are a monophyletic group that is a sister to the eumetazoans, a large group that includes most or all other animals (Fig. 42.1).

Cnidarians are the architects of life's largest constructions: coral reefs.

Most animal diversity falls within the **Eumetazoa**, a group that includes animals as varied as jellyfish and elephants. Despite their tremendous anatomical diversity, eumetazoans are united by a number of synapomorphies, including the development of diverse cell types, distinct tissues, and gastrulation, which is the inward movement of cells of the ball-like blastula to form distinct layers during early development (Chapter 33).

Cnidaria form one major branch of the Eumetazoa. Many of us have encountered at least a few cnidarians. Jellyfish flourish in marine environments from coastlines to the deep sea; sea anemones cling to the seafloor, extending their tentacles upward to gather food and deter predators; corals secrete massive skeletons of calcium carbonate, forming reefs that fringe continents and islands in tropical oceans.

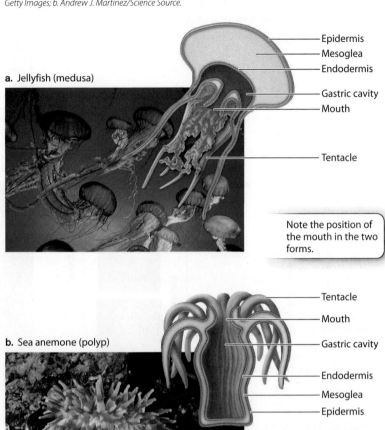

FIG. 42.3 Two cnidarians. (a) Jellyfish; (b) sea anemone. *Photo sources: a. Image Source/Getty Images; b. Andrew J. Martinez/Science Source.*

Jellyfish and sea anemones look strikingly different, but they share a body plan common to all cnidarians (**Fig. 42.3**). Synapomorphies of cnidarians include radial symmetry (Chapter 33) and a mouth surrounded by tentacles armed with stinging cells that subdue prey and defend against enemies. An anemone is like a jellyfish stuck upside down onto the seafloor with its tentacles and mouth facing upward. We call a free-floating jellyfish a medusa, and the stationary (sessile) form of an anemone a polyp. Some cnidarians develop into one form or the other, but many have life cycles that alternate between the two.

Rather than individual cells metabolizing food particles as in sponges, eumetazoans have a digestive system in which food is taken into a specialized compartment, where it is digested using enzymes secreted by cells that line its walls. In cnidarians, the gastric cavity is closed: food and waste enter and exit through the same opening. Cnidarians can digest large food items, such as a whole fish, whose dissolved nutrients can then be absorbed

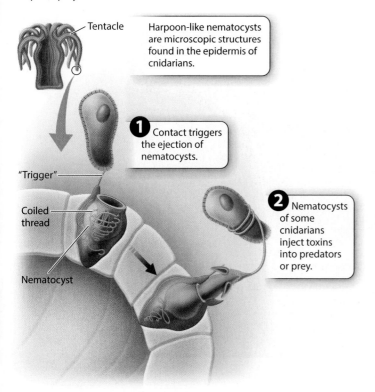

FIG. 42.4 Nematocysts. Nematocysts are harpoonlike organelles that capture prey.

- Tentacle
- Harpoon-like nematocysts are microscopic structures found in the epidermis of cnidarians.
- "Trigger"
- Coiled thread
- Nematocyst
- **1** Contact triggers the ejection of nematocysts.
- **2** Nematocysts of some cnidarians inject toxins into predators or prey.

through the walls of the cavity. As a consequence, cnidarians can consume many types of food that are unavailable to sponges.

The cnidarian body develops from a diploblastic embryo (one with two germ layers; Chapter 33). The embryo's outer layer, the **epidermis**, develops from the outer germ layer, or ectoderm; its inner lining of cells, the **endodermis**, derives from the inner germ layer, or endoderm (Chapter 40). These tissues enclose a gelatinous mass called the **mesoglea** (the "jelly" of jellyfish). This organization sounds a bit like the sponge body plan, but there are important differences, many of which involve characters shared between cnidarians and their sister group, bilaterian animals.

First, in cnidarians and bilaterians, the cells that form the epidermis and endodermis occur as closely packed layers of cells embedded in a protein-rich matrix, forming an epithelium (Chapter 10). The epithelium comprises an unusually tight layer of specialized cells, which are bounded by a matrix composed of protein and connected to one another by junctions that regulate the passage of ions or other molecules. These epithelial layers line compartments within eumetazoan animals, often absorbing or secreting substances, and thereby making possible tissues and organs such as those of the digestive system (Chapter 38). For many years, most biologists accepted that sponges do not form epithelia. More recently, careful anatomical studies have shown that more complex sponges can form epithelia, although these are somewhat simpler in terms of cellular connections than the epithelia of other animals.

Cnidarians also have a wider array of cell types than sponges do, permitting more sophisticated tissue function. Muscle cells allow jellyfish to swim through the ocean, and a simple network of nerve cells permits cnidarians to sense their environment and respond to it through directional movement. No such cells occur in sponges. Although cnidarians lack a brain, a few have light-sensitive cells that function as simple eyes.

Moreover, whereas sponges filter water to gain food, cnidarians are predators, capturing prey with their tentacles and digesting it in the gastric cavity. Specialized cells on the tentacles contain a tiny harpoonlike organelle called a nematocyst, often tipped with a powerful neurotoxin that aids in capturing prey and defending the cnidarians against other predators (**Fig. 42.4**). Other specialized cells lining the gastric cavity of cnidarians secrete digestive enzymes that break down ingested food into molecules that can be taken up by the cells lining the cavity by endocytosis. There is no specialized passage for waste removal. Instead, waste is excreted back into the gastric cavity and leaves by way of the mouth. Oxygen uptake and carbon dioxide release occur by diffusion.

Many cnidarians reproduce asexually to form colonies. For example, corals form extensive rounded, fan-shaped, or hornlike colonies by budding; over time, they can build massive structures on the seafloor (**Fig. 42.5a**). In the Portuguese Man-of-War, different individuals in the same colony develop distinct morphologies: some are specialized for flotation, others for prey capture, and still others for reproduction (**Fig. 42.5b**). Among animals without complex organs, such colonies represent the height of morphological complexity.

Cnidarians exhibit a number of features that differentiate them from sponges, although they lack the complex organ systems that characterize their sister group, the bilaterian animals. Approximately 9000 species of cnidarians live in the oceans. Coral reefs are remarkable not only for the size of their skeletal colonies, but also for their mode of nutrition. Many have lost the ability to capture prey, but instead obtain nutrition from the symbiotic algae found in their surface tissues.

Ctenophores and placozoans represent the extremes of body organization among phyla that branch from early nodes on the animal tree.

Our exploration of nonbilaterian animals concludes with ctenophores (also called comb-jellies) and placozoans. Neither of these groups is diverse (just 100–150 species of comb-jellies have been named, and so far only one species of placozoan), but they may shed light on the evolutionary relationships among sponges, cnidarians, and bilaterians.

Ctenophores resemble cnidarians in body plan, and for many years most biologists thought that ctenophores and cnidarians were close relatives. Like cnidarians, ctenophores

FIG. 42.5 Colony formation in cnidarians. (a) In soft corals, the individual members of the colony are similar in form and function. (b) In the Portuguese Man-of-War, the members of the colony differentiate to serve distinct functions, such as floating, feeding, and reproduction. *Sources: a. ifish/Getty Images; b. George G. Lower/Science Source.*

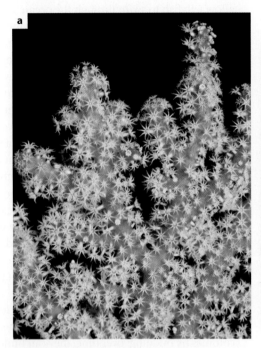

have radial symmetry and tissues, with an outer epithelium and an inner endodermis that enclose a gelatinous interior (**Fig. 42.6**). Ctenophores also have a simple nerve net, as well as rudimentary gonads. These predators feed by ingestion, digesting prey within their gastric cavity through the action of enzymes secreted from the cells lining the gut cavity. As in cnidarians, gas exchange occurs by diffusion. However, there are important differences between cnidarians and ctenophores. Ctenophores propel themselves through the oceans by the coordinated beating of cilia that extend from their epidermal cells. These cilia are usually arranged in comblike groups, which explains the common name for this phylum. Unlike in cnidarians, the digestive wastes generated by ctenophores move through a simple stomach for elimination through an anal pore opposite the mouth.

If ctenophores are the most complex animals discussed so far, **placozoans** are the simplest. Each of these tiny (millimeter-scale) animals contains only a few thousand cells arranged into upper and lower epithelia that sandwich an interior fluid crisscrossed by a network of fiber cells with multiple nuclei (**Fig. 42.7**). Placozoans have no specialized tissues and few differentiated cell types. Cilia on their cell surfaces allow movement, and gas exchange occurs by diffusion. Placozoans can absorb dissolved organic molecules, but commonly feed by taking in food particles and secreting digestive enzymes to break those particles down. Despite their morphological simplicity, placozoans have a genome that contains many of the genes for transcription factors and signaling molecules that are present in cnidarians and bilaterian animals. The roles that these genes play in placozoan biology remain unclear.

FIG. 42.6 Ctenophores, or comb-jellies. *Photo source: Andrey Nekrasov/Alamy.*

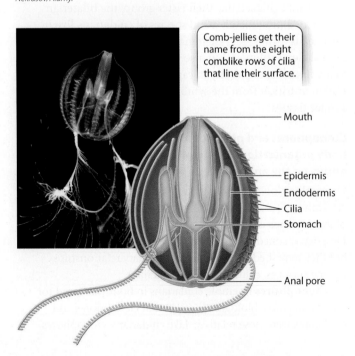

Comb-jellies get their name from the eight comblike rows of cilia that line their surface.

- Mouth
- Epidermis
- Endodermis
- Cilia
- Stomach
- Anal pore

FIG. 42.7 The placozoan *Trichoplax adhaerens*. *Photo source: Reprinted by permission from Macmillan Publishers Ltd. M. Srivastava et al. (August 21, 2008), "The Trichoplax Genome and the Nature of Placozoans," Nature 454:955–960. Copyright 2008. Image courtesy of Ana Signorovitch.*

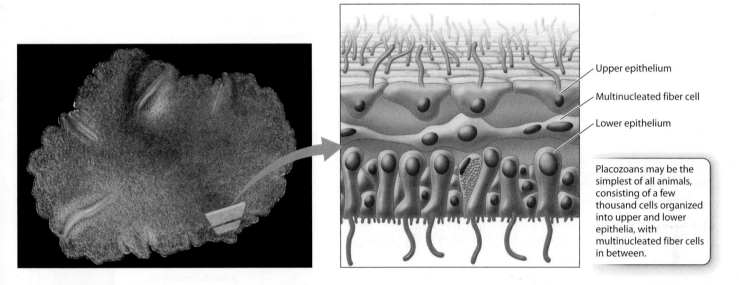

Placozoans may be the simplest of all animals, consisting of a few thousand cells organized into upper and lower epithelia, with multinucleated fiber cells in between.

Branching relationships among early nodes on the animal tree remain uncertain.

As we saw in Chapter 33, we can infer relationships among different groups from comparisons of observable morphological traits, from DNA sequence data, and even from similarities and differences in how their embryos develop (for example, the geometry of the first cell divisions and two versus three germ layers in early development). Molecular, anatomical, and embryological data suggest that bilaterally symmetrical animals form a monophyletic group whose sister is the radially symmetrical (and also monophyletic) Cnidaria. Increasingly, DNA sequence data place placozoans as the sister group to cnidarians and bilaterian animals. This placement reinforces the view that early animals fed by the ingestion of particles and the absorption of organic molecules by individual cells, as observed in choanoflagellates and sponges. Only later did increasing complexity of tissues result in digestive systems specialized for this function. Note that many of the earliest animals found as fossils from the Ediacaran Period (see Fig. 33.14) also seem to have little more than upper and lower epithelia sandwiching a fluid-filled cavity; they fed by phagocytosis and absorption of organic molecules and exchanged gases by diffusion. Living placozoans, therefore, may help us to understand the phylogenetic relationships of the oldest known animal fossils.

Where to place ctenophores in this tree is less clear (**Fig. 42.8**). Traditionally, ctenophores were thought to branch along with cnidarians and bilaterians, as sister to the sponges, and some molecular data support this view (Fig. 42.8a). In

FIG. 42.8 Phylogenetic trees of animals, showing the placement of placozoans and two hypotheses for the placement ctenophores in red. (a) The traditional hypothesis is that Porifera are sister to all other animals, with uncertainty in the branching relationships of placozoans and ctenophores. (b) The alternative hypothesis is that ctenophores are the sister group to all other animals.

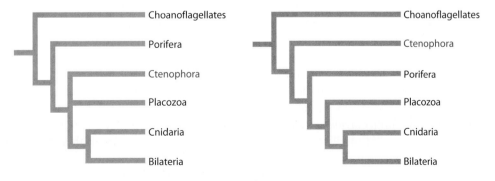

a rather surprising twist, DNA sequence comparisons have recently been cited as evidence that ctenophores are the sister group to all other animals (Fig. 42.8b). If this second hypothesis is correct, it means either that the complex tissues of ctenophores evolved independently of those found in cnidarians and bilaterians or that the morphological simplicity of living sponges reflects the loss of features retained in other animals. DNA sequences and biochemical data indicate that wherever ctenophores belong on the animal tree of life, their nervous system evolved independently of those found in cnidarians and bilaterian animals. At present, we simply don't know the correct branching order of the animal tree.

> **Self-Assessment Questions**
>
> 1. How do muscle cells enable cnidarians to function in ways that sponges cannot?
> 2. Which synapomorphies unite sponges and eumetazoans as sister groups?
> 3. The hypothesis that ctenophores are sister to all other animals runs counter to the traditional view that sponges and a group containing all other animals diverged from the earliest node on the animal tree. What are the two possible ways of reconciling this new phylogenetic hypothesis with anatomical and morphological observations?

42.2 PROTOSTOME ANIMALS

The phyla discussed in the preceding section include approximately 18,000 species, a tiny fraction of all animal diversity. All other animals described to date reside on the bilaterian branch, the final stop in this broad tour. Synapomorphies of bilaterian animals include bilateral symmetry and complex organs that develop from a triploblastic (three germ layers) embryo. The anatomical complexity of bilaterian animals makes possible types of locomotion, feeding, gas exchange, behavior, and reproduction that are unknown in the other groups. These capabilities underpin the remarkable diversity of bilaterian animals.

As biologists discovered in the nineteenth century, bilaterians can be divided into two groups on the basis of early embryological development. In the **Protostomia**, the blastopore—the first opening into the embryo interior that is formed during gastrulation—usually develops into the organism's mouth. In contrast, in the **Deuterostomia**, the blastopore develops into the anus (**Fig. 42.9**). Molecular sequence data support the protostome–deuterostome groupings of bilaterian animals and show that protostome animals can be further divided into two groups: the **Lophotrochozoa** (which include mollusks and annelid worms) and the **Ecdysozoa** (which include insects and other arthropods).

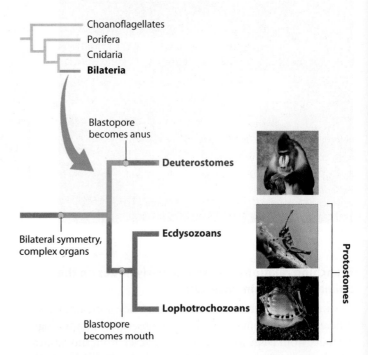

FIG. 42.9 A simple phylogenetic tree of bilaterian animals. Bilaterian animals can be divided into two major groups, the deuterosomes and the protostomes. Protostomes can be further subdivided into lophotrochozoans and ecdysozoans. *Photo sources: (top to bottom) Glenn Nagel/iStockphoto; Lal/Getty Images; Franco Banfi/WaterF/AGE Fotostock.*

Lophotrochozoans account for nearly half of all animal phyla, including the diverse and ecologically important annelids and mollusks.

United largely by their molecular synapomorphies, the Lophotrochozoa include several very different types of animals. On the one hand, there are the brachiopods, bryozoans, and related phyla that have a tentacle-lined organ for filter feeding called a lophophore. On the other hand, there are the mollusks, annelid worms, and their close relatives, which develop from a type of larva called a trochophore. The odd name "lophotrochozoan" is a contraction of the terms "lophophore" and "trochophore," with "zoan" added to indicate that these organisms are animals (*zoon* is Greek for "animal"). The majority of lophotrochozoan species also have a distinctive form of spiral cleavage early in development.

Seventeen of the 36 recognized animal phyla branch from this limb of the animal tree (Table 42.1). Most lophotrochozoans are small animals that live on or within sediments or swim in the sea. A few, including the brachiopods, were once ecologically important but are abundant only locally today. And a few have diversified over time to become diverse and ecologically important animals in both marine and terrestrial environments. In this section, we focus on two phyla of lophotrochozoans: **annelid worms**, well known to most of us as earthworms but far more diverse in the oceans, and **mollusks**, which include clams, snails, and squid (**Fig. 42.10**). These two phyla are closely related but have evolved quite different anatomies in the half billion years since they diverged from one another.

The annelids most familiar to us are the earthworms, which, as their name suggests, are commonly found in soil (**Fig. 42.11a** and **42.11b**). Most of the 15,000 known annelid species, however, live in the oceans (**Fig. 42.11c**). All annelids have a cylindrical body with distinct segments, and they illustrate the advantages of a bilaterian body plan, including a distinct head region (cephalization) and segments modified to become different structures (Chapter 33). In the case of annelids, the head at the front end has a well-developed mouth and, internally, a cerebral ganglion (a collection of nerve cells) that connects to an extensive nervous system. A digestive system extends through the body from the head to an anus, with a sequence of specialized organs for crushing ingested food, then digesting it, and finally excreting wastes, much like the digestive system in the human body. Working together, these features enable annelids to move through their environment, actively searching for food and digesting it efficiently after ingestion.

Aquatic annelids have external gill-like organs for gas exchange, whereas terrestrial earthworms exchange gases through their skin. In both cases, a closed circulatory system moves oxygen and other dissolved gases through the body. Annelids have waste-filtering organs called nephridia, gonads (repeated in most segments), and a fluid-filled coelom, or body cavity. Fluids in the coelom form a hydrostatic skeleton that works in coordination with paired muscles in each segment to direct the organism's movement. Earthworms have a glandular ring in their epidermis called a clitellum that secretes fluids in which eggs are deposited.

Annelids all share some basic features of anatomy and development, but they have evolved impressive functional diversity, which is well illustrated by the ways in which they feed. Many annelids are predators that capture and ingest prey. Others, such as the earthworms, ingest sediment, digesting the organic matter it contains and excreting mineral particles. A few marine annelids have evolved tentacles that enable them to filter food particles from water, while leeches, a specialized group of freshwater annelids, attach themselves to vertebrates to suck out meals of blood (**Fig. 42.11d**). Remarkably, the meter-long vestimentiferan worms that live around hydrothermal vents in the oceans have given up ingestion altogether (**Fig. 42.11e**). These enormous worms, which lack mouths, gain nutrition from chemosynthetic bacteria that live within a collar of specialized tissue found in the worm. Discovered in the 1970s, vestimentiferan worms were originally placed in their own phylum, but molecular sequence comparisons indicate that they are annelids.

Mollusks are the second major phylum of the Lophotrochozoa, and include such familiar groups as snails, clams, mussels, squids, and octopuses. Mollusks are distinguished from other lophotrochozoans by a unique structure called the mantle, which plays a major role in breathing and excretion, and which forms their shells (when shells are present). The mantle lines the visible body of a squid as well as what we commonly (but mistakenly) view as the head of an octopus. Mollusks develop from a distinctive larva called a trochophore that has a tuft of cilia at its top and additional cilia bands around its middle. Marine annelids develop from trochophores as well, and the presence of this unique form of larva in both phyla suggests a close evolutionary relationship between the two. This hypothesis is supported by molecular sequence comparisons. Nonetheless, mollusks and annelids share very few features in their adult body plans.

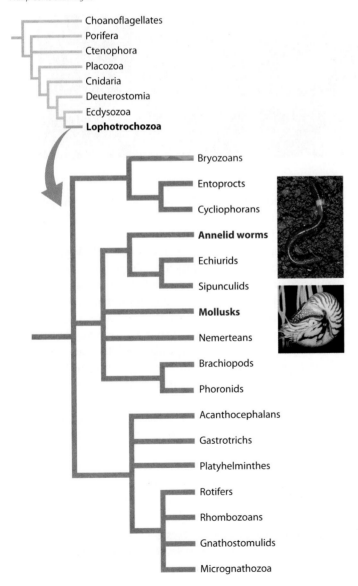

FIG. 42.10 A phylogenetic tree of lophotrochozoan animals, showing all 17 phyla in this group. Phyla discussed in the text are shown in red. Uncertainties in branching order are shown by nodes with more than two branches. *Photo sources: (top) Valerie Giles/Science Source; (bottom) Phillip Colla/Oceanlight.*

FIG. 42.11 Annelids. (a and b) Earthworms typify the segmented body plan found throughout the Annelida. (c) The most diverse annelids are polychaete worms, which live in the oceans. (d) Leeches are specialized for ingesting blood from mammalian hosts. (e) Vestimentiferan worms live along hydrothermal rift vents in the oceans, gaining nutrition from chemosynthetic bacteria in specialized organs (red in the image). *Photo sources: a. Valerie Giles/ Science Source; c. Alexander Semenov/Science Source; d. Volk, F/AGE Fotostock; e. Susan Dabritz/SeaPics.com.*

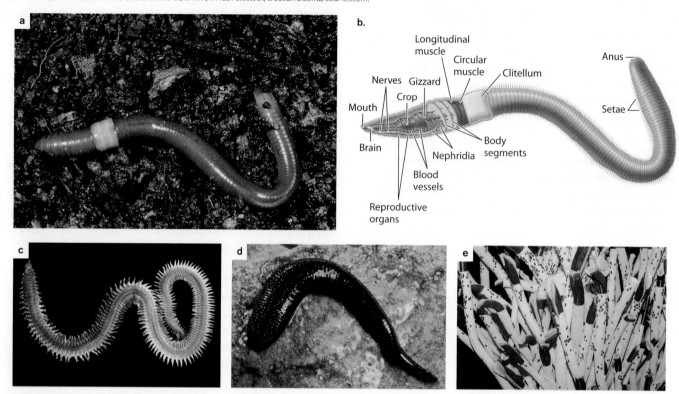

As is true of annelid worms, mollusks share a number of anatomical and developmental features but have diversified in terms of body plan, enabling them to thrive in a number of ecological roles and environmental settings. More than 80,000 species of mollusks have been described, most of them gastropods.

Gastropods (snails and slugs) provide an illuminating example of the versatile mollusk body plan (**Fig. 42.12**). Gastropods have a head with a well-developed mouth that contains a toothlike structure called a radula for feeding. The mouth connects to a gut cavity that extends to an anus. Featherlike gills facilitate gas exchange, and a muscular foot is used for locomotion. A neural ganglion in the head coordinates a nervous system that extends through the body; the body also contains a well-developed circulatory system, gonads, and nephridia. Gastropods have a coelom, but the body cavity is generally reduced to small pouches that surround the heart and other organs. The outer surface of the body consists of the mantle. In many gastropods, the mantle tissues secrete an external skeleton of calcium carbonate, which forms a shell.

Some gastropods eat algae, but many others are predators. Approximately half dwell in the ocean, and the other half are freshwater or terrestrial species that primarily feed on plants. In land snails and slugs (the only terrestrial mollusks), the gills have been lost, and gas exchange occurs in an internal cavity that has been modified to function as a lung.

The second great class of mollusks is the **cephalopods**—about 700 species of squid, cuttlefish, octopus, and the chambered nautilus (**Fig. 42.13**). Cephalopods share a number of features with gastropods, including much of the internal anatomy, featherlike gills, and mantle. At the same time, they have distinctive adaptations for their unique modes of life. Most obviously, cephalopods have muscular tentacles that capture prey and sense the environment. These leglike appendages are found on the head region rather than on the sides of the body as in other animals; this structure explains their distinctive name, which is derived from the Greek *kephale,* meaning "head," and *podos,* meaning "possession of a foot."

Cephalopods are superb swimmers, able to dart through the water by means of a jet propulsion system that forces water through mantle tissue that is fused to form a siphon. The combination of tentacles and rapid locomotion helps to make cephalopods important predators in the oceans. Moreover, cephalopods have exceptional eyesight and exhibit the most complex behavior of any invertebrate animals: they are able to learn visual patterns and solve puzzles to gain food. The

FIG. 42.12 A gastropod, showing a body plan typical of the phylum Mollusca. Photo source: ARCO/W Rolfes/AGE Fotostock.

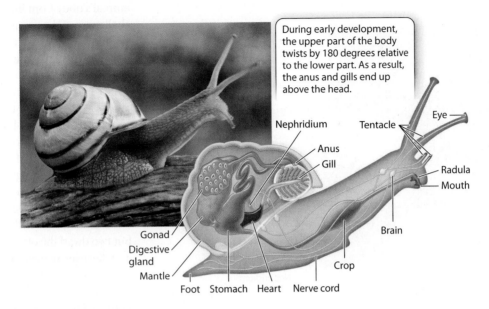

chambered nautilus secretes a coiled shell of calcium carbonate (Figs. 42.13a and 42.13b). In contrast, in other cephalopods, such as squid and octopus, the shell is reduced or absent (Fig. 42.13c). In oceans of the Mesozoic Era (252–66 million years ago), shelled cephalopods called ammonites were among the most abundant and diverse predators.

FIG. 42.13 Cephalopods. (a and b) Nautilus, (c) octopus, and squid (not shown) are ecologically important predators in the oceans. Photo sources: a. Phillip Colla/Oceanlight; c. Michael Carey/SeaPics.com.

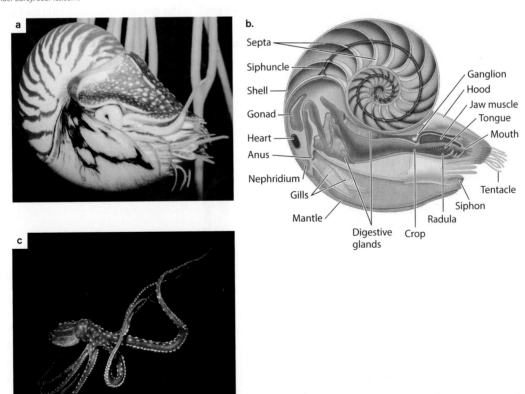

FIG. 42.14 *Mercenaria mercenaria*, a bivalve mollusk commonly harvested from coastal waters of eastern North America. *Photo source: Glenn Price/Dreamstime.com.*

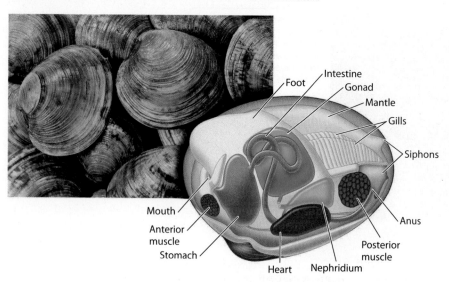

A third major group of mollusks, the **bivalves**, includes the familiar clams, oysters, and mussels. These animals also have a mantle and other anatomical features that point to their close relationship to snails and squids, but have some distinctive features that set them apart (**Fig. 42.14**). Notably, bivalves have (literally) lost their heads. As their name implies, bivalves have evolved a skeleton in which two hard shells (called valves) are connected by a flexible hinge.

Most bivalves obtain food by filtering particles from seawater. However, many bivalves live within marine sediments, burrowing into sand or mud with their muscular foot. How can bivalves filter food from seawater while buried in the sediment? Once again, natural selection has modified the mantle in these mollusks, making possible a new function. Flaps of mantle tissue have fused to form a pair of siphons that extend upward from the bivalve's body to the surface of the seafloor above it. One siphon draws water containing food and oxygen into the body. The second siphon returns water and waste materials to the environment.

Approximately 9000 species of bivalves have been described, mostly from the oceans, but also many from fresh water. Bivalves include the longest-lived animals in the world, the mahogany clams of the North Atlantic, for which the record is more than 400 years of age. These bivalves would make excellent subjects for study of rates of mutation and DNA repair (Chapter 21).

Ecdysozoans are animals that episodically molt their external cuticle during growth.

The second major group of protostome animals is the Ecdysozoa (**Fig. 42.15**), which gets its name from the process of ecdysis, or molting. Uniquely among animals, ecdysozoans secrete a protein- or polysaccharide-rich cuticle that covers their bodies. This tough cuticle acts like a suit of flexible, lightweight armor, protecting the animal's body from injury and from physical challenges of the environment such as drying. This "armor" has a special property: the hard cuticle can be used to form appendages that function as tools, weapons, or even wings. Like a suit of armor, the cuticle is always a perfect fit for the wearer but does not stretch, so it must be exchanged episodically during growth for a larger size to fit a larger, growing body. Ecdysozoans molt their external cuticle during growth, revealing a very soft, larger replacement underneath that soon hardens into a new protective covering.

The Ecdysozoa include eight phyla, but two dwarf the others in abundance and diversity: nematodes and arthropods (see Table 42.1). The most numerous of all animals are the small roundworms known as **nematodes** (**Fig. 42.16**). Most nematodes are so tiny that you've probably never noticed them, but they occur by the billions all around you and may well be inside your body as you read this. Approximately 20,000

FIG. 42.15 A phylogenetic tree of ecdysozoan animals, showing all eight phyla in this group. Phyla discussed in the text are shown in red. Uncertainties in phylogenetic position are shown by nodes with more than two branches. *Photo sources: (top) London School of Hygiene & Tropical Medicine/Science Source; (bottom) Cindy Singleton/iStockphoto.*

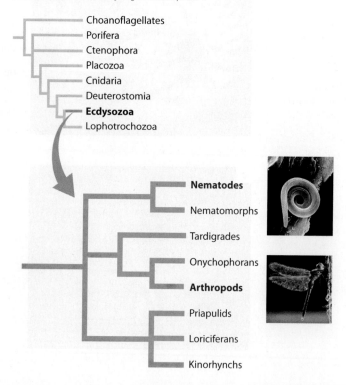

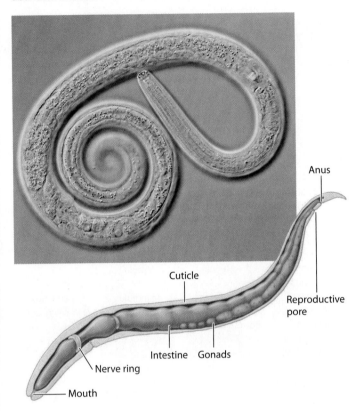

FIG. 42.16 A nematode. Nematodes are small but remarkably abundant animals. *Photo source: Gerd Guenther/Science Source.*

nematode species have been described, but estimates of their true diversity run as high as half a million. Most nematodes are microscopic, preying on microorganisms in soils and marine sediments. Their elongated cylindrical bodies have a mouth at one end, an intestine for digestion, and an anus in the rear. Other organs are quite simple: nerve cords extend along the main axis from a ring in back of the mouth, rudimentary muscles line the body, and gonads expel sperm and eggs for reproduction. As in other ecdysozoans, the nematode body is covered by a tough cuticle that molts during growth. Parasitic nematodes are common and well known to doctors; hookworms, in particular, are major sources of illness in the tropics.

The second major group of ecdysozoan animals is the **arthropods**, whose members include insects, which we discuss in the next section.

Self-Assessment Questions

4. Of the morphological traits of present-day snails, squids, and clams, which features do you think were present in the common ancestor of all three?

5. The protostomes form one major branch of bilaterian animals. Which features do they share with other bilaterians, and which features distinguish them as a separate group within the bilaterians?

42.3 ARTHROPODS

If nematodes are the most numerous animals, arthropods are the most diverse, accounting for a majority of all known animal species. Which features have led to the remarkable evolutionary success of this phylum?

The Arthropoda is the most diverse animal phylum.

Perhaps the most obvious feature of arthropods is the one for which they are named: jointed legs (*arthro* means "jointed," and *pod* means "foot" or "leg"). The jointed leg is the most versatile part of the arthropod body; it has been modified through evolution into structures that function as paddles, spears, stilts, pincers, needles, hammers, and more. The jointed leg allowed arthropods to colonize land by providing a multipoint, independent suspension—much like that of an all-terrain vehicle—able to support a relatively large body and move over the most uneven ground. So well adapted for locomotion is the arthropod leg that NASA engineers use it as a model for planetary exploration vehicles. Researchers have conducted many studies to understand how this one type of limb could give rise to so many diverse structures with different functions (**Fig. 42.17**).

The arthropods' other defining characteristic is the material that forms their hard external skeleton: a strong, lightweight and nearly indestructible polysaccharide called **chitin**. Chitin, as you may recall from Chapter 32, also forms the wall of fungal cells. Next to the cellulose that forms the wall of plant cells, chitin may be the most abundant biomolecule on Earth.

There are four main groups of arthropods (**Fig. 42.18**), of which the most diverse group by far is the **insects**. The other three groups of arthropods are **chelicerates**, which include spiders, scorpions, and their relatives; **myriapods**, which include centipedes and millipedes; and **crustaceans**, which include lobsters, shrimp, crabs, and their relatives.

Chelicerates, named for their pincerlike claws called chelicerae, are the only arthropods that lack antennae. The chelicerates are mostly carnivores, except for plant-eating mites. Scorpions and spiders use venom to subdue prey. In spiders, the venom is injected through chelicerae modified to form fanglike appendages. In scorpions, the venom is injected through the curved tail segment of their abdomen. Other familiar chelicerates include the harvestmen (often called daddy long-legs in the United States), ticks, and horseshoe crabs (which are not crustaceans despite their name). Giant sea scorpions called eurypterids, up to 2 m long, were dominant predators in warm shallow seas 450–252 million years ago. The chelicerates were among the very first animals to invade land, more than 400 million years ago, aided by their protective cuticle and supported by their arrays of jointed legs.

Myriapods are named for their many pairs of legs, which make them easy to recognize. Centipedes ("hundred legs") may have as many as 300 pairs of legs and are fast-moving predators with front legs modified into venomous, fanglike organs.

HOW DO WE KNOW?

FIG. 42.17

How did the diverse feeding structures of arthropods arise?

BACKGROUND Arthropod species diversity is matched by the diversity of form and function of their appendages, including legs, wings, jaws, and feelers. One challenge for scientists has been recognizing and relating the equivalent parts of the very different front ends of arthropods as varied as insects, crabs, spiders, and centipedes.

HYPOTHESIS Researchers hypothesized that the diverse structures in the head region of different arthropods develop from similar limb buds clearly visible in developing embryos. This hypothesis was based in part on the observation that the same limb buds in different vertebrate embryos develop into a wide range of structures, such as bird wings, whale flippers, and human arms.

EXPERIMENT The early stages of arthropod embryos have a definite head region and several body segments. Each segment has a pair of appendages. *Hox* genes specify the identity of each body segment (Chapter 19). Researchers can visualize the expression patterns of different *Hox* genes by using labeled probes on whole embryos. In this way, researchers can compare the development of appendages on the same segment in different arthropods.

The dark bands show the expression of a *Hox* gene in the paired limb buds that can be seen in the embryo of the spider *Cupiennius salei*.

Photo source: Damen, W. G. M., et al. (1998), "A Conserved Mode of Head Segmentation in Arthropods Revealed by the Expression Pattern of Hox Genes in a Spider," Proceedings of the National Academy of Sciences, USA 95(18):10665–10670. Copyright 1998 National Academy of Sciences, U.S.A.

RESULTS Studies of the embryos reveal the various feeding structures that develop from each segment of different arthropods. For example, the maxillae of insects and crustaceans form from the same segment of the developing embryo. By contrast, in chelicerates (such as spiders and scorpions), this segment develops into legs. Moving one segment forward (toward the head), the jawlike mandibles (top row of photos) of insects and crustaceans form from the same segment. In chelicerates, this segment develops into pedipalps (sensory organs in spiders, pincers in scorpions). Moving another segment forward, the second pair of antennae in crustaceans forms from the segment that gives rise to chelicerae (fangs in spiders and their counterparts in scorpions; bottom row of photos) in chelicerates.

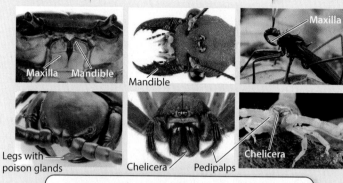

Feeding structures in the head regions of arthropods develop from paired appendages in the embryo.

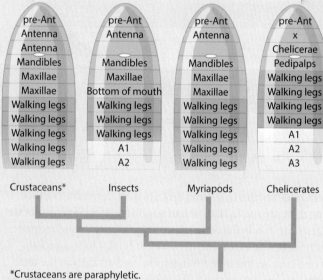

Blue segments in the head region of the developing arthropod embryos develop into various feeding structures. The white ovals indicate the position of the mouth.

*Crustaceans are paraphyletic.

Photo source: Piotr Naskrecki.

CONCLUSIONS The diverse feeding structures of arthropods arise from limb buds in the first segments of the developing embryo. In closely related groups, such as insects and crustaceans, the same segments develop into similar structures. In more distantly related groups, such as insects and spiders, the same segments develop into quite different structures. Evolution has modified simple limb buds into a remarkable diversity of forms, enabling structures as specialized as the grinding mandibles of crabs and the sharp fangs of spiders.

FOLLOW-UP WORK There has been recent controversy over the front appendages of sea spiders. Researchers have questioned whether they are equivalent to large clawlike appendages—called "great appendages," aptly enough—on some Cambrian arthropod fossils.

SOURCE Damen, W. G. M., et al. 1998. "A Conserved Mode of Head Segmentation in Arthropod Revealed by the Expression Pattern of *HOX* Genes in a Spider." Proceedings of the National Academy of Sciences USA 95:10665–10670.

FIG. 42.18 Arthropods. This diverse phylum includes four main groups: chelicerates (spiders, scorpions, and ticks), myriapods (centipedes and millipedes), crustaceans (including lobsters, shrimp, and crabs), and insects. *Photo sources: (top to bottom) B. G. Thomson/Science Source; Tom McHugh/Science Source; Michael S. Nolan/SeaPics.com; Christian Hütter/imageBROKER/AGE Fotostock.*

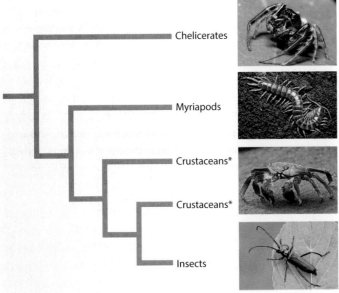

*Crustaceans are paraphyletic. All other lineages are monophyletic.

Millipedes ("thousand legs") may have several hundred pairs of legs (though never a thousand) and are slow-moving herbivores and detritivores (eaters of dead organisms). Millipedes often have glands lining their long bodies that produce a substance containing cyanide, which they secrete to defend against predators.

Crustaceans also have distinctive appendages, notably their branched legs. Although crustaceans share a number of anatomical synapomorphies, they are a paraphyletic group: the descendants of their common ancestor also include insects and their closest relatives, which diverged from their crustacean sister group more than 400 million years ago.

Crustaceans navigate their world with the aid of two pairs of sensory antennae, the larger of which has two branches. The unique crustacean larva, called a nauplius, swims with the aid of the smaller antenna and has a single eye that typically disappears by adulthood. Most crustaceans are aquatic, but a few, including land crabs and woodlice, live on land. In the sea, these versatile arthropods fill many of the ecological roles that insects play on land, eating plants, other animals, or detritus. The tiny shrimplike plankton known as krill occur in vast numbers in polar seas, supporting ecosystems of fish, birds, and mammals such as whales. One Antarctic krill species is estimated to total more than 500 million tons of biomass, exceeding the global total for humans. Many lobsters and crabs are detritivores, scouring the ocean bottoms for dead and dying animals. Some crustaceans are even parasites, attaching themselves to the skin or inside the mouths of fish or sea mammals so that they can feed. The strange creatures known as barnacles begin life resembling shrimp but soon settle on a surface for a life of filter-feeding. As barnacles mature, their bodies become covered with thickened plates of armor and their legs elongate and develop long brushes that sweep food into their mouths.

Insects make up the majority of all known animal species and have adaptations that allow them to live in diverse habitats.

Insects are extraordinarily diverse; the approximately 1 million described insect species account for 80% or more of all known animal species. Insects' success may be a function of four adaptations that allow them to live in diverse habitats: desiccation-resistant eggs; wings; specialized respiratory systems; and **metamorphosis**, a major change in form from one developmental stage to another.

Unlike other terrestrial arthropods, all insects have highly specialized eggshells that can withstand desiccation while still allowing gas exchange. The ability of insect eggs to resist drying out means that they can be laid on almost any surface. Insects can therefore occupy habitats that are difficult for other terrestrial arthropods to access. For example, a butterfly or beetle can lay a single egg on a leaf far above the forest floor, where winds would quickly dry the eggs of a scorpion or spider.

The wings of insects allow them to reach habitats and feed in ways other animals cannot. Insects were the first animals—and the only arthropods—to evolve wings, nearly 350 million years ago. Modern-day mayflies and dragonflies still bear structural similarities to these first fliers. The earliest-branching members of this group, the silverfish and rockhoppers, are referred to as primitively wingless because they evolved before the origin of wings. Some other insects, such as fleas, have lost the wings that were present in their common ancestor, presumably because their current parasitic lifestyles are better served by jumping than flying. Like all arthropods, insects repeatedly shed their exoskeleton as they grow into adulthood. No insect can molt wings, however, so only adult insects can fly.

Insects have evolved another critical adaptation for life on land: specialized respiratory systems. Aquatic arthropods obtain oxygen through gills, but gills don't work well in air because their large surface area dries out. Some terrestrial chelicerates have evolved simple lungs, called book lungs. These structures are filled with hemolymph and have a large surface area for gas exchange, but are kept moist inside a protective pocket. Some crabs are able to make extended forays on land by keeping their gills moist underneath their shell. In contrast, insects exchange gases through small pores in their exoskeletons called **spiracles**. The spiracles connect to an internal system of tubes, the **tracheae**, that directs oxygen to and removes carbon dioxide from respiring tissues, much like air exchange ducts in a building (Chapter 37).

FIG. 42.19 Metamorphosis in butterflies. Butterflies are a group of insects that undergo a major developmental change called metamorphosis. *Source: Ralph A. Clevenger/Getty Images.*

Complete metamorphosis involves a pupal stage (center) in which the tissues of the larva (caterpillar on left) are reorganized into a very different adult form (butterfly on right).

Within the insects, the main evolutionary dividing line is between those that undergo metamorphosis, such as butterflies and fruit flies, and those that do not, such as grasshoppers and water bugs. Insects that do not undergo metamorphosis look much like miniature adults when they hatch, and the main change in body form involves the appearance of wings. Dragonflies, which spend much of their early lives in lakes or ponds, also may change the form of their legs or eyes as they leave the aquatic environment.

Some insects do undergo metamorphosis, a remarkable transformation from a wormlike larval form specialized for feeding to an adult form specialized for mating and reproduction (**Fig. 42.19**). During the pupa stage, the body tissues undergo a dramatic change from the relatively simple larva to a very different-looking adult, such as a fly, butterfly, wasp, or beetle. Insects undergo a nearly complete dissolution of their muscles and other body parts during metamorphosis. The pupa must be immobile within its chitinous exoskeleton and so is sometimes (in moths) encased in a cocoon spun from silk. Hidden from view, it transforms from something as simple as a maggot, which lacks even an obvious head or tail, into an animal as sophisticated as a fly, which is able to walk on water, hover motionless in front of a flower, or fly at 60 mph. This evolutionary step has little parallel elsewhere in life and is one of the keys to insect diversification.

Self-Assessment Question

6. Which four adaptations account at least in part for the success of insects? Why?

42.4 DEUTEROSTOME ANIMALS

Having introduced diversity within protostome animals, we now turn our attention to the other great bilaterian group, the Deuterostomia. As noted previously, one character that unites deuterostomes and separates them from protostomes is a feature of early development: the blastopore develops into the mouth instead of the anus. Deuterostome animals (**Fig. 42.20**) include three major phyla: **Hemichordata** (acorn worms), **Echinodermata** (sea urchins and sea stars), and **Chordata** (vertebrates and closely related invertebrate animals such as sea squirts).

FIG. 42.20 A phylogenetic tree of deuterostome animals, showing all phyla (and the major subdivisions of the chordates) in this group. Phyla discussed in the text are shown in red. *Photo sources: (top to bottom) Darlyne A. Murawski/Getty Images; Azure Computer & Photo Services/Animals Animals; © www.konig-photo.com; Reinhard Dirscherl/WaterFrame/Getty Images/Dwight Kuhn.*

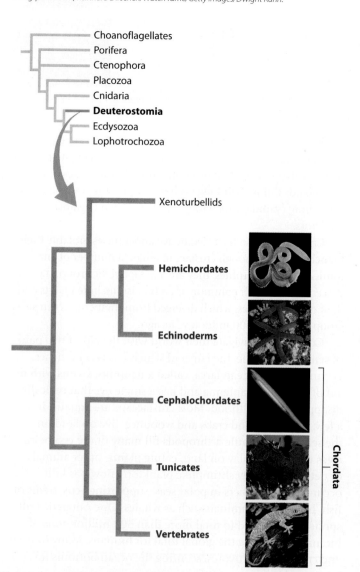

Hemichordates include acorn worms and pterobranchs, and echinoderms include sea stars and sea urchins.

We begin with the deuterostomes most distantly related to the vertebrates: the hemichordates and echinoderms. Hemichordates include acorn worms, approximately 75 species of wormlike animals that move through seafloor sediments in search of food particles. Pterobranchs, which are also hemichordates, consist of about a dozen species of animals that attach to the seafloor and use tentacles to filter food from seawater (**Fig. 42.21**). All hemichordates have a mouth on an elongated protuberance called a proboscis that connects to the digestive tract by a tube called the **pharynx**. The pharynx contains a number of vertical openings called **pharyngeal slits** separated by stiff rods of protein. Hemichordates also have a **dorsal nerve cord** that forms the basis of a nervous system. These characters are found, as well, in chordate species, at least at some stage of their development. Additional characters that relate hemichordates to other deuterostomes are apparent in embryological and larval development. Nonetheless, adult hemichordates have body plans quite distinct from those of chordates and echinoderms.

Another major group of deuterostomes is the echinoderms, among the most distinctive of all animal phyla. Echinoderms include sea stars and sea urchins (including sand dollars), as well as brittle stars, sea cucumbers, and sea lilies. All of these organisms share a unique fivefold symmetry on top of their basic bilaterian organization (**Fig. 42.22**). Echinoderms form distinctive skeletons made of interlocking plates of porous calcite, a form of calcium carbonate. This structure makes most of them hard and stony and protects them from predators. Another unique feature of echinoderms is their **water vascular system**, a series of fluid-filled canals that permit bulk flow of oxygen and nutrients. **Tube feet**, small projections of the water vascular system that extend outward from the body surface, allow echinoderms to crawl, sense their environment, capture food, and breathe. Sea stars can use the tube feet on their powerful arms to pull open the valves of clams. Approximately 7000 echinoderm species reside in present-day oceans, but the fossil record suggests that this phylum was more diverse in the past.

Zoologists have long known that echinoderm and hemichordate larvae share many features. Molecular sequence data now confirm that echinoderms are the closest relatives of hemichordates, with chordates being a sister group to the group formed by echinoderms and hemichordates (see Fig. 42.20).

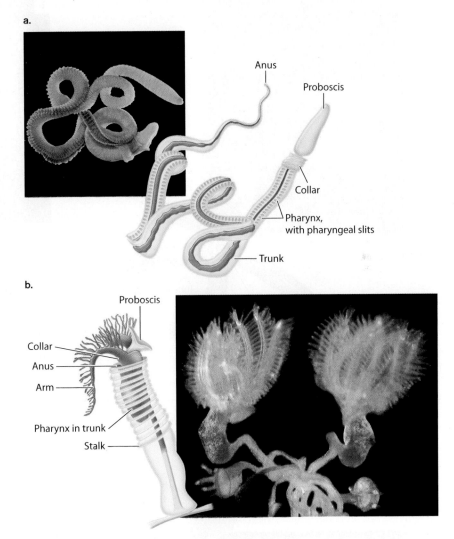

FIG. 42.21 Hemichordates. Hemichordates include (a) acorn worms that move through sediments in search of food and (b) pterobranchs, which have filter-feeding tentacles. *Photo sources: a. Darlyne A. Murawski/Getty Images; b. Johanna T. Cannon.*

Chordates include vertebrates, cephalochordates, and tunicates.

The other great branch of the deuterostome tree is the chordates, the phylum that includes vertebrate animals (Table 42.1). Within the phylum Chordata, there are three subphyla: the **cephalochordates**, the **tunicates**, and the **vertebrates** (also called **craniates**). Like the hemichordates, all chordates have a pharynx with pharyngeal slits (illustrated by the invertebrate chordate amphioxus in **Fig. 42.23**). In fish, the pharyngeal slits form the gills, but in terrestrial animals such as humans, these slits can be seen only in developing embryos. The **notochord**, a unique feature of chordates, is a stiff rod of collagen and other proteins that runs along the back, providing resistance for the muscles along each side in some chordates. In vertebrates, the notochord is apparent only during early embryogenesis and is replaced functionally by a **vertebral column** later in

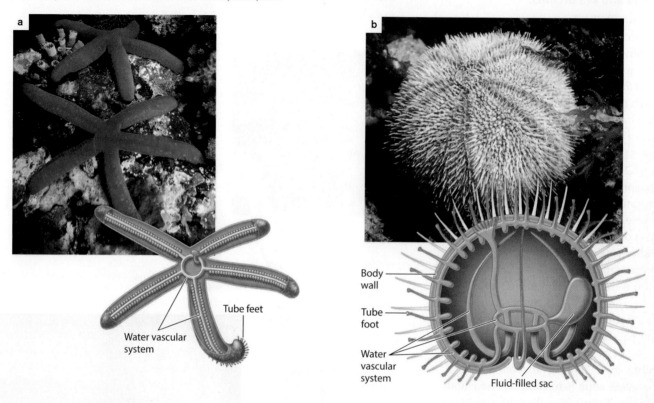

FIG. 42.22 Echinoderms. (a) Sea stars are familiar sights on the shallow seafloor. (b) Sea urchins are motile echinoderms found on most seafloors. *Photo sources: a. Azure Computer & Photo Services/Animals Animals; b. Sue Daly/naturepl.com.*

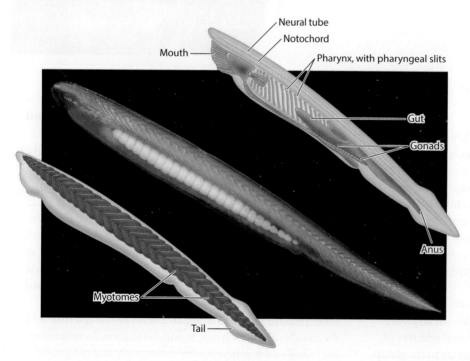

FIG. 42.23 Amphioxus, a cephalochordate. Amphioxus has many features in common with vertebrates, including those labeled here, but it lacks a well-differentiated head and has no cranium or vertebrae. *Photo source: © www.konig-photo.com.*

development. Also forming during early development is the **neural tube**, a cylinder of embryological tissue that develops into a dorsal nerve cord. Body musculature is organized into a series of segments called **myotomes**, most visible in fish as the flakes of muscle that become separated when the fish is cooked. Chordates have a tail (reduced but still internally present in humans and other apes) that extends posterior to the anus, and some chordates have muscularized appendages that include the fins of fish and limbs of terrestrial vertebrates.

Amphioxus (Fig. 42.23) is a cephalochordate. Cephalochordates share key features of body organization with vertebrates but lack a well-developed brain and eyes, have no lateral appendages, and do not form a mineralized skeleton. Nonetheless, their many similarities to vertebrates suggest they are closely related, and molecular sequence comparisons confirm this hypothesis.

FIG. 42.24 Tunicates. (a and b) Adult form; (c) larval form. *Photo source: a. Reinhard Dirscherl/Getty Images.*

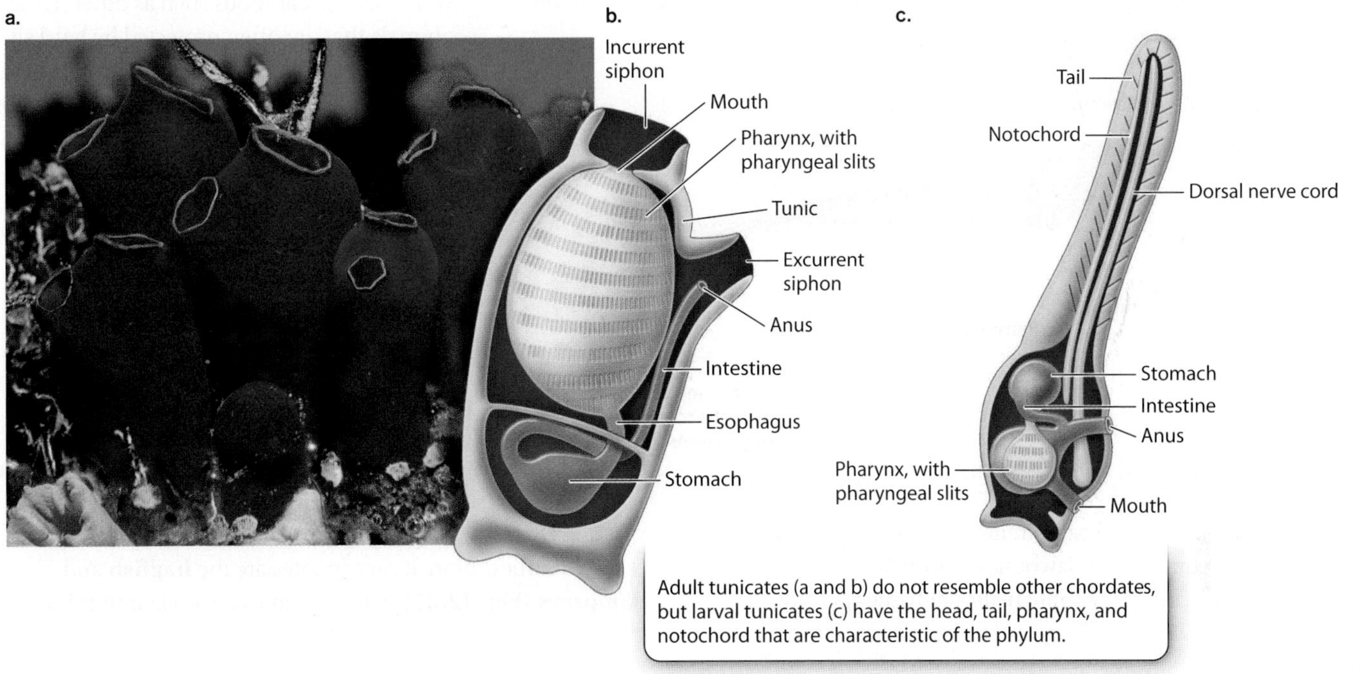

Adult tunicates (a and b) do not resemble other chordates, but larval tunicates (c) have the head, tail, pharynx, and notochord that are characteristic of the phylum.

The tunicates also share key features of the chordate body plan during early development, such as the notochord, neural tube, and a long tail with muscles arranged in myotomes (**Fig. 42.24**). However, tunicates have a unique adult form, with the pharynx and its pharyngeal slits being the only obvious similarities to other chordates. Tunicates include approximately 3000 species of filter-feeding marine animals, such as sea squirts that are anchored to the seafloor and salps that float in the sea. The adult tunicate body is a basketlike structure that is highly modified for filter-feeding, not unlike the shapes of barnacles, which are also adapted to this feeding mode.

A tunicate might be mistaken for a sponge, but its complex internal structure shows that it lies far from sponges on the animal tree. The tunicate's body wall, or tunic, has a siphonlike mouth at one end that draws water through an expanded pharynx; this pharynx captures food particles and exchanges gases. Water and wastes are expelled through an anal siphon.

For many years, zoologists considered cephalochordates to be the closest relatives of vertebrate animals. Molecular data now suggest that, in fact, the tunicates are our closest invertebrate relatives (**Fig. 42.20**). The common ancestor of tunicates, cephalochordates, and vertebrate animals was a small animal much like amphioxus in organization. After the divergence of the three major groups of chordates, however, the tunicates' adaptation for filter-feeding led to the unique anatomy they have today.

Self-Assessment Questions

7. Draw a phylogeny of deuterostomes, indicating the most likely relationships among chordates, hemichordates, and echinoderms, with common examples of each group.

8. You do not have a notochord, a long tail, or pharyngeal slits. How do you know that you are a chordate?

42.5 VERTEBRATES

In many ways, the last stop in our exploration of animal diversity is the most familiar because it includes our own species, *Homo sapiens*. The animals known as vertebrates are named for their jointed skeletons, which run along the main axis of the body and form a series of hard segments collectively termed **vertebrae** (singular, **vertebra**). In addition to features shared with other chordates, vertebrate animals are distinguished by a cranium that protects a well-developed brain, a pair of eyes, a distinctive mouth for food capture and ingestion, and an internal skeleton commonly mineralized by calcium phosphate (**Fig. 42.25**). Vertebrates also have a coelom in which the organs are suspended and a closed circulatory system. Like the other chordates, vertebrates have pharyngeal slits in at least their early embryonic stages.

Many of the features that separate vertebrates from invertebrate chordates can be found in the head, including a bony

FIG. 42.25 Vertebrate characters. Anatomical features of vertebrates that are characteristic of chordates are shown in red; those that are unique to vertebrates are shown in blue. *Source: Stockbroker xtra/AGE Fotostock.*

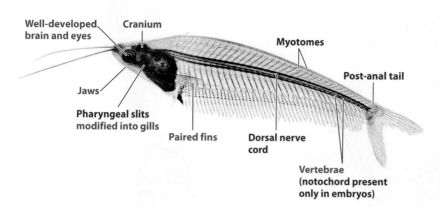

cranium that protects the brain; in most vertebrates, the cranium joins with the mandible (or jawbone) to form a skull. This protective helmet for the delicate neuronal tissues permits the development of a larger brain than would otherwise be possible. The mandible permits vertebrates to eat foods such as other living animals that are often protected by hard shells and plants containing tough cellulose and lignin. Because hagfish, which branched from an early node on the vertebrate tree, have a cranium but not vertebrae (or a mandible), some biologists use the name "Craniata" for this group, rather than the traditional "Vertebrata."

Fish are the most diverse vertebrate animals.

We sometimes think of fish as one group, but the aquatic animals we commonly call "fish" include four distinct groups of aquatic vertebrates, not all of them monophyletic (**Fig. 42.26**).

The earliest-branching craniates are the **hagfish** and **lampreys** (**Fig. 42.27**). These animals have a cranium built of

FIG. 42.26 A phylogeny of the vertebrates. This phylogeny is favored by recent molecular data, but, like all phylogenies, it is a hypothesis. The placements of hagfish relative to other vertebrates and turtles relative to other sauropsids have provoked debate for many years.

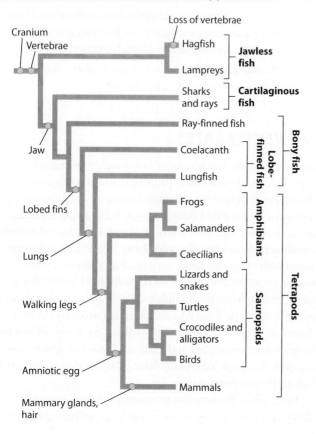

FIG. 42.27 Hagfish and lampreys. Commonly grouped together as jawless fishes, (a) hagfish and (b) lampreys are the earliest-branching vertebrates still alive today. *Sources: a. Tom McHugh/Science Source; b. F Hecker/AGE Fotostock.*

FIG. 42.28 Cartilaginous fish. The skeletons of sharks are made of cartilage. *Source: Michael Patrick O'Neill/Alamy.*

cartilage but lack jaws. Lampreys also have a vertebral column built of cartilage. Hagfish lack vertebrae, although current research suggests that this trait existed in a common ancestor of hagfish and lampreys but was lost in hagfish. These eel-like organisms eat soft foods without the aid of jaws: they diverged before jaws evolved from pharyngeal slits. Hagfish feed on marine worms and on dead and dying sea animals, while lampreys live parasitically, sucking body fluids from their fish prey. Both hagfish and lampreys have a series of gill slits through which water enters to bring oxygen to the gills.

Biologists have long debated the phylogenetic relationships among hagfish, lampreys, and other vertebrate animals. The presence of a vertebral column in lampreys and the absence of vertebrae in hagfish convinced many biologists that lampreys are the sister group to other vertebrates, and that hagfish are the sister group to *all* vertebrates. However, several lines of molecular data now favor the view shown in Fig. 42.26: hagfish and lampreys *together* form the sister group to all other vertebrates. Approximately 65 species of lampreys and hagfish live in fresh water and coastal oceans. Fossils record the existence of many extinct species of jawless fishes. When we include fossils, the jawless fishes as a whole are paraphyletic because one of their groups gave rise to fishes with jaws.

Cartilaginous fish, formally known as Chondrichthyes, diverged from the next deepest node on the vertebrate tree.

This monophyletic group includes approximately 800 species of sharks, rays, and chimaeras, all of which have jaws and a skeleton made of cartilage (**Fig. 42.28**). Cartilaginous fishes form deposits of hard calcium phosphate minerals like those in human bones, but only in their teeth and in small toothlike structures called denticles embedded in the skin. The best-known cartilaginous fishes are the sharks, many of which are predators that put those mineralized teeth to good use. The same group also includes Whale Sharks, filter-feeding animals that, at a length up to 12 m (40 feet), are gentle giants of the sea. Rays are closely related to sharks but have a flattened body adapted for life on the seafloor, where they feed on clams and crabs. The cartilaginous fish can retain levels of urea in their tissues that would poison other vertebrates, a physiological adaptation that helps them to maintain their salt balance without well-developed kidneys (Chapter 39). Nearly all cartilaginous fishes occur in the oceans, but Bull Sharks (*Carcharhinus leucas*) commonly swim into rivers—they've been recorded well up the Mississippi—and several species of rays live in fresh water.

Moving up one more branch in the vertebrate tree, we encounter the **bony fishes**, or Osteichthyes (see Fig. 42.26). Bony fish have a cranium, jaws, and bones that consist of mineralized calcium phosphate. Numbering some 28,000 freshwater and marine species, these are the fish that humans

FIG. 42.29 Bony fish. The char, well known to those who fish in fresh water, illustrates the principal anatomical features of bony fishes: a cranium, jaws, and mineralized bones. *Source: JOEL SARTORE/National Geographic Creative.*

most commonly encounter (**Fig. 42.29**). Bony fish are by far the most diverse group of vertebrates, possessing several unique features that facilitate their occupation of diverse habitats. First, they have a system of movable elements in their jaws that allows them to specialize and diversify their feeding on many different types of food. Second, they possess a unique gas-filled sac called a swim bladder that permits exquisite control over their position in the water column through changes in buoyancy. Third, bony fish have kidneys that allow them to regulate water balance, so that they can occupy waters over a wide range of salinities.

The swim bladder plays a special role in vertebrate evolution. Early vertebrates evolved a gut sac that enabled them to gulp air to obtain additional oxygen. In some fishes, this sac evolved into the swim bladder, but in one group the sac was modified to become a lung. Charles Darwin was among the first to recognize that the swim bladder and the vertebrate lung are homologous organs.

The common ancestor of tetrapods had four limbs.

Most animal phyla occur in the oceans, and many are exclusively marine. A subset of animal phyla has successfully colonized fresh water, and a smaller subset has radiated onto the land. Eleven groups on the animal phylogenetic tree contain both aquatic *and* terrestrial species: nematodes, flatworms, annelids (earthworms and leeches), snails, tardigrades (microscopic animals with eight legs, sometimes called water bears), onychophorans (velvet worms), four groups of arthropods (millipedes and centipedes, scorpions and spiders, land crabs, and insects), and vertebrates. In each of these groups, the aquatic species branch from the earliest nodes, and the terrestrial species diverge from later nodes. No two of these groups share a last common ancestor that lived on land; thus, they all made the transition independently. Earlier, we outlined the extraordinary diversity of terrestrial arthropods. Here, we focus on vertebrates as the other great animal colonists of the land.

Most bony fish have fins supported by a raylike array of thin bones. About half a dozen closely related species, however, have pectoral and pelvic fins that extend from a muscularized bony stalk. These **lobe-finned fish** include the **coelacanth** and **lungfish** (**Fig. 42.30**). Although these animals resemble other fish, the coelacanth and lungfish are the closest relatives of **tetrapods** (four-legged animals). Lungfish can survive periods when their watery habitat dries up by burying themselves in moist mud and breathing air. Coelacanths first appeared more than 400 million years ago and were thought to have gone extinct more than 80 million years ago, until a specimen was discovered in a fisherman's catch off South Africa in 1938. Hence, they are sometimes referred to as living fossils. There are two living species of coelacanths and three species of lungfish, one each in South America, Africa, and Australia. As shown in Fig. 42.26, tetrapods descended from the same ancestor that is common to all four fish groups. Thus, there is no monophyletic group that corresponds to our popular notion of "fish."

Fossils document the anatomical transition in vertebrate animals as they moved from water to land, showing changes in the form of the limbs, rib cage, and skull (Fig. 22.22 discusses

FIG. 42.30 Lobe-finned fish. The fish most closely related to tetrapods are the lobe-finned fishes: (a) the coelacanth and (b) the lungfish. These fishes have paired fins with an anatomical structure similar to that of tetrapod legs. *Sources: a. Gerard Lacz/Animals Animals–Earth Scenes; b. D. R. Schrichte/SeaPics.com.*

Tiktaalik and other fossils that connect tetrapods to lobe-finned fishes). Living tetrapods include amphibians, such as frogs and salamanders; lizards, turtles, crocodilians, and birds; and mammals. The last common ancestor of all these animals had four limbs, which is the source of the name **Tetrapoda** ("four legs"). Some, like snakes and a few amphibians and lizards, lost their legs in the course of evolution.

The more than 4500 species of **Amphibia** ("double life") range in size from tiny frogs a few millimeters in length to the Chinese Giant Salamander, which is more than 1 m long. Their name reflects their distinctive life cycle (**Fig. 42.31**). Most amphibian species have an aquatic larval form with gills that permit breathing under water and a terrestrial adult form that has lungs for breathing air. Because their eggs dry out easily, amphibians must reproduce in the water or in moist habitats, so they are not completely terrestrial. Whereas many amphibian larvae graze on algae, the adults are predators and often have a muscular tongue for capturing prey. Many amphibians are protected by toxins they secrete from glands in their skin, sometimes advertised by their brilliant color patterns. Most amphibian adults have simple lungs and so require moist skin as a secondary means of obtaining oxygen, with toads and red efts being notable exceptions. This is why frogs in particular have fallen victim to a fungus that affects the skin's ability to breathe (Chapter 48).

Amniotes evolved terrestrial eggs.

Like insects, some vertebrates evolved an egg adapted to tolerate dry conditions that accompany life on land. The **amniotic egg** has a desiccation-resistant shell and four membranes that permit gas exchange and management of waste products produced by the embryo (Chapter 40). These eggs must be fertilized internally before the eggshell is produced by the female because sperm cannot penetrate the shell. The amniotic egg can exchange gases while retaining water, which enables the group of vertebrates known as **amniotes** to live in dry terrestrial habitats that amphibian eggs cannot tolerate. Amniotic eggs permit long development times, with the embryo's nutrition being supplied by a large yolk or a placenta. They keep wastes separated from the embryo so they do not poison it, giving the embryo time to build more complex bodies than otherwise would be possible.

FIG. 42.31 The amphibian life cycle.

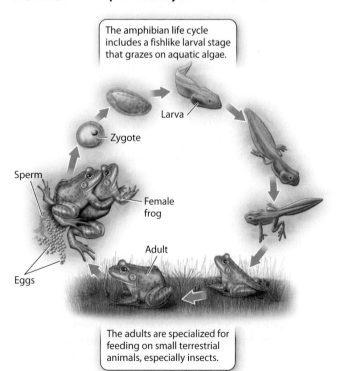

FIG. 42.32 Amniotes. (a) Turtles, (b) lizards, (c) mammals, and (d) birds. *Sources: a. Gary Meszaros/Science Source.; b. François Dorothé/Getty Images; c. Sarah Peters/Getty Images; d. NASA/Jim Grossmann.*

Amniotes include lizards, snakes, turtles, and crocodilians (approximately 6000 species of scaly animals commonly referred to as reptiles), as well as birds (at least 10,000 species) and mammals (5500 species; **Fig. 42.32**). Most mammals, many lizards, and some snakes have evolved live birth rather than laying eggs. Nonetheless, their eggs retain the specialized membranes characteristic of all amniotes. Instead of being wrapped around the embryo inside a tough egg shell, these membranes surround and protect the embryo inside the womb.

As shown in Fig. 42.26, amniotes form a monophyletic group with two major branches. One consists of the mammals, whereas the other branch contains turtles, birds, and the amniotes traditionally grouped together as reptiles. Like fish, however, reptiles are not monophyletic. As discussed in Chapter 22, fossils show that birds diverged from dinosaur ancestors, and molecular sequence comparisons indicate that crocodiles are the closest living relatives of birds. Unlike their closest living and fossil relatives, birds lack teeth, most of their scales have been modified into feathers (except on the legs and toes), and they are generally adapted for flight. Adaptations for flight include hollow bones that are light but remain strong, and a method of breathing that extracts much oxygen from each breath, permitting high-performance flight (Chapter 37).

One consequence of this streamlined, lightweight bird body is a lack of the heavy jaws that many other tetrapods use to grind up plant leaves or other low-nutrition foods. Instead, birds have a grinding organ, the gizzard (also found in alligators and crocodiles), inside their bodies near their center of gravity, where it interferes less with flight.

All **mammals** are covered with hair and feed their young milk from the mammary glands for which the class Mammalia is named. Like many other groups of animals and plants, the early-branching mammals show intermediate stages in the evolution of the body plans that dominate on Earth today.

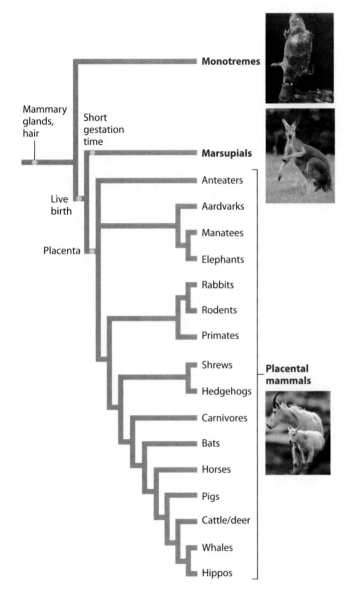

FIG. 42.33 A phylogenetic tree of mammals. There are three major groups of mammals: monotremes, marsupials, and placental mammals. *Photo sources: (top to bottom) Dave Watts/NHPA/Photoshot; C. Wermter/AGE Fotostock; David Macias/Science Source.*

The first fossils that clearly show evidence of hair appeared approximately 210 million years ago, during the Triassic Period. Like feathers in birds, hair permits retention of body heat. The earliest-branching living mammals, the monotremes, lay eggs, like birds or lizards (**Fig. 42.33**), but their hatched young drink milk secreted from pores in the skin of the mother's belly. Living monotremes, the platypus and four species of echidna, or spiny anteater, are found only in Australia and New Guinea.

The first mammals that gave birth to live young appeared in the Jurassic Period, about 160 million years ago. These animals gave rise to the two major living groups, marsupial and placental mammals. **Marsupials** include kangaroos, koalas, and related groups native to Australia, as well as the opossums found in the Americas. Their young are born at an early stage of development, and the tiny babies complete development while attached to mammary glands equipped with nipples that provide them with milk. In some marsupials, notably kangaroos, this development occurs inside a pouch. Only later do fully formed juveniles venture outside to begin life on the ground.

The **placental mammals** are named for a temporary organ called the **placenta** that develops in the uterus along with the embryo, providing nutrition that enables the offspring to be larger and more quickly independent when born (Chapter 40). Most living mammals fall into this group. Placental mammals include the carnivores, such as lions and weasels; the primates, including monkeys, apes, and humans; and the hooved mammals, which include cattle, pigs, deer, and—perhaps surprisingly—their marine relatives, the whales.

The most diverse mammals belong to two groups of mostly small animals, the bats and the rodents. Bats, which evolved flight and a form of sonar called echolocation to find food at night, include more than 1000 species. The largely plant-feeding rodents are as diverse as the large mammals and bats combined, numbering more than 2000 species. Beavers, gophers, rats, mice, hamsters, and squirrels are all rodents. Rodent evolutionary success is often attributed to innovations in tooth development. Their teeth are unusual in that they grow continually throughout life. This constant growth permits constant wear, allowing these animals to gnaw through hard protective coverings to obtain nutrition from a remarkably wide variety of otherwise hard-to-access sources such as the contents of seeds.

Fig. 42.34 shows animal phylogeny in the context of life's broader diversity. Clearly, the diversity that surrounds us today has evolved over the immense time scale of Earth history, on a planet that has changed markedly through time. The dynamic interplay of life and environment underpins ecology as well as diversity, and it provides a fundamental challenge for sustaining biological diversity in the changing world of the twenty-first century (Chapter 48).

Self-Assessment Questions

9. Why do bird feet have scales like those found on snakes and lizards?
10. Which features are common to bony fishes, amphibians, amniotes, and mammals?
11. Which features underpin the ecological success of tetrapods on land?
12. Name two features that are unique to mammals and suggest their importance for mammal evolution.

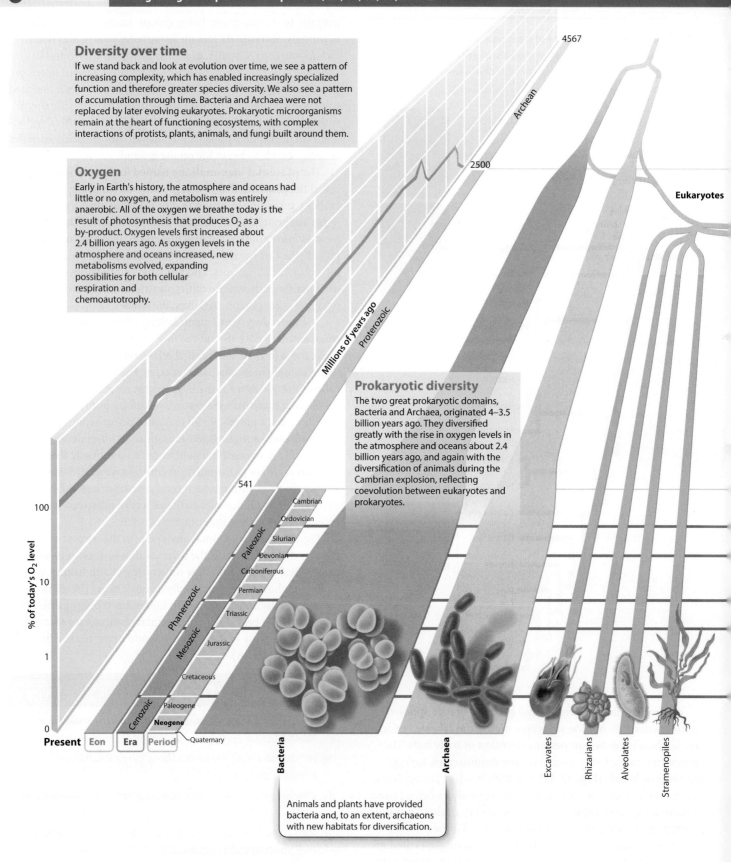

VISUAL SYNTHESIS
FIG. 42.34

Diversity Through Time
Integrating concepts from Chapters 24, 25, 31, 32, 33, and 42

Diversity over time
If we stand back and look at evolution over time, we see a pattern of increasing complexity, which has enabled increasingly specialized function and therefore greater species diversity. We also see a pattern of accumulation through time. Bacteria and Archaea were not replaced by later evolving eukaryotes. Prokaryotic microorganisms remain at the heart of functioning ecosystems, with complex interactions of protists, plants, animals, and fungi built around them.

Oxygen
Early in Earth's history, the atmosphere and oceans had little or no oxygen, and metabolism was entirely anaerobic. All of the oxygen we breathe today is the result of photosynthesis that produces O_2 as a by-product. Oxygen levels first increased about 2.4 billion years ago. As oxygen levels in the atmosphere and oceans increased, new metabolisms evolved, expanding possibilities for both cellular respiration and chemoautotrophy.

Prokaryotic diversity
The two great prokaryotic domains, Bacteria and Archaea, originated 4–3.5 billion years ago. They diversified greatly with the rise in oxygen levels in the atmosphere and oceans about 2.4 billion years ago, and again with the diversification of animals during the Cambrian explosion, reflecting coevolution between eukaryotes and prokaryotes.

Animals and plants have provided bacteria and, to an extent, archaeons with new habitats for diversification.

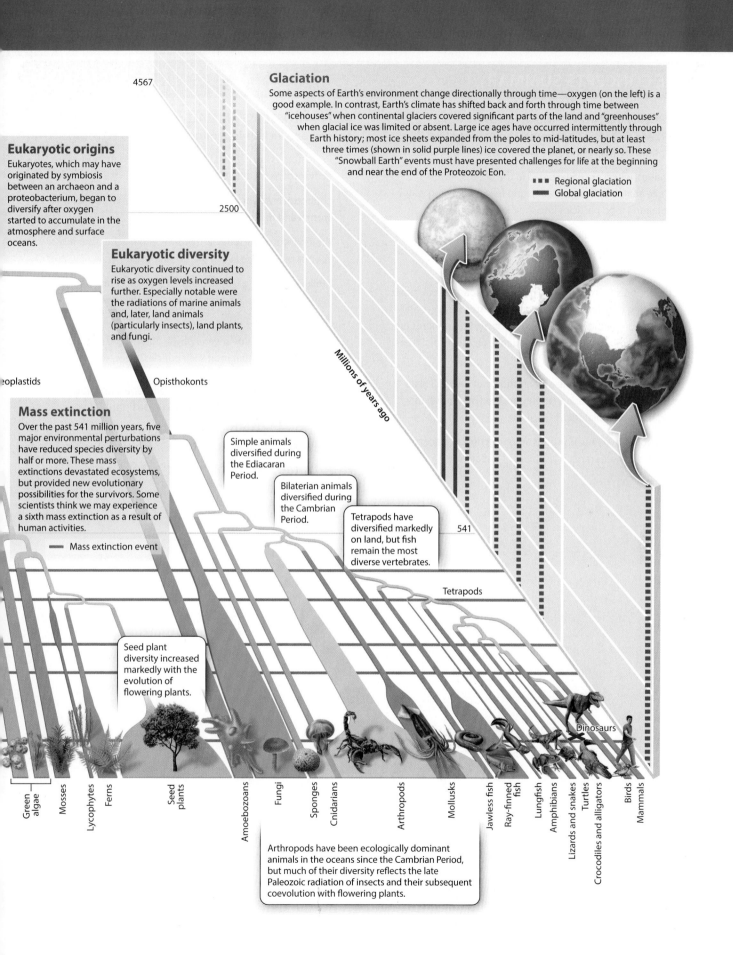

CORE CONCEPTS SUMMARY

42.1 SPONGES, CNIDARIANS, CTENOPHORES, AND PLACOZOANS: The simplest animals evolved multicellularity from single-celled ancestors.

Sponges, which are widespread in the oceans, have a simple anatomical organization and feed by drawing water containing food particles into their interiors. page 946

Cnidarians form multiple tissues, differentiate nerve and muscle cells, and feed as predators by means of specialized stinging cells. page 948

Ctenophores, or comb-jellies, resemble cnidarians but move by beating cilia, have an anal pore for waste excretion, and have a rudimentary mesoderm germ layer. Ctenophores are a difficult limb to place with confidence on the animal tree of life. page 949

Placozoans have a simple organization, but genomic data support the hypothesis that they are the sister group to cnidarians and bilaterians. page 950

The evolutionary relationships among sponges, placozoans, and ctenophores remain a subject of research. page 951

42.2 PROTOSTOME ANIMALS: Protostomes are bilaterian animals and include lophotrochozoans and ecdysozoans.

Protostomes are divided into lophotrochozoans and ecdysozoans. page 952

Lophotrochozoans consist of 17 phyla, including annelid worms and mollusks. page 952

Annelid worms include earthworms and leeches, but most species of annelids live in the oceans. page 952

Mollusks include gastropods (snails, slugs), cephalopods (squid, octopuses), and bivalve mollusks (clams, oysters). page 953

Ecdysozoans include several phyla, notably the nematodes and arthropods, that molt their cuticles. page 956

42.3 ARTHROPODS: Arthropods, bilaterian animals named for their jointed legs, are extraordinarily diverse.

Arthropods can be organized into four main groups: insects, chelicerates (spiders, scorpions), myriapods (centipedes, millipedes), and crustaceans (lobsters, shrimp). page 957

The jointed legs of arthropods have been modified to include a remarkable diversity of structures. page 957

The most diverse group of arthropods is the insects. page 959

All insects have desiccation-resistant eggs. page 959

Insects were the first animals to evolve wings. page 959

Insects efficiently exchange gases through spiracles and tracheae. page 959

Some insects undergo metamorphosis, a dramatic developmental change from a pupa to an adult. page 960

42.4 DEUTEROSTOME ANIMALS: Deuterostomes are bilaterian animals and include humans and other chordates, as well as acorn worms and sea stars.

Deuterostomes include three main phyla—Chordata, Hemichordata, and Echinodermata. page 960

Well-known echinoderms include sea stars and sea urchins, which have a fivefold symmetry superimposed on their basic bilateral body plan. page 961

Chordates include vertebrates, cephalochordates, and tunicates. page 961

42.5 VERTEBRATES: Vertebrates are deuterostomes with a bony cranium and typically a vertebral column, enabling them to become ecologically important in the water and on land.

The group commonly known as "fish" consists of four distinct groups of aquatic vertebrates: hagfish and lampreys, cartilaginous fishes, bony fishes, and lobe-finned fishes. page 964

Lungfish are the closest relatives of tetrapods, which include amphibians, lizards, turtles, crocodilians, birds, and mammals. page 966

Amphibians have an aquatic larval form and a terrestrial adult form. page 967

Amniotes, such as lizards, snakes, crocodiles, birds, and mammals, have an amniotic egg, permitting these animals to colonize dry, terrestrial habitats. page 967

Mammals are covered with hair and feed their young milk from mammary glands. page 968

Log in to **LaunchPad** to check your answers to the Self-Assessment Questions and to access additional learning tools.

CASE 8

Conserving Biodiversity

Rainforest and Coral Reef Hotspots

When we think about mass extinctions, the demise of the dinosaurs is often the first example that springs to mind. But many conservationists believe we are on the brink of a mass extinction today. The International Union for the Conservation of Nature has estimated that more than 800 plant and animal species in habitats around the world have died out in the last 500 years, and more than 20,000 are currently at risk of extinction.

The cause of this die-off isn't an asteroid impact or a cataclysmic volcanic event. Instead, human activity is dramatically altering the planet and its biological diversity.

We cannot save every single species threatened with extinction. Conservationists must make difficult choices, prioritizing which species to protect. To get the most bang for every conservation buck, ecologists have identified 25 priority areas, or biodiversity hotspots.

These hotspots are areas with disproportionately large concentrations of endemic species—that is, species found nowhere else. For example, scientists estimate that as many as 44% of all vascular plants and 35% of vertebrate species are confined to these 25 hotspots. Those figures are impressive, considering that the hotspots make up just 1.4% of Earth's land surface.

The islands of the Caribbean—the West Indies—constitute one such biodiversity hotspot. The Caribbean Sea contains more than 7000 islands rich in biological diversity, including an estimated 6550 native plant species. All of the Caribbean islands are flush with flora and fauna, but one island in particular epitomizes the concept of a biodiversity hotspot. Hispaniola, the second largest of the Caribbean islands, was the site of the first European colonies in the New World, founded by Christopher Columbus. Today, Hispaniola is shared by two countries, Haiti to the west and the Dominican Republic to the east.

Hispaniola is roughly the same size as Ireland, but it contains a striking assortment of habitats, including desert, savanna, rainforests, dry forests, pine forests, coastal estuaries, and interior mountains. With such a rich abundance of habitats, Hispaniola is, in some ways, a microcosm of the planet's many ecosystems. And like the Earth's other natural places, Hispaniola faces many challenges to its native biological diversity.

Of all the habitats on Hispaniola, the forests contain the greatest variety of plants, animals, and other organisms. Ferns, orchids, and bromeliads (spiky relatives of the pineapple) grow in treetops far above the forest floor. These plants, known as epiphytes, absorb water from dew, rainwater, and even the air. Their roots never reach the soil.

The forest is alive with activity. Ants march silently. Crickets and cicadas buzz and whine in the treetops. Vertebrate vocalists such as frogs and birds fill the air

> *Conservationists must make difficult choices, prioritizing which species to protect. To get the most bang for every conservation buck, ecologists have identified 25 priority areas, or biodiversity hotspots.*

with chirps and whistles, making themselves known to both potential mates and rivals. Bats swoop through the darkened skies, using ultrasonic calls to locate insect prey. An estimated 30,000 species of multicellular organisms are found on Hispaniola, and a third of those are endemic. One example is the Hispaniolan solenodon, a long-nosed mammal that resembles an overgrown shrew and secretes toxic saliva. Listed as endangered, the solenodon has flirted with extinction for decades.

How did all these species come to live on the island? Over tens of millions of years, organisms traveled to Hispaniola by air and by sea. Geographically separated from their ancestors on the mainland, many of these organisms evolved to become distinct species. As they adapted to fill distinctive niches on the island, animals such as *Anolis* lizards, frogs, and crickets evolved an impressive diversity of forms.

In some cases, scientists have been able to explore Hispaniola's evolutionary history directly. They have found the remains of insects and other species that became entombed in amber, which is fossilized tree resin, 30–23 million years ago. Many of these preserved specimens resemble species still living on Hispaniola. Others are unlike any creatures found on the island today.

The stunning diversity displayed on Hispaniola doesn't exist just on land. Conservationists consider the warm, shallow reefs of the Caribbean to be one of the world's marine biodiversity hotspots. Coral reefs in general are among the world's most biologically diverse ecosystems, with species from all recognized phyla of animals populating the reefs.

Around the world, reefs support many thousands of species, including snails, corals, lobsters, and fish. Many, but not all, of these creatures spend their entire life in this habitat. Reefs also act as vital nursery grounds for animals that live in other parts of the ocean, from sea turtles to sharks. Reef biodiversity is also critical to maintaining fisheries that provide food and income for millions of people.

Hispaniola's rainforests and reefs—indeed, all the biodiversity hotspots identified by conservationists—are notable for more than just their biological richness. They have also been given priority status because they face serious threats.

Reefs around the world are being damaged by human activities. Overfishing and pollution take their toll on reef ecosystems. Climate change is increasing seawater temperature, and increasing CO_2 is altering ocean chemistry.

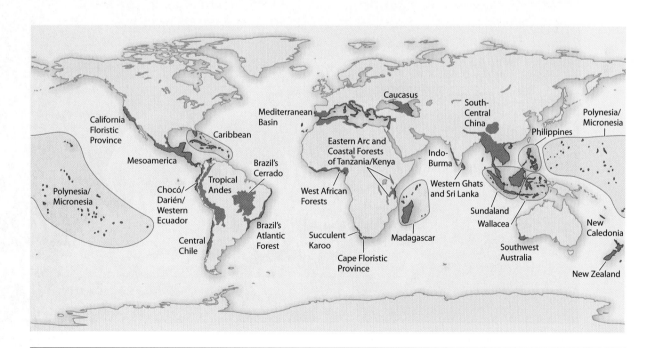

Biodiversity hotspots. The areas highlighted in orange have a particularly high diversity of species found nowhere else.

Data from Macmillan Publishers Ltd: N. Meyers et al., 2000, "Biodiversity Hotspots for Conservation Priorities," Nature 403:853–858.

CASE 8

Those changes threaten the coral organisms that form the backbone of the reefs.

Likewise, ecosystems on land face significant challenges. The Caribbean islands now have only approximately 11% of their original vegetation. Humans have cleared great swaths of land for agriculture and chopped trees to use as fuel. Humans have also introduced non-native, or invasive, species to the islands, both intentionally and accidentally. On Hispaniola, Europeans introduced the Javanese mongoose to control rats in sugarcane fields. The mongoose has devastated local populations of reptiles, amphibians, and birds.

If Hispaniola serves as a microcosm of the planet's ecosystems, it also serves as a cautionary tale. Two divergent paths have emerged on the single island. Haiti has historically suffered greater poverty than the Dominican Republic, and it has much greater population density. Those factors have put more pressure on the environment in Haiti, contributing to much greater deforestation on the western half of the island.

The case of Hispaniola illustrates the complex challenges facing conservationists. Solutions to environmental challenges must take into consideration the effects of poverty, culture, and other social pressures. Balancing the needs of humans with the needs of natural ecosystems is no easy task. But as a relatively small island, Hispaniola may actually be in a better position to turn things around than larger, more politically complex countries. The world's biodiversity hotspots might yet lead the charge in finding ways for humans and the rest of the planet's inhabitants to coexist peacefully.

Coral reefs. Coral reefs support many species but are threatened by warming, changing seawater chemistry, overfishing, and pollution. *Source: Jeff Hunter/Getty Images.*

DELVING DEEPER INTO CASE 8

For each of the Cases, we provide a set of questions to pique your curiosity. These questions cannot be answered directly from the Case itself, but instead introduce issues that will be addressed in chapters to come. We revisit this Case in Chapters 43–48 on the pages indicated below.

1. How do populations colonize islands? See page 1015.
2. Can competition drive species diversification? See page 1023.
3. How is biodiversity measured? See page 1032.
4. How do evolutionary and ecological history explain biodiversity? See page 1090.
5. How has global environmental change affected coral reefs around the world? See page 1100.
6. What are our conservation priorities? See page 1111.

CHAPTER 43 Behavior and Behavioral Ecology

CORE CONCEPTS

43.1 TINBERGEN'S QUESTIONS: For any behavior, we can ask what adaptive function it serves, what causes it, how it develops, and how it evolved.

43.2 DISSECTING BEHAVIOR: Animal behavior is shaped in part by genes acting through the nervous and endocrine systems.

43.3 LEARNING: Learning is a change of behavior as a result of experience.

43.4 INFORMATION PROCESSING: Orientation, navigation, and biological clocks all require information processing.

43.5 COMMUNICATION: Communication involves an interaction between a sender and a receiver.

43.6 SOCIAL BEHAVIOR: Social behavior is shaped by natural selection.

There is no Nobel Prize for biology. Instead, for biological research, the Nobel Committee awards the Prize for Physiology or Medicine. Biologists specializing in other areas, such as behavior and ecology, are not even in the running for science's ultimate prize. However, in 1973, in an unprecedented move, the Nobel Committee bent its own rules and awarded the Prize for Physiology or Medicine to three scientists with no physiological or medical credentials: Niko Tinbergen, Konrad Lorenz, and Karl von Frisch. The award was given in recognition of their roles in pioneering the study of animal behavior under natural conditions.

What exactly does the expression "under natural conditions" mean and why is it important to emphasize? In the early 1900s, Carl von Hess, a prominent visual physiologist, claimed that bees are color blind, a conclusion based on experiments in laboratory settings. However, by training bees to go to food sources associated with differently colored cards in a natural context, the Austrian behavioral biologist Karl von Frisch demonstrated in 1914 that, in fact, bees have excellent color vision. Von Frisch's work illustrated that behavior—in this case, foraging for food—cannot be studied in isolation, but can be fully understood only when studied in a natural context. Von Frisch emphasized this point at a scientific meeting where he showed how he had trained bees to associate blue cards with food. Von Frisch's presentation caused quite a stir for more than just scientific reasons. By chance, the audience was wearing blue name tags, which proved highly popular with the blue-fixated bees!

In this chapter, we explore animal behavior: how animals learn, communicate with one another, form social groups, and choose mates. In each case, we can consider animal behavior on multiple levels. At the individual level, behavior is determined in part by a complex interplay of the nervous and endocrine systems. Behavior is also influenced by the environment, as von Frisch's bee studies illustrate. Finally, behaviors, like all traits, are shaped by natural selection.

43.1 TINBERGEN'S QUESTIONS

We begin our exploration of animal behavior by asking a simple question: why does an animal exhibit a particular behavior? In 1963, the Dutch behavioral biologist Niko Tinbergen divided this overarching question into four separate questions, each one focusing on a different aspect of the behavior (**Table 43.1**).

TABLE 43.1	Tinbergen's Four Questions		
		PROXIMATE: Short-term processes acting within an individual's lifetime affecting the behavior	**ULTIMATE:** Long-term evolutionary processes affecting the behavior
ORIGIN OF A BEHAVIOR		**Developmental:** How is the behavior acquired over an individual's lifetime?	**Evolutionary:** Where and how did the behavior arise in the past?
IMPLEMENTATION OF A BEHAVIOR		**Mechanistic:** How is the behavior caused through neural, muscular, and other processes?	**Adaptive:** How does the behavior enhance the survival and/or reproduction of individuals?

Tinbergen asked proximate and ultimate questions about behavior.

Let's consider an example: why does a bird sing? While apparently straightforward, this question can be broken down into more specific questions using Tinbergen's approach and then answered in a number of different ways:

1. Adaptive function. What is the behavior for? How does the behavior promote the individual's ability to survive and reproduce? In this case, the answer to the question might be that a male bird sings to attract a mate and then reproduce.

2. Mechanistic causation. What physiological mechanisms cause the behavior? This question can have multiple answers. A bird sings because its hormone levels have changed in response to changes in day length. Or, more immediately, a bird sings because air passing through its specialized singing organ, the syrinx, causes membranes to vibrate rhythmically.

3. Development. How did the behavior develop? Here, the focus is on the role of genes and the environment in shaping the development of the behavior. In birds, typically the male sings, and he has learned the song from his father or other males.

4. Evolutionary history. How did the behavior evolve over time? Complex bird songs may have evolved from vocalizations made by ancestors that became increasingly standardized over time, so much so that we can often identify a bird species simply by hearing it sing. A behavior may have originated to fulfill a function different from the one it currently serves. For example, the song of a particular species may have first evolved to claim a territory but now is used to attract mates.

Tinbergen's analysis allows us to see that different answers to the question "Why does a bird sing?" can all be correct. The first two categories, mechanistic causation and development, provide *proximate* explanations of behavior because they deal with *how* the behavior came to be. The second two categories, adaptive function and evolutionary history, provide evolutionary explanations of how natural selection has shaped a behavior over time; these *ultimate* explanations of behavior deal with *why* the behavior came to be. Tinbergen's four questions are complementary ways of looking at the same problem, and are a good starting point for the analysis of behavior.

The answers to Tinbergen's questions rely on an interplay between genes and the environment. The influence of genes is especially clear in **innate** behaviors, which are instinctive and carried out regardless of earlier experience. Male silkworm moths of the genus *Bombyx* exhibit an innate behavior when

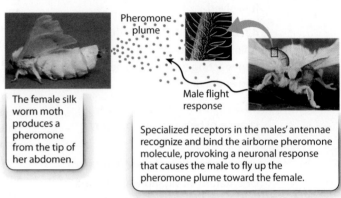

FIG. 43.1 An example of innate behavior. A male moth flies toward the source of a female-produced pheromone. *Photo sources: (left to right) Leal Lab, Department of Molecular and Cellular Biology/UC Davis/Regents of the University of California; Courtesy of Prof. Dr. R. A. Steinbrecht. Cell Tissue Research (1995) 282:203–217; Fabio Pupin/FLPA/Science Source.*

they fly upwind toward the source of a female-produced pheromone, an airborne chemical signal to members of the same species; in this case, the communication acts to attract members of the opposite sex (**Fig. 43.1**). Pheromones released from a female's abdominal gland are sensed by small hairs on the male's antennae. In the presence of pheromones, the antennal sensory hairs fire action potentials (Chapter 34). When approximately 200 hairs are being activated per second, a male flies upwind distances of a kilometer or more, tracking the increasing pheromone concentration until he finds the female. The male moth does not need to learn this behavior; he performs it spontaneously. His genes encode molecular receptors to which the pheromone binds, and this binding triggers a cascade of events that result in the moth's heading up the pheromone's concentration gradient.

In contrast to an innate behavior, which is instinctive, a **learned** behavior depends on an individual's experience. As we will see throughout this chapter, even neurologically simple organisms have a considerable capacity to learn. Fruit flies, for example, learn to avoid certain locations or substances if they associate them with an unpleasant experience.

We can consider, then, the extent to which a particular behavior is genetically encoded (part of the animal's nature) and the extent to which it is conditioned by the environment in which the animal develops (part of the animal's nurture). However, as we discussed in Chapter 17, nature and nurture are inextricably linked. For example, most human behaviors are the product of an interaction between nature and nurture. Consider language in the context of Tinbergen's second question concerning the development of behavior: humans have an innate ability to acquire language, but the specific language that we acquire depends entirely on our nurture. A baby in Italy learns Italian, and one in Finland learns Finnish.

> **Self-Assessment Questions**
>
> 1. Analyze the behavior of the male silkworm moth described in this section in terms of the four questions that Tinbergen might have asked about it.
> 2. Explain the difference between innate and learned behaviors, and provide one example of each type of behavior.

43.2 DISSECTING BEHAVIOR

Behavior can be influenced by a large number of factors, both genetic and environmental. For that reason, complex behaviors can sometimes defy analysis. By choosing simple behaviors that can be more easily studied and understood, scientists can identify, isolate, and investigate the separate components of more complex behaviors.

The fixed action pattern is a stereotyped behavior.

Some of the first behaviors to be carefully analyzed were **displays**. Displays are species-specific patterns of behavior that tend to follow the same sequence of actions whenever they are repeated, in a way that is similar from one individual to the next. Because these behaviors are so similar whenever they occur, they are considered "stereotyped." Presumably, natural selection favors display behaviors that send unmistakable signals. Consider, for example, courtship displays in birds that are preparing to mate (**Fig. 43.2**). Experiments have shown that, in some species, a bird raised in complete isolation from other members of its species still performs the stereotyped courtship display with great precision, suggesting that the behavior is genetically encoded; that is, it is instinctual rather than learned.

Many displays are **fixed action patterns (FAPs)**: a sequence of behaviors that, once triggered, is followed through to completion. A classic example, originally studied by Tinbergen, is the response of a goose to an egg that has fallen from its nest (**Fig. 43.3**). The **key stimulus** initiates the behavior. In the goose's case, the key stimulus is the sight of the misplaced egg. This sight provokes in the goose an egg-retrieval FAP, which consists of rolling the egg back to the nest with the underside of its beak (Fig. 43.3a). This response cannot be broken down into smaller subunits and is always carried out to the end, even if it is interrupted. In fact, even if the researcher ties a string around the misplaced egg and removes it while the goose is in mid-action, the goose will continue the task of rolling the now-absent egg (Fig. 43.3b).

To better understand this behavior, the researcher can vary attributes of the key stimulus (in this case, the misplaced egg). A remarkable finding is that many geese respond most strongly to the largest round object provided, even a soccer ball, as illustrated in Fig. 43.3c. A soccer ball not only elicits the egg-retrieval FAP, but does so even more strongly than a normal egg. The soccer ball is considered a **supernormal stimulus** because it is larger than any egg the goose would naturally encounter and elicits an exaggerated response. Natural selection has likely favored geese that recognize and respond to large eggs that have rolled outside the nest. However, because eggs never grow to the size of soccer balls, selection has not shaped an appropriate response to unrealistically large egg-shaped objects.

The nervous system processes stimuli and evokes behaviors.

How do animals recognize the stimulus that leads to a particular behavior? In the example of the goose retrieving an egg outside its nest, how does the goose recognize that the object in question is sufficiently egg-like to elicit the behavior? At its root, this is an extraordinarily difficult problem. The world is full of stimuli and many things look like an egg. Thus, the challenge for the animal is to filter out the correct signal (the egg) from the surrounding noise (everything else). Animals have to process many different kinds of information in three dimensions.

We know that stimulus recognition is often carried out by **feature detectors**, specialized sensory receptors or groups

FIG. 43.2 Courtship display in the Raggiana Bird-of-Paradise. The male (top) is performing a highly stereotyped set of behaviors to entice the female to mate with him. *Source: Bruce Beehler/AGE Fotostock.*

FIG. 43.3 A fixed action pattern. A goose displays and completes the behavior to retrieve an egg (a), even if the behavior is interrupted (b) or the stimulus altered (c).

of sensory receptors that respond to important signals in the environment. Frogs provide a good example. Male frogs call to attract females and warn off other males. A single pond may contain many species of frogs, each with its specific call. A heavily populated frog pond can be a noisy place, but it is critical that an individual frog is able to distinguish the call of other members of its own species from those of other species. A further complication is that different species of frogs can have similar calls, which are often virtually indistinguishable to the human ear.

One approach to the problem of understanding how frogs recognize their species-specific call is experimenting with recordings of frog calls. By recording frog calls, altering aspects of the call in specific ways, playing them back, and then monitoring the responses of frogs, we can identify the impact of different components of the call—pitch, duration, and pulse frequency, for example. Such studies have demonstrated that the frog's auditory nerves act as a feature detector, with each one "tuned" to a specific component of the call. The combination of different feature detector neurons stimulated by different components adds up to the frog's recognition of a specific call, as shown in **Fig. 43.4**. Once the sound is correctly identified, the appropriate behavior follows.

Hormones can trigger certain behaviors.

The effect of a neuron is typically short lived and local: a neuron may, for example, connect to another neuron or cause a muscle to contract (Chapter 34). One way in which a stimulus may provoke a more widespread and prolonged effect is through the production of hormones. Hormones can affect multiple cells in target organs simultaneously (Chapter 36).

Anolis lizards demonstrate the important role of hormones in sexual behavior, showing both how social stimuli can affect the release of hormones and how hormones can affect behavior (**Fig. 43.5**). If females of *Anolis carolinensis* are isolated from males during the breeding season, approximately 80% of individuals will be found to have active egg follicles in their ovaries (Fig. 43.5a). If one male is added to a group of females, that figure increases to 100% (Fig. 43.5b). The courting behavior of the male lizard stimulates the females to produce hormones that cause the full development of the ovaries, making all the females reproductively active. Now consider what happens if a group of males, instead of a single male, is added to an all-female population: only 40% of the females undergo ovarian development (Fig. 43.5c). This unexpected result seems to occur because the males fight among themselves rather than court the females; for some of the females, then, the courtship

FIG. 43.4 A feature detector. In its natural environment, a frog is subjected to a barrage of sound stimuli. Each species has a feature detector tuned to its species-specific call, allowing it to distinguish the call of its own species (red arrow) from the calls of other frogs (blue arrows).

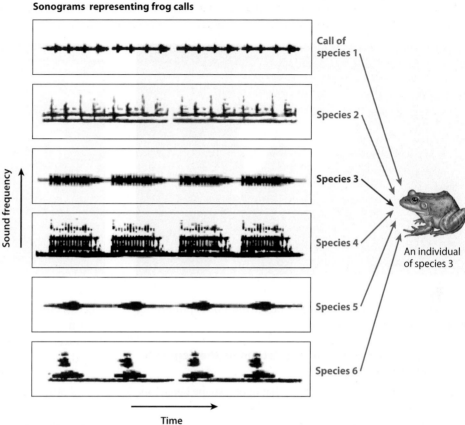

stimulus is lacking. Castrated males (which do not produce testosterone) that are added to a group of females have no significant effect on rates of ovarian development, which remains approximately 80% (Fig. 43.5d). They fail to court because the courting behavior is caused in part by testosterone. In contrast, a castrated male that is injected with testosterone does display courting behavior; his presence has the same effect as the presence of a single male, inducing all the females to undergo ovarian development (Fig. 43.5e).

These simple laboratory experiments demonstrate the complex interplay between hormones and behavior. Testosterone mediates both the male–male interactions and the male–female

---

FIG. 43.5 Role of hormones in behavior, demonstrated by groups of *Anolis* lizards.

a. + 0 males

In a group of all female *Anolis carolinensis* collected in the spring, 80% of the females are prepared for reproduction with egg follicles in their ovaries.

d. + castrated male

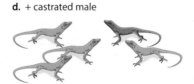

Castrated males do not display courtship behavior, and so do not affect the females' reproductive readiness.

b. + 1 male

A male displaying courtship behavior added to the all-female group causes 100% of females to become reproductively active.

e. + castrated male injected with testosterone

An injection of testosterone causes a castrated male to display courtship behavior, affecting the group of females in the same way as a non-castrated male.

c. + more than one male

More than one male added to a group of females fight with each other instead of courting females. Now only 40% of the females are reproductively active.

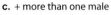 Reproductive   Nonreproductive

FIG. 43.6 Artificial selection. Artificial selection can alter behavior, as shown by the remarkable differences in behavior we see among breeds of dogs, such as (a) a Pointer and (b) a Border Collie. *Sources: a. Adriano Bacchella/naturepl.com; b. Fred Lord/Alamy.*

interactions that cause females to produce hormones promoting their reproductive development.

Breeding experiments can help determine the degree to which a behavior is genetic.

The nervous and endocrine systems that in part shape behaviors are the product of genetic instructions. Thus, on the one hand, all behaviors have a genetic component. After all, it is the instructions in the genome that build the nervous system, muscles, and glands whose actions lead to a behavior. On the other hand, all behaviors can be considered environmental since, without the appropriate environment, an organism would not develop normally and would therefore lose the ability to perform the behavior. On a very basic level, an organism in an environment without appropriate nutrients would lack the necessary energy to perform a given behavior. How, then, do we study the genetic basis of behavior? Modern approaches harness the power of molecular genetics, but more traditional analyses are also highly informative.

Artificial selection (Chapter 20), in which humans breed animals and plants for particular traits, provides strong evidence of the role of genetics in influencing behavior. Dogs were domesticated from wolves approximately 10,000 years ago. Today, dogs display great diversity, although they are all the same species. Obviously, there has been extensive selection on physical traits: a Dachshund, for example, looks quite different from a German Shepherd or a Great Dane. Selection has also been applied to behavior: a Pointer has extraordinary "pointing" behavior that indicates the location of a hunter's prey, and a Border Collie is an excellent herder (**Fig. 43.6**).

In the late 1950s, William Dilger studied the nest-building behavior of lovebirds, which are members of the genus *Agapornis*. Some species transport their nesting material in their beaks, whereas others tuck pieces of the material into their tail feathers. Does the nesting behavior of these birds have a genetic basis? To answer this question, Dilger set up crosses between species with different nest-building techniques to produce hybrids. He then observed how the hybrids built their nests. Interestingly, the hybrid offspring showed nest-building behavior intermediate between that of the parents: they tried to tuck material under their feathers, but were not successful. These experiments suggest that this behavior has a genetic basis.

It was not possible to do a true Mendelian analysis of the lovebirds' behavior (Chapter 15). Such an analysis would require additional crosses, which were not possible because the hybrids were sterile. Like most behaviors, the lovebirds' nest-building behavior is probably controlled by a large number of genes, each of which has a relatively small effect (Chapter 17). Given their multiple influences, it is difficult to identify which genes are responsible for the behavior. Even with molecular methods, mapping and identifying the underlying controlling genes in cases where many different genes govern a particular trait is a daunting task.

Molecular techniques provide new ways of testing the role of genes in behavior.

Molecular biology is changing the way we approach behavioral genetics. Some studies have identified complex behaviors that are strongly influenced by a single gene. One well-understood instance of a single gene's effect on behavior comes from the fruit fly *Drosophila*.

In *Drosophila*, different alleles of the *foraging* (*for*) gene, for^s and for^R, are present in populations. The two alleles have different effects on the behavior of *Drosophila* larvae. The gene encodes an enzyme expressed in the brain that affects neuronal activity and alters behavior. In the absence of food, both "sitter" (for^s) and "rover" (for^R) larvae move about in search of food. In the presence of food, however, sitters barely move, feeding on the patches on which they find themselves, whereas rovers move extensively both within a patch of food and between patches (**Fig 43.7a**).

Both alleles are present in natural populations (**Fig. 43.7b**); typically, 70% are rovers, 30% are sitters. That both strategies are found in populations suggests that both are adaptive. Otherwise, if one strategy were markedly inferior, we would expect natural selection to eliminate it. Furthermore, studies have shown that rovers are selected for in crowded environments in which seeking new food sources carries a survival advantage, and sitters are selected for in less crowded environments because sitters take maximum advantage of their current food source. In this case, variation at a single gene affects a complex behavior in fruit flies.

FIG. 43.7 The effect of genotype at the foraging locus on the feeding behavior of *Drosophila melanogaster* larvae. (a) When food is present, rover larvae travel more extensively than sitter larvae. (b) Both types of larvae exist in natural populations. *Source: M. B. Sokolowski, 2001, "Drosophila: Genetics Meets Behaviour," Nature Reviews Genetics 2:879–890, doi:10.1038/35098592.*

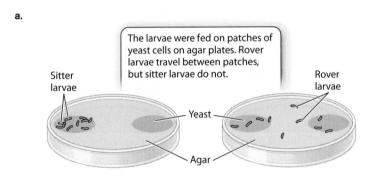

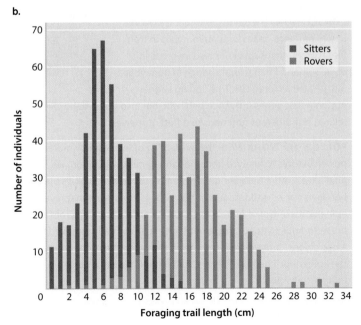

The *foraging* gene is present in other insects as well, including honeybees, suggesting that the gene has been evolutionary conserved. Honeybees with low levels of *for* expression in their brain tend to stay in the hive, whereas those with high levels of *for* expression are likely to be foragers (**Fig. 43.8**).

Studies of the North American vole, in the genus *Microtus*, have also revealed how genes influence behavior (**Fig. 43.9**). The Prairie Vole is monogamous: couples pair-bond for life and mate only with each other. By contrast, the Prairie Vole's close relative, the Montane Vole, is promiscuous: a male mates and moves on, and a female over her lifetime produces litters by several different males. What differences underlie these different behaviors?

Hormones and their receptors provide an answer. In all mammals, antidiuretic hormone (ADH) controls the concentration of urine (Chapter 39). The receptor for ADH is expressed not only in the kidney, but also in the brain, where it affects behavior. Both the monogamous Prairie Vole and the promiscuous Montane Vole produce ADH, but the hormone receptors in the two species are different. In the 1990s, the American neuroscientist Thomas Insel analyzed and compared the genes that encode the ADH receptor in the two vole species. He found a difference in the region of the gene that determines under which conditions and in which cells it is expressed. Because of this difference, the distribution of ADH receptors in the brain differs in the Prairie Vole and in the Montane Vole.

Does the difference in gene expression explain the different sexual behaviors of the two species? The answer seems to be yes. Insel and his colleague Larry Young inserted the Prairie Vole's ADH receptor gene, complete with its regulatory region, into a laboratory mouse (a member of a promiscuous species, like the Montane Vole). Interestingly, Insel and Young observed a marked change in the mouse's behavior. Rather than mating with a female and then moving on to another mate, the transgenic male mouse pair-bonded with the female. In short, the addition of a single gene affected a complex behavior such as pair bonding, making the otherwise promiscuous mouse more likely to pair bond.

Self-Assessment Questions

3. What is fixed about a fixed action pattern?
4. Why are feature detectors a key aspect of sensory systems?
5. Why are long-term changes in behavioral state—for example, from sexually immature to mature—typically mediated by hormones?
6. Why is it typically difficult to identify individual genes that govern particular behaviors?
7. Explain, with an example, how crosses between closely related species can help us understand the genetic basis of behavior.

HOW DO WE KNOW?

FIG. 43.8
Can the same gene influence behavior differently in different species?

BACKGROUND In fruit flies (*Drosophila melanogaster*), variation in the *foraging* (*for*) gene affects feeding behavior of larvae. Two alleles of the *for* gene exist in natural populations: *for*^{s} larvae ("sitters") tend to stay on a patch of food and *for*^{R} larvae ("rovers") tend to move from patch to patch. Honeybees (*Apis mellifera*) also forage for food, but, in contrast to fruit flies, bee foraging behavior changes with age. Young honeybees ("nurses") stay at the hive, and older ones ("foragers") forage for nectar. The *for* gene is present in honeybees, raising the possibility that this gene is involved in the developmental change in foraging in honeybees.

HYPOTHESIS In 2002, Yehuda Ben-Shahar and colleagues at the University of Illinois hypothesized that foraging in honeybees is influenced by the expression of the *for* gene, with nurses having lower expression levels than foragers.

EXPERIMENT 1 Levels of *for* mRNA were measured in honeybee nurses and foragers in two different colonies.

RESULTS Levels of *for* mRNA are significantly higher in foragers than in nurses (Fig. 43.8a). However, foragers are older than nurses. Thus, based on the results of just the initial measurements, it was unclear whether the increased gene expression in foragers compared with nurses is associated with differences in age or differences in behavior. To answer this question, the researchers repeated the experiment and measured levels of *for* mRNA in honeybees that forage at the age at which bees are normally nurses. These young foragers were also found to have higher *for* mRNA levels than nurses (Fig. 43.8b).

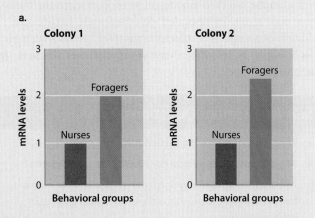

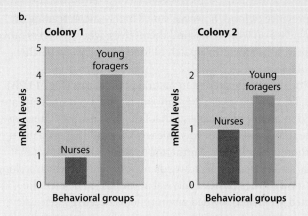

EXPERIMENT 2 The *for* gene encodes a cGMP-dependent kinase that phosphorylates other proteins (Chapter 9). In a second experiment, the researchers treated honeybee nurses with cGMP to activate the kinase and monitored their subsequent behavior. They used a related compound, cAMP, which has a chemical structure similar to cGMP but does not affect the kinase, as a control to ensure that any effect observed was specific for the cGMP pathway and not, say, a generalized stress response to being experimentally manipulated.

RESULTS Treatment with cGMP changed the behavior of nurses, causing them to forage (Fig. 43.8c). Furthermore, the higher the dose, the more foraging behavior was observed. No effect on foraging was seen in the control treatment (Fig. 43.8d).

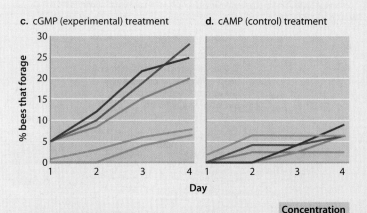

CONCLUSION The same gene that is involved in two different behavioral phenotypes in fruit flies (sitters versus rovers) is also involved in a developmental behavioral change in honeybees (nurses versus foragers). This finding suggests not only that genes influence behavior, but also that a gene can have related but different functions in different species.

FOLLOW-UP WORK With the availability of increasingly more powerful tools to analyze genomes and genetic regulation, we now have a more detailed understanding of the behavioral switch between nurses and foragers in honey bees. It appears that reversible epigenetic control of gene expression (Chapter 18) plays a role in the nurse/forager switch. If nurse bees are experimentally removed from a bee colony, foragers revert to being nurses, in the process reversing the epigenetic changes that resulted in their previous nurse-to-forager transition.

SOURCE Ben-Shahar, Y., et al. 2002. "Influence of Gene Action Across Different Time Scales on Behavior." *Science* 296:741–744.

FIG. 43.9 Differing mating behavior in voles. The different behaviors in (a) monogamous Prairie Voles and (b) promiscuous Montane Voles results from variation at a single gene. *Sources: a. Todd Ahern/Quinnipiac University; b. Rick & Nora Bowers/Alamy.*

43.3 LEARNING

As we have noted, genes are not the only determinants of behavior. We all know that experience often leads to changes in behavior, a process we call **learning**. Humans are extraordinary learners, but we are not alone in this ability. Even many relatively simple organisms respond to experience. Researchers have identified several types of learning. The categories are not strictly delimited, and there is some overlap among them, but they are a useful way to organize our thinking and give us deeper insights into the subject.

Non-associative learning occurs without linking two events.

Non-associative learning is learning that occurs in the absence of any particular outcome, such as a reward or punishment. One type of non-associative learning is **habituation**, the reduction or elimination of a behavioral response to a repeatedly presented stimulus. Chicks presented with silhouettes of predators flying overhead provide an example. Initially, any overhead silhouette provokes a defensive crouching posture, but eventually chicks habituate to overhead silhouettes that have proved to be benign. In the absence of any consequences, chicks no longer crouch in response to a harmless silhouette passing repeatedly overhead.

Sensitization is another form of non-associative learning. Sensitization is the enhancement of a response to a stimulus that is achieved by presenting a prior strong or novel stimulus. This initial stimulus makes the animal more alert and responsive to the next stimulus. The sea slug, *Aplysia*, a model organism among scientists interested in the neuronal basis of behavior, exhibits sensitization. *Aplysia* withdraws its gills in response to a touch on the siphon (the tube through which it draws water over its gills). Interestingly, if the animal is given a weak electric shock first, its response to a touch on its siphon is much more rapid, suggesting that the shock has made the slug more sensitive to later stimuli.

Associative learning occurs when two events are linked.

Associative learning (also called **conditioning**) occurs when an animal learns to link (or associate) two events. Perhaps the most famous example of conditioning is the response of Ivan Pavlov's dogs. In the course of experiments on animal digestion during the 1890s, Pavlov, a Russian physiologist, first presented dogs with meat powder, and they salivated in response. He then presented the dogs with an additional cue, a ringing bell, whenever he presented the dogs with meat powder. After repeatedly experiencing the two stimuli together, the dogs salivated at the sound of the bell alone, in expectation of the meat reward. This form of conditioning, in which two stimuli are paired, is called **classical conditioning**.

A second form of associative learning, which links a behavior with a reward or punishment, is called **operant conditioning**. Operant conditioning involves a novel, undirected behavior that becomes more or less likely over time. If the behavior is rewarded (positively reinforced), it is more likely to occur the next time around. It is not just out of the goodness of her heart that the trainer at Sea World gives a sea lion or dolphin a fish reward at the end of each trick that the animal performs. In contrast, if behavior is punished (negatively reinforced), the response becomes less likely.

In classical conditioning, an association is made between a stimulus and a behavior, whereas in operant conditioning, an association is made between a behavior and a response. In classical conditioning, one stimulus is exchanged for another (in the case of Pavlov's dogs, the smell of food was exchanged for the sound of a bell). As a result, a novel stimulus evokes a behavior. In operant conditioning, a behavior is associated with a reward or punishment, so the behavior becomes more likely (with a reward) or less likely (with a punishment).

HOW DO WE KNOW?

FIG. 43.10

To what extent are insects capable of learning?

BACKGROUND European digger wasps, *Philanthus triangulum*, live in the sand. These wasps are sometimes called "bee wolves" because they hunt honeybees to feed their developing young. After mating, each female digs a long burrow with a few chambers at the end where she lays her eggs. She then forages for honeybees that she brings back to these chambers for her larvae to eat. The wasp faces a navigational challenge: having captured her prey, how can she find her way back to and recognize her nest? As part of his PhD studies in the early 1930s, Niko Tinbergen noticed that wasps lingered briefly near a new nest before heading off to hunt; he hypothesized that they were learning local landmarks associated with the nest.

HYPOTHESIS Wasps learn visual cues around their nests to help locate the nest upon their return.

METHOD Tinbergen recognized that a good test of the learning abilities of an insect should take place in its natural environment. He combined his skills as a naturalist and as an experimentalist to devise an elegant demonstration of the way in which female wasps learn landmarks for navigation. He placed a ring of pine cones around the nest of a wasp (Fig. 43.10a); then, once she had left to hunt, he shifted them to a new location away from the nest entrance (Fig. 43.10b). If visual cues are key to the wasp's ability to locate the nest, the wasp should return to the displaced pine cone ring. In contrast, if cues are, for example, olfactory, the wasp should return directly to the nest.

RESULTS Female digger wasps carried out a brief landmark-learning flight on departure from the nest. When the landmarks were displaced, the females returned to the wrong location (Fig. 43.10b).

a.

b.

CONCLUSION Female digger wasps learn and use local landmarks, such as the experimental ring of pine cones, as cues to the nest location.

SOURCE Tinbergen, N. 1958. *Curious Naturalists*. New York: Basic Books.

Learning is an adaptation.

The capacity to learn often has adaptive functions. Much learning, including human learning, is based on **imitation**, in which one individual copies another. Learning by imitation was famously observed in the days when people in Britain had milk deliveries left outside on the doorstep. As we saw in Chapter 23, a number of birds, including Great Tits and Blue Tits, learned to open foil-topped milk bottles, and this behavior spread rapidly through populations of birds as other birds copied the successful birds' approach. Individuals can also learn by imitating individuals of a different species. An octopus can learn to open a jar with a reward inside by watching a human or another octopus do it.

The capacity to learn a particular task seems in many instances to be innate. Consider female digger wasps, which are foragers. Tinbergen showed that a digger wasp learns the landmarks around her nest and then uses this information to find her way back to it from hunting. In his experiment, illustrated in **Fig. 43.10**, Tinbergen placed a ring of pine cones around a wasp's nest and, after she had flown out and returned a few times, moved the circle of pine cones to the side of the nest. The returning wasp then flew past her own nest and went directly to the center of the ring. This innate ability to learn landmarks is adaptive: it improves the survival or reproductive success of the individual.

This adaptive aspect of animal learning is also revealed by taste aversion experiments, in which an animal typically learns to avoid certain flavors associated with a negative outcome. Rats learn to avoid flavored water if consuming it is associated with nausea. However, they do *not* learn to avoid flavored water if it is associated with a different kind of negative reinforcement, such as a mild electric shock. This result indicates that rats can make some associations, but not others. The ones they can learn are the biologically meaningful ones, those that

FIG. 43.11 Imprinting on (a) Konrad Lorenz and (b) an ultralight aircraft. By manipulating newly hatched birds' first sight of a moving object, it is possible to induce powerful attachments in the birds. *Sources: a. Nina Leen/Time Life Pictures/Getty Images; b. Jeffrey Phelps/Getty Images.*

favor survival. In the course of evolution, the rat's ancestors encountered poisoned food that resulted in nausea and the aversion response evolved. Until humans started doing experiments on them, however, rats had never encountered a meal that resulted in an electric shock. It is not surprising, then, that the ability to pair these two phenomena never evolved. These experiments also show that the specific evolutionary history of the species in question matters when analyzing animal behavior.

In addition to adaptive predispositions for *what* can be learned and not learned, many species exhibit predispositions for *when* learning takes place. This is particularly evident in imprinting, a form of learning typically seen in young animals in which they acquire a certain behavior in response to key experiences during a critical period of development. In 1935, Austrian ethologist Konrad Lorenz made imprinting famous by exploiting the observation that newly hatched goslings and ducklings rapidly learn to treat any sufficiently large moving object they happen to see shortly after hatching as their mother. Lorenz found that, if he was the first person the hatchlings saw, they would follow him as though he were their mother (**Fig. 43.11**).

This behavior is adaptive because the first being that a hatchling usually sees is a parent; the resulting close parent–offspring association ensures that the parent can provide care and protection. This **filial imprinting** is most common in bird species whose offspring leave the nest and walk around while still young, such as ducklings. By comparison, filial imprinting is rare in species of birds whose young stay in the nest until they are able to fly away.

Experiments have shown that filial imprinting typically occurs during a specific, sensitive period in the animal's life and that the results are usually irreversible. After Lorenz's baby ducks had imprinted on him, they would not change their minds about who their parent was even when presented with their real mother duck. The timing of the sensitive period varies from species to species. In some cliff-nesting sea birds, imprinting on auditory stimuli (such as the call of the parents) begins while the chick is still in the egg.

Self-Assessment Questions

8. Both classical and operant conditioning involve learning to associate two events. How do the two kinds of conditioning differ?
9. Distinguish between associative and non-associative learning, and provide an example of each form of learning.
10. What is the adaptive value of imprinting in goslings?

43.4 INFORMATION PROCESSING

Learning is a form of information processing in which experience shapes behavior. At a cellular level, learning typically involves a change in the strength of connections between neurons. The central nervous system of even relatively simple animals is capable of extraordinary feats of information processing. The ways in which animals' nervous and endocrine systems and the input of experience are integrated to generate adaptive behaviors are remarkable. Some of these adaptive behaviors include ways of moving, navigating, and keeping time, which we examine in this section.

Orientation involves a directed response to a stimulus.

Even the simplest bacteria and protozoa are capable of moving in response to stimuli. A *Paramecium* that finds itself in an unfavorable environment, such as water that is too warm or too salty, increases its speed and begins to make random turns. When it encounters favorable conditions, such as cooler water, it slows and reduces its turning rate. These random, undirected movements are termed **kineses** (singular, **kinesis**).

In contrast, **taxes** (singular, **taxis**) are movements in a specific direction in response to a stimulus. An interesting example of a taxis is movement oriented to a magnetic field. In

1975, American microbiologist Richard Blakemore was the first to use the term *magnetotaxis* to describe this behavior. Blakemore found that anaerobic bacteria in the genus *Aquaspirillum*, which swim by means of flagella, tend to swim toward magnetic north. These bacteria can be attracted to the side of a dish with a bar magnet. Little bits of magnetized iron oxide, arranged in a row inside the *Aquaspirillum* bacterial cell, allow the bacteria to sense the magnetic field. Blakemore hypothesized that the bacteria swim north in an effort to swim deeper. In the Northern Hemisphere, the north magnetic pole is inclined downward. At Woods Hole, Massachusetts, where these bacteria were found, the magnetic pole is at approximately a 70-degree incline. *Aquaspirillum* are anaerobic and must stay buried in sediment to survive; the bacteria remain buried by moving downward along the magnetic gradient.

To test his hypothesis, Blakemore looked for and found bacteria in New Zealand that exhibited similar behavior to the Woods Hole bacteria, except that they swam toward the *south* magnetic pole, bringing them downward in the Southern Hemisphere. Finally, he took New Zealand cultures back to Woods Hole, where the bacteria responded inappropriately to the unfamiliar magnetic field of the Northern Hemisphere, swimming up into the oxygen and dying. These experiments demonstrated that bacteria are able to sense a magnetic field and move in a directed fashion relative to that field.

Navigation is demonstrated by the remarkable ability of homing in birds.

Many animals use environmental cues to migrate long distances. The navigational achievements of homing pigeons are legendary. They can home—that is, find their way back to the place where they are housed and fed—over extremely long distances, even of more than a thousand miles. These pigeons use a wealth of cues when homing. A compass may tell you which way is north, but you also need **map information**—which tells you where you are with respect to your goal—if the compass is to be useful for finding a particular location. Homing pigeons, therefore, must have both compass and map senses. That map sense is presumably based on landmarks. Given the long distances for which the pigeons travel, these landmarks must vary in their type and probably include olfactory as well as visual cues.

The pigeon compass relies on different kinds of information. For example, pigeons can navigate during the day using the sun as a compass, and at night using the stars. And, like *Aquaspirillum*, pigeons can detect Earth's magnetic field. This magnetic navigation system may be important to pigeons on cloudy days. Researchers have performed experiments with pigeons wearing magnetic helmets, which disrupt the birds' ability to detect Earth's magnetic field. On cloudy days, the pigeons wearing the helmets are unable to navigate home. On sunny days, however, the pigeons' sun compass overrides the erroneous magnetic information of the helmets.

The sun compass requires information about time as well. Every hour, the sun moves 360/24 = 15 degrees through the sky. To determine where north is, a pigeon needs information from the sun as well as some way to keep time. It turns out that pigeons have a clock—a biological one.

Biological clocks provide important time cues for many behaviors.

Like us, other animals live in space *and* time. For many species, time is a life-or-death matter. When to migrate or mate is a critical decision in a seasonal environment. Researchers are beginning to unravel the neural and genetic underpinnings of the clocks in some species. In model organisms such as *Drosophila*, for example, researchers have identified a number of genes that, when mutated, cause the organism's biological clocks to run slow or fast.

A biological clock is produced by a set of interacting proteins that cycles on its own to provide a regular rhythm. Different clocks work on different timescales. Daily cycles are governed by a **circadian clock**. Circadian clocks regulate many daily rhythms in animals, such as those of feeding, sleeping, hormone production, and core body temperature. Some species, especially seacoast species living in habitats where the tides are important, time their activities by a **lunar** (moon-based) **clock**. There are also **annual** (yearly) **clocks**. For example, periodical cicadas are insects in the genus *Magicicada* that have a generation time of either 13 or 17 years: in a given location, a cicada outbreak occurs every 13 or 17 years.

Circadian clocks are observed in many organisms, including plants, fungi, and animals. Humans also have circadian clocks, as jet lag never fails to remind us. The circadian clock is based on a set of "clock genes" whose protein products oscillate through a series of feedback loops in a roughly 24-hour cycle. Thus, when animals that, like humans, are active during the day and inactive at night are placed in artificial conditions that are always lit, they continue to follow a basic day–night, active–inactive cycle. The clocks are not perfect, though: as the period spent in constant light is prolonged, the circadian clock drifts slowly until the animal is eventually no longer synchronized with the true day–night cycle.

Biological clocks remain synchronized with the day–night cycle because they are often reset by external inputs. For the circadian clock, light is the primary input. The natural light–dark cycle keeps the clock from drifting. For clocks related to the seasons, day length (known as **photoperiod**) is the critical input because it serves as a good indicator of the time of the year (Chapter 28). For example, many mammals produce their offspring in the spring so that they can grow over the summer, when resources are most abundant. The ability to synchronize reproduction to a time when young have the greatest chance of survival has been selected for over many generations. Photoperiod governs the timing of many kinds of behaviors, including migration, development, and reproduction.

The importance of the sun compass coupled with the biological clock in homing pigeons can be demonstrated by experimentally disturbing the birds' clock, as shown in **Fig. 43.12**.

HOW DO WE KNOW?

FIG. 43.12

Does a biological clock play a role in birds' ability to orient?

BACKGROUND One suggestion for how pigeons home is that they use a sun compass. A sun compass requires information about time as well as direction. For example, if you are in the Northern Hemisphere and the time is 12 noon, then the sun is due south of you. But orienting yourself by this method is possible only if you know the time, so the question arises as to whether homing birds have the ability to tell the time. One way to answer this question is to "clockshift" the birds. That is, researchers can raise birds in an artificial day–night cycle that is out of sync with the actual one. In this way, the birds' sense of time can be shifted by a set number of hours.

Event	Dawn	Dusk
Actual time	6 am	6 pm
Clockshifted time	12 noon	12 midnight

HYPOTHESIS If a bird's ability to home is dependent on an internal clock, clockshifting should affect the bird's homing ability in a predictable way. Given that the sun travels 360 degrees in 24 hours, a 6-hour clockshift will result in a 90-degree error in homing direction because $360/(24/6) = 90$.

EXPERIMENT In 1969, American biologist William Keeton clockshifted pigeons by raising them in a chamber under an artificial light. Two groups of birds, one clockshifted and one not, from Ithaca, New York, were released on a sunny day at Marathon, New York, approximately 30 km east of Ithaca. Release on a sunny day made it possible for the birds to use the sun's position to navigate.

RESULTS As expected, the control birds (those that were not clockshifted) were usually good at picking the direction of their home loft, heading approximately westward toward Ithaca. A minority did fly in the wrong the direction, suggesting that other factors sometimes override a pigeon's clock-based sense of direction. The clockshifted birds, in contrast to the non-clockshifted ones, miscalculated the appropriate direction more consistently, tending to head northward, as shown by the positions of the red triangles on the compass in the figure.

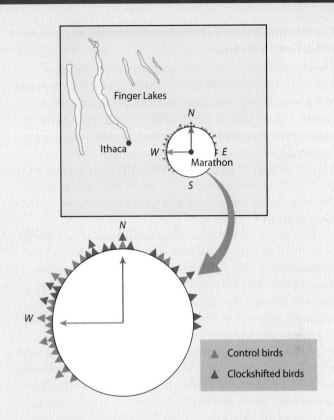

INTERPRETATION Assume the birds are released at 12 noon, when the sun is due south. The control birds know to fly in a direction 90 degrees clockwise from the direction of the sun, but the clockshifted birds "think" it is 6 p.m., so they expect the sun to be in the west. Their 90-degree clockwise correction results in them flying due north.

CONCLUSION The clear difference between control and clockshifted birds in this experiment shows that an internal time-based sun compass is an important component of the birds' homing abilities. However, the scatter of points (for both the experiment and control groups) suggests that other factors are important as well. This conclusion is reinforced by the observation that birds home well on cloudy days, when they cannot use a sun compass, suggesting that birds use multiple cues and navigational systems when they are homing.

SOURCE Keeton, W. T. 1969. "Orientation by Pigeons: Is the Sun Necessary?" *Science* 165:922–928.

Self-Assessment Questions

11. Differentiate between a kinesis and a taxis.
12. Why is photoperiod a widely used time cue?
13. Animals that migrate over long distances using the sun for directional information typically also have the ability to estimate the time of day. Why?

43.5 COMMUNICATION

Up to this point, this chapter has focused on individual behaviors. Behaviors often depend on environmental cues, as we saw with the examples of navigation and keeping time. In addition, behaviors are often shaped by interactions with other individuals. Learning sometimes depends on communication between a teacher and student, and, as anyone knows who has ever watched birds tending their young, communication is central to parenting.

Communication is the transfer of information between a sender and a receiver.

The sophistication of animal communication varies enormously, as can be seen by the various types of signal illustrated in **Fig. 43.13**. Even our closest relatives, the chimpanzees, cannot rival the human facility with language, but a number of species have evolved forms of communication that convey remarkably complex and specific information. For example, a Vervet Monkey that perceives a threat to its group utters an alarm call that not only warns of a threat, but actually specifies the nature of the threat—whether it is a hawk, leopard, snake, or some other type of predator.

The simplest definition of **communication** is the transfer of information between two individuals, the **sender** and the **receiver**. The sender supplies a signal that elicits a response from the receiver. For example, the bright petals of a flower signal to an insect that nectar and pollen are available. This definition, however, has its problems. An owl hears the rustling of a mouse and responds accordingly. Most people would agree that the owl is not really communicating with the mouse, or vice versa. For this reason, some biologists prefer to define communication as a transfer of information between two individuals in which the sender attempts to manipulate in some way the behavior of the receiver.

How has communication evolved? Some communication may have evolved through co-opting and modifying behaviors used in another context. This process, called **ritualization**, involves (1) increasing the conspicuousness of the behavior, (2) reducing the amount of variation in the behavior so that it can be immediately recognized, and (3) increasing the separation of the behavior from its original function. The scent markings with which mammals mark their territory provide an example. The original function was simply the elimination of waste, but this function has been modified through evolution: strategically placed marks communicate with other individuals, indicating, for example, the extent of an individual's territory. The communication advantage is that an intruder detecting a territorial scent may be less likely to invade, thereby avoiding a potentially dangerous fight.

Many forms of communication have evolved that prevent animals from coming to harm through a "limited-war strategy." Instead of battling it out, two males may engage in elaborate displays to size each other up, either literally by standing side by side or through displays of physical prowess such as roaring, in an attempt to determine who is dominant. Fighting would be disadvantageous to both because even the winner might be seriously injured. In this case, communication may have evolved as a way for individuals to assess one another.

In some cases, communication in the natural world can be deceitful. A male may attempt to convince other males (or a female) that he is bigger (and stronger) than he really is. Possibly one reason why dogs circling in a fight raise the hair

---

FIG. 43.13 Types of signals in different forms of communications. Signals can be (a) visual, as in frilled lizard's threat display; (b) auditory, such as bird song; (c) electrical, as in the species-specific electric pulses emitted by African electric fish; (d) chemical, such as the pheromones with which a queen termite controls her colony; and (e) mechanical, as when spiders "twang" webs to interact. *Sources: a. DWI YULIANTO/Shutterstock; b. Nataba/Getty Images; c. © Matthew E. Arnegard; d. Dan L. Perlman/EcoLibrary.org; e. Ron Rowan Photography/Shutterstock.*

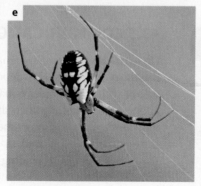
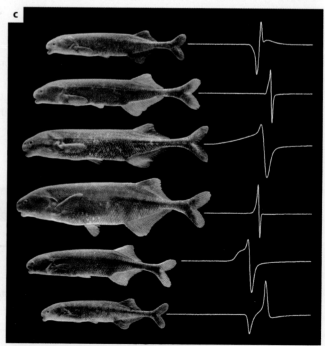

along their backs is to inflate their apparent size. A potential prey may attempt to convince a predator that it is not, in fact, prey. This form of deceit is especially evident in some species of butterfly that mimic leaves or bark when their wings are at rest. Alternatively, predators may emit deceitful signals to entice their prey. Females of one species of firefly, for instance, mimic the flashes of the females of another species in an effort to lure males of the second species, which they then eat.

Some forms of communication are complex and learned during a sensitive period.

Bird song is one of the best-known and richest forms of animal communication. Because of the clear connection between what birds hear and how they respond, songbirds often serve as a model system in the study of learning and communication. Bird songs are complex sequences of sounds, often repeated over and over. Like cricket and frog calls, bird songs are **advertisement displays**, behaviors by which individuals draw attention to their status. For example, the displays might indicate that the birds are sexually available or are holding territory. Bird songs are typically produced by males during the breeding season.

This sex difference in patterns of communication often results from intersexual selection (Chapter 20), and can extend to other traits. For example, male birds usually have bright plumage and elaborate displays, and females do not. Why do we see such differences between males and females? The key is the level of investment that an individual makes in an offspring. The eggs a female bird produces are metabolically expensive, whereas the sperm a male bird produces are metabolically cheap. Because of this and other costs associated with reproduction, a female typically has only one opportunity or just a few opportunities to reproduce. Given this reality, it is in her best interest to ensure that she has chosen a high-quality male, one that sings well or has especially fine plumage. By contrast, it is in the male's interest to maximize his access to females—that is, to be as sexually attractive as possible. Sexual selection therefore tends to produce showy, attractive males and females that choose males on the basis of their performance or appearance.

In many species of birds that sing, some or all of the song is learned, often during a specific, sensitive period. Detailed studies of the White-Crowned Sparrow, shown in **Fig. 43.14**, pioneered in the 1960s by neurobiologist and ethologist Peter Marler, and extended by Luis Baptista, have yielded the following general picture of the process.

Song in White-Crowned Sparrows is learned by imprinting: the young male hears adult song during a sensitive period, 10 to 50 days after hatching. Shortly after this period, young male White-Crowned Sparrows produce unstructured twittering sounds, known as a subsong, comparable to the babbling of human babies. If deprived of hearing adult song, perhaps because of isolation during the sensitive period, the bird sings for the rest of his life a song not much different from unstructured twittering, even if he hears another male sing both before and after the sensitive period.

Between about 250 and 300 days, the male sings an imperfect copy of the song (known as plastic song). By about 300 to 350 days after hatching, song acquisition is complete (this song is known as structured song). At this point, even if the bird is deafened in the lab, he will still sing correctly. The song he produces is typically a precise copy of the male's song he heard during the sensitive period, complete with most of its individual as well as species-specific characteristics.

White-Crowned Sparrows, then, have a programmed predisposition for *when* song learning takes place. Even more interesting, *what* can be learned is similarly constrained. If a tape of another species' song is played during the sensitive period, even the song of closely related Swamp Sparrows, the White-Crowned Sparrow male cannot learn that song; his adult song ends up not much different from unstructured twittering. However, if he is provided with a live tutor, such as a live male Swamp Sparrow rather than a tape, his ability to learn the other bird's songs is greatly improved. Furthermore, if he hears tapes of the song of his own species and that of a closely related species played together, he can pick out the correct elements specific to his own species and sing a perfect White-Crowned Sparrow song. Thus, birds of this species preferentially learn their species-specific song, but they cannot

---

FIG. 43.14 Song acquisition in the White-Crowned Sparrow. A sonogram shows differences between the unstructured twittering of early life and the final, structured song. *After T. J. Carew, 2000, Behavioral Neurobiology, Sunderland, MA: Sinauer Associates, p. 244; Photo source: Paul Sutherland/Getty Images.*

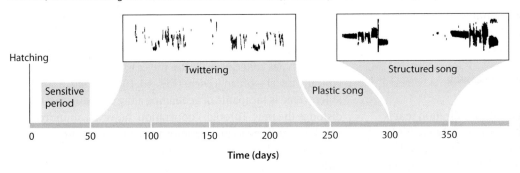

FIG. 43.15 The waggle dance of honeybees. This form of communication provides information to other foragers about the distance to and direction of a food source. *After M. B. Sokolowski, 2001, "Drosophila: Genetics Meets Behaviour," Nature Reviews Genetics 2:879–890, doi:10.1038/35098592.*

The distance to the food source is communicated by the length of time it takes to move up the middle of the circle (the "waggle").

The direction of the food source relative to the sun is communicated by the angle of the waggle run relative to vertical.

sing a normal song without learning it from another member of their own species.

Song learning in White-Crowned Sparrows has been well studied, and provides just one example of song learning in birds. Some species, such as mockingbirds, starlings, and canaries, learn a wider range of songs than do White-Crowned Sparrows, and have a longer period of song learning, in some cases probably lifelong. By contrast, other species, such as many New World flycatchers, apparently have hard-wired (innate) songs: they produce the appropriate song even if raised in isolation from other birds. Researchers are just beginning to decode and understand other sound-based forms of communication, such as those produced by dolphins and whales.

Various forms of communication can convey specific information.

Honeybees have an elaborate means of communication that is quite different from the sound-based communication of birds. Bees travel far and wide to forage for food for their hive and, upon returning home, use a variety of cues to inform outgoing foragers about the location and nature of the food resources they have found.

Returning worker bees pass samples of the collected food to other workers that gather around the foragers when they return to the hive. The returning foragers begin a series of movements in specific patterns that encode information about the direction and distance of food sources. This dance language of honeybees was originally discovered by Karl von Frisch in the 1920s.

If the food source is more than about 50 m away, the incoming forager does a waggle dance (**Fig. 43.15**). The forager moves quickly back and forth ("waggles") as it moves forward, circles back and repeats the waggle, and then circles back the other way, and so on. This dance usually goes on for several minutes and may take much longer. The distance to the food source is conveyed by the length of time it takes to waggle up the middle of the circle. The direction, with reference to the position of the sun, is conveyed by the angle of the line of the waggle run: if the food source is in a direct line with the sun, the dancer's waggle line points straight up; if the food is at an angle of 45 degrees to the left of the sun, the dancer's line is 45 degrees to the left of vertical. The bees' dance language is among the natural world's most extraordinary and effective methods of communication.

Self-Assessment Questions

14. Explain why many biologists do not consider just any transfer of information between a sender and a receiver to be a form of communication.

15. Why do we hear regional "dialects" in the songs of some bird species but not in the songs of other species?

43.6 SOCIAL BEHAVIOR

Returning to Tinbergen's four questions, we can consider further the adaptive function and evolutionary history of animal behavior. The evolutionary advantages of most behaviors are readily apparent. The imprinting of Lorenz's goslings on the first moving object they encountered makes sense when you consider that this object is usually a parent (not Konrad Lorenz!) and that sticking close to a parent is an excellent survival strategy in a world filled with predators.

Social behaviors also have evolutionary advantages. Consider the cooperative hunting seen in killer whales. Killer whales participate in a behavior called wave washing, in which a pod of whales collectively creates waves to wash a prey species—typically a seal—off an iceberg (**Fig. 43.16**). By itself, a single killer whale is incapable of generating a big enough wave to dislodge the prey. Through cooperation, however, an individual can contribute to prey capture and gain some portion of the prey. This behavior is collaborative, and it is clear why each individual participates.

FIG. 43.16 Mutually beneficial cooperation. Killer whales cooperate to create a wave to wash prey off an iceberg. *Sources: (top) John Durban/NOAA; (bottom) R. Pitman/NOAA SWFSC.*

Sometimes how a behavior enhances survival and reproduction is less clear. Perhaps the most prominent examples of this kind of behavior are sacrifices made by individuals apparently for the good of others. Formally, we describe an act of self-sacrifice that helps another as **altruistic**. Such acts seem to fly in the face of natural selection, which selects for traits that increase fitness and selects against traits that decrease fitness. Altruistic acts, by definition, decrease fitness and therefore should be selected against. This contradiction is particularly stark in the case of species like honeybees: honeybee workers, which are females, do not reproduce but instead assist the reproduction of the queen. Darwin confessed that such behaviors caused him sleepless nights. How, Darwin wondered, could altruism arise?

Group selection is a weak explanation of altruistic behavior.

Darwin eventually suggested a solution to the altruism problem that today we call **group selection**. The idea is that natural selection operating on individuals is a less powerful force than another form of selection that operates on groups. Consider two groups: in one group, individuals demonstrate altruism, helping one another out; in the other group, individuals act entirely selfishly. Darwin argued that, in a conflict between the two groups, the first group would defeat the second because the first group works better as a unit. In this example, selection in favor of altruism benefits the group and trumps selection in favor of selfish behavior.

Although an attractive idea, group selection is probably not important in the evolution of altruism. Altruism under group selection is typically not an **evolutionarily stable strategy**, meaning that this kind of behavior can be readily driven to extinction by an alternative strategy. As an example, consider lemmings, a species of Scandinavian rodent. It was once thought lemmings committed mass suicide by plunging into the ocean when the population became too large to sustain. This self-sacrificial, altruistic act would ensure that the population did not crash after exhausting available resources. This is an appealing story, but untrue. Lemmings do not actually commit suicide; they take to the water to swim to new territories during migration.

We can also see the evolutionary logic prohibiting the evolution of such altruistic behavior. An individual that "cheats" by not performing the altruistic act will survive and breed, and its altruistic companions will not. Now consider that this lemming behavior—to kill oneself or not—is genetically encoded. A mutation that arises in a lemming population that causes the individual to "cheat" and not kill itself will spread rapidly through the population by natural selection until, after many generations, it will completely replace the altruistic allele that causes the individual to kill itself. The altruistic lemming behavior, then, is not an evolutionarily stable strategy because natural selection favors an alternative strategy.

This description is oversimplified because such traits are unlikely to have a simple genetic basis, but the point stands: group selection is in general not an evolutionarily stable strategy because it can readily be outcompeted by "selfish" strategies.

Reciprocal altruism is one way that altruism can evolve.

If group selection does not provide a strong explanation for the evolution of altruism, how can we explain altruistic behavior? **Reciprocal altruism**, whereby individuals exchange favors, is one way that altruism can evolve. This idea is perhaps best summarized as "You scratch my back, and I'll scratch yours."

For his kind of behavior to work, individuals must be able to recognize each other and remember previous interactions. In other words, if I did you a favor, I need to be able to recall who you are so that I can be sure I get the expected return.

A famous example of reciprocal altruism comes from vampire bats. Adult female vampire bats live in groups of 8 to 12, and often remain in these groups for years. In one study, two tagged females shared the same roost for 12 years. The female groups are occasionally joined by unrelated females, so not all of the bats are closely related. Because they have long-term associations, there is opportunity for reciprocal interactions.

Feeding for these bats is a "boom or bust" activity: either a bat has found and successfully fed on a blood source (boom), or a bat has failed to feed (bust). As a result, bats are often either overfed or underfed. Underfed bats are at risk of dying by starvation. Bats returning from a successful feeding expedition often regurgitate blood to unsuccessful ones. In turn, they will benefit on future occasions when they have been unsuccessful in the hunt for food, when those they have fed in the past have food to spare.

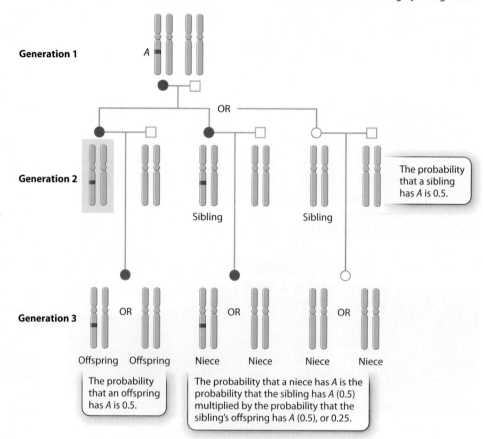

FIG. 43.17 Computing relatedness. One individual's relatedness (r) to another individual is the probability that one shares, through inheritance, a given segment of DNA with that individual. In this illustration, relatedness is calculated with reference to the individual shown on the gray background.

Kin selection is based on the idea that it is possible to contribute genetically to future generations by helping close relatives.

In the 1960s, researchers explored the idea that the relatedness of participating individuals plays an important role in the evolution of altruism. The key insight here is that there are at least two ways to contribute genetically to the next generation: either an individual can produce its own offspring, or it can produce no offspring and help close relatives reproduce instead. The latter strategy can become established in a population through a process called **kin selection**, a form of natural selection that favors the spread of alleles that promote behaviors that help close relatives (kin).

What relationships qualify as kin? Consider the family depicted in **Fig. 43.17**. One animal (indicated by gray shading in the figure) has an allele A at a particular locus. If the animal reproduces, the probability of each offspring inheriting an A allele is 0.5. But what if the animal does not reproduce? According to Mendel's laws, there is a 50% chance that the animal's parents passed on the same A allele to the animal's sibling (Chapter 15). If the sibling does, in fact, inherit allele A, the probability that it passes allele A to each of its offspring is also 0.5. Given the two probabilities (that the sibling has an A, and that the allele passed on to an offspring), the probability that the animal's niece or nephew has A is 0.25. In other words, a diploid animal (such as humans) is twice as closely related to its own offspring as it is to its nephew or niece. Therefore, if the animal does not have its own offspring but instead helps its sibling have two offspring, the animal will make the same net genetic contribution to the next generation as it would by having one of its own offspring.

This logic was formalized in 1964 in a famous paper by British evolutionary biologist William D. Hamilton. Hamilton defined three key variables: B, the benefit of the behavior to the recipient (how many individuals does the behavior save?); C, the cost of the behavior to the donor (how many offspring are lost?); and r, the degree of relatedness between the recipient and donor. Hamilton asserted that as long as rB exceeds C, altruism can evolve.

How does kin selection work in practice? Let's start with an extreme case of relatedness: clonal organisms that are genetically identical to each other (meaning that the probability of a sibling having allele *A* is 1 or 100%). Many aphid species have a phase of their life cycle in which a "mother" aphid reproduces parthenogenetically to produce a small colony of clones of herself (Chapter 40). Any aphid in the colony is capable of reproducing in this way, but, in a case of reproductive altruism, some individuals do not reproduce at all. These "soldier" aphids instead defend the colony against predators, typically other insects. As specialized defenders, they are unable to reproduce. In this case, the Hamilton calculation is straightforward. By helping their clone-mates survive, the sterile soldier aphids are ensuring that their own genes survive to be passed on to the next generation.

The most famous cases of reproductive altruism occur in the insect group Hymenoptera, which includes ants, bees, and wasps. Many species of Hymenoptera are **eusocial**, meaning that they have overlapping generations in a nest, provide cooperative care of the young, and feature clear and consistent division of labor between reproducers (the queen of a honeybee colony) and nonreproducers (the workers). Often called social insects, these species are one of evolution's most extraordinary success stories.

Which features of Hymenoptera predisposed these insects to eusociality? Ant expert E. O. Wilson argues that natural selection has strongly favored group behavior in these insects through the group benefits derived from living in and defending a nest. For example, they can easily communicate with one another about the location of food resources such as nectar.

William Hamilton also noted that most Hymenoptera have an unusual mode of sex determination (called haplodiploidy): females (the queen and workers) are diploid, whereas males (drones) are haploid, produced from unfertilized eggs. As a result, the degree of relatedness, *r*, is higher for the sister–sister relationship (0.75) than for the mother–daughter one (0.5), as shown in **Fig. 43.18**. Put another way, a haplodiploid female is more closely related to her sister than she is to her daughter. Therefore, it makes more evolutionary sense, in terms of genetic representation in the next generation, for a haplodiploid worker to help her mother, the queen, to produce more sisters than for her to produce her own offspring.

This form of social organization can result in a huge colony of related individuals all working collectively, much as different cells work collectively within a body. Often members of these colonies specialize into castes: some ants are soldiers, defending the colony; others are foragers; and so on. If we extend the analogy of individuals acting like cells in a body, the multiple castes are equivalent to organs, all coordinating and working together. The term "superorganism" is sometimes used to describe a single social insect colony.

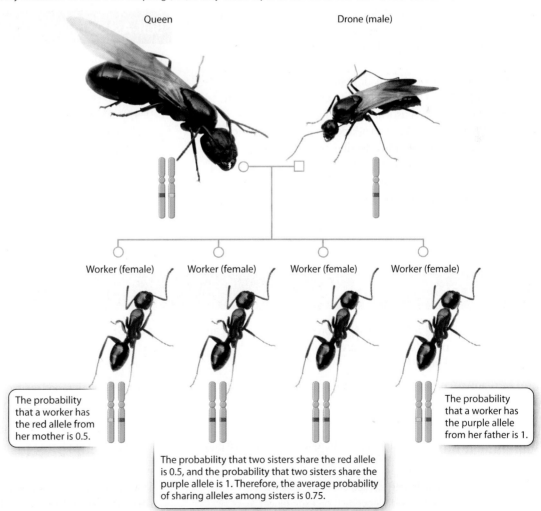

FIG. 43.18 Relatedness in haplodiploid insects, such as ants. Sisters are more related to each other (0.75) than they would be to their own offspring (0.5) if they could reproduce. *Photos source: Alex Wild/alexanderwild.com.*

FIG. 43.19 Social behavior in non-insects. (a) Naked mole rats are eusocial but not haplodiploid. (b) Florida Scrub Jays display altruistic behavior but are not eusocial. *Photos source: a. Joel Sartore/Getty Images, b. Alamy Stock Photo.*

Some species are eusocial but not haplodiploid, including the naked mole rat (**Fig. 43.19a**). These examples suggest that, while haplodiploidy may predispose a species to eusociality, it is not required.

Other species display altruistic behavior toward close relatives, as predicted by Hamilton, but are not as extreme in their social behavior as eusocial species. Florida Scrub Jays (**Fig. 43.19b**) live in habitats that lack enough territories for all adult birds. As a result, the offspring of a breeding pair on a territory may remain in that territory and help its parents raise more young. If a young bird were to strike out on its own, it would likely fail to reproduce, so, in terms of making a genetic contribution to the next generation, it is better to help its parents raise more siblings.

Self-Assessment Questions

16. It's sometimes said that an animal does something "for the good of the species." Why is this argument incorrect?

17. Reciprocal altruism is unlikely to evolve in species with short life-spans. Why?

18. Describe how an altruistic behavior might evolve by kin selection.

19. How many cousins are genetically equivalent to a single offspring?

CORE CONCEPTS SUMMARY

43.1 TINBERGEN'S QUESTIONS: For any behavior, we can ask what adaptive function it serves, what causes it, how it develops, and how it evolved.

Tinbergen suggested that we ask questions about animal behavior at four different levels of analysis, focusing on function, cause, development, and evolution. page 979

Tinbergen's four questions have different answers, and all may be correct explanations of the behavior. page 980

Behavior can be innate (carried out in the absence of experience) or learned (dependent on experience). page 980

43.2 DISSECTING BEHAVIOR: Animal behavior is shaped in part by genes acting through the nervous and endocrine systems.

Most behavior is produced by the interaction between genes (nature) and environment (nurture). page 981

A fixed action pattern is a stereotyped sequence of behaviors that, once initiated by a key stimulus, always goes to completion. page 981

Stimuli are recognized by feature detectors, which are specialized sensory receptors of the nervous system. page 981

Hormones can affect behavior through their wide-ranging and often long-term effects on multiple body systems. page 982

The extent to which genes influence a behavior can be determined by crossing closely related species having different behaviors and analyzing the behavior of the offspring. page 984

Most behaviors are influenced by many genes, but a few are strongly influenced by a single gene. page 984

Molecular studies are providing new ways to understand the role of genes in behavior in laboratory animals. page 984

43.3 LEARNING: Learning is a change of behavior as a result of experience.

Non-associative learning is a form of learning that occurs without linking two events. page 987

Habituation, a form of non-associative learning, is the reduction or elimination of a response to a repeated stimulus. page 987

Sensitization, another form of non-associative learning, is the enhancement of a response to a stimulus that follows a priming stimulus. page 987

Associative learning occurs when an animal learns to link two separate events. page 987

Classical conditioning is a form of associative learning in which a novel association is made between a formerly neutral stimulus and a behavior. page 987

Operant conditioning is a form of associative learning in which rewarding or punishing a behavior makes that behavior more likely or less likely to occur, respectively. page 987

Imitation and imprinting are forms of learning. page 988

43.4 INFORMATION PROCESSING: Orientation, navigation, and biological clocks all require information processing.

Kineses are undirected movements, whereas taxes are directed movements. page 989

Long-distance navigation may require the processing of many external cues, including the position of the sun, Earth's magnetic field, and information about time. page 990

Biological clocks control some physiological and behavioral aspects of animal activity. page 990

43.5 COMMUNICATION: Communication involves an interaction between a sender and a receiver.

Communication is an interaction between two individuals, the sender and the receiver, and often involves attempts by the sender to manipulate the behavior of the receiver. page 992

Communication can evolve by the modification of noncommunicative aspects of an animal's behavior. page 992

Bird song is a rich form of communication that is often learned during a sensitive period. Acquisition of bird song proceeds through distinct stages. page 993

The dance language of honeybees communicates information about the location and direction of a food source. page 994

43.6 SOCIAL BEHAVIOR: Social behavior is shaped by natural selection.

One of the most perplexing behaviors from the standpoint of natural selection is altruism, in which an organism reduces its own fitness by acts that benefit others. page 995

Group selection is the idea that altruistic behaviors can evolve if they benefit the group, not the individual; this idea is not widely accepted. page 995

Reciprocal altruism requires that individuals interact regularly and remember each other so that they can "pay back" past favors. page 995

Kin selection is the concept that altruistic behaviors can evolve if they benefit close relatives. page 996

Log in to LaunchPad to check your answers to the Self-Assessment Questions and to access additional learning tools.

CHAPTER 44 Population Ecology

CORE CONCEPTS

44.1 POPULATIONS AND THEIR PROPERTIES: A population consists of all of the individuals of a given species that live and reproduce in a particular place and time.

44.2 POPULATION GROWTH AND DECLINE: Population size can respond to changes in the physical and biological environment.

44.3 AGE-STRUCTURED POPULATION GROWTH: The age structure of a population helps ecologists understand past changes and predict future changes in population size.

44.4 METAPOPULATION DYNAMICS: The dynamics of populations are influenced by the colonization and extinction of smaller, interconnected populations that make up a metapopulation.

In 2004, undergraduates on a field trip to the Dominican Republic, on the island of Hispaniola, swept up an unfamiliar black-and-white butterfly in their nets. That butterfly was a Lime Swallowtail (*Papilio demoleus*), which is native to Asia. The students' specimen was the very first of this species to be captured in the New World.

Soon, the Lime Swallowtail spread to all corners of Hispaniola. This species posed a serious concern on the island because their caterpillars strip the leaves off young lime and orange trees. Within three years, the Lime Swallowtail had also established itself in nearby Puerto Rico and Jamaica, and it has now reached Cuba. From there, it is just a short flight to Florida and the multibillion-dollar citrus industry of the United States. Can we enlist basic principles of biology to help understand this butterfly's Caribbean expansion and predict its future as it continues to spread?

Ecology is the study of the relationships of organisms to one another and to the environment. Ecological relationships sustain a flow of energy and materials from the sun, Earth, oceans, and atmosphere through organisms, with carbon and other elements eventually returning to the environment. It is these interactions that determine the shifting ranges of Lime Swallowtails, as well as all species on Earth.

In previous chapters, we discussed the anatomical, physiological, reproductive, and behavioral characteristics of organisms. These attributes are the products of evolution, and they determine the ways in which individual organisms interact with their physical and biological environment. Physical, or **abiotic**, factors are nonliving aspects of the environment that affect organisms, such as temperature, water, nutrients, and toxins. Biological, or **biotic**, factors are living organisms, including competitors, predators, parasites, and prey, as well as organisms that provide shelter or food.

Ecology and evolution are closely intertwined. In the following chapters, we develop the ideas of ecology in the context of interactions between members of the same and different species, and we explore the different strategies species can follow that maximize their survival and reproduction in their environment. We trace the processes that govern interactions from the evolved traits of individual organisms to continent-spanning patterns of diversity and biomes.

44.1 POPULATIONS AND THEIR PROPERTIES

When the first U.S. census was completed in 1790, the British scholar Thomas Malthus supposed that the remarkable growth of the American population, which had doubled in the previous 25 years, was a consequence of ample food supply and the active encouragement of marriages. In his 1798 book *An Essay on the Principle of Population*, Malthus argued that a similar periodic doubling of population in the British Isles would outstrip the more slowly increasing food supply in 50 years and lead to starvation. That is, increasing the number of births per year would eventually result in an increased number of deaths because the amount of food the land could produce was limited. This view of the human condition influenced the thinking of Charles Darwin, who as an inquisitive scholar read the most exciting works of the day. Darwin realized that the tension between population growth and resource limitation applies to all species, leading to a struggle for existence among and within species that results in adaptations that enhance survival (Chapter 20).

A population includes all the individuals of a species in a particular place.

In Chapter 20, we introduced the concept of population as a fundamental unit of evolution: it is populations, not individuals, that evolve. In this chapter, we return to populations, this time as fundamental units of ecology. A population is all the individuals of a given species that live and reproduce in a particular place. For example, populations of the American Red Squirrel (*Tamiasciurus hudsonicus*) occur in patches of conifer (spruce, pine, or hemlock) forests across the northern United States and Canada. Each group of American Red Squirrels that lives in a distinct area is considered a population. Collectively, all of the American Red Squirrels in all of the populations constitute the species.

Natural selection enables populations to adapt to the abiotic and biotic components of their environment. The various traits that arise in individuals and evolve to be prominent in populations under natural selection ultimately determine whether a population will grow or shrink (or even disappear) under a given set of physical and biological conditions in a local environment.

How are populations characterized? What controls changes in population size? In this chapter, we first consider key features of populations. Then we look at what happens when the number of births exceeds the number of deaths, and what happens when resources such as food start to run out. We also consider how individuals of different ages contribute to population growth, and how local populations are connected to one another. In Chapter 45, we introduce the idea of interactions with other species and examine how those interactions shape the growth of populations.

Three key features of a population are its size, range, and density.

Populations are defined by three key features: size, range, and density (**Fig. 44.1**). Population size is the number of individuals of all ages alive at a particular time in a particular place. For example, the total size of the population of Antarctic krill (*Euphausia superba*) in the cold waters around Antarctica is estimated at 800 trillion, making them by mass the most abundant animals on Earth (Fig. 44.1a).

A second key feature of a population is its geographic range: the area over which that population is spread. For example, the Antarctic krill population lives in an area of approximately 19 million square kilometers around Antarctica (Fig. 44.1b). The range of any species reflects in part the range of climates a population can tolerate and determines how many other species the population encounters. The range of a plant population, for example, influences how many kinds of insect eat it, and the range of an insect population determines how many kinds of plants it encounters.

FIG. 44.1 (a) Size, (b) range, and (c) density of a population. The places where populations occur can be mapped, and the numbers of individuals within those populations can be estimated. *Sources: Data from A. Atkinson, V. Siegel, E. A. Pakhomov, M. J. Jessopp, and V. Loeb, 2009, "A Reappraisal of the Total Biomass and Annual Production of Antarctic Krill," Deep-Sea Research I 56:727–740. Photos: Jean Paul Ferrero/ARDEA (inset) British Antarctic Survey/Getty Images.*

a. Size

The size of the Antarctic population of krill is estimated to be 800 trillion animals.

b. Range

c. Density

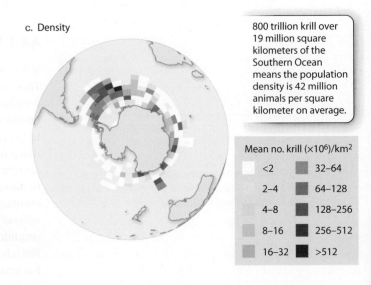

Their range is approximately 19 million square kilometers in the Southern Ocean.

800 trillion krill over 19 million square kilometers of the Southern Ocean means the population density is 42 million animals per square kilometer on average.

Mean no. krill ($\times 10^6$)/km^2

<2	32–64
2–4	64–128
4–8	128–256
8–16	256–512
16–32	>512

Taken together, population size and range contribute to a third key attribute of a population: density. Population density is defined as a population's size divided by its range. In effect, density tells us how crowded or dispersed the individuals are that make up the population. For the Antarctic krill population, 800 trillion animals spread over 19 million square kilometers yields a population density of 42 million animals per square kilometer (Fig. 44.1c). This number describes the average density because krill, like most populations of organisms, vary in local density and individuals are not evenly spread out over the environment. As we will see, population density can affect population size.

This way of characterizing population density, as average density across the range of a population, is often useful, but it can be misleading depending on how the organisms are distributed (**Fig. 44.2**). For example, individuals within a population may be distributed randomly. In such a case, a new individual has an equal chance of occupying any position within the range, and the location of one individual has no influence on where the next individual will be found (Fig. 44.2a). Sometimes, however, resources are distributed patchily within the range, or the chances of a new individual's survival are enhanced by the presence of other individuals. In these circumstances, the population may be clumped, with individuals clustered together (Fig. 44.2b). In still other cases, one individual of a population might prevent another from settling nearby, such as when all individuals depend on the same set of limited resources. Or predators, having located one prey individual, may search for other members of the same species nearby. Under these conditions, individuals may be distributed more evenly than would be predicted by chance. Ecologists call such populations uniform (Fig. 44.2c).

Ecologists estimate population size by sampling.

Imagine trying to calculate the size of a population of Lime Swallowtails by counting every individual in the Dominican Republic. The task is hopeless. Like the Lime Swallowtail population, the populations of most species are either too large or too widely dispersed for every individual to be counted. Instead, ecologists estimate population size by taking repeated samples of a population and, based on these samples, estimating the total number of individuals. With an estimate of population size and range, ecologists can then determine population density.

For sessile, or sedentary, organisms such as milkweeds in a field or barnacles on a rocky shoreline, ecologists may lay a rope across the area of interest and count every individual that touches it, or they may throw a hoop and count the individuals inside it. Each such count is one sample. Based on the data for a number of samples, the ecologists can estimate the number and distribution of individuals in the population. For example, suppose 1-m² hoops are placed in three different spots in a field,

FIG. 44.2 Distribution of individuals within populations. A population may be (a) randomly distributed, (b) clumped, or (c) uniform.

a.

The distributions of individual trees or other organisms can appear to be random, with no clear pattern to where they occur.

b.

If resources are clustered or spatial proximity to other individuals enhances fitness, individuals may be clustered.

c.

When resources are limited or predators target a single species, an individual might be better off if it is as far from others as possible, producing a uniform pattern of distribution.

encircling 4, 5, and 3 individual milkweeds, respectively. We can multiply the average number of individuals (in this case, 4) per 1-m² sample by the total square meters in the area of interest (in this case, the size of the entire field) to estimate the milkweed population size.

It is relatively easy to count sedentary organisms along a straight line or within a grid, but many species move around. How do ecologists estimate population size for turtles in a pond, or for mice in a forest? In these cases, a method called **mark-and-recapture** is useful. Biologists capture individuals,

HOW DO WE KNOW?

FIG. 44.3

How many butterflies are there in a given population?

Photo source: Doug Wechsler/Nature Picture Library.

BACKGROUND Individuals in many populations of butterflies (and other animals) often live and die within the patch of habitat where they were born. Biologists estimate population sizes of such animals using mark-and-recapture, a technique that involves capturing and releasing organisms twice, on two different days. The assumption is that the population size is approximately the same on the second day as the first day. Biologists also assume that the captured and released animals mix with others in the population and do not intentionally avoid recapture on the second day. This technique was used to estimate the population size of Monarch Butterflies (*Danaus plexippus*) in Florida during their spring migration north from the butterflies' wintering grounds in Mexico.

METHOD Biologists captured the butterflies in a field of milkweeds, their host plant; marked the butterflies with an identification sticker on the wing as shown in the photo; and then released them. The number of marked butterflies was recorded. The next day, butterflies were caught and the number of marked and unmarked butterflies recorded. The butterflies that were already marked are called recaptures.

To determine the population size (N), the total number of marked butterflies and unmarked butterflies caught on the second day (C) is divided by the number of recaptures (R). This number is then multiplied by the number of butterflies marked on the first day (M), as follows:

$$N = (C/R) \times M$$

RESULTS Let's say 100 butterflies are captured, marked, and released on the first day. On the second day, 120 butterflies are captured, 30 of which are recaptures. The rest are unmarked. We conclude that the population size is $(120/30) \times 100$, which in this example is 400 butterflies total in the sampled population.

RELATED METHODS Mark-and-recapture techniques are extremely useful in estimating population sizes of organisms ranging from polar bears to disease victims. In addition, some refinements to the mark-and-recapture approach take into account the movement of individuals between populations. These methods involve taking multiple samples and adding variables to the equation. Many animals have distinctive color markings, such as the blotches on humpback whale tailfins and skin-spotting patterns on frogs; these animals can be recorded without having to be marked.

SOURCE Knight, A., L. P. Brower, and E. H. Williams. 1999. "Spring Remigration of the Monarch Butterfly *Danaus plexippus* (Lepidoptera: Nymphalidae) in North-Central Florida: Estimating Population Parameters Using Mark–Recapture." *Biological Journal of the Linnean Society* 68:531–556.

mark them in a way that doesn't affect function or behavior, release them back into the wild, and then capture a second set of individuals, some of which were previously marked. **Fig. 44.3** illustrates how the mark-and-recapture method can be used to estimate the number of butterflies in an area. If a large number of butterflies are caught, marked, released, and then recaptured, it is possible to estimate the total population size by extrapolating from the proportion of the second sample that is marked.

These methods are simple, and they make a number of assumptions about the population. For example, we assume that the population has not changed in size between our first and second samples. Nonetheless, mark-and-recapture methods have proved effective in estimating the sizes of populations in nature. More sophisticated analyses are available for more complicated situations, as when more than two samples are caught, or when immigration or emigration occurs during the time between samples. For example, researchers tracking invasive populations of pythons in the Florida Everglades have placed radio-transmitters inside small samples of pythons caught and released back to the Everglades. Even though pythons are nearly impossible to recapture, mathematical models using radio-tracked python movements, coupled with counts of road-crossings by these and uncaptured pythons, help the researchers estimate their population sizes.

Self-Assessment Questions

1. What is the difference between a population and a species?
2. What basic facts about populations do we need to know to describe them?

44.2 POPULATION GROWTH AND DECLINE

The size of a population may increase, decrease, or stay the same over time. Trends in population size are a key focus of ecological research and its application to conservation biology. A decrease in size through time, for example, might signal declining health in a local population. This trend can be a matter of practical concern when the population in question is a commercially harvested fish or an endangered species. In the 1990s, for example, catastrophic declines in cod populations led to a ban on commercial cod fishing on the Grand Banks of Newfoundland in northeastern Canada. The declines affected both the cod and fishing communities, and led to changes in the Grand Banks ecosystems: crab and lobster populations have increased, as there are fewer cod to prey on them (see Chapter 48).

Population size is affected by birth, death, immigration, and emigration.

Many factors affect population size. An obvious process that contributes to population increase is birth. Another is immigration: individuals may arrive in a population from elsewhere. In contrast, mortality (death) and emigration, the departure of individuals from a population, both act to decrease population size (**Fig. 44.4**).

We can describe the changes in a population's size mathematically, defining population size as N and the change in population size over a given time interval as ΔN, where the Greek letter Δ (delta) means "change in." ΔN is the number of individuals at a given time (time 1) minus the number of individuals at an earlier time (time 0), which is notated as $N_1 - N_0$:

$$\Delta N = N_1 - N_0$$

The factors leading to a change in population size over time are births, deaths, immigration, and emigration. If B = number of births, D = number of deaths, I = number of immigrants, and E = number of emigrants, we can also quantify ΔN as follows:

$$\Delta N = (B - D) + (I - E)$$

Commonly, ecologists want to know not just whether population size is increasing or decreasing, but also the rate at which population size is changing—that is, the change in population size (ΔN) over a given period of time (Δt), or $\Delta N/\Delta t$. Let's say a population starts with 80 individuals and after two years has grown to 120 individuals. The population, then, has gained 40 individuals in two years, which gives us a rate of $40/2 = 20$ individuals per year.

The importance we place on such a number strongly depends on the population size. An increase of 40 individuals in two years is significant if the starting population had 80 individuals, but it is less significant if the starting population included 10,000 individuals. Therefore, we are usually interested in the proportional increase or decrease over time, especially the rate of population growth per individual, known as the **per capita growth rate** (*capita* is a Latin term that means "heads"). The per

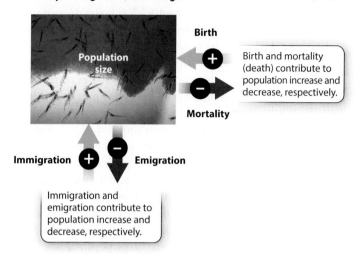

FIG. 44.4 Factors affecting the size of a population: birth, mortality, immigration, and emigration. *Photo source: Jean Paul Ferrero/ARDEA.*

capita growth rate, symbolized by r, is the change in population size ΔN in a given period of time Δt, divided by the size of the population N:

$$r = (\Delta N/\Delta t)/N$$

The per capita growth rate is the average number of offspring per individual (that is, *per capita*). For example, an adult American woman has, on average, 2.3 children; this is the average for the population as a whole, since no one individual can actually have 2.3 children.

When r is positive, the population increases in size; when r is negative, it decreases in size; and when r equals 0, the size of the population does not change. In addition, r can change from one generation to the next. In our example, r equals 40 individuals divided by two years, or 20 individuals per year, divided by 80 individuals (the starting population size), for an average per capita growth rate of 0.25 per year.

Population size increases rapidly when the per capita growth rate is constant over time.

As we just discussed, we can use mathematical models, including equations and graphs of those equations, to describe natural phenomena. These models not only explain what we observe, but also help us make predictions. Ecologists have developed a variety of models to describe and predict how populations will change over time. These models often describe idealized populations and make certain assumptions. For example, many population growth models assume that the population is isolated, with no immigration into or emigration out of the population. These assumptions allow us to focus on the main factors that influence population growth, birth and death, without having to adjust for factors (such as migration of individuals) that are more difficult to estimate but also generally have less influence on populations.

Let's consider a model of population growth for species such as bacteria and humans, in which life cycles are not synchronized. In other words, organisms are born and develop at different times, and any individual that reaches sexual maturity in the population can reproduce immediately. The result is that the population increases or decreases continuously. This type of population growth is known as **continuous growth**.

One of the simplest models of continuous population growth assumes that there are no constraints on population size and the growth rate of the population is constant. Even though the growth rate is constant, the actual number of individuals added per unit of time increases as the population grows. This is because each new individual also reproduces, so over time more and more individuals contribute to the growing population. As a result, the population grows extremely rapidly and exhibits **exponential growth**. Exponential growth occurs when the growth rate is constant, meaning the population grows in proportion to the number of individuals.

The characteristic shape of an exponential growth curve is shown in **Fig. 44.5**. The size of the population increases slowly at first, but then increases rapidly. Compound interest in a bank account works the same way; as you save more, you earn more even when the interest rate remains the same.

We can describe the curve in Fig. 44.5 mathematically. Let's start with the formula for per capita growth we calculated previously, but rearrange it to solve for the change in population size over time, as follows:

$$\Delta N/\Delta t = rN$$

Because growth is continuous, we are interested in describing population growth at any instant of time, rather than the average growth over an interval of time. Therefore, we use the symbol "d" instead of Δ:

$$dN/dt = rN$$

Focusing on a single instant of time is useful for describing populations that reproduce continuously with overlapping generations because choosing an interval in which to measure the population, such as a year or a decade, would be arbitrary. Similarly, in this equation, r is not the average per capita growth rate, but instead the per capita growth at an instant of time, called the **intrinsic rate of increase**.

Exponential growth is typical of small populations when resources are abundant, such as with the initial growth of bacteria in a laboratory culture or algae in a pond. The human population has also grown exponentially, a topic we return to in Chapter 48. When there are no environmental factors that limit population growth, r reaches its maximal value, symbolized as r_{max}. Every species has a characteristic value for r_{max}.

We can calculate the population size N at any time t using the following equation:

$$N_t = N_0 e^{rt}$$

where e is Euler's number, a mathematical constant approximately equal to 2.718. Euler's number is the base of natural logarithms.

Population growth is not continuous for all species. Consider an annual plant, in which each generation of organisms germinates at the same time, grows together to sexual maturity, reproduces, and then dies. The number of germinating seeds produced by organisms in any one generation determines the number of organisms in the succeeding generation. This type of population growth is known as **discrete growth** because the population size increases (or decreases) in one discrete step at the beginning of each new generation. It is typical of organisms that breed seasonally or annually.

For discrete growth, the growth rate in one generation, symbolized by the Greek letter lambda (λ), is determined by dividing the population at one point in time (N_1) by the population size one generation earlier (N_0):

$$\lambda = N_1/N_0$$

For example, let's say the population size of annual plants is 804 in one year and 992 in the next year. The growth rate λ is $992/804 = 1.23$, which represents a 0.23 or 23% rate of growth in one year.

If we know the starting population size (N_0) and the growth rate of the population (λ), we can rearrange the formula to determine the population size one generation later (N_1) as follows:

$$N_1 = N_0 \lambda$$

As with continuous growth, we can consider a situation in which the growth rate (in this case λ) is constant in each

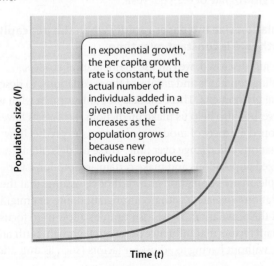

FIG. 44.5 Exponential growth. When the per capita growth rate, r, is constant, the curve shows ever-larger increases in population size over time.

generation and the population grows in proportion to the number of individuals. In this case, the population grows geometrically. In both geometric and exponential growth, the growth rate is constant; however, in geometric growth the population increases or decreases in steps (discretely), whereas in exponential growth it increases or decreases continuously.

If the growth rate is constant and we want to calculate the population size at any time t (N_t), we can use the following formula:

$$N_t = N_0 \lambda^t$$

A chessboard example drives home the remarkable consequences of geometric growth. Suppose we place one grain of wheat on the first square of an 8×8 chessboard that has 64 squares, two grains on the second square, four grains on the third, and so on, doubling the number on each successive square. In this example, $N_0 = 1$, $\lambda = 2$, and $t = 64$. By the time we reach the 64th square, the total number of grains on the board (N_t) is 2^{64} or approximately 1.8×10^{19}, which is roughly equal to the number of grains of sand on Earth.

Malthus recognized that rapid population growth cannot continue for long in nature because the number of individuals eventually outstrips the resources available to support the population. Darwin, in turn, recognized that when the number of young produced at birth exceeds the number of adults that can be supported by available resources, individuals within the population compete to obtain the resources needed for growth and reproduction. This **intraspecific** (within-species) **competition** can result in natural selection when some individuals have genetic mutations that give them a competitive advantage over other individuals (Chapter 20).

Individuals can also compete for resources with individuals from different populations of species. Such **interspecific** (between-species) **competition** can result in either an increase or a decrease in population size. We focus for the most part on intraspecific competition in this chapter, reserving a detailed discussion of the effects of interspecific competition for Chapter 45.

Carrying capacity is the maximum number of individuals a habitat can support.

As a population grows, crowding commonly results in a decrease in the availability or quality of resources required for continuing growth. As a consequence, growth rate declines. On Hispaniola, for example, Lime Swallowtails have probably reached the limit of available citrus groves. Individual females have to search longer to find suitable plants on which no other females have already laid eggs. With that delay, and the pressure of eggs that are continuing to mature inside their bodies, females may end up laying too many eggs in one place or on poor-quality food, so there is not enough food for all larvae to survive. All of these factors lead to fewer births per female in the Lime Swallowtail population than occurred immediately after they first arrived in Hispaniola, when resources were not a limiting factor.

For many organisms, crowding affects the death rate as well as the birth rate. For example, as the fish population in a pond increases, there is less food available for each individual, and oxygen is used up by microbes that proliferate because of nutrients in the fish waste.

In other words, birth and death rates are themselves affected by population density, which we defined earlier as the number of individuals per unit of area or (for aquatic species) volume of habitat. Increasing density is not always detrimental to a population. For example, salmon or sturgeon eggs are more likely to be fertilized in rivers or streams swarming with their species than when relatively few reproducing adults are present. Generally, however, at some point increasing population density causes the environment to deteriorate because of a scarcity of resources. When the density of a growing population reaches this point, the rate of growth slows and eventually levels off as the death rate increases to equal the birth rate.

The maximum number of individuals of a given species that a habitat can support is its **carrying capacity (K)**. K represents the interplay between the requirements of individuals for growth and reproduction and the environmental resources such as food and space available to support these needs.

Many factors can keep a population below K, including predation and parasitism. For example, increasing numbers of Lime Swallowtails on Hispaniola have no doubt attracted predators and parasites that feed on the butterflies. Similarly, a high population density of fish in a pond increases the likelihood of infection by viruses, bacteria, and fungi because these agents are transmitted from fish to fish. Predation and parasitism can reduce the population size below K, but they do not change K itself.

Logistic growth produces an S-shaped curve and describes the growth of many natural populations.

When a population is small, resources are not a limiting factor and the population can grow at or near its maximum intrinsic rate of increase, r_{max}. But as the population increases, drawing closer to its carrying capacity, worsening conditions due to crowding can slow the rate of growth. For any given population at any given moment, such as deer in New England forests or striped bass in Chesapeake Bay, we may wish to know what fraction of the carrying capacity is available for further population growth. The answer, expressed mathematically, is $(K - N)/K$.

We can again model population growth, this time taking into account the carrying capacity. In this case, we get a curve like the one shown in **Fig. 44.6**. Growth is exponential at first, but then slows as the population size approaches its maximum sustainable size, K. The resulting pattern of population growth,

FIG. 44.6 Logistic growth. The graph shows the characteristic S-shaped logistic growth curve, which is determined by r and K.

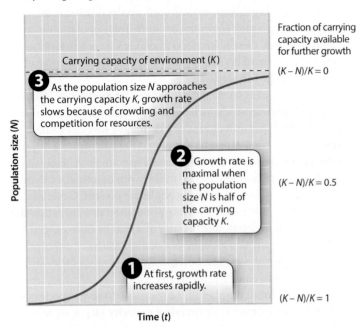

with its characteristic S-shaped curve, is termed **logistic growth**. This curve can be described mathematically as follows:

$$dN/dt = r[(K-N)/K]N$$

Logistic growth curves are S-shaped because they reflect two sets of processes, each related to population size. At first, growth reflects only the rate at which individuals can reproduce (when N is very small relative to K), so the curve rises steeply. The second part of the curve, where it starts to level off, indicates a slower growth rate. This portion of the curve reflects the onset of additional factors that start to have a significant effect on survival and reproduction above a certain population size, such as decreasing availability of food and space.

Keep in mind that logistic growth is a model of population growth, not a graph of real data points from an existing population. Natural populations often show variations from this idealized S-shaped curve. Nevertheless, logistic curves apply to many other phenomena in biology, as well as in such diverse fields as chemistry, sociology, political science, and economics. For example, a logistic curve describes the rate at which oxygen binds to a population of hemoglobin molecules, as seen in Chapter 37.

Factors that influence population growth can be dependent on or independent of population density.

In the preceding discussion, the factors that limited population size depended on the density of the population. At low density, competition for food and other resources does not limit growth.

However, as density increases, competition exerts more and more influence on population growth, as resources become more limited. Similarly, at high population density, individuals may be more vulnerable to predation or to infection. Factors such as competition and predation are described **density-dependent** factors.

Not all changes in population growth rates result from population density. **Density-independent** factors influence population size without regard for the population's density. These factors include events such as a severe drought or a prolonged cold period, either of which can cause widespread mortality independent of population density. A graph of density-independent effects on populations often shows instantaneous, rather than gradual, drops in population size, and these changes can occur at different starting densities.

An example of a density-independent factor can be seen in a population of Scottish Red Deer (*Cervus elaphus scoticus*) (**Fig. 44.7**). Researchers found that the deer population's growth rate is influenced by the weather in the spring, when deer are born. A cold, wet spring reduces both birth weights and survival of newborn deer. The females that are born in such years are smaller in the following breeding seasons, reducing their average breeding success, and their survival as adults is lower, reducing their contributions to the population size. A factor such as weather acts on all individuals in a

FIG. 44.7 Density-independent effects on population size. Independent of population density, cold and rainy weather have an effect on birth weight and survival of newborn Scottish Red Deer. *Data from S. D. Albon, T. H. Clutton-Brock, and F. E. Guinness, 1987, "Early Development and Population Dynamics in Red Deer. II. Density-Independent Effects and Cohort Variation," Journal of Animal Ecology 56(1):69–81.*

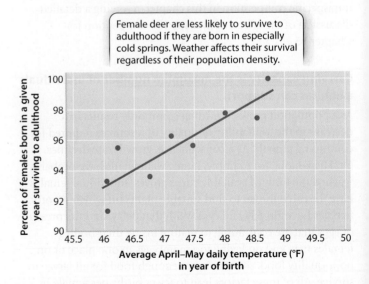

population regardless of density and can have lasting effects on population growth.

> **Self-Assessment Questions**
>
> 3. What four processes affect population size?
> 4. If a population triples in size in a year, what is the per capita growth rate?
> 5. Would you rather have a million dollars today, or start with one cent and double the amount each day for a month? Why?
> 6. Draw an exponential growth curve and a logistic growth curve. What accounts for their different shapes?
> 7. Name two density-dependent and two density-independent factors that can limit the size of a population.

44.3 AGE-STRUCTURED POPULATION GROWTH

Ecologists use direct counting and mark-and-recapture methods to assess the size and density of populations in a particular place. However, not all organisms in a population contribute equally to population growth because some reproduce more than others. Therefore, it is also useful to take into account individuals that have differing reproductive capacities. Variations in reproductive capacities most often depend on differences in age.

Birth and death rates vary with age and environment.

To understand past changes and predict future changes in populations, ecologists consider birth rates, expected longevity of individuals, and the proportion of individuals in a population that is able to reproduce. Insights into these parameters come from the **age structure** of a population, the number of individuals within each age group of the population. A population in which most individuals are past the age of greatest reproduction or a population mostly composed of individuals that have not yet reached peak reproductive age will not increase as rapidly as one dominated by individuals capable of the greatest reproductive output.

To reconstruct the age structure of a population, researchers commonly divide the population, or a sample of the population, into age classes. Such classes may include, for example, individuals born within a single year or those born within an interval of several years. Age may be determined by measuring an aspect of the organism's anatomy (for example, counting growth rings in fish scales or a tree trunk) or estimated by overall size (for example, fish length or tree height).

Estimating the number of individuals of differing ages enables ecologists to predict whether the size of a population will increase or decrease. A growing population usually shows a pyramid-shaped age distribution, with the youngest classes much more abundant than the older classes, whereas a stable population shows a more even distribution of age classes. **Fig. 44.8** shows the difference in age structure between France, a country with a stable population, and India, a country with a growing population.

In humans, differences in age structure may reflect, in part, socioeconomic variation among countries, but differences in the age structure of other species reveal threats from various sources.

FIG. 44.8 Age structure of two populations. These data from 2010 are depicted as histograms, bar graphs in which the length of each bar indicates the number of individuals in the group (in this example, age group). *Data from United Nations Department of Economic and Social Affairs, Population Pyramids (France and India), http://esa.un.org/wpp/populationpyramids/population-pyramids.htm.*

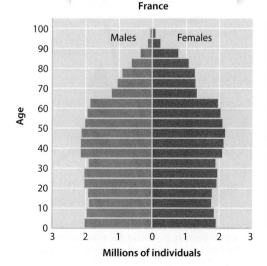
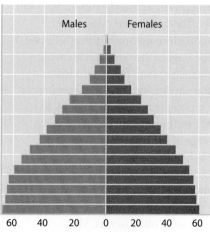

Fig. 44.9 shows the effects of overfishing on the age structure of a species of long-lived Pacific rockfish, determined by comparing the ages of fish sampled in unfished waters with those in heavily fished waters. In the unfished population, the curve shows many individuals close to and past the age of 10 years, when reproductive ability starts to climb. In the fished populations, the curve shows a sharp decline in the number of individuals older than 6 years because mostly older and larger fish are caught. Because older fish produce most of the eggs and young fish, the annual reproductive effort of fished populations is severely affected by the removal of the older fish.

Survivorship curves record changes in survival probability over an organism's life-span.

The age structure of a population reflects death rates as well as birth rates. **Demography** is the study of the size, structure, and distribution of populations over time, including changes in populations in response to birth and death rates, aging, and migration. The demography of the invasive Lime Swallowtail butterfly has been studied in Southeast Asia, where it is native. This investigation may be helpful in predicting the population dynamics of this species in coming years as it spreads from Hispaniola across the Americas.

The factors that cause insect mortality commonly differ from one life stage to the next, and knowing their effects can help us understand the physical and biological influences on population growth. **Fig. 44.10** shows the different dangers in each stage of the Lime Swallowtail's life. Each butterfly lays 30 or more eggs, but only a fraction of the eggs survive; most are lost to parasitism, drying out, falling from leaves, or predation by ants or other insects. The larvae have many defenses against predators, among them camouflage

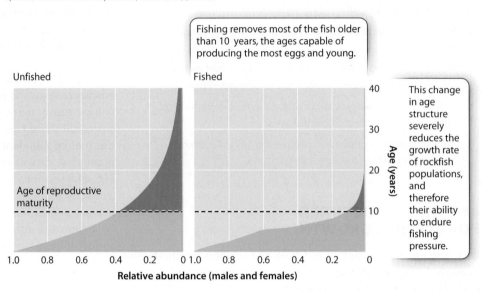

FIG. 44.9 Age structure in Pacific rockfish populations in fished and unfished waters. Harvesting larger and older individuals reduces the population's capacity for growth, jeopardizing commercial stocks. *Data from S. A. Berkeley, M. A. Hixon, R. J. Larson, and M. S. Love, 2004, "Fisheries Sustainability via Protection of Age Structure and Spatial Distribution of Fish Populations," Fisheries 29(8):23–32.*

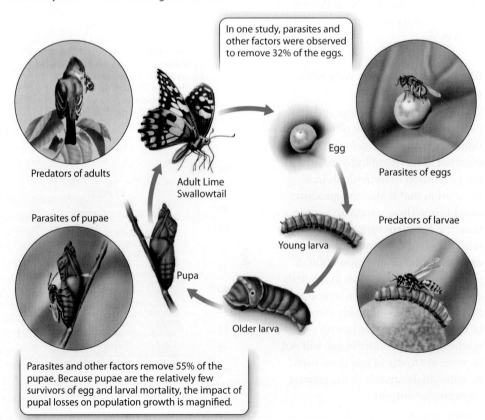

FIG. 44.10 Lime Swallowtail life cycle and natural enemies. The factors that cause insect mortality commonly differ from one life stage to the next.

and chemical toxins, but they still fall prey to wasps, as well as to lizards and other animals. Thus, the faster they develop into adults, the more likely they are to escape discovery by their enemies. The pupae, which are large and immobile, are perhaps the most vulnerable stage in the Lime Swallowtail life cycle, and pupae are often highly parasitized by a particular kind of wasp.

Fig. 44.11 presents data on the duration of each stage in the Lime Swallowtail life cycle and the numbers killed in that stage. The individual eggs that begin our study form a **cohort**, a group defined as the individuals born at a given time, and we can trace their survival as they age. The information about how many individuals of a cohort are alive at different points in time can be summarized in a life table, from which an ecologist can identify patterns of survival in a population. The proportion of individuals from the initial cohort that survive to each successive stage of the life cycle is called survivorship. When survivorship curves for the Lime Swallowtail are plotted through time, they show an ever-declining fraction of the starting number of eggs (Fig. 44.11).

FIG. 44.12 Survivorship curves. A Type I curve shows most mortality late in life; a Type II curve shows even mortality throughout life; a Type III curve shows most mortality early in life.

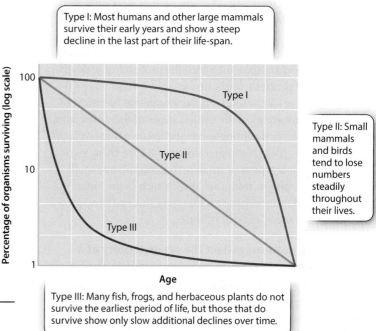

FIG. 44.11 Life table and survivorship curve for a population of Lime Swallowtail butterflies. (a) A life table summarizes the number of individuals alive in each cohort of a population and can be represented graphically by (b) a survivorship curve, which plots the proportion of the original cohort alive at the end of each stage or time period. *Data from M. Pathak, and P. Q. Rivsi, 2002, "Age Specific Life Tables of Papilio demoleus on Different Hosts," Annals of Plant Protection Sciences 10(2):375–376.*

a.

Stage	Approximate duration	Number alive	Survivorship
Egg	5 days	250	0.68
Larva	18 days	170	0.93
Pupa	15 days	158	0.45
Adult	5 days	71	

b.

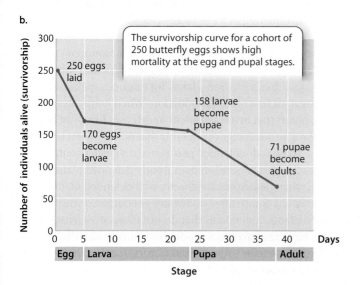

We can see in the life table and the survivorship curve for this example that survival is lowest among eggs (68% survive to the next stage) and pupae (only 45% survive to the next stage), which cannot move to escape harm. If we divide the number lost at each stage by the number of days at that stage, we see that the egg stage losses are 16 per day, the pupal stage losses are 6 per day. By contrast, the larval stage losses are only 0.6 per day. Given these data, there is strong selective pressure to decrease the length of each stage, especially those stages most vulnerable to mortality. Nevertheless, if we remember that each adult female can lay 30 or more eggs, only a very few eggs need to survive to the adult egg-laying stage to ensure the growth of the population. In warm climates such as the Caribbean, these butterflies may reproduce eight or nine times in a year. Most of their eggs will not survive to become adults. If they did, available resources would soon be depleted.

Patterns of survivorship vary among organisms.

Populations differ in the ages at which they experience the most mortality. In some species, many offspring are produced, but high mortality early in life means that few survive to reproductive age. In other species, mortality is highest in older individuals. In still other species, mortality occurs with equal frequency at all ages. **Fig. 44.12** illustrates the three principal types of survivorship curve—Type I, Type II, and Type III— which correspond to three different patterns of mortality:

whether most mortality occurs late in life, is even throughout life, or is concentrated early in life, respectively.

Human populations in affluent countries experience most mortality late in life, so they typically show a Type I survivorship curve (Fig. 44.12). Other large animals, such as elephants and whales, show similar survivorship curves. In contrast, birds and many small mammals commonly experience consistent levels of mortality throughout life, which can be graphed as a declining straight line. These animals, for which the probability of death does not change much through the life cycle, exhibit a Type II survivorship curve (Fig. 44.12). Finally, organisms like butterflies show high mortality at the earliest stages of their life cycles, resulting in a sharply declining survivorship curve that levels off (Fig. 44.12). This pattern is characteristic of a Type III survivorship curve, which is typical of small herbaceous plants, small animals such as mice or minnows, insects such as the Lime Swallowtail, and other organisms that grow fast and produce a large number of seeds, eggs, or offspring.

Reproductive patterns reflect the predictability of a species' environment.

In the preceding sections, we discussed population size and growth curves, and we outlined age structure and survivorship. These topics converge when we consider patterns of reproduction. Some species produce large numbers of offspring but provide few resources for their support. Such species are described as **r-strategists** (Chapter 40). In contrast, other species produce relatively few young but invest considerable resources in their support. These are termed **K-strategists**. How do these concepts relate to the intrinsic rate of increase r and carrying capacity K in population growth?

In general, species that live in unpredictable environments produce many young. When the environment is full of predators watching for vulnerable prey, such as gulls that watch for young turtles on a beach, or when resources are patchily distributed and abundant for only a short time, the chances of a young organism surviving to adulthood are low. Laying an egg, then, is like buying a lottery ticket: there is a limited chance of success. It is more advantageous to buy many tickets cheaply (that is, to have many offspring and make little parental investment in any of them) than to pay a lot for one ticket. Thus, salmon and other fish produce many thousands of young, each with an extremely small chance of survival.

As we saw earlier, organisms like insects and plants that display Type III survivor curves also experience high mortality in the first stages of life. Because their reproductive success depends on the small chance that a few of their young will survive the hazards they encounter, parents produce many offspring and invest relatively little in any of them. We refer to species like the fish we just discussed as r-strategists because they reproduce at rates approaching r_{max}, the maximum intrinsic rate of increase.

In contrast, species living where resources are predictable often produce a lower number of offspring and invest their reproductive resources in just a few, often larger and better-cared-for, offspring. These offspring are better protected from predators, able to compete with others when they are young, and more likely to survive to adulthood. Animals such as humans and bears, which have a Type I survivorship curve, produce only a few young and are K-strategists because their population densities commonly lie near the carrying capacity K.

These two strategies reflect an evolutionary trade-off. A reproductive individual has access to a finite amount of resources and can invest in either many inexpensive young or a few well-provisioned offspring. Depending on the combination of physiology and environment, natural selection in a given set of circumstances may favor one strategy over the other. However, r- and K-strategists represent the opposite ends of a continuum in reproductive pattern. Many species lie between these extremes.

The life history of an organism shows trade-offs among physiological functions.

As noted earlier, population growth can be limited by resource availability. How an organism makes use of the resources that are available can also affect population growth. Each individual must devote some of its total food and energy intake to growth, some to the maintenance of cells and tissues, and some to reproduction. This allocation is very much like a household budget, in which incoming resources are divided among the day-to-day expenses of acquiring food and paying for housing and transportation, meeting present and future medical and educational expenses, and savings held in reserve for emergencies. When one category goes up in cost—it requires more money, or more resources—another must come down.

In nature, individual organisms strike a balance in how they allocate resources to the basic needs of growth, maintenance, and reproduction. This balance has evolved to minimize energy loss while retaining the flexibility to use resources in different ways depending on environmental circumstances. These kinds of trade-offs evolve through natural selection and are the rule in ecology, explaining in part why species tend to produce either many small offspring or a few large ones.

The typical pattern of resource investment in each stage of a given species' lifetime is its **life history**. The amount of energy available for growth, maintenance, and reproduction vary over an individual's lifetime. Plants, for example, make trade-offs between defense and reproduction. When herbivores are around, plants that invest resources in defenses like poisons or thorns have an advantage over plants that do not invest in such defenses. These well-defended plants make a greater contribution to the next generation than poorly defended plants. In contrast, when herbivores are absent, well-defended plants are commonly at a competitive disadvantage compared with poorly defended plants. Individuals that do not invest in defenses can allocate more resources to reproduction. Such trade-offs in plant defense are especially important in nutrient-poor soils, where the cost of replacing valuable tissues lost to herbivores is higher than in nutrient-rich soils (Chapter 30).

The availability of nutrients and other resources may vary from place to place as well as over time, whether in different seasons or in different years. To take advantage of different conditions in their environment, the life history of many species includes some degree of control over their allocation of resources. Galápagos Lava Lizards, for example, show considerable flexibility in allocating resources of fat and protein that can be used for building eggs. As is the case for many other animals, availability of resources such as insects for food in the Lava Lizards' habitat varies from year to year due to variation in rainfall. This variation in resources determines how much energy is available for both growth and reproduction.

In dry years, the lizards show a physiological trade-off between the number of eggs produced and the average size of each egg. Larger eggs require more resources to build, leaving less for additional eggs, and larger eggs hatch into larger offspring that are better able to fend for themselves. If the environment carries a high predation risk, however, then laying more eggs scattered in different places may be a better strategy than laying a few larger eggs in one place. Lizards may not reproduce at all if they are stressed by lack of resources for their other physiological needs, such as growth and development. Does this flexibility make these lizards K-strategists or r-strategists? The answer is neither. Which strategy is optimal depends on the nature of the risks faced by the Lava Lizards' young.

The key point is that food and other resources are typically limiting factors. The various ways in which animals and plants allocate such resources to growth and reproduction are rooted in their physiology, and their physiological traits are, themselves, evolved traits. Thus, ecology both reflects and determines how organisms function in nature.

Self-Assessment Questions

8. Draw a graph showing the age structure of a population that is growing rapidly and a graph of a population that is stable.

9. Plot a survivorship curve for a species with high rates of predation early in life and one for a species with high mortality late in life. Name the type of survivorship curve displayed in each graph.

10. How do r and K strategies relate to the predictability of the environment? In which kinds of environment is each strategist more successful?

11. What is a trade-off in physiological functions? Give an example.

44.4 METAPOPULATION DYNAMICS

Up to this point, we have focused for the most part on single populations: how they are characterized, how they change in size over time, and what the distributions of ages are within them. In this section, we consider how multiple populations of a single species influence one another. In nature, the populations of many, if not most, species are unevenly distributed across the landscape. Deer occur only where there are forest patches of suitable size; milkweeds grow only in open fields; and ducks need to be near water. As a consequence of these preferences, local populations are usually separated from one another in space and are connected only occasionally by individuals that migrate between habitat patches. Moreover, some habitat areas that could, in principle, support populations of a given species are unoccupied, as invasive species, such as the Lime Swallowtail in Hispaniola, illustrate. How does this kind of distribution influence the dynamics—that is, the change in abundances through time—of populations in nature?

A metapopulation is a group of populations linked by immigrants.

Groves of trees, separated from other groves by fields, are patches of habitat for the birds, insects, and fungi that rely on them for food and shelter. Similarly, humans are patches of habitat for infectious viruses and certain bacteria; coral reefs are patches for fish; and mountaintops are patches for mountain goats and other high-altitude species. For birds such as nuthatches or insects such as bark beetles, crossing between groves of trees is a risky business: predators abound and the possibility looms of not finding another suitable grove. Similarly, populations of reef corals are often separated by open waters full of predators. In effect, all of these habitats are islands separated by uninhabitable territory. We can define a patch as a bit of habitat that is separated from other bits of habitat by an inhospitable environment that may be difficult or risky for individuals to cross.

The nuthatches that occupy a particular tree grove interact with one another more frequently than they do individuals of other tree groves, and each group is considered a local population. Occasionally, a few migrants move from grove to grove, enabling gene flow between local populations. We can think of local populations of nuthatches in separate tree groves as parts of a larger, more inclusive group called a metapopulation. A **metapopulation** is a large population made up of smaller populations linked by occasional movements of individuals among them (**Fig. 44.13**).

To understand the dynamics of populations, we must consider not only the birth and death of individuals, but also the colonization and extinction of local populations. Local populations within a metapopulation have independent fates and commonly become extinct because of their relatively small sizes. In general, small populations become extinct more often than large populations because they are more susceptible to random factors that increase mortality, such as extreme weather, natural disasters, or predation. For example, a nest of predators can eliminate a prey species locally. Or, if population density falls too low in a small local population because of disease or lack of resources, the survivors may not be able to find each other for mating. Nevertheless, as long as some of the local populations remain—that is, if all individuals do not all become extinct at the same time—these populations can recolonize old

FIG. 44.13 Metapopulations. A metapopulation is a group of independent populations connected by occasional immigrants, illustrated here by populations of nuthatches in groves of trees. *Photo source: Charles O'Rear/Corbis.*

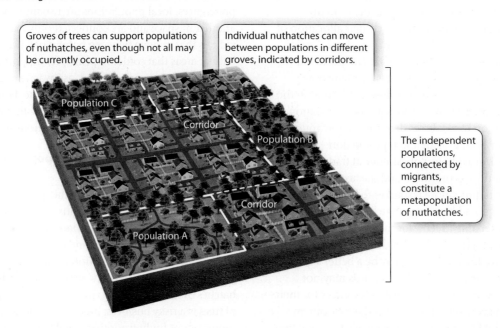

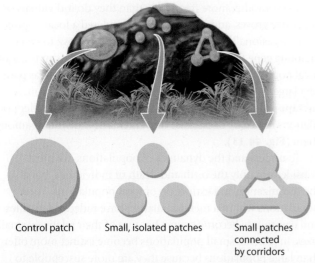

FIG. 44.14 Experimental patches of habitat. This study investigated the effects of patch size and connections in moss and the rates of colonization and extinction of the tiny arthropods that live there.

habitats or find new habitat patches, thus ensuring survival of the metapopulation.

Experiments with patches of moss growing on boulders show that the size and connectedness of the patches of moss affect the probability that species of arthropods inhabiting the moss will become extinct (**Fig. 44.14**). Researchers cut large patches of moss into several small fragments and sampled the number of insect species in each patch over time. As expected, isolated patches experienced decreases in species diversity, as much as 40% in one year; some species became extinct. However, when nearly isolated moss patches were left with a narrow corridor of moss connecting them, the rate of extinction was lower. Even small corridors of suitable habitat can allow species to move between patches of habitat and prevent extinction, or serve to reintroduce species that become locally extinct.

Studies of metapopulations not only help us understand the dynamics of populations, but also have relevance to conservation biology (Chapter 48). For example, researchers have monitored nine forest patches near Manaus, Brazil, for more than 30 years as part of the Biological Dynamics of Forest Fragments Project, one of the largest and longest-running ecological studies ever undertaken (**Fig. 44.15**). The researchers learned that fragmentation strongly affects forest biodiversity, providing clues to the design of effective refuges. Species survival depends on both the health of local populations and the ability of individuals to colonize new habitat patches. As the simple experiment in Fig. 44.14 with mosses shows, corridors that connect endangered populations can dramatically increase the probability that a species will survive. However, corridors

FIG. 44.15 Habitat patches. The Biological Dynamics of Forest Fragments Project in Brazil monitors forest patches, including this one, to better understand how fragmentation affects population health and species diversity. *Source: Rob Bierregaard.*

are important only to species that use them to cross barriers between patches. For species such as ferns, which produce spores that are able to float between patches on breezes, corridors may not be as important as they are to patch colonization. Dispersal ability differs among species, strongly influencing their capacity to colonize distant patches.

CASE 8 CONSERVING BIODIVERSITY: RAINFOREST AND CORAL REEF HOTSPOTS

How do populations colonize islands?

Islands represent extremes in habitat patchiness for colonizing species: organisms that live on a mainland must successfully cross a barrier of water to establish a new population on an island. For this reason, ecologists commonly use the term "island" to refer to any habitat patch that is surrounded by a substantial expanse of inhospitable environment. While this definition certainly includes parcels of land surrounded by water (what we usually think of as islands), it can also apply to a body of water surrounded by land, an expanse of forest surrounded by grasslands, or a mountaintop alpine field surrounded by lower-elevation forests.

Anolis lizards provide a classic and dramatic example of evolutionary diversification on the island of Hispaniola, where many dozens of closely related species coexist. Phylogenies suggest that *Anolis* lizards colonized Hispaniola in five or six separate events, but the dozens of species present today indicate that the populations diversified markedly after colonization. For flightless animals such as lizards, the ocean that surrounds Hispaniola is a barrier that prevents easy crossing, yet the island is clearly not unreachable. The distance of Hispaniola from other Caribbean islands and from the mainland is such that lizards and frogs and other small flightless animals have some chance of reaching it, perhaps by riding natural rafts of vegetation set adrift following mainland storms.

When the first lizards landed on the island, there may have been few natural predators and/or an abundance of small prey, allowing the population to grow rapidly. At the same time, Hispaniola featured unique types of habitats not available on the mainland, allowing the lizards to diversify over time into new species by adaptive radiation (**Fig. 44.16**; Chapter 21).

Not all species are equally able to colonize a new island. For example, as you might expect, flying animals such as bats and birds more frequently arrive on oceanic islands than nonflying animals such as rodents, lizards, and frogs. However, the nonflying animals that do colonize islands successfully more often evolve into multiple new species. If the new, closely related species do not themselves colonize another area, they will be found nowhere else in the world; these are called endemic species. Endemic species are particularly vulnerable to invasive species, such as the Lime Swallowtail of Hispaniola.

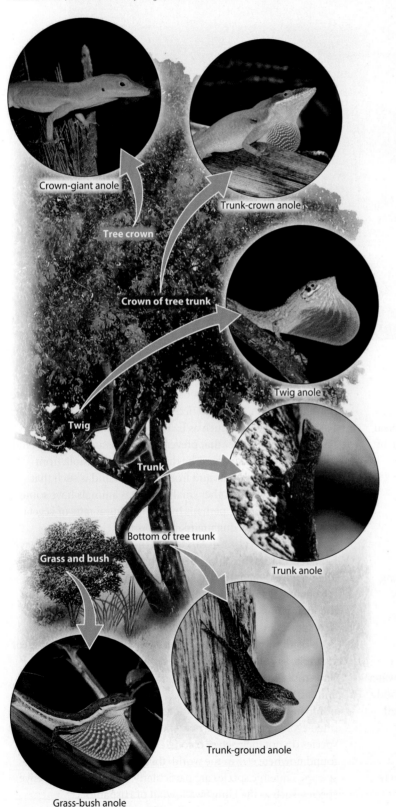

FIG. 44.16 *Anolis* lizards in Hispaniola. The lizards arrived as colonists from elsewhere in the Caribbean and subsequently diversified to produce a variety of species. *Photo sources: (top to bottom) Eladio M. Fernandez; Eladio M. Fernandez; Eladio M. Fernandez; © 2013 Jake Scott; Stan Osolinski/Getty Images; Eladio M. Fernandez.*

Islands are natural laboratories for studies of evolution, ecology, and conservation biology. On islands, we can study how populations grow after the initial colonization event, and then how populations adapt to their new environment, which can be quite different from the mainland where they originated. We can also look at how different species interact with one another, the subject of the next chapter. Finally, in a world increasingly marked by habitat fragmentation caused by human activities, studies of islands are playing an important role in conservation biology (Chapter 48) by helping us better understand the impacts of fragmentation.

Self-Assessment Questions

12. What is the difference between a population and a metapopulation?
13. Why are corridors important to the health of local populations?
14. List three examples of a habitat island. What makes each an island?

CORE CONCEPTS SUMMARY

44.1 POPULATIONS AND THEIR PROPERTIES:
A population consists of all the individuals of a given species that live and reproduce in a particular place.

- Population size is the number of individuals at a particular time and place. page 1002

- A population is characterized by its size, range, and density. page 1002

- Population range is the total area in which a population lives. page 1002

- Population density is the population size divided by its range. page 1003

- Population size can be estimated by sampling and mark-and-recapture techniques. page 1003

44.2 POPULATION GROWTH AND DECLINE:
Population size can increase or decrease over time.

- Population size is determined by births, deaths, immigration, and emigration. page 1005

- The population growth rate is the change in population size over time and is commonly expressed as the per capita growth rate, the rate of growth (or decline) per individual. page 1005

- Populations grow continuously when life cycles are not synchronized, and discretely when organisms reproduce seasonally so that population size changes at one step in each generation. page 1006

- Exponential growth is characterized by a constant per capita growth rate, so that the population grows in proportion to the number of individuals. page 1006

- Population density often limits population growth because of decreasing resources. This limit defines the carrying capacity (K) of the environment. page 1007

- The growth of natural populations is often logistic, producing an S-shaped curve in which growth is exponential at first, but then tapers off as population size approaches the carrying capacity of the environment. page 1007

- Population density can be limited by density-dependent factors, such as competition for resources and predation, and by density-independent factors, such as drought and fire. page 1008

44.3 AGE-STRUCTURED POPULATION GROWTH:
The age structure of a population helps ecologists understand past changes and predict future changes in population size.

- The age structure of a population is the number of individuals of each age group and is often depicted as a histogram. page 1009

- A growing population is characterized by a pyramid-shaped age distribution, with the youngest classes more abundant than older classes, whereas a stable population shows an even distribution of age classes. page 1009

- Survivorship is the proportion of individuals from an initial cohort that survives to each successive age or stage of the life cycle. page 1010

- Type I, Type II, and Type III survivorship curves correspond to mortality that occurs predominantly late in life, is even throughout life, or is concentrated early in life, respectively. page 1011

- r-Strategists are organisms that live in unpredictable environments and produce many young, but invest relatively little in each offspring. page 1012

- K-Strategists live where resources are predictable; they produce few young but invest heavily in each offspring. page 1012

- The life history of an organism is the evolved pattern of resource investment in each stage of a given species' lifetime. page 1012

- Organisms often exhibit trade-offs between reproduction and other physiological functions. page 1012

44.4 METAPOPULATION DYNAMICS:
The dynamics of populations are influenced by the colonization and extinction of smaller, interconnected populations that make up a metapopulation.

- The size of local populations (patches) and the degree of connectedness between patches determines in part whether a given patch will persist. page 1014

- An ecological island may be a true island surrounded by water or any habitat "island" surrounded by uninhabitable areas. page 1015

Log in to **LaunchPad** to check your answers to the Self-Assessment Questions and to access additional learning tools.

CHAPTER 45
Species Interactions and Communities

CORE CONCEPTS

45.1 THE NICHE: The niche encompasses the physical habitat and ecological role of a species.

45.2 ANTAGONISTIC INTERACTIONS: Competition, predation, parasitism, and herbivory are interactions in which at least one species is harmed.

45.3 MUTUALISTIC INTERACTIONS: Mutualisms are interactions in which both species benefit.

45.4 COMMUNITIES: Communities are composed of local populations of multiple species that may interact with one another.

The chocolate tree, *Theobroma cacao*, known simply as cacao, is native to tropical America, where the environment is consistently warm and humid. Physical environment alone, however, does not determine where cacao lives. Cacao reproduction, and ultimately its geographic range, is limited by the requirement that tiny flies called midges pollinate the plants (**Fig. 45.1**). The midges' eggs and young need deep shade and a thick layer of moist rotting vegetation that forms either from the natural fall of leaves onto the forest floor or from banana plants intentionally left to decay in cacao plantations. Because of the needs of the midges, cacao cannot reproduce—at least not without human help—in areas that experience full sun or on clean ground. The natural distribution of cacao plants, therefore, depends on interactions with both the physical environment and other species.

The reliance of chocolate trees on midges is not unusual. Every species interacts with other species in ways that affect growth and reproduction. These other species may be the insects that help a plant pollinate its flowers or ones that eat its leaves. They may be a fungus that helps a tree gain nutrients or another that attacks it. They may be other species that compete for resources, predators and pathogens that reduce population size, prey species that provide a specialized food resource, or microbes that provide or consume nutrients while living inside their hosts (Case 5 The Human Microbiome). Associations among species, whether positive or negative, form a complex web of interactions in which species influence the population size and distribution of other species, whether they are partners, enemies, or food. In this chapter, we explore where and how species live and how they interact with one another in a community.

45.1 THE NICHE

Why do species live where they do? As discussed in Chapter 44, a population persists only in places where its members can tolerate the physical environment. Plants, for example, are sensitive to rainfall, temperature, and nutrient levels in the soil. The habitat occupied by a species usually contains other species, and the distribution of each species reflect interactions with the other species, as the example of cacao illustrates. Competition between species can occur when different species use the same resources in the same place. Distinct species can coexist within shared habitat only if they use different resources, if they are active at different times, or if the local populations of one or both species have so few individuals that resources do not serve to limit their survival and reproduction. In other cases, such as cacao, species persist only when other species that benefit their growth or reproduction are also present.

The niche is a species' place in nature.

The combination of a species' physical habitat and its ecological role in that habitat defines a species' **niche**. The meaning of this term has changed over the years. Nearly a century ago, the American ecologist Joseph Grinnell defined the niche as the sum of the habitat requirements needed for a species' survival and reproduction. Later, ecologist Charles Elton redefined the niche as the role a species plays in a community, shifting the emphasis from the habitat to the species itself. By the middle of the twentieth century, ecologist G. Evelyn Hutchinson had combined these ideas, popularizing the concept of the niche as a multidimensional habitat that allows a species to practice its way of life.

The niche is determined by physical (abiotic) factors such as climate and soil chemistry, as well as by biological (biotic) factors based on interactions with other species. Although species are commonly associated with a specific habitat, the terms "niche" and "habitat" are not interchangeable. Niches have a dual nature: they reflect both where organisms occur and what they do there. Furthermore, niches encompass the ways that organisms respond to, and also affect, the resources and other species found in the habitat.

The realized niche of a species is more restricted than its fundamental niche.

The **fundamental niche** of a species includes the full range of climate conditions and food resources that permit the individuals in a species to live. In nature, however, many species do not occupy all the habitats permitted by their physiology because of interactions with other species, such as competition and predation, that can limit the range actually occupied. The actual range of habitats occupied by a species is its **realized niche**.

For example, Red-Winged Blackbirds (*Agelaius phoeniceus*) nest in marshes (**Fig. 45.2**). Their survival is most successful

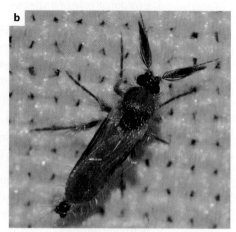

FIG. 45.1 A mutually beneficial interaction between two species. Development of (a) seed-rich cacao fruits depends on pollination of the cacao flower by (b) midges. *Sources: a. Exotica im/AGE Fotostock; b. Brian V. Brown.*

when they nest over deep water that provides protection from predators. In the western United States, Yellow-Headed Blackbirds (*Xanthocephalus xanthocephalus*) compete for territory with Red-Winged Blackbirds. In this region, the larger Yellow-Headed Blackbirds occupy the deep water parts of the marshes, while the Red-Winged Blackbirds nest at the water's edge. In this example, the entire marsh is the fundamental niche for Red-Winged Blackbirds (Fig. 45.2a). By contrast, the shallow water sites are the realized niche for the Red-Winged Blackbirds, where they co-occur with Yellow-Headed Blackbirds (Fig. 45.2b).

FIG. 45.2 Fundamental and realized niches. (a) The fundamental niche includes all the habitat potentially available to a population. (b) The realized niche is the subset of that habitat actually occupied by the population. *Photo sources: Ken Canning/Getty Images; All Canada Photos/Alamy.*

a. Red-Winged Blackbirds only

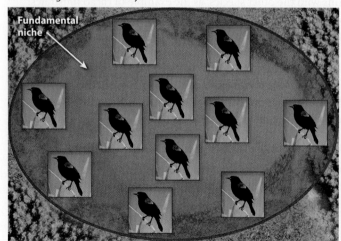

Red-Winged Blackbirds are found throughout marshes when Yellow-Headed Blackbirds are absent.

b. Red-Winged Blackbirds and Yellow-Headed Blackbirds

When Yellow-Headed Blackbirds are present, they compete successfully for space in the middle of the marsh. Because of competition for space, the realized niche of Red-Winged Blackbirds is smaller than the fundamental niche.

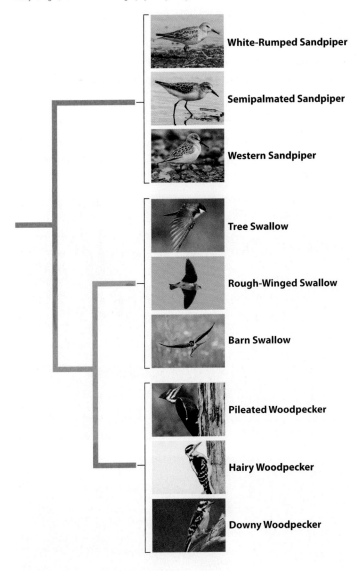

FIG. 45.3 Phylogenetic niche conservatism. Semipalmated, White-Rumped, and Western Sandpipers run along the water line on beaches in search of places to probe for tiny invertebrates buried in the sand. Tree, Barn, and Rough-Winged Swallows have long wings and broad bills that help catch flying insects. Downy, Hairy, and Pileated Woodpeckers drill holes in trees to reach insects. *Photo sources: (top to bottom) Paul Reeves Photography/Getty Images; impr2003/Getty Images; BirdImages/Getty Images; Janet Forjan/Alamy; Hal Beral/VWPics/Alamy; imageBROKER/Alamy; Canon_Bob/Getty Images; rpbirdman/Getty Images; Paul Reeves Photography/Getty Images.*

Some of the best evidence that organisms can live beyond the ranges they actually occupy comes from studies of invasive species (Chapter 48). For example, cane toads were introduced into Australia in 1935 and now occupy a wider range of habitats than they do in their native South America. Similarly, many familiar European weedy plants are much more widespread in North America than they are in Europe, hinting at the limits imposed by competitors and predators in their native ranges.

Invasive species show us that the realized niche is often much smaller than the fundamental niche.

Niches are shaped by evolutionary history.

The niches of nearly all species are narrow in at least one way, and often in several ways. For example, among woodland birds, warblers hunt insects on leaves, thrushes seek insects in soil, and woodpeckers find insects that live inside the wood of trees. These specializations may be explained at least in part by the niches of the ancestors from which these different woodland birds evolved. The tendency of species to retain aspects of their ancestral niches, and therefore for closely related species to resemble each other in niches, is called **phylogenetic niche conservatism**. While the conservatism of morphological traits has long been appreciated and studied, study of the importance of niche conservatism is a relatively new focus in evolutionary biology and speaks to our increasing awareness of the strong influence of evolutionary history in shaping ecological niches of organisms.

For example, the ecological niches of birds often remain relatively constant even as new species form, with the result that closely related species are similar in their ways of life, including the kinds of nests they build and the kinds of food they eat. Nearly all sandpipers have long bills they use to probe wet sandy beaches for the tiny animals they consume; different swallow species all hunt insects as they fly over fields, rivers, and ponds; and different species of woodpecker feed on insects living inside the wood of tree trunks and branches (**Fig. 45.3**).

The evolutionary ancestors of sandpipers, swallows, and woodpecker species were adapted for their ways of finding food. Food gathering is enabled by the long legs and bills of sandpipers that allow them to probe along long sandy beaches. Similarly, the long wings and broad mouths of swallows are particular well adapted for catching flying insects. Woodpeckers have stiff tailfeathers that allow them to brace vertically on trees and use their chisel-like bills to bore holes that they probe in search of insects with their extraordinarily long, barbed tongues. These examples illustrate that closely related species (in this case, different species of sandpipers, swallows, and woodpeckers) share similar niches (feeding on insects on beaches, in the air, and in wood, respectively), supporting the concept of phylogenetic niche conservatism.

Self-Assessment Questions

1. What defines the niche of a species?
2. Choose an organism, such as an oak tree, and explain the types of parameters that define its niche.

45.2 ANTAGONISTIC INTERACTIONS

Comparing the realized and fundamental niches of Red-Winged Blackbirds makes it clear that interactions with other species play key roles in the distribution and abundance of populations (see Fig. 45.2). Virtually every organism interacts with other species: think of plants and their pollinators or predators and their prey. These interactions often narrow the geographic distributions of species within the larger boundaries of their physical requirements, and help shape their realized niches.

All interactions involve costs (often in resources) and benefits (usually increases in reproduction). Some interactions are direct, where species physically interact (such as plants and their pollinators). Others are indirect, where species influence one another through competition for a shared resource, such as food. In many interactions, the benefits are one-sided: one organism actually consumes another or successfully competes with it for food or some other resource. In this section, we focus on **antagonisms**, interactions in which at least one participant loses more than it gains.

Limited resources foster competition.

Whenever there are fewer resources than individuals seeking them, competition can arise. **Competition** is an interaction in which the use of a mutually needed resource by one individual or group of individuals lowers the availability of that resource to another individual or group. For example, different species of sea anemones compete for space in rocky shallows and male dung beetles compete with each other for animal droppings. In many animal species, males compete with each other for female mates.

Recall from Chapter 44 that in many populations the tension between the intrinsic rate of increase and the environmental carrying capacity results in **intraspecific competition**, competition among individuals of a single species. The increase in intraspecific competition that accompanies increased population density is a major reason why population growth slows as the population approaches the environment's carrying capacity. Competition among individuals of different species, called **interspecific competition**, is also common.

Whether competition occurs among individuals of one species or two, it is generally a lose–lose situation: each side loses access to resources or spends energy it would not expend in the absence of the other. Indeed, Darwin recognized that the "struggle for existence," as he called it, is a primary driver of natural selection, leading to the evolution of traits that aid survival and reproduction by reducing competition in one way or another.

Species compete for resources other than food.

Food is the resource we most often think of as the limiting factor for population growth and reproduction. Many species, though, just require a place to stand, so to speak, or suitable conditions for a nest. Barnacle and mussel larvae compete for a place to attach on submerged rocks. The food these organisms consume, tiny plankton, is almost always abundant. Thus, food is not as likely to limit population growth for these species as attachment space is.

Similarly, plants compete for space in fields and forests so they can collect sunlight and obtain water and nutrients from the soil. Of course, sunlight cannot be used up, but room to collect sunlight on the ground below can be filled. Nesting sites for birds that lay their eggs in tree hollows or on cliff faces can also be strongly limited by the availability of space, as can the settling of corals on reef surfaces (**Fig. 45.4**).

Competitive exclusion prevents two species from occupying the same niche at the same time.

In general, when two species have overlapping niches (if, for example, they share a physical habitat and overlap strongly in resource use), one will either become extinct in that place or change its niche. Coyotes and foxes eat the same small prey, but when they meet, a coyote wastes energy by chasing the fox, and

FIG. 45.4 Competition for space. (a) Herons, egrets, and cormorants compete for nest sites in wetland trees. (b) Corals compete for space on reef surfaces. *Sources: a. Courtesy of Jim Maddox, Galloway, NJ; b. Franco Banfi/Getty Images.*

FIG. 45.5 Competitive exclusion. European Red Squirrels (a) and American Grey Squirrels (b) share the same niche, and the Grey Squirrels outcompete the Red Squirrels where they co-occur. *Sources: a. Sarah Peters/Getty Images; b. Zen Rial/Getty Images.*

the fox loses energy by running away from the coyote. In regions where foxes and coyotes can potentially share a habitat, the foxes avoid interactions with the larger coyotes by leaving the area, placing their dens in locations where coyotes do not roam.

The idea that two species cannot occupy the exact same niche at the same time is called **competitive exclusion**. Although the general idea is discussed in Darwin's *On the Origin of Species*, this principle is often attributed to the Russian biologist Georgii Gause. A consequence of competitive exclusion can be as simple as one species moving its territory when the other is present, as in the example of foxes and coyotes. A similar outcome occurs when Red-Winged and Yellow-Headed Blackbirds occupy the same marsh. Competitive exclusion is one way in which a fundamental niche can be reduced to a smaller realized niche.

More dramatically, competitive exclusion can lead to the local extinction of one of the two species. Squirrels provide a well-known example (**Fig. 45.5**). For millennia, European Red Squirrels (*Sciurus vulgaris*, different from the American Red Squirrel discussed in Chapter 44) thrived in woodland habitats throughout the British Isles. In 1876, however, the American Grey Squirrel (*Sciurus carolinensis*) was introduced to Great Britain and spread rapidly. Grey and Red Squirrels have similar dietary preferences, but the larger Grey Squirrels are better at gathering these resources. For this reason, as Grey Squirrels expanded, Red Squirrels declined to near extinction, excluded by their competitors.

While Grey Squirrels excluded Red Squirrels in Great Britain, nine closely related squirrel species live together in the tropical forests of Malaysia. They are able to coexist because the different species move through distinct areas of the forest, prefer different foods, and are active at different hours of the day. All of these traits were likely shaped by competition in the past. **Resource partitioning**, the division of resources by different species living in the same habitat, can minimize competition, allowing two or more species to coexist. Competitive exclusion is what keeps species that depend on the same resource from living at the same place at the same time, and resource partitioning is what allows similar species to coexist.

Even if two species overlap substantially in resource use, competitive exclusion may not always take place. It is more likely to occur when the population sizes of both species are large enough to encounter each other or reduce the resources available. As discussed later in the chapter, predation and parasitism can keep population sizes low.

 CASE 8 CONSERVING BIODIVERSITY: RAINFOREST AND CORAL REEF HOTSPOTS

Can competition drive species diversification?

In Chapter 44, we discussed the *Anolis* lizards on the island of Hispaniola and other islands of the Caribbean. Many closely related species of *Anolis* lizards live in similar areas and they all

FIG. 45.6 **Adaptations of *Anolis* lizards.** (a) *Anolis* species that feed on twigs have short limbs and a prehensile tail that they use to support themselves when they sleep; (b) those that feed on trunks have long limbs. *Source: (a and b) Eladio M. Fernandez.*

feed on insects and other invertebrates (see Fig. 44.16). How do they coexist?

Within their fundamental niche of tropical forests, different *Anolis* species differ in how and where they search for prey. Lizards with long legs run on the ground, while those with shorter legs and long tails climb on vegetation. Still others are large and robust and live in the high canopies of trees. Altogether, *Anolis* lizards have evolved at least six different feeding strategies with corresponding morphological traits (**Fig. 45.6**). In other words, they exhibit resource partitioning, allowing them to coexist in one place.

In many cases, resource partitioning reflects an evolutionary history in which competition for similar resources promoted the use of different niches by different groups, and eventually the diversification of species in this group. That is, competitive exclusion, played out over time, resulted in divergence in habitat use and feeding strategies, and ultimately speciation. The result is that the overlap between closely related groups is minimized. Among animals, resource partitioning usually entails behavioral shifts (and eventually often divergence in morphology) that enable different groups of individuals to use resources beyond those that the two groups share. For example, if two groups of finches normally feed on the same type of seed, resource partitioning may result in the groups accumulating physiological differences that allow each to feed more effectively on a different size of seed (Chapter 21).

Predation, parasitism, and herbivory are interactions in which one species benefits at the expense of another.

In contrast to competition, some interactions benefit one participant and harm the other. **Predation** is a type of interaction in which one organism, a predator, consumes another, its prey. In this interaction, the predator benefits at the expense of the prey. Early experiments by Georgii Gause showed that a simple system with one predator population and one prey population is inherently unstable. In such a scenario, the predator will overexploit the prey, driving it to extinction, and then become extinct itself.

In 1958, the American scientist Carl Huffaker demonstrated that if the prey had refuges where some individuals could escape from predators, the prey species could persist while predator populations declined. Through time, the prey population would recover. Population density would then rise to a point where predators would again expand and cause prey abundance to decline—and then decline again themselves (**Fig. 45.7**). Huffaker showed that predators and prey cycle repeatedly through periods of increasing and then decreasing density, as predators track their prey and some prey escape predation. A long-term, stable oscillation pattern can be achieved when prey are able to find temporary escape in refuges. In natural environments, predators may switch to other more common prey species when a particular prey becomes scarce, allowing prey populations time to recover.

Predation can affect population dynamics beyond just those of the predator and its prey. For example, predators can limit the population sizes of their prey to the extent that competitive exclusion does not occur between the prey and a competing population.

For example, in ponds throughout the southern United States, tadpoles of the large Southern Toad, the Eastern Spadefoot, and the tiny Spring Peeper all depend on the same food source; that is, they all graze on algae. In the presence of the two larger species' tadpoles, Spring Peeper tadpoles compete poorly for food (**Fig. 45.8**). As a result, they have low survival rates, and those individuals that do make it to maturity are often undersized because as tadpoles they did not have access to adequate nutrition. In the presence of a predatory Red-Spotted Newt, however, the tables are turned. The newt prefers to eat the larger tadpole species, and its voraciousness prevents the larger species from becoming sufficiently abundant to outcompete the

HOW DO WE KNOW?

FIG. 45.7

Can predators and prey coexist stably in certain environments?

BACKGROUND In the 1950s, it was not clear whether predators and prey could and exist in the same habitat indefinitely or whether both groups of species would become extinct as predators consumed all available prey. The results of experiments up to that time gave no concrete conclusions, and ecologists knew of examples of predators introduced to islands that hunted their prey to extinction.

HYPOTHESIS Ecologist Carl Huffaker hypothesized that predators and prey could stably coexist if temporary refuges were available for the prey.

EXPERIMENT Huffaker studied two kinds of mites: one that eats orange peels, and another that preys on the orange peel–eating mite. Huffaker put two sets of oranges on a table and separated them with barriers of petroleum jelly, essentially making each orange its own habitat patch (Fig. 45.7a). He added toothpicks to one set of oranges so that the peel-eating mites could let out silk and float over the petroleum jelly barriers and, at least temporarily, escape predation.

RESULTS In the setup in which prey could not escape, a single cycle of increase and decline was observed (graphed in Fig. 45.7b): the size of the prey population (red line) increased but was closely tracked by rising populations of the predator (blue line), and eventually both declined to extinction.

In the environments where toothpicks were provided, the orange-feeding mites climbed the toothpicks and fled on the silken threads they produced, escaping predators and reaching oranges without mites. These populations of predators and prey went through three cycles of increase and decline (graphed in Fig. 45.7c), and cycles would probably have continued if oranges continued to be supplied.

CONCLUSION Predator–prey systems can be stable if there are sufficient areas available where prey can escape predators, at least temporarily.

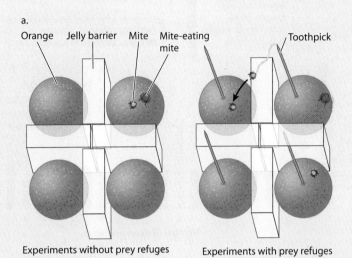

a. Experiments without prey refuges | Experiments with prey refuges

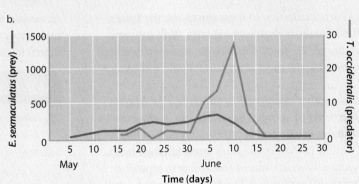

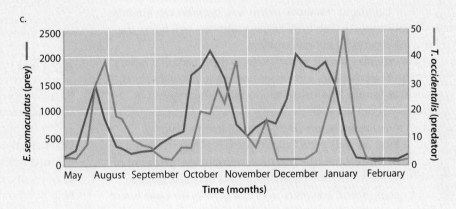

SOURCE Information from Huffaker, C. B. 1958. "Experimental Studies on Predation: Dispersion Factors and Predator–Prey Oscillations." *Hilgardia: A Journal of Agricultural Science* 27:795–834.

FIG. 45.8 Effect of predators on the outcome of competition. Newt predators in ponds prefer to eat bigger tadpoles that compete well for food, enabling smaller tadpoles that are poorer competitors to survive to adulthood. *Data from P. J. Morin, 1981, "Predatory Salamanders Reverse Outcome of Competition Among Three Species of Anuran Tadpoles," Science 212:1284–1286.*

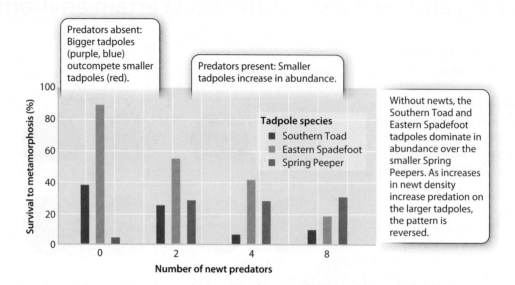

smaller Spring Peeper tadpoles. In these situations, the Spring Peeper tadpoles have the highest survival rates. With competition for food reduced, the small tadpoles grow to their maximal body size, and become the most abundant tadpole species in the ponds.

Parasitism is another interaction in which one participant benefits at the expense of another. **Parasites** live in close association with another species, gaining nutrition by consuming their hosts' tissues. Unlike predators, parasites rarely kill their hosts, but they can reduce the host's fitness and ability to reproduce or increase the host's susceptibility to harsh environmental conditions by tapping its resources. In this way, parasites can limit the population size of their host, keeping numbers well below the carrying capacity of the environment.

An extreme example of parasitism is the fungus that infects the American chestnut, a tree once dominant in eastern North American forests. This fungus attacks the vascular system of young chestnut trees, which usually succumb when they reach 2 to 3 m in height, before reaching reproductive maturity. Today, there are only a few isolated patches of large chestnut trees that have escaped the fungus in Vermont and a handful of other places. As chestnut populations have declined, populations of oak, beech, and other trees that grow in the same forest have increased.

Herbivory, the consumption of plant parts, benefits herbivorous animals by providing nutrients, and harms plants by directly affecting the products of photosynthesis. Plants are generally not as nutritious for animals as other animals are, so herbivorous animals must eat greater amounts of plant material to obtain the nutrients they need for survival and reproduction.

As we saw in Chapter 30, plants are not passive victims of herbivores; most are well defended with chemical or physical deterrents to herbivory.

Self-Assessment Questions

3. What is the difference between competitive exclusion and resource partitioning?

4. How can predators influence competition and affect the niches of other species?

5. What do competition, predation, parasitism, and herbivory all have in common?

45.3 MUTUALISTIC INTERACTIONS

As our chapter-opening example of cacao and midges shows, some interactions benefit both participants, though usually each participant's benefits are associated with some costs. When the benefits for each participant outweigh their costs, the interaction is a **mutualism**. Typical benefits include access to nutrients, shelter from predators or weather, and direct help in reproduction. Ultimately, all benefits are measured by natural selection in terms of reproductive output. Some typical costs include the proteins, fats, and carbohydrates that are invested in building structures such as flowers that attract pollinators or specialized tissues that house bacteria or algae. Costs can also take the form of energy-consuming activities such as the transport of pollen or seeds or the loss of food resources consumed by a partner.

Mutualisms are interactions between species that benefit both participants.

The midges in the chapter-opening example obtain the benefit of food from the cacao blossoms at the cost of pollen transport between flowers. For its part, the chocolate tree obtains the benefit of pollination at the cost of producing the sugars and amino acids in the flowers' nectar.

The nitrogen-fixing bacteria that live in nodules on the roots of soybeans (Chapters 24 and 27) provide another illuminating example of a mutualism. The bacteria occur naturally in the soil and are attracted by molecular signals to the soy plant roots, where they take up residence in nodules produced by the plant. The bacteria provide their host plant with nitrogen in a biologically useful form, and the plant provides the bacteria with food and a stable environment. In response to signals from the bacteria, the plant builds the bacteria's home, the nodules, out of root tissue. The cost involved in this construction is outweighed by the benefit of enhanced access to nitrogen, which is measured in greater growth and reproductive output for the plant. In the interaction between plant and bacteria, both sides win.

Despite both partners in a mutualism benefiting from the interaction, it is important to remember that each side is acting in its own self-interest and bears costs that are weighed against benefits in terms of growth and reproduction. Associations that, overall, are beneficial to one partner are often also beneficial to the other partner and, therefore, more often represented in the next generation. For example, nitrogen-fixing bacteria are ubiquitous. Likewise, their principal hosts, members of the plant family to which beans belong, are unrivaled in their diversity of species, dominating tropical forests all over the world. Nitrogen-fixing bacteria are equally important in the sandy shallows of tropical seas, where they form associations with vast beds of sea grasses, which in turn serve as nurseries for tropical reef fish.

In other words, mutualisms are subject to natural selection, just like any other adaptation (Chapter 20), and are at the heart of some of the most important biological interactions on Earth.

Mutualisms may evolve increasing interdependence.

Close interactions between species, either antagonisms or mutualisms, that have evolved over long periods of time are called **symbioses** (singular, **symbiosis**). One of the best-studied symbiotic mutualisms involves aphids (and their insect relatives) and closely associated bacteria. The insects suck plant sap for food, and the bacteria live in tissues that develop within the insects adjacent to their digestive system. The insects provide a home for the bacteria at some physiological cost, but also gain an important benefit from their residence: plant sap contains relatively few nutrients, and the bacteria provide their hosts with essential amino acids. The bacteria benefit from a stable and favorable environment at the cost of some nutrients supplied to the aphids. The aphids pass the bacteria from mother to daughter in egg cells, guaranteeing that offspring have bacteria.

When the interaction between species drives reciprocal adaptations in both participants, there is a possibility for long-term coevolution because the descendants of each side are associated with each other, just as were past generations. Any mutations that arise can affect both members of the partnership, because neither aphids nor bacteria are found without the other.

DNA evidence suggests that aphids and their bacteria have been associated for 100 million years or more, and that the bacterial genomes have been gradually losing genes they would need for life outside the insects (**Fig. 45.9**). In fact, some bacterial genes have been acquired by the insects. We saw a similar process of coevolution in Chapter 24 in the example of mitochondria and chloroplasts, which is probably nature's most dramatic example of a coevolved mutualism.

While some mutualisms involve close interactions between specific species, others are less particular. For example, sweet-tasting fruits evolved in many flowering plants, attracting mammals and birds that disperse their seeds, but not in response to any particular bird or mammal species. Similarly, flowers evolved in response to insect pollinators such as bees and flies, whose own adaptations for visiting flowers evolved in response to flower availability. General associations such as these can be beneficial, especially when the population of one partner undergoes declines. For example, *Freycinetia arborea*, a vine found only in the Hawaiian Islands, is pollinated exclusively by birds introduced from Japan in 1929. How was *F. arborea* pollinated before the twentieth century? Pollen grains on the pelts of extinct Hawaiian bird species preserved in museum collections show that the vine was once pollinated by birds that are no longer found in Hawaii. The vines persist because they were able to form a mutualism with the newly introduced Japanese species.

Digestive symbioses recycle plant material.

In Case 5 The Human Microbiome, we saw that diverse microorganisms in the human digestive tract provide important services to our own well-being. Humans are not alone in hosting microbial symbionts; most animals have microbiomes. All herbivorous animals studied to date, from cows and sheep to ants, beetles, and termites, are able to feed on plant tissues only with the help of microbial symbionts. Much of Earth's biomass consists of wood and other plant tissues rich in cellulose and lignin. Few organisms have the ability to digest these molecules, but some microorganisms, mainly bacteria and fungi, can. When these microorganisms reside in an animal gut, such as the rumen or large intestine of a cow, they enable their host to benefit from plant tissues that the animal itself cannot break down. In this way, both host and symbiont benefit from the association.

Not all plant-digesting symbionts live within the digestive tracts of their hosts. Consider leaf-cutting ants. Colonies of these ants in a Central American forest can consume as much as 14% of the leaves of nearby trees, hauling each leaf in pieces down

HOW DO WE KNOW?

Fig. 45.9

Have aphids and their symbiotic bacteria coevolved?

BACKGROUND Aphids are small insects that are common pests of garden plants. Specialized cells in the aphids harbor populations of the bacterium *Buchnera*, which provide their aphid hosts with amino acids essential for growth. Mother aphids pass the bacteria through their eggs to their daughters.

HYPOTHESIS Aphids and *Buchnera* bacteria have coevolved for millions of years.

EXPERIMENT American evolutionary ecologist Nancy Moran sequenced the DNA of the *Buchnera* bacteria and their host aphids to establish the phylogenetic relationships of aphid species and their symbionts. Moran reconstructed the phylogeny of both groups and compared their relationships. She reasoned that if the two phylogenies matched each other, the hypothesis of coevolution is supported.

RESULTS The phylogenetic trees of aphids and their associated bacteria match perfectly, just as if the bacteria were a gene of the aphids instead of a separate organism. This matching of phylogenetic trees is seen both among aphid species in different genera and for aphid populations within a single species. A DNA-based molecular clock (Chapter 20) further shows that aphids, and the insects related to them, have all coevolved with these bacteria over nearly 200 million years.

Photo source: Nigel Cattlin/Science Source.

CONCLUSION A single group of *Buchnera* has been passed down through generations of its aphid hosts for nearly 200 million years. Host–symbiont systems can be stable over long time intervals and show long-term coevolution.

FOLLOW-UP WORK Researchers are studying the changes in the genomes of bacteria living inside aphids to see how they lost the genes necessary for living independently of their aphid hosts.

SOURCE Moran, N. A., et al. 1993. "A Molecular Clock in Endosymbiotic Bacteria Is Calibrated Using the Insect Hosts." *Proceedings of the Royal Society. London, Series B*, 253:167–171.

---

into their subterranean nests. A single colony of leaf-cutting ants can consume as many kilograms of leaves per hectare as an herbivore the size of a sheep over the same amount of time. The ants do not feed on the leaves directly, but rather use them to feed gardens of fungi that the ants tend carefully (**Fig. 45.10**). The fungi, in turn, serve as food for the ant colony. The fungi benefit from being fed, and the ants benefit from the nutrients that the fungi provide.

These insect–fungi symbioses can be thought of as a form of agriculture that, like human agriculture, has a large impact on the biosphere through their efficient breakdown of plant materials. Insect–fungus mutualisms are certainly dramatic, but they are more the rule than the exception: nearly all eukaryotes depend on microbial symbionts with which they have coevolved (Chapter 25 and Case 5 The Human Microbiome).

FIG. 45.10 Plant-digesting symbioses. Leaf-cutter ants (a) feed leaf fragments to gardens of fungi (b), which digest the leaves and ultimately become food for the ants. *Sources: a. Redmond Durrell/Alamy; b. Martin Dohrn/naturepl.com.*

In some cases, animal–bacterial mutualisms have been implicated in resource partitioning and speciation, providing another link between ecology and evolution. Snails of the genus *Alviniconcha* occur widely along hydrothermal ridge systems in the western Pacific Ocean, gaining nutrition from chemoautotrophic bacteria within their tissues. The types and amounts of gases and ions emanating from the ridge vary along the ridges, as do the species of snails. Three closely related snail species are each found in association with distinct bacterial symbionts. Each group of bacterial symbionts facilitates growth under a different range of ridge chemistries. Resource partitioning and speciation are evidently based not solely on either the snails or their bacterial symbionts, but rather on the close mutualistic association of snail and bacterium.

Mutualisms may be obligate or facultative.

When one or both sides of a mutualism cannot survive without the other, the association is said to be **obligate**. The association of aphids and bacteria is obligatory for both sides: bacteria cannot live without the shelter the aphids provide, and aphids cannot live without the nutrients provided by the bacteria.

Many other associations are not so tightly intertwined, such that one or both participants can survive without the other. These interactions are termed **facultative**. For the midges that pollinate cacao, the association is facultative because the midges have other sources of food, including other flowers, in the wet forests they inhabit. For cacao, however, the relationship is obligate because bees and other insects do not usually pollinate them. Thus, their reproduction is more or less dependent on visits by midges.

Many obligate mutualisms are thought to have begun as facultative relationships that became reinforced over time by natural selection. For example, the mutation that originally made a cacao flower sprout from the tree trunk close to the ground rather than high out on a branch made the flower easier for low-flying midges to pollinate while decreasing the chances of bees discovering and pollinating the flowers. If midges were consistently more reliable pollinators than bees, the mutation would have been selected and then spread throughout the population of cacao (Chapter 20).

In some interactions, one partner is unaffected by the interaction.

In addition to interactions that benefit both partners and those in which at least one partner is harmed, there are interactions in which one partner benefits and there is no apparent effect on the other. These interactions are called **commensalisms**. For example, Gray Whales in the Pacific Ocean are commonly covered with barnacles. The barnacles benefit from the association, obtaining both a substrate for growth and a free ride through waters rich in planktonic food. By comparison, the presence or absence of barnacles doesn't seem to affect the whales one way or the other.

Some ecologists also recognize another class of interaction, called amensalism, in which one species in an interacting pair is harmed with no apparent effect on the other. For example, as they forage for food, elephants in African forests and woodlands commonly destroy trees. This behavior has a harmful effect on birds and insects living in the trees, not to mention the trees themselves, without apparently benefiting or harming the elephant populations. Similarly, wild pigs destroy plants as they plow up acres of soil in forests in their search for roots and small subterranean animals. The spectrum of costs and benefits associated with interactions among organisms is summarized in **Table 45.1**.

TABLE 45.1 Major Types of Species Interactions

INTERACTION	EFFECT ON EACH SPECIES	EXAMPLE
Competition	−/−	A grass and a wildflower: Each species loses the water, nutrients, and access to sunlight that the other takes.
Predation	+/−	Arctic Foxes and lemmings: Foxes benefit from eating lemmings; lemmings lose opportunities to reproduce.
Herbivory	+/−	Bison and grass: Bison benefit from eating grass; grass loses biomass that is eaten.
Parasitism	+/−	Tapeworms and humans: Tapeworms benefit from absorbing nutrients in human intestine; humans lose nutrients.
Mutualism	+/+	Flowers and bees: Flowers gain pollination; bees gain nectar and some pollen.
Commensalism	+/0	Egrets and cattle: Egrets benefit from insects stirred up by cattle; cattle are unaffected by egrets.

Facilitation occurs when one species indirectly benefits another.

In the mutualisms and commensalisms we have considered, species affect each other's fitness through direct interactions. Species can also benefit through indirect interactions. For example, trees may alter the amount of sunlight that reaches the forest floor, creating conditions that benefit particular understory plants. Similarly, kelp forests in the oceans can create suitable habitat for other species. When one species creates an environment that helps another, the interaction is called **facilitation**. Facilitation benefits one species, with the other at least not harmed, so it can include both mutualisms and commensalisms.

This type of interaction can be more complex, as, for example, when two different species attack a third. Both fungi and beetles attack pine tree phloem, and the attack of each benefits the other. Bark beetles facilitate fungi that attack pine trees by providing entryways through their tunnels into the tree, while fungi facilitate bark beetles by lowering the amount of resinous defenses the pines produce. Like insects, bacteria and other pathogens usually attack only one particular type of host tissue, such as the gastrointestinal, respiratory, nervous, or circulatory system. Severe infections can sometimes render the hosts more susceptible to infection by other pathogens.

The costs and benefits of species interactions can change over time.

Associations are not fixed: they can change over time. A mutualism can, in some cases, become antagonistic if one of the partners "cheats" by imposing a larger cost than benefit on the other. In fact, mutualisms that are loose associations among changing partners can become one sided rather quickly. For example, many plants have evolved tubular flowers that guide bees past their anthers or stigma on the way to the nectar at the base. Navigating this route takes just a little more time for the bees, which try to visit as many flowers as possible. Some bees have short-circuited this plant mechanism by nipping the flower base from the outside and then drinking the nectar without pollinating the flower. Most plants have enough successful pollination to ensure seed production, but losses from cheating can still be costly.

Some interactions begin as a benefit to one species, with no benefit or cost to the other species. For example, cattle egrets follow water buffalo to pick up the insects stirred up by the buffalo as they pass. The egrets gain food, but the buffalo are not affected either way—an example of commensalism (**Fig. 45.11**).

FIG. 45.11 Changing costs and benefits. Egrets feed on insects stirred up by water buffalo, which are usually not affected by the egrets—a commensalism. If, however, the egrets spot lions and the poor-sighted buffalo take warning, the relationship becomes a mutualism.
Source: Biosphoto/Gérard Lacz.

However, cattle–egret commensalism can become mutualism over time if the egrets give early warnings of nearby predators such as lions or eat insects like tsetse flies that carry diseases harmful to the buffalo.

Self-Assessment Questions

6. What are some of the costs and benefits to an apple tree and to the honeybee that pollinates it? Which kind of interaction is this?

7. Give an example of an obligate mutualism. Why is it considered obligate?

45.4 COMMUNITIES

We have been discussing how species interact and how these interactions can shape the distributions of populations, limiting distributions in one way or another. Now we look collectively at all the species that occur together in the same place. How, and to what extent, do species interact? What are the properties of these interacting sets of species?

Species that live in the same place make up communities.

A **community** is the set of all populations of two or more different species found in a given place at a particular time. Populations in a community, whether a salt marsh, mountaintop, or desert, are connected by their various interactions as well as by their physical location. The populations in a local community share the same rainfall, climate, and soil, and often encounter one another physically as they grow, reproduce, and die. Some provide food or shelter for others, and all end up being recycled by the many small decomposers in the soil. We often characterize communities by the principal plants and animals they harbor, such as the cacti and lizards of deserts, the tiny, rock-loving mosses and marmots of a mountaintop, or the grasses and mosquitoes of a coastal salt marsh. Indeed, communities often seem quite distinctive to us and easily recognized. At first glance, it would even appear that the plants and animals and fungi that live in one kind of place might be competing for the same local resources.

When we look at the details of where particular species occur, almost no two species have exactly the same geographic distribution. Consider the bogs of northern New England, as shown in **Fig. 45.12**. This community is made up of Black Spruce, Bog Club Moss, and other bog-loving species of plants. They all occur in this area because of their mutual preference for the water-saturated, acidic, sandy soils found in northern bogs. But when we look more closely, we find that these different species of plants have only partially overlapping geographic ranges. Black Spruce populations occur from New England north through Canada, while Bog Club Moss populations occur as far south as Virginia. Even in one location, such as Victory Bog in northern Vermont, the Black Spruce and Bog Club Moss are not distributed in exactly the same way across the wet bog and surrounding woods. Nevertheless, we focus on the assemblage of species that occur in a particular place to discuss their interactions with each other.

The activities of different populations in a community vary in time as well as space. For example, flies are active during the day, pollinating flowers that open in the morning. Likewise, predators of the flies, such as dragonflies and birds, are active during the day. In contrast, moths are active at night, as are the spiders and bats that prey on moths. Similarly, in coastal marine communities, oysters, crabs, and many other animals follow the daily rise and fall of tides. And, of course, the activities of many species change seasonally in responses to sunlight, temperature, and rainfall. The protein-rich insects that birds prey on to feed their young are most abundant in the spring, and fruits and seeds rich in the oils that migratory birds need to supply fat reserves are mostly available in late summer and fall.

FIG. 45.12 A bog community. This community in northern New England consists of bog-loving species of plants, such as Black Spruce and Bog Club Moss. *Source: Erin Paul Donovan/Alamy.*

CASE 8 CONSERVING BIODIVERSITY: RAINFOREST AND CORAL REEF HOTSPOTS

How is biodiversity measured?

A key feature of any community is its **biodiversity**, or biological diversity. Biodiversity is often taken to mean the number of species, but we can also think of it more broadly and at many levels, including the variety of genetic sequences, cell types, metabolisms, species, life histories, phylogenetic groups, and communities. We can also consider biodiversity for a particular place, such as a single tropical rainforest on the island of Hispaniola, or the whole island of Hispaniola, or the entire planet.

Biodiversity is usually first measured by counting the number of species present in a particular area, known as the species richness. Because of limitations in time and expertise, scientists often focus on a subset of the taxonomic groups present. One approach is to start with the number of plant species. In terrestrial environments, the number of plant species is often a good predictor of total species diversity. For example, there are approximately four times as many animal species as plant species known on Earth. Therefore, scientists estimate the numbers of animal species by multiplying the number of plant species by 4. These animal species include arthropods and other small animals that live on or near the plants, birds and other vertebrates that prey on the insects and plants, and larger predators that prey on these birds and other vertebrates.

Another aspect of biodiversity is the relative proportions of individual species in a community, known as species evenness. Consider two communities of 100 individuals distributed across 10 different species of plants. The number of individuals of each species can differ greatly between the two communities. The first community might have 91 individuals of a single plant species and one individual each of 9 other plant species, while the second community might have 10 individuals of each of the 10 species of plants. The two communities have the same number of species, or species richness, but the second community is more even.

This idea of differences in evenness among communities can be measured and used to compare communities. Species evenness is an important measure because a community that is dominated by a single species may be more vulnerable to pathogens or unusual environmental conditions than a community in which every species is equally abundant. Of course, when determining evenness, the samples must be taken in the same way in the two places to avoid possible biases. Such measures of evenness, therefore, are most feasible and appropriate for local communities that can be sampled in a repeatable, consistent way in different places.

When we are comparing the biodiversity of regions or countries, however, we usually have only the most basic information on the total numbers of species, and for most kinds of organisms, we don't even have that. For most communities on the planet, the richness of their flowering plant species is among the best-known elements of biodiversity, rivaled only by knowledge of the birds or mammals. In contrast, the number of species of the main plant pollinators, consumers, and decomposers, and the insects and fungi in communities around the world, is poorly known. On the island of Hispaniola, for example, nearly every flowering plant species, approximately 6000 in all, is known (and so are the birds and mammals). Because scientists estimate that there are about two to three times as many kinds of insects as flowering plants at any given place on Earth, the number of insect species on Hispaniola will likely total 12,000 to 18,000 once all have been identified. Indeed, each year new studies of different insect groups are published that report a doubling or tripling of the known species. For example, recent intensive studies of fireflies on Hispaniola reported a new total of 69 species, up from the 17 previously known.

Tropical islands such as Hispaniola, Hawaii, and the Galápagos are also missing elements of biodiversity because certain species have not been able to migrate there from the nearest mainland. For example, only a few of the many South American frog and lizard groups are found on Hispaniola, and only large iguanas made it to the Galápagos, which are even farther away from the mainland. Thus, the ability of organisms to disperse from one place to another influences biological diversity by limiting the kinds of immigrants.

Species influence each other in a complex web of interactions.

The interactions among all the populations in a given habitat define the nature of that community. In habitats with many different predator and prey species, identifying the ways that each species affects others can be challenging. For this reason, much research on how predators and prey influence community structure has been carried out in the Arctic, where communities tend to have relatively few species.

One well-studied community is on Bylot Island in the Canadian territory of Nunavut. There, lemmings and Snow Geese are the main herbivores, eating grasses and other low plants that grow on the island. The main predators of lemmings and Snow Geese are Arctic Foxes and Snowy Owls. Lemmings, foxes, owls, geese, and plants are linked by a set of interactions illustrated in **Fig. 45.13**. Lemming populations rise and fall every 4 years or so. When lemming populations decline, foxes turn to other prey, including Snow Geese. The population density of Snow Geese, therefore, is indirectly influenced by the population density of lemmings. When the foxes focus on the Snow Geese as a food source, the now harder-to-find lemmings escape, allowing lemming populations to increase again. Snowy Owls chase the foxes away, which in turn indirectly benefits snow geese, which are preyed upon by foxes when lemming populations are low.

FIG. 45.13 Predator–prey relationships in the Arctic grassland community on Bylot Island, Canada. Species have either a positive effect (+) or negative effect (–) on other species.

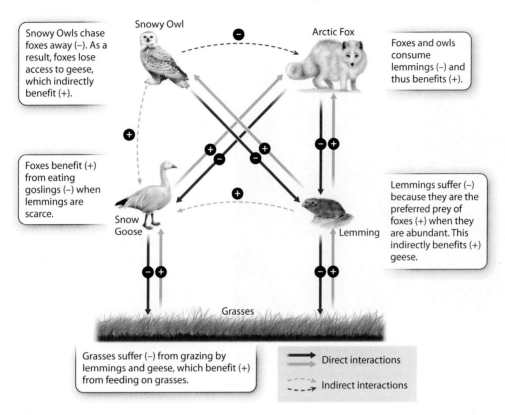

Keystone species have disproportionate effects on communities.

While all or nearly all species in a given place have some effect on one another, the integrity of a community may depend on a single species. This situation can occur when one species influences the interactions of species in the community downstream of its own, or when the species modifies its physical environment, as beavers do when they make ponds. Pivotal populations are called **keystone species** because they support a community in much the same way that an architectural keystone supports an arch. Although all species contribute to the structure of a community, a keystone species affects other members of the community in ways that are disproportionate to its abundance or biomass.

A classic experiment by American zoologist Robert Paine in the 1960s gave birth to the concept of keystone species. Along the northwestern coast of the United States, mussels adhere to rocks in the intertidal zone, where they coexist with about two dozen other species of invertebrate animals and seaweeds. Mussel populations are regulated by their principal predator, a sea star called *Pisaster ochraceus*. When Paine removed the sea star from experimental plots, mussel populations exploded, outcompeting other animals and seaweeds for space and so eliminating the other animals and plants from the rocky shoreline. Control plots in which the sea star remained present showed no such change. From these results, Paine concluded that the activities of the predatory sea star had a singularly strong influence on the diversity and distribution of other populations in the rocky intertidal community. *Pisaster ochraceus* is a keystone species.

Another experiment, this time unintended, drives home the point. Along the west coast of the United States and Canada, giant kelp grow more than 50 m tall, providing a home for a remarkable diversity of fish and invertebrates (**Fig. 45.14**). The kelp community depends strongly on a single predator: the sea otter. Sea otters prey on sea urchins, which graze on the rootlike holdfasts that anchor kelp to the seafloor. When the otters were removed by overhunting decades ago, sea urchin abundance soared and the kelp nearly disappeared, along with many other populations that live in kelp forests. Now, in areas where otters are protected, they again keep urchin populations in check, enabling the kelp to return, and along with it, diverse fish and animal species.

Keystone species called ecosystem engineers actively shape the physical environment, creating habitat for others. The classic example is beavers, which produce ponds by damming streams in forests, creating habitat that would not otherwise exist (**Fig. 45.15**). Hippopotamuses provide another only recently

FIG. 45.14 Keystone species. The activities of keystone species, such as otters in Pacific kelp forests shown here, strongly influence the communities in which those species live. *Source: Carl Gwinn, blackcormorant.net.*

Sea otter predation is key to the health of a kelp forest. Notice the many fish beneath the kelp canopy.

When otters are absent, their sea urchin prey rise in abundance and consume most of the kelp holdfasts. Without kelp forests, many animals decline in abundance.

FIG. 45.15 An ecosystem architect's effect on the environment. This satellite image shows the world's largest beaver dam in Alberta, Canada, which has flooded the surrounding area, creating a new habitat. *Source: DigitalGlobe/ScapeWare3d/Getty Images.*

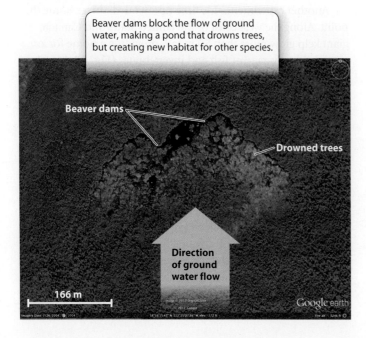

Beaver dams block the flow of ground water, making a pond that drowns trees, but creating new habitat for other species.

recognized example of this role. In East Africa, hippos loll about in rivers and lakes during the day, but venture onto the land at night to feed on grasses and other low-lying plants. The hippos return to the water before the sun rises and—importantly—defecate in their watery home. In this way, hippos transfer a substantial amount of nutrients from land to river, and these nutrients support a community that is more abundant and diverse than would be possible in the absence of hippos.

Disturbance can modify community composition.

Earlier, we discussed how predation, herbivory, and parasitism can reduce the effects of competition (section 45.2). Severe physical impacts on a habitat, such as those caused by storms, earthquakes, or road building, can also dramatically lower the abundances of some species. For example, fire may remove the grasses and trees in a woodland, with the species that depend on them, such as deer and birds, then disappearing. Such events are known as disturbances, and they have effects on populations of interacting species that are independent of their densities, which we discussed in Chapter 44. Disturbances often affect multiple species in the same community, thereby exerting a strong influence over community composition. We usually think of disturbances as abiotic, but the line is not so clear when we consider road building by humans, for example, or dam building by beavers.

The impact of severe weather in forests demonstrates how disturbance can affect a community. When wind takes down tall trees in a forest, it also takes away the shade that had prevented some species from establishing new seedlings. Fires are frequent sources of disturbance in regions with a seasonally dry climate, such as parts of western North America and southwestern Australia. Like storms, fires remove existing plants and animals, opening up habitat for recolonization. Some species have adapted to frequent fires. For example, several pine species have evolved thick bark and high branches, traits that help protect them from fire. Other pines, which have evolved cones that open only in the wake of fire, have adapted to use the habitat space opened up in the aftermath of the fire and the nutrients that dead plant matter releases into the soil.

Noting the influence of storms on community composition in forests and coral reefs, American ecologist Joseph Connell proposed in the 1970s that the diversity of species in a given community reflects, at least in part, the frequency and intensity of disturbance. If disturbance is intense, few species can tolerate the physical conditions of the habitat and diversity is low. By comparison, if disturbance is rare or weak, competition may take over and only a few, stronger, competitors may remain in the community. Species diversity tends to be highest when disturbance is frequent or intense enough to inhibit competition, but not so strong as to limit the number of species that can tolerate the environment.

Succession describes the community response to new habitats or disturbance.

Following disturbance or the appearance of a new habitat, a community goes through a series of changes. These changes involve a predictable sequence of colonizations by species that transform the community, in what can appear to be a linear process of maturation. Each species modifies the habitat by affecting the environment or changing the community structure, creating opportunities for other kinds of species to colonize. The pace of this transformation may be rapid, occurring over a few decades, or it may be much slower, depending on the environment.

This process of species replacing each other in time in a predictable sequence is called **succession** (Fig. 45.16). Ecologists sometimes differentiate between the colonization of a new and previously uninhabited area (such as land exposed by a retreating glacier, an important factor following the last ice age, and again today), known as primary succession, and the reestablishment of communities following disturbance, known as secondary succession. Whether succession is primary or secondary, the succession of species reflects the adaptations of each species to the changes in sunlight, soil, water, and space that occur as the first organisms arrive, mature, and die. These organisms influence their habitat in both life and death, whether aquatic or terrestrial.

Succession over thousands of years can turn bare rock into forest. Lichens are often the first organisms to colonize the bare rock newly formed by cooling lava or exposed by landslides. Initially, only a few mites and other tiny invertebrates may live in the lichen. As the stones weather, and microorganisms feed on the lichen and on one another, they form particles of soil that accumulate in cracks and crevices. Eventually, small plants, such as sedges and grasses, take root, and the growing plant community begins to gain height and diversity as other plants establish themselves.

Soon, larger insects, spiders, and other invertebrates join the community, as well as the first small mammals and birds. Many of these early colonizing species are r-strategists (Chapter 44), growing and reproducing quickly, and spreading their young far and wide. Birds carry the seeds of shrubs and other plants in their droppings, and the gradually deepening soil eventually permits the establishment of shrubs that in turn provide habitats for larger mammals and birds, and perhaps snakes and salamanders.

The first sun-loving trees, such as birches and pines, appear next, providing shade for the seedlings of later colonizing trees such as oaks and maples. In many forests, stable, long-lived K-strategists (Chapter 44) make up a final stage in the succession, forming a mature assembly commonly called a climax community. A **climax community** is one in which there is little further change in species composition.

The Pin Cherry, a resident of New England forests, illustrates how the life histories of individual species can contribute to succession. If you walk through a forest in New Hampshire, you will see oak trees and beeches, pine and hemlock. A mature forest may contain few or no Pin Cherry trees. However, if you examine seeds lying dormant within the forest soil, you will see numerous Pin Cherry seeds. How did they get there?

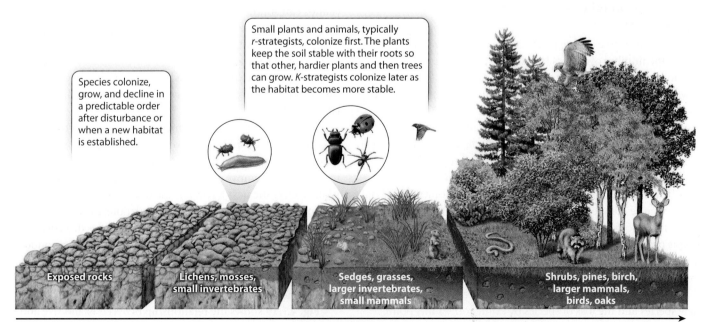

FIG. 45.16 Succession. In new habitats or following disturbance, species colonize and transform the habitat in a predictable sequence that reflects the adaptations of participating species.

Years earlier, disturbance from a storm or fire opened up the forest floor in these areas. Pin Cherries established themselves in the now-sunlit soils, growing from seeds left in the droppings of birds that had consumed the cherries elsewhere. The seedlings of Pin Cherries grow well in full sunlight, so the trees grew rapidly to reproductive maturity. Pin Cherries are considered r-strategists because they thrive in unpredictable openings in the forest, with few competitors, and reproduce early in life. Eventually, though, Pin Cherries and other pioneering species such as birches are shaded out by K-strategists such as oaks and beeches.

Once a forest canopy is established, Pin Cherries disappear because their seeds do not germinate in the shade, although the seeds remain on the forest floor. In contrast, oak seedlings prosper in shade. Over time, oaks and other shade-loving species come to dominate this patch of forest, forming a climax community that other tree seedlings cannot enter. They will continue to dominate it until the next disturbance, when the removal of mature trees will once more expose the forest floor to bright sunlight, literally giving Pin Cherries another day in the sun.

Each stage of plant succession is accompanied by changes in the animals (and fungi) associated with them and the habitat they create. The pioneering Pin Cherry is colonized quickly by tent caterpillars that cannot feed on the later climax species of oaks or beeches. Later, acorns and beech nuts support rodents such as squirrels and mice, as well as nut weevils, and the large trunks of these trees provide homes for woodpeckers and other species that cannot nest in small Pin Cherry trees. Underground, the fungi associated with the roots of these different trees also undergo succession in parallel, as do the insects that feed on them. Entire assemblages of species turn over as sunlight and other physical factors change along with the succession of species that depend on them and each other.

Island biogeography explains species diversity on habitat islands.

Islands allow us to examine patterns of ecological succession as well as the number of species (the species richness) in a particular area. Species diversity of an island reflects both the rate at which new species arrive on the island and the rate at which species already on the island become extinct (**Fig. 45.17**). As species arrive on an initially uninhabited island, the colonization rate is high because there are no predators yet and the small number of colonists means that there is little competition for resources. Over time, as more species arrive and successfully establish populations on the island, the rate of colonization slows because competition and predation increase, making successful colonization of newly arrived species more difficult. At this point, there is also a smaller number of species left in the pool of potential colonists. In contrast, the extinction rate increases as newly arrived species compete.

At some point, a balance is achieved between the arrival of new species and the loss of species by extinction. At this point, diversity is said to be at equilibrium (Fig. 45.17). This does not

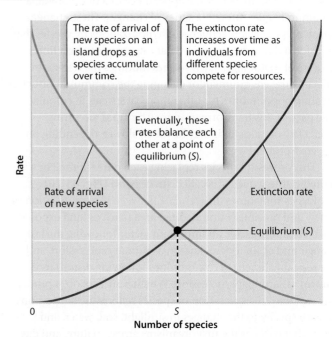

FIG. 45.17 Island species diversity, a balance of colonization and extinction.

mean that no new species will colonize the island, but instead that the establishment of a new colonist species will be matched by the extinction of one already in place. There is generally turnover in species composition through time.

The **theory of island biogeography**, articulated in 1967 by the American biologists Robert MacArthur and Edward O. Wilson, states that the number of species that can occupy an island depends on two factors (**Fig. 45.18**). The first factor is the size of the island. Because of their size, larger islands receive more colonists than smaller ones. Furthermore, larger islands can support more species than smaller ones can. For these reasons, the equilibrium number of species for larger islands is greater than that for smaller islands (Fig. 45.18a). The second factor is the distance of the island from a source of colonists. More distant islands have lower rates of colonization, reducing the number of species at equilibrium (Fig. 45.18b).

Wilson and his student Daniel Simberloff tested the theory of island biogeography in a novel way. Working among small mangrove islands off the Florida Keys, they used short-term insecticides to remove all the arthropods living on individual islands. Over the next few years, the researchers followed ecological succession as new arthropod populations established themselves on the islands. They found that the species diversity of islands differing in size and distance from shore corresponded to those predicted by the theory. Larger and closer islands had more successful colonists, whereas smaller and more distant islands had fewer colonists.

Fig. 45.19 shows the relationship between island size and species number. Similar relationships between island size and

FIG. 45.18 Effect of (a) island size and (b) distance from source populations on the equilibrium number of species on an island.

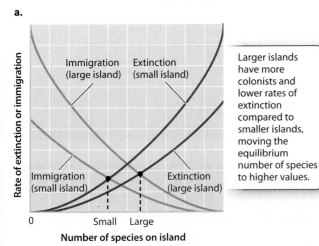

a.

Larger islands have more colonists and lower rates of extinction compared to smaller islands, moving the equilibrium number of species to higher values.

b.

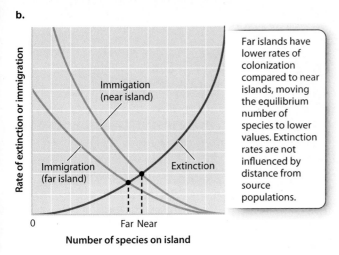

Far islands have lower rates of colonization compared to near islands, moving the equilibrium number of species to lower values. Extinction rates are not influenced by distance from source populations.

FIG. 45.19 Island size and number of species. The diversity of amphibian and reptile species on Caribbean islands fits the predictions of island biogeography theory. *After R. H. MacArthur and E. O. Wilson, 1967, The Theory of Island Biogeography, Princeton, NJ: Princeton University Press.*

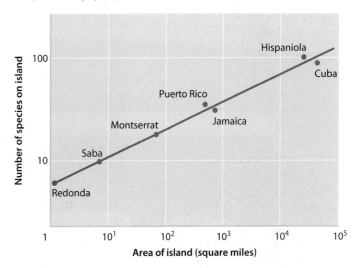

equilibrium species diversity have been observed repeatedly in various locations. The relationship is described mathematically as a **species–area relationship**:

$$S = cA^x$$

where S is the number of species at equilibrium, c is a mathematical constant that depends on the species in question, A is the habitat area, and x is an experimentally determined exponent that relates island size (A) to the number of species (S) on the island and is between 0.1 and 0.4. The observation that the value of x falls in this narrow range indicates not only that larger islands support more species than smaller ones, but also that there is a quantitative, consistent relationship between island size and species number.

This mathematical formula translates into a useful back-of-the-envelope relationship: a ten-fold increase in island size results in a doubling of species diversity. Perhaps more relevant for conservation, a 90% reduction of habitat area results in a 50% loss of species diversity if no other factors are taken into account. The species–area relationship applies to habitat islands as well as true islands, and it holds broadly for areas as large as continents. This relationship has many applications in conservation biology. For example, it is used to determine how large a protected area must be to support some species and how close reserves must be to one another to ensure the movement of at least some individuals between them.

Many oceanic islands, such as Hawaii and the Galápagos, are volcanic in origin, so they began with no organisms at all. Their histories provide models both for habitat islands and for ecological succession. We follow changes over time in an aquatic and terrestrial habitat to summarize key concepts about population growth, species interactions, and communities in **Fig. 45.20**. We also introduce ways in which humans affect these processes, a topic that we revisit in Chapter 48.

Self-Assessment Questions

8. Lemmings affect many other species in the Bylot Island community. Why are lemmings not considered a keystone species?

9. What are three factors that help determine the species composition of a community?

10. How can a physical disturbance, such as a drought, affect community composition?

11. What is meant by "ecological succession"? Give an example.

12. Which factors determine the diversity of species on a habitat island? What is the relevance of these factors in managing the habitat of an endangered species?

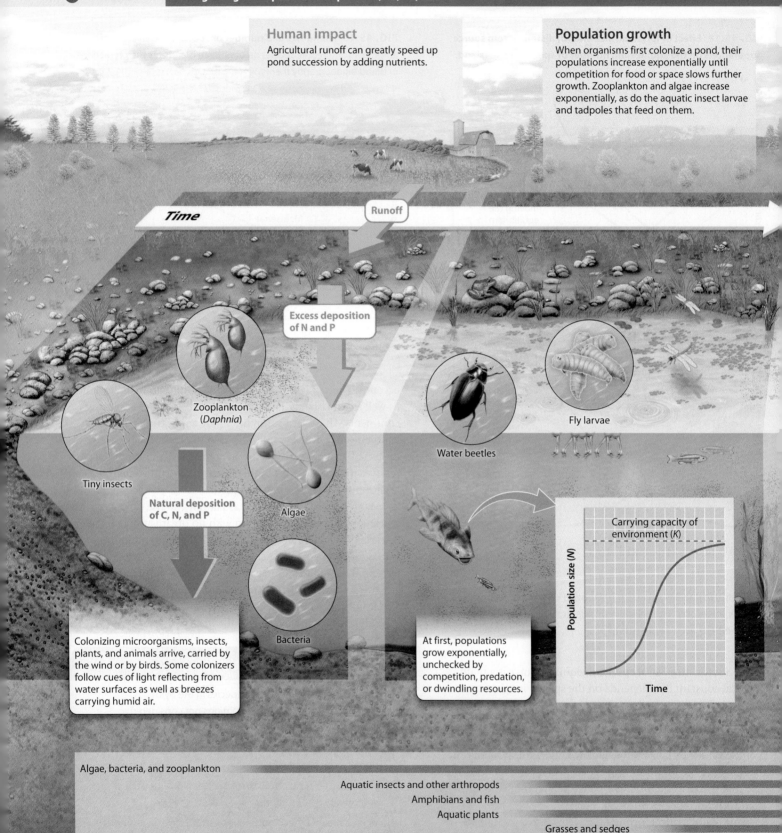

Succession: Ecology in Microcosm
VISUAL SYNTHESIS — FIG. 45.20
Integrating concepts from Chapters 44, 45, 46, and 48

Human impact
Agricultural runoff can greatly speed up pond succession by adding nutrients.

Population growth
When organisms first colonize a pond, their populations increase exponentially until competition for food or space slows further growth. Zooplankton and algae increase exponentially, as do the aquatic insect larvae and tadpoles that feed on them.

Colonizing microorganisms, insects, plants, and animals arrive, carried by the wind or by birds. Some colonizers follow cues of light reflecting from water surfaces as well as breezes carrying humid air.

At first, populations grow exponentially, unchecked by competition, predation, or dwindling resources.

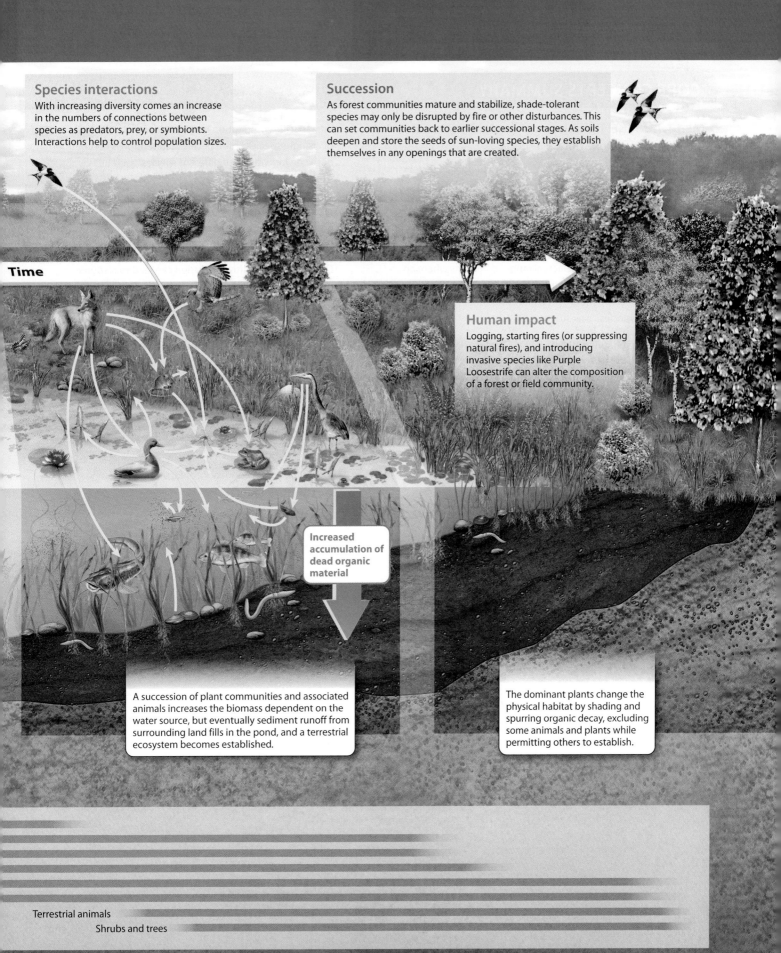

CORE CONCEPTS SUMMARY

45.1 THE NICHE: The niche encompasses the physical habitat and ecological role of a species.

The niche reflects the biological and nonbiological factors that determine where an organism lives and how it functions. page 1019

The fundamental niche is the full range of physical conditions and food resources that permit an organism to live. page 1020

The realized niche is the actual range of habitats where an organism lives. It is often smaller than the fundamental niche because of competition, predation, and other interactions between species. page 1020

Many closely related species have broadly similar niches as a result of their shared evolutionary history. page 1021

45.2 ANTAGONISTIC INTERACTIONS: Competition, predation, parasitism, and herbivory are interactions in which at least one species is harmed.

Competition is a form of interaction in which two individuals of the same or different species use the same limited resource. page 1022

Intraspecific competition is competition between individuals of a single species; interspecific competition is competition between individuals of different species. page 1022

Individuals can compete for food, space, mates, or other resources. page 1022

Competitive exclusion prevents two species from occupying the same niche at the same time. page 1022

Over time, competitive exclusion can lead to resource partitioning, in which two species evolve in such a way that they occupy different niches. page 1023

Predation, parasitism, and herbivory are interactions in which one species benefits and one is harmed. page 1024

45.3 MUTUALISTIC INTERACTIONS: Mutualisms are interactions in which both species benefit.

In mutualism, the benefits to both species outweigh the costs of participation. page 1026

Mutualisms can be obligate, meaning that they are required for survival or reproduction, or facultative, meaning that they are not required. page 1029

Obligate mutualisms can strongly influence the evolution of both partners, whose lineages may remain associated through their respective joint speciation and diversification. page 1029

Interactions in which one species benefits and the other is unaffected are commensalisms. page 1029

Facilitation is a type of indirect interaction where one species creates a suitable habitat for another. page 1030

Species interactions can change over time. For example, partners can become parasites. page 1030

45.4 COMMUNITIES: Communities are composed of local populations of multiple species that may interact with each other.

An ecological community consists of all the organisms living in one place at a given time. page 1031

Biodiversity, or biological diversity, includes the diversity of genetic sequences, cell types, metabolisms, species, life histories, phylogenetic groups, and communities. page 1032

Biodiversity is often measured by the number of species present in a given area. page 1032

Populations of consumers and producers may affect each other directly through the act of consumption or indirectly by influencing competition and other kinds of interactions. page 1032

Keystone species have a large influence on the composition of a community disproportionate to their numbers. page 1033

Physical disturbances, such as a storm or drought, can reshape communities. page 1034

Succession is the predictable order of species colonization and replacement in a new or newly disturbed patch of habitat. page 1035

Species diversity on habitat islands reflects both the rate at which new species arrive and the rate at which colonist species become extinct. page 1036

The theory of island biogeography proposes that the number of species that can occupy an island depends on the size of the island and its distance from the mainland. page 1036

Log in to **LaunchPad** to check your answers to the Self-Assessment Questions and to access additional learning tools.

CHAPTER 46　Ecosystem Ecology

CORE CONCEPTS

46.1 THE SHORT-TERM CARBON CYCLE: Photosynthesis and respiration are the key biochemical pathways for the biological, or short-term, carbon cycle.

46.2 THE LONG-TERM CARBON CYCLE: Physical processes govern the carbon cycle on longer timescales.

46.3 FOOD WEBS AND TROPHIC PYRAMIDS: Species interactions cycle carbon and transfer energy through ecosystems.

46.4 OTHER BIOGEOCHEMICAL CYCLES: Nitrogen and phosphorus also cycle through ecosystems.

46.5 THE ECOLOGICAL FRAMEWORK OF BIODIVERSITY: Biogeochemical cycles help us understand the evolution of biological diversity.

In previous chapters, we have seen how the functional characteristics of different species link them together in ecological interactions. Species compete with one another, prey on each other, and respond in varying ways to disturbance. All of these interactions help to shape communities, the associations of species that live together at a given time in a given place. But populations do not interact only with other populations; they also interact with their physical environment in intimate ways. An **ecosystem** consists of the organisms in a community and their physical environment, with each influencing the other through time. Ecosystem ecology, then, explores the processes that link the physical and biological Earth, both the ways that physical environments shape communities and the ways that organisms affect the physical environment. Biogeochemical cycles, the pathways by which biologically important elements move back and forth between Earth and life, provide a fundamental illustration of ecosystems at work.

46.1 THE SHORT-TERM CARBON CYCLE

Let's imagine that we could tag a carbon atom at its moment of origin and then follow its odyssey through time and space. Formed in a nuclear blast furnace deep within an ancient star and then ejected into space as the star exploded in death, our atom was eventually swept up with other materials to form the Earth, a small planet orbiting the newer star we now call the sun. Volcanoes introduced our carbon atom into the early atmosphere as carbon dioxide (CO_2). Slowly, over millions of years, this CO_2 reacted chemically with water and rocks, transferring the carbon to limestone that accumulated on the seafloor. Here, our atom sat for many millions of years, until earthquakes, erosion, or other geologic activities returned it to the atmosphere as, once again, CO_2. Slowly but surely, geologic processes on the early Earth cycled carbon from atmosphere to rocks, and back again.

Sometime between 4 and 3.5 billion years ago, our carbon atom began to cycle more rapidly—much more rapidly. Communities of interacting microorganisms began to cycle carbon through metabolism, with photosynthetic microorganisms converting CO_2 into organic molecules, and respiring microorganisms returning CO_2 to the environment. To this day, the biological carbon cycle endlessly propels our atomic wayfarer from atmosphere and oceans to cells and back again, while the slow burial of organic matter and limestone in sedimentary rocks continues to bring carbon atoms to the solid Earth, returning them to the atmosphere only slowly, on geologic timescales.

The **carbon cycle** is the intricately linked network of biological and physical processes that shuttles carbon among rocks, soil, oceans, air, and organisms. Why focus on carbon? Because the carbon cycle provides an organizing principle for understanding life on Earth. The chemistry of life is, in no small part, the chemistry of carbon. When we discuss primary producers and consumers, predators and prey, or decomposers, we are really talking about how carbon moves from one species to another, and between organisms and their surrounding environment. Photosynthetic organisms convert the energy of the sun to chemical energy in carbon molecules, and consumers gain energy by eating those molecules. Thus, the carbon cycle also traces the transfer of energy through ecosystems. In short, the cycling of carbon and other biologically important elements underpins both the efficient functioning of ecosystems and their persistence over an immense span of time. The carbon cycle, therefore, focuses our attention on the ways that physical and biological processes together determine the properties of ecosystems. In addition, it provides a basis for assessing the role that humans play in our environmental present and future.

The Keeling curve records changing levels of carbon dioxide in the atmosphere over time.

In 1958, American chemist Charles David Keeling began a novel program to monitor the atmosphere. At the time he began his study, scientists had only a vague notion of how CO_2 behaves in air. Many thought that the CO_2 concentration might vary unpredictably from time to time, and from place to place. Keeling decided to find out if, in fact, it did.

From five towers set high on Mauna Loa in Hawaii, 3400 m above sea level, Keeling sampled the atmosphere every hour and measured its composition with an infrared gas analyzer. Within the first few years of his project, a pattern of seasonal oscillation became apparent: CO_2 concentration in the air reached its annual high point in spring, but then declined by about 6 parts per million (ppm, by volume) to a minimum in early fall (**Fig. 46.1**). Such regular variation was unexpected: what could be causing it?

Perhaps the pattern was local, the researchers thought. Airplane traffic to Hawaii peaked in winter, and maybe that was affecting the air around Mauna Loa. As measurements continued through a number of years, however, the pattern persisted, even as trends in air traffic changed. Moreover, monitoring stations in many other parts of the globe gave similar results. Knowing that the Mauna Loa measurements are representative of the atmosphere as a whole allows us to appreciate what an annual variation of 6 ppm really means. With calendar-like regularity, approximately 47 billion metric tons of CO_2 enters and leaves the atmosphere annually—that's 13 billion metric tons of the element carbon (the rest of the mass is oxygen). To explain such a pattern, we must consider processes that affect the planet as a whole.

A second pattern is evident in Fig. 46.1. Summertime removal of CO_2 from the atmosphere does not quite balance winter increase, with the result that the amount of CO_2 in the atmosphere on any given date is greater than it was a year earlier. Atmospheric CO_2 levels have increased steadily through the period of monitoring. For example, in 1958 the maximum monthly value recorded for CO_2 was 317 ppm; in April 2018 the CO_2 level reached 410 ppm. This is an increase of about 30%, and the rise does not appear poised to slow or stop any time soon.

What causes atmospheric CO_2 to vary through the year and over a timescale of decades? To answer this question, we need to ask which processes introduce CO_2 into the atmosphere and which processes remove it. Carbon dioxide is added to the atmosphere by (1) geologic inputs, mainly from volcanoes and mid-ocean ridges, where CO_2 is emitted as new seafloor forms; (2) biological inputs, especially respiration; and (3) human activities, including deforestation and the burning of fossil fuels.

Processes that remove CO_2 from the atmosphere include (1) geologic removal, especially by chemical weathering, in which CO_2 in rainwater reacts with exposed rocks, and (2) biological removal, mainly through photosynthesis, in which CO_2 is turned into organic molecules used for growth and reproduction. As we will see in the next section, geologic processes play key roles in the carbon cycle on timescales of centuries and longer. Only biology, however, can account for the annual rhythm of the Keeling curve.

Photosynthesis and respiration are key processes in short-term carbon cycling.

To grow, all organisms require both carbon and energy. As we saw in Chapter 8, photosynthetic organisms use energy from the sun to form ATP and NADPH, which reduce CO_2 to carbohydrates. As photosynthesis pulls CO_2 out of the atmosphere (or water, for aquatic organisms), the carbon is transferred to carbohydrates in the Calvin cycle, and oxygen is given off as a by-product. The overall chemical reaction is shown here:

$$6CO_2 + 6H_2O \rightarrow C_6H_{12}O_6 + 6O_2$$

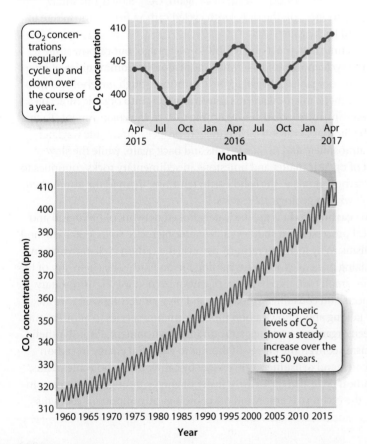

FIG. 46.1 The Keeling curve. The Keeling curve provides a record of atmospheric CO_2 concentrations over half a century, measured from an observatory at Mauna Loa, Hawaii. *Data from Dr. Pieter Tans, NOAA/ESRL (www.esrl.noaa.gov/gmd/ccgg/trends/) and Dr. Ralph Keeling, Scripps Institution of Oceanography (scrippsco2.ucsd.edu/).*

It is estimated that photosynthetic organisms remove approximately 210 billion metric tons of carbon (in nearly 770 billion metric tons of CO_2) from the atmosphere each year, with land plants removing approximately half of this mass and phytoplankton and seaweeds in the oceans removing the rest. This annual photosynthetic removal of CO_2 equals approximately 25% of the total CO_2 in the atmosphere. In the absence of processes that restore CO_2 to the atmosphere, high rates of photosynthesis would use up the atmospheric reservoir of CO_2 in just a few years. But the Keeling curve tells us that atmospheric CO_2 is not decreasing rapidly; instead, it is rising. Moreover, the geologic record indicates that photosynthesis has fueled ecosystems for billions of years without causing catastrophic CO_2 depletion.

Which processes return CO_2 to the atmosphere, balancing photosynthetic removal? In Chapter 7, we discussed aerobic respiration. Humans and many other organisms gain both the energy and carbon needed for growth from organic molecules (that is, by consuming food). As explained in Chapter 7, aerobic respiration uses oxygen to oxidize organic molecules to CO_2, converting chemical energy in the organic molecules to ATP for use in cellular processes. Aerobic respiration can be summarized by this chemical equation:

$$C_6H_{12}O_6 + 6O_2 \rightarrow 6CO_2 + 6H_2O$$

Plants respire, too, thereby converting the chemical energy in carbohydrates into useful work. On the present-day Earth, the amount of CO_2 returned annually to the atmosphere by aerobic respiration and related processes is about equal to the amount that plants and algae remove by photosynthesis. That is why photosynthesis does not deplete the CO_2 in air.

As we discussed in Chapter 8, the equations for oxygenic photosynthesis and aerobic respiration mirror each other. Photosynthesis continually converts CO_2 to organic molecules, using energy from the sun, and respiration harvests the chemical energy stored in organic molecules, releasing CO_2 and water in the process. Photosynthesis and aerobic respiration are complementary metabolic pathways: each uses the products of the other and generates a new supply of substrates for its complement. Anoxygenic photosynthesis and anaerobic respiration also form complementary metabolisms. The result is a cycle, called the short-term, or biological, carbon cycle (**Fig. 46.2**).

The regular oscillation of CO_2 reflects the seasonality of photosynthesis in the Northern Hemisphere.

Now that we have identified the major processes in the short-term (days to decades) carbon cycle, we can examine what makes atmospheric CO_2 level oscillate seasonally. Except in a narrow band around the equator, photosynthesis is seasonal: it occurs at higher rates in the summer and at lower rates in the winter, when many plants go dormant and deciduous trees shed their leaves. Respiration, in contrast, remains more or less constant through the year.

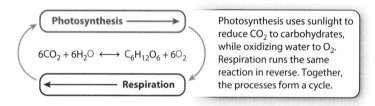

FIG. 46.2 The short-term carbon cycle. The complementary metabolic processes of photosynthesis and respiration drive the short-term cycling of carbon through the biosphere.

Photosynthesis uses sunlight to reduce CO_2 to carbohydrates, while oxidizing water to O_2. Respiration runs the same reaction in reverse. Together, the processes form a cycle.

Even a quick look at the globe shows that land is distributed asymmetrically, with more land—and hence more plants—in the Northern Hemisphere than in the Southern Hemisphere. For this reason, global atmospheric CO_2 declines through the northern summer, when rates of photosynthesis are highest relative to respiration, and then increases through fall and winter, when the ratio of photosynthesis to respiration is reversed. The result is the seasonal oscillation of atmospheric CO_2 levels documented by Keeling.

Human activities play an important role in the modern carbon cycle.

Now let's consider the second major pattern evident in the Keeling curve. Although CO_2 levels oscillate on an annual basis, the overall pattern is one of sustained increase. During the 1960s, the observed increase was less than 1 ppm each year. In contrast, in 2016, atmospheric CO_2 increased by 3 ppm. The current level of atmospheric CO_2 is 30% higher than it was 60 years ago. Why has CO_2 input into the atmosphere outstripped removal for the past half century? Should we consider this pattern unusual?

Answering the second question helps us to address the first. The question of whether it is unusual for atmospheric CO_2 to increase 30% in 60 years can be answered only if we can compare the Keeling curve with longer records of atmospheric CO_2 levels. Before Keeling began his monitoring, researchers had not systematically collected air samples. Fortunately, however, nature did it for us.

When glacial ice forms, it traps tiny bubbles of air. Each year, snowfall gives rise to a new layer of ice, and as layers accumulate through time, their bubbles preserve a history of the atmosphere. The graph in **Fig. 46.3** shows the amount of CO_2 in air bubbles trapped in Antarctic ice that has accumulated over the past 1000 years. Ice samples show that CO_2 levels actually began to increase slowly in the 1800s, the time of major transformations in mining, manufacturing, and transportation we call the Industrial Revolution. Before that period, atmospheric CO_2 levels had varied little since the Middle Ages, staying at 270–280 ppm for centuries on end. On a 1000-year timescale, then, big changes in CO_2 abundance happened only over the last 200 years. Atmospheric CO_2 inputs and outputs were approximately in balance until the Industrial Revolution.

HOW DO WE KNOW?

FIG. 46.3

How much CO_2 was in the atmosphere 1000 years ago?

BACKGROUND Direct measurements of the atmosphere show that CO_2 has increased by nearly 30% over the past 60 years. Is this pattern representative of atmospheric change over the past millennium, or is the pattern recorded by the Keeling and his successors unusual?

OBSERVATION Scientists recognized that the snow that accumulates on glaciers to form new layers of ice is initially full of small spaces that are in contact with the air. Through time, as snow changes to ice, these spaces become sealed off from the surrounding environment and form bubbles, preserving a sample of air at the time of ice formation. Cores drilled through glacial ice, then, provide an opportunity to view the Keeling curve within a longer framework of time.

Ice core drilled from a glacier in Antarctica

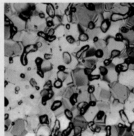
Viewed under the microscope, glacial ice contains small bubbles that trap samples of air.

Photo sources: (left) Courtesy of Ted Scambos and Rob Bauer, National Snow and Ice Data Center, University of Colorado, Boulder; (right) Courtesy of Vin Morgan.

MEASUREMENTS By identifying chemical indicators that vary on an annual basis in successive layers of ice, scientists were able to assign ages to ice samples taken from cores drilled through the Law Dome, a large ice dome in Antarctica. The youngest ice samples overlap in age with the first part of the Keeling data, providing a check that air samples trapped in ice have CO_2 concentrations similar to those in air measured directly over the same interval.

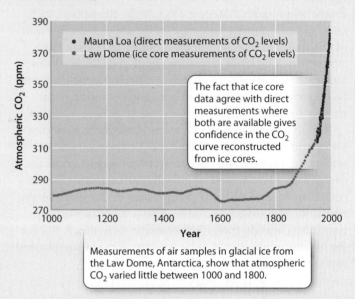

The fact that ice core data agree with direct measurements where both are available gives confidence in the CO_2 curve reconstructed from ice cores.

Measurements of air samples in glacial ice from the Law Dome, Antarctica, show that atmospheric CO_2 varied little between 1000 and 1800.

CONCLUSIONS Before the Industrial Revolution, atmospheric CO_2 levels had varied little over 1000 years, generally ranging between 270 and 280 ppm. From this, scientists have concluded that current changes in atmospheric CO_2 are unusual on the timescale of the past millennium.

SOURCE Etheridge, D. M., et al. 1996. "Natural and Anthropogenic Changes in Atmospheric CO_2 over the Last 1000 Years from Air in Antarctic Ice and Firn." *Journal of Geophysical Research* 101:4115–4128.

Fig. 46.3 shows that there is a **correlation** between the increasing CO_2 content of the atmosphere and human activities. A correlation simply indicates that two events or processes occur together. But can we actually demonstrate that humans have played a role in recent atmospheric change? By itself, correlation does not establish **causation**, a relationship in which one event leads to another. We might propose instead that recent CO_2 increase reflects natural processes and matches the period of industrialization just by coincidence.

Carbon isotopes show that much of the CO_2 added to air over the past half century comes from burning fossil fuels.

We can test the hypothesis that human activities have contributed to the increases in atmospheric CO_2 measured at Mauna Loa and in ice cores by making careful measurements of a chemical detail: the isotopic composition of atmospheric CO_2. As discussed in Chapter 2, many elements have several isotopes, atoms of the element that vary in atomic mass because they have different numbers of neutrons. Carbon has three isotopes: ^{12}C (with six neutrons, accounting for about 99% of all carbon atoms), ^{13}C (with seven neutrons, accounting for most of the remaining 1%), and the rare ^{14}C (with eight neutrons, about one part per trillion of atmospheric carbon).

More than 40 years ago, the Austrian-born chemist Hans Suess began to measure the relative abundances of the three isotopes of carbon in atmospheric CO_2, a program that continues today. He observed that the proportion of ^{13}C in atmospheric CO_2 has declined as the total amount of CO_2 has increased (**Fig. 46.4**). This subtle change in isotopic composition has occurred because the CO_2 now being added to the atmosphere has less ^{13}C than the CO_2 already in the air.

HOW DO WE KNOW?

FIG. 46.4

What is the major source of the CO_2 that has accumulated in Earth's atmosphere over the past two centuries?

BACKGROUND Chemical analyses show that CO_2 levels in the current atmosphere are 46% higher than they were at the time of the American Revolution. This rise coincides with major advances in manufacturing and transportation, which are powered by the burning of fossil fuels. These coincidences in timing suggest that human activities are responsible for increasing CO_2 levels.

HYPOTHESIS The burning of fossil fuels, a development central to the emergence of modern industrial societies, has been the principal cause of the measured increase of CO_2 in the atmosphere.

EXPERIMENT AND RESULTS Hans Suess measured the abundance of the two stable isotopes of carbon, ^{13}C and ^{12}C, in air samples and demonstrated that the amount of ^{13}C in atmospheric CO_2 has decreased relative to the amount of ^{12}C over the second half of the twentieth century. Additional measurements of coral skeletons and wood show that the ratio of ^{13}C to ^{12}C in air has been decreasing over the entire period during which atmospheric CO_2 levels have been increasing.

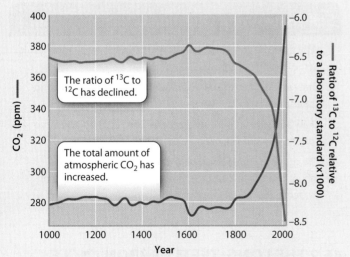

DISCUSSION The ratio of ^{13}C to ^{12}C in CO_2 added to the atmosphere over the past 200 years is lower than that in CO_2 that was already in the air 250 years ago. The ratio of ^{13}C to ^{12}C in CO_2 emitted by volcanoes or other geologic processes is too high to account for these data. In contrast, organic matter formed by photosynthesis has just the right ratio of ^{13}C to ^{12}C to account for Suess's measurements. By itself, the changing abundance of ^{13}C in the air could reflect the conversion of plant carbon to CO_2 by burning, or it could reflect the burning of fossil fuels formed by the burial of plant and algal materials in the geologic past.

FOLLOW-UP WORK Further studies of ^{14}C, the third, unstable isotope of C, also show a strong decrease in the abundance of ^{14}C in atmospheric CO_2 over time, relative to other isotopes. While the burning of vegetation to clear land for agriculture results in CO_2 with the right ratio of ^{13}C to ^{12}C, modern vegetation has far too much ^{14}C to account for the decrease in the ^{14}C composition of air. Only fossil fuels have the right ratios of all three carbon isotopes—^{12}C, ^{13}C, and ^{14}C—to account for the pattern of isotopic change in atmospheric CO_2 measured by Suess and others. In fact, ^{14}C data from the high Rocky Mountains indicate that if fossil fuels were the principal source of added CO_2, CO_2 levels should have increased by a little less than 2 ppm per year between 2004 and 2008—which is very close to the increase actually measured.

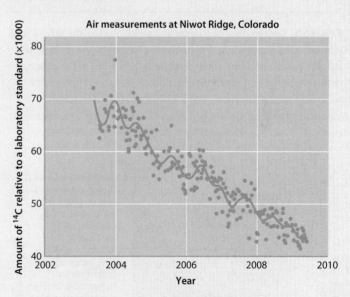

CONCLUSION Burning of fossil fuels by industrialized societies has been, and continues to be, a principal source of CO_2 buildup in Earth's atmosphere.

SOURCES Revelle, R., and H. E. Suess. 1957. "Carbon Dioxide Exchange between Atmosphere and Ocean and the Question of an Increase of Atmospheric CO_2 during the Past Decades." *Tellus* 9:18–27; Rubino, M., et al. 2013. "A Revised 1000 Year Atmospheric δ^{13}C-CO_2 record from Law Dome and South Pole, Antarctica." *Journal of Geophysical Research Atmospheres* 118:8482–8499; Turnbull, J. C., et al. 2007. "A New High Precision $^{14}CO_2$ Time Series for North American Continental Air." *Journal of Geophysical Research: Atmospheres* 112(D11):D11310. doi: 10.1029/2006JD008184.

What does this change in isotopic composition tell us about the source of carbon in the atmosphere? In photosynthesis, enzymes preferentially incorporate CO_2 containing the lighter isotope ^{12}C into biomolecules rather than CO_2 containing ^{13}C. Carbon in volcanic gases and dissolved in seawater has a different ratio of carbon isotopes. Only the isotope ratio contained in the organic matter of organisms matches the shift in carbon isotope ratios in Suess's measurements of the atmosphere. Therefore, processes that cause a net conversion of organic matter to CO_2, such as burning of forests to clear land or fossil fuel combustion, must be adding CO_2 to the atmosphere.

Here's where the other isotope of carbon, ^{14}C, comes in. Measurements also show that the sources of added CO_2 are depleted in ^{14}C relative to the CO_2 already in the air. Modern organic matter contains too much ^{14}C to account for the observed pattern, but ancient organic matter—the coal, petroleum, and natural gas burned as fossil fuels—is isotopically just right. Chemical analyses, therefore, support the hypothesis that human activities cycle carbon in amounts high enough to affect the chemical composition of the atmosphere.

That atmospheric CO_2 is increasing is not a hypothesis; it is a measurement. That the burning of fossil fuels plays an important role in this increase is also unambiguously documented by chemical analyses. Current debate focuses not on these observations, but instead on the more difficult problem of understanding how increasing CO_2 will affect climate. We discuss this problem in Chapter 48. For now, however, we can use some relevant numbers to reveal another important aspect of the carbon cycle.

As shown in the top graph in **Fig. 46.5**, humans now add almost 9 billion metric tons of carbon to the atmosphere each year, nearly all of it as CO_2. Much of this material comes from power plants, airplanes, ships, and automobiles, which oxidize ancient organic matter in the form of fossil fuels to CO_2 on a grand scale. Other important inputs come from the conversion of forests to agriculture or pastureland. Clearing and burning forest trees also oxidizes organic carbon to CO_2.

Interestingly, if we compare our best estimates of human CO_2 generation since 1958 (not including CO_2 added through respiration) with the measured increase in atmospheric CO_2, we find that roughly half of the CO_2 released by human activities did not end up in the air. Where did it go? This has been a major question for years. It now appears that much of the "missing carbon" has been stored as inorganic carbon in the form of CO_2, bicarbonate (HCO_3^-), and carbonate (CO_3^{2-}) ions dissolved in the oceans. Some has also been taken up by vegetation or stored in soils. The bottom graph in Fig. 46.5 shows an estimate of the various places, or sinks, where human-produced CO_2 ends up. These observations indicate that Earth's carbon cycle is more complicated than a simple exchange of carbon between photosynthesis and respiration. There are additional storehouses of carbon and additional processes that move carbon from one store to another, topics we discuss next.

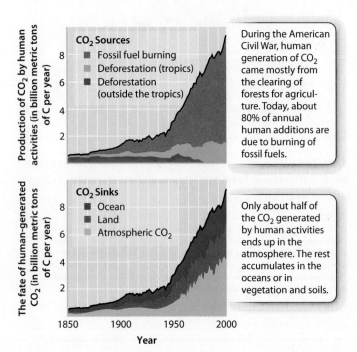

FIG. 46.5 Carbon dioxide sources and sinks. Much of the CO_2 introduced by humans into the atmosphere has not stayed there. About half of it has accumulated as dissolved inorganic carbon in the oceans or as new biomass on the continents. *Data from J. G. Canadell, C. Le Quéré, M. R. Raupach, C. B. Field, E. T. Buitenhuis, P. Ciais, T. J. Conway, N. P. Gillett, R. A. Houghton, and G. Marland, 2007, "Contributions to Accelerating Atmospheric CO_2 Growth from Economic Activity, Carbon Intensity, and Efficiency of Natural Sinks," Proceedings of the National Academy of Sciences USA 104:18866–18870.*

Self-Assessment Questions

1. Why does the coupling of photosynthesis and respiration result in the cycling of carbon?
2. Draw and explain the curve representing the changing atmospheric levels of CO_2 over the course of a year.
3. Draw and explain the curve representing the changing atmospheric levels of CO_2 over the past 150 years.
4. The graph in Fig. 46.3 suggests that human activities have influenced CO_2 levels in the atmosphere. What is one line of evidence that lends support to this hypothesis?

46.2 THE LONG-TERM CARBON CYCLE

Up to this point, we have considered only the short-term carbon cycle: exchanges over days, years, and decades driven by the biological processes of photosynthesis and respiration, and altered in recent times by human activities. Geologic evidence, however, tells us that on timescales of thousands to millions of years,

FIG. 46.6 Atmospheric CO_2 content over the past 400,000 years. (a) The graphs show measurements made from ice (lower graph) and air bubbles (upper graph) trapped in glacial ice at Vostok, Antarctica. (b) Twenty thousand years ago during Earth's most recent ice age, glacial ice (shown in white) covered much of North America and Europe, as well as mountain ranges such as the Andes and Himalayas. Sea ice (light blue) expanded markedly in both northern and southern oceans. *Data from P. Rekacewicz, "Temperature and CO_2 Concentration in the Atmosphere over the Past 400,000 Years," from Vital Climate Graphics, UNEP/GRID–Arendal Maps & Graphics Library, JPG file, 2005, http://www.grida.no/publications/vg/climate/page/3057.aspx. Based on J. R. Petit et al., 1999, "Climate and Atmospheric History of the Past 420,000 Years from the Vostok Ice Core, Antarctica," Nature 399:429–436. Plate tectonic maps and continental drift animations: Last Glacial Maximum by C. R. Scotese, PALEOMAP Project.*

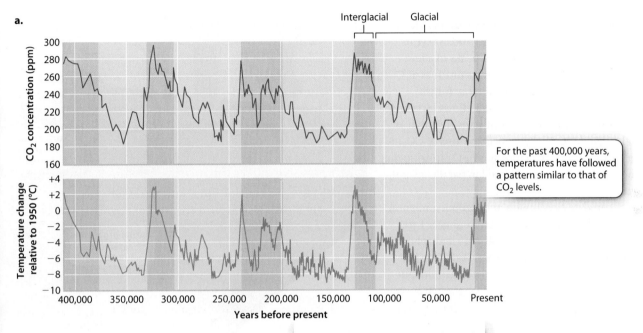

For the past 400,000 years, temperatures have followed a pattern similar to that of CO_2 levels.

Cold glacial and warm interglacial periods correlate with minimum and maximum atmospheric CO_2 levels, respectively.

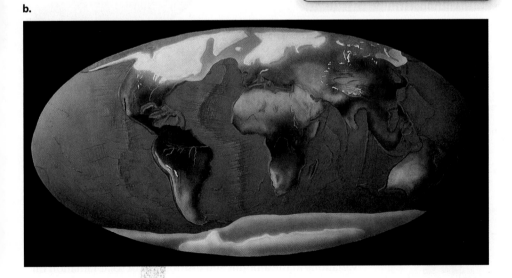

Records of atmospheric composition over 400,000 years show periodic shifts in CO_2 levels.

In the preceding section, we discussed evidence that before the Industrial Revolution, atmospheric CO_2 levels had not changed appreciably for 1000 years or more. Longer-term records, however, show that the CO_2 levels in air can change substantially through time. For example, at Vostok, high on the Antarctic ice sheet, glacial ice records more than 400,000 years of environmental history (**Fig. 46.6**). As shown in the upper graph of Fig. 46.6a, the youngest samples at Vostok show

CO_2 levels in air have changed even more dramatically. These longer-term changes mean that we must consider additional contributions to the carbon cycle: physical processes, including volcanism and climate change. Indeed, the complete carbon cycle links Earth's physical and biological processes, providing a foundation for understanding the interconnected histories of life and environment through our planet's long history.

about 285 ppm CO_2 in the atmosphere, consistent with direct measurements of air, including the first years of the Keeling curve. By comparison, CO_2 levels were much lower 20,000 years ago—about 180 ppm. In fact, the Vostok ice core in its entirety shows that atmospheric CO_2 has oscillated between 285 ppm and 180 ppm for at least 400,000 years.

The lower graph in Fig. 46.6a shows an estimate of surface temperature obtained by chemical analysis of oxygen isotopes in ice from the same glacier—specifically, how it differed from the average temperature in 1950. Over the last 400,000 years, Antarctic temperature has oscillated between peaks of a few degrees warmer than today and temperatures as much as 6°C to 8°C colder than the present. As discussed more fully in Chapter 48, CO_2 is a potent greenhouse gas: it allows incoming solar radiation to reach Earth's surface, but traps heat that is re-emitted from land and sea. Higher concentrations of CO_2 result in warmer temperatures. Therefore, it is not surprising that climate and atmospheric CO_2 levels changed in parallel over the past 400,000 years.

The two curves correlate closely with another phenomenon, the periodic growth and decay of continental ice sheets. Large glaciers expanded in the Northern and Southern Hemispheres a few million years ago, ushering in an ice age. Today, we live in an interglacial interval, when climate is relatively mild, but 20,000 years ago thick sheets of ice extended far enough away from the poles to cover the present site of Boston (Fig. 46.6b).

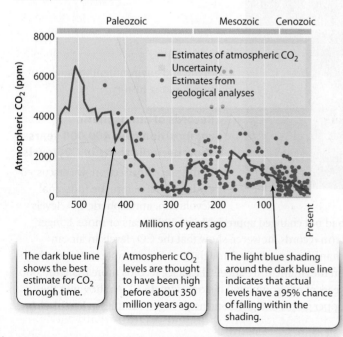

FIG. 46.7 Atmospheric CO_2 content for the past 541 million years. Models and indirect estimates from geological measurements suggest that CO_2 levels have varied substantially through Earth's history. *Data from Fig 3.4 in K. Wallmann and G. Aloisi, 2012, "The Global Carbon Cycle: Geological Processes," in Fundamentals of Geobiology, edited by A. H. Knoll, D. E. Canfield, and K. O. Konhauser, pp. 20–35, Chichester, UK: Wiley-Blackwell.*

The dark blue line shows the best estimate for CO_2 through time.

Atmospheric CO_2 levels are thought to have been high before about 350 million years ago.

The light blue shading around the dark blue line indicates that actual levels have a 95% chance of falling within the shading.

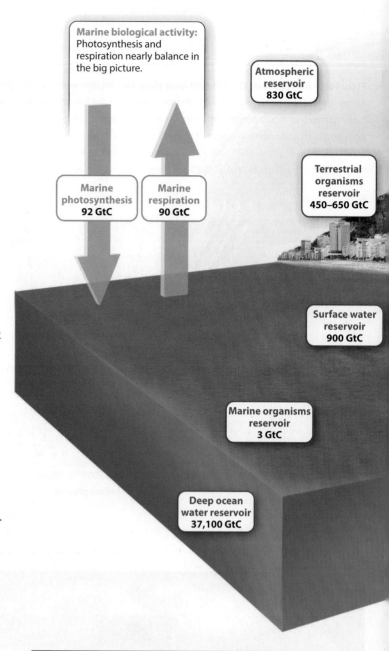

FIG. 46.8 Movement of carbon between reservoirs. The numbers for each reservoir are gigatons of carbon (GtC), and those for fluxes are gigatons of carbon per year. (1 gigaton = 1 billion metric tons.) *Photo source: Fotosearch Stock Images; Data from P. Rekacewicz, "The Carbon Cycle," from Vital Climate Graphics, UNEP/GRID–Arendal Maps & Graphics Library, JPG file, 2005, http://www.grida.no/graphicslib/detail/the-carbon-cycle_a224.*

The repeated climatic shifts recorded in ice cores reflect periodic variations in the amount and distribution of solar radiation on Earth's surface, caused by oscillating changes in Earth's orbit around the sun. Scientists continue to debate why atmospheric CO_2 oscillates in parallel with glacial expansion and retreat, but the proposed mechanisms commonly suggest interactions involving the ocean's capacity to store large amounts of inorganic carbon, especially during glacial times. Whatever the explanation, the historical record of the past 400,000 years shows that climate can and does change without any input from

humans, something we must take into account when considering our climatic present and future (Chapter 48).

On still longer timescales, atmospheric levels of CO_2 have varied even more widely. The atmospheric record trapped in glacial ice extends back in time less than 1 million years. To estimate atmospheric CO_2 levels for earlier intervals of Earth history, we must rely on computer models and measurements of chemical or paleontological features of ancient rocks that are thought to reflect the levels of atmospheric CO_2 at their time of formation. Such indirect observations and analyses come with a large degree of uncertainty, but most Earth scientists accept at least the broad pattern of atmospheric history shown in **Fig. 46.7**. On this timescale, atmospheric CO_2, and hence climate, are determined in large part by geologic processes, especially the chemical weathering of continental rocks uplifted into mountains.

Reservoirs and fluxes are key in long-term carbon cycling.

To understand the long-term carbon cycle better, we must look at how carbon is distributed among its various **reservoirs**, the places where carbon is found on the Earth. As shown in the purple boxes in **Fig. 46.8**, reservoirs of carbon include organisms, the atmosphere, soil, the oceans, and sedimentary rocks. Both biological and geologic processes cycle carbon back and forth among these pools.

How much carbon is stored in each of Earth's major reservoirs? If we add up all the carbon contained in the total mass of organisms living on land, that amount of carbon is just a bit smaller than the amount of carbon stored as CO_2 in the atmosphere. Soil, by comparison, stores as much carbon as do land organisms and the atmosphere combined, mostly as slowly decaying organic compounds. In the oceans, the amount of carbon contained in living organisms is actually very small. Much more resides as inorganic carbon dissolved in the water—most of it in the deep oceans as CO_2 and bicarbonate and carbonate ions.

The biggest carbon reservoir of all lies beneath our feet, within sediments and sedimentary rocks. Calcium carbonate minerals ($CaCO_3$), which form limestone, and organic matter preserved in sedimentary rocks dwarf all other carbon reservoirs combined by three orders of magnitude. Coal, petroleum, and natural gas make up only a small part of the organic matter stored in sedimentary rocks, but as we have seen, releasing the carbon in this reservoir by combustion plays an important role in the modern carbon cycle.

Fluxes are the rates at which carbon flows from one reservoir to another. In addition to the biological processes of photosynthesis and respiration described earlier, physical processes add and remove CO_2 from the atmosphere. Volcanoes and mid-ocean ridges forming new seafloor release an estimated 0.1 billion metric tons of carbon (as CO_2) into the atmosphere each year. More CO_2, approximately 0.05 billion metric tons of carbon per year, is released by the slow oxidation of coal, oil, and other ancient organic material in sedimentary rocks exposed at the Earth's surface. In nature, bacteria and fungi accomplish most of this oxidation by respiring old organic molecules. By burning fossil fuels, humans accelerate this process dramatically, increasing by a hundredfold the rate at which sedimentary organic carbon is oxidized.

Approximately 0.43 billion metric tons of CO_2 are removed from the atmosphere each year by chemical reactions involving air, water, and exposed rocks, a process called chemical weathering. Ions released during this weathering eventually precipitate as $CaCO_3$ in the oceans, contributing to the large sedimentary reservoir of carbon.

One final set of geological processes completes the long-term carbon cycle. Earth's crust is constantly in motion, propelled by heat within the Earth's interior. Plate tectonics is the name given to this dynamic movement of our planet's outer layer. Plate tectonics provides geology's best explanation for the formation of mountains and ocean basins, and it plays an important role in the long-term carbon cycle. New crust that forms the seafloor and the sediments that accumulate on top of it are returned to the mantle by subduction, in which one plate sinks beneath

FIG. 46.9 Geological processes that drive the long-term carbon cycle. Numbers are estimates for annual fluxes of carbon in gigatons (GtC). Red boxes and arrows indicate processes that move CO_2 from geologic reservoirs into the atmosphere. Blue boxes and arrows indicate processes that move CO_2 from the atmosphere into geologic reservoirs. *Data from J. Gaillardet and A. Galy, 2008, "Himalaya—Carbon Sink or Source?" Science 320:1728, doi:10.1126/science.1159279.*

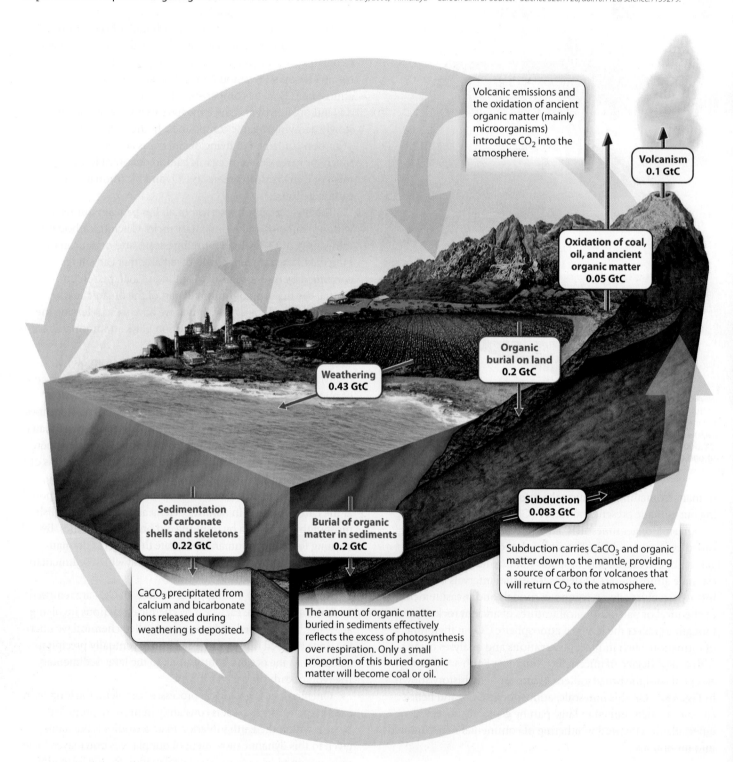

another. $CaCO_3$ and organic carbon deposited as part of those sediments return to the mantle in this way, eventually to be recycled to the surface as CO_2 emitted from volcanoes and at mid-ocean ridges where new seafloor forms. **Fig. 46.9** shows the physical processes at work in Earth's long-term carbon cycle.

In essence, carbon moves from the atmosphere to sediments, where it can be stored for millions of years before returning to the surface environment through physical processes of volcanism and seafloor formation. The sensitivity of different reservoirs to change depends on the relative sizes of the reservoir and of the amount of movement of material into and out of it. When fluxes are large relative to the size of the reservoir, reservoir size can change rapidly. In particular, the amount of carbon stored as CO_2 in air is not that much larger than the annual fluxes into and out of the atmosphere. For this reason, atmospheric CO_2 abundance can be influenced by a number of processes at work in the carbon cycle.

Self-Assessment Questions

5. Draw and explain the curve representing changing atmospheric levels of CO_2 over the last 400,000 years.
6. How do geologic processes participate in the long-term carbon cycle?
7. If plate tectonic processes form a chain of high mountains, would you expect atmospheric CO_2 to increase or decrease?
8. Rates of respiration are known to vary positively with temperature. As increasing CO_2 causes temperatures at the Earth's surface to rise, would changing rates of respiration tend to exacerbate or dampen this rise?

46.3 FOOD WEBS AND TROPHIC PYRAMIDS

So far, we have outlined the biological and physical pathways that cycle carbon through Earth's surface environments. The cycling of carbon and other elements depends on and reflects interactions among species as well. Moreover, the physical and biological components of ecosystems are linked by the processes that cycle carbon and other elements and transfer energy through the system.

Food webs trace carbon and other elements through communities and ecosystems.

As you sit quietly by a woodland pond, the carbon cycle is at work all around you. In the forest, plants generate carbohydrates as they photosynthesize, and algae and photosynthetic bacteria do the same in the pond. In Chapter 6, we called photosynthetic organisms autotrophs ("self feeders") because they synthesize the organic molecules needed for growth and reproduction from CO_2. Ecologists commonly refer to these organisms as **primary producers**, reflecting the fact that they fix inorganic carbon into organic molecules that feed other organisms in the community.

An insect grazes on leaves, and ducks feed on both insects and algae. Ducks and insects obtain the carbon they need for growth and reproduction from the foods they eat, and they also gain energy by respiring food molecules. In Chapter 6, we called such organisms heterotrophs ("other feeders"); ecologists refer to them as **consumers**. All the animals in both pond and forest are consumers. Beneath our feet, similar interactions play out on a microscopic scale, especially among the fungi, protists, and bacteria that consume the dead remains of other organisms.

Primary consumers consume primary producers. As we saw in Chapter 45, this kind of interaction is used to classify consumers as herbivores or grazers, whether they are eating plants on land or algae in aquatic environments. Primary consumers transfer carbon and energy drawn by the primary producers from the environment to a second biological reservoir and, through respiration, return some of the carbon to surrounding air or water.

Secondary consumers, in turn, are predators or scavengers that feed on primary consumers. These organisms may provide food for still other consumers, generating successive tiers of organisms that feed upon one another. Like primary consumers, secondary consumers continue the transfer of carbon and energy from reservoir to reservoir, eating the primary consumers and returning some of the carbon stored in that reservoir to the environment through respiration.

Eventually, having passed from one consumer to another, the carbon originally fixed by photosynthesis is returned to the atmosphere by the respiration of fungi, bacteria, and other **decomposers** that break down dead tissues. The same is true of other biologically important elements such as nitrogen and phosphorus (discussed in the next section).

The transfer of carbon and energy from one organism to another forms a food chain. Because most heterotrophs within a community can consume or be consumed by a number of other species, biologists often prefer to speak of **food webs,** a term that provides a better sense of the complexity of biological interactions within the carbon cycle (**Fig. 46.10**). Food webs define the interactions among organisms in ponds, forests, and many other habitats and trace the passage of carbon atoms through the biological carbon cycle. Put another way, biogeochemical cycles underpin the ecological structure of biological communities.

An organism's typical place in a food web is its **trophic level.** The first trophic level consists of primary producers; consumers occupy the levels above the first trophic level. At the second trophic level are primary consumers; they comprise the herbivores, organisms that eat plants or other primary producers. Secondary consumers are found at the third

FIG. 46.10 A food web. A food web traces the cycling of carbon and other elements from the environment through a succession of organisms and back to the environment. Energy is transferred from primary producers to consumers, eventually to be dissipated through work and the production of heat.

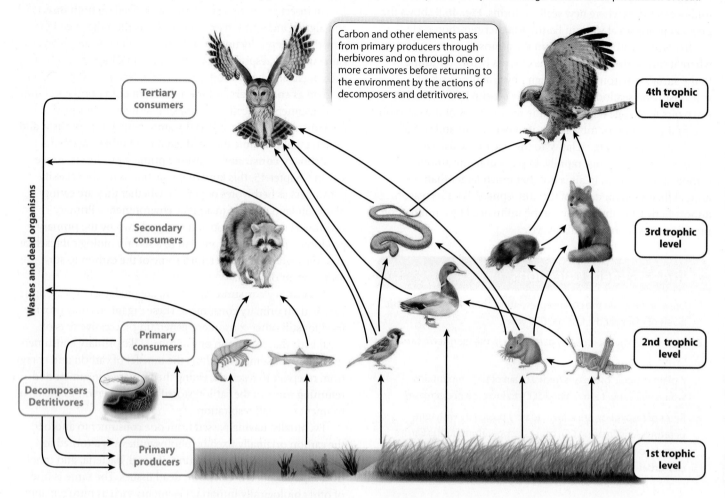

trophic level and potentially higher levels as well. Top or apex consumers, which have few, if any, natural predators, occupy the highest trophic level. Parasites exploit all levels of the system, obtaining carbon and other nutrients from primary producers and consumers alike. Finally, the all-important decomposers and detritivores include fungi, bacteria, protists, and certain animals. Decomposers feed on the dead cells or bodies of other organisms, and detritivores do much the same, consuming partially decomposed plants or animals in soil or sediment, as well as organic compounds in feces. These organisms both provide food for microbe-eating consumers and return carbon dioxide and other inorganic compounds to the environment, completing the cycle of elements through the ecosystem.

Energy as well as carbon is transferred through ecosystems.

As the organisms in food webs move carbon and other elements from one organism to another, they also transfer energy. Carbon and energy are closely intertwined in food webs for the simple reason that the energy sources for consumers are the carbon-rich organic molecules in the organisms they eat. Unlike carbon, energy does not cycle through an ecosystem, so new energy must continually be harvested from the environment to sustain the community. In essentially all ecosystems where sunlight is available, photosynthesis is necessary for the species interactions that define a community and ecosystem. Plants, algae, and photosynthetic bacteria capture energy from the sun and use it to synthesize energy-rich organic molecules. Where sunlight is absent, especially in the vast depths of ocean, bacteria and archaeons drive primary production through chemical reactions.

As plants capture energy from sunlight and carbon from the atmosphere, they build chemical bonds that store energy in carbohydrates. The organisms that consume these organic molecules store some of the energy in other molecules such as fat and starch, but they also use much of their food energy

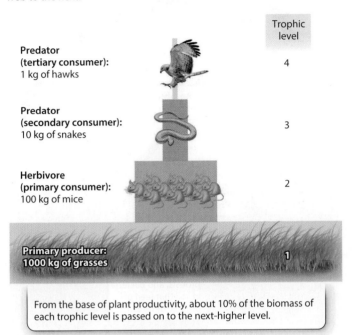

FIG. 46.11 A trophic pyramid. A trophic pyramid graphically shows the consequences of inefficient energy transfer from one level in a food web to the next.

From the base of plant productivity, about 10% of the biomass of each trophic level is passed on to the next-higher level.

to do work (such as moving, growing, or reproducing) or generate heat. When carbon is released during respiration by consumers, it can be reincorporated into organisms by photosynthesis; in contrast, energy expended as heat or work cannot be captured again by other organisms. Therefore, while carbon and other elements are recycled in ecosystems, energy transfer is unidirectional.

Because organisms at one trophic level rarely consume all the resources at the level beneath them, and because energy is lost through wastes, work, and heat dissipation, only approximately 10% of the energy available at one trophic level in an ecosystem is incorporated into the next level. The consequences of this inefficient transfer of energy through food webs can be illustrated by a **trophic pyramid** (**Fig. 46.11**), which shows the biomass supported at each level by the biomass and energy available in the level beneath it. For example, 1000 kg of grass and herb biomass will support approximately 100 kg of mice (a little more than 3000 animals), and that, in turn, will support 10 kg of small predators like snakes. At the apex of the pyramid, we find a single kilogram of hawk or owl tissue. As the trophic pyramid suggests, primary production exerts a powerful influence over the rest of the community. To get more foxes and eagles, an ecosystem would need more photosynthesis.

On land, trophic pyramids generally have a wide base and a narrow apex. In aquatic habitats, however, trophic webs may be inverted, at least with respect to biomass. Planktonic algae at the base of the pyramid are tiny and may have high rates of primary production, resulting in high rates of energy generation but low total biomass. The herbivores that consume these algae and the predators that eat them in turn are successively larger, are longer lived, and forage for food over a larger area. This trend can result in increasing biomass as we move upward through the pyramid, leading to its inverted shape. But note that while the trophic pyramids for aquatic communities may be inverted with respect to biomass, they are not inverted with respect to energy: because of energy loss through inefficient feeding, work, and heat dissipation, communities always require more energy to be available at lower trophic levels to support higher ones.

In summary, the complementary metabolic processes of photosynthesis and respiration cycle carbon and transfer energy through terrestrial and aquatic ecosystems (**Fig. 46.12**). On land, primary production is dominated by a single group of photosynthetic organisms, vascular plants such as the tree illustrated in Fig. 46.12a. In the oceans, many groups of algae and cyanobacteria convert carbon dioxide to organic molecules (Fig. 46.12b). And in the dark waters of the deep sea, primary production by chemoautotrophic Bacteria and Archaea can contribute in important ways to the carbon cycle. Consumers, from animals to bacteria, live on organic matter originally generated by primary producers. Plants and animals may be the most conspicuous components of the biological carbon cycle, but it is the microorganisms—fungi, bacteria, and archaeons—that complete the cycle, generating CO_2 by respiration or fermentation from organic materials that leak into soils, sediments, and deep seawaters.

Complementary metabolic processes also cycle nitrogen, phosphorus, and other elements required for life, as we discuss in the next section. By continually recycling materials, biogeochemical cycles, which involve both biological and physical processes, sustain life over long intervals. In their absence, life could hardly have persisted for 4 billion years.

Self-Assessment Questions

9. How does a food web that diagrams the movement of carbon through an ecosystem also show how energy is transferred from one organism to the next?

10. Biomass generally declines substantially as we move upward from one trophic level to the next in a trophic pyramid. This pattern was discussed in this section in terms of energy loss from one level to the next, but carbon must be lost as well. What happens to it?

11. Describe how herbivores can affect the abundances of organisms at higher and lower trophic levels.

FIG. 46.12 The biological carbon cycle. Carbon cycles, both (a) terrestrial and (b) marine, link metabolism and ecology into functioning ecosystems.

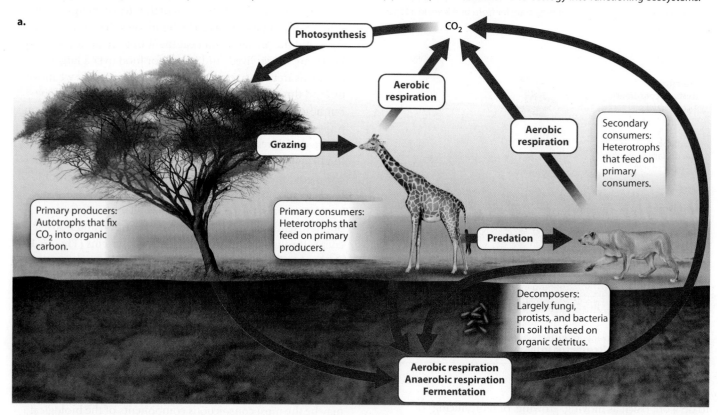

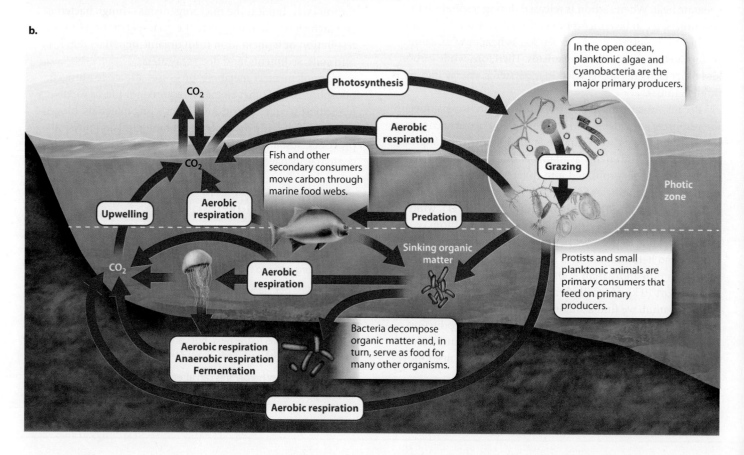

FIG. 46.13 The nitrogen cycle. The nitrogen cycle also links metabolism and ecology, governing the availability of nitrogen for primary producers and thereby exerting a strong influence on the biological carbon cycle.

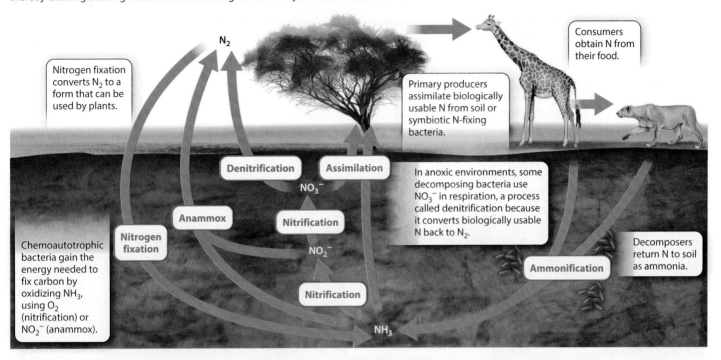

46.4 OTHER BIOGEOCHEMICAL CYCLES

Until now, we have focused on carbon, and for good reason. Carbon atoms form the backbone of most molecules in cells other than water, and carbon compounds such as fatty acids and sugars are the principal means through which energy is transferred from consumed to consumer. However, other nutrients, such as nitrogen (N) and phosphorus (P), are critical for building amino acids and nucleic acids that play key roles in the synthesis of proteins, DNA, and RNA, among other organic molecules important for the growth, maintenance, and reproduction of all species. Thus, as organisms cycle carbon through ecosystems, they also cycle nitrogen, phosphorus, and other elements. The primary producers that fix CO_2 into carbohydrates also take up nitrogen and phosphorus from the environment, and use them to synthesize amino acids, nucleic acids, and membranes. Consumers depend on primary producers not only for their supply of carbon and energy, but also for the nitrogen and phosphorus they need to synthesize their own organic molecules. Along the way in these biogeochemical cycles, metabolic wastes, including excess nitrogen and phosphorus, as well as carbon dioxide, are passed back to the environment through excretion and gas exchange.

For this reason, the long-term maintenance and functioning of ecosystems require that organisms actively cycle not only carbon, but also other elements critical for life. In the following discussion, we focus on the nitrogen and phosphorus cycles.

The nitrogen cycle is closely linked to the carbon cycle.

Fig. 46.13 illustrates the same terrestrial ecosystem as shown in Fig. 46.12a, but highlights the cycling of nitrogen. The primary producers, consumers, and decomposers linked in the carbon cycle also cycle nitrogen. Indeed, the biological carbon and nitrogen cycles are closely connected by both ecology and metabolism.

Primary producers need nitrogen as well as carbon. While they fix CO_2 into organic molecules by photosynthesis or chemosynthesis, primary producers generally obtain the nitrogen they require by taking up biologically usable nitrogen found as nitrate (NO_3^-) or ammonia (NH_3) in soil or water, a process called **assimilation.** Primary consumers obtain nitrogen from the primary producers they consume, just as they obtain carbon from these organisms. Through such consumption, both nitrogen and carbon travel across food webs. Decomposers return nitrogen to the environment as ammonia, a process termed **ammonification.**

As shown in Fig. 46.13, a number of bacteria and archaeons use nitrate or ammonia to generate energy, releasing nitrogen gas (N_2) to the atmosphere. These processes include denitrification (a form of anaerobic respiration in which nitrate, rather than oxygen, serves as the terminal electron acceptor), nitrification (a chemoautotrophic process that uses energy gained from the oxidation of ammonia or nitrite, NO_2^-, by oxygen), and anammox (a form of chemoautotrophy, with energy gleaned from the reaction of ammonia and nitrite) (Chapter 24). Because of these processes, the amount of biologically usable nitrogen in communities would decline through time to very low levels were it not for **nitrogen fixation,** the process by which some bacteria and archaeons reduce nitrogen gas (N_2) to biologically usable ammonia (Chapter 24). Nitrogen fixation, then, is the process by which nitrogen enters food webs.

In the oceans, nitrogen cycling works much as outlined in the preceding paragraph: microorganisms fix N_2, decomposers

FIG. 46.14 The phosphorus cycle. Phosphorus, another key nutrient, is weathered from rocks and eventually returns to its geological reservoir through deposition in sediments. Before it does, organisms repeatedly cycle phosphorus through ecosystems.

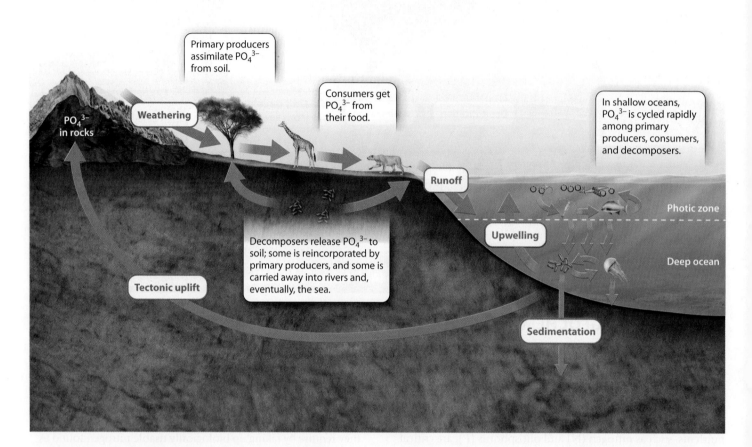

return biologically usable nitrogen compounds to the water, primary producers assimilate this nitrogen, and consumers cycle nitrogen even as they move carbon and energy through ecosystems. Eventually, as on land, microbes convert biologically usable nitrogen compounds back to N_2.

The continual generation of biologically usable nitrogen compounds by nitrogen-fixing bacteria is fundamental not only to the nitrogen cycle but also to the carbon cycle, as it supplies critical nutrients for the primary producers at the base of an ecosystem's trophic pyramid and food web. As discussed in Chapter 48, humans now fix a great deal of nitrogen through industrial processes, altering the balance of nutrients in many ecosystems.

Phosphorus cycles through ecosystems, supporting primary production.

Like nitrogen, phosphorus is a key element needed for growth; this nutrient is incorporated into the nucleic acids and membranes found in all organisms as well as into ATP, the universal currency of life. The phosphorus cycle, however, is quite different from the nitrogen cycle. Whereas nitrogen is found largely in the atmosphere, phosphorus is mostly present in rocks. In addition, although nitrogen and phosphorus are both assimilated by primary producers, phosphorus does not provide electron donors or electron acceptors for energy metabolism. It simply enters the food web as phosphate ions (PO_4^{3-}) released from rocks by chemical weathering; is taken up, or assimilated, by primary producers; and then is transferred from one organism to another through cycles of consumption and decomposition, eventually returning to its geological reservoir by accumulation in sediments (**Fig. 46.14**).

In the long term, phosphorus-containing sediments will be uplifted by tectonic processes to form mountains, making the phosphorus available again for release by chemical weathering. As is the case for nitrogen, humans have greatly accelerated phosphorus cycling through the mining of phosphate ores for fertilizer (Chapter 48).

Self-Assessment Questions

12. Describe two ways that the biological nitrogen cycle interacts with the biological carbon cycle.

13. What are two ways in which the nitrogen and phosphorus cycles differ?

46.5 THE ECOLOGICAL FRAMEWORK OF BIODIVERSITY

Among the most remarkable features of life is its astonishing diversity. In previous chapters, we have discussed biological diversity, or biodiversity, in terms of the remarkable variety of form and function displayed across the tree of life. Here, we revisit biodiversity within the framework of ecology. While biodiversity is the product of evolution, it is shaped and sustained by ecological interactions among organisms and between organisms and the physical environment.

Biological diversity reflects the many ways that organisms participate in biogeochemical cycles.

An estimated 500,000 species of photosynthetic organisms lie at the base of trophic pyramids in all but a few deep-sea and subterranean ecosystems. In terms of nutrient metabolism and their function in an ecosystem's food web, all these species do pretty much the same thing. Why, then, is there such a diversity of photosynthetic organisms? Why don't just a few species dominate the world's photosynthesis?

The example of a forest pond helps to make the basis of photosynthetic diversity clear. In the forest community, the leaves of several different tree species form a photosynthetic canopy above the forest floor. Below, shrubs grow, making use of light not absorbed by the leaves above them. And below the shrubs are grasses, herbs, ferns, and mosses that can grow in the reduced light levels of the forest floor. In the nearby pond, a few species of aquatic plants line the water's edge, but beyond that algae and photosynthetic bacteria dominate photosynthesis, with some species anchored to the pond bottom and others floating in the water column.

Within even a single ecosystem, then, the different photosynthetic species that transfer carbon atoms from CO_2 to organic molecules subdivide the forest on the basis of light, water, and nutrient availability—that is, they occupy distinct niches (Chapter 45). On a larger scale, climate and physical features (mountains, valleys, rivers, lakes) vary tremendously from one region to another, and these features also help explain how primary producers can build and maintain diversity even though they all perform the same basic function in ecosystems. In the forests of New England, the tree species that thrive in wet regions differ from those found on well-drained hillsides. In southern California, seasonally dry woodlands support yet another set of plant species. In general, plants of varying size, shape, and physiology inhabit physically and biologically distinct environments, and the same is true of photosynthetic organisms in lakes and oceans.

Thus, the immense diversity of photosynthetic organisms found today does not reflect differences in how energy-rich molecules are produced for a community of organisms (although some of that occurs; see Chapter 24) so much as it does structural and physiological adaptations. These adaptations allow the effective gathering of light, nutrients, and—critical to life on land—water, in widely varying local environments. Natural selection, acting on local populations, links the diversity of photosynthetic organisms to the carbon cycle.

If half a million species function as primary producers, an estimated 10 million species help to transfer energy and cycle carbon and other elements onward through food webs. These include plants, algae, and bacteria that respire as well as conduct photosynthesis, and animals, fungi, and microorganisms that obtain their carbon, nitrogen, phosphorus, and energy from organic compounds in photosynthetic organisms or the consumers that eat them. These organisms are essential to the completion of the short-term carbon cycle, returning carbon atoms to the environment as CO_2.

Heterotrophic bacteria, amoebas, and humans may use essentially the same biochemical pathway to respire organic molecules, but they differ markedly in how they feed and, therefore, in what they can eat. Bacteria (and also fungi) absorb molecules from their environment, but amoebas and many other eukaryotic microorganisms can capture and ingest cells—they are capable of predation. Animals capture prey as well, but commonly feed on organisms far too large for an amoeba to eat. As photosynthetic organisms have adapted structurally and physiologically to local environments across the globe, consumers have adapted by means of locomotion, mouth and limb specialization, perception, and behavior to obtain their food.

Biological diversity can influence primary production and therefore the biological carbon cycle.

The biodiversity of a given region reflects its level of primary productivity. We have seen that predators' population sizes depend on the population sizes of their prey, which in turn depend on what those prey consume, going all the way down the trophic pyramid to primary production. Greater primary production also means that a greater variety of plants of different sizes and kinds can coexist, which in turn supports a greater variety of herbivorous animals. While this pattern suggests that high productivity can foster high diversity, recent experiments show that the reverse may also be true: biodiversity may itself play an important role in the productivity of ecosystems (**Fig. 46.15**). This role reflects the many different ways that different plants and animals may use the resources provided by soil, water, and sunlight in a particular place. Because no species is adapted for all functions in a given set of conditions, a diverse collection of species with different adaptations will collectively be more productive than a single species.

Biogeochemical cycles weave together biological evolution and environmental change through Earth history.

As we have seen, the short-term carbon cycle and associated biological cycles of nitrogen, phosphorus, and other elements provide a framework for understanding ecology. It is the

HOW DO WE KNOW?

FIG. 46.15

Does species diversity promote primary productivity?

BACKGROUND The different plant species found in a community have combinations of leaves, stems, and roots that tap different sources of water and nutrients in the soil and capture light at different levels of sun and shade. Do the differences among plant species result in higher levels of primary production than would be possible with fewer species?

HYPOTHESIS Plant diversity promotes primary production within a community.

EXPERIMENT Ecologist David Tilman conducted a long-term experiment in Minnesota, seeding each of 11 plots with 1, 2, 4, or 16 grassland species. He controlled each plot's amount of nutrients, water, and carbon dioxide, and also manipulated the plots' exposure to herbivory and disturbance by fire. The primary production of each plot was measured annually for up to 23 years.

RESULTS The graph shows that the addition of nitrogen fertilizer [quantified as kilograms of nitrogen (kg of N) per hectare (ha)] increased primary production significantly, and additional water, higher CO_2, herbivory, and disturbance also resulted in at least a modest increase in productivity. The difference in primary production between high-diversity (16 species) and low-diversity (1, 2, or 4 species) plots was at least as great as the largest additions of fertilizer—and much greater than any of the other treatments.

CONCLUSION The hypothesis was supported: in the grassland community under study, species diversity promoted primary production.

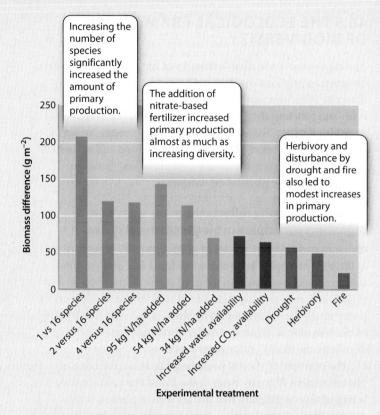

FOLLOW-UP WORK Researchers continue to study the ecosystem effects of species diversity. Steadily accumulating data show the importance of conserving the species diversity found in nature.

SOURCE Tilman, D., P. B. Reich, and F. Isbell. 2012. "Biodiversity Impacts Ecosystem Productivity as Much as Resources, Disturbance, or Herbivory." *Proceedings of the National Academy of Sciences USA*, 109:10394–10397.

---

long-term cycles of carbon and other elements, however, that establish the context of evolution and environmental change through Earth's history.

The world did not always support the biological diversity we see today. Indeed, 2 billion years ago, the carbon cycle included mostly photosynthetic bacteria and microbial heterotrophs because there were no plants or animals (**Fig. 46.16**). Subsequently, algae gained an ecological foothold as nutrients became more widely available. Single-celled eukaryotic heterotrophs expanded their reach by gathering food in ways not possible for bacteria, ingesting cells and other types of particulate food. With the evolution of multicellularity, animals and seaweeds added further complexity to carbon cycling in the sea. Eventually, some aquatic green algae evolved the capacity to live on land, and animals soon followed, building the carbon cycle on land and thereby generating unprecedented levels of diversity on our planet.

The history of life is one of accumulating variety and complexity, based mostly on the great biological components of the carbon cycle: photosynthesis and respiration. Indeed, the carbon cycle did more than provide a framework for accumulating diversity: it changed the very nature of Earth's surface environments in ways that enabled new types of organism to evolve.

When life began, Earth's atmosphere and oceans contained little or no oxygen gas. How did our present oxygen-rich environment come to be? The key lies in the formulas for photosynthesis and respiration, shown in Fig. 46.2. Photosynthesis and respiration cycle not only carbon, but oxygen (and water) as well. For this reason, the history of atmospheric oxygen is closely tied to the workings of the carbon cycle through time. When oxygen production by photosynthesis and oxygen consumption by respiration are in balance, oxygen levels do not change. However, organic carbon generated by photosynthesis can be buried in sediments, thereby avoiding its usual

FIG. 46.16 Microbial communities in hot springs in New Zealand. This environment and others that few eukaryotes tolerate provide a glimpse of how the carbon cycle worked before plants and animals evolved. *Source: Albert Aanensen/naturepl.com*

FIG. 46.17 Oxygen levels in Earth's atmosphere during the history of life.

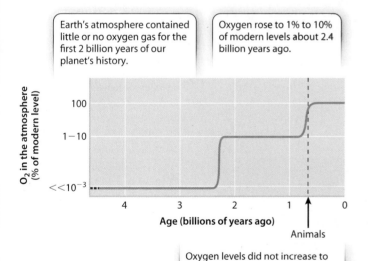

Earth's atmosphere contained little or no oxygen gas for the first 2 billion years of our planet's history.

Oxygen rose to 1% to 10% of modern levels about 2.4 billion years ago.

Oxygen levels did not increase to levels that humans could breathe until about 580 million years ago, coinciding with the first appearance of animals in the fossil record.

return to the environment through respiration. Some of the O_2 also generated by primary producers can then accumulate in the atmosphere and oceans. The geologic history of oxygen, shown in **Fig. 46.17**, results from the interactions through time between the carbon and other biogeochemical cycles with

plate tectonics. Not until Earth was 2 billion years old did the photosynthetic production of O_2 come to exceed oxygen consumption owing to reduced gases and minerals, as well as by respiring organisms. And only about 580 million years ago did increasing nutrient levels result in an atmosphere much like our own. Our oxygen-rich world is a latecomer in Earth history, but its emergence was key to the diversity of animals that require high levels of O_2 to live.

But life does not just respond passively to changing environments: it can also drive environmental change through time. For example, in Fig. 46.7, notice the large drop in the line representing atmospheric CO_2 in the mid-Paleozoic Era, 400–350 million years ago. This sharp decline is thought to reflect the evolution of a new player in the carbon cycle: woody plants. The evolution of trees increased the size of the carbon reservoir on land and ushered in an important new mechanism for removing carbon from the air and ultimately transferring it to sedimentary rocks. That mechanism was the burial of plant material on land, forming peat and, eventually, coal.

In the current century, a single animal species is influencing the carbon cycle in new ways: humans. As noted in our discussion of the Keeling curve, we are adding CO_2 to the atmosphere at a high rate, warming Earth's surface by the greenhouse effect. The consequences of human activities for Earth and its living organisms are discussed more fully in Chapter 48. For now, just bear in mind that modern humans are not simply observers of the carbon cycle; we are major participants in it.

Fig. 46.18 summarizes the global flow of carbon and energy through ecosystems, including their interactions with the nitrogen and phosphorus cycles and human impacts.

Self-Assessment Questions

14. If photosynthetic organisms all gain carbon and energy in much the same way, how do diverse photoautotrophs persist in ecosystems?

15. How can plant diversity influence rates of primary production in terrestrial ecosystems?

16. Compare and contrast a modern marine food web with food webs that existed two billion years ago.

VISUAL SYNTHESIS
FIG. 46.18
Flow of Matter and Energy Through Ecosystems
Integrating concepts from Chapters 45–48

Terrestrial carbon cycle
Carbon (C) enters the cycle through photosynthesis and is transferred through food webs by consumers. Decomposers complete the return of carbon to the atmosphere as CO_2.

Terrestrial phosphorus cycle
Phosphate ions released from rocks by weathering are incorporated into biomass by plants. Phosphorus (P) is transferred through food webs by consumers, eventually to be deposited in sediments or transferred to the oceans by rivers.

Terrestrial nitrogen cycle
Nitrogen (N) enters the cycle through nitrogen fixation, as bacteria in soils reduce N_2 to the biologically available form of ammonia. Plants assimilate biologically available nitrogen from soils. In turn, nitrogen is transferred through food webs by consumers, and decomposers return nitrogen compounds to soils, where ammonia can be oxidized to nitrate by chemoautotrophic bacteria, and nitrate can be reduced to N_2 (which returns to the atmosphere) by bacteria that respire anaerobically. **Go to the Visual Synthesis Map on LaunchPad to explore how the nitrogen cycle interacts with the cycles shown here.**

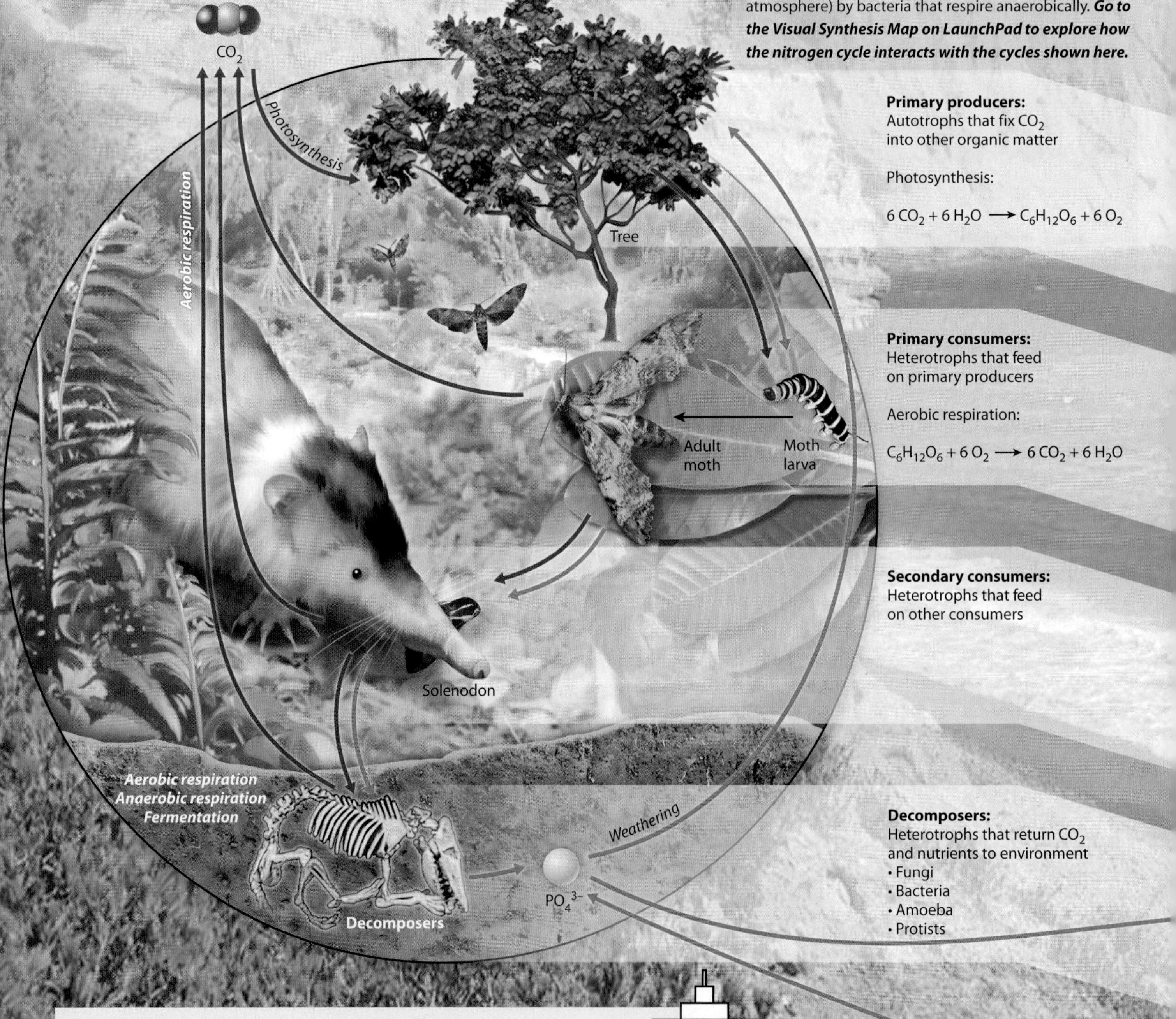

CO_2

Aerobic respiration

photosynthesis

Tree

Adult moth

Moth larva

Solenodon

Aerobic respiration
Anaerobic respiration
Fermentation

Weathering

PO_4^{3-}

Decomposers

Primary producers:
Autotrophs that fix CO_2 into other organic matter

Photosynthesis:
$$6\ CO_2 + 6\ H_2O \longrightarrow C_6H_{12}O_6 + 6\ O_2$$

Primary consumers:
Heterotrophs that feed on primary producers

Aerobic respiration:
$$C_6H_{12}O_6 + 6\ O_2 \longrightarrow 6\ CO_2 + 6\ H_2O$$

Secondary consumers:
Heterotrophs that feed on other consumers

Decomposers:
Heterotrophs that return CO_2 and nutrients to environment
- Fungi
- Bacteria
- Amoeba
- Protists

Trophic pyramids
As organisms cycle carbon, they also transfer energy, as illustrated by trophic pyramids. Some of the energy stored chemically in the sugars synthesized by primary producers and the biomolecules consumers eat is converted to heat and work by respiration. For this reason, only a portion of the energy taken in is available to consumers further up the food web. **Go to the Visual Synthesis Map on LaunchPad to explore how trophic pyramids reflect the carbon cycle.**

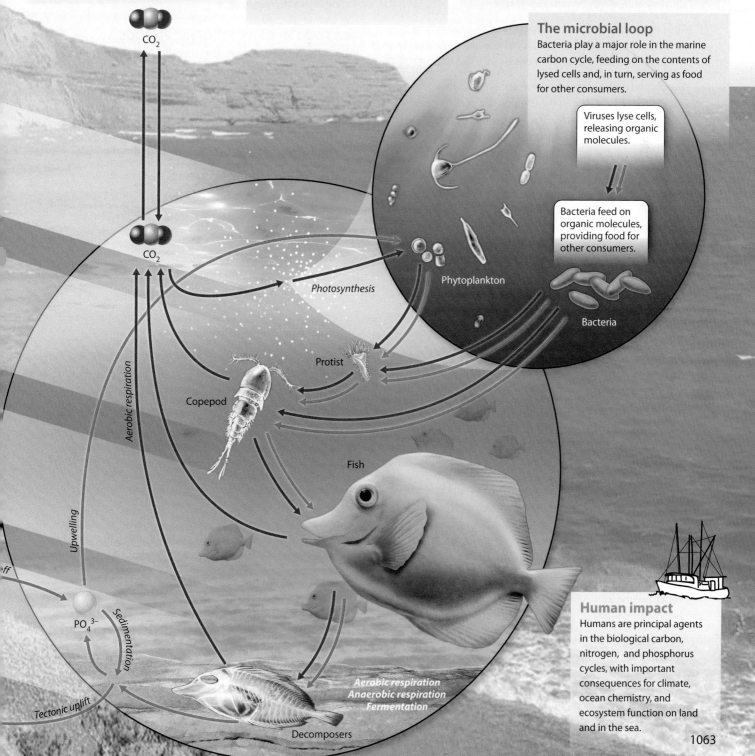

CORE CONCEPTS SUMMARY

46.1 THE SHORT-TERM CARBON CYCLE: Photosynthesis and respiration are the key biochemical pathways of the biological, or short-term, carbon cycle.

- Keeling's program to monitor CO_2 levels in the atmosphere revealed two patterns: (1) CO_2 levels oscillate on a yearly basis and (2) CO_2 levels have increased annually to the present day since measurements began. page 1044

- The first pattern (seasonal oscillation) can be explained by the global imbalance between photosynthesis and respiration in the summer and in the winter months. page 1045

- The second pattern (annual increase) can be extended further back in time by measuring the composition of air trapped in gas bubbles within polar ice. These measurements indicate that CO_2 levels in the atmosphere remained steady for more than a thousand years, but then started to increase in the mid-1800s, during the Industrial Revolution. page 1045

- The ratio of different isotopes of carbon in the atmosphere indicates that most of the carbon added to the atmosphere in recent decades comes from human activities, particularly the burning of fossil fuels. page 1046

46.2 THE LONG-TERM CARBON CYCLE: Physical processes govern the carbon cycle on longer timescales.

- Physical processes, which act at slower rates and over longer timescales, are the major drivers of the long-term carbon cycle. page 1048

- Over the last 400,000 years, CO_2 levels in the atmosphere have gone through periodic shifts that coincide with repeated cycles of glaciation. page 1050

- CO_2 levels in the atmosphere 500 million years ago may have been as much as 15 to 20 times higher than present-day levels (Fig. 46.7). page 1050

- In following carbon through its cycle, it is important to consider where carbon is stored (reservoirs) and the rates at which it is added to or removed from these stores (fluxes). page 1051

- Important reservoirs of carbon include organisms, the atmosphere, soils, the oceans, and sedimentary rocks. Sedimentary rocks are by far the largest carbon reservoir. page 1051

- Physical processes that remove CO_2 from the atmosphere include the weathering of rocks; processes that add CO_2 to the atmosphere include volcanoes and the oxidation of ancient organic matter in sedimentary rocks. page 1051

- The slow leakage of organic matter from biological communities to sediments accumulating on the seafloor links the short-term carbon cycle populated by organisms to the long-term carbon cycle driven by the Earth. page 1051

- Over time spans of hundreds of millions of years, movements of Earth's tectonic plates become major influences on the carbon cycle. page 1051

46.3 FOOD WEBS AND TROPHIC PERIODS: Species interactions cycle carbon and transfer energy through ecosystems.

- An ecosystem is the community and the physical habitat in which the organisms live. page 1053

- Photosynthetic and other autotrophic organisms are primary producers, converting CO_2 into organic molecules. page 1053

- Heterotrophic organisms that eat primary producers are called primary consumers. page 1053

- Heterotrophs that prey on grazers are called secondary consumers. page 1053

- Primary producers gain energy, carbon, nitrogen, phosphorus, and other elements from the environment. Consumers gain all of these resources from their food. page 1053

- Interacting networks of species, linked by predator–prey interactions, are called food webs. page 1053

- Trophic pyramids show that losses of energy and carbon from one level in a food web to the next result in ever decreasing biomass at higher trophic levels. page 1055

- Because of energy lost to heat, work, and the inefficiencies of consumers, each trophic level in an ecosystem generally has approximately 10% of the biomass found at the next-lower level. page 1055

- Carbon cycles through ecosystems, but energy is used and dissipated. page 1055

46.4 OTHER BIOGEOCHEMICAL CYCLES: Nitrogen and phosphorus also cycle through ecosystems.

- Nitrogen in the atmosphere is fixed by bacteria and archaeons, and then cycles through primary consumers, consumers, and decomposers; thus, the nitrogen cycle is closely linked to the carbon cycle. page 1057

- Phosphorus is found mostly in rocks, enters the food web as phosphate ion by chemical weathering, and cycles through ecosystems, supporting primary production. page 1058

46.5 THE ECOLOGICAL FRAMEWORK OF BIODIVERSITY: Biogeochemical cycles help us understand the evolution of biological diversity.

Primary producers are well adapted to obtaining sunlight, water, and nutrients in different environments. page 1059

Primary and secondary consumers are adapted to obtaining different sources and sizes of food. page 1059

The evolution of life on Earth can be considered a long history during which organisms have evolved different ways of obtaining energy and carbon from their environment. page 1059

The interactions between biological and geologic processes at work in the carbon cycle are also responsible for Earth's oxygen-rich atmosphere and oceans. page 1060

Log in to LaunchPad to check your answers to the Self-Assessment Questions and to access additional learning tools.

CHAPTER 47   Climate and Biomes

CORE CONCEPTS

47.1 CLIMATE: Solar radiation, wind and ocean currents, and topography determine the distribution of major climatic zones on Earth.

47.2 BIOMES: Biomes are broad, ecologically uniform areas whose characteristic species reflect regional climate.

47.3 GLOBAL PATTERNS: Global patterns of primary production and biodiversity reflect climate, nutrient availability, and evolutionary history.

Watching a movie or video, you often don't need a map to figure out where the action is taking place. Cactus and scraggly shrubs suggest the deserts of Mexico or the American Southwest; dense stands of low-growing conifers bring to mind central Canada or Siberia. The deciduous forests of eastern North America and Europe are equally distinctive, as are prairies and tropical rainforests. Plants conspicuously show that biological communities vary from place to place, reflecting the adaptations to local conditions of temperature, rainfall, and nutrient availability of the species that live there. If we take the time to survey the animals in these diverse communities, we will find that they are distinctive as well.

Biomes are broad geographic areas with similar communities. Climate exerts a major influence on the nature and distribution of biomes on land and in the sea. Before we consider this interaction between Earth and life, however, we must first ask what climate is and why it varies from place to place.

47.1 CLIMATE

Climate can be thought of as long-term average weather. In Boston, any given day in March may be warm or cold, dry or rainy, or even snowy. Nevertheless, records compiled over decades show that, on average, March days in Boston vary from about 42°F (5°C) at the beginning of the month to 47°F (9°C) at its end. March brings an average of 4.3 inches (110 mm) of precipitation in Boston, at least some of it typically snow. Montreal is colder and snowier in March; Phoenix is much warmer and drier.

What controls this broadly predictable distribution of average weather across the planet? Climate in Boston, and around the world, is determined by solar radiation, global patterns of wind and ocean circulation, and Earth's varying topography.

The principal control on Earth's surface temperature is the angle at which solar radiation strikes the surface.

In our solar system, Earth stands out as the planet with life. Our nearest neighbors, Venus and Mars, are, respectively, too hot for liquid water to persist and too cold. Earth is the Goldilocks planet: just the right distance from the sun so that, in conjunction with the greenhouse effect of the atmosphere, solar radiation maintains surface temperatures that can keep water in its liquid form. As we look across Earth, we see that climate varies markedly from one region to another. This variation reflects the relationship between Earth and the sun, modified by the physical features of Earth and the circulation of air and seawater.

Fig. 47.1 shows how mean annual temperature is distributed across Earth's surface. Earth is hot near the equator (the red colors in Fig. 47.1) and cold at the poles (blue and purple). This temperature distribution results from incoming solar radiation. Sunlight strikes the equatorial regions directly, but at higher latitudes (closer to the poles), the curvature of the Earth means that its surface is oriented at an angle to incoming radiation. To see what a difference this makes, shine a flashlight directly on a sheet of paper placed perpendicular to the beam of light. Now, tilt the paper at 45 degrees to the light. You'll notice that the same amount of light (and, therefore, the same amount of energy) is distributed over a

FIG. 47.1 Solar radiation and surface temperature. The angle at which solar radiation strikes our planet's curved surface is the primary determinant of Earth's surface temperatures.

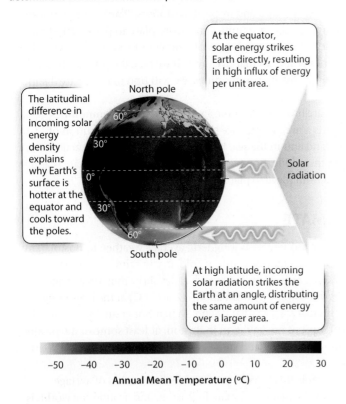

At the equator, solar energy strikes Earth directly, resulting in high influx of energy per unit area.

The latitudinal difference in incoming solar energy density explains why Earth's surface is hotter at the equator and cools toward the poles.

At high latitude, incoming solar radiation strikes the Earth at an angle, distributing the same amount of energy over a larger area.

Annual Mean Temperature (°C)

FIG. 47.2 Seasonality. Because Earth's axis of rotation tilts at an angle of 23.5 degrees, most parts of the globe experience seasonal variations in temperature.

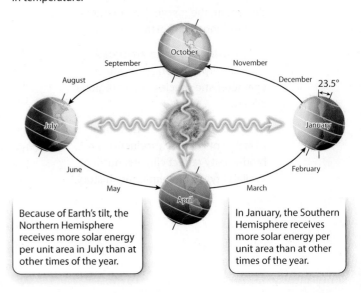

Because of Earth's tilt, the Northern Hemisphere receives more solar energy per unit area in July than at other times of the year.

In January, the Southern Hemisphere receives more solar energy per unit area than at other times of the year.

larger area. On Earth, the amount of solar energy striking a given area in the tropics is about twice that found near the poles, so tropical regions are warmer than polar regions.

At higher latitudes, temperature is not only lower, but also exhibits greater variation throughout the year. Seasonality reflects a second feature of Earth as a planet: Earth's axis of rotation is not oriented perpendicular to incoming sunlight but rather tilts at an angle of 23.5 degrees (**Fig. 47.2**). Because of this tilt, solar radiation strikes North America and Europe more directly in July than it does in January—and so these areas are warmer in summer and colder in winter. At the extreme, above the Arctic Circle, incoming solar radiation can be absent for days to months during the winter. A similar pattern of seasonality characterizes the Southern Hemisphere, but winter is in July and summer is in January. These patterns of temperature variation through the year go a long way toward explaining the geographic distribution of biomes across the globe.

Topography—that is, the physical features of Earth's surface—also contributes to global temperature patterns. Even at low latitudes, mountaintops can have glaciers, reflecting the fact that temperature declines with increasing elevation. In general, temperature drops about 3.6°F for every 1000 feet in elevation (6.5°C per kilometer). For this reason, climate and biomes on mountains change as elevation increases, mirroring the spatial pattern observed from lower to higher latitudes.

Heat is transported toward the poles by wind and ocean currents.

To further understand how temperature is distributed across the Earth, we need to think about another set of processes: those that transport heat energy from the equator toward the poles. Evidence that heat moves from low latitudes to high can be found in the distribution of temperatures at the Earth's surface. Temperatures at the equator are actually a bit cooler than we might predict based solely on how directly solar radiation strikes Earth at that latitude; the poles, in turn, are somewhat warmer than a diagram of solar radiation like Fig. 47.1 would predict. To reconcile these predicted and measured temperatures, we need to consider how heat is transported from low to high latitudes. This heat transport is carried out by wind and ocean currents.

The gas molecules that make up air are constantly in motion. When they are heated, they move faster, and (unless confined) the volume of the air expands. For this reason, warm air is less dense than colder air and, therefore, rises through the atmosphere. Not surprisingly, the rise of warm air is particularly strong at the equator (**Fig. 47.3**). As shown in the inset in Fig. 47.3, air cools as it rises. Once it reaches an altitude of 10 to 15 km above Earth's surface, the air no longer continues to rise but instead spreads toward the poles, continuing to cool as it goes. Eventually, at about 23 to 30 degrees north and south latitude, the cooling air becomes dense enough to sink back to the surface. This air continues to move along Earth's surface: some of it moves toward a pole, while the rest moves back toward the equator. In either case, the air warms again as it moves across Earth's surface, rising once more at the equator or at about 60 degrees north and south latitude. Thus, as shown in Fig. 47.3, Earth's lower atmosphere is organized into cells of rising and

FIG. 47.3 Wind currents. The rise of warm air through the atmosphere and the descent of colder air govern patterns of prevailing winds, transporting heat from the equator toward the poles.

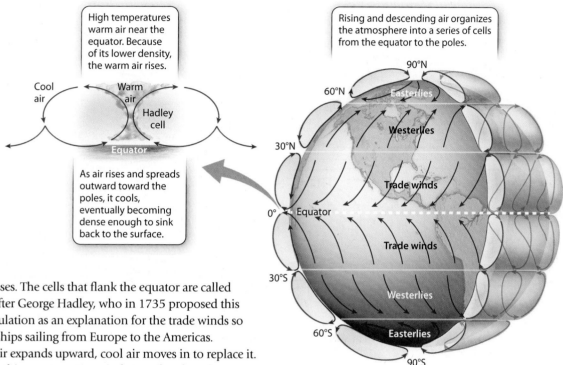

falling air masses. The cells that flank the equator are called Hadley cells after George Hadley, who in 1735 proposed this pattern of circulation as an explanation for the trade winds so important to ships sailing from Europe to the Americas.

As warm air expands upward, cool air moves in to replace it. We experience this movement as wind. As sailors have known for thousands of years, the prevailing direction of the wind differs from one latitude to another. Prevailing winds reflect those atmospheric cells of rising and falling air. Warm air moves toward the equator in some cells and away from the equator in others, with a net transport of heat occurring from the equator to the poles. The winds don't move strictly in a north or south direction; instead, prevailing wind direction has a strong east–west component (as, for example, with the westerlies and trade winds shown in Fig. 47.3). To understand this, we have to consider one additional aspect of our planet: its rotation about an axis.

Earth rotates in a counterclockwise direction, moving from west to east. In the course of one daily rotation, a spot at the equator moves through a distance equivalent to our planet's circumference—nearly 25,000 miles, or 40,000 km. Points at higher latitudes travel a shorter distance in a single rotation; the distance traveled eventually is zero at the poles. So, in a period of 24 hours, a point at the equator has traveled farther than a point at higher latitude; that is, the point at the equator moves more quickly than the point closer to the poles. As the cooling air moves north and south from the equator as shown in Fig. 47.3, the land beneath it rotates to the east, but at a slower speed than the land at the equator. Because of this, winds moving north or south appear to deflect to the right in the Northern Hemisphere and to the left in the Southern Hemisphere (**Fig. 47.4**). This phenomenon is called the **Coriolis effect**; in conjunction with the cells of rising and

FIG. 47.4 The Coriolis effect. Because of Earth's counterclockwise rotation about its axis, winds in the Northern Hemisphere deflect to the right; those in the Southern Hemisphere deflect to the left.

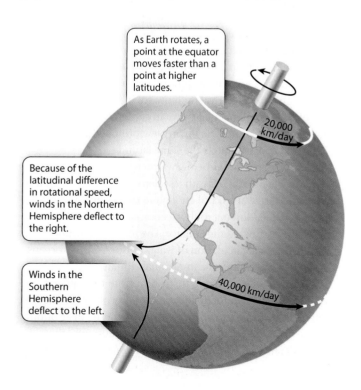

FIG. 47.5 Ocean currents. Prevailing ocean currents reflect the winds that drive the circulation of water at the ocean's surface and transport a large amount of heat from the equator toward the poles.

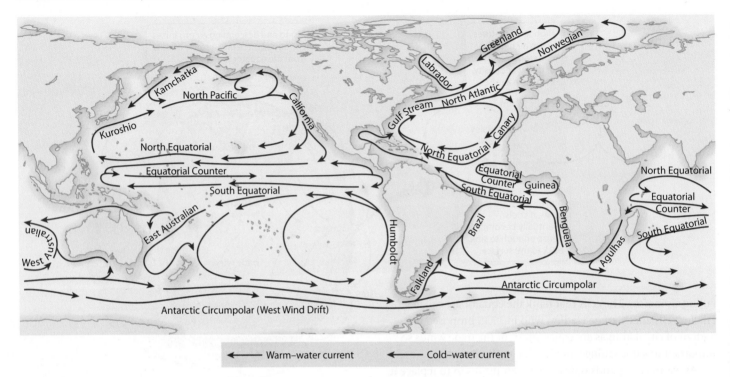

descending air masses, it explains the pattern of prevailing winds illustrated in Fig. 47.3.

Winds, in turn, push on water in the oceans, directing surface currents as shown in **Fig. 47.5**. Water can carry much more heat than air can carry, so ocean currents transport a great deal of heat toward the poles. For example, the ocean current known as the Gulf Stream (and its northeasterly extension, the North Atlantic Drift) transports warm water north along the eastern margin of North America and then east toward Europe. The heat it carries helps to keep northwestern Europe much warmer than land at equivalent latitudes in North America. Thus, subtropical plants that thrive in gardens of coastal Scotland would be unthinkable in Newfoundland, Canada, which lies at approximately the same latitude.

Just as colder air sinks below warmer and less dense air, cold dense waters sink beneath less dense water masses. At high latitudes, surface waters cool enough to sink beneath warmer, less dense waters, sinking to the deep ocean where they begin to move slowly toward the equator. The result is a complex three-dimensional circulation of water through Earth's oceans that plays an important role in regional climates.

In sum, wind and water transfer heat from the equator toward the poles, modifying the direct effects of solar radiation. Together, these processes result in the temperature distribution observed across the planet, explaining why palm trees that cannot tolerate freezing are confined to low latitudes, while deciduous trees that lose their leaves every fall occur mostly in the seasonal climates found at mid-latitudes. However, if you drive due west from New York to northern California, you will pass sequentially through deciduous forest, grass-dominated prairies, and stands of conifers on mountains. Clearly, there is more to the distribution of biomes than temperature alone.

Global circulation patterns determine patterns of rainfall, but topography also matters.

To fully understand the features and distribution of biomes around Earth, we must introduce one more physical principle: warm air can carry more water vapor than cold air can carry. Armed with this insight as well as the observation from the previous section that on a global scale, heat is transferred from the equator to the poles in cells of rising and falling air, we return to the global circulation of air. Especially at the equator, warm air carries a great deal of water vapor, but, as it rises and the air cools, the amount of water vapor the air can carry decreases. As a result, the vapor condenses, and it rains. Conversely, the air descending at about 23 to 30 degrees latitude north and south warms as it sinks; thus it takes up water vapor rather than releasing it, leading to aridity (**Fig. 47.6**). Zones of high annual rainfall (polar and subtropical highs in Fig. 47.6) and low annual rainfall (subpolar lows and ITCZ equatorial low) follow this broad pattern of rising and descending masses, with the highest rainfall occurring near the equator and the

FIG. 47.6 Global rainfall patterns. The rise and fall of air masses govern the global distribution of rainfall.

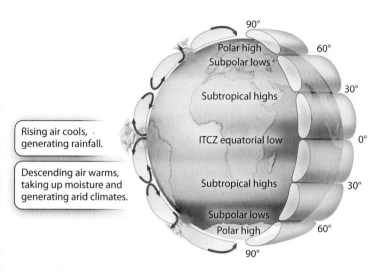

FIG. 47.7 Topography and rainfall. Mountains impose a regional pattern on climate because of the rain shadow, which results in differing amounts of rain on opposite sides of mountain ranges. *Data for top map from PRISM Group and Oregon Climate Service, Oregon State University.*

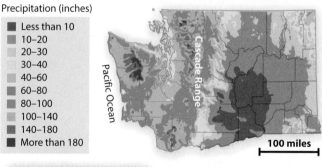

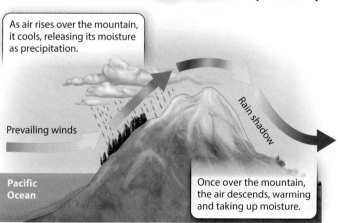

lowest rainfall between 23 and 30 degrees latitude north and south.

Regional precipitation (rain or snow) patterns are actually more complicated than this simple picture would suggest, as seen in the precipitation map for Washington state (**Fig. 47.7**). From the coast eastward into the Cascade Mountains, Washington and neighboring British Columbia are wet all year long, supporting magnificent conifer forests. However, east of the Cascades and the Coast Mountains that extend to the north, the land is dry. As this observation suggests, the mountains themselves impose a regional pattern of rainfall. This effect, called a rain shadow, reflects the same general principles of rising and descending air masses. As shown in Fig. 47.7, as wet air moves from the ocean up into the mountains, it cools, releasing its moisture as precipitation. Moving past the mountains, air masses descend, warming as they go and taking up water vapor. For this reason, lands in the rain shadow of the mountains are arid.

In general, regional climate reflects the interactions among solar radiation, global patterns of circulation, and Earth's varying topography. Biomes, in turn, reflect these variations in climate.

Self-Assessment Questions

1. Explain how incoming solar radiation exerts a direct influence on the distribution of temperature across Earth's surface.
2. How would climates be different if Earth's axis of rotation were perpendicular to incoming solar radiation rather than tilted at 23.5 degrees?
3. Why are equatorial latitudes generally wet, but land masses at about the 23 to 30 degrees north and south latitudes generally dry?
4. Why are the sides of tall mountains that face oncoming wind generally wetter than their opposite slopes?

47.2 BIOMES

Most of us are familiar with the dense canopies of tropical rainforests and the rolling grasslands of prairies. Likewise, we know that coral reefs occur in tropical to subtropical waters. As we noted at the beginning of the chapter, we can recognize these and other biomes based on their characteristic communities. On land, the most conspicuous feature of any biome is its vegetation, which reflects the evolutionary adaptation of plant form and physiology to climate. But how close is this relationship between biome and climate?

In 1884, Wladimir Köppen published a global climate map that, with relatively few changes, is still used today. You might wonder just how much information was available to Köppen about temperature and rainfall in, for example, the

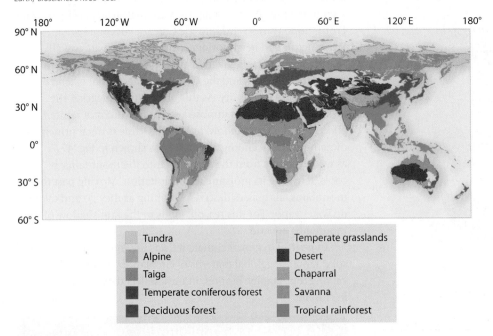

FIG. 47.8 Global distribution of terrestrial biomes. Distributions of terrestrial biomes around the world reflect regional climates. *Data Source: D. M. Olson et al., 2001, "Terrestrial Ecoregions of the World: New Map of Life on Earth," Bioscience 51:933–938.*

Legend: Tundra, Alpine, Taiga, Temperate coniferous forest, Deciduous forest, Temperate grasslands, Desert, Chaparral, Savanna, Tropical rainforest

Amazon rainforest or central Asian steppes. The answer: not much. In fact, Köppen's diagram actually charts global vegetation; the reason it works so well as a climate map is that plant form strongly correlates with global patterns of temperature and rainfall (**Fig. 47.8**). In designating major climate zones according to their characteristic vegetation, Köppen mapped out Earth's terrestrial biomes.

Terrestrial biomes reflect the distribution of climate.

In Chapter 27, we saw that plants perform a physiological balancing act, opening their stomata to allow carbon dioxide into leaves, but losing water vapor through transpiration (evaporation from leaves) as a result. **Evapotranspiration** is the sum of evaporation directly from soils and water bodies plus the amount transpired by plants (**Fig. 47.9**). In many ecological communities, transpiration returns a large amount of water vapor to the atmosphere. For example, a single acre of corn can transpire 3000–4000 gallons (10,000–15,000 L) of water per day.

As we've seen, temperature varies with latitude and altitude, as do humidity and wind. Temperature, humidity, and wind all determine the rate at which evapotranspiration removes water from plants and soils. The water within plants is replenished mostly by roots that take up water from soil. This water, in turn, must be replenished by precipitation and ground water flow. As a consequence, rates of transpiration by plants reflect both rate of photosynthesis and the availability of soil and ground water. Where water resources are limited, transpiration is limited as well. Plants conserve water by closing their stomata, which restricts intake of carbon dioxide and photosynthesis.

In deserts, which have low annual precipitation, the moisture in soils evaporates rapidly. Therefore, to conserve water, desert plants must keep transpiration (and therefore photosynthetic rate) to a minimum. The plants' ability to store water in their tissues provides a defense against drought, so succulents—plants such as cactus, which store water in their stems—are common. At the other extreme, rainforests experience high rates of precipitation, which permit nearly unlimited transpiration (and therefore high photosynthetic rates). The result is wet forests with high biomass. Intermediate levels of water availability and evapotranspiration

FIG. 47.9. Evapotranspiration. The supply (precipitation and groundwater flow) and demand (evapotranspiration) for water determine the nature of vegetation in terrestrial biomes.

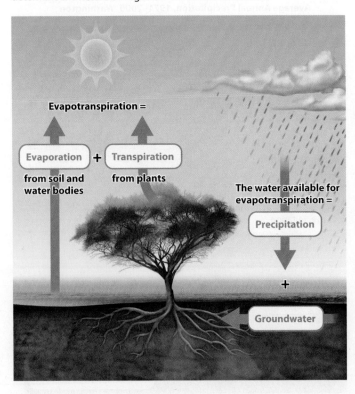

characterize the temperate zones that encompass much of the United States, Europe, and parts of eastern Asia, as well as mid-altitudes, below the highest mountain peaks but well above the coastline. Both latitude and altitude influence evapotranspiration through their effects on temperature and precipitation, and the combined effects of these factors are reflected in the distribution of biomes around the world (**Fig. 47.10**).

In similar biomes on different continents, dominant plants generally look much the same, but commonly they are not closely related. Plants have independently evolved adaptations to particular climatic regimes by convergence (Chapter 22). For example, the cacti that grow in deserts of western North America look very similar to plants in arid regions of Africa, but the African plants belong to an entirely different family (**Fig. 47.11**). Convergent evolution is also seen in the trees of rainforests in different regions of the world. Distantly related trees have evolved leaves with long tips that allow excess water to drip away, a reddish color in new leaves that protects them from intense sunlight, and roots with high buttresses that support tall trees atop thin soil.

In terrestrial biomes, vascular plants dominate primary production, although mosses, lichens, algae, and cyanobacteria contribute as well. Plants also provide a physical structure for biomes, with high canopies in wet temperate forests and tropical forests, but shorter vegetation where water or nutrients are scarce, or where the growing season (the part of the year when temperature and precipitation permit plant growth) is short.

Primary consumers include a variety of animals. These consumers have adaptations that facilitate feeding on specific plant resources, and—like the plants they eat—consumers commonly show evidence

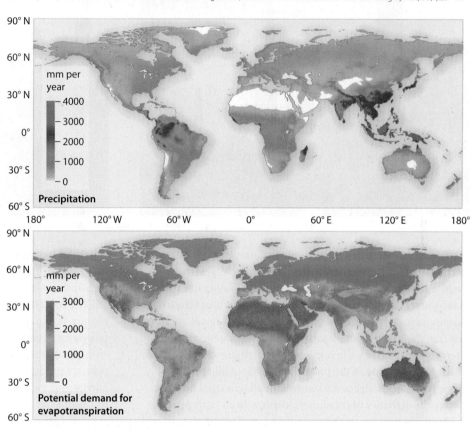

FIG. 47.10 Evapotranspiration, precipitation, and terrestrial biomes. Precipitation and the potential for evapotranspiration change with latitude and altitude. The ratio between precipitation and evapotranspiration acts as a primary determinant of vegetation type. *Data from International Water Management Institute Global Monthly Dataset. Relationship of biomes to precipitation and temperature from J. R. Mather and G. A. Yoshioka, 1968, "The Role of Climate in the Distribution of Vegetation," Annals of the Association of American Geographers, 58(1):29–41.*

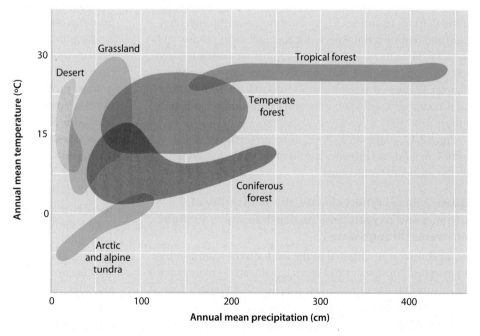

FIG. 47.11 Convergence of plant form. (a) Cacti in the deserts of North America and (b) euphorb plants in Africa look similar because they are both adapted to conserve and store water, but they are not closely related. *Sources: a. McDonald Wildlife Photography/Animals Animals; b. Michele Westmorland/naturepl.com*

of convergent evolution. In many environments, mammals and insects are the principal herbivores, grazing on leaves, fruits, seeds, and stems. Insects and vertebrates also play major roles as secondary consumers. Predatory insects in various terrestrial biomes include hornets, the praying mantis, and larval ant lions that trap their prey in small sandy pits. Terrestrial biomes are often characterized by the carnivorous vertebrates that live there, which include some of the most charismatic of all animals— for example, lions, tigers, and wolves. In all terrestrial biomes, despite the activities of consumers, detritus from both plants and animals ends up on the ground and may accumulate in soil.

Decomposers, especially fungi, protists, and bacteria, feed on this detritus, returning CO_2 to the atmosphere through respiration and returning nutrients to the soil. Still more protists and animals feed on these decomposers, populating a detritus-based food web within the soil. Soil animals include tiny nematodes (Chapter 42) as well as larger earthworms that eat both detritus and decomposers, ventilating the soil as they tunnel in search of food.

Fig. 47.12, pages 1075–1082, summarizes the principal biomes on land and provides graphs of their average annual precipitation and temperature.

Aquatic biomes reflect climate, the availability of nutrients and oxygen, and the depth to which sunlight penetrates through water.

Life, of course, is not limited to the land; diverse communities thrive in both freshwater and marine (saltwater) environments. Organisms that live in water face functional challenges distinct from the challenges on land, and this helps to explain the nature of aquatic biomes. Terrestrial organisms evolved from ancestors living in water, but to understand global ecology, it is helpful to begin with familiar terrestrial biomes and then move on to aquatic examples, which we generally experience only indirectly.

The most obvious distinction between terrestrial and aquatic biomes is that water is denser than air. For primary producers, buoyancy in water can provide mechanical support, reducing or eliminating the need for structural tissues such as wood. Also, primary producers on land are tethered to the ground both by gravity and the need to extract nutrients and water from soil, but this is not the case in aquatic biomes. In aquatic environments, nutrients are dissolved in the water, allowing primary producers to live as buoyant cells within surface waters, taking in dissolved nutrients and harvesting sunlight, often far from shore.

As it the case on land, the structure and distribution of primary producers governs the occurrence and abundance of consumers. Aquatic plants and macroscopic algae are important primary producers in shallow aquatic environments, where they can take advantage of nutrients transported from land while remaining anchored to the riverbed, lakebed, or seafloor. In larger lakes and the open ocean, however, phytoplankton— photosynthetic microorganisms that float in sunlit waters—are the predominant primary producers, combining high rates of primary production with low biomass.

In shallow aquatic environments such as lake margins and coastal marine settings, much plant, algal, and animal debris accumulates in sediments, supporting abundant growth by bacteria and various invertebrates. These decomposers provide food for many larger consumers, including the larvae and juveniles of commercially important fish and invertebrate species. Fungi, so important in terrestrial biomes as decomposers, can also have important actions in decomposing plant litter in streams and lakes. Fungi are generally thought to play only

FIG. 47.12 Common land biomes. Sources: (Tundra) Helge Schulz/Picture Press/Getty Images; (Alpine) Rolf E. Kunz/AGE Fotostock; (Taiga) Uwe-Bergwitz/Getty Images; (Temperate coniferous forest) Charlie Ott/Science Source; (Deciduous forest) NatureByDarrellYoung/Alamy; (Temperate grassland) Robert and Jean Pollock/Science Source; (Desert) Wayne Lynch/Getty Images; (Chaparral) Amit Basu Photography/Getty Images; (Savanna) age fotostock/Alamy; (Rainforest) Berndt Fischer/Getty Images. Climograph data sources: (Tundra) Michael E. Ritter, 2006, The Physical Environment: An Introduction to Physical Geography; (Alpine) climate-data.org; (Taiga) https://borealshieldandplains.weebly.com/; (Temperate coniferous forest) Michael E. Ritter, 2006, The Physical Environment: An Introduction to Physical Geography; (Deciduous forest) http://drought.unl.edu/archive/climographs/images/CharlestonWVMetric.jpg; (Temperate grassland) climate-data.org; (Desert) http://media.diercke.net/omeda/800/12664E_1_Klimadia_In_Salah.jpg; (Chaparral) National Climatic Data Center; (Savanna) http://www.geoforcxc.com/natural-systems/weather-and-climate/ecosystems/tropical-continental-climate/; (Rainforest) World Meteorological Association.

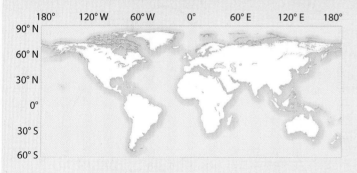

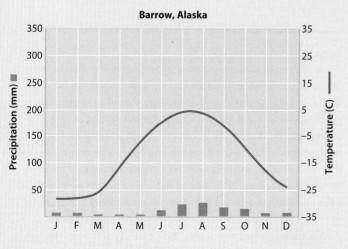

Tundra

Tundra is the coldest biome, and short days in winter limit the growing season. Tundra occurs close to the North Pole, above 65° N. The South Pole is largely surrounded by the ice and seas of Antarctica, so there is very little area with plants. The climate diagram from Barrow, Alaska, shows that in tundra biomes, both temperatures and precipitation are low, with highs for both occurring during the summer. Although precipitation is low, rates of evaporation are also low. Because drainage is commonly poor, the ground is usually waterlogged and permanent ice occurs below a few centimeters of soil. Primary producers are mostly mosses, lichens, herbs, and low shrubs. Grasses and sedges occur in drier places, as do other flowering plants. Plant diversity is low, and most plants are small. Caribou are conspicuous grazers, but other primary consumers, including musk oxen, rabbits, birds, and insects, occur as well. Wolves and foxes are key predators. Fungi and bacteria decompose organic detritus, but low temperatures and waterlogged soils limit rates of respiration.

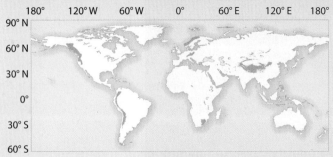
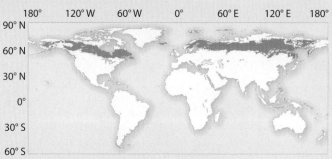

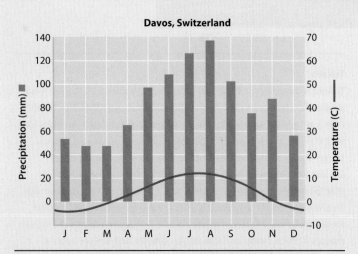
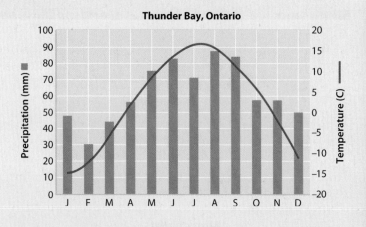

Alpine

The Alpine biome is similar to tundra but lacks permanent ice below the soil, and the temperatures vary more widely. Alpine areas occur throughout the world, often at approximately 10,000 feet (3000 m) at lower latitudes, but always just below the elevation where snow cover persists throughout the year. Because of their altitude, they are windy, cold places. The thin atmosphere provides only limited protection from UV radiation, so many alpine plants are low and slow growing. Grasses are abundant, as are herbs that can provide spectacular wildflower displays during the short growing season. Alpine grazers include mountain goats, llamas, yaks, and marmots, as well as seasonally active insects. Predators include wolves and cats such as the Himalayan snow leopard.

Taiga

These cool, moist forests occur from 50° N to 65° N. The short summer brings rain, and most of the plants are low-growing conifers like spruce, fir, larch, and pine, with an understory dominated by shrubs in the blueberry and rose families. Mosses commonly form an extensive and soggy groundcover. The soils are deep with accumulated organic matter because the low temperatures result in slow decomposition, but they are also acidic and poor in nutrients, helping to explain the short stature of most plants. Birds, insects, and other invertebrate animals are more abundant and diverse than those found in tundra biomes, but mammals are the most conspicuous consumers. Grazers include elk, moose, caribou, porcupines, hares, and a diversity of rodents. Bears, lynx, wolverines, weasels, mink, wolves, and foxes are well-known predators.

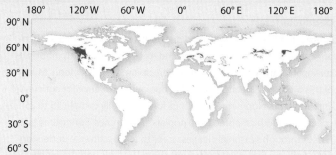

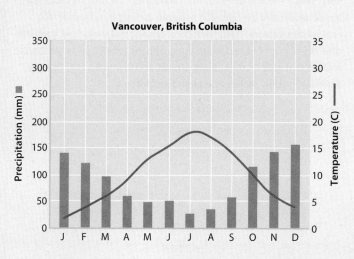

Temperate coniferous forest

Two broad areas of temperate coniferous forest occur below 50° N in North America, northern Japan, and parts of Europe and continental Asia. Along the Pacific coast of North America, warm summers and mild winters, along with abundant precipitation in the winter, permit growth of enormous conifers such as Douglas Fir, Red Cedar, Sitka Spruce, and redwoods. Much of the undergrowth consists of grasses, ferns, and members of the blueberry family. Understory shrubs are uncommon. In the interior of North America, much less precipitation and colder winter temperatures support drought-resistant conifers such as Ponderosa and Lodgepole Pines and Englemann Spruce. Diverse invertebrate species live in the soils, supported by active microbial decomposers. Insect and vertebrate diversity are all higher in temperature coniferous forests than in taiga communities.

Deciduous forest

Forests dominated by hardwood deciduous trees occur across much of eastern North America, Europe, and Asia in climates with moderate seasonal temperature variation and precipitation all year long. Much of this biome has been subjected to human disturbance for agriculture and urban development. Deciduous forests usually contain 15 to 25 tree species, including maples, oaks, poplars, and birches. Springtime sun passes through seasonally leafless trees to reach a diverse understory flora. Soils are rich in nutrients from annual leaf fall, and the moderate temperatures and precipitation promote decomposition, while the cool winters promote accumulation of organic materials. Soils contain a diversity of microbes and fungi, along with the protists and invertebrate animals that feed on them. Insects, birds, and mammals are relatively diverse, joined by snakes, lizards, and amphibians.

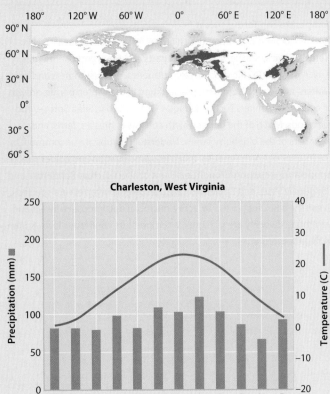

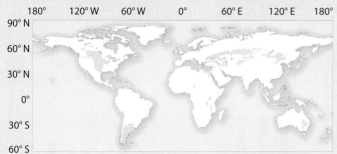

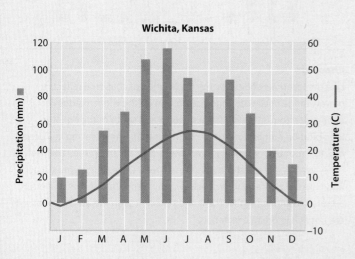

Temperate grassland

Before human settlement, this biome occupied most of the midwestern United States and south central Canada, dominated by blue-stem and buffalo grasses. Disturbance by fire helps to maintain grass populations in this biome. Lack of precipitation also prevents many species of trees from surviving in this region, and those trees that do grow successfully are usually found in low areas that have more available moisture. Because of the presence of active decomposers, soils in temperature grasslands accumulate nutrients, providing some of the most productive agricultural lands in the world. Before colonization by humans, temperate grasslands of North America supported abundant and diverse grazers, including bison, pronghorns, horses, and mammoths. Wild horses and mammoths became extinct in North America about 11,000 years ago, and herds of bison and pronghorn dwindled in the nineteenth century. Burrowing grazers such as prairie dogs remain relatively common, although their population sizes have fallen; these species support predators such as ferrets, badgers, foxes, and birds of prey.

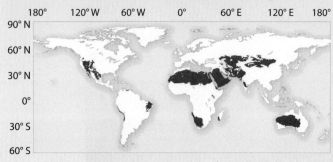
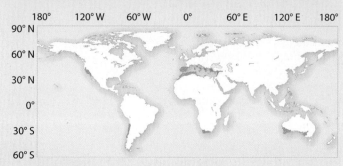
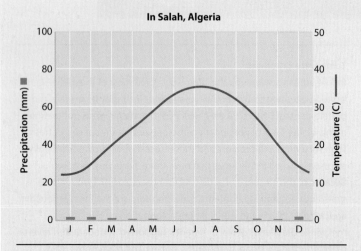
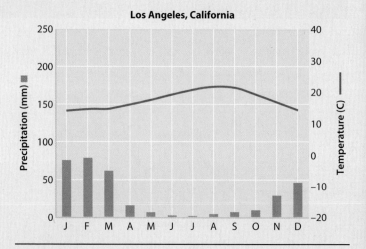

Desert

Desert occurs in continental interiors north and south of the equator from 15° to 35° latitude (Fig. 47.6). Wind patterns prevent this biome from receiving more than a few centimeters of precipitation annually. Plants commonly have deep roots, and many, including cactus and euphorb plants, have stems adapted to store water. Primary production is low, and soils are poor in nutrients but may have high surface salt due to evaporation. Given the low rates of primary production, primary consumers tend to be small, and include diverse lizards as well as rodents. Predators include snakes, cat species, coyotes, and birds such as owls, hawks, and eagles.

Chaparral

Like the locations of deserts, the distribution of chaparral reflects a narrow range of climate conditions found on the western edge of continents from 32° to 40° latitude north and south of the equator. Precipitation ranges from 30 to 75 cm per year, usually falling in a period of 2 to 4 months. Typical plants are herbs that die and reseed themselves every year, evergreen shrubs, and small trees. Olives, eucalyptus, acacia, and oaks are typical woody species, always drought resistant and often adapted to withstand fire. The limited precipitation means chaparral soils are not rich in organic materials.

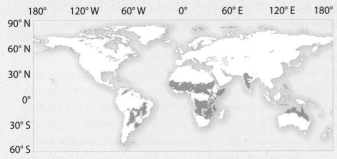

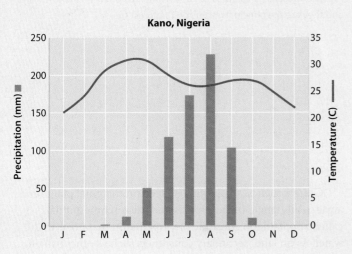

Savanna

Tall, perennial grasses dominate this biome, which occurs in warm, relatively dry regions of eastern Africa, South America, and Australia. Rain is seasonal and ranges from 75 to 150 cm per year. Scattered trees and shrubs usually drop their leaves in the dry season to conserve moisture. As in temperate grasslands, fire plays a key role in maintaining this biome. Large mammalian grazers are abundant and diverse; they include the migrating antelopes, zebras, and giraffes well known from Africa, but also kangaroos and other marsupials in Australia and large rodents in South America. Predator diversity can also be high, as exemplified by lions and other cat species, hyenas, and wild dogs. Dingoes are important predators in Australian savannas, but were introduced only a few thousand years ago from southeastern Asia. Before that time, the top carnivores included large, now-extinct lizards, as well as snakes and smaller lizards still present in the biome.

Tropical rainforest

The most diverse of all terrestrial biomes, tropical rainforests extend north and south of the equator from 10° N to 10° S. Temperatures are warm, and annual precipitation commonly exceeds 250 cm. Tropical rainforests harbor a great diversity of species; tree diversity alone often exceeds 300 species per hectare. Trees grow tall, and many have buttressed roots for support. Vines and other epiphytic plants are common. Most leaves are evergreen and leathery, and many have long pointed tips that facilitate drainage of excess moisture. Because of the high temperatures and heavy rains, decomposition by fungi and bacteria is rapid, preventing the accumulation of organic materials in clay-rich or sandy soils. Few large grazers are found in rainforests, but smaller mammals such as primates, bats, and rodents are highly diverse, as are birds, snakes, and lizards. Insects are especially abundant and diverse. Ants alone make up as much as 30% of animal biomass in rainforests, and they are the principal grazers on rainforest trees.

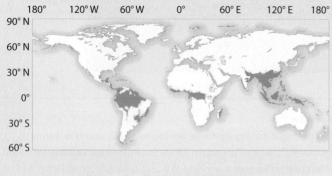

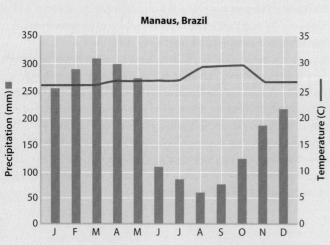

a minor role in the oceans, although recent surveys of marine sediments show that in areas with abundant organic matter and little oxygen, microscopic fungi capable of fermentation can be prominent decomposers. Bacteria also play a critical role in cycling the remains of phytoplankton and other open-water organisms, both within the water and on deep lakebeds or seafloors.

Grazers such as snails and fish feed on marginal plants and algae, while single-celled protists and microscopic animals called zooplankton consume phytoplankton that floats in the water. As on land, secondary consumers such as other fish or squid feed on the grazers, adding complexity to food webs. Plankton, whether photosynthetic or heterotrophic, generally have only limited powers of locomotion, but rather drift with

FIG. 47.13 Common freshwater biomes. *Sources: (Lakes) Ken Gillespie/Getty Images; (Rivers) Ignacio Palacios/Getty Images.*

Lakes
Lakes and ponds range in size from tiny temporary pools to the Great Lakes of North America, which contain more than 20% of all freshwater habitats in the world. Lakes range geographically from the equator to above 80° N. The ions dissolved in the waters, pH, and salinity vary with the types of terrain in which lakes are found, and nutrient levels can range from low to extremely high. Aquatic plants and macroscopic green algae live along the shallow margins of lakes, and both algae and cyanobacteria are primary producers throughout the photic zone. Zooplankton include small arthropods and rotifers, whereas the nekton is dominated by fish. Turtles and birds also play important roles as grazers and predators in many lakes.

Rivers
Rivers are freshwater biomes characterized by moving water. Because of their currents, streams and rivers erode, transport, and deposit sediment, so that their shape continually changes. Rivers and streams, like lakes, vary tremendously in size and chemistry. Rivers are generally well oxygenated, although oxygen levels may be lower in slowly moving rivers on floodplains. As with lakes, plants and large algae grow along river margins, while phytoplankton photosynthesize throughout the water column. Insects are important consumers, but other invertebrates, fish, turtles, and birds can all be abundant and diverse. Salmon and some other fish spend much of their lives in the oceans but swim upriver to reproduce.

the current. In contrast, some consumers swim actively through their watery surroundings; called nekton, they include fish, squid, and whales. Because oxygen gas dissolves in water only to a limited extent, surface waters directly beneath the atmosphere have an O_2 concentration about 20 times lower than that of the air. This compels larger animals to move constantly to keep their gill systems well oxygenated, or to live in places such as cold running streams and rocky marine coastlines where currents naturally produce well-oxygenated water.

The main freshwater biomes are lakes and rivers (**Fig. 47.13**). These vary tremendously in climate as well as in nutrient input and oxygen availability. Commonly, stratification reflects temperature differences between surface and deeper waters, since cool water is denser than warm water. In general, the turbulent flow of water ensures that rivers are well oxygenated. In contrast, some lakes are physically stratified, meaning that the density of bottom waters keeps them from mixing upward with shallow waters. Organic debris that sinks to deep waters provides food for respiring bacteria, which in turn draws down oxygen levels. Where nutrient influx is high, oxygen levels in such lakes can become severely depleted in their deeper parts; this outcome is particularly common where human activities have increased the nutrient loads entering the lake (Chapter 48). Seasonal cooling of surface waters can eliminate temperature differences from top to bottom, causing previously stratified water masses to mix and bringing oxygen to deep waters and nutrients to the surface. The seasonal influx of nutrients to surface waters commonly results in seasonal variations in both primary producers and rates of photosynthesis

Marine biomes cover most of our planet's surface.

Only approximately 2.5% of Earth's water is fresh, and most of this water occurs in glaciers, permafrost, and groundwater within soils. Overwhelmingly, the water on our planet is seawater. Covering 71% of Earth's surface and reaching a depth of nearly 11,000 m (almost 7 miles), the oceans constitute Earth's largest biomes. In fact, the oceans are so large and so diverse in their biology that ecologists commonly subdivide them into distinct zones based on depth and proximity to the shoreline (**Fig. 47.14**).

FIG. 47.14 Biomes in the sea. Marine biomes (those discussed in Fig. 47.15 are noted in red) reflect distance from the shore and the depth of ocean basins.

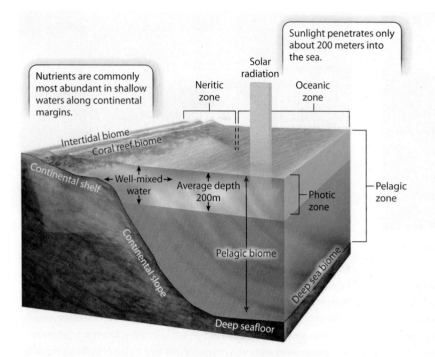

The dimensions of the oceans determine the distribution of nutrients, oxygen, and light. Runoff from the continents (supplemented by nitrogen fixation by marine cyanobacteria) provides most of the nutrients required by primary producers in the sea, but because sunlight rarely penetrates much deeper than about 200 m (called the **photic zone**), most of the ocean's volume is off limits for photosynthetic organisms. The **neritic zone** encompasses environments that are near the shore, with a shallow seafloor, relatively high availability of nutrients, and persistent sunlight. Distinct biomes within the neritic zone include intertidal and coral reef biomes (Case 8 Conserving Biodiversity Hotspots). Deeper waters beyond the continental shelf make up the **oceanic zone**.

Within the oceanic zone, biologists commonly distinguish between the pelagic biome in open ocean water and the deep sea biome on the deep seafloor (Fig. 47.14). In the open ocean, nutrients are commonly less abundant, and waters are deeper, so much of the pelagic biome does not receive sunlight. Life in the deep sea biome is largely sustained by organic debris that rains downward from surface waters, although chemoautotrophic bacteria and archaeons contribute as well. Particle- and filtering-feeding animals, as well as microbes, respire this rain of organic matter, which returns phosphorus and nitrogen to the water. Nutrient-rich deep waters are eventually transported back to the surface, in a process called upwelling. Upwelling waters occur mostly near the shores of continents, where they support locally high rates of photosynthesis. Upwelling is commonly seasonal, so primary producers in this zone change from one part of the year to another. Upwelling regions make up only about 1% of the ocean's surface area, but by some estimates, they support more than 20% of all commercial fish populations.

Many different groups of algae contribute to primary production on continental shelves, with diatoms, coccolithophorids, and dinoflagellates dominating in most regions. These algae turn over rapidly: most are eaten by grazers or lysed by viruses within a few days. As a consequence, their total biomass is commonly lower than that of zooplankton that graze on them, inverting the trophic pyramid characteristic of terrestrial ecosystems (Chapter 46). In the middle of large ocean basins, broad areas called gyres, which are bounded by major currents, lie far from continental and upwelling sources of nutrients; their low primary productivity is dominated by cyanobacteria and tiny algae discovered only recently.

Viruses play a particularly large role in marine biomes, lysing as much as 30% of phytoplankton daily in coastal ecosystems and releasing large amounts of organic molecules into the water column, the water that overlies the seafloor. The organic matter supports large populations of respiring bacteria and archaeons, which play an important role in returning carbon to the environment as carbon dioxide (Chapter 24). These microbial populations also support diverse populations of larger consumers.

Earlier in this chapter, we noted that winds drive ocean circulation, but winds affect only the uppermost waters of the sea. The surface layer that is mixed by the wind, broadly equivalent to the photic zone, is well oxygenated. In contrast, in deeper waters, oxygen can become depleted by respiration as microbes feed on sinking organic matter. In a few places where primary production is high on the surface, the oxygen level in water masses just beneath the mixed layer declines to essentially zero. In these areas, an oxygen-minimum zone forms that supports mostly anaerobically respiring microbial populations and features a limited presence of zooplankton and animals. On the deep seafloor and in the waters above the seafloor, oxygen is present but food is sparse, consisting only of a slow rain of particles from shallower depths. Lifeforms consist of scattered populations of animals (especially fish, brittle stars, and polychaete worms), heterotrophic protists, and microbes. One notable exception occurs at hydrothermal vents, where high concentrations of sulfide, hydrogen, and methane support primary production by chemosynthetic bacteria and archaeons. These, in turn, support locally dense populations of animals.

Key marine biomes are summarized in **Fig. 47.15**, pages 1085–1086.

FIG. 47.15 Common marine biomes. *Sources: (Intertidal) Dan L. Perlman/EcoLibrary; (Coral reefs) DIRSCHERL Reinhard/hemispicture.com/Getty Images; (Pelagic realm) Michael Aw/SeaPics.com; (Deep sea) Barrett Lia/SeaPics.com.*

Intertidal
The part of the neritic (shallow-water) zone closest to shore is the intertidal biome, the part of coastlines between the high and low tide marks. Intertidal organisms are exposed to the atmosphere on a daily basis; thus they are challenged by periodic desiccation as well as by temperature extremes that accompany exposure to direct sunlight. As a consequence, the intertidal biome has distinctive species associations. In sandy intertidal zones, or tidal flats, many animals burrow into the wet sand to escape exposure at low tide. In rocky intertidal zones, organisms such as mussels and barnacles may close their shells tightly when the tide is out. Waves can sweep organisms away; to avoid this fate, especially along rocky coastlines, both algae and animals securely attach themselves to the substrate. Nutrient levels can be high, favoring strong algal growth. Consumers include a variety of sea stars, sea urchins, mollusks, barnacles, and polychaete worms. Intertidal biomes exhibit the greatest diversity where tidal exposure is relatively short. Tidal pools harbor diverse animal and algal species.

Coral reefs
Coral reefs form another neritic biome—indeed, the most diverse biome in the oceans. The best-known reefs occur in shallow, tropical to subtropical environments where the skeletons of corals form physical structures that rise above the seafloor. Nutrient levels are commonly low, but primary production is high and mostly tied to dinoflagellate algae that live within the tissues of corals. Rapid recycling of nutrients between the corals and their algal symbionts ensures high levels of primary production. Free-living algae occur in reefs, but are kept at low abundance by grazing fish. Many species of invertebrates live within the reefs, often in nooks and crannies among coral heads that provide shelter from predators. Fish diversity can be especially high. Coral reefs also occur in the deep sea, but build gradually there, as coral growth occurs only slowly in the absence of symbiotic algae.

Pelagic
The pelagic biome—the part of the ocean that is neither close to shore nor close to the seafloor—forms the bulk of the oceanic system. Both plankton and nekton live within the water column. In the upper 200 m or so of the pelagic biome, sunlight permits photosynthesis, with diverse algae and cyanobacteria acting as primary producers. In deeper waters, sunlight is absent and life is sustained by the rain of organic particles from surface waters. Zooplankton are mostly small arthropods, while fish and cephalopods are key components of the nekton. Although this is true in both surface and deep waters, animals are less abundant in deep waters. Fish and squids are conspicuous consumers, but recent research shows that, in the upper water column, heterotrophic protists may actually be more abundant and diverse than fishes. Bacteria and archaeons occur throughout the pelagic biome.

Deep sea

Kilometers below the surface, deep seawaters are cold and dark, though not sterile. Sinking detritus supports sparse populations of animal consumers, as well as bacteria and archaeons that feed on organic particles or generate energy and organic molecules by chemosynthetic pathways tied to the nitrogen cycle (Figure 24.20). Despite their limited biomass, deep-sea biomes exhibit species diversity as high as those of shallow marine communities. Nevertheless, the composition of communities in deep-sea biomes differs relatively little from place to place around the world, so the total diversity of these environments is lower than that found within the photic zone globally. Dense animal populations occur locally on the deep seafloor, where hydrothermal vents expel fluids containing high abundances of hydrogen, hydrogen sulfide, and methane. These compounds support locally high levels of primary production by chemosynthetic bacteria, many of them present as symbionts within the tissues of vent animals (Chapter 24).

Self-Assessment Questions

5. Why can wet soils occur both near the equator, where rainfall is high, and at high latitudes, where rainfall is low?
6. Choose five terrestrial biomes, describe their climate and vegetation, and explain why they differ from one another.
7. How do marine biomes differ between the surface ocean and the deep seafloor?

47.3 GLOBAL PATTERNS

Biomes show how the physical environment influences the abundance, form, and diversity of species in different parts of the planet. In Chapter 46, we discussed another aspect of the interactions between organisms and their environments: the ways that ecosystems cycle energy and nutrients. What do we see when we look at these patterns globally, rather than at the ecosystem or biome level? In this section, we focus on global patterns of primary production and biodiversity, both of which depend on interactions between Earth and life.

Global patterns of primary production reflect variations in climate and nutrient availability.

The biomes we just discussed all depend on a foundation of primary production. This brings up an important question: what controls rates of photosynthesis? Sunlight plays a role. Although all ecosystems, from polar deserts to equatorial forests, receive the same total number of hours of sunlight annually, the intensity of solar radiation declines from the equator to the poles because of Earth's curvature (see Fig. 47.1). Seasonality also becomes more pronounced toward the poles, influencing the ability of primary producers to sustain growth throughout the year at higher latitudes (see Fig. 47.2).

In addition to sunlight, water and nutrients govern the rate of primary production in most ecosystems. In Chapter 27, we saw that plants perform a physiological balancing act, opening their stomata to allow carbon dioxide into their leaves but losing water vapor as a result. Because photosynthesis requires both carbon dioxide and water, the photosynthetic rate, defined as the amount of carbon dioxide fixed into organic matter in a given time interval, is closely tied to water availability. Lack of regular access to water limits the productivity of photosynthetic organisms in deserts and seasonally arid landscapes, explaining why irrigation increases crop yields in arid places such as southern California, northern Africa, and the Middle East.

Even where water is not a limiting factor, rates of photosynthesis are constrained by the availability of nutrients. For this reason, application of fertilizer rich in nutrients generally increases crop yields. Plants, algae, and photosynthetic bacteria all need nitrogen, phosphorus, sulfur, iron, and an array of trace elements to grow, and a shortage of any one of these nutrients can limit primary production. In forests, nitrogen is commonly the soil nutrient that is least abundant relative to the needs of plants. In the open ocean, nitrogen can also be limiting, although more commonly a scarcity of phosphorus limits photosynthesis. In other parts of the ocean, nitrogen and phosphorus are both present in abundance, but primary production is limited by a deficiency of iron, which is needed for many enzymes and electron transport molecules (**Fig. 47.16**).

The idea that primary production is limited by the nutrient that is least available relative to the needs of primary producers is called **Liebig's Law of the Minimum**, after Justus von Liebig, a nineteenth-century German scientist who first popularized the concept. Climate and nutrient limitation therefore play important roles in determining the primary productivity of individual ecosystems. It stands to reason, then, that global patterns of primary production reflect planetary-scale variations in climate and nutrient availability.

HOW DO WE KNOW?

FIG. 47.16

Does iron limit primary production in some parts of the oceans?

BACKGROUND Plants, algae, and other primary producers require nitrogen and phosphorus to make proteins, nucleic acids, lipids, and other biomolecules. In turn, the availability of these nutrients is thought to limit primary production in the oceans. Photosynthetic organisms generally photosynthesize until their environment becomes depleted in a nutrient required for growth. In many parts of the oceans, nitrogen (most commonly in the form of nitrate ion, NO_3^-) is found in extremely low levels in the photic zone, identifying it as a limiting nutrient. Phosphorus (as phosphate ion, PO_4^{3-}) is also depleted in most surface waters, indicating that it, too, can limit rates of photosynthesis. Regionally, however, nitrogen and phosphorus concentrations can both be high. This indicates that primary producers are not making use of these available nutrients and strongly suggests that some other factor is limiting primary production.

HYPOTHESIS In some parts of the oceans, iron—required for the functioning of enzymes critical to photosynthesis and nitrogen uptake—is the nutrient in lowest supply relative to the needs of primary producers. In such cases, it limits rates of photosynthesis.

EXPERIMENT In the late 1980s, American oceanographers John Martin and Steve Fitzwater became intrigued by subarctic waters of the northeastern Pacific Ocean, where high nitrogen and phosphorus concentrations suggested that some other nutrient was limiting primary production. They collected samples of seawater, along with its photosynthetic phytoplankton, and added small amounts of iron to different samples: 1, 5, or 10 nanomoles per kilogram of sample (in the control sample, no iron was added). As photosynthesis proceeded over the course of a week, the researchers made daily measurements of chlorophyll concentration (an indicator of photosynthetic rate), as well as of the concentrations of nitrogen (NO_3^-, the nitrate ion in Fig. 47.16b) and phosphorus (PO_4^{3-}, the phosphate ion in Fig. 47.16c) in the sample treatments.

RESULTS As the graphs show, over the course of the experiments, samples to which iron (Fe) was added showed increased chlorophyll concentrations (Fig. 47.16a) and decreased levels of nitrogen (Fig. 47.16b) and phosphorus (Fig. 47.16c), with higher additions of iron generally showing more pronounced effects. In contrast, controls with no added iron showed little change over the course of the experiment.

CONCLUSION In samples from the northeastern Pacific Ocean, addition of iron increased rates of photosynthesis and resulted in primary producers taking up more nitrogen and phosphorus. This supports the hypothesis that, in these waters, iron—not nitrogen or phosphorus—limits primary production.

FOLLOW-UP WORK Since this initial experiment, many researchers have conducted similar studies with consistent results in areas of the surface ocean containing relatively high levels of nitrogen and phosphorus. Some researchers have suggested that seeding these ocean waters with iron could increase primary production and lead to the sinking of organic carbon into the deep sea. This would remove carbon dioxide from the atmosphere, helping to alleviate global climate change associated with increasing human-generated CO_2 emissions. To date, when this approach has been tried, results have been mixed, with experiments showing a spike of increased photosynthesis after the addition of iron, followed by an interval of enhanced respiration, with little carbon ultimately sinking into the deep sea.

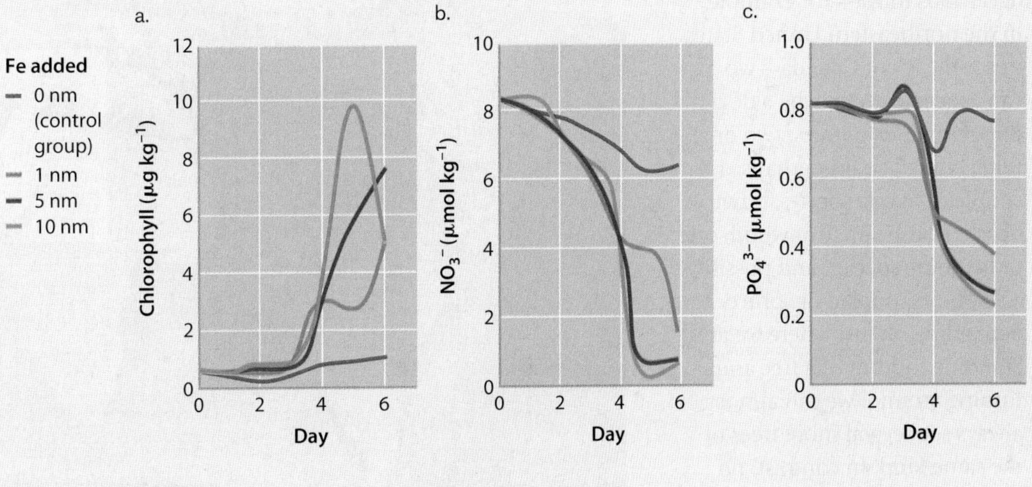

Fe added
— 0 nm (control group)
— 1 nm
— 5 nm
— 10 nm

a. Chlorophyll ($\mu g\ kg^{-1}$) vs Day
b. NO_3^- ($\mu mol\ kg^{-1}$) vs Day
c. PO_4^{3-} ($\mu mol\ kg^{-1}$) vs Day

SOURCE Martin, J. H., and S. E. Fitzwater. 1988. "Iron-Deficiency Limits Phytoplankton Growth in the Northeast Pacific Subarctic." *Nature* 331:341–343.

Fig. 47.17 illustrates the global distribution of primary production on land. The similarity of this figure to maps of global evapotranspiration and precipitation (see Fig. 47.10) shows the strong control exerted by climate on patterns of terrestrial primary production. Warm and wet climates near the equator support highly productive forests, whereas dry habitats and cold environments with limited growing seasons have lower levels of primary productivity.

By contrast, **Fig. 47.18** illustrates how rates of photosynthesis vary in large lakes and the ocean. In the sea, the relationship to global climatic zones is not so clear, which is not surprising because, in aquatic ecosystems, water is not a limiting factor for photosynthesis. The principal pattern shown in Fig. 47.18 is one of high photosynthetic rates along the margins of continents and especially low rates in the central gyres of circulating oceans. Rates of primary production are high near continents because runoff of nitrogen and phosphorus from the continents supplies abundant nutrients to the sea. Upwelling zones, mostly coastal areas where nutrient-rich deep waters return to the surface, are also regions of high primary production. In contrast, the interiors of large ocean basins contain little biologically available nitrogen or phosphorus, so rates of primary production are low in these regions.

Global patterns of primary production help to determine the properties of biomes. Moreover, as we will see in the next section, primary production helps determine global patterns of biological diversity.

Biodiversity is highest at the equator and lowest toward the poles.

On a half-hour walk through a deciduous forest—for example, in the northeastern United States or southeastern Canada—we can expect to see maybe a dozen different kinds of trees: two or three kinds of oaks perhaps, a couple of maple species, birch, beech, hornbeam, ironwood, one or two pine species, and possibly a sycamore, poplar, or spruce, depending on just where we are. Standing in front of a tree and turning around, we can almost always see several more trees of the same kind. In contrast, on a similar walk through a forest in the Amazon or in Borneo, we would pass 300 or 400 different tree species, with almost no two trees of the same kind right next to each other. The great Victorian naturalist Alfred Russel Wallace took note of the low population density of trees in tropical forests and observed, "If the traveller ... notices a particular species and wishes to find more like it, he may often turn his eyes in vain in every direction."

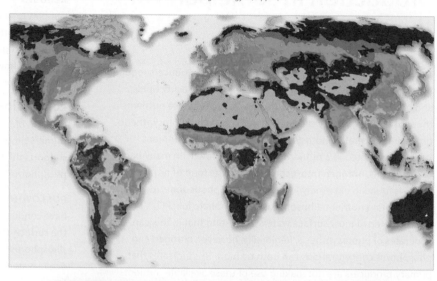

FIG. 47.17 Primary production on land. High rates of primary production occur where climates are warm and wet. *Data from W. Cramer et al., 1999, "Comparing Global Models of Terrestrial Net Primary Productivity (NPP): Overview and Key Results," Global Change Biology 5 (suppl 1): 1–15.*

<100 200 300 400 500 600 700 800 900 1000 1100 1200 >1200
Annual primary production (g C/m²)

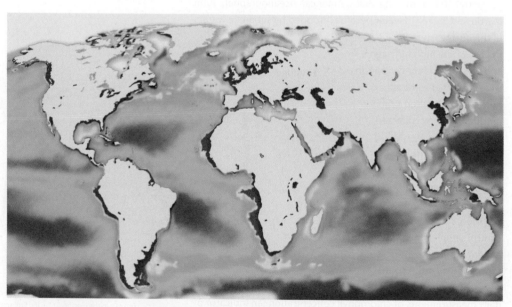

FIG. 47.18 Primary production in the oceans. In the sea, nutrient availability governs rates of primary production. *Data from K. Sherman, P. Celone, and S. Adams, 2004, NOAA Fisheries Service's Large Marine Ecosystems Program: Status Report. NOAA Tech Memo NMFS NE 183.*

0   45   90   135   180   225   270   315   360   405   450
Annual primary production (g C/m²)

Our intuitive feel for biodiversity is reflected in the global distribution of plant species richness shown in **Fig. 47.19**. Equatorial rainforests harbor the highest number of plant species, and in general the amount of plant diversity decreases as we move from the equator toward the poles. This change in species diversity from the equator to the poles is called the **latitudinal diversity gradient**. Deserts have particularly low plant diversity, a pattern linked to their low primary productivity. In contrast, chaparral climates—temperate regions with dry summers and wet winters, often called Mediterranean climates because they characterize lands around the Mediterranean Sea—can support large numbers of plant species. This pattern is seen, for example, along the southern tip of Africa, southwestern Australia, and Mediterranean Europe. In these regions, additional factors only indirectly linked to climate influence regional plant diversity, including habitat diversity (in Mediterranean regions, a mixture of woodland, savannahs, and grassland) and frequent fires (disturbance, discussed in Chapter 45).

Mammals, birds, insects, and many other kinds of animal show similar patterns of species diversity, with the highest species diversity found near the equator and the lowest near the poles. As an example, **Fig. 47.20** shows the global distribution of mammal species. To get a sense of animal diversity in tropical rainforests, consider that one survey found more than 700 species of leaf beetle in a single small set of trees in the Amazon, a number equal to the total diversity of leaf beetles in all of eastern North America. A single tree in this same forest harbored 43 species of ants, exceeding the total for all of Great Britain.

What causes the latitudinal diversity gradient, and what are its consequences? Does this global pattern reflect the ecological processes that characterize communities? Ideas abound on why the tropics contain so many species. One

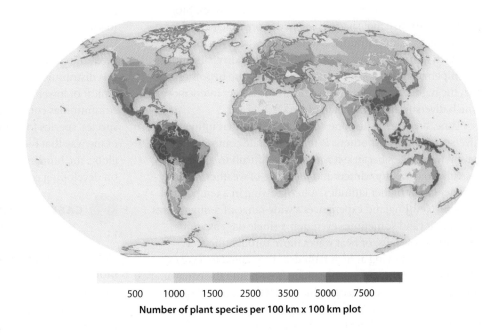

FIG. 47.19 Latitudinal diversity gradient in vascular plants. Species diversity is highest in the wet tropics and declines toward the poles. *Data from Jens Mutke and Wilhelm Barthlott, 2008, "Biodiversität und ihre Veränderung im Rahmen des Globalen Umweltwandels: Biologische Aspekte" (Biodiversity and its change in the context of global environmental change: biological aspects), in Dirk Lanzerath, Jens Mutke, Wilhelm Barthlott, Stefan Baumgärtner, Christian Becker, and Tade M. Spranger, 2008, Biodiversität (Biodiversity), (Ethik in den Biowissenschaften–Sachstandsberichte des DRZE, Bd.5) (Series: Ethics in the life sciences–DRZE expert reports, vol. 5), Freiburg i.B.: Alber: 63.*

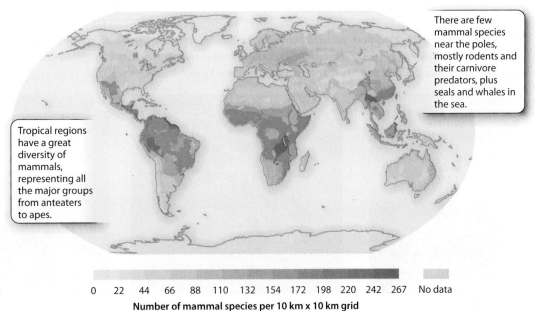

FIG. 47.20 Latitudinal diversity gradient in mammals. The diversity of mammal species in regional ecosystems decreases from the equator to the poles. *Source: After D. M. Kaufman, 1995, "Diversity of New World Mammals: Universality of the Latitudinal Gradients of Species and Bauplans," Journal of Mammalogy 76:322–334.*

hypothesis that has garnered broad support looks to the relative ages of tropical and higher-latitude biomes. Tropical biomes tend to be older, having evolved over tens of millions of years, whereas biomes at higher latitudes have changed identity as the climate changed over the past few million years. For instance, parts of North America that were covered by glaciers 20,000 years ago now support forests and grasslands. Because tropical habitats have existed much longer than temperate ones, biologists hypothesize that species have had more time to evolve and diversify at low latitudes. Several studies of diversification rate, such as in frogs, show comparable rates of diversification in temperate and tropical biomes, supporting the hypothesis that the higher tropical diversity reflects the longer interval over which diversification has taken place.

Time, however, is just one of several factors that influence diversity. Another hypothesis is that temperate communities have fewer species because it is relatively difficult to adapt successfully to cold, dry winters and the range of weather typically experienced at higher latitudes. A tree growing in a temperate area like New England experiences a wide range of temperatures over a year (and sometimes even over a single day) as winds blow and snow falls, or heat waves move through the area. Because species living at high latitudes are adapted to a greater range of conditions, on average, than those nearer the equator, they tend to have larger geographic ranges than tropical species, which are adapted to the narrow range of environmental variation experienced in any particular place.

One hypothesis suggests that the many insects and fungi specialized to attack particular kinds of trees are generally more abundant in wet tropical forests, so trees of a given species tend to be as far apart as they can be, thus reducing the chance of being found by their enemies. This spacing leaves more room for different species in any given place. In addition, more land surface area is available at low latitudes, so the species–area relationship (Chapter 44) may play a role in the latitudinal diversity gradient.

Very likely, the latitudinal diversity gradient reflects all these influences, and perhaps more not yet studied. One lesson is clear, though: where there are more species of plants, there are also more species of animals and other consumers. Tropical temperatures and high humidity have permitted a wide range of plant species to evolve over the past 100 million years, creating many niches for the animals that depend on them.

Both on land and in the sea, biodiversity hotspots occur predominantly in tropical to subtropical environments. In part, this distribution simply reflects the higher species diversity typical of lower latitudes, but it also highlights that tropical communities contain a relatively high proportion of endemic species, species found in one place and nowhere else. Because of the way that biodiversity hotspots are distributed around the globe, the burden of conservation efforts falls disproportionately on developing countries.

CASE 8 CONSERVING BIODIVERSITY: RAINFOREST AND CORAL REEF HOTSPOTS

How do evolutionary and ecological history explain biodiversity?

Earth's biomes are the result of natural selection and environmental change through time. For this reason, the fossil record can shed light on how current patterns of diversity came to be. In general, fossils tell us little about the origins of tropical forests. On the island of Hispaniola, however, there is a remarkable exception: nuggets of amber that provide an extraordinary window on life in the island's rainforest 30 to 23 million years ago.

Amber is fossilized tree resin, and in Hispaniola it comes from the tree *Hymenea*, a genus of plants within the pea family. *Hymenea* trees grew in the Hispaniola rainforest 30 to 23 million

---

FIG. 47.21 Fossils in 30–23-million-year-old Dominican amber. Tree resins (a) trap insects and other organisms and harden into amber, preserving their contents as fossils. Fossils from Dominican amber document the evolutionary history of (b) orchid-pollinating stingless bees, now extinct on Hispaniola, and (c) *Anolis* lizards that, like their modern descendants, hunted insects in trees. *Source: a. Maria Arts/Shutterstock; b and c. Courtesy George Poinar.*

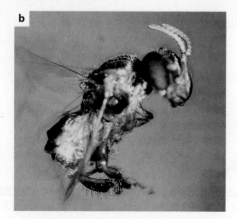

years ago, much as they do today. Lizards, tree frogs, insects, plants, and fungi were all captured in their sticky sap, which hardened into amber, preserving a record of forest diversity and ecology (**Fig. 47.21**).

Studies of these fossils reveal ecological relationships similar to some observed on Hispaniola today. Not surprisingly, some of the most abundant animals in the amber are insects whose modern counterparts have ecological associations with *Hymenea*. Ancient ants that patrolled the bark of ancient *Hymenea* and other tropical trees in search of food, beetles that fed on trees that secrete resin as a defense mechanism, and stingless bees that gathered resins for their nests are all preserved intact in the amber. Many of the insects seen in this amber, perhaps even the majority, are practically indistinguishable from species that continue to exist today on Hispaniola.

Vertebrate fossils are rare in Dominican amber, but they include *Anolis* lizards closely related to lizards found in Hispaniola today that are specialized for life in the canopies and high on the trunks of forest trees. In other words, resource partitioning (Chapter 45) began long ago in the Hispaniola rainforest.

Studies of these kinds of fossils add a deeper dimension of time to our understanding of how ecosystems have come to be. Whether forest, reef, or grassland, every ecosystem contains the present-day survivors of groups representing a broad range of origins in evolutionary time and geographic space. The amber fossils remind us that what G. Evelyn Hutchinson called "the ecological theater and the evolutionary play" have been intertwined throughout our planet's history, producing the diversity of species we see today.

Self-Assessment Questions

8. Why do patterns of global primary production on land show strong latitudinal variation, whereas primary production in the oceans varies more strongly with distance from shore?

9. Describe the general pattern of diversity from the equator to the poles. Provide two hypotheses to explain this pattern of diversity.

10. From what you know about how species diversity changes with latitude, predict how species diversity changes with altitude.

CORE CONCEPTS SUMMARY

47.1 CLIMATE: Solar radiation, wind and ocean currents, and topography determine the distribution of major climactic zones on Earth.

Equatorial climates are warm and polar climates are cold because solar radiation strikes Earth's curved surface head-on at the equator and at an increasing angle at increasingly higher latitudes. page 1067

Because of the tilt of Earth on its axis, solar radiation strikes the Northern and Southern Hemispheres unevenly at different parts of Earth's orbit, causing seasonality. page 1068

Rising warm air near the equator and sinking cool air drives winds that transport heat from the equator toward the poles. page 1068

Prevailing winds drive ocean currents that transport heat from the equator toward the poles. page 1069

Topographical features, such as mountains, regionally modify global circulation patterns. page 1071

47.2 BIOMES: Biomes are broad, ecologically uniform areas whose characteristic species reflect regional climate.

The distribution of the plants that characterize terrestrial biomes reflects factors such as climate and soil. page 1072

Regional climate determines the ratio of potential evapotranspiration to precipitation, a key determinant of vegetation type, primary production, and biodiversity in terrestrial biomes. page 1073

Aquatic biomes reflect the spatial distribution of nutrients and the depth to which sunlight penetrates water. page 1074

47.3 GLOBAL PATTERNS: Global patterns of primary production and biological diversity reflect climate, nutrient availability, and evolutionary history.

Climate and nutrient availability determine global patterns of primary production. page 1086

In most terrestrial ecosystems, water and nutrients limit rates of primary production. page 1086

In the sea, nutrients commonly limit primary production. page 1088

Species diversity generally declines from the equator toward the poles, a pattern known as the latitudinal diversity gradient. page 1088

Earth's biomes are the historical outcome of environmental change through time and natural selection. page 1090

Log in to **LaunchPad** to check your answers to the Self-Assessment Questions and to access additional learning tools.

CHAPTER 48

The Anthropocene
Humans as a Planetary Force

CORE CONCEPTS

48.1 THE ANTHROPOCENE EPOCH: Some scientists call the time in which we are living the Anthropocene Epoch to reflect humans' significant impact on the planet.

48.2 HUMAN INFLUENCE ON THE CARBON CYCLE: Humans have a major impact on the carbon cycle, primarily through the burning of fossil fuels, which returns carbon dioxide to the atmosphere at high rates.

48.3 HUMAN INFLUENCE ON THE NITROGEN AND PHOSPHORUS CYCLES: Humans have an important impact on the nitrogen and phosphorus cycles, primarily through the use of fertilizer in agriculture.

48.4 HUMAN INFLUENCE ON EVOLUTION: The impact of humans on the environment is changing the setting in which evolution acts.

48.5 CONSERVATION BIOLOGY: Conservation biology aims to safeguard biodiversity for future generations.

48.6 SCIENTISTS AND CITIZENS IN THE TWENTY-FIRST CENTURY: In the twenty-first century, biologists, doctors, engineers, teachers, and informed citizens have vital roles to play in understanding our changing planet and making wise choices for our future.

By almost any measure, you were born into humanity's golden age. Beginning with the invention of the steam engine, humans have mastered the ability to harness Earth's store of chemical energy, dramatically extending the power of our own and other animals' muscles. You probably don't find it remarkable to drive across a continent in less than a week or fly across an ocean overnight, but your great-great-grandparents would have been astonished. At the same time, advances in antibiotics and public health have drastically reduced infectious disease in many parts of the world. Smallpox, long one of life's deadliest realities, has been eliminated, and polio may eventually follow the same path. In addition, the Green Revolution of the last 70 years has culminated in today's unprecedented amounts of food for a hungry planet (Case 6 Agriculture).

One consequence of this success is an increasing human population: our numbers have doubled since 1970, exceeding 7.6 billion as of 2018. Another ramification is that our actions increasingly affect the world around us directly or indirectly. In this chapter, we look at some of these impacts, and what we can do about them. Can we sustain or control current patterns of population growth and resource use? Can we ensure that future generations will be able to enjoy and benefit from the diversity of life we see today? As discussed in the following sections, there is reason for both concern and optimism. Together, scientific insight, technical ingenuity, and wise citizenship provide our best hope of conserving our natural world while sustaining hard-won gains in food, energy, and health through the twenty-first century and beyond.

48.1 THE ANTHROPOCENE EPOCH

Increasingly, scientists refer to the modern time as the **Anthropocene Epoch**, from the Greek *anthropos*, meaning "human," and *kainos*, meaning "recent." This name emphasizes the dominant impact of humans on the present day Earth. Our collective impact is partly a consequence of our sheer numbers, which in turn drive our energy and land use, but it also reflects the individual and societal choices we make.

Humans are a major force on the planet.

More humans are alive today than ever before, and our numbers are climbing rapidly. A world that had approximately 500 million inhabitants when Christopher Columbus set sail in 1492 now supports more than 7.6 billion people (**Fig. 48.1**). By the time you retire, our numbers will probably exceed 9 billion.

Of course, population size doesn't tell the whole story. Whether driving a car, living in a heated apartment, plugging in a computer, or eating fruit shipped from another hemisphere, you use a great deal more energy than your ancestors did. **Fig. 48.2** shows daily energy use per person in selected countries. As you can see in this bar chart, energy use by individuals varies greatly from one part of the world to another. Americans, Canadians, Europeans, and Australians live energy-intensive lives, whereas populations in rural Africa and parts of Asia consume relatively little energy per person.

In truth, our direct use of energy reflects only part of our impact on the planet. Nearly everything humans use requires both land and energy. All of us consume food grown in fields, fertilized and harvested through energy, and shipped to a local store. We live in houses built of wood,

FIG. 48.1 Growth through time of the human population. Data from "Historical Estimates of World Population," U.S. Census Bureau, International Data Base, last modified June 2012, http://www.census.gov/population/international/data/worldpop/table_history.php.

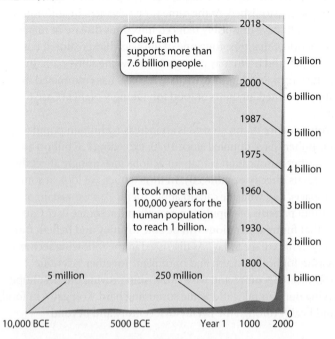

FIG. 48.3 Our ecological footprint. The graph shows the number of hectares needed to support an average person in different countries compared with their ranking on the Human Development Index, a measure of standard of living. Source: European Environmental Agency, http://www.eea.europa.eu/data-and-maps/figures/correlation-of-ecological-footprint-2008. Data from Global Footprint Network, 2008; United Nations Development Programme, 2012.

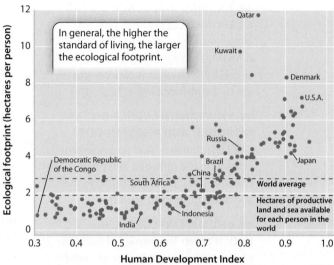

brick, and stone, and use clothing, furniture, and appliances fashioned from plant materials grown in fields and forests or from minerals extracted from Earth by mining. It takes land to grow plants, and energy to sow seeds, make and distribute fertilizer, harvest useful materials, and manufacture, ship, and clean the finished products.

The **ecological footprint** concept represents an attempt to quantify our individual claims on global resources by adding up all the energy, food, materials, and services we use and estimating how much land is required to provide those resources. **Fig. 48.3** plots the average ecological footprint per person against the Human Development Index, a measure of standard of living maintained by the United Nations, for many countries. Not surprisingly, developed countries tend to have large ecological footprints: it takes more than 7 hectares of land to support an average American. By contrast, in many parts of Asia and Africa, the average citizen is supported by only a single hectare.

In many countries, living standards have risen over the past 50 years, nearly always increasing the mean ecological footprint of their citizens. By some estimates, the total ecological footprint of humanity now exceeds the actual surface area of Earth. Taken together, then, trends in population size and the ecological footprint highlight both the remarkable technological successes of the past century and the challenges we face as citizens, policy makers, scientists, and engineers in the decades to come.

FIG. 48.2 Energy use by individuals among countries. Energy use is distributed unevenly among countries. Data from United Nations Department of Economic and Social Affairs, 2009, World Economic and Social Survey 2009: Promoting Development, Saving the Planet, New York: United Nations, p. 44, http://www.un.org/en/development/desa/news/policy/wess-2009.shtml.

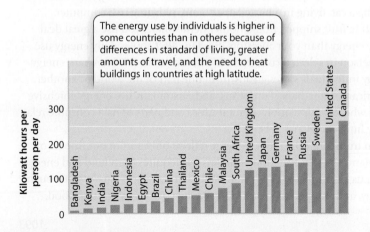

Self-Assessment Questions

1. Take a look at Fig. 48.1. How long did it take the human population to double from 1 billion to 2 billion? How about from 2 billion to 4 billion? What does this pattern say about the rate of population growth?

2. Why does the mean "ecological footprint" of a country's citizens increase as its standard of living increases?

48.2 HUMAN INFLUENCE ON THE CARBON CYCLE

In Chapter 46, we discussed the carbon cycle, explaining how photosynthesis removes carbon dioxide (CO_2) from the atmosphere and respiration replaces it. The Keeling curve (see Fig. 46.1) shows that atmospheric CO_2 levels have increased annually over the past 60 years, and analyses of air bubbles trapped in glacial ice indicate that this pattern of increase began even earlier, with the onset of the Industrial Revolution in the nineteenth century (see Fig. 46.3).

Clearly, humans have become major contributors to the carbon cycle, particularly the part of the cycle in which ancient organic matter, or fossil fuel, is oxidized to CO_2, returning CO_2 to the atmosphere (**Fig. 48.4**). As discussed in Chapter 46, the amount of CO_2 we add to the atmosphere each year by burning fossil fuel is about 100 times that produced by all Earth's volcanoes taken together. And, as we also saw in Chapter 46, measurements over the past several decades of the relative abundances of different isotopes of carbon in atmospheric CO_2 indicate that the CO_2 now being added to our air comes primarily from the combustion of fossil fuels (see Fig. 46.4). Clearing forests for agriculture generates still more CO_2 (Fig. 48.4). When preparing land for agriculture, the existing vegetation is generally burned off, converting much of the organic carbon in biomass and soil organic matter to CO_2. No counteracting process removes CO_2 at comparable rates.

As atmospheric carbon dioxide levels have increased, so has mean surface temperature.

For the past century, scientists, sailors, and interested citizens have monitored temperature at weather stations around the world. More recently, satellites have enabled us to measure temperature in places as remote as the high Arctic and the middle of the ocean. The results are clear: in most parts of the world, mean annual temperature during the decade 1999–2008 was warmer than the average temperatures for 1940–1980 (**Fig. 48.5**). In some places, the temperature change has been slight; in others, especially at high latitudes, the increase has been as much as 2°C.

We can measure CO_2 levels in the atmosphere, and they are increasing. We can measure global temperature, and it is increasing. Is increasing CO_2 responsible for the observed temperature changes? To address this question, we must understand that CO_2 is a **greenhouse gas**, a gas that absorbs heat energy and then

FIG. 48.4 Human-generated sources of carbon dioxide. Every year, humans convert organic matter to carbon dioxide at rates far higher than those found in nature, predominantly by (a) burning fossil fuels and (b) biomass burning, as seen in this Amazon forest. *Sources: a. Tim McCaig/Getty Images; b. Stockbyte/Getty Images.*

FIG. 48.5 Global warming. *Data from R. A. Rohde, Global Warming Art, http://www.globalwarmingart.com/. Based on data from the Hadley Centre Coupled Model version 3 (HadCM3), http://badc.nerc.ac.uk/view/badc.nerc.ac.uk_ATOM_dpt_1162913571289262.*

Overall, Earth's mean surface temperature has increased by almost 1°C.

Higher latitudes have warmed more than environments nearer the equator.

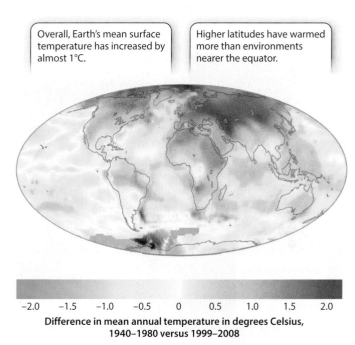

Difference in mean annual temperature in degrees Celsius, 1940–1980 versus 1999–2008

FIG. 48.6 The greenhouse effect. Carbon dioxide is a greenhouse gas: it traps heat from the sun, warming the planet. *Photo source: Fotosearch Stock Images.*

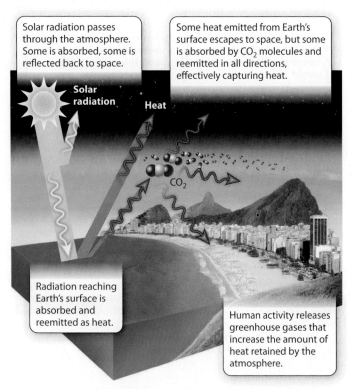

emits it in all directions. As shown in **Fig. 48.6**, solar radiation (yellow in Fig. 48.6) passes freely through the atmosphere, from top to bottom. Some incoming radiation is reflected from Earth's surface, and the rest is absorbed by the land and sea. In turn, some of the energy absorbed by Earth's surface is radiated back again as infrared radiation, or heat (red in Fig. 48.6). Greenhouse gases in the atmosphere absorb the infrared radiation reflected up from Earth's surface and emit it in all directions. Some is directed upward, out of the atmosphere, but half of the trapped heat is directed downward, toward Earth. The net effect is like that of the glass panes of a greenhouse, which allow sunlight to enter but prevent heat from leaving.

Carbon dioxide is just one of several important greenhouse gases in the atmosphere: water vapor is another, and methane a third. Without these gases absorbing and trapping heat, average surface temperatures would fall below freezing and life would not be possible. However, because CO_2 is increasing rapidly, its greenhouse effect is also increasing. Likewise, methane levels are rising rapidly, in large part because of increasing food production. Most of the methane added to the atmosphere each year is generated by methane-producing archaeons (Chapter 24), which thrive in the guts of cattle and in the waterlogged paddies where rice is cultivated. More beef and expanded rice paddies result in higher rates of methane production. The thawing of permafrost at high latitudes releases additional methane that was trapped in frozen soils when the ice formed long ago.

Physics can help us determine how much atmospheric warming is due to increases in greenhouse gases. Each molecule of greenhouse gas absorbs and emits a specific amount of heat, and calculations show that the increases in atmospheric greenhouse gases measured over the past 50 years have increased the difference between incoming (solar) and outgoing radiation by approximately 2.5 watts per square meter. This difference winds up adding heat to the oceans and atmosphere. The scientific consensus, as reflected in reports from the Intergovernmental Panel on Climate Change, is that this greenhouse effect is the principal cause of the observed temperature change over the course of the twentieth century. Scientific consensus also holds that global temperature will continue to rise as atmospheric CO_2 continues to increase in the twenty-first century.

If this view is correct, human activities are changing the world. But can we really eliminate the possibility that the observed increases in greenhouse gases and temperature have natural causes? After all, as we saw in Chapter 46, the long-term geologic record indicates that climate and atmospheric composition have changed dramatically and repeatedly throughout our planet's history, long before humans walked the Earth. Perhaps volcanic emissions have increased, driving the observed changes. As it turns out, volcanic activity cannot explain current global warming; measurements of the isotopic composition of atmospheric CO_2 effectively eliminate volcanic eruptions as a principal source of the increasing CO_2 level (see Fig. 46.4).

Moreover, we can monitor the effects of volcanoes as they occur, and we can gauge the effects of past eruptions because volcanic ash accumulates along with the ice in continental glaciers. Historically, the major effect of large volcanic eruptions has been to *decrease* temperature because volcanic ash and aerosols reflect incoming solar radiation back into space. Furthermore, the impact of volcanic eruptions lasts only a few years; it doesn't drive the kind of century-long temperature increase that we are currently observing.

What about the possibility that the amount of solar radiation entering the atmosphere has varied through time? The sun's output oscillates, and we know from direct measurements how solar radiation has varied during the past 70 years. The effect of this variation can be calculated, and it is small relative to greenhouse effects. It does not seem that variable solar radiation can account for the amount of temperature change that we have observed. In short, natural processes alone cannot explain the temperature increases observed in recent decades (**Fig. 48.7**).

Scientists now overwhelmingly agree that atmospheric CO_2 levels are rising, that temperature is increasing, and that the physics of greenhouse gases relates the two effects. Beyond these facts, however, it is difficult to say with certainty how human-induced global trends in climate will affect a given area.

FIG. 48.7 Changes in monthly mean global temperature over the past 70 years, with model results for human and natural contributions to the observed temperature increase. The models match the temperature data only when the influence of human activities is taken into account.

Source: K. Haustein et al., 2017, "A Real-Time Global Warming Index," Scientific Reports 7:15417.

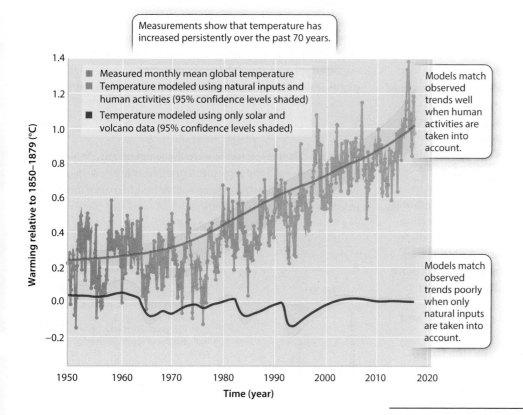

reflecting radiation back into space? And what about feedbacks: will the consequences of climate change lead to further climate change? For example, will methane and carbon dioxide released by human activities melt permafrost at high latitudes, contributing to further temperature increases? How will oceanic and atmospheric circulation patterns change, and how will these changes affect climate?

We still seek definitive answers to all these questions. Nonetheless, most climate models suggest that mean global temperature will increase by 1°C to 4°C during the twenty-first century, depending on our own actions. That increase is not expected to occur uniformly throughout the globe (**Fig. 48.8**). As has been true during the past 50 years, changing circulation patterns may cool some places and warm others, with particularly pronounced warming occurring at high latitudes. Likewise, rainfall is likely to increase in some areas, and decline in others.

Why don't we know for certain? Modeling future climatic conditions is challenging because of the many complex interactions that contribute to climate. Climate models attempt to understand how climate works by applying equations that relate a simplified set of variables and interactions. Using mathematical models that accurately approximate current climate, scientists can change the inputs into a climate model—adding CO_2 to the atmosphere, for example, or changing the extent of forest cover. The model then generates results that can be used to predict future climate. All models must be checked against actual observations, and most are sensitive to assumptions made in constructing the model. That said, climate models do a relatively good job of explaining global-scale features of climate and climatic change.

Nevertheless, questions remain. How, in detail, might cloud cover change over the next century, and how would such a change affect temperature and precipitation in, for example, midwestern North America? What will be the effect of adding more air pollutants, such as the black carbon particles released into the air when coal or wood burns? Will black carbon increase warming by absorbing solar radiation, or decrease it by

FIG. 48.8 A prediction for global warming by the end of the twenty-first century based on current rates of CO_2 increase. Source: Adapted from R. A. Rohde, "Global Warming Predictions," from Global Warming Art, JPEG image, http://www.globalwarmingart.com/wiki/File:Global_Warming_Predictions_Map.jpg. Based on data from the Hadley Centre Coupled Model version 3 (HadCM3), http://badc.nerc.ac.uk/view/badc.nerc.ac.uk_ATOM_dpt_1162913571289262.

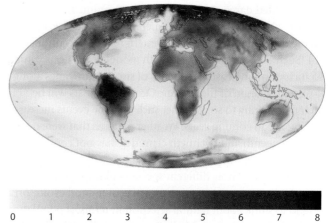

As discussed in Chapter 47, climate can be defined as average weather over a long time interval. From one year to the next, however, weather varies tremendously: there will be cold years and warm years, wet ones and dry ones. Because weather is so variable, evidence for climate change comes not from individual weather events but rather from records kept over decades. We can draw few conclusions from the single observation that global temperature for June 2018 was 0.75°C (1.4°F) higher than the June average for the twentieth century. However, the fact that June 2018 was the 402nd month in a row that was warmer than its twentieth-century average tells us that the world is warming.

Models also predict how rainfall patterns may change as Earth warms. To test their accuracy on this front, oceanographers analyzed 1.7 million measurements of seawater salinity taken over the past century. The oceans were chosen because they contain 97% of our planet's water and receive 80% of its rainfall. The salinity of surface seawater reflects both the addition of fresh water by rain, which decreases salinity, and evaporation, which increases it. For this reason, changing salinity can indicate whether the balance of rainfall and evaporation is shifting over broad regions of Earth. Consistent with many model predictions, wet areas of Earth are becoming wetter and dry regions drier.

Changing environments affect species distribution and community composition.

What will be the consequences of contemporary climate change? Without question, some locations will benefit. For example, temperature increase in New England and Scandinavia will mean longer growing seasons. Other regions will suffer. As precipitation patterns change, many places will become drier, including already water-limited areas of the southwestern United States. A number of climate models predict that some of the strongest declines in rainfall will occur in regions that currently produce much of the corn and wheat that feed the world. Already, farmers in southeastern Australia have experienced the worst droughts in a century, and with them unprecedented damage from brushfires. Over much of North America, 2012 was the driest summer in many decades, and extreme drought returned in the western half of the continent in 2014. Such events are consistent with model predictions, which also include an increased frequency of strong hurricanes along the western margin of the Atlantic Ocean. Scientists will be watching carefully to see whether such extreme events increase in frequency in coming years. Weather patterns that were normal in the nineteenth and twentieth centuries may not continue to hold in future decades.

To understand how different species will respond to climate change, we return to basic principles of ecology and evolution. We tend to think about natural selection in terms of populations adapting to a constant or slowly changing environment. In the twenty-first century, however, the environment will probably be a rapidly moving target, so genotypes that conferred high fitness in the past may not prove advantageous in the future.

Plants illustrate the complicated ways in which organisms respond to environmental change. Meticulous notes taken during the 1840s by Henry David Thoreau provide an inventory of plant species diversity in woodlands west of Boston, Massachusetts. Climate records show that mean annual temperature in this area has increased by about 2.5°C since Thoreau began his survey; in response, many, but not all, species now flower a week earlier than they did in the nineteenth century (**Fig. 48.9**). A number of species have declined in abundance or even disappeared since Thoreau's time, and these tend to be the plants least able to change their flowering time (Chapter 28).

FIG. 48.9 Plant responses to warming temperatures. Observations over a century and a half show that (a) the time of first flowering has become earlier as (b) regional temperatures have become warmer in the woods of Concord, Massachusetts. *Data from A. J. Miller-Rushing and R. B. Primack, 2008, "Global Warming and Flowering Times in Thoreau's Concord: A Community Perspective," Ecology 89(2):332–341, doi: 10.1890/07-0068.1.*

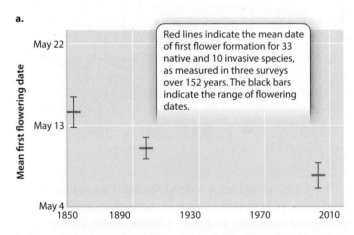

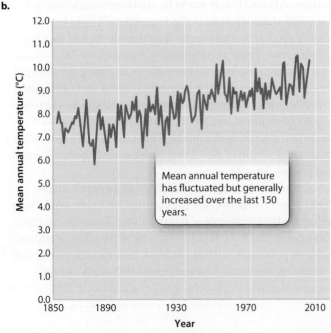

This example highlights the simple but important observation that different plant species respond in distinct ways to changing climate, which favors some populations over others. Earlier onsets of leaves, flowers, and fruit have also been documented throughout the temperate zones of North America, Europe, and Asia.

The example of Thoreau's woods shows that as climate changes, some plants can modify their flowering time. Experimental evidence indicates that many plants also show a direct growth response to elevated levels of CO_2. Where nitrogen in the soil is abundant, increased carbon dioxide levels tend to stimulate growth and, therefore, increase removal of CO_2 from the atmosphere by photosynthesis. One might hope that increased rates of photosynthesis would help to draw CO_2 from the atmosphere, but in most terrestrial environments, soil nitrogen levels are low, limiting any growth response to elevated CO_2. Moreover, increased drought stress and more frequent outbreaks of insect pests and wild fires may counteract any capacity for accelerated growth and CO_2 removal.

Plants grown in air with CO_2 levels like those predicted for the end of this century often change the way that they distribute the products of photosynthesis to different organs: the amount of nitrogen in plant tissues commonly declines, as does the proportion of resources devoted to reproduction. This difference doesn't reflect genetic change, but rather an individual plant's physiological response to altered environmental conditions. But if environmental change persists for a long time, the varying abilities of different plants to respond physiologically will result in changing allele frequencies, as some variants survive and reproduce better than others. That is, plant populations may evolve by natural selection in response to climate change.

Evolution is one response to climate change, but, given the pace at which atmospheric composition and climate are changing, many populations may not have time to adapt. In this case, populations will either migrate or become extinct. Fossils deposited as Earth's climate warmed at the end of the last ice age show that many plant species dramatically changed their geographic distributions in response to changing climate. As discussed in Chapter 28, trees don't move, but they can migrate by dispersing seeds to new areas. Between 18,000 and 8000 years ago, Red Pine and Jack Pine migrated thousands of kilometers to the north from the Gulf of Mexico, where they had lived during the ice age, to their current distributions in the northern United States and Canada (**Fig. 48.10a**). Hickory also migrated northward, expanding its range rather than simply shifting from one place to another as ice retreated and climate warmed (**Fig. 48.10b**).

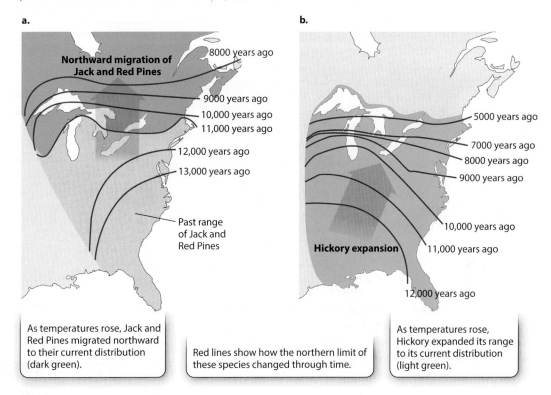

FIG. 48.10 Plant migration in response to past climate change. Records of fossil pollen over the 13,000 years since the last ice age document (a) the northward migration or (b) northward range expansion of selected tree species. *Source: After M. B. Davis, 1976, "Pleistocene Biogeography of Temperate Deciduous Forests," Geoscience and Man 13:13–26.*

Migration is a potentially important response to environmental changes projected for the twenty-first century, but it requires a continuous route to get from one place to another, and it can be a challenge to find these direct paths on continents broken up by agricultural lands and cities. For this reason, some biologists advocate "assisted migration": the deliberate transplantation of plant populations from existing habitats to new ones more favorable to growth. To make sure that assisted migration does more good than harm, we need to do further research to understand how plants grow and compete in new ecological situations and under different environmental conditions.

Fig. 48.11 shows an estimate of the sensitivity of plants around the world to projected twenty-first-century climate. Areas in red are especially vulnerable: plants at high latitudes

FIG. 48.11 Estimated sensitivity of global vegetation to twenty-first-century climate change. Red and orange colors indicate high vulnerability of regional vegetation. *Source: Alistair W. R. Seddon, Marc Macias-Fauria, Peter R. Long, David Benz, and Kathy J. Willis, 2016, "Sensitivity of Global Terrestrial Ecosystems to Climate Variability," Nature 531:229–232.*

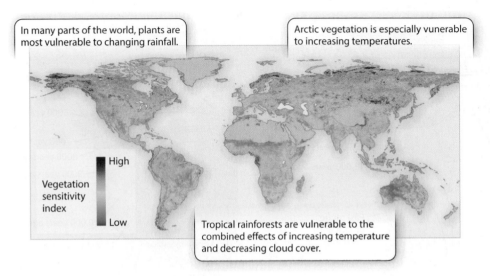

are particularly sensitive to warming, whereas tropical rainforests are also vulnerable to predicted decreases in cloud cover. In other areas, plants are most sensitive to predicted changes in water availability. Species in these regions could face diminished population size, decreased growth rates, or even extinction.

CASE 8 CONSERVING BIODIVERSITY: RAINFOREST AND CORAL REEF HOTSPOTS

How has global environmental change affected coral reefs around the world?

Increasing atmospheric CO_2 is not just influencing life on land. It is also having an observable impact on organisms in the oceans, not least on the coral reefs that provide hotspots of biological diversity in the marine realm. The Great Barrier Reef is visible from space, its wave-washed necklace of limestone stretching more than 2000 km along the northeastern coast of Australia (**Fig 48.12**). A global diversity hotspot, the reef's 1500 fish species (a tenth of all known fishes), more than 4000 species of clams and snails, and myriad other species prompt basic questions about ecology and diversity, many of which we have addressed in previous chapters. Lately, however, one question has come to dominate scientific discussion of the Great Barrier Reef: will it exist a century from now?

Corals are the principal architects of the Great Barrier and most other reefs (Chapter 42), so coral biology is the key to understanding both the reef's unusual diversity and its vulnerability in the face of environmental change. Most reef corals obtain nutrition from unicellular algae that live within their tissues. The algal symbionts in reef corals provide food for their host, in return receiving nutrients, a means of waste disposal, and a stable environment. The efficient cycling of carbon, nitrogen, and phosphorus between corals and their symbionts underpins the biological richness of coral reefs. Reef corals accomplish another noteworthy feat: they make skeletons of calcium carbonate ($CaCO_3$) that over time build the reef's three-dimensional framework.

The very processes that facilitate reef growth—biomineralization and microbial symbiosis—are exposing their Achilles' heel in the twenty-first century. Today, on the Great Barrier Reef, throughout the Caribbean, and elsewhere, corals are dying. Bleaching, indicated by the white skeletons of dead

FIG. 48.12 The Great Barrier Reef. The world's largest coral reef, lying off the coast of Australia, is visible from space. *Source: NASA.*

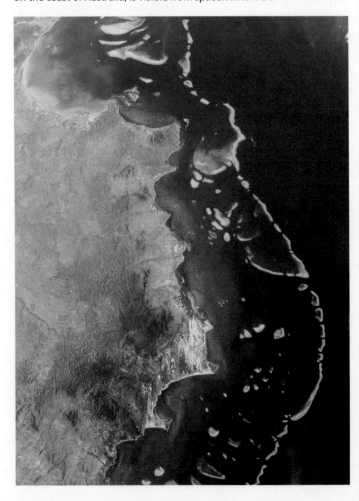

corals, occurs when the symbiotic algae that feed the corals abandon their hosts, thereby sentencing them to death (**Fig. 48.13**). Corals and their symbionts are sensitive to warming, and algae cannot remain in corals when temperatures exceed a critical threshold. Increasing seawater temperature also facilitates the spread of infections that can kill coral. In 2016, the northern part of the Great Barrier Reef experienced a record-breaking heat wave, and the associated bleaching and disease reduced coral cover by as much as 60%. Historically, major bleaching has hit the Great Barrier Reef only sporadically, perhaps once or twice per decade, and at least some of the corals affected in these events have survived and regained their symbiotic algae. In 2017, however, massive bleaching visited the reef again, contributing to continued stress and massive coral mortality. Scientists expect that as the world continues to warm, temperature stress will intensify, dramatically changing the ecology of the Great Barrier ecosystem.

Another long-term threat comes from increasing CO_2 levels in the ocean. As noted in Chapter 46, not all the CO_2 produced by humans during the past century has accumulated in the atmosphere: one-fourth to one-third has been absorbed by the ocean. That's good news for slowing climate change, because it means there is less CO_2 in the atmosphere to trap heat. Conversely, it is potentially bad news for marine habitats, because CO_2 dissolves in seawater to form carbonic acid (H_2CO_3) (Chapters 6 and 37). Unlike climate change, the effects of increasing CO_2 on ocean pH are straightforward: increasing the abundance of CO_2 in the oceans causes the pH of seawater to decrease, a phenomenon known as **ocean acidification**.

In the past 50 years, as the concentration of CO_2 in the atmosphere has increased from approximately 315 ppm to as much as 410 ppm, the pH of surface oceans has

FIG. 48.13 Coral bleaching. This coral shows a pattern of whitening that occurs when corals lose their algal symbionts. *Source: Scubazoo/Science Source.*

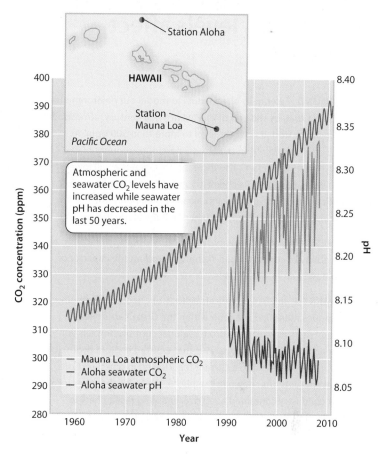

FIG. 48.14 Trends in atmospheric CO_2 levels (red), seawater CO_2 levels (blue), and seawater pH (purple) in the central Pacific Ocean. *Data from S. C. Doney et al., 2009, "Ocean Acidification: The Other CO2 Problem," Annual Review of Marine Science 1:169–192.*

correspondingly dropped by 0.1 unit (**Fig. 48.14**). That may not seem like a lot, but recall that the pH scale is logarithmic. Thus, a 0.1-unit drop represents an increase in hydrogen ions of about 30%. In turn, ocean acidification causes the levels of carbonate ions in seawater to decrease, making it more difficult for some marine algae and corals to build their $CaCO_3$ skeletons (**Fig. 48.15**).

Some corals can survive this change. For example, experiments on two Mediterranean corals showed that at a pH level 0.8 unit below that of present-day seawater, the corals grew naked—essentially as sea anemones—and recovered their skeletons upon return to normal seawater. But, if ocean acidification persists over timescales much longer than coral generations, the structural framework for tropical reefs may be irrevocably lost.

What can be done?

From the preceding discussion, it is clear that rising CO_2 has both direct effects on the physiology of land and sea organisms and indirect effects related to climate change and ocean chemistry. To ensure our environmental future, scientists must

HOW DO WE KNOW?

FIG. 48.15

How will rising twenty-first-century CO_2 levels affect coral reefs?

BACKGROUND Essentially all attempts to model our environmental future predict that, by the end of the twenty-first century, atmospheric CO_2 will increase from its current level to more than 500 ppm. As CO_2 levels continue to rise in the atmosphere, the amount of CO_2 absorbed by the oceans will also increase. This increase will lower the pH of seawater, changing its chemistry in a way that might make it more difficult for some animals and algae to precipitate the calcium carbonate skeletons that are key to their biological function.

HYPOTHESIS Increasing twenty-first-century ocean acidification will limit the ability of corals to build skeletons, eroding the capacity of corals to build and maintain reefs.

EXPERIMENT 1 In 2014, American biologist Rebecca Albright and her colleagues identified a lagoon within the Great Barrier Reef where coral reefs were isolated from regional currents. In their first experiment, they set up an apparatus that emitted sodium hydroxide (NaOH) for three weeks into a current that flowed in one direction over adjacent reefs. (As shown in the photo, the fluids that alter local seawater chemistry contained a red dye, enabling researchers to monitor the flow of the fluid across the reef.) NaOH is a base, so it had the effect of increasing the pH of water flowing over the reef to values that more closely resembled the pH of seawater before the Industrial Revolution. Albright and her coworkers measured rates of community calcification (net increase or decrease in the skeletal mass of reef corals and coralline algae) in the affected reef both before and after the NaOH treatment, and did the same for an adjacent control area not influenced by the NaOH.

EXPERIMENT 2 In a second, complementary experiment, Albright and her colleagues added CO_2 to waters flowing over one reef area, in an effort to simulate the projected pH of late twenty-first-century seawater. Again, they measured rates of community calcification and compared those to a control area in which no additional CO_2 was added to the water.

RESULTS In the first experiment, when NaOH increased pH to pre–Industrial Revolution levels, net community calcification increased by approximately 7%. No change in calcification was observed in the control area (Fig. 48.15a; error bars reflect natural day-to-day variability in the reefs). In contrast, in the second experiment, the addition of CO_2 reduced net community calcification by 34%; again, no change was observed in the control area (Fig. 48.15b).

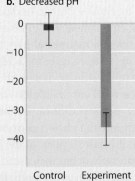

CONCLUSION Corals and other reef-building organisms in the Great Barrier Reef are sensitive to ocean acidification, showing decreased ability to form $CaCO_3$ skeletons both relative to pre–Industrial Revolution seawater chemistry and under conditions likely to occur over the twenty-first century.

FOLLOW-UP WORK Reef corals show a consistent vulnerability to pH decline, although some populations appear to be more resistant than others. Resistant corals might provide a means to sustain reefs in the future, but their ecology will differ from those we see today.

SOURCES Albright, R., et al. 2016. "Reversal of Ocean Acidification Enhances Net Coral Reef Calcification." *Nature* 531:362–365; Albright, R., et al. 2018. "Carbon Dioxide Addition to Coral Reef Waters Suppresses Net Community Calcification." *Nature* 555:51–519. Photo by Aaron Takeo Ninokawa, University of California, Davis.

understand the climate system and biological responses to climate change much better than they do now. Because the principal source of rising CO_2 is the burning of fossil fuels, the solutions we develop as citizens will clearly focus on energy, including both how we obtain it and how we use it. As the global population continues to increase and people aspire to higher standards of living, human impact on the carbon cycle will continue to grow, at least if we continue business as usual. Business as usual, however, may not be a viable option for the long term because population growth may simply outstrip resource availability, especially for petroleum. Ultimately, much of our success in both slowing rates of climate change and sustaining energy supplies will depend on our ability to use energy more efficiently and to generate new forms of power, most likely from the sun, wind, and nuclear sources.

Half a century ago, geologist M. King Hubbard predicted from patterns of oil use and the rate at which new oil fields are discovered that petroleum production in the United States would peak in the 1970s and decline thereafter. Global production, he suggested, would peak later, in the early decades of the twenty-first century. As King predicted, conventional petroleum production in the United States peaked in 1972 and has fallen since then to about half its peak rate of production. Nevertheless, unconventional petroleum resources, made available in no small part by hydraulic fracturing, or fracking, is expected to restore total petroleum production in the United States to near-1970s levels by 2020. (Fracking is a technique in which large volumes of fluid and sand are injected into shale rock, fracturing the rock and releasing trapped oil or gas.)

Another change not anticipated by Hubbard is that fossil fuel demand is expected to peak before mid-century, as alternative energy sources become more widespread. Such changes in both supply and demand suggest petroleum production will not decline as early as Hubbard predicted. Even so, it will still decline before the end of the twenty-first century, both in the United States and globally. Coal, natural gas, and unconventional reserves such as the oil sands found widely in western Canada can quench the world's thirst for fossil fuels for another century or more, and biofuels (fuels generated from plant biomass) may also help conserve petroleum. Nevertheless, the energy produced in this way will be expensive, and climate change will continue.

Natural gas (methane, CH_4) looms large in estimates of twenty-first-century energy generation. Fracking has revolutionized natural gas extraction; in the United States alone, shale gas production has increased by 1300% since 2007. Of all fossil fuels, combustion of natural gas generates the most energy for each mole of CO_2 released to the environment; thus, generating electricity by natural gas has advantages over burning coal. Combusting natural gas leads to CO_2 emissions, however, and gas leaks from expanding wells and pipelines add more methane, a strong greenhouse gas, to the atmosphere. Scientists and citizens continue to debate whether the toxic fluids used in fracking will contaminate groundwater in regions where shale gas is produced. Another issue in the debate of costs versus benefits is earthquakes: notable upticks in the frequency of moderate earthquake activity accompany the injection of fluids into the shale rock during fracking.

Like fossil fuels, alternative energy sources—wind, the sun, tidal energy, and nuclear power (**Fig. 48.16**)—all have important pros and cons, not the least of which, in the case of nuclear power, is the problem of long-term storage of

FIG. 48.16 Alternative energy sources, including (a) solar and (b) wind energy. *Sources: a. euroluftbild.de/picture-alliance/Newscom; b. Travel Pix/Robert Harding.*

radioactive waste. At the same time, all of these sources have the potential to benefit humanity by reducing the amount of CO_2 emitted per kilowatt of energy generated, conserving finite natural resources, and helping to contain the cost of energy production. The development of more fuel-efficient cars, airplanes, and buildings will also help us maintain our energy-intensive lifestyle while decreasing its environmental consequences. As individuals, we can choose energy-efficient transportation and appliances and develop habits that decrease energy waste. When one person turns off the light in an empty room, the energy saved is tiny, but when millions of people do it, the benefits add up.

Wise choices by citizens and consumers can slow the increase in atmospheric CO_2 levels, but the CO_2 already in the atmosphere will not go away quickly. As noted earlier, we have dramatically increased the rate at which CO_2 is added to the atmosphere, but we have not changed the rate at which it is removed. Therefore, even if we could cut CO_2 emissions by 90% tomorrow, CO_2 would not return to nineteenth-century levels for hundreds or even thousands of years.

Another possible solution to this problem is to actively remove CO_2 from the air. Reforestation of previously cleared landscapes removes carbon from the atmosphere because plants build biomass from CO_2 during photosynthesis (Chapter 8). It is also possible to capture CO_2 as it rises upward through smokestacks, although it is currently expensive to do so. Importantly, several pilot projects have provided a proof-of-concept for scrubbing CO_2 directly from the atmosphere. Both the scale of operation needed to make a difference in global CO_2 and the costs of doing so remain challenging, but if we could develop this technology at reasonable cost, it would apply a significant brake to climate change.

Clearly, forestry and twenty-first-century technology have potentially important roles to play in undoing the consequences of humans' actions on the carbon cycle over the past hundred years. For the most part, however, gaining control over our environmental future will entail wise personal and governmental choices about how we obtain and use energy.

Self-Assessment Questions

3. Scientists have pointed out that if humans stopped emitting CO_2 into the air today, atmospheric CO_2 levels would not return to pre–Industrial Revolution levels for hundreds of years. Why is this so?
4. What is the difference between global warming and the greenhouse effect?
5. What are two ways organisms can respond to changes in CO_2 levels and temperatures?
6. What are several possible solutions to the problem of increased atmospheric CO_2?

48.3 HUMAN INFLUENCE ON THE NITROGEN AND PHOSPHORUS CYCLES

Human activities influence the environment not because we operate outside the carbon cycle, but because we have become an important part of it. Our interest in improved crop production has also resulted in unprecedented human impact on the nitrogen and phosphorus cycles. Once again, environmental changes tied to human nitrogen and phosphorus use are altering the ecological landscape for species in lakes and the ocean.

Nitrogen fertilizer transported to lakes and the sea causes eutrophication.

As discussed in Chapter 24, all organisms require nitrogen to synthesize proteins, nucleic acids, and other molecules. Although nitrogen is plentiful in the atmosphere as N_2 gas, this form is not biologically available to animals and plants. Nitrogen-fixing bacteria and archaeons convert N_2 to biologically available forms of nitrogen, called fixed nitrogen, such as ammonia (NH_3) and nitrate (NO_3^-). Plants then obtain nitrogen from ammonia and nitrate in the soil, and animals get the nitrogen from the food they eat. Crops, harvested year after year in the same fields, deplete available nitrogen, so sustaining high yields requires that farmers add biologically available nitrogen to the soil as fertilizer.

Through the use of fertilizer and other activities, humans add approximately 240 million tons of fixed nitrogen to the biosphere each year, similar to the amount fixed naturally by microbes. The majority of this human-generated fixed nitrogen is created industrially by reacting nitrogen gas and hydrogen in the presence of a catalyst to produce ammonia. Most commercially produced ammonia is further processed into nitrate, which is spread across fields as fertilizer to increase crop production. Success in feeding the world's people owes much to fertilization with nitrates, but only about one-third of the nitrogen added to croplands actually ends up in food. That is, much of the nitrate fertilizer leaves fields as surface runoff and travels by rivers to lakes or the ocean. In addition, denitrifying bacteria use some of the added nitrate for respiration, returning N_2 to the atmosphere—and also generating N_2O, yet another potent greenhouse gas.

In lakes and oceans, the fertilizing effects of agricultural runoff are finally realized, albeit in unintended ways. The added nutrients lead to a great increase in primary production by algae and cyanobacteria in a process called **eutrophication**. These algal and cyanobacterial masses eventually sink to the bottom, where heterotrophic organisms, mostly bacteria, feed on them, fueling high rates of aerobic respiration. The bacteria's demand for oxygen can completely deplete O_2 in bottom waters, with potentially catastrophic consequences for animal life on the seafloor or lakebed.

In the Gulf of Mexico, nutrients leached from croplands and transported down the Mississippi River have created an ominously named dead zone of oxygen-depleted bottom water. This dead zone expands seasonally from the mouth of the river west toward Texas; by 2017, it covered an area about the size of New Jersey (**Fig. 48.17**). Within the dead zone, fish and seafloor invertebrates die in masses. Several hundred dead zones have been identified in coastal oceans around the world, and most are expanding year by year. Those in the Gulf of Mexico and the Baltic Sea are among the largest.

Nitrogen applied as fertilizer but spread by runoff and streamflow can affect the full range of ecological interactions discussed in Chapters 43 through 47. Through its use, we benefit tremendously from increased crop production but pay a price in the eutrophication of lakes and oceans. We pay another price as well: as much as 1% to 2% of all fossil fuel consumption goes to make nitrogen fertilizer.

Phosphate fertilizer is also used in agriculture, but has finite sources.

The other major nutrient used to fertilize fields is phosphate (PO_4^{3-}), which creates a somewhat different problem. Like nitrate, much of the phosphate added to fields as fertilizer leaves as runoff, and phosphate runoff can be an important contributor to eutrophication. The Everglades of south Florida provide a good example. This ecosystem is built of plants that have adapted to the low phosphate levels characteristic of the Everglades. However, in recent decades phosphate levels in parts of the Everglades have increased up to a hundredfold. Chemical tracers show that these increased phosphate levels stem from phosphate fertilizer used on agricultural lands to the north. Bolstered by the presence of this nutrient, phosphate-loving cattails and other introduced plants are expanding at the expense of native plants, decreasing the diversity of this remarkable ecosystem.

Overuse of phosphates undoubtedly has detrimental effects on certain natural ecosystems, yet a key concern is the risk that we will eventually run out of phosphates to use as fertilizers. Unlike the case with nitrate, phosphate is not produced by an industrial process; that is, the only source of phosphate fertilizer is mining. Only a few places in the world have sedimentary rock with phosphates in high enough concentrations to be worth mining, and these stores are being depleted.

Globally, phosphate production peaked about 1990, raising serious questions about how the world's need for phosphate fertilizer will be satisfied in the future. In many ways, the problem of phosphate availability parallels that of petroleum production. As abundant sources become depleted, we will have to mine poorer and poorer resources, raising production costs and ultimately calling into question our capacity to sustain agricultural production.

Human encroachment on the nitrogen cycle and long-term phosphate availability, then, converge on a single question: can we manage future agricultural production in ways that feed the growing human population in a sustainable fashion?

What can be done?

The Green Revolution in agriculture began in the 1940s with the introduction of high-yield, disease-resistant varieties of wheat to Mexico (Case 6 Agriculture). In subsequent decades, new strains of corn, wheat, and rice; widespread application of industrial fertilizer; and the spread of farm machinery increased global crop production to levels that could scarcely have been imagined at the beginning of the twentieth century. As population growth has accelerated, however, the need to grow ever more food remains strong. Other factors, notably rising meat consumption in many countries and the increasing use of corn to generate biofuel, are also placing ever greater demands on global agriculture. It is estimated that, over the next decade alone, global corn and wheat production will need to increase by 15% to meet the world's demand for food.

How can we meet the needs of a growing population? Conventionally, agronomists see two options: we can devote more land to crops, or we can increase the yield of fields already in place. As shown in **Fig. 48.18**, globally both the amount of land devoted to growing corn and corn yield have increased steadily over the past 50 years. It is possible to increase yield further, as the yield per hectare in the United States, Canada, China, and western Europe far exceeds that of most other regions. Increasing yields, however, will require increasing the use of fertilizer and the use of fossil fuels to power farm machinery, with all the attendant problems of climate change and eutrophication. Increasing crop area also comes at a cost, as forests and natural grasslands go under the plow, decreasing the biological storage of carbon and threatening biodiversity.

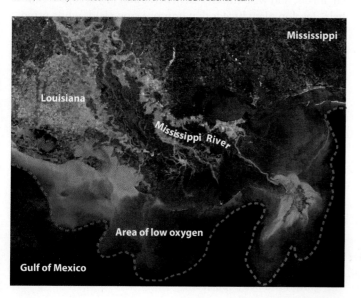

FIG. 48.17 A dead zone, visible in this NASA satellite image as billows of green. *Source: Image courtesy of Liam Gumley, Space Science and Engineering Center, University of Wisconsin–Madison and the MODIS Science Team.*

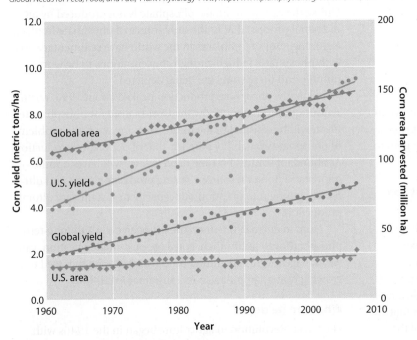

FIG. 48.18 Corn yield and area in the United States and globally from 1961 to 2007. Global corn yield and area have increased at about the same rate, whereas U.S. corn yield has risen much more quickly than corn area because of energy-intensive agriculture and use of fertilizer. *Data from M. D. Edgerton, 2009, "Increasing Crop Productivity to Meet Global Needs for Feed, Food, and Fuel," Plant Physiology 149:9, http://www.plantphysiol.org/content/149/1/7.*

> In the United States, the area devoted to cornfields has increased only a small amount over the last 50 years, but the total corn harvest has grown steadily because of increasing yield.

> Globally, corn yields remain much lower than in the United States and so increasing area under cultivation continues to play a major role in increasing the corn harvest.

For more than a century, biologists have developed improved varieties of crop plants by breeding, the highly effective practice that Darwin called "variation under domestication" and used to frame his arguments about natural selection. Today, we have another tool available: genetic engineering. Biologists can now introduce genes for desirable traits into crop plants, improving their yield, disease resistance, or nutritional value (Chapter 30). The potential benefits of genetically modified crops are obvious, but some biologists see risks as well, many of which are detailed in Chapter 12.

Conventional breeding and genetic modification may improve crop yields, but, in a world of changing climate and evolving pests, these gains may have something of a "Red Queen" flavor to them (Chapter 40). In Lewis Carroll's *Through the Looking-Glass*, the Red Queen tells Alice, "Now, here, you see, it takes all the running you can do, to keep in the same place. If you want to get somewhere else, you must run at least twice as fast as that!" Much of evolution appears to work this way: as the physical or biological environment changes, continued adaptation is needed just to maintain fitness. Strains of wheat, modified to improve drought tolerance, may simply maintain high yields in a changing environment, not increase them. Increases in yields through genetic modification may be possible, but limited. Thus, massive fertilization by nitrate and phosphate fertilizers will remain a key component of agriculture.

Another type of biological intervention, the reduction of pests, may help increase some crop yields. We are not the only species with a nutritional interest in croplands. Fungi, bacteria, protists, nematodes, insects, birds, and mice all target cornfields, sharply reducing yield. In the United States alone, such unwanted guests cause an estimated $1.5 billion worth of damage per year. Conventional breeding and genetic modification have produced improved resistance to pests, but fungus-resistant plants provide strong selective pressure for fungi that can circumvent the mechanisms of resistance, thereby gaining access to otherwise unavailable food.

After harvest, food spoilage also reduces the effective yield of agricultural lands. An estimated one-third of the food grown in the United States is never eaten, and figures for Europe are similar. Fungi and bacteria spoil food during storage and transportation, on supermarket shelves, and in refrigerators. The problem also exists in less developed countries. The United Nations estimates that in some parts of Africa, one-fourth of all crops spoil before they ever reach the market. More efficient storage, transportation, and packaging could sharply increase the effective yield of fields and pastures.

As with energy, achieving a sustainable future for agriculture will require the combined efforts of creative biologists, ingenious engineers, and wise citizens. Geneticists and physiologists can help develop plant breeds that improve crop yields while using less fertilizer. Working with microbiologists, engineers can discover new ways of storing and transporting food, which will increase the likelihood that it will be eaten before it spoils. As individuals, we make daily choices about the food we eat. We can choose to eat more plant products, for example. The basic ecological logic of the trophic pyramid (Chapter 46) tells us that a hectare of grain fed directly to humans feeds more people than the meat of cattle fed on the same crops. We can also choose to eat locally grown products where possible, saving energy while decreasing the likelihood of spoilage.

Self-Assessment Questions

7. What are the causes and consequences of eutrophication?

8. What are several possible solutions to the problem of feeding a growing human population?

48.4 HUMAN INFLUENCE ON EVOLUTION

It has sometimes been claimed that, as humans exert increasing influence on nature, evolution will grind to a halt. Nothing could be further from the truth. As examples in the preceding sections show, what is changing, and what will continue to change as the human footprint grows, is the selective landscape within which evolution operates. Some species, such as rats and cockroaches, thrive in the urban environments we have constructed for our own benefit. Others will expand or decline as we replace prairies and forests with cropland and pastureland. What are our conscious and inadvertent effects on evolution and biodiversity?

Human activities have reduced the quality and size of many habitats, decreasing the number of species they can support.

As noted in Chapter 1, humans make use of nearly 25% of the entire photosynthetic output on land through the crops and livestock we harvest, the biomass we burn, and the primary production we eliminate when we clear-cut forests. Ecologists also estimate that crops cover 10% to 15% of Earth's land surface, and pastureland another 6% to 8% (**Fig. 48.19**).

The conversion of natural ecosystems to cropland decreases the area of natural habitats, as does the expansion of cities and towns. As discussed in Chapter 45, the theory of island biogeography predicts that as the size of a habitable area decreases, species diversity within the habitat will decline. Habitat loss in tropical rainforests poses a particularly important threat to biodiversity (**Fig. 48.20**). These forests represent the greatest concentrations of species diversity on land, but clear-cutting to make room for grazing and croplands has destroyed more than half of their pre-Industrial Revolution area. It is estimated that deforestation over the next century could eliminate 40% to 50% of all tree species in the Amazon forest.

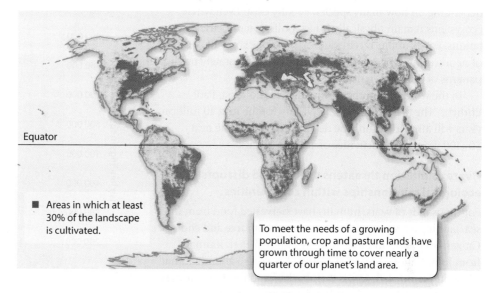

FIG. 48.19 Global agriculture land. Through time, land under cultivation has expanded, diminishing natural landscapes and the species diversity they support. *Source: After Millennium Ecosystem Assessment, 2005, Ecosystems and Human Well-being: Synthesis, Washington, DC: Island Press.*

- Areas in which at least 30% of the landscape is cultivated.

To meet the needs of a growing population, crop and pasture lands have grown through time to cover nearly a quarter of our planet's land area.

FIG. 48.20 Forests under threat. (a) The map shows the extent of deforestation worldwide. (b) This desolate scene in Amazonia was once an expanse of tropical rainforest. Nearly 20% of the Amazon rainforest has been cleared for crops, pasture, or mining. *Photo source: Andre Penner/AP Images.*

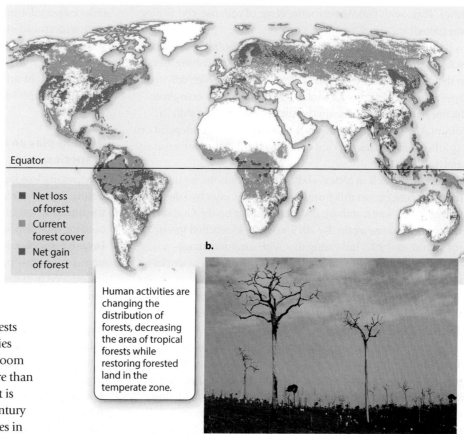

- Net loss of forest
- Current forest cover
- Net gain of forest

Human activities are changing the distribution of forests, decreasing the area of tropical forests while restoring forested land in the temperate zone.

As the forest disappears, so, too, will untold numbers of insect, mite, and other animal species. The renowned biologist E. O. Wilson has estimated that Earth is losing 0.25% of its species annually—that is, 5000 to 25,000 species per year—depending on how many species actually exist. Even where ecosystems remain physically intact, pollution can erode populations. For example, recent studies in Europe show that levels of neonicotinoid pesticides in soil and water accurately predict patterns of population decline in bees and birds.

In the words of another distinguished biologist, Paul Ehrlich, "The fate of biological diversity for the next 10 million years will almost certainly be determined during the next 50–100 years by the activities of a single species."

Overexploitation threatens species and disrupts ecological relationships within communities.

For thousands of years, humans have harvested food from the sea, tapping a resource that long seemed both free and endless. Citizens of Newfoundland, however, know better. Journals from John Cabot's voyage of 1498 tell of cod so abundant that a person "could walk across their backs," and for two centuries cod supported the Newfoundland economy. The amount of cod caught each year increased markedly in the 1960s; surveys estimated that as much as 60% of adult cod was being harvested. By the late 1980s, however, harvests had plummeted (**Fig. 48.21**). By 1992, cod stocks were estimated at 1% of 1960 levels (which were themselves well below those in Cabot's time). Thus, with fish stocks on the verge of collapse, cod fishing was halted.

Unfortunately, the tragedy of cod in Newfoundland is not an isolated story. By 2003, yields in nearly 30% of all fishing areas in the open ocean had declined by 90% or more. With collapsing fish stocks comes a loss in food production, economic decline, and, hidden beneath the surface of the sea, a shift in community structure, as species that compete with or depend on exploited fish species respond to changed conditions of natural selection.

Recent research in Newfoundland offers reasons for both hope and concern on this front. Around 2010, after two decades of persistently low numbers, cod populations on the Grand Banks began to rise again. By 2015, cod stocks reached levels not seen since 1992, indicating that with careful management, damaged fisheries might recover. Unfortunately, stocks plummeted again in 2017, underscoring the need for continuing vigilance. On a positive note, through careful management, the percentage of overexploited fishing areas in waters off the United States has declined by 50% since 2000.

Overexploitation occurs on land as well, perhaps most conspicuously affecting the majestic mammal species that have graced the African savanna for millennia. A global appetite for ivory has encouraged poaching, contributing to the reduction of elephant populations from an estimated 3–5 million animals in the 1930s to less than 10% of that today. Fueled by

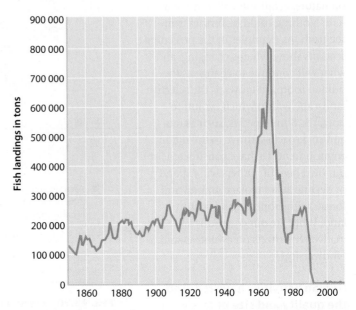

FIG. 48.21 Collapse of cod fishing in Newfoundland. Persistent harvesting of cod at rates higher than those at which stocks could be replenished led to the collapse of cod stocks. *Data from Millennium Ecosystem Assessment.*

a (biologically unfounded) demand in parts of Asia for rhino horn as an aphrodisiac or folk medicine, poaching has similarly reduced populations of most rhinoceros species, limiting these once widespread animals to protected areas in both Africa and southeastern Asia. Note that habitat loss and overexploitation do not occur independently in these ecosystems; their effects are additive and, in some cases, synergistic, meaning that overexploitation amplifies the harmful effects of habitat loss, and vice versa.

Humans play an important role in the dispersal of species.

Biodiversity loss has many causes. In addition to habitat destruction and overexploitation, one of the most insidious is the introduction of non-native species. Non-native species that become established in new ecosystems are commonly termed **invasive species**. Freed from the natural constraints on their population growth, these species can expand dramatically when introduced into new areas, sometimes with negative consequences for native species and ecosystems.

Species can sometimes be transported long distances by natural processes. For example, the Gulf Stream and North Atlantic Drift occasionally carry coconuts from the Caribbean to the coast of Ireland. Migratory birds can transport microorganisms across continents. In the Anthropocene Epoch, however, humans have dramatically increased dispersal. Ships fill their ballast tanks with seawater in Indonesia and empty it into Los Angeles Harbor, carrying organisms across the Pacific

Ocean. Insects in fruit from South America wind up on a dock in London. Humans have become major agents of dispersal, adding new complexity to twenty-first-century ecology.

It has been estimated that 49,000 non-native species have been introduced into the United States (**Fig. 48.22**). Many of these provide food or pleasure (for example, ornamental plants), and only a few have escaped into natural ecosystems. Some of the invasive species that have become established in their new surroundings do little beyond increasing community diversity, but others put strong pressure on native species.

Kudzu, for example, is a plant that was originally imported from Japan to slow soil erosion. It now covers more than 3 million hectares in the southeastern United States, displacing native plants through competition for space and resources (Fig. 48.22a). The Zebra Mussel, a European bivalve that hitchhiked to the Great Lakes in the ballast water of cargo ships, has also multiplied dramatically in its new habitat, displacing native species and clogging intake pipes to power plants (Fig. 48.22b). Nearly 500 introduced plant species contribute disproportionately to crops lost each year to competition from weeds. And in Australia, more than 10% of all indigenous mammal species have become extinct in the past 200 years, largely because European colonists introduced cats and red foxes into their new home.

Invasive species have particularly devastating effects on islands because island species have commonly evolved with relatively few competitors or predators. Bones in Hawaiian lava caves record more than 40 bird species that existed 1000 years ago but are now extinct, eliminated by the rats and pigs introduced by Polynesian colonists, as well as by landscape alteration. Introduced predators have had a similar impact on the island of Guam. Brown tree snakes introduced from Australasia since World War II have reduced native bird and reptile diversity by 75% (Fig. 48.22c).

FIG. 48.22 Invasive species. (a) Kudzu (*Pueraria lobata*), a Japanese vine, was introduced to the United States in 1876 and now covers more than 3 million hectares. (b) The Zebra Mussel (*Dreissena polymorpha*), introduced into the Great Lakes from Europe in 1985, disrupts lake ecosystems, threatening native biological diversity. (c) The Brown Tree Snake (*Boiga irregularis*), introduced into Guam from Asia and Australia shortly after World War II, has decimated indigenous bird populations across the island. *Sources: a. James K. York/Dreamstime.com; b. Dr. David J. Jude, School of Natural Resources and Environment, University of Michigan; c. John Mitchell/Science Source.*

Such introductions contribute to the changing diversity and ecology of twenty-first-century landscapes, but it may be the spread of modern diseases that should impress (and concern) us most. In a world filled with airplanes and oceangoing ships, bacteria and viruses spread much more rapidly than they did in our pre-industrial past. The H1N1 (swine) flu, a global concern in 2009, is a notable example. The first diagnosed case was identified on April 13, 2009, in Mexico, and by mid-May of that year the flu had been reported on five continents.

Humans have altered the selective landscape for many pathogens.

From the Black Death of the fourteenth century to the flu pandemic of 1918, disease has played a major role in human history. In the Anthropocene Epoch, we have unprecedented opportunities to control or even eradicate some of the great diseases of the past and present. At the same time, the strong selective pressures placed on pathogens by antibiotics pose a key challenge for twenty-first-century medicine.

Malaria, discussed in Case 4, provides an example. In any given year, malaria infects an estimated 500 million people, causing 1 million or more deaths. Infection is particularly severe in parts of Africa, where the direct and indirect effects of malaria account for almost half of all childhood mortality. Insecticide-soaked bed nets have proved a highly effective means of protection against the disease in affected areas, reducing transmission by as much as 90%. Natural selection, however, has begun to chip away at success in controlling the disease: mosquito strains have begun to emerge that are resistant to some insecticides.

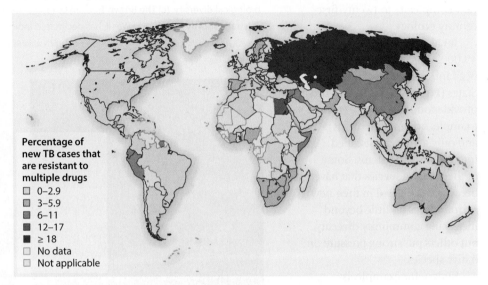

FIG. 48.23 Distribution of extremely drug-resistant tuberculosis. By 2016, 44 countries had reported at least 1000 cases of extremely drug-resistant tuberculosis. *Data from World Health Organization, Global Tuberculosis Report 2017.*

Quinine, a chemical extracted from the bark of the cinchona tree, was the first effective treatment for malaria, and remained in widespread use until the 1940s. Since that time, many new antibiotics have been developed. Over time, their usefulness has declined as resistance to each of these drugs has evolved in populations of *Plasmodium*, the infectious agent of malaria. More recently, artemisinins, antibiotics based on a traditional Chinese plant remedy, have proved effective in treatment. Again, however, resistant strains of malaria have been identified.

The arms race between antibiotics and pathogens is not limited to malaria, but rather characterizes nearly the whole range of diseases caused by viruses, bacteria, protists, and fungi that infect humans. For example, tuberculosis, a lung infection spread by inhalation of the bacterium *Mycobacterium tuberculosis*, causes 1.8 million deaths annually, mostly in sub-Saharan Africa, where many people are also infected with HIV (Chapter 41). Antibiotic treatment is effective against tuberculosis, but many people who are infected do not or cannot comply with the strict regimen of daily treatment for 6 months, hastening the evolution of drug-resistant *M. tuberculosis* strains. An incomplete course of antibiotics does not eradicate all of the bacteria, thereby selecting for those bacteria that are least susceptible to the antibiotic. In 2016, approximately 600,000 new cases of antibiotic-resistant tuberculosis were reported. More concerning, a small percentage of these infections has been characterized as "extremely drug resistant," meaning that the pathogens resist both standard treatments and alternative antibiotics developed over the past 25 years (**Fig. 48.23**). In effect, extremely drug-resistant tuberculosis is untreatable.

Because bacteria can pass genes horizontally (Chapter 24), antibiotic resistance that evolves in one species can jump to another species. Thus, the bacterial incubators for drug resistance can arise anywhere and spread rapidly. And the problem is getting worse: between 2000 and 2015, daily doses of antibiotics increased globally by 65%, mainly in developing countries.

Clearly, despite the remarkable gains of the twentieth century, public health will remain a key issue in the decades ahead. Improved sanitation, in the form of clean drinking water and safe food, can do much to reduce transmission of infectious diseases. In addition, the search continues for vaccines that can further reduce or even eliminate some historic scourges. Even so, the ongoing evolution of drug resistance requires the continuing efforts of scientists to stay one step ahead of pathogens.

Are amphibians ecology's "canary in the coal mine"?

In the nineteenth century and well into the twentieth century, British, American, and Canadian coal miners sometimes took canaries with them down long mine shafts. Canaries are more sensitive than humans to carbon monoxide and other toxic gases that can accumulate in mines. When the canaries passed out, the miners knew it was time to leave. Because many amphibian species have narrow environmental tolerances and exchange gases through their skin, they can be especially sensitive to environmental disruption. For this reason, some biologists see them as "canaries" for the global environment.

If amphibians are indeed our early-warning system, we should be concerned. Fully one-third of all amphibian species are threatened with extinction, and more than 40% have undergone significant population decline over the past

FIG. 48.24 Amphibians under threat. The Harlequin Toad (*Atelopus varius*), native to Costa Rica and Panama, has been classified as critically endangered. *Source: Marie Read/Science Source.*

decade (**Fig. 48.24**). What accounts for this pattern? Nearly all of the environmental issues discussed in this chapter come into play. Amphibian habitat is being destroyed in many parts of the world, especially old-growth forests that provide damp, food-rich environments in abundance. Pesticides and other toxins of human manufacture take their toll as well. For example, the common herbicide atrazine has been shown experimentally to impair development in frogs. Studies of placental cells grown in the laboratory suggest that atrazine in our food may also affect human reproduction. When atrazine was applied to human placental cells, it caused overexpression of genes associated with abnormal fetal development. Because of concerns about public health, atrazine has been banned in several European countries.

Recently, infection has also been implicated in amphibian decline. The fungus *Batrachochytrium dendrobatidis* (*Bd*) infects the skin of frogs and impairs many of the skin's functions. Recent studies of *Bd* show the classic features of epidemics—rapid geographic spread and high mortality—along the expanding front of fungal populations. How *Bd* spreads and whether it acts with other aspects of global change remain controversial, but there can be no doubt that disease has joined habitat destruction and pollution as a major cause of amphibian decline.

Self-Assessment Questions

9. How do invasive species affect species diversity of communities and ecosystems?

10. How does the Red Queen hypothesis apply to interactions of pathogens and antibiotics?

11. What are three examples of species that have benefited and three examples of species that have been harmed by human activity?

48.5 CONSERVATION BIOLOGY

Biological diversity, or **biodiversity**, refers to the summed variation of life on Earth (Chapter 45). It encompasses genetic and population variation within species; the number of species in local communities, biomes, or the planet as a whole; and the variations in species composition among ecosystems across the planet. We don't know how many species exist on our planet, although the number certainly runs into the millions and may exceed 10 million. Whether local or global, patterns of biodiversity reflect all of the processes discussed in earlier chapters: natural selection, physiology and other aspects of biological function, ecological interactions among species, and the distribution of physical environments. Increasingly, biodiversity also reflects human activities.

As discussed earlier in this chapter, habitat destruction, overexploitation, pollution, and invasive species all pose threats to biodiversity, and in the twenty-first century, climate change threatens biodiversity as well. A 2016 census of population size for more than 3700 species of mammals, birds, fishes, and amphibians found that the total number of vertebrate animals in nature has declined by 60% over the past 50 years, prompting concern about mass extinction comparable to the great biological cataclysms of the past. Unlike past mass extinctions, however, current threats come not from meteorites or massive volcanism, but rather from humans. Importantly, as human activities lie at the heart of threats to biodiversity, we can make choices that will help us to preserve and protect the natural world. **Conservation biology** addresses the challenge of sustaining biodiversity in a changing world crowded with people.

 CASE 8 CONSERVING BIODIVERSITY: RAIN-FOREST AND CORAL REEF HOTSPOTS

What are our conservation priorities?

Before examining how biologists and policy makers are working to conserve biodiversity, we should take a moment to consider what we wish to preserve. At the broadest level, most biologists would hope to sustain our shared heritage of global species richness—a tall order on a planet increasingly marked by human activities. Not all ecosystems are equal in terms of biodiversity. Indeed, following the lead of British conservationist Norman Myers, biologists have identified about two dozen **biodiversity hotspots** on land. Hotspots are relatively small areas that have unusually high numbers of endemic species and that are under threat from human activities. Endemic species are species found in one place and nowhere else. By definition, hotspots contain at least 1500 endemic plant species and have lost 70% of their original habitat area. Biodiversity hotspots represent only about 2% of Earth's land surface, but they support more than half of the world's endemic plant species and nearly 43% of its endemic bird, mammal, reptile, and amphibian species. Therefore, biodiversity hotspots are high priorities for conservation efforts.

Biodiversity reflects both species richness within local communities and the variations in species composition from one community to another, and from one biome to the next. While biodiversity hotspots have garnered widespread interest, many conservation biologists wish to conserve species richness in local communities and regional ecosystems throughout the world. Good evidence suggests that biodiversity provides important **ecosystem services**—that is, benefits to humans—including cleaner air and water, greater primary productivity (Chapter 46), improved resilience to environmental disruption (Chapter 45), and untapped sources of food and molecular compounds for use in medicine and agriculture.

As discussed in Chapter 14, genetic variation exists within as well as among species, and this variation is likely to play an important role in conserving species in a time of climate change. Genetic variants in plants provide the raw materials for breeding crops that can thrive in habitats that may become warmer or drier than at present. And within marine species vulnerable to global change, geneticists are beginning to find local populations with enhanced capacity to evolve tolerance to warmer temperature or decreased pH. Such variants may help to sustain the commercial shellfishing in the twenty-first century and provide robust coral populations able to sustain tropical reefs.

Conservation biologists have a diverse toolkit for confronting threats to biodiversity.

As shown in **Fig. 48.25**, thousands of plant and animal species face the threat of extinction. Among mammals, for example, nearly 10% of all species are endangered or critically endangered, meaning that they are unlikely to persist without human intervention. Declines in population size and geographic range make another 10% of mammal species vulnerable to extinction.

The scope of the problem is large, requiring a variety of conservation strategies to address it. Some programs focus on individual species, targeting specific threats that have reduced or could reduce populations below sustainable levels. Widely publicized efforts to protect elephants and rhinos provide examples. By protecting and restoring the habitats where these animals thrive, expanding efforts to curb poaching, and working to shut down trade in elephant ivory and rhinoceros horn, conservationists hope to build and maintain sustainable populations. In some cases, such as with the Arabian Oryx, long extinct in the wild but recently reintroduced to the deserts of the Arabian Peninsula, zoos and captive breeding programs have played a critical role in rebuilding populations. As the Arabian Oryx illustrates, species-specific programs are often successful. Similarly, the Southern White Rhino, known from a single local population a century ago, has prospered in protected reserves and now numbers more than 20,000

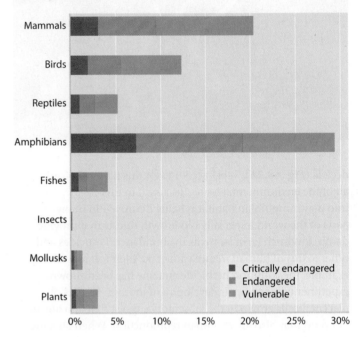

FIG. 48.25 Biodiversity at risk. A substantial proportion of vertebrate species has been classified as vulnerable (likely to become endangered in the absence of intervention), endangered (likely to become extinct in the absence of intervention), or critically endangered (very high risk of extinction in the wild). In the twenty-first century, these proportions are increasing year by year. *Data from C. Hogan, 2014, "Endangered Species," http://www.eoearth.org/view/article/51cbedac7896bb431f69323a.*

individuals. Whooping Cranes, majestic Northern American birds reduced to just 23 known individuals in 1940, now number more than 600.

While conservation strategies focused on specific species are appropriate and necessary, they are simply not practical for the greater diversity of species at risk. Fortunately, one key component of targeted conservation efforts helps to conserve many species at once: protecting and restoring ecosystems.

In 1872, Yellowstone National Park was established as the world's first national park. The idea was to set aside a wilderness region of outstanding beauty and preserve it for all time. Since then, national parks and other reserves have been established throughout the world, and they have played a key role in preserving our natural heritage. Reserves speak to the three major threats to biodiversity introduced earlier in this chapter. First, they protect habitat area and, in many cases, provide land to restore degraded habitats. Second, they provide managed areas within which overexploitation and pollution can be eliminated or minimized. Third, they provide a buffer against invasive species. Notably, although we may value reserves because they protect charismatic animals such as lions, rhinos, or bison, they actually help to conserve entire ecosystems.

A great deal of research has shown that some designs for reserves work better than others. Not surprisingly, bigger is better. Species differ in the area needed to support sustainable populations, with top carnivores requiring particularly large territories to thrive. Thus, one large reserve functions better than four smaller reserves of the same total size. Also, because conservation is commonly more challenging along the edges of reserves, round or square areas function better than elongated or fragmented areas because they minimize edge effects. Importantly, successful plans for reserves include **corridors** that provide routes for migration from one reserve to another.

Climate change provides new challenges for conservation biology in the twenty-first century.

Since 1985, the area of Australia's Great Barrier Reef covered by coral has declined by more than 50%, driven by three main processes: physical destruction by tropical cyclones, predation by Crown-of-Thorns Sea Stars, and bleaching. Over the same interval, reefs in the Caribbean Sea have similarly declined at a rate of about 1.5% per year, mostly because of storms, disease, bleaching, and the spread of algae due to overexploitation of the fish that keep algal populations in check.

The declining health, size, and diversity of coral reefs reflect some of the same factors that challenge biodiversity on land. Outbreaks of Crown-of-Thorns Sea Stars may reflect both the overexploitation of predators on the sea stars and increased nutrient influx that appears to increase the survival of sea star larvae. Increased storm frequency and intensity, as well as bleaching, reflect the impacts of another factor: climate change. Human activities such as agriculture and fishing will continue into the future, but warming temperatures and ocean acidification will pose increasingly dire threats to reefs in coming decades. Thus, even as policies are put in place to protect reefs from local disturbances such as pollution and overexploitation, reef conservation and management will require strategies to mitigate the effects of global change.

The same is true on land. It is critical to maintain reserves that protect natural habitats and the species they support, but, on the timescale of the next century, the climate of protected areas will change, and with it the capacity of a reserve to support its native biodiversity. Under these conditions, conserving or creating corridors between habitats (Chapter 44) will play a particularly important role in enabling populations to adapt to climate change, aiding their move from an inhospitable reserve to another with local climate more amenable for growth and reproduction. Many ecologists think that natural corridors to promote migration will not be sufficient; that is, they believe that conservation biologists will need to implement a carefully considered program of assisted migration to match populations with appropriate environments.

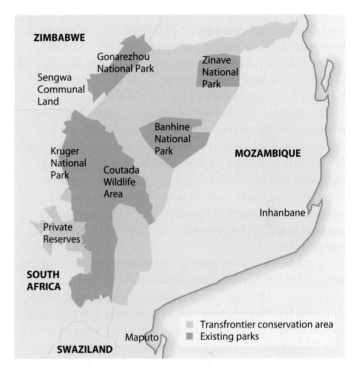

FIG. 48.26 Linking nature reserves in southern Africa. In a multinational effort, national parks and other reserves in South Africa, Zimbabwe, and Mozambique are being linked by a broad conservation area that provides corridors for migration. Within this area, animals and people coexist. *Data from http://www.krugerpark.co.za/great-limpopo-transfrontier-park-south-africa.html.*

In southern Africa, conservation efforts have benefited substantially by grouping reserves across national borders and connecting them with corridors. For example, a broad conservation area now links national parks and other reserves in South Africa, Mozambique, and Zimbabwe, providing an enlarged area for conservation corridors that connect established reserves and permit seasonal migration (**Fig. 48.26**). A key point is that, within this conservation area, animals and people coexist. Many corridors, in Africa and elsewhere, necessarily pass through populated areas, so local support is essential. In southern Africa, people within the conservation area have been persuaded that the preservation of biodiversity is in their best economic interests.

Sustainable development provides a strategy for conserving biodiversity while meeting the needs of the human population.

By now it should be clear that resource use and conservation biology are closely intertwined: the fate of many species will depend on the decisions we make about food and energy for a burgeoning human population. With this point in mind,

sustainable development has emerged as an important principle for trying to balance priorities for humans and the natural world. In principle, the concept is simple: the rate at which we use resources should be no higher than the rate at which they can be replenished.

Sustainable development is most easily understood in terms of species that grow and are harvested by humans. Thus, logging in the forests of western North America or the Amazon basin is sustainable only if tree harvest is matched by growth, keeping the forests intact. Similarly, fisheries are sustainable only if annual catch is no larger than the capacity of fish populations to grow and produce. By changing the way that fishermen are distributed across these areas and the species they fish, the harvest of fish along the Pacific coast of the United States has become more predictable (as has the income of fishermen) and stocks of commercial species are increasing. Expanding marine reserves is also proving successful as a means to rebuild fish populations. At the same time, decisions by both fishermen and consumers are shifting the targets of commercial fishing away from vulnerable species and toward those that can be harvested sustainably.

It can be difficult to quantify sustainable yields for agriculture, as the cost of producing food includes the costs of fuel and fertilizer. Moreover, where episodic floods or droughts influence food supply, sustainable yields may vary from year to year. Despite these difficulties, the broader target of sustainable development—reducing our ecological footprint—has begun to influence agricultural practice and consumer choices. The task is formidable, but the goal is important: to ensure that meeting the needs of humans does minimal damage to Earth's biodiversity.

Without question, there will be fewer species when the twenty-first century ends than when it began. Lost species mean lost opportunities for the discovery of novel compounds for medical research. Diminished biodiversity may make ecological communities less productive and less resilient to fires, hurricanes, or other environmental events. There is a pressing need to feed the world, but there are also many good reasons to conserve the biodiversity that has evolved over 4 billion years. Balancing these two imperatives will entail creative approaches to both science and policy, including thoughtful ways of using land that serve both people and conservation.

Self-Assessment Questions

12. In what ways do habitat loss, overexploitation, and invasive species threaten biodiversity?
13. How do reserves and corridors help to conserve biodiversity?

48.6 SCIENTISTS AND CITIZENS IN THE TWENTY-FIRST CENTURY

The decades ahead will be a time of great opportunity, but they will also present increasing challenges as humans' ecological footprint continues to grow around the world. How we will meet those challenges is, in large measure, up to you.

If you become a biologist, you will be able to help improve our planet's health in many different ways (**Fig. 48.27**). Field biologists are needed to document Earth's true biodiversity, even as much of it comes under threat. Physiologists are needed to help us understand how species respond to environmental change.

Ecologists and population geneticists are needed to discover the principles that govern species interactions in communities so that we can understand how physiological responses to global change in one species will affect others. Only then will we be able to design effective ways to preserve the world's biodiversity. Molecular biologists will help us combat disease in new ways, and geneticists will develop crops that yield more food. Perhaps most important, biologists will work with economists, lawmakers, and others to ensure that government policies reflect what we know about the natural world on which we all depend.

If you become a doctor or a public health worker, you will have many opportunities to improve the health of people in countries around the world. Through laboratory research, physicians will discover new ways to treat cancer, malaria, AIDS, and other diseases. And, by treating patients in new ways, doctors will reduce the threat of antibiotic-resistant pathogens. Working with government and private agencies, public health workers will help ensure supplies of clean water and adequate food are available for people in parts of Africa, Asia, and wherever else the basic needs of life pose an ongoing challenge.

If you become an engineer, you will have the opportunity to design energy-efficient machines or establish improved ways of slowing the spoilage of food. Novel solutions for managing resources and balancing resource use with conservation will benefit all people.

If you become a teacher, you will have a particularly important role in making sure that our children have the knowledge they need to meet the challenges of the Anthropocene Epoch. And, by choosing to be a responsible citizen, you can help to ensure that your family, workplace, and government make wise choices. In an age of environmental change, knowledge of Earth and life provides a key framework for personal choice. As citizens, our votes shape governments, and biological education can help to ensure wise leadership.

FIG. 48.27 Human impact. Scientists, doctors, teachers, and informed citizens can make a difference. *Sources: (Doctor) AP Photo/Sebastian John; (Science teacher) LWA/Dann Tardif/Getty Images; (Biochemist underwater) Alexis Rosenfeld/Science Source; (Testifying Department of Health and Human Services Assistant Secretary) Matthew Cavanaugh/EPA/REX/Shutterstock; (Earth Day tree planting) Chas Metivier/The Orange County Register/ZUMAPRESS.com.*

We cannot change the world we've been given, but we can and will influence the one our children inherit. It is easy to despair about the future, but as the novelist Philip Pullman put it, "Hope is not the name of a temperament, or of an emotion, but of a virtue." Hope inspires action, and our collective capacity for positive change is great. Knowledge of biology will not by itself ensure that humanity's golden age can be sustained in coming years, but it will surely be an important part of our solutions.

CORE CONCEPTS SUMMARY

48.1 THE ANTHROPOCENE EPOCH: Some scientists call the time in which we are living the Anthropocene to reflect humans' significant impact on the planet.

Today, more than 7.6 billion people live on our planet. page 1093

Ecological footprint is the amount of land required to support an individual at an average standard of living for a community, country, or the world as a whole. page 1094

Our ecological footprint reflects all the activities that support our daily life, from growing and transporting the foods that we eat to supplying the materials and energy for our homes and the vehicles that provide our transportation. page 1094

48.2 HUMAN INFLUENCE ON THE CARBON CYCLE: Humans have a major impact on the carbon cycle, primarily through the burning of fossil fuels, which returns carbon dioxide to the atmosphere at high rates.

Carbon dioxide is a greenhouse gas, trapping heat and thereby keeping the surface of Earth warm. page 1095

Greenhouse gases play an important role in maintaining temperatures on Earth compatible with life, but increasing levels of greenhouse gases are resulting in higher mean surface temperatures. page 1096

Changing climate has already affected the distribution of species and the composition of communities, and will continue to do so in the twenty-first century. page 1098

Increasing levels of CO_2 are increasing the temperature and decreasing the pH of the oceans, affecting marine organisms, including corals. page 1100

We can reduce CO_2 emissions, conserve finite natural resources, and contain the cost of energy by developing energy sources other than fossil fuels and improving energy efficiency. page 1103

48.3 HUMAN INFLUENCE ON THE NITROGEN AND PHOSPHORUS CYCLES: Humans have an important impact on the nitrogen and phosphorus cycles, primarily through the use of fertilizer in agriculture.

Humans add large amounts of industrially fixed nitrogen to the environment, which enters the biosphere as fertilizer to increase food production. page 1104

Approximately one-third of the nitrogen in fertilizer is taken up by crops; the rest is returned to the atmosphere by soil microorganisms or ends up in lakes and oceans, where it results in eutrophication, a boost in primary production that depletes oxygen and can lead to the death of fish and other aquatic organisms. page 1104

A second major nutrient in fertilizer is phosphorus, which is mined rather than produced industrially. The finite supply of phosphorus raises questions about our ability to sustain agricultural production in the future. page 1105

We have increased agricultural yields through the use of fertilizer, high-yield and disease-resistant plants, and farm machinery. page 1105

To feed a growing population, we can devote more land to agriculture or increase productivity further, but each option brings with it additional costs. page 1105

48.4 HUMAN INFLUENCE ON EVOLUTION: The impact of humans on the environment is changing the setting in which evolution acts.

Humans have altered the landscape in dramatic ways, leading to the expansion of some species, such as corn and wheat, and the decrease and extinction of others. page 1107

Habitat degradation and loss may cause extinction in some of Earth's most diverse ecosystems, such as tropical rainforests. page 1107

Overexploitation of species by hunting, fishing, and logging threatens exploited species and could change the ecology of ecosystems where these species live. page 1108

Invasive species, spread by human activity, are major players in biodiversity loss. page 1108

The use of antibiotics is one of the success stories of twentieth-century medicine, but has led to the spread of antibiotic resistance, a major challenge for twenty-first-century biologists and physicians. page 1110

Amphibian populations have declined dramatically because of a combination of habitat loss, pesticide use, and the spread of infection. page 1110

48.5 CONSERVATION BIOLOGY: Conservation biology aims to safeguard biodiversity for future generations.

Declining population sizes indicate that many species are vulnerable to extinction. Preserving the world's biological heritage will require the sustained efforts of scientists and citizens. page 1111

Reserves provide strong defense against habitat loss, overexploitation, and invasive species. page 1112

Corridors that allow species to move between reserves strengthen conservation and will be particularly important as global change affects reserves. page 1113

Sustainable development strives to match the rate at which we use resources to the rate at which they can be replenished. page 1113

48.6 SCIENTISTS AND CITIZENS IN THE TWENTY-FIRST CENTURY: In the twenty-first century, biologists, doctors, engineers, teachers, and informed citizens have vital roles to play in understanding our changing planet and making wise choices for our future.

Human impact can be positive as well as negative. page 1114

Log in to **LaunchPad** to check your answers to the Self-Assessment Questions and to access additional learning tools.

Glossary

10-nm fiber A relaxed 30-nm chromatin fiber, the state of the chromatin fiber in regions of the nucleus where transcription is currently taking place.

3′ end The end of a nucleic acid strand that carries a free 3′ hydroxyl.

30-nm fiber A chromosomal conformation created by the folding of the nucleosome fiber of DNA and histones.

3-phosphoglycerate (3-PGA) A 3-carbon molecule; during the Calvin cycle, two molecules of 3-PGA are the first stable products following the addition of carbon dioxide.

5′ cap The modification of the 5′ end of the primary transcript by the addition of a special nucleotide attached in an unusual chemical linkage.

5′ end The end of a nucleic acid strand containing a free 5′ phosphate group.

ABC model A model of floral development that invokes three activities—A, B, and C; each activity represents the function of a protein or proteins that specify organ identity in each whorl.

abiotic factors Aspects of the physical environment, such as temperature, water availability, or wind, that affect an organism or population.

abomasum The fourth chamber in the stomach of ruminants, where protein digestion occurs.

abscisic acid A plant hormone that triggers stomatal closure, stimulates root elongation, and maintains seed dormancy.

absolute temperature (T) Temperature measured on the Kelvin scale.

absorption The direct uptake of molecules by organisms, commonly to obtain food. In vertebrate digestion, it is the process by which breakdown products are taken up into the bloodstream.

accessory pigment A light-absorbing pigment other than chlorophyll in the photosynthetic membrane; carotenoids are important accessory pigments.

acclimatization An adaptive change in body function to a new environment.

acidic Describes a solution in which the concentration of protons is higher than that of hydroxide ions (the pH is lower than 7).

acrosome An organelle that surrounds the head of the sperm containing enzymes that enable sperm to transverse the outer coating of the egg.

actin A protein that forms microfilaments; it associates with the motor protein myosin in muscles to contract and generate force.

action potential A brief membrane electrical signal transmitted from the nerve cell body along one or more axons.

activation energy (E_A) The energy input necessary to reach the transition state.

activator A compound that increases the activity of an enzyme.

active site The portion of an enzyme that binds substrate and converts it to product.

active transport The "uphill" movement of substances against a concentration gradient requiring an input of energy.

adaptive (acquired) immunity The part of the vertebrate immune system that is specific to a given pathogen and that responds more strongly to subsequent infections.

adaptive radiation A period of unusually rapid evolutionary diversification in which natural selection accelerates the rate of speciation within a group, with new species adapted for specific niches.

addition rule The principle that the probability of either of two mutually exclusive outcomes occurring is given by the sum of their individual probabilities.

adenine (A) A purine base.

adenosine triphosphate (ATP) The molecule that provides energy in a form that all cells can readily use to perform the work of the cell.

adherens junction A beltlike junctional complex composed of cadherins that attaches a band of actin to the plasma membrane.

adrenal glands Paired glands located adjacent to the kidneys that secrete cortisol in times of stress.

adrenal medulla The inner part of the adrenal gland, which is stimulated by the sympathetic nervous system to secrete norepinephrine.

advantageous mutations Genetic changes that improve their carriers' chances of survival or reproduction.

advertisement display Behavior by which individuals draw attention to themselves or their status.

aerobic Utilizing oxygen.

aerobic metabolism Energy metabolism that uses oxygen gas to oxidize organic compounds, generating ATP; found in Bacteria, Archaea, and the mitochondria of eukaryotic organisms.

age structure The number of individuals within each age group of a population.

agonist muscles Muscle pairs that combine to produce similar motions.

aldosterone A hormone produced by the cortex of the adrenal glands that stimulates the distal convoluted tubule and collecting ducts to take up more salt and water.

alga (plural, algae) A photosynthetic protist.

alkaloid Any one of a group of nitrogen-bearing compounds that can affect the nervous system of animals, produced by some plants as a defense against herbivores.

allantois In the amniotic egg, a membrane that encloses a space where metabolic wastes collect.

allele frequency The proportion of a specified allele among all the alleles of a gene in a population.

alleles The different forms of a gene, corresponding to different DNA sequences in each different form.

allometry An increase in body size accompanied by a change in shape; literally, "different measure."

allopatric Describes populations that are geographically separated from each other.

allosteric effect A change in the activity or affinity of a protein as the result of binding of a molecule to a site other than the active site.

allosteric enzyme An enzyme whose activity is affected by binding a molecule at a site other than the active site. Typically, allosteric enzymes change their shape on binding an activator or inhibitor.

alpha (α) carbon The central carbon atom of each amino acid.

alpha (α) helix One of the two principal types of secondary structure found in proteins.

alternation of generations A sexual life cycle in which haploid and diploid phases alternate; especially in plants, the alternation of generations involving a multicellular haploid gametophyte and a multicellular diploid sporophyte.

alternative splicing A process in which primary transcripts from the same gene can be spliced in different ways to yield different mRNAs and therefore different protein products.

altruistic Describes self-sacrificial behavior in which an individual's actions decrease its own fitness while increasing that of another individual.

Alveolata A eukaryotic superkingdom, defined by the presence of cortical alveoli, small vesicles that, in some species, store calcium ions. Includes ciliates, dinoflagellates, and apicomplexans.

alveoli (singular, alveolus) Clusters of tiny thin-walled sacs in the lungs where gas exchange by diffusion takes place; found at the ends of very fine respiratory bronchioles.

amine hormone A hormone that is derived from a single aromatic amino acid, such as tyrosine.

amino acid An organic molecule containing a central carbon atom attached to a carboxyl group, an amino group, a hydrogen atom, and a side chain. Amino acids are the building blocks of proteins.

amino acid replacement A change in the identity of an amino acid at a particular site in a protein resulting from a mutation in a gene.

amino end The end of a polypeptide chain that has a free amino group.

amino group NH_2; a nitrogen atom bonded to two hydrogen atoms. An amino group is present covalently linked to the central carbon atom of an amino acid.

aminoacyl (A) site One of three binding sites for tRNA on the large subunit of a ribosome.

aminoacyl tRNA synthetase An enzyme that attaches a specific amino acid to a specific tRNA molecule.

ammonification The return of biological nitrogen to the environment as ammonia during the breakdown of organic matter by decomposers.

amnion In the amniotic egg, a membrane surrounding a fluid-filled cavity that allows the embryo to develop in a watery environment.

amniotes The group of vertebrate animals that produces amniotic eggs; this group includes lizards, snakes, turtles, crocodilians, birds, and mammals.

amniotic egg An egg that can exchange gases while retaining water, permitting reproduction in dry terrestrial habitats.

Amoebozoa A superkingdom of eukaryotes with amoeba-like cells that move and gather food by means of pseudopodia.

Amphibia A monophyletic group of vertebrates, including frogs and salamanders, with an aquatic larval form with gills and an adult terrestrial form that usually has lungs.

amphipathic Having both hydrophilic and hydrophobic regions.

amplified In PCR technology, an alternative term for "replicated."

amylase An enzyme that breaks down carbohydrates.

anabolism The set of chemical reactions that build molecules from smaller units utilizing an input of energy, usually in the form of ATP. Anabolic reactions result in net energy storage within cells and the organism.

anaerobic metabolism Energy metabolism in the absence of oxygen.

analogous Describes similar characters that evolved independently in different groups as a result of similar selection pressures; the noun form is *analogy*.

anammox Anaerobic ammonia oxidation; energy metabolism found in some bacteria in which ammonium ions are oxidized by nitrite, yielding nitrogen gas as a by-product.

anaphase The stage of mitosis in which the sister chromatids separate.

anaphase I The stage of meiosis I in which the two homologous chromosomes of each bivalent separate as they are pulled in opposite directions, but the sister chromatids remained joined at the centromere.

anaphase II The stage of meiosis II in which the centromere of each chromosome splits and the separated chromatids are pulled toward opposite poles of the spindle.

anchor A membrane protein that attaches to other proteins and helps to maintain cell structure and shape.

angiosperms The flowering plants; a monophyletic group of seed plants characterized by flowers, double fertilization, and fruits.

angiotensin II A hormone that causes the smooth muscles of arterioles throughout the body to constrict, which increases blood pressure as well as stimulates the release of the hormone aldosterone, increasing reabsorption of water and electrolytes by the kidneys.

annealing The coming together of complementary strands of single-stranded nucleic acids by base pairing.

annelid worms A phylum of worms (Annelida) that have a cylindrical body with distinct segments and a bilaterian body plan.

annual clock A biological clock that corresponds closely to a solar year.

anoxygenic Not producing oxygen; anoxygenic photosynthetic bacteria do not use water as an electron donor and so do not generate oxygen gas.

antagonisms Interactions in which at least one participant loses more than it gains.

antagonist muscles Muscle pairs that pull in opposite directions at a joint to produce opposing motions.

anterior pituitary gland An endocrine organ that forms the anterior lobe of the pituitary gland; it responds to hormones from the hypothalamus that stimulate it to release hormones in turn.

anther In flowering plants, a structure at the top of a stamen consisting of several sporangia in which pollen is produced.

Anthropocene Epoch The modern era, so named to reflect the dominant impact of humans on Earth.

antibody A large protein produced by plasma cells that binds to molecules called antigens.

anticodon The sequence of three nucleotides in a tRNA molecule that base pairs with the corresponding codon in an mRNA molecule.

antidiuretic hormone (ADH) A posterior pituitary gland hormone that acts on the kidneys and controls the water permeability of the collecting ducts, thus regulating the concentration of urine that an animal excretes; also known as *vasopressin*.

antigen Any molecule that leads to the production of antibodies.

antigenic drift The gradual process by which mutation leads to changes in the amino acid sequences of antigens. For example, antigenic drift allows a population of viruses to evolve over time and evade memory B and T cells.

antigenic shift Reassortment of RNA strands in a viral genome, leading to sudden changes in cell-surface proteins.

antigenic variation The encoding of a protein at different times by any one of a number of different genes.

antigen-presenting cell A type of cell (including macrophages, dendritic cells, and B cells) that takes up an antigen and returns portions of it to the cell surface bound to MHC class II proteins.

antiparallel Oriented in opposite directions; the strands in a DNA duplex are antiparallel.

aorta A large artery through which oxygenated blood flows from the left ventricle of the heart to the head and rest of the body.

aortic body A sensory structure of the vertebrate aorta that monitors the levels of oxygen and carbon dioxide and the pressure of blood moving through the body.

aortic valve A valve between the left ventricle and the aorta.

apical dominance The suppression of growth of axillary buds by the shoot apical meristem.

apomixis A process in some species of flowering plants in which seeds develop in the absence of fertilization.

appendicular Describes the part of the vertebrate skeleton that consists of the bones of the shoulder, pelvis, and limbs.

appendix A narrow, tubelike structure that extends from the cecum. The appendix is a vestigial structure that has no clear function in nonherbivorous animals.

aquaporin A protein channel that allows water to cross the plasma membrane more readily than by diffusing through the lipid bilayer.

aqueous Watery.

Archaea One of the three domains of life, consisting of single-celled organisms with a single circular chromosome and no nucleus that divide by binary fission and differ from bacteria in many aspects of their cell and molecular biology.

Archaeplastida A eukaryotic superkingdom of photosynthetic organisms; includes the land plants.

Ardi A specimen of *Ardipithecus ramidus*, an early hominin, dating from about 4.4 million years ago.

arteriole A small branch of an artery.

artery A large, high-pressure vessel that moves blood flow away from the heart to the tissues.

arthropods A phylum of animals that includes insects, crabs and shrimp; contains more than half of all known animal species; distinguished by their segmented bodies and jointed legs.

artificial selection A form of directional selection similar to natural selection, but with selection done intentionally by humans, usually with a specific goal in mind, such as increased milk yield in cattle.

ascomycetes A monophyletic fungal subgroup of the Dikarya in which nuclear fusion and meiosis take place in an elongated saclike cell called an ascus; also called *sac fungi*.

asexual reproduction The reproduction of organisms by mitotic cell division or fragmentation. Offspring are clones of the parent.

assimilation The process by which organisms incorporate nutrients obtained from the environment into more complex molecules.

associative learning (conditioning) Learning that two events are correlated.

atom The basic unit of matter.

atomic mass The mass of the atom determined by the number of protons and neutrons.

ATP synthase An enzyme that couples the movement of protons through the enzyme with the synthesis of ATP.

atrioventricular (AV) node A specialized region of the heart containing pacemaker cells that transmit action potentials from the sinoatrial node to the ventricles of the heart.

atrioventricular (AV) valve A valve between the right atrium and the right ventricle and between the left ventricle and left atrium.

atrium (plural, atria) A heart chamber that receives blood from the lungs or the rest of the body.

autocrine signaling Signaling between different parts of one cell; the signaling cell and the responding cell are one and the same.

autoimmune disease A disease in which tolerance is lost and the immune system becomes active against antigens of the host.

autonomic nervous system The involuntary component of the peripheral nervous system, which controls internal functions of the body such as heart rate, blood flow, digestion, excretion, and temperature.

autosome Any chromosome other than the sex chromosomes.

autotroph An organism that is able to synthesize its own food using energy from sunlight or inorganic chemicals; a primary producer.

auxin A plant hormone that plays a major role in plant development through the establishment of concentration gradients that guide patterns of cell growth and differentiation.

avirulent In plants, describes pathogens that damage only a small part of the plant because the host plant is able to contain the infection. In animals, describes nonpathogenic microorganisms.

axial Describes the part of the vertebrate skeleton that consists of the skull and jaws of the head, the vertebrae of the spinal column, and the ribs.

axillary bud A meristem that forms at the junction between a leaf and the stem.

axon The fiberlike extension from the cell body of a neuron that transmits signals away from the nerve's cell body; the output end of a nerve cell.

axon hillock The junction of the nerve cell body and its axon, which initiates an action potential.

B lymphocyte (B cell) A cell of the immune system that matures in the bone marrow of vertebrates and produces antibodies.

Bacteria One of the three domains of life, consisting of single-celled organisms with a single circular chromosome but no nucleus that divide by binary fission and differ from archaeons in many aspects of their cell and molecular biology.

bacteriochlorophyll A light-harvesting pigment found in photosynthetic bacteria; closely related to the chlorophyll found in plants, algae, and cyanobacteria.

bacteriophage A virus that infects bacterial cells.

balancing selection Natural selection that acts to maintain two or more alleles of a given gene in a population.

ball-and-socket joint A joint that allows rotation in many axes, like the hip and shoulder.

basal lamina A specialized form of extracellular matrix that underlies and supports all epithelial tissues; also called a basement membrane.

base A nitrogen-containing compound that makes up part of a nucleotide.

base excision repair A specialized repair system in which an incorrect DNA base and its sugar are both removed and the resulting gap is repaired.

base stacking Stabilizing hydrophobic interactions between bases in the same strand of DNA.

basic Describes a solution in which the concentration of protons is lower than that of hydroxide ions (the pH is higher than 7).

basidiomycetes A monophyletic fungal subgroup of the Dikarya, including smuts, rusts, and mushrooms, in which nuclear fusion and meiosis take place in a club-shaped cell called a basidium; also called club fungi.

basophil A type of granulocyte that, along with eosinophils, defends against parasitic infections but also contributes to allergies.

behaviorally isolated Describes individuals that only mate with other individuals on the basis of specific courtship rituals, songs, and other behaviors.

beta (β) sheet One of the two principal types of secondary structure found in proteins.

beta-(β-)oxidation The process of shortening fatty acids by a series of reactions that sequentially remove two carbon units from their ends.

Bilateria The monophyletic group of animals with bilateral symmetry.

bilayer A two-layered structure of the cell membrane with hydrophilic "heads" pointing outward toward the aqueous environment and hydrophobic "tails" oriented inward away from water.

bile A fluid produced by the liver that aids in fat digestion by breaking large clusters of fats into smaller lipid droplets.

binary fission The process by which cells of bacteria or archaeons divide to form two daughter cells.

binding affinity The tightness of the binding between the receptor and the signaling molecule.

biodiversity Biological diversity; the total number of species, or, more broadly, the diversity of genetic sequences, cell types, metabolism, life history, phylogenetic groups, communities, and ecosystems.

biodiversity hotspots Relatively small areas that have unusually high numbers of endemic species and that are under threat from human activities.

biological species concept (BSC) As described by Ernst Mayr, the concept that "species are groups of actually or potentially interbreeding populations that are reproductively isolated from other such groups." The BSC is the most widely used and accepted definition of a species, but cannot be applied to asexual organisms.

biologist A scientist who studies life.

biology The science of life and how it works.

biome The distinctive and stable assemblage of species found over a broad region of Earth; terrestrial biomes are each recognized by their distinctive vegetation.

biotic factors Any aspect of organisms that affects another organism.

biotrophic pathogen A plant pathogen that obtains resources from living cells.

biparental inheritance A type of inheritance in which the organelles in the offspring cells derive from those in both parents.

bipedal Habitually walking upright.

bivalent A four-stranded chromosomal structure formed in prophase I of meiosis composed of a pair of homologous chromosomes, each consisting of two chromatids, paired side by side and held together by one or more chiasmata.

bivalves A group of mollusks that includes clams, oysters, and mussels; they have an enclosing skeleton made up of two mineralized shells (valves) connected by a flexible hinge.

bladder A hollow organ in tetrapods (mammals, birds, reptiles, and amphibians) for the storage and elimination of urine; in fishes, a swim bladder controls air volume and buoyancy.

blastocyst A hollow, fluid-filled ball of cells with an inner cell mass that develops from the morula and is an early stage of embryo development in mammals.

blastula A hollow, fluid-filled ball of cells that develops from the morula and is an early stage of embryo development in animals.

blending inheritance The now-discredited model in which heredity factors transmitted by the parents become intermingled in the offspring instead of retaining their individual genetic identities.

blood The circulatory fluid in vertebrates.

bone marrow A fatty tissue between trabeculae and within the central cavity of a bone that contains many important cell populations.

bony fish Osteichthyes; fish that have a cranium, jaws, and mineralized bones. This group includes about 20,000–25,000 species.

bottleneck An extreme, usually temporary, reduction in population size that may result in marked loss of genetic diversity and, in the process, genetic drift.

Bowman's capsule A membranous sac that encases the glomerulus in the kidney.

brain The centralized concentration of neurons in an organ located in the head that processes complex sensory stimuli from the environment or from anywhere in the body.

brainstem Part of the vertebrate brain, formed from the midbrain, which activates the forebrain by relaying information from lower spinal levels.

bronchiole Any one of the fine branches of secondary bronchi.

bud scale Small modifed leaves that protect shoot apical meristems from desiccation and damage due to cold.

budding A form of asexual reproduction in fungi, plants, and some animals in which a bud forms on the organism and eventually breaks off to form a new organism that is smaller than its parent.

bulbourethral glands Glands below the prostate gland that produce a clear fluid that lubricates the urethra for passage of the sperm.

bulk flow The movement of fluids over long distances through organisms due to pressure differences.

C_3 plant A plant that does not use 4-carbon organic acids to supply the Calvin cycle with carbon dioxide.

C_4 plant A plant in which carbon dioxide is incorporated into 4-carbon organic acids in mesophyll cells that are then used to supply the Calvin cycle in bundle sheath cells; results in the suppression of photorespiration.

cadherin A calcium-dependent adherence protein, important in the adhesion of cells to other cells in animals.

calmodulin A protein that binds with Ca^{2+} and activates the enzyme myosin kinase.

Calvin cycle The process in which carbon dioxide is reduced to synthesize carbohydrates, with ATP and NADPH as the energy sources.

Cambrian explosion A transition period in geologic time during which the body plans characteristic of most bilaterian phyla diversified.

canine One of the teeth in carnivores specialized for piercing the body of prey.

capacitation A series of physiological changes that allow the sperm to fertilize the egg.

capillary A very small blood vessel, arranged in finely branched networks connecting arterioles to venules, where gases, nutrients and waste products are exchanged by diffusion with surrounding tissues.

capsid The protein coat that surrounds the nucleic acid genome of a virus.

carbohydrate An organic molecule containing C, H, and O atoms that provides a source of energy for metabolism and that forms the starting point for the synthesis of all other organic molecules.

carbon cycle The intricately linked network of biological and physical processes that shuttles carbon among rocks, soil, oceans, air, and organisms.

carboxyl end The end of a polypeptide chain that has a free carboxyl group.

carboxyl group COOH; a carbon atom with a double bond to oxygen and a single bond to a hydroxyl group.

carboxylation A chemical reaction in which carbon dioxide is added to another molecule and a carboxylic acid group is produced. It occurs, for example, in the Calvin cycle when carbon dioxide is added to a 5-carbon molecule.

cardiac cycle The contraction of the two atria of the heart followed by contraction of the two ventricles.

cardiac muscle Muscle cells that make up the walls of the atria and ventricles and contract to pump blood through the heart.

cardiac output (CO) The volume of blood pumped by the heart over a given interval of time; the key measure of heart function.

carotid body A chemosensory structure of the carotid artery that monitors the levels of carbon dioxide and oxygen, and the pH of blood moving to the brain.

carpel The ovule-producing organ in flowers.

carrier A transporter that facilitates movement of molecules.

carrying capacity (K) The maximum number of individuals a habitat can support.

cartilage A type of connective tissue found, for example, in the walls of intervertebral discs and the joint surfaces between adjacent bones.

cartilaginous fish Chondrichthyes; a monophyletic group of fish that includes about 800 species of sharks, rays, and chimaeras.

Casparian strip A band of hydrophobic material that encircles each cell of the endodermis of a root, forcing water and solutes to reach the xylem by passing through the cytoplasm of endodermal cells.

catabolism The set of chemical reactions that break down molecules into smaller units and, in the process, produce ATP to meet the energy needs of the cell.

causation A relationship in which one event leads to another.

cavitation The abrupt replacement of liquid water in a xylem conduit by water vapor, which prevents water from being pulled through the xylem by transpiration.

cecum A chamber that branches off the large intestine; along with the colon, the site of hindgut fermentation.

cell The simplest self-reproducing entity that can exist as an independent unit of life.

cell adhesion molecule A cell-surface protein that attaches cells to one another and to the extracellular matrix.

cell cycle The collective name for the steps that make up eukaryotic cell division.

cell division The process by which a parent cell gives rise to two daughter cells.

cell junction A complex of proteins in the plasma membrane where a cell makes contact with another cell or the extracellular matrix.

cell-mediated immunity A form of immunity in which T cells recognize and act against pathogens directly.

cell wall A structural layer external to the plasma membrane that helps maintain the shape and internal composition of the cell; present in plants, fungi, many protists, archaeons, and bacteria.

cellular blastoderm In *Drosophila* development, the structure formed by the nuclei in the single-cell embryo when they migrate to the periphery of the embryo and each nucleus becomes enclosed in its own cell membrane.

cellular respiration A series of chemical reactions in which organic molecules are oxidized to carbon dioxide, converting the energy stored in organic molecules to ATP.

cellulase An enzyme that breaks down cellulose.

central dogma Originally, the idea that information flows from nucleic acids to proteins, but not in the opposite direction. More generally, the view that information transfer in a cell usually goes from DNA to RNA to protein.

central nervous system (CNS) In vertebrates, the brain and spinal cord; in invertebrates, centralized information-processing ganglia.

centromere A constriction that physically holds sister chromatids together; the site of the attachment of the spindle fibers that move the chromosome in cell division.

centrosome A compact structure that is the microtubule organizing center for animal cells.

cephalization The concentration of nervous system components at one end of the body.

cephalochordates A subphylum of Chordata that shares key features of body organization with vertebrates but lacks a well-developed brain and eyes, has no lateral appendages, and does not have a mineralized skeleton. Amphioxus, or the lancelet, is the best known example.

cephalopods A monophyletic group of mollusks, including squid, cuttlefish, octopus, and chambered nautilus, with distinctive adaptations such as well-developed eyes, a siphon derived from fused mantle tissue that directs movement by jet propulsion, and muscular tentacles that capture prey and sense the environment.

cerebellum Part of the vertebrate brain, formed from the hindbrain, which coordinates complex motor tasks by integrating motor and sensory information.

cerebral cortex Part of the vertebrate brain formed from a portion of the forebrain that is greatly expanded in mammals, particularly primates.

cerebrum The outer left and right hemispheres of the cerebral cortex.

cervix The end, or neck, of the uterus.

chain terminator A term for a dideoxynucleotide, which if incorporated into a growing daughter strand stops strand growth because there is no hydroxyl group to attack the incoming nucleotide.

channel A transporter with a passage that allows the movement of molecules through it.

chaperone A protein that helps shield a slow-folding protein until it can attain its proper three-dimensional structure.

character In the discipline of systematics, an anatomical, physiological, or molecular feature of an organism that varies among taxa.

character state The observed condition of a character, such as presence or absence of lungs or the arrangement of petals.

checkpoint One of multiple regulatory mechanisms that coordinate the temporal sequence of events in the cell cycle.

chelicerates One of the four main groups of arthropods, including spiders and scorpions, chelicerates have pincer-like claws and are the only arthropods that lack antennae.

chemical bond Any form of attraction between atoms that holds them together.

chemical energy A form of potential energy held in the chemical bonds between pairs of atoms in a molecule.

chemical reaction The process by which molecules are transformed into different molecules.

chemiosmotic hypothesis The hypothesis that the gradient of protons across a membrane provides a source of potential energy that is converted into chemical energy stored in ATP.

chemoautotroph A microorganism that obtains its carbon by the fixation of carbon dioxide, using energy derived from chemical reactions, not from sunlight.

chemoreceptor A receptor that responds to molecules that bind to specific protein receptors on the cell membrane of the sensory receptor.

chemotroph An organism that derives its energy directly from organic molecules such as glucose.

chiasma (plural, chiasmata) A crosslike structure within a bivalent constituting a physical manifestation of crossing over.

chitin A modified polysaccharide containing nitrogen that makes up the cell walls of fungi and the hard exoskeletons of arthropods.

chlorophyll The major photosynthetic pigment contained in the photosynthetic membranes of eukaryotes and cyanobacteria; it plays a key role in the cell's ability to capture energy from sunlight. Chlorophyll appears green because it is poor at absorbing green wavelengths.

chloroplast An organelle in photosynthetic eukaryotes that converts energy of sunlight into chemical energy by synthesizing simple sugars.

chloroplast genome In photosynthetic eukaryotes, the genome of the chloroplast.

choanocyte A type of cell that lines the interior surface of a sponge; choanocytes have flagella and function in nutrition and gas exchange.

choanoflagellate The closest protistan relatives of animals; mostly unicellular protists characterized by a ring of microvilli around the cell's single flagellum.

cholecystokinin (CCK) A peptide hormone that causes the gallbladder to contract and thus release bile into the duodenum.

cholesterol An amphipathic lipid that is a major component of animal cell membranes.

Chordata One of the three major phyla of deuterostomes, this group includes vertebrates and closely related invertebrate animals such as sea squirts and lancelets.

chorion In the amniotic egg, a membrane that surrounds the entire embryo along with its yolk and allantoic sac.

chromatin A complex of DNA, RNA, and proteins that gives chromosomes their structure; chromatin fibers are either 30 nm in diameter or, in a relaxed state, 10 nm.

chromatin remodeling The process in which the nucleosomes are repositioned to expose different stretches of DNA to the nuclear environment.

chromosome condensation The progressive coiling of the chromatin fiber, an active, energy-consuming process requiring the participation of several types of proteins.

chytrid Primarily aquatic fungi that have simple, often unicellular bodies and that produce flagellated gametes and spores.

cilium (plural, cilia) Hairlike structures that extend from the surface of cells. Nonmotile cilia don't move and often have a sensory function; motile cilia propel the movement of cells or fluid surrounding the cell.

circadian clock A biological clock, on a near-daily cycle, that can be set by external cues and regulates many daily rhythms. In animals, the circadian clock may affect feeding, sleeping, hormone production, and core body temperature.

circular muscle Smooth muscle that encircles the body or an organ; in the digestive tract, a circular muscle layer contracts to reduce the size of the lumen. A circular muscle layer contracts alternately with longitudinal muscle to move contents through the digestive tract and to enable locomotion in animals with hydrostatic skeletons.

circulation The movement of a specialized body fluid in animals that carries oxygen and carbon dioxide, and nutrients and waste products, through the body.

cis-regulatory element A short DNA sequence adjacent to a gene, usually at the 5' end, that interacts with transcription factors.

cisternae (singular, cisterna) The series of flattened membrane sacs that make up the Golgi apparatus.

citric acid cycle The third stage of cellular respiration, in which acetyl-CoA is broken down and carbon dioxide is released.

cladistics Phylogenetic reconstruction on the basis of shared evolutionary changes in characters, often called synapomorphies.

class In the Linnaean system of classification, a group of closely related orders.

class I A class of gene of the major histocompatibility complex that is expressed on the surface of all nucleated cells.

class II A class of gene of the major histocompatibility complex that is expressed on the surface of macrophages.

class III A class of gene of the major histocompatibility complex that encodes several proteins of the complement system and proteins involved in inflammation.

classical conditioning Associative learning in which two stimuli are paired.

cleavage The successive mitotic divisions of the zygote after fertilization, in which the single large egg is divided into many smaller cells.

climax community A mature assembly, a final stage in succession, in which, in the absence of environmental disturbance, there is little further change in species composition.

clitoris The female homolog of the glans penis.

clonal selection A hypothesis proposing that the antigen instructs the antibody to fold in a particular way so that the two interact in a specific manner; now a central principle of immunology.

clone An individual that carries an exact copy of the nuclear genome of another individual; clones are genetically identical cells or individuals.

closed circulatory system A circulatory system made up of a set of internal vessels and a heart that functions as a pump to move blood to different regions of the body.

Cnidaria A phylum characterized by radial symmetry, two germ tissues in the embryo, a closed internal gastric cavity, and well-developed tissues but not organs; includes jellyfish, sea anemones, and corals.

cochlea A coiled chamber within the skull containing hair cells that convert sound (pressure) waves into an electrical impulse that is sent to the brain.

codon A group of three adjacent nucleotides in RNA that specifies an amino acid in a protein or that terminates polypeptide synthesis.

coelacanth Two species of lobe-finned fish found long thought to have been extinct for 80 million years but which still live today; along with lungfish, the nearest relatives of tetrapods.

coelom A body cavity surrounding the gut.

coenzyme Q (CoQ) In respiration, a mobile electron acceptor that transports electrons from complexes I and II to complex III in the electron transport chain and moves protons from the mitochondrial matrix to the intermembrane space.

coenocytic Containing many nuclei within one giant cell; the nucleus divides multiple times, but the nuclei are not partitioned into individual cells.

coevolution The process in which species evolve together, each responding to selective pressures from the other.

cognition The ability of the brain to process and integrate complex sources of information, interpret and remember past events, solve problems, reason, and form ideas.

cohesion Attraction between molecules; one consequence of cohesion is surface tension.

cohort A group of the individuals born at a given time.

collagen A strong protein fiber found in bone, artery walls, tendons, and other connective tissues.

collecting duct A type of duct in the vertebrate kidney where urine collects.

colon Part of the hindgut and the site of reabsorption of water and minerals; also known as the *large intestine*.

combinatorial control Regulation of gene transcription by means of multiple transcription factors acting together.

commensalism An interaction between species in which one partner benefits with no apparent effect on the other.

communication The transfer of information between two individuals, the sender and the receiver.

community The set of all populations found in a given place.

compact bone Dense, mineralized bone tissue that forms the walls of a bone's shaft.

companion cell In angiosperms, a cell that carries out cellular functions such as protein synthesis for sieve tubes.

comparative genomics The analysis of the similarities and differences in protein-coding genes and other types of sequence in the genomes of different species.

competition An interaction in which the use of a mutually needed resource by one individual or group of individuals lowers the availability of the resource for another individual or group.

competitive exclusion The result of a competitive interaction in which one species is prevented from occupying a particular habitat or niche.

complement system The collective name for certain proteins circulating in the blood that participate in innate immune function and thus support other parts of the immune system.

complementary Describes the relationship of purine and pyrimidine bases, in which the base A pairs only with T, and G pairs only with C.

complementary DNA (cDNA) A DNA molecule produced by reverse transcriptase from an RNA template.

complex carbohydrate A long, branched chain of monosaccharides.

complex trait A trait that is influenced by multiple genes as well as by the environment.

compound eye An eye structure found in insects and other arthropods that consists of a number of ommatidia, individual light-focusing elements.

concordance The percentage of cases in which both members of a pair of twins show the trait when it is known that at least one member shows it.

cone cell A type of photoreceptor cell on the vertebrate retina that detects color.

conjugation The direct cell-to-cell transfer of DNA, usually in the form of a plasmid.

connective tissue A type of animal tissue that underlies epithelial tissues and is found elsewhere as well; it has an extensive extracellular matrix and few cells.

conservation biology The efforts by biologists and policymakers to address the challenge of sustaining biodiversity in a changing world crowded with people.

conserved Describes sequences that are similar in different organisms.

constant (C) region Describes an unchanging region of the H and L chains of an antibody.

constitutive Describes expression of a gene that occurs continuously.

consumer An organism that obtains the carbon it needs for growth and reproduction from the foods it eats and gains energy by respiring food molecules; heterotrophic organisms of all kinds that directly consume primary producers or consume those that do.

continuous growth A type of population growth in which the population increases or decreases continuously.

contractile ring In animal cells, a ring of actin filaments that forms at the equator of the cell perpendicular to the axis of what was the spindle at the beginning of cytokinesis.

contractile vacuole A type of cellular compartment that takes up excess water and waste products from inside the cell and expels them into the external environment.

control group The group that is not exposed to the variable in an experiment.

cooperative binding The increase in binding affinity with additional binding of oxygen.

copy-number variation (CNV) Differences among individuals in the number of copies of a region of the genome.

cork cambium Lateral meristem that renews and maintains an outer layer that protects the stem against herbivores, mechanical damage, desiccation, and fire.

corpus luteum A temporary endocrine structure in the ovaries that secretes progesterone.

correlation The co-occurrence of two events or processes; correlation does not imply causation.

Coriolis effect The phenomenon in which, because of Earth's counterclockwise rotation about its axis, winds and oceanic currents in the Northern Hemisphere deflect to the right, and those in the Southern Hemisphere deflect to the left.

cortex In a stem, the region between the epidermis and the vascular bundles, composed of parenchyma cells. In the mammalian brain, the highly folded outer layer of gray matter, about 4 mm thick, made up of densely packed neuron cell bodies and their dendrites. In the mammalian renal system, the outer layer of the kidney.

co-speciation A process in which two groups of organisms speciate in response to each other and at the same time, producing matching phylogenies.

countercurrent exchange A mechanism in which two fluids flow in opposite directions, exchanging properties.

countercurrent multiplier A system that generates a concentration gradient as two fluids move in parallel but in opposite directions, as occurs in the loops of Henle in the vertebrate kidney.

covalent bond A chemical bond formed by a shared pair of electrons holding two different atoms together.

CpG island A cluster of CpG sites on a DNA strand where cytosine (C) is adjacent to guanosine (G); the "p" represents the phosphate in the backbone.

cranial nerve In vertebrates, a nerve that links specialized sensory organs to the brain; most contain axons of both sensory and motor neurons.

craniates A subphylum of Chordata, distinguished by a bony cranium that protects the brain; also known as *vertebrates*.

crassulacean acid metabolism (CAM) A mechanism in plants that helps balance carbon dioxide gain and water loss by capturing carbon dioxide into 4-carbon organic acids at night, when transpiration rates are low, and then using it to supply the Calvin cycle during the day while stomata remain closed.

CRISPR (clustered regularly interspaced short palindromic repeats) DNA sequences in Bacteria and Archaea that contain viral sequences and are used for defense; the basis of a DNA editing technique that allows researchers to modify any sequence in the genome.

crisscross inheritance A pattern in which an *X* chromosome present in a male in one generation is transmitted to a female in the next generation, and in the generation after that can be transmitted back to a male.

Cro-Magnon The first known population of *Homo sapiens* in Europe, named for the site in France where specimens were first described.

crop In many birds, an expanded part of the esophagus, which serves as an initial storage chamber.

cross-bridge The binding of the head of a myosin molecule to actin at a specific site between the myosin and actin filaments.

cross-bridge cycle Repeated sequential interactions between myosin heads that bind to and release from sites on actin filaments, forming and unforming cross-bridges, that cause a muscle fiber to contract.

crosscurrent Running across another current at 90 degrees.

crossover The physical breakage, exchange of parts, and reunion between non-sister chromatids.

crustaceans One of the main groups of arthropods, including lobsters, shrimp, and crabs; distinguished by two pairs of antennae and their branched legs or other appendages.

ctenophores Comb-jellies; species in this phylum have a radial body plan but a flow-through gut; they propel themselves by cilia arranged like a comb along the long axis of the body.

cultural transmission The transfer of information among individuals through learning or imitation.

cuticle In plants, a protective layer of a waxy substance secreted by epidermal cells that limits water loss; in animals, an exoskeleton that covers the bodies of invertebrates such as nematodes and arthropods.

C-value paradox The disconnect between genome size and organismal complexity (the C-value is the amount of DNA in a reproductive cell).

cyanobacteria A monophyletic group of bacteria capable of oxygenic photosynthesis.

cyclic electron transport An alternative pathway for electrons during the photosynthetic electron transport chain that increases the production of ATP.

cyclin A regulatory protein whose levels rise and fall with each round of the cell cycle.

cyclin-dependent kinase (CDK) A kinase that is always present within the cell but active only when bound to the appropriate cyclin.

cytochrome–b_6f complex (Cyt) Part of the photosynthetic electron transport chain, through which electrons pass between photosystem II and photosystem I.

cytochrome c The enzyme to which electrons are transferred in complex III of the respiratory electron transport chain.

cytokine A chemical messenger released by phagocytes that recruits other immune cells to the site of injury or infection.

cytokinesis In eukaryotic cells, the division of the cytoplasm into two separate cells.

cytokinin A plant hormone that stimulates cell division and delays leaf senescence.

cytoplasm The contents of the cell other than the nucleus.

cytosine (C) A pyrimidine base.

cytoskeleton In eukaryotes, an internal protein scaffold that helps cells to maintain their shape and serves as a network of tracks for the movement of substances within cells.

cytosol The region of the cell inside the plasma membrane but outside the organelles; the jelly-like internal environment that surrounds the organelles.

cytotoxic T cell A type of T cell that is activated by cytokines released from helper T cells and that can kill other cells.

daughter strand In DNA replication, the strand synthesized from a parental template strand.

day-neutral plant A plant that flowers independently of any change in day length.

decomposer An organism that breaks down dead tissues, feeding on the dead cells or bodies of other organisms.

delayed hypersensitivity reaction A reaction initiated by helper T cells, which release cytokines that attract macrophages to the site of exposure, such as the skin.

deleterious mutations Genetic changes that are harmful to an organism.

deletion A missing region of a gene or chromosome.

demography The study of the size, structure, and distribution of populations over time, including changes in response to birth, aging, migration, and death.

denaturation The unfolding of proteins by chemical treatment or high temperature; the separation of paired, complementary strands of nucleic acid.

dendrite A fiberlike extension from the cell body of a neuron that receives signals from other nerve cells or from specialized sensory endings; the input end of a nerve cell.

dendritic cell A type of cell with long cellular projections that is typically part of the natural defenses found in the skin and mucous membranes.

denitrification The process in which some bacteria use nitrate as an electron acceptor in respiration, releasing N_2 to the environment.

density-dependent Describes processes affecting populations that are influenced by the number of individual organisms, such as the use of resources or susceptibility to predation or parasitism.

density-independent Describes factors such as severe drought that influence population size without regard for the density of the population.

deoxyribonucleic acid (DNA) A linear polymer of four subunits; the information archive in all organisms.

deoxyribose The sugar in DNA.

depolarization An increase in membrane potential from a negative resting potential.

dermis The layer of skin beneath the epidermis, consisting of connective tissue, hair follicles, blood and lymphatic vessels, and glands. It supports the epidermis both physically and by supplying it with nutrients and provides a cushion surrounding the body.

desiccation Excessive water loss; drying out.

desiccation tolerance A suite of biochemical traits that allows cells to survive extreme dehydration by minimizing damage to membranes and macromolecules.

desmosome A buttonlike point of adhesion that holds the plasma membranes of adjacent cells together.

deuterostome A bilaterian in which the blastopore, the first opening to the internal cavity of the developing embryo, becomes the anus. The taxonomic name is Deuterostomia and includes humans and other chordates.

Deuterostomia Bilaterians in which the blastopore develops into the anus. Includes echinoderms, hemichordates, and chordates.

development The process in which a fertilized egg undergoes multiple rounds of cell division to become a multicellular organism, often with specialized tissues and organs.

diaphragm A domed sheet of muscle at the base of the lungs in mammals that separates the thoracic and abdominal cavities and contracts to drive inhalation.

diaphysis The middle region of a bone.

diastole The relaxation of the ventricles, allowing the heart to fill with blood.

dideoxynucleotide A nucleotide lacking both the 2 and 3 hydroxyl groups on the sugar ring.

dietary mineral In animals, a chemical element other than carbon, hydrogen, oxygen, or nitrogen that is required in the diet and must be obtained in food.

differentiation The process in which cells become progressively more specialized as a result of gene regulation.

diffusion The random motion of molecules, with net movement occurring from areas of higher to lower concentration of the molecules.

digestion The process of breaking down food into smaller molecules that can be absorbed.

digestive tract Collectively, the passages that connect the digestive organs; also known as the *gut*.

Dikarya A fungal group that includes about 98% of all described fungal species and in which dikaryotic cells are formed.

dikaryotic ($n + n$) Having two haploid nuclei, one from each parent, in each cell.

diploblastic Describes animals in which the embryo has two germ layers, the endoderm and the ectoderm, from which the adult tissues develop.

diploid Describes a cell with two complete sets of chromosomes.

directional selection A form of selection that results in a shift of the mean value of a trait in a population over time.

discrete growth A type of population growth in which the population size increases (or decreases) in one discrete step at the beginning of each new generation. It is typical of organisms that breed seasonally or annually.

dispersal The process in which some individuals colonize a distant place far from the main source population.

display A pattern of behavior used in communication that is species specific and tends to be highly repeatable and similar from one individual to the next.

disruptive selection A form of selection that operates in favor of extremes and against intermediate forms, selecting against the mean.

distal convoluted tubule The third portion of the renal tubule, in which urea is the principal solute and into which other wastes from the bloodstream are secreted.

dizygotic (DZ) Describes twins that arise when two separate eggs, produced by double ovulation, are fertilized by two different sperm.

DNA editing Techniques used to "rewrite" a nucleotide sequence so that specific mutations can be introduced into genes to better understand their function, or to correct mutant versions of genes to restore normal function.

DNA fingerprinting The analysis of a small quantity of DNA to uniquely identify an individual; also known as *DNA typing*.

DNA ligase An enzyme that uses the energy in ATP to close a nick in a DNA strand, joining the 3′ hydroxyl of one end to the 5′ phosphate of the other end.

DNA polymerase An enzyme that is a critical component of a large protein complex that carries out DNA replication.

DNA replication The process of duplicating a DNA molecule, during which the parental strands separate and new partner strands are made.

DNA transposons Repeated DNA sequences that replicate and can move from one location to another in the genome by DNA replication and repair.

DNA typing The analysis of a small quantity of DNA to uniquely identify an individual; also known as *DNA fingerprinting*.

domain One of the three largest limbs of the tree of life: Eukarya, Bacteria, or Archaea.

dominant Describes an allele or trait that is expressed in heterozygotes. Only one dominant allele is needed to express the phenotype.

dormancy A time period in the life of an organism when growth, development, and metabolism slow down or stop; in plants, a state in which seeds are prevented from germinating even though environmental conditions are favorable.

dorsal nerve cord A nerve cord that develops in a location dorsal to the notochord; this embryonic feature is unique to chordates.

dosage compensation The differential regulation of *X*-chromosomal genes in females and in males.

double bond A covalent bond in which covalently joined atoms share two pairs of electrons.

double fertilization In angiosperms, the process in which two sperm from a single pollen tube fuse with (1) the egg and (2) the two haploid nuclei of the central cell. The first fusion forms the zygote; the second develops into the nutritive endosperm.

double helix The structure formed by two strands of complementary nucleotides that coil around each other.

Down syndrome A condition resulting from the presence of an extra copy of chromosome 21; also known as *trisomy 21*.

downstream gene A gene that functions later than another in development.

duodenum The initial section of the small intestine, into which food enters from the stomach.

duplication A region of a chromosome that is present twice instead of once.

duplication and divergence The process of creating new genes by duplication followed by change in sequence over evolutionary time.

dynamic instability Cycles of polymerization and depolymerization in microtubules.

dynein A motor protein that carries cargo away from the plasma membrane toward the minus ends of microtubules.

Ecdysozoa A subclade of the protostome animals, characterized by the episodic molting of an exoskeleton during growth.

Echinodermata One of the three major phyla of deuterostomes, defined by five-part symmetry; this group includes sea urchins and sea stars.

ecological footprint The quantification of individual human claims on global resources by adding up all the energy, food, materials, and services used and estimating how much land is required to provide those resources.

ecological isolation Pre-zygotic isolation between individuals that specialize ecologically in different ways.

ecological species concept (ESC) The concept that there is a one-to-one correspondence between a species and its niche.

ecology The study of how organisms interact with one another and with their physical environment.

ecosystem A community of organisms and the physical environment it occupies.

ecosystem services Benefits to humans provided by biodiversity, such as cleaner air and water, greater primary productivity, improved resilience to environmental disruption, and untapped sources of food and molecular compounds for use in medicine and agriculture.

ectoderm In animals, the outer germ layer, which differentiates into epithelial cells, pigment cells in the skin, nerve cells in the brain, and, in animals with eyes, the cornea and lens.

ectomycorrhizae One of the two main types of mycorrhizae; ectomycorrhizae produce a thick sheath of fungal cells (hyphae) that surround the root tip, as well as hyphae that grow between, but do not penetrate, cells in the interior of the root.

ectotherm An animal that obtains most of its heat to warm its body from the environment.

egg The larger, female gamete; also call an *ovum*.

ejaculatory duct The duct through which sperm travel from the vas deferens to the urethra.

elastin A protein fiber found in artery walls that provides elasticity.

electrochemical gradient A gradient that combines the charge gradient and the chemical gradient of protons and other ions.

electromagnetic receptor A receptor that responds to electrical, magnetic, or light stimuli.

electron A negatively charged particle that moves around the atomic nucleus.

electron carrier A molecule that carries electrons (and energy) from one set of reactions to another.

electron transport chain The system that transfers electrons along a series of membrane-associated proteins from an electron donor molecule to a final electron acceptor.

electronegativity The ability of atoms to attract electrons.

element A pure substance, such as oxygen, copper, gold, or sodium, that cannot be further broken down by the methods of chemistry.

elongation The process in protein translation in which successive amino acids are added one by one to the growing polypeptide chain.

elongation factor A protein that breaks the high-energy bonds of the molecule GTP to provide energy for ribosome movement and elongation of a growing polypeptide chain.

embryo An early stage of multicellular development that results from successive mitotic divisions of the zygote.

endergonic Describes reactions with a positive ΔG that are not spontaneous and so require an input of energy.

endocrine system A system of cells and glands that secretes hormones into the bloodstream and works with the nervous system to regulate an animal's internal physiological functions.

endocrine signaling Signaling by molecules that travel through the bloodstream.

endocytosis The process in which a vesicle buds off from the plasma membrane, bringing material from outside the cell into that vesicle, which can then fuse with other membranes.

endoderm In animals, the inner germ layer, which differentiates into cells of the lining of the digestive tract and associated organs.

endodermis In plants, a layer of cells surrounding the xylem and phloem at the center of the root that controls the movement of nutrients into the xylem. Also, the inner lining of the cnidarian body.

endomembrane system A cellular system that includes the nuclear envelope, the endoplasmic reticulum, the Golgi apparatus, lysosomes, the plasma membrane, and the vesicles that move between them.

endomycorrhizae One of the two main types of mycorrhizae; endomycorrhizal hyphae penetrate into root cells, where they produce highly branched structures (arbuscules) that provide a large surface area for nutrient exchange.

endophyte A fungus that lives within leaves and that may help the host plant by producing chemicals that deter pathogens and herbivores.

endoplasmic reticulum (ER) An organelle composed of a network of membranes that is involved in the synthesis of proteins and lipids.

endoskeleton A skeleton that develops within the body, as the bones of vertebrates.

endosperm A triploid tissue within angiosperm seeds that supplies nutrition to the embryo.

endosymbiosis A symbiosis in which one partner lives within the other.

endotherm An animal that produces most of its own heat as a by-product of metabolic reactions to maintain a warm body temperature.

energetic coupling The driving of a non-spontaneous reaction by a spontaneous reaction.

energy A property of objects that makes work possible; energy can be transferred from one object to another, and cannot be created or destroyed.

energy balance A form of homeostasis in which the amount of energy calories from food taken in equals the amount of calories used over time to meet metabolic needs.

energy intake Sources of energy taken into the body to support metabolism.

energy use The ways in which energy is expended by an organism.

enhancer A specific DNA sequence necessary for transcription.

enthalpy (H) The total amount of energy in a system.

entropy (S) The amount of disorder (or the number of possible positions and motions of molecules) in a system.

envelope A lipid structure that surrounds the capsids of some viruses; also used to describe the plasma membrane and cell wall of bacteria and polysaccharide structures that encompass cyanobacterial cells.

environmental risk factor A characteristic in a person's surroundings that increases the likelihood of developing a particular disease.

environmental variation Variation among individuals that is due to differences in the environment.

enzyme A protein that functions as a catalyst to accelerate the rate of a chemical reaction; enzymes are critical in determining which chemical reactions take place in a cell.

eosinophil A type of granulocyte that defends against parasitic infections but also contributes to allergies.

epidermis In animals, the outer layer of skin, which serves as a water-resistant, protective barrier. In plants, the outermost layer of cells in leaves, young stems (lacking secondary growth), and roots. In general, the outer layer of the body.

epididymis An organ that lies above the testes where sperm become motile and are stored prior to ejaculation.

epigenetic Describes effects on gene expression due to differences in DNA packaging, such as modifications in histones or chromatin structure.

epiglottis A flap of tissue at the bottom of the pharynx that prevents food from entering the trachea and lungs.

epiphysis The end region of a bone; blood vessels invading the epiphysis and diaphysis trigger the transformation of cartilage into bone.

epiphyte A plant that grows on branches or trunks of trees, without contact with the soil.

epistasis Interaction between genes that modifies the phenotypic expression of genotypes.

epithelial tissue A type of animal tissue that provides a lining for all of the spaces inside and outside the body; it is composed of closely packed cells.

equational division Another name for meiosis II because cells in meiosis II have the same number of chromosomes at the beginning and at the end of the process.

esophagus Part of the foregut; the passage from the mouth to the stomach.

essential amino acid An amino acid that cannot be synthesized by cellular biochemical pathways and instead must be ingested.

estrogen A hormone secreted by the ovaries that stimulates the development of female secondary sexual characteristics.

estrus cycle A cycle in placental mammals other than humans and chimpanzees characterized by phases in which females are sexually receptive.

ethanol fermentation The fermentation pathway in plants and fungi during which pyruvate releases carbon dioxide to form acetaldehyde and electrons from NADH are transferred to acetaldehyde to produce ethanol and NAD^+.

Eukarya The eukaryotes; one of the three domains of life, in which cells have a true nucleus and divide by mitosis.

eukaryote An organism whose cells have a true nucleus.

Eumetazoa A group of anatomically diverse animals that are united by a number of synapomorphies, including the development of diverse cell types, distinct tissues, and gastrulation.

eusocial Describes behavior most commonly observed in species of Hymenopteran insects, in which they have overlapping generations in a nest, cooperative care of the young, and clear and consistent division of labor between reproducers (the queen of a honeybee colony) and nonreproducers (the workers).

eutrophication The process in which the addition of nutrients to a body of water greatly stimulates the growth of plants, algae, and cyanobacteria, often leading to a decrease in dissolved oxygen concentration and therefore a decline in species diversity.

evapotranspiration The sum of water evaporated directly from soils and water bodies plus the amount transpired by plants.

evo-devo The short name for evolutionary developmental biology, a field of study that compares the genetic programs for growth and development in different species in order to understand how diversity has evolved within and between groups.

evolutionarily conserved Characteristics that persist relatively unchanged through diversification of a group of organisms and therefore remain similar in related species.

evolutionarily stable strategy A type of behavior that cannot readily be driven to extinction by an alternative strategy.

excitation–contraction coupling The process that produces muscle force and movement, by coupling the excitation of the muscle cell by a motor neuron to contraction of the muscle.

excitatory postsynaptic potential (EPSP) A positive change in the postsynaptic membrane potential.

excretory tubule In renal systems, a type of tube that drains waste products and connects to the outside of the body.

exergonic Describes reactions with a negative ΔG that release energy and proceed spontaneously.

exhalation The expelling of oxygen-poor air by the elastic recoil of the lungs and chest wall.

exit (E) site One of three binding sites for tRNA on the large subunit of a ribosome.

exocytosis The process in which a vesicle fuses with the plasma membrane and empties its contents into the extracellular space or delivers proteins to the plasma membrane.

exon A sequence that is left intact in mRNA after RNA splicing.

exoskeleton A rigid skeletal system that lies external to the animal's soft tissues.

experimentation A disciplined and controlled way of learning about the world and testing hypotheses in an unbiased manner.

exponential growth The pattern of population increase that results when r (the per capita growth rate) is constant through time.

expressed Turned on or activated, as a gene or protein.

extension (joint) The joint motion in which bone segments move apart.

extension (PCR) A step in the polymerase chain reaction (PCR) for producing new DNA fragments in which the reaction mixture is heated to the optimal temperature for DNA polymerase, and each primer is elongated by means of deoxynucleoside triphosphates.

external fertilization Fertilization that takes place outside the body of the female; in some aquatic organisms, for example, eggs and sperm are released into the water.

extracellular matrix A meshwork of proteins and polysaccharides outside the cell; the main constituent of connective tissue.

extraembryonic membrane In the amniotic egg, one of several sheets of cells that extend out from the developing embryo and form the yolk sac, amnion, allantois, and chorion.

extravasation The process in which phagocytes that travel in the blood move from a blood vessel to the site of infection.

eyecup An eye structure found in flatworms that contains photoreceptors that point up and to the left or right.

F_1 generation The first filial, or offspring, generation.

F_2 generation The second filial generation; the offspring of the F_1 generation.

facilitated diffusion Diffusion through a membrane protein, bypassing the lipid bilayer.

facilitation An indirect interaction between two species in which one species changes the environment in such a way as to benefit another species.

facultative In mutualisms, describes one in which one or both sides can survive without the other. In metabolism, describes a means of obtaining energy that is sometimes but not always used.

fallopian tube A tube from each ovary, through one of which a released oocyte passes; also called *oviduct*.

family A group of closely related genera.

fast-twitch Describes muscle fibers that generate force quickly, producing rapid movements, but that also consume more energy and fatigue more quickly than slow-twitch fibers.

fatty acid A long chain of carbons attached to a carboxyl group; three fatty acid chains attached to glycerol form a triacylglycerol, a lipid used for energy storage.

feature detector A specialized sensory receptor or group of sensory receptors that respond to important signals in the environment.

fermentation A variety of metabolic pathways that produce ATP from the partial oxidation of organic molecules without oxidative phosphorylation or an electron acceptor, such as oxygen.

ferns and horsetails A monophyletic group of vascular plants in which both the gametophyte and sporophyte generations are free living and that disperse by spores.

fertilization The union of gametes to produce a diploid zygote.

fetus An embryo in mammals.

fiber In angiosperms, a narrow cell with thick walls that provides mechanical support in wood. In animals, a muscle cell, which produces forces within an animal's body and exerts forces on the environment.

filament In animals, a thin thread of proteins that interacts with other filaments to cause muscles to shorten. In plants, the part of the stamen that supports the anther.

filial imprinting Imprinting in which newborn offspring rapidly learn to treat any animal they see shortly after birth as their mother.

filtration The separation of solids from fluids, as when circulatory pressure pushes fluid containing wastes through specialized filters into an extracellular space.

first-division nondisjunction Failure of chromosome separation in meiosis I.

first law of thermodynamics The law of conservation of energy: energy can neither be created nor destroyed—it can only be transformed from one form into another.

fitness A measure of the extent to which an individual's genotype is represented in the next generation.

fixed In genetics, describes the situation in which all individuals in a population are homozygous for the same allele of a particular gene; the noun form is *fixation*. In metabolism, refers to the processes by which simple inorganic molecules are converted into biologically available forms, especially carbon fixation and nitrogen fixation.

fixed action pattern (FAP) A sequence of behaviors that, once triggered, is followed through to completion.

flagellum (plural, flagella) An organelle that propels the movement of cells; short flagella are also called cilia.

flexion (joint) The joint motion in which bone segments rotate closer together.

fluid When describing membranes, the ability of membrane lipids to move in the plane of the membrane.

fluid mosaic model A model that proposes that the lipid bilayer is a fluid structure that allows molecules to move laterally within the membrane and is a mosaic of two types of molecules, lipids and proteins.

flux The rate at which a substance—for example, carbon—flows from one reservoir to another.

folding domain A region of a protein that folds in a similar way across a protein family relatively independently of the rest of the protein.

follicle A type of cell that makes up the shell of cells surrounding an oocyte that supports the developing oocyte.

follicle-stimulating hormone (FSH) A hormone secreted by the anterior pituitary gland that stimulates the male and female gonads to secrete testosterone in males and estrogen and progesterone in females.

follicular phase The phase of the menstrual cycle during which FSH acts on granulosa cells, resulting in the maturation of several oocytes, of which, usually, only one becomes completely mature.

food web A map of the interactions that connect consumer and producer organisms; the movement of carbon through the biological components of an ecosystem.

force An interaction that changes the movement of an object, such as a push or pull by one object interacting with another object.

forebrain The region of the vertebrate brain that governs cognitive functions.

foregut The first part of an animal's digestive tract, including the mouth, esophagus, and stomach.

founder event A type of genetic drift that occurs when only a few individuals establish a new population.

fragmentation A form of asexual reproduction in which new individuals arise by the splitting of one organism into pieces, each of which develops into a new individual.

frameshift mutation A mutation in which an insertion or deletion of some number of nucleotides that is not a multiple of three causes a shift in the reading frame of the mRNA, changing all following codons.

frequency of recombination The proportion of recombinant chromosomes among the total number of chromosomes observed.

frontal lobe The region of the brain located in the anterior region of the cerebral cortex, important in decision making and planning.

fruit In angiosperms, the structure that develops from the ovary, sometimes united with adjacent tissues, and serves to protect immature seeds and enhance dispersal once the seeds are mature.

fruiting body A multicellular structure in some fungi that facilitates the dispersal of sexually produced spores.

functional groups Groups of one or more atoms that have particular chemical properties of their own, regardless of what they are attached to.

fundamental niche The full range of climate conditions and food resources in which individuals in a species may live.

fungi An abundant and diverse group of heterotrophic eukaryotic organisms, principally responsible for the decomposition of plant and animal tissues.

G protein A protein that binds to the guanine nucleotides GTP and GDP.

G protein-coupled receptor A receptor that couples to G proteins, which bind to the guanine nucleotides GTP and GDP.

G_0 phase The gap phase in which cells pause in the cell cycle between M phase and S phase; it may last for periods ranging from days to more than a year.

G_1 phase The gap phase in which the size and protein content of the cell increase and specific regulatory proteins are made and activated in preparation for S-phase DNA synthesis.

G_2 phase The gap phase in which the size and protein content of the cell increase in preparation for M-phase mitosis and cytokinesis.

gain-of-function mutation Any mutation in which a gene is expressed in the wrong place or at the wrong time.

gallbladder The digestive organ in which bile produced by the liver is stored.

gamete A reproductive haploid cell; gametes fuse in pairs to form a diploid zygote. In many species, there are two types of gametes: eggs in females, sperm in males.

gametic isolation Incompatibility between the gametes of different individuals (typically belonging to different species).

gametogenesis The formation of gametes.

gametophyte In alternation of generations, describes the haploid multicellular generation that gives rise to gametes.

ganglion (plural, ganglia) A group of nerve cell bodies that processes sensory information received from a local, nearby region, resulting in a signal to motor neurons that control some physiological function of the animal.

gap junction A type of connection between the plasma membranes of adjacent animal cells that permits materials to pass directly from the cytoplasm of one cell to the cytoplasm of another.

gas exchange The transport of gases such as oxygen and carbon dioxide between an organism and its environment.

gastrin A peptide hormone produced in the stomach that stimulates cells lining the stomach to increase their production of HCl.

gastropods A group of mollusks consisting of snails and slugs.

gastrula A stage of early embryo development in animals when cells of the hollow blastula migrate and reorganize to form a multilayered structure in which different cell types begin to differentiate.

gastrulation A highly coordinated set of cell movements in animals in which the cells of the blastula migrate inward, creating germ layers of cells within the embryo.

gel electrophoresis A technique that is used to separate DNA, RNA, or proteins based on size and/or charge, using an electric current passed through a jelly-like substance.

gene The unit of heredity; the stretch of DNA that affects one or more traits in an organism, usually through an encoded protein or noncoding RNA.

gene expression The production of a functional gene product.

gene family A group of genes with related functions, usually resulting from multiple rounds of duplication and divergence.

gene flow The movement of alleles from one population to another through interbreeding between members of each population.

gene pool All the alleles present in all individuals in a population or species.

gene regulation The various ways in which cells control gene expression.

general transcription factors A set of proteins that bind to the promoter of a gene whose combined action is necessary for transcription.

genetic code The correspondence between codons and amino acids, in which 20 amino acids are specified by 64 codons.

genetic drift A random change in the frequency of an allele due to the statistical effects of finite population size.

genetic incompatibility Genetic dissimilarity between two organisms, such as different numbers of chromosomes, that is sufficient to act as a post-zygotic isolating factor.

genetic information Information carried in DNA, organized in the form of genes.

genetic map A diagram showing the relative positions of genes along a chromosome.

genetic risk factor Any mutation that increases the risk of a given disease in an individual.

genetic test A method of identifying the genotype of an individual.

genetic variation Differences in genotype among individuals in a population.

genetically modified organism (GMO) An organism that has been genetically engineered, such as modified viruses and bacteria, laboratory organisms, agricultural crops, and domestic animals; also known as a *transgenic organism*.

genome The genetic material transmitted from a parental cell or organism to its offspring.

genome annotation The process by which researchers identify the various types of sequence present in genomes.

genomic rearrangement In immunology, the process of joining different gene segments as a B cell differentiates to produce a specific antibody. In genetics, evolutionary changes in gene order or organization.

genotype The genetic makeup of a cell or organism; the particular combination of alleles present in an individual.

genotype-by-environment interaction Unequal effects of the environment on different genotypes, resulting in different phenotypes.

genotype frequency The proportion of a specified genotype among all the genotypes for a particular gene or set of genes in a population.

genus (plural, genera) A group of closely related species.

geographic isolation Spatial segregation of individuals.

geologic timescale The series of time divisions that mark Earth's long history.

germ cells The reproductive cells that produce sperm or eggs and the cells that give rise to them.

germ layers Three sheets of cells in animals—the ectoderm, mesoderm, and endoderm—formed by migrating cells of the gastrula that differentiate further into specialized cells and tissues.

germ-line mutation A mutation that occurs in eggs and sperm or in the cells that give rise to these reproductive cells and therefore is passed on to the next generation.

gibberellic acid A plant hormone that stimulates the elongation of stems.

Gibbs free energy (*G*) The amount of energy available to do work.

gills Highly folded delicate structures in aquatic animals that facilitate gas exchange with the surrounding water.

gizzard In birds, alligators, crocodiles, and earthworms, a compartment with thick muscular walls in the digestive tract where food mixed with ingested rock or sediment is broken down into smaller pieces.

glans penis The head of the human penis.

glial cell A type of cell that surrounds neurons and provides them with nutrition and physical support.

glomeromycetes A monophyletic fungal group of apparently low diversity but tremendous ecological importance that occurs in association with plant roots.

glomerulus A tufted loop of porous capillaries in the vertebrate kidney that filters blood.

glycerol A 3-carbon molecule with OH groups attached to each carbon.

glycogen The form in which glucose is stored in animals.

glycolysis The breakdown of glucose to pyruvate; the first stage of cellular respiration.

glycosidic bond A covalent bond that attaches one monosaccharide to another.

Golgi apparatus The organelle that modifies proteins and lipids produced by the endoplasmic reticulum and acts as a sorting station as they move to their final destinations.

gonad In mammals, the part of the reproductive system where haploid gametes are produced. Male gonads are testes, where sperm are produced. Female gonads are ovaries, where eggs are produced.

gonadotropin-releasing hormone (GnRH) A hormone released by the hypothalamus that stimulates the anterior pituitary gland to secrete luteinizing hormone and follicle-stimulating hormone.

gram-positive bacteria Bacteria that retain, in their thick peptidoglycan walls, the diagnostic dye developed by Hans Christian Gram. (Bacteria with thin walls, which do not retain the dye, are said to be gram negative.)

grana (singular, granum) Interlinked structures resembling flattened sacs that form the thylakoid membrane.

granulocyte A type of phagocytic cell that contains granules in its cytoplasm.

granuloma A structure formed by lymphocytes surrounding infected macrophages that helps to prevent the spread of an infection and aids in killing infected cells.

gravitropic Orientation of growth in response to gravity. A negative gravitropic response, as in stems, is growth upward against the force of gravity; a positive gravitropic response, as in roots, is with the force of gravity.

gray matter Densely packed neuron cell bodies and dendrites that make up the cortex, a highly folded outer layer of the mammalian brain about 4 mm thick.

greenhouse gas A gas in the atmosphere that allows incoming solar radiation to reach the Earth's surface, but absorbs radiation re-emitted as heat, trapping it in the atmosphere and causing the temperature to rise.

ground tissues In plants, all tissues other than the epidermis and vascular tissues.

group selection Selection caused by the differential success of groups rather than individuals.

growth factor In animals, any one of a group of small, soluble molecules, usually the signal in paracrine signaling, that affect cell growth, cell division, and changes in gene expression.

growth plate A region of cartilage near the end of a bone where growth in bone length occurs.

growth ring One of the many rings apparent in the cross section of the trunk of a tree, produced by decreases in the size of secondary xylem cells at the end of the growing season, that make it possible to determine the tree's age.

guanine (G) A purine base.

guard cell One of two cells surrounding the central pore of a stoma.

gustation The sense of taste.

gut Collectively, the passages that connect digestive organs; also known as the *digestive tract*.

gymnosperms Seed plants whose ovules are not enclosed in a carpel; gymnosperms include pine trees and other conifers, as well as cycads, gnetophytes, and ginkgo.

habituation The reduction or elimination of a behavioral response to a repeatedly presented stimulus.

hagfish A vertebrate animal with a cranium built of cartilage but no jaws; hagfish feed on marine worms and dead and dying sea animals.

hair cell A specialized mechanoreceptor that senses movement and vibration.

half-life The time it takes for an amount of a substance to reach half its original value. Radioactive half-life is the time it takes for half of the atoms in a given sample of a substance to decay.

haploid Describes a cell with one complete set of chromosomes.

haplotype The particular combination of alleles present in any defined region of a chromosome.

Hardy–Weinberg equilibrium A state in which allele and genotype frequencies do not change over time, implying the absence of evolutionary forces. It also specifies a mathematical relationship between allele frequencies and genotype frequencies.

heart The pump of the circulatory system, which moves blood to different regions of the body.

heart rate (HR) The number of heartbeats per unit time.

heartwood The center of the stem in long-lived trees, which does not conduct water.

heavy (H) chains Two of the four polypeptide chains that make up an antibody molecule.

helicase A protein that unwinds the parental double helix at the replication fork.

helper T cell One type of T cell that assists other cells of the immune system by secreting cytokines, thus activating B cells to secrete antibodies, for example.

hematocrit The fraction of red blood cells within the blood of vertebrates.

Hemichordata One of the three major phyla of deuterostomes, this group includes acorn worms and pterobranchs.

hemidesmosome A type of desmosome in which integrins are the prominent cell adhesion molecules.

hemoglobin An iron-containing molecule specialized for oxygen transport.

hemolymph The circulatory fluid in invertebrates.

heterokaryotic Describes a stage in the life cycle of some fungi, in which plasmogamy is not followed immediately by karyogamy and the cells have unfused haploid nuclei from both parents.

hemophilia A trait characterized by excessive bleeding that results from a recessive mutation in a gene encoding a protein necessary for blood clotting.

herbivory The consumption of plant tissues.

heritability In a population, the proportion of the total variation in a trait that is due to genetic differences among individuals.

heterotroph An organism that obtains its carbon from organic molecules synthesized by other organisms.

heterozygote advantage A form of balancing selection in which the heterozygote's fitness is higher than that of either of the homozygotes, resulting in selection that ensures that both alleles remain in the population at intermediate frequencies.

heterozygous Describes an individual who inherits different types of alleles from the parents, or genotypes in which the two alleles for a given gene are different.

hierarchical Describes gene regulation during development, in which the genes expressed at each stage in the process control the expression of genes that act later.

hindbrain Along with the midbrain, the region of the vertebrate brain that controls basic body functions and behaviors.

hindgut The last part of an animal's digestive tract, including the large intestine and rectum.

hinge joint A simple joint that allows one axis of rotation, like the elbow and knee.

hippocampus A posterior region of the limbic system within the mammalian brain involved in spatial cognition and long-term memory formation.

histamine A chemical messenger released by mast cells and basophils; an important contributor to allergic reactions and inflammation.

histone code The pattern of modifications of the histone tails that affects the chromatin structure and gene transcription.

histone tail A string of amino acids that protrudes from a histone protein in the nucleosome.

homeodomain The DNA-binding domain in homeotic proteins, whose sequences are very similar from one homeotic protein to the next.

homeostasis The active regulation and maintenance, in organisms, organs, or cells, of a stable internal physiological state in the face of a changing external environment.

homeotic (*Hox*) gene A gene that specifies the identity of a body part or segment during embryonic development.

hominins A member of one of the different species in the group leading to humans.

homologous Describes characters that are similar in different species because of descent from a common ancestor; the noun form is *homology*.

homologous chromosomes Pairs of chromosomes, matching in size and appearance, that carry the same set of genes; one of each pair was received from the mother, the other from the father.

homozygous Describes an individual who inherits an allele of the same type from each parent, or a genotype in which both alleles for a given gene are of the same type.

horizontal gene transfer The transfer of genetic material between organisms that are not parent and offspring.

hormone A chemical signal that influences physiology and development in both plants and animals; in animals, hormones are released into the bloodstream and circulate throughout the body.

host cell A cell in which viral replication occurs.

host plant A plant species that can be infected by a given pathogen.

hotspot In genetics, a site in the genome that is especially mutable. In ecology, a region with high levels of biodiversity, often with large numbers of endemic species.

human chorionic gonadotropin (hCG) A hormone released by the developing embryo that maintains the corpus luteum.

hybrid offspring The offspring produced by a cross; sometimes applied specifically to interbreeding between two closely related species.

hybridization Interbreeding between two different varieties or species.

hydrogen bond A bond between a hydrogen atom in one molecule and an electronegative atom in another molecule.

hydrophilic "Water loving"; describes a class of molecules with which water can undergo hydrogen bonding.

hydrophobic "Water fearing"; describes a class of molecules poorly able to undergo hydrogen bonding with water.

hydrophobic effect The exclusion of nonpolar molecules by polar molecules, which drives biological processes such as the formation of cell membranes and the folding of proteins.

hydrostatic skeleton A skeletal system in which fluid contained within a body cavity is the supporting element.

hydroxyapatite The calcium phosphate mineral found in bone.

hypersensitive response A type of plant defense against infection in which uninfected cells surrounding the site of infection rapidly produce large numbers of reactive oxygen species, triggering cell wall reinforcement and causing the cells to die, thus creating a barrier of dead tissue.

hyperthermophile An organism that requires an environment with high temperature.

hyphae In fungi, highly branched multicellular filaments that provide a large surface area for absorbing nutrients.

hypothalamus The underlying brain region of the forebrain, which interacts with the autonomic and endocrine systems to regulate the general physiological state of the body.

hypothesis A tentative explanation for one or more observations that makes predictions that can be tested by experiments or additional observations.

IgA One of the five antibody classes, IgA is typically a dimer and the major antibody on mucosal surfaces.

IgD One of the five antibody classes, IgD is a monomer and is typically found on the surface of B cells; it helps initiate inflammation.

IgE One of the five antibody classes, IgE is a monomer that plays a central role in allergies, asthma, and other immediate hypersensitivity reactions.

IgG The most abundant of the five antibody classes, IgG is a monomer that circulates in the blood and is particularly effective against bacteria and viruses.

IgM One of the five antibody classes, IgM is a pentamer in mammals and a tetramer in fish and is particularly important in the early response to infection, activating the complement system and stimulating an immune response.

ileum A section of the small intestine that, with the jejunum, carries out most nutrient absorption.

imitation Observing and copying the behavior of another.

immediate hypersensitivity reaction A reaction characterized by a heightened or an inappropriate immune response to common antigens.

immunodeficiency Any disease in which part of the immune system does not function properly.

imprinting (genetics) Silencing of gene expression by chemical modification of DNA or chromatin that differs according to the parent of origin.

in vitro fertilization (IVF) A process in which eggs and sperm are brought together in a petri dish, where fertilization and early cell divisions occur.

inbred line A true-breeding, homozygous strain.

inbreeding depression A reduction in fitness resulting from breeding among relatives causing homozygosity of deleterious recessive mutations.

incisor One of the teeth in the front of the mouth, used for biting.

incomplete dominance Inheritance in which the phenotype of the heterozygous genotype is intermediate between those of homozygous genotypes.

incomplete penetrance The phenomenon in which some individuals with a genotype corresponding to a trait do not show the phenotype, either because of environmental effects or because of interactions with other genes.

incus A small bone in the vertebrate middle ear that helps amplify the sound waves that strike the tympanic membrane.

induced pluripotent stem cell (iPS cell) A cell that has been reprogrammed to become pluripotent by activation of certain genes, most of them encoding transcription factors or chromatin proteins.

inducer A small molecule that elicits gene expression.

inflammation A physiological response of the body to injury that removes the inciting agent if present and begins the healing process.

inhalation The drawing of oxygen-rich air into the lungs by the expansion of the thoracic cavity.

inhibitor A synthesized compound that decreases the activity of an enzyme.

inhibitory postsynaptic potential (IPSP) A negative change in the postsynaptic membrane potential.

initiation The stage of translation in which methionine is established as the first amino acid in a new polypeptide chain.

initiation factor A protein that binds to mRNA to initiate translation.

innate Describes behaviors that are instinctive and carried out regardless of earlier experience.

innate (natural) immunity The part of the immune system that provides protection in a nonspecific manner against all kinds of infection; it does not depend on exposure to a pathogen.

inner cell mass A mass of cells in one region of the inner wall of the blastocyst from which the body of the embryo develops.

insects The most diverse of the four main groups of arthropods, with a head, thorax, and abdomen, and six walking legs.

instantaneous speciation Speciation that occurs in a single generation.

integral membrane protein A protein that is permanently associated with the cell membrane and cannot be separated from the membrane experimentally without destroying the membrane itself.

integrin A transmembrane protein, present on the surface of virtually every animal cell, that enables cells to adhere to the extracellular matrix.

intercostal muscles Muscles of vertebrates attached to adjacent pairs of ribs that assist the diaphragm by elevating the ribs on inhalation and depressing them during exhalation.

intermediate filament A polymer of proteins, which vary according to cell type, that combine to form strong, cable-like filaments that provide animal cells with mechanical strength.

intermembrane space The space between the inner and outer mitochondrial membranes.

internal fertilization Fertilization that takes place inside the body of the female.

interneuron A neuron that processes information received by sensory neurons and transmits it to motor neurons in different body regions.

internode The segment between two nodes on a shoot.

interphase The time between two successive M phases.

intersexual selection A form of sexual selection involving interaction between males and females, as when females choose from among males.

interspecific competition Competition between individuals of different species.

intervertebral disc A gel-filled support structure found between the bony vertebrae of the backbone that enables flexibility and provides cushioning of loads.

intrasexual selection A form of sexual selection involving interactions between individuals of one sex, as when members of one sex compete with one another for access to the other sex.

intraspecific competition Competition between individuals of the same species.

intrinsic rate of increase The per capita growth rate; the maximum rate of growth when no environmental factors limit population increase.

intron A sequence that is excised from the primary transcript and degraded during RNA splicing.

invasive species Non-native species; since they are removed from natural constraints on population growth, invasive species can expand dramatically when introduced into new areas, sometimes with devastating consequences for native species and ecosystems.

inversion The reversal of the normal order of a block of genes.

involuntary component The part of the peripheral nervous system that controls unconscious reactions.

ion An electrically charged atom or molecule.

ion channels Channel-like membrane proteins that open and close, thereby altering the flow of ions across the plasma membrane.

ionic bond A chemical bond in which two ions with opposite charges associate with each other due to their difference in electronegativity.

island population An isolated population; may apply to true islands (land surrounded by water) or to ecological islands that similarly isolate their inhabitants (for example ponds, for fish).

isomers Molecules that have the same chemical formula but different structures.

isometric Describes the generation of muscle force without a change in muscle length.

isometry An increase in body size while keeping the same overall shape; literally, "same measure."

isotopes Atoms of the same element that have different numbers of neutrons.

jejunum A section of the vertebrate small intestine that, with the ileum, carries out most nutrient absorption.

juxtaglomerular apparatus The structure formed by specialized cells of the efferent arteriole leaving the glomerulus of each nephron, which secretes the hormone rennin into the bloodstream.

karyogamy The fusion of two nuclei following plasmogamy.

karyotype A standard arrangement of chromosomes, showing the number and shapes of the chromosomes representative of a species.

key stimulus A stimulus that initiates a fixed action pattern.

keystone species Pivotal populations that affect other members of the community in ways that are disproportionate to their abundance or biomass.

kidneys In vertebrates, paired organs of the renal system that remove waste products and excess fluid; their action contributes to homeostasis.

kin selection A form of natural selection that favors the spread of alleles promoting behaviors that help close relatives.

kinesin A motor protein, similar in structure to myosin, that transports cargo toward the plus end of microtubules.

kinesis (plural, kineses) A random, undirected movement in response to a stimulus.

kinetic energy The energy of motion.

kinetochore The protein complexes on a chromatid where spindle fibers attach.

kingdom A group of closely related phyla.

Klinefelter syndrome A sex-chromosomal abnormality in which an individual has 47 chromosomes, including two X chromosomes and one Y chromosome.

knee-extension reflex A reflex commonly tested by physicians to evaluate peripheral nervous and muscular system function.

K-strategist A species that produces relatively few young but invests considerable resources into their support.

labia majora Outer folds of skin in the vulva.

labia minora Inner folds of skin in the vulva that meet at the clitoris.

lactic acid fermentation The fermentation pathway in animals and bacteria during which electrons from NADH are transferred to pyruvate to produce lactic acid and NAD^+.

lagging strand A daughter strand that has its $5'$ end pointed toward the replication fork, so as the parental double helix unwinds, a new DNA piece is initiated at intervals, and each new piece is elongated at its $3'$ end until it reaches the piece in front of it.

lamellae (singular, lamella) A thin layer. In animals, the many thin, sheetlike structures spread along the length of each gill filament, giving gills a large surface area relative to their size. In plants, the middle lamella is the thin pectin layer between cells.

lamprey A vertebrate animal with a cranium and vertebral column built of cartilage but no jaws; many lampreys live parasitically, sucking body fluids from fish prey.

large intestine Part of the hindgut and the site of reabsorption of water and minerals; also known as the *colon*.

lariat A loop and tail of RNA formed after RNA splicing.

larynx The structure, above the trachea, that contains the vocal cords.

lateral inhibition Inhibition of a process in nerve cells adjacent to the cell receiving a signal inducing that process, enhancing the strength of a signal locally but diminishing it peripherally.

lateral meristem The source of new cells that allows plants to grow in diameter.

latex A white sticky liquid produced in some plants.

latitudinal diversity gradient The increase in species diversity from the poles to the equator.

leading strand A daughter strand that has its 3′ end pointed toward the replication fork, so as the parental double helix unwinds, this daughter strand can be synthesized as one long, continuous polymer.

learned Describes a behavior that depends on an individual's experience.

learning The process in which experience leads to changes in behavior.

lengthening contraction The contraction of a muscle against a load greater than the muscle's force output, leading to a lengthening of the muscle.

lenticel A region of less tightly packed cells in the outer bark that allows oxygen to diffuse into the stem.

leptosporangium A distinctive sporangium, found in some ferns, whose outer wall is only a single cell thick and in which drying produces a catapult-like mechanism to release spores.

Leydig cell A type of cell in the testes that secretes testosterone.

Liebig's Law of the Minimum The principle that primary production is limited by the nutrient that is least available relative to its use by primary producers.

lichens Stable associations between a fungus and a photosynthetic microorganism, usually a green alga but sometimes a cyanobacterium.

life history The typical pattern of resource investment in each stage of a given species' lifetime.

ligand Alternative term for a signaling molecule that binds with a receptor, usually a protein.

ligand-binding site The specific location on the receptor protein where a signaling molecule binds.

ligand-gated ion channel A receptor that alters the flow of ions across the plasma membrane when bound by its ligand.

light (L) chains Two of the four polypeptide chains that make up the simplest antibody molecule.

limbic system Inner components of the vertebrate forebrain that control physiological drives, instincts, emotions, motivation, spatial memory, and the sense of reward.

linked Describes genes that are sufficiently close together in the same chromosome that they do not assort independently.

lipase A type of enzyme produced by the pancreas that breaks apart lipids, thus enabling their more effective digestion.

lipid A hydrocarbon molecule that is soluble in non-polar solvents but not water; lipids store energy, act as signaling molecules, and are a component of cell membranes.

lipid raft Lipids assembled in a defined patch in the cell membrane.

liposome An enclosed bilayer structure spontaneously formed by phospholipids in environments with neutral pH, like water.

liver A vertebrate organ that aids in the digestion of proteins, carbohydrates, and fats in the duodenum by producing bile, which breaks down fat.

lobe-finned fish Species of fish with paired pectoral and pelvic fins that have a bone structure similar to that of tetrapod limbs.

logistic growth The pattern of population growth that results as growth potential slows down as the population size approaches K, its maximum sustainable size.

long-day plant A plant that flowers only when the light period exceeds a critical value.

longitudinal muscle Smooth muscle that runs lengthwise along a body or organ; in the digestive tract, a longitudinal muscle layer contracts to shorten small sections of the gut.

loop of Henle The middle portion of the vertebrate renal tubule, which creates a concentration gradient that allows water passing through the collecting duct to be reabsorbed.

Lophotrochozoa A group of bilaterians that includes mollusks and annelid worms.

loss-of-function mutation A genetic change that inactivates the normal function of a gene.

Lucy An unusually complete specimen of an early hominin fossil, *Australopithecus afarensis*, found in 1974 in Ethiopia and dating to 3.2 million years ago.

lumen In eukaryotes, the continuous interior of the endoplasmic reticulum; in plants, a fluid-filled compartment enclosed by the thylakoid membrane; generally, the interior of any tubelike structure.

lunar clock A moon-based biological clock that times activities in some species, especially those living in habitats where tides are important.

lungfish Several species of lobe-finned fish that use a simple lung to survive periods when their watery habitat dries by burying themselves in moist mud and breathing air; along with coelacanths, the closest relatives of tetrapods.

lungs The internal organs for gas exchange in many terrestrial animals.

luteal phase The phase of the menstrual cycle of mammals beginning with ovulation.

luteinizing hormone (LH) A hormone secreted by the anterior pituitary gland of vertebrates that stimulates the gonads to secrete testosterone in males and estrogen and progesterone in females.

lycophytes A monophyletic group of spore-dispersing vascular plants that are the sister group to all other vascular plants.

lymph The fluid in the vertebrate lymphatic system in which T and B cells circulate.

lymphatic system A network of vessels distributed through the vertebrate body with important functions in the immune system.

lysis The breakage, or bursting, of a cell.

lysogeny The integration of viral DNA into a host cell's DNA, which is then transmitted to offspring cells.

lysosome A vesicle derived from the Golgi apparatus that contains enzymes that break down macromolecules such as proteins, nucleic acids, lipids, and complex carbohydrates.

lytic pathway The alternative to the lysogenic pathway; in the lytic pathway, a virus bursts, or lyses, the cell it infects, releasing new virus particles.

M phase The stage of the cell cycle consisting of mitosis and cytokinesis, in which the parent cell divides into two daughter cells.

macrophage A type of large phagocytic cell that is able to engulf other cells.

mainland population The central and largest population of a species.

major groove The larger of two uneven grooves on the outside of a DNA duplex.

major histocompatibility complex (MHC) A group of genes and the proteins they encode that appear on the surface of most mammalian cells; only antigens associated with MHC proteins are recognized by T cell receptors.

malleus A small bone in the vertebrate middle ear that helps amplify the waves that strike the tympanic membrane.

Malpighian tubule One of the tubes in the main body cavity of insects and other terrestrial arthropods through which fluid passes and which empties into the hindgut.

mammals A class of vertebrates distinguished by body hair and mammary glands from which they feed their young.

mandibles Paired mouthparts found in some arthropods.

map information The knowledge of where an individual is with respect to its goal.

map unit A unit of distance in a genetic map equal to the distance between genes resulting in 1% recombination.

mark-and-recapture A method in which individuals are captured, marked in way that doesn't affect their function or behavior, and then released. The percentage of marked individuals captured later enables ecologists to estimate population size.

marsupials A group of mammals that includes kangaroos, koalas, and opossums; their young are born at an early stage of development and must crawl to abdominal mammary glands that provide them with milk.

mass extinction A catastrophic drop in recorded diversity, which has occurred five or more times in the past 541 million years.

mast cell An immune cell that releases histamine.

maternal-effect gene A gene that is expressed by the mother that affects the phenotype of the offspring, typically through the composition or organization of the oocyte.

maternal inheritance A type of inheritance in which the organelles in the offspring cells derive from those in the mother.

mating types Genetically distinct forms of individuals of a fungus species that, by enabling fertilization only between different types, prevent self-fertilization and promote out-crossing.

mean The average value of a set of numbers.

mechanical incompatibility Structural configuration of the genitalia that prevents mating with another individual (typically of another species).

mechanoreceptor A sensory receptor that responds to physical deformations of its membrane produced by touch, stretch, pressure, motion, or sound.

mediator complex A complex of proteins that interacts with the Pol II complex and allows transcription to begin.

medulla A part of the brainstem; also, the inner layer of the mammalian kidney and adrenal gland.

meiosis I Reductional division; the first stage of meiotic cell division, in which the number of chromosomes is halved.

meiosis II Equational division; the second stage of meiotic cell division, in which the number of chromosomes is unchanged.

meiotic cell division A form of cell division that includes only one round of DNA replication but two rounds of nuclear division; meiotic cell division makes sexual reproduction possible.

membrane attack complex (MAC) A complex of complement proteins that makes holes in bacterial cells, leading to cell lysis.

membrane potential A difference in electrical charge across the plasma membrane.

memory cell A type of long-lived immune cell produced after exposure to a pathogen that contains membrane-bound antibodies having the same specificity as the parent cell.

menopause The cessation of menstrual cycles resulting from decreased production of estradiol and progesterone by the mammalian ovaries.

menstrual cycle A monthly cycle in females of some mammals in which oocytes mature and are released from the ovary under the influence of hormones.

menstruation The monthly shedding of the uterine lining in some mammals.

meristem A discrete population of actively dividing, totipotent plant cells; apical meristems are located at the tip of stems and roots and produce cells that allow plants to grow in length, while lateral meristems surround stems and roots and produce cells that allow growth in diameter.

meristem identity gene A gene that contributes to meristem stability and function.

mesentery A membrane in the vertebrate abdominal cavity through which blood vessels, nerves, and lymph travel to supply the gut.

mesoderm In animals, the intermediate germ layer, which differentiates into cells that make up connective tissue, muscle cells, red blood cells, bone cells, kidney cells, and gonad cells.

mesoglea In cnidarians, a gelatinous mass enclosed by the epidermis and endodermis.

mesohyl A gelatinous mass that lies between the interior and exterior cell layers of a sponge that contains some amoeba-like cells that function in skeleton formation and the dispersal of nutrients.

mesophyll A leaf tissue of loosely packed photosynthetic cells.

messenger RNA (mRNA) The RNA molecule that combines with a ribosome to direct protein synthesis; it carries the genetic "message" from the DNA to the ribosome.

metabolic rate An animal's overall rate of energy use.

metabolism The chemical reactions occurring within cells that convert one molecule into another and transfer energy in living organisms.

metamorphosis The process in some animals in which the body changes dramatically at key stages in development.

metanephridia A pair of excretory organs in each body segment of annelid worms that filters the body fluid.

metaphase The stage of mitosis in which the chromosomes are aligned in the middle of the dividing cell.

metaphase I The stage of meiosis I in which the meiotic spindle is completed and the bivalents move to lie on an imaginary plane cutting transversely across the spindle.

metaphase II The stage of meiosis II in which the chromosomes line up so that their centromeres lie on an imaginary plane cutting across the spindle.

metapopulation A large population made up of smaller populations linked by migration.

methanogens Archaea that generate natural gas (methane, CH_4) as a by-product of anaerobic energy metabolism.

MHC class I (genes and proteins) MHC genes and proteins in vertebrates that are expressed on the surface of all nucleated cells.

MHC class II (genes and proteins) MHC genes and proteins in vertebrates that are expressed on the surface of macrophages, dendritic cells, and B cells.

MHC class III (genes and proteins) MHC genes and proteins in vertebrates that encode several proteins of the complement system and proteins involved in inflammation.

micelle A spherical structure in which lipids with bulky heads and a single hydrophobic tail are packed.

microfilament A helical polymer of actin monomers, present in various locations in the cytoplasm, that helps make up the cytoskeleton.

microfossil A microscopic fossil, including fossils of bacteria and protists.

microRNA (miRNA) Small, regulatory RNA molecules that can cleave or destabilize RNA or inhibit its translation.

microtubule A hollow, tubelike polymer of tubulin dimers that helps make up the cytoskeleton.

microvilli Highly folded surfaces of villi, formed by fingerlike projections on the surfaces of epithelial cells.

midbrain Along with the hindbrain, the region of the vertebrate brain that controls basic body functions and behaviors; it is a part of the brainstem.

midgut The middle part of an animal's digestive tract, including the small intestine.

migration The movement of organisms from one place to another, including the movement of individuals from one population to another.

minor groove The smaller of two unequal grooves on the outside of a DNA duplex.

mismatch repair A mechanism for repairing nucleotide mismatches in newly replicated DNA.

mitochondria (singular, mitochondrion) Specialized organelles that are the site of respiration in eukaryotic cells, oxidizing chemical compounds such as sugars to carbon dioxide and transferring their chemical energy to ATP.

mitochondrial DNA (mtDNA) A small circle of DNA, about 17,000 base pairs long in humans, found in every mitochondrion.

mitochondrial genome In eukaryotic cells, the DNA in the mitochondria.

mitochondrial matrix The space enclosed by the inner membrane of the mitochondria.

mitosis In eukaryotic cells, the division of the nucleus, in which the chromosomes are separated into two nuclei.

mitotic spindle A structure in the cytosol made up predominantly of microtubules that pull the chromosomes into separate daughter cells.

Modern Synthesis The current theory of evolution, which combines Darwin's theory of natural selection and Mendelian genetics.

molar One of the teeth in the back of the mouth of mammals specialized for crushing and shredding tough foods such as meat and fibrous plant material.

molecular clock The observation of rate constancy in molecular evolution. The extent of genetic divergence at a gene in two taxa is thus a reflection of the time since the taxa last shared a common ancestor.

molecular evolution Evolution at the level of DNA, which in time results in the genetic divergence of populations.

molecular fossils Sterols, bacterial lipids, and some pigment molecules, which are relatively resistant to decomposition, that can be preserved in sedimentary rocks, documenting organisms that rarely form conventional fossils.

molecular orbital A merged orbital traversed by a pair of shared electrons.

molecular self-assembly The process by which, when conditions and relative amounts are suitable, viral components spontaneously interact and assemble into mature virus particles.

molecule A substance made up of two or more atoms.

mollusks A monophyletic invertebrate group distinguished by a mantle, which plays a major role in movement, skeleton-building, breathing, and excretion; includes clams, snails, and squid.

molting Periodic shedding, as of an exoskeleton.

monophyletic Describes groupings in which all members share a single common ancestor not shared with any other species or group of species.

monosaccharide A simple sugar.

monozygotic (MZ) Describes twins that arise from a single fertilized egg, which after several rounds of cell division separates into two distinct, but genetically identical, embryos.

morphospecies concept The idea that members of the same species usually look like each other more than like other species.

morula In animals, the solid ball of cells resulting from early cell divisions of the fertilized egg.

motor endplate The post-synaptic region on a muscle cell where acetylcholine binds with receptors.

motor neuron A neuron that signals a muscle or gland to cause a response in the body.

motor protein Any of various proteins that are involved in intracellular transport or cause muscle contraction by moving the actin microfilaments inside muscle cells.

motor unit A vertebrate motor neuron and the population of muscle fibers that it innervates.

mouth The first part of the foregut, which receives food.

mucosa An inner tissue layer with secretory and absorptive functions surrounding the lumen of the digestive tract.

multiple alleles Two or more different alleles of the same gene, occurring in a population of organisms.

multiplication rule The principle that the probability of two independent events occurring together is the product of their respective probabilities.

multipotent Describes cells that can form a limited number of types of specialized cell.

multiregional hypothesis The idea that modern humans derive from the *Homo ergaster* populations that spread around the world starting 2 million years ago.

muscle tissue A type of animal tissue made up of cells (fibers) that are able to shorten or contract, resulting in movement.

mutagen An agent that increases the probability of mutation.

mutation Any heritable change in the genetic material, usually a change in the nucleotide sequence of a gene.

mutualism An interaction between two or more species that benefits all.

mycelium (plural, mycelia) A network of branching hyphae.

mycorrhizae Fungi that live symbiotically among roots, enhancing nutrient uptake by the plants.

myelin Fatty insulating layers surrounding the axons of vertebrate neurons, produced by glial cells, that speed up the propagation of action potentials.

myofibril A long rodlike structure in muscle fibers that contains parallel arrays of actin and myosin filaments.

myoglobin An oxygen-binding protein in the cells of vertebrate muscles, related to hemoglobin, that facilitates oxygen delivery to mitochondria.

myosin A motor protein found in cells that carries cargo to the plus ends of microfilaments and is also used by both striated and smooth muscles to contract and generate force.

myotome In chordates, any one of a series of segments that organizes the body musculature.

myriapods One of the four main groups of arthropods, including centipedes and millipedes; distinguished by their many pairs of legs.

natural killer cell A cell type of the innate immune system that recognizes and kills host cells that are infected by a virus or have become cancerous or otherwise abnormal.

natural selection The process in which, when there is genetic variation in a population of organisms, the variants best suited for growth and reproduction in a given environment contribute disproportionately to future generations. Of all the evolutionary mechanisms, natural selection is the only one that leads to adaptations.

Neanderthal *Homo neanderthalensis*, a species similar to humans, but with thicker bones and flatter heads; present in the fossil record 600,000–30,000 years ago.

necrotrophic pathogen A plant pathogen that kills cells before drawing resources from them.

negative feedback The effect in which the final product of a biochemical pathway inhibits the first step; the process in which a stimulus acts on a sensor that communicates with an effector, producing a response that opposes the initial stimulus. Negative feedback is used to maintain steady conditions, or homeostasis.

negative regulation The process in which a regulatory molecule must bind to the DNA at a site near the gene to prevent transcription.

negative selection Natural selection that reduces the frequency of a deleterious allele.

negatively selected In immunology, describes B and T cells that react too strongly to self antigens in association with MHC and are eliminated through cell death.

nematodes Roundworms, the most numerous of all animals; a phylum of the Ecdysozoa.

neoteny The process in which the timing of development is altered so that a sexually mature organism retains the physical characteristics of the juvenile form.

nephron The functional unit of the vertebrate kidney, consisting of the glomerulus, Bowman's capsule, renal tubules, and collecting ducts.

neritic zone Within the oceans, environments that are near the shore, with a shallow seafloor, relatively high availability of nutrients, and persistent sunlight. Distinct biomes within the neritic zone include intertidal and coral reef biomes.

nerve A bundle of long fiberlike extensions (axons) from multiple nerve cells.

nerve cord A bundle of long fiberlike extensions from multiple nerve cells that serves as the central nervous system of invertebrates, such as flatworms and earthworms, and, together with the brain, as the central nervous system of vertebrates.

nervous system A network of many interconnected nerve cells.

nervous tissue A type of animal tissue found in the nerve nets of cnidarians and in the brain, spinal cord, and peripheral nerves of vertebrates; it takes in sensory information from the environment, processes information, and sends signals to target organs to elicit a response.

neural tube In chordates, a cylinder of embryological tissue that develops into a dorsal nerve cord.

neuron Nerve cell; the basic fundamental unit of nervous systems.

neurosecretory cell A neuron in the vertebrate hypothalamus and the posterior pituitary gland that secretes hormones into the bloodstream.

neurotransmitter A molecule that conveys a signal from the end of an axon to the postsynaptic target cell, such as another neuron or a muscle fiber.

neutral mutations Genetic changes that have no effect or negligible effects on the organism, or whose effects are not associated with differences in survival or reproduction.

neutron An electrically neutral particle in the atomic nucleus.

neutrophil A type of phagocytic cell that is very abundant in the blood and is often one of the first cells to respond to infection.

niche A complete description of the role a species plays in its environment, and of its requirements, both abiotic and biotic.

nicotinamide adenine dinucleotide phosphate (NADPH) An important cofactor in many biosynthetic reactions; the reducing agent used in the Calvin cycle.

nitrification The process by which chemoautotrophic bacteria oxidize ammonia (NH_3) to nitrite (NO_2^-) and then nitrate (NO_3^-).

nitrogen fixation The process by which some Bacteria and Archaea convert nitrogen gas (N_2) into ammonia (NH_3), a form biologically useful to primary producers.

nitrogenous waste Waste in the form of ammonia, urea, and uric acid, which are toxic to organisms in varying degrees.

node In phylogenetic trees, the point where a branch splits, representing the common ancestor from which the descendant species diverged. In plants, the point on a shoot where one or more leaves are attached.

nodes of Ranvier Sites on an axon of vertebrate nerve cells that lie between adjacent myelin-wrapped segments, where the axon membrane is exposed.

non-associative learning Learning that occurs in the absence of any particular outcome, such as a reward or punishment.

nondisjunction The failure of a pair of chromosomes to separate normally during anaphase of cell division.

nonpolar Describes compounds that do not have regions of positive and negative charge.

nonpolar covalent bond A covalent bond between atoms that have the same, or nearly the same, electronegativity.

nonrandom mating Mate selection biased by genotype or relatedness.

nonrecombinants Progeny in which the alleles are present in the same combination as that present in a parent.

nonself From the point of view of a given organism, describes molecules and cells that are foreign.

nonsense mutation A mutation that creates a stop codon, terminating translation.

non-sister chromatids Chromatids of different members of a pair of homologous chromosomes; although they carry the same complement of genes, they are not genetically identical.

nonsynonymous (missense) mutation A point mutation (nucleotide substitution) that causes an amino acid replacement.

norm of reaction A graphical depiction of the change in phenotype across a range of environments.

normal distribution A distribution whose plot is a bell-shaped curve.

notochord In chordates, a stiff rod of cartilage-like tissue that runs along the back and provides support for the axis of the body.

nuclear envelope The cell structure, composed of two membranes, inner and outer, that defines the boundary of the nucleus.

nuclear genome In eukaryotic cells, the DNA in the chromosomes.

nuclear localization signal The signal sequence for the nucleus that enables proteins to move through pores in the nuclear envelope.

nuclear pore One of many protein channels in the nuclear envelope that act as gateways that allow molecules to move into and out of the nucleus and are thus essential for the nucleus to communicate with the rest of the cell.

nuclear transfer A procedure in which a hollow glass needle is used to insert the nucleus of a cell into the cytoplasm of an egg whose own nucleus has been destroyed or removed.

nucleic acid A polymer of nucleotides that encodes and transmits genetic information.

nucleoid In prokaryotes, a cell structure with multiple loops formed from supercoils of DNA.

nucleolus A distinct, dense, non–membrane-bound spherical structure within the nucleus that contains the genes and transcripts for ribosomal RNA.

nucleoside A molecule consisting of a 5-carbon sugar and a base.

nucleosome A beadlike repeating unit of histone proteins wrapped with DNA making up the 10-nm chromatin fiber.

nucleotide A constituent of nucleic acids, consisting of a 5-carbon sugar, a nitrogen-containing base, and one or more phosphate groups.

nucleotide excision repair The repair of multiple mismatched or damaged bases across a region; a process similar to mismatch repair, but over a much longer piece of DNA, sometimes thousands of nucleotides.

nucleus (of an atom) The dense central part of an atom containing protons and neutrons.

nucleus (of a cell) The compartment of the cell that houses the DNA in chromosomes.

obligate In ecology, a type of mutualism in which one participant requires the other to survive. In metabolism, a means of gaining energy or carbon that is required by the organism; that is, it has no other options, as in an obligate photoautroph.

observation The act of viewing the world around us.

occipital lobe The region of the brain that processes visual information from the eyes.

ocean acidification A decrease in the pH of seawater caused by an increase in the amount of carbon dioxide in the oceans.

oceanic zone Deeper waters beyond the continental shelf; it includes the pelagic biome in open ocean water and the deep sea biome on the deep seafloor.

Okazaki fragment In DNA replication, any of the many short DNA pieces in the lagging strand.

olfaction The sense of smell.

oligonucleotides Short stretches of nucleic acids; in the polymerase chain reaction (PCR), the primer sequences produced by chemical synthesis, which are typically 20–30 nucleotides long.

omasum The third chamber in the stomach of ruminants, into which the mixture of food and bacteria passes and where water is reabsorbed.

ommatidia (singular, ommatidium) Individual light-focusing elements that make up the compound eyes of insects and other arthropods; the number of ommatidia determines the resolution of the image.

oncogene A cancer-causing gene.

oocyte An unfertilized egg cell in animals produced by the mother; a developing female gamete.

oogenesis The formation of ova or eggs.

open circulatory system A circulatory system found in many invertebrate animals in which most of the circulating fluid is contained within the animal's body cavity.

open reading frame (ORF) A stretch of DNA or RNA consisting of codons for amino acids uninterrupted by a stop codon. In genome annotation, this sequence motif identifies the region as potentially protein coding.

operant conditioning Associative learning in which a behavior that was initially undirected has become paired with a particular stimulus through reward/punishment reinforcement.

operator The binding site for a repressor protein.

operon A group of functionally related genes located in tandem along the DNA and transcribed as a single unit from one promoter; the region of DNA consisting of the promoter, the operator, and the coding sequence for the structural genes.

Opisthokonta A superkingdom that encompasses animals, fungi, and related protists.

opsin A photosensitive protein that converts the energy of light photons into electrical signals in the receptor cell.

opsonization The binding of a molecule to a pathogen to facilitate uptake by a phagocyte.

orbital A region in space where an electron is present most of the time.

organ Two or more tissues that combine and function together.

organ system A group of organs that have related functions. For example, the intestine is one organ of the digestive system, which also includes the stomach, liver, pancreas, and other organs.

organelle Any one of several compartments in eukaryotes that divide the cell contents into smaller spaces specialized for different functions.

organic molecule A molecule containing carbon atoms, usually in chains or rings.

organogenesis The transformation of the three germ layers into all the organ systems of the body in animals.

origin of replication Any point on a DNA molecule at which DNA synthesis is initiated.

osmoconformer An animal that matches its internal osmotic pressure to that of its external environment.

osmoregulation The regulation of water and solute levels to control osmotic pressure.

osmoregulator An animal that maintains internal solute concentrations that differ from that of its environment.

osmosis The net movement of a solvent, such as water, across a selectively permeable membrane toward the side of higher solute concentration.

osmotic pressure The pressure needed to prevent water from moving from one solution into another by osmosis.

osteoblast A type of cell that forms bone tissue.

osteoclast A type of cell that digests and removes bone tissue.

out-of-Africa hypothesis The idea that modern humans arose from *Homo heidelbergensis* descendants in Africa before dispersing beyond Africa around 60,000 years ago.

ovary In angiosperms, a hollow structure at the base of the carpel in which the ovules develop and which protects the ovules from being eaten or damaged by animals; in animals, the female gonad where eggs are produced.

oviduct A tube from each ovary, through one of which a released oocyte passes; also called *fallopian tube*.

oviparity Laying eggs.

ovoviviparity Giving birth to live young, with nutritional support of the embryo from the yolk.

ovulate cone A reproductive shoot in gymnosperms that produces ovules.

ovule A reproductive structure in seed plants consisting of the female gametophyte developing within a sporangium and surrounded by outer protective layers. Ovules, when fertilized, develop into seeds.

ovum (plural, ova) The larger, female gamete; also called an *egg*.

oxidation A reaction in which a molecule loses electrons.

oxidation–reduction reaction A reaction involving the loss and gain of electrons between reactants. In biological systems these reactions are often used to store or release chemical energy.

oxidative phosphorylation A set of metabolic reactions that occurs by passing electrons along an electron transport chain to the final electron acceptor, oxygen, pumping protons across a membrane, and using the proton electrochemical gradient to drive synthesis of ATP.

oxygen dissociation curve The curve that results when blood pO_2 is plotted against the percentage of O_2 bound to hemoglobin.

oxygenic Producing oxygen.

oxytocin A hormone released by the posterior pituitary gland that causes uterine contraction during labor and stimulates the release of milk during breastfeeding.

P_1 generation The parental generation in a series of crosses.

pacemaker Describes cardiac muscle cells that function as a regulator of heart rhythm.

palindromic Reading the same in both directions; describes sequence identity along the paired strands of a duplex DNA molecule; a symmetry typical of restriction sites.

pancreas A secretory gland that has both endocrine function, secreting hormones, including insulin, directly into the blood, and exocrine function, aiding the digestion of proteins, carbohydrates, and fats by secreting digestive enzymes into ducts that connect to the duodenum.

paracrine signaling Signaling by a molecule that travels a short distance to the nearest neighboring cell to bind its receptor and deliver its message.

paraphyletic Describes groupings that include some, but not all, the descendants of a common ancestor.

parasexual Describes asexual species that generate genetic diversity by the crossing over of DNA during mitosis.

parasites Organisms that live in close association with another species, consuming or gaining nutrition from their hosts' tissues, generally without killing them.

parasympathetic division The division of the vertebrate autonomic nervous system that slows the heart and stimulates digestion and metabolic processes that store energy, enabling the body to "rest and digest."

parathyroid gland A gland adjacent to the thyroid gland that secretes parathyroid hormone (PTH), which, with calcitonin, regulates the actions of bone cells and levels of calcium in the blood.

parenchyma In plants, describes thin-walled cells that carry out a variety of functions, including photosynthesis and storage; capable of further cell division if stimulated.

parietal lobe The vertebrate brain region, posterior to the frontal lobe, that controls body awareness and the ability to perform complex tasks.

parsimony Choosing the simplest hypothesis to account for a given set of observations. In phylogenetic reconstruction, opting for the tree requiring the fewest evolutionary steps.

parthenogenesis A form of asexual reproduction in which females produce eggs that are not fertilized by males but divide by mitosis and develop into new individuals.

partial pressure (p) The fractional concentration of a gas relative to other gases present multiplied by the atmospheric pressure exerted on the gases.

partially reproductively isolated Describes populations that have not yet diverged into separate species but whose genetic differences are extensive enough for the hybrid offspring they produce to have reduced fertility or viability compared with offspring produced by crosses between individuals within each population.

paternal inheritance A type of inheritance in which the organelles in the offspring cells derive from those in the father.

pathogens Organisms and other agents, such as viruses, that cause disease.

peat bog Wetland in which dead organic matter accumulates.

pedigree A diagram of family history that summarizes the record of the ancestor–descendant relationships among individuals.

penis The male copulatory organ.

pepsin An enzyme produced in the stomach that breaks down proteins.

peptide bond A covalent bond that links the carbon atom in the carboxyl group of one amino acid to the nitrogen atom in the amino group of another amino acid.

peptide hormone A hormone that is a short chain of linked amino acids.

peptidoglycan A complex polymer of sugars and amino acids that makes up the cell wall in bacteria.

peptidyl (P) site One of three binding sites for tRNA on the large subunit of a ribosome.

per capita growth rate (r) The rate of population growth per individual; it is the change in population size ΔN in a given period of time Δt, divided by the size of the population N.

pericycle In roots, a single layer of cells just to the inside of the endodermis from which new root apical meristems are formed, allowing roots to branch.

periodic selection The episodic loss of diversity as a successful variant outcompetes others.

periodic table of the elements The arrangement of the chemical elements in tabular form, organized by their chemical properties.

peripatric speciation A specific kind of allopatric speciation in which a few individuals from a mainland population disperse to a new location remote from the original population and evolve separately.

peripheral nervous system (PNS) Collectively, the sensory and motor nerves, including the cranial and spinal nerves, neurons of the autonomic nervous system, and ganglia.

peripheral membrane protein A protein that is temporarily associated with the lipid bilayer or with integral membrane proteins through weak noncovalent interactions.

peristalsis Waves of muscular contraction that move food toward the base of the stomach.

personalized medicine An approach in which the treatment is matched to the patient, not the disease; examination of an individual's genome sequence, by revealing his or her disease susceptibilities and drug sensitivities, allows treatments to be tailored to that individual; also called *precision medicine*.

petal A structure, often brightly colored and distinctively shaped, occurring in the next-to-outermost whorl of a flower; petals attract and orient animal pollinators.

phagocyte A type of immune cell that engulfs and destroys foreign cells or particles.

phagocytosis The engulfing of a cell or particle by another cell.

pharyngeal slit A vertical opening separated from other slits by stiff rods of protein in the pharynx of chordates.

pharynx In chordates, the region of the throat that connects the mouth and nasal cavities to the digestive system and, in vertebrates, to the lungs.

phenol Any one of a class of chemical compounds, produced by some plants as a defensive mechanism.

phenotype The expressed physical, behavioral, and biochemical traits of an individual, including height, weight, eye color, and so forth.

pheromone A water- or airborne chemical compound released by animals into the environment that signals and influences the behavior of other members of their species.

phloem The vascular tissue that transports carbohydrates from leaves to the rest of the plant body.

phloem sap The sugar-rich solution that flows through both the lumen of the sieve tubes and the sieve plate pores.

phosphatase An enzyme that removes a phosphate group from another molecule.

phosphate group A chemical group consisting of a phosphorus atom bonded to four oxygen atoms.

phosphodiester bond A bond that forms when a phosphate group in one nucleotide is covalently joined to the sugar unit in another nucleotide. Phosphodiester bonds are relatively stable and form the backbone of a DNA strand.

phospholipid A type of lipid and a major component of the cell membrane.

photic zone The surface layer of the ocean through which enough sunlight penetrates to enable photosynthesis.

photoheterotroph An organism that uses the energy from sunlight to make ATP and relies on organic molecules obtained from the environment as the source of carbon for growth and other vital functions.

photoperiod Day length.

photoperiodism The effect of the photoperiod, or day length, on physiological or developmental processes; in plants, the effect of day length on flowering.

photoreceptor A molecule whose chemical properties are altered when it absorbs light; also, photoreceptors are the sensory receptors in the eye.

photorespiration A process in which rubisco acts as an oxygenase, resulting in release of carbon dioxide and a net loss of energy.

photosynthesis The biochemical process in which carbohydrates are built from carbon dioxide and water using the energy of sunlight. In anoxygenic photosynthesis, electrons are donated by compounds other than water, especially H_2S.

photosynthetic electron transport chain A series of redox reactions in which light energy absorbed by chlorophyll is used to power the movement of electrons; in oxygenic photosynthesis, the electrons ultimately come from water and the terminal electron acceptor is $NADP^+$.

photosystem A protein–pigment complex that absorbs light energy to drive redox reactions and thereby sets the photosynthetic electron transport chain in motion.

phototroph An organism that captures energy from sunlight.

phototropic Bending in response to light. A positive phototropic response, as in stems, is toward light; a negative phototropic response, as in roots, is away from light.

phragmoplast In dividing plant cells, a structure formed by overlapping microtubules that guide vesicles containing cell wall components to the middle of the cell.

phylogenetic niche conservatism The observed similarity in closely related species of some aspect of their niches, indicating its presence in their common ancestor.

phylogenetic species concept (PSC) The idea that members of a species all share a common ancestry and a common fate.

phylogenetic tree A tree-like diagram representing a hypothesis about the evolutionary relationships among populations or species.

phylogeny The history of descent with modification and the accumulation of change over time.

phylum (plural, phyla) A group of closely related classes, defined by having a distinct body plan.

phytochrome A photoreceptor for red and far-red light that switches back and forth between two stable forms, active and inactive, depending on its exposure to light.

pilus (plural, pili) A hollow, threadlike structure that connects bacteria, drawing two cells together so that plasmids can be transferred between them in the process of conjugation.

pineal gland A gland located in the thalamic region of the brain that responds to autonomic nervous system input by secreting melatonin, which controls wakefulness.

pinna (plural, pinnae) Small unit of the photosynthetic surface on fern leaves; also, the external structure of mammalian ears that enhance the reception of sound waves contacting the ear.

pith In a stem, the region inside the ring of vascular bundles.

pituitary gland An endocrine gland located at the base of the vertebrate brain that produces a number of different hormones and controls many of the other organs of the endocrine system.

placenta In placental mammals, an organ formed by the fusion of the chorion and allantois that allows the embryo to obtain nutrients directly from the mother.

placental mammal A mammal that provides nutrition to the embryo through the placenta, a temporary organ that develops in the uterus.

placozoans Possibly the simplest of all animals; each contains only a few thousand cells arranged into upper and lower epithelia that sandwich an interior fluid crisscrossed by a network of multinucleate fiber cells.

plasma (effector) cell An immune cell that secretes antibodies.

plasma membrane The membrane that surrounds the cytoplasm of the cell, separating the inside of the cell from the outside of the cell; also called the cell membrane.

plasmid In bacteria, a small circular molecule of DNA carrying a small number of genes that replicate independently of the DNA in the bacterium's circular chromosome.

plasmodesmata (singular, plasmodesma) Connections between the plasma membranes of adjacent plant cells that permit molecules to pass directly from the cytoplasm of one cell to the cytoplasm of another.

plasmogamy The cytoplasmic union of two cells.

pleiotropy The phenomenon in which a single gene has multiple effects on seemingly unrelated traits.

pluripotent A cell that can give rise to any type of cell in an organism. In animals, it describes embryonic stem cells (cells of the inner mass), which can give rise to any of the three germ layers and therefore to any cell of the body.

podocytes Cells with footlike processes that make up one of the three layers of the filtration barrier of the vertebrate kidney glomerulus.

point mutation A mutation in which a base pair is replaced by a different base pair; this is the most frequent type of mutation; also known as a *nucleotide substitution*.

Pol II The RNA polymerase complex responsible for transcription of protein-coding genes.

polar (molecule) A molecule that has regions of positive and negative charge.

polar body A small cell produced by the asymmetric first meiotic division of a primary oocyte.

polar covalent bond Bonds that do not share electrons equally.

polar transport The coordinated movement of auxin across many cells.

polarity An asymmetry such that one end of a structure differs from the other.

polarized Having opposite properties in opposite parts; describes a resting membrane potential in which there is a buildup of negatively charged ions on the inside surface of the cell's plasma membrane and positively charged ions on its outer surface.

pollen In seed plants, the multicellular male gametophyte surrounded by a sporopollenin-containing outer wall.

pollen cone A reproductive shoot in gymnosperms that produces pollen.

pollination The process in which pollen is carried to an ovule.

poly(A) tail The nucleotides added to the 3′ end of the primary transcript by polyadenylation.

polyadenylation The addition of a long string of consecutive A-bearing ribonucleotides to the 3′ end of the primary transcript.

polycistronic mRNA A single molecule of messenger RNA that is formed by the transcription of a group of functionally related genes located next to one another along bacterial DNA.

polymer A complex organic molecule made up of repeated simpler units connected by covalent bonds.

polymerase chain reaction (PCR) A selective and highly sensitive method for making copies of a piece of DNA, which allows a targeted region of a DNA molecule to be replicated into as many copies as desired.

polymorphism Any genetic difference among individuals that is present in multiple individuals in a population.

polypeptide A polymer of amino acids connected by peptide bonds.

polyphyletic Describes groupings that do not include the last common ancestor of all members.

polyploidy The condition of having more than two complete sets of chromosomes in the genome.

polysaccharide A polymer of simple sugars. Polysaccharides provide long-term energy storage or structural support.

polyspermy In animals, fertilization by more than one sperm.

pons A part of the brainstem.

population All the individuals of a given species that live and reproduce in a particular place; one of several interbreeding groups of organisms of the same species living in the same geographical area.

population density The size of a population divided by its range.

population size The number of individuals of all ages alive at a particular time in a particular place.

Porifera Sponges; an animal phylum characterized by few cell types, no well-defined tissues, and no clearly defined plane of symmetry.

positive feedback In the nervous and endocrine systems, a type of feedback in which a stimulus causes a response that leads to an enhancement of the original stimulus that leads to a larger response in the same direction. The process reinforces itself until interrupted.

positive regulation The process in which a regulatory molecule must bind to the DNA at a site near the gene in order for transcription to take place.

positive selection Natural selection that increases the frequency of a favorable allele.

positively selected In immunology, describes B and T cells that recognize self MHC molecules on epithelium cells and continue to mature.

posterior pituitary gland An endocrine organ that forms the posterior lobe of the pituitary gland; it develops from neural tissue at the base of the brain and contains axons of neurosecretory cells from the hypothalamus that secrete releasing hormones.

posttranslational modification The modification, after translation, of proteins in ways that regulate their structure and function.

post-zygotic isolation Describes factors that cause the failure of the fertilized egg to develop into a fertile individual.

potential energy Stored energy that is released by a change in an object's structure or position.

power stroke The stage in the muscle cross-bridge cycle in which the myosin head pivots and generates a force, causing the myosin and actin filaments to slide relative to each other.

precision medicine An approach in which the treatment is matched to the patient, not the disease; examination of an individual's genome sequence, by revealing his or her disease susceptibilities and drug sensitivities, allows treatments to be tailored to that individual; also called *personalized medicine*.

predation An interaction between organisms in which one (the predator) consumes the other (the prey).

pregnancy The carrying of one or more embryos in the mammalian uterus.

premolar One of the teeth of mammals between the canines and molars that are specialized for shearing tough foods.

pre-zygotic isolation Describes factors that prevent the fertilization of an egg.

primary active transport Active transport that uses the energy of ATP directly.

primary bronchi The two divided airways from the vertebrate tetrapod trachea, supported by cartilage rings, each airway leading to a lung.

primary consumers Herbivores, which consume primary (plant, algae, or cyanobacterial) producers. Sometimes called grazers.

primary growth In plants, growth originating in the apical meristems of shoots and roots; resulting in the production of new leaves and an increase in length.

primary motor cortex The part of the frontal lobe of the vertebrate brain that produces complex coordinated behaviors by controlling skeletal muscle movements.

primary oocyte A diploid cell formed by mitotic division of oogonia during fetal development.

primary producer An organism that takes up inorganic carbon, nitrogen, phosphorus, and other compounds from the environment and converts them into organic compounds; a source of food for heterotrophic organisms in the local environment.

primary response The response to the first encounter with an antigen, during which there is a short lag before antibody is produced.

primary somatosensory cortex The part of the vertebrate parietal lobe that integrates tactile information from specific body regions and relays it to the motor cortex.

primary spermatocyte A diploid cell formed by mitotic division of spermatogonia at the beginning of spermatogenesis.

primary structure The sequence of amino acids in a protein.

primary transcript The initial RNA transcript that comes off the template DNA strand.

primate A member of the order of mammals that includes humans, distinguished by a number of molecular and morphological features, including relatively large brains, nails rather than claws, front-facing eyes, and, in some species, an opposable thumb.

primer A short stretch of RNA at the beginning of each new DNA strand that serves as a starter for DNA synthesis; an oligonucleotide that serves as a starter in the polymerase chain reaction.

primordia (singular, primordium) An organ in its earliest stage of development; in plants, leaf primordia form near the tips of shoot apical meristems and develop into leaves.

principle of independent assortment The principle that segregation of one set of alleles of a gene pair is independent of the segregation of another set of alleles of a different gene pair.

principle of segregation The principle by which half the gametes receive one allele of a gene and half receive the other allele.

probability Among a very large number of observations, the expected proportion of observations that are of a specified type.

probe A labeled DNA fragment that can be tracked in a procedure such as a Southern blot.

procambial cell A plant cell that retains the capacity for cell division and gives rise to both xylem and phloem.

product Any one of the transformed molecules that result from a chemical reaction.

progesterone A hormone secreted by the vertebrate ovaries that maintains the thickened and vascularized uterine lining.

prokaryote An organism whose cell or cells do not have a membrane-bound nucleus; sometimes used to refer collectively to archaeons and bacteria.

prometaphase The stage of mitosis in which the nuclear envelope breaks down and the microtubules of the mitotic spindle attach to chromosomes.

prometaphase I The stage of meiosis I in which the nuclear envelope breaks down and the meiotic spindles attach to kinetochores on chromosomes.

prometaphase II The stage of meiosis II in which the meiotic spindles attach to kinetochores on chromosomes.

promoter A regulatory region where RNA polymerase and associated proteins bind to the DNA duplex.

proofreading The process in which DNA polymerases can immediately correct their own errors by excising and replacing a mismatched base.

prophase The stage of mitosis characterized by the appearance of visible chromosomes.

prophase I The beginning of meiosis I, marked by the visible manifestation of chromosome condensation.

prophase II The stage of meiosis II in which the chromosomes in the now-haploid nuclei recondense to their maximum extent.

prostate gland A vertebrate exocrine gland that produces a thin, slightly alkaline fluid that helps maintain sperm motility and counteracts the acidity of the female reproductive tract.

protease inhibitor A molecule that blocks the action of enzymes that break down proteins. Manufactured by some plants as a defense against herbivores.

protein family A group of proteins that are structurally and functionally related.

protein sorting The process by which proteins end up where they need to be in the cell to perform their function.

proteins The key structural and functional molecules that do the work of the cell, providing structural support and catalyzing chemical reactions. The term "protein" is often used as a synonym for "polypeptide."

proteobacteria The most diverse bacterial group, defined largely by similarities in rRNA gene sequences; it includes many of the organisms that populate the expanded carbon cycle and other biogeochemical cycles.

protist An organism with cells having a nucleus but lacking other features specific to plants, animals, or fungi.

proton A positively charged particle in the atomic nucleus.

protonephridia Excretory organs in flatworms that isolate waste from the body cavity.

proto-oncogene The normal cellular gene counterpart to an oncogene, which is similar to a viral oncogene but can cause cancer only when mutated.

protostome A bilaterian in which the blastopore, the first opening to the internal cavity of the developing embryo, becomes the mouth. The taxonomic name is Protostomia, and the group includes mollusks and arthropods.

Protostomia Bilaterians in which the blastopore—the first opening into the embryo interior that is formed during gastrulation—usually develops into the organism's mouth.

protozoan (plural, protozoa) A heterotrophic protist.

proximal convoluted tubule The first portion of the vertebrate renal tubule from which electrolytes and other nutrients are reabsorbed into the blood.

pseudogene A gene that is no longer functional.

pulmonary artery One of two arteries, left and right, that carries deoxygenated blood from the right ventricle of the vertebrate heart to the lungs.

pulmonary capillary A small blood vessel that supplies the alveolar wall of tetrapod lungs.

pulmonary circulation Circulation of the blood to the lungs of amphibians, reptiles, birds, and mammals.

pulmonary valve A valve between the common ventricle of the heart of amphibians and reptiles, or the left ventricle of mammals, and the pulmonary trunk.

pulmonary vein A vein that returns oxygenated blood from the lungs of amphibians, reptiles and mammals to the heart, or to the left atrium of reptiles and mammals.

Punnett square A worksheet in the form of a checkerboard used to predict the consequences of a random union of gametes.

purine In nucleic acids, either of the bases adenine and guanine, which have a double-ring structure.

pyloric sphincter A band of muscle at the base of the vertebrate stomach that opens to allow small amounts of digested food to enter the small intestine.

pyrimidine In nucleic acids, any of the bases thymine, cytosine, and uracil, which have a single-ring structure.

quantitative trait A complex trait that is influenced by many genes and the environment and that is measurable along a continuum, such as height and weight.

quaternary structure The structure that results from the interactions of several polypeptide chains.

R gene Any one of the group of genes that express the R proteins in plants.

R group A chemical group attached to the central carbon atom of an amino acid, whose structure and composition determine the identity of the amino acid; also known as a *side chain*.

R protein Any one of a group of receptors in plant cells, each expressed by a different gene, that function as part of the plant's immune system by each binding to a specific pathogen-derived protein.

radiometric dating Dating ancient materials using the decay of radioisotopes as a yardstick, including the decay of radioactive ^{14}C to nitrogen for time intervals up to a few tens of thousands of years, and the decay of radioactive uranium to lead for most of Earth history.

reabsorption In renal systems, an active or passive process in which substances that are important for an animal to retain are taken up by cells of the excretory tubule and returned to the bloodstream.

reactant Any of the starting molecules in a chemical reaction.

reaction center Specially configured chlorophyll molecules where light energy is converted into electron transport.

reactive oxygen species Highly reactive forms of oxygen.

reading frame Following a start codon, a consecutive sequence of codons for amino acids.

realized niche The actual resources and habitats used by a species.

receiver The individual who, during communication, receives from the sender a signal designed to elicit a response.

receptor A molecule on cell membranes that detects critical features of the environment. Receptors detecting signals that easily cross the cell membrane are sometimes found in the cytoplasm.

receptor activation The "turning on" of a receptor, which often occurs when a signaling molecule binds to a receptor on a responding cell.

receptor kinase A receptor that is an enzyme that adds a phosphate group to another molecule.

receptor protein The molecule on the responding cell that binds to the signaling molecule.

recessive Describes an allele or trait that is only expressed in homozygotes, and not expressed in heterozygotes. Two recessive alleles are needed to express the phenotype.

reciprocal altruism The exchange of favors between individuals.

reciprocal cross A cross in which the female and male parents are interchanged.

reciprocal inhibition The activation of opposing sets of muscles so that one set is inhibited as the other is activated, allowing the movement of joints such as the knee.

reciprocal translocation Interchange of parts between nonhomologous chromosomes.

recombinant An offspring with a different combination of alleles from that of either parent, resulting from one or more crossovers in prophase I of meiosis.

recombinant DNA DNA molecules from two (or more) different sources combined into a single molecule.

recovery metabolism An animal's elevated consumption of oxygen following activity.

rectum The part of the hindgut where feces are stored until elimination.

reduction A reaction in which a molecule gains electrons.

reductional division An alternative name for meiosis I because this division reduces the number of chromosomes by half.

refractory period The period following an action potential during which the inner membrane voltage falls below and then returns to the resting potential.

regeneration In the context of the Calvin cycle, the reactions that produce the 5-carbon molecule needed for carboxylation.

regenerative medicine A discipline that aims to use the natural processes of cell growth and development to replace diseased or damaged tissues.

regression toward the mean With regard to complex traits, the principle that offspring exhibit an average phenotype that is intermediate between that of the parents and that of the population as a whole.

regulatory transcription factor A protein that recruits the components of the transcription complex to the gene.

release factor A protein that causes a finished polypeptide chain to be freed from the ribosome.

renal tubules Tubes in the vertebrate kidney that process the filtrate from the glomerulus by reabsorption and secretion.

releasing hormone A peptide hormone that binds to receptors on cells in the anterior pituitary gland, leading that organ to release a much larger amount of the associated hormones.

renaturation The base pairing of complementary single-stranded nucleic acids to form a duplex; also known as *hybridization*, it is the opposite of denaturation.

repetitive DNA The collective term for repeated sequences of various types in eukaryotic genomes.

replication The process of copying DNA so genetic information can be passed from cell to cell or from an organism to its progeny.

replication bubble A region formed by the opening of a DNA duplex at an origin of replication, which has a replication fork at each end.

replication fork The site where the parental DNA strands separate as the DNA duplex unwinds.

repressor A protein that, when bound with a sequence in DNA, can inhibit transcription.

reservoir A supply or source of a substance. Reservoirs of carbon, for example, include organisms, the atmosphere, soil, the oceans, and sedimentary rocks.

residue In the context of protein synthesis, any of the amino acids that is incorporated into a protein.

resource partitioning A pattern in which species whose niches overlap may diverge to minimize the overlap.

responding cell The cell that receives information from the signaling molecule.

response A change in cellular behavior, such as activation of enzymes or genes, following a signal.

resting membrane potential The negative voltage across a cell membrane at rest.

restriction enzyme Any one of a class of enzymes that recognizes specific, short nucleotide sequences in double-stranded DNA and cleaves DNA at or near these sites.

restriction site A recognition sequence in DNA cutting, which is typically four or six base pairs long; most restriction enzymes cleave double-stranded DNA at or near these restriction sites.

reticulum The second chamber in the stomach of ruminants, which, along with the rumen, harbors large populations of anaerobic bacteria that break down cellulose.

retina A thin tissue in the posterior of the vertebrate eye that contains the photoreceptors and other nerve cells that sense and initially process light stimuli.

retinal A derivative of vitamin A that absorbs light and binds to rhodopsin, a transmembrane protein in the photosensitive cells of vertebrates.

retrotransposons Transposable elements in DNA sequences in which RNA is used as a template to synthesize complementary strands of DNA, a reversal of the usual flow of genetic information from DNA into RNA.

reverse transcriptase An RNA-dependent DNA polymerase that uses a single-stranded RNA as a template to synthesize a DNA strand that is complementary in sequence to the RNA.

rhizosphere The soil layer that surrounds actively growing roots.

ribonucleic acid (RNA) A molecule chemically related to DNA that is synthesized by proteins from a DNA template.

ribose The sugar in RNA.

ribosomal RNA (rRNA) Noncoding RNA found in all ribosomes that aid in translation.

ribosome A complex structure of RNA and protein, bound to the cytosolic face of the rough endoplasmic reticulum in the cytoplasm, on which proteins are synthesized.

ribulose bisphosphate carboxylase oxygenase (rubisco) The enzyme that catalyzes the carboxylation reaction in the Calvin cycle.

ribulose 1,5-bisphosphate (RuBP) The 5-carbon sugar to which carbon dioxide is added by the enzyme rubisco.

RISC (RNA-induced silencing complex) A protein complex that is targeted to specific mRNA molecules by base pairing with short regions on the target mRNA, inhibiting translation or degrading the RNA.

ritualization The process of co-opting and modifying behaviors used in another context by increasing the conspicuousness of the behavior, reducing the amount of variation in the behavior so that it can be immediately recognized, and increasing its separation from the original function.

RNA editing The process in which some RNA molecules become a substrate for enzymes that modify particular bases in the RNA, thereby changing its sequence and sometimes what it codes for.

RNA polymerase The enzyme that carries out polymerization of ribonucleoside triphosphates from a DNA template to produce an RNA transcript.

RNA polymerase complex An aggregate of proteins that synthesizes the RNA transcript complementary to the template strand of DNA.

RNA primase An RNA polymerase that synthesizes a short piece of RNA complementary to the DNA template and does not require a primer.

RNA processing Chemical modification that converts the primary transcript into finished mRNA, enabling the RNA molecule to be transported to the cytoplasm and recognized by the translational machinery.

RNA splicing The process of intron removal from the primary transcript.

RNA transcript The RNA sequence synthesized from a DNA template.

RNA world hypothesis The hypothesis that the earliest organisms relied on RNA for both catalysis and information storage.

rod cell A type of photoreceptor cell on the vertebrate retina that detects light and shades ranging from white to shades of gray and black, but not color.

root apical meristem A group of totipotent cells near the tip of a root that is the source of new root cells.

root cap A structure that covers and protects the root apical meristem as it grows through the soil.

root hair A slender outgrowth produced by epidermal cells that greatly increases the surface area of the root.

root nodule A structure, formed by dividing root cells, in which nitrogen-fixing bacteria live.

rough endoplasmic reticulum (RER) The part of the endoplasmic reticulum with attached ribosomes.

***r*-strategist** A species that produces large numbers of offspring but provides few resources for their support.

rumen The first chamber in the stomach of ruminants, which, along with the reticulum, harbors large populations of anaerobic bacteria that break down cellulose.

S phase The phase of interphase in which the entire DNA content of the nucleus is replicated.

saccharide The simplest carbohydrate molecule; also called a *sugar*.

saltatory propagation The movement of an action potential along a vertebrate myelinated axon, "jumping" from node to node.

Sanger sequencing A procedure in which chemical termination of daughter strands help in determining the DNA sequence.

sapwood In long-lived trees, the layer adjacent to the vascular cambium that contains the functional xylem.

sarcomere The region of myofilaments located within muscle myofibrils that span from one Z disc to the next; the basic contractile unit of a muscle.

sarcoplasmic reticulum (SR) A modified form of the endoplasmic reticulum surrounding the myofibrils of muscle cells.

saturated Describes fatty acids that do not contain double bonds; the maximum number of hydrogen atoms is attached to each carbon atom, "saturating" the carbons with hydrogen atoms.

scaffold A supporting protein structure in a metaphase chromosome.

scientific inquiry A deliberate, systematic, careful, and unbiased way of learning about the natural world.

scrotum A sac outside the abdominal cavity of the male that holds the testes.

second-division nondisjunction Disjunction in the second meiotic division.

second law of thermodynamics The principle that the transformation of energy is associated with an increase in the degree of disorder in the universe.

second messenger An intermediate cytosolic signaling molecule that transmits signals from a receptor to a target within the cell. (First messengers transmit signals from outside the cell to a receptor.)

secondary active transport Active transport that uses the energy of an electrochemical gradient to drive the movement of molecules.

secondary consumers Predators or scavengers that feed on primary consumers.

secondary growth In plants, growth originating in lateral meristems; resulting in larger diameter stems and roots.

secondary oocyte A large cell produced by the asymmetric first meiotic division of the primary oocyte.

secondary phloem New phloem cells produced by the vascular cambium, which are located to the outside of the vascular cambium.

secondary response The response to re-exposure to an antigen, which is quicker, stronger, and longer than the primary response.

secondary sexual characteristic A trait that characterizes and differentiates the two sexes but that does not relate directly to reproduction.

secondary spermatocyte A diploid cell formed during the first meiotic division of the primary spermatocyte.

secondary structure The structure formed by interactions between stretches of amino acids in a protein.

secondary xylem New xylem cells produced by vascular cambium, which are located to the inside of the vascular cambium.

secretin A hormone released by cells lining the duodenum in response to the acidic pH of the stomach contents entering the small intestine and that stimulates the pancreas to secrete bicarbonate ions.

secretion In renal systems, an active process that eliminates substances that were not previously filtered from the blood.

seed A mature fertilized ovule consisting of a seed coat, a nutritive tissue, and the embryo; seeds are multicellular structures that allow offspring to disperse away from the parent plant.

seed coat A protective outer structure surrounding the seed.

segmentation The formation of discrete repeating parts or segments in the developing body of many animals.

segregate Separate; applies to chromosomes or members of a gene pair moving into different gametes.

selection The retention or elimination of mutations in a population of organisms.

selectively permeable Describes the properties of some membranes, including the plasma membrane, which lets some molecules in and out freely, lets others in and out only under certain conditions, and prevents other molecules from passing through at all.

self Describes an organism's own molecules and cells.

self-compatible Describes species in which pollen and ovules produced by flowers on the same plant can produce viable offspring.

self-incompatible Describes species in which pollination by the same or a closely related individual does not lead to fertilization.

self-propagating Continuing without input from an outside source; action potentials are self-propagating in that they move along axons by sequentially opening and closing adjacent ion channels.

semen A fluid that nourishes and sustains sperm as they travel in the male and then the female reproductive tracts.

semicircular canal One of three connected fluid-filled tubes in the mammalian inner ear that contains hair cells that sense angular motions of the head in three perpendicular planes.

semiconservative replication The mechanism of DNA replication in which each strand of a parental DNA duplex serves as a template for the synthesis of a new daughter strand.

seminal vesicles Two glands at the junction of the vas deferens and the prostate gland that secrete a protein- and sugar-rich fluid that makes up most of the semen and provides energy for sperm motility.

seminiferous tubules A series of tubes in the testes where sperm are produced.

sender The individual who, during communication, supplies a signal designed to elicit a response from the receiver.

sensitization The enhancement of a response to a stimulus that is achieved by first presenting a strong or novel stimulus.

sensory neuron A neuron that receives and transmits information about an animal's environment or its internal physiological state.

sensory organ A group of sensory receptors that converts particular physical and chemical stimuli into nerve impulses that are processed by a nervous system and sent to a brain.

sensory receptor cell A sensory neuron with specialized membranes in which receptor proteins are embedded.

sensory transduction The conversion of physical or chemical stimuli into nerve impulses.

sepal A structure, often green, that forms the outermost whorl of a flower and, with other sepals, encases and protects the flower during its development.

septum (plural, septa) In fungal hyphae, a wall that partially divides the cytoplasm into separate cells.

sequence assembly The process in which short nucleotide sequences of a long DNA molecule are arranged in the correct order to generate the complete sequence.

sequence motif Any of a number of sequences or sequence arrangements that indicate the likely function of a segment of DNA.

serosa An outer layer of cells and connective tissue that covers and protects the gut.

Sertoli cell A type of cell in the seminiferous tubules of vertebrate testes that supports sperm production.

set point A steady-state value in homeostatic regulation.

sex chromosome Any of the chromosomes associated with sex determination.

sexual reproduction The process of producing offspring that receive genetic material from two parents; in eukaryotes, the process occurs through meiosis and fertilization.

sexual selection A form of selection that promotes traits that increase an individual's access to reproductive opportunities.

shell (of an atom) One or more orbital zones within an atom where electrons occur. Different shells have distinct energy levels.

shoot The collective name for the leaves, stems, and reproductive organs; the major aboveground organ systems of vascular plants.

shoot apical meristem A group of totipotent cells at the tip of a stem or branch that gives rise to new shoot tissues in plants.

short-day plant A plant that flowers only when the day length is less than a critical value.

shotgun sequencing DNA sequencing method in which the sequenced fragments do not originate from a particular gene or region, but from sites scattered randomly across the molecule.

side chain A chemical group attached to the central carbon atom of an amino acid, whose structure and composition determine the identity of the amino acid; also known as an *R group*.

sieve plate In angiosperm phloem, a modified end wall with large pores that links sieve elements.

sieve tube A multicellular unit composed of sieve elements that are connected end to end, through which phloem transport takes place.

sigma factor A protein that associates with RNA polymerase that facilitates its binding to specific promoters.

signal sequence An amino acid sequence that directs a protein to its proper cellular compartment.

signal transduction The process in which an extracellular molecule acts as a signal to activate a receptor, which transmits information through the cytoplasm.

signal-anchor sequence In protein sorting, an amino acid sequence in a polypeptide chain that embeds the chain in the membrane.

signaling cell The source of a signaling molecule.

signaling molecule The carrier of information transmitted when the signaling molecule binds to a receptor; also referred to as a *ligand*.

signal-recognition particle (SRP) An RNA–protein complex that binds with part of a polypeptide chain and marks the molecule for incorporation into the endoplasmic reticulum (eukaryotes) or the plasma membrane (prokaryotes).

silencers DNA sequences that bind with regulatory transcription factors and repress transcription.

single-gene trait A trait determined by Mendelian alleles of a single gene with little influence from the environment.

single-lens eye An eye structure found in vertebrates and cephalopod mollusks that works like a camera to produce a sharply defined image of the animal's visual field.

single-nucleotide polymorphism (SNP) A site in the genome where the base pair that is present differs among individuals in a population.

single-strand binding protein A protein that binds single-stranded nucleic acids.

sink In plants, any portion of the plant that imports carbohydrates to fuel growth and respiration, such as a root, young leaf, or developing fruit.

sinoatrial (SA) node A specialized region of the vertebrate heart containing pacemaker cells where the heartbeat is initiated.

sister chromatids The two copies of a chromosome produced by DNA replication.

sister groups Groups that are more closely related to each other than either of them is to any other group.

skeletal muscle Striated muscle that connects to the body skeleton to move an animal's limbs and torso.

sliding filament model The hypothesis that striated muscles produce force and change length by the sliding of actin filaments relative to myosin filaments.

slow-twitch Describes muscle fibers that contract slowly, resist fatigue, and consume less ATP than do fast-twitch fibers to produce force.

small interfering RNA (siRNA) A type of small double-stranded regulatory RNA that becomes part of a complex able to cleave and destroy single-stranded RNA with a complementary sequence.

small intestine Part of the midgut; the site of the last part of digestion and most nutrient absorption.

small nuclear RNA (snRNA) Noncoding RNA found in eukaryotes and involved in splicing, polyadenylation, and other processes in the nucleus.

small regulatory RNA A short RNA molecule that can block transcription, cleave or destabilize RNAA, or inhibit mRNA translation.

smooth endoplasmic reticulum (SER) The portion of the endoplasmic reticulum that lacks ribosomes.

smooth muscle The muscle in the walls of arteries, the respiratory system, and the digestive and excretory systems; smooth muscle appears uniform under the light microscope.

solubility The ability of a substance to dissolve.

solute A dissolved molecule such as the electrolytes, amino acids, and sugars often found in a solvent such as water.

solvent A liquid capable of dissolving a substance.

somatic cell A nonreproductive cell, the most common type of cell in the body of a multicellular organism.

somatic mutation A mutation that occurs in somatic cells.

somatic nervous system The voluntary component of the peripheral nervous system, which is made up of sensory neurons that respond to external stimuli and motor neurons that synapse with voluntary muscles.

source In plants, a region that supplies carbohydrates to other parts of the plant.

Southern blot A method for determining the size and number of copies of a DNA sequence of interest by means of a labeled probe hybridized to DNA fragments separated by size by means of gel electrophoresis.

spatial summation The converging of multiple receptors onto a neighboring neuron, increasing its firing rate proportionally to the number of signals received.

speciation The process whereby new species are produced.

species A group of individuals that can exchange genetic material through interbreeding to produce fertile offspring.

species–area relationship A mathematical relationship between the number of species and the size of an island.

spermatogenesis The formation of sperm.

spermatozoa (sperm) The smaller, male gametes.

spinal cord In vertebrates, a central tract of neurons that passes through the vertebrae to transmit information between the brain and the periphery of the body.

spinal nerve In vertebrates, a nerve running from the spinal cord to the periphery containing axons of both sensory and motor neurons.

spiracle An opening in the exoskeleton on either side of an insect's abdomen through which gases are exchanged.

spliceosome A complex of RNA and protein that catalyzes RNA splicing.

sponges Porifera; a phylum characterized by few cell types, no complex tissues or organs, and an irregular form, with no clearly defined plane of symmetry.

spongy bone Vertebrate bone tissue consisting of trabeculae, and thus lighter than compact bone, found in the ends of limb bones and within vertebrae.

spontaneous Occurring in the absence of any assignable cause; most mutations are spontaneous.

sporangium In plants, a multicellular structure in which haploid spores are formed by meiosis.

spore A reproductive cell capable of growing into a new individual without fusion with another cell.

sporophyte Describes the diploid multicellular generation in plants that produces spores.

sporopollenin A complex mixture of polymers that is remarkably resistant to environmental stresses such as ultraviolet radiation and desiccation. In plants, found in the walls of pollen and spores.

stabilizing selection A form of selection that selects against extremes and therefore maintains the status quo.

stamen A pollen-producing floral organ.

stapes A small bone in the vertebrate middle ear that helps amplify the waves that strike the tympanic membrane; the stapes connects to the oval window of the cochlea.

starch The storage form of carbohydrates in plants.

Starling's Law The correspondence between change in stroke volume and change in the volume of blood filling the vertebrate heart.

statocyst A type of gravity-sensing organ found in most invertebrates.

statolith In plants, a large starch-filled organelle in the root cap that senses gravity; in animals, a dense particle that moves freely within a statocyst, enabling it to sense gravity.

stem cell An undifferentiated cell that can undergo an unlimited number of mitotic divisions and differentiate into any of a large number of specialized cell types.

stereocilia Nonmotile cell-surface projections on hair cells whose movement causes a depolarization of the cell's membrane.

steroid A type of lipid.

steroid hormone A hormone that is derived from cholesterol.

stigma The surface at the top of the carpel, to which pollen adheres.

stomach The last part of the foregut, which serves as a storage and digestive chamber.

stomata (singular, stoma) Pores in the epidermis of a leaf that regulate the diffusion of gases between the interior of the leaf and the atmosphere.

Stramenopila A eukaryotic superkingdom including unicellular organisms, giant kelps, algae, protozoa, free-living cells, and parasites; distinguished by a flagellum with two rows of stiff hairs and, usually, a second, smooth flagellum.

striated muscle Skeletal muscle and cardiac muscle, which appear striped under a light microscope due to regular thick (myosin) and thin (actin) myofilament spacing.

strigolactone A hormone, produced in roots and transported upward in the xylem, that inhibits the outgrowth of axillary buds.

stroke volume (SV) The volume of blood pumped during each heartbeat.

stroma The region of the chloroplast that surrounds the thylakoid, where the Calvin cycle takes place.

stromatolite A layered structure that records sediment accumulation by microbial communities.

structural gene A gene that codes for the sequence of amino acids in a polypeptide chain.

style A cylindrical stalk between the ovary and the stigma, through which the pollen tube grows.

suberin In plants, a waxy compound found in the walls of cork cells that protects against mechanical damage, the entry of pathogens, and water loss.

submucosa A tissue layer surrounding the mucosa that contains blood vessels, lymph vessels, and nerves.

subspecies Allopatric populations that have yet to evolve even partial reproductive isolation but which have acquired population-specific traits.

substrate (S) A molecule acted upon by an enzyme.

substrate-level phosphorylation A way of generating ATP in which a phosphate group is transferred to ADP from an organic molecule, which acts as a phosphate donor or substrate.

succession In ecology, the process of species replacing one another in time in a predictable sequence; each species modifies the habitat by affecting the environment or changing the community structure, creating opportunities for other species.

sugar The simplest carbohydrate molecule; also called a *saccharide*.

sulci (singular, sulcus) Deep crevices in the brain that separate the lobes of the cerebral hemispheres.

supercoil A coil of coils; a circular molecule of DNA can coil upon itself to form a supercoil.

superkingdom One of seven major groups of eukaryotic organisms, classified by molecular sequence comparisons.

supernormal stimulus An exaggerated stimulus that elicits a response more strongly than the normal stimulus.

surfactant A compound that reduces the surface tension of a fluid film.

suspension filter feeding The most common form of food capture by animals, in which water with food suspended in it passes through a sievelike structure.

sustainable development Use of natural resources at rates no higher than the rate at which they can be replenished.

symbiont An organism that lives in closely evolved association with another species.

symbiosis (plural, symbioses) A close interaction that has evolved between species that live together, often interdependently.

sympathetic division The division of the vertebrate autonomic nervous system that generally produces arousal and increased activity; active in the fight-or-flight response.

sympatric Describes populations that are in the same geographic location.

synapomorphy A shared derived character; the basis of cladistic phylogenetic reconstruction.

synapse A junction through which the axon terminal of a nerve cell communicates with a neighboring cell.

synapsis The gene-for-gene pairing of homologous chromosomes in prophase I of meiosis.

synaptic cleft The space between the axon of the presynaptic nerve cell and the neighboring postsynaptic cell.

synonymous (silent) mutation A mutation resulting in a codon that does not alter the corresponding amino acid in the polypeptide.

systemic acquired resistance (SAR) The ability of a plant to resist future infections, occurring in response to a wide range of pathogens.

systemic circulation Circulation of the blood to the body, excluding the lungs.

systole The phase of the vertebrate cardiac cycle when the ventricles contract to eject blood from the heart.

T cell receptor (TCR) A protein receptor on a T cell that recognizes and binds to an antigen.

T lymphocyte (T cell) A cell of the immune system that matures in the thymus and includes helper and cytotoxic cells.

tandem repeat A region along a DNA molecule in which many duplicated copies of a short sequence of nucleotides are adjacent to one another.

tannin Any one of a group of phenols found widely in plant tissues that bind with proteins and reduce their digestibility.

taste bud One of the sensory organs for taste.

TATA box A DNA sequence present in many promoters in eukaryotes and archaeons that serves as a protein-binding site for a key general transcription factor.

taxis (plural, taxes) Movement in a specific direction in response to a stimulus.

taxon (plural, taxa) A named taxonomic group at any rank, such as a species, a genus, or a family.

telomerase An enzyme containing an RNA template from which complementary telomere repeats are synthesized.

telomere A repeating sequence at each end of a eukaryotic chromosome.

telophase The stage of mitosis in which the nuclei of the daughter cells are formed and the chromosomes uncoil to their original state.

telophase I The stage of meiosis I in which the chromosomes uncoil slightly, a nuclear envelope briefly reappears, and in many species the cytoplasm divides, producing two separate cells.

telophase II The stage of meiosis II in which the chromosomes uncoil and become diffuse, a nuclear envelope forms around each set of chromosomes, and the cytoplasm divides by cytokinesis.

template strand In DNA replication, the parental strand whose sequence is used to synthesize a complementary daughter strand.

temporal isolation Pre-zygotic isolation between individuals that are reproductively active at different times.

temporal lobe The region of the brain involved in the processing of sound, language and reading, and object identification and naming.

temporal summation The summation of postsynaptic potentials over time, which determines whether the postsynaptic cell fires an action potential.

temporomandibular joint A specialized jaw joint in mammals that allows the teeth of the lower and upper jaws to fit together precisely.

tendon A collagen structure that attaches vertebrate muscles to the skeleton and transmits muscle forces over a wide range of joint motion.

termination In protein translation, the time at which the addition of amino acids stops and the completed polypeptide chain is released from the ribosome. In cell communication, the stopping of a signal.

terminator A DNA sequence at which transcription stops and the transcript is released.

terpene Any one of a group of compounds that do not contain nitrogen and are produced by some plants as a defensive mechanism.

tertiary structure The overall three-dimensional shape of a protein, formed by interactions between secondary structures.

test (of a protist) A "house" constructed of organic molecules with or without precipitated minerals or adhering sediment grains that shelters a protist.

test group The experimental group that is exposed to the variable in an experiment.

testcross Any cross of an unknown genotype with a homozygous recessive genotype.

testis (plural, testes) The male gonad, where sperm are produced.

testosterone A steroid hormone, secreted by the vertebrate testes, that plays key roles in male growth, development, and reproduction.

tetanus A muscle contraction of sustained force resulting from multiple action potentials.

Tetrapoda A monophyletic group of animals whose last common ancestor had four limbs; this group includes amphibians, lizards, turtles, crocodilians, birds, and mammals; some tetrapods, like snakes, have lost their legs in the course of evolution.

tetrapods Four-legged vertebrate animals.

thalamus The inner brain region of the forebrain, which acts as a relay station for sensory information sent to the cerebrum.

thallus A flattened, often photosynthetic structure lacking differentiated vegetative organs; produced by the body of the gametophyte generation in hornworts and some liverworts.

theory A general explanation of a natural phenomenon supported by a large body of experiments and observations.

theory of island biogeography A theory that states that the number of species that can occupy a habitat island depends on two factors: the size of the island and the distance of the island from a source of colonists.

thick filament A parallel grouping of myosin molecules that form the myosin filament within muscle sarcomeres.

thin filament Two helically arranged actin polymers twisted together that form the actin filament within muscle sarcomeres.

threshold potential The critical depolarization voltage of −50 mV required for an action potential.

thylakoid membrane A highly folded membrane in the center of the chloroplast that contains light-collecting pigments and that is the site of the photosynthetic electron transport chain.

thymine (T) A pyrimidine base.

thyroid gland A gland located in the front of the vertebrate neck that leads to the release of two peptide hormones, thyroxine and triiodothyronine.

Ti plasmid A small circular DNA molecule in virulent strains of *R. radiobacter* containing genes that can be integrated into the host cell's genome, as well as the genes needed to make this transfer.

tidal ventilation A breathing technique in most land vertebrates in which air is drawn into the lungs during inhalation and moved out during exhalation.

tidal volume The amount of air inhaled and exhaled in a cycle; in humans, tidal volume is 0.5 liter when breathing at rest.

tight junction A junctional complex that establishes a seal between animal cells so that the only way a substance can travel from one side of a sheet of epithelial cells to the other is by moving through the cells by a cellular transport mechanism.

tissue A collection of cells that work together to perform a specific function.

tolerance In immunology, the ability of B and T cells not to respond to self antigens even though the immune system functions normally otherwise.

toll-like receptors (TLRs) A family of transmembrane receptors on phagocytes that recognize and bind to molecules on the surface of microorganisms, providing an early signal that an infection is present.

topography The physical features of Earth.

topoisomerase II An enzyme that breaks a DNA double helix, rotates the ends, and seals the break.

totipotent Describes cells that have the potential to give rise to a complete organism; a fertilized egg is a totipotent cell.

trabeculae Small plates and rods with spaces between them, found in spongy bone.

trace fossil A track or trail, such as a dinosaur track or the feeding trails of snails and trilobites, left by an animal as it moves about or burrows into sediments.

trachea The central airway leading to the lungs of air-breathing animals, supported by cartilage rings.

tracheae An internal system of tubes in insects that branch from openings along the abdominal surface into smaller airways, directing oxygen to and removing carbon dioxide from respiring tissues.

tracheid A unicellular xylem conduit.

trade-off An exchange in which something is gained at the expense of something lost.

trait A characteristic of an individual.

transcription The synthesis of RNA from a DNA template.

transcriptional activator protein A protein that binds to a sequence in DNA to enable transcription to begin.

transcriptional regulation The mechanisms that collectively regulate whether transcription occurs.

transduction Horizontal gene transfer by means of viruses.

transfer RNA (tRNA) Noncoding RNA that carries individual amino acids for use in translation.

transformation The conversion of cells from one state to another, as from nonvirulent to virulent, when DNA released to the environment by cell breakdown is taken up by recipient cells. In recombinant DNA technology, the introduction of recombinant DNA into a recipient cell.

transgenic organisms An alternative term for genetically modified organisms.

transition state The brief time in a chemical reaction in which chemical bonds in the reactants are broken and new bonds in the product are formed.

translation Synthesis of a polypeptide chain corresponding to the coding sequence present in a molecule of messenger RNA.

transmembrane proteins Proteins that span the entire lipid bilayer; most integral membrane proteins are transmembrane proteins

transmission genetics The discipline that deals with the manner in which genetic material is passed from generation to generation.

transpiration The loss of water vapor from leaves.

transporters Membrane proteins that move ions or other molecules across the cell membrane.

transposable element (TE) A DNA sequence that can replicate and move from one location to another in a DNA molecule; also known as a *transposon*.

transposon A DNA sequence that can replicate and move from one location to another in a DNA molecule; also known as *transposable element (TE)*.

triacylglycerol A lipid composed of a glycerol backbone and three fatty acids.

trimesters The three periods of pregnancy, each lasting about 3 months.

triose phosphate A 3-carbon carbohydrate molecule, produced by the Calvin cycle and exported from the chloroplast.

triploblastic Describes animals in which the embryo has three germ layers, with the mesoderm between the endoderm and ectoderm.

trisomy 21 A condition resulting from the presence of three, rather than two, copies of chromosome 21; also known as *Down syndrome*.

trophic level An organism's typical place in a food web as a producer or consumer.

trophic pyramid A diagram that traces the flow of energy through communities, showing the amount of energy available at each level to feed the next. The pyramid shape results because biomass and the energy it represents generally decrease from one trophic level to the next.

tropic hormone A hormone that controls the release of other hormones.

tropism The bending or turning of an organism in response to an external signal such as light or gravity.

tropomyosin A protein that runs in the grooves formed by the actin helices and blocks the myosin-binding sites on the actin filament.

troponin A protein that moves tropomyosin away from myosin-binding sites, allowing cross-bridges between actin and myosin to form and the muscle to contract.

true breeding Describes a trait whose physical appearance in each successive generation is identical to that in the previous one.

trypsin A digestive enzyme produced by the vertebrate pancreas that breaks down proteins.

tube feet In echinoderms, small projections of the water vascular system that extend outward from the body surface and facilitate locomotion, sensory perception, food capture, and gas exchange.

tubulin Dimers (composed of an α tubulin and a β tubulin) that assemble into microfilaments.

tumor suppressor One of a family of genes that encode proteins whose normal activities inhibit cell division.

tunicates A subphylum of Chordata that includes about 3000 species of filter-feeding marine animals, such as sea squirts and salps.

turgor pressure Pressure within a cell resulting from the movement of water into the cell by osmosis and the tendency of the cell wall to resist deformation.

Turner syndrome A sex-chromosomal abnormality in which an individual has 45 chromosomes, including only one *X* chromosome.

twitch A muscle contraction that results from a single action potential.

twofold cost of sex The observation that population size can increase more rapidly in asexually reproducing organisms than in sexually reproducing organisms because only females produce offspring in sexually reproducing organisms.

unsaturated Describes fatty acids that contain carbon–carbon double bonds.

uracil (U) A pyrimidine base in RNA, where it replaces the thymine found in DNA.

ureter A large tube in the vertebrate kidney that brings urine from the kidneys to the bladder.

urethra A tube from the mammalian urinary bladder that in males carries semen as well as urine from the body.

uterus A hollow organ within the reproductive tract of female mammals with thick, muscular walls that is adapted to support the developing embryo if fertilization occurs and to deliver the baby during birth.

vacuole A membrane-bound organelle present in some cells, including plant and fungal cells, that contains fluid, ions, and other molecules; in some cases, it absorbs water and contributes to turgor pressure.

vagina A tubular channel connecting the mammalian uterus to the exterior of the body; also known as the birth canal.

valence electrons The electrons farthest from the nucleus, which are at the highest energy level.

van der Waals interactions The binding of temporarily polarized molecules because of the attraction of opposite charges.

variable (in experimentation) The feature of an experiment that is changed by the experimenter from one treatment to the next.

variable expressivity The phenomenon in which a particular phenotype is expressed with a different degree of severity in different individuals.

variable (V) region A region of the heavy (H) and light (L) chains of an antibody; the variable region distinguishes a given antibody from all others.

vas deferens A long, muscular tube that carries sperm from the testes to the ejaculatory duct of amniote vertebrates.

vasa recta The blood vessels in the vertebrate kidneys.

vascular cambium (plural, cambia) Lateral meristem that is the source of new xylem and phloem.

vascular tissues A plant tissue that extends from leaves to roots and allows water and nutrients to move by bulk flow.

vasoconstriction The narrowing of a vessel by the contraction of smooth muscle, which in turn increases resistance and decreases blood flow.

vasodilation The widening of a vessel by the relaxation of smooth muscle, which in turn decreases resistance and increases blood flow.

vasopressin In vertebrates, a posterior pituitary gland hormone that acts on the kidneys and controls the water permeability of the collecting ducts, thus regulating the concentration of urine that an animal excretes; also known as *antidiuretic hormone (ADH)*.

vector In recombinant DNA, a carrier of the donor fragment, usually a plasmid.

vegetative reproduction Asexual reproduction by growth and fragmentation.

veins In plants, the system of vascular conduits within the leaf; in animals, the large, low-pressure vessels that return blood to the heart.

vena cava (plural, venae cavae) One of two large veins in the vertebrate body that drain blood from the head and body into the heart.

ventilation The movement of an animal's respiratory medium—water or air—past a specialized respiratory surface.

ventricle In vertebrates and some invertebrates, a heart chamber that pumps blood to the lungs or the rest of the body.

venule A blood vessel into which capillaries drain as blood is returned to the heart.

vernalization A prolonged period of exposure to cold temperatures necessary to induce flowering in some plants.

vertebrae (singular, vertebra) The series of hard bony segments making up the jointed skeleton that runs along the main axis of the body in vertebrates.

vertebral column A skeletal structure in vertebrates that functionally replaces the embryonic notochord that supports the body.

vertebrates A subphylum of Chordata, distinguished by a bony cranium that protects the brain and (unless lost through evolution), a vertebral column; also known as *craniates*.

vesicle A small membrane-enclosed sac that transports substances within the cell.

vessel A multicellular xylem conduit.

vessel element An individual cell that is part of a xylem vessel for water transport in plants.

vestibular system A system in the vertebrate inner ear made up of two statocyst chambers and three semicircular canals.

vestigial structure A structure that has lost its original function over time and is now much reduced in size.

vicariance The process in which a geographic barrier arises within a single population, separating it into two or more isolated populations that are isolated from each other.

villi Highly folded inner surfaces of the jejunum and ileum of the small intestine.

virulent Describes pathogens that are able to overcome a host's defenses and lead to disease.

virus A small infectious agent that contains a nucleic acid genome packaged inside a protein coat called a capsid.

visible light The portion of the electromagnetic spectrum apparent to our eyes.

vitamin An organic molecule that is required in very small amounts in the diet.

viviparity Giving birth to live young, with nutritional support of the embryo from the mother.

vocal cords Twin organs in the vertebrate (amniote) larynx that vibrate as air passes over them, enabling sound production.

voltage-gated channels Ion membrane channels that open and close in response to changes in membrane potential.

voluntary component The part of the peripheral nervous system that controls conscious reactions.

vulva The external genitalia of the female.

water potential A parameter that combines all of the factors that influence the movement of water, such as pressure, osmosis, and gravity.

water vascular system A series of fluid-filled canals that permit bulk transport of oxygen and nutrients in echinoderms.

white blood cell (leukocyte) A type of cell in the immune system that arises by differentiation from stem cells in the bone marrow.

white matter Collectively, the axons of cortical neurons in the interior of the vertebrate brain and spinal cord; it is the fatty myelin produced by glial cells surrounding the axons that makes these regions white.

wild type The most common allele, genotype, or phenotype present in a population; nonmutant.

X chromosome One of the sex chromosomes; a normal human female has two copies of the X chromosome; a normal male has one X chromosome and one Y chromosome.

xanthophyll Any one of several yellow-orange pigments that slow the formation of reactive oxygen species by reducing excess light energy; these pigments accept absorbed light energy directly from chlorophyll and convert this energy to heat.

X-inactivation The process in female mammals in which dosage compensation occurs through the inactivation of one X chromosome in each cell.

X-linked genes A gene in the X chromosome.

xylem Vascular tissue consisting of lignified conduits that transports water and nutrients from the roots to the leaves.

Y chromosome One of the sex chromosomes; a normal human male has one X chromosome and one Y chromosome.

yeast A single-celled fungus found in moist, nutrient-rich environments. Many yeasts can metabolize sugar by means of fermentation.

Y-linked gene A gene that is present in the region of the Y chromosome that shares no homology with the X chromosome.

yolk A substance in the eggs of animals with external fertilization that provides all the nutrients that the developing embryo needs until it hatches.

Z disc A protein backbone found regularly spaced along the length of a myofibril.

Z scheme Another name for the photosynthetic electron transport chain, so called because the overall energy trajectory resembles a "Z."

zygomycetes Fungi that produce hyphae undivided by septa and do not form multicellular fruiting bodies; they make up less than 1% of known fungal diversity.

zygote The diploid cell formed by the fusion of two gametes.

Index

bold face indicates a definition
italics indicate a figure
t indicates a table

A

A. *See* Adenine (A)
ABC model, 412–414, *413*
Abiotic factors, **1001**
ABO blood system, 429–430, 430*t*
Abomasum, **874**, *874*
Abscisic acid, 645, 645*t*, **659**
Absolute temperature (T), **125**
Absorption, **867**, 871–872, *872*
Accessory pigments, **164**–165
Acclimatization, **823**
Acetylation, **378**, *379*
Acetyl-coenzyme A (acetyl-CoA), **138**, *138*, *139*, 141
  citric acid cycle and, 142–143, *142–143*
Acidic solutions, **37**
  Archaea and, 537, *537*
Acquired immune deficiency syndrome (AIDS), **478**
Acrosome, **905**, *906*
ACTH (Adrenocorticotropic hormone), 817, 818*t*
Actin, **205**, *206*, 210, 210*t*
  in muscles, 788–795, *789–793*
  slime molds and, 557
Action potentials, **759**, *760*, 762–763, *763*
  propagation of, 764, 764–766, *765*
Activation energy (E_A), **127**–128, *128*
Activators, **129**, **387**, *387*
Active sites, **128**, *129*
Active transport, **101**, *103*
  primary, 102, *102*
  secondary, 102, 102–103
Adalia bipunctata (ladybug), 430, *430*
Adaptations
  natural selection and, 434–436
  reproduction and, 903–905, *903–905*
Adaptive immunity, **925**, 925*t*
  B cells and, *926*, 930–936, *931–933*, *935*
  complement system and, *929*, 929–930
  T cells and, *926*, 930, 936–939, *937*, *938*, 938*t*
  See also Antibodies
Adaptive radiations, *456*, **457**
Addition rule, **332**, *332*
Adenine (A), **41**, *41*, 54
Adenosine diphosphate (ADP), 122

Adenosine monophosphate (AMP), **122**
  cyclic, **194**, *194*, *195*
Adenosine triphosphate (ATP), **14**, **119**, **135**, **856**
  active transport and, *102*, 102–103
  Calvin cycle and, 160, *161*, 163
  cellular energy and, 122–123
  cellular respiration and, 135–136, *136*, 146*t*, 152–153, *153*
  citric acid cycle production of, *142*, 142–143
  cyclic electron transport and, 169, *169*
  exercise and, 153–154
  hydrolysis of, 126, *126*, *127*
  metabolism and, 856, *856*
  phosphorylation and, 192–193, *193*
  structure of, *122*
  synthesis of, 144, *146*, 146–147
Adenovirus, 298, *298*
ADH. *See* Antidiuretic hormone (ADH)
Adherens junctions, **214**, *215*, 216*t*
ADP. *See* Adenosine diphosphate (ADP)
Adrenal cortex, 819*t*
Adrenal glands, *820*, **822**
Adrenaline, **816**
  endocrine signaling and, 188
  G protein-coupled receptor amplification and, *194*, *194*, *195*
Adrenal medulla, 819*t*, **823**
Adrenocorticotropic hormone (ACTH), 817, 818*t*
Advantageous (beneficial) mutations, *303*, **303**, *304*, **429**
Advertisement displays (animal communication), **993**
Aerobic metabolism, **855**–857, *856*, *856*
Aerobic reactions, **525**
Aerobic respiration, **135**–136
Age structure, of populations, *1009–1011*, **1009**–1013
Aging, telomeres and, 266
Agonists, **794**
Agriculture
  artificial selection and, 439, 587, 635
  environmental risk factors and, *363*, 363
  eutrophication and, 1104
  evolution and, 21–22, 508
  fertilizers and, 613
  fungi and, 731–732, *732*

  genetic diversity and, 706–707, *707*
  genetic engineering and, 275, *276*, 588, 670–671, 680–682, 1106
  history of, 586–589
  integrated pest management and, 681
  plant defenses and, 680–682, *682*
  plant hormones and, 646–647
  polyploidy and, 290, *291*
  population growth challenges for, 1105–1106, *1106*
  price of, 586
  See also Green Revolution; Wheat
Agrodiaetus (butterfly), 449, *449*
AIDS (Acquired immune deficiency syndrome), **478**
Aizenberg, Joanna, 736
Albinism, 337, *339*
Albright, Rebecca, 1102
Aldosterone, **894**
Algae, **554**, 1084
  brown, 560, *560*, *571*
  green, 558, *559*, 559–560, 578, 583, *583*, 591–592
  life cycles of, *618*, **618**
  red, 558, *559*, *571*
Alkaloids, **673**, 674*t*
Allantois, **904**, *905*
Allard, Henry, 660
Allele frequencies, **429**
  genetic drift and, 440–441
  natural selection and, 436
  nonrandom mating and, 442
Alleles, **302**, *304*, **327**
  multiple, 338–339
  segregation of, 330–331, *331*
Allometry, **753**
Allopatric speciation, **452**–453, *456*, *459*
Allosteric effect, **387**
Allosteric enzymes, **129**, **131**, *131*
α carbons, *39*, *40*, 71–72, *72*
α helices, **75**, *75*–76
Alpine biomes, 1076
Alternation of generations, *617–621*, 617–622, **619**
  evolution of, 685, *686*
Alternative energy sources, *1103*, 1103–1104
Alternative splicing, *67*, **67**, *382*, *383*
  genome size and, 289
Altitude, oxygen and, 842
Altruism, **995**–998, *998*
Alvarez, Walter, 6–7
Alveolates, **561**, *561*, *563*
Alveoli, **835**, *836*

Amine hormones, *815*, **815**–816
Amino acid replacement, **310**
Amino acids, **39**, **862**–863, 863*t*
  hydrophilic, 73
  hydrophobic, 72, *73*
  in proteins, 39–40, *40*
  side chains and, 71–73, *72*
  structure of, 72–73, *73*
Aminoacyl (A) sites, **80**, *80*
Aminoacyl tRNA synthetases, **81**, *81*
Amino ends, **74**, *74*
Amino groups, **39**–40, *40*, 71–72, *72*
Ammonia, **882**, 882–883
  See also Fertilizers
Ammonification, **1057**, *1057*
Amnions, **904**, *904*–905, *905*
Amniotes, 752, **967**–969, *968*
Amniotic eggs, *904*, **904**–905, *905*, **967**
Amoebas, **546**, *546*, 556, *557*, 558
  diseases and, 557
  photosynthetic, 550, *551*
Amoebozoans, 556, *556*–558, *557*
AMP. *See* Adenosine monophosphate
Amphibia, **967**
Amphibians
  as ecological warning, 1110–1111, *1111*
  heart and circulatory system of, *847*, 847–848
  life cycle of, *967*, **967**
  waste excretion by, 886
Amphioxus, *962*, **962**
Amphipathic molecules, **94**, *94*
Amplification, **194**, *194*, *195*, **267**, *268*, 816–817, *817*
Amylase, **868**, 868*t*
Anabolism, **121**, *121*, 124, **856**
  energy in, 127
Anaerobic ammonia oxidation (anammox) reactions, **530**
Anaerobic metabolism, **148**–150, *149*, 855–857, *856*, *856*
Anaerobic respiration, **135**–136
Analogy, **472**, *472*
Anammox (anaerobic ammonia oxidation) reactions, **530**
Anaphase (mitosis), **229**, *230*, *249*
Anaphase I (meiosis I), *233*, **233**
Anaphase II (meiosis II), *234*, **235**
Anatomy, ecological systems and, 19–20
Anchoring junctions, *213*, **214**, *215*, 216*t*
Anchors, **97**, *97*
Androgen, 812*t*
Anfinsen, Christian, 77, 78

I-1

Angiosperms, **594**, *594*, 626–629, 626–635, *633*, *634*
   biodiversity of, 688t, 701–707, *702*, *703*, *706*
   coevolution of pollination and, 702–703
   evolution of, 688–689, *689*
   genomes of, 289, *290*
   life cycles of, *632*
   phylogenetic tree for, *702*
   visual synthesis of, *708–709*
   See also Flowers
Angiotensin II, **894**
Animal rights movement, 509
Animals
   biology-inspired design from form and function of, 745, *745*
   body plans of, 739–745
   breeding and behavior of, 984, *984*
   cells of, 106, *106*
   classification of, 740–741, *741*
   colonization, 752
   communication of, 991–994, *992–994*
   culture and, *508*, 508–509
   as deuterostomes, 742–743, *743*, *744*
   digestion by, 867, 867–874, 868t, *869–874*
   diploblastic, 742, *743*
   as ectotherms, 858–859, *861*
   as endotherms, 858–859, *860*
   evolution of, 580–582, *581*, 749–754
   extinction threatened, 1112, *1112*
   extracellular matrix of, 217–218, *217–219*
   eye development in, 409–411, *409–411*
   features of, 739–740
   feeding by, 864–867, *865–867*
   fossil record of, 749–750, *750*
   germ layers of, 742–743
   homeostasis and, 748–749, *749*
   language and, 509, *509*
   mass extinctions and, *751*, 751–752
   metabolic rate in, 857–859, *858*
   metabolism in, 855–859, *856–858*
   molecular sequencing of, 744, *744–745*
   multicellularity in plants compared to, 578–580
   nutrition and diet of, 862–864, *863*, 863t, 864t
   organs of, 747, *747–748*
   phylogenetic tree for, 740, *740*, *743*, *744*, *758*, *945*, *947*, *951*, *951–952*
   phylogeny and, 946, *947*
   as protostomes, 742–743, *743*, *744*
   seed dispersal and, 634, *634*
   segmentation and, 742, *742*
   sensory systems of, 773–780, *774–777*, *779*, *780*
   size of, 752–754
   skeletons of, 799–806
   tissues of, 745–747, *746*, *747*
   visual synthesis of, *971*
   weight of, 365–366, *366*
   See also Circulatory systems; Immune systems (animals); Muscles; Nervous systems, animal; Renal systems; Reproduction; Respiratory systems
Annealing, **267**, *268*
Annelid worms, 952–954, *953*, *954*
Annual clocks, **990**
Anolis carolinensis (lizards), 982–984, *983*
Anoxygenic reactions, 525–526, *526*
Antagonisms, 1022–1026
Antagonist muscles, **794**, *794*
Anterior pituitary gland, 818t, *820*, **820–821**, *821*
Anthers, 627, **628**
   crossing plants and, *326*, 326–327
Anthropocene Epoch, **1093–1094**
Antibiotics, 542
Antibodies, **931**
   binding of, 931–932
   classes of, 932, *932*
   clonal selection and, 932–933, *933*
   diversity of, 934, 935–936
   genomic rearrangement and, *935*, 935–936
   structure of, 931, *931*
Anticodons, **81**, *81*
Antidiuretic hormone (ADH), 818t, **821**, **892**
   urine concentration control by, 892, *892*
Antigenic drift, **940**, *940*
Antigenic shift, **940**, *940*
Antigenic variation, **942**
Antigen-presenting cells, **937**, *937*–938
Antigens, **931**, *931*
Antioxidants, **170**, *171*
Antiparallel strands, **56**
Antirrhinum majus (snapdragons), *331*, *331*
Ants
   fungal infection of, 728–729
   pheromones and, *825*, 825–826
   plant defenses and, 674–675, *675*
   plant symbiosis with, 675, *675*
   relatedness in, 997, *997*

Aorta, **847**, *847*
Aortic bodies, **837**, *837*
Aortic valve, **848**, *849*
Apes, phylogenetic trees for, 491–493, *492*
Aphids, 517, 997
   mutualisms and, 1027–1028
Apical dominance, 647–648, *648*
Apomixis, **635**
Appendicular skeletons, **802**, *802*
Appendix, **874**
Aquaporins, **101**, **878**
Aquatic biomes. *See* Freshwater biomes; Marine biomes
Aqueous solutions, 35–37
Arabidopsis thaliana (mustard plant)
   circadian clock and, 661
   floral development of, *412*, 412–414, *413*
   genes of, 550
   genome of, 288, 288t
   growth of, 659
Arbacia punctulata (sea urchin), 242
Archaea, **14**, **519**
   acidic solutions and, 537, *537*
   Bacteria and Eukarya compared to, 523, 524t
   biodiversity of, 535–539
   genomes of, 289, *290*
   nitrogen cycle and, *529*, 529–530
   phylogenetic tree for, 17, *17*, 523, *523*, 532, *532*, 535–536, *536*
   in sulfur cycle, *528*, 528–529
Archaeopteryx lithographica (feathered dinosaur), 484, *485*
Archaeplastida, **558**, 558–560
Ardi (hominin), 494, *495*
Ardipithecus ramidus (hominin), 494
Aristotle, 324, 724
Arnold, William, 166
Arrhenius, Svante, 130
Arteries, **843**, 843–845
Arterioles, **843**, 844
Arthropods, **957**–960, *959*, *960*
Artificial heart valves, 849–850
Artificial joints, 805
Artificial selection, **15**, *15*, 368, 438, **439**
   agriculture and, 439, 587, 635
   behavior and, 984, *984*
Asclepias species (milkweeds), 672, *672*
Ascomycetes, 726, **726**–728, *727*
Asexual reproduction, **225**, **897**
   biological species concept and, 449
   evolution and, *900*, 900–902, *901*
   forms of, 897–899, *898*
   in plants, 635–636, *636*
Asian elephant (*Elephas maximus*), 448, *448*

A (aminoacyl) sites, **80**, *80*
Aspen tree (*Populus tremuloides*), 636, *636*
Assimilation, **528**, *528*, **1057**, *1057*
Associative learning, **987**
Assortment. *See* Independent assortment
Astrocytes, **761**
Atomic mass, 29–30
Atoms, **29**–31, *30*
ATP. *See* Adenosine triphosphate (ATP)
ATP synthase, 144, **146**, *146*
Atrioventricular (AV) nodes, **850**, *850*–851
Atrioventricular (AV) valve, **848**, *849*
Atrium, **847**, *847*
Auditory cortex, **782**, *782*
Australopithecus afarensis (hominin), 494–495, *495*
Autocrine signaling, **189**, *189*
Autoimmune diseases, **924**
Autonomic nervous system, **770**–771, *771*, 851
Autosomes, **344**
Autotrophs, **120**, *120*
Auxin, 645, 645t, *646*, *647*
Avery, Oswald, 52–53
Avirulent pathogens, **667**, *667*
AV (atrioventricular) nodes, **850**, *850*–851
AV (atrioventricular) valve, **848**, *849*
Axial skeleton, **802**, *802*
Axillary buds, **642**, *642*
   apical dominance and, 647–648, *648*
Axon hillock, **759**, *760*, 768, *768*
Axons, **759**–760, *760*, 767

B

Bacillus thuringiensis (bacterium), 682, *682*
Bacteria, **14**, **519**
   Archaea and Eukarya compared to, 523, 524t
   biodiversity of, 530–535, *535*
   carbon cycle and, 524–528, *525–527*
   cell size of, 519–521, *520–522*
   chlorophyll in, 525, *526*
   cyano-, 534, *535*
   evolution experiments with, 18–19
   fermentation in, 527
   genetic diversity in, *522*, 522–523
   genomes of, 289, *290*
   gram-positive, 534
   intestinal, 541, 541–542
   myxo-, 521, *521*
   nitrogen cycle and, *529*, 529–530
   nitrogen fixing, *529*, 529–530

as pathogens, 923–924, *924*
photosynthesis and, 534–535, *535*
phylogenetic tree for, 17, 523, *523*, 530, 532, 532–533
plant genomes and, 670–671, *671*
proteo-, *533*, 534
in sulfur cycle, 528, 528–529
symbiotic, 553, *554*
Bacteriochlorophyll, **525**, *526*
Bacteriophages, **298**, *298*, *391*, *392*
Baker's yeast. *See Saccharomyces cerevisiae* (baker's yeast)
Balancing selection, **437**
Ball-and-socket joints, 805–806, *806*
Baltimore, David, 296–297
Baltimore system, 296–297, *297*
Barcode of Life, 449
Bark, *649*, 649–650, *650*
innate immunity and, 925, *925*
visual synthesis of, *709*
Barnacles, 959
Barrier junctions, 214, *215*, *216t*
Basal lamina, **204**, *205*, *218*, *219*
Basal resistance, **667**, *668*
Base excision repair, **319**, *319*
Base pairing
hydrogen bonds and, 56–57, *57*
translation and, 81–82, *82*
Bases, **40**, *53*, *54*
complementary, **42**, *42*, 56
purine, **41**, *41*, *54*
pyridamine, **41**, *41*, *54*
Base stacking, **57**
Basic solutions, **37**
Basidiomycetes, **726**, *726*, 728–731, *729*, *730*
Basophils, **926**, *927*
Basswood (*Tilia americana*), *650*
Bats, 996
B cells, **926**, **930**, 930–936, *931–933*, *935*
See also Antibodies
BCG vaccine, 940–941
Bdelloid rotifers, 901–902
Bees
genetic influence on behavior of, 986
language and, 509
pheromones and, 824–825, *825*
as pollinators, *19*, *19*
waggle dance and communication of, *994*, 994
Behavior
artificial selection and, 984, *984*
communication and, 991–994, *992–994*
displays and, *981*, *981*
ecological systems and, 19–20
fixed action pattern as stereotyped, *981*, 982

genetic influence on, 984–986, *985*, *987*
hormones and, 982–984, *983*
information processing and, 989–991
innate, **980**, *980*
learned, **980**
learning and, 987–989, *989*
nervous system and, 981–982, *983*
social, 994–998, *995–998*
Tinbergen's questions on, 979–980, *979t*, *994*
Behavioral isolation, **451**
Beneficial (advantageous) mutations, **303**, *303*, *304*, **429**
Bennett, J. Claude, 934, *935*
Ben-Shahar, Yehuda, 986
Benson, Andrew, 162
β barrels, 86, *87*
β-galactosidase, 388
β-globin gene family, *314*, *315*
β-oxidation, **152**, *152*
β sheets, **75–76**, *76*
Betula occidentalis (water birch), *650*
Bilateral symmetry, **581**
Bilateria, **740**, *741*
Bilaterians, **952**, *952*
Bilayers, **94–95**, *95*
Bile, **871**
Binary fission, **225–226**, *226*
Binding affinity, **196**
Biodiversity
of angiosperms, *688t*, 701–707, *702*, *703*, *706*
of Archaea, 535–539
of Bacteria, 530–535, *535*
of bryophytes, 689–692, *690–692*
carbon cycle and, 1055, *1056*
climate change impact on coral reefs and, 1100, 1100–1102, *1101*
competition and, 1023–1024, *1024*
complex multicellularity and, 583
conserving, 974–976, *975*, *976*, 1111–1114
ecosystems and, 1059–1061, *1061*
of eukaryotes, 554–564
of ferns, *696*, 696–697, *697*
of fish, **964**–**966**, *964*–*966*
of fungi, 724–731, *725*–*727*, *729*, *730*
geologic timescale and, 565
global patterns of, 1088–1090, *1089*
of gymnosperms, *688t*, 697–700, *698–701*
of horsetails, *696*, 696–697
latitudinal diversity gradient and, *1089*, 1089–1090
of lycophytes, *693*, 693–696, *694*

measurement of, 1032
nature reserves protecting, 1112–1113
of plants, 679–682, *681*, *682*, 685–689, *686*, *687*, *688t*, *689*
primary production and, 1059–1060
sustainable development for conservation of, 1113–1114
of vascular plants, 692–697, *693–697*
of vertebrates, 963–969, *964–969*
visual synthesis of, *970–971*
See also Genetic diversity
Biodiversity hotspots, 974–976, *975*, *976*, **1111**–**1112**
Biological clocks, 990–991
Biological Dynamics of Forest Fragments Project, 1014–1015, *1015*
Biological species concept (BSC), 447–450, **448**
Biologists, 3, 1114, *1115*
Biology, **3**
Biology-inspired design. *See* Case 7: Biology-Inspired Design
Biomes, **1067**, 1071
freshwater, 1074, 1083, *1083*
marine, 1083–1084, *1083–1086*
terrestrial, 1072–1074, *1072–1082*
Biomimetic Dexterous Manipulation Laboratory, 736
Biotic factors, **1001**
Biotrophic pathogens, **666**
Biparental inheritance, **357**
Bipedalism, **494**, **501**, *501*
Birds, *968*, *968*
asexual reproduction and, 900, *900*
communication of, 992, *993*, 993–994
consciousness and, 509–510
culture and, *508*, *508*
digestion by, 868
dinosaurs and, 484, *485*
Galápagos finches, 438, *438*, 455, *456*, *457*
heart and circulatory system of, 848–849
learning in, 989, *989*
lungs of, *836*, 837
navigation of, 990–991
salt excretion from glands in, *881*, *881*
seed dispersal and, 634, *634*
towhees, 449, *450*
Bird song, 992, *993*, 993–994
Birth anomalies, 371–372, *372*
Birth rates, 1005, *1005*
Bissell, Mina, 221
Bivalents, **232**, *232*
Bivalves, **956**, *956*
Björkman, Olle, 600

Bladder, **886**, *886*
Blakemore, Richard, 990
Blastocysts, **400**, *400*, *915*, *915*
Blastoderm, cellular, **403**, *404*
Blastulas, **579**, **914**
Blending inheritance, **324**, 324–325, 332–333
Blight, potatoes and, *665*, *666*
Blood, **830**
composition of, 838, *838*
glucose levels in, 150
kidney regulation of volume of, 893, 893–894
pH of, 37
vertebrate filtration of, 885–886, *886*
Blood cells. *See* Red blood cells
Blood groups, 429–430, *430t*
Blood pressure, **845**, 845–846
kidney regulation of, 893, 893–894
B lymphocytes. *See* B cells
BMPs. *See* Bone morphogenetic proteins (BMPs)
Bog communities, 1031, *1031*
Bohr effect, 842
Bonds, chemical, **32**
of carbon, *37*, 37–38
covalent, **32**, 32–33
double, 32
energy and, 122
glycosidic, **43**, *43*
hydrogen, **35**, 35–36, 56–57, *57*, 74–76, *75*, *76*
ionic, **33**–**34**, *34*
nonpolar covalent, **33**
peptide, **40**, *40*, 73–74, *74*
phosphodiester, **41**, *41*, *55*, *55*
polar covalent, **33**, *33*
Bone marrow, **804**, *804*
Bone morphogenetic proteins (BMPs), 824
Bones
compact, **803**, *804*
endoskeletons and, **802**, 802–803
formation of, 803, *804*
growth of, 804
repair of, 805
spongy, **803**–**804**, *804*
Bony fish, **965–966**, *966*
Borlaug, Norman, 635, 681, 731–732
Borthwick, Harry, 657
Bottleneck, population, **440**
Bowman, William, 886
Bowman's capsule, **886**, *886*, **888**, *888*
Brachiopods, **751**, 751–752
Brachydactyly, *337*, *338*
Bracken fern (*Pteridium aquilinum*), *621*, 621–622
Brains, **758**, *759*
evolution of, 758

Brains (*Continued*)
  organization and function of, 780–783, *781–783*
  size of, *502*, 502–503
Brainstem, **780**, *781*
Brakefield, Paul, 584
Branching
  of roots, *653*, 653, 658–659
  of shoots, 641–642, *642*
BRCA1 mutations, 252–254, *253*, 284
BRCA2 mutations, 252, *253*, 254
Breathing
  citric acid cycle and, 143
  homeostasis and, *837*, 837–838
Breeding, animal, 984, *984*
  See also Artificial selection
Brenner, Sydney, 60
Bridges, Calvin B., 348–349
Bronchioles, **835**, *836*
Brown algae, 560, *560*, 571
Bryophytes, 592, *593*
  alternation of generations in, 619–620, *620*
  biodiversity of, 689–692, *690–692*
  desiccation tolerance in, 593
  vascular plants convergent evolution with, 690–691
  See also Mosses
BSC (biological species concept), 447–450, **448**
Budding, 898, *898*
Bud scales, **643**, *643*
Bulbourethral glands, **906**, *906*
Bulk flow, **573**, 578, 830
  multicellularity and, 573–574, *574*
  steps of, 830–831, *831*
Bundle sheath cells, 599, *599*
Burgess Shale, 481, *481*
Burs, 734, *734*
Butterflies, 449, *449*
  color pattern in wings of, 584
  life cycle of, *1010*, 1010–1011
  metamorphosis in, 960, *960*
Bylot Island, 1032, *1033*

C

C. *See* Cytosine (C)
C_3 plants, **599**
C_4 plants, 599, **599**–600
Cabot, John, 1108
Cadherins, **213**, *213*, 214, 216*t*
  complex multicellularity and, 575
Caenorhabditis elegans (nematode worm), 956–957, *957*
  development of, 414–417, *415*, *416*
  dosage compensation in, 380
  genome of, 288, 288*t*
  mechanoreceptors in, 775–776

Calcitonin, 818*t*
Calmette, Albert, 940
Calmodulin, **793**–794
Calvin, Melvin, 162, 163
Calvin cycle, **158**, **159**, 160–163, *161*
  evolution of, 171
  visual synthesis of, *176*
CAM. *See* Crassulacean acid metabolism (CAM)
Cambrian explosion, **750**–751
Cambrian Period, *750*, 750–751
cAMP (cyclic adenosine monophosphate), **194**, *194*, *195*
Cancer
  cell-cycle checkpoints and, 245
  human papillomavirus (HPV) and, 180–183, *181*
  metastasis of, *219*, 219–220, 247
  mutations and, 247, *247*, 306–307, *307*
  oncogenes and, 245–247
  signaling errors and, 198–199
  telomerase and, 266
  vaccination for, 180–183
  viruses causing, 246
  visual synthesis of, *418*
Canines, 866, *866*
Cann, Rebecca, 498–499, 504
Capacitation, **912**–913
Capillaries, **843**, *844*, 845, *845*, 847, *847*
Capsids, **287**
Carbohydrates, **39**, 41
  Calvin cycle and, 163
  carbon dioxide incorporated into, 162
  complex, 42
  complex, structure of, 41–43, *42*, *43*
  glycolysis and, 150, *151*
  starch storage of, 163, *163*
  transport of, in phloem, 606, 606–608, *607*
Carbon
  atoms of, 29–30, *30*
  bonds of, 37, 37–38
  ecosystem transfer of energy and, 1054–1055, *1055*
  metabolism and, 119–121, *120*
  molecules containing, 38–39, *38–39*
Carbon cycle, 519, **1043**
  bacteria and, 524–528, *525–527*
  biodiversity and, 1055, *1056*
  evolution of, 540, 1060–1061, *1061*
  food webs and, 1053–1055, *1054*
  human activities and, 1045–1046, *1095–1103*, 1095–1104
  long-term, 1048–1053, *1049–1052*

  short-term, 1043–1048, *1045*
  visual synthesis of, *1062–1063*
Carbon dioxide
  in Calvin cycle, 160–163, *161*
  in carbohydrates, 162
  coral reefs and levels of, 1102
  decreasing emissions of, 1103–1104
  in Earth's atmosphere, 1043–1053, *1044*, *1045*, *1048–1052*, *1095–1096*, *1096*
  from fossil fuels, 1046–1048, *1048*, 1095, *1095*
  Keeling curve and, 1044, *1044*, 1045–1047
  leaves and, 595, *596*
  respiratory system and, 829–831, *831*
  seasonal oscillation of atmospheric, 1045
Carboxylation, **160**, *161*
Carboxyl ends, **74**, *74*
Carboxyl groups, **39**–40, *40*, **71**–72, *72*
Cardiac cycle, **849**, *849*
Cardiac muscles, **788**, *788*–790, *850*, 850–851
Cardiac output (CO), **851**
Cardiovascular systems. *See* Circulatory systems
Carnivores, evolution of, 583
Caron, Joan, 221
Carotenoids, 165
Carotid bodies, *837*, **837**
Carpels, **626**–627, *627*
Carriers, **100**, **101**
Carroll, Lewis, 901, 1106
Carrying capacity (*K*), **1007**
Carsonella rudii, 26
Cartilage, **800**, *801*, 803
Cartilaginous fish, **965**, *965*
Case 1: Life's Origins, 25–27, *26*, *27*
  cell membranes and, 95
  cellular respiration and, 149–150, *150*
  chemical reactions and, 131–132
  citric acid cycle and, 143, *143*
  genetic code and, 85–86
  molecules of life, 45–47
  nucleic acids and, 60–61
  photosynthesis and, 173
Case 2: Cancer, 180–183, *181*
  metastasis of, *219*, 219–220
  oncogenes and, 245–247
  signaling errors and, 198–199
Case 3: Your Personal Genome, 252–255, *253*
  ancestry and, 356, *356*, *357*, 358
  genetic risk factors and, 307
  genetic tests and, 339–340
  lifestyle choices and, 386

  personalized medicine and, 284, 373–374
  regenerative medicine and, 403
  sequencing of, 273
Case 4: Malaria, 422–425, *423*
  coevolution and, 437, 457–458
  genetic variation in response to, 506–507
Case 5: The Human Microbiome, 514–517, *515*
  chloroplast symbiosis and, 548–551, *551*
  eukaryotic cell origins and, *552*, 553
  intestinal bacteria and, *541*, 541–542
  mitochondrial symbiosis and, *551*, 551–553, *552*
Case 6: Agriculture, 586–589
  fertilizers and, 613
  fungi and, 731–732, *732*
  genetic diversity and, 706–707, *707*
  Green Revolution and, 635
  plant defenses and, 680–682, *682*
  plant hormones and, 646–647
Case 7: Biology-Inspired Design, 734–736, *734–737*
  animal form and function for, 745, *745*
  bone repair and, 805
  cochlear implants and, 778
  dialysis and, 894
  heart valve replacements and, 849–850
  for sepsis treatment, 930
Case 8: Conserving Biodiversity, 974–976, *975*, *976*
  biodiversity measurement and, 1032
  climate change challenges for, 1113
  climate change impact on coral reefs and, *1100*, 1100–1102, *1101*
  competition and, 1023–1024, *1024*
  fossil record and, *1090*, 1090–1091
  island population colonization and, 1015–1016, *1016*
  priorities for, 1111–1112
  strategies for, 1112–1113
  sustainable development and, 1113–1114
Casparian strips, **609**, *610*
Catabolism, **121**, *121*, 124, **856**
  energy in, 127
Cats
  calico, *380*, 381
  cloned, *402*, 402–403
  pheromones and, 825

Cattle, rumen of, 515–516
Causation, **1046**
Cavitation, **605**, *605*
CCK (cholecystokinin), 816, 819*t*, 871
CDKs (cyclin-dependent kinases), **241**, *241*–243, *243*
cDNA (complementary DNA), **274**
Cecum, **874**, *874*
Cell adhesion molecules, 212, **213**, *213*, 216*t*, 249
 multicellularity and, 574–576
Cell communication, 185–187, *186*, *187*
 complex multicellularity and, 576, 576–577, *577*
 in development, 414–417, *415*, *416*
 distance in, 188–191
 endocrine system and, 824–826, *824*–826
 visual synthesis of, *249*
Cell cycle, **227**, *227*
 checkpoints, 243, *244*, 245
 regulation of, 241–244
Cell division, **225**
 cytoplasmic, 230, *231*
 eukaryotes and mitotic, 226
 meiotic, 231–238, *232*–234, 235*t*, *236*, *237*
 mitotic, 226–230, *229*–231
 prokaryotes, binary fission and, 225–226, *226*
 visual synthesis of, *248*
 See also Meiosis I; Meiosis II; Mitosis
Cell junctions, **204**, 212–214, 212–215, 216*t*
Cell membranes, 94–99, *98*, 191, 191–193
Cells, 11–15, *12*
 active transport and, 103, *103*
 animal, 106, *106*
 antigen-presenting, 937, *937*–938
 B, **926**, 930–936, *931*–933, *935*
 bundle sheath, 599, *599*
 chloride, 880, *880*
 coenocytic, 557, *557*
 companion, 606, *606*
 cone, 780, *780*
 cytotoxic T, **926**, 936, 938, 938*t*
 dendritic, **926**, 927
 dikaryotic (*n* + *n*), 722, *722*–723
 diploid, 228
 distance between, 188–191
 epithelial, 209, *209*
 extracellular matrix influencing shape of, 220, *220*
 first observation of, 93, *93*
 flagella of, 209, *209*, 235
 germ, 265, 306
 glial, 761, 765
 guard, 596–597, *597*

hair, 776, *776*–777, *777*
haploid, 228
helper T, **926**, 936, 938, 938*t*
hepatocytes, *204*, 221
heterokaryotic, 722, *722*
hormone-binding, 815
host, 297–298
induced pluripotent stem, 403
internal organization of, 104–107
Leydig, 909
mast, **926**, 927
memory, **926**, 932
microvilli of, *204*, *206*
multipotent, 400
natural killer, **926**, 927
neurosecretory, 812
parenchyma, 595, 601
pH of, 40
plant, 106, *107*
plasma, **926**, 932
pluripotent, 400
procambial, **643**
red blood, 103, *103*, *204*
responding, 186, *186*
rod, 780, *780*
sensory receptor, 773
Sertoli, 909
signaling, 186, *186*
somatic, 306
stem, 265, 400–401
T, 930, 936–939, *937*, *938*, 938*t*
totipotent, 399–400
types of, 203, *204*
white blood, **926**, 926–927
Cell-surface receptors, 191–193, *192*
Cell theory, **93**
Cellular blastoderm, **403**, *404*
Cellular matrix, **204**
Cellular respiration, **135**
 ATP and, 135–136, *136*, 146*t*, 152–153, *153*
 chemical energy and, 135–136, *136*
 energy flow in, 146–147, *148*
 evolution of, 149–150, *150*
 first cells and, 149–150, *150*
 glucose and, 138, *139*
 oxidation-reduction reactions and, 136–137, *137*
 photosynthesis compared to, 175, *176*–177
 stages of, 137–138, *138*
 visual synthesis of, *176*–177
Cellulase, **873**
Cellulose, 216–217, *217*
Cell walls, 99, *107*, **107**
 cell shape and, 103–104, *104*
 in plants, 578–579
Centipedes, 957
Central dogma of molecular biology, **13**, *13*, **59**, *59*, **80**, *80*

Central nervous system (CNS), 769–770, *770*
Centromeres, **228**, *228*, 230, 315
Centrosomes, **205**, *206*, **229**, *229*
Cephalization, **741**, *741*–742
Cephalochordates, **961**, *962*
Cephalopods, 954–955, *955*
Cerebellum, **780**, *781*
Cerebral cortex, **780**, *781*
Cerebrum, **781**, *781*
Cervix, **907**, *908*
CFTR (cystic fibrosis transmembrane conductance regulator), 311, *312*
Chain terminators, **272**, *272*
Channels, 100, **100**–101
Chaparral biomes, **1080**
Chaperones, 77, **79**
Chara corollina (alga), 618, *618*
Characters, **472**
Character states, **472**
Chargaff, Erwin, 55
Chase, Martha, 53
Checkpoints, 243, *244*, 245
Chelicerates, **957**, *959*
Chemical bonds, 32–35
 See also Bonds, chemical
Chemical energy, **122**
 cellular respiration and, 135–136, *136*
Chemical reactions, 34, **34**–35
 aerobic, 525
 anammox (anaerobic ammonia oxidation), 530
 anoxygenic, 525–526, *526*
 energetically coupled, 126, *126*
 enzymes and, 127–132
 exergonic *vs.* endergonic, 125, *125*
 first cells and, 131–132
 heat and, 128
 oxidation-reduction, 136–137, *137*, 158, 158–160, 524–525
 oxygenic, 525
 process of, 124, *124*–125
 thermodynamics and, 125–126
Chemiosmotic hypothesis, **146**, *147*
Chemistry, 8–11
Chemoautotrophs, **526**, *526*–527
Chemoreceptors, **773**–775, *775*
Chemotrophs, **120**, *120*
Chiasmata, **232**, *232*–233
Childbirth, **916**, *917*
Chimaeras, 965
Chimpanzees
 culture and, 508, *508*–509
 human lineage and, 493–494, *494*
 human traits compared to, 501–503, *502*, *503*
 language and, 509, *509*
Chinese Giant Salamander, 967, *967*

Chitin, **714**, 801, 957
Chlamydomonas (algae), 209, *209*, 548, *559*, 560, 578
Chlorella (algae), 163
Chloride cells, 880, *880*
Chlorophyll, **115**, *164*, **164**–166, *165*
 in bacteria, 525, *526*
 photosynthesis evolution and, 173
Chloroplast genome, 295–296
Chloroplasts, 106–107, *107*, **115**, *115*, 163
 endosymbiosis and, *174*, 174–175, 517, 548–551, *551*
 inheritance and, 357
 metabolism and, 547
Choanocytes, **947**, *947*
Choanoflagellates, 555, *556*, 578, 945
 cell adhesion and, 575
 sponges compared to, 946–947
Cholecystokinin (CCK), 816, 819*t*, 871
Cholesterol, **96**
 cell membranes and, 96, 96–97
 genes affecting, 372, *372*
 structure of, 44–45, *45*
Chomsky, Noam, 509
Chordates, 960–963, *961*–963
Chorion, **904**, *905*
Chromatids
 non-sister, 232, *232*
 sister, **228**, *228*–230
Chromatin, 228, **293**, *293*, 378
Chromatin remodeling, **378**
 development and, 408
 vernalization and, 662
Chromatotropins, 812*t*
Chromosome condensation, 293, **294**, *294*
Chromosomes
 abnormalities of sex, 240, 240–241
 allele segregation and, 330–331, *331*
 gene expression regulation and, 380, 380–381
 genes and, 348, 348–349
 homologous, **228**, *228*, 232, 294
 of human genome, 294–295, *295*
 independent assortment and, 334, *336*
 meiotic cell division and, 231–235, *232*–234, 235*t*, *236*
 mitotic cell division and, 228–230, *229*–231
 mutations in, 314–317, *314*–317
 in nondisjunction, 238, 238–239
 nondisjunction and, 348, 348–349
 nonhomologous, 317, *317*
 replication of, 264–266, *264*–266

Cichlid fish, 826, *826*
Cigarette smoke, 836–837
Cilia, **208–209**, *209*, **926**
Circadian clocks, 661–662, **990**
Circular muscle layer, **873**, *873*
Circulation, 830–831, *831*
　pulmonary, 847
　systemic, 847
Circulatory systems, 830–831, *831*
　arteries in, *843*, 843–845
　blood pressure and, 845, 845–846
　bulk flow and, 573, *574*
　capillaries in, *843*, 844, 845, *845*, 847, *847*
　closed, *842*, 842–843
　evolution of, 843
　heart in, 842, *842*, 846–850, 846–851
　open, *842*, 842–843
　veins in, *843*, 843–845
　vessel sizes in, *843*, 843–844
Cis-regulatory elements, **411**
Cisternae, **110**, *110*
Citric acid cycle, **138**, *138*, 142–143, *142–143*, **856**, *856*
　visual synthesis of, *177*
Cladistics, **473**
　See also Statistics
Clams, 956
Classes, **471**, *471*
Classical conditioning, **987**
Classification, 470–471, *471*
　See also Phylogenetic trees; Taxonomy
Class I/II/III genes, **937**
Claudins, 215, 216*t*
Cleavage, **914**, *914*, *915*
　visual synthesis of, *918*
Climate, 1067–1071, *1068–1071*
Climate change
　biodiversity conservation challenges with, 1113
　communities and, 1098–1100
　evolution and, 1099, *1099*
　global warming and, *1095–1097*, 1095–1098
　plant responses to, 1098–1100, *1098–1100*
　responding to, 1101, 1103–1104
Climate models, *1097*, 1097–1098
Climax communities, **1035**–1036
Clitoris, **908**, *908*
Clocks
　annual, 990
　biological, 990–991
　circadian, 661–662, 990
　lunar, 990
　molecular, 443–444, *444*, 477
Clonal selection, **932–935**, *933*
Clones, **402**, *402–403*, 897–899
Closed circulatory systems, **842**, 842–843

Clustered regularly interspaced short palindromic repeats (CRISPR), **276–278**, *277*
Cnidarians, **740**, *741*, **948**
　anatomy of, *948*, 948–949, *949*
　colony formation in, 949, *949*
CNS (central nervous system), **769–770**, *770*
CNVs (copy-number variations), **315–316**, *316*
CO (cardiac output), **851**
Coal, 715
　lycophytes and, 694–696
　See also Fossil fuels
Coccolithophorids, 565–566, *566*
Cochlea, **777**, *777*–778
Cochlear implants, 778
Codons, **80–84**, *81*
Coelacanth, **966**, *967*
Coelom, **744**, *744–745*
Coenocytic cells, **557**, *557*
Coenzyme Q (CoQ), **144**
Coevolution, **540**
　of angiosperms and pollination, 702–703
　malaria and, 437, 457–458
　of plants and ants, 675
　of prokaryotes and eukaryotes, 540
Cofactors, **131–132**
Cognition, **783**
Cohesion, **36**
Cohorts, **1011**
Collagen, 218, *218*, **803**, 844–845
Collecting ducts, **886**, *886*
Colon, 872
Colonization, animal, 752
　island populations and, 1015–1016, *1016*
Color blindness, 349, *349*
Combinatorial control, **382**, *412–414*, **412–414**
Comb-jellies (ctenophores), 949–952, *950*, *951*
Commensalisms, **1029–1030**, *1030*, 1030*t*
Common ancestors, 468–469, *469*
　of humans, 492–496, 498–500
Communicating junctions, **214**, 215, 216*t*
Communication (animal), 991–994, *992*, *992–994*
Communication (cell). *See* Cell communication
Communities, **1031**
　bog, 1031, *1031*
　climate change and, 1098–1100
　climax, 1035–1036
　disturbance of, *1034*, 1034–1036, *1035*
　island species diversity and, *1036*, 1036–1037, *1037*

　keystone species and, 1033–1034, *1034*
　predation and, 1032, *1032*
　succession and, *1035*, 1035–1036
　visual synthesis of, *1038–1039*
　See also Populations
Compact bones, **803**, *804*
Companion cells, **606**, *606*
Comparative genomics, **287**
Competition, **1022**, 1030*t*
　biodiversity and, 1023–1024, *1024*
　costs and benefits of, *1030*, 1030–1031
　fitness and, 436
　herbivory and, 1026
　interspecific and intraspecific, 1007, 1022
　limited resources fostering, 1022
　parasites and, 1026
　predation and, 1024–1026, *1026*
　sexual selection and, 439–440
　for space, 1022, *1022*
Competitive exclusion, 1022–**1023**, *1023*
Complementary bases, **42**, *42*, 56
Complementary DNA (cDNA), **274**
Complement system, **929**, *929*–930
Complex carbohydrates, **42**
Complex traits, **361**
　concordance of, 369–370, 370*t*
　diseases and disorders and, 362, 362–363, 371–374, *372*
　environmental risk factors and, 363, *363*
　heritability and, 367–368
　multiple genes and, 363–365, *364*
　Punnett squares and, 364, *364*
　regression toward the mean and, 366–367, *367*
Compound eyes, **778–779**, *779*
Concentration gradients, **100**, *100*, 103
Concordance, **369–370**, 370*t*
Concurrent exchange, **833**, *833*
Conditioning, **987**
Cone cells, **780**, *780*
Conformational changes
　ATP synthesis and, 146
　membrane proteins and, 101
　receptors and, 191–192
Conifers, **700**, *700*
Conjugation, *522*, **523**
Connective tissue, 218, **746**
Connell, Joseph, 680, 1034
Connexins, 215, 216*t*
Consciousness, 509–510
Conservation biology, **974–976**, *975*, *976*, **1111–1114**
　See also Case 8: Conserving Biodiversity
Conserved sequences, **287**

Constant (C) regions, **931**, *931*
Constitutive phenotypes, **390**, *391*
Consumers, **1053**, *1054*
Contact-dependent signaling, **189**, *189*, 191
Continuous growth, **1006–1007**
Contractile rings, **230**, *231*
Contractile vacuoles, **103**, *104*, **883**, *883*
Contrasting traits, 325–327, *326*
Control groups, **5**
Convergent evolution, **472**, *472*
　of bryophytes with vascular plants, 690–691
Cooperative binding, **840**
Cope, Edward, 753
CopyCat, **402**, *402–403*
Copy-number variations (CNVs), **315**, *315–316*
CoQ (coenzyme Q), **144**
Coral bleaching, 1100–1101, *1101*, 1113
Coral reefs
　as biodiversity hotspots, **974–976**, *975*, *976*
　as biomes, *1085*
　carbon dioxide levels and, 1102
　climate change impact on biodiversity and, *1100*, 1100–1102, *1101*
Corey, Robert, 75–76
Coriolis effect, *1069*, **1069–1070**
Cork cambia, **648–650**, *649*, *650*
Corn (*Zea mays*), 312–313
Corpus luteum, **909**, *910*
Correlation, **1046**
Corridors, **1113**
Cortex (kidney), **887**, *887–888*, *889*, *890*
Cortex (plant), **609**, *610*, **644**, *644*
Cortisol, 819*t*
Co-speciation, **457**, *457*
Countercurrent exchange, **833**, *833*
Countercurrent multiplier, **890**
Covalent bonds, **32**, *32–33*
Coyote tobacco (*Nicotiana attenuata*), 676–677, *677*
CpG islands, **379**, *379*
Crabs, 959
Cranial nerves, **770**, *770*
Craniates, **961**
　See also Vertebrates
Crassulacean acid metabolism (CAM), **597–598**, *598*
C (constant) regions, **931**, *931*
Crick, Francis H. C., 53, 55–56, 258, 839
CRISPR (clustered regularly interspaced short palindromic repeats), **276–278**, *277*
Crisscross inheritance, **347**, *347–348*
Cro-Magnons, 500

Crop, **868**
Cross-bridge cycle, **791**–**792**, *792*
Cross-bridges, **791**–**792**, *792*
Crosscurrent flow, **837**
Crosses
　crisscross inheritance and, *347*, 347–348
　flowers and, 630–631
　of plants, *326*, 326–327
　reciprocal, 327
　test, 330, *330*, 330*t*
　with two traits, *333*, 333–334
　See also Artificial selection
Crossover, **232**, *232*, **344**
　linked genes and, 351–352, *352*
Crown gall disease, 670–671, *671*
Crown-of-Thorns Sea Stars, 1113
CRP–cAMP complex, 390–391, *392*
Crustaceans, 900, *900*, **957**, 959, *959*
Cryptic species, 449
Ctenophores (comb-jellies), 949–952, *950*, *951*
Cultural transmission, **368**
Culture, **507**, 507–509, *508*
Cuticle (animal), **801**, *801*
Cuticle (plant), **596**, *596*
　innate immunity and, **925**, *925*
C-value paradox, **289**–290
Cyanobacteria, **534**, *535*
　evolution of, 150
　photosynthesis evolution and, 173–175, *174*
　photosystems in, 174, *174*
　See also Bacteria
Cycads, 698–699, *699*
Cyclic adenosine monophosphate (cAMP), **194**, *194*, *195*
Cyclic electron transport, **169**, *169*
Cyclin-dependent kinases (CDKs), **241**, 241–243, *243*
Cyclins, **241**, 241–243, *243*
Cysteine, 73, *73*
Cystic fibrosis
　genetic mapping and, 353
　mutations and, 311, *312*
Cystic fibrosis transmembrane conductance regulator (CFTR), 311, *312*
Cytochrome-b_6f complex (Cyt), **167**, *168*
Cytochrome *c*, **144**
Cytokines, **928**
Cytokinesis, **226**, 230, 231, 249
Cytokinins, **645**, 645*t*, 647–648
Cytoplasm, **14**, **106**
　in meiotic cell division, 235, *237*
Cytosine (C), **41**, *41*, *54*
Cytoskeletons, **106**, *106*, **204**, 205–211, *206*, 210*t*, **546**, 546–547
Cytosol, **107**
　in signal amplification, 194, *194*, *195*

Cytotoxic T cells, **926**, **936**, 938, 938*t*

D

Daeschler, Edward, 486
Daphnia (crustacean), 900, *900*
Darwin, Charles, 361, 435, 1001
　altruism and, 995
　appendix and, 874
　artificial selection and, 15, *15*, 368
　blending inheritance and, 324–325, 332–333
　competition and, 1007
　competitive exclusion and, 1023
　evolution and, 17, 47, 324
　Galápagos finches and, 438, *438*, 455, *456*, 457, 460, *461*
　human evolution and, 491, 494, 510
　natural selection and, 434–436, 460–461
　nectar spurs and, 630
　phototropism and, 654–655
　phylogeny and, 467
　scientific inquiry and, 4
　sexual selection and, 439, 506
　speciation, 460–461
　species and, 447
　swim bladder and, 966
　variation and, 427
Darwin, Francis, 654–655
Daughter strands, **257**–259, *258*
Day-neutral plants, **660**
Deaf community, 778
Deciduous forest biomes, *1078*
Decomposers, **1053**, *1054*
　fungi as, 715, *716*
Deep sea biomes, *1086*
Deer, 1008, *1008*
Delayed hypersensitivity reactions, **938**
Deleterious (harmful) mutations, 302, *303*, **429**
Deletions, **314**, 314–315
Delta proteins, 191
de Mestral, George, 734
Demography, **1010**
Denaturation, **76**–77, 267, *268*
Dendrites, **759**–760, *760*
Dendritic cells, **926**, **927**
Denisovans, **497**, 500
Density, of populations, *1002*, 1003, *1003*, *1008*, 1008–1009
Density-dependent factors, **1008**
Density-dependent mortality, 680, *681*, 703
Density-independent factors, **1008**, *1008*
Denitrification, **529**, 529–530
Deoxyribonucleic acid (DNA), **13**, *13*, **40**, **51**
　complementary, 274

　composition and structure of, 40, 40–41, 42, 52–57, *54–57*
　damage to, 317–318, *318*
　of eukaryotes, 228, *228*, 293, *293*–294
　function of, 58–59
　isolation and sequencing of, 266–273
　mitochondrial, 498–500
　of organelles, 295–296
　origins of life and, 25–26
　repair of to, 318–320, *319*, *320*
　replication of, 257–263, *258*–*263*
　RNA as intermediary between protein and, 58–59, *59*
　RNA compared to, 60, *60*
　in transcription, 59–61, *61*
　See also Recombinant DNA; Repetitive DNA
Deoxyribose, **54**
Depolarization, **762**
Dermis, **204**, *205*
Descent of Man, and Selection in Relation to Sex, The (Darwin, C.), 439, 491
Desert biomes, *1072*, *1080*
Desiccation, **591**
Desiccation tolerance, **593**
Desmosomes, **214**, **215**, 216*t*
Deuterostomes, **742**–743, *743*, *744*, *952*, 952, 960–963, *960*–*963*
Development, embryonic, **399**
　cell communication in, 414–417, *415*, *416*
　combinatorial control in, 412–414, *412*–*414*
　complex multicellularity and, 577–578
　contact-dependent signaling and, 189, 191
　endocrine system and, *810*, 810–812
　evolutionary conservation of transcription factors and, 409–411, 503, 503–504
　female reproductive system and, 907–908
　genetic basis of, 399–403
　hemoglobin and, 841, 841–842
　hierarchical control of, 403–408
　neoteny and, 502
　pregnancy and, 914–917, *915*
　visual synthesis of, *419*, 918–919
Devonian Period, 752
Diabetes, 814
Dialysis, 894
Diaphragm, **834**, *835*
Diaphysis, **804**, *804*
Diastole phase, **849**, *849*
Diatoms, 560, *561*
Dideoxynucleotides, **272**, *272*, *273*
Diet, 862–864, *863*, 863*t*, 864*t*
Dietary minerals, **863**, *863*

Differentiation, **399**
　complex multicellularity and, 577–578
Diffusion, **100**, **829**
　bacteria cell size and, 520–521, *521*, *522*
　facilitated, 100, *100*
　gas exchange and, 829–830, *830*
　multicellularity and, 572–573, *573*
　oxygen and, 572–573
Digestion, **867**, 867–874, 868*t*, 869–874
Digestive symbioses, 1027–1029, *1029*
Digestive tract, **867**, 867–868, *873*, 873–874, *874*
Digger wasps (*Philanthus triangulum*), 988
Dikarya, **726**, *726*
Dikaryotic (*n* + *n*) cells, **722**, 722–723
Dilger, William, 984
Dinoflagellates, 561, *561*, 563
Dinosaurs
　birds and, 484, *485*
　fossil record of, 479, *479*, 484, *485*
　mass extinction of, 6–7, 479
　size of, 752
Diploblastic animals, **742**, *743*
Diploids, **228**, **548**, *548*
Dipodomys ingens (giant kangaroo rat), 887, *887*
Directional selection, **437**–438, *438*
Direct-to-consumer (DTC) testing, 253–254, 340
Disaccharides, 150–151, *151*
Discrete growth, **1006**
Diseases and disorders
　acquired immune deficiency syndrome (AIDS), 478
　allele frequency and, 436
　amoebas and, 557
　autoimmune, 924
　complex traits and, *362*, 362–363, 371–374, *372*
　crown gall disease, 670–671, *671*
　cystic fibrosis, 311, *312*, 353
　diabetes, 814
　Down syndrome (trisomy 21), 239, 239–240, 371, *372*
　emphysema, 302, *303*, 339
　epidemics and, 21, 1109–1110, *1110*
　epidermolysis bullosa, 210, *211*
　genetic engineering for correction of, 278
　genetic risk factors and, 302, *303*, 307
　hemophilia, 349–350, *350*
　human immunodeficiency virus (HIV), 287–288, *288*, 303, *303*, 478

Diseases and disorders (*Continued*)
human microbiome and, 514–517, *541*, 541–542
human papillomavirus (HPV), 180–183, *181*
Huntington's disease, 252
Klinefelter syndrome, 240, *240*
mitochondrial inheritance and, 358, *358*
from nondisjunction, 239, 239–240
phenylketonuria (PKU), 338–339
sickle-cell anemia, 304–305, 437, 506–507
tobacco mosaic virus, 297, *298*, 298–299, 668, *668*, 669, 679
tuberculosis (TB), 940–941, *941*, 1110, *1110*
Turner syndrome, 240, *240*
xeroderma pigmentosum (XP), 319
See also Malaria
Dispersal (life cycle)
of seeds, *633*, 633–634, *634*
of spores, 620–622, *621*
visual synthesis of, *708*
Dispersal (speciation), **453**–457, *455*, *456*
Displays, **981**, *981*
Disruptive selection, **439**, *439*
sympatric speciation and, 458, *459*
Distal convoluted tubule, **888**, *889*, *890*, 891
Divergence, **315**
speciation and, 452, *453*
Dizygotic (fraternal) twins, **369**, *369*
DNA. *See* Deoxyribonucleic acid (DNA)
DNA editing, **276**–278, *277*
DNA fingerprinting, **316**, *316*
DNA ligase, **261**, *261*, 318
DNA polymerase, **260**, 263, *263*
DNA replication, **257**–263, *258–263*
DNA sequencing, **272**–273, *273*
genetic variation and, 430
of genomes, 281–284, *283*
human evolution and, 493–494
species and, 449
DNA transposons, **291**, *291*
DNA typing, **316**, *316*, 338
Dobzhansky, Theodosius, 17, 510
Doctors, 1114, *1115*
Dogs
breeding and behavior of, 984, *984*
pheromones and, 825, *825*
Domains, 14, 17, **471**, *471*
Dominant traits, **327**, *327*, 330t
incomplete, **331**, *331*
in pedigrees, 337, *338*
Donohue, Jerry, 55

Dormancy, **625**
phytochrome and, 657, *658*
Dorsal nerve cord, **961**
Dosage compensation, **380**
Double bonds, **32**
Double fertilization, **632**, *633*, *633*
visual synthesis of, *708*
Double helix, **41**, *42*, *51*, 55–57, *56*, *57*
Downstream genes, **411**, 414
Down syndrome (trisomy 21), *239*, 239–240, 371, *372*
DPANN group, 535–536, *536*
Dreyer, William, 934, 935
Drosophila melanogaster (fruit fly)
development of, 403–408, *404–408*
dosage compensation in, 380
genetically engineered, 410–411, *411*
genetic influence on behavior of, 984–986, *985*
genome of, 282–283, 288t, 289t
life cycle of, 403–404, *404*
linked genes in, 350–352, *351*
X-linked genes in, 345–348, 345–350
Drugs
antibiotics and, 542
for human immunodeficiency virus (HIV), 288
penicillin, 924
personalized medicine and, 374
from plants, 673–674, 674t
resistance to, 18, 87
See also Diseases and disorders
DTC (direct-to-consumer) testing, 253–254, 340
Duodenum, **870**, *870*
Duplications, **314**, 314–315
Dynamic instability, **207**, *207*
Dynein, **208**, *208*

E
E_A (activation energy), **127**–128, *128*
Ear, **777**, 777–778
Earth
atmospheric carbon dioxide in, 1043–1053, *1044*, *1045*, *1048–1052*, 1095–1096, *1096*
atmospheric oxygen history in, 149–150, 174, 484, 540, 582, 1060–1061, *1061*
composition of, 9, *9*
geological processes and, 436
glaciation and, *971*
history of, 483–484, *484*
life on, water and, 36
seasonality of, 1068, *1068*
See also Climate change
Earthworms, 759, *759*, 952–954, *953*, *954*

digestion by, 868
hydrostatic skeletons in, 800, *800*
waste excretion by, 884, 885
Ecdysone, 812t
Ecdysozoans, **952**, *952*, *956*, 956–957, *957*
Echinoderms, **960**–961, *962*
Ecological footprint, **1094**, *1094*
Ecological isolation, *451*, 451–452
Ecological niches, **450**
Ecological species concept (ESC), **450**
Ecological systems, 19–20, *20*
Ecological Theater and the Evolutionary Play, The (Hutchinson), 20
Ecologists, 1114
Ecology, 19–20, **1001**
See also Populations
Ecosystems, **1043**
biodiversity and, 1059–1061, *1061*
carbon and energy transfer in, 1054–1055, *1055*
food webs and, 1053–1055, *1054*
long-term carbon cycle and, 1048–1053, *1049–1052*
nitrogen cycle and, *1057*, 1057–1058
phosphorus cycle and, 1058, *1058*
short-term carbon cycle and, 1043–1048, *1045*
trophic pyramids and, 1055, *1055*
visual synthesis of, *1062–1063*
Ecosystem services, 1112
Ectoderm, **400**, *400*, *915*, 916
Ectomycorrhizae, **611**, 611–612, **717**
Ectotherms, **858**–859, *861*
Eggs, 235, 237, **899**
amniotic, **904**, 904–905, *905*, 967
as totipotent cells, 399–400, *400*
Ehrlich, Paul, 680, 926
Eiffel Tower, 734–735, *735*
Ejaculatory duct, **906**, *906*
Elastin, **217**, *217*, 844–845
Electrochemical gradients, *102*, **103**
Electrolytes, **877**–881
Electromagnetic receptors, **778**–780, *779*, *780*
Electromagnetic spectrum, 164
Electron carriers, **136**, *136*
Electron donors, 158, 173–174
Electronegativity, **33**
Electrons, **29**
orbitals and, *30*, 30–31, *32*
valence, 32
Electron transport chain, **136**, *136*, 144–147, *145*, *148*, *150*, 856, 856–857
cyclic, 169, *169*

photosynthetic, **159**–160, *160*, 177
visual synthesis of, *177*
Electrophoresis, gel, **269**, *269*
Elements, **29**
periodic table of, 31, *31*, *32*
Elephas maximus (Asian elephant), 448, *448*
Elongation factors, **85**
Elongation stage, of transcription, **61**
Elongation stage, of translation, **84**, 84–85
Elton, Charles, 1019
Embryos, **899**, 915, *915*
visual synthesis of, *918*
See also Development, embryonic
Emerson, Robert, 166
Emigration, **1005**, *1005*
Emphysema, 302, *303*, 339
Encyclopedia of Life, 477
Endangered species, 22, *22*
Endergonic reactions, **125**, *125*
Endocrine signaling, **188**, *189*
Endocrine system, **809**
cell communication and, 824–826, *824–826*
function of, 809–814, *810*, *813*, *814*
homeostasis and, 812–814, *813*, *814*
vertebrate, **820**, 820–823, *821*
Endocytosis, **108**, *108*, 546, *546*
Endoderm, **400**, *400*, *915*, 916
Endodermis, **609**, *610*, 949
Endomembrane system, 107–114, *108*, *108*, 545
Endomycorrhizae, **611**, 611–612, 717
visual synthesis of, *709*
Endophytes, **717**–718
Endoplasmic reticulum (ER), **106**, *106*, 109, **109**–110, 545, *546*
Endoskeletons, **799**, *802*, 802–803
Endosperms, **632**, *633*, *633*
Endosymbiosis, 174–175, **548**–553, *551*, *552*
evolution and, *174*, 517
origin of photosynthesis and, 562–564, *563*
Endosymbiotic hypothesis, 175
Endotherms, **858**–859, 860
Endurance Crater, Mars, 25, *26*
Energetic coupling, *126*, **127**
Energy, **121**
activation (E_A), **127**–128, *128*
alternative, *1103*, 1103–1104
in anabolic reactions, 127
in catabolic reactions, 127
cellular respiration and, 146–147, *148*
chemical, 122, 135–136, *136*
ecosystem transfer of carbon and, 1054–1055, *1055*

kinetic, *121*, 121–122
metabolism and, 14–15
nutrient uptake and, 610–611
potential, *121*, 122, 144, 146, *146*
solar, 172, 1103, *1103*
thermodynamics and, 9–10, *123*, 123–124
wind, 1103, *1103*
Energy balance, **862**
Energy intake, **862**
Energy levels, *30*, *31*, *32*
Energy use, **862**
by humans, 1093–1094, *1094*
Engineers, 1114
Enhancers, **62**, *63*, **382**, *382*
Enthalpy (H), **125**
Entropy (S), 9–10, *10*, 123–124, **125**
Envelopes, **297**
Environmental risk factors, **363**, *363*
genetic risk factors and, *365*, 365–366, *366*
Environmental variation, 16
Enzymes, **39**, **97**, *97*, **127**
activation energy and, 127–128, *128*
allosteric, 129, 131, *131*
chemical reactions and, 127–132
influences on, 129, *131*
restriction, 269–270, *270t*, *271*
specificity of, 129
Eosinophils, *926*, *927*
Epidemics, 21, 1109–1110, *1110*
Epidermis (animal), **204**, *205*, *211*, **949**
Epidermis (plant), **594**, *594*
Epidermolysis bullosa, 210, *211*
Epididymis, **905**, *907*
Epigenetic mechanisms of gene regulation, **379**
Epiglottis, **868**, *869*
Epinephrine, 816, 819*t*
Epiphysis, **804**, *804*
Epiphytes, **598**, *598*, **690**
Epistasis, **336**, *336*–337, 372–373
Epithelial tissue, **746**
Epithelium, **204**, *205*, **209**, 209–210
cell junctions and, *214*, 215
EPSPs. *See* Excitatory postsynaptic potentials (EPSPs)
Equational division, **235**
See also Meiosis II
ER. *See* Endoplasmic reticulum (ER)
ESC (ecological species concept), **450**
Escherichia coli (bacterium), **521**, *521*
DNA repair and, 318
evolution experiments and, 18–19

genome of, 288*t*, *292*, *292*
transcriptional regulation in, 388–389, *389*
transcription and, 60
translation and, 85
viral infection of, 391, *392*
E (exit) sites, **80**, *80*
Esophagus, **867**, *867*–868, *869*
Essay on the Principle of Population, An (Malthus), 435, 586, 1001
Essential amino acids, **862**–863, 863*t*
Estrogen, 816, 818*t*, **909**, *909*
Estrus cycle, **911**
Ethanol fermentation, **148**–149, *149*
Ethics
genetic engineering and, 275, 278
genetic tests and, 340
personalized medicine and, 284
visual synthesis of, *418*
Ethnicity
genetic variation and, 505–506
haplotype and, 356, *356*, *357*
Ethylene, **645**, 645*t*, 659
Eudicots, **705**–706, *706*
Eukarya, **14**, **554**
Archaea and Bacteria compared to, 523, 524*t*
photosynthesis evolution and, *174*, 174–175
phylogenetic tree for, *17*, *17*, *523*, *523*, *532*
Eukaryotes, **14**, **105**
biodiversity of, 554–564
coevolution of prokaryotes and, 540
DNA of, **228**, *228*, *293*, 293–294
gene regulation in, **378**, *378*
genetic diversity, sexual reproduction and, 547–549, *548*
genomes of, 289–290, 290*t*, *291*
internal organization of, 105–107, *105–107*, 545–547, *546*
meiosis and, 231–238
metabolism and, 547
mitosis and, 226–230
multicellular, 569–571, *570*
origins of, 549–553, *551*, *552*, *554*
phylogenetic tree for, *554*, *555*
replication and, 258, *259*
symbiotic bacteria and, 553, *554*
visual synthesis of, *971*
Euryarchaeota, **535**–537, *536*
Eusocial species, **997**–998
Eutrophication, **1104**–1105
Evaporative pumps, 602–603
Evapotranspiration, **1072**, *1072*–1073, *1073*

Evo-devo, **399**, **584**
Hox genes and, 408
master regulatory genes and, 411, 584
Evolution, **4**, **15**–18
agriculture and, 21–22, 508
of alternation of generations, 685, *686*
of altruism, 995–996
of angiosperms, 688–689, *689*
of animals, **580**–582, *581*, 749–754
asexual reproduction and, *900*, 900–902, *901*
of brain, 758
of Calvin cycle, 171
of carbon cycle, 540, 1060–1061, *1061*
of carnivores, 583
of cellular respiration, 149–150, *150*
of circulatory systems, 843
of citric acid cycle, 143
climate change and, 1099, *1099*
of complex multicellularity, 580–584, *581–583*
convergent, **472**, *472*
cultural change and, 507–508
of cyanobacteria, 150
of cytoskeleton, 210–211
Darwin, C., and, 17, 47, 324
ecological systems and, 20, *20*
endosymbiosis and, *174*, 174–175, 517
Escherichia coli experiments and, 18–19
evidence for, 488, *488*
experimental, 18–19
of fungi, **583**, *583*, 724
of gene families, *314*, 315
genetic variation and, 427–431, *428*, *430*
of glycolysis, 149
of gymnosperms, 688
Hardy-Weinberg equilibrium and, 432–434, *433*
of hormones, 817–819
human activities as agents of, 21, 21–22, 1107–1111, *1107–1111*
of human immunodeficiency virus (HIV), 478
of leaves, **642**, *642*
of mammals, 968–969
mass extinctions and, 486–487, *487*
Modern Synthesis and, 436
molecular, 442–444, 475–477, *476*
multicellularity and, 570–571
mutations and, 13
niches and, 1021
non-adaptive mechanisms of, 440–442, *441*
of phloem, 686, *686*

of photosynthesis, 173–175, *174*
of photosynthetic electron transport chain, 173–174
phylogenetic trees and, 468–469, *469*
of plant life cycles, 624, *624*
of plants, 290, *291*, **618**, *618*
of plants biodiversity, 685–689, *686*, *687*, 688*t*, *689*
of pollen, 686, *686*
of populations, 432
of predation, 565
of prokaryotes, **539**, *539*–542, *540*
of proteins, 86–87
rate of, 477, *477*
of reproduction, 897–902
of seeds, 686, *686*
sexual reproduction and, 238
transcriptional regulation and, 393
variation and, 15–16
of vascular plants, 692–693
of wood, 695
of xylem, 686, *686*
See also Biodiversity; Genetic diversity; Human evolution; Mutations; Natural selection
Evolutionary conservation, of developmental transcription factors, 409–411, **503**, 503–504
Evolutionary stable strategy, **995**
Excitation–contraction coupling, **792**–793, *793*
Excitatory postsynaptic potentials (EPSPs), **768**, *768*–769, 774
Excretory organs, **883**–885, *883–885*
Excretory tubules, **883**, *883*–884
Exercise, 153–154
oxygen consumption during, 857, *857*
Exergonic reactions, **125**, *125*
Exhalation, **834**, *834*–835
Exit (e) sites, **80**, *80*
Exocytosis, **108**, *108*, **546**, *546*
synapses and, 767, *767*
Exons, **66**, *67*, **382**
gene structure and, 286, *286*–287
Exoskeletons, **799**, **801**, *801*–802
innate immunity and, 925, *925*
Experimentation, **4**
controlled, 5
DNA structure and, 55–57
evolution and, 18–19
fluorescent dyes, 295, *295*
fluorescent recovery after photobleaching (FRAP), 98–99
gel electrophoresis and, 269, *269*, 429–430
isotopes and, 159, 258
origin of organic molecules and, 46

Experimentation (*Continued*)
  replica plating and, 308–309
  Sanger sequencing and, 272–273, *273*
  X-ray crystallography and, *75*, 75–76, *76*
  See also How Do We Know?
Exponential growth, **1006**, *1006*
Extension, **267**, *268*, **794**, *794*
External fertilization, **903**, *903*
Extinctions
  animals under threat of, 1112, *1112*
  biological species concept and, 449
  fossil record and, 479, 486–487, *487*
  human influences on, 22, *22*
  See also Mass extinctions
Extracellular matrix, **204**, **803**
  of animals, 217–218, *217–219*
  cancer metastasis and, *219*, 219–220
  cell shape influenced by, 220, *220*
  gene expression influenced by, 220, 221
  of plants, 216–217, *217*
Extraembryonic membranes, **904**, *905*
Extravasation, 928–929, *929*
Extremophiles, 27
Eyecups, 778–779, *779*
Eyes
  development of, 409–411, *409–411*
  photoreceptors and, 778–780, *779*, *780*
  types of, 778–779, *779*

F

F_1 generation, *327*, **327**–328, *328*
F_2 generation, **328**, *328*, 328t
Facilitated diffusion, **100**, *100*
Facilitation, **1030**
Facultative mutualisms, **1029**
FAD/FADH$_2$. *See* Flavin adenine dinucleotide (FAD, FADH$_2$)
Fallopian tubes, *907*, **908**
Families, **471**, *471*
FAPs (fixed action patterns), **981**, *982*
Fast-twitch fibers and, *798*, 798–799
Fatty acids, **39**
  dietary, 151–152
  saturated compared to unsaturated, 43, *44*, 96
Feathered dinosaur (*Archaeopteryx lithographica*), 484, *485*
Feature detectors, 981–982, *983*
Feedback. *See* Negative feedback; Positive feedback

Feeding
  active swimming and, *865*, *865*
  arthropods and, 958
  jaws and teeth in, 865–867, *866*, *867*
  suction, 864–865, *865*
  suspension filter, 864, *865*
Female reproductive system, *907*, 907–908, *908*
Fermentation, **148**–149, *149*, **527**
  in bacteria, 527
  in ruminants, 873–874, *874*
Ferns, **594**, *594*, 621–622
  biodiversity of, *696*, 696–697, *697*
  life cycles of, *621*
  phylogenetic tree for, *696*, *696*
Fertilization, 237–238, **899**, *899*
  double, *632*, **633**, *633*, *708*
  external, **903**, *903*
  flowers and, 631–633, *632*, *633*
  human reproduction and, 912–914, *913*
  internal, **903**
  visual synthesis of, *918*
  in vitro (IVF), 914
Fertilizers
  agriculture and, 613
  eutrophication and, 1104
  Green Revolution and, 635
  nitrogen fixation and, 613
Fetus, **915**
  visual synthesis of, *919*
FGFs. *See* Fibroblast growth factors (FGFs)
Fibers (muscle), **788**
Fibers (wood), **650**, *651*
Fibroblast growth factors (FGFs), 824
Fibroblasts, *217*, *218*, *220*
Fight-or-flight responses, 186, 816, 823
Filial imprinting, **989**
Filtration, **883**, 883–886, *886*
First-division nondisjunction, *238*, *239*
First law of thermodynamics, 9, *9*, **123**, *123*
Fish
  age structure in, 1010, *1010*
  biodiversity of, 964–966, *964–966*
  blood composition in, 838, *838*
  bony, 965–966, *966*
  cartilaginous, 965, *965*
  gills and, 831–833, *832*, *833*
  heart and circulatory system of, 846, *847*
  lobe-finned, 966, *967*
  as osmoregulators, 880, *880*
  pheromones and, 826, *826*
  suction feeding and, 864–865, *865*

  swim bladder and, 966
  tetrapod link to, 484, 486
  waste excretion by, 886
Fisher, Ronald, 436
Fitness, **436**
  inbreeding and, 442
  spore dispersal and, 620
Fitzwater, Steve, 1087
5' caps, **65**–66, *66*
5' ends, 55–56
  replication and, 258–260, *260*
  transcription and, *61*, *61*
Fixed action patterns (FAPs), **981**, *982*
Fixed populations, 429
Flagella, **209**, *209*, **905**, *906*
  in opisthokonts, *555*, 555–556
  in sperm, 235
Flatworms, 758, *759*
  gas exchange in, 830, *830*
  waste excretion by, *884*, 884–885
Flavin adenine dinucleotide (FAD, FADH$_2$), 137
  citric acid cycle and, *142*, 142–143
  electron transport chain and, 144–147, *145*, *148*
Fleming, Alexander, 924
Flexion, **794**, *794*
Flies. *See Drosophila melanogaster* (fruit fly)
Flowers, **626**, *628*, *635*
  development of, 412–414, *412–414*, 644
  diversity of, 414, *414*, *628*, *629*, *629*
  fertilization and, 631–633, *632*, *633*
  fruit transformation from, *633*, 633–634, *634*
  organization of, 626–627, *627*
  outcrossing and, 630–631
  pheromones and, *629*, *629*
  pollination of, 628–630, *629*, *630*
  visual synthesis of, *708*
  See also Angiosperms
Fluid mosaic model, **98**
Fluids, **95**
Fluorescent dyes, *295*, *295*
Fluorescent recovery after photobleaching (FRAP), 98–99
Flu virus, 939–940, *940*
Fluxes, **1051**, *1052*
Folding domains, **86**, *87*
Follicles, *909*, **910**
Follicle-stimulating hormone (FSH), 818t, **908**, *909*, *910*, 911
Follicular phase, of menstrual cycle, **909**, *910*
Food webs, 1053–1055, *1054*

Foraminifera, 565, *565*
Forces, **788**
  isometric, *796*, **796**
  muscle generation of, 795–798, *796*
  van der Waals, **44**, *44*, 72, 95–96
Forebrain, 780–781, *781*
Foregut, 867–868, 873–874, *874*
Fossil fuels
  alternatives to, *1103*, 1103–1104
  carbon dioxide from, 1046–1048, *1048*, *1095*, *1095*
  origin of, 157
  supply and demand changes for, 1103
Fossilization, 479
  molecular fossils and, 480–481
  trace fossils and, *480*, *480*
Fossil record
  of animals, 749–750, *750*
  biodiversity and, *1090*, 1090–1091
  of Cambrian Period, *750*, 750–751
  complex multicellularity and, 580–582, *581*
  of dinosaurs, 479, *479*, 484, *485*
  fossilization and, 479–480, *480*
  geologic data and, 481–484
  human evolution and, 494–496, *495–497*
  mass extinctions and, 479, 486–487, *487*
  microfossils and, *564*, 564–565
  molecular clock and, 443
  phylogeny and, 487–488, *488*
  of protists, 564–566, *564–566*
  in Rhynie chert, 692–693, *693*
  tree of life and, 17
  vascular plants and, 651
Founder events, **440**
Fox, George, 523
Fragmentation, **898**, 898–899
Frameshift mutations, **312**, *312*
Franklin, Rosalind, 55
FRAP (fluorescent recovery after photobleaching), 98–99
Fraternal (dizygotic) twins, **369**, *369*
Frequency of recombination, 352–353
Freshwater biomes, 1074, **1083**, *1083*
Frisch, Karl von, 979
Frontal lobe, *781*, 782
Fruit flies. *See Drosophila melanogaster* (fruit fly)
Fruiting bodies, 720–721, *721*
Fruits, **633**, 633–634, *634*
  visual synthesis of, *708*
FSH. *See* Follicle-stimulating hormone (FSH)
Functional groups, 39

Fundamental niches, **1020**, *1020*
Fungi, **571**, **713**
  agriculture and, 731–732, *732*
  ants infected by, 728–729
  biodiversity of, 724–731, *725–727*, *729*, *730*
  as decomposers, *715*, *716*
  dikaryotic, *726*, **726**
  evolution of, *583*, **583**, 724
  growth and nutrition of, 713–719, *714–719*
  life cycles of, 721–723, *722*
  mycorrhizae and, *611*, 611–612
  as opisthokonts, 724
  as pathogens, *716*, 716–717, *717*, 728–729, *924*, **924**
  phylogenetic tree for, *724*, **724**
  plants symbiosis with, *718*, 718–719, *719*
  reproduction of, 719–724, *720–723*
  spores of, 719–721, *720*

G

G. *See* Gibbs free energy (G); Guanine (G)
G_0 phase, **227**, *227*
G_1 phase, **227**, *227*
G_2 phase, **227**, *227*
Gain-of-function mutations, 411
Galápagos finches, 438, *438*, 455
  adaptive radiation of, *456*, 457
  speciation by hybridization and, *460*, 461
Gallbladder, *870*, **871**
Galton, Francis, 361, 366, 368, 370
Gametes, **225**, 237, **328**–329, *329*, **899**
Gametic isolation, **451**
Gametogenesis, 911–912, *912*
Gametophytes, *619*, 619–622, *621*, 690, 691
Ganglia, **758**
Gap genes, **406**, *406*
Gap junctions, *214*, **215**, 216t, *576*, **576**
Garner, Wightman, 660
Gas exchange, **829**
  diffusion and, 829–830, *830*
  respiratory, 831–838, *832–834*, *836*, *837*
Gastrin, 816, 819t, **868**
Gastropods, **954**, *955*
Gastrulas, **400**, *400*, *579*, *580*
Gastrulation, **403**–404, *404*, *915*, *916*
  visual synthesis of, *918*
Gatorade, 877
Gause, Georgii, 1024
Geckos, *736*, **736**
Gehring, Walter, 410–411

Gel electrophoresis, **269**, *269*
  population genetics and, 429–431
GenBank, 477
Gene expression, **51**, **377**
  extracellular matrix influencing, 220, *221*
  influences on, 378, *378*–380, *379*
  regulation of, *380*, 380–381
  transcription and, 381–382, *382*
  visual synthesis of, *88–89*
Gene families, *314*, **315**
Gene flow, **441**
Gene pools, **427**
Genera, *471*, **471**
General transcription factors, **62**, *63*, 381–382, *382*
Gene regulation, **51**, **377**
  epigenetic mechanisms of, 379
  in eukaryotes, 378, *378*
  in prokaryotes, 386–393
  RNA processing and, *382*, 382–383, *383*
Genes, **13**, **51**
  behavior influenced by, 984–986, *985*, *987*
  cholesterol and, 372, *372*
  chromosomes and, *348*, 348–349
  class I/II/III, 937
  complex traits and multiple, 363–365, *364*
  downstream, 411, 414
  gap, **406**, *406*
  homeotic (*Hox*), **407**–408, *407–409*
  housekeeping, 62
  imprinted, 902
  linked, 350–355, *351*, *352*
  maternal-effect, 405
  meristem identity, 641
  number of, 288–289, *288t*
  oncogenes, 245–247
  pair-rule, **406**, 406–407
  proto-oncogenes, **245**
  pseudo-, 443
  R, 668
  structural, 388–389, *389*
  structure of, *286*, 286–287
  X-linked, 345–350, *345–350*
  Y-linked, 355–356, *355–357*
Genetically modified organisms (GMOs), **275**, *276*
  See also Genetic engineering
Genetic code, **82**, *82*–86
Genetic diversity
  agriculture and, 706–707, *707*
  in bacteria, **522**, 522–523
  cross over and, 232
  eukaryotes, sexual reproduction and, *547*–549, *548*
  first cells and, 85–86
  mass extinctions and, 486
  of roots, 653–654, *654*
  of seeds, 625, *625*

sexual reproduction and, 237–238
  See also Biodiversity; Evolution; Mutations; Natural selection
Genetic drift, **432**, 440–441, *441*
  visual synthesis of, *462*
Genetic engineering
  agriculture and, 275, *276*, 588, 680–682, 1106
  disease and disorder correction with, 278
  DNA editing and, 276–278, *277*
  ethics and, 275, 278
  of fruit flies, 410–411, *411*
  genetically modified organisms (GMOs) and, **275**, *276*
  mosquitoes and, 422–425
  plasmids and, 274
  recombinant DNA and, 274–275, *275*
Genetic incompatibility, **452**
Genetic information, **51**
  flow of, 52–53
  nucleic acids and, 40–41
Genetic Information Nondiscrimination Act (GINA), 340
Genetic maps, 353–355, *355*
Genetic risk factors, 252–255, *253*, **302**, *303*
  diseases and disorders and, 307
  environmental risk factors and, 365, 365–366, *366*
  genetic maps and, 353–355, *355*
Genetic tests, 339–340
  visual synthesis of, *418*
Genetic variation, **16**, **301**
  copy-number variations and, *315*, 315–316
  DNA sequencing and, 430
  ethnicity and, 505–506
  evolution and, 427–431, *428*, *430*
  genotypes and, 301–305
  human, 504–507, *505*
  malaria and, 506–507
  measuring, 429–430, *430*
  migration and, 441–442
  mutations and, **16**, *428*, 428–429, 442
  phenotypes and, 301–305
  population genetics and, 427–431
  recombinants and, **428**, 428–429
  segregation and, 332–333
  single-gene, **430**, *430*
  visual synthesis of, *418*
Genome annotation, 284–288, **285**, *285*–288
Genomes, **281**
  annotation of, 284–288, *285–288*
  of archaea, 289, *290*

of bacteria, 289, *290*
  chloroplast, 295–296
  of eukaryotes, 289–290, *290t*, *291*
  mitochondrial, 295–296
  mutation rates in, *305*, 305–306
  nuclear, 295–296
  organization of, *292*–295, *292–296*, 547
  of plants, bacteria and, 670–671, *671*
  point mutations and, 310–311, *311*
  sequencing of, 281–284, *283*, 494
  size of, 289–290, *289t*, *290*
  of viruses, 289, *290*, 296–299, *297*, *298*
  See also Case 3: Your Personal Genome
Genomic rearrangement, **935**, 935–936
Genomics, comparative, 287
Genotype-by-environment interaction, *365*, 365–366, *366*
  visual synthesis of, *419*
Genotype frequency, **429**
Genotypes, 301–305, **302**, **327**
  Punnett squares and, *329*, **329**
Genus, *471*, **471**
Geographic isolation, *451*, **451**–452
Geographic range, of populations, *1002*, **1002**
Geologic timescale, *482*, **482**–483
  biodiversity and, 565
Germ cells, **265**, **306**
  mutations in, 306–307, *307*
Germination, *630*
Germ layers, **400**, *400*, *915*, *916*
Germ-line mutations, **306**–307, *307*, **428**
GH. *See* Growth hormone (GH)
Giant kangaroo rat (*Dipodomys ingens*), *887*, **887**
Giant sea scorpions, 957
Gibberellic acid, **645**, 645t, 647
Gibbs free energy (G), **125**, *125*
Giemsa stains, **295**, *295*
Gills, **831**–833, *832*, *833*
Gilmore, Claudia, 252, 252–255, 284
GINA (Genetic Information Nondiscrimination Act), 340
Ginkgos, **699**, 699–700
Gizzard, **868**
Glaciation, visual synthesis of, *971*
Glands, salt excretion from, *881*, **881**
Glans penis, **907**
Glaucocystophytes, *558*, 558–559
Glial cells, **761**, 765
Global warming, *1095–1097*, 1095–1098
  See also Climate change

Glomeromycetes, 726
Glomerulus, 885–886, *886*, **888**, *888*
Glucagon, 814, *814*, 819*t*
Glucose, 42, *42*, 813–814, *814*, 856
  in blood, 150
  cellular respiration and, 138, *139*
  small intestine absorption of, 871–872, *872*
  transcriptional regulation and, 390–391, *392*
Glycerol, 43, *43*–44
Glycine, 72, *73*
Glycogen, 150, *151*, 814, *814*
Glycolysis, 138, *138*, *139*, *140*, 856, *856*
  carbohydrates and, 150, *151*
  evolution of, 149
  visual synthesis of, *177*
Glycosidic bonds, 43, *43*
GMOs (genetically modified organisms), 275, *276*
  See also Genetic engineering
Gnetophytes, 700, *701*
GnRH (gonadotropin-releasing hormone), **908**, *909*, *910*, 911
Golgi apparatus, **106**, *106*, *110*, 110–111, 545, *546*
Gonadotropin-releasing hormone (GnRH), **908**, *909*, *910*, 911
Gonads, **905**, *906*
Gordon, Albert, 795
Gottschalk, Carl W., 891
G protein-coupled receptors, **192**, *192*
  activation of, 193–194, *194*
  amplification of, 194, *194*, *195*
  responses to, 194–195
  termination of, 196, *196*
G proteins, **192**, *192*, 194
Gram, Hans Christian, 534
Gram-positive bacteria, 534
Grana, **160**
Granulocytes, **926**, *927*
Granulomas, **940**, *941*
Grasses, 704–705
Gravitropism, **654**, *656*
Gray matter, **781**, *782*
Gray whales, 1029
Great Barrier Reef, *1100*, 1100–1102, 1113
Green algae, 558, 559, 559–560, 578, 583, *583*, 591–592
  evolution of plants and, 618, *618*
Greenhouse effect, 1096, *1096*
Greenhouse gases, **1095**–1096, *1096*
Green Revolution, 635, 1105
  pesticides and herbicides and, 681
  wheat stem rusts and, 731
  See also Agriculture
Griffith, John, 55
Griffiths, Frederick, 52–53
Grinnell, Joseph, 1019

Ground tissue, **594**, *594*
Group selection, **995**
Growth factors, **188**–190
Growth hormone (GH), 818*t*
Growth plate, **804**, *804*
Growth rings, **649**, *649*
Guanine (G), **41**, *41*, 54
Guard cells, **596**–597, *597*
Guérin, Camille, 940
Gurdon, John, 401–402
Gustation, **773**
Gut, **867**, *867*–868
Gymnosperms, **594**, *594*
  biodiversity of, 688*t*, 697–700, *698–701*
  evolution of, 688
  life cycle of, 622–624, *623*, *624*
  phylogenetic tree for, *698*
Gyres, 1084

H

H (enthalpy), **125**
Habitat
  disturbance of, *1034*, 1034–1036, *1035*
  human activities influencing, 1015–1016, *1016*, *1107*, 1107–1108
  island species diversity and, 1036, 1036–1037, *1037*
  niches and, *1020*, 1020–1021
  patches, *1014*, 1014–1015, *1015*
Habituation, **987**
Hagfish, **964**, *964*–965
Hair cells, **776**, *776*–777, *777*
Haldane, J. B. S., 754
Half-life, **483**
Haloquadratum walsbyi, 521
Hamilton, William D., 996–998
Hanson, Jean, 790
Haploids, **228**, 548, *548*
Haplotypes, **356**, *356*, *357*
Hardy, G. H., 432
Hardy-Weinberg equilibrium, **432**–434, *433*
Harmful (deleterious) mutations, 302, *303*, 429
Hatch, Marshall, 162
hCG (human chorionic gonadotropin), **911**
H (heavy) chains, **931**, *931*
HCl (hydrochloric acid), 868
Hearing, mechanoreceptors for, 777, *777*–778
Heart, **842**, *842*
  structure and function of, 846–850, *846*–*851*
Heart rate (HR), **851**
Heart valve replacements, 849–850
Heartwood, **649**
Heat
  chemical reactions and, 128
  thermodynamics and, 124

Heavy (H) chains, **931**, *931*
Height, of humans, 366–367, *367*, 373
Helianthus annuus (sunflower), 459, *460*
Helianthus anomalus (sunflower), 459, *460*
Helianthus petiolaris (sunflower), 459, *460*
Helicase, **261**, *262*
Helicobacter pylori (bacterium), 515, 541
Helper T cells, **926**, *936*, 938, 938*t*
Hematocrit, **838**, *838*
Hemichordates, **960**–961, *961*
Hemidesmosomes, **214**, **215**, 216*t*
Hemoglobin, **838**
  development and, *841*, 841–842
  oxygen dissociation curve for, 839–842, *840*, *841*
  oxygen transport by, 838–842, *840*, *841*
  sickle-cell anemia and, 304, 437, 506–507
  structure of, 839
Hemolymph, **830**
Hemophilia, **349**–350, *350*
Hendricks, Sterling, 657
Hepatocytes, **204**, 221
Herbicides, 681
Herbivory, **1026**, 1030*t*
  competition and, 1026
  plant defenses against, 671–676, *672*, *673*, 674*t*, *675*, *676*
Heritability, **367**–368
Hermaphrodites, **415**
Hershey, Alfred, 53
Hess, Carl von, 979
Heterokaryotic cells, **722**, *722*
Heterotrophs, **120**, *120*
Heterozygosity, **302**, 304–305, *329*
Heterozygote advantage, **437**
Hierarchical control, of embryonic development, 403–408
Hill, A. V., 796
Hindbrain, **780**, *781*
Hindgut, **867**, *867*–868, 872, *874*, *874*
Hinge joints, **805**–806, *806*
Hippocampus, **781**, *781*, 783
Hippocrates, 323–324
Hispaniola, 1001
  as biodiversity hotspot, 974–976, *975*
  biodiversity measurement on, 1032
  carrying capacity in, 1007
  fossil record and biodiversity in, *1090*, 1090–1091
  population colonizations on, 1015, *1016*
  resource partitioning on, 1023–1024
Histamine, **927**

Histone code, **379**
Histones, 228, **294**, *294*
  molecular clocks and, 443, *444*
Histone tails, **378**–379
HIV. *See* Human immunodeficiency virus (HIV)
Hodges, Scott, 630
Hodgkin, Alan, 765–766
Holley, Robert W., 83
Holtfreter, Johannes, 213
Homeodomain, **408**
Homeostasis, **26**, **99**, **748**, **877**
  animals and, 748–749, *749*
  breathing and, *837*, 837–838
  endocrine system and, 812–814, *813*, *814*
  energy balance and, 862
  negative feedback and, 131
  plasma membranes maintaining, 99–100
  set points and, 823
  visual synthesis of thermoregulation and, 860–861
Homeotic (*Hox*) genes, **407**–408, *407*–409
Homing pigeons, 990–991
Hominins, **494**–496, *494*–*497*
Homo genus, 495–496, *495*–*497*
Homologous chromosomes, **228**, *228*, *232*, *294*
Homology, **472**, *472*, *473*
Homozygosity, **302**, 304, 328, *329*
Hooke, Robert, 93, 104
Horizontal gene transfer, **522**, *522*–523, 532
Hormones, **645**, **809**
  amine, **815**, *815*–816
  behavior and, 982–984, *983*
  cells binding, 815
  evolution of, 817–819
  major types of, 812, 812*t*, 818*t*–819*t*
  peptide, **815**, *815*–816, *816*, 818*t*–819*t*
  in plants, 644–648, 645*t*, *647*
  releasing, **817**, *817*, 818*t*
  reproductive system and, 908–911, *909*, *910*
  signal amplification of, 816–817, *817*
  as signaling molecules, 186, 188
  steroid, **815**, *815*–816, *816*, 818*t*–819*t*
  tropic, **821**
Hornworts, 689, *690*
Horsetails, **594**, *594*
  biodiversity of, *696*, 696–697
  phylogenetic tree for, *696*, 696
Host cells, **297**–298
Host plants, **666**–667
Hosts, co-speciation and, 457, *457*
  See also Parasites

Hotspots (genome), **306**
Housekeeping genes, 62
How Do We Know?
  action potentials, 766
  animal size increases, 753
  antibody diversity, 934
  arthropod feeding structures, 958
  artificial selection, 438
  ATP synthesis, 147
  bdelloid rotifer asexual reproduction, 902
  biodiversity and primary production, 1060
  butterfly wings, 584
  carbon dioxide incorporated into carbohydrates, 162
  carbon dioxide in Earth's atmosphere, 1046–1047
  cell cycle regulation, 242
  chlorophyll molecules in groups, 166
  climate change impact on coral reefs, 1102
  complex traits, 370–371
  convergent evolution, 695
  dinosaur mass extinction, 6–7
  DNA replication, 258–259
  enzymes in chemical reactions, 130
  evolution demonstrated through experiments, 18–19
  evolution of AIDS virus, 478
  extracellular matrix influencing gene expression, 220, 221
  fish-tetrapod link, 486
  fungal infection of ants, 728–729
  fungal spore shape, 721
  genes influence behavior, 986
  genetic code, 83
  genetic maps, 354
  genetic material, 52–53
  genetic variation, 431
  genome sequencing, 282–283
  grassland expansion, 704–705
  growth factors, 190
  hemoglobin structure, 839
  human lineage, 493
  human origins, 498–499
  inheritance of single-gene traits, 335
  insect growth and development, 811
  iron as limiting nutrient, 1087
  learning in insects, 988
  leaves pull water from soil, 603
  mammalian kidney concentrated urine production, 891
  metabolic rate, 859
  muscle contraction, 795
  mutation randomness, 308–309
  myoglobin structure, 839
  navigation of birds, 991
  oceanic Archaea, 538
  oceanic bacteria, 531
  origin of cell adhesion, 575
  origin of chloroplasts, 550
  origin of organic molecules, 46
  origin of photosynthesis, 562
  photorespiration, 600
  photosynthetic oxygen, 159
  phototropism, 655
  plant communication, 678
  plant immune systems, 669
  pollinator shifts, 630–631
  population size, 1004
  predator-prey coexistence, 1025
  proteins move in plane of membranes, 98–99
  protein structure, 78–79
  seed germination, 657
  speciation and vicariance, 454
  spontaneous generation, 10–11
  stem cells, 401
  transcriptional regulation, 388
  transposons, 313
  viruses and cancer, 246
  See also Experimentation
Hox (homeotic) genes, **407**–408, *407–409*
Hozumi, Nobumichi, 934
HPV (human papillomavirus), 180–183, *181*
HR (heart rate), **851**
Hubbard, M. King, 1103
Huffaker, Carl, 1024–1025
Human activities
  carbon cycle and, 1045–1046, *1095–1103*, 1095–1104
  as evolution agents, *21*, 21–22, *1107–1111*, 1107–1111
  extinction and, *22*, 22
  habitat influenced by, 1015–1016, *1016, 1107*, 1107–1108
  invasive species and, 1108–1109, *1109*
  nitrogen cycle and, 1104–1105, *1105*
  overexploitation and, 1108, *1108*
  pathogens and, 1109–1110, *1110*
  phosphorus cycle and, 1105
  red tides and, 561, 563
  solar energy and, 172
  visual synthesis of, *1038–1039, 1063*
Human chorionic gonadotropin (hCG), **911**
Human Development Index, 1094, *1094*
Human evolution
  African origins of, 497–500
  fossil record and, 494–496, *495–497*
  natural selection and, 506
  phylogenetic tree for, 491–493, *492*
  sexual selection and, 506
  uniqueness of, 510
  See also Evolution
Human genome, 281
  chromosomes of, 294–295, *295*
  repetitive DNA in, 290–291, *291*
  size of, 289t
Human immunodeficiency virus (HIV)
  drugs for, 288
  evolution of, 478
  genome of, 287–288, *288*
  polymorphisms and, 303, *303*
Human papillomavirus (HPV), 180–183, *181*
Humans
  agriculture challenges for population growth of, 1105–1106, *1106*
  chemical composition of, *37*, 37
  chimpanzee traits compared to, 501–503, *502, 503*
  common ancestors of, 492–496, 498–500
  consciousness and, 509–510
  culture and, 507–508, *508*
  energy use by, 1093–1094, *1094*
  ethnic origins and, 356, *356*
  eyes of, *409*, 410
  genetic variation in, 504–507, *505*
  height of, 366–367, *367*, 373
  language and, 509
  lungs of, 835–836
  population growth of, 1093, *1094*
  prehistory of, 504–505, *505*
  reproduction anatomy and physiology in, 905–911, *906–910*
  traits of, 500–504, *501–503*
Hummingbirds, 4, *4*
Hunt, Tim, 242
Huntington's disease, 252
Hutchinson, G. Evelyn, 20, 1019–1020
Huxley, Andrew, 765–766, 790, 795
Hybridization, **325**–327, *327*, **449**
  biological species concept and, 449–450, *450*
  instantaneous speciation and, 459–460, *460, 461*
Hybrid offspring, **449**
Hydrochloric acid (HCl), 868
Hydrogen bonds, **35**, 35–36
  base pairing and, 56–57, *57*
  in proteins, 74–76, *75, 76*
Hydrogenosomes, 551, *551*
Hydrolysis, of ATP, *126, 126, 127*
Hydrophilic molecules, **35**
  amino acids, 72, *73*
  cell membranes and, 94–95, *95*
Hydrophobic effect, **35**
Hydrophobic molecules, **35**
  amino acids, 72, *73*
  cell membranes and, 94–95, *95*
Hydrostatic skeletons, **799**–801, *800*
Hydrothermal vents, 27, *27*
  chemoautotrophs and, *526*, 526–527
  eukaryote origins and, 553
  hyperthermophiles and, 536–537
Hydroxyapatite, **803**
Hydroxyl groups, 60, *60*
Hymenoptera, 997
Hypersensitive response, **668**, *668*
Hyperthermophiles, **536**–537
Hyphae, **713**–715, *714*
Hypothalamus, **781**, *781*, 818t, *820*, 820–822, *821*, 892
Hypotheses, 4–7

I
Identical (monozygotic) twins, 369, *369*
IgA, **932**, *932*
IgD, **932**, *932*
IgE, **932**, *932*
IgG, **932**, *932*
IgM, **932**, *932*
Ileum, **870**, *870*
Imitation, **988**
Immediate hypersensitivity reactions, **932**
Immigration, 1005, *1005*, 1013–1015, *1014*
Immune systems (animals)
  adaptive, 925, 925t, *926*, 929, 929–939, *931–933, 935, 937, 938*, 938t
  flu virus and, 939–940, *940*
  innate, 925–929, 925–930, 925t
  malaria and, *941*, 941–942
  overview of, 923–925
  tuberculosis (TB) and, 940–941, *941*
Immune systems (plants), 666–668, *667*
Immunity. *See* Adaptive immunity; Innate immunity
Immunodeficiency, **924**
Immunoglobulins, **932**, *932*
Implantation, visual synthesis of, *918*
Imprinted genes, 902
Imprinting, **379**–380
Inbred lines, 363
Inbreeding depression, **442**
Incisors, **866**, *866*
Incomplete dominance, **331**, *331*
Incomplete penetrance, **339**
Incus, **777**, *777*
Independent assortment, **333**
  chromosomes and, 334, *336*
  phenotypes and, 334, *336, 336*
  Punnett squares and, 334, *334*, 335
  segregation and, *333*, 333–334

Induced pluripotent stem cells (iPS cells), 403
Inducers, 387, 389, *390*
Inflammation, *928*, **928**-929
Information processing, 989-991
Inhalation, *834*, **834**-835
Inheritance
  biparental, 357
  blending, *324*, 324-325, 332-333
  chloroplasts and, 357
  crisscross, *347*, 347-348
  early theories of, 323-325
  maternal, 357-358, *358*
  in mitochondria, 357-358, *358*
  paternal, 357
  of sex chromosomes, *345*, *346*, 357
  X-linked, 345-350, *345-350*
  Y-linked, 355-356, *355-357*
  See also Transmission genetics
Inhibitors, **129**, *131*
Inhibitory postsynaptic potentials (IPSPs), *768*, **768**-769
Initiation codons, 82
Initiation factors, **84**, *84*
Initiation stage, of transcription, 61
Initiation stage, of translation, *84*, **84**-85, *85*
Innate behavior, **980**, *980*
Innate immunity, **925**, 925t
  complement system and, *929*, 929-930
  extravasation and, 928-929, *929*
  inflammation and, *928*, 928-929
  phagocytosis and, 926-927, *927*
  skin and, *925*, 925-926
  TLRs and, 927-928
  white bloods cells and, *926*, 926-927
Inner cell masses, **400**, *400*
Insects, *957*, *959*
  compound eyes in, 778-779, *779*
  learning in, 988
  mandibles in, 866-867, *867*
  metamorphosis in, 959-960, *960*
  relatedness in, *997*, 997
  respiratory systems of, *832*, 834
  waste excretion by, *884*, 884-885, *885*
Instantaneous speciation, 459-460, *460*, *461*
Insulin, 814, *814*, 819t
Integral membrane proteins, **97**, *98*
Integrated pest management, 681
Integrins, **213**, *213*, *214*, 216t
  complex multicellularity and, 575
Intercostal muscles, **835**
Intermediate filaments, **205**, *206*, *210*, 210-211, 210t, *211*
Intermembrane space, **141**, *141*
Internal fertilization, 903

Interneurons, **757**
Internodes, **639**, *640*, *641*, *641*
Interphase, **227**, *227*
Intersexual selection, **439**-440, *440*
Interspecific competition, **1007**, **1022**
Intertidal biomes, *1085*
Intervertebral discs, *800*, *801*
Intestinal bacteria, *541*, 541-542
Intracellular receptors, *191*
Intrasexual selection, **439**, *440*
Intraspecific competition, **1007**, **1022**
Intrinsic rate of increase, **1006**
Introns, 66, *67*, **382**
  gene structure and, *286*, 286-287
Invasive species, **1108**-1109, *1109*
Inversions, **316**-317, *317*
In vitro fertilization (IVF), **914**
Involuntary components, **770**
Ion channels, *192*, *193*, **767**
Ionic bonds, **33**-34, *34*
Ions, **30**
iPS cells (induced pluripotent stem cells), 403
IPSPs (inhibitory postsynaptic potentials), *768*, **768**-769
Iridium, 6-7
Iron, 1086-1087
Island biogeography, theory of, **1036**-1037, *1037*
Island populations, *453*, *455*, **455**
  colonization and, 1015-1016, *1016*
  diversity of, *1036*, 1036-1037, *1037*
Isomers, **38**-39, *39*
Isometric forces, **796**, *796*
Isometry, **753**
Isotopes, **29**, *258*
  in radiometric dating, **483**, *483*
IVF (in vitro fertilization), **914**

J

Jacob, François, 60, 388, *390*, 736
Janzen, Daniel, 680
Jaws, 865-867, *866*, *867*
Jejunum, **870**, *870*
Jellyfish, 747
  anatomy of, *948*, 948-949
  diffusion and, *573*, 573
  eyes of, *409*, 410
Jenkins, Farish, 486
Joints, 805-806, *806*
Jolie, Angelina, 254
Joyce, Gerald, 8
Julian, Fred, 795
Juvenile hormone, 812t
Juxtaglomerular apparatus, *893*, **893**-894

K

K (Carrying capacity), **1007**
Kacelnik, Alex, 509
Karlson, Peter, 812

Karyogamy, **722**, *722*
Karyotype, **228**, *228*, *294*, *295*
Keeling, Charles David, 1044
Keeling curve, 1044, *1044*, 1045-1047
Keeton, William, 991
Kelp, 560, *560*
Kendrew, John, 839
Keratinocytes, 204, *205*
Key stimulus, **981**, *982*
Keystone species, **1033**-1034, *1034*
Khorana, Har Gobind, 82, *83*
Kidneys, **877**
  blood filtration in vertebrate, 885-886, *886*
  blood pressure and blood volume regulation by, *893*, 893-894
  dialysis and, 894
  in mammals, 887-893, *887-894*
  See also Renal systems
Killer whales, 994, *995*
Kineses, **989**
Kinesin, **208**, *208*
Kinetic energy, *121*, **121**-122
Kinetochores, **230**, *230*
King, Marie-Claire, 493, 501-502
King, Nicole, 575
Kingdoms, **471**, *471*
Kingfisher (*Tanysiptera galatea*), *455*, 455
Kin selection, **996**, 996-997, *997*
Kit receptor kinase, *197*, 197-198
Klinefelter syndrome, **240**, *240*
Knee-extension reflex, *772*, *772*
Knowlton, Nancy, 454
Kohler, Nancy, 188, 190
Köppen, Wladimir, 1071-1072
Krebs cycle. *See* Citric acid cycle
Krings, Matthias, 500
K-strategists, 903-904, *904*, **1012**
Kuhn, Alfred, 810-811
Kuhn, Werner, 891

L

Labia majora, **908**, *908*
Labia minora, **908**, *908*
Lactase, 870-871
Lactic acid fermentation, **148**-149, *149*
Lactose intolerance, 871
Lactose operons, **389**-391, *389-391*
Ladybug (*Adalia bipunctata*), 430, *430*
Lagging strands, *260*-262, **261**-263
Lake biomes, **1083**, *1083*
Lamarck, Jean-Baptiste, 324
Lamellae, *832*, *833*
Laminin, **217**, 220, *220*, 221
Lampreys, **964**, 964-965
Language, 509, *509*
Large intestine, *867*, **868**
  digestion in, 872-873
Lariats, **66**, *67*

Larynx, **835**, *836*
Lateral inhibition, **415**, *415*
Lateral meristems, **648**
Latex, **672**, *672*
Latitudinal diversity gradient, *1089*, **1089**-1090
L (light) chains, *931*, *931*
Leading strands, *260*-262, **261**-263
Learned behavior, **980**
Learning, 782-783, **987**-989, *989*
Leaves, 595-600, *597-598*, 601-610, *604*
  carbon dioxide and, *595*, *596*
  evolution of, 642, *642*
  growth of, 642-644
  production and arrangement of, 642-643, *643*
  visual synthesis of, *709*
Lederberg, Esther, 308-309
Lederberg, Joshua, 308-309
Leeuwenhoek, Anton van, 93
Lengthening contraction, **796**, *796*
Lenski, Richard, 18
Lenticels, **650**, *650*
Lentiviruses, *287*, 287-288
Leptosporangium, **697**
Leukemia, 245
Leydig cells, **909**
LH. *See* Luteinizing hormone (LH)
Lichens, **718**, **718**-719, *719*
Liebig's Law of the Minimum, **1086**
Life
  cell membrane origins and, 95
  characteristics of, 8
  chemical and physical principles of, 8-11
  chemical reactions leading to, 131-132
  eukaryotes origins and, 549-553, *551*, *552*, *554*
  genetic code and, 85-86
  origin of, 10-11, 13, 25-27, *539*, 539-540
  origin of molecules for, 45-47, 60
  water and, on Earth, 36
Life cycles
  of algae, *618*, *618*
  alternation of generations and, 617-622, *618*, *619*, *621*
  of amphibians, *967*, *967*
  of angiosperms, *632*
  of ascomycetes, *727*, 727-728
  of basidiomycetes, *730*
  of butterflies, *1010*, 1010-1011
  dispersal in, 620-622, *621*, *633*, 633-634, *634*, *708*
  of ferns, *621*
  of fungi, 721-723, *722*
  of gymnosperms, 622-624, *623*, *624*
  of land plants, 624, *624*
  of mosses, 618-619, *619*
Life histories, **1012**-1013

Ligand-binding sites, **191**
Ligand-gated ion channels, *192*, *193*, **767**
Ligands, **191**, 414–417, *415*, *416*
Light, visible, **164**, 164–165
Light (L) chains, **931**, *931*
Lignin, 216, *217*
 in xylem, 605
Limbic system, **781**, *781*
"Limited-war strategy," 992
Linked genes, 350–355, *351*, *352*
Linnaeus, Carolus, 467
Lipase, 868, 868*t*
Lipid rafts, **97**
Lipids, **39**
 cell membranes and, 94–95, *95*
 structure of, 43–45, *43–45*
Liposomes, **95**, *95*
Lipton, Allan, 188, 190
Liver, **870**, *870*
Liverworts, 689, *690*
Lizards, 968, *968*, 982–984, *983*, 1013, 1015–1016, *1016*, 1023–1024, *1024*
Lobe-finned fish, 966, *967*
Lobsters, 959
Logistic growth, 1007–1008, *1008*
Long-day plants, **660**, *660*, *661*
Longitudinal muscle layer, **873**, *873*
Long-term carbon cycle, 1048–1053, *1049–1052*
Loop of Henle, 888–891, *889*, *890*
Lophotrochozoans, *952–956*, 952–956
Lorenz, Konrad, 989, *989*
Loss-of-function mutations, **410**
Lucy (hominin), 494–495, *495*
Lumen, **110**, **160**, **873**, *873*
Lunar clocks, 990
Lungfish, 966, *967*
Lungs, 831–832, *832*, **834**, 834–837, *836*
Luteal phase, of menstrual cycle, **909**, *910*
Luteinizing hormone (LH), 818*t*, 908–909, *909*, *910*, 911
Lycophytes, **594**, *594*
 biodiversity of, *693*, 693–696, *694*
 coal and, 694–696
 life cycle of, *621*, 621–622
 phylogenetic tree for, 694, *694*
Lymph, **845**
Lymphatic system, **845**
Lyon, Mary F., 380
Lysis, **391**, *392*
Lysogeny, **392**, *392*
Lysosomes, **106**, *106*, **111**, *111*
Lytechinus pictus (sea urchin), 242
Lytic pathways, **391**, *392*

M

MAC (membrane attack complex), **930**
MacArthur, Robert, 1036
MacLeod, Colin, 52–53
Macrophages, **926**, *927*
Maggot flies (*Rhagotletis pomonella*), **439**, *439*
Magnaporthe grisea (rice blast), 666
Mainland populations, 453, *455*
Major grooves, **56**
Major histocompatibility complex (MHC), **937**, 937–939, *938*
Malaria, 422–423, *423*, *425*, **941**, 941–942
 coevolution and, 437, 457–458
 genetic variation in response to, 506–507
 vaccination for, 424, 1110
Male reproductive system, 905–907, *906*, *907*
Malleus, **777**, *777*
Malpighian tubules, **885**, *885*
Malthus, Thomas, 4, 435, 586, 1001
Mammals, **968**
 blood composition in, 838, *838*
 evolution of, 968–969
 fermentation in ruminants, 873–874, *874*
 heart and circulatory system of, 848, 848–849, *849*
 kidneys in, 887–893, 887–894
 latitudinal diversity gradient in, *1089*, 1089
 phylogenetic tree for, 969
 placental, 969
 respiratory systems of, *832*
 waste excretion by, 886
Mandibles, 866–867, *867*
Mannose-binding lectin (MBL), **930**
Map information, **990**
MAP (mitogen-activated protein) kinase pathway, 198–200, *199*
Map units, **353**
Margulis, Lynn, 549–550
Marine biomes, 1083–1084, *1083–1086*
Mark-and-recapture, 1003–1004
Mars, 25, *26*
Marsupials, 969
Martin, John, 1087
Maryland Mammoth tobacco plant, 660
Mass extinctions, **486**, **751**
 animals and, *751*, 751–752
 of dinosaurs, 6–7, 479
 evolution and, 486–487, *487*
 fossil record and, 479, 486–487, *487*
 genetic diversity and, 486
 visual synthesis of, *971*
Mast cells, **926**, *927*
Maternal-effect genes, **405**
Maternal inheritance, 357–358, *358*
Mating types, **723**
Matzinger, Polly, 924–925

Mayr, Ernst, 448, 533
MBL (mannose-binding lectin), **930**
McCarty, Maclyn, 52–53
McClintock, Barbara, 313
Mean, **366**
Mechanical incompatibility, **451**
Mechanoreceptors, **775**
 for gravity and motion, *776*, 776–777
 for hearing, *777*, 777–778
Mediator complex, **62**, *63*
Medicine
 dialysis and, 894
 personalized, 284, 373–374
 regenerative, 403
 See also Drugs; Vaccination
Medulla (brain), **780**, *781*, 819*t*
Medulla (kidney), **887**, 887–888, *889*, *890*
Meiosis I, **231**–233, *232*, *233*, 235*t*, *236*, 899
Meiosis II, **231**, *234*, 234–235, 235*t*, *236*, 899
Meiotic cell division, 231–238, *232–234*, 235*t*, *236*, *237*, **899**, *899*
Melanin, 208, *209*
Melanocytes, 204, *205*
Melanocyte-stimulating hormone, 812*t*, 818*t*
Melanophores, 208, *209*
Melatonin, 819*t*
Membrane attack complex (MAC), **930**
Membrane potentials, **761**, 913
Membranes
 cell, 94–98
 extraembryonic, **904**, *905*
 plasma, 13–14, *14*, 98–104, *106*, *106*, *545*, *546*
Memory, 782–783
Memory cells, **926**, **932**
Mendel, Gregor, 361
 independent assortment and, 333–337
 laws of, 335, *419*
 linked genes and, 350
 Modern Synthesis and, 436
 multiple genes and, 363–364
 segregation and, 328–334
 sex chromosomes and, 343–344, *357*
 transmission genetics and, 323–327
Mendeleev, Dmitri, 31
Menopause, 911
Menstrual cycle, **909**, *910*, 911
Menstruation, *910*, **911**
Menthol, 674*t*
Merezhkovsky, Konstantin Sergeevic, 549–550
Meristem identity genes, **641**

Meristems, **579**, *579*, **639**
 lateral, 648
 root apical, 651–652, *652*
 shoot apical, 639–643, *640–642*
Merriman, Curtis, 370
Meselson, Matthew, 60, 258, 902
Mesentery, **873**, *873*
Mesoderm, **400**, *400*, **915**, *916*
Mesoglea, **949**
Mesohyl, **947**, *947*
Mesophyll, **595**, *596*
Messel Shale, 481, *481*
Messenger RNA (mRNA), **65–67**, *65–67*, **80**, *80*
 gene regulation and, *382*, 382–383, *383*
 gene structure and, *286*, 286–287
 polycistronic, 65, **85**, *85*, **389**, *389*
Metabolic rate, **857**–859, *858*
Metabolism, 14–15, 119–121, *120*, *121*
 adenosine triphosphate (ATP) and, **856**, *856*
 aerobic, 855–857, *856*
 anaerobic, 148–150, *149*, 855–857, *856*
 in animals, 855–859, *856–858*
 chloroplasts and, 547
 eukaryotes and, 547
 integration of, 150–154
 mitochondria and, 547
 recovery, **857**, 857–858
Metamorphosis, **810**, *810*, 959–960, *960*
Metanephridia, **884**, *885*
Metaphase (mitosis), **229**, *230*, 249
Metaphase I (meiosis I), *233*, 233
Metaphase II (meiosis II), *234*, 235
Metapopulations, **1013**–1015, *1014–1015*
Metastasis, cancer, *219*, 219–220, 247
Methane, 1096–1097
Methanogens, **537**
Methylation, **379**, *379*
MHC (major histocompatibility complex), **937**, 937–939, *938*
Micelles, **94**–95, *95*
Microbial loop, visual synthesis of, *1063*
Microbial mats, 525–527, *526*
Microbiomes. *See* Case 5: The Human Microbiome
Microfilaments, 205, *206*, *207*, 207–208, 210*t*
Microfossils, **564**, 564–565
Microraptor gui (feathered dinosaur), 479
MicroRNA (miRNA), **68**, **384**, *384*
Microsporidia, 556
Microtubules, 205, *206–208*, *207*–209, 210*t*
Microvilli, *204*, *206*, **871**, *871*, 888–889

Midbrain, **780**, *781*
Midgut, **867**, *867–868*
Migration, **441**
 genetic variation and, 441–442
 Hardy–Weinberg equilibrium and, 432
Milkweeds (*Asclepias* species), 672, *672*
Miller, Stanley, 45, 46
Millipedes, 959
Minor grooves, **56**
miRNA. *See* MicroRNA (miRNA)
Mismatch repair, **318**, *319*
Missense (nonsynonymous) mutations, 310, *311*
Mistletoe, 666, *667*
Mitchell, Peter, 144, 146
Mitochondria, **106**, *106*, **114**, *114–115*
 endosymbiosis and, *174*, 174–175, 517, *551*, 551–553, *552*
 inheritance in, 357–358, *358*
 metabolism and, 547
 pyruvate oxidation and, 141, *141*
Mitochondrial DNA (mtDNA), 498–500
Mitochondrial genome, 295–296
Mitochondrial matrix, **141**, *141*
Mitogen-activated protein (MAP) kinase pathway, 198–200, *199*
Mitosis, **226**–230, *229–231*, 235t, *236*, 248
Mitotic spindles, *229*, **229**–230
Modern Synthesis, **436**
Molars, **866**, *866*
Molecular clocks, **443**–444, *444*, 477
Molecular evolution, 442–444
 phylogenetic trees and, 475–477, *476*
Molecular fossils, 480–481
Molecular orbital, **32**
Molecular self-assembly, **298**–299
Molecular sequencing, of animals, **744**, *744–745*
 See also DNA sequencing
Molecules, **32**–34, *33*
 amphipathic, **94**, *94*
 carbon-based, *38–39*
 cell adhesion, *212*, 213, *213*, 216t, *249*, 574–576
 hydrophilic, 35, 94–95, *95*
 hydrophobic, 35, 94–95, *95*
 nonpolar, 35
 organic, 37, 39–47
 origin of life and, 45–47, 60
 polar, 35
 signaling, 186, *186*, 188
Mollusks, **953**, *953*, 956, *956*
Molting, **802**, *810*, *810*
Monkeys, 491–492, *492*
 evolution of, 443
Monocots, **703**, *703–705*

Monod, Jacques, 388, 390
Monophyletic groups, **470**, *470*
Monosaccharides, **42**–43, *43*
 dietary, 150–151, *151*
Monosiga brevicollis (choanoflagellate), 578
Monozygotic (identical) twins, **369**, *369*
Moran, Nancy, 1028
Morgan, Thomas Hunt, 345–348, *354*
Morphine, 674t
Morphospecies concept, 448–449
Mortality rates, 1005, *1005*, 1010–1012, *1011*
Morulas, **400**, *400*, **914**, *914*
Mosquitoes, malaria and, 422–425, *423*, 458, 941, *941–942*
Mosses
 biodiversity of, 689–691, *690*, *691*
 habitat patches and, 1014, *1014*
 life cycle of, 618–619, *619*
Moths
 development of, *810*, 810–811
 pheromones and, 980, *980*
Motor endplates, **792**, *793*
Motor neurons, **757**
Motor proteins, **208**, *208*
Motor units, **797**, *797–798*
Mouth, **867**, *867–868*
M phase, **227**, *227*
mRNA. *See* Messenger RNA (mRNA)
mtDNA (mitochondrial DNA), 498–500
Mucosa, **873**, *873*
Multicellularity
 biodiversity and complex, 583
 bulk flow and, 573–574, *574*
 cell adhesion and, 574–576
 cell communication and, *576*, 576–577, *577*
 coenocytic cells and, 570, *570*
 complex, *571*, 571–572, *572*
 development and, 577–578
 differentiation and, 577–578
 diffusion and, 572–573, *573*
 eukaryotes and, 569–571, *570*
 evolution and, 570–571
 evolution of complex, 580–584, *581–583*
 fossil record and complex, 580–582, *581*
 oxygen and complex, *582*, 582–583
 phylogenetic tree for, *572*, 580, *581*
 in plants compared to animals, 578–580, *579*, *580*
Multicellular organisms, 12
Multiple alleles, **338**–339
Multiplication rule, **332**, *332*

Multipotent cells, **400**
Multiregional hypothesis, **497**
Muscles
 actin in, 788–795, *789–793*
 antagonist, 794, *794*
 cardiac, *788*, 788–790, *850*, 850–851
 contractile properties of, 794, *794–799*, *796–798*
 fast-twitch fibers and, *798*, 798–799
 force generation of, 795–798, *796*
 function of, 787–794
 intercostal, 835
 lengthening contraction of, 796, *796*
 myosin in, 788–795, *789–793*
 skeletal, 788–790, *788–790*
 slow-twitch fibers and, *798*, 798–799
 smooth, 788, 788–789
 speed and, 799
 striated, 788, *788*
Muscle tissue, **746**–747, *747*
Mushrooms, 719–720, 729, *729*
 See also Fungi
Mussels, 956, 1033
Mustard plant. *See Arabidopsis thaliana* (mustard plant)
Mutagens, **317**–318, *318*
Mutations, 13, 58, 87, **301**
 beneficial (advantageous), 303, *303*, 304, 429
 BRCA1, 252–254, *253*, 284
 BRCA2, 252, *253*, 254
 cancer and, 247, *247*, 306–307, *307*
 in chromosomes, 314–317, *314–317*
 deleterious (harmful), 302, *303*, 429
 frameshift, 312, *312*
 gain-of-function, 411
 genetic variation and, 16, *428*, 428–429, 442
 genotype and environment impact on, 303–305, *304*
 germ-line, 306–307, *307*, 428
 Hardy–Weinberg equilibrium and, 432–433
 loss-of-function, 410
 neutral, 302, 429
 nonsense, 311, *311*
 nonsynonymous (missense), 310, *311*
 nucleotide deletion or insertion in, 311–312, *312*
 Pax6, 410, 410–411, *411*
 point, 309–311, *310*, *311*
 proofreading and, 306
 protein evolution and, 87
 randomness of, 308–309

 rates of, *305*, 305–306
 receptor kinases and, *197*, 197–198
 small-scale, 309–313
 somatic, *306*, 306–307, 428
 spontaneous, 305
 synonymous (silent), 310, *311*
 transposons and, *312*, 312–313
 visual synthesis of, *418*
 See also Evolution; Genetic diversity; Natural selection
Mutualisms, **1027**, 1030t
 aphids and, 1027–1028
 costs and benefits of, *1030*, 1030–1031
 facultative, 1029
 obligate, 1029
 symbioses and, 1027–1028
 trees and, 1019, *1020*, 1026–1027
Mycelia, **714**
Mycoplasma genitalium (bacterium), 289
Mycorrhizae, **611**, *611–612*
Myelin, **765**, *765*
Myofibrils, **789**, *789*
Myoglobin, **798**, **840**
 oxygen dissociation curve for, 840, *840*
 oxygen storage and, 840, *841*
 structure of, 839
Myosin, **209**
 in muscles, 788–795, *789–793*
 slime molds and, 557
Myotomes, **962**, *962*
Myriapods, **957**, 959, *959*
Myxobacteria, 521, *521*

N

NADH. *See* Nicotinamide adenine dinucleotide (NADH)
Nadler, Jason, 735
NADPH. *See* Nicotinamide adenine dinucleotide phosphate (NADPH)
Naked mole rats, 998, *998*
Natural gas, 1103, *1103*
Natural killer cells, **926**, *927*
Natural selection, 4, **15**
 adaptations and, 434–436
 allele frequencies and, 436
 directional, 437–438, *438*
 disruptive, 439, *439*, 458, *459*
 human evolution and, 506
 speciation and, 460–461
 stabilizing, 437, *437*
 visual synthesis of, *463*
 See also Evolution; Genetic diversity; Mutations
Nature reserves, 1112–1113
Nautilus, 954–955, *955*
Navigation, 990–991
Neanderthals, *496*, 496–497, 500
Necrotrophic pathogens, **666**

Nectar spurs, 630–631
Negative feedback, **131**, **748**, *748*, 813
  endocrine system and, 813, *813*, *814*
Negative regulation, 386–387, *387*
Negative selection, **436**, 443, *444*, 939
Nematocysts, **949**, *949*
Nematodes, 956–957, *957*
  See also Caenorhabditis elegans (nematode worm)
Neoteny, **502**
Nephrons, **886**, *886*, 889, 890
Neritic zone, **1084**, *1084*
Nerve cords, **759**, *759*
Nerve growth factors (NGFs), 824
Nerve nets, **758**, *759*
Nerves, **759**, *759*
  cranial, 770, *770*
  spinal, 770, *770*
Nervous systems, animal, 740–742, *741*, **757**
  behavior and, 981–982, *983*
  central, 769–770, *770*
  evolution of, 757–759
  neuron organization and structure in, 759–761, *760*
  neuron signal transmission in, 761–769, *762–765*, *767*, *768*
  neuron types in, 757–758
  organization of, 769–773
  peripheral, 769–772, *770*
  simple to complex, 758–759, *759*
Nervous tissue, **747**, *747*
Neufeld, Fred, 52
Neural tube, **962**, *962*
Neurons, **757**
  organization and structure of, 759–761, *760*
  signal transmission and function of, 761–769, *762–765*, *767*, *768*
  synapses and communication of, 766–769, *767*, *768*
  types of, 757–758
Neurosecretory cells, **812**
Neurotransmitters, 189, **760**, *760*
  synapses and, 767, *767*
Neutral mutations, **302**, 429
Neutrons, **29**
Neutrophils, **926**, *927*
NGF (nerve growth factors), 824
Niches, **1019**–1021, *1020*, *1021*
Nicolson, Garth, 98
Nicotiana attenuata (coyote tobacco), 676–677, *677*
Nicotinamide adenine dinucleotide (NADH), 137, 141
  citric acid cycle and, *142*, 142–143
  electron transport chain and, 144–147, *145*, *148*
  fermentation and, 148–149, *149*

Nicotinamide adenine dinucleotide phosphate (NADPH), **159**, 161, *161*
  in photosystems, 167, *167*, 168
Niedergerke, Rolf, 790
Nilsson-Ehle, Herman, 364
Nirenberg, Marshall E., 83
Nitrification, **529**, *529*
Nitrogen cycle, **529**, 529–530, *1057*, *1057*–1058
  human activities influencing, 1104–1105, *1105*
  visual synthesis of, *1062–1063*
Nitrogen fixation, **529**, 529–530, 532, **612**, **1057**, *1057*
  fertilizers and, 613
Nitrogenous wastes, **882**, *882*–883
n + n (dikaryotic) cells, *722*, 722–723
Nodes (phylogeny), **467**, *468*
Nodes (plant morphology), **639**, *640*
Nodes of Ranvier, **765**, *765*
Non-associative learning, **987**
Nondisjunction, **238**, 348
  chromosomes in, *238*, 238–239
  disorders from, *239*, 239–240
  genes in chromosomes and, *348*, 348–349
Nonhomologous chromosomes, **317**, *317*
Nonpolar covalent bonds, **33**
Nonpolar molecules, **35**
Nonrandom mating, **433**, 442
Nonrecombinants, 351–352, *352*
Nonself, **924**, **938**, 938–939
Nonsense mutations, **311**, *311*
Non-sister chromatids, **232**, *232*
Nonsynonymous (missense) mutations, **310**, *311*
Nontemplate strands, **61**, *61*
Noradrenaline, 816
Norepinephrine, 816, *819t*
Normal distribution, **364**, *364*
Norm of reaction, **365**, *365*
Notch receptors, 191
Notochord, **961**, *962*
Nuclear envelope, **108**, *109*, 229
Nuclear genome, **295**–296
Nuclear localization signals, **112**, *112*
Nuclear pores, **108**, *109*
Nuclear transfer, 401–402, *402*
Nuclei (atomic), **29**, *30*
Nuclei (cellular), **14**, **105**, **106**, *106*
Nucleic acids, **13**, **39**
  first cells and, 60–61
  genetic information and, 40–41
  See also Deoxyribonucleic acid (DNA); Ribonucleic acid (RNA)
Nucleoids, **105**, **292**, *292*–293

Nucleolus, **67**
Nucleosides, **54**, *54*
Nucleosomes, **293**, *294*
Nucleotide excision repair, **319**, *320*
Nucleotides, **39**, **53**–54, *54*
  mutations with deletion or insertion of, 311–312, *312*
Nucleotide substitutions, 310
  See also Point mutations
Nusslein-Volhard, Christiane, 404–406
Nutrients, plants and, 608–613, *609t*
Nutrition, 862–864, *863*, *863t*, *864t*

O

Obesity, 365–366, *366*, 862
Obligate mutualisms, **1029**
Observation, 4
Occipital lobe, **782**, *782*
Occludins, 215, *216t*
Ocean acidification, **1101**, *1101*
Ocean currents, 1070, *1070*, 1084
Oceanic zone, **1084**, *1084*
Octet rule, 32
Octobot, **745**, *745*
Octopus, 745, 954–955, *955*
Okazaki, Reiji, 261
Okazaki fragments, **261**, *262*, 265
Olfaction, **773**, *775*
Oligonucleotides, **267**
Omasum, **874**, *874*
Ommatidia, **779**, *779*
Oncogenes, **245**–247
On the Origin of Species (Darwin, C.), 4, 47, 427, 434–436, 447, 1023
Oocytes, 235, *237*, 404–405, **907**, *908*, *909*, **911**
  development of, 911–912, *912*
  fertilization and, 913, *913*
Oogenesis, **911**–912, *912*
Oomycetes, **560**, **665**, *666*
Open circulatory systems, **842**, 842–843
Open reading frame (ORF), **285**–286, *286*
Operant conditioning, **987**
Operators, **389**–391, *389–392*
Operons, **85**, *389–391*, **389**–391
Opisthokonts, **555**, *555*–556
  fungi as, 724
Opsins, **778**, *779*
Opsonization, **930**
Orbitals, **30**
  electrons and, *30*, 30–31, *32*
  molecular, 32
Orders, **471**, *471*
Ordovician Period, 751, *752*
ORF (open reading frame), **285**–286, *286*
Organelles, **106**–107, *106–110*, *110*
  DNA of, 295–296

Organic molecules, **37**, 39–44
  origin of, 45–47
Organogenesis, **916**
  visual synthesis of, *919*
Organs, 203–204, **746**
  of animals, 747, *747*–748
  See also Cells; Tissues
Organ system, **747**
Orgasms, 913
Orgel, Leslie, 47
Orientation, **989**–991
Origin of replication, **264**, *264*
Osmoconformers, **879**
Osmoregulation, **879**–881, *880*
Osmoregulators, **880**, *880*
Osmosis, **101**, *101*
  red blood cells and, 103, *103*
  renal systems and, 877–878, *878*
  reverse, 603
  water loss and, 597, *597*
Osmotic pressure, **101**, **878**, *878*
Osmotic pumps, **607**, *607*
Osteoblasts, **803**
Osteoclasts, **803**
Otters, 1033, *1034*
Outcrossing, flowers and, 630–631
Out-of-Africa hypothesis, **497**–500
Ova. *See* Eggs
Ovaries (flower), **627**, *627*–628
Ovaries (human), *818t*, **820**, **822**, **907**, *907*
Overexploitation, human activities and, **1108**, *1108*
Oviducts, **907**, *908*
Oviparity, **904**
Ovoviviparity, **905**
Ovulate cones, *623*, **623**–624
Ovulation, 909, *910*
Ovules, **622**, *623*
  flower fertilization and, 631–633, *632*, *633*
Oxidation-reduction reactions, **136**–137, *137*, **158**, 524–525
  photosynthesis as, *158*, 158–160
Oxidative phosphorylation, **136**, *136*, **138**, *138*, 150
Oxygen
  altitude and, 842
  complex multicellularity and, 582, *582*–583
  diffusion and, 572–573
  in Earth's atmosphere, 149–150, 174, 484, 540, 582, 1060–1061, *1061*
  exercise and consumption of, 857, *857*
  hemoglobin and transport of, 838–842, *840*, *841*
  myoglobin and storage of, 840, *841*
  photosynthesis and, 159

Oxygen (Continued)
  respiratory system and, 829–831, 831
  visual synthesis of, 970
Oxygen dissociation curves, 839–842, 840, 841
Oxygenic reactions, **525**
Oxytocin, 818t, **821**
Oysters, 956

P

P₁ generation, **327**, 327
p53 proteins, 244, 244–245, 307
Pääbo, Svante, 500
Pacemakers, **850**, 850–851
Pacific rockfish, 1010, 1010
Paine, Robert, 1033
Pair-rule genes, **406**, 406–407
Palindromic symmetry, **270**, 271
Pancreas, 819t, **870**, 870
Pangaea, 484, 484
Paracrine signaling, **188**, 189
Paramecium (algae), 209, 209
Paraphyletic groups, **470**, 470
Parasexuality, **723**, 723–724
  in glomeromycetes, 726
Parasites, **1026**, 1030t
  competition and, 1026
  co-speciation and, 457, 457
  plants as, 666, 667
Parasympathetic division, **771**, 771
Parathyroid glands, 819t, 822–823
Parathyroid hormones (PTH), 819t, 822–823
Parenchyma cells, **595**, 601
Parietal lobe, **781**, 782
Parsimony, **475**
Parthenogenesis, **898**, 899
Partial pressure, **830**, 830, 838
Partial reproductive isolation, 452
Passive transport, plasma membranes and, 100–101
Pasteur, Louis, 10–11
Patches, habitat, **1014**, 1014–1015, 1015
Paternal inheritance, 357
Pathogens
  avirulent, **667**, 667
  bacteria as, 923–924, 924
  biotrophic, 666
  fungi as, 716, 716–717, 717, 728–729, 924, 924
  human activities and, 1109–1110, 1110
  necrotrophic, 666
  plant protection against, 665–671, 666–671
  protists as, 924, 924
  virulent, **667**, 667
  viruses as, 923–924, 924
  worms as, 924, 924
  See also Diseases and disorders

Paulinella chromatophora (amoeba), 550–551, 551
Pauling, Linus, 75–76
Pax6 mutations, 410, 410–411, 411
Payne, Jonathan, 753
PCRs (polymerase chain reactions), 267, 268, 269
PDGF. See Platelet-derived growth factor (PDGF)
Peas. See Pisum sativum
Peat bogs, 691–692, 692
Pedigrees, **337–339**, 337–339
  X-linked genes and, 349, 349–350, 350
Pelagic biomes, 1085
Penicillin, 924
Penis, **906**, 906–907
Pepsin, 868t, **869**
Peptide bonds, **40**, 40, 73–74, 74
Peptide hormones, **815**, 815–816, 816, 818t–819t
Peptidoglycans, **520**, 534
Peptidyl (P) sites, **80**, 80
Per capita growth rates, **1005**–1007, 1006
Pericles, **653**, 653
Periodic selection, 534
Periodic table of the elements, **31**, 31, 32
Peripatric speciation, **453**, 455, 455
Peripheral membrane proteins, **97**–98, 98
Peripheral nervous system (PNS), **769**–772, 770
Peristalsis, **868**
Permeability, selective, **99**
Peroxisomes, **106**, 106
Personalized medicine, **284**, 373–374
  See also Case 3: Your Personal Genome
Perutz, Max, 839
Pesticides, 681
Petals, **626**, 627
PFK-1 (phosphofructokinase-1), 153, 154
3-PGA (3-phosphoglycerate), **161**, 161
pH
  of blood, 37
  of cells, 40
  lysosomes and, 111
  ocean acidification and, 1101, 1101
  of water, 36–37
Phagocytes, **926**–928, 927
Phagocytosis, **546**, 546, **926**–927, 927
Pharyngeal slits, **961**, 961
Pharynx, 868, **869**, **961**, 961
Phenols, 673, 674t
Phenotypes, **301**–305, **302**, 327
  constitutive, **390**, 391

environmental risk factors and, 363, 363
epistasis and, **336**, 336–337
independent assortment and, 334, 336, 336
Punnett squares and, 329, 329
Phenylketonuria (PKU), 338–339
Pheromones, **824**–826, 825
  flowers and, 629, 629
  moths and, 980, 980
Philanthus triangulum (digger wasps), 988
Phloem, **594**, 601, 601, 610, 611
  carbohydrate transport in, 606, 606–608, 607
  evolution of, 686, 686
  secondary, 649, 649, 650
  visual synthesis of, 709
Phloem sap, 606–607, 607
Phosphatases, **193**, 193
Phosphate groups, 53–54, 54
Phosphodiester bonds, **41**, 41, 55, 55
Phosphofructokinase-1 (PFK-1), 153, 154
3-phosphoglycerate (3-PGA), **161**, 161
Phospholipids, **45**
Phosphorus cycle, **1058**, 1058
  human activities influencing, 1105
  visual synthesis of, 1062–1063
Phosphorylation
  ATP and, 192–193, 193
  oxidative, 136, 136, 138, 138, 150
  protein, cell cycle and, 241–242
  substrate-level, 136, 136
Photic zone, 157, **1084**, 1084
Photoheterotrophs, **527**
Photoperiodism, **660**, 660, 661, 990
Photoreceptors, **778**–780, 779, 780
Photorespiration, **171**–172, 172, 598–600, 599
Photosynthesis, **115**, **157**
  bacteria and, 534–535, 535
  CAM and, 597–598, 598
  cellular respiration compared to, 175, 176–177
  challenges with, 169–172
  efficiency of, 172–173, 173
  evolution of, 173–175, 174
  in extreme environments, 157–158, 158
  first cells and, 173
  on land, 591–595
  mechanism of, 164–169
  origin of, 562–564, 563
  as oxidation-reduction reaction, 158, 158–160, 524–525
  oxygen and, 159
  photorespiration and, 598–600, 599

  short-term carbon cycle and, 1044–1045, 1045
  visual synthesis of, 176–177, 709
Photosynthetic electron transport chain, **159**–160, 160
  evolution of, 173–174
  photosystems and, 166–167, 167, 168
  visual synthesis of, 176
Photosystems, **164**
  in cyanobacteria, 174, 174
  photosynthetic electron transport chain and, 166–167, 167, 168
Phototrophs, **119**–120, 120
Phototropism, **654**–656, 656
Photovoltaic cells, 172
Phragmoplast, **230**
Phyla, **471**, 471
Phylogenetic niche conservatism, **1021**, 1021
Phylogenetic species concept (PSC), **450**
Phylogenetic trees, **467**
  for angiosperms, 702
  for animals, 740, 740, 743, 744, 758, 945, 947, 951, 951–952
  for apes, 491–493, 492
  for Archaea, 17, 17, 523, 523, 532, 532, 535–536, 536
  for arthropods, 959
  for Bacteria, 17, 17, 523, 523, 530, 532, 532–533
  for bilaterians, 952, 952
  classification and, 470–471, 471
  for complex multicellularity, 572, 580, 581
  construction of, 472–478, 474
  for ecdysozoans, 956
  for eudicots, 706
  for Eukarya, 17, 17, 523, 523, 532
  for eukaryotes, 554, 555
  for euterostomes, 960
  evolution and, 468–469, 469
  for ferns, 696, 696
  for fungi, 724, 724
  for gymnosperms, 698
  for horsetails, 696, 696
  for land plants, 591–592, 592, 617, 617, 685, 686
  for lophotrochozoans, 953
  for lycophytes, 694, 694
  for mammals, 969
  molecular evolution and, 475–477, 476
  for monocots, 703
  for primates, 16, 16–17, 491–493, 492
  problem solving with, 477–478
  reading, 467–471
  speciation and, 467, 468
  for vertebrates, 468–469, 469, 866, 964, 964

Phylogeny, **467**, *468*
  animals and, 946, *947*
  fossil record and, 487–488, *488*
Phylum, **471**, *471*
Physical activity, oxygen consumption during, 857, *857*
Physics, 8–11
Physiology, ecological systems and, 19–20
Phytochrome, **657**, *658*
Phytophthora infestans (oomycete), 665, *666*
Pigeons, homing, 990–991
Pigmentation, in skin, 204, 208, *209*
Pigments
  accessory, 164–165
  melanin, 208, *209*
Pili, **105**, 523
Pineal gland, 819t, **823**
Pine tree (*Pinus taeda*), *623*, 623–624
Pinnae, **696**
Pinus taeda (pine tree), *623*, 623–624
Pipilo maculatus (towhee bird), 449, *450*
Pipilo ocai (towhee bird), 449, *450*
Pisum sativum (pea)
  in Mendel's experiments, 325–335
  as model organism, 325, *325*
Pith, **644**, *644*
Pits, **601**, 601–602
Pituitary gland, **812**, 818t, *820*, 820–822, *821*
PKU (phenylketonuria), 338–339
Placenta, **905**, 915–916, **969**
Placental mammals, **969**
Placozoans, **950**, *951*
Planarian flatworm, eyes of, 409, *410*
Plant reproduction
  alternation of generations and, **617**–621, *617–622*
  asexual, 635–636, *636*
  flowers and, 626–635, *627*, *628*, *632*, *633*
  seed plants and, 622–625, *623–625*
Plants
  ants and defense of, 674–675, *675*
  ants and symbiosis with, 675, *675*
  bacteria and genomes of, 670–671, *671*
  biodiversity of, 679–682, *681*, *682*
  burs on, 734, *734*
  C_3, 599
  C_4, 599, 599–600
  cells of, 106, *107*
  cell walls in, 578–579
  climate change and, 1098–1100, *1098–1100*
  composition of, 609t
  crossing of, *326*, 326–327
  day-neutral, 660
  defense resources and, 676–679
  development of, 639–644, *640–644*
  drugs from, 673–674, 674t
  environmental factors affecting, 654–659, *656*, *658–659*
  evolution of, 290, *291*, 618, *618*
  extracellular matrix of, 216–217, *217*
  flowering, *626–629*, 626–635, *632–634*
  fungi symbiosis with, 718, 718–719, *719*
  herbivore defenses of, 671–676, *672*, *673*, 674t, *675*, *676*
  hormones in, 644–648, 645t, *647*
  host, 666–667
  immune systems of, 666–668, *667*
  as invasive species, 1109, *1109*
  leaves of, 595–600, *596–598*, *601–610*, *604*
  life cycle of land, *624*, 624
  long-day, 660, *660*, *661*
  multicellularity in animals compared to, 578–580
  nutrients and, 608–613, 609t
  as parasites, 666, *667*
  pathogen defense of, 665–671, *666–671*
  photoperiodism and, 660, *660*, *661*, 990
  phylogenetic tree for land, 591–592, *592*, 617, *617*, 685, *686*
  polyploidy in, 460, *460*
  primary growth of, 648, *708*
  as primary producers, 528
  roots of, 594, *595*, 601–606, 608–613, *610–612*
  secondary growth of, 648–651, *649–651*, *708*
  seed, 622–625, *623–625*, 686
  short-day, 660, *660*, *661*
  soil and, 608, 609t
  stems of, 594, *595*, 601, *601*, 608
  sulfur cycle and, 528, *528*
  visual synthesis of, *970–971*
  See also Vascular plants
Plasma cells, **926**, **932**
Plasma membranes, **13**–14, **98**, **106**, *106*, 545, *546*
  active transport and, 101–103, *102*
  homeostasis maintained by, 99–100
  passive transport and, 100–101

Plasmids, **105**, **274**, **520**, *520*
  genetic engineering and, 274
  Ti, 671
Plasmodesmata, **107**, *107*, **214**, **215**, 216t, 577
Plasmodium falciparum, 422–425, *423*, 457–458, 1110
Plasmogamy, **722**, *722*
Platelet-derived growth factor (PDGF), 188, 190
  leukemia and, 245
Pleiotropy, **372**
Ploidy, 228
Pluripotent cells, **400**
PNS. *See* Peripheral nervous system (PNS)
Podocytes, **888**, *888*
Point mutations, **309**–311, *310*, *310*, *311*
Polar bodies, **235**, *237*, **912**, *912*
Polar covalent bonds, **33**, *33*
Polarity, **55**
Polarization, **761**
Polar molecules, **35**
Polar transport, **646**, *647*
Pol II, **62**, *63*
Pollen, **622**, *623*
  evolution of, 686, *686*
Pollen cones, **623**, *623*
Pollination, **622**, *623*
  coevolution of angiosperms and, 702–703
  of flowers, 628–630, *629*, *630*
  visual synthesis of, *708*
  wind, 629
Pollinator shifts, 630–631
Polyadenylation, **66**
Poly(A) tails, **66**, *66*
Polycistronic mRNA, **65**, **85**, *85*, **389**, *389*
Polymerase chain reactions (PCRs), **267**, *268*, 269
Polymerization, 47, *47*, 62, 64, *64*
Polymers, **39**
Polymorphisms, **302**
  beneficial, 303, *303*, *304*
  harmful, 302, *303*
  neutral, 302
  single-nucleotide, 310, 338, 354–355, *355*
Polypeptides, **74**, *74*
Polyphyletic groups, **470**, *470*
Polyploidy, 228, **290**
  agriculture and, 290, *291*
  instantaneous speciation and, 460, *460*
  in plants, 460, *460*
Polysaccharides, **42**
Polyspermy, **913**
Polytrichum commune (moss), 618–619, *619*
Pons, **780**, *781*
Population bottleneck, **440**

Population genetics, 427–431
  human origins and, 499–500
Populations, **427**–428
  age structure of, *1009–1011*, 1009–1013
  agriculture challenges for growth of, 1105–1106, *1106*
  colonization of, 1015–1016, *1016*
  density of, *1002*, 1003, *1003*, 1008, 1008–1009
  evolution of, 432
  features of, 1001–1004, *1002*, *1003*
  fixed, 429
  geographic range of, 1002, *1002*
  growth and decline of, *1005*, 1005–1009, *1006*, *1008*
  island, 453, 455, *455*
  mainland, 453, 455
  meta-, 1013–1015, *1014–1015*
  sampling, 1003–1004
  size of, 1002, *1002*, 1004–1009
  survivorship and, 1010–1012, *1011*
  visual synthesis of, *1038*
  See also Communities
Populus tremuloides (aspen tree), 636, *636*
Porifera, **740**, *741*
Positive feedback, **762**, *763*, **813**
  endocrine system and, 813, *814*
Positive regulation, **386**–387, *387*
Positive selection, **436**, **939**
Posterior pituitary gland, 818t, *820*, 820–821, *821*
Posttranslational modification, **385**–386
Post-zygotic isolating factors, *451*, 452
Potatoes, blight and, 665, *666*
Potential energy, *121*, **122**, 144
  ATP synthase and, 146, *146*
Power strokes, **791**, *792*
Precision medicine, **373**
  See also Personalized medicine
Predation, 547, **1024**, 1030t
  butterfly wings and, 584
  communities and, 1032, *1032*
  competition and, 1024–1026, *1026*
  ecological systems and, 20
  evolution of, 565
Predator–prey coexistence, 1025
Pregnancy, 914–917, *915*
Premolars, **866**, *866*
Pressure
  blood, 845, 845–846
  osmotic, 101, 878, *878*
  partial, 830, *830*, 838
  See also Turgor pressure
Prey. *See* Predation
Pre-zygotic isolating factors, *451*, 451–452

Primary active transport, **102**, *102*
Primary bronchi, **835**, *836*
Primary consumers, **1053**, *1054*
Primary growth of plants, **648**, *708*
Primary motor cortex, **782**, *782*
Primary oocytes, **911**, *912*
Primary producers, **528**, **1053**, *1054*, 1059–1060
  global patterns of, 1086–1088, *1088*
Primary responses, **933**, *933*
Primary somatosensory cortex, *782*, **782**–783, *783*
Primary spermatocytes, **911**, *912*
Primary structure of proteins, **74**, *75*
Primary transcripts, **65**, *65*, **382**, *382*–383
Primates, **491**
  phylogenetic tree for, *16*, 16–17, 491–493, *492*
Primers, **261**, *261*
Primordia, **643**
Principle of independent assortment, **333**, 335
  See also Independent assortment
Principle of segregation, **328**–330, *329*, 335
Probability, **331**
  normal distribution and, 364, *364*
  transmission genetics and, 331–332, *332*
Probes, **271**
Procambial cells, **643**
Products, **34**, **125**, 128–129
Progesterone, 816, 818*t*, **909**, *909*
Prokaryotes, **14**, **105**
  cell division, binary fission and, 225–226, *226*
  coevolution of eukaryotes and, 540
  evolution of, *539*, 539–542, *540*
  gene regulation in, 386–393
  internal organization of, **105**, *105*
  visual synthesis of, *970*
Prolactin, 818*t*
Proline, *72, 73*
Prometaphase (mitosis), **229**, *229*–230, *249*
Prometaphase I (meiosis I), **233**, *233*
Prometaphase II (meiosis II), *234*, **235**
Promoters, **61**, *62*
  sequence motifs and, 285
  in transcriptional regulation, 387, *387*
Proofreading, **263**, *263*
  DNA repair and, 318
  mutations and, 306
Prophase (mitosis), **228**–229, *229*, *249*

Prophase I (meiosis I), **231**–233, *232, 233*
Prophase II (meiosis II), **234**, *234*
Prostate gland, **906**, *906*
Protease inhibitors, **674**
Protein families, **86**
Proteins, **13**, *13*, **39**, **74**
  amino acids in composition of, 39–40, *40*
  anchor, 97
  bone morphogenetic, 824
  cell cycle and phosphorylation of, 241–242
  in cell membranes, 97–99, *98*, *191*, 191–193
  conformational changes in, 101
  Delta, 191
  denatured, 76–77
  dietary, 152
  evolution of, 86–87
  folding of, 74–79, 86, *87*
  G, **192**, *192*, *194*
  hydrogen bonds in, 74–76, *75*, *76*
  integral membrane, 97, *98*
  motor, *208*, **208**
  p53, **244**, 244–245, 307
  peripheral membrane, 97–98, *98*
  posttranslational modification of, 385–386
  R, 668
  RNA as intermediary between DNA and, 58–59, *59*
  single-strand binding, 261, *262*
  structure of, 71–79, *73, 75, 77–79*
  synthesis of, 80–86
  transcriptional activator, **62**, *63*
  transmembrane, 97, *98*, 213, *214*
  transporter, 97
  See also Receptors
Protein sorting, **112**–114
Proteobacteria, *533*, **534**
Prothoracicotropic hormone (PTTH), **812**, 812*t*, 817
Protists, **554**
  fossil record of, 564–566, *564–566*
  as pathogens, 924, *924*
Protonephridia, **884**, *884*
Proton pumps, 144, *145*, *168*
Protons, **29**–30, *30*
Proto-oncogenes, **245**
Protostomes, **742**–743, *743, 744*, 952–957
Protozoa, **554**
Proximal convoluted tubule, **888**–889, *889, 890*
PSC (phylogenetic species concept), **450**
Pseudogenes, **443**
P (peptidyl) sites, **80**, *80*
Pteridium aquilinum (bracken fern), 621, 621–622

PTH. *See* Parathyroid hormones (PTH)
PTTH. *See* Prothoracicotropic hormone (PTTH)
Public health workers, 1114
Puccinia species (wheat leaf rust), 666
Pullman, Philip, 1115
Pulmonary arteries, *848*, **849**
Pulmonary capillaries, **835**, *836*
Pulmonary circulation, **847**
Pulmonary valve, *848*, **849**
Pulmonary veins, *848*, **849**
Pumps
  evaporative, 602–603
  osmotic, **607**, *607*
  proton, 144, *145*, *168*
Punnett, Reginald, 329, *330*
Punnett squares, **329**, *329*
  complex traits and, 364, *364*
  epistasis and, *336*
  Hardy–Weinberg equilibrium and, *433*, 433–434
  incomplete dominance and, 331
  independent assortment and, 334, *334*, 335
  principle of segregation and, **329**, 329–330, 335
  X-linkage and, *346*, 346–347
Purine bases, **41**, *41*, **54**
Pyloric sphincter, **870**
Pyrimidine bases, **41**, *41*, **54**
Pyruvate oxidation, 138, *138*, *139*, *141*, **141**, 856, *856*

Q

Quantitative traits, **363**
Quaternary structure of proteins, **74**, *75, 77, 78*
Quorum sensing, **186**–187

R

Race, 505–506
Racker, Efraim, 147
Radiolaria, *565*, **565**
Radiometric dating, **483**, *483*
Rainfall
  global warming and, 1098
  in terrestrial biomes, 1072–1073, *1073*
  topography and, 1070–1071, *1071*
Rainforests
  as biodiversity hotspots, 974, *975*, **975**
  as biomes, *1082*
  rainfall in, 1072
Rana pipiens (leopard frog), 402
Rats, 988–989
Raven, Peter, 680
Rays, 965
Reabsorption, *883*, **883**–884
Reactants, **34**, **125**, 128–129
Reaction center, *165*, **165**–166, 173

Reactive oxygen species, **169**–171, *170*
Reading frames, **80**–81
  open, **285**–286, *286*
Realized niches, *1020*, **1020**–1021
Receivers (communication), **992**
Receptor activation, **187**, *187*
  in G protein-coupled receptors, 193–194, *194*
  of receptor kinases, 198, *198*
Receptor kinases, **192**, **192**–193
  activation and signaling of, 198, *198*
  MAP kinase pathway and, 198–200, *199*
  mutations and, **197**, 197–198
  responses to, 198
  termination of, 198
Receptors, **97**, *97*, **186**, *186*, 414–417, *416*
  cell-surface, 191–193, *192*
  chemo-, 773–775, *775*
  conformational changes and, 191–192
  electromagnetic, 778–780, *779, 780*
  G protein-coupled, *192*, **192**–196, *194–196*
  intracellular, 191
  mechano-, 775–778, *776, 777*
  notch, 191
  photo-, 778–780, *779, 780*
  signaling, 191–193, *191–193*
Recessive traits, **327**, *327*
  in pedigrees, 337–338, *339*
Reciprocal altruism, **995**–996
Reciprocal crosses, **327**
Reciprocal inhibition, **772**, *772*–773
Reciprocal translocation, **317**, *317*
Recombinant DNA, **274**–275, *275*
  See also Genetic engineering
Recombinants, **351**
  crossover and, 351–352, *352*
  frequency of, 352–353
  genetic variation and, *428*, 428–429
Recovery metabolism, **857**, 857–858
Rectum, **868**
Red algae, *558, 559*, **571**
Red blood cells, *204*
  active transport and, 103, *103*
Redi, Francesco, 10–11
Red Queen hypothesis, 901
Red tides, 561, 563
Reduction, **160**, *161*
Reductional division, **233**
  See also Meiosis I
Reefs. *See* Coral reefs
Reflexes, **772**, *772*–773
Refractory periods, **763**, *764*
Regeneration, **160**, *161*

Regenerative medicine, **403**
Regression toward the mean, 366–367, *367*
  See also Statistics
Regulatory transcription factors, **381**–382, *382*
Relatedness, **996**, 996–998, *997*
Release factors, **85**, *85*
Releasing hormones, **817**, *817*, 818t
Remoras, 735, *735*
Renal systems
  dialysis and, 894
  excretory organs and, 883–885, *883–885*
  mammalian kidneys and, **887**–893, *887–894*
  osmoregulation and, 879–881, *880*
  osmosis and, 877–878, *878*
  waste excretion and, 881–886, *882–886*
  See also Kidneys
Renal tubules, **886**, *886*
Renaturation, **270**–272
Renner, Otto, 603
Repetitive DNA, **283**, *283*
  in human genome, 290–291, *291*
Replica plating, 308–309
Replication, **13**, **58**, **257**
  of chromosomes, 264–266, *264–266*
  of DNA, 257–263, *258–263*
  DNA structure and, 58, *58*
  eukaryotic, 258, *259*
  5′ ends and, 258–260, *260*
  origin of, **264**, *264*
  semiconservative, **257**–260, *258*
  3′ ends and, 258–260, *260*
Replication bubbles, **264**, *264*
Replication forks, **258**, 258–259, *262*
Repressors, **387**, *387*, 389, 390
Reproduction
  adaptations and, 903–905, *903–905*
  childbirth and, *916*, 917
  evolution of, 897–902
  fertilization and, 912–914, *913*
  of fungi, 719–724, *720–723*
  gametogenesis and, 911–912, *912*
  human, anatomy and physiology of, 905–911, *906–910*
  menstrual cycle and, 909, *910*, 911
  pregnancy and, 914–917, *915*
  vegetative, **636**, *636*
  visual synthesis of, *918–919*
  See also Asexual reproduction; Plant reproduction; Sexual reproduction
Reproductive isolation, **447**, 447–448
  partial, 452
  post-zygotic factors of, 451, 452
  pre-zygotic factors of, *451*, 451–452
Reproductive system
  female, *907*, 907–908, *908*
  hormonal regulation of, 908–911, *909*, *910*
  male, 905–907, *906*, *907*
  menstrual cycle and, 909, *910*, 911
Reptiles
  heart and circulatory system of, 847–848
  waste excretion by, 886
Reserves, nature, 1112–1113
Reservoirs, *1050–1051*, **1051**, *1052*
Residues, **74**
Resistance
  basal, **667**, 668
  to drugs, 18, 87
  specific, 668
  systemic acquired, **668**, 670
Resource partitioning, **1023**–1024
Respiration
  aerobic compared to anaerobic, 135–136
  short-term carbon cycle and, 1044–1045, *1045*
  See also Cellular respiration
Respiratory gas exchange, 831–838, *832–834*, *836*, *837*
Respiratory systems
  carbon dioxide and, 829–831, *831*
  gills and, 831–833, *832*, *833*
  lungs and, 831–832, *832*, *834*, 834–837, *836*
  oxygen and, 829–831, *831*
  tracheae and, **831**–832, *832*, 834
Responding cells, **186**, *186*
Responses, *187*, **187**–188
  to G protein-coupled receptors, 194–195
  primary, *933*, **933**
  to receptor kinases, 198
  secondary, *933*, **933**
Resting membrane potentials, **761**, *762*
Restriction enzymes, **269**–270, 270t, *271*
Restriction sites, **270**, 270t
Reticulum, **873**, *874*
Retina, **779**, **780**
Retinal, **778**, *779*
Retrotransposons, **291**, *291*
Reverse osmosis, 603
Reverse transcriptase, **297**, *297*
R genes, **668**
R groups, **39**–40, *40*, **72**, *72*
Rhagotletis pomonella (maggot flies), *439*, 439
Rhizobium radiobacter (bacterium), 670–671, *671*, 681
Rhizosphere, **608**
Rhodnius (insect), 810–812, *813*
Rhynie chert, 692–693, *693*
Ribonucleic acid (RNA), **13**, **40**, **58**
  composition and structure of, *40*, 40–41
  as DNA and protein intermediary, 58–59, *59*
  DNA compared to, 60, *60*
  messenger, 65–67, *65–67*
  noncoding types of, 67–68
  processing, **65**–67, *65–68*
  small regulatory, **384**, *384*
  splicing, 67–68, *68*
  in transcription, 59–61, *61*
  world hypothesis, 60, 86
Ribose, **60**, *60*
Ribosomal RNA (rRNA), **67**
  bacterial diversity and, 530–532
Ribosomes, **59**, **80**, *80*, **109**
Ribulose 1,5-bisphosphate (RuBP), **160**–161, *161*, 163
Ribulose bisphosphate carboxylase oxygenase (rubisco), **160**–161
  photorespiration and, 171–172, *172*
Rice blast (*Magnaporthe grisea*), 666
RISC. *See* RNA-induced silencing complex (RISC)
Ritualization, **992**
River biomes, *1083*, **1083**
RNA. *See* Ribonucleic acid (RNA)
RNA editing, **383**, *383*
RNA-induced silencing complex (RISC), **384**, *384*
RNA polymerase, **61**
  transcription and, 62, 64, *64*
RNA polymerase complex, **381**–382, *382*
RNA primase, **261**, *261*
RNA processing, **65**–67, 65–68, **382**
  gene regulation and, *382*, 382–383, *383*
  visual synthesis of, *88*
RNA splicing, 66–67, *67*, **382**–383
RNA transcripts, **61**, *61*, 65–67, *65–68*
RNA world hypothesis, 60, 86
RoboTuna, 745
Rod cells, **780**, *780*
Root apical meristems, **651**–652, *652*
Root caps, **652**, *652*
Root hairs, **609**, *610*
Root nodules, **612**, *612*
Roots, 594, *595*, 601–606, 608–613, *610–612*
  branching of, 653, *653*, 658–659
  genetic diversity of, 653–654, *654*
  growth and development of, 651–654, *652–654*
  phototropism and, 655–656
  visual synthesis of, *709*

Roper, Marcus, 721
Ross, A. F., 668
Rough endoplasmic reticulum, **109**, **110**
Rous, Peyton, 245, 246
R proteins, **668**
rRNA. *See* Ribosomal RNA (rRNA)
r-strategists, **903**, *904*, **1012**
Rubisco. *See* Ribulose bisphosphate carboxylase oxygenase (rubisco)
RuBP. *See* Ribulose 1,5-bisphosphate (RuBP)
Rumbaugh, Susan Savage, 509
Rumen, **873**, *874*
Ruminants, fermentation in, 873–874, *874*
Rusts, 728–729, *729*, 731
Rutherford, Ernest, 31

S

S. *See* Entropy (*S*)
Saccharides, 41–42
Saccharomyces cerevisiae (baker's yeast), 715
  budding of, *898*, **898**
  genome of, 288t
Sahelanthropus tchadensis (hominin), *494*, **494**
Saint-Hilaire, Étienne Geoffroy, 502
Salicylic acid, 674t
Saltatory propagation, **765**, *765*
Salt excretion, from glands, 881, *881*
Sanger, Frederick, 272
Sanger sequencing, **272**–273, *273*
SA (sinoatrial) nodes, **850**, *850*
Sapwood, **649**
SAR (systemic acquired resistance), **668**, 670
Sarcomeres, **789**–791, *790*
Sarcoplasmic reticulum (SR), **792**–793, *793*
Sarich, Vince, 443
Saturated fatty acids, *43*, **44**, *96*
Savanna biomes, *1081*
Scaffolds, **294**, *294*
Schlieden, Matthias, 93
Scholander, Fredrick, 603
Schwann, Theodor, 93
Schwendener, Simon, 718
Scientific inquiry, 4–8, *5*, **6**
Scientists, in twenty-first century, 1114–1115, *1115*
Scorpions, 957
Scrotum, **905**, *906*
Scrub Jays, *998*, **998**
Sea anemones, 758, *759*
  anatomy of, *948*, 948–949
  hydrostatic skeletons in, 799–800, *800*
Sea urchin (*Arbacia punctulata*), 242

Seaweeds, 559–560
Secondary active transport, *102*, 102–**103**
Secondary consumers, **1053**, *1054*
Secondary growth of plants, 648–651, *649–651*, 708
Secondary oocytes, **912**, *912*
Secondary phloem, **649**, *649*, *650*
Secondary responses, **933**, *933*
Secondary sexual characteristics, **909**
Secondary spermatocytes, **911**, *912*
Secondary structure of proteins, 74–76, *75*, 78–79
Secondary xylem, **649**, *649*, *650*
Second-division nondisjunction, *238*, **239**
Second law of thermodynamics, 9–10, *10*, *123*, **123**–124
Second messengers, **194**
Secretin, 819*t*, **871**
Secretion, **883**, *883*–884
Seed coats, 624–625
Seed plants, 622–625, *623–625*, 686
See also Angiosperms; Gymnosperms
Seeds, **622**
  dispersal of, 633, *633–634*, *634*
  evolution of, *686*, **686**
  genetic diversity of, **625**, *625*
  germination and, 656–657, *657*
  structure of, *624*, 624–625
Segmentation, **404**, *404*, **742**, *742*
Segregation, 328–334
Selection, **87**, **432**
  balancing, 437
  directional, 437–438, *438*
  disruptive, **439**, *439*, *458*, *459*
  group, 995
  Hardy–Weinberg equilibrium and, 432
  intersexual, 439–440, *440*
  intrasexual, **439**, *440*
  kin, **996**, *996–997*, *997*
  negative, 436, 443, 444, 939
  periodic, 534
  positive, 436, 939
  sexual, 439–440, *440*, 506
  stabilizing, **437**, *437*
  See also Artificial selection; Natural selection
Selective permeability, **99**, **878**, *878*
Self, **924**, **938**, *938–939*
Self-compatibility, **630**
Self-incompatibility, **631**
Self-propagation, **765**
Semen, **906**
Semicircular canals, **776**, *777*
Semiconservative replication, 257–260, *258*
Seminal vesicles, **906**, *906*
Seminiferous tubules, **905**–906, *907*

Senders (communication), **992**
Sensitization, **987**
Sensory neurons, **757**
Sensory organs, **773**
Sensory receptor cells, **773**
Sensory systems, 773–780, *774–777*, *779*, *780*
Sensory transduction, **773**, *774*
Sepals, **626**, *627*
Sepkoski, Jack, 486, 751
Sepsis, 930
Septa, *714*, **714**–715
Sequence assembly, 282–283, *283*
Sequence motifs, 285–286, *286*
Serosa, **873**, *873*
Sertoli cells, **909**
Set points, **749**
  homeostasis and, 823
Sex chromosomes, **344**
  abnormalities of, *240*, 240–241
  inheritance of, *345*, *346*, *357*
  segregation of, 344–345, *345*
  transmission genetics and, 343–344, *344*
Sex-determining region in the Y chromosome (*SRY*), 355
Sexual reproduction, **225**, **899**
  eukaryotes, genetic diversity and, 547–549, *548*
  evolution and, 238
  genetic diversity and, 237–238
  meiotic cell division and, 237–238
  processes of, *899*, **899**
  twofold cost of, **901**, *901*
Sexual selection, 439–440, *440*
  human evolution and, 506
Sharks, **965**, *965*
Shells (egg), **925**, *925*
Shells (electrons), *30*, *31*, *32*
Shine-Dalgarno sequence, *85*, **85**
Shoot apical meristems, 639–643, *640–642*
Shoots, **594**, *594*
  growth and development of, 639–644, *640–644*
  phototropism and, 654–656, *656*
  reproductive, **626**, *626–629*, *627*
  shade and, 658, *659*
Short-day plants, **660**, *660*, *661*
Short-term carbon cycle, 1043–1048, *1045*
Shotgun sequencing, **282**
Shrimp, 959
Shubin, Neil, 486
Sickle-cell anemia, 304–305, 437, 506–507
Side chains, 39–40, *40*, *72*, **72**
Siemens, Hermann, 370
Sieve plates, **606**, *606*
Sieve tubes, **606**, *606–607*, *607*
Sigma factor, **62**

Signal amplification, **194**, *194*, *195*, *267*, *268*
  of hormones, 816–817, *817*
Signal-anchor sequence, **113**, *113*–114
Signaling cells, **186**, *186*
Signaling molecules, **186**, *186*, *188*
Signaling receptors, 191–193, *191–193*
Signal-recognition particle (SRP), **113**, *113*
Signal sequences, **112**, *112*
Signal transduction, **187**, *187*, 414–417, *416*
Silencers, **382**, *382*
Silent (synonymous) mutations, **310**, *311*
Simberloff, Daniel, 1036
Singer, S. Jonathan, 98
Single-gene traits, **361**
Single-gene variation, **430**, *430*
Single-lens eyes, 778–780, *779*
Single-nucleotide polymorphism (SNP), **310**, 338
  genetic maps and, 354–355, *355*
Single-strand binding proteins, **261**, *262*
Sinks, **606**–607, *607*
Sinoatrial (SA) nodes, **850**, *850*
siRNA. See Small interfering RNA (siRNA)
Sister chromatids, **228**, *228*–230
Sister groups, **469**, *469*–470
  synapomorphies and, 473
  taxonomy and, 471
Size, of populations, *1002*, **1002**, 1004–1009
Skeletal muscles, **747**, *747*, 788–790, *788–790*
Skeletons
  of animals, 799–806
  appendicular, **802**, *802*
  axial, **802**, *802*
  endo-, **799**, 802, *802*–803
  exo-, **799**, *801*, 801–802
  hydrostatic, **799**–801, *800*
  vertebrate, **802**, *802*–806, *804*, *805*
Skin
  innate immunity and, *925*, 925–926
  intermediate filaments in, 210, *211*
  layers of, *746*, **746**
  pigmentation in, 204, 208, *209*
  structure of, 204–205, *205*
Sliding filament model, **790**–791, *791*
Slime molds, **557**, *557*–558
Slow-twitch fibers, **798**, *798*–799
Slugs, 954, *955*, 987
Small interfering RNA (siRNA), **68**, **384**, *384*, 670
  plant defenses and, *670*, **670**

Small intestine, 819*t*, **867**, **868**
  digestion in, 870–872, *870–872*
Small nuclear RNA (snRNA), **67**
Small regulatory RNAs, **384**, *384*
Smell, 773–775, *775*
Smith, John Maynard, 901
Smoking, cigarette, 836–837
Smooth endoplasmic reticulum, *109*, **110**
Smooth muscles, **788**, *788*–789
Smuts, **728**, *729*
Snails, 954, *955*
Snakes, digestion by, *869*, **869**
Snapdragons (*Antirrhinum majus*), *331*, **331**
SNP. See Single-nucleotide polymorphism (SNP)
snRNA. See Small nuclear RNA (snRNA)
Social behavior, 994–998, *995–998*
Sogin, Mitchell, 531
Soil, plants and, 608, 609*t*
Solar energy, 172, **1103**, *1103*
Solar radiation, global warming and, 1096
Solubility, **838**
Solutes, **877**–878, *878*
Solvents, **35**
Somatic cells, **306**
Somatic mutations, *306*, **306**–307, **428**
Somatic nervous system, 770–771
Somatostatin, 819*t*
Sources, **606**
Southern, Edwin M., 271
Southern blots, **271**, *271*–272
Soybeans, nitrogen fixation and, *529*, **529**
Spatial summation, **768**, *769*
Speciation, **447**
  allopatric, 452–453, *456*, *459*
  co-, **457**, *457*
  dispersal and vicariance and, 453–457, *455*, *456*
  divergence and, *452*, *453*
  instantaneous, 459–460, *460*
  natural selection and, 460–461
  peripatric, **453**, *455*, **455**
  phylogenetic trees and, 467, *468*
  sympatric, **458**–459, *459*
  visual synthesis of, 462–463
Species, **427**, **447**, *447*
  biological species concept (BSC), 447–450
  cryptic, 449
  DNA sequencing and, 449
  ecological species concept (ESC) and, 450
  endangered, 22
  invasive, 1108–1109, *1109*
  keystone, **1033**–1034, *1034*
  morphospecies concept and, 448–449

phylogenetic species concept (PSC) and, 450
sub-, 452–453
Species-area relationships, **1037**
Specific resistance, 668
Speed, muscles and, 799
Sperm, 209, *209*, **899**, **905**, *906*, 907
  development of, 911, *912*
  fertilization and, 912–913, *913*
  flagella in, 235
  mutation rates in, 306
Spermatogenesis, 911–912, *912*
Sphagnum (moss), 691, *692*
S phase, **227**, *227*
Spiders, 957
Spinal cord, **769**, *770*
Spinal nerves, **770**, *770*
Spiracles, **834**, *959*
Spisula solidissima (surf clam), 242
Spliceosomes, **66**
Sponges, 747, 758, **946**
  anatomy of, *947*, 947–948
  choanoflagellates compared to, 946–947
  diffusion and, 573, *573*
Spongy bones, **803**–804, *804*
Spontaneous generation, 10–11
Spontaneous mutations, **305**
Sporangia, **619**
  in bryophytes, 619–620, *620*
  in fungi, 720, *720*
  in slime molds, 557, *557*
Spores, **899**
  of fungi, 719–721, *720*
  See also Ferns; Fungi; Horsetails; Lycophytes
Sporophytes, **619**, 619–622, *621*, 690, *691*
Sporopollenin, **620**
Squids, 954–955
  eyes of, *409*, 410
  luminescence in, *516*, 517
Squirrels, 1002, 1023, *1023*
SR (sarcoplasmic reticulum), **792**–793, *793*
SRP (signal-recognition particle), **113**, *113*
SRY. *See* Sex-determining region in the Y chromosome (*SRY*)
Stabilizing selection, **437**, *437*
Stahl, Franklin W., 258
Stamens, **626**, *627*
Stapes, **777**, *777*
Starch, 39, 43, **150**, *151*
  carbohydrate storage in, 163, *163*
Starling's law, 796, **851**
Statistics
  cladistics and, 473
  genetic drift and, 440–441, *441*
  normal distribution and, 364, *364*
  regression toward the mean and, 366–367, *367*

transmission genetics and, 331–332, *332*
Statocysts, **776**, **777**
Statoliths, **656**, *656*, **776**, **777**
Stem cells, **265**, **400**–401
  induced pluripotent, 403
Stems, 594, *595*, 601, *601*, 608
  branching of, 641–642, *642*
  growth of, 639–642, *640*–*642*
  wind exposure and, 659
  See also Phloem; Xylem
Stereocilia, **776**
Stern, Kurt, 130
Steroid hormones, **815**, 815–816, *816*, 818*t*–819*t*
Steroids, 44–45
Stigma, **627**, **628**
  crossing plants and, *326*, 327
Stoeckenius, Walther, 147
Stomach, 819*t*
  digestion in, **867**, 867–870
Stomata, **596**–597, *597*
Stop codons, 82
  gene structure and, 286, 286–287
Stramenopila, 560, **560**–561, *561*, 563
Streptococcus pneumoniae, 52–53, *521*
Streptomyces, 521
Striated muscles, **788**, *788*
Strigolactone, 646, **648**
Stroke volume (SV), **851**
Stroma, **160**
Stromatolites, **539**, *539*
Structural genes, **388**–389, *389*
Sturtevant, Alfred H., 350, 354
Styles, **627**, **628**
Suberin, **650**
Submucosa, **873**, *873*
Subspecies, 452–453
Substrate-level phosphorylation, **136**, *136*
Substrates, 128–130, *129*
Succession, *1035*, **1035**–1036
  visual synthesis of, *1038*–*1039*
Suction feeding, 864–865, *865*
Suess, Hans, 1047
Sugars, 39, 53–54, *54*
  dietary, 150–151
Sulci, **781**, *782*
Sulfur cycle, 528, *528*–529
Sunflowers, 459, *460*
Supercoils, **292**, 292–293
Superkingdoms, 556
Supernormal stimulus, **981**, *982*
Superorganisms, 997
Surfactant, **835**
Surf clam (*Spisula solidissima*), 242
Survivorship, 1010–1012, *1011*
Suspension filter feeding, **864**, *865*
Sustainable development, 1113–**1114**
Sutherland, John, 47

SV (stroke volume), **851**
Swallowing, 868, *869*
Swim bladder, 966
Swimming, feeding and active, 865, *865*
Symbionts, **549**
Symbioses, **549**–551, *551*
  digestive, 1027–1029, *1029*
  mutualisms and, 1027–1028
  between plants and ants, 675, *675*
  between plants and fungi, 718, 718–719, *719*
  root nodules and, 612, *612*
Sympathetic division, **771**, *771*
Sympatric speciation, **458**–459, *459*
Synapomorphies, **473**, *473*, **476**, *476*
Synapses, **759**, *760*
  neuron communication and, 766–769, *767*, *768*
Synapsis, **231**–232, *232*
Synaptic cleft, **759**–760, *760*
Synonymous (silent) mutations, **310**, *311*
Systemic acquired resistance (SAR), **668**, *670*
Systemic circulation, **847**
Systole phase, **849**, *849*
Szostak, Jack W., 60

T

T. *See* Absolute temperature (T); Thymine (T)
TACK group, 535–536, *536*
Taiga biomes, *1076*
Tandem repeats, **316**, *316*
Tannins, 673–674, **674***t*
Tanysiptera galatea (kingfisher), 455, *455*
Taste, 773, 775
Taste buds, **775**
TATA boxes, **61**, *62*, 381
Taxa, **470**
Taxes, **989**–990
Taxonomy, 468, **470**–471, *471*, 477
Taylor, C. Richard, 859
TB. *See* Tuberculosis (TB)
T cell receptors (TCRs), **936**–937, *937*
T cells, 926, 930, **930**, 936–939, *937*, *938*, 938*t*
TCRs (T cell receptors), **936**–937, *937*
TDF (testis-determining factor), 916
Teachers, 1114, *1115*
Teeth, 865–867, *866*, *867*
Teleosts, **880**, *880*
Telomerase, 264–266, **265**, *266*
Telomeres, **265**, *266*
Telophase (mitosis), **229**, *230*, 249

Telophase I (meiosis I), **233**, *233*
Telophase II (meiosis II), **234**, **235**
Temin, Howard, 297
Temperate coniferous forest biomes, *1077*
Temperate grassland biomes, *1079*
Temperature
  absolute, 125
  climate and, 1068–1070, *1068*–*1070*
  metabolic rate and body, 858–859
  seasonality of, 1068, *1068*
Template strands, **257**, *258*
Temporal isolation, **451**
Temporal lobe, **781**–782, *782*
Temporal summation, **768**, *769*
Temporomandibular joint, **866**, *866*
10-nm fibers, **293**, *294*
Tendons, **802**
Termination (cellular communication), **187**, *187*
  of G protein-coupled receptors, 196, *196*
  of receptor kinases, 198
Termination stage, of transcription, **61**
Termination stage, of translation, 84–85, *85*
Terminator, **62**, *62*
Terpenes, 673, 674*t*
Terrestrial biomes, 1072–1074, *1072*–*1082*
Tertiary structure of proteins, **74**, *75*, 76–79, *77*
Testcrosses, **330**, *330*, 330*t*
Testes, 819*t*, **820**, **822**, **905**, *906*
Test groups, 5
Testis-determining factor (TDF), 916
Testosterone, 816, 819*t*, **909**, *909*
Tests of protists, **554**
Tetanus, **797**, *797*
Tetrapods, 484, 486, **966**–967, *967*
Thalamus, **781**, *781*
Thallus, **689**
Thaumarchaeota, 536, 537, *538*
Theories, **7**–8
Theory of island biogeography, **1036**–1037, *1037*
Thermodynamics, laws of, **9**, *10*
  chemical reactions and, 125–126
  energy and, 9–10, **123**, 123–124
  heat and, 124
  work and, 123
Thermoregulation, **858**–859, *860*–*861*
Thick filaments, **789**, *789*
Thin filaments, **789**, *789*
Thiomargarita namibiensis (bacterium), 521, *522*
30-nm fibers, **293**, *294*

Thomas, Eugen, 718
Thoreau, Henry David, 1098–1099
3' ends, 55–56
  replication and, 258–260, 260
  transcription and, 61, 61
Threshold potentials, 762
Through the Looking-Glass (Carroll), 901, 1106
Thylakoid, 115
Thylakoid membrane, 160, 160
Thymine (T), 41, 41, 54
Thyroid gland, 818t, 820, 822
Thyroid hormones, 818t
Thyroid-stimulating hormone (TSH), 817, 818t
Tidal ventilation, 834, 834–835
Tidal volume, 834, 835
Tight junctions, 214, 215, 216t
Tiktaalik roseae (fish-tetrapod fossil), 484, 486
Tilia americana (basswood), 650
TIM barrels, 86, 87
Tinbergen, Niko, 979–980, 979t, 988, 994
Ti plasmids, 671
Tissues, 203–204, 745
  of animals, 745–747, 746, 747
  connective, 746
  epithelial, 746
  muscle, 746–747, 747
  nervous, 747, 747
TLRs (toll-like receptors), 927–928
T lymphocytes. *See* T cells
Toad (*Xenopus laevis*), 401
Tobacco mosaic virus, 297, 298, 298–299, 668, 669, 679
Tolerance, 939
Toll-like receptors (TLRs), 927–928
Tonegawa, Susumu, 934
Topography, 1068
  rainfall and, 1070–1071, 1071
Topoisomerase II, 261, 262, 292
Totipotent cells, 399–400
Towhees, 449, 450
Trabeculae, 803
Trace fossils, 480, 480
Trachea, 835, 836
Tracheae, 831–832, 832, 834, 959
Tracheids, 601, 601
Trade-offs, 678–679, 679
Traits, 323–324
  contrasting, 325–327, 326
  crosses with two, 333, 333–334
  dominant, 327, 327, 330t, 331, 331, 337, 338
  quantitative, 363
  recessive, 327, 327, 337–338, 339
  single-gene, 361
  See also Complex traits
Transcription, 13, 13, 59
  DNA and RNA in, 59–61, 61
  DNA structure and, 59, 59

Escherichia coli and, 60
gene expression and, 381–382, 382
process of, 61–64, 61–64
regulation of, 386–387, 387
RNA polymerase and, 62, 64, 64
RNA transcripts and, 65–67, 65–68
visual synthesis of, 88
Transcriptional activator proteins, 62, 63
Transcriptional regulation, 381–382, 382
  from chromatin to mRNA, 377–383
  epigenetic mechanisms of, 379
  evolution and, 393
  hierarchical, 403–408
  from mRNA to phenotype, 384–386
  in prokaryotes, 386–393
  visual synthesis of, 394
Transcription factors
  evolutionary conservation of developmental, 409–411, 503, 503–504
  general, 62, 63, 381–382, 382
  gradients in, 405, 405–406
  regulatory, 381–382, 382
Transduction, 522, 523
Transfer RNA (tRNA), 67
Transformation, 275, 522, 523
Transgenic organisms, 275, 276
  See also Genetic engineering
Transition states, 128
Translation, 13, 13, 59, 80, 83, 86
  base pairing and, 81–82, 82
  components of, 80–81
  DNA structure and, 59, 59
  elongation stage of, 84, 84–85
  Escherichia coli and, 85
  initiation stage of, 84, 84–85, 85
  regulation of, 384–385, 385
  termination stage of, 84–85, 85
  visual synthesis of, 89
Translocation, reciprocal, 317, 317
Transmembrane proteins, 97, 98, 213, 214
Transmission, cultural, 368
Transmission genetics, 323
  in family histories, 337–339, 337–340
  foundations of, 325–327
  independent assortment and, 333–337
  Mendel and, 323–327
  probability and, 331–332, 332
  segregation and, 328–334
  sex chromosomes and, 343–344, 344
  statistics and, 331–332, 332
Transpiration, 595–596
Transporters, 97, 97

Transposable elements (transposons), 291, 291, 312, 312–313
Tree of life, 17, 17, 468, 477
  See also Phylogenetic trees
Trees
  density-dependent mortality and, 680, 681
  mutualisms and, 1019, 1020, 1026–1027
  succession and, 1035–1036
  See also Angiosperms; Gymnosperms
Triacylglycerol, 43, 43
Trilobites, eyes of, 409, 410
Trimesters, 914–917
Triose phosphates, 161
Triploblastic animals, 742, 743
Trisomy 21 (Down syndrome), 239, 239–240, 371, 372
tRNA (transfer RNA), 67
Trophic levels, 1053–1054, 1054
Trophic pyramids, 1055, 1055
  visual synthesis of, 1062
Tropical rainforests. *See* Rainforests
Tropic hormones, 821
Tropism, 654
Tropomyosin, 789, 789
Troponin, 793, 793
True breeding, 325–327, 327
  See also Artificial selection
Truffles, 728
Trypsin, 868t, 871
TSH (thyroid-stimulating hormone), 817, 818t
Tube feet, 961, 962
Tuberculosis (TB), 940–941, 941, 1110, 1110
Tubulin, 205, 206, 210, 210t
Tumor suppressors, 245, 247, 247
Tundra biomes, 1075
Tunicates, 961, 963, 963
Turgor pressure, 104
  in hyphae, 714
  phloem sap and, 607, 607
Turner syndrome, 240, 240
Turtles, 968, 968
Twins, identical compared to fraternal, 369, 369
Twin studies, 369–370, 370t
Twitch, 797, 797
Twofold cost of sex, 901, 901

U
U. *See* Uracil (U)
Ug99, 731–732, 732
Unicellular organisms, 12
Unsaturated fatty acids, 43, 44, 96
Uracil (U), 41, 41, 60, 60
Urea, 879, 882, 882–883
Ureter, 886, 886
Urethra, 906, 906
Urey, Harold, 45

Uric acid, 882, 882–883
Urine, 891, 892, 892–893
Uterus, 907, 908

V
Vaccination, 933–935
  for cancer, 180–183
  for malaria, 424, 1110
  for tuberculosis (TB), 940–941
  See also Drugs
Vacuoles, 104, 107, 107
Vagina, 907, 908
Valence electrons, 32
Vampire bats, 996
van der Waals forces, 44, 44, 72, 95–96
Van Helmont, Jan Baptist, 157
Variable expressivity, 339
Variable (V) regions, 931, 931
Variables, 5
Variation
  antigenic, 942
  environmental, 16
  evolution and, 15–16
  genetic, 16
  mutations and, 13
  See also Genetic variation
Vasa recta, 890, 890–891
Vascular cambia, 648–649, 649, 650
  convergent evolution of, 695
Vascular plants
  biodiversity of, 692–697, 693–697
  bryophytes convergent evolution with, 690–691
  development of, 644
  dispersal in, 620–622, 621
  evolution of, 692–693
  fossil record and, 651
  latitudinal diversity gradient in, 1089, 1089
  organs and tissues of, 594–595, 595
  types of, 594, 594
  See also Angiosperms; Ferns; Gymnosperms; Horsetails; Lycophytes; Plants
Vascular systems, bulk flow and, 573, 574
Vascular tissues, 594–595, 595
Vascular wilt disease, 668, 669
Vas deferens, 906, 907
Vasoconstriction, 845
Vasodilation, 845
Vasopressin, 821, 892
Vavilov, Nicolai, 707
Vectors, 274, 275
Vegetative reproduction, 636, 636
Veins (human), 843, 843–845
Veins (plant), 595, 596
Velcro, 734, 734
Vena cavae, 845
Ventilation, 830, 831
  tidal, 834, 834–835

Ventricles, **847**, *847*
Venules, *843*, **844**
Vernalization, *661*, **662**
Vertebrae, **963**, *964*
Vertebral column, **961**
Vertebrates, **961**
  biodiversity of, 963–969, *964–969*
  endocrine system of, *820*, 820–823, *821*
  features of, 963–964, *964*
  filtration of blood in, 885–886, *886*
  lymphatic system in, 845
  phylogenetic tree for, 468–469, *469*, *866*, *964*, *964*
  respiratory systems of, *832*, *834*, 834–837, *836*
  skeletons of, *802*, 802–806, *804*, *805*
Vesicles, **107**, 545–546, *546*
Vessel elements, **601**, *601*, 650–651, *651*
Vessels, **601**, *601*
Vestibular system, **776**, *777*
Vestigial structures, **874**
Vicariance (speciation), 453–457, *455*, *456*
Villi, **871**, *871*
Virchow, Rudolf, 93
Virulent pathogens, **667**, *667*
Viruses, 15, **287**
  cancer caused by, 246
  classification of, 296–297, *297*
  flu, 939–940, *940*
  genomes of, 289, *290*, 296–299, *297*, *298*

host range of, *297*, 297–298
as pathogens, 923–924, *924*
transcriptional regulation and, 391–393, *392*, *393*
visual synthesis of, 394–395
Visible light, *164*, **164**–165
Vitamins, **863**–864, 864t
Viviparity, **905**
Vocal cords, **835**, *836*
Volcanism, global warming and, 1096
Voles, *985*, *987*
Voltage-gated channels, **762**, *763*
Voluntary components, **770**
Volvox (algae), *559*, 560, 578
V (variable) regions, **931**, *931*
Vulva, **908**

W

Waggle dance, of bees, 994, *994*
Wallace, Alfred Russel, 434–435, *435*, 1088
Wasps, 988
Waste excretion, 881–886, *882–886*
Water
  balance of, 877–881, *878*, *880*
  as greenhouse gas, 1096
  hydrogen bonds in, 35, 35–36, *36*
  life on Earth and, 36
  pH of solution of, 36–37
  as polar molecule, 35
  properties of, *36*, 36
  transport of, in xylem, 602–604, *604*
Water birch (*Betula occidentalis*), 650
Water potential, **604**

Water vascular system, **961**, *962*
Watson, James D., 53, 55–56, 258, 839
Weight, of animals, 365–366, *366*
Weinberg, Wilhelm, 432
Welch, Mark, 902
Went, Frits, 655
Whales, 994, *995*, 1029
Whale sharks, 965
Wheat
  breeding of, 635, 681
  domestication of, *587*, 587–588
  stem rusts in, 731
Wheat leaf rust (*Puccinia* species), 666
White blood cells, *926*, **926**–927
White-Crowned Sparrows, *993*, 993–994
White matter, **781**, *782*
Whittall, Justen, 630
Wieschaus, Eric F., 404–406
Wigglesworth, Vincent, 810–811
Wild types, **345**, *345*
Wilkins, Maurice, 55
Wilson, Allan, 443, 493, 498, 501–502
Wilson, Edward O., 997, 1036
Wilson, H. V., 212–213
Wind
  currents, 1068–1069, *1069*, 1084
  energy, 1103, *1103*
  pollination, 629
  stems and exposure to, 659
Woese, Carl, 523
Wood, 649–651, *649–651*
  decomposition of, 715, *716*
  evolution of, 695

Worms, as pathogens, 924, *924*

X

Xanthophylls, *170*, **171**
X chromosomes, 343–345, **344**, *344*
Xenopus laevis (toad), 401
Xeroderma pigmentosum (XP), 319
X-inactivation, *380*, **380**–381, *381*
X-linked genes, *345–350*, **345–350**
XP (Xeroderma pigmentosum), 319
X-ray crystallography, **75**, 75–76, *76*
Xylem, **594**, 601, *601*, *602*, 605
  evolution of, 686, *686*
  lignin in, 605
  secondary, 649, *649*, *650*
  visual synthesis of, *709*
  water transport in, 602–604, *604*

Y

Y chromosomes, 343–345, **344**, *344*
  human origins and, 499–500
Yeasts, **715**, *715*
Yellowstone National Park, 1112
Y-linked genes, **355**–356, *355–357*
Yolk, **904**

Z

Z discs, **789**, *790*, *791*
Zea mays (corn), 312–313
Z scheme, **167**, *167*
Zygomycetes, *725*, **725**–726
Zygotes, **237**, **329**, **400**, *400*, 548, *548*, **899**